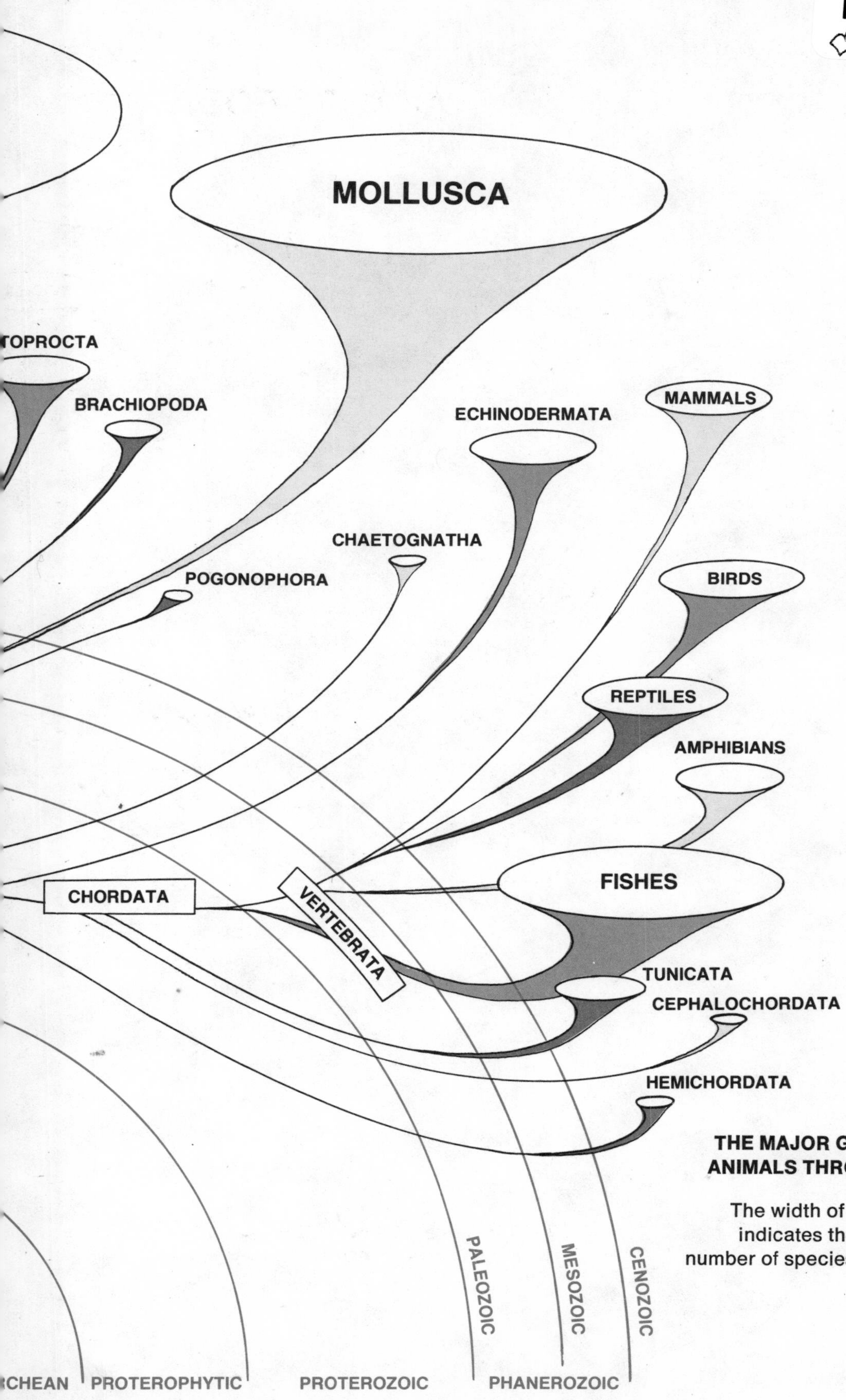

MOLLUSCA

TOPROCTA

BRACHIOPODA

ECHINODERMATA

MAMMALS

CHAETOGNATHA

POGONOPHORA

BIRDS

REPTILES

AMPHIBIANS

CHORDATA

VERTEBRATA

FISHES

TUNICATA

CEPHALOCHORDATA

HEMICHORDATA

PALEOZOIC

MESOZOIC

CENOZOIC

CHEAN PROTEROPHYTIC PROTEROZOIC PHANEROZOIC

THE MAJOR GROUPS OF ANIMALS THROUGH TIME.

The width of each oval indicates the relative number of species in that group.

INTEGRATED PRINCIPLES OF

ZOOLOGY

INTEGRATED PRINCIPLES OF

ZOOLOGY

CLEVELAND P. HICKMAN, Sr.

Late Professor Emeritus,
Department of Zoology,
DePauw University,
Greencastle, Indiana

CLEVELAND P. HICKMAN, Jr.

Department of Biology,
Washington and Lee University,
Lexington, Virginia

FRANCES M. HICKMAN

Formerly of the Department of Zoology,
DePauw University,
Greencastle, Indiana

In association with

LARRY S. ROBERTS

Department of Zoology,
University of Massachusetts,
Amherst, Massachusetts

SIXTH EDITION

With 1205 illustrations, including original drawings by

WILLIAM C. OBER, M.D.

Crozet, Virginia

THE C. V. MOSBY COMPANY

ST. LOUIS • TORONTO • LONDON 1979

Cover photographs by Cleveland P. Hickman, Jr.

FRONT COVER

Panama treefrog *Smilisca phaeota*
Mudskipper *Periophthalmus* sp.
Indian Ocean soft coral *Alcyonium* sp.
Eastern box turtle *Terrapene carolina carolina*

SPINE

Parrot snake *Leptophis ahaetulla*

BACK COVER

Galápagos crab *Grapsus grapsus*
Marine iguana *Amblyrhynchus cristatus*
Red-footed booby *Sula sula*
White grunts *Pomadasys* sp.

SIXTH EDITION

Printed in the United States of America

The C. V. Mosby Company
11830 Westline Industrial Drive, St. Louis, Missouri 63141

Library of Congress Cataloging in Publication Data

Hickman, Cleveland Pendleton, 1895-1978
Integrated principles of zoology.

Includes bibliographies and index.
1. Zoology. I. Hickman, Cleveland P., joint author. II. Hickman, Frances Miller, joint author. III. Title.
QL47.2.H54 1979 591 78-27064
ISBN 0-8016-2172-0

GW/VH/VH 9 8 7 6 5 4 3 2 1 02/C/221

In memoriam

Cleveland P. Hickman, Sr.

It is with a deep sense of personal loss that we report the death of our senior author. Far more than an author to us, he was not only our beloved husband and father, he was our teacher, our leader, our inspiration, and our challenge. This book was his dream. It was his vision that planned it, his genius that executed it. Into its various editions he put 25 years of labor, but it was always a labor of love, and into the fabric of its pages are woven his love for his students, his love of teaching, his love of all nature, and finally his love of writing. For us, it was a joy and a privilege to help him with the earlier editions and finally to take an active part in the revision of the later editions. We have tried to follow his example, to strive for the goals he set, and to maintain the high quality of his work. We grieve that he is not here to see the publication of this sixth edition, but we rejoice that his spirit lives on, not only in our hearts but in the hearts of his students and colleagues and in the pages of this book.

Frances M. Hickman
Cleveland P. Hickman, Jr.

PREFACE

The sixth edition of *Integrated Principles of Zoology* follows the fifth after five years of exciting and important discoveries in the biologic sciences. So much new information is a challenge to instructors attempting to relate the knowledge of zoology fully and interestingly to their students. In some respects, the growing complexity of biology renders effective communication more difficult, since specialists in the different biologic disciplines tend to view the life process from their own particular vantage points. Nevertheless, advances in each area provide more knowledge about the basic mechanisms of life and thus strengthen the bridges interconnecting all zoologic disciplines. As exciting and important as recent discoveries are, they must all be viewed against the background of a relatively few unifying concepts and principles that can be readily understood by the beginning student in zoology. These common denominators of biologic inquiry form the central theme of this undergraduate textbook of zoology.

The objectives of this book were established in the first edition by Cleveland P. Hickman, Sr. Our aim is to provide students of zoology with a knowledge and appreciation of animals and the nature of animal life as it is presently understood. Zoology is a discipline that deals with an animal in all its aspects. Considered purely as a living organism, an animal is a self-perpetuating, self-regulating physiochemical system that is in continuous adjustment with its environment. This admittedly colorless description of an animal nevertheless serves to emphasize that understanding what an animal is, what it does, and where it comes from requires the fusion of numerous disciplines, ranging from the burgeoning field of molecular biology to evolution and population biology. The zoologist must be concerned with an animal's morphology, physiology, behavior, environmental relationships, development, and evolutionary history.

The amount of information that has been accumulated in any one of these areas is enormous; combined, it is beyond the comprehension of any one person. Furthermore, many aspects of life itself remain unsolved. Fortunately, despite their complexity and diversity, animals share a basic organization of form and function, and all have arrived where they are today through the same evolutionary process. Consequently, it is possible to order and integrate our present knowledge about animals into a logically developed presentation, as we have attempted to do in this textbook.

In this edition, several chapters have been completely rewritten to reorder emphasis, improve clarity, and incorporate new information. Other chapters have been revised to remove errors and to introduce new material.

Chapter 1, which as in previous editions introduces several fundamental concepts and principles that underlie zoologic study, has been wholly reorganized and rewritten. A new section explaining the concept of evolution has been added, and the biologic principles that appear on pp. 9 to 16 have been updated, restated, and grouped into natural subdivisions of biologic knowledge to improve their accessibility and usefulness. Well-formulated principles are indispensable in helping us interrelate facts and apply biologic knowledge, and we encourage the reader to refer frequently to these principles while progressing through the book. Chapter 2 contains new sections on colloidal systems, lipids, and the properties of water that relate to water's essential role in living systems. Chapter 3, which traces the evolution of life on earth from its origins some 3 billion years ago to the appearance of eukaryotic organisms toward the close of the Precambrian era, has been revised and serves to introduce the eukaryotic cell, the subject of Chapter 4. Chapter 5 contains a revised treatment of enzyme function, and the section on cellular metabolism has been thoroughly recast and reillustrated. Chapter 6 on animal architecture has been completely reillustrated and contains revised sections on animal complexity, body organization, homology and analogy, and body size. The section on animal phylogeny, treated in Chapter 7, has been rewritten to embrace more recent theories of the relationship of the animal kingdom to other forms of life.

Part two on the diversity of animal life has been updated throughout and reorganized for better flow and clarity. Sections dealing with invertebrate parasites have been rewritten to reflect current knowledge, and many new illustrations have been added. Behavioral biology and physiologic adaptations received greater emphasis throughout this section. In each of the invertebrate chapters, the section on phylogeny and adaptive radiation has been moved to the chapter's end where the reader, once acquainted with the phylum, is in a better position to understand the group's relationship to other phyla.

Part three, dealing with animal activity, contains new and revised sections on the immune system (Chapter 29), nutritional requirements of animals (Chapter 31), synapse physiology and functional specialization of the cerebral hemispheres (Chapter 32), and mechanisms of hormone action (Chapter 33). Part three closes with a new chapter that emphasizes stereotyped and social behavior of animals.

The five chapters of Part four center on the continuity and evolution of animal life. The discussions of homology of sex organs and origin of germ cells have been rewritten (Chapter 35), and a new section on aging has been added to Chapter 36. Chapters 37 (genetics) and 38 (evolution) have been wholly revised and largely reillustrated. Chapter 37 is cast in a historical sequence that begins with the laws of Mendel, progresses through chromosomal theory and gene theory, and ends with a discussion of the role of nucleic acids in the storage and transmission of genetic information. Chapter 38, too, is organized historically. It contains a new section on the significance of Charles Darwin's voyage on the Beagle and reworked sections on natural selection, the Hardy-Weinberg equilibrium, and speciation. The final chapter of Part four has been rewritten to describe the recent and exciting harvest of hominid fossils and the current thinking of physical anthropologists regarding the descent of the human species.

The environment of animals is the theme of the three chapters of Part five. The chapter on the biosphere has been expanded to include a new section on biome diversity. The next chapter centers on the ecosystem and ecologic principles. The final chapter departs from the traditional exposition of biologic facts and principles to explore the turbulent relationship between humans and wildlife and the causes of animal extinction—issues that affect our values and culture as well as our natural environment.

A unique feature of this book is a section (the Appendix) on the development of zoology. It contains a listing of key discoveries in zoology and an annotated list of books and publications. We have carefully reviewed the entire section, deleting entries that we felt were of lesser significance, correcting and rewriting many of the existing descriptions, and adding a number of more recent discoveries and publications. We have also devised a system of key symbols to make easier the task of locating important discoveries in a given area. The list is somewhat shorter than before but we believe it is even more useful as a reference source. The historical background of zoology is too often neglected in presenting zoologic information to students, yet most instructors realize that every discovery rests on foundations assembled by others.

The reference lists at the ends of the chapters have been updated, and we have again separated the popular *Scientific American* articles from the general references. Throughout, we have made every effort to define terms where they are first introduced in the text, but the reader may at any time look up a new term in the glossary, which has been thoroughly revised and expanded. The derivation of each term is provided, as well as its pronunciation and definition, since it has always been our belief that biologic terms and scientific names of animals are remembered more easily and are more meaningful when they are learned in conjunction with their derivations.

Many people gave us valuable advice about the organization and content of this book. We are especially indebted to Professor William S. Hoar of the University of British Columbia, who read all of Part three (animal physiology and behavior) and furnished numerous suggestions for improvements. Michael D.

Johnson of DePauw University reviewed and commented on the three chapters dealing with the phylum Arthropoda. The second author's colleagues at Washington and Lee University were especially responsive to his pleas for assistance. John J. Wielgus read the sections on cell physiology, L. Randlett Emmons reviewed the chapters on genetics and evolution, John H. McDaniel reviewed the chapter on human evolution, and Gary H. Dobbs offered constructive suggestions on various sections. Dean Foster (Virginia Military Institute) reviewed the chapters on animal behavior, the biosphere, and the final chapter on animals in the human environment. We appreciate the helpful comments on the Appendix made by Mindaugas S. Kanlenas, Theodore D. Sargent, and David J. Klingener of the University of Massachusetts. Rufus Rickenbacher of Maplewood, N.J., read most of the manuscript and galley proofs, ferreting out errors and questioning us on exposition. We are also indebted to Maggie Ober of Crozet, Virginia for checking portions of the manuscript.

Finally, we thank the many users of previous editions who have written with their comments and suggestions. We are especially grateful for the detailed comments on many sections of the fifth edition that we received from Dwight L. Ryerson (Pomona College), Barbra A. Roller (Harcum Junior College), and Charles J. Ellis (Iowa State University). Others who contributed comments include: G. M. Tvrdik, California State University, Sacramento; D. M. Smith, Northwest Missouri State University; C. E. Vasey, SUNY College of Arts and Science; E. S. Benes, California State University, Sacramento; E. R. Decker, Mississippi Gulf Coast Junior College (Jeff Davis Campus); J. B. Abram, Hampton Institute; R. Feldman, Community College of Allegheny County; L. Niblock, Livingstone College; J. H. Crowe, University of Wisconsin, Eau Claire; D. L. G. Noakes, University of Guelph; C. L. Pritchett, Brigham Young University; R. W. Flournoy, Louisiana Tech University; P. B. Ahrens, Iowa State University; L. Anderson, North Carolina State University; H. L. Willis, University of Wisconsin, Platteville; K. M. Standing, California State University, Fresno; W. G. Bennett, Iowa Central Community College; E. S. Hunter, Tidewater Community College; R. C. Wall, Lake Sumter Community College; E. Waldorf, Louisiana State University; C. A. Gathman, Palm Beach Junior College; H. Barker, St. Cloud State College; T. S. Robinson, University of Louisville, J. E. Trainer, Jacksonville University; T. M. Berra, Ohio State University; P. Castor, California Polytechnic State University; S. B. Castle, Arizona State University; J. Pfleegor, Corning Community College; J. W. Merker, Texas A & M University; H. R. Krear, Michigan Technological University; M. Duffy, Allegheny County Community College; H. N. Cunningham, Jr., Behrend College; A. Feldherr, Santa Fe Community College; L. X. Washington, Sr., Southern University; S. L. Cave, Niagara County Community College; J. Bast, Allegheny Community College; T. A. Woolley, Colorado State University; S. T. Grant, Holyoke Community College; B. Johnson, Eastern Michigan University; F. A. Fraembs, Eastern Illinois University; C. M. Chandler, Middle Tennessee State University; R. F. Denoncourt, York College of Pennsylvania; C. H. F. Rowell, University of California, Berkeley; V. V. Lyons, Trinity College; S. J. Grabowski, Allentown College; J. Benson, Ohio University; Y. V. Amrein, Pomona College; W. G. Walther, Bates College; M. S. James, University of Utah; R. E. Molnar, North Dakota State University; T. Ferretti, Delta State University; H. Perlman, Delgado Junior College; J. Munro, Université Laval; C. E. Waldrop, Gadsden State Junior College; W. Chizinsky, Briarcliff College; W. H. Volker, Thomas More College; W. A. Dreyer, University of Cincinnati, T. L. Pearce, Colgate University; R. B. Brown, West Chester State College; R. B. Loomis, California State University, Long Beach; M. L. Overton, Cheyney State College; C. G. Wilber, Colorado State University; A. J. Mia, Bishop College; F. R. Munter, University of the Pacific; R. Reinhard, Mt. San Antonio College. Their suggestions and those of many others were of great assistance in the preparation of this edition.

• • •

We welcome the association of Larry S. Roberts, who contributed extensively to the revision of many of the invertebrate chapters as well as to Chapter 1 and the historical Appendix. His expertise in invertebrate zoology and his broad experience in general zoology bring new talent to this venture. We look forward to his continued collaboration with us in future editions.

Cleveland P. Hickman, Jr.
Frances M. Hickman

CONTENTS

PART ONE

INTRODUCTION TO THE LIVING ANIMAL

Barren-ground caribou bull. (Photograph by L. L. Rue, III.)

In these seven chapters, we discuss the basic nature of life as we understand it today. So far as can be determined, the same materials and physical and chemical laws of the nonliving world also apply to the living. The essential difference appears to lie in the organization of the elementary materials of the nonliving into the highly specific architecture of the living. In its simplest form, life is associated with a heterogeneous substance constructed of organic macromolecules. From this substance has evolved an organizational hierarchy of cells, tissues, organs, organisms, and, finally, the vastly complex community of life, the ecosystem. The living animal's activity, growth, reproduction, and use of its environment require a still further organization of energy processes beyond that encountered in the strictly nonliving world.

CHAPTER 1

LIFE

General considerations and biologic principles

A woodchuck surveys its portion of the earth environment that it shares with human beings.

Photograph by L. L. Rue, III.

LIFE ON EARTH

What is life?

We could hardly begin our consideration of zoology with a more difficult question, one that many biologists have considered almost unanswerable. Even with the benefit of today's scientific sophistication and understanding, it is difficult to define the basic substance of life precisely and simply.

Nevertheless, there are several biologic manifestations that clearly distinguish the living from the nonliving. The essential differences between the two appear to be organization, metabolism, growth, adaptability, irritability, and reproduction. As we have learned more about the structural and organizational properties of protoplasm—living matter—it has become increasingly apparent that both the living and the nonliving share the same kind of chemical elements and that both obey the law of conservation of energy (Principles 1 and 2, p. 9). Few biologists today hold to the vitalistic view, prevalent not so many years ago, that living things are endowed with an inexplicable vital force. The accumulating weight of scientific evidence indicates that life is governed by physical and chemical laws, even though we still may be far from understanding the complex fabric of life as a whole.

In the living organism, atoms and molecules are combined into patterns that have no counterparts in the nonliving. Combinations of certain large molecules (macromolecules) are unique to life: the proteins, fats, carbohydrates, and nucleic acids. Such combinations of common elements, which are highly improbable on thermodynamic grounds, provide dynamic systems of coordinated chemical and physical activities that, taken as a whole, distinguish the living from the nonliving. It is these unique macromolecules of life, organized as they are into complexes—some of which capture, store, and transmit energy and others of which provide continuity by self-replication—that make the more obvious manifestations of life possible. Thus, living things maintain and repair themselves, respond to stimulation, and reproduce (Principles 10, p. 11, and 25, p. 15). Many of them undergo development and growth, finally to become senescent and die (Principles 21 and 22, p. 14). It is true that almost any single criterion of life has its counterpart in the nonliving world. Only living things, however, have combined these properties into unique structural and functional patterns.

A further important characteristic of living organisms is the intimate, reciprocal relationship between them and their environment (Principle 27, p. 15). Life and the environment are inseparable. The evolutionary history of the organism has placed it in a specific environment that has determined the structural, functional, and behavioral properties of the organism. The environment is fit for the organism, and the organism is fitted to the environment and responds to its changes. The interaction between the living and the nonliving has determined the character of life on earth and the character of the earth itself. Each has deeply marked the other.

Nature and origin of matter

Matter is any object or entity that occupies space. The chief properties of matter are gravitation and inertia. Mass, which all material bodies have, is a measure of inertia. Matter and energy make up the universe, and these two components are interconvertible according to Einstein's theory of relativity, which is expressed in the equation Energy $= mc^2$, where m equals the mass and c equals the speed of light. This equation represents the theoretic basis for converting matter to energy. As we commonly understand matter and energy, they are separate. Matter in whatever form (solid, liquid, or gas) occupies space and has mass. Energy has the ability to produce change or motion, that is, the ability to perform work.

The 92 naturally occurring elements of which the living and nonliving are made are the result of a cosmic evolution. According to present theory, the nuclei of helium and all the heavier elements are composed ultimately of hydrogen nuclei and neutrons. All elements heavier than hydrogen were formed by supernovas that occur at the end of the lives of large stars.

Throughout most of the life of a star, energy is produced by the fusion of hydrogen nuclei into helium. When the hydrogen is mostly depleted, the star, deprived of its source of nuclear energy, collapses under the force of its own weight, generating central temperatures of hundreds of millions of degrees. As the temperature rises during collapse, carbon nuclei are formed by fusion of helium nuclei. With continued collapse and still higher temperatures all of the elements are produced by nuclear fusion. The final catastrophic end comes when central temperatures reach several billion degrees, releasing so much energy that the star is blown apart. All 92 elements are sprayed into space.

In the universe, helium makes up about one-fifth and hydrogen about four-fifths of all matter. The heavier elements are thus very rare in the universe and represent not more than 1% or 2% by mass of all matter. Therefore, the earth and the other planets of the solar system are composed of rare material.

Living matter is even more selective. Of the 92 naturally occurring elements, only 24 are essential for life and, of these, six (carbon, hydrogen, nitrogen, oxygen, phosphorus, and sulfur) play especially important roles in living systems. But as emphasized already, the presence or absence of specific elements alone cannot explain the fundamental difference between the living and the nonliving. The chief difference is the arrangement of atoms into organic macromolecules that can perform the basic functions of life.

Is life found only on earth?

At present there is no positive evidence that life flourishes elsewhere than on earth. Yet earth is a minor planet circling an ordinary star in one galaxy among billions. In the immensity of the universe with its countless billions of stars (more than 10^{20} stars have been revealed by powerful telescopes) there must be many stars accompanied by planetary bodies. Unfortunately a planet the size of earth circling even the nearest star cannot be observed with our telescopes. However, astronomers predict that most star systems include planets and of these, about 10% have lifetimes in excess of 2 billion years—time enough for life to originate and evolve. A *conservative* estimate is that there are at least 100,000 possibilities for life *in our galaxy alone*. Perhaps all of these planets are sterile bodies of rock washed by lifeless seas. But this is unlikely. The elements required for life are found throughout the universe and obey the same laws of chemistry and physics that apply on earth. Also, the primitive conditions on our earth that spawned life some 3 billion years ago may occur on other planets. Indeed in our galaxy, estimated to be about 10 billion years old, there are many stars (and planets, presumably) much older than earth's 4.6 billion years. Life may have evolved and existed on their planetary systems for billions of years longer than on earth. Can we still maintain that life on earth is unique in the universe and that the human species is at the center of all things?

All of this speculation is exciting and alluring, but it is speculation nonetheless. There is not a shred of hard evidence for the existence of life anywhere but on earth. Clearly, life is not a necessary component of the universe, as are matter and energy. Life is the historic outcome of certain molecular combinations. At the moment ours is the only planet we know of that is supporting life, and until we discover life forms elsewhere, or until life elsewhere reveals itself to us, humans may be forgiven for considering themselves truly unique.

There is an old theory that life on earth originated from spores, bacteria, or macromolecules carried to this planet on meteorites or similar bodies. Even allowing that such structures could survive cosmic radiation and the searing temperatures of entry into the earth's atmosphere, such a theory is simply "passing the buck" to another planet for an explanation of the origin of life.

Nevertheless, there exists one particularly intriguing class of meteorites called carbonaceous chondrites. This rare form of meteorite contains carbon, mostly in the form of an insoluble polymer. About 1% of the total carbon, however, is composed of soluble, high–molecular weight substances including hydrocarbons, fatty and aromatic acids, and many kinds of amino acids, some of which occur in biologic systems.

Such findings do not prove the existence of extraterrestrial life for two reasons. First, the amino acids found in such meteorites are mixtures of both optic isomers (D and L forms), whereas amino acids in living systems are predominantly of the L optic isomer form. Second, many of the meteorite amino acids are very rare or nonexistent in the life forms that we know. Such amino acids are strikingly similar to those produced by electric discharges acting on mixtures of methane, nitrogen, ammonia, and water in recent prebiotic synthesis experiments (Chapter 3). Since we do not know where meteorites originate, it is difficult to assess the significance of such findings. At present it seems wiser to assume that the organic substances present in carbonaceous meteorites were synthesized under abiotic conditions somewhere in space than to suggest that they are evidence of life forms on another planet.

Protoplasm as the life substance

The living substance of organisms is described by the general term **protoplasm.** The term is used less by biologists today than it once was because it has not been possible to define it specifically. Structurally, protoplasm resembles a complicated colloid system of many phases. These appear as numerous localized areas of varying sizes, states of aggregation, and physical and chemical natures suspended in a continuous phase possessing qualities of a gel. It has been compared to a netlike fibrous structure, but the idea of a permanent molecular framework is not consistent with the dynamic nature of protoplasmic organization that distinguishes it from nonliving structures. It is often extremely fluid, revealing flow and movement. It maintains itself by continually building up and breaking down substances that compose it. It is not a homogeneous matter and it is not possible to collect a drop of it that can be called representative of the whole. Consequently protoplasm is often now regarded as a vague and nebulous term. Nevertheless many biologists find it a useful one for describing the indispensable groundwork—or true living matter—of cells, even though the intricate organization and peculiar properties of protoplasm are quite complex.

Levels of complexity

The biomass of the earth is composed of a **hierarchy of structure and function** ranging all the way from atoms and subatomic particles to the highly complex ecosystems of the total earth environment. This hierarchy, encompassing the nonliving as well as the living, is commonly divided into the following levels of interacting components: atom, molecule, organelle, cell, tissue, organ, organism, population, community, and ecosystem. Beginning with the atom it is evident that each level in the hierarchy furnishes the components, or building stones, for the units of the next higher level. Atoms and molecules are integrated into cellular structures and cells, cells into tissues, different tissues into organs and organ systems, these into multicellular organisms, and organisms into populations, which finally are united into communities and ecosystems.

Note that any given level contains all the lower levels as components. For instance, there are fewer tissues than cells and fewer cells than organelles. In addition to its own complexity, each level also includes the complexities of all the lower levels. Energy is expended in the shift from one organizational level to another, and when a higher level has been attained, energy is necessary to maintain that level.

Atoms and subatomic particles are usually the domain of physicists, while molecules are the focus in the study of chemistry. The laws of physics and chemistry as they apply to atoms and molecules also apply to living materials, which are considered to be within the domain of biology.

We can view the biologic sciences themselves as a variety of disciplines: biophysics, biochemistry, molecular biology, cell biology, organismal physiology, animal behavior, and ecology. Each of these is associated with the study of a particular level in the biologic hierarchy, and the investigators in each of these disciplines tend to view the biomass from their own particular vantage point. Nonetheless, these disciplines are mutually interdependent, not mutually exclusive. For example, the biophysicist may be interested in a mathematical description of energy and mass transfer in cell respiration, the biochemist might concentrate on the chemistry of the pertinent reactions, and the molecular biologist or cell biologist might ask questions related to the spatial location of the reaction, as in the mitochondria. None of these investigators would feel constrained by artificial limits, however. The cell biologist must be concerned with the mode of entry of nutrient molecules into the cell and with the chemical reactions; the biochemist must be concerned with the location of the reactions, and so on. Similarly, the organismal physiologist may study the action of a hormone on a whole organ or organism but must recognize the fact that the hormone may act by regulating entry of food molecules into cells; and the actions of an animal, studied by the behaviorist, may be determined by the hormonal and nutritional state of the organism. Finally, the ecologist may study the energy budget or energy flow in an entire community or ecosystem, but this is the sum of the behavior and actions of the organisms within the system, which in turn depend on their hormonal and nutritional states, and ultimately on the chemical and physical events in the cells that comprise the organisms.

It is clear that a unified view of the earth's biomass will emerge only as it becomes increasingly possible to make such bridges between the adjacent disciplines of biology. The aim is not to explain an organism or an ecosystem in terms of the laws of physics, but ultimately to develop a synthesis of the *interactions* of the different levels of complexity.

Concept of the ecosystem

Animals may be examined at several different levels of organization within the biosphere. The individuals of an interbreeding group of organisms of the same species make up a **population.** When a population coexists with other populations of different species that are all adapted to a certain set of environmental conditions, a **community** is formed. A community is a localized aggregation of closely interdependent populations, organized so that they are more or less independent of other communities. Each community is a self-sufficient entity with a distinctive unity, such as a coniferous forest, a grassland, or a coral reef. Within its organization are different strata of populations, each adapted for a particular role in the maintenance of the community. One stratum may be the producers (plants), which can trap solar energy and convert it into food energy. A second stratum is the primary consumers, which eat plants and are themselves consumed by secondary consumers (the third stratum). A fourth stratum contains the decomposers (chiefly bacteria and fungi), which disassemble dead organisms, releasing basic nutrients to be used again by plants. The community and the nonliving environment interact to form an ecologic system, or **ecosystem.** An ecosystem therefore is the sum total of the physical and biologic factors operating within an area that includes living organisms and abiotic substances. The exchange of materials between the living and nonliving occurs in a circular path. Such a system usually contains a few species with large populations and many species with small populations.

The ecosystem is the basic functional unit in ecology and the highest level of organization in the biologic hierarchy. All other levels of organization are parts or fractions of the ecosystem. It may be large or small, provided that it meets the requirements of major components that operate together to form a functional stability. Life cycles result from the interactions between the biotic and abiotic components of the system. A forest region, a meadow, a lake, a desert, or even an ocean are examples of ecosystems. If the system is well balanced, no supporting materials are ever exhausted. An ecosystem is characterized by self-regulation, balanced energetics, functional and interspecies diversity, and independence except for a solar energy source.

Energy and life

All life requires energy. The steady flow of solar energy together with the appropriate kinds of matter provided on earth were responsible for the origin of life, as well as for its continued existence. The sun, the earth, and outer space form a **steady-state system** in which the total amount of energy reaching earth from the sun is exactly balanced by an equal amount of energy lost to outer space. However, in its passage through the atmosphere, light energy is degraded, since the light emitted from the earth to outer space is of longer wavelength and thus of lower energy than the shorter wavelength, higher energy quanta received from the sun. Thus there is no violation of the second law of thermodynamics, which states that an isolated system must decay spontaneously toward maximum entropy or disorder (Principle 2, p. 9).

But on earth, the presence of living systems appears to be inconsistent with the second law because the evolution of life has seen the emergence of organisms of ever-increasing complexity and order and requiring ever more energy for their sustenance. Animals, plants, and the ecosystems containing them, sustained as they are by the continuous flow of solar energy through the system, remain as steady-state systems of high internal order that oppose the tendency for entropic decay, at least for the period of time that life exists on earth. The sun–earth–outer space system as a whole, however, is always losing energy.

To be of use to living organisms, light energy must be absorbed. Nearly all of the energy that enters the living ecosystem is first absorbed by chlorophyll molecules in green plants, which transform a portion of this energy into potential or food energy. The energy accumulated by plants is called net productivity, and it supports all the rest of life on earth. Herbivorous animals eat plants and convert a small part of the potential chemical energy into animal tissue and a large part into heat that is invisibly lost from the earth's surface. Carnivorous animals eat the herbivores, and they in turn are eaten by other carnivores. At each step in the transfer of energy in such a food chain, a large part of the energy is lost as heat. It is the steady flow of solar energy, year in and year out, trapped initially as chemical energy in green plants, that maintains order in the ecosystem. Without energy transfer, there could be no growth, no reproduction, no organic evolution, and no ecosystems.

At the cellular level, animals and plants use chemical energy to synthesize hundreds of different kinds of molecules, perform mechanical work, and drive organic compounds and inorganic ions across membranes against concentration gradients. All of this is achieved through the **chemical bond energy** of adenosine triphosphate (ATP), the universal energy coinage of biological systems. Bond energy is a form of potential energy, or energy of position, as opposed to kinetic energy, or the energy of motion. Potential energy becomes kinetic energy when the animal utilizes the chemical-bond energy stored in food to carry out its energy-requiring activities. Thus animals are totally dependent on plants, which fortunately use only a portion of the energy that they accumulate by photosynthesis for their own growth and maintenance. We see then that plant photosynthesis and animal respiration are inextricably woven together in a vast cycle that is driven by the constant flux of energy from the sun.

Concept of organic evolution

Organic evolution explains the diversity of modern organisms, their characteristics, and their distribution as the historic outcomes of gradual, continuous change from previously existing forms. The doctrine of organic evolution holds that all existing forms of life have descended over an enormous span of time from a single, simple, cellular, protoplasmic mass that arose spontaneously, probably in the sea. The possibility that life has arisen more than once is not excluded, however. Organic evolution refers to the change of life forms over many generations through time and is an aspect of the larger view of **inorganic evolution,** which is concerned with the development of the physical universe from unorganized matter. The principal alternative interpretation for the diversity of life is that of special creation, which holds that all life forms were divinely created much as we find them today. However, the study of biology must account for the physical evidence that life has evolved and changed over time. Biologists today consider the evidence for organic evolution to be so overwhelming that its existence cannot be denied except by abandoning reason. Since the late nineteenth century, the principal concern of scientists has not been whether evolution occurs but how it occurs.

Evolutionary theory is strongly identified with Charles Robert Darwin, one of the most important figures in the history of human culture. Darwin and Alfred Russell Wallace simultaneously presented the first credible explanation of the occurrence of evolution, the **principle of natural selection** (Principle 4, p. 10). The supporting evidence for the principle, so lucidly and so forcibly presented by Darwin in his book *On The Origin of Species* (1859), established his undisputed preeminence in evolutionary theory. Darwin's theory generated a storm of controversy between those who supported his ideas and those who opposed them. The major weakness in the theory—ignorance of the mechanisms that underlie hereditary change—was removed at the beginning of the twentieth century, when our knowledge of genetics expanded and scientists began to learn about the mechanisms of inheritance and mutations (Principles 13 to 16, pp. 12 to 13).

Natural selection is the guiding force of evolution. The principle is founded on the observed facts that no two organisms are exactly alike; that some variations, at least, are inheritable; that all groups tend to overproduce their kind; and that because more individuals are reproduced than can survive, there is a struggle for existence among them. In this struggle, according to Darwin, organisms possessing those variations that make them better adapted to the environment will survive. If, because of their survival, they have more offspring than do those with less favorable variations, their inheritable characteristics will appear in greater proportion in the next generation. Continuation of this natural selection will produce continuous evolutionary change. Natural selection, therefore, is based on **adaptations**—those characteristics possessed by an organism that enable it to survive and reproduce in its environment (Principle 5, p. 10). The idea of *fitness* is embodied in the well-known nineteenth century phrase, "survival of the fittest."

With the discovery of genes and chromosome systems at the turn of the century and the recognition that genes were self-reproducing, operative units of heredity, the mechanism of inheritable variation became understandable. Genes can undergo sudden, spontaneous, and random changes called **mutations.** Mutations are the ultimate source of all new variation. Natural selection acts to encourage the spread of some mutations while repressing the spread of others. *Mutations are the creative force of evolution; natural selection provides direction.*

The essence of natural selection is reproductive success. Because of genetic differences between individuals within a population, some have more progeny that live to reproduce than do others. This means that the "winning" genes will become common in the population, while others are selected against and become rare or absent. Evolution is a continuing process in which chance variations are constantly interacting with the environment to determine individual success. This leads to the evolution of individual adaptations and to the gradual modification of a group of organisms as a whole (phyletic evolution). The evolutionary process is all-pervasive. Every feature of life as we know it today —the vast diversity of life on earth; every adaptation; every aspect of structure, function, and behavior of every organism—has been shaped by the unrelenting pressure of natural selection. Organic evolution is the keystone for all aspects of biologic knowledge.

SOME IMPORTANT BIOLOGIC PRINCIPLES AND CONCEPTS

A principle or generalization is a statement of fact that has a wide application and can be used to formulate other principles and concepts. Like the other sciences, biology has a number of fundamental principles and assumptions, though they may not lend themselves to the mathematic exactness of the physical sciences. Well-formulated principles are indispensable to help us organize our thinking, see important relationships, and form basic conclusions from the awesome mass of factual material with which we are confronted. Principles in biology, as in other sciences, are based on observation and experimentation and have been tested by many workers over long periods of time. Inevitably, some will be subject to revision and new interpretation in the light of new knowledge.

Inevitably also, any such list compiled by other zoologists might be subject to other organization or emphasis, or even to addition or deletion of certain points, depending on the background and training of those individuals. Nonetheless, there would be a significant degree of agreement on the principles that demonstrate the essential unity of the organism and the integration of all biologic systems.

Statements of principles are necessarily condensed, and many terms used here, while defined in the glossary, will be fully explained in later chapters. *The student is not expected to memorize the principles at this point, but rather should assimilate them gradually while progressing through the book and beginning to interrelate information.* References to some of these principles have already been made, and further references to them will be made in later chapters, where they will have significant application and will be better understood. Some are so well-accepted that they form a part of the conceptual scheme of virtually all biologists. Others are of more recent development and have yet to be thoroughly integrated into the science.

The application of physical and chemical principles

1. Living systems and their constituents obey physical and chemical laws. All living systems are subject to the same physical and chemical laws as are nonliving systems. Within the cells of any organism, the living substance, or **protoplasm,** is itself comprised of a multitude of nonliving constituents: proteins, nucleic acids, fats, carbohydrates, waste metabolites, crystalline aggregates, pigments, and many others, all of which are composed of molecules and their constituent atoms. The protoplasm is alive because of the highly complex organization of these nonliving substances and the way they interact with one another, just as a watch is a timepiece only when all of its gears, springs, and bearings are organized in a particular way and interact with one another. Neither the gears of a watch nor the molecules in protoplasm can interact in any way that is contrary to universal physical laws. Consequently, the more completely we can understand the functioning of protoplasm and its constituents on the basis of chemical principles, the more completely we can understand the phenomena of life (Chapter 2).

2. All organisms capture, store, and transmit energy. Of the many physicochemical principles pertinent to the maintenance of life, the laws governing energy and its transformations (thermodynamics) are central. The first law of thermodynamics is the **law of conservation of energy.** This says that *energy can be neither created nor destroyed, though it may be transformed from one form to another.* All aspects of life require energy in some form, and the energy to support life on earth flows from the fusion reactions in our sun and reaches the earth in the form of light and heat. Sunlight is captured by green plants and transformed by the process of photosynthesis into the form of chemical bond energy. Chemical bond energy is a form of potential energy that can be released when the bond is broken; the energy is then used to perform electric, mechanical, and osmotic tasks in the cell. Energy made and stored in plants is used by animals, and these animals may be eaten by other animals.

Ultimately, all the energy transformed and stored by the plants will have again been transformed, little by little, and dissipated as heat. This is in accord with the second law of thermodynamics, which otherwise might seem to be violated by living systems. *The second law states that there is a tendency in nature to proceed toward a state of greater molecular disorder* or **entropy.** Thus, the high degree of molecular organization in protoplasm is only attained and maintained as long as energy fuels the organization. The ultimate fate of that protoplasm is degradation and dissipation of its chemical bond energy as heat, thereby approaching greater molecular disorder, or randomness (Chapters 5 and 41).

3. There is a biochemical and molecular unity of living systems. Given that living systems are subject to universal physicochemical laws and that a particular molecule or element has the same characteristics, whether it is in a bacterial or an elephant cell, it should not be surprising that living systems possess a high degree of molecular and biochemical unity. All cells require chemical constituents to serve the biochemical functions of food utilization and energy transformation and the hereditary functions of genetic replication. Despite the occurrence of more than 170 amino acids in nature, only 20 are used almost ubiquitously as building blocks of active proteins in living organisms. Deoxyribonucleic acid, the hereditary material of life, is constructed from only four particular nitrogenous bases, in addition to certain sugar phosphate units. The macromolecules used in energy transformations often show remarkable consistency in structure. The same sequences of metabolic reactions to release the stored energy in chemical bonds occur over an enormous variety of unrelated organisms, and the reactions often differ only in detail (Chapters 2 and 5).

The concept of evolution

4. Animals evolve by the mechanism of natural selection. That existing organisms have evolved by organic evolution from preexisting organisms through the mechanism of natural selection constitutes one of the great unifying principles of all biology. This concept has already been introduced and will be discussed further in Chapter 38. Evolution by means of natural selection allows us to ask two fundamental questions about any biologic phenomenon or characteristic: "What is its function?" and "How did it evolve?"

The idea of function is unique to biology among the sciences. By function we do not mean "purpose" in the sense of consciously directed objective, but the way in which a particular characteristic helps an organism or a population to survive and reproduce. For example, the function of legs or limbs in particular animals may be locomotion, food gathering, food handling, copulation, perhaps respiration, or even a combination of these. The legs thus help the animal to survive and reproduce; hence they have **adaptive value,** that is, they are an **adaptation.** By the same token, the function of a particular enzyme may be to help derive the energy from a food molecule, or an entire metabolic sequence of reactions may have such a function, and they can thus be viewed as adaptations. A living organism is a bundle of adaptations, some of which may be very widely possessed by a great variety of organisms, and some of which are narrow, or specialized, adaptations, helping the organism to survive in its particular habitat (Chapter 38).

5. All organisms are adapted to their habitat. Environmental conditions vary from one habitat to another, and these conditions constitute **selective factors** or **pressures,** which favor the survival of organisms with certain characteristics over others, that is, they favor the survival of those organisms that are better adapted to a specific environment. Thus, over the generations, organisms with adaptations suitable for a particular habitat (special adaptations) evolve because of persistent selection by environmental pressures (Chapters 7 to 27 and 38).

6. Animals that have many morphologic characters in common share a common descent. Since existing animals have evolved from preexisting ones, it is highly probable that those having many morphologic characteristics in common share a common ancestor. The more characters organisms have in common, the more closely they are related, and the closer they are to their common ancestor. The phylogenetic scheme forms the basis for modern classification of animals. A few common characters shared by two groups may have limited significance because of the possibility of **convergent evolution;** similar selective pressures may have selected for the evolution of similar adaptations in unrelated groups. However, when common characters are homologous—similar in origin—the evidence for relationship is strong (Chapters 7 to 27).

7. The embryos of animals tend to resemble the embryos of their ancestors. Studies on the embryogenesis of animals in the context of evolution resulted in the observation that embryos often go through stages resembling their ancestors, which led to the formulation of the **biogenetic law.** As originally conceived, it was believed that the embryonic stages of an animal were similar to the adult stages of its phylogenetic ancestors: the **principle of recapitulation,** or the idea that ontogeny (the life history) repeats phylogeny (the ancestral history). This viewpoint assumed that all evolutionary advancements were added to the terminal stages of the life histories of organisms. Though subsequent observation has shown that this interpretation is incorrect, there is, nevertheless, a tendency for early developmental patterns to become more or less stabilized in successive ontogenies of later descendants **(paleogenesis).** Thus, a modern restatement of the biogenetic law would be that *embryos of animals tend to go through stages resembling the embryos of their ancestors.* This tendency is not well demonstrated, or evident at all, in *all* animals for two reasons: embryos are just as subject to natural selection as are adults and so may show embryonic adaptations, and adult adaptations often begin their development in the embryonic stages. Therefore, any similarity to ancestral embryos may be completely obscured. Paleogenesis is important, nevertheless, because study of the less highly specialized or modified members of a group may provide evidence of phylogenetic relationships with other groups (Chapters 7 and 36).

8. Evolution is irreversible. An evolutionary generalization of wide applicability is sometimes known as **Dollo's law:** that with the possible exception of short intervals and in a very restricted sense, evolution is a one-way process, or is irreversible. An evolutionary path once taken cannot be retraced, though to be sure, the organisms following the path had no choice in the matter. A consequence of this fact is that the possible range of adaptations possessed by descendant groups is predetermined and limited by the array of adaptations possessed by their ancestors. Accordingly, each major group of animals usually has a basic adaptive pattern that may have determined its evolutionary divergence. For example, in the phylum Mollusca—a very large group of animals with over 100,000 living species—there is a certain array of characteristics that, if figuratively extrapolated backward, suggest a hypothetical ancestral mollusc. Most investigators of molluscs agree that the hypothetical ancestral mollusc must be very much like the actual ancestral mollusc, and all living molluscs, from the tiny snail to the giant squid, are but elaborations on the ancestral pattern. Thus, the adaptive radiation of present-day molluscs, though enormous, is strictly limited by the characteristics of their ancient ancestor (Chapters 8 to 27, 38).

Biogenesis

9. Life comes from life. Certainly a basic principle of the widest possible applicability is the concept that all organisms come from preexisting organisms, or the **principle of biogenesis.** This is in contrast to the notion that life can spring from inanimate matter, or the theory of abiogenesis or spontaneous generation—a theory that is now discredited. That living organisms could be spontaneously generated was widely accepted until the experiments of several workers, including those of Pasteur were carried out. Pasteur showed that if nutrient broth was boiled in flasks with S-shaped necks, thus sterilizing the medium, no bacterial growth subsequently occurred in the broth because the S-shaped neck prevented entry of bacteria on dust particles from the air. If the S-shaped neck was broken off, then bacterial colonies rapidly developed.

Modern biologists have recognized, however, that while the principle of biogenesis applies now, it was not always so. Life must have originated on earth from nonliving matter at least once, perhaps several times. But of course, conditions were far different in the primeval seas than they are now (Chapter 3).

10. Reproduction is a property of living systems. A corollary of the principle of biogenesis is the ability of living organisms to reproduce. Reproduction is a unique and ubiquitous property of living systems. This principle becomes self-evident when one considers the alternatives (Chapter 35).

The cell theory

11. Cells are the fundamental functional units of life. The cell theory, or **cell doctrine** as it is often called, is another of the great unifying concepts of biology. As understood currently, the doctrine states that the cell is the fundamental structural and functional unit of all living things. All animals and plants are comprised of cells and cell products. New cells come from division of preexisting cells, and the activity of a multicellular organism as a whole is the sum of the activities of its constituent cells and their interactions.

Despite great variation in cell size (though most are between 0.5 and 40 μm in diameter), shape, and function, all are basically similar in structure. It is a remarkable fact that living forms, from amebas and unicellular algae to whales and giant redwood trees, are formed from this single type of building unit (Chapter 4).

12. Cells contain differentiated and functionally interdependent structures. The cell is surrounded by a membrane, and in plants, by a cell wall secreted by the cell. Within the cell are various structures referred to as **organelles.** In plant and animal cells there are a number of organelles bounded by membranes, including a **nucleus,** a relatively large structure containing the hereditary material. The material and structures outside the nucleus are collectively called the **cytoplasm.** In addition to the membranous organelles in the cytoplasm, there are particulate structures, proteins in colloidal suspension, and smaller molecules in solution; there also may be other substances, such as lipid (fat). The cells of bacteria and blue-green algae lack membrane-bound organelles, and their nuclear area is not surrounded by a membrane (Chapter 4).

The gene theory

13. The principle of hereditary transmission is that all organisms inherit a structural and functional organization from their progenitors. This generalization involves the laws of heredity and applies to all living things. The highest and lowest types of animals have the capacity for reproducing their own kind and transmitting their characteristics to their offspring. What is inherited by an offspring is not necessarily an exact copy of the parent because heredity is not as simple as this. What is inherited is a certain type of organization that, under the influence of developmental and environmental forces, gives rise to a certain manifested result. Many potentialities may be inherited, but those that are actually expressed are much more limited (Chapters 35 and 37).

14. The gene is the fundamental unit of inheritance, and inheritance obeys Mendel's laws. The gene is a bit of coded information that confers the potential for expression of a certain characteristic on the cell or organism. The genes are borne on bodies called **chromosomes;** each chromosome bears many genes. With certain notable exceptions, the nucleus of each cell has two sets of chromosomes **(diploid condition),** with pairs of chromosomes bearing the genes for the same set of characteristics. The two genes for a particular characteristic, one on each member of a pair of homologous chromosomes, are described as allelic genes, or **alleles.** Only one of the alleles may result in the production of a detectable characteristic in the organism (the dominant allele), though both are present in each cell, and either may be passed on to the progeny.

During the production of the sex cells (**gametes: sperm** and **ova**), the number of chromosomes is halved **(haploid condition),** with each gamete receiving one of each of the homologous pairs of chromosomes. Thus, each gamete has only one of each kind of gene. This is the **law of segregation,** first formulated by Gregor Mendel in 1866.

Mendel's second law, the **law of independent assortment,** implies that the likelihood of an individual progeny inheriting a particular allele is the result of chance alone. Though each gamete has a full haploid set of chromosomes, which chromosome of each of the homologous pairs that is actually passed to a particular gamete is completely random, that is, it is independent of the other homologues passed to that gamete. Of course, all of the genes on the same chromosome tend to be inherited together and do not assort independently (Chapter 37).

15. Gene frequencies in a population can remain stable under certain conditions. When the female gamete (ovum) is fertilized by the male (sperm), a fusion of their nuclei restores the number of chromosomes to the diploid condition. Thus, the progeny receive one set of chromosomes from each parent and, consequently, have two genes for each trait. However, in the population as a whole, there may be several other alleles for the same trait, and the frequency of occurrence of a given gene can be expressed as a fraction of one (one equals the whole population, or the frequencies of all the alleles added together). It might be expected that if the frequency of a gene in a population were very low, then the allele would eventually disappear. However, it can be shown that, provided certain conditions are met, the gene frequencies of all alleles in a population remain constant. This is the **Hardy-Weinberg law.** The conditions required for the Hardy-Weinberg law to operate are an *absence* of natural selection, mutation, and genetic drift (Chapter 38).

16. Gene frequencies in a population can change. The presence of numerous alleles, or combinations of alleles, in a population provides variability in the characteristics possessed by the individuals in that population; and clearly, if some of these characteristics are helpful to survival and reproduction of those individuals, they will provide a disproportionate share of their genes to the next generation. If selective pressures are strong, then gene frequencies in a population may change markedly over time. Thus, *natural selection influences gene frequencies.*

Second, the genes themselves may change, that is, the coded information in a gene undergoes a molecular alteration so that the trait produced by that gene is different. This is a **mutation.** Mutations are the ultimate source of biologic variation. Mutations may produce genetic effects that are helpful in terms of natural selection, or they may be detrimental. They may be induced by artificial means, such as x-rays or radioactivity, or by a variety of chemical agents, but their natural causes are largely obscure. Mutations may occur in somatic (body) cells, but only mutations in germ cells, which give rise to gametes, will be inherited by the next generation. The frequencies of genes with detrimental effects may be reduced over time, or they may be constant if the rate of mutation offsets the negative selective value of its effect.

Genetic drift is a change in gene frequency by chance alone. This can occur when the population is small, and "sampling errors" can accumulate, thus changing gene frequencies in the population over time (Chapters 37 and 38).

17. DNA is the substance that contains the inherited information. We have referred to the gene as a "bit of coded information." The discovery of the nature of the code and how the code is translated into the expression of a characteristic is among the great triumphs of modern biology. It has been found that the genetic material on the chromosomes is **deoxyribonucleic acid (DNA).** This is a high–molecular weight substance composed of many units of a sugar phosphate (deoxyribose phosphate) and many units of four nitrogenous bases (adenine, guanine, thymine, and cytosine, abbreviated A, G, T, and C, respectively). In 1953 James Watson and Francis Crick suggested a **double helix model** for DNA, for which abundant evidence has subsequently been accumulated. According to the Watson-Crick model, there are two very long strands of sugar phosphate units connected together, and the strands twist around each other in a double helix form. Projecting toward each other between the two strands are the nitrogenous bases. When a precisely scaled molecular model was constructed, it was found that adenine would fit opposite thymine and cytosine would fit opposite guanine, but other combinations would not fit in the available space. This suggested a hypothesis for DNA replication: the strands could unwind, and each could serve as a template for the synthesis of the complementary strand, resulting in two double helices exactly the same as the preexisting one (Chapters 2 and 37).

18. The genetic code is in the linear of bases on the DNA strand. The Watson-Crick model implied that the genetic code would lie in the linear order of the nitrogenous bases: A, G, T, and C. It was noted previously that proteins are constructed of 20 different amino acid building blocks. Assuming that the order of bases in DNA represented a code for the order of amino acids in a protein, it was clear that some combination of bases must code for one protein, since there are only four bases but 20 amino acids. The minimum number of bases to provide a sufficient number of different combinations to specify each of the amino acids is three, as suggested by Francis Crick in 1961. Subsequent research has substantiated Crick's hypothesis. The genetic code is based on triplets of nitrogenous bases along the DNA molecule. Each triplet, called a **codon,** codes for a single amino acid, and the codons do not overlap. The genetic code is apparently universal, that is, the same codon or sequence of three nitrogenous bases codes for the same amino acid in all organisms (Chapter 37).

19. Transcription and translation of genetic information are mediated by RNA. Though the transmission of genetic information and the nature of the genetic code now seem to us at once ingenious and astonishingly simple, they fall far short of explaining how the cell reads the code and translates it into cell products and structures. Nevertheless, much progress has been made in elucidating these critical processes. They depend on molecules of **ribonucleic acid (RNA),** a substance somewhat similar to DNA except that ribose is present in its strand instead of deoxyribose, and uracil is present instead of thymine. There are three principal types of RNA: **messenger-RNA (mRNA), transfer-RNA (tRNA),** and **ribosomal-RNA (rRNA).** Molecules of mRNA are synthesized alongside strands of DNA in the chromosome, using the DNA as a template and thus transcribing the code from the DNA to the mRNA. The mRNA moves out of the nucleus into the cytoplasm where it interacts with submicroscopic particles called **ribosomes.** Ribosomes are constructed of about 50% protein and 50% rRNA. The molecules of tRNA provide the translation function. The various types bind with specific amino acids. At a particular position in a specific tRNA molecule there is a triplet sequence of nitrogenous bases, the **anticodon,** which is complementary to the codon sequence in the mRNA. Ribosomes move along the strand of mRNA, positioning the molecules of tRNA according to the codon-anticodon specificity, and the amino acid attached to the tRNA is reattached to the chain of amino acids being assembled into a protein. The specificity or particular kind of protein is determined by the sequence of amino acids in it, and this depends on the sequence of codons in DNA as transcribed and translated by the RNA system (Chapter 37).

20. The "one gene–one enzyme" principle is that a gene corresponds to the information to synthesize an enzyme. A gene, then, may be redefined as a series of codons of DNA that result in the production of a particular protein. The physical and chemical properties of the protein depend on the ordinal sequence of its amino acids. The properties of many proteins allow them to function as **enzymes.** Enzymes are catalysts for the myriad chemical reactions that go on in a cell; that is, they facilitate reactions by participating in them but are not "used up" in the reactions—they are not substrates or products of the reactions. The hundreds of chemical reactions, collectively called **metabolism** or **metabolic reactions,** are each catalyzed by specific enzymes. These reactions would not occur, or would occur very slowly, if they were not mediated by enzymes. The reactions to derive energy from food molecules and synthesize cell structures and products, even the synthesis of DNA itself, require the participation of enzymes. In 1941 G. W. Beadle and E. L. Tatum suggested the one gene—one enzyme hypothesis—that a gene would constitute the genetic information necessary to synthesize one enzyme. The validity of the hypothesis is now widely accepted in a slightly modified form.

Since it is now known that a particular protein may be made of several chains of amino acids (polypeptides), the Beadle and Tatum hypothesis is expressed as **one gene–one polypeptide.** Thus a mutation is a change in the order of the nitrogenous bases in the DNA, which codes for the order of amino acids in an enzyme, producing an enzyme with different properties. Mutations often involve a change in several nitrogenous bases in the gene but in some cases may result from the alteration of a single pair. When the properties of an enzyme are changed, the metabolic reaction it catalyzes is altered; therefore, *metabolic reactions are under genic control* (Chapters 5 and 37).

Growth, life cycle, and differentiation

21. Growth is a fundamental characteristic of life. Its manifestation as the synthesis of protoplasm in every living organism has been referred to as the **growth law.** True growth is characterized by an increase in tissue or protoplasmic mass with cell division, or cell enlargement, or both. It thus depends on the incorporation of new materials from the environment and is subject to the availability of those materials (Chapters 6 and 36).

22. All organisms have a characteristic life cycle that they pass through as they grow. In the case of unicellular organisms, this life cycle may be as simple as growth and cell division; but the interspersion of resting stages, sexual reproduction, cyclic changes in body form, and other such events are common. In multicellular organisms, development begins with a single cell, usually a fertilized egg, or **zygote.** This initially undergoes division **(cleavage)** into smaller and smaller cells with no increase in total mass and no true growth, but the preliminary cell division increases the total surface area of cell membranes, which facilitates the incorporation of nutrients required for actual growth. True developmental growth is accompanied by shaping of the embryo, called **morphogenesis,** and a progressive specialization of the cells, tissues, and structures into the organization of the whole organism, called **differentiation.** Thus begins the life cycle of the organism, which may be quite complex, including embryonic and juvenile development, adolescence, adult equilibrium, and senescence (Chapters 5, 6, 35, 36).

23. All organisms develop a characteristic body plan. The patterns of growth are genetically directed and are usually constant for each stage of the life cycle from generation to generation. The inherited body plan may be described in terms of broadly inclusive characteristics, such as presence and type of internal body cavities, symmetry, type of nervous system, and others—all of which would be possessed by organisms with a common ancestor—or the plan may be described in terms of narrow characteristics possessed only by individuals of a certain species or population (Chapter 6).

24. Nuclear equivalence and cell differentiation are characteristics of cells in multicellular organisms. Since the body plan and all organization of the body are inherited, it is clear that *all* genetic information (the **genome**) for the attainment of that body plan must be present in the zygote, which replicates its DNA and confers that information on all its progeny cells. Thus, the cell nuclei of a multicellular organism, with few exceptions, are genetically alike. This is the **principle of nuclear equivalence.** However, it is also clear that *differentiated* cells are not all alike. Therefore, only some of the information is transcribed and translated in a given cell. Cell differentiation results from the differential activity of the same set of genes in different cells. In the fully differentiated cell, most of the genome is fully repressed and silent. A muscle cell, for example, never translates the parts of its DNA instructions that would produce proteins unique to a nerve cell. During embryonic development, nuclei in different regions of the embryo provide different genetic information, resulting in regional differentiation of cells and tissue. The control of differentiation remains one of the great unsolved problems of biology. Whatever the molecular mechanisms, there is abundant evidence that differentiation is one of the responses to conditions or substances in the environment of a cell (Chapter 37).

Responses and relations to the environment of a cell or organism

25. Irritability is a characteristic of life. One of the fundamental characteristics of life is irritability, that is, the ability to react to an environmental stimulus. This is perhaps the most widely occurring of all adaptations. Broadly speaking, it can be interpreted as the ability to react to a stimulus in such a way as to promote the continued life of a cell or organism. The reaction may be to consume a food molecule or particle when such is present, to seek or avoid light, to pursue a prey organism, to avoid a noxious substance, and many others. When such reactions can no longer be accomplished, the organism perishes. Throughout the life of an animal, matter and energy pass through the body, providing perturbations of the internal physiologic state. Many physiologic and metabolic mechanisms exist that function to compensate for such disturbances and to maintain conditions that are compatible with continued life within the organism. This tendency toward internal stabilization was recognized by Claude Bernard in the nineteenth century and was developed in the twentieth century by Walter B. Cannon, who called it the **principle of homeostasis.** Homeostasis is accomplished at the cell, organ, and system levels by material and energy transport and in many instances is controlled by feedback mechanisms (Chapters 28 to 34).

26. Cells in a multicellular animal communicate with and affect each other. Reaction to stimuli and maintenance of homeostasis require mechanisms for communication between cells and organs in multicellular animals. Increasingly complex levels of organization in animals require correspondingly increasing complexities in communication between cells and organs. Communication between parts of an animal is by two primary means: **neural** and **hormonal.** Relatively rapid communication is by neural mechanisms and involves propagated electrochemical changes in cell membranes. Highly organized and complex nervous systems carry out this function in higher animals. Relatively less rapid or long-term adjustments in an animal are by hormonal or endocrine mechanisms. Hormones are substances produced by cells in one part of the body that regulate cell processes elsewhere in the organism when carried there by body fluids. Sometimes the hormones are produced by nerve cells, or cells derived from nerve cells, and in those cases the system is described as **neuroendocrine.** Many instances are now known in which substances produced by some individual animals reach other individuals in the population, there to function as hormones in the other individuals, producing adjustments in their physiologic processes or behavior. Such substances are called **pheromones** (Chapters 32 to 34).

27. All animals interact with their environment. Whether able to react to a given environmental stimulus or not, all organisms in a given area interact with both the biologic and physical factors in their environment. The earth's biomass is organized into a hierarchy of interacting units: the individual organism, the population, the community, and the ecosystem. The concepts of the community and food chains (producers; primary and secondary consumers; decomposers) have been mentioned already. Other important generalizations have developed in ecology. Among these is the fact that all organisms have **habitats** and **niches.** The habitat is the spatial location where an animal lives.

It is always physically circumscribed, though the space may be very large, such as an entire ocean, or it may be tiny, such as the intestine of an insect. What the organism does—its role—in its habitat is its niche. The niche of an animal must be described in terms of the effects it has on other organisms in its habitat, and vice versa, as well as the effects on and by nonliving resources in the habitat. The possible effects on other animals include an array, such as whether the organism interacts with them as a predator, a prey, a mutual, a parasite, a competitor, or others. Insofar as individuals or populations compete with each other for food, space, or other resources in a habitat, their niches overlap. If the competition is too strong, one of the populations will perish, be driven out of the habitat, or be forced to utilize other resources—to occupy a different niche. In fact, it has been found that no two species can occupy the same ecologic niche **(principle of competitive exclusion; Gause's rule)** (Chapters 38 and 41).

The unity of biologic science

At this point the strong interrelationships of these biologic principles should be evident. All the cellular constituents, like the cells themselves, are governed by natural laws. Living organisms can come only from other living organisms, just as new cells can be produced only from preexisting cells. The nucleus of the cell carries the hereditary material, which confers a certain set of characteristics on its cellular progeny, which themselves may become new organisms on which natural selection may act. Living organisms have mechanisms that tend to compensate through internal and external changes to keep conditions within limits compatible with continued life.

CHAPTER 2

MATTER AND LIFE

A DNA molecule is an extremely long helical chain, and its molecular architecture is reflected in its gross structure. This photograph shows the final step in a common DNA isolation scheme, in which many long, viscous strands of concentrated and purified DNA are being removed from an ice-cold alcohol suspension by winding them on a glass rod.

Courtesy Ted Lane.

BASIC STRUCTURE OF MATTER

Nature of atoms and molecules

Although the ancient Greeks had certain conceptions of the composition of matter, such as the universe being composed of the four elements—fire, air, earth, and water—our present concepts of the nature of matter have originated within the past two centuries. A. L. Lavoisier, the great French scientist of the eighteenth century, compiled the first list of elements, a total of 28, and explained the precise nature of respiration (1778). J. Dalton, the English chemist, conceived that matter was composed of atoms, which combine in definite proportions to form chemical compounds (1808). The Italian investigator A. Avogadro showed how many atoms of each kind make up each compounded atom and gave us the concept of the molecule (1811). By 1869 no fewer than 92 kinds of atoms had been ascertained, and D. I. Mendeléeff worked out his periodic table of the elements. He arranged the elements into 8 groups on the basis that a relation of the chemical elements can be expressed by their properties, which are periodic functions of their atomic weights. If the elements are in a group of elements that have similar properties and relations, they follow a regular progression in the individual differences of their members.

For many years the atom was considered solid and indivisible. In 1911 Lord Rutherford showed that every atom consists of a positively charged nucleus surrounded by a negatively charged planetary system of electrons (Fig. 2-1). An electron is a negatively charged particle with almost no mass. The nucleus, containing most of the atom's mass, is made up of two kinds of particles, **protons** and **neutrons.** These two particles have about the same mass, each being about 2,000 times heavier than an electron. The protons bear positive charges, and the neutrons are uncharged (neutral).

The positively charged protons, which would normally repel each other because they bear similar charges, are held together by powerful short-range nuclear forces. When an atomic nucleus is torn apart in a fission reaction, such as during the explosion of an atomic bomb, enormous amounts of energy are released. This energy stabilizes the structure of atoms but it plays no part in the physical and chemical interactions that occur in the living body.

Although there is the same number of protons in the nucleus as there are electrons revolving around the nucleus, the number of neutrons may vary. For every positively charged proton in the nucleus, there is a negatively charged electron. The total charge of the atom is thus neutral.

The number of protons in the nucleus is the **atomic number** of the atom. Thus hydrogen, helium, and lithium, containing respectively 1, 2, and 3 protons in their nuclei, have atomic numbers of 1, 2, and 3, respectively. Since atoms are electrically neutral, the atomic number is also equal to the number of electrons revolving around the nucleus.

The **mass number** of an atom is the total number of protons and neutrons in its nucleus. The nucleus of an oxygen atom contains 8 protons and 8 neutrons. It therefore has a mass number of 16. The heaviest natural element, uranium, has a nucleus of 92 protons (its atomic number is thus 92) and 146 neutrons, and so its mass number is 238. The elements are designated by convenient symbols that show the atomic number as a subscript, and the total number of protons and neutrons (mass number) as a superscript. Thus oxygen is designated ${}_8O^{16}$, hydrogen ${}_1H^1$ (1 proton, no neutrons), helium ${}_2He^4$ (2 protons, 2 neutrons), and so on (Fig. 2-2).

The **atomic weight** of an atom is nearly the same as its mass number. However, a quick examination of a periodic table shows that none of the elements has an atomic weight of an exact integer. How are atomic weights derived? Obviously an atom weighs far too little to serve as a useful index for weight comparison. A hydrogen atom, for example, weighs 1.67×10^{-24} g. Consequently, physicists have assigned a set of

FIG. 2-1 Structure of carbon atom. A planetary system of six negatively charged electrons revolves around a dense nucleus of six positive protons and six uncharged neutrons.

meaningful relative weights to the elements, using carbon as a base for comparison. Carbon, $_6C^{12}$, with 6 protons and 6 neutrons was assigned the integral value 12. By this scale, protons and neutrons have masses of 1.0073 and 1.0087, respectively—masses close to, but not exactly, 1. (Electrons have an almost negligible mass of 0.00055.) Thus no element, except carbon 12, has an atomic weight of an exact integer: hydrogen weighs 1.008; helium, 4.0026; lithium, 6.939; and so on.

The **gram atomic weight** of an element is its atomic weight expressed in grams. The number of atoms in a gram atomic weight is 6.02×10^{23} (Avogadro's number). This means that the gram atomic weight of all elements has the same number of atoms. One gram of hydrogen (its gram atomic weight) has the same number of atoms as does 16 g of oxygen or 12 g of carbon, their gram atomic weights.

The relative weight of a molecule, composed of 2 or more atoms, is known as the **gram molecular weight.** It is the sum of the gram atomic weights of all the atoms in a molecule. One gram molecular weight of carbon dioxide (CO_2) is 44 g (12 + 16 + 16); of water (H_2O) is 18 g; and of oxygen (O_2), 32 g. As with gram atomic weight, defined previously, the gram molecular weight of any compound contains the same number of molecules (6.02×10^{23}) as the gram molecular weight of any other compound.

Still another expression of quantity of a substance is the **mole.** This unit combines the molecular weight and the number of molecules present. The number of moles of a compound present in a particular instance is equal to the weight of the compound present in that instance divided by its molecular weight:

$$\text{moles} = \frac{\text{weight in grams}}{\text{molecular weight}}$$

For example, 18 g of water, having a molecular weight of 18, is 1 mole of water; 36 g of water is 2 moles. Since most compounds in living systems are present in much less than 1-mole quantities, the **millimole** (mmole, 0.001 mole) and the **micromole** (μmole, 0.000001 mole) are more commonly used in biology to express molecular quantities than is the mole.

Isotopes. It is possible for 2 atoms of the same element to have the same number of protons in their nuclei, but a different number of neutrons. For example hydrogen exists in nature primarily as $_1H^1$, that is, it contains 1 proton, but no neutron. However, there are also trace amounts of two other forms of hydrogen: $_1H^2$, which has 1 proton and 1 neutron and is called deuterium, and $_1H^3$, which has 1 proton and 2 neutrons and is called tritium. These three varieties differ only in the number of neutrons in the nucleus (Fig. 2-3). Such forms of an element, having the same charge but different atomic weights, are called **isotopes.**

Radioactive isotopes. Although most of the naturally occurring elements are stable, all elements have at least one radioactive isotope. These isotopes undergo spontaneous disintegration, with the emission of one or more of three types of particles, or rays—**gamma rays** (a form of electromagnetic radiation), **beta rays** (elec-

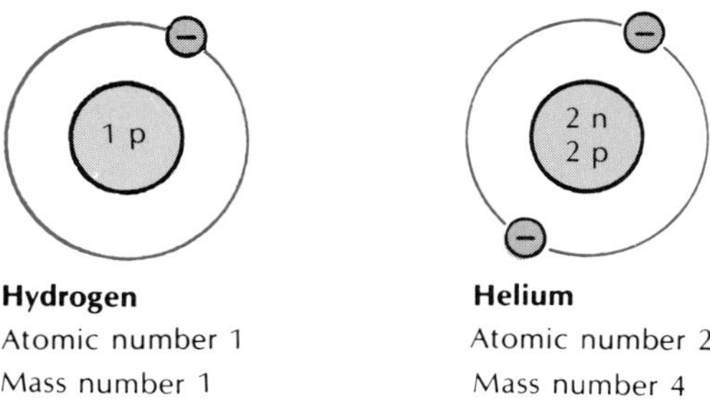

FIG. 2-2 Two lightest atoms. Since first shell closest to atomic nucleus can hold only 2 electrons, helium shell is closed so that helium is chemically inactive.

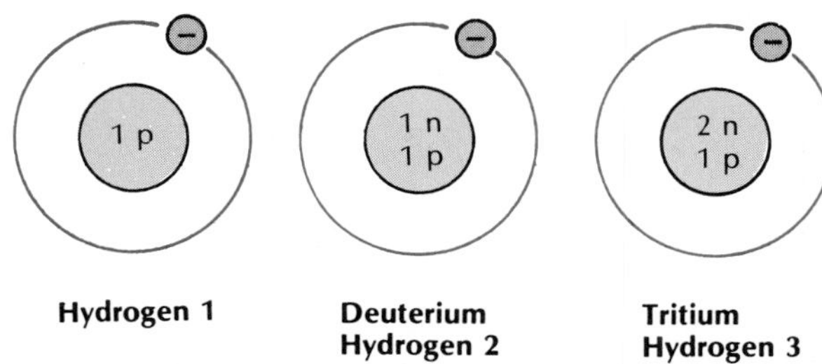

FIG. 2-3 Three isotopes of hydrogen. Of the three isotopes, hydrogen 1 makes up about 99.98% of all hydrogen, and deuterium (heavy hydrogen) makes up about 0.02%. Tritium is radioactive and is found only in traces in water. Numbers indicate approximate atomic weights. Most elements are mixtures of isotopes. Some elements (for example, tin) have as many as ten isotopes.

trons), and **alpha rays** (positively charged helium nuclei stripped of their electrons). Most of the isotopes of greatest use in biologic tracer studies are beta and gamma emitters. Virtually all are prepared synthetically in nuclear reactors and cyclotrons. Among the commonly used radioisotopes are carbon 14 ($_6C^{14}$), tritium ($_1H^3$), and phosphorus 32 ($_{15}P^{32}$). Using radioisotopes, biologists are able to trace movements of elements and tagged compounds through organisms. Our present understanding of metabolic pathways in animals and plants is in very large part the result of this powerful analytic tool. Radioisotopes are also used to great advantage in the diagnosis of disease in humans, such as cancer of the thyroid gland.

Electron "shells" of atoms

According to Niels Bohr's planetary model of the atom, the electrons revolve around the nucleus of an atom in precise orbits, or shells. This simplified picture of the atom has been greatly modified by recent experimental evidence.

According to the quantum theory, the electrons surrounding the nucleus exist at discrete energy levels, called quantum levels. This theory replaces the older idea that electrons revolve around the nucleus in definite shells, or orbit patterns. A quantum level represents a discrete energy value. These energies, and hence the energies of electrons in these quantum levels, increase as the distance from the nucleus increases. Electrons tend to move as close to the nucleus as possible. However, there is a physical maximum to the number of electrons that can occupy each quantum level. Thus as the inner level becomes filled, additional electrons are forced into more distant quantum levels. These outer electrons are more excited and have a higher energy content.

Although the picture of the atom described by the quantum theory provides a much better basis for understanding the atom, the old planetary model is still useful in interpreting chemical phenomena. The number of concentric "shells," or the paths of the electrons in their orbits, varies with the element. Each shell can hold a maximum number of electrons. The first shell next to the atomic nucleus can hold a maximum of 2 electrons (hydrogen has only 1), and the second shell can hold 8; other shells also have a maximum number, but no atom can have more than 8 electrons in its outermost shell. Inner shells are filled first, and if there are not enough electrons to fill all the shells, the outer shell is left incomplete. Hydrogen has 1 proton in its nucleus and 1 electron in its single orbit but no neutron. Since its shell can hold 2 electrons, it has an incomplete shell. Helium has 2 electrons in its single shell, and its nucleus is made up of 2 protons and 2 neutrons. Since the 2-electron arrangement in helium's shell is the maximum number for this shell, the shell is closed and precludes all chemical activity. There is no known compound of helium. Neon is another inert gas (chemically inactive) because its outer shell contains 8 electrons, the maximum number (Fig. 2-4). However, stable compounds of xenon (an inert gas) with fluorine and oxygen are formed under special conditions. Oxygen has an atomic number of 8. Its 8 electrons are arranged with 2 in the first shell and 6 in the second

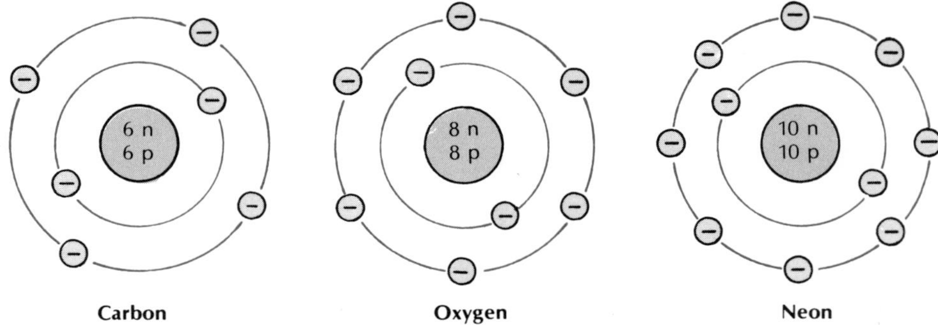

FIG. 2-4 Electron shells of three common atoms. Since no atom can have more than 8 electrons in its outermost shell and 2 electrons in its innermost shell, neon is chemically inactive. However, the second shells of carbon and oxygen, with 4 and 6 electrons, respectively, are open so that these elements are electronically unstable and react chemically whenever appropriate atoms come into contact. Chemical properties of atoms are determined by their outermost electron shells.

shell (Fig. 2-4). It is active chemically, forming compounds with almost all the elements except the inert gases.

Molecules and chemical bonds

A molecule is a combination of 2 or more atoms joined by chemical bonds. A molecule may be composed of just one element, such as molecules of oxygen (O_2) and of hydrogen (H_2), or of different elements, such as carbon dioxide (CO_2) and methane (CH_4). The chemical bond joining 2 atoms in a molecule contains stored potential energy that is released when the bond is broken and is reestablished between atoms during chemical reactions. All chemical bonds involve the sharing of electrons by 2 atoms.

Ions and oxidation states. Elements react in such a way as to gain a stable configuration of electrons in their outer shells. The number of electrons in the outer shell varies from 0 to 8. With either 0 or 8 in this shell, the element is chemically inactive. When there are fewer than 8 electrons in the outer shell, the atom will tend to lose or gain electrons to have an outer shell of 8, which will result in a charged ion. Atoms with 1 to 3 electrons in the outer shell tend to lose them to other atoms and to become positively charged ions because of the excess protons in the nucleus. Atoms with 5 to 7 electrons in the outer orbit tend to gain electrons from other atoms and to become negatively charged ions because of excess electrons over the protons. Positive and negative ions tend to unite.

Every atom has a tendency to complete its outer shell to increase its stability. Let us examine how 2 atoms with incomplete outer shells, sodium and chloride, can interact to fill their outer shells. Sodium, with 11 electrons, has 2 electrons in its first shell, 8 in its second shell, and only 1 in the third shell. The third shell is highly incomplete; if this third-shell electron were lost, the second shell would be the outermost shell and would produce a stable atom. Chlorine, with 17 electrons, has 2 in the first shell, 8 in the second, and 7 in the incomplete third shell. Chlorine must gain an electron to fill the outer shell and become a stable atom. Clearly, the transfer of the third-shell sodium electron to the incomplete chlorine third shell would yield simultaneous stability to both atoms.

Sodium, now with 11 protons but only 10 electrons, becomes electropositive (Na^+). In gaining an electron from sodium, chlorine contains 18 electrons but only 17 protons, and thus becomes an electronegative chloride ion (Cl^-). Since unlike charges attract, a chemical bond is formed, called an **ionic bond** (Fig. 2-5). The ionic compound formed, sodium chloride, can be represented in electron dot notation (''fly-speck formulas'') as:

$$Na\cdot + \cdot\ddot{\underset{\cdot\cdot}{Cl}}: \rightarrow Na^+ + (:\ddot{\underset{\cdot\cdot}{Cl}}:)^-$$

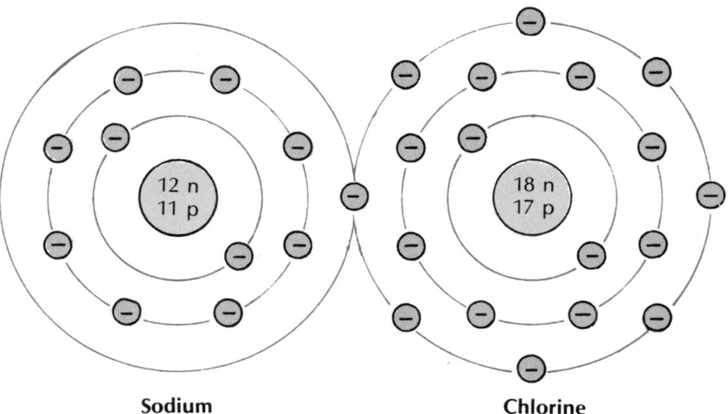

FIG. 2-5 Ionic bond. When 1 atom of sodium and 1 of chlorine react to form a molecule, a single electron in the outer shell of sodium is transferred to the outer shell of chlorine. This causes the outer or second shell (third shell is empty) of sodium to have 8 electrons and also chlorine to have 8 electrons in its outer or third shell. The compound thus formed is called sodium chloride (NaCl). By losing 1 electron, sodium becomes a positive ion, and by gaining 1 electron, chlorine (chloride) becomes a negative ion. This ionic bond is held together by a strong electrostatic force.

Processes that involve the **loss of electrons** are **oxidation** reactions; those that involve the **gain of electrons** are called **reduction** reactions. Since oxidation and reduction always occur simultaneously, each of these processes is really a "half-reaction." The entire reaction is called an **oxidation-reduction** reaction, or simply **redox** reactions. The terminology is confusing because oxidation-reduction reactions involve electron transfers, rather than (necessarily) any reaction with oxygen. However, it is easier to learn the system than to try to change accepted usage.

We now need to introduce the concept of **oxidation number.** This term refers to the charge an atom would have if the bonding electrons were arbitrarily assigned to the more electronegative of two interacting elements. For example, in the sodium chloride reaction, we consider that the bonding electron has been transferred to the chlorine atom. Consequently sodium, having lost its electron, becomes electropositive and is said to have an oxidation number of +1. Chlorine, with its newly acquired electron, becomes electronegative, and takes an oxidation number of −1. Some elements always exist in compounds in the same oxidation number. For example, oxygen almost always has an oxidation number of −2; sodium, +1; magnesium, +2; potassium, +1; and calcium, +2. However, most metals have two or more oxidation numbers. For example, iron may exist as +2 or +3, chromium as +3 or +6, manganese as +2, +4, or +7. Other elements, such as hydrogen, can take either positive or negative oxidation numbers.

Covalent bonds. Stability can also be achieved when 2 atoms **share** electrons. Let us again consider the chlorine atom, which, as we have seen, has an incomplete 7-electron outer shell. Stability is attained by gaining an electron. One way this can be done is for 2 chlorine atoms to share one pair of electrons (Fig. 2-6). To do this, the 2 chlorine atoms must **overlap** their third shells, so that the electrons in these shells can now spread themselves over both orbits. Many other elements can form covalent (or electron-pair) bonds. For example: hydrogen (H_2)

$$\mathrm{H\cdot} + \mathrm{H\cdot} \rightarrow \mathrm{H{:}H}$$

and oxygen (O_2)

$$\mathrm{\ddot{\underset{..}{O}}{:}} + \mathrm{{:}\ddot{\underset{..}{O}}} \rightarrow \mathrm{\ddot{\underset{..}{O}}{::}\ddot{\underset{..}{O}}}$$

In this case, oxygen with an oxidation number of −2 must share two pairs of electrons to achieve stability. Each atom now has 8 electrons available to its outer shell, the stable number.

Covalent bonds are of great significance to living systems, since the major elements of protoplasm (carbon, oxygen, nitrogen, hydrogen) almost always share electrons. Carbon, which usually has an oxidation number of either +4 or −4 (its outer shell contains 4 electrons), can share its electrons with hydrogen to form methane:

$$\mathrm{\cdot\dot{\underset{\cdot}{C}}\cdot} + 4\,\mathrm{H\cdot} \rightarrow \begin{array}{c}\mathrm{H}\\ \mathrm{H{:}\ddot{\underset{..}{C}}{:}H}\\ \mathrm{H}\end{array}$$

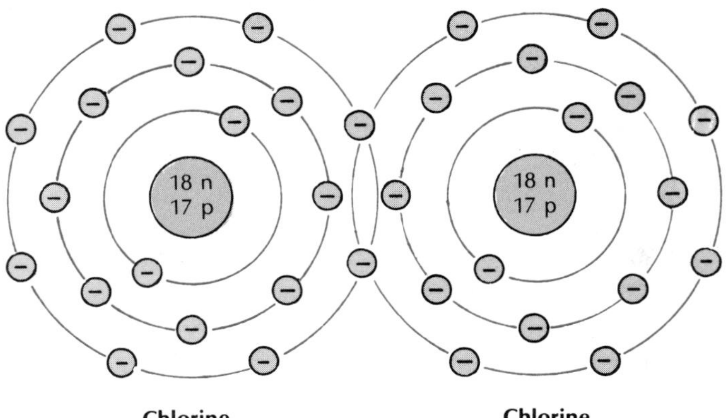

FIG. 2-6 Covalent bond. Each chlorine atom has 7 electrons in its outer shell, and by sharing one pair of electrons, each atom acquires a complete outer shell of 8 electrons, thus forming a molecule of chlorine (Cl_2). Such a reaction is called a molecular reaction, and such bonds are called covalent bonds.

Carbon now achieves stability with 8 electrons, and each hydrogen atom becomes stable with 2 electrons. Carbon also forms covalent bonds with oxygen:

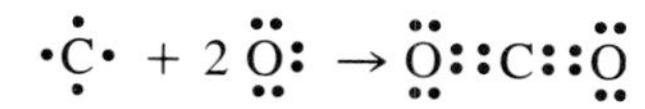

$$\cdot\dot{\underset{\cdot}{C}}\cdot + 2\,\ddot{\underset{\cdot\cdot}{O}}\!: \rightarrow \ddot{\underset{\cdot\cdot}{O}}\!::\!C\!::\!\ddot{\underset{\cdot\cdot}{O}}$$

Carbon can also bond with itself (and hydrogen) to form, for example, ethane:

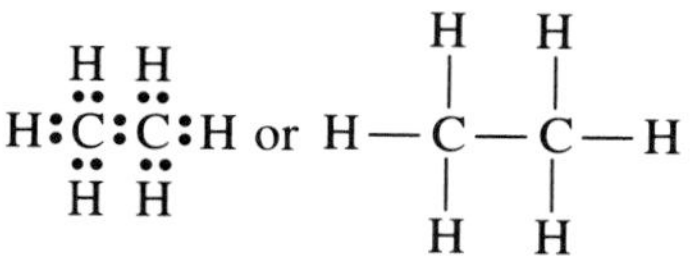

$$\begin{matrix} & H & H & \\ H: & \ddot{\underset{\cdot\cdot}{C}} : & \ddot{\underset{\cdot\cdot}{C}} : & H \\ & H & H & \end{matrix} \quad \text{or} \quad \begin{matrix} & H & & H & \\ & | & & | & \\ H- & C & - & C & -H \\ & | & & | & \\ & H & & H & \end{matrix}$$

Carbon can join in "double bond" configuration; for example, ethylene:

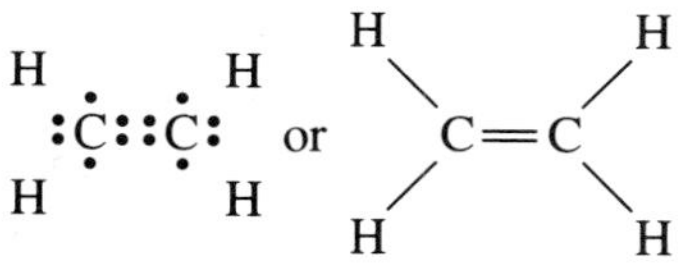

$$\begin{matrix} H & & H \\ & :\dot{\underset{\cdot}{C}}::\dot{\underset{\cdot}{C}}: & \\ H & & H \end{matrix} \quad \text{or} \quad \begin{matrix} H & & & & H \\ & \diagdown & & \diagup & \\ & & C{=}C & & \\ & \diagup & & \diagdown & \\ H & & & & H \end{matrix}$$

Or even in "triple bond" configuration; for example, acetylene:

$$H:C:::C:H \text{ or } H-C{\equiv}C-H$$

These examples only begin to illustrate adequately the amazing versatility of carbon. It is a part of virtually all compounds comprising living substance, and without carbon life as we know it would not exist.

Acids, bases, and salts

The hydrogen ion (H^+) is one of the most important ions in living organisms. The hydrogen atom contains a single electron. When this is completely transferred to another atom (not just shared with another atom as in the covalent bonds with carbon), only the hydrogen nucleus with its positive proton remains. Any molecule that dissociates in solution and gives rise to a hydrogen ion, also called simply a proton, is an **acid.** An acid is classified as strong or weak, depending on the extent to which the acid molecule is dissociated in solution. Those that dissociate completely in water (H_2SO_4, HNO_3, and HCl) are called strong acids. Weak acids such as acetic acid (CH_3COOH) dissociate only weakly. A solution of acetic acid is mostly undissociated acetic acid molecules with only a small number of acetate (CH_3COO^-) and hydrogen ions (H^+) present.

A **base** contains negative ions called hydroxyl ions and may be defined as a molecule or ion that will accept a proton. Bases are produced when compounds containing them are dissolved in water. NaOH (sodium hydroxide) is a strong base because it will dissociate completely in water into sodium (Na^+) and hydroxyl (OH^-) ions. Among the characteristics of bases is the ability to combine with hydrogen ions, thus decreasing their concentration. Like acids, bases vary in the extent to which they dissociate in aqueous solutions into hydroxyl ions.

A **salt** is a compound resulting from the chemical interaction of an acid and a base. Common salt, sodium chloride (NaCl), is formed by the interaction of hydrochloric acid (HCl) and sodium hydroxide (NaOH). In water the HCl is dissociated into H^+ and Cl^- ions. The hydrogen and hydroxyl ions combine to form water (H_2O), and the sodium and chloride ions combine to form salt (NaCl):

$$\underset{\textbf{Acid}}{H^+Cl^-} + \underset{\textbf{Base}}{Na^+OH^-} \rightarrow \underset{\textbf{Salt}}{Na^+Cl^-} + H_2O$$

Organic acids are usually characterized by having in their molecule the carboxyl group (—COOH). They are weak acids because a relatively small proportion of the H^+ reversibly dissociates from the carboxyl:

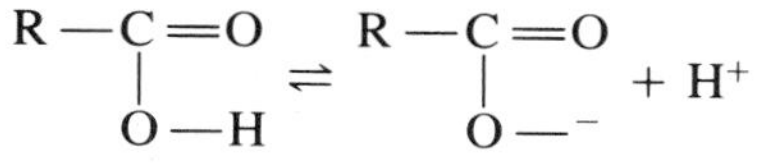

$$\begin{matrix} R-C{=}O \\ | \\ O-H \end{matrix} \rightleftharpoons \begin{matrix} R-C{=}O \\ | \\ O-^{-} \end{matrix} + H^+$$

R refers to an atomic grouping unique to the molecule. In water, the COO— group will behave as a weak acid. Some common organic acids are acetic, citric, formic, lactic, and oxalic. Many more of these will be encountered later in discussions of cellular metabolism.

Hydrogen ion concentration (pH)

Solutions are classified as acid, base, or neutral, according to the proportion of hydrogen (H^+) and hydroxyl (OH^-) ions they possess. In acid solutions there is an excess of hydrogen ions; in alkaline, or basic, solutions the hydroxyl ion is more common; and in neutral solutions both hydrogen and hydroxyl ions are present in equal numbers.

To express the acidity or alkalinity of a substance, a logarithmic scale, a type of mathematical shorthand, is employed that uses the numbers 1 to 14. This is the pH scale, and the numbers refer to the negative logarithm of the hydrogen ion concentration. Thus H^+ concentration *increases* with *decreasing* pH numbers (*negative* log). Numbers below 7 indicate an acid range. The number 7 indicates neutrality, that is, the presence of equal numbers of H^+ and OH^- ions. According to this logarithmic scale, a pH of 3 is ten times more acid than one of 4; a pH of 9 is ten times more alkaline than one of 8.

In protoplasmic systems pH plays an important role, for, in general, deviations from the normal usually result in severe damage. Most substances and fluids in the body are close to the point of neutrality, that is, pH of around 7. Blood, for instance, has a pH of about 7.35 in most terrestrial vertebrates, or just slightly on the alkaline side. Lymph is slightly more alkaline than blood. Saliva has a pH of 6.8, on the acid side. Gastric juice is the most acid substance in the body, about pH 1.6. The regulation of the pH in the body tissue fluids involves many important physiologic mechanisms; one of the most important is the buffer action of certain salts.

Buffer action

The hydrogen ion concentration in the extracellular fluids must be regulated so that metabolic reactions within the cell will not be adversely affected by a constantly changing hydrogen ion concentration. A change of only 0.2 pH unit from the normal blood pH of 7.35 can cause serious metabolic disturbances. To maintain pH within physiologic limits, there are certain substances in cells and organisms that tend to compensate for any change in the pH when acids or alkalies are produced in metabolic reactions or are added to the body fluids. These are called **buffers.** The hydrogen ion concentration within the cells is probably greater (pH is lower) than the hydrogen ion concentration in the extracellular fluids because of the metabolic production of CO_2, which reacts with the cellular water to form carbonic acid (H_2CO_3). Certain phosphates, sulfates, and organic acid radicals also add to the acidic nature of the intracellular fluids. Within the cells the high content of protein serves as a buffer and thus tends to keep the pH from going too low.

The buffer function of blood is dependent on both plasma and red blood corpuscle buffer mechanisms. The chief buffer of plasma and tissue fluid is sodium bicarbonate ($NaHCO_3$). This salt dissociates into sodium ions (Na^+) and bicarbonate ions (HCO_3^-). When a strong acid (for example, HCl) is added to the fluid, the H^+ ions of the dissociated acid will react with the bicarbonate ion (HCO_3^-) to form a very weak acid, carbonic acid, which dissociates only slightly. Thus the H^+ ions from the HCl are removed and the pH is little altered.

Mixtures and their properties

Whenever masses of different kinds are thrown together, we have what is called a mixture. All the different states of matter (solids, liquids, gases) may be involved in these mixtures. The mixtures we are mainly interested in here are those in which water or other fluid is one of the states of matter. When something is mixed with a liquid, any one of three kinds of mixtures is formed.

Molecular solutions. If crystals of salts or sugars are added to water, the molecules (or ions in the case of salts) are uniformly dispersed through the water, forming a **true solution.** Such solutions are transparent. In such a case the water is the **solvent** and the dissolved salt or sugar the **solute.** The freezing point of solutions is lower and the boiling point is higher than those of pure water.

Suspensions. If solids added to water remain in masses larger than molecules, the mixture is a suspension. Muddy water is a good example. When allowed to stand, the particles in suspension will settle out to the bottom. Suspensions have a turbid appearance and have the same boiling and freezing points as pure water.

Colloidal solutions. Whenever the dispersed particles are intermediate in size between the molecular state and the suspension, a third mixture is the result—the colloidal solution. The colloidal state is a condition in which one substance, such as a protein or other macromolecule, is dispersed in another substance to form many small phases suspended in one continuous phase. Thus every colloid system consists of two

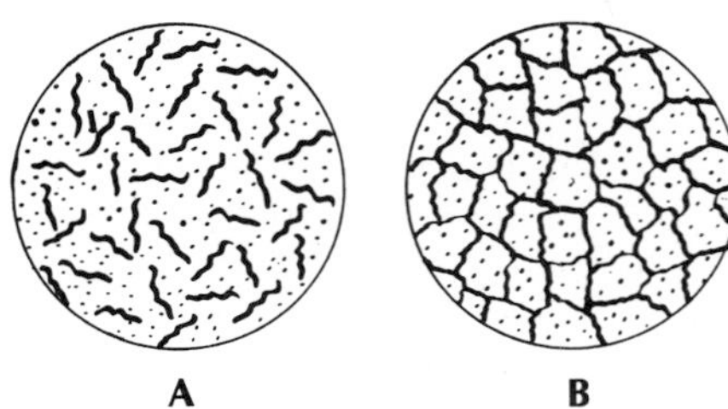

FIG. 2-7 Sol and gel. **A,** Sol condition in which gelatin particles are the discontinuous phase, water the continuous phase. **B,** Gel condition in which gelatin particles form continuous phase (network), enclosing water as discontinuous phase.

phases: a discontinuous, or dispersed, phase and a continuous, or dispersion, phase (Fig. 2-7). Colloidal systems are common not only in living systems but in many materials that we encounter in our daily lives, as the examples in the list below illustrate.

Colloidal system	Example
Solid dispersed in a liquid	Ink
Liquid dispersed in a liquid	Emulsions, such as cream
Liquid dispersed in a solid	Jellies, gelatin
Solid dispersed in a solid	Stained glass
Gas dispersed in a solid	Floating soap
Gas dispersed in a liquid	Foam, carbonated water
Liquid dispersed in a gas	Fog, clouds
Solid dispersed in a gas	Smoke, fine dust

All protoplasm is a colloidal solution, because the main molecular components of protoplasm—proteins—show all the characteristics of the colloidal state. Proteins are stable colloids because, first, they are charged ions in solution that repel each other, and second, each protein molecule attracts water molecules around it in definite layers. Immediately adjacent to the protein the water is closely oriented in a highly structured manner. This is referred to as "bound" water. Farther from the protein surface, the water molecules gradually lose their order, becoming completely random in the suspending fluid itself. Each protein molecule therefore becomes a much-enlarged colloid particle as a result of surface forces that will orient water in a regular structured form around it. The water forms the continuous phase of this colloidal system, as well as contributing to the discontinuous phase of hydrated protein particles. In fact, the protein more or less binds itself to the solvent, forming a rigid, stable system. Since the living cell is filled with proteins and other biopolymers, most of the cellular water becomes definitely oriented and structured by the presence of these colloids. This is an important characteristic that provides structure to the aqueous medium of the cell in which all metabolic activities proceed.

Many, though not all, colloidal systems show phase reversal. For example, when gelatin is poured into hot water, the gelatin particles (discontinuous phase) are dispersed through the water (continuous phase) in a thin consistency that is freely shakable (Fig. 2-7). Such a condition is called a **sol.** When the solution cools, gelatin now becomes the continuous phase and the water is in the discontinuous phase. Moreover, the solution has stiffened and become semisolid and is called a **gel.** Heating the solution will cause it to become a sol again, and the phases are reversed.

Some proteins also exist in sol and gel states. The sol state is the normal fluid state of cellular proteins. The gel state occurs when proteins form numerous stronger linkages between molecules, producing a meshwork of intertwining fibers. The proteins then become the continuous phase that entraps water. The gelation of heated egg albumin and the formation of a blood fibrin clot in shed blood are examples of permanent phase reversals of protein caused by slight changes in their properties. Reversible gelation has been observed in the living cell and is believed to accompany several physiologic processes in normal cellular activity.

Why do colloids play such an important role in the structure of protoplasm? There are several reasons, among which may be mentioned the following:

1. Great surface exposure, which allows for many chemical reactions
2. The property of phase reversal, which helps explain how protoplasm can carry on diverse functions and change its appearance during metabolic activities
3. The inability of colloids to pass through membranes, which promotes the stability and organization of the cellular system, such as cell and nuclear membranes and cytoplasmic inclusions
4. The selective absorption or permeability of the cell membrane, which is largely dependent on the phase reversal of its colloidal structure

CHEMISTRY OF LIFE

Fitness of earth for life

In his classic book *Fitness of the Environment* (1913), the biochemist L. J. Henderson maintained that earth possesses "the best of all possible environments for life." Although we no longer maintain that only the earth in all the universe harbors life, we do believe that only planets very similar to earth will have the special conditions that permit the evolution of life. Why is the earth so fit for life?

The earth is large enough to have a surface density that permits molecules to collect and align properly. Protoplasm is in a colloidal state, an intermediate between conditions too solid to allow change and conditions too fluid or gaseous to permit molecular organization. The earth's gravity is strong enough to hold an extensive gaseous atmosphere but not so strong that more than a trace of free hydrogen remains.

The temperature on earth is suitable for life, meaning practically within the range of −50° C to +100° C. This is a narrow range of tolerance for life in a universe in which temperatures extend over millions of degrees. Above 100° C life as we know it is impossible because water boils and is driven off at that temperature. Temperatures much below the freezing point prevent the growth of organisms by slowing chemical processes. Nevertheless, some bacteria and algae survive in and around hot springs at temperatures approaching 90° C. At the other extreme, many organisms flourish in the Arctic and Antarctic, where air temperatures may drop below −70° C. Thus life on earth presses against the limits imposed by the thermal requirements for life processes based on protoplasm.

There are other characteristics of the earth that make it an especially fit environment for life as we know it. The earth contains a suitable array of major and minor elements in the right proportions for the synthesis of organic compounds. And, finally, earth contains an abundance of water.

Water and life

The evolution of complex biochemical systems based on carbon depended on the presence of water. Without water, life could not have evolved on earth, no matter how suitable the earth's environment might have been in other respects. Fanciful systems of life based on ammonia and silicates rather than on water and carbon have been suggested but such speculations are totally hypothetical and in any case would require environmental conditions vastly different from those on earth.

Water is the most abundant of all protoplasmic compounds, making up about 60% to 90% of most living organisms. The maintenance of a constant aqueous internal environment is a major physiologic task for all organisms, both terrestrial and aquatic.

There are several extraordinary properties of water that make it especially fit for its essential role in protoplasmic systems. It is the most stable yet versatile of all solvents. Water is the only substance that occurs in nature in the three phases of solid, liquid, and vapor within the ordinary range of earth's temperatures. Water has a **high specific heat:** it requires 1 calorie to elevate the temperature of 1 g of water 1° C (such as from 15° to 16° C). Every other liquid but ammonia requires less heat to accomplish the same temperature increase. Water has twice the specific heat of alcohol or oil, four times the specific heat of aluminum, and ten times the specific heat of iron. Thus the high thermal capacity of water has a great moderating effect on environmental temperature changes and is a great protective agent for all life.

Water also has a **high heat of vaporization.** It requires more than 500 calories to change 1 g of liquid water into water vapor. This is twice as much heat as is required to vaporize methyl alcohol. For terrestrial animals (and plants), cooling produced by the evaporation of water is an important means of getting rid of excess heat.

Another important property of water from a biologic standpoint is its **unique density behavior** during changes of temperature. Most liquids become continually more dense with cooling. Water, however, reaches its maximum density at 4° C and then becomes lighter with further cooling. Therefore ice *floats* rather than forming on the bottom of lakes and ponds. If it were not for this property, bodies of water would freeze solid from the bottom up in winter, and, except in warmer climates, would not necessarily completely melt in summer. Under these conditions, aquatic life would be severely limited.

Water has a **high surface tension,** greater than any other liquid but mercury. This property, caused by the great cohesiveness of water molecules, is important in the maintenance of protoplasmic form and movement. It also creates a unique ecologic niche for insect forms, such as water striders and water boatmen, that skate on the surfaces of ponds. Despite its high surface tension, water has **low viscosity,** a property that favors the movement of blood through minute capillaries and of cytoplasm inside cellular boundaries.

From what we have said, clearly it is the great stability of water that makes it such a fine medium for life. Actually, the great resistance of water to potential changes in state is surprising because water is composed of oxygen and hydrogen, two of the most reactive elements known. We now know that the remarkable properties of water can be explained in large part on the basis of its structure. Water is a hydride of oxygen. The water molecule is shaped like an isosceles triangle: the legs of the triangle are two O—H bonds that subtend an angle of approximately 105°.

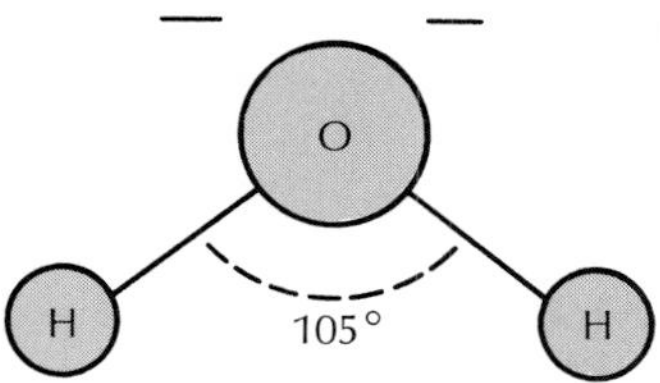

The water molecule is electrically neutral because it has an equal number of protons and electrons. However, the oxygen is so strongly electronegative that it tends to draw electrons away from the hydrogen nuclei. Consequently, each of the hydrogen atoms carries a net positive charge, and the oxygen atom carries two local negative charges. Thus the water molecule is a permanent dipole with zones of negative and positive charge.

Water molecules tend to orient themselves so that the positively charged zone of one molecule faces the negatively charged zone of another. Every oxygen atom, therefore, becomes the center of a tetrahedron of other

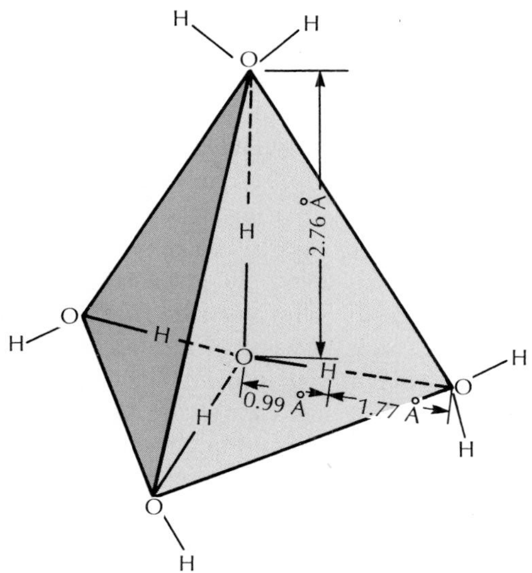

oxygen atoms. The kind of linkage indicated in the diagram above by a dashed line is a **hydrogen bond.** Although it is much weaker than the O—H bonds within each water molecule, it nevertheless requires substantial energy for breakage (approximately 5 kcal/mole). Hydrogen bonds are of great importance in the chemistry of life because they serve as atomic bridges that participate in the formation of macromolecules, especially proteins.

Hydrogen bonds also give water most of its special properties. The attraction of water molecules to each other through hydrogen bonds is responsible for the high surface tension of water. The high specific heat of water is a consequence of attractive forces between hydrogen bonds. When water is heated, much of the heat energy is used to rupture some of the hydrogen bonds, leaving a relatively small amount of heat to increase the kinetic energy (molecular movement), and thus the temperature, of the water. Similarly, the heat of vaporization of water is high because all of the hydrogen bonds between a water molecule and its neighbors must be ruptured before water can escape the surface and enter the air.

Finally, water is an excellent **solvent** for the ions of salts, which are so important to life processes. Salts dissociate to a much greater degree in water than they do in any other solvent. This results from the dipolar nature of water, which causes it to orient itself around charged particles dissolved in it. When, for example, NaCl is dissolved in water, it rapidly ionizes into Na^+ and Cl^- ions. The negative zones of the water dipoles align themselves around the Na^+ ions while the positive zones align themselves around the Cl^- ions. This keeps the ions separated, thus promoting a high degree of dissociation.

Water and life are part and parcel of each other. The special conditions on earth resulting from its ideal size, element composition, and nearly circular orbit at a perfect distance from a long-lived star, the sun, made possible the accumulation of water on the earth's surface. It is difficult even to imagine the origin of life without it.

The chemical elements of life

Of the 92 naturally occurring elements, perhaps 46 are found in protoplasm. Twenty-four of these are considered essential for life, while others are present in protoplasm only because they exist in the environment with which the organism interacts. Of the 24 essential elements, six play especially important roles in living systems. These major elements, shown in red in Fig. 2-8, are carbon (C), hydrogen (H), nitrogen (N), oxygen (O), phosphorus (P), and sulfur (S). Most organic molecules are built with these six elements. Another five essential elements found in less abundance in living systems and shown in light red in

1 **H** 1.01																	2 He 4.00
3 Li 6.94	4 Be 9.01											5 B 10.8	6 C 12.0	7 N 14.0	8 O 16.0	9 **F** 19.0	10 Ne 20.2
11 Na 23.0	12 Mg 24.3											13 Al 27.0	14 **Si** 28.1	15 P 31.0	16 S 32.1	17 Cl 35.5	18 Ar 39.9
19 K 39.1	20 Ca 40.1	21 Sc 45.0	22 Ti 47.9	23 **V** 50.9	24 **Cr** 52.0	25 **Mn** 54.9	26 **Fe** 55.8	27 **Co** 58.9	28 **Ni** 58.7	29 **Cu** 63.5	30 **Zn** 65.4	31 Ga 69.7	32 Ge 72.6	33 As 74.9	34 **Se** 79.0	35 Br 79.9	36 Kr 83.8
37 Rb 85.5	38 Sr 87.6	39 Y 88.9	40 Zr 91.2	41 Nb 92.9	42 **Mo** 95.9	43 Tc (99)	44 Ru 101.	45 Rh 103.	46 Pd 106.	47 Ag 108.	48 Cd 112.	49 In 115.	50 **Sn** 119.	51 Sb 122.	52 Te 128.	53 **I** 127.	54 Xe 131.
55 Cs 133.	56 Ba 137.	57 La 139.	72 Hf 178.	73 Ta 181.	74 W 184.	75 Re 186.	76 Os 190.	77 Ir 192.	78 Pt 195.	79 Au 197.	80 Hg 201.	81 Tl 204.	82 Pb 207.	83 Bi 209.	84 Po (210)	85 At (210)	86 Rn (222)
87 Fr (223)	88 Ra (226)	89 Ac (227)															
				58 Ce 140.	59 Pr 141.	60 Nd 144.	61 Pm (147)	62 Sm 150.	63 Eu 152.	64 Gd 157.	65 Tb 159.	66 Dy 162.	67 Ho 165.	68 Er 167.	69 Tm 169.	70 Yb 173.	71 Lu 175.
				90 Th 232.	91 Pa (231)	92 U 238.	93 Np (237)	94 Pu (242)	95 Am (243)	96 Cm (247)	97 Bk (247)	98 Cf (251)	99 Es (254)	100 Fm (253)	101 Md (256)	102 No (254)	103 Lr (257)

FIG. 2-8 Periodic table of elements showing those essential to life. Six major elements in living systems are set in bright red; five essential minor elements are set in light red. Fourteen trace elements essential to life are set in boldface black. Other elements, notably aluminum, antimony, mercury, cadmium, lead, silver, and gold, are usually present in living systems in trace amounts but are not dietary essentials. The mass number is shown above, and the atomic weight (approximate) below, each element.

Fig. 2-8 are calcium (Ca), potassium (K), sodium (Na), chlorine (Cl), and magnesium (Mg). Several other elements, called trace elements, are found in minute amounts in animals and plants, but are nevertheless indispensable for life. These are manganese (Mn), iron (Fe), iodine (I), molybdenum (Mo), cobalt (Co), zinc (Zn), selenium (Se), copper (Cu), chromium (Cr), tin (Sn), vanadium (V), silicon (Si), nickel (Ni), and fluorine (F). They are shown in black boldface type in Fig. 2-8.

Most trace elements are necessary components of enzyme systems. Some trace elements are also important constituents of organic compounds, such as the iron of hemoglobin and myoglobin. Iodine, the heaviest essential trace element, is a vital part of the thyroid hormones, thyroxine and triiodothyronine. The discovery of necessary trace elements requires ingenious and elaborate dietary studies to determine whether or not a particular element promotes growth or prevents deficiency diseases in test animals. At the present time, tin, vanadium, silicon, and fluorine appear to be essential for the growth of young animals.

Many elements are extremely toxic to life, even in small amounts. The most dangerous appear to be certain metals, normally present in trace quantities in the environment, but which may be concentrated by industrial activities. Trace pollutants of beryllium, nickel, and chromium have been linked with cancer; antimony and lead with heart disease; and cadmium and tungsten with enzyme impairment, leading to reduced life expectancy.

Mercury is being found with increasing frequency in both fresh and salt waters. Mercury from paper mills, which until recently used phenyl mercuric acetate as a fungicide to prevent cellulose rot, and from spillage and leakage from plants engaged in chlorine formation, settles to the bottom of freshwater streams, where it is converted by bacteria into soluble and highly toxic metalloorganic compounds. The half-life of these mercuric compounds in fish (which obtain it from the water and from invertebrate food organisms) may be as much as 1,000 days. In humans who eat the fish, the half-life is only about 70 days.

Despite recent concern with the harmful effects of metal pollutants, it is not exclusively a problem of our technologic society. Poisoning from lead salts was known among the ancient Romans, who preferred lead utensils and drank water conducted through lead pipes.

It has even been suggested that lead poisoning may account for the decline of Rome's ruling dynastic families.

It is not yet known just how our chemical milieu affects the structure of the animal body, or how necessary certain elements found in the body may be. Since living matter has come from the same elements that are found on this planet it is evident that the composition of the earth restricts what elements go into the composition of the organism. But evolution has been very selective in choosing certain elements as the building stones of life. It is clear from the previous listing that only a select number of the earth's elements are found in protoplasm. Their interactions pose many problems about which little is known.

Chemical complexity of living matter

The basis of biologic activities is chemical reactions. These reactions involve chemical elements that are found in all organisms as well as in the nonliving world. Life must have had its beginning in combinations and reactions of chemicals. At first, only a small number of combinations and reactions were necessary for the initiation of life. As time went on, more complex substances or compounds with successively higher levels of organization and reactions appeared. Despite the incredible complexity of living matter, all biologic phenomena operate according to the physical laws of chemistry and physics (Principle 1, p. 9).

Analysis of typical living matter reveals that it is composed of about 60% to 90% water (higher animals are usually about 60% to 70% water), 15% protein, 10% to 15% fats and other lipids, 1% carbohydrates, and 5% inorganic ions of various kinds (Na^+, K^+, Cl^-, $SO_4^=$, and others) (Fig. 2-9).

Many different categories of biomolecules are found in every cell of the organism. Most of them are dissolved or suspended in cellular water, either in ionized or nonionized form. Some organic substances (molecules containing carbon) form a part of the hard substances (horn, claws, hoofs, keratin, and others) of the body. Besides being in the cells and tissues of the body, organic components are also found in the environment, especially that of aquatic animals. Surprisingly, lake water may contain a far higher organic content than does ocean water. Some organic substances found in water are organic phosphorus and organic nitrogen compounds, amino acids, carotenoid substances, and vitamins.

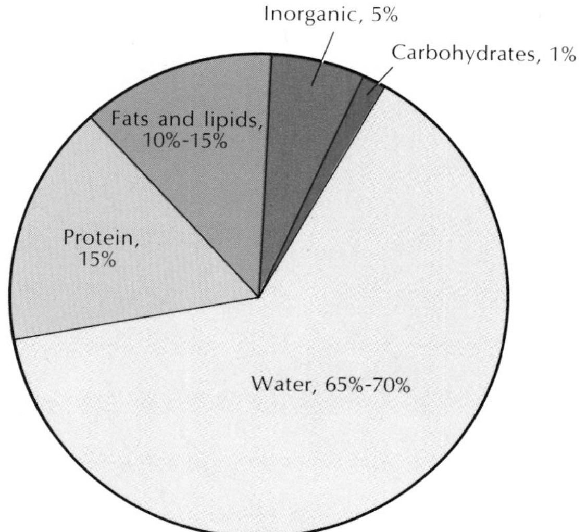

FIG. 2-9 Relative composition of major constituents of cells of higher animals.

Inorganic matter actually outweighs the organic part of living matter. The inorganic portion includes (1) water; (2) mineral solids consisting of crystals, secreted precipitates, bone, and shells; and (3) cellular minerals, either free or combined with organic compounds. Most of the cellular minerals are in the form of ions (H^+, Ca^{++}, Na^+, and Mg^{++}, OH^-, $CO_3^=$, HCO_3^-, $PO_4^{\equiv}$, Cl^-, and $SO_4^=$ ions). Animals get these constituents from the water in which the minerals are dissolved from rocks and the soil.

Organic molecules

The term "organic compounds" has been applied to substances derived from plants and animals. All organic compounds contain carbon, but many also contain hydrogen and oxygen as well as nitrogen, sulfur, phosphorus, salts, and other elements. Organic compounds in a specific way are those carbon compounds in which the principal bonds are carbon-to-carbon and carbon-to-hydrogen.

Carbon with its valence of four has a great ability to bond with other carbon atoms in chains of varying lengths and configurations. More than a million organic compounds have been identified; more are being added daily. Carbon-to-carbon combinations introduce the possibility of enormous complexity and variety into molecular structure. As a chain it can combine with hydrogen to form an **aliphatic** compound:

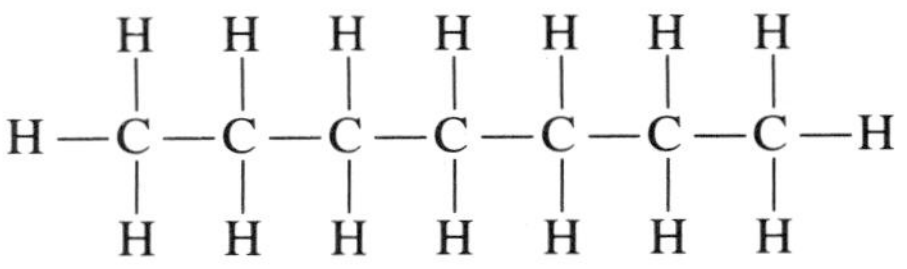

```
    H   H   H   H   H   H   H
    |   |   |   |   |   |   |
H — C — C — C — C — C — C — C — H
    |   |   |   |   |   |   |
    H   H   H   H   H   H   H
```

Heptane

or a ring structure (**aromatic** compound):

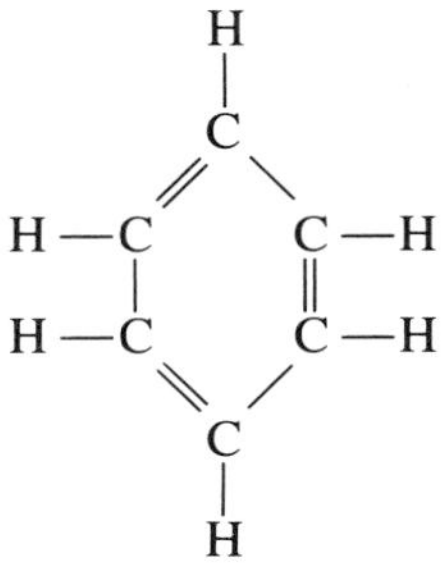

```
          H
          |
          C
        //  \
  H — C        C — H
      |        ||
  H — C        C — H
        \\  /
          C
          |
          H
```

Benzene

Other types of configurations include rings and chains joined to each other, multiple branches of chains, helix arrangements, and others. The diversity of carbon molecules has made possible the complex kinds of macromolecules that form the essence of life.

Carbohydrates: nature's most abundant organic substance

Carbohydrates are compounds made of carbon, hydrogen, and oxygen. They are usually present in the ratio CH_2O and are grouped as H—C—OH. Familiar examples of carbohydrates are sugars, starches, and cellulose. There is more cellulose (the woody structure of plants) on earth than all other organic materials combined. Carbohydrates are made synthetically from water and carbon dioxide by green plants, with the aid of the sun's energy. This process, called **photosynthesis,** is a reaction on which all life depends, for it is the starting point in the formation of food.

Carbohydrates are usually divided into the following three classes: (1) **monosaccharides,** or simple sugars; (2) **disaccharides,** or double sugars; and (3) **polysaccharides,** or complex sugars. Simple sugars are composed of carbon chains containing 4 carbons (tetroses), 5 carbons (pentoses), or 6 carbons (hexoses). Other simple sugars have up to 10 carbons, but these are not biologically important. Simple sugars, such as glucose, galactose, and fructose, all contain a free sugar group,

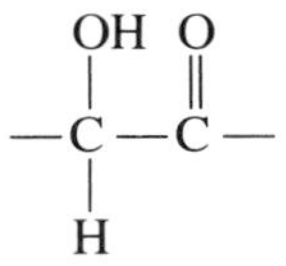

```
   OH  O
   |   ||
 — C — C —
   |
   H
```

in which the double-bonded O may be attached to the terminal C of a chain (an aldehyde) or to a nonterminal C (a ketone). The hexose **glucose** (also called dextrose) is the most important carbohydrate in the living world. Glucose is often shown as a straight-chain aldehyde.

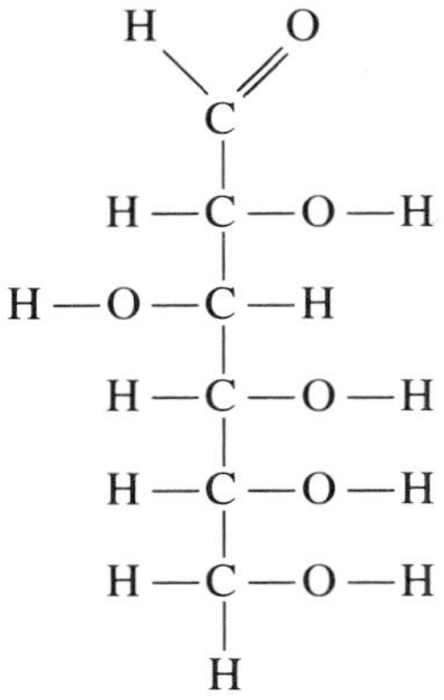

```
        H     O
         \  //
           C
           |
       H — C — O — H
           |
   H — O — C — H
           |
       H — C — O — H
           |
       H — C — O — H
           |
       H — C — O — H
           |
           H
```

but in fact it tends to form a cyclic compound:

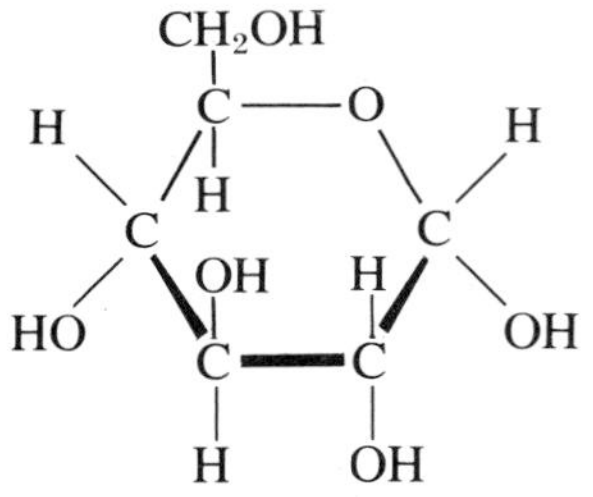

```
            CH2OH
             |
   H         C ——— O         H
     \      /|       \      /
       \  /  H         \  /
         C               C
       /  \ OH       H /  \
    HO      \|       |/    OH
             C ━━━━━ C
             |       |
             H       OH
```

This formula shows the ring structure of glucose, but it is misleading because it obscures the three-dimensional form of the molecule. Rather than lying in a flat plane, the glucose molecule is "puckered," because the 4 atoms bonded to any single carbon molecule occupy the corners of a regular tetrahedron, four-cornered, pyramid-like figure. To better represent the structure of glucose than in the flat-plane ring structure above, organic chemists have devised a "chair" model to show the configuration of glucose and other hexose molecules:

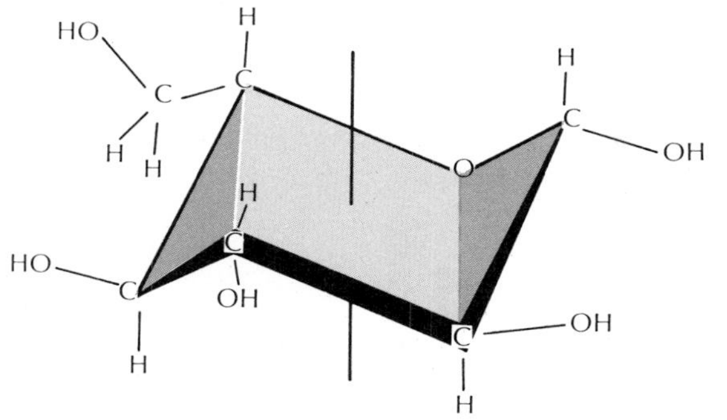

Although the three-dimensional chair conformation is the most accurate way to represent the simple sugars, we must remember that all forms of glucose, however represented, are the same molecule.

Other hexoses of biologic significance are galactose and fructose. Their straight-chain structures are compared with glucose in Fig. 2-10.

Disaccharides are double sugars formed by the bonding together of two simple sugars. An example is maltose (malt sugar) composed of 2 glucose molecules. As shown in Fig. 2-11, the 2 glucose molecules are condensed together by the removal of a molecule of water. This dehydration reaction, with the sharing of an oxygen atom by the two sugars, characterizes the formation of all disaccharides. Two other common disaccharides are sucrose (ordinary cane, or table, sugar), formed by the linkage of glucose and fructose, and lactose (milk sugar), comprised of glucose and galactose.

Polysaccharides are made up of many molecules of simple sugars (usually glucose) and are referred to by the chemist as polymers. Their empirical formula is usually written $(C_6H_{10}O_5)_n$, where n stands for the unknown number of simple sugar molecules of which they are composed. Starch is common in most plants

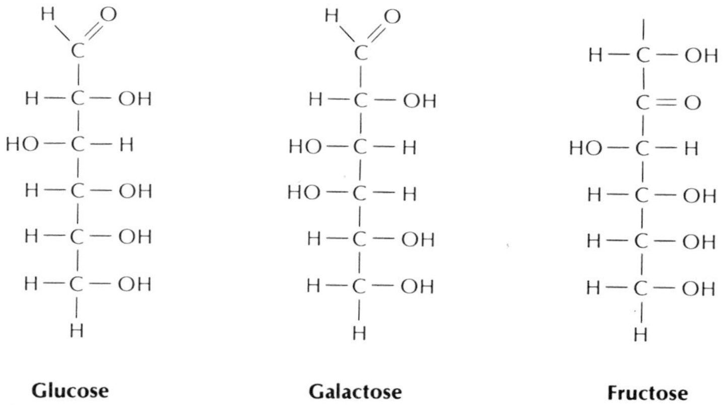

Glucose

Galactose

Fructose

FIG. 2-10 These three hexoses are the most common monosaccharides. Glucose and galactose are aldehyde sugars; fructose is ketone sugar.

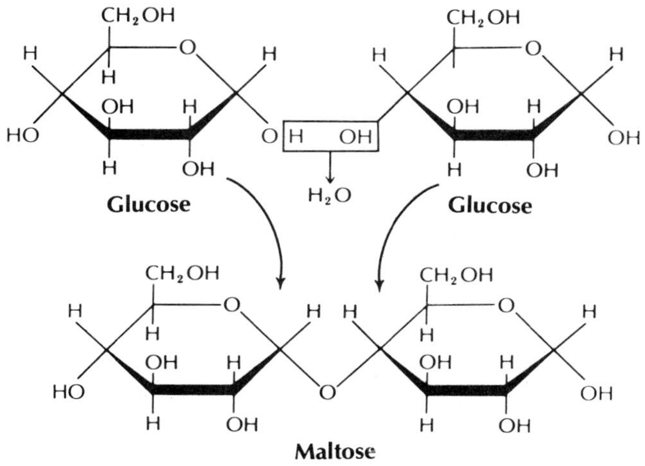

Glucose

Glucose

Maltose

FIG. 2-11 Formation of a double sugar (disaccharide maltose) from two glucose molecules with the removal of a molecule of water.

and is an important food constituent. **Glycogen** (Fig. 2-12), or animal starch, is found mainly in liver and muscle cells in vertebrates. When needed, glycogen is converted into glucose and is delivered by the blood to the tissues. Another polymer is **cellulose,** which is an important part of the cell walls of plants (Fig. 2-13). Cellulose cannot be digested by humans, but some other animals, such as the herbivores, with the aid of bacteria, and termites, with the aid of flagellates, can do so.

The main role of carbohydrates in protoplasm is to serve as a source of chemical energy. Glucose is the most important of these energy carbohydrates. Some carbohydrates become basic components of protoplasmic structure, such as the pentoses that form constituent groups of nucleic acids and of nucleotides.

Proteins: foundation substance of protoplasm

Proteins are large, complex molecules characterized by a high nitrogen content. They are the predomi-

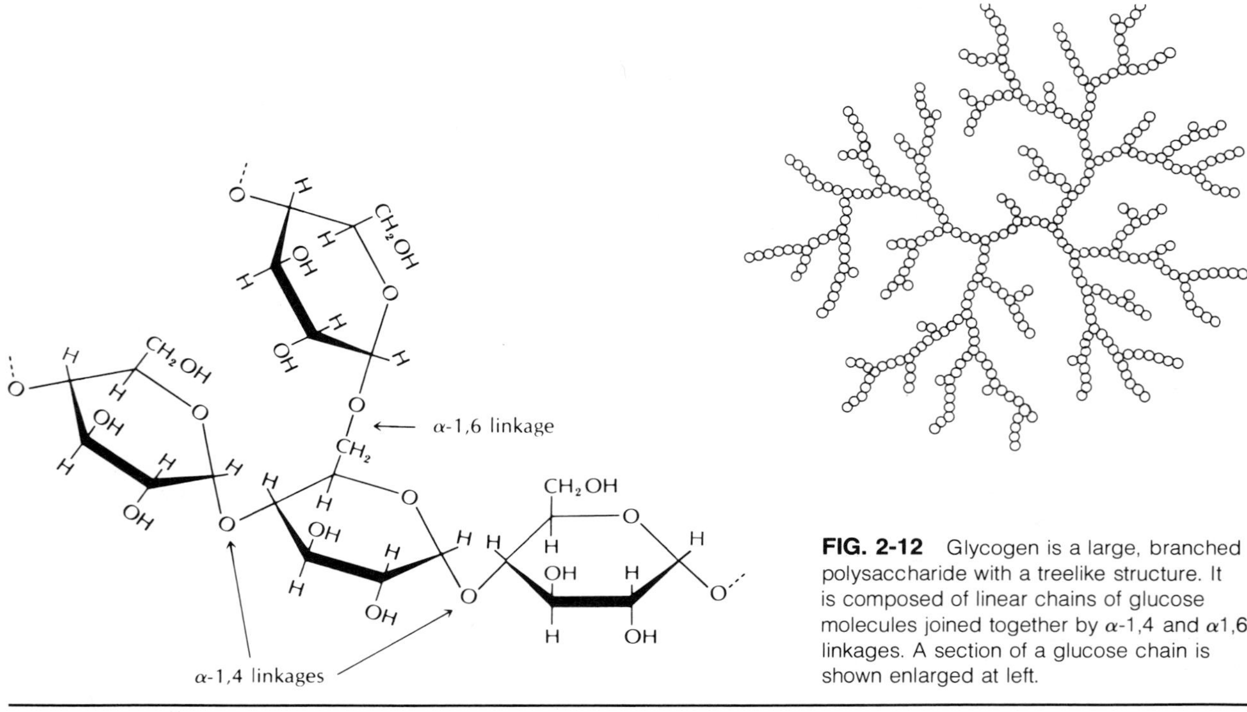

FIG. 2-12 Glycogen is a large, branched polysaccharide with a treelike structure. It is composed of linear chains of glucose molecules joined together by α-1,4 and α1,6 linkages. A section of a glucose chain is shown enlarged at left.

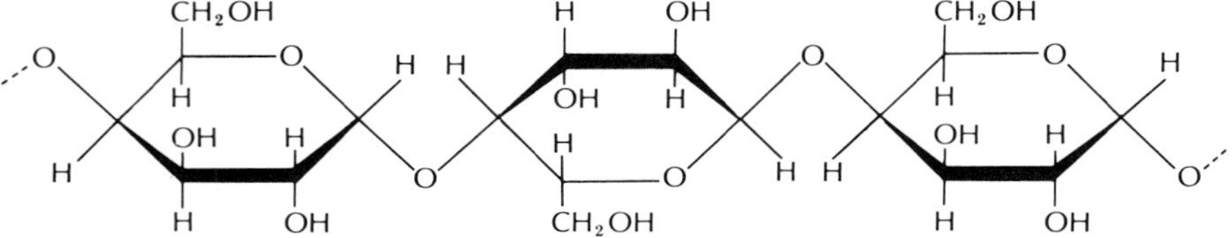

FIG. 2-13 Cellulose, the structural carbohydrate of plants, and the most abundant of all organic compounds. Cellulose differs chemically from glycogen in lacking molecular branching. Its glucose molecules are joined together in straight chains by β-1,4 linkages.

nant type of macromolecule in cells, constituting more than 50% of the cell's dry weight.

Proteins serve as the chief structural material of protoplasm and play numerous other essential roles in living systems. They form enzymes—globular proteins specialized to serve as catalysts in virtually all biochemical activities of cells. Other proteins are antibodies, transport proteins, storage proteins, contractile proteins, and some hormones. In every living organism there are thousands of different proteins, each fitted to perform a specific functional or structural role.

A typical protein is composed of perhaps 200 amino acids arranged into a long, chainlike molecule. However, there is great variation in protein size, ranging from small proteins consisting of approximately 50 amino acids to complex enzymes containing more than 20,000 amino acids divided among several distinct chains.

Each of the numerous amino acids in a protein molecule could be any one of 20 different kinds of amino acids that commonly occur in living organisms (Fig. 2-14). There are many other amino acids in nature—

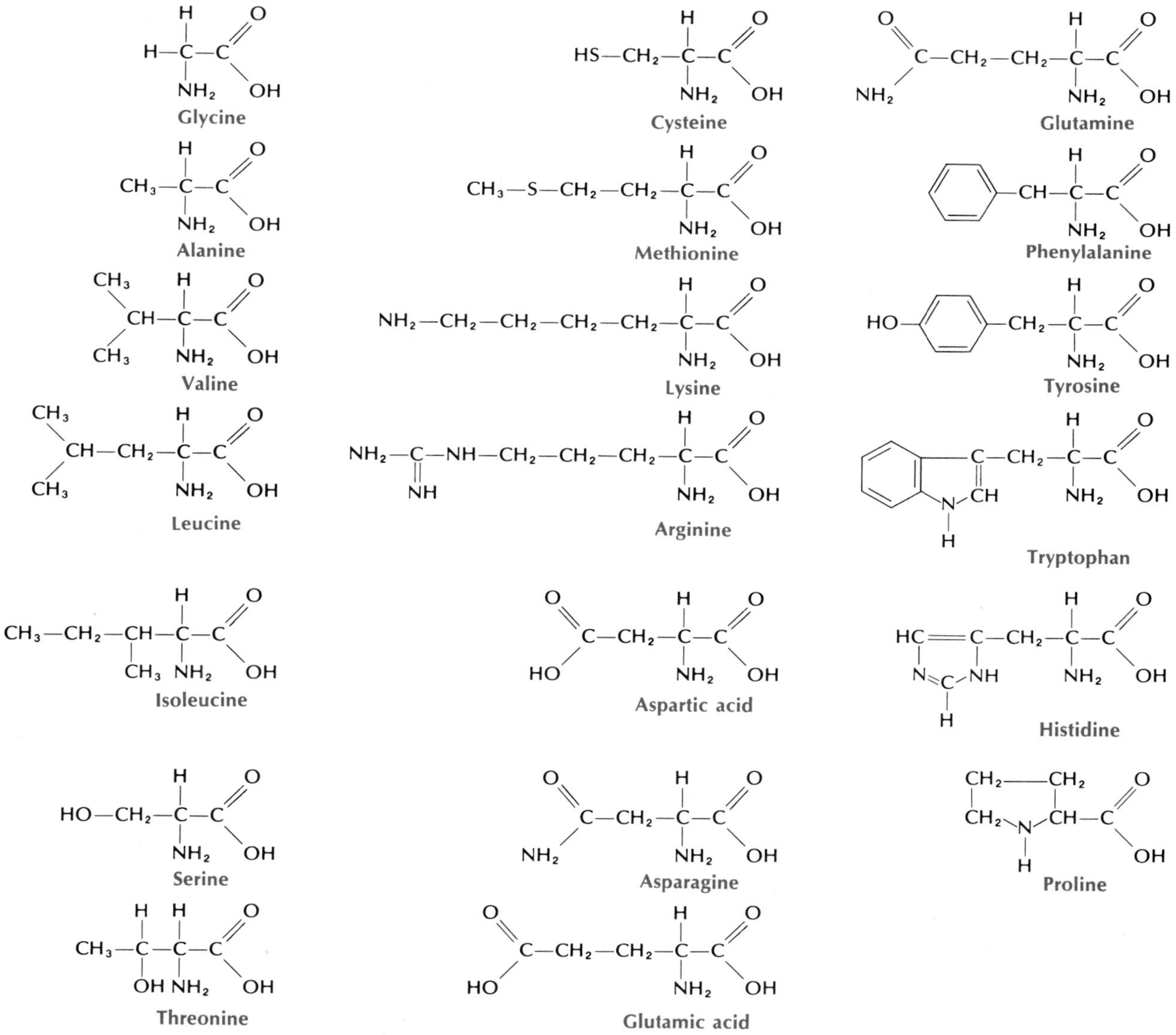

Glycine

Cysteine

Glutamine

Alanine

Methionine

Phenylalanine

Valine

Lysine

Tyrosine

Leucine

Arginine

Tryptophan

Isoleucine

Aspartic acid

Histidine

Serine

Asparagine

Proline

Threonine

Glutamic acid

FIG. 2-14 The 20 naturally occurring amino acids.

more than 170 different amino acids are known to occur (usually rarely) in various animal and plant tissues—but only the standard 20 are coded in the cell's genes (Principle 18, p. 13). The remaining amino acids, when they occur, are derivatives produced by chemical modification of the standard 20 after the protein is synthesized.

Since the 20 amino acids may be arranged in all possible combinations, the number of different amino acid sequences even in the smallest proteins is incomprehensibly vast. For a protein consisting of 200 amino acids, there are 200^{20}, or 10^{260}, possible combinations, a figure so large that it can be appreciated only by comparing it with an equally incomprehensible quantity: the number of electrons in the universe, estimated by the astronomer Sir Arthur Eddington to be 10^{256}. Since proteins determine the biochemical uniqueness of each living organism, there is obviously no practical limit to biologic variation. Yet proteins are anything but the products of random shuffling of amino acid building blocks. The sequence of amino acids in every protein is precisely determined by the genetic code in the cell's genes. Even a single alteration in this sequence may destroy its function.

Each of the 20 amino acids contains one amino group ($-NH_2$) and one carboxyl group ($-COOH$) attached to the same carbon atom. The remainder of the molecule is unique for each amino acid. The general formula for an amino acid is as follows:

$$R-\underset{NH_2}{\overset{H}{\underset{|}{\overset{|}{C}}}}-C\begin{matrix}\diagup\!\!\diagup O\\ \diagdown OH\end{matrix}$$

The symbol *R* represents a "side group" unique for each acid. The amino acids range in complexity from glycine with a simple hydrogen atom as its side group to tryptophan with a complex aromatic ring side group.

In forming polypeptide chains, amino acids are linked head to tail by a bond between the amino group of one amino acid to the carboxyl group of another. Water is eliminated in the condensation, and a covalent **peptide bond** is formed.

The bonding of two amino acids, as shown below, forms a **dipeptide;** the addition of a third amino acid forms a **tripeptide.** In this manner—sequential addition of amino acids through peptide bonds—long **polypeptide** chains are built:

The polypeptide (shown at top of p. 35) consists of five different amino acids linked by peptide bonds. When the chain exceeds about 50 amino acids, the molecule is called a **protein** rather than a polypeptide. Thus proteins are built up of a backbone involving the amino and carboxyl groups. The various side groups and the folding of the peptide chains are responsible for the chemical and biologic properties of the protein.

For convenience, biochemists have recognized four levels of protein organization called primary, secondary, tertiary, and quaternary. The **primary** structure of a protein is determined by the kind and sequence of amino acids making up the polypeptide chain. The polypeptide chain or chains tend to spiral into a definite helical pattern, like the turns of a screw. This precise coiling, known as the **secondary** structure of the protein, most commonly takes a clockwise direction called an **alpha-helix** (Fig. 2-15). The spirals of the chains are stabilized by weak **hydrogen bonds,** usually between a hydrogen atom of one amino acid and the peptide-bond oxygen of another amino acid in an adjacent turn of the helix. Hydrogen bonds provide definite spacing to the helix: there is 1 amino acid every 1.5 Å along the axis, and there are an average 3.6 amino acids per turn.

The polypeptide chain (primary structure) not only spirals into helical configurations (secondary structure), but also the helices themselves bend and fold, giving the protein its complex, yet stable, three-dimensional **tertiary** structure (Fig. 2-16). The folded chains are stabilized by the interactions between side groups of amino acids. One of these interactions is the **disulfide bond,** a strong covalent bond between pairs of cysteine (sis′tee-in) molecules that are brought together by folds in the polypeptide chain. Other kinds of bonds that help to stabilize the tertiary structure of proteins are hydrogen bonds, ionic bonds, and hydrophobic bonds.

The term **quaternary** structure describes those pro-

$$\begin{matrix}H\\ \diagdown\\ \end{matrix}\!N-\underset{R}{\underset{|}{CH}}-C\begin{matrix}\diagup\!\!\diagup O\\ \diagdown \boxed{OH\ \ H}\end{matrix}\!\!-N-\underset{R}{\underset{|}{CH}}-C\begin{matrix}\diagup\!\!\diagup O\\ \diagdown OH\end{matrix} \rightarrow \begin{matrix}H\\ \diagdown\\ \end{matrix}\!N-\underset{R}{\underset{|}{CH}}-\boxed{\overset{O}{\overset{\|}{C}}-\overset{H}{\overset{|}{N}}}-\underset{R}{\underset{|}{CH}}-C\begin{matrix}\diagup\!\!\diagup O\\ \diagdown OH\end{matrix} + H_2O$$

Peptide bond

TERMINAL AMINO GROUP **TERMINAL CARBOXYL GROUP**

OH (on benzene ring) | H_3C CH_3 | N—H, C=CH (indole ring)

H | CH_2 | CH_3 | CH | CH_2

NH_2—CH—C—NH—CH—C—NH—CH—C—NH—CH—C—NH—CH—C—OH

O | O | O | O | O

Glycine **Tyrosine** **Alanine** **Valine** **Tryptophan**

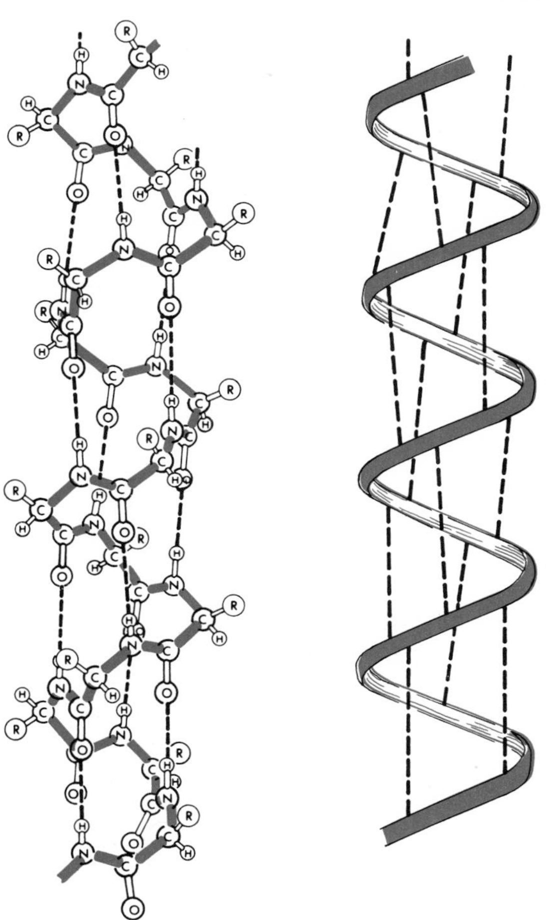

FIG. 2-15 Alpha-helix pattern of a polypeptide chain. *Dashed lines,* Hydrogen bonds that stabilize adjacent turns of the helix. *R,* Amino acid side chains. (Adapted from Green, D. 1956. Currents of biochemical research. New York, Interscience Publishers, Inc.)

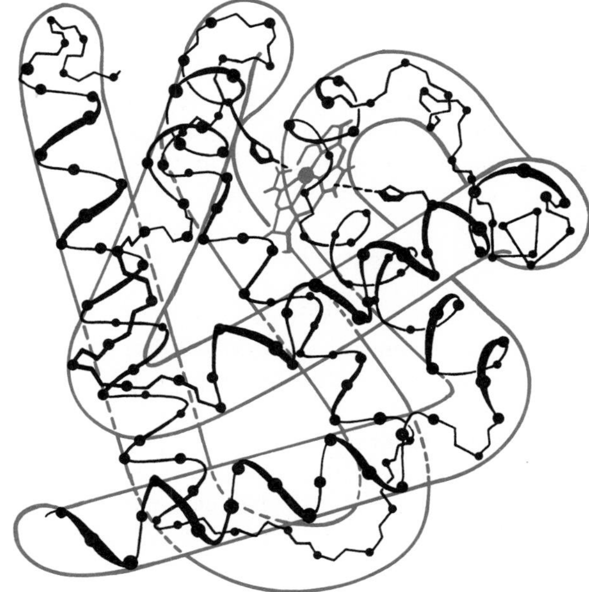

FIG. 2-16 Three-dimensional tertiary structure of the protein myoglobin. Adjacent folds of the polypeptide chain are held together by disulfide bonds that form between pairs of cysteine molecules. In the upper center of the molecule is the heme group, which combines with oxygen. (From Neurath, H. 1964. The proteins, vol. 2, ed. 2. New York, Academic Press, Inc.)

teins that contain more than one polypeptide chain unit. For example, hemoglobin of higher vertebrates is composed of four polypeptide subunits nested together into a single protein molecule.

The complete structure of several proteins has now been thoroughly worked out. This has proved to be a monumental task, for not only must the correct amino acid sequence be determined, but also the complete three-dimensional configuration, that is, the exact way the polypeptide chains are folded and bonded together. Insulin, the pancreatic hormone that governs glucose metabolism, was the first protein to have its amino acid sequence determined. Insulin is a small protein (molecular weight 5,700), consisting of two polypeptide chains containing 51 amino acids. The amino acid sequence was worked out by Frederick Sanger and his colleagues at Cambridge University in 1953. Sanger could not determine the three-dimensional configuration of insulin by the laborious techniques available at that time. By using x-ray diffraction, subsequent workers have constructed what are believed to be fairly accurate pictures of the shape of many native proteins.

Nucleic acids: genetic apparatus of the cell nucleus

Nucleic acids are complex substances of high molecular weight that represent a basic manifestation of life. The genetic information necessary for all aspects of biologic inheritance is encoded in the sequence of these polymeric molecules (Principles 17 and 18, p. 13). Nucleic acids direct the synthesis of proteins, including all of the enzymes that in turn carry out the numerous routine synthetic and functional activities of the cell. Most important, nucleic acids are the only molecules that have the power (with the help of the right enzymes) to replicate themselves. This is the essential process that guarantees the accuracy of reproduction. To understand how nucleic acids perform these roles, we need to look rather closely at their chemical structures.

There are two kinds of nucleic acids in all cells—deoxyribonucleic acid (DNA) and ribonucleic acid (RNA). DNA is a polymer built of repeated units called **nucleotides.** Each nucleotide contains three parts—a **sugar,** a **phosphate group,** and a **nitrogenous base.** The sugar in DNA is a pentose (5-carbon) sugar called **deoxyribose,** with the structural formula:

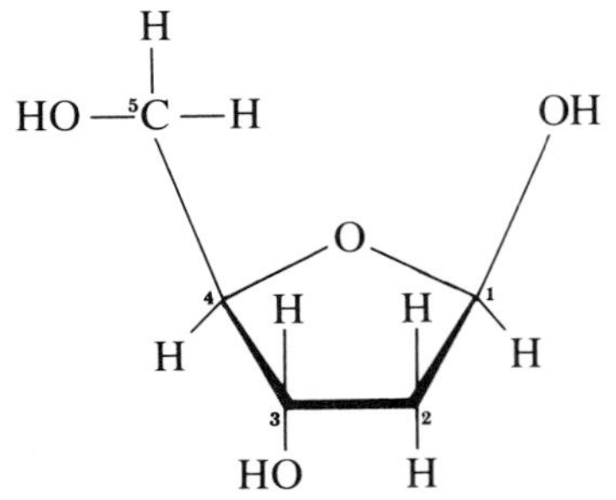

Deoxyribose sugar

The **phosphoric acid** has the structural formula:

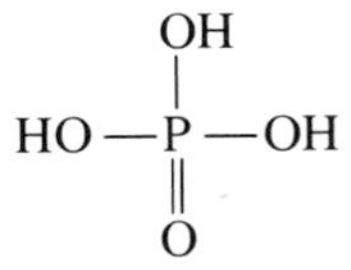

Four **nitrogenous bases** are found in DNA. Two of them are organic compounds composed of nine-membered double rings, classed as **purines.** They are **adenine** and **guanine:**

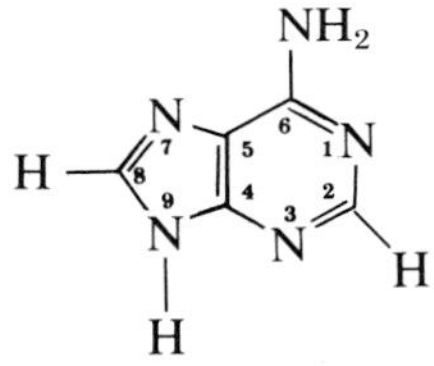

Adenine 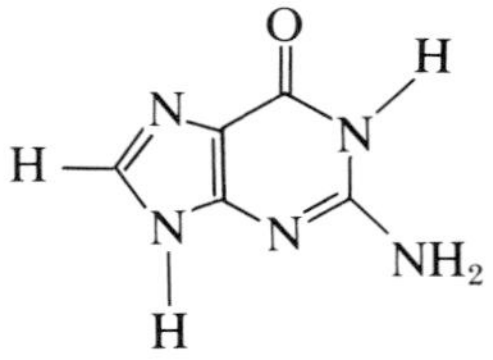 **Guanine**

The other two nitrogenous bases belong to a different class of organic compounds called **pyrimidines,** consisting of six-membered rings. The two pyrimidines found in DNA are **thymine** and **cytosine:**

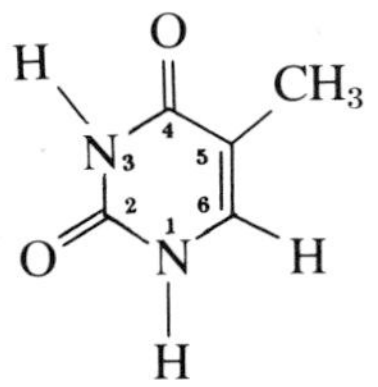

Thymine 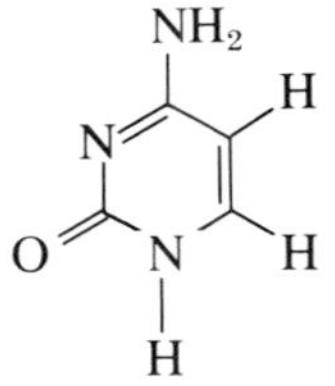 **Cytosine**

The sugar, phosphate group, and nitrogenous base are linked as shown in the generalized scheme (p. 37) for a **nucleotide:**

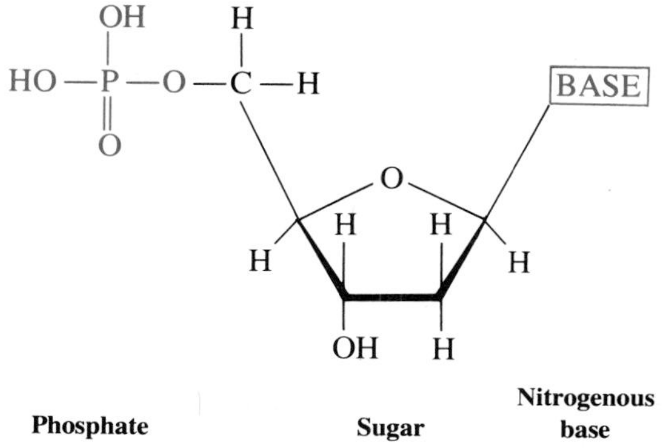

In DNA, the backbone of the molecule is built of phosphoric acid and deoxyribose sugar; to this backbone are attached the nitrogenous bases (Fig. 2-17). However, one of the most interesting and important discoveries about the nucleic acids is that DNA is not a single polynucleotide chain but rather consists of *two* complementary chains that are precisely cross-linked by specific hydrogen bonding of purine and pyrimidine bases. It was found that the number of adenines is equal to the number of thymines, and the number of guanines equals the number of cytosines. This fact suggested a pairing of bases: adenine with thymine (AT) and guanine with cytosine (GC). The larger adenine (a purine) always attaches to the smaller thymine (a pyrimidine) by two hydrogen bonds; and the larger guanine (a purine) always attaches to the smaller cytosine (a pyrimidine) by three hydrogen bonds:

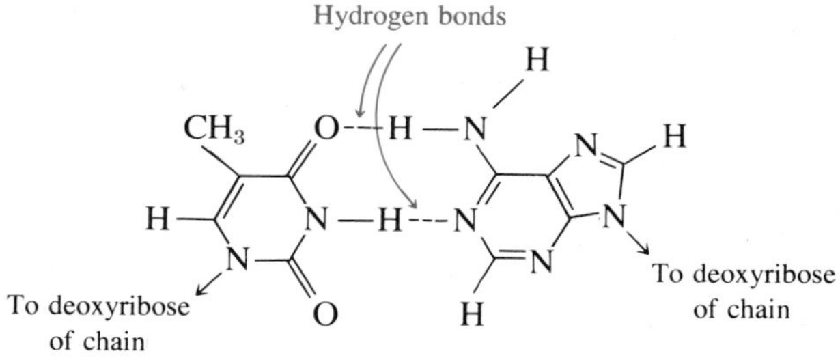

Thymine—adenine

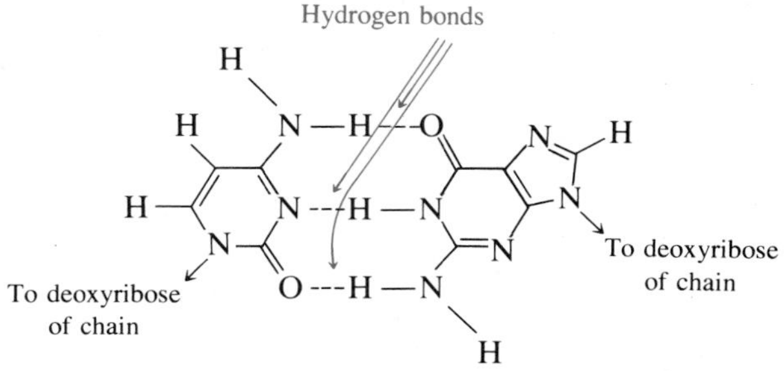

Cytosine—guanine

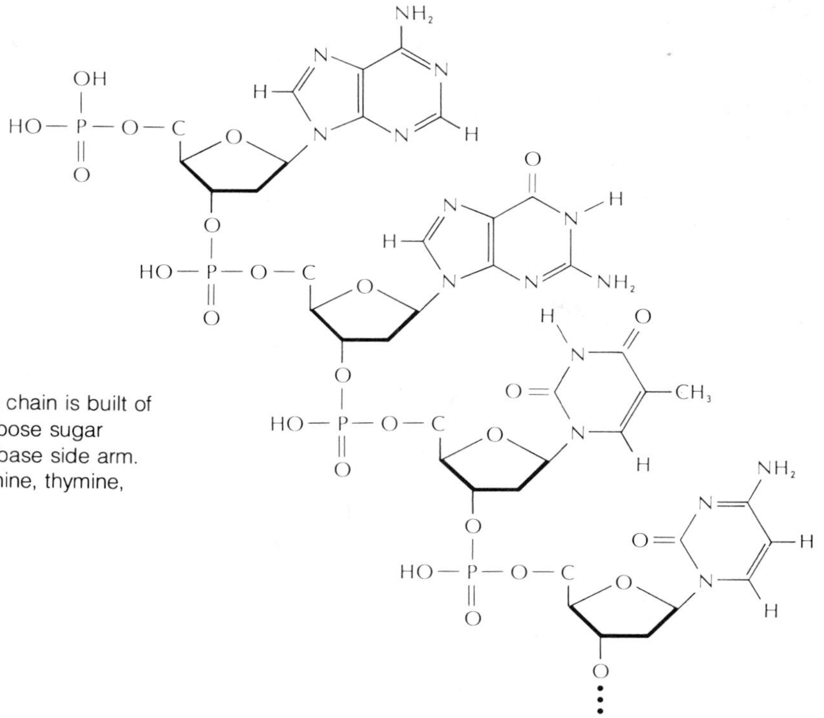

FIG. 2-17 Section of DNA. Polynucleotide chain is built of a backbone of phosphoric acid and deoxyribose sugar molecules. Each sugar holds a nitrogenous base side arm. Shown from top to bottom are adenine, guanine, thymine, and cytosine.

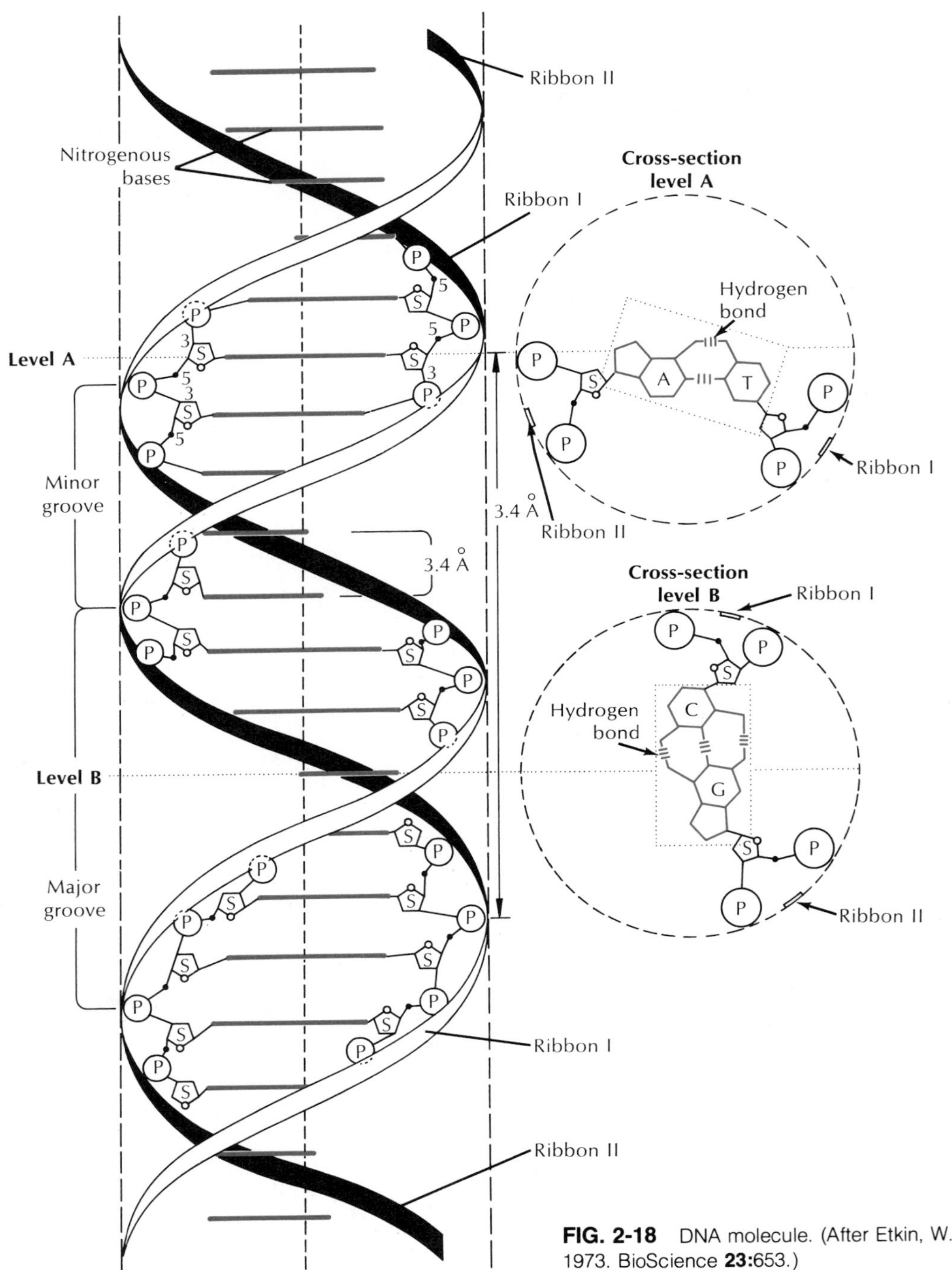

FIG. 2-18 DNA molecule. (After Etkin, W. 1973. BioScience **23:**653.)

The result is a ladder structure. The uprights are the sugar-phosphate backbones, and the connecting rungs are the paired nitrogenous bases, AT or GC.

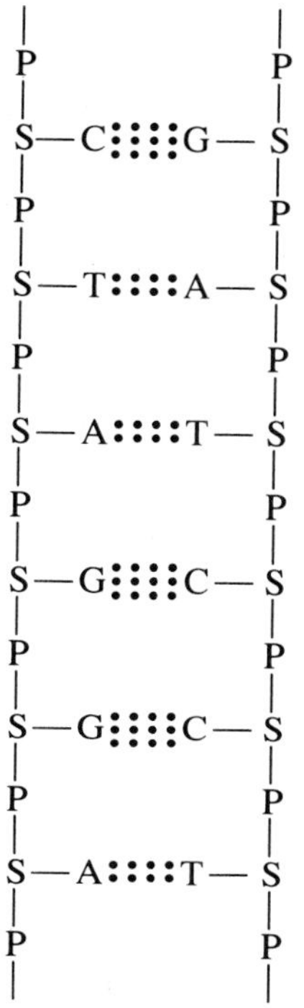

```
|           |
P           P
|           |
S—C::::G—S
|           |
P           P
|           |
S—T::::A—S
|           |
P           P
|           |
S—A::::T—S
|           |
P           P
|           |
S—G::::C—S
|           |
P           P
|           |
S—G::::C—S
|           |
P           P
|           |
S—A::::T—S
|           |
P           P
|           |
```

However, the ladder is twisted into a **double helix,** with about 10 base pairs for each complete turn of the helix (Fig. 2-18).

The determination of the structure of DNA has been widely acclaimed as the single most important biologic discovery of this century. It was based on the x-ray diffraction studies of Maurice H. F. Wilkins and the ingenious proposals of Francis H. C. Crick and James D. Watson, published in 1953. Watson, Crick, and Wilkins were later awarded the Nobel Prize for Medicine and Physiology for their momentous work.

Ribonucleic acid (RNA) is very similar to DNA in structure except that it consists of a *single* polynucleotide chain. In RNA one of the four bases, thymine (T), is replaced by uracil (U), and the sugar deoxyribose is replaced by ribose. In other respects the single RNA chain is joined together like each of the two DNA chains. RNA functions in the transcription and translation of genetic information (Principle 19, p. 13).

Lipids: fuel storage and building material

Lipids are fats and fatlike substances. They are composed of molecules of low polarity; consequently they are virtually insoluble in water, but are soluble in organic solvents such as acetone, alcohol, and ether. The three principal groups of lipids are neutral fats, phospholipids, and steroids.

Neutral fats. The neutral or "true" fats are the major fuels of animals. The average American male at age 25 contains 7.7 kg (17 lb) of stored fat; women have even more: 10 kg (22 lb). This represents approximately 40 days of stored fuel. Obese individuals may carry enough fat to supply the equivalent of a year of normal metabolism. Many species make much better use of this stored fat than do humans. When Atlantic salmon enter fresh water, they stop feeding and may remain active for as long as a year, utilizing stored fat exclusively.

Most fat in higher vertebrates is stored as **adipose tissue,** which accumulates especially under the skin and in the abdominal cavity. In fishes, lipids are stored principally in the liver and between the fibers of skeletal muscle. Stored fats represent food ingested in excess of the amount required to meet ongoing metabolic demands. Stored fat may be derived directly from dietary fat or indirectly from dietary carbohydrates (especially glucose) that are converted to fat for storage. Fats are oxidized and released into the bloodstream as needed to meet tissue demands, especially for muscles.

Fats are triglycerides, which are molecules consisting of glycerol and 3 molecules of fatty acids. True fats are therefore esters, that is, a combination of an alcohol (glycerol) and an acid. Fatty acids in triglycerides are simply long-chain monocarboxylic acids; they vary in size, but the common ones are 14 to 24 carbons long. The production of a typical fat by the union of glycerol and stearic acid is shown by the following reaction:

$$
\begin{array}{llll}
C_{17}H_{35}CO\,\boxed{OH\quad H}\,O-CH_2 & & C_{17}H_{35}OCO-CH_2 & \\
C_{17}H_{35}CO\,\boxed{OH + H}\,O-CH & \rightarrow & C_{17}H_{35}OCO-CH & + 3H_2O \\
C_{17}H_{35}CO\,\boxed{OH\quad H}\,O-CH_2 & & C_{17}H_{35}OCO-CH_2 & \\
\textbf{Stearic acid (3 mol)} \quad \textbf{Glycerol (1 mol)} & & \textbf{Stearin (1 mol)} &
\end{array}
$$

In this reaction it is seen that the 3 fatty acid molecules have united with the OH group of the glycerol to form stearin (a neutral fat), with the production of 3 molecules of water.

Most triglycerides contain two or three different fatty acids attached to glycerol, bearing ponderous names such as myristoyl palmitoyl stearoyl glycerol:

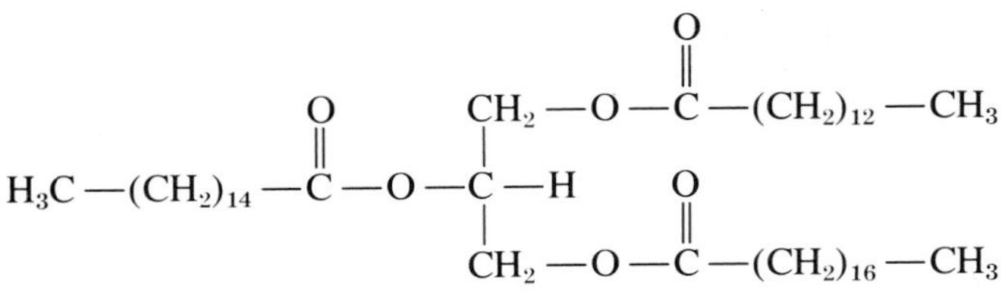

$$
\begin{array}{l}
\qquad\qquad\qquad\qquad\qquad\qquad CH_2-O-\overset{O}{\overset{\|}{C}}-(CH_2)_{12}-CH_3 \\
H_3C-(CH_2)_{14}-\overset{O}{\overset{\|}{C}}-O-\overset{|}{\underset{|}{C}}-H \\
\qquad\qquad\qquad\qquad\qquad\qquad CH_2-O-\overset{O}{\overset{\|}{C}}-(CH_2)_{16}-CH_3
\end{array}
$$

The fatty acids in this triglyceride are **saturated,** that is, every carbon within the chain holds two hydrogen atoms. Saturated fats, more common in animals than in plants, are usually solid at room temperature. **Unsaturated** fatty acids, typical of plant oils, have carbon atoms joined by double bonds; that is, the carbons are not "saturated" with hydrogen atoms and are able to form additional bonds with other atoms. Plant fats such as peanut oil and corn oil tend to be liquid at room temperature.

Phospholipids. Unlike the fats that are fuels and serve no structural roles in the cell, phospholipids are important components of the molecular organization of tissues, especially membranes. They resemble triglycerides in structure, except that one of the three fatty acids is replaced by phosphoric acid and an organic base. An example is lecithin, an important phospholipid of nerve membrane.

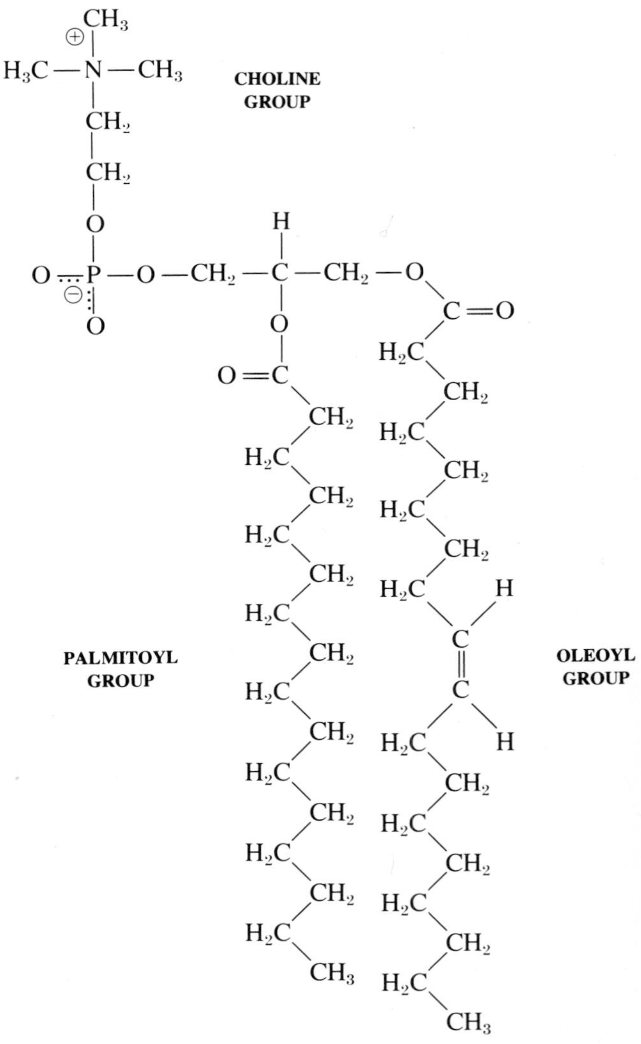

Although phospholipids are related structurally to fats, they differ from them in one important respect: the phosphate group is charged and polar and therefore soluble in water. Since the remainder of the molecule is nonpolar, phospholipids can bridge two environments and serve to bind water-soluble molecules such as proteins to water-insoluble materials. In membranes, phospholipids tend to align themselves into bilayers with the insoluble fatty acid chains orientated toward the center of the "sandwich" and the polar phosphate groups directed toward the outer aqueous phases. The significance of this configuration is discussed in Chapter 5.

Steroids. Steroids are complex alcohols unrelated chemically to fats but bearing fatlike properties. The steroids are a large group of biologically important molecules, including cholesterol (Fig. 33-10, p. 768), vitamin D, many adrenal cortical hormones, and the sex hormones.

CHOLINE GROUP

$$
\begin{array}{l}
\quad\; \oplus\; CH_3 \\
H_3C-\overset{|}{\underset{|}{N}}-CH_3 \\
\quad\; CH_2 \\
\quad\; CH_2 \\
\quad\; O \qquad\qquad\qquad\quad H \\
O \overset{\ominus}{=} \overset{|}{\underset{|}{P}}-O-CH_2-\overset{|}{\underset{|}{C}}-CH_2-O-\overset{O}{\overset{\|}{C}}-(CH_2)_7-CH=CH-(CH_2)_7-CH_3 \\
\quad\; O \qquad\qquad\qquad\quad O-\overset{O}{\overset{\|}{C}}-(CH_2)_{14}-CH_3
\end{array}
$$

PALMITOYL GROUP

OLEOYL GROUP

Annotated references

Selected general readings

Standard biochemistry and cellular biology texts are ideal sources of additonal information on the topics in this chapter. The following selection is by no means exhaustive.

De Robertis, E. D. P., W. W. Nowinski, and F. A. Saez. 1970. Cell biology, ed. 5. Philadelphia, W. B. Saunders Co. *This intermediate level text emphasizes the living cell.*

Dowben, R. M. 1969. General physiology. New York, Harper & Row, Publishers. *Advanced text focusing on physiochemical aspects of biological systems.*

Lehninger, A. L. 1975. Biochemistry: the molecular basis of cell structure and functions, ed. 2. New York, Worth Publishers, Inc. *Clearly presented advanced text.*

McGilvery, R. W. 1970. Biochemistry: a functional approach. W. B. Saunders Co. *Advanced biomedical approach with fine summaries of general biochemical principles.*

White, A., P. Handler, and E. L. Smith. 1973. Principles of biochemistry, ed. 5. New York, McGraw-Hill Book Co. *Comprehensive medical biochemistry with detailed considerations of organic molecules and their biologic roles.*

Yost, H. T. 1972. Cellular physiology. Englewood Cliffs, N.J., Prentice-Hall, Inc. *Advanced treatment of molecular physiology and biochemistry.*

Selected *Scientific American* articles

The collection of *Scientific American* readings edited by Hanawalt and Haynes (listed below) is an excellent source of articles on biochemical aspects of cell biology and should be consulted for articles that appeared before 1973. A selection of more recent articles appears here.

Capaldi, R. 1974. A dynamic model of cell membranes. **230:**26-33 (Mar.). *Describes the membrane lipid bilayer and associated protein molecules.*

Hanawalt, P. C., and R. H. Haynes [Eds.]. 1973. The chemical basis of life. Readings from *Scientific American*, San Francisco, W. H. Freeman and Co., Publishers.

Sharon, N. 1974. Glycoproteins. **230:**78-86 (May). *These proteins with linked sugar side chains play several important functional roles.*

Stroud, R. M. 1974. A family of protein-cutting proteins. **231:**74-88 (July). *The serine proteases perform diverse functions despite their common descent from a single ancestral enzyme.*

CHAPTER 3
ORIGIN OF LIFE

Since life's beginnings more than 3 billion years ago, life and the environment have been inseparable. Their interactions have determined the character of life on earth and the character of the earth itself.

Photograph by C. P. Hickman, Jr.

Where did life on earth come from? This is an ancient personal question that we must suppose has aroused human curiosity since the dawn of cultural development. Although the great religions have sought to satisfy this curiosity, they have not provided nor were ever intended to provide a scientific explanation of the detailed sequence of events that culminated in the first appearance of life. To most biologists the question of life's beginnings is one of profound interest. The biologist is struck by the remarkable unity of nature (Principle 3, p. 10). As more concerning the identifiable components of life is learned, an evolutionary pattern in the structure and function of living things can be seen.

All organisms, from humans to the smallest microbes that transcend the rather arbitrary boundary between life and nonlife, share two kinds of basic biomolecules—nucleic acid and protein. Both are large and complex in form. As we have seen in the previous chapter, the nucleic acids DNA and RNA are composed of nitrogenous bases, sugars, and phosphoric acid. They are the basic genetic polymers of all living things and carry the informational blueprint that determines the cell's activities and directs the synthesis of proteins. Proteins are composed of 20 different amino acids joined together with peptide linkages. Thus all organisms use the same simple building blocks: 20 amino acids, 5 bases (adenine, guanine, cytosine, uracil, and thymine) 2 sugars (ribose and deoxyribose), and 1 phosphate.

The remarkable uniformity of life extends also to cell function. The metabolic processes that convert foodstuffs into a usable form of energy consistently occur in the simplest organisms to the most complex. This example, along with other examples of molecular and functional identity, consequently suggests that all life must have had a common beginning.

Even though we acknowledge the kinship of living things, we must admit at the beginning that we do not know how life on earth originated. Until recently the study of life's origins was not considered worthy of serious speculation by biologists because, it was argued, the absence of a geologic record made the course of events resulting in the appearance of life unknowable. This situation has changed.

Since 1950 several laboratories around the world have been devoting full-time research efforts to origin-of-life studies. It is a multidisciplinary effort that requires the contributions of scientists of several specialties—biologists, chemists, physicists, geologists, and astronomers. From such studies it has been possible to reconstruct a scenario of ancient events in which simple single-celled living organisms evolved more than 3 billion years B.P. (before the present) from inorganic constituents present on the surface of the earth. These studies are not attempts to prove or disprove any existing belief, religious or philosophic, but rather they are endeavors to solve a great cosmic mystery. Although in some ways a historic reconstruction, origin-of-life studies are buoyed by the recent successful simulation of the "primordial broth" in which life began, the discovery of amino acids in extraterrestrial meteorites, and a vast amount of supportive geochemical information.

HISTORIC PERSPECTIVE

People have always been awed and mystified by the question of how life originated on the earth. From ancient times it was commonly believed that life could arise by spontaneous generation from dead material, in addition to arising from parental organisms by reproduction (biogenesis). Frogs appeared to arise from damp earth, mice from putrified matter, insects from dew, maggots from decaying meat, and so on. Warmth, moisture, sunlight, and even starlight were often mentioned as beneficial factors that encouraged spontaneous generation.

These ideas were developed into an elaborate theory by the Greeks and reappeared frequently in the writings of Aristotle (384 to 322 B.C.). They became firmly entrenched in virtually all cultures, including those of the Far East, and were unquestioningly accepted by even relatively recent great figures such as Copernicus, Bacon, Galileo, Harvey, Descartes, Goethe, and Schelling. Spontaneous generation was also supported by Christian philosophers who pointed out that, according to the first chapter of Genesis, God did not create plants and animals directly but bade the waters to bring them forth.

Inevitably the question of spontaneous generation fell under the scrutiny of experimental science (sixteenth and seventeeth centuries). At first such studies were ill-conceived efforts to supplement natural observations of spontaneous generation by artificially producing various organisms in the laboratory. A typical recipe is one for making mice, given by the Belgian plant nutritionist Jean Baptiste van Helmont. "If you press a piece of underwear soiled with sweat together with some wheat in an open jar, after about 21 days the odor changes and the ferment. . . . changes the wheat into mice. But what is most remarkable is that the mice which came out of the wheat and underwear were not small mice, not even miniature adults or aborted mice, but adult mice emerge!''

The first attack on the doctrine of spontaneous generation occurred in 1668 when the Italian physician Francesco Redi exposed meat in jars, some of which were uncovered while some were covered with parchment and some with wire gauze. The meat in all three kinds of vessels spoiled, but only the open vessels had maggots, and he noticed that flies were constantly entering and leaving these vessels. He concluded that, if flies had no access to the meat, no worms would be found.

Although Redi's refutation of spontaneous generation became widely known, the doctrine was too firmly entrenched to be disbelieved. In 1748 the English Jesuit priest John T. Needham boiled mutton broth and put it in corked containers. After a few days the medium was swarming with microscopic organisms. He concluded that spontaneous generation was real because he believed that he had killed all living organisms by boiling the broth and that he had excluded the access of others by the precautions he took in sealing the tubes.

However, an Italian investigator, Abbé Lazzaro Spallanzani (1767), was critical of Needham's experiments and conducted experiments that dealt another blow against the theory of spontaneous generation. He thoroughly boiled extracts of vegetables and meat, placed these extracts in clean vessels, and sealed the necks of the flasks hermetically in flame. He then immersed the sealed flasks in boiling water for several minutes to make sure that all germs were destroyed. As controls, he left some tubes open to the air. At the end of 2 days he found the open flasks swarming with organisms; the others contained none.

This experiment still did not settle the issue, for the advocates of spontaneous generation maintained either that air, which Spallanzani had excluded, was necessary for the production of new organisms or that the method he used had destroyed the vegetative power of the medium. When oxygen was discovered (1774), the opponents of Spallanzani seized on this as the vital principle that he had destroyed in his experiments.

It remained for the great French scientist Louis Pasteur to silence all but the most stubborn proponents of spontaneous generation with an elegant series of experiments with his famous ''swan-neck'' flasks. Pasteur (1861) answered the objection to the lack of air by introducing fermentable material into a flask with a long S-shaped neck that was open to the air (Fig. 3-1). The flask and its contents were then boiled for a long time. Afterwards the flask was cooled and left undisturbed. No fermentation occurred, for all organisms that entered the open end were deposited on the floor of the neck and did not reach the flask contents. When the neck of the flask was cut off, the organisms in the air could fall directly on the fermentable mass and fermentation occurred within it in a short time. Pasteur concluded that, if suitable precautions were taken to keep out the germs and their reproductive elements

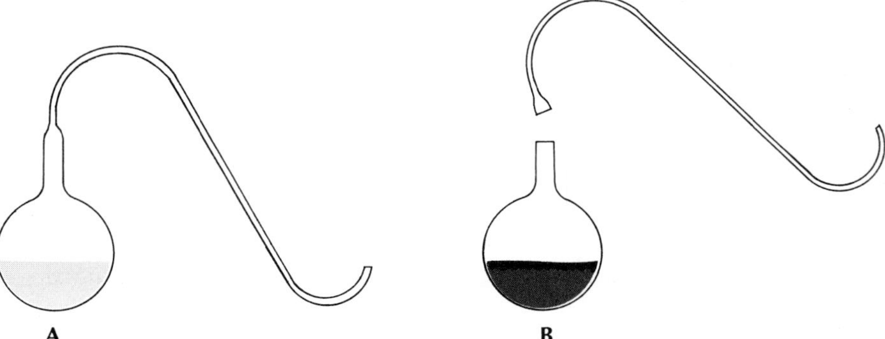

FIG. 3-1 Louis Pasteur's swan-neck flask experiment. **A,** Sugared yeast water boiled in swan-neck flask remains sterile until neck is broken. **B,** Within 48 hours, flask is swarming with life.

(such as eggs and spores), no fermentation or putrefaction could take place.

Pasteur brought an end to the long and tenacious career of the concept of spontaneous generation. Pasteur's work showed that no living organisms come into existence except as descendants of similar organisms (Principle 9, p. 11). In announcing his results before the French Academy, Pasteur proclaimed, "Never will the doctrine of spontaneous generation arise from this mortal blow." Paradoxically, in showing that spontaneous generation did not occur as previously claimed (production of mice, maggots, frogs, and others), Pasteur also ended for a time further inquiry into the spontaneous origins of life. A lengthy period of philosophic speculation followed, but virtually no experimentation on life's origins was performed for 60 years.

Renewal of inquiry: Haldane-Oparin hypothesis

The rebirth of interest into the origins of life occurred in the 1920s. In this decade the Russian biochemist Alexander I. Oparin and the British biologist J. B. S. Haldane independently proposed that life originated on earth after an inconceivably long period of "abiogenic molecular evolution." Rather than arguing that the first living organisms miraculously originated all at once, a notion that had constrained fresh thinking for so long, Haldane and Oparin suggested that the simplest living units (for example, bacteria) came into being very gradually by the progressive assembly of inorganic molecules into more complex organic molecules. These molecules would react with each other to form living microorganisms.

Haldane proposed that the earth's primitive atmosphere consisted of water, carbon dioxide, and ammonia. When ultraviolet light shines on such a gas mixture, many organic substances such as sugars and amino acids are formed.

Today an ozone layer, produced by short-wave ultraviolet rays acting on atmospheric oxygen to produce ozone (O_3), serves as a protective screen to prevent deadly ultraviolet light from reaching the earth's surface. But on the primitive earth, long before oxygen began to accumulate in the atmosphere, ultraviolet radiation must have been very intense. Haldane believed that the early organic molecules could accumulate in the primitive oceans to form "a hot dilute soup." In this primordial broth carbohydrates, fats, proteins, and nucleic acids might have been assembled to form the earliest microorganisms.

Oparin, too, suggested that the earth's primitive atmosphere lacked oxygen and instead contained hydrogen, methane, ammonia, and other reducing compounds. He proposed that the organic compounds required for life were formed spontaneously in such a reducing atmosphere under the influence of sunlight, lightning, and the intense heat of volcanos.

The Haldane-Oparin hypothesis greatly influenced theoretic speculation on the origins of life during the 1930s and 1940s. Finally in 1953 Stanley Miller, working with Harold Urey in Chicago, made the first attempt to simulate with laboratory apparatus the conditions thought to prevail on the primitive earth. This strikingly successful experiment, which is described in more detail later in this chapter, demonstrated that important biomolecules are formed in surprisingly large amounts when an electric discharge is passed through a reducing atmosphere of the kind proposed by Haldane and Oparin. With the realization that it was possible to simulate successfully a prebiotic milieu in the laboratory, a new era in origin-of-life studies was ushered in. It coincided with the dawn of the space age and a new public interest in the question of life's origins.

PRIMITIVE EARTH

Formation of the solar system

Astronomers have recently placed the age of the universe at 20 billion years. The sun and the planets are believed to have been formed approximately 4.6 billion years ago out of a spheric cloud of cosmic dust and gases that had some angular momentum. The cloud collapsed under the influence of its own gravity into a rotating disc. As the material in the central part of the disc condensed to form the sun, a substantial amount of gravitational energy was released as radiation. The pressure of this outwardly directed radiation prevented the complete collapse of the nebula into the sun. The material left behind began to cool and eventually gave rise to the planets (Fig. 3-2).

The inner planets (Mercury, Venus, Earth, and Mars) are small and composed mostly of nonvolatile elements because the high temperatures of the cooling planets drove off the volatile elements. The outer planets are composed largely of volatile compounds such as methane, ammonia, water, hydrogen, and helium because these materials were blown outward into this re-

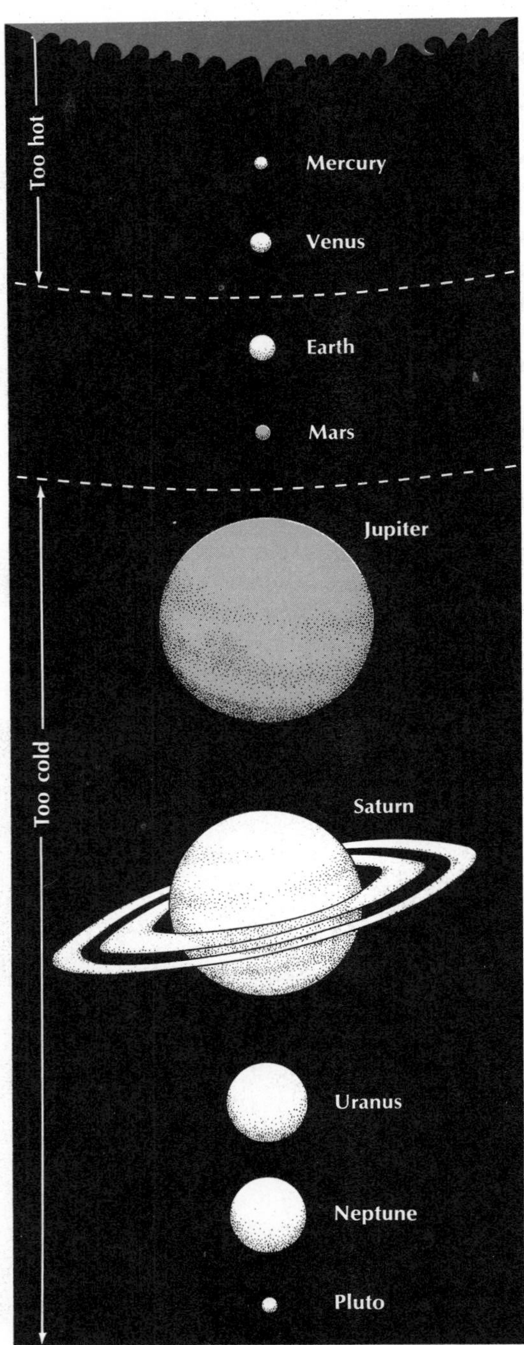

FIG. 3-2 Solar system showing narrow range of conditions suitable for life.

gion of the collapsing disc by the sun's radiation pressure. In the region occupied by the planets Jupiter and Saturn, massive bodies of hydrogen and helium condensed. The low temperatures and enormous gravitational fields of these bodies prevented their gases from escaping. It is believed that the condensation of the solar system was 95% complete within the relatively short time of 50,000 years.

Origin of the earth's atmosphere

Evolution of the primeval reducing atmosphere. While the earth was still more or less gaseous, the various atoms were sorted out according to weight. Heavy elements (silicon, aluminum, nickel, and iron) gravitated toward the center, while the lighter elements (hydrogen, oxygen, carbon, and nitrogen) remained in the surface gas. Hydrogen and helium, because of their volatility, continued to escape as the earth condensed, and these elements became severely depleted from the primeval atmosphere. Neon and argon were almost completely lost from the atmosphere. Oxygen, nitrogen, and water, which are the major constituents of our present oxidizing atmosphere, could not escape from the earth because they were present in nonvolatile, chemically combined forms trapped in dust particles. Later as the earth condensed, water, carbon compounds, and nitrogen or ammonia were released into the atmosphere from the earth's interior by volcanic activity, which was much more extensive then than it is now.

It is generally agreed that the earth's primeval atmosphere contained no more than a trace of oxygen. It was a **reducing atmosphere;** that is, it contained a predominance of molecules having less oxygen than hydrogen. Methane (CH_4), ammonia (NH_3), and water (H_2O) are examples of fully reduced compounds; such compounds are believed to have composed the early atmosphere of the earth. Carbon dioxide (CO_2), nitrogen (N_2), and traces of hydrogen (H_2) may also have been present.

The character of the primeval atmosphere is very important in any discussion of the origins of life because the organic compounds of which living organisms are made are not stable in an oxidizing atmosphere. Organic compounds are not synthesized in our oxidizing atmosphere today, and they are not stable if they are introduced into it. Obviously life could not have originated without the basic organic building blocks from which the first organisms were assembled. Consequently modern origin-of-life theories assume that the primeval atmosphere was reducing because the synthesis of compounds of biologic importance occurs only under reducing conditions. Fortunately geochemical evidences tend to support the belief that the early atmosphere was a reducing one that arose by degassing of the earth's interior.

Appearance of oxygen. As water, nitrogen, and carbon compounds continued to enter the atmosphere from volcanos, the atmosphere became saturated with water vapor, and rain began to fall. Small lakes formed, enlarging into oceans. The oceans gradually became salted as rocks weathered. During this early period, which lasted 1.5 billion years, the atmosphere was still reducing well after the first living protocells such as algae and bacteria had evolved.

Our atmosphere today is strongly oxidizing. It contains 79% molecular nitrogen, approximately 21% free oxygen, and a small amount of carbon dioxide. Although the time course for its development is much disputed, at some point oxygen began to appear in significant amounts in the atmosphere. In the primitive reducing atmosphere oxygen was formed by the decomposition of water in either of two ways.

The first is the photolytic action of ultraviolet light from the sun on water in the upper atmosphere.

$$2\,H_2O \xrightarrow{\text{UV light}} 2\,H_2 + O_2$$

The hydrogen produced escapes from the earth's gravitational field, leaving free oxygen to accumulate in the atmosphere. The amount produced by the photodissociation of water is now quite small, although it may have been significant over the immense span of time in which it occurred.

The second and probably more important source of oxygen is photosynthesis. Almost all oxygen produced at the present time is produced by algae and land plants. Each day the earth's plants combine approximately 400 million tons of carbon with 70 million tons of hydrogen to set free 1.1 billion tons of oxygen. Oceans are the major source of oxygen. Almost all oxygen produced today is consumed by organisms oxidizing their food to carbon dioxide; if this did not occur, the amount of oxygen in the atmosphere would double in approximately 3,000 years. Since Precambrian fossil algae resemble modern algae and since algae today are the greatest producers of oxygen, it seems probable that most of the oxygen in the early atmosphere was produced by photosynthesis.

CHEMICAL EVOLUTION

Sources of energy

According to the Haldane-Oparin hypothesis, a variety of carbon compounds gradually accumulated on the surface of the earth during a lengthy period of prebiotic chemical evolution. The primitive reducing atmosphere contained simple gaseous compounds of carbon, nitrogen, oxygen, and hydrogen, such as carbon dioxide, molecular nitrogen, water vapor, methane, and ammonia. These were the starting materials from which organic compounds were made. However, if these gaseous compounds are mixed together in a closed glass system and allowed to stand at room temperature, they never chemically react with each other. To promote a chemical reaction a continuous source of **free energy** sufficient to overcome reaction-activation barriers must be supplied.

For example, one of the simplest and most important prebiotic chemical reactions is the formation of hydrogen cyanide (HCN) from nitrogen (N_2) and methane (CH_4). When an electric discharge is passed through an atmosphere of nitrogen, some of the molecules absorb enough energy to dissociate into atoms.

$$N_2 \rightarrow 2\,N$$

Dissociated nitrogen atoms are highly reactive. If methane is present in the atmosphere, nitrogen atoms react to form hydrogen cyanide and hydrogen.

$$N + CH_4 \rightarrow HCN + \frac{3}{2}H_2$$

Other sources of energy such as ultraviolet light or heat could equally well cause the formation of hydrogen cyanide from nitrogen and methane. Once formed, hydrogen cyanide dissolves in rain and is carried into lakes and oceans where it and other reactive molecules can form organic compounds.

The sun is by far the most powerful source of free energy for the earth. Each year each square centimeter of the earth receives an average 260,000 calories of radiant energy. The way in which this energy is distributed over the infrared, visible, and ultraviolet re-

TABLE 3-1 Present sources of energy averaged over the earth*

Source	Energy (cal/cm²/year)
Total radiation from sun	260,000
Infrared (above 7,000 Å)	143,000
Visible (3,500-7,000 Å)	113,600
Ultraviolet	
2,500-3,500 Å	2837
2,000-2,500 Å	522
1,500-2,000 Å	39
<1,500 Å	1.7
Electric discharges	4
Shock waves	1.1
Radioactivity (to 1.0 km depth)	0.8
Volcanos	0.13
Cosmic rays	0.0015

*Modified from Miller, S. L., and L. E. Orgel. 1974. The origins of life on the earth. Englewood Cliffs, N.J., Prentice-Hall, Inc.

gions of the spectrum is shown in Table 3-1. The gases present in the primitive atmosphere would not have absorbed visible or infrared energy. Since the solar energy available falls off rapidly in the ultraviolet region, only approximately 0.2% of the total energy is at wavelengths shorter than 2,000 Å, where it can be absorbed by molecules such as methane, water, and ammonia. Still, ultraviolet radiation must have been an important source of energy for photochemical reactions in the primitive atmosphere.

Electric discharges must have provided another source of energy for chemical evolution. Although the total amount of electric energy released by lightning is small compared to solar energy, nearly all of the electric energy of lightning is effective in synthesizing organic compounds in a reducing atmosphere. A single flash of lightning through a reducing atmosphere generates a large amount of organic matter. Thunderstorms are thought to have been more prevalent on the earth than they are today and may have been one of the most important sources of energy for organic synthesis.

Volcanos and hot springs were available on the primitive earth as sources of energy, but it is doubtful that they were a major site of prebiotic organic synthesis. Cosmic rays, radioactivity, and sonic energy generated by ocean waves were also available sources of energy, but their contribution cannot be evaluated and was probably small in any case. Of the other sources of energy, only shock waves generated by meteorites passing through the primitive atmosphere may have produced a large amount of organic matter. Meteorites generate intense temperatures as high as 20,000° C and pressures exceeding 15,000 atmospheres. A large meteorite could generate millions of tons of organic matter in its wake.

It is not known which energy source was most important for prebiotic synthesis on earth. Most authorities believe both electric discharges and ultraviolet radiation made major contributions, but the quantity and quality of prebiotic energy sources remain a subject of debate and inquiry.

Prebiotic synthesis of small organic molecules

The first steps leading toward life on earth happened when simple organic molecules were generated in the earth's reducing atmosphere by the action of lightning, ultraviolet radiation, and perhaps other kinds of available energy. These simple molecules were washed into lakes and oceans where they formed a "primordial soup." Our understanding of what happened to produce the first reactive organic molecules comes largely from a young branch of chemistry called prebiotic chemistry.

Earlier in this chapter we referred to Stanley Miller's pioneering simulation of primitive earth conditions in the laboratory. Miller built an apparatus designed to circulate a mixture of methane, hydrogen, ammonia, and water past an electric spark (Fig. 3-3). Water in the flask was boiled to produce steam that helped to circulate the gases. The products formed in the electric discharge (representing lightning) were condensed in the condenser and collected in the U-tube and small flask (representing ocean).

After a week of continuous sparking, the water containing the products was analyzed. The results were surprising. Approximately 15% of the carbon that was originally in the reducing "atmosphere" had been converted into organic compounds that collected in the "ocean." The most striking finding was that many compounds related to life were synthesized. These included four amino acids commonly found in proteins, urea, and several simple fatty acids.

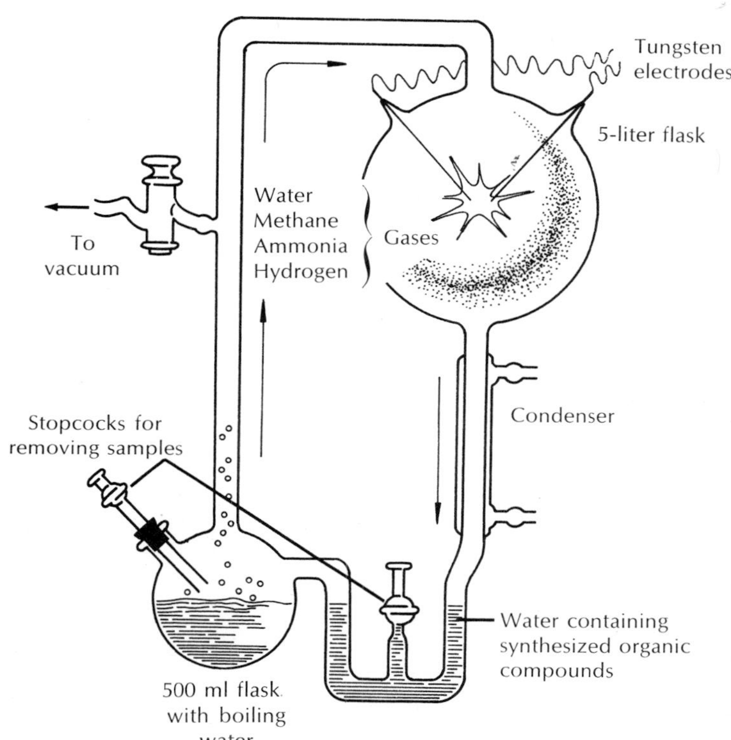

FIG. 3-3 Dr. S. L. Miller and apparatus used in experiment on the synthesis of amino acids with an electric spark in a reducing atmosphere. (Courtesy S. L. Miller.)

We can appreciate the astonishing nature of this synthesis when we consider that there are thousands of known organic compounds with structures that are no more complex than those of the amino acids formed. Yet in Miller's synthesis most of the relatively few substances formed were compounds found in living organisms. This was surely no coincidence, and it suggests that prebiotic synthesis on the primitive earth occurred under conditions that were not greatly different from those that Miller chose to simulate.

Miller's work stimulated many other investigators to repeat and extend his experiment. It was soon found that amino acids could be synthesized in many different kinds of gas mixtures that were heated (volcanic heat), irradiated with ultraviolet light (solar radiation), or subjected to electric discharge (lightning). All that was required to produce amino acids was that the gas mixture be reducing and that it be subjected violently to some energy source. It was also confirmed that no amino acids could be produced in an atmosphere containing oxygen.

Miller discovered that amino acids were not formed directly in the spark, but rather were produced by the condensation of certain reactive intermediates, especially hydrogen cyanide and formaldehyde (HCHO) reacting with ammonia. It was found that hydrogen cyanide would react with ammonia under prebiotic conditions to form adenine, one of the four bases found in nucleic acids and a component of ATP, the universal energy intermediate of living systems. Adenine is a purine, a chemically complex molecule (p. 36). The ease with which it was produced under prebiotic conditions suggests that it came to occupy a central position in biochemistry because it was abundant on the primitive earth. Other nucleic acid bases and sugars have been synthesized under conditions thought to have prevailed on the primitive earth.

Thus the experiments of many scientists have shown that highly reactive intermediate molecules such as hydrogen cyanide, formaldehyde, and cyanoacetylene are formed when a reducing mixture of gases is subjected to a violent energy source. These react with water and ammonia to form more complex organic molecules, including amino acids, fatty acids, urea, aldehydes, sugars, several purine and pyrimidine bases, ATP, porphyrins—indeed all the building blocks required for the synthesis of the most complex organic compounds of living matter.

Formation of polymers

Need for concentration. The next stage in chemical evolution involved the condensation of amino acids, purines, pyrimidines, and sugars to yield larger molecules that resulted in proteins and nucleic acids. Such condensations do not occur easily in dilute solutions because the presence of excess water tends to drive reactions toward decomposition. Although the primitive ocean has been called a primordial soup, it was probably a rather dilute one containing organic material that was approximately one-tenth to one-third as concentrated as chicken bouillon.

Prebiotic synthesis must have occurred in restricted regions where concentrations were higher. The primordial soup might have been concentrated by evaporation in lakes, ponds, or tidepools. Dilute aqueous solutions could also have been concentrated effectively by freezing. As ice freezes, organic solutes are concentrated in the solution that separates from the pure ice. This technique is employed to produce applejack from cider in the northern United States and Canada. When a barrel of cider is allowed to freeze, a liquid residue remains that contains most of the alcohol and flavoring materials.

The Haldane-Oparin hypothesis suggests that the primitive ocean was a *warm* primordial soup. However, there is increasing evidence that prebiotic synthesis may have occurred in a cold rather than in a warm ocean. In a cold ocean, not unlike the ocean today with an average temperature of 4° C, much of the polar water would be frozen. This would be favorable for prebiotic synthesis for two reasons: (1) freezing of the water would concentrate the organic molecules in the residue, and (2) the low temperatures would slow the spontaneous decomposition of organic compounds and polymers.

Prebiotic molecules also might have been concentrated in the foam of breaking waves or by adsorption on the surface of clays and other minerals in the mud of estuaries. Clay has the capacity to concentrate and condense large amounts of organic molecules from an aqueous solution.

An attractive related model for prebiotic condensation was proposed by Oparin. He suggested that colloidal droplets assembled from two or more polymers of opposite charge, which he called "coacervates" (from Latin *coacervare,* to assemble or cluster together), provided an important site for the adsorption of soluble organic molecules. Coacervate droplets were formed initially when two or more polymers of opposite charges were mixed. Even though the coacervate droplets were floating in the ocean or in a lake, they would provide a locally nonaqueous environment that was favorable for condensation reactions. Oparin and his Russian associates have prepared coacervate droplets from many different biologic compounds (for example, soaps, serum albumin, gelatin, and gum arabic); however, it has not yet been shown that important prebiotic synthesis occurred within coacervates.

Thermal condensations. Most biologic polymerizations are dehydration reactions; that is, monomers are linked together by the removal of water. The peptide bond is a familiar example. In living systems dehydration reactions always take place in an aqueous (cellular) environment. This apparent paradox—the removal of water with water all around—is possible because in living systems the appropriate dehydrating enzymes are present. Without enzymes and energy supplied by ATP, the macromolecules (proteins and nucleic acids) of living systems soon break down into their constituent monomers.

One of the most critical problems is explaining how dehydrations occurred in primitive earth conditions before enzymes appeared. The simplest dehydration is accomplished by driving off water from solids by direct heating. For example, if a mixture of all 20 amino acids is heated to 180° C, a good yield of polypeptides is obtained.

The thermal synthesis of polypeptides to form "proteinoids" has been studied extensively by the American scientist Sidney Fox. When proteinoids are boiled in water they form enormous numbers of hard, minute spherules called microspheres. Proteinoid micro-

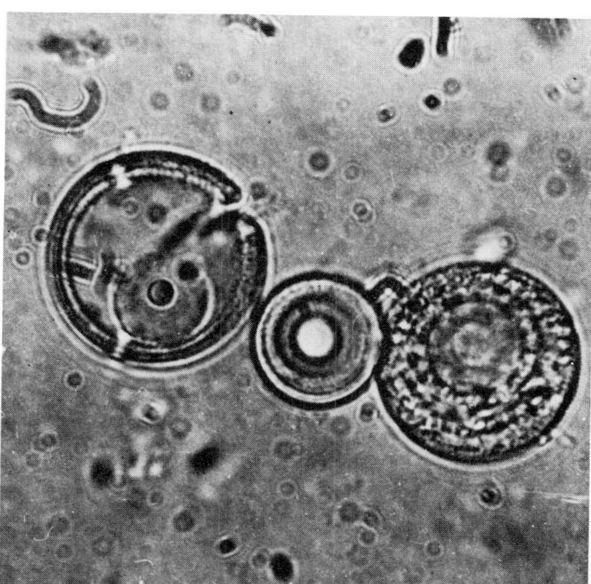

FIG. 3-4 Electron micrograph of proteinoid microspheres. These proteinlike bodies can be produced in the laboratory from polyamino acids and may represent precellular forms. They have definite internal ultrastructure. (×1700.) (Courtesy S. W. Fox, Institute of Molecular Evolution, University of Miami, Coral Gables, Fla.)

spheres (Fig. 3-4) possess certain characteristics of living systems. Each is not more than 2 μm in diameter and is comparable in size and shape to spheric bacteria. Their outer walls appear to have a double layer, and they show osmotic and selective diffusion properties. They may grow by accretion or proliferate by budding. There is no way to know whether proteinoids may have been the ancestors of the first cells or whether they are just interesting creations of the chemist's laboratory. They must be formed under conditions that would have been found only in volcanos. Possibly organic polymers might have condensed on or in volcanos and then, wetted by rain or dew, reacted further in solution to form polypeptides or polynucleotides.

ORIGIN OF LIVING SYSTEMS

The first living organisms were cells—autonomous membrane-bound units with a complex functional organization that permitted the essential activities of assimilation, metabolism, excretion, reproduction, irritability, and other characteristic cellular functions. The primitive chemical systems we have described lack all of these properties. The principal problem in understanding the origin of life is explaining how primitive chemical systems could have become organized into living autonomous cells.

As we have seen, a lengthy chemical evolution on the primitive earth produced several molecular components of living forms. In a later stage of evolution, nucleic acids began to behave as simple genetic systems that directed the synthesis of other polymers, especially proteins. We are now confronted with a troublesome chicken-egg question: (1) How could nucleic acids have appeared without enzymes? (2) How could enzymes have evolved without nucleic acids to direct their synthesis? Certainly nucleic acids are necessary for replication, and no great evolutionary progress could have been made without them. But, if nucleic acids (or polymers resembling nucleic acids) evolved before proteins, they must have been able to catalyze their own synthesis, as well as direct the synthesis of other types of compounds.

Proponents of the theory that nucleic acids arose first argue that they behaved as "naked" genes that possessed the means for expressing information as well as storing it. Thus far, however, it has not been possible to synthesize nucleic acids with natural linkages under prebiotic conditions. Furthermore no known nucleic acids of living organisms today have any enzymatic capabilities: they are pure repositories of information that cannot be expressed in the absence of proteins. This does not mean that they never did have catalytic powers, but it seems unlikely that nucleic acids would have evolved to code for proteins that did not already exist.

On the other hand, there is some evidence that proteins with rudimentary enzymatic and metabolic functions may have formed primitive cell-like structures with limited reproductive capabilities. The transmission of properties to subsequent generations would be less exact in the absence of nucleic acids, but more exact self-replication would not be an important requirement for early life forms on the primitive earth because lack of environmental competition would encourage, rather than prevent, diverse individuality.

Several authors favor the idea that proteins arose before nucleic acids. This idea is appealing for several reasons. First of all, proteins are the basis of the cell life cycle; the cell is built mainly of and functions principally on proteins. No cell could live on nucleic acids alone. Second, reactant amino acid molecules exhibit a high degree of internal ordering when combined under prebiotic conditions. They link together in nonrandom arrangements to form polypeptides of very limited diversity. Furthermore, artificial peptides exhibit a natural tendency to coil and form a typical alpha-helix as soon as the polypeptide chain exceeds 8 or 10 amino acids in length. This suggests that nucleic acids would not have been absolutely essential in specifying sequences in primitive proteins. Finally, it is a premise of evolutionary theory that something must exist (for example, a protein) before its existence can be directed (by nucleic acids).

These reasons establish a basis for understanding how the first ''organisms'' may have been self-ordered and self-assembling microsystems of proteins. What did these first creatures look like? They may have consisted of nothing more than grains of clay or mineral onto which were adsorbed amino acids and other micromolecules. These organic molecules, being concentrated in a microenvironment that was segregated from the water around it, were then able to form primitive proteins. The mineral may also have acted as a primitive catalyst to help polymerize amino acids and nucleotides. Only later did the nucleic acids assume the role of administrators that controlled the synthesis of polypeptides and protoenzymes.

Once this stage of organization was reached, natural selection began acting on these primitive self-replicating systems. This was a critical point. Before this stage, biogenesis was shaped by the favorable environmental conditions on the primitive earth and by the nature of the reacting elements themselves. When self-replicating systems became responsive to the forces of natural selection, their subsequent evolution became directed. The more rapidly replicating and more successful systems were favored and they replicated even faster. In short, the most efficient forms survived. From this evolved the genetic code and fully directed protein synthesis. The system was a protocell and it could be called a living organism.

Origin of metabolism

Living cells today are organized systems that possess complex and highly ordered sequences of enzyme-mediated reactions. Some cells trap solar energy and convert it into chemical bond energy, which is stored in glucose, ATP, and other molecules. Other cells are able to utilize these sources of bond energy to grow, divide, and maintain their internal integrity. Indeed, the attributes of life, involving energy conversion, assimilation, secretion, excretion, responsiveness to stimuli, and capacity to reproduce, all depend on the complex metabolic patterns characteristic of contemporary cells. How did such vastly complex metabolic schemes develop?

The earliest microorganisms were most likely anaerobic heterotrophs, probably bacteria* that obtained all their nutrients directly from the environment. They would not need and doubtless did not possess any biosynthetic capacity. Chemical evolution had already supplied generous stores of nutrients in the prebiotic soup. There would be neither advantage nor need for the earliest organisms to synthesize their own compounds, as long as they were freely available from the environment. But as soon as deficiencies of certain essential nutrients occurred, alternative sources had to be found. At this point, those microorganisms that could synthesize these essential compounds from other accessible compounds would clearly have a greater advantage for surviving than those that could not.

For example, ATP is the immediate energy coinage of all living organisms. Since it has been formed in simulated prebiotic experiments, there is good reason to believe that it was present in the primitive environment and available to and used by protocells. Thus early organisms would have depended on the environmental supply for their ATP requirements.

Once the supply was exhausted or became precarious, perhaps because of an increase in the numbers of organisms using it, those protocells able to convert a precursor such as glycerate-2-phosphate to ATP would have had a tremendous advantage over those that lacked this capability. They would be selected for survival and would thrive. In the same way, when the supply of glycerate-2-phosphate became limiting it would be necessary to synthesize it from another precursor supplied by the environment. Again, the most successful organisms would have been those that chanced to develop this metabolic capability. Long se-

*The smallest and simplest contemporary cells are found in a group of bacteria, the pleuropneumonia-like organisms, also known as mycoplasmas. Of these, *Mycoplasma laidlawii,* a free-living bacterium of sewage is the smallest, only 0.1 μm in diameter. By comparison most bacteria are 1 to 5 μm in diameter, and the average eukaryotic cell is 20 to 50 μm in diameter. Mycoplasmas are enclosed by a simple ''unit'' membrane (p. 75) and the only organelles are ribosomes and double-stranded loops of DNA.

quences of reactions could have arisen in this manner.

It is important to realize that an enzyme is required to catalyze each of these reactions. So, when we say that early protocells developed a reaction sequence as we have described (A made from B, B from C, and so on), we are really saying that the appropriate enzymes appeared to catalyze these reactions. The numerous enzymes of cellular metabolism appeared when cells became able to utilize proteins for catalytic functions and thereby gained a selective advantage. No planning was required; the results were achieved through natural selection.

Appearance of photosynthesis and oxidative metabolism

Eventually, almost all utilizable energy-rich nutrients of the prebiotic soup were consumed. This ushered in the next stage of biochemical evolution, the use of readily available solar radiation to provide metabolic energy. Photosynthesis, the production of organic compounds from sunlight and atmospheric carbon dioxide, is the only process that restores free energy to the biosphere. Plant photosynthesis makes possible the richness of life on earth as we know it today. The self-nourishing photoautotrophic organisms, mainly green plants, capture solar energy and use it to convert simple inorganic substances into organic materials. The energy-rich compounds they produce provide not only for their own functioning but for the heterotrophic organisms, mainly animals, that feed on autotrophs. The heterotrophs in turn release important raw materials for autotrophs. This is the energy cycle of the biosphere that is powered by a steady supply of energy from the sun (Principle 2, p. 9).

The appearance of photosynthesis was of enormous consequence for evolution, but like other metabolic events it did not appear all at once. In plant photosynthesis, water is the source of the hydrogen that is used to reduce carbon dioxide to sugars. Oxygen is liberated into the atmosphere.

$$6\ CO_2 + 6\ H_2O \xrightarrow{\text{light}} C_6H_{12}O_6 + 6\ O_2$$

However, the first steps in the development of photosynthesis almost certainly did not involve the splitting of water because a large input of energy is required. Hydrogen sulfide is thought to have been abundant in the primitive earth and was probably the first reducing agent used in photosynthesis.

$$6\ CO_2 + 12\ H_2S \xrightarrow{\text{light}} C_6H_{12}O_6 + 12\ S + 6\ H_2O$$

Later, as hydrogen sulfide and other reducing agents except water were used up, oxygen-evolving photosynthesis appeared. This is thought to have occurred approximately 3 billion years B.P. Gradually oxygen began to accumulate in the atmosphere. The accumulation was probably very slow at first because until an ozone screen appeared as ultraviolet shielding—and this first required the appearance of oxygen—photosynthesizing organisms were restricted to deeper and poorly illuminated photic zones in the ocean where the surface waters would protect them from lethal ultraviolet radiation. When atmospheric oxygen reached approximately 1% of its present level, ozone began to accumulate and ultraviolet radiation was screened out. Now land and surface waters could be occupied, and oxygen production probably increased sharply.

But at this important juncture, accumulating atmospheric oxygen began to interfere with cellular metabolism, which up to this point had evolved under strictly reducing conditions. As the atmosphere slowly changed from a reducing to an oxidizing one, a new and highly efficient kind of energy metabolism appeared: **oxidative (aerobic) metabolism.** By using the available oxygen to oxidize glucose to carbon dioxide and water, 36 molecules of ATP are obtained from each molecule of glucose. By comparison, anaerobic fermentation of glucose to lactic acid releases only 2 molecules of ATP from each glucose molecule. Thus oxidative metabolism allowed organisms to recover much of the bond energy stored by photosynthesis. Most living forms became wholly dependent on oxidative metabolism, and oxygen-evolving photosynthesis became essential for the continuation of life on earth.

The final phase in life's evolution followed. Although a vast span of time—more than 2 billion years—passed before multicellular organisms appeared, living cells, very much as we know them today, surrounded by semipermeable membranes and supporting an efficient oxygen-consuming form of metabolism, were flourishing on earth.

PRECAMBRIAN LIFE

As depicted on the inside back cover of this book, the Precambrian period spanned the geologic time before the beginning of the Cambrian period 600 million years B.P. At the beginning of the Cambrian period, most of the major phyla of invertebrate animals made their appearance within a few million years. This has been called the "Cambrian explosion" because before this time fossil deposits were rare and almost devoid of anything more complex than single-celled algae. What forms of life existed on earth before the burst of evolutionary activity in the early Cambrian world?

Prokaryotes and the age of blue-green algae

The earliest microorganisms were probably **anaerobic bacteria.** Their food was organic matter produced abiotically in the primitive oceans. The bacteria proliferated, giving rise to a great variety of bacterial forms, some of which were capable of photosynthesis. From these arose the oxygen-evolving **blue-green algae** (cyanophytes) some 3 billion years B.P.

Bacteria and blue-green algae are called **prokaryotes,** meaning literally "before nucleus." They lack not only the nuclei typical of higher unicellular organisms but other organelles as well; they have no mitochondria, plastids, chromosomes, centrioles, or vacuoles. They contain the genetic material DNA, but it is not complexed with RNA or proteins as it is in higher forms of life. They reproduce primarily by fission or budding, never by true mitotic cell division.

Bacteria and especially blue-green algae ruled the earth's oceans unchallenged for some 1½ to 2 billion years. The blue-green algae reached the zenith of their success approximately 1 billion years B.P., when filamentous forms produced great floating mats on the ocean surface. This long period of blue-green algae dominance, encompassing approximately two-thirds of the history of life, has been called with justification the "age of blue-green algae."* Bacteria and blue-green algae are so completely different from forms of life that evolved later that they are placed in a separate kingdom, Monera (p. 126).

Appearance of the eukaryotes

Approximately 1 billion years B.P., or perhaps even earlier, organisms with nuclei appeared. These **eukaryotes** (true nucleus), as they are called, have cells with membrane-bound nuclei containing chromosomes in which the DNA is complexed with RNA and proteins. Eukaroytes are larger than prokaryotes, contain at least 1,000 times more DNA, and divide by some form of mitosis. They contain organelles, including mitochondria in which the enzymes for oxidative metabolism are packaged. All of the familiar forms of life (protozoans, fungi, nucleated algae, green plants, and multicellular animals) are composed of eukaryotic cells.

Prokaryotes and eukaryotes are profoundly different from each other and clearly represent a marked dichotomy in the evolution of life. The ascendency of the eukaryotes resulted in a rapid decline in the dominance of blue-green algae since the eukaryotes proliferated and quickly captured rule of the seas.

Why were the eukaryotes immediately so successful? Because they developed that most important prerequisite for rapid evolution—sex. Sex promotes almost limitless genetic variability by mixing the genes of two individuals. By preserving favorable genetic variants, natural selection encourages rapid evolutionary change (see Principle 4, p. 10). Prokaryotes, on the other hand, are asexual and can only stamp out carbon copies of parental cells. The prokaryotes propagate themselves effectively and efficiently, but change occurs only when a genetic mutation intervenes.

The first eukaryotes were no doubt unicellular planktonic forms. Some were photosynthetic autotrophs; others were heterotrophs—grazing herbivores that fed on the prokaryotes. As the blue-green algae were cropped, their dense filamentous mats began to thin, providing space for other species. Carnivores appeared to feed on the herbivores. Soon a balanced ecosystem of carnivores, herbivores, and primary producers appeared. This was ideal for evolutionary diversity. By freeing space, cropping herbivores encouraged a greater diversity of producers, which in turn promoted the evolution of new and more specialized croppers. An ecologic pyramid appeared with carnivores at the top.

The burst of evolutionary activity that followed at the end of the Precambrian period and beginning of the Cambrian period was unprecedented, and nothing approaching it has occurred since. Nearly all animal and plant phyla appeared and established themselves within a relatively brief period of a few million years. The eukaryotic cell made possible the richness and diversity of life on earth today.

*Schopf, J. W. 1974. Paleobiology of the precambrian: the age of blue-green algae. Evolutionary Biology **7**:1-43.

Annotated references

Selected general readings

Blum, H. F. 1968. Time's arrow and evolution, ed. 3. Princeton, N.J., Princeton University Press. *Advanced undergraduate level emphasizing the chemical problems of evolution.*

Cairns-Smith, A. G. 1971. The life puzzle. Toronto, University of Toronto Press. *An unorthodox treatment emphasizing the physical forces that lead to order.*

Dose, K., S. W. Fox, G. A. Deborin, and T. E. Pavlovskaya, editors. 1974. The origin of life and evolutionary biochemistry. New York, Plenum Publishing Corp. *This edited volume of 41 chapters by specialists in the origin of life commemorates the fiftieth anniversary of the appearance of Aleksandr Oparin's first published work in 1924.*

Fox, S. W., and K. Dose, 1977. Molecular evolution and the origin of life, ed. 2. San Francisco, W. H. Freeman & Co., Publishers. *An advanced survey of wide scope. The author's experiments with proteinoids are detailed.*

Kenyon, D. H. and G. Steinman. 1969. Biochemical predestination. New York, McGraw-Hill Book Co. *One of the best books on origin of life. Advanced and detailed but highly readable and with excellent summaries.*

Lemmon, R. M. 1970. Chemical evolution. Chemical Reviews **70:**95-109. *Excellent advanced review of this subject.*

Miller, S. L. 1953. A production of amino acids under possible primitive earth conditions. Science **117:**528. *The first report of the synthesis of amino acids in a reducing atmosphere.*

Miller, S. L. and L. E. Orgel. 1974. The origins of life. Englewood Cliffs, N.J., Prentice-Hall, Inc. *Advanced undergraduate level emphasizing the chemical problems of life's origins.*

Oparin, A. I. 1953. The origin of life. New York, Dover Publications, Inc. *This is the paperback edition of Oparin's most important work first published by Macmillan, Inc., in 1938. Later but in some ways less satisfactory versions have been published by Academic Press, Inc.*

Orgel, L. E. 1973. The origins of life: molecules and natural selection. New York, John Wiley & Sons, Inc. *Well-balanced treatment at the intermediate level.*

Ponnamperuma, C. 1972. The origins of life. New York, E. P. Dutton & Co., Inc. *Popularized account. Well-illustrated.*

Selected *Scientific American* articles

Barghoorn, E. S. 1971. The oldest fossils. **224:**30-42 (May). *Describes and illustrates fossils of algae and bacteria, some more than 3 billion years old.*

Cameron, A. G. W. 1975. The origin and evolution of the solar system. **233:**32-41 (Sept.).

Glaessner, M. F. 1961. Precambrian animals. **204:**72-78 (Mar.).

Huang, S. 1960. Life outside the solar system. **202:**55-63 (Apr.).

Lawless, J. G., C. E. Folsome, and K. A. Kvenvolden. 1972. Organic matter in meteorites. **226:**38-46 (June). *Organic compounds found in meteorites appear not to be of biologic origin.*

Sagan, C., and F. Drake. 1975. The search for extraterrestrial life. **232:**80-89 (May). *There is little doubt that civilizations even more advanced than ours exist elsewhere in the universe, but locating them will require great effort.*

Wald, G. 1954. The origin of life. **191:**44-53 (Aug.). *A popular, concise account.*

CHAPTER 4

THE CELL AS THE UNIT OF LIFE

Dividing cells in developing embryos of the common urchin *Echinus esculentus*. Present are an unfertilized egg (*top center,* shown without a surrounding membrane), fertilized eggs *(right),* 2-cell and 4-cell stages. Cytokinesis is nearly complete in the two embryos at bottom.

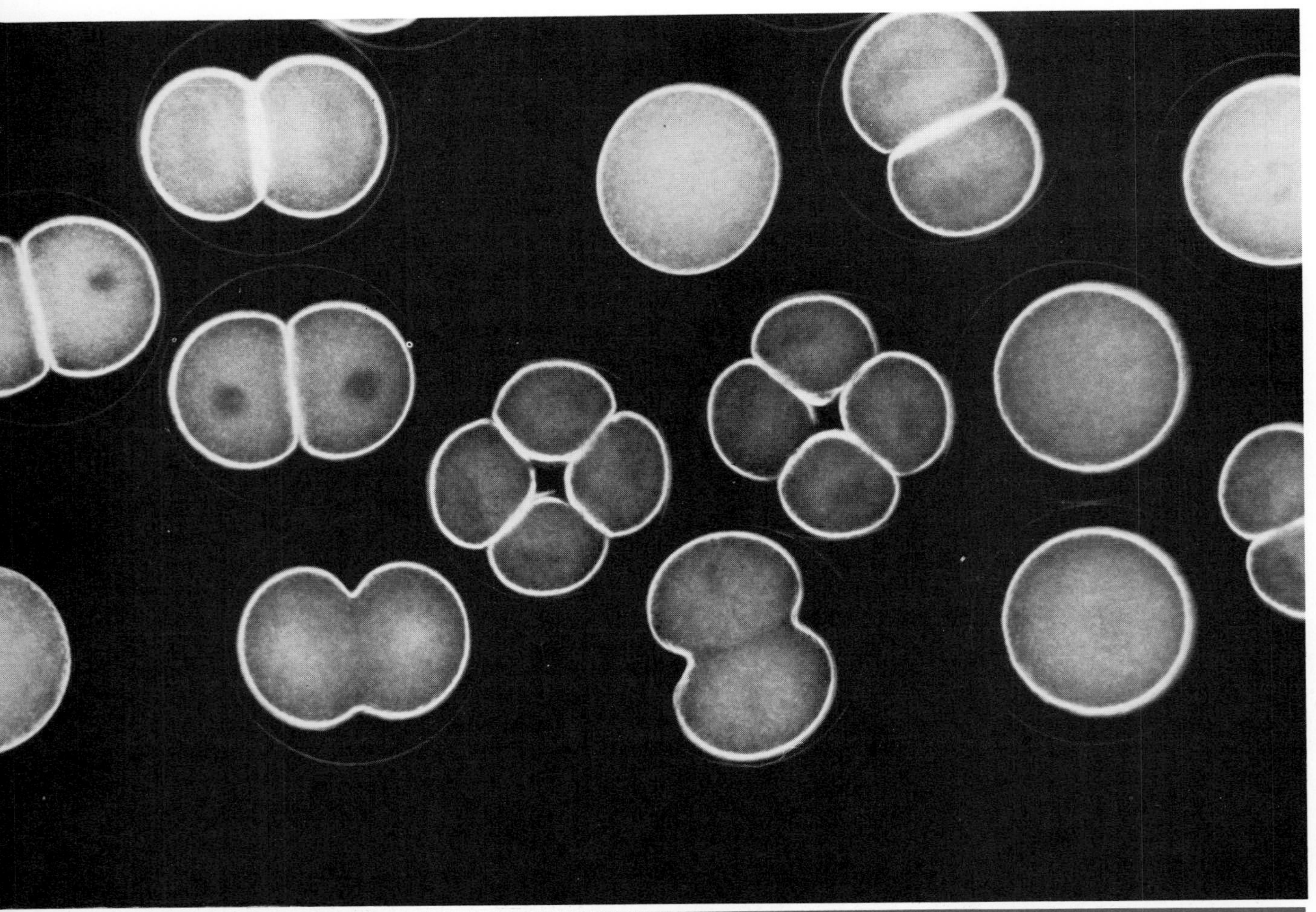

Photograph by D. P. Wilson.

CELL CONCEPT

More than 300 years ago the English scientist and inventor Robert Hooke, using a primitive compound microscope, observed boxlike cavities in slices of cork and leaves. He called these compartments "little boxes or cells." In the years that followed Hooke's first demonstration of the remarkable powers of the microscope before the Royal Society of London in 1663, biologists gradually began to realize that cells were far more than simple containers filled with "juices."

Cells are the fabric of life. Even the most primitive cells are enormously complex structures that form the basic units of all living matter. All tissues and organs are composed of cells. In a human being an estimated 40 trillion cells interact, each performing its specialized role in an organized community. In single-celled organisms, all the functions of life are performed within the confines of one microscopic package. There is no life without cells. The idea that the cell represents the basic structural and functional unit of life is the most important unifying concept of biology (Principle 11, p. 11).

With the exception of eggs, which are the largest cells (in volume) known, cells are small and mostly invisible to the unaided eye. Consequently, our understanding of cells paralleled technical advances in the resolving power of microscopes. The Dutch microscopist A. van Leeuwenhoek, using high-quality single lenses that he had made, sent letters to the Royal Society of London containing detailed descriptions of the numerous organisms he had observed (1673 to 1723). In the early nineteenth century, the improved design of the microscope permitted biologists to see separate objects only 1 μm* apart. This advance was quickly followed by new discoveries that laid the groundwork for **modern cell theory**—a theory that states that all living organisms are composed of cells.

In 1838 Matthias Schleiden, a German botanist, announced that all plant tissue was composed of cells. A year later one of his countrymen, Theodor Schwann, described animal cells as being similar to plant cells, which is an understanding that had been long delayed because the animal cell is bounded only by a nearly invisible plasma membrane and because it lacks the distinct cell wall characteristic of the plant cell. Schleiden and Schwann are thus credited with the unifying cell theory that ushered in a new era of productive exploration in cell biology.

In 1840 J. Purkinje introduced the term **protoplasm** to describe the cell contents. Protoplasm was at first thought to be a granular, gel-like mixture with special and elusive life properties of its own; the cell was thus viewed as a bag of thick soup containing a nucleus. Later the interior of the cell became increasingly visible, as microscopes were improved and better tissue-sectioning and staining techniques were introduced. Rather than being a uniform granular soup, the cell interior is composed of numerous **cell organelles,** each performing a specific function in the life of the cell (Principle 12, p. 12).

How cells are studied

Because cells are submicroscopic structures with complex and delicate internal organization, new technical approaches were required for the visual exploration of cell structure and function. The light microscope, with all its variations and modifications, has contributed more to biologic investigation than any other instrument. It has been a powerful exploratory tool for 300 years and continues to be so more than 45 years after the invention of the electron microscope. But, until the electron microscope was perfected, our concept of the cell was limited to that which could be seen with magnifications of 1,000 to 2,000 diameters. This is the practical limit of the light microscope.

The electron microscope, invented in Germany in approximately 1931 and developed in the 1940s, employs high voltages to drive a beam of electrons through a vacuum to magnify the object being studied. The wavelength of electrons is approximately 0.00001 times that of ordinary white light, thus permitting far greater magnification (compare *A* and *B* of Fig. 4-1). Electron microscopes can detect objects only 0.001 the size of objects discernible with the best light microscope. Even large molecules such as DNA and proteins can actually be seen with the electron microscope.

The source of ordinary light for the light microscope is replaced in the electron microscope with a beam of electrons emitted from a tungsten filament; the glass lenses of the light microscope are replaced with magnets for shaping the electron beam; and the human eye is replaced with a fluorescent screen or a camera (Fig. 4-2). The system must be completely evacuated,

*Units of measurement commonly used in microscopic study are micrometers, nanometers, and angstroms: 1 micrometer (μm) = 0.000001 meter (about $^{1}/_{25,000}$ inch); 1 nanometer (nm) = 0.000000001 meter; 1 angstrom (Å) = 0.0000000001 meter. Thus 1 m = 10^3 mm = 10^6 μm = 10^9 nm = 10^{10} Å.

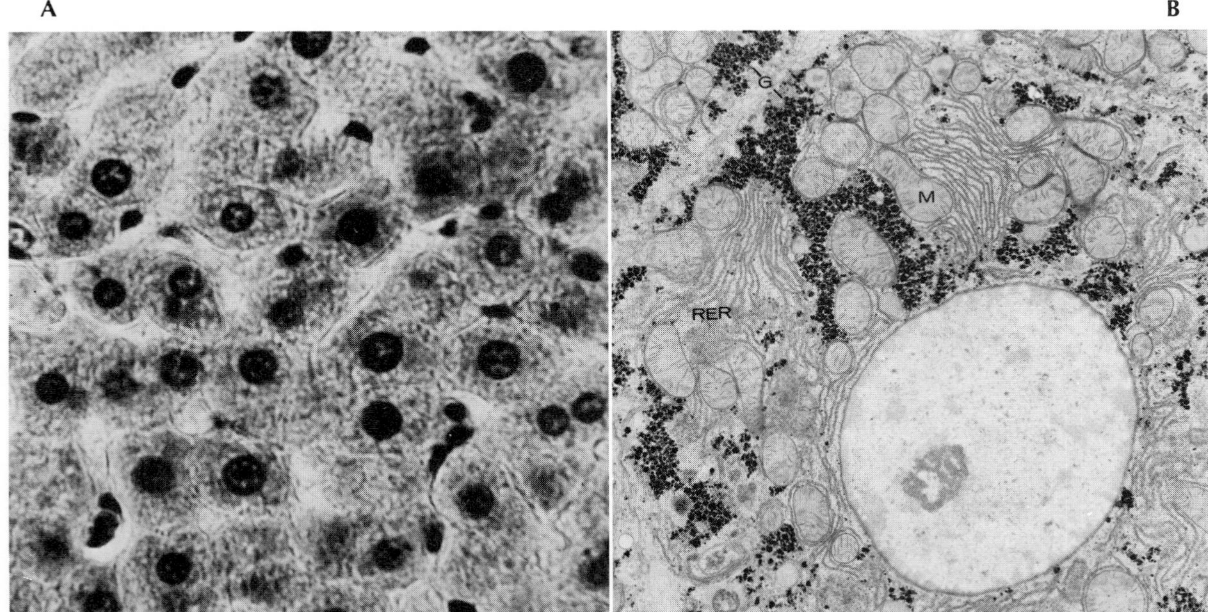

FIG. 4-1 Liver cells of rat. **A,** Magnified about 700 times. Note prominently stained nucleus in each polyhedral cell. **B,** Portion of single liver cell, magnified about 6,000 times. Single large nucleus dominates field; mitochondrion *(M),* rough endoplasmic reticulum *(RER),* and glycogen granules *(G)* are also seen. (From Morgan, C. R., and R. A. Jersild. 1970. Anat. Record **166:** 575-586.)

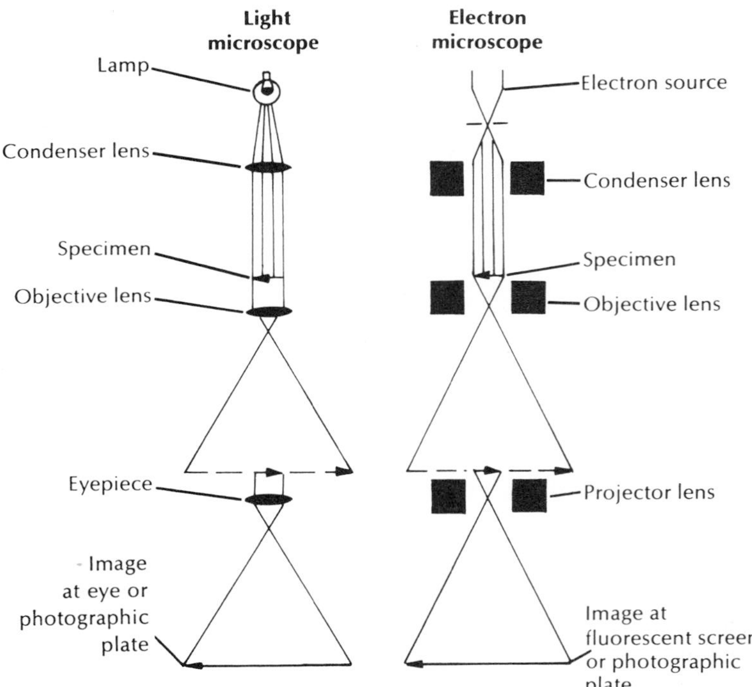

FIG. 4-2 Comparison of optical paths of light and electron microscopes. Note that to facilitate comparison, the scheme of the light microscope has been inverted from its usual orientation with light source below and image above.

and samples must be dry, nonvolatile, and cut into extremely thin sections. Image formation depends on differences in electron scattering, which in turn depend on the density of objects in the electron beam.

Since the atomic nuclei of such elements as oxygen, hydrogen, carbon, and phosphorus, of which the cell is largely composed, have "equivalent mass," the tissue must be treated with "electron stains" containing ions such as osmium, lead, or uranium whose large nuclei block electrons more completely. These ions react with different chemical groups in the cell and are precipitated during these reactions. The electron micrograph shows these locations as dark, electron-opaque parts of the cell.

More recently, the **scanning electron microscope** has been introduced. Because of its great depth of field, this instrument has been used to obtain strikingly beautiful photographs of the surfaces of animals and epithelial tissues. Example of scanning electron micrographs is shown on p. 74.

Advances in the techniques for cell study were not limited to improvements in microscopes. Because most tissues were almost impossible to study in their living state, biologists of the last century found it necessary to kill, preserve, stain, and section tissues to make them optically suitable for examination. Histologists (scientists who specialize in the study of tissue structure) experimented patiently for many years to discover fixatives that arrested life processes almost instantly and preserved the cells and their organelles in the same structural relationships they possessed when alive. The microtome, invented in 1870, allowed controlled sectioning of extremely thin slices of tissues. Organic chemists developed stains and aniline dyes that were applied to provide contrast to cellular structures. The search continues for even more precise physiochemical methods that will reveal specific entities within cells. Special stains, fluorescent antibody techniques, and radioactive tracers have contributed enormously to our present understanding of cell structure and function.

ORGANIZATION OF THE CELL

If we were to restrict our study of cells to fixed and sectioned tissues, we would be left with the erroneous impression that cells are static, quiescent, rigid structures. In fact, the cell interior is in a constant state of upheaval. Most cells are continually changing shape, pulsing and heaving; their organelles twist and regroup in a cytoplasm teeming with starch granules, fat globules, and membrane fragments. This description is derived from studies with time-lapse photography of living cell cultures. If we could see the swift shuttling of molecular traffic through gates in the cell membrane and the metabolic energy transformations within cell organelles, we would have an even stronger impression of internal turmoil. But the cell is anything but a bundle of disorganized activity. There is order and harmony in the cell's functioning that represents the elusive phenomenon we call life. Studying this dynamic miracle of evolution through the microscope, we realize that as we gradually comprehend more and more about this unit of life and how it operates, we are gaining a greater understanding of the nature of life itself.

Prokaryotic and eukaryotic cells

The radically different cell plan of prokaryotes and eukaryotes was previously described (p. 54). A fundamental distinction is that prokaryotes lack the membrane-bound nucleus present in all eukaryotic cells. Other major differences are summarized in Table 4-1.

Despite these differences, which are of paramount importance in cell studies, there is much that prokaryotes and eukaryotes have in common. Both have DNA, use the same genetic code, and synthesize proteins. Many specific molecules such as ATP perform similar roles. These fundamental similarities imply common ancestry (Principle 3, p. 10).

Prokaryotic cells are microbes—bacteria and blue-green algae of the kingdom Monera. The most complex of these are the filamentous forms of blue-green algae and some bacteria. All other organisms are eukaryotes distributed among four kingdoms: the unicellular kingdom Protista (protozoans and nucleated algae) and three multicellular kingdoms—Plantae (green plants), Fungi (true fungi), and Animalia (multicellular animals). The kingdom classifications are discussed in Chapter 7 (p. 126). The following discussion is restricted to eukaryotic cells of which all animals are composed.

Components of the eukaryotic cell and their functions

If the inside of the cheek is gently scraped with a blunt instrument and if the scrapings are put on a slide in a drop of physiologic salt solution and examined unstained with a microscope, living cells can be seen.

TABLE 4-1 Comparison of prokaryotic and eukaryotic cells

Characteristic	Prokaryotic cell	Eukaryotic cell
Cell size	Mostly small (1-10 μm)	Mostly large (10-100 μm)
Genetic system	DNA not associated with proteins; no chromosomes	DNA complexed with proteins in chromosomes
Cell division	Direct by binary fission or budding; no mitosis	Some form of mitosis; centrioles, mitotic spindle present
Sexual system	Absent in most; highly modified if present	Present in most; male and female partners; gametes that fuse
Nutrition	Absorption by most; photosynthesis by some	Absorption, ingestion, photosynthesis
Energy metabolism	Mitochondria absent; oxidative enzymes bound to cell membrane, not packaged separately; great variation in metabolic pattern	Mitochondria present; oxidative enzymes packaged therein; unified pattern of oxidative metabolism throughout
Intracellular movement	None	Cytoplasmic streaming, phagocytosis, pinocytosis

The flat circular cells with small nuclei are the squamous epithelial cells that line the mouth region.

Flat epithelial cells are only one variety of many different shapes assumed by cells. Although many cells, because of surface tension forces, assume a spheric shape when freed from restraining influences, others retain their shape under most conditions because of their characteristic cytoskeleton, or framework of microtubules (Fig. 4-3).

A eukaryotic cell includes both its outer wall, or membranes, and its contents. Typically, it is a semifluid mass of microscopic dimensions, completely enclosed within a thin, differentially permeable **plasma membrane.** It usually contains two distinct regions—the nucleus and the cytoplasm. The **nucleus** is enclosed by a **nuclear membrane** and contains **chromatin** and one **nucleolus** or more. Within the cytoplasm are many organelles such as mitochondria, Golgi complex, centrioles, and endoplasmic reticulum. Plant cells may also contain plastids, especially chloroplasts.

All structures, or organelles, of the cell have separate, important functions. The **nucleus** (Figs. 4-4 and 4-5) has two important roles: (1) to store and carry hereditary information as chromosomal DNA from generation to generation of cells and individuals and (2) to synthesize messenger RNA, which migrates into the cytoplasm where it translates genetic information into the kind of protein characteristic of the cell and thus determines the cell's specific role in the life process.

One or more **nucleoli** (sing., nucleolus) usually occur within the nucleus. These globular bodies are rich in RNA and perform an essential role in synthesizing **ribosomes,** which are required for protein synthesis. In the resting (interphase) nucleus the chromosomes are present as a deeply staining, diffuse material called **chromatin;** individual chromosomes are not visible until the cell begins to divide. The remainder of the nucleus is filled with **nucleoplasm** (Fig. 4-5).

A **plasma membrane,** or plasmalemma, surrounds the cell. It is a sturdy envelope that encloses the cell and behaves as a selective "gatekeeper" that determines what can and what cannot enter or leave the cell. Membranes similar to, if not identical to, the cytoplasmic membranes also surround the organelles within the cell. Membranes thus serve as partitions to subdivide the cell space into many self-contained compartments in which biochemical reactions may proceed. With the electron microscope, the cell membrane appears as two dark lines, each approximately 2.5 to 3.0 nm thick at each side of a light zone. The entire membrane is 7.5 to 10.0 nm thick.

The **endoplasmic reticulum,** or **ER,** is a complex of membranes that separates some of the products of the cell from the synthetic machinery that produces them. This separation facilitates storage and secretion and ensures that the synthetic machinery is retained (Figs. 4-5 and 4-6). The nuclear membrane is formed from parts of this membrane system, and it is so arranged that there is direct continuity between the nucleus and the synthetic part of the cytoplasm because of openings in the nuclear membrane (Fig. 4-5). There are two types of endoplasmic reticulum: rough-surfaced and smooth-surfaced. The rough-surfaced type has on its

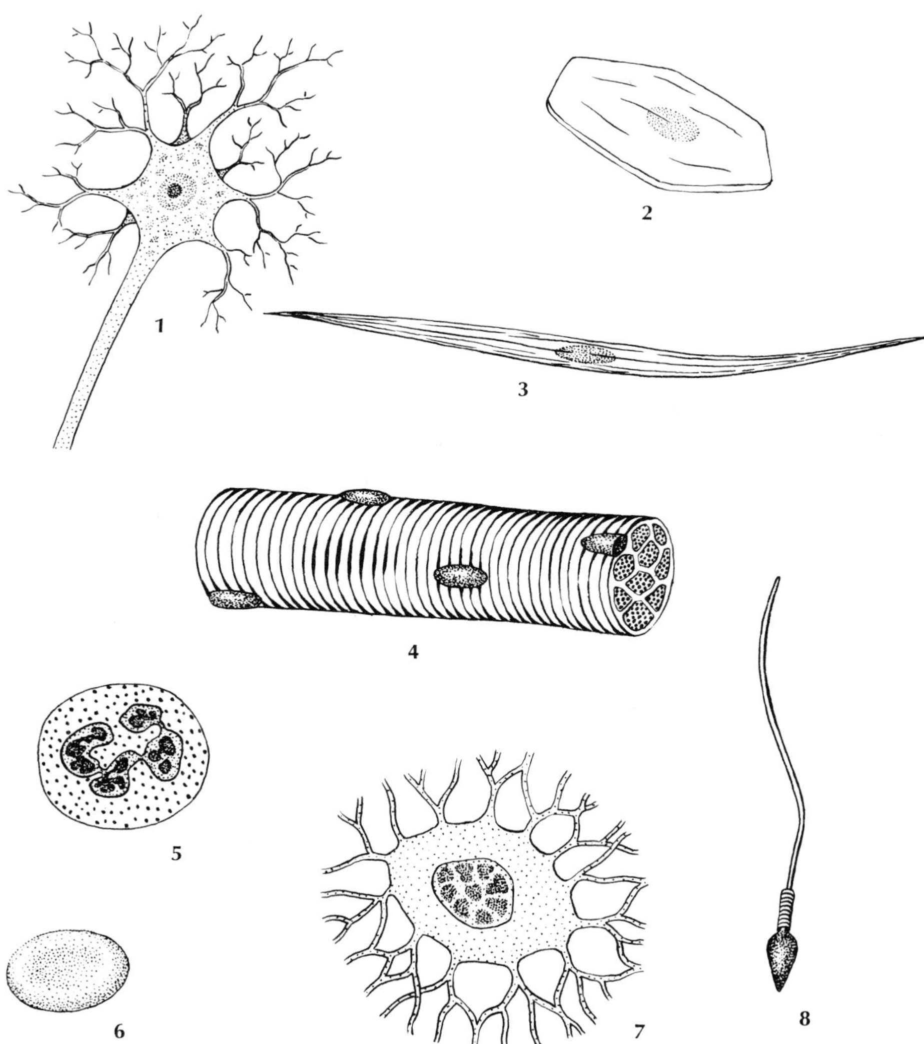

FIG. 4-3 Types of cells. **1,** Nerve cell, showing cell body (soma) surrounded by numerous dendritic extensions and a portion of the axon extending below; **2,** epithelial cell from lining of the mouth; **3,** smooth muscle cell from intestinal wall; **4,** striated muscle cell from skeletal muscle; **5,** white blood corpuscle; **6,** red blood corpuscle (erythrocyte); **7,** bone cell; **8,** human spermatozoon. (Not drawn to the same scale.)

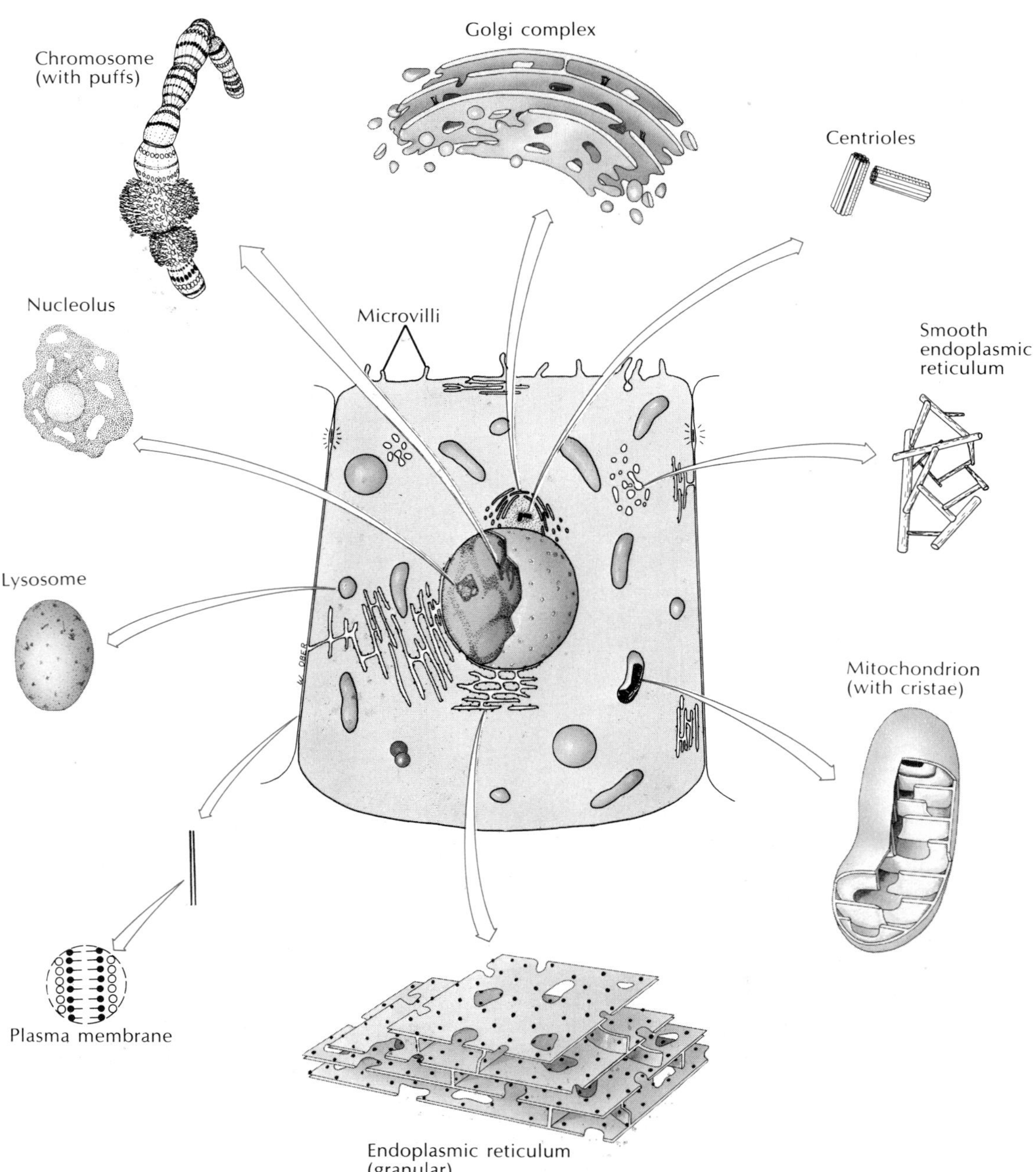

FIG. 4-4 Generalized cell with principal organelles, as might be seen with the electron microscope. Each of the major organelles is shown enlarged. Membranes of organelles are believed to be continuous with, or derived from, the plasma membrane by an infolding process. Structure of other membranes (of nucleus, endoplasmic reticulum, mitochondria, and others) is probably similar to that of plasma membrane, shown enlarged at lower left.

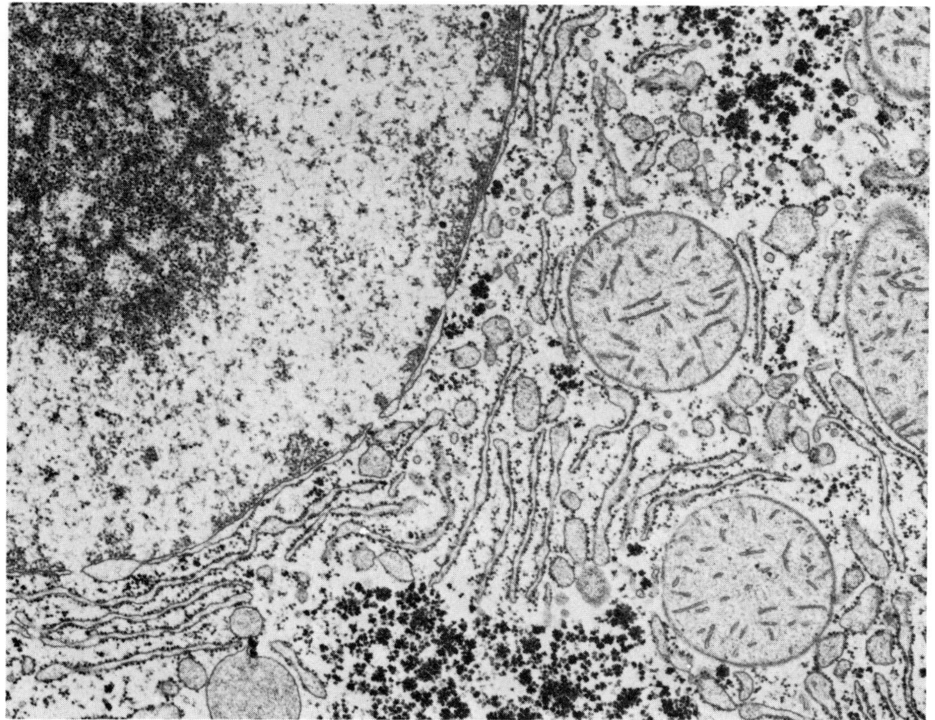

FIG. 4-5 Electron micrograph of part of hepatic cell of rat showing portion of nucleus (left) and surrounding cytoplasm. Endoplasmic reticulum and mitochondria are visible in cytoplasm, and pores are seen in nuclear membrane. (×14,000.) (Courtesy G. E. Palade, The Rockefeller University, New York.)

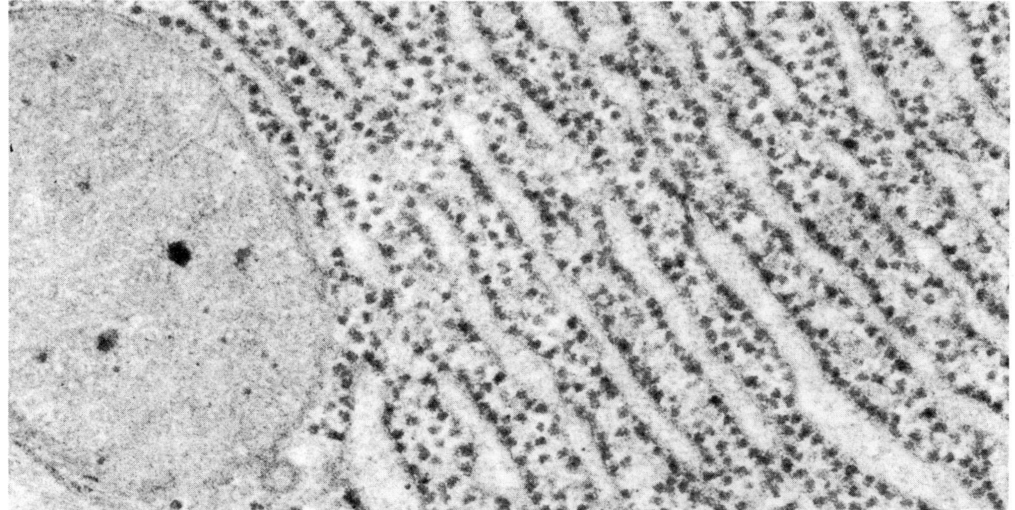

FIG. 4-6 Electron micrograph of portion of pancreatic exocrine cell from guinea pig showing endoplasmic reticulum with ribosomes (small dark granules). Oval body (left) is mitochondrion. (×66,000.) (Courtesy G. E. Palade, The Rockefeller University, New York.)

outer surface the small granules called ribosomes (Fig. 4-6). The ribosomes are important sites of protein synthesis.

The **Golgi complex** (shown enlarged at the top of Fig. 4-4)—one part of the smooth endoplasmic reticulum—is the primary site for processing and packaging many secretory products destined to be exported from glandular cells. The Golgi complex also secretes enzyme-rich droplets that remain in the same cell that produces them. These membrane-enclosed droplets are called **lysosomes** (literally "loosening body," a body capable of causing lysis, or disintegration). The enzymes that they contain are involved in the breakdown of foreign material, including bacteria engulfed by the cell. Lysosomes also are capable of breaking down injured or diseased cells, since the enzymes they contain are so powerful that they kill the cell that formed them if the lysosome membrane ruptures. In normal cells the enzymes remain safely enclosed within the protective membrane.

The **mitochondrion** (Fig. 4-7) is a conspicuous organelle present in nearly all eukaryotic cells. Mitochondria are diverse in shape, size, and number; some are rodlike, and others are more or less spheric in shape. They may be scattered more or less uniformly through the cytoplasm, or they may be localized near cell surfaces and other regions where there is unusual metabolic activity. The mitochondrion is composed of a double membrane. The outer membrane is smooth, whereas the inner membrane is folded into numerous platelike projections called **cristae** (Figs. 4-4 and 4-7). These characteristic features serve to make mitochondria easy to identify among the organelles. Mitochondria are often called "powerhouses of the cell" because enzymes located on the cristae carry out the energy-yielding steps of aerobic metabolism. ATP, the most important energy storage molecule of all cells, is produced in this organelle. Mitochondria contain DNA and synthesize some of their own proteins that contribute to their structure, although they cannot synthesize the energy-producing enzymes.

All of these cellular entities (except ribosomes) are composed of or enclosed within membranes. Eukaryotic cells also contain a variety of nonmembranous elements. **Centrioles** (shown in the upper right in Fig. 4-4) determine the orientation of the plane of cell division. An epithelial cell may have several hundred **cilia** on its surface. Each grows out from a **basal body** that is thought to arise by repeated replication of the original pair of centrioles in the cell; the basal body is identical to the centriole in structure. **Microtubules** and **microfilaments** are associated with cellular movement phenomena, including cytoplasmic streaming and the movement of chromsomes during cell division. They also form the main supportive and propulsive machinery of cilia and flagella.

Surfaces of cells and their specializations

The free surface of epithelial cells (cells that cover the surface of a structure or line a tube or cavity) frequently bear either **cilia** or **flagella.** These are vibratile, locomotory extensions of the cell surface that serve to sweep materials past the cell. In single-celled animals (many of the protozoans) and some primitive multicellular forms, they propel the entire animal through a liquid medium.

Cilia occur in large numbers on each cell and are relatively short (5 to 10 μm). Flagella typically, though not always, occur singly or in small numbers per cell and are long whiplike structures that may reach 150 μm in length. Both cilia and flagella form from basal bodies that are anchored in the cell cytoplasm. Most cilia and flagella have the same internal structure: nine fibrils surrounding a pair of central fibrils.

The surface of continuous cells, or cells packed together, have junction complexes between them. There are several types of these specializations. The adjoining surfaces of cells are sealed only in restricted areas. Nearest the free surface, the two opposing membranes appear to fuse to form a **tight junction** (Fig. 4-8). Below this is a slightly widened **intermediate junction** and then **desmosomes,** small ellipsoid discs scattered between the epithelial cells. Desmosomes act as adhesion sites between apposing plasma membranes. They measure approximately 250 to 410 nm in their greatest diameter. From the cell cytoplasm, tufts of fine filaments converge onto the desmosomes. Between the two apposed plates of a desmosome is a narrow intercellular space (20 to 24 nm wide). All of these special junctional complexes produce the **terminal bar,** which is found at the distal junctions of adjacent columnar epithelial cells. They form a complete beltlike junction just beyond the luminal or apical portion of the plasma membrane. It is believed that cell junctions, in addition to serving as points of attachment, serve as avenues of chemical communication between adjacent cells.

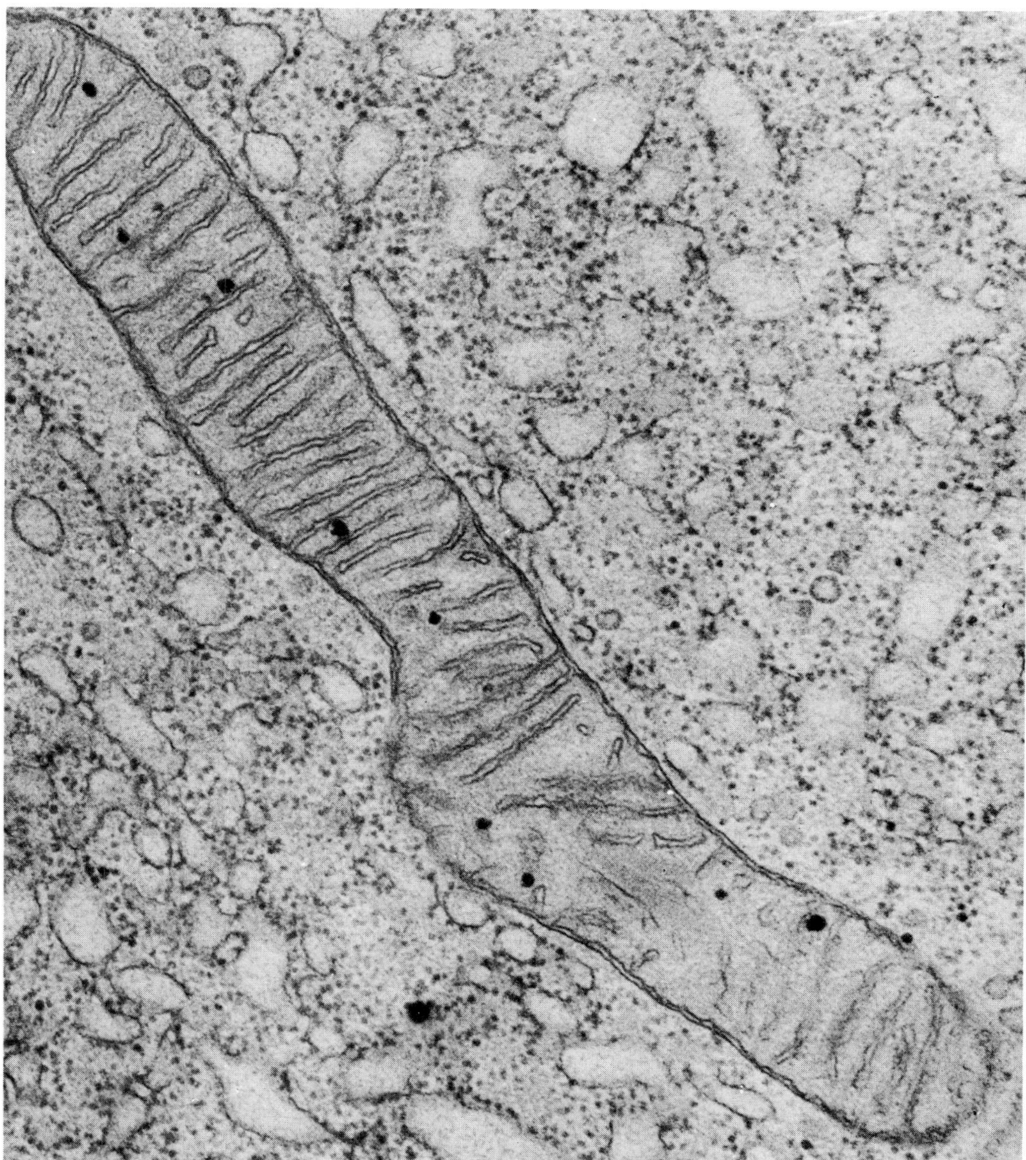

FIG. 4-7 Electron micrograph of elongated mitochondrion in pancreatic exocrine cell of guinea pig. The mitochondrion has two membranes, an outer smooth membrane and an inner membrane that is folded inward into invaginations called cristae. These appear as incomplete transverse septa inside the mitochondrion. (×50,000.) (Courtesy G. E. Palade, The Rockefeller University, New York.)

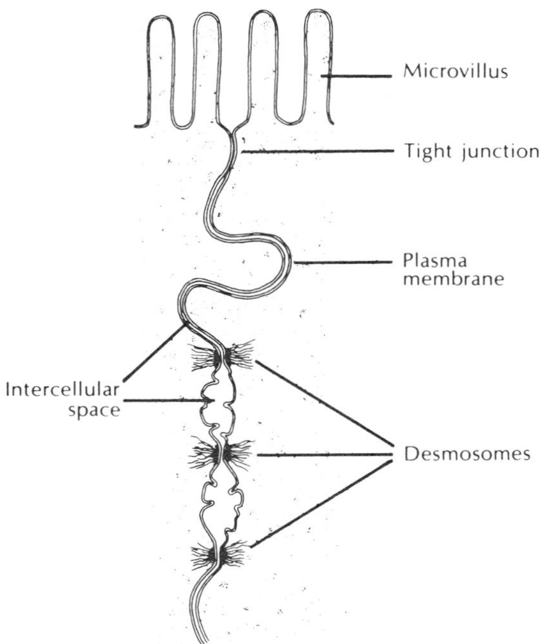

FIG. 4-8 Two opposing plasma membranes forming the boundary between two epithelial cells. Various kinds of junctional complexes are found. Tight junction is a firm, adhesive band completely encircling the cell. Desmosomes are isolated "spot welds" between cells that serve as sites of intercellular communication. Intercellular space may be greatly expanded in epithelial cells of some tissues.

Other specializations of the cell surface are the interdigitations of confronted cell surfaces where the plasma membranes of the cells infold and interdigitate very much like a zipper. They are especially common in the epithelium of kidney tubules. The distal or apical boundaries of some epithelial cells, as seen with the electron microscope, show regularly arranged **microvilli.** They are small, fingerlike projections consisting of tubelike evaginations of the plasma membrane with a core of cytoplasm (Fig. 4-8). They are seen clearly in the lining of the intestine where they greatly increase the absorptive and digestive surface. Such specializations appear as brush borders by the light microscope. The spaces between the microvilli are continuous with tubules of the endoplasmic reticulum, which may facilitate the movement of materials into the cells.

CELL DIVISION (MITOSIS)

All cells of the body arise from the division of preexisting cells (Principle 11, p. 11). All the cells found in most multicellular organisms have originated from the division of a single cell, the **zygote,** which is formed from the union (fertilization) of an **egg** and a **sperm.** Cell division provides the basis for one form of growth, for both sexual and asexual reproduction, and for the transmission of hereditary qualities from one cell generation to another cell generation.

In the formation of **body cells** (somatic cells) the process of nuclear division is referred to as **mitosis.** By mitosis each "daughter cell" is assured of receiving a complete set of genetic instructions. Mitosis is a delivery system for distributing the chromosomes and the DNA they contain to continuing cell generations. Thus all somatic cells, which number hundreds of billions in large animals, have the same genetic content, since all are descended by faithful reproduction of the original fertilized egg. As an animal grows, its somatic cells differentiate and assume different functions and appearances because of differential gene action. Even though most of the genes in specialized cells remain silent and unexpressed throughout the lives of those cells, every cell possesses a complete genetic complement. Mitosis ensures equality of genetic potential; later, other processes direct the orderly expression of genes during embryonic development by selecting from the genetic instructions that each cell contains. (These fundamental properties of cells of multicellular organisms are stated in Principles 21, 22, and 24, pp. 14 to 15, and will be discussed further in Chapter 36.)

In animals that reproduce **asexually,** mitosis is the only mechanism for the faithful transfer of genetic information from parent to progeny. In animals that reproduce **sexually** the parents must produce **sex cells** (gametes or germ cells) that contain only half the usual number of chromosomes, so that the offspring formed by the union of the gametes will not contain double the number of parental chromosomes. This requires a special type of *reductional* division called **meiosis.** The gametes (eggs and sperm) formed by meiosis contain half the parental number of chromosomes. When the gametic nuclei fuse to form a zygote, the full chromosomal complement is restored. Within the zygote lies all the information for future expansion and propagation. Meiosis is described further in Chapter 35.

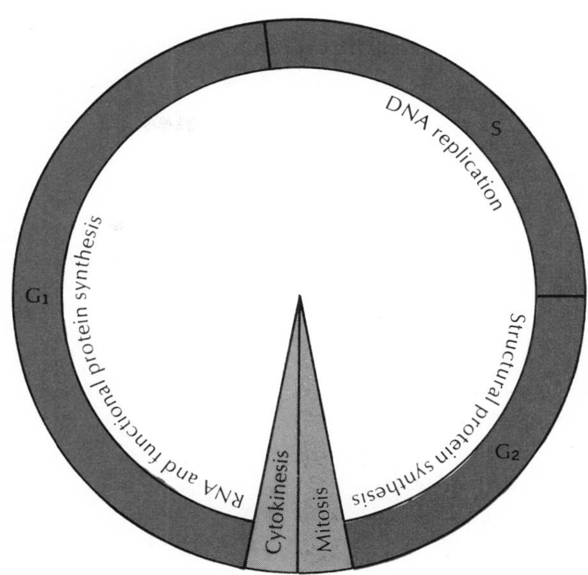

FIG. 4-9 The cell cycle, showing relative duration of recognized periods. *S*, G_1, and G_2 are periods within interphase: *S*, synthesis of DNA; G_1, presynthetic period; G_2, postsynthetic period. Actual duration of the cycle and the different periods varies considerably in different cell types.

The cell cycle

Cycles are conspicuous attributes of life. The descent of a species through time is in a very real sense a sequence of life cycles (Principle 22, p. 14). Similarly, cells undergo cycles of growth and replication as they repeatedly divide. A cell cycle is a mitosis-to-mitosis cycle, that is, the interval between one cell generation and the next (Fig. 4-9).

The most visible event in the cell cycle is mitosis: the condensation of chromosomes, their separation to opposite poles of the cell, and their transformation into diffuse chromatin of the daughter nuclei. Mitosis (nuclear division) is immediately followed by **cytokinesis,** the division of the cytoplasmic contents to complete the division of the cell into two daughter cells.

Actual nuclear division occupies only about 5% of the cell cycle; the rest of the cell's time is spent in **interphase,** the stage between nuclear divisions. For many years it was thought that interphase was a period of rest because the nucleus appeared inactive as observed with the ordinary light microscope. In the early 1950s new techniques for revealing DNA replication in nuclei were introduced at the same time that biologists came to appreciate fully the significance of DNA as the genetic material. It was then discovered that DNA replication occurred during the interphase stage. Further studies revealed that many other protein and nucleic acid components essential to normal cell growth and division were synthesized during the seemingly silent interphase period.

The most vital event in the cell cycle—the replication of DNA—occurs during a phase called the S period (period of synthesis). In mammalian cells in tissue culture, the S period lasts about 6 of the 18 to 24 hours required to complete one cell cycle. In this period both strands of DNA must replicate, that is, new complementary partners are synthesized for both strands so that two identical molecules are produced from the original strand. In human beings, each chromosome contains approximately 175 million nucleotide pairs arranged in a double helix that makes one turn for every 10 nucleotide pairs (Fig. 2-18, p. 38). At an estimated rate of DNA synthesis of approximately 6,000 base pairs per minute it would take 486 hours to complete the replication of one DNA molecule. In fact, replication is completed in 6 hours. To explain how this could happen, it was suggested that DNA replication must proceed simultaneously at numerous points of origin along the molecule. Electron micrographs of replicating DNA isolated from chromosomes have indeed confirmed this hypothesis. At multiple points of origin, DNA replicates bidirectionally, forming replicating loops **(replicons)** that eventually unite (Fig. 4-10). The result of replication is two complete DNA molecules, each carrying the same coded message as the parent molecule.

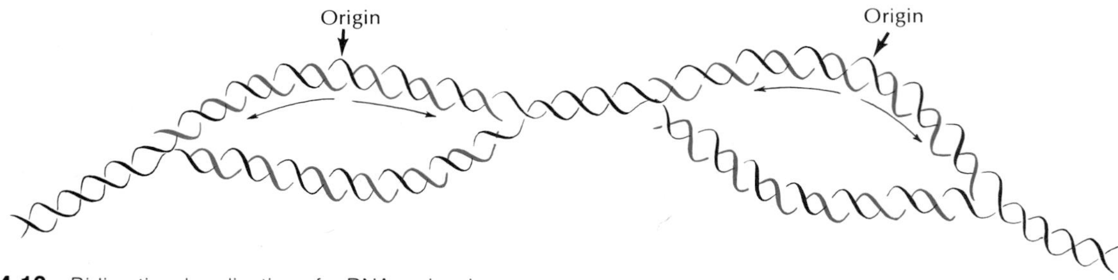

FIG. 4-10 Bidirectional replication of a DNA molecule.

The S period is preceded and succeeded by G_1 and G_2 periods respectively (G stands for "gap"), during which no DNA synthesis is occurring. For most cells, G_1 is an important preparatory stage for the replication of DNA that follows. During G_1, transfer RNA, ribosomes, messenger RNA, and several enzymes are synthesized. During G_2, spindle and aster proteins are synthesized in preparation for chromosome separation during mitosis. G_1 is typically of longer duration than G_2, although there is much variation in different cell types. Embryonic cells divide very rapidly since there is no cell growth between divisions, only subdivision of mass. DNA synthesis may proceed a hundred times more rapidly in embryonic cells than in adult cells, and the G_1 period is almost completely omitted. Cancer cells not only divide rapidly but also grow between divisions. The usual constraints that operate to switch off normal cell division when tissue growth is complete are lacking in uncontrolled cancer cell growth. Normal cells "know" when to stop dividing; cancer cells do not. Much of the current research on cancer is focused on how the cell division switch is controlled and why it becomes overridden by internal and external influences during tumor growth.

Structure of chromosomes

The genetic material DNA does not occur freely in the nucleus but is combined with special stabilizing proteins called **histones** to form nucleoproteins. These, together with other structural proteins and enzymes, form **chromatin.** In cells that are not dividing the chromatin is loosely organized into irregular clumps of dispersed material, but in dividing cells the chromatin condenses into the elongate **chromosomes** (color bodies), so named because they stain deeply with biologic dyes.

Chromosomes are of varied lengths and shapes, some bent, some rodlike. Their number is constant for the species, and every body cell (but not the germ cells) has the same number of chromosomes regardless of the cell's function. A human being, for example, has 46 chromosomes in each body (somatic) cell.

During mitosis (nuclear division) the chromosomes shorten and become increasingly condensed and distinct, and each assumes a characteristic shape. At some point on the chromosome is a **centromere,** or constriction, to which are attached several spindle fibers that pull the chromosome toward the pole during mitosis.

Chromosomes are always arranged in pairs, or two of each kind. Of each pair, one comes from one parent and the other from the other parent. Thus there are 23 pairs in the human species. Each pair usually has certain characteristics of shape and form that aid in identification. A biparental organism begins with the union of two gametes, each of which furnishes a **haploid** set of chromosomes (23 in humans) to produce a somatic or **diploid** number of chromosomes (46 in humans). The chromosomes of a haploid set are also called a **genome.** Thus a fertilized egg consists of a paternal genome (chromosomes contributed by the father) and a maternal genome (chromosomes contributed by the mother).

How is DNA packaged in chromosomes? Packaging is a formidable problem because, by some estimates, the DNA in a human cell is nearly 4 m in length, yet is packed into 46 chromosomes having an aggregate length of only 200 μm. The DNA must obviously be coiled or folded up in some way to fit into a small space. Yet the packaging must be arranged so that the genetic instructions are accessible for reading during the transcription process (the formation of messenger RNA from nuclear DNA, described in Chapter 37).

In 1974 it was discovered that the chromatin is composed of repeating subunits, called **nucleosomes.** Each nucleosome is a narrow ''spool'' or disc of histone proteins around which two turns of double-helical DNA are wound to form a superhelix. The nucleosomes are linked together by the continuous DNA strand much like beads on a string. This arrangement is thought to explain the knobby appearance of chromatin fibers as revealed by high-resolution electron micrographs. Since the DNA is wound around the *outside* of the nucleosome beads, all base pairs of the molecule are accessible for transcription.

Stages in cell division

There are two distinct phases of cell division: the division of the nuclear chromosomes **(mitosis)** and the division of the cytoplasm **(cytokinesis).** Mitosis, the division of the nucleus (that is, chromosomal segregation), is certainly the most obvious and complex part of cell division and that of greatest interest to the cytologist. Cytokinesis normally immediately follows mitosis, although there are occasions when the nucleus may divide a number of times without a corresponding division of the cytoplasm. In such a case the resulting mass of protoplasm containing many nuclei is referred to as a **multinucleate cell.** An example is the giant resorptive cell type of bone (osteoclast) that may contain 15 to 20 nuclei. Sometimes a multinucleate mass is formed by cell fusion rather than nuclear proliferation. This arrangement is called a **syncytium.** An example is vertebrate skeletal muscle, which is composed of multinucleate fibers that have arisen by the fusion of numerous embryonic cells.

The process of mitosis is arbitrarily divided for convenience into four successive stages or phases, although one stage merges into the next without sharp lines of transition. These phases are prophase, metaphase, anaphase, and telophase (Fig. 4-11). When the cell is not actively dividing, it is in the ''resting'' stage or **interphase.** However, as discussed previously, the cell is not really ''resting'' at this stage because the DNA content of the nucleus is being duplicated between divisions. Thus when the cell begins ''active'' mitosis, it already has a double set of chromosomes. One requirement for division of animal cells is the presence of **centrioles**—permanent, self-duplicating, rodlike bodies usually found in pairs. Each cell inherits one set of centrioles and produces another set.

Prophase. At the start of prophase the centrioles migrate toward opposite sides of the nucleus. At the same time, fine fibers appear between the centriole complexes to form a football-shaped **spindle,** so named because of its resemblance to nineteenth century wooden spindles, used to twist thread together in spinning. Other fibers radiate outward from each centriole to form **asters.** The entire structure is the **mitotic apparatus,** and it increases in size and prominence as the centrioles move farther apart (Fig. 4-12).

At the same time, the diffuse nuclear chromatin condenses to form visible chromosomes. These actually consist of two identical sister **chromatids** formed during the S period of interphase. The sister chromatids are joined together at their centromere. At the end of prophase, the nuclear envelope quickly disappears.

Metaphase. During metaphase the condensed sister chromatids rapidly migrate to the middle of the nuclear region to form a **metaphasic plate** (Fig. 4-13). The centromeres line up precisely on the plate with the arms of the chromatids trailing off randomly in various directions. Spindle fibers now attach to each centromere.

Anaphase. The two chromatids of each double chromosome thicken and separate. The single centromere that has held the two chromatids together now splits so that two independent chromosomes, each with its own centromere, are formed. The chromosomes part more, evidently pulled by the spindle fibers attached to the centromeres. The arms of each chromosome trail along behind as though the chromosome were being dragged through a resisting medium. The spindle fibers are actually microtubules that contain bundles of **actin,** one of the contractile proteins present in muscle cells. Recent evidence suggests that the actin bundles, powered by ATP, somehow contract to drag the chromosomes toward their respective poles. However, the fibers maintain the same thickness throughout anaphase, neither thickening nor thinning as the chromosomes separate.

Telophase. When the daughter chromosomes reach their respective poles, telophase has begun. The daughter chromosomes are crowded together and stain intensely with histologic stains. Two other events also occur: the appearance of a **cleavage furrow** encircling the surface of the cell (Fig. 4-11) and a **cell plate** (in plants) that originates from the central portions of the interpolar spindle fibers. Eventually the cleavage furrow deepens and constricts the cell into two daughter cells. Other changes that terminate the telophase period are the disappearance of the spindle fibrils that revert

FIG. 4-11 Stages of mitosis, showing division of a cell with two pairs of chromosomes. One chromosome of each pair is shown in red.

Interphase

Chromatin material appears granular; each chromosome reaches its maximum length and minimum thickness; duplication of chromosome occurs at this stage

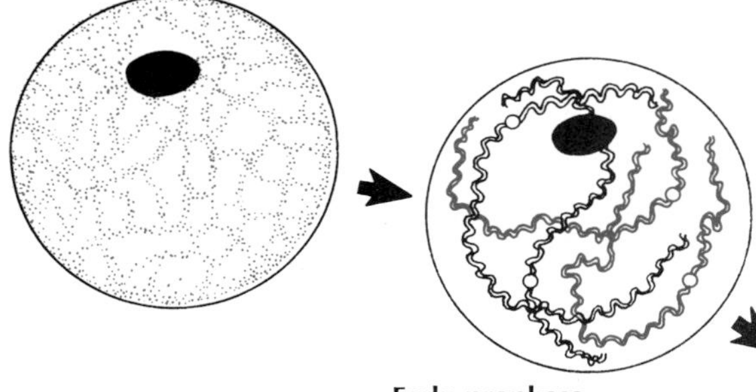

Early prophase

Each elongated chromosome now consists of 2 chromatids attached to single centromere; centrosome divides and spindle starts development

Late prophase

Double nature of short, thick chromosome more apparent; each chromosome made up of 2 half-chromosomes or sister chromatids; nucleolus usually disappears; nuclear envelope disintegrates

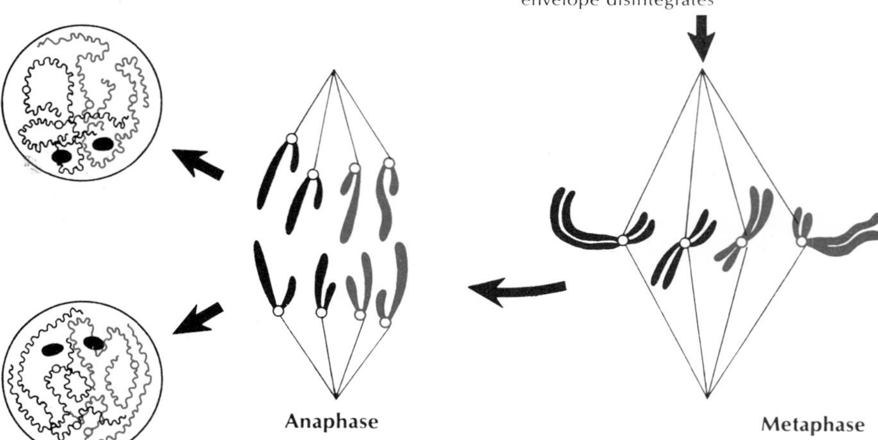

Metaphase

Chromosomes arranged on equatorial plate; centromeres (not yet divided) anchored to equator of spindle

Anaphase

Chromatids, now called daughter chromosomes are in 2 distinct groups; daughter centromeres, which may be attached at various points on different chromosomes, move apart and drag daughter chromosomes toward respective poles

Telophase

Chromosomes become longer and thinner; chromosomes may lose identity; nuclear membrane reappears and spindle-astral fibers fade away; cell body divides into 2 daughter cells, each of which now enters interphase

FIG. 4-12 Fine structure of mitotic apparatus. At each pole of the spindles is a clear zone occupied by a pair of centrioles and surrounded by short microtubules. Other microtubules form spindle fibers, some of which extend from pole to pole, while others attach to chromosomes. (From DuPraw, E. J. 1968. Cell and molecular biology. New York, Academic Press, Inc.)

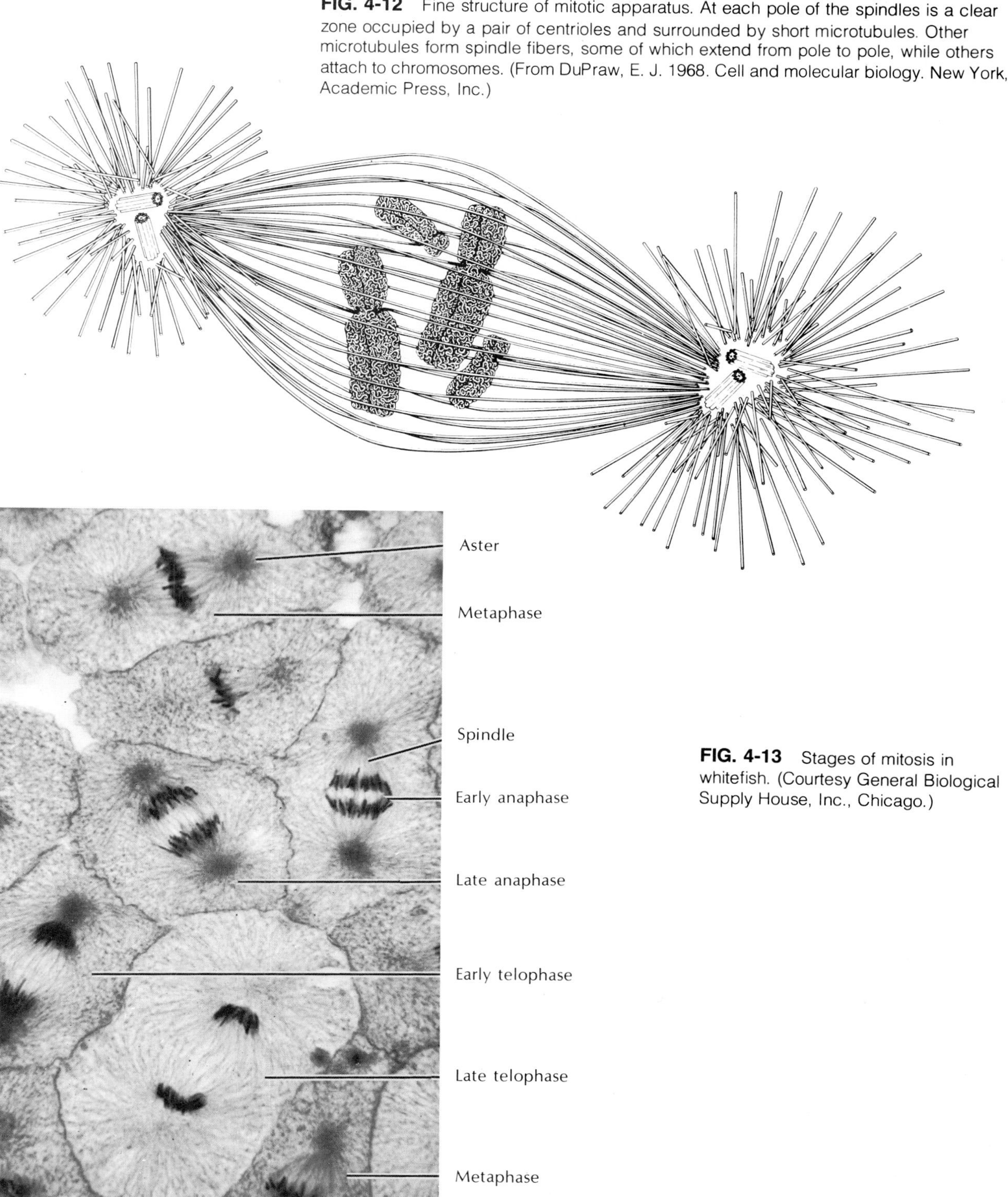

FIG. 4-13 Stages of mitosis in whitefish. (Courtesy General Biological Supply House, Inc., Chicago.)

to a sol from a gel condition, the gradual assumption of a chromatin network as the chromosomes lose their identity, the formation of a new nuclear membrane, and the manufacture of new nucleoli by the chromosomes.

The flux of cells

Cell division is important for growth, for replacement of cells lost to natural attrition and wear and tear, and for wound healing. Cell division is especially rapid during the early development of the organism. At birth the human child has about 2 trillion cells from repeated division of a single fertilized egg. This immense number could be attained by just 42 cell divisions, with each generation dividing once every 6 to 7 days. With only five more cell divisions, the cell number would increase to approximately 60 trillion, the number of cells in a mature man weighing 75 kg. But of course no organism develops in this machine-like manner. Cell division is rapid during early embryonic development, then slows with age. Furthermore, different cell populations divide at widely different rates. In some, the average period between divisions is measured in hours while in others it is measured in days, months, or even years. Cells in the central nervous system stop dividing altogether after the early months of fetal development and persist without further division for the life of the individual. Muscle cells, too, stop dividing during the third month of fetal development and future growth depends on enlargement of fibers already present.

In other tissues that are subject to wear and tear there must be a constant replacement of cells that are lost. It has been estimated that in humans about 1% to 2% of all body cells—a total of 100 billion—are shed daily. Mechanical rubbing wears away the outer cells of the skin, and emotional stress can result in physical stress that affects many cells. Food in the alimentary canal rubs off lining cells, the restricted life cycle of blood corpuscles must involve enormous numbers of replacements, and during active sex life many millions of sperm are produced each day. Such losses of cells are made up by mitosis.

Cells undergo a senescence with aging. At some point in the life cycle of most cells there is a breakdown of cell substance, the formation of inert material, a slowing down of metabolic processes, and a decrease in the synthetic power of enzymes. These factors lead eventually to the death of the cell. In certain cases, parts of the cell such as scales, feathers, and bony structures may persist after the death of the cell.

Annotated references

Selected general readings

Avers, C. J. 1976. Cell biology. New York, D. Van Nostrand Co. *Textbook of cell structure and function; strong on organelle genetics and cytogenetics.*

Jensen, W. A., and R. B. Park. 1967. Cell ultrastructure. Belmont, Calif., Wadsworth Publishing Co., Inc. *A collection of electron micrographs and illustrated cells and their components from both plants and animals.*

Kennedy, D. 1974. Cellular and organismal biology: Readings from Scientific American. San Francisco, W. H. Freeman & Co., Publishers. *An anthology of Scientific American articles on cell structure, regulation, and functional roles.*

Nilsson, L., and J. Linberg. 1973. Behold man. Boston, Little, Brown & Co. *Collection of superb photographs and scanning electron micrographs of organs, tissues, and embryos by an internationally recognized scientific photographer. Excellent explanatory legends accompany illustrations.*

Novikoff, A. B., and E. Holtzman. 1976. Cells and organelles, ed. 2. New York, Holt, Rinehart & Winston, Inc. *Clearly written and well-illustrated description of structure of cells and cell components and their function.*

Pfeiffer, J. 1972. The cell. New York, Time-Life Books. *Structure and function of cells with excellent picture essays.*

Swanson, C. P. 1969. The cell, ed. 3. Englewood Cliffs, N.J., Prentice-Hall, Inc. *One of a series of biologic monographs. A concise, well-illustrated account of the modern concept of the cell.*

Trumbore, R. H. 1966. The cell: chemistry and function. St. Louis, The C. V. Mosby Co. *A good basic, well-organized text, with summaries, on all aspects of the cell.*

Selected *Scientific American* articles

Albrecht-Buehler, G. 1978. The tracks of moving cells. **238:** 68-76 (Apr.). *Many biologic phenomena involve cell movements. A method for following cell migration in culture is described.*

Capaldi, R. A. 1974. A dynamic model of cell membranes. **230:**26-33 (Mar.). *Structure and function of the cell membrane are described.*

de Duve, C. 1963. The lysosome. **208:**64-72 (May).

Fox, C. F. 1972. The structure of cell membranes. **226:**30-38 (Feb.).

Koshland, D. E., Jr. 1973. Protein shape and biological control. **229:**52-64 (Oct.). *The ability of enzymes to bend un-*

der external influences can explain how they control life processes.

Margulis, L. 1971. Symbiosis and evolution. **225:**48-53 (Aug.). *The author offers evidence suggesting that mitochondria and chloroplasts were once independent organisms.*

Mazia, D. 1974. The cell cycle. **230:**54-64 (Jan.).

Neutra, M., and C. P. Leblond. 1969. The Golgi apparatus. **220:**100-107 (Feb.).

Racker, E. 1968. The membrane of the mitochondrion. **218:** 32-39 (Feb.).

Staehelin, L. A., and B. E. Hull. 1978. Junctions between living cells. **238:**140-152 (May).

Stent, G. S. 1972. Cellular communication. **227:**42-51 (Sept.). *Cells communicate by means of hormones and nerve impulses.*

Wessells, N. K. 1971. How living cells change shape. **225:** 76-82 (Oct.). *Microtubules and microfilaments act as a skeleton and muscle for the cell.*

CHAPTER 5

THE PHYSIOLOGY OF THE CELL

Scanning electron micrograph of a kidney tubule (from a marine teleost fish, *Micrometrus* sp.) sectioned to reveal eight epithelial cells. Note the numerous small microvilli bordering the tubular canal and the less numerous, but much larger, cilia. (×4,100.)

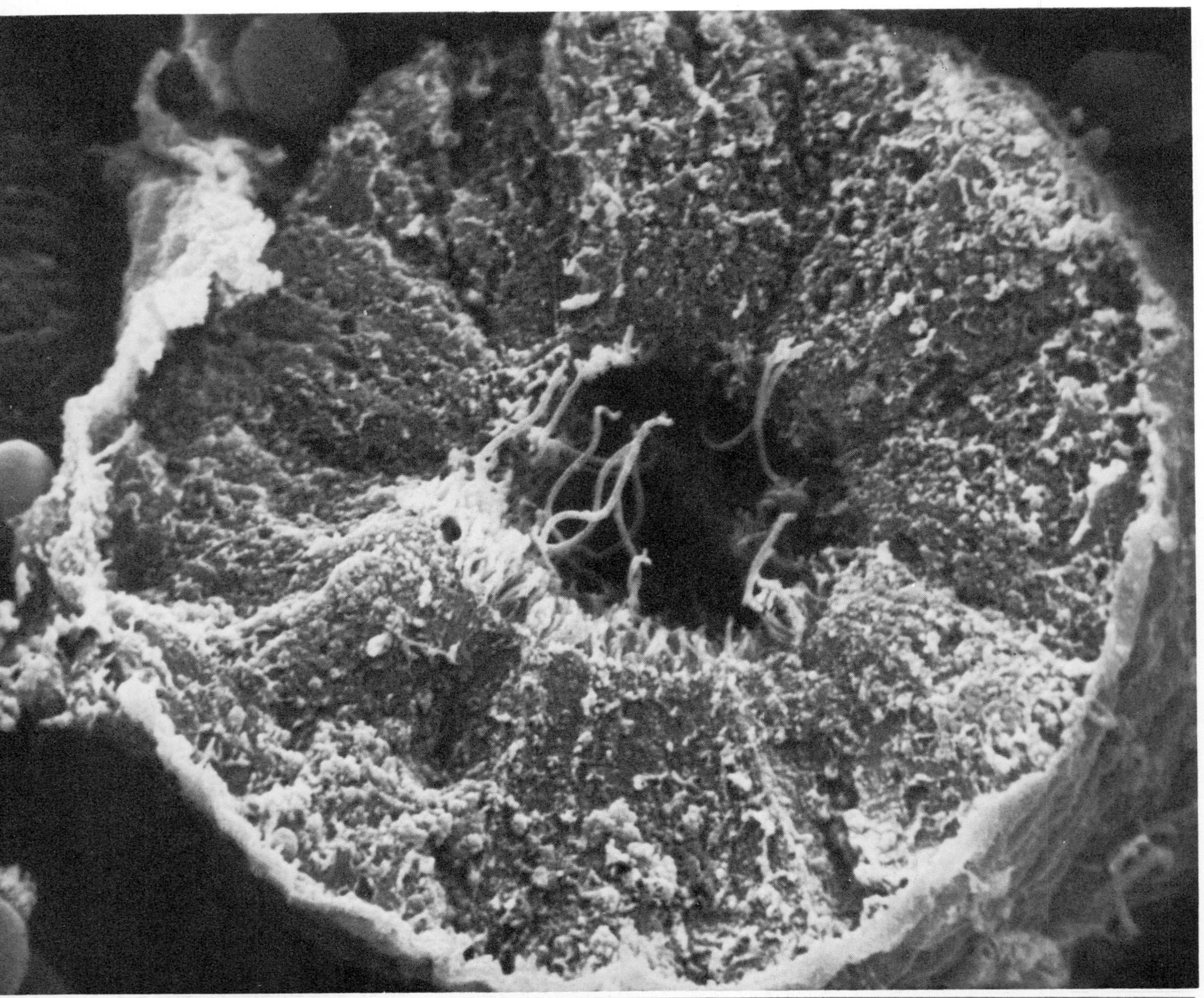

Courtesy G. H. Dobbs, III, Washington and Lee University.

DIFFERENTIATION OF CELL FUNCTIONS

In the previous chapter, we described the generalized blueprint of an animal cell. We pointed out that a unicellular organism contains within its cell boundaries all the equipment required for life and its propagation. These basic functions are the metabolic production of energy, the biosynthesis of cellular constituents, the regulation of metabolism, the maintenance of a protective and selective membrane boundary, movement, and reproduction.

These are attributes of all animal life—of advanced multicellular as well as the simplest unicellular forms. Whereas in protozoans all life processes are performed within the confines of a single cell, in metazoans they are distributed among groups of specialized cells that have specific functions, such as nervous coordination, digestion, and excretion, that benefit the whole organism. The advantages of the "division of labor" that characterizes multicellular forms are evidenced by their undisputed evolutionary success. But cellular differentiation is progress that carries a price, the loss of cellular independence. The specialized cell can no longer perform *all* the functional activities of an organism; its life role is to perform a single function. It is not unlike the city dweller who knows and performs a useful service for other people and who in turn is profoundly dependent on the numerous services others render.

Nevertheless, certain activities are performed by every animal cell, whether it be a liver cell, a nerve cell, or a muscle cell. All cells must produce energy, manufacture their own internal structures, control much of their own activity, and guard their boundaries. It is these basic shared activities that will be considered in this chapter.

MEMBRANE STRUCTURE AND FUNCTION

The incredibly thin, yet sturdy, plasma membrane that encloses every cell is vitally important in maintaining cellular integrity. Once believed to be a rather static entity that defined cell boundaries and kept cell contents from spilling out, the plasma membrane has proved to be a dynamic structure having remarkable activity and selectivity. It is a permeability barrier that separates the interior from the external environment of the cell, regulates the vital flow of molecular traffic into and out of the cell, and provides many of the unique functional properties of specialized cells.

Membranes inside the cell—those surrounding the mitochondria, Golgi apparatus, endoplasmic reticulum, lysosomes, and other organelles—also have vital functions and share many of the structural features of the plasma membrane. Internal membranes provide the site for many, perhaps most, of the enzymatic reactions required by the cell.

Present concept of membrane structure

The plasma membrane is composed almost entirely of proteins and lipids. The lipids that give the membrane its strength and provide other gross structural properties belong to a class of compounds called **phospholipids.** These are **amphipathic** molecules; that is, one end is insoluble in water (hydrophobic), while the opposite end is water soluble (hydrophilic) and polar, carrying an ionic charge. The nonpolar end consists of hydrocarbon chains of **fatty acids,** and the polar (charged) end consists of **glycerol** attached to **phosphate** and other groups (Fig. 5-1).

In 1934 Danielli proposed that the phospholipids were arranged in a bimolecular layer sandwiched between two layers of protein. This remarkably insightful theory was developed from extensive studies of the chemical and biologic properties of the plasma membrane long before it was possible to view the membrane with the electron microscope. Only recently (1971) was it possible to confirm the Danielli theory.

High-resolution electron microscopy reveals a very thin double line at the cell surface approximately 7.5 nm thick. The two narrow dark lines are believed to represent the protein layers and the polar ends of the phospholipids; the light middle interspace represents the fatty acid chain bilayer. This three-layered structure—two dark lines separated by a light interspace—is also characteristic of the membranes that surround the organelles within the cell. This observation resulted in the **unit-membrane hypothesis,** which states that all membranes are of similar structure and composed of lipid and protein layers as originally proposed by Danielli.

Although all membranes are built of lipids and proteins, the proteins are not found arranged in an orderly, static array on the lipid bilayer. Some proteins penetrate into the lipid core of the bilayer; others extend all the way through it (Fig. 5-1). Furthermore the

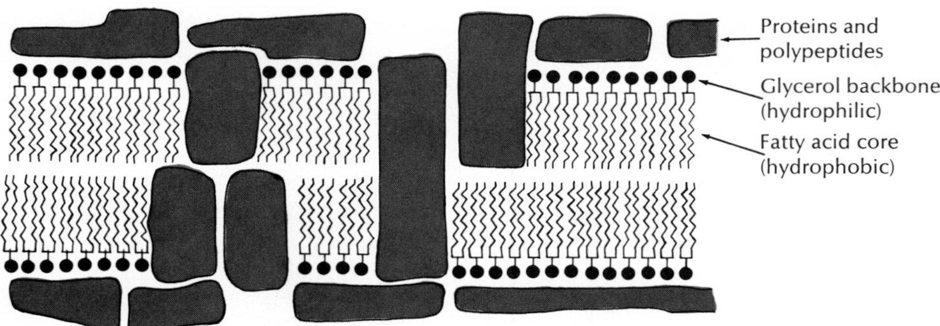

FIG. 5-1 Diagram of plasma membrane. Membrane is a lipid bilayer sandwiched between layers of protein molecules that penetrate and may even extend through the membrane in places. The proteins serve as routes for solute penetration.

membrane appears to be remarkably restless and fluid, with proteins constantly moving and reorganizing their molecular configuration. Some of the proteins that extend through the membrane are thought to behave as "channels" through which small molecules such as sodium, chloride, and potassium are allowed to pass.

Membrane permeability

The plasma membrane acts as a gatekeeper for the entrance and exit of the many substances involved in cell metabolism. Some substances can pass through with ease, others enter slowly and with difficulty, and still others cannot enter at all. This is called the **selective behavior** of the cell membrane. Because conditions outside the cell are different from and more variable than conditions within the cell, it is necessary that the passage of substances across the membrane be rigorously controlled.

We recognize three principal ways that a substance may traverse the cell membrane: (1) by **free diffusion** along a concentration gradient; (2) by a **mediated-transport system,** in which the substance binds to a specific site that in some way assists it across the membrane; and (3) by **endocytosis,** in which the substance is enclosed within a vesicle that forms on and detaches from the membrane surface to enter the cell.

Free diffusion and osmosis. If a living cell surrounded by a membrane is immersed in a solution having more solute molecules than the fluid inside the cell, a **concentration gradient** instantly exists between the two fluids. More solute molecules strike the membrane from the outside than from inside. Assuming the membrane is **permeable** to the solute, there is a net movement of solute toward the inside, the side having the lower concentration. The solute diffuses "downhill" across the membrane until its concentrations on each side are equal.

Most cell membranes are **semipermeable,** that is, permeable to water but selectively permeable or impermeable to solutes. In free diffusion it is this selectiveness that regulates molecular traffic. As a rule, gases (such as O_2 and CO_2), urea, and lipids (such as hydrocarbons and alcohol) are the only solutes that can diffuse through biologic membranes with any degree of freedom. This happens because of the lipid nature of membranes that form a natural barrier to most biologically important molecules that are not lipid soluble. Since many water-soluble molecules readily pass through membranes, such movements cannot be explained by simple diffusion. Instead sugars, as well as many electrolytes and macromolecules, are moved across membranes by carrier-mediated processes, described in the next section.

If a membrane is placed between two unequal concentrations of solutes to which the membrane is impermeable or only weakly permeable, water flows through the membrane from the more dilute to the more concentrated solution. In effect the water molecules move down a concentration gradient from an area where the water molecules are more concentrated to an area where they are less concentrated. This is **osmosis.** To understand why this happens we must view the system from the standpoint of the state of the water on each side.

Water exists as a mixture of complexes in equilibrium with free water. The solutes bind water or exaggerate the formation of complexes, leaving less of the water free to diffuse through the membrane. Thus there is a net flow of water across the membrane to the more concentrated solution. Unbound water, like any other material, tends to diffuse "downhill," that is, from an area of higher concentration to an area of lower concentration.

Osmosis differs from unrestricted diffusion in that in osmosis only the water can diffuse; the solute is restricted by the selectively permeable membrane. Another difference is that in osmosis the movement of water creates a volume change. This can be demonstrated by a familiar experiment in which a selectively permeable membrane such as a collodion membrane is tied over the end of a funnel. The funnel is filled with a sugar solution and placed in a beaker of pure water so that the water levels inside and outside the funnel are equal. In a short time the water level in the glass tube rises, indicating that water is passing through the collodion membrane into the sugar solution (Fig. 5-2).

Inside the funnel are sugar molecules, as well as water molecules. In the beaker outside the funnel are only water molecules. Thus the concentration of water is greater on the outside because some of the water on the inside is complexed with sugar molecules. The water therefore goes from the greater concentration of water (outside) to the lesser concentration (inside).

As the fluid rises in the tube against the force of gravity, it exerts a hydrostatic pressure on the collodion membrane and glass tubing (small arrows in Fig. 5-2). This hydrostatic pressure opposes the movement of water molecules into the funnel. Eventually the hydrostatic pressure becomes so great that there is no further *net* movement of water from the beaker into the funnel, and the fluid level in the glass tube stabilizes. We see then that *osmosis can perform work*. A diffusion gradient (large arrows in Fig. 5-2) drives water through the membrane into the solution and creates an opposing **hydrostatic pressure** head. When the hydrostatic pressure (measured by the height of the fluid column) equals the osmotic pressure, no more water enters the osmometer. (Actually water molecules continue to traverse the membrane, but the movement inward is matched by movement outward.) Osmotic pressure can thus be expressed in terms of the height of the fluid column, which in turn depends on the concentration of the sugar solution.

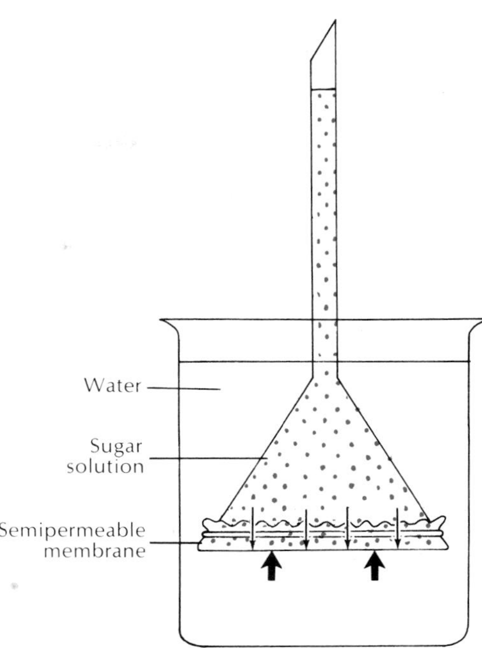

FIG. 5-2 Simple membrane osmometer.

The *direct* measurement of osmotic pressure in biologic solutions is seldom done today. This is because the osmotic pressures of most biologic solutions are so great that it would be impractical, if not impossible, to measure them with the simple membrane osmometer described. The osmotic pressure of human blood plasma would lift a fluid column more than 250 feet—if we could construct such a long, vertical tube and find a membrane that would not rupture from the pressure.

Indirect methods of measuring osmotic pressure are more practical. By far the most widely used measurement is the **freezing point depression.** This is a much faster and more accurate determination than is the direct measurement of osmotic pressure by the collodion membrane osmometer. Pure water freezes at exactly 0° C. As solutes are added, the freezing point is lowered; the greater the concentration of solutes, the lower the freezing point. Human blood plasma freezes at approximately −0.56° C; seawater freezes at approximately −1.80° C. Although the lowering of the freezing point of water by the presence of solutes is small, great accuracy of measurement is possible because the instruments used by biologists can detect differences of as little as 0.001° C.

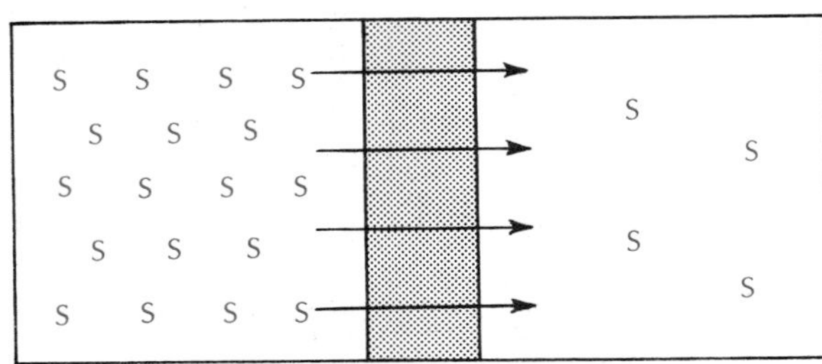

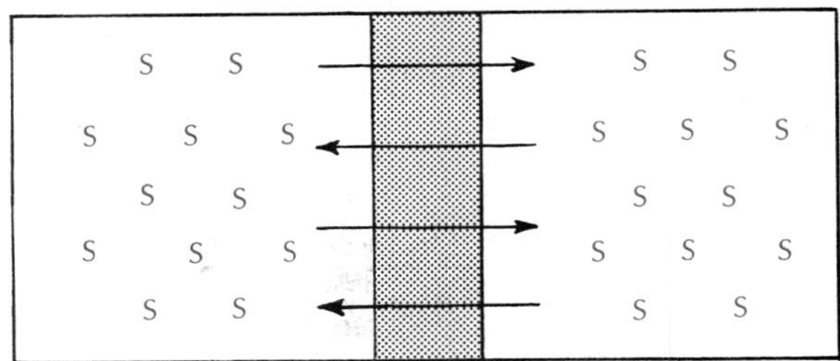

A Passive diffusion

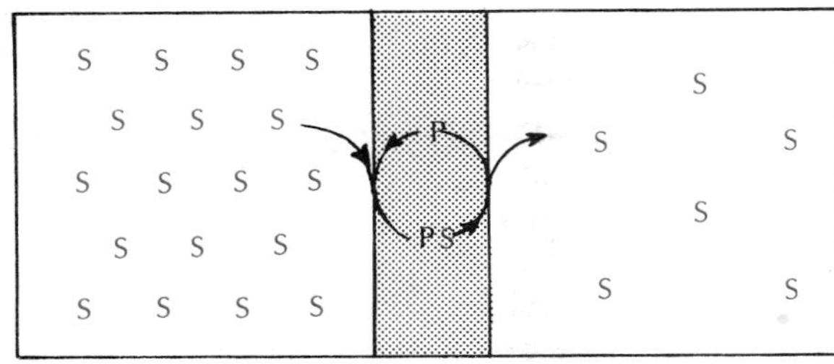

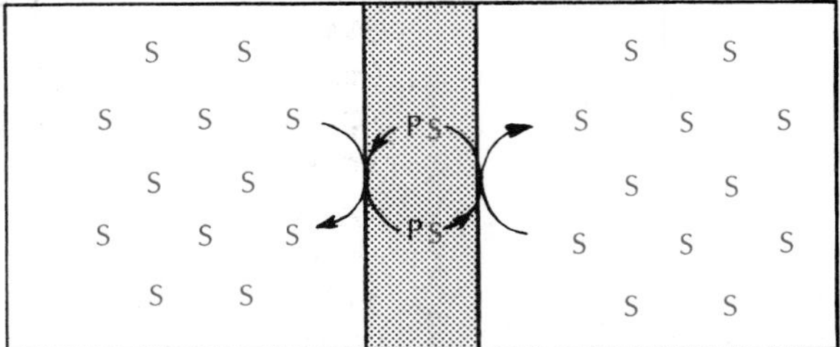

B Facilitated transport

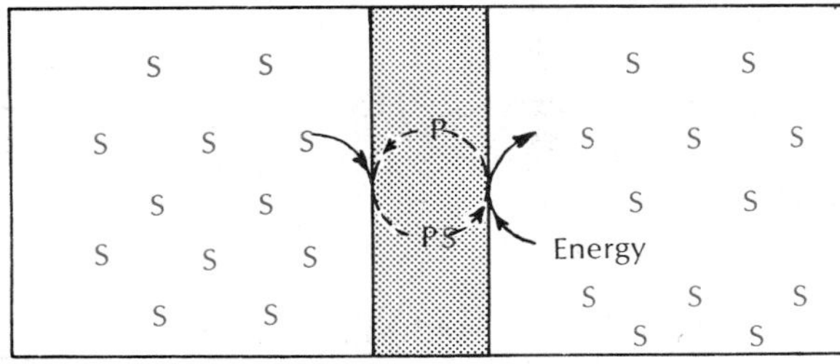

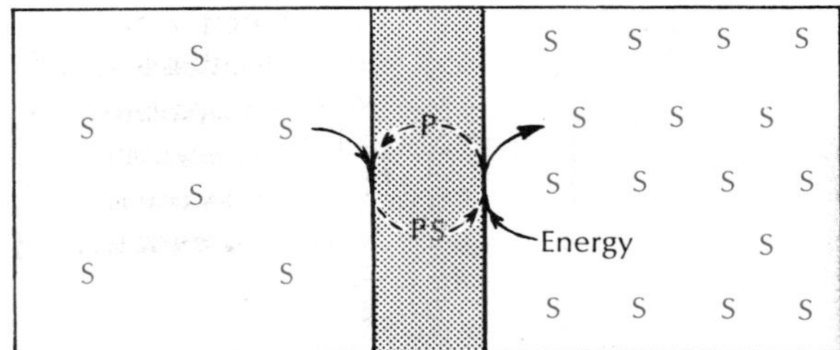

C Active transport

FIG. 5-3 Schematic representation of passive diffusion, facilitated transport, and active transport. In passive diffusion, movement of substance *S* is always down a concentration gradient and at equilibrium there is no further net movement. In facilitated transport, a carrier protein *(P)* assists the substance across a membrane it cannot otherwise penetrate, but no energy is required, and movement is always down a concentration gradient. In active transport, energy is supplied by ATP. Starting from equilibrium, one compartment loses solute and the other gains solute.

Mediated transport. We have seen that the cell membrane is a very effective barrier to the free diffusion of most molecules of biologic significance. Yet it is essential that such materials enter and leave the cell. Nutrients such as sugars and materials for growth such as amino acids must enter the cell, and the wastes of metabolism must leave. Such molecules are moved across the membrane by special mechanisms built into the structure of the membrane. This is called **mediated transport,** meaning that a specific transport mechanism mediates, or facilitates, transfer across the membrane barrier.

Experimental evidence suggests that mediated transport of sugars, amino acids, and other solutes involves the reversible combination with membrane proteins. Such proteins are called **carriers:** protein molecules positioned within the membrane and capable of shuttling from one membrane surface to the other. It is assumed that the carrier molecule captures a solute molecule to be transported, forming a solute-carrier complex. It moves or rotates to the opposite surface with its fare, where the solute detaches and leaves the membrane. The carrier moves back, again presenting its attachment site for the pickup and transport of another solute molecule. Protein carriers are usually quite specific, recognizing and transporting only a limited group of chemical substances or perhaps even a single substance.

At least two distinctly different kinds of carrier-mediated transport mechanisms are recognized: (1) **facilitated transport,** in which the carrier assists a molecule to diffuse through the membrane that it cannot otherwise penetrate, and (2) **active transport,** in which energy is supplied to the carrier systems to transport molecules in the direction opposite to the gradient (Fig. 5-3). Facilitated transport therefore differs from active transport in that it sponsors movement in a downhill direction (in the direction of the concentration gradient) only and requires no metabolic energy to drive the carrier system.

In higher animals facilitated transport is important for the transport of glucose (blood sugar) into body cells that burn it as a principal energy source for the synthesis of ATP. The concentration of glucose is greater in the blood than in the cells that consume it, favoring inward diffusion, but glucose is a large, polar molecule that does not, by itself, penetrate the membrane rapidly enough to support the metabolism of many cells; the carrier system increases the inward flow of glucose.

In active transport, molecules are moved uphill against the forces of passive diffusion. Active transport always involves the expenditure of energy (from ATP) because materials are pumped against a concentration gradient. Among the most important active transport systems in all animals are those that maintain sodium and potassium gradients between cells and the surrounding extracellular fluid or external environment. Most animal cells require a high internal concentration of potassium for protein synthesis at the ribosome and for certain enzymatic functions. The potassium concentration may be 20 to 50 times greater inside the cell than outside. Sodium, on the other hand, may be 10 times more concentrated outside the cell than inside. Both of these electrolyte gradients are maintained by the active transport of sodium into and potassium out of the cell. It is known that in many cells the outward pumping of sodium is linked to the inward pumping of potassium; the same carrier molecule is used for both. As much as 10% to 40% of all the energy produced by the cell is used by the **sodium-potassium exchange pump.**

Endocytosis. The ingestion of solid or fluid material by cells was observed by microscopists nearly 100 years before phrases like "active transport" and "protein carrier mechanism" were a part of the biologist's vocabulary. Endocytosis is a collective term that describes two similar processes, **phagocytosis** and **pinocytosis.**

Phagocytosis, which literally means "cell eating," is a common method of feeding among the Protozoa and lower Metazoa. It is also the way in which white blood cells (leukocytes) engulf cellular debris and uninvited microbes in the blood. By phagocytosis, the cell membrane forms a pocket that engulfs the solid material. The membrane-enclosed vesicle then detaches from the cell surface and moves into the cytoplasm where its contents are digested by intracellular enzymes (Fig. 31-5, p. 715).

Pinocytosis, or "cell drinking," is similar to phagocytosis except that drops of fluid are sucked discontinuously through tubular channels into cells to form tiny vesicles. These may combine to form larger vacuoles. Both processes require metabolic energy, and in this respect they are forms of active transport.

CENTRAL ROLE OF ENZYMES IN LIFE

The life process involves thousands of chemical reactions occurring within cells. However, the chemical breakdown of large molecules and the release of energy by cellular activities would not proceed at any meaningful rate without catalysts. As every chemist knows, catalysts are chemical substances that accelerate reaction rates without affecting the products of the reaction and without being altered or destroyed as a result of the reaction. A catalyst cannot make an energetically impossible reaction happen; it simply accelerates a reaction that would proceed at a very slow rate in its absence.

Enzymes are the catalysts of the living world. They are involved in every aspect of life. Approximately 1,000 different enzymes in an average cell regulate the release of energy used in respiration and promote the synthesis of materials for growth and for replace-

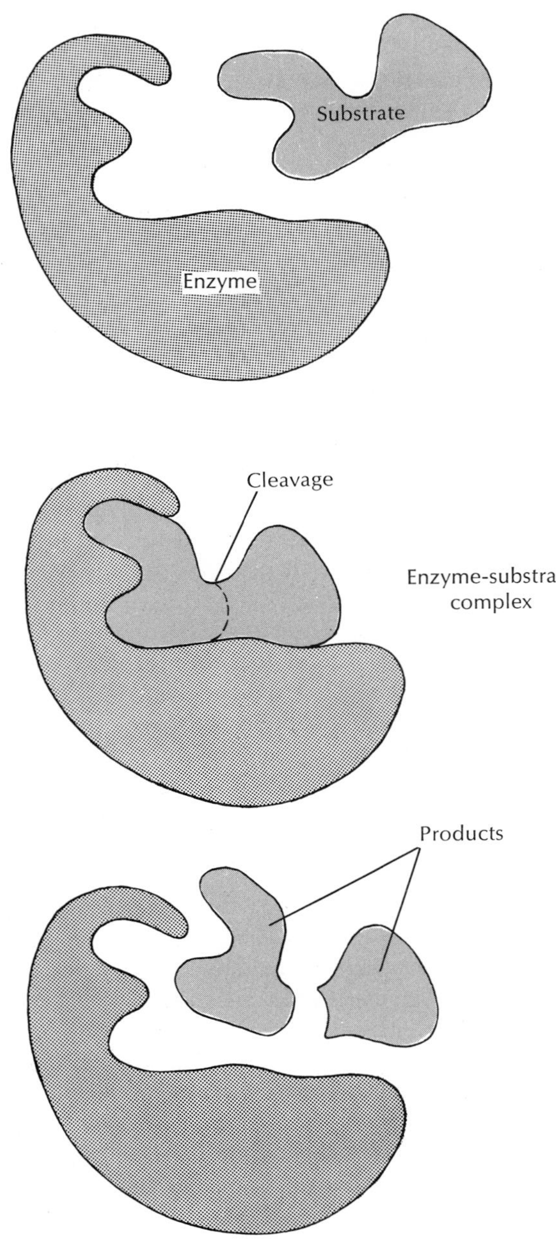

FIG. 5-4 Enzyme action and specificity. Substrate must match the geometry of the enzyme's active site. Once the substrate is precisely aligned to form a complex, certain molecules in the enzyme bond covalently to specific substrate molecules, forming an intermediate chemical state. One bond in the substrate then breaks, and the products are expelled from the enzyme.

ment of losses resulting from structural wear and tear. They regulate muscle contraction, determine molecular movement across membranes, and are involved in mental activities. Other enzymes control the digestion and absorption of food. Enzymes bring control and order to the intricacy of life activities.

Nature of enzymes. Enzymes are complex molecules varying in size from small, simple proteins with a molecular weight of 10,000 to highly complex molecules with molecular weights up to 1 million. Many enzymes are pure proteins—delicately folded and interlinked chains of amino acids. There is nothing novel about the structure of such enzymes that distinguishes them from other proteins; they contain the same kinds of amino acids and the same chemical bonds found in other proteins. It is curious that among the many kinds of chemical reactions catalyzed by pure protein enzymes, some enzymes serve to split apart other proteins. The digestive enzyme pepsin, which initiates the breakdown of protein foodstuffs, is an example.

There are many biochemical reactions that cannot be catalyzed by enzymes built entirely of amino acids. For these, effective catalysis requires the presence of special active chemical groups called **coenzymes** that are attached to or built into the protein enzymes. A protein-coenzyme complex is called a **holoenzyme;** and when the coenzyme is removed by biochemical techniques, the remaining protein is called an **apoenzyme.** The apoenzyme is inactive by itself.

Coenzymes often have unique structures, quite different from any of the metabolites on which cellular enzymes act. Many coenzymes are formed from **vitamins,** compounds that must be supplied in the diet. Not all coenzymes require vitamin precursors, but all of the B complex vitamins and probably vitamin C are essential components of various coenzymes. Since animals have lost the ability to synthesize the vitamin components of coenzymes, it is obvious that a vitamin deficiency can be serious. However, unlike dietary fuels and nutrients that must be replaced, once burned or assembled into structural materials, vitamins are recovered in their original form and used repeatedly. Therefore, only small amounts of vitamins are required in the diet.

Naming of enzymes. In the past, many enzymes have been named by adding the suffix *-ase* to the name of the **substrate,** the molecule on which the enzyme acts. Thus **sucrase** catalyzes the breakdown of **sucrose** to glucose and fructose, **urease** splits **urea** to ammonia and carbon dioxide, and **phosphatase** catalyzes the hydrolysis of **phosphate esters.** Other enzymes such as trypsin and pepsin were named before the convention of adding *-ase* to the substrate was adopted. Because the number of newly discovered enzymes is rapidly increasing, biochemists more recently have attempted to adopt a systematic classification that specifies the reaction catalyzed. Thus the reaction

$$\text{ATP} + \text{Creatine} \rightleftharpoons \text{ADP} + \text{Phosphocreatine}$$

is catalyzed by an enzyme called **ATP:creatine phosphotransferase.** This is the recommended systematic name but in practice the older and less ponderous trivial name **creatine kinase** is usually used.

Action of enzymes. An enzyme functions by combining in a highly specific way with its substrate. According to the classic **lock-and-key theory,** each enzyme contains an **active site,** which is a unique molecular configuration that is exactly complementary to at least a portion of the specific substrate molecule (Fig. 5-4). Most substrates are in fact much smaller than a large protein enzyme. By fitting onto or around the substrate, the enzyme provides a unique chemical environment that increases the probability of a chemical reaction at specific sites on the substrate. The special catalytic talent of an enzyme resides in its power to reduce the high internal energy barrier through which substrate molecules must pass to be transformed into products. In effect, an enzyme steers the reaction through one or more intermediate steps, each of which requires much less activation energy than that required for a single-step reaction (Fig. 5-5). The enzyme combines with its substrate to form a precisely aligned enzyme-substrate complex in which the substrate is secured by covalent bonds to several points in the active site of the enzyme. The substrate is split, and its products are liberated from the enzyme, which is restored to its active form.

The lock-and-key theory is still largely accepted by biochemists. However, the active site of the enzyme may be a flexible surface that infolds, and conforms to, the substrate, rather than a fixed and nonyielding template, as the original theory held. This newer **conformational theory** has not altered a firmly held principle of enzyme action: the enzyme and substrate must combine so that active groups on the enzyme come into precise alignment with reactive sites on the substrate. Only then can the substrate be altered chemically. The necessity of correct alignment explains the high specificity of enzymes.

Enzymes that engage in important main-line sequences—such as the crucial energy-providing reactions of the cell that go on constantly—seem to operate in enzyme sets rather than in isolation. For example, the conversion of glucose to carbon dioxide and water proceeds through 19 reactions, each requiring a specific enzyme. Main-line enzymes are found in relatively high concentrations in the cell, and they may implement quite complex and highly integrated enzymatic sequences. One enzyme carries out one step, then passes the product to another enzyme that catalyzes another step, and so on.

Specificity of enzymes. One of the most distinctive attributes of enzymes is their high specificity. This is a consequence of the exact molecular fit that is required between enzyme and substrate. Furthermore, an enzyme catalyzes only one reaction; unlike reactions carried out in the organic chemist's laboratory, no side reactions or by-products result. Specificity of both substrate and reaction is obviously essential to prevent a cell from being overwhelmed with useless by-products.

However, there is some variation in degree of specificity. Some enzymes, such as succinic dehydrogenase, will catalyze the oxidation (dehydrogenation) of one substrate only, succinic acid. Others, such as proteases (for example, pepsin and trypsin), will act on almost any protein. Usually an enzyme will take on one substrate molecule at a time, catalyze its chemical change, release the product, and then repeat the process with another substrate molecule. The enzyme may repeat this process billions of times until it is finally worn out (a few hours to several years) and is broken down by scavenger enzymes in the cell. Some enzymes are able to undergo successive catalytic cycles at dizzying speeds of up to a million cycles per minute; most operate at slower rates. The digestive enzymes, which are secreted into the digestive tract to degrade food materials are "one-shot" enzymes. After breaking down their substrate, they are themselves digested and lost to the body.

Enzyme-catalyzed reactions. Enzyme-catalyzed reactions are theoretically reversible. This is signified by the double arrows between substrate and products.

$$\text{Succinic acid} + H_2O \rightleftharpoons \text{Fumaric acid}$$

However, for various reasons the reactions catalyzed by most enzymes tend to go only in one direction. For

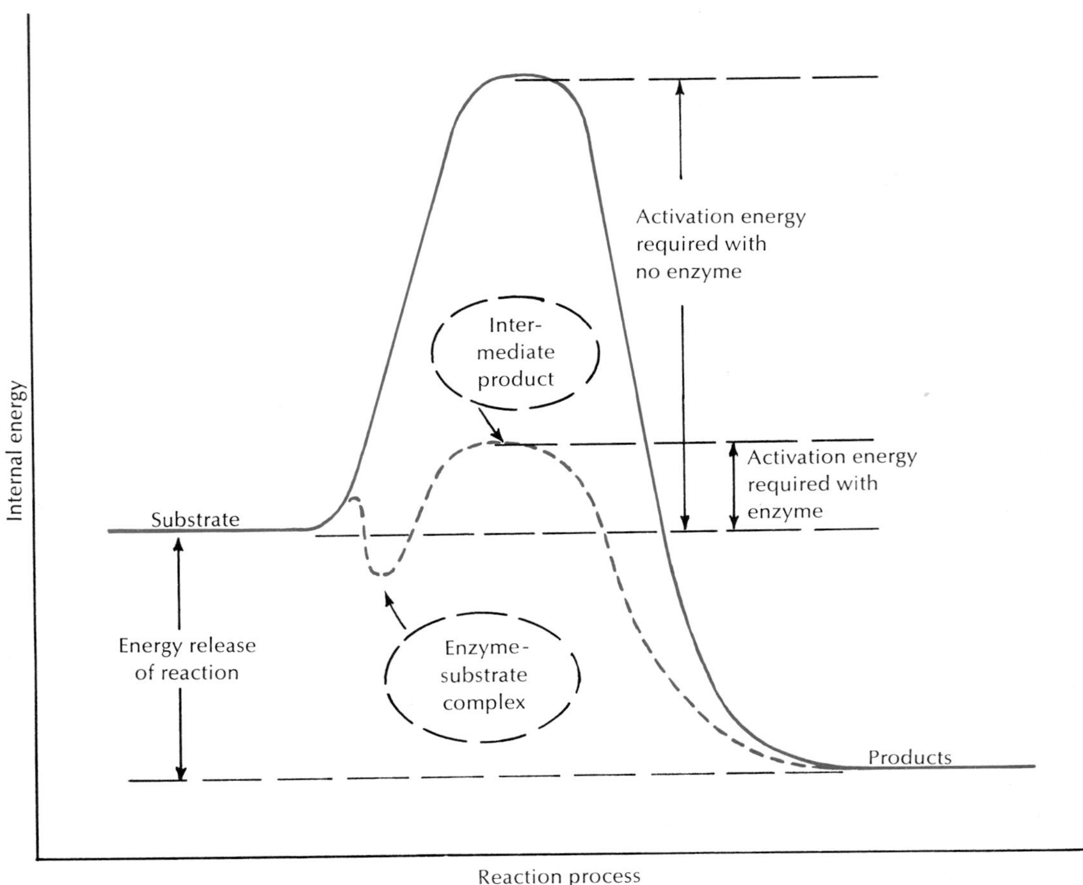

FIG. 5-5 Energy changes during enzyme catalysis of a substrate. The overall reaction proceeds with a net release of energy. In absence of enzyme, substrate is stable because of the large amount of activation energy needed to disrupt strong chemical bonds. The enzyme reduces the energy barrier by forming a chemical intermediate with much lower internal energy state.

example, the proteolytic enzyme pepsin can degrade proteins into amino acids, but it cannot accelerate the rebuilding of amino acids into any significant amount of protein. The same is true of most enzymes that catalyze the hydrolysis of large molecules such as nucleic acids, polysaccharides, lipids, and proteins. There is usually one set of reactions and enzymes that break them down, but they must be resynthesized by a different set of reactions that are catalyzed by different enzymes. This apparent irreversibility exists because the chemical equilibrium usually favors the formation of the smaller degradation products.

The net **direction** of any chemical reaction is dependent on the relative energy contents of the substances involved. If there is little change in the chemical-bond energy of the substrate and the products, the reaction is more easily reversible. However, if large quantities of energy are released, more energy must be provided in some way to drive the reaction in the reverse direction. Thus many, if not most, enzyme-catalyzed reactions are practically irreversible, unless the reaction is coupled to another that makes energy available. In the cell, both reversible and irreversible reactions are combined in complex ways to make possible both synthesis and degradation.

Sensitivity of enzymes. Enzyme activity is sensitive to temperature and pH. As a general rule enzymes work faster with increasing temperature, but will do so only within certain limits. Moreover, this increase in velocity is not proportional to the rise in temperature. Usually the rate is doubled with each 10° C rise, but a change from 20° to 30° C has a greater effect than one from 30° to 40° C. The optimum temperature for enzymes of endothermic animals is about body temperature. Above 40° to 50° C most enzymes become unstable and may be inactivated altogether because the enzyme is denatured, or cooked.

Each enzyme usually works best within a certain range of acidity or alkalinity. Pepsin of the acid gastric juice is most active at about pH 1.8; trypsin of the alkaline pancreatic juice is most active at about pH 8.2. Most cellular enzymes work best when the pH is around neutral. In strong acid or alkaline solutions, most enzymes irreversibly lose their catalytic power.

CELLULAR METABOLISM

Cellular metabolism refers to the collective total of chemical processes that occur within living cells. It is often called **intermediary metabolism** because the exchange of matter and energy between the cell and its environment proceeds in a stepwise manner through chemical pathways composed of numerous intermediates, or **metabolites.** Although intermediary metabolism appears hopelessly complex as depicted in detailed metabolic charts that often grace the walls of biochemist's laboratories, the central metabolic routes through which matter and energy are channeled are not difficult to understand. Biochemists are vastly furthered in their research by the fact that the same kinds of reaction sequences in metabolism occur in a great variety of life forms from bacteria to humans (Principle 3, p. 10).

Metabolism consists of two major classes of reactions: **catabolism,** the fragmentation of nutrient molecules into smaller and simpler parts, and **anabolism,** the putting together, or biosynthesis, of larger molecules from molecular fragments. Every metabolic reaction is catalyzed by a specific enzyme and is accompanied by an exchange of energy. Energy that is present in the chemical structure of organic molecules is released in catabolism, whereas anabolism requires the input of chemical energy. Catabolism and anabolism proceed simultaneously in cells (Fig. 5-6).

Living cells have two major requirements from the environment: (1) **nutrients** as a source of carbon and inorganic materials that are used to build all of their or-

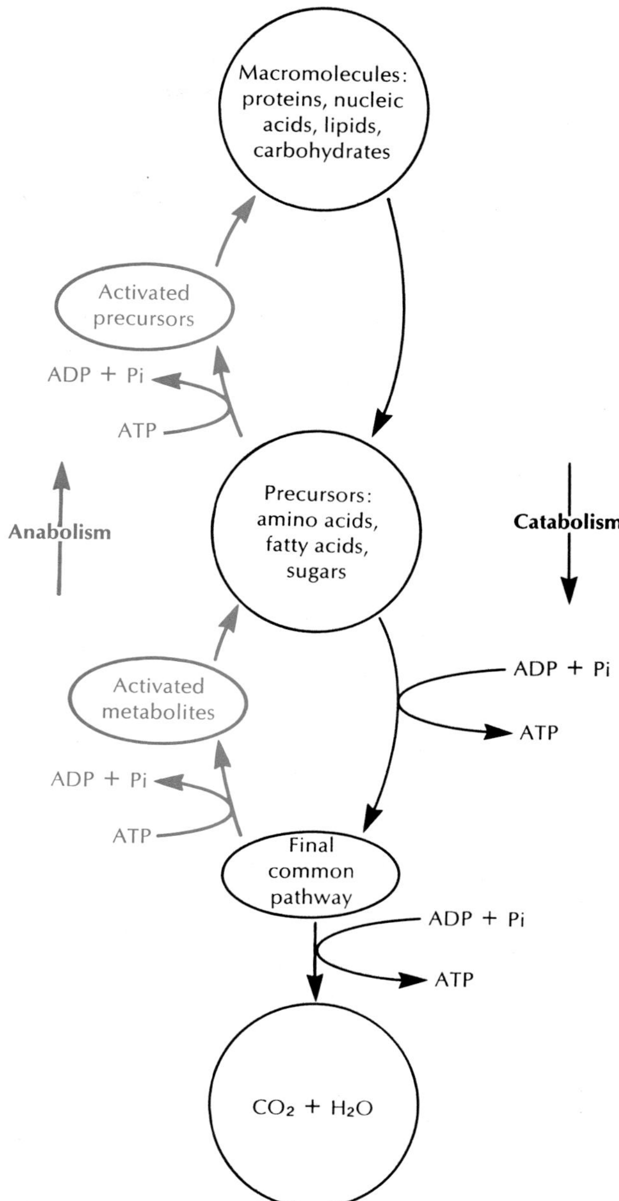

FIG. 5-6 Principal features of cellular metabolism. In catabolic pathways (in black, leading downward) large nutrient molecules are progressively degraded and channeled into a final common pathway in which the products are oxidized to carbon dioxide and water. Catabolism yields free energy from metabolites, which is used to generate ATP, the "energy currency" of life. In anabolism (in red, leading upward) metabolites are converted into precursor building blocks, which are assembled into macromolecules. Anabolism is energetically uphill and requires the input of ATP.

ganic structure and (2) **fuels** as a source of energy. Plants (**autotrophic** or "self-feeding" organisms) can utilize carbon dioxide as a source of carbon for constructing organic molecules. Animals (**heterotrophic** organisms that feed on others) must obtain carbon from the relatively complex organic compounds elaborated by plants. The photosynthetic plants use light as their energy source and consequently are relatively self-sufficient. Animals, of course, are totally dependent on plants for ready-made fuels.

Animal cells tap the stored energy of organic fuels (for example, simple sugars, fatty acids, amino acids) through a series of controlled degradative steps. This process makes use of molecular oxygen from the atmosphere. In return animal cells give off carbon dioxide as an end product, which is used by plant cells in making glucose and the more complex molecules. In this way the cellular energy cycle of life involves the harnessing of sunlight energy by green plants directly and by animal cells indirectly (Principle 2, p. 9).

Cellular energy transfer

There are certain limited parallels between the combustion of fuel in a fire and metabolic combustion of fuel in a living cell. In both, oxygen is consumed and energy is liberated as heat (exergonic reactions). If the fuel is burned in an internal combustion engine so that work is performed, the parallel is even better. But here the similarity ends. The burning of gasoline in a cylinder is an explosive event that promotes just one function, the rapid expansion of gas. Many chemical bonds are broken simultaneously, and much energy is lost as heat.

In contrast, metabolic oxidations are flameless and of low-temperature. They proceed gradually instead of explosively, and the energy liberated is coupled to a great variety of energy-consuming reactions. Although metabolic energy exchanges proceed with impressive efficiency, heat is inevitably liberated. Heat can be put to some use, of course; the endothermic vertebrates (birds and mammals) use it to elevate and maintain a constant internal body temperature, just as the heat of gasoline combustion is made to warm the occupants of a car. But for the most part, heat is a useless commodity to a cell, since it is a nonspecific form of energy that cannot be captured and redistributed to power metabolic processes. There is actually only one way in which the oxidative release of energy is made available for use by cells: it is coupled to the production of high-energy phosphate bonds, usually in the form of **ATP (adenosine triphosphate)** by addition of inorganic phosphate to **ADP (adenosine diphosphate).** The structure of ATP is shown at the bottom of the page.

The ATP molecule consists of a purine (adenine), a 5-carbon sugar (ribose), and 3 molecules of phosphoric acid linked together by two pyrophosphate bonds to form a triphosphate group. The pyrophosphate bonds are called **high-energy bonds** because they are repositories of a great deal of chemical energy. This energy has been transferred to ATP from other low-energy bonds in the respiratory process. Respiration, by the stepwise oxidation of fuel substrates, redistributes bond energies so that a few high-energy bonds are created and stored in ATP. Obviously this energy is gained at the expense of fuel energy; the end products of cellular respiration, carbon dioxide and water, contain much

Adenosine triphosphate (ATP)

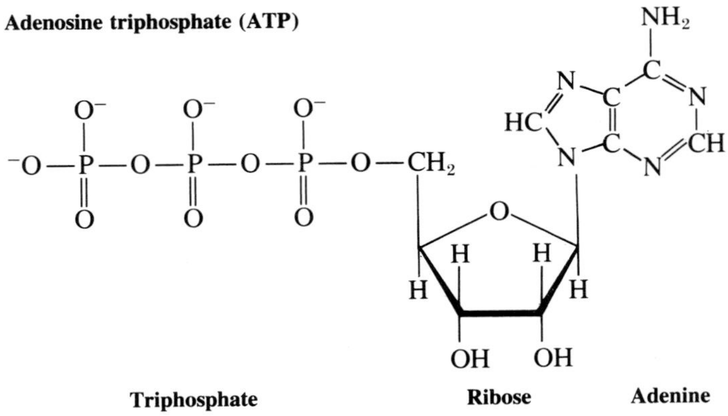

Triphosphate **Ribose** **Adenine**

less bond energy than do the fuel substrates (for example, glucose) that entered the oxidative pathway.

The high-energy pyrophosphate bonds of ADP and ATP are designated by the "tilde" symbol ~. Thus a low-energy phosphate bond is shown as —P, and a high-energy one as ~ P. ADP can be represented as A—P ~ P and ATP as A—P ~ P ~ P.

The trapping of energy by ATP can be shown as follows:

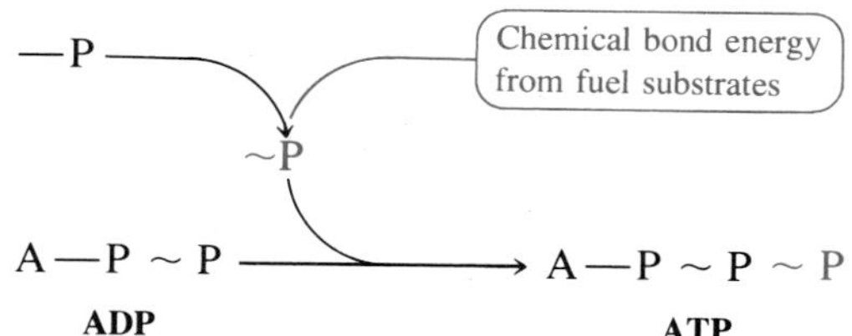

A low-energy phosphate group containing bond energy of 2,000 to 3,000 calories/mole is converted into a high-energy phosphate group containing bond energy of 8,000 to 10,000 calories/mole. Where does the low-energy phosphate group (—P) come from? Ultimately it comes from the diet. But in cellular respiration, ATP itself is the immediate source of the phosphate group necessary to start the oxidation process, donating the —P to the fuel substrate molecule (usually glucose).

ATP ADP

Fuel substrate → Fuel substrate —P

The fuel molecule is phosphorylated with a low-energy phosphate group, and the phosphorylated fuel can then be oxidized to yield energy. Quite obviously, the fuel molecule must release more energy—actually it provides much more energy—than is loaned to it by ATP at the start.

This is a kind of initial deficit financing that is required for an ultimate energy return many times greater than the original energy investment. We will return later to this subject of energy budgeting.

The amount of ATP produced in respiration is totally dependent on its rate of utilization. In other words, ATP is produced by one set of reactions and immediately consumed by another. ATP is a great **energy-coupling** molecule, used to transport energy from one reaction to another. For example, ATP formed from the oxidation of glucose is used to synthesize proteins or lipids or to provide power for some other process, such as the contraction of skeletal muscles. The point is, living organisms do not produce and put aside vast amounts of ATP, hoarded against some future energy need. What they do store is the fuel itself, in the form of carbohydrates and lipids especially. ATP is formed as it is needed, primarily by oxidative processes in the mitochondria. Oxygen is not consumed unless ADP and phosphate molecules are available, and these do not become available until ATP is hydrolyzed by some energy-consuming process. *Metabolism is therefore mostly self-regulating.*

Although ATP is a carrier rather than a reservoir of energy, muscle and nerve cells have a phosphate compound especially adapted for energy storage. This is phosphocreatine:

H
\
N ~ PO_3^-
/
H_2N—C
\
N—CH_3
|
CH_2
|
COO^-

The phosphocreatine reservoir contains a high-energy bond that can provide instant power to the muscle contractile machinery, which often has sudden energy needs. Phosphocreatine is formed from creatine and ATP and is in direct chemical equilibrium with ATP. A burst of muscle activity will deplete the available ATP within seconds, raising the concentration of ADP. High-energy phosphate groups are transferred to ADP from phosphocreatine, providing more ATP for contraction. This system can provide chemical energy for several minutes of muscle contractions. Creatine phosphate is found in the muscles of all vertebrate animals, but other phosphate compounds, especially arginine phosphate, are present among the invertebrates.

Respiration: generating ATP in the presence of oxygen

How electron transport is used to trap chemical bond energy. Having seen that ATP is the one common energy denominator by which all cellular machines are powered, we are in a position to ask how this energy is captured from fuel substrates. This question directs us to an important generalization: *all cells obtain their chemical energy requirements from oxidation-reduction reactions*. This means simply that in the degradation of fuel molecules, hydrogen atoms (electrons and protons) are passed from reducing agents to oxidizing agents with a release of energy. A portion of this energy is trapped and used to form the high-energy bonds of ATP. The release of energy during electron transfer and its conservation as ATP is the mainspring of cell activity and was a crucially important evolutionary achievement.

Before proceeding further and to avoid later confusion, let us consider what we mean by oxidation-reduction ("redox") reactions. In these reactions there is a transfer of electrons from an electron donor (the reducing agent) to an electron acceptor (the oxidizing agent). As soon as the electron donor loses its electrons, it becomes oxidized. As soon as the electron acceptor accepts electrons, it becomes reduced. In other words, a reducing agent becomes oxidized when it reduces another compound, and an oxidizing agent becomes reduced when it oxidizes another compound. Thus for every oxidation there must be a corresponding reduction.

In the cell, the electrons flow through a series of carriers. Each carrier is reduced by accepting electrons and then is reoxidized by passing electrons to the next carrier in the series. By transferring electrons stepwise in this manner, energy is gradually released, and a maximum yield of ATP is realized.

Aerobic and anaerobic metabolism. Ultimately, the electrons are transferred to a **final electron acceptor.** The nature of this final acceptor is the key that determines the overall efficiency of cellular metabolism. The heterotrophs can be divided into two great groups: **aerobes,** those that use molecular oxygen as the final electron acceptor, and **anaerobes,** those that employ some other molecule as the final electron acceptor.

Because humans and all of the other animals with which we are most familiar are dependent on oxygen, it may seem surprising that many primitive organisms live successfully under oxygen-free conditions. But we must recall that life on earth had its origins under highly reducing conditions; there was no oxygen in the primitive atmosphere for perhaps the first billion years of life on the primeval earth. During this formative period, the basic skeleton of cellular metabolism evolved under strictly anaerobic conditions. Only later as evolving photosynthetic organisms began to produce oxygen did metabolic reactions appear in which the generation of ATP was coupled to the utilization of oxygen.

Today most animals utilize oxygen to generate high-energy phosphate. Strictly anaerobic organisms still exist and indeed occupy some important ecologic niches. These organisms for the most part are bacteria that are restricted to the relatively few oxygen-free environments that remain on earth. For the majority of animals, oxygen is a necessity of life. Evolution has favored aerobic metabolism, not only because oxygen became available, but also because aerobic metabolism is vastly more efficient than anaerobic metabolism. In the absence of oxygen, only a very small fraction of the bond energy present in foodstuffs can be released. For example, when an anaerobic microorganism degrades glucose, the final electron acceptor (lactic acid) still contains most of the energy of the original glucose molecule. On the other hand, an aerobic organism, using oxygen as the final electron acceptor, can completely oxidize glucose to carbon dioxide and water. Almost 20 times as much energy is released when glucose is completely oxidized as when it is degraded only to the stage of lactic acid. A very obvious advantage of aerobic metabolism is that a much smaller quantity of foodstuff is required to maintain a given rate of metabolism.

General description of respiration. Aerobic metabolism is more familiarly known as **respiration,** defined as the oxidation of fuel molecules by molecular oxygen. Let us look at the process in general before considering it in more detail. Respiration proceeds through three stages that result in the complete oxidation of foodstuff molecules to carbon dioxide and water. The three stages are: preliminary degradation of fuel molecules, the Krebs cycle, and oxidative phosphorylation. Stage I occurs in the cytoplasm and stages II and III in the mitochondrion.

In stage I, a fuel molecule such as glucose is broken up into a pair of 3-carbon molecules (pyruvate). In stage II, pyruvate is degraded into a 2-carbon acetyl group, which is coupled with a coenzyme to form **acetyl coenzyme A (acetyl CoA).** Acetyl CoA is channeled into the Krebs cycle where the acetyl group of acetyl CoA is degraded to form 2 molecules of carbon dioxide. In both stages I and II, hydrogen atoms are removed from the carbon skeleton of the fuel molecule and passed to electron acceptor compounds in the electron transport chain of stage III. Here, electron energy is "trapped" and transferred to high-energy ATP molecules. At the end of the chain the electrons (and the protons that accompany them) are oxidized by molecular oxygen, the final electron acceptor, to form water.

We will now examine each of these stages in more detail, beginning with the *last* stage, the electron transport chain. We will proceed then to stages II and I, in that order, and then rebuild the entire respiratory scheme in a summary.

Oxidative phosphorylation

The transfer of electrons from fuel substrates to molecular oxygen through a series of enzymes is referred to as **oxidative phosphorylation.** This electron transport chain is composed of a series of large molecules localized in the inner mitochondrial membranes. The electron carriers are compounds that can be reduced by accepting electrons from the previous carrier, and then be oxidized again by passing electrons to the next carrier. Each successive carrier is a somewhat stronger oxidizing agent than the one before; that is, *successive carriers are increasingly stronger electron acceptors.* Finally, the electrons, as well as the hydrogen protons that accompany them, are transferred to molecular oxygen to form water:

$$\text{Fuel-(2H)} \curvearrowright \text{A} \curvearrowright \text{B} \curvearrowright \text{C} \curvearrowright \text{D} \curvearrowright \text{Oxygen} \rightarrow H_2O$$

It is conventional to represent electron transfer through the electron carrier system as the transfer of hydrogen atoms, although we must emphasize that it is the energized electrons and not the protons of the hydrogen that are the important energy packets. The proton of each hydrogen atom simply takes a free ride during this electron shuttle until, at the end, it bonds with reduced oxygen and forms water.

The whole function of the electron transport chain is the capture of energy from the original fuel substrate and the transformation of it into a form the cell can use. To do this, the large chemical potential of food molecules is drawn off in small steps (rather than in one explosive burst as in ordinary combustion) by successive electron carriers. At three points along the chain, ATP production takes place by the phosphorylation of ADP. This method of energy capture is called oxidative phosphorylation because the formation of high-energy phosphate is coupled to oxygen consumption, and this depends, as we have seen, on the demand for ATP by other metabolic activities within the cell. The actual mechanism of ATP formation by oxidative phosphorylation is not known with certainty; we can only say that the transfer of electrons does something that is translated into the production of high-energy phosphate bonds. Currently, the most widely accepted explanation of this mechanism is the chemiosmotic theory (Hinkle and McCarty, 1978).

Localization and function of electron carriers. Oxidative phosphorylation, a complex process, would be unable to function efficiently, if at all, were the enzymes just floating freely in the cytoplasm of the cell. There is now abundant evidence that the oxidative enzymes and electron carriers are arranged in a highly ordered state on the membranes of the mitochondria.

It will be recalled that mitochondria are composed of two membranes. The outside membrane forms a smooth sac enclosing the inner membrane that is turned into numerous ridges called **cristae** (Fig. 4-7, p. 65). The inner membrane is studded with enormous numbers of minute, stalked particles called **inner-membrane spheres.** The electron carriers of the respiratory chain are restricted to the inner membrane where, presumably, they are located on the inner-membrane spheres. A section of mitochondrion as it might appear under the high-resolution electron microscope is shown (highly diagrammatically) in Fig. 5-7.

Pairs of electrons donated initially from food substrates flow along the electron carriers in succession (Fig. 5-8). For most food substrates the initial electron acceptor is NAD (nicotinamide adenine dinucleotide, a derivative of the vitamin niacin). The substrate is oxidized in the process (because it loses electrons), and NAD is reduced (because it gains electrons).

Next flavoprotein (a riboflavin derivative) oxidizes the reduced NAD by accepting its electrons. Flavoprotein becomes reduced (having gained electrons), and NAD is returned to its original oxidized state. In the same way the pair of electrons is passed sequentially to coenzyme Q (chemically related to vitamin K) and then through a series of electron acceptors called **cytochromes.** The cytochromes are large molecules that

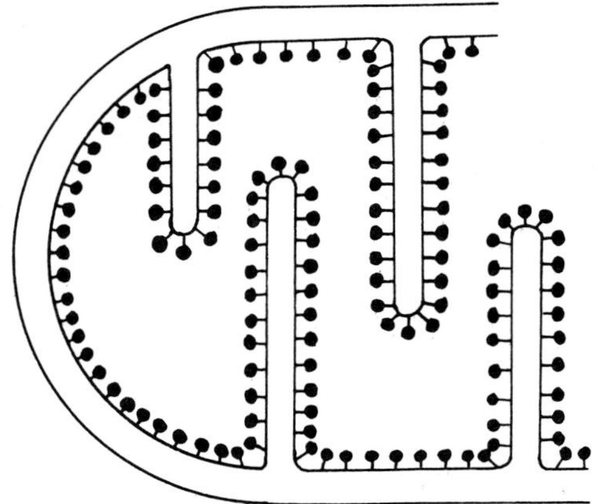

FIG. 5-7 Representation of section of mitochondrion as seen through high-resolution electron microscope, showing the inner membrane spheres that bear enzymes of the respiratory chain. The density of the spheres is actually many times greater than that depicted in this diagram.

belong to a class of proteins called chromoproteins because they contain colored coenzyme groups. The coenzyme group of a cytochrome, like hemoglobin, which is closely related to it, is an iron-bearing group that can be reversibly reduced.

$$Fe^{+++} + e^- \underset{\text{Oxidation}}{\overset{\text{Reduction}}{\rightleftarrows}} Fe^{++}$$

As the electrons are passed from cytochrome to cytochrome, each is successively reduced and then oxidized. Finally the electrons are passed to molecular oxygen. The transfer of electrons and the points of high-energy phosphate bond formation are shown in Fig. 5-8. Thus for every pair of electrons moved along the carriers to oxygen, a total of 3 ATP molecules is formed.

We have seen that ATP production is coupled to electron transfer, which in turn is completely dependent on oxygen, the final hydrogen and electron acceptor. Without oxygen the process stops because the electrons have nowhere to go. The components of the electron transfer chain would quickly become fully reduced and remain so, in the absence of the electron sink, molecular oxygen.

Acetyl CoA: strategic intermediate in respiration

In respiration, most fuel molecules are progressively stripped of their carbon atoms until those carbons are finally converted into molecules of carbon dioxide. During this degradation the hydrogens and their electrons are removed and passed into the important energy-yielding electron transport chain we have already described. But to reach this final sequence, most carbon atoms appear in the 2-carbon group, acetyl coenzyme A (acetyl CoA). This is a critically important compound. Some two-thirds of all the carbon atoms in foods eaten by animals appear as acetyl CoA at some stage. The strategic metabolic position of acetyl CoA is illustrated in Fig. 5-9. It is the final oxidation of

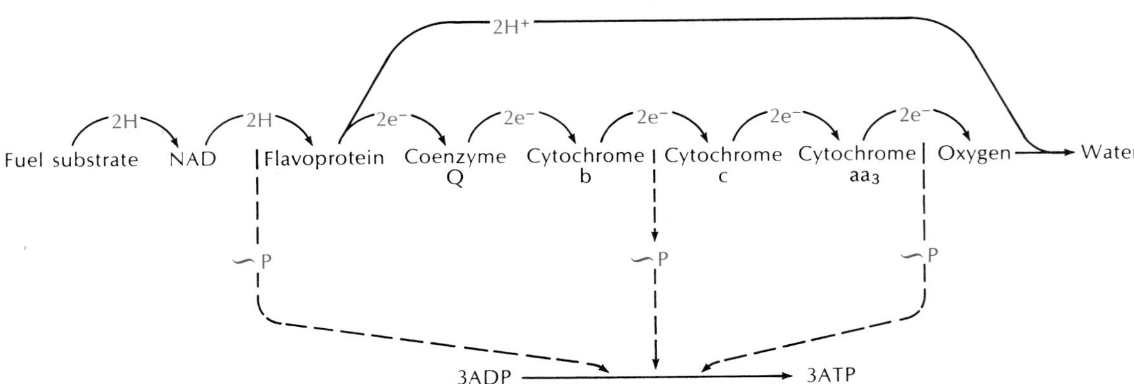

FIG. 5-8 Electron transport chain. Electrons are transferred from one carrier to the next, terminating with molecular oxygen to form water. A carrier is reduced by accepting electrons and then is reoxidized by donating its electrons to the next carrier. ATP is generated at three points in the chain. These electron carriers are located on inner membrane spheres of mitochondria.

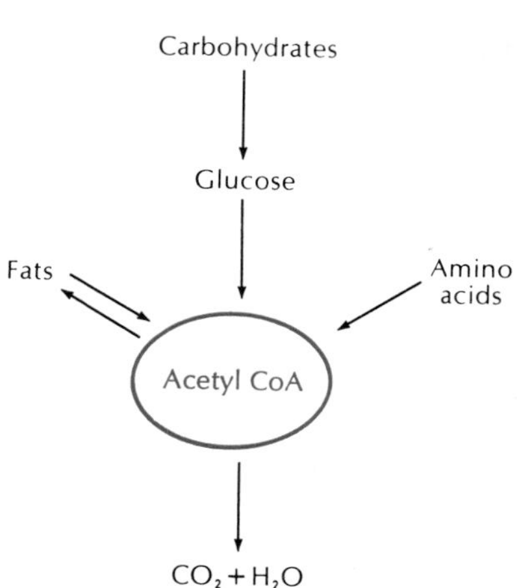

FIG. 5-9 Acetyl CoA is an important intermediate in oxidation of carbohydrates, proteins (amino acids), and fats.

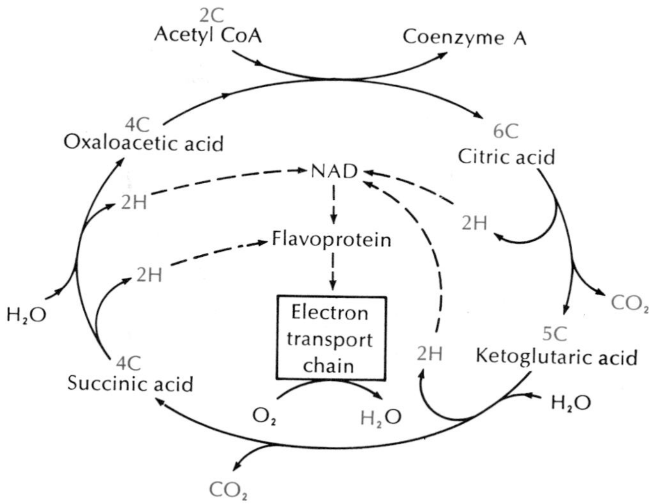

FIG. 5-10 Krebs cycle. (See text for explanation.)

acetyl CoA that provides the energized electrons used to generate ATP. Acetyl CoA is also the source of nearly all the carbon atoms found in the body's fats, as the reverse arrow in Fig. 5-9 indicates. The structure of acetyl CoA can be shown in abbreviated form as:

$$\underbrace{CH_3-\overset{\displaystyle O}{\overset{\|}{C}}}_{\text{Acetyl group}} \;\vdots\; \underbrace{S-CoA}_{\text{Coenzyme A group}}$$

The right hand side of the molecule is a coenzyme containing the vitamin **pantothenic acid,** another example of how vitamins play important structural roles in critical cellular functions.

Krebs cycle: oxidation of acetyl CoA

The degradation (oxidation) of the 2-carbon acetyl group of acetyl CoA occurs in a cyclic sequence called the Krebs cycle after its British discoverer Sir Hans A. Krebs. In this cycle the acetyl group is enzymatically broken down to form 2 molecules of carbon dioxide and four pairs of hydrogen atoms. The latter (or actually their electrons) then enter the electron transport system described previously.

The Krebs cycle begins with the condensation of the 2-carbon acetyl group of acetyl CoA with the 4-carbon compound oxaloacetic acid to form the 6-carbon compound citric acid. Citric acid is then degraded through a series of reactions that yields 2 molecules of carbon dioxide from the original acetyl group. The 4-carbon oxaloacetic acid is regenerated and ready to begin another turn of the cycle. Oxaloacetic acid therefore serves as a carrier for the 2 carbons of the acetyl group; it does not disappear in the cyclic process but will continue to accept an indefinite number of acetyl groups.

As the acetyl group is oxidized, carbon atom by carbon atom, four pairs of electrons and 4 protons (shown as four pairs of hydrogen atoms in Fig. 5-10) are transferred to the electron transfer chain (shown in the center of the cycle in Fig. 5-10). Three pairs of electrons are passed to NAD; the remaining pair is passed directly to flavoprotein. Each pair of electrons then shuttles down the electron transport chain to molecular oxygen, a process that is accompanied by the generation of ATP as already described.

Three molecules of ATP are generated for *each* molecule of NAD receiving electrons; this yields a total of 9 ATP molecules per acetyl group. Two more molecules of ATP are generated from the electrons passed directly to flavoprotein. One more high-energy bond is generated at another point in the cycle; it forms a compound called GTP (guanosine-5′-triphosphate), which has the same energy yield as ATP, and for sim-

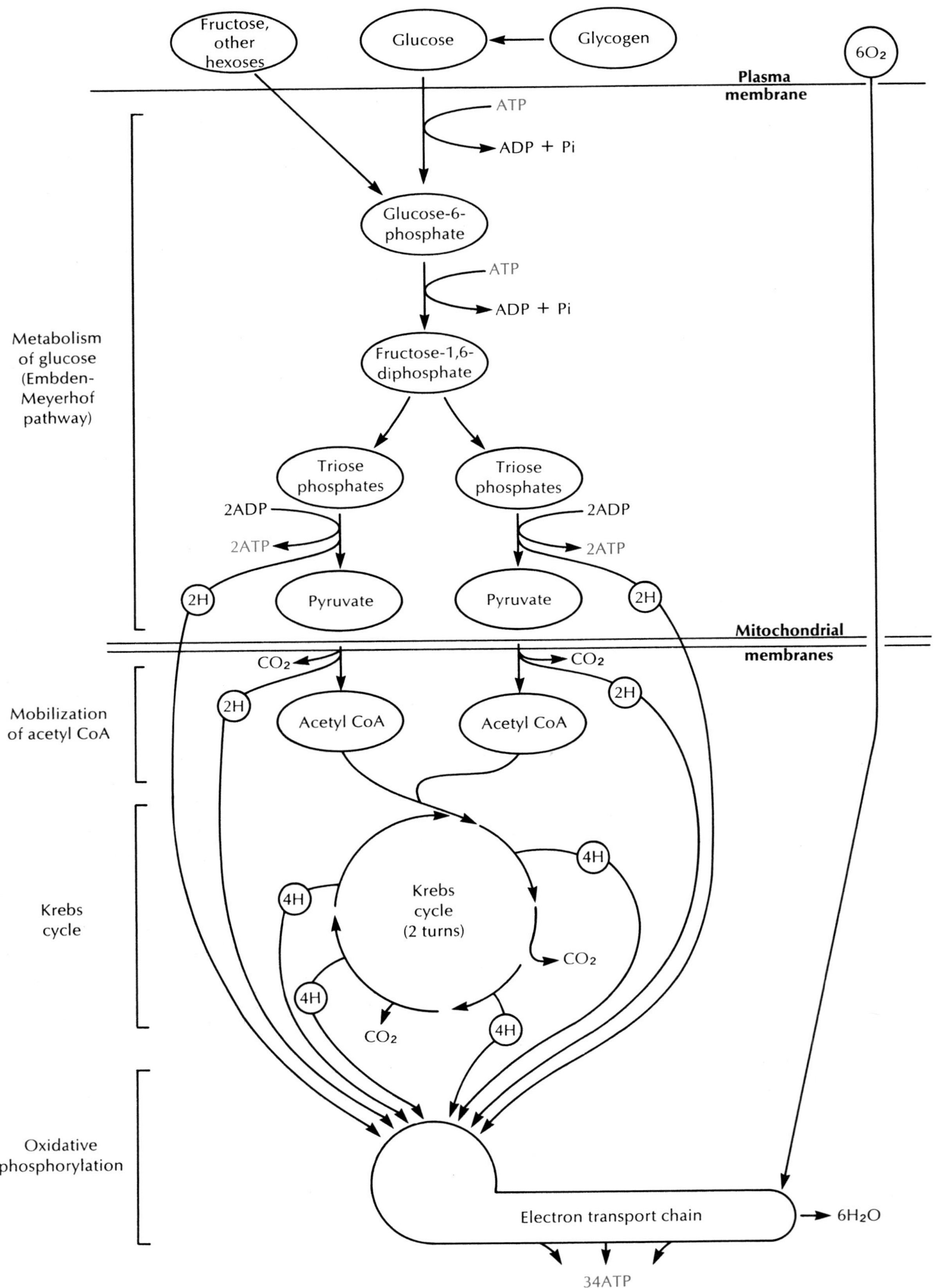
Fructose, other hexoses
Glucose
Glycogen
6O2
Plasma membrane
ATP
ADP + Pi
Glucose-6-phosphate
ATP
ADP + Pi
Fructose-1,6-diphosphate
Metabolism of glucose (Embden-Meyerhof pathway)
Triose phosphates
Triose phosphates
2ADP
2ATP
2ADP
2ATP
2H
Pyruvate
Pyruvate
2H
Mitochondrial membranes
CO2
CO2
2H
Acetyl CoA
Acetyl CoA
2H
Mobilization of acetyl CoA
4H
Krebs cycle
4H
Krebs cycle (2 turns)
CO2
4H
CO2
4H
Oxidative phosphorylation
Electron transport chain
6H2O
34ATP

FIG. 5-11 Pathway for oxidation of glucose and other carbohydrates. Glucose is degraded to pyruvate by cytoplasmic enzymes (Emden-Meyerhof pathway). Acetyl CoA is formed from pyruvate and is fed into the Krebs cycle. An acetyl group (2 carbons) is oxidized to two molecules of CO_2 with each turn of the cycle. Pairs of electrons (2H) are removed from carbon skeleton of the substrate at several points in the pathway and are carried by oxidizing agents (NAD or flavoprotein, not shown) to the electron transport chain where 34 molecules of ATP are generated. Four molecules of ATP are also generated by substrate phosphorylation in the Emden-Meyerhof pathway, yielding a total of 38 molecules of ATP per glucose molecule. Note that molecular oxygen is involved only at the very end of a basically anaerobic pathway.

plicity's sake, we will call it ATP. Thus the net yield is 12 molecules of ATP for the single acetyl group fed into the cycle. We must keep firmly in mind that 11 of these 12 high-energy phosphate bonds are generated by oxidative phosphorylation in the electron transport chain (Fig. 5-8) and not in the Krebs cycle itself. The Krebs cycle simply provides a means for the release of energized electrons during the oxidation of the acetyl group. All of these reactions occur in mitochondria. But electron release through the Krebs cycle is believed to occur in the matrix, the gel-like inner compartment of the mitochondrion, whereas the electron carriers and the coupling to oxidative phosphorylation occurs in the inner mitochondrial membrane. Thus there is a spatial as well as functional separation of these processes.

Glucose: major source of acetyl CoA

All the major fuels (glucose, fats, amino acids) serve as sources of acetyl CoA. Glucose, however, is a particularly important fuel for most tissues, especially the brain. Glucose is first converted to a 3-carbon compound, **pyruvate** (pyruvic acid), through a series of reactions that are called the Embden-Meyerhof pathway. Pyruvic acid is then enzymatically stripped of a carbon atom to form acetyl CoA. The general outline for this sequence is shown in Fig. 5-11. Again, we shall simplify the biochemical story by condensing this glucose metabolism pathway, which actually consists of ten consecutive enzymic reactions, into four major steps.

The metabolism of glucose begins with its phosphorylation by ATP to form **glucose-6-phosphate** (as shown at the bottom of the page).

Glucose-6-phosphate is an important intermediate because it is a "stem" compound that can lead into any of several metabolic pathways. However, the predominant metabolic fate for glucose-6-phosphate is entry into the Embden-Meyerhof sequence. Following a conversion to fructose and another phosphorylation, fructose-1,6-diphosphate is split into two 3-carbon sugars, called **triose phosphates.** Each triose phosphate is oxidized and rearranged to form **pyruvate,** resulting in a yield of high-energy phosphate as ATP. A pair of hydrogen atoms and a molecule of carbon dioxide are then removed from pyruvate, forming acetyl CoA.

Let us now summarize the entire oxidation of glucose (Fig. 5-11). Glucose first enters the cells of tissues, passing through the plasma membrane by a transport process that requires the presence of the pancreatic hormone **insulin.** Glucose is then phosphorylated and enters the Embden-Meyerhof pathway in the cytoplasm. Through this sequence it is split in the middle to form two 3-carbon sugars (triose phosphates) that are converted to pyruvate. Pyruvate is decarboxylated to form acetyl CoA. This sets the stage for entry into the Krebs cycle located in the mitochondrial matrix. After condensing with oxaloacetic acid to form citric acid, the 2-carbon acetyl fragment is oxidized—yielding four pairs of electrons and 4 protons that are passed along the electron transport chain located on the inner

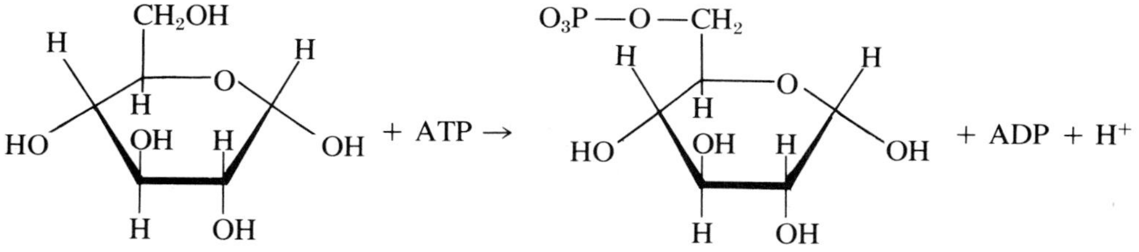

CH_2OH $O_3P—O—CH_2$

H, H, O, H, HO, OH, H, OH, H, OH + ATP → H, H, O, H, HO, OH, H, OH, H, OH + ADP + H^+

Glucose   **Glucose-6-phosphate**

mitochondrial membrane. The electrons finally arrive at oxygen, the ultimate electron acceptor.

What has been accomplished? A molecule of glucose has been completely oxidized to carbon dioxide and water. ATP has been generated at several points along the way. Let us now balance up the yield. First of all, 2 molecules of ATP were consumed in the initial phosphorylation of glucose. Then 4 molecules of ATP were produced by direct transfer of high-energy phosphates to ADP from the triose phosphates. Next we add 10 molecules of ATP generated in the electron transport chain for the 8 hydrogens released in the formation of acetyl CoA from the triose phosphates. Then we add 12 molecules of ATP generated in the complete oxidation of *each* of the acetyl CoA molecules in the Krebs cycle. Our total is $-2 + 4 + 10 + 12 + 12 = 36$ moles of ATP. The whole sequence can be summarized as follows:

$$\text{Glucose } (C_6H_{12}O_6) + 6O_2 + 36ADP + 36\text{—P} \rightarrow 6\ CO_2 + 6\ H_2O + 36ATP$$

Efficiency of oxidative phosphorylation

It is probably obvious that no energy-transforming process can be 100% efficient, not even the remarkable cellular oxidative machinery produced by organic evolution. If we burn a mole of glucose (180 g) in a bomb calorimeter, it releases about 686,000 calories. This is the *potential* energy for forming ATP. It has been determined that 8,000 to 12,000 calories are required to synthesize 1 mole of ATP. Consequently glucose theoretically *could* provide enough energy to generate 50 to 85 moles of ATP from ADP. It actually turns out 38 moles of ATP. If we assume that each ATP mole represents an average energy equivalent of 10,000 calories, then 38 moles of ATP represents 380,000 calories. Thus the efficiency of glucose oxidation is 380:686, or about 55%. Engineers would be delighted if they could build machines that could do as well.

Glycolysis: generating ATP without oxygen

Up to this point we have been describing aerobic metabolism, or respiration. The complete oxidation of glucose involves the formation of pyruvate by cytoplasmic enzymes followed by the oxidation of pyruvate in the mitochondria. It is in the latter mitochondrial stage of respiration that the energies of electrons are tapped and used to generate ATP by oxidative phosphorylation. The important feature of aerobic metabolism is that oxygen is the final electron acceptor, and the complete oxidation of fuel substrates is possible only in the presence of oxygen.

We will now consider how animals generate ATP without oxygen, that is, anaerobically. Anaerobic organisms break down carbon compounds by **fermentation,** an ancient metabolic device for obtaining energy in an atmosphere devoid of oxygen. The term "fermentation," meaning "cause to rise," was originally used to describe the action of yeasts that break down

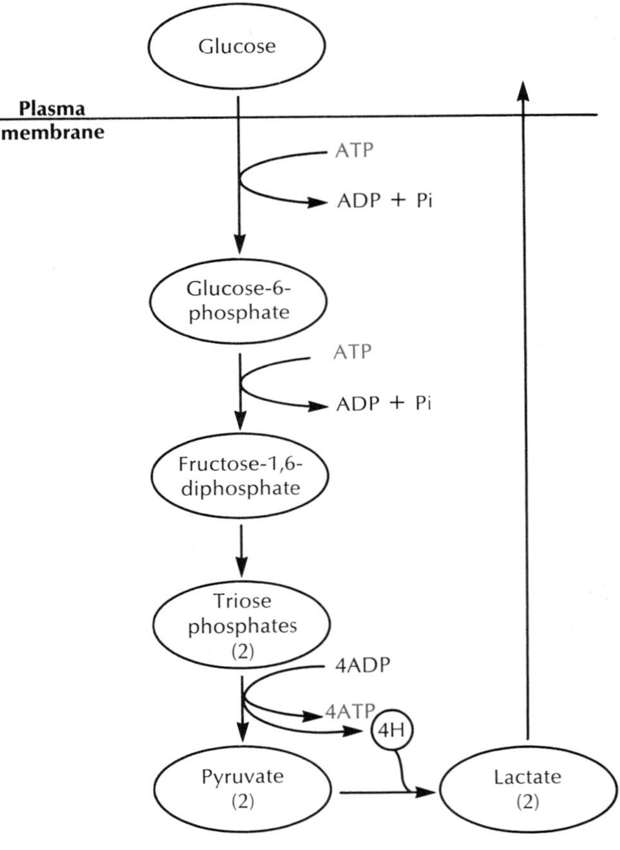

FIG. 5-12 Glycolysis, an anaerobic process that proceeds in the absence of oxygen. Glucose is broken down to 2 molecules of pyruvate, generating 4 molecules of ATP and yielding 2, since 2 molecules of ATP are used to phosphorylate fructose-1,6-diphosphate. Pyruvate is converted to lactate, the final electron acceptor for the hydrogen atoms and electrons released during pyruvate formation.

glucose into alcohol (ethanol) and carbon dioxide, end products that long ago created occupations for both brewers and bakers. **Alcoholic fermentation** is one of two major fermentative sequences that exist in organisms today. The other is **glycolytic fermentation,** or simply **glycolysis,** defined as the anaerobic degradation of glucose to yield lactic acid.

In glycolysis, glucose and other 6-carbon sugars are first broken down stepwise to a pair of 3-carbon pyruvate molecules, yielding 2 molecules of ATP and 2 atoms of hydrogen. This pathway, shown in Fig. 5-12, is precisely the same Embden-Meyerhof pathway that in aerobic metabolism directs glucose into the Krebs cycle (compare Fig. 5-11 with Fig. 5-12). But in the absence of molecular oxygen, further oxidation of pyruvate cannot occur. Both pyruvate and hydrogen accumulate in the cytoplasm because neither can proceed in oxidative channels without oxygen. The problem is neatly solved by forming lactate from pyruvate. Lactate becomes the final electron acceptor and the end product of glycolysis.

Glycolysis is a primitive and inefficient metabolic pathway that yields only 2 moles of ATP per mole of glucose; by comparison oxidative phosphorylation yields 36 moles. Despite its inefficiency, it has not been discarded by evolution—its key virtue being that it provides *some* high-energy phosphate in situations where oxygen is absent or in short supply. Many microorganisms have the opportunistic capacity to metabolize anaerobically when the oxygen supply is limited, then to switch to aerobic metabolism when oxygen is plentiful. Most animals have retained the glycolytic pathway that serves as an important backup system capable of providing short-term generation of ATP during brief periods of heavy energy expenditure, when the slow rate of oxygen delivery would be a limiting factor. Vertebrate skeletal muscle may rely heavily on glycolysis during short bursts of activity when contraction is too rapid and too powerful to be sustained by oxidative phosphorylation. This may mean the difference between survival and death in emergency situations. The lactate that accumulates in muscle during glycolysis diffuses out into the blood where it is later disposed of in the liver.

Some animals rely heavily on glycolysis during normal activities. For example, diving birds and mammals fall back on glycolysis almost entirely to give them the needed energy to sustain a long dive. And salmon would never reach their spawning grounds were it not for glycolysis that provides almost all the ATP used in the powerful muscular bursts needed to carry them up rapids and falls.

Metabolism of lipids

Animal fats are **triglycerides** (neutral fats), molecules composed of glycerol and 3 molecules of fatty acids. These fuels are important sources of energy for many metabolic processes in all animals.

The first step in the breakdown of a triglyceride is the splitting of glycerol from the 3 fatty acid molecules. Glycerol and the fatty acids then proceed through separate pathways. Glycerol, a 3-carbon carbohydrate, is phosphorylated and enters the Embden-Meyerhof pathway as a triose phosphate. From this point it is oxidized in the aerobic pathway like any other carbohydrate.

The remainder of the triglyceride molecule is fatty acids, carboxylic acids with long hydrocarbon chains. One of the abundant naturally occurring fatty acids is **stearic acid** (see below).

We know that fats enter the mitochondrial metabolic processes through acetyl CoA (Fig. 5-9). What happens in brief is that the long hydrocarbon chain of a fatty acid is sliced up by oxidation, 2 carbons at a time; these are released from the end of the molecule as acetyl CoA. The process is repeated until the entire chain has been reduced to several 2-carbon acetyl units. The oxidation of a fatty acid is diagrammed in Fig. 5-13, using a shorthand representation in which each jog in the chain symbolizes a saturated carbon ($—CH_2—$) of the fatty acid stearic acid.

The oxidation of a fatty acid occurs in a repeated three-step sequence. In the first step, coenzyme A is combined with the fatty acid. In step two, the third car-

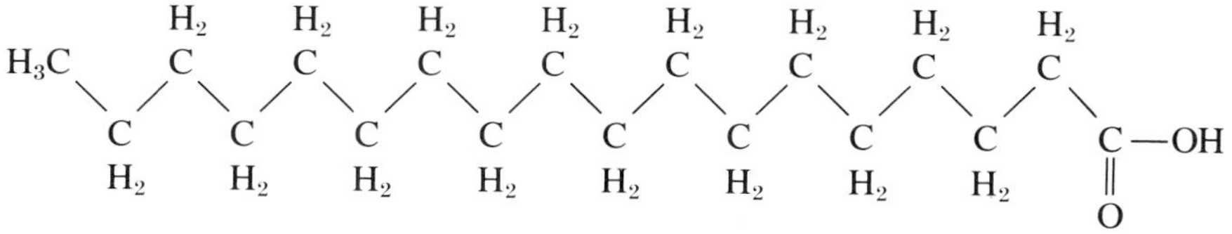

$$H_3C-\underset{H_2}{C}-\overset{H_2}{C}-\underset{H_2}{C}-\overset{H_2}{C}-\underset{H_2}{C}-\overset{H_2}{C}-\underset{H_2}{C}-\overset{H_2}{C}-\underset{H_2}{C}-\overset{H_2}{C}-\underset{H_2}{C}-\overset{H_2}{C}-\underset{H_2}{C}-\overset{H_2}{C}-\underset{H_2}{C}-\overset{H_2}{C}-\underset{\underset{O}{\|}}{C}-OH$$

Stearic acid

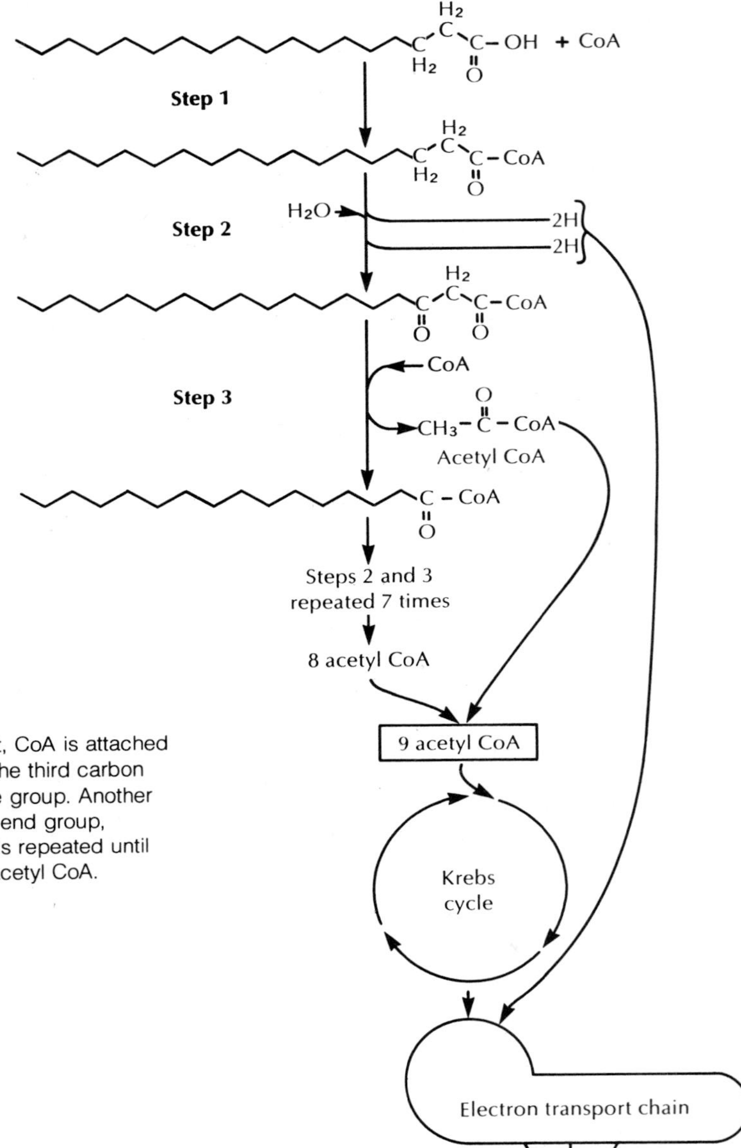

FIG. 5-13 Oxidation of a fatty acid. First, CoA is attached to the carboxylate end of the acid. Then, the third carbon from the end is oxidized, yielding a ketone group. Another molecule of CoA cleaves off the 2-carbon end group, liberating acetyl CoA. The whole process is repeated until the chain has been entirely converted to acetyl CoA.

bon from the end is stripped of its hydrogens; these are accepted by carriers (NAD and flavoprotein) and transferred to the electron transport chain. In step three a molecule of acetyl CoA is sliced off the end by another molecule of coenzyme A, which then adds itself to the chain. The hydrocarbon chain, now 2 carbons shorter, is left with coenzyme A on its end. Steps two and three are repeated until the whole chain has been converted into acetyl CoA. All 9 molecules of acetyl CoA are fed into the Krebs cycle as they are formed, to be broken down into carbon dioxide and water, and their electron energy is used to generate ATP.

How much ATP is gained by fatty acid oxidation? It will be noted that ATP is generated in the electron transport chain by the hydrogens stripped from the carbon chain during fatty acid breakdown, as well as by metabolism of the 9 acetyl CoA units. With allowance for the ATP expended to attach the first coenzyme A molecule, it has been calculated that the complete oxidation of 18-carbon stearic acid will yield 147 ATP molecules. By comparison, 3 molecules of glucose (also totaling 18 carbons) yield 108 ATP molecules. Since there are three fatty acids in each triglyceride molecule, a total of 441 ATP molecules are formed. An additional 22 molecules of ATP are generated in the breakdown of glycerol, giving a grand total of 463 molecules of ATP. Little wonder that fat is considered the king of animal fuels! Fats are more concentrated fuels than carbohydrates because fats are almost pure hydrocarbons; they contain more hydrogen per carbon atom than sugars do, and it is the energized electrons of hydrogen that generate high-energy bonds, when they are carried through the mitochondrial electron transport system.

Stored fats are the greatest reserve fuel in the body. Most of the utilizable fat resides in **adipose tissue** composed of specialized cells packed with globules of triglycerides. Adipose tissue is widely distributed in the abdominal cavity, in muscles, around deep blood vessels, and especially under the skin. Women possess about 30% more fat than men on the average, and this is responsible in no small measure for the curved contours of the female figure. However, its aesthetic contribution is strictly subsidiary to its principal function as an internal fuel depot. Indeed, human beings can only too easily deposit large quantities of fat, generating personal unhappiness and hazards to their health. Perhaps there are redeeming qualities to obesity. Since fat is a poor conductor of heat fat people may be more comfortable in winter, and in time of famine they are able to metabolize their internal reserves while their formerly trim companions suffer. The physiological and psychological aspects of obesity are now being investigated by many researchers.

Fat stores are derived principally from surplus fats and carbohydrates in the diet. Acetyl CoA is the source of carbon atoms used to build fatty acids (Fig. 5-9). Since all major classes of organic molecules (carbohydrates, fats, and proteins) can be degraded to acetyl CoA, all can be converted into stored fat. The biosynthetic pathway for fatty acids resembles a reversal of the catabolic pathway we have already described, but requires an entirely different set of enzymes. From acetyl CoA, the fatty acid chain is assembled two carbons at a time. Since fatty acids release energy when they are oxidized, they obviously require an input of energy for their synthesis. This is provided principally by electron energy transferred from the Embden-Meyerhof pathway during glucose degradation.

Metabolism of proteins

We end our discussion of cellular metabolism by examining the metabolism of proteins. Since proteins are composed of amino acids, 20 in all (Fig. 2-14, p. 33), the central topic of our consideration is amino acid metabolism. Amino acid metabolism is complex. For one thing each of the 20 amino acids requires a separate pathway of biosynthesis and degradation. For another, amino acids are precursors to tissue proteins, enzymes, nucleic acids and other nitrogenous constituents that form the very fabric of the cell. The central purpose of carbohydrate and fat oxidation is to provide energy needed to construct and maintain these vital macromolecules.

Let us begin with the **amino acid pool** in the blood and extracellular fluid from which the tissues draw their requirements. When animals eat proteins, these are digested in the gut, releasing the constituent amino acids, which are then absorbed. Tissue proteins also are hydrolyzed during normal growth, repair, and tissue restructuring; their amino acids join those derived from protein foodstuffs to enter the amino acid pool. A portion of the amino acid pool is used to rebuild tissue proteins, but most animals ingest a protein surplus. Since amino acids are not excreted as such in any significant amounts, they must be disposed of in some way. In fact, amino acids can be and are metabolized through oxidative pathways to yield high-energy phosphate. In short, proteins serve as fuel as do carbohydrates and fats. Their importance as fuel obviously depends on the nature of the diet. In carnivores that ingest a diet of almost pure protein and fat, nearly half of their high-energy phosphate is derived from amino acid oxidation.

Before entering the fuel depot, nitrogen must be removed from the amino acid molecule. This is done in either of two ways. Some amino acids are **deaminated** (their amino group splits off) to yield ammonia and a keto acid:

$$\underset{\textbf{Amino acid}}{R-\underset{\displaystyle NH_2}{\overset{|}{CH}}-COOH} + H_2O \rightarrow \underset{\textbf{Keto acid}}{R-\overset{\displaystyle O}{\overset{\|}{C}}-COOH} + \underset{\textbf{Ammonia}}{NH_3} + H_2$$

Most amino acids, however, undergo **transamination** in which the amino group is transferred to a keto acid to yield a new amino acid, often glutamic acid:

$$\underset{\textbf{Amino acid}}{R-\underset{\underset{NH_2}{|}}{CH}-COOH} + \underset{\alpha\textbf{-Ketoglutaric acid}}{HOOC-CH_2-CH_2-\overset{\overset{O}{\|}}{C}-COOH} \rightleftharpoons$$

$$\underset{\textbf{Keto acid}}{R-\overset{\overset{O}{\|}}{C}-COOH} + \underset{\textbf{Glutamic acid}}{HOOC-CH_2-CH_2-\underset{\underset{NH_2}{|}}{CH}-COOH}$$

Glutamic acid can then be oxidized to liberate ammonia. Thus amino acid degradation yields two main products, ammonia and carbon skeletons, which are handled in different ways.

Once the nitrogen atoms are removed, the carbon skeletons of amino acids can be completely oxidized, usually to either pyruvate or acetate. These residues then enter regular routes utilized by carbohydrate and fat metabolism. This brings us finally to the disposal of ammonia, a complicated story both biochemically and evolutionarily that we can examine only in a general way in this chapter.

Ammonia is potentially a highly toxic waste product. Its disposal offers little problem to aquatic animals because it is soluble and readily diffuses into the surrounding medium through the respiratory surfaces. Terrestrial forms cannot get rid of ammonia so conveniently and must detoxify it by converting it to a relatively nontoxic compound. The two principal compounds formed are **urea** and **uric acid,** although a variety of other detoxified forms of ammonia are excreted by different invertebrate and vertebrate groups. Among the vertebrates, amphibians and especially mammals produce urea. Reptiles and birds, as well as many terrestrial invertebrates, produce uric acid.

The key feature that seems to determine the choice of nitrogenous waste is the availability of water in the environment. When water is abundant the chief nitrogenous waste is ammonia. When water is restricted, it is urea. And for animals living in truly arid habitats (for example, insects, pulmonate snails, birds, and lizards) it is uric acid. The distinguishing attribute of uric acid is that it is highly insoluble and easily precipitates from solution, allowing its removal in solid form. It is not so much the adults that benefit from uric acid's insolubility (since many forms excreting urea also live successfully in dry habitats) as it is their embryos. Birds and reptiles lay eggs enclosed in water-impermeable shells containing not only the embryo but also all of the supportive supplies that it will require for its development: nutrients, fuel, and a small amount of water that must be used with the greatest economy. Furthermore, the embryo has no way to jettison its wastes. As the embryo develops, amino acids are metabolized, yielding ammonia that must be immediately detoxified. Urea is an unsuitable product because it requires too much water for its storage. The solution is uric acid, which is retained in harmless, solid form in the allantois, one of the embryonic membranes (p. 826). When the hatchling emerges into its new world, the accumulated uric acid, along with the shell and the membranes that supported development, is discarded.

Annotated references

Selected general readings

Barker, G. R. 1968. Understanding the chemistry of the cell. Institute of Biology's Studies in biology, no. 13. New York, St. Martin's Press. *This paperback contains a concise account of cell chemistry, energy and metabolism, and research approaches.*

Giese, A. C. 1973. Cell physiology, ed. 4. Philadelphia, W. B. Saunders Co. *An excellent cell physiology test for students who have had good introductory courses in chemistry and physics.*

Hochachka, P. A., and G. N. Somero. 1973. Strategies of biochemical adaptation. Philadelphia, W. B. Saunders Co. *Important treatment of comparative biochemistry.*

Howland, J. L. 1968. Introduction to cell physiology: information and control. New York, The Macmillan Co. *Brief though moderately advanced treatment of biochemistry, enzyme action, and cell physiology.*

Lehninger, A. L. 1975. Biochemistry: the molecular basis of cell structure and function, ed. 2. New York, Worth Publishers. *A very lucidly written and amply illustrated undergraduate biochemistry text, particularly suitable for the student who leans more toward chemistry than biology.*

McGilvery, R. W. 1970. Biochemistry: a functional approach. Philadelphia, W. B. Saunders Co. *Well-written mammalian biochemistry.*

Stryer, L. 1975. Biochemistry. San Francisco, W. H. Freeman Co. *Clear explanations and good diagrams.*

White, A., P. Handler, and E. L. Smith. 1973. Principles of biochemistry, ed. 5. New York, McGraw-Hill Book Co. *Advanced mammalian biochemistry.*

Yost, H. T. 1972. Cellular physiology. Englewood Cliffs, N.J., Prentice-Hall, Inc. *Advanced treatment of molecular physiology and biochemistry.*

Selected *Scientific American* articles

Capaldi, R. A. 1974. A dynamic model of cell membranes. **230:**26-33 (Mar.)

Changeux, J. R. 1965. The control of biochemical reactions. **212:**36-45 (Apr.)

Frieden, E. 1959. The enzyme-substrate complex. **201:**119-125 (Aug.)

Green, D. C. 1960. The synthesis of fat. **202:**46-51 (Feb.).

Hinckle, P. C., and R. E. McCarty. 1978. How cells make ATP. **238:**104-123 (Mar.).

Koshland, D. E., Jr. 1973. Protein shape and biological control. **229:**52-64 (Oct.).

Lehninger, A. L. 1960. Energy transformation. **202:**102-114 (May).

Loewenstein, W. R. 1970. Intercellular communication. **222:**79-86 (May).

Margaria, R. 1972. The sources of muscular energy. **226:**84-91 (Mar).

Neurath, H. 1964. Protein-digesting enzymes. **211:**68-79 (Dec.).

Phillips, D. C. 1966. The three-dimensional structure of an enzyme molecule. **215:**78-90 (Nov.).

Rustad, R. C. 1961. Pinocytosis. **204:**121-130 (Apr.).

Solomon, A. K. 1960. Pores in the cell membrane. **203:**146-156 (Dec.).

Solomon, A. K. 1962. Pumps in the living cell. **207:**100-108 (Aug.).

Stroud, R. M. 1974. A family of protein-cutting proteins. **231:**74-88 (July).

CHAPTER 6
ARCHITECTURAL PATTERN OF AN ANIMAL

Cerianthus membranaceus, an organism exhibiting radial symmetry.

Photograph by C. P. Hickman, Jr.

The architecture of most animals conforms to a well-defined plan. The basic uniformity of biologic organization derives from the supposed common ancestry of animals and from their basic cellular construction. Despite the vast differences of structural complexity of animals ranging from the simplest protozoa to humans, all share an intrinsic material design and fundamental functional plan. In the last analysis, whatever unity we see among animal organization is explained by just one fact: all animals live on Earth, a planet bearing a unique set of physical properties that has molded the nature of life on it. Yet even as the biologist takes some comfort from the belief in the basic unity of life—and many areas of biologic endeavor, such as general physiology and molecular biology, are grounded on this faith—we must admit that animals exhibit an incredible diversity of specific structural and functional adaptations.

Animals are shaped by their particular habitat: its physical nature and its biotic community. Such adaptations are not *caused* by the environment; rather they arose when, by natural selection, beneficial variations were preserved. This concept will be clearer when evolution is discussed in Chapter 38. In this chapter we deal with the fundamental uniformity of animal structure that is recognizable despite the numerous specializations, both subtle and prominent, that have modified the basic plan.

EVOLUTION OF ANIMAL COMPLEXITY

How has increased animal complexity, so evident in animal phylogeny, arisen? The **unicellular** forms are complete organisms and carry on all the functions of higher forms. Within the confines of their cell, they often show remarkable organization and division of labor, such as skeletal elements, locomotor devices, fibrils, and beginnings of sense organs. The **metazoan** or multicellular animal, on the other hand, has cells differentiated into tissues and organs that are specialized for specific functions. The metazoan cell is not the equivalent of a protozoan cell; it is only a specialized part of the whole organism and is incapable of independent existence.

Size and complexity

It is a demonstrable fact that animals (and plants for that matter) have increased in size in the course of evolution. There are exceptions of course; many advanced groups contain small species as well as large. But generally, groups that have appeared more recently in geologic history have larger members than the ancestral groups from which they sprang. With increased size has come increased complexity. Complexity has increased as the result of specialization and division of labor within body tissues. An ameba can move without muscles, digest food without a gut, excrete its wastes without a kidney, coordinate its activities without a brain, and breathe without gills. Larger and more advanced animals have specialized organs for these functions. An alimentary canal is not merely an epithelial tube. It contains muscles for movement, glands for secretion, nerves and endocrine glands for control, and blood and lymph vessels to move materials about. Large size offers many advantages (a point we will return to at the end of this chapter), but it requires elaborate machinery and more energy for construction and maintenance.

Does this mean that life is progressing toward ever more advanced and more complex types? And is complexity the same thing as success? It would be ridiculous to argue that there has been no progress in evolution. Progress can be measured in several ways, and certainly increased complexity of organization is one of them. From the minute and relatively simple organisms that comprised life's beginnings, geologic and fossil records lead us through a history of ever more intricately organized forms. And while it satisfies our egos to consider ourselves superior to all other forms of life, we cannot dismiss our position as simply ego gratification. We are the dominant species today, and we alone can and probably will determine the survival of all species, including ourselves.

But we must not confuse progress with success. Bacteria and blue-green algae still survive and flourish after 3 billion years of competition with "higher" forms of life. Every species is superior to all others in its own ecologic niche; if it were not, it would be driven out by another. Survival, not complexity, is the measure of success. Extinction is the price of failure.

Concept of individuality

The question, "What is an individual animal?" may seem trifling and obvious. We have little difficulty recognizing a worm, a bird, or a dog as an individual and certainly all human beings are individuals. But in unicellular animals that reproduce by asexual fission or budding or fragmentation, the distinction of individuality begins to blur. Where is the "individual" in an organism that merely divides generation after generation? Is a flatworm an individual when it can be cut into numerous pieces, each of which will grow into a new organism? Are colonial forms individuals? We can appreciate that these questions are difficult to answer. Individualization also includes death, since the individual must have a limited existence. Clearly, the concept of individuality does not apply readily to many primitive forms that divide repeatedly and are, in a sense, imperishable.

An individual is traditionally identified as an adult. When we refer to a worm, a bird, or a dog we instantly and automatically picture an adult worm, bird, or dog. But a dog is a dog from fertilized egg to death; as zygote, embryo, fetus, puppy, adolescent, and adult. Zoologists feel more comfortable considering an individual to be a brief time-slice out of a **life cycle** of an organism (Principle 22, p. 14). The idea that the whole life cycle is the central unit of biology on which evolution acts is finding increasing favor among zoologists (Bonner, 1965).

Grades of organization

An animal is an organization of subordinate units that are united at successive levels, and the whole is integrated for carrying on the life processes. Organelles are integrated into cells, cells into tissues, tissues into organs, and organs into systems. Each level is more complex than the one before, and, as a general rule, it is more advanced and a more recent evolutionary product.

Protoplasmic grade of organization. Protoplasmic organization is found in protozoans and other unicellular organisms. All life functions are confined within the boundaries of a single cell, the fundamental unit of life. Within the cell, protoplasm is differentiated into organelles capable of carrying on specialized functions.

Cellular grade of organization. Cellular organization is an aggregation of cells that are functionally differentiated. A division of labor is evident, so that some cells are concerned with, for example, reproduction, others with nutrition. Such cells have little tendency to become organized into tissues. Some protozoan colonial forms having somatic and reproductive cells might be placed in this category. Many authorities also place the sponges at this level.

Cell-tissue grade of organization. A step beyond the preceding is the aggregation of similar cells into definite patterns or layers, thus becoming a tissue. Sponges are considered by some authorities to belong to this grade, although the jellyfish and their relatives (Cnidaria) are usually referred to as the beginning of the tissue plan. Both groups are still largely of the cellular grade of organization because most of the cells are scattered and not organized into tissues. An excellent example of a tissue in cnidarians is the **nerve net,** in which the nerve cells and their processes form a definite tissue structure, with the function of coordination.

Tissue-organ grade of organization. The aggregation of tissues into organs is a further step in advancement. Organs are usually made up of more than one kind of tissue and have a more specialized function than tissues. The first appearance of this level is in the flatworms (Platyhelminthes), in which there are a number of well-defined organs such as eyespots, proboscis, and reproductive organs. In fact, the reproductive organs are well organized into a reproductive system.

Organ-system grade of organization. When organs work together to perform some function we have the highest level of organization—the organ system. The systems are associated with the basic bodily functions—circulation, respiration, digestion, and the others. Typical of all the higher forms, this type of organization is first seen in the nemertean worms in which a complete digestive system, distinct from the circulatory system, is present.

PRELIMINARY SURVEY OF ANIMAL EMBRYOLOGY

Embryology is the study of the progressive growth and differentiation that occurs during the early development of an organism. A brief summary of embryology is necessary for understanding the pattern of an animal and also for understanding some of the basic concepts used in describing animal groups and their classification. The developmental process and its control are discussed in greater detail in Chapter 36.

General pattern of development

Let us begin by restating Principle 22 (p. 14) introduced in the preceding section. All animals have a characteristic life cycle. Admittedly the life cycle is somewhat indistinct in many protozoans (such as the ameba) that are potentially immortal because they are part of ancestral lines that have never experienced natural senescence and death. The parental organism reproduces by simply dividing into two daughter cells, each essentially a continuation of the parent. In these forms there is no true individual, only progressive individualization that ''ends'' with division.

Early in the life history of all metazoans, however, there occurs a differentiation of the germ cell from the body or soma cells. It is the uniting of the germ cells (male sperm and female ova) that gives rise to a new generation (sexual reproduction), while the body (soma) cells die. The real life cycle of a metazoan starts with the union of an ovum (egg) with a spermatozoan, a process called **fertilization.**

The fertilized egg, called a **zygote,** is really a one-celled organism, and from it develops a complete animal by the process of **differentiation.** How this occurs is only partly known. All the information necessary for development is contained within the fertilized egg, principally in the genes of the egg's nucleus. The actual blueprint for differentiation is coded within the DNA molecules of the genes. The heredity of the organism stabilizes the pattern of development; the variation that makes evolutionary changes possible is contributed by the gene segregation that occurs during the formation of gametes and recombination of maternal and paternal genes at fertilization.

Every species develops a characteristic body plan (Principle 23, p. 14). As the body plan takes shape during development, certain basic characteristics of the phylum appear before the specific qualities of the species appear. Such basic qualities may be symmetry, a longitudinal axis, and (if a vertebrate) a notochord, dorsal tubular nerve cord, three major pairs of sensory organs, paired pharyngeal pouches, a chambered heart, a liver, paired kidneys, paired pectoral and pelvic appendages, and others. There are overlappings of some of these characteristics in both vertebrates and invertebrates. As development continues, the individual acquires the morphologic characters of its lower taxa (class, order, family, and genus) and finally of its own species. This indicates that development proceeds from the general to the specific in gross morphologic characters.

The stages of embryogenesis are as follows:

1. **Fertilization.** The activation of the egg by a sperm in biparental reproduction. The union of these male and female gametes forms a zygote; this is the starting point for development.
2. **Cleavage and blastulation.** The division of the zygote into smaller and smaller cells (cleavage) to form a hollow ball of tiny cells (blastula).

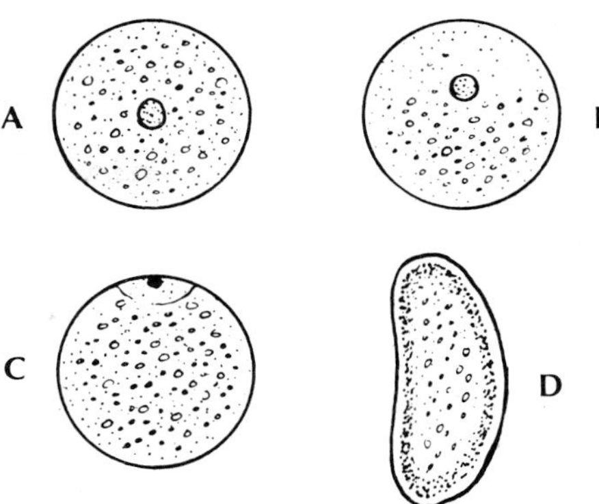

FIG. 6-1 Types of eggs. **A,** Isolecithal (echinoderms and amphioxus); small yolk evenly distributed. **B,** Telolecithal with holoblastic cleavage (amphibians and bony fishes); yolk concentrated at vegetal pole. **C,** Telolecithal with meroblastic cleavage (birds and reptiles); protoplasm concentrated in germinal disc at animal pole. **D,** Centrolecithal (insects and other arthropods); protoplasm centered but migrates out at cleavage, leaving yolk centered.

3. **Gastrulation.** The sorting out of cells of the blastula into layers (ectoderm, mesoderm, endoderm) that become committed to the formation of future body organs.
4. **Differentiation.** The formation of body organs and tissues, which take on their specialized functions. The basic body plan of the animal becomes established.
5. **Growth.** Increase in size of the animal by cell division or cell enlargement. Growth depends on the intake of food to supply material for the synthesis of protoplasm.

Types of eggs

Eggs may be classified as follows with respect to yolk distribution (Fig. 6-1):

The **isolecithal** egg (also called homolecithal) is small, and the small amount of yolk (deutoplasm) and cytoplasm is uniformly distributed through the egg, with the nucleus near the center. Cleavage is typically **holoblastic** (that is, the cleavage planes pass completely through the embryo), and the cells are completely separated by cell membranes at each division. The cells (blastomeres) are nearly equal in size. Such eggs are found in the protochordates (such as sea squirts and *Amphioxus*) and the echinoderms (sea stars and sea urchins).

In the **telolecithal egg** the large amount of yolk (50%

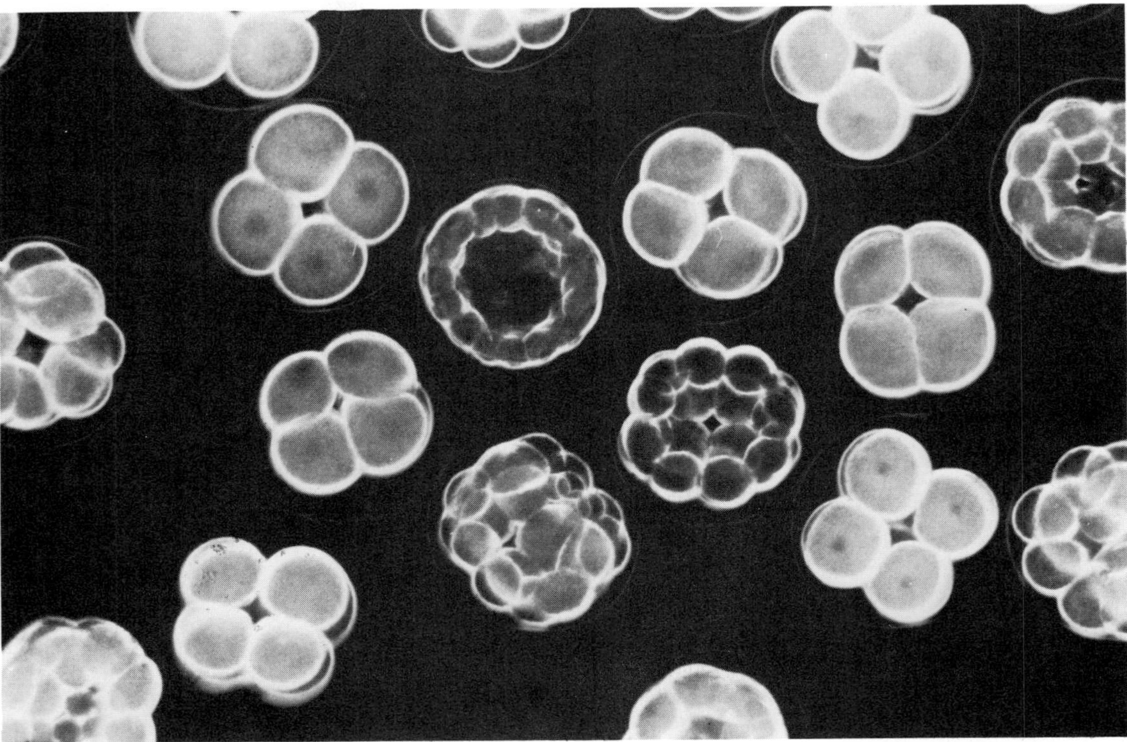

FIG. 6-2 Cell division in developing embryos of *Echinus esculentus,* the common sea urchin. Note how cells become progressively smaller with no growth between divisions. Present are four- and eight-cell stages, many-celled stage, and early blastula. (× 165.) (Photograph by D. P. Wilson.)

to 90%) tends to be concentrated toward one pole (the vegetal pole) where metabolism is lower. The protoplasm and nucleus are found mainly at the opposite (animal) pole, where metabolic activity is greater. There are two classes of telolecithal eggs. The first is yolked eggs with holoblastic cleavage, in which the later cleavages produce unequal cells—small cells, or micromeres, at the animal pole and large cells, or macromeres, at the vegetal pole. Amphibians and bony fish have this kind of egg. The second is yolked eggs (megalecithal) with **meroblastic** (partial or discoidal) cleavage, in which the small amount of protoplasm is concentrated at the animal pole in the germinal disk, or blastoderm, where cleavage occurs. These eggs are usually large, contain a great deal of albumin (egg white) derived from the oviducts, and have a hard or soft shell. Bird and reptile eggs are good examples (Fig. 6-4).

In the **centrolecithal egg** the nucleus and the surrounding layer of protoplasm are at first in the center of the egg; but as cleavage occurs, most of the nucleated masses of cytoplasm migrate to the periphery and form a cellular layer (blastoderm), leaving the yolk in the center of the egg. From the blastoderm the embryo develops. These eggs are characteristic of advanced arthropods such as insects.

In addition to the plasma membrane universally present in eggs, there are in the various animal groups many kinds of protective membranes or envelopes around the eggs. These include the **vitelline membrane,** or fertilization membrane, secreted by the egg, the **zona pellucida** formed by ovarian follicle cells of mammals, the **egg jelly** from the oviducts of bony fish and amphibians, and the **chitinous shell** (chorion) from the ovarian tubules of insects.

Fertilization and formation of the zygote

The fusion of the pronuclei of sperm and egg to form a zygote restores the diploid number of chromosomes, combines the maternal and paternal genetic traits, and

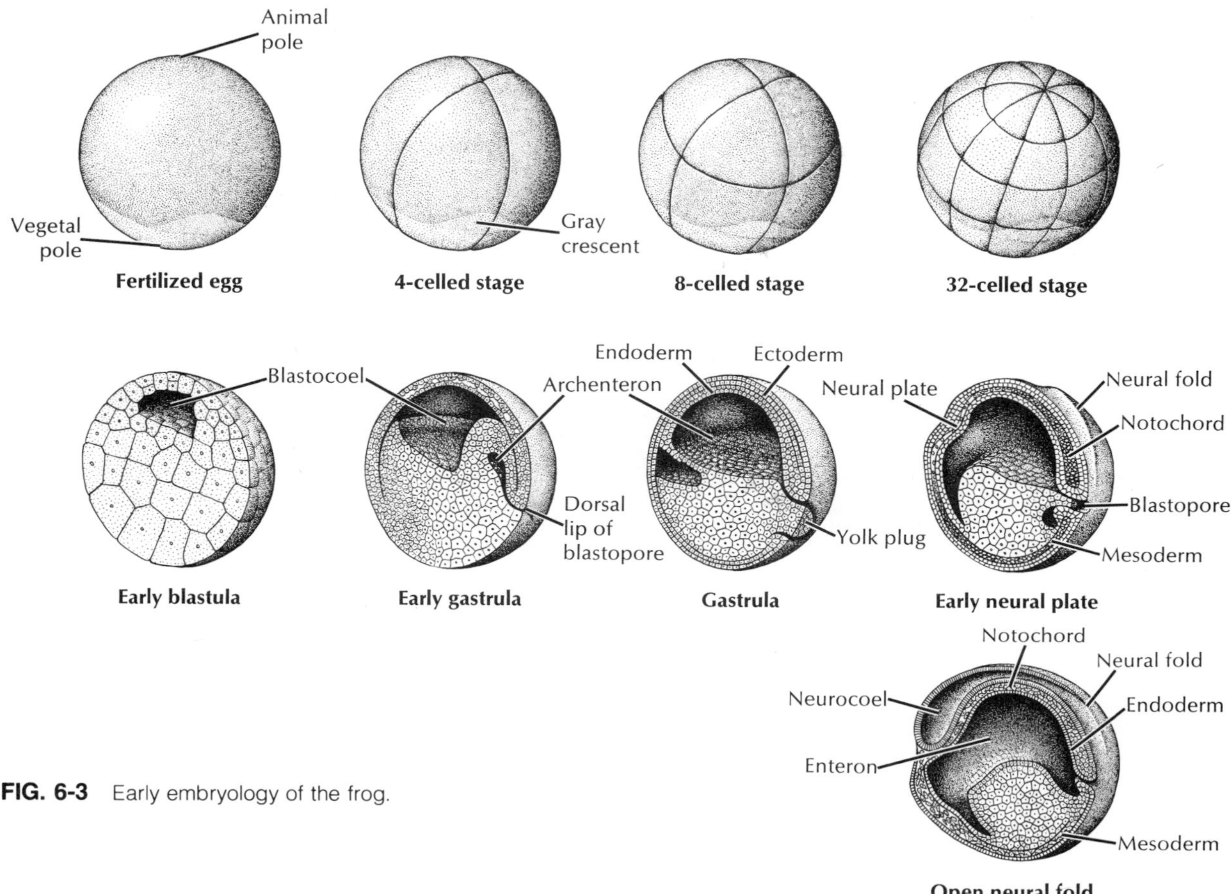

FIG. 6-3 Early embryology of the frog.

activates the egg to develop. This process is treated in more detail in Chapter 36.

Cleavage and blastulation

The unicellular zygote now begins to divide, first into two cells, those two into four cells, those four into eight. Repeated again and again, these cell divisions soon convert the zygote into a ball of cells. This process, called **cleavage,** occurs by mitosis. But unlike ordinary body-cell mitosis, there is no true growth (that is, increase in protoplasmic mass) (Fig. 6-2). With each subsequent division, the cells are reduced in size by one half. The cleavage process converts a single, very large, unwieldy egg into many small, more maneuverable, ordinary-sized cells.

Cleavage patterns are much affected by the amount of yolk in the egg. In eggs with very little yolk, such as those of mammals, the cytoplasm is uniformly distributed through the egg, and the nucleus is in, or near, the egg center. In such eggs, cleavage is complete **(holoblastic),** and the daughter cells formed at each division are of approximately equal size. The eggs of frogs and other amphibians are richly supplied with yolk that tends to be massed in the so-called vegetal pole of the egg. The opposite, or animal, pole contains the egg cytoplasm and the nucleus. The early cleavage divisions tend to be displaced toward the animal pole because the mass of relatively inactive yolk in the vegetal pole retards the rate of cleavage in that region (Fig. 6-3). Birds and reptiles produce the largest eggs of all animals. Nearly all of this comparatively enormous size is storage food—the part of the egg we commonly call yolk and the investment of albumin, or egg "white." The active cytoplasm, containing the nucleus, is but a tiny disc (the blastodisc) resting on top of the ball of yolk. Cleavage (called **meroblastic)**

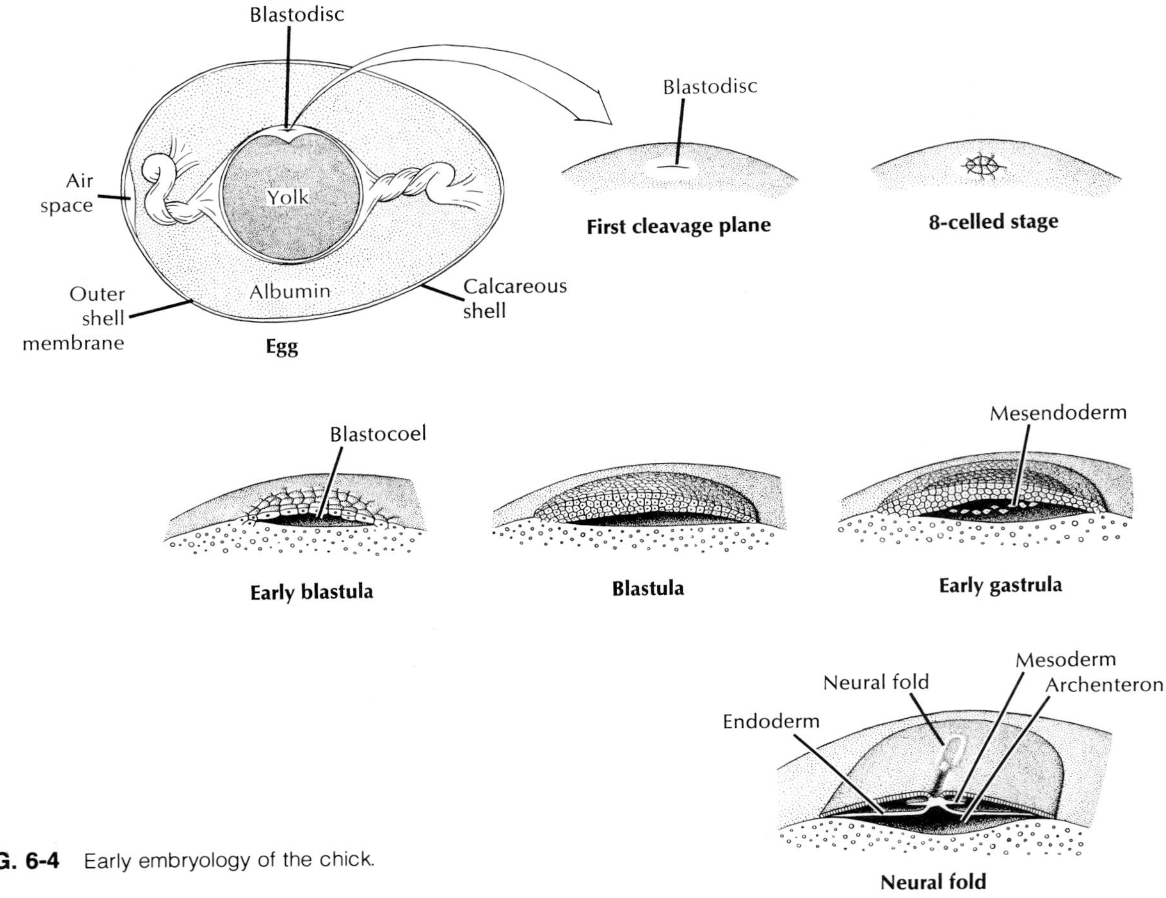

FIG. 6-4 Early embryology of the chick.

is confined to this area of cytoplasm, since the cell divisions cannot possibly cut through the vast bulk of inert yolk (Fig. 6-4).

Radial and spiral cleavage. Holoblastic cleavage is also classified on the basis of radial and spiral types. In **radial cleavage** the cleavage planes are symmetrical to the polar axis and produce tiers, or layers, of cells on top of each other. Radial cleavage is **indeterminate,** that is, there is no definite relation between the position of any blastomere and the specific tissue it will form in the embryo. Very early blastomeres, if separated, may each be capable of giving rise to a complete embryo. In **spiral cleavage** the cleavage planes are diagonal to the polar axis and produce alternate clockwise and counterclockwise quartets of unequal cells around the axis of polarity. Spiral cleavage is **determinate,** and the fate of each blastomere is fixed very early in embryonic development. All blastomeres must be present to form a whole embryo. With few exceptions, radial cleavage is found in the deuterostomes and spiral cleavage in the protostomes. We pursue this distinction in more detail in Chapter 36.

Cleavage, however modified by the presence of varying amounts of yolk, results in a cluster of cells, called a **blastula** (Figs. 6-3 to 6-5). In many animals, such as the amphibian and mammalian embryos, the cells rearrange themselves around a central fluid-filled cavity called the **blastocoel.** We have seen that cleavage has resulted in the proliferation of several thousand maneuverable cells poised for further development. There has been a great increase in total DNA content, since each of the many daughter cell nuclei, by chro-

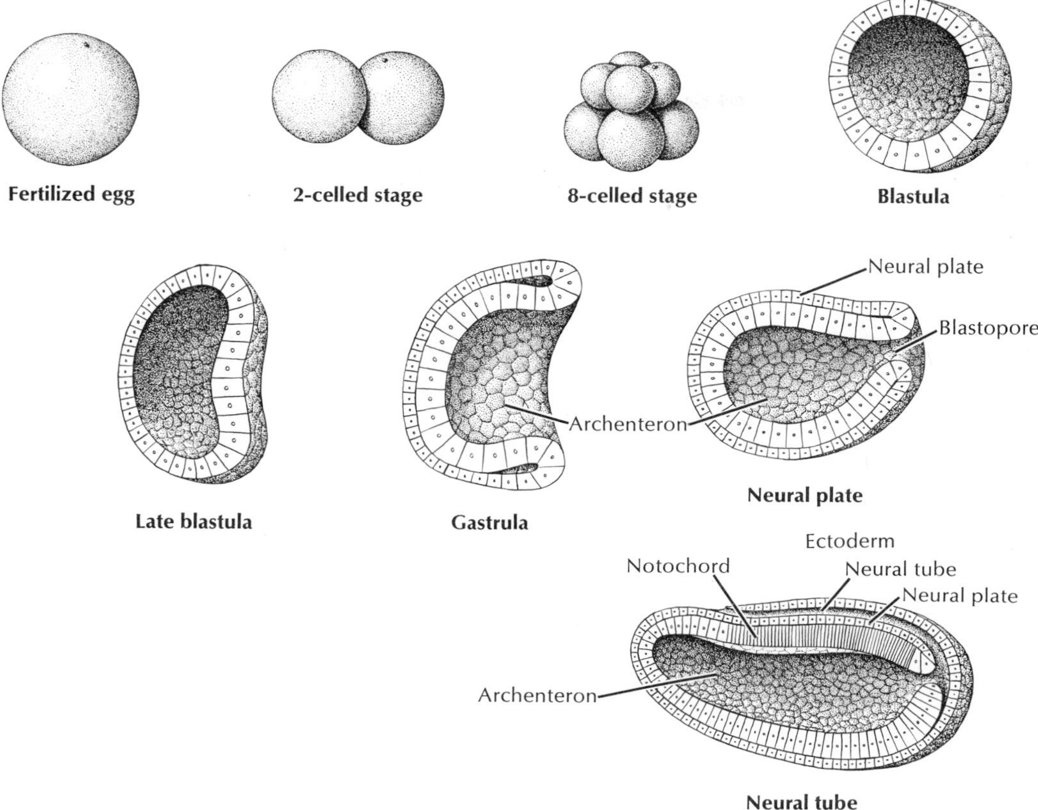

FIG. 6-5 Early embryology of amphioxus.

mosomal replication at mitosis, contains as much DNA as the original zygote nucleus.

Gastrulation

Gastrulation is a regrouping process in which new and important cell associations are formed. Up to this point the embryo has divided itself up into a multicellular complex; the cytoplasm of these numerous cells is nearly in the same position as in the original undivided egg. In other words, there has been no significant movement or displacement of the cells from their place of origin. As gastrulation begins, the cells become rearranged in an orderly way by morphogenetic movements.

In **amphioxus** (Fig. 6-5) the blastoderm of the vegetal pole bends inward so that the whole embryo becomes converted into a double-walled, cup-shaped structure. The structure lining the new cavity thus formed is the **archenteron** (primitive gut), and its opening to the outside is the **blastopore.** In amphibians, the type of gastrulation that occurs in amphioxus is impossible. The cleavage divisions at the lower or vegetal pole are slowed down by the inert yolk so that the resulting blastula consists of many small cells at the animal pole and a few large cells at the vegetal pole. Cells on the surface begin to sink inward (invaginate) at one point, the blastopore. Through the curved groove of the blastopore, surface cells move as a sheet to the interior to form a two-layered embryo (Fig. 6-3). A rodlike **notochord** forms at this time, growing forward to run lengthwise along the dorsal side of the embryo. Continued rearrangements of cells

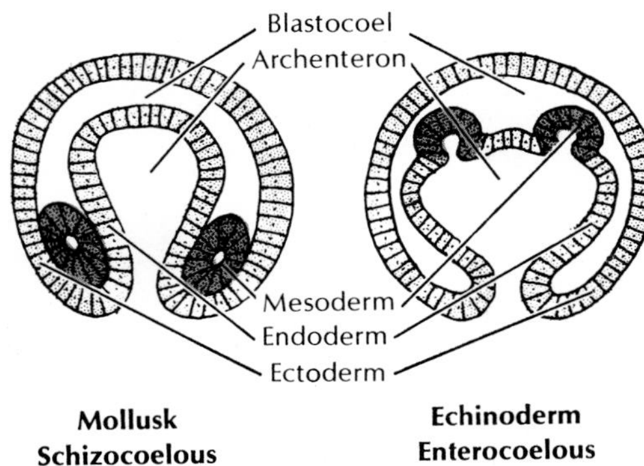

FIG. 6-6 Two types of mesoderm and coelom formation. Schizocoelous, in which mesoderm originates from wall of archenteron near lips of blastopore, and enterocoelous, in which mesoderm and coelom develop from endodermal pouches.

form a third layer; these three layers, called **germ layers,** are the primary structural layers that play crucial roles in the further differentiation of the embryo. The outer layer, or **ectoderm,** will give rise to the nervous system and outer epithelium of the body. The middle layer, or **mesoderm,** will give rise to the circulatory, skeletal, and muscular structures. The inner layer, or **endoderm** (also called entoderm), will develop into the digestive tube and its associated structures.

In certain simple metazoans, only two germ layers are formed, the endoderm and ectoderm. These animals are called **diploblastic.** In all higher forms, the mesoderm also appears, either from pouches of the archenteron or from other cells. This three-layered condition is called **triploblastic.**

Formation of the coelom

The coelom, or true body cavity that contains the viscera, may be formed by one of two methods—**schizocoelous** or **enterocoelous** (Fig. 6-6), or by modification of these methods. (The two terms are descriptive, for *schizo* comes from the Greek *schizein,* to split; *entero* is a Greek form from *enteron,* gut; coelous comes from the Greek *koilos,* hollow or cavity.) In schizocoelous formation the coelom arises, as the word implies, from the splitting of mesodermal bands that originate from the blastopore region and grow between the ectoderm and endoderm; in enterocoelous formation the coelom comes from pouches of the archenteron, or primitive gut.

Since coelom formation occurs very early in embryonic development, the appearance of two quite different methods of formation among animals is believed to signal a fundamental division in metazoan evolution. The **deuterostome** division of the Metazoa (echinoderms, chaetognaths, protochordates, and chordates) mostly follows the enterocoelous method of coelom formation. The **protostome** division (almost all other **Metazoa**) follow the schizocoelous method. The chordates are exceptions to this distinction because their coelom is formed by mesodermal splitting (schizocoelous), although in other respects they develop as deuterostomes, the division to which they are assigned. Other characteristics that distinguish these two phylogenetic divisions of bilateral animals are radial, indeterminate cleavage and spiral, determinate cleavage, mentioned before.

Differentiation

With formation of the three primary germ layers, cells continue to regroup and rearrange themselves into primordial cell masses. As masses develop, they become increasingly committed to specific directions of differentiation. Cells that previously had the potential to develop into a variety of structures now lose this diverse potential and assume commitments to become, for example, kidney cells, intestinal cells, or brain cells. Differentiation is discussed in more detail in Chapter 36.

Fate of germ layers

Following are some of the vertebrate structures that normally arise from the three germ layers:

Ectoderm

Epidermis of skin
Lining of mouth, anus, nostrils
Sweat and sebaceous glands
Epidermal coverings such as hair, nails, feathers, horns, epidermal scales, enamel of teeth
Nervous system, including sensory parts of eye, nose, ear

Endoderm

Lining of alimentary canal
Lining of respiratory passages and lungs
Secretory parts of liver and pancreas
Thyroid, parathyroid, thymus
Urinary bladder
Lining of urethra

Mesoderm

Skeleton and muscles
Dermis of skin
Dermal scales and dentin
Excretory and reproductive systems
Connective tissue
Blood and blood vessels
Mesenteries
Lining of coelomic cavity

The assignment of early embryonic layers to specific "germ layers" (a somewhat unfortunate term that is not to be confused with "germ cells," which are the eggs and sperm) is for the convenience of embryologists and of no concern to the embryo. The idea that each germ layer can give rise to certain tissues and organs only, and to no others, is no longer held. It is now known that the interactions of cells play a part in determining their differentiation in vertebrate animals. The precise position of a cell with relation to other cells and tissues during early development often controls the real fate of that cell. Under some conditions a certain germ layer may give rise to structures normally arising from a different germ layer. Experiments have demonstrated that a presumptive ectodermal structure, when grafted into appropriate regions, will form organs that normally come from a different germ layer. Therefore, the topographic position of cells in their development must play a significant role in their final fate and destiny.

Embryologic development must be considered quite flexible. The biologic system cannot be restricted to a definite pattern even though normally it appears to be. Are there any organs or structures that do not come from any germ layer? In a strict sense the germ cells do not originate from any of the three germ layers because they come directly from cells that were segregated in the early cleavage stages of the fertilized egg.

ORGANIZATION OF THE BODY

The body consists of three elements: (1) cells (discussed earlier), (2) body fluids, and (3) extracellular structural elements.

Body fluids. Water permeates all tissues and spaces in the body but it is naturally separated into certain fluid "compartments." In all metazoans, the two major body fluid compartments are the **intracellular space,** within the body's cells, and the **extracellular space,** outside the cells. In animals with closed vascular systems (such as the vertebrates), the extracellular space can be further subdivided into the **blood plasma** (blood corpuscles are really part of the intracellular compartment) and the **interstitial fluid.** The interstitial fluid, or tissue fluid, occupies the spaces surrounding the cells. There is much variation in body fluid compartmentalization among the invertebrates, but all share with the vertebrates the basic subdivision of fluid between the intracellular and extracellular compartments.

Extracellular structural elements. Extracellular, or intercellular,* substance is the supporting material of the organism such as connective tissue, cartilage, and bone. It provides mechanical stability, protection, a depot of materials for exchange, and a mediator for intercellular reactions. It is responsible for tissue firmness and cellular support. Two types of intercellular tissue are recognized—**formed** and **amorphous.** The formed type includes collagen (white fibrous tissue). This is the most abundant protein in the animal kingdom and makes up the major part of the fibrous constituent of the skin, tendon, ligaments, cartilage, and bone. Elastin, which gives elasticity to the tissues, also belongs to the formed type. Amorphous intercellular substance (ground substance) is composed of mucopolysaccharides arranged in long chain polymers.

Strictly speaking, extracellular substances are lifeless secretions of cells, which alone are living, according to classic cell theory. However, they are protoplasmic materials, or transformations of protoplasm, that are part of the wholeness of the living organism. They are subject to aging and disease just as cells are. The extracellular substances are products of cells, and whether they should be considered living or dead can only be resolved by one's definition of a living mass—as a collection of cells or as an organization of tissues and systems. Certainly tissues and systems are alive, and most biologists today would accept the unitary view that the "living mass" includes both cells and extracellular substances.

Tissues

A **tissue** is a group of similar cells (together with associated cell products) specialized for the performance of a common function. The study of tissues is called **histology.** All cells in metazoan animals take part in the formation of tissues. Sometimes the cells of

*The term intercellular, meaning "between cells," should not be confused with the term intracellular, meaning "within cells."

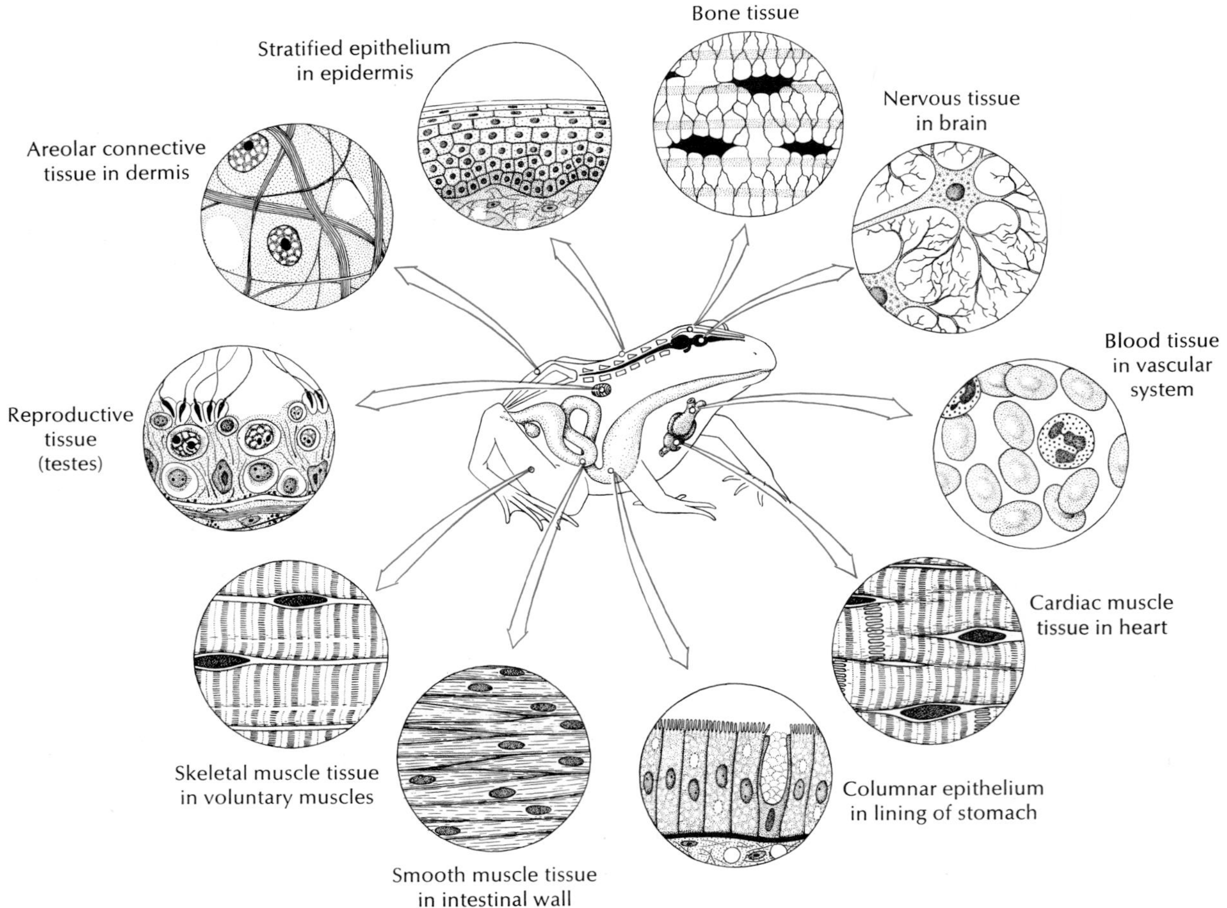

FIG. 6-7 Diagram of a frog showing various types of tissue.

a tissue may be of several kinds, and some tissues have a great many intercellular materials.

The different types of tissues originate from the basic properties of protoplasm. These properties, and the tissues that are the manifestations of these properties, are irritability and conductivity (nervous tissue), contractility (muscle tissue), support and adhesion (connective tissue), absorption and secretion (epithelial tissue), and fluidity and conductivity (vascular tissue) (Fig. 6-7). This is a surprisingly short list of basic tissue types that are able to meet the requirements of the diverse morphologic patterns of all animals. Each of the basic tissues can be subdivided into several types that are specialized for many different functions. Some of these are depicted in Fig. 6-7.

During embryonic development, the germ layers differentiate into the five major tissues by a process called **histogenesis.** Their origins are as follows: (1) epithelial tissue, from all three germ layers; (2) connective or supporting tissue, from mesoderm; (3) muscular or contractile tissue, from mesoderm; (4) nervous tissue, from ectoderm; and (5) vascular tissue, from mesoderm.

Epithelial tissue. An **epithelium** is a tissue that covers an external or internal surface. It also includes hollow or solid derivatives from this tissue. Epithelial tissues (Figs. 6-8 and 6-9) are made up of closely associated cells, with some intercellular material between the cells. Some cells are bound together by **intercellular bridges** of cytoplasm. Most of them have one sur-

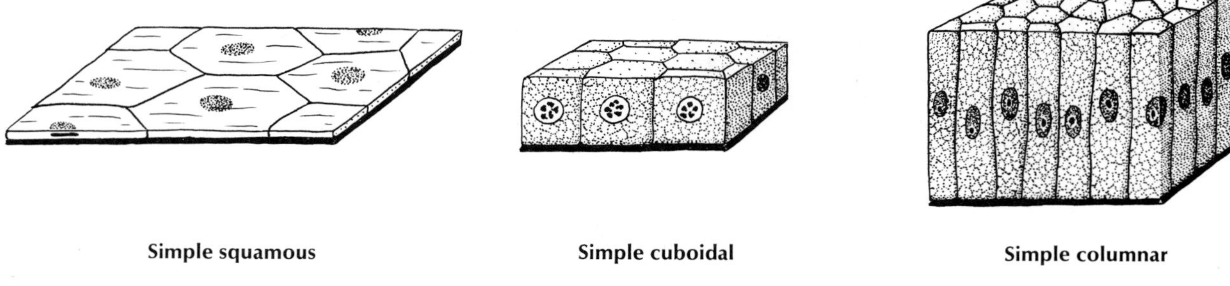

FIG. 6-8 Types of simple epithelium.

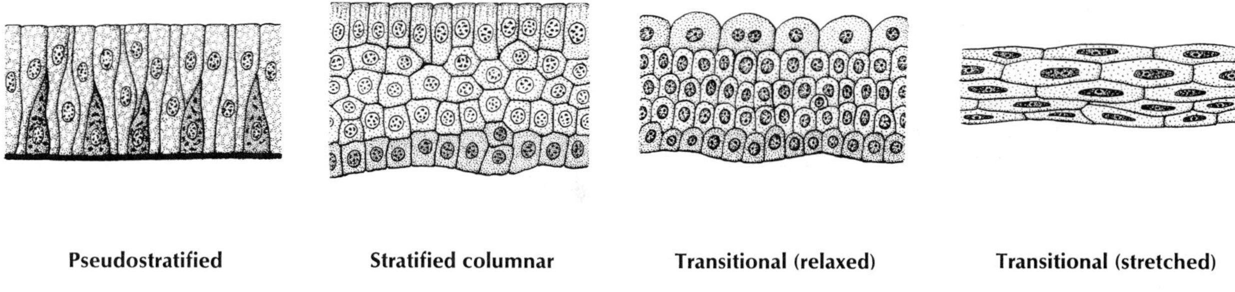

FIG. 6-9 Types of stratified and transitional epithelial tissue.

face free and the other surface lying on vascular connective tissue. A noncellular **basement membrane** is often attached to the basal cells. Epithelial cells are often modified to produce secretory glands that may be unicellular or multicellular. Some free surfaces (joint cavities, bursae, brain cavity) are not lined with typical epithelium.

Epithelia are classified on the basis of cell form and number of cell layers. **Simple epithelium** is one layer thick (Fig. 6-8), and its cells may be **squamous** (flat), as in endothelium of blood vessels, **cuboidal** (short prisms), as in glands and ducts, and **columnar** (tall), as in stomach and intestine. Any of these three forms of cells may occur in several layers as a **stratified epithelium** (as in skin, sweat glands, urethra) (Fig. 6-9). Some stratified epithelia can change the number of their cell layers by movement (**transitional,** such as the bladder). Others have cells of different heights and give the appearance of stratified epithelia (**pseudostratified,** such as the trachea). Many epithelia may be **ciliated** at their free surfaces (such as the oviduct). Epithelia serve to protect, secrete, excrete, lubricate, and perform other functions.

Connective tissue. Connective tissues bind together and support other structures. They are so common that the removal of all other tissues from the body would still leave the complete form of the body clearly apparent. Connective tissue is made up of **scattered cells,** a great deal of formed materials such as **fibers** and ground substance **(matrix)** secreted by the cells. There

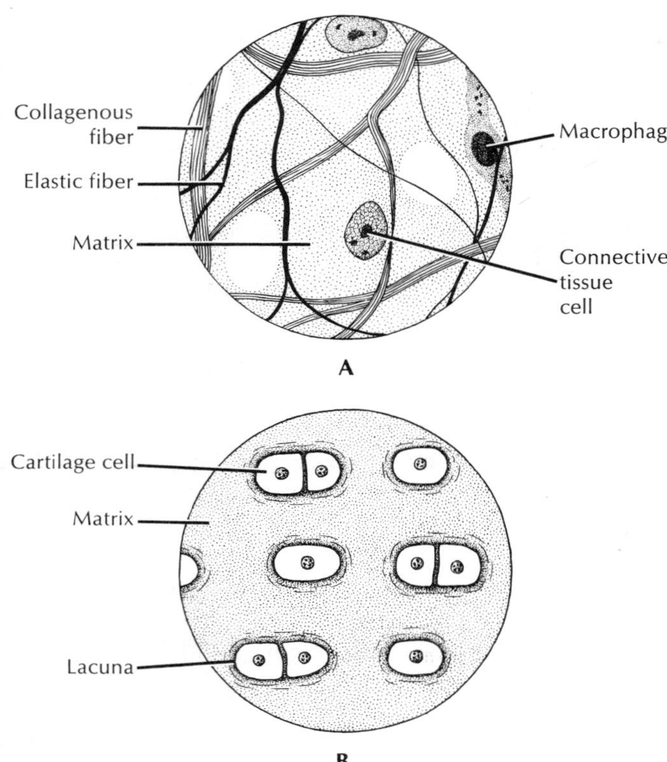

FIG. 6-10 **A,** Areolar, a type of loose connective tissue. **B,** Hyaline cartilage, most common form of cartilage in body and a type of dense connective tissue.

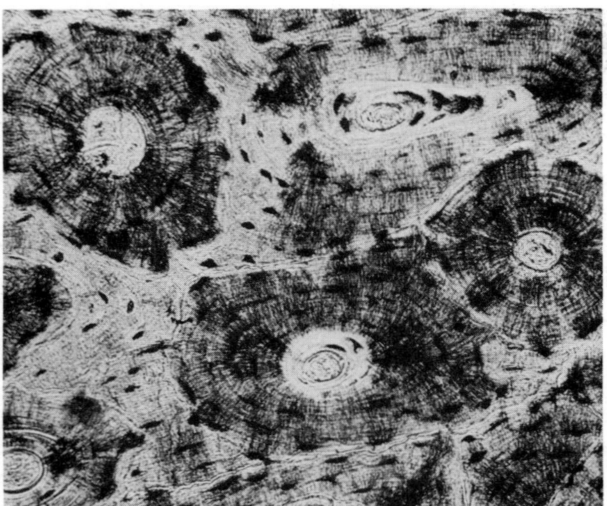

FIG. 6-11 Section of bone, a type of dense connective tissue, showing several cylindric haversian systems typical of bone. (×180.)

are three types of fibers: white or collagenous (collagen is the most common structural protein in the body), yellow or elastic, and branching or reticular. Connective tissue may be classified in various ways, but all the types fall under either **loose connective tissue** (reticular, areolar, adipose) or **dense connective tissue** (sheaths, ligaments, tendons, cartilage, bone) (Figs. 6-10 and 6-11). Adipose stresses cells, ligaments stress fibers, and cartilage stresses ground substance (matrix).

Connective tissues are derived from the **mesenchyme,** a generalized embryonic tissue that can differentiate also into vascular tissue and smooth muscle. Mesenchyme may also be considered the most primitive connective tissue. When its cells are closely packed together, it is called **parenchyma;** when loosely arranged with gelatinous material, it is called **collenchyma.**

Muscular tissue. Muscle is the most common tissue in the body of most animals. It is made up of elongated cells or fibers specialized for contraction. It originates (with few exceptions) from the mesoderm, and its unit is the cell or **muscle fiber.** The unspecialized cytoplasm of muscles is called **sarcoplasm,** and the contractile elements within the fiber (cell) are the **myofibrils.** Functionally, muscles are either **voluntary** (under control of will) or **involuntary.** Structurally, they are either **smooth** (fibers unstriped) or **striated** (fibers cross-striped).

The three kinds of muscular tissue (Figs. 6-12 and 6-13) are **smooth involuntary** (walls of viscera and walls of blood vessels), **striated involuntary** or **cardiac** (heart), and **striated voluntary or skeletal** (limb and trunk). Another type of muscular tissue is made up of **myoepithelial** cells (Fig. 6-14) found in sweat, salivary, and mammary glands between the epithelium and connective tissue. They extend branching processes around the secretory cells of the glands. Their function may be to squeeze secretions from the acini toward the surface openings of the larger ducts.

Nervous tissue. Nervous tissue is specialized for the detection and conduction of stimuli. The structural and functional unit of the nervous system is the **neuron** (Fig. 6-15), a nerve cell made up of a body containing the nucleus and its processes or fibers. It originates from an embryonic ectodermal cell called a **neuroblast.** (Part of the nervous system of echinoderms may be mesodermal in origin.) In most animals the bodies of nerve cells are restricted to the central nervous system and ganglia, but the fibers may be very long

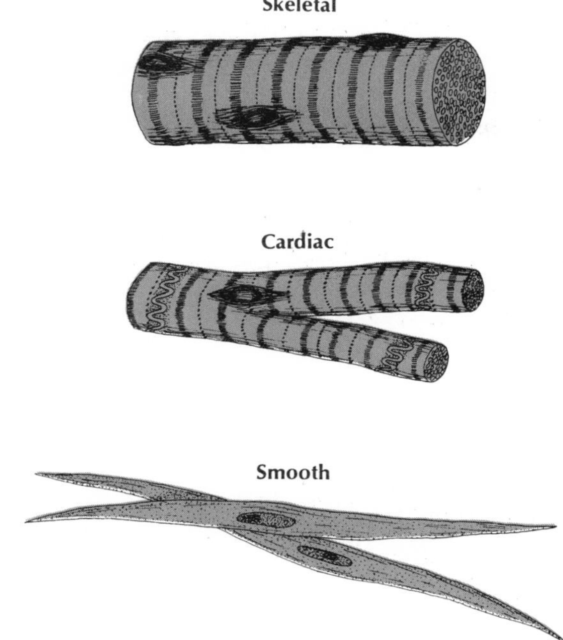

FIG. 6-12 Three kinds of vertebrate muscle fibers, as they appear when viewed with the light microscope.

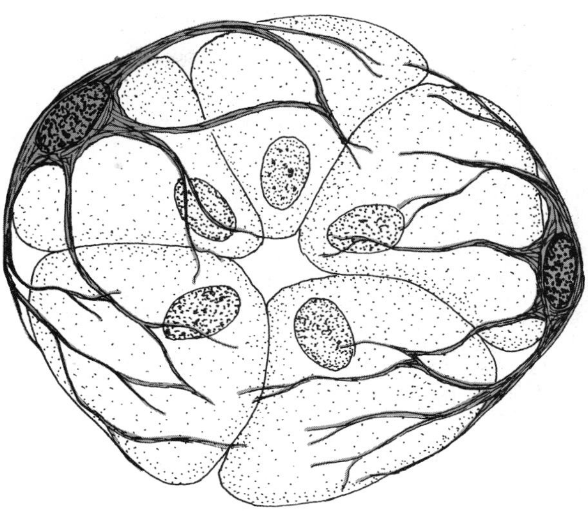

FIG. 6-14 Two myoepithelial basket cells surrounding salivary secretory cells. Each myoepithelial cell is made up of a central body with long cytoplasmic processes and may be considered a fourth type of muscular tissue. Myoepithelial cells resemble smooth muscle cells.

FIG. 6-13 Photomicrograph of skeletal muscle showing several striated fibers lying side by side. (×600.) (Photograph by J. W. Bamberger.)

and ramify through the body. Neurons are arranged in chains, and the point of contact between neurons is a **synapse.** Some of the fibers bear a sheath (medullated, or myelinated); in others the sheath is absent (nonmedullated).

Sensory neurons are concerned with picking up impulses from sensory **receptors** in the skin or sense organs and transmitting them to nerve centers (brain or spinal cord). **Motor neurons** carry impulses from the nerve centers to muscles or glands **(effectors)** that are thus stimulated to act. **Association neurons** may form various connections between other neurons.

Vascular tissue. Blood, lymph and tissue fluids are types of vascular tissues. Blood is a fluid tissue composed of **white blood cells, red blood cells, platelets,** and a liquid—**plasma.** Traveling through blood vessels, the blood carries to the tissue cells the materials necessary for their life processes. **Lymph** and **tissue fluids** arise from blood by filtration and serve in the exchange between cells and blood.

Organs and systems

An organ is a group of tissues organized into a larger functional unit. In higher forms an organ may have most or all of the five basic tissue types. For example, the heart (Fig. 6-16) has epithelial tissue for covering and lining, connective tissue for framework, muscular walls for contraction, nervous elements for coordination, and vascular tissue for transportation.

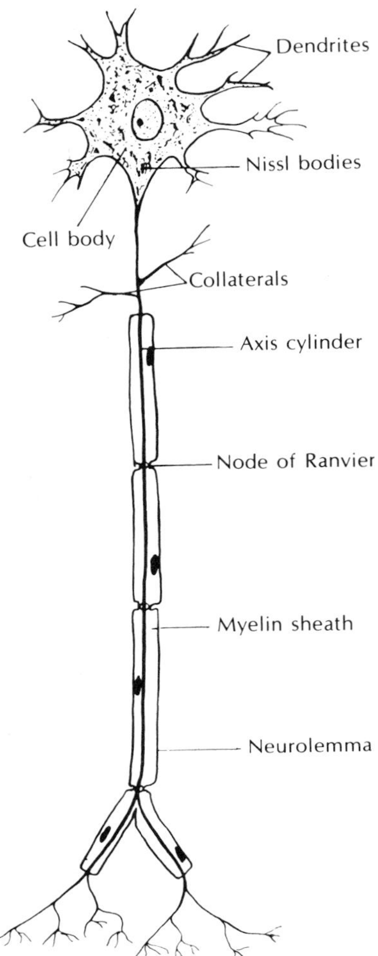

FIG. 6-15 Neuron.

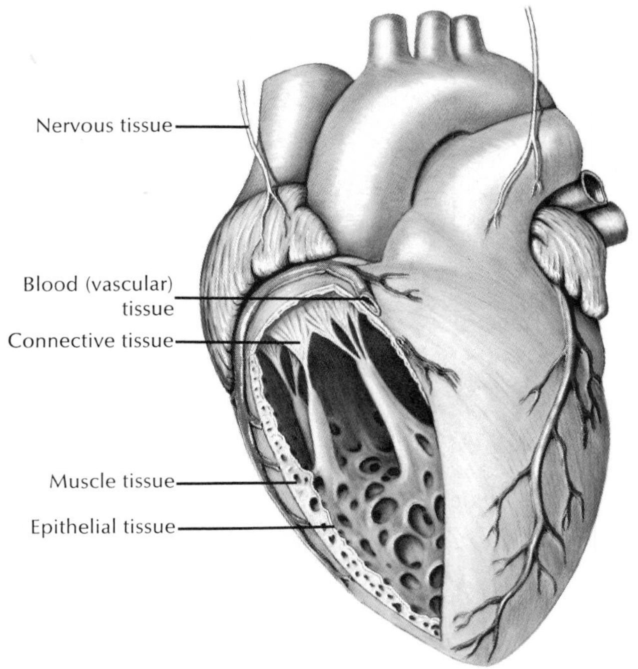

FIG. 6-16 Heart showing various types of tissue in its structure.

All organs have a characteristic structural plan. Usually one tissue carries the burden of the organ's chief function, as muscle does in the heart; the other tissues perform supportive roles. The chief functional cells of an organ are called its **parenchyma;** the supporting tissues are its **stroma.** For instance, in the pancreas the secreting cells are the parenchyma; the capsule and connective tissue framework represent the stroma.

Organs are, in turn, associated in groups to form **systems,** with each system concerned with one of the basic functions. The higher metazoans have 11 organ systems: skeletal, muscular, integumentary, digestive, respiratory, circulatory, excretory, nervous, special sensory, endocrine, and reproductive. However, all living organisms perform the same basic functions. The need for procuring and utilizing food and for movement, protection, perception, and reproduction are equally basic to an ameba, a clam, an insect, or a human being. Obviously, because of differences in size, structure, and environment, each must meet these problems in a different manner.

ANIMAL BODY PLANS

Thus far in this chapter we have considered those characteristics of body design and development that animals share. We have seen that animals show great diversity in grade of organization and in other characteristics, such as the presence or absence of a body cavity and overall size and complexity. We are ready now to consider the various architectural plans that distinguish major groups of animals. These appear enormously diverse as even a cursory examination of the different invertebrate and vertebrate groups will reveal. However, it is possible to resolve them into four ''master plans'' (Gardiner, 1972). These are the unicellular plan, the cell-aggregate plan, the blind-sac

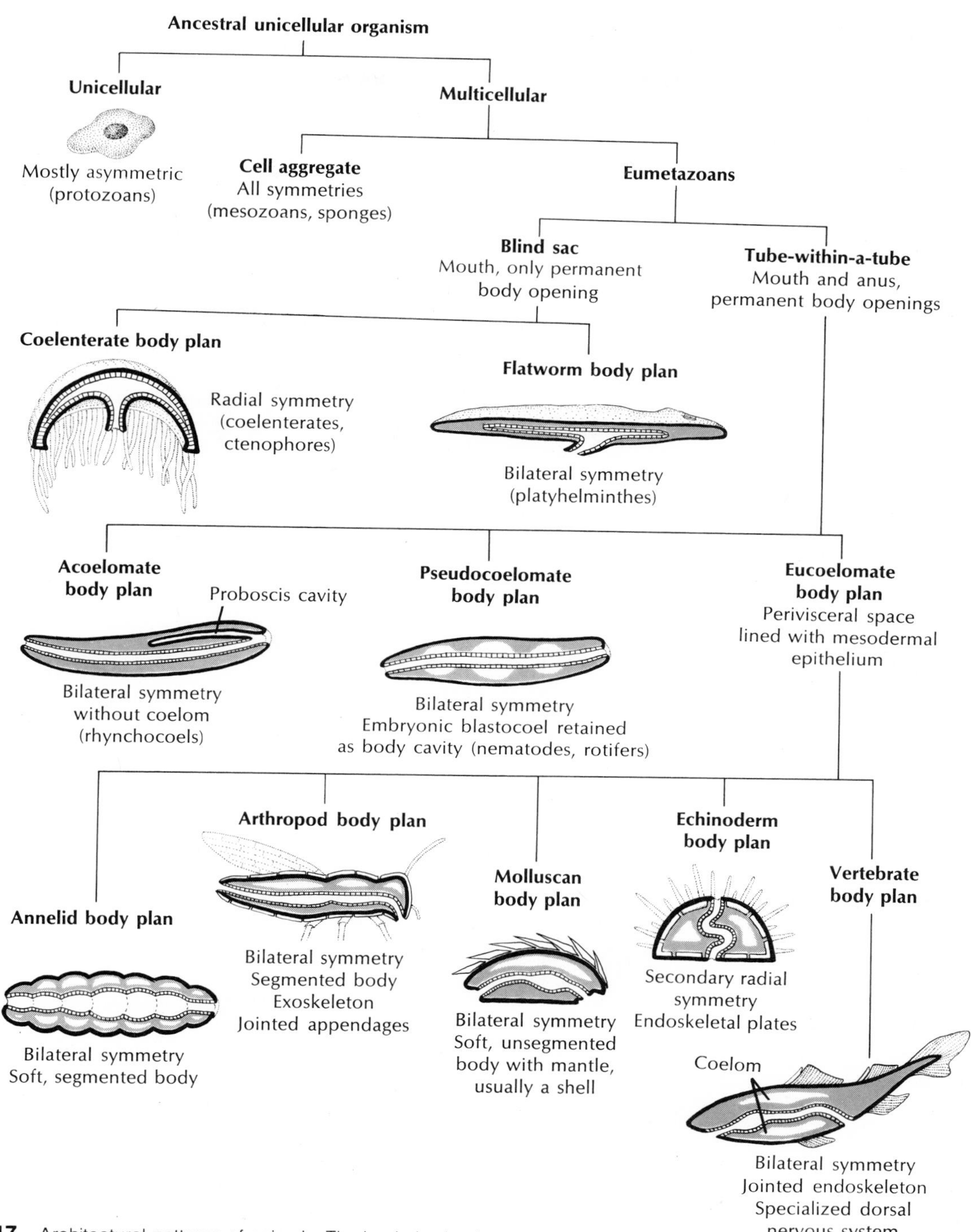

FIG. 6-17 Architectural patterns of animals. The basic body plans are unicellular (or acellular), cell aggregate, blind sac, and tube-within-a-tube. These, especially the last, have been variously modified during evolutionary descent to fit animals to a great variety of life-styles. Ectoderm is shown in solid black, endoderm as open blocks, mesoderm in red.

plan, and the tube-within-a-tube plan. These, together with some of their most important modifications, are shown in Fig. 6-17. All but the first are multicellular plans. Four of the most important determinants of multicellular body plans are symmetry, presence or absence of a body cavity, presence or absence of segmentation, and cephalization. We will discuss these in turn.

Animal symmetry

Symmetry refers to balanced proportions, or the correspondence in size and shape of parts on opposite sides of a median plane. **Spheric symmetry** means that any plane passing through the center divides the body into equivalent, or mirrored, halves. This type of symmetry is found chiefly among some of the protozoans and is rare in other groups of animals. Spheric forms are best suited for floating and rolling.

Radial symmetry applies to forms that can be divided into similar halves by any plane passing through the longitudinal axis. These are the tubular, vase, or bowl shapes found in some sponges and in the hydras, jellyfish, sea urchins, and the like, in which one end of the longitudinal axis is usually the mouth (see photograph, p. 98). A variant form is **biradial symmetry** in which, because of some part that is single or paired rather than radial, only one or two planes passing through the longitudinal axis produce mirrored halves. Sea walnuts, which are more or less globular in form but have a pair of tentacles, are an example. Radial and biradial animals are usually sessile or slow moving. The two phyla that are primarily radial, Cnidaria and Ctenophora, are called the **Radiata.**

In **bilateral symmetry** only a sagittal plane can divide the animal into two mirrored portions—right and left halves (Fig. 6-18). Bilateral animals make up all of the higher phyla and are collectively called the **Bilateria.** They are better fitted for forward movement than radially symmetric animals.

Let us review some of the convenient terms used for locating regions of animal bodies (Fig. 6-18). **Anterior** is used to designate the head end; **posterior,** the opposite or tail end; **dorsal,** the back side; and **ventral,** the front or belly side. **Medial** refers to the midline of the body, **lateral** to the sides. **Distal** parts are farther from a point of reference; **proximal** parts are nearer. **Pectoral** refers to the chest region or the area supporting the forelegs, and **pelvic** refers to the hip region or the area supporting the hind legs. A **frontal plane** divides a bilateral body into dorsal and ventral halves by running through the anteroposterior axis and the right-left axis at right angles to the **sagittal plane,** the plane

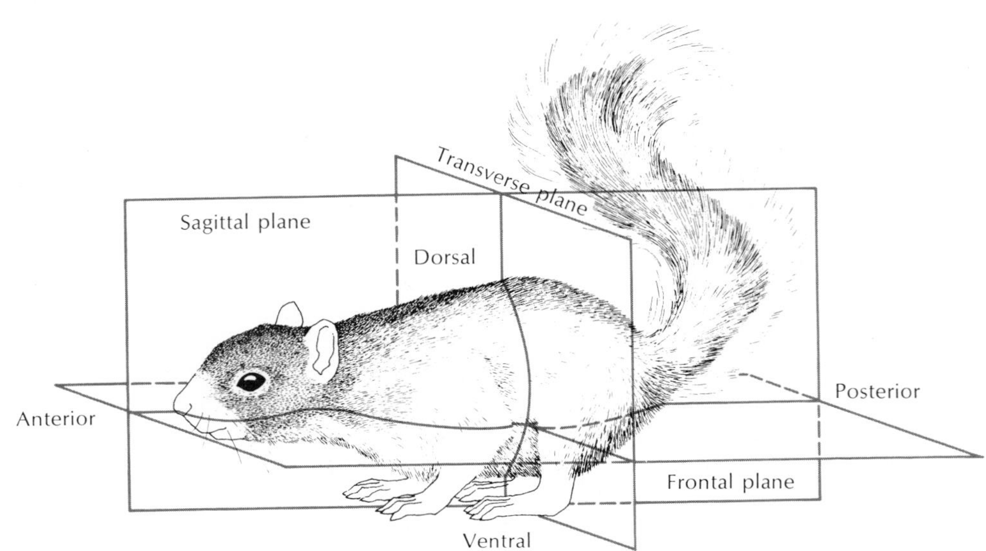

FIG. 6-18 The planes of symmetry as illustrated by a bilateral animal.

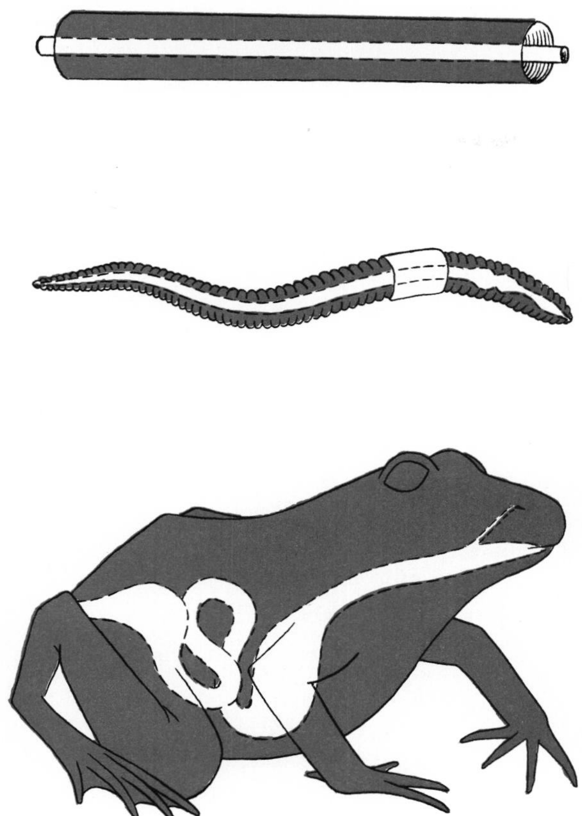

FIG. 6-19 Tube-within-a-tube arrangement.

dividing an animal into right and left halves. A **transverse plane** would cut through a dorsoventral and a right-left axis at right angles to both the sagittal and frontal planes and would result in anterior and posterior portions.

Body cavities

The bilateral animals can be grouped according to their body-cavity type or lack of body cavity (Fig. 6-17). In higher animals the main body cavity is the **coelom,** a fluid-filled space that surrounds the gut. This provides coelomic animals with a "tube-within-a-tube" arrangement (Fig. 6-19). The two methods of coelom formation—schizocoelous and enterocoelous—were described earlier (p. 106).

The coelom is of great significance in animal evolution. It provides increased body flexibility and space for visceral organs and permits greater size and complexity by exposing more cells to surface exchange. The fluid-filled space also serves as a hydrostatic skeleton in some forms, aiding them in such functions as movement and burrowing.

Acoelomate Bilateria. The more primitive bilateral animals do not have a true coelom. In fact, the flatworms and a few others have no body cavity surrounding the gut. The region between the ectodermal epidermis and the endodermal digestive tract is completely filled with mesoderm in the form of parenchyma.

Pseudocoelomate Bilateria. Nematodes and related phyla have a cavity surrounding the gut, but it is not lined with mesodermal peritoneum. It is derived from the blastocoel of the embryo; it represents a persistent blastocoel. This type of body cavity is called a **pseudocoel.**

Eucoelomate Bilateria. Animals possessing a true coelom lined with mesodermal peritoneum include the remainder of the bilateral animals.

Metamerism (segmentation)

Metamerism is the serial repetition of similar body segments along the longitudinal axis of the body. Each segment is called a metamere, or **somite.** In forms such as the earthworm and other annelids, in which metamerism is best represented, the segmental arrangement includes both external and internal structures of several systems. There is repetition of muscles, blood vessels, nerves, and the setae of locomotion. Some other organs, such as those of sex, are repeated in only a few somites. In higher animals much of the segmental arrangement has become obscure.

When the somites are similar, as in the earthworm, the condition is called **homonomous metamerism;** if they are dissimilar, as in the lobster and insect; it is called **heteronomous metamerism.**

Segmentation often appears in embryonic stages but becomes obscure in the adult. For example, muscles of vertebrate animals show a decided metamerism in the embryo but little in the adult. On the other hand, the arrangement of the vertebrae is clearly metameric in adult vertebrates.

True metamerism is found in only three phyla: Annelida, Arthropoda, and Chordata (Fig. 6-20), although superficial segmentation of the ectoderm and the body wall may be found among many diverse groups of animals.

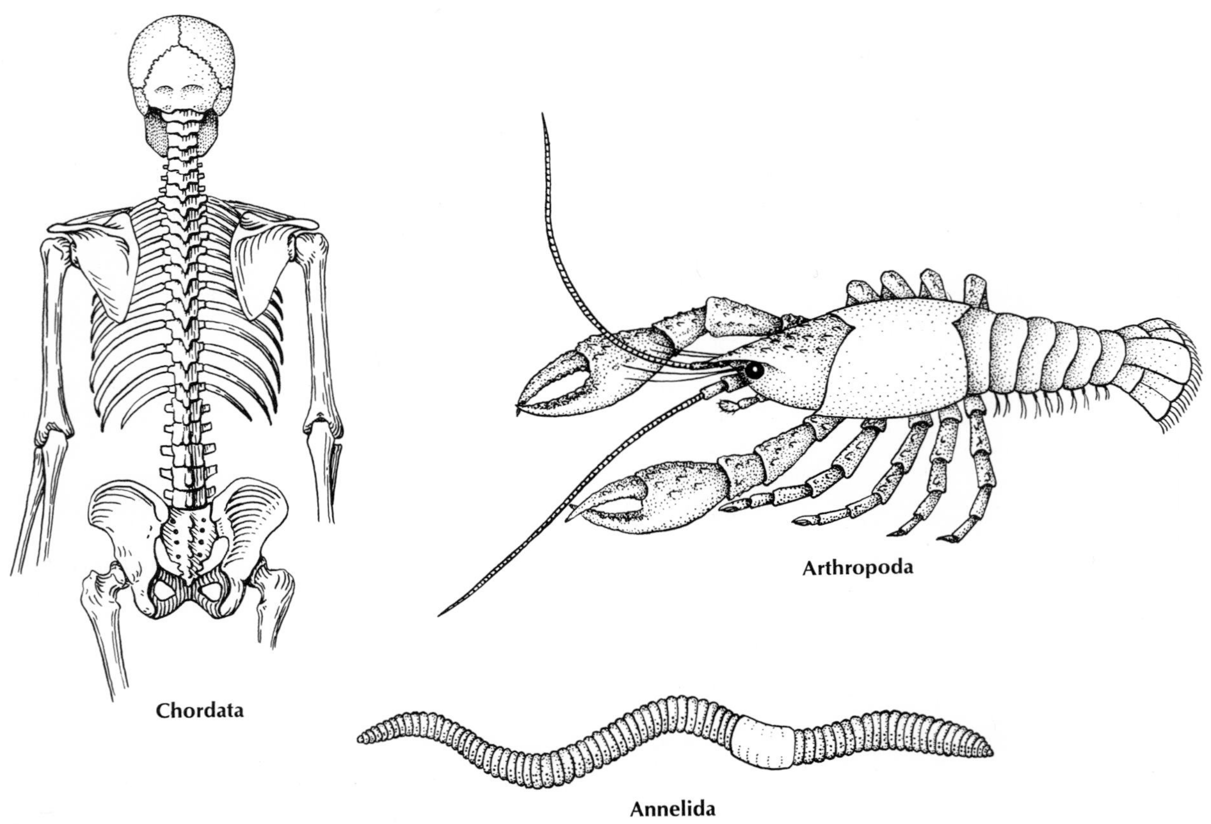

FIG. 6-20 Segmented phyla. These three phyla have all made use of an important principle in nature—metamerism, or repetition of structural units. Annelids and arthropods are definitely related, but chordates have derived their segmentation independently. Segmentation brings more varied specialization because segments, especially in arthropods, have become modified for different functions.

In the more advanced annelids and arthropods, the segments may be united into functional groups of two or more somites, each group being structurally separated from other groups and specialized to perform a certain function for the whole animal. This is called **tagmosis.** The primitive annelids and arthropods, however, adhere to the primitive pattern of having a serial succession of nearly identical somites. As will be seen later, tagmosis is more common among the arthropods than the annelids because the soft-bodied annelid requires the numerous fluid-filled body somites in its locomotion. A good example of tagmosis is the familiar insect that has three tagmata: head, thorax, and abdomen. Each tagma in this animal represents a grouping of a certain number of segments, and each is specialized for a specific function or functions.

Cephalization

The differentiation of a definite head end is called **cephalization** and is found chiefly in bilaterally symmetric animals. The concentration of nervous tissue and sense organs in the head bestows obvious advantages to an animal moving through its environment head first. This is the most efficient positioning of instruments for sensing the environment and responding to it. Usually the mouth of the animal is located on the head as well, since so much of an animal's activity is concerned with procuring food. Cephalization is always accompanied by differentiation along an anteroposterior axis **(polarity).** Polarity usually involves gradients of activities between limits, such as between the anterior and the posterior ends.

HOMOLOGY AND ANALOGY

In comparative studies of animals the concepts of **homology** (similarity in origin) and **analogy** (similarity in function or appearance, but not in origin) are frequently used to express the mechanisms responsible for similar patterns of morphology or of function in different animals. For example, the bones in a whale flipper are homologous to the bones of a human arm and to those of a bat wing (Fig. 38-7, p. 886), although in appearance and function they are quite different. They are homologous because all share a common ancestry and embryonic origin. All develop similarly from a limb bud within which forms a single proximal bone (humerus), two parrallel bones located more distally (radius and ulna), and several small bones more distal to the latter (bones of wrist and phalanges). Later in development, significant modifications appear that shape the limb for the totally different function each will perform in these three different mammals.

Analogy denotes similarity of function and often of appearance as well. Just as different functions make homologous structures dissimilar in form, similar functions can make nonhomologous structures resemble each other. The wings of a bird and a butterfly obviously perform similar flight functions, but they are not homologous because they have totally different origins. Their resemblance is only superficial. The skeletal framework of the bird's wing with its feathered surface corresponds to the basic vertebrate pentadactyl limb plan. The wing of a butterfly is supported by cuticular thickenings (''veins''), and the surface is a cuticular membrane composed of two cell layers. They are analogous structures.

Homology is a relative term. If one always insists on similarity of structure *and* development for strict homology, as is sometimes done, then many structures that might be considered homologous do not meet these requirements. For instance, the wing of a bird and the wing of a bat are homologous only insofar as both are derived from pentadactyl forelimbs. They are analogous, however, for they have the same function. In examining homologous structures it is best to keep in mind the following question: to what extent or degree do they share a common origin? Of course, even this question contains a nebulous element, since all animals are related, having presumably descended from a common ancestor in the long distant past. Homology implies a close evolutionary kinship. Convergent evolution, or the independent origin of two similar structures, is caused by independent mutations that are favored under similar environments. Horns have appeared independently many times in mammals, but should they be considered homologous? Perhaps the best criterion for homology would be homologous genes, but this is impossibly strict at our present state of knowledge. Embryologists deal with phenotypes, or the visible expressions, which is not the same thing as the genotypes, or hereditary constitution.

The principle of homology, although difficult to define strictly, has a wide application in zoology. It is used as an argument for evolution because it is based on the idea of inheritance from common ancestors. Homology is also important in classifying animals. As more is known about animals, chemical identity may become an important aspect of homology among organisms. A plan of archetypes with which the structures of animals are compared may thus be formulated, not only from gross morphologic structures, but also from their biochemical resemblances.

SIZE OF THE ORGANISM

Throughout our survey of animal architectures we have repeatedly encountered a central theme: progression of animal complexity. From the dawn of life to the present, animals have tended toward more intricate organization, greater division of labor, and larger body size. Large animals have not displaced the small; indeed there is reason to suspect that small organisms are as plentiful today as they ever were; the reason is simply that small animals occupy other niches. Small organisms long ago occupied all the available small ecologic niches, and the only way a new species can succeed is to displace an organism from an existing niche or adapt to a new and larger one. This may well be why there has been an evolutionary trend toward increasing body size.

Large size offers certain advantages. Perhaps the most obvious is that in the predator-prey contest, predators are almost always larger than their prey. Exceptions are few and usually are related to an especially aggressive behavior that compensates for small size. Thus food chains are composed of a series of animals of increasing size, each preying on a smaller creature of optimum size for eating, and each being preyed on in turn by a larger predator. At the bottom of the chain are the producers of small body size and large biomass; at the top are the carnivores of large body size and small biomass. (Food chains are described in more detail on p. 949.)

Larger animals prey on smaller ones not only because they have bigger mouths and appetites, but also because they usually are faster. Since the predator must first catch its prey, and the prey must escape its predator if it is to survive, the advantage of larger size is clear. Another advantage that accompanies large size is greater internal stability, that is, the capacity to regulate the internal environment of the body. (This is known in physiologic language as **homeostasis** and is discussed in Chapter 30.) The ability to maintain internal stability despite changes in the external environment frees organisms to invade otherwise hostile habitats. For example, only the homeotherms (warm-blooded animals such as birds and mammals) can remain active through all seasons in regions with severe winter climates.

We must concede in closing that the evident advantages of large size derive at least in part from our perspective as large animals ourselves. Every animal, whether large or small, is a success in its ecologic niche. In fact, the most abundant groups of animals on earth today are those with short generation times and, consequently, small size. To these countless creatures, "small is good."

Annotated references

Selected general readings

Balinsky, B. I. 1975. An introduction to embryology, ed. 4. Philadelphia, W. B. Saunders Co. *Authoritative, stressing mechanisms of development.*

Berrill, N. J., and G. Karp. 1976. Development. New York, McGraw-Hill Book Co. *Balanced developmental biology text: comprehensive yet readable and well ordered.*

von Bertalanffy, L. 1952. Problems of life. London, Watts & Co. *Levels of organization are among several topics discussed in this classic book dealing with biologic concepts.*

Bloom, W., and D. W. Fawcett. 1975. A textbook of histology, ed. 10. Philadelphia, W. B. Saunders Co. *Authoritative and superbly illustrated advanced treatise on human histology.*

Bonner, J. T. 1965. Size and cycle. Princeton, Princeton University Press. *An essay on the significance of size of the organism in the life cycle.*

Gardiner, M. S. 1972. The biology of invertebrates. New York, McGraw-Hill Book Co.

Lorenz, K. Z. 1974. Analogy as a source of knowledge. Science **185**:229-234.

Montagna, W. 1956. The structure and function of skin. New York, Academic Press, Inc. *A good account of the most versatile organ in the body.*

Romer, A. S., and T. S. Parsons. 1977. The vertebrate body. Philadelphia, W. B. Saunders Co. *An exemplary comparative anatomy text with good descriptions of body tissues and organs.*

Webster, D., and M. Webster. 1974. Comparative vertebrate morphology. New York, Academic Press, Inc.

Selected *Scientific American* articles

Fraser, R. D. B. 1969. Keratins. **221**:86-96 (Aug.).

Gross, J. 1961. Collagen. **204**:120-130 (May).

McLean, F. C. 1955. Bone. **192**:84-91 (Feb.).

Ross, R., and P. Bornstein. 1971. Elastic fibers in the body. **224**:44-52 (June).

CHAPTER 7

CLASSIFICATION AND PHYLOGENY OF ANIMALS

Carolus Linnaeus, the great Swedish naturalist who founded our modern system of classification in the mid-eighteenth century. This statue of a youthful Linnaeus stands before his home in the old university town of Uppsala, Sweden.

Photograph by C. P. Hickman, Jr.

ANIMAL CLASSIFICATION

Zoologists have named more than 1.5 million species of animals, and thousands more are added to the list each year. Yet some evolutionists believe that species named so far make up less than 20% of all living animals and less than 1% of all those that have existed in the past.

To communicate with each other about the diversity of life, biologists have found it a practical necessity not only to name living organisms but to classify them. It is not just that the desire to put things into some kind of order is a fundamental activity of the human mind. A system of classification is a storage, retrieval, and communication mechanism for biologic information. The science of **systematics** embraces the studies of speciation, classification, and phylogeny and is concerned with the identification and naming of each kind of organism by a uniformly adopted system that best expresses the degree of similarity between organisms.

Linnaeus and the development of classification

Although the history of human efforts to distinguish and name plants and animals must have been rooted in the beginnings of language, the great Greek philosopher and biologist Aristotle was the first to attempt seriously the classification of organisms on the basis of structural similarities. Following the Dark Ages in Europe, the English naturalist John Ray (1627-1705) introduced a more comprehensive classification system and a modern concept of species. The flowering of systematics in the eighteenth century culminated in the work of Carolus Linnaeus (1707-1778), who gave us our modern scheme of classification.

Linnaeus was a Swedish botanist at the University of Uppsala. He had a great talent for collecting and classifying objects, especially flowers. Linnaeus worked out a fairly extensive system of classification for both plants and animals. His scheme of classification, published in his great work *Systema Naturae,* emphasized morphologic characters as a basis for arranging specimens in collections. Actually his classification was largely arbitrary and artificial, and he believed strongly in the constancy of species. Linnaeus wrote, "The Author of Nature, when He created species, imposed on His Creations an eternal law of reproduction and multiplication within the limits of their proper kinds. He did indeed in many instances allow them the power of sporting in their outward appearances, *but never that of passing from one species to another*" (emphasis ours).

Linnaeus divided the animal kingdom down to species, and according to his scheme each species was given a distinctive name. He recognized four classes of vertebrates and two classes of invertebrates. These classes were divided into orders, the orders into genera, and the genera into species. Since his knowledge of animals was limited, his lower groups, such as the genera, were very broad and included animals that are now placed in several orders or families. As a result, much of his classification has been drastically altered, yet the basic principle of his scheme is followed at the present time.

Linnaeus's scheme of arranging organisms into an ascending series of groups of ever-increasing inclusiveness is the **hierarchic system** of classification. Species were grouped into genera, genera into orders, and orders into classes. This taxonomic hierarchy has been considerably expanded since Linnaeus's time. The major categories, or **taxa** (sing., **taxon**), now used are as follows, in descending series: kingdom, phylum, class, order, family, genus, and species. This hierarchy of seven ranks can be subdivided into finer categories, such as superclass, subclass, infraclass, superorder, suborder, and so on. In all, more than 30 taxa are recognized. For very large and complex groups, such as the fishes and insects, these additional ranks are required to express recognized degrees of evolutionary divergence. Unfortunately they also contribute complexity to the system.

Linnaeus' system for naming species is known as **binomial nomenclature.** Each species has a Latinized name composed of two words (hence binomial). The first word is the genus, written with a capital initial letter; the second word is the specific name that is peculiar to the species and is written with a small initial letter (Table 7-1). The genus name is always a noun, and the specific name is usually an adjective that must agree in gender with the genus. For instance, the scientific name of the common robin is *Turdus migratorius* (L., *turdus,* thrush; *migratorius,* of the migratory habit).

There are times when a species is divided into subspecies, in which case a **trinomial nomenclature** is employed (see katydid example, Table 7-1). Thus to distinguish the southern form of the robin from the eastern robin, the scientific term *Turdus migratorius achrustera* (duller color) is employed for the southern

TABLE 7-1 Examples of classification of animals

	Human	Gorilla	Southern leopard frog	Katydid
Phylum	Chordata	Chordata	Chordata	Arthropoda
Subphylum	Vertebrata	Vertebrata	Vertebrata	
Class	Mammalia	Mammalia	Amphibia	Insecta
Subclass	Eutheria	Eutheria		
Order	Primates	Primates	Salientia	Orthoptera
Suborder	Anthropoidea	Anthropoidea		
Family	Hominidae	Pongidae	Ranidae	Tettigoniidae
Subfamily			Raninae	
Genus	*Homo*	*Gorilla*	*Rana*	*Scudderia*
Species	*Homo sapiens*	*Gorilla gorilla*	*Rana pipiens*	*Scudderia furcata*
Subspecies			*Rana pipiens sphenocephala*	*Scudderia furcata furcata*

type. Taxa lower than subspecies are sometimes employed when four words are used in the scientific name, the last one usually standing for **variety.** In this latter case the nomenclature is **quadrinomial.** The trinomial and quadrinomial nomenclatures are really additions to the Linnaean system, which is basically binomial.

It is important to recognize that *only* the species is binomial. All ranks above the species are uninomial nouns, written with a capital initial letter. We must also note that the second word of a species is an epithet that has no meaning by itself. The scientific name of the white-breasted nuthatch is *Sitta carolinensis*. The "carolinensis" may be and is used in combination with other genera to mean "of Carolina," as for instance *Parus carolinensis* (Carolina chickadee) and *Dumetella carolinensis* (catbird). The genus name, on the other hand, may stand alone to designate a taxon that may include several species.

Species

Despite the central importance of the species concept in biology, biologists do not agree on a single rigid definition that applies to all cases. Before Darwin's time, the species was considered a primeval pattern, or archetype, divinely created. With gradual acceptance of the concept of organic evolution, scientists realized that species were not fixed, immutable units but have evolved one from another. Sometimes the gaps between species are so subtle that they can be distingiushed only by the most careful examination. In other instances, a species is so distinctive in every way that it is clearly unique and only remotely related to other species. Consequently the criteria of taxonomy have undergone gradual changes.

At first each species was supposed to have been represented by a **type** that was used as a fixed standard. The **typologic specimen** was duly labeled and deposited in some prestigious center such as a museum. Anyone classifying a particular group would always take the pains to compare his specimens with the available typologic specimens. Since variations from the type specimen nearly always occurred, these differences were supposed to be attributable to imperfections during embryonic development and were considered of minor significance.

The **typologic** (or **morphologic) species concept** of classifying persisted for a long period (and still does to some extent). During this time, though, the idea that species represent lineages in evolutionary descent was becoming more firmly established. Gradually taxonomists began to think of a species as **groups of interbreeding natural populations that are reproductively isolated from other such groups** (Mayr, 1969). This concept of **genetic incompatibility** between different species therefore replaced the earlier idea of **character discontinuity** for species separation. Taxonomists began to think of species as populations in which every individual is unique and in which every individual may change to a greater or lesser extent when placed in a different environment. This is the **biologic species concept.**

This modern viewpoint is, of course, the antithesis of that of the typologist to whom variations from the type specimen are illusions caused by small mistakes during embryonic development. In population studies, the type is considered an abstract average of *real* variations that occur within the interbreeding population. Thus the species must be regarded as an **interbreeding population** made up of individuals of common descent and sharing intergrading characteristics. A species is set apart from all other species by a distinct evolutionary role.

The biologic species concept is not without difficulties. For one thing, it does not apply to organisms that reproduce asexually because there is no way to test the interbreeding criterion in uniparental species. For another, it is a "static" concept that does not flex easily with the slow, usually unmeasurable, stages that occur in a species through time. Furthermore, even in species that reproduce sexually, reproductive isolation may be only partial. Not infrequently populations are found that are in an intermediate stage of differentiation between races, which can interbreed, and species, which cannot interbreed. They cannot be classed as definite races or as definite species, so are sometimes called semispecies.

Despite these problems, the biologic species concept works satisfactorily most of the time. Reproductive barriers do exist. Sometimes pairs of species can be made to hybridize, but produce sterile offspring. A well-known example is the sterile mule, produced by a cross between a horse and an ass. Even when fertile hybrids are possible, they are normally prevented by various external barriers, such as anatomic incompatibilities and, especially, behavioral patterns that create aversions to mating with the wrong species.

Basis for formation of taxa

Classification emphasizes the natural relationships of animals. Descent from a common ancestor explains similarity in character; the more recent the descent, the more closely the animals are grouped in taxonomic units (Principle 6, p. 10). The genera of a particular family show less diversity than do the families of an order. This is because the common ancestor of families within an order is more remote than the common ancestor of different genera within a family. The same applies to higher categories. The common ancestor of the various vertebrate classes, for example, must be much older than the common ancestor of orders of mammals within the class Mammalia.

It is apparent that this criterion of evolutionary ancestry is not a very definite one for use in setting up taxa. There is no way to define a class that does not apply equally well to an order or a family. The genera of certain ancient groups, such as the molluscs, may actually be much older than orders and classes of other groups. The taxonomist must arrange the classifications so that all members of a taxon resemble one another more closely than they do the members of any other taxon of the same rank. Obviously the assignment of taxonomic rank depends on the opinion of the taxonomist making the study, and this is one reason classification has been called an art and not a science (G. G. Simpson, 1961). As more is learned about animals and their relationships, changes in classification are required. This brings instability to biologic nomenclature and reduces its efficiency as a reference system.

The **law of priority** also brings about frequent changes. The first name proposed for a taxonomic unit that is published and meets other proper specifications has priority over all subsequent names proposed. The rejected duplicate names are called **synonyms.** It is disturbing sometimes to find that a species that has been well established for years must undergo a change in terminology when some industrious systematist discovers that on the basis of priority or for some other reason the species is, according to this "law," misnamed.

Despite such difficulties, the hierarchic system of classification is the only accepted system we have. To reduce confusion in the field of taxonomy and to lay down a uniform code of rules for the classification of animals, the International Commission on Zoological Nomenclature was established in 1898. This commission meets from time to time to formulate rules and to make decisions in connection with taxonomic work.

Recent trends in taxonomy

Since the recognition of the concept of organic evolution, biologic units have been classified on the basis of common descent, using morphologic characters. But it is not always clear just what this evolutionary relationship is. Only the fossil record can supply convincing evidence, and this is lacking more often than not. There is an inherent danger of circular reasoning in taxonomic work: hypotheses about taxonomic relationships are used as evidence of phylogenetic relationships, and these in turn are used to make judgments about taxonomic relationships. Clearly, there is a duality to the science of systematics. On the one hand it is charged with naming organisms and placing them in some kind of order. On the other, systematics is supposed to help unravel the course of evolution. It is an unsteady mix of the practical and the theoretical. Most biologists firmly believe that taxonomy should legitimately concern itself with both these functions, but the difficulty of doing so has placed stresses on established taxonomic procedures and stimulated the search for new approaches.

Numerical taxonomy

In systematics it is necessary to recognize the characters that have taxonomic value and to make a proper analysis of the data (characters) before assigning an animal to the proper taxon. Often the analysis of the data involves a statistical approach. In 1763 the Frenchman M. Adanson proposed a scheme of classification that involved the grouping of individuals into a particular species according to the number of shared characteristics. Thus each member of a species would have a majority of the total characteristics of the taxon, even though some of its characters are not shared with others of the same taxon. Such a classification has quantitative rather than qualitative significance. The classification lacks the phyletic relationship of evolutionary taxonomy (Adanson lived long before Darwin's time). However, this scheme has been revived (especially by plant taxonomists) in recent years and has given rise to **numerical taxonomy,** which makes use of the computer method for ascertaining calculations of similarity. Similar and dissimilar characteristics are simply fed into a computer and an analysis is made of its calculations for determining the taxon of a group. The **phenetic species concept,** as this approach is called, makes no claim of inferring evolutionary relationship but aims simply at producing meaningful groupings of organisms.

Other approaches

Other techniques are now being developed in systematics that offer great promise in solving problems of phylogenetic relationships and evolution. Research is now being carried out in animal behavior, comparative biochemistry, serology, cytology, genetic homology, and comparative physiology with this aim in view.

For example, it has been shown that there are certain homologies among polynucleotide sequences in the DNA molecules of such different forms as fish and humans. These sequences appear to be genes that have been retained with little change throughout vertebrate evolution. Possible phenotypic expressions of these homologous sequences are bilateral symmetry, notochord, and hemoglobin, as well as others. By using a single strand of DNA from one species and short radioactive pieces of a DNA strand from another species and mixing the strands together, it was found that some of the smaller strands paired with similar regions on the large strand, indicating that the paired parts had common genes.

Another recent technique involves the recognition of RNA codons by transfer RNA of another species. These new biochemical methods of classification are used to complement, rather than replace, the older, more established methods. In general, molecular evidences have not agreed very well with the fossil evidence. Nevertheless, the new molecular approach provides a potentially powerful tool for the systematist.

ANIMAL PHYLOGENY

Phylogeny (''tribe origin'') is the science of genealogic relationships among lineages of life forms. The phylogenist's attempts to reconstruct lines of descent resulting in living organisms are not unlike the efforts to ferret out the ancestral history (genealogy) of one's own family. Neither plant or animal species nor members of one's family arose spontaneously, but rather they represent branching of ancestral forms. The phylogeny of an animal group, then, represents our concept of the path its evolution has taken. It is the evolutionary history of the group.

Unfortunately, the evolutionary origin of most animal phyla is shrouded in the obscurity of Precambrian times. Because fossil records are fragmentary, our reconstructions of patterns of evolutionary relationships must rely to a large extent on evidence from comparative morphology and embryology. Relationships are naturally more easily established within the smaller taxonomic units (such as species, genera, orders) than in the classes and phyla.

Although the representation of ancestral relationships of animals in the form of a ''family tree'' seems obvious to us today, it was not apparent to biologists before classification became founded on evolutionary theory. Even Darwin never attempted a pictorial diagram of animal relationships. Yet, despite all the shortcomings that any phylogenetic tree must possess, especially the danger of depicting and accepting highly tentative relationships as dogmatic certainty, such a scheme is a valuable tool. A family tree serves to tie the taxa together in an evolutionary blueprint. Constructing a family tree is a creative activity based on judgment and experience and as such is always subject to modification as new information becomes available.

In each of the following chapters on animal phyla, we include a summary of that group's origins and the relationships within the group. The student is encouraged to make frequent reference to the geologic time scale on the inside back cover of this book and to the phylogenetic tree of multicellular animals on the inside front cover. We have tried to base conclusions about group histories on recent informed opinion. Although we recognize that family trees can give false impressions, we have nonetheless used them in the absence of a better alternative. They may be viewed as educated speculations and as such with a certain measure of skepticism; at the same time they are not science fiction. They are derived from close morphologic reasoning and a thorough understanding of general biologic principles by scientists who have devoted their lives to this form of detective work. Evolution, with its idea of life transforming itself through the ages, is supported by a vast wealth of fossil and living evidence. It is after all the framework of biology.

Some helpful definitions

In the discussions on evolutionary relationships in this and ensuing chapters, terms such as lower, higher, primitive, advanced, specialized, generalized, adaptation, and fitness will be used frequently, and an understanding of their meaning may be critical to an understanding of the discussion to follow.

The terms **lower** and **higher** usually refer to a group's relative position on a phylogenetic tree; that is, the level at which it is believed to have branched off a main stem of evolution. For example, sponges and jellyfish are usually considered to be "lower" phyla because they are believed to have been among the earliest metazoans to have originated. In other words, they originated near the base of the family tree of the animal kingdom.

The terms **primitive** and **advanced** are often used when discussing relationships within a particular group. A primitive species is one that possesses a great many of the same characteristics thought to have belonged to the ancestral stock from which it evolved. An advanced species is one that has undergone considerable change from the primitive condition, usually because of adaptation to a changed environment or to a different mode of living. Among the molluscs, for example, the chitons are considered to be more primitive (that is, more like the hypothetical mollusc ancestor) than the snails, clams, or octopuses. A primitive species is not "less perfect" than an advanced species, for it may be as well adapted to its own type of environment as the advanced one is to the environment for which it has become especially adapted. Also, a species or group may be primitive in some respects and advanced in others.

Specialized might refer either to an organism or to one or more of its parts that has become adapted to a particular ecologic niche or to a particular function. A more **generalized** species or structure may share the characteristics of two or more distinct groups or structures. For example, many of the smaller aquatic crustaceans have a number of similar feathery trunk appendages, all serving several functions, such as swimming, respiration, filter feeding, or egg bearing. Such multipurpose appendages would be considered generalized in comparison with the highly specialized defensive chelipeds or sensory antennae of the crayfish, the pollen-collecting legs of the honeybee, or the digging forelegs of the mole cricket. Each of these specialized appendages is adapted for one primary function.

An **adaptation** is any characteristic of an organism taken in the context of how the characteristic helps the organism survive and reproduce. Many adaptations are possessed in common by a wide variety of animals, but the word is most often used in reference to special adaptations for a particular habitat or environment. Indeed, the more *special adaptations* the animal has, the more *specialized* the animal is. Any characteristic or any change in a characteristic is said to have **adaptive value,** if it tends toward greater fitness of the organism to existence in a particular niche or habitat.

An organism is **fitted** to its environment when it is adjusted or adapted to it. Its **fitness** for any particular place or mode of living is the degree of its adjustment, suitability, or adaptation to that particular environment or niche.

Origin of Metazoa

Unraveling the origin of the multicellular animals (metazoans) has presented many problems for zoologists. It is generally believed that they evolved from unicellular organisms, but there is much disagreement over which group of unicellular organisms gave rise to them, and how. The three prevalent theories in current use are (1) that the metazoans arose from a syncytial (multinucleate) ciliate in which cell boundaries later evolved, (2) that they arose from a colonial flagellate in which the cells gradually became more specialized and interdependent, and (3) that the origin of metazoans was polyphyletic, or derived from more than one group of unicellular organisms.

Proponents of the **syncytial ciliate theory** believe that metazoans arose from primitive single-celled ciliates, which at first were multinucleated (having more than one nucleus) but later, by acquiring cell membranes around the nuclei, became cellularized into the multicellular condition. The ancestral ciliates, like many modern ones, tended toward bilateral symmetry, so that the earliest metazoans were bilateral and similar to the present primitive flatworms. There are several objections to this theory. It ignores the embryology of the flatworms in which nothing similar to cellularization occurs; it does not explain the presence of flagellated sperm in the metazoans; and, perhaps more importantly, it infers that the radially symmetric cnidarians are derived from the flatworms, thus making bilateral symmetry more primitive than radial symmetry.

The **colonial flagellate theory**—first proposed by Haeckel, a German evolutionist, in 1874—is the classic theory, which, with various revisions, still has many followers. According to this theory, the metazoans are derived from a hollow, spherical, colonial flagellate (protozoan). Many flagellates form colonies of from a few to many cells (zooids), such as *Gonium* (4 to 16 cells), *Eudorina* (32 cells), *Pleodorina* (32 to 128 cells), or *Volvox* (a spherical, hollow colony of thousands of flagellated cells). In *Volvox,* the cells are usually differentiated into a few reproductive cells and a great many somatic (body) cells, which cannot reproduce except to divide and form more somatic cells. A further differentiation of function in such a colonial ancestral form could, according to the theory, have led to the metazoan condition. According to Haeckel, this hollow, spherical ancestral stage was the **blastea,** and the hollow blastula stage of the metazoan embryo was a recapitulation of this ancestral stage. Simple invagination of the blastea would have produced the **gastrea,** corresponding to the gastrula embryonic stage. Metschnikoff (1877) claimed that the hypothetical ancestral organism was a solid structure, or **stereo gastrea,** rather than a hollow one, and was similar to the free-swimming planula larva of the cnidarians (jellyfishes and others), a larva that is radially symmetric and has no mouth. The lower metazoans, according to this theory, derived from this planuloid (planula-like) ancestor, and thus accounted for the radial symmetry of the cnidarians. Bilateral symmetry could have evolved later. According to Hyman (1951) some of these planuloid ancestors became adapted to life on the ocean floor, where they acquired a creeping method of locomotion in which the dorsal and ventral surfaces became differentiated, developed a ventral mouth, and evolved a tendency toward cephalization (a concentration of neurons and sensory structures at the anterior). This would have led to bilateral symmetry and the evolution of the primitive flatworms, which later gave rise to all other bilateral forms.

Some zoologists prefer the theory that the metazoans had a **polyphyletic origin** and suggest that the sponges, cnidarians, ctenophores, and flatworms each evolved independently—the sponges and cnidarians probably evolving from colonial flagellates, and the ctenophores and flatworms from ciliates. This is, of course, a compromise between the other two theories.

In any case, by whatever theory, the primitive flatworms (acoel) are believed to be the most primitive of the bilateral animals.

Ontogeny and phylogeny

Phylogeny, as we have seen, represents the evolutionary history of any taxon. **Ontogeny,** on the other hand, refers to the history of the development of an individual from zygote to maturity. All metazoans pass through certain common developmental stages. The stages, in succession, are the zygote, cleavage, blastulation, and gastrulation. Haeckel believed that each of these successive stages in the development of an individual represented one of the adult forms that appeared in its evolutionary history. As he phrased it, "Ontogeny is the short and rapid recapitulation of phylogeny. During its own development . . . an individual repeats the most important changes in form evolved by its ancestors during their long and slow paleontological development." According to this theory, the zygote of a developing individual represented the protozoan or protistan stage of its evolutionary history; the blastula represented the hollow colonial protozoans; and the gastrula, the adult cnidarians. The human embryo with gill depressions in the neck was believed to represent the stage when our adult ancestors were aquatic. On this basis he gave his generalization: *ontogeny (individual development) recapitulates (repeats) phylogeny (evolutionary descent)*. This notion later became known as the **biogenetic law** (Principle 7, p. 11).

Haeckel's hypothesis of recapitulation is often called the **gastrea hypothesis.** It was based primarily on cnidarian development, which, however, occurs in a different way from what Haeckel supposed. The type of gastrulation that he considered to be primitive is now known to be far from universal among the cnidarians. Haeckel's theory also was based on the premise that evolutionary change occurs by the successive addition of stages to the end of an unaltered ancestral ontogeny. This was an outgrowth of his belief in Lamarck's theory of the inheritance of acquired characters (p. 875).

K. E. von Baer, a nineteenth century embryologist, had noticed long before Haeckel the general similarity between the embryonic stages and the adults of certain animals, but arrived at a different and sounder interpretation. According to his view, the earlier stages of all embryos tend to look alike, but as development proceeds, the embryos become more dissimilar. In other words, he believed that the embryos of higher and lower forms resemble each other more the earlier in their development that they are compared and *not* that embryos of higher forms resemble the adults of lower forms.

The fascination with and debates over recapitulation gradually waned as scientists became more absorbed with Mendelian genetics, experimental embryology, and cell and molecular biology; and the whole question fell into a mild state of disrepute. S. J. Gould (1977) has brought the matter again into the limelight. He contends that there *are* parallels between the stages of ontogeny and phylogeny, but that features appearing at one stage of an ancestral ontogeny may be shifted to earlier or later stages in descendants. If a feature that appears at a specific point in the ontogeny of an ancestor arises earlier and earlier in descendants, we have one parallel producing **recapitulation** (the descendant's ontogeny repeats an evolutionary sequence of stages that characterized its ancestors at that specific point). If a feature that appears at a specific point in the ontogeny of an ancestor arises later and later in the descendants, we have an inverse parallel producing **paedomorphosis** (which is the retention of ancestral juvenile characters by later stages in the ontogeny of descendants). In other words, the rate of development of a feature in a descendant's ontogeny may be faster or slower than the rate of development of that same feature in an ancestor's ontogeny. This phyletic change in timing of development Gould calls **heterochrony.**

Kingdoms of life

Since Aristotle's time, it has been traditional to assign every living organism to one of two kingdoms—plant and animal. However, the two-kingdom system has outlived its usefulness. Although it is easy to place rooted, photosynthetic organisms such as trees, flowers, mosses, and ferns among the plants and to place food-ingesting, motile forms such as worms, fishes, and mammals among the animals, unicellular organisms present difficulties. Some forms are claimed both for the plant kingdom by botanists and for the animal kingdom by zoologists. An example is *Euglena* (p. 150) and its phytoflagellate kin, which are motile, like animals, but have chlorophyll and photosynthesize, like plants. Other groups such as the bacteria were rather arbitrarily assigned to the plant kingdom.

It was inevitable that biologists would try to resolve these problems by separating problem groups into new kingdoms. This was first done in 1866 by Ernst Haeckel who proposed the new kingdom Protista to include all single-celled organisms. At first the bacteria and blue-green algae, forms that lack nuclei bounded by a membrane, were included with nucleated unicellular organisms. Finally the important differences between the anucleate bacteria and blue-green algae **(prokaryotes)** and all other organisms that have cells with membrane-bound nuclei **(eukaryotes)** were recognized. The prokaryote-eukaryote distinction is actually much more profound than the plant-animal distinction of the traditional system. The many differences between prokaryotes and eukaryotes are summarized on p. 60.

In 1969 R. H. Whittaker proposed a five-kingdom system that incorporated the basic prokaryote-eukaryote distinction (Fig. 7-1). The kingdom Monera contains the prokaryotes; the eukaryotes are divided among the remaining four kingdoms. The kingdom Protista contains the unicellular eukaryotic organisms (protozoans and eukaryotic algae). The multicellular organisms are split into three kingdoms on the basis of mode of nutrition and other fundamental differences in organization. The kingdom Plantae includes multicel-

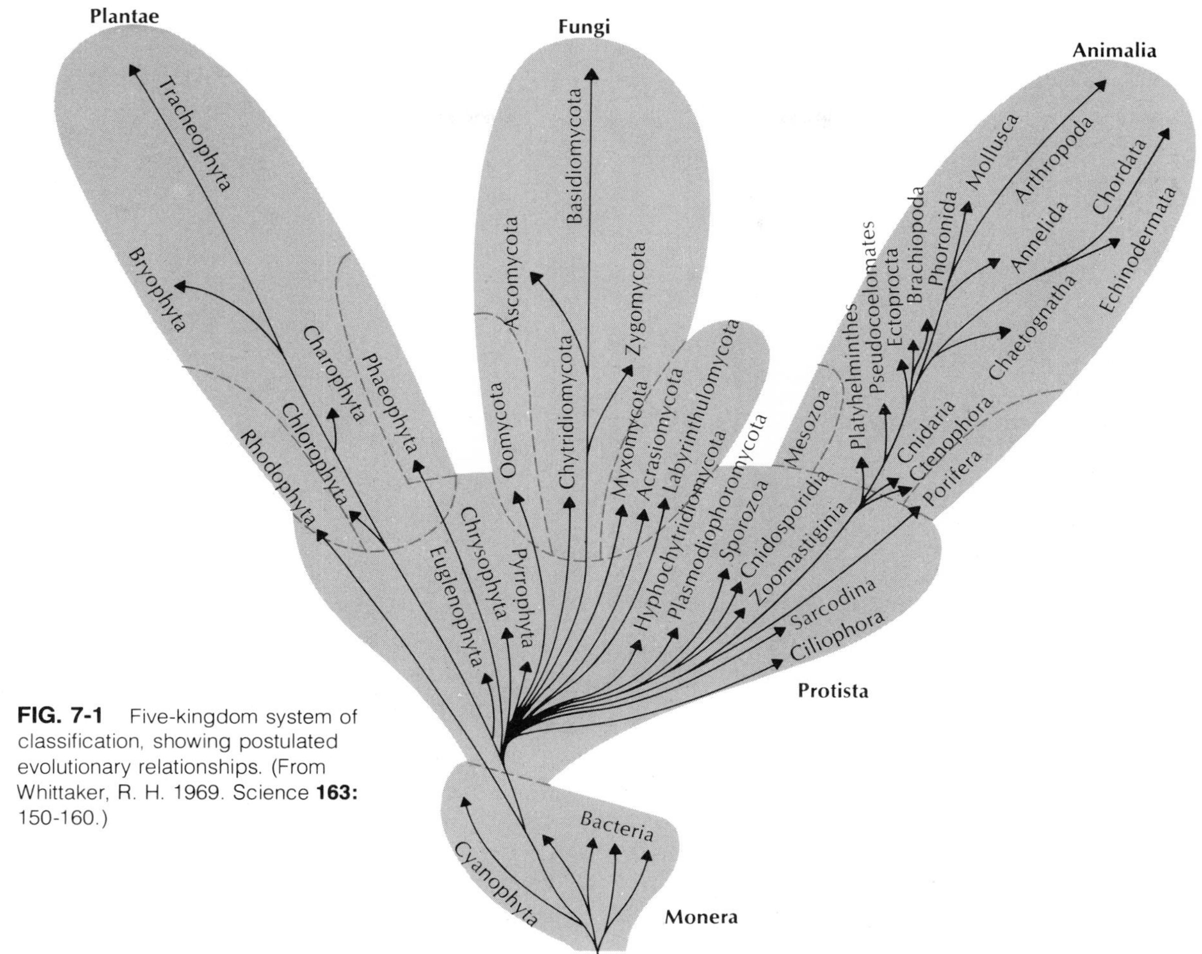

FIG. 7-1 Five-kingdom system of classification, showing postulated evolutionary relationships. (From Whittaker, R. H. 1969. Science **163:** 150-160.)

lular photosynthesizing organisms, higher plants, and multicellular algae. Kingdom Fungi contains the molds, yeasts, and fungi that obtain their food by absorption. The invertebrates (except the Protozoa) and the vertebrates comprise the kingdom Animalia. Most of these forms ingest their food and digest it internally, although some parasitic forms are absorptive. The supposed evolutionary relationships of the five kingdoms are shown in Fig. 7-1. The Protista are believed to have given rise to all three multicellular kingdoms, which therefore have evolved independently.

Exclusion of the unicellular protozoans from the animal kingdom, in which they have been traditionally included, presents a didactic problem for the zoologist. In the five-kingdom system, phylum Protozoa has been divided into seven groups that are each elevated to phylum status. This change emphasizes the fluid nature of the hierarchic system of classification pointed out earlier in this chapter. Since many protozoologists had already advocated splitting the protozoans into even more phyla on the basis of fundamental differences between groups, taxonomic inflation of the protozoans, as treated in the five-kingdom system, may be legitimate and defensible. On the other hand, the protozoans share many animal-like characteristics. Most ingest their food; many have specialized organelles and advanced locomotory systems, portending tissue differentiation in multicellular forms; many reproduce sexually; and some flagellate forms are colo-

TABLE 7-2 Diphyletic theory of phylogeny*

Protostomes	Deuterostomes
Mouth from, at, or near blastopore	New mouth from stomodeum
Anus new formation	Anus from or near blastopore
Cleavage mostly spiral and involving all organ systems	Cleavage mostly radial and involving only dorsal myotomes
Embryology mostly determinate (mosaic)	Embryology usually indeterminate
Coelom forms as split in mesodermal bands; schizocoelous	Coelom from fusion of enterocoelous pouches (except chordates, which are schizocoelous)
Endomesoderm usually from blastomere 4d	Endomesoderm from enterocoelous pouching (except chordates)
Includes phyla Platyhelminthes, Rhynchocoela, Annelida, Mollusca, Arthropoda, minor phyla	Includes phyla Echinodermata, Hemichordata, Chaetognatha, and Chordata

*Embryos of many of the advanced members of each group may have some or all of these characteristics obscured or lost.

nial with division of labor among cell types, again suggestive of a metazoan pattern.

Whether or not the protozoans should be classified as a single animal phylum, as their name suggests, or split into several phyla is a matter of opinion. At the present time it is still reasonable and workable to retain the protozoans as an animal phylum. This relationship is depicted in the family tree of animals on the inside front cover of this book.

Larger divisions of the animal kingdom

Although the phylum is often considered to be the largest and most distinctive taxonomic unit, zoologists often find it convenient to combine phyla under a few large groups because of certain common embryologic and anatomic features. Such large divisions may have a logical basis, for the members of some of these arbitrary groups are not only united by common traits, but evidence also indicates some relationship in phylogenetic descent.

- SUBKINGDOM PROTOZOA (UNICELLULAR)—phylum Protozoa
- SUBKINGDOM METAZOA (MULTICELLULAR)—all other phyla
  - BRANCH A (MESOZOA)—phylum Mesozoa, the mesozoans
  - BRANCH B (PARAZOA)—phylum Porifera, the sponges
  - BRANCH C (EUMETAZOA)—all other phyla
    - GRADE I (RADIATA)—phyla Cnidaria (Coelenterata), Ctenophora
    - GRADE II (BILATERIA)—all other phyla
      - DIVISION A (PROTOSTOMIA)—bilateral animals with spiral and determinate cleavage and mouth arising from or near the blastopore
        - ACOELOMATES—phyla Platyhelminthes, Rhynchocoela (Nemertina)
        - PSEUDOCOELOMATES—phyla Rotifera, Gastrotricha, Kinorhyncha, Gnathostomulida, Nematoda, Nematomorpha, Acanthocephala, Entoprocta
        - EUCOELOMATES—phyla Mollusca, Annelida, Arthropoda, Priapulida, Echiurida, Sipunculida, Tardigrada, Pentastomida, Onychophora, Pogonophora, Phoronida, Ectoprocta, Brachiopoda
      - DIVISION B (DEUTEROSTOMIA)—with radial and indeterminate cleavage and mouth arising some distance from the blastopore; all coelomates
        - Phyla Echinodermata, Chaetognatha, Hemichordata, Chordata

As in the outline, the bilateral animals are customarily divided into protostomes and deuterostomes on the basis of their embryologic development (Table 7-2). However, some of the phyla are very difficult to place into one of these two categories because they possess some of the characteristics of each group. Ivanov (1976), for example, objects to the use of the term Protostomia as a taxon. He suggests dividing the Bilateria into Scolecida (the lower worms) and Coelomata (all of the coelomates). Coelomata would then consist of five superphyla: Trochozoa (phyla Annelida, Mollusca, Arthropoda, and some minor phyla), Tentaculata (the lophophorate phyla Phoronida, Ectoprocta, and Brachiopoda), Chaetognatha (phylum Chaetognatha), Pogonophora (phylum Pogonophora), and Deuterostomia (phyla Echinodermata, Hemichordata, and Chordata). His view reflects a belief in a monophyletic origin for all coelomates, including the annelids, molluscs, and arthropods, rather than the diphyletic origin for coelomates shown in Table 7-2.

Annotated references

Selected general readings

Anfinson, C. B. 1959. The molecular basis of evolution. New York, John Wiley & Sons, Inc. *This book has made a great impact on all serious students of the life sciences. Chapter 1, although brief, will give the student a good phylogenetic orientation.*

Clifford, H. T., and W. Stephenson. 1975. An introduction to numerical classification. New York, Academic Press, Inc. *Methods of numerical approaches to classification are clearly described and compared to classic taxonomy. Less detailed than the Sneath and Sokal book.*

Crowson, R. A. 1970. Classification and biology. New York, Atherton Press. *A thorough treatment of the methods and problems of zoologic, botanic and paleontologic classification.*

de Beer, G. R. 1940. Embryos and ancestors. New York, Oxford University Press. *An excellent evaluation of the biogenetic law.*

Dougherty, E. C., and others (eds.). 1963. The lower metazoa: comparative biology and phylogeny. Los Angeles, University of California Press. *Papers from a symposium on the phylogeny and comparative biology of the lower metazoa.*

Florkin, M. 1966. A molecular approach to phylogeny. New York, Elsevier Publishing Co.

Gould, S. J. 1977. Ontogeny and phylogeny. Cambridge, Mass., The Belknap Press of Harvard University Press. *A review of the historical development and collapse of the theory of recapitulation, and a discussion of modern concepts, such as heterochrony—changes in developmental timing, producing parallels between ontogeny and phylogeny—and neoteny, which Gould terms the most important determinant of human evolution.*

Grant, V. 1977. Organismic evolution. San Francisco, W. H. Freeman and Co.

Hadzi, J. 1963. The evolution of the Metazoa. New York, Pergamon Press, Inc.

Hyman, L. H. 1940-1967. The invertebrates, vols. 1-6. New York, McGraw-Hill Book Co. *Informative discussions on the phylogenies of most of the invertebrates are treated in this outstanding series of monographs. Vol. 1 contains a discussion of the colonial theory of the origin of metazoans, and vol. 2 a discussion of the origin of bilateral animals, body cavities, and metamerism.*

Ivanov, A. V. 1976. Correlation between the Protostomia and the Deuterostomia and the classification of the animal kingdom. Zool. Zh. **55**(8):1125-1137. *In Russian with Russian and English summaries.*

Jeffrey, C. 1973. Biological nomenclature. London, Edward Arnold. *A concise, practical guide to the principles and practice of biologic nomenclature and a useful interpretation of the Codes of Nomenclature.*

Leone, C. A. (ed.). 1964. Taxonomic biochemistry and serology. New York. The Ronald Press Co. *Collection of 47 papers treating the principles of systematics, molecular taxonomy, and taxonomic biochemistry and serology of animals, plants, and microorganisms.*

Margulis, L. 1970. Origin of eukaryotic cells. New Haven, Yale University Press. *Detailed consideration of theories of origins and evolution of microbial, plant, and animal cells.*

Margulis, L. 1974. Five-kingdom classification and the origin and evolution of cells. Evol. Biol. **7:**45-78. *Describes a modified version of the Whittaker five-kingdom system.*

Mayr, E. (ed.). 1957. The species problem. Washington, D. C., American Association for the Advancement of Science. *This is a symposium by many authorities on the problems of species.*

Mayr, E. 1969. The biological meaning of species. Biol. J. Linn. Soc. **1:**311-320.

Mayr, E., E. G. Linsley, and R. C. Usinger. 1953. Methods and principles of systematic zoology. New York, McGraw-Hill Book Co. *The present status of the rules and regulations of taxonomy is well discussed.*

Morescalchi, A. 1970. Karyology and vertebrate phylogeny. Bull. Zool. **37:**1-20.

Raven, P. H., B. Berlin, and D. E. Breedlove. 1971. The origins of taxonomy. Science **174:**1210-1213.

Ross, H. H. 1974. Biological systematics. Reading, Mass., Addison-Wesley Publishing Co., Inc. *Theory and practice of systematics is presented and exemplified from animals, plants, and microorganisms. Very comprehensive and useful.*

Simpson, G. G. 1961. Principles of animal taxonomy. New York, Columbia University Press. *An outstanding contribution of the basic principles of systematics and morphologic diversity.*

Slobodchikoff, C. N. (ed.). 1976. Concepts of species. Benchmark papers in systematic and evolutionary biology, vol. 3. Stroudsburg, Pa., Dowden, Hutchinson and Ross, Inc. *A selection of papers on the historic and philosophic development of the species concept with helpful introductions by the editor.*

Sneath, P. H. A., and R. R. Sokal. 1973. Numerical taxonomy. San Francisco, W. H. Freeman & Co. Publishers. *Extensive treatment of taxonomic theory, methods of classification, and the implications and application of numerical taxonomy.*

Whittaker, R. H. 1969. New concepts of the kingdoms of organisms. Science **163:**150-160. *The author's five-kingdom system.*

Selected *Scientific American* article

Sokal, R. R. 1966. Numerical taxonomy. **215:**106-116 (Dec.).

PART TWO

THE DIVERSITY OF ANIMAL LIFE

Marine tubeworms and (below) sea squirts. (Photograph by C. P. Hickman, Jr.)

The 1.5 million named species of animals are witness to the enormous diversity of animal life believed to have descended from a common ancestor by means of a single evolutionary process (discussed in Part four). Classification of animals is based on evolutionary relationships—the only way, biologically, that the diversity of life can be understood. Through progressive adaptation to varied and changing environmental opportunities, animals have expanded into practically every environmental niche that can support life.

The following chapters deal with the adaptive radiation that has occurred among animals, the relationships of the various groups to each other, the evolutionary advances that have pointed the way toward higher phyla, and the distinctive characteristics of each of the types that make up the varied fauna of our earth.

CHAPTER 8

THE UNICELLULAR ANIMALS—PHYLUM PROTOZOA

A radiolarian—one of the ocean-dwelling protozoans.

Courtesy American Museum of Natural History.

Position in the animal kingdom

The protozoan is a complete organism in which all life activities are carried on within the limits of a single plasma membrane. As its protoplasmic mass is not subdivided into cells, the protozoan is sometimes termed "acellular," but many people prefer "unicellular" to emphasize the many structural similarities to the cells of multicellular animals (see Principle 11, p. 11). Protozoa may be said to belong to a protoplasmic level of organization.

The term "Protista" is used by some biologists to refer to all unicellular organisms, whether they are plant or animal. When applied in this way, the category includes the unicellular algae, protozoa, bacteria, yeasts, and so on. Most biologists prefer to use "Protista" to denote only those unicellular organisms whose nucleus is bounded by a nuclear membrane (eukaryotic) and the term "Monera" to include those unicellular organisms without a nuclear membrane (prokaryotic), such as the bacteria and the blue-green algae.

The protozoa are eukaryotic protistans. The eukaryote-prokaryote distinction was summarized in Chapter 4 (p. 60) and mentioned again in Chapter 7 (p. 126).

Biologic contributions

1. **Protoplasmic specialization** (division of labor within the cell) involves the organization of protoplasm into functional organelles.
2. The earliest indication of **division of labor between cells** is seen in certain colonial protozoans that have both somatic and reproductive zooids (individuals) in the colony.
3. **Asexual reproduction** by mitotic division is first developed in the protists.
4. **True sexual reproduction** with zygote formation is found in some protozoans.
5. The responses (taxes) of protozoans to stimuli represent the **beginning of reflexes and instincts** as we know them in metazoans.
6. The most primitive animals with an **exoskeleton** are certain shelled protozoans.
7. **All types of nutrition** are developed in the protozoans: autotrophic, saprozoic, and holozoic.

THE PROTOZOA

Though protozoa are unicellular, it is erroneous to think of them as simple. They are functionally complete organisms and have many complicated structures. Physiologically they are quite complex. Within the cytoplasm of a protozoan there is specialization and division of labor. The specialized structures within the cytoplasm are called **organelles,** and each is fitted for a specific function, just as specialized organs or groups of cells perform a special function in the metazoan (many-celled) animal (see Principle 12, p. 12). Particular organelles may perform as skeletons, sensory systems, conducting mechanisms, contractile systems, organs of locomotion, defense mechanisms, and so on.

Among the protozoa, the dividing line between animals and plants becomes quite vague and arbitrary. Many of the flagellates are **autotrophic** (holophytic), that is, they contain chlorophyll and, like plants, can manufacture their own carbohydrates by photosynthesis. They may be considered plants by botanists and animals by zoologists, each to suit the convenience of their taxonomic systems. Perhaps in the early evolution of animals some of these autotrophic flagellates may have lost their green chloroplasts to become colorless animals that must either absorb nutrients from their environment (saprozoic) or feed on other plant or animal matter (heterotrophic, or holozoic).

Most protozoa live singly, but many, particularly among the flagellates, live in distinct colonies of from several to hundreds of individual zooids. The distinctions between such protozoan colonies and the simplest metazoans are also vague and arbitrary. In a protozoan colony, some cells may be specialized for reproduction, but the rest can perform all other body functions independently of each other. In the simplest metazoa, there may be only slightly more interdependence, but in the higher and more complex metazoa, most cells become highly specialized and depend strongly on the integrity of the organism for survival.

Size

Most protozoa are microscopic, usually from 3 to 300 μm long. The largest are among the Foraminifera, some of which have shells 100 to 125 mm in diameter, and some extinct protozoa reached a size of several centimeters. Certain amebas may be 4 to 5 mm in diameter.

Number of species

The number of named species of protozoans is close to 50,000, of which over 20,000 are fossils (mostly Foraminifera), and of the remainder, about one-third are parasitic. These figures may represent only a fraction of the total number of species. Some protozoologists think that there may be more protozoan species than all other species together because each species of the higher phyla may have several species of parasitic protozoa, and many protozoa bear parasites themselves. In reality, there may be more parasitic species than free-living ones.

Ecologic relationships

Protozoa are found wherever life exists. They are highly adaptable and easily distributed from place to place. Moisture is necessary, whether they live in marine or freshwater habitats, soil, decaying organic matter, or plants and animals. Protozoa may be sessile or free-swimming, and they form a large part of the floating plankton. The same species are often found widely separated in time as well as space. Some species may have spanned geologic eras of more than 100 million years.

Despite their wide distribution, many protozoans can live successfully only within narrow environmental ranges. Species adaptations vary greatly, and successions of species frequently occur as environmental conditions change. These changes may be brought about by physical factors, as in the drying up of a pond, seasonal changes in temperature, and so on, or they may be biologic changes, such as predator pressure.

Protozoa play an enoromous role in the economy of nature. Their fantastic numbers are attested by the gigantic ocean and soil deposits formed by their skeletons. As mutuals, they may be indispensable to certain forms of life—for example, the termites—but as parasites they may give rise to serious diseases of humans and some animals. They also may be important sources of water contamination.

Characteristics

1. **Unicellular;** some colonial
2. **Mostly microscopic,** although some large enough to be seen with the unaided eye
3. All symmetries represented in the group; shape variable or constant (oval, spherical, or other)
4. **No germ layer present**
5. No organs or tissues, but **specialized organelles** found; nucleus single or multiple
6. Free living, mutualism, commensalism, parasitism all represented in the group
7. Locomotion by **pseudopodia, flagella, cilia,** and direct cell movements; some sessile
8. Some provided with a **simple protective exoskeleton,** but mostly naked
9. **Nutrition of all types:** autotrophic (manufacturing own nutrients by photosynthesis), heterotrophic (depending on other plants or animals for food), saprozoic (using nutrients dissolved in the surrounding medium)
10. Habitat: aquatic, terrestrial, or parasitic
11. Reproduction **asexually** by fission, budding, and cysts and **sexually** by conjugation or by syngamy (union of male and female gametes to form a zygote)

Brief classification

Traditionally, four main groups of protozoa have been recognized: the flagellates, the amebas, the sporeformers, and the ciliates. More recent studies have shown that the classification based on those groups was unnatural, and the system that follows places these organisms in a more phylogenetic arrangement. The student should realize that the phylum Protozoa encompasses a very heterogeneous assemblage, and the subphyla would probably each be considered phyla in their own right if they were metazoan groups.

Subphylum Sarcomastigophora (sar'ko-mas-ti-gof' o-ra) (Gr., *sarkos,* flesh, + *mastix,* whip, + *phora,* bearing). Flagella, pseudopodia, or both types of locomotory organelles; usually with only one type of nucleus; typically no spore formation; sexuality, when present, essentially syngamy.

Superclass Mastigophora (mas-ti-gof'o-ra) (Gr., *mastix,* whip, + *phora,* bearing). One or more flagella typically present in adult stages; autotrophic or heterotrophic, or both; reproduction usually asexual by fission.

Class Phytomastigophorea (fi'to-mas-ti-go-for'e-a) (Gr., *phyton,* plant, + *mastix,* whip, + *phora,* bearing). Plantlike flagellates, usually bearing chromoplasts, which contain chlorophyll. Examples: *Chilomonas, Euglena, Volvox, Ceratium, Peranema.*

Class Zoomastigophorea (zo'o-mas-ti-go-for'e-a) (Gr., *zoon,* animal, + *mastix,* whip, + *phora,* bearing). Flagellates without chromoplasts; one to many flagella; ameboid forms with or without flagella in some groups; species predominately symbiotic. Examples: *Trichomonas, Trichonympha, Trypanosoma, Leishmania, Proterospongia, Dientamoeba.*

Superclass Opalinata (o'pa-lin-a'ta) (NF, *opaline,* like opal in appearance, + *-ata,* group suffix). Body covered with longitudinal rows of cilium-like organelles; parasitic; cytostome (cell mouth) lacking; two to many nuclei of one type. Examples: *Opalina, Protoopalina.*

Superclass Sarcodina (sar-ko-di'na) (Gr., *sarkos,* flesh, + *ina,* belonging to) **(Rhizopoda).** Pseudopodia typically present; flagella present in developmental stages of some; cortical zone of cytoplasm relatively undifferentiated compared with other major taxa; body naked or with external or internal skeletons; free-living or parasitic.

Class Actinopodea (ak'ti-no-po'de-a) (Gr., *aktis, aktinos,* ray, + *pous, podos,* foot). With pseudopodia radiating from a spherical body. Examples: *Actinosphaerium, Actinophrys, Thalassicola.*

Class Rhizopodea (ri-zo-po'de-a) (Gr., *rhiza,* root, + *pous, podos,* foot). Variety of pseudopodia. Examples: *Amoeba, Entamoeba, Arcella.*

Subphylum Apicomplexa (a'pi-com-plex'a) (L., *apex,* tip or summit, + *complex,* twisted around, + *a,* suffix). Characteristic set of organelles (apical complex) associated with anterior end present in some developmental stages; cilia and flagella absent except for flagellated microgametes in some groups; cysts often present; all species parasitic.

Class Sporozoa (spor-o-zo'a) (Gr., *sporos,* seed, + *zoon,* animal). Spores typically present, without polar filaments and with one to many sporozoites. Examples: *Monocystis, Gregarina, Eimeria, Plasmodium, Toxoplasma.*

Class Piroplasmea (pir-o-plaz'me-a) (L., *pirum,* pear, + Gr., *plasma,* a thing molded). Spores absent; no flagella or cilia at any stage; locomotion by body flexion or gliding; two-host life cycles, known hosts are vertebrates (in red blood cells) and ticks; apical complex present, but reduced. Example: *Babesia.*

Subphylum Myxospora (mix-os'por-a) (Gr., *myxa,* slime, mucus, + *sporos,* seed). Spores of multicellular origin; one or more sporoplasms; spores enclosed by two or three (rarely one) valves; parasites of lower vertebrates, especially fishes, and invertebrates. Examples: *Myxosoma, Henneguya.*

Subphylum Microspora (mi-cros'por-a) (Gr., *micro,* small, + *sporos,* seed). Spore of unicellular origin; single sporoplasm; single valve; parasites of invertebrates, especially arthropods, and lower vertebrates. Examples: *Nosema, Glugea, Plistophora.*

Subphylum Ciliophora (sil-i-of'or-a) (L., *cilium,* eyelash, + Gr., *phora,* bearing). Cilia or ciliary organelles in at least one stage of life cycle; two types of nucleus (except in a few forms); asexual reproduction by binary fission across rows of cilia; sexuality involving conjugation, autogamy, and cytogamy.

Class Ciliata (sil-i-ah'ta) (L., *cilium,* eyelash, + *-ata,* group suffix). Characteristics of subphylum.

Subclass Holotrichia (ho-lo-trik'e-a) (Gr., *holos,* entire, + *thrix, trichos,* hair). Ciliation usually uniform over body; specialized buccal cilature inconspicuous or absent. Examples: *Paramecium, Colpoda, Tetrahymena, Spirochona, Balantidium.*

Subclass Spirotrichia (spir-o-trik'e-a) (L., *spiro,* coil + Gr. *thrix, trichos,* hair). Adoral zone of membranelles winding clockwise toward cytostome; cilia sparse; some with cirri on ventral surface. Examples: *Stentor, Blepharisma, Halteria, Epidinium, Euplotes.*

Subclass Peritrichia (per-i-trik'e-a) (Gr., *peri,* around, + *thrix, trichos,* hair). Conspicuous oral ciliature winding counterclockwise around apical end; body cilia usually lacking in adults; mostly attached stalked ciliates. Examples: *Vorticella, Carchesium, Trichodina.*

Subclass Suctoria (suk-tor'e-a) (L., *sugere, suctum,* to suck). Stalked ciliates with tentacles; no external ciliature in adult. Examples: *Podophyra, Ephelota.*

Protozoan fauna of plankton

Plankton is a general term for those organisms that passively float and drift with the wind, tides, and currents of both freshwater and marine water. It is composed mostly of microscopic animals and plants, of which protozoans form an important part. Sarcodina, particularly the foraminiferans and radiolarians, and Mastigophora make up most of the protozoan part of plankton, but the ciliates are also represented, and even some parasitic sporozoans are found within other plankton animals.

Plankton is a very important source of food for other marine animals. It forms a large part of the base of the food pyramid in aquatic ecosystems (producers and primary consumers, p. 950). Not only invertebrates feed on plankton, but many fish and even the huge whalebone whales feed directly on it. As the organisms of the shallow-water plankton die, they sink to deeper levels to serve as food for other animals there.

Symbiotic relationships

The term **symbiosis** refers to an intimate relationship between two organisms of different species in which one or both species derives benefit. Used in the strict sense, one of the individuals (the **symbiont**) lives in or on the body of the other (the **host**). The association may be beneficial to both species **(mutualism),** beneficial to one partner but not harmful to the other **(commensalism),** or beneficial to the symbiont and harmful to the host **(parasitism).** These relationships are not always easy to distinguish, but clear cases of all three are shown by protozoans.

Mutualism. *Paramecium bursaria,* a ciliate, harbors green algae (zoochlorellae) that manufacture carbohydrates (by photosynthesis), which benefit the paramecium. The algae receive a safe shelter and probably other physiologic benefits in return. Zoochlorellae are also found in other protozoans, such as *Stentor,* certain amebas, and heliozoans.

Several species of flagellates live in the intestines of termites and wood-feeding roaches, where they secrete enzymes for digesting the cellulose that is thus made available to their hosts (Fig. 8-1). The protozoa ingest the wood after the insects have chewed it up into small

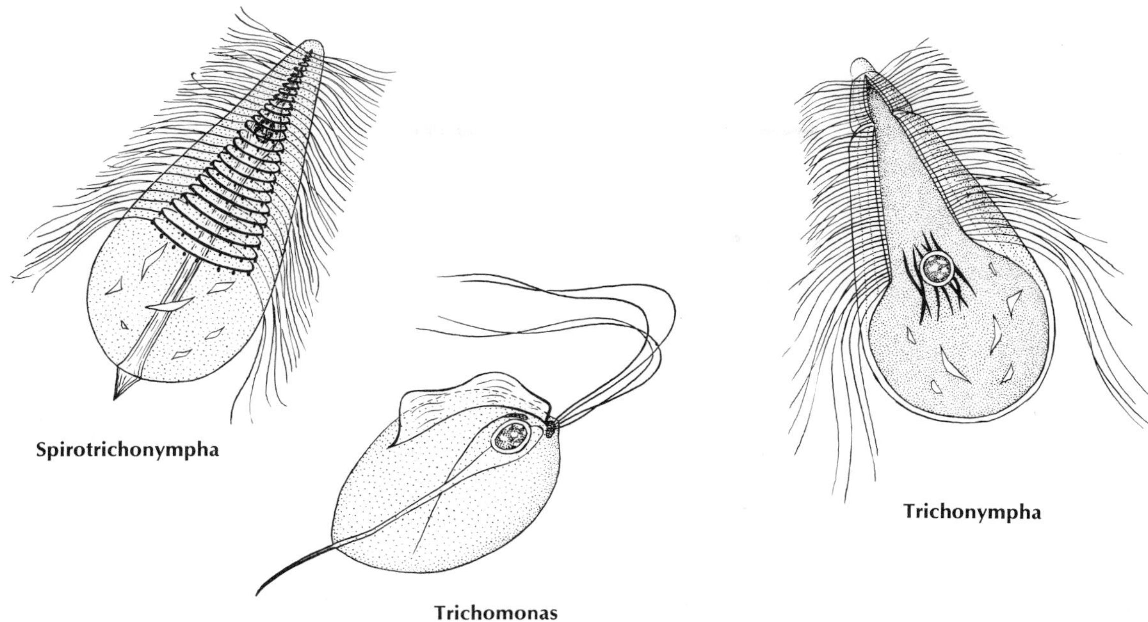

FIG. 8-1 Three flagellates that are common symbionts in the gut of termites and wood roaches.

bits, and the insects are nourished by the metabolic products excreted by the protozoa. The flagellates cannot live outside the host, and when termites lose their protozoan fauna (by high oxygen exposure or high temperature or during molting), they starve to death in a short time, even when fed an abundance of wood, unless they are reinfected with the flagellates.

Commensalism. Commensal protozoa may live on the outside (ectocommensals) or, more commonly, on the inside (endocommensals) of another organism. A variety of species, including *Vorticella* and some suctorians, have been reported as ectocommensals on hydroids, crustaceans, annelids, and other aquatic invertebrates. Endocommensals are common in the digestive tracts of invertebrates and vertebrates.

Herbivorous mammals contain great numbers of ciliates and also a few flagellates and amebas. In ruminants such as cattle and sheep they are found in the first two stomach compartments, where they digest bacteria and foodstuffs in the host's food. Eventually they pass into the other compartments of the stomach and intestine, where they are destroyed and digested so that the host gets all the nutrition after all. Estimates are given of as many as 100,000 to 1 million ciliates per cubic millimeter of gut contents and a total number in a mature cow of 10 to 50 billion. Apparently protozoa are not essential to the cattle, which can live and grow normally when the commensals are removed. Most of the essential digestion of cellulose in ruminants is accomplished by anaerobic bacteria.

Parasitism. Parasitism is a most common form of symbiosis among protozoa. Ectoparasites live on the outside of the body and endoparasites live within the host. Most animals, especially higher ones, have one or more kinds of protozoan parasites. Protozoans themselves, even protozoan parasites, are often parasitized by other protozoans. For instance, the opalinid that lives in the frog's intestine is parasitized by a certain ameba.

Parasitic species are found among all classes of protozoa; several groups are entirely parasitic. Every vertebrate species probably harbors a parasitic or endocommensal ameba. Protozoan parasites have different ways of infecting the host. Some are transferred by fecal contamination *(Entamoeba histolytica);* by arthropods or other carriers (*Trypanosoma* and *Plasmodium*); by placenta (blood parasites); and by invasion of ovary or egg *(Babesia)*.

Protozoan parasites may differ little in structure from free-living forms, but they have physiologic adapta-

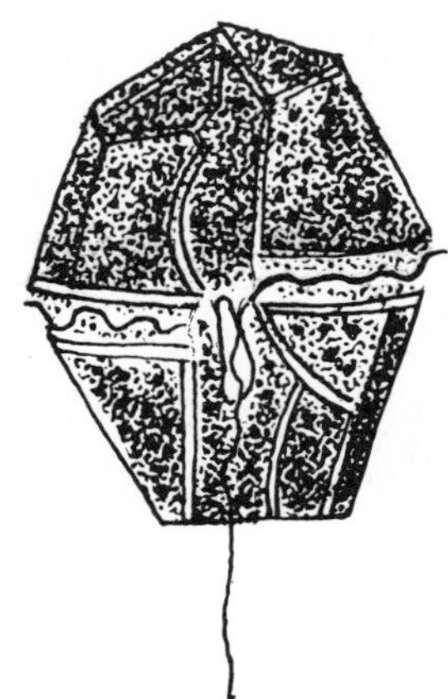

FIG. 8-2 *Gonyaulax polyhedra,* one of the dinoflagellates responsible for "red tides" along the coast of southern California. This organism produces a toxic alkaloidal substance that is very destructive to fish. Similar dinoflagellates cause frequent "red tides" along the Florida and New England coasts. Shellfish that feed on these organisms may be a source of food poisoning in humans.

tions. Some protozoan parasites are adapted to a wide range of hosts and parasitize many species; others are restricted to a few species.

Contamination of water

Many protozoan parasites are transmitted in water and may give rise to serious diseases such as amebic dysentery. Protozoans also affect the taste and odor of water and often determine its drinking qualities. Water derived from surface sources and stored in large reservoirs is most likely to suffer. The chief cause of fishy, aromatic, or other odors is the production of aromatic oils by the disintegration of microscopic organisms. Certain flagellates, such as *Dinobryon* and *Uroglena,* impart to water a pronounced fishy odor, not unlike cod-liver oil. *Synura,* another flagellate, gives a bitter and spicy taste to water. Of course other factors, such as algae and decomposing organic matter, also cause bad odors and taste.

Red tides and toxins

When physical conditions are favorable, a species of phytomastigophoran or an alga may reproduce in great profusion, producing a "bloom." Blooms of certain dinoflagellates in coastal areas sometimes produce "red tides," so-called because the organisms are present in such abundance that they color the water. Often the color is not red, but yellow or brown. Some people believe that a dinoflagellate bloom is the explanation of the incident in which Moses turned the Nile to "blood." Substances produced by the dinoflagellates may be highly toxic to fish and other sea life, and they may be fatal to humans or other animals drinking the waters or feeding on fish or shellfish. Red tides in Florida have been associated with *Gymnodinium brevis* and in California and in New England with species of *Gonyaulax* (Fig. 8-2). There may be little overt effect on sea life in some instances, but the clams and mussels feed on the dinoflagellates, concentrating the toxin, and become quite poisonous to humans. Initial symptoms of the neurotoxin are numbness of the extremities and loss of coordination, and at the beginnings of red tide episodes, unsuspecting physicians have been known to diagnose alcoholic drunkenness—with very unfortunate consequences.

Red tides may occur every 2 to 3 years or more irregularly, but the economic loss to the fish and shellfish industry is very large.

Locomotion

Pseudopodia

Pseudopodia are the chief means of locomotion of members of the Sarcodina, and they can be formed by some flagellates and by various other cells in many invertebrates and vertebrates. Typically, they may form at any point on the cell surface and extend in any direction, but to be effective in locomotion the cell membrane must be in contact with a substratum. The characteristic movement with pseudopodia is called **ameboid movement.**

In many amebas, a peripheral, nongranular layer of cytoplasm, the **ectoplasm,** can be distinguished, which encloses the central, granular **endoplasm.** The ectoplasm is in the gel state of a colloid, and the endoplasm is in the sol state (p. 25). When a pseudopodium is beginning to form, an ectoplasmic projection called the **hyaline cap** appears, and a stream of endoplasm begins to flow into it (Fig. 8-3). At the advancing end of the forming pseudopodium, the endoplasmic stream fountains out toward the periphery. Here it converts to a gel stage, thus forming a stiff, and continually lengthening, ectoplasmic tube around the inner core of fluid endoplasm. At some point the tube becomes anchored, through its plasmalemma, to the substratum, and the animal is drawn forward. At the tem-

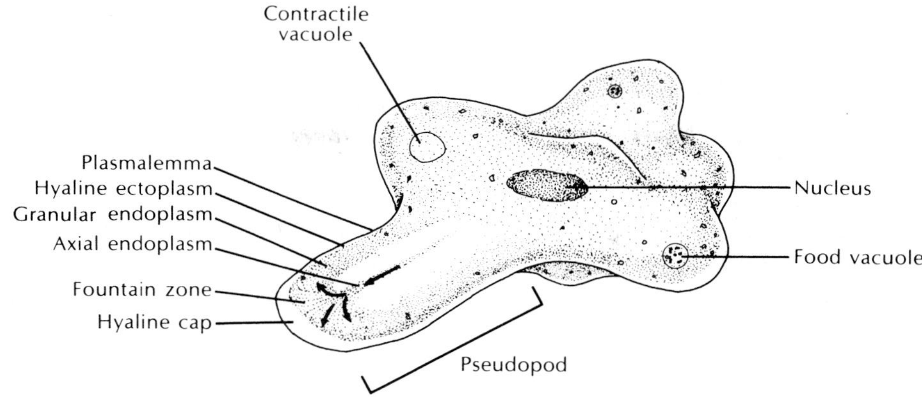

FIG. 8-3 Structure of *Amoeba* in active locomotion. Arrows indicate direction of streaming plasmasol. First sign of new pseudopodium is thickening of ectoplasm to form clear hyaline cap. Into this hyaline region flow granules from fluid endoplasm (plasmasol), forming a type of tube with walls of plasmagel and core of plasmasol. As plasmasol flows forward it fountains out and is converted into plasmagel. Substratum is necessary for ameboid movement, but only tips of pseudopodia touch it. (See text for explanation of ameboid movement.)

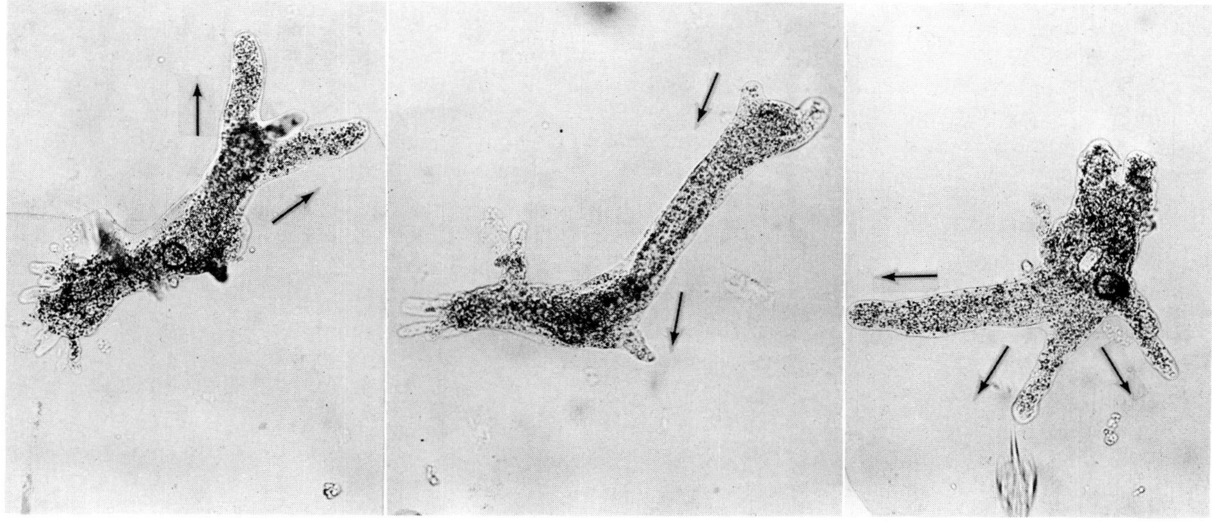

FIG. 8-4 Ameboid movement. Series photographed at intervals of about half a minute. *Right,* Pseudopodium is extending toward escaping rotifer. (Photographs by F. M. Hickman.)

porary "tail end" of the pseudopodium the ectoplasmic tube is being converted back to streaming protoplasm, thus replenishing the forward flow. In essence, then, the ameba creates a tube, anchors it, and flows through it as it moves forward. At any time the direction of flow can be reversed, its length shortened, and new pseudopodia started (Fig. 8-4).

Several theories have been advanced to explain the mechanism of ameboid movement, two of which depend on the active *contraction of molecules* of cellular proteins, either in the fountain zone or in the "tail end." If this contraction occurs in the tail region, the animal is being pushed forward (rear-contraction theory); if it is in the fountain zone, the animal is being pulled along (front-contraction theory). More recent research has brought both of these theories into ques-

FIG. 8-5 Cross section of flagella. Each flagellum has nine peripheral and two central fibrils, each made up of two microfibrils enclosed in a sheath. Cilia have the same construction. (Electron micrograph, courtesy I. R. Gibbons, Harvard University.)

tion. Ameboid movement, like the movement of flagella and cilia, and even that of muscle cells, may be effected by tiny filaments moving past each other. Presumably, fibrils in the gel phase would pull particles or fibrils in the solated endoplasm toward the direction of pseudopod extension. Support of this theory comes from experiments in which *extracts* of amebas are treated with ATP and show gelation, contractility, streaming of particles, and extrusion of water. Oriented fibrils have been shown in the gel phase by electron microscopy, and, recently, myosin filaments have been identified in *Amoeba proteus*.

Though well exemplified and most easily studied in the ameba, ameboid movement is of very wide occurrence and of extreme importance in multicellular animals. Much of the defense against disease in the human body depends on ameboid white blood cells, and ameboid cells in many other animals, vertebrate and invertebrate, play similar roles.

Cilia and flagella

Cilia and **flagella** comprise the other major locomotory mechanisms of protozoans, and they are no less important than pseudopodia for multicellular animals. Not only do many small metazoans use cilia for locomotion, the cilia of many create water currents for their feeding and respiration. Ciliary movement is vital to many species in such functions as handling food, reproduction, excretion, osmoregulation, and so on.

It is now known that the distinction between cilia and flagella is purely an arbitrary one based on the fact that cilia are generally shorter and more numerous than flagella. Electron microscopy has shown that they are structurally the same. Each contains nine pairs of longitudinal fibrils arranged in a circle around a central pair (Fig. 8-5), and this is true for all flagella and cilia in the animal kingdom, with a few notable exceptions. This ''9 + 2'' tube of fibrils in the flagellum or cilium is its **axoneme;** the axoneme is covered by a membrane continuous with the cell membrane covering the rest of the organism. At about the point where the axoneme enters the cell proper, the central pair of fibrils ends at a small plate within the circle of nine pairs (Fig. 8-6). Also at about that point, another fibril joins each of the nine pairs, so that these form a short tube extending from the base of the flagellum into the cell and consisting of nine *triplets* of fibrils. The short tube of nine triplets is the **kinetosome** and is exactly the same in structure as the **centriole** (see p. 64). The centrioles of some flagellates may give rise to the kinetosomes, or the kinetosomes may function as centrioles. All typical flagella and cilia have a kinetosome at their base, regardless of whether they are borne by a protozoan or metazoan cell. The kinetosomes of protozoa have older, traditional names **(blepharoplast, basal body, basal granule)** that are still in common usage.

The current theory to account for ciliary and flagellar movement is the **sliding-microtubule hypothesis.** The movement is powered by the release of chemical bond energy in ATP (p. 84). Two little arms are visible in electron micrographs on each of the pairs of peripheral tubules in the axoneme (Figs. 8-5 and 8-6), and these bear the enzyme to cleave the ATP, adenosine triphosphatase (ATPase). When the bond energy in ATP is released, the arms ''walk along'' one of the filaments in the adjacent pair, causing it to slide relative to the other filament in the pair. Shear resistance, causing the axoneme to bend when the filaments slide past each other, is provided by ''spokes'' from each doublet to the central pair of fibrils. These are also visible in electron micrographs.

Reproduction and life cycles

Reproduction

Reproduction in most protozoa is primarily by asexual cell division. It is comparable in many respects to cell division in the multicellular animals. The protozoa, however, often have certain structural specializa-

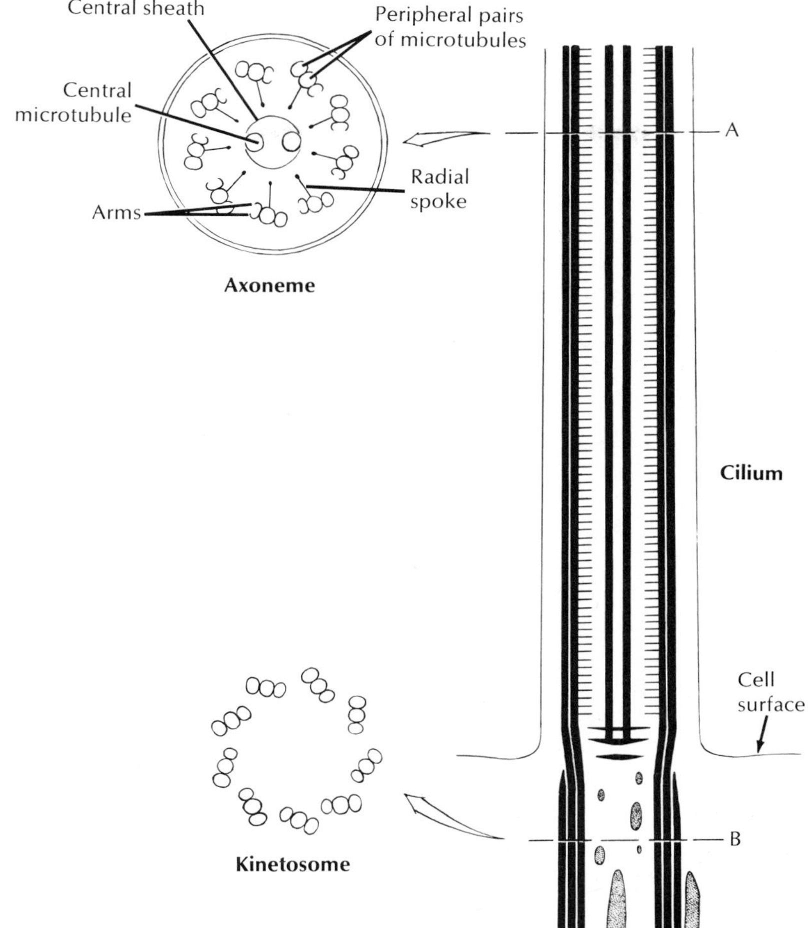

FIG. 8-6 Structure of a cilium showing at *A*, a section through the axoneme within the cell membrane, and at *B*, a section through the kinetosome. The nine pairs of filaments plus the central pair make up the axoneme. The central pair ends at about the level of the cell surface in a basal plate (axosome). The peripheral filaments continue inward for a short distance to comprise two of each of the triplets in the kinetosome.

tions (organelles)—such as flagella, cilia, contractile vacuole, and gullet—that may be divided equally or unequally between the two daughter cells, so that a certain amount of differentiation or regeneration may be necessary to make the new animal complete. Some of these organelles are self-reproducing, but others are lost by resorption (dedifferentiation), then differentiated anew in each of the daughter organisms.

Three general modes of cell division are recognized: binary fission, multiple fission, and budding. All of these are basically asexual processes. In many species, however, asexual reproduction is often followed at certain periods by some form of sexual reproduction that may or may not be necessary for the continued existence of the organisms.

Binary fission. Binary fission, the most common type of reproduction among protozoans, involves the division of the organism—both its nucleus and cytoplasm—into two essentially equal daughter organisms. Binary fission may be transverse (across ciliary rows, as in most ciliates) or between flagellar groups or rows (as in Mastigophora). The nucleus divides by mito-

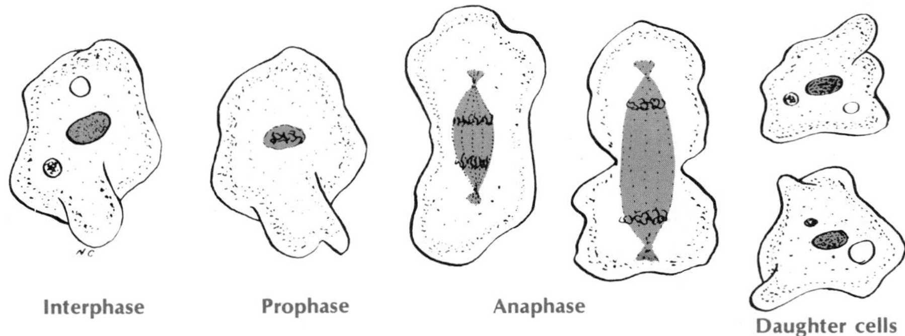

FIG. 8-7 Mitosis in *Amoeba.*

sis (Fig. 8-7), but the nuclear membrane may persist through the division in some forms. In many forms the chromosomes are similar in structure and behavior to metazoan chromosomes; in others the chromosomes are granular and highly atypical.

Budding. Budding is quite similar to binary fission except that the division products are unequal. The daughter cell is usually much smaller than the parent.

Multiple fission (schizogony, sporogony). In multiple fission, the nucleus divides a number of times, followed by the division of the cytoplasm (cytokinesis) to form as many cells as there are nuclei. It occurs in some species of flagellates and several Sarcodina and is characteristic of the Sporozoa. Sporogony is distinguished from schizogony in that sporogony takes place at some time following fertilization of gametes (zygote formation) and often involves spore formation (p. 153).

Protozoan colonies. Protozoan colonies are formed when the daughter zooids remain associated instead of moving apart and living a separate existence (Fig. 8-8). Protozoan colonies vary from individuals embedded together in a gelatinous substance to those that have protoplasmic connections among them. The arrangement of the individuals results in certain types of colonies, such as **linear** (daughter cells attached endwise), **spherical** (grouped in a ball shape), **discoid** (platelike arrangement), and **arboroid** (treelike branches). Simple colonies may have only a few zooids *(Pandorina)* (Fig. 8-8), or they may have thousands of zooids *(Volvox)* (Fig. 8-18). Usually the individuals of a colony are structurally and physiologically the same, although there may be some division of labor, such as differentiation of reproductive and somatic zooids. Division of labor, however, may be carried so far that it is difficult to distinguish between a protozoan colony and a metazoan individual.

Sexual phenomena. Sexual reproduction involves that special kind of nuclear division called **meiosis** (pp. 805-809 and Chapter 35), in which the number of chromosomes is reduced by half (diploid number to haploid number). The union of gamete nuclei (zygote formation) restores the diploid condition. In multicellular animals, reduction division occurs during the formation of gametes and is therefore **gametic.** This kind of meiosis is found in protozoans among some Heliozoa and flagellates and in the Opalinata and the ciliates. If meiosis occurs in the first divisions after zygote formation, and all intervening stages from that point to gamete formation are haploid, the meiosis is called **zygotic.** Zygotic meiosis is found in some flagellates and all Sporozoa. The gametes may be similar in appearance **(isogametes)** or unlike **(anisogametes).**

The fertilization of one individual gamete by another is called **syngamy,** but some sexual phenomena in protozoans do not involve that process. Examples are **autogamy,** in which gametic nuclei arise by meiosis and fuse to form a zygote within the same organism that produced them; **parthenogenesis,** the development of an organism from a gamete without fertilization; and **conjugation,** in which there is an exchange of gametic nuclei between paired organisms (conjugants). Some of these processes are described further under the discussion of the paramecium.

Life cycles

Many protozoans have very complex life cycles; others have simple ones. A simple life cycle may consist of an active phase and a cyst. In some cases the cyst may be lacking. *Amoeba* has a relatively simple

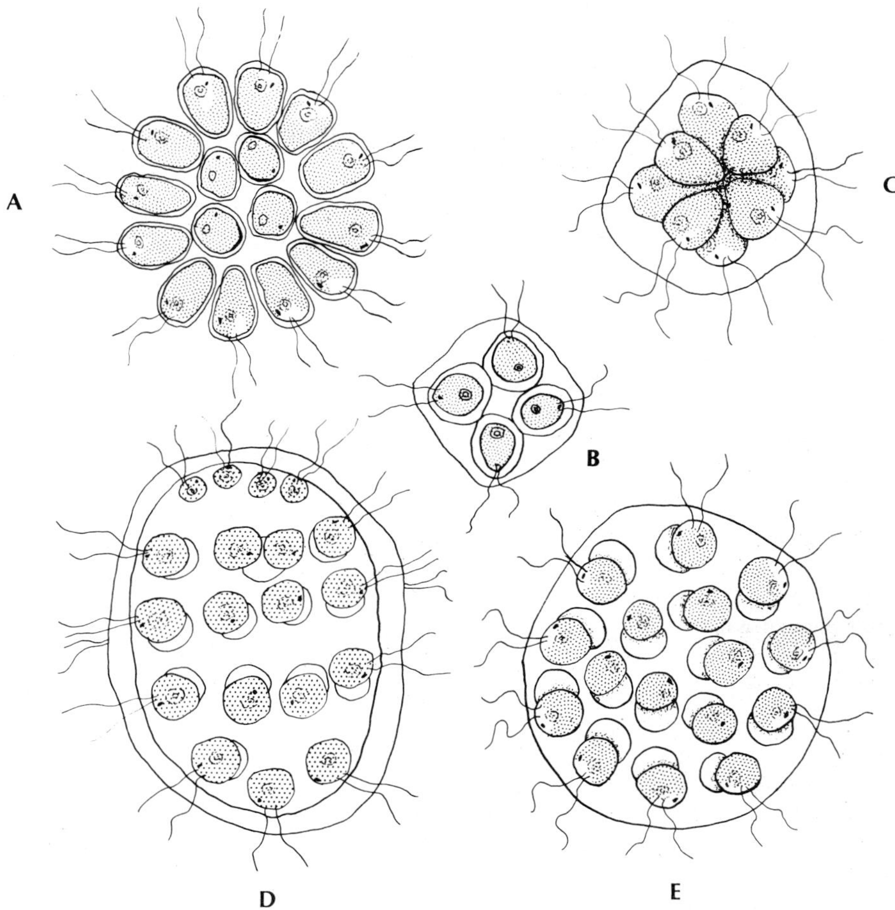

FIG. 8-8 Some colonial protozoans. Such colonies may have been the forerunners of the metazoans. **A,** *Gonium pectorale,* usually of 16 zooids. **B,** *G. sociale,* four zooids, each of which may give rise to a daughter colony by cell division or may serve as an isogamete. **C,** *Pandorina morum,* 16 zooids, each of which divides four times to form new colony. **D,** *Pleodorina illinoisensis,* 32 zooids, of which the four anterior are sterile and the others are capable of both sexual and asexual reproduction. **E,** *Eudorina elegans,* 32 zooids, each of which forms a new colony.

life cycle. The more complex life cycles include two or more stages in the active phase and a reproductive phase that may include sexual as well as asexual phenomena. For example, some protozoans have both a ciliated and a nonciliated stage; others have ameboid and flagellate stages; and still others, free-swimming and sessile stages. The most complex protozoan life cycles are found in the parasitic class Sporozoa—a good example of which is *Plasmodium* (Fig. 8-20), the malarial parasite (p. 154).

Encystment is common among protozoans, helping them withstand unfavorable conditions. There is usually a complex series of events that occurs when the organism encysts. It becomes quiescent, and many organelles, such as cilia, flagella, and contractile vacuole, may dedifferentiate and disappear. A cyst wall is secreted over the surface so that the animal can withstand desiccation, temperature changes, and other harsh conditions. Reproductive cycles, such as budding, fission, and syngamy, may also occur in the encysted condition. The cysts of some species may be viable for many years.

REPRESENTATIVE TYPES

A fairly representative member of each large group of protozoans will be described to give the student a basis for comparison of the groups. Forms such as *Amoeba* and *Paramecium,* however, although large and easy to obtain for study, are not wholly representative because their life histories are somewhat simple compared with other members of their respective groups.

Subphylum Sarcomastigophora

Sarcomastigophora includes both those protozoa that move by flagella (Mastigophora) and those that move by means of pseudopodia (Sarcodina). As a rule the nuclei are all of one type (monomorphic) although there may be more than one nucleus in an individual.

Superclass Sarcodina

Amoeba proteus

HABITAT. *Amoeba proteus* is widely distributed. They live in slow streams and ponds of clear water, often in shallow water on the underside of lily pads and other aquatic vegetation, or on the sides of dams, in watering troughs, and in the sides of ledges where the water runs slowly from a brook or spring. They are rarely found free in water, for they require a substratum on which to glide.

STRUCTURE. The shape of the ameba is irregular and continuously changing because of its power to thrust out **pseudopodia,** or false feet, at any point on its body (Fig. 8-4). It is colorless and about 250 to 600 μm in its greatest diameter. Sometimes its shape is almost spherical—when all its pseudopodia are withdrawn. The thin delicate outer membrane is called the **plasmalemma** (Fig. 8-3). Just beneath this is a nongranular layer, the **ectoplasm,** which encloses the granular **endoplasm.** The ectoplasm is in the gel state of a colloid, and the endoplasm is in the sol state (p. 25).

Most organelles are found within the endoplasm. The granular, disc-shaped **nucleus** is found there. Another organelle is the water expulsion vacuole **(contractile vacuole)**—a bubblelike body that grows to a maximum size, then contracts to expel its fluid contents to the outside. Scattered through the endoplasm are **food vacuoles,** which enclose food particles. There are also other vacuoles, **crystals,** and **granules** of various shapes and forms.

NUTRITION. The ameba lives on algae, protozoa, rotifers, and even other amebas. It shows some selection in its food, for it will not ingest everything that comes its way. Food may be taken in at any part of the body surface. When the ameba engulfs food, it thrusts out pseudopodia to enclose the food particle completely (phagocytosis) (Fig. 8-9). Along with the food, some water in which the food is suspended may also be taken in. These food vacuoles are carried around by the streaming movements of the endoplasm. Lysosomes with enzymes fuse with each food vacuole, and digestion proceeds within the vacuole. As digestion proceeds, the vacuoles decrease in size because of loss of water and the passage of the digested material into the surrounding cytoplasm. Finally, indigestible material is eliminated by passing out through the plasmalemma as the animal flows away.

The ameba is able to live for many days without food but decreases in volume during this process. The time necessary for the digestion of a food vacuole varies with the kind of food, but it is usually around 15 to 30 hours.

OSMOREGULATION AND METABOLISM. The ameba needs and utilizes energy just as multicellular animals do. It gets this energy by oxidation, which results in waste products such as carbon dioxide, water, and ammonia. Some of these waste substances are eliminated through the body surface, but some are discharged through the contractile vacuole, which also gets rid of excess water that the ameba is continually taking in. It is thus responsible for regulating the osmotic pressure of the body. The ameba has solutes in its protoplasm that make it hypertonic to the surrounding fresh water. Water will therefore enter the ameba by osmosis through its plasmalemma. It is interesting to note that marine amebas do not have contractile vacuoles because they are immersed in isotonic sea water (but when placed in fresh water they will form them).

As in other protozoa, exchange of respiratory gases occurs directly through the body surface by diffusion. Oxygen is dissolved in the water, which is in contact with every part of the cell membrane so that the gas is easily accessible and diffuses into the ameba.

REPRODUCTION. When the ameba reaches full size, it divides into two animals by the process of **binary fission.** Typical mitosis (Fig. 8-7) occurs, taking about 30 minutes. During the process of division the shape of the ameba is spherical, with a number of small pseudopodia. The nuclear membrane disappears during the metaphase, and the body elongates and separates by

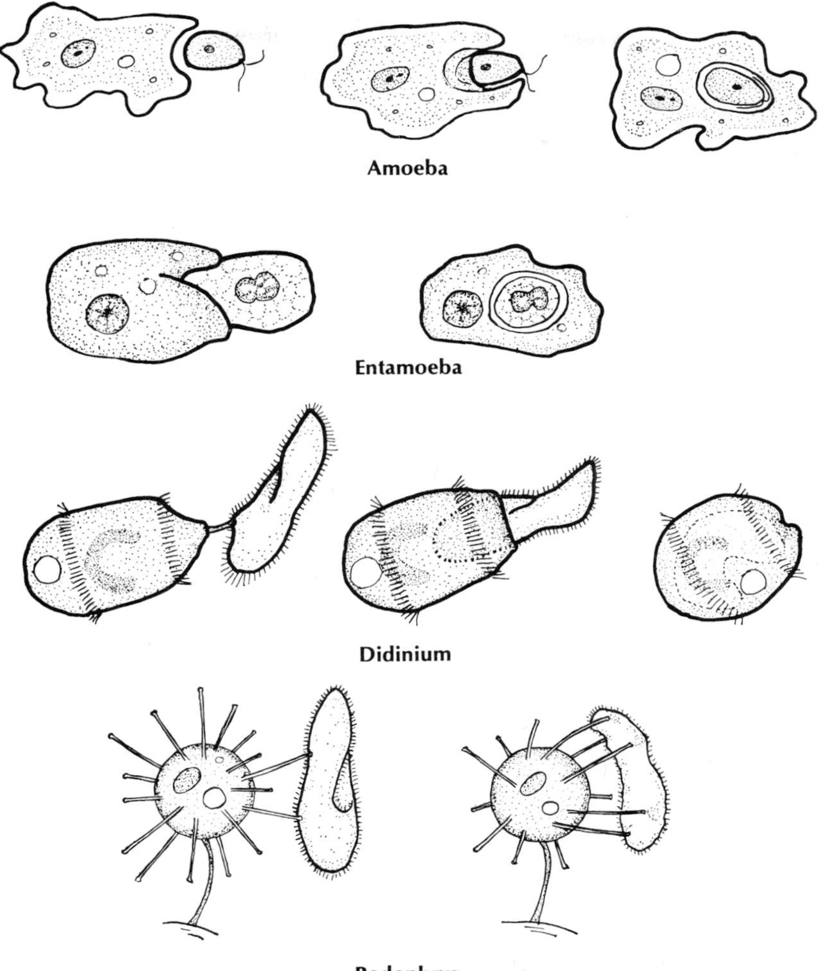

FIG. 8-9 Some typical methods of ingestion among protozoans. *Amoeba* ingests a small flagellate. *Entamoeba,* a parasite, engulfs a leukocyte. *Didinium,* a holotrich, eats only paramecia. It pierces prey before swallowing it whole. Suctorians such as *Podophrya* have protoplasmic tentacles with knobbed ends through which protoplasm from the prey is ingested.

fission into two daughter cells. Under ordinary conditions the ameba attains a size for division about every 3 days.

Multiple fission and budding have been reported to occur in the ameba, but binary fission seems to be the only regular method employed.

BEHAVIOR. The ameba reacts to physical and chemical stimuli as do other animals. Its reactions center around food-gathering, locomotion, changes in shape, avoidance of unfavorable environments, and other stimuli. Its responses to these different forms of stimuli vary. In a positive reaction the ameba goes toward the stimulus; in a negative reaction it moves away. If touched with a needle it will draw back and move away, but when floating it will respond in a positive way to a solid object. It moves away from a strong light and may change its direction a number of times to avoid it, but it may react positively to a weak light. The ameba's rate of locomotion is lessened by colder temperatures and may cease entirely near the freezing

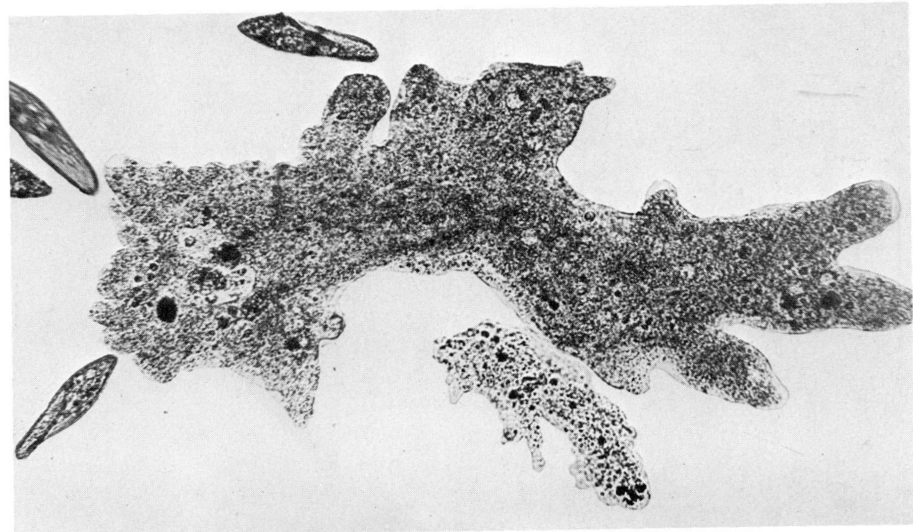

FIG. 8-10 Comparison in size of *Pelomyxa* (larger) and *Amoeba*. The former may attain length of 5 mm. Several paramecia also shown. (Courtesy Carolina Biological Supply Co., Burlington, N.C.)

point. Its rate increases up to 32° C (90° F), but it ceases to move at temperatures higher than this.

Its response to chemicals varies with the nature of the chemical. Although indifferent to most normal constituents in its medium, the ameba will react positively toward substances of a food character.

Other members of Sarcodina

RHIZOPODEA. The amebas belong to a class of sarcodines called Rhizopodea, all of which have pseudopodia similar to those of *Amoeba proteus*. There are many species of *Amoeba;* for example, *A. verrucosa* has short pseudopodia; *Pelomyxa carolinensis (Chaos chaos)* (Fig. 8-10) is several times as large as *A. proteus;* and *A. radiosa* has many slender pseudopodia.

There are many entozoic amebas, most of which live in the intestines of humans or other animals. Two common genera are *Endamoeba* and *Entamoeba*. *Endamoeba blattae* is the endocommensal in the intestine of cockroaches, and related species are found in termites. *Entamoeba histolytica* is the most important rhizopod parasite of humans. It lives in the large intestine and on occasion can invade the intestinal wall by secreting enzymes that attack the intestinal lining. If this occurs, a serious amebic dysentery may result, and the outcome is sometimes fatal. The organisms may be carried by the blood to the liver and other organs and cause abscesses there. Many infected persons show few or no symptoms but are carriers, passing cysts in their feces. Infection is spread by contaminated water or food containing the cysts.

Other species of *Entamoeba* found in humans are *E. coli,* in the intestine, and *E. gingivalis,* in the mouth. Neither of these species is known to cause disease.

Not all rhizopods are "naked" as are the amebas. Some have their delicate plasma membrane covered with a protective **test** or shell. *Arcella* and *Diffugia* (Fig. 8-11) are common sarcodines of the order Arcellinida (Testacida), which have a test of secreted siliceous or chitinoid material that may be reinforced with grains of sand. They move by means of pseudopodia that project from openings in the shell.

The **foraminiferans** (order Foraminiferida) are an ancient group of shelled rhizopods found in all oceans, with a few in fresh and brackish water. They are mostly bottom-living, but a few live in open water. Their tests are of numerous types (Figs. 8-11 and 8-12). Most tests are many-chambered and are made of calcium carbonate, although silica, silt, and other foreign materials are sometimes used. Slender pseudopodia extend through openings in the test, then branch and run together to form a protoplasmic net in which they ensnare their prey. Here the captured prey is digested, and the digested products are carried into the interior by the flowing protoplasm. The life cycles of foraminiferans are complex, for they have multiple division

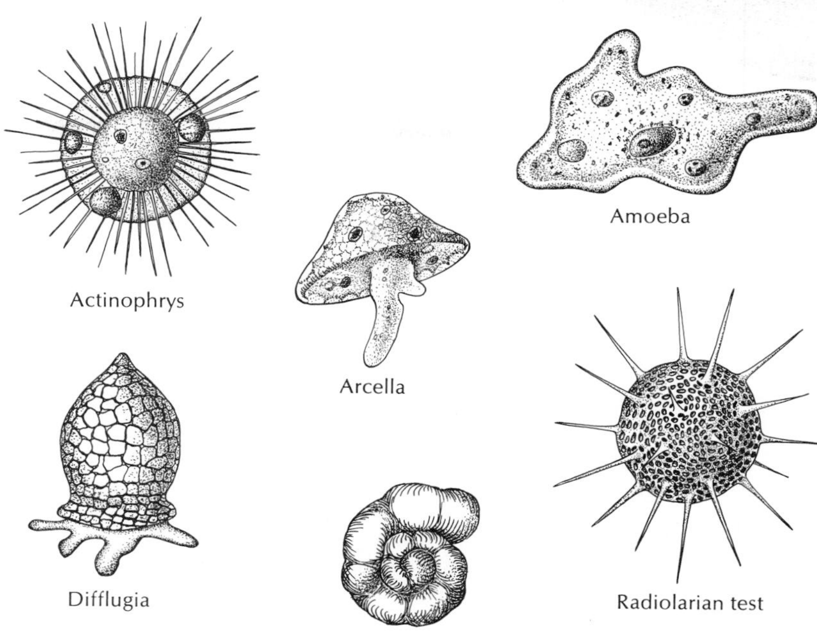

FIG. 8-11 A group of Sarcodina. *Amoeba, Arcella, Difflugia,* and foraminiferans are rhizopods. *Actinophrys* and radiolarians are actinopods.

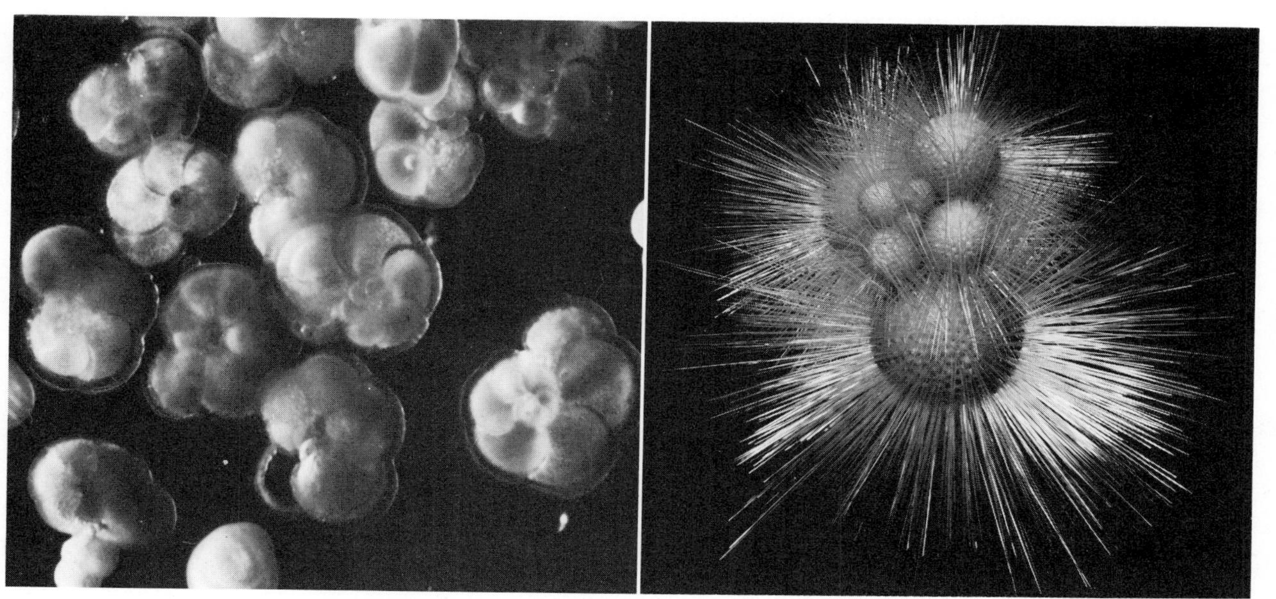

FIG. 8-12 **A,** Living foraminiferans. **B,** Glass model of *Globigerina bulloides.* Foraminiferans are ameboid marine protozoans that secrete a calcareous, many-chambered shell in which to live and then extrude protoplasm through pores to form a layer over the outside. The animal begins with one chamber, and as it grows, it secretes a succession of new and larger chambers, continuing this process throughout life. Foraminiferans are planktonic animals, and when they die, their shells are added to the ooze on the ocean's bottom. (**A** courtesy Roman Vishniac. **B** courtesy American Museum of Natural History, New York.)

FIG. 8-13 Types of radiolarian shells. In his study of these beautiful forms collected on the famous *Challenger* expedition, Haeckel worked out our present concepts of symmetry. (Courtesy General Biological Supply House, Inc., Chicago.)

and alternation of haploid and diploid generations. The slime molds belong to the rhizopod group.

ACTINOPODEA. In the class Actinopodea of sarcodines, the pseudopodia are slender and usually radiate from the central test. Some forms have the slender pseudopods stiffened by an axial rod running down their center. These protozoans are beautiful little animals. Included in this group are the heliozoans, which are mostly freshwater forms. They include *Actinosphaerium,* which is about a millimeter in diameter and can be seen with the naked eye, and *Actinophrys* (Fig. 8-11), only 50 μm in diameter; neither has a test. *Clathrulina* secretes a latticed test.

The radiolarians are the oldest known group of animals. They are all marine and nearly all pelagic (live in open water). Most of them are planktonic in shallow water, though some live in deep water. Their highly specialized siliceous skeletons are intricate in form and of great beauty (Fig. 8-13). The body is divided by a central capsule that separates inner and outer zones of cytoplasm. The central capsule, which may be spherical, ovoid, or branched, is perforated to allow cytoplasmic continuity. The skeleton is made of silica or strontium sulfate and usually has a radial arrangement of spines that extend through the capsule from the center of the body. At the surface a shell may be fused with the spines. Around the capsule is a frothy mass of cytoplasm from which stiff pseudopodia arise (p. 138). These are sticky for catching the prey that are carried by the streaming protoplasm to the central capsule to be digested. It is here that both the rear-contraction and the front-contraction theories of pseudopodial movement (p. 139) meet difficulties: the ectoplasm on one side of the axial rod moves outward, or toward the tip, while on the other side it moves inward, or toward the test.

Radiolarians may have one or many nuclei. Their life history is not completely known, but binary fission, budding, and sporulation have been observed in them.

Role in building earth deposits. Two orders, Foraminifera and Radiolaria, which have existed since Precambrian times, have left excellent fossil records, for their hard shells have been preserved unaltered. Many of the extinct species are identical to present ones. They were especially abundant during the Cretaceous and Tertiary periods. Some of them were among the largest of protozoans, measuring up to 100 mm or more in diameter.

For untold millions of years the tests of dead Foraminifera have been sinking to the bottom of the ocean, building up a characteristic ooze rich in lime and silica. Most of this ooze is made up of the shells of the genus *Globigerina* (Fig. 8-12, *B*). About one-third of all the sea bottom is covered with *Globigerina* ooze. This ooze is especially abundant in the Atlantic Ocean.

The Radiolaria (Figs. 8-11 and 8-13), with their less soluble siliceous shells, are usually found at greater depths (15,000 to 20,000 feet), mainly in the Pacific and Indian Oceans. Radiolarian ooze probably covers about 2 to 3 million square miles. Under certain conditions, radiolarian ooze forms rocks (chert). Many fossil Radiolaria are found in the Tertiary rocks of California.

The thickness of these deep-sea sediments has been estimated at from 700 to 4,000 m. Although the average rate of sedimentation must vary greatly, it is always very slow. *Globigerina* ooze has probably increased 1 to 12.5 mm in 1,000 years. As many as 50,000 shells of Foraminifera may be found in a single gram of sediment, which gives some idea of the magnitude of numbers of these microorganisms and the length of time it has taken them to form the sediment carpet on the ocean floor.

Of equal interest and of greater practical importance are the limestone and chalk deposits that were laid down when sea covered what is now land. Later, through a rise in the ocean floor and other geologic changes, this sedimentary rock emerged as dry land. The chalk deposits of many areas of England, including the White Cliffs of Dover, were laid down by the accumulation of these small microorganisms. The great pyramids of Egypt were made from stone quarried from limestone beds that were formed by a very large foraminiferan that flourished during the early Tertiary period.

Since fossil foraminiferans and radiolarians can be brought up in well drillings, their identification is often important to oil geologists for correlation of rock strata.

Superclass Mastigophora, the flagellated protozoans

The superclass Mastigophora, commonly called the flagellates, includes protozoa that typically have one or more flagella, and that organelle is normally their means of locomotion. What are probably the most

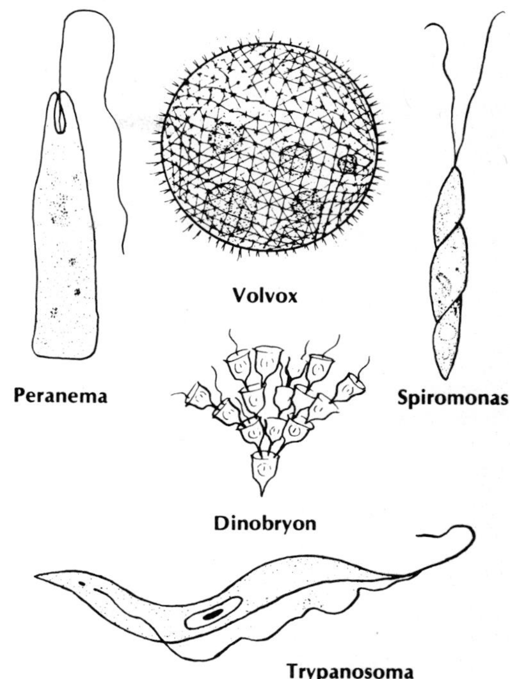

FIG. 8-14 Some common types of Mastigophora, both solitary and colonial.

primitive protozoa are found in this superclass. The group is divided into the phytoflagellates (class Phytomastigophorea), which usually have chlorophyll and are thus plantlike, and the zooflagellates (class Zoomastigophorea), which do not have chlorophyll but are either holozoic or saprozoic and thus are animal-like.

The **phytoflagellates** usually have one or two flagella (sometimes four) and chromatophores (also called chromoplasts or chloroplasts), which contain the pigments used in photosynthesis. They are mostly free-living and include such familiar forms as *Euglena, Chlamydomonas, Peranema* (Fig. 8-14), *Volvox,* and the dinoflagellates that cause the red tides. *Peranema* is related to *Euglena* but is a colorless phytoflagellate that is holozoic in nutrition. *Chilomonas* is another common form that is an important food item for amebas. *Noctiluca,* a marine species, is luminescent and produces a striking greenish light. *Ceratium,* which is found in both fresh water and salt water, has a body covering of cellulose plates and horns. Some of these flagellates are colonial, living in groups of individuals **(zooids),** and the number of zooids per colony are characteristic for the species (Fig. 8-8). *Volvox* has

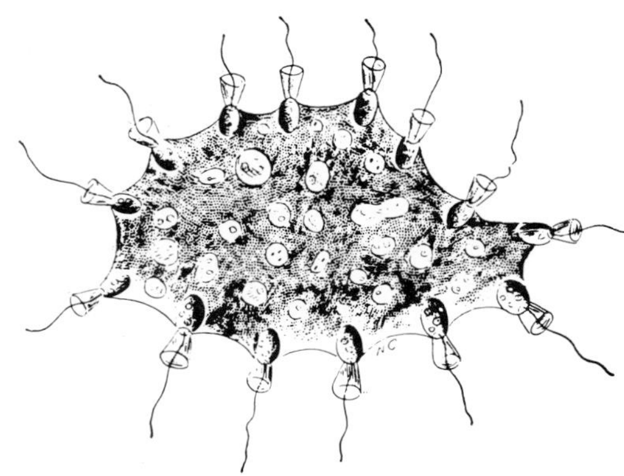

FIG. 8-15 *Proterospongia,* a colonial choanoflagellate. In a gelatinous mass, collared zooids are embedded on the outside and collarless ameboid zooids on the inside. Collared cells resemble choanocytes of sponges. Only choanoflagellates and sponges have these peculiar cells.

thousands of zooids per colony (Figs. 8-14 and 8-18).

The **zooflagellates** are all colorless, lacking chromatophores, and have holozoic or saprozoic nutrition. Many are parasitic.

Some protozoans have pseudopodia as well as flagella. One example is *Proterospongia,* which grows as a colony of collared zooids that has been suggested as a forerunner of the sponges, in which collared cells are typical (Fig. 8-15).

Some of the most important protozoan parasites are zooflagellates. Many of them belong to the genus *Trypanosoma* (Fig. 8-14) and live in the blood of fish, amphibians, reptiles, birds, and mammals. Some are nonpathogenic, but others produce severe diseases in humans and domestic animals. *T. brucei gambiense* and *T. brucei rhodesiense* cause African sleeping sickness in humans, and *T. brucei brucei* causes a related disease in domestic animals. They are transmitted by the tsetse fly *(Glossina). T. b. rhodesiense,* the more virulent of the sleeping sickness trypanosomes, and *T. b. brucei* have natural reservoirs (antelope and other wild mammals) that are apparently not harmed by the parasites. *T. cruzi* causes Chagas' disease in humans in Central America and South America. It is transmitted by the bite of the "kissing bug" *(Triatoma).* Three species of *Leishmania* cause severe disease in humans. One is a visceral disease, especially affecting the liver and spleen; one causes lesions in the mucous membranes of the nose and throat; and the least serious causes a type of skin lesion. They are transmitted by sandflies. Visceral leishmaniasis and cutaneous leishmaniasis are common in parts of Africa and Asia, and the mucocutaneous form is found in Central America and South America.

Several species of *Trichomonas* (Fig. 8-1) are symbiotic. *T. hominis* is found in the cecum and colon of humans and apparently causes no disease. *T. vaginalis* inhabits the urogenital tract of humans, is transmitted venereally, and may cause vaginitis. Other species of *Trichomonas* are widely distributed through all classes of vertebrates and many invertebrates. *Giardia lamblia* often causes no disease in the intestine of humans but may sometimes produce severe diarrhea. It is transmitted through fecal contamination.

Euglena viridis, a phytoflagellate

HABITAT. The normal habitat of *Euglena viridis* (order Euglenida) (Fig. 8-16) is freshwater streams and ponds where there is considerable vegetation. They are sometimes so numerous as to give a distinctly greenish color to the water. Although light is necessary for their photosynthesis, they are often found at various depths below the surface of water, for they are fairly active forms.

STRUCTURE. The spindle-shaped body of *E. viridis* is about 60 μm (0.06 mm) long, but some species are smaller and some larger (*E. oxyuris* is 500 μm long). It is covered by a **pellicle** flexible enough to permit movement (Fig. 8-17, *A*). Inside the pellicle is a thin layer of clear **ectoplasm** surrounding the mass of **endoplasm.** From a flask-shaped **reservoir** in the anterior end a **flagellum** extends. There is another, short flagellum that extends only into the reservoir. This shorter flagellum gives the mistaken impression that the long, locomotory flagellum has a bifurcated root. Each of the flagella terminates in a **kinetosome (basal body, blepharoplast),** located in the cytoplasm at the base of the reservoir.

A large **contractile vacuole** which is formed by fusion of smaller vacuoles, empties wastes and excess water into the reservoir, the anterior opening of which is an exit.

Near the reservoir is a red **eyespot,** or **stigma** (Fig. 8-16). This is a shallow cup of pigment that allows light from only one direction to strike a light-sensitive receptor located as a swelling near the base of the large flagellum. When the euglena is moving toward the light, the receptor is illuminated; when it changes direction, the shadow of the pigment falls on the recep-

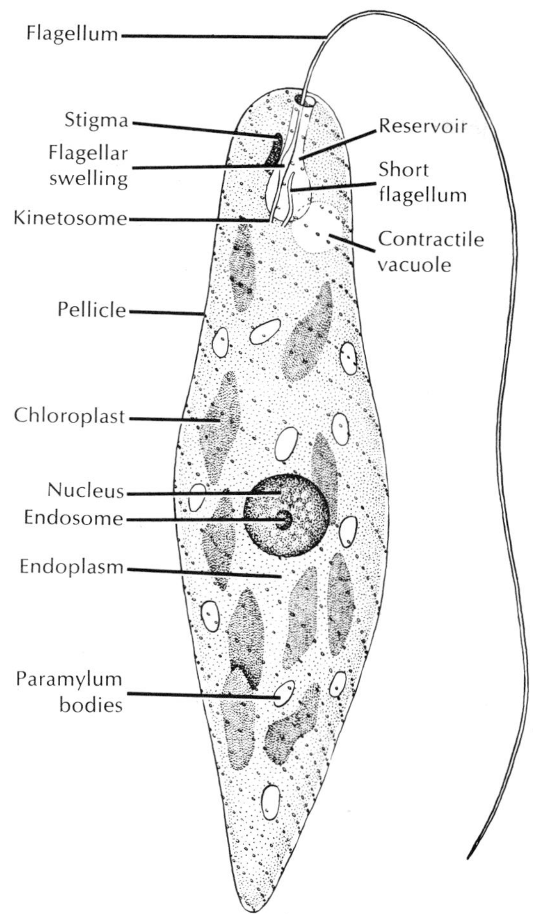

FIG. 8-16 *Euglena.* Features shown are a combination of those visible in living and stained preparations.

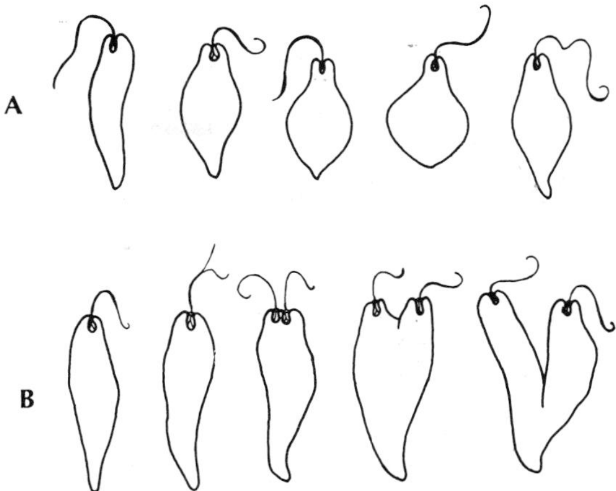

FIG. 8-17 A, Changes of shape in mastigophorans such as *Euglena* are called "euglenoid movement." **B,** Reproduction in *Euglena* occurs by longitudinal fission, beginning at anterior end.

tor. Thus the organism, which depends on sunlight for its photosynthesis, can orient itself toward the light. If given the choice, it will avoid both shady areas and regions of very bright light.

Within the cytoplasm are oval **chromoplasts** (chloroplasts) that bear chlorophyll and give the euglena its greenish color. **Paramylum bodies** of various shapes are masses of a starchlike food storage material.

NUTRITION. The euglena normally derives its food through **autotrophic (holophytic)** nutrition, using solar energy captured in its chromoplasts to reduce carbon dioxide and produce carbohydrates. If kept in the dark, the organism makes use of **saprozoic** nutrition, absorbing nutrients through its body surface, and the chlorophyll in its chromoplasts gradually fades and disappears. When returned to the light, the color returns, but mutants of *Euglena* can be produced that have permanently lost their photosynthetic ability. *Euglena* does not ingest solid food at all, nor does the opening of the reservoir function as a cytostome (cell mouth).

LOCOMOTION. The euglena swims freely by the movement of its flagellum, which moves in a whiplike manner or with rotary motion, with the undulation passing from base to tip. This motion pulls the animal forward in a straight course while the body rotates spirally. Flagellates can travel from a few tenths to 1 mm/per second, according to their size, with the speed increasing with the larger flagellates. Some flagellates use the flagella to push rather than to pull themselves along. *Euglena* can also change its shape by peristalsis-like "euglenoid" movement (Fig. 8-17, *A*).

REPRODUCTION. *Euglena* reproduces through **binary fission.** The nucleus undergoes mitotic division, while the body, beginning at the anterior end, divides lengthwise (Fig. 8-17, *B*). During inactive periods and for protection, the euglena assumes a spherical shape surrounded with a gelatinous covering, thus becoming **encysted.** In this condition it can withstand drought and can become active when it is in water again. Euglenas usually divide while encysted, thus each cyst may contain two or more euglenas.

Volvox globator. The group to which *Volvox* belongs (order Volvocida) includes many freshwater fla-

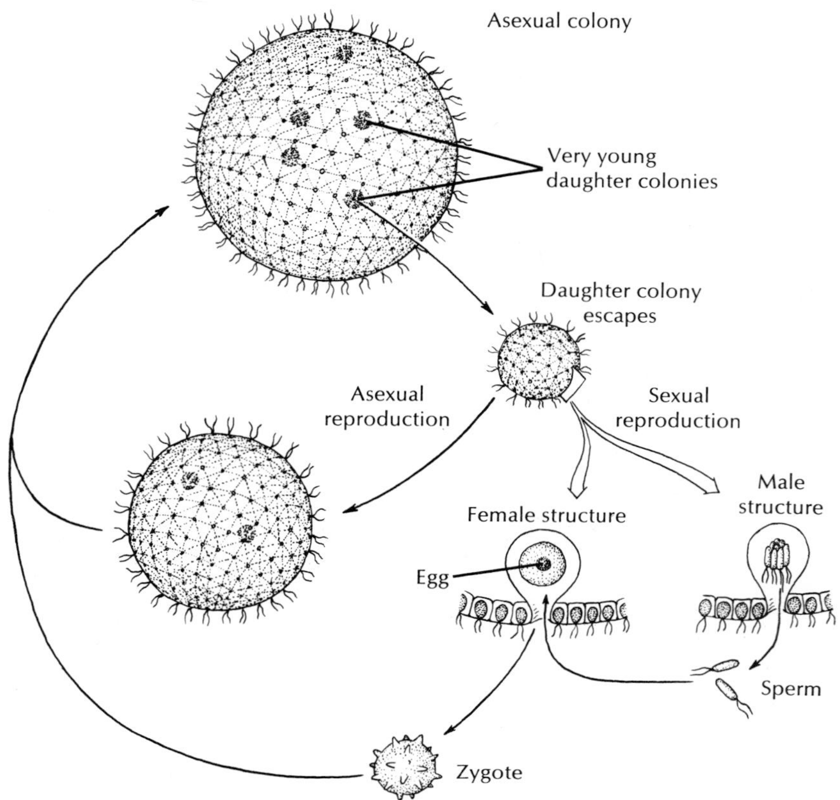

FIG. 8-18 Life cycle of *Volvox*. Asexual reproduction occurs in spring and summer when specialized diploid reproductive cells divide to form young colonies that remain in the mother colony until large enough to escape. Sexual reproduction occurs largely in autumn when haploid sex cells develop. The fertilized ova may encyst and so survive the winter, developing into mature asexual colonies in the spring. In some species the colonies have separate sexes; in others both eggs and sperm are produced in the same colony.

gellates, mostly green, that have a close resemblance to algae. The bodies of most species are enclosed in a cellulose cell wall through which two short flagella project. Most of them are provided with green chromoplasts. Many are colonial forms (Fig. 8-8).

Volvox (Fig. 8-18) is a green hollow sphere that may reach a diameter of 0.5 to 1 mm. It is a colony of many thousands of zooids (up to 50,000) embedded in the gelatinous surface of a jelly ball. Each cell is much like a euglena, with a nucleus, a pair of flagella, a large chromoplast, and a red stigma. Adjacent cells are connected with each other by cytoplasmic strands. At one pole (usually in front as the colony moves), the stigmata are a little larger. Coordinated action of the flagella causes the colony to move by rolling over and over.

Here we have the beginning of division of labor to the extent that most of the zooids are somatic cells concerned with nutrition and locomotion, and a few germ cells located in the posterior half are responsible for reproduction. Reproduction is asexual or sexual. In either case only certain zooids located around the equator or in the posterior half take part.

Asexual reproduction in *Volvox* occurs by the repeated mitotic division of one of the germ cells, to form a hollow sphere of cells, with the flagellated ends of the cells inside. The sphere then invaginates, or turns itself inside out, to form a daughter colony similar to the parent colony. Several daughter colonies are formed inside the parent colony before they escape by rupture of the parent (Fig. 8-18).

In **sexual reproduction** some of the zooids differentiate into **macrogametes** (ova) or **microgametes** (sperm) (Fig. 8-18). The macrogametes are fewer and larger and are loaded with food for nourishment of the young colony. The microgametes, by repeated division, form bundles or balls of small flagellated sperm that, when mature, leave the mother colony and swim about to find a mature ovum. When a sperm fertilizes an egg, the zygote so formed secretes a hard, spiny, protective shell around itself. When released by the breaking up of the parent colony, the zygote remains quiescent during the winter. Within the shell the zygote undergoes repeated division until a small colony is produced that is released in the spring. A number of asexual generations may follow, during the summer, before sexual reproduction occurs again.

Subphylum Apicomplexa

All apicomplexans are endoparasites, and their hosts are found in many animal phyla. The presence of a certain combination of organelles, the **apical complex,** distinguishes this subphylum (Fig. 8-19). The apical complex is usually present only in certain developmental stages of the organisms; for example, **merozoites** and **sporozoites** (Fig. 8-20). Some of the structures, especially the **rhoptries** and **micronemes,** apparently are of aid in penetration of the host's cells or tissues.

Locomotor organelles are not as obvious in this group as they are in other protozoa. Pseudopodia occur in some intracellular stages, and gametes of some species are flagellated. Tiny contractile fibrils can form waves of contraction across the body surfaces to propel the organism through a liquid medium.

The life cycle usually includes both asexual and sexual reproduction, and there is sometimes an invertebrate intermediate host. At some point in the life cycle, the organisms develop a **spore (oocyst),** which is infective for the next host and is often protected by a resistant coat.

Class Sporozoa

Coccidia. The most important class of the subphylum Apicomplexa, the Sporozoa, contains two subclasses, the Gregarinia and the Coccidia. The gregarines are common parasites of invertebrates, but they are of little economic significance.

The Coccidia are intracellular parasites in invertebrates and vertebrates, and the group contains species of very great medical and veterinary importance.

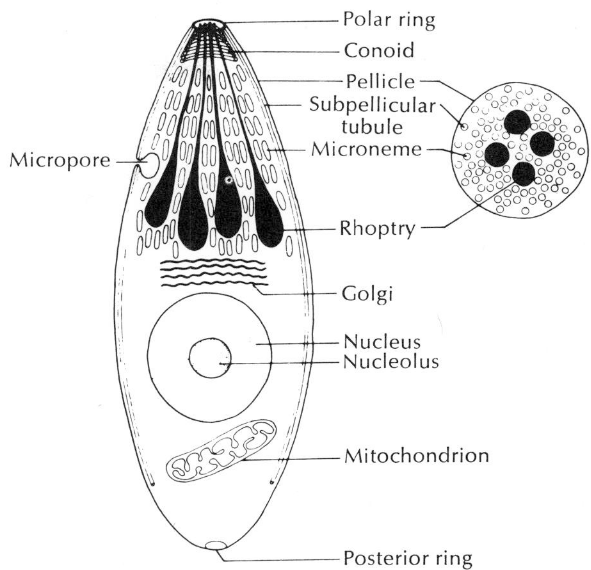

FIG. 8-19 Diagram of an apicomplexan sporozoite or merozoite at the electron microscope level, illustrating the apical complex. The polar ring, conoid, micronemes, rhoptries, subpellicular microtubules, and micropore (cytostome) are all considered components of the apical complex. (From Levine, N. D. 1973. In D. M. Hammond and P. L. Long. The Coccidia. Baltimore, University Park Press.)

EIMERIA. The name "coccidiosis" is generally applied only to infections with *Eimeria* or *Isospora*. Humans are occasionally infected with species of *Isospora,* but apparently there is little disease. Some species of *Eimeria* may cause serious disease in some domestic animals. The symptoms are usually severe diarrhea or dysentery. *E. magna* and *E. stiedai* infect the intestine and liver of rabbits.

E. tenella is often fatal to young fowl, producing severe pathogenesis in the intestine. The organisms undergo schizogony (p. 142) in the intestinal cells, finally producing gametes. After fertilization, the zygote forms an oocyst that passes out of the host in the feces. Sporogony occurs within the oocyst outside the host, producing eight sporozoites in each oocyst. Infection occurs when a new host accidentally ingests a sporulated oocyst, and the sporozoites are released by digestive enzymes.

TOXOPLASMA. A similar life cycle occurs in *Toxoplasma gondii,* a parasite of cats, but this species produces extraintestinal stages as well. The extraintestinal stages can develop in a wide variety of animals other

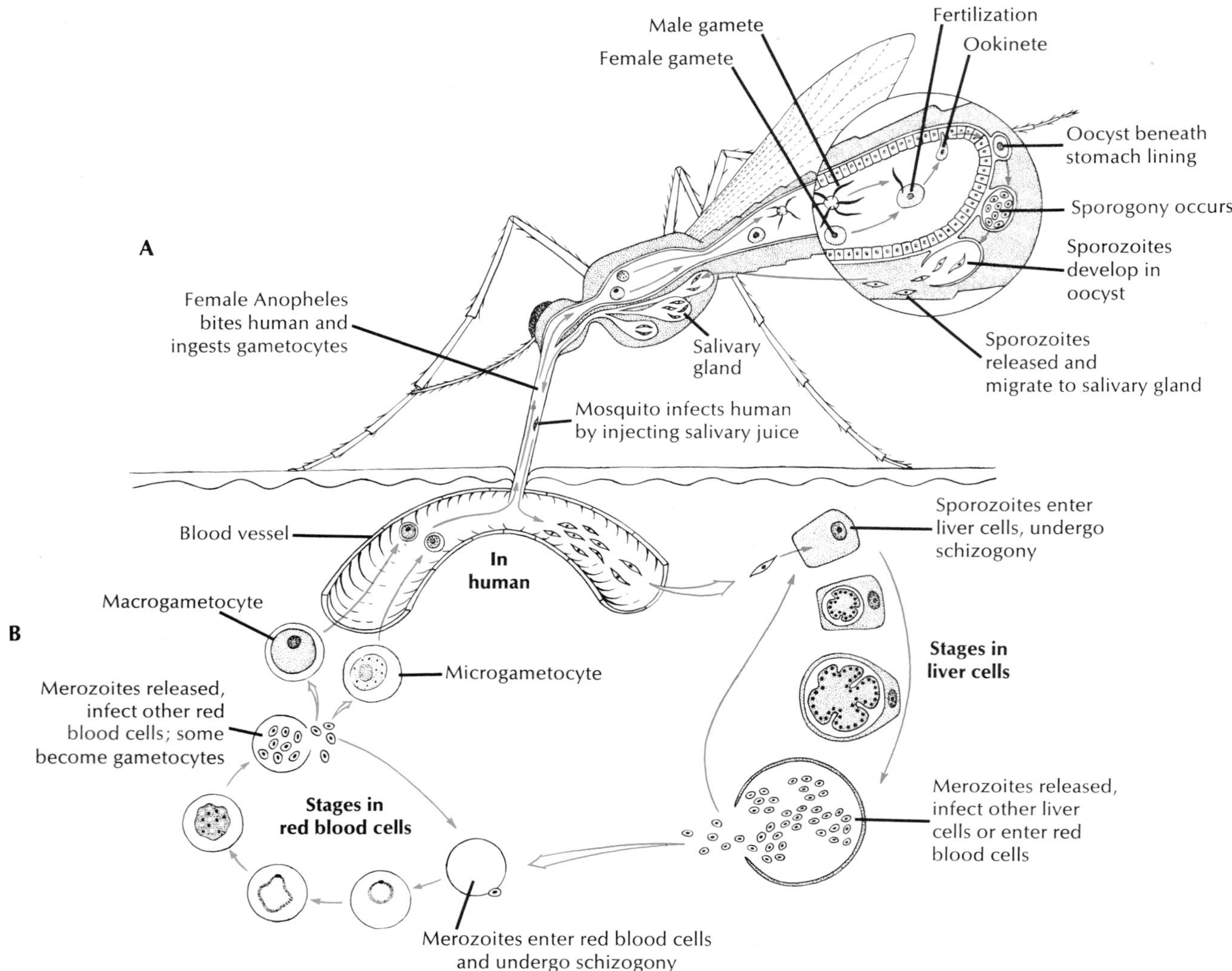

FIG. 8-20 Life cycle of *Plasmodium vivax,* one of the protozoa (class Sporozoa) that causes malaria in humans. **A.** Sexual cycle produces sporozoites in body of mosquito. **B,** Sporozoites infect human and reproduce asexually, first in liver cells then in red blood cells. Malaria is spread by *Anopheles* mosquito, which sucks up gametocytes along with human blood, then, when biting another victim, leaves sporozoites in new wound.

than cats—for example, rodents, cattle, and humans. Gametes and oocysts are not produced by the extraintestinal forms, but they can initiate the intestinal cycle in a cat if the cat eats infected prey. In humans *Toxoplasma* causes little or no ill effects except in a woman infected during pregnancy, particularly in the first trimester. Such infection greatly increases the chances of a birth defect in the baby; it is now believed that 2% of all mental retardation in the United States is a result of congenital toxoplasmosis. The normal route of infection for humans is apparently the consumption of infected beef that is insufficiently cooked.

PLASMODIUM, THE MALARIAL ORGANISM. The best known of the coccidians is *Plasmodium,* the causative organism of the most important infectious disease of humans: **malaria.** Malaria is a very serious disease, difficult to control and very widespread, particularly in tropical and subtropical countries. Four species of *Plasmodium* infect humans, and, though each produces its own peculiar clinical picture, all four have essentially the same cycles of development in their hosts (Fig. 8-20).

The parasite is carried by mosquitoes *(Anopheles),* and sporozoites are injected into the human with the insect's saliva during its bite. The sporozoites penetrate liver cells and initiate schizogony. The products of this division then enter other liver cells to repeat the schizogonous cycle, or in *P. falciparum* they penetrate the red blood cells after only one cycle in the liver. The period when the parasites are in the liver is the **incubation period,** and it lasts from 6 to 15 days, depending on the species of *Plasmodium.*

Merozoites released from the liver schizogony enter red blood cells and begin a series of schizogonous cycles in the red blood cells. When they enter the cells, they become ameboid **trophozoites,** feeding on hemoglobin. The end product of the parasite's digestion of hemoglobin is a dark, insoluble pigment: **hemozoin.** The hemozoin accumulates in the host cell, then is released when the next generation of merozoites is produced, and eventually accumulates in the liver, spleen, or other organs. The trophozoite within the cell grows and undergoes schizogony, producing six to 36 merozoites, depending on the species, which burst forth to infect new red cells. When the red blood cell containing the merozoites bursts, it releases the parasite's metabolic products, which have accumulated there. The release of these foreign substances into the patient's circulation causes the chills and fever characteristic of malaria.

Since the populations of schizonts maturing in the red blood cells are synchronized to some degree, the episodes of chills and fever have a periodicity characteristic of the particular species of *Plasmodium.* In *P. vivax* (benign tertian malaria), the episodes occur every 48 hours; in *P. malariae* (quartan malaria), every 72 hours; in *P. ovale,* every 48 hours; and in *P. falciparum* (malignant tertian), about every 48 hours, though the synchrony is less well defined in this species. People usually recover from infections with the first three species, but mortality may be high in untreated cases of *P. falciparum* infection. Unfortunately, *P. falciparum* is the most common species, accounting for 50% of all malaria in the world.

After some cycles of schizogony in the red blood cells, infection of new cells by some of the merozoites results in the production of **microgametocytes** (male) and **macrogametocytes** (female) rather than another generation of merozoites. When the gametocytes are ingested by a mosquito feeding on the patient's blood, they mature into **gametes,** and fertilization occurs. The zygote becomes a motile **ookinete,** which penetrates the stomach wall of the mosquito and becomes the **oocyst.** Within the oocyst sporogony occurs, and thousands of **sporozoites** are produced. The oocyst ruptures, and the sporozoites migrate to the salivary glands, from which they are transferred to a human by the bite of the mosquito. Development in the mosquito requires from 7 to 18 days but may be longer in cool weather.

The elimination of mosquitoes and their breeding places by insecticides, drainage, and other methods has been effective in controlling malaria in some areas. However, the difficulties in carrying out such activities in remote areas and the acquisition of resistance to insecticides by mosquitoes and to antimalarial drugs by *Plasmodium* (especially *P. falciparum*) mean that malaria will be a serious disease of humans for a long time to come.

Other species of *Plasmodium* parasitize birds, reptiles, and mammals. Those of birds are transmitted chiefly by the *Culex* mosquito.

Subphylum Ciliophora

Class Ciliata

The ciliates are a large and interesting group, with a great variety of forms living in all types of freshwater and marine habitats. They are the most structurally complex and diversely specialized of all the protozoa. The majority are free-living, but some are commensal or parasitic. They are usually solitary and motile, but some are sessile and some colonial. There is great diversity of shape and size. In general they are larger than most other protozoans, but they range from very small (10 to 12 μm) up to 3 mm long. All have cilia that beat in a coordinated rhythmic manner, though the arrangement of the cilia may vary.

Ciliates are always multinucleate, possessing at least one **macronucleus** and one **micronucleus,** but varying from one to many of either type. The macronuclei are apparently responsible for metabolic and developmental functions and for maintaining all the visible traits, such as the pellicular apparatus. Macronuclei

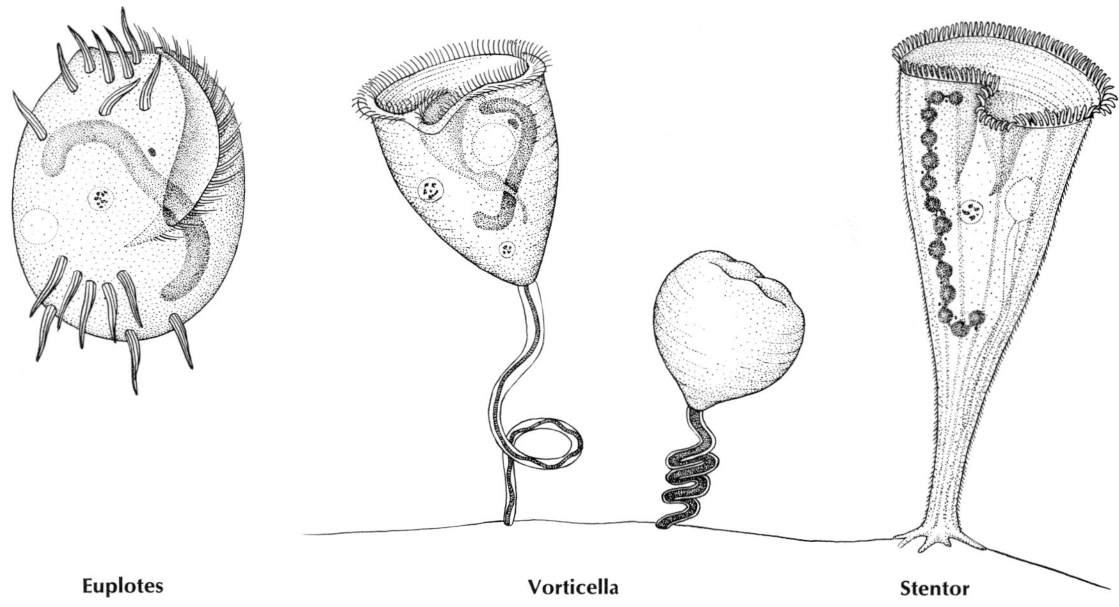

FIG. 8-21 Some representative ciliates, *Euplotes,* have stiff cirri used for crawling about. Contractile myonemes in ectoplasm of *Stentor* and in stalks of *Vorticella* allow great expansion and contraction. Note the macronuclei, long and curved in *Euplotes* and *Vorticella,* shaped like a string of beads in *Stentor.*

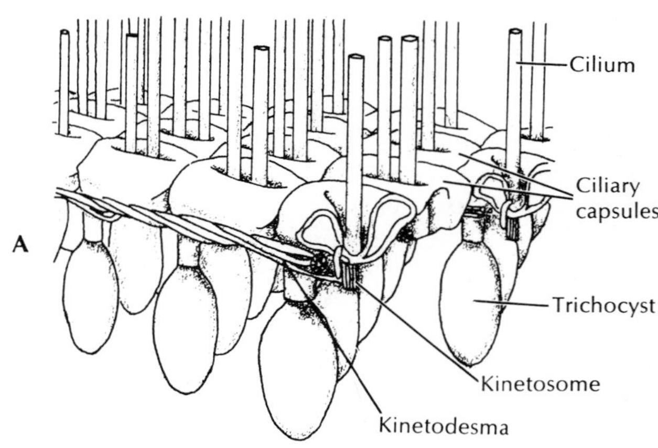

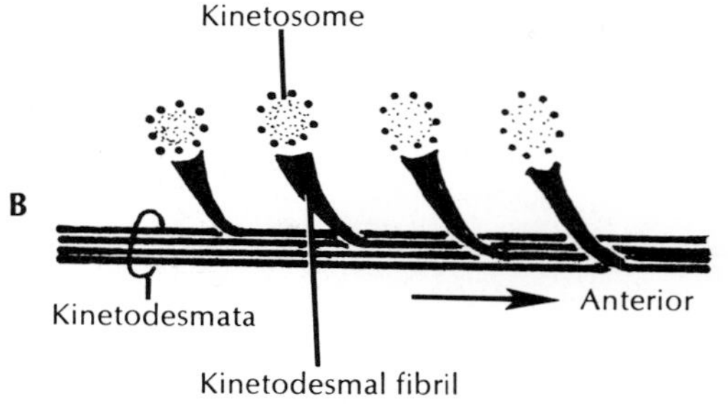

FIG. 8-22 The infraciliature of a ciliate. **A,** Structure of the pellicle and its relation to the neuromotor system. **B,** Portion of a kinety, made up of the kinetosomes (basal bodies of the cilia) and their kinetodesmata (fibrils).

are varied in shape among the different species (Figs. 8-21 and 8-24). The micronuclei participate in sexual reproduction and give rise to macronuclei after exchange of micronuclear material between individuals. The micronuclei divide mitotically, and the macronuclei divide amitotically.

Ciliates are covered by a cell membrane or **pellicle,** which may be very thin or, in some species, may form a thickened armor. The cilia are short and usually arranged in longitudinal or diagonal rows. Their structure is exactly comparable to flagella, with axoneme and kinetosome, except that cilia are shorter (Figs. 8-5 and 8-6). Cilia may cover the surface of the animal or may be restricted to the oral region or to certain bands.

In some forms the cilia are fused into a sheet called an **undulating membrane** or into smaller **membranelles,** both used to propel food into the **cytopharynx** (gullet). In other forms there may be fused cilia forming stiffened tufts called **cirri,** often used in locomotion by the creeping ciliates (Fig. 8-21).

An apparently structural system of fibers, in addition to the kinetosomes, makes up the **infraciliature,** just beneath the pellicle (Fig. 8-22). Each cilium terminates beneath the pellicle in its kinetosome, and from each kinetosome a **kinetodesmal fibril** arises and passes along beneath the row of cilia, joining with the other fibrils of that row. The kinetosomes and fibrils **(kinetodesmata)** of that row make up what is known as a **kinety** (Fig. 8-22). All ciliates seem to have kinety systems, even those that lack cilia at some stage. The infraciliature apparently does not coordinate the ciliary beat, as formerly thought. Coordination of the ciliary movement seems to be by waves of depolarization of the cell membrane moving down the animal, similar to the phenomena in a nerve impulse (p. 729).

As was pointed out before for ameboid movement, ciliary movement is extremely important for multicellular animals. Not only do many small metazoa use cilia for locomotion, the cilia of many others create water currents for their feeding and respiration. Ciliary movement is vital to many species in such functions as food-handling, reproduction, excretion, osmoregulation, and so on.

Many ciliates have contractile fibrils that early investigators called **myonemes.** With the aid of the light microscope, these could be seen to run in rows parallel with the rows of kinetosomes. The electron microscope reveals that there are actually two systems of fibrils, the **km fibers** (composed of stacks of microtubules) and the **M bands** (bundles of tubular microfilaments) lying beneath the km fibers. Both km fibers and M bands are attached to the kinetosomes. Recent studies indicate that the M bands may be the ones responsible for contraction. In *Stentor,* mechanical stimuli cause waves of contractions to spread in both oral and aboral directions; an electric stimulus, however, causes contraction simultaneously in all areas of the body.

Most ciliates are **holozoic.** Most of them possess a cytostome (mouth) that in some forms is a simple opening and in others is connected to a gullet or ciliated groove. The mouth in some is strengthened with stiff, rodlike trichites for swallowing larger prey; in others ciliary water currents carry microscopic food particles toward the mouth, as in the paramecia. *Didinium* has a proboscis for engulfing the paramecia on which it feeds (Fig. 8-9). Suctorians paralyze their prey and then ingest their contents through tubelike tentacles by a complex feeding mechanism that apparently combines phagocytosis with a sliding filament action of microtubules in the tentacles (Fig. 8-9). In any case the food is digested within food vacuoles.

In one of the subclasses, Holotrichia, small bodies called **trichocysts** are located in the ectoplasm (Figs. 8-22 and 8-24). When discharged, they form long threadlike filaments that pass through the pellicle and harden, except for the tips, which are sticky for attachment. In some ciliates the discharged trichocysts seem to help anchor the animal while feeding; in others they are apparently used to paralyze other small organisms for defense or capture of prey.

Among the more striking and familiar of the ciliates are the following: *Stentor,* trumpet-shaped and solitary, with a bead-shaped macronucleus (Fig. 8-21); *Vorticella,* bell-shaped and attached by a contractile stalk (Fig. 8-21); *Euplotes* and *Stylonychia,* with flattened bodies and groups of fused cilia (cirri) that function as legs (Fig. 8-21); *Blepharisma,* slender and pink in color; and *Spirostomum,* very long and wormlike in appearance.

Paramecium, a representative ciliate. Paramecia are usually abundant in ponds or sluggish streams containing aquatic plants and decaying organic matter.

STRUCTURE. The paramecium is often described as slipper-shaped. *Paramecium caudatum* is from 150 to 300 μm (0.15 to 0.3 mm) in length and is blunt anteriorly and somewhat pointed posteriorly (Fig. 8-23). The animal has an asymmetric appearance because of the **oral groove,** a depression that runs obliquely backward on the ventral side.

The **pellicle** is a clear, elastic membrane divided

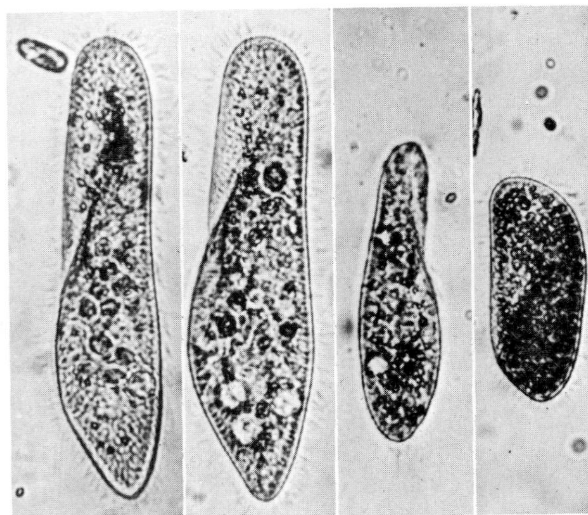

FIG. 8-23 Comparison of four common species of *Paramecium* photographed at the same magnification. Left to right: *P. multimicronucleatum, P. caudatum, P. aurelia,* and *P. bursaria.* (Courtesy Carolina Biological Supply House Co., Burlington, N.C.)

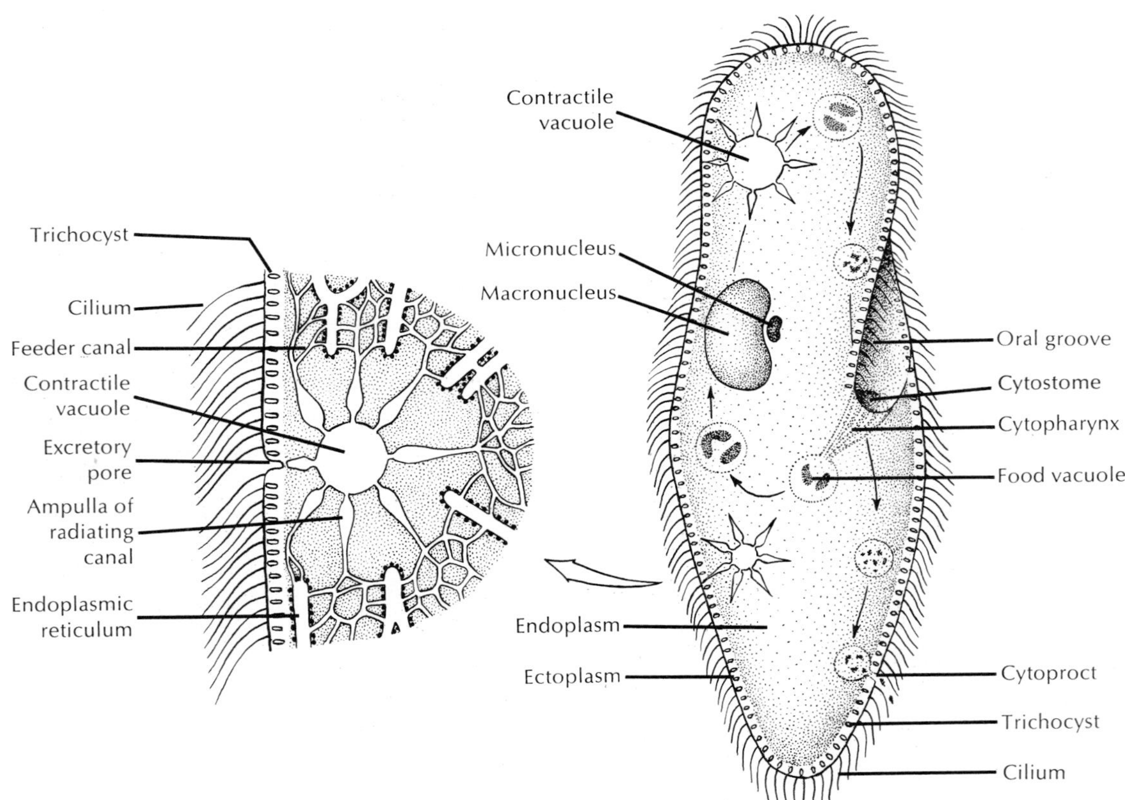

FIG. 8-24 *Left,* enlarged section of a contractile vacuole (water expulsion vesicle) of *Paramecium.* Water is believed to be collected by endoplasmic reticulum, emptied into feeder canals and then into the vesicle. The vesicle contracts to empty its contents to the outside, thus serving as an osmoregulatory organelle. *Right, Paramecium,* showing gullet, food vacuoles, and nuclei.

into hexagonal areas by tiny elevated ridges (Fig. 8-22) and is covered over its entire surface with cilia arranged in lengthwise rows. Just below the pellicle is the thin clear **ectoplasm** that surrounds the larger mass of granular **endoplasm** (Fig. 8-24). Embedded in the ectoplasm just below the surface are the spindle-shaped **trichocysts** which alternate with the bases of the cilia. The infraciliature can be seen only by special fixing and staining methods.

The **cytostome** at the end of the oral groove leads into a tubular **cytopharynx, or gullet.** Along the gullet an undulating membrane of modified cilia keeps food moving. Fecal material is discharged through an **anal pore (cytoproct)** posterior to the oral groove (Fig. 8-24). The endoplasm contains food vacuoles containing food in various stages of digestion. There are two **contractile vacuoles,** each consisting of a central space surrounded by several **radiating canals** that collect fluid and empty it into the central vacuole.

P. caudatum has two nuclei: a large kidney-shaped **macronucleus** and a smaller **micronucleus** fitted into the depression of the former. These can usually be seen only in stained specimens. The number of micronuclei varies in different species. *P. multimicronucleatum* may have as many as seven.

NUTRITION. Paramecia are holozoic, living on bacteria, algae, and other small organisms. They are selective in choosing their food, for some items are taken in and others rejected. The cilia in the oral groove sweep food particles in the water into the cytostome from which point they are carried into the cytopharynx by the undulating membrane. From the cytopharynx the food is collected into a food vacuole that is constricted into the endoplasm. The food vacuoles circulate in a definite course through the protoplasm (cyclosis) while the food is being digested by enzymes from the endoplasm. The indigestible part of the food is ejected through the anal pore (cytoproct). Digestion in ciliates is fairly rapid. A *Didinium* can digest a whole paramecium in about 20 minutes.

EXCRETION AND OSMOREGULATION. The two contractile vacuoles regulate the water content of the body and help eliminate nitrogenous waste. They lie close to the dorsal surface and drain fluid from the cytoplasm by means of radiating canals that connect the vesicle with the endoplasmic reticulum (Fig. 8-24). When the vacuole reaches a certain size, it discharges to the outside through a pore. The two vacuoles contract alternately. They contract more frequently at higher temperatures and in the mature animal. Since the contents of the paramecium are hypertonic to the surrounding fresh water, osmotic pressure would cause water to diffuse into the cell, and thus one of the main functions of the vacuoles is to get rid of the excess water.

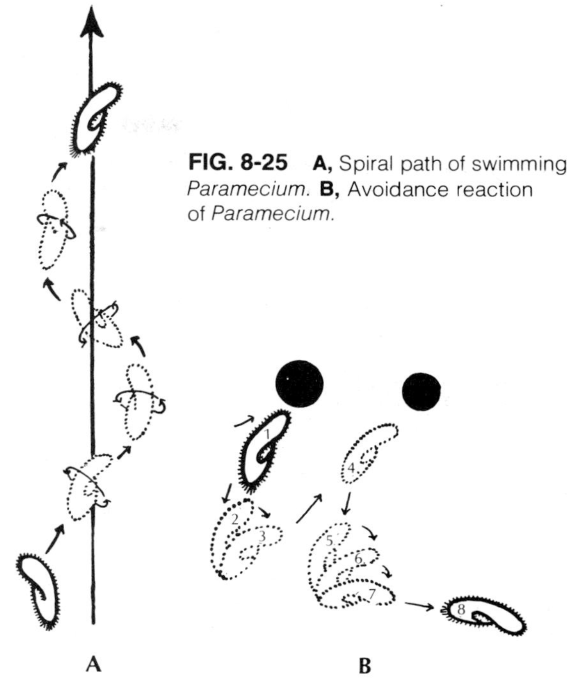

FIG. 8-25 **A,** Spiral path of swimming *Paramecium.* **B,** Avoidance reaction of *Paramecium.*

LOCOMOTION. The body of the paramecium is elastic, allowing it to bend and squeeze its way through narrow places. Its cilia can beat either forward or backward, so that the animal can swim in either direction. The cilia beat obliquely, thus causing the animal to rotate on its long axis. In the oral groove the cilia are longer and beat more vigorously than the others so that the anterior end swerves aborally. As a result of these factors, the animal follows a spiral path moving forward (Fig. 8-25, *A*). In swimming backward the beat and path of rotation are reversed.

BEHAVIOR. When a ciliate, such as a paramecium, comes in contact with a barrier or a disturbing chemical stimulus, it reverses its cilia, backs up a short distance, and swerves the anterior end as it pivots on its posterior end. This is called an **avoiding reaction** (Fig. 8-25, *B*). While it is doing this, samples of the surrounding medium are brought into the oral groove. When the sample no longer contains the unfavorable stimulus, the animal moves forward. In this "trial-and-

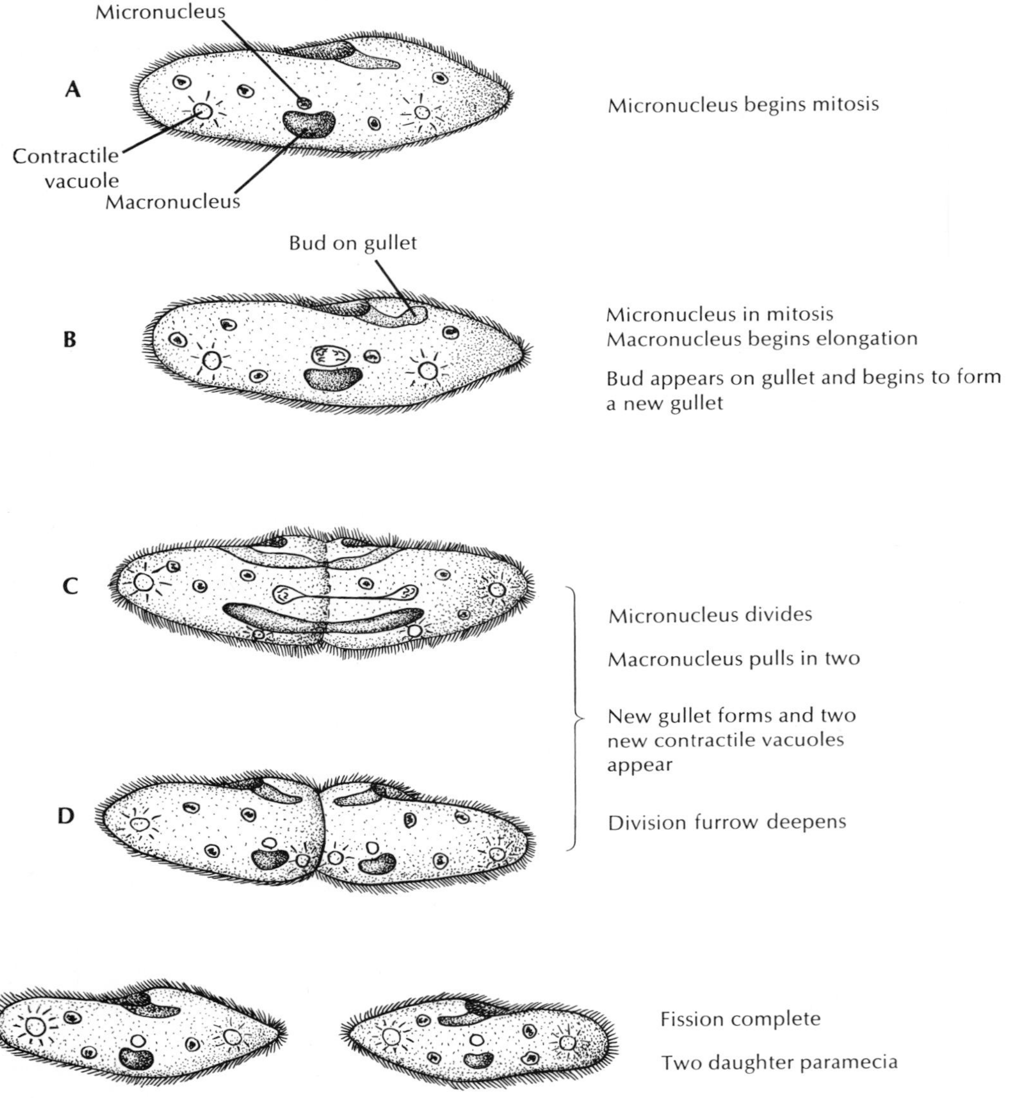

FIG. 8-26 Binary fission in *Paramecium*, showing the sequence of stages.

error'' method the animal attempts many directions until it finds one that is favorable.

Paramecia do not always respond in the same manner to the same stimuli. Their physiologic states vary with conditions. A hungry animal will react in a different way from one that is well fed. In general its behavior is conditioned by factors that favor or hinder the normal life processes.

Locomotor responses, by which an animal orients itself toward or away from particular stimuli, are called **taxes.** If the response is movement toward the stimulus, it is a positive response; an avoiding reaction is a negative response. With respect to the type of stimulus, a taxis or tropism might be classified as one of the following: thermotaxis, response to heat; phototaxis, response to light rays; thigmotaxis, response to contact; chemotaxis, response to chemical substances; rheotaxis, response to currents of air or water; galvanotaxis, response to constant electric current; or geotaxis, response to gravity.

Since no nervous system is found in protozoans (except perhaps in the infraciliature), these responses must be attributable to the innate irritability of protoplasm. The complex responses of higher forms are believed to have developed from these simple mechanical responses.

REPRODUCTION. Paramecia reproduce only by binary fission across kineties but have certain forms of sexual phenomena called conjugation and autogamy.

In **binary fission** the micronucleus divides mitotically into two daughter micronuclei, which move to opposite ends of the cell (Fig. 8-26). The macronucleus elongates and divides amitotically. Another cytopharynx is budded off and two new contractile vacuoles appear. In the meantime a constriction furrow appears near the middle of the body and deepens until the cytoplasm is completely divided. The process of binary fission requires from ½ to 2 hours.

The division rate usually varies from one to four times each day. About 600 generations can be produced in a year.

Conjugation occurs at intervals in ciliates. Conjugation is the temporary union of two individuals for the purpose of exchanging chromosomal material (Fig. 8-27). During the union the macronucleus disintegrates and the micronucleus of each individual undergoes meiosis, giving rise to four haploid micronuclei, three of which degenerate (Fig. 8-27, *A* to *D*). The remaining micronucleus then divides into two haploid pronuclei, one of which crosses over into the conjugant partner. When the exchanged pronucleus unites with the pronucleus of the partner, the diploid number of chromosomes is restored (Fig. 8-27, *E* and *F*).

The two paramecia now separate, and in each the fused micronucleus, which is comparable to a zygote in higher forms, divides by mitosis into two, four, and eight micronuclei. Four of these enlarge and become macronuclei, and three of the other four disappear (Fig. 8-27, *G* to *J*). Now the paramecium itself divides twice, resulting in four paramecia, each with one micronucleus and one macronucleus. After this complicated process, the animals may continue to reproduce by binary fission without the necessity of conjugation.

The result of conjugation is similar to that of zygote formation, for each exconjugant contains hereditary material from two individuals. The advantage of sexual reproduction is that it permits gene recombinations, thus stimulating evolutionary change. Although ciliates in clone cultures can apparently reproduce repeatedly and indefinitely without conjugation, the stock seems eventually to lose vigor. Conjugation restores vitality to a stock. Seasonal changes or a deteriorating environment will usually stimulate sexual reproduction.

In 1937 it was discovered that not every paramecium would conjugate with any other paramecium of the same species. Sonneborn found that there were physiologic differences between individuals that set them off into **mating types.** Ordinarily conjugation will not occur between individuals of the same mating type but only with an individual of another (complementary) mating type. It was also found that within a single species there are a number of varieties,* each of which has mating types that conjugate among themselves but not with the mating types of other varieties. In *Paramecium aurelia,* for instance, each of six varieties has two mating types; conjugation, however, will occur only between members of opposite or complementary mating types within their own variety. With few exceptions, each variety has only two interbreeding mating types. There is no morphologic basis for distinguishing mating types within a variety; such differences that exist must be physiologic. Some varieties, however, can be distinguished from each other morphologically.

Mating types are also found in other species of paramecia and in other ciliates.

Autogamy is a process of self-fertilization that is similar to that of conjugation except that there is no exchange of nuclei. After the disintegration of the macronucleus and the meiotic divisions of the micronucleus, the two haploid pronuclei fuse to form a synkaryon that is, however, homozygous rather than heterozygous (as in the case of exconjugants).

Parasitic ciliates. Many symbiotic ciliates live as commensals, but some can be harmful to their hosts. *Balantidium coli* lives in the large intestine of humans, pigs, rats, and many other mammals (Fig. 8-28). There seem to be host-specific strains, and the organism is

*Within each species of *Paramecium* the individuals exhibit morphologic and physiologic differences. Since these differences are usually more minor and more superficial than those that distinguish species, the groups within a species are referred to as strains, biotypes, or varieties. Most species of Protozoa can be divided into a number of these groups.

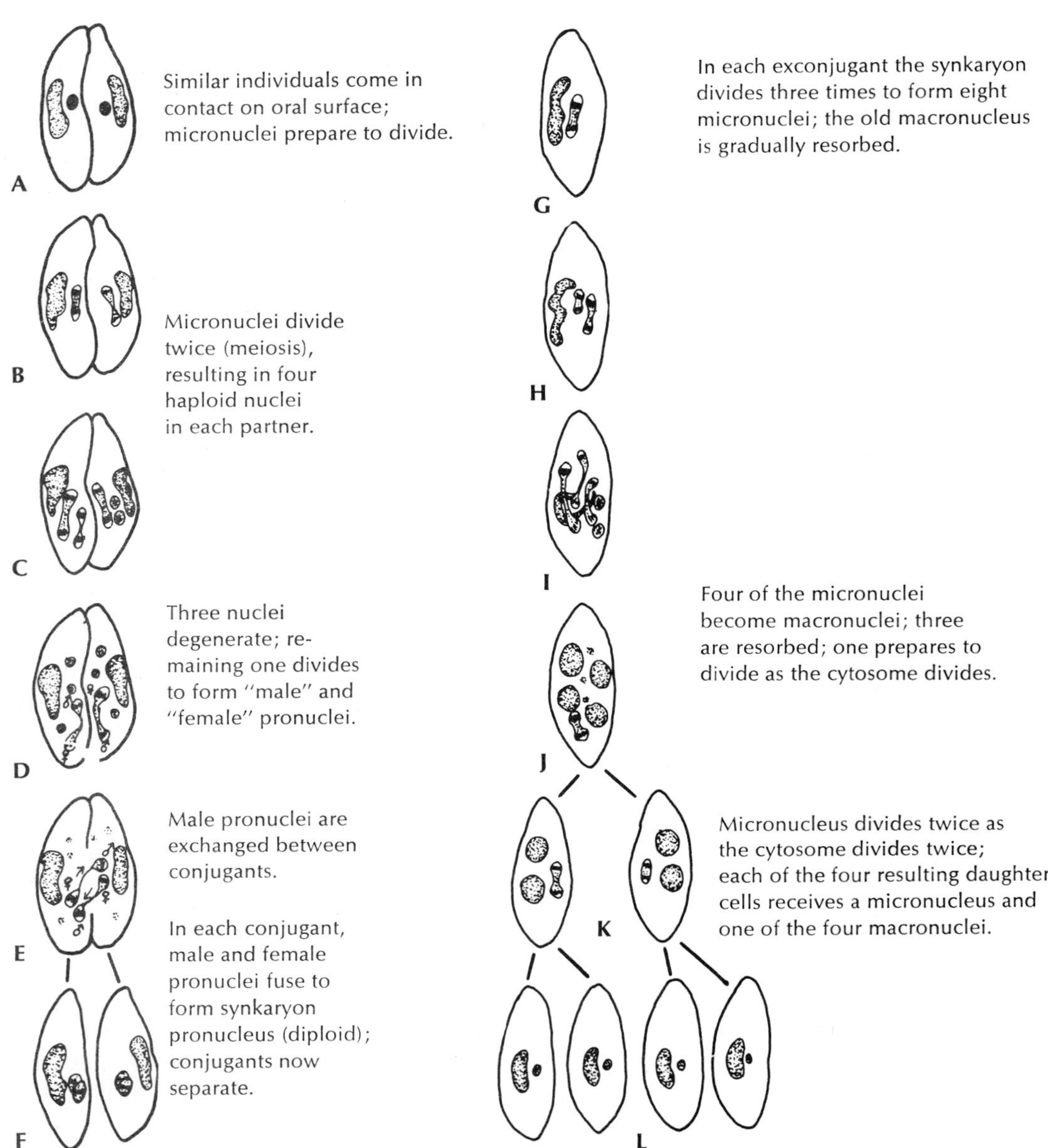

FIG. 8-27 Sexual reproduction, or conjugation, in *Paramecium caudatum*.

not easily transmitted from one species to another. Transmission is by fecal contamination of food or water. Usually the organisms are not pathogenic, but in humans they sometimes invade the intestinal lining and cause a dysentery similar to that caused by *Entamoeba histolytica*. The disease can be serious and even fatal.

Other species of ciliates live in other hosts. *Epidinium* (Fig. 8-29) is a commensal in the rumen of cattle, and *Nyctotherus* occurs in the colon of frogs and toads. In aquarium and wild freshwater fish, *Ichthyophthirius* causes a disease known to many fish culturists as "ick."

Suctorians. Suctorians are ciliates in which the young possess cilia and are free-swimming, and the adults grow a stalk for attachment, become sessile, and lose their cilia. They have no cytostome but feed by

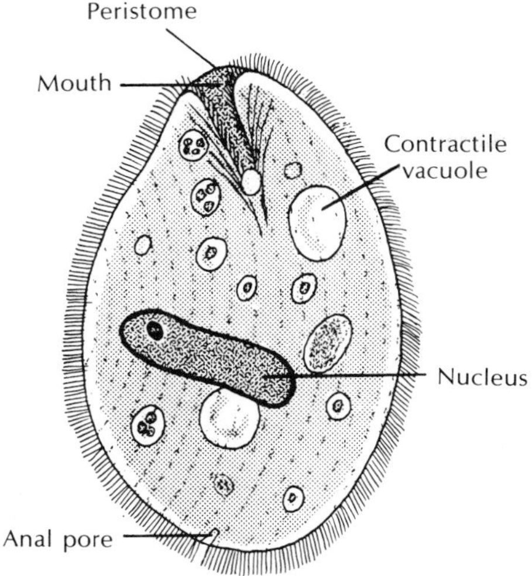

FIG. 8-28 *Balantidium coli,* a ciliate parasitic in humans and other mammals. This ciliate is common in pigs, in which it does little damage. In humans it may produce intestinal ulcers and severe chronic dysentery. Infections are common in parts in Europe, Asia, and Africa, but are rare in the United States.

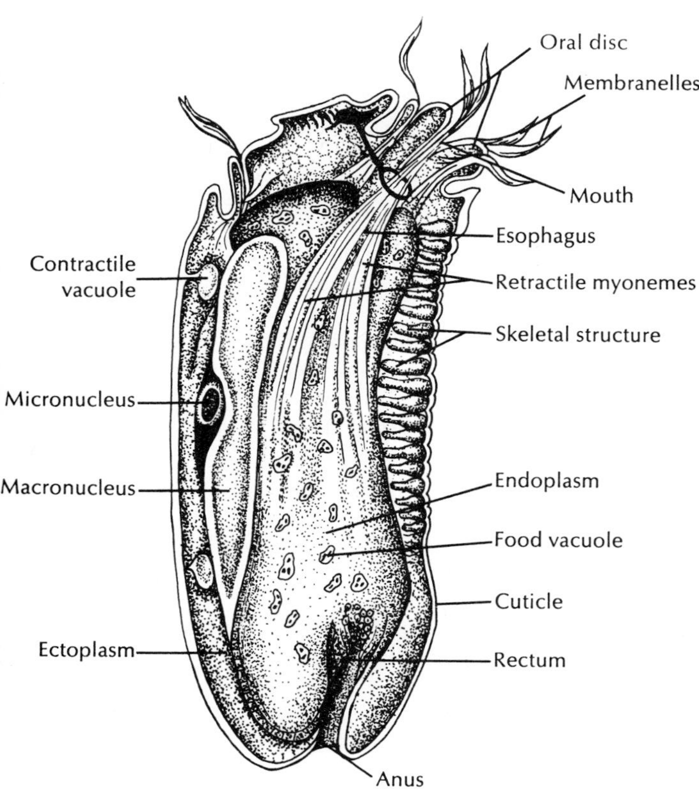

long, slender, tubelike tentacles. The suctorian captures living prey, usually a ciliate, by the tip of one or more tentacles and paralyzes it. The cytoplasm of the prey then flows through the attached tentacles, forming food vacuoles in the feeding suctorian (Fig. 8-9).

One of the best places to find freshwater suctorians is in the algae that grows on the carapace of turtles. Common genera of suctorians found there are *Anarma* (without stalk or test) and *Squalorophyra* (with stalk and test). Other freshwater representatives are *Podophyra* (Fig. 8-9) and *Dendrosoma*. *Acinetopsis* and *Ephelota* are saltwater forms.

Suctorian parasites include the *Trichophyra,* which is parasitic on the gills of the small-mouthed black bass and may cause serious damage to the fish; *Allantosoma,* which occurs in the intestine of certain mammals; and *Sphaerophyra,* which is found in *Stentor*.

PHYLOGENY AND ADAPTIVE RADIATION

Phylogeny. Protozoa are usually placed at the base of the phylogenetic tree. Doubtless, multicellular animals were derived from a protozoan or protozoan-like ancestor, perhaps more than once. Sponges, for example, may well have been derived separately from other metazoa. Colonial protozoa, particularly among flagellates (Fig. 8-8), show various degrees of cell aggregation and some differentiation that suggest the body plans of early metazoa.

With the exception of certain shell-bearing Sarcodina, such as Foraminifera and Radiolaria, protozoans have left no fossil records. Mastigophorans may be the oldest of all protozoa, perhaps having arisen from bacteria and spirochetes, but the group is probably polyphyletic. Most of the phytoflagellates are more closely related to the green algae than they are to zooflagellates. Some colorless phytoflagellates have chlorophyll-bearing relatives, and some autotrophic forms are facultatively saprophytic in darkness. Hence, the common origin of animals and plants may lie in the Phytomastigophora. That the amebas were derived

FIG. 8-29 *Epidinium,* which lives in the digestive tract of ruminants, shows how complex a protozoan can be. It has specialized organelles for coordination, ingestion and digestion, support, contraction, food storage, egestion, and hereditary continuity. (Modified from Sharp, R. G. 1914. Univ. Calif. Publ. Zool. **13:**43-122.)

from the flagellates is indicated by the ameboid stages in some flagellates and flagellated stages of some amebas. However, the different orders of Sarcodina may have arisen independently from different kinds of flagellates. Sporozoa, which are all specialized parasites, probably came from flagellated ancestors: they often have ameboid feeding stages and flagellated gametes. The origin of the ciliates is somewhat obscure, but the basic structural similarity of the flagellum to the cilium is undeniable.

Adaptive radiation. Some of the wide range of adaptations of protozoa have been illustrated in the previous pages. The Sarcodina range from bottom-dwelling, naked species to planktonic forms such as the foraminiferans and radiolarians with beautiful, intricate tests. There are many symbiotic species of amebas. Flagellates likewise show adaptations for a similarly wide range of habitats, with the added variation of photosynthetic ability in many species of Phytomastigophorea. The fine line between plants and animals at this level is shown by our ability to turn a ''plant'' into an ''animal'' by experimentally destroying the chloroplasts of *Euglena*.

Within a single-cell body plan, the division of labor and specialization of organelles is carried furthest by the ciliates. These have become the most complex of all protozoans. Specializations for intracellular parasitism have been adopted by the Sporozoa.

Derivation and meaning of names

Amoeba (Gr., *amoibē,* change) Genus of Sarcodina.

Arcella (L., diminutive of *arca,* box) Genus of Sarcodina; with a boxlike test.

Coccidia (Gr., *kokkos,* kernel or berry) Order of Sporozoa.

Difflugia (L. *diffluo,* flow apart) Genus of Sarcodina; refers to the flowing out of the pseudopodia.

Eimeria (after Eimer, German zoologist) Genus of the order Coccidia of Sporozoa.

Entamoeba (Gr., *entos,* within, + *amoibē,* change) **histolytica** (Gr., *histos,* tissue, + *lysis,* a loosing) Genus and species in Sarcodina.

Euglena (Gr., *eu,* true, good, + *glēnē,* eyeball or eye pupil) Refers to the stigma, or eyespot.

Euplotes (Gr., *eu,* true, good, + *plōtēr,* swimmer) Genus of Ciliata.

Foraminifera (L., *foramen,* hole, + *fero,* bear) The tests are frequently perforated.

Leishmania (after Leishman, who discovered it) Parasitic genus of Zoomastigophorea.

Noctiluca (L., the moon, from *nox,* night, + *lucere,* to shine) Genus of Mastigophora. Their luminescence is most apparent at night.

Opalina (L., *opalus,* opal) Genus of superclass Opalinata.

Paramecium (Gr., *paramekes,* oblong) Genus of Ciliata.

Pelomyxa (Gr., *pelos,* mud, + *myxa,* nasal mucus) Refers to the black and brown inclusions in body. Genus of Sarcodina.

Plasmodium (Gr., *plasma,* something molded) Genus of Sporozoa causing malaria.

Radiolaria (L., *radiolus,* small ray) Order of Sarcodina. The axopodia radiate in all directions.

Rhizopodea (Gr., *rhiza,* root, + *peus, podos,* foot) Many pseudopodia may be extended at one time.

Stentor (Grecian herald with loud voice) Genus of Ciliata. Shaped like a megaphone.

Toxoplasma (Gr., *toxo,* a bow, + *plasma,* molded) Genus of Coccidia; sexual reproduction in cats; may cause birth defects in humans.

Trichomonas (Gr., *thrix,* hair, + *monas,* single) Genus of Mastigophora.

Trypanosoma (Gr., *trypanon,* auger, + *sōma,* body) Genus of Mastigophora. The body with undulating membrane is twisted.

Volvox (L., *volvere,* to roll) Genus of Mastigophora. Refers to their characteristic movement.

Vorticella (L., diminutive of *vortex,* a whirlpool) Genus of Ciliata. Movement of their cilia creates a spiral movement of water.

Annotated references

Selected general readings

Allen, R. D., D. Francis, and R. Zeh. 1971. Direct test of the positive pressure gradient theory of pseudopod extension and retraction in amoebae. Science **174:**1237-1240. *Provides evidence that front-contraction and rear-contraction theories of pseudopodial movement probably are not correct.*

Bannister, L. H., and E. C. Tatchell. 1968. Contractility and the fibre systems of *Stentor caeruleus*. J. Cell. Sci. **3:**295-308.

Blake, J. H., and M. A. Sleigh. 1974. Mechanics of ciliary locomotion. Biol. Rev. **49:**85-125.

Brokaw, C. J. 1972. Flagellar movement: a sliding filament model. Science **178:**455-461.

Christianson, R. G., and J. M. Marshall. 1965. A study of phagocytosis in the ameba *Chaos chaos*. J. Cell Biol. **25:**443-457.

Corliss, J. O. 1961. The ciliated Protozoa: characterization, classification and guide to the literature. New York, Pergamon Press, Inc.

Cushman, J. A. 1948. Foraminifera, their classification and economic use, ed. 4. Cambridge, Mass., Harvard University Press.

D'Haese, J., and H. Hinssen. 1974. Structure of synthetic and native myosin filaments from *Amoeba proteus*. Cell Tissue Res. **151:**323-335.

Edmondson, W. T. (ed.). 1959. Ward and Whipple's fresh-water biology, ed. 2. New York, John Wiley & Sons, Inc. *In this handbook there are useful keys to the major groups of Protozoa.*

Gojdies, Mary. 1953. The genus *Euglena*. Madison, University of Wisconsin Press. *A comprehensive account of this group. Much attention is given to taxonomy, but there are also good descriptions of morphology.*

Grell, K. G. 1973. Protozoology. Heidelberg, Springer-Verlag.

Hedley, R. H., and C. G. Adams. 1974. Foraminifera, vol. 1. New York, Academic Press, Inc.

Hickman, C. P. 1972. Biology of the invertebrates, ed. 2. St. Louis, The C. V. Mosby Co.

Honigberg, B. M., W. Balamuth, E. C. Bovee, and others. 1964. A revised classification of the Phylum Protozoa. J. Protozool. **11:**7-20. *Authoritative revision of the higher taxonomy of the protozoa by a committee of the Society of Protozoologists. A few important changes have been made since its publication, for example, the creation of the Apicomplexa and the placement of the Piroplasmea, Myxospora, and Microspora.*

Hutner, S. H. (ed.). 1964. Biochemistry and physiology of protozoa, vol. 3. New York, Academic Press, Inc.

Hutner, S. H., and A. Lwoff (eds.). 1955. Biochemistry and physiology of protozoa, vol. 2. New York, Academic Press, Inc.

Hyman, L. H. 1940. The invertebrates: Protozoa through Ctenophora, vol. 1. New York, McGraw-Hill Book Co. *An extensive and exhaustive section is devoted to the morphology and physiology of protozoa.*

Jahn, T. L., and F. F. Jahn. 1949. How to know the Protozoa. Dubuque, Iowa, William C. Brown Co., Publishers. *A manual on the identification and description of protozoan forms.*

Jennings, H. S. 1906. Behavior of the lower organisms. New York, Columbia University Press. *A classic work on the tropisms (taxes) of protozoan forms. This treatise had a profound influence on all subsequent investigations along this line.*

Jennings, R. K., and R. F. Acker. 1970. The protistan kingdom. New York, Van Nostrand Reinhold Co.

Jones, A. R. 1974. The ciliates. New York, St. Martin's Press, Inc.

Kidder, G. W. (ed.). 1967. Chemical zoology, vol. 1. New York, Academic Press, Inc.

Kudo, R. R. 1966. Protozoology, ed. 5. Springfield, Ill., Charles C Thomas, Publisher.

Leedale, G. F. 1967. Euglenoid flagellates. Englewood Cliffs, N.J., Prentice-Hall, Inc.

Levine, N. D. 1973. Protozoan parasites of domestic animals and man. Minneapolis, Burgess Publishing Co.

Mackinnon, D. L., and R. S. J. Hawes. 1961. An introduction to the study of Protozoa. Oxford, The Clarendon Press.

Manwell, R. D. 1961. Introduction to protozoology. New York, St. Martin's Press, Inc.

Marshall, J. M., and V. Nachmias. 1965. Cell surface and pinocytosis. J. Histochem. Cytochem. **13:**92-104.

Murray, J. W. 1973. Distribution and ecology of living benthic foraminiferids. New York, Crane, Russak & Co.

Newman, E. 1972. Contraction in *Stentor coeruleus:* a cinematic analysis. Science **177:**447-449.

Papas, G. D., and P. W. Brandt. 1958. The fine structure of the contractile vacuole in *Amoeba*. J. Biophys. Biochem. Cytol. **4:**485-488.

Pennak, R. W. 1953. Fresh-water invertebrates of the United States. New York, The Ronald Press Co. *A reference work with considerable attention devoted to Protozoa.*

Pettersson, H. 1954. The ocean floor, New Haven, Conn., Yale University Press. *The role the Foraminifera and Radiolaria have played in building up the sediment carpet of the ocean floor is vividly described in this little book.*

Pitelka, D. R. 1963. Electron-microscopic structure of Protozoa. New York, The Macmillan Co. *A fine account of the ultrastructure of Protozoa.*

Rudzinska, M. A. 1973. Do suctoria really feed by suction? BioScience **23**(2):87-94.

Sarjeant, W. A. S. 1974. Fossil and living dinoflagellates. New York, Academic Press, Inc.

Satir, P. 1968. Studies on cilia, III. Further studies on the cilia tip and a "sliding filament" model of ciliary motility. J. Cell. Biol. **39:**77-94.

Schmidt, G. D., and L. S. Roberts. 1977. Foundations of parasitology. St. Louis, The C. V. Mosby Co. *A parasitology text containing a good account of protozoan parasites.*

Sleigh, M. A. 1962. The biology of cilia and flagella. New York, The Macmillan Co. *An excellent summary.*

Sleigh, M. A. 1973. The biology of protozoa. New York, American Elsevier. *Well-written text, integrated form and function with many micrographs and diagrams of cell cycles.*

Sonneborn, T. M. 1950. *Paramecium* in modern biology. BioScience **21:**31-43.

Sonneborn, T. M. 1957. Breeding systems, reproductive methods, and species problems in Protozoa. In Mayr, E. (ed.). The species problem. Washington, D.C., American Association for the Advancement of Science, pp. 155-324. *A good analysis of mating types and the concept of syngen as applied to the varieties of Protozoa.*

Tartar, V. 1961. The biology of *Stentor*. New York, Pergamon Press, Inc.

Vickerman, K., and F. E. G. Cox. 1967. The Protozoa. Boston, Houghton Mifflin Co.

Wichterman, R. 1953. The biology of *Paramecium*. New York, McGraw-Hill Book Co. *A well-written treatise on one genus of protozoans.*

Selected *Scientific American* articles

Allen, R. D. 1962. Amoeboid movement. **206:**112-122 (Feb.)

Alvarado, C. A., and L. J. Bruce-Chevatt. 1962. Malaria. **206:**86-98 (May).

Bonner, J. T. 1949. The social amoebae. **180:**44-47 (June).

Bonner, J. T. 1950. *Volvox:* A colony of cells. **182:**52-55 (May).

Bonner, J. T. 1969. Hormones in social amoebae and mammals. **220:**78-91 (June).

Hawking, F. 1970. The clock of the malaria parasite. **222:** 123-131 (June).

Hutner, S. H., and J. A. McLaughlin. 1958. Poisonous tides. **199:**92-98 (Aug.).

Satir, P. 1974. How cilia move. **231:**45-52 (Nov.).

CHAPTER 9

THE LOWEST METAZOANS

Phylum Mesozoa
Phylum Porifera—sponges

The breadcrumb sponge *Halichondria panicea* growing on a rock under seaweed at low tide. The pores that dot the surface of the sponge are the openings through which seawater circulates through the animal.

Photograph by D. P. Wilson.

Position in animal kingdom

The multicellular animals, or Metazoa, are typically divided into three branches—Mesozoa (a single phylum), Parazoa (phylum Porifera, the sponges), and Eumetazoa (all other phyla).

Although Mesozoa and Parazoa are multicellular, neither fits into the general plan of organization of the other phyla. Such cellular layers as they possess are not homologous to the germ layers of the Eumetazoa, and neither group has developmental patterns in line with the other metazoans. The poriferans are considered to be aberrant, that is, deviating widely from standard patterns. This is the reason for the name Parazoa, which means the "beside-animals."

Biologic contributions

Although the simplest in organization of all the metazoans, the mesozoans and sponges do comprise a higher level of morphologic and physiologic integration than that found in protozoan colonies. Both groups may be said to belong to a **cellular level of organization.**

The mesozoans, although composed simply of an outer layer of somatic cells and an inner layer of reproductive cells, nevertheless have a very complex reproductive cycle somewhat suggestive of that of the trematodes (flukes). Mesozoans are entirely parasitic.

The poriferans are more complex, with several types of cells differentiated for various functions, some of which are organized into **incipient tissues** of a low level of integration.

The developmental patterns of both phyla are different from those of other phyla, and their embryonic layers are not homologous to the germ layers of other phyla.

The sponges have developed a unique system of **water currents** on which they depend for food and oxygen.

PHYLUM MESOZOA

The name Mesozoa (mes-o-zo'a) (Gr., *mesos,* in the middle, + *zoia,* animals) was coined by an early investigator (van Beneden, 1876) who believed the group to be a "missing link" between protozoans and metazoans. These minute, ciliated, wormlike animals represent the simplest level of organization of any known metazoan. Mesozoans all live as parasites in marine invertebrates and the majority of them are only 0.5 to 7 mm in length. Most of them are made up of only 20 to 30 cells arranged basically in two layers. The layers, however, are not homologous to the germ layers of higher metazoans.

There are two classes of mesozoans, the Dicyemida and the Orthonectida, but they differ so much from each other that some authorities believe they should be placed in separate phyla.

Dicyemida

The dicyemids live in the kidneys of benthic cephalopods (bottom-dwelling octopuses, cuttlefish, and their relatives). The adults, called **vermiforms** (or nematogens), are long and slender. They have an outer layer of 20 to 30 ciliated somatic cells (the number is constant within each species) surrounding a long, slender axial cell in which there are a number of small reproductive cells (Fig. 9-1, *A*). The reproductive cells by repeated division give rise to **vermiform embryos** (also called **primary nematogens**) similar to the parent. These are released from the parent's body into the urine of the host, where they grow and eventually attach to the spongy tissue of the kidney and begin to reproduce. The kidney lining may eventually be covered with thousands of these wormlike parasites.

As the population in the kidney becomes overcrowded, some of the adult members begin another type of reproduction by which dispersal of the species can occur. Instead of developing into vermiform embryos, the reproductive cells develop into a gonadlike structure that produces male and female gametes that fuse in fertilization. The zygotes develop into minute (0.04 mm) ciliated **infusoriform larvae** (Fig. 9-1, *B*) quite unlike the parent. These are shed with the urine into the seawater and sink to the bottom. Here they may be ingested by bottom-crawling cephalopods. When the infusoriforms reach the kidney of the new host (perhaps by way of the circulatory system) they develop into **stem vermiforms** (stem nematogens), which develop into adult vermiforms and produce vermiform larvae. This starts a new population, which is increased by more vermiform larvae until overpopulation again brings about the development of the infusoriform dispersion larvae. It was formerly believed

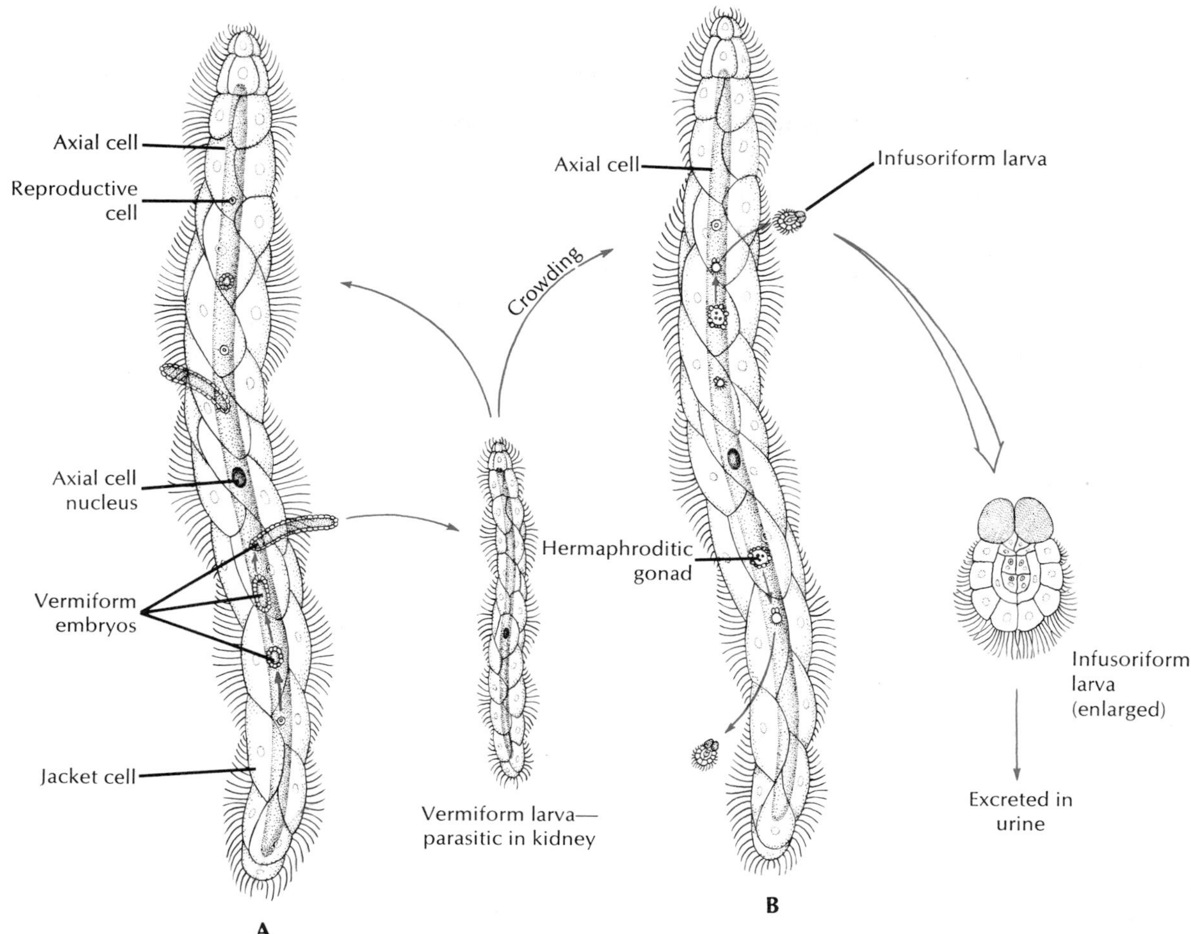

FIG. 9-1 Two methods of reproduction by mesozoans. **A,** Asexual development of vermiform larvae from reproductive cells in the axial cell of adult. **B,** Under crowded conditions in the host kidney, reproductive cells develop into gonads with gametes that produce infusoriform dispersal larvae that are shed in the host urine. (Modified from Lapan, E. A., and H. Morowitz. 1972. Sci. Am. **227:**94-101 [Dec.].)

that the infusoriforms were picked up by an intermediate host, but recent evidence indicates that no intermediate host is needed.

Orthonectida

Orthonectids parasitize a variety of invertebrates, such as brittle stars, bivalve molluscs, polychaetes, and nemerteans. The best known orthonectid is *Rhopalura,* a parasite of brittle stars. The life cycles involve sexual and asexual phases. The asexual stage is quite different from that of the dicyemids. It consists of a multinucleated mass of protoplasm called a **plasmodium,** which by division ultimately gives rise to males and females. These are composed of an outer layer of ciliated somatic cells surrounding an inner mass of germ cells (Fig. 9-2). Large numbers of the adults are shed into the seawater where fertilization occurs.

Phylogeny of mesozoans

There is still much to learn about these mysterious little parasites, but probably one of the most intriguing questions is the place of mesozoans in the evolutionary picture. Some investigators believe them to be primitive or degenerate flatworms and even believe they

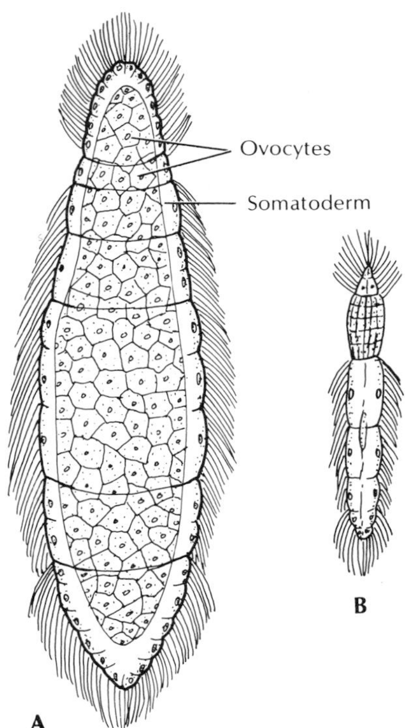

FIG. 9-2 **A,** Female and, **B,** male orthonectid *(Rhopalura).* This mesozoan parasitizes such forms as flatworms, molluscs, annelids, and brittle stars. Note that the structure consists of a single layer of ciliated epithelial cells surrounding an inner mass of sex cells.

should be classed with the Platyhelminthes. Others place them close to the protozoans as truly primitive forms, possibly related to the ciliates. Whether metazoans and mesozoans derived independently from protozoan beginnings or whether mesozoans are indeed degenerate flatworms is still an enigma.

PHYLUM PORIFERA—SPONGES

Sponges belong to phylum Porifera (po-rif′er-a) (L., *porus,* pore, + *fera,* bearing). The bodies of sponges bear myriads of tiny pores and canals that comprise a filter-feeding system adequate for their inactive life-style, for they are sessile animals. They depend on the water currents carried through their unique canal systems to bring them food and oxygen and to carry away their body wastes. Their bodies are little more than masses of cells embedded in a gelatinous matrix and stiffened by a skeleton of minute **spicules** of calcium or silica or by "spongy" fibers of a keratose substance called **spongin.** They have no organs or true tissues, and even their cells show a certain degree of independence. As sessile animals with only negligible body movement, they have not evolved a nervous system or sense organs and have only the simplest of contractile elements.

So, although they are multicellular, sponges share few of the characteristics of other metazoan phyla. They seem to be outside the line of evolution leading from the protozoans to the other metazoans—a dead-end branch. It is for this reason that they are often called the Parazoa (Gr. *para,* beside or along side of, + *zoia,* animals).

Most sponges are colonial, and they vary in size from a few millimeters to the great loggerhead sponges, which may reach 2 m or more across. Many sponge species are brightly colored because of pigments in the dermal cells. Red, yellow, orange, green, and purple sponges are not uncommon. However, the color fades quickly when the sponges are removed from the water. Some sponges, including the simplest and most primitive, are radially symmetric, but many are quite irregular in shape. Some stand erect, some are branched or lobed, others are low, even encrusting, in form. Some bore holes into shells or rocks.

The sponges are an ancient group, with an abundant fossil record extending back to the Cambrian period and even, according to some claims, the Precambrian.

Ecologic relationships

Most of the 5,000 or more sponge species are marine, although some 150 species live in fresh water. Marine sponges are abundant in all seas and at all depths, and a few even exist in brackish water. Although the embryos are free-swimming, the adults are always attached, usually to rocks, shells, corals, or other submerged objects. Some benthic forms even grow on sand or mud bottoms. Their growth patterns often depend on the shape of the substratum, the direction and speed of the water currents, and the availability of space, so that the same species may differ markedly in appearance under different environmental circumstances. Sponges in calm waters may grow taller and straighter than those in rapidly moving waters.

Many animals (crabs, nudibranchs, mites, bryozoans) live as commensals or parasites in or on sponges. The larger sponges particularly tend to harbor a large variety of invertebrate commensals. On the other hand, sponges grow on many other living animals, such as molluscs, barnacles, brachiopods, corals, or hydroids. Some crabs attach pieces of sponge to their carapaces for camouflage and for protection, as most predators seem to find sponges distasteful. Some reef fishes, however, are known to graze on shallow-water sponges.

Characteristics

1. Multicellular; body a loose aggregation of cells of mesenchymal origin
2. Body with pores (ostia), canals, and chambers that serve for passage of water
3. Mostly marine; all aquatic
4. Radial symmetry or none
5. Epidermis of flat pinacocytes; most interior surfaces lined with flagellated collar cells (choanocytes) that create water currents; a gelatinous protein matrix called mesohyl (mesoglea) contains amebocytes, collencytes and skeletal elements
6. Skeleton of calcareous or siliceous crystalline spicules, protein spongin, or a combination
7. No organs or true tissues; digestion intracellular; excretion and respiration by diffusion
8. Reactions to stimuli apparently local and independent; nervous system probably absent
9. All adults sessile and attached to substratum
10. Asexual reproduction by buds or gemmules and sexual reproduction by eggs and sperm; free-swimming ciliated larvae

Classes

There are three classes of sponges, classified mainly by the kinds of skeletons they possess.

Class Calcispongiae (cal-si-spun'je-e) (L., *calcis,* lime, + Gr., *spongos,* sponge) **(Calcarea).** Have spicules of calcium carbonate that often form a fringe around the osculum (main water outlet). Spicules are needle-shaped or three- or four-rayed. All three types of canal systems (asconoid, syconoid, leuconoid) represented. All marine. Examples: *Scypha, Leucosolenia.*

Class Hyalospongiae (hy'a-lo-spun'je-e) (Gr., *hyalos,* glass, + *spongos,* sponge) **(Hexactinellida).** Have three dimensional, six-rayed, siliceous spicules extending at right angles from a central point; spicules often united to form network; body often cylindric or funnel-shaped. Flagellated chambers in simple syconoid or leuconoid arrangement. Habitat mostly deep water; all marine. Examples: Venus's flower basket *(Euplectella), Hyalonema.*

Class Demospongiae (de-mo-spun'je-e) (Gr., *demos,* people, + *spongos,* sponge). Have siliceous spicules that are not six-rayed, or spongin, or both. Leuconoid-type canal systems. One family found in fresh water; all others marine. Examples: *Thenea, Cliona, Spongilla, Myenia,* and all bath sponges.

Form and function

The only body openings of these unusual animals are pores, usually many tiny ones called **ostia** for incoming water, and a few large ones called **oscula** (sing., **osculum**) for water outlet. These openings are connected by a system of canals, some of which are lined with peculiar flagellated collar cells called **choanocytes,** whose flagella maintain a current of environmental water through the canals. Water enters the canals through a multitude of tiny incurrent pores **(dermal ostia)** and leaves by way of one or more large oscula. The choanocytes not only keep the water moving but also trap and phagocytize food particles that are carried in the water. The cells lining the passageways are very loosely organized. Collapse of the canals is prevented by the skeleton, which, depending on the species, may be made up of needlelike calcareous or siliceous spicules, a meshwork of organic spongin fibers, or a combination of the two.

Sessile animals make few movements and therefore need little in the way of nervous, sensory, or locomotor parts. Sponges apparently have lived as sessile animals from their earliest appearance and have never acquired specialized nervous or sensory structures, and they have only the very simplest of contractile systems.

Types of canal systems

Most sponges fall into one of three types of canal systems—asconoid, syconoid, or leuconoid (Fig. 9-3).

Asconoids—flagellated spongocoels. The asconoid sponges have the simplest type of organization. They are small and tube-shaped. Water enters through microscopic dermal pores into a large cavity called the **spongocoel,** which is lined with choanocytes. The choanocyte flagella pull the water through the pores and expel it through a single large osculum (Fig. 9-3). *Leucosolenia* is an asconoid type of sponge. Its slender, tubular individuals grow in groups attached by a common stolon, or stem, to objects in shallow seawater (Fig. 9-4). Asconoids are found only in class Calcispongiae.

Syconoids—flagellated canals. Syconoid sponges look somewhat like larger editions of asconoids, from which they were derived. They have the tubular body and single osculum, but the body wall, which is thicker and more complex than that of asconoids, contains choanocyte-lined radial canals that empty into the spongocoel (Fig. 9-3). The spongocoel in syconoids is lined with epithelial-type cells rather than the flagellated cells as in asconoids. Water enters through a large number of dermal ostia into **incurrent canals** and then filters through tiny openings called **prosopyles** into the choanocyte-lined **radial canals** (Fig. 9-5). Here food is ingested by the choanocytes, whose flagella force the water on through internal pores **(apo-**

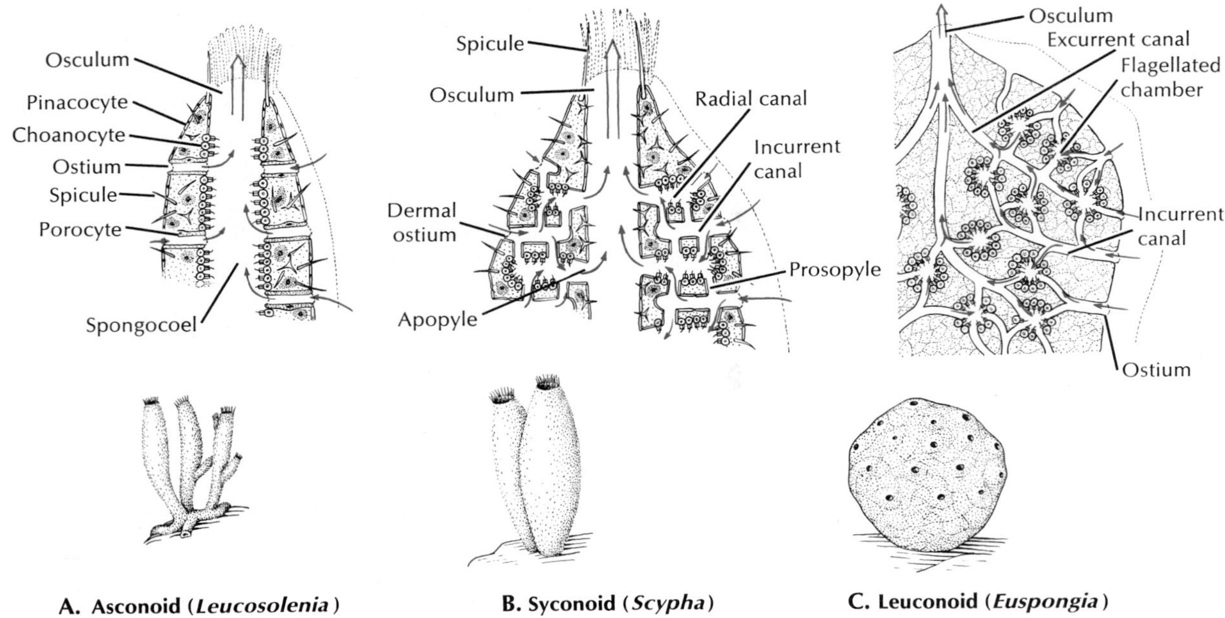

FIG. 9-3 Three types of sponge structure. The degree of complexity from simple asconoid to complex leuconoid type has involved mainly the water-canal and skeletal systems, accompanied by outfolding and branching of the collar cell layer. The leuconoid type is considered the major plan for sponges, for it permits greater size and more efficient water circulation.

pyles) into the spongocoel. From there it emerges through the osculum. Syconoids do not usually form highly branched colonies as do the asconoids. During development, syconoid sponges pass through an asconoid stage; the flagellated canals form by evagination of the body wall. This is one evidence that syconoid sponges were derived from asconoid ancestral stock. Syconoids are found in classes Calcispongiae and Hyalospongiae. *Sypha* is a commonly studied example of the syconoid type of sponge (Fig. 9-4).

Leuconoids—flagellated chambers. Leuconoid organization is the most complex of the sponge types and the best adapted for increase in sponge size. Most leuconoids form large colonial masses, each member of the mass having its own osculum, but individual members are poorly defined and often impossible to distinguish (Fig. 9-4 and p. 167). Clusters of flagellated chambers are filled from incurrent canals and discharge water into excurrent canals that eventually lead to the osculum (Fig. 9-3). Most sponges are of the leuconoid type, which occurs in both class Calcispongiae and class Demospongiae.

These three types of canal systems—asconoid, syconoid, and leuconoid—are correlated with the evolution of sponges, from simple to complex forms. Evolutionary changes involved increasing the flagellated surfaces in proportion to the volume, thus providing more collar cells to meet the food demands. This was achieved by the outpushing of the spongocoel of a simple sponge such as the asconoid type to form the radial canals (lined with flagellated cells) of the syconoid type. Further folding of the body wall produced the complex canals and chambers of the leuconoid type.

Types of cells

Sponge cells are loosely arranged in a gelatinous matrix called **mesohyl** (mesoglea, mesenchyme) (Fig. 9-6). The mesohyl is the "connective tissue" of the sponges; in it are found various ameboid cells, fibrils, and skeletal elements.

There are several types of cells in sponges.

Pinacocytes. The nearest approach to a true tissue in sponges is found in the arrangement of the **pinacocyte** cells of the epidermis (Fig. 9-6). These are thin, flat,

FIG. 9-4 A cluster of sponges removed from a dock where it was attached by a slender stalk, perhaps a bit of algae. Around the stalk (above) is a bit of breadcrumb sponge *Halichondria,* a leuconoid type sponge. To the right is a mass of branching *Leucosolenia,* an asconoid sponge, and at left are three large and two small *Scypha,* syconoid sponges. (Photograph by D. P. Wilson.)

epithelial-type cells that cover the exterior surface and some interior surfaces. Pinacocytes are somewhat contractile and help regulate the surface area of the sponge. Some of the pinacocytes are modified as contractile **myocytes,** which are usually arranged in circular bands around the oscula or pores, where they help regulate the rate of water flow.

Choanocytes. The choanocytes, which line the flagellated canals and chambers, are ovoid cells with one end embedded in the mesohyl and the other exposed. The exposed end bears a flagellum surrounded by a collar (Figs. 9-6 and 9-7). The electron microscope shows the collar to be made up of adjacent protoplasmic processes, or fibrils, connected to each other

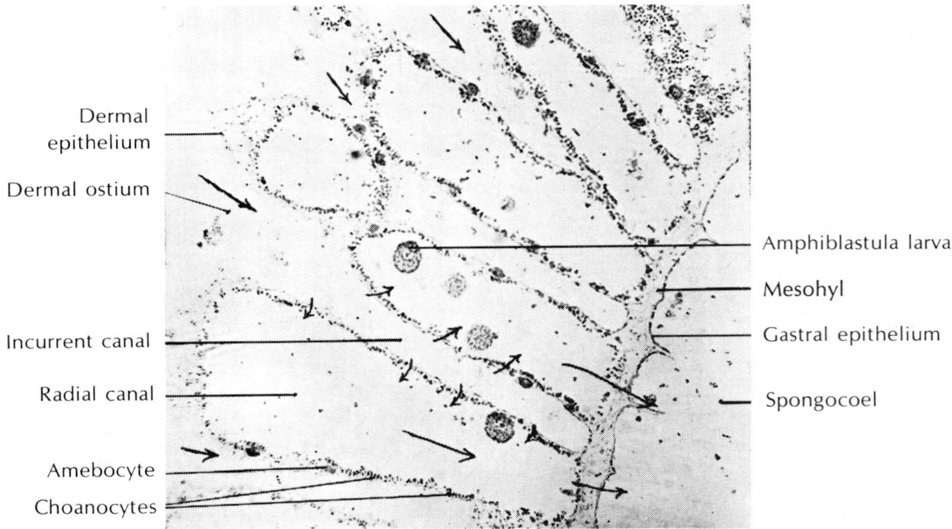

FIG. 9-5 Cross section through wall of sponge *Scypha*, showing canal system. Photomicrograph of stained slide. (Micrograph by F. M. Hickman.)

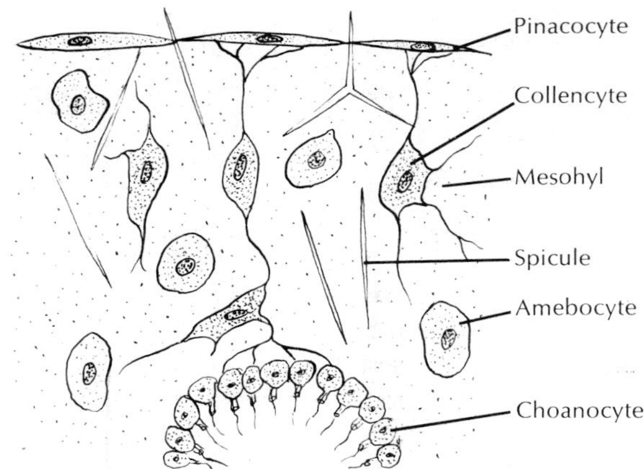

FIG. 9-6 Small section through sponge wall showing four types of sponge cells. Pinacocytes are protective and contractile; choanocytes create water currents and engulf food particles; amebocytes have a variety of functions; collencytes appear to have a contractile function.

by delicate microvilli, so that the collar forms a fine filtering device for straining food particles from the water (Fig. 9-7, *B* and *C*). The beat of the flagellum pulls water through the sievelike collar and forces it out through the open top of the collar. Particles too large to enter the collar become trapped in secreted mucus and slide down the collar to the base where they are phagocytized by the cell body. Larger particles have already been screened out by the small size of the dermal pores and prosopyles. The food engulfed by the cells may be digested in food vacuoles within the choanocyte or passed on to a neighboring amebocyte for digestion.

Amebocytes. Various ameboid cells, called amebocytes, move about in the mesohyl (Fig. 9-6) and carry out a number of functions. Some are filled with food reserves, aid in digestion, or carry pigments; others, **archaeocytes,** form reproductive cells. Some, called **scleroblasts,** secrete the spicules, and others, called **spongioblasts,** secrete the spongin fibers of the skeleton. Some star-shaped cells called **collencytes** appear to be contractile and to secrete collagenous fibrils into the mesohyl. **Lophocytes** are a highly motile type of amebocyte and function in secretion of bundles of fibrils. Some authorities have suggested a nervous func-

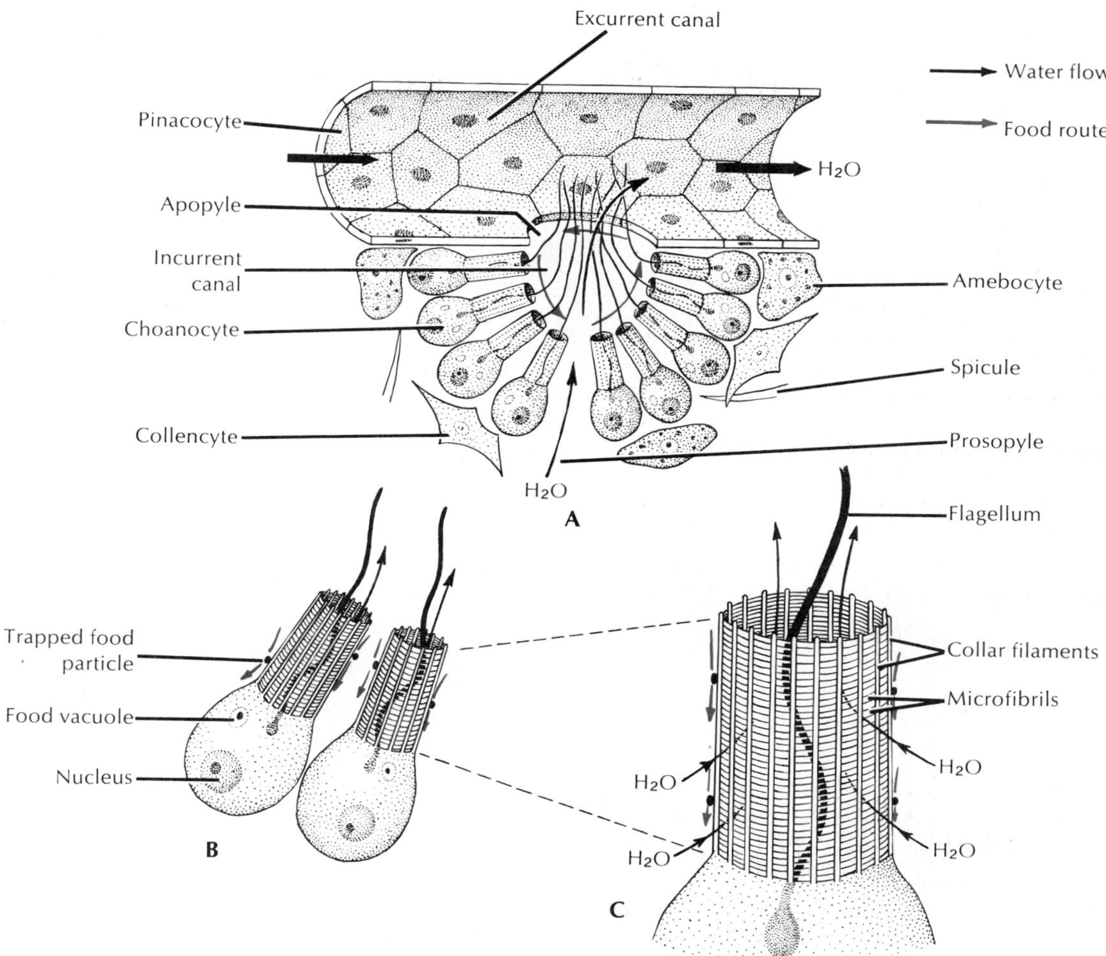

FIG. 9-7 Food trapping by sponge cells. **A,** Cutaway section of canals showing cellular structure and direction of water flow. **B,** Two choanocytes and, **C,** structure of the collar. Small red arrows indicate movement of food particles.

tion for the collencytes or lophocytes, but as yet there is no conclusive evidence to support this claim.

Types of skeletons

The skeleton gives support to the sponge, preventing collapse of the canals and chambers. In calcareous sponges the spicules are composed mostly of crystalline calcium carbonate and have one, three, or four rays (Fig. 9-8). The Demospongiae have siliceous spicules (with one, two, or four rays), spongin fibers, or both spicules and spongin. Glass sponges have siliceous spicules with six rays arranged in three planes at right angles to each other. In *Monorhaphis,* a glass sponge, the spicules are composed of an axial fiber of a protein called spiculin, surrounded by cylindrical layers of noncrystalline hydrated silica, and finally by another layer of spiculin. There are many variations in the shape of spicules, and these structural variations are of taxonomic importance.

The protein spongin forms a tough branching fibrous network that gives support to the soft lining tissues alone or in combination with spicules.

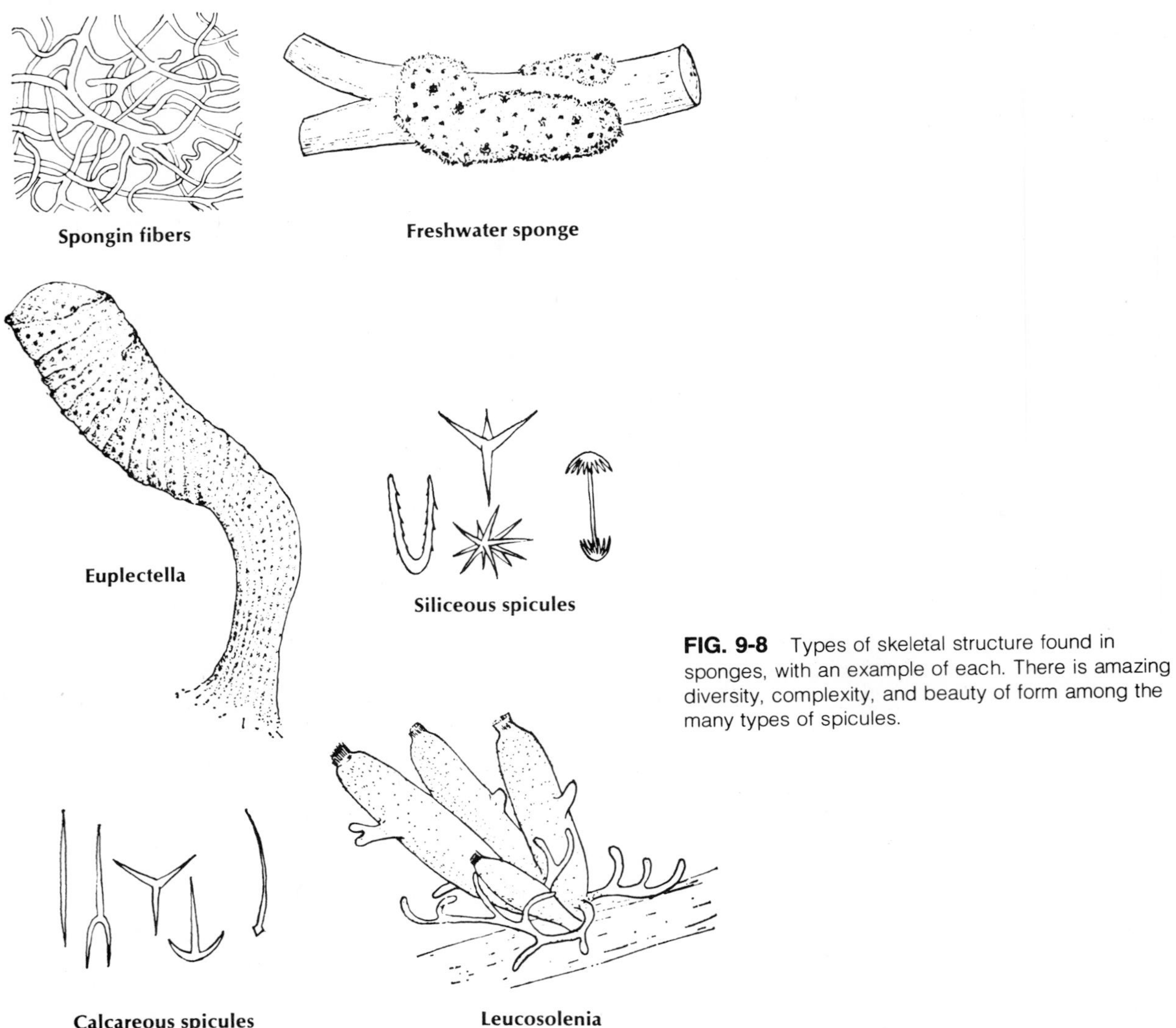

FIG. 9-8 Types of skeletal structure found in sponges, with an example of each. There is amazing diversity, complexity, and beauty of form among the many types of spicules.

Sponge physiology

Sponges apparently feed on fine detritus particles, planktonic organisms, and bacteria that are screened by the dermal pores and prosopyles and again by the choanocyte collars. Choanocytes ingest most of the food, but pinacocytes and amebocytes also can phagocytize food particles. Food is often passed from cell to cell. Digestion, which is **intracellular** (occurs within the cells), may be started in a collar cell and completed in an amebocyte or transferred to a third cell for the final stage. Thus the wandering amebocytes perform a carrier service, act as digesters, and serve as storage warehouses. Although the choanocyte-lined chambers are the chief areas for feeding activity, they are not in any sense comparable to the digestive tracts of other metazoans.

There are no respiratory or excretory organs; both functions are apparently carried out by diffusion in individual cells. Contractile vacuoles have been found in amebocytes and choanocytes of freshwater sponges.

All the life activities of the sponge depend on the current of water flowing through the body. A sponge pumps a remarkable amount of water. *Leuconia,* for example, is a small leuconoid sponge about 10 cm tall and 1 cm in diameter. It is estimated that water enters through some 81,000 incurrent canals at a velocity of 0.1 cm/second. However, *Leuconia* has over 2 million flagellated chambers whose combined diameter is much greater than that of the canals, so that in the chambers the water slows down to 0.001 cm/second, thus allowing ample opportunity for food capture by the collar cells. All of the water is expelled through a single osculum at a velocity of 8.5 cm/second—a jet force capable of carrying waste products some distance away from the sponge. Some large sponges have been found to filter 1,500 liters of water a day.

Only the finest food particles reach the choanocytes because of the screening action of the ostia, the prosopyles, and the collars. Ostia average 50 μm in diameter and prosopyles about 5 μm. The tiny spaces between the cytoplasmic tentacles that make up a collar are only about 0.1 μm wide. Thus the choanocytes capture the smallest particles, and slightly larger bits may be engulfed by amebocytes or pinacocytes.

The only visible activities and responses, other than the propulsion of water, are slight alterations in shape and the closing and opening of the incurrent and excurrent pores, and these movements are very slow. The most common response is closure of the oscula. Apparently excitation spreads from cell to cell, although some zoologists point to the possibility of coordination by means of substances carried in the water currents, and some have tried, not very successfully, to demonstrate the presence of nerve cells.

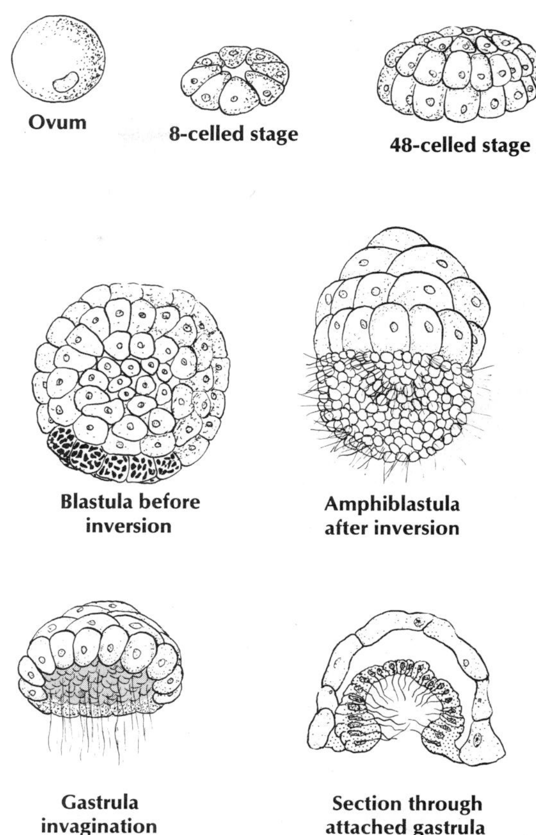

FIG. 9-9 Development of the sponge *Scypha.*

Reproduction

Sponges reproduce both asexually and sexually. **Asexual reproduction** occurs by means of bud formation. **External buds,** after reaching a certain size, may become detached from the parent and float away to form new sponges, or they may remain to form colonies. **Internal buds,** or **gemmules,** are formed in freshwater sponges and some marine sponges. Here, archaeocytes are collected together in the mesohyl and become surrounded by a siliceous shell or by a cluster of spicules. When the parent animal dies, the gemmules survive and remain dormant, thus preserving the life of the species during periods of freezing or severe drought. Later the cells in the gemmules escape through a special opening, the **micropyle,** and develop into new sponges. Gemmulation in freshwater sponges (Spongillidae) is thus an adaptation to the changing seasons. Gemmules are also a means of colonizing new habitats, since they can be spread by streams or by animal carriers. What prevents the gemmules from hatching during the season of formation rather than remaining dormant? It has been shown that some species secrete a substance that inhibits early germination of the gemmules. Gemmule formation is apparently influenced by both light and temperature, although sponges have no known photoreceptors.

It is interesting to note that investigators working on *Ephydatia fluviatilis,* a freshwater sponge, have discovered that there are several strains of the species, and that sponges hatched from gemmules of one strain can fuse with others of the same strain, but not with products of gemmules of any other strain. The differences in strains are assumed to be genetic.

In **sexual reproduction** ova and sperm develop from archaeocytes or from choanocytes. The ova are fertilized in the mesohyl, the embryos develop there, and they finally break out into the canals and spongocoel to be carried out the osculum. Embryogenesis follows a very unusual path, which is found in no other metazoan group. In the Calcispongiae and primitive Demospongiae a hollow blastula develops, with flagellated cells toward the interior. The blastula then turns *inside out* **(inversion),** the flagellated ends of the cells becoming directed to the outside! The larva is now called an **amphiblastula** (Figs. 9-5 and 9-9). Its flagellated cells **(micromeres)** are at one end, and the larger, nonflagellated cells **(macromeres)** are at the other. In contrast to other metazoan embryos, the micromeres invaginate into or are overgrown by the macromeres. The flagellated micromeres become the choanocytes of the new sponge (Fig. 9-9). The free-swimming larva of other sponges is a solid-bodied **parenchymula.** The outwardly directed, flagellated cells migrate to the interior after the larva settles and become the choanocytes in the flagellated chambers.

Some sponges are **monoecious** (have both male and female sex cells in one individual) others are **dioecious** (have separate sexes).

Regeneration and somatic embryogenesis

Sponges have a tremendous ability to repair injuries and to restore lost parts, a process called **regeneration.** Regeneration does not imply a reorganization of the entire animal, but only of the wounded portion.

On the other hand, if a sponge is cut into small fragments, or if the cells of a sponge are entirely dissociated and are allowed to fall into small groups, or aggregates, entire new sponges can develop from these fragments or aggregates of cells. This process has been termed **somatic embryogenesis.** Somatic embryogenesis involves a complete reorganization of the structure and functions of the participating cells or bits of tissue. Isolated from the influence of adjoining cells, they can realize their own potential to change in shape or function as they develop into a new organism.

A great deal of experimental work is currently being done in this field. The process of reorganization appears to differ in sponges of differing complexity. There is still some controversy concerning just what mechanisms lead to the adhesion of the cells and the share that each type of cell plays in the formative process.

Class Calcispongiae (Calcarea)

The Calcispongiae are the calcareous sponges, so-called because their spicules are composed of calcium carbonate. The spicules are straight (monaxons) or three- or four-rayed. These sponges tend to be small—10 cm or less in height—and tubular or vase-shaped. They may be asconoid, syconoid, or leuconoid in structure. Though many are drab in color, some are bright yellow, red, green, or lavender. *Leucosolenia* and *Scypha* are marine shallow-water forms commonly used in the laboratory. *Leucosolenia* is a small asconoid sponge that grows in branching colonies, usually arising from a network of horizontal, stolonlike tubes (Fig. 9-4). *Scypha* is a solitary sponge that may live singly or form clusters by budding (Fig. 9-4). The vase-shaped, typically syconoid animal is 1 to 3 cm long, with a fringe of straight spicules around the osculum that discourages small animals from entering. Spicules projecting from the body wall give the animal a bristly appearance. *Sypha (Sycon),* the form studied in North America, is often confused with a European genus, *Grantia,* but differs from it through lacking an outer body covering, or cortex.

Class Hyalospongiae (Hexactinellida)—glass sponges

The glass sponges make up the class Hyalospongiae. They are nearly all deep-sea forms that are collected by dredging. Most of them are radially symmetric, with vase- or funnel-shaped bodies that are usually attached by stalks of root spicules to a substratum (Fig. 9-8, *Euplectella*). In size they range from 7.5 to 10 cm to more than 1.3 m in length. Their distinguishing features are the skeleton of six-rayed siliceous spicules that are commonly bound together into a network forming a glasslike structure, and the **trabecular net** of living tissue produced by the fusion of the pseudopodia of many types of amebocytes. Within the trabecular net are elongated finger-shaped chambers lined with choanocytes and opening into the spongocoel. The osculum is unusually large and may be covered over by a seivelike plate of silica. There is no epidermis or gelatinous mesohyl, and both the external surface and spongocoel are lined with the trabecular net. The skeleton is rigid, and muscular elements (myocytes) appear to be absent. The general arrangement of the chambers fits glass sponges into both syconoid and leuconoid types. Their structure is adapted to the slow constant currents of sea bottoms, for the channels and pores of the sponge wall are relatively large and uncomplicated and permit an easy flow of water. Little, however, is known about their physiology.

The latticelike network of spicules found in many

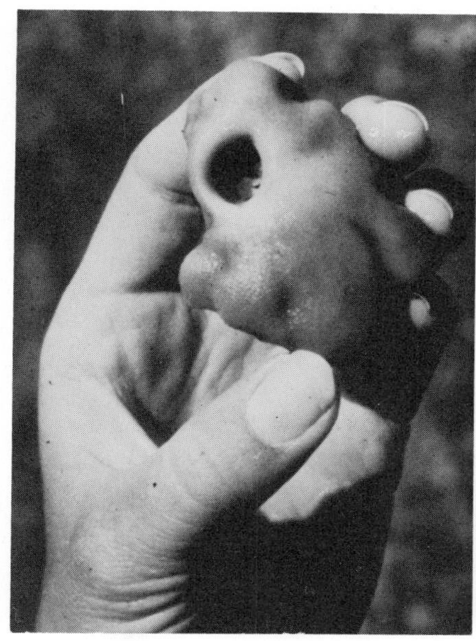
A

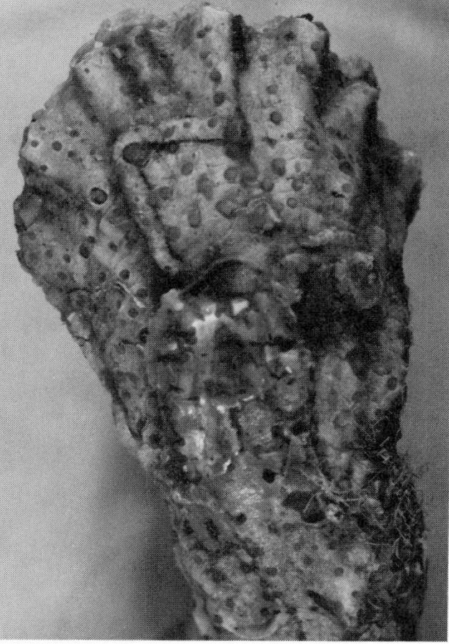
B

FIG. 9-10 **A,** *Suberites ficus* containing a hermit crab. The sponge grows over and dissolves the snail shell in which the crab lives, leaving the crab encased in sponge for shelter. **B,** *Cliona celata,* the sulfur sponge, bores into mollusc shells. This one is an oyster shell. Class Demospongiae. (**A,** Courtesy G. J. Bakus. **B,** Photograph by F. M. Hickman.)

glass sponges is of exquisite beauty, such as that of *Euplectella,* or Venus's flower basket (Fig. 9-8), a classic example of the Hyalospongiae.

Class Demospongiae

Class Demospongiae contains more than 80% of the sponge species, including most of the larger sponges. The spicules are siliceous but are not six-rayed, and they may be bound together by spongin or may be absent altogether. All members of the class are leuconoid and all are marine except one family, the Spongillidae, or freshwater sponges.

Freshwater sponges are widely distributed in well-oxygenated ponds and streams, where they are found encrusting plant stems and old pieces of submerged wood. They may resemble a bit of wrinkled scum, be pitted with pores, and be brownish or greenish in color. Common genera are *Spongilla* and *Myenia.* Freshwater sponges are most common in midsummer, although some are more easily found in the fall. They die and disintegrate in late autumn, leaving the gemmules (already described) to produce the next year's population. They also reproduce sexually. When examined closely, freshwater sponges reveal a thin dermis overlying large subdermal spaces (separated by columns of spicules) with many water channels in the interior. There are usually several oscula, each of which (at least in *Myenia*) is mounted on a small chimneylike tube. Their spiculation also includes a spongin network.

The marine Demospongiae are quite varied in both color and shape. Some are encrusting (Fig. 9-10, *A*); some are tall and fingerlike; some are low and spreading (p. 167); some bore into shells (Fig. 9-10, *B*); and some are shaped like fans, vases, cushions, or balls

FIG. 9-11 Two specimens of a football-shaped sponge *Geodia baretti,* which occur in great numbers in Scandinavian waters below 60 m. (Photograph by T. Lundälv, Kristinebergs Zoological Station, Sweden.)

(Fig. 9-11). Loggerhead sponges may grow several meters in diameter.

The so-called bath sponges *(Spongia, Hippospongia)* belong to the group called horny sponges, which have spongin skeletons and lack spicules entirely.

Commercial sponges are collected by hooks, by dredging or trawling, or by divers. After their collection they are exposed out of water to kill them and are then placed in shallow water, where they are squeezed or treaded on to remove the softened animal matter until only the horny, spongin skeleton remains. After being cleaned and bleached, they are trimmed and sorted for the market.

Sponges are often cultured by cutting out pieces of the individual animals, fastening them to concrete or rocks, and dropping them into the proper water conditions. It takes many years for sponges to grow to market size. Most of the household sponges now on the market are synthetic.

Phylogeny and adaptive radiation

Phylogeny. The origin of sponges dates back before the Cambrian period. Two groups of calcareous spongelike organisms occupied early Paleozoic reefs. The Devonian period saw the rapid development of many glass sponges. That sponges are related to the protozoans is shown in their phagocytic method of nutrition and the resemblance of their flagellated larvae to colonial protozoans. The theory that sponges arose from choanoflagellates (protozoans that bear collars and flagella), similar perhaps to the colonial *Proterospongia* (Fig. 8-15), earned support for a time. However, many zoologists object to the theory because sponges do not acquire collars until late in their embryologic development. The outer cells of the larvae are flagellated but not collared, and they do not become collar cells until they become internal. Also, collar cells are found in certain corals and echinoderms, so they are not unique to the sponges.

Another theory is that sponges derived from a hollow, free-swimming colonial flagellate, such as gave rise to the ancestral stocks of other metazoans. Certainly the sponge larvae resemble such flagellate colonies. The curious process of inversion occurs in the colonial phytoflagellate *Volvox*. The development of the water canals and the movement of the flagellated cells to the interior to become choanocytes may have occurred as the sponges began to assume a sessile existence. Whatever the origin, it is obvious that the sponges diverged early from the main line leading to other metazoans. That they are remote, phylogenetically, from other metazoans is shown by their low level of organization, the independent nature of their cells, the absence of organs, and their body structure built around a system of water canals. They became a "dead-end" phylum.

Adaptive radiation. The Porifera have been a highly successful group that has branched out into several thousand species and a variety of marine and freshwater habitats. Their diversification centers largely around their unique water-current system and its various degrees of complexity. The proliferation of the flagellated chambers in the leuconoid sponges was more favorable to an increase in body size than that of the asconoid and syconoid sponges because facilities for feeding and gaseous exchange were greatly enlarged.

Derivation and meaning of names

Dicyemida (Gr., *di-*, two, + *kyēma,* embryo) An order of the Mesozoa.

Euplectella (NL from Gr., *euplektos,* well-plaited) Genus of glass sponge having a basketlike skeleton of interwoven siliceous spicules (Hyalospongiae).

Hippospongia (Gr., *hippos,* horse, + L., *spongia,* sponge) Genus of sponge (Demospongiae).

Leucosolenia (Gr., *leukos,* white, + *sōlen,* pipe) Genus of sponges (Calcispongiae).

Orthonectida (Gr., *orthos,* straight, + *nektōs,* swimming) An order of the Mesozoa.

Parazoa (Gr., *para,* beside, + *zōon,* animal) Sponges are so called because they do not appear to be closely related to any group of the Metazoa.

Scypha (Gr., *skyphos,* cup) This genus is often incorrectly called *Grantia* or *Sycon* (Calcispongiae).

Spongilla (L., *spongia* from Gr., *spongos,* sponge) Genus of freshwater sponge (Demospongiae).

Annotated references
Selected general readings

Barnes, R. D. 1974. Invertebrate zoology, ed. 3. Philadelphia, W. B. Saunders Co.

Brien, P. 1968. The sponges, or Porifera. In M. Florkin and B. T. Sheer (eds.). Chemical zoology, vol. 2. New York, Academic Press, Inc.

Brown, C. H. 1975. Structural materials in animals. New York, John Wiley & Sons, Inc.

Bullock, T. H., and G. A. Horridge. 1965. Structure and function of the nervous system of invertebrates, vol. 1. San Francisco, W. H. Freeman Co., pp. 450-453.

Elvin, D. W. 1976. Seasonal growth and reproduction of an intertidal sponge, *Haliclona permollis* (Bowerbank). Biol. Bull. (Woods Hole) **151**(1):108-125.

Florkin, M., and B. T. Scheer. 1968. Chemical zoology, vol. 2. New York, Academic Press, Inc. *Section I contains six papers on Porifera.*

Fry, W. G. 1970. Biology of the Porifera. New York, Academic Press, Inc. *A collection of papers presented at a symposium of the Zoological Society of London.*

Garrone, R., and J. Pottu-Boumendil. 1976. Cell movements, collagen biosynthesis and organization in the Porifera. Bull. Soc. Zool. Fr. **101**(1):23-29. *In French with English summary.*

Giese, A. C., and S. S. Pearse (eds.) 1974. Reproduction of marine invertebrates, vol. 1. New York, Academic Press, Inc.

Gilbert, J. J., and T. L. Simpson. 1976. Gemmule polymorphism in the freshwater sponge *Spongilla lacustris*. Arch. Hydrobiol. **78**(2):268-277. *Describes two types of gemmules and discusses their adaptive significance.*

Gilbert, J. J., and T. L. Simpson. 1976. Sex reversal in a freshwater sponge. J. Exp. Zool. **195**(1):145-151.

Gosner, K. L. 1971. Guide to identification of marine and estuarine invertebrates: Cape Hatteras to the Bay of Fundy. New York, John Wiley & Sons, Inc., pp. 51-65.

Hickman, C. P. 1973. Biology of the invertebrates, ed. 2. St. Louis, The C. V. Mosby Co.

Hyman, L. H. 1940. The invertebrates: Protozoa through Ctenophora, vol. 1. New York, McGraw-Hill Book Co. *Contains chapters on Mesozoa and Porifera.*

Jewel, M. 1959. Porifera. In W. T. Edmondson, H. B. Ward, and G. C. Whipple (eds.). Freshwater biology, ed. 2. New York, John Wiley & Sons, Inc. *Contains key to common freshwater sponges.*

Jones, W. C. 1962. Is there a nervous system in sponges? Biol. Rev. **37:**1-150. *The author refutes the idea that sponges have sensory and ganglionic cells.*

Jørgensen, C. B. 1966. The biology of suspension feeding. New York, Pergamon Press.

Kaestner, A. 1967. Invertebrate zoology, vol. 1. New York, John Wiley & Sons, Inc.

Korotkova, G. P. 1970. Regeneration and somatic embryogenesis in sponges. In W. G. Fry (ed.). Biology of the Porifera. New York, Academic Press, Inc., pp. 423-436.

Light, S. F., R. I. Smith, F. A. Pitelka, D. P. Abbott, and F. M. Weesner. 1967. Intertidal invertebrates of the central California coast. Berkeley, University of California Press. *Contains keys to common West Coast sponges.*

Long, M. E., and D. Doubelet. 1977. Consider the sponge. Nat. Geogr. **151**(3):392-407. *Beautiful color photographs of sponges.*

McConnaughey, B. H. 1963. The Mesozoa. In E. C. Dougherty (ed.). The lower Metazoa: comparative biology and phylogeny. Berkeley, University of California Press. *The author stresses the resemblance between the Mesozoa and the parasitic flatworms.*

McConnaughey, B. H. 1968. The Mesozoa. In M. Florkin and B. T. Scheer. Chemical zoology, vol. 2. New York, Academic Press, Inc.

Pennak, R. W. 1953. Freshwater invertebrates of the United States. New York, The Ronald Press Co. *Has chapter on freshwater sponges and a key to North American species.*

Rasmont, R. 1968. Chemical aspects of hibernation. In M. Florkin and B. T. Scheer. Chemical zoology, vol. 2. New York, Academic Press, Inc.

Rasmont, R. 1970. Some new aspects of the physiology of freshwater sponges. In W. G. Fry (ed.). Biology of the Porifera. New York, Academic Press, Inc. *He believes that the gemmules of some species, or at least of some strains of sponges, undergo a true diapause, similar to that of insects.*

Schmidt, G. D., and L. S. Roberts. 1977. Foundations of parasitology. St. Louis, The C. V. Mosby Co., pp. 182-189. *Has a chapter on the Mesozoa.*

Stunkard, H. W. 1972. Clarification of taxonomy in Mesozoa. Syst. Zool. **21**(2):210-214. *Contains also a good literature review.*

Vacelet, J. 1976. Electron microscope study of the association between bacteria and sponges of the genus *Verongia* (Dictyoceratida). J. Micros. Biol. Cell **23**(3):271-288. *Discusses the physiologic importance of a bacterial population, mainly in nutrition.*

Van Beneden, E. 1876. Recherces sur les dicyémides. Bruxelles Acad. Roy. Belg. Bull. Cl. Sci. **41:**1160-1205, **42:**35-97. *One of the first investigations on the group that was discovered in 1839. He considered them intermediate between the Protozoa and the Metazoa.*

Wilson, H. V. 1907. On some phenomena of coalescence and regeneration in sponges. J. Exp. Zool. **5:**245-258. *This classic experimental work on siliceous sponges first showed the phenomenon of regeneration after dissociation. A new sponge is formed by aggregation and fusion out of the cells of an old sponge, which have been separated by squeezing through a piece of gauze. This phenomenon also occurs in forms other than the Porifera.*

Selected *Scientific American* articles

Lapan, E. A., and H. J. Morowitz. 1972. The Mesozoa. **227:**94-101 (Dec.). *The author presents evidence that the Dicyemida (the Orthonectida may represent a different group because of their complexity) are not simplified flatworms but are at the simplest known level of multicellular organization and may help in understanding the processes of cell differentiation.*

Newell, N. D. 1972. The evolution of reefs. **226**(6):54-65.

CHAPTER 10

THE RADIATE ANIMALS

Phylum Cnidaria
Phylum Ctenophora

Sea anemones, the "flowers of the sea," are marine cnidarians and are radially symmetric. They are carnivores; they use myriads of tiny stinging cells on their tentacles to paralyze their prey and then bend the tentacles to carry the food to the mouth.

Courtesy R. C. Hermes.

Position in animal kingdom

The two phyla Cnidaria and Ctenophora make up the radiate animals, which are characterized by **primary radial** or **biradial symmetry,** and represent the most primitive of the eumetazoans. Radial symmetry, in which the body parts are arranged concentrically around the oral-aboral axis, is particularly suitable for **sessile** or sedentary animals. Biradial symmetry is basically a type of radial symmetry in which only two planes through the oral-aboral axis divide the animal into mirror images because of the presence of some part that is single or paired. All other eumetazoans have a primary bilateral symmetry, that is, they are bilateral or were derived from an ancestor that was bilateral.

Neither phylum has advanced generally beyond the **tissue level of organization,** although a few organs occur. In general the ctenophores have a more complex structural grade than that of the cnidarians.

Biologic contributions

1. Both phyla have developed two well-defined **germ layers,** ectoderm and endoderm; a third, or mesodermal, layer, which is derived embryologically from the ectoderm, is present in some. The body plan is saclike, and the body wall is composed of two distinct layers, epidermis and gastrodermis, derived from the ectoderm and endoderm, respectively. The gelatinous matrix, mesoglea, between these layers may be structureless, may contain a few cells and fibers, or may be composed largely of mesodermal connective tissue and muscle fibers.
2. An internal body cavity, the **gastrovascular cavity,** is lined by the gastrodermis and has a single opening, the mouth, which also serves as the anus.
3. **Extracellular digestion** occurs in the gastrovascular cavity, as does intracellular digestion in the gastrodermal cells.
4. Most radiates have **tentacles,** or extensible projections around the oral end, that aid in food capture.
5. The first true **nerve cells** (protoneurons) occur in the radiates, but the nerves are arranged as a nerve net, with no central nervous system.
6. **Sense organs** appear first in the radiates and include well-developed statocysts (organs of equilibrium) and ocelli (photosensitive organs).
7. Locomotion in the free-moving forms is achieved either by **muscular contractions** (cnidarians) or **ciliary comb plates** (ctenophores). However both groups are still better adapted to floating or being carried by currents than to strong swimming.
8. **Polymorphism** in the cnidarians has widened their ecologic possibilities. In many species the presence of both a polyp (sessile and attached) stage and a medusa (free-swimming) stage permits occupation of a benthic (bottom) and a pelagic (open-water) habitat by the same species.
9. Some unique features are found in these phyla, such as **nematocysts** (stinging organoids) in cnidarians and **colloblasts** (adhesive organoids) and **ciliary** comb plates in ctenophores.

PHYLUM CNIDARIA

The phylum Cnidaria (ny-dar′e-a) (Gr., *knide,* nettle, + L., *-aria,* [pl. suffix] like or connected with) is a large and interesting group of more than 9,000 species. It takes its name from the cells called **cnidocytes,** which contain the stinging organoids **(nematocysts)** that are so characteristic of the phylum. Nematocysts are *formed and used* only by cnidarians and by one species of ctenophore. Another name for the phylum, Coelenterata (se-len′te-ra′ta) (Gr., *koilos,* hollow, + *enteron,* gut, + L., *-ata* [pl. suffix] characterized by), is used less frequently than formerly, and it sometimes is now employed to refer to both radiate phyla, since its meaning is equally applicable to both.

The cnidarians are generally regarded as being close to the basic stock of the metazoan line. Although their organization has a structural and functional simplicity not found in other Metazoa, they are a rather successful phylum, forming a significant proportion of the biomass in some locations. They are widespread in marine habitats, and there are a few in fresh water. Though they are sessile or, at best, fairly slow-moving or slow-swimming, they are quite efficient predators of organisms that are much more complex and swift. The phylum includes some of nature's strangest and loveliest creatures—the branching, plantlike hydroids; the flowerlike sea anemones; the jellyfish; and those architects of the ocean floor, the corals, sea whips, sea fans, sea pansies, and all the hard corals whose thousands of years of calcareous house-building have produced great reefs and coral islands (Fig. 10-1).

Ecologic relationships

Cnidarians are found most abundantly in shallow marine habitats, especially in warm temperate and

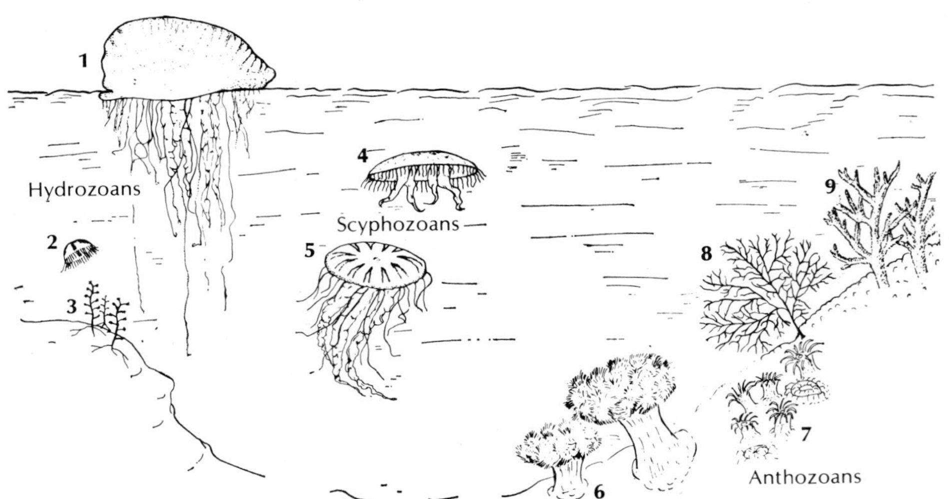

FIG. 10-1 Representative cnidarians. Class Hydrozoa: **1,** *Physalia;* **2,** *Gonionemus;* **3,** *Obelia.* Class Scyphozoa: **4,** *Aurelia;* **5,** *Chrysaora.* Class Anthozoa: **6,** *Metridium;* **7,** *Astrangia;* **8,** *Gorgonia;* **9,** staghorn coral *(Acropora).* (Not shown to scale.)

tropical regions. There are no terrestrial species. Colonial hydroids are found usually in shallow coastal water attached to mollusc shells, rocks, wharves, and other animals, but some species are found at great depths. Floating and free-swimming medusae are found in open seas and lakes, often a long distance from the shore. Floating colonies such as the Portuguese man-of-war and *Velella* have floats or sails by which the wind carries them.

Some molluscs and flatworms eat hydroids bearing nematocysts and utilize these stinging cells for their own defense. Some other animals feed on cnidarians, though cnidarians rarely serve as food for humans.

Cnidarians sometimes live symbiotically with other animals, often as commensals on the shell or other surface of their host. Algae may live as mutuals in the tissues of cnidarians, notably in some freshwater hydras and in reef-building corals. The presence of the algae in reef-building corals limits the occurrence of coral reefs to relatively shallow, clear water where there is sufficient light for the photosynthetic requirements of the algae. These kinds of corals form an essential component of coral reefs, and reefs are extremely important habitats in tropical waters. Additional comments on coral reefs are given later in the chapter.

Economic importance

Though many cnidarians have little economic importance, reef-building corals are an important exception. Fish and other animals associated with reefs provide substantial amounts of food for humans, and reefs are of economic value as tourist attractions. Precious coral serves for jewelry and ornaments, and coral rock serves for building purposes.

Planktonic medusae may be of some importance as food for fish that are of commercial value; on the other hand, the reverse is also true—the young of the fish fall prey to cnidarians.

Characteristics

1. Entirely aquatic, some in fresh water but mostly marine
2. **Radial symmetry** or biradial symmetry around a longitudinal axis with **oral** and **aboral** ends; no definite head
3. Two basic types of individuals: **polyps** and **medusae**
4. Exoskeleton or endoskeleton of chitinous, calcareous, or protein components in some
5. Body with two layers, epidermis and gastrodermis, with mesoglea **(diploblastic);** mesoglea with cells and connective tissue (ectomesoderm) in some **(triploblastic)**
6. **Gastrovascular cavity** (often branched or divided with septa) with a single opening that serves as both mouth and anus; extensible tentacles usually encircling the mouth or oral region

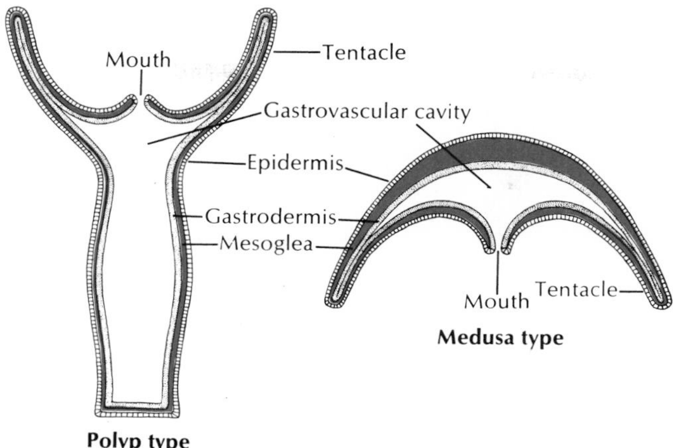

FIG. 10-2 Comparison between the polyp and medusa types of individuals.

7. Special stinging cell organoids called **nematocysts** in either or both epidermis and gastrodermis; nematocysts abundant on tentacles, where they may form batteries or rings
8. **Nerve net** with symmetric and asymmetric synapses; with some sensory organs; diffuse conduction
9. Muscular system (epitheliomuscular type) of an outer layer of longitudinal fibers at base of epidermis and an inner one of circular fibers at base of gastrodermis; modifications of this plan in higher cnidarians, such as separate bundles of independent fibers in the mesoglea
10. Reproduction by asexual budding (in polyps) or sexual reproduction by gametes (in all medusae and some polyps); sexual forms monoecious or dioecious; **planula larva;** holoblastic cleavage
11. No excretory or respiratory systems
12. No coelomic cavity

Classes

Class Hydrozoa (hy-dro-zo'a) (Gr., *hydra,* water serpent, + *zoon,* animal). Solitary or colonial; asexual polyps and sexual medusae, although one type may be suppressed; hydranths with no mesenteries; medusae (when present) with a velum; both fresh-water and marine. Examples: *Hydra, Obelia, Physalia, Tubularia.*

Class Scyphozoa (sy-fo-zo'a) (Gr., *skyphos,* cup, + *zōon,* animal). Solitary; polyp stage reduced or absent; bell-shaped medusae without velum; gelatinous mesoglea much enlarged; margin of bell or umbrella typically with eight notches that are provided with sense organs; all marine. Examples: *Aurelia, Cassiopeia, Rhizostoma.*

Class Anthozoa (an-tho-zo'a) (Gr., *anthos,* flower, + *zoon,* animal). All polyps; no medusae; solitary or colonial; enteron subdivided by at least eight mesenteries or septa bearing nematocysts; gonads endodermal; all marine.

Subclass Zoantharia (zo'an-tha'ri-a) (NL). With simple unbranched tentacles. Sea anemones and hard corals. Examples: *Metridium, Adamsia, Astrangia, Cerianthus.*

Subclass Alcyonaria (al'cy-o-na'ri-a) (NL). With eight pinnate tentacles. Soft corals. Examples: *Tubipora, Alcyonium, Gorgonia, Renilla.*

Dimorphism in cnidarians

Cnidarians may live singly or in colonies. Two basic morphologic types of individuals are recognized in the group (Fig. 10-2).

1. **Polyps** have tubular bodies. A mouth surrounded by tentacles is located at one end; the other end is blind and usually attached by a pedal disc or other device to a substratum. Sometimes there is more than one type of polyp, each specialized for a certain function, such as feeding or reproduction.
2. **Medusae,** or free-swimming jellyfish, have bell- or umbrella-shaped bodies with a mouth located centrally on a projection of the concave side. Around the margin of the umbrella are the tentacles, which are provided with stinging cells.

Some species have both types of individuals in their life history (*Obelia* and other hydroids); others have

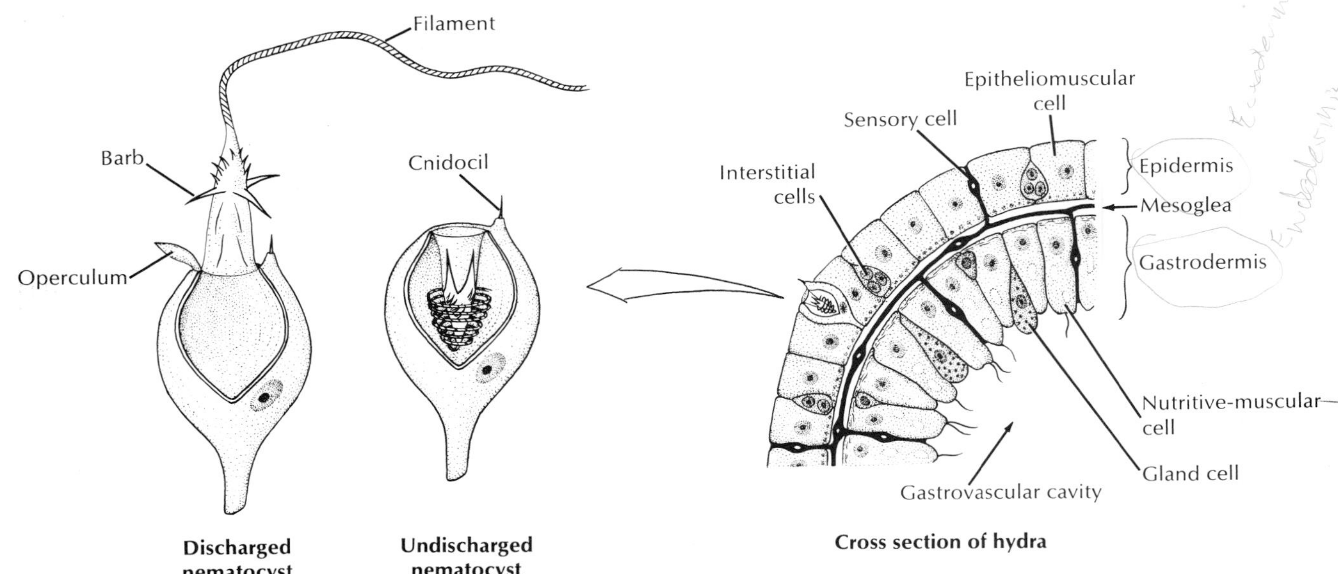

FIG. 10-3 At left, structure of a stinging cell. At right, portion of the body wall of a hydra. Cnidocytes, which contain the nematocysts, arise in the epidermis from interstitial cells.

only the polyp stage (hydras and Anthozoa); and still others have only the jellyfish, or medusa, stage (certain Scyphozoa). Many of the colonial hydroids are polymorphic, having a medusa stage and two or more types of polyps in the polyp stage.

Though polyps and medusae seem superficially to be different from each other, actually this difference is not pronounced. If a polyp form such as that of the hydra were inverted and broadened out laterally to shorten the oral-aboral axis and the mesoglea greatly increased, the result would be a structure similar to a medusa, or jellyfish (Fig. 10-2). The great amount of mesoglea in the medusa gives a neutral buoyancy to at least some species to aid in swimming.

In the evolution of the two types, polyps and medusae, probably the jellyfish represents the complete and typical cnidarian whereas the polyp is a persistent larval, or juvenile, stage.

Nematocysts—the stinging organoids

One of the most characteristic structures in the entire cnidarian group is the stinging organoid called the **nematocyst** (Fig. 10-3). Over 20 different types of nematocysts have been described in the cnidarians so far; they are important in taxonomic determinations. The nematocyst is a tiny capsule composed of material similar to chitin and containing a coiled tubular "thread" or filament, which is a continuation of the narrowed end of the capsule. This end of the capsule is covered by a little lid, or **operculum.** The inside of the undischarged thread may bear little barbs, or spines.

The nematocyst is enclosed in the cell that has secreted it, the **cnidocyte** (during its development, the cnidocyte is properly called the **cnidoblast**). Most cnidocytes are provided with a triggerlike **cnidocil,** which is a modified flagellum with a kinetosome at its base. Contact of the cnidocil with an object such as prey provides tactile stimulation for the nematocyst to discharge. Nematocysts used in defense and food capture require chemical stimulation (presence of animal fluids) to discharge, as well as tactile stimulation. Adhesive nematocysts usually do not discharge in food capture.

The mechanism of nematocyst discharge is remarkable. Inside the capsule, there is an osmotic pressure of 140 atm. When stimulated to discharge, the nematocyst membrane becomes permeable to water, and the high internal osmotic pressure causes water to rush into the capsule. The operculum opens, the increase in *hydrostatic pressure* within the capsule pushes the thread

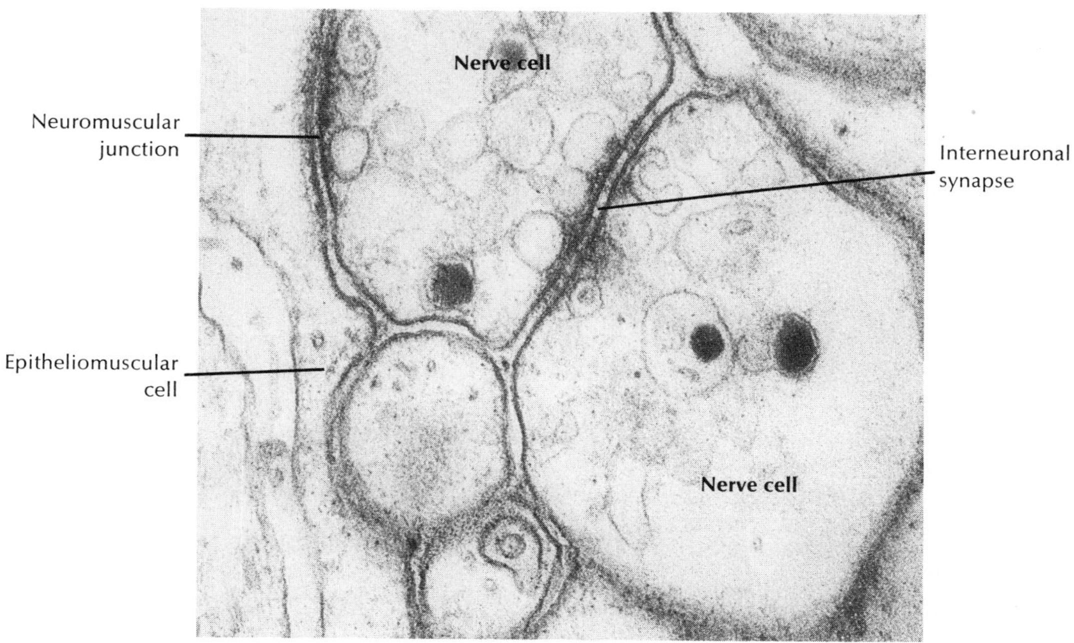

FIG. 10-4 Transmission electron micrograph of neuronal synapse and neuromuscular junction in the scyphistoma of the jellyfish *Aurelia.* Contrast the symmetric, two-way interneuronal synapse (vesicles on both sides) to the asymmetric junction of the nerve with the epitheliomuscular cell. (×67,300.) (From J. A. Westfall. 1973. Am. Zool. **13:**237.)

out with great force, and the thread turns inside out as it goes. At the everting end of the thread, the barbs flick to the outside like tiny switchblades. This minute but awesome weapon then injects poison when it penetrates the prey.

Nematocysts are considered **independent effectors,** that is, their discharge is not caused by neural stimulation. However, the nervous system plays some role in nematocyst discharge, probably by affecting the threshold of stimulation: a well-fed hydra ceases to discharge nematocysts at prey.

Cnidocytes may occur singly or in batteries consisting of one large and many small ones. They are borne in invaginations of ectodermal cells and, in some forms, in gastrodermal cells, and they are especially abundant on the tentacles. When a nematocyst is discharged, its cnidocyte is absorbed and a new one replaces it.

The nematocysts of most cnidarians are not harmful to humans, but the stings of the Portuguese man-of-war (Fig. 10-15) and certain jellyfish are quite painful and in some cases may be dangerous.

The nerve net

The nerve net of the cnidarians is one of the best examples in the animal kingdom of a diffuse nervous system. This plexus of nerve cells is found both at the base of the epidermis and at the base of the gastrodermis, forming two interconnected nerve nets. **Nerve processes (axons)** end on other nerve cells at synapses or at junctions with sensory cells or effector organs (nematocysts or epitheliomuscular cells). Nerve impulses are transmitted from one cell to another by release of a neurotransmitter from small vesicles on one side of the synapse or junction (p. 731). One-way transmission between nerve cells in higher animals is assured because the vesicles are located only on one side of the synapse. However, cnidarian nerve nets are peculiar in that many of the synapses have vesicles of neurotransmitters on both sides, allowing transmission across the synapse in either direction (Fig. 10-4). Another peculiarity of cnidarian nerves is the absence of any sheathing material (myelin) on the axons; these are apparently the only truly naked nerve fibers in the animal kingdom.

A

B

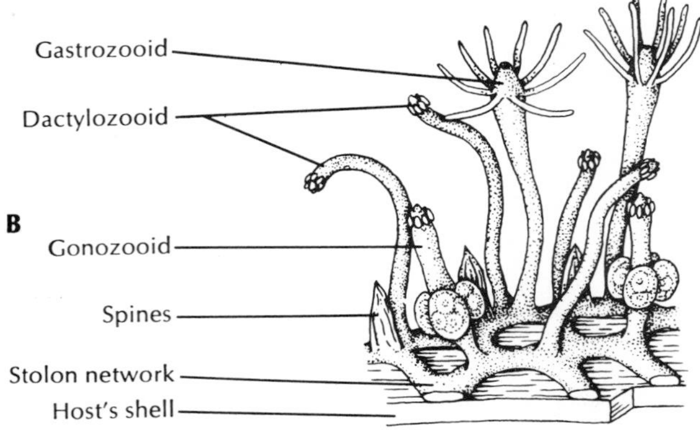

FIG. 10-5 *Hydractinia,* a colonial hydrozoan. **A,** Hermit crab, *Pagurus floridanus,* with its shell covered with a velvety colony of *Hydractinia,* and with a large anemone perched on top. **B,** Portion of a colony of *Hydractinia,* showing the types of zooids and the stolon (hydrorhiza) from which they grow. The hermit crab itself seeks out a suitable anemone, detaches it with its pincers, then holds it against its shell until the anemone attaches there. The crab is camouflaged by the cnidarians and protected by their nematocysts. The cnidarians get a free ride and bits of food from their host's meals. (**A** courtesy R. C. Hermes.)

There is no concentrated grouping of nerve cells to suggest a "central nervous system." Nerves are grouped, however, in the "ring nerves" of hydrozoan medusae and in the marginal sense organs of scyphozoan medusae. In some cnidarians the nerve nets form two or more systems: in Scyphozoa there is a fast conducting system to coordinate swimming movements and a slower one to coordinate movements of tentacles.

The nerve cells of the net have synapses with slender sensory cells that receive external stimuli, and the nerve cells have junctions with epitheliomuscular cells and nematocysts. Together with the contractile fibers of the epitheliomuscular cells, the sensory-nerve cell net combination is often referred to as a **neuromuscular system,** an important landmark in the evolution of the nervous system. The nerve net is never completely lost from higher forms. Annelids have it in their digestive systems. In the human digestive system it is represented by the plexus of Auerbach and the plexus of Meissner. The rhythmic peristaltic movements of the stomach and intestine are coordinated by this counterpart of the cnidarian nerve net.

Another type of conduction is found in at least some cnidarians. It is referred to as **neuroid conduction** and is defined as a propagation of electric events in nonnervous, nonmuscular cells. Conduction is generally through an epithelial layer and is involved in protective or locomotory responses. Local or complex responses are mediated by nerve cells. Some investigators believe that specialized nerve and muscle tissues arose from myoepithelial cells with neuroid conduction.

Class Hydrozoa

The majority of hydrozoans are marine and colonial in form, and the typical life cycle includes both the asexual polyp and the sexual medusa stages. Some, however, such as the freshwater hydra, have no medusa stage. The colonial marine hydroid *Hydractinia* (Fig. 10-5), which grows profusely on the shells of certain hermit crabs, also lacks a medusa stage. The colonial hydroid *Tubularia* develops medusa buds (gonophores) that never become free medusae but shed their gametes while still attached to the parent polyp (Fig. 10-6).

The marine jellyfish *Liriope* has no hydroid stage; the larvae develop directly into medusae. Some medusae, such as *Sarsia,* not only reproduce sexually but bud young medusae from the manubrium, a projection that hangs down around the mouth, or from the base of the tentacles.

The hydra, although exceptional, has, because of its size and ready availability, become a favorite as an in-

FIG. 10-6 A colony of *Tubularia* growing on a bit of ocean rock. Medusa buds do not detach from *Tubularia* polyps, but shed their gametes while in place. (Photograph: colony by D. P. Wilson; inset by B. Tallmark.)

troduction to this phylum. Combining its study with that of a representative colonial marine hydroid such as *Obelia* gives an excellent idea of the class Hydrozoa.

Hydra, a freshwater hydrozoan

The common freshwater hydra is a solitary polyp and one of the few cnidarians found in fresh water. Its normal habitat is the underside of aquatic leaves and lily pads in cool, clean fresh water of pools and streams. The hydra family is found throughout the world, with ten species occurring in the United States. Common species are the green hydra *(Chlorohydra viridissima),* which owes its color to symbiotic algae (zoochlorella) in its cells, and the brown hydra *(Pelmatohydra oligactis).*

Body plan. The body of the hydra can extend to a length of 25 to 30 mm or can contract to a tiny, jellylike mass. It is a cylindric tube with the lower (aboral) end drawn out into a slender stalk, ending in a basal or **pedal disc** for attachment. This pedal disc is provided with gland cells to enable the hydra to adhere to a substratum and also to secrete a gas bubble for floating. In the center of the disc there may be an excretory pore. The **mouth,** located on a conical elevation called the **hypostome,** is encircled by six to ten hollow tentacles that, like the body, can be greatly extended when the animal is hungry.

The mouth opens into the **gastrovascular cavity,** which communicates with the cavities in the tentacles. In some individuals **buds** may project from the sides, each with a mouth and tentacles like the parent. Testes or ovaries, when present, appear as rounded projections on the surface of the body (Fig. 10-9).

Body wall. The body wall surrounding the gastrovascular cavity consists of an outer **epidermis** (ectodermal) and an inner **gastrodermis** (endodermal) with **mesoglea** between them (Fig. 10-3).

EPIDERMIS. The epidermis is made up of small cubical cells and is covered with a delicate cuticle. This layer contains several types of cells—epitheliomuscular, interstitial, gland, cnidocyte, and sensory and nerve cells.

Epitheliomuscular cells make up most of the epidermis and serve both for covering and for muscular contraction. The bases of most of these cells are extended parallel to the tentacle or body axis and contain myofibrils (myonemes), thus forming a layer of longitudinal muscle next to the mesoglea. Contraction of these fibrils shortens the body or tentacles.

Interstitial cells are the undifferentiated stem cells found among the bases of the epitheliomuscular cells. Differentiation of the interstitial cells gives rise to cnidoblasts, sex cells, buds, nerve cells, and others, but generally not to epitheliomuscular cells (which reproduce themselves).

Gland cells are tall cells around the pedal disc and mouth that secrete an adhesive substance for attachment and sometimes a gas bubble for floating.

Cnidocytes (nematocytes) are found throughout the epidermis, especially on the tentacles. They contain the **nematocysts,** which they have secreted during their development as **cnidoblasts.** Three functional kinds of nematocysts are found in the hydra.

1. The **penetrant** (Fig. 10-3) is long and threadlike and bears spines along its length. When discharged, it is capable of piercing the bodies of small animals that happen to touch the tentacles, paralyzing them with the poison it injects.
2. The **volvent** is threadlike with some spines but with a closed tip. After discharge, the volvent quickly recoils, entangling the prey.
3. The **glutinant** produces an adhesive secretion and is used in locomotion and attachment.

Sensory cells are scattered among the other epidermal cells, especially around the mouth and the tentacles and on the pedal disc. The free end of each sensory cell bears a flagellum, which is the sensory receptor for chemical and tactile stimuli. The other end branches into fine processes, which synapse with the nerve cells.

Nerve cells of the epidermis are generally multipolar (have many processes), though in more highly organized cnidarians the cells may be bipolar (with two processes). Their processes (axons) form synapses with sensory cells and other nerve cells and junctions with epitheliomuscular cells and cnidocytes. There are both one-way (morphologically asymmetric) and two-way synapses with other nerve cells (Fig. 10-4).

GASTRODERMIS. The gastrodermis, a layer of cells lining the gastrovascular cavity, is made up chiefly of large, flagellated, columnar epithelial cells with irregular flat bases. The cells of the gastrodermis include nutritive-muscular, interstitial, and gland cells.

Nutritive-muscular cells are usually tall columnar cells and have laterally extended bases containing myonemes. The myofibrils run at right angles to the body or tentacle axis and so form a circular muscle layer. However, this muscle layer in hydras is very weak, and longitudinal extension of the body and tentacles is brought about mostly by increasing the volume of water in the gastrovascular cavity. The water is brought in through the mouth by the beating of the flagella on the nutritive-muscular cells. Thus, the water in the gastrovascular cavity serves as a **hydraulic skeleton.** The two flagella on the free end of each cell also serve to circulate food and fluids in the digestive cavity. The cells often contain large numbers of food vacuoles. Gastrodermal cells in the green hydra *(Chlorohydra)* bear green algae (zoochlorella), which give the hydras their color. This is probably a case of symbiotic mutualism, for the algae utilize the carbon dioxide to form organic compounds useful to the host and secrete oxygen as a by-product of their photosynthesis. They receive shelter and probably other physiologic requirements in return.

Interstitial cells are scattered among the bases of the nutritive cells. They may transform into other types of cells when the need arises.

Gland cells in the hypostome and in the column secrete digestive enzymes. Mucous glands about the mouth aid in ingestion.

Cnidocytes are not found in the gastrodermis, for nematocysts are lacking in this layer.

MESOGLEA. The mesoglea lies between the epidermis and gastrodermis and is attached to both layers. It is gelatinous, or jellylike, and has no fibers or cellular elements. It is a continuous layer that extends over both body and tentacles, thickest in the stalk portion and thinnest on the tentacles. This arrangement allows the pedal region to withstand great mechanical strain and gives the tentacles more flexibility. The mesoglea helps to support the body and acts as a type of elastic skeleton.

Locomotion. A hydra has several ways of moving from one location to another (Fig. 10-7). One is by gliding on the basal disc, aided by secretions from mucous glands. In a ''measuring-worm'' type of movement the hydra bends over and attaches its tentacles, slides its basal disc up close to the tentacles, and then releases its tentacles and straightens up. Another meth-

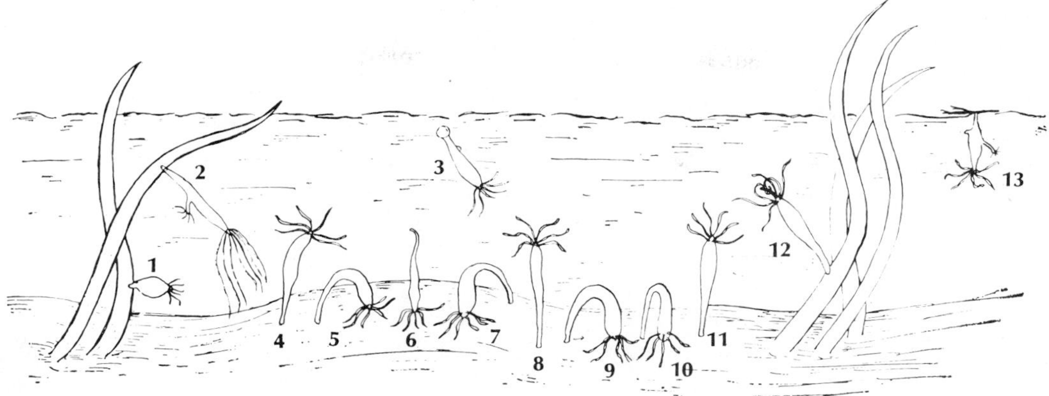

FIG. 10-7 Methods of behavior in hydra. **1,** Contracted; **2,** extended; **3,** rising to surface by bubble; **4** to **8,** steps in "somersaulting"; **9** to **11,** steps in "measuring-worm" movements; **12,** ingesting food by aid of tentacles; **13,** floating while suspended by air bubble.

od is like a handspring, in which the animal attaches its tentacles, flips its basal disc completely over, and attaches it to a new position. The hydra also may move from one place to another in an inverted position by using its tentacles as legs. To rise to the surface of the water it often forms a gas bubble on its basal disc and floats up.

Feeding and digestion. The hydra feeds on a variety of small crustaceans, insect larvae, and annelid worms. A hungry hydra waits for its food to come to it. It may, if necessary, shift to a more favorable location, but once attached to its chosen substratum, it waits with its tentacles fully extended. Any small organism that brushes against one of its tentacles is immediately stopped, and held by dozens of tiny nematocyst threads; the penetrants pierce the prey's tissues to inject a poisonous paralyzing fluid, while volvents coil themselves about bristles, hairs, or spines. The hapless prey may be several times larger than its captor. Now the tentacles, some of them attached to the prey, move slowly toward the hydra's mouth. The mouth slowly opens, and the prey slides in. It is not swallowed by muscular action; the mouth simply widens and, well moistened with mucus, glides over and around the prey (Fig. 10-8).

The activator that actually causes the mouth to open is a reduced form of glutathione, which is found to some extent in all living cells. Glutathione is released from the prey through the wounds made by the nematocysts, but only those animals that release enough of the chemical to activate the feeding response are eaten by the hydra. This explains how a hydra distinguishes between *Daphnia,* which it relishes, and some other forms that it refuses. When glutathione is placed in water containing hydras, each hydra will go through the motions of feeding even though no prey is present.

Inside the gastrovascular cavity, contraction of the body wall forces the food downward. Gland cells in the gastrodermis discharge enzymes on the food. The digestion started in the gastrovascular cavity is called **extracellular digestion,** but many of the food particles are drawn by pseudopodia into the nutritive-muscular cells of the gastrodermis, where **intracellular digestion** occurs. Indigestible particles are forced back out of the mouth by contraction of the body. Ameboid cells may carry undigested particles to the gastrovascular cavity, where they are eventually expelled with other indigestible matter. Digested food products may be stored in the gastrodermis or distributed by diffusion to other cells, including the epidermis.

Respiration and excretion. Special organs for respiratory exchange or excretion are not necessary in hydras because the tissues are so thin that such functions can be accomplished by direct diffusion into the surrounding water or into the gastrovascular cavity. Nitrogenous wastes are largely in the form of ammonia, which diffuses through the body surface. Excess water taken in with food or by osmosis can be expelled by periodic body contractions that force the fluid out of the mouth.

Reproduction. The hydra can reproduce both sexually and asexually. **Asexual reproduction** is by **bud-**

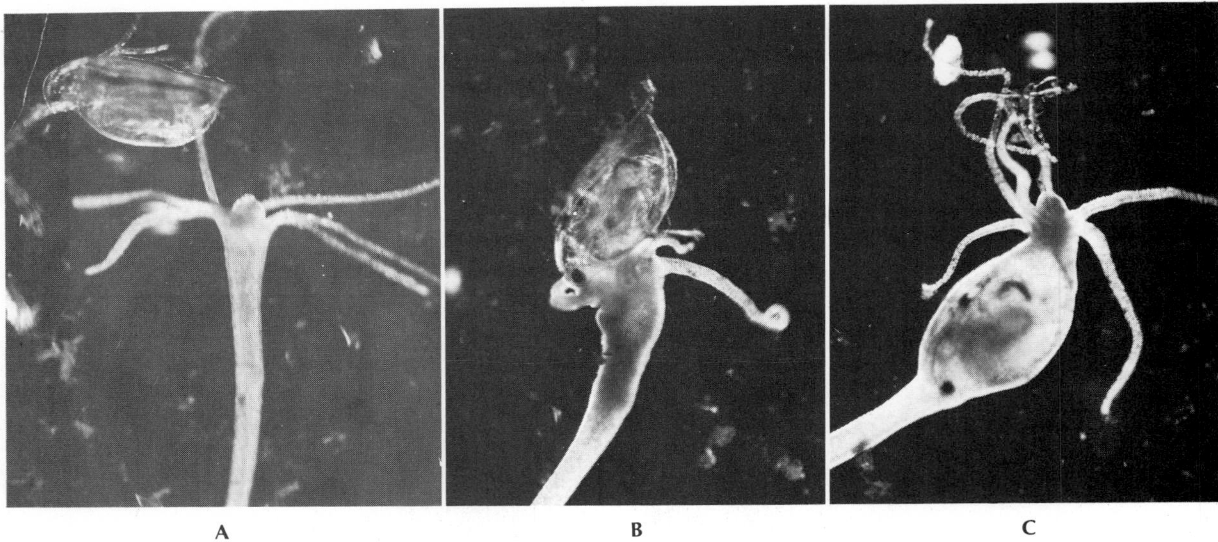

FIG. 10-8 **A,** Hungry hydra catches an unwary water flea with the stinging cells of its tentacles and, **B,** swallows it whole. **C,** Hydra is full, but not too full to capture a protozoan for dessert. (Photographs by F. M. Hickman.)

ding (Fig. 10-9, *A*). Buds form as outpocketings of the body wall, with the gastrovascular cavity of the bud being in communication with the cavity of the parent. A hypostome with a mouth and a ring of tentacles differentiates on the bud, and eventually it constricts at its base and detaches to lead a separate existence. Several buds may be formed on the same parent.

In **sexual reproduction** most hydra species are **dioecious** (have separate sexes); others are **monoecious** (hermaphroditic). Sexual reproduction involves the formation of gonads, which are more common in the autumn. Reduction of water temperature will promote their formation, but other adverse environmental conditions, such as increased carbon dioxide concentration and reduced aeration in stagnant water, may be factors in induction of sexuality in the hydra. The adaptive value in these phenomena is apparent when one realizes that the overwintering stage of the hydra is its encapsulated egg.

Gonads develop as temporary structures formed from differentiated interstitial cells that have aggregated at certain locations along the stalk (Fig. 10-9, *B* and *C*). These multiply and undergo **gametogenesis** with meiosis. Haploid sperm are set free into the water. In the development of the egg, one centrally located egg cell enlarges by the union of other interstitial cells and eventually occupies most of the space in the ovary. In some species the eggs ripen one at a time, in succession; in others, several may ripen at once.

The sperm fertilize the eggs while they are still in the ovary, and embryogenesis begins there. The zygote undergoes holoblastic **cleavage** to form a hollow **blastula.** The inner portions of the cells forming the blastula divide off (gastrulation by delamination) and fill the blastocoel to form a solid-bodied **gastrula.** The surface cells are now the **ectoderm,** which will become the epidermis, and the inner mass of cells is the **endoderm,** which will become the gastrodermis. The **coelenteron** is later formed within the gastrodermal mass, and the mesoglea is laid down between the ectoderm and the endoderm. At about this time a chitinous shell is secreted about the embryo, which breaks loose from the parent. The embryo may pass the winter in the encysted condition. When weather conditions are more favorable, the embryo completes its development, the shell ruptures, and a young hydra with tentacles hatches.

Regeneration. Over 230 years ago, Abraham Trembley was astonished to discover that isolated sections of the stalk of hydra could regenerate and each become a complete animal. Since then, over 2,000 investigations of hydra have been published, and it has become a classic model for the study of morphogenesis. However, only in recent years with modern tech-

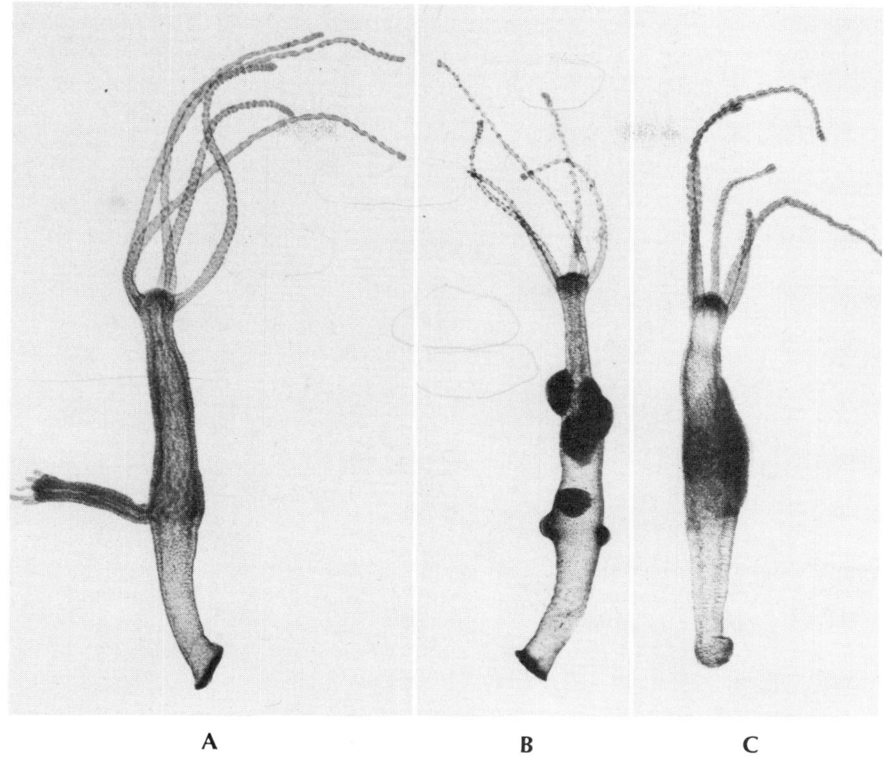

FIG. 10-9 Stained specimens of hydras in reproductive phases. **A,** Budding. The bud forms as an outpocketing of the body wall, and its gastrovascular cavity is continuous with that of the parent. **B,** Male hydra with testes. **C,** Female with ovary. Testes and ovaries develop from the epithelial layer. (Courtesy Carolina Biological Supply Co., Burlington, N.C.)

niques has some insight been gained into the mechanisms involved. The mechanisms governing morphogenesis (development of the structural form of an animal) are of more than academic interest and have great practical importance; the simplicity of hydra lends itself to these investigations.

Though simple, hydra is the most complex animal known whose cells can be entirely disaggregated, then reaggregate and regenerate a complete individual. When sections are simply cut from the column and then allowed to regenerate, they show a morphologic polarity. That is, the end of the section originally closer to the head (oral) end always regenerates a hypostome and tentacles, and the other end forms a basal disc. It is now known that such a result occurs because a chemical gradient of a substance controlling differentiation of cells **(morphogen)** exists in the column, with the highest concentration nearest the hypostome end. The morphogen in this case is called "head activator substance" and has been extracted from hydras and partially purified. It is apparently a peptide with a molecular weight of about 900 and is active in extremely low concentration (10^{-10} M). When the substance is extracted and put into the water with hydras undergoing regeneration, it increases the regeneration rate. In the water with unregenerating hydras, the substance increases the rate of bud formation. Within the hydra, the morphogen is produced and stored in nerve cells. Thus, head activator substance is apparently analogous to neurosecretory hormones (hormones synthesized and secreted by nerve cells), a number of which are very important in controlling development in higher invertebrates and vertebrates.

Behavior. The hydra responds to various stimuli, both internal and external. Spontaneous movements of body and tentacles occur while the animal is attached.

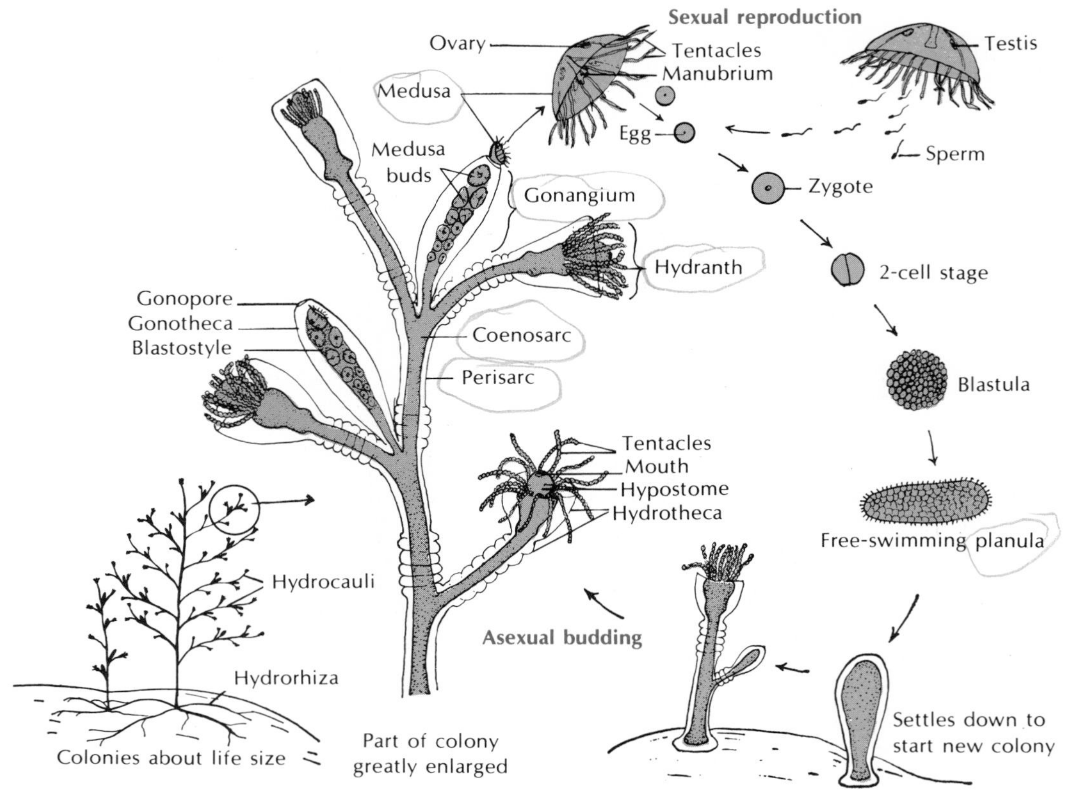

FIG. 10-10 Life cycle of *Obelia,* showing alternation of polyp (asexual) and medusa (sexual) stages. *Obelia* is a calyptoblastic hydroid, that is, its polyps as well as its stems are protected by continuations of the perisarc. Contrast with *Eudendrium* (Fig. 10-12).

If the individual is well fed, its movements are slow, but they increase whenever it becomes hungry. These movements are produced by the contractile fibers in the wall when they are stimulated through the nerve net.

How the hydra reacts to stimuli depends on the intensity and kind of the stimuli and on the physiologic state of the hydra. A slight jar will cause the whole animal to contract rapidly. A localized stimulus, such as touching one of the tentacles with a sharp needle, may produce the same effect. The explanation for this probably lies in the widespread transmission of the nerve impulse over all the nerve net. If such a localized stimulus is mild, there may be a more or less localized response, such as the contraction of a single tentacle or the pulling away of that part of the body touched.

A unique behavior pattern called the **contraction burst** has been described in some hydras under constant conditions. At 5- to 10-minute intervals a hydra suddenly assumes a ball shape by contracting its longitudinal muscles. After a few seconds the individual usually extends itself again. It is believed that this behavior pattern enables the hydra to sample its environment intermittently, and the pattern may play a part in movement and light orientation. A neuronal pacemaker located in the subhypostome region may initiate the impulse for the action. Body contraction is also associated with the periodic expulsion of excess water through the mouth.

Hydras respond to light stimuli in an optimum way, tending to avoid very strong light but seeking moderately lighted regions. By trial and error they find the situation that best suits them. When subjected to a weak constant electric current, they orient the oral end toward the anode and the basal end toward the cathode.

FIG. 10-11 A colony of *Obelia*, a calyptoblastic hydrozoan. The tentacled zooids are the feeding hydranths. The club-shaped zooids growing from the bases of some of the hydranths are reproductive gonangia. Note the attachment area (hydrorhiza) at the base of the stems. Stalked protozoans are attached along the stems of the hydroids. (Photograph by D. P. Wilson.)

Obelia and Gonionemus as examples of hydroid and medusa stages

Both *Obelia* and *Gonionemus* are marine forms that have polyp and medusa stages in their life histories. *Obelia* has a prominent hydroid (juvenile) stage, but an inconspicuous medusa (adult) stage (Fig. 10-10). The reverse conditions are found in *Gonionemus,* in which the medusa is large and the hydroid small.

Obelia may be considered a typical colonial hydroid (Fig. 10-11). It is found on both the Atlantic and Pacific coasts. It attaches to stones and other objects by a rootlike base called **hydrorhiza,** from which arise branching stems **(hydrocauli).** On these stems are large numbers of polyps, which are of two types: **hydranths,** which are nutritive, and **gonangia,** which are reproductive. The hydranths furnish nutrition for the colony; the gonangia produce young medusae asexually by budding. The medusae are sexual, giving rise to sperm and eggs. When zygotes are formed, they develop through a series of stages, terminating in a polyp

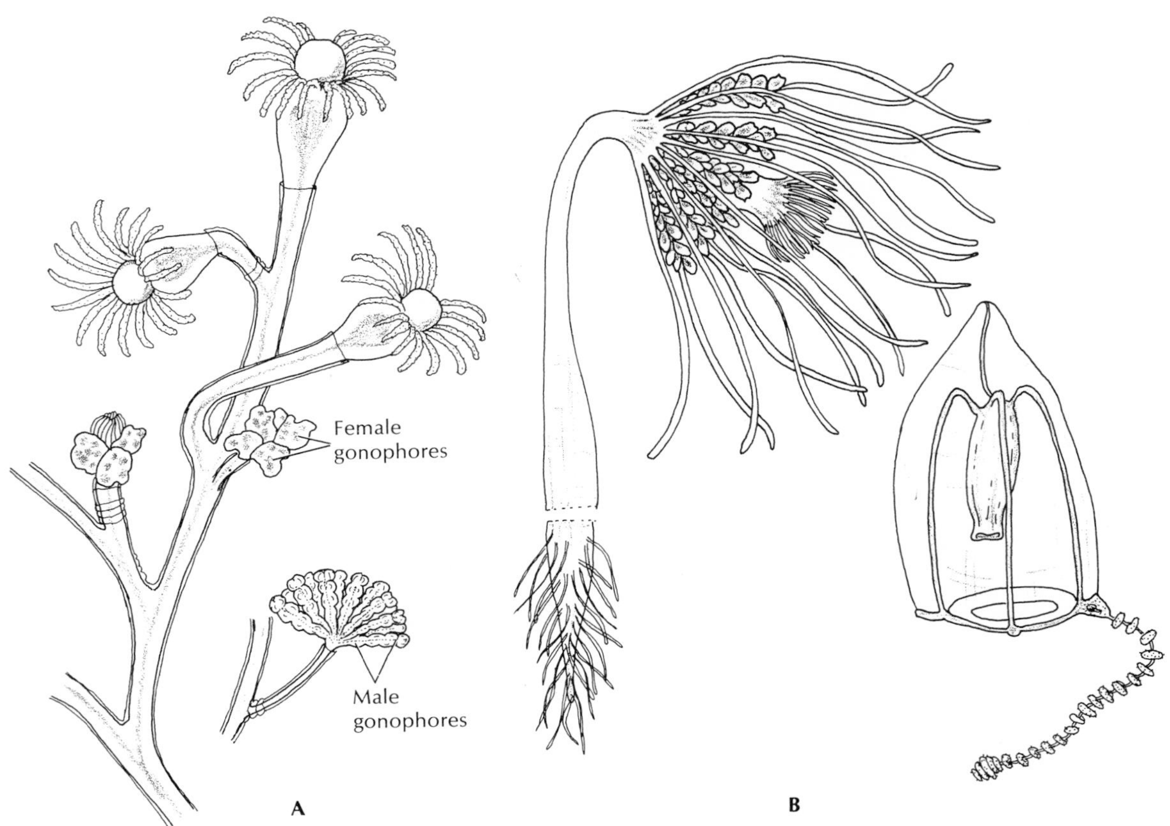

FIG. 10-12 Gymnoblastic hydroids. **A,** *Eudendrium* forms a bushy colony with naked hydranths and gonophores. There are no free medusae. **B,** *Corymorpha* is a solitary hydroid. Its gonophores produce free-swimming medusae, each with a single training tentacle.

form, thus completing the life cycle (Fig. 10-10).

The **hydrocaulus,** or stem that bears the polyps, is a hollow tube composed of a cellular body wall **(coenosarc)** surrounding the **gastrovascular cavity** and covered by a transparent chitinous **perisarc.** The coenosarc, like the body of the hydra, has an outer epidermis, an inner gastrodermis, and mesoglea between them. The gastrovascular cavity is continuous throughout the colony so that nourishment can be distributed from polyps to hydrorhiza. The protective perisarc is also continuous, being modified to cover the polyps.

The nutritive polyp, or **hydranth,** is much like a miniature hydra, with a **hypostome** and **mouth** surrounded by many **tentacles.** By means of the tentacles and **nematocysts,** these feeding polyps capture small crustaceans, worms, or other invertebrates that come within their reach. Within the gastrovascular cavity food is reduced by digestive enzymes to a broth of small particles. This is driven throughout the colony by flagellary movement and by contractions of the hydranth. Cells of the gastrodermis pick up the food and complete digestion in food vacuoles; so digestion is both extracellular and intracellular. As far as is known, starches, cellulose, and chitin are not digested by hydroids. The hydranth is protected by a cuplike **hydrotheca,** a continuation of the perisarc, into which the tentacles can contract.

In the reproductive **gonangium** the medusae develop as lateral buds called **gonophores.** The gonangium is surrounded by a protective sheath **(gonotheca)** with an opening through which young medusae escape (Fig. 10-10).

Not all hydroids have their polyps protected by this continuation of the perisarc as do *Obelia, Campanu-*

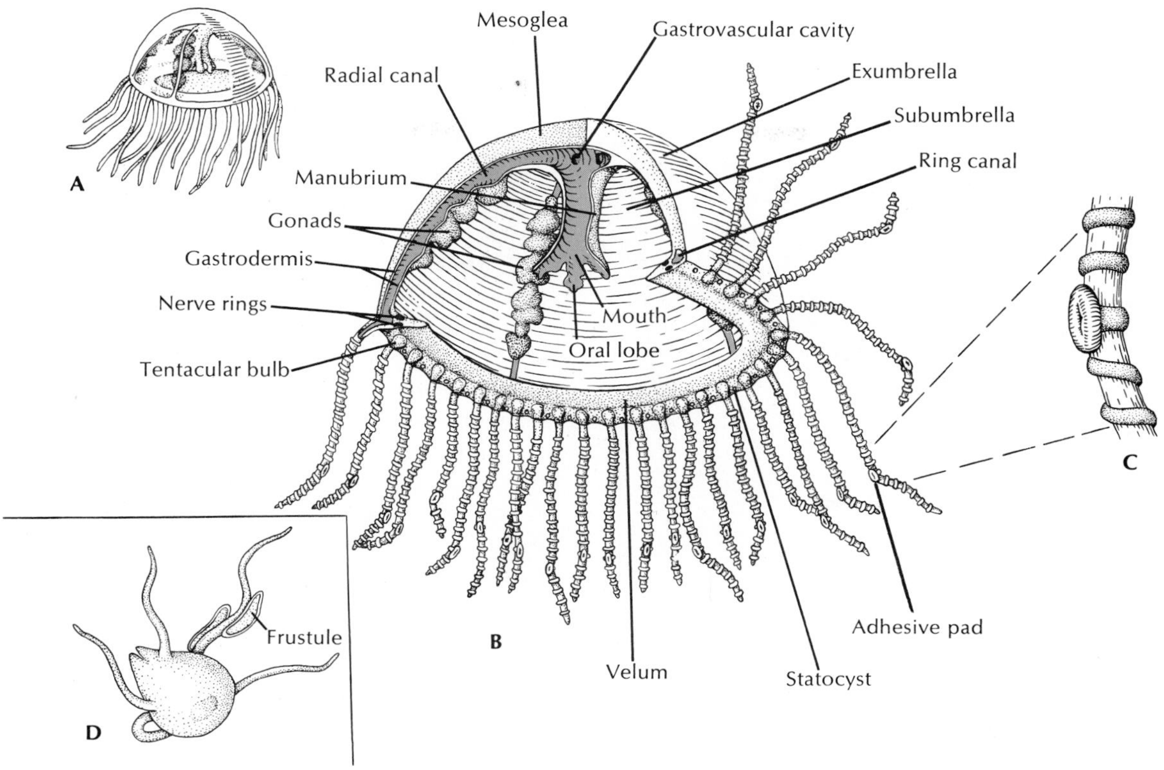

FIG. 10-13 *Gonionemus,* a marine hydrozoan medusa. **A,** Medusa is typically tetramerous in arrangement and bears a velum. **B,** Cutaway to show morphology of the medusa. **C,** Portion of tentacle with its adhesive pad and ridges of nematocysts. **D,** The tiny polyp stage, or hydroid, that develops from the planula larva. It can either bud off frustules, which become more polyps, or can bud off young medusae.

laria, and others (known collectively as the Calyptoblastea). Many other hydroids, such as *Hydractinia* (Fig. 10-5), *Tubularia* (Fig. 10-6), and *Corymorpha* and *Eudendrium* (Fig. 10-12) (the Gymnoblastea) have naked polyps.

The cellular structure of the colony is much like that of the individual hydra. Myonemes from the epitheliomuscular and nutritive-muscular layers provide movement, stimulated through a **nerve net. Sensory cells** are most abundant around the mouth and tentacles.

Gonionemus is frequently studied as an example of hydrozoan medusae, since its medusa stage is much larger than that of *Obelia* and is fairly typical of the class (Fig. 10-13). The polyp of *Gonionemus* is extremely small. The medusa is bell-shaped and 1 to 3 cm in diameter. The convex, or aboral, side is called the **exumbrella,** whereas the concave, or oral, side is the **subumbrella.** Around the margin of the bell are a score or more of **tentacles** with nematocysts, each tentacle with a bend near the tip bearing an **adhesive pad.**

Inside the margin is a thin muscular membrane, the **velum,** which partly closes the open side of the bell. The velum distinguishes the hydrozoan medusa from the scyphozoan medusa, which has no velum. The velum is used in swimming movements. Contractions in the velum and body wall bring about a pulsating movement that alternately fills and empties the subumbrellar cavity. As the animal contracts, forcing water out of the cavity, it is propelled forward, aboral side first, with a sort of "jet propulsion." By constricting the opening through which water passes from the subumbrellar space, the velum increases water velocity, thus improving swimming efficiency. The animal swims upward, turns over, and floats lazily

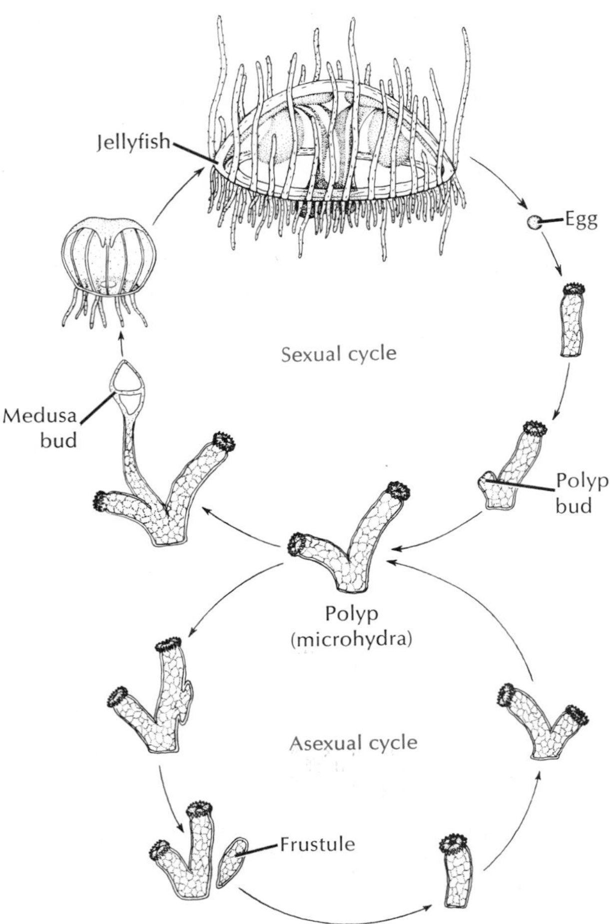

FIG. 10-14 *Craspedacusta,* a freshwater hydrozoan—life cycle. The planula develops into a minute polyp, called microhydra *(center),* which, by budding and by production of frustules that give rise to more polyps, form a small colony (asexual cycle). The polyps may then bud off medusa buds, which become sexually mature medusae and produce eggs or sperm (sexual cycle). Fertilized eggs develop into planula larvae, which settle down and become tiny polyps (microhydra).

downward, tentacles outspread to capture unwary prey. It rests while attached to vegetation by its adhesive pads.

Hanging down inside the bell is the **manubrium,** at the tip of which is the **mouth** surrounded by four **oral lobes.** From the mouth a **gullet** leads to the **stomach** at the base of the manubrium. Four **radial canals** lead out from the stomach to a **ring canal** around the margin, which connects with all the tentacles. The entire system from the gullet to the tips of the tentacles makes up the **gastrovascular cavity,** in which food is partly digested by enzymes and is distributed to all parts of the body. Digestion is completed in the cells of the gastrodermis. Worms, crustaceans, and small fish are favorite foods.

Medusae are dioecious. In *Gonionemus* the **gonads** are suspended under each of the radial canals (Fig. 10-13). The eggs or sperm are shed into the water outside. The fertilized egg develops into a ciliated **planula larva,** which swims about for a time, then settles down, attaches to some object, loses its cilia, and develops into a minute polyp. The cycle begins again with the young polyp budding off additional polyps that finally produce new medusae by asexual budding.

A more elaborate nervous system is of greater adaptive value in the active, free-swimming medusae than in the sessile polyps. In addition to the **nerve net,** neural cells are concentrated in two **nerve rings** at the base of the velum, one in the exumbrellar and one in the subumbrellar epithelium. **Statocysts** are organs of equilibrium that are found around the umbrellar margin in many medusae. Each statocyst is a small sac containing a movable, hard concretion, like a tiny pebble, called a **statolith.** As the animal moves, the statolith stimulates surrounding sensory endings, allowing the organism to orient with respect to gravity.

Tentacular bulbs are enlargements located at the base of the tentacles. Within the bulbs nematocysts are formed and migrate out to the batteries on the tentacles. The bulbs may also help in intracellular digestion. The entire animal seems to be photosensitive.

Freshwater medusae

The freshwater medusa *Craspedacusta sowerbyi* (Fig. 10-14) (class Hydrozoa, order Hydroida) may have evolved from marine ancestors in the Yangtze River of China. Probably introduced with shipments of aquatic plants, this interesting form has now been found in many parts of Europe, all over the United States, and in parts of Canada.

The polyp phase of this animal is tiny (2 mm) and appears to be more or less degenerate, for it has no perisarc and no tentacles. It occurs in colonies of a few polyps. For a long time its relation to the medusa was not recognized, and thus the polyp was given a name of its own, *Microhydra ryderi.* On the basis of its relationship to the jellyfish and the law of priority, both the polyp and the medusa should be called *Craspedacusta.*

The polyp has three methods of asexual reproduc-

FIG. 10-15 Portuguese man-of-war *Physalia physalis* (order Siphonophora, class Hydrozoa) eating a fish. This colony of medusa and polyp types is integrated to act as one individual. As many as a thousand zooids may be found in one colony. They often drift on ocean beaches in the southern United States, where they are a hazard to bathers. Although a drifter, the colony has restricted directional movement. Their stinging organoids secrete a powerful neurotoxin. (Courtesy New York Zoological Society.)

tion: (1) by budding off new individuals, which may remain attached to the parent (colony formation); (2) by constricting off nonciliated planula-like larvae (frustules), which can move around and give rise to new polyps; and (3) by producing medusa buds, which develop into sexual jellyfish. Although the medusae are dioecious, usually all jellyfish in a particular habitat are of the same sex. It is assumed that the population arose from a single polyp introduced into that habitat.

The jellyfish, which may attain a diameter of 20 mm when mature, has some odd features. The tentacles are numerous, are unequal in length, and are not provided with adhesive pads. Only one kind of nematocyst is found. The gonads are enlarged sacs that hang down inside the subumbrella, and the manubrium extends down almost to the level of the velum.

Order Siphonophora

The siphonophorans are highly specialized hydrozoans. They are polymorphic swimming or floating colonies made up of several types of modified medusae and polyps. This is a very ancient group, judging by fossils and other evidence.

Physalia, or the Portuguese man-of-war (Fig. 10-15), is one such colony with a rainbow-hued float of blues and pinks that carries it along on the surface waters of tropical seas. Many are blown to shore on our eastern coast. The long graceful tentacles, actually zooids, are laden with nematocysts and are capable of inflicting painful stings. The float, called a **pneumatophore,** is believed to have expanded from the original larval polyp. It contains an air sac arising from the body wall and filled with a gas similar to air. The float acts as a type of nurse-carrier for future generations of individuals that bud from it and hang suspended in the water. Some of the siphonophores, such as *Stephalia* and *Nectalia,* possess swimming bells as well as a float.

There are several types of polyp individuals. The **gastrozooids** are feeding polyps with a single long tentacle arising from the base of each. Some of these long stinging tentacles become separated from the feeding polyp and are called **dactylozooids,** or fishing tentacles. These sting the prey and lift them to the lips of the feeding polyps. Among the modified medusoid individuals are the **gonophores,** which are little more than sacs containing either ovaries or testes.

An interesting mutualistic relationship exists between *Physalia* and a small fish called *Nomeus* that swims among the tentacles with perfect safety. Other larger fish trying to catch *Nomeus* are caught by the deadly tentacles. *Nomeus* probably feeds on bits of *Physalia's* prey and certainly gains a measure of protection from predation by its refuge among the *Physalia* tentacles. Why the fish is not stung to death by its host's nematocysts is unclear, but like the anemone fish to be discussed later, *Nomeus* is probably protected by a skin mucus that does not stimulate nematocyst discharge.

Class Scyphozoa

Class Scyphozoa contains "true" jellyfish (Fig. 10-16). Some of these medusae, such as *Cyanea* (Fig. 10-16, *B*), are a meter or more in diameter, with tentacles more than 23 m (75 feet) long. Others, however, are quite small. Most are found floating free in the open sea, but the members of one order are sessile and at-

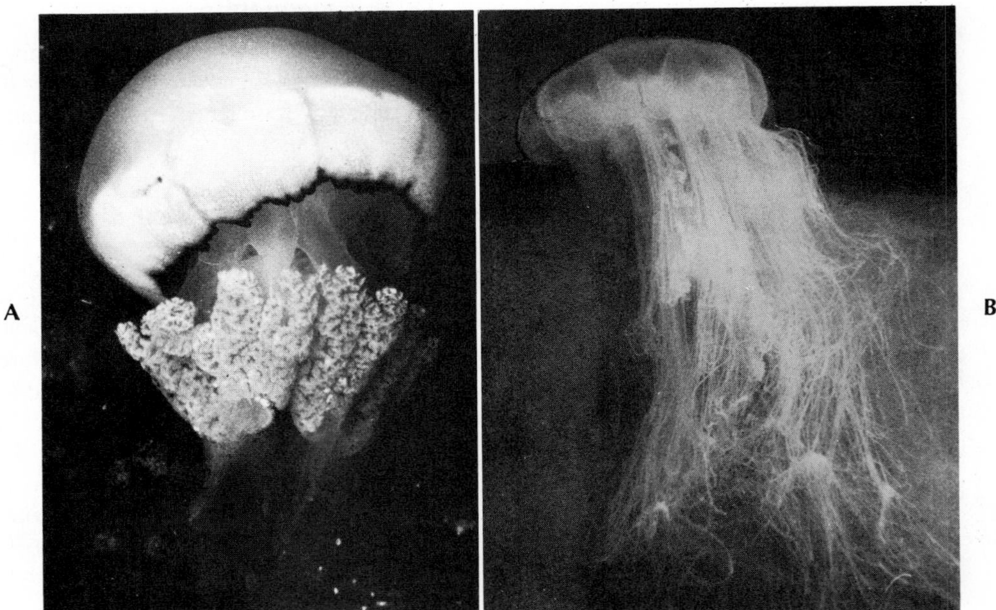

FIG. 10-16 Scyphozoan jellyfishes. **A,** *Rhizostoma pulmo,* the "lung medusa," from the north Atlantic. It is large (50 cm across), heavy bodied, and bluish white with violet edge coloration. **B,** Large jellyfish *Cyanea.* Some Arctic forms attain diameter of more than 2 m. Its many hundred tentacles may reach a length of 25 m or more. (**A,** Courtesy B. Tallmark, Uppsala University, Sweden; **B,** photograph by C. P. Hickman, Jr.).

tach to seaweed, stones, and the like. The polypoid stage is lacking or is limited to a small larval stage.

The bells of different species vary in depth from a shallow saucer shape to a deep helmet or goblet shape. The jelly (mesoglea) layers are unusually thick, giving the bell a fairly firm consistency. The jelly is 95% to 96% water. Unlike the hydromedusae, this layer in the scyphomedusae also contains ameboid cells and fibers, so that it is now called a **collenchyme.** Movement is by rhythmic pulsations of the umbrella. There is no velum as in the hydromedusae. There may be many tentacles or few, and they may be short as in *Aurelia* or long as in *Cyanea. Aurelia* (Fig. 10-17) is a familiar species 7 to 10 cm in diameter, commonly found in the waters off both our east and west coasts, and is widely used for the study of Scyphozoa.

The margin of the umbrella is scalloped, usually with each indentation bearing a pair of **lappets,** and between them is a sense organ called a **rhopalium** (tentaculocyst). *Aurelia* has eight such notches. Some scyphozoans have four, others 16. Each rhopalium is club-shaped and contains a hollow statocyst for equilibrium and one or two pits lined with sensory epithelium. In some species the rhopalia also bear **ocelli.** Cubomedusae, one of the order of Scyphozoa, have some surprisingly complex eyes near the bell margin, and they may well be able to form images.

The mouth is centered on the subumbrella side. The manubrium is usually drawn out to form four frilly **oral arms** that are used in capturing and ingesting prey.

The tentacles, manubrium, and often the entire body surface are well supplied with nematocysts that can give painful stings. In fact among the scyphozoans are the so-called stinging nettles, so dreaded by swimmers, and also the deadly sea wasp *Chironex fleckeri,* the stings of which are considered quite dangerous and sometimes lethal, with death occurring rapidly (usually within a few minutes). Most of the fatalities resulting from stings have been reported from tropical Australian waters. However, the primary function of scyphozoan nematocysts is not to attack humans but to paralyze prey animals, which are conveyed to the mouth lobes with the help of the other tentacles or by the bending of the umbrella margin.

Aurelia, which has comparatively short tentacles,

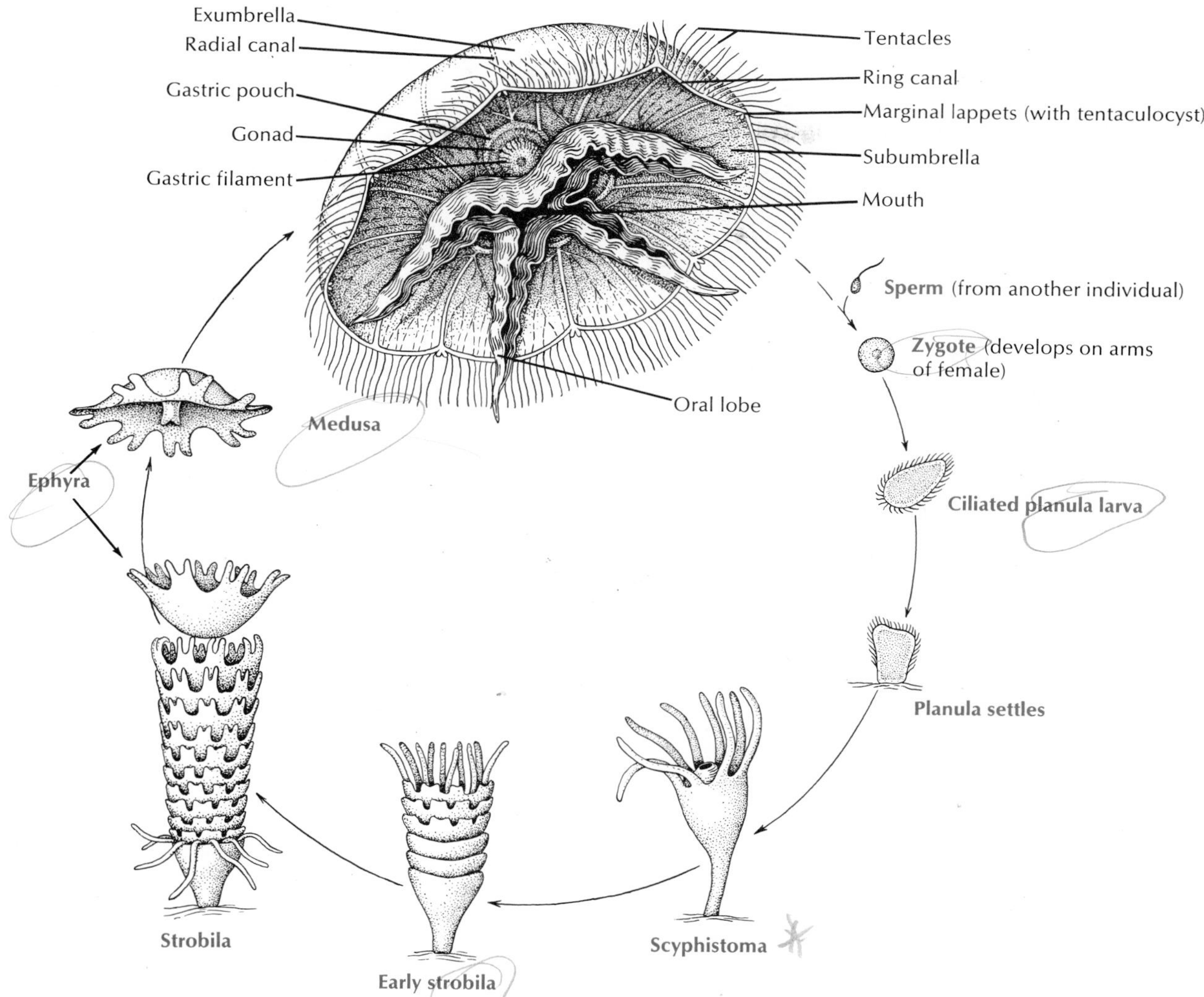

FIG. 10-17 Life cycle of *Aurelia*, a marine scyphozoan medusa.

feeds on small planktonic animals. These are caught in the mucus of the umbrella surface, are carried to "food pockets" on the umbrella margin by cilia, and are picked up from the pockets by the oral lobes whose cilia carry the food to the gastrovascular cavity. Flagella in the gastrodermis layer keep a current of water moving to bring food and oxygen into the stomach and carry out wastes.

Cassiopeia, a jellyfish common to Florida waters, and *Rhizostoma* (Fig. 10-16, *A*), which can be found in colder waters, belong to a group differing from that of *Aurelia* both in their lack of tentacles on the umbrella margin and in the structure of the oral arms. During development, the edges of the oral lobes fold over and fuse, forming canals (**arms** or **brachial canals**) that become highly branched. These open to the surface at frequent intervals by pores called "mouths"; the original mouth is obliterated in the fusion of the oral lobes. Planktonic organisms caught in the mucus of the frilly oral arms are transported to the mouths and then up the brachial canals to the gastric cavity by cilia. In contrast to the usual swimming habit of medusae, *Cassiopeia* is usually found lying on its "back" in shallow lagoons. Its umbrella margin contracts about 20 times

FIG. 10-18 **A,** Scyphistoma (polyp stage) of the large jellyfish *Cassiopeia.* The individual at left shows the mouth. To the left of the bell are two buds which will detach and develop into new scyphistomas. In the spring young ephyrae bud off, one at a time, from the oral side to become medusae. **B,** Scyphistoma *(right)* and strobilas *(left)* of *Aurelia aurita* hanging down from a rock. The strabila at left has an ephyra nearly ready to separate. (**A,** Photograph by F. M. Hickman; **B,** photograph by D. P. Wilson.)

per minute, creating water currents to bring plankton into contact with the mucus and nematocysts of its oral lobes.

Internally, extending out from the stomach of scyphozoans are four **gastric pouches** in which gastrodermis extends down in little tentacle-like projections called **gastric filaments.** These are covered with nematocysts to further quiet any prey that may still be struggling. Gastric filaments are lacking in the hydromedusae. A complex system of **radial canals** branch out from the pouches to a **ring canal** in the margin and make up a part of the gastrovascular cavity.

The **nervous system** in scyphozoans seems to be of a synaptic nerve net type, with a subumbrella net that controls bell pulsations, and another more diffuse net that controls local reactions such as feeding.

The sexes are separate, with the gonads located in the gastric pouches. Fertilization is internal, with the sperm being carried by ciliary currents into the gastric pouch of the female. The zygotes may develop in the seawater or may be brooded in folds of the oral arms. The ciliated planula larva becomes attached and develops into a **scyphistoma,** a hydralike form (Fig. 10-17). By a process of **strobilation** the scyphistoma of *Aurelia* forms a series of saucerlike buds, **ephyrae,** and is now called a **strobila** (Figs. 10-17 and 10-18, *B*). When the ephyrae break loose, they grow into mature jellyfish.

In *Cassiopeia,* the scyphistoma does not form a strobila, as in *Aurelia,* but buds off young medusae one at a time (monodisc strobilation) (Fig. 10-18, *A*). The scyphistoma can also constrict off small planula-like ciliated buds that swim about and then settle down to become other scyphistomas.

Class Anthozoa

The anthozoans, or "flower animals," are polyps with a flowerlike appearance. There is no medusa stage. Anthozoans are all marine and are found in both deep and shallow water and in polar seas as well as tropical seas. They vary greatly in size and may be solitary or colonial. Many of the forms are supported by skeletons.

The class has two subclasses—**Zoantharia,** made up of the sea anemones and stony corals, and **Alcyonaria,** which includes the sea fans, sea plumes, sea pansies, and other soft corals. The zoantharians have a **hexamerous** plan (of six or multiples of six) or polymerous symmetry and have simple tubular tentacles arranged

FIG. 10-19 *Actinia equina,* a sea anemone of the family Actiniidae, whose members are common in temperate waters. (Photograph by D. P. Wilson.)

in one or more circlets on the oral disc. The alcyonarians are **octamerous** (built on a plan of eight) and always have eight pinnate (featherlike) tentacles arranged around the margin of the oral disc.

The gastrovascular cavity is large and partitioned by mesenteries, or septa, that are inward extensions of the body wall. The walls and mesenteries contain both circular and longitudinal muscle fibers.

The mesoglea is a mesenchyme containing ameboid cells. There is a general tendency toward biradial symmetry in the septal arrangement and in the shape of the mouth and pharynx. There are no special organs for respiration or excretion.

Sea anemones

Sea anemone polyps are larger and heavier than hydrozoan polyps (Fig. 10-19 and p. 182). Most of them range from 5 mm or less to 100 mm in diameter, and from 5 mm to 200 mm long, but some grow much larger. Some of them are quite colorful. Anemones are found in coastal areas all over the world, especially in the warmer waters, and they attach by means of their pedal discs to shells, rocks, timber, or whatever submerged substrata they can find. Some burrow in the bottom mud or sand.

Sea anemones are cylindric in form with a crown of tentacles arranged in one or more circles around the mouth on the flat **oral disc** (Fig. 10-20). The slit-shaped mouth leads into a **pharynx.** At one or both ends of the mouth is a ciliated groove called a **siphonoglyph,** which extends into the pharynx. The siphonoglyphs create water currents directed into the pharynx. The cilia elsewhere on the pharynx direct water outward. The currents thus created carry in oxygen and remove wastes. They also help maintain an internal fluid pressure or a hydrostatic skeleton that serves in lieu of a true skeleton as a support for opposing muscles.

The pharynx leads into a large **gastrovascular cavity** that is divided into six radial chambers by means

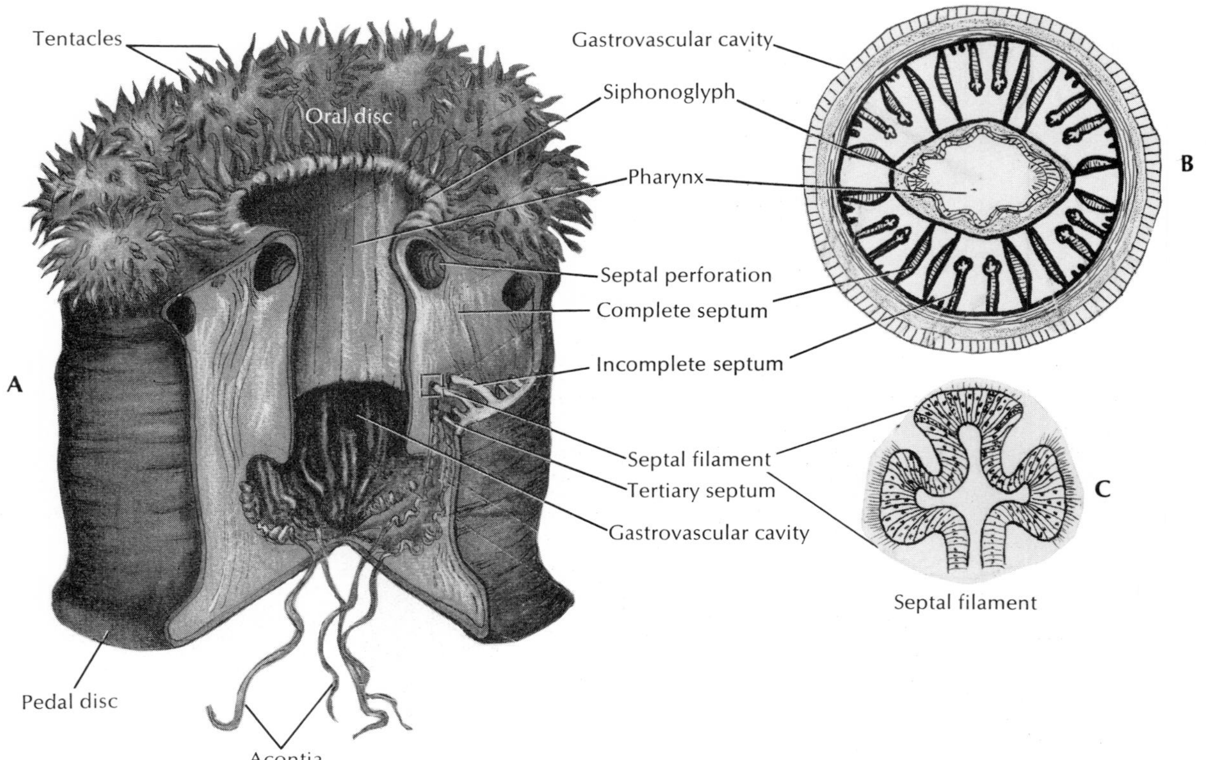

FIG. 10-20 Structure of sea anemone *Metridium*. **A,** Cutaway view of pharynx and gastrovascular cavity. **B,** Transverse section through the pharynx. **C,** Enlargement of one of the septal filaments that are located along the edges of the incomplete septa (cross section).

of six pairs of **primary (complete) septa,** or **mesenteries,** that extend vertically from the body wall to the pharynx (Fig. 10-20). These chambers communicate with each other and are open below the pharynx. Smaller **(incomplete)** mesenteries partially subdivide the larger chambers and provide a means of increasing the surface area of the gastrovascular cavity. The free edge of each incomplete septum forms a type of sinuous cord called a **septal filament** that is provided with nematocysts and with gland cells for digestion. In many anemones the lower ends of the septal filaments are prolonged into **acontia threads,** also provided with nematocysts and gland cells, that can be protruded through the mouth or through pores in the body wall to help overcome prey or provide defense. The pores also aid in the rapid discharge of water from the body when the animal is endangered and contracts to a small size.

Sea anemones are carnivorous, feeding on fish or almost any live animals of suitable size. Some species live on minute forms caught by ciliary currents.

Feeding behavior in many zoantharians is under chemical control. Some respond to reduced glutathione. In certain others two compounds are involved: asparagine, the feeding activator, causes a bending of the tentacles toward the mouth; then reduced glutathione induces swallowing of the food.

Sea anemones are muscular, having muscle fibers not only in the epidermis and gastrodermis but also in the collenchyme as well. There are definite muscle bands in the mesenteries. Most anemones can glide along slowly on their pedal discs. They can expand and stretch out their tentacles in search of small vertebrates and invertebrates, which they overpower with tentacles and nematocysts and carry to the mouth.

When disturbed, sea anemones contract and draw in their tentacles and oral discs. Some anemones are able to swim, to a limited extent, by rhythmic bending movements, which may be a mechanism for escape from enemies such as sea stars and nudibranchs. *Stomphia,* for example, at the touch of a predatory sea star, will detach its pedal disc and make creeping or swimming movements to escape (Fig. 10-21). This escape reaction is caused not only by the touch of the star but also by exposure to drippings exuded by the star or to crude extracts made from its tissues. The sea star drippings contain steroid saponins that are toxic and irritating to most invertebrates. Extracts given off by nudibranchs can also provoke this reaction in sea anemones.

Anemones form some interesting mutualistic relationships with other animals. Some species habitually attach to the shells occupied by certain hermit crabs (Fig. 10-5). The hermit encourages the relationship and, finding its favorite species, which it recognizes by touch, it massages the anemone until it detaches. The hermit crab holds the anemone against its own shell until the anemone is firmly attached. When the crab moves to a larger shell, it will move its anemone to the new shell, too. The hermit is not only immune to its passenger's stings but is protected by them from its enemies. The anemone gets free transportation and particles of food dropped by the hermit crab.

Certain damselfish (family Pomacentridae) form associations with large anemones, especially in tropical Indo-Pacific waters. An unknown property of the skin mucus of the fish causes the anemone's nematocysts not to discharge, but if some other fish is so unfortunate as to brush the anemone's tentacles, it is likely to become a meal. The anemone obviously provides shelter for the anemone fish, and the fish may help venti-

FIG. 10-21 Reaction of *Stomphia* to a predatory sea star *Dermasterias.* On contact with an arm of the star, **A,** the anemone contracts and, **B,** withdraws its tentacles; then, **C,** it loosens its pedal disc and, **D,** using a combination of tentacular movement and muscular body-wall movements enhanced by shifting of fluids in the gastrovascular cavity, it may roll, creep, or swim to a safer location. (Courtesy Dr. D. M. Ross; from Feder, H. 1972. Sci. Am. **227:**93-100 [July].)

FIG. 10-22 Group of polyps of the stony coral *Astrangia danae* protruding from their shallow calcareous cups. When not feeding, or when disturbed, they can retract into the cups. This is a common zoantharian coral of the Atlantic coast. (Courtesy General Biological Supply House, Inc., Chicago.)

FIG. 10-23 Stony coral *Oculina* showing cups in which polyps once lived. A zoantharian coral. Note the sclerosepta in each cup, which in life project up into the polyp between the mesenteries.

late the anemone by its movements, keep the anemone free of sediment, and even lure an unwary victim to seek the same shelter.

The sexes are separate in sea anemones, and the gonads are arranged on the margins of the mesenteries. The zygote develops into a ciliated larva. Asexual reproduction commonly occurs by **pedal laceration** or by longitudinal fission, occasionally by transverse fission or by budding. In pedal laceration, small pieces of the pedal disc break off as the animal moves, and these each regenerate a small anemone.

Zoantharian corals

The zoantharian corals include the true stony corals (order Scleractinia) and the black corals (order Antipatharia). The stony corals might be described as miniature sea anemones that live in calcareous cups they themselves have secreted (Figs. 10-22 and 10-23). Like that of the anemones, the coral polyp's gastrovascular cavity is subdivided by mesenteries arranged in multiples of six (hexamerous) and its hollow tentacles surround the mouth, but there is no siphonoglyph.

Instead of a pedal disc, the epidermis at the base of the column secretes the limy skeletal cup, including the sclerosepta, which project up into the polyp between the true mesenterial septa (Fig. 10-23). The living polyp can retract into the safety of the cup when not feeding. Since the skeleton is secreted below the living tissue rather than within it, the calcareous material is an exoskeleton. In the colonial corals, the skeleton may become very large and massive, building up over many years, with the living coral forming a sheet of tissue over the surface. The gastrovascular cavities of the polyps are all connected through this sheet of tissue.

The black, or thorny, corals (order Antipatharia) usually form slender branching colonies several feet tall. Their polyps have six nonretractile tentacles.

Several other small orders of Zoantharia are recognized.

Alcyonarian corals

Alcyonarians are often referred to as octocorals because of their strict octomerous symmetry, with eight pinnate tentacles and eight unpaired, complete mesenteries (Fig. 10-24). They are all colonial, and the gastrovascular cavities of the polyps communicate through a system of gastrodermal tubes called **solenia.** The solenia run through an extensive mesoglea **(coenenchyme)** in most alcyonarians, and the surface of the colony is covered by epidermis. The skeleton is secreted in the coenenchyme and consists of limey spicules, fused spicules, or a horny protein, often in combination. Thus, the skeletal support of most alcyonarians is an endoskeleton. The variation in pattern among the species of alcyonarians lends great variety to the

FIG. 10-24 *Eunicella verrucosa,* an alcyonarian coral of the group that contains the sea fans, sea rods, and sea whips. Inset shows the pinnate tentacles, arranged eight to each polyp. (Photograph by D. P. Wilson.)

form of the colonies: from the soft coral *Alcyonium* (Fig. 10-25), with its spicules scattered through the coenenchyme, to the tough, axial supports of the sea fans and other gorgonian corals (Fig. 10-26, *B* and *C*), to the fused spicules of the organ-pipe coral (Fig. 10-26, *A*). *Renilla,* the sea pansy, is a colony reminiscent of a pansy flower. Its polyps are embedded in the fleshy upper side and a short stalk that supports the colony is embedded in the sea floor (Fig. 10-25, *B*).

The graceful beauty of the alcyonarians—in hues of yellow, red, orange, and purple—helps create the "submarine gardens" of the coral reefs.

Coral reefs

Coral reefs are among the most productive of any ecosystem, and they have a diversity of life forms rivaled only by the tropical rain forest. They are large formations of calcium carbonate (limestone) in shallow tropical seas laid down by living organisms over thousands of years; living plants and animals are confined to the top layer of reefs where they add more and more calcium carbonate to that deposited by their predecessors. The most important organisms that precipitate calcium carbonate from sea water to form reefs are the **hermatypic** (reef-building) **corals** and **coralline algae.** Not only do the coralline algae contribute to the total mass of calcium carbonate, but their precipitation of the substance helps to hold the reef together. Some alcyonarians and hydrozoans (especially *Millepora*) contribute in some measure to the calcareous material, and an enormous variety of other organisms contribute small amounts. However, hermatypic corals seem essential to the formation of large reefs, since such reefs do not occur where these corals cannot live.

Hermatypic corals require warmth, light, and the salinity of undiluted sea water. This limits coral reefs to shallow waters between 30 degrees north and 30 degrees south latitude and excludes them from areas with upwelling of cold water or areas near major river outflows with attendant low salinity and high turbidity. These corals require light because they have mutualistic algae (zooxanthellae) living in their tissues. The

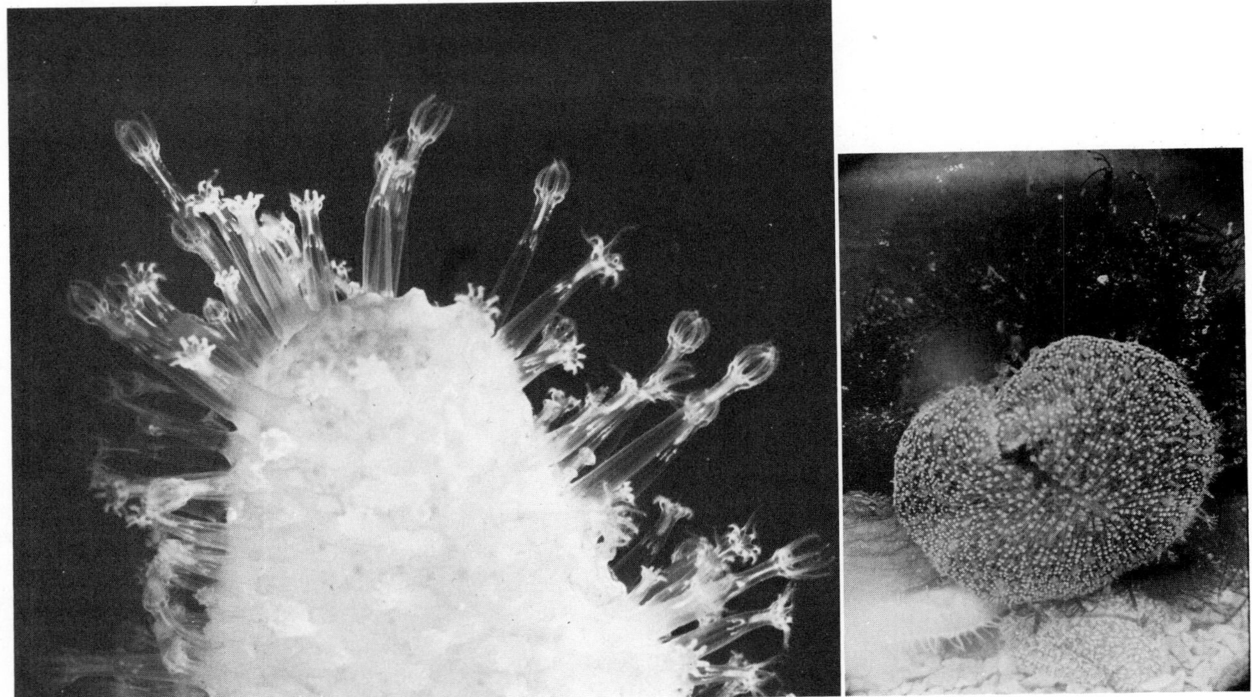

FIG. 10-25 **A,** *Alcyonium,* a soft coral of the North Atlantic. **B,** *Renilla,* the sea pansy, in which the polyps grow on the upper side of a leaflike rachis attached to the mud bottom by a stalk. (**A,** Photograph by D. P. Wilson; **B,** photograph by F. M. Hickman.)

microscopic zooxanthellae are very important to the corals: their photosynthesis and fixation of carbon dioxide furnish food molecules for their hosts, they recycle phosphorus and nitrogenous waste compounds that otherwise would be lost, and they enhance the ability of the coral to deposit calcium carbonate.

Several types of reefs are commonly recognized. The **fringing reef** is close to a land mass with either no lagoon or a narrow lagoon between it and the shore. A **barrier reef** runs roughly parallel to shore and has a wider and deeper lagoon than does a fringing reef. **Atolls** are reefs that encircle a lagoon but not an island. These types of reefs typically slope rather steeply into deep water at their seaward edge. **Patch** or **bank reefs** occur some distance back from the steep, seaward slope in the lagoons of barrier reefs or atolls. The so-called Great Barrier Reef, extending 1,200 miles long and up to 90 miles from the shore off the northeast coast of Australia, is actually a complex of reef types.

Several theories to account for the structure of reefs have been proposed, beginning with that of Darwin. He suggested that all reefs began as fringing reefs, which then became barrier reefs and then atolls, as the land mass slowly sank or subsided, and the growth of the coral kept up with the subsidence. The most acceptable theory at present is the **karst mechanism,** proposed by Purdy and others and named for a well-known terrestrial feature called **karst topography.** This is caused when rain, along with carbon dioxide from the air dissolved in it to form a weak carbonic acid, falls on exposed limestone rock formations. As the weak acid percolates through the rock, the limestone unevenly dissolves, eventually forming subterranean caverns and underground rivers. On the surface, areas of greater solution or places where a cavern roof caves in, form cup-shaped **sinks.** Areas of lesser solution are left as **prominences.** During the most recent glacial period about 10,000 years ago (Wisconsin glaciation), so much water was tied up in glaciers that the sea level

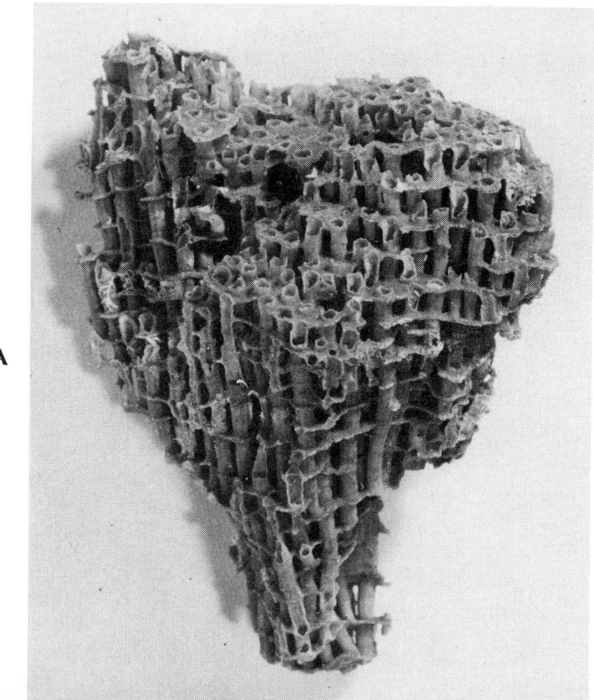

FIG. 10-26 Dried skeletons of alcyonarian corals. **A,** Organpipe coral *Tubipora.* The skeleton of this species is of fused spicules that are red in color because of iron salts. **B** and **C,** Gorgonians. **B,** Sea fan *Gorgonia.* **C,** *Pseudopterogorgia.* These and other gorgonians are prominent on Caribbean reefs. Their skeleton consists of a central rod of hornlike protein, gorgonin, surrounded by limy spicules secreted by the coenenchyme.

was about 100 m lower than it is now. The reefs that had been formed up to that time would have been exposed to the air, high above sea level, and subjected to the effects of rainwater. They would have constituted precisely the type of geologic formation that results in terrestrial karst topography. When the sea level rose again, hermatypic corals again colonized their old areas, so that what we see now is a thin veneer of recent coral whose reef form reflects the underlying karst. For example, Purdy showed experimentally that the rim of a limestone platform exposed to acid rain dissolves more slowly than the central area; thus, the barrier reef and atoll would represent the rim of the platform, and the lagoons are the central areas. Reef passes were formed as drainage breaches in the rim as the water ran seaward, much like a present-day river. "Blue holes," which are very deep, cup-shaped areas in lagoons, represent ancient sinks, and patch reefs are presumed to be recolonized karst prominences.

PHYLUM CTENOPHORA

General relationships

Ctenophora (te-nof′o-ra) (Gr., *kteis, ktenos,* comb, + *phora,* pl. of bearing) comprise a small group of fewer than 100 species. All are marine forms. They take their name from the eight rows of comblike plates they bear for locomotion. Common names for ctenophores are "sea walnuts" and "comb jellies." Ctenophores, along with cnidarians, represent the only two phyla having primary radial symmetry, in contrast to other metazoans, which have primary bilateral symmetry.

Ctenophores do not have nematocysts, except in one species *(Euchlora rubra)* that is provided with nematocysts on certain regions of its tentacles but lacks colloblasts. These nematocysts are a part of this ctenophore and are not "appropriated" by eating a cnidarian.

In common with the cnidarians, ctenophores have not advanced beyond the tissue grade of organization. There are no definite organ systems in the strict meaning of the term.

Ecologic relationships

The ctenophores are strictly marine animals and are all free-swimming except for a few creeping forms. They occur in all seas, but especially in warm waters.

Although feeble swimmers and more common in surface waters, ctenophores are sometimes found at considerable depths. They are often at the mercy of tides and strong currents, but they avoid storms by swimming downward in the water. In calm water they may rest vertically with little movement, but when moving, they use their ciliated comb plates to propel themselves mouth-end forward. Highly modified forms such as *Cestum* use sinuous body movements as well as their comb plates in locomotion.

The fragile, transparent bodies of ctenophores are easily seen at night when they emit light (luminesce).

Characteristics

1. Symmetry **biradial;** arrangement of internal canals and the position of the paired tentacles change the radial symmetry into a combination of the two **(radial + bilateral)**
2. Usually ellipsoidal or spherical in shape, **with radially arranged rows of comb plates for swimming**
3. Ectoderm, endoderm, and a mesoglea (ectomesoderm) with scattered cells and muscle fibers; may be considered **triploblastic**
4. Nematocysts absent (except in one species), but **adhesive cells (colloblasts)** present
5. Digestive system consisting of mouth, pharynx, stomach, and a series of canals
6. Nervous system consisting of a subepidermal plexus concentrated around the mouth and beneath the comb plate rows; an **aboral sense organ** (statocyst)
7. No polymorphism or attached stages
8. Reproduction monoecious; gonads (endodermal origin) on the walls of the digestive canals, which are under the rows of comb plates; determinate cleavage; cydippid larva
9. Luminescence common

Comparison with Cnidaria

Ctenophores resemble the cnidarians in the following ways:

1. Form of radial symmetry; together with the cnidarians, they form the group Radiata
2. Aboral-oral axis around which the parts are arranged
3. Well-developed gelatinous ectomesoderm (collenchyme)
4. No coelomic cavity
5. Diffuse nerve plexus
6. Lack of organ systems

They differ from the cnidarians in the following ways:

1. No nematocysts except in *Euchlora*
2. Development of muscle cells from mesenchyme
3. Presence of comb plates and colloblasts
4. Mosaic, or determinate type of development
5. Presence of pharynx generally
6. No polymorphism

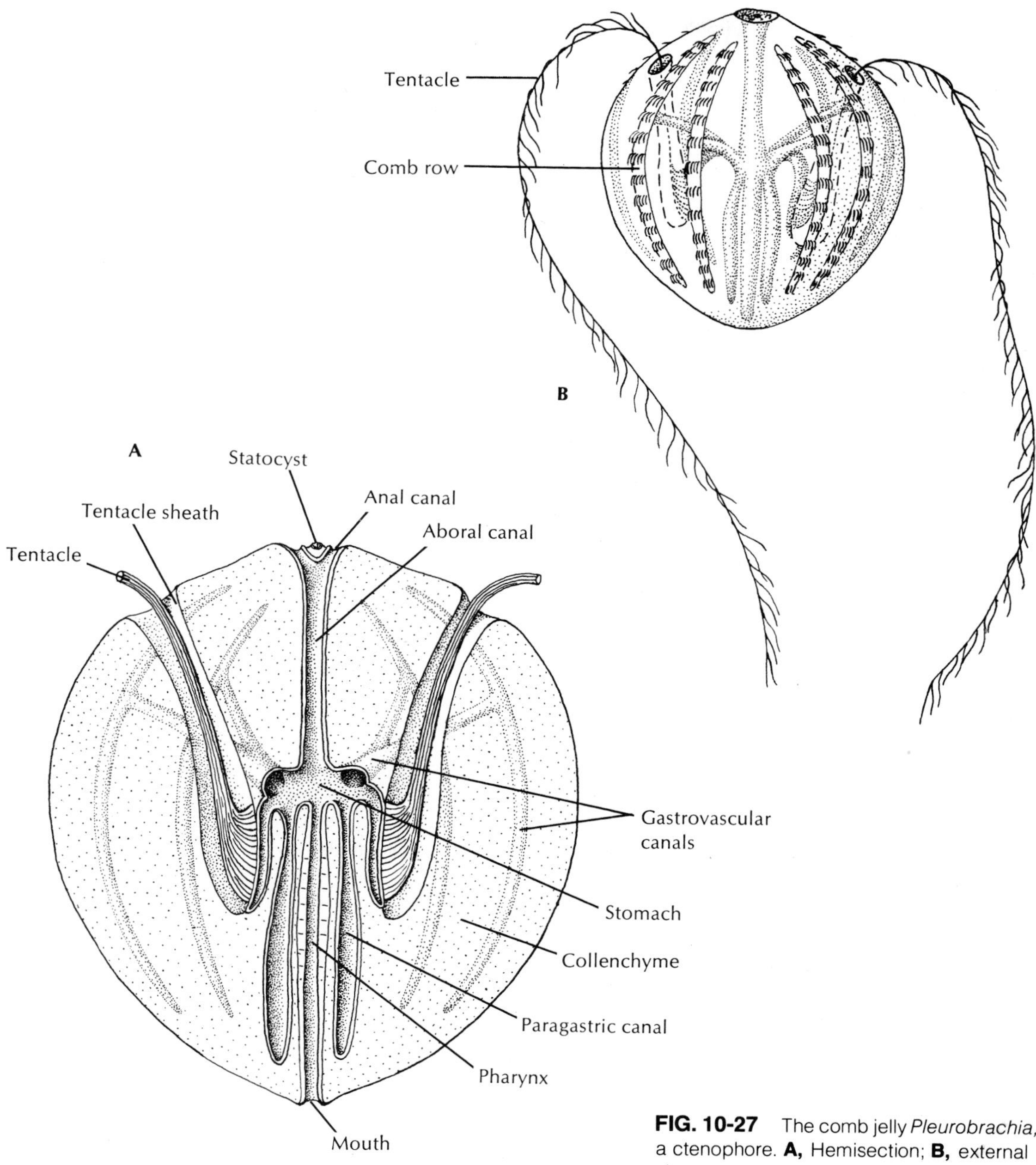

FIG. 10-27 The comb jelly *Pleurobrachia,* a ctenophore. **A,** Hemisection; **B,** external view.

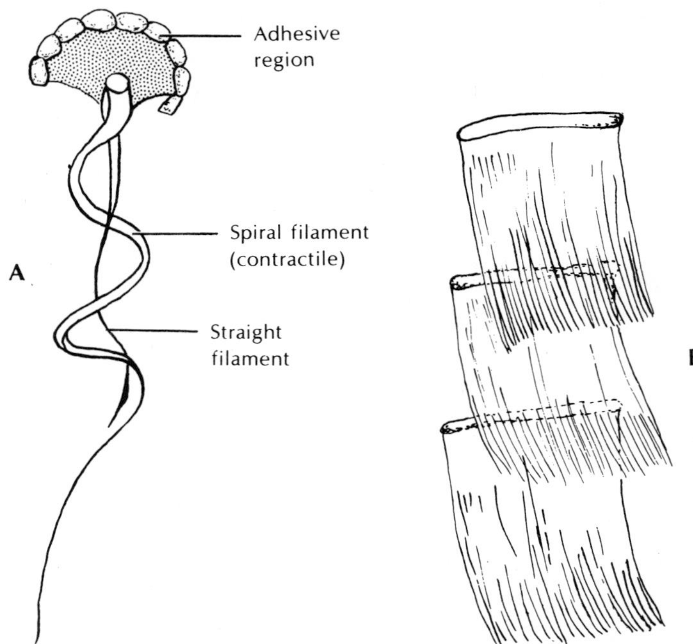

FIG. 10-28 **A,** Colloblast, an adhesive cell characteristic of ctenophores. **B,** Portion of a comb row showing three comb plates, each composed of transverse rows of long fused cilia.

Classes

Class Tentaculata (ten-tak-yu-la′ta) (L., *tentaculum,* feeler, + *-ata,* group suffix). With tentacles. Tentacles may or may not have sheaths into which they retract. Some types flattened for creeping; others compressed to a bandlike form. In some the comb plates may be confined to the larval form. Examples: *Pleurobrachia, Cestum.*

Class Nuda (nu′da) (L., *nudus,* naked). Without tentacles; conical form; wide mouth and pharynx; gastrovascular canals much branched. Example: *Beroë.*

Class Tentaculata

Representative type—Pleurobrachia

Pleurobrachia is often used as a type of this group of ctenophores. Its transparent body is about 1.5 to 2 cm in diameter (Fig. 10-27). The oral pole bears the mouth opening, and the aboral pole has a sensory organ, the **statocyst.**

Comb plates. On the surface are eight equally spaced bands called **comb rows** that extend as meridians from the aboral pole and end before reaching the oral pole (Fig. 10-27). Each band is made up of transverse plates of long fused cilia called **comb plates** (Fig. 10-28, *B*). Ctenophores are propelled by the beating of the cilia on the comb plates. The beat in each row starts at the aboral end and proceeds successively along the combs to the oral end. All eight rows normally beat in unison. The animal is thus driven forward with the mouth in advance. The animal can swim backward by reversing the direction of the wave.

Tentacles. The two **tentacles** are long and solid and very extensible, and they can be retracted into a pair of **tentacle sheaths.** When completely extended, they may be 15 cm long. The surface of the tentacles bears **colloblasts,** or glue cells (Fig. 10-28, *A*), which secrete a sticky substance for catching and holding small animals.

Body wall. The cellular layers of ctenophores are similar to those of the cnidarians. Between the epidermis and gastrodermis is a jellylike **collenchyme,** which makes up most of the interior of the body and contains muscle fibers and ameboid cells.

Digestive system and feeding. The **gastrovascular system** consists of a mouth, a pharynx, a stomach, and a system of gastrovascular canals that branch through the jelly to run to the comb plates, tentacular sheaths, and elsewhere (Fig. 10-27). Two blind canals terminate near the mouth, and an aboral canal passes near the statocyst and divides into two small **anal canals** through which undigested material is expelled.

Ctenophores live on small planktonic organisms such as copepods. The glue cells on the tentacles enable the ctenophore to adhere to the small prey and carry it to the mouth. Digestion is both extracellular and intracellular.

Respiration and excretion. Respiration and excretion occur through the body surface.

Nervous and sensory systems. Ctenophores have a nervous system similar to that of the cnidarians. It is made up of a subepidermal plexus, which is concentrated under each comb plate, but there is no central control as is found in higher animals.

The **sense organ** at the aboral pole is a statocyst. Tufts of cilia support a calcareous statolith, with the whole being enclosed in a bell-like container. Alterations in the position of the animal change the pressure of the statolith on the tufts of cilia. The sense organ is also concerned in coordinating the beating of the comb rows, but does not trigger their beat.

The epidermis of ctenophores is abundantly supplied with sensory cells, so the animals are sensitive to chemical and other forms of stimuli. When a ctenophore comes in contact with an unfavorable stimulus, it often reverses the beat of its comb plates and backs up. The comb plates are very sensitive to touch, which often causes them to be withdrawn into the jelly. Neuroid conduction is present in the epithelial cells.

Reproduction. *Pleurobrachia,* in common with other ctenophores, is monoecious. The gonads are located on the lining of the gastrovascular canals under the comb plates. Fertilized eggs are discharged through the epidermis into the water.

Cleavage in the ctenophores is determinate (mosaic), since the various parts of the animal that will be formed by each cleavage cell are determined early in embryogenesis. If one of the cells is removed in the early stages, the resulting embryo will be deficient. This type of development is just the opposite of that of cnidarians. The free-swimming cydippid larva is superficially similar to the adult ctenophore and develops directly into an adult.

Other ctenophores

Ctenophores are fragile and beautiful creatures. Their transparent bodies glisten like fine glass, brilliantly iridescent during the day and luminescent at night.

One of the most striking ctenophores is *Beroë,* which may be more than 100 mm in length and 50 mm in breadth. Its shape is conical or ovoid, and it is provided with a large mouth but no tentacles. It is pink-colored and the body wall is covered with an extensive network of canals formed by the union of the paragastric and meridional canals. Venus's girdle *(Cestum)* is compressed and bandlike, may be more than a meter long, and presents a graceful appearance as it swims. The highly modified *Ctenoplana* and *Coeloplana* are very rare but are interesting because they have flattened disc-shaped bodies and are adapted for creeping rather than swimming. Both have unusually long tentacles. A common ctenophore along the Atlantic and Gulf coasts is *Mnemiopsis,* which has a laterally compressed body with two large oral lobes and unsheathed tentacles.

Nearly all ctenophores give off flashes of luminescence at night, especially such forms as *Mnemiopsis* (Fig. 10-29). The vivid flashes of light seen at night in southern seas are often caused by members of this phylum.

PHYLOGENY AND ADAPTIVE RADIATION

Phylogeny. The origin of the cnidarians and ctenophores is obscure, though the most widely supported theory today is that the radiate phyla arose from a radially symmetric, planula-like ancestor. It may well have been that such an ancestor could have been common to the radiates and to the higher metazoa, the latter having been derived from a branch whose members habitually crept about on the sea bottom. Such a habit would select for bilateral symmetry. Others became sessile or free-floating, conditions for which radial symmetry is a selective advantage. A planula larva in which an invagination formed to become the gastrovascular cavity would correspond roughly to a cnidarian with an ectoderm and an endoderm.

The trachyline medusae (an order of class Hydrozoa) are sometimes considered the most primitive of modern cnidarians because of their direct development from the planula and actinula larvae to the medusa. Such a group

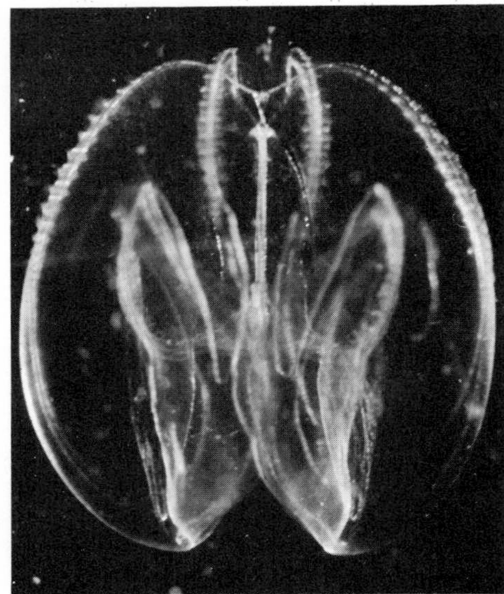

FIG. 10-29 The sea walnut *Mnemiopsis,* a ctenophore. Its fragile beauty is particularly apparent at night, when large numbers can be seen pulsating in the water. Common in Atlantic and Gulf of Mexico waters. (Photograph by Roman Vishniac; courtesy Encyclopaedia Britannica Films, Inc.)

could have given rise to the three classes of cnidarians, with further development of the polyp and subsequent loss of the medusa occurring in the anthozoan line. Though some zoologists have argued differently, that the Cnidaria were derived from a bilateral ancestor via the most "bilateral" of the cnidarians (Anthozoa), the preponderance of evidence is that the Hydrozoa is the most primitive class and the Anthozoa is the most specialized.

In light of their many resemblances to the cnidarians, the ctenophores may have originated from the same ancestral stock, such as the trachyline medusae, but conclusive evidence is not available.

Adaptive radiation. In their evolution neither phylum has deviated very far from its basic plan of structure. In the Cnidaria, both the polyp and medusa are constructed on the same scheme. Likewise, the ctenophores have adhered to the arrangement of the comb plates and their biradial symmetry.

Nonetheless, the cnidarians are a successful phylum in terms of numbers of individuals and species, demonstrating a surprising degree of diversity considering the simplicity of their basic body plan. They are efficient predators, many feeding on prey quite large compared to themselves. Some are adapted for feeding on small particles. The colonial form of life is well-explored, with some colonies growing to great size among the corals, and others, such as the siphonophores, showing astonishing polymorphism and specialization of individuals within the colony.

Derivation and meaning of names

Aurelia (L., *aurum,* gold) Genus (Scyphozoa).

Calyptoblastea (Gr., *kalyptos,* covered, + *blastos,* bud, sprout) Order of Hydroida having the gonophores in a gonotheca; also called Leptomedusae.

Cestum (Gr., *kestos,* girdle) Genus (Tentaculata). The body is compressed into a ribbonlike form.

Chlorohydra (Gr., *chlōros,* green, + hydra) Genus (Hydrozoa).

Craspedacusta (Gr., *kraspedon,* border, velum, + *kystis,* bladder) Genus (Hydrozoa).

Gonionemus (Gr., *gōnia,* angle, + *nema,* thread) Genus (Hydrozoa).

Gymnoblastea (Gr., *gymnos,* naked, bare, + *blastos,* bud, sprout) Order of Hydroida having naked medusa buds; also called Anthomedusae.

Hydra (Greek mythology, water serpent)

Metridium (Gr., *mētridios,* fruitful) Genus (Anthozoa).

Nomeus (Gr., herdsman) Refers to the habits of this commensal fish in inducing larger fish to chase it into the grasping tentacles of the Portuguese man-of-war.

Obelia (Gr., *obelias,* round cake) Genus (Hydrozoa).

Pelmatohydra (Gr., *pelma, pelmatos,* stalk or sole, + hydra) Genus (Hydrozoa). These hydras have a stalk at the basal end of body.

Physalia (Gr., *physallis,* bladder) Genus (Hydrozoa). Has a bladderlike gas-filled float.

Pleurobrachia (Gr., *pleuron,* side, + L., *brachia,* arms) Genus (Tentaculata). Refers to the tentacles, one on each side.

Siphonophora (Gr., siphon + *phora,* pl. of bearing) An order of Hydrozoa. Refers to the gastrozooid polyp with a single hollow tentacle in these polymorphic hydrozoan colonies.

Trachylina (Gr., *trachys,* rough, + L., *linum,* flax) An order of Hydrozoa. Refers to the appearance produced by the long-haired sensory cells and lithostyles or sense clubs.

Annotated references

Selected general readings

Barnes, D. J. 1970. Coral skeletons: an explanation of their growth and structure. Science **170:**1305-1308 (Dec.).

Bayer, F. M. 1961. The shallow-water Octocorallia of the West Indian region. The Hague, Martinus Nijhoff.

Bayer, F. M. 1971. Coral. Natural History **80**(3):42-47 (Mar.). *Excellent color photographs of reef corals.*

Bayer, F. M., and H. B. Owre. 1968. The free-living lower invertebrates. New York, The Macmillan Co.

Bloom, A. L. 1974. Geomorphology of reef complexes. In L. F. Laporte (ed.). Reefs in time and space. Special publication no. 18. Tulsa, Society of Economic Paleontologists and Mineralogists.

Bourne, G. C. 1900. Ctenophora. In E. R. Lankester. A treatise on zoology. London, A. & C. Black, Ltd. *Good detailed descriptions of the morphology of ctenophores.*

Brown, F. A., Jr. 1950. Selected invertebrate types. New York, John Wiley & Sons, Inc. *Good for certain representative marine forms.*

Buchsbaum, R. 1976. Animals without backbones: an introduction to the invertebrates, ed. 2. Chicago, University of Chicago Press. *Many excellent illustrations of cnidarians.*

Buck, J. 1973. Bioluminescent behavior in *Renilla*. Biol. Bull. **144**(1):19-37 (Feb.).

Bullough, W. S. 1950. Practical invertebrate anatomy. New York, The Macmillan Co. *A practical manual of certain selected types, but taxonomy is outdated.*

Burnett, A. L. 1973. Biology of *Hydra*. New York, Academic Press, Inc.

Bushnell, J. H., Jr., and T. W. Porter. 1967. The occurrence and prey of *Craspedacusta sowerbyi* (particularly polyp stages) in Michigan. Trans. Am. Microsc. Soc. **86:**22-27.

Compton, G. 1970. What is the world's deadliest animal? Sci. Dig. **68**(2):24-28 (Aug.).

Crowell, S. (ed.). 1965. Behavioral physiology of coelenterates. Am. Zool. **5:**335-389.

Dougherty, E. C. (ed.). 1963. The lower Metazoa. Berkeley, University of California Press. *Deals with comparative biology and phylogeny of the lower metazoan phyla.*

Fraser, C. M. 1937. Hydroids of the Pacific coast of Canada and the United States. Toronto, University of Toronto Press. *This monograph and a similar one on the hydroids of the Atlantic coast are the most comprehensive taxonomic studies yet made on the group in America.*

Goreau, T. 1961. Problems of growth and calcium deposition in reef corals. Endeavor **20:**32-39. *A summary of the recent concept involving the role of zooxanthellae in coral formation.*

Goreau, T. F., and L. S. Land. 1974. Fore-reef morphology and depositional processes, North Jamaica. In L. F. Laporte (ed.). Reefs in time and space. Special publication no. 18. Tulsa, Society of Economic Paleontologists and Mineralogists.

Gosner, K. L. 1971. Guide to identification of marine and estuarine invertebrates. New York, Interscience Publishers. *An essential aid for students of the invertebrates found along the northeastern coast of the United States.*

Hardy, A. C. 1956. The open sea. Boston, Houghton Mifflin Co. *Many beautiful plates of medusae and other cnidarians.*

Hickman, C. P. 1973. Biology of the invertebrates, ed. 2. St. Louis, The C. V. Mosby Co.

Hyman, L. H. 1940. The invertebrates: Protozoa through Ctenophora. New York, McGraw-Hill Book Co. *Extensive accounts are given of the Cnidaria and the ctenophores in the last two chapters of this authoritative work.*

Jha, R. K., and G. O. Mackie. 1967. The recognition, distribution, and ultrastructure of hydrozoan nerve elements. J. Morphol. **123:**43-61.

Kaestner, A. 1967. Invertebrate zoology, vol. 1. New York, Interscience Publishers.

Lenhoff, H. M., and W. F. Loomis. 1961. The biology of hydra and some other coelenterates. Coral Gables, Fla., University of Miami Press.

Lenhoff, H. M., L. Muscatine, and L. V. Davis (eds.). 1971. Experimental coelenterate biology. Honolulu, University of Hawaii Press.

Mackie, G. O. 1970. Neuroid conduction and the evolution of conducting tissues. Q. Rev. Biol. **45:**319-332.

Macklin, M., T. Roma, and K. Drake. 1972. Water excretion by hydra. Science **179:**194-195 (Jan.).

Mariscal, R. N. 1971. Experimental studies on the protection of anemone fishes from sea anemones. In T. C. Cheng (ed.). Aspects of the biology of symbiosis. Baltimore, University Park Press.

McCullough, C. B. 1963. The contraction burst pacemaker system in hydras. Proc. Sixteenth Int. Congr. Zool. (Washington) **1:**25.

Morin, J. C., and J. W. Hastings. 1971. Energy transfer in a bioluminescent system. J. Cell. Physiol. **77:**313-318.

Muscatine, L., and H. M. Lenhoff (eds.). 1974. Coelenterate biology. Reviews and new perspectives. New York, Academic Press, Inc. *Highly recommended for further reading; it has chapters on histology, nematocysts, neurobiology, development, symbioses, bioluminescence, and Ctenophora.*

Pearse, J. S., and V. B. Pearse. 1978. Vision in cubomedusan jellyfishes. Science **199:**458.

Pennak, R. W. 1953. Freshwater invertebrates of the United States. New York, The Ronald Press Co. *Chapter 4 is devoted to the freshwater cnidarians.*

Purdy, E. G. 1974. Reef configurations: cause and effect. In L. F. Laporte (ed.). Reefs in time and space. Special publication no. 18. Tulsa, Society of Economic Paleontologists and Mineralogists. *Extensive consideration of karst theory of coral reef morphology, with experimental evidence.*

Rees, W. J. (ed.). 1966. The Cnidaria and their evolution. New York, Academic Press, Inc. *A symposium on various aspects of cnidarian evolution, plus articles on other topics.*

Russell, F. S. 1953. The medusae of the British Isles. Cambridge, Cambridge University Press. *In this fine monograph the many species of British medusae are described in text and by beautiful plates, many in color.*

Schaller, H. C. 1976. Action of the head activator as a growth hormone in hydra. Cell Differ. **5:**1-11.

Schaller, H. and A. Gierer. 1973. Distribution of the head-activating substance in hydra and its localization in membranous particles in nerve cells. J. Embryol. Exp. Morphol. **29:**39-52.

Smith, F. G. W. 1948. Atlantic reef corals. Miami University of Miami Press. *This is a valuable aid for identification of Atlantic hermatypic corals.*

Totton, A. K., and G. O. Mackie. 1960. Studies on *Physalia physalis* (L.). Discovery Reports **30:**301-407. Cambridge, Cambridge University Press. *Monograph on the natural history and morphology (Totton) and the behavior and histology (Mackie) of the Portuguese man-of-war.*

Uchida, T. 1963. Two phylogenetic lines in the coelenterates from the viewpoint of their symmetry. Proceedings of the Sixteenth International Congress of Zoology (Washington), vol. 1, p. 24.

Warshofasky, F. 1966. Portuguese man-of-war. Nat. Wildlife. **4**(4):37-40 (June-July).

Webster, G. 1971. Morphogenesis and pattern formation in hydroids. Biol. Rev. **46**(1):107.

Westfall, J. A. 1965. Nematocysts of the sea anemone, *Metridium*. Am. Zool. **5:**373-393.

Westfall, J. A. 1973. Ultrastructural evidence for neuromuscular systems in coelenterates. Am. Zool. **13:**237-246.

Selected *Scientific American* articles

Berrill, N. J. 1957. The indestructible hydra. **197:**118-125 (Dec.). *Describes the remarkable regenerative powers of the hydra.*

Feder, H. A. 1972. Escape responses in marine invertebrates. **227:**92-100 (July).

Gierer, A. 1974. Hydra as a model for the development of biological form. **231:**44-54. (Dec.)

Lane, C. E. 1960. The Portuguese man-of-war. **202:**158-168 (Mar.). *This beautiful but dangerous hydrozoan represents a remarkable colonial development.*

Loomis, W. F. 1959. The sex gas of hydra. **200:**145-156 (Apr.). *The carbon dioxide tension of the water appears to play an important role in controlling sexual reproduction in hydras.*

Newell, N. D. 1972. The evolution of reefs. **226:**54-65 (June).

CHAPTER 11

THE ACOELOMATE ANIMALS

Phylum Platyhelminthes
Phylum Rhynchocoela

Scanning electron micrograph of male and female *Schistosoma mansoni.* Most flukes are hermaphroditic, but this medically important species is dioecious. The female is held by the male, clasped in a large groove (gynocophoric canal) in the male's ventral surface.

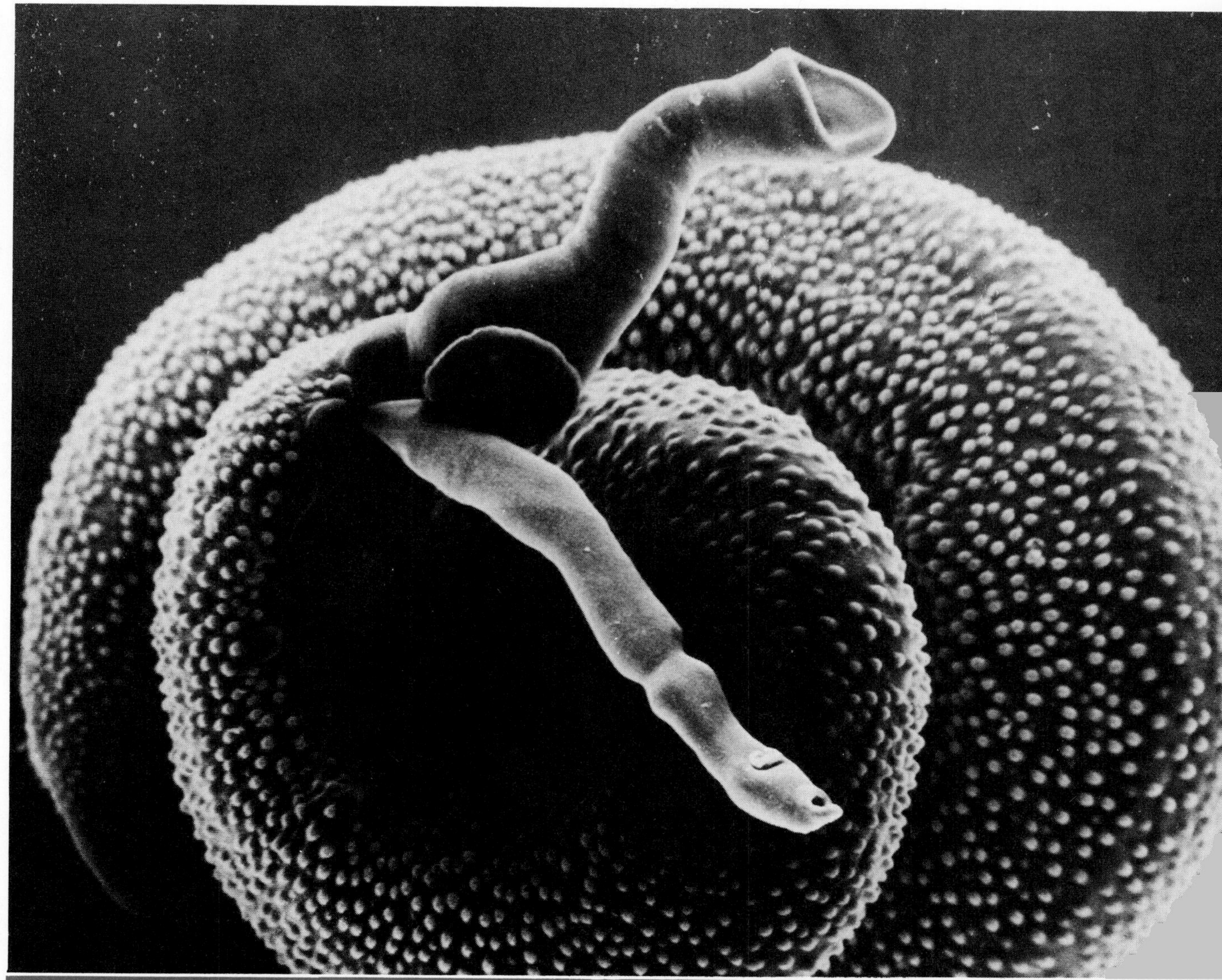

Courtesy A. Sue Carlisle, University of Pennsylvania School of Medicine, Philadelphia.

Position in animal kingdom

1. The Platyhelminthes, or flatworms, and the Rhynchocoela, or ribbon worms, are the most primitive groups of animals to have **primary bilateral symmetry,** the type of symmetry assumed by all higher animals.
2. These phyla have only one internal space, the digestive cavity, with the region between the ectoderm and endoderm filled with mesoderm in the form of muscle fibers and mesenchyme (parenchyma). Since they lack a coelom or a pseudocoelom, they are termed **acoelomate animals,** and because they have three well-defined germ layers, they are termed triploblastic.
3. Acoelomates show more specialization and division of labor among their organs than do the radiate animals because the mesoderm makes more elaborate organs possible. Thus the acoelomates are said to have reached the **organ-system level of organization.**
4. They belong to the protostome division of the Bilateria and have spiral and determinate cleavage.

Biologic contributions

1. The acoelomates have developed the basic **bilateral** plan of organization that has been widely exploited in the animal kingdom.
2. The **mesoderm** is now developed into a well-defined embryonic germ layer **(triploblastic),** thus making available a great source of tissues, organs, and systems.
3. Along with bilateral symmetry, **cephalization** has been established. There is some centralization of the nervous system evident in the **ladder type of system** found in flatworms.
4. Along with the subepidermal musculature, there is also a mesenchymal system of muscle fibers.
5. An **excretory system** appears for the first time.
6. The rhynchocoels have developed the **first circulatory system** with blood and the **first one-way alimentary canal.** Although not stressed by zoologists, the rhynchocoel cavity in ribbon worms is technically a true coelom, but as it is merely a part of the proboscis mechanism, it is not of evolutionary significance.
7. Unique and specialized structures occur in both phyla. The parasitic habit of many flatworms has led to many specialized adaptations, such as organs of adhesion.

PHYLUM PLATYHELMINTHES

General relations

The term ''worm'' has been loosely applied to elongated, bilateral invertebrate animals without appendages. At one time zoologists considered worms (Vermes) to be a group in their own right. Such a group included a highly diverse assortment of forms. Modern classification has broken up this group into phyla and reclassified them. By tradition, however, zoologists still refer to the various groups of these animals as ''flatworms,'' ''ribbon worms,'' ''roundworms,'' segmented worms,'' and the like.

The term Platyhelminthes (plat'y-hel-min'theez) (Gr., *platys,* flat, + *helmins,* worm), was first proposed by Gegenbaur (1859) and was applied to the animals now included under that heading. At first rhynchocoels and some others were included, but later they were removed to other groups.

The flatworms are bilaterally symmetric, a basic plan for all animals that have advanced very far in complexity of organization. They were derived from an ancestor that probably had many cnidarian-like characteristics, perhaps an ancestor shared by the cnidarians. Nonetheless, replacement of the jellylike mesoglea of cnidarians with a cellular mesodermal (endomesodermal) mesenchyme (parenchyma) laid the basis for a more complex organization. Parenchyma is a form of connective tissue containing more cells and fibers than the mesoglea of the cnidarians.

Flatworms range in size from a millimeter or less to some of the tapeworms that are many meters in length. Their flattened bodies may be slender, broadly leaflike, or long and ribbonlike.

Ecologic relationships

The flatworms include both free-living and parasitic forms, but the free-living members are found exclusively in the class Turbellaria. A few turbellarians are symbiotic or parasitic, but the majority are adapted as bottom dwellers in marine (Fig. 11-1) or fresh water or live in moist places on land. Many, especially of the larger species, are found on the underside of stones and other hard objects in freshwater streams or in the littoral zones of the ocean. Some may actually be found

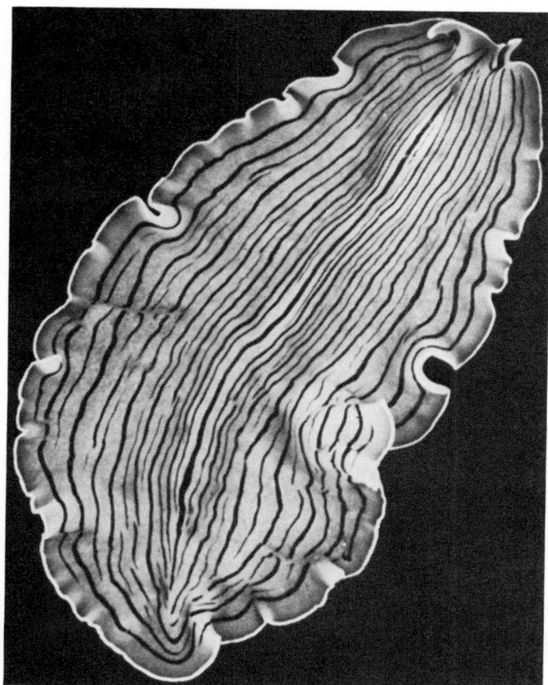

FIG. 11-1 *Prostheceraeus vittatus;* a marine polyclad turbellarian. Note sensory tentacles at upper right. (Photograph by D. P. Wilson.)

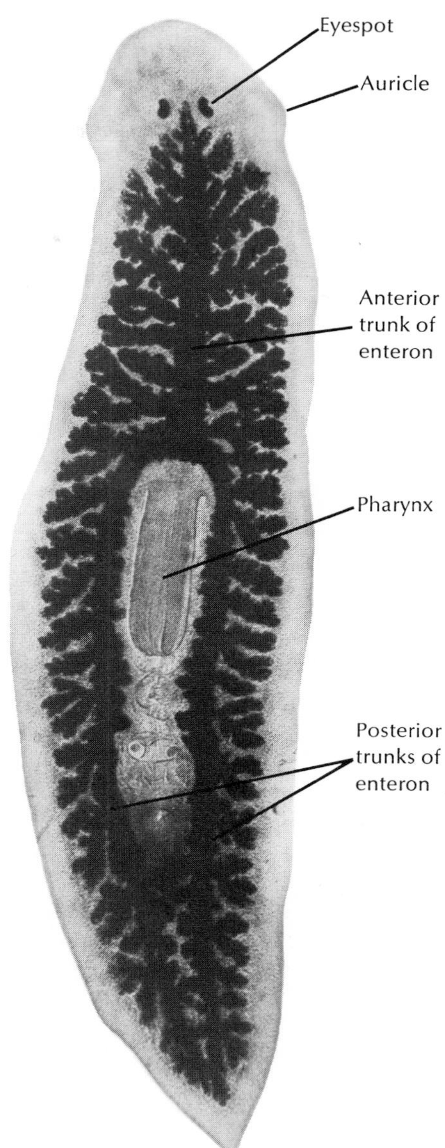

FIG. 11-2 Stained planarian. (Courtesy Carolina Biological Supply Co., Burlington, N.C.)

at considerable depths in the ocean. Pelagic existence is usually restricted to larval forms, although some adults are pelagic in tropic or subtropic seas. Those that live in the tidal zones of shores can adapt themselves to a wide range of temperature and salinity.

Relatively few turbellarians live in fresh water. Planarians (Fig. 11-2) and some others frequent streams and spring pools; others prefer flowing water of mountain streams. Some species occur in fairly hot springs.

Terrestrial turbellarians are found in fairly moist places under stones and logs. Land planarians, such as *Bipalium,* occur frequently in greenhouses, where temperature conditions are favorable. Some of the turbellarians harbor symbiotic chlorellae within their mesenchyme. An example is the genus *Convoluta,* which may, at certain stages, digest their chlorellae.

All members of the classes Monogenea and Trematoda (the flukes) and the class Cestoda (the tapeworms) are parasitic. Most of the Monogenea are ectoparasites, but all the trematodes and cestodes are endoparasitic. Many species have indirect life cycles with more than one host; the first host is often an invertebrate, and the final host is usually a vertebrate. Humans serve as hosts for a number of species. Certain larval stages may be free-living. Both flukes and tapeworms have undergone many adaptive changes in their evolution as specialized parasites. Such adaptations involve, for example, loss of sense organs, the appearance of adhesive organs or suckers, loss of digestive system, and greatly increased reproductive capacity.

Characteristics

1. Three germ layers **(triploblastic)**
2. **Bilateral symmetry;** definite polarity of anterior and posterior ends
3. **Body flattened dorsoventrally;** oral and genital apertures mostly on ventral surface
4. Body segmented in one class (Cestoda)
5. Epidermis may be cellular or syncytial (ciliated in some); **rhabdites** in epidermis of most Turbellaria; epidermis a synctial **tegument** in Monogenea, Trematoda, and Cestoda
6. Muscular system of a sheath form and of mesodermal origin; layers of circular, longitudinal, and oblique fibers beneath the epidermis
7. No internal body space other than digestive tube (acoelomate); spaces between organs filled with parenchyma, a form of connective tissue or mesenchyme
8. Digestive system incomplete (gastrovascular type); absent in some
9. **Nervous system consisting of a pair of anterior ganglia with longitudinal nerve cords connected by transverse nerves and located in the mesenchyme** in most forms; similar to cnidarians in primitive forms
10. Simple sense organs; eyespots in some
11. Excretory system of two lateral canals with branches bearing **flame cells (protonephridia);** lacking in some primitive forms
12. Respiratory, circulatory, and skeletal systems lacking; lymph channels with free cells in some trematodes
13. Most forms monoecious; reproductive system complex, usually with well-developed gonads, ducts, and accessory organs; internal fertilization; development direct in free-swimming forms and those with a single host in the life cycle; usually indirect in internal parasites in which there may be a complicated life cycle often involving several hosts
14. Class Turbellaria mostly free-living; classes Monogenea, Trematoda, and Cestoda entirely parasitic

Classes

Class Turbellaria (tur'bel-lar'e-a) (L., *turbellae* [pl.], stir, bustle, + *-aria,* like or connected with). The turbellarians. Usually free-living forms with soft flattened bodies; covered with ciliated epidermis containing secreting cells and rodlike bodies (rhabdites); mouth usually on ventral surface sometimes near center of body; no body cavity except intercellular lacunae in parenchyma; mostly hermaphroditic, but some have asexual fission. Examples: *Dugesia* (planaria), *Microstomum, Planocera.*

Class Monogenea (mon'o-gen'e-a) (Gr., *mono,* single, + *gene,* origin, birth). The monogenetic flukes. Body covered with a syncytial tegument without cilia; body usually leaflike to cylindric in shape; posterior attachment organ with hooks, suckers, or clamps, usually in combination; monoecious; development direct, with single host and usually with free-swimming, ciliated larva; all parasitic, mostly on skin or gills of fish. Examples: *Dactylogyrus, Polystoma, Gyrodactylus.*

Class Trematoda (trem'a-to'da) (Gr., *trematodes,* with holes, + *eidos,* form). The digenetic flukes. Body covered with a syncytial tegument without cilia; leaflike or cylindric in shape; usually with oral and ventral suckers, no hooks; alimentary canal usually with two main branches; nervous system similar to turbellarians; mostly monoecious; development indirect, with first host a mollusc, final host usually a vertebrate; parasitic in all classes of vertebrates. Examples: *Fasciola, Clonorchis, Schistosoma.*

Class Cestoda (ses-to'da) (Gr., *kestos,* girdle, + *eidos,* form). The tapeworms. Body covered with nonciliated, syncytial tegument; general form of body tapelike; scolex with suckers or hooks, sometimes both, for attachment; body usually divided into series of proglottids; no digestive organs; usually monoecious; parasitic in digestive tract of all classes of vertebrates; development indirect with two or more hosts; first host may be vertebrate or invertebrate. Examples: *Diphyllobothrium, Hymenolepis, Taenia.*

Class Turbellaria

Turbellarians are mostly free-living worms that range in length from 5 mm or less to 50 cm. Usually covered with ciliated epidermis, these are mostly creeping worms that combine muscular with ciliary movements to achieve locomotion. The mouth is on the ventral side. Unlike the trematodes and cestodes, they have simple life cycles.

The orders of Turbellaria can be divided into two groups based on specialization of the female reproductive system and embryogenesis. In the more primitive group the yolk for nutrition of the developing embryo is contained in the egg cell itself **(endolecithal),** and the embryogenesis shows the spiral determinate cleavage typical of protostomes (p. 128). In the more advanced group the egg cell contains little or no yolk, and the yolk is contributed by cells released from organs called **vitellaria (ectolecithal).** Usually a number of yolk cells are combined with the zygote within the eggshell, affecting cleavage in such a way that the spiral pattern cannot be distinguished.

Of the several orders in the more primitive group, the order Acoela is often regarded as having changed least from the ancestral form. Its members are small and have a mouth but no gastrovascular cavity or excretory system. Food is merely passed through the mouth or pharynx into temporary spaces that are surrounded by a syncytial mesenchyme where gastrodermal phagocytic cells digest the food intracellularly. The order has a syncytial epidermis and a diffuse nervous system. An example is *Otocelis,* which lives as a commensal in the digestive system of sea urchins and sea cucumbers.

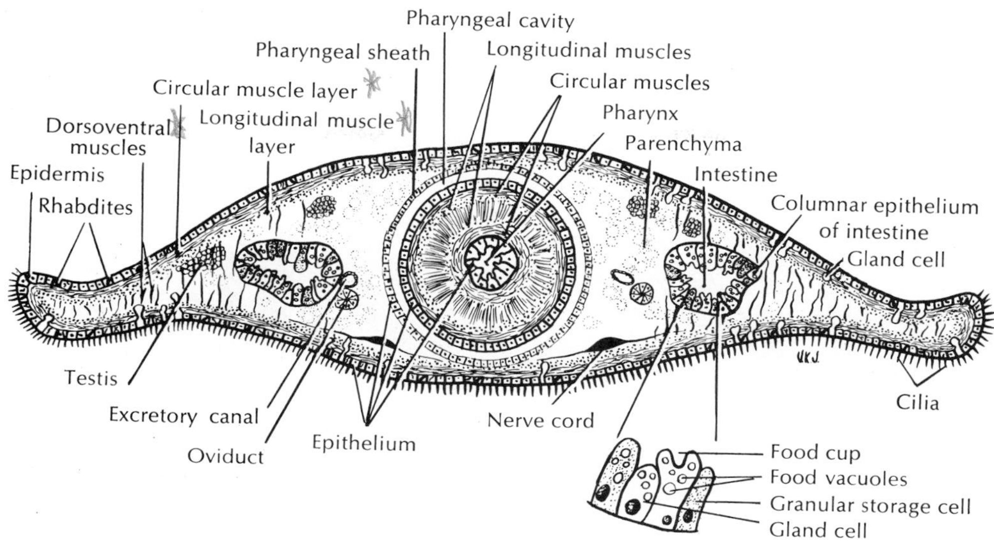

FIG. 11-3 Cross section of planarian through pharyngeal region showing relations of body structures.

Another of the primitive orders, the Polycladida, is comprised of leaflike marine forms that have many intestinal branches from the central digestive cavity. They are often brightly colored and have numerous eyes. *Prostheceraeus* has sensory tentacles, each of which bears several eyes (Fig. 11-1). A feature of some polyclads is the presence of a ciliated, free-swimming, larval form. One of these, called Müller's larva, is provided with ciliated projecting lobes and eyespots.

Of the more advanced orders, Neorhabdocoela and Seriata contain many marine and freshwater species. Members of the suborder Tricladida (order Seriata) have a three-branched digestive tract. The common freshwater planarians belong to this group (Fig. 11-2). Some members of the suborder are marine, such as *Bdelloura* (Fig. 11-5), an ectoparasite on the gills of *Limulus,* the horseshoe crab. *Bipalium* is a common large terrestrial form often found in greenhouses.

Representative type—*Dugesia,* the common brown planarian

Habitat. A well-known representative of the triclad turbellarians (suborder Tricladida) is the common freshwater brown planarian *(Dugesia)* (Fig. 11-2). Several species of *Dugesia* are found on the underside of rocks and debris in brooks, ponds, and springs of the United States. They are small and flat and their dark mottled color blends perfectly with the rocks or plants to which they cling. Other common genera are *Dendrocelopsis* and *Cura,* both black planarians, and *Phagocata,* a white variety.

Structure. *Dugesia* is flat and slender and about 2 cm or less in length. The head region is triangular, with two lateral lobes known as **auricles.** These are not ears but olfactory organs. Two **eyespots** on the dorsal side of the head near the midventral line give the animal a cross-eyed appearance. Its background color may range from a dark yellow to olive, or dark brown. The ventral side is lighter. Near the center of the ventral side is the **mouth,** through which the muscular **pharynx (proboscis)** can be extended for the capture of prey (Fig. 11-4, *C*). The **genital pore** opens posterior to the mouth.

The outer covering is ciliated epidermis resting on a basement membrane (Fig. 11-3). It contains rod-shaped **rhabdites** that, when discharged with water, swell and form a protective mucous sheath around the body. Single-cell mucous glands open on the surface of the epidermis. In the body wall below the basement membrane are layers of **muscle fibers** that run circularly, longitudinally, and diagonally. A meshwork of **parenchyma** cells, developed from mesoderm, fills the spaces between muscles and visceral organs.

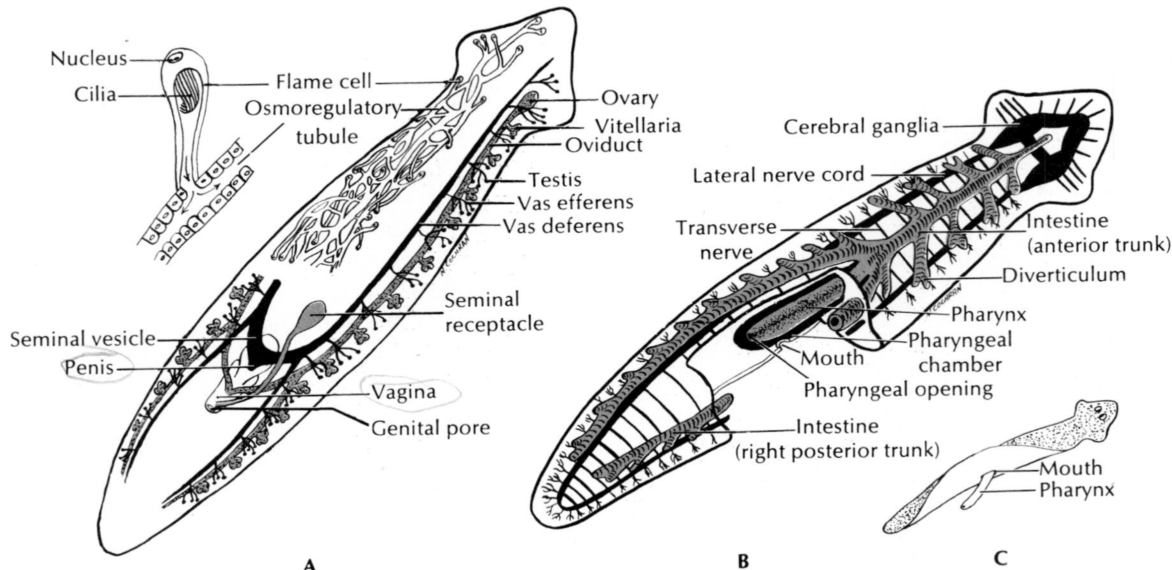

FIG. 11-4 **A,** Reproductive and osmoregulatory systems in planaria. Portions of male and female organs omitted to show part of osmoregulatory system. Inset at left is enlargement of flame cell. **B,** Diagrammatic view of digestive system and ladder-type nervous system of planaria. Cut section shows relation of pharynx, in resting position, to digestive system and mouth on ventral surface. **C,** Pharynx extended through ventral mouth.

Locomotion. Freshwater planarians move by gliding, head slightly raised, over a slime tract secreted by the marginal adhesive glands. The beating of the epidermal cilia in the slime tract drives the animal along. Rhythmic muscular waves can also be seen passing backward from the head as it glides. A less common method is crawling. The worm lengthens, anchors its anterior end, and by contracting its longitudinal muscles pulls up the rest of its body.

Digestive system. The digestive system includes a mouth, pharynx, and intestine. The pharynx, enclosed in a **pharyngeal sheath** (Fig. 11-3), opens posteriorly just inside the mouth, through which it can extend (Fig. 11-4, *B*). The intestine has three many-branched trunks, one anterior and two posterior. The whole forms a **gastrovascular cavity** lined with columnar epithelium (Figs. 11-3 and 11-4, *B*).

Planarians are mainly carnivorous, feeding largely on small crustaceans, nematodes, rotifers, and insects. They can detect food from some distance by means of chemoreceptors. They entangle their prey in mucous secretions from the mucous glands and rhabdites. The planarian grips its prey with its anterior end, wraps its body around the prey, extends its proboscis, and sucks up minute bits of the food. Intestinal secretions contain proteolytic enzymes for some **extracellular digestion.** Bits of food are sucked up into the intestine, where the phagocytic cells of the gastrodermis complete the digestion **(intracellular).** The gastrovascular cavity ramifies to most parts of the body, and food is absorbed through its walls into the body cells. Undigested food is egested through the pharynx. Planarians can go a long time without feeding; during starvation the cells of the mesenchyme and reproductive systems are resorbed, and the body volume may be reduced by as much as 300 times.

Excretion and osmoregulation. The osmoregulatory system consists of two longitudinal canals with a complex network of tubules that branch to all parts of the body and end in **flame cells (protonephridia)** (Fig. 11-4, *A*). The flame cell surrounds a small space into which a tuft of cilia projects. The space is continuous with the tubule complex, and the beat of the cilia provides force to drive fluids through the system. The tubules lead into collecting ducts that finally open dorsally by two pores. That this system is mainly os-

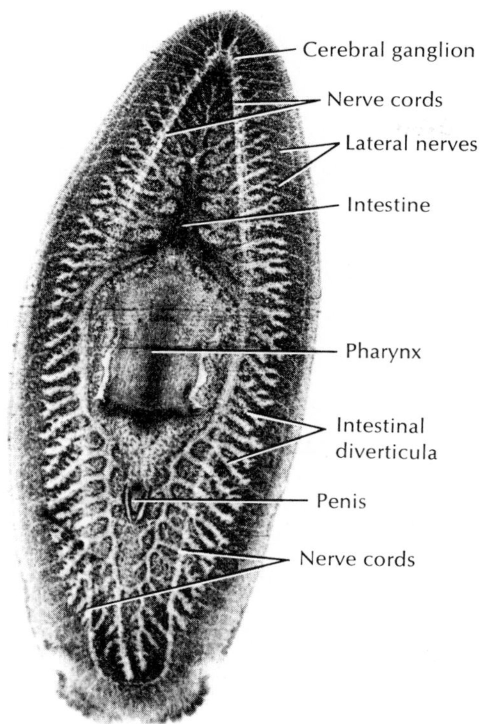

FIG. 11-5 Ladder type of nervous system shows up clearly in this photomicrograph of stained preparation of *Bdelloura*, a marine triclad. (Photograph by F. M. Hickman.)

moregulatory is indicated by the observation that it is reduced or absent in marine turbellarians, which do not have to expel excess water.

Metabolic wastes are largely removed by diffusion through the body wall.

Respiration. There are no respiratory organs. Exchange of gases takes place through the body surface.

Nervous system. Two **cerebral ganglia** beneath the eyespots serve as the **"brain."** Two ventral **nerve cords** extending from the brain are connected by transverse nerves to form a "ladder type" of nervous system (Figs. 11-4, *B*, and 11-5). The more primitive turbellarians may have three, four, or five pairs of longitudinal nerve trunks.

The **eyespots** (Fig. 11-2) are pigment cups into which retinal cells extend from the brain, with the photosensitive ends of the cells inside of the cups. They are sensitive to light intensities and the direction of a source but can form no images. Since light must enter the opening of the cup to stimulate the sensory cells, the direction of the light source is detected (Fig. 11-6). Planaria are negatively phototactic and are most active at night. Receptors for taste, smell, and touch are distributed over the body but are concentrated on small lobes at the anterior end, the **auricles.**

Reproduction. Triclad turbellarians reproduce both sexually and asexually. Asexually the animal merely constricts behind the pharyngeal region and separates into two animals. Each new animal regenerates its

FIG. 11-6 Ocelli, or eyes, of planarian. Pigment cup lets light enter open side, parallel to long axis of retinal (light-sensitive) cells. Planarian determines light direction from stimulation of light-sensitive cells.

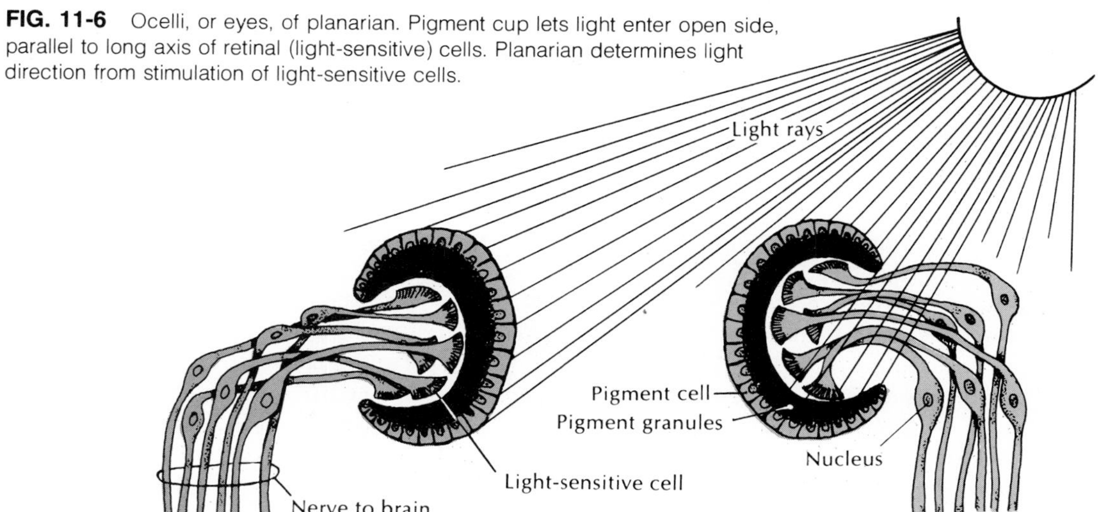

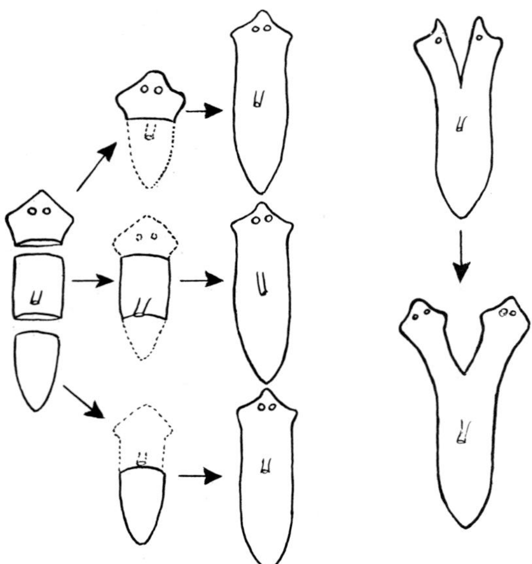

FIG. 11-7 Regeneration in planarian. When cut transversely into three separate pieces *(left)*, each piece develops into planarian. If head is split *(right)*, a double-headed animal results.

missing parts. Sexually the worm is **monoecious;** each individual is provided with both male and female organs (Fig. 11-4, *A*).

In the **male system** (Fig. 11-4, *A*) a row of **testes** on each side empty sperm by way of tiny tubes into a common duct, the **vas deferens,** which enlarges posteriorly to form a **seminal vesicle** where sperm are stored until discharged through the muscular **penis** (Fig. 11-5) during copulation (the mating act).

The **female system** contains two anterior **ovaries** that discharge eggs into tubular **oviducts,** which in turn join to form the **vagina.** Vitelline ducts empty into the oviducts. Connected to the vagina is the seminal receptacle, which receives and stores sperm from the mating partner.

Although hermaphroditic, turbellarians practice cross-fertilization, mutually exchanging sperm during copulation. At copulation each worm inserts its penis into the genital pore of the mate and in each worm sperm pass through the penis from the male seminal vesicle to the female seminal receptacle of the mate. After the worms separate, sperm pass up the oviducts to fertilize eggs as they leave the ovaries. Several eggs together with yolk cells from vitellaria become enclosed in small spherical egg capsules. The egg capsules, when laid, are attached by little stalks to the underside of stones or plants. Embryos finally emerge as little planarians (juveniles).

Regeneration. Planarians have great power to regenerate lost parts (Fig. 11-7). When an individual is cut in two, the anterior end will grow a new tail and the posterior end a new head. Planarians may also be grafted or cut in ways that produce freakish designs such as two heads or two tails.

Experiments have shown that when a planarian is cut across, free cells **(neoblasts)** from the mesenchyme migrate to the cut surface and aggregate there to form a blastema, which develops into the new part. Irradiation of a worm will destroy these neoblasts, and no regeneration will occur. It has been suggested that the neoblasts are attracted to the cut region by chemical emanations, which cease when the blastema is formed.

Behavior. Planarians respond to the same types of stimuli as do other animals. Their ventral surfaces are positively thigmotactic, whereas their dorsal surfaces are negatively thigmotactic. Planarians respond in a positive way to the juices of foods. They avoid strong light and seek out dark or dimly lighted regions. They are more active at night than during the day. The auricular lobes appear to be sensitive to both water currents and chemical stimuli. In general their reactions to water currents depend on their normal habitats. Those from flowing water are usually positively rheotactic; those from still water will not react or else are positively rheotactic to weak currents only.

The behavior patterns of flatworms have been the subject of numerous investigations all over the world, for animal behaviorists have considered them to be the lowest group that show any capacity for learning in response to simple conditioned changes.

Class Monogenea

The monogenetic flukes traditionally have been placed as an order of the Trematoda, but they are sufficiently different to deserve a separate class. They are all parasites, primarily of the gills and external surfaces of fish. A few are found in the urinary bladders of frogs and turtles, and one has been reported from the eye of a hippopotamus. Though widespread and common, monogeneans seem to cause little damage to their hosts under natural conditions. However, like numerous other fish pathogens, they become a serious threat when their hosts are crowded together, as in fish farming.

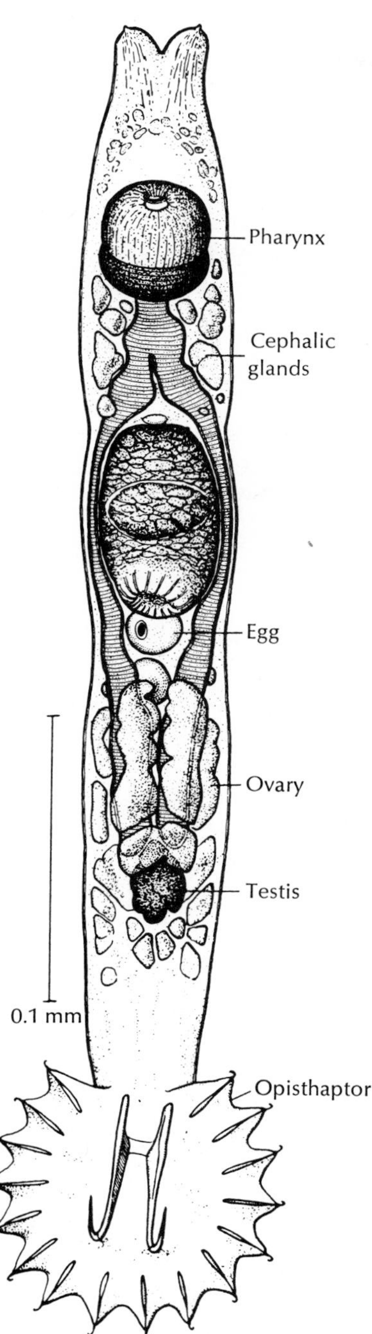

FIG. 11-8 *Gyrodactylus cylindriformis,* ventral view. (From Mueller, J. F., and H. J. Van Cleave. 1932. Roosevelt Wildlife Annals.)

The life cycles of monogeneans are direct, with a single host. The egg hatches a ciliated larva, the **oncomiracidium,** that attaches to the host or swims around awhile before attachment. The oncomiracidium bears hooks on its posterior, which in many species become the hooks on the large posterior attachment organ (the **opisthaptor**) of the adult. Because the monogenean must cling to the host and withstand the force of water flow over the gills or skin, adaptive radiation has produced a wide array of opisthaptors in different species. Opisthaptors may bear large and small hooks, suckers, and clamps, often in combination with each other.

Common genera are *Gyrodactylus* (Fig. 11-8) and *Dactylogyrus,* both of economic importance to fish culturists, and *Polystoma,* found in the urinary bladder of frogs.

Class Trematoda

Trematodes are all parasitic flukes, and as adults they are almost all found as endoparasites of vertebrates. They are chiefly leaflike in form and are structurally similar in many respects to the more advanced Turbellaria. A major difference is found in the body covering, or **tegument,** which does not bear cilia in the adult. Furthermore, in common with Monogenea and Cestoda, the nuclei of the tegument are found in cell bodies sunk beneath the outer layer (Fig. 11-9). The cell bodies are internal to the superficial muscle layers and communicate with the outer layer (distal cytoplasm) by processes extending between the muscles. The cell bodies have mitochondria, Golgi complex, endoplasmic reticulum, and a variety of vesicles in their cytoplasm. The distal cytoplasm is crowded with vesicles and some mitochondria. Because the distal cytoplasm is continuous, with no intervening cell membranes, the tegument is syncytial. This peculiar epidermal arrangement is probably related to adaptations for parasitism in ways that are still unclear.

Other structural adaptations for parasitism are more apparent: various penetration glands or glands to produce cyst material; organs for adhesion such as suckers, hooks, and so on; and increased reproductive capacity. Otherwise, trematodes retain several turbellarian characteristics, such as a well-developed alimentary canal (but with the mouth at the anterior, or cephalic, end) and similar reproductive, excretory, and nervous systems, as well as a musculature and mesen-

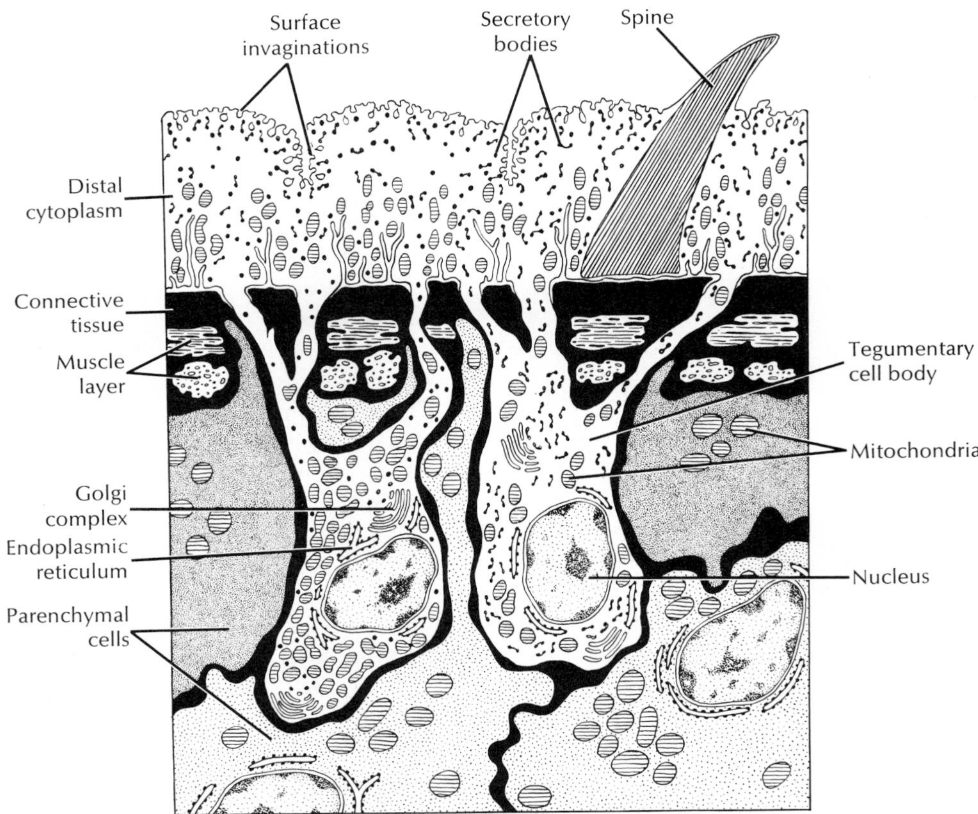

FIG. 11-9 Diagrammatic drawing of the structure of the tegument of a trematode *Fasciola hepatica*. (Drawing by L. T. Threadgold, from Schmidt, G. D. and L. S. Roberts. 1977. Foundations of Parasitology. St. Louis, The C. V. Mosby Co.)

chyme that are only slightly modified from those of the Turbellaria. Sense organs are poorly developed.

Of the subclasses of Trematoda, Aspidogastrea and Didymozoidea are small and poorly known groups, but Digenea is a large group with many species of medical and economic importance.

Digenea

With rare exceptions, digenetic trematodes have an indirect life cycle, the first **(intermediate)** host being a mollusc and the final **(definitive)** host being a vertebrate. In some species a second, and sometimes even a third, intermediate host intervenes. Some investigators believe that the ancestors of digeneans were parasites of molluscs as larvae and free-living as adults; then, when vertebrates evolved, some of the ancestral forms were able to exploit that habitat and gave rise to digenetic trematodes. The group has been very successful, and they inhabit, according to species, a wide variety of sites in their hosts: all parts of the digestive tract, the respiratory tract, the circulatory system, the urinary tract, and the reproductive tract.

One of the world's most amazing biologic phenomena is the digenean life cycle. Although the cycles of different species vary widely in detail, a typical example would include the adult, egg, miracidium, sporocyst, redia, cercaria, and metacercaria stages (Fig. 11-10). The **egg** usually passes from the definitive host in the excreta and must reach water to develop further. There, it hatches to a free-swimming, ciliated larva, the **miracidium.** The miracidium penetrates the tissues of a snail, where it transforms into a **sporocyst.** The sporocyst reproduces asexually to yield either more sporocysts or a number of **rediae.** The rediae, in turn, reproduce asexually to produce more rediae or to produce **cercariae.** In this way a single egg can give

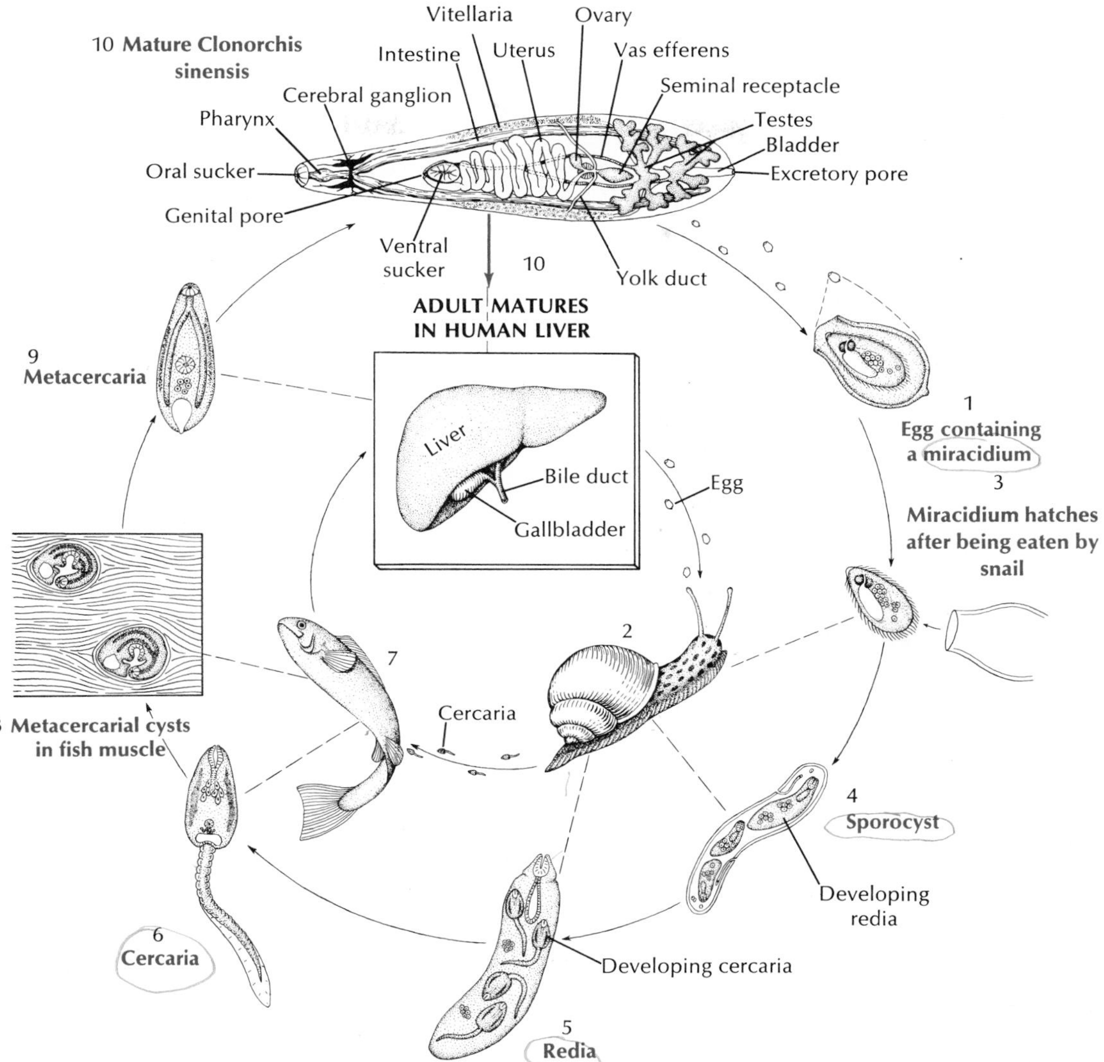

FIG. 11-10 Life cycle of human liver fluke *Clonorchis sinensis.* Egg, *1,* shed from adult trematode, *10,* in bile ducts of human, is carried out of body in feces and is ingested by snail *(Parafossarulus), 2,* in which miracidium, *3,* hatches and becomes mother sporocyst. *4,* Young rediae are produced in sporocyst, grow, *5,* and in turn produce young cercariae. Cercariae now leave snail, *6,* find a fish host, *7,* and burrow under scales to encyst in muscle, *8.* When raw or improperly cooked fish containing cysts is eaten by humans, metacercaria is released, *9,* and enters bile duct, where it matures, *10,* to shed eggs into feces, *1,* thus starting another cycle.

rise to an enormous number of progeny. The cercariae emerge from the snail and penetrate a second intermediate host or encyst on vegetation or other objects to become **metacercariae,** which are juvenile flukes. The adult grows from the metacercaria when that stage is eaten by the definitive host.

Some of the most serious parasites of humans and domestic animals belong to the Digenea. The first digenetic life cycle to be worked out was that of *Fasciola hepatica,* which causes "liver rot" in sheep and other ruminants. The adult fluke lives in the bile passages of the liver, and the eggs are passed in the feces. After hatching, the miracidium penetrates a snail to become a sporocyst. There are two generations of rediae, and the cercaria encysts on vegetation. When the infested vegetation is eaten by the sheep or other ruminant, the metacercariae excyst and grow into young flukes.

Clonorchis sinensis—liver fluke of humans

Clonorchis (Fig. 11-10) is the most important liver fluke of humans and is common in many regions of the Orient, especially in China, southern Asia, and Japan. Cats, dogs, and pigs are also often infected.

Structure. The worms vary from 10 to 20 mm in length (Fig. 11-10). They have an **oral sucker** and a **ventral sucker** and are covered externally by a tegument. The **digestive system** consists of a pharynx, a muscular esophagus, and two long, unbranched intestinal ceca. The **excretory system** consists of two protonephridial tubules, with branches provided with flame cells or bulbs. The two tubules unite to form a single median tubule that opens to the outside. The **nervous system,** like that of turbellarians, is made up of two cerebral ganglia connected to longitudinal cords that have transverse connectives. The **muscular system** also is of the planarian type.

The **reproductive system** is hermaphroditic and complex. The **male system** is made up of two branched **testes** and two **vasa efferentia** that unite to form a single **vas deferens,** which widens into a **seminal vesicle.** The seminal vesicle leads into an **ejaculatory duct,** which terminates at the genital opening. Unlike most trematodes, *Clonorchis* does *not* have a protrusible copulatory organ, the cirrus. The female system contains a branched **ovary** with a short **oviduct,** which is joined by ducts from the **seminal receptacle** and the **vitellaria.** The oviduct is surrounded by a glandular mass, **Mehlis' gland,** of uncertain function. From Mehlis' gland the much-convoluted **uterus** runs to the genital pore. Cross-fertilization between individuals is usual, and sperm are stored in the seminal receptacle. When an ovum is released from the ovary, it is joined by a sperm and a group of vitelline cells and is fertilized. The vitelline cells release a proteinaceous shell material, which is stabilized by a chemical reaction, the Mehlis' gland secretions are added, and the egg passes into the uterus.

Life cycle. The normal habitat of the adults is in the bile passageways of humans and other fish-eating mammals (Fig. 11-10). The eggs, each containing a complete **miracidium,** are shed into the water with the feces but do not hatch until they are ingested by the snail *Parafossarulus* or related genera. The eggs, however, may live for some weeks in water. In the snail the miracidium enters the tissues and transforms into the **sporocyst** (a baglike structure with embryonic germ cells), which produces one generation of **rediae.** The redia is elongated, with an alimentary canal, a nervous system, an excretory system, and many germ cells in the process of development. The rediae pass into the liver of the snail where, by a process of internal budding, they give rise to the tadpolelike **cercariae.**

The cercariae escape into the water, swim about until they meet with fish of the family Cyprinidae, and then bore into the muscles or under the scales. Here the cercariae lose their tails and encyst as **metacercariae.** If a mammal eats raw infected fish, the metacercarial cyst dissolves in the intestine, and the metacercariae apparently migrate up the bile duct, where they become adults. Here the flukes may live for 15 to 30 years.

The effect of the flukes on humans depends mainly on the extent of the infection. A heavy infection may cause a pronounced cirrhosis of the liver and may result in death. Cases are diagnosed through fecal examinations. To avoid infection, all fish used as food should be thoroughly cooked. Destruction of the snails that carry larval stages would be a method of control.

Schistosoma—blood flukes

Schistosomiasis, infection with blood flukes of the genus *Schistosoma,* ranks as one of the major infectious diseases in the world, with 200 million people infected. The disease is widely prevalent over much of Africa and parts of South America, the West Indies, the Middle East, and the Far East. The old generic name for the worms was *Bilharzia,* and the infection was called *bilharziasis,* a name still used in many areas.

The blood flukes differ from most other flukes in being dioecious and having the two branches of the digestive tube united into a single tube in the posterior part of the body. The male is broader and heavier and has a large, ventral groove, the **gynecophoric canal,** posterior to the ventral sucker. The gynecophoric canal embraces the long, slender female (Fig. 11-11).

Three species account for most of the schistosomiasis in humans: *S. mansoni,* which lives primarily in the venules draining the large intestine; *S. japonicum,*

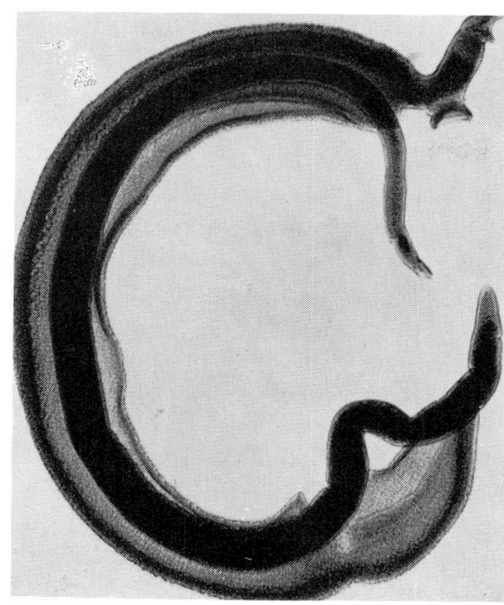

FIG. 11-11 Adult male and female *Schistosoma mansoni* in copulation. Male has long sex canal that holds female (the darkly stained individual) during insemination and oviposition. Humans are usually hosts of adult parasites, found mainly in Africa but also in South America and elsewhere. Humans become infected by wading or bathing in cercaria-infested waters. (AFIP No. 56-3334.)

which is found mostly in the venules of the small intestine; and *S. haematobium,* which lives in the venules of the urinary bladder. *S. mansoni* is common in parts of Africa, Brazil, northern South America, and the West Indies; and species of *Biomphalaria* are the principal snail intermediate hosts. *S. haematobium* is widely prevalent in Africa, using snails of the genera *Bulinus* and *Physopsis* as the main intermediate hosts. *S. japonicum* is confined to the Far East, and its hosts are several species of *Oncomelania*.

The plan of the life history of blood flukes is similar in all species. Eggs are discharged in human feces or urine; if they get into water, they hatch out as ciliated **miracidia,** which must contact the required kind of snail within a few hours to survive. In the snail, they transform into **sporocysts,** which produce another generation of sporocysts. The daughter sporocysts give rise to **cercariae** directly, without the formation of rediae. The cercariae escape from the snail and swim about until they come in contact with the bare skin of a human. They penetrate through the skin, shedding their tails in the process, and reach a blood vessel where they enter the circulatory system. There is no metacercarial stage. The young schistosomes make their way to the hepatic portal system of blood vessels and undergo a period of development in the liver before migrating to their characteristic sites. As eggs are released by the adult female, they are somehow extruded through the wall of the venule and through the gut or bladder lining, to be voided with the feces or urine, according to species. Many eggs do not make this difficult transit and are swept by the blood flow back to the liver or other areas, where they become centers of inflammation and tissue reaction.

The main ill effects of schistosomiasis result from the eggs. With *S. mansoni* and *S. japonicum,* eggs in the intestinal wall cause ulceration, abscesses, and bloody diarrhea with abdominal pain. Similarly, *S. haematobium* causes ulceration of the bladder wall with bloody urine and pain on urination. Eggs swept to the liver or other sites cause symptoms associated with the organs where they lodge. When they are caught in the capillary bed of the liver, they impede circulation and cause cirrhosis, a fibrotic reaction that interferes with liver function. Of the three species, *S. haematobium* is considered the least serious and *S. japonicum* the most severe. The prognosis is poor in heavy infestations of *S. japonicum* without early treatment.

Control is best carried out by educating people to dispose of their body wastes hygienically, a difficult problem with poor people under primitive conditions. Unfortunately, some projects intended to raise the standard of living in some countries, such as the Aswan High Dam in Egypt, have increased the prevalence of schistosomiasis by creating more habitats for the snail intermediate hosts.

Schistosome dermatitis (swimmer's itch). Various species of schistosomes in several genera are known to cause a rash or dermatitis when their cercariae penetrate hosts that are unsuitable for further development. The cercariae of several genera whose normal hosts are North American birds cause dermatitis in bathers in northern lakes. The severity of the rash increases with an increasing number of contacts with the organisms, or sensitization. After penetration, the cercariae are attacked and killed by the host's immune mechanisms, and they release allergenic substances, causing itching. The condition is more an annoyance than a serious threat to health, but there may be economic losses to persons depending on vacation trade around infested lakes.

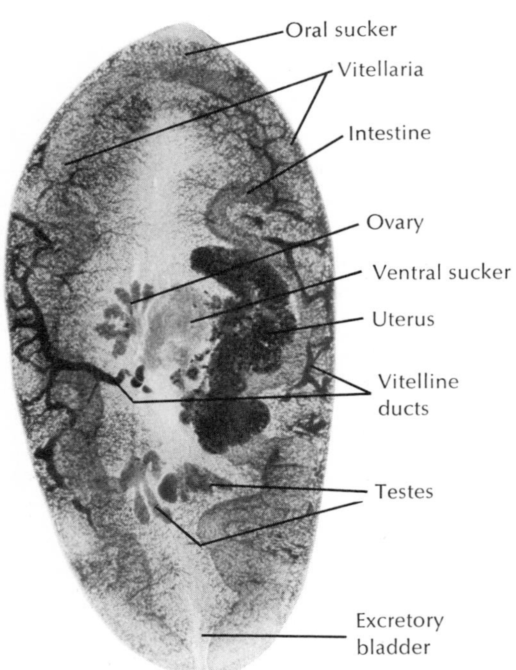

FIG. 11-12 Pulmonary fluke *Paragonimus westermani* infects human lung, producing paragonimiasis. Adults are up to 2 cm long. Eggs discharged in sputum or feces hatch into free-swimming miracidia that enter snails. Cercariae from snail enter freshwater crabs and encyst in soft tissues. Humans are infected by eating poorly cooked crabs or by drinking water containing larvae freed from dead crabs. (Photograph by R. E. Kuntz and J. A. Moore; from Schmidt, G. D., and L. S. Roberts. 1977. Foundations of Parasitology. St. Louis, The C. V. Mosby Co.)

Paragonimus—lung flukes

Several species of *Paragonimus,* a fluke that lives in the lungs of its host, are known from a variety of mammals. *Paragonimus westermani* (Fig. 11-12), found in the Orient, Southwest Pacific, and some parts of South America, parasitizes a number of wild carnivores, humans, pigs, and rodents. Its eggs are coughed up in the sputum, swallowed, then eliminated with the feces. The metacercariae are found in freshwater crabs, and the infection is acquired by eating uncooked crab meat. The infection causes respiratory symptoms, with breathing difficulties and chronic cough. Fatal cases are common. A closely related species, *P. kellicotti,* is found in mink and similar animals in North America, but only one human case has been recorded. Its metacercariae are in crayfish.

Some other trematodes

Fasciolopsis buski (intestinal fluke of humans) parasitizes humans and pigs in India and China. Larval stages occur in several species of planorbid snails, and the cercariae encyst on water chestnuts, an aquatic vegetation eaten raw by humans and pigs.

Leucochloridium is noted for its remarkable sporocysts. Snails *(Succinea)* eat vegetation infected with eggs from bird droppings. The sporocysts become much enlarged and branched, and the cercariae encyst within the sporocyst. The sporocysts enter the snail's head and tentacles, become brightly striped with orange and green bands, and pulsate at frequent intervals. Birds are attracted by the enlarged and pulsating tentacles, eat the snails, and so complete the life cycle.

Class Cestoda

Cestoda, or tapeworms, differ in many respects from the preceding classes: they usually have long flat bodies made up of many segments, or **proglottids,** and there is a complete lack of a digestive system. As in Monogenea and Trematoda, there are no external, motile cilia in the adult, and the tegument is of a distal cytoplasm with sunken cell bodies beneath the superficial muscle layer (Fig. 11-9). In contrast to the monogenes and trematodes, however, their entire surface is covered with minute projections called **microtriches.** The microtriches greatly amplify the surface area of the tegument, which is a vital adaptation of the tapeworm since it must absorb all its nutrients across the tegument.

Tapeworms are nearly all monoecious. They have well-developed muscles, and their excretory system and nervous system are somewhat similar to those of other flatworms. They have no special sense organs but do have sensory endings in the tegument that are modified cilia. One of their most specialized structures is the **scolex,** or holdfast, which is the organ of attachment. It is usually provided with suckers or suckerlike organs and often with hooks or spiny tentacles (Fig. 11-13).

With rare exceptions, all cestodes require at least two hosts, and the adult is a parasite in the digestive tract of vertebrates. Often one of the intermediate hosts is an invertebrate.

The class Cestoda is divided into two subclasses:

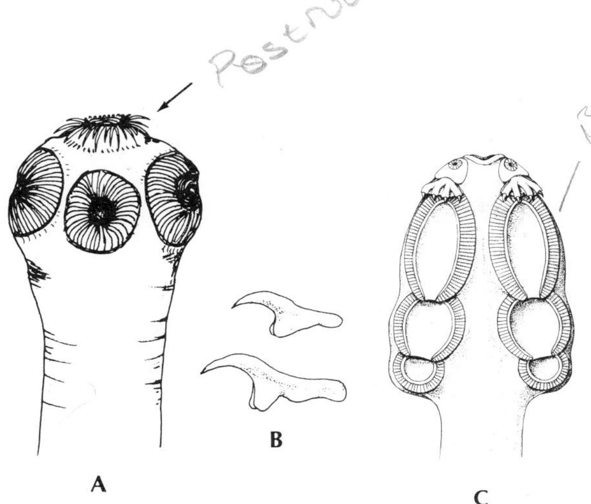

FIG. 11-13 Two tapeworm scolices. **A,** Scolex of *Taenia solium* (pork tapeworm) with apical hooks and suckers. (Scolex of *Taeniarhynchus saginatus* is similar, but without hooks). **B,** Hooks of *T. solium.* **C,** Scolex of *Acanthobothrium coronatum,* a tapeworm of sharks. This species has large leaflike sucker organs divided into chambers with apical suckers and hooks.

Cestodaria and Eucestoda. The Cestodaria is a small group found in a few lower vertebrates. Their bodies are not divided into separate proglottids and have only one set of reproductive organs **(monozoic).** Their larvae have ten hooks.

The Eucestoda contains the great majority of species in the class. With the exception of two small orders, the members of this subclass have the body divided into a series of proglottids, each of which contains a set of reproductive organs **(polyzoic).** Their larval forms all have six hooks, rather than ten as in the Cestodaria. The main body of the worms, the chain of proglottids, is called the **strobila.** Typically, there is a **germinative zone** just behind the scolex where new proglottids are budded off. As new proglottids are differentiated in front of it, each individual segment moves posteriorly in the strobila, and its gonads mature. The proglottid is usually fertilized by another proglottid in the same or a different strobila. The shelled embryos form in the uterus of the proglottid, and they are either expelled through a uterine pore, or the entire proglottid is shed from the worm as it reaches the posterior end.

Some zoologists have maintained that the proglottid formation of cestodes represents ''true'' segmentation (metamerism), but we do not support this view. Segmentation of tapeworms is best considered a replication of sex organs to increase reproductive capacity and is not related to the metamerism found in Annelida, Arthropoda, and Chordata (p. 115).

More than 1,000 species of tapeworms are known to parasitologists. Almost all vertebrate species are infected. Normally, adult tapeworms do little harm to their hosts. The most common tapeworms found in humans are given in Table 11-1.

Taeniarhynchus saginatus—beef tapeworm

Structure. The beef tapeworm lives as an adult in the alimentary canal of humans, whereas the larval form is found primarily in the intermuscular tissue of cattle. The mature adult may reach a length of 10 m or more. Its **scolex** has four **suckers** for attachment to the intesti-

TABLE 11-1 Common cestodes of humans

Common and scientific name	Means of infection; prevalence in humans
Beef tapeworm *(Taeniarhynchus saginatus)*	Eating rare beef; most common of all tapeworms in humans
Pork tapeworm *(Taenia solium)*	Eating rare pork
Fish tapeworm *(Diphyllobothrium* [*Dibothriocephalus*] *latum)*	Eating rare or poorly cooked fish; fairly common in Great Lakes region of United States, and other areas of world where raw fish is eaten
Dog tapeworm *(Dipylidium caninum)*	Unhygienic habits of children (larvae in flea and louse); moderate frequency
Dwarf tapeworm *(Hymenolepis nana)*	Larvae in flour beetles; common
Unilocular hydatid *(Echinococcus granulosus)*	Larval cysts in humans; infection by contact with dogs; common wherever humans are in close relationship with dogs and ruminants
Multilocular hydatid *(Echinococcus multilocularis)*	Larval cysts in humans; less common than unilocular hydatid

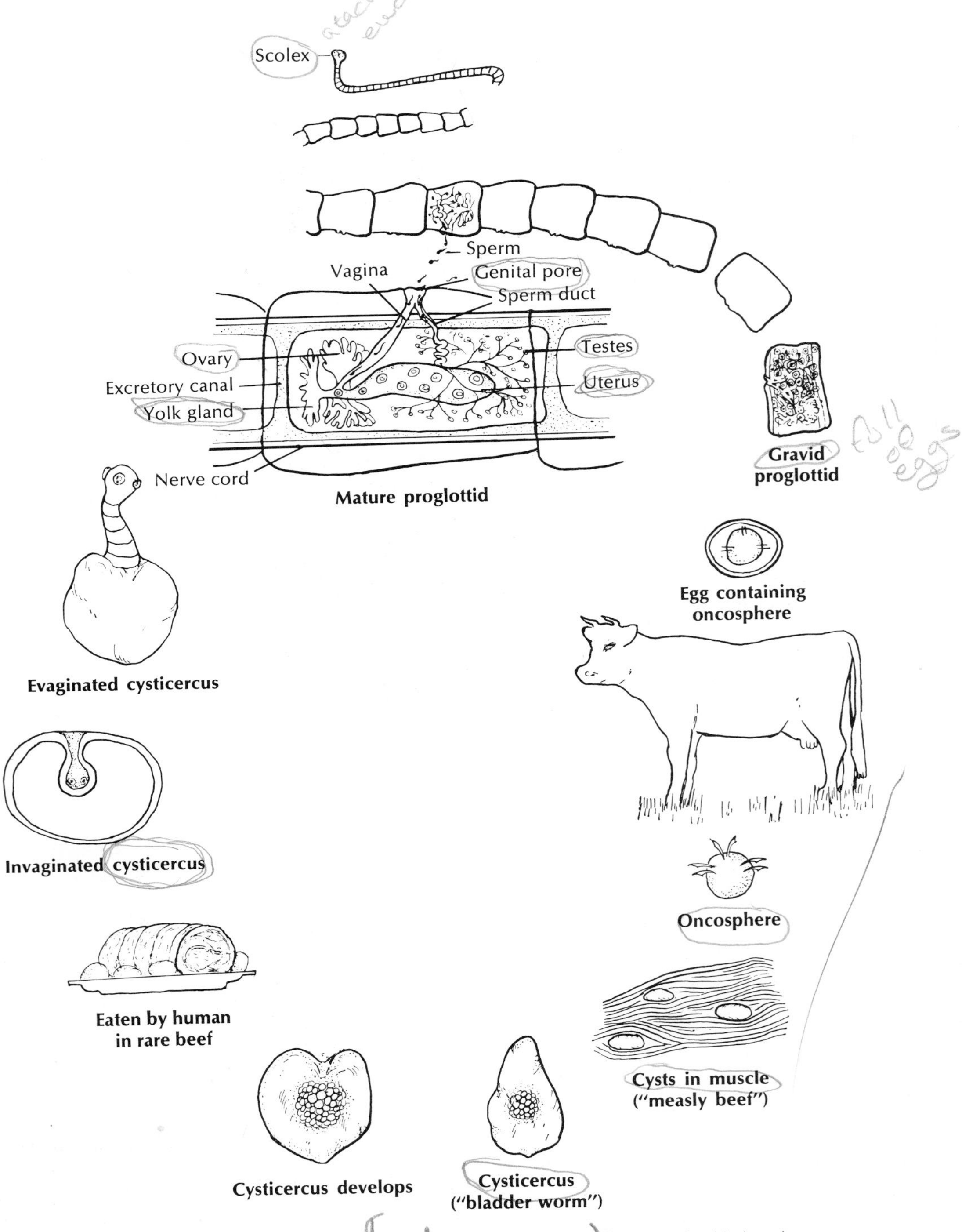

FIG. 11-14 Life cycle of beef tapeworm. *Taeniarhynchus saginatus.* Ripe proglottids break off in human intestine, pass out in feces, and are ingested by cattle. Eggs hatch in cow's intestine, freeing oncospheres, which penetrate into muscles and encyst, developing into "bladder worms." Human eats infected rare beef and cysticercus is freed in intestine where it develops, forms a scolex, attaches to intestine wall, and matures.

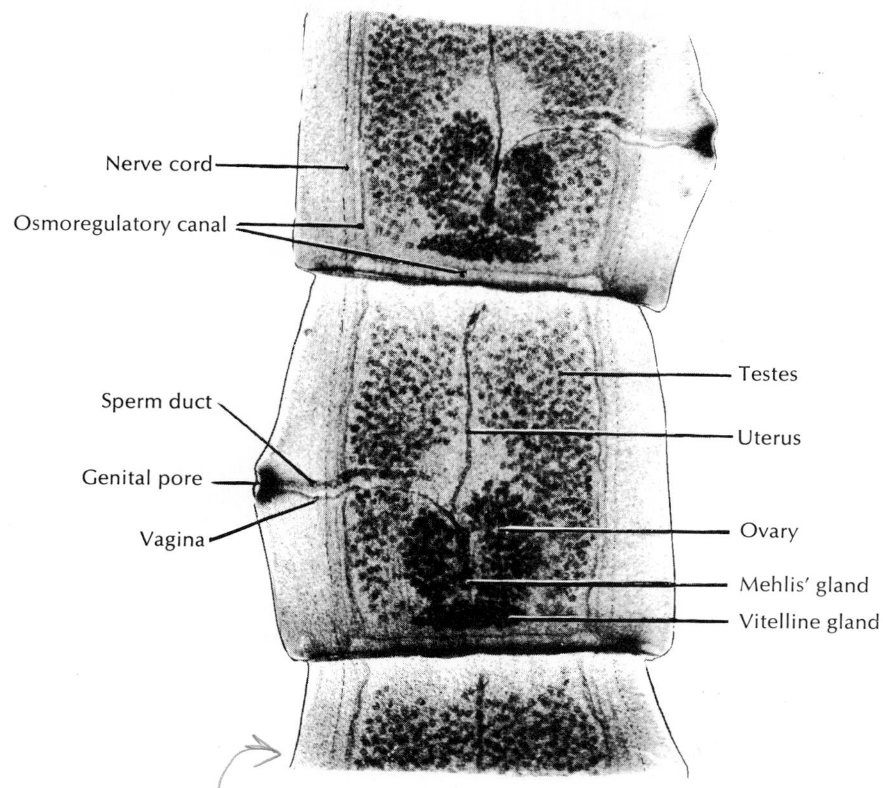

FIG. 11-15 Photomicrograph of mature proglottid of *Taenia pisiformis,* dog tapeworm. Portions of two other proglottids also shown. (Courtesy General Biological Supply House, Inc., Chicago.)

nal wall, but no hooks (Fig. 11-14). A short neck connects the scolex to the **strobila,** which may be made up of as many as 2,000 **proglottids.** New proglottids are formed by **transverse budding** of the neck region. As they mature and move backward, the proglottids increase in size, so that the proglottids are narrow near the scolex and broader and larger toward the posterior end. The terminal **gravid proglottids** (Fig. 11-14) are finally detached and shed in the feces.

The tapeworm shows some unity in its organization, for **excretory canals** in the scolex are also connected to the canals, two on each side, in the proglottids, and two longitudinal **nerve cords** from a **nerve ring** in the scolex run back into the proglottids also (Fig. 11-15). Attached to the excretory ducts are the **flame cells.** Each mature proglottid also contains **muscles** and **parenchyma** as well as a complete set of **male** and **female organs** similar to those of a trematode.

In the order to which this species belongs, however, the vitellaria are typically a single, compact **vitelline gland** located just posterior to the ovaries. When the gravid proglottids break off and pass out with the feces, they usually crawl out of the fecal mass and onto vegetation nearby. Here they may be picked up by grazing cattle. The proglottid ruptures as it dries up, further scattering the embryos on soil and grass. The embryos may remain viable on grass for as long as 159 days.

Life cycle. When cattle swallow the eggs, the eggs hatch, and the six-hooked larvae **(oncospheres)** burrow through the intestinal wall into the blood or lymph vessels and finally reach voluntary muscle, where they encyst to become **bladder worms (cysticerci).** Here the larvae develop an invaginated scolex but remain quiescent. When infected "measly" meat (Fig. 11-16) is eaten by a suitable host, the cyst wall dissolves, the scolex evaginates and attaches to the intestinal mu-

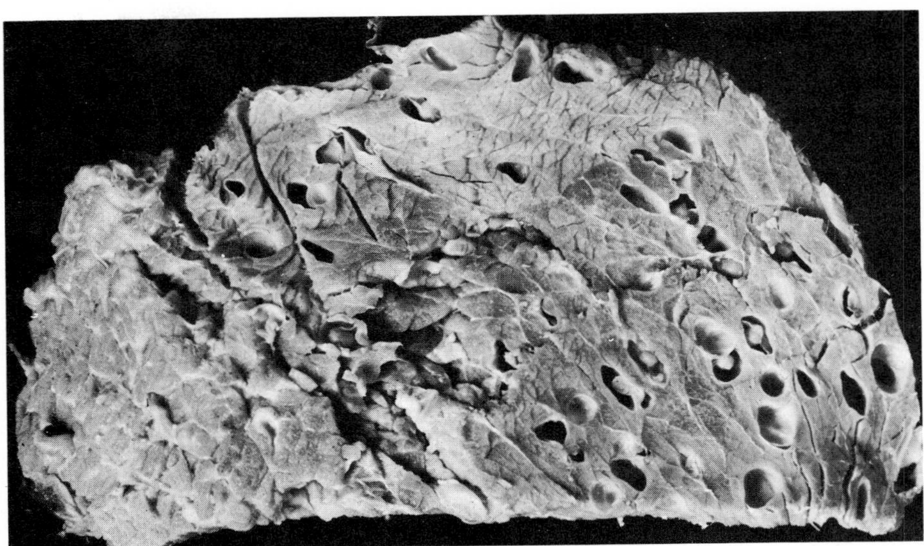

FIG. 11-16 "Measly" pork showing cysts of bladder worms, *Taenia solium*. Beef heavily infected with beef tapeworm has similar appearance but lighter infections may be much less obvious. (Photograph by F. M. Hickman.)

cosa, and new proglottids begin to develop. It takes 2 to 3 weeks for a mature worm to form. When a person is infected with one of these tapeworms, numerous gravid proglottids are expelled daily, sometimes crawling out the anus by themselves. Humans become infected by eating rare roast beef, steaks, and barbecues. Considering that about 1% of American cattle are infected, that 20% of all cattle slaughtered are not federally inspected, and that even when inspected one-fourth of infections are missed, it is not surprising that tapeworm infection is fairly common. Infection is precluded when meat is thoroughly cooked.

Some other tapeworms

Taenia solium (pork tapeworm). The adult *Taenia solium* lives in the small intestine of humans, whereas the larvae live in the muscles of pigs. Adults may be 7 m or longer. The scolex has both suckers and hooks arranged on its tip (Fig. 11-13, *A* and *B*), the **rostellum.** The life history of this tapeworm is similar to that of the beef tapeworm. Humans become infected by eating improperly cooked pork. The incidence of infection is much lower than that of the beef tapeworm, probably because rare pork is less popular.

The pork tapeworm *(T. solium)* is considered more dangerous than *T. saginatus* because the cysticerci, as well as the adults, can develop in humans. If eggs or proglottids are accidentally ingested by a human, or proglottids shed from an adult worm are carried back to the stomach by reverse peristalsis, the liberated embryos migrate to any of several organs and form cysticerci. The condition is called **cysticercosis.** Common sites are the eye or brain, and infection in them results in blindness or serious neurologic symptoms.

Diphyllobothrium latum (fish tapeworm). The adult tapeworm is found in the intestine of humans, dogs, and cats; the immature stages are in crustaceans and fish. This tapeworm, often called the broad tapeworm, is the largest of the cestodes that infect humans. It sometimes reaches a length of 20 m and may have more than 3,000 proglottids. Eggs discharged in water by the human host hatch and may be swallowed by the first intermediate host, a tiny crustacean *(Cyclops)*. When the crustacean is eaten by a fish, the young larva migrates to the muscles where it may grow to as much as 25 mm long. Usually it encysts in the muscle. When raw or poorly cooked fish is eaten by a suitable host, the larva is liberated and grows into adult form. It has been known to live in a person for many years. Broad tapeworm infections are found all over the world; in the United States infections are most common in the Great Lakes region. In Finland, but apparently not other areas, the worm may cause a serious anemia.

Dipylidium caninum (dog tapeworm). *Dipylidium*

is common in pet dogs and cats and sometimes in children. It may be 30 cm or more in length and may have about 200 proglottids. The larva is found in the louse and flea. The dog or cat becomes infected by licking or biting these ectoparasites. It takes about 2 weeks for the worm to mature.

Hymenolepis nana (dwarf tapeworm). The dwarf tapeworm is the smallest and perhaps the most common human tapeworm in the world. Its life cycle is unique in that an intermediate host is optional. The adults are 20 to 40 mm long and have 100 to 200 proglottids. The same species, though perhaps a different strain, is very common in mice. When eaten by a human or rodent, the larval forms are liberated and penetrate the intestinal mucosa, where they grow into a **cysticercoid** juvenile. After a few days they reenter the lumen of the intestine and mature into adult worms. Alternatively, if a grain beetle or any one of numerous other insects eats the eggs along with rodent feces, the embryo penetrates to the insect's hemocoel and forms the cysticercoid there. The human or rodent becomes infected by ingesting the grain beetle.

The common tapeworm of rats, *Hymenolepis diminuta,* is occasionally found in humans, but it is important for other reasons. Because its hosts are easily maintained in the laboratory, and because the worms are moderately large (up to 1 m long), *H. diminuta* is a very convenient form for experimental parasitology. Consequently, parasitologists know more about this species than any other cestode.

Echinococcus granulosus (unilocular hydatid). The adult *E. granulosus* (Fig. 11-17, *A*) is found in dogs and other canines; the larvae are found in more than 40 species of mammals, including humans, monkeys, sheep, reindeer, and cattle. Thus humans serve as an intermediate host in the case of this tapeworm. The adults are only 5 to 10 mm long and have just three or four proglottids. When ingested by a suitable intermediate host, the embryo migrates to the liver, lung, or some other organ and slowly begins to form a special kind of cysticercus called a **hydatid cyst.** The hydatid grows very slowly, taking 5 months to reach a size of 1 cm, but it can grow for a very long time—up to 20 years—reaching the size of a basketball in an unrestricted site such as the liver. If the hydatid grows in a critical site such as the heart or central nervous system, serious symptoms may be felt in a much shorter time. The main cyst maintains a single or unilocular chamber, but within the main cyst, daughter cysts bud off, and each contains thousands of scolices. Each scolex will produce a worm when eaten by a canine. The only treatment is surgical removal of the hydatid.

Echinococcus multilocularis (multilocular hydatid). *Echinococcus multilocularis* is closely related to *E. granulosus,* except that the main definitive hosts are foxes and the main intermediate hosts are rodents. Humans can also serve as intermediate hosts. Instead of developing a thick, laminated layer and growing into a large, single cyst, like *E. granulosus, E. multilocularis* has a thin outer wall that grows and infiltrates processes into the surrounding host tissues like a cancer. Each process may have several small, fluid-filled pockets containing scolices; thus, the hydatid is **multilocular.** *E. multilocularis* is more rare in humans than is *E. granulosus,* and that is fortunate, since surgery is much more difficult because of the invasive nature of the cyst.

Moniezia expansa (sheep tapeworm). The adult is found in sheep and goats (Fig. 11-18); the larva grows in a small mite (family Oribatidae).

Taenia pisiformis (dog tapeworm). *Taenia pisiformis* is widely used as a type for study in the laboratory (Fig. 11-15). The larvae occur in the mesenteries and liver of rabbits; the adults are found in cats and dogs. Its occurrence in humans is rare.

PHYLUM RHYNCHOCOELA (NEMERTINA)

General relations

Rhynchocoela (ring′ko-se′la) (Gr., *rhynchos,* beak, + *koilos,* hollow) are often called the ribbon worms. Their name refers to the proboscis, a long muscular tube that can be thrust out swiftly to grasp the prey. The phylum was formerly called Nemertea or Nemertina (nem′er-ti′na) (Gr., *Nemertes,* one of the Nereids, unerring one, + *-ina,* belonging to), with both names referring to the unerring aim of the proboscis. These worms are still often spoken of as the nemertean or nemertine worms. They are thread-shaped or ribbon-shaped worms; nearly all of them are marine. Some live in secreted gelatinous tubes. There are about 650 species in the group.

Nemertine worms are usually less than 20 cm long, though a few are several meters in length. *Lineus longissimus* is said to reach 30 m. Their colors are often bright, though most are dull or pallid. In the odd genus *Gorgonorhynchus* the proboscis is divided into

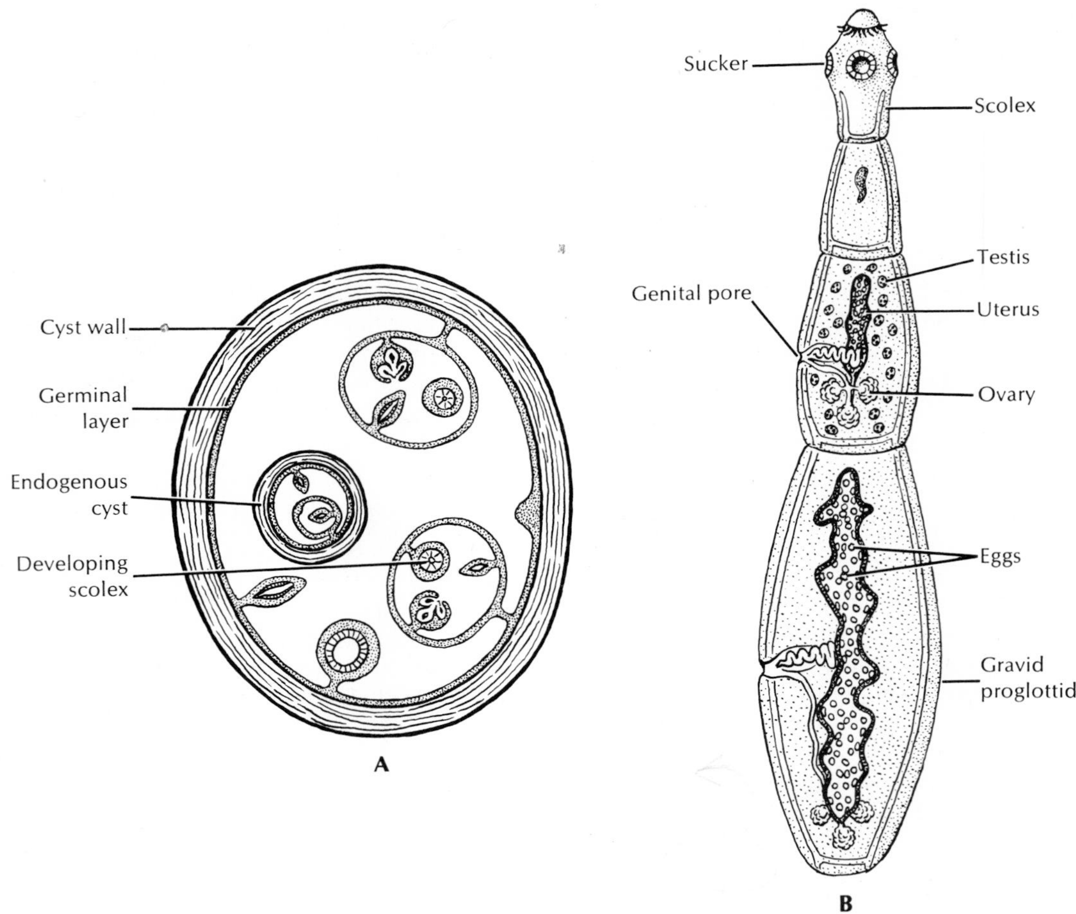

FIG. 11-17 *Echinococcus granulosus,* dog tapeworm, which may be dangerous to humans. **A,** Early hydatid cyst or bladder-worm stage found in cattle, sheep, hogs, and sometimes humans produces hydatid disease. Humans acquire disease by unsanitary habits in association with dogs. When eggs are ingested, liberated larvae usually encyst in liver, lungs, or other organs. Brood capsules containing scolices are formed from inner layer of each cyst. Cyst enlarges, developing other cysts with brood pouches. May grow for years, to size of basketball, necessitating surgery. **B,** Adult tapeworm lives in intestine of dog or other carnivore.

many proboscides, which appear as a mass of wormlike structures when everted.

With few exceptions, the general body plan of the nemertines is similar to that of Turbellaria. Like the latter, their epidermis is ciliated and has many gland cells. Another striking similarity is the presence of flame cells in the excretory system. Recently rhabdites have been found in several nemertines, including *Lineus*. However, nemertines differ from flatworms in their reproductive system. They are mostly dioecious. In the marine forms there is a ciliated **pilidium larva.** This helmet-shaped larva has a ventral mouth but no anus—another flatworm characteristic. It also has some resemblance to the trochophore larva that is found in annelids and molluscs. Other flatworm characteristics are the presence of bilateral symmetry and a mesoderm and the lack of a coelom. All in all, the present evidence seems to indicate that the nemertines came from an ancestral form closely related to Platyhelminthes.

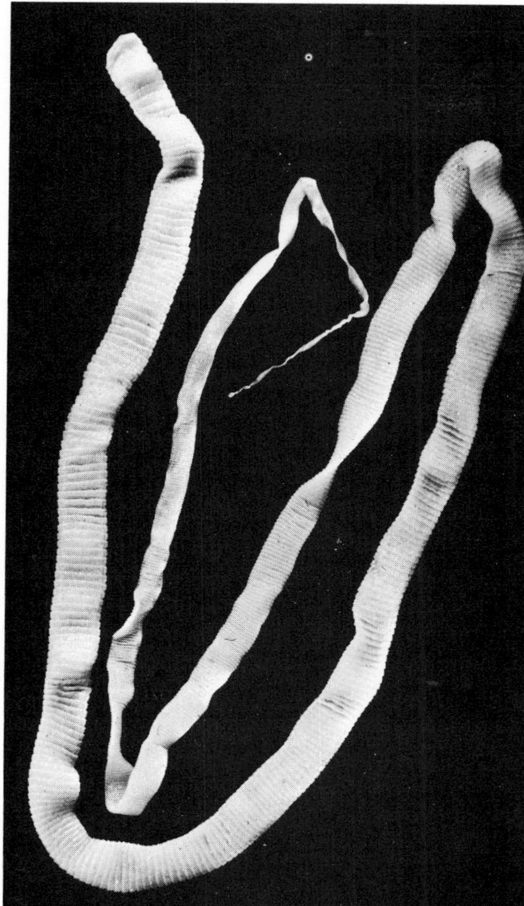

FIG. 11-18 Sheep tapeworm *Moniezia expansa.* Note progressive increase in size from young proglottids budded from neck *(center)* to oldest (gravid) proglottids, shown at upper left. (Photograph by F. M. Hickman.)

The nemertines show some advances over the flatworms. One of these is the eversible **proboscis** and its sheath, for which there are no counterparts among Platyhelminthes. Another difference is the presence of an **anus** in the adult. These forms have a **complete digestive system,** the first to be found in the animal kingdom. They are also the simplest animals to have a **blood vascular** system.

Characteristics

1. Bilateral symmetry; highly contractile body that is cylindric anteriorly and flattened posteriorly
2. Three germ layers
3. Epidermis with cilia and gland cells; rhabdites in some
4. Body spaces with parenchyma, which is partly connective tissue and partly gelatinous
5. An **eversible proboscis,** which lies free in a cavity (rhynchocoel) above the alimentary canal
6. **Complete digestive system** (mouth to anus)
7. Body-wall musculature of outer circular and inner longitudinal layers with diagonal fibers between the two; sometimes another circular layer inside the longitudinal layer
8. **Blood vascular system with two or three longitudinal trunks**
9. Acoelomate, though rhynchocoel technically may be considered a true coelom
10. Nervous system usually a four-lobed brain connected to paired longitudinal nerve trunks or, in some, middorsal and midventral trunks
11. Excretory system of two coiled canals, which are branched with **flame cells**
12. Sexes separate with simple gonads; asexual reproduction by fragmentation; few hermaphrodites; **pilidium larva** in some
13. No respiratory system
14. Sensory **ciliated pits** or **head slits** on each side of head, which communicate between the outside and the brain; tactile organs and ocelli (in some)
15. In contrast to Platyhelminthes, few parasitic nemertines

Ecologic relationships

A few of the nemertines are found in moist soil and fresh water, but by far the larger number are marine. At low tide they are often coiled up under stones. It seems probable that they are active at high tide and quiescent at low tide. Some nemertines such as *Cerebratulus* often live in empty mollusc shells. The small species live among seaweed, or they may be found swimming near the surface of the water. Nemertines are often secured by dredging at depths of 5 to 8 m or deeper. A few are commensals or parasites. *Prostoma rubrum,* which is 20 mm or less in length, is a well-known freshwater species.

Classes

Class Enopla (en′op-la) (Gr., *enoplos,* armed). Proboscis usually armed with stylets; mouth opens in front of brain. Example: *Amphiporus, Prostoma.*

Class Anopla (an′o-pla) (Gr., *anoplos,* unarmed). Proboscis lacks stylets; mouth opens behind brain. Example: *Cerebratulus, Tubulanus, Lineus.*

Class Enopla

Most nemertines have a close resemblance to each other, although some are very long and difficult to

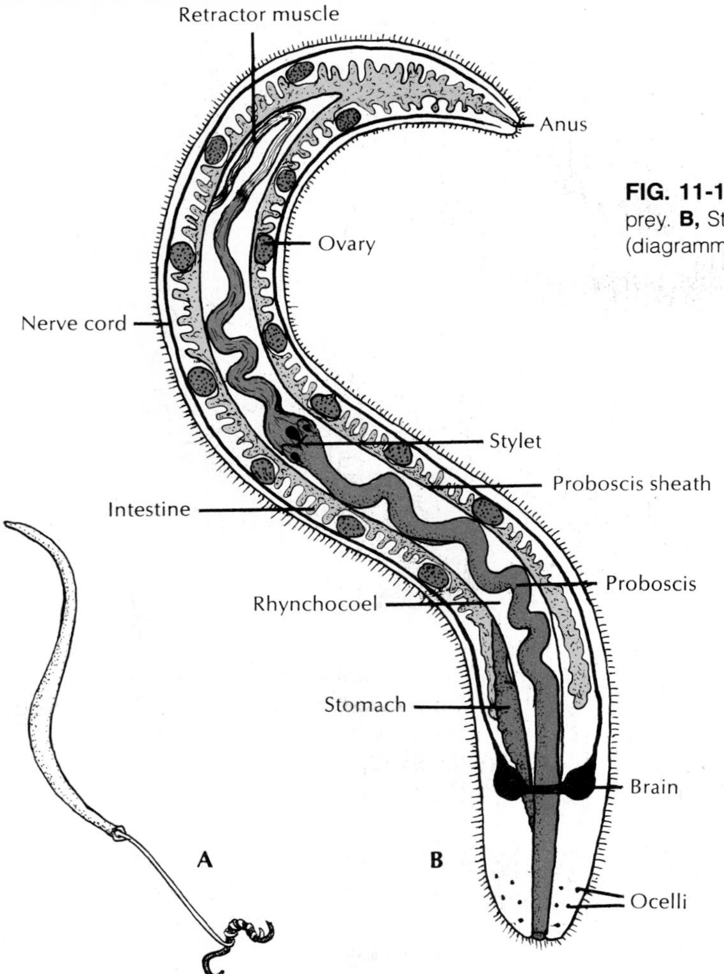

FIG. 11-19 **A,** *Amphiporus,* with proboscis extended to catch prey. **B,** Structure of female nemertine worm *Amphiporus* (diagrammatic). Dorsal view to show proboscis.

study in the laboratory because their internal organs are not easily seen. Nemertines are slender worms and very fragile, with a great diversity in size. *Amphiphorus,* which is taken here as a representative type, is one of the smaller ones.

Representative type—Amphiporus ocraceus, a ribbon worm

Amphiporus (Fig. 11-19) is from 20 to 80 mm long and about 2.5 mm wide. It is dorsoventrally flattened and has rounded ends. The body wall comprises an epidermis of ciliated columnar cells and layers of circular and longitudinal muscles (Fig. 11-20, *A*). A partly gelatinous parenchyma fills the space around the visceral organs. Ocelli are located at the anterior end. The thick-lipped mouth is anteroventral, with the opening of the proboscis just above it.

The **proboscis** is not connected with the digestive tract but is an eversible organ that can be protruded from its cavity, the **rhynchocoel,** and used for defense and catching prey (Fig. 11-19). It lies within a sheath to which it is attached by muscles. The rhynchocoel is filled with fluid, and by muscular pressure on this fluid the anterior part of the tubular proboscis is everted, or turned inside out. The proboscis apparatus is an invagination of the anterior body wall, and its structure therefore duplicates that of the body wall. The retractor muscles attached at the end are used to retract the everted proboscis, much like inverting the tip of a finger of a glove by a string attached inside at its tip. The proboscis is armed with a sharp-pointed stylet. A frontal gland also opens at the anterior end by a pore.

Locomotion. *Amphiporus* can move with considerable rapidity by the combined action of its well-

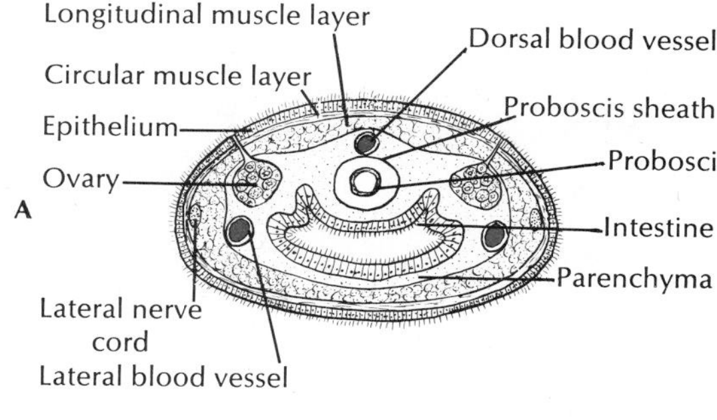

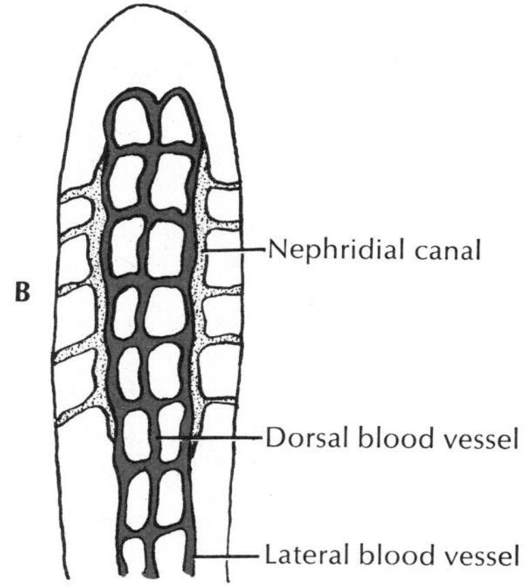

FIG. 11-20 **A,** Diagrammatic cross section of female nemertine worm. **B,** Excretory and circulatory systems of anterior region of nemertine worm. Flame bulbs along nephridial canal are closely associated with lateral blood vessels.

developed musculature and its cilia. It glides mainly against a substratum; some species make use of muscular waves in crawling. Some nemertines have the interesting method of protruding the proboscis, attaching themselves by means of the stylet, and then drawing the body up to the attached position.

Feeding and digestion. The nemertines are carnivorous and voracious, eating either dead or living prey. In seizing their prey they thrust out the slime-covered proboscis, which quickly ensnares the prey by wrapping around it (Fig. 11-19, *A*). The stylet also pierces and holds the prey. Then retracting the proboscis, the nemertine draws the prey near the mouth, where it is engulfed by the esophagus that is thrust out to meet it.

The **digestive system** is complete and extends straight through the length of the body to the terminal **anus,** lying ventral to the proboscis sheath. The **esophagus** is straight and opens into a dilated part of the tract, the **stomach.** The blind anterior end of the intestine as well as the main intestine is provided with paired **lateral ceca.** The alimentary tract is lined with ciliated epithelium, and in the wall of the esophagus there are glandular cells.

Digestion is largely extracellular in the intestinal tube, and when the food is ready for absorption, it passes through the cellular lining of the intestinal tract into the blood-vascular system. The indigestible material passes out the anus (Fig. 11-19, *B*), in contrast to Platyhelminthes in which it leaves by the mouth.

Circulation. The blood-vascular system is simple and enclosed with a single dorsal vessel and two lateral vessels (Fig. 11-20, *B*) connected by transverse vessels. All three longitudinal vessels join together anteriorly to form a type of collar. The blood is usually colorless, containing nucleated corpuscles. However, in some nemertines the blood is red, green, yellow, or orange from the presence of pigments whose function is unknown. There is no heart, and the blood is propelled by the muscular walls of the blood vessels and by bodily movements.

Excretion and respiration. The excretory system contains a pair of lateral tubes with many branches and flame cells (Fig. 11-20, *B*). Each lateral tube opens to the outside by one or more pores. Waste is picked up from the parenchymal spaces and blood by the flame cells and carried by the excretory ducts to the outside. Many of the protonephridia are so closely associated with the circulatory system that their function appears truly excretory, in contrast to their apparently osmoregulatory function in Platyhelminthes.

Respiration occurs through the body surface.

Nervous system. The nervous system includes a brain composed of four fused ganglia, one pair dorsal and one pair ventral, united by commissures. Five longitudinal nerves extend from the brain posteriorly—a large lateral trunk on each side of the body, paired dorsolateral trunks, and one middorsal trunk. These are connected by a network of nerve fibers. From the brain, nerves run to the proboscis, to the ocelli and other sense organs, and to the mouth and esophagus. In addition to the ocelli, there are other sense organs, such as tactile papillae, sensory pits and grooves, and probably auditory organs.

Reproduction and development. The reproductive system in *Amphiporus* is dioecious. The gonads in either sex lie between the intestinal ceca (Fig. 11-9). From each gonad a short duct (gonopore) runs to the dorsolateral body surface. Eggs and sperm are discharged into the water, where fertilization occurs. Egg production in the females is usually accompanied by degeneration of the other visceral organs.

Nemertines have a spiral, mosaic cleavage and a hollow coeloblastula. Gastrulation is by invagination, and the mesoderm is partly from the endoderm and partly from the ectoderm. The rhyncocoel develops as a cavity in the mesoderm and is, therefore, technically a coelomic cavity, but it is not homologous to the coelom in higher forms.

A pilidium larva develops, which is helmet-shaped and bears a dorsal spike of fused cilia and a pair of lateral lobes. The entire larva is covered with cilia and has a mouth and alimentary canal but no anus. In some nemertines the zygote develops directly without undergoing metamorphosis. The freshwater species, *Prostoma rubrum,* is hermaphroditic. A few nemertines are viviparous.

Regeneration. Nemertines have great powers of regeneration. At certain seasons some of them fragment by autotomy, and from each fragment a new individual develops. This is especially noteworthy in the genus *Lineus*. Fragments from the anterior region will produce a new individual more quickly than will one from the posterior part. Sometimes the proboscis is shot out with such force that it is broken off from the body. In such a case a new proboscis is developed within a short time.

PHYLOGENY AND ADAPTIVE RADIATION

Phylogeny. Platyhelminthes and Rhynchocoela are apparently closely related, with the flatworms being the more primitive. There can be little doubt that the bilaterally symmetric flatworms were derived from a radial ancestor, perhaps one very similar to the planula larva of the cnidarians. Some investigators believe that this **planuloid ancestor** may have given rise to one branch of descendants that were sessile or free-floating and radial, which became the Cnidaria, and another that acquired a creeping habit and bilateral symmetry, which became the Platyhelminthes. Bilateral symmetry is a selective advantage for creeping or swimming animals because sensory structures are concentrated on the anterior, the end which first encounters environmental stimuli (cephalization).

The transformation from a planuloid ancestor to an early platyhelminth involved a number of body modifications, such as an oral-aboral flattening, with the oral end becoming the ventral surface and the ventral surface adapting for locomotion with the aid of cilia and muscles. The small flatworms of the order Acoela seem to meet many of the requirements of an early ancestor of Platyhelminthes. They have several characteristics in common with the planula larva of the cnidarians, such as no epidermal basement membrane, no digestive system, a nerve plexus under the epidermis, and no distinct gonads. The acoeloid ancestor gave rise to the other orders of Turbellaria and the other classes in the phylum. It may be that the ancestral cestodes never had a digestive tract and therefore did not lose it in their adaptation to parasitism.

The Rhynchocoela probably arose from flatworm stock: the body construction of ciliated epidermis, muscles, and mesenchyme-filled spaces are similar in both groups. The rhynchocoels are more advanced than the flatworms in having a complete digestive system, a vascular system, and a more highly organized nervous system.

Adaptive radiation. The flatworm body plan, with its creeping adaptation, placed a selective advantage on bilateral symmetry and further development of cephalization, ventrodorsal regions, and caudal differentiation. Because of their body shape and metabolic requirements, early flatworms must have been well preadapted for parasitism and gave rise to symbiotic lines on numerous occasions. These lines produced descendants that were extremely successful as parasites, and many flatworms became very highly specialized for that mode of existence.

The ribbon worms have stressed the proboscis apparatus in their evolutionary diversity. Its use in capturing prey may have been secondarily evolved from its original function as a highly sensitive organ for exploring the environment. Though the ribbon worms have advanced beyond the flatworms in their complexity of organization, they have been dramatically less successful as a group. Perhaps the possession of a proboscis was a deterrent to a parasitic habit but highly efficient as a predator tool, or perhaps some critical preadaptations were simply not present.

Derivation and meaning of names

Acoela (Gr., *a,* without, + *koilos,* hollow) Order (Turbellaria). These worms have no enteron.

Amphiporus (Gr., *amphi,* on both sides, + *poros,* pore) Genus (Enopla). Refers to the mouth and proboscis pores at the anterior end.

Cerebratulus (L., *cerebrum,* brain, + *ulus,* dim) Genus (Anopla). Refers to the relatively prominent cerebral ganglia.

Clonorchis (Gr., *clon,* branch, + *orchis,* testis) Genus (Trematoda). Testes are branched.

Digenea (Gr., *dis,* double, + *genos,* race) Subclass (Trematoda) These flukes require two or more hosts for their complete development.

Diphyllobothrium (Gr., *dis,* double, + *phyllon,* leaf, + *bothrion,* small hole) Genus (Cestoda). The scolex of this tapeworm has only two suckers instead of four, which is the number commonly found.

Dugesia (formerly called *Euplanaria* but changed by priority to *Dugesia* after Dugès, who first described the form in 1830) Genus (Turbellaria)

Echinococcus (Gr., *echinos,* spiny, + *coccus,* berry) Genus (Cestoda). The multiple scolices give a spiny and berry-like appearance to the dangerous tapeworm larval cysts.

Hymenolepis (Gr., *hymen,* membrane, + *lepis,* scale) Genus (Cestoda).

Moniezia (*Moniez,* a surname, patronymic, + *ia,* suffix) Genus (Cestoda). Genus named for Romain-Louis Moniez, a famous parasitologist of the latter nineteenth century.

Monogenea (Gr., *monas,* single, + *genos,* race) Class (Platyhelminthes). Only one host required for development.

Neorhabdocoele (Gr., *neo,* new, recent, + *rhabdos,* red, + *koilos,* cavity) Order (Turbellaria). The more advanced group of turbellarians with a straight intestine of smooth contour.

Paragonimus (Gr., *para,* beside, + *gonimos,* generative) Genus (Trematoda). Refers to the position of the reproductive organs. Testes lie side by side and the ovary lies opposite the uterus.

Polycladida (Gr., *poly,* many, + *klados,* branch) Order (Turbellaria) have intestines of many branches.

Schistosoma (Gr., *schistos,* divided, + *soma,* body) Genus (Trematoda). Male canal in which the female is held gives a split appearance to the body.

Taenia (Gr., *tainia,* band, ribbon) Genus (Cestoda).

Taeniarhynchus (Gr., *tainia,* band, ribbon, + *rhynchus,* snout, beak) Genus (Cestoda). Differs from *Taenia* in having no hooks on its rostellum.

Tricladida (Gr., *tri-,* three, + *klados,* branch) Suborder (order Seriata, class Turbellaria). With three-branched intestines.

Annotated references
Selected general readings

Baer, J. C. 1952. Ecology of animal parasites. Urbana, Ill., University of Illinois Press.

Böhmig, L. 1929. Artikel. Nemertini. In W. Kükenthal and T. Krumbach (eds.). Handbuch der Zoologie, vol. 2, part 1, sec. 3. Berlin, Walter de Gruyter & Co.

von Brand, T. 1966. Biochemistry of parasites. New York, Academic Press, Inc.

Burger, O. 1890. Anatomie und Histologie des Nemertinen. Z. Wiss. Zool. **50:**1-279. *A classic study of the nemertines.*

Cable, R. M. 1965. Thereby hangs a tail. J. Parasitol. **51:** 3-12. *This reference and the following one have interesting discussions on the nature of the trematode life cycle.*

Cable, R. M. 1971. Parthenogenesis in parasitic helminths. Am. Zool. **11:**267-272.

Canning, E. U., and C. A. Wright (eds.). 1972. Behavioural aspects of parasite transmission. Zool. J. Linnean Society **51**(Suppl. 1).

Coe, W. R. 1943. Biology of the nemerteans of the Atlantic Coast of North America. Hartford, Trans. Connecticut Acad. Arts Science **35:**129. *This is only one of many valuable papers on Rhynchocoela by this author.*

Cort, W. W., D. J. Ameel, and A. Van der Woude. 1954. Parasitological reviews—germinal development in the sporocysts and radiae of the digenetic trematodes. Exp. Parasitol. **3:**185-225.

Dawes, B. 1946. The Trematoda with special reference to British and other European forms. New York, Cambridge University Press.

Erasmus, D. A. 1972. The biology of trematodes. London, Edward Arnold (Publishers) Ltd.

Fallis, A. M. (ed.). 1971. Ecology and physiology of parasites. Toronto, University of Toronto Press.

Goodrich, E. S. 1945. The study of nephridia and genital ducts since 1895. Q. J. Microsc. Sci. **86:**113-392.

Grassé, P. P. (ed.). 1961. Traité de zoologie, anatomie, systématique, biologie, vol. IV. Platyhelminthes, mésozoaires, acanthocéphales, némertiens (first fascicule). Paris, Masson & Cie. *An authoritative treatise on these groups.*

Hickman, C. P. 1973. Biology of the invertebrates, ed. 2. St. Louis, The C. V. Mosby Co.

Hunter, G. W., J. C. Swartzwelder, and D. F. Clyde. 1976. Tropical medicine, ed. 5. Philadelphia, W. B. Saunders Co. *A valuable source of information on parasites of medical importance.*

Hyman, L. H. 1951. The invertebrates: Platyhelminthes and Rhynchocoela, vol. 2. New York, McGraw-Hill Book Co. *A comprehensive account of this group, with excellent figures and bibliography.*

Jenkins, M. M. 1966. Note on stalk formation in cocoons

of *Dugesia dorotocephala* (Woodworth, 1897). Trans. Am. Microsc. Soc. **85:**168.

Jenkins, M. M., and H. P. Brown. 1964. Copulation activity and behavior in the planarian *Dugesia dorotocephala* (Woodworth, 1897). Trans. Am. Microsc. Soc. **83:**32-40.

Jennings, J. B. 1963. Some aspects of nutrition in the Turbellaria, Trematoda, and Rhynchocoela. In E. C. Dougherty (ed.). The lower Metazoa: comparative biology and phylogeny. Berkeley, University of California Press.

Karling, T. G. 1963. Some evolutionary trends in turbellarian morphology. In E. C. Dougherty (ed.). The lower Metazoa: comparative biology and phylogeny. Berkeley, University of California Press.

Lumsden, R. D. 1975. Surface ultrastructure and cytochemistry of parasitic helminths. Exp. Parasitol. **37:**267-339.

Mueller, J. R. 1965. Helminth life cycles. Am. Zool. **5:**131-139.

Olsen, O. W. 1974. Animal parasites: their life cycles and ecology. Baltimore, University Park Press.

Oschman, J. L. 1967. Microtubules in the subepidermal glands of *Convoluta roscoffensis* (Acoela, Turbellaria). Trans. Am. Microsc. Soc. **86:**159-162.

Pappas, P. W., and C. P. Read. 1975. Membrane transport in helminth parasites: a review. Exp. Parasitol. **37:**469-530.

Pennak, R. W. 1953. Fresh-water invertebrates of the United States. New York, The Ronald Press Co. *Includes an excellent description of the structure and life history of the freshwater nemertine* Prostoma rubrum.

Schell, S. C. How to know the trematodes. Dubuque, Iowa, William C. Brown, Publishers.

Schmidt, G. D. 1970. How to know the tapeworms. Dubuque, Iowa, William C. Brown, Publishers.

Schmidt, G. D., and L. S. Roberts. 1977. Foundations of parasitology. St. Louis, The C. V. Mosby Co.

Smyth, J. D. 1966. The physiology of trematodes. San Francisco, W. H. Freeman & Co.

Smyth, J. D. 1969. The physiology of cestodes. San Francisco, W. H. Freeman & Co.

Thomas, A. P. 1883. The life history of the liver fluke *(Fasciola hepatica)*. Q. J. Microsc. Sci. (ser. 2) **23:**99-133. *This classic work represents the first life history of a digenetic trematode to be worked out. It gave a great impetus to work in the field of parasitology.*

Wardle, R. A., and J. A. McLeod. 1952. The zoology of tapeworms. Minneapolis, University of Minnesota Press. *A comprehensive account.*

Wardle, R. A., J. A. McLeod, and S. Radinovsky. 1974. Advances in the zoology of tapeworms, 1950-1970. Minneapolis, University of Minnesota Press.

Wells, M. 1968. Lower animals. New York, McGraw-Hill Book Co.

Yamaguti, S. 1958. Systema helminthum, vol. I. (two parts). The digenetic trematodes of vertebrates. New York, Interscience Publishers.

Yamaguti, S. 1959. Systema helminthum, vol. II. The cestodes of vertebrates. New York, Interscience Publishers.

CHAPTER 12

THE PSEUDOCOELOMATE ANIMALS

Phylum Rotifera
Phylum Gastrotricha
Phylum Kinorhyncha
Phylum Nematoda
Phylum Nematomorpha
Phylum Acanthocephala
Phylum Entoprocta
Phylum Gnathostomulida

Anterior end of a hookworm *Necator americanus.* A parasite of humans, the common name is derived from the characteristic dorsal flexure, shown here. The worms attach themselves to the host's intestinal wall and suck blood, and the buccal capsule with its large teeth *(at right)* and the muscular esophagus *(running to left)* are very effective in this function.

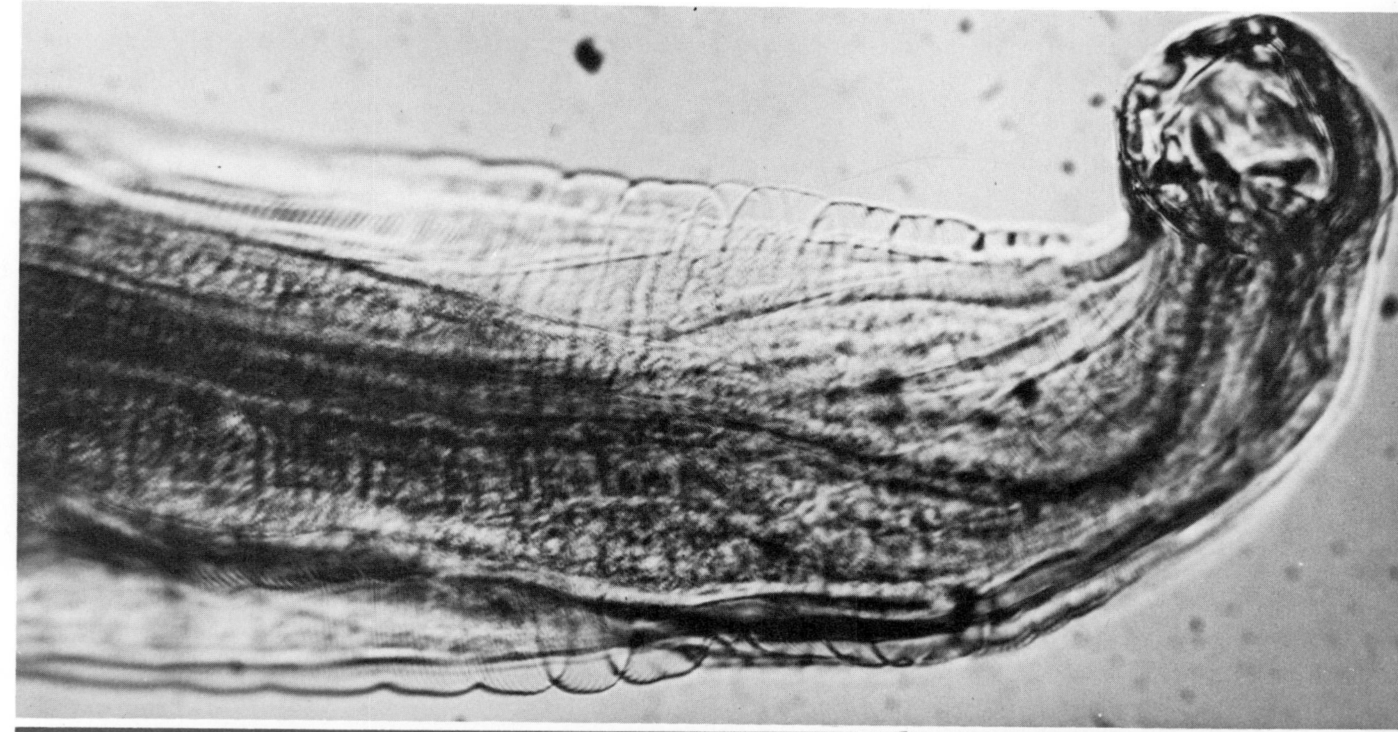

Photograph by L. S. Roberts.

Position in animal kingdom

In the first seven phyla above, the original blastocoel of the embryo persists as a space, or body cavity, between the enteron and the body wall. Because this cavity lacks the peritoneal lining found in the true coelomates, it is called a **pseudocoel,** and the animals possessing it are called **pseudocoelomates.**

Gnathostomulida is a newly described phylum whose relationship with other phyla is still quite unclear. Although the animals in it do not have pseudocoel spaces, certain other features suggest affinities with some pseudocoelomates.

Pseudocoelomates belong to the Protostomia division of the bilateral animals.

Biologic contributions

The pseudocoel is a distinct advancement over the solid body structure of the acoelomates. It may be filled with fluid or may contain a gelatinous substance with some mesenchyme cells. In common with a true coelom, the pseudocoel presents certain adaptive potentials, though these are by no means realized in all members: (1) greater freedom of movement, (2) space for the development and differentiation of digestive, excretory, and reproductive systems, (3) a simple means of circulation or distribution of materials throughout the body, (4) a storage place for waste products to be discharged to the outside by excretory ducts, and (5) a hydrostatic organ. Since most pseudocoelomates are quite small, the most important functions of the pseudocoel are probably in circulation and as a means to maintain a high internal hydrostatic pressure.

The complete mouth-to-anus digestive tract is now well established and found in these phyla (except Gnathostomulida) and in all higher phyla.

THE PSEUDOCOELOMATES

Vertebrates and higher invertebrates have a true **coelom,** or peritoneal cavity, which is formed in the mesoderm during embryonic development and is, therefore, lined with a layer of mesodermal epithelium, the **peritoneum.** The pseudocoelomate phyla have a pseudocoel rather than a true coelom. It is derived from the embryonic blastocoel rather than from a secondary cavity within the mesoderm. It is a space not lined with peritoneum, between the gut and the mesodermal and ectodermal components of the body wall.

Seven distinct groups of animals belong to the pseudocoelomate category. These are Rotifera, Gastrotricha, Kinorhyncha, Nematoda, Nematomorpha, Acanthocephala, and Entoprocta. The first five of these groups have certain similarities that have led some authorities to place them as classes in a phylum called Aschelminthes (as'kel-min'theez) (Gr., *askos,* bladder, + *helmins,* worm). However, they differ so much that any phyletic relationship is highly debatable at least, and other authorities prefer to consider them as separate phyla. Some group the five together loosely as individual phyla under a superphylum Aschelminthes. The Entoprocta have sometimes been grouped with the Ectoprocta, together called the Bryozoa (moss animals). However, because the ectoprocts have a true coelom, they are usually considered as a separate phylum, and the term "bryozoans" is generally taken to exclude the entoprocts in current usage.

An eighth phylum, the Gnathostomulida, is included because it has some resemblances to the others in spite of its lack of pseudocoleomate spaces.

However one classifies them, the pseudocoelomates are a heterogeneous assemblage of animals that seem to have little in common except a pseudocoel. Most of them are small; some are microscopic; some are fairly large. Some, such as the nematodes, are found in freshwater, marine, terrestrial, and parasitic habitats; others, such as the Acanthocephala, are strictly parasitic. Some have unique characteristics such as the lacunar system of the acanthocephalans, the ciliary corona of the rotifers, or the zonites of the kinorhynchs.

Even in such a diversified grouping some characteristics are shared. In all there is a body wall of epidermis (often syncytial), a dermis, and muscles surrounding the pseudocoel. The digestive tract is complete (except in Gnathostomulida), and it, along with the gonads and excretory organs, is within the pseudocoel and bathed in perivisceral fluid. The epidermis in many secretes a nonliving cuticle with some specializations such as bristles, spines, and the like.

A constant number of cells or nuclei in the individuals of a species, a condition known as **eutely,** is common to several of the groups, and in most of them there is an emphasis on the longitudinal muscle layer.

Characteristics of pseudocoelomates

1. Symmetry bilateral; unsegmented; triploblastic (three germ layers)

2. Body cavity an unlined **pseudocoel**
3. Size mostly small; some microscopic; a few a meter or more in length
4. Body vermiform; body wall a **syncytial** or cellular epidermis with thickened cuticle, sometimes molted; muscular layers mostly of **longitudinal fibers;** cilia mostly absent
5. Digestive system (lacking in acanthocephalans) complete with mouth, enteron, and anus; pharynx muscular and well developed: **tube-within-a-tube arrangement;** digestive tract usually only an epithelial tube with **no definite muscle layer**
6. Circulatory and respiratory organs lacking
7. Excretory system of canals and protonephridia in some; cloaca that receives excretory, reproductive, and digestive products may be present
8. Nervous system of cerebral ganglia or of a circumenteric nerve ring connected to anterior and posterior nerves; sense organs of **ciliated pits,** papillae, bristles, and some eyespots
9. Reproductive system of gonads and ducts that may be single or double; sexes nearly always separate, with the male usually smaller than the female; eggs microscopic with shell often containing chitin
10. Development may be direct or with a complicated life history; cleavage mostly determinate; **cell or nuclear constancy common**

PHYLUM ROTIFERA

Rotifera (Ro-tif′e-ra) (L., *rota,* wheel, + *-fera,* those that bear) derive their name from the characteristic ciliated crown, or **corona,** that, when beating, often gives the impression of rotating wheels. Rotifers range in size from 40 μm to 3 mm in length, but most are between 100 and 500 μm long. Some have beautiful colors, though most are transparent, and some have odd and bizarre shapes. Their shapes are often correlated with their mode of life. The floaters are usually globular and saclike; the creepers and swimmers are somewhat elongated and wormlike; and the sessile types are commonly vaselike, with a cuticular envelope (lorica). Some are colonial. One of the best known genera is *Philodina* (Fig. 12-1), which is often used for study.

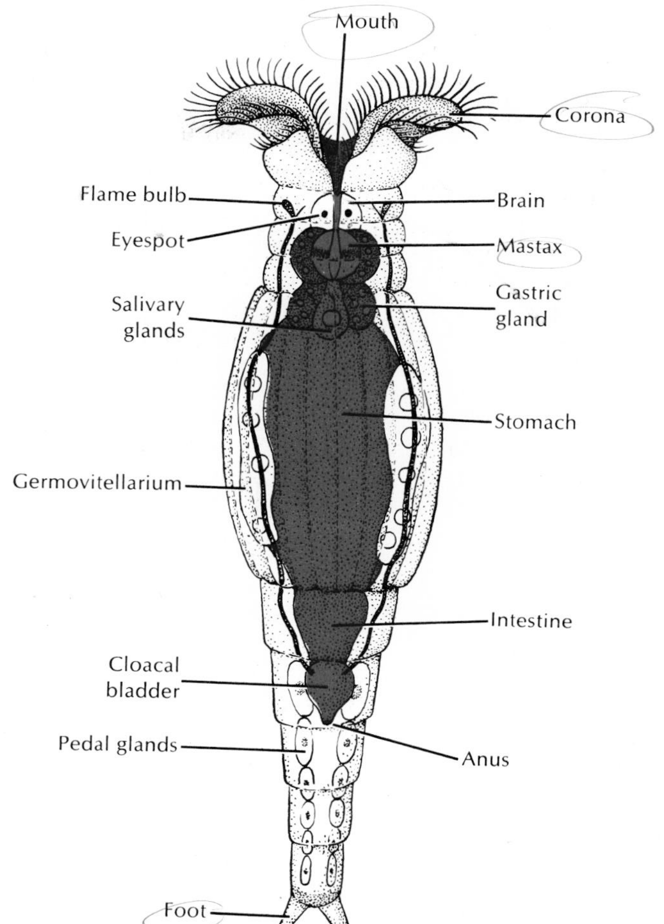

FIG. 12-1 Structure of *Philodina* rotifer.

Ecologic relationships

Rotifers are a cosmopolitan group of about 2,000 species, some of which are found throughout the world. Most of the species are freshwater inhabitants, but a few are marine; some are terrestrial, and some are epizoic or parasitic.

Rotifers are adapted to many kinds of ecologic conditions. Some can endure wide ranges of pH and temperature, though some are more specific in their requirements. Many are found only in acid waters with a pH of 4 to 6, but more are found in alkaline waters. Sessile rotifers are especially sensitive to chemical conditions.

Most species are benthic, occurring on the bottom or in vegetation of ponds or along the shores of large freshwater lakes where they swim or creep about on the vegetation. Some terrestrial species frequent mosses and other plants where they often select a special kind of plant. A large proportion of the species that live in the water film between sand grains of sandy

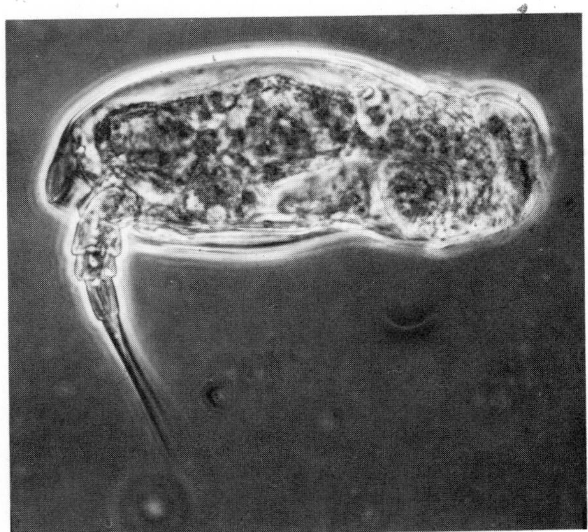

FIG. 12-2 Rotifer. Note the foot with its ringed joints and the two long toes. The toes contain pedal glands for attachment and, in swimming forms, the foot is often ventral and used as a rudder. (Courtesy R. P. Higgins.)

beaches (psammolittoral) are rotifers. Pelagic forms are common in the surface waters of freshwater lakes and ponds. These forms may secrete oil drops for buoyancy. Some planktonic rotifers may exhibit cyclomorphosis, that is, seasonal variations in body form.

Many species of rotifers can endure long periods of desiccation, during which they resemble grains of sand. While in a desiccated condition, rotifers are very tolerant of temperature variations. This is especially true of moss-dwelling rotifers. True encystment in which a protective cyst is formed occurs in only a few rotifers. On addition of water, desiccated rotifers resume their activity.

Strictly marine species of rotifers are rather few in number. Some of the littoral species of the sea may be freshwater ones that are able to adapt to saltwater.

Classification

Rotifers are usually divided into three orders (or classes):

SEISONACEA—epizoic marine rotifers; elongated form; corona poorly developed; sexes similar in size and form. Example: *Seison*.

BDELLOIDEA—swimming or creeping forms; anterior end retractile; corona usually with pair of trochal discs; no males; parthenogenesis; two germovitellaria. Examples: *Philodina* (Fig. 12-1), *Rotaria*.

MONOGONONTA—swimming or sessile forms, with single germovitellarium; males reduced in size; eggs of three types (amictic, mictic, dormant). Example: *Asplanchna, Epiphanes*

Morphology and physiology

External features. The body of the rotifer comprises a head bearing a ciliated corona, a trunk, and a posterior tail, or foot. It is covered with a cuticle and is nonciliated except for the corona.

The ciliated **corona,** or crown, surrounds a nonciliated central area of the head called the **apical field,** which may bear sensory bristles or papillae. The appearance of the head end depends on which of the several types of corona it has—usually a circlet of some sort, or a pair of trochal (coronal) discs (the term trochal comes from a Greek word meaning wheel). The cilia on the corona beat in succession, giving the appearance of a revolving wheel or pair of wheels. The **mouth** is located in the corona on the midventral side. The coronal cilia are used in both locomotion and feeding.

The **trunk** may be elongated, as in *Philodina* (Fig. 12-1), or saccular in shape (Fig. 12-2). It contains the visceral organs and often bears sensory antennae. It is covered by a transparent cuticle that in *Philodina* and others is superficially ringed so as to simulate segmentation, but in many other forms is much thickened to form an outer case or **lorica,** often arranged in plates or rings.

The **foot** is narrower and usually bears one to four toes. Its cuticle may be ringed so that it is telescopically retractile. It is tapered gradually in some forms (Fig. 12-1) and sharply set off in others (Fig. 12-2). The foot is an attachment organ and contains pedal glands that secrete an adhesive material used by both sessile and creeping forms. In swimming pelagic forms, the foot is usually reduced. Rotifers move either by creeping with leechlike movements aided by the foot, or by swimming with the coronal cilia, or both.

Internal features. Underneath the cuticle is the **syncytial epidermis,** which secretes the cuticle, and bands of **subepidermal muscles,** some circular, some longitudinal, and some running through the pseudocoel to the visceral organs. The **pseudocoel** is large, occupying the space between the body wall and the viscera. It is filled with fluid, some of the muscle bands, and a network of mesenchymal ameboid cells.

The **digestive system** is complete. Some rotifers

feed by sweeping minute organic particles or algae toward the mouth by the beating of the coronal cilia. The cilia are able to sort out and dispose of the larger unsuitable particles. The **pharynx** is fitted with a muscular portion **(mastax)** that is equipped with hard jaws **(trophi)** for sucking in and grinding up the food particles. The constantly chewing pharynx is often a distinguishing feature of these tiny animals. Carnivorous species feed on protozoans and small metazoans, which they capture by trapping or grasping. The trappers have a funnel-shaped area around the mouth. When small prey swim into the funnel, the lobes fold inward to capture and hold them till they are drawn into the mouth and pharynx. The hunters have trophi that can be projected and used like forceps to seize the prey, bring it back into the pharynx, and then pierce it or break it up so the edible parts can be sucked out and the rest discarded. There are both salivary and gastric glands that are believed to secrete enzymes for extracellular digestion. Absorption occurs in the stomach.

The **excretory system** typically consists of a pair of **protonephridial tubules,** each with several **flame cells,** that empty into a common bladder. The bladder, by pulsating, empties into the **cloaca**—into which the intestine and oviducts also empty. The fact that the rate of pulsation is fairly rapid—1 to 4 times per minute—would indicate that the protonephridia are important osmoregulatory organs. The water apparently enters through the mouth, rather than across the cuticle; even marine species empty their bladder at frequent intervals.

The **nervous system** consists of a bilobed brain, dorsal to the mastax, that sends paired nerves to the sense organs, mastax, muscles, and viscera. Sensory organs include paired eyespots (in some species such as *Philodina*), sensory bristles and papillae, and ciliated pits and dorsal antennae.

Reproduction. Rotifers are dioecious, but the males are usually smaller than the females. In the order Bdelloidea males are entirely unknown, and in Monogononta they seem to occur only for a few weeks of the year.

The female reproductive system consists of combined ovaries and yolk glands **(germovitellaria)** and oviducts that open into the cloaca. Yolk is supplied to the developing ova by way of flow through cytoplasmic bridges, rather than as separate yolk cells as in many Platyhelminthes.

In the Bdelloidea (*Philodina,* for example) all females are parthenogenetic and produce diploid eggs that hatch into diploid females. These reach maturity in a few days. In the order Seisonacea the females produce haploid eggs that must be fertilized and that develop into either males or females. In the Monogononta, however, females produce two kinds of eggs. During most of the year diploid females produce **amictic eggs** that are thin-shelled and diploid. These develop parthenogenetically into diploid amictic females. However such rotifers often live in temporary ponds or streams and are cyclic in their reproductive patterns. Any one of several environmental factors—for example, crowding, diet, or photoperiod (according to species)—may induce the amictic eggs to develop into diploid mictic females that will produce thin-shelled haploid **mictic eggs.** If these eggs are not fertilized, they develop into haploid males. But if fertilized, the eggs develop a thick, resistant shell and become dormant. They survive over winter ("winter eggs") or until environmental conditions are again suitable, at which time they hatch into diploid females. Dormant eggs are often dispersed by winds or birds, a fact that may account for the peculiar distribution patterns of rotifers.

The male reproductive system includes a single testis and a ciliated sperm duct that runs to a genital pore (males usually lack a cloaca). The end of the sperm duct is specialized as a copulatory organ. Copulation is usually by hypodermic impregnation, that is, the penis can penetrate any part of the female body wall and inject the sperm directly into the pseudocoel.

Females hatch with adult features, needing only a few days growth to reach maturity. Males often do not grow and are sexually mature at hatching.

Cell or nuclear constancy. Most structures in rotifers are syncytial, but the nuclei in the various organs are said to show a remarkable constancy in numbers in any given species **(eutely).** For example, Martini reported that in one species of rotifer he always found 183 nuclei in the brain, 39 in the stomach, 172 in the corona epithelium, and so on. Not all zoologists are convinced that cell or nuclear constancy is always that absolute.

PHYLUM GASTROTRICHA

Gastrotricha (gas-trot'ri-ka) (N.L., Gr., *gaster, gastros,* stomach or belly, + *thrix, trichos,* hair) includes small, ventrally flattened animals about 65 to 500 μm

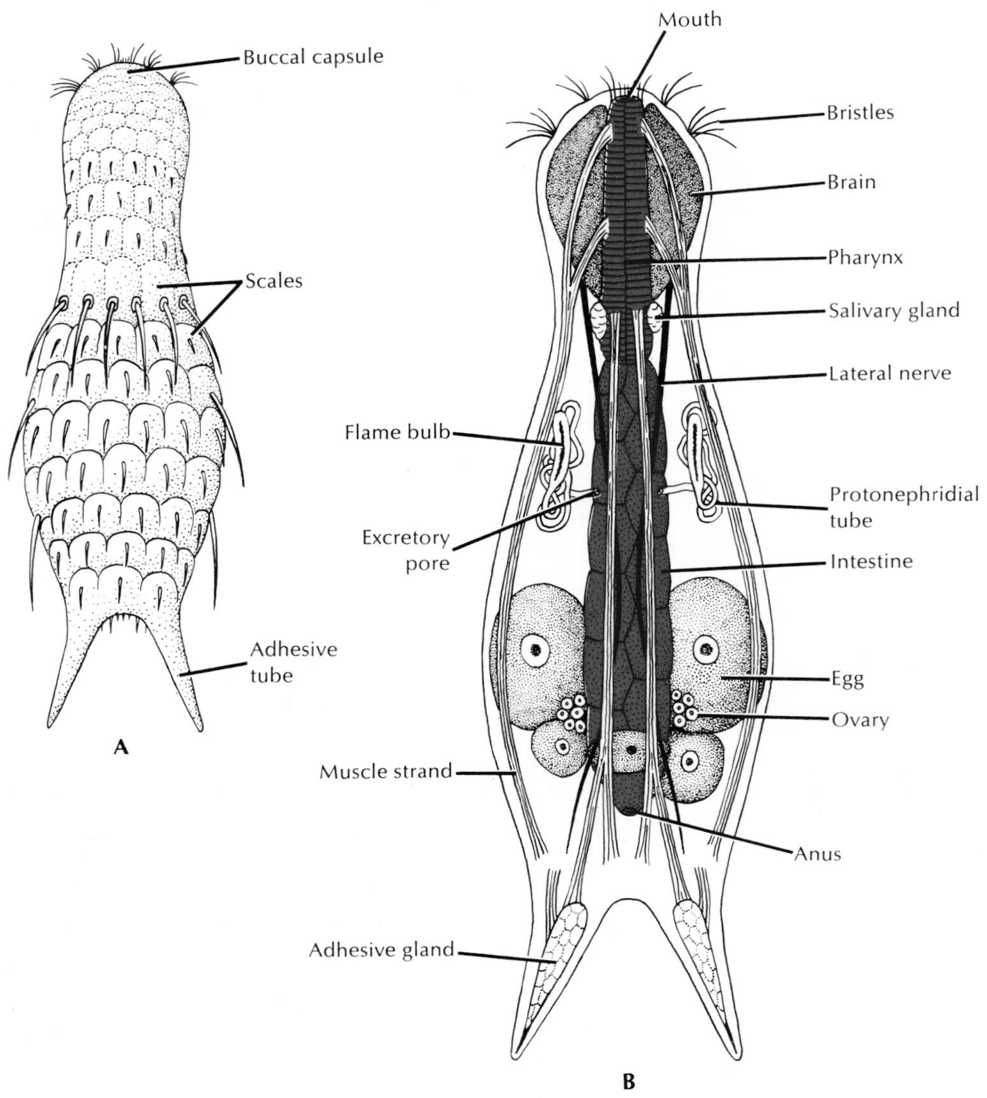

FIG. 12-3 *Chaetonotus,* a gastrotrich. **A,** Dorsal surface. **B,** Internal structure, ventral view.

long, somewhat like rotifers but lacking the corona and mastax and having a characteristically bristly or scaly body. They are usually found gliding on the bottom, or on an aquatic plant or animal substrate, by means of their ventral cilia, or they can be found in the interstitial spaces between bottom particles. They may be closely related to the rotifers, which they resemble in having cilia, protonephridia, and a similar muscle pattern. On the other hand, some characteristics are more similar to those of the nematodes.

Ecologic relationships

Gastrotrichs are found in both fresh and salt water. The 500 or so species are about equally divided between the two media. Freshwater forms have benthic habits and occur chiefly among the vegetation of ponds and lakes. They usually retain contact with a substratum of some sort. Many of their species are cosmopolitan, but only a few occur in both fresh water and the sea. Much is yet to be learned about their distribution. Gastrotrichs make up part of the psammolittoral populations found between the sand grains of sandy beaches. Like the rotifers, chaetonotid gastrotrichs produce thin-walled rapidly developing eggs and thick-shelled, dormant eggs. They can withstand low oxygen concentrations. Their thick-shelled eggs can withstand harsh environmental conditions and may survive dormancy for some years.

Structure

The gastrotrich (Fig. 12-3) is usually elongated, with a convex dorsal surface bearing a pattern of bristles, spines, or scales, and a flattened ciliated ventral surface. The head is often lobed and ciliated and the tail end may be forked.

A syncytial epidermis is found beneath the cuticle. Longitudinal muscles are better developed than are circular ones, and in most cases they are unstriped. Adhesive tubes secrete a substance for attachment. The pseudocoel is somewhat reduced and contains no amebocytes.

The digestive system is complete and is made up of a mouth, muscular pharynx, stomach-intestine, and anus (Fig. 12-3, *B*). Their food is largely algae, protozoans, and detritus, which is directed to the mouth by the head cilia. Digestion appears to be extracellular. Protonephridia are restricted to certain species.

The nervous system contains a brain near the pharynx and a pair of lateral nerve trunks. Sensory structures are similar to those in rotifers, except that the eyespots are generally lacking.

Gastrotrichs are hermaphroditic, though the male system of the chaetonotids is so poorly developed that they are functionally parthenogenetic females. The female reproductive system consists of one or two ovaries, a uterus, an oviduct, and a gonophore, which may open anteriorly to, or in common with, the anus. Eggs are laid on some substratum such as weeds and hatch in a few days. Development is direct, and the juveniles have the same form as the adults. Species of *Chaetonotus* are common freshwater gastrotrichs (Fig. 12-3).

PHYLUM KINORHYNCHA

Kinorhyncha (kin′o-ring′ka) are marine worms a little larger than rotifers and gastrotrichs but usually not over 1 mm long. Their name comes from the Greek words *kinein,* to move, and *rhynchos,* beak, and refers to their retractile proboscis. The phylum has also been called Echinodera, meaning spiny-skinned.

Ecologic relationships

Kinorhynchs seem to be most abundant in shallow mud bottoms of ocean shores, but their general distribution is spotty. Some live among algae where they feed on diatoms. The presence of algae in many species may account somewhat for their color. About a hundred species have been reported.

Structure

The body of the kinorhynch is divided into 13 or 14 rings (zonites), which bear spines, but they have no external cilia (Fig. 12-4). The retractile head has a circlet of spines with a small retractile proboscis. The body is flat underneath and arched above. Their body wall is made up of a cuticle, a syncytial epidermis, and longitudinal epidermal cords, much like those of nematodes. The arrangement of the muscles is correlated with the zonites, and circular, longitudinal, and diagonal muscle bands are all represented.

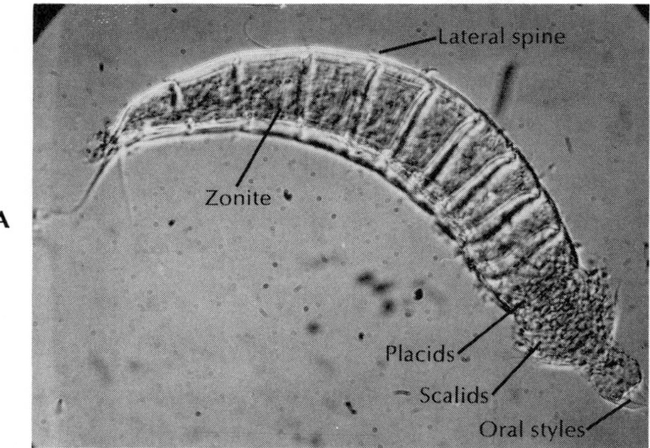

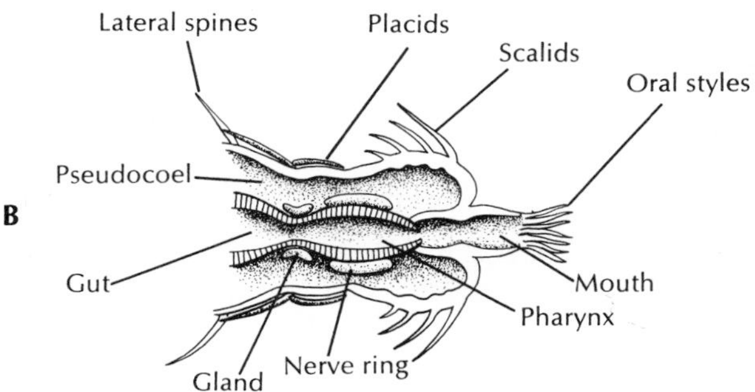

FIG. 12-4 *Echinoderes,* a kinorhynch, is a minute marine worm. Segmentation is superficial. Head with its circle of spines is retractile. Note detail of internal structure in **B.** (**A,** Photograph by L. S. Roberts.)

A kinorhynch cannot swim. In the silt and mud where it commonly lives, it burrows by extending the head into the mud and anchoring it with spines. It then draws the body forward until the head is retracted into the body. When disturbed, the kinorhynch draws in the head and protects it with a closing apparatus of cuticular plates (Fig. 12-4, *A*).

The digestive system is complete, with a mouth at the tip of the proboscis (Fig. 12-4, *B*), a pharynx, an esophagus, a stomach-intestine, and an anus. Kinorhynchs feed on diatoms or on organic material in the mud where they burrow.

The pseudocoel is filled with fluid-bearing amebocytes. The excretory system is made up of a multinucleated **solenocyte** (protonephridium) on each side of the eleventh zonite. Each solenocyte has one long and one short flagellum.

The nervous system is in contact with the epidermis, with a nerve ring encircling the pharynx, and with a ventral ganglionated nerve cord extending throughout the body. Sense organs are represented by eyespots in some and by the sensory bristles.

Sexes are separate, with paired gonads and gonoducts. The young larvae have only the first three zonites, but as they grow, new zonites are added to the others.

Among the best known genera of the Kinoryncha are *Echinoderes, Echinoderella, Pycnophyes,* and *Trachydemus*.

PHYLUM NEMATODA

The phylum Nematoda (nem'a-to'da) (Gr., *nema,* thread, + *eidos,* form) already numbers more than 12,000 species, and when all the species have been classified, they may even outnumber the arthropods. They are worldwide in distribution.

Distinguishing characteristics of this large group of animals are their cylindric shape; their flexible, nonliving cuticle; their lack of motile cilia or flagella; and the muscles of their body wall, which have several unusual features, among these being the fact that the muscles run in a longitudinal direction only. Correlated with their lack of cilia, nematodes do not have protonephridia; their excretory system consists of either one or more large gland cells (the **renette**) opening by an excretory pore, or a canal system without renette cells, or both renette and canals together. Their pharynx is characteristically muscular with a triradiate lumen and resembles the pharynx of the gastrotrichs and of the kinorhynchs. Use of the pseudocoel as a hydrostatic organ, is highly developed in the nematodes, and much of the functional morphology of the nematodes can be best understood in the context of the high **hydrostatic pressure** (turgor) in the pseudocoel.

Most nematode worms are under 5 cm long, and many are microscopic, but some parasitic nematodes are over a meter in length.

Ecologic relationships

Free-living nematodes occur in almost every conceivable kind of ecologic niche and habitat and are probably the most widespread of all metazoans. They have been found from the arctic regions to the tropics and occur in the sea, fresh water, and soil. They are present in high mountains, deserts, hot springs, and great ocean depths. Wherever found, their numbers may be enormous. Many thousands may be found in a single rotting apple. A fistful of soil may yield millions. Several billion per acre may occur in good farmland. The ability of many to survive extreme environmental conditions probably exceeds that of any other group of animals. Vinegar eels, *Turbatrix aceti,* can withstand a pH of 1.5, and most marine species can endure a wide range of salinity. Some species can endure extremes of temperature and dessication by passing into a dormant **(cryptobiotic)** state and reviving when conditions are more favorable.

Free-living nematodes feed on bacteria, yeasts, fungal hyphae, and algae. They may be saprozoic, saprophytic, or coprozoic (live in fecal material). Predatory species may eat rotifers, tardigrades, small annelids, and other nematodes. Many species feed on plant juices from higher plants, which they penetrate, sometimes causing agricultural damage of great proportions. Nematodes themselves may serve as prey for mites, insect larvae, and even nematode-capturing fungi.

Virtually every species of vertebrate and many invertebrates serve as hosts for one or more types of parasitic nematodes. Nematode parasites in humans cause much discomfort, disease, and death, and in domestic animals they are the source of much economic loss. The most common nematode parasites of humans in the United States are listed in Table 12-1.

Classes

Classification of the nematodes is somewhat more satisfactory at the order and superfamily level; the division into

TABLE 12-1 Common parasitic nematodes of humans in North America

Common and scientific names	Mode of infection; prevalence
Hookworm (*Ancylostoma duodenale* and *Necator americanus*)	Contact with larvae in soil that burrow into skin; common in southern states
Pinworm *(Enterobius vermicularis)*	Inhalation of dust with ova and by contamination with fingers; most common parasite in United States
Intestinal roundworm *(Ascaris lumbricoides)*	Ingestion of embryonated ova in contaminated food; common in rural areas of Appalachia and southeastern states
Trichina worm *(Trichinella spiralis)*	Ingestion of infected pork muscle; occasional in humans throughout North America
Whipworm *(Trichuris trichiura)*	Ingestion of contaminated food or by unhygienic habits; usually common wherever *Ascaris* is found

classes relies on characteristics that are not very striking and that are difficult for the novice to distinguish. Two classes are usually recognized.

Class Phasmidia (faz-mid'e-a) (Gr., *phasm,* phantom, + -idia, diminutive) **(Secernentea).** Body with a pair of minute sensory pouches (phasmids) near posterior tip; similar pair of sense organs at anterior end (amphids) poorly developed; excretory system with one or two lateral canals, with or without associated glandular cells; both free-living and parasitic forms. Examples: *Rhabditis, Ascaris, Enterobius.*

Class Aphasmidia (a'faz-mid'e-a) (Gr., *a,* without, + *phasm,* phantom, + *-idia,* diminutive) **(Adenophorea)**—phasmids lacking, amphids usually well developed; excretory system of one or more renette cells (glandular); caudal and hypodermal glands common; mostly free-living, but includes some parasites. Examples: *Dioctophyme, Trichinella, Plectus.*

Classes Phasmidia and Aphasmidia

Representative types—Ascaris lumbricoides and A. suum, the large roundworms of humans and pigs

Because of its size and availability, *Ascaris* is usually selected as a type for study in zoology, as well as in experimental work. Thus, it is probable that more is known about the structure, physiology, and biochemistry of *Ascaris* than of any other nematode. There are several species in this genus. One of the most common, *A. megalocephala,* is found in the intestine of horses. The large roundworm of pigs, *A. suum,* is morphologically close to *A. lumbricoides,* found in humans, and they were long considered the same species.

A. lumbricoides (Fig. 12-5) is one of the most common parasites found in humans; recent surveys have shown a prevalence of up to 64% in some areas of the southeastern United States. Infection occurs by swallowing embryonated eggs. Unsanitary defecation habits "seed" the soil, and infection usually occurs

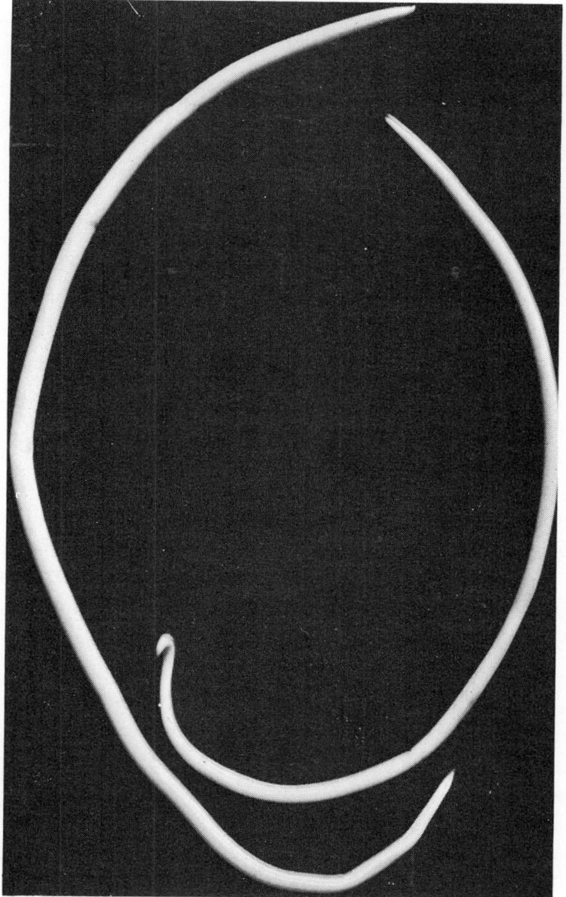

FIG. 12-5 Intestinal roundworm, *Ascaris lumbricoides,* male and female. Male *(right)* has characteristic sharp kink in end of tail.

when eggs are ingested with uncooked vegetables or when children put soiled fingers or toys in their mouths. The eggs are very resistant and can remain viable for many months or even years in the soil.

Morphology and physiology

The females of these species are about 20 to 30 cm long. The whitish yellow worms are pointed at both ends, and four pale lines can be discerned extending the length of the body. These lines are the **epidermal cords:** thickened areas of epidermis that bear the nuclei of the syncytial epidermis, as well as the **dorsal** and **ventral nerve cords** and the lateral **excretory canals.** Three **lips,** provided with papillae, surround the mouth. The male can be distinguished by its smaller size and by its sharply curved posterior end bearing two **copulatory spicules** which often protrude from the cloaca.

The outer body covering is the thick, noncellular **cuticle,** secreted by the epidermis. The cuticle is of great functional importance to the worm, serving to contain the high hydrostatic pressure exerted by the fluid in the pseudocoel. The several layers of the cuticle are primarily of **collagen,** a structural protein also abundant in vertebrate connective tissue. Three of the layers are comprised of crisscrossing fibers, which confer some longitudinal elasticity on the worm but severely limit its capacity for lateral expansion.

Beneath the epidermis is a layer of **longitudinal muscles.** There are no circular muscles in the body wall. The muscles are arranged in four bands, or quadrants, marked off by the four epidermal cords. Each muscle cell is large and spindle-shaped, with a striated, contractile process abutting the epidermis on one side and another process running to the dorsal or ventral nerve on the other. Nematode muscles are very unusual in that they have a muscle process running to the nerve instead of a nerve process innervating the muscle.

The fluid-filled **pseudocoel,** in which the internal organs lie, constitutes a **hydrostatic skeleton.** Hydrostatic skeletons, found in many invertebrates, lend support and transmit the force of muscle contraction because the volume of the enclosed fluid is constant. Normally, muscles are arranged antagonistically, so that movement is effected by contraction of one group of muscles and relaxation of the other. However, nematodes do not have circular body wall muscles to antagonize the longitudinal muscles; therefore, the cuticle must serve that function. Compression of the cuticle on the side of muscular contraction and stretching on the opposite side are the forces that return the body to resting position when the muscles relax; this produces the characteristic thrashing motion seen in nematode movement. An increase in efficiency of this system can only be achieved by an increase in hydrostatic pressure, and the hydrostatic pressure in the nematode pseudocoel is much higher than is usually found in animals that have hydrostatic skeletons but that also have antagonistic muscle groups.

The alimentary canal of the nemtaode consists of a **mouth** (Fig. 12-6), a muscular **pharynx,** a long nonmuscular **intestine,** a short **rectum,** and a terminal **anus.** Food material is sucked into the pharynx when the muscles in its anterior portion contract rapidly and open the lumen. Relaxation of the muscles anterior to the food mass closes the lumen of the pharynx, forcing the food posteriorly toward the intestine. The intestine is one cell layer thick and is normally collapsed by the surrounding hydrostatic pressure in the pseudocoel. Food matter is moved posteriorly by body movements and by additional food being passed into the intestine from the pharynx. Defecation is accomplished by muscles that simply pull the anus open, and the expulsive force is provided by the pseudocoelomic pressure.

The excretory system of *Ascaris* consists of two lateral **excretory canals,** with a connecting transverse network, emptying through a pore just behind the mouth. The ''excretory'' system of nematodes has important osmoregulatory functions and, in some species, secretory functions as well. Also, excretion of waste products occurs through the cuticle and through the intestinal wall and anus. The major nitrogenous waste products are ammonia and urea, and the main waste products of energy metabolism are short-chain fatty acids, but many other compounds also are excreted.

Adult *Ascaris* and many other parasitic nematodes have an anaerobic energy metabolism; thus a Krebs cycle and classic cytochrome system of electron transport are absent. Energy is derived through glycolysis and probably through some incompletely known electron transport sequences. Interestingly, some free-living nematodes and free-living stages of parasitic nematodes are obligate aerobes and have a Krebs cycle and cytochrome system. For example, the eggs of *Ascaris* require oxygen for embryonation, and development stops in the absence of oxygen. However, after the eggs hatch and the young worms begin growing

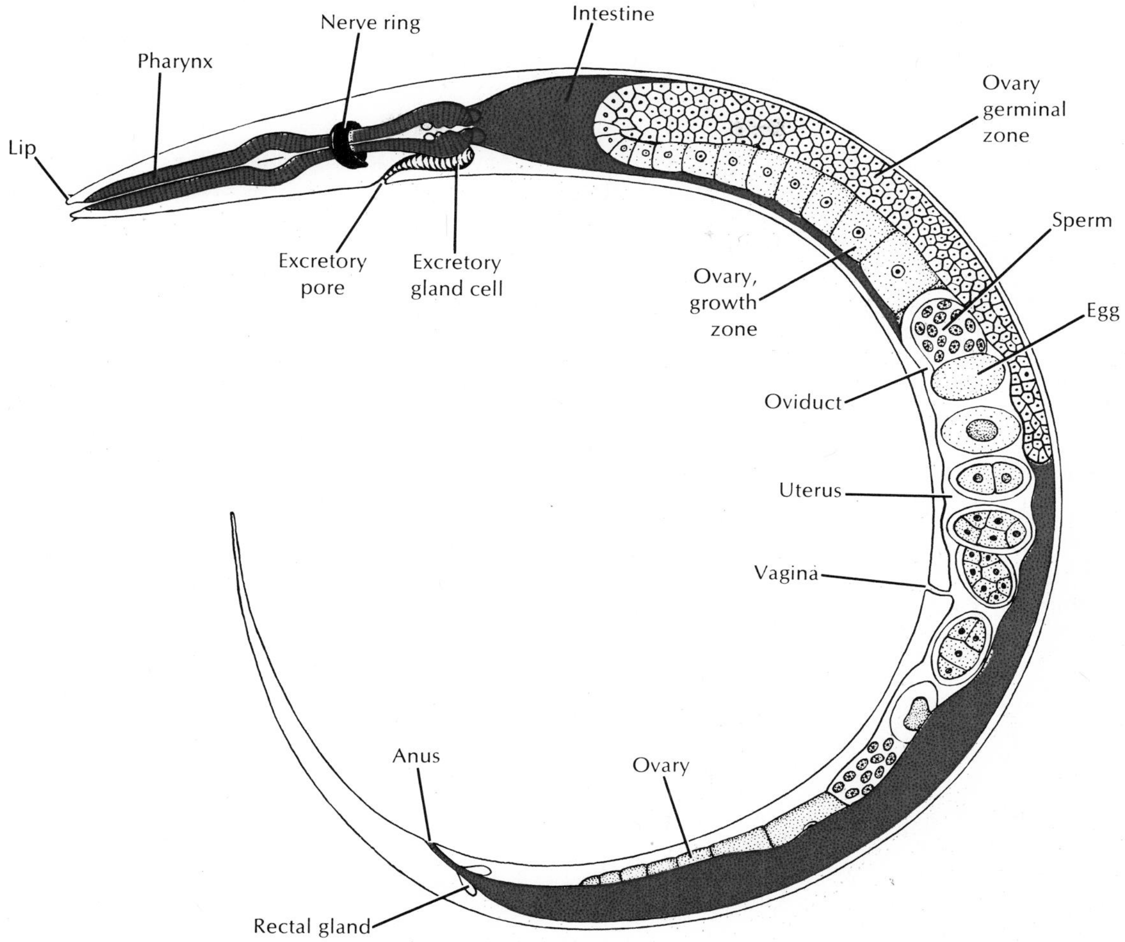

FIG. 12-6 *Rhabditis,* a common free-living nematode that feeds on decaying plant and animal matter. Some species that feed on manure undergo a developmental arrest at the third-stage juvenile until they "hitch-hike" to a new food supply on a dung beetle. In this drawing of a female, the intestine overlays and hides the germinal zone of the posterior ovary. As in most nematodes, the proximal end of the oviduct serves as a sperm storage area; the oocytes are penetrated by ameboid sperm as they pass through and then complete meiosis. (After Hirschman, modified from Sasser, J. N., and W. R. Jenkins. 1960. Nematology. Chapel Hill, University of North Carolina Press.)

in their host, their energy metabolism becomes anaerobic, and oxygen in atmospheric concentrations even becomes toxic to them.

A **ring of nerve tissue** and ganglia around the pharynx gives rise to small nerves to the anterior end and to two **nerve cords,** one dorsal and one ventral. The **papillae** are tactoreceptor organs, which are elaborately developed around the lips and around the posterior end of the male—the latter being important in copulation. A pair of more complex sensory organs, the **amphids,** are found near the anterior end, and a pair of similar organs, the **phasmids,** may be found near the posterior. These are pitlike chemoreceptors, and in some species they may have secretory function. Amphids are most developed in free-living nematodes, being reduced in parasitic species; phasmids are found only in the class Phasmidia. The sensory endings in amphids and phasmids, and in at least some papillae, are modified cilia and are the only cilia that occur in nematodes.

Reproduction and life cycle

The male reproductive system consists of a long tubular **testis, vas deferens, seminal vesicle, ejaculatory duct,** and two **spicules.** The spicules do not conduct sperm into the female's vagina, but they serve to hold the vulva open against the surrounding hydrostatic pressure of the female's pseudocoel, so that the ejaculatory duct can discharge the sperm into the female tract. The female reproductive system consists of **ovaries, oviducts, uterus,** and a **vagina** that empties through the **genital pore,** or **vulva.**

A female *Ascaris* may lay 200,000 eggs a day; they are passed out in the host's feces. Given suitable soil conditions, embryonation is complete within 2 weeks. Direct sunlight and high temperatures are rapidly lethal, but the eggs have an amazing tolerance to other adverse conditions, such as desiccation or lack of oxygen. When embryonated eggs are swallowed by a host, the tiny larvae hatch. They burrow through the intestinal wall into the veins or lymph vessels and are carried through the heart to the lungs. There they break out into the alveoli and are carried up the bronchi to the trachea. If the infection is large, they may cause a serious pneumonia at this stage. On reaching the pharynx, the larvae are swallowed, pass through the stomach, and finally mature about 2 months after the eggs were ingested. In the intestine the worms cause abdominal symptoms and allergic reactions, and in large numbers they may cause intestinal blockage. Perforation of the intestine with resultant peritonitis is not uncommon, and wandering worms may occasionally emerge from the anus or throat or may enter the trachea or eustachian tubes and middle ears.

Visceral larva migrans

Infection of dogs and cats with species of ascarids *(Toxocara)* is extremely common in the United States. Surveys indicate that 20% of adult dogs and 98% of puppies are infected. The life cycles of these species are quite similar to that of *Ascaris lumbricoides,* with soil being seeded with eggs from dog and cat feces. Humans, especially children, may become infected, but in this abnormal host the worms do not mature in the intestine. Their development is arrested, and the larval worms continue to wander in the tissues, causing a condition called **visceral larva migrans.** Visceral larva migrans is much more common than reported cases suggest because most cases go undetected unless the larvae wander into some vital spot, such as the eye, where they may cause blindness in the affected eye.

Other nematode parasites

Nearly all vertebrates and many invertebrates are parasitized by nematodes. Some of the more important parasites of humans are described subsequently.

Hookworm. The hookworms commonly found in humans are *Necator americanus* and *Ancylostoma duodenale*. Their common name refers to the fact that their anterior end curves dorsally, suggesting a hook (p. 243). *N. americanus* was discovered in the New World, but it was probably introduced there through the slave trade. It is the more common of the two in humans and accounts for 95% of hookworm infection in the United States. Formerly, this parasite was a major disease agent in the southeast United States, but through sanitation efforts and control campaigns, the prevalence there was reduced from over 40% in 1910 to around 4% in 1963.

Hookworms have a large buccal cavity armed with cutting plates or teeth (Fig. 12-7). They attach to the wall of the small intestine and feed on blood. For reasons that are not clear, they suck much more blood from the gut wall then they use for food, simply passing it through their alimentary tract, mostly unchanged. Hence, **hookworm disease** is a disease of **blood loss** and is largely manifested by symptoms of **anemia** (low hemoglobin and red cells in the blood). The severity of the disease, or even whether any disease at all is produced, is determined by several factors, the most important of which are the number of worms present and the general nutritional state of the host. Larger numbers of worms cause more blood loss, and even a small number of worms can produce disease in a person who has a diet deficient in protein.

The life cycle differs from that of *Ascaris,* though as with *Ascaris* only one host is involved. The eggs pass out with the feces of the host, embryonate in the soil, and hatch spontaneously. The juveniles are free-living, feeding on microorganisms and fecal matter. They grow and molt their cuticle twice to become infective third-stage juveniles. Development is then arrested until they reach a new host. When a host comes in contact with the third-stage juvenile, the worm penetrates the host skin, is carried by the blood to the lungs,

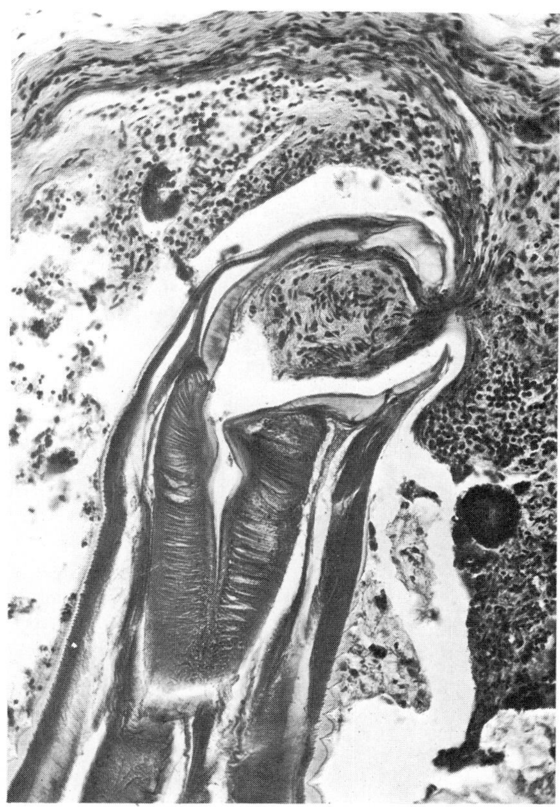

FIG. 12-7 Section through anterior end of hookworm attached to human intestine. Note the cutting plates of the mouth pinching off a bit of mucosa from which the thick muscular pharynx sucks blood. Esophageal glands secrete an anticoagulant to prevent blood clotting. (AFIP No. 33810.)

breaks through the alveoli, goes up the trachea to the pharynx, and then is swallowed and finally reaches the small intestine. Since the eggs and juveniles require shade, moisture, and warmth for survival, hookworm occurs mainly in tropical and subtropical areas.

Trichina worm. *Trichinella spiralis* is the tiny nematode responsible for the potentially lethal disease **trichinosis.** Adult worms burrow in the mucosa of the small intestine where the female produces living larvae. The larvae penetrate into blood vessels and are carried to the skeletal muscles where they coil up and form a cyst that eventually becomes calcified (Fig. 12-8). The worms may live in the cysts for years. When meat containing live cysts is swallowed, the larvae are liberated into the intestine where they mature and produce living larvae.

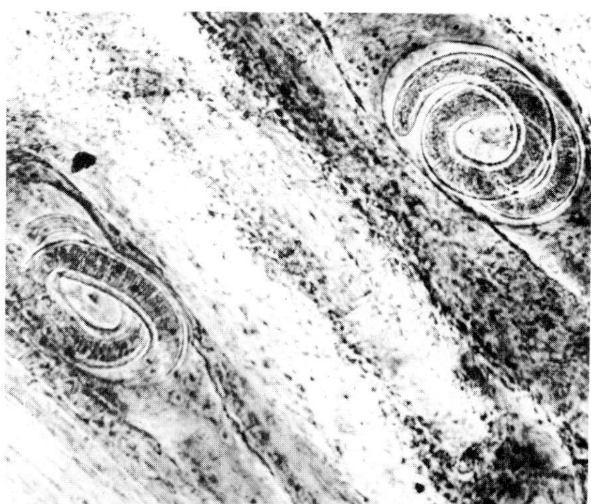

FIG. 12-8 Muscle infected with trichina worm *Trichinella spiralis.* Larvae may live 10 to 20 years in these cysts. If eaten in poorly cooked meat, the larvae are liberated in the intestine. They quickly mature and release many larvae into the blood of the host.

In addition to humans, *T. spiralis* can infect many other mammals, including hogs, rats, cats, and dogs. Humans most often acquire the parasite by eating improperly cooked pork. Hogs become infected by eating garbage containing pork scraps with cysts or by eating infected rats.

Heavy infections are very serious; about 20 deaths have been caused by *T. spiralis* in the United States during the last 10 years. Lighter infections are much more common, and about 2.4% of the population of the United States is infected. The simplest preventive measure is the thorough cooking of all pork.

Pinworms. The pinworm, *Enterobius vermicularis,* is the most common helminth parasite of humans in the United States. Its prevalence is estimated conservatively at 30% in children and 16% in adults, and it is found among people in all socioeconomic groups, from the deep South to Alaska. Fortunately, it is relatively nonpathogenic. The adults (Fig. 12-9, *A*) live in the large intestine and cecum. The females, up to about 12 mm in length, migrate to the anal region at night to lay their eggs (Fig. 12-9, *B*). Scratching the resultant itch effectively contaminates the hands and bed clothes. Eggs embryonate rapidly and become infective within 6 hours at body temperature. When they are swallowed, they hatch in the duodenum and ma-

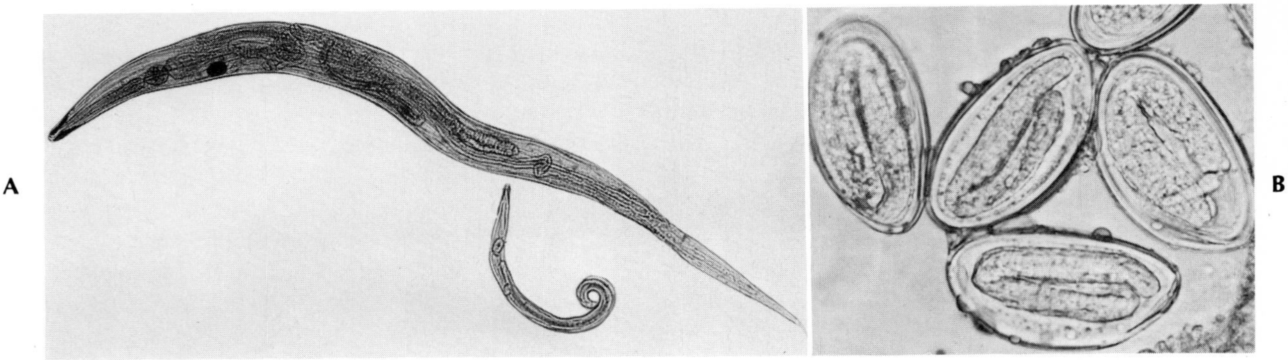

FIG. 12-9 Pinworms, *Enterobius vermicularis.* **A,** Adult pinworms; the male is much smaller. **B,** Group of pinworm eggs, which are usually discharged at night around the anus of the host, who, by scratching during sleep, may get fingernails and clothing contaminated. This may be the most common and widespread of all human helminth parasites. (Courtesy Indiana University School of Medicine, Indianapolis.)

ture in the large intestine. Diagnosis is made by applying the sticky side of cellulose tape around the anus to collect eggs and then examining the tape under the microscope. Several drugs are very effective against pinworm, but all members of a family should be treated at the same time, since the worm easily spreads through a household. Small numbers of pinworms are usually asymptomatic, but larger numbers may cause intestinal disturbance, perianal itching, and even nervousness and irritability in children.

Whipworms. *Trichuris trichiura* is probably the most common nematode of humans after *Enterobius* and *Ascaris,* especially in children in warmer locales. The eggs are passed in the feces and embryonate in the soil. Adults live in the lower ileum and cecum, where damage to the intestinal mucosa in heavy infections can result in serious disease, and occasionally even death.

Filarial worms. At least eight different species of filarial nematodes infect humans, and some of these worms are major causes of disease. Some 250 million people in tropical countries are infected with *Wuchereria bancrofti* or *Brugia malayi,* which places these species among the scourges of humanity. The worms live in the lymphatic system, and the females are as

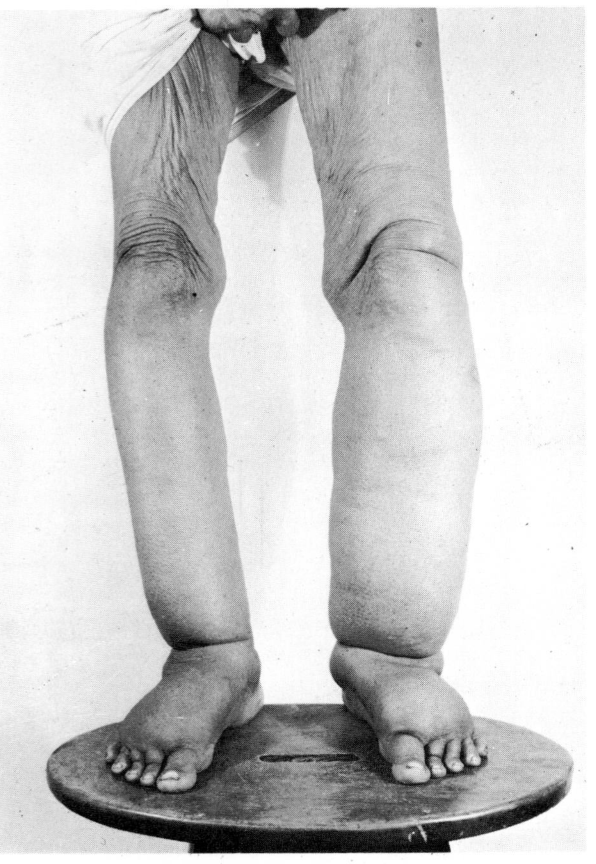

FIG. 12-10 Elephantiasis of leg caused by adult filarial worms of *Wuchereria bancrofti,* which live in lymph passages and block the flow of lymph. Larval microfilariae are transmitted by mosquitoes. (AFIP No. 44430-1.)

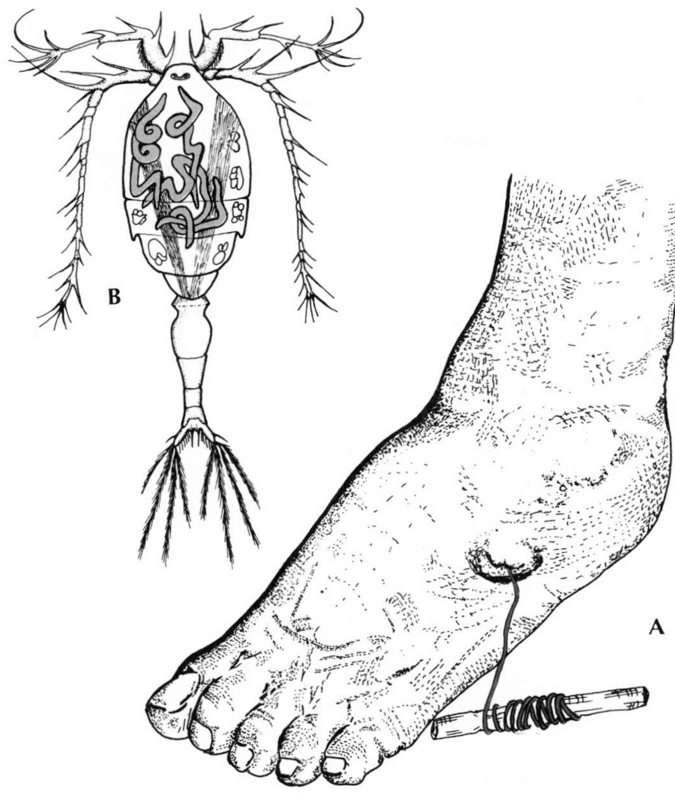

FIG. 12-11 **A,** Guinea worm *Dracunculus* being extracted from host by winding one turn of the stick each day to prevent rupturing worm and causing infection. A more modern method is surgery. **B,** *Cyclops,* a copepod, containing *Dracunculus* larvae in body cavity.

long as 100 mm. The disease symptoms are associated with inflammation and obstruction of the lymphatic system. The females release live young, the tiny **microfilariae,** into the blood and lymph. When the microfilariae are ingested by the appropriate species of mosquito, along with its blood meal, they penetrate the mosquito's gut and develop to third-stage, infective juveniles. They escape from the proboscis sheath of the mosquito when it is feeding again on a human and penetrate the wound made by the mosquito bite. Though the infection may be asymptomatic, various inflammatory responses usually occur, and the dramatic manifestations of **elephantiasis** are occasionally produced. This condition is associated with long and repeated exposure to the worms and is marked by an excessive growth of connective tissue and enormous swelling of affected parts, such as the scrotum, legs, arms, and more rarely, the vulva and breasts (Fig. 12-10).

Another filarial worm, *Oncocerca volvulus,* infects more than 30 million people in parts of Africa, Arabia, Central America, and South America. It causes **river blindness** and is carried by the blackfly, *Simulium* spp.

Guinea worms. *Dracunculus medinensis,* the guinea worm, is found in certain areas of Africa, India, and the Middle East. The male is small, but the female is up to 80 cm long and lies just under the surface of the skin of its host, where it appears much like a varicose vein. The worm causes an ulcer in the skin through which it discharges living larvae on contact with water. The larvae are picked up by the small aquatic crustacean *Cyclops,* where they undergo further development (Fig. 12-11, *B*). Infections occur from drinking water containing the *Cyclops*. The worms can be removed surgically, but the time-honored method of removing the adult worm has been to wind it out on a small stick, a little each day (Fig. 12-11, *A*). It is interesting to note that the "fiery serpent" of Biblical times was probably the guinea worm.

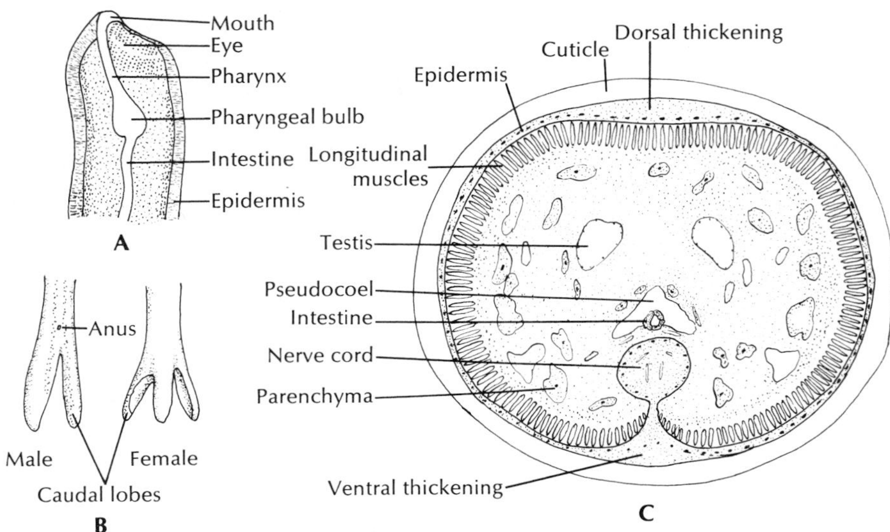

FIG. 12-12 Structure of *Paragordius,* a nematomorph. **A,** Longitudinal section through the anterior end. **B,** Posterior end of male and female worms. **C,** Transverse section. Nematomorphs, or "horsehair worms," are very long and very thin. Their pharynx is usually a solid cord of cells and is nonfunctional. *Paragordius,* whose pharynx opens through to the intestine, is unusual in this respect and also in the possession of a photosensory organ ("eye").

PHYLUM NEMATOMORPHA

The popular name for the Nematomorpha (nem′a-to-mor′fa) (Gr., *néma, nēmatos,* thread, + *morphe,* form) is "horsehair worms," based on an old superstition that the worms arise from horsehairs that happen to fall into the water. They were long included with the nematodes, with which they share the structure of the cuticle, presence of epidermal cords, longitudinal muscles only, and pattern of nervous system. However, the early larval form of some species has a striking resemblance to the Priapulida so that it is impossible to say to what group the nematomorphs are most closely related.

Ecologic relationships

The horsehair worms are free-living as adults and parasitic in arthropods as juveniles. As a group they have worldwide distribution and a variety of aquatic habitats and may be found in both running and standing water. Adults do not feed but will live almost anywhere in wet or moist surroundings if the oxygen is adequate. Juveniles do not emerge from the arthropod host unless there is water nearby. Adults are often seen wriggling slowly about in ponds or streams, with males being more active than females. The female discharges her eggs in water in long strings. Some juveniles, such as *Gordius,* a cosmopolitan genus, are believed to encyst on vegetation that may later serve as food for a grasshopper or other arthropod. In the marine form *Nectonema* juveniles occur in hermit crabs and other crabs.

Morphology

Horsehair worms are extremely long and slender, with a cylindric body. They may reach a length of 1.5 m with a diameter of only 3 mm, but most are smaller with a diameter of 1 to 2 mm. The anterior ends are usually rounded, and the posterior ends are rounded or have two or three caudal lobes (Fig. 12-12).

The body wall is much like that of nematodes: a secreted **cuticle,** an **epidermis,** and musculature of **longitudinal muscles** only. Ventral or dorsal and ventral, but not lateral, **epidermal cords** are present. The **pseudocoel** in many species is largely filled with parenchymal cells.

The **digestive system** is vestigial. The pharynx is a solid cord of cells, and the intestine does not open to the cloaca. The larval forms absorb food from their arthropod hosts through the body wall, and the adults apparently live on stored nutrients.

Circulatory, respiratory, and excretory systems are lacking. There is a nerve ring around the pharynx and a midventral nerve cord. Each sex has a pair of gonads and a pair of gonoducts that empty into the cloaca. The young hatch from the eggs and somehow gain entry into the arthropod host. After several months in the hemocoel of the host, the matured worm emerges into the water. Curiously, if the host is a terrestrial insect, it is stimulated by an unknown mechanism to seek water.

PHYLUM ACANTHOCEPHALA

The members of the phylum Acanthocephala (a-kan'tho-sef'a-la) (Gr., *akantha,* spine or thorn, + *kephale,* head) are commonly known as "spiny-headed worms." The phylum derives its name from one of its most distinctive features, a cylindric invaginable proboscis bearing rows of recurved spines, by which it attaches itself to the intestine of its host. All acanthocephalans are endoparasitic, living as adults in the intestines of vertebrates.

Various species range in size from less than 2 mm to over 1 m in length, with the females of a species usually being larger than the males. In life, the body is usually bilaterally flattened, with numerous transverse wrinkles. The worms are usually cream-color but may be yellowish or brown from absorption of pigments from the intestinal contents.

Ecologic relationships

Acanthocephalans inflict traumatic damage from penetration of the intestinal wall by the spiny proboscis. In many cases there is remarkably little inflammation, but in some species the inflammatory response of the host is intense. Great pain can be produced, particularly if the gut wall is completely perforated.

Over 500 species are known, most of which parasitize fish, birds, and mammals, and the phylum is worldwide in distribution. However, no species is normally a parasite of humans, though rarely humans are infected with species that usually occur in other hosts.

Larvae of spiny-headed worms develop in arthropods, either crustaceans or insects, depending on the species.

Phylogenetic relationships

Acanthocephalans are highly specialized parasites with a unique structure and have doubtless been so for millions of years. Any ancestral or other related group that would shed a clue to the phyletic relationships of the Acanthocephala is probably long since extinct. Like the cestodes, acanthocephalans have no digestive tract and must absorb all nutrients across the tegument, but the tegument of the two groups is quite different in structure. Also, acanthocephalans are pseudocoelomate and show eutely, as in the nematodes, though here, too, the structural and developmental differences are great. Thus, the Acanthocephala are an isolated phylum, not closely related to any known form.

Characteristics

1. Body cylindric (often flattened in life); anterior end with **spiny retractile proboscis** and sheath
2. **Tegument** syncytial, containing fluid-filled channels (lacunae); body wall with circular and longitudinal muscles
3. Fluid-filled pseudocoel
4. No digestive, respiratory, or circulatory systems
5. Excretory system, when present, with two branched ciliated protonephridia connected to a common excretory duct
6. Nervous system with a central ganglion within the proboscis receptacle and nerves to proboscis and body; sensory endings on proboscis and genital bursa
7. Sexes separate; shelled larvae embryonate in pseudocoel of female; special selector apparatus in female system
8. Adults parasitic in intestine of vertebrates; juveniles develop in arthropods; other animals may serve as transport hosts

Representative type—Macracanthorhynchus hirudinaceus, intestinal spiny-headed worm of pigs

Macracanthorhynchus hirudinaceus occurs throughout the world in the small intestine of pigs and occasionally in other mammals. Its juveniles are parasitic in certain beetle larvae. Its common occurrence and large size makes it suitable for study, and its life cycle has been known for a long time.

Morphology and physiology

In life the body is somewhat flattened, though it is usual for specimens to be treated with tap water before fixation so that fixed specimens are turgid and cylindric (Fig. 12-13, *C*). The female varies from 10 to 65 cm in length and is about 5 mm in diameter; the male is 5 to 9 cm in length.

The body wall is syncytial, and its surface is punctured by minute crypts 4 to 6 μm deep, which greatly increase the surface area of the tegument. About 80%

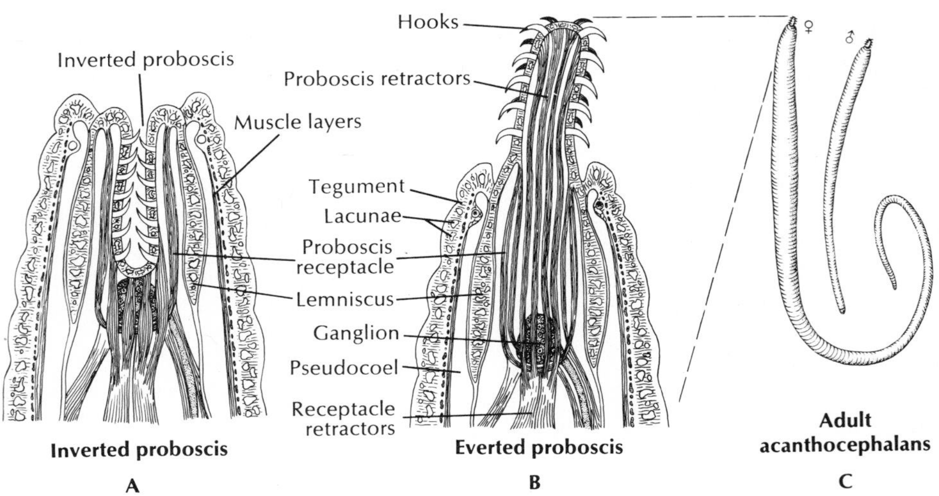

FIG. 12-13 Structure of a spiny-headed worm (phylum Acanthocephala). **A** and **B,** Eversible spiny proboscis by which the parasite attaches to the intestine of the host, often doing great damage. Since they lack a digestive tract, food is absorbed through the tegument. **C,** Male is typically smaller than female.

of the thickness of the tegument is the radial fiber zone, which contains a **lacunar system** of ramifying fluid-filled canals (Fig. 12-13, *A* and *B*). The function of the lacunar system is unclear, but it may serve in distribution of nutrients. Attached to the neck region are two elongated **lemnisci** (extensions of the epidermis and lacunar system) that may serve as reservoirs of the lacunar fluid from the proboscis when that organ is invaginated.

The **proboscis,** bearing six rows of recurved hooks, is attached to the neck region (Fig. 12-13) and can be inverted into a **proboscis receptacle** by retractor muscles. The proboscis is a forbidding organ and can cause serious damage to the intestine of its host.

In *M. hirudinaceus* and other members of its family, there are a pair of **protonephridia** with flame cells that unite to form a common tube that opens into the sperm duct or uterus.

Since acanthocephalans have no digestive tract, they must absorb all nutrients through their tegument. They can absorb various molecules by specific membrane transport mechanisms, and their tegument can carry out pinocytosis. The tegument bears some enzymes, such as peptidases, which can cleave several dipeptides, and the amino acids are then absorbed by the worm. Like cestodes, acanthocephalans have a requirement for host dietary carbohydrate, but their mechanism for absorption of glucose is different. As glucose is absorbed, it is rapidly phosphorylated and compartmentalized, so that a metabolic "sink" is created into which glucose in the surrounding medium can flow. Glucose can diffuse down a concentration gradient into the worm because it is constantly removed as soon as it enters.

A pair of tubular **genital ligaments,** or **ligament sacs,** extend posteriorly from the end of the proboscis receptacle. The male has a pair of **testes,** each with a **vas deferens,** and a common **ejaculatory duct** that ends in a small **penis.** During copulation sperm are ejected into the vagina, travel up the genital duct, and escape into the pseudocoel.

In the female, the ovarian tissue in the ligament sac breaks up into **ovarian balls** that rupture the ligament sacs and float free in the pseudocoel. One of the ligament sacs leads to a funnel-shaped **uterine bell** that receives the developing shelled embryos and passes them on to the uterus (Fig. 12-14). An interesting and unique **selective apparatus** operates here. Fully developed embryos are slightly longer than immature ones, and they are passed on into the uterus, while immature eggs are retained for further maturation.

The shelled embryos, which are discharged in the feces of the vertebrate host, do not hatch until eaten by the intermediate host. For *M. hirudinaceus* this is

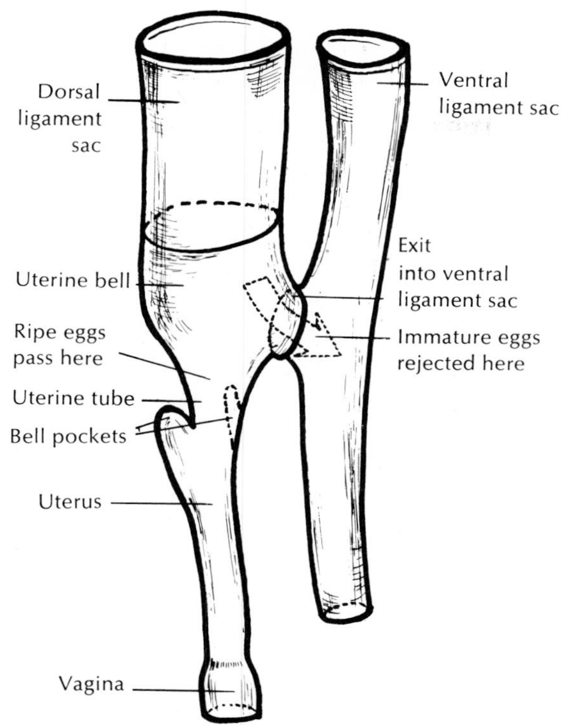

FIG. 12-14 Scheme of the genital selective apparatus of a female acanthocephalan. It is a unique device for separating immature from mature fertilized eggs. Eggs containing larvae enter the uterine bell and pass on to the uterus and exterior. Immature eggs are shunted into the ventral ligament sac or into the pseudocoel to undergo further development.

any of several species of soil-inhabiting beetle larvae, especially scarabeids. Grubs of the June beetle *(Phyllophaga)* are frequent hosts. Here the larva **(acanthor)** burrows through the intestine and develops to the juvenile **(cystacanth).** Pigs are infected by eating the grubs. Multiple infections may do considerable damage to the pig's intestine, and perforations may occur.

PHYLUM ENTOPROCTA

Entoprocta (en'to-prok'ta) (Gr., *entos,* + *proktos,* anus) is a small phylum of less than a hundred species of tiny, sessile animals that, superficially, look much like hydroid cnidarians, but their tentacles are ciliated and tend to roll inward (Fig. 12-15). Most entoprocts are microscopic, and none is more than 5 mm long. They are all stalked and sessile forms; some are colonial, and some are solitary. All are ciliary feeders.

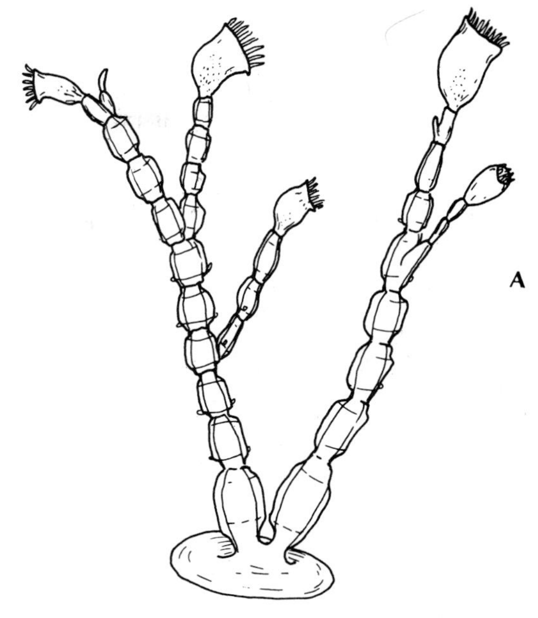

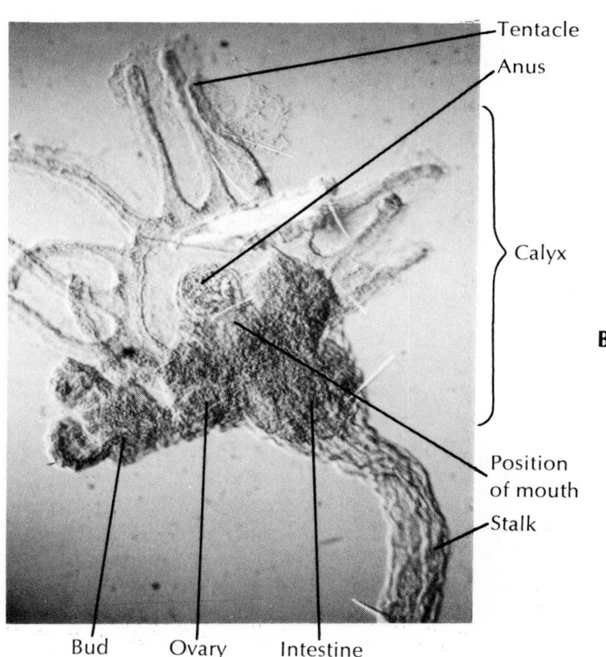

FIG. 12-15 **A,** *Urnatella,* a freshwater entoproct, forms small colonies of two or three stalks from a basal plate. **B,** *Loxosomella,* a solitary entoproct. Both solitary and colonial entoprocts can reproduce asexually by budding, as well as sexually. (**A,** Modified from Cori, 1936; **B,** photograph by L. S. Roberts.)

Phylogenetic relationships

The entoprocts were once included with the phylum Ectoprocta in a phylum called Bryozoa, but the ectoprocts are true coelomate animals, and many zoologists prefer to place them in a separate group. Ectoprocts are still often referred to as bryozoans. The Entoprocta may be distantly related to the Ectoprocta, but there is little evidence of close relationship. The entoprocts may have arisen as an early offshoot of the same line that led to the ectoprocts.

Ecologic relationships

With the exception of the genus *Urnatella,* all entoprocts are marine forms that have a wide distribution from the polar regions to the tropics. Most marine species are restricted to coastal and brackish waters and often grow on shells and algae. Some are commensals on marine annelid worms. Freshwater entoprocts occur on the underside of rocks in running water. *U. gracilis* is the only common freshwater species in North America (Fig. 12-15, *A*).

Morphology and physiology

The body, or **calyx,** of the entoproct is cup-shaped, bears a **crown,** or circle, of ciliated **tentacles,** and may be attached to a substratum by a single **stalk** and an attachment disc with adhesive glands as in the solitary *Loxosomella* (Fig. 12-15, *B*) or by two or more stalks in colonial forms. Both tentacles and stalk are continuations of the body wall. The 8 to 30 tentacles making up the crown are ciliated on their lateral and inner surfaces and each can move individually. The tentacles can roll inward to cover and protect the mouth and anus but cannot be retracted into the calyx.

Movement is usually restricted in entoprocts, but *Loxosoma,* which lives in the tubes of marine annelids, is described as quite active, moving over the annelid and its tube freely.

The gut is U-shaped and ciliated, and both the mouth and anus open within the circle of tentacles. Entoprocts are **ciliary filter feeders.** Long cilia on the sides of the tentacles keep a current of water containing protozoans, diatoms, and particles of detritus moving in between the tentacles. Short cilia on the inner surfaces of the tentacles capture the food and direct it downward toward the mouth.

The **body wall** consists of a cuticle, cellular epidermis, and longitudinal muscles. The **pseudocoel** is largely filled with a gelatinous parenchyma in which is embedded a pair of **protonephridia** and their ducts, which unite and empty near the mouth. There is a well-developed **nerve ganglion** on the ventral side of the stomach, and the body surface bears sensory bristles and pits. Circulatory and respiratory organs are absent. Exchange of gases occurs through the body surface, probably much of it through the tentacles.

Some species are monoecious, some dioecious, and some appear to be **protandric**—that is, the gonad at first produces sperm and later eggs. The gonoducts open within the circle of tentacles.

Fertilized eggs develop in a depression, or brood pouch, between the gonopore and the anus. Entoprocts have a modified spiral cleavage pattern with mosaic blastomeres. The coeloblastula is formed by invagination. The trochophore-like larva is ciliated and free-swimming. It has an apical tuft of cilia at the anterior end and a ciliated girdle around the ventral margin of the body. Eventually the larva settles to the substratum and inverts to form the adult.

PHYLUM GNATHOSTOMULIDA

The first species of the Gnathostomulida (nath'o-sto-myu'lid-a) (Gr., *gnathos,* jaw, + *stoma,* mouth, + L. *-ulus,* diminutive) was observed in 1928 in the Baltic, but its description was not published until 1956. Since then these animals have been found in many parts of the world, including the Atlantic coast of the United States, and over 80 species in 18 genera have been described.

Gnathostomulids are delicate wormlike animals and are 0.5 to 1 mm long. They live in the interstitial spaces of very fine sandy coastal sediments and silt and can endure conditions of very low oxygen. They are often found in large numbers and frequently in association with gastrotrichs, nematodes, ciliates, tardigrades, and other small forms.

Lacking a pseudocoel, a circulatory system, and an anus, the gnathostomulids show some similarities to the turbellarians and were at first included in that group. However, their mesenchyme is greatly reduced, and their pharynx is reminiscent of the rotifer mastax. The pharynx is armed with a pair of lateral jaws used to scrape fungi and bacteria off the substratum. And, though the epidermis is ciliated, each epidermal cell has but one cilium, a condition rarely found except in the sponges and cnidarians. The phyletic relationships of the Gnathostomulida are still obscure.

Gnathostomulids can glide, swim in loops and spirals, and bend the head from side to side. Sexual stages may include males, females, and hermaphrodites. Fertilization is internal.

PHYLOGENY AND ADAPTIVE RADIATION OF THE PSEUDOCOELOMATES

Phylogeny. The problematic phylogenetic relationships of the Acanthocephala, the Entoprocta, and the Gnathostomulida have been discussed previously. The remaining pseudocoelomate phyla were grouped into a single phylum (Aschelminthes) by Hyman. All of these phyla share a certain combination of characteristics, including the fact that they are usually wormlike, have a cuticle secreted by an epidermis that is underlain by muscles not arranged in regular circular and longitudinal layers, have a brain that is a circumenteric nerve ring, have determinate cleavage and eutely, and lack a muscle layer in the intestine. Hyman contended that the evidences of relationships were so concrete and specific that they could not be disregarded. Nevertheless, most authors now consider that differences between the groups are sufficient to merit phylum status for each, though some accept the concept of the Aschelminthes as a superphylum. These phyla may well have been derived originally from the protostome line via an acoelomate common ancestor resembling the rhabdocoel flatworms.

Adaptive radiation. Certainly, the most impressive adaptive radiation in this group of phyla is shown by the nematodes. They are by far the most numerous in terms of both individuals and species, and they have been able to adapt to almost every habitat available to animal life. Their basic pseudocoelomate body plan, with the cuticle, hydrostatic skeleton, and longitudinal muscles, has proved generalized and plastic enough to adapt to an enormous variety of physical conditions. Free-living lines gave rise to parasitic forms on at least several occasions, and virtually all potential hosts have been exploited. All types of life cycle occur—from the simple and direct to the complex, with intermediate hosts; from normal dioecious reproduction to parthenogenesis, hermaphroditism, and alternation of free-living and parasitic generations. A major factor contributing to the evolutionary opportunism of the nematodes has been their extraordinary capacity to survive conditions suboptimal for viability, for example, the developmental arrests in many free-living and animal parasitic species and the ability to undergo cryptobiosis in many free-living and plant parasitic species.

Derivation and meaning of names

Ascaris (Gr., *askaris,* intestinal worm). Genus of Nematoda

Dioctophyma (Gr., *diokto-* [irreg.] from *dionkoun,* to distend, + *phyma,* swelling). Refers to the papillae around the mouth, 6 to 18 in number. Giant kidney worm; genus of Nematoda

Dracunculus (L., *draco,* dragon, + *-unculus,* small; from Gr., *drakontion,* guinea worm). Guinea worm; genus of Nematoda

Enterobius (Gr., *enteron,* intestine, + *bios,* life). Pinworm; genus of Nematoda

Gordius (Greek mythologic king who tied an intricate knot). Genus of Nematomorpha

Macracanthorhynchus (Gr., *makros,* long, + *akantha,* thorn, + *rhynchos,* beak). Genus of Acanthocephala

Necator (L., *necator,* killer). Hookworm; genus of Nematoda

Trichinella (Gr., *trichinos,* of hair, + *-ella,* diminutive). Trichina worm; genus of Nematoda

Annotated references

Selected general readings

Baer, J. G. 1952. Ecology of animal parasites. Urbana, Ill., University of Illinois Press. *Excellent account of the ways parasites have adapted themselves.*

Behme, R., and others. 1972. Biology of nematodes. New York, MSS Information Corp.

Bird, A. F. 1971. The structure of nematodes. New York, Academic Press, Inc.

Chitwood, B. G., and M. B. Chitwood. 1974. An introduction to nematology. Baltimore, University Park Press. *A revision and consolidation of earlier editions of an authoratitive work.*

Cori, C. 1936. Kamptozoa. In H. G. Bronn (ed.). Klassen und Ordnungen des Tier-Reichs, vol. 4, part 2. Leipzig, Akademische Verlagsgesellschaft.

Croll, N. A. 1976. The organization of nematodes. New York, Academic Press, Inc.

Croll, N. A., and B. E. Matthews. 1977. Biology of nematodes. New York, John Wiley & Sons.

Donner, J. 1966. Rotifers, New York, Frederick Warne & Co., Inc.

Dougherty, E. C. (ed.). 1963. The lower Metazoa. Berkeley, University of California Press. *This is a comparative phylogeny, morphology, and physiology of the invertebrates below the annelids. More than 30 specialists contributed monographs to this important work.*

Edgar, R. S., and W. B. Wood. 1977. The nematode *Caenorhabditis elegans:* a new organism for intensive biological study. Science **198:**1285-1286.

Giese, A. C., and J. S. Pearce. 1974. Reproduction of marine invertebrates, vol. 1. New York, Academic Press, Inc. *Includes chapters on Nematoda, Rotifera, Gastrotricha, Kinorhyncha, and Gnathostomulida.*

Higgins, R. P. 1965. The homalorhagid Kinorhyncha of northeastern U.S. coastal waters. Trans. Am. Microsc. Soc. **84:**65-72.

Hyman, L. H. 1951. The invertebrates: Acanthocephala, Aschelminthes, and Entoprocta, vol. 3. New York, McGraw-Hill Book Co. *Describes with great accuracy the many types of these phyla.*

Jenkins, W. R., and D. P. Taylor. 1967. Plant nematology. New York, Van Nostrand Reinhold Co.

Kaestner, A. 1967. Invertebrate zoology, vol. 1. New York, Interscience Publishers, Inc.

Lee, D. L., and H. J. Atkinson. 1977. The physiology of nematodes, ed. 2. New York, Columbia University Press.

Mariscal, R. N. 1965. The adult and larval morphology and life history of the entoproct *Barentsia gracilis* (M. Sars, 1835). J. Morphol. **116:**311-388.

Martini, E. 1912. Studien über die Konstanz histologischer Elemente. III. Hydatina Senta. Zeitsch. Wiss. Zool. **102:**425-645.

Nickerson, W. 1901. On *Loxosoma davenporti*. J. Morphol. **17:**357-376.

Pennak, R. W. 1953. Fresh-water invertebrates of the United States. New York, The Ronald Press Co. *This excellent work deals with the free-living, freshwater invertebrates and omits the parasitic forms. Many of the sections describe various types of Aschelminthes.*

Reidl, R. I. 1969. Gnathostomulida from America. Science **163:**445-452.

Roggen, D. R., D. J. Rask, and N. O. Jones. 1966. Cilia in nematode sensory organs. Science **152:**515-516.

Rogick, M. D. 1948. Studies on marine Bryozoa, part 2. Biol. Bull. **94:**128-142. *The author considers the Entoprocta as a class under the Bryozoa.*

Sacks, M. 1964. Life history of an aquatic gastrotrich. Trans. Am. Microsc. Soc. **83:**358-362.

Sasser, J. N., and W. R. Jenkins (ed.). 1960. Nematology. Chapel Hill, N.C., University of North Carolina Press. *An important work by more than a score of eminent authorities in this difficult field.*

Schmidt, G. D., and L. S. Roberts. 1977. Foundations of parasitology. St. Louis, The C. V. Mosby Co.

Van Cleve, H. J. 1941. Relationships of Acanthocephala. Am. Naturalist **75:**1-20. *In this and many other articles this investigator attempts to appraise the position of the Acanthocephala in the animal kingdom.*

Warren, K. S. 1974. Helminthic diseases endemic in the United States. Am. J. Trop. Med. Hyg. **23:**723-730.

Selected *Scientific American* articles

Crowe, J. H., and A. F. Cooper, Jr. 1971. Cryptobiosis. **225:**30-36 (Dec.).

Edwards, C. A. 1969. Soil pollutants and soil animals. **220:** 88-99 (Apr.).

Hawking, P. 1958. Filariasis. **199:**94-100 (July).

CHAPTER 13
THE MOLLUSCS
Phylum Mollusca

The common whelk *Buccinum undatum,* a large marine snail, shown crawling on its muscular foot. Its head and tentacles are at the right and above them is the siphon which maintains a flow of respiratory water.

Photograph by D. P. Wilson.

Position in animal kingdom

1. The molluscs are one of the major groups of true **coelomate** animals.
2. They belong to the **protostome** branch, or schizocoelomous coelomates, and have spiral and determinate cleavage.
3. **All the organ systems** are present and well developed.
4. Many molluscs have a **trocophore larva** similar to the trocophore larva of marine annelids, marine turbellarians, and some others.

Biologic contributions

1. In molluscs gaseous exchange occurs not only through the body surface as in lower invertebrates, but also in specialized **respiratory organs** in the form of **gills or lungs.**
2. They have an open **circulatory system** in most classes, with pumping **heart,** vessels, and blood sinuses. In most cephalopods the circulatory system is closed.
3. The efficiency of the respiratory and circulatory systems in the cephalopods has made greater body size possible. Invertebrates reach their largest size in some of the cephalopods.
4. They introduce for the first time a fleshy **mantle** that, in most cases, secretes a shell and is variously modified for a number of functions.
5. Features unique to the phylum are the **radula** and the muscular **foot.**
6. The highly developed direct **eye** of higher molluscs is similar to the indirect eye of the vertebrates but arises as a skin derivative in contrast to the brain eye of vertebrates (convergent evolution, Principle no. 6, p. 10).

THE MOLLUSCS

Next to Arthropoda the phylum Mollusca (mollus'ka) (L., *molluscus,* soft) has the most named species in the animal kingdom—probably more than 100,000 living species, not to mention some 35,000 fossil species discovered to date. The name Mollusca indicates one of their distinctive characteristics, a soft body.

This very diverse group includes the chitons, tooth shells, snails, slugs, nudibranchs, sea butterflies, clams, mussels, oysters, squid, octopuses, and nautiluses. The group ranges from fairly simple organisms to some of the most complex of invertebrates, and in size from almost microscopic to the giant squid *Architeuthis harveyi*. The body of this huge species may grow up to 12 m long, and the two longest of its tentacles may stretch as much as another 50 m. It may weigh between 225 and 270 kg (500 and 600 pounds). The shells of some of the giant clams *Tridacna gigas,* which inhabit the Indo-Pacific coral reefs, reach 1.5 m in length and weigh over 225 kg. These are extremes, however, for probably 80% of all molluscs are less than 5 cm in maximum shell size. The phylum includes some of the most sluggish and some of the swiftest and most active of the invertebrates. It includes herbivorous grazers, predaceous carnivores, and ciliary filter feeders.

The enormous variety, great beauty, and easy availability of the shells of molluscs has made shell collecting a popular pasttime. However, many amateur shell collectors, even though able to name hundreds of the shells that grace our beaches, know very little about the living animals that created those shells and once lived in them. Just what is a mollusc? Reduced to its simplest dimensions, a mollusc might be described as consisting of the main body mass, called the **visceral hump,** which is covered by a **shell** and from which extends a sensory and feeding area, the **head,** and a locomotor area, the **foot.** It is the various adaptations and combinations of these basic components that produce the great confusion of different patterns making up this major group of animals.

Ecologic relationships

Molluscs are found in a great range of habitats, from the tropics to polar seas, at altitudes exceeding 7,000 m, in ponds, lakes, and streams, on mud flats, in pounding surf, and in open ocean from the surface to the abyssal depths. Most of them live in the sea, and they represent a variety of life-styles, including bottom feeders, burrowers, borers, and pelagic forms.

According to the fossil evidence, the molluscs originated in the sea, and most of them have remained there. Much of their evolution occurred along the shores, where food was abundant and habitats were varied. Only the bivalves and gastropods moved on to brackish and freshwater habitats. As filter feeders, the bivalves were unable to leave aquatic surroundings. Only the snails (gastropods) actually invaded the land. Terrestrial snails are limited in their range by their need for humidity, shelter, and the presence of calcium in the soil.

Life cycles are generally short, although some molluscs live for several years. Nudibranchs probably live only a year; oysters may live up to 10 years; *Unio,* 12 years; freshwater snails, 4 to 5 years; squid *(Loligo)* up to 2 years; and giant squid 10 years.

Economic importance

A group as large as the molluscs would naturally affect humans in some way. A wide variety of molluscs are used as food. Pearl buttons are obtained from shells of bivalves. The Mississippi and Missouri river basins have furnished material for most of this industry in the United States; however, supplies are becoming so depleted that attempts are being made to propagate bivalves artificially. Pearls, both natural and cultured, are produced in the shells of clams and oysters, most of them in a marine oyster, *Meleagrina,* found around eastern Asia (Fig. 13-2, *B*).

Some molluscs are destructive. The burrowing shipworms, chiefly *Teredo* (Fig. 13-22), do great damage to wooden ships and wharves. To prevent the ravages of shipworms, wharves must either be creosoted or built of cement. Snails and slugs are known to damage garden and other vegetation. In addition, snails often serve as intermediate hosts for serious parasites. The boring snail *Urosalpinx* is second only to the sea star in destroying oysters.

Evolution of molluscs

The molluscs, which date from the Precambrian era, have left a continuous fossil record since Cambrian times. The primitive larva of the mollusc, which many still retain, is the ciliated **trocophore** type (Fig. 13-21) similar to that of marine annelids, and the type of egg cleavage is also similar in molluscs and annelids, which indicates a relationship between the two phyla. The ladderlike nervous system of some molluscs resembles that of the turbellarians. It is conceivable that a flatworm type of ancestor gave rise to the two main protostome groups—the nonsegmented molluscs and the segmentally arranged annelid-arthropod stem.

Significance of the coelom

Heretofore we have been studying animals that lacked a true coelom. These follow several patterns. In the radiates with gelatinous mesoglea between the body surface and the enteron, diffusion of substances is a simple matter. In flatworms body spaces are filled with cellular parenchyma of endomesoderm; here the need for a better transport method is met by the extensive branching of the gastrovascular cavity and of the protonephridial system throughout the body. In the nemertines this is aided by a system of blood vessels.

In the pseudocoelomates the parenchyma is replaced by spongy or open spaces—the pseudocoel. Fluid in the pseudocoel bathes the organs, thus providing a means of internal transport serviceable enough for small animals. Lacking mesenteries, the organs lie loose in the body cavity. The pseudocoelomates are all small—obviously such an arrangement would be unsuitable for more sizable forms.

In the coelomates the coelom develops as a secondary cavity within the mesoderm. The coelomic cavity is completely surrounded by mesodermal epithelium, called **parietal peritoneum.** There is not only ample room in the coelom for organs, but the organs are held in place by **mesenteries,** which are continuations of the peritoneum. The organs are themselves covered with **visceral peritoneum.** This ensures a more stable arrangement of organs with less crowding. The alimentary canal can become more muscular, more highly specialized, and more diversified without interfering with other organs, such as the heart, liver, or lungs.

The coelom performs other important functions. It is filled with **coelomic fluid,** and its lining is often ciliated to keep the fluid moving. Thus it aids in the movement of materials, such as absorbed foods and metabolic wastes, from one place to another. In many smaller coelomates no other transport system is necessary. In animals with a vascular system, the mesenteries provide an ideal location for the network of blood vessels necessary to deliver blood to every body organ.

The coelom can also serve as a hydrostatic skeleton. Circular and longitudinal body wall muscles, acting as antagonists, can contract or relax to vary the force exerted on the coelomic fluid and thus produce a variety of body movements.

Altogether the development of the coelom must be considered an important stepping stone in the evolution of larger and more complex forms. The three major phyla of coelomate protostomes are the molluscs, the annelids, and the arthropods. A number of smaller invertebrate phyla, the echinoderms, and the vertebrates are also coelomates.

Characteristics

1. Body bilaterally symmetric (bilateral asymmetry in some); unsegmented; usually with definite head

2. Ventral body wall specialized as a muscular **foot,** variously modified but used chiefly for locomotion
3. Dorsal body wall forms pair of folds called the **mantle,** which encloses the **mantle cavity,** is modified into **gills** or **lungs,** and secretes the **shell** (shell absent in some)
4. Surface epithelium usually ciliated and bearing mucous glands and sensory nerve endings
5. Coelom mainly limited to area around heart and perhaps lumen of gonads
6. Complex digestive system; rasping organ **(radula)** usually present; anus usually emptying into mantle cavity
7. **Open circulatory system** (mostly closed in cephalopods) of heart (usually three-chambered), blood vessels, and sinuses; respiratory pigments in blood
8. Gaseous exchange by **gills, lungs, mantle,** or **body surface**
9. One or two kidneys **(metanephridia)** opening into the pericardial cavity and usually emptying into the mantle cavity
10. Nervous system of paired cerebral, pleural, pedal, and visceral ganglia, with nerve cords and subepidermal plexus; ganglia centralized in nerve ring in gastropods and cephalopods
11. Sensory organs of touch, smell, taste, equilibrium, and vision (in some); eyes highly developed in cephalopods

Classes

The classes of molluscs are based on such features as type of shell, type of foot, and shape of shell.

Class Monoplacophora (mon'o-pla-kof'o-ra) (Gr., *monos,* one, + *plax,* plate, + *phora,* bearing) (Fig. 13-4). Body bilaterally symmetric with a broad flat foot; a single limpetlike shell; mantle cavity with five or six pairs of gills; large coelomic cavities; radula present; six pairs of nephridia, two of which are gonoducts; separate sexes. Example: *Neopilina.*

Class Polyplacophora (pol'y-pla-kof'o-ra) (Gr., *polys,* many, several, + *plax,* plate, + *phora,* bearing)—**chitons.** Elongated, dorsoventrally flattened body with reduced head; bilaterally symmetric; radula present; shell of eight dorsal plates; foot broad and flat; gills multiple, along sides of body between foot and mantle edge; sexes usually separate, with a trochophore but no veliger larva. Examples: *Mopalia, Chaetopleura* (Fig. 13-5).

Class Aplacophora (a'pla-kof'o-ra) (Gr., *a-,* not, without, + *plax,* plate, + *phora,* bearing)—**solenogasters.** Shell, mantle, and foot absent, but radula present; calcareous spicules embedded in integument; gills absent in some; some hermaphroditic; trochophore present; formerly united with chitons in class Amphineura. Examples: *Neomenia, Chaetoderma.*

Class Scaphopoda (ska-fop'o-da) (Gr., *skaphe,* trough, boat, + *pous, podos,* foot)—**elephant tusk shells.** Body enclosed in a one-piece tubular shell open at both ends; conical foot; mouth with radula and tentacles; head absent; mantle for respiration; sexes separate; trochophore larva. Example: *Dentalium* (Fig. 13-6).

Class Gastropoda (gas-trop'o-da) (Gr., *gaster,* belly, + *pous, podos,* foot)—**snails and others.** Body asymmetric, usually in a coiled shell (shell uncoiled or absent in some); head well developed, with radula; foot large and flat; one or two gills, or with mantle modified into secondary gills or a lung; most with single auricle and single nephridium; nervous system with cerebral, pleural, pedal, and visceral ganglia; dioecious or monoecious, some with trochophore, typically with veliger, some without pelagic larva. Examples: *Busycon* (Fig. 13-13), *Physa, Helix, Aplysia* (Fig. 13-14).

Class Bivalvia (bi-val've-a) (L., *bi-,* two, + *valva,* folding door, valve) **(Pelecypoda)—bivalves.** Body enclosed in a two-lobed mantle; shell of two lateral valves of variable size and form, with dorsal hinge; head greatly reduced, but mouth with labial palps; no radula; no cephalic eyes, a few with eyes on mantle margin; foot usually wedge-shaped; gills platelike; sexes usually separate, typically with trochophore and veliger larvae. Examples: *Anodonta, Venus, Tagelus* (Fig. 13-15), *Teredo* (Fig. 13-22).

Class Cephalopoda (sef'a-lop'o-da) (Gr., *kephale,* head, + *pous, podos,* foot)—**squids and octopuses.** Shell often reduced or absent; head well developed with eyes and a radula; foot modified into arms or tentacles; siphon present; nervous system of well-developed ganglia, centralized to form a brain; sexes separate, with direct development. Examples: *Loligo* (Fig. 13-24), *Octopus, Sepia* (Fig. 13-25).

Body plan of the mollusc

The typical mollusc has a **head** with a tonguelike rasping organ called a **radula,** a **visceral mass** (visceral hump) covered with soft skin, and a ventral muscular **foot** usually used in locomotion. Two folds of skin, outgrowths of the dorsal body wall, make up a protective **mantle,** or **pallium,** which encloses a space between the mantle and body wall called the **mantle cavity** (pallial cavity). The mantle cavity houses the gills or lungs, and in most molluscs the mantle secretes a protective **shell.** These structures are modified in various ways among the different groups, and in some the head, the shell, or the mantle may be lacking. The epidermis is soft and glandular, and much of it is ciliated.

The mantle and mantle cavity. The mantle is a sheath of skin extending from the visceral hump that hangs down on each side of the body, protecting the soft parts and creating between itself and the visceral mass the space called the mantle cavity. The outer surface of the mantle secretes the shell.

The mantle cavity plays an enormous role in the life of the mollusc. It usually houses the respiratory organs (gills or lung), which develop from the mantle, and the mantle's own exposed surface serves also for gaseous exchange. Into the mantle cavity are emptied products from the digestive, excretory, and reproductive systems. In aquatic molluscs a continuous current of water, kept moving by surface cilia or by muscular pumping, brings in oxygen, and in some forms, food; flushes out wastes; and carries reproductive products out to the environment. In aquatic forms the mantle is usually equipped with sensory receptors for sampling the environmental water. In cephalopods (squid and octopuses) the muscular mantle and its cavity create the jet propulsion used in locomotion. Many molluscs can withdraw the head or foot into the mantle cavity, which is surrounded by the shell, for protection.

The foot. The molluscan foot may be adapted for locomotion, for attachment to a substratum, for food capture (in cephalopods), or for a combination of these functions. It is usually a ventral, solelike structure in which waves of muscular contraction effect a creeping locomotion (p. 265). However, there are many modifications, such as the attachment disc of the limpets, the laterally compressed "hatchet foot" of the bivalves, or the division of the foot into the suckered arms and tentacles of the squid and octopuses. Secreted mucus is often used as an aid to adhesion or as a slime track by small molluscs that glide on cilia.

The foot is manipulated by a combination of muscles and hydrostatic skeleton. Some molluscs glide by rhythmic waves of muscular action. Burrowing forms can extend the foot into the mud by muscular action and then anchor it by enlargement with blood pressure while the body is drawn forward. In pelagic forms the foot may be modified into winglike parapodia, or thin, mobile fins for swimming.

The radula. The radula is a rasping, protrusible, tonguelike organ found in all molluscs except the bivalves. It is a ribbonlike membrane on which are mounted rows of tiny teeth that point backward (Fig. 13-1). Complex muscles move the radula and its supporting cartilages **(odontophore)** in and out while the membrane is partly rotated over the tips of the cartilages. There may be a few or as many as 250,000 teeth, which, when protruded, can scrape, pierce, tear, or cut. The function of the radula is twofold: to rasp off fine particles of food material and to serve as a conveyor belt for carrying the particles in a continuous stream toward the digestive tract. As the radula wears away anteriorly, new rows of teeth are continuously replaced by secretion at its posterior end. The pattern and number of teeth in a row are specific for each species and are used in the classification of molluscs.

The shell. The shell of the mollusc, when present, is secreted by the mantle and is lined by it. There are usually three typical layers (Fig. 13-2). The **periostracum** is the outer horny layer, composed of an organic substance called conchiolin, which consists of quinone-tanned protein. It helps to protect the underlying calcareous layers from erosion by boring organisms. It is secreted by a fold of the mantle edge and growth occurs only at the margin of the shell. On the older parts of the shell the periostracum often becomes worn away. The middle **prismatic layer** is composed of densely packed prisms of calcium carbonate, laid down in a protein matrix, and deposited as either calcite or aragonite. It is secreted by the glandular margin of the mantle, and increase in shell size occurs at the shell margin as the animal grows. The inner **nacreous layer** of the shell lies next to the mantle and is secreted continuously by the mantle surface, so that it increases in thickness during the life of the animal. The calcareous nacre is laid down in thin layers. Very thin and wavy layers produce the iridescent mother-of-pearl found in the abalones *(Haliotis)*, the chambered nautilus *(Nautilus)*, and many bivalves. Such shells may have 450 to 5,000 fine parallel layers of aragonite for each centimeter of thickness.

Freshwater molluscs usually have a thick periostra-

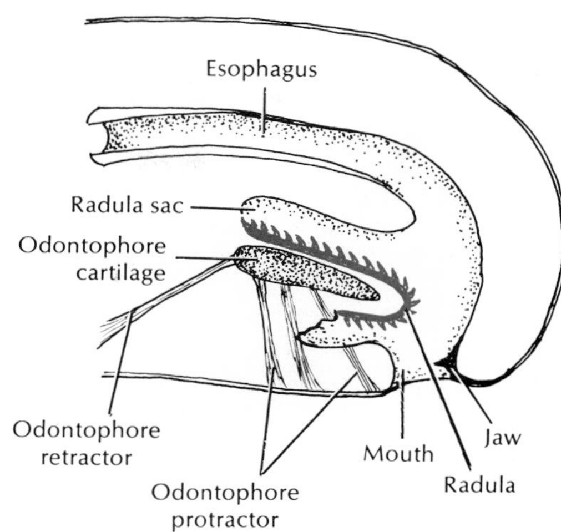

FIG. 13-1 Diagrammatic longitudinal section of gastropod head showing the radula and radula sac. The radula moves back and forth over the odontophore cartilage. As the animal grazes, the mouth opens, the odontophore is thrust forward, the radula gives a strong scrape backward bringing food into the pharynx, and the mouth closes. The sequence is repeated rhythmically. As the radula ribbon wears out anteriorly it is continually replaced posteriorly.

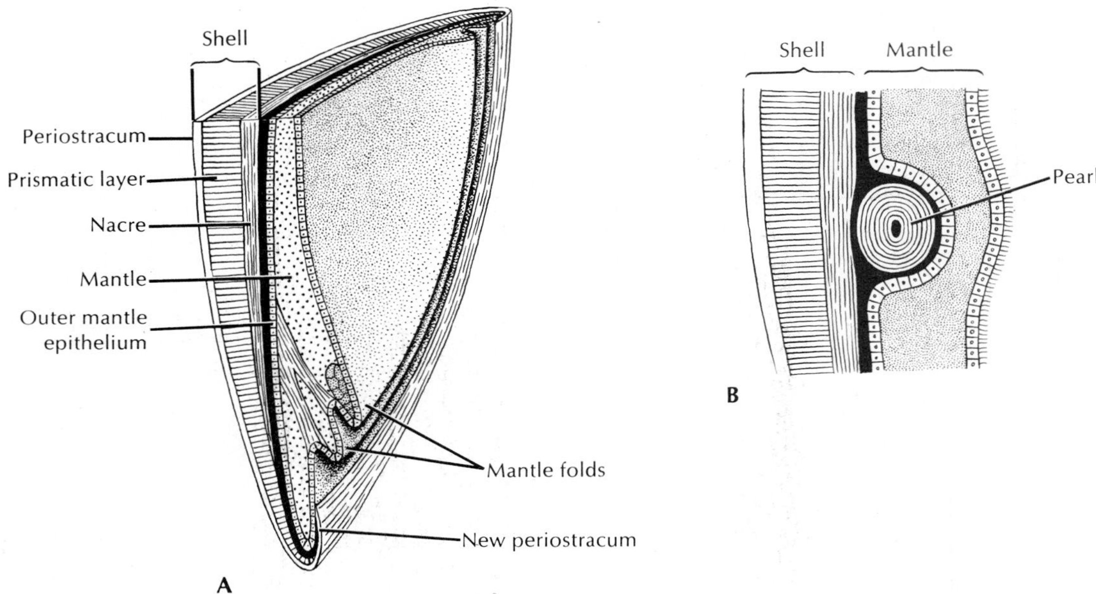

FIG. 13-2 A, Diagrammatic vertical section of shell and mantle of a bivalve. The outer mantle epithelium secretes the shell; the inner epithelium is usually ciliated. **B,** Formation of pearl between mantle and shell as a parasite or bit of sand under the mantle becomes covered with nacre.

cum that gives some protection against the acids produced in the water by the decay of leaf litter. In many marine molluscs the periostracum is relatively thin, and in cowries it is absent. There is a great range in variation in shell structure. Calcium for the shell comes from the environmental water or soil or from food. The first shell appears during the larval period and grows continuously throughout life.

Internal structure and function. In the molluscs, oxygen–carbon dioxide exchange occurs not only through the body surface, particularly that of the **mantle,** but in specialized respiratory organs such as gills or lungs, which are derivatives of the mantle. There is an **open circulatory system** with a pumping **heart,** blood vessels, and blood sinuses. Most cephalopods have a closed blood system with heart, vessels, and capillaries. The digestive tract is complex and highly specialized, according to the feeding habits of the various molluscs. Most molluscs have a pair of **kidneys** (nephridia), which connect with the coelom; the ducts of the kidneys in many forms serve also for the discharge of eggs and sperm. The **nervous system,** consisting of several pairs of ganglia with connecting nerve cords, is in general simpler than that of the annelids and arthropods. Neurosecretory cells have been identified in the nervous system that, at least in certain air-breathing snails, produce a growth hormone and function in osmoregulation. There are a number of types of highly specialized sense organs.

Most molluscs are dioecious, although some of the gastropods are hermaphroditic. Many aquatic molluscs pass through free-swimming trocophore and veliger larval stages. The **veliger** is the free-swimming larva of most marine snails, tusk shells, and bivalves. It develops from the trochophore and has the beginning of a foot, shell, and mantle (Fig. 13-3).

Class Monoplacophora

Until 1952 it was thought that the Monoplacophora (mon-o-pla-kof′o-ra) consisted only of Paleozoic shells. However, in that year living specimens of *Neopilina* were dredged up from the ocean bottom near the west coast of Costa Rica. These molluscs are small and have a low, rounded shell and a creeping foot (Fig. 13-4). They have a superficial resemblance to the limpets, but unlike most other molluscs, a number of organs are serially repeated. Such serial repetition occurs to a more limited extent in the chitons. *Neopilina*

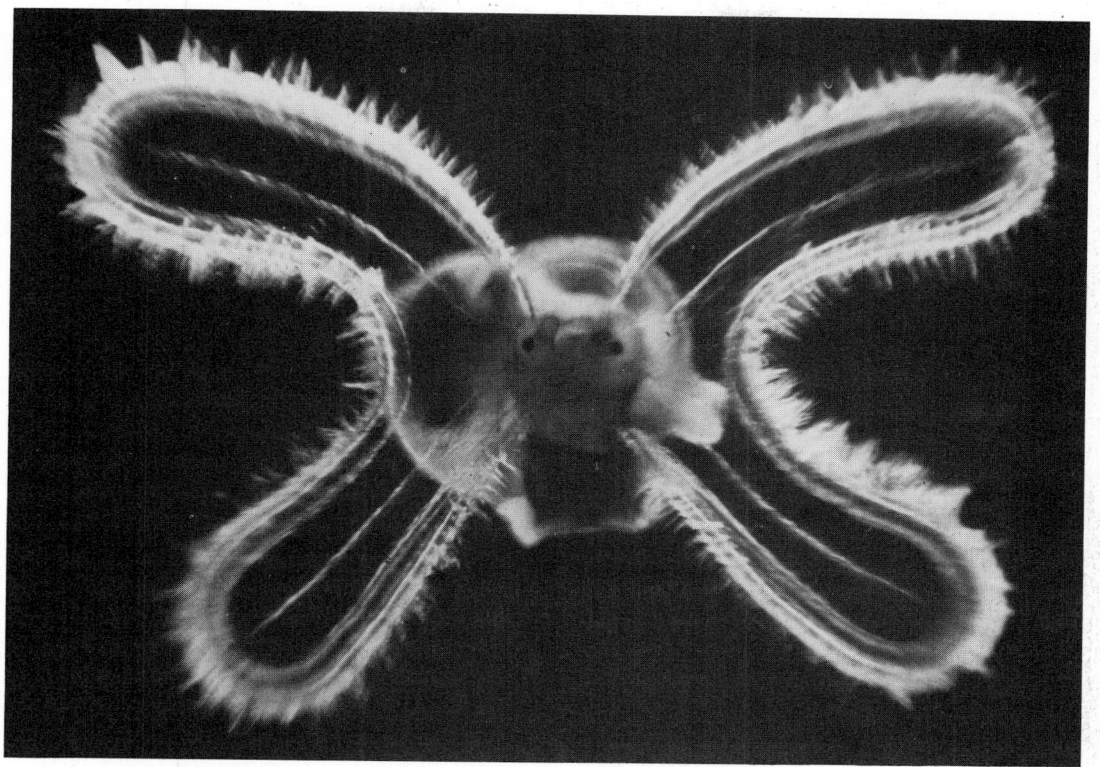

FIG. 13-3 Veliger larva of the thick-lipped dog whelk *Nassarius incrassatus*, swimming. (Photograph by D. P. Wilson.)

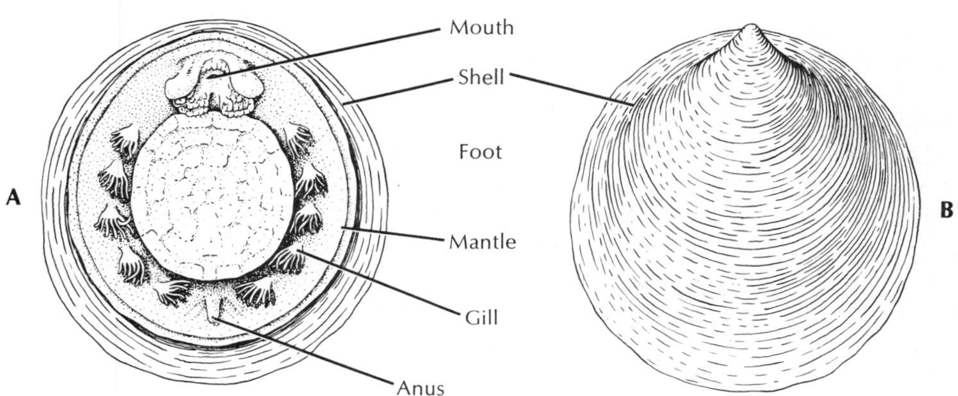

FIG. 13-4 *Neopilina*, class Monoplacophora. **A,** Ventral view; **B,** dorsal view.

has five pairs of gills, two pairs of auricles, six pairs of nephridia, one or two pairs of gonads, and a ladder-like nervous system with ten pairs of pedal nerves. The mouth bears the characteristic radula. Some authors (Lemche and Wingstrand; Hyman) have considered the monoplacophorans truly metameric and constituting evidence that the molluscs were descended from a metameric, annelid-like ancestor. Others (Clark; Russel-Hunter) believe that *Neopilina* shows only pseudometamerism, though the phylogenetic relationship of the annelids and molluscs, based on embryologic evidence, is unquestioned.

Class Polyplacophora—chitons

The chitons are somewhat flattened and have a convex dorsal surface that bears eight articulating limy **plates,** or **valves,** which give them their name (Fig. 13-5). The term Polyplacophora means "bearing many plates" in contrast to the Monoplacophora, which bear one shell (*mono,* single), and the Aplacophora, which bear none (*a,* without). The plates overlap posteriorly and are usually dull-colored to match the rocks to which the chitons cling.

Most chitons are small (2 to 5 cm); the largest, *Cryptochiton,* rarely exceeds 30 cm. They prefer rocky surfaces in intertidal regions, though some live at great depths. Chitons are stay-at-home organisms, straying only very short distances for feeding. In feeding, a sensory subradular organ protrudes from the mouth to explore for algae. When some are found, the radula is then projected to scrape them off. The chiton clings tenaciously to its rock with the broad flat foot. If detached, it can roll up like an armadillo for protection.

The mantle forms a girdle around the margin of the plates, and in some species mantle folds cover part or all of the plates. On each side of the broad ventral foot and lying between the foot and the mantle is a row of gills suspended from the roof of the mantle cavity. With the foot and the mantle margin adhering tightly to the substrate, these grooves become closed chambers, open only at the ends. Water enters the grooves anteriorly, flows across the gills, and leaves posteriorly, thus bringing to the gills a continuous supply of oxygen.

Blood pumped by the three-chambered heart reaches the gills by way of an aorta and sinuses. A pair of kidneys (metanephridia) carry waste from the pericardial cavity to the exterior. Two pairs of longitudinal nerve cords are connected in the buccal region. Sense organs include shell eyes on the surface of the shell (in some) and a pair of **osphradia** (sense organs for sampling water).

Sexes are separate in chitons. Sperm shed by males in the exhalant currents enter the gill grooves of the females by inhalant currents. Eggs are shed into the sea singly or in strings or masses of jelly.

Class Aplacophora

The class Aplacophora, or solenogasters, includes less than 100 species of strange aberrant molluscs. They are small wormlike creatures with a terminal mouth and anus (Fig. 13-27). Most are approximately 2.5 cm long, but some range up to 30 cm. There is no shell, but the skin is embedded with spicules. The integument encloses the body except for a ventral groove in the nonburrowing forms; burrowing forms lack the groove. The foot is vestigial. A pair of gills is located posteriorly in nonburrowing forms. Some have a radula. They live in fairly deep water where some are scavengers or carnivores and others feed on hydroids and corals. *Neomenia* and *Chaetoderma* are common species along the Atlantic coast.

Class Scaphopoda

The Scaphopoda (ska-fop′o-da), commonly called the tusk shells or tooth shells, are sedentary marine molluscs that have a slender body covered with a mantle and a tubular shell open at both ends. Here the molluscan body plan has taken a new direction, with the mantle wrapped around the viscera and fused to form a tube. Most scaphopods are 2.5 to 5 cm long, though they range from 4 mm to 25 cm long. *Dentalium* is a common Atlantic genus.

The foot, which protrudes through the larger end of the shell, is used to burrow into mud or sand, always leaving the small end of the shell exposed to the water above (Fig. 13-6, *A*). Respiratory water is circulated through the mantle cavity both by movements of the foot and by ciliary action (Fig. 13-6, *B*). Gaseous exchange occurs in the mantle. Food particles suspended in the water are caught in ciliary mantle ridges and are conveyed to the foot. Most of the food, however, is detritus and protozoans from the substrate. It is caught on the cilia of the foot or on the mucous-covered, ciliated knobs of long tentacles, called captacula, and conveyed to the nearby mouth. The radula carries the food to a crushing gizzard.

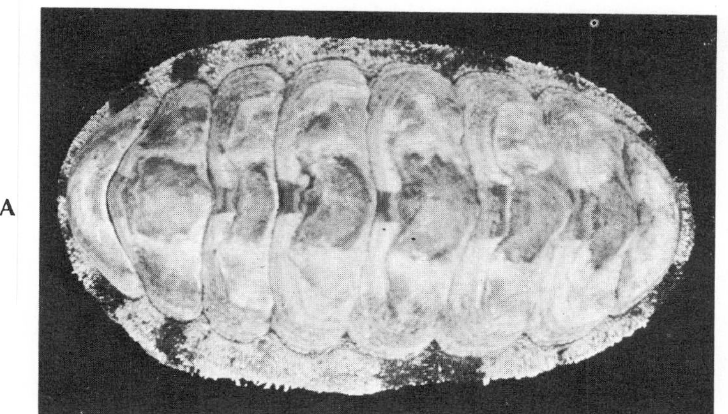

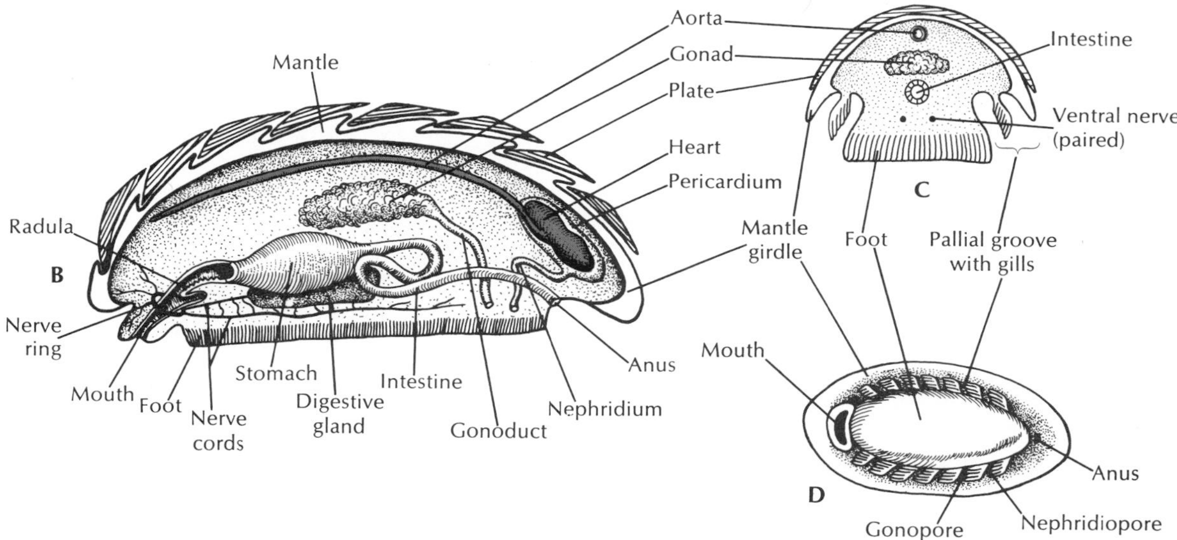

FIG. 13-5 Anatomy of a chiton (class Polyplacophora). **A,** Dorsal view. Note overlapping plates and surrounding marginal girdle. **B,** Longitudinal section. **C,** Transverse section. **D,** External ventral view. (**A,** Photograph by F. M. Hickman.)

The sexes are separate, and the larva is a trochophore.

Class Gastropoda

Among the molluscs the class Gastropoda is by far the largest and most successful, containing about 35,000 living and 15,000 fossil species. It is made up of members of such diversity that there is no single general term in our language that can apply to them as a group. They include snails, limpets, slugs, whelks, conchs, periwinkles, sea slugs, sea hares, sea butterflies, and others. They range from some of the most primitive of marine molluscs to the highly evolved terrestrial air-breathing snails and slugs. These animals are basically bilaterally symmetric, but because of **torsion,** a twisting process that occurs in the early larval stage, the visceral mass has become asymmetric.

The shell, when present, is always of one piece **(univalve)** and may be coiled or uncoiled. Starting at the apex, which contains the oldest and smallest whorl, the whorls become successively larger and spiral about the central axis, or **columella** (Fig. 13-7). The shell may

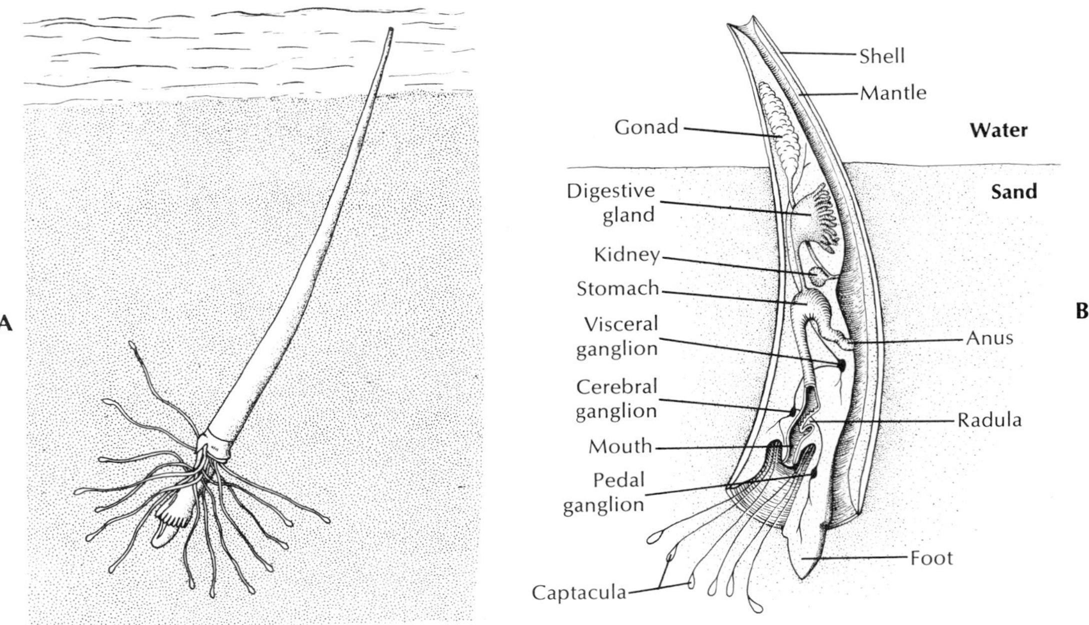

FIG. 13-6 Scaphopoda. **A,** *Dentalium* (class Scaphopoda), with its tubular shell, burrows into soft mud or sand and feeds by means of its prehensile tentacles. Water enters and leaves by way of the posterior aperture. **B,** Internal anatomy of *Dentalium*.

be right-handed **(dextral)** or left-handed **(sinistral),** depending on the direction of coiling. Dextral shells are far more common. The direction of coiling is genetically determined.

Gastropods range from microscopic forms to giant marine forms such as *Hemifusus proboscifera,* a snail with a shell up to 60 cm long, and the sea hare *Aplysia,* some species of which have grown up to a meter in length. Most of them, however, are between 1 and 8 cm in length. Some fossil gastropods were as much as 2 m long. Gastropod life-spans are not well known, but some gastropods live from 5 to 15 years.

Natural history. The range of gastropod habitats is large. In the sea, gastropods are common both in the littoral zones and at great depths, and some are even pelagic (free-swimming). Some are adpated to brackish water and some to fresh water. On land they are restricted by such factors as the mineral content of the soil and extremes of temperature, dryness, and acidity. Even so, they are widespread, and some have been found at great altitudes and some even in polar regions. Snails have all kinds of habitats—small pools or large bodies of water, in woodlands, in pastures, under rocks, in mosses, on cliffs, in trees, and underground. They have successfully undertaken every mode of life except aerial locomotion.

Gastropods are usually sluggish, sedentary animals because most of them have heavy shells and slow locomotor organs. Some are specialized for climbing, swimming, or burrowing. Shells are their chief defense, although they are also protected by coloration and by secretive habits. Some snails have an **operculum,** a horny plate that covers the shell aperture when the body is withdrawn into the shell. Some lack shells altogether. Some are distasteful to other animals, and a few such as *Strombus* can deal an active blow with the foot, which bears a sharp operculum. Nevertheless, they are eaten by birds, beetles, small mammals, fish, and other predators. Serving as intermediate hosts for many kinds of parasites, snails are often harmed by the larval stages of parasites.

Torsion. Of all the molluscs, only gastropods undergo torsion. Torsion is a peculiar phenomenon that moves the mantle cavity to the front of the body and

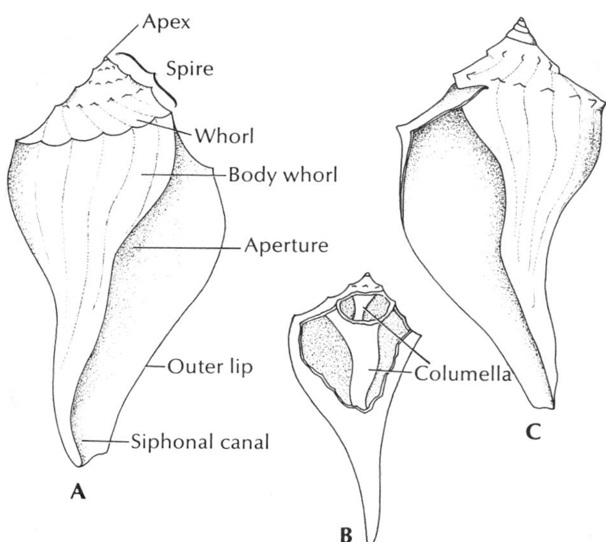

FIG. 13-7 The shell of the whelk *Busycon*. **A** and **B,** *Busycon carica,* a dextral, or right-handed, shell. **C,** *B. contrarium,* a sinistral, or left-handed, shell.

then twists the visceral organs and mantle cavity in a 180-degree rotation to produce the typical gastropod asymmetry.

Torsion occurs very early in the embryo (veliger stage), and in some species the entire action may take only a few moments. Before torsion occurs, the embryo is bilaterally symmetric with an anterior mouth and a posterior anus and mantle cavity (Fig. 13-8, *A*). Torsion is brought about by an uneven growth of the right and left muscles that attach the shell to the head-foot.

During the first stage of torsion there is a ventral flexure, bringing the anal region downward and then forward (Fig. 13-8, *B*), so that the anus opens anteriorly below (ventral to) the head and mouth. In the second phase the mantle cavity and associated viscera rotate 180 degrees, so that the structures that were ventral shift up the right side to a dorsal position and the dorsal structures shift down the left side to a ventral position (Fig. 13-8, *C*). This also shifts the originally left gill, kidney, and other organs to the right side and the originally right organs to the left side, bends the digestive tract, and twists the nerve cords into a figure eight. The mantle cavity with its anal opening now faces forward and lies above the head and mouth. Such a torsion allows the sensitive head end of the animal to be drawn

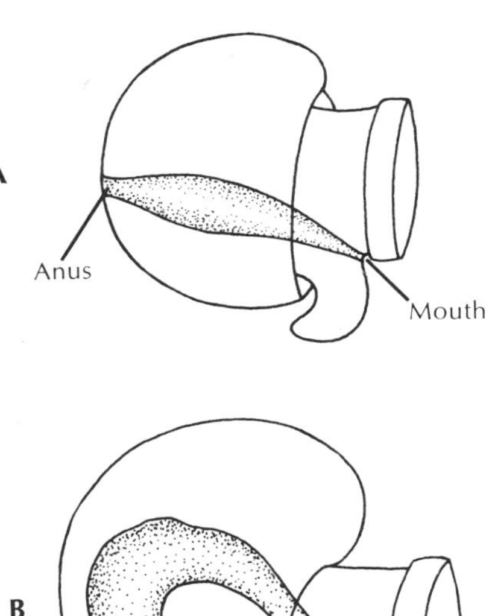

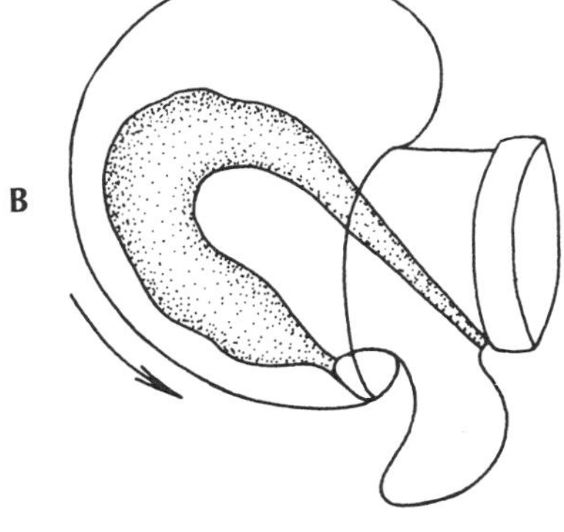

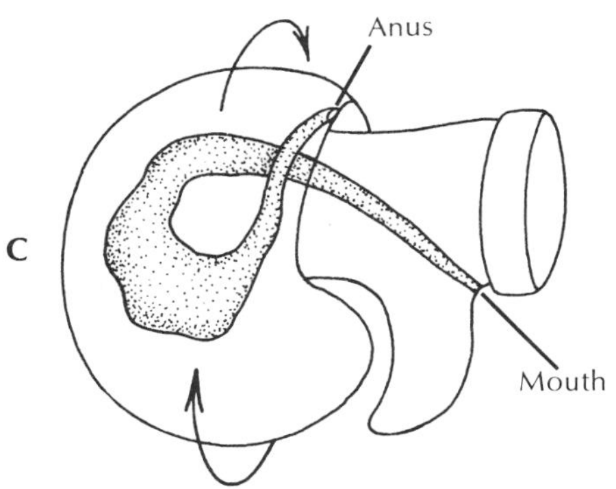

FIG. 13-8 Torsion in a gastropod larva. **A,** Anus and mantle cavity are posterior. **B,** Anus and mantle cavity move downward and forward to an anterior and ventral location. **C,** Rotation of visceral mass 180 degrees counterclockwise (when viewed from the dorsal side) brings the anus and mantle cavity above the head.

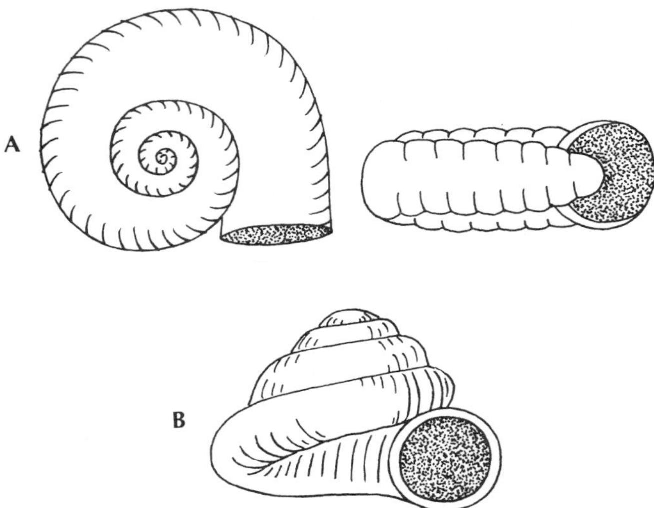

FIG. 13-9 Coiling of the shell, which is independent of torsion. **A,** Two views of planospiral coiling, or coiling that occurs in a single plane. **B,** Conispiral coiling, which produces a cone-shaped shell.

into the protection of the mantle cavity, with the tougher foot forming a barrier to the outside.

There are varying degrees of torsion in the different groups of gastropods. In some gastropods, such as the sea hare, *Aplysia,* a reverse action, or **detorsion,** follows torsion, partially restoring the bilateral symmetry.

The curious arrangement that results from torsion poses a serious sanitation problem by creating the possibility of wastes being washed back over the gills **(fouling)** and causes us to wonder what strong evolutionary pressures selected for such a strange realignment of body structures. We can only speculate. Perhaps crawling on a surface of loose sediment would have raised a trail of disturbed particles that could contaminate the respiratory water drawn into posterior gills and mantle cavity, but would not so affect gills in a mantle cavity facing forward. Perhaps torsion was originally beneficial to the larvae in their adaptation to a free-swimming and plankton-feeding existence, then proved to be of preadaptive value to the adults as they exploited bottom-dwelling habits. Other explanations have been proposed, but certainly the consequences of torsion and the resulting need to avoid fouling have been very important in the subsequent evolution of gastropods. These consequences cannot be explored, however, until another unique aspect of gastropods—coiling—has been described.

Coiling. The coiling, or spiral winding, of the shell and visceral hump is not the same as torsion. Coiling may occur in the larval stage at the same time as torsion, but the fossil record shows that coiling was a separate evolutionary event and originated in gastropods earlier than torsion did. Nevertheless, all living gastropods have descended from coiled torted ancestors, whether or not they now show these characteristics.

Early gastropods had a bilaterally symmetric **planospiral** shell, that is, all the whorls lay in a single plane (Fig. 13-9, *A*). Such a shell was not very compact since each whorl had to lie completely outside the preceding one. Curiously, a few modern species have secondarily returned to the planospiral form. The compactness problem of the planospiral shell was solved by the **conispiral** shape, in which each succeeding whorl was at the side of the preceding one (Fig. 13-9, *B*). However, this shape was clearly unbalanced, hanging as it was with much weight over to one side. Better weight distribution was achieved by shifting the shell upward and posteriorly, with the shell axis oblique to the longitudinal axis of the foot (p. 265). The weight and bulk of the main body whorl, the largest whorl of the shell, pressed on the right side of the mantle cavity, however, and apparently interfered with the organs on that side. Accordingly, the gill, auricle, and kidney of the right side have been lost in all except primitive living gastropods, leading to a condition of *bilateral asymmetry*.

Though the loss of the right gill was probably an adaptation to the mechanics of carrying the coiled shell, that condition made possible a way to avoid the problem of torsion—fouling—that is displayed in most modern prosobranchs. Water is brought into the left side of the mantle cavity and out the right side, carrying with it the wastes from the anus and nephridiopore, which lie near the right side. Ways in which fouling is avoided in other gastropods will be mentioned subsequently.

Feeding habits. Feeding habits of gastropods are as varied as their shapes and habitats, but all include the use of some adaptation of the radula. The majority of gastropods are herbivorous, rasping off particles of algae. Some herbivores are grazers, some are browsers, some are planktonic feeders. *Haliotis,* the abalone, holds seaweed with the foot and breaks off pieces with

FIG. 13-10 Purple sea-snail *Janthina janthina* floating at the surface with its raft of air bubbles. (Class Gastropoda) (Photograph by D. P. Wilson.)

the radula. Land snails forage at night for green vegetation.

Some snails, such as *Bullia* and *Buccinum* (p. 265) are scavengers, living on dead and decaying flesh; others are carnivores, which tear their prey with the radular teeth. *Melongena* feeds on clams, especially *Tagelus,* the razor clam, thrusting its proboscis between the gaping shell valves. *Fasciolaria* and *Busycon* (Fig. 13-13) feed on a variety of molluscs, preferably oysters. The snail slips the outer edge of its shell between the shell valves of its prey to hold them apart before inserting its proboscis. *Urosalpinx cinerea,* the oyster borer, or "tingle," drills holes through the shell of the oyster. Its radula, bearing three longitudinal rows of teeth, is used first to begin the drilling action, then the animal glides forward, everts an accessory boring organ through a pore in the anterior sole of its foot, and holds it against the shell, using a chemical agent to soften the shell. Short periods of rasping alternate with long periods of chemical activity until a neat round hole is completed. With its proboscis inserted through the hole, the snail may feed continuously for hours or days, using the radula to tear away the soft flesh. *Urosalpinx* is attracted to its prey at some distance by sensing some chemical, probably one released in the metabolic wastes of the prey.

Certain species of *Conus* feed on other gastropods. When *Conus* senses the presence of its prey, its proboscis fills with poisonous venom, and a single radular tooth is slid into position at the tip of the proboscis. When the proboscis strikes the prey, the tooth, followed by a cloud of poison, is shot at the prey, quieting it at once. *Conus* then puts its mouth over the orifice of the shell and devours the prey. Some species of *Conus* can deliver very painful stings, and in several species the sting is lethal to humans. The venom is apparently a neurotoxin.

The purple sea snail *Janthina* is a drifter that has no eyes and cannot swim, but builds a float of air bubbles to keep it at the surface (Fig. 13-10). It feeds on floating cnidarians with which it happens to come in contact. *Janthina* secretes a purple ink that may paralyze

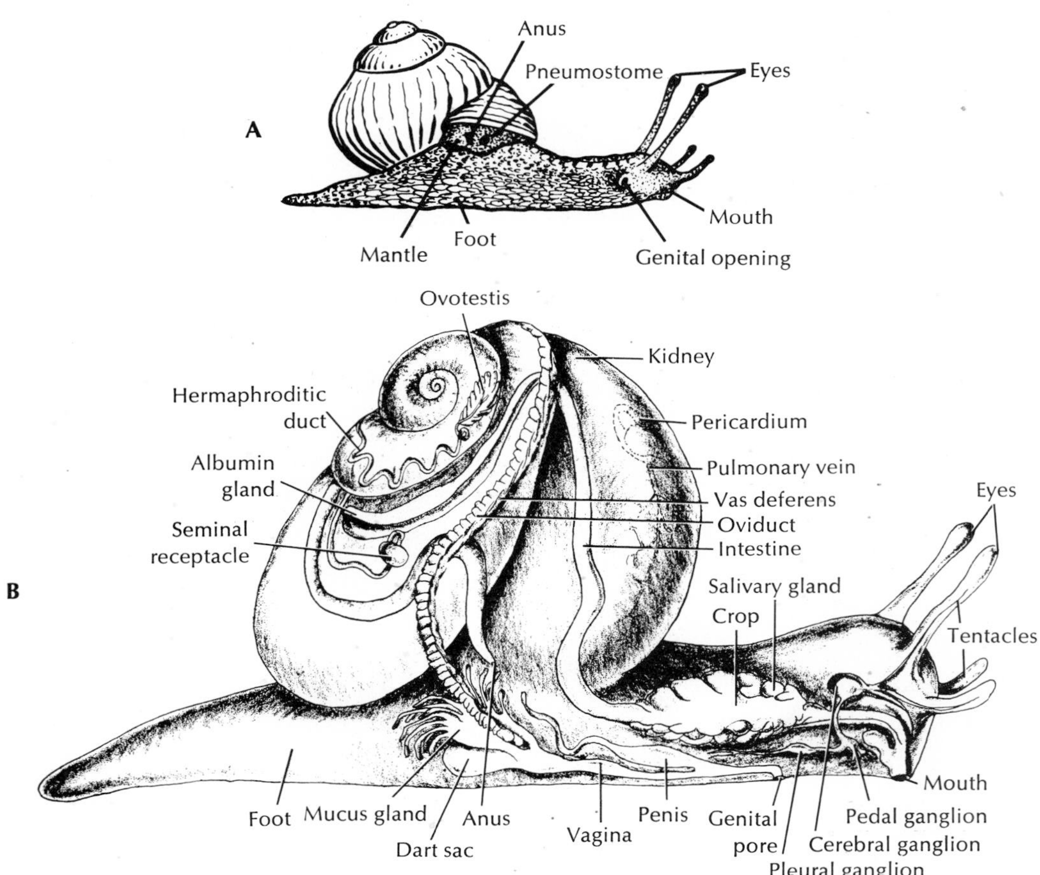

FIG. 13-11 **A,** *Helix,* a common land snail; **B,** internal structure.

its victim or that may be used as a defense mechanism.

Some gastropods feed on organic deposits on the sand or mud. Others collect the same sort of organic debris but can digest only the microorganisms contained in it. Some of the sessile gastropods, such as the limpets, are ciliary feeders that use the gill cilia to draw in particulate matter that is rolled into a mucous ball and is carried to the mouth. Some of the sea butterflies secrete a mucous net to catch small planktonic forms; then they draw the web into the mouth.

After maceration by the radula or by some grinding device, such as the gizzard in the sea hare *Aplysia* (Fig. 13-14, *A*), digestion is usually extracellular in the lumen of the stomach or digestive glands. In ciliary feeders the stomachs are sorting regions, and most of the digestion is intracellular in the digestive glands.

Internal form and function. Respiration in most gastropods is carried out by a gill (two gills in primitive prosobranchs) located in the mantle cavity, although some aquatic forms, lacking gills, depend on the mantle and skin. The pulmonates have a highly vascular area in the mantle that serves as a **lung** (Fig. 13-11). Many aquatic pulmonates must surface in order to expel a bubble of gas from the lung. To take in air they curl the edge of the mantle around the pneumostome (a pulmonary opening in the mantle cavity) to form a siphon.

Most gastropods have a single nephridium (kidney). The circulatory and nervous systems are well developed (Fig. 13-11). The latter includes three pairs of ganglia connected by nerves. Sense organs include eyes or simple photoreceptors, statocysts, tactile or-

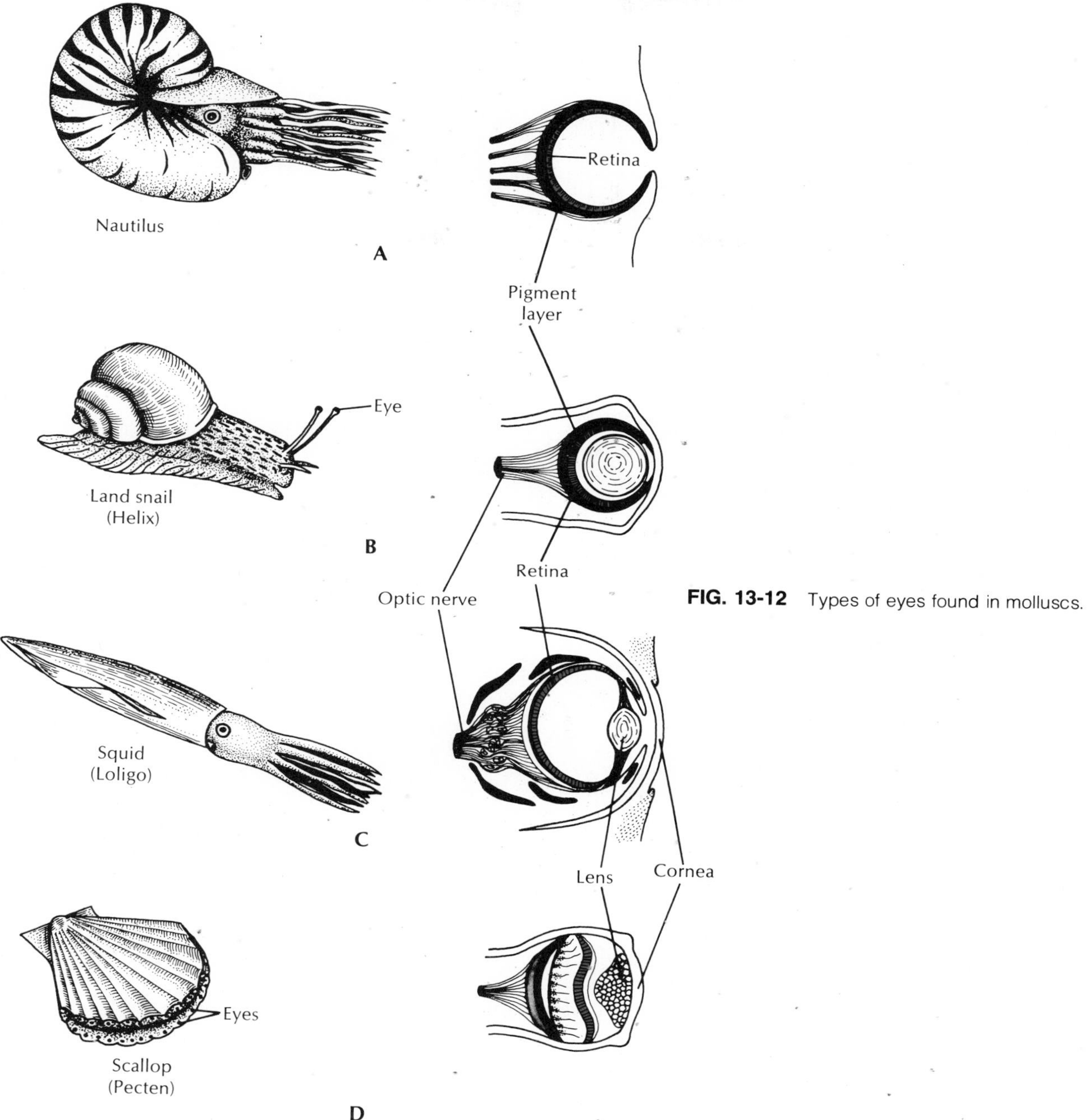

FIG. 13-12 Types of eyes found in molluscs.

gans, and chemoreceptors. The simplest type of gastropod eye is simply a cuplike indentation in the skin lined with pigmented photoreceptor cells. In many gastropods the eyecup contains a lens and is covered with a cornea (Fig. 13-12, *B*). A sensory area called the *osphradium,* located in the incurrent siphon of most gastropods, is known to be chemosensory in some forms, although its function may be mechanoreceptive in some and is still unknown in others.

There are both dioecious and monoecious gastropods. Many gastropods perform courtship ceremonies. During copulation in monoecious species there is an exchange of spermatophores (bundles of sperm). Many terrestrial pulmonates eject a dart from a dart sac (Fig.

FIG. 13-13 The whelk *Busycon.* **A,** Ventral view showing the operculum closed to protect the soft body. **B,** Egg laying by *Busycon.* Eggs are laid in a string of double-edged, parchmentlike discs, one end of which will be fastened to the substrate when completed. The string of discs may be a meter long and have as many as 100 capsules in which the eggs develop into minute snails. (Class Gastropoda) (Photographs by C. P. Hickman, Jr.)

13-11) into the partner's body to heighten excitement prior to copulation. After copulation each partner deposits its eggs in shallow burrows in the ground. In the most primitive gastropods ova and sperm are discharged into the sea water where fertilization occurs, and the embryo soon hatches as a free-swimming trochophore larva. In most gastropods fertilization is internal.

Fertilized eggs encased in transparent shells may be emitted singly to float among the plankton or may be laid in gelatinous layers attached to the substratum. In some marine forms, eggs are enclosed, either in small groups or in large numbers, in tough egg capsules, in a wide variety of parchment egg cases (Fig. 13-13, *B*), or in "collars." The young generally emerge as veliger larvae (Fig. 13-3), or they may spend the veliger stage in the case or capsule and emerge as young snails. The cases or capsules may be attached to the substratum or deposited on shells of other members of the species; some are kept covered by the parental shell. Some species, including many freshwater snails, are ovoviviparous, brooding their eggs and young in the pallial oviduct.

Major groups of gastropods

There are three subclasses of gastropods, the Prosobranchia, Opisthobranchia, and Pulmonata.

Prosobranchia. The mantle cavity is anterior as a result of torsion, with the gill or gills lying in front of the heart. Water enters the left side and exits from the right side, and the right edge of the mantle is often extended into a long siphon to further separate exhalant from inhalant flow. In the primitive prosobranchs with two gills (for example, the abalone *Haliotis*), fouling is avoided by having the exhalant water go up and out through one or more holes in the shell above the mantle cavity.

Prosobranchs have one pair of tentacles. The sexes are usually separate. An operculum is often present.

This group contains most of the marine snails and some of the freshwater and terrestrial gastropods. They range in size from the periwinkles and small limpets (*Patella* and *Fissurella*) to the giant conch *(Strombus),* the largest univalve in America. Familiar examples of prosobranchs are the abalone *(Haliotis),* which has an ear-shaped shell; the whelk *(Busycon)* (Fig. 13-13), which lays its eggs in double-edged, disc-shaped capsules attached to a cord a meter long (Fig. 13-13, *B*); the common periwinkle *(Littorina);* the slipper or boot shell *(Crepidula);* the oyster borer *(Urosalpinx),* which bores into oysters and sucks out their juices; the rock shell *(Murex),* of which a European species was used for making the royal purple of the ancient Romans; and the freshwater forms (*Goniobasis* and *Viviparus*).

Opisthobranchia. The opisthobranchs show partial or complete detorsion; thus, the anus and gill (if present) are displaced to the right side or rear of the body. Clearly, the fouling problem is obviated if the anus is moved to the posterior. Two pairs of tentacles are usually found, and the shell is typically reduced or absent. All are monoecious. The opisthobranchs are an odd assemblage of molluscs that include sea slugs, sea hares, sea butterflies, canoe shells, and others. They are all marine; most of them are shallow-water forms, hiding under stones and seaweed; a few are pelagic. Currently 8 to 12 or more orders of opisthobranchs are recognized, but for convenience they can be divided into two classic groups: tectibranchs, with gill and shell usually present, and nudibranchs, in which there is no shell or true gill, but in which secondary gills are present along the dorsal side or around the anus.

Among the tectibranchs is the large sea hare *Aplysia* (Fig. 13-14, *A*), which has large earlike anterior tentacles and a vestigial shell; and the pteropods, or sea butterflies (*Carolina* and *Clione*). In pteropods the foot is modified into fins for swimming; thus they are pelagic and form a part of the plankton fauna.

The nudibranchs are represented by the sea slugs, which are often brightly colored and carnivorous. The plumed sea slug *Aeolis,* which lives on sea anemones and hydroids, often draws the color of its prey into the elongated papillae (cerata), which cover its back. It also salvages the nematocysts of the hydroids for its own use. The frilled sea slug *Tridachia* (Fig. 13-14, *C*) is a lovely little green and white nudibranch common in Florida waters. *Dendrodoris* is one of the common West coast nudibranchs (Fig. 13-14, *B*).

Pulmonata. The pulmonates show torsion and include the land and freshwater snails and slugs (and a few brackish and saltwater forms). They lack gills, but the vascularized mantle wall has become a **lung,** which fills with air by contraction of the mantle floor. The anus and nephridiopore open near the opening of the lung to the outside **(pneumostome),** and waste is expelled forcibly with air or water from the lung. They have single, monoecious gonads. The aquatic species have one pair of nonretractile tentacles, at the base of which are the eyes; land forms have two pairs of tentacles, with the posterior pair bearing the eyes (Fig. 13-12, *B*). Among the thousands of land species some of the most familiar American forms are *Helix, Polygyra, Succinea, Anguispira, Zonitoides, Limax,* and *Agriolimax*. Aquatic forms are represented by *Helisoma, Lymnaea,* and *Physa*. *Physa* is a left-handed (sinistral) snail.

Class Bivalvia (Pelecypoda)

The Bivalvia (bi-val've-a) are also known as Pelecypoda (pel-e-sip'o-da), or "hatchet-footed" ani-

A

B

C

FIG. 13-14 Opisthobranch gastropods. **A,** *Aplysia,* the sea hare, photographed against a glass surface. One small black eye is visible just beneath the second pair of tentacles. This species is much used in neurologic studies because of unusually large ganglia. **B,** *Dendrodoris,* a Pacific coast nudibranch, is a rich yellow and has a dorsal cluster of gills. **C,** The frilled sea slug *Tridachia* is a dainty little green and white nudibranch familiar in Florida waters. (Photographs: **A,** by B. Tallmark, Upsalla University; **B,** by F. M. Hickman; **C,** by R. C. Hermes.)

mals, as their name implies (Gr., *pelekus,* hatchet, + *pous, podus,* foot). They are the bivalved molluscs that include the mussels, clams, scallops, oysters, and shipworms and that range in size from tiny seed shells 1 to 2 mm in length to giant South Pacific clams *Tridacna,* which may reach more than 1 m in length and as much as 225 kg (500 pounds) in weight. Most bivalves are sedentary **filter feeders** that depend on ciliary currents produced by the gills to bring in food materials. Unlike the gastropods, they have no head, no radula, and very little cephalization.

Most pelecypods are marine, but many live in brackish water and in streams, ponds, and lakes.

Shell. Bivalves are laterally compressed, and their

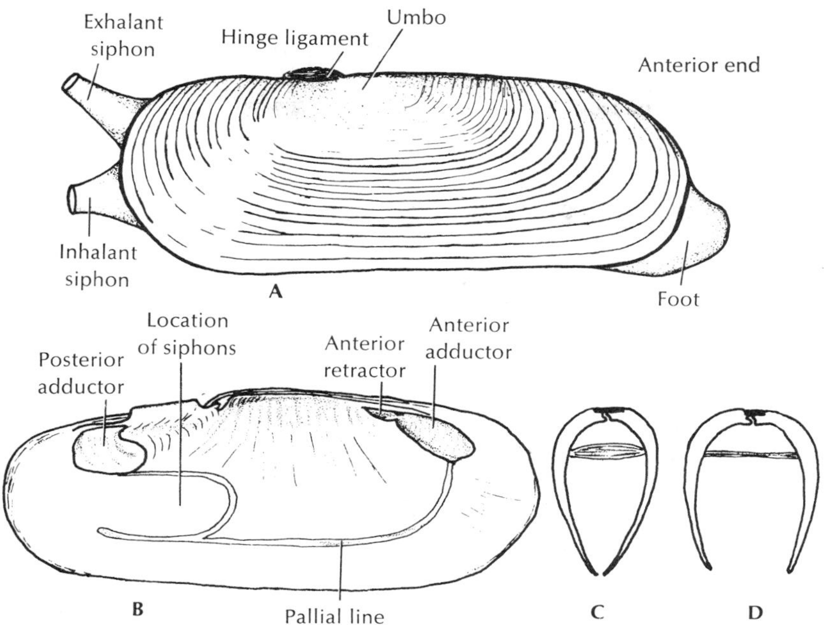

FIG. 13-15 *Tagelus gibbus,* the stubby razor clam (class Bivalvia). **A,** External view of right valve. **B,** Inside of left shell showing scars where muscles were attached. The mantle was attached at the pallial line. **C** and **D,** Sections showing function of adductor muscles and hinge ligament. In **C** adductor muscle is contracted, pulling valves together. In **D** adductor muscle is relaxed, allowing the hinge ligament to pull the valves apart.

two shells **(valves)** are held together dorsally by a hinge ligament that causes the valves to gape ventrally. The valves are drawn together by adductor muscles that work in opposition to the hinge ligament (Fig. 13-15, *B* and *C*). The valves function largely for protection, but those of the shipworms *(Teredo)* have microscopic teeth for rasping wood, and the rock borers *(Pholas)* use spiny valves for boring into rock. A few bivalves such as scallops use their shells for locomotion by clapping the valves together so that they move in spurts.

The umbo is the oldest part of the shell, and growth occurs in concentric lines around it. (Fig. 13-15, *A*).

Pearl production is the by-product of a protective device used by the animal when a foreign object (grain of sand, parasite, or other) becomes lodged between the shell and mantle. The mantle secretes many layers of nacre around the irritating object (Fig. 13-2). Pearls are cultured by inserting particles of nacre, usually taken from the shells of freshwater clams, between the shell and mantle of a certain species of oyster and by keeping the oysters in enclosures for several years. *Meleagrina* is an oyster used extensively by the Japanese for pearl culture.

The body. The body is made up of the **visceral mass** suspended from the dorsal midline, a muscular **foot** attached to the visceral mass anteroventrally, and a pair of **gills** (ctenidia) on each side, each covered by a fold of the **mantle.** Some bivalves have a single gill rather than a double gill on each side.

Mantle. The mantle hangs down on each side of the visceral mass, adhering to the valves. The posterior edges of the mantle folds are modified to form dorsal **excurrent (exhalant)** and ventral **incurrent (inhalant) apertures** for regulating water flow (Fig. 13-16, *A*). In some marine bivalves the mantle is drawn out into long muscular siphons that allow the clam to burrow into the mud or sand and extend the siphons to the water above (Fig. 13-16, *B* to *D*). Cilia on the gills and inner surface of the mantle direct the flow of water over the gills.

Locomotion. Pelecypods move by extending a slender muscular foot between the valves (Fig. 13-16, *D*). Blood swells the end of the foot to anchor it in mud or

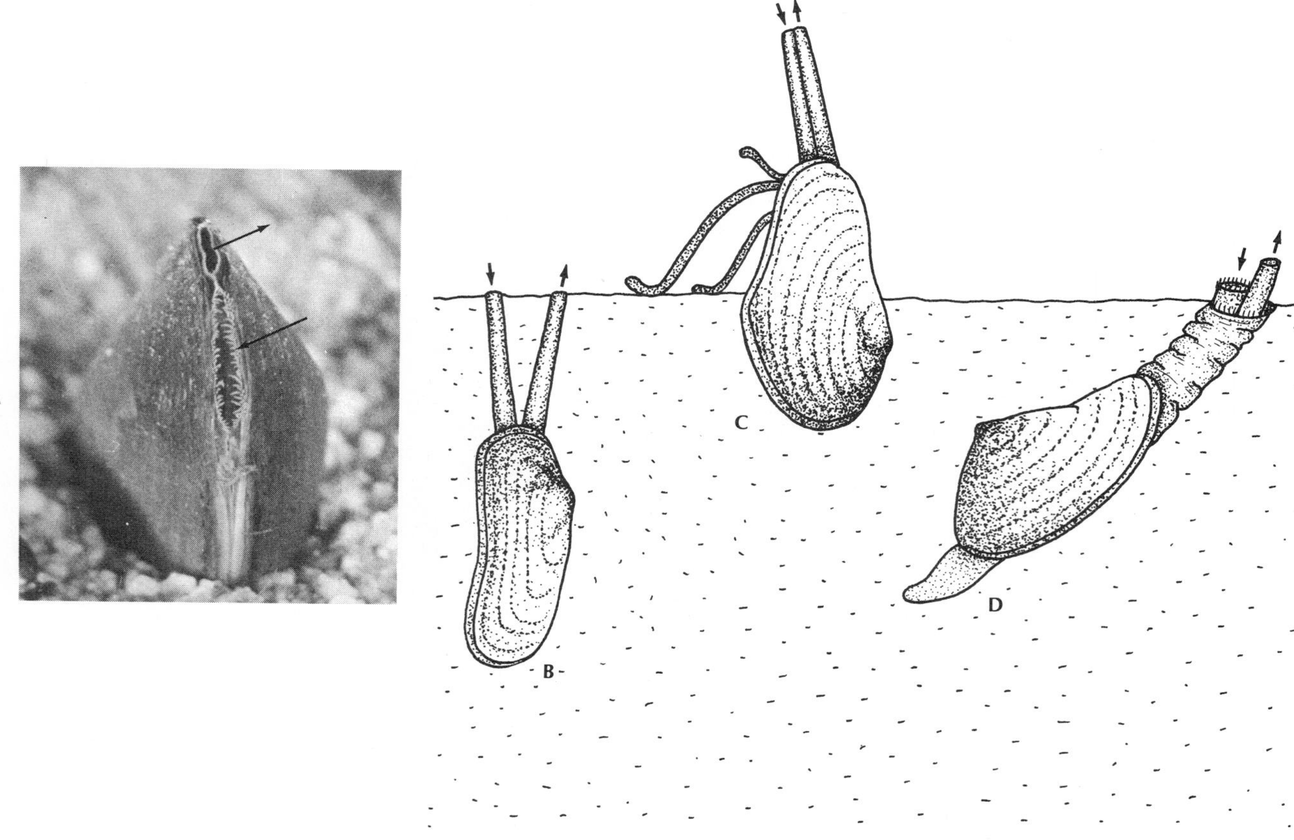

FIG. 13-16 Adaptations of bivalves for inhalant and exhalant water currents. **A,** In the freshwater clam the mantle edges form apertures. **B** to **D,** In many marine forms the mantle is drawn out into long siphons. **B,** *Tagelus,* the stubby razor clam; **C,** *Yoldia;* **D,** *Mya.* In **A, B,** and **D,** the inhalant current brings in both food and oxygen. **C,** In *Yoldia* the siphons are largely respiratory; long ciliated palps feel about over the mud surface and convey food to the mouth. (**A** from *Adaptive Radiation—the Mollusks,* an Encyclopaedia Britannica film.)

sand, and then longitudinal muscles contract to shorten the foot and pull the animal forward. In most bivalves the foot is used for burrowing, such as in *Venus, Mya,* and the razor clams.

Probably the most primitive form of locomotion is creeping, represented in a few forms (*Solemya* and *Lepton*). Some, such as *Yoldia* (Fig. 13-16, *C*), can leap over surfaces; *Kellia* is able to climb up a surface; and the scallops and file shells are able to swim jerkily by clapping their valves together to create a sort of jet propulsion (Fig. 13-17).

Gills. Gaseous exchange is carried on by both the mantle and the gills. In most bivalves each gill is formed of two walls (lamellae) joined together at their ventral margins. Each lamella is made up of many vertical gill filaments strengthened by chitinous rods. Water enters the gills through innumerable small pores in the walls and is propelled by ciliary action. Partitions between the walls divide the gill internally into many vertical **water tubes** that carry water upward (dorsally) into a common **suprabranchial chamber** (Fig. 13-18) and through it to the excurrent aperture. Blood vessels or spaces in the interlamellar partitions are used for exchange of gases.

The water tubes in the female often double as brood pouches for eggs and larvae during the breeding season. The various orders of Bivalvia are classified on the basis of their gill type and hinge structure.

FIG. 13-17 **A,** The file shell *Lima hians*. **B,** Scallops react vigorously to the presence of their enemy, the sea star, by clapping their valves to swim away. The dark spots along their mantle edges are bright blue eyes. (Class Bivalvia) (Photographs by D. P. Wilson.)

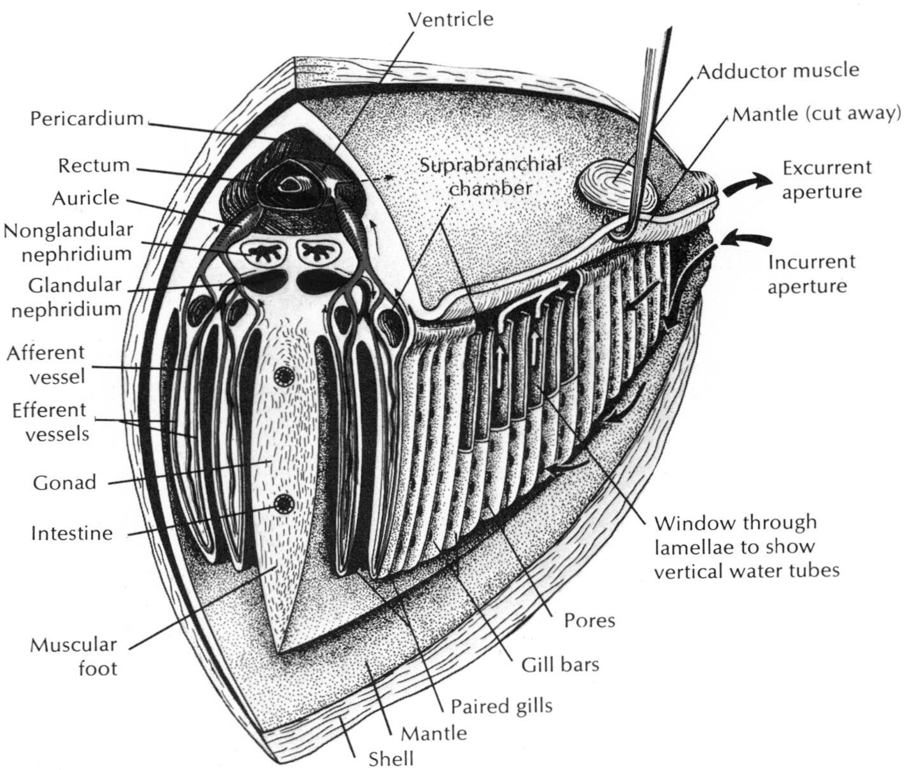

FIG. 13-18 Section through heart region of a freshwater clam to show relation of circulatory and respiratory systems. Respiratory water currents: water is drawn in by cilia, enters gill pores, and then passes up water tubes to suprabranchial chambers and out excurrent aperture. Blood in gills exchanges carbon dioxide for oxygen. Blood circulation: ventricle pumps blood forward to sinuses of foot and viscera, and posteriorly to mantle sinuses. Blood returns from mantle to auricles; it returns from viscera to the kidney, and then goes to the gills, and finally to the auricles.

Feeding and digestion. Most bivalves are filter feeders. The respiratory currents bring both oxygen and organic materials to the gills where ciliary tracts direct them to the tiny pores of the gills. Gland cells on the gills and labial palps secrete copious amounts of mucus, which entangles particles too large to enter the gill pores. These mucous masses slide down the outside of the gills toward food grooves at the lower edge of the gills (Fig. 13-19). Heavier particles of sediment drop off the gills as a result of gravitational pull, but smaller particles travel along the food grooves toward the labial palps. The palps, being also grooved and ciliated, direct the mucous mass into the mouth. The palps have long been thought to serve as sorting areas for particles, but more recent research (Bernard, 1974) indicates that by the time the mucus reaches the palps, the separation and rejection of particles has already occurred through gravitational settling. This is aided by the contrast between the size of the incurrent and excurrent apertures and that of the gill pores. For example, in the oyster an inhalant aperture measuring 0.51 cm^2 is one-third as large as the exhalant aperture, but only 1% as large as the *total combined area* of the gill pores. Thus water that enters the mantle cavity moving rather fast slows down considerably as it enters the gill passages. Heavier particles then tend to fall and only the small ones enter the gill pores to be carried away in the exhalant current or else become enmeshed in the mucus to be carried to the palps and mouth. The palps, when undisturbed, are usually pressed together and serve to turn away excess amounts of empty mucus and to direct the more concentrated food-mucous masses to the mouth.

Some bivalves, such as *Nucula* and *Yoldia,* are de-

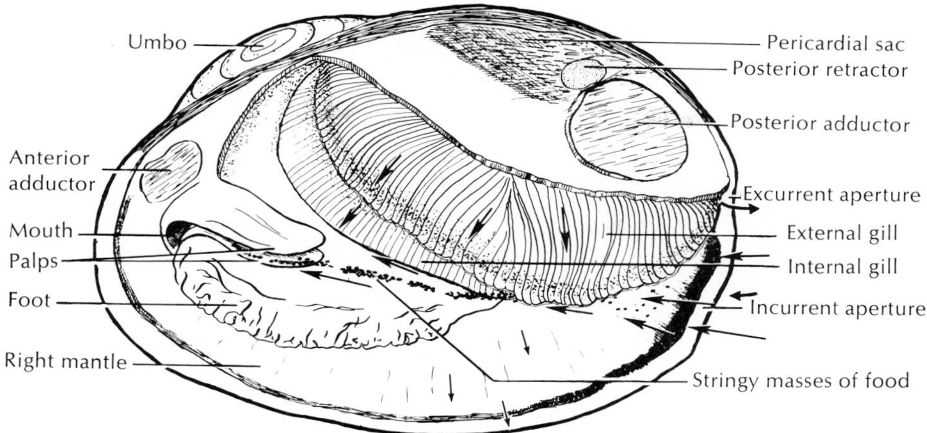

FIG. 13-19 Feeding mechanism of freshwater clam. Left valve and mantle are removed. Water enters the mantle cavity posteriorly and is drawn forward by ciliary action to the gills and palps. As water enters the tiny openings of the gills, food particles are sieved out and caught up in strings of mucus that are carried by cilia to the palps and directed to the mouth. Sand and debris drop into the mantle cavity and are removed by cilia.

posit feeders and have long proboscides attached to the labial palps (Fig. 13-16, *C*). These can be protruded onto the sand or mud to collect food particles, in addition to the particles attracted by the gill currents.

Another group of bivalves (septibranchs) draw small crustaceans or bits of organic debris into the mantle cavity by sudden inflow of water created by the pumping action of a muscular septum in the mantle cavity.

The floor of the stomach of filter-feeding bivalves is folded into ciliary tracts for sorting the continuous stream of particles. A cylindric style sac opening into the stomach secretes a gelatinous rod called the **crystalline style,** which projects into the stomach and is kept whirling by means of cilia in the style sac (Fig. 13-20). The crystalline style is composed of mucoproteins and some enzymes (amylase, for example). Surface layers of the rotating style dissolve in the stomach fluids, freeing enzymes for extracellular digestion. The end of the rotating style becomes attached to the mucous food mass, causing it to rotate too. As the mass spins, food particles detach from it and land on the ridged and ciliated sorting surface of the stomach floor. Here large or unsuitable particles are directed to the intestine for elimination. Smaller or partially digested nutritive particles are taken into the digestive gland or may be picked up by amebocytes. In either case digestion is completed intracellularly. Molluscs as a group have a great variety of digestive enzymes, although there are some differences in the distribution of the enzymes among the various classes.

The stomach of the carnivorous septibranch is muscular and serves as a crushing gizzard. Digestible material, after being broken up, passes to the digestive diverticula and is phagocytized and digested intracellularly.

Circulation. The three-chambered heart, which lies in the pericardial cavity (Fig. 13-18), is made up of two auricles and a ventricle and beats at the rate of about six times a minute. Blood is pumped through an anterior aorta to the foot and viscera and through a posterior aorta to the rectum and mantle. Part of the blood is oxygenated in the mantle and is returned to the ventricle through the auricles; the other part circulates through sinuses and passes in a vein to the kidneys, from there to the gills for oxygenation, and back to the auricles. Carbon dioxide and other wastes are carried to the gills and kidneys for elimination. The blood is colorless and contains nucleated ameboid corpuscles.

Excretion. A pair of U-shaped kidneys (nephridial tubules) lie just below the heart (Fig. 13-18). The glandular portion of each tubule opens into the pericardium; the bladder portion empties into the suprabranchial chamber. The kidneys remove waste from the blood that circulates through its network and from the pericardial cavity.

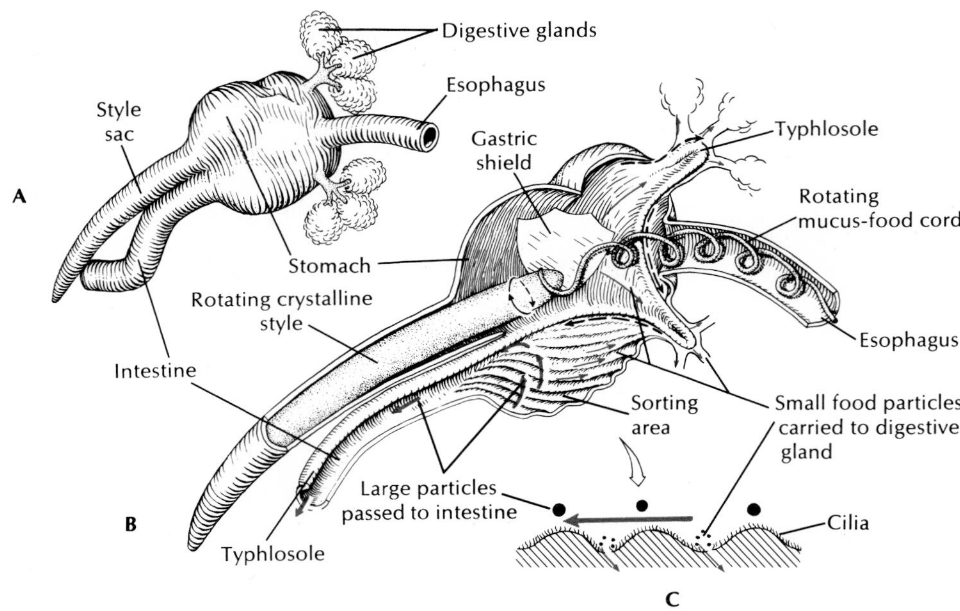

FIG. 13-20 Stomach and crystalline style of ciliary-feeding clam. **A,** External view of stomach and style sac. **B,** Transverse section showing direction of food movements. Food particles in incoming water are caught in a cord of mucus that is kept rotating by the crystalline style. Ridged sorting areas direct large particles to the intestine, and small food particles to digestive glands. **C,** Sorting action of cilia. (**B,** Modified from J. E. Morton, 1967. Mollusca, ed. 4, London, Hutchinson & Co.)

Nervous and sensory systems. The nervous system consists of three pairs of widely separated ganglia connected by commissures and a system of nerves. The ganglia are the **cerebropleural ganglia** near the mouth, the **pedal ganglia** in the foot, and the **visceral ganglia** below the posterior adductor muscle. Neurosecretory cells comparable to those of arthropods and vertebrates have been found in some molluscs.

Sense organs are poorly developed. There is a pair of statocysts in the foot and a pair of osphradia of uncertain function in the exhalant area of the mantle cavity. Tactile cells are abundant on exposed surfaces and are concentrated on the mantle border near the inhalant and exhalant openings. Ocelli are common in larval forms but are often lost in metamorphosis. Some forms have simple pigment cells on the mantle. Scallops have a row of small blue eyes along each mantle edge (Fig. 13-17, *B*). Each is made up of cornea, lens, retina, and pigmented layer (Fig. 13-12, *D,* and p. 726).

Reproduction and development. Sexes are usually separate. Gonads are located in the visceral mass above the foot (Fig. 13-18), with the sperm ducts or oviducts discharging their gametes into the suprabranchial chamber to be carried out with the exhalant current. An oyster may produce 50 million eggs in a single season. In most bivalves fertilization is external. The embryo develops into trochophore, veliger, and spat stages (Fig. 13-21). Oyster spats swim about for about 2 weeks before settling down for attachment. It takes about 4 years for oysters to grow to commercial size.

In most freshwater clams, fertilization is internal. Eggs drop into the water tubes of the gills where they are fertilized by sperm entering with the inhalant current. They develop there into a bivalved **glochidium larva** stage, which is a specialized veliger. When discharged, the glochidia are carried by water currents, and if they come in contact with a passing fish, they attach to its gills or skin and live as parasites for several weeks. Then they sink to the bottom to begin independent lives. Larval ''hitchhiking'' helps distribute a form whose locomotion is very limited.

Boring bivalves. Many pelecypods are able to burrow into mud or sand, but some have evolved a mechanism for burrowing into much harder substances, such as wood or stone.

Teredo and *Bankia* are well-known wood-boring

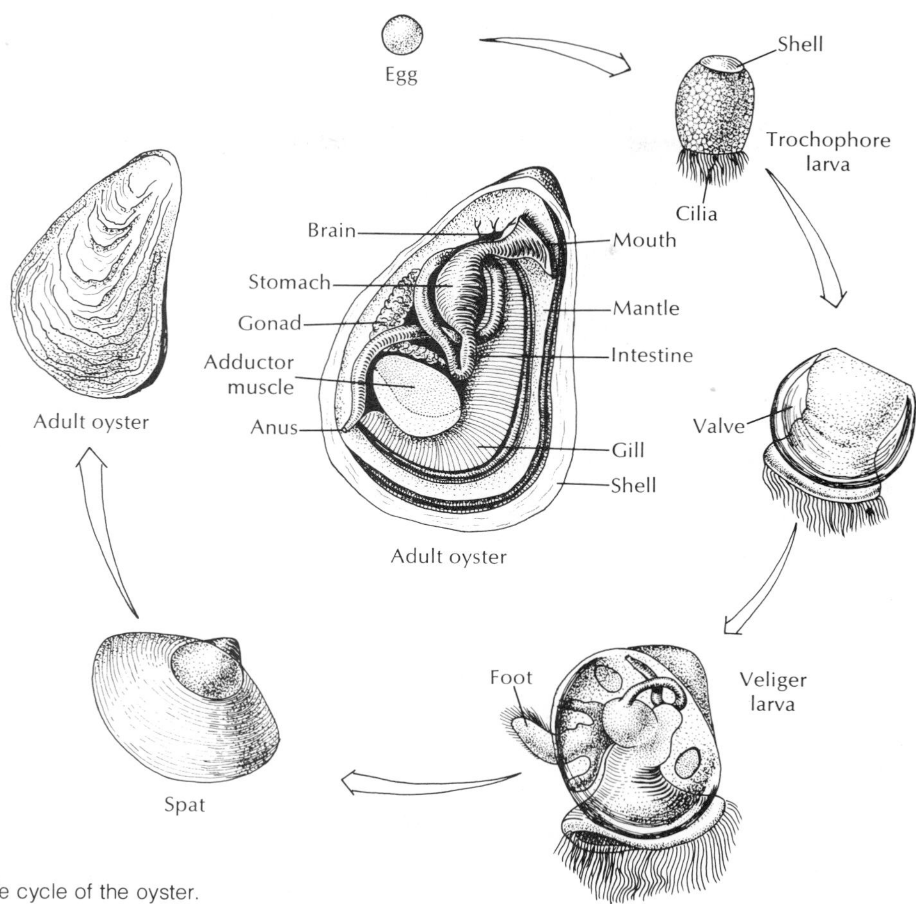

FIG. 13-21 Life cycle of the oyster.

molluscs that can be very destructive to wooden ships and wharves. These strange little clams have a long, wormlike appearance, with a pair of slender siphons on the posterior end that keep water flowing over the gills, and a pair of small globular valves on the anterior end with which they burrow (Fig. 13-22). The valves have microscopic teeth so that they function as very effective wood rasps. The animals extend their burrows with an unceasing rasping motion of the valves. This sends a continuous flow of fine wood particles into the digestive tract where they are attacked intracellularly by the enzyme cellulase. In addition to wood as nourishment they also collect incoming plankton particles with their gills and palps.

Fertilization is internal, with the zygotes becoming implanted in the gill tissue. The young *Teredo* are liberated as globular ciliated larvae 0.5 μm in diameter that swim about looking for a likely spot to penetrate wood and metamorphose. Clinging to the new burrow with its foot and boring forward as it grows, the young *Teredo* may grow to a length of 100 to 125 mm and a diameter of 5 mm, always extending its siphons backward toward the opening to pump in the oxygen- and food-bearing water. One species of *Bankia* makes burrows up to a meter long and 12 mm in diameter.

Some clams bore into rock. The piddock *(Pholas)* bores into limestone, shale, sandstone, and sometimes wood or peat. It has strong valves that bear spines by which it gradually cuts away the rock while anchoring itself by its foot. *Pholas* may grow to 15 cm long and make rock burrows up to 30 cm long.

Class Cephalopoda

The Cephalopoda (sef-a-lop′o-da) are the most advanced of the molluscs—in fact, in some respects they

FIG. 13-22 Shipworms. Wood from a jetty split open to show living shipworms *Teredo navalis* boring in it. (Class Bivalvia) (Photograph by D. P. Wilson.)

are the most advanced of all the invertebrates. They include the squid, octopuses, nautiluses, devilfish, and cuttlefish. All are marine, and all are active predators.

Cephalopods are the "head-footed" molluscs (Gr., *kephalē,* head, + *pous, podos,* foot) in which the modified foot is concentrated in the head region. The edges of the foot are drawn out into arms and tentacles that bear sucking discs for seizing prey; also, part of the foot is modified to form a funnel for expelling water from the mantle cavity.

Cephalopods range upward in size from 2 to 3 cm. The squid *Loligo* is about 30 cm long. The giant squid *Architeuthis* is the largest invertebrate known.

Fossil records of cephalopods go back to Cambrian times. The earliest shells were straight cones; others were curved or coiled, culminating in the coiled shell similar to that of the modern *Nautilus*—the only remaining member of the once flourishing nautiloids (Fig. 13-23). Cephalopods without shells or with internal shells (such as octopuses and squid) are believed to have evolved from some early straight-shelled nautiloid.

The natural history of the cephalopods is known only in part. They are saltwater animals and appear sensitive to the degree of salinity. Few are found in the Baltic Sea, where the water has a low salt content. Cephalopods are found at various depths. The octopus is often seen in the intertidal zone, lurking among rocks and crevices, but occasionally is found at great depths. The more active squid are rarely found in shal-

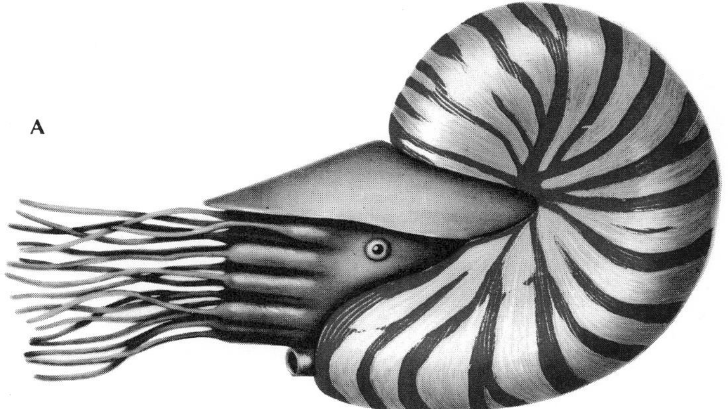

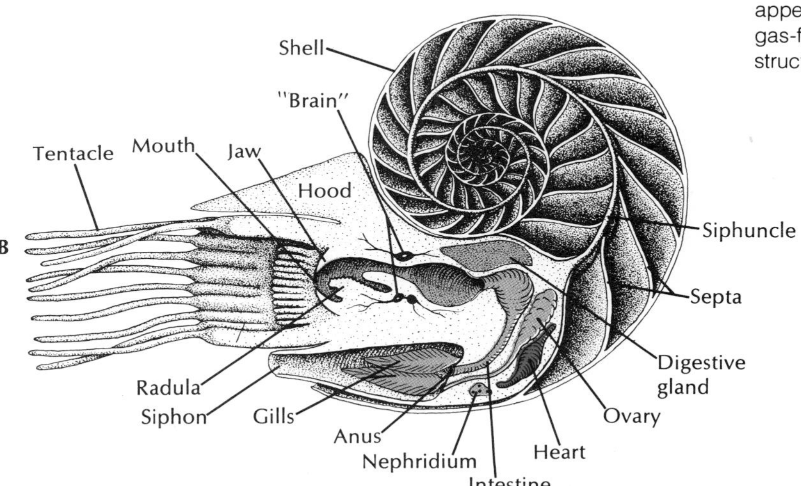

FIG. 13-23 *Nautilus,* a cephalopod. **A,** External appearance. **B,** Longitudinal section, showing gas-filled chambers of shell, and diagram of body structure.

low water, and some have been taken at depths of 5,000 m. *Nautilus* is usually taken from the ocean floor near islands (southwestern Pacific) where the water is several hundred meters deep.

Major groups of cephalopods

There are three subclasses of cephalopods: the Nautiloidea, which have two pairs of gills; the entirely extinct Ammonoidea; and the Coleoidea, which have one pair of gills. The Nautiloidea populated the Paleozoic and Mesozoic seas, but there survives only one genus, *Nautilus* (Fig. 13-23), of which there are three living species. The Ammonoidea were widely prevalent in the Mesozoic era but became extinct by the end of the Cretaceous period.

The subclass Coleoidea includes all living cephalopods except *Nautilus*. There are two main orders: Decapoda (squid), with ten arms and a relatively large coelom, and Octopoda (octopuses), with eight arms and reduced coelom.

Two of the decapod's arms are modified into tentacular arms for seizing prey. They can either be retracted into special pouches or doubled back on themselves. They bear suckers only at the ends and are situated between the third and fourth arms on each side of the head (Fig. 13-24, *A*). The suckers in squid are stalked (pedunculated), with horny rims bearing teeth; in octopuses the suckers are sessile and have no horny rims.

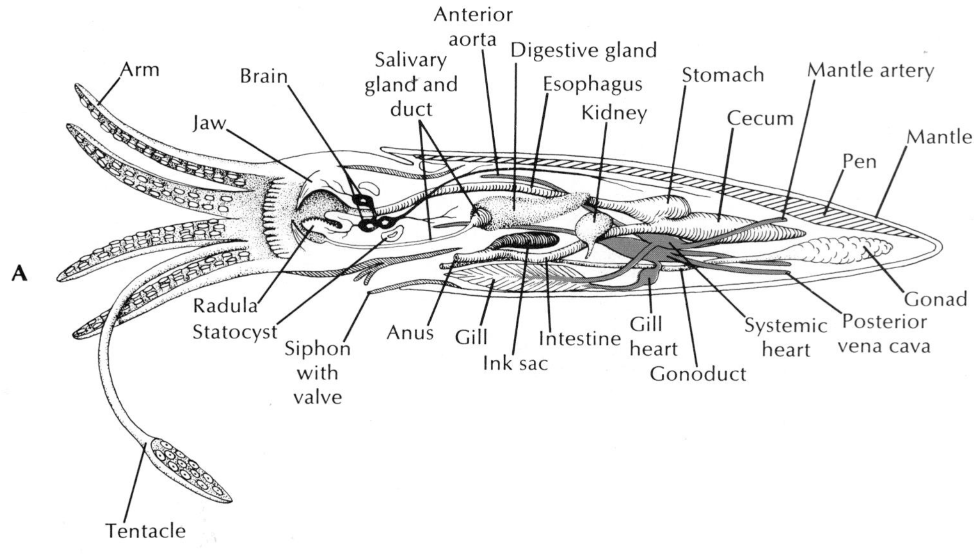

FIG. 13-24 **A,** Lateral view of squid anatomy; left half of mantle removed. **B,** School of young squid, swimming. (Class Cephalopoda) (Photograph by C. P. Hickman, Jr.)

Form and function

Shell. Early nautiloid shells were heavy, but they were made buoyant by a series of **gas chambers,** as is that of *Nautilus* (Fig. 13-23, *B*). The shell of *Nautilus,* although coiled, is quite different from that of a gastropod. Its shell is divided by transverse septa into internal chambers. The living animal inhabits only the last chamber. As it grows, it moves forward, secreting behind it a new septum. The chambers are connected by a tube called the **siphuncle,** which extends from the visceral mass and secretes gas into the empty chambers. The resulting buoyancy allows the animal to swim.

Cuttlefish also have a small coiled or curved shell, but it is entirely enclosed by the mantle. In the squid most of the shell has disappeared, leaving only a thin, horny strip called a pen, which is enclosed by the mantle. In the octopus the shell has disappeared entirely.

Locomotion. Most cephalopods swim by forcefully expelling water from the mantle cavity through a ventral **funnel**—a sort of jet-propulsion method. The funnel is mobile and can be pointed forward or backward

FIG. 13-25 **A,** *Sepia,* the cuttlefish. **B,** *Octopus.* (Class Cephalopoda) (Photograph by D. P. Wilson.)

to control direction; speed is controlled by the force with which water is expelled.

Squid and cuttlefish are excellent swimmers. The squid body is streamlined and built for speed (Fig. 13-24). The cuttlefish (Fig. 13-25, *A*) is slower but can regulate its buoyancy by regulating the relative amounts of fluid and gas in the narrow spaces of its shell. Both squid and cuttlefish have lateral fins that can serve as stabilizers, but they are held close to the body for rapid swimming.

Nautilus is active at night; its gas-filled chambers keep the shell upright. Though not as fast as the squid, it moves surprisingly well.

Octopus has a globular body and no fins (Fig. 13-25, *B*). The octopus can swim backward by spurting jets of water from its funnel, but it is better adapted to crawling about over the rocks and coral, using the suction discs on its arms to pull or to anchor itself. Some deep-water octopods have arms webbed like an umbrella and swim in a sort of medusa fashion.

External features. During the larval development of the cephalopod, the head and foot become indistinguishable. The ring around the mouth, which bears the arms, or tentacles, is considered to be derived from the foot.

In *Nautilus* the head with its 60 to 90 or more tentacles can be extruded from the opening of the body compartment of the shell (Fig. 13-23). Its tentacles have no suckers but are made adhesive by secretions. They are used in searching for, sensing, and grasping food. Beneath the head is the funnel. The mantle, mantle cavity, and visceral mass are sheltered by the shell. Two pairs of gills are located in the mantle cavity.

Cephalopods other than nautiloids have either eight (octopods) or ten (cuttlefish) appendages around the mouth, no external shell, and only one pair of gills. Octopods have eight arms; squid and cuttlefish (decapods) have ten arms, one pair of which are long retractile tentacles. The thick mantle covering the trunk fits loosely at the neck region, allowing intake of water into the mantle cavity. When the mantle edges contract closely about the neck, water is expelled through the funnel. The water current thus created provides oxygenation for the gills in the mantle cavity, jet power for locomotion, and a means of carrying wastes and gametes away from the body.

Feeding and nutrition. Cephalopods are predaceous, feeding chiefly on small fishes, molluscs, crustaceans, and worms. Their arms, which are used in food capture and handling, have a complex musculature and are capable of delicately controlled movements. Except in *Nautilus,* whose tentacles secrete a sticky substance, the inner surface of cephalopod arms bear powerful suction cups. The longer tentacles of squid and cuttlefish are suckered only at the ends. They are highly mobile and are used for swiftly seizing the prey and bringing it to the mouth. Strong, beaklike **jaws** can bite or tear off pieces of flesh, which are then pulled into the mouth by the tonguelike action of the **radula.** Octopods and cuttlefish have salivary glands that secrete a poison for immobilizing prey. Digestion is extracellular and occurs in the stomach and cecum.

Nervous and sensory systems. The nervous and sensory systems are more highly advanced in cephalopods than in other molluscs. The brain, the largest of any of the invertebrates, consists of many ganglia concentrated chiefly in the head and coordinated by millions of nerve cells. Squid have giant nerve fibers (among the largest known in the animal kingdom), which are activated when the animal is alarmed and which initiate maximal contractions of the mantle muscles for a speedy escape.

Some sense organs are well developed. The statocysts are larger and more complex than in other molluscs. Eyes are highly advanced and except in *Nautilus,* which has relatively simple eyes, they are able to form images. They are remarkably like vertebrate eyes, with cornea, lens, chambers, and retina (Fig. 13-13). One striking difference is that they are controlled by the statocysts and are held in a constant relation to gravity, so that the slit-shaped pupils are always in a horizontal position. Octopods can be taught to discriminate between colors and between shapes—for example, a square and a rectangle—and to remember such a discrimination for a considerable time. Experimenters find it easy to modify their behavior patterns by devices of reward and punishment. Octopods use their arms for tactile exploration, and with their tactile sense can discriminate between textures but apparently not between shapes. The arms are well supplied with both tactile and chemoreceptor cells. They seem to lack a sense of hearing.

A neurosecretory system is thought to supply a hormone that maintains the tonus of the body.

Communication. Little is known of the social behavior of nautiloids or of the deep-water coleoids, but inshore and littoral forms such as *Sepia, Sepioteuthis, Loligo,* and *Octopus* have been extensively studied. Although their tactile sense is well developed and they have some chemical sensitivity, visual signals predominate as a means of communication. These consist of a host of movements of the arms, fins, and body, as well as many color changes. The movements may range from minor body motions to exaggerated spreading, curling, raising, or lowering of some or all of the arms. Color changes are brought about by cells in the skin called chromatophores, which contain pigment granules. Each elastic chromatophore is surrounded by tiny muscle cells whose contractions pull the cell boundary of the chromatophore outward, causing it to expand greatly in size. As the cell expands the pigment becomes dispersed, changing the color pattern of the animal. When the muscles relax, the chromatophores return to their original size, and the pigment becomes concentrated again. By means of the chromatophores, which are under nervous and probably hormonal control, an elaborate system of changes in color and pattern is possible, including general darkening or lightening; flushes of pink, yellow, or lavender; and the formation of bars, stripes, spots, or irregular blotches. These may be used variously as danger signals, as protective coloring, in courtship rituals, and probably in other ways. Although visual signals cannot be seen far in aquatic environments, they work well for cephalopods because cephalopods are such gregarious animals, and because they seem to prefer the clearer waters.

By assuming different color patterns on different parts of the body, one member of a group of squid can transmit three or four different messages *simultaneously* to different individuals and in different directions, and it can, in an instant if necessary, change any or all of the messages. Probably no other system of communication can convey so much information so rapidly, unless it would be the acoustic repertoires of some of the primates and song birds.

Deepwater cephalopods may have to depend more on chemical or tactile senses than their littoral or surface cousins, but they also produce their own type of visual signals, for they have evolved many elaborate luminescent organs.

Cephalopods other than nautiloids have another protective device. An ink sac that empties into the rectum contains an **ink gland** that secretes into the sac **sepia,** a dark fluid containing the pigment melanin. When the animal is alarmed, it releases a cloud of ink, which may hang in the water as a blob, or be contorted by water currents. The animal quickly leaves the scene, leaving the ink to serve as a decoy to the predator.

Reproduction. Sexes are separate in cephalopods. In the male seminal vesicle the spermatozoa are encased in spermatophores and stored in a sac that opens into the mantle cavity. One arm of the adult male is modified as an intromittent organ, which, during copulation, plucks a spermatophore from his own mantle cavity and inserts it into the mantle cavity of the female near the oviduct opening (Fig. 13-26). Before copulation males often undergo color displays, apparently directed against rival males. Eggs are fertilized as they leave the oviduct and are usually attached to stones or other objects to develop. Some octopods tend their eggs, and *Argonauta,* the paper nautilus, secretes a fluted "shell," or capsule in which she broods her eggs.

PHYLOGENY AND ADAPTIVE RADIATION

Phylogeny. There is a striking similarity between the annelids and the molluscs in the formation and development of the embryo. As has been pointed out, both have spiral and determinate cleavage, and the trochophore larvae of many marine molluscs is almost identical with those of marine annelids. This is strong evidence for a common origin of the two groups. There are good reasons for believing that both groups diverged from some flatworm stock that became coelomate (developed a coelom).

When did this divergence occur? The dorsal plates of the chitons seem to suggest an external metamerism, but the internal organs are not metamerically arranged as in the annelids. However, some authors believe that the monoplacophorans show some internal metameric organization. The question is, then, did the molluscs diverge before metamerism appeared in the ancestral line? Or did the common ancestor possess a metamerism, which was retained and enlarged upon by the annelids but was lost by the molluscs except for the traces left in the Monoplacophora? If the latter theory is correct and metamerism is primitive in molluscs, then one might expect the trait to show up in molluscan larvae, but there is no characteristic suggesting metamerism in the development of any known molluscan larva.

Some zoologists suggest that the replication of body parts found in the ancestral stock was pseudometamerism, which developed into true segmentation in the annelids, but in the molluscs gained expression only in the chitons and monoplacophorans.

C. R. Stacek (1972) suggests that the first step in the evolution of the molluscan body plan from that of the turbellarian ancestor (which possessed some pseudometamerism) was a complete digestive system and the ability to secrete a protective mucous covering. This, he says, is the stage at which the annelid and

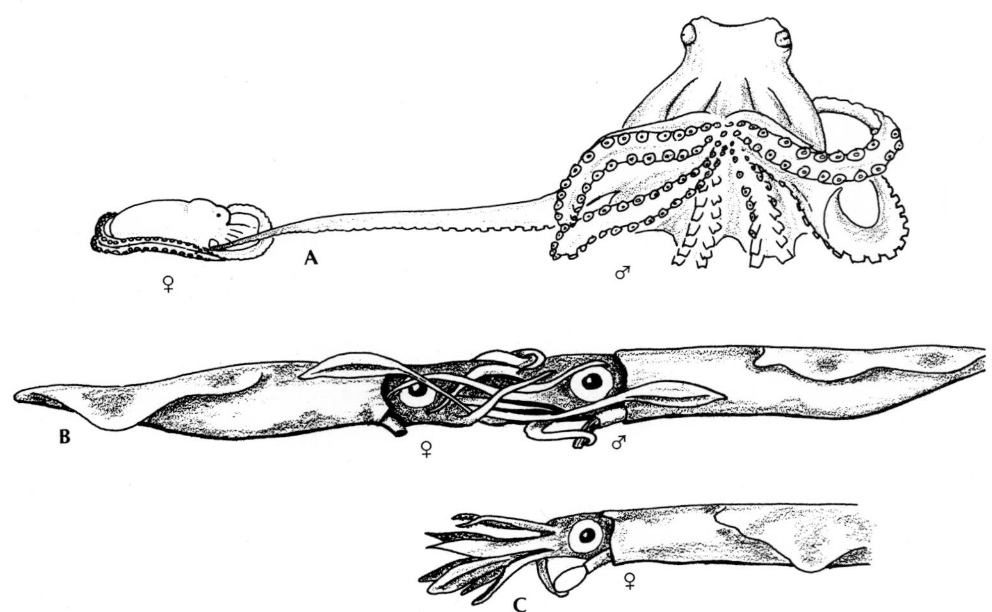

FIG. 13-26 Copulation in cephalopods. **A,** Large male octopus uses modified arm to deposit spermatophores in female mantle cavity to fertilize her eggs. Octopuses often tend their eggs during development. **B,** Male squid grasps spermatophore as it emerges from his funnel and will thrust it between the ventral arms of the female near her sperm reservoir. **C,** After copulation, the female reaches for a string of fertilized eggs emerging from her funnel and will attach them to a rock or other base.

mollusc stems diverged. Next came the radula and the development of a dorsal cuticle containing calcareous spicules. Here one evolutionary line led to the aplacophorans and another to the polyplacophorans, in which some pseudometamerism is found. Further advancements were the development of a mantle, mantle cavity, and gills beneath the cuticle. Layers of calcium carbonate secreted by the mantle formed the characteristic molluscan shell, and the original cuticle became the periostracum. The locomotory part of the body was the foot. This development led to the Monoplacophora. The remaining four classes—Gastropoda, Scaphopoda, Bivalvia, and Cephalopoda—evolved from the Monoplacophora. According to this theory, the initial factor that led to the evolution of the molluscan body plan and to the establishment of the group as a phylum was the dorsal cuticle.

Adaptive radiation. Zoologists have hypothesized a generalized type of ancestral mollusc (Fig. 13-27) that might have had a soft body, a dorsally arched calcareous shell, a ventral muscular foot, a mantle and mantle cavity, a complete digestive tract and radula, and a pair of gills near the anus. The primitive body plan was that of a ventral head-foot complex with an anteroposterior axis. Later as the body became more bulky, a visceral mass took shape above the foot, adding a dorsoventral axis. Adaptive radiation has brought modifications of this plan; for example, in the gastropods with their asymmetric growth and torsion; in the cephalopods, which have lengthened in the original dorsoventral axis and in which the foot has moved forward and become transformed into prehensile tentacles around the mouth; and in the bivalves in which the foot is modified for burrowing, the shell is divided into two parts, the head is lost, and the gills are modified for feeding.

Most of the diversity among molluscs is related to their adaptation to different habitats and modes of life and to a wide variety of feeding methods, ranging from sedentary filter feeding to active predation. There are many adaptations for food-gathering within the phylum and an enormous variety in

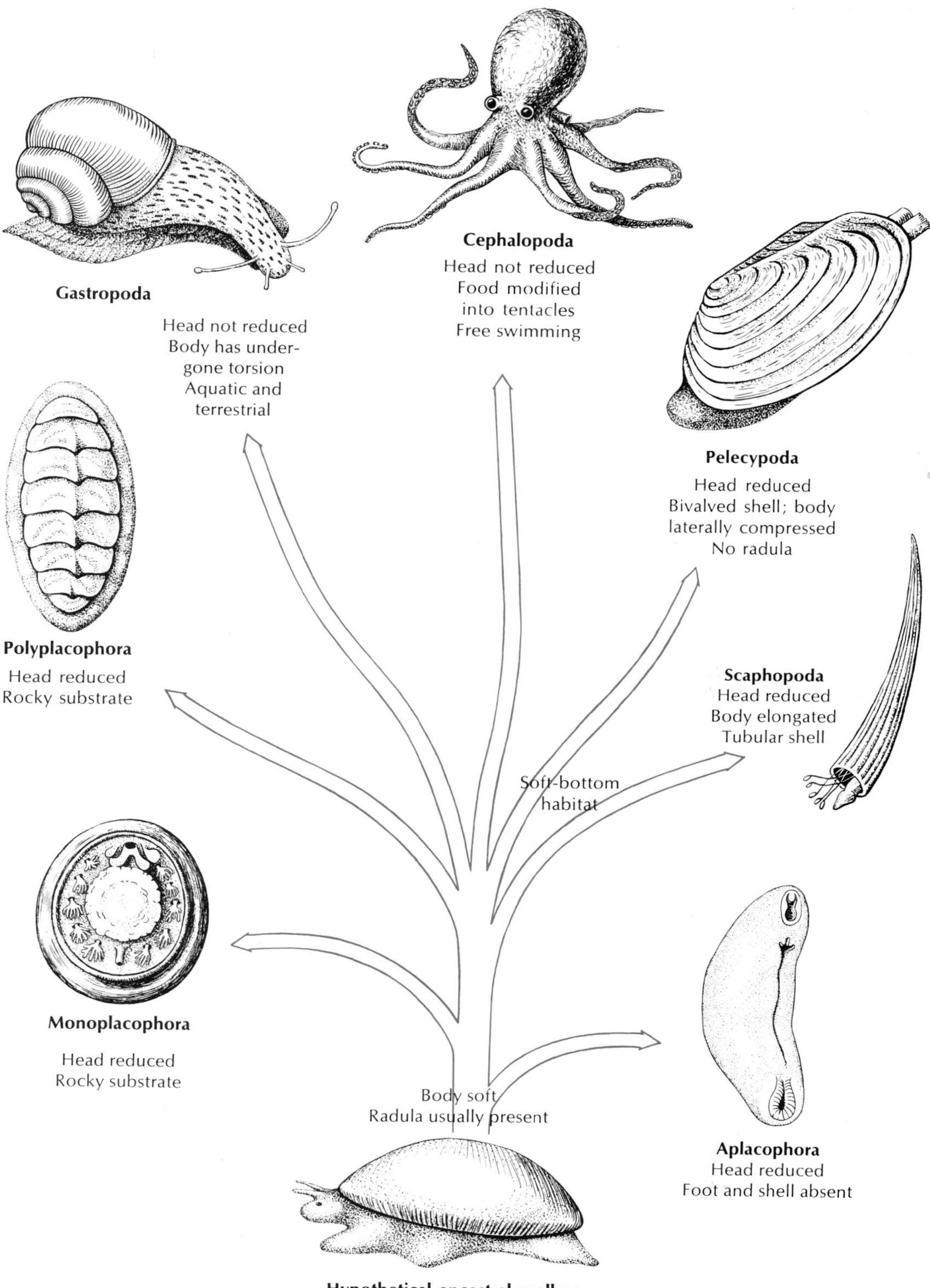

FIG. 13-27 The classes of Mollusca, showing their adaptive radiation.

radular structure and function, particularly among the gastropods.

The versatile glandular mantle has probably shown more plastic adaptive capacity than any other molluscan structure. Besides secreting the shell and forming the mantle cavity, it is variously modified into gills, lungs, siphons, and apertures, and it sometimes functions in locomotion, in the feeding processes, or in a sensory capacity. The shell, too, has undergone a variety of evolutionary adaptations.

Derivation and meaning of names

Anodonta (Gr., *an-*, without, + *odous, odontos,* tooth) A common genus of freshwater clams.

Busycon (Gr., *bous,* ox, + *sykon,* fig) Giant whelk.

Chiton (Gr., coat of mail, tunic). Refers to plates that cover this primitive mollusc.

Coleoidea (Gr., *coleo,* sheath, + *-idea,* group suffix) Subclass of cephalopods in which the shell is absent or internal (sheathed).

Dentalium (L., *dentalis,* dental) Genus of Scaphopoda.

Helix (Gr., twisted; a spiral) Genus of land snails introduced from Europe.

Nautilus (Gr., *nautilos,* sailor) This sole survivor of the most ancient molluscs is an active swimmer.

Nudibranchia (L., *nudus,* naked, + Gr., *branchia,* gills) These opisthobranch gastropods have no shells.

Opisthobranchia (Gr., *opisthe,* behind, + *branchia,* gills) Gills are located posteriorly in these gastropods.

Physa (Gr., bellows, air stream) Refers to bubble of air carried by snail when it submerges.

Prosobranchia (Gr., *prosō,* forward, + *branchia,* gills) Gastropods with gills located anteriorly.

Pulmonata (L. *pulmo,* lung, + *-ata,* group suffix) Gastropods having a lung.

Solenogasters (Gr., *sōlēn,* pipe, + *gastēr,* belly) Synonym of Aplacophora.

Strombus (L., a kind of spiral snail) Giant conch.

Annotated references

Selected general readings

Abbott, R. T. 1968. Seashells of North America, a guide to field identification. New York, Golden Press. *Excellent illustrations, and only slightly less information than his more expensive* American seashells.

Abbott, R. T. 1974. American seashells, ed. 2. New York, Van Nostrand Reinhold Co.

Barnes, R. D. 1974. Invertebrate zoology, ed. 3. Philadelphia, W. B. Saunders Co.

Bernard, F. R. 1974. Particle sorting and labial palp function in the Pacific oyster *Crassostrea gigas* Thunberg, 1975. Biol. Bull. **146:**1-10.

Buchsbaum, R. M., and L. J. Milne. 1960. The lower animals: living invertebrates of the world. Garden City, N.Y., Doubleday & Co., Inc. *Superb photographs (many in color) and concise accounts of the invertebrate phyla.*

Burch, J. B. 1962. How to know the eastern land snails. Dubuque, Iowa, William C. Brown Co., Publishers.

Carriker, M. R., and D. V. Zandt. 1972. Predatory behavior of a shell-boring muricid gastropod. In H. E. Winn, and B. L. Olla (eds.). Behavior of marine animals, vol. 1. New York, Plenum Press, pp. 157-244.

Clark, R. B. 1964. Dynamics in metazoan evolution. The origin of the coelom and segments. Oxford, Clarendon Press.

Corning, W. C., J. A. Dyal, and A. O. D. Willows (eds.). 1973, 1975. Invertebrate learning, vols. 2 and 3. New York, Plenum Publishing Corp.

Cousteau, J., and P. Diolé. 1973. Octopus and squid, the soft intelligence. Garden City, N.Y., Doubleday & Co. Inc. *Popular interest; beautiful color plates; illustrated glossary.*

Edmondson, W. T. (ed.). 1959. Ward and Whipple's freshwater biology, ed. 2. New York, John Wiley & Sons, Inc. *A taxonomic key to the freshwater families of molluscs is included.*

Fretter, V. (ed.). 1968. Studies in the structure, physiology, and ecology of molluscs. New York, Academic Press, Inc.

Fretter, V., and A. Graham. 1962. British prosobranch molluscs: their functional anatomy and ecology. London, The Ray Society.

Fretter, V., and J. Peake (eds.). 1975. Pulmonates: functional anatomy and physiology. New York, Academic Press, Inc.

Galtsoff, P. S. 1964. The American oyster *Crassostrea virginica* Gmelin. U.S. Fish Wildlife Serv. Fish. Bull. **64:**1-480.

Gregoire, C. 1972. Structure of the molluscan shell. In M. Florkin and B. Scheer (eds.). Chemical zoology, vol. 7, New York, Academic Press, Inc., pp. 45-210.

Hickman, C. P. 1973. Biology of the invertebrates, ed. 2. St. Louis, The C. V. Mosby Co.

Hiscock, I. D. 1972. Phylum Mollusca. In A. J. Marshal and W. D. Williams (eds.). Textbook of zoology: invertebrates. New York, American Elsevier Publishing Co., Inc.

Hyman, L. H. 1967. The invertebrates: Mollusca, vol. 6. New York, McGraw-Hill Book Co. *This volume covers four groups of the molluscs—Aplacophora, Polyplaco-*

phora, Monoplacophora, and Gastropoda—and upholds the fine traditions of the other volumes in this series.

Jorgensen, C. B. 1966. Biology of suspension feeding. New York, Pergamon Press, Inc.

Kaestner, A. 1964. Invertebrate zoology, vol. 1. New York, John Wiley & Sons, Inc.

Keen, A. M. 1971. Marine molluscan genera of western North America, ed. 2. Stanford, Calif., Stanford University Press.

Keen, A. M., and J. H. McLean. 1971. Sea shells of tropical west America. Stanford, Calif., Stanford University Press.

Lemche, H. 1957. A new living deep-sea mollusk of the Cambro-Devonian class Monoplacophora. Nature **179:** 413. *An account of Neopilina.*

Lemche, H., and K. G. Wingstrand. 1959. The anatomy of *Neopilina galatheae*. Galathea Reports **3:**9-71.

Light, S. F., R. I. Smith, F. A. Pitelka, D. P. Abbott, and F. M. Weesner. 1967. Intertidal invertebrates of the central California coast. Berkeley, University of California Press.

MacGinitie, G. E., and N. MacGinitie. 1967. Natural history of marine animals, ed. 2. New York, McGraw-Hill Book Co. *A section is devoted to Mollusca.*

McLean, J. H. 1969. Marine shells of Southern California. Los Angeles, Los Angeles County Museum of Natural History. *Handbook with black-and-white photographs.*

Meglitsch, P. A. 1972. Invertebrate zoology, ed. 2. New York, Oxford University Press.

Morris, P. A. 1973. A field guide to shells of the Atlantic and Gulf coasts and the West Indies, ed. 3. In W. J. Clench (ed.). Boston, Houghton Mifflin Co. *An excellent revision of a popular handbook.*

Morton, J. E. 1967. Molluscs, ed. 4. London, Hutchinson & Co.

Moynihan, M. H., and A. F. Rodaniche. 1977. Communication, crypsis, and mimicry among cephalopods. In T. A. Sebeok (ed.). How animals communicate. Bloomington, Ind., Indiana University Press, pp. 293-302.

Packard, A. 1972. Cephalopods and fish: the limits of convergence. Biol. Rev. Cambridge Phil. Soc. **47**(2):241-307. *Thought-provoking review with extensive bibliography.*

Pennak, R. W. 1953. Fresh-water invertebrates of the United States. New York, The Ronald Press Co., pp. 667-726.

Potts, W. T. W. 1967. Excretion in the mollusks. Biol. Rev. **42:**1-41.

Purchon, R. D. 1968. The biology of the Mollusca. New York, Pergamon Press, Inc. *A readable book that reviews in eight essays the research of the past 40 to 50 years, stressing the fascinating adaptive radiation within the classes.*

Rice, T. 1971. Marine shells of the Pacific Northwest. Edmonds, Wash., Ellison Industries. *Handy field guide with color plates.*

Runham, N. W., and P. J. Hunter. 1970. Terrestrial slugs. London, Hutchinson University Library.

Russell-Hunter, W. D. 1968. A biology of lower invertebrates. New York, The Macmillan Co.

Solem, A. 1974. The shell makers: introducing mollusks. New York, John Wiley & Sons, Inc. *Good for the shell collector who wants to know more about the living molluscs.*

Thompson, T. E. 1976. Biology of the Opisthobranch molluscs, vol. 1. London, The Ray Society. *Basically an identification guide; beautifully illustrated.*

Thompson, T. E. 1976. Nudibranchs. Neptune, N.J., T.F.H. Publications, Inc. *A slender handbook filled with superb color photographs.*

Wells, M. J. 1962. Brain and behavior in cephalopods. Stanford, Calif., Stanford University Press.

Wilbur, K. M., and C. M. Yonge (eds.). 1964, 1966. Physiology of Mollusca, vols. 1 and 2. New York, Academic Press, Inc. *A monograph that summarizes much of the research work on molluscs up to that time.*

Yonge, C. M. 1960. Oysters. London, William Collins Sons and Co., Ltd. *Excellent general account.*

Yonge, C. M., and T. E. Thompson. 1976. Living marine molluscs. London, William Collins Sons and Co., Ltd. *Interesting book about both the more common molluscs and those with particularly interesting adaptive features.*

Young, J. Z. 1961. Learning and form discrimination by octopus. Biol. Rev. Cambridge Phil. Soc. **36:**31-96.

Selected *Scientific American* articles

Boycott, B. B. 1965. Learning in the octopus. **212:**42-50 (Mar.).

Feder, H. M. 1972. Escape responses in marine invertebrates. **227:**92-100 (July).

Korringa, P. 1953. Oysters. **189:**86-91 (Nov.). *Their life history is described.*

Lane, C. E. 1961. The teredo. **204:**132-142 (Feb.). *Biology of the shipworm.*

Willows, H. O. D. 1971. Giant brain cells in mollusks. **224:**68-75 (Feb.).

Yonge, C. M. 1975. Giant clams. **232:**96-105 (Apr.).

CHAPTER 14

THE SEGMENTED WORMS

Phylum Annelida

A beautiful twin-fan worm *Bispira volutacornis* extends its feathery plume from the mouth of its tube to trap minute food particles on its delicate radioles. It has built its tube in a hole in the rock.

Photograph by D. P. Wilson.

Position in animal kingdom

1. Annelids belong to the **protostome** branch of the animal kingdom and have spiral and determinate cleavage.
2. Annelids have a true **coelom** (body cavity).
3. Annelids as a group show a primitive metamerism with comparatively few differences between the different somites.
4. All organ systems are present and well developed.

Biologic contributions

1. The introduction of **metamerism** by the group represents the greatest advancement of this phylum and lays the groundwork for the more highly specialized metamerism of the arthropods.
2. A true coelomic cavity reaches a high stage of development in this group.
3. Specialization of the head region into differentiated organs, such as the tentacles, palps, and eyespots of the polychaetes, is carried further in some annelids than in other invertebrates so far considered.
4. The tendency toward **centralization of the nervous system** is more developed, with cerebral ganglia (brain), two closely fused ventral nerve cords with giant fibers running the length of the body, and various ganglia with their lateral branches.
5. The circulatory system is much more complex than any we have so far considered. It is a closed system with muscular blood vessels and aortic arches (''hearts'') for propelling the blood.
6. The appearance of the fleshy **parapodia,** with their respiratory and locomotor functions, introduces a suggestion of the paired appendages and specialized gills found in the more highly organized arthropods.
7. The well-developed **nephridia** in most of the somites have reached a differentiation that involves a removal of waste from the blood as well as from the coelom.
8. Annelids are the most highly organized animals capable of complete regeneration. However, this ability varies greatly within the group.

PHYLUM ANNELIDA

Annelida (an-nel'i-da) (L., *annellus,* little ring, + *-ida,* pl. suffix) consists of the segmented worms. It is a large phylum, numbering approximately 9,000 species, the most familiar of which are the earthworms and freshwater worms (oligochaetes) and the leeches (hirudineans). However, approximately two-thirds of the phylum is comprised of the marine worms (polychaetes), which are less familiar to most people. Among the latter are many curious members; some are strange, even grotesque, whereas others are graceful and beautiful (see lead photograph). They include the clamworms, plumed worms, parchment worms, scaleworms, lugworms, and many others. The annelids are true coelomates and belong to the protostome branch, with spiral and determinate cleavage. They are a highly developed group in which the nervous system is more centralized and the circulatory system more complex than those of the phyla we have studied thus far.

The Annelida are worms whose bodies are divided into similar rings, or **segments,** arranged in linear series, and externally marked by circular grooves called **annuli;** the name of the phylum is descriptive of this characteristic. Body segmentation, or **metamerism,** in the annelids is not merely an external feature but is also seen internally in the repetitive arrangement of organs and systems and in the partitioning off of segments (also called metameres or somites) by septa. Metamerism, however, is not limited to annelids; it is shared by the arthropods (insects, crustaceans, and others), which are related to the annelids, and also by the vertebrates, in which it evolved independently.

Annelids are sometimes called ''bristle worms'' because, with the exception of the leeches, most annelids bear tiny chitinous bristles called **setae** (Gr., *chaite,* hair or bristle). Short needlelike setae help anchor the somites during locomotion to prevent backward slipping; long, hairlike setae aid aquatic forms in swimming. Since many annelids are either burrowers or live in secreted tubes, the stiff setae also aid in preventing the worm from being pulled out or washed out of its home. Robins know from experience how effective the earthworms' setae are.

ECOLOGIC RELATIONSHIPS

Annelids have a worldwide distribution and occur in marine and brackish waters, fresh water, and terrestrial soils. A few of the species may be called cosmopolitan.

Polychaetes are chiefly marine forms and make up

about two-thirds of the annelid worms. Most of them are benthic, but some live free in the open sea. They are usually divided into two groups—the sedentary polychaetes, or Sedentaria, and the free-moving polychaetes, or Errantia. Sedentary polychaetes are mainly tubicolous, that is, they spend all or much of their time in tubes or permanent burrows. Many of them, especially those that live in tubes, have elaborate devices for feeding and respiration. In sabellids and serpulids the cirri or tentacles around the mouth give rise to great featherlike "branchial crowns" that are involved in both feeding and respiration (p. 300). By means of ciliary currents water passes between the tentacles and carries food entangled in mucus to the mouth. The errant, or free-moving, polychaetes have various habitats; some are strictly pelagic and others live in crevices or under rocks or shells, never straying far in open water.

Many polychaetes are euryhaline (can tolerate a wide range of environmental salinity) and occur in brackish water. Freshwater polychaete fauna is more diversified in warmer regions than in the temperate zones.

Clitellates (oligochaetes and leeches) (Fig. 14-2) occur predominately in freshwater or terrestrial soils. Some freshwater species burrow in the bottom mud and sand and others among submerged vegetation. The swimming species usually have long setae, whereas the common earthworms, the most familiar example of the oligochaetes, have short setae.

Many of the leeches (class Hirudinea) are predators, and many are specialized for piercing their prey and feeding on blood or soft tissues. A few leeches are marine, but most of them live in fresh water or in damp regions. Suckers are typically found at both ends of the body for attachment to the substratum or to their prey. Some are adapted for forcing their pharynx or proboscis into soft tissues as in the gills of fish. The most specialized leeches, however, have sawlike chitinous jaws with which they can cut through tough skin. Many leeches live as carnivores on small invertebrates; some are temporary parasites; and some are permanent parasites, never leaving their host.

Characteristics

1. Body **metamerically segmented;** symmetry bilateral
2. Body wall with outer circular and inner longitudinal muscle layers; outer transparent moist cuticle secreted by epithelium
3. **Chitinous setae,** often present on fleshy appendages called **parapodia;** setae absent in leeches
4. Coelom (schizocoel) well developed and divided by septa, except in leeches; coelomic fluid supplies turgidity and functions as hydrostatic skeleton
5. **Blood system closed** and segmentally arranged; respiratory pigments (hemoglobin, hemerythrin or chlorocruorin) often present; amebocytes in blood plasma
6. Digestive system complete and not metamerically arranged
7. Respiratory gas exchange through skin, **gills,** or **parapodia**
8. Excretory system typically a **pair of nephridia for each metamere**
9. Nervous system with a double ventral nerve cord and a pair of ganglia with lateral nerves in each metamere; brain a pair of dorsal cerebral ganglia with connectives to cord
10. Sensory system of tactile organs, taste buds, statocysts (in some), photoreceptor cells, and eyes with lenses (in some)
11. Hermaphroditic or separate sexes; larvae, if present, are trochophore type; asexual reproduction by budding in some; spiral and determinate cleavage

Classes

The annelids are classified primarily on the basis of the presence or absence of parapodia, setae, metameres, and other morphologic features. Because both the oligochaetes and the hirudineans (leeches) bear a saddlelike enlargement, called a **clitellum** (L., *clitellae,* packsaddle) (Fig. 14-3, *B*), that is involved in reproduction, these two groups are often placed under the heading Clitellata (cli-tel-la'ta) and members are called clitellates. On the other hand, because both the Oligochaeta and the Polychaeta possess setae, some authorities place them together in a group called Chaetopoda (ke-top'o-da) (Gr., *chaitē,* long hair, + *pous, podos,* foot).

Class Polychaeta (pol"e-ke'ta) (Gr., *polys,* many, + *chaitē,* long hair). Mostly marine; head distinct and bearing eyes and tentacles; most segments with parapodia (lateral appendages) bearing tufts of many setae; clitellum absent; sexes usually separate; gonads transitory; asexual budding in some; trochophore larva usually; mostly marine.

Subclass Errantia (er-ran'she-a) (L., *errare,* to wander, + *-ia,* pl. suffix). Segments usually similar except in head and anal regions; parapodia alike and with acicula (long, stout setae); pharynx usually protrusible; head appendages usually present; free-living, tube-dwelling, and pelagic species, mostly marine. Examples: *Neanthes, Aphrodite, Glycera.*

Subclass Sedentaria (sed-en-ta're-a) (L., *sedere,* to sit, + *-aria,* like or connected with). Body with unlike segments, and parapodia and with regional differentation; prostomium small or indistinct; head appendages modified or absent; pharynx without jaws and mostly nonprotrusible; parapodia reduced and without acicula; gills anterior or absent; tube-dwelling or in burrows. Examples: *Arenicola, Chaetopterus, Amphitrite.*

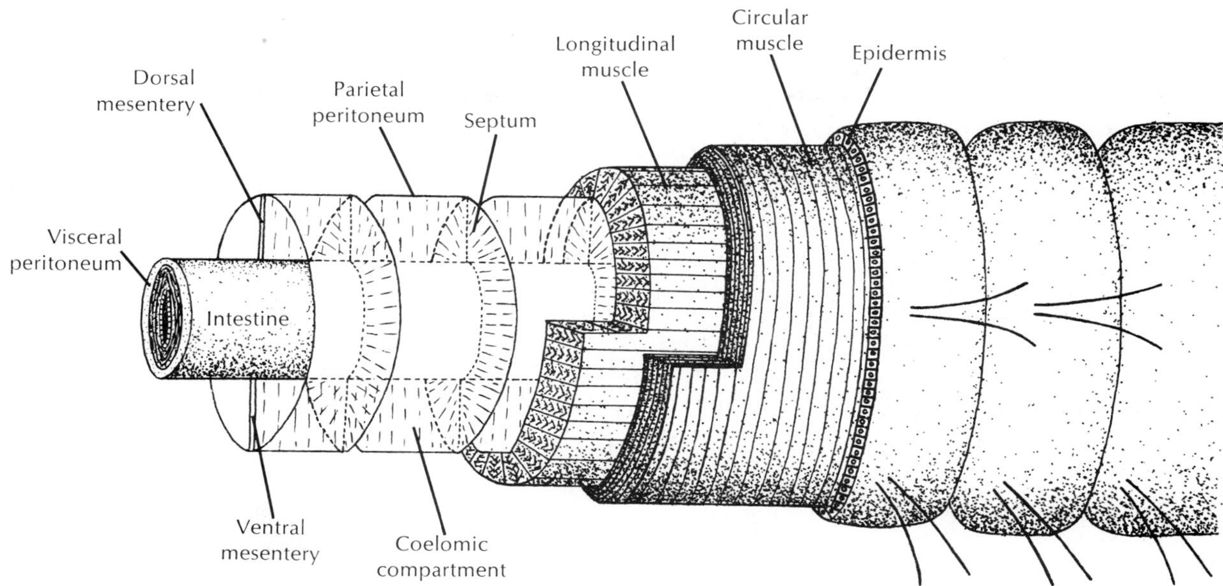

FIG. 14-1 Annelid body plan. The body wall and intestine, a "tube-within-a-tube," are separated by the coelomic cavity, which forms a pair of coelomic spaces in each somite. The peritoneum of each coelomic space lines the body wall (parietal peritoneum), covers the intestine and viscera (visceral peritoneum), and lies against the peritoneum of adjoining coelomic spaces to form the mesenteries and septa.

Class Oligochaeta* (ol'i-go-ke'ta) (Gr., *oligos,* few, + *chaitē,* long hair). Body with conspicuous segmentation; number of segments variable, setae few per metamere; no parapodia; head absent; coelom spacious and usually divided by intersegmental septa; hermaphroditic; development direct, no larva; chiefly terrestrial and freshwater. Examples: *Lumbricus, Stylaria, Aeolosoma, Tubifex.*

Class Hirudinea (hir'u-din'e-a) (L., *hirudo,* leech, + *-ea,* characterized by)—**leeches.** Body with fixed number of segments (34) with many annuli; body with anterior and posterior suckers usually; clitellum present; no parapodia; setae absent (except *Acanthobdella*†); coelom closely packed with connective tissue and muscle; development direct; hermaphroditic; terrestrial, freshwater, and marine. Examples: *Hirudo, Placobdella, Macrobdella.*

*The Branchiobdellida, a group of small annelids that are parasitic or commensal on crayfish and show similarities to both oligochaetes and leeches are here placed with the oligochaetes, but they are considered by some authorities to be a separate class. They have 14 or 15 segments and bear a head sucker.

†One genus of leech, *Acanthobdella,* is a primitive type, with some characteristics of leeches and some of oligochaetes; it is sometimes separated from the other leeches into a special class, Acanthobdellida, that characteristically has 27 somites, setae on the first five segments, and the anterior sucker absent.

BODY PLAN

The annelid body typically has a head or prostomium, a segmented body, and an anal segment called the pygidium. New segments form just in front of the pygidium; thus, the oldest segments are at the anterior end and the youngest segments are at the posterior end. The body wall is made up of strong circular and longitudinal muscles adapted for swimming, crawling, and burrowing, and is covered with epidermis and a thin, outer layer of nonchitinous cuticle (Fig. 14-1).

In most annelids the coelom develops as a pair of coelomic compartments in each segment. Each compartment is surrounded with **peritoneum** (a layer of mesodermal epithelium), which lines the body wall, forms dorsal and ventral **mesenteries,** and covers all the organs (Fig. 14-1). Where the peritonea of adjacent segments meet, the **septa** are formed. These are perforated by the gut and longitudinal blood vessels. Not only is the coelom metamerically arranged, but practically every body system is affected in some way by this segmental arrangement.

Except in the leeches, the coelom is filled with fluid, which serves as a hydrostatic skeleton that provides the rigidity and resistance necessary for effective muscular movement. Because a segmented body is compartmentalized by septa, the coelomic fluid skeleton is localized, and widening and elongation can occur in restricted areas. Crawling motions are effected by alternating waves of contraction by longitudinal and circular muscles (peristaltic contraction) passing down the body. Segments in which longitudinal muscles are contracted widen and anchor themselves against burrow walls or other substratum while other segments, in which circular muscles are contracted, elongate and stretch forward. Forces powerful enough for burrowing as well as locomotion can thus be generated. Swimming forms use undulatory rather than peristaltic movements in locomotion.

CLASS OLIGOCHAETA

The more than 3,000 species of oligochaetes are found in a great variety of sizes and habitats. They include the familiar earthworms and many species that live in fresh water. Most are terrestrial or freshwater forms, but some are parasitic, and a few live in marine or brackish water.

Oligochaetes, with few exceptions, bear setae, which may be long or short, straight or curved, blunt or needlelike, arranged singly or in bundles. Whatever the type, they are less numerous in oligochaetes than in polychaetes, as is implied by the class name, which means "few setae." Aquatic forms usually have longer setae than do earthworms.

Some of the more common oligochaetes, other than earthworms, are the small freshwater forms such as *Aeolosoma* (Fig. 14-2) (1 mm long), which contains red or green pigments, has bundles of setae, and is often found in hay cultures; *Nais* (2 to 4 mm long), which is brownish in color, with two or three bundles of setae on each segment; *Stylaria* (10 to 25 mm long), which has two bunches of setae on each segment, with the prostomium extended into a long process, and black eyespots; *Dero* (5 to 10 mm long), which is reddish in color, lives in tubes, and has three tail gills; *Tubifex* (30 to 40 mm long), which is reddish in color and lives with its head in mud at the bottom of ponds and its tail waving in the water; *Chaetogaster* (10 to 15 mm long), which has only ventral bundles of setae; and *Enchytraeus,* small whitish worms that live in both moist soil and in water. Some oligochaetes, such as *Aeolosoma,* may form chains of zooids asexually by transverse fission (Fig. 14-2, *B*).

Earthworms

The most familiar of the oligochaetes are the earthworms ("night crawlers"), which burrow in moist, rich soil, emerging at night to explore their surroundings. In damp, rainy weather they stay near the surface, often with mouth or anus protruding from the burrow. In very dry weather they may burrow several feet underground, coil up in a slime chamber, and become dormant. *Lumbricus terrestris,* the form commonly studied in school laboratories, is approximately 12 to 30 cm long (Fig. 14-3). *Eisenia,* a small genus, is usually abundant in manure piles. Giant tropical earthworms may have from 150 to 250 or more segments and may grow to as much as 4 m in length. They usually live in branched and interconnected tunnels.

Aristotle called earthworms the "intestines of the soil." Some fifteen centuries later Charles Darwin published his observations in his classic *The Formation of Vegetable Mould Through the Action of Worms.* He showed how worms enrich the soil by bringing subsoil to the surface and mixing it with the topsoil. An earthworm can ingest its own weight in soil every 24 hours, and Darwin estimated that from 10 to 18 tons of dry earth per acre pass through their intestines annually, thus bringing up potassium and phosphorus from the subsoil and also adding to the soil nitrogenous products from their own metabolism. They expose the mold to the air and sift it into small particles. They also drag leaves, twigs, and organic substances into their burrows closer to the roots of plants. Their activities are important in aerating the soil.

Form and function

In earthworms the mouth is overhung by a fleshy prostomium at the anterior end and the anus on the last segment (Fig. 14-3, *B*). *Lumbricus* may have 100 to 175 segments. Few new segments are added after hatching.

In most earthworms, each segment, with the exception of the first and last, bears four pairs of chitinous setae (Fig. 14-3, *C*) located on the ventral and lateral surfaces. Each seta is a bristlelike rod set in a sac within the body wall and moved by tiny muscles (Fig. 14-4). The setae project through small pores in the cuticle to the outside. In locomotion and burrowing setae anchor parts of the body to prevent slipping. When a seta

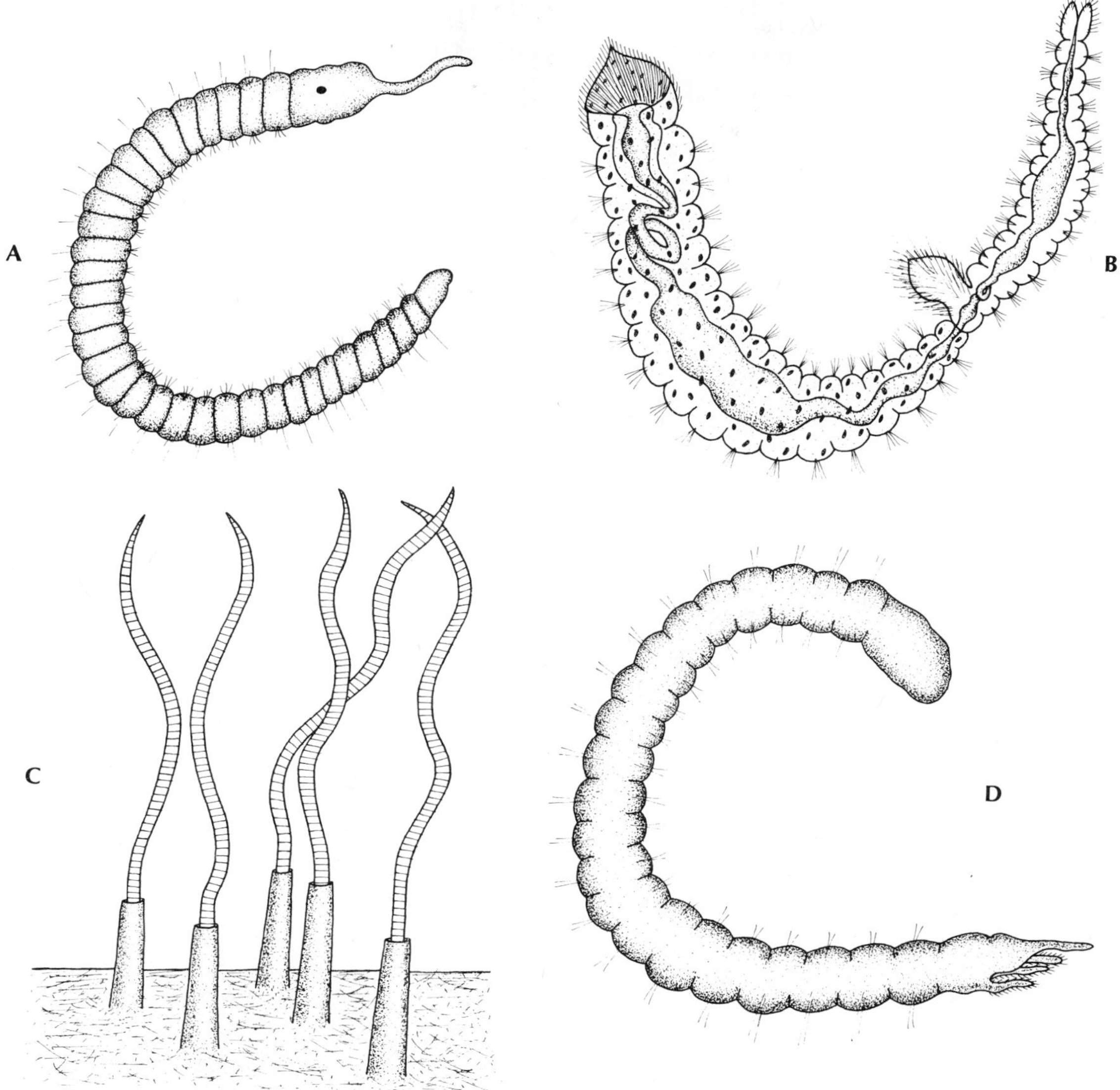

FIG. 14-2 Some freshwater oligochaetes. **A,** *Stylaria* has the prostomium drawn out into a long snout. **B,** *Aeolosoma* uses cilia around the mouth to sweep in food particles, and it buds off new individuals asexually. **C,** *Tubifex* lives head down in long tubes. **D,** *Aulophorus* is provided with ciliated anal gills.

is lost, a new one is formed in a reserve follicle to replace it.

Locomotion. Earthworms move by peristaltic movement. Contraction of circular muscles in the anterior end lengthen the body, thus pushing the anterior end forward where it is anchored by setae; contractions of longitudinal muscles then shorten the body, thus pulling the posterior end forward. As these waves of contraction pass along the entire body, it is gradually moved forward. Setae are rarely used as levers in crawling, but rather as anchors to prevent slipping. The fluid pressure in the individual coelomic compartments is a significant factor in the forward thrust of the body. Such pressure has been calculated to be equivalent to forces of 1.5 to 8 g. When burrowing in soft soil, the earthworm thrusts its anterior end into crevices

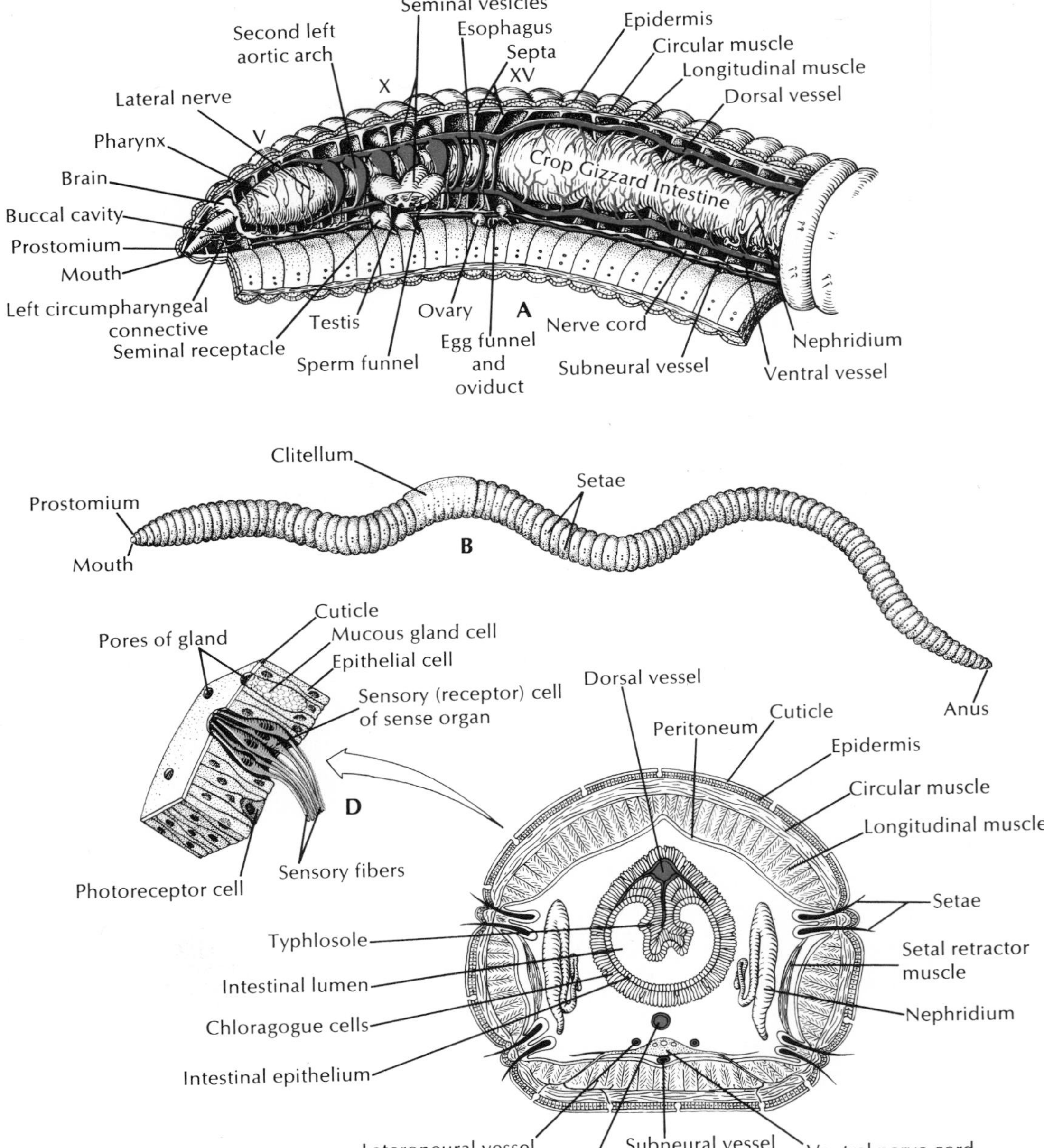

FIG. 14-3 Earthworm anatomy. **A,** Internal structure of anterior portion of worm. **B,** External features, lateral view. **C,** Generalized transverse section through region posterior to clitellum. **D,** Portion of epidermis showing sensory, glandular, and epithelial cells.

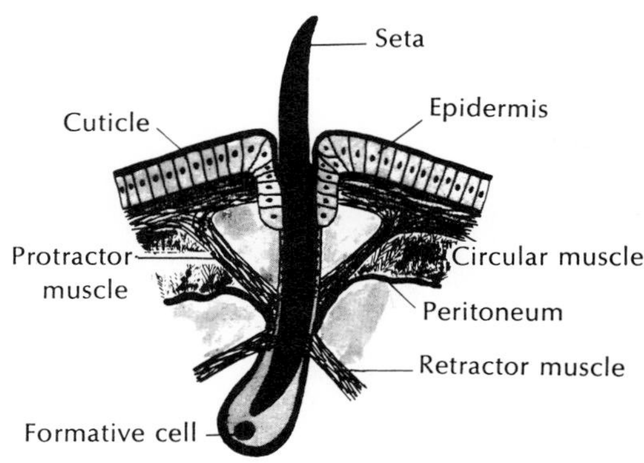

FIG. 14-4 Seta with its muscle attachments showing relation to adjacent structures. Setae lost by wear and tear are replaced by new ones, which develop from formative cells. (Modified from J. Stephenson, 1930, and others.)

between soil particles and then enlarges its pharynx to push the soil aside. In firmer soil it burrows by literally eating its way through the soil, which accounts for the numerous castings.

Nutrition. Most oligochaetes are scavengers. Earthworms feed mainly on decayed organic matter, bits of leaves and vegetation, refuse, and animal matter. After being moistened by secretions from the **mouth,** food is drawn in by the sucking action of the muscular **pharynx.** The liplike prostomium aids in manipulating the food into position. The calcium from the soil swallowed with the food tends to produce a high blood calcium level. **Calciferous glands** along the **esophagus** secrete calcium ions into the gut and so reduce the calcium-ion concentration of the blood. Calciferous glands are really excretory, rather than digestive, organs. They also function in regulating the acid-base balance of the body fluids, maintaining the pH at a fairly stable value.

Leaving the **esophagus,** food is stored temporarily in the thin-walled **crop** before being passed on into the **gizzard,** which grinds the food into small pieces. Digestion and absorption take place in the **intestine.** Along the dorsal side, the wall of the intestine is infolded to form a **typhlosole,** which greatly increases the absorptive and digestive surface (Figs. 14-3, *C*, and 14-5, *B*). The digestive system secretes various enzymes to break down the food: pepsin, which acts on proteins; amylase, which acts on carbohydrates; cellulase, which acts on cellulose; and lipase, which acts on fats. The indigestible residue is discharged through the **anus.**

Surrounding the intestine and dorsal vessel and filling much of the typhlosole is a layer of yellowish **chlorogogue tissue** derived from the peritoneum. This tissue serves as a center for the synthesis of glycogen and fat, a function roughly equivalent to that of liver cells. The chlorogogue cells when ripe (full of fat) are released into the coelom where they float free as cells called **eleocytes,** which transport materials to the body tissues. They apparently can pass from segment to segment, and have been found to accumulate around wounds and regenerating areas, where they break down and release their contents into the coelom. Chlorogogue cells also function in excretion.

Earthworms take in a great deal of soil, sand, and other indigestible matter along with their food. The food products are absorbed into the blood, which carries them to the various parts of the body for assimilation. Some of the food is absorbed into the coelomic fluid, which also aids in food distribution.

Circulation. Annelids have a double transport system—the coelomic fluid and the circulatory system. Food, wastes, and respiratory gases are carried by both coelomic fluid and blood in varying degrees. The blood is carried in a closed system of blood vessels, including capillary systems in the tissues. There are five main blood trunks, all running lengthwise through the body (Fig. 14-5).

The **dorsal vessel** (single) runs above the alimentary canal from the pharynx to the anus. It is a pumping organ, provided with valves, and functions as the true heart. This vessel receives blood from vessels of the body wall and digestive tract and pumps it anteriorly into the five pairs of aortic arches. The chief function of the aortic arches is to maintain a steady pressure of blood into the ventral vessel.

The **ventral vessel** (single) serves as the aorta. It receives blood from the aortic arches and delivers it to the brain and rest of the body, giving off segmental vessels to the walls, nephridia, and digestive tract.

The **lateral neural vessels** (paired) lie one on each side of the nerve cord. These receive the blood from the segmental vessels and carry it posteriorly with many branches to the nerve cord.

The **subneural vessel** (single), under the nerve cord, is the main vein. It receives blood from the nerve cord, nephridia, and body wall and passes it backward and upward into paired parietal vessels, which return it to

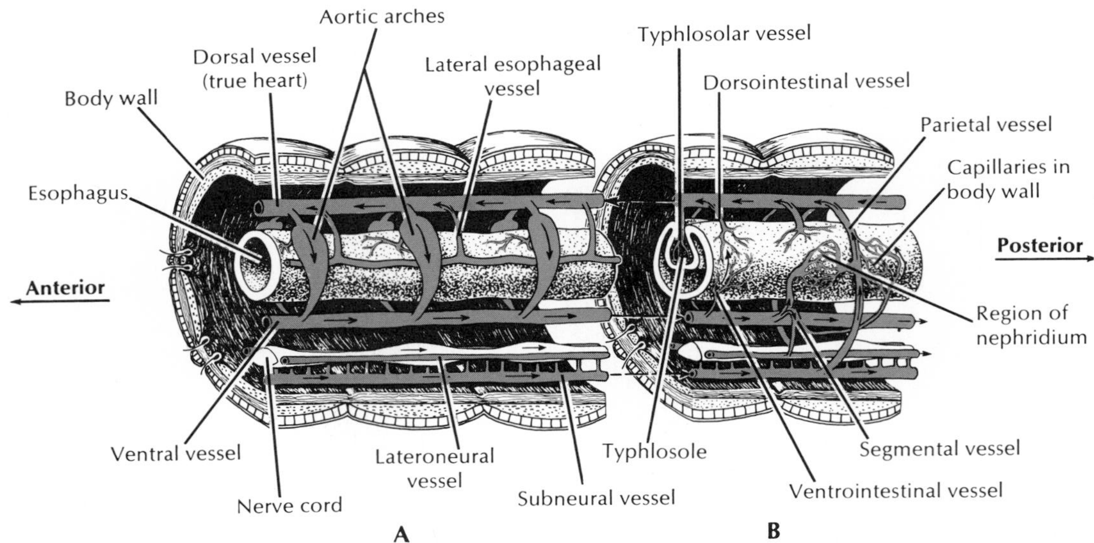

FIG. 14-5 Scheme of circulation in the earthworm. **A,** Somites 9, 10, and 11, showing three of the five aortic arches that receive blood from the dorsal vessel and pass it down to the ventral vessel to be carried posteriorly and distributed to the body tissues. **B,** Any somite posterior to the esophagus, showing distribution of blood from ventral vessel to intestine, nerve cord, and body wall and return of blood to dorsal vessel by the parietal and intestinal vessels. The dorsal vessel is the chief pumping organ.

the dorsal vessel. The parietals also drain blood from the nephridia and body wall.

The propulsion necessary to force the blood along is provided by the peristaltic or milking action of the muscular walls of the blood vessels, particularly the dorsal vessel. Valves in the vessels prevent backflow.

The blood of the earthworm is made up of a **liquid plasma** in which are colorless ameboid cells, or **corpuscles.** Dissolved in the blood plasma is the respiratory pigment **hemoglobin,** an iron-containing protein. This gives a red color to the blood and aids in the transportation of oxygen for respiration.

Excretion. The organs of excretion are the **nephridia,** a pair of which is found in each somite except the first three and the last one. Each one occupies parts of two successive somites (Fig. 14-6). A ciliated funnel, known as the **nephrostome,** lies just anterior to an intersegmental septum and leads by a small ciliated tubule through the septum into the somite behind, where it connects with the main part of the nephridium. This part of the nephridium is made up of several complex loops of increasing size, which finally terminate in a bladderlike structure leading to an aperture, the **nephridiopore;** this opens to the outside near the ventral row of setae. By means of cilia, wastes from the coelom are drawn into the nephrostome and tubule, where they are joined by organic wastes filtered from blood capillaries in the glandular part of the nephridium. All the waste is discharged to the outside through the nephridiopore.

Aquatic oligochaetes excrete ammonia; terrestrial oligochaetes excrete the much less toxic urea. *Lumbricus* produces both, the level of urea depending somewhat on environmental conditions. Both urea and ammonia are produced by chlorogogue cells, which may break off and enter the nephridia directly, or their products may be carried by the blood. Some nitrogenous waste is also eliminated through the body surface.

Oligochaetes are largely freshwater animals, and even such terrestrial forms as earthworms must exist in a moist environment. Osmoregulation is a function of the body surface and the nephridia, as well as the gut and the dorsal pores. *Lumbricus* will gain weight when placed in tap water and lose it when returned to the soil. Salts as well as water can pass across the integument, apparently by active transport.

Respiration. The earthworm has no special respiratory organs, but gaseous exchange is made in the

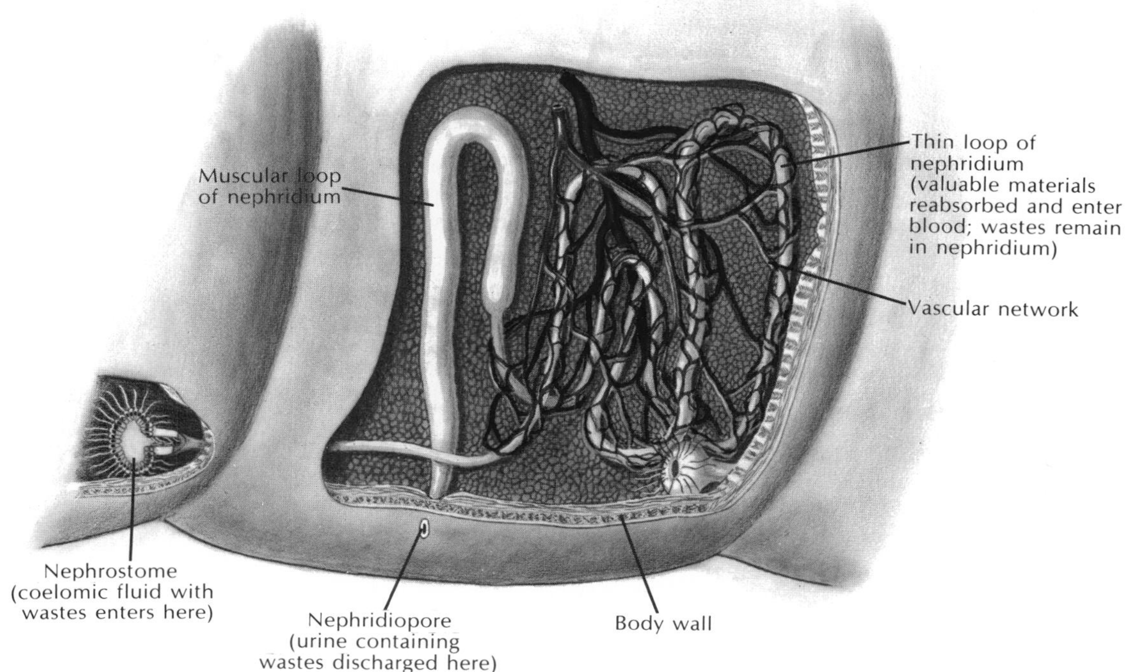

FIG. 14-6 Nephridium of earthworm. Wastes are drawn into the ciliated nephrostome in one segment, then passed through the loops of the nephridium, and are expelled through the nephridiopore of the next segment.

moist skin, where oxygen is picked up and carbon dioxide is given off. Blood capillaries are fairly numerous just below the cuticle, and the oxygen combines with the hemoglobin of the plasma and is carried to the various tissues.

Nervous system and sense organs. The nervous system in earthworms (Fig. 14-7) consists of a **central system** and **peripheral nerves.** The central system is made up of a pair of cerebral ganglia (the brain) above the pharynx and a pair of connectives passing around the pharynx connecting the brain with the first pair of ganglia in the nerve cord; a ventral nerve cord, really double, running along the floor of the coelom to the last somite; and a pair of fused ganglia on the nerve cord in each somite. Each pair of fused ganglia gives off nerves to the body structures, which contain both sensory and motor fibers. As in all higher forms, the sensory neurons carry impulses from special sensory cells in the epidermis to the nerve cord. Motor neurons run to muscles or glands.

Neurosecretory cells have been found in the brain and ganglia of annelids, both oligochaetes and polychaetes. They are endocrine in function and secrete neurohormones concerned with the regulation of reproduction, secondary sex characteristics, and regeneration.

For rapid escape movements most annelids are provided with from one to several very large axons commonly called **giant axons,** or giant fibers, located in the ventral nerve cord. Their large diameter increases the rate of conduction and makes possible simultaneous contractions of muscles in many segments. In most earthworms the nerve cord contains three quite large giant axons located dorsally (Fig. 14-8) and a pair of smaller ones located ventrally. Each giant axon connects to a cell body in each segment. The middorsal fiber receives impulses from sense organs in the anterior end of the body. The dorsolateral ones, which are connected by transverse fibers, are fired by stimulation posterior to the clitellum. Little is known of the function of the ventral pair. In the dorsal median giant fiber of *Lumbricus,* which is 90 to 160 μm in diameter,

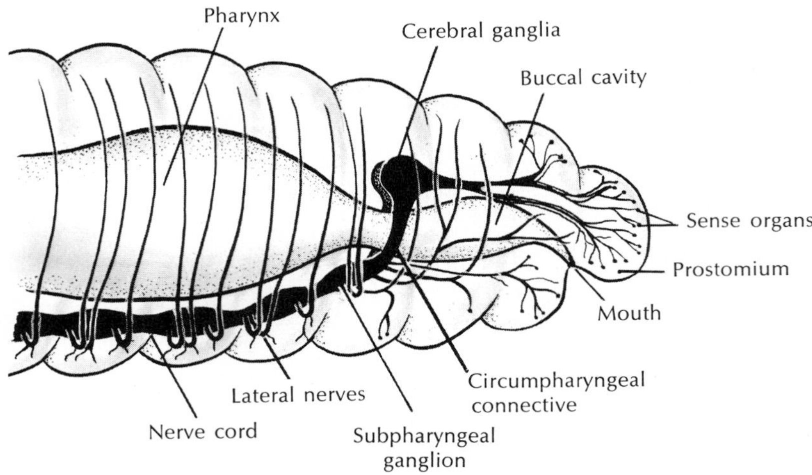

FIG. 14-7 Anterior portion of earthworm and nervous system. Note concentration of sense organs in this region.

the speed of conduction has been estimated at 20 to 45 m/second, several times faster than in ordinary neurons. This is also much faster than in polychaete giant fibers, probably because in the earthworms the giant fibers are enclosed in myelinated sheaths. The speed of conduction may be altered by changes in temperature.

Simple **sense organs** are distributed all over the body. Earthworms have no eyes but do have many photoreceptors in the epidermis. Each lens-shaped photoreceptor has a basal process that extends to the nerve plexus beneath the epidermis. Most oligochaetes are negatively phototactic to strong light but positively phototactic to weak light. Many single-celled sense organs are widely distributed in the epidermis; it has been estimated that there are 700 per square millimeter on the dorsal side of the prostomium of *Lumbricus*. Some sensory cells are found in clusters, forming tubercles from which sensory processes extend above the cuticle. These are presumably chemoreceptors and are most numerous on the prostomium. There are many free nerve endings in the integument, which are probably tactile.

General behavior. Earthworms are among the most defenseless of creatures, yet their abundance and wide distribution indicate their ability to survive. Although they have no specialized sense organs, they are sensitive to many stimuli, such as **mechanical,** to which they are positive when it is moderate; **vibratory** (such as a footfall near them), which causes them to retire quickly into their burrows; and **light,** which they avoid unless it is very weak. Chemical responses aid them in the choice of food.

Chemical as well as tactile responses are very important to the worm. It must be able to sample not only the organic content of the soil to find food, but also must sense its texture, acidity, and calcium content. Earthworms can discriminate between different kinds of leaves and between acids, alkalies, and sugars. *Lumbricus* will avoid soil with a pH below 4.1, for acid soils are deficient in calcium and some calcium is necessary for their well-being.

When irritated, earthworms eject coelomic fluid through their **dorsal pores,** one of which is located in each segment.

Experiments show that earthworms have some learning ability. They can be taught to avoid an electric shock, and thus an association reflex can be built up in them. Darwin credited earthworms with a great deal of intelligence in pulling leaves into their burrows, for they apparently seized the leaves by the narrow end, which is the easiest way for drawing a leaf-shaped object into a small hole. Darwin assumed that the seizure of the leaves by the worms did not result from random handling or from chance but was purposeful in its mechanism. However, investigations since Darwin's time have shown that the process is mainly one of trial and error, for they often seize a leaf several times before attaining the right position.

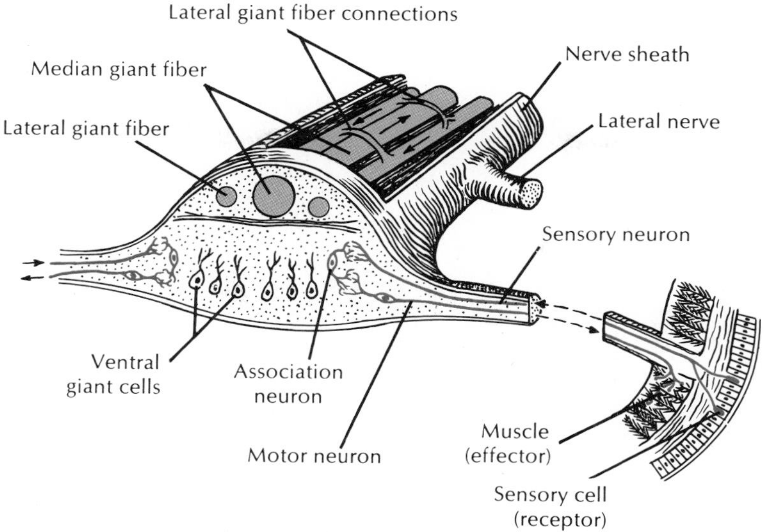

FIG. 14-8 Portion of nerve cord of earthworm showing arrangement of simple reflex arc *(in foreground)* and the three dorsal giant fibers that are adapted for rapid reflexes and escape movements. Ordinary crawling involves a succession of reflex acts, the stretching of one somite stimulating the next to stretch, and so on. Impulses are transmitted much faster in giant fibers than in regular nerves so that all segments can contract simultaneously when quick withdrawal into a burrow is necessary.

Reproduction and development. Earthworms are monoecious (hermaphroditic); that is, both male and female organs are found in the same animal (Fig. 14-3, *A*). In *Lumbricus* the reproductive systems are found in somites 9 to 15. Two pairs of small testes and two pairs of sperm funnels are surrounded by three pairs of large seminal vesicles. Immature sperm from the testes mature in the seminal vesicles, then pass into the sperm funnels and down sperm ducts to the male genital pores in somite 15, where they are expelled during copulation. Eggs are discharged by a pair of small ovaries into the coelomic cavity, where they are picked up by ciliated funnels and are carried by oviducts to the outside through female genital pores on somite 14. Two pairs of seminal receptacles in somites 9 and 10 receive and store sperm from the mate during copulation.

Reproduction in earthworms may occur at any season, but it is most common in warm moist weather. They exchange sperm during **copulation,** which usually occurs at night. When two worms mate, they extend their anterior ends from their burrows and bring their ventral surfaces together with their anterior ends pointed in opposite directions (Fig. 14-9). This arrangement of the bodies places the seminal receptacle openings of one worm in opposition to the clitellum of the other worm (Fig. 14-10, *A*). The worms are held together by mucus secreted by the clitellum and by special ventral setae, which penetrate each other's bodies in the regions of contact. Each worm secretes a slime tube about itself from somites 9 to 36. Sperm discharged from the sperm ducts of each worm travel by seminal grooves on the ventral surface to the openings of the seminal receptacles of the other worm (Fig. 14-10, *A*). After this reciprocal exchange of sperm is made, the worms separate. Copulation requires about 2 hours.

After copulation each worm secretes around its clitellum, first a mucous tube and then a tough, chitinlike band that forms a cocoon. As the cocoon passes forward, eggs from the oviducts, albumin from the skin glands, and sperm from the mate (stored in the seminal receptacles) are poured into it. Fertilization of the eggs now takes place within the cocoon (Fig. 14-10).

When the cocoon leaves the worm, its ends close, producing a lemon-shaped body (Fig. 14-10, *E*). The size of the cocoon varies with different species of

FIG. 14-9 Two earthworms in copulation. Their anterior ends point in opposite direction as their ventral surfaces are held together by mucous bands secreted by the clitella. Mutual insemination occurs during copulation. After separation each worm secretes a cocoon to receive its eggs and sperm. (Courtesy G. Carter.)

earthworms; those in *Lumbricus terrestris* are about 7 by 5 mm. In this form only one of several fertilized eggs develops into a worm, the others acting as nurse cells. Cocoons are commonly deposited in the earth, although they may be found at the surface near the entrance of burrows. Between copulations the earthworm continues to form cocoons so long as there are sperm in the seminal receptacles.

Cleavage is holoblastic, unequal, and spiral. The embryo passes through the blastula and gastrula stages and forms the three germ layers typical of the development of higher forms. The form that hatches from the egg is a young worm similar to the adult (Fig. 14-10, *F*) that escapes from the cocoon in 2 to 3 weeks. It does not develop a clitellum until it is sexually mature.

Regeneration. Annelid worms are perhaps the most highly organized animals that have the power of complete regeneration. Not all annelids have this capacity, and in most of them there are limitations. Earthworms vary; some species can form two complete worms when cut in two, but other species cannot. In the common earthworm *(Lumbricus)* a posterior piece may regenerate a new head of three to five segments, and an anterior piece may form a new tail, with the level of the cut determining the number of segments regenerated. A worm cut at segment 50 will regenerate ten fewer segments than one cut at the level of segment 40. Earthworms can also be grafted, and pieces of several worms have been grafted end to end to form long worms.

CLASS HIRUDINEA—THE LEECHES

Leeches are found predominantly in freshwater habitats, but a few are marine, and some have even adapted

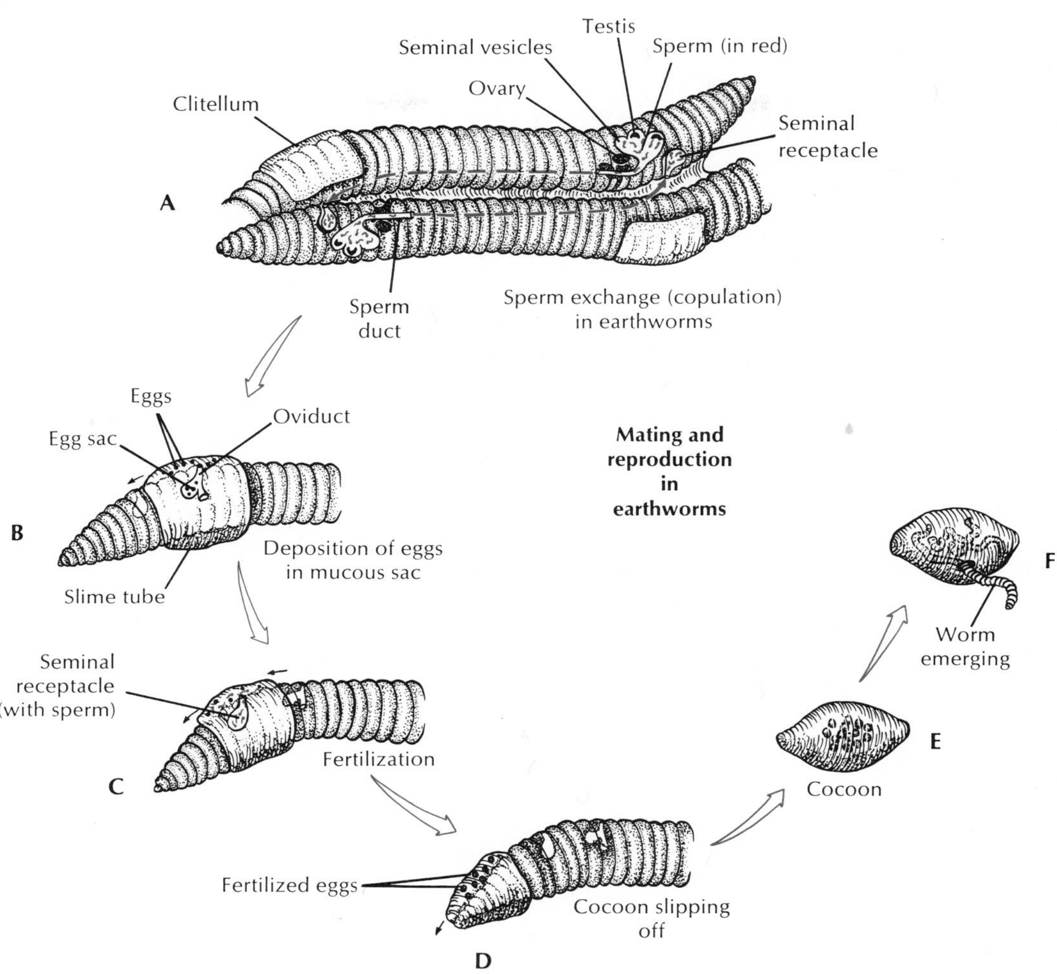

FIG. 14-10 Earthworm copulation and formation of egg cocoons. **A,** Mutual insemination occurs during copulation; sperm from genital pore (somite 15) pass along seminal grooves to seminal receptacles (somites 9 and 10) of each mate. **B** and **C,** After worms separate, a slime tube formed over the clitellum passes forward to receive eggs from oviducts and sperm from seminal receptacles. **D,** As cocoon slips off over anterior end, its ends close and seal. **E,** Cocoon is deposited near burrow entrance. **F,** Young worms emerge in 2 to 3 weeks.

to terrestrial life in warm, moist places. They are more abundant in tropical countries than in temperate zones. Some tropical leeches attack human beings and are a nuisance. In Egypt and nearby areas a small aquatic leech is often swallowed in drinking water and becomes a serious pest by fastening on the pharynx and epiglottis.

Most leeches are between 2 and 6 cm in length, but some are smaller; some, including the ''medicinal'' leech, reach 20 cm, but the giant of all is the Amazonian *Haementeria,* which reaches 30 cm.

Leeches are found in a variety of patterns and colors —black, brown, red, or olive green. They are usually flattened dorsoventrally.

Like the oligochaetes, leeches have a **clitellum** during the breeding season. The clitellum secretes a cocoon for the reception of eggs. Both groups are hermaphroditic, but leeches are more highly specialized. As fluid feeders and blood suckers, they have lost the setae used by the oligochaetes in locomotion and have developed **suckers** for attachment while sucking blood; their gut is specialized for storage of large quantities of blood.

Form and function

Unlike other annelids, leeches have a **fixed number of somites** (34), but they appear to have many more because each somite is marked by transverse grooves to form from two to 16 superficial rings called **annuli** (Fig. 14-11).

The **coelom** represents another difference between leeches and other annelids; leeches lack distinct coelomic compartments. In all but one species the septa have disappeared, and the coelomic cavity is filled with connective tissue and a system of spaces called **sinuses.** The coelomic sinuses form a regular system of channels filled with coelomic fluid, which in some leeches serves as an auxiliary circulatory system.

Locomotion. Most leeches creep with looping movements of the body, by attaching first one sucker and then the other and pulling up the body. Aquatic leeches can also swim with a graceful undulatory movement.

Nutrition. Leeches are popularly considered to be parasitic, but it would be more accurate to call them predaceous. Even the true bloodsuckers rarely are host specific or remain on the host for a long period of time. Most freshwater leeches are active predators or scavengers equipped with a proboscis that can be extended to draw in small invertebrates or to take blood from cold-blooded vertebrates. Some freshwater leeches are true bloodsuckers, preying on cattle, horses, humans, and others. Some terrestrial leeches feed on insect larvae, earthworms, and slugs, which they hold by an anterior sucker while using a strong sucking pharynx to draw in the food. Other terrestrial forms climb bushes or trees to reach warm-blooded vertebrates such as birds or mammals.

Most leeches are fluid feeders. Many prefer to feed on tissue fluids and blood pumped from wounds already open. The true bloodsuckers, which include the so-called medicinal leech *Hirudo,* have cutting plates, or ''jaws,'' for cutting through tissues. Some species are in fact true parasites, living permanently on their hosts.

Respiration and excretion. Gas exchange occurs only through the skin except in some of the fish leeches, which have gills. There are ten to 17 pairs of nephridia, in addition to coelomocytes and certain other specialized cells that may also be involved in excretory functions.

Nervous and sensory systems. Leeches have two ''brains,'' one in the head, composed of six pairs of fused ganglia forming a ring around the pharynx, and one in the tail, composed of seven pairs of fused ganglia. The additional 21 pairs of ganglia are segmentally arranged along the double nerve cord. In addition to free sensory nerve endings and photoreceptor cells in the epidermis, there is a row of sense organs, called sensillae, in the central annulus of each segment; there are also a number of pigment cup ocelli.

Reproduction. Leeches are hermaphroditic but practice cross-fertilization during copulation. Sperm are transferred by a penis or by hypodermic impregnation. After copulation the clitellum secretes a cocoon that receives the eggs and sperm. Cocoons are buried in bottom mud, attached to submerged objects or, in terrestrial species, placed in damp soil. Development is similar to that of oligochaetes.

Circulation. In leeches the coelom has been reduced, by the invasion of connective tissue and in some by a proliferation of chlorogogue tissue, to a system of coelomic sinuses and channels. Some orders of leeches retain the typical oligochaete circulatory system and in these the coelomic sinuses act as an auxiliary blood-vascular system. In other orders the traditional blood vessels have disappeared and the system of coelomic sinuses forms the only blood-vascular system. In those the blood (the equivalent of coelomic fluid) is propelled by contractions of certain longitudinal channels.

The ''medicinal leech''—Hirudo medicinalis

For centuries the ''medicinal leech'' was employed in medical practice for bloodletting because of the mistaken idea that bodily disorders and fevers were caused by a plethora of blood. Since the leech is 10 to 12 cm long and can extend to a much greater length when distended with blood, the amount of blood it can suck out of a patient is considerable. Leech collecting and leech culture in ponds were practiced in Europe on a commercial scale during the nineteenth century.

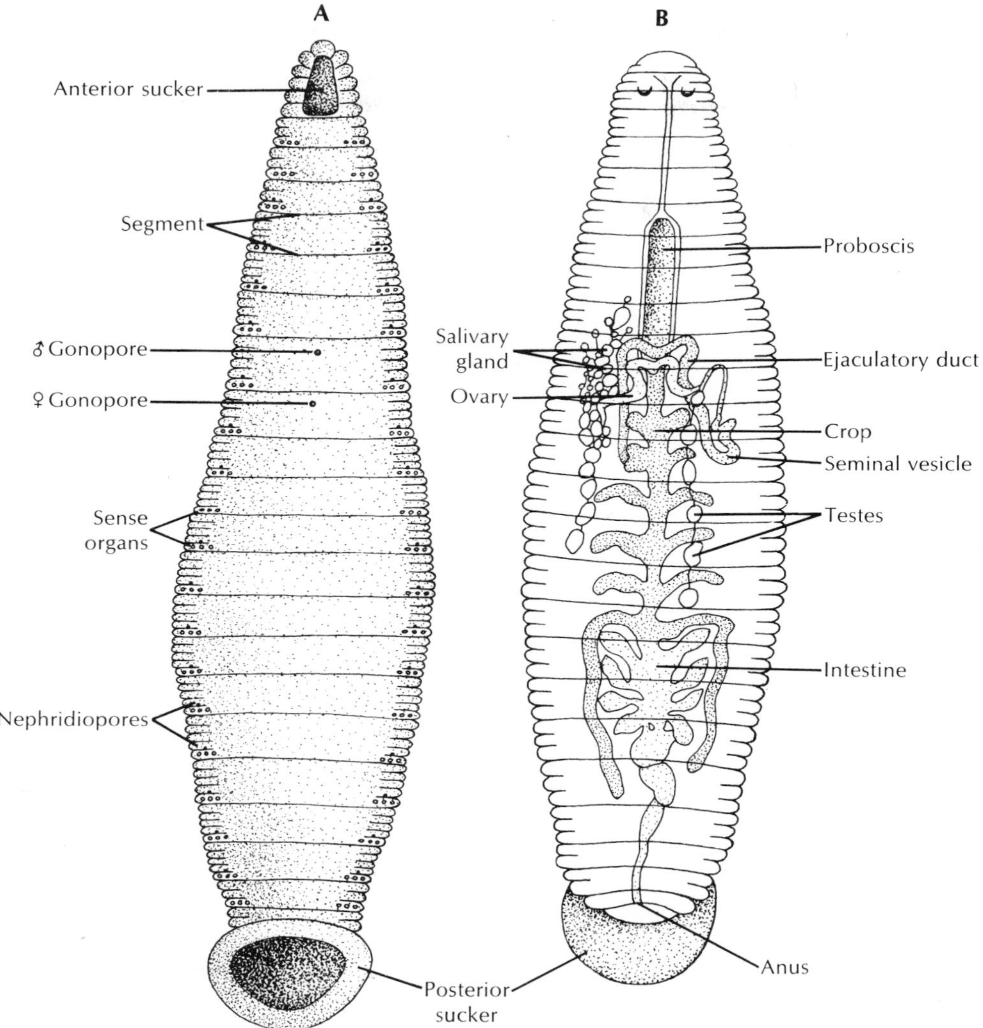

FIG. 14-11 Structure of the leech. **A,** Ventral view of *Hirudo.* **B,** Internal structure of *Helobdella. Hirudo* is an aquatic bloodsucker with five annuli on most of its segments. *Helobdella* is an aquatic leech with three annuli to a segment.

Wordworth's interesting poem "The Leech-Gatherer" was based on this use of the leech.

Structural and behavioral characteristics. In *Hirudo medicinalis* and other true blood suckers, the anterior sucker surrounds the mouth. The body is covered with a cuticle, which is secreted by an epidermis beneath, and there are many mucous glands that open on the surface. The body is divided by transverse furrows into 102 annuli. The midbody segments each comprise five annuli, with the number decreasing toward the extremities. A **clitellum** is found on segments 10 to 12 in the breeding season. The muscular system is well developed, with circular, longitudinal, and oblique bands.

The **alimentary canal** (Fig. 14-11, *B*) is made up of a mouth with three jaws and chitinous teeth, a muscular pharynx with salivary glands, a crop with 11 pairs of lateral ceca, a slender intestine, a rectum, and an anus; the anus opens anterior to the posterior sucker.

The bloodsucking leech attaches itself to the prey by

the posterior sucker and then moves the anterior end about to locate a vulnerable spot to which to attach the anterior sucker that surrounds the mouth and jaws. The jaws make a typical triradiate incision. Secretions from the leech's salivary glands contain an anticoagulant called **hirudin,** an inhibitor of thrombokinase, that prevents clotting of the host's blood as it flows from the incision. The glands also seem to secrete a histamine-like substance that causes the capillaries around the wound to dilate. The food is sucked up by the muscular pumping pharynx, is stored in the large crop, and is digested slowly. *Hirudo* can ingest as much as three to five times its body weight at a single feeding and then may not feed again for 3 to 6 months; *Haemadipsa* can take in up to ten times its weight. This is a useful adaptation, for leeches may not often have the chance to eat.

The **male organs** (Fig. 14-11, *B*) are the paired testes beneath the crop, a pair of vasa deferentia with some glands, and a median penis, which opens into the male pore. **Female organs** (Fig. 14-11, *B*) include a pair of ovaries with oviducts, an albumin gland, and a vagina, which opens near the male pore. The sperm are transferred in little packets (spermatophores) by the filiform penis, which penetrates the vagina of the mate in mutual copulation. The fertilized eggs are deposited in cocoons formed by glandular secretions. These cocoons may be attached to stones or other objects or even to the leech itself. Leeches have little or no regenerative ability.

CLASS POLYCHAETA

The polychaetes are the largest and most primitive class of annelids, with more than 35,000 species, most of them marine. Though the majority of them are from 5 to 10 cm long, some are less than a millimeter, and others may be as long as 3 m. Some are brightly colored in reds and greens; others are dull or iridescent. Some are picturesque, such as the "feather-duster" worms (Fig. 14-12).

Polychaetes live under rocks, in coral crevices, or in abandoned shells, or they burrow into mud or sand; some build their own tubes on submerged objects or in bottom material; some adopt the tubes or homes of other animals; some are pelagic, making up a part of the planktonic population. They are extremely abundant in some areas; for example, a square meter of mud flat may contain thousands of polychaetes. They play a significant part in marine food chains, since they are eaten by fish, crustaceans, hydroids, and many others.

Polychaetes differ from other annelids in having a well-differentiated head with specialized sense organs; paired, paddlelike appendages, called parapodia, on most segments; and no clitellum (Fig. 14-18). As their name implies, they have many setae, usually arranged in bundles on the parapodia. They show a pronounced differentiation of some body somites and a specialization of sensory organs practically unknown among clitellates.

In contrast to clitellates, polychaetes have no permanent sex organs, possess no permanent ducts for their sex cells, and usually have separate sexes. Their development is indirect, for they undergo a form of metamorphosis that involves a trochophore larva.

Polychaetes are usually divided into two subclasses —Errantia and Sedentaria. These subclasses are convenient but probably artificial. The **Errantia,** or errant worms (L., *errare,* to wander), include the free-moving pelagic forms, active burrowers, crawlers, and the tube worms that leave their tubes for feeding or breeding. Most of these, like *Nereis,* the clam worm (Fig. 14-18), are predatory forms equipped with jaws or teeth. They have a muscular eversible pharynx armed with teeth that can be thrust out with surprising speed and dexterity for capturing prey. The **Sedentaria** (L., *sedere,* to sit) are largely sedentary worms that rarely expose more than the head end from the tubes or burrows in which they live (p. 300 and Fig. 14-12).

Most sedentary tube and burrow dwellers are particle feeders, using ciliary or mucoid methods of obtaining food. The principal food source is plankton and detritus. Some, like *Amphitrite* (Fig. 14-13), with head peeping out of the mud, send out long extensible tentacles over the surface. Cilia and mucus on the tentacles entrap particles found on the sea bottom and move them toward the mouth.

The fanworms, or "featherduster" worms, are beautiful tubeworms, fascinating to watch as they emerge from their secreted tubes and unfurl their lovely tentacular crowns to feed. A slight disturbance, sometimes even a passing shadow, causes them to duck quickly into the safety of the homes they have built. Food attracted to the feathery arms, or radioles, by ciliary action is trapped in mucus and is carried down ciliated food grooves to the mouth (Fig. 14-14, *B* and *C*). Particles too large for the food grooves are carried along the margins and dropped off. Further sorting may

FIG. 14-12 Polychaete tubeworms. **A,** *Protula* builds a calcareous tube (family Serpulidae). **B,** *Spirographis spallanzani* builds a noncalcareous tube (family Sabellidae). (Photographs by C. P. Hickman, Jr.)

occur near the mouth where only the small particles of food enter the mouth, and sand grains are stored in a sac to be used later in enlarging the tube.

Some worms, such as *Chaetopterus,* secrete mucous filters through which they pump water to collect edible particles (Fig. 14-19). The lugworm *Arenicola* lives in an L-shaped burrow in which, by peristaltic movements, it keeps water filtering down through the sand and out the open end of the burrow. It ingests the food-laden sand brought by the water current (Fig. 14-15).

Tube dwellers secrete many types of tubes. Some are parchmentlike (Fig. 14-19); some are firm, calcareous tubes attached to rocks or other surfaces (Fig. 14-12, *A*); and some are simply grains of sand or bits of shell or seaweed cemented together with mucous secretions (Fig. 14-14). Many burrowers in sand and mud flats simply line their burrows with mucus (Fig. 14-15).

The polychaete typically has a head, or **prostomium,** which may or may not be retractile and which often bears eyes, antennae, and sensory palps (Figs.

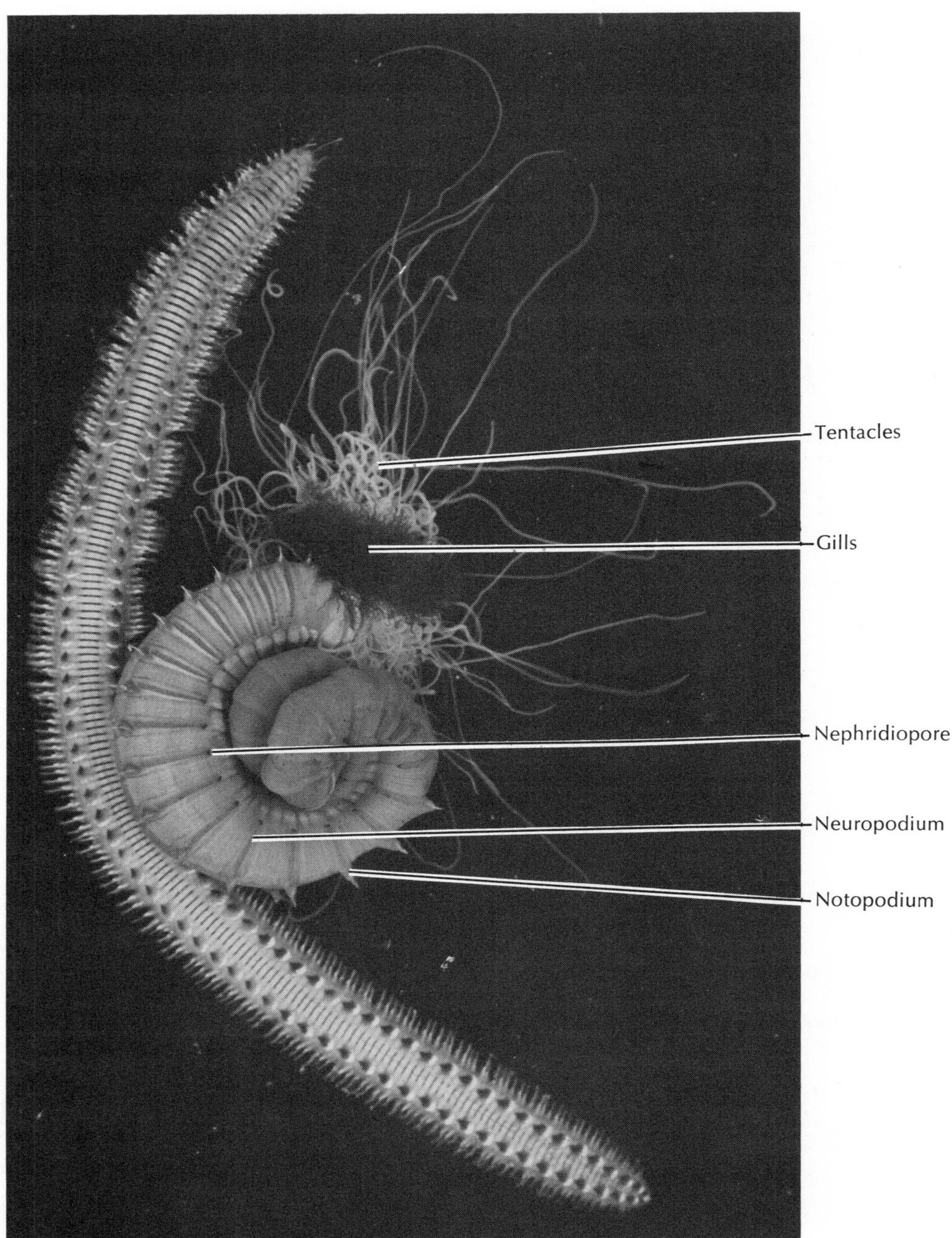

FIG. 14-13 Commensal polychaetes. The coiled worm is *Amphitrite,* (family Terebellidae), which lives in a membranous tube buried in the mud. It extends long, grooved tentacles over the surface to feed and has red gills. It shares its tube with the errant scaleworm *Lepidometria (Lepidasthenia),* (family Polynoidae). (Photograph by D. P. Wilson.)

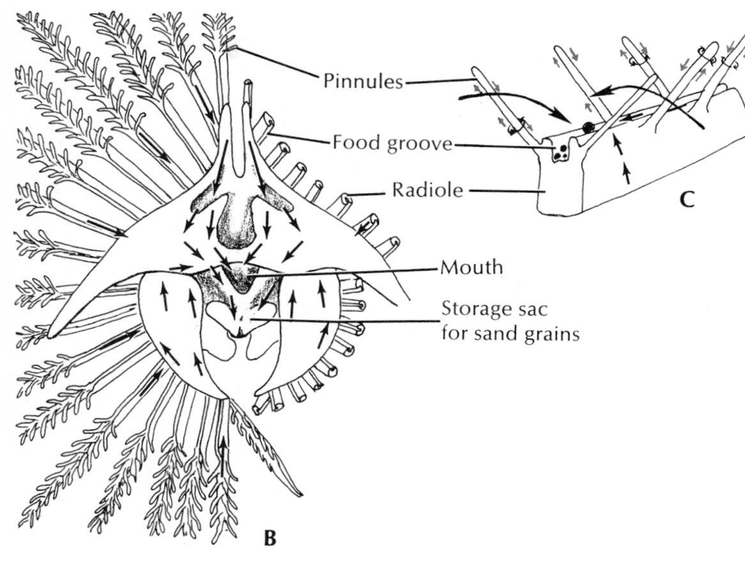

FIG. 14-14 *Sabella,* a polychaete ciliary feeder. **A,** *Sabella pavonia* with its crown of feeding radioles extended from its leathery secreted tube, reinforced with sand and debris. **B,** Anterior view of crown. Cilia direct small food particles along grooved radioles to mouth and discard larger particles. Sand grains are directed to storage sacs and later are used in tube building. **C,** Distal portion of radiole showing ciliary tracts of pinnules and food grooves. (**A,** Photograph by T. Lundälv, Kristinebergs Zoological Station, Sweden; **B,** Adapted from several sources.)

14-17 and 14-18). The first segment **(peristomium)** surrounds the mouth and may bear setae, palps, or, in predatory forms, chitinous jaws. Ciliary feeders may bear a tentacular crown that may be opened like a fan or withdrawn into the tube.

The trunk is segmented and most segments bear fleshy appendages called **parapodia,** which may have lobes, cirri, setae, and other parts on them (Fig. 14-17). The parapodia are used in crawling, swimming, or anchoring in tubes. They usually serve as the chief respiratory organs, although some polychaetes may also have gills. *Amphitrite,* for example, has three pairs of branched gills and long extensible tentacles (Fig. 14-13). *Arenicola,* the lugworm (Fig. 14-15), which burrows through the sand leaving characteristic castings at the entrance to its burrow, has paired gills on certain somites.

Sense organs are more highly developed in polychaetes than in oligochaetes and include eyes, nuchal organs, and statocysts. Eyes, when present, may range from simple eyespots to well-developed organs. They are most conspicuous in errant worms. Usually the eyes are retinal cups, with rodlike photoreceptor cells lining the cup wall and directed toward the lumen of the cup. The highest degree of development is found in the family Alciopidae, which has large, image-resolving eyes similar in structure to those of the squid (Fig. 13-12, *C*), with cornea, lens, retina, and retinal pigment. The alciopid eye also has accessory retinas, a characteristic shared by deep-sea fishes and some deep-sea cephalopods. The function of the accessory retinas may be to serve as a depth gauge for these pelagic animals. Recent studies with electroencephalograms show these eyes to be especially well adapted to utilizing the dim

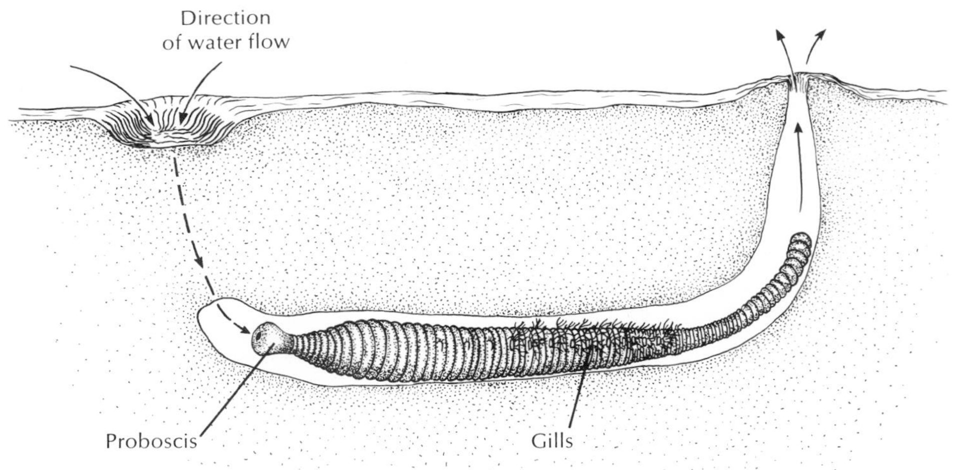

FIG. 14-15 *Arenicola,* the lugworm, lives in an L-shaped burrow in intertidal mud flats. It burrows by successive eversions and retractions of its proboscis. By peristaltic movements it keeps water filtering down through the sand and out the open end of the burrow. The worm then ingests the food-laden sand.

light of the deep sea. Nuchal organs are ciliated sensory pits or slits that appear to be chemoreceptive, an important factor in food-gathering. Some burrowing and tube-building polychaetes have statocysts that function in body orientation.

Reproductive systems are simple. Gonads appear as temporary swellings of the peritoneum and shed their gametes into the coelom. They are carried outside through gonoducts, through nephridia, or by rupture of the body wall. Fertilization is external, and the early larva is a trochophore.

Some polychaetes live most of the year as sexually unripe animals called **atokes,** but during the breeding season a portion of the body develops into a sexually ripe form called an **epitoke,** which is swollen with gametes (Fig. 14-16). An example is the palolo worm *Eunice viridis,* which lives in burrows among the coral reefs of the South Seas. During the reproductive cycle, the posterior somites become swollen with gametes. During the swarming period, which occurs at the beginning of the last quarter of the October-November moon, these epitokes break off and float to the surface. Just before sunrise the sea is literally covered with them, and at sunrise they burst, freeing the eggs and sperm for fertilization. The anterior portions of the worms regenerate new posterior sections. A related form, *Leodice,* swarms in the Atlantic in the third quarter of the June-July moon.

Some interesting polychaetes

Clam worms—Nereis (Neanthes). The clam worms, or sand worms, as they are sometimes called, are errant polychaetes that live in mucus-lined burrows in or near low tide (Fig. 14-17). Sometimes they are found in temporary hiding places such as under stones, where they stay, bodies covered and heads protruding. They are most active at night, when they wiggle out of their hiding places and swim about or crawl over the sand in search of food.

The body, containing about 200 somites, may grow to 30 or 40 cm in length. The head is made up of a prostomium and a peristomium. The prostomium bears a pair of stubby **palps,** sensitive to touch and taste; a pair of short sensory **tentacles;** and two pairs of small dorsal **eyes** that are light-sensitive. The peristomium bears the ventral mouth, a pair of chitinous **jaws,** and four pairs of sensory tentacles (Fig. 14-18, *A*).

Each parapodium is formed of two lobes—a dorsal **notopodium** and a ventral **neuropodium** (Fig. 14-18, *C*). Each lobe is supported by one or more chitinous spines (acicula). The parapodia bear setae and are abundantly supplied with blood vessels. The parapodia

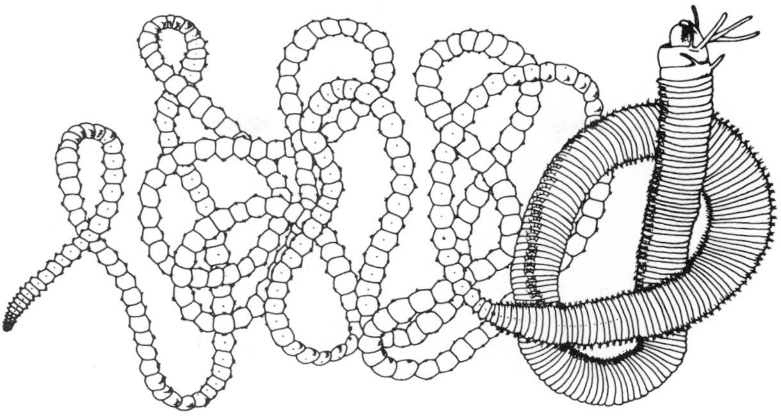

FIG. 14-16 *Eunice viridis,* the Samoan palolo worm. The posterior segments make up the epitokal region, consisting of segments packed with gametes. Each segment has an eyespot on the ventral side. Once a year the worms swarm and the epitokes detach, rise to the surface, and discharge their ripe gametes, leaving the water milky. By the next breeding season, the epitokes are regenerated. (Modified from W. M. Woodworth, 1907.)

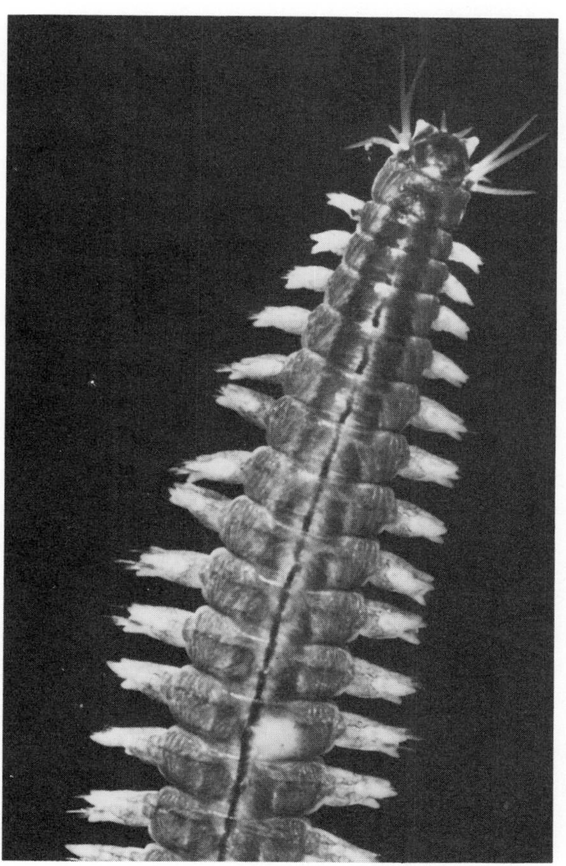

FIG. 14-17 *Nereis diversicolor,* head and anterior segments. Note the well-defined segments, the lobed parapodia, and the prostomium with tentacles. (Photograph by D. P. Wilson.)

are used for both creeping and swimming and are manipulated by oblique muscles that run from the midventral line to the parapodia in each somite. The worm swims by lateral undulatory wriggling of the body—unlike the peristaltic movement of the earthworms. It can dart through the water with considerable speed. These undulatory movements can be used in the burrow to suck water into or pump it out of the burrow. The worm will usually adapt some kind of burrow if it can find one. When a worm is placed near a glass tube, it will wriggle in without hesitation.

The clam worm feeds on small animals, other worms, larval forms, and the like. It seizes them with its chitinous jaws, which are protruded through the mouth when the pharynx is everted. When the pharynx is withdrawn, the food is swallowed. Movement of the food through the alimentary canal is by peristalsis.

Sexes are separate, but the reproductive organs are not permanent, for the sex cells are budded off from the coelomic lining and are carried to the outside by the nephridia and by bursting through the body wall. Fertilization is external, and the zygote develops into a free-swimming trochophore larva, which later metamorphoses into the adult form.

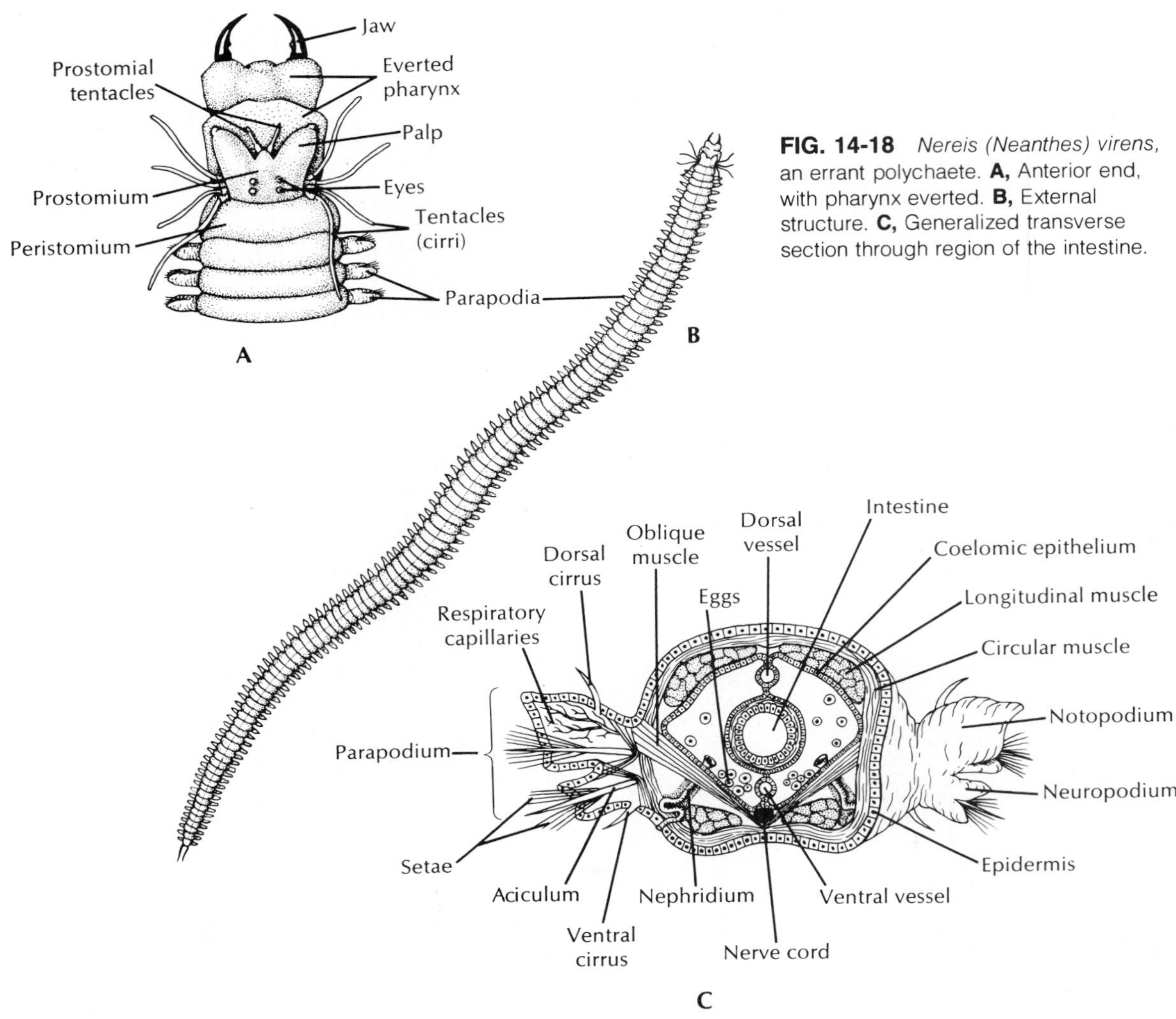

FIG. 14-18 *Nereis (Neanthes) virens,* an errant polychaete. **A,** Anterior end, with pharynx everted. **B,** External structure. **C,** Generalized transverse section through region of the intestine.

Parchment worms—Chaetopterus

Chaetopterus lives in a U-shaped parchment tube buried, except for the tapered ends, in sand or mud along the shore (Fig. 14-19). The worm attaches to the side of the tube by ventral suckers. Fans (modified parapodia) on segments 14 to 19 pump water through the tube by rhythmic movements. A pair of enlarged parapodia in the tenth segment secretes a long mucous bag that reaches back to a small food cup just in front of the fans. All the water passing through the tube is filtered through this mucous bag, the end of which is rolled up into a ball by cilia in the cup. When the ball is about the size of a BB shot, the fans stop beating and the ball of food and mucus is rolled forward by ciliary action to the mouth and is swallowed.

SIGNIFICANCE OF METAMERISM

No truly satisfactory theory has yet been given for the origins of metamerism and the coelom, though the subject has stimulated much speculation and debate over the years. One theory for the origin of metamerism is that chains of subzooids formed by asexual fission in flatworms may, instead of separating into distinct individuals, have held together and developed structural and functional unity with the passage of time.

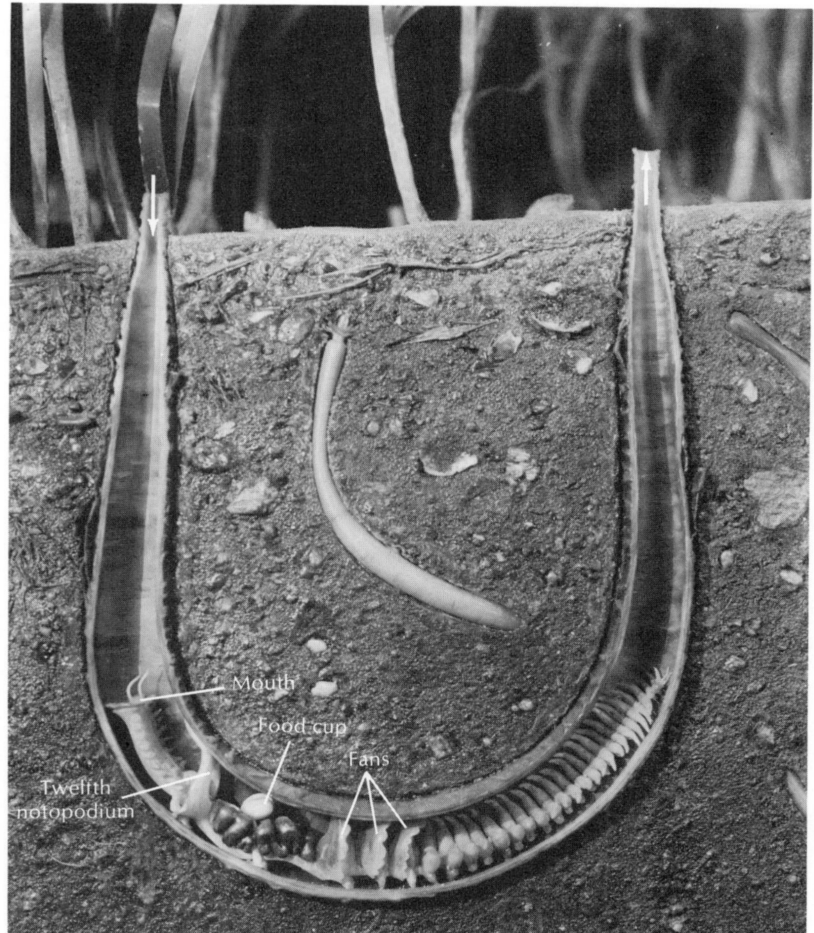

FIG. 14-19 *Chaetopterus*, a sedentery polychaete (in U-tube) and *Phascolosoma*, a sipunculan worm (in center). *Chaetopterus* lives in a parchment tube through which it pumps water with its three pistonlike fans. The fans beat 60 times per minute to keep water currents moving. The winglike notopodia of the twelfth segment continuously secrete a mucus net that strains out food particles. As the net fills with food, the food cup rolls it into a ball and, when the ball is large enough (about 3 mm), the food cup bends forward and deposits the ball in a ciliated groove to be carried by cilia to the mouth and swallowed. (Courtesy The American Museum of Natural History, New York.)

Another theory, which might be called the pseudometamerism theory, stresses the secondary origin of metamerism by the repetition of certain body parts. Serially arranged plates and muscles are found in the body wall of many elongated organisms, including some of the pseudocoelomates. In many flatworms, gonads and excretory structures are arranged in regular series. True segmentation could have been selected for as an adaptation for swimming in such a pseudometameric animal. Both of these theories separate the origin of metamerism from that of the coelom, that is, the ancestor that experienced incomplete fission or that was pseudometameric could have already been coelomate. Other theories, such as the gonocoel theory, imply a simultaneous origin of the coelom and metamerism. The gonocoel theory holds that a series of coelomic cavities (therefore, metameric) arose in an ancestor from gonadal cavities that persisted after the gametes were expelled. Like other theories requiring a simultaneous origin of the coelom and metamerism, the gonocoel theory has difficulty explaining the condition in groups such as the molluscs, which are coelomates, but for which there is little or no evidence that their ancestors were ever metameric.

These and all the other classic theories of the origin of metamerism and the coelom have had important arguments leveled against them, and more than one may be correct, or none, as suggested by R. B. Clark. The coelom and metamerism may have evolved independently in more than one group of animals, as, for example, in the chordates and in the protostome line. Clark stressed the functional and evolutionary significance of these features to the earliest animals that possessed them. He argued forcefully that the adaptive value of the coelom in the protostomes, at least, was as a **hydrostatic skeleton** in a burrowing animal. Thus, contraction of muscles in one part of the animal could act antagonistically on muscles in another part by transmission of the force of contraction through the enclosed constant volume of fluid in the coelom.

Though burrowing in the substrate may have selected for the coelom, certain other advantages accrued to its possessors. The coelomic fluid would have acted as a circulatory fluid for nutrients and wastes, making large numbers of flame cells distributed throughout the tissues unnecessary. Gametes could be stored in the spacious coelom for release simultaneously with other individuals in the population, thus enhancing chances of fertilization, and this would have selected for greater nervous and endocrine control. Finally, separation of the coelom into a series of compartments by septa, resulting in metamerism, would have greatly increased the capacity for powerful burrowing. The force exerted by muscle contraction on the fluid in one coelomic compartment would not be transmitted to the fluid in other compartments, making possible independent and separate movements by the separate metameres. Independent movements of metameres in different parts of the body would have placed selective value on a more sophisticated nervous system for control of the movements and led to elaboration of the central nervous system.

PHYLOGENY AND ADAPTIVE RADIATION

Phylogeny. There are so many similarities in the early development of the molluscs, annelids, and primitive arthropods that there seems little doubt about their close relationship. It is thought that the common ancestor of the three phyla was some type of flatworm. Many marine annelids and molluscs have an early embryogenesis typical of protostomes, in common with some marine flatworms, suggesting a real, if remote, relationship. Annelids share with the arthropods an outer secreted cuticle and a similar nervous system, and there is a similarity between the lateral appendages (parapodia) of many marine annelids and the appendages of certain primitive arthropods. The most important resemblance, however, probably lies in the segmented plan of the annelid and the arthropod body structure.

Which of the annelids came first? This has been the subject of a long and continuing debate. It was long assumed that the archiannelids, a small group of minute polychaetes, were the most primitive of the annelids. This view is no longer widely held. Rather than primitive, they appear to be polychaetes that have become secondarily simplified as a result of their small size and their adaptation to a psammolittoral life—that is, living in the interstitial spaces between the grains of sand or mud bottom. R. B. Clark (1969) suggests that the origin of annelids needs to be viewed in terms of the origin of the metameric arrangement of worms. He emphasizes that the only consistent correlation between the structure and function of a segmented coelom is a mechanical one. The division of the coelom, a hydrostatic organ, into units is a mechanical device limiting the transfer of hydrostatic pressure from one part of the coelom to another. There is an advantage in such a locomotor device to macroscopic animals, but not to microscopic ones, so there is little reason to assume it evolved first in the microscopic archiannelids. Most theories of annelid origin have assumed that the polychaetes were the earliest segmented worms and that metamerism arose in connection with the development of lateral appendages (parapodia). However, the oligochaete body is adapted to vagrant burrowing in the substratum with a peristaltic movement that is highly benefited by a metameric coelom. Polychaetes, on the other hand, are adapted to swimming and crawling in a medium too fluid for effective peristaltic locomotion. Although the parapodia do not prevent such locomotion, they do little to further it, and it seems unlikely that they evolved as an adaptation for swimming. Though the polychaetes are more primitive in some respects, for example, in their reproductive system, it would appear more plausible to consider the ancestral annelids as having been similar to the oligochaetes in overall body plan and as having given rise to the more specialized leeches and polychaetes. The leeches are closely related to the oligochaetes and probably evolved from them in connection with a swimming existence and the abandonment of a burrowing mode of life.

Adaptive radiation. Annelids are an ancient group that has undergone extensive adaptive radiation. The basic body structure, particularly of the polychaetes, lends itself to almost endless modification. As marine worms, polychaetes have a wide range of habitats in an environment that is not physically or physiologically demanding. Unlike the earthworms, whose environment imposes strict physical and physiologic selective pressure, the polychaetes have been free to experiment and thus to have achieved a wide range of adaptive features.

A basic adaptive feature in the evolution of annelids is their septal arrangement, resulting in fluid-filled coelomic compartments. Fluid pressure in these compartments is used as a hydrostatic skeleton in precise movements such as burrowing and swimming. Powerful circular and longitudinal muscles have been adapted for flexing, shortening, and lengthening the body.

There is a wide variation in feeding adaptations, from the sucking pharynx of the oligochaetes and the chitinous jaws of carnivorous polychaetes to the specialized tentacles and cirri of the ciliary feeders.

In polychaetes the parapodia have been adapted in many ways and for many functions, chiefly locomotion and respiration.

In leeches many of their adaptations, such as suckers, cutting jaws, pumping pharynx, distensible gut, and the production of hirudin, are related to their predatory and bloodsucking habits.

Derivation and meaning of names

Amphitrite (Greek mythology, sea goddess) A common genus of marine polychaete annelids.

Arenicola (L., *arena,* sand, + *colere,* to inhabit) A genus of polychaete worms.

Chaetopterus (Gr., *chaitè,* hair or mane, + *pteron,* wing) Genus of polychaete worm.

Enchytraeus (Gr., *enchytraeus,* living in an earthen pot) Genus of small white oligochaete.

Lumbricus (L., intestinal worm) Genus of common earthworm.

Nereis (Greek mythology, a Nereid, or daughter of Nereus, ancient sea god) A genus of polychaete worms.

Sabella (L., *sabulum,* sand) Genus of polychaete tubeworm.

Tubifex (L., *tubus,* a tube, + *faciens,* to make or do) Genus of aquatic oligochaete.

Annotated references

Selected general readings

Barnes, R. D. 1965. Tube-building and feeding in chaetopterid polychaetes. Biol. Bull. **129:**217-233.

Barnes, R. D. 1974. Invertebrate zoology, ed. 3. Philadelphia, W. B. Saunders Co., pp. 233-316.

Brinkhurst, R. O., and Jamieson, B. G. 1972. Aquatic oligochaetes of the world. Toronto, Toronto University Press.

Brown, S. C. 1975. Biomechanics of water pumping by *Chaetopterus variopedatus* Renier. Skeletomusculature and kinetics. Biol. Rev. **149:**136-156 (Aug.).

Clark, L. B., and W. N. Hess. 1940. Swarming of the Atlantic palolo worm, *Leodice fucata.* Tortugas Lab. Papers **332:**21-70.

Clark, R. B., 1964. Dynamics in metazoan evolution. The origin of the coelom and segments. Oxford, England. The Clarendon Press.

Clark, R. B. 1969. Systematics and phylogeny: Annelida, Echiura, Sipuncula. In K. Florkin, and B. T. Scheer (eds.). Chemical zoology, vol. 4. New York, Academic Press, Inc.

Corning, W. C., J. A. Dyal, and A. O. D. Willows (eds.). 1973. Invertebrate learning, vol. 1. New York, Plenum Publishing Corp.

Dales, R. P. 1967. Annelids, ed. 2. London, The Hutchinson Publishing Group Ltd. *A concise account of the annelids.*

Darwin, C. R. 1881. The formation of vegetable mould through the action of worms. *A classic account of the way in which earthworms improve and transform the surface of the soil.*

Edmondson, W. T. (ed.). 1959. Fresh-water biology, ed. 2. New York, John Wiley & Sons, Inc. *Contains a guide to identification of freshwater annelids, with keys.*

Gardner, C. R. 1976. The neuronal control of locomotion in the earthworm. Biol. Rev. **51:**25-52.

Giese, A. C., and J. S. Pearse (eds.). Repoduction of marine invertebrates, vol. 3. New York, Academic Press, Inc.

Gosner, K. L. 1971. Guide to identification of marine and estuarine invertebrates: Cape Hatteras to the Bay of Fundy. New York, Interscience Publishers, pp. 326-387.

Hartman, O., 1968. Atlas of the errantiate polychaetous annelids from California. Los Angeles, University of Southern California, Allan Hancock Foundation. *Extensive keys for identification.*

Hickman, C. P. 1973. Biology of the invertebrates, ed. 2. St. Louis, The C. V. Mosby Co.

Kaestner, A. 1967. Invertebrate zoology, vol. 1. New York, Interscience Publishers, pp. 454-566.

Laverack, M. S. 1963. The physiology of earthworms. New York, The Macmillan Co. *A concise account.*

Light, S. F., R. I. Smith, F. A. Pitelka, D. P. Abbott, and F. M. Weesner. 1967. Intertidal invertebrates of the central California coast. Berkeley, University of California Press, pp. 63-108.

Mangum, C. 1970. Respiratory physiology in annelids. Am. Sci. **58**(6):641-647.

Mann, K. H. 1962. Leeches *(Hirudinea),* their structure, physiology, ecology, and embryology. New York, Pergamon Press, Inc.

Meglitsch, P. A. 1972. Invertebrate zoology, ed. 2. New York, Oxford University Press, Inc.

Mulloney, B. 1970. Structure of the giant fibers of earthworms. Science **168:**994-996 (May).

Pennak, R. W. 1953. Freshwater invertebrates of the United States. New York, The Ronald Press Co., pp. 278-320. *A brief account with keys to families and genera of freshwater annelids.*

Sawyer, R. T. 1972. North American freshwater leeches, exclusive of the Piscicolodae, with a key to all species. Ill. Biol. Monogr. No. 46.

Stephenson, J. 1972. The Oligochaeta. New York, Wheldon & Wesley Ltd., Stechert-Hafner Service Agency, Inc. *A reprint of the classic 1930 edition. Out of date in some areas, but still valuable.*

Wald, G., and S. Rayport. 1977. Vision in annelid worms. Science **196:**1434-1439 (June).

Selected *Scientific American* articles

Nicholls, J. C., and D. Van Essen. 1974. The nervous system of the leech. **230:**38-48 (Jan.).

Wells, P. 1959. Worm autobiographies. **200:**132-141 (June).

CHAPTER 15

THE ARTHROPODS

Phylum Arthropoda
Subphylum Trilobita
Subphylum Chelicerata

A eurypterid, or sea scorpion, attacks a swimming crustacean, as depicted in this Silurian diorama at the American Museum of Natural History. Arthropods are an ancient group that were already thriving in Cambrian times.

Courtesy The American Museum of Natural History, New York.

Position in animal kingdom

1. Arthropods belong to the **protostome** branch, or schizocoelous coelomates, of the animal kingdom, and primitively they have spiral and determinate cleavage.
2. Arthropods have much of the characteristic structure of higher forms—bilateral symmetry, triploblastic, coelomic cavity, and organ systems.
3. They share with the annelids the property of conspicuous segmentation, but in their somites they emphasize greater variety and more grouping for specialized purposes, and to the somites have been added the jointed appendages, with pronounced division of labor, resulting in greater variety of action.

Biologic contributions

1. **Cephalization** makes additional advancements, with centralization of fused ganglia and sensory organs in the head.
2. The **somites** have gone beyond the sameness of the annelid type and are now **specialized** for a variety of purposes, forming functional groups of somites **(tagmosis).**
3. The presence of paired **jointed appendages** diversified for numerous uses makes for greater adaptability.
4. Locomotion is by extrinsic limb muscles, in contrast to the body musculature of annelids. **Striated muscles** are emphasized, thus ensuring rapidity of movement.
5. Although **chitin** is found in a few other forms below arthropods, its use is better developed in the arthropods. The **cuticular exoskeleton,** containing chitin, is a great advance over that of the annelids, making possible a wide range of adaptations.
6. The gills, and especially the **tracheae,** represent a breathing mechanism more efficient than that of most invertebrates.
7. The alimentary canal shows greater specialization by having chitinous teeth, compartments, and gastric ossicles.
8. Behavior patterns have advanced far beyond those of most invertebrates, with a higher development of primitive intelligence and **social** organization.
9. **Metamorphosis** is common in development.
10. Many arthropods have well-developed protective coloration and protective resemblances.

PHYLUM ARTHROPODA

Phylum Arthropoda (ar-throp′o-da) (Gr., *arthron,* joint, + *pous, podos,* foot) is the most extensive phylum in the animal kingdom, making up more than three-fourths of all known species. Approximately 900,000 species have been recorded, and probably as many more remain to be classified. Arthropods include the spiders, scorpions, ticks, mites, crustaceans, millipedes, centipedes, insects, and some others. In addition there is a rich fossil record extending to the very late Precambrian period (p. 327).

Arthropods are eucoelomate protostomes with well-developed organ systems, and they share with the annelids the property of conspicuous segmentation.

Arthropods have an exoskeleton containing chitin, and their primitive pattern is that of a linear series of similar somites, each with a pair of jointed appendages. However, the pattern of somites and appendages varies greatly in the phylum. There is a tendency for the somites to be combined or fused into functional groups, called **tagmata,** for specialized purposes; the appendages are frequently differentiated and specialized for pronounced division of labor.

Few arthropods exceed 60 cm in length, and most are far below this size. The largest is the Japanese crab *Macrocheira,* which has approximately a 3.7 m span; the smallest is the parasitic mite *Demodex,* which is less than 0.1 mm long.

Arthropods are usually active, energetic animals. Judged by their great diversity and their wide ecologic distribution, as well as by the vast numbers of species, their success is surpassed by no other group of animals.

Although arthropods compete with humans for food supplies and spread serious diseases, they are essential in pollination of many food plants, and they also serve as food, yield useful drugs and dyes, and produce useful products such as silk, honey, and beeswax.

Ecologic relationships

The arthropods are more widely and more densely distributed throughout all regions of the earth than are members of any other phylum. They are found in all types of environment from low ocean depths to very high altitudes, and from the tropics far into both north and south polar regions. Different species are adapted for life in the air; on land; in fresh, brackish, and marine waters; and in or on the bodies of plants and other animals. Some species live in places where no other form could survive.

Although all types—carnivorous, omnivorous, and

symbiotic—occur in this vast group, the majority are herbivorous. Most aquatic arthropods depend on algae for their nourishment, and the majority of land forms live chiefly on plants. In diversity of ecologic distribution the arthropods have no rivals.

Characteristics

1. Bilateral symmetry; **metameric body,** often divided into head and trunk; head, thorax, and abdomen; or cephalothorax and abdomen
2. **Jointed appendages;** primitively, one pair to each somite, but number often reduced; appendages often modified for specialized functions
3. **Exoskeleton of cuticle** containing protein, lipid, chitin, and often calcium carbonate secreted by underlying epidermis and shed (molted) at intervals
4. **Complex muscular system,** with exoskeleton for attachment; **striated muscles** for rapid action; smooth muscles for visceral organs; no cilia
5. **Reduced coelom** in adult; most of body cavity consisting of hemocoel (sinuses, or spaces, in the tissues) filled with blood
6. Complete digestive system; mouthparts modified from appendages and adapted for different methods of feeding
7. Open circulatory system, with dorsal **contractile heart,** arteries, and hemocoel (blood sinuses)
8. Respiration by **body surface, gills, tracheae** (air tubes), or **book lungs**
9. Paired excretory glands called coxal, antennal, or maxillary glands present in some, homologous to metameric nephridial system of annelids; some with other excretory organs, called **malpighian tubules**
10. Nervous system of annelid plan, with dorsal brain connected by a ring around the gullet to a double nerve chain of ventral ganglia; fusion of ganglia in some species; well-developed sensory organs
11. Sexes usually separate, with paired reproductive organs and ducts; usually internal fertilization; oviparous or ovoviviparous; often with **metamorphosis;** parthenogenesis in a few forms

Classification

Subphylum Trilobita (tri'lo-bi'ta) (Gr., *tri-,* three, + *lobos,* lobe) **trilobites.** All extinct forms; Cambrian to Carboniferous; body divided by two longitudinal furrows into three lobes; distinct head, thorax, and abdomen; biramous (two-branched) appendages.

Subphylum Chelicerata (ke-lis'e-ra'ta) (Gr., *chēlē,* claw, + *keras,* horn, + *ata,* group suffix)—**eurypterids, horseshoe crabs, spiders, ticks.** First pair of appendages modified to form chelicerae; pair of pedipalps and four pairs of legs; no antennae, no mandibles; cephalothorax and abdomen usually unsegmented.

Class Merostomata (mer'o-sto'-ma-ta) (Gr., *mēros,* thigh, + *stoma,* mouth, + *ata,* group suffix)—**aquatic chelicerates.** Cephalothorax and abdomen; compound lateral eyes; appendages with gills; sharp telson.

Subclass Eurypterida (yu-rip-ter'i-da) (Gr., *eurys,* broad, + *pteryx,* wing or fin, + *-ida,* pl. suffix)—**giant water scorpions.** Extinct; cephalothorax covered by dorsal carapace; abdomen with 12 segments and postanal telson; pair of simple ocelli and pair of compound eyes. Example: *Eurypterus.*

Subclass Xiphosurida (zif-o-su'ri-da) (Gr., *xiphos,* sword, + *oura,* tail)—**horseshoe crabs.** Cephalothorax convex- and horseshoe-shaped; abdomen unsegmented and terminated by long spine; three-jointed chelicerae and six-jointed walking legs; pair of simple eyes and pair of compound eyes; book gills. Example: *Limulus.*

Class Pycnogonida (pik'no-gon'i-da) (Gr., *pyknos,* compact, + *gony,* knee, angle)—**sea spiders.** Small (3 to 4 mm), but some reach 500 mm; body chiefly cephalothorax; tiny abdomen; usually eight pairs of long walking legs (some with 10 to 12 pairs); one pair of subsidiary legs (ovigers) for egg-bearing; mouth on long proboscis; four simple eyes; no respiratory or excretory system. Example: *Pycnogonum.*

Class Arachnida (ar-ack'ni-da) (Gr., *arachnē,* spider)—**scorpions, spiders, mites, ticks, harvestmen.** Four pairs of legs; segmented or unsegmented abdomen with or without appendages and generally distinct from cephalothorax; respiration by gills, tracheae, or book lungs; excretion by malpighian tubules or coxal glands; dorsal bilobed brain connected to ventral ganglionic mass with nerves; simple eyes; sexes separate; chiefly oviparous; no true metamorphosis. Examples: *Argiope, Centruroides.*

Subphylum Mandibulata* (man-dib'u-la'ta) (L., *mandibula,* mandible, + *ata,* group suffix). Head appendages consisting of one or two pairs of antennae, one pair of mandibles, and one or two pairs of maxillae.

Class Crustacea (crus-ta'she-a) (L., *crusta,* shell, + *acea,* group suffix)—**crustaceans.** Mostly aquatic, with gills; hard exoskeleton; cephalothorax usually with dorsal carapace; biramous appendages, modified for various functions; two pairs of antennae. Sexes usually separate. Development primitively with nauplius stage (see resumé of subclasses, p. 363).

Class Diplopoda (di-plop'o-da) (Gr., *diploos,* double, + *pous, podos,* foot)—**millipedes.** Subcylindric body; head with short antennae and simple eyes; body with variable number of somites; short legs, usually two pairs of legs to a somite; separate sexes; oviparous. Examples: *Julus, Spirobolus.*

Class Chilopoda (ki-lop'-o-da) (Gr., *cheilos,* lip, + *pous, podos,* foot)—**centipedes.** Dorsoventrally flattened body; variable number of somites, each with one pair of legs; one pair of long antennae; separate sexes; oviparous. Examples: *Cermatia, Lithobius, Geophilus.*

*Subphylum Mandibulata has traditionally contained those arthropods possessing mandibles and antennae—crustaceans, myriapods, and insects. However, many authorities now believe that crustacean mandibles and insect-myriapod mandibles have resulted from convergent evolution rather than a natural close relationship, and they therefore believe that although the term "mandibulates" is a convenient common term of inclusion for those groups having mandibles, the term "Mandibulata" should not be used to imply evolutionary significance.

Class Pauropoda (pau-rop'o-da) (Gr., *pauros,* small, + *pous, podos,* foot)—**pauropods.** Minute (1 to 1.5 mm); cylindric body consisting of double segments and bearing nine or ten pairs of legs; no eyes. Example: *Pauropus.*

Class Symphyla (sym'fy-la) (Gr., *syn,* together, + *phylon,* tribe)—**garden centipedes.** Slender (1 to 8 mm) with long, filiform antennae; body consisting of 15 to 22 segments with ten to 12 pairs of legs; no eyes. Example: *Scutigerella.*

Class Insecta (in-sek'ta) (L., *insectus,* cut into)—**insects.** Body with distinct head, thorax, and abdomen and usually a marked constriction between thorax and abdomen; pair of antennae; mouthparts modified for different food habits; head of six fused somites; thorax of three somites; abdomen with variable number, usually 11 somites; thorax with two pairs of wings (sometimes one pair or none) and three pairs of jointed legs; separate sexes; usually oviparous; gradual or abrupt metamorphosis. (A brief description of insect orders is given on pp. 408 to 412.)

Comparison of Arthropoda with Annelida

Similarities between Arthropoda and Annelida are as follows:

1. External segmentation
2. Segmental arrangement of muscles
3. Metamerically arranged nervous system with dorsal cerebral ganglia

Arthropods differ from Annelids in having the following:

1. Fixed number of segments
2. Usually a lack of intersegmental septa
3. Tagmatization
4. Coelomic cavity reduced; main body cavity a hemocoel
5. Open (lacunar) circulatory system
6. Special mechanisms (gills, tracheae, book lungs) for respiration
7. Exoskeleton containing chitin
8. Jointed appendages
9. Compound eye and other well-developed sense organs
10. Absence of cilia
11. Metamorphosis in many cases

Why have arthropods been so successful?

The success of the arthropods is attested to by their diversity, number of species, wide distribution, variety of habitats and feeding habits, and power of adaptation to changing conditions. Some of the structural and physiologic patterns that have been helpful to them are briefly summarized in the following discussion.

1. A versatile exoskeleton. The arthropods possess an exoskeleton that is highly protective without sacrificing mobility. This skeleton is the **cuticle,** an outer covering secreted by the underlying epidermis. The cuticle is made up of an inner and usually thicker **endocuticle** and an outer, relatively thin **epicuticle.** The endocuticle contains **chitin** bound with protein. Chitin is a tough, resistant, nitrogenous polysaccharide that is insoluble in water, alkalis, and weak acids. Thus the endocuticle is not only flexible and light-weight but also affords protection, particularly against dehydration. In some crustaceans the chitin may make up as much as 60% to 80% of the endocuticle, but in insects it is probably not more than 40%. In most crustaceans, the endocuticle in some areas is also impregnated with **calcium salts,** which reduces its flexibility. In the hard shells of lobsters and crabs, for instance, this calcification is extreme. The outer epicuticle is composed of protein and lipid. The protein is stabilized and hardened by tanning, adding further protection. Both the endocuticle and epicuticle are laminated, that is, composed of several layers each (Fig. 28-1, p. 647).

The cuticle may be soft and permeable or may form a veritable coat of armor. Between body segments and between the segments of appendages it is thin and flexible, creating movable joints and permitting free movements. In crustaceans and insects the cuticle forms ingrowths (apodemes) that serve for muscle attachment. It may also line the fore- and hindgut; line and support the trachea; and be adapted for biting mouthparts, sensory organs, copulatory organs, and ornamental purposes. It is indeed a versatile material.

The nonexpansible cuticular exoskeleton does, however, impose important conditions on growth. To grow, an arthropod must shed its outer covering at intervals and grow a larger one—a process called **ecdysis,** or **molting.** Arthropods molt from four to seven times before reaching adulthood, and some continue to molt after that. An exoskeleton is also relatively heavy and becomes proportionately heavier with increasing size. This tends to limit the ultimate body size.

2. Segmentation and appendages for more efficiency and better locomotion. Typically each somite is provided with a pair of jointed appendages, but this arrangement is often modified, with both segments and appendages specialized for adaptive functions. The limb segments are essentially hollow levers that are moved by internal muscles, most of which are striated for rapid action. The jointed appendages are equipped with sensory hairs and have been modified and adapted

FIG. 15-1 Fossils of early arthropods. **A,** Trilobite, dorsal view; plaster-cast impression. These animals were abundant in mid-Cambrian period. **B,** A eurypterid fossil; eurypterids flourished in Europe and North America from Ordovician to Permian periods. (Photographs by F. M. Hickman.)

for sensory functions, food handling, swift and efficient walking legs, and swimming appendages. This affords greater efficiency and a wider capacity for adjustment to varied habitats.

3. Air piped directly to cells. Most land arthropods have the highly efficient tracheal system of air tubes, which delivers oxygen directly to the tissues and cells and makes a high metabolic rate possible. Aquatic arthropods breathe mainly by some form of gill that is quite efficient.

4. Highly developed sensory organs. Sensory organs are found in great variety, from the compound (mosaic) eye to those simpler senses that have to do with touch, smell, hearing, balancing, chemical reception, and so on. Arthropods are keenly alert to what goes on in their environment.

5. Complex behavior patterns. Arthropods exceed most other invertebrates in the complexity and organization of their activities. Innate (unlearned) behavior unquestionably controls much of what they do, but learning also plays an important part in the lives of many of them.

6. Reduced competition through metamorphosis. Many arthropods pass through metamorphic changes, including a larval form quite different from the adult in structure. The larval form is often adapted for eating a different kind of food from that of the adult, resulting in less competition within a species.

Economic importance

Arthropods serve as an important source of food for many animals. Lobsters, crabs, shrimp, and crayfish are eaten by humans all over the world. Plankton contains many crustaceans and is food for fish and other aquatic animals. Insects are important as food for many birds and animals, and some are relished by humans in various areas of the world.

They are useful in other ways than as food. Some insects are predators that live on other insects, helping to keep the harmful ones in check. Spiders probably stand at the top of the list as enemies of insects. Arthropod products include shellac produced by scale insects, cochineal (a dye) from other scale insects, silk spun by silkworm larvae, honey and beeswax from bees, and so on. They are also essential in pollination of trees and other plants, including many food plants of humans.

Many arthropods are harmful. Insects destroy millions of dollars worth of food each year. Arthropods carry devastating diseases: mosquitoes carry malaria; copepods carry larval stages of the guinea worm and fish tapeworm; mites and ticks carry diseases and also live as ectoparasites; some spiders and scorpions deliver poisonous bites; barnacles foul ship bottoms; sow bugs and a variety of insects damage gardens and greenhouse crops.

SUBPHYLUM TRILOBITA

The trilobites probably had their beginnings a million or more years before the Cambrian period in which they flourished. They have been extinct some 200 million years, but were abundant during the Cambrian and Ordovician periods. Their name refers to the trilobed shape of the body, caused by a pair of longitudinal grooves. They were bottom dwellers, probably scavengers (Fig. 15-1, *A*). Most of them could roll up like pill bugs, and they ranged from 2 to 67 cm in length.

The exoskeleton contained chitin, strengthened in

FIG. 15-2 Horseshoe crabs mate in shallow water at high tide. The larger female burrows into the sand to lay her many eggs while the male covers them with sperm. As the tide recedes, she covers her eggs with sand, and they return to the sea. The eggs, warmed by the sun, hatch in about 2 weeks. (Photograph by L. L. Rue, III.)

some areas by calcium carbonate. The body comprised three tagmata: head, thorax, and pygidium. The head was one piece but showed signs of former segmentation; the thorax had a variable number of somites; and the somites of the pygidium, at the posterior end, were fused into a plate. The head bore a pair of antennae, compound eyes, mouth, and four pairs of jointed appendages. Each body somite except the last also bore a pair of appendages each with a branch bearing a fringe of filaments that probably served as gills.

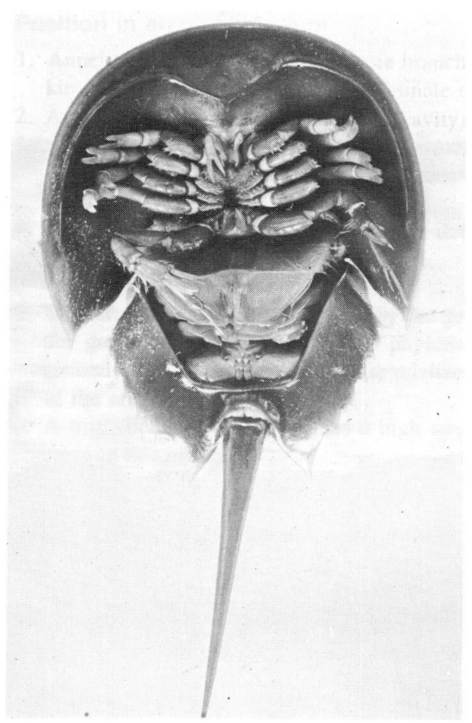

FIG. 15-3 Ventral view of horseshoe crab *Limulus* (class Merostomata). They grow up to 0.5 m in length. (Photograph by F. M. Hickman.)

SUBPHYLUM CHELICERATA

The chelicerate arthropods make up a very ancient group that includes the eurypterids (extinct), horseshoe crabs, spiders, ticks and mites, scorpions, and sea spiders. They are characterized by having six pairs of appendages that include a pair of **chelicerae,** a pair of **pedipalps,** and **four pairs of walking legs.** They have **no mandibles** and **no antennae.** Most chelicerates suck up liquid food from their prey.

Class Merostomata

Class Merostomata is represented by the eurypterids, all now extinct, and by the xiphosurids, or horseshoe crabs, an ancient group sometimes referred to as ''living fossils.''

Eurypterids—subclass Eurypterida. The eurypterids, or giant water scorpions (Fig. 15-1, *B,* and p. 327) were largest of all fossil arthropods, with some of them reaching a length of 3 m. Their fossils are found in rocks from the Ordovician to the Permian periods. They had many resemblances to the marine horseshoe crabs (Figs. 15-2 and 15-3) and also to the scorpions, their land counterparts. The head had six fused segments and bore both simple and compound eyes and six pairs of appendages. The abdomen had 12 segments and a spikelike tail.

Theories of their early habitats differ. Some believe the eurypterids evolved mainly in fresh water, along with the ostracoderms; others hold that they arose in brackish lagoons.

Horseshoe crabs—subclass Xiphosurida. The xiphosurids are an ancient marine group that dates from the Cambrian period. Our common horseshoe crab *Limulus (Xiphosura)* (Fig. 15-2) goes back practically unchanged to the Triassic period. There are only three genera (five species) living today: *Limulus,* which lives in shallow water along the North American Atlantic coast; *Carcinoscorpius,* along the southern shore of Japan; and *Tachypleus,* in the East Indies and along the coast of southern Asia. They usually live in shallow water.

Xiphosurids have an unsegmented, horseshoe-shaped **carapace** (hard dorsal shield) and a broad abdomen, which has a long **telson,** or tailpiece. The cephalothorax bears five pairs of walking legs and a pair of chelicerae, whereas the abdomen has six pairs of broad thin appendages that are fused in the median line (Fig. 15-3). On some of the abdominal appendages, **book gills** (flat, leaflike gills) are exposed. There are two compound and two simple eyes on the carapace. The horseshoe crab swims by means of its abdominal plates and can walk with its walking legs. It feeds at night on worms and small molluscs, which it seizes with its chelicerae.

During the mating season the horseshoe crabs come to shore at high tide to mate (Fig. 15-2). The female burrows into the sand where she lays her eggs, with one or more smaller males following her closely to add their sperm to the nest before she covers the eggs with sand. Here the eggs are warmed by the sun and protected from the waves until the young larvae hatch and return to the sea by another high tide. The larvae are segmented and are often called ''trilobite larvae'' because they resemble the trilobites, which are often considered to have been their ancestors.

Class Pycnogonida—sea spiders

Some sea spiders are only a few millimeters long, but others are much larger. They have small, thin

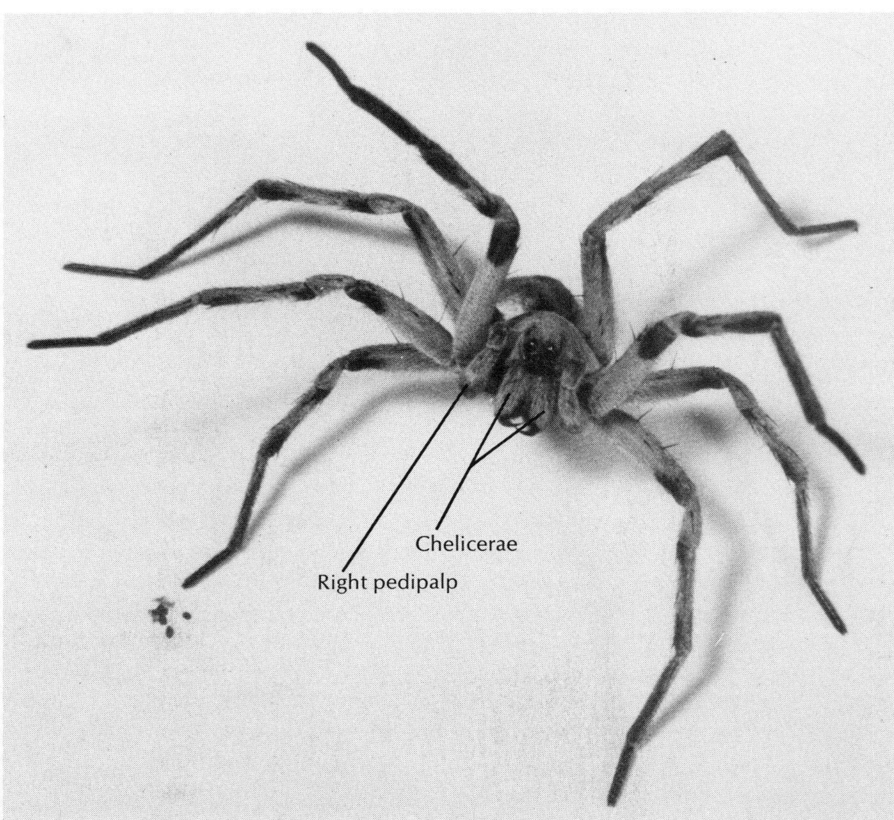

FIG. 15-4 A wolf spider strikes a defensive pose that shows its appendages, chelicerae, and eyes to good advantage. (Photograph by F. M. Hickman.)

bodies and usually four pairs of very long, thin walking legs. Some species are provided with chelicerae and palps. One unique feature is the subsidiary pair of egg-bearing legs **(ovigers)** on which the males carry the developing eggs. The mouth is located at the tip of a long suctorial proboscis used to suck juices from cnidarians and soft-bodied animals. Most of them have four simple eyes. They have a heart for slowly circulating the blood in the hemocoel, but there are no small blood vessels. Excretory and respiratory systems are absent. The long, thin body and legs provide a large surface, in proportion to volume, that is evidently sufficient for diffusion of gases and wastes. Because of the small size of the body, the digestive system sends branches into the legs, and most of the gonads are also found there.

Sea spiders are found in all oceans, but they are most abundant in polar waters. *Pycnogonum* is a common intertidal genus found on both the Atlantic and Pacific coasts.

Some authorities believe that the pycnogonids are more closely related to the crustaceans than to other arthropods; others have placed them closer to the arachnids. However, they are so highly modified and their fossil record is so scarce that any assertions made about their origins and relationships must be purely speculative.

Class Arachnida

The arachnids (Gr., *arachnē,* spider) are not as closely knit a group as the insects. In addition to spiders, the group includes scorpions, pseudoscorpions, whip scorpions, ticks, mites, daddy longlegs (harvestmen), and others. There are many differences among these with respect to form and appendages. Most of them are free-living and are far more common in warm, dry regions than elsewhere.

The arachnid body is made up of a cephalothorax and abdomen and usually bears a pair of chelicerae, a pair of pedipalps, and four pairs of walking legs (Fig.

15-4). Antennae and mandibles are lacking. Most arachnids are predaceous and may be provided with claws, fangs, poison glands, or stingers. They usually have sucking mouthparts or a strong sucking pharynx with which they suck the fluids and soft tissues from the bodies of their prey. Among their interesting adaptations are the spinning glands of the spiders.

Arachnids have been extremely successful. More than 50,000 species have been described so far. They were the first of the arthropods to move into terrestrial habitats. Scorpions are found among Silurian fossils, and by the end of the Paleozoic period mites, spiders, and solpugids had appeared.

Most arachnids are harmless to humans, and actually do much good by destroying injurious insects. A few, such as the black widow and the brown recluse spiders, can give painful or even dangerous bites. The sting of the scorpion may be quite painful. Some ticks and mites are vectors of diseases as well as causes of annoyance and painful irritations. Certain mites damage a number of important food and ornamental plants by sucking their juices.

Spiders—order Araneae. The spiders are a large group of 35,000 species, distributed all over the world. The spider body is compact—a **cephalothorax (prosoma)** and **abdomen (opisthosoma),** both unsegmented and joined by a slender pedicel.

The anterior appendages are a pair of **chelicerae** (Fig. 15-4), which have terminal **fangs** provided with ducts from poison glands, and a pair of **pedipalps** with basal parts with which they chew (Fig. 15-4). Four pairs of **walking legs** terminate in claws.

All spiders are predaceous and feed largely on insects. Hunting spiders include wolf spiders, jumping spiders, fishing spiders, crab spiders, tarantulas, trapdoor spiders, and others. Some of these stalk their prey. Some lay a dragline. Jumping spiders leap on their prey. Most other spiders spin some sort of web to snare and entangle the prey, the most familiar being the geometric designs of the orb-weavers (Fig. 15-5). The lasso spider *Mastophora* throws a thread with a sticky ball attached to the end at passing insects. Another lasso spider *Cladomelea* uses the sticky ball and thread as a pendulum, which it swings back and forth with one leg until it catches some luckless passerby. *Scytodes,* the spitting spider, squirts viscid threads of gum from its chelicerae to entangle its prey. The prey of most spiders is limited to small invertebrates, but cases are known of mice and fledgling birds being caught by larger spiders, and the fishing spider *Dolomedes* is known to catch small fish (Fig. 15-6).

The spider seizes its prey with chelicerae and pedipalps, injecting into it venom from the poison glands. Web-weavers wrap up the captured victim with more silk before biting them. Orb-weavers, however, have evolved a means of dealing with a moth or butterfly, which, because of the loose scales that are shed into the web, can escape more easily than other insects. The spider applies a long bite to quiet such prey before wrapping it up in the usual manner.

Spiders with toothless chelicerae puncture the prey with the fangs, allow digestive enzymes to flow from the mouth into the wound to liquefy the tissues, then suck up the resulting broth into the stomach. Wolf spiders, tarantulas, and others with toothed chelicerae crush or chew up the prey, aiding digestion by enzymes from the mouth. The digestive system fills most of the abdomen and the posterior half of the cephalothorax (Fig. 15-7).

Spiders breathe by means of book lungs or tracheae or both. Book lungs, which are unique in spiders, consist of many parallel air pockets extending into a blood-filled chamber (Fig. 15-7). Air enters the chamber by a slit in the body wall. Because these air pockets are flattened and leaflike, the whole structure is called a book lung.

The tracheae make up a system of air tubes that carry air directly to the tissues from openings in the body wall called **spiracles.** The tracheae are similar to those in insects (p. 384) but are much less extensive. Oxygen is distributed by the blood in spiders, whereas it is distributed directly to the body cells by the tracheal system of insects. The circulatory system in spiders includes a dorsal heart, vessels, and sinuses.

Spiders and insects have a unique **excretory system of malpighian tubules** (Fig. 15-7), which work in conjunction with specialized rectal glands. Potassium and other solutes and waste materials are secreted into the tubules, which drain the fluid, or "urine," into the intestine. The rectal glands reabsorb most of the potassium and water, leaving behind such wastes as uric acid. By this cycling of water and potassium, species living in dry environments may conserve body fluids, producing a nearly dry mixture of urine and feces. Many spiders also have **coxal glands,** which are modified nephridia, that open at the coxa, or base, of the first and third walking legs.

Spiders usually have eight **simple eyes,** each provided with a lens, optic rods, and a retina (Fig. 15-7,

FIG. 15-5 *Argiope aurantia,* the common yellow and black garden spider, builds its orb web in gardens or tall grasses and then hangs there, head down, waiting for the arrival of fresh meat in the form of an unwary insect. (Photograph by J. H. Gerard.)

B). They are used chiefly for perception of moving objects, but some, such as those of the hunting and jumping spiders, may form images. Since vision is usually poor, a spider's awareness of its environment depends a great deal on its **sensory hairs.** Every hair or bristle on its surface, whether or not it is actually connected to receptor cells, is useful in communicating some information about the surroundings, the air currents, or the changing tensions in the spider's web. By sensing the vibrations of its web, the spider can judge the size and activity of its entangled prey or can receive the message tapped out by a prospective mate.

Before mating, the male spins a tiny web, about 3 mm long, taps it gently with his genital opening to deposit a drop of sperm on it, then picks up the sperm into special cavities of his pedipalps. Then he is ready to search for a female. Finding her, he may perform a courtship ritual, according to his species. In web-building spiders, this may be a sort of Morse code tapped out on her web in vibrations. The hunting spider, whose vision is better, may wave a "semaphore" message with his legs or pedipalps or may display certain of his male decorations. Eventually he seizes her to prevent her escape, inserts his pedipalps into her genital opening, and stores the sperm in her seminal receptacles. Soon after mating he recharges his palps and seeks another female.

When she is ready to lay her eggs, the female spins a shallow silk bag to receive them, then extrudes the mass of eggs, fertilizes them with the stored sperm, and

FIG. 15-6 A fishing spider *Dolomedes sexpunctatus* is sucking out the body juices of a minnow it has captured. Fishing spiders usually live near the water and are often seen running about over aquatic vegetation. They are hunters and do not use the web to ensnare prey. (Photograph by J. H. Gerard.)

covers the eggs with more silk, forming an egg cocoon. Some mothers guard the eggs till they hatch. Wolf spiders and some others carry their egg sacs attached to their spinnerets and will fiercely defend them (Fig. 15-8, *A*). The jumping spider *Ballus* attaches hers to a branch, then spins two sheets over it, leaving a space between them for her to take up a guarding position. Some spider egg cases are covered with a tough silk, perhaps camouflaged with mud, and are then attached to a substratum and abandoned. The number of eggs in a case varies with the species but may be close to a thousand in the garden spider, or as few as two in the tiny six-eyed *Oonops*. Spiders usually lay more than one batch of eggs, with those that lay a few at a time usually producing more batches. Spiderlings hatch in about 2 weeks but may remain in the egg case for a few

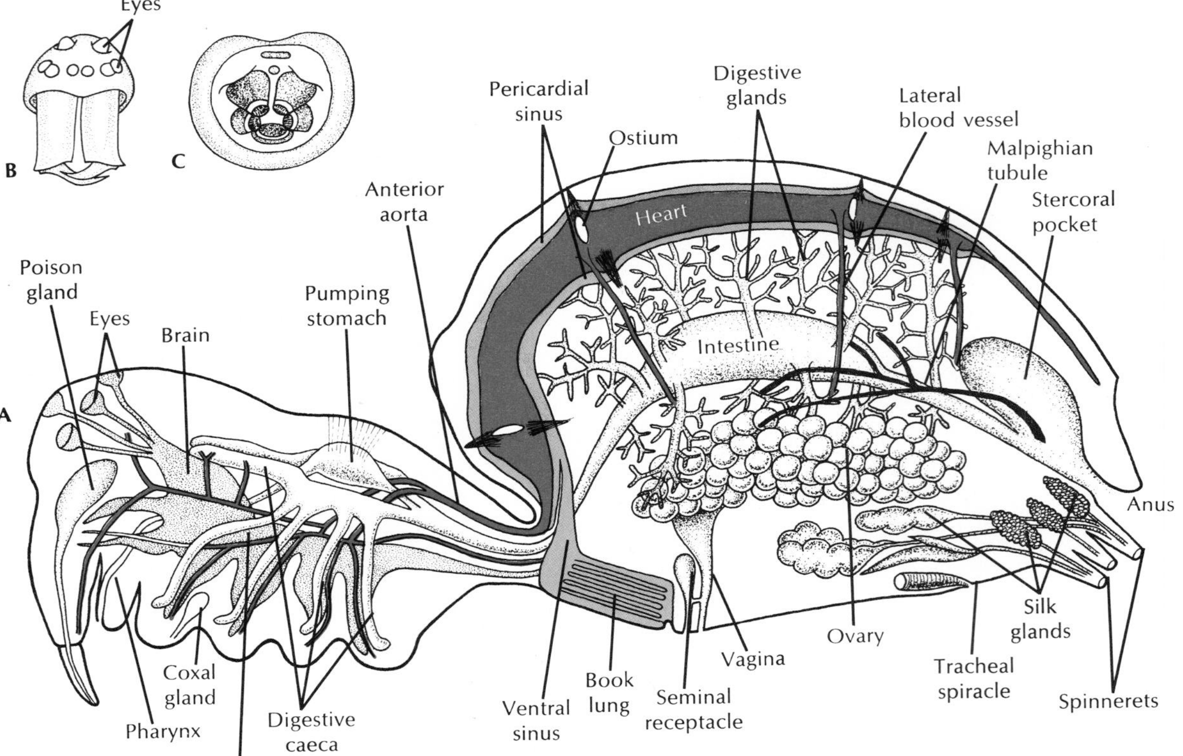

FIG. 15-7 **A,** Internal structure of the spider. **B,** Anterior view of head, showing eyes and chelicerae with fangs. **C,** Ventral view of spinnerets. The type of chelicerae and the number, arrangement, and/or locations of eyes, lung slits, tracheal spiracles, and spinnerets are all identifying characteristics used in classifying spiders.

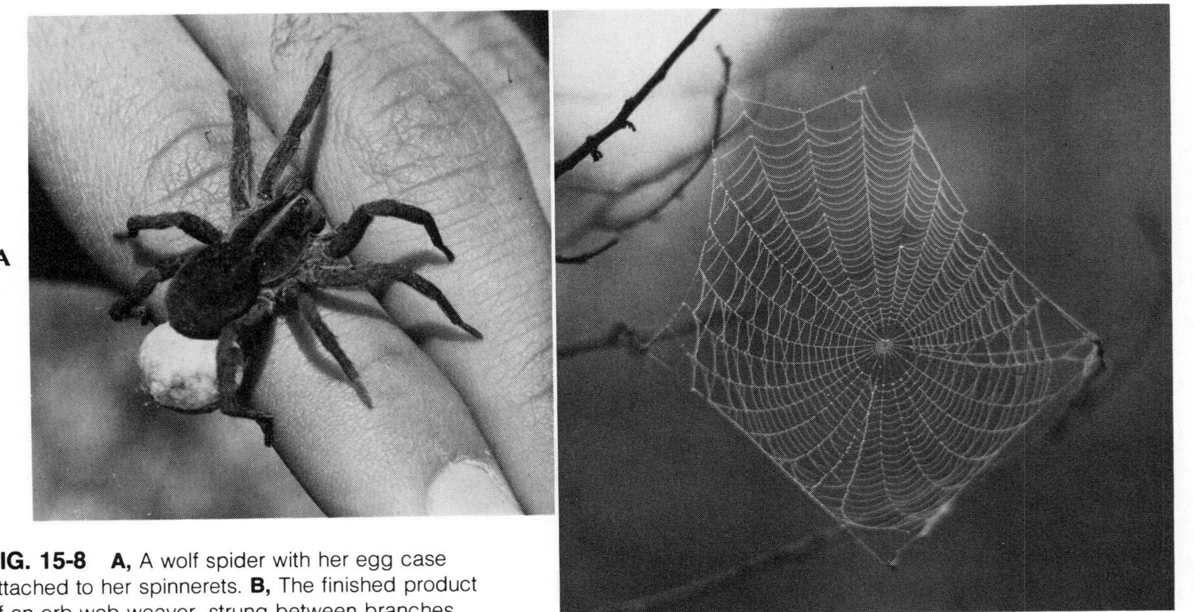

FIG. 15-8 **A,** A wolf spider with her egg case attached to her spinnerets. **B,** The finished product of an orb-web weaver, strung between branches. Note, however, that wolf spiders are not orb weavers. (Photography by C. P. Hickman, Jr.)

weeks longer and molt once before leaving the sac. Some spiders spend the winter months as eggs or young still in the egg cases and emerge in the spring. Spiderlings will molt several times before they mature.

Some spiders care for the juveniles for a time. On hatching, young wolf spiders of the family Lycosidae climb on the mother's back where they spend several days holding on to the spiny knobbed hairs that are peculiar to female lycosids. In another family of wolf spiders, the Pisauridae, the female carries the sac until just about time for hatching, then fastens it to a blade of grass, weaves a tent over it, and stands guard outside. The juveniles remain in the tent until shortly before the second molt. The trap-door spider, which spends its entire life in its burrow, raises its young there (Fig. 15-10).

WEB-SPINNING HABITS. The ability to spin silk is an important factor in the lives of spiders. Two or three pairs of spinnerets containing hundreds of microscopic tubes run to special abdominal **silk glands** (Fig. 15-7, *A* and *C*). A scleroprotein secretion emitted as a liquid hardens on contact with air to form the silk thread. Spiders' silk threads are stronger than steel threads of the same diameter and are said to be second in strength only to fused quartz fibers. The threads will stretch one-fifth of their length before breaking.

Spiders use silk threads for many purposes besides web-making. They use them to line their nests; form sperm webs or egg sacs; build draglines; make bridge lines, warning threads, molting threads, attachment discs, or nursery webs; or trap and entangle their prey and wrap them up securely. The dragline is the lifeline of the spider. It is used as a bridge between two objects or across a stream or as a trapline to detect the presence of an insect. A spider may use it to escape a predator by dropping off an object and remaining suspended in midair until danger is past, then climbing back up the line. When a spider feels the urge to "fly," it climbs to the top of some object, faces the wind, raises its abdomen, and spins out several threads into the breeze. When there are enough threads to form a parachute that will support its weight, it lets go and sails away, not entirely at the mercy of the wind for it can exert some control over its "balloon" by spinning more threads or by pulling some in. Newly hatched spiders often use ballooning to find new homes away from their birthplace and so reduce competition with their siblings.

The orb-weaver often establishes its bridge thread (Fig. 15-9, *A*) by spinning a long thread into the breeze until it sticks on some neighboring object. If it fails to stick, the spider pulls it in and tries again. As the spider moves along the bridge it has made, it bites the thread in two, holds both ends with its legs, and collects the thread ahead of it as it adds strengthening substance to the thread behind it. Of the several kinds of silk spun by spiders, two in particular are used in web-making. A tough, dry, elastic thread is used for the frame and radii of the web and for a temporary spiral. Then beginning at the outside, an adhesive thread is used to construct the sticky spiral on which the prey is caught (Fig. 15-9, *F*). Two adaptations aid the spider in moving about the web with great agility without sticking in its own viscid spirals. One is a special hook-shaped claw, which gives it a firm grip; the other is a coating of oil on the legs that protects them from the viscid fluid.

Spider webs are snares for trapping insects. The kind of net varies with the species. Some are primitive and consist merely of a few strands of silk radiating out from a spider's burrow or place of retreat. The domestic spider *Theridion tepidariorum* builds an irregular tangle of threads that we recognize as the familiar cobweb. The hammock spider *Linyphia phrygiana* makes a sheet web closely woven in a single plane. The grass spider *Agelenopsis naevia* builds a funnel web, which is a sheet web with a tube extending from it to a retreat. Orb webs are the conspicuous and beautiful geometric webs often seen on vegetation (Fig. 15-8, *B*); the garden spider *Argiope aurantia* (Fig. 15-5) builds such a web, as do the members of the subfamily Araneinae.

Not all spiders spin webs for traps. Some, such as the wolf spiders, simply chase and catch their prey. Some spin an anchor thread to warn them of the prey's approach. The trap-door spider found in the western and southern United States excavates a burrow 15 to 25 cm deep and about 3 cm wide, lines it with silk, and then, to fit the opening of the burrow, constructs a trap door of silk mingled with soil and camouflages it with vegetation (Fig. 15-10). The spider can close the door to keep intruders out or to keep the burrow rainproof, or it can hold the door ajar to attract insects and small crustaceans, which it seizes and drags into its burrow.

ARE SPIDERS REALLY DANGEROUS? It is truly amazing that such small and helpless creatures as the spiders have generated so much unreasoning fear in the human heart. Spiders are timid creatures, which, rather than being dangerous enemies to humans, are actually allies in the continuing battle with insects. The venom pro-

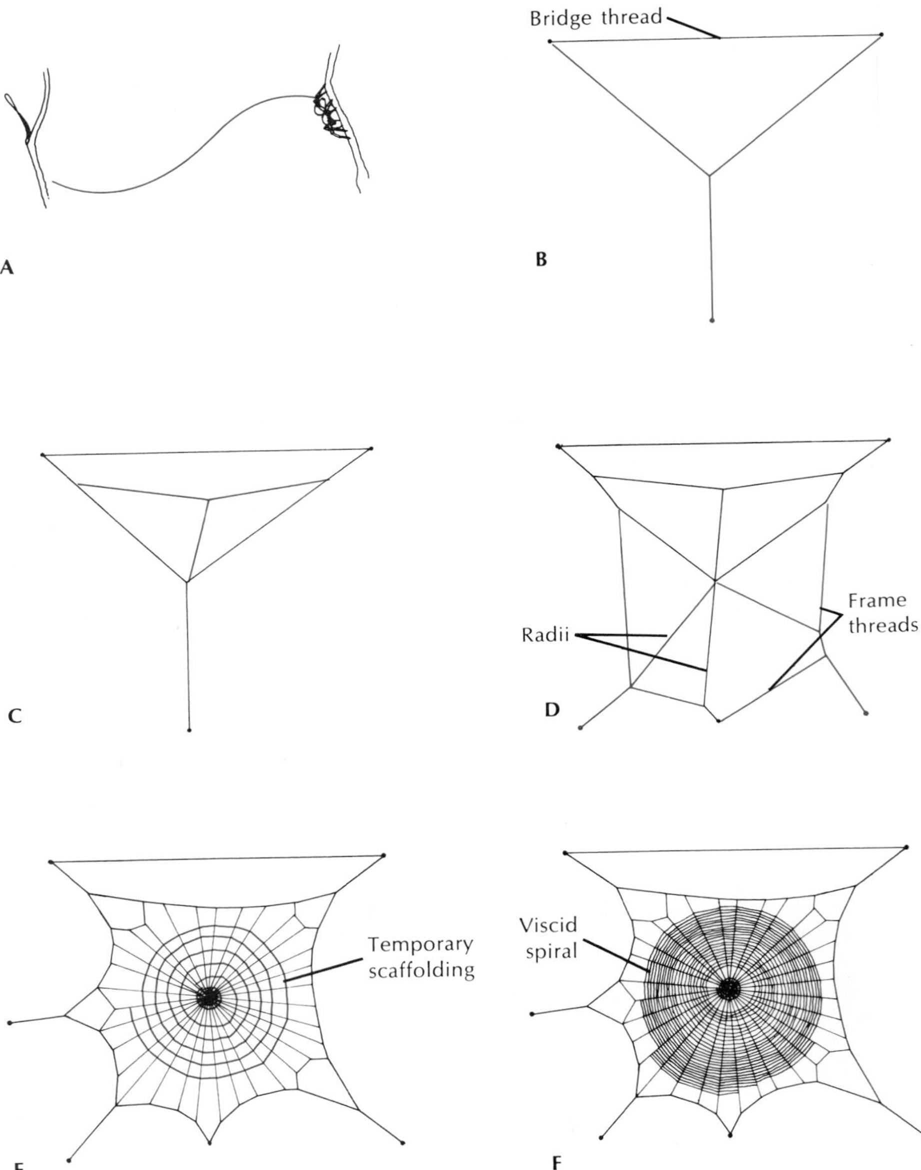

FIG. 15-9 Construction of typical orb web. **A,** Spider establishes a bridge thread, often assisted by a breeze. **B,** A second horizontal line is loosely fixed, then drawn tight by a vertical line fastened below. The juncture of these lines is the central hub, staging point for further progress. **C** and **D,** Frame threads and radii are added. **E,** These are followed by a temporary spiral of nonsticky thread. **F,** Finally the spider lays down a viscid spiral, starting at the outside and working inward, removing and eating the temporary scaffolding as it goes. (Modified from Savory, T. H. 1960. Scientific American **202:**115.)

FIG. 15-10 Burrow of trap-door spider, with lid thrown open and its owner just retreating within. (Photograph by F. M. Hickman.)

FIG. 15-11 A tarantula spider *Dugesiella lentzi* just captured. These interesting spiders are not dangerous; they make good pets and live for years. (Photograph by J. H. Gerard.)

duced to kill the prey is usually harmless to human beings. Even the most poisonous spiders bite only when tormented or when defending their eggs or young. The American tarantulas (Fig. 15-11), in spite of their fearsome size, are *not* dangerous. They rarely bite, and their bite is not considered serious.

There are, however, two species in the United States that can give severe or even fatal bites—the **black widow** and the **brown recluse.** The black widow is found worldwide, but in the United States it is more common in the south and west. The brown recluse is more common in the south central United States. The black widow, *Latrodectus mactans,* is small to moderate in size and shiny black, with a bright orange or red "hour-glass" on the underside of the abdomen (Fig. 15-12, *A*). The web is an irregular mesh with a funnel-shaped retreat and is built close to the ground under logs or stones, in holes in the ground, or in barns and outbuildings. Black widow venom is neurotoxic; that is, it acts on the nervous system. Approximately four or five out of each 1,000 bites reported have proved fatal.

The brown recluse, *Loxosceles reclusa,* is smaller than the black widow, is brown, has six eyes instead

FIG. 15-12 **A,** This black widow spider *Latrodectus* suspended on her web, has just eaten a large cockroach. Note "hourglass" marking (orange-colored) on ventral side of abdomen. Its venom is neurotoxic. **B,** The brown recluse *Loxosceles reclusa* is a small brown venomous spider. Note the small violin-shaped marking on its cephalothorax. Its venom is hemolytic. A bite from either spider can be serious. (**A,** Photograph by F. M. Hickman; **B,** courtesy J. H. Gerard.)

of the usual eight, and bears a violin-shaped dorsal stripe (Fig. 15-12, *B*). Its venom is hemolytic rather than neurotoxic, producing death of the tissues surrounding the bite. Reactions to its bite range from mild to severe and are occasionally fatal. The brown recluse lives under rocks and bark and is found around buildings, indoors as well as outdoors. It is nocturnal in its habits.

BEHAVIOR AND COMMUNICATION. Much of the behavior of arachnids is stereotyped and innate. The orb-weaver *Araneae diademata* spins an unrecognizable pattern soon after hatching, producing its artistic web only after the first molt. If the young spiderling is placed in a glass tube so small that it cannot string threads and is released after its first molt, it builds its characteristic web just as perfectly as other members of its species not so confined, indicating that the pattern is genetically programmed, not learned. If, as the spider weaves its geometric web, someone cuts some of the threads, it is incapable of going back and replacing those threads but continues on through the preprogrammed series of operations. When undisturbed, however, its instincts lead to an amazingly precise and perfect pattern. In ethologic language the orb-building behavior of spiders is a fixed action pattern (p. 774).

Communication by means of chemical, mechanical, acoustic, and visual signals has been observed in many different spiders.

Chemical communication. This is probably the most primitive form of spider communication. Contact chemoreception seems to be involved in inducing courtship behavior. In some spiders this occurs when the male accidentally touches a female, in others when the male contacts a female's dragline or the web of a female. Contact chemoreception is also a means of recognition in social spiders. A pheromone that acts as an airborne sex attractant is emitted by the females of *Crytophora* and possibly by some of the jumping spiders.

Mechanical communication. Messages may be con-

FIG. 15-13 The striped scorpion *Centruroides vittatus* found in the southwestern United States. Its sting is painful, but in this particular species is not considered lethal. (Photograph by J. H. Gerard.)

veyed by a vibrating web. The males of different species drum on the female's web, or pluck certain threads, or attach new threads to the web and pluck or shake them. The female may indicate by tapping her own signal that she is in a receptive rather than a predatory mood, an important difference to the male, which is usually smaller and in danger of being devoured. Tactile stimuli are involved in mating, when the male may tap or stroke the female's body. In the brood case of wolf spiders the young respond to the knobby hairs of the mother's abdomen but will refuse to settle on her if the hair has been shaved or covered up.

Some spiders have evolved a means of breaking the code of tactile stimuli of other spiders. *Ero,* one of the Mimetidae, which do not build webs but are predatory on other spiders, plucks on the web of an orb-weaving spider in the manner of an approaching male. The female, lured to the edge of her web by the signal, is devoured by her enemy rather than wooed by a suitor.

Acoustic communication. Some spiders produce sounds by drumming the pedipalps on a substratum or by stridulatory structures on the body. Certain theridiid (comb-footed) spiders, by twisting the abdomen up and down, rub hairy tubercles on the abdomen against certain ridges on the cephalothorax, producing a sound used in courtship. Males of the wolf spiders *Lycosa* and *Schizocosa* produce sound by rapidly oscillating their pedipalps at the tibiotarsal joints. Stridulation is also used in threat display between males.

Visual communication. Wolf spiders and jumping spiders have better vision than most spiders. During courtship the male, having found a female, may signal her with a distinctive pattern of foreleg or pedipalp waving or by certain stepping movements—each species with its own pattern. Visual displays have also been observed in threat behavior of wolf spiders (Fig. 15-4).

SPIDER ENEMIES. In spite of their expertise at predation, the spiders are also the source of food for many other animals. Among vertebrate enemies are birds, frogs and toads, lizards, and shrews. Among the invertebrates, spiders themselves are probably their own greatest enemy. One family, the Mimetidae, or pirate spiders, preys entirely on other spiders, and the young of many species, if unable to find food, will feed on their siblings. Next in order would no doubt come certain hymenopteran insects, such as the Ichneumons (Fig. 17-20); the pompilid wasps (spider wasps); and trypoxylonid wasps (organ-pipe mud-daubers), which provision their young with spiders that they have paralyzed.

Scorpions—order Scorpionida. Although scorpions are more common in tropical and subtropical regions, some also occur in temperate zones. Scorpions are generally secretive, hiding in burrows or under objects by day and feeding at night. They feed largely on insects and spiders, which they seize with the pedipalps and tear up with the chelicerae.

Sand-dwelling scorpions apparently locate their prey by sensing surface waves generated by movements of insects on or in the sand (Brownell, 1977). These waves are picked up by compound slit sensilla located on the basitarsal segments of the legs. The scorpion can locate a burrowing cockroach 50 cm away and reach it in three or four quick orientation movements.

The scorpion body consists of a rather short cephalothorax, which bears the appendages; a pair of large median eyes and two to five pairs of small lateral eyes; a preabdomen of seven segments; and a long slender postabdomen, or tail, of five segments, which ends in a stinging apparatus (Fig. 15-13). The chelicerae are small and three-jointed; the pedipalps are large, chelate (pincerlike), and six-jointed; and the four pairs of walking legs are eight-jointed.

On the ventral side of the abdomen are curious comblike pectines, which are tactile organs used for exploring the ground and for sex recognition. The stinger on the last segment consists of a bulbous base and a curved barb that injects the venom. The venom of most species is not harmful to humans but may produce a painful swelling. However, certain species of *Androctonus* in Africa and *Centruroides* in Mexico can be fatal unless antivenom is available.

Scorpions perform a complex mating dance, the male holding the female's chelae and stepping back and forth. He taps her genital area with his forelegs and stings her pedipalp. Finally he deposits a spermatophore and pulls the female over it until the sperm mass is taken up in the female orifice. Scorpions are either ovoviviparous or truly viviparous, that is, the females brood their young within the female reproductive tract. After several months or a year of development, anywhere from six to 90 young are produced, depending on the species. The young, only a few millimeters long, crawl up on the mother's back until after the first molt. They mature in about a year.

Pseudoscorpions—order Pseudoscorpionida. Pseudoscorpions, or false scorpions, are tiny, usually less than 8 mm long (Fig. 15-14). They live under bark and stones and in leaf mold, and a few species live in caves and in houses. They feed chiefly on small insects and mites. In winter some hibernate deep in the ground, where they construct silken chambers from silk glands that open on the chelicerae. The pedipalps are similar to those of scorpions except that they bear the poison glands used to paralyze or kill the prey.

Ticks and mites—order Acarina. Ticks and mites are arachnids in which the cephalothorax and abdomen are fused into an unsegmented ovoid body with eight legs. They are found almost everywhere—in both fresh and salt water, on vegetation, on the ground, and parasitic in animals. There are about 15,000 species, many of which are of importance to humans.

FIG. 15-14 Pseudoscorpion, or false scorpion. Order Chelonethida (Pseudoscorpionida). Most members of this group do not exceed 5 to 8 mm in length. They live under stones, bark of trees, and sometimes between pages of books. Their food is chiefly small insects and mites. In winter they construct cocoons from silk glands, which open on the chelicerae. Note large pedipalps, which resemble those of true scorpions. (Courtesy R. Weber and W. Vesey.)

Ticks are usually larger than mites. They pierce the skin of vertebrates and suck up the blood until enormously distended; then they drop off and digest the meal. After molting, they are ready for another meal. Some ticks are important disease vectors. Texas cattle fever is caused by a protozoan parasite transmitted by the tick *Boophilus annulatus*. The wood tick *Dermacentor* (Fig. 15-15) is the vector for Rocky Mountain spotted fever, caused by a rickettsial organism. Tularemia (rabbit fever) is also carried by a species of *Dermacentor*.

Most mites are less than 2 mm long, and they are varied in their habits and habitats. The spider mites, or red spiders, are destructive to plants. The mange mite *Demodex* is responsible for mange. Chiggers, *Eutrombicula,* lay their eggs on the ground. The larvae have only six legs until after the first molt. When the larvae find a suitable host, they attach to the skin and, with the aid of digestive secretions, form a tiny crater where they feed on partially digested tissues. The adults are not parasitic.

Phylogeny and adaptive radiation

Phylogeny. The similarities between the annelids and the arthropods give strong support to the theory that both phyla originated from a line of coelomate segmented protostomes, which in time diverged to form a protoannelid line with laterally located parapodia and a protoarthropod line with more ventrally located parapodia. The protoannelid line gave rise eventually to the polychaetes. The protoarthropod line apparently diverged further into three branches: (1) the trilobites and the chelicerates (spiders and horseshoe crabs), (2) the crustaceans, and (3) the insects, millipedes, and centipedes. The phylum Onychophora also may have come from some part of this protoarthropod line.

Some biologists consider the arthropods to be three separate phyla, each organized at the arthropod level. This is in accord with the three-stem origin of the arthropods, since a phylum is supposed to be monophyletic.

Adaptive radiation. Annelids show little specialization or fusion of somites and little differentiation of appendages. However, in arthropods the adaptive trend has been toward tagmatization of the body by differentiation or fusion of somites, giving rise in more advanced groups to such tagma as head and trunk; head, thorax, and abdomen; or cephalothorax (fused head and thorax) and abdomen. Primitive arthropods tend to have similar appendages, whereas the more advanced forms have appendages specialized for specific functions, or some appendages may be lost entirely.

Much of the amazing diversity in arthropods seems to have developed because of modification and specialization of their cuticular exoskeleton and their jointed appendages, thus resulting in a wide variety of locomotor and feeding adaptations.

W. S. Bristowe estimated that at certain seasons a Sussex field that had been undisturbed for several years had a population of 2 million spiders to the acre. He concluded that so many could not successfully compete except for the many specialized adaptations they had evolved. There are among them adaptations to cold and heat, wet and dry conditions, and light and darkness.

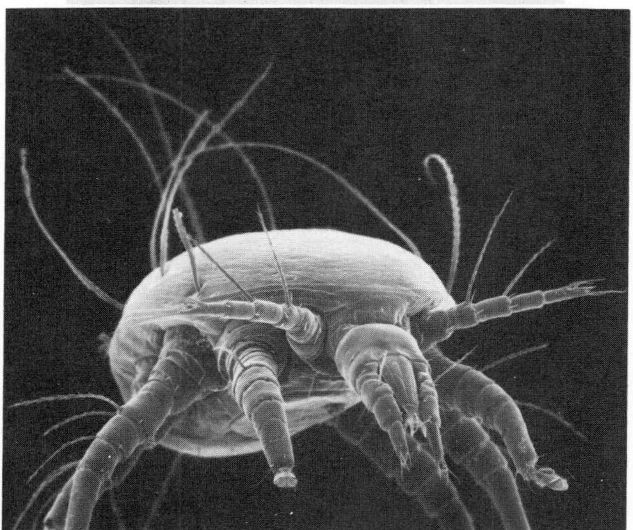

FIG. 15-15 **A,** The wood tick *Dermacentor,* one species of which transmits Rocky Mountain spotted fever and tularemia, as well as produces tick paralysis in pets. **B,** The mite *Diarthrophallus quercus.* (×114.) (**A,** Photograph by F. M. Hickman; **B,** Courtesy R. Schuster.)

Some capture large insects, some only small ones; web-builders snare mostly flying insects, whereas hunters seek those that live on the ground. Some lay eggs in the spring, others in the late summer. Some feed by day, others by night, and some have developed flavors that are distasteful to birds or to certain predatory insects. As it is with the spiders, so has it been with other arthropods; their adaptations are many and diverse and contribute in no small way to their long success.

Annotated references

Selected general readings

Arthur, D. R. 1960. Ticks. Cambridge, England, Cambridge University Press.

Arthur, D. R. 1962. Ticks and disease, rev. ed. Elmsford, N.Y., Pergamon Press, Inc.

Baerg, W. J. 1958. The tarantula. Lawrence, University of Kansas Press. *A monograph on the habits and natural history of a greatly misunderstood member of an animal group.*

Bristowe, W. S. 1971. The world of spiders. London, William Collins Sons & Co., Ltd. *The author is a supreme spider-watcher and tells of spider activities in a style that should convince any reader that spiders can be fun. Well-illustrated.*

Brownell, P. H. 1977. Compressional and surface waves in sand: used by desert scorpions to locate prey. Science **197:**477-479 (July).

Carthy, J. D. 1965. The behavior of arthropods. Edinburgh, Oliver and Boyd Ltd. *A slim book and not too difficult.*

Carthy, J. D. 1968. The pectines of scorpions, In J. D. Carthy and G. E. Newell (eds.). Invertebrate receptors. Symposium of the Zoological Society of London, no. 23. New York, Academic Press, Inc., pp. 251-261.

Clarke, K. U. 1973. The biology of the Arthropoda. New York, American Elsevier Publishing Co. *Gives a basis of facts and some theoretical material suitable for first- and second-year students.*

Cloudsley-Thompson, J. L. 1968. Spiders, scorpions, centipedes and mites. New York, Pergamon Press, Inc. *Deals with seven orders, emphasizing their general behavior, feeding habits, and mating behavior.*

Corning, W. C., J. A. Dyal, and A. O. D. Willows (eds.). 1973. Invertebrate learning, vol. 2. Arthropods and gastropod mollusks. New York, Plenum Publishing Corp.

Daniel, M., and B. Rosicky (eds.). 1973. Proceedings of the Third International Congress of Acarology, Prague, 1971. The Hague, Holland, Dr. W. Junk B. V., Publishers.

Fabre, J. H. 1919. The life of the spider. New York, Dodd, Mead and Co., Inc. *A classic account.*

Gertsch, W. J. 1949. American spiders. New York, Van Nostrand Reinhold Co. *Chapters on life-cycles, spinning, courtship, and evolution, in addition to accounts of the various orders; 32 color plates and 32 in gravure.*

Headstrom, R. 1973. Spiders of the United States. New York, A. S. Barnes & Co., Ltd. *Both for information and for identification of the spiders (omitting the tarantulas and trap-door spiders).*

Kaestner, A. 1968. Invertebrate zoology, vol. 2. Arthropod

relatives, Chelicerata, Myriapoda. New York, Interscience Publishers.

Kaston, B. J. 1972. How to know the spiders, ed. 2. Dubuque, William C. Brown Co., Publishers. *Spiral-bound identification manual.*

King, P. E. 1973. Pycnogonids. New York, St. Martin's Press, Inc.

Moore, R. C., C. G. Lalicker, and A. G. Fisher. 1952. Invertebrate fossils. New York, McGraw-Hill Book Co. *Has two chapters dealing with trilobites and chelicerates.*

Naegle, J. A. (ed.). 1963. Advances in acarology, vol. I. Ithaca, N.Y., Cornell University Press. *For the specialist.*

Sankey, J. H. P., and T. H. Savory. 1974. The British harvestmen, Linnean Society's Synopsis, 2nd Series. Synopses Br. Fauna, N. S. no. **4:**1-76.

Savory, T. H. 1952. The spider's web. New York, Frederick Warne & Co., Ltd. *Deals with the nature of silk, the use of the silk glands, and the types of webs produced.*

Savory, T. H. 1961. Spiders, men and scorpions. London, *An historical account of the science of arachnology from Aristotle to about 1950.*

Savory, T H. 1971. Spiders. Boston, Ginn & Co. *Intended for beginners, with attractive illustrations and an emphasis on simple experiment.*

Savory, T. H. 1974. Introduction to arachnology. London, Frederick Muller, Ltd. *A slim book for the beginning student.*

Savory, T. H. 1977. Arachnida. ed. 2. New York, Academic Press, Inc. *Fairly advanced account, covering morphology, physiology, embryology, ethology, and more about the group. Devotes a chapter to each of the orders.*

Sebeok, T. A. (ed.). 1977. How animals communicate. Bloomington, Indiana University Press, pp. 315-325.

Snow, K. R. 1970. The Arachnida: an introduction. New York, Columbia University Press. *A short account of the class. Emphasis is on mites, ticks, and spiders; other orders given more briefly; includes* Limulus.

Tuttle, D. M., and E. W. Baker. 1968. Spider mites of the southwestern United States and a revision of the Tetranychidae. Tucson, University of Arizona Press.

Weygoldt, P. 1969. The biology of pseudoscorpions. Cambridge, Mass., Harvard University Press. *An excellent treatise on the false scorpions.*

Witt, P. N. (ed.). 1969. Web-building spiders. Am. Zool. **9:**70-238.

Wolken, J. J. 1971. Invertebrate photoreceptors. New York, Academic Press, Inc.

Selected *Scientific American* articles

Burgess, J. W. 1976. Social spiders **234:**100-107 (Mar.). *Gregarious spiders are rare but some form communal webs for catching prey.*

Savory, T. H. 1962. Daddy longlegs. **207:**119-128 (Oct.).

Savory, T. H. 1968. Hidden lives. **219:**108-114 (July). *Almost half of the cryptozoic species are arachnids.*

CHAPTER 16

THE AQUATIC MANDIBULATES

Phylum Arthropoda
Class Crustacea

One of the most abundant organisms in the world, *Calanus finmarchicus,* along with other planktonic copepods, forms an extremely important segment of the marine ecosystem. They are the primary consumers of marine phytoplankton, a crop five times greater annually than all land vegetation. They are in turn fed on by large numbers of small fishes, such as sardines, herring, menhaden, and others. Some investigators believe that "copepods may hold one key to our future survival" (Russell-Hunter, 1969).

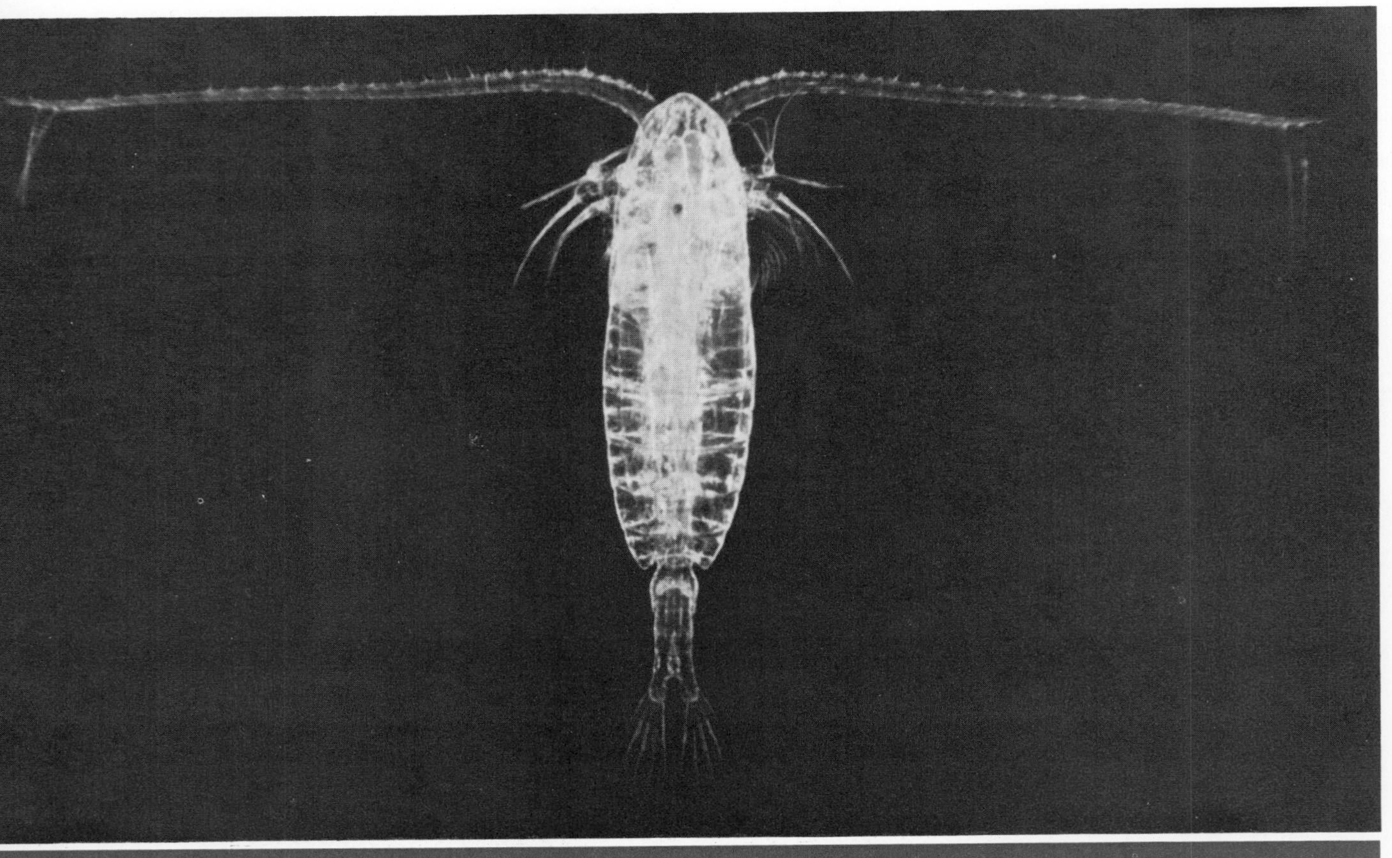

Photograph by D. P. Wilson.

Mandibulates are those arthropods that possess mandibles (jaws), including the crustaceans, insects, and myriapods (centipedes, millipedes, pauropods, and symphylans). Although these groups were formerly included in a subphylum of the Arthropoda, it is now widely held that the mandibles of the crustaceans and those of the insects and myriapods are not homologous in development or phylogeny. Therefore, the term "mandibulates" is now used only for convenience to refer to those arthropods with certain types of feeding appendages. Nonetheless, all of these forms have, at least, a pair of **antennae,** a pair of **mandibles,** and a pair of **maxillae** on the head. These appendages perform sensory, mastigatory, and food-handling functions, respectively. The body may consist of a head and trunk, but in the more advanced forms, a high degree of tagmosis (p. 328) has occurred so that there is a well-defined head, thorax, and abdomen. In most Crustacea one or more thoracic segments have become fused with the head to form a **cephalothorax.** The thoracic and abdominal appendages are mainly for walking or swimming, but in some groups they are highly specialized in function. Among the mandibulates, the Crustacea are mainly marine; however, there are many freshwater and a few terrestrial species, while the insects and myriapods are mainly terrestrial. There are numerous species of insects in freshwater habitats, and a few in marine. The myriapods and insects are dealt with in Chapter 17.

CLASS CRUSTACEA

The members of the class Crustacea (L., *crusta,* shell) get their name from the hard shells most of them bear. The 30,000 or more species in this class include lobsters, crayfish, shrimp, crabs, water fleas, copepods, barnacles, and some others. It is the only arthropod class that is primarily aquatic. The majority are free-living, but many are sessile, commensal, or parasitic. Though crustaceans differ from most other arthropods in a variety of ways, the only truly diagnostic characteristic is that crustaceans are the only arthropods with **two pairs of antennae.**

General nature of a crustacean

In addition to the two pairs of antennae and mandibles, crustaceans have two pairs of maxillae on the head, followed by a pair of appendages on each body segment that are modified for various functions. All appendages, except perhaps the first antennae, are primitively **biramous** (two main branches), and at least some of the appendages of present-day adults show that condition. Organs specialized for respiration, if present, are in the form of **gills.** Crustaceans lack malpighian tubules.

Most crustaceans have between 16 and 20 segments, but some primitive forms have 60 segments or more. The more advanced crustaceans tend to have fewer segments and increased tagmatization. The major tagmata are the **head, thorax,** and **abdomen,** but these are not homologous throughout the class (or even within some subclasses) because of varying degrees of fusion of segments (somites); for example, as in the cephalothorax.

The most advanced and by far the largest group of crustaceans is the subclass Malacostraca, which includes the lobsters, crabs, shrimps, beach fleas, sow bugs, and many others. These show a surprisingly constant arrangement of body segments and tagmata that is often referred to as the **caridoid facies*** and is considered the ancestral plan of the group (Fig. 16-1). This typical body plan has a head of five (six embryonically) fused somites, a thorax of eight somites, and an abdomen of six somites (seven in a few species). At the anterior end is the nonsegmented **rostrum** and at the posterior end is the nonsegmented **telson,** which with the last abdominal somite and its **uropods** form the tail fan in many forms.

In many crustaceans, the dorsal cuticle of the head may extend posteriorly and around the sides of the animal to cover or be fused with some or all of the thoracic and abdominal somites. This covering is called the **carapace.** In some groups the carapace forms clamshell-like valves that cover most or all of the body. In the decapods (including lobsters, shrimp, crabs, and others) the carapace covers the entire cephalothorax, but not the abdomen.

Relationships and origin of crustaceans

The relationship of the crustaceans to other arthropods has long been a puzzle. According to a widely held theory, the trilobites are considered the ancestors of the crustaceans, but some zoologists have proposed that both these groups evolved independently from different nonarthropod ancestors.

*"Caridoid" derives from the scientific name of a group of crustaceans; "facies" means face or general appearance.

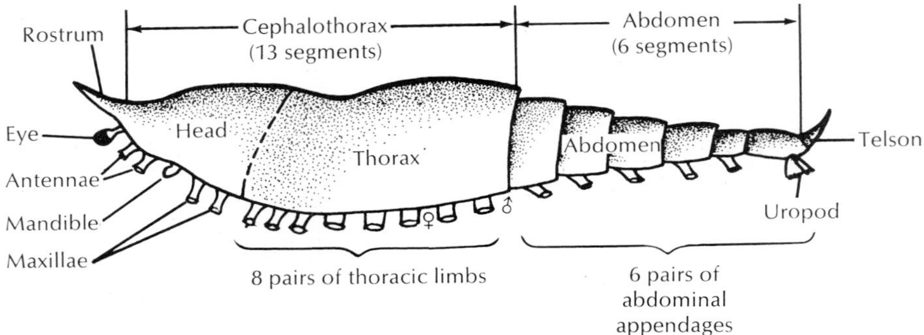

FIG. 16-1 The archetypic plan (caridoid facies) of the Malacostraca. In the general plan found today in the decapods—lobsters, crayfish, crabs, and others—the anterior three pairs of thoracic legs have become maxillipeds.

Some light was shed on the problem when the primitive crustacean *Hutchinsoniella macracantha* was discovered in 1954 in Long Island Sound. This form was assigned to a new subclass of its own (Cephalocarida) because of its several very primitive characteristics. These include thoracic appendages that are similar to each other and second maxillae that are similar to the thoracic appendages. Except for the eighth thoracic legs, which lack the innermost branch, they are all essentially three-branched, or **triramous.** Many authorities see similarities between these appendages and those of the trilobites, and the many specialized limbs of other crustaceans are derivable from the triramous cephalocarid appendages. The characteristics of this little form (about 4 mm long) support the idea that crustaceans were derived from primitive trilobites or that the two groups were derived from a common ancestor.

Representative type—crayfish

Though the crayfish is far from the most primitive crustacean, it is easily available and is good for beginning study of this class. It is a member of the subclass Malacostraca, order Decapoda.

The American lobster *Homarus americanus* is very similar to the crayfish but is much larger and is marine. The crayfish are freshwater forms, and more than 130 species in several genera have been described in the United States. *Orconectes* and *Cambarus* are common genera in North America east of the Rocky Mountains, and *Astacus* occurs west of the Rocky Mountains and also in Europe.

Habitat. Crayfish are among the most common animals in fresh water. They are found in streams, ponds, and swamps over most of the world. Some are found in slow-moving water, others prefer swift streams, and some blind ones dwell in caves. They are primarily bottom dwellers and spend the day under rocks and vegetation, coming out at night in search of food. Some are semiterrestrial in their habits and make characteristic chimneylike burrows in the soil, if the water level is not too far under the surface.

Behavior. In their burrows crayfish face outward so that they can move out for food. While stationary, they keep currents of water moving by waving the bailer and swimmerets back and forth. They walk with most of the weight supported by the fourth pair of legs. When escaping from danger, they move rapidly backward by extending and flexing the abdomen, uropods, and telson, thus producing a series of backward darts.

Crayfish are positively thigmotactic and try to get most of the body in contact with a surface. Many chemical substances, unless concentrated, will attract them. Most light sources will cause them to retreat, although they apparently are attracted to red.

The behavior of the crayfish is mostly instinctive, but they and several other decapods can be taught certain acts, such as running simple mazes, through reinforcement of success with food.

External features. As with other crustaceans, the body of the crayfish is covered with an exoskeleton composed of chitin, protein, and lime. This hard protective covering is soft and thin at the joints between the somites, allowing flexibility of movement. The tagmata (**cephalothorax** and **abdomen**) and somites are typical caridoid facies.

The cephalothorax, enclosed dorsally and laterally

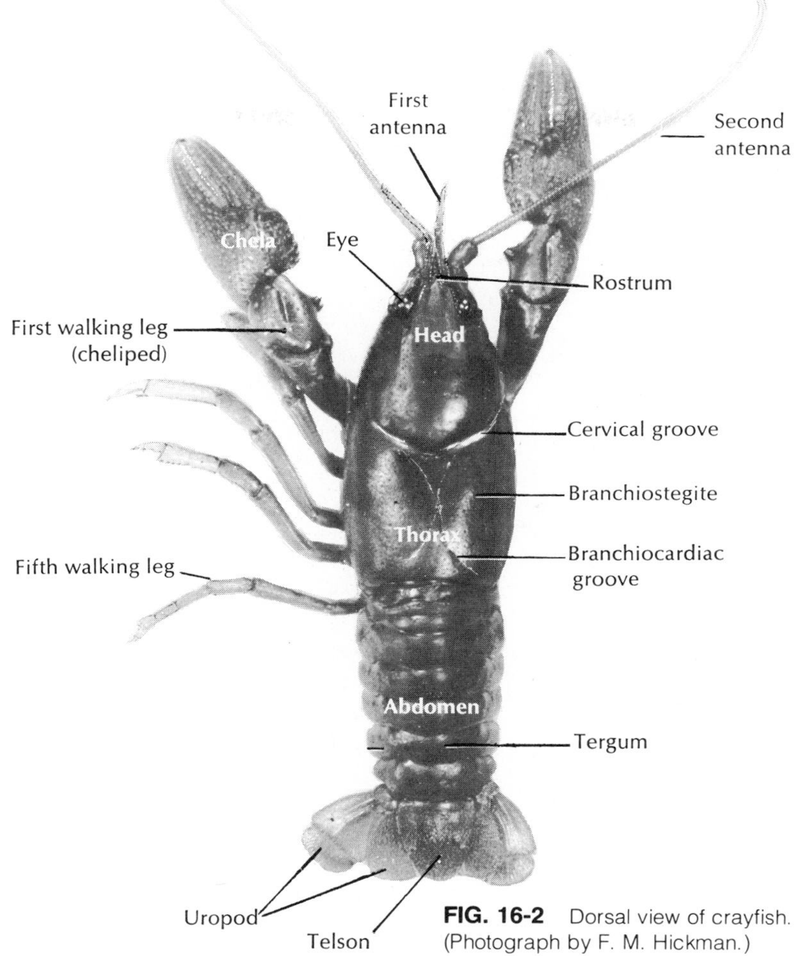

FIG. 16-2 Dorsal view of crayfish. (Photograph by F. M. Hickman.)

by the **carapace,** appears unsegmented except for a groove between the head and thorax (Fig. 16-2). The anterior tip of the carapace is the **rostrum.** The **compound eyes,** which are stalked and movable, lie beneath and on each side of the rostrum.

Each somite of the abdomen is covered by a dorsal cuticular plate, or **tergum** (Fig. 16-2), and a ventral transverse bar, the **sternum,** lies between the swimmerets (Fig. 16-3). Lateral to the base of each swimmeret is the **pleural** area. The abdomen terminates in the broad, flaplike **telson,** which is not considered a somite. On the ventral side of the telson is the **anus.**

The openings of the paired **vasa deferentia** are on the median side at the base of the fifth pair of walking legs, and those of the paired **oviducts** are at the base of the third pair. In the female the opening to the seminal receptacle is located in the midventral line between the fourth and fifth pairs of walking legs.

APPENDAGES. The crayfish typically has one pair of jointed appendages on each somite (Fig. 16-3). These appendages differ from each other, depending on their functions. All, however, are variations of the basic, biramous plan, most easily illustrated by the abdominal appendages, the **swimmerets** (Fig. 16-4). The basal portion, the **protopodite,** bears a lateral **exopodite** and a medial **endopodite.** The protopodite is made up of two joints (**basipodite** and **coxopodite**), whereas the exopodite and endopodite have from one to several segments each. All the other appendages can be viewed as modifications of the basic plan, though the walking legs have become secondarily uniramous.

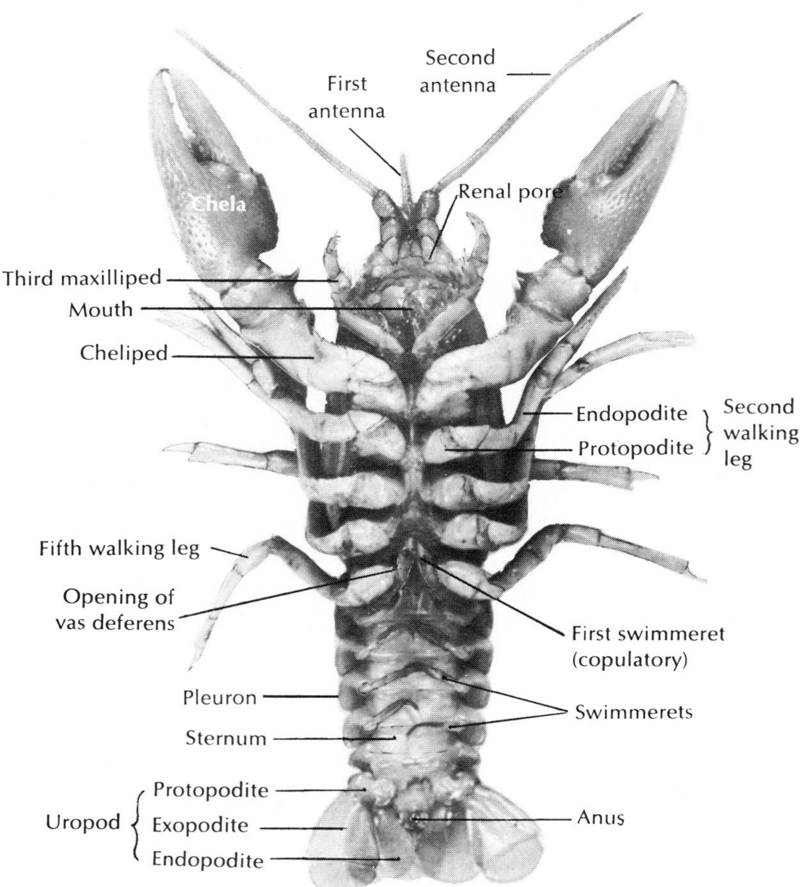

FIG. 16-3 Ventral view of male crayfish. (Photograph by F. M. Hickman.)

Structures that have a similar basic plan and have descended from a common form are said to be **homologous,** whether they have the same function or not. Since the specialized walking legs, mouthparts, chelipeds, and swimmerets have all developed from a common biramous type but have become modified to perform different functions, they are all homologous to each other (serially homologous). During the evolution of this structural modification, some branches have been reduced, some lost, some greatly altered, and some new parts added. The crayfish and their allies possess the best examples of **serial homology** in the animal kingdom.

Table 16-1 shows how the various appendages have become modified from the biramous plan to fit specific functions.

Regeneration and autotomy. Crayfish have the power to regenerate certain lost parts. In general, any of the appendages and the eyes will be renewed when removed, although regeneration is faster in young animals. A lost part is partially renewed at the next molting, and after several moltings it is completely restored, though the new structure is not always identical to the one lost.

The power of self-amputation is called **autotomy.** It refers to the breaking off of an injured leg or chela at a definite preformed breakage plane by means of a reflex act. The breaking point is near the base of the legs and is marked by an encircling line on the basal segment of the chelae and at the third joint on the walking legs. If one of these appendages is injured, all parts terminal to the breaking point are cast off. The process is effected

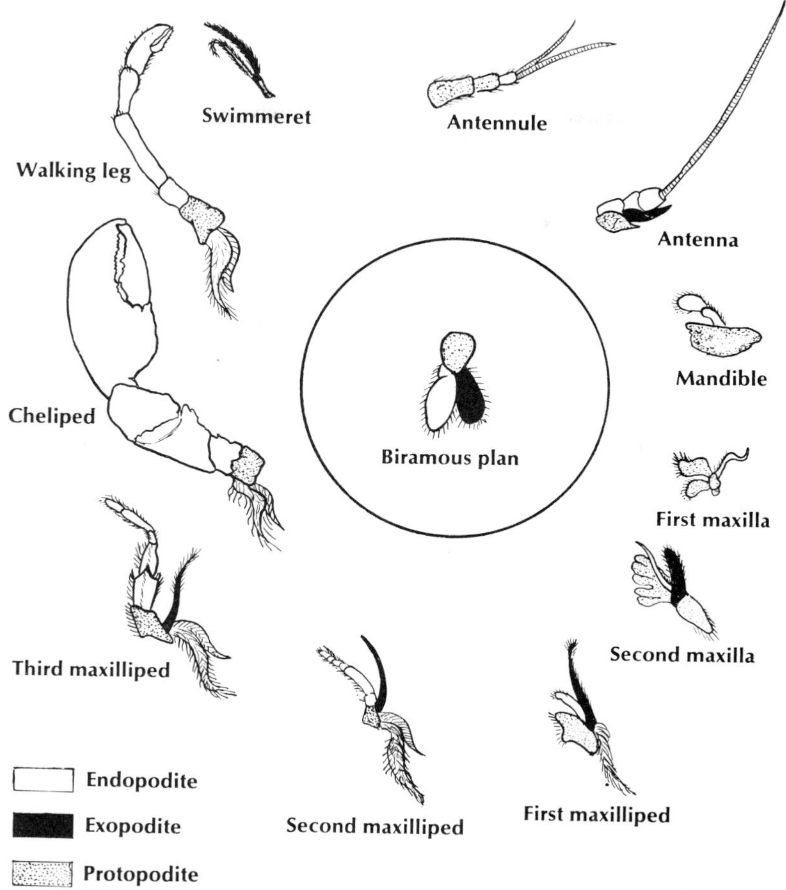

FIG. 16-4 Appendages of the crayfish showing how they have become modified from the basic biramous plan, as found in a swimmeret.

by a special autotomizer muscle, which contracts very strongly to cause extreme flexion of the leg, putting pressure on the breakage plane until it ruptures (Fig. 16-5). After a part is cast off in this manner, a replacement regenerates in the regular way.

Internal features. Some systems, such as the muscular and nervous, and the entire segmented abdomen clearly show the metamerism inherited from the annelid ancestors, but there are marked modifications in other systems. Most of the changes involve concentration of parts in a particular region or else reduction or complete loss of parts, for example, the intersepta.

HEMOCOEL. The major body space in the arthropods is not a true coelom but a blood-filled **hemocoel.** The hemocoel is a *primary* body cavity that is actually a persistent blastocoel. A true coelom, on the other hand, is a *secondary* body cavity that forms within the mesoderm. The only remaining coelomic compartments in the crustaceans are the end sacs of the excretory organs and the space around the gonads.

MUSCULAR SYSTEM. Striated muscles make up a considerable part of the body of a crayfish. It uses them for body movements and for manipulation of the appendages. The muscles are usually arranged in antagonistic groups: **flexors,** which draw a part toward the body, and **extensors,** which straighten it out. The abdomen has powerful flexors (Fig. 16-6), which are used when the animal swims backward—its best means of escape. Strong muscles on either side of the stomach manipulate the mandibles.

RESPIRATORY SYSTEM. Respiratory gas exchange in

TABLE 16-1 Crayfish appendages

Appendage	Protopodite	Endopodite	Exopodite	Function
Antennule	3 segments, statocyst in base	Many-jointed feeler	Many-jointed feeler	Touch, taste, equilibrium
Antenna	2 segments, excretory pore in base	Long, many-jointed feeler	Thin, pointed blade	Touch, taste
Mandible	2 segments, heavy jaw and base of palp	2 distal segments of palp	Absent	Crushing food
First maxilla	2 thin medial lamellae	Small unjointed lamella	Absent	Food-handling
Second maxilla	2 bilateral lamellae, extra plate, epipodite	1 small pointed segment	Dorsal plate, the scaphognathite (bailer)	Draws currents of water into gills
First maxilliped	2 medial plates and epipodite	2 small segments	1 basal segment, plus many-jointed filament	Touch, taste, food-handling
Second maxilliped	2 segments plus gill	5 short segments	2 slender segments	Touch, taste, food-handling
Third maxilliped	2 segments plus gill	5 larger segments	2 slender segments	Touch, taste, food-handling
First walking leg (cheliped)	2 segments plus gill	5 segments with heavy pincer	Absent	Offense and defense
Second walking leg	2 segments plus gill	5 segments plus small pincer	Absent	Walking and prehension
Third walking leg	2 segments plus gill; genital pore in female	5 segments plus small pincer	Absent	Walking and prehension
Fourth walking leg	2 segments plus gill	5 segments, no pincer	Absent	Walking
Fifth walking leg	2 segments; genital pore in male; no gill	5 segments, no pincer	Absent	Walking
First swimmeret	In female reduced or absent; in male fused with endopodite to form tube			In male, transfers sperm to female
Second swimmeret				
Male	Structure modified for transfer of sperm to female			
Female	2 segments	Jointed filament	Jointed filament	Creates water currents; carries eggs and young
Third, fourth, and fifth swimmerets	2 short segments	Jointed filament	Jointed filament	Create current of water; in female carry eggs and young
Uropod	1 short, broad segment	Flat, oval plate	Flat, oval plate; divided into 2 parts with hinge	Swimming; egg protection in female

the smaller crustaceans takes place over thinner areas of the cuticle (for example, in the legs) and specialized structures may be absent. The larger crustaceans have gills, which are delicate, featherlike projections with very thin cuticle. In the decapods, the sides of the carapace enclose the gill cavity, which is open anteriorly and ventrally (Fig. 16-7). Gills may project from the pleural wall into the gill cavity, from the articulation of the thoracic legs with the body, or from the thoracic coxopodites. The latter two types are typical of crayfish. The "bailer," a part of the second maxilla, draws water over the gill filaments, into the gill cavity at the bases of the legs, and out of the gill cavity at the anterior.

FEEDING AND DIGESTION. Crayfish and lobsters will eat almost anything they can get, dead or alive. Before

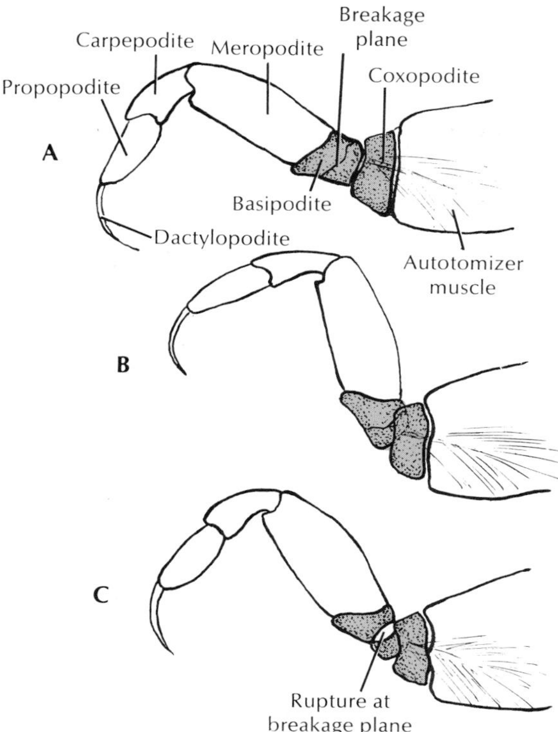

FIG. 16-5 Autotomy in the crab is a reflex act in which extreme flexion causes the leg to rupture at a preformed breakage plane. The stimulus is injury to the leg distal to the breakage plane. **A,** Leg in normal resting position showing location of preformed breakage plane. **B,** Extreme flexion of the leg from contraction of autotomizer muscle causes basipodite to press against rim of coxopodite. **C,** Continued contraction and pressure causes rupture at breakage plane. A double membrane at this location pinches off the nerve and blood vessels at autotomy and prevents blood loss and tissue damage at stump. Regeneration occurs faster at this location than elsewhere. (Modified from Wood, F. D., and H. E. Wood. 1932. J. Exp. Zool. **62:**1-55.)

reaching the mouth, the food is torn into small bites and is kneaded into a usable size and shape by the maxillipeds, maxillae, and mandibles. Although the mandibles can crush, they do not really chew. Chewing is a function of the stomach. The mouth opens on the ventral side between the mandibles and leads to a short esophagus (Fig. 16-6).

The crayfish stomach is large and thin-walled, contains two compartments, and is manipulated by a group of gastric muscles. When food reaches the anterior, or cardiac, portion of the stomach, it is chewed by a **gastric mill** (composed of three movable chitinous teeth, one dorsal and two lateral) (Fig. 16-8). Then it passes through a filtering device in the posterior (pyloric) part of the stomach. A special sieve of fine bristles allows only the finest particles to enter a filter pouch through which the food moves in a fluid stream into the ducts of the large digestive glands. Food particles too large for the filter pouch are funneled into the intestine for elimination or else are regurgitated. Digestion is completed in the tubules of the digestive glands, and the nutrients are absorbed into the blood, which carries them to the tissue cells to be utilized. The contents of the digestive tract are propelled by peristalsis.

The gastric mill enables these somewhat sedentary animals to consume their food rapidly and chew it later while they are safely hidden from enemies.

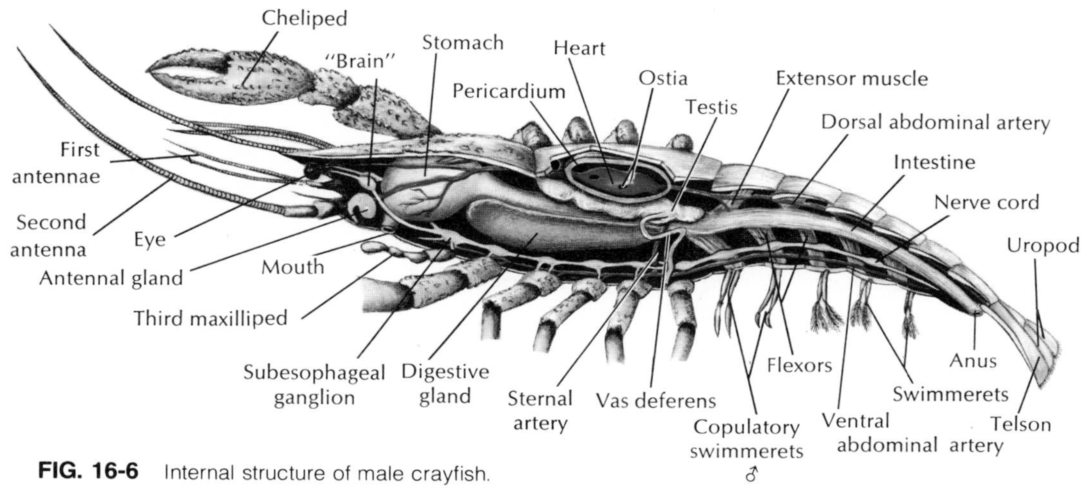

FIG. 16-6 Internal structure of male crayfish.

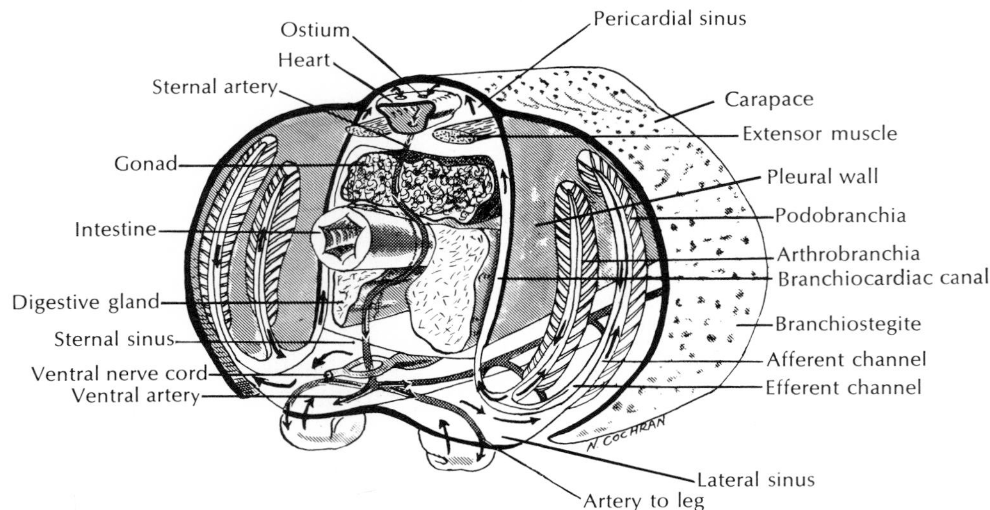

FIG. 16-7 Diagrammatic cross section through heart region of crayfish showing direction of blood flow in this "open" blood system. Blood is pumped from heart to body tissues through arteries, which empty into tissue sinuses. Returning blood enters sternal sinus, is carried to gills for gas exchange and then back to pericardial sinus by branchiocardiac canals. Note absence of veins.

The esophagus and hindgut are lined with cuticle, which is periodically shed during ecdysis (molting).

CIRCULATION. Crayfish and other arthropods have an "open" or lacunar type of blood system. This means that there are no veins, but that the blood leaves the heart by way of arteries and returns to venous sinuses, or spaces, instead of veins before it reenters the heart. Recall that the annelids have a closed system, as do the vertebrates.

A dorsal heart is the chief propulsive organ. It is a single-chambered sac of striated muscle. Blood enters the heart from the surrounding pericardial sinus through three pairs of ostia, with valves that prevent backflow into the sinus (Fig. 16-6). From the heart the blood enters a system of arteries and capillaries. Valves in the arteries prevent a backflow of blood. Small arteries or capillaries empty into the tissue sinuses, which in turn discharge into the large sternal sinus (Fig. 16-7).

From there, afferent sinus channels carry blood to the gills where oxygen and carbon dioxide are exchanged. The blood is now returned to the pericardial sinus by efferent channels (Fig. 16-7).

Blood in arthropods is largely colorless. It includes ameboid blood cells of at least two types. Hemocyanin, a copper-containing respiratory pigment, or hemoglobin, an iron-containing pigment, may be dissolved in the blood. The blood has the property of clotting, which prevents loss of blood in minor injuries. The blood corpuscles release a thrombinlike coagulant that precipitates the clotting.

EXCRETION. The excretory organs of adult crustaceans are usually a single pair of tubular structures located in the ventral part of the head anterior to the esophagus (Fig. 16-6). They are called **antennal glands** or **maxillary glands,** depending on whether they open at the base of the antennae or of the second maxillae. The excretory organs of the decapods are antennal glands, also called **green glands** in this group.

The **end sac** of the antennal gland, which is derived from an embryonic coelomic compartment, consists of a small vesicle **(saccule)** and a spongy mass called a **labyrinth.** The labyrinth is connected by an **excretory tubule** to a dorsal **bladder,** which opens to the exterior by a pore on the ventral surface of the basal antennal segment (Fig. 16-9). Hydrostatic pressure within the hemocoel provides the force for filtration of fluid into the end sac, and as the filtrate passes through the excretory tubule, it is modified by resorption of salts and water and finally excreted as urine.

Excretion of nitrogenous wastes (mostly ammonia) takes place by diffusion across thin areas of cuticle, especially the gills, and the so-called excretory organs

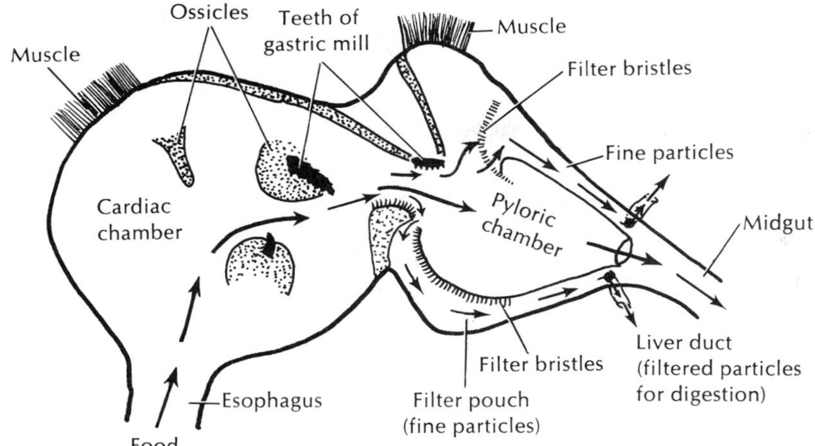

FIG. 16-8 Malacostracan stomach showing gastric "mill" and directions of food movements. Note that mill is provided with chitinous ridges, or teeth for mastication, and setae for straining the food before it passes into the pyloric stomach.

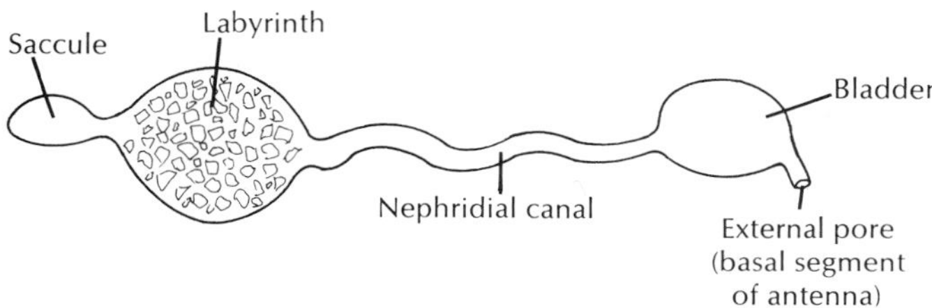

FIG. 16-9 Scheme of antennal gland (green gland) of crayfish. (In natural position organ is much folded.) Most selective resorption takes place in the labyrinth, a complicated spongy mass. Some crustaceans lack a labyrinth, and the excretory tubule (nephridial canal) is a much-coiled tube.

function principally to regulate the ionic and osmotic composition of the body fluids. Freshwater crustaceans, such as the crayfish, are constantly threatened with overdilution of the blood by water, which diffuses across the gills and other water-permeable surfaces. The green glands, by forming a dilute, low-salt urine, act as an effective flood-control device. In marine crustaceans such as lobsters and crabs the kidney functions to adjust the salt composition of the blood by selective modification of the salt content of the tubular urine. In these forms the urine remains isosmotic to the blood.

NERVOUS SYSTEM. The nervous systems of the crayfish and earthworm have much in common, although that of the crayfish is somewhat larger and has more fusion of ganglia (Fig. 16-6). The **brain** is a pair of **supraesophageal ganglia** that supply nerves to the eyes and the two pairs of antennae. It is joined by connectives to the **subesophageal ganglion,** a fusion of at least five pairs of ganglia that supply nerves to the mouth, appendages, esophagus, and antennal glands. The double ventral nerve cord has a pair of ganglia for each somite (from the eighth to the nineteenth) and gives off nerves to the appendages, muscles, and other parts (Fig. 16-10, *B*). In addition to this central system, there is also a sympathetic nervous system associated with the digestive tract.

In the primitive branchiopods the ventral nerve cord has the ladder arrangement characteristic of flatworms and annelids, for the two parts of the cord are separate and are connected by transverse commissures (Fig. 16-10, *A*).

SENSORY SYSTEM. Crustaceans have better developed sense organs than do the annelids. The largest sense organs of the crayfish are the eyes and the statocysts. Tactile organs are widely distributed over the body in the form of **tactile hairs,** delicate projections of the cuticle, which are especially abundant on the chelae, mouthparts, and telson. The chemical senses of **taste** and **smell** are found in hairs on the antennae, mouthparts, and other places.

A saclike **statocyst** is found on the basal segment of

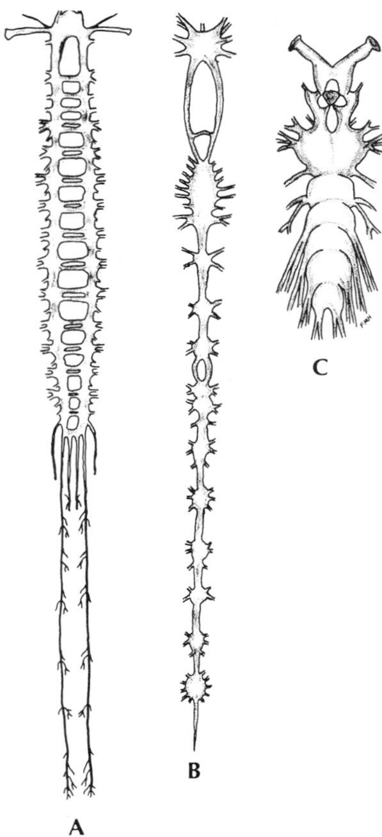

FIG. 16-10 Types of crustacean nervous systems. **A,** *Branchinecta,* a fairy shrimp. **B,** *Astacus,* a crayfish. **C,** *Argulus,* a branchiuran. Note tendency toward concentration and fusion of ganglia. Ladderlike arrangement in **A** may be considered a primitive condition, whereas that in **B** is similar to nervous system of annelid type.

each first antenna and opens to the surface by a dorsal pore. The statocyst contains a ridge that bears sensory hairs formed from the chitinous lining and grains of sand that serve as **statoliths.** Whenever the animal changes its position, corresponding changes in the position of the grains on the sensory hairs are relayed as stimuli to the brain, and the animal can adjust itself accordingly. The cuticular lining of the statocyst is shed at each molting (ecdysis), and with it the sand grains are also lost, but new grains are picked up through the dorsal pore after ecdysis.

The eyes in crayfish are **compound,** since they are made up of many units called **ommatidia** (Fig. 16-11). Covering the rounded surface of each eye is a transparent area of the cuticle, the **cornea,** which is divided into some 2,500 small squares known as **facets.** These are the outer ends of the ommatidia. Each ommatidium, starting at the surface, consists of a **corneal facet;** two **corneagen cells,** which form the cornea; a **crystalline cone** of four **cone cells** (vitrellae); a pair of **retinular cells** around the crystalline cone; several retinular cells, which form a central **rhabdome;** and black **pigment cells,** which separate the retinulae of adjacent ommatidia. The inner ends of the retinular cells connect with sensory nerve fibers that pass through optic ganglia to form the optic nerve to the brain.

The movement of pigment in the arthropod compound eye permits it to adjust for different amounts of light. In each ommatidium are three sets of pigment cells: distal retinal, proximal retinal, and reflecting; these are so arranged that they can form a more or less complete collar or sleeve around each ommatidium. For strong light or day adaptation the distal retinal pigment moves inward and meets the outward moving proximal retinal pigment so that a complete pigment sleeve is formed around the ommatidium (Fig. 16-11). In this condition only those rays that strike the cornea directly will reach the retinular cells, for each ommatidium is shielded from the others. Thus each ommatidium will see only a limited area of the field of vision (a mosaic, or apposition, image). In dim light the distal and proximal pigments separate so that the light rays, with the aid of the reflecting pigment cells, have a chance to spread to adjacent ommatidia and to form a continuous, or superposition, image. This second type of vision is less precise but takes maximum advantage of the limited amount of light received.

Reproduction. The crayfish is dioecious, and there is some sexual dimorphism, for the female has a broader abdomen than the male and lacks the modified swimmerets, which serve as copulatory organs in the male. The **male organs** consist of paired **testes** and a pair of **vasa deferentia,** which pass over the digestive glands and down to **genital pores** located at the base of the fifth walking legs (Fig. 16-6). The **female organs** are paired **ovaries** and short **oviducts,** which pass over the digestive glands and down to the genital pores at the base of the third walking legs.

Development. After fertilization, the eggs are attached to the swimmerets and embryonate there (Fig. 16-12). On hatching, the juvenile resembles the adult except for size and degree of gonad development; it

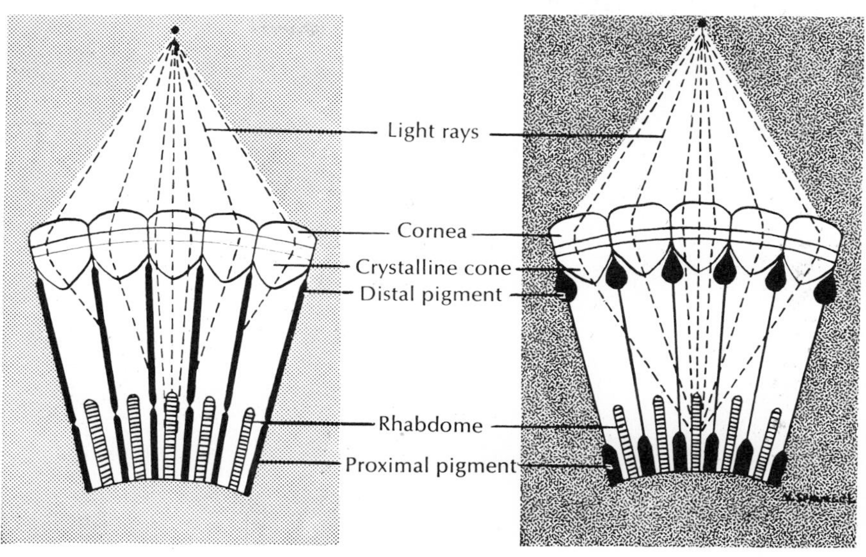

FIG. 16-11 Portion of compound eye of arthropod showing migration of pigment in ommatidia for day and night vision. Five ommatidia represented in each diagram. In daytime each ommatidium is surrounded by a dark pigment collar so that each ommatidium is stimulated only by light rays that enter its own cornea (mosaic vision); in nighttime, pigment forms incomplete collars, and light rays can spread to adjacent ommatidia (continuous, or superposition, image). (Redrawn from Moment, G. B. 1967. General zoology. Boston, Houghton Mifflin Co.)

FIG. 16-12 Female crayfish with eggs attached to swimmerets. Crayfish carrying eggs are said to be "in berry." The young, when hatched, also cling to swimmerets, protected by the tail fan. (Photograph by F. M. Hickman.)

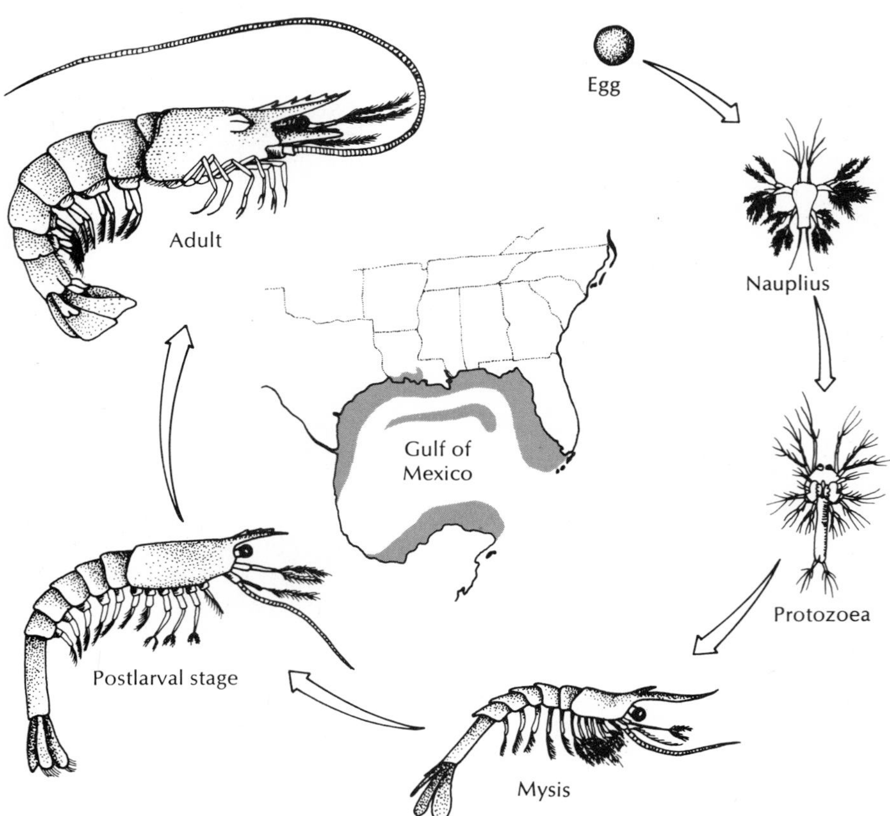

FIG. 16-13 Life cycle and distribution *(in red)* of the Gulf shrimp *Pennaeus*. Pennaeids spawn at depths of 20 to 50 fathoms. The young larval forms make up part of the plankton fauna. Older shrimp spend their days hidden in the loose deposits on the bottom, coming up at night to feed. Development is considered slightly metamorphic.

then grows and matures as it goes through a series of molts.

The type of development seen in the crayfish is described as **epimorphic:** the juvenile that hatches from the egg resembles a miniature adult, and there is no larval form. In the majority of crustaceans, however, a larva quite unlike the adult in structure and appearance hatches from the egg. The primitive and most widely occurring larva in the Crustacea is the **nauplius** (Figs. 16-13 and 16-16). The nauplius bears only three pairs of appendages: uniramous first antennae, biramous second antennae, and biramous mandibles. They all function as swimming appendages at this stage. Through a series of molts, the organism grows and adds somites and appendages, slowly becoming more like the adult. Such development is described as **anamorphic.** In many crustaceans the development is neither epimorphic nor anamorphic, but **metamorphic,** in which assumption of the adult form involves radical changes. For example, the metamorphosis of a barnacle proceeds from a free-swimming nauplius to a larva with a bivalve carapace called a cypris, and finally to the sessile adult with calcareous plates.

Ecdysis. Ecdysis (ek′duh-sis), or molting, is necessary for the body to increase in size because the exoskeleton is nonliving and does not grow as the animal grows. Much of the crustacean's functioning, including its reproduction, behavior, and many metabolic processes, are directly affected by the physiology of the molting cycle.

The **cuticle,** which is secreted by an underlying cell layer, the epidermis, is composed of several layers. The outermost is the **epicuticle,** a very thin layer of

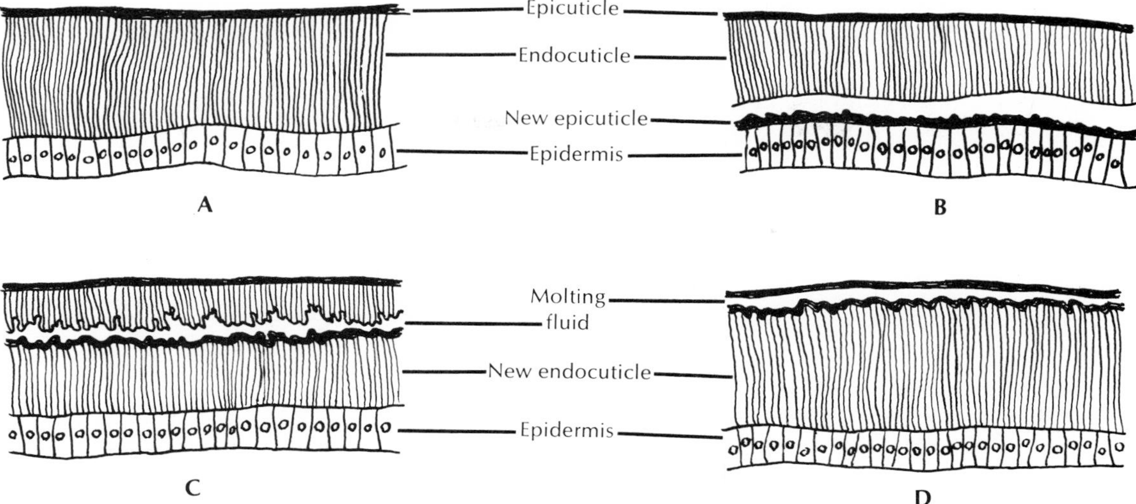

FIG. 16-14 Cuticle secretion and resorption in preecdysis. **A,** Interecdysis condition. **B,** Old endocuticle separates from epidermis, which secretes new epicuticle. **C,** As new endocuticle is secreted, molting fluid dissolves old endocuticle, and the solution products are resorbed. **D,** At ecdysis, little more than the old epicuticle is left to discard. In postecdysis, new cuticle is stretched and unfolded, and more endocuticle is secreted. (Modified from Schmidt, G. D., and L. S. Roberts. 1977. Foundations of Parasitology. St. Louis, The C. V. Mosby Co.)

lipid-impregnated protein. The bulk of the cuticle is made up of the several layers of the **endocuticle:** (1) the **pigmented layer,** which is just beneath the epicuticle and contains protein, calcium salts, and chitin; (2) the **calcified layer,** which contains more chitin and less protein and is heavily mineralized with calcium salts; and (3) the **uncalcified layer,** which is a relatively thin layer of chitin and protein.

Some time before the actual ecdysis, the epidermis separates from the uncalcified layer and secretes a new epicuticle (Fig. 16-14). Enzymes are released through canals to the area above the new epicuticle. The enzymes begin to dissolve the old endocuticle, and the soluble products are resorbed and stored within the body of the crustacean. Some of the calcium salts are stored as gastroliths in the walls of the stomach. Finally, little more than the epicuticle is left. The animal swallows water, which it absorbs through its gut, and its blood volume increases greatly. The internal pressure causes the cuticle to split, and the animal pulls itself out of its old exoskeleton (Fig. 16-15). Then follows a stretching of the still soft cuticle, redeposition of the salvaged inorganic salts and other constituents, hardening of the new cuticle, and deposition of more endocuticular material. During the period of molting, the animal is defenseless and remains hidden away.

When the crustacean is young, ecdysis must occur frequently to allow growth, and the molting cycle is relatively short. As the animal approaches maturity, intermolt periods become progressively longer, and in some species, molting ceases altogether. During intermolt periods, increase in tissue mass occurs, that is, water is replaced by living tissue.

HORMONAL CONTROL OF THE ECDYSIS CYCLE. Though ecdysis is hormonally controlled, the cycle is often initiated by an environmental stimulus perceived by the central nervous system. Such stimuli may include temperature, day length, and humidity (in the case of land crabs). The action of the signal from the central nervous system is to decrease the production of a **molt-inhibiting hormone** by the **X-organ.** The X-organ is a group of neurosecretory cells in the medulla terminalis of the brain. In the crayfish and other decapods, the medulla terminalis is found in the eye stalk. The hormone is carried in the axons of the X-organ to the **sinus gland** (which is probably not glandular in function, however), also in the eye stalk, where it is released into the blood.

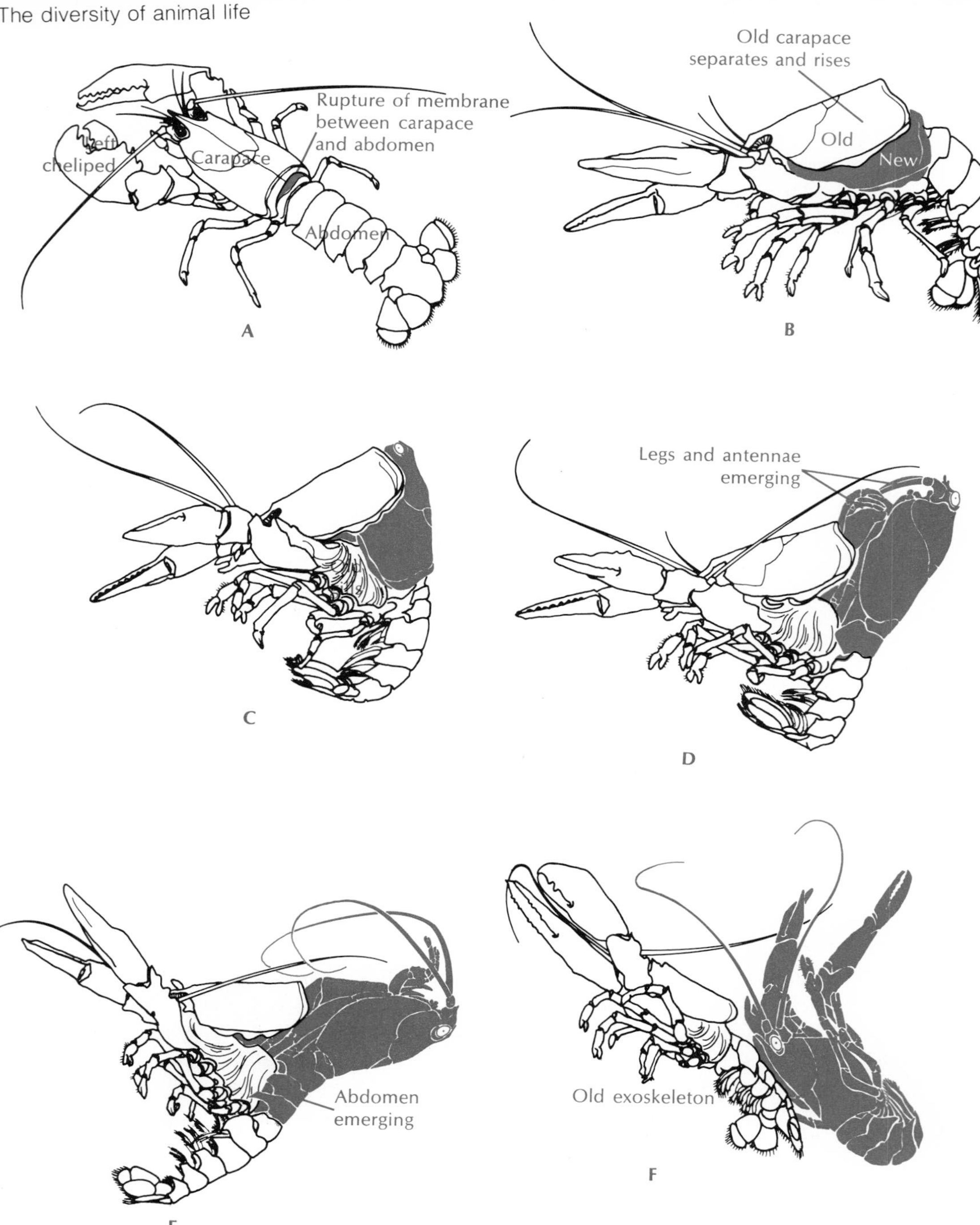

FIG. 16-15 Molting sequence in the lobster *Homarus americanus*. **A,** Membrane between carapace and abdomen ruptures, and carapace begins slow elevation. This step may take up to 2 hours. **B** to **E,** Head, thorax, and finally abdomen are withdrawn. This process usually takes no more than 15 minutes. **F,** Immediately after ecdysis, chelipeds are desiccated and body is very soft. Lobster continues rapid absorption of water so that within 12 hours body increases about 20% in length and 50% in weight. Tissue water will be replaced by protein in succeeding weeks. (Drawn from photographs by D. E. Aiken.)

When the level of molt-inhibiting hormone drops, release of a **molting hormone** from the **Y-organs** is promoted. The Y-organs are located beneath the adductor muscles of the mandibles and are homologus to the prothoracic glands of insects, which produce the hormone ecdysone. The action of the molting hormone is to initiate the processes leading to ecdysis (proecdysis). Once initiated, the cycle proceeds automatically, without further action of hormones from either the X- or Y-organs.

Other endocrine functions. Not only does removal of eyestalks accelerate molting, it was also found over 100 years ago that such crustaceans can no longer adjust body coloration to background conditions. About 50 years later it was discovered that the defect was not caused by loss of vision but by loss of hormones in the eyestalks. Body color of crustaceans is largely a result of pigments in special branched cells (chromatophores) in the epidermis. Color change is achieved by concentration of the pigment granules in the center of the cells, which causes a lightening effect, or by dispersal of the pigment throughout the cells, which causes a darkening effect. The pigment behavior is controlled by hormones from neurosecretory cells in the eyestalk, as is migration of retinal pigment for light and dark adaptation in the eyes (Fig. 16-11).

Release of neurosecretory material from the pericardial organs in the wall of the pericardium causes increase in the rate and amplitude of the heartbeat.

Androgenic glands, which were found in an amphipod (*Orchestia,* the common beach flea), probably occur in all male malacostracans. These are not neurosecretory organs. Their secretion stimulates the expression of male sexual characteristics. Young malacostracans have rudimentary androgenic glands, but in females these glands fail to develop. If they are artificially implanted in a female, her ovaries transform to testes and begin to produce sperm, and her appendages begin to take on male characteristics at the next molt. In isopods the androgenic glands are found in the testes; in all other malacostracans they are between the muscles of the coxopodites of the last thoracic legs and partly attached near the ends of the vasa deferentia. Though females do not possess organs similar to androgenic glands, their ovaries produce one or two hormones that influence secondary sexual characteristics.

Hormones that influence other body processes in Crustacea may be present, and evidence is good for a neurosecretory substance produced in the eyestalk that regulates blood sugar level.

Brief résumé of the crustaceans

The crustaceans are an extensive group with many subdivisions. There are many patterns of structure, habitat, and mode of living among them. Some are much larger than the crayfish; others are smaller, even microscopic. Some are highly developed and specialized; others have simpler organizations.

The reader should realize that the following summary of crustacean groups is misleadingly brief. Though all subclasses are mentioned, a complete presentation would require coverage of a surprisingly large number of taxa in the hierarchy below the subclass level. Among the Malacostraca, for example, some very unusual taxa such as ''series'' and ''section'' are used.

Subclass Cephalocarida

Cephalocarida (sef′a-lo-kar′i-da) (Gr., *kephalē,* head, + *karis,* shrimp, + *-ida,* pl. suffix) is a small group, with only four species described so far, that is found along both coasts of the United States, in the West Indies, and in Japan. They are 2 to 3 mm long and have been reported from bottom sediments from the intertidal zone to a depth of 300 m. They are even more primitive than the oldest known fossil crustacean, dating from the Devonian period. The similarity of the thoracic limbs to each other and to the second maxillae was mentioned before. The serial similarity of the body somites and the musculature is primitive. The second maxillae and the first seven thoracic legs may be considered triramous, bearing a pseudepipodite in addition to the exopodite and endopodites. The eighth legs lack the endopodite. Cephalocarids do not have eyes, carapace, or abdominal appendages. True hermaphrodites, they are unique among the Arthropoda in discharging both eggs and sperm through a common duct.

Subclass Branchiopoda

Branchiopoda (bran′kee-op′o-da) (Gr., *branchia,* gills, + *pous, podos,* foot) also represents a primitive crustacean type. Four orders are recognized: **Anostraca** (fairy shrimp and brine shrimp, Fig. 16-16, *G*), which lack a carapace; **Notostraca** (tadpole shrimp such as *Triops,* Fig. 16-16, *D*), whose carapace forms a large dorsal shield covering most of the trunk somites; **Conchostraca** (clam shrimp such as *Lynceus*), whose carapace is bivalve and usually encloses the entire

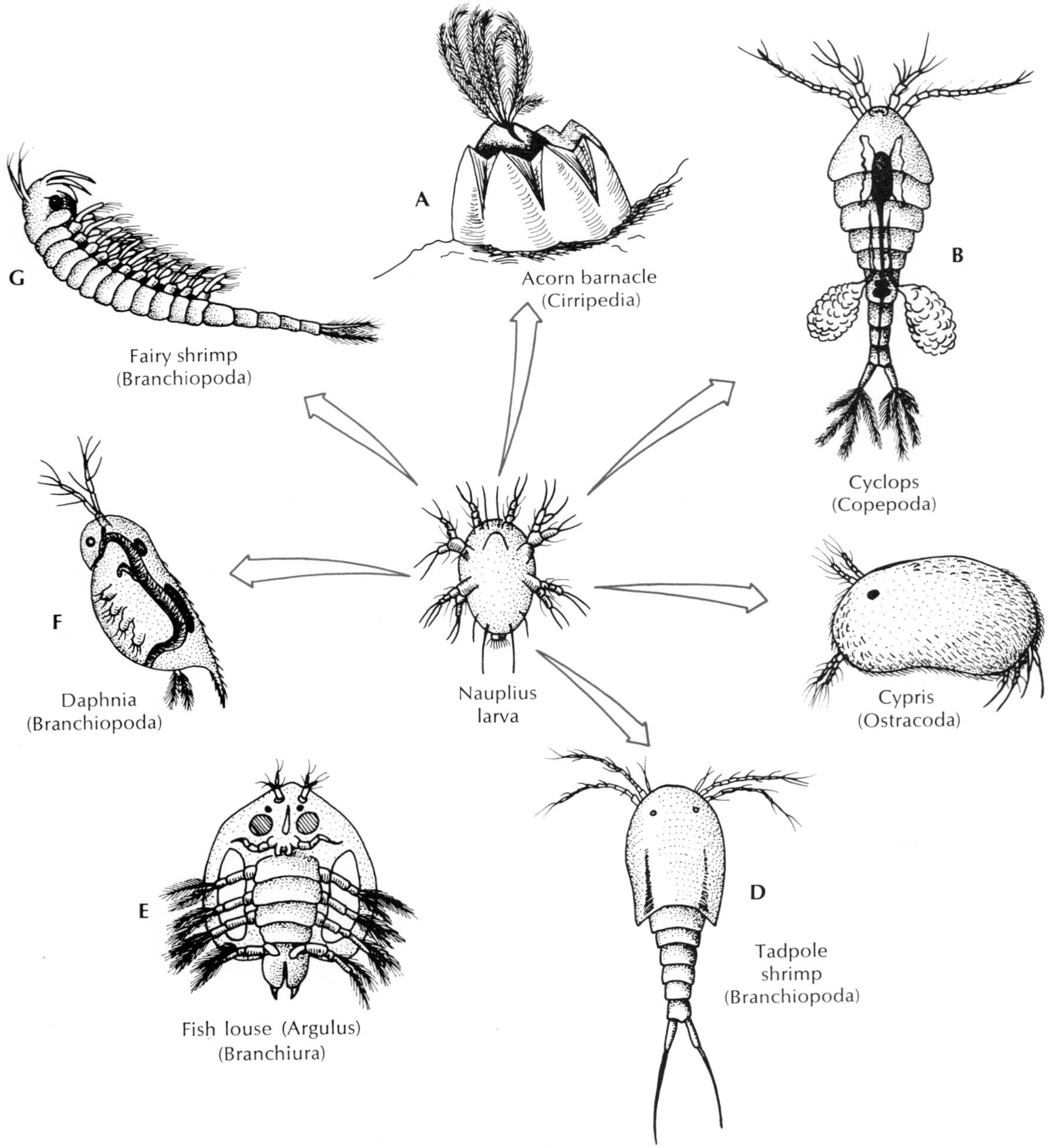

A
Acorn barnacle
(Cirripedia)

B
Cyclops
(Copepoda)

C
Cypris
(Ostracoda)

D
Tadpole
shrimp
(Branchiopoda)

E
Fish louse (Argulus)
(Branchiura)

F
Daphnia
(Branchiopoda)

G
Fairy shrimp
(Branchiopoda)

Nauplius
larva

FIG. 16-16 Group of smaller crustaceans. A nauplius hatches from the egg in most cases, except in the Branchiura, in which the nauplius is passed embryonically, and a postlarva hatches.

body; and **Cladocera** (water fleas such as *Daphnia*, Fig. 16-16, *F*), with a carapace typically covering the entire body but not the head. Branchiopods have reduced first antennae and second maxillae. Their legs are flattened and leaflike **(phyllopodia)** and are the chief respiratory organs (hence, the name branchipods). The legs also are used in filter feeding in most branchipods, and in groups other than the cladocerans, they are used for locomotion as well.

Most branchiopods are freshwater forms, and the most important and successful order is the Cladocera, which often forms a large segment of the freshwater zooplankton. Their reproduction is most interesting and is reminiscent of that found in some rotifers (Chapter 12). During the summer they often produce only females, by parthenogenesis, rapidly increasing the population. With the onset of unfavorable conditions, some males are produced, and eggs that must be fertilized are produced by normal meiosis. The fertilized eggs are very resistant to cold and desiccation, and they are very important for survival of the species over winter and for passive transfer to new habitats. Cladocera are mostly epimorphic, while other branchiopods are mainly anamorphic.

Subclass Ostracoda

Members of Ostracoda (os-trak'o-da) (Gr., *ostrakodes,* testaceous, that is, having a shell) are, like the conchostracans, enclosed in a bivalve carapace and resemble tiny clams, 0.25 to 8 mm long (Fig. 16-16, *C*). Ostracods show considerable fusion of trunk somites, and numbers of thoracic appendages are reduced to two or none. Feeding and locomotion are principally by use of the head appendages. Most ostracods live on the bottom or climb on plants, but some are planktonic, burrowing, or parasitic. Feeding habits are diverse; there are particle, plant, and carrion feeders and predators. They are widespread in both marine and freshwater habitats. Development is anamorphic.

Subclass Mystacocarida

The Mystacocarida (mis-tak'o-kar'i-da) (Gr., *mystax,* mustache, + *karis,* shrimp, + *-ida,* pl. suffix) is a subclass of tiny crustaceans (less than 0.5 mm long), which live in the interstitial water between sand grains of marine beaches (psammolittoral habitat). Only three species have been described, but they are widely distributed through many parts of the world. They are primitive in several characteristics and are believed to be related to the Copepoda.

FIG. 16-17 Group of acorn barnacles *Balanus* (subclass Cirripedia) and sea mussels *Mytilus,* attached to rock in intertidal zone. The mussels are molluscs. (Photograph by F. M. Hickman.)

Subclass Copepoda

The Copepoda (ko-pep'o-da) (Gr., *kōpē,* oar, + *pous, podos,* foot) is an important subclass of Crustacea, second only to the Malacostraca in numbers of species. The copepods are small (usually a few millimeters or less in length), rather elongate, tapering toward the posterior, lacking a carapace, and retaining the simple, median, nauplius eye in the adult (Fig. 16-16, *B*). They have a single pair of uniramous maxillipeds and four pairs of rather flattened, biramous, thoracic swimming appendages. The fifth pair of legs is reduced. The posterior part of the body is usually separated from the anterior, appendage-bearing portion by a major articulation. The first antennae are often longer than the other appendages. The Copepoda have been very successful and evolutionarily enterprising, with large numbers of symbiotic as well as free-living species. Many of the parasites are highly modified, and the adults may be so highly modified (and may depart so far from the description just given) that they can hardly be recognized as arthropods.

Ecologically, the free-living copepods are of extreme importance, often dominating the primary consumer level (p. 7) in aquatic communities. In many marine localities, the copepod *Calanus* is the most numerous organism in the zooplankton and has the greatest proportion of the total biomass (p. 6). In other localities it may only be surpassed in the biomass by euphausids (p. 367). *Calanus* forms a major por-

FIG. 16-18 Gooseneck barnacles *Lepas fascicularis* attach by long stalks. The barnacle is enclosed in a soft mantle that secretes the calcareous plates. It sweeps in food particles with its thoracic legs, the cirri. Adductor muscles can pull the plates together when the animal is not feeding. (Photograph by R. C. Hermes.)

tion of the diet of such economically and ecologically important fish as herring, menhaden, sardines, and the larvae of larger fish and (along with euphausids) is an important food item for whales and sharks. Other genera commonly occur in the marine zooplankton, and some forms such as *Cyclops* and *Diaptomus* may form an important segment of the freshwater plankton. Many species of copepods are parasites of a wide variety of other marine invertebrates and marine and freshwater fish, and some of the latter are of economic importance. Some species of free-living copepods serve as intermediate hosts of parasites of humans, such as *Diphyllobothrium* (a tapeworm) and *Dracunculus* (a nematode), and of other animals.

Development in the free-living species is commonly anamorphic or slightly metamorphic, but in some of the highly modified parasites, it is strongly metamorphic.

Subclass Branchiura

Branchiura (brank-i-ur′a) (Gr., *branchia,* gills, + *ura,* tail) is a small group of primarily fish parasites, which, despite its name, has no gills (Fig. 16-16, *E*). Members of this group are usually between 5 and 10 mm long and may be found on marine or freshwater fish. They typically have a broad, shieldlike carapace, compound eyes, four biramous thoracic appendages for swimming, and a short, unsegmented abdomen. The second maxillae have become modified as suction cups, and they can move about on their fish host or even from fish to fish. Development is almost epimorphic: there is no nauplius, and the young resemble the adults except in size and degree of development of the appendages.

Subclass Cirripedia

The Cirripedia (sir-ri-ped′di-a) (L., *cirrus,* curl of hair, + *ped, pedis,* foot) includes the barnacles, which are usually enclosed in a shell of calcareous plates, as well as three smaller orders of burrowing or parasitic forms. Barnacles are sessile as adults and may be attached to the substrate by a stalk (goose barnacles) (Fig. 16-18) or directly (acorn barnacles) (Figs. 16-16, *A,* and 16-17). Typically, the carapace (mantle) surrounds the body and secretes a shell of calcareous plates. The head is reduced, the abdomen absent, and the thoracic legs are long, many-jointed cirri with hairlike setae. The cirri are extended through an opening between the calcareous plates to filter from the water the small particles on which the animal feeds (Fig. 16-18). Though all barnacles are marine, they are often found in the intertidal zone and are, therefore, exposed to drying and sometimes fresh water for some periods of time. During these periods the aperture between the plates can be closed to a very narrow slit.

Barnacles are hermaphroditic and undergo a striking metamorphosis during development. Most hatch as nauplii, which soon become cypris larvae, so called because of their resemblance to the ostracod genus *Cypris.* They have a bivalve carapace and compound eyes. The cypris attaches to the substrate by means of its first antennae, which have adhesive glands associated with them, and begins its metamorphosis. This involves several dramatic changes, including secretion of the calcareous plates, loss of the eyes, and transformation of the swimming appendages to cirri.

Barnacles frequently foul ship bottoms by settling and growing there. So great may be their number that the speed of ships may be reduced 30% to 40%, necessitating drydocking the ship to clean them off.

Sacculina (Fig. 16-19) (order Rhizocephala) is a highly modified parasitic cirripede. It is a parasite of crabs, growing within them and sending a system of rootlike tubules throughout their bodies to absorb nutrients.

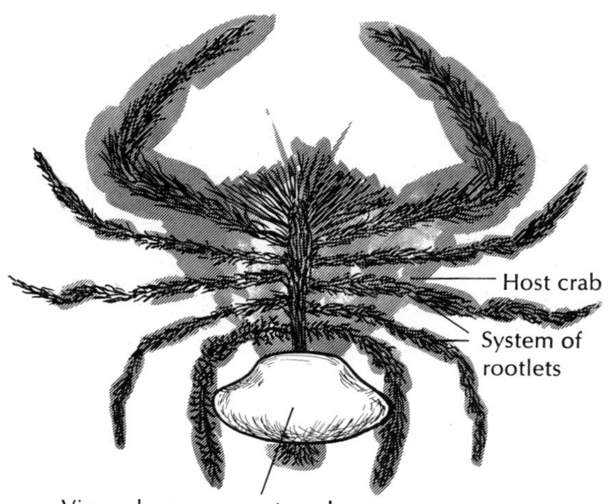

FIG. 16-19 *Sacculina,* an unusual crustacean (subclass Cirripedia) that parasitizes crabs. *Sacculina* larva attaches to the host and discharges cells into the crab's blood through a thin area of cuticle. At the junction of the crab's stomach and intestine, these cells become attached and form a mass, which sends branching rootlets to all parts of the host's body to absorb nutrients. Eventually *Sacculina* forms an opening on the ventral side of the host and extrudes a soft sac filled with eggs. The parasite destroys the crab's androgenic glands, altering its sex hormones so much that the host assumes a female form at its next molt.

Subclass Malacostraca

The Malacostraca (mal-a-kos′tra-ka) (Gr., *malakos,* soft, + *ostrakon,* shell) is the largest subclass of Crustacea and shows great diversity. The diversity is indicated by the higher classification of the group, which includes two series, four superorders, 12 or 13 orders, 16 suborders, plus numerous sections, subsections, and superfamilies. We will confine our coverage to mentioning a few of the most important orders. The characteristic caridoid facies of the malacostracans is described on p. 349.

Order Isopoda. The Isopoda (i-sop′o-da) (Gr., *isos,* equal, + *pous, podos,* foot) is one of the few crustacean groups to have successfully invaded terrestrial habitats in addition to fresh- and seawater habitats.

Isopods are commonly dorsoventrally flattened, lack a carapace, and have sessile compound eyes; their first pair of thoracic limbs are maxillipeds. The remaining thoracic limbs lack exopodites and are similar, while the abdominal appendages bear the gills and, except the uropods, also are similar to each other (hence the name isopods).

Common land forms are the sow bugs, or pill bugs (*Porcellio* and *Armadillidium*), which live under stones and in damp places. Though they are terrestrial, they do not have the efficient cuticular covering and other adaptations possessed by insects to conserve water; therefore, they must live in moist conditions. *Asellus* (Fig. 16-20, *F*) is a common freshwater form found under rocks and among aquatic plants. *Ligia* is a common marine form that scurries about on the beach or rocky shore. Some isopods are parasites of fish or crustaceans and may be highly modified.

Development is essentially epimorphic but may be highly metamorphic in the specialized parasites.

Order Amphipoda. Amphipoda (am-fip′o-da) (Gr., *amphis,* on both sides, + *pous, podos,* foot) resembles isopods in that the members have no carapace and have sessile compound eyes and one pair of maxillipeds. However, they are usually compressed laterally, and their gills are in the typical thoracic position. Further, their thoracic and abdominal limbs are each arranged in two or more groups that differ in form and function. For example, one group of abdominal legs may be for swimming and another for jumping. There are many marine amphipods, including some beach-dwelling forms (for example, the beach flea, *Orchestia*), numerous freshwater species (*Hyalella* and *Gammarus,* Fig. 16-20, *D*), and a few parasites. Development is epimorphic.

Order Euphausiacea. The Euphausiacea (u-faws-i-a′si-a) (Gr., *eu,* well, + *phausi,* shining bright, + *-acea,* L., suffix, pertaining to) is a group of only about 90 species, but they are important as the oceanic plankton known as "krill." They are about 3 to 6 cm long; have a carapace fused with all the thoracic segments, but which does not entirely enclose the gills; have no maxillipeds; and have all thoracic limbs with exopodites (Fig. 16-20, *C*). Most are bioluminescent, with a light-producing substance in an organ called a **photophore.** Some species may occur in enormous swarms, up to 45 m^2 and extending 100 to 500 m in one direction. They form a major portion of the diet of baleen whales and many fishes. Eggs hatch as nauplii, and development is anamorphic.

Order Decapoda. Decapoda (de-cap′o-da) (Gr., *deka,* ten, + *pous, podos,* foot) have three pairs of maxillipeds and five pairs of walking legs, of which the

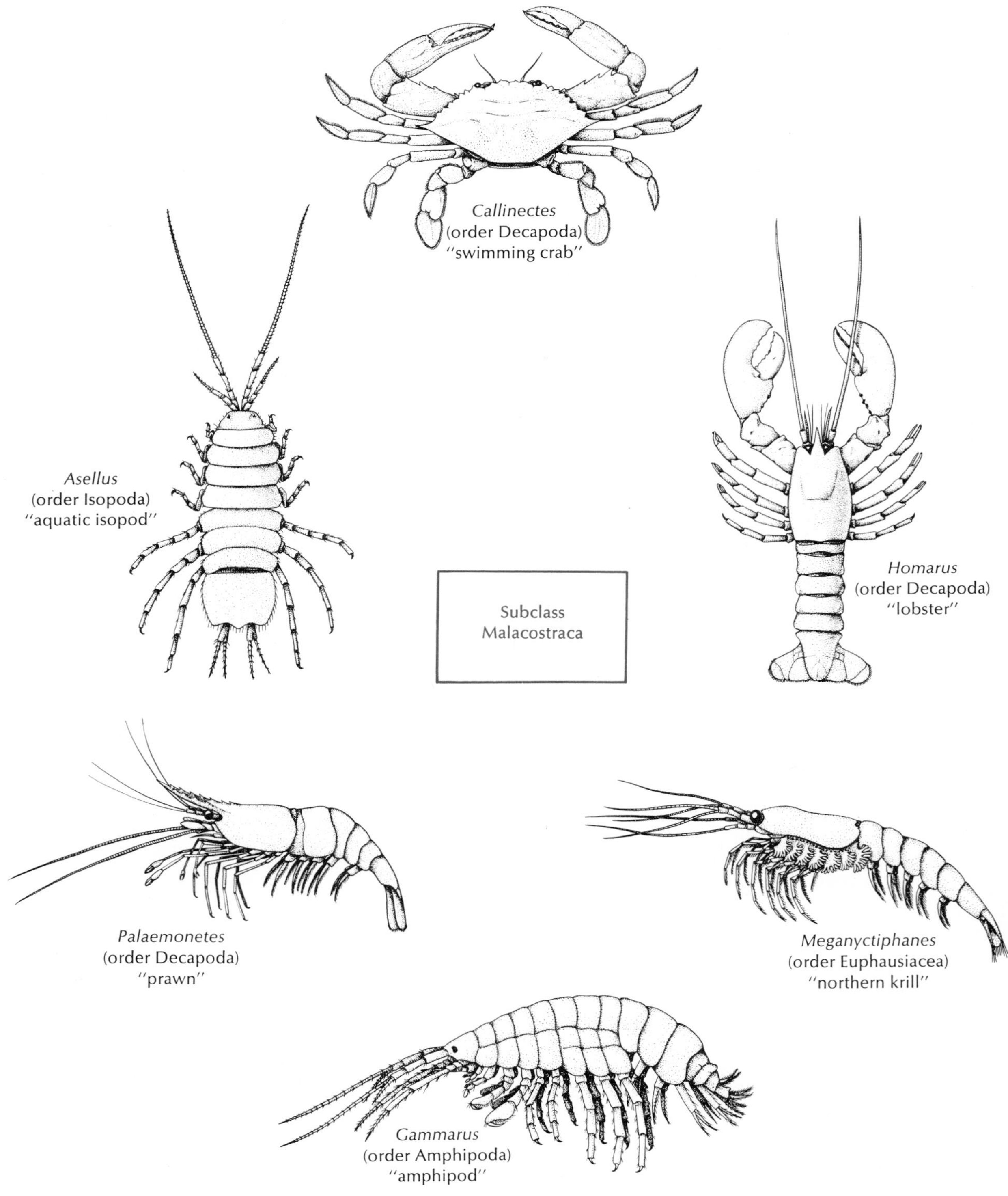

FIG. 16-20 Some crustaceans of the subclass Malacostraca.

FIG. 16-21 Marine hermit crab, *Eupagurus bernhardus,* has chosen an empty snail shell (Buccinum) as its home. The thin cuticle on the crab's abdomen makes a protective home a necessity. As it grows, it will select larger shells. Because of the antics of hermit crabs, they might be called the clowns of the sea. (Photograph by B. Tallmark, Uppsala University, Sweden.)

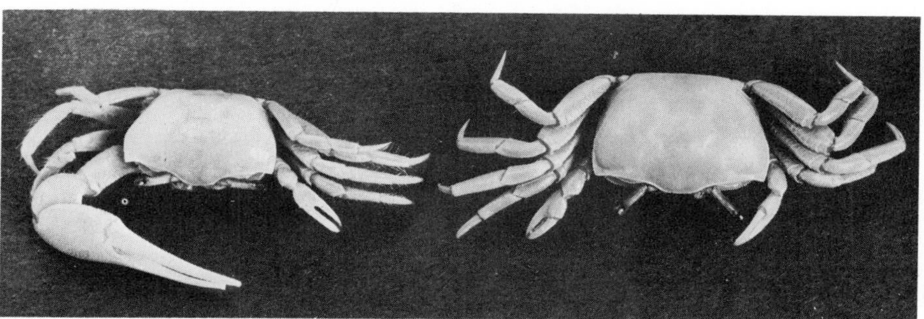

FIG. 16-22 Fiddler crab *Uca* (subclass Malacostraca, order Decapoda). Male *(left)* waves its large chela back and forth in the presence of a female; hence the name "fiddler." (Photograph by F. M. Hickman.)

A

B

FIG. 16-23 "Decorator crab" *Stenocionops furcata* obtaining a sea anemone *Calliactis tricolor* to place on its back. **A,** The crab prods the anemone near its base until it detaches its pedal disc. **B,** The crab then raises the anemone toward its own carapace. The anemone usually attaches to the carapace with its tentacles first and then later turns and attaches its pedal disc. It took this crab 13 minutes to loosen and attach the anemone. Decorator crabs also attach sponges to their backs as part of their camouflage. (Courtesy D. M. Ross, from Cutress, C., D. M. Ross, and L. Sutton. 1970. Canad. J. Zool. **48** (2):371-376.)

first is modified to form pincers (chelae). They range in size from a few millimeters to the largest of all arthropods, the Japanese spider crab, whose chelae span 4 m. The crayfish, which was described earlier in the chapter, and the lobsters, crabs, and "true" shrimp belong in this group (Figs. 16-13 and 16-20). There are over 8,500 species of decapods, and the order is extremely diverse. They are very important ecologically and economically, and numerous species are items of food for humans.

The crabs, especially, exist in a great variety of forms. Although resembling the pattern of crayfish, they differ from the latter in having a broader cephalothorax and a reduced abdomen. Familiar examples along the seashore are the hermit crabs (Fig. 16-21), which live in snail shells because their abdomens are not protected by the same heavy exoskeleton as the anterior parts are; the fiddler crabs, *Uca* (Fig. 16-22), which burrow in the sand just below the high-tide level and come out to run about over the sand while the tide is out; and the spider crabs such as *Libinia* and the interesting decorator crabs *Stenocionops,* which cover their carapaces with sponges and sea anemones for protective camouflage (Fig. 16-23).

Annotated references
Selected general readings

Adiyodi, K. G., and R. G. Adiyodi. 1970. Endocrine control of reproduction in decapod Crustacea. Biol. Rev. **45:** 121-165.

Barnard, J. L., R. J. Menzies, and M. C. Bacescu. 1962. Abyssal Crustacea. New York, Columbia University Press. *A series of three papers based on the findings of the Columbia University research vessel Vema. Emphasis is on systematics, including new species.*

Barnes, R. D. 1974. Invertebrate zoology, ed. 3. Philadelphia, W. B. Saunders Co.

Barnwell, F. H. 1966. Daily and tidal patterns of activity in individuals of fiddler crab (genus *Uca*) from the Woods Hole region. Biol. Bull. **130:**1-17.

Barrington, E. J. W. 1967. Invertebrate structure and function. Boston, Houghton Mifflin Co.

Bliss, D. E., and L. H. Mantel. 1968. Terrestrial adaptations in Crustacea. Am. Zool. **8:**307-700.

Caspari, E. (ed.). 1964. Behavior genetics. Am. Zool. **4:** 97-173.

Clarke, K. U. 1973. The biology of the Arthropoda. New York, American Elsevier Publishing Co., Inc.

Copeland, E. 1966. Salt transport organelle in *Artemia salenis* (brine shrimp). Science **151:**470-471.

Edmondson, W. T. (ed.). 1959. Ward and Whipple's freshwater biology, ed. 2. New York, John Wiley & Sons, Inc.

Gosner, K. L. 1971. Guide to identification of marine and estuarine invertebrates. New York, Interscience Publishers.

Green, J. 1961. A biology of Crustacea. London, H. F. & G. Witherby, Ltd. *A concise account of the structure and physiology of this group.*

Jackson, R. M., and F. Raw. 1966. Life in the soil. New York, St. Martin's Press, Inc. *Shows how organisms form an integral part of soil and their importance in soil formation and plant growth.*

Kabata, Z. 1970. Diseases of fishes, book 1. Crustacea as enemies of fishes. In S. F. Snieszko and H. R. Axelrod (eds.). Jersey City, N.J., T. F. H. Publications, Inc. *An excellent account of the crustacean parasites of fishes, from the little modified to the bizarre; includes pathology and treatment of affected hosts.*

Kaestner, A. 1970. Invertebrate zoology, vol. 3. New York, Interscience Publishers. *The best and fullest account of the class in one volume in English; a must for serious students of Crustacea.*

Lockwood, A. P. M. 1968. Aspects of the physiology of Crustacea. Edinburg, Oliver & Boyd, Ltd.

Manton, S. M. 1964. Mandibular mechanisms and the evolution of arthropods. Philosoph. Trans. R. Soc. Lond. (Biol.) **247:**1-183.

Marshall, S. M. 1973. Respiration and feeding in copepods. In F. S. Russell and M. Yonge (eds.). Advances in marine biology, vol. 11, New York, Academic Press, Inc., pp. 57-120.

Marshall, S. M., and A. P. Orr. 1955. The biology of a marine copepod *Calanus finmarchicus* (Gunnerus). Edinburgh, Oliver & Boyd, Inc.

Mauchline, J., and L. R. Fisher. 1969. The biology of euphausiids. In F. S. Russell and M. Yonge (eds.). Advances in marine biology, vol. 7. New York, Academic Press, Inc.

Newell, R. C. 1970. Biology of intertidal animals. London, Logos Press, Ltd.

Passano, L. M. 1961. The regulation of crustacean metamorphosis. Am. Zool. **1:**89-95.

Pennak, R. W. 1953. Fresh-water invertebrates of the

United States. New York, The Ronald Press Co. *Unusually clear taxonomic keys and illustrative drawings.*

Russell-Hunter, W. D. 1969. A biology of the higher invertebrates. New York, The Macmillan Co.

Sanders, H. L. 1957. The Cephalocarida and crustacean phylogeny. Syst. Zool. **6:**112-128.

Schmidt, G. D., and L. S. Roberts. 1977. Foundations of parasitology. St. Louis, The C. V. Mosby Co. *Good treatment in a parasitology text of crustacean parasites of vertebrates and invertebrates.*

Schmitt, W. L. 1965. Crustaceans. Ann Arbor, The University of Michigan Press. *A good little book by one of the most eminent students of the group; very interesting reading and easy to understand.*

Southward, A. J. 1955. Feeding of barnacles. Nature **175:** 1124-1125.

Street, P. 1966. The crab and its relatives. London, Faber and Faber, Ltd.

Tombes, A. S. 1970. An introduction to invertebrate endocrinology. New York, Academic Press, Inc.

Waterman, T. H. (ed.). 1960, 1961. The physiology of Crustacea, vols. 1 and 2. New York, Academic Press, Inc. *Becoming increasingly outdated, but still a valuable compendium of information.*

Whittington, H. B., and W. D. I. Rolfe (eds.). 1963. Phylogeny and evolution of Crustacea. Special publication. Cambridge, Mass., Museum of Comparative Zoology.

Wood, F. D., and H. E. Wood. 1932. Autotomy in decapod Crustacea. J. Exp. Zool. **62:**1-55.

Wulff, V. J. 1956. Physiology of the compound eye. Physiol. Rev. **36:**145-163.

Selected *Scientific American* articles

Benson, A. A., and R. F. Lee. 1975. The role of wax in oceanic food chains. **232:**77-86 (Mar.).

Brown, F. A., Jr. 1954. Biological clocks and the fiddler crab. **190:**34-37 (Apr.).

Caldwell, R. L., and H. Dingle. 1976. Stomatopods. **234:**81-89. (Jan.).

Murphy, R. C. 1962. The oceanic life of the Antarctic. **207:** 187-210. (Sept.).

Palmer, J. D. 1975. Biological clocks of the tidal zone. **232:** 70-79 (Feb.).

CHAPTER 17

THE TERRESTRIAL MANDIBULATES—THE MYRIAPODS AND INSECTS

Phylum Arthropoda
Classes Chilopoda, Diplopoda, Pauropoda, Symphyla, Insecta

The Dynastinae, which are the giants among beetles, include the elephant beetles, rhinoceros beetles, and Hercules beetles. The males are larger than the females and have extraordinarily long horns. This is a male elephant beetle *Megasoma sp.*

Courtesy Smithsonian Institution.

FIG. 17-1 Centipede *Scolopendra* (class Chilopoda). Most segments have one pair of appendages each. First segment bears pair of poison claws, which in some species can inflict serious wounds. Centipedes are carnivorous. (Photograph by F. M. Hickman.)

Along with the arachnids, the insects and myriapods are primarily terrestrial arthropods. Only a few of them have returned to aquatic life, usually in fresh water.

The term ''myriapod,'' meaning ''many-footed,'' is commonly used for a group of several classes of mandibulate arthropods that have evolved a pattern of two tagmata—head and trunk—with paired appendages on all trunk somites except the last. The myriapods include the Chilopoda (centipedes), Diplopoda (millipedes), Pauropoda (pauropods), and Symphyla (symphylans).

The insects have evolved a pattern of three tagmata—head, thorax, and abdomen—with appendages on the head and thorax but greatly reduced or absent from the abdomen. Insects may have arisen from an early myriapod or protomyriapod form.

The terrestrial mandibulates have no antennules and their appendages are unlike those of the crustaceans in being always uniramous, never biramous. And although some insect young are aquatic and have gills, the gills are not homologous to those of the crustaceans.

The insects and myriapods share with the onychophorans and some of the arachnids the use of tracheae for carrying the respiratory gases directly to and from all body cells.

Excretion is usually by malpighian tubules.

Class Chilopoda

The Chilopoda (ki-lop'o-da), or centipedes, are land forms with somewhat flattened bodies that may contain from a few to 177 somites (Fig. 17-1). Each somite, except the one behind the head and the last two in the body, bears a pair of jointed appendages. The appendages of the first body segment are modified to form poison claws.

The head appendages are similar to those of an insect. There is a pair of antennae, a pair of mandibles, and one or two pairs of maxillae. A pair of eyes on the dorsal side of the head consist of groups of ocelli.

The digestive system is a straight tube into which salivary glands empty at the anterior end. Two pairs of malpighian tubules empty into the hind part of the intestine. There is an elongated heart with a pair of arteries to each somite. Respiration is by means of a tracheal system of branched air tubes that come from a pair of spiracles in each somite. The nervous system is typically arthropod, and there is also a visceral nervous system.

Sexes are separate, with unpaired gonads and paired ducts. Some centipedes lay eggs and others are viviparous. The young are similar to the adults.

Centipedes prefer moist places such as under logs, bark, and stones. They are very agile and are carnivorous in their eating habits, living on earthworms, cockroaches, and other insects. They kill their prey with their poison claws and then chew it with their mandibles. The common house centipede, *Scutigera forceps,* with 15 pairs of legs is often seen scurrying around bathrooms and damp cellars, where they catch insects. Most species are harmless to humans. Some of the tropical centipedes may reach a length of 30 cm.

Class Diplopoda

The diplopods are called millipedes, which literally means ''thousand feet'' (Fig. 17-2). Even though they do not have that many legs, they do have a large number of appendages since each abdominal somite has two

FIG. 17-2 Millipede *Narceus americanus* (class Diplopoda.) Note the typical doubling of appendages on most segments. In contrast to centipedes, millipedes have cylindric rather than flattened bodies and are usually vegetarians. (Photograph by C. P. Hickman, Jr.)

pairs of appendages, a condition that may have arisen from the fusion of pairs of somites. Their cylindric bodies are made up of 25 to 100 somites. The short thorax consists of four somites, each bearing one pair of legs.

The head bears two clumps of simple eyes, a pair each of antennae, mandibles, and maxillae. The general body structures are similar to those of centipedes, with a few variations here and there. Two pairs of spiracles on each abdominal somite open into air chambers that give off the tracheal air tubes. Two genital apertures are found toward the anterior end.

In most millipedes the appendages of the seventh somite are specialized for copulatory organs. After copulation the eggs are laid in a nest and are carefully guarded by the mother. The larval forms have only one pair of legs to each somite.

Millipedes are not so active as centipedes. They walk with a slow, graceful motion, not wriggling as the centipedes do. They prefer dark moist places under logs or stones. They are herbivorous, feeding on decayed matter, whether plant or animal, although sometimes they eat living plants. When disturbed, they often roll up into a coil. Common examples of this class are *Spirobolus* and *Julus,* both of which have wide distribution.

Class Pauropoda

The pauropods are a group of minute (2 mm or less), soft-bodied myriapods, of which only 60 or 70 species have been named so far. They have a small head with branched antennae and no eyes, but they have a pair of sense organs that have the appearance of eyes (Fig. 17-3, *A*). Their 12 trunk segments usually bear nine pairs of legs (none on the first or the last two segments). They have only one tergal plate covering each two segments.

Tracheae, spiracles, and circulatory system are lacking. Pauropods are probably most closely related to the diplopods but are considered to be more primitive.

Although widely distributed, the pauropods are the least well known of the myriapods. They live in moist soil, leaf litter, or decaying vegetation and under bark and debris. Representative genera are *Pauropus* and *Allopauropus*.

Class Symphyla

Symphylans are small (2 to 10 mm) and have centipede-like bodies (Fig. 17-3, *B*). They live in humus, leafy mold, and debris. *Scutigerella immaculata* is often a pest on vegetables and flowers, particularly in greenhouses. They are soft-bodied, with 14 segments, 12 of which bear legs and one a pair of spinnerets. The antennae are long and unbranched.

Symphylans are eyeless but have sensory pits at the bases of the antennae. The tracheal system is limited to a pair of spiracles on the head and tracheal tubes to the anterior segments only.

The mating behavior of *Scutigerella* is unusual. The male places a spermatophore at the end of a stalk. When the female finds it, she takes it into her mouth, storing the sperm in special buccal pouches. Then she removes the eggs from her gonopore with her mouth and attaches them to moss or lichen, or to the walls of crevices, smearing them during the handling with some of the semen and so fertilizing them. The young at first have only six or seven pairs of legs.

Class Insecta

The insects are the most successful biologically of all the groups of arthropods. There are more species of insects than species in all the other classes of animals combined. The recorded number of insect species has been estimated to be close to 800,000, with thou-

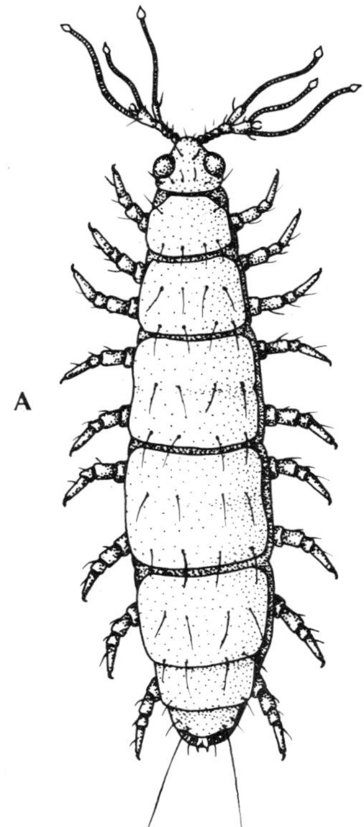

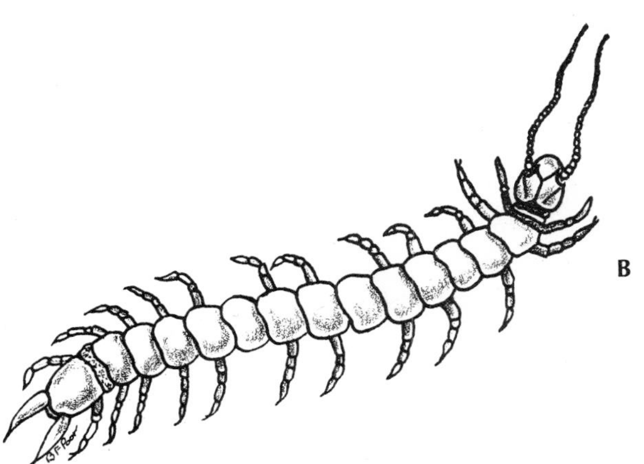

FIG. 17-3 **A,** A pauropod. Pauropods are minute, whitish myriapods with three-branched antennae and nine pairs of legs. They live in leaf litter and under stones. They are eyeless but have sense organs that resemble eyes. **B,** *Scutigerella,* a symphylan, is a minute whitish myriapod that is sometimes a greenhouse pest. (**B,** Modified from Snodgrass, R. E., 1952.)

sands of other species yet to be discovered and classified. There is also striking evidence that evolution is continuing among insects at the present time, even though the group as a whole is considered to be stable, according to the fossil record.

It is difficult to appreciate fully the significance of this extensive group and its role in the biologic pattern of animal life. The study of insects **(entomology)** occupies the time and resources of skilled men and women all over the world. The struggle between humans and their insect competitors seems to be an endless one, yet paradoxically insects have so interwoven themselves into the economy of nature in so many useful roles that we would have a difficult time without them.

Insects differ from other arthropods in having **three pairs of legs** and usually **two pairs of wings** on the thoracic region of the body, although some have one pair of wings or none. In size insects range from less than 1 mm to 20 cm in length—the majority being less than 2.5 cm long. Generally, the largest insects live in tropical areas.

Distribution

Insects are among the most abundant and widespread of all land animals. They have spread into practically all habitats that will support life except the deeper waters of the sea. Only a relatively few are marine. The marine water striders *(Halobates)*, which live on the surface of the ocean, are the only marine invertebrates that live on the sea-air interface. Insects are common in brackish water, in salt marshes, and on sandy beaches. They are abundant in fresh water, in soils, in forests, and in plants, and they are found even in deserts and wastelands, on mountain tops, and as parasites in and on the bodies of plants and animals.

Their wide distribution is made possible by their powers of flight and their highly adaptable nature. In most cases they can easily surmount barriers that are well nigh impassable to many other animals. Their small size allows them to be carried by currents of both wind and water to far regions. Their well-protected eggs can withstand rigorous conditions and can be carried long distances by birds and animals. Their

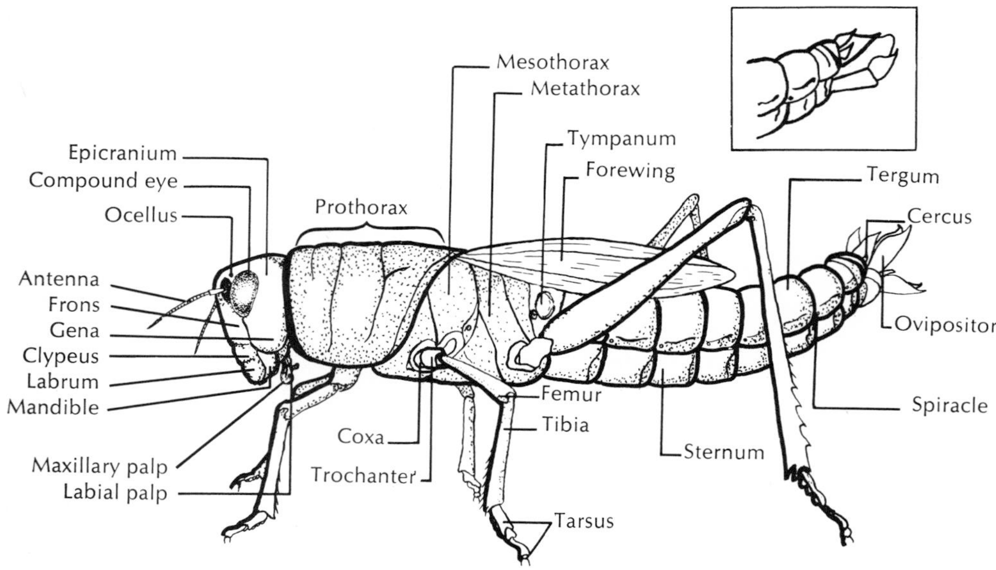

FIG. 17-4 External features of female grasshopper. Terminal segments of male shown in inset.

agility and aggressiveness enable them to fight for every possible niche in a habitat. No single pattern of biologic adaptation can be applied to them.

Adaptability

Insects, during their evolution, have shown an amazing adaptability, as is evidenced by their wide distribution and enormous diversity of species. Most of their structural modifications center around the wings, legs, antennae, mouthparts, and alimentary canal. Such wide diversity enables this vigorous group to take advantage of all available resources of food and shelter. Some are parasitic, some suck the sap of plants, some chew up the foliage of plants, some are predaceous, and some live on the blood of various animals. Within these different groups, specialization occurs, so that a particular kind of insect will eat, for instance, the leaves of only one kind of plant. This specificity of eating habits lessens competition and to a great extent accounts for their biologic success.

Insects are well-adapted to dry and desert regions. The hard and protective exoskeleton helps prevent evaporation, but some insects also extract the utmost in fluid from food and fecal material, as well as moisture from the water by-product of bodily metabolism.

The exoskeleton is made up of a complex system of plates known as **sclerites** connected to one another by concealed, flexible hinge joints. The muscles between the sclerites enable the insect to make precise movement. The rigidity of its exoskeleton is attributable to the unique scleroproteins and not to its chitin component, and its lightness makes flying possible. By contrast, the cuticle of crustaceans is stiffened by mineral matter and that of the arachnids by organic materials.

External features

Insects show a remarkable variety of morphologic characters. Some are fairly generalized in body structure, some are highly specialized. The grasshopper, or locust, is a generalized type that is usually used in laboratories to demonstrate the general features of insects (Fig. 17-4).

The insect body is made up of the head, thorax, and abdomen. The cuticle of each body segment is typically composed of four plates (sclerites), a dorsal notum (tergum), a ventral sternum, and a pair of lateral pleura. The pleura of abdominal segments is membranous rather than sclerotized.

The head usually bears a pair of relatively large compound eyes, a pair of antennae, and usually three ocelli. The antennae, which vary greatly in size and form (Fig. 17-5), act as tactile organs, organs of smell, and in some cases organs of hearing. Mouthparts, formed

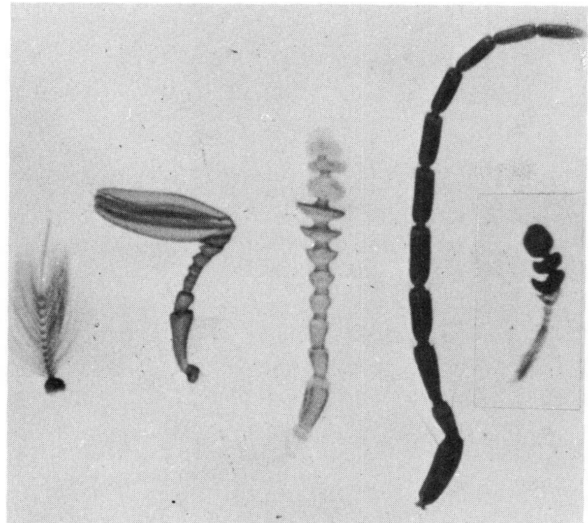

FIG. 17-5 A few of the various types of insect antennae. *Left to right:* mosquito (plumose); May beetle (laminate); click beetle (serrate); tenebrionid beetle (moniliform); and water scavenger beetle (clavate). (Photomicrograph by F. M. Hickman.)

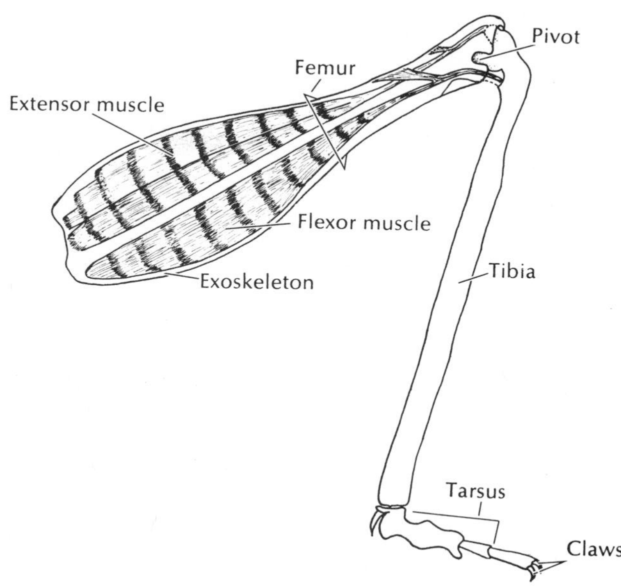

FIG. 17-6 Hind leg of grasshopper. Muscles that operate the leg are found within a hollow cylinder of exoskeleton. Here they are attached to the internal wall, from which they manipulate segments of limb on the principle of a lever. Note pivot joint and attachment of tendons of extensor and flexor muscles, which act reciprocally to extend and flex the limb.

from specially hardened cuticle, typically consist of a labrum, a pair each of mandibles and maxillae, a labium, and a tonguelike hypopharynx. The type of mouthparts an insect possesses determines how it feeds. Some of these modifications will be discussed later.

The thorax is composed of the prothorax, mesothorax, and metathorax, each bearing a pair of legs (Fig. 17-4). In most insects the mesothorax and metathorax each bear a pair of wings. The wings are cuticular extensions formed by the epidermis. They consist of a double membrane that contains veins of thicker cuticle that serve to strengthen the wing. Although these veins vary in their patterns among the different species, they are constant within a species and serve as one means of classification and identification.

Legs of insects are often modified for special purposes. Terrestrial forms have walking legs with terminal pads and claws as in beetles. These pads may be sticky for walking upside down, as in houseflies. The hind legs of the grasshoppers and crickets are adapted for jumping (Fig. 17-6). The mole cricket has the first pair of legs modified for burrowing in the ground (Fig. 17-8, *D*). Water bugs and many beetles have paddle-shaped appendages for swimming. For grasping its prey, the forelegs of the praying mantis are long and strong.

The **honeybee** is a good example of how an insect's legs are developed for special purposes (Fig. 17-7). The first pair of legs in this insect is suited to collect pollen by having a feathery **pollen brush** on the metatarsus and a fringe of hairs on the medial edge of the tibia for cleaning the head and the compound eye. A semicircular indentation lined with teeth is found in the metatarsus, and this is covered over with a spine, the **velum,** from the tibia. As the antenna is pulled through this notch, it is cleaned of pollen; hence this structure is called the **antennae cleaner.** The middle leg also has a **pollen brush** on the metatarsus and a **spur** on the tibia for removing wax from the wax glands on the abdomen. The hind limb is the most specialized of all, for it bears the **pollen basket,** the **pollen packer,** and **pollen brushes.** The pollen basket is a concavity on the tibia, with hairs along both edges that are kept moist with secretions from the mouth. The pollen packer consists of a row of stout bristles on the tibia and the auricle, a smooth plate on the metatarsus. The pecten removes the pollen from the pollen brush of the opposite leg onto the auricle. When the leg is flexed, the auricle presses against the end of the tibia, compressing the pollen. Pollen brushes (combs) occur

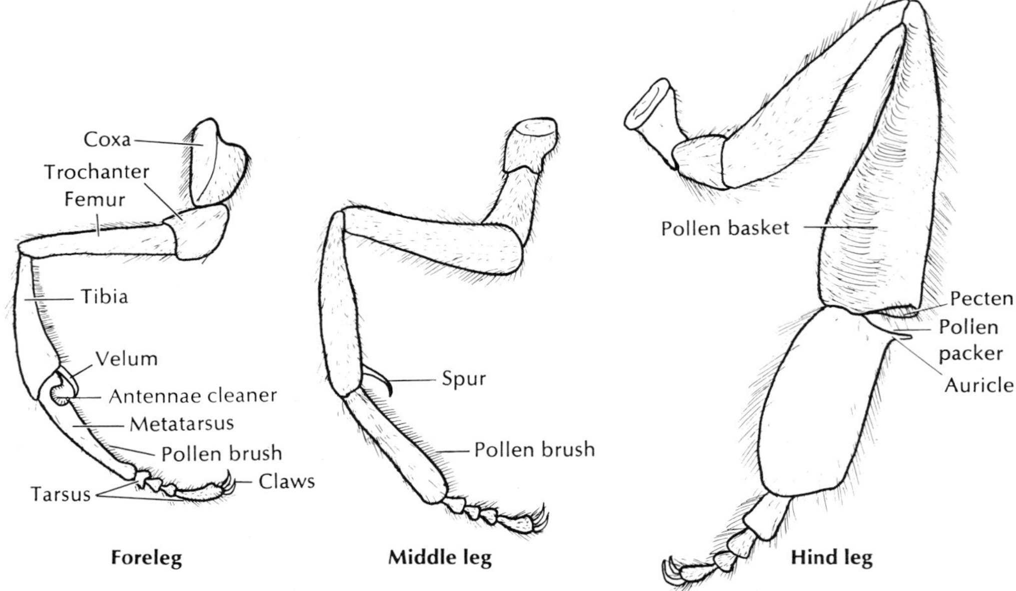

FIG. 17-7 Adaptive legs of worker honeybee *(left side).* In the foreleg, toothed indentation covered with velum is used to comb out antennae. Spur on the middle leg removes wax from wax glands on the abdomen. Pollen picked up on body hairs is combed off by pollen brushes on the front and middle legs and deposited on pollen brushes of the hind legs. Long hairs of pecten on the hind leg remove pollen from the brush of the opposite leg; then the auricle is used to press it into a pollen basket when the leg joint is flexed back. A bee carries her load in both baskets to the hive and pushes pollen into a cell, to be cared for by other workers.

on the inner surface of the metatarsus and consist of rows of stout spines.

The abdomen of insects is composed of nine to 11 segments—the eleventh, when present, is reduced to a pair of cerci (appendages at the posterior end). Larval or nymphal forms have a variety of abdominal appendages, but these are lacking in the adults. The end of the abdomen bears the external genitalia (Fig. 17-4).

There are innumerable variations in body form among the insects. Beetles are usually thick and plump; damselflies, crane flies, and walking sticks (Fig. 17-9, *A*) are long and slender; aquatic beetles are streamlined (Fig. 17-8, *C*); and cockroaches are flat, adapted to living in crevices. The female ovipositor of the ichneumon wasp is extremely long (Fig. 17-8, *A*). The cerci form horny forceps in the earwigs and are long and many jointed in stoneflies and mayflies (Fig. 17-34, *B* and *G*). Antennae are long in cockroaches and katydids, short in dragonflies and most beetles, knobbed in butterflies, and plumed in most moths. Dobsonflies have greatly enlarged mandibles, so long in males that they resemble long thin forceps (Fig. 17-9, *B*).

Locomotion in insects

Walking. Most insects, when walking, use a triangle of legs involving the first and last leg of one side together with the middle leg of the opposite side. In this way, insects keep three of their six legs on the ground, a tripod arrangement for stability. A slow-moving insect alternates its movement first on one side and then on the other. When it goes faster, the two phases tend to overlap and one side may begin before the other side has finished. In some insects all six legs may be used simultaneously and not alternately.

Some insects, such as the water strider *(Gerris),* are able to walk on the surface of water. The water strider has on its footpads nonwetting hairs that do not break the surface film of water but merely indent it. As it skates along, *Gerris* uses only the two posterior pairs of legs and steers with the anterior pair. The body of the marine water strider *Halobates,* an excellent surfer on rough ocean waves, is further protected by a water-repellent coat of close-set hairs shaped like thick hooks.

Power of flight. Insects share the power of flight with birds and flying mammals. However, their wings have evolved in a different manner from that of the

FIG. 17-8 **A,** Female ichneumon wasp (order Hymenoptera) uses long ovipositor to bore into a tree and lay an egg near a wood-boring beetle larva, which will become food for the ichneumon larva. This specimen had an overall length of over 15 cm (see also Fig. 17-20). **B,** Ichneumon head showing compound eyes, antennae, and some mouthparts. **C,** Giant diving beetle *Dytiscus* (order Coleoptera), about 3 cm long, is streamlined for life as an active aquatic predator. **D,** Mole cricket *Gryllotalpa* (order Orthoptera) has forelegs adapted for digging. (Photographs by F. M. Hickman.)

limb buds of birds and mammals and are not homologous to them. Insect wings are formed by outgrowths from the body wall of the mesothoracic and metathoracic segments and are composed of cuticle.

Most insects have two pairs of wings, but the Diptera (true flies) have only one pair, the hind wings being represented by a pair of tiny **halteres** (balancers) that vibrate and are responsible for equilibrium during flight. Males of order Strepsiptera have only the hind pair of wings and an anterior pair of halteres. The males of the scale insects also have one pair of wings but no halteres. Some insects are wingless. Ants and termites, for example, have wings only on males, and on females during certain periods; workers are always wingless. Lice and fleas are always wingless.

Wings may be thin and membranous, as in flies and many others (Fig. 17-8, *A*); thick and horny, as in the front wings of beetles (Fig. 17-8, *C*); parchment-like, as in the front wings of grasshoppers; covered with fine scales, as in butterflies and moths; or with hairs, as in caddis flies.

Wing movements are controlled by a complex of muscles in the thorax. These include some of the largest and strongest of all insect muscles. Dragonflies, orthopterans, moths, butterflies, and some other insects have **direct flight muscles** that insert directly on the base of the wings and control usually the tilting, or "feathering," of the wing. However, in the horizontal wings of the dragonflies they also provide the up and down movements (Fig. 17-10, *A*). Most of the more

A

B

FIG. 17-9 **A,** Walking stick (order Orthoptera). Its resemblance to the twigs it lives among is a protective device. **B,** Dobsonfly (order Neuroptera) has aquatic naiad, the hellgrammite, often used by fishermen as bait. (Photographs by C. P. Hickman, Jr.)

highly specialized insects have the more powerful **indirect flight muscles.** These include both longitudinal muscles and dorsoventral muscles, which attach to the walls of the thorax instead of directly on the wings. These raise or lower the wings by changing the shape of the thoracic walls (Fig. 17-10, *B*). Contraction of the longitudinal muscles causes the notum to bulge upward and thus lowers the wings. Contraction of the dorsoventral muscles depresses the notum and forces the wings upward. The wing behaves as a lever, pivoting on the pleural process, and the base of the wing, which projects inward from the pivot, moves as the notum moves but in the opposite direction.

Obviously flying entails more than a simple flapping of the wings; a forward thrust is necessary. As the indirect flight muscles alternate rhythmically to raise and lower the wings, the direct flight muscles alter the angle of the wings so that they act as lifting airfoils during both the upstroke and the downstroke, twisting the leading edge of the wings downward during the downstroke and upward during the upstroke. This produces a figure-eight movement that aids in spilling air from the trailing edges of the wings. The quality of the forward thrust depends, of course, on several factors, such as variations in wing venation, how much the wings are tilted, how they are feathered, and so on.

Flight speeds vary. The fastest flyers usually have narrow, fast-moving wings with a strong tilt and a

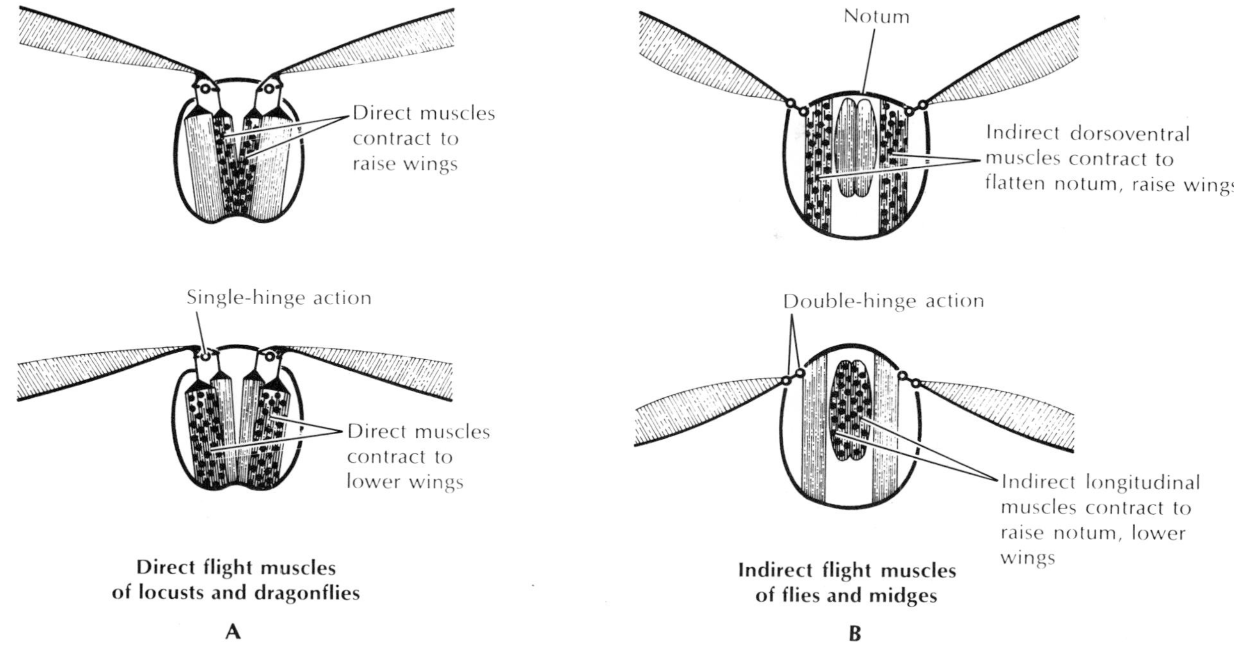

FIG. 17-10 Insect flight muscles. **A,** Direct muscles are attached to the base of the wings at their articulation and raise, lower, or tilt the wings by direct action. **B,** Indirect muscles are attached to the thoracic walls and exert a lever action upon the wings by changing the shape of the thorax.

strong figure-eight component. Sphinx moths and horseflies are said to achieve approximately 48 km (30 miles) per hour and dragonflies approximately 40 km (25 miles) per hour. Some insects are capable of long continuous flights. The migrating monarch butterfly *Danaus plexippus* travels south for hundreds of miles in the fall, flying at a speed of approximately 10 km (6 miles) per hour.

The frequency of wing beat also varies. Light-bodied insects with large wings, such as butterflies, may beat as few as 4 times per second. The small wings of heavy insects, such as flies and bees, may vibrate at 100 beats per second or more. The fruit fly *Drosophila* can fly at 300 beats per second, and midges have been clocked at more than 1,000 beats per second.

Internal form and function

Nutrition. The digestive system (Fig. 17-11) consists of a foregut (mouth with salivary glands, esophagus, crop for storage, and gizzard for grinding); a midgut (stomach and gastric ceca); and a hindgut (intestine, rectum, and anus). The fore- and hindguts are lined with cuticle, so absorption of food is confined largely to the midgut, though some absorption may take place in all sections. The majority of insects feed on plant juices and plant tissues. Such a food habit is called **phytophagous.** Some insects feed on specific plants; others, such as grasshoppers, will eat almost any plant. The caterpillars of many moths and butterflies eat the foliage of only certain plants. Certain species of ants and termites cultivate fungous gardens as a source of food.

Many beetles and the larvae of many insects live on dead animals **(saprophagous).** A number of insects are **predaceous,** catching and eating other insects as well as other types of animals. The so-called predaceous diving beetle *Cybister fimbriolatus,* however, has been found not to be as predaceous as supposed but is largely a scavenger.

Many insects, adults as well as larvae, are **parasitic.** Fleas, for instance, live on the blood of mammals, and the larvae of many varieties of wasps live on spiders and caterpillars (Fig. 17-12). In turn, many are parasitized by other insects. Some of the latter are

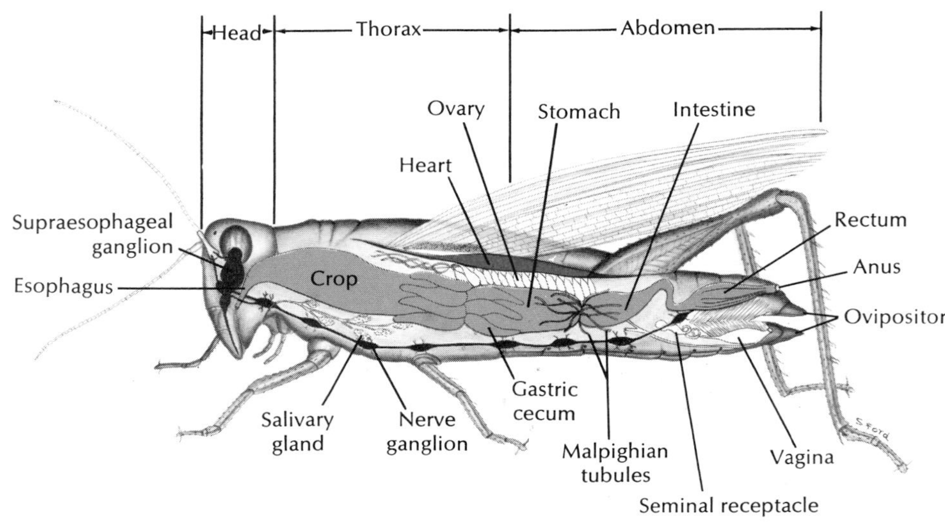

FIG. 17-11 Internal structure of female grasshopper.

FIG. 17-12 **A,** Hornworm, larval stage of a sphinx moth (order Lepidoptera). The more than 100 species of North American sphinx moths are strong fliers and mostly nocturnal feeders. Their larvae, called hornworms because of the large, fleshy posterior spine, are often pests of tomatoes, tobacco, and other plants. **B,** Hornworm parasitized by a tiny wasp *Apanteles,* which laid its egg inside of it. The wasp larvae have emerged, and their pupae are on the catepillar's skin. Young wasps emerge in 5 to 10 days, but the caterpillar usually dies. (**A,** Photograph by C. P. Hickman, Jr.: **B,** Courtesy O. W. Olsen.)

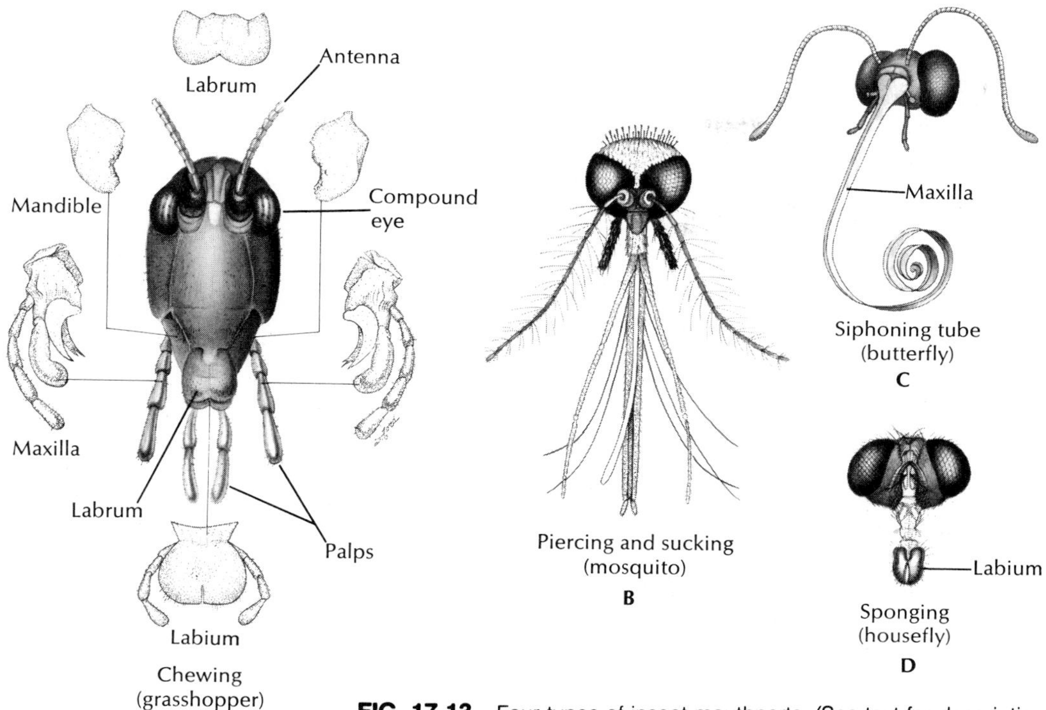

FIG. 17-13 Four types of insect mouthparts. (See text for description of types and examples.)

beneficial to humans by controlling the numbers of injurious insects. When parasitic insects are themselves parasitized by other insects, the condition is known as **hyperparasitism,** which often becomes quite involved.

For each type of feeding, the mouthparts are adapted in a specialized way. The **sucking mouthparts** are usually arranged in the form of a tube and can pierce the tissues of plants or animals. This arrangement is well-shown in the water scorpion *(Ranatra fusca),* a member of the order Hemiptera. This elongated, stick-like insect with a slender caudal respiratory tube has a beak in which there are four piercing, needlelike stylets made up of two mandibles and two maxillae. These parts are fitted together to form two tubes, a salivary tube for injecting saliva into the prey and a food tube for drawing out the body fluid of the prey. The mosquito also combines piercing with needle-like stylets and sucking through a food channel (Fig. 17-13, *B*). In honeybees the labium forms a flexible and contractile "tongue" covered with many hairs. When the bee plunges its proboscis into nectar, the tip of the tongue bends upward and moves back and forth rapidly. Liquid enters the tube by capillarity and is drawn up continuously by a pumping pharynx. In butterflies and moths, mandibles are usually absent, and the maxillae are modified into a long sucking proboscis (Fig. 17-13, *C*) for drawing nectar from flowers. At rest the proboscis is coiled up into a flat spiral. In feeding it is extended, and fluid is pumped up by pharyngeal muscles.

Houseflies, blowflies, and fruitflies have **sponging and lapping mouthparts** (Fig. 17-13, *D*). At the apex of the labium are a pair of large, soft lobes with grooves on the lower surface that serve as food channels. These flies lap up liquid food, or liquefy it first with salivary secretions. Horseflies, however, are fitted not only to sponge up surface liquids but to bite into the skin with slender, tapering mandibles and then sponge up blood.

Biting mouthparts such as those of the grasshopper and many other herbivorous insects are adapted for seizing and crushing food (Fig. 17-13, *A*); those of most carnivorous insects are sharp and pointed for piercing their prey. The mandibles of chewing insects are strong, toothed plates whose edges can bite or tear

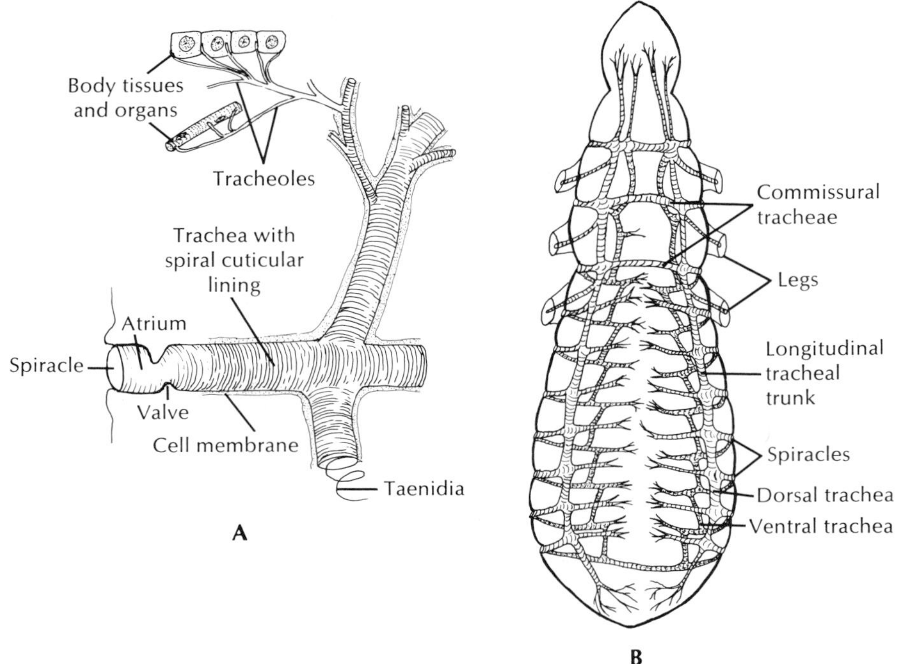

FIG. 17-14 **A,** Relationship of spiracle, tracheae, taenidia (chitinous bands that strengthen the tracheae), and tracheoles (diagrammatic). **B,** Generalized arrangement of insect tracheal system (diagrammatic). Air sacs and tracheoles not shown.

while the maxillae hold the food and pass it toward the mouth. Enzymes secreted by the salivary glands add chemical action to the chewing process.

The kind of mouthparts an insect has may determine the type of insecticide used in destroying it. For those that bite and chew their food the poison can be applied directly to the food; those that suck may be smothered with gaseous mixtures that interfere with their respiration.

Circulation. A tubular heart in the pericardial cavity (Fig. 17-11) moves blood forward through the only blood vessel, a dorsal aorta. The heartbeat is a peristaltic wave. Accessory pulsatory organs help move the blood into the wings and legs, and blood flow is also facilitated by various body movements. The blood consists of plasma and amebocytes. The blood apparently has little to do with oxygen transport.

Gas exchange. Terrestrial animals require efficient respiratory systems that permit rapid oxygen–carbon dioxide exchange but at the same time restrict water loss. In insects this is the function of the **tracheal system,** an extensive network of thin-walled tubes that branch into every part of the body (Fig. 17-14). The tracheal trunks open to the outside by paired **spiracles,** usually two on the thorax and seven or eight on the abdomen. A spiracle may be merely a hole in the integument, as in primary wingless insects, but it is usually provided with a valve or some sort of closing mechanism that cuts down water loss. The evolution of such a device must have been very important in enabling insects to move into drier habitats. The spiracle may also possess a filtering device such as a sieve plate or a set of interlocking bristles that may prevent entrance of water, parasites, or dust into the tracheae. The **tracheae** are composed of a single layer of cells and are lined with cuticle that is shed, along with the outer cuticle, during the molt. They are supported by spiral thickenings of the cuticle (called taenidia) that prevent their collapse. The tracheae branch out into smaller tubes, ending in very fine, fluid-filled tubules called **tracheoles** (not lined with cuticle), which branch into a fine network over the cells. In large insects the largest tracheae may be several millimeters in diameter but taper down to 1 to 2 μm. The tracheoles then taper to 0.5 to 0.1 μm in diameter. In one of the stages of the silkworm larva it is estimated that there are 1.5

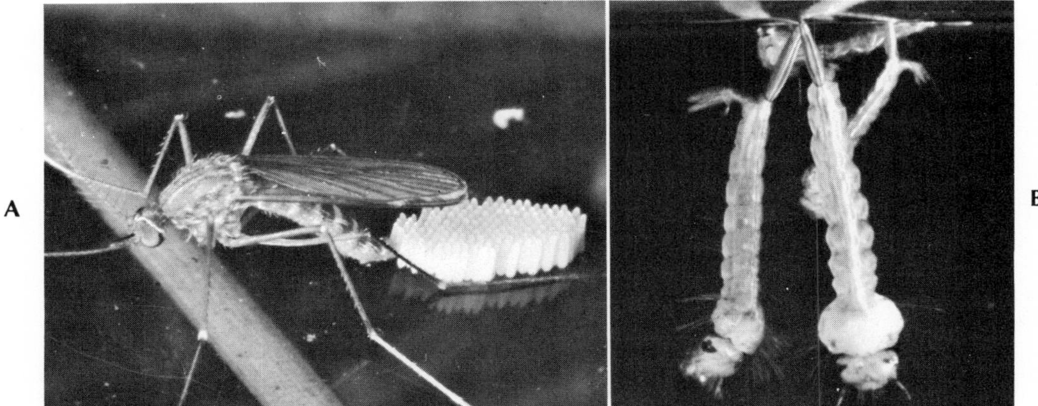

FIG. 17-15 **A,** Mosquito *Culex* (order Diptera) lays her eggs in small packets or rafts on the surface of standing or slowing moving water. **B,** Mosquito larvae are the familiar wrigglers of ponds and ditches. To breathe they hang head down, with respiratory tubes projecting through the surface film of water. Motion of vibratile tufts of fine hairs on the head brings a constant supply of food. (From *Mosquito,* an Encyclopaedia Britannica film.)

million tracheoles! Scarcely any living cell is located more than a few micrometers away from a tracheole. In fact, the ends of some tracheoles actually indent the membranes of the cells they supply, so that they terminate close to the mitochondria. The tracheal system affords an efficient system of transport without the use of oxygen-carrying pigments in the blood.

The tracheal system may also include **air sacs,** which appear to be dilated tracheae without taenidia (Fig. 17-16, *A*). They are thin-walled and flexible and are located largely in the body cavity but also in appendages. In some insects the sacs may have functions other than respiratory. For example, they may allow internal organs to change in volume during instar stages without changing the shape of the larvae; they reduce the weight of large insects; or they may act as sound resonators or heat insulators.

In some very small insects gas transport occurs entirely by diffusion along a concentration gradient. As oxygen is used, a partial vacuum develops in the tracheae, and air is sucked in through the spiracles. Larger or more active insects employ some ventilation device for moving air in and out of the tubes. Usually muscular movements in the abdomen perform the pumping action that draws air in or expels it. In some insects—locusts, for example—additional pumping is provided by telescoping the abdomen, pumping with the prothorax, or thrusting the head forward and backward.

The tracheal system is primarily adapted for air breathing, but many insects (nymphs, larvae, and adults) live in water. In small, soft-bodied aquatic nymphs, the gaseous exchange may occur by diffusion through the body wall, usually into and out of a tracheal network just under the integument. The aquatic nymphs of stoneflies and mayflies are quipped with **tracheal gills,** which are thin extensions of the body wall containing a rich tracheal supply. The gills of dragonfly nymphs are ridges in the rectum (rectal gills) where gas exchange occurs as water moves in and out.

Although the diving beetle *Dytiscus* (Fig. 17-8, *C*) can fly, it spends most of its life in the water as an excellent swimmer. It uses an "artificial gill" in the form of a bubble of air held under its wing covers. The bubble is kept stable by a layer of hairs on top of the abdomen and is in contact with the spiracles on the abdomen. Oxygen from the bubble diffuses into the tracheae and is replaced by diffusion of oxygen from the water. Thus the bubble can last for several hours before the beetle must surface to replace it. Mosquito larvae are not good swimmers but live just below the surface, putting out short breathing tubes to the surface for air (Fig. 17-15). Spreading oil on the water, a favorite method of mosquito control, clogs the tracheae with oil and so suffocates the larvae. "Rat-tailed maggots" of the syrphid flies have an extensible tail that can stretch as much as 15 cm to the water surface.

Excretion and water balance. Insects and spiders

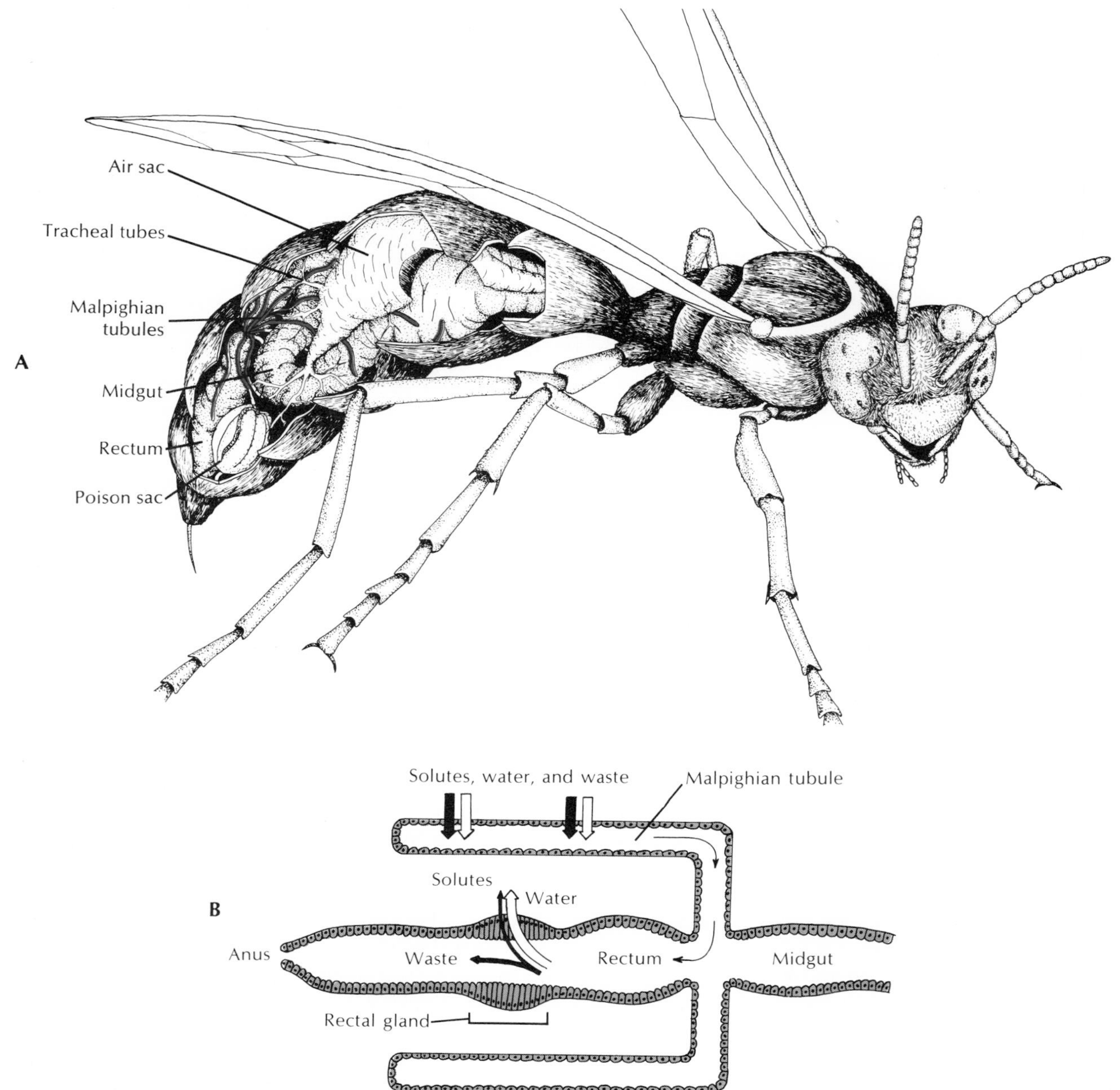

FIG. 17-16 Malpighian tubules of insect. **A,** Malpighian tubules are located at the juncture of the midgut and hindgut (rectum) as shown in the cutaway of a wasp. **B,** Function of malpighian tubules is shown. Solutes, especially potassium, are actively secreted into the tubules. Water and wastes follow. This fluid moves into the rectum where solutes and water are actively reabsorbed, leaving wastes to be excreted.

have a unique excretory system consisting of **malpighian tubules** that operate in conjunction with specialized glands in the wall of the rectum. The malpighian tubules, variable in number, are thin, elastic, blind tubules attached to the juncture between the midgut and hindgut (Fig. 17-11 and 17-16, *A*). The free ends of the tubules lie free in the hemocoel and are bathed in blood (hemolymph).

Since the malpighian tubules are closed and lack an arterial supply, urine formation cannot be initiated by blood ultrafiltration as in the crustaceans and vertebrates. Instead potassium is actively secreted into the tubules (Fig. 17-16, *B*). This primary secretion of ions pulls water along with it by osmosis to produce a potassium-rich fluid. Other solutes and waste materials also are secreted or diffuse into the tubule. The fluid, or "urine," then drains from the tubules into the intestine. In the rectum, specialized rectal glands actively resorb potassium and water as needed, leaving behind wastes such as uric acid.

Since water requirements vary among different types of insects, this ability to cycle water and salts is very important. Insects living in dry environments may resorb nearly all water from the rectum, producing a nearly dry mixture of urine and feces. Leaf-feeding insects take in and excrete quantities of fluid. Freshwater larvae need to excrete water and conserve salts. Insects that feed on dry grains need to conserve water and excrete salt.

Nervous system. The nervous system in general resembles that of the larger crustaceans, with a similar tendency toward fusion of ganglia (Fig. 17-11). A giant fiber system has been demonstrated in a number of insects. There is also a stomodeal nervous system that corresponds in function with the autonomic nervous system of vertebrates. Neurosecretory cells located in various parts of the brain have an endocrine function, but, except for their role in molting and metamorphosis, little is known of their activity.

Insect sense organs. Along with their neuromuscular coordination, the sensory perceptions of insects are unusually keen. Most of their sense organs are microscopic and located chiefly in the body wall. Each type usually responds to a specific stimulus. The various organs are receptive to mechanical, auditory, chemical, visual, and other stimuli.

MECHANORECEPTION. Mechanical stimuli, or those dealing with touch, pressure, vibration, and the like, are picked up by sensilla. A sensillum may be simply a seta, or hair, connected with a nerve cell process (Fig. 17-17, *B*), a nerve ending just under the cuticle and lacking a seta, or a more complex organ (scolopophorous organ) consisting of sensory cells with their endings attached to the body wall. Such organs are widely distributed over the antennae, legs, and body.

AUDITORY RECEPTION. Airborne sounds may be detected by very sensitive hairs (hair sensilla) or by tympanal organs. In tympanal organs a number of sensory cells (ranging from a few to hundreds) extend to a very thin tympanic membrane that encloses an air space in which vibrations can be detected. Tympanal organs are found in certain Orthoptera (Fig. 17-4), Homoptera, and Lepidoptera. Some insects are fairly insensitive to airborne sounds but can detect vibrations reaching them through the substrate. Vibrations of the substrate are detected by organs usually on the legs.

CHEMORECEPTION. Chemoreceptors (for taste or smell) are usually bundles of sensory cell processes that are often located in sensory pits. These are usually on mouthparts, but in ants, bees, and wasps are also found on the antennae, and butterflies, moths, and flies also have them on the legs. The chemical sense is generally keen and some insects can detect certain odors for several kilometers. Many of the patterns of insect behavior such as feeding, mating, habitat selection, and host-parasite relations are mediated through the chemical senses. These senses are also involved in the responses of insects to man-made repellents and attractants.

VISUAL RECEPTION. Insect eyes are of two types, simple and compound. Simple eyes are found in some nymphs and larvae and in many adults. Most insects have three ocelli on the head. Recent experiments indicate that the honeybee uses them to monitor light intensity, but not in the formation of images.

Compound eyes are found in most adults and may cover most of the head and consist of thousands of ommatidia—6,300 in the eye of a honeybee, for example. The structure of the compound eye is similar to that of crustaceans (Figs. 16-11 and 17-17, *B*). An insect such as a honeybee can see simultaneously in almost all directions around its body, but it is more myopic than the human, and images, even of nearby objects, are fuzzy. However, most flying insects rate much higher than humans in flicker-fusion tests. Flickers of light become fused in the human eye at a frequency of 45 to 55/second, but bees and blowflies can

FIG. 17-17 **A,** Portion of mosquito antennae. Note the sensory hairs and the segmented nature of the antennae. **B,** Female Hercules beetle *Dynastes tityus* showing the sensitive hairs on the underside of the body. This is one of the largest of the beetles (50 to 60 mm). The male has a longer horn extending forward over the head. (**A,** Courtesy P. P. C. Graziadei, Florida State University; **B,** Photograph by C. P. Hickman, Jr.)

distinguish as many as 200 to 300 separate flashes of light per second. This should be an advantage in analyzing a fast-changing landscape during flight.

A bee can distinguish colors, but its sensitivity begins in the ultraviolet range, which the human eye cannot see, and stops in the orange; the honeybee cannot distinguish shades of red from shades of gray.

OTHER SENSES. Insects also have well-developed senses for temperature, especially on the antennae and legs, and for humidity, proprioreception, gravity, and others.

Neuromuscular coordination. Insects are active creatures with excellent neuromuscular coordination. Arthropod muscles are typically cross-striated, just as vertebrate skeletal muscles are. A flea can leap a distance of 100 times its own length, and an ant can carry in its jaws a load greater than its own weight. This sounds as though insect muscle were stronger than that of other animals. Actually, however, the force a particular muscle can exert is related directly to its cross-sectional area, not its length. Based on maximum load moved per square centimeter of cross-section, their

FIG. 17-18 Crane flies (order Diptera) mating. They live in damp places with abundant vegetation. Their larvae are aquatic or semiaquatic. (Photograph by C. P. Hickman, Jr.)

FIG. 17-19 Tiger moth and eggs. She always lays her eggs on the weed that serves as food for her larvae. (Courtesy C. G. Hampton, University of Alberta, Edmonton, Alberta.)

strength is relatively the same as that of vertebrate muscle.

Reproduction. Sexes are separate in insects, and fertilization is usually internal. Insects have various means of attracting mates. The female moth gives off a powerful pheromone that can be detected for a great distance by the male. Fireflies use flashes of light; some insects find each other by means of sounds or color signals and by various kinds of courtship behavior.

Sperm are usually deposited in the vagina of the female at the time of copulation (Fig. 17-18). In some orders the sperm are encased in spermatophores that may be transferred at copulation or deposited on the substratum to be picked up by the female. The silverfish deposits a spermatophore on the ground, then spins signal threads to guide the female to it. During evolutionary transition from aquatic to terrestrial life, spermatophores were widely used, with copulation evolving much later.

Usually the sperm are stored in the spermatheca of the female in numbers sufficient to fertilize more than one batch of eggs. Many insects mate only once during their lifetime, and none mates more than a few times.

Insects usually lay a great many eggs. The queen honeybee, for example, may lay more than 1 million eggs during her lifetime. On the other hand, some flies are viviparous and bring forth only a single offspring at a time. Forms that make no provision for the care of the young may lay many more eggs than those that provide for the young or those that have a very short life cycle.

Most species normally lay their eggs in a particular type of place to which they are guided by visual, chemical, or other clues. Butterflies and moths lay their eggs on the specific kind of plant on which the caterpillar must feed (Fig. 17-19). The tiger moth may look for a pigweed and the sphinx moth for a tomato or tobacco plant and the monarch butterfly for a milkweed plant (Fig. 17-21). Insects whose immature stages are aquatic lay their eggs in water (Fig. 17-15, *A*). A tiny braconid wasp lays her eggs on the caterpillar of the sphinx moth where they will feed and pupate in tiny white cocoons (Fig. 17-12). The ichneumon wasp, with unerring accuracy, seeks out a certain kind of

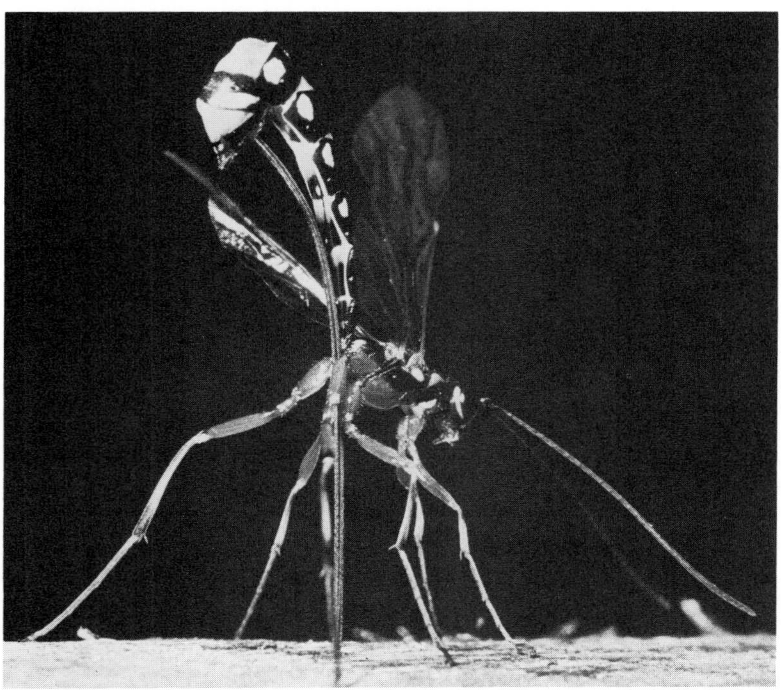

FIG. 17-20 An ichneumon wasp with the end of the abdomen raised to thrust her long ovipositor into wood to find a tunnel made by the larva of a wood wasp or wood-boring beetle. She can bore 13 mm or more into the wood to lay her eggs in the larva of the wood-boring beetle, which will become host for her own larvae. Other ichneumon species attack spiders, moths, flies, crickets, caterpillars, and others. (See also Fig. 17-8.) (Photograph by L. L. Rue, III.)

larva in which her young will live as internal parasites. Her long ovipositors may have to penetrate 1 to 2 cm of wood to find the larva of a wood wasp or a wood-boring beetle in which she will deposit her eggs (Fig. 17-20).

Metamorphosis and growth

Early development occurs within the egg, and the hatching young escape from the egg in various ways. During the postembryonic development most insects change in form—that is, they undergo **metamorphosis** (Fig. 17-21). During this period in order to grow they must undergo a number of molts, and each stage of the insect between molts is called an **instar** (Fig. 17-22).

Although metamorphosis is not limited to insects, insects illustrate it more dramatically than any other group. The transformation, for instance, of the hickory horned devil caterpillar into the beautiful royal walnut moth represents an astonishing morphologic change in development. In insects metamorphosis is associated with the evolution of wings, which are restricted to the reproductive stage where they can be of the most benefit. Not all insects undergo metamorphosis, but most of them do in some form or other.

Complete metamorphosis. Approximately 88% of insects undergo a complete metamorphosis, which separates the physiologic processes of growth **(larva)** from those of differentiation **(pupa)** and reproduction **(adult)** (Fig. 17-21). Each stage functions efficiently without competition with the other stages, for the larvae often live in entirely different surroundings and eat different foods from the adults. The wormlike larvae, which usually have chewing mouthparts, are known as caterpillars, maggots, bagworms, fuzzy worms, grubs, and so on. After a series of instar stages during which the wings are developing internally, the larva forms a case or cocoon about itself and becomes a pupa, or chrysalis, a nonfeeding stage in which many insects pass the winter. When the final molt occurs over winter the full-grown adult emerges, pale and with wings wrinkled. In a short time the wings expand and harden, and the insect is on its way. The stages, then, are egg, larva (several instars), pupa, and adult. The adult undergoes no further molting. Insects that undergo complete metamorphosis are said to be **holometabolous** (Gr., *holo,* complete, + *metabolē,* change).

Gradual metamorphosis. Some insects undergo a type of gradual, or incomplete, metamorphosis. These

FIG. 17-21 Egg laying and metamorphosis of monarch butterfly *Danaus plexippus*: **A,** Adult lays eggs on milkweed plant and hatched larvae feed on milkweed leaves. **B,** Larva hangs on milkweed as it prepares to pupate. At this stage wings develop internally but are not everted until last larval instar. They have chewing mouthparts but no compound eyes. **C,** Larva has transformed into chrysalis, or pupa, an inactive stage that does not feed and is covered by a cocoon or protective covering. **D,** Adult has emerged, with short, wrinkled wings. Wings will expand and harden and pigmentation will develop, and the butterfly will go on its way. (Photograph by J. H. Gerard.)

include the grasshoppers, cicadas and mantids, which have terrestrial young, and mayflies, stoneflies, and dragonflies, which lay their eggs in water. The young are called **nymphs** (or **naiads,** if aquatic), and their wings develop externally as budlike outgrowths in the early instars and increase in size as the animal grows by successive molts and becomes a winged adult (Figs. 17-22 and 17-23). The aquatic naiads have tracheal gills or other modifications for aquatic life (Fig. 17-24). The stages are egg, nymph (several instars), and adult. Insects that undergo gradual metamorphosis are called **hemimetabolous** (Gr., *hemi,* half, + *metabolē,* change).

Epimorphosis. A few insects, such as silverfish and springtails, are said to undergo epimorphosis. The young, or juveniles, are similar to the adults except in size. The stages are egg, juvenile, and adult. Such insects include the wingless insects (apterygote orders).

Physiology of metamorphosis. Metamorphosis in insects is controlled and regulated by hormones. There

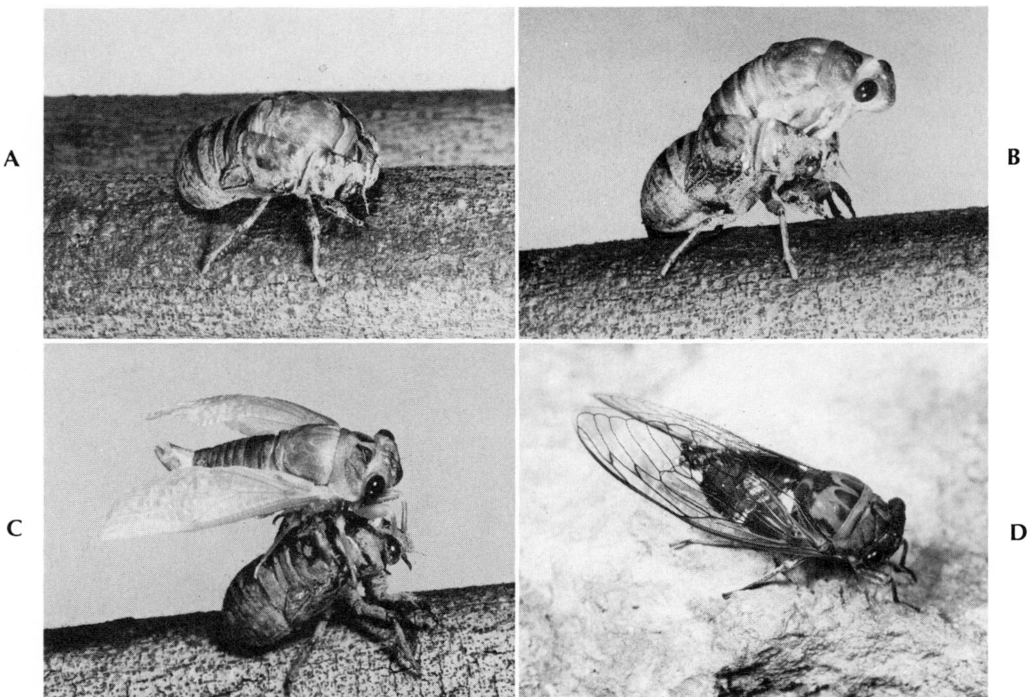

FIG. 17-22 Ecdysis in the dog-day cicada *Tibicen.* Before the old cuticle is shed a new one forms underneath. **A,** Old cuticle splits in dorsal midline as result of blood pressure and of air forced into thorax by muscle contraction. **B,** Emerging insect is pale, and its new cuticle is soft. **C,** The wings begin to expand as blood is forced into the veins, and insect enlarges by taking in air. **D,** Within an hour or two the cuticle begins to darken and harden, and the cicada is ready for flight. (Photograph by J. H. Gerard.)

are three major endocrine organs involved in development through the larval stages to the pupa and eventually to the emergence of the adult. These organs are the **brain,** the **ecdysial (prothoracic) glands,** and the **corpora allata** (Fig. 33-4, p. 757).

The intercerebral part of the brain and the ganglia of the nerve cord contain several groups of neurosecretory cells that produce an endocrine substance called the **brain hormone (activation hormone).** These neurosecretory cells may send their axons to paired organs behind the brain, the **corpora cardiaca,** which serves as a storage place for the brain hormone. The corpora cardiaca are of nervous origin, similar to the neurohypophysis of vertebrates. The brain hormone is carried in the blood to the ecdysial gland, a glandular organ in the head or the prothorax that is stimulated to produce the **molting hormone,** or **ecdysone** (ek′ duh-sone). This hormone sets in motion certain processes that lead to the casting off of the old skin (ecdysis) by proliferation of the epidermal cells.

If the larval form is retained at the end of this process, it is called simple molting; if the insect undergoes changes into pupa or adult, it may be referred to as metamorphosis. Simple molting persists as long as a certain **juvenile hormone** (neotenine) is present in sufficient amounts, along with the molting hormone in the blood, and each molting simply produces a larger larva.

The juvenile hormone is produced by a pair of tiny glands **(corpora allata)** located near the corpora cardiaca. Even the kind of cuticle produced depends on the amount of juvenile hormone present. If only a small amount of this hormone is present, a pupal cuticle is the result. When the corpora allata cease to produce the juvenile hormone, the molting hormone alone is secreted into the blood and the adult emerges (metamorphosis). It is thus seen that the molting hormone is necessary for each molt but is modified by the juvenile hormone, whose action is to maintain larval characters in the young insect.

A

B

FIG. 17-23 **A,** Young praying mantids (nymphs) emerging from their egg capsule. Egg capsules (oothecae) are glued to shrubbery and other objects in late summer and fall. When eggs hatch in spring, enormous swarms of wingless nymphs emerge from a single capsule. **B,** Praying mantis (order Orthoptera), about life size. It gets this name from the way it holds its forelimbs but is far more interested in preying on other insects than in pious devotions. (**A,** Photograph by F. M. Hickman; **B,** courtesy J. W. Bamberger.)

Experimental evidence shows that when the corpora allata (and thus the juvenile hormone) are removed surgically from the larva, the following molt will result in metamorphosis into the adult. Conversely, if the corpora allata from a young larva are transplanted into an old larva, the latter can be converted into a giant larva, because metamorphosis to the pupa or adult stage cannot occur. Many other experimental modifications on this theme have been performed. Progress has also been made in determining the chemical nature of the hormones (p. 758).

The mechanism of molting and metamorphosis just described is that found in holometabolous insects, but the same factors also apply in general to the molting nymphal stages of hemimetabolous insects, in which there are no pupal stages. A recent method of insect control involves the use of compounds that mimic the juvenile hormones. These prevent insects from becoming sexually competent when treated just before they become adults.

What factors initiate the sequence of secretion and role of these three different hormones? How are they correlated with cyclic events in the life histories of insects? Experimentally it has been shown that in some insects low temperature activates the neurosecretory cells of the intercerebral gland of the brain, which then sets in motion the sequence of events already related. The chilling of the brain seems to be all important in the initiation of metamorphosis. Adults do not molt and grow since the ecdysial glands degenerate after the last ecdysis. Many aspects of the control mechanism of these interesting processes have not yet been worked out.

Diapause

Many animals, including many types of insects, undergo a period of dormancy in their annual life cycle. In temperate zones there may be a period of winter dormancy, called hibernation, or a period of summer dormancy, called aestivation, or both. It is well-known

FIG. 17-24 Gradual metamorphosis. A young dragonfly (order Odonata) has just emerged from its larval case. The aquatic naiad, after several molts, leaves the water for its final molt to become an adult. While it is aquatic, the naiad breathes by means of anal gills. (Photograph by L. L. Rue, III.)

that there are periods in the life cycle of many insects when eggs, larvae, pupae, or even adults remain dormant for a long time because external conditions of climate, moisture, and the like are too harsh or unfavorable for survival under states of normal activity. Thus the life cycle is synchronized with periods of suitable environmental conditions and abundance of food. Most insects enter such a state when some factor of the environment, such as temperature, becomes unfavorable, and the state continues until conditions again become favorable.

Some species, however, become dormant at a certain time *before* conditions become unsuitable and remain so for a definite period, even if conditions do not change. This type of dormancy is called **diapause** (di′a-poz) (Gr., *dia,* through, dividing into two parts, + *pausis,* a stopping). Diapause, or the capacity to anticipate periods of stressful conditions, is genetically determined in each species but is set off by some certain signal. The most constant and reliable indicator of extensive temperature changes to come is the lengthening or shortening of the days. It is not surprising, then, that photoperiod, or day length, is the signal that initiates diapause. The arrival of a critical length of day starts the physiologic machinery for establishing diapause, which continues until the proper day-length or other signal is received. In the American silkworm *Hyalophora* the pupa goes into diapause. Activity is resumed *only* if the pupa is chilled for approximately 10 weeks at 3° to 5° C, after which the final metamorphosis occurs. In this instance chilling seems to be the necessary stimulus for the resumption of neurosecretory action and ecdysone production; no juvenile hormone is produced at the end of diapause.

Sometimes in insects a distinction is made between this obligatory type of dormancy, known as **primary**

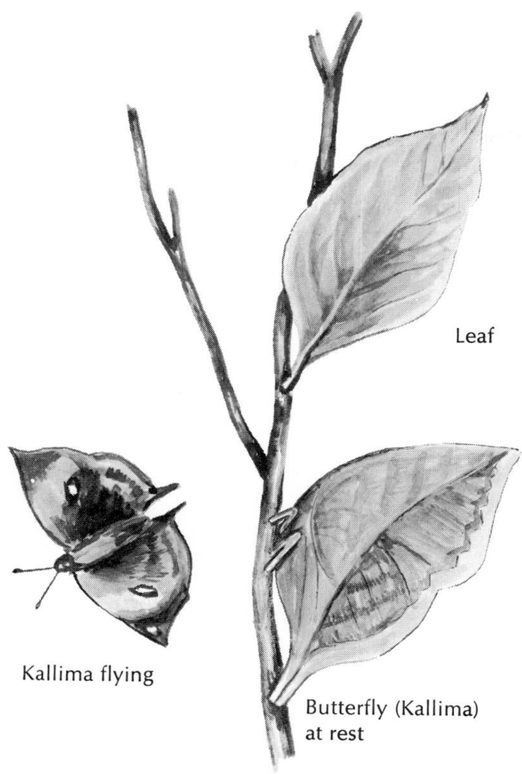

FIG. 17-25 Striking case of protective resemblance in butterfly *Kallima*, which mimics a leaf when perched on a twig. This butterfly is a native of East Indies and was first described by the famous English naturalist Alfred Russel Wallace.

or **obligatory diapause,** and the dormancy that is initiated directly by stressful environmental conditions, which is then called **facultative diapause.**

Diapause always occurs at the end of an active growth stage of the molting cycle so that, when the diapause period is over, the insect is ready for another molt. One species of the ant *Myrmica* reaches the third instar stage in late summer. Many of the larvae do not develop beyond this point until the following spring, even if temperatures are mild or if the larvae are kept in a warm laboratory.

Protection and coloration

Insects as a group display many colors. This is especially true of butterflies, moths, and beetles. Even in the same species the color pattern may vary in a seasonal way, and there also may be color differences between males and females. Some of the color patterns in insects are probably highly adaptive, such as those for **protective coloration, warning coloration, mimicry** (Fig. 17-25), and others.

Besides color, insects have other methods of protecting themselves. The cuticular exoskeleton affords a good protection for many of them; some, such as stinkbugs, have repulsive odors and tastes; others protect themselves by a good offense, for many are very aggressive and can put up a good fight (for example, bees and ants); and still others are swift in running for cover when danger threatens.

Monarch butterflies (Fig. 17-21) are known to be poisonous to birds because their caterpillars assimilate cardiac glycosides from certain species of milkweed *(Asclepiadaceae)*. This substance confers unpalatability on the butterflies after metamorphosis and induces vomiting in their predators. Certain leaf beetles that are toxic or unpalatable to their predators have been found to produce defensive secretions containing cardiac glycosides, which are not sequestered from their foods but are synthesized by the beetles themselves. Bats detect their insect prey by echolocation but certain moths have evolved special ultrasonic ears (tympanic membranes) by which they can pick up the bat chirps and thus evade capture.

Chemical repellents. Many insects practice chemical warfare in a variety of ingenious ways. Some use chemical exudates that either repel an assault because of their bad taste or odor or poisonous properties or that mechanically prevent a predator from attacking. When certain aphids are touched by a parasitic wasp, they secrete a waxy substance that hardens in air and may fatally entangle the wasp. A social wasp of the American tropics hangs its comb from the end of a very thin stalk, then secretes onto the stalk a repellent from its abdominal gland. Its comb is then safe from marauding ants, which will not walk down the treated stalk.

Some repellents are injected directly into the enemy. The stings of bees, wasps, and ants are good examples. The bombardier beetle, on the other hand, produces an irritating spray that it aims accurately at attacking ants or other enemies.

Migrations

Some insect species practice seasonal mass movements somewhat similar to the migrations of birds, except that the migrating individuals do not usually make the return flight; this is made by members of the next generation. The monarch butterfly *Danaus plexip-*

FIG. 17-26 Tumblebugs, or dung beetles. *Canthon pilularis* chew off a bit of dung, roll it into a ball, and then roll it to where they wish to bury it in soil. One beetle pushes while the other pulls. Eggs are laid in the ball and the larvae feed on the dung. Tumblebugs are black, 2.5 cm or less in length, and common in pasture fields. (Photograph by J. H. Gerard.)

pus is among the best known of the migratory insects. In late summer they start the southern flight from all over the United States and Canada. The farther they go the more numerous they become, with huge swarms of them congregating in some places. They winter in the southern United States and Mexico and reproduce there or on the spring flight north. Most of the adult monarchs that drift southward in the fall have developed during the preceding summer. Those that go northward in the spring reproduce on milkweeds along the way and give rise to the fall migrants. By tagging the wings of thousands of monarchs, much has been learned about their routes, and it has been determined that some of them make flights of close to 3,200 km (about 2,000 miles). The painted lady *Cynthia cardui* is another butterfly that breeds in the United States and overwinters in Mexico. Their larvae feed on thistles. The actual flight of these insects is not as directional as that of birds, as they are more apt to be carried along by wind currents and so find themselves in places where they do not usually resort.

In California the ladybug beetles, which have been of great help in conquering the scale insect, have been found to migrate from the valleys to the mountains for a part of their life cycle. Some insects breed in the south then migrate north in the spring and summer. Some breed in the north but do not winter there. Among such migrating insects are certain leafhoppers and noctuid moths. Mass migrations of grasshoppers have been known since biblical times, and they used to occur in the United States, but have been uncommon in the past century. Such mass migrations have been known to involve billions of individuals and to cover a territory of thousands of square miles, with the grasshoppers eating everything in sight wherever they landed.

Innate and learned behavior

The keen sensory perceptions of insects make them extremely responsive to many stimuli. The behavior of insects, as of other animals, consists of their conduct in response to stimulation. The stimuli may be internal (physiologic) or external (environmental). The responses are governed by both the physiologic state of the animal and the pattern of nerve pathways traveled by the impulses. The nerve pathways are largely hereditary, so the responses are largely automatic. The response may be orientation toward or away from the stimulus, as, for example, the attraction of a moth to light, the avoidance of light by a cockroach, the attraction of carrion flies to the odor of dead flesh, or the orientation of the caddis fly larva toward a water current or of a dragonfly to an air current. The insect may react positively to certain chemicals in the location of food or negatively to being touched. A response to a specific stimulus may be modified by other stimuli. For instance, honeybees respond to bright light by leaving the hive *if* the temperature is high but not if the temperature is low.

Much of the behavior of insects, however, is not a simple matter of orientation but involves a complex series of responses. A pair of tumblebugs, or dung beetles, chew off a bit of dung, roll it into a ball, and roll the ball laboriously to where they intend to bury it, after laying their eggs in it (Fig. 17-26). The cicada slits the bark of a twig and then lays an egg in each of the slits. The female potter wasp *Eumenes* scoops up clay into pellets, carries them one by one to her building site, and fashions them into dainty little narrow-necked clay pots, into each of which she lays an egg. Then she hunts and paralyzes a number of caterpillars, pokes them into the opening of a pot, and closes up the opening with clay. Each egg, in its own protective pot, hatches to find a well-stocked larder of food awaiting it.

Much of such behavior is "innate," that is, entire sequences of actions apparently have been genetically programmed. However, a great deal more learning is involved than was once believed. The potter wasp, for example, must learn where she has left her pots if she is to return to fill them with caterpillars one at a time. In fact, in many respects insects have a surprising capacity for learning. Training experiments with honeybees, for example, show them capable of learning signals in every known area of sensory perception, and learning them quickly in most cases. Tasks involving multiple signals in various sensory areas can be memorized and performed in sequence. Worker bees have been trained to walk through mazes that involved five turns in sequence, using such clues as the color of a marker, the distance between two spots, or the angle of a turn. The same is true of ants. Workers of one species of *Formica* learned a six-point maze at a rate only two or three times slower than laboratory rats. The foraging trips of ants and bees often wind and loop about in a circuitous route, but once the forager has found food the return trip is relatively direct. One investigator suggests that the continuous series of calculations necessary to figure the angles, directions, distance, and speed of the trip and to convert it into a direct return could involve a stop watch, a compass, and integral vector calculus. How the insect does it is not known.

Social insects, which have been studied extensively, have been found capable of most of the basic forms of learning used by mammals. The exception is insight learning. Apparently insects cannot, when faced with a new problem, reorganize their memories to construct a new response. For example, the female digger wasp *Ammophila* simultaneously cares for up to fifteen nests and not only learns their locations, but to each larva she brings the proper number of caterpillars each day. To a recently hatched larva she brings one to three caterpillars a day and to older ones she brings three to seven. A full-grown larva is not fed but is sealed off and left to mature. Before she begins her day's work she inspects the contents of each active nest and what she finds in this one inspection determines her behavior for the day. When an investigator (G. P. Baerends) substituted a large for a small larva *after* the inspection she still brought only one to three caterpillars. If the substitution was made *before* the day's inspection she brought three to seven. She was unable to learn anything new about the nests after the one learning trip. However, this is no disadvantage to the wasp under normal circumstances. Normally the only changes that occur in those nests result from her own activities, and in one inspection she is able to learn all she really needs to know about her work for the day.

Communication

Insects, of both social and nonsocial species, communicate with other members of their species by means of olfactory, visual, auditory, and tactile signals. Perhaps the most important of these, and certainly the one that has created the most interest and research in recent years, is the olfactory, or chemical communication, which involves the secretion of pheromones.

Olfactory communication. Pheromones are substances secreted by an animal that act as hormones when picked up or encountered by other individual animals. They usually affect the behavior or physiologic processes of other individuals of the same species, although some alarm pheromones are interspecific. Pheromones include sex attractants, releasers of certain behavior patterns, trail markers, alarm signals, territorial markers, and the like. Like hormones, pheromones are effective in minute quantities.

When a female pine sawfly *Diprion* was caged and placed in a field, its sex pheromone attracted more than 11,000 males. Queen honeybees produce a sex attractant in the mandibular glands that attracts males from a considerable distance. Female sex attractants also have been demonstrated in both myrmicine and ponerine ants.

Pheromones may attract individuals of the same sex, as well as those of the opposite sex and are often used for recruitment. A queen honeybee removed from her swarm can attract workers to the new location in a short time. In fact, the crushed heads of several queens on a bit of filter paper can attract swarms that have lost their queens. A fire ant worker heading home after having discovered a food source draws the tip of its sting along the ground surface, extruding a pheromone. It is dispensed in exceedingly minute quantities, but it induces workers that encounter the trail to follow it outward from the nest toward the food. When concentrated amounts of the pheromone are collected and allowed to diffuse from a glass rod held in the air near the nest, worker ants congregate beneath it and can be led about by the vapor alone if the rod is moved along slowly.

Odor trails are laid by ants and termites. The "harvesting termites" of the tropics go forth in enormous foraging armies at night to gather humus, leaf litter, and lichen. The armies, which might contain as many as 300,000 individuals, would be led by ten or so workers marching abreast and flanked regularly by soldiers facing outward. The columns are guided by odor trails, produced by the sternal gland of the abdomen. The American termite *Zootermopsis,* which never leaves the nest to forage, uses trail odor to summon help when a breach occurs in the wall of the nest. Termite nymphs disturbed by increased light or air current run back to the nest interior, leaving an odor trail that attracts other nymphs and guides them to the breach to repair it. If their number is too small to make the repair, they too lay trails to attract others until a sufficient repair crew is assembled and the repair is completed.

Social insects, such as bees, ants, wasps, and termites, can recognize a nestmate—or an alien in the nest—by means of identification pheromones. An intruder from another species is violently attacked at once. If from the same species but another colony there may be a variety of responses. It may be attacked and killed; it may be investigated but finally accepted; or it may be accepted but given less food than the others until it has had time to acquire the colony odor. Caste determination in termites, and to some extent in ants and bees, is determined by pheromones. In fact, pheromones are probably a primary integrating force in populations of social insects. Many insect pheromones have been extracted and chemically identified.

Auditory communication. Sound production and reception (phonoproduction and phonoreception) in insects have been studied extensively, and it is evident that, although a sense of hearing is not present in all insects, this means of communication is meaningful to insects that use it. Sounds serve as warning devices, as advertisement of territorial claims, or as courtship songs.

The sounds of crickets and grasshoppers seem to be concerned with courtship and aggression. Grasshoppers rub the femur of the third pair of legs over the ridges of the forewings. Male crickets scrape the rough edges of the forewings together to produce their characteristic chirping. The hum of mosquitoes is caused by the rapid vibration of their wings. The long, drawn-out sound of the male cicada, a recruitment call, is produced by the vibrating membranes in a pair of organs located on the ventral side of the basal abdominal segment. The humming of bees varies with the temperament of the hive. When excited, the more rapid vibration of the wings produces a difference in sound that is readily detected by those familiar with their ways. Water striders "create" vibrations on the water that are used as signals. Whirligig beetles appear to be the only insects that use echos to detect obstacles—a type of sonar behavior found in bats, porpoises, oil birds, and others.

Visual communication. Certain kinds of flies, springtails, and beetles manufacture their own visual signals in the form of **bioluminescence.** The best known of the luminescent beetles are the fireflies, or lightning bugs, in which the flash of light is a means of locating a prospective mate. Each species has its own characteristic flashing rhythm produced on the ventral side of the last abdominal segments. The male

FIG. 17-27 A fatal embrace. The female firefly *Photuris* (order Coleoptera) *(left)* has seized and begun to feed on a male of another species, which she had lured by mimicking his species' lighting response. (Courtesy J. E. Lloyd, from Science, Feb. 7, 1975, cover. Copyright 1975 by the American Association for the Advancement of Science.)

flashes his species-specific pattern while flying. If a female flashes the proper answer after the appropriate interval, he flies toward her, giving his signal again. The dialogue usually culminates in copulation.

An interesting instance of mimicry of light signals has been observed in females of several species of *Photuris,* which prey on male fireflies of other species by mimicking the female mating signals of the prey species and then by capturing and devouring the luckless males that court them (Fig. 17-27).

Tactile communication. There are many forms of tactile communication, such as tapping, stroking, grasping, and antennae touching, which evoke responses varying from recognition to recruitment and alarm. The food-begging behavior of the larvae of the ant *Formica* is to tap the mandibles of the worker with their own mouthparts, thus triggering the regurgitation of liquid food by the worker. When an adult *Formica* worker taps another worker with its antennae, the signal is to stop moving about.

The dances of worker honeybees, when they return to the hive after successful foraging, are performed in the darkness of a crowded hive where visual perception is impossible. However, it is not known whether the bees that follow the dancing worker about sense its movements by the touch of antennae or by the sound of the air currents made during the dance—or perhaps by both. The dances themselves will be described in more detail in Chapter 34.

Social behavior

Insects rank very high in the animal kingdom in their organization of social groups. Social communities are not all as complex as those of the honeybees, however. Some community groups are temporary and uncoordinated, as the hibernating associations of carpenter bees or the feeding gatherings of aphids. Some are coordinated for only brief periods, such as the mating swarms of mosquitoes or mayflies. Others cooperate more fully, such as the tent caterpillars *Malacosoma,* which not only gather in sleeping and feeding communities but join in building a home web and a feeding net. However, even these are still open communities, and their social behavior is limited to the larval stage of the life cycle.

In the true societies of the higher orders, such as honeybees, ants, and termites, a complex social life is necessary for the perpetuation of the species. Such societies are closed. In them all stages of the life cycle are involved, the communities are usually permanent, all activities are collective, and there is reciprocal communication. There is a high degree of efficiency in the division of labor. Such a society is essentially a family group in which the mother or perhaps both parents remain with the young, sharing the duties of the group in a cooperative manner. The society is usually characterized by polymorphism, or **caste** differentiation, along with differences in behavior that are associated with the division of labor.

Honeybees. The honeybees have one of the most complex organizations in the insect world. Instead of lasting one season, their organization continues for a more or less indefinite period. As many as 60,000 to 70,000 honeybees may be found in a single hive. Of these, there are three castes—a single sexually mature female, or **queen,** a few hundred **drones,** which are sexually mature males, and the **workers,** which are sexually inactive genetic females (Fig. 17-28).

The workers take care of the young, secrete wax with which they build the six-sided cells of the honey-

FIG. 17-28 **A,** A wild bee-tree colony. **B,** Portion of colony with workers at work on the hive. The pair at left seem to be engaged in mutual feeding. (Photograph by C. P. Hickman, Jr.)

comb, gather the nectar from flowers, manufacture honey, collect pollen, and ventilate and guard the hive. Each worker appears to be responsible for a specific task, depending on its age, but during its lifetime of a few weeks it performs all of the various tasks.

One drone, sometimes more, fertilizes the queen during the mating flight, at which time enough sperm is stored in her spermatheca to last her a lifetime. Drones have no stings and are usually driven out or killed by the workers at the end of the summer. A queen may live as long as five seasons, laying thousands of eggs in that time. She is responsible for keeping the hive going through the winter, and only one reigning queen is tolerated in a hive at one time.

Castes are determined partly by fertilization and partly by what is fed to the larvae. Drones develop from unfertilized eggs (and consequently are haploid); queens and workers develop from fertilized eggs (and thus are diploid). Female larvae that are destined to become queens are fed royal jelly, a secretion from the salivary glands of the nurse workers. Royal jelly differs from the "worker jelly" fed to ordinary larvae, but the components in it that are essential for queen determination have not yet been identified. Honey and pollen are added to the worker diet about the third day of larval life. Female workers are prevented from maturing sexually by pheromones in the "queen substance," which is produced by the queen's mandibular glands. Royal jelly is produced by the workers only when the level of "queen substance" pheromone in the colony drops. This occurs when the queen becomes too old, dies, or is removed. Then the workers' ovaries develop, and they start enlarging a larval cell and feeding the larva the royal jelly that produces a new queen.

There is an interesting temporal division of labor among the workers. During the early days of adulthood the hypopharyngeal and mandibular glands, which are the principal source of larval food, reach the peak of their development. The wax glands also develop then. Thus for the first 2 or 3 weeks of adulthood young workers act as nurses, feeding the larvae and queen, and they also work at the construction and sealing of the cells of the honeycomb. After this time the glands shrink, and the workers gradually become field bees, patrolling and guarding the hive and foraging for nectar. Work schedules are flexible, and the bees seem to respond to the needs of the hive as they arise. Workers also spend a great deal of time just resting or moving about the hive so that there is always a reserve force readily available for major emergencies such as invasion by a predator or the overheating of the nest. Even their glandular functions are flexible, for if a shortage of nurses or of wax-producing bees arises, the glands of some of the older workers redevelop and become functional again.

Bees collect nectar from many kinds of flowering plants, many of which give a distinctive flavor to the honey. But the bees make some changes in the nectar before making it into honey. Nectar has as its chief sugar the 12-carbon sugar, sucrose. With the enzyme invertase, bees convert sucrose into fructose and glucose (6-carbon sugars); so honey contains little sucrose. To prevent fermentation, bees remove part of the moisture from the nectar by spreading it in the various cells of the honeycomb and fanning it with their wings to evaporate the excess water.

Social vespids. The social vespids, or paper wasps, such as the bald-faced hornets (Fig. 17-29) and the yellow jackets, also have a caste system, but it has not evolved as far as those of the other major groups of social insects. The bald-faced hornet *Vespula* constructs a nest of papery material consisting of wood or foliage chewed up and elaborated by the wasp (Fig. 17-29, *B*). The nest is usually hung in a tree or other sheltered place.

In the spring a queen, which has mated in the fall and hibernated over the winter, emerges and begins to build her nest. She chews wood and vegetable fibers into a pulp to build the first of her papery cells, into each of which she lays an egg. She feeds the first of the larvae which, as they mature, become workers. The workers add new cells and feed and care for more hatching larvae, each new crop of workers enlarging the nest and caring for more young. They gather soft-bodied insects, nectar, and fruit pulp to feed the larvae. Toward the end of the summer several hundred queens and males begin to hatch. The determination of queens and workers seems to be the result of a difference in the amount of food given them by the nurse workers, although the possibility of special additions being secreted into the food cannot be ruled out. Workers are kept subordinate to the queens by aggressive dominance behavior on the part of the queens, and social hierarchies are often established among the young queens. The mother queen dies and the virgin queens and males leave the nest and mate. Males and workers perish during the winter, but the mated queens hibernate under bark, in cracks, or in insect burrows where they await the spring and the establishment of new colonies.

Ants and termites. Ants and termites also have complicated social lives. In both groups winged and fertile males and females are produced in large numbers at certain seasons. They leave the nest in swarms and engage in mating flights. After mating they shed their wings and start their colonies.

Termites (often called "white ants") differ from ants in being soft-bodied and usually light-colored and having a broad joint between the thorax and abdomen. Ants are dark and hard bodied and have a narrow constriction between the thorax and abdomen. At rest, winged stages of termites hold their wings flat over the

FIG. 17-29 **A,** Bald-faced hornet *Vespula maculata* (order Hymenoptera), one of the paper wasps noted for their globular, papery nests, in which larvae are reared. **B,** Paper nest of bald-faced hornet, lower side removed to show tiers of cells. Cells are open on the lower side while the larvae are growing and sealed when the larvae pupate. Nests are attached to bushes or trees and are composed of fibers of weatherworn wood. (Photographs by F. M. Hickman.)

abdomen while ants usually hold them above the body.

Termite colonies contain two main castes, the reproductives, both males and females, and the sterile individuals (Fig. 17-30). Some of the fertile individuals may have wings and may leave the colony, mate, lose their wings, and as **king** and **queen** start a new colony. Wingless fertile individuals may under certain conditions substitute for the king or queen. Sterile members are wingless and become **workers** and **soldiers.** Soldiers have large heads and mandibles and serve for the defense of the colony. As in bees and ants, caste differentiation is caused by extrinsic factors. Reproductive individuals and soldiers secrete inhibiting pheromones that are passed throughout the colony to the nymphs through a mutual feeding process, called **trophallaxis,** so that they become sterile workers. Workers also produce pheromones, and if the level of "worker substance" or "soldier substance" falls, as might happen after an attack by marauding predators, for example, compensating proportions of the appropriate caste are produced in the next generation.

The phenomenon of trophallaxis, or exchange of nutrients, appears to be common among all social insects because it integrates the colony by passage of pheromones. The process involves feeding of the young by the queen and workers, which in turn may receive a drop of saliva from the young. It may also involve mutual licking, grooming, and the like.

Termites feed on cellulose but some species cannot digest it themselves, depending instead on myriads of symbionts (flagellates and bacteria) in the digestive tract to perform that function for them—one of the best examples of mutualistic symbiosis. The termites live in galleries constructed in wood or soil. Some of the colonies are huge and quite complex, with a ventilating system that regulates the oxygen supply and temperature. In one African species the termites maintain fungous gardens in the center of the nest, which they utilize for food. A system of channels brings in fresh air and carries out stale air.

In ant colonies the males die soon after mating and the queen either starts her own new colony or joins some established colony and does the egg-laying. The sterile females are wingless workers and soldiers that do the work of the colony—gather food, care for the young, and protect the colony. In many larger colonies there may be two or three types of individuals within each caste.

Ants have a varied diet. In some the larvae are the real food digesters for the colony, as they can digest solid food and the adults feed on liquid foods. Nutrients are distributed among the members by means of trophallaxis.

Ants have evolved some striking patterns of "economic" behavior, such as making slaves, farming fungi, herding "ant cows" (aphids), sewing their nests together with silk, and using tools.

Some 35 species of ants—mostly living in cold cli-

FIG. 17-30 Termites. **A,** Reproductive adult. After the mating flight, adults shed their wings and then go in pairs to start a new colony. **B,** Workers are wingless sterile adults that tend the nest, care for the young, and perform other tasks. Termites are pale, soft bodied, and broad waisted in contrast to ants, which are dark, hard bodied, and narrow waisted. (Courtesy J. H. Gerard.)

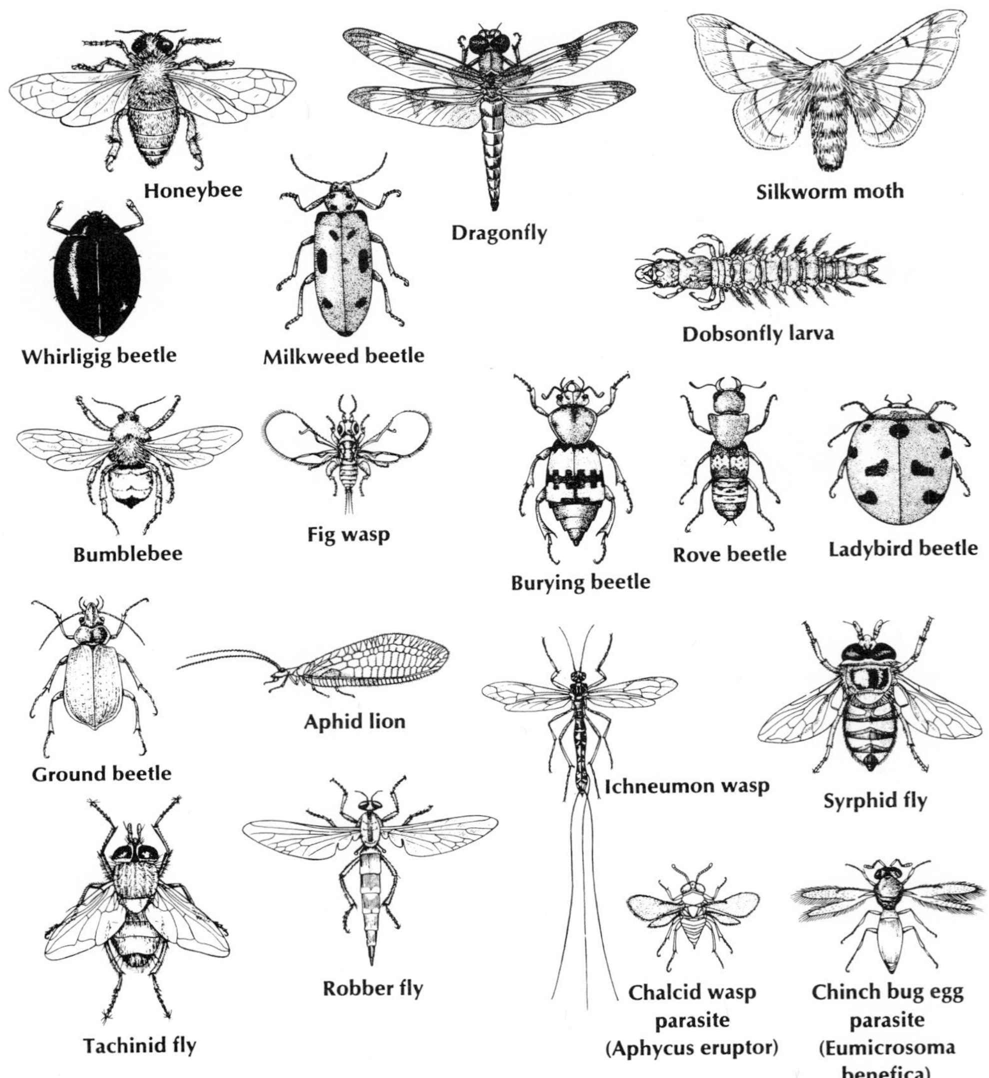

FIG. 17-31 Beneficial insects (not shown to scale). (Courtesy General Biological Supply House, Inc., Chicago.)

mates—depend on slave labor for their existence. Most slave-making species are so highly specialized for raiding that they are poorly suited to carry out the other work of the colony. The Amazon ants *Polyergus* have saber-shaped mandibles. They raid the nest of a slave species, pierce the bodies of the defending ants, and carry the cocoons containing pupae back to their own nest. The pupae hatch into nonreproductive workers that do all the work for the colony. Some slave-makers would actually starve to death without the presence of their slaves.

Slave-makers usually enslave members of a different species somewhat similar to their own. However, an interesting adaptation of slavery warfare between neighboring colonies of the *same* species has been observed in the honeypot ants *(Myrmecocystus mimicus)*. The honeypot ants have a special caste, called "repletes," that engorge honey, store it in their greatly expandable abdomens, and during the off-season regurgitate it to feed the colony.

The unusually thin cuticle of this species, which allows the abdomen to expand, would be a liability to the ants in the deadly territorial warfare carried on by other species; however, this species has developed a nonlethal type of territorial display in the form of elaborate tournaments between neighboring colonies. When invaded, hundreds of workers rush out. Opponents approach each other on stilt legs, turn sideways, and drum on each other's abdomens with their antennae until the weaker partners yield. If the invaded colony is too small to defend itself, the invaders may rush in and carry off larvae, pupae, workers, and repletes as slaves for their own colony.

Insects and human welfare

Beneficial insects. Some insects are highly beneficial to human interests (Fig. 17-31). Some of them produce useful materials, such as honey and beeswax from bees, silk from silkworms, and shellac from a wax secreted by the lac insects. Lac insects are small but commercially important scale insects (family Lacciferidae), common in Indochina and the Phillipines, that secretes so much wax that the twigs of host plants become coated with a layer 6 to 13 mm thick. The wax is collected and used in making shellac, a multimillion-dollar business.

Insects are necessary for the cross-fertilization of many fruits and other crops. Bees, for example, are indispensable in raising fruits, clover, and other crops, and the Smyrna fig will not grow in California without the help of a small fig wasp, *Blastophaga,* which carries pollen from the nonedible caprifig.

Insects and higher plants very early in their evolution formed a relationship of mutual adaptations that have been to each other's advantage. Insects exploit flowers for food, and flowers exploit insects for pollination. Each floral development of petal and sepal arrangement is correlated with the sensory adjustment of certain pollinating insects. Among these mutual adaptations are amazing devices of allurements, traps, specialized structures, precise timing, and so on.

Many predaceous insects, such as tiger beetles, aphid lions, ant lions, praying mantids, and ladybird beetles, destroy harmful insects. Some insects control harmful ones by parasitising them or by laying their eggs where their young, when hatched, may devour the host. Dead animals are quickly taken care of by maggots hatched from eggs laid in carcasses.

Insects and their larvae serve as an important source of food for birds, fish, and other animals.

In Australia, cattle, first introduced in 1788, have increased to over 30 million, but Australian dung beetles prefer the drier marsupial dung. Consequently pastures become literally covered with cow pads that breed insect pests and limit the grazing space. Since 1967 dung beetle eggs from Africa have been collected, placed in balls of dung, and buried in Australian soil to hatch. Their rapid spread is resulting in increased available pasture, some control of the buffalo fly, and increased soil fertility, since the beetles break up the dung and carry it underground.

Harmful insects. Harmful insects include those that eat and destroy plants and fruits, such as grasshoppers, chinch bugs, corn borers, boll weevils, grain weevils, San Jose scale, and scores of others (Fig. 17-32). Practically every cultivated crop has some insect pest. Lice, blood-sucking flies, warble flies, botflies, and many others attack domestic animals. Malaria, carried by the *Anopheles* mosquito, is still one of the world's killers; yellow fever and filariasis are also transmitted by mosquitoes. Fleas carry the plague, which at many times in history has almost wiped out whole populations. The housefly is the vector for typhoid and the louse for typhus fever; the tsetse fly carries African sleeping sickness; and a blood-sucking bug, *Rhodnius,* is a carrier of Chagas fever. In addition there is a tremendous destruction of food, clothing, and property by weevils, cockroaches, ants, clothes moths, termites, and carpet beetles. Not the least of the insect pests is the bedbug, *Cimex,* a blood-sucking hemipterous insect that humans contracted, probably early in their evolution, from bats that shared their caves.

Control of insects. Because all insects are an integral part of the ecologic communities to which they belong, their total destruction would probably do more harm than good. Food chains would be disturbed, some of our most loved birds would disappear, and the biologic cycles by which dead animal and plant matter disintegrate and return to enrich the soil would be seriously impeded. The beneficial role of insects in our environment has often been overlooked, and in our zeal to control the pests we have indiscriminately sprayed the landscape with extremely effective "broad-spectrum" insecticides that eradicate the good, as well as the harmful, insects. We have also found, to our chagrin, that many of the chemicals we have used persist in the environment and accumulate as residues in the bodies of animals higher up in the food chains. Also many strains of insects have developed a resistance to the insecticides in common use.

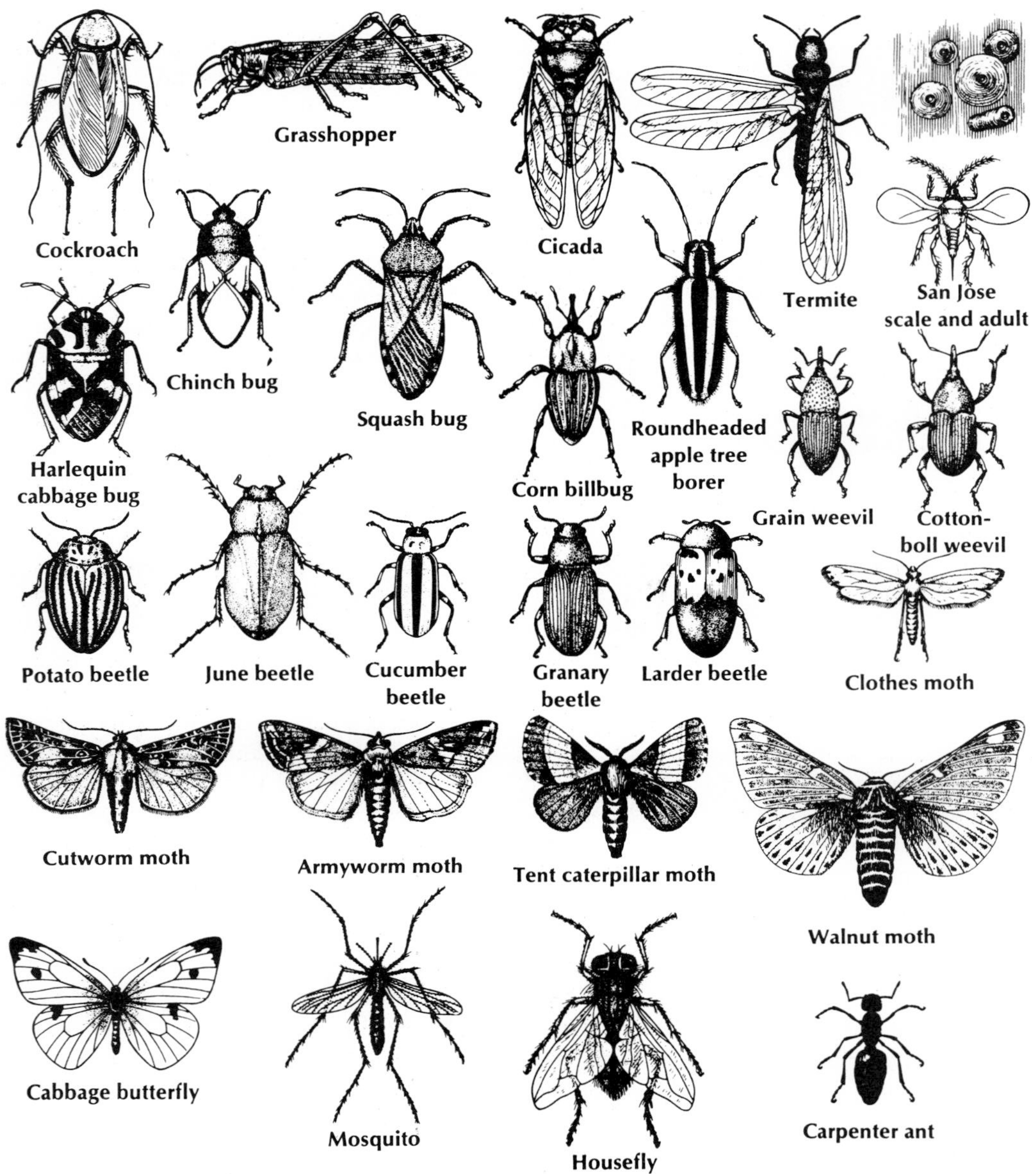

FIG. 17-32 Harmful insects (not shown to scale). (Courtesy General Biological Supply House, Inc., Chicago.)

In recent years an effort has been made to be more selective in the choice and use of pesticides that are specific in their targets. In addition to chemical control, other methods of control have been under intense investigation and experimentation.

The development of **insect-resistant crops** is one area of investigation. So many factors, such as yield and quality, are involved in developing resistant crops that the teamwork of specialists from many related fields is required. Some progress, however, has been made.

BIOLOGIC CONTROLS. Several types of biologic controls have been developed by the U. S. Department of Agriculture and others. All of these areas present problems but also show great possibilities. One is the use of **pathogens** such as *Bacillus thuringensis,* which is used to control several lepidopteran pests (cabbage looper, imported cabbage worm, tomato worm). This spore-forming bacterium forms a protein crystal that is poisonous to lepidopteran larvae. However, it attacks all lepidopterans (butterflies and moths), not just the specific pest.

A second type of control is the use of various **viruses** that are natural enemies of insects and could be cultivated in large numbers and applied at the most opportune time. Many viruses have been isolated that seem to have potential as insecticides. However, specific viruses are difficult to rear and could be expensive to put into commercial production.

A third method is to interfere with the metabolism or reproduction of the insect pests, for example by introducing natural predators or by using the sterile male approach. There have already been some successes with **natural predators** such as the vedalia beetle brought from Australia to counteract the work of the cottony-cushion scale on citrus plants and the parasites introduced from Europe for control of the alfalfa weevil.

The **sterile male approach** has been used effectively in eradicating screwworm flies, a livestock pest. Large numbers of male insects, sterilized by irradiation, are introduced into the natural population; females that mate with the sterile flies lay infertile eggs.

Still another method is the control of insect behavior by the use of **naturally occurring organic compounds** that act as hormones or pheromones, including sex attractants, repellents, stimulants, deterrents, and arrestants. Such research, although very promising, is slow because of our limited understanding of insect behavior and the problems of isolating and identifying complex compounds that are produced in such minute amounts. Successful pheromone research and promising field trapping has been done with the leaf-roller moth in fruit tree orchards, the cabbage looper in cabbage fields, the pink bollworm in cotton fields, the gypsy moth in forest and shade trees, and others. In the future, pheromones will probably play an important role in the concerted efforts of biologic pest control.

Phylogeny and adaptive radiation

Insect fossils, while not abundant, have been found in numbers sufficient to give a general idea of their evolutionary history. Although a variety of marine arthropods, such as trilobites, crustaceans, and xiphosurans, was present in the Cambrian period, the first terrestrial arthropods—the scorpions and millipedes—did not appear until the Silurian period. The first insects, which were wingless, date from the Devonian period. By the Carboniferous period, several orders of winged insects, most of which are now extinct, had appeared.

Not all zoologists have put the same interpretation on the comparative data that are available, but certain general relationships are evident. It is believed that the insects arose from a myriapod ancestor that had paired leglike appendages. Both myriapods and insects have clearly defined heads provided with antennae and mandibles. However, the evolution of insects involved specialization of the next three segments to become the locomotor segments (thorax) and a loss or reduction of appendages on the rest of the body (abdomen). The primitively wingless apterygotes are undoubtedly the most primitive of the insects, and traits similar to those of the myriapods are found in them. Probably some ancestral form similar to the Protura or Thysanura gave rise to two major lines of winged insects, which differed in their ability to flex their wings. One of these led to the Odonata and Ephemeroptera, which have outspread wings that cannot be folded back over the abdomen. The other line branched into three groups, all of which were present by the Permian period. One group with gradual metamorphosis, chewing mouthparts, and cerci includes the Orthoptera, Dermaptera, Isoptera, and Embioptera; another with gradual metamorphosis and a tendency toward sucking mouthparts includes the Thysanoptera, Hemiptera, and Homoptera and perhaps also the Psocoptera, Zoraptera, Mallophaga, and Anoplura, although there is some disagreement among authorities about the last group. Insects with complete metamorphosis are the most specialized, and the Neuroptera, which were probably the earliest of these, may have given rise to the other endopterygote orders, with the social insects being the most advanced.

The adaptive nature of the insects has been stressed throughout this chapter. The direction and range of their adaptive radiation, both structurally and physiologically, have been amazingly varied. Whether it be in the area of habitat, feeding adaptations, means of locomotion, reproduction, or general mode of living, the adaptive achievements of the insects are truly remarkable.

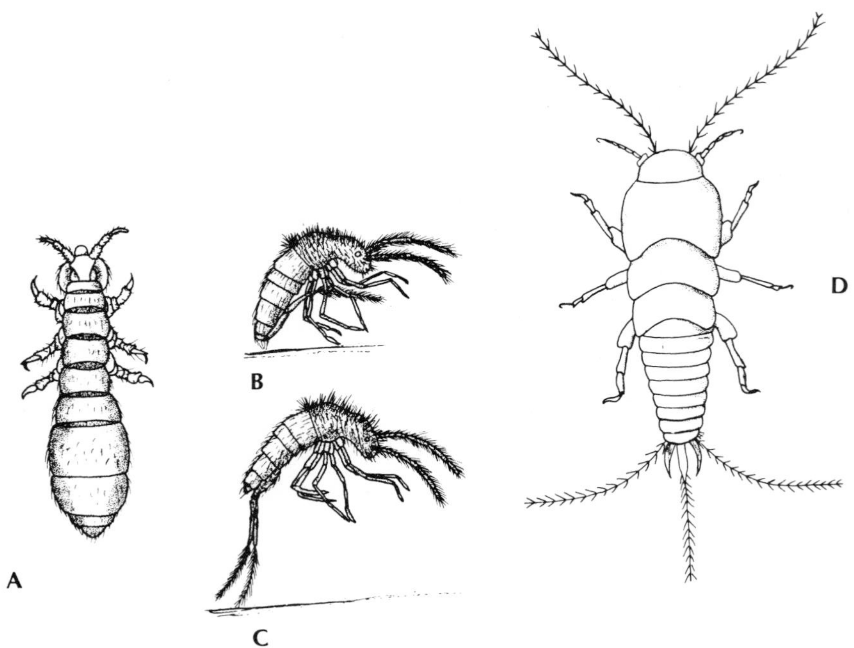

FIG. 17-33 A to C, Springtails (order Collembola). A, *Anurida*. B and C, *Orchesetta* in resting and leaping positions. D, Silverfish *Lepisma* (order Thysanura) is often found in homes.

Brief review of insect orders

Insects are divided into orders on the basis of wing structure, mouthparts, metamorphosis, and so on. Entomologists do not all agree on the names of the orders or on the limits of each order. Some tend to combine and others to divide the groups. However, the following synopsis of the orders is one that is rather widely accepted.

Subclass Apterygòta (ap-ter-y-go' ta) (Gr., *a,* not, + *pterygōtos,* winged) **(Ametabola).** Primitive **wingless** insects, anamorphic or epimorphic.

Order Protura (pro-tu' ra) (Gr., *protos,* first + *oura,* tail). Minute (1 to 1.5 mm); no eyes or antennae; appendages on abdomen as well as thorax; live in soil and dark, humid places; anamorphic.

Order Diplura (dip-lu'ra) (Gr., *dis,* double, + *oura,* tail)—**japygids.** Usually under 10 mm; pale, eyeless; a pair of long terminal filaments or pair of caudal forceps; live in damp humus or rotting logs; epimorphic.

Order Collembola (col-lem'bo-la) (Gr., *kolla,* glue, + *embolon,* peg, wedge)—**springtails and snow fleas.** Small (5 mm or less); no eyes; respiration by trachea or body surface; a springing organ folded under the abdomen for leaping (Fig. 17-33, *A* to *C*); abundant in soil; sometimes swarm on pond surface film or on snow banks in spring; epimorphic.

Order Thysanura (thy-sa-nu'ra) (Gr., *thysanos,* tassel, + *oura,* tail)—**silverfish and bristletails** (Fig. 17-33, *D*). Small to medium size; large eyes, long antennae; three long terminal cerci; live under stones and leaves and around human habitations; epimorphic.

Subclass Pterygòta (ter-y-go' ta) (Gr., *pterygōtos,* winged) **(Metabola). Winged insects** (some secondarily wingless) **with metamorphosis;** includes 97% of all insects.

Superorder Exopterygòta (ek-sop-ter-i-go'ta) (Gr., *exo,* outside + *pterygotos,* winged) **(Hemimetabola). Metamorphosis gradual; wings develop externally** on larvae; compound eyes present on larvae; larvae called **nymphs** (or **naiads,** if aquatic).

Order Ephemeroptera (e-fem-er-op'ter-a) (Gr., *ephēmeros,* lasting but a day, + *pteron,* wing)—**mayflies** (Figs. 17-34, *F* and *G*). Wings membranous; forewings larger than hindwings; adult mouthparts vestigial; nymphs aquatic, with lateral tracheal gills.

Order Odonata (o-do-na' ta) (Gr., *odontos,* tooth, + *ata,* characterized by)—**dragonflies, damselflies** (Figs. 17-24 and 17-34, *C, D,* and *E*). Large; membranous wings are long, narrow, net-veined, and similar in size; long and slender body; aquatic nymphs with aquatic gills and prehensile labium for capture of prey.

Order Orthoptera (or-thop'ter-a) (Gr., *orthos,* straight, + *pteron,* wing)—**grasshoppers, locusts, crickets, cockroaches, walkingsticks** (Fig. 17-9, *A*). **praying mantids** (Fig. 17-23). Wings when present, with forewings thickened and hindwings folded like a fan under forewings; chewing mouthparts.

Order Dermaptera (der-map' ter-a) (Gr., *derma,* skin, + *pteron,* wing)—**earwigs.** (Fig. 17-35). Very short forewings; large and membranous hind wings folded under forewings when at rest; biting mouthparts; forcepslike cerci.

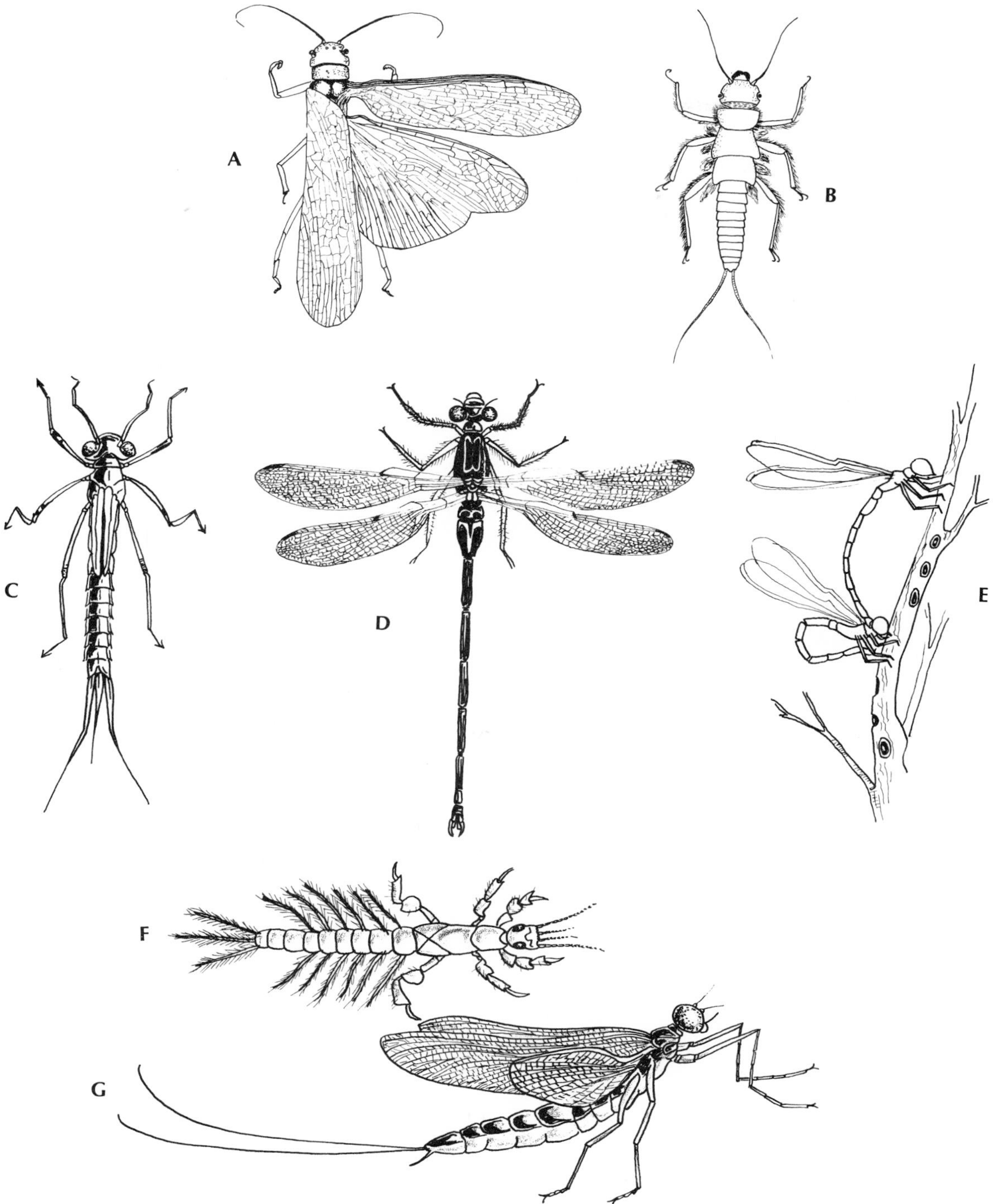

FIG. 17-34 **A** and **B,** Stonefly adult and naiad (order Plecoptera.) **C** to **E,** Damselfly (order Odonata) naiad, adult, and mating pair. **F** and **G,** Mayfly (order Ephemeroptera) naiad and adult. All have gradual metamorphosis and aquatic larvae.

FIG. 17-35 Earwig (order Dermaptera). Forcepslike cerci at posterior end are usually better developed in male and are used as organs for defense and offense. (Stained preparation, greatly enlarged.) (Photograph by F. M. Hickman.)

FIG. 17-36 Box elder bug *Leptocoris* (order Hemiptera). These often become a nuisance in the fall when they enter houses, seeking place to hibernate. However, they do no damage to house contents. (Photograph by F. M. Hickman.)

Order Plecoptera (ple-kop′ter-a) (Gr., *plekein,* to twist, + *pteron,* wing)—**stoneflies** (Fig. 17-34, *A* and *B*). Membranous wings; larger and fanlike hind wings; aquatic nymph with tufts of tracheal gills.

Order Isoptera (i-sop′ter-a) (Gr., *isos,* equal, + *pteron,* wing)—**termites** (Fig. 17-30). Small; membranous, narrow wings similar in size with few veins; wings shed at maturity; erroneously called "white ants"; distinguishable from true ants by broad union of thorax and abdomen; complex social organization.

Order Embioptera (em-bi-op′ter-a) (Gr., *embios,* lively, + *pteron,* wing)—**webspinners.** Small; male wings membranous, narrow, and similar in size; wingless females; chewing mouthparts; colonial; make silk-lined channels in tropical soil.

Order Psocoptera (so-cop′ter-a) (Gr., *psoco,* rub small, + *pteron,* wing) **(Corrodentia)—psocids, "book lice," "bark lice,"** Body usually small, may be as large as 10 mm; membranous, narrow wings with few veins, usually held rooflike over abdomen when at rest; some wingless species; found in books, bark, birdnests, on foliage.

Order Zoraptera (zo-rap′ter-a) (Gr., *zōros,* pure, + *apterygos,* wingless)—**zorapterans.** As large as 2.5 mm; membranous, narrow wings usually shed at maturity; colonial and termitelike.

Order Mallophaga (mal-lof′a-ga) (Gr., *mallos,* wool, + *phagein,* to eat)—**biting lice.** As large as 6 mm; wingless; chewing mouthparts; legs adapted for clinging to host; live on birds and mammals.

Order Anoplura (an-o-plu′ ra) (Gr., *anoplos,* unarmed, + *oura,* tail)—**sucking lice.** Depressed body; as large as 6 mm; wingless; mouthparts for piercing and sucking; adapted for clinging to warm-blooded host; includes the head louse, body louse, crab louse, others.

Order Thysanoptera (thy-sa-nop′ter-a) (Gr., *thysanos,* tassel, + *pteron,* wing)—**thrips.** Length, 0.5 to 5 mm (a few longer); wings, if present, long, very narrow, with few veins, and fringed with long hairs; sucking mouthparts; destructive planteaters, but some feed on insects.

Order Hemiptera (he-mip′ter-a) (Gr., *hemi,* half, + *pteron,* wing) (Heteroptera)—**true bugs.** Size 2 to 100 mm; wings present or absent; forewings with basal portion leathery, apical portion membranous (Fig. 17-36); hindwings membranous; at rest, wings held flat over abdomen; piercing-sucking mouthparts; many with odorous scent glands; include water scorpions, water striders, bedbugs, squash bugs, assassin bugs, chinch bugs, stinkbugs, plant bugs, lace bugs, others.

Order Homoptera (ho-mop′ter-a) (Gr., *homos,* same, + *pteron,* wing)—**cicadas** (Fig. 17-22), **aphids, scale insects, leafhoppers.** (Often included as suborder under Hemiptera.) If winged, either membranous or thickened front wings and membranous hindwings; wings held rooflike over body; piercing-sucking mouthparts; all plant-eaters; some destructive; a few serving as source of shellac, dyes, etc.; some with complex life histories.

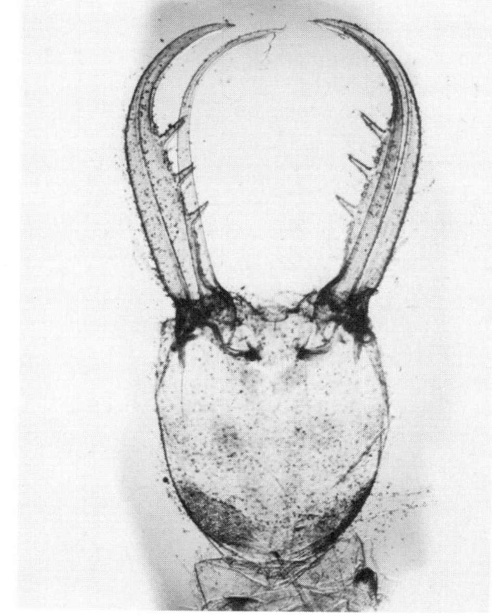

FIG. 17-37 **A,** Conical crater pits of ant lion larvae or "doodlebugs," designed to trap ants. When ant starts to slide into a sandy pit, the ant lion, which is concealed in pit, helps by undermining sand beneath the ant. **B,** Head of ant lion larva *Myrmeleon* (order Neuroptera), (greatly enlarged), showing large mandibles used for seizing prey. (**A,** Photograph by C. P. Hickman, Jr.; **B,** by F. M. Hickman.)

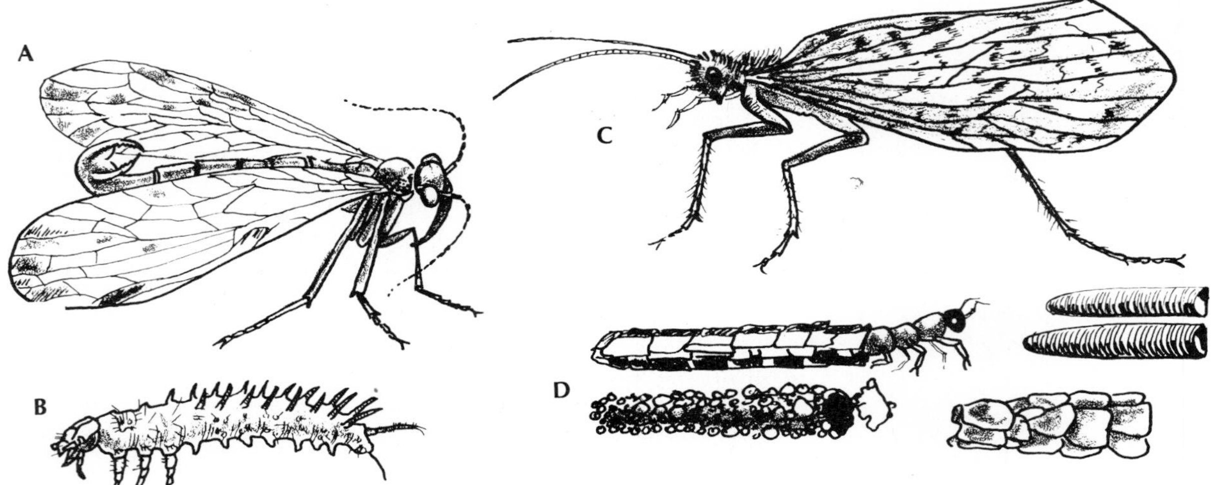

FIG. 17-38 **A,** Male scorpionfly *Panorpa* (order Mecoptera) has recurved abdomen with scorpion-like claspers. **B,** Scorpionfly nymph. **C,** Caddis fly (order Trichoptera). **D,** Several types of larval cases built by aquatic caddis fly larvae, often on the underside of stones.

Superorder Endopterygota (en-dop-ter-y-go′ ta) (Gr., *endon,* inside, + *pterygotos,* winged) **(Holometabola). Metamorphosis complete; wings develop internally;** larvae without compound eyes.

Order Neuroptera (neu-rop′ter-a) (Gr., *neuron,* nerve, + *pteron,* wing)—**dobsonflies** (Fig. 17-9, *B*), **antlions,** (Fig. 17-37), **lacewings.** Medium to large size; similar, membranous wings with many cross veins; chewing mouthparts; dobsonflies with greatly enlarged mandibles in males, and with aquatic larvae; ant lion larvae (doodlebugs) make craters in sand to trap ants.

Order Coleoptera (ko-le-op′ter-a) (Gr., *koleos,* sheath, + *pteron,* wing)—**beetles** (Figs. 17-8, *C,* 17-17 and 17-26), **fireflies** (Fig. 17-27), **weevils.** The largest order of animals in the world; front wings (elytra) thick, hard, opaque; membranous hindwings folded under front wings at rest; mouthparts for biting and chewing; includes ground beetles, carrion beetles, whirligig beetles, darkling beetles, stag beetles, dung beetles (Fig. 17-26), diving beetles, boll weevils, others.

Order Strepsiptera (strep-sip′ter-a) (Gr., *strepsis,* a turning, + *pteron,* wing)—**stylops.** Minute; females with no wings, eyes, or antennae; males with vestigial forewings and fan-shaped hindwings; females and larvae parasitic in bees, wasps, and other insects.

Order Mecoptera (me-kop′ter-a) (Gr., *mekos,* length, + *pteron,* wing)—**scorpionflies** (Fig. 17-38). Small to medium size, wings long, slender, with many veins; at rest, wings held rooflike over back; scorpionlike male clasping organ at end of abdomen; carnivorous; live in moist woodlands.

Order Lepidoptera (lep-i-dop′ter-a) (Gr., *lepidos,* scale, + *pteron,* wing)—**butterflies and moths.** Membranous wings covered with overlapping scales, wings coupled at base; mouthparts a sucking tube, coiled when not in use; larvae (caterpillars) with chewing mandibles for plant eating, stubby prolegs on the abdomen, and silk glands for spinning cocoons; antennae knobbed in butterflies and usually plumed in moths.

Order Diptera (dip′ter-a) (Gr., *dis,* two, + *pteron,* wing)—**true flies.** Single pair of wings, membranous and narrow; hindwings reduced to inconspicuous balancers (halteres); sucking mouthparts or adapted for sponging or lapping or piercing; legless larvae called maggots or, when aquatic, called wigglers (Fig. 17-15); includes crane flies (Fig. 17-18), mosquitoes, moth flies, midges, fruit flies, flesh flies, houseflies, horseflies, botflies, blowflies, and many others.

Order Trichoptera (tri-kop′ter-a) (Gr., *trichos,* hair, + *pteron,* wing)—**caddis flies** (Fig. 17-38, *C* and *D*). Small, soft-bodied; wings, well-veined and hairy, folded rooflike over hairy body; chewing mouthparts; aquatic larvae construct cases of leaves, sand, gravel, bits of shell, or plant matter, bound together with secreted silk or cement; some make silk feeding nets attached to rocks in stream.

Order Siphonaptera (si-fon-ap′ter-a) (Gr., *siphon,* a siphon, + *apteros,* wingless)—**fleas.** Small; wingless; bodies laterally compressed; legs adapted for leaping; no eyes; ectoparasitic on birds and mammals; larvae legless and scavengers.

Order Hymenoptera (hi-men-op′ ter-a) (Gr., *hymen,* membrane, + *pteron,* wing)—**ants, bees, wasps.** Very small to large; membranous, narrow wings coupled distally; subordinate hindwings; mouthparts for biting and lapping up liquids; ovipositor sometimes modified into stinger, piercer, or saw (Fig. 17-20); both social and solitary species; most larvae legless, blind, and maggotlike.

Annotated references
Selected general readings

Askew, R. R. 1971. Parasitic insects. New York, American Elsevier Publishing Co., Inc.

Beck, S. D. 1968. Insect photoperiodism. New York, Academic Press, Inc. *Discusses behavioral photoperiodism, such as locomotion, feeding, mating, and swarming.*

Birch, M. C. (ed.). 1974. Pheromones. New York, American Elsevier Publishing Co., Inc.

Blower, J. G. (ed.). 1974. Myriapoda. Symposia of the Zoological Society of London, no. 32. New York, Academic Press, Inc.

Borror, D. J., D. M. Delong, and C. A. Triplehorn. 1976. An introduction to the study of insects, ed. 4. New York, Holt, Rinehart and Winston, Inc.

Butler, C. G. 1975. The world of the honeybee, rev. ed. Collins, London. *Updated to add queen substances, pheromones, communication and so on.*

Chu, H. F. 1949. How to know the immature insects. Dubuque, Iowa, William C. Brown Co., Publishers. *An identification guide.*

Corning, W. C., J. A. Dyal, and A. O. D. Willows. 1973. Invertebrate learning, vol. 2. New York, Plenum Publishing Corp.

Debach, P. 1974. Biological control by natural enemies. New York, Cambridge University Press.

Dethier, V. G. 1963. The physiology of insect senses. New York, John Wiley & Sons, Inc.

Dethier, V. G. 1976. The hungry fly. Cambridge, Mass., Harvard University Press.

Evans, H. E., and M. J. Eberhard. 1970. The wasps. Ann Arbor, University of Michigan Press.

Fox, R. M., and J. W. Fox. 1964. Introduction to comparative entomology. New York, Reinhold Publishing Co.

von Frisch, K. 1967. The dance language and orientation of bees. Cambridge, Mass., The Belknap Press of Harvard University.

von Frisch, K. 1971. Bees: their vision, chemical senses, and language, rev. ed. Ithaca, N.Y., Cornell University Press. *An outstanding work on the way bees communicate with each other and reveal the sources of food supplies.*

von Frisch, K. 1974. Decoding the language of the bee. Science **185:**663-668 (Aug.). *A Nobel lecture.*

Goetsch, W. 1957. The ants. Ann Arbor, The University of Michigan Press. *A concise account of ants and their ways.*

Holland, W. J. 1968. The moth book: a guide to the moths of North America (revised by E. A. Brower). New York, Dover Publications.

Hölldobler, B. 1976. Tournaments and slavery in a desert ant. Science **192:**912-914 (May).

Hölldobler, B., and C. P. Haskins. 1977. Sexual calling behavior in primitive ants. Science **195:**793-794 (Feb.).

Jacobson, M. 1972. Insect sex pheromones. New York, Academic Press, Inc.

Jaques, H. E. 1947. How to know the insects, ed. 2. Dubuque, Iowa, William C. Brown Co., Publishers. *A good key to the families of insects.*

Jaques, H. E. 1951. How to know the beetles. Dubuque, Iowa, William C. Brown Co., Publishers. *A useful and compact manual for the coleopterist.*

Jonson, C. G. 1969. Migration and dispersal of insects by flight. London, Methuen & Co. Ltd.

Krisha, K., and F. M. Weesner (eds.). 1969, 1970. Biology of termites, vols. 1 and 2. New York, Academic Press, Inc.

Lindauer, M. 1971. Communication among social bees. Cambridge, Mass., Harvard University Press.

Little, V. A. 1972. General and applied entomology, ed. 3. New York, Harper & Row, Publishers.

Markl, H., and M. Lindauer, 1965. Physiology of insect behavior. M. Rockstein (ed.). In Physiology of insecta, vol. 2. New York

Menn, J. J., and M. Beroza (eds.). 1972. Insect juvenile hormones, chemistry and action. New York, Academic Press, Inc. *A symposium on the chemistry and action of insect juvenile hormones.*

Michener, C. D. 1974. The social behavior of the bees. Cambridge, Mass., Harvard University Press.

Michener, C. D., and M. H. Michener. 1951. American social insects. New York, Van Nostrand Reinhold Co.

Novák, V. J. A. 1975. Insect hormones, ed. 2. New York, John Wiley & Sons, Inc.

Richards, O. W. 1971. Biology of the social wasps (Hymenoptera, Vespidae). Biol. Rev. **46:**483-528.

Robinson, A. S. 1976. Progress in the use of chromosomal translocations for the control of insect pests. Biol. Rev. **51:**1-24.

Rockstein, M. 1973, 1974. The physiology of insecta, vols. 1 to 6, ed. 2, New York, Academic Press, Inc. *An extensive review of this large group.*

Ross, H. H. 1965. A textbook of entomology, ed. 3. New York, John Wiley & Sons, Inc.

Schneirla, T. C. 1971. Army ants: a study in social organization. In H. R. Topoff (ed.). San Francisco, W. H. Freeman & Co.

Skaife, S. H. 1961. The study of ants. New York, Longman, Inc.

Spoczynski, J. O. I. 1975. The world of the wasp. New York, Crane, Russak & Co. *An interesting little book.*

Sudd, J. H. 1967. An introduction to the behavior of ants. New York, St. Martin's Press, Inc.

Swan, L. A., and C. S. Papp. 1972. The common insects of North America, New York, Harper & Row, Publishers.

Wigglesworth, V. B. 1972. Principles of insect physiology,

ed. 7. New York, John Wiley & Sons, Inc. *A comprehensive account.*

Wigglesworth, V. B. 1974. Insect physiology, ed. 7. New York, John Wiley & Sons, Inc. *A brief and easily understandable summary.*

Wilson, E. O. 1971. The insect societies. Cambridge, Mass., Harvard University Press.

Selected *Scientific American* articles

Bartholomew, G. A. 1972. Temperature control in flying moths. **226:**70-77 (June).

Batra, S. W. T., and L. R. Batra. 1967. The fungus gardens of insects. **217:**112-120 (Nov.).

Bennet-Clark, H. C., and A. W. Ewing. 1970. The love song of the fruit fly. **223:**85-92 (July).

Bentley, D., and R. R. Hoy, 1974. The neurobiology of cricket song. **231:**34-44 (Aug.). *The songs give clues linking genetic information, development, the nervous system, and behavior.*

Bishop, J. A., and L. M. Cook. 1975. Moths, melanism and clean air. **232:**90-99 (Jan.). *Light moths that got darker in industrial areas are now getting lighter as pollution is alleviated.*

Buck, J., and E. Buck. 1976. Synchronous fireflies. **234:** 74-85 (May). *Discusses the evolutionary role of some fireflies flashing in unison and some being unsyncronized.*

Cambi, J. M. 1971. Flight orientation in locusts. **225:**74-81 (Aug.).

Evans, H. E. 1963. Predatory wasps. **208:**144-154 (Apr.).

Evans, H. E., and R. W. Matthews. 1975. The sand wasps of Australia. **233:**108-115 (Dec.).

von Frisch, K. 1962. Dialects in the language of the bees. **207:**78-86 (Aug.).

Heinrich, B. 1973. The energetics of the bumblebee. **228:** 96-102 (Apr.).

Hinton, H. E. 1970. Insect eggshells. **223:**84-91 (Aug.).

Hölldobler, B. 1971. Communication between ants and their guests. **224:**86-93 (Mar.).

Hölldobler, B. K., and E. O. Wilson. 1977. Weaver ants. **237:**146-154 (Dec.). *Weaver ants use their larvae as shuttles to weave nests in the tropical rain forest.*

Horridge, G. A. 1977. The compound eye of insects. **237:** 108-120 (July).

Johansson, G. 1975. Visual motion perception. **232:**76-88 (June).

Johnson, C. G. 1963. The aerial migration of insects. **209:** 132-138 (Dec.).

Jones, J. C. 1968. The sexual life of a mosquito. **218:**108-116 (Apr.).

Milne, L. J., and M. Milne. 1976. The social behavior of burying beetles. **235:**84-89 (Aug.).

Morse, R. A. 1972. Environmental control in the beehive. **226:**93-98 (Apr.).

Roeder, K. D. 1965. Moths and ultrasound. **212:**94-102 (Apr.).

Rothschild, M., and others. 1973. The flying leap of the flea. **222:**92-101 (Nov.).

Saunders, D. S. 1976. The biological clock of insects. **234:** 114-121 (Feb.).

Savory, T. H. 1968. Hidden lives. **219:**108-114 (July).

Schneider, D. 1974. The sex-attractant receptor of moths. **231:**28-35 (July).

Topoff, H. R. 1972. The social behavior of army ants. **227:** 71-79 (Nov.).

Waterhouse, D. F. 1974. The biological control of dung. **230:**100-109 (Apr.).

Wehner, R. 1976. Polarized-light navigation by insects. **235:** 106-115 (July).

Weis-Fogh, T. 1975. Unusual mechanisms for the generation of lift in flying. **233:**80-87 (Nov.).

Wenner, A. M. 1964. Sound communication in honeybees. **210:**116-124 (Apr.).

Williams, C. M. 1967. Third-generation pesticides. **217:** 13-17 (July).

Wilson, D. M. 1968. The flight-control system of the locust. **218:**83-90 (May).

Wilson, E. O. 1972. Animal communication. **227:**52-71 (Sept.).

Wilson, E. O. 1975. Slavery in ants. **232:**32-49 (June).

Wright, R. H. 1975. Why mosquito repellents repel. **233:** 104-111 (July).

CHAPTER 18

THE LESSER PROTOSTOMES

Phylum Sipuncula
Phylum Echiura
Phylum Pogonophora
Phylum Priapulida
Phylum Pentastomida
Phylum Onychophora
Phylum Tardigrada

This little creature, which resembles some prehistoric monster but is only 300 to 500 μm long, is *Echiniscus maucci,* one of the "water bears" of phylum Tardigrada. Unable to swim, it clings to moss or water plants with its claws, and if the environment dries up, it goes into a state of suspended animation and "sleeps away" the drought.

Courtesy D. R. Nelson and R. O. Schuster.

Position in animal kingdom

The phyla discussed in this chapter are all coelomate protostomes, although some also have some deuterostome characteristics in their embryologic development. Their relationship to each other and to the major protostome phyla is often puzzling, but they have all probably digressed at different times from the annelid-arthropod stem line. Possible phylogenetic relationships will be mentioned in the discussion of each group.

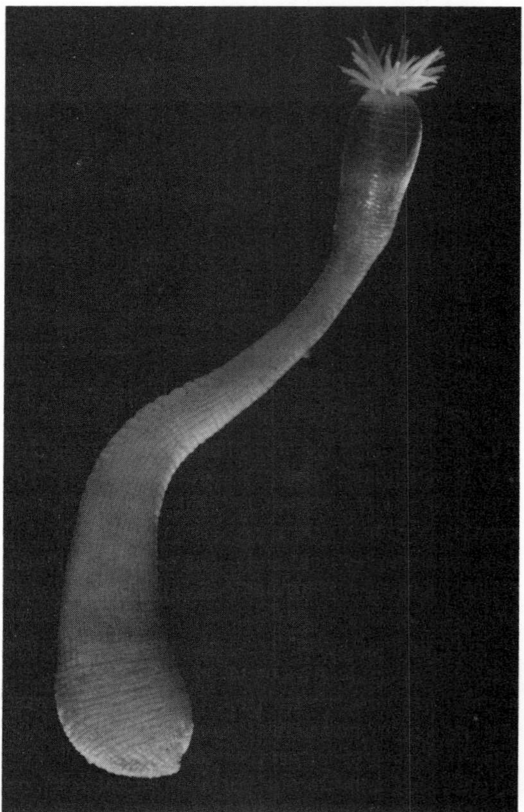

FIG. 18-1 A sipunculan worm *Themiste lageniformis* from oyster beds in Fort Pierce, Florida. (Courtesy M. Rice, Smithsonian Institution.)

THE LESSER PROTOSTOMES

This chapter includes a brief discussion of seven phyla whose positions in the phylogenetic lines of the animal kingdom are somewhat problematic, as are their relationships to each other. The great evolutionary flow that began with the appearance of the coelom and led to the three major phyla—Mollusca, Annelida, and Arthropoda—also produced a number of other lines. Some are now extinct, whereas others, though small in number of species and marked by very little evolutionary divergence within each phylum, have survived. The seven phyla grouped here probably all stemmed from the annelid-arthropod line, all following different adaptive trends.

Four of the phyla, Sipuncula, Echiura, Priapulida, and Pogonophora, are benthic (bottom-dwelling) marine worms that seem to have some affinity with the annelids. The first three have a variety of proboscis devices used in burrowing and food-gathering. The pogonophores live in tubes, mostly in deep sea mud, have long anterior tentacles, and lack a digestive tract. The Pentastomida, Onychophora, and Tardigrada have sometimes been grouped together and called the pararthropods because they have unjointed limbs with claws (at some stage) and a cuticle that undergoes molting and thus show a relationship with the arthropods. The Pentastomida are entirely parasitic; the Onychophora are terrestrial but are limited to damp areas; the Tardigrada are found in marine, freshwater, and terrestrial habitats.

PHYLUM SIPUNCULA

The phylum Sipuncula (sigh-pun'kyu-la) (L., *sipunculus,* little siphon, + *ida,* pl. suffix) consists of benthic marine worms, predominantly littoral or sublittoral. They live sedentary lives in burrows in mud or sand (Fig. 14-19), occupy borrowed snail shells, or live in coral crevices or among vegetation. Some species construct their own rock burrows by chemical and perhaps mechanical means. More than half the species are restricted to tropical zones. Some are tiny, slender worms, but the majority range from 15 to 30 cm in length. Some of them are commonly known as "peanut worms" because, when disturbed, they can contract to a peanut shape (Fig. 18-1).

Sipunculans have no segmentation or setae. They are most easily recognized by a slender retractile introvert, or proboscis, that is continually and rapidly being run in and out of the anterior end. The walls of the trunk are muscular. When the introvert is everted, the mouth can be seen at its tip surrounded by a crown of ciliated tentacles. Undisturbed sipunculans usually extend the anterior end from the burrow or hiding place and stretch out the tentacles to explore and feed. They are largely deposit feeders living on organic matter collected in mucus on the tentacles and moved to the

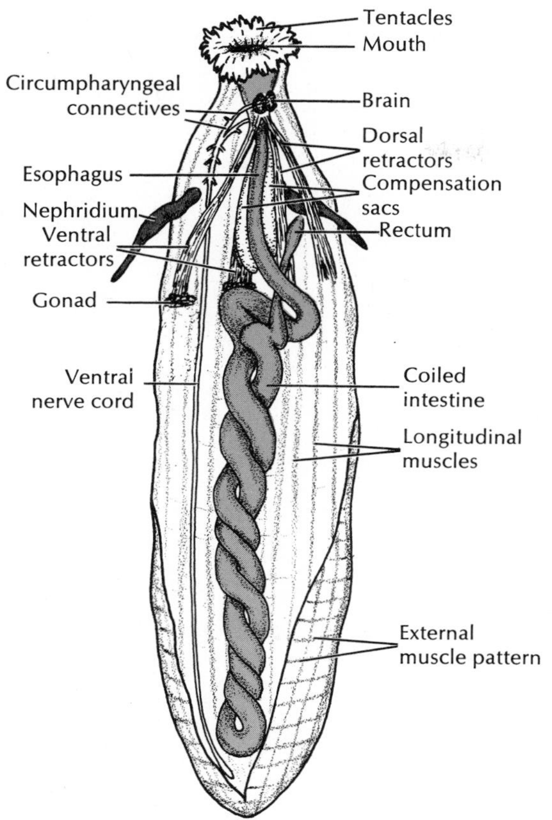

FIG. 18-2 Internal structure of *Sipunculus.*

mouth by ciliary action. The introvert is extended by hydrostatic pressure produced by contraction of the body wall muscles against the coelomic fluid. It is retracted by special retractor muscles. Its surface is often rough because of surface spines, hooks, or papillae.

There is a large fluid-filled coelom traversed by muscle and connective tissue fibers. The digestive tract is a long tube that doubles back on itself to end in the anus near the base of the introvert (Fig. 18-2). A pair of large nephridia open to the outside to expel waste-filled coelomic amebocytes; they also serve as gonoducts. Circulatory and respiratory systems are lacking, but the coelomic fluid contains red corpuscles that bear a respiratory pigment, hemerythrin, used in the transportation of oxygen. The nervous system has a bilobed cerebral ganglion just behind the tentacles and a ventral nerve cord extending the length of the body. The sexes are separate. Permanent gonads are lacking, and ovaries or testes develop seasonally in the connective tissue covering the origins of one or more of the retractor muscles. Sex cells are released through the nephridia. Asexual reproduction also occurs by transverse fission, the posterior one-fifth of the parent constricting off to become the new individual. The larval form is usually a trochophore.

There are approximately 330 species and 14 genera, which are placed by some authorities into four families. The best known genera are probably *Sipunculus, Phascolosoma, Aspidosiphon,* and *Golfingia.*

The early embryologic development of sipunculans, echiurans, and annelids is almost identical, showing a very close relationship among the three. It is also similar to the molluscan development. The four phyla are grouped together by some authors into a supraphyletic assemblage called the ''Trochozoa'' because of the common possession of a trochophore larva. There are other similarities, too, that point to the close relationship of the sipunculans to the echiurans and annelids. The sipunculans and echiurans are simpler animals than the annelids and probably represent collateral evolutionary lines that have retained more primitive features than the annelids have.

PHYLUM ECHIURA

The phylum Echiura (ek-ee-yur′a) (Gr., *echis,* adder, + *oura,* tail, + *-ida,* pl. suffix) consists of marine worms that burrow into mud or sand or live in empty snail shells or sand dollar tests, rocky crevices, and so on. They are found in all oceans—most commonly in littoral zones of warm waters—but some have been found in polar waters and some dredged from depths of 2,000 m. They vary in length from a few millimeters to 40 or 50 cm.

The echiurans have only a third as many species as the sipunculans (about 100), but there is much more diversity, and they are found in greater densities. There are two classes, Echiurida and Sactosomatida, with Echiurida being much larger and containing two orders and five families.

The body of the echiuran is cylindric and somewhat sausage-shaped (Fig. 18-3). Anterior to the mouth is a flattened, extensible proboscis (introvert), which, unlike that of the sipunculids, cannot be retracted into the trunk. Echiurids are often called ''spoonworms'' because of the shape of the contracted proboscis in some worms. The proboscis, which contains the brain, is actually a cephalic lobe, probably homologous to the annelid prostomium. The proboscis has a ciliated groove leading to the mouth. While the animal lies

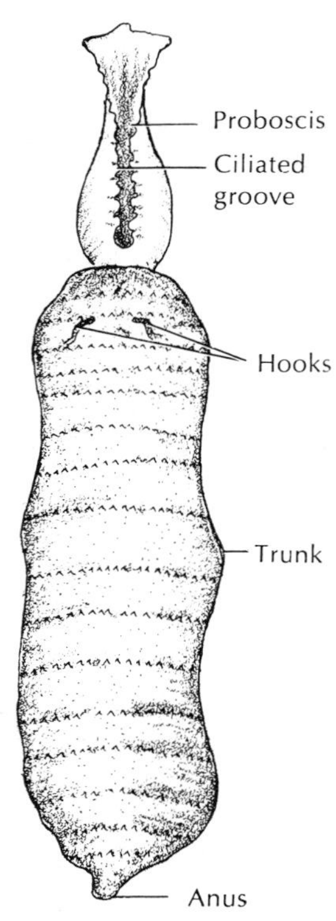

FIG. 18-3 *Echiurus,* an echiurid common on both Atlantic and Pacific coasts. The shape of the proboscis lends them the common name of "spoon worms."

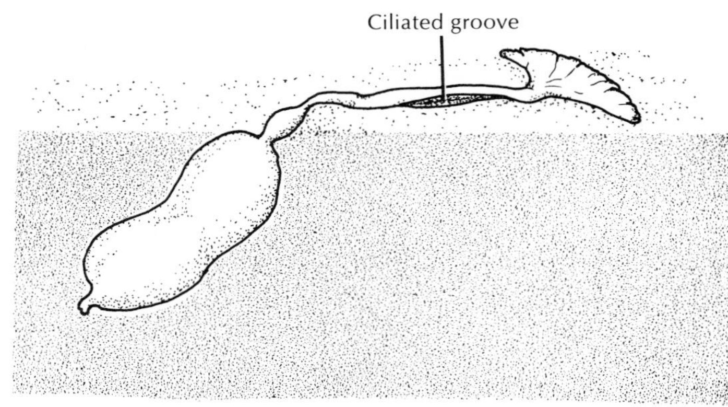

FIG. 18-4 *Tatjanellia,* an echiurid, is a marine detritus feeder that burrows in mud or sand. It extends its long proboscis to explore the bottom surface. Organic particles are picked up by the proboscis and carried along a ciliated food groove to the mouth. (After Zenkevitch; modified from Dawydoff, C. 1959.)

buried, the proboscis can extend out over the mud for exploration and deposit feeding (Fig. 18-4). In *Bonellia viridis* very small particles are picked up and moved along the proboscis by cilia; larger particles are moved by a combination of cilia and muscular action or by muscular action alone. Unwanted particles can be rejected along the route to the mouth. The proboscis in some forms is short and in others long. *Bonellia,* which is only 8 cm long, can extend its proboscis to a meter in length.

One common form, *Urechis,* lives in a U-shaped burrow in which it secretes a funnel-shaped mucous net. It pumps water through the net, capturing bacteria and fine particulate material in it. When loaded with food, the net is swallowed. *Lissomyena* lives in empty gastropod shells in which it constructs galleries irrigated by rhythmic pumping of water and feeds on sand and mud drawn in by the irrigation process.

The muscular body wall is covered with cuticle and epithelium, which may be smooth or ornamented with papillae. There may be a pair of anterior setae or a row of bristles around the posterior end. The coelom is large. The digestive tract is long and coiled and terminates at the posterior end. A pair of anal vesicles may have an excretory and osmoregulatory function. Most echiurans have a closed circulatory system with colorless blood but contain hemoglobin in coelomic corpuscles and certain body cells. Two or three pairs of nephridia serve mainly as gonoducts. A nerve ring runs around the pharynx and forward into the proboscis, and there is a ventral nerve cord. There are no specialized sense organs.

The sexes are separate, with a single gonad in each sex. The mature sex cells break loose from the gonads and leave the body cavity by way of the nephridia, and fertilization is usually external.

In some species sexual dimorphism is pronounced, with the female being much the larger of the two. *Bonellia* is noteworthy both for its extreme sexual dimorphism and for the way sex is determined in this genus. At first the freeswimming larvae are sexually undifferentiated. Those larvae that come into contact with the proboscis of a female become tiny males (1 to 3 mm long) that migrate to the female uterus. About 20 males are usually found in a single female. Larvae that do not contact a female proboscis metamorphose into females. It is not known whether the stimulus for male development is a chemical from the female proboscis, a matter of the chemical content of the

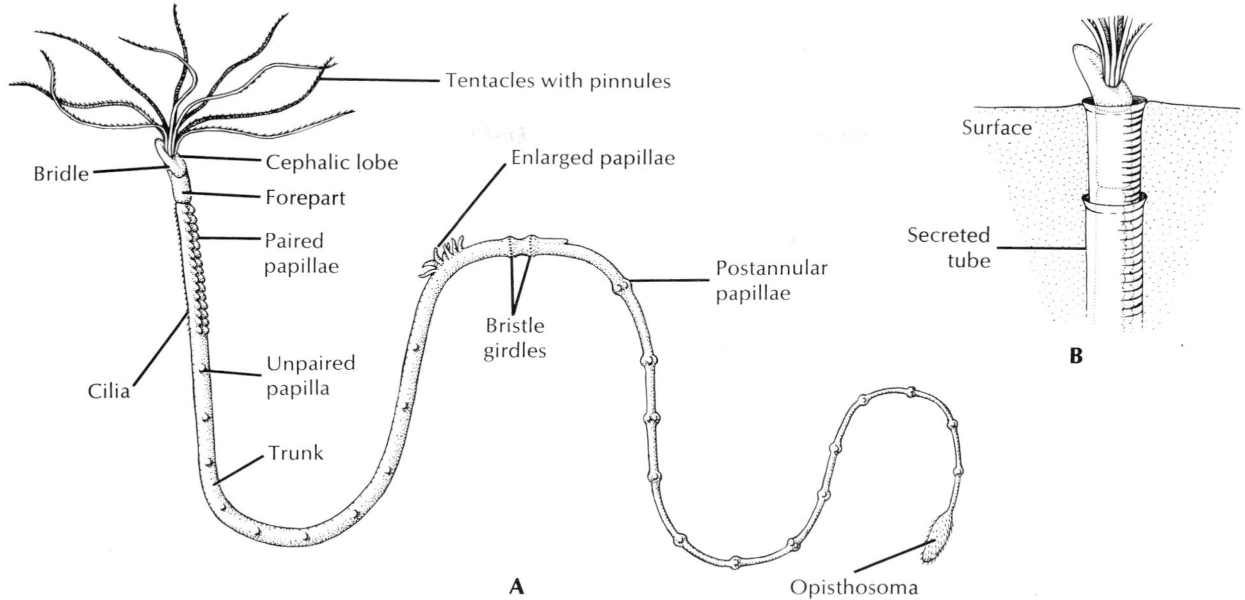

FIG. 18-5 A typical pogonophoran, diagrammatic. **A,** External features. The body, in life, is much more elongated than shown in this diagram. **B,** Position in tube.

environmental water, or a dimorphism in the eggs.

Early cleavage and trochophore stages are very similar to those of annelids and sipunculans. The trochophore stage, which may last from a few days to 3 months, according to the species, is followed by gradual metamorphosis to the wormlike adult.

The relationship of the echiurans to annelids and sipunculans has already been mentioned. Echiurans show a transitory metamerism during their embryogenesis and so may have descended from a metameric ancestor. Sipunculans, however, show no trace of metamerism.

PHYLUM POGONOPHORA

The phylum Pogonophora (po′-go-nof′ e-ra) (Gr., *pōgōn,* beard, + *phora,* bearing), or beardworms, was entirely unknown before the twentieth century. The first specimens to be described were collected from deep-sea dredgings in 1900 off the coast of Indonesia. They have since been discovered in several seas, including the western Atlantic off the U.S. eastern coast. Some 80 species have been described so far; they have been divided into two orders: Athecanephria and Thecanephria.

These elongated tube-dwelling forms have left no known fossil record. Their closest affinity seems to be to the annelids.

Most pogonophores live in the bottom ooze on the ocean floor, always below the intertidal zone and usually at depths of more than 200 m. This accounts for their delayed discovery, for they are obtained only by dredging. Their length varies from 5 to 85 cm, with a diameter usually of a fraction of a millimeter. They are sessile animals that secrete very long chitinous tubes in which they live, probably extending the anterior end only for feeding. The tubes are usually oriented upright in the bottom ooze. The tube is usually about the same length as the animal, which can move up or down inside the tube but cannot turn around.

External features. The beardworm has a long, cylindric body, covered with cuticle. The body is divided into a short anterior forepart; a long, very slender trunk; and a small, segmented opisthosoma (Fig. 18-5). It is difficult to determine which is the ventral and which is the dorsal side of the animal because the pogonophore lives in a vertical position, and if removed, simply curls up and never crawls. Therefore, rather than use the terms dorsal and ventral, some authors call the side of the body containing the main

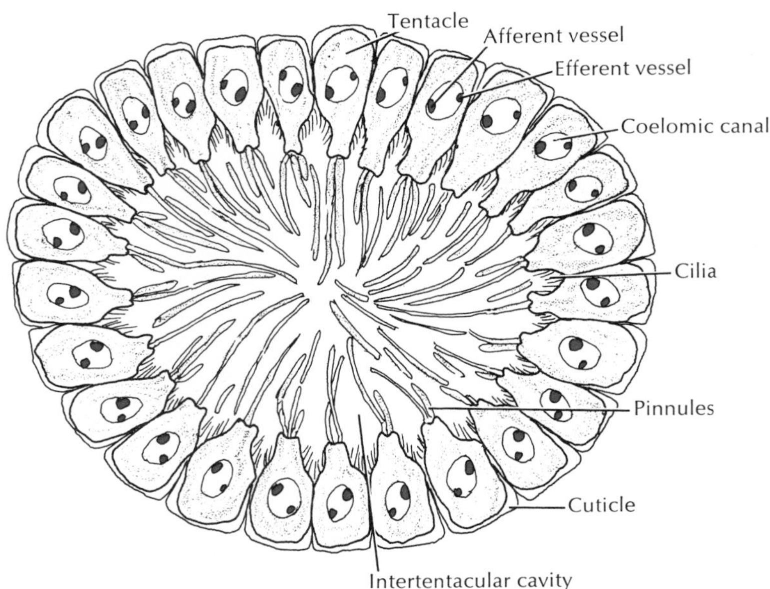

FIG. 18-6 Cross section of tentacular crown of pogonophore *Lamellisabella*. Tentacles arise from ventral side of forepart at base of cephalic lobe. Tentacles (which vary in number in different species) enclose a cylindric space, with the pinnules forming a kind of food-catching network. Food may be digested in this pinnular meshwork and absorbed into the blood supply of tentacles and pinnules.

nerve trunk the neural side, and the other side the antineural. The **forepart** is made up of the anterior cephalic lobe, which contains the central nervous ganglion and a short muscular region following the lobe. On the antineural side the **cephalic lobe** bears some long **tentacles** (the "beard" that gives the phylum its name). There may be from one to 260 or more tentacles depending on the species. They are hollow extensions of the coelom and bear minute pinnules. For a part or all of their length the tentacles lie parallel with each other, enclosing a cylindric intertentacular space into which the pinnules project (Fig. 18-6). Each tentacle contains an afferent and an efferent blood vessel. The forepart also bears a pair of slanting ridges consisting of thickened cuticle and called the **bridle,** which may help the animal cling to the inner surface of the tube as it moves up and down, or it may be the means of distributing the glandular exudate that forms the tube.

The long **trunk** bears **papillae** and, about midway back, two rings of short toothed setae called **girdles,** or **belts,** which are used to grip the wall of the tube, thus allowing the two halves of the body to contract or extend independently in the tube. Anterior to the girdles is the preannular region of the trunk, which usually bears longitudinal bands of cilia on the neural side and papillae on the opposite side. The postannular part of the trunk is very thin and is easily broken when the animals are collected. In fact, the segmented tail end of the animal, or **opisthoma,** was not found and described until after 1963. It is thicker than the trunk, is divided into five to 23 short segments which bear setae, and ends in a small depression.

The mucous-coated **cuticle** that covers the body is made up of criss-cross layers of fibers with microvilli protruding through the mesh. Its structure is similar to that of the cuticle of annelids and sipunculans. There are characteristic thickenings of the pogonophoran cuticle that seem to help the animal grip the tube wall. The **setae** are chitinous and are structurally and chemically similar to annelid setae.

Internal features. The body wall is composed of cuticle, epidermis, and circular and longitudinal muscles.

There is a coelomic compartment in each of the body divisions, and septa between the segments of the opisthosoma also divide it into coelomic compartments.

Pogonophores are remarkable in having **no mouth or digestive tract,** making their mode of nutrition a rather puzzling matter. It is assumed that the tentacles extend from the tube and gather suspended organic detritus. In species that have many tentacles, the tentacles form the cylindric arrangement mentioned previously (Fig. 18-5). In species with a single tentacle *(Siboglinum)* or a group of tentacles *(Spirobrachia)* the tentacle or tentacles coil to form a cylindric funnel. Water is driven into the cylinder by cilia, and particles are trapped in mucus on the pinnules. Enzymes from gland cells near the pinnules may digest the food particles externally. The nutrients could then be absorbed directly into the tentacles and distributed by the blood. It has been demonstrated that amino acids can be absorbed anywhere on the body surface, and this may be one means of gaining nourishment. Pogonophora are the *only nonparasitic animals* that lack a digestive system and have external digestion.

Gas exchange probably also occurs principally in the tentacles. There is a well-developed **closed blood vascular system.**

Sexes are separate in pogonophores, with a pair of gonads and a pair of coelomoducts in the trunk coelom. Pogonophores are not mobile enough to migrate for spawning, but they often live in dense populations, apparently close enough together for reproduction. As a male *Siboglinum* ejects spermatophores from the genital openings, it reaches into the tube with its tentacles, picks up the spermatophores, and draws them toward the top of the tube. As more spermatophores are drawn forward, they push the earlier ones out. Presumably the waterborne spermatophores are captured by the tentacles of nearby females or are caught in their tubes. Fertilization occurs in the female's tube, and the zygotes accumulate in the top part of the tube. Different species produce from about five eggs *(Siboglinum)* to more than 100 eggs *(Oligobrachia)*. The embryos are brooded in the tubes in several species of the order Athecanephria, but this has not been observed in the order Thecanephria. The cleavage is unequal but atypical. It seems to be closer to radial than to spiral. Gastrulation is by epiboly; there is no blastocoel or blastopore. The development of the coelom is schizocoelic, not enterocoelic as was originally described. The worm-shaped embryo is ciliated, but a poor swimmer. It is probably swept along by water currents until it settles.

The tube. The tubes of pogonophores are almost cylindric, ranging from 15 to 20 cm long in smaller species to 150 cm in *Zenkevitchiana longissima*. Diameters vary from 0.1 to 2.8 mm. The tube may be soft and pliable *(Siboglinum)*, parchmentlike and whitish *(Zenkevitchiana)*, or thick, dark, and rigid *(Spirobrachia)*. It may be uniform or composed of regular or irregular rings that may or may not have frilled edges. The tube is apparently secreted by glandular papillae on the preannular trunk. It is thought that after the glandular exudate is produced it is smeared over the inside of the tube as the animal moves up and down in the tube. How the tube is started by the larva is not known. It has been theorized that after leaving the maternal tube the wormlike larva buries itself upright in the mud, produces the exudate, and spreads it about itself by rotating on its axis. As the animal elongates and the tube is enlarged, the bridle may be used in the spreading process.

Phylogeny and adaptive radiation. Because the earlier specimens of Pogonophora that were dredged up lacked the segmented opisthosoma, Ivanov and other early workers, who believed they were working with whole specimens, described the coelom as trimeric (composed of three parts), like that of the hemichordates, and assumed that they were deuterostomes. Ivanov also described the larval coelom as being trimeric. The later discovery of the segmented posterior end brought about some revision of theory. The adult coelom has proved to be polymeric, not trimeric. That fact and the schizocoelic development of the larva both point toward a protostome affinity rather than deuterostome. The pogonophore tubes were originally thought to resemble those of the hemichordate pterobranchs, but analysis of their amino acid and chitin content shows no relationship to the pterobranchs. Pogonophores have photoreceptor cells very similar to those of annelids (oligochaetes and leeches), and the structure of the cuticle, the makeup of the setae, and the segmentation of the opisthosoma all point strongly toward a close relationship with the annelids.

Adaptive radiation has not been extensive. The chief areas of diversity are in the structure of the tentacular crown and in the tube structure.

PHYLUM PRIAPULIDA

The Priapulida (pri'a-pyu'li-da) (Gr., *priapos,* phallus, + *-ida,* pl. suffix) are a small group (only nine species) of marine worms found chiefly in the colder waters of both hemispheres. They have been reported along the U.S. Atlantic coast from Massachusetts to

Greenland and on the Pacific coast from California to Alaska. They live in the bottom mud and sand of the sea floor and range from intertidal zones to depths of several thousand meters.

Their cylindric bodies are rarely more than 12 to 15 cm long. Most of them are burrowing predaceous animals that usually orient themselves upright in the mud with the mouth at the surface. They are adapted for burrowing by body contractions. *Tubiluchus* is a minute detritus feeder adapted to interstitial life in warm coralline sediments. *Maccabeus* is a tiny tube-dweller discovered in muddy Mediterranean bottoms.

External features. The body includes a proboscis, trunk, and usually one or two caudal appendages (Fig. 18-7). The eversible proboscis is ornamented with papillae and ends with rows of curved spines that surround the mouth. When the proboscis is everted, the spines point forward; when invaginated, they are directed posteriorly (Fig. 18-8). The proboscis is used in sampling the surroundings as well as for the capture of small soft-bodied prey. One genus, *Maccabeus,* has a crown of branchial tentacles around the mouth.

The trunk is not truly segmented but is superficially divided into 30 to 100 rings and is covered with tubercles and spines. The tubercles are probably sensory in function. The anus and urogenital pores are located at the posterior end of the trunk. The caudal appendages (one in *Priapulus caudatus* and two in *Priapulopsis bicaudatus)* are hollow stems bearing many hollow vesicles that communicate with the coelom. They are believed to be respiratory and probably chemoreceptive in function. *Tubiluchus* bears a long terminal tail instead of caudal appendages. The body is covered with a chitinous cuticle that is molted periodically throughout life.

Internal features. The muscular body wall contains a single layer of epithelium beneath the cuticle, layers of circular and longitudinal muscles, and a thin peritoneum.

The digestive system consists of the proboscis, a muscular pharynx, and a straight intestine and rectum. There is a nerve ring around the pharynx and a midventral nerve cord. The coelom contains amebocytes and, at least in *Priapulus caudatus,* corpuscles bearing hemerythrin.

Sexes are separate. The paired urogenital organs are each made up of a gonad and clusters of solenocytes, both connected to a protonephridial tubule that carries both gametes and urinary products to the outside. Fertilization is external. Spawning has been observed in *Priapulus caudatus,* in which the male usually spawns first, the female's spawning apparently being stimulated by that of the male. Sexual dimorphism is absent in priapulids except in *Tubiluchus,* in which the male abdomen is covered with setae.

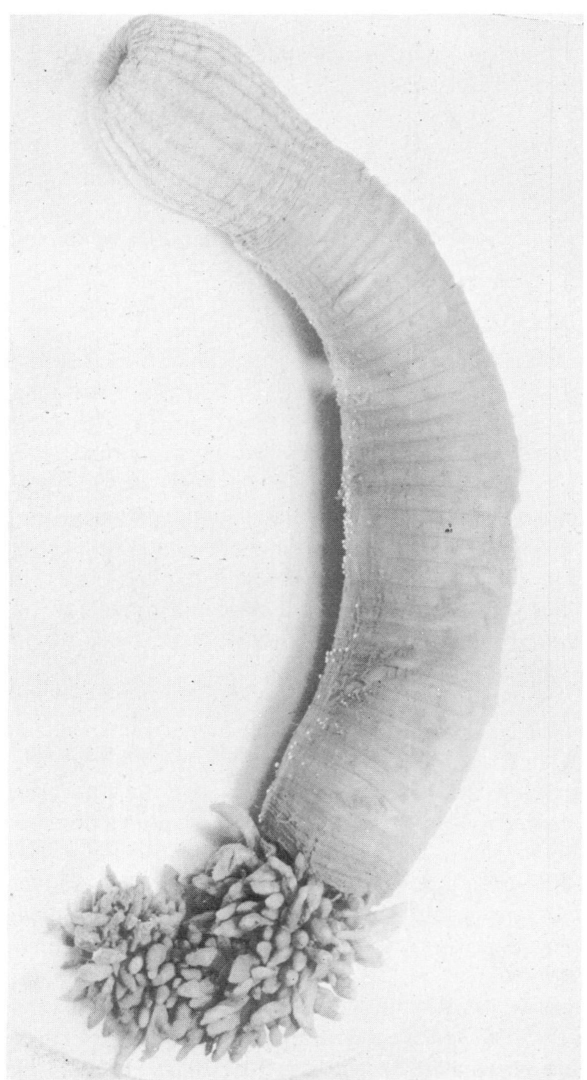

FIG. 18-7 *Priapulus caudatus* Lamarck. Proboscis (top) is partially withdrawn. Note rows of papillae on proboscis, superficial segmentation along trunk, and bushy caudal appendage. When fully expanded, some specimens may attain length of 145 mm. (Courtesy W. L. Shapeero, University of Washington, Seattle.)

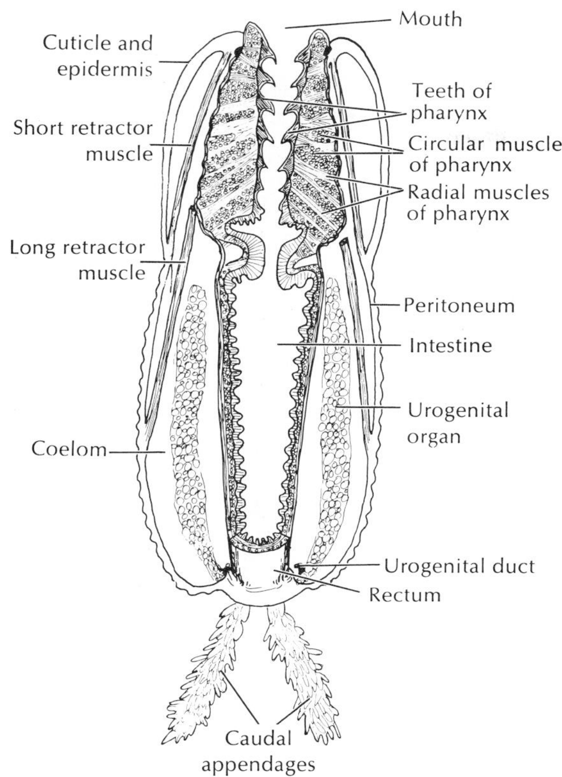

FIG. 18-8 Major internal structures of *Priapulus.*

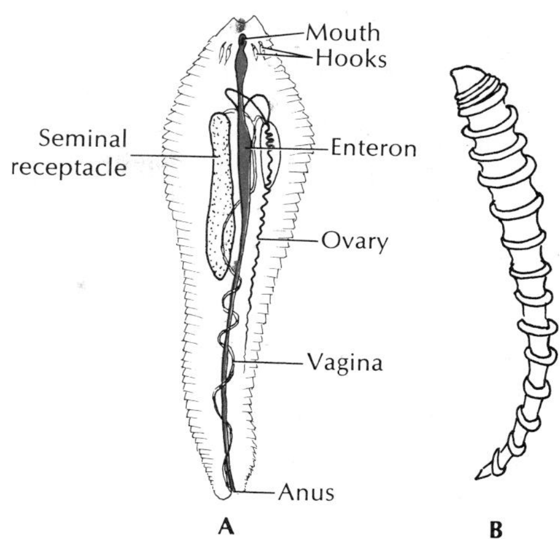

FIG. 18-9 Two pentastomids. **A,** *Pentastomum,* found in lungs of snakes and other vertebrates. Female is shown with some internal structures. **B,** Female *Armillifer,* a pentastomid with pronounced body rings. In parts of Africa and Asia, humans are parasitized by immature stages; adults (10 cm long or more) live in lungs of snakes. Human infection may occur from eating snakes or from contaminated food and water.

The embryology of the Priapulida is poorly known. In *Priapulus* and *Halicryptus* the egg undergoes radial cleavage and develops into a stereogastrula. The larvae develop a strong cuticle, forming a lorica, much like that of rotifers, into which the animal can wholly withdraw. The larvae of *Priapulus* dig into the mud and become detritus feeders. The cuticle is molted between successive stages, and development may take as long as 2 years. Metamorphosis occurs at the last molt.

Phylogeny and adaptive radiation. Priapulids were formerly placed with the pseudocoelomates by some authorities, but a peritoneum has been identified, placing them among the true coelomates. However, their relationship to other coelomates is obscure. Some authors believe them to be remnants of an ancient group because they have retained several primitive features. They bear a fairly close resemblance to the Kinorhyncha. A better knowledge of their embryogeny is needed to understand their phylogeny.

The small number of species indicates either that the priapulids have undergone little adaptive radiation or that only remnants remain of groups that were once more widely distributed. Their stable ecologic niche has given them little opportunity or necessity for wide evolutionary diversification within the group.

PHYLUM PENTASTOMIDA

The Pentastomida (pen-ta-stom′i-da) (Gr., *pente,* five, + *stoma,* mouth), or tongue worms, are a phylum of about 60 to 70 species of wormlike parasites of the respiratory system of vertebrates. The adults live mostly in the lungs of reptiles, such as snakes, lizards, and crocodiles, but one species, *Reighardia sternae,* lives in the air sacs of terns and gulls; and another, *Linguatula serrata,* lives in the nasopharynx of canines and felines (and occasionally humans). Though more common in tropical areas, they are also found in North America, Europe, and Australia.

Morphology. The adults range from 1 to 13 cm in length. Transverse rings give their bodies a segmented appearance (Fig. 18-9). The body is covered with a chitinous cuticle that is molted periodically during larval stages. The anterior end may bear five short protuberances (from which the name Pentastomida is derived). Four of these bear claws. The fifth bears the mouth and two pairs of sclerotized hooks for attachment to the host tissues. There is a simple straight digestive system, adapted for sucking. The nervous system, similar to that of annelids and arthropods, has paired ganglia along the ventral nerve cord. The only sense organs appear to be papillae. There are no circulatory, excretory, or respiratory organs.

Reproduction and life cycles. Sexes are separate, and the females are usually larger than the males. Fertilization is internal. A female may produce several million eggs, which pass up the trachea of the host, are swallowed, and pass out with the feces. The larvae hatch out as oval, tailed creatures with four stumpy legs, each with one or two retractile claws and a penetration apparatus at the anterior end of the body.

Most pentastomid life cycles require an intermediate vertebrate host such as a fish, a reptile, or, rarely, a mammal, which is eaten by the definitive vertebrate host. After ingestion by the intermediate host, the larva penetrates the intestine, migrates randomly in the body, and finally metamorphoses into a nymph, which lacks the legs and penetration apparatus. After several molts the nymph develops mouth hooks and a segmented appearance and grows in size. It finally becomes encapsulated and dormant. When eaten by the final host, the nymph finds its way to the lung, feeds on blood and tissue, and matures. Females of *Linguatula serrata* live 2 years or more in the nasopharyngeal mucosa. Their eggs leave the host via nasal secretions or feces.

Several species have been found encysted in humans, the most common being *Armillifer armillatus,* but usually they cause few symptoms. *Linguatula serrata* is the most common cause of nasopharyngeal pentastomiasis, or ''halzoun,'' a disease of humans in the Middle East and India.

Phylogeny. The phylogenetic affinities of the Pentastomida are uncertain. They have some similarities to the Annelida, and some workers have believed they are offshoots from the polychaetes. Their larval appendages and molting cuticle, however, are arthropod characters. Their larvae resemble tardigrade larvae. Most modern taxonomists align them with the arthropods, but there is little agreement as to where they fit in that phylum. Some authors suggest that they are related to the myriapods; others see an affinity with the Acarina because the larvae resemble mite larvae. Wingstrand (1972) says pentastomid spermatozoa are identical with those of the crustacean subclass Branchiura and suggests that they should be classed with the Branchiura. Because their modifications for parasitic life make it difficult to do more than guess at the free-living forms from which they arose, it seems best to keep them in a separate phylum.

PHYLUM ONYCHOPHORA

Members of the phylum Onychophora (on-y-kof′o-ra) (Gr., *onyx,* claw, + *pherein,* to bear) are commonly called the ''velvet worms,'' or ''walking worms.'' They comprise approximately 70 species of caterpillar-like animals, ranging from 1.4 to 15 cm in length. They live in rain forests and other moist, leafy habitats in tropical and subtropical regions and in some temperate regions of the Southern Hemisphere.

The fossil record of the onychophorans shows that they have changed little in their 500-million-year history. A fossil form, *Aysheaia,* discovered in the Burgess shale deposit of British Columbia and dating back to mid-Cambrian times, is very much like the modern onychophorans. Onychophorans have been of unusual interest to zoologists because they share so many characteristics with both the annelids and the arthropods. They have been called, a bit too hopefully perhaps, the ''missing link'' between the two phyla. According to the fossil record, onychophorans were originally marine animals and were probably far more common at one time than they are now. Today they are all terrestrial and are extremely retiring, coming out only at night or when the air is nearly saturated with moisture.

These curious animals have been dubbed velvet worms because they are covered with a velvety skin, and walking worms because they have from 14 to 43 pairs of stumpy, unjointed legs (Fig. 18-10), each ending with a flexible pad and a pair of claws. In fact, it was for this characteristic that the first onychophorans, discovered in 1825, were given the generic name *Peripatus,* from the Greek *peripatos,* meaning ''a walking about.''

External features. The onychophoran body is more or less cylindric and shows no external segmentation except for the paired appendages. The skin is soft and velvety and is covered with a thin, flexible cuticle that contains protein and chitin. In structure and chemical composition it resembles arthropod cuticle; however, it never hardens like arthopod cuticle, and it is molted in patches rather than all at one time. The body is studded with tiny **tubercles,** some of which bear sen-

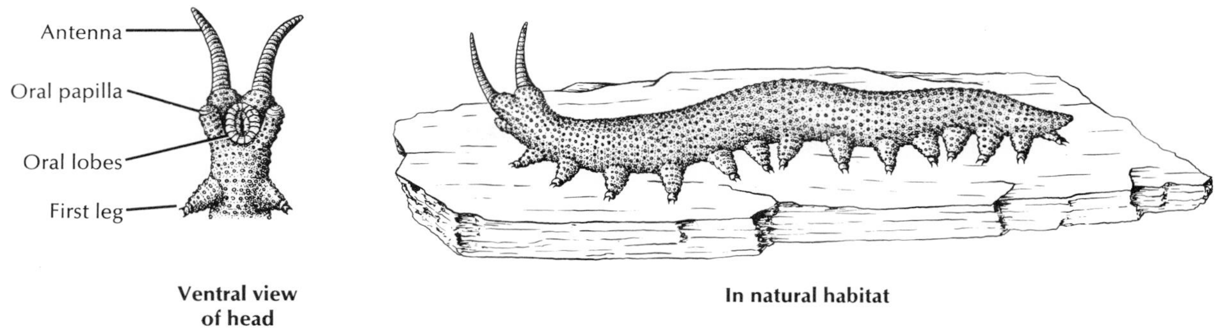

FIG. 18-10 *Peripatus,* a caterpillar-like onychophoran that has both annelid and arthropod characteristics.

sory bristles. The color may be green, blue, orange, dark gray, or black; and minute scales on the tubercles give it an iridescent and velvety appearance. The head bears a pair of large **antennae,** each with an annelid-like **eye** at the base (Fig. 18-10). The ventral mouth has a pair of clawlike **mandibles** and is flanked by a pair of **oral papillae** from which a defensive secretion can be expelled.

The **unjointed legs** are short, stubby, and clawed. Locomotion is achieved by waves of contraction passing from anterior to posterior. When a segment is extended, the legs are lifted up and moved forward. The legs are more ventrally located than are the parapodia of annelids.

Internal features. The body wall is muscular like that of the annelids. The body cavity is a **hemocoel,** imperfectly divided into compartments, or sinuses, much like those of the arthropods. **Slime glands** on each side of the body cavity open on the oral papillae. When disturbed by a predator, the animal can eject from the slime glands two streams of a sticky substance that rapidly hardens.

The mouth, surrounded by lobes of skin, contains a dorsal tooth and a pair of lateral mandibles used for grasping and cutting prey. There is a muscular pharynx and a straight digestive tract. Most velvet worms are predaceous, feeding on caterpillars, insects, snails, worms, and the like. Some onychophorans live in termite nests and feed on termites. Prey discovered by the tentacles is held by the lips and sucking pharynx while the mandibles slit open the body wall. A pair of **salivary glands** secretes salivary enzymes into the wound and the digested tissues are sucked into the mouth.

Each segment contains a pair of **nephridia,** each nephridium with a vesicle, ciliated funnel and duct, and nephridiopore opening at the base of a leg. There is some evidence that the chief excretory product is uric acid, a water-salvaging device used by birds, lizards, land snails, and most insects. Absorptive cells in the midgut excrete crystalline uric acid; and certain pericardial cells function as nephrocytes, storing excretory products taken from the blood.

For respiration there is a **tracheal system** that ramifies to all parts of the body and communicates with the outside by many openings, or **spiracles,** scattered all over the body. The spiracles cannot be closed to prevent water loss; so although the tracheae are efficient, the animals are restricted to moist habitats. The tracheal system is somewhat different from that of arthropods and probably has evolved independently.

The open circulatory system has, in the pericardial sinus, a dorsal, tubular heart with a pair of ostia in each segment.

There are a pair of cerebral ganglia with connectives and a pair of widely separated nerve cords with connecting commissures. The brain gives off nerves to the antennae and head region, and the cords send nerves to the legs and body wall. Sense organs include the pigment cup ocelli, taste spines around the mouth, tactile papillae on the integument, and hygroscopic receptors that orient the animal toward water vapor.

Onychophorans are dioecious, with paired reproductive organs. The males usually deposit their sperm in spermatophores in the female seminal receptacle. The South African *Peripatopsis,* however, has no seminal receptacles. The male deposits the spermatophores on the back of the female, which may accumulate a number of them. White blood cells dissolve the skin beneath the spermatophores. The sperm can then enter the body cavity and migrate in the blood to the ovaries to fertilize the eggs. Onychophorans may be oviparous, ovoviparous, or viviparous. Only the Australian genera *Ooperipatus* and *Symparipatus* are oviparous and lay shell-covered eggs in moist places. In all other onychophorans the eggs develop in the uterus, and living young are produced. Female *Peripatus acacioi,* raised in the laboratory (Lavallard and Campiglia, 1975), were able to give birth by 15 to 23 months of age. They usually had an annual brood of one to eight young. Parturition took about half an hour, with the young freeing themselves gradually, head first. In some species there is a placental attachment between mother and young (viviparous); in others the young develop in the uterus without attachment (ovoviviparous).

Phylogeny. Onychophorans resemble the annelids with their soft body, nonjointed appendages, segmentally arranged nephridia, muscular body wall, pigment-cup ocelli, and ciliated reproductive ducts. Arthropod characteristics are the cuticle, the tubular heart and open circulatory system, the presence of tracheae, a hemocoel for a body cavity, and the large size of the brain. They differ from either phylum in their scanty metamerism, structure of the mandibles, and the separate arrangement of the nerve cords. They are more primitive than insects and are somewhat like the centipedes in the arrangement of internal metamerism.

Some authors believe the onychophorans should be included with the arthropods, but that would involve redefining the phylum Arthropoda. Manton (1972) recommends placing the Onychophora, myriapods, and insects in a phylum called Uniramia. In spite of their obvious relationship to the myriapods and insects, however, most authors believe that the differences seem to warrant keeping the onychophorans in a separate phylum.

PHYLUM TARDIGRADA

Tardigrada (tar-di-gray′da) (L., *tardus,* slow, + *gradus,* step), or "water bears," are minute forms usually less than a millimeter in length. Most of the 300 to 400 species are terrestrial forms that live in the water film that surrounds mosses and lichens. Some live in freshwater algae or mosses or in the bottom debris; and a few are marine, inhabiting the interstitial spaces between sand grains, in both deep and shallow seawater. They share many characteristics with the arthropods.

FIG. 18-11 *Echiniscus (H.) gladiator* (phylum Tardigrada) as seen under a scanning electron microscope. (×1250.) Its claws and slow, lumbering movement have earned it the name "water bear." Members of this species are 100 to 150 μm long. (Courtesy D. R. Nelson and R. O. Schuster.)

External features. The body is elongated, cylindric, or a long oval and is unsegmented. The head is merely the anterior part of the trunk. The trunk bears four pairs of short, stubby, unjointed legs, each armed with four to eight claws, which may be unequal in size and are used for clinging to its substrate (Fig. 18-11). The last pair of appendages lies at the posterior end of the body. The body is covered by a nonchitinous cuticle, secreted by the epidermis, that is molted along with the claws and buccal apparatus four or more times in its life history. Cilia are absent. Common American species are *Macrobiotus, Echiniscus,* and *Hypsibius.*

Internal features. The mouth is at the anterior end and opens into a buccal tube that empties into a muscular pharynx that is adapted for sucking. Two needlelike stylets flanking the buccal tube can be protruded through the mouth. The stylets are used for piercing the cellulose walls of plant cells, and the liquid contents are sucked in by the pharynx. Some tardigrades suck the body juices of nematodes, rotifers, and other small animals. Glands empty into both the pharynx and the digestive tract. Some tardigrades, such as

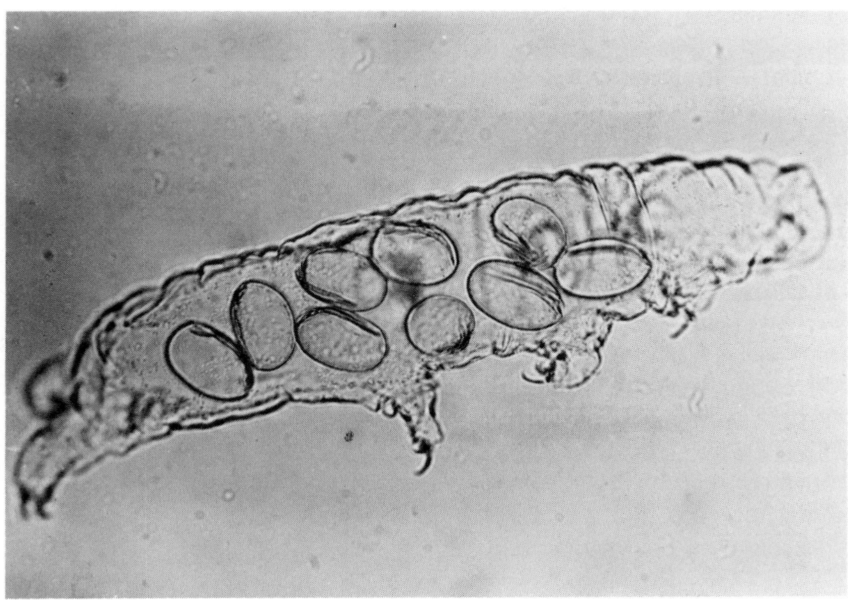

FIG. 18-12 Molted cuticle of a tardigrade, containing a number of fertilized eggs. (From Sayre, R. M. 1969. Trans. Am. Microsc. Soc. **88:**266-274.)

Echiniscus, expel feces when molting, leaving the feces in the discarded cuticle.

At the junction of the stomach and rectum, three glands, often called malpighian tubules, empty into the digestive system. They are thought to be excretory in function.

Most of the body cavity is a hemocoel, with the true coelom restricted to the gonadal cavity. There are no circulatory or respiratory systems, as fluids freely circulate through the body spaces, and gaseous exchange can occur through the body surface.

The muscular system consists of a number of long muscle bands, each composed of a single smooth muscle cell. Most have their origins and insertions on the body wall. Circular muscles are absent, but the hydrostatic pressure of the body fluid may act as a type of skeleton. Being unable to swim, the water bear creeps about awkwardly, clinging to the substrate with its claws.

The brain is large and covers most of the dorsal surface of the pharynx. It connects by way of circumpharyngeal connectives to the subpharyngeal ganglion, from which the double ventral nerve cord extends posteriorly as a chain of four ganglia. Sense organs usually consist of a pair of pigmented eyespots (in most species) and various sensory spines and bristles.

Sexes are separate in tardigrades. In some freshwater and moss-dwelling species, males are unknown and parthenogenesis seems to be the rule. In marine species, however, males and females occur with approximately equal frequency. The gonad is single, and the gonoduct may end at a gonopore or may join the rectum. Year round reproduction is typical of marine species. Egg-laying, like defecation, apparently occurs only at molting, when the volume of coelomic fluid is reduced. Females of some species deposit the eggs in the molted cuticle (Fig. 18-12). Males gather round the old cuticle and shed sperm into it. Other species are fertilized internally, but only at time of molting.

Cleavage is holoblastic but atypical, and a stereogastrula is formed. Six pairs of coelomic pouches arise from the gut, but all except the last pair disaggregate to form the buccal apparatus, pharynx, and body musculature. The last pair fuses to form the gonad. Thus the gonocoel (which is enterocoelic) is the only true coelom left in the adult. Development is direct, and the young juveniles use stylets and claws to open the eggs. Further growth and development occur through periodic molting. The life span of marine tardigrades is thought to be about 3 to 4 months.

Cryptobiosis. One of the most intriguing features of terrestrial tardigrades is their capacity to enter a state of suspended animation, called cryptobiosis (formerly called anabiosis), during which metabolism is virtually imperceptible; the organism can withstand harsh environmental conditions. Under gradual drying conditions, the water content of the body is reduced from 85% to only 3%, movement ceases, and the body becomes barrel-shaped. In a cryptobiotic state tardigrades can resist temperature extremes, ionizing radiations, oxygen deficiency, and other adverse conditions, and may survive for years. Activity resumes when moisture is again available.

Phylogeny. The affinities of tardigrades are among the most puzzling of all animal groups. They have some similarities to the rotifers, particularly in their reproduction and their cryptobiotic tendencies, and some authors call them pseudocoelomates. Their embryology, however, would seem to put them among the coelomates. The nervous system indicates a relationship to the annelids and arthropods. Some authors place them close to the arthropods, particularly the mites. Their enterocoelic origin of the mesoderm is a deuterostome characteristic. For the present, at least, their status is quite uncertain.

Annotated references

Selected general readings

Barnes, R. D. 1974. Invertebrate zoology, ed. 3. Philadelphia, W. B. Saunders Co.

Clark, R. B. 1972. Systematics and phylogeny: Annelida, Echiura, Sipuncula. In M. Florkin and B. T. Scheer (eds.). Chemical zoology, vol. 4. New York, Academic Press, Inc.

Fisher, W. K. 1946. Echiurid worms of the North Pacific Ocean. U.S. Nat. Museum Proc. **96:**213-292.

Fisher, W. K. 1952. The sipunculid worms of California and Baja California. Proc. U.S. Nat. Museum. **102:**371-450.

Florkin, M., and B. T. Scheer (eds.). 1972. Chemical zoology, vol. 4. New York, Academic Press, Inc.

George, J. D., and E. C. Southward. 1973. A comparative study of the setae of Pogonophora and polychaetous Annelida. J. Mar. Biol. Assoc. U.K. **53:**403-424.

Gosner, K. L. 1971. Guide to identification of marine and estuarine invertebrates: Cape Hatteras to the Bay of Fundy. New York, Interscience Publishers.

Gould-Somero, M. 1974. Echiura. In A. C. Giese and J. S. Pearse (eds.). Reproduction of marine invertebrates, vol 3. New York, Academic Press, Inc.

Hackman, R. H., and M. Goldberg. 1975. Peripatus: its affinities and its cuticle. Science **190:**582-583 (Nov.).

Hickman, C. P. 1973. Biology of the invertebrates, ed. 2. St. Louis, The C. V. Mosby Co. *General treatment and descriptions of all the minor coelomate phyla.*

Hyman, L. H. 1959. The invertebrates, vol. 5. New York, McGraw-Hill Book Co.

Ivanov, A. V. 1963. Pogonophora. New York, Consultants Bureau.

Jaccarini, V., and P. J. Schembri. 1977a. Feeding and particle selection in the echiurian worm *Bonellia viridis* Rolando (Echiura: Bonellidae). J. Exp. Mar. Biol. Ecol. **28** (2): 163-182.

Jaccarini, V., and P. J. Schembri. 1977b. Locomotory and other movements of the proboscis of *Bonellia viridis* (Echiura, Bonnelidae). J. Zool. (Lond.) **182**(4):467-476.

Kaestner, A. 1967, 1968. Invertebrate zoology, vols 1 and 2. New York, Interscience Publishers.

Lavallard, R., and S. Campiglia. 1975. Contribution to the biology of *Peripatus acacioi* Marcus and Marcus (Onychophora). V. Studies of the breeding in a laboratory culture. Zool. Anz. **195**(5/6):338-350. *In French with French and English summary.*

Light, S. F., and others. 1967. Intertidal invertebrates of the central California coast. Berkeley, University of California Press.

MacGinitie, G. E., and N. MacGinitie. 1967. Natural history of marine animals, ed. 2. New York, McGraw-Hill Book Co.

Manton, S. M. 1950. The locomotion of *Peripatus*. J. Linn. Soc. Zool. **41:**529-539.

Manton, S. M. 1972. The evolution of arthropodan locomotory mechanisms. Part 10. Zool. J. Linn. Soc. **51:**203-400. *A new taxon, the phylum Uniramia, was proposed to include the Onychophora, Myriapoda, and Hexapoda* (Insecta) *as subphyla.*

Marcus, E. 1959. Tardigrada. In W. T. Edmondson, H. B. Ward, and G. C. Whipple (eds.). Freshwater biology, ed. 2. New York, John Wiley & Sons, Inc., pp. 508-521. *Keys to freshwater species.*

Meglitsch, P. A. 1972. Invertebrate zoology, ed. 2. New York, Oxford University Press.

Nelson, D. R. 1975. The hundred-year hibernation of the water bear. Nat. Hist. **84**(7):62-65 (Aug.-Sept.).

Newby, W. W. 1940. The embryology of the echiuroid worm, *Urechis caupo*. Am. Phil. Soc. Mem., vol. 16.

Peck, S. B. 1975. A review of the New World Onychophora with the description of a new cavernicolous genus and species from Jamaica. Psyche **82**(3/4):341-358. *A key to the two families and eight genera of New World onychophorans.*

Pennak, R. W. 1953. Fresh-water invertebrates of the United States. New York, The Ronald Press Co., pp. 240-255. *An account of the tardigrades, with keys to the common species.*

Pollack, L. W. 1975. Tardigrada. In A. C. Giese and J. S. Pearse (eds.). Reproduction of marine invertebrates, vol. 2. New York, Academic Press, Inc., pp. 43-54.

Rice, M. E. 1970. Asexual reproduction in a sipunculan worm. Science **167:**1618-1620 (Mar.).

Rice, M. E. 1973. Morphology, behavior and histogenesis of the pelagosphera larva of *Phascolosoma agassizii* (Sipuncula). Smithson. Contrib. Zool. **132:**1-51.

Rice, M. E. 1975. Sipuncula. In A. C. Giese and J. S. Pearse (eds.). Reproduction of marine invertebrates, vol. 2. New York, Academic Press, Inc., pp. 67-128.

Riggen, G. T. 1962. Tardigrada of southwest Virginia with the addition of a new species from Florida. Blacksburg, Va., Virginia Agricultural Experiment Station.

Shapeero, W. 1961. Phylogeny of Priapulida. Science **133:** 879-880.

Schembri, P. J., and V. Jaccarini. 1977. Locomotory and other movements of the trunk of Bonellia viridis (Echiura, Bonnelidae). J. Zool. (Lond.) **182**(4):477-494.

Schmidt, G. D., and L. S. Roberts. 1977. Foundations of parasitology. St. Louis, The C. V. Mosby Co.

Southward, E. C. 1971. Recent researches on the Pogonophora. Oceanogr. Mar. Biol. Ann. Rev. **9:**193-220.

Southward, E. C. 1975a. Fine structure and phylogeny of the Pogonophora. In E. J. W. Barrington and R. P. S. Jefferies (eds.). Protochordates. London, Zoological Society of London, no. 36.

Southward, E. C. 1975b. Pogonophora. In A. C. Giese and J. S. Pearse (eds.). Reproduction of marine invertebrates, vol. 2. New York, Academic Press, Inc., pp. 129-156.

Stephen, A. C., and S. J. Edmonds. 1972. The phyla Sipuncula and Echiura. London, British Museum.

van der Land, J. 1970. Systematics, zoogeography, and ecology of the Priapulida, no. 112. Zoologische verhandelingen uitgegeven door het Rijksmuseum van Natuurlijke Historie te Leiden. Leiden, E. J. Brill. *An up-to-date monograph.*

van der Land, J. 1975. Priapulida. In A. C. Giese and J. S. Pearse (eds.). Reproduction of marine invertebrates, vol. 2. New York, Academic Press, Inc., pp. 55-66.

Webb, M. 1964. The posterior extremity of *Siboglinum fiordicum* (Pogonophora). Sarsia **15:**33-35.

Webb, M. 1965. Additional notes on the adult and larva of *Siboglinum fiordicum* and on the possible mode of tube formation. Sarsia **20:**21-43.

Wingstrand, K. G. 1972. Comparative spermatology of a pentastomid, *Raillietiella hemidactyli,* and a branchiuran crustacean, *Argulus foliaceus,* with a discussion of pentastomic relationships. Kongelige Danske Videnskabernes Selskab Biol. Skrifter **19**(4):1-72.

Selected *Scientific American* article

Crowe, J. H., and A. F. Cooper, Jr. 1971. Cryptobiosis. **225**(6):30-36 (Dec.).

CHAPTER 19

THE LOPHOPHORATE ANIMALS

Phylum Phoronida
Phylum Ectoprocta (Bryozoa)
Phylum Brachiopoda

The lophophorate animals are distinguished by a crown of delicate ciliated tentacles surrounding the mouth and used for particle feeding. This is the U-shaped lophophore of one of the microscopic individuals of a colony of the ectoproct *Plumatella*.

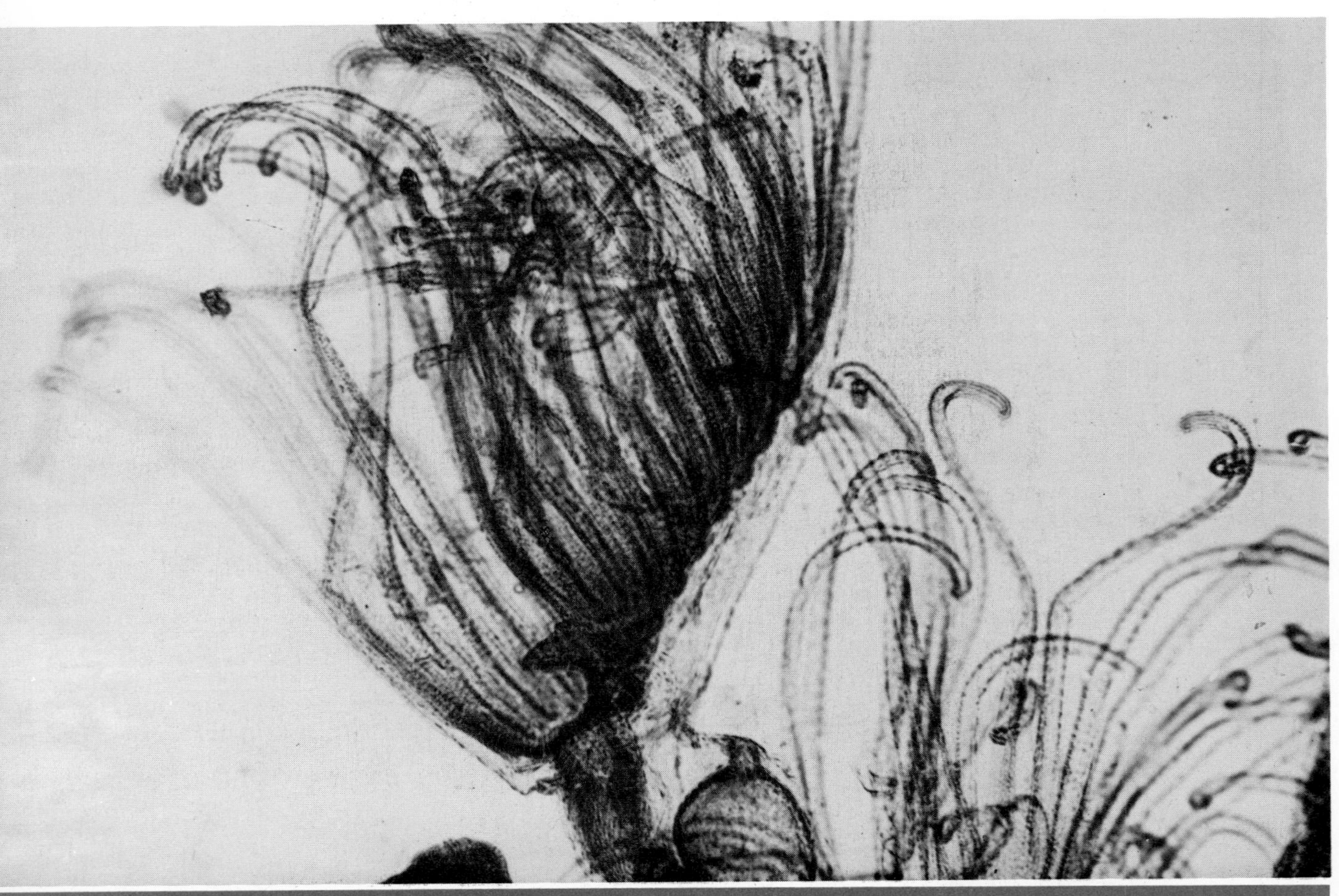

Photograph by F. M. Hickman.

Position in animal kingdom

1. The lophophorate phyla possess a **true coelom,** that is, a body cavity that is lined with a layer of mesodermal epithelium called the peritoneum.
2. They belong to the **protostome** branch of the **bilateral** animals, but they have some of the characteristics of the deuterostomes.
3. The three phyla are usually grouped together because they all possess the crown of tentacles called a **lophophore,** which is specialized for sedentary filter feeding. The lophophore **surrounds the mouth but not the anus,** thus differing from the tentacular crown of Entoprocta.

Biologic contributions

1. The lophophore, a unique ridge that bears hollow, ciliated tentacles, is an efficient specialized filter-feeding device, forming a ciliated route, or trough, for trapping and directing food particles to the mouth.
2. The brachiopods and phoronids possess vascular systems for circulation of food nutrients and other materials.
3. The blood in phoronids possesses red blood corpuscles that contain hemoglobin for carrying oxygen.

THE LOPHOPHORATES

The three lophophorate phyla might appear on superficial examination to have nothing in common except that they are all aquatic invertebrates, mostly marine. The **phoronids** (phylum Phoronida) are wormlike marine forms that live in secreted tubes in sand or mud or attached to rocks or shells. The **ectoprocts** (phylum Ectoprocta) are minute forms, mostly colonial, whose protective cases often form encrusting masses on rocks, shells, or plants. The **brachiopods** (phylum Brachiopoda) are bottom-dwelling marine forms that superficially resemble molluscs because of their bivalved shells.

One might wonder why these three apparently different types of animals are lumped together in a group called Lophophorata. Actually they have more in common than first appears. They are all eucoelomates; all have some protostome characteristics; all are sessile; and none has a distinct head. But these characteristics are also shared by other phyla. What really sets them apart from other phyla is the common possession of a **ciliary feeding device** called a **lophophore,** a term that means crest bearer (Gr., *lophos,* crest or tuft, + *phorein,* to bear).

A lophophore is a unique arrangement of ciliated tentacles borne on a ridge (a fold of the body wall), which surrounds the mouth but not the anus. The lophophore with its crown of tentacles contains within it an extension of the coelom, and the thin, ciliated walls of the tentacles not only comprise an efficient feeding device but also serve as a respiratory surface for exchange of gases between the environmental water and the coelomic fluid. The lophophore can usually be extended for feeding or withdrawn for protection.

In addition, all three phyla have a **U-shaped alimentary canal,** with the anus placed near the mouth but **outside the lophophore.** All have a **free-swimming larval stage** but are **sessile as adults.**

PHYLUM PHORONIDA

The phylum Phoronida (fo-ron′i-da) (Gr., *phoros,* bearing, + L., *nidus,* nest) comprises approximately 15 species of small wormlike animals that live on the bottom of shallow coastal waters, especially in temperate seas. They range from a few millimeters to 30 cm in length. Each worm secretes a leathery or chitinous tube in which it lies free, but which it never leaves (Fig. 19-1). The tubes may be anchored singly or in a tangled mass on rocks, shells, or pilings or buried in the sand. The tentacles on the lophophore are thrust out for feeding, but if the animal is disturbed it can withdraw completely into its tube. The phylum name means "nest-bearing," the nest, of course, referring to the tentacled lophophore.

The lophophore is made up of two parallel ridges curved in a horseshoe shape, the bend located ventrally and the mouth lying between the two ridges. The horns of the ridges are often coiled into twin spirals. Each ridge carries hollow ciliated tentacles, which, like the ridges themselves, are extensions of the body wall.

The cilia on the tentacles direct a water current toward a groove between the two ridges, which leads toward the mouth. Plankton and detritus caught in this current become entangled in mucus and are carried by the cilia to the mouth. The anus lies dorsal to the mouth, outside the lophophore, flanked on each side by a nephridiopore. Water leaving the lophophore passes over the anus and nephridiopores, carrying

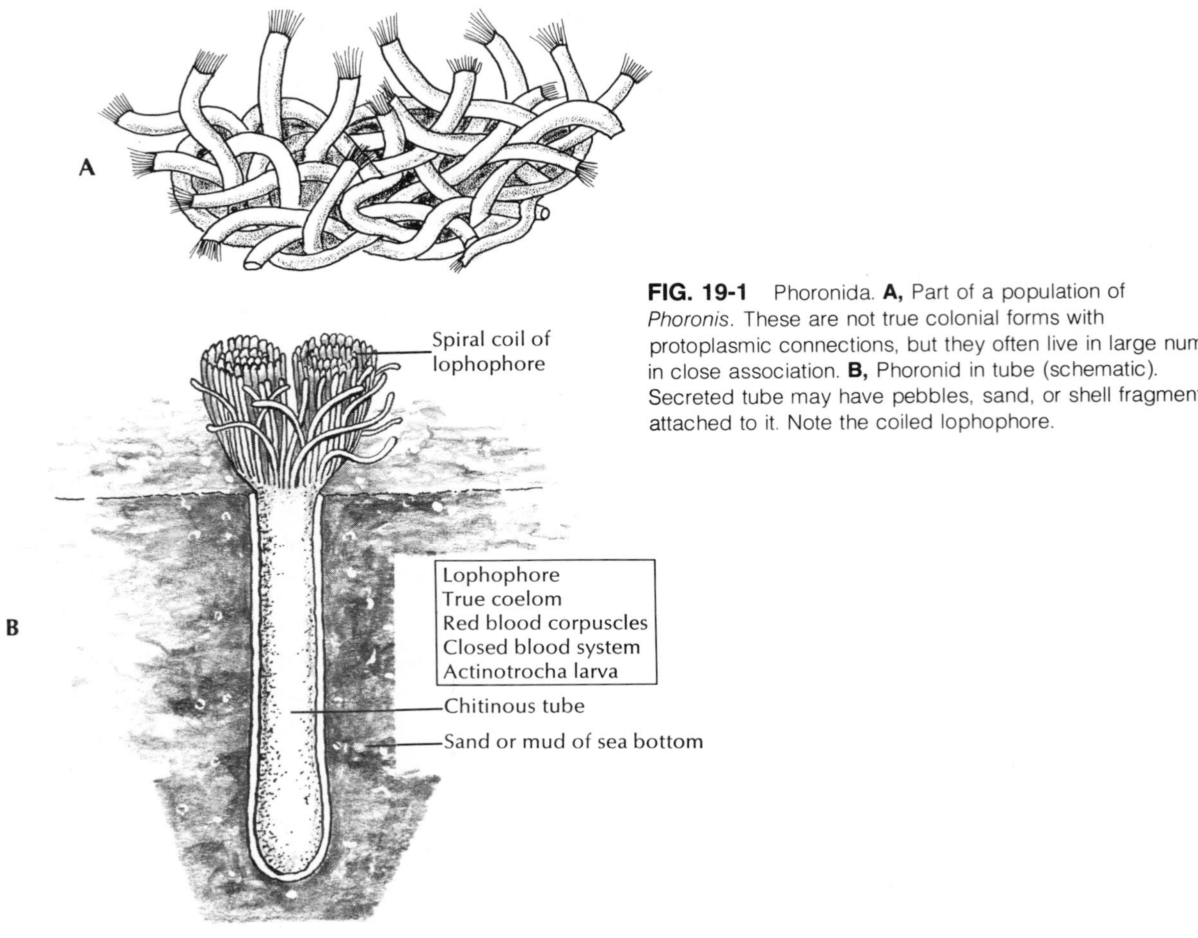

FIG. 19-1 Phoronida. **A,** Part of a population of *Phoronis*. These are not true colonial forms with protoplasmic connections, but they often live in large numbers in close association. **B,** Phoronid in tube (schematic). Secreted tube may have pebbles, sand, or shell fragments attached to it. Note the coiled lophophore.

away the wastes. Cilia in the stomach area of the U-shaped gut aid in food movement.

The body wall is made of cuticle, epidermis, and both longitudinal and circular muscles. The coelomic cavity is subdivided by mesenteric partitions into compartments similar to those of the deuterostomes. The phoronids have a closed system of contractile blood vessels but no heart (Fig. 19-2); the red blood contains hemoglobin. There is a pair of metanephridia. A nerve ring sends nerves to the tentacles and body wall; a single giant motor fiber lies in the epidermis; and an epidermal nerve plexus supplies the body wall and epidermis.

There are both monoecious (the majority) and dioecious species of Phoronida, and at least one species reproduces asexually. Cleavage seems to be related to both the spiral and the radial types. The free-swimming, ciliated larva, called an actinotroch, metamorphoses into the adult, which sinks to the bottom, secretes a tube, and becomes sessile.

Phoronopsis californica is a large, orange-colored form approximately 30 cm long found along the west coast of the United States. *Phoronis architecta* is a smaller Atlantic coast species (approximately 12 cm long) that has a very wide distribution.

PHYLUM ECTOPROCTA (BRYOZOA)

The Ectoprocta (ek-to-prok′ta) (Gr., *ektos,* outside, + *proktos,* anus) have long been called bryozoans, or moss animals (Gr., *bryon,* moss, + *zōon,* animal), a term that originally included the Entoprocta also. However, because the entoprocts are pseudocoelomates and have the anus located within the tentacular crown, they are no longer classed with the ectoprocts, which, like the other lophophorates, are eucoelomate and have

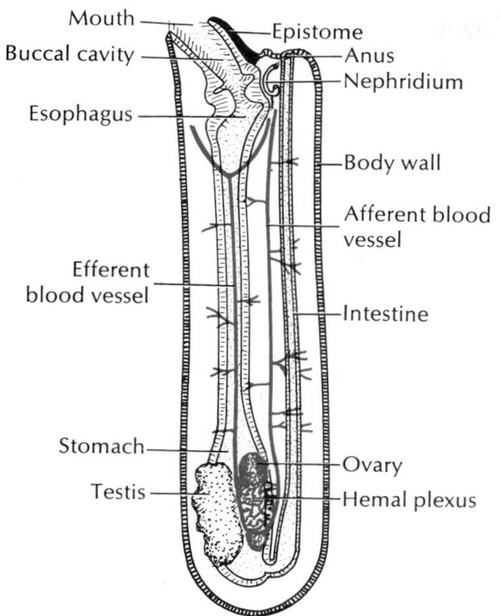

FIG. 19-2 Chief internal structures of a phoronid. Lophophore has been omitted.

the anus outside the circle of tentacles. Some authors continue to use the name "Bryozoa" but now exclude the entoprocts from the group.

Of the 4,000 or so species of ectoprocts, few are more than 0.5 mm long; all are aquatic, both freshwater and marine, but are largely found in shallow waters; and most, with very few exceptions, are colony builders. Ectoprocts have been very successful. They have left a rich fossil record since the Ordovician era. Marine forms today exploit all kinds of firm surfaces, such as shells, rock, large brown algae, mangrove roots, and ship bottoms.

Each member of a colony lives in a tiny chamber, called a **zooecium,** which is secreted by its epidermis (Fig. 19-3). Each individual, or **zooid,** consists of a feeding polypide and a case-forming cystid. The **polypide** includes the lophophore, digestive tract, muscles, and nerve centers. The **cystid** is the body wall of the animal, together with its secreted exoskeleton. The exoskeleton, or zooecium, may, according to the species, be gelatinous, chitinous, or stiffened with calcium and possibly also impregnated with sand. The shape may be boxlike, vaselike, oval, or tubular.

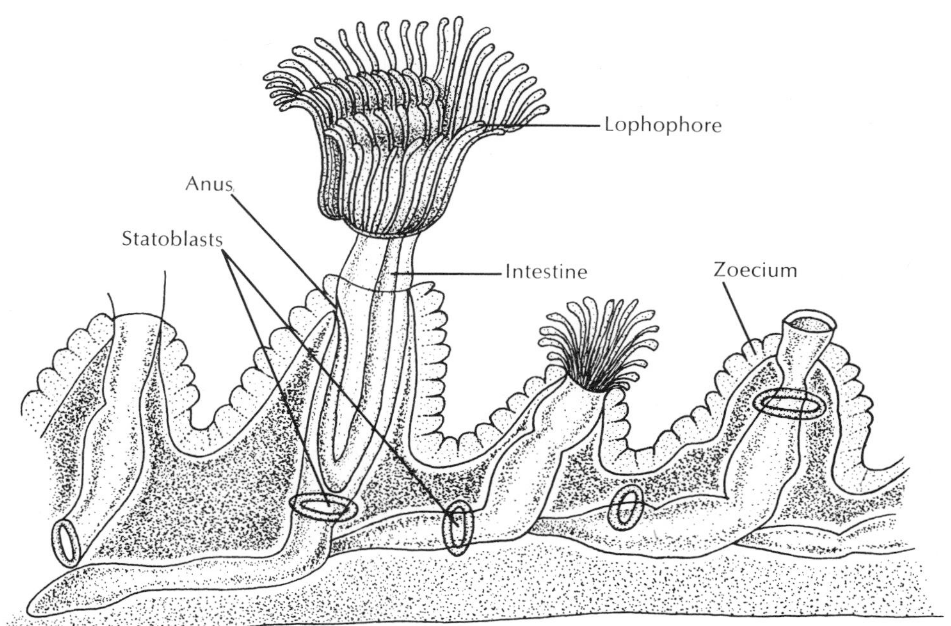

FIG. 19-3 Small portion of freshwater colony of *Plumatella* (phylum Ectoprocta), which grows on the underside of rocks. The tiny individuals disappear into their chitinous zoecia when disturbed.

FIG. 19-4 **A,** Ciliated lophophore of *Flustrella,* a marine ectoproct. **B,** *Plumatella repens,* a freshwater bryozoan (phylum Ectoprocta). It grows in branching, threadlike colonies on the underside of rocks and vegetation in lakes, ponds, and streams. (**A,** Courtesy J. A. Cooke, Museum of Natural History; **B,** photograph by R. Vishniac.)

Some colonies form limy encrustations on seaweed, shells, and rocks; others form fuzzy or shrubby growths or erect, branching colonies that look like seaweed. Some ectoprocts might easily be mistaken for hydroids but can be distinguished under a microscope by the fact that their tentacles are ciliated (Fig. 19-4). In some freshwater forms the individuals are borne on finely branching stolons that form delicate tracings on the underside of rocks or plants. Other freshwater ectoprocts are embedded in large masses of gelatinous material. Although the zooids are minute, the colonies may be several centimeters in diameter; some encrusting colonies may be a meter or more in width (Fig. 19-5), and erect forms may reach 30 cm or more in height. Freshwater ectoprocts may form mosslike colonies on the stems of plants or on rocks, usually in shallow ponds or pools. They may be able to slide along slowly on the object that supports them.

The polypide lives a type of jack-in-the-box existence, popping up to feed and then quickly withdrawing into its little chamber, which often has a tiny trapdoor (operculum) that shuts to conceal its inhabitant. To extend the tentacular crown, certain muscles contract, which increases the hydraulic pressure within the body cavity and pushes the lophophore out. Other muscles can contract to withdraw the crown to safety with great speed.

The lophophore ridge tends to be circular in marine ectoprocts (Fig. 19-4, *A*) and U-shaped in freshwater species (Fig. 19-4, *B*). When feeding, the animal extends the lophophore and spreads the tentacles out into a funnel. Cilia on the tentacles draw water into the funnel and out between the tentacles. Food particles trapped in mucus in the funnel are drawn into the mouth, both by the pumping action of the muscular pharynx and by the action of cilia in the pharynx. Undesirable particles can be rejected by reversing the ciliary action, by drawing the tentacles close together, or by retracting the whole lophophore into the zoecium. Digestion in the ciliated, U-shaped digestive tract ap-

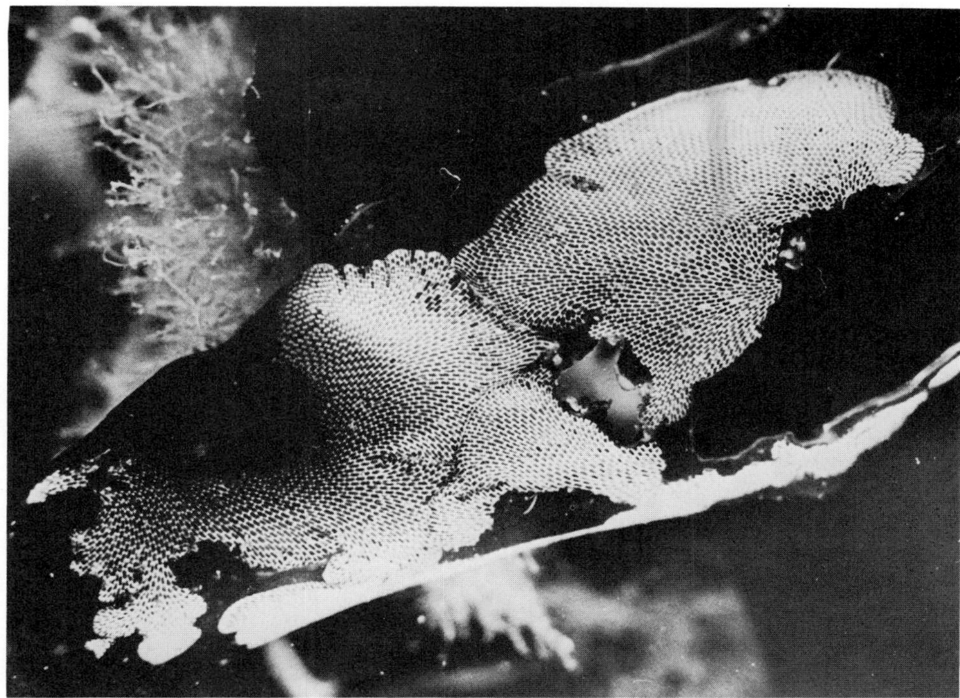

FIG. 19-5 Skeletal remains of a colony of *Membranipora,* a marine encrusting form of Ectoprocta. Each little oblong zoecium is the calcareous former home of a tiny ectoproct. Some cnidarian hydroids are visible at lower right. (Photograph by B. Tallmark, Uppsala University, Sweden.)

pears to be extracellular for protein and starches and intracellular for fats.

Respiratory, vascular, and excretory organs are absent. Gaseous exchange occurs through the body surface, and, since the ectoprocts are small, the coelomic fluid is adequate for internal transport. Coelomocytes engulf and store waste materials. There is a ganglionic mass and a nerve ring around the pharynx, but no sense organs are present. The coelom is divided by a septum into an anterior portion in the lophophore and a larger posterior portion. Pores in the walls between adjoining zooids permit exchange of materials by way of the coelomic fluid.

Most colonies are made up of feeding individuals, but polymorphism also occurs. One type of modified zooid resembles a bird beak that snaps at small invading organisms that might foul a colony. Another type has a long bristle that sweeps away foreign particles.

Most ectoprocts are hermaphroditic. Some species shed eggs into the seawater, but most brood their eggs, some within the coelom and some externally in a special ovicell, which is a modified zooecium in which the embryo develops. Marine species have radial cleavage but a highly modified trochophore larva with a vibratile plume of sensory cilia, an adhesive sac, and a piriform organ. After swimming about for a time the larva uses the vibratile plume to select a suitable site for settling, then attaches temporarily with a sticky acid mucopolysaccharide secreted from the piriform organ. Later the adhesive sac produces acid mucopolysaccharide and protein secretions that effect a permanent attachment.

Freshwater species reproduce both sexually and asexually. Asexual reproduction is by budding or by means of **statoblasts,** which are hard, resistant capsules containing a mass of germinative cells that are formed during the summer and fall (Fig. 19-3). When the colony dies in late autumn, the statoblasts are released, and in spring they can give rise to new polypides and eventually to new colonies.

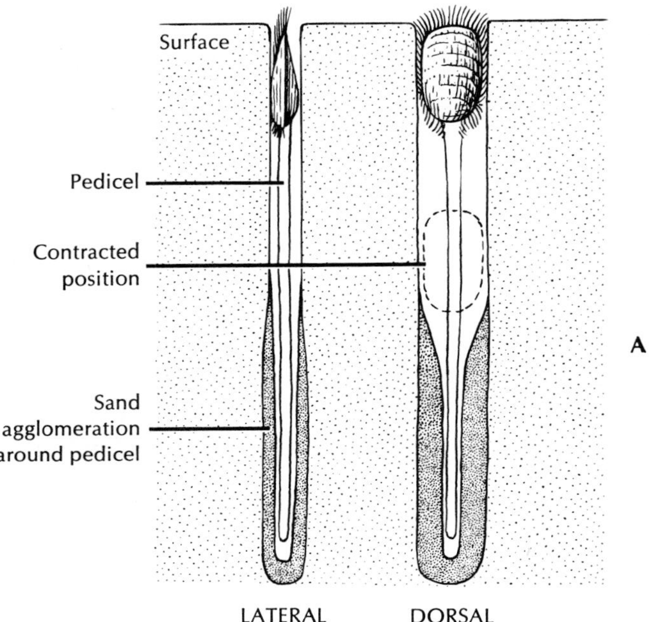

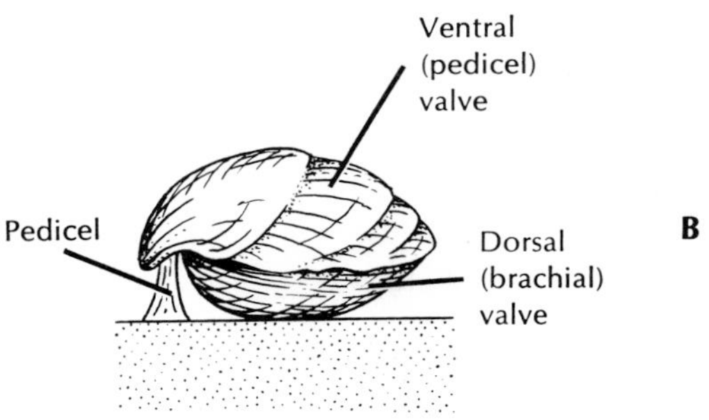

FIG. 19-6 Brachiopods. **A,** *Lingula* (two views) in feeding position within its mud burrow. The contractile pedicel can draw the body down to the position indicated by the dotted line. The valves are held together by muscles. Bundles of setae channel the water in and out over the lophophore. **B,** *Hemithyris*. The valves have a tooth and socket articulation and a short pedicel projects through the pedicel valve to attach to the substratum.

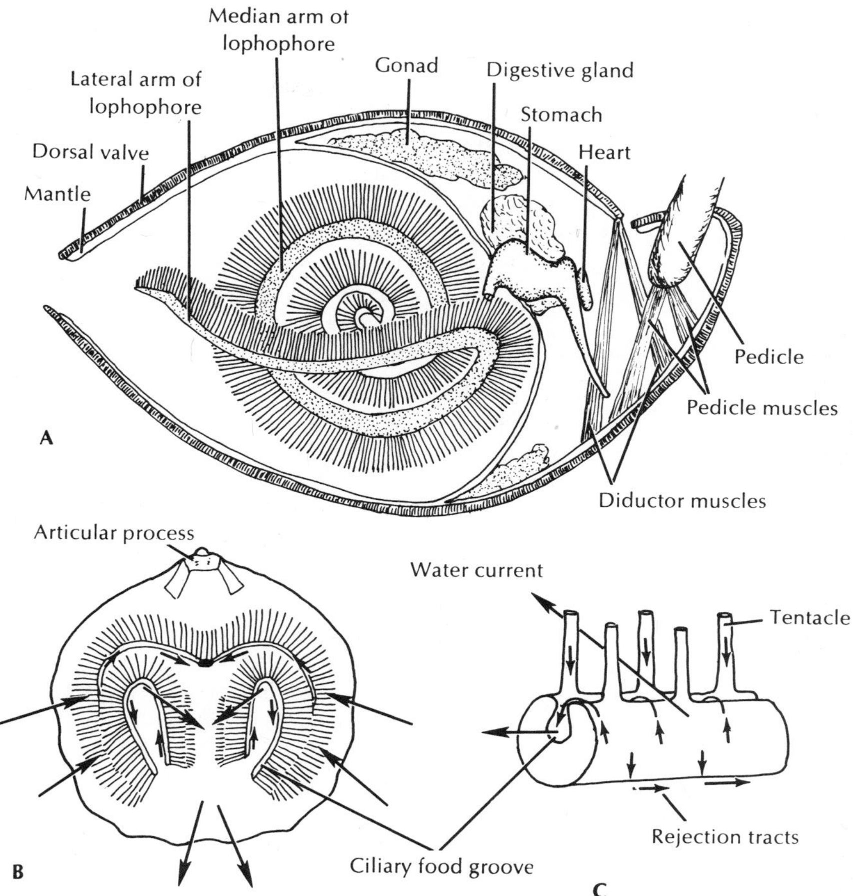

FIG. 19-7 Phylum Brachiopoda. **A,** An articulate brachiopod (longitudinal section). **B** and **C,** Feeding and respiratory currents. **B,** Large arrows show water flow over lophophore; small arrows indicate food movement toward mouth in ciliated food groove. **C,** Portion of lophophore arm showing current direction in feeding tracts and rejection tracts. (**B** and **C,** modified from Russell-Hunter, W. D. 1969. A biology of higher invertebrates. New York, The Macmillan Co.)

Brooding is often accompanied by degeneration of the lophophore and gut of the adults, the remains of which contract into minute dark balls, or **brown bodies.** Later, new internal organs may be regenerated in the old chambers. The brown bodies may remain passive or may be taken up and eliminated by the new digestive tract—an unusual kind of storage excretion.

PHYLUM BRACHIOPODA

The Brachiopoda (brak-i-op′o-da) (Gr., *brachiōn,* arm, + *pous, podos,* foot), or lamp shells, are an ancient group. Compared with the fewer than 300 species now living, some 30,000 fossil species, which once flourished in the Paleozoic and Mesozoic seas, have been described. Modern forms have changed little from the early ones. *Lingula* (Fig. 19-6, *A*) is probably the most ancient of these "living fossils," having existed virtually unchanged since Ordovician times. Most modern brachiopod shells range between 5 and 80 mm in length, but some fossil forms reached 30 cm.

Brachiopods are all attached, bottom-dwelling, marine forms that mostly prefer shallow water. Their name, which means "arm-footed," refers to the arms of the **lophophore.** Externally brachiopods resemble the bivalved molluscs in having two calcareous shell valves secreted by the mantle. They were, in fact, classed with the molluscs until the middle of the nineteenth century. Brachiopods, however, have **dorsal** and **ventral valves** instead of right and left lateral valves as do the bivalve molluscs and, unlike the bivalves, most of them are attached to a substrate either directly or by means of a fleshy stalk called a **pedicel** (or pedicle). Some, such as *Lingula,* live in vertical burrows in sand or mud. Muscles open and close the valves and provide movement for the stalk and tentacles.

In most brachiopods the ventral (pedicel) valve is slightly larger than the dorsal (brachial) valve, and one end projects in the form of a short pointed beak that is perforated where the fleshy stalk passes through (Fig. 19-6, *B*). In many the shape of the pedicel valve is that of the classic oil lamp of Greek and Roman times, so that the brachiopods came to be known as the "lamp shells."

There are two classes of brachiopods based on shell structure. The shell valves of Articulata are connected by a hinge with an interlocking tooth-and-socket arrangement, as in *Terebratella;* those of Inarticulata lack the hinge and are held together by muscles only, as in *Lingula* and *Glottidia*.

The body occupies only the posterior part of the space between the valves (Fig. 19-7), and extensions of the body wall form mantle lobes that line and secrete the shell. The large horseshoe-shaped lophophore in the anterior mantle cavity bears long ciliated tentacles used in respiration and feeding. Ciliary water currents carry food particles between the gaping valves and over the lophophore. Food is caught in mucus on the tentacles and carried in a ciliated food groove along the arm of the lophophore to the mouth. Unwanted particles are carried down rejection tracts to the mantle lobe and carried out in ciliary currents. Organic detritus and some algae are apparently primary food sources. The brachiopod lophophore not only can create food currents, as do other lophophorates, but also seems to be able to absorb dissolved nutrients directly from the environmental seawater.

The coelom, like that of other lophophorates, has an anterior mesocoel and a posterior metacoel. One or two pairs of nephridia open into the coelom and empty into the mantle cavity. Coelomocytes, which ingest particulate wastes, are carried out by the nephridia. There is an open circulatory system with a contractile heart. The lophophore and mantle are probably the chief sites of gaseous exchange. There is a nerve ring with a small dorsal and a larger ventral ganglion.

Sexes are separate and paired gonads discharge gametes through the nephridia. Most fertilization is external, but a few species brood their eggs and young.

The development of brachiopods is similar in some ways to that of the deuterostomes, with radial, mostly equal, holoblastic cleavage and the coelom forming enterocoelically in the articulates. The free-swimming larva resembles the trochophore. In the articulates, metamorphosis occurs after the larva has attached by a pedicel. In inarticulates the larva resembles a minute brachiopod with a coiled pedicel in the mantle cavity. There is no metamorphosis. As the larva settles, the pedicel attaches to the substratum, and adult existence is begun.

PHYLOGENY AND ADAPTIVE RADIATION

The possession of a lophophore by all three phyla is considered evidence of their close relationship, but each phylum has specialized along its own lines and developed its own life-style. As a group they seem to occupy a phylogenetic position somewhere between the protostomes and the deuterostomes. The coelom is divided into three regions as in deuterostomes, although the anterior portion, or protocoel, is repressed, since there is no head present. The lack of a head may be correlated with their ciliary method of feeding. In their embryology they display both protostome and deuterostome characteristics, with radial cleavage and, in the brachiopods, enterocoelic development of the coelom. All have a trochophore type of larva. Some authorities consider them the most primitive of the deuterostomes or at least ancestral to the other deuterostomes.

All lophophorates are **filter feeders** and most of their evolutionary diversification has been guided by this function. The tubes of phoronids vary according to their habitats. Various ectoprocts tend to build their protective exoskeletons of chitin or gelatin, which may or may not be impregnated with calcium and sand. Brachiopod variations occur largely in their shells and lophophores.

Annotated references

Selected general readings

American Society of Zoology. 1977. Biology of lophophorates. Am. Zool. **17**(1):3-150. *A collection of 13 papers.*

Barnes, R. D. 1974. Invertebrate zoology, ed. 3. Philadelphia, W. B. Saunders Co.

de Beauchamp, P., and J. Roger. 1960. Brachiopoda. In P. Grasse (ed.). Traite de zoologie, vol. 5, pp. 1380-1499.

Emig, C. C. 1977a. The lophophore: significant structures in Lophophorata (Brachiopoda, Bryozoa, Phoronida). Zool. Ser. **5**(3/4):133-138.

Emig, C. C. 1977b. Embryology of Phoronida. Am. Zool. **17**:21-37.

Gosner, K. L. 1971. Guide to identification of marine and estuarine invertebrates: Cape Hatteras to the Bay of Fundy. New York, Interscience Publishers, pp. 222-248.

Hickman, C. P. 1973. Biology of the invertebrates, ed. 2. St. Louis, The C. V. Mosby Co.

Hyman, L. H. 1959. The invertebrates: smaller coelomate groups, vol. 5. New York, McGraw-Hill Book Co. *Phoronida, Ectoprocta, Brachiopoda, and Sipunculida are covered in this volume.*

Larwood, G. P. 1973. Living and fossil bryozoa: recent advances in research. New York, Academic Press, Inc.

Light, S. F., and others. 1967. Intertidal invertebrates of the central California coast. Berkeley, University of California Press.

MacGinitie, G. E., and N. MacGinitie. 1967. Natural history of marine animals, ed. 2. New York, McGraw-Hill Book Co.

McCammon, H. M. 1969. The food of articulate brachiopods. J. Paleontol. **43**(4):976-985.

McCammon, H. M., and W. A. Reynolds. 1976. Experimental evidence for direct nutrient assimilation by the lophophore of articulate brachiopods. Mar. Biol. **34**(1): 41-51.

Meglitsch, P. A. 1972. Invertebrate zoology, ed. 2. New York, Oxford University Press.

Nielson, C. 1977. The relationships of Entoprocta, Ectoprocta and Phoronida. Am. Zool. 17:149-150.

Pennak, R. W. 1953. Fresh-water invertebrates of the United States. New York, The Ronald Press Co., pp. 256-277.

Rogick, M. D. 1959. Bryozoa. In W. T. Edmondson (ed.). Ward and Whipple's freshwater biology, ed. 2. New York, John Wiley & Sons, Inc. *Keys to both freshwater entoprocts and ectoprocts.*

Rudwick, M. J. S. 1970. Living and fossil brachiopods. London, Hutchinson University Library, Hutchinson & Co.

Ryland, J. S. 1970. Bryozoans. London, Hutchinson University Library, Hutchinson & Co.

Steele-Petrovic, H. M. 1976. Brachiopod food and feeding processes. Palaeontology **19**(3):417-436.

CHAPTER 20

THE ECHINODERMS

Phylum Echinodermata

Sea urchins, the "pin cushions" of the sea, are bottom dwellers that attach themselves by hydraulically operated tube feet and feed by scraping the rocks with a peculiar five-toothed mechanism called "Aristotle's lantern." This is a group of *Echinus acubus*, photographed at a depth of 25 m off the west coast of Norway.

Courtesy T. Lundälv, Kristinebergs Zoological Station, Sweden.

Position in animal kingdom

Phylum Echinodermata (e-ki'no-der'ma-ta) (Gr., *echinos,* sea urchin, hedgehog, + *derma,* skin, + *-ata,* characterized by) belongs to the **Deuterostomia** branch of the animal kingdom and the members of it are enterocoelous coelomates. The other phyla of this group are Chaetognatha, Hemichordata, and Chordata. Primitively, deuterostomes have the following embryologic features in common: anus developing from or near the blastopore, and mouth developing elsewhere; coelom budded off from the archenteron (enterocoel); radial and indeterminate cleavage; and endomesoderm (mesoderm derived from or with the endoderm) from enterocoelic pouches. Though only distantly related, the Echinodermata is the only major invertebrate group showing affinities with the vertebrates. The typical deuterostome embryogenesis is shown only by some of the lower vertebrates, but the similarity probably has real evolutionary meaning. Thus, the echinoderms, chordates, and the lesser deuterostome phyla presumably have been derived from a common ancestor. Nevertheless, their evolutionary history has taken the echinoderms to the point where they are very much unlike any other animal group.

Biologic contributions

There is one word that best describes the echinoderms: strange. They are a major group, but they occupy the end of a side branch of the phylogenetic tree, so none of their characteristics can be said to presage those of any more advanced group. They have a unique constellation of characteristics that are found in no other phylum. Though an enormous amount of research has been devoted to the echinoderms, we are still far from a satisfactory understanding of many aspects of their biology. As Libbie Hyman (1955) said, echinoderms are a "noble group especially designed to puzzle the zoologist."

Among the more striking of the features shown by the echinoderms are the system of coelomic channels comprising the **water-vascular system,** the calcareous **dermal endoskeleton,** the **hemal system,** and the **metamorphosis** from bilateral larva to radial adult.

THE ECHINODERMS

The echinoderms are marine forms and include the sea stars, brittle stars, sea urchins, sea cucumbers, and sea lilies. They represent a bizarre group sharply distinguished from all other members of the animal kingdom. Their name is derived from their external spines or protuberances. A calcareous endoskeleton is found in all members of the phylum, either in the form of plates or represented by scattered tiny ossicles.

The most noticeable characteristics of the echinoderms are (1) the spiny endoskeleton of plates, (2) the water-vascular system, (3) the pedicellariae, (4) the dermal branchiae, and (5) radial or biradial symmetry. Radial symmetry is not limited to echinoderms, but no other group with such complex organ systems has radial symmetry.

Echinoderms are an ancient group of animals extending back to the Cambrian period. Despite the excellent fossil record, the origin and early evolution of the echinoderms are still obscure. It seems clear that they descended from bilateral ancestors because their larvae are bilateral but become radially symmetric later in their development. Many zoologists believe that early echinoderms were sessile and evolved radiality as an adaptation to the sessile existence. Bilaterality is of adaptive value to animals that travel through their environment, while radiality is of value to animals whose environment meets them on all sides equally. Hence, the body plan of present day echinoderms seems to have been derived from one that was attached to the bottom by a stalk, had radial symmetry and radiating grooves (ambulacra) for food-gathering, and whose oral side faced upward. Attached forms were once plentiful, but only about 80 species, all in the class Crinoidea, still survive. Oddly, conditions have favored the survival of their free-moving descendants, though they are still quite radial, and among them are some of the most abundant marine animals. Nevertheless, in the exception that proves the rule, at least three groups of echinoderms (two groups of echinoids and the holothuroids) seem to be evolving back toward bilaterality.

Characteristics

1. Body unsegmented (nonmetameric) with **radial, pentamerous symmetry;** body rounded, cylindric, or star-shaped, with five or more radiating areas, or ambulacra, alternating with interambulacral areas
2. **No head or brain;** few specialized sensory organs; sensory system of tactile and chemoreceptors, podia, terminal tentacles, photoreceptors, and statocysts

3. Nervous system with circumoral ring and radial nerves; usually two or three systems of networks located at different levels in the body, varying in degree of development according to group
4. **Endoskeleton** of **dermal calcareous ossicles** with **spines** or of calcareous **spicules** in dermis; covered by an epidermis (ciliated in most); **pedicellariae** (in some)
5. A unique **water-vascular system** of coelomic origin that extends from the body surface as a series of tentacle-like projections (**podia,** or **tube feet**) that are protracted by increase of fluid pressure within them; an opening to the exterior **(madreporite** or **hydropore)** usually present
6. Locomotion by tube feet, which project from the ambulacral areas, or by movement of spines, or by movement of arms, which project from central disc of body
7. Digestive system usually complete; axial or coiled; anus absent in ophiuroids
8. Coelom extensive, forming the perivisceral cavity and the cavity of the water-vascular system; coelom of enterocoelous type; coelomic fluid with amebocytes.
9. Blood-vascular system **(hemal system)** much reduced, playing little, if any, role in circulation, and surrounded by extensions of coelom **(perihemal sinuses);** main circulation of body fluids (coelomic fluids) by peritoneal cilia
10. Respiration by **dermal branchiae,** by **tube feet,** by **respiratory tree** (holothuroids), and by **bursae** (ophiuroids)
11. **Excretory organs absent**
12. Sexes separate (except a few hermaphroditic) with large gonads, single in holothuroids but multiple in most; simple ducts, with no elaborate copulatory apparatus or secondary sexual structures; fertilization usually external; eggs brooded in some
13. Development through **free-swimming, bilateral, larval stages** (some with direct development); metamorphosis to radial adult or subadult form
14. Autotomy and regeneration of lost parts conspicuous

Classification

There are about 6,000 living and 20,000 extinct or fossil species of Echinodermata. The traditional classification of the echinoderms placed all the free-moving forms oriented with oral side down in the subphylum Eleutherozoa, containing most of the living species. The other subphylum, Pelmatozoa, contained mostly forms with stems and oral side up; most of the extinct classes and the living Crinoidea belong to this group. There has been growing recognition that these subphyla were each polyphyletic, and most students of echinoderms have adopted the following classification (proposed by Fell and used in Moore, 1966, 1967); characteristics for strictly fossil classes are omitted here:

Subphylum Homalozoa (ho-mal′o-zo′a) (Gr., *homalos,* level, even, + *zoion,* animal). Ancient fossil echinoderms called carpoids, which were not radially symmetric; fossil classes Homostelea, Homoiostelea, Stylophora, and Ctenocystoidea.

Subphylum Crinozoa (krin′o-zo′a) (Gr., *krinon,* lily, + *zoion,* animal). Radially symmetric, with rounded or cup-shaped theca and brachioles or arms. Attached during part or all of life by stem. Oral surface directed upward. Fossil classes: Eocrinoidea, Paracrinoidea, Cystoidea, and Blastoidea.

Class Crinoidea (krin-oi′de-a) (Gr., *krinon,* lily, + *eidos,* form, + *-ea,* characterized by)—**sea lilies and feather stars.** Aboral attachment stalk of dermal ossicles; mouth and anus on oral surface; five arms branching at base and bearing pinnules; ciliated ambulacral grooves on oral surface with tentacle-like tube feet for food-gathering; spines, madreporite, and pedicellariae absent. Examples: *Antedon, Florometra* (Fig. 20-16).

Subphylum Asterozoa (as′ter-o-zo′a) (Gr., *aster,* star, + *zoion,* animal). Radially symmetric, star-shaped echinoderms, unattached as adults.

Class Stelleroidea (stel′ler-oi′de-a) (L., *stella,* star, + Gr., *eidos,* form). With characteristics of subphylum, body of a central disc with radially arranged rays or arms.

Subclass Somasteroidea (som′ast-er-oi′de-a) (Gr., *soma,* body, + *aster,* star, + *eidos,* form). Mostly extinct sea stars with primitive skeletal structure; single living species, *Platasterias latiradiata,* from deep water off western coast of Mexico.

Subclass Asteroidea (as′ter-oi′de-a) (Gr., *aster,* star, + *eidos,* form, + *-ea,* characterized by)—**sea stars.** Star-shaped echinoderms, with the arms not sharply marked off from the central disc; ambulacral grooves open, with tube feet on oral side; tube feet often with suckers; anus and madreporite aboral; pedicellariae present. Example: *Asterias* (Fig. 20-1, *E*).

Subclass Ophiuroidea (o′fe-u-roi′de-a) (Gr., *ophis,* snake, + *oura,* tail, + *eidos,* form)—**brittle stars and basket stars.** Star-shaped, with the arms sharply marked off from the central disc; ambulacral grooves closed, covered by ossicles; tube feet without suckers and not used for locomotion; pedicellariae absent. Examples: *Ophiura, Gorgonocephalus* (Fig. 20-6).

Subphylum Echinozoa (ek′in-o-zo′a) (Gr., *echinos,* sea urchin, hedgehog, + *zoion,* animal). Mostly unattached, globoid or discoid echinoderms without arms. Fossil classes: Helicoplacoidea, Edrioasteroidea, Ophiocistioidea.

Class Echinoidea (ek′i-noi′de-a) (Gr., *echinos,* sea urchin, hedgehog, + *eidos,* form)—**sea urchins, sea biscuits, and sand dollars.** More or less globular or disc-shaped echinoderms with no arms; compact skeleton or test with closely fitting plates; movable spines; ambulacral grooves covered by ossicles, closed; tube feet with suckers; pedicellariae present. Examples: *Arbacia, Strongylocentrotus, Lytechinus* (Fig. 20-9), *Mellita.*

Class Holothuroidea (hol′o-thu-roi′de-a) (Gr., *holothourion,* sea cucumber + *eidos,* form)—**sea cucumbers.** Cucumber-shaped echinoderms with no arms; spines absent; microscopic ossicles embedded in thick muscular wall; anus present; ambulacral grooves closed; tube feet with suckers; circumoral tentacles (modified tube feet); pedicellariae absent; madreporite plate internal. Examples: *Thyone, Stichopus, Cucumaria* (Fig. 20-14).

Ecologic relationships

Echinoderms are all marine; they have no ability to osmoregulate and are thus rarely found in waters that are brackish. They are found in all oceans of the world and at all depths, from the intertidal to the abyssal regions. Often the most common animals in the deep ocean are echinoderms. The most abundant species found in the Philippine Trench (10,540 m) was a holothurian. They are virtually all bottom dwellers, though there are a few pelagic species.

No parasitic echinoderms are known, but a few are commensals. On the other hand, a wide variety of other animals make their homes in or on echinoderms, including parasitic or commensal algae, protozoa, ctenophores, turbellarians, cirripedes, copepods, decapods, snails, clams, polychaetes, fish; and other echinoderms.

The asteroids, or sea stars (Fig. 20-1), are commonly found in various types of bottom habitats, often on hard, rocky surfaces, but numerous species are at home on sandy or soft bottoms. Some species are particle feeders, but many are predators, feeding particularly on sedentary or sessile prey, since the sea stars themselves are relatively slow moving.

Ophiuroids — brittle stars, or serpent stars (Fig. 20-6) — are by far the most active echinoderms, moving by their arms rather than by tube feet. A few species are reported to have swimming ability, and some burrow. They may be scavengers, browsers, or deposit or filter feeders. Some are commensal in large sponges, in whose water canals they may live in great numbers.

Holothurians, or sea cucumbers (Fig. 20-14), are widely prevalent in all seas. Many are found on sandy or mucky bottoms where they lie concealed. Compared to other echinoderms, holothurians are greatly extended in the oral-aboral axis and are oriented with that axis more or less parallel to the substrate and lying on one side. Most are suspension or deposit feeders.

Echinoids, or sea urchins (Figs. 20-9, 20-10, and p. 440), are adapted for living on the ocean bottom and always keep their oral surface in contact with a substratum. The ''regular'' sea urchins prefer hard bottoms, but the sand dollars and heart urchins (''irregular'' urchins) are usually found on sand. The regular urchins, which are radially symmetric, feed chiefly on algae or detritus, while the irregulars, which are secondarily bilateral, feed on small particles.

Crinoids (Fig. 20-16) stretch their arms out and up like a flower's petals and feed on plankton and suspended particles. Most living species become detached from their stems as adults, but they nevertheless spend most of their time on the substrate, holding on by means of aboral appendages called cirri.

Importance to humans

The zoologist who admires the fascinating structure and function of echinoderms can share with the layperson an admiration of the beauty of their symmetry, often enhanced by bright colors. Many species are rather drab, but others may be orange, red, purple, blue, and often bicolor.

Because of the spiny nature of their structure, echinoderms are not often the prey of other animals — except other echinoderms (sea stars). In scattered parts of the world, however, sea urchin gonads are relished food items of humans, either raw or roasted on the half shell. Trepang, the cured body wall of certain large holothurians, is considered a delicacy, particularly in some Oriental countries. It is highly nutritious, containing over 50% of highly digestible protein, and it is said to impart a delicate flavor to soups.

Sea stars feed on a variety of molluscs, crustaceans, and other invertebrates, but their chief economic impact is on clams and oysters. A single star may eat as many as a dozen oysters or clams in a day. To rid shellfish beds of these pests, large rope ''mops'' in which the sea stars become entangled are sometimes dragged over oyster beds, and the collected sea stars are destroyed in hot water vats. A more effective method is to distribute lime over areas where they abound. Lime damages the delicate epidermal membrane, destroying the dermal branchiae and ultimately the animal itself. Meanwhile, the oysters remain with their shells tightly closed until the lime dissipates.

Echinoderms have been widely used in experimental embryology, for their gametes are usually abundant and easy to collect and handle in the laboratory. The investigator can follow the embryonic developmental stages with great accuracy. Artificial parthenogenesis was first discovered in sea urchin eggs, when it was found that, by treating the eggs with hypertonic sea water or subjecting them to a variety of other stimuli, development would take place without the presence of sperm.

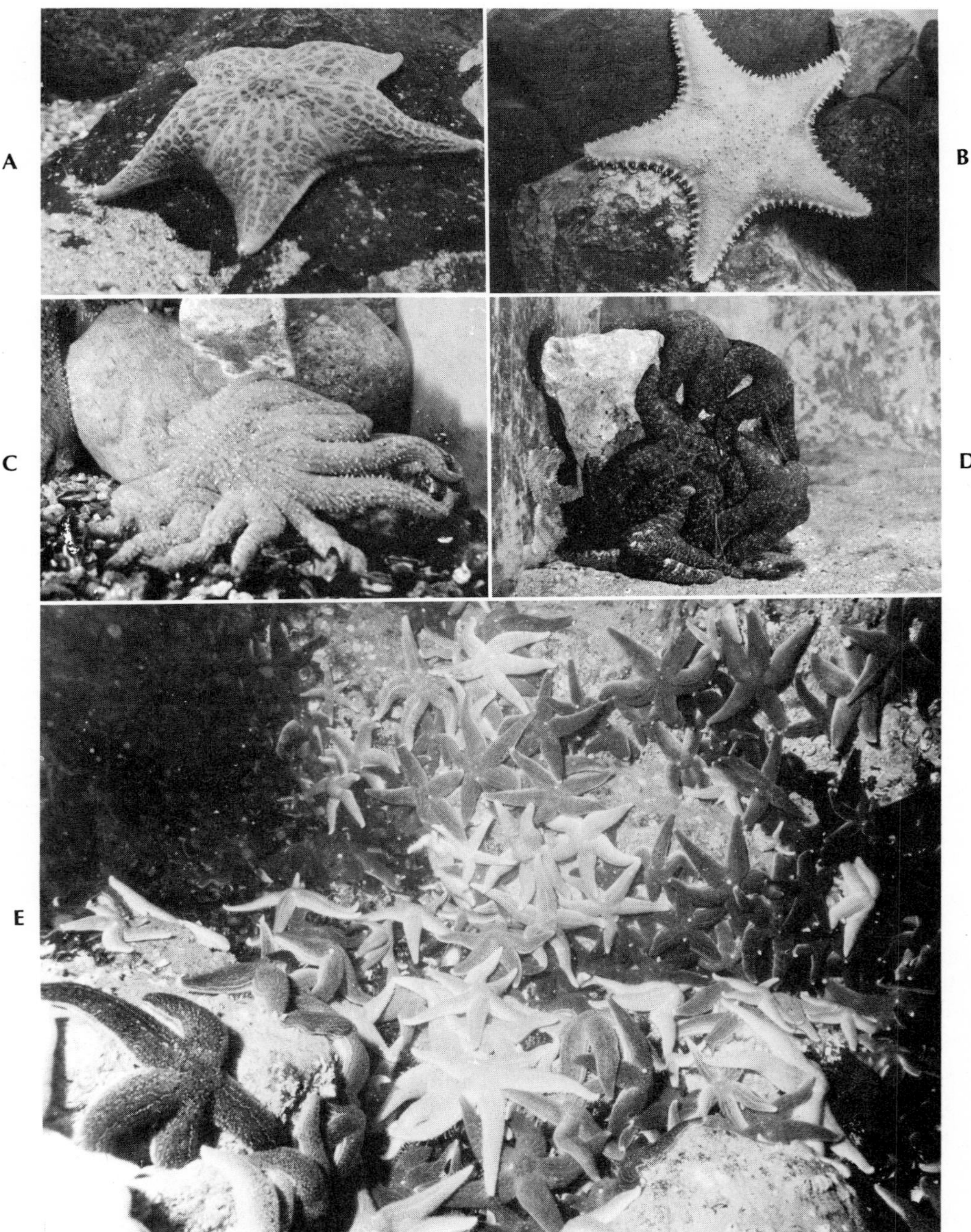

FIG. 20-1 Some sea stars (subclass Asteroidea) **A** to **D** are found on the West coast of the United States. **A,** *Dermasterias;* **B,** *Hippasteria;* **C,** *Pycnopodia;* **D,** *Pisaster;* **E,** group of *Asterias rubens* that has aggregated on a mussel bed to feed on mussels. Some of the mussels that were too close to the surface were washed down by heavy wave action to a rock shelf below, and the stars are feasting on them. (**A, C, D,** Photographs by C. P. Hickman, Jr.; **B,** courtesy M. Newman, Vancouver Public Aquarium, British Columbia; **E,** courtesy T. Lundälv.)

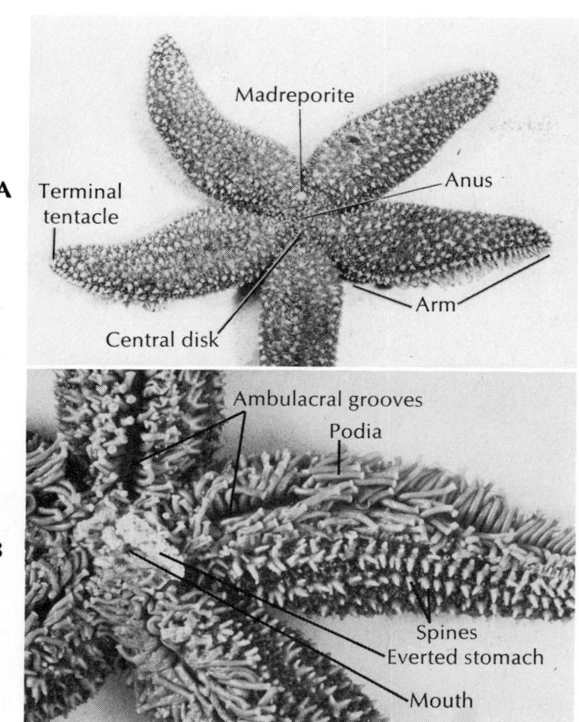

FIG. 20-2 *Asterias.* **A,** Aboral view; **B,** oral view. (Photographs by F. M. Hickman.)

Class Stelleroidea

Sea stars, subclass Asteroidea

Sea stars are familiar along the shore lines where sometimes large numbers of them may aggregate on the rocks (Fig. 20-1, *E*). Sometimes they cling so tenaciously that they are difficult to dislodge without tearing off some of their tube feet. They also live in muddy or sandy bottoms and among coral reefs. They are often brightly colored and range in size from a centimeter in greatest diameter to about a meter across. *Asterias* (Fig. 20-2) is one of the common genera of the east coast of the United States and is used in many zoology laboratories. *Pisaster* (Fig. 20-1, *D*) is common on the California coast, as is *Dermasterias,* the leather star (Fig. 20-1, *A*).

External features. Sea stars are composed of a central disc that merges gradually with the tapering arms (rays). They tend to be pentamerous and typically have five arms, but there may be more (Fig. 20-1, *C*). The body is flattened and flexible, and covered with a ciliated pigmented epidermis. The mouth is centered on the under, or oral, side, surrounded by a soft peristomial membrane. An **ambulacral groove** along the oral side of each arm is bordered by movable **spines** that protect the rows of **tube feet** (podia) projecting from the groove. Viewed from the oral side, the large **radial nerve** can be seen in the center of each ambulacral groove (Fig. 20-4, *C*), between the rows of tube feet. The nerve is very superficially located, covered only by thin epidermis. Under the nerve is an extension of the coelom and the radial canal of the water-vascular system, all of which are external to the underlying ossicles (Fig. 20-4, *C*). In all other classes of living echinoderms except the crinoids, these structures are covered over by ossicles or other dermal tissue; thus, the ambulacral grooves in asteroids and crinoids are said to be **open,** and those of the other groups are **closed.**

The aboral surface is usually rough and spiny, though the spines of many species are flattened, so that the surface appears smooth. Around the bases of the spine are groups of minute pincerlike **pedicellariae,** bearing tiny jaws manipulated by muscles (Fig. 20-3). These help keep the body surface free of debris, protect the papulae, and sometimes aid in food capture. The **papulae (dermal branchiae** or **skin gills)** are soft delicate projections of the coelomic cavity, covered only with epidermis and lined internally with peritoneum; they extend out through spaces between the ossicles and are concerned with respiration (Fig. 20-4, *C*). Also on the aboral side are the inconspicuous **anus** and the circular **madreporite** (Fig. 20-2, *A*), a calcareous sieve leading to the water-vascular system.

Endoskeleton. Beneath the epidermis of the sea star is a mesodermal endoskeleton of small calcareous plates, or **ossicles,** bound together with connective tissue. From these ossicles project the spines and tubercles that are responsible for the spiny surface. Muscles in the body wall move the rays and can partially close the ambulacral grooves by drawing their margins together.

Coelom, excretion, and respiration. The coelomic compartments of the larval echinoderm give rise to several structures in the adult, one of which is a spacious body **coelom** filled with fluid. The fluid contains amebocytes **(coelomocytes),** bathes the internal organs, and projects into the papulae. The coelomic fluid is circulated around the body cavity and into the papulae by the ciliated peritoneal lining. Exchange of respiratory gases and excretion of nitrogenous

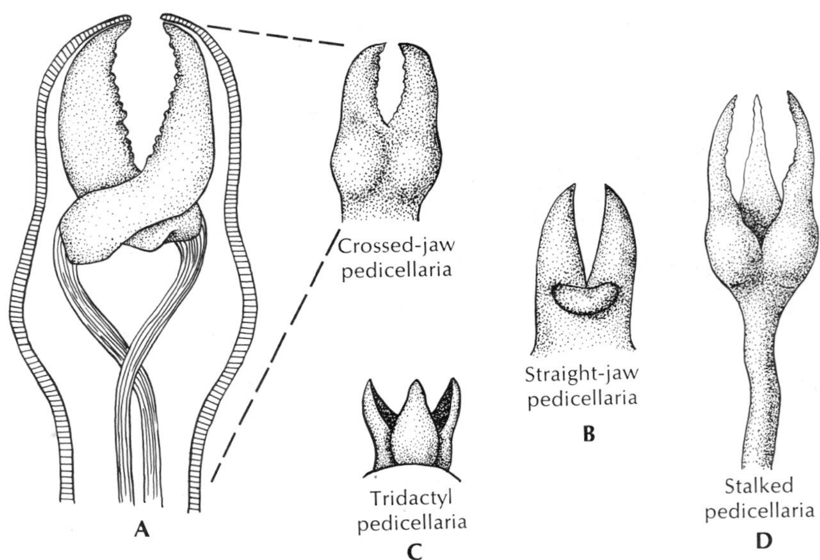

FIG. 20-3 Pedicellariae are tiny devices for cleaning the epidermis of echinoderms. The calcareous "jaws" are manipulated by muscles. **A** shows the abductor muscles (in optical section) used to open the jaws. The two-jawed varieties, **A** and **B,** are common in asteroids; the three-jawed varieties, **C** and **D,** are found in echinoids.

waste, principally ammonia, take place by diffusion through the thin walls of the papulae and tube feet. Some wastes may be picked up by the coelomocytes, which migrate through the epithelium of the papulae or tube feet to the exterior, or the tips of papulae containing the waste-laden coelomocytes may be pinched off.

Water-vascular system. The water-vascular system is another coelomic compartment and is unique to the echinoderms. Showing exploitation of hydraulic mechanisms to a greater degree than in any other animal group, it is a system of canals and specialized tube feet that, together with the dermal ossicles, has determined the evolutionary potential and limitations of this phylum. In sea stars the primary functions of the water-vascular system are locomotion and food-gathering, as well as those of respiration and excretion.

Structurally, the water-vascular system opens to the outside through small pores in the **madreporite.** The function of the madreporite seems to be to allow rapid adjustment of hydrostatic pressure within the water-vascular system in response to changes in external hydrostatic pressure resulting from depth changes, as in tidal fluctuations. The madreporite of asteroids is on the aboral surface (Fig. 20-2, *A*), and leads into the **stone canal,** which descends toward the **ring canal** around the mouth (Fig. 20-4, *B*). **Radial canals** diverge from the ring canal, one into the ambulacral groove of each ray. Also attached to the ring canal are four or five pairs of folded, pouchlike **Tiedemann's bodies** and from one to five **polian vesicles** (polian vesicles are absent in some sea stars; for example, *Asterias*). The Tiedemann's bodies are thought to produce coelomocytes, and the polian vesicles are apparently for fluid storage.

A series of small **lateral canals,** each with a one-way valve, connects the radial canal to the cylindric **podia,** or **tube feet,** along the sides of the ambulacral groove in each ray. Each podium is a hollow, muscular tube, the inner end of which is a muscular sac, the **ampulla,** that lies within the body coelom (Fig. 20-4, *A* and *C*), and the outer end of which usually bears a **sucker.** Some species lack the suckers. The podia pass to the outside between the ossicles in the ambulacral groove.

The water-vascular system operates hydraulically and is an effective locomotor mechanism. The valves in the lateral canals prevent backflow of fluid into the radial canals. The tube foot has in its walls connective tissue that maintains the cylinder at a relatively con-

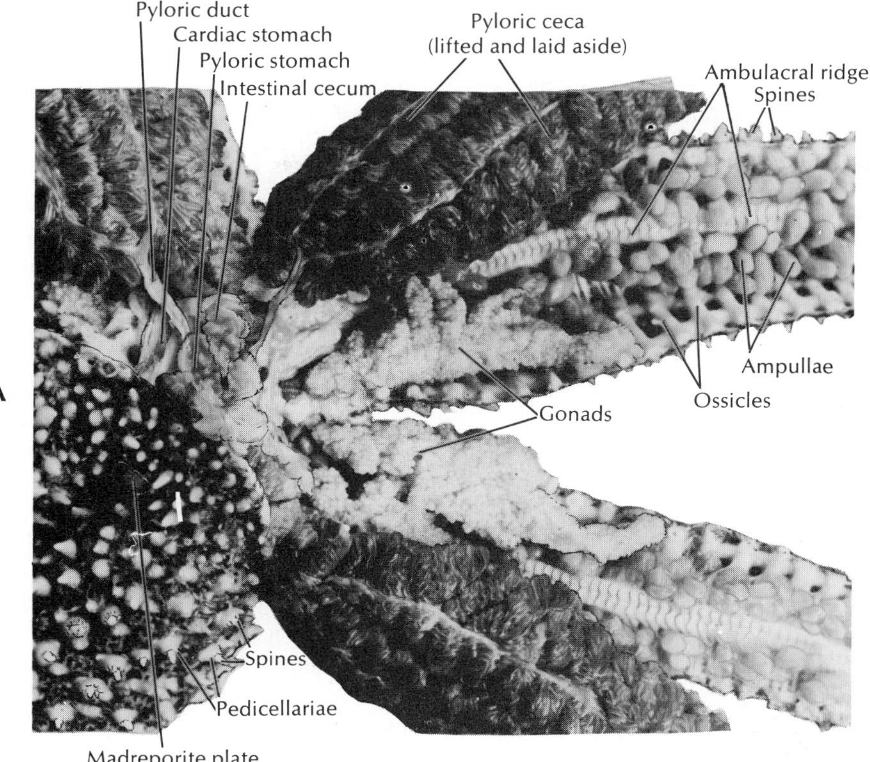

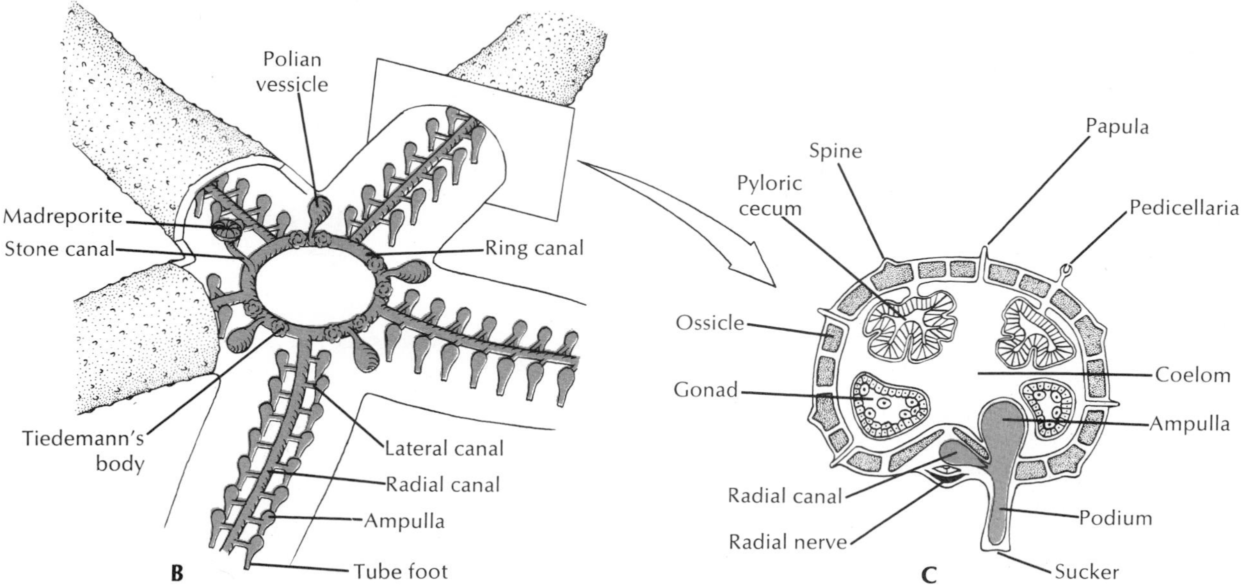

FIG. 20-4 A, Sea-star dissection, aboral view; at right, portions of two arms with digestive glands lifted aside to expose gonads and ampullae. At lower left, two undissected arms show madreporite and spines. **B,** Water-vascular system. Podia penetrate between ossicles. **C,** Cross section of arm at level of gonads, illustrating open ambulacral grooves. (**A,** Photograph by F. M. Hickman.)

stant diameter. On contraction of muscles in the ampulla, fluid is forced into the podium, extending it. Conversely, contraction of the longitudinal muscles in the tube foot retracts the podium, forcing fluid back into the ampulla. Contraction of muscles in one side of the podium bends the organ toward that side. Small muscles at the end of the tube foot can raise the middle of the disclike end, thus creating a suction-cup effect when the end is applied to the substrate. It has been estimated that by combining mucous adhesion with suction, a single podium can exert a pull equal to 25 to 30 g. Coordinated action of all or many of the tube feet is sufficient to draw the animal up a vertical surface or over rocks.

On a soft surface, such as muck or sand, the suckers are ineffective (and numerous sand-dwelling species have no suckers), so the tube feet are employed as legs. Locomotion now becomes mainly a stepping process. Most sea stars can move only a few centimeters per minute, but some very active ones can move 75 to 100 cm per minute; for example, *Pycnopodia* (Fig. 20-1, *C*). When inverted, the sea star bends its rays until some of the tubes reach the substratum and attach as an anchor; then the sea star slowly rolls over.

Tube feet are innervated by the central nervous system rather than through the diffuse subepidermal plexus. Nervous coordination enables the tube feet to move in a single direction, though not in unison, so that the sea star may progress. If the radial nerve in an arm is cut, the podia in that arm lose coordination, though the podia can still function. If the circumoral nerve ring is cut, the podia in all arms become uncoordinated, and movement ceases.

Feeding and digestive system. The **mouth** on the oral side leads through a short **esophagus** to a large **stomach** in the central disc. The lower **(cardiac)** part of the stomach can be everted through the mouth during feeding (Fig. 20-2, *B*), and excessive eversion is prevented by gastric ligaments. The upper **(pyloric)** part is smaller and connects by ducts to a pair of large **pyloric ceca (digestive glands)** in each arm (Fig. 20-4, *A*). Digestion is mostly extracellular, although some intracellular digestion may occur in the ceca. A short **intestine** leads aborally from the pyloric stomach, and there are usually a few small, saclike **intestinal ceca** (Fig. 20-4, *A*). The **anus** is inconspicuous, and some sea stars lack an intestine and anus.

Many sea stars are carnivorous and feed on molluscs, crustaceans, polychaetes, echinoderms, other invertebrates, and sometimes small fish. Sea stars will often consume a rather wide range of food items, but many show particular preferences. Some like brittle stars, sea urchins, or sand dollars, swallowing them whole and later regurgitating undigestible ossicles and spines. Some attack other sea stars, and if they are small compared to their prey, they may attack and begin eating at the end of one arm.

Asteroids feed heavily on molluscs, and *Asterias* is a significant predator on commercially important clams and oysters. When feeding on a bivalve, a sea star will hump over its prey, attaching its podia to the valves, and then exert a steady pull, using its feet in relays. A force of some 1,300 g can thus be exerted. In half an hour or so the adductor muscles of the bivalve fatigue and relax. With a very small gap available, the star inserts its soft everted stomach into the space between the valves and wraps it around the soft parts of the shellfish. After feeding, the sea star draws in its stomach by contraction of the stomach muscles and relaxation of body wall muscles.

Some sea stars feed on small particles, either entirely or in addition to carnivorous feeding. Plankton or other organic particles coming in contact with the animal's surface are carried by the epidermal cilia to the ambulacral grooves and then to the mouth.

Hemal system. The so-called hemal system is not very well developed in asteroids, and its function in all echinoderms is unclear. The hemal system has little or nothing to do with circulation of body fluids. It is a system of tissue strands enclosing unlined sinuses and which is enclosed in another coelomic compartment, the **perihemal channels.** The main channel is the **axial sinus,** which extends up through a spongy mass of tissue (the **axial gland**) lying along the stone canal. At its oral end it connects to a **periesophageal hemal ring,** which in turn gives rise to **radial hemal lacunae** that run out into each arm between the radial nerve and the radial water canal. The axial gland also connects to a **gastric hemal ring** with branches to each pyloric cecum and to an **aboral hemal ring** with branches to the gonads. Coelomocytes are present in the hemal fluid, and they may be produced in the hemal system. The hemal system may be useful in distributing digested products, but its specific functions are not really known.

Nervous system. The nervous system consists of three units placed at different levels in the disc and arm. The chief of these systems is the **oral (ectoneu-**

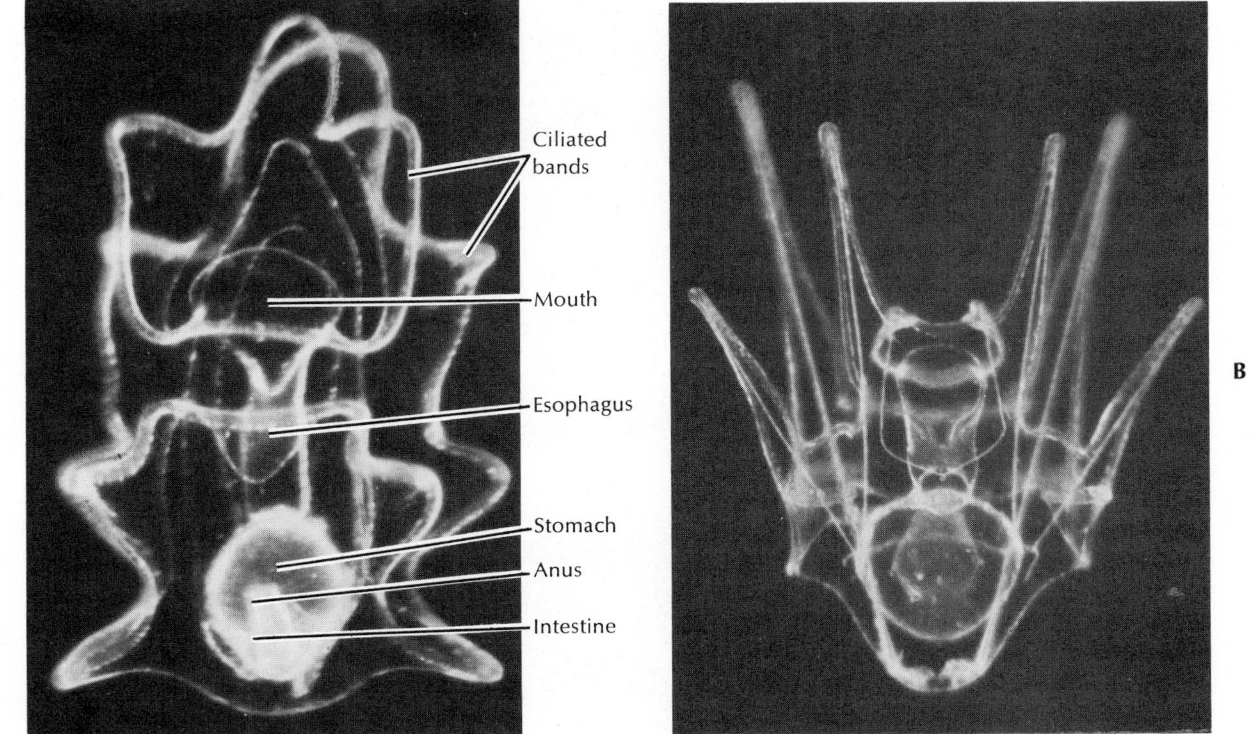

FIG. 20-5 Bilaterally symmetric larvae of asteroids and echinoids. **A,** Bipinnaria of *Asterias rubens*. **B,** Echinopluteus of *Psammechinus miliaris;* as in the ophiopluteus, the ciliated bands extend onto the long, graceful arms. (Photograph by D. P. Wilson.)

ral) system composed of a **nerve ring** around the mouth and a main **radial nerve** into each arm. It appears to coordinate the tube feet. A **deep (hyponeural)** system lies aboral to the oral system, and an **aboral** system consists of a ring around the anus and radial nerves along the roof of the rays. An **epidermal nerve plexus** or nerve net freely connects these systems with the body wall and related structures. The epidermal plexus has been shown to coordinate the responses of the dermal branchiae to tactile stimulation—the only instance known in echinoderms in which coordination occurs through a nerve net.

The sense organs are not well developed. There are tactile organs and other sensory cells scattered over the surface and an ocellus at the tip of each arm. Their reactions are mainly to touch, temperature, chemicals, and differences in light intensity. Sea stars are usually more active at night.

Reproductive system and regeneration and autotomy. Most sea stars have separate sexes. A pair of gonads lie in each interradial space (Fig. 20-4, *A*). Fertilization is external and occurs in early summer when the eggs and sperm are shed into the water. It has been shown that the maturation and shedding of sea star eggs are stimulated by a secretion from neurosecretory cells located on the radial nerves (Kanatani and others).

Echinoderms can regenerate lost parts. Sea star arms can regenerate readily, even if all are lost. Stars also have the power of autotomy and can cast off an injured arm near the base. It may take months to regenerate a new arm.

If an arm is broken off or removed, and it contains a part of the central disc (about one-fifth), the arm can regenerate a complete new sea star! In former times fishermen used to dispatch sea stars they collected from their oyster beds by chopping them in half with a hatchet—a worse-than-futile activity. Some sea stars reproduce asexually under normal conditions by cleaving the central disc, each part regenerating the rest of the disc and missing arms.

Development. In some species the liberated eggs are brooded, either under the oral side of the animal or

FIG. 20-6 **A,** Brittle star *Ophiura albida*. **B,** Basket star *Gorgonocephalus caryi*. Ophiuroids have sharply marked off arms and are very agile. They are bottom dwellers. (**A,** Photograph by B. Tallmark, Uppsala University, Sweden; **B,** courtesy M. Newman, Vancouver Public Aquarium, British Columbia.)

in specialized aboral structures, and development is direct; but in most species the embryonating eggs are free in the water and hatch to free-swimming larvae.

Early embryogenesis shows the typical primitive deuterostome pattern. Gastrulation is by invagination, and the anterior end of the archenteron pinches off to become the coelomic cavity, which expands in a U shape to fill the blastocoel. Each of the legs of the U, at the posterior, constricts to become a separate vesicle, and these eventually give rise to the main coelomic compartments of the body **(somatocoels).** The anterior portion of the U gives rise to the water-vascular system and the perihemal channels. The free-swimming larva has cilia arranged in bands and is called a **bipinnaria** (Fig. 20-5, *A*). The ciliated tracts become extended into larval arms. Soon the larva grows three adhesive arms and a sucker at its anterior end and is now called a **brachiolaria.** At this time it attaches to the substratum, forms a temporary attachment stalk, and undergoes metamorphosis.

Metamorphosis involves a dramatic reorganization of a bilateral larva into a radial juvenile. The anteroposterior axis of the larva is lost, and *what was the left side becomes the oral surface, and the larval right side becomes the aboral surface*. Correspondingly, the larval mouth and anus disappear, and a new mouth and anus form on what was originally the left and right sides, respectively. The portion of the anterior coelomic compartment from the left side expands to form the ring canal of the water-vascular system around the mouth, and then it grows branches to form the radial canals. As the short, stubby arms and the first podia appear, the animal detaches from its stalk and begins life as a young sea star.

Crown-of-thorns star. Since 1963 there have been numerous reports of increasing numbers of the crown-of-thorns sea star *(Acanthaster planci)* that were damaging large areas of coral reef in the Pacific Ocean. The crown-of-thorns star feeds on coral polyps, and it sometimes occurs in large aggregations, or "herds." Some of the suggested reasons for its increase were human destruction of the giant triton, a gastropod that feeds on sea stars, or that dredging, blasting, and other activities had destroyed the creatures that feed on the sea star larvae. Various attempts have been made to correct the problem. Indications at present are that the crown-of-thorns epidemic is tapering off and that recovery of the damaged reefs may be more rapid than was formerly believed.

Brittle stars, subclass Ophiuroidea

The brittle stars are the largest of the major groups of echinoderms in numbers of species, and they are probably the most abundant also. They abound in all types of benthic marine habitats, even carpeting the abyssal sea bottom in many areas.

Apart from the typical possession of five arms, the brittle stars are surprisingly different from the asteroids. The arms of brittle stars are slender and sharply set off from the central disc (Fig. 20-6). They have no pedicellariae or papulae, and their ambulacral grooves are closed, covered with arm ossicles. The tube feet are without suckers; they aid in feeding but are of limited use in locomotion. In contrast to that in the asteroids, the madreporite of the ophiuroids is located on the oral surface, on one of the oral shield ossicles (Fig. 20-7). Ampullae on the podia are absent, and force for protrusion of the podium is generated by a proximal muscular portion of the podium.

Each of the jointed arms consists of a column of articulated ossicles (the so-called **vertebrae**), connected by muscles and covered by plates. Locomotion is by arm movement. The arms are moved forward in pairs and are placed against the substratum, while one (any one) is extended forward or trailed behind, and the animal is pulled or pushed along in a jerky fashion.

Structure. The mouth is surrounded by five movable plates that serve as **jaws** (Fig. 20-7). There is no anus. The skin is leathery, with dermal plates and spines arranged in characteristic patterns. Surface cilia are mostly lacking.

The visceral organs are confined to the central disc since the rays are too slender to contain them (Fig. 20-8). The **stomach** is saclike and there is no intestine. Indigestible material is cast out of the mouth.

Five pairs of **bursae** (peculiar to ophiuroids) open toward the oral surface by **genital slits** at the bases of the arms. Water circulates in and out of these sacs for exchange of gases. On the coelomic wall of each bursa are small **gonads** that discharge into the bursa their ripe sex cells, which pass through the genital slits into the water for fertilization. Sexes are usually separate; a few ophiuroids are hermaphroditic. Some brood their young in the bursae; the young escape through the genital slits or by rupturing the aboral disc. The larva is called the **ophiopluteus,** and its ciliated bands extend onto delicate, beautiful larval arms, like those of the echinopluteus of the echinoids (Fig. 20-5, *B*). During

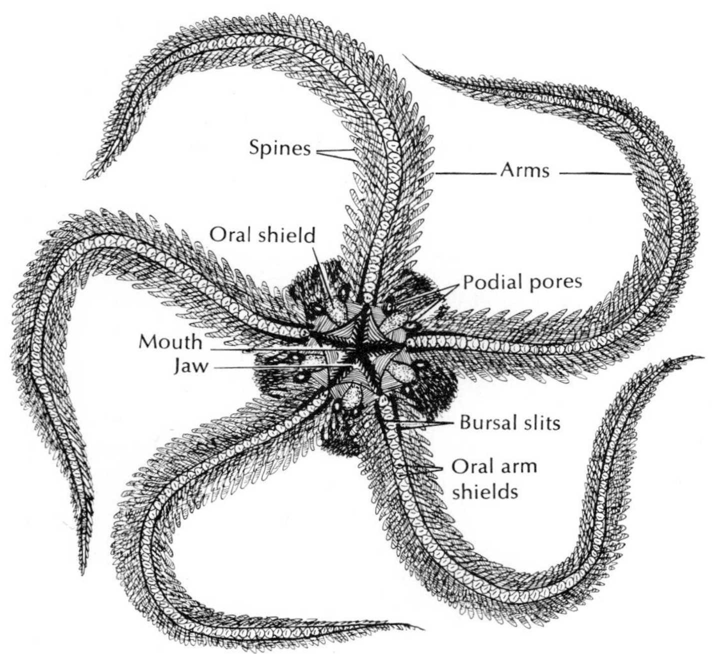

FIG. 20-7 Oral view of spiny brittle star *Ophiothrix*.

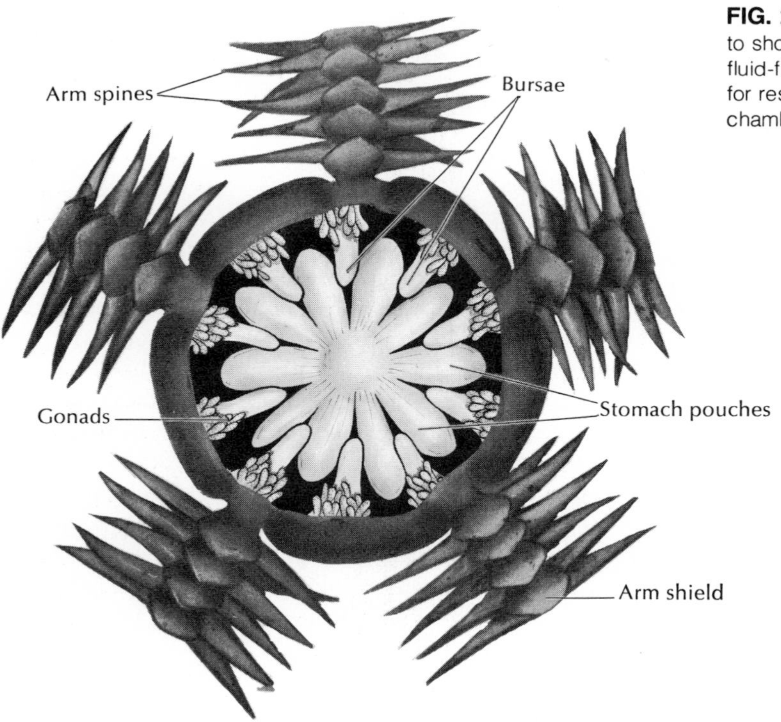

FIG. 20-8 Ophiuroid with aboral disc wall cut away to show principal internal structures. The bursae are fluid-filled sacs in which water constantly circulates for respiration. They also serve as brood chambers. Only bases of arms are shown.

the metamorphosis to the juvenile, there is no temporarily attached phase, as in the asteroids.

Water-vascular, nervous, and hemal systems are similar to those of the sea stars. In each arm there is a small coelom, a radial nerve, and a radial canal of the water-vascular system.

Biology. Brittle stars tend to be secretive, living on hard bottoms where light penetrates. They are generally negatively phototropic and insinuate themselves into small crevices between rocks, becoming more active at night. They are commonly fully exposed on the bottom in the permanent darkness of the deep sea. Ophiuroids feed on a variety of small particles, either browsing food from the bottom or filter feeding. Podia are important in transferring food to the mouth. Some brittle stars extend arms into the water and catch suspended particles in mucous strands between the arm spines.

Regeneration and autotomy are even more pronounced in brittle stars than in sea stars. Many seem very fragile, releasing an arm or even part of the disc at the slightest provocation. Some can reproduce asexually by cleaving the disc; each progeny then regenerates the missing parts.

Some common ophiuroids along the Atlantic coast of the United States are *Amphipholis* (viviparous and hermaphroditic), *Ophioderma, Ophiothrix,* and *Ophiura* (Fig. 20-6, *A*). Along the Pacific coast are *Amphipholis, Amphiodia* with long delicate arms, the ultraspiny *Ophiothrix,* and the sand-colored *Ophioplocus* (viviparous). The basket star *Gorgonocephalus* (Fig. 20-6, *B*) has arms that branch repeatedly. Most ophiuroids are drab in color, but some are attractive, with variegated color patterns.

Sea urchins, sand dollars, and heart urchins, class Echinoidea

The echinoids have a compact body enclosed in an endoskeletal **test,** or shell. The dermal ossicles, which have become closely fitting plates, make up the test. Echinoids lack arms, but their tests reflect the typical pentamerous plan of the echinoderms in their five ambulacral areas. The most notable modification of the ancestral body plan is that the oral surface has become expanded over the sides and top, so that the ambulacral areas extend up to the area around the anus **(periproct).** The majority of living species of sea urchins are referred to as ''regular''; they are hemispheric in shape, radially symmetric, and have medium to long spines (Fig. 20-9 and p. 440). Sand dollars (Fig. 20-11) and heart urchins (Fig. 20-10) are ''irregular'' because the orders to which they belong have become secondarily bilateral; their spines are usually very short. Regular urchins move by means of their tube feet, with some assistance from their spines, and irregular urchins move chiefly by their spines (Fig. 20-11). Some echinoids are quite colorful.

Echinoids have a wide distribution in all seas, from the intertidal regions to the deep oceans. Regular urchins often prefer rocky or hard bottoms, while sand dollars and heart urchins like to burrow into a sandy substrate. Distributed along one or both coasts of North America are common genera of regular urchins (*Arbacia, Strongylocentrotus, Lytechinus* (Fig. 20-9), and sand dollars *(Mellita* [Fig. 20-11], *Dendraster, Echinarachnius).* The West Indies–Florida region is rich in echinoderms, including echinoids, of which *Diadema,* with its long, needle-sharp spines, is a prominent example.

Structure. The echinoid **test** is a compact skeleton of ten double rows of plates that bear movable, stiff spines (Fig. 20-12). The plates are firmly sutured. The five pairs of ambulacral rows are homologous to the five arms of the sea star and have pores (Fig. 20-12, *B*) through which the long tube feet extend. The plates bear small tubercles on which the round ends of the spines articulate as ball-and-socket joints. The spines are moved by small muscles around the bases.

There are several kinds of **pedicellariae,** the most common of which are three-jawed and are mounted on long stalks (Figs. 20-3, *D,* and 20-9). Pedicellariae help keep the body clean and capture small organisms. The pedicellariae of many species bear poison glands, and the toxin paralyzes small prey.

The mouth of regular urchins is surrounded by five converging teeth. In some sea urchins branched **gills** (modified podia) encircle the peristome. The **anus, genital openings,** and **madreporite** are located aborally in the periproct region (Fig. 20-12). The sand dollars also have teeth, and the mouth is located at about the center of the oral side, but the anus has shifted to the posterior margin or even the oral side of the disc, so that an anteroposterior axis and bilateral symmetry can be recognized. Bilateral symmetry is even more accentuated in the heart urchins, with the anus near the posterior on the oral side and the mouth moved away from the oral pole toward the anterior (Fig. 20-10).

FIG. 20-9 Sea urchin *Lytechinus*. Note the slender suckered tube feet. They often attach to bits of shell, seaweed, or other things for camouflage. Stalked pedicellariae can be seen between the spines. This is a common Atlantic form from the Carolinas to the West Indies. (Photograph by R. C. Hermes.)

Inside the test (Fig. 20-12) is the coiled digestive system and a complex chewing mechanism (in the regular urchins and in sand dollars), called **Aristotle's lantern** (Fig. 20-13), to which the teeth are attached. A ciliated siphon connects the esophagus to the intestine and enables the water to bypass the stomach to concentrate the food for digestion in the intestine. Sea urchins eat algae and other organic material, which they graze with their teeth. Sand dollars have short club-shaped spines that move the sand and its organic contents over the aboral surface and down the sides. Fine food particles drop between the spines and are carried by ciliated tracts on the oral side to the mouth.

The hemal and nervous systems are basically similar to those of the asteroids. The ambulacral grooves are closed, and the radial canals of the water-vascular system run just beneath the test, one in each of the ambulacral radii (Fig. 20-12). The podia are supplied with ampullae within the test, each of which usually communicates with its podium by *two* canals through pores in the ambulacral plate; consequently, such pores in the plates are in pairs. The peristomial gills, where present, are of little or no importance in respiratory gas exchange, this function being carried out principally by the other podia. In the irregular urchins, the respiratory podia are thin-walled, flattened, or lobulate and are arranged in ambulacral fields called **petaloids** on the aboral surface. The irregular urchins also have

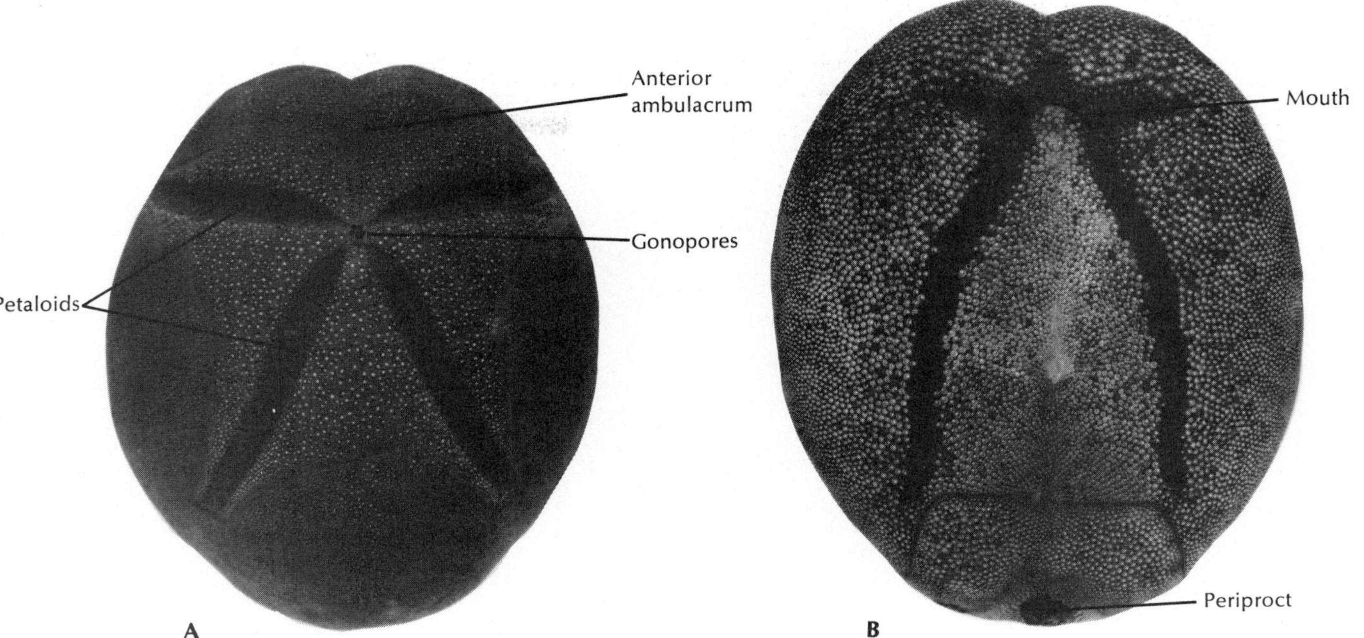

FIG. 20-10 The test of an irregular echinoid *Meoma,* one of the largest heart urchins (test up to 18 cm). *Meoma* occurs in the West Indies and from the Gulf of California to the Galápagos Islands. **A,** Aboral view. Note two rows of paired pores for respiratory podia in the petaloids. Anterior ambulacral area is not modified as a petaloid in the heart urchins, though it is in sand dollars. **B,** Oral view. Note curved mouth at anterior and periproct at posterior ends. In life, periproct bears numerous small ossicles surrounding anus. (Photographs by L. S. Roberts.)

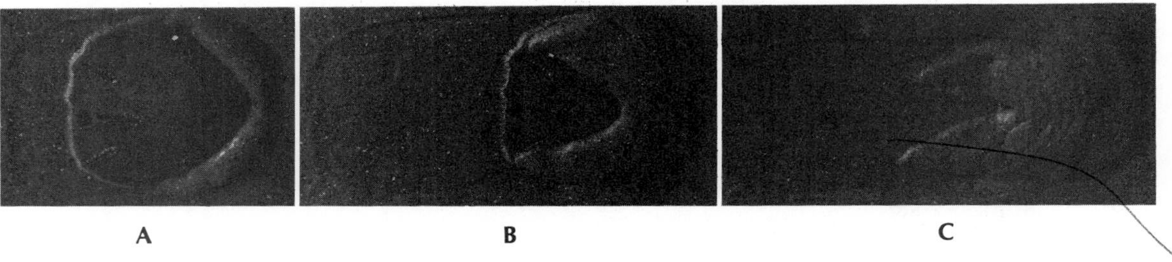

FIG. 20-11 Sand dollars *Mellita* bury themselves in the wet sand by moving the many short spines on their flat oral side. They scour the sand for edible particles, using tube feet to pass food to the mouth, which is centered on the underside. (Photographs by C. P. Hickman, Jr.)

short, suckered, single-pored podia in the ambulacral and sometimes interambulacral areas; these function in food-handling.

Sexes are separate, and both eggs and sperm are shed into the sea for external fertilization. Some, such as the slate pencil urchins, brood their young in depressions between the spines. The **echinopluteus larvae** (Fig. 20-5, *B*) of nonbrooding echinoids may live a planktonic existence for several months and then metamorphose quickly into young urchins.

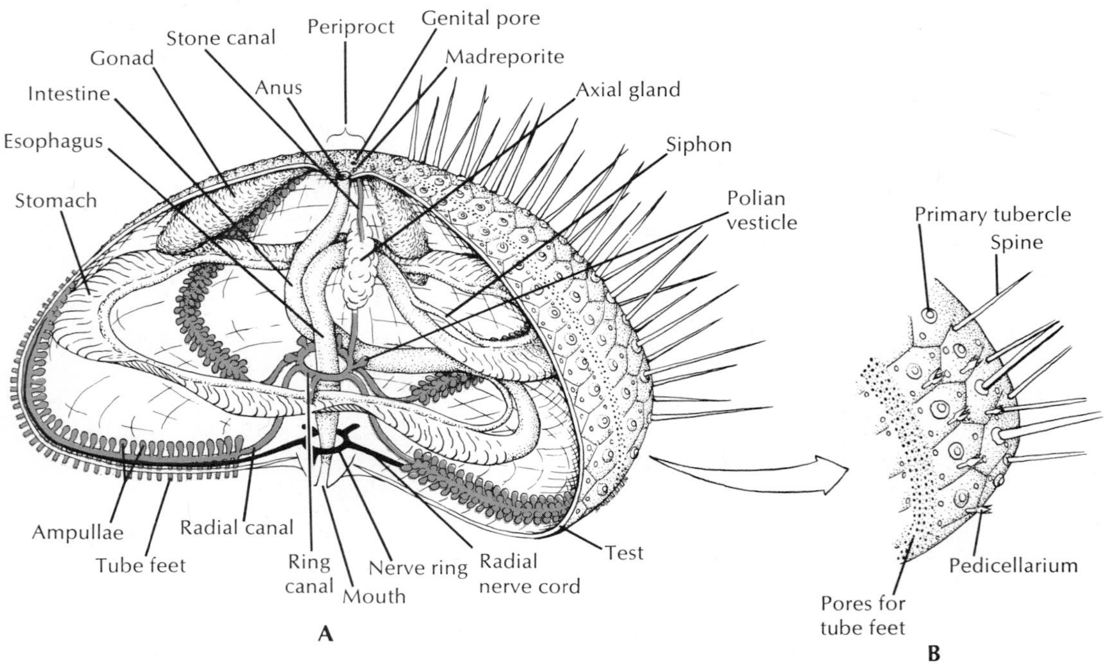

FIG. 20-12 **A,** Internal structure of the sea urchin; water-vascular system in red. **B,** Detail of portion of endoskeleton.

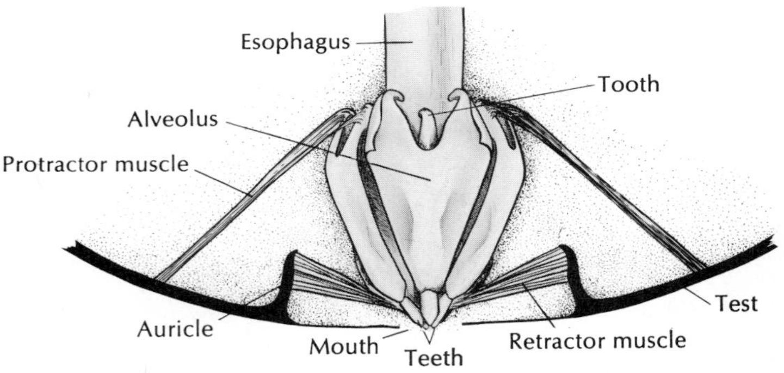

FIG. 20-13 Aristotle's lantern, the complex mechanism used by the sea urchin for masticating its food. Five pairs of retractor muscles draw the lantern and teeth up into the test; five pairs of protractors push the lantern down and expose the teeth. Other muscles produce a variety of movements. Only major skeletal parts and muscles are shown in the diagram.

Sea cucumbers, class Holothuroidea

In a phylum characterized by odd animals, class Holothuroidea contains members that both structurally and physiologically are among the strangest of all. These animals have a remarkable resemblance to the vegetable after which they are named (Fig. 20-14). Compared to the other echinoderms, the holothurians are greatly elongated in the oral-aboral axis, and the ossicles are much reduced in most, so that the animals are soft-bodied. Some species characteristically crawl on the surface of the sea bottom, others are found beneath rocks, and some are burrowers.

Common species along the east coast of North America are *Cucumaria frondosa* (Fig. 20-14), *Thyone briareus* (Fig. 20-15), and the transluscent, burrowing *Leptosynapta*. Along the Pacific coast there are sev-

FIG. 20-14 The sea cucumber *Cucumaria frondosa,* with tentacles extended for feeding. In the background is the sea anemone *Tealia telina. Cucumaria* is a suspension feeder; minute plankton organisms adhere to the tentacles, which when loaded bend over and wipe off the food into the pharynx. (Photograph by T. Lundälv.)

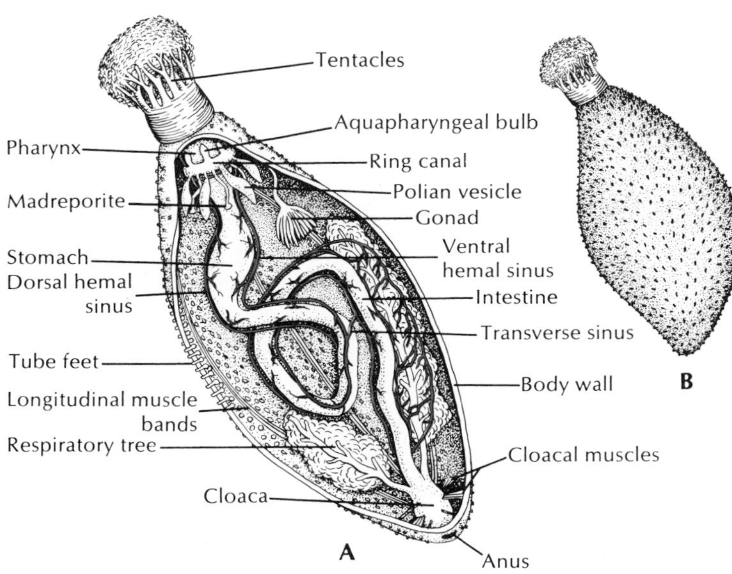

FIG. 20-15 Anatomy of the sea cucumber *Thyone.* **A,** Internal. **B,** External. *Red,* Hemal system.

FIG. 20-16 The feather star *Florometra serratissima,* a crinoid. Feather stars can crawl by holding on to objects with the adhesive ends of the pinnules and pulling along by contracting arms. They can also swim by raising and lowering alternate sets of arms. (Courtesy M. Newman, Vancouver Public Aquarium, British Columbia.)

eral species of *Cucumaria* and the striking reddish brown *Stichopus,* with very large papillae.

Structure. The body wall is usually leathery, with tiny ossicles embedded in it, though a few species have large ossicles forming a dermal armor. Because of the elongate body form of the sea cucumbers, they characteristically lie on one side. In some species the locomotory tube feet are equally distributed to the five ambulacral areas, or all over the body; but most have tube feet well developed only in the ambulacra normally applied to the substratum. Thus, a secondary bilaterality is present, albeit of quite different origin from that of the irregular urchins. The side applied to the substratum has three ambulacra and is called the **sole;** the tube feet in the dorsal ambulacral areas, if present, are usually without suckers and may be modified as sensory papillae. All tube feet, except oral tentacles, may be absent in burrowing forms.

The **oral tentacles** are ten to 30 retractile, modified tube feet around the mouth. The body wall contains circular and longitudinal muscles along the ambulacra.

The **coelomic cavity** is spacious and fluid-filled and has many coelomocytes. The digestive system empties posteriorly into a muscular **cloaca** (Fig. 20-15). A **respiratory tree** composed of two long, many-branched tubes also empties into the cloaca, which pumps seawater into it. The respiratory tree serves both for respiration and excretion and is not found in any other group of living echinoderms. Gas exchange also occurs through the skin and tube feet.

The hemal system is more well developed in holothurians than in other echinoderms. The **water-vascular system** is peculiar in that the madreporite lies free in the coelom.

The sexes are separate, but some holothurians are hermaphroditic. Among the echinoderms, only the sea cucumbers have a single gonad, and this is considered a primitive character. The gonad is usually in the form of one or two clusters of tubules that join at the gonoduct. Fertilization is external, and the free-swimming larva is called an **auricularia.** Some species brood the young either inside the body or somewhere on the body surface.

Biology. Sea cucumbers are sluggish, moving partly by means of their ventral tube feet and partly by waves of contraction in the muscular body wall. The more sedentary species trap suspended food particles in the mucus of their outstretched oral tentacles or pick up particles from the surrounding bottom. They then stuff the tentacles into their pharynx, one by one, sucking off the food material. Others crawl along, grazing the bottom with their tentacles.

Sea cucumbers have a peculiar power of what appears to be self-mutilation, but is really a mode of defense. Some, when irritated, may cast out a part of their viscera by a strong muscular contraction that may either rupture the body wall or evert its contents through the anus. The lost parts are soon regenerated.

There is an interesting commensal relationship between some sea cucumbers and a small fish, *Carapus,* that uses the cloaca and respiratory tree of the sea cucumber as shelter.

Sea lilies and feather stars, class Crinoidea

The crinoids are the most primitive of the living echinoderms. As fossil records reveal, crinoids were once far more numerous than now. They differ from

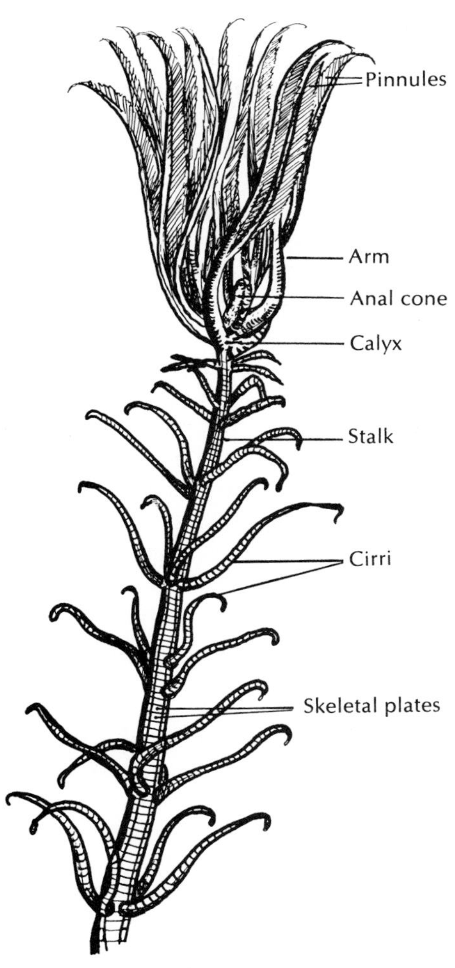

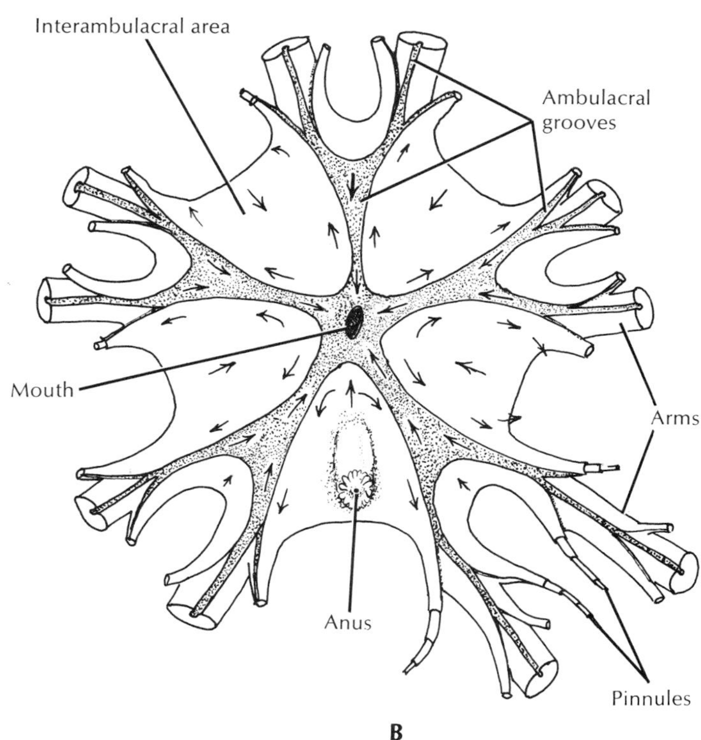

FIG. 20-17 Crinoid structure. **A,** Feather star (stalked crinoid) with portion of stalk. Some crinoids have stalks up to 60 cm long. **B,** Oral view of calyx of the crinoid *Antedon,* showing direction of ciliary food currents. Ambulacral grooves with podia extend from mouth along arms and branching pinnules. Food particles touching podia are tossed into ambulacral grooves and carried, tangled in mucus, by strong ciliary currents toward mouth. Particles falling on interambulacral areas are carried by cilia first toward mouth and then outward and finally dropped off the edge, thus keeping the oral disc clean.

other echinoderms by being attached during a substantial part of their lives. Sea lilies have a flower-shaped body that is placed at the tip of an attached stalk (Fig. 20-17, *A*). The feather stars have long, many-branched arms, and the adults are free-swimming, though they may remain in the same spot for long periods (Fig. 20-16). During metamorphosis feather stars become sessile and stalked, but after several months they detach and become free-moving. Many crinoids are deep-water forms, but feather stars may inhabit shallow waters, especially in the Indo-Pacific and the West Indian–Caribbean regions, where the largest numbers of species are found.

Structure. The body disc, or **calyx,** is covered with a leathery skin **(tegmen)** containing calcareous plates. The epidermis is poorly developed. Five flexible arms branch to form many more arms, each with many lateral **pinnules** arranged like barbs on a feather (Fig. 20-17). Calyx and arms together are called the **crown.** Sessile forms have a long, jointed **stalk** attached to the aboral side of the body. This stalk is made up of plates, appears jointed, and may bear **cirri.** Madreporite, spines, and pedicellariae are absent.

The upper (oral) surface bears the mouth, which opens into a short esophagus, from which the long **intestine** with diverticula proceeds aborally for a distance and then makes a complete turn to the **anus,** which may be on a raised cone (Fig. 20-17, *A*). With the aid of tube feet and mucous nets, crinoids feed on small organisms that are caught in the ambulacral grooves. The **ambulacral grooves** are open and ciliated and serve to carry food to the mouth (Fig. 20-17, *B*). Tube feet in the form of tentacles are also found in the grooves.

The **water-vascular system** has the echinoderm plan. The nervous system is made up of an **oral ring** and a **radial nerve** that run to each arm. The aboral or entoneural system is more highly developed in crinoids, in contrast to most other echinoderms. Sense organs are scanty and primitive.

The sexes are separate. The gonads are simply masses of cells in the genital cavity of the arms and pinnules. The gametes escape without ducts through a rupture in the pinnule wall. The **doliolaria** larvae are free-swimming for a time before they become attached and metamorphose. Most living crinoids are from 15 to 30 cm long, but some fossil species had stalks 25 m in length.

Phylogeny and adaptive radiation

Phylogeny. Based on the embryologic evidence of the bilateral larvae of echinoderms, there can be little doubt that their ancestors were bilateral and that their radial symmetry is secondary. Some recent investigators believe that the radial symmetry arose in a free-moving echinoderm ancestor and that sessile groups were derived several times independently from the free-moving ancestors. However, this view does not account for the adaptive significance of the radial symmetry, that is, as an adaptation for the sessile existence. The more traditional view is that the first echinoderms were sessile, became radial as an adaptation to that existence, and then gave rise to the free-moving groups. Certainly, the most primitive living echinoderms are the crinoids; and the existence of a transitory stalked phase in the asteroids, which also have some primitive characteristics such as open ambulacra, supports the traditional sequence. It is believed also that the endoskeleton was an adaptation for a sessile existence, and that the original function of the water-vascular system was in feeding.

The nature of the hypothetical preechinoderm has also been subject to debate. Some have held that it was a **dipleurula** ancestor, which was a creeping, soft-bodied animal with three, paired coelomic compartments. Hyman and others have suggested that it was a **pentacula** ancestor, more like a lophophorate animal, with tentacles around the mouth. According to the latter theory, the water-vascular system was derived from the tentacles around one side, which were five in number and contained extensions of the middle coelomic compartment of the ancestor.

Adaptive radiation. The radiation of the echinoderms has been determined by the limitations and potentials of their most important characteristics: radial symmetry, the water-vascular system, and their dermal endoskeleton. If their ancestors had a brain and specialized sense organs, these were lost in the adoption of radial symmetry. Thus, it is unsurprising that there are large numbers of creeping, benthic forms with filter feeding, deposit feeding, scavenging, and herbivorous habits, comparatively few predators, and very rare pelagic forms. In this light, the relative success of the asteroids as predators is impressive and probably attributable to the extent to which they have exploited the hydraulic mechanism of the tube feet.

The basic body plan of echinoderms has severely limited their evolutionary opportunities to become parasites. Indeed, the most mobile of the echinoderms, the ophiuroids, which are also the ones most able to insert their bodies into small spaces, are the only group with significant numbers of commensal species.

Derivation and meaning of names

Antedon (Gr., *Anthēdōn,* name of a nymph) A sea lily, genus of class Crinoidea.

Arbacia (Gr., *Arbakes,* ancient king) A prevalent genus of sea urchin, class Echinoidea.

Asterias (Gr., *asterias,* starred, fr. *astēr,* star) Genus of sea star, subclass Asteroidea.

Echinarachnius (Gr., *echinos,* sea urchin, hedgehog, + *rachis,* spine) Genus of sand dollar, class Echinoidea.

Ophiothrix (Gr., *ophis,* serpent, + *thrix,* hair) Genus of brittle star, subclass Ophiuroidea.

Ophiura (Gr., *ophis,* serpent, + *oura,* tail) Genus of brittle star, subclass Ophiuroidea.

Stichopus (Gr., *stichos,* a row or line, + *pous,* footed) Genus of sea cucumber, class Holothuroidea.

Annotated references

Selected general readings

Barnes, R. D. 1974. Invertebrate zoology, ed. 3. Philadelphia, W. B. Saunders Co.

Barrington, E. J. W. 1967. Invertebrate structure and function. Boston, Houghton Mifflin Co.

Binyon, J. 1972. Physiology of echinoderms. Elmsford, N.Y., Pergamon Press, Inc.

Boolootian, R. A. (ed.). 1966. Physiology of Echinodermata. New York, Interscience Publishers. *A valuable account of the physiology of this group.*

Boolootian, R. A., and A. C. Giese. 1959. Clotting of echinoderm coelomic fluid. J. Exp. Zool. **140:**207-229.

Brown, T., and K. Willey. 1972. Crown of thorns: the death of the Great Barrier Reef? Sydney, Angus & Roberts, Ltd.

Buchsbaum, R., and L. Milne. 1960. The lower animals, living invertebrates of the world. Garden City, N.Y., Doubleday & Co., Inc. *Excellent color photographs of echinoderms.*

Burnett, A. L. 1960. The mechanism employed by the starfish *Asterias forbesi* to gain access to the interior of the bivalve *Venus mercenaria.* Ecology **41:**583-584. *This investigator and others have shown that the starfish uses no narcotic agent to produce a small gap between the clam's valves through which the starfish can squeeze its stomach.*

Clark, A. M. 1962. Starfishes and their relations. London, British Museum.

Coe, W. R. 1972. Starfishes, serpent stars, sea urchins and sea cucumbers of the Northeast. New York, Dover Publications, Inc.

Collins, A. R. S. 1975. Biochemical investigation of two responses involved in the feeding behaviour of *Acanthaster planci* (L.). III. Food preferences. J. Exp. Mar. Biol. **17:**87-94.

Corning, W. C., J. A. Dyal, and A. O. D. Willows (eds.). 1975. Invertebrate learning, vol. 3. New York, Plenum Publishing Corp.

Fell, H. B. 1972. Phylum Echinodermata. In A. J. Marshall and W. D. Williams (ed.). Textbook of zoology: invertebrates. New York, American Elsevier Publishing Co., Inc.

Fontaine, A. R. 1965. Feeding mechanisms of the ophiuroid *Ophiocomina nigra.* J. Mar. Biol. Assoc. U. K. **45:**373.

Goodbody, I. 1960. The feeding mechanism in the sand dollar *Mellita sexiesperforata.* Biol. Bull. **119**(1):80-86.

Gosner, K. L. 1971. Guide to identification of marine and estuarine invertebrates. New York, Interscience Publishers. *A comprehensive identification aid for invertebrates of the east coast of North America, especially from Cape Hatteras to the Bay of Fundy.*

Grassé, P. P. (ed.). 1948. Traité de zoologie XI. Paris, Masson & Cie.

Harvey, L. B. 1956. The American *Arbacia* and other sea urchins. Princeton, N. J., Princeton University Press. *A comprehensive monograph on this interesting group, which have furnished so many basic concepts in the field of cytology and development.*

Hyman, L. H. 1955. The invertebrates: Echinodermata, vol. 4. New York, McGraw-Hill Book Co. *The most comprehensive work yet published on the echinoderms, an invaluable reference.*

Jennings, H. S. 1907. Behavior of the starfish *Asterias forreri* De Loriol. Univ. Calif. Pub. Zool. (Berkeley) **4:**339-411. *An account of the reactions of the starfish, including the righting movement.*

Kanatani, H. and H. Shirai. 1969. Mechanism of a starfish spawning. II. Some aspects of action of a neural substance obtained from radial nerves. Biol. Bull. **137:**297-311.

MacGinitie, G. E., and N. MacGinitie. 1968. Natural history of marine animals, ed. 2. New York, McGraw-Hill Book Co. *The section on echinoderms includes some good descriptions and photographs of representative forms, especially those of the Pacific coast.*

Millott, N. (ed.). 1967. Echinoderm biology. New York, Academic Press, Inc.

Moore, R. C. (ed.). 1966, 1967. Treatise on invertebrate paleontology: Echinodermata, Parts U and S. Lawrence, Kan., Geological Society of America and University of Kansas Press.

Nichols, D. 1966. Echinoderms. London, Hutchinson & Co., Ltd.

Nicol, J. A. C. 1969. The biology of marine animals, ed. 2. New York, Pitman Publishing Corp.

Ormond, R. F. G., and others. 1973. Formation and breakdown of aggregations of the crown-of-thorns starfish, *Acanthaster planci* (L.). Nature **246:**167-169.

Ricketts, E. F., and J. Calvin. 1968. Between Pacific tides, ed. 4 (revised by J. W. Hedgpeth). Stanford, Calif., Stanford University Press. *In many ways this is a unique book of seashore life. It stresses the habits and habitats of the Pacific coast invertebrates (including echinoderms), and the illustrations are revealing. It includes an excellent systematic index and annotated bibliography.*

Russell-Hunter, W. D. 1969. A biology of higher invertebrates. New York, The Macmillan Co.

Smith, R. I., and others (ed.). 1957. Intertidal invertebrates of the central California coast. Berkeley, University of California Press. *This is a revision of S. F. Light's Laboratory and Field Text in Invertebrate Zoology. Consists mainly of taxonomic keys to the forms found in the intertidal zone.*

Selected *Scientific American* articles

Benson, A. A., and R. F. Lee. 1975. The role of wax in oceanic food chains. **232**:77-86 (Mar.). *The possession of enzymes to digest the wax in corals may account for the ability of* Acanthaster planci *to feed on coral polyps.*

Feder, H. M. 1972. Escape responses in marine invertebrates. **227**:93-100 (July). *Slow-moving molluscs and other echinoderms can undergo some remarkable antics in efforts to escape predatory sea stars.*

CHAPTER 21

THE LESSER DEUTEROSTOMES

Phylum Chaetognatha
Phylum Hemichordata

Living arrowworms, *Sagitta elegans*. They make up an important part of the marine plankton in both littoral and open seas. They are rarely more than 5 to 7 cm long and belong to phylum Chaetognatha.

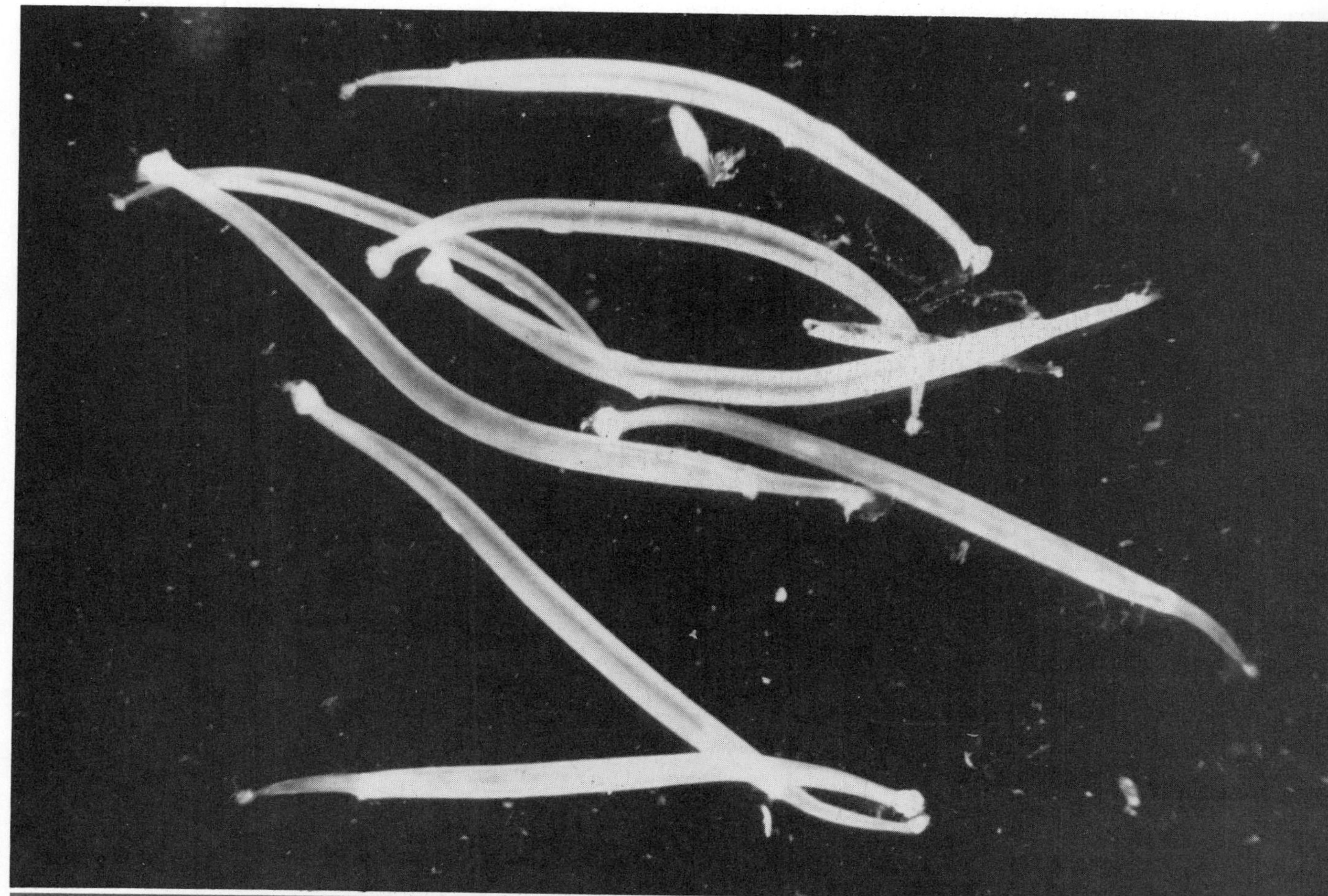

Photograph by D. P. Wilson, Plymouth, England.

The deuterostomes include, along with the Echinodermata, three other phyla—Chaetognatha, Hemichordata, and Chordata. Two of the chordate subphyla—Urochordata and Cephalochordata—are also invertebrate groups.

The term lesser (or minor) deuterostomes is usually used in reference to the hemichordates and chaetognaths, but only in the sense that they have a relatively small number of species. However, they are widespread and contain some commonly found invertebrate forms. Often smaller groups deserve much more attention than is usually given to them, for they contribute much to our understanding of evolutionary diversity and relationships.

These phyla have enterocoelous development of the coelom and some form of radial cleavage. The hemichordates were formerly included as a subphylum of the Chordata, but they are probably more closely related to the Echinodermata. The Chaetognatha apparently are not closely related to any other group.

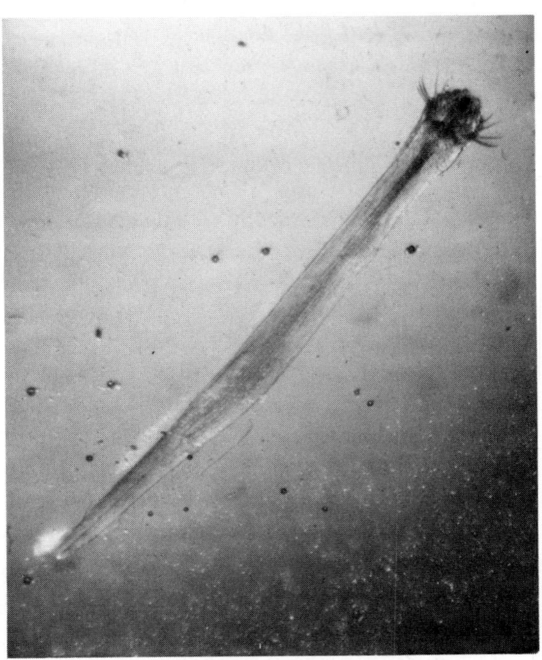

FIG. 21-1 Arrowworm *Sagitta*. Head *(top)* is largely covered with hood formed from epidermis. When worm is engaged in catching its prey, hood is retracted to neck region. (Preserved specimen, photograph by L. S. Roberts.)

PHYLUM CHAETOGNATHA

A common name for the chaetognaths is arrowworms. They are all marine animals and are considered by some to be related to the nematodes and by others to be related to the annelids. However, they actually seem to be aberrant and show no distinct relations to any other group. Only their embryology indicates their position as deuterostomes.

The name Chaetognatha (ke-tog′na-tha) (Gr., *chaitē,* long flowing hair, + *gnathos,* jaw) refers to the sickle-shaped bristles on each side of the mouth. This is not a large group, for there are only some 65 known species. Their small, straight bodies resemble miniature torpedoes, or darts, ranging from 2.5 to 10 cm in length.

The arrowworms are all adapted for a planktonic existence, except for *Spadella,* a benthic genus. They usually swim to the surface at night and descend during the day. Much of the time they drift passively, but they can dart forward in swift spurts, using the caudal fin and longitudinal muscles—a fact that no doubt contributes to their success as planktonic predators. Horizontal fins bordering the trunk are used in flotation rather than in active swimming.

External features. The body of the arrowworm is unsegmented and is made up of head, trunk, and postanal tail (Figs. 21-1 and 21-2). On the underside of the head is a large vestibule leading to the mouth. There are teeth in the vestibule, and flanking it on both sides are curved chitinous spines used in seizing the prey. There is a pair of eyes on the dorsal side. A peculiar hood formed from a fold of the neck can be drawn forward over the head and spines. When the animal captures prey, the hood is retracted, and the teeth and raptorial spines spread apart and then snap shut with startling speed. Arrowworms are voracious feeders, living on planktonic forms, especially copepods, and even small fish.

The body is covered with a thin cuticle and a layer of epidermis that is single-layered except along the sides of the body, where it is stratified in a thick layer. These are the only invertebrates with a many-layered epidermis.

Internal features. Arrowworms are fairly advanced worms in that they have a complete digestive system, a well-developed coelom, and a nervous system with a nerve ring containing large dorsal and ventral ganglia and a number of lateral ganglia. Sense organs include the eyes, sensory bristles, and a U-shaped ciliary loop extending over the neck from the back of the head that is believed to detect water currents and chemical nature. Vascular, respiratory, and excretory systems, however, are entirely lacking.

Arrowworms are hermaphroditic with either cross- or self-fertilization. The eggs of *Sagitta* are coated with jelly and are planktonic. Eggs of other arrowworms may be attached to the body and carried about for a time. The juveniles develop directly without metamorphosis. Chaetognath embryology differs from that of other deuterostomes in that the coelom is formed by a backward extension from the archenteron rather than by pinched-off coelomic sacs. There is no true peritoneum lining the coelom. Cleavage is radial, complete, and equal.

The best known species is *Sagitta,* the common arrowworm (Figs. 21-1 and 21-2).

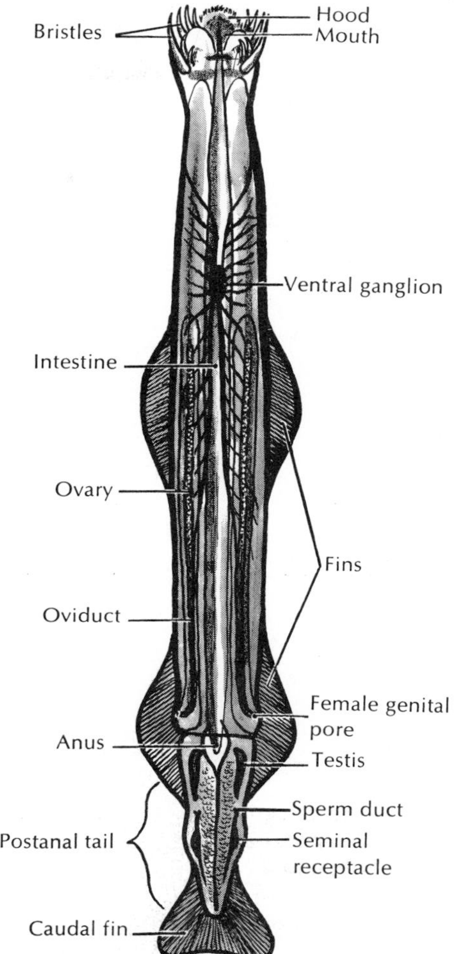

FIG. 21-2 Arrowworm *Sagitta,* ventral view. Arrowworms have many resemblances to certain pseudocoelomates, and some authorities hesitate to call them coelomate animals, although they are enterocoelous and are placed in Deuterostomia. Among their features are postanal tail, hood, fins, and stratified epidermis.

PHYLUM HEMICHORDATA—THE ACORN WORMS

Position in animal kingdom

1. Hemichordates belong to the deuterostome branch of the animal kingdom and are enterocoelous coelomates with radial cleavage.
2. Hemichordates show some of both echinoderm and chordate characteristics.
3. A chordate plan of structure is suggested by gill slits and a restricted dorsal tubular nerve cord.
4. Similarity to the echinoderms is shown in larval characteristics.

Biologic contributions

1. A **tubular dorsal nerve cord** in the collar zone may represent an early stage of the condition in chordates; a diffused net of nerve cells is similar to the uncentralized, subepithelial plexus of echinoderms.
2. The **gill slits** in the pharynx, which are also characteristic of chordates are used primarily for filter feeding and only secondarily for breathing and are thus comparable to those in the protochordates.

The hemichordates

The Hemichordata (hem′i-kor-da′ta) (Gr., *hemi,* half, + *chorda,* string, cord) are marine animals that were formerly considered a subphylum of the chordates, based on their possession of gill slits and a rudimentary notochord. However, it is now generally agreed that the so-called hemichordate "notochord" is really a stomochord and not homologous with the chordate notochord, so the hemichordates are given the rank of a separate phylum.

Hemichordates are vermiform bottom dwellers, living usually in shallow waters. Some are colonial and live in secreted tubes. Most are sedentary or sessile. Their distribution is fairly worldwide, but their secretive habits and fragile bodies make collecting them difficult.

Members of class Enteropneusta (acorn worms) range from 20 mm to 2.5 m in length and 3 to 200 mm in breadth. Members of class Pterobranchia are smaller, usually from 5 to 14 mm, not including the stalk. About 70 species of enteropneusts and three small genera of pterobranchs are recognized.

Hemichordates have the typical tricoelomate structure of deuterostomes.

Characteristics

1. Soft-bodied; wormlike, or short and compact with stalk for attachment
2. Body divided into proboscis, collar, and trunk; coelomic pouch single in proboscis, but paired in other two; buccal diverticulum in posterior part of proboscis
3. Enteropneusta free-moving and of burrowing habits; pterobranchs sessile, mostly colonial, living in secreted tubes
4. Circulatory system of dorsal and ventral vessels and dorsal heart
5. Respiratory system of gill slits (few or none in pterobranchs) connecting the pharynx with outside as in chordates
6. No nephridia; a single glomerulus connected to blood vessels may have excretory function
7. A subepidermal nerve plexus thickened to form dorsal and ventral nerve cords, with a ring connective in the collar; dorsal nerve cord of collar hollow in some
8. Sexes separate in Enteropneusta, with gonads projecting into body cavity; in pterobranchs reproduction may be sexual or asexual (in some) by budding; tornaria larva in some Enteropneusta

Class Enteropneusta

The enteropneusts, or acorn worms, are sluggish wormlike animals that live in burrows or under stones, usually in mud or sand flats of intertidal zones. *Balanoglossus* and *Saccoglossus* (Fig. 21-3) are common genera.

The mucus-covered body is divided into a tongue-like **proboscis,** a short **collar,** and a long **trunk** (protosome, mesosome, and metasome).

Proboscis. The proboscis is the active part of the animal. It probes about in the mud, examining its surroundings and collecting food in mucous strands on its surface. These are carried by cilia to the groove at the edge of the collar, are directed to the mouth on the underside, and are swallowed. Large particles can be rejected by covering the mouth with the edge of the collar (Fig. 21-4).

Burrow dwellers use the proboscis to excavate, thrusting it into the mud or sand and allowing cilia and mucus to move the sand backward. Or they may eat the sand and mud as they go, extracting from it its organic contents. They build U-shaped mucus-lined burrows, usually with two openings 10 to 30 cm apart and with the base of the U 50 to 75 cm below the surface. They can thrust the proboscis out the front opening for feeding. Defecation at the back opening builds characteristic spiral fecal mounds that leave a telltale clue to the location of their burrows.

In the posterior end of the proboscis is a small coelomic sac (protocoel) into which extends the **buccal diverticulum,** a slender, blindly ending pouch of the gut that reaches forward into the buccal region and was formerly believed to be a notochord. A slender canal connects the protocoel with a **proboscis pore** to the outside (Fig. 21-3, *B*). The paired coelomic cavities in the collar also open by pores. By taking in water through the pores into the coelomic sacs, the proboscis and collar can be stiffened to aid in burrowing. Contraction of the body musculature then forces the excess water out through the gill slits, reducing the hydrostatic pressure and allowing the animal to move forward.

Branchial system. A row of **gill pores** is located dorsolaterally on each side of the trunk just behind the collar (Fig. 21-4, *A*). These open from a series of gill chambers that in turn connect with a series of **gill slits** in the sides of the pharynx. No gills are attached to the gill slits, but some respiratory gaseous exchange occurs in the vascular branchial epithelium, as well as in the body surface. Ciliary currents keep a fresh supply of water moving from the mouth through the pharynx and out the gill slits and branchial chambers to the outside.

Feeding and the digestive system. Hemichordates are largely ciliary-mucus feeders. Behind the buccal cavity lies the large pharynx containing in its dorsal part the U-shaped gill slits (Fig. 21-3, *B*). Since there are no gills, it is assumed that the primary function of the branchial mechanism of the pharynx is food-gathering. Food particles caught in mucus and brought to the mouth by ciliary action on the proboscis and collar are strained out of the branchial water that leaves through the gill slits and are directed along the ventral part of the pharynx and esophagus to the intestine, where digestion and absorption occur (Fig. 21-4).

Circulatory and excretory systems. A middorsal vessel carries blood forward above the gut. In the collar the vessel expands into a sinus and a heart vesicle above the buccal diverticulum. Blood is then driven

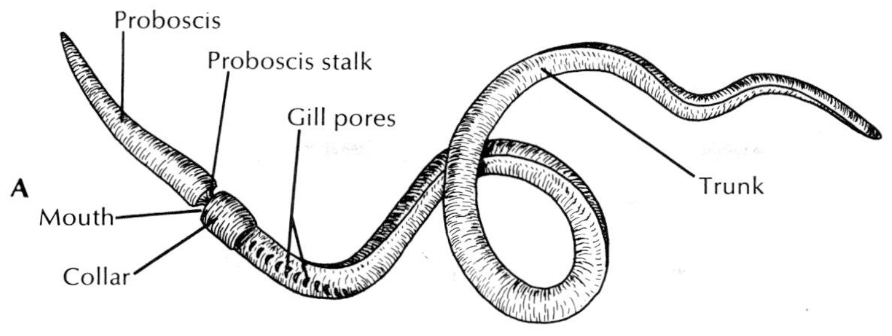

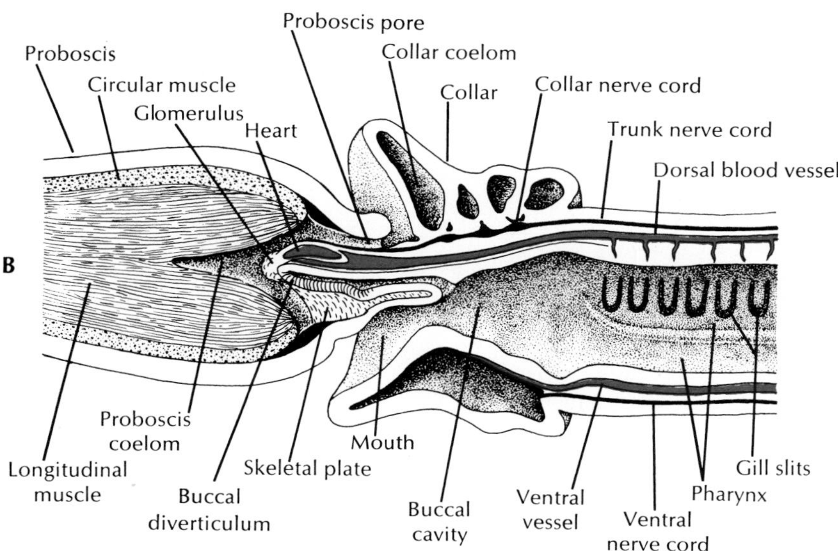

FIG. 21-3 *Saccoglossus,* an acorn worm. **A,** External lateral view. **B,** Longitudinal section through anterior end. (Hemichordata, class Enteropneusta.)

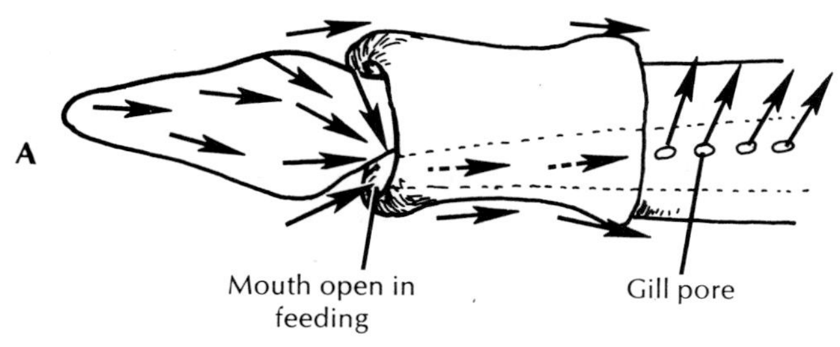

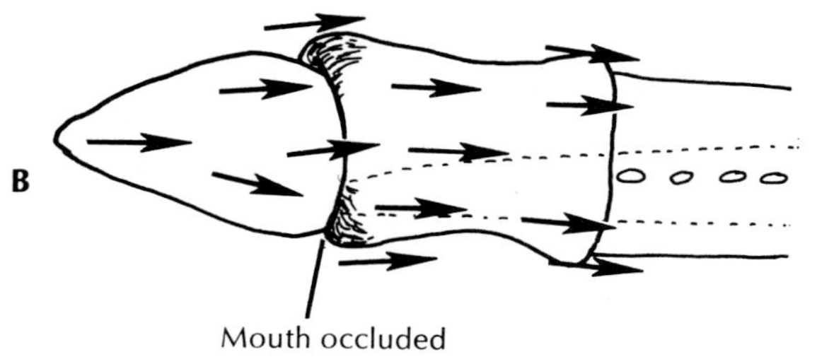

FIG. 21-4 Food currents of enteropneust hemichordate. **A,** Side view of acorn worm with mouth open, showing direction of currents created by cilia on proboscis and collar. Food particles are directed toward mouth and digestive tract. Rejected particles move toward outside of collar. Water leaves through gill pores. **B,** When mouth is occluded, all particles are rejected and passed onto the collar. Nonburrowing and some burrowing hemichordates utilize this feeding method. (Modified from Russell-Hunter, W. D. 1969. A biology of the higher invertebrates. New York, The Macmillan Co.)

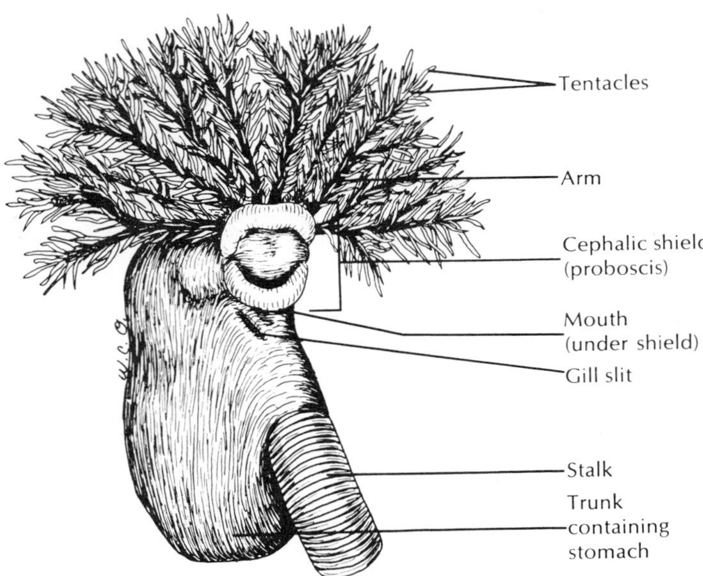

FIG. 21-5 *Cephalodiscus*, a pterobranch hemichordate. These tiny (5 to 7 mm) forms live in coenecium tubes in which they can move about. Ciliated tentacles and arms direct currents of food and water toward mouth. These deep-sea animals may be close to the ancestral stock of echinoderms and chordates.

into a network of blood sinuses called the **glomerulus,** which partially surrounds these structures. The glomerulus is assumed to have an excretory function (Fig. 21-3, *B*). Blood travels posteriorly through a ventral vessel below the gut, passing through extensive sinuses to the gut and body wall. The blood is colorless.

Nervous and sensory systems. The nervous system consists mostly of a subepithelial network, or plexus, of nerve cells and fibers to which processes of epithelial cells are attached. Thickenings of this net form dorsal and ventral nerve cords that are united posterior to the collar by a ring connective. The dorsal cord continues into the collar and furnishes many fibers to the plexus of the proboscis. The collar cord is hollow in some species and contains giant nerve cells with processes running to the nerve trunks. This primitive nerve plexus system is highly reminiscent of that of the cnidarians and echinoderms.

Sensory receptors include neurosensory cells throughout the epidermis (especially in the proboscis, a preoral ciliary organ that may be chemoreceptive) and photoreceptor cells.

Reproductive system and development. Sexes are separate in enteropneusts. Gonads are arranged in a dorsolateral row on each side of the anterior part of the trunk. Fertilization is external, and in some species a ciliated **tornaria** larva develops that at certain stages is so similar to the echinoderm bipinnaria that it was once believed to be an echinoderm larva. The familiar *Saccoglossus* of American waters has direct development without a tornaria stage.

Class Pterobranchia

The basic plan of the class Pterobranchia is similar to that of the Enteropneusta, but certain structural differences are correlated with the sedentary mode of life of pterobranchs. The first pterobranch ever reported was obtained by the famed "Challenger" expedition of 1872 to 1876. Although first placed among the Polyzoa (Entoprocta and Ectoprocta), its affinities to the hemichordates were later recognized. Only two genera (*Cephalodiscus* and *Rhabdopleura*) are known in any detail.

Pterobranchs are small animals, usually within the range of 1 to 7 mm in length, although the stalk may be longer. Many individuals of *Cephalodiscus* (Fig. 21-5) live together in gelatinous tubes, which often form an anastomosing system. The zooids are not connected, however, and live independently in the tubes. Through apertures in these tubes, they extend their crown of tentacles. They are attached to the walls of the tubes by extensible stalks that can jerk the owners back into the tubes when necessary.

The body of *Cephalodiscus* is divided into the three regions—proboscis, collar, and trunk—characteristic of the hemichordates. There is only one pair of gill slits, and the alimentary canal is U-shaped, with the anus near the mouth. The proboscis is shield-shaped.

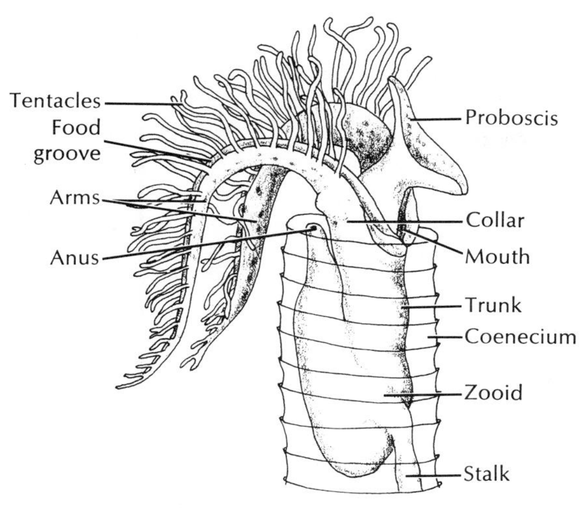

FIG. 21-6 *Rhabdopleura,* a pterobranch hemichordate in its tube. Individuals live in branching tubes, connected by stolons, and protrude the tentacled lophophore for feeding.

At the base of the proboscis there are five to nine pairs of branching arms with tentacles containing an extension of the coelomic compartment of the mesosome, as in a lophophore. Ciliated grooves on the tentacles and arms collect food. Some species are dioecious, and others are monoecious. Asexual reproduction by budding may also occur.

In *Rhabdopleura,* which is smaller in size than *Cephalodiscus,* the members remain together to form a colony of zooids connected by a stolon and enclosed in coenecium tubes (Fig. 21-6). The collar in these forms bears two branching arms or lophophores. No gill clefts or glomeruli are present. New individuals are reproduced by budding from a creeping basal stolon, which branches on a substratum. In none of the pterobranchs is there a tubular nerve cord in the collar, but otherwise their nervous system is similar to that of the Enteropneusta.

Atubaria, a little known genus, has no tube but attaches its stalk to colonial hydroids.

The fossil graptolites of the middle Paleozoic era are often placed as an extinct class under Hemichordata. Their tubular chitinous skeleton and colonial habits indicate an affinity with *Rhabdopleura.* They are considered important index fossils of the Ordovician and Silurian geologic strata.

Phylogeny and adaptive radiation

Phylogeny. Hemichordate phylogeny has long been puzzling. Hemichordates share characteristics with both the echinoderms and the chordates. They share with the chordates the gill slits, which are used primarily for filter feeding and secondarily for breathing as in some of the protochordates. A short dorsal, somewhat hollow nerve cord in the collar zone foreshadows the nerve cord of the chordates. The buccal diverticulum in the hemichordate mouth cavity, which was long believed to be a rudimentary notochord homologous with the notochord of chordates, is now considered of no phylogenetic importance. The relationship to the echinoderms is striking. The early embryology is remarkably like that of echinoderms, and the early tornaria larva is almost identical with the bipinnaria larva of asteroids. The similarity between the hydraulic action of the coelomic pouches and that of the water-vascular system in echinoderms and the similarity in plan of the subepithelial nerve plexus of the two groups are further evidence of their relationship.

Within the phylum, the class Pterobranchia is considered more primitive than the class Enteropneusta and shows affinities with the Ectoprocta, Brachiopoda, and others because of its lophophore and sessile habits. Some believe that the pterobranchs may be similar to the common ancestors of both the hemichordates and the echinoderms.

Adaptive radiation. The pterobranchs, because of their sessile lives and being found largely in secreted tubes in ocean bottoms, where conditions are fairly stable, have undergone little adaptive divergence. They have retained a tentacular type of ciliary feeding. The enteropneusts, on the other hand, although sluggish, are more active than the pterobranchs. Having lost the tentaculated arms, they use a proboscis to trap small organisms in mucus, or they eat sand as they burrow and digest organic sediments from the sand. Their evolutionary divergence, though greater than that of the pterobranchs, is still modest.

Derivation and meaning of names

Balanoglossus (Gr., *balanos,* acorn, + *glōssa,* tongue) Refers to the shape of the proboscis in this enteropneust genus.

Saccoglossus (Gr., *sakkos,* sac, + *glōssa,* tongue) The most familiar genus of enteropneusts.

Sagitta (L., lalen root arrow) Refers to the shape of the body in this common genus of chaetognaths.

Annotated references
Selected general readings

Alvarino, A. 1965. Chaetognaths. Oceanogr. Mar. Biol. Ann. Rev. **3:**115-194.

Barrington, E. J. W. 1965. The biology of Hemichordata and Protochordata. San Francisco, W. H. Freeman & Co., Publishers. *Concise account of behavior, physiology, and reproduction of hemichordates, urochordates, and cephalochordates.*

Barrington, E. J. W. 1967. Invertebrate structure and function. Boston, Houghton Mifflin Co.

Bieri, R. 1959. The distribution of planktonic Chaetognatha in the Pacific and their relationship to the water masses. Limnol. Oceanogr. **4:**1-28.

Bullock, T. H. 1940. Functional organization of the nervous system of Enteropneusta. Biol. Bull. **79:**91-113.

Dilly, P. N. 1975. The pterobranch *Rhabdopleura compacta:* its nervous system and phylogenetic position. Symp. Zool. Soc. Lond. **36:**1-16.

Edmondson, W. T. (ed.). 1966. Marine biology III. New York, The New York Academy of Sciences.

Ghirardelli, E. 1968. Some aspects of the biology of the chaetognaths. Adv. Mar. Biol. **6:**271-375.

Gosner, K. L. 1971. Guide to the identification of marine and estuarine invertebrates. New York, Interscience Publishers, Inc.

Hyman, L. H. 1959. The invertebrates: smaller coelomate groups, vol. 5. New York, McGraw-Hill Book Co. *The best appraisal in English of this strange group.*

Knight-Jones, E. W. 1952. On the nervous system of *Saccoglossus cambrensis.* Phil. Trans. R. Soc. London B. **236:**315-354.

Kowalevsky, A. 1866. Anatomie des *Balanoglossus.* Mémoires de l'Académie impériale des sciences de St. Pétersburg, ser. 7, vol. 10, no. 3. *A classic paper and the first accurate description of this group.*

Newell, G. E. 1951. The stomochord of Enteropneusta. Proc. Zool. Soc. **121:**741. *An appraisal of the status of the stomochord in comparison with a true notochord.*

Parry, D. A. 1944. Structure and function of the gut in *Spadella* and *Sagitta.* J. Marine Biol. Assoc. U. K. **26:**16-36.

Russell-Hunter, W. D. 1969. A biology of higher invertebrates. New York, The Macmillan Co.

Stebbing, A. R. D., and P. N. Dilly. 1972. Some observations on living *Rhabdopleura compacta* (Hemichordata). J. Mar. Biol. Assoc. U. K. **52:**443-449.

CHAPTER 22

THE CHORDATES

Ancestry and evolution, general characteristics, protochordates

Tube sea squirts *Ciona intestinalis.* These sessile marine protochordates are ascidians—one of three classes of the subphylum Urochordata. Although adult ascidians are highly specialized and modified chordates, their free-swimming "tadpole" larvae bear all the right chordate hallmarks—notochord, gill slits, dorsal nerve cord, and postanal tail—and occupy an important position in theories of chordate ancestry.

Photograph by D. P. Wilson.

Position in animal kingdom

1. All available evidence indicates that chordates have evolved from the invertebrates, but it has been impossible to establish the exact relationship.
2. Two possible lines of ancestry have been proposed in the phylogenetic background of the chordates. One of these is the annelid-arthropod-mollusc group; the other is the echinoderm-protochordate group.
3. The echinoderms as a group have certain characteristics that are shared with the chordates, such as **indeterminate cleavage, same type of mesoderm and coelom formation,** and **anus derivation from blastopore with mouth of secondary origin.** Thus the echinoderms appear to have a close kinship to the chordate phylum.
4. Taking the phylum as a whole, there is more fundamental unity of plan throughout all the organs and systems of this group than there is in any of the invertebrate phyla.
5. From gill filter-feeding ancestors to the highest vertebrates, the evolution of chordates has been guided by the specialized basic adaptations of the living endoskeleton, paired limbs, and advanced nervous system.

Biologic contributions

1. A **living endoskeleton** is characteristic of the entire phylum. Two endoskeletons are present in the group as a whole. One is the rodlike **notochord,** present in all members of the phylum at some time; the other is the **vertebrate skeleton** with its axial **vertebral column** that largely replaces the notochord in higher chordates.
2. The living endoskeleton permits **continuous growth** and provides a light but exceptionally sturdy structural **framework for muscular attachment.** The muscular and skeletal systems together comprise most of the body mass, and other body systems have become specialized to serve locomotory demands.
3. The endoskeleton of vertebrates consists of **axial, appendicular,** and **visceral** divisions. The more primitive axial skeleton is made up of the vertebral column, cranium, and rib cage. The appendicular skeleton, composed of the pectoral and pelvic girdles, developed to provide support for stabilizing fins and later for weight-bearing **paired limbs.** The visceral skeleton is the framework for the gill region.
4. A **postanal tail,** present at some stage in most chordates, provided increased propulsive efficiency for primitive aquatic chordates. It was later modified to serve various locomotory, balance, and prehensile functions.
5. A high-pressure **closed circulatory system** with **ventral heart** became a highly efficient internal transport system that performs nutritive, respiratory, excretory, regulatory, and communicative roles. The blood contains **erythrocytes with hemoglobin,** and there are several classes of leukocytes.
6. The **perforated pharynx** (gill slits) was introduced in primitive chordates to serve respiratory and feeding functions. **Jaws** and **internal gills** later evolved from the pharyngeal region.
7. The higher chordates (all vertebrates) have a **pituitary complex** that controls most endocrine functions in the body and is itself regulated by the brain.
8. A **dorsal hollow nerve cord** that expands anteriorly into a brain is universally present at some stage. The system is divided into a sensory-motor **central nervous system** and an **autonomic nervous system** that governs involuntary functions. A cluster of keen sensory receptors connects to the brain.

THE CHORDATES

Animals most familiar to the student belong to the great phylum Chordata (kor-da′ta) (L., *chorda,* cord). Humans themselves are members and share one of the common characteristics from which the phylum derives its name—the **notochord** (Gr., *nōton,* back, + L., *chorda,* cord) (Fig. 22-1). This structure is possessed by all members of the phylum, in either the larval or the embryonic stages or throughout life. The notochord is a rodlike, semirigid body of vacuolated cells, which extends, in most cases, the length of the body between the enteric canal and the central nervous system. Its primary purpose is to support and to stiffen the body, that is, to act as a skeletal axis.

The structural plan of chordates retains many of the features of invertebrate animals, such as bilateral symmetry, anteroposterior axis, coelom tube-within-a-tube arrangement, metamerism, and cephalization. Yet, whereas the kinship of chordates and invertebrates is obvious, it has not been possible to establish the exact relationship with certainty.

Ecologically the phylum has been very successful in the animal kingdom. Chordates are among the most adaptable of organic forms and are able to occupy most kinds of habitats. From a purely biologic viewpoint, chordates are of primary interest because they illustrate so well the broad biologic principles of evolution, development, and relationship. They represent as a group the background of human beings.

Characteristics

1. Bilateral symmetry; head and tail regions; segmented body; three germ layers; coelom well-developed
2. **Notochord** (a skeletal rod) present at some stage in life cycle
3. **Nerve cord dorsal and tubular;** anterior end of cord usually enlarged to form brain
4. **Pharyngeal gill slits present at some stage in life cycle** and may or may not be functional
5. A **postanal tail** usually projecting beyond the anus at some stage and may or may not persist
6. **Heart ventral,** with dorsal and ventral blood vessels; closed circulatory system
7. Complete digestive system
8. Exoskeleton often present; well-developed in fishes, reptiles, and birds
9. A cartilaginous or bony **endoskeleton** present in the majority of members (vertebrates)

Four chordate hallmarks

The four distinctive characteristics that set chordates apart from all other phyla are the **notochord, dorsal tubular nerve cord, pharyngeal gill slits,** and **postanal tail.** These characteristics are always found in the early embryo, although they may be altered or may disappear altogether in later stages of the life cycle.

These four features are so important that each merits an introduction of its own.

Notochord. This rodlike body develops in the embryo as a longitudinal outfolding of the dorsal side of the alimentary canal. It is endodermal in origin, although in some forms there is a possibility that the other germ layers have contributed to its formation. In most it is a semirigid rod extending the length of the body. It is the first part of the endoskeleton to appear in the embryo. As an axis on which the muscles can act, it permits undulatory movements of the body. In most of the protochordates and in primitive vertebrates the notochord persists throughout life. In all vertebrates a series of cartilaginous or bony vertebrae are formed from the connective tissue sheath around the embryonic notochord and replace it as the chief mechanical axis of the body.

Dorsal tubular nerve cord. In the invertebrate phyla the nerve cord (often paired) is ventral to the alimentary canal and is solid, but in the chordates the cord is dorsal to the alimentary canal and is formed as a tube. The anterior end of this tube in vertebrates becomes enlarged to form the brain. The hollow cord is produced by the infolding of ectodermal cells on the dorsal side of the body above the notochord (Fig. 6-3, p. 103). Among the vertebrates the nerve cord lies in the neural arches of the vertebrae, and the brain is surrounded by a bony or cartilaginous cranium.

Pharyngeal gill slits. Pharyngeal gill slits are perforated slitlike openings that lead from the pharyngeal cavity to the outside. They are formed by the invagination of the outside ectoderm and the evagination of the endodermal lining of the pharynx. The two pockets break through when they meet, to form the slit. In higher vertebrates these pockets may not break through and only grooves are formed instead of slits; all traces of them usually disappear. In forms that use the slits for breathing, gills with blood vessels develop at the margins of the slits; here gaseous exchange (oxygen and carbon dioxide) proceeds between the blood and the water that flows over the gills. The slits have in their walls supporting frameworks of gill bars. Primitive forms such as amphioxus have a large number of slits, but only six or seven are the rule in the fishes. The transitory appearance of the slits in land vertebrates is often used as evidence for evolution.

Postanal tail. This chordate innovation, together with somatic musculature and the stiffening notochord, provides the motility that larval tunicates and amphioxus need for their free-swimming existence. As a structure added to the body behind the end of the digestive tract, it has evolved specifically for propulsion in water. Its efficiency is later increased in fishes with the addition of fins.

What advances do chordates show over other phyla?

Living endoskeleton. One distinguishing characteristic of the chordates is the **endoskeleton,** which, as we have seen, is first found in the echinoderms. An endoskeleton is an internal structure that provides support and serves as a framework for the body. Most chordates possess two types of axial endoskeletons in their life cycle. The first is the **notochord** (Fig. 22-1), possessed at some stage by all chordates. The second is the **vertebral column** and accessory structures such as the appendages. This second type of endoskeleton, which is more specialized and more adaptable for evolutionary growth, is possessed by only part, although the greater part, of the chordate phylum.

The endoskeleton was probably composed initially of cartilage that later gave way to bone. Cartilage

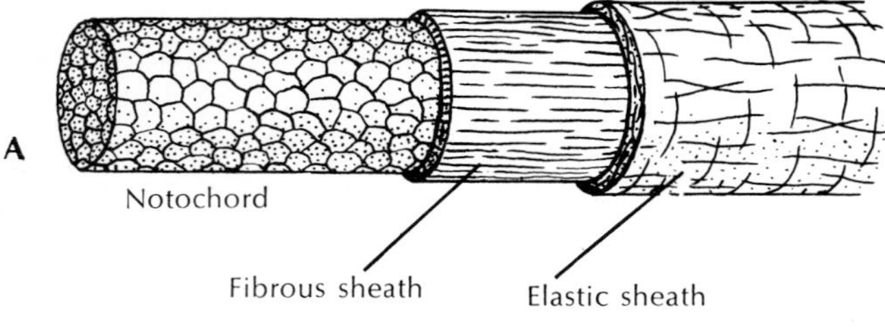

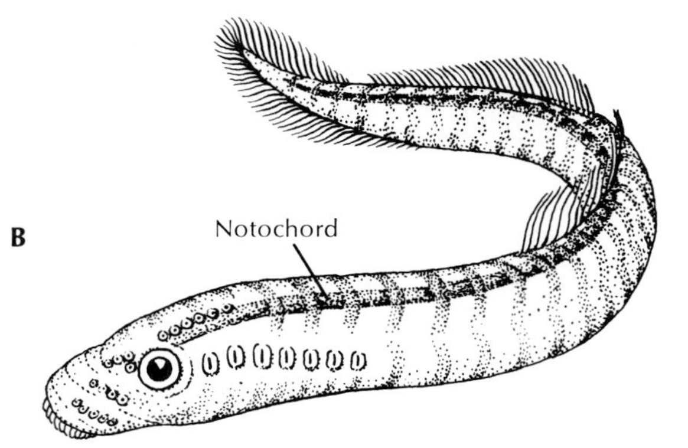

FIG. 22-1 **A,** Structure of the notochord and its surrounding sheaths. Cells of the notochord proper are thick-walled, pressed together closely, and filled with semifluid. Stiffness is caused mainly by turgidity of fluid-filled cells and surrounding connective tissue sheaths. This primitive type of endoskeleton is characteristic of all chordates at some stage of their life cycle. The notochord provides longitudinal stiffening of the main body axis, a base for myomeric muscles, and an axis around which the vertebral column develops. In hagfishes, **B,** and lampreys it persists throughout life, but in higher vertebrates it is largely replaced by the vertebrae. In humans, slight remnants are found in the nuclei pulposi of the intervertebral discs. Its method of formation is different in the various groups of animals. In amphioxus it originates from the endoderm; in birds and mammals it arises as an anterior outgrowth of the primitive streak.

forms a perfectly suitable endoskeleton, especially for an aquatic existence, and still is the first skeleton to appear in all larval and embryonic vertebrates. In the agnathans (hagfish and lampreys), the sharks and their kin, and even some primitive bony fishes such as sturgeons, the adult endoskeleton is composed strictly of cartilage. Bone appears in the endoskelton of more advanced vertebrates; it offers two clear advantages over cartilage, which are probable reasons for its evolution. First, it serves as a reservoir for phosphate, an indispensable component of ATP and an intermediate in several mainline enzymatic pathways in all cells. Stored phosphate provides an important resource for aquatic vertebrates during periods when low environmental phosphate would otherwise be a limiting factor to growth and maintenance. Second, only bone could provide the structural strength required for life on land, where mechanical stresses on the endoskeleton are far greater than they are in water.

The vertebrate endoskeleton, composed of either bone or cartilage, is a **living tissue,** whereas the invertebrate exoskeleton is dead noncellular material. Still, the exoskeleton has not been totally discarded; many vertebrates have a keratinoid exoskeleton, although here the exoskeleton serves mainly for protection rather than for the attachment of muscles. The endoskeleton has the advantage of allowing continuous growth, without the necessity of shedding. For this reason vertebrate animals can attain great size; some of them are the most massive in the animal kingdom. Endoskeletons provide much surface for muscle attachment, and size differences between animals result mainly from the amount of muscle tissue they possess. More muscle tissue necessitates greater development of body systems, such as circulatory, digestive, respiratory, and excretory. Thus it is seen that the endoskelton is the chief basic factor in the development and specialization of the higher animals.

From an evolutionary viewpoint the function of the skeleton, as represented by the exoskeleton and the endoskeleton, has shifted more from a protective to a supportive one. The limy shells of clams and other molluscs and the chitinous armor of arthropods are excellent defensive armors, even though they also serve for attachment of muscles and support of bodily structures. However, endoskeletons have their protective functions as well as supporting ones, as revealed by such excellent protective boxes as the cranium for the brain and the thoracic rib cage for important visceral organs.

Pharynx and efficient respiration. The perforated pharynx (gill slits), present in all chordates at some stage in their life cycle, evolved as a filter-feeding apparatus. In primitive chordates (such as amphioxus), water with suspended food particles is drawn through the mouth by ciliary action and flows out through the gill slits where the food is trapped in mucus. A later improvement was the replacement of cilia with a muscular feeding apparatus that pumps water through the pharynx by expanding and contracting the pharyngeal cavity. This adaptation led to the development of **internal gills.** Many invertebrate groups have gills, but they are typically feathery projections from the body surface and all lack the pharyngeal suction pump that makes the vertebrate gill the exceedingly effective device that it is. Some fish gills can remove 85% of the oxygen from water in its single pass across the gill surface. No invertebrate aquatic respiratory device can approach this efficiency. The tracheal system of arthropods is a highly efficient aerial respiratory device but is effective only in small animals. The chordate paired gill clefts and the internal gills that evolve from them are one of the most distinctive chordate features.

Advanced nervous system. No single system in the body is more strongly associated with functional and structural advancement than is the nervous system. Longitudinal nerve cords appear in various invertebrate groups but are typically paired and ventrally situated and contain nerve fibers mixed with clusters of nerve cells (ganglia). Only the chordates have a single, hollow, dorsal nerve cord, with clear separation of **brain,** wherein are concentrated nerve cells, from the **spinal cord,** which contains principally nerve fibers. Such a system permits the greatest possible utilization of space for well-integrated nervous patterns. Sense organs are developed well beyond those of any invertebrate group: paired eyes with lens and inverted retinas; pressure receptors, such as paired ears designed for equilibrium and later redesigned to include sound reception; and chemical receptors, including taste and the exquisitely sensitive olfactory organs of many vertebrates.

Classification of living chordates

There are three subphyla under phylum Chordata. Two of these subphyla are small, lack a vertebral column, and are of interest primarily as borderline or first chordates (protochordates). The third subphylum, Vertebrata, is subdivided into eight classes.

Phylum Chordata

Group Protochordata (Acrania)

Subphylum Urochordata (u'ro-kor-da'ta) (Gr., *oura,* tail, + L., *chorda,* cord, + *-ata,* characterized by) **(Tunicata)—tunicates.** Notochord and nerve cord in free-swimming larva only; sessile adults encased in tunic.

Subphylum Cephalochordata (sef'a-lo-kor-da'ta) (Gr., *kephalē,* head, + L., *chorda,* cord)—**lancelets** *(Amphioxus).* Notochord and nerve cord found along entire length of body and persist throughout life; fishlike in form.

Group Craniata

Subphylum Vertebrata (ver'te-bra'ta) (L., *vertebratus,* backboned). Bony or cartilaginous vertebrae surrounding spinal cord; notochord in all embryonic stages, persisting in some of the fish; also may be divided into two great groups (superclasses) according to presence of jaws.

Superclass Agnatha (ag'na-tha) (Gr., *a,* without, + *gnathos,* jaw) **(Cyclostomata)—hagfishes, lampreys.** Without true jaws or appendages.

Class Petromyzontes (pet'ro-my-zon'teez) (Gr., *petros,* stone, + *myzon,* sucking)—**lampreys.** Suctorial mouth with horny teeth; nasal sac not connected to mouth; seven pairs of gill pouches.

Class Myxini (mik-sy'ny) (Gr., *myxa,* slime)—**hagfishes.** Terminal mouth with four pairs of tentacles; buccal funnel absent; nasal sac with duct to pharynx; five to 15 pairs of gill pouches; partially hermaphroditic.

Superclass Gnathostomata (na'tho-sto'ma-ta) (Gr., *gnathos,* jaw, + *stoma,* mouth)—**jawed fishes, all tetrapods.** With jaws and (usually) paired appendages.

Class Chondrichthyes (kon-drik'thee-eez) (Gr., *chondros,* cartilage, + *ichthys,* a fish)—**sharks, skates, rays, chimaeras.** Streamlined body with heterocercal tail; cartilaginous skeleton; five to seven gills with separate openings, no operculum, no swim bladder.

Class Osteichthyes (os'te-ik'thee-eez) (Gr., *osteon,* bone, + *ichthys,* a fish) **(Teleostomi)—bony fishes.** Primitively fusiform body but variously modified; mostly ossified skeleton; single gill opening on each side covered with operculum; usually swim bladder or lung.

Class Amphibia (am-fib'e-a) (Gr., *amphi,* both or double, + *bios,* life)—**amphibians.** Poikilothermic tetrapods; respiration by lungs, gills, or skin; development through larval stage; skin moist, containing mucous glands, and lacking scales.

Class Reptilia (rep-til'e-a) (L., *repere,* to creep)—**reptiles.** Poikilothermic (some endothermic) tetrapods possessing lungs; embryo develops within shelled egg; no larval stage; skin dry, lacking mucous glands, and covered by epidermal scales.

Class Aves (ay'veez) (L., pl. of *avis,* bird)—**birds.** Endothermic vertebrates with front limbs modified for flight; body covered with feathers; scales on feet.

Class Mammalia (ma-may'lee-a) (L., *mamma,* breast)—**mammals.** Endothermic vertebrates possessing mammary glands; body covered with hair; well-developed brain.

TABLE 22-1 Divisions of the phylum Chordata

<table>
<tr><th>Urochordata (tunicates)</th><th>Cephalochordata (lancelets)</th><th>Petromyzontes (lampreys)</th><th>Myxini (hagfishes)</th><th>Chondrichthyes (sharks)</th><th>Osteichthyes (bony fishes)</th><th>Amphibia (amphibians)</th><th>Reptilia (reptiles)</th><th>Aves (birds)</th><th>Mammalia (mammals)</th></tr>
<tr><td colspan="10">← Chordata →</td></tr>
<tr><td colspan="2">← Protochordata →</td><td colspan="8">← Vertebrata →</td></tr>
<tr><td colspan="2">← Acrania →</td><td colspan="8">← Craniata →</td></tr>
<tr><td colspan="2"></td><td colspan="2">← Agnatha →</td><td colspan="6">← Gnathostomata →</td></tr>
<tr><td colspan="4"></td><td colspan="2">← Pisces →</td><td colspan="4">← Tetrapoda →</td></tr>
<tr><td colspan="2"></td><td colspan="5">← Anamniota →</td><td colspan="3">← Amniota →</td></tr>
</table>

The basic taxonomic division of the phylum Chordata is only one of several systems used to characterize different groups within the phylum. These are summarized in Table 22-1. A fundamental separation is the Protochordata from the Vertebrata. Since the former lack a well-developed head, they are also called Acrania. All vertebrates have a well-developed skull case enclosing the brain and are called Craniata.

Two basic branches of the vertebrates are the Agnatha, forms lacking jaws (lampreys and hagfish), and Gnathostomata, forms having jaws (all other vertebrates). These two groups represent early separations within the Vertebrata. The Gnathostomata in turn can be subdivided into Pisces, jawed aquatic vertebrates with limbs, if any, in the shape of fins; and the Tetrapoda, jawed vertebrates with two pairs of limbs. Still another fundamental separation among the vertebrates is based on embryologic patterns. The embryos of reptiles, birds, and mammals develop within a special fluid-filled membranous sac, the amnion. These are called the Amniota, and their shelled eggs can be laid on land. Fishes and amphibians, the Anamniota, lack this important adaptation.

ANCESTRY AND EVOLUTION

Since the early nineteenth century when the theory of organic evolution became the focal point for ferreting out relationships between groups of living organisms, zoologists have debated the question of chordate origins. Nearly every major invertebrate group has at one time or another been advanced as a candidate for the chordate ancestral group, and all, for one reason or another, have failed to meet the qualification requirements for undisputed election. The great chasm between chordates and invertebrates seems nearly as wide today as it did 150 years ago when zoologists began serious speculation on the issue.

The most serious obstacle to progress is the almost total absence of fossil material for reconstructing lines of descent. The earliest protochordates were in all probability soft-bodied forms that stood little chance of being preserved even under the most ideal conditions. Consequently, speculations must come from the study of living organisms, especially from an analysis of how they develop. But well over 600 million years have elapsed since the chordates had their beginnings, and even the highly conservative early developmental stages of animals cannot be expected to remain unchanged over such an enormous span of time. Nevertheless zoologists continue the attempt to reconstruct chordate origins by careful study of the available data, perhaps because, as one biologist puts it, it is an "interesting and enjoyable venture to speculate concerning the Cambrian and Precambrian happenings that have led to my own existence."*

Early theories of chordate evolution. The earliest speculations about chordate origins understandably focused on the most successful, and in many respects most advanced, of invertebrate groups, the Arthropoda. It was quickly recognized that if one took an arthropod with its segmented body, ventral nerve cord, and dorsal heart and turned it over, one has the basic plan of a vertebrate. The idea was first proposed by St. Hilaire in 1818 and received detailed support by subsequent zoologists. In 1875 Semper and Dohrn independently transferred the ancestral award to the Annelida because this group shares a basic body plan with the arthropods and in addition has an excretory system

*Berrill, N. J. 1955. The origins of vertebrates. Oxford, Oxford University Press.

that strikingly resembles that of primitive vertebrates. Although the annelid-vertebrate theory continued to receive support as late as 1922, it contained unresolvable difficulties. An inverted annelid has its brain and mouth in the wrong relative positions. The annelid's nerve cord is ventral but connects to a dorsal brain via circumpharyngeal connectives through which the digestive tube passes (Fig. 14-3, p. 306). When an annelid is inverted, as was required for the annelid-vertebrate theory, the mouth ends up on top of the head and the brain below. Most proponents of the theory devoted their efforts to explaining away this discrepancy. The annelid theory was eventually discarded, like the arthropod theory before it, when zoologists began to base their speculations on developmental patterns of animals rather than on adult forms, which are the highly differentiated and specialized end products of development.

Ribbonworm (nemertine) theory. Toward the end of the last century, the Dutch embryologist Hubrecht suggested that the vertebrates originated from the ribbonworms (nemertine worms, phylum Rhynchocoela). The argument rests on the presence in ribbonworms of a number of features that, in the course of evolution, could ''easily'' be modified into primitive vertebrate characteristics. For example, the nemertine proboscis could be modified into the stiffening notochord to permit fishlike movement through the water. The pharyngeal structures of certain present-day nemertines, if serially perforated, would closely resemble the typical chordate gill slits. Several other potential vertebrate features have been noted. Although ribbonworm embryology differs significantly from that of the vertebrates, and the group is a very primitive one whose own origins are presumably separated by a vast span of time from vertebrate origins, the ribbonworm theory has several appealing features that explain its recent revival (Willmer, 1974).

Echinoderm theory. When, early in this century, further theorizing became rooted in developmental patterns of animals, it immediately became apparent that the echinoderms deserved serious consideration as the chordates' ancestor. Echinoderms and chordates belong to the deuterostome branch of the animal kingdom in which the mouth is formed as a secondary opening and the blastopore of the gastrula becomes the anus. The coelom of these two phyla is primitively enterocoelous: it is budded off from the archenteron of the embryo. Furthermore, there is a great resemblance between the bipinnaria larvae of certain echinoderms and tornaria larvae of the hemichordates, a phylum bearing some chordate characteristics. Both have similar ciliated bands in loops, sensory cilia at the anterior end, and a complete digestive system with ventral mouth and posterior anus. Both echinoderm and chordate embryos show indeterminate cleavage; that is, each of the early blastomeres has equivalent potentiality for supporting full development of a complete embryo. These characteristics are shared by brachiopods and pterobranchs (a hemichordate group) as well as by echinoderms, protochordates, amphioxus, and vertebrates. This is probably a natural grouping and almost certainly indicates interrelationships, although remote.

Efforts also have been made to relate primitive fossil echinoderms (stem-bearing forms similar to sea lilies today) with the vertebrate ostracoderms, a group of primitive, jawless fishes. Both groups have excellent fossil records and both coexisted during the Ordovician period, some 500 million years ago. But this approach again is an attempt to pin together affinities by looking at differentiated adult forms and is the same flawed approach that misled earlier zoologists to relate annelids and vertebrates.

However, the uncertainty of fossil evidence in no way invalidates the embryologic evidence for a remote kinship between echinoderms and chordates. It remains the best guess we have (Fig. 22-2). At the same time, we must acknowledge that there is no compelling reason to designate *any* invertebrate group as chordate ancestral stock. The vertebrates may have arisen independently from a simple unicellular protist stage, as the Canadian biologist J. R. Nursall (1962) has suggested, with all intermediate types having been lost.

Unable to narrow the search for an invertebrate-chordate connecting link any further, zoologists are presently focusing on groups within the chordate phylum itself, which share enough anatomic and developmental features to make unraveling the evolutionary past a more profitable effort.

Candidates for vertebrate ancestral stock

As discussed previously, all members of the phylum Chordata share four anatomic features at some time in their life histories: a notochord, forming a stiff mechanical axis for the body; a dorsal tubular nerve cord; pharyngeal gill slits used for breathing and feeding; and a postanal tail for larval motility. The three

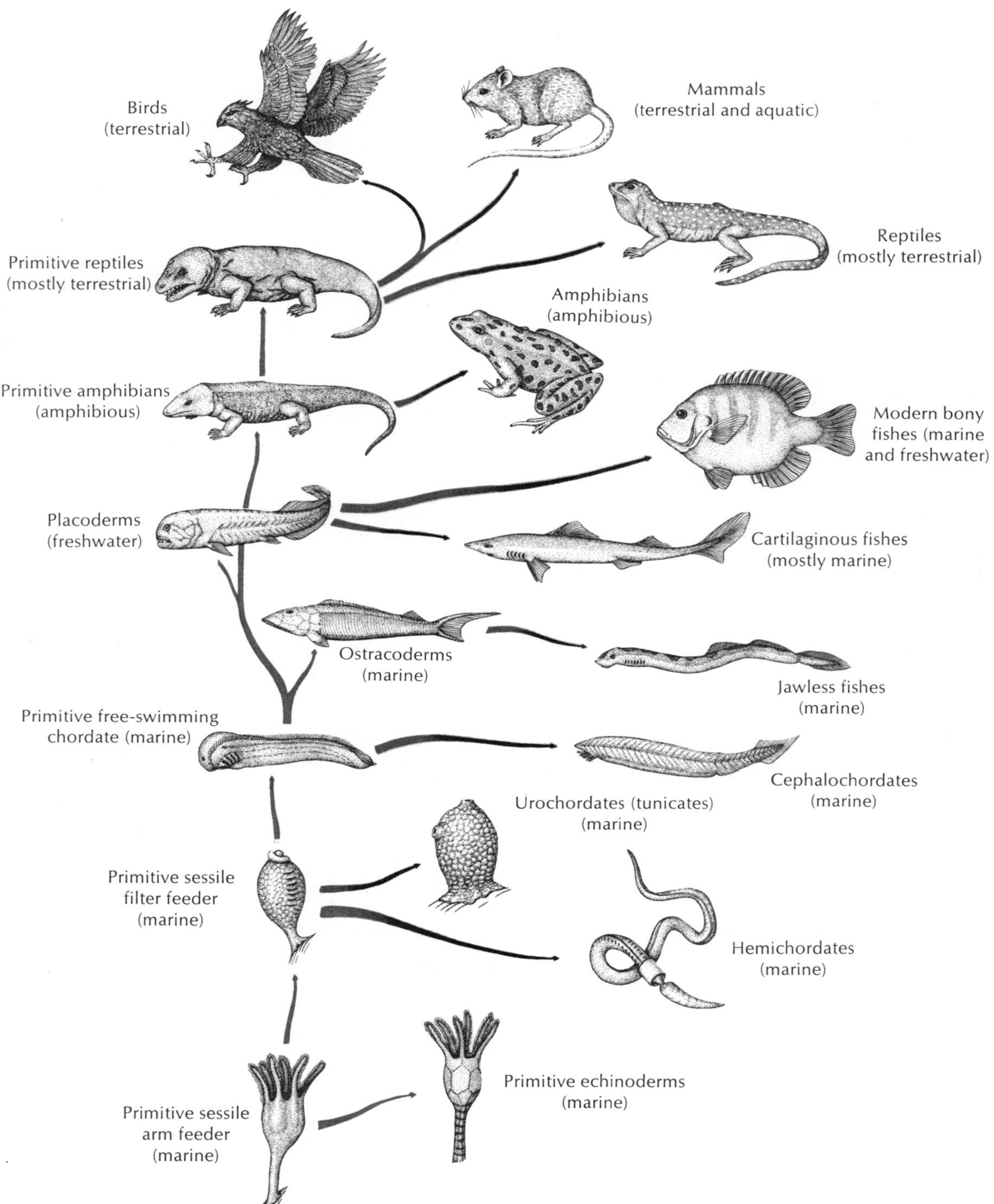

FIG. 22-2 Hypothetical family tree of the chordates, suggesting probable origin and relationships. *Red,* Extinct stem groups. *Black,* Living groups.

subphyla that bear these characteristics are the Cephalochordata (of which the lancelet amphioxus is its famous representative), the Urochordata (tunicates, or sea squirts), and the Vertebrata. The vertebrates comprise by far the greatest number of the chordates, so the terms "vertebrate" and "chordate" are frequently used (somewhat incorrectly) as synonyms. The backboneless members of the phylum—cephalochordates and urochordates—are usually referred to as protochordates or prevertebrates and have long been considered good candidates for the vertebrate ancestral stock.

The Hemichordata, described in the preceding chapter, previously were also assigned to the phylum Chordata since they possess a perforated pharynx (gill slits) and a middorsal thickening of nervous tissue, which may be a forerunner of the vertebrate tubular nerve cord. However the stomochord of the hemichordates (see Fig. 21-3) cannot be homologized to the notochord as was once believed. Furthermore the gill slits and dorsal nerve center may represent a convergent resemblance to similar chordate structures rather than a true homology. Consequently the hemichordates are usually excluded from chordate membership although most biologists agree that a real, though remote, relationship exists.

The other group formerly included with the chordates are the pogonophores. This is a highly aberrant and specialized group that possesses almost none of the diagnostic chordate features. Yet they are deuterostomes, and their tripartite body organization and certain other characteristics indicate a relationship to the hemichordates. If we are to admit a remote affinity between hemichordates and chordates, we must do the same for the pogonophores.

Position of amphioxus. The problems of sorting out lines of descent from this assemblage of candidates are enormous because most of the prechordates and protochordates are highly specialized remnants of once successful and well-represented groups. The first approach was to place the chordate and prechordate groups in an order of increasing morphologic complexity leading to the vertebrates. By this analysis, amphioxus becomes the logical structural ancestor because it possesses as an adult all four chordate hallmarks plus five vertebrate characteristics: segmented musculature, the beginning of optic and olfactory sense organs, a liver diverticulum, beginnings of a ventral heart, and separation of dorsal and ventral spinal roots in the vertebrate style (Figs. 22-10 and 22-11). Little wonder that amphioxus once attained a pinnacled position among zoologists searching for their vertebrate ancestor.

But on closer scrutiny, amphioxus failed to meet the qualifications for a generalized ancestral type. Its notochord is overdeveloped into a forward extension for its specialized burrowing mode of life and it effectively prevents the development of a proper brain. Its kidney is a solenocyte type that bears little resemblance to the vertebrate glomerular-tubular nephron. Its unique atrium has no vertebrate counterpart and there is a non–vertebrate-like proliferation of gill slits. Amphioxus today is usually regarded as a highly specialized and degenerative member of the chordate family: it lies as an offshoot, rather than in the main line, of chordate descent.

Urochordata and recapitulation. Attention then became focused on the alternative protochordate group, the Urochordata (tunicates). This group is composed of three groups of which the ascidians (sea squirts) are the most common and the simplest. At first glance, more unlikely candidates for vertebrate ancestor could hardly be imagined. As adults, ascidians are virtually immobile forms surrounded by a tough, cellulose-containing tunic of variable color. Their adult life is spent in one spot attached to some submarine surface, filtering vast amounts of seawater from which they extract their planktonic food. As adults (Fig. 22-6) they lack notochord, tubular nerve cord, postanal tail, sense organs, and segmental musculature. Superficially they resemble sponges far more than they resemble any known vertebrate. However, the chordate nature of the ascidians is abundantly evident in their tadpole larvae. These tiny, active, site-seeking forms (Fig. 22-9) have all the right qualifications for membership in the prevertebrate club: notochord, hollow dorsal nerve cord, gill slits, postanal tail, brain, and sense organs (otolith balance organ and an eye complete with lens).

The discovery of this form in 1869 not only placed the urochordates squarely in the vertebrate camp but greatly influenced E. Haeckel in formulating his theory of recapitulation (biogenetic law; see Principle 7 on p. 11). According to this theory, adult stages of ancestors are repeated during the embryonic development of their descendants; in other words the development of an organism is an accurate record of past evolutionary history. We recognize now that this record is very slurred and telescoped and must be interpreted

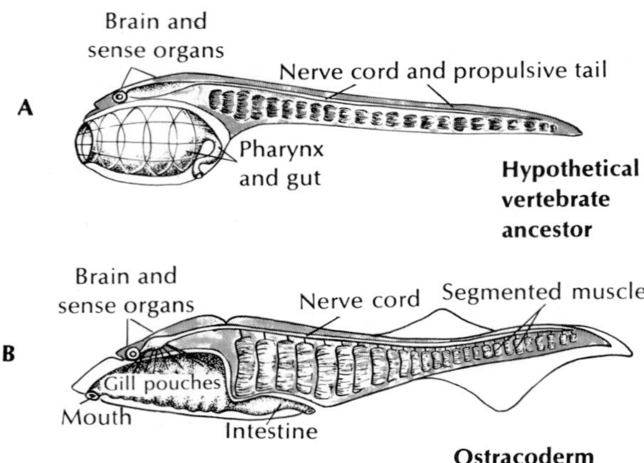

FIG. 22-3 Hypothetical vertebrate ancestor compared with an ostracoderm, earliest fossil vertebrate. According to the Garstang hypothesis, the vertebrate common ancestor was an ascidian tadpole larva that became neotenous and evolved into a new free-swimming species. Continued evolution led to the jawless ostracoderms and jawed placoderms. (Redrawn from Roe, A., and Simpson, G. G. 1958. Behavior and evolution. New Haven, Yale University Press.)

with caution. But at the time the true nature of the ascidian tadpole larva was first understood, it was considered to be a relic of an ancient free-swimming chordate ancestor of the ascidians. Adult ascidians then came to be regarded as degenerate, sessile descendants of the ancient chordate form.

Garstang's theory of chordate larval evolution. It remained for W. Garstang in England (1928) to introduce totally fresh thinking to the vertebrate ancestor debate. In effect, Garstang turned the sequence around: rather than the ancestral tadpole larva giving rise to a degenerative sessile ascidian adult, he suggested that the ancestral chordate stock was primarily a filter-feeding sessile marine group not unlike modern ascidians. The free-swimming tadpole larva was evolved from these attached, bottom-dwelling forms to meet the need for seeking out new habitats. Thus the tadpole larva was visualized as an ascidian creation, evolved within the group to enhance site-seeking capabilities. Garstang next suggested that at some point the tadpole larva became paedogenetic, that is, became capable of maturing gonads and reproducing in the larval stage. With continued larval evolution, a new group of free-swimming animals would appear. The best evidences for this theory are found in the living tunicates today, especially among the two planktonic groups, the thaliaceans and the larvaceans. In the latter group (Fig. 22-8, *A*), the basic larval form is retained throughout life; they are in effect paedogenetic tunicates, although extremely specialized.

Garstang departed from previous thinking by suggesting evolution may occur in the larval stages of animals. This idea received slow acceptance by zoologists accustomed to thinking of developmental stages as being largely insulated from change, as embodied in the "biogenetic law." Yet, in all likelihood an evolutionary sequence similar to that proposed by Garstang occurred. Garstang's theory has received detailed support from the zoologist N. J. Berrill, who, however, disagrees with Garstang in the way the evolutionary sequence proceeded.

The ascidian tadpole larva with its propulsive tail, stiffening notochord, and dorsal nerve cord that integrates sensory information and motor activity clearly suggests and foreshadows the vertebrate line. The resemblance of an early vertebrate ostracoderm to this hypothetical ancestor is suggested in Fig. 22-3.

Summary. Let us summarize our reconstruction of possible vertebrate origins. We have hypothesized that the primitive chordate ancestor evolved from the echinoderms, which share several embryologic affinities with the chordates. In our judgment, these evidences make them stronger candidates for chordate ancestry than the nemertine worms, despite the recent revival of interest in the latter group as possible candidates. The ancient echinoderm is thought to have been a sessile form that trapped food on waving, tentacular arms. It may have *resembled* present-day crinoids, but we are not inferring direct descendance from crinoids.

The next step was improvement of the feeding apparatus by replacing arms with an internal filtering system, a stage that is partly preserved in modern adult ascidians. The larvae of these immobile forms were probably bilaterally symmetric planktonic forms similar to larvae of echinoderms and certain hemichordates. Then followed a crucial step: the evolution of a free-swimming tadpolelike larva, better able to seek out new habitats and promote rapid spread of progeny. The tadpole larva next became reproductively active by paedogenesis, a well-known phenomenon among modern vertebrates. Then, by dropping the sessile adult stage from this independent, free-swimming

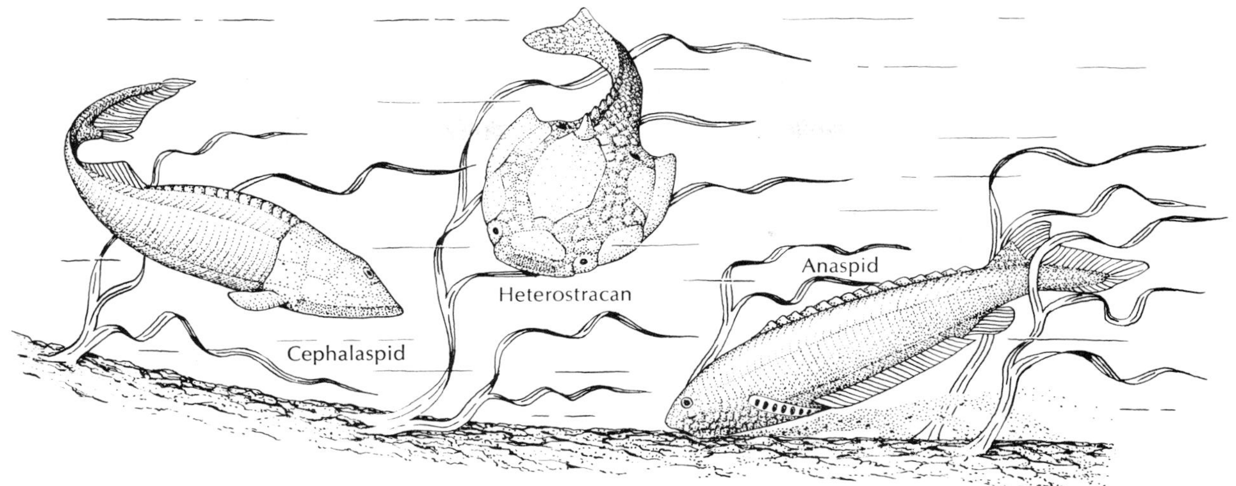

FIG. 22-4 Three ostracoderms, jawless fishes of Silurian and Devonian times. Representatives of three of the best known ostracoderm groups are illustrated as they might have appeared while searching for food on the floor of a Devonian sea. All were filter feeders, drawing water and organic debris into the mouth, straining out the organic matter, and expelling the water through the gill openings and between the ventral head plates.

form, the larva began its own evolutionary line, bearing the basic vertebrate framework.

Jawless ostracoderms—earliest vertebrates

The earliest vertebrate fossils are fragments of bony armor discovered in Ordovician rock in Russia and in the United States. They were small, jawless creatures collectively called **ostracoderms** (os-trak′o-derm) (Gr., *ostrakon,* shell, + *derma,* skin), which belong to the Agnatha division of the vertebrates. These earliest ostracoderms, called **heterostracans** (Fig. 22-4), lacked paired lateral fins that subsequent fishes found so important for stability. Their swimming movements must have been inefficient and clumsy, although sufficient to propel them from one mud bed to another where they practiced their mud-grubbing search for nutrients in the form of organic debris on the sea bottom. Later in the Silurian period and especially in the Devonian period the heterostracan ostracoderms underwent a major radiation, resulting in the appearance of several peculiar-looking forms varying in shape and length of the snout, dorsal spines, and dermal plates. Most continued their mud-loving existence, although at least one species became a surface feeder. Without ever evolving paired fins or jaws, the heterostracans dominated the early Devonian period until eclipsed by another ostracoderm group, the **cephalaspids.**

The cephalaspids improved the efficiency of a benthic life by evolving paired fins. These fins, located just behind the head shield, provided control over pitch and yaw that ensured well-directed forward movement. The best known genus in this group is *Cephalaspis* (Fig. 22-4), the subject of a brilliant and classic series of studies by the Swedish paleozoologist E. A. Stensiö. *Cephalaspis* was a small animal, seldom exceeding 30 cm in length, and was covered by a well-developed armor, the head by a solid shield (rounded anteriorly) and the body by bony plates. It had no axial skeleton or vertebrae. The mouth was ventral and anterior, and it was jawless and toothless. Its paired eyes were located close to the middorsal line. A pineal eye was also present, in front of which was a single nasal opening. At the lateroposterior corners of the head shield were a pair of flaplike fins. The trunk and tail were apparently adapted for active swimming. Between the margin of the head shield and the ventral plates there were ten gill openings on each side. They also had a lateral line system. They were adapted for filter feeding, which may explain the large expanded size of the head, made up as it is of a large pharyngeal gill-slit filtering apparatus.

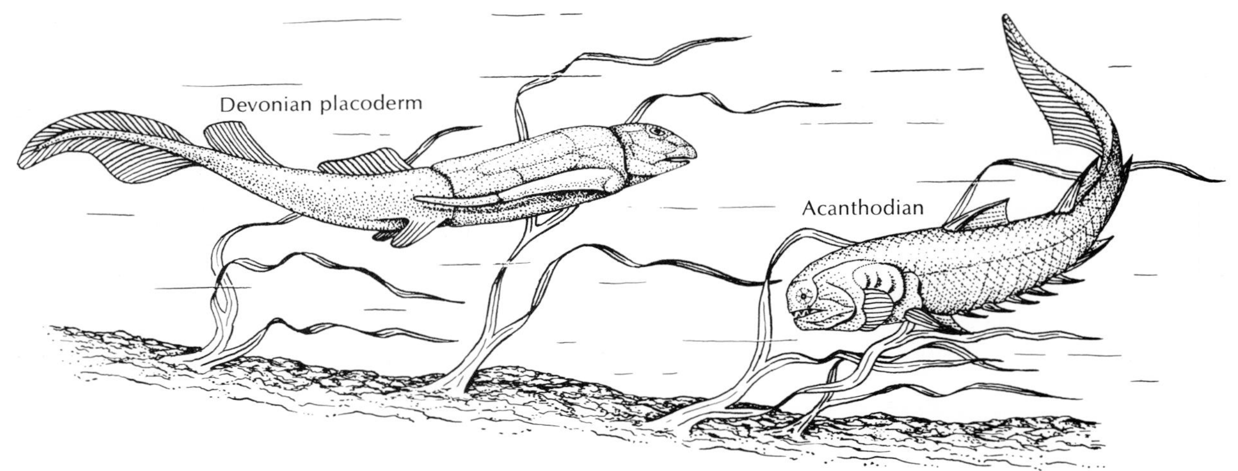

FIG. 22-5 Early jawed fishes of the Devonian period, 400 million years ago. The placoderm *(left)* and a related acanthodian *(right)* were highly mobile and voracious forms. Though more successful than the less maneuverable ostracoderms, they eventually failed in competition with their successors, the bony and cartilaginous fishes.

A third major group of ostracoderms, the **anaspids,** were clearly related to the cephalaspids although they looked very different (Fig. 22-4). Like the cephalaspids, they had paired fins, but these were peculiar, elongate, posterior stabilizers that formed a part of the caudal (tail) fin. The anaspids were much more streamlined than other ostracoderms and are believed to have practiced an active mode of life. Some paleontologists believe that one of the two groups of living agnathans, the lampreys, may have evolved from the anaspids (Fig. 23-2, p. 498).

As a group, the ostracoderms were basically fitted for a simple, bottom-feeding life. Yet despite their anatomic limitations, they enjoyed a respectable radiation in the Silurian and Devonian periods. Their overall contribution was enormous, for they provided a blueprint for subsequent vertebrate evolution. But they could not survive the competition of the more advanced jawed fishes that began to dominate the Devonian period, and in the end they disappeared.

Early jawed vertebrates

All jawed vertebrates, whether extinct or living, are collectively called gnathostomes (''jaw mouth''), in contrast to jawless vertebrates, the agnathans (''without jaw''). The latter are also often referred to as cyclostomes (''circle mouth'').

The first jawed vertebrates to appear in the fossil record were the **placoderms** (plak′o-derm) (Gr., *plax,* plate, + *derma,* skin) (Fig. 22-5). The advantages of jaws are obvious, since they allowed predation on large and active forms of food. The possessors of jaws would enjoy a tremendous advantage over jawless vertebrates, which were restricted to a wormlike existence of sifting out organic debris and small organisms in the bottom mud. Jaws arose through modifications of the first two of the serially repeated cartilaginous gill arches. The beginnings of this trend can, in fact, be seen in some of the jawless ostracoderms where the mouth became bordered by strong dermal plates that could be manipulated somewhat as jaws with the gill arch musculature. The more anterior arches continued a gradual modification to permit more efficient seizing, and the skin surrounding the mouth was modified into teeth. Eventually the anterior gill arches became bent into the characteristic position of vertebrate jaws, as seen in the placoderms.

The early jawed fishes had well-developed lateral fins. These may have arisen as lateral folds in the body wall, which broke up into a series of fins, such as is found in the **acanthodians,** a primitive group contemporary with the placoderms (Fig. 22-5). In these, however, the anterior and posterior pairs were larger than the others and eventually became the pectoral and pelvic fins; the intermediate ones disappeared. Some paleontologists believe, however, that these numerous paired fins may have arisen independently as separate structures. Their internal skeleton so far as is known was composed of bone.

Placoderms evolved into a great variety of forms, some large and grotesque in appearnce. They were armored fishes, covered with diamond-shaped scales or with large plates of bone. All became extinct by the end of the Paleozoic era.

Evolution of modern fishes and tetrapods

Reconstruction of the origins of the vast and varied assemblage of modern living vertebrates is, as we have seen, based in large part on fossil evidence. When paleozoologists sit down with an array of fossil forms before them, many of them represented by more or less incomplete exoskeletal remains, they immediately encounter a conflict in their attempts to prove or disprove relationships between two distinct fossil groups. They see on the one hand a number of structural features that are shared by both groups: this suggests relationship. They also see features that are widely different for the two groups: this suggests early separation of lines of evolution. Their task then is to decide whether shared features are **primitive** ones, that is, features that have been inherited from a common ancestor, or whether shared features have arisen independently because of similar habitats or some kind of retrogressive development. As for widely differing characteristics, they must decide whether these are latter-day specializations of little consequence or whether they represent fundamental and ancient departures in evolutionary trends. These problems obviously are complex, but presumably they are what keep paleontologists excited with their chosen profession. The outcome is that paleontologists usually can only suggest group memberships; they seldom can make the decisive, indisputable judgments that students (and zoology professors too, for that matter), who are not especially interested in the details, would like to have.

Although the earliest vertebrates tell us much less than we would like about subsequent trends in evolution, affinities become much easier to establish as the fossil record improves in more recent geologic times. For instance, the descent of birds and mammals from reptilian ancestors has been worked out in a highly convincing manner from the relatively abundant fossil record available. By contrast, the ancestry of modern fishes is shrouded in uncertainty. The Swedish paleontologist E. Jarvik has emphasized that the main vertebrate stem groups (such as agnathans, lungfishes, sharks, bony fishes, and stem tetrapods) became anatomically specialized some 400 to 500 million years ago and have changed relatively little since then. Thus main evolutionary lines as seen in the fossil record run back almost in parallel; if extended backward to their illogical extreme, they would hardly ever meet. Obviously they must meet at some point in the distant past, but this exercise reveals that the crucial separations in vertebrate evolution occurred in the Cambrian period, perhaps even the Precambrian period, long before the fossil record became established for the convenience of paleozoologists.

Despite the difficulties of establishing early lines of descent for the vertebrates, they are a natural, monophyletic group, distinguished by a great number of common characters. They have almost certainly descended from a common ancestor, the nature of which we have already discussed. Very early in their evolution, the vertebrates divided into two great stems, the agnathans and the gnathostomes. These two groups differ from each other in many fundamental ways, in addition to the obvious lack of jaws in the former group and their presence in the latter. Thus both groups are very old and of about the same age. On this basis we cannot say that agnathans are more "primitive" than gnathostomes even though the latter have continued on a marvelous evolutionary advance that produced most of the modern fishes, all of the tetrapods, and the reader of this book. Although the agnathans are represented today only by the hagfishes and the lampreys, these creatures too are successful in their own way.

CHORDATE SUBPHYLA

Subphylum Urochordata (Tunicata)

The tunicates (Fig. 22-6) are widely distributed in all seas from near the shoreline to great depths. Most of them are sessile, at least as adults, although some are free-living. The name "tunicate" is suggested by the nonliving tunic (test) that surrounds them and contains cellulose. A common name for them is sea squirt because some of them discharge water through the excurrent siphon when irritated. They vary in size from microscopic forms to several centimeters in length.

As a group they may be considered as degenerative

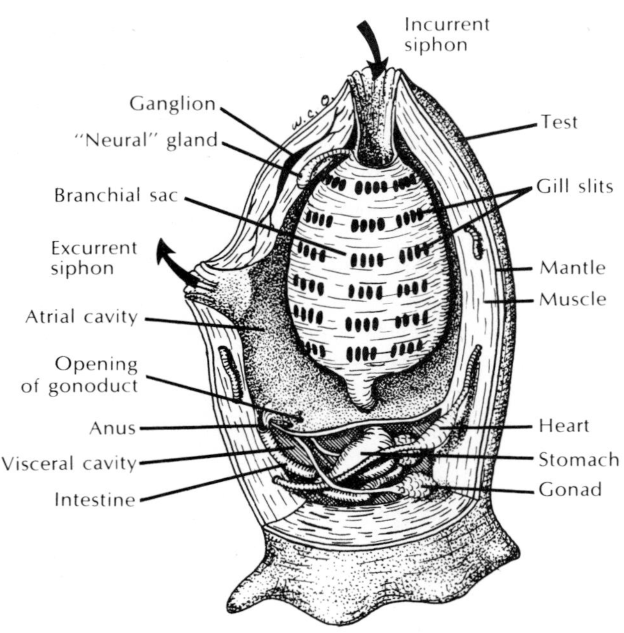

FIG. 22-6 Structure of adult solitary, simple ascidian. *Arrows* indicate direction of water currents.

FIG. 22-7 A compound sea squirt *Botryllus schlosseri,* common in shallow coastal waters and rock tidepools. Each of the star-shaped patterns represents a colonial arrangement in which the arms of the star are individuals, each with its own incurrent siphon at the end of the arm. All are united centrally where they share a common test, forming a compound ascidian. (Approximately ×2.) (Photograph by D. P. Wilson.)

or specialized chordates, for the adults lack many of the common characteristics of chordates. For a long time they were classified among the molluscs. In 1866 Kowalevsky, the Russian embryologist, worked out their true position.

Urochordata is divided into three classes: Ascidiacea, Larvacea, and Thaliacea. Of these, the members of **Ascidiacea,** commonly known as the ascidians, are by far the most common and the best known. One of this group is described later as a representative tunicate. Ascidians may be solitary, colonial, or compound (Fig. 22-7). Each of the solitary and colonial forms has its own test, but among the compound forms many individuals may share the same test. In some of these compound ascidians each member has its own **incurrent siphon,** but the **excurrent opening** is common to the group. Most ascidians are monoecious, but some are dioecious, and they can also reproduce asexually by budding or gemmation. The larvae may develop outside or in the atrium of the parent. Many ascidians are among the most vividly colored invertebrates.

The **Larvacea** and **Thaliacea** are pelagic (open sea) forms and are not often found in the intertidal zones where ascidians are common. The members of Larvacea are small, tadpolelike forms under 5 mm in length and resemble ascidian tadpoles (Fig. 22-8, *A*). They perhaps represent persistent larval forms that have become paedogenetic. They secrete around themselves cellulose tunics and are filter feeders. *Oikopleura* is a larvacean form often collected in tow-net samplings of ocean plankton.

The class Thaliacea is made up of members that may reach a length of 8 to 10 cm. Their transparent body is spindle-shaped or cylindric and is surrounded by bands of circular muscles, with their incurrent and excurrent siphons at opposite ends (Fig. 22-8, *B*). They are mostly carried along by currents, although by contracting their circular muscle bands they can force water out of their excurrent siphons and can move by jet propulsion. Many are provided with luminous organs and give a brilliant light at night. Most of the body is hollow, with the viscera forming a compact mass on the ventral side. They appear to have come from attached ancestors as did the ascidians. Some of them have complex life histories. In forms like *Doliolum* there is alternation of generations between sexual and asexual forms. After hatching from the egg, the larval tadpole changes into a barrel-shaped nurse or asexual stage, which produces small buds on a ventral stolon. These buds break free, become attached to another part of the parent, and develop into three kinds of individuals; one kind breaks free to become the sexual stage. *Salpa* also has alternation of generations and is a common form along the Atlantic coast.

Adult ascidian—Molgula

Molgula is globose in form and is attached by its base to piles and stones. Lining the test or tunic is a membrane or **mantle.** On the outside are two projections: the **incurrent** and **excurrent siphons** (Fig. 22-6). Water enters the incurrent siphon and passes into the pharynx (branchial sac) through the mouth. On the midventral side of the pharynx is a groove, the **endostyle,** which is ciliated and secretes mucus. Food material in the water is entangled by the mucus in this endostyle and carried into the esophagus and stomach. The intestine leads to the anus near the excurrent siphon. The water passes through the pharyngeal slits in the walls of the pharynx into the atrial cavity. As the water passes through the slits, respiration occurs.

The circulatory system contains a ventral **heart** near the stomach and two large vessels, one connected to each end of the heart. The action of the heart is peculiar in that it drives the blood first in one direction and then in the other. This reversal of blood flow is found in no other animal. The excretory system is a type of nephridium near the intestine. The nervous system is restricted to a nerve ganglion and a few nerves that lie on the dorsal side of the pharynx. A notochord is lacking. The animals are hermaphroditic, for both ovaries and testes are found in the same animal. Ducts lead from the gonads close to the intestines and empty near the anus. The germ cells are carried out the excurrent siphon into the surrounding water, where cross-fertilization occurs.

Of the four chief characteristics of chordates, adult tunicates have only one: the pharyngeal gill slits. However, the larval form gives away the secret of their true relationship. The tadpole larva (Fig. 22-9) is an elongate, transparent form with all four chordate characteristics: a notochord, a hollow dorsal nerve cord, a propulsive postanal tail, and a large pharynx with endostyle and gill slits. The larva does not feed but swims about for some hours before fastening itself vertically by its adhesive papillae to some solid object. It then undergoes a retrograde metamorphosis to become the sessile adult.

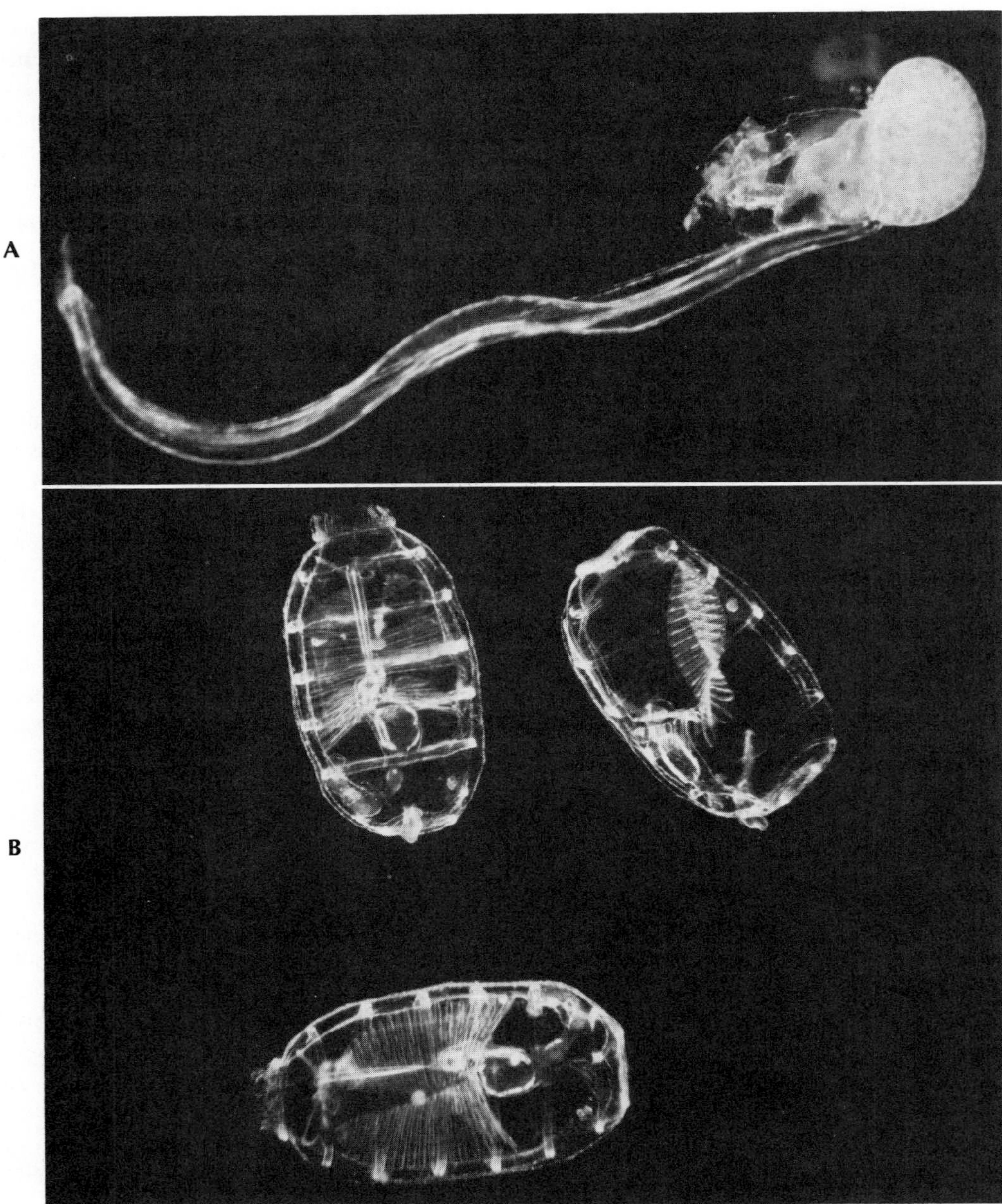

FIG. 22-8 **A,** A larvacean *Oikopleura dioca*—a typical tadpolelike member of the class Larvacea. (Approximately ×55.) **B,** A thaliacean *Doliolum nationalis*—a member of the class Thaliacea. (Approximately ×25.) (Photographs by D. P. Wilson.)

Subphylum Cephalochordata

Amphioxus

Subphylum Cephalochordata is the most interesting of all the protochordates, for one of its members is the lancelet *Branchiostoma (Amphioxus)* (Fig. 22-10), one of the classic animals in zoology. Cephalochordates are found mainly on the sandy marine beaches of southern waters, where they burrow in the sand, with the anterior end projecting out. One American species, *B. virginiae,* is found from Florida to the Chesapeake

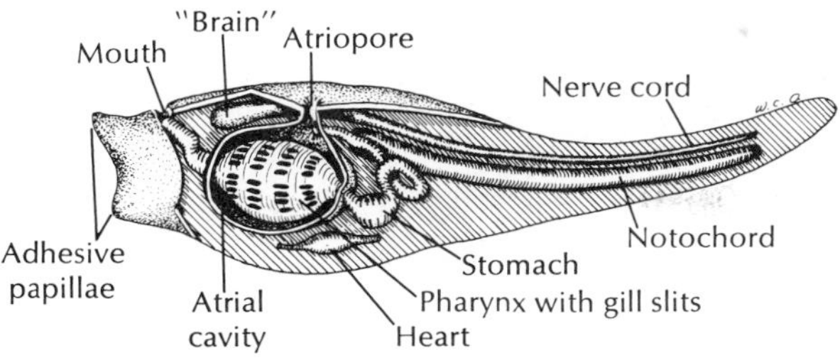

FIG. 22-9 Structure of tunicate (ascidian) tadpole larva. This larva is believed to closely resemble the ancestor of all vertebrates and shows all four principal chordate characteristics—notochord, dorsal nerve cord, pharyngeal gill slits, and postanal tail.

FIG. 22-10 Amphioxus. This interesting bottom-dwelling cephalochordate possesses the four distinctive chordate characteristics (notochord, dorsal nerve cord, pharyngeal gill slits, and postanal tail) that once made it the prime candidate for our vertebrate ancestor. However, because it also bears many specialized and degenerate features, zoologists now consider it a divergent offshoot from the main line of chordate evolution. (Courtesy B. Tallmark.)

Bay. Altogether there are 28 species, of which four are North American, scattered over the world. They can swim in open water by swift lateral movements of the body.

Amphioxus is especially interesting, for it has the four distinctive characteristics of chordates in simple form, and in other ways it may be considered a blueprint of the phylum. It has a slender, laterally compressed body, 5 to 6 cm long (Figs. 22-10 and 22-11), with both ends pointed. There is a long **dorsal fin,** which passes around the tail end to form the **caudal fin.** A short **ventral fin** is also found. These fins are reinforced by **fin rays** of connective tissue. The ventral side of the body is flattened and bears along each side a **metapleural fold.** There are three openings to the outside: the ventral anterior **mouth,** the **anus** near the base of the caudal fin, and the **atriopore** just anterior to the ventral fin.

The body is covered with a soft **epithelium,** one layer thick, resting on some connective tissue. The **notochord,** which extends almost the entire length of the body, is made up of cells and gelatinous substance enclosed in a sheath of connective tissue. Above the notochord is the tubular dorsal **nerve cord,** with a slight dilation at the anterior end known as the **cerebral vesicle.** Along each side of the body and tail are the numerous <-shaped **myotomes,** or muscles, which have a metameric arrangement (Fig. 22-11). The myotomes are separated from each other by **myosepta** of connective tissue. The myotomes of the two sides alternate with each other.

The anterior end of the body is called the **rostrum.** Just back of this and slightly below is a median opening surrounded by a membrane, the **oral hood,** which bears some twenty **oral tentacles (buccal cirri).** The oral hood encloses the chamber known as the **vestibule,** at the bottom of which lies the true **mouth** with a membrane, the **velum,** around it. Around the mouth are 12 **velar tentacles.** The cirri and tentacles serve to strain out large particles and have sensory functions. Ciliated patches on the walls of the buccal cavity in front of the velum produce a rotating effect and are

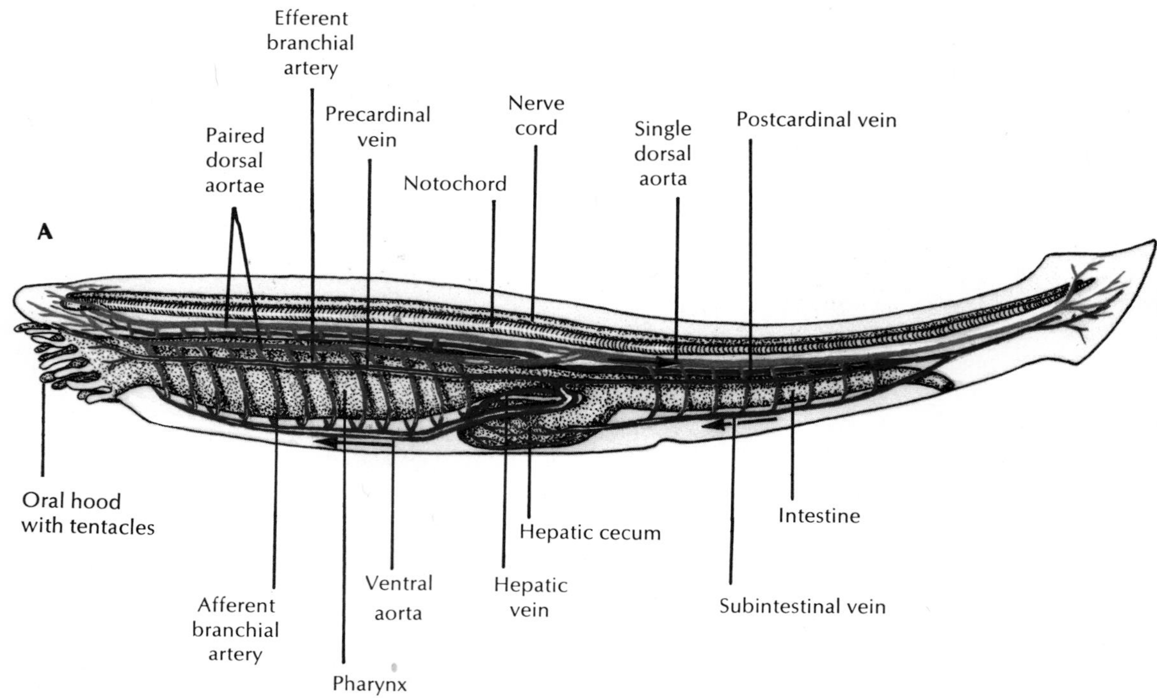

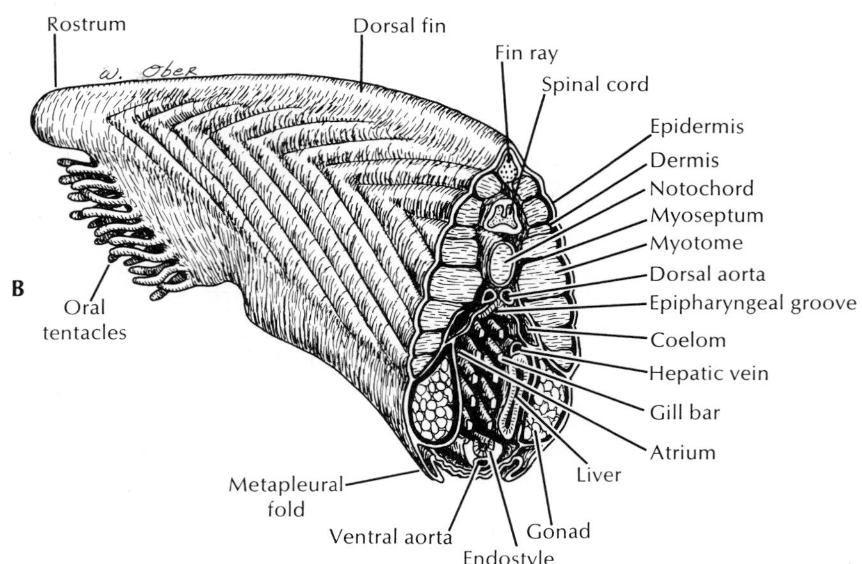

FIG. 22-11 **A,** Structure of amphioxus showing scheme of circulation. **B,** Cross section of anterior end of amphioxus.

called the **wheel organ,** which propels water currents. On the dorsal side of the oral hood is Hatschek's groove and pit, an embryonic relic from the first coelomic sac on the left. It may be homologous with the pituitary of vertebrates.

Just behind the mouth is the large compressed pharynx with more than a hundred pairs of **gill slits,** which act as strainers in filter feeding as well as in respiration. From the pharynx the narrow tubular **intestine** extends backward to the anus. On the ventral side of the intestine is a large diverticulum, the **hepatic cecum.** The **coelom** is reduced and is confined to the region above the pharynx and around the intestine (Fig. 22-11). Connecting the coelom to the atrium are about a hundred pairs of ciliated **nephridia** of the solenocyte type, a modified kind of flame cell. The big cavity around the pharynx is the **atrium;** it is lined with ectoderm and is therefore not a coelom. The pharynx has a middorsal groove, the **hyperbranchial** (epipharyngeal) **groove,** and a midventral one, known as the **endostyle.** Both of these grooves are lined with cilia and gland cells (Fig. 22-11). Food is entangled by the mucus of the endostyle and is carried to the intestine; the water passes through the gill slits into the atrium and gives up oxygen to the blood vessels in the **gill bars.**

Although there is no heart, the **blood system** is similar to that of higher chordates (Fig. 22-11). The blood moves posteriorly in the **dorsal aorta** and anteriorly in the **ventral aorta;** a **hepatic portal** vein leads from the intestines to the liver. Blood is propelled by contractions of the ventral aorta and is carried to the dorsal aorta by vessels in the gill bars. The blood is almost colorless, with a few red corpuscles.

The **nervous system** is above the notochord and consists of a single dorsal **nerve cord,** which is hollow. This nerve cord gives off a pair of ''nerves'' alternately to each body segment, or myotome, as dorsal and ventral roots. Although this arrangement resembles the vertebrate spinal cord, the ventral root actually does not contain nerves but rather the drawn-out extensions of myotomic muscle fibers. Ciliated cells with **sensory** functions are found in various parts of the body.

Sexes are separate, and each sex has about 25 pairs of gonads located on the wall of the atrium. The sex cells are set free in the atrial cavity and pass out the atriopore to the outside, where fertilization occurs. Cleavage is total (holoblastic), and a gastrula is formed by invagination. The larva hatches soon after deposition and gradually assumes the shape of the adult.

No other chordate shows so well the basic diagnostic chordate characteristics as does the amphioxus. Not only are the four chief characters of chordates — dorsal nerve cord, notochord, pharyngeal gill slits, and postanal tail — well-represented but so are secondary characteristics, such as liver diverticulum, hepatic portal system, and the beginning of a ventral heart. Indicative also of the condition in vertebrates is the much thicker dorsal portion of the muscular layer. This is in contrast to the invertebrate phyla in which the muscular layer is about the same thickness around the body cavity. The metameric arrangement of the muscles is suggestive of a similar plan in the embryos of vertebrates.

Just where amphioxus is placed in the evolutionary blueprint of the chordates and the vertebrates is a controversial point. It is placed by many authorities near the primitive fishes, ostracoderms, but whether it comes before or after these fishes in the evolutionary line is not settled. Many regard the amphioxus as a highly specialized or degenerate member of the early chordates and believe that the overdeveloped notochord was developed in them as a correlation to their burrowing habits. The forward extension of the notochord into the tip of the snout may be one of the reasons for the small development of the brain of the amphioxus. Many authorities therefore assign amphioxus to a divergent side branch of some stage intermediate between the early filter-feeding prevertebrates and the vertebrates (Fig. 22-2).

Subphylum Vertebrata

The third subphylum of the chordates is the large and eminently successful Vertebrata, the subject of the next five chapters of this book. The subphylum Vertebrata shares the basic chordate characteristics with the other two subphyla, but in addition it has a number of features that the others do not share. The characteristics that give the members of this group the name ''Vertebrata'' or ''Craniata'' are a braincase, or **cranium,** and a **spinal column of vertebrae,** which forms the chief skeletal axis of the body.

Characteristics

1. Chief diagnostic features of chordates — **notochord, dorsal nerve cord, pharyngeal gill slits,** and **postanal tail** — all present at some stage of the life cycle
2. **Integument** basically of two divisions: an outer **epidermis** of stratified epithelium from the ectoderm and an inner **dermis** of connective tissue derived from the

mesoderm; many modifications of skin among the various classes, such as glands, scales, feathers, claws, horns, and hair.

3. Notochord more or less replaced by the spinal column of **vertebrae** composed of cartilage or bone or both; distinctive **endoskeleton** consisting of the vertebral column with the cranium, visceral arches, limb girdles, and two pairs of jointed appendages
4. **Many muscles** attached to the skeleton to provide for movement
5. **Complete digestive system** ventral to the spinal column and provided with large digestive glands, **liver,** and **pancreas**
6. Circulatory system consisting of the **ventral heart** of two to four chambers; a closed vascular system of arteries, veins, and capillaries; blood containing red corpuscles, with hemoglobin, and white corpuscles; paired aortic arches connecting the ventral and dorsal aortas and giving off branches to the gills among the aquatic vertebrates; in the terrestrial types aortic arch plan modified into **pulmonary** and **systemic** systems.
7. Well-developed **coelom** largely filled with the visceral systems
8. **Excretory system** made up of paired kidneys (mesonephric or metanephric type in adults) provided with ducts to drain the waste to the cloaca or anal region
9. Brain typically divided into five vesicles
10. Ten or 12 pairs of cranial nerves with both motor and sensory functions is the rule; a pair of spinal nerves supplying each primitive myotome; an autonomic nervous system in control of involuntary functions of internal organs
11. **Endocrine system** of ductless glands scattered through the body
12. Nearly always separate sexes, each sex containing paired gonads with ducts that discharge their products either into the cloaca or into special openings near the anus
13. **Body plan** consisting typically of **head, trunk,** and **postanal tail; neck** present in some, especially terrestrial forms; two pairs of appendages generally, although entirely absent in some; coelom divided into a pericardial space and a general body cavity; thoracic cavity in mammals

AMMOCOETE LARVA OF LAMPREY AS CHORDATE ARCHETYPE

The ammocoete larva of lampreys possesses many of the basic structures one would expect to find in a chordate archetype. Many of its structures are simple in form and similar to those in higher vertebrates. It has a heart, ear, eye, thyroid gland, and pituitary gland, which are characteristic of vertebrates but are lacking in amphioxus. This larva is so different from the adult lamprey that it was for a long time considered to be a separate species; not until it was shown to metamorphose into the adult lamprey was the exact relationship explained. This eel-like larva spends several years buried in the sand and mud of shallow streams, until it finally emerges as an adult that may continue to live in fresh water (freshwater lampreys) or else may migrate to the sea (marine lampreys).

Since Stensiö's important work on ostracoderms in 1927, the similarity of this ammocoete larva to the cephalaspids of that ancient group of fishes has become more and more apparent, and many zoologists are substituting it for amphioxus as a basic ancestral type. It is true that the ammocoete has some degenerative specializations of its own, for it lacks the bony exoskeleton, an important feature in ostracoderms. Stensiö, Romer, and other paleontologists have emphasized that a hard or bony exoskeleton is characteristic of ancestral vertebrates and that cartilaginous structures in the adult represent a specialized embryonic condition that has been retained. Romer has shown that cartilage serves a real purpose in the embryo. Cartilage is not present in dermal bones such as certain skull bones that are laid down directly in membrane and have simple growth, but only in internal bones where it is necessary to maintain complicated relationships with blood vessels, muscles, and other bones throughout the entire growth period. Bone grows only by accretion and does not have the power to expand, as cartilage, and thus the latter represents an ideal embryonic material before the adult elements are fully formed.

Some of the generalized characteristics of the ammocoete larva will be pointed out in the following summary.

General chordate features. The ammocoete has a long, slender body, with the front end broader and blunter than the tail end (Fig. 22-12). A median membranous fin fold extends along most of the posterior dorsal border, passes around the caudal end, where the fin is broader, and then continues forward on the ventral side. The **notochord** is large and extends from the very tip of the tail to a region near the posterior end of the brain. The **dorsal nerve cord,** unlike that of the amphioxus, is enlarged anteriorly to form a complete brain. Instead of the numerous gill slits of the amphioxus, there are only seven pairs of gill pouches and slits in the ammocoetes (there are six pairs in shark embryos). Muscular segmentation is also found in the

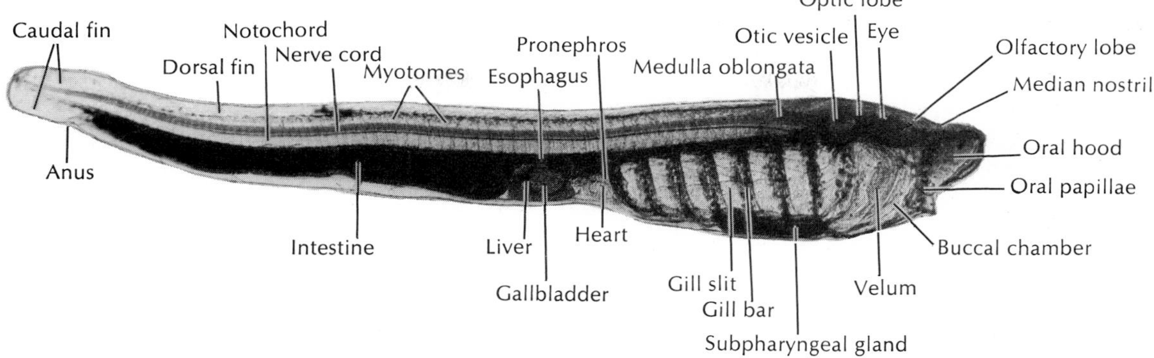

FIG. 22-12 Ammocoete larva. Photograph of stained slide, showing larval structure (Photograph by F. M. Hickman.)

form of myotomes along the dorsal part of the body. The skeleton is meager and in a degenerate condition. Such parts as are found are entirely cartilaginous, for example, the gill bars of the branchial basket, the scattered plates of the braincase, and the small vertebrae near the notochord. The well-developed notochord is the chief supporting skeleton. There are no paired fins or jaws.

Digestive and respiratory systems. At the ventral anterior end of the larva is a cup-shaped **oral hood,** which encloses the **buccal cavity.** Numerous **oral papillae,** or branched projections, are attached to the sides and roof of the oral hood and surround the mouth cavity. They have a sensory function. Between the buccal cavity and the **pharynx** is the **velum,** consisting of a pair of flaps that help create currents of water entering the mouth. The expanded pharynx makes up a large part of the alimentary canal and bears in its lateral walls the seven pairs of **gill pouches.** Each of these pouches opens to the exterior by a **gill slit.** Each gill slit has a fold or **gill** on both its anterior and posterior surfaces, and the wall of the pharynx between adjacent gill slits contains supporting rods of cartilage **(gill bars).** The gills are richly supplied with blood capillaries and are bathed by currents of water that enter the mouth and pass to the exterior through the gill slits. Oxygen from the water enters the blood in the gills and carbon dioxide is given off by the blood in exchange.

The **endostyle (subpharyngeal gland),** a closed furrow or tube extending the length of four gill pouches, is found in the floor of the pharynx. It secretes mucus, which is discharged into the pharynx by a small duct. This sticky mucus entangles the food particles brought in by water currents produced by muscular contractions of the pharynx. Cords of food thus formed are carried into the intestine by the ciliated groove in the floor of the pharynx. During metamorphosis a portion of the endostyle is converted into the **thyroid gland,** which secretes the thyroid hormone containing iodine.

The pharynx narrows at its posterior end to form a short **esophagus,** which opens into the straight **intestine.** Opening into the intestine by the **bile duct** is the **liver,** with which is associated a large and conspicuous **gallbladder. Pancreatic cells** are found in the wall of the anterior part of the intestine but do not form a distinct gland. The intestine opens posteriorly into the cloaca, which also receives the **kidney ducts.** The **anus** is found a short distance in front of the postanal tail.

The generalized vertebrate features of the ammocoetes are thus seen to be its jawless filter feeding, its relatively undifferentiated alimentary canal, its gill arrangement, and its endostyle characteristic of primitive feeding organisms. The development of the thyroid gland from part of the endostyle, as well as the muscular branchial movement, also represents a plan that higher vertebrates have followed.

Circulatory system. The hypothetical, primitive chordate plan of four major longitudinal blood vessels—**dorsal aorta, subintestinal artery, right cardinal vein,** and **left cardinal vein**—and two major connections between these blood vessels—**right** and **left ducts of Cuvier** and the **aortic arches**—is generally followed in the ammocoetes with certain modifications. The posterior end of the subintestinal vein has become modified to form the **hepatic portal vein,** the anterior portion to form the **heart** of one **auricle** (atrium) and one **ventricle** arranged in tandem. From the ventricle the short **ventral aorta** runs forward to give off eight pairs of **aortic arches** to the gill pouches. Each arch is composed of an **afferent branchial artery** carrying blood to the capillaries of the gills and an **efferent branchial artery** carrying aerated blood from the gill capillaries to the **dorsal aorta.** The dorsal aorta gives off many branches to the body tissues, and a large posterior branch, the **intestinal,** to the intestine. The **cardinal veins** return blood from the tissues to the right and left **ducts of Cuvier,** which empty into the **sinus venosus,** a thin-walled chamber that empties into the atrium, and then to the ventricle of the heart. Each cardinal vein has an **anterior cardinal** and a **posterior cardinal branch.** The hepatic portal vein picks up blood laden with nutrients from the intestine and carries it to the liver. The **hepatic vein** carries blood from the liver to the **sinus venosus,** and so back to the heart.

Excretory system. The ammocoete has a **pronephric kidney** that functions to remove nitrogenous wastes, certain ions, and excess water. The pronephric kidney is the most primitive of three types of vertebrate kidneys that develop from the **nephrotomic plate,** an area of segmented mesoderm that extends along each side of the vertebrate embryo. The three generations of kidney—pronephros, mesonephros, and metanephros—are described in Chapter 30. Thus the ammocoete kidney conforms to the basic chordate plan, whereas the solenocyte-type of flame cell found in amphioxus is altogether different.

Reproductive system. The **gonads** are paired ridge-like structures on the dorsal side of the coelom. Each gonad appears to have been formed by the fusion of a number of units. Since they lack genital ducts, the adult lampreys shed their gametes into the coelom, where an opening into the **mesonephric duct** allows the gametes to escape to the outside through the **urogenital papillae.**

Nervous and sensory systems. Both brain and spinal cord conform to the basic chordate plan. The **brain** has the typical three divisions of **forebrain, midbrain,** and **hindbrain.** Each of these divisions is associated with an important sense organ—olfaction, vision, and hearing, respectively. From the dorsal side of the forebrain are two outgrowths or stalks, each of which bears a vestigial **median eye,** the only instance among vertebrates of two median eyes. Other vertebrates may have two outgrowths, but the anterior one is the parietal body, which appears to have been a median eye, and the posterior one (epiphysis) is the pineal gland. In no living vertebrate does a median eye function as such. In some vertebrates only the pineal gland or body is present.

The **pituitary** is a double gland composed of (1) the **neurohypophysis,** formed from an evagination **(infundibulum)** of the forebrain, and (2) the **adenohypophysis,** formed from the **hypophyseal pouch** of the nasohypophyseal sac (the ammocoete differs from all other vertebrates in that the adenohypophysis does not develop from the primitive mouth cavity). The functional **eyes** of the ammocoetes are small and develop from the forebrain. The spinal cord gives off a pair of **dorsal roots** (mainly sensory) and a pair of **ventral roots** (motor) in every muscle segment. The dorsal and ventral roots do not join as they do in higher vertebrates, nor do the nerves have myelin sheaths.

Annotated references
Selected general readings

Barrington, E. J. W. 1965. The biology of Hemichordata and Protochordata. San Francisco, W. H. Freeman & Co. *A synthesis of recent work on these deuterostomes most closely related to vertebrates. Discusses the possible homologies between the endostyle and the vertebrate thyroid gland.*

Berrill, N. J. 1955. The origin of vertebrates. New York, Oxford University Press. *The author stresses the tunicates as the basic stock from which other protochordates and vertebrates arose. He believes that such a sessile filter feeder was really the most primitive animal and was not a mere degenerate side branch of chordate evolution.*

Colbert, E. H. 1969. Evolution of the vertebrates, ed. 2. New York, John Wiley & Sons, Inc. *A clear and well-written presentation.*

Garstang, W. 1928. The morphology of the Tunicata, and its bearings on the phylogeny of the Chordata. Quat. J. Micr. Sci. **72:**51.

Halstead, L. B. 1968. The pattern of vertebrate evolution. San Francisco, W. H. Freeman & Co. *In this thoughtful interpretation of fossil evidence, the author considers physiologic and ecologic aspects of evolution.*

Halstead, L. B. 1973. The heterostracan fishes. Biol. Rev. **48:**279-332. *Thorough review of the earliest known vertebrates.*

Jarvik, E. 1967. Aspects of vertebrate phylogeny. In T. Ørvig (ed.). Nobel Symposium 4. Stockholm, Almqvist & Wiksell.

Nursall, J. R. 1962. On the origins of the major groups of animals. Evolution **16:**118-123.

Romer, A. S. 1959. The vertebrate story. Chicago, University of Chicago Press. *A comprehensive background of the evolutionary trends and relationships of the various vertebrate groups leading up to humans.*

Romer, A. S. 1966. Vertebrate paleontology, ed. 3. Chicago, University of Chicago Press. *An authoritative work by a distinguished paleontologist.*

Romer, A. S., and T. S. Parsons. 1977. The vertebrate body, ed. 5. Philadelphia, W. B. Saunders Co. *Early chapters of this popular comparative anatomy text deal with vertebrate adaptations, ancestry, and body plan.*

Stahl, B. J. 1974. Vertebrate history: problems in evolution. New York, McGraw-Hill Book Co. *Vertebrate evolution and paleontology.*

Wessells, N. K. 1974. Vertebrate structures and functions. Readings from Scientific American. San Francisco, W. H. Freeman & Co. *An anthology of Scientific American articles with excellent introductions by N. K. Wessells.*

Willmer, E. N. 1974. Nemertines as possible ancestors of the vertebrates. Biol. Rev. **49:**321-363. *The author resurrects and strengthens a nineteenth-century theory of vertebrate origins.*

Young, J. Z. 1962. The life of vertebrates, ed. 2. New York, Oxford University Press. *A classic.*

CHAPTER 23

THE FISHES

Phylum Chordata
Classes Petromyzontes, Myxini, Chondrichthyes, and Osteichthyes

Migrating Pacific salmon jumping at Brooks Falls, Alaska.

Courtesy G. B. Kelez, United States Fish and Wildlife Service.

Fishes are the undisputed masters of the aquatic environment. Because they live in a habitat that is basically hostile to humans, we have not always found it easy to appreciate the incredible success of these vertebrates. Plato considered fishes "senseless beings . . . which have received the most remote habitations as a punishment for their extreme ignorance." And average North Americans today are probably unconscious of and uninformed about fishes unless they happen to be sports fishermen or tropical fish enthusiasts.

Nevertheless the world's fishes have enjoyed an adaptive radiation easily as spectacular as that of all the land vertebrates, with the possible exception of the mammals (the latter having succeeded in the air and in the sea as well as on land). Their numerous structural adaptations have produced a great variety of forms ranging from gracefully streamlined trout to grotesque creatures that dwell in the blackness of the ocean's abyssal depths. Considered either in numbers of species (more than 20,000 named species) or in numbers of individuals (countless billions), they easily outnumber the four terrestrial vertebrate classes combined.

Although fishes are the oldest vertebrate group, there is not the slightest evidence that, like their amphibians and reptile successors, they are declining from a period of earlier glory; certain groups of ancient fishes have become extinct but have been replaced by successful modern fishes. There are indeed more bony fishes today than ever before, and no other group threatens their domination of the seas.

Their success can be attributed to one thing: they are perfectly adapted to their dense medium. A trout or pike can hang motionless in the water at any depth, varying its neutral buoyancy by adding or removing air from the swim bladder, or it can dart forward or at angles, using its fins as brakes and tilting rudders. Fish have excellent olfactory and visual senses and a unique lateral line system, which with its exquisite sensitivity to water currents and vibrations, provides a "distance touch" in water. Their gills are the most effective respiratory devices in the animal kingdom for extracting oxygen from water. With highly developed organs of salt and water exchange, bony fishes are excellent osmotic regulators, capable of fine-tuning their body fluid composition in their chosen freshwater or seawater environment. Fishes have evolved complex behavioral mechanisms for dealing with emergencies, and many have evolved elaborate reproductive behavior concerned with courtship, nest building, and care of the young. These are only a few examples of many such adaptations evident in this varied phylogenetic assemblage, which includes four of the eight vertebrate classes.

ANCESTRY AND RELATIONSHIPS OF MAJOR GROUPS OF FISHES

Theories of chordate and vertebrate origins were described in the preceding chapter. The fishes like all other vertebrates take their origin from an unknown common ancestor that may have arisen from a free-swimming larval form of an ancient ascidian or ascidian-like stock. During the Cambrian period, or perhaps even in the Precambrian, the earliest fishlike vertebrates branched into the jawless agnathans and the jawed gnathostomes (Fig. 23-1). All subsequent vertebrates descended from one or the other of these two great stems.

The agnathans include the ostracoderms, which are now extinct, and the living hagfishes (Myxini) and lampreys (Petromyzontes). Biologists generally agree that the living agnathans are successors of the ostracoderms, but there is no agreement about which of the three ostracoderm groups (heterostracans, cephalaspids, anaspids) they are most closely related to. The weight of evidence points to an affinity between the anaspids and the living lampreys, but the heritage of the hagfishes is unknown. Both hagfishes and lampreys are highly specialized, and aside from both lacking jaws, they bear little resemblance to the ostracoderms. Moreover it has long been recognized that hagfishes and lampreys only superficially resemble each other. There are numerous anatomic and physiologic contrasts, as well as important differences in their mode of life and life cycles. This suggests that the hagfishes and lampreys have been phylogenetically independent for a very long time and should be considered as belonging to entirely separate classes, as ichthyologists now regard them.

All the rest of the fishes are gnathostomes that have descended from one or more early jawed ancestors. The placoderms have often been suggested, but the fossil evidence is actually so fragmentary that it is impossible to pick out ancestral groups with certainty. Whatever the early lines of descent, ichthyologists now recognize several natural groups of living jawed fishes, which are depicted in Fig. 23-1.

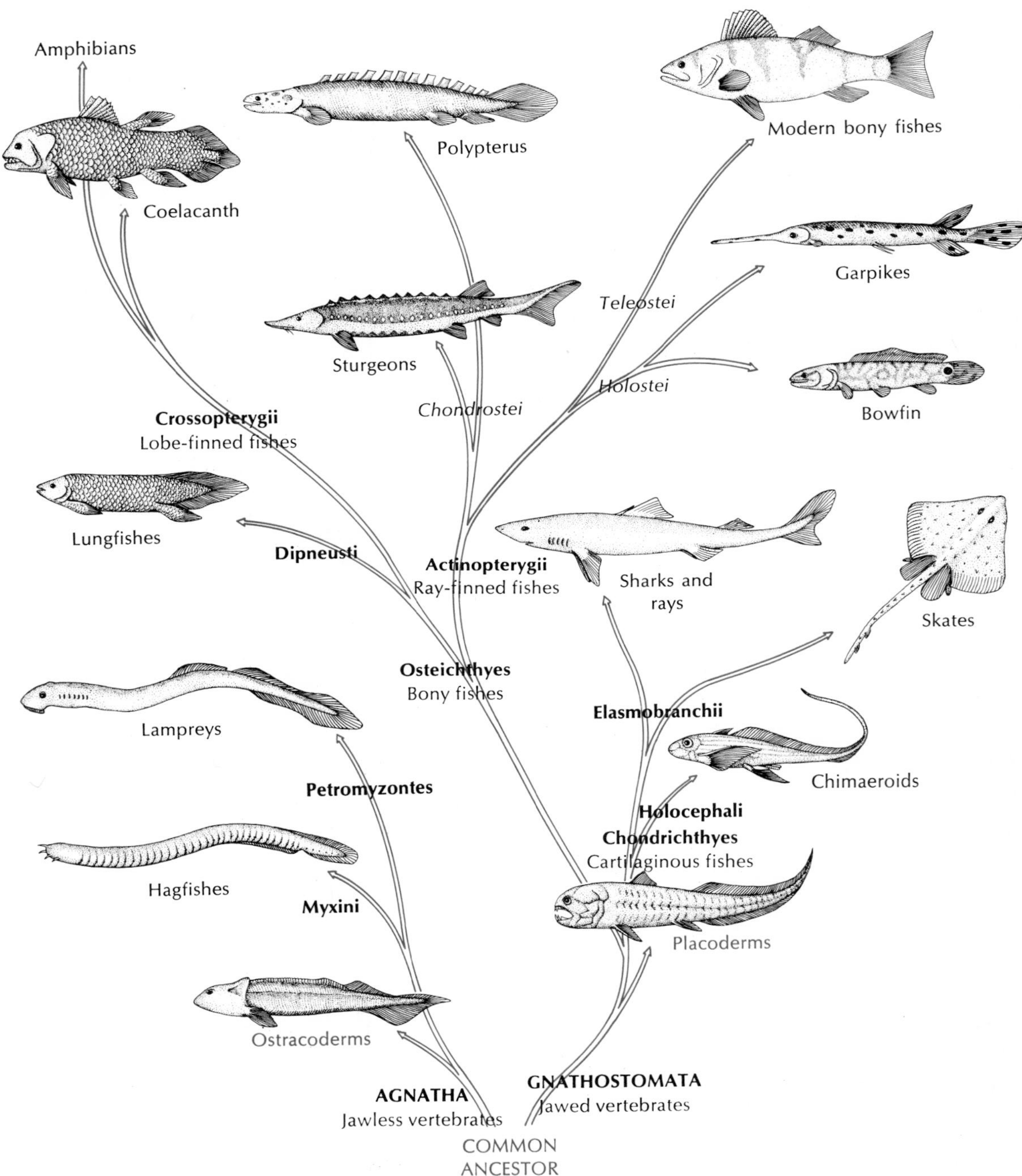

FIG. 23-1 Family tree of the fishes. Lines suggest probable relationships. *Red,* Extinct groups.

The sharks, skates, and rays comprise one natural, compact group, the subclass Elasmobranchii of the class Chondrichthyes. These animals lost the heavy armor of the early placoderms and adopted cartilage instead of bone for the skeleton (thought to be a secondary degeneration), an active and predatory habit, and a sharklike body form that has undergone only minor changes over the ages. The elasmobranchs flourished during the Devonian and Carboniferous eras but declined dangerously close to extinction in the Permian period. They staged a recovery in the early Mesozoic period and radiated into the modest but thoroughly successful assemblage of modern sharks.

Obviously sharing a remote relationship to the elasmobranchs are the bizarre chimaeras of the subclass Holocephali. They appeared in the Jurassic period, apparently as an offshoot of some now extinct elasmobranch stock. These peculiar yet strangely appealing fishes have many internal elasmobranch structures and consequently have been included with the latter in the class Chondrichthyes (cartilaginous fishes).

The bony fishes (class Osteichthyes) are the dominant fishes today. This diverse assemblage offers perplexing classification problems. We can recognize three great stems of descent: first there are the actinopterygians, or **ray-finned fishes,** which radiated into the modern bony fishes; second are the crossopterygians, the **lobe-finned fishes,** from which the amphibians are descended; and third are the dipneusts, or **lungfishes.** Ichthyologists formerly classified the last two of these three groups together as Choanichthyes ("funnel-fish") in the belief that they shared internal nostrils, that is, nostrils that open into the mouth cavity in the style of higher vertebrates. But more thorough studies of fossil forms revealed that the internal nostril of the lungfish has a completely different origin; this and other differences in the skeleton have now set the lungfishes apart from the mainline of vertebrate descent.

The distinction of being ancestors to the tetrapods now rests with the crossopterygians, which are represented today by a single species, the coelacanth *Latimeria chalumnae*. Both the crossopterygians, with one species, and the dipneusts, with six, are relic groups. These seven survivors are meager evidence of stocks that flourished in the Devonian period. Having remained mostly unchanged over the subsequent 400 million years to present, they are in a sense "living fossils." The lobe-finned fishes and lungfishes are now grouped with the ray-finned fishes under the class Osteichthyes (Teleostomi in some classifications), to produce a highly unbalanced living representation: one lobe-fin (the coelocanth), six species (three genera) of lungfishes, and some 18,000 species of ray-finned fishes. The last are divided into three superorders as shown in Fig. 23-1. The evolution of these groups is discussed later in this chapter.

One further point of evolutionary importance must be mentioned. The great weight of paleontologic opinion now favors a marine origin for the vertebrates. But at some later time, probably in the Silurian or Lower Devonian period, most of the major groups of fishes penetrated into estuaries and then into freshwater rivers. This invasion was of momentous importance to subsequent vertebrate evolution. It compelled the development of physiologic mechanisms to maintain osmotic and ionic balance in the highly dilute freshwater environment. These Devonian invaders not only met the challenge but also established a body fluid concentration (about one-third that of seawater) and ionic composition that fixed the body fluid pattern for all vertebrates to evolve later, whether aquatic, terrestrial, or aerial. Thus it was that when the early bony fishes (Osteichthyes) and elasmobranchs *returned* to the sea in the Triassic period, they encountered a new osmotic challenge: they were now much more dilute than their surroundings. Rather than sliding back into osmotic equilibrium with seawater they became osmotic and ionic regulators. In bony fishes and sharks this physiologic problem is solved in different ways that are described later. Our interest at this point is that both the marine elasmobranchs and marine bony fishes are descendants of freshwater stocks: they have come full circle by returning to their ancestral home in the sea.

Classification of living fishes

The following broad classification mostly follows that of Nelson (1976). No one scheme is accepted by even the majority of ichthyologists. When we contemplate the incredible difficulty of ferreting out relationships among some 20,000 living species and a vast number of fossils of varying ages, we can appreciate why fish classification has been, and will continue to be, undergoing continuous change.

Subphylum Vertebrata

Superclass Agnatha (ag′na-tha) (Gr., *a,* not, + *gnathos,* jaw) **(Cyclostomata).** No jaws; cartilaginous skeleton; ventral fins absent; two semicircular canals; notochord persistent.

Class Petromyzontes (pet′ro-my-zon′teez) (Gr., *petros,* stone, + *myzon,* sucking)—**lampreys.** Mouth suctorial with horny teeth; nasal sac not connected to mouth; gill pouches, seven pairs. Examples: *Petromyzon, Lampetra*.

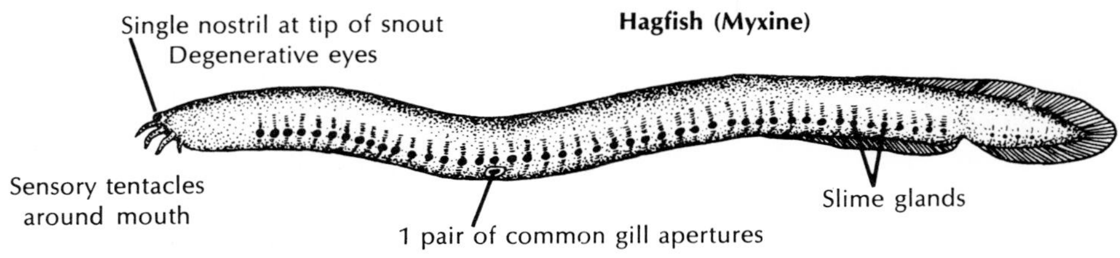

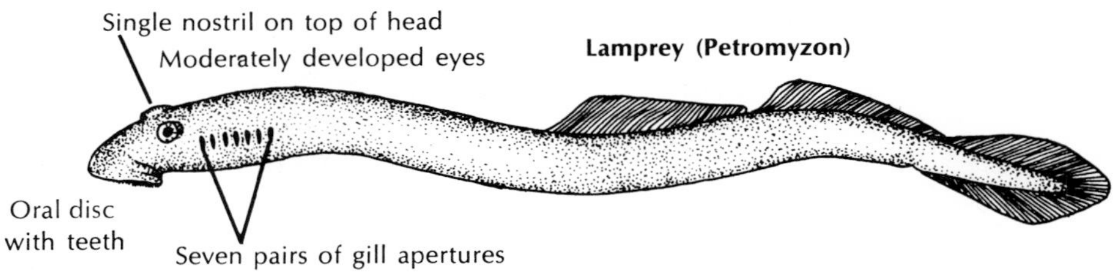

FIG. 23-2 Comparison of hagfish (class Myxini) and lamprey (class Petromyzontes), representatives of the superclass Agnatha.

Class Myxini (mik-sy'ny) (Gr., *myxa,* slime)—**hagfishes.** Mouth terminal with four pairs of tentacles; buccal funnel absent; nasal sac with duct to pharynx; gill pouches, five to 15 pairs; partially hermaphroditic. Examples: *Myxine, Bdellostoma.*

Superclass Gnathostomata (na'tho-sto'ma-ta) (Gr., *gnathos,* jaw, + *stoma,* mouth). Jaws present; usually paired limbs; three semicircular canals; notochord persistent or replaced by vertebral centra.

Class Chondrichthyes (kon-drik'thee-eez) (Gr., *chondros,* cartilage, + *ichthys,* fish)—**cartilaginous fishes.** Cartilaginous skeleton; teeth not fused to jaws; no swim bladder.

Subclass Elasmobranchii (e-laz'mo-bran'kee-i) (Gr., *elasmos,* a metal plate, + *branchia,* gills)—**sharks, skates, rays.** Placoid scales or no scales; five to seven gill arches and gills in separate clefts along pharynx. Examples: *Squalus, Raja.*

Subclass Holocephali (hol'o-sef'a-li) (Gr., *holos,* entire, + *kephalē,* head)—**chimaeras,** or **ghostfish.** Gill slits covered with operculum; jaws with tooth plates; single nasal opening; without scales; accessory clasping organs in male; lateral line an open groove. Example: *Chimaera.*

Class Osteichthyes (os'te-ik'thee-eez) (Gr., *osteon,* bone, + *ichthys,* a fish) **(Teleostomi)—bony fishes.** Body primitively fusiform but variously modified; skeleton mostly ossified; single gill opening on each side covered with operculum; usually swim bladder or lung.

Subclass Crossopterygii (cros-sop-te-rij'ee-i) (Gr., *krossoi,* fringe or tassels, + *pteryx,* fin, wing)—**lobe-finned fishes.** Heavy bodied; paired fins lobed with internal skeleton of basic tetrapod type; premaxillae, maxillae present; scales large with tubercles and heavily overlapped; three-lobed diphycercal tail; skeleton with much cartilage; bony spines hollow; air bladder vestigial; gills hard with teeth; intestine with spiral valve; spiracle present. Example: *Latimeria.*

Subclass Dipneusti (dip-nyu'sti) (Gr., *di-,* two, + *pneustikos,* of breathing)—**lungfishes.** All median fins fused to form diphycercal tail; fins lobed or of filaments; scales of cycloid bony type; teeth of grinding plates; no premaxillae or maxillae; air bladder of single or paired lobes and specialized for breathing; intestine with spiral valve; spiracle absent. Examples: *Neoceratodus, Protopterus, Lepidosiren.*

Subclass Actinopterygii (ak'ti-nop-te-rij'ee-i) (Gr., *aktis,* ray, + *pteryx,* fin, wing)—**ray-finned fishes.** Paired fins supported by dermal rays and without basal lobed portions; nasal sacs open only to outside. Examples: *Salmo, Perca.*

JAWLESS FISHES—SUPERCLASS AGNATHA

The living members of the Agnatha are represented by some 60 species almost equally divided between two classes: Petromyzontes (lampreys) and Myxini, (hagfishes) (Fig. 23-2). They have in common the absence of jaws, internal ossification, scales, and paired fins, and both share porelike gill openings and an eel-like body form. At the same time there are so many important differences, some of which are indicated in the following listing, that they have been assigned to separate vertebrate classes.

Characteristics

1. Body slender, **eel-like,** rounded, with **soft skin** containing **mucous glands** but **no scales**
2. Median fins with cartilaginous fin rays, but **no paired appendages**

3. **Fibrous** and **cartilaginous** skeleton; notochord persistent
4. Suckerlike oral disc with well-developed teeth in lampreys; mouth with two rows of eversible teeth in hagfishes
5. Heart with one auricle and one ventricle; aortic arches in gill region; blood with erythrocytes and leukocytes
6. Seven pairs of gills in lampreys; five to 15 pairs of gills in hagfishes
7. **Mesonephric kidney** in lampreys; **pronephric kidney** anteriorly and mesonephric kidney posteriorly in hagfishes
8. Dorsal nerve cord with differentiated brain; eight to ten pairs of cranial nerves
9. Digestive system **without stomach;** intestine with spiral fold in lampreys; spiral fold absent in hagfishes
10. Sense organs of taste, smell, hearing; eyes moderately developed in lampreys but highly degenerate in hagfishes
11. External fertilization; gonad single without duct; sexes separate and long larval stage (ammocoete) in lampreys; hermaphroditic and direct development with no larval stage in hagfishes

Lampreys—class Petromyzontes

All the lampreys of the northern hemisphere belong to the family Petromyzontidae. The destructive marine lamprey *Petromyzon marinus* is found on both sides of the Atlantic Ocean (in America and Europe) and may attain a length of 1 m. *Lampetra* also has a wide distribution in North America and Eurasia and ranges from 15 to 60 cm long. There are 19 species of lampreys in North America. About half of these belong to the nonparasitic brook type; the others are parasitic. The nonparasitic species have probably descended from the parasitic forms by degeneration of the teeth, alimentary canal, and so on. The genus *Ichthyomyzon,* which contains three parasitic and three nonparasitic species, is restricted to eastern North America. On the west coast of North America the chief marine form is *Lampetra (Entosphenus) tridentatus*.

All lampreys, marine as well as freshwater forms, spawn in the spring in North America (autumn in some European species) in shallow gravel and sand in freshwater streams. The males begin nest-building and are joined later by females. Using their oral discs to lift stones and pebbles and vigorous body vibrations to sweep away light debris, they form an oval depression (Fig. 23-3). At spawning, with the female attached to a rock to maintain position over the nest, the male attaches to the dorsal side of her head. He curls the posterior part of his body tightly about that of the female so that their cloacal openings are together. As the eggs are shed into the nest, they are fertilized by the male. Only small numbers of eggs are shed at each mating, and the process may be repeated numerous times over a period of 1 to 2 days. The sticky eggs adhere to pebbles in the nest and soon become covered with sand. The adults die soon after spawning.

FIG. 23-3 Brook lampreys *Lampetra lamattei* clearing pebbles for a spawning nest in a small creek. (Courtesy J. W. Jordan, Jr.)

The eggs hatch in about 2 weeks, releasing small larvae (ammocoetes), which are so unlike their parents that early biologists were deceived into giving them distinct scientific names (such as *Ammocoetes branchioles*). The larva bears a remarkable resemblance to amphioxus and possesses the basic chordate characters in such simplified and easily visualized form that it has been considered a chordate archetype (p. 490). After absorbing the remainder of its yolk supply, the young ammocoete, now about 7 mm long, leaves the nest gravel and drifts downstream to burrow in some suitable sandy, low-current area. Here it remains for an extraordinarily long time, 3 to 7 years, feeding on minute organisms and debris in the mud. During this time the ammocoete grows and then in the fall rapidly metamorphoses into an adult. This change involves the development of larger eyes, the replacement of the hood by the oral disc with teeth, a shifting of the nostril to the top of the head, and the development of a rounder but shorter body.

Parasitic lampreys either migrate to the sea, if marine, or else remain in fresh water, where they attach themselves by their suckerlike mouth to fish and, with their sharp horny teeth, rasp away the flesh and suck

A

B

C

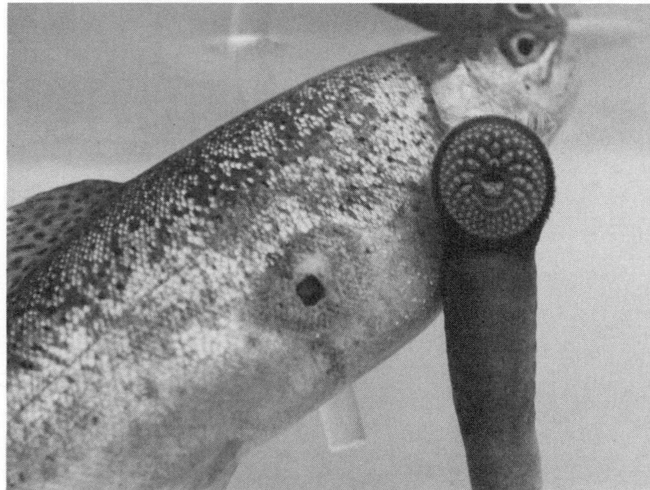

FIG. 23-4 Sea lampreys *Petromyzon marinus* attacking trout. **A,** Recently metamorphosed sea lampreys, 15 to 18 cm long, attack 20 cm long brook trout *Salvelinus fontinalis* in experimental aquarium. **B,** Head of 38 cm long sea lamprey feeding on a rainbow trout *Salmo gairdneri*. Note the single nostril on top of head and the eyes and gill apertures. **C,** Lamprey detached from rainbow trout to show feeding wound that had penetrated body cavity and perforated gut. Trout died from wound. Note chitinous teeth on underside of lamprey head. (Courtesy United States Bureau of Sport Fisheries and Wildlife, Fish Control Laboratory, La Crosse, Wis.; **A,** by R. E. Lennon; **B** and **C,** by L. L. Marking.)

out the blood (Fig. 23-4). To promote the flow of blood, the lamprey injects an anticoagulant into the wound. When gorged, the lamprey releases its hold but leaves the fish with a large gaping wound that may prove fatal. The parasitic freshwater adults live a year or more before spawning and then die; the marine forms live longer.

The nonparasitic lampreys do not feed after emerging as adults, for their alimentary canal degenerates to a nonfunctional strand of tissue. Within a few months they also spawn and die.

The invasion of the Great Lakes by the landlocked sea lamprey *Petromyzon marinus* in this century has had a devastating effect on the fisheries there. No lampreys were present in the Great Lakes west of Niagara Falls until the Welland Ship Canal was built in 1829. Even then nearly 100 years elapsed before sea lampreys were first seen in Lake Erie. After that the spread was rapid, and the sea lamprey was causing extraordinary damage in all the Great Lakes by the late 1940s. No fish species was immune from attack, but the lampreys preferred lake trout, and this multimillion dollar fishing industry was brought to total collapse in the early 1950s. Lampreys then turned to rainbow trout, whitefish, turbot, yellow perch, and lake herring, all important commercial species. These stocks were decimated in turn. The lampreys then began attacking chubs and suckers. Coincident with the decline in attacked species, the sea lampreys themselves began to decline after reaching a peak abundance in 1951 in Lakes Huron and Michigan and in 1961 in Lake Superior. The fall has been attributed both to depletion of food and to the effectiveness of control measures (electric barriers and use of chemical larvicides in selected spawning streams). Lake trout are presently recovering slowly, but wounding rates are still high. Fishery organizations are experimenting with the introduction into the Great Lakes of species that appear to be more resistant to lamprey attack, such as kokanee salmon (landlocked Pacific sockeye salmon).

Hagfishes—class Myxini

The hagfishes are an entirely marine group that feed on dead or dying fishes, annelids, molluscs, and crustaceans. Thus they are neither parasitic like lampreys nor predaceous, but are scavengers. There are only 32 described species of hagfishes, of which the best known in North America are the Atlantic hagfish *Myxine glutinosa* (Fig. 23-2) and the Pacific hagfish *Bdellostoma stouti*.

Hagfishes have long been of interest to comparative physiologists. Unlike any other vertebrate, they are isosmotic to sea water like marine invertebrates. They are the only vertebrates to have both pronephric and mesonephric kidneys in the adult, although only the latter forms urine. They have no less than four sets of hearts positioned at different places in the body to boost blood flow through their low-pressure circulatory system.

Despite other unique anatomic and physiologic features of interest to biologists, they probably have fewer human admirers than any other group of fishes. The sports fisherman who catches one discovers that the hook is so deeply swallowed that retrieval is impossible. To cap the sportsman's misfortune, the animal secretes enormous quantities of slimy mucus from large and small mucous glands located all over the body and from special slime glands positioned along its sides (Fig. 23-2). A single hagfish is said to be capable of converting a bucket of water into a mass of whitish jelly in minutes. Their habit of biting into and entering the bodies of gill-netted fish has not endeared them to commercial fishermen either. Entering through the anus or gills, hagfishes set about eating out the contents of the body, leaving behind a loose sack of skin and bones. But as fishing methods have passed from the use of drift nets and set lines to large and efficient otter trawls, hagfishes have ceased to be an important pest.

The yolk-filled egg of the hagfish may be 14 to 25 mm in diameter and is enclosed in a horny shell. There is no larval stage and growth is direct. They are hermaphroditic but can produce only one kind of gamete at a time; a single individual may produce sperm at one season and eggs the next.

CARTILAGINOUS FISHES—CLASS CHONDRICHTHYES

The approximately 625 species of sharks, skates, rays, and chimaeras form an ancient but compact and highly developed group. Although a much smaller and less diverse assemblage than the bony fishes, their impressive combination of well-developed sense organs, powerful jaws and swimming musculature, and predaceous habits ensure them of a secure and lasting niche in the aquatic community. One of their distinctive features is their cartilaginous skeleton, which must be considered degenerate instead of primitive. Although there is some calcification here and there, bone is entirely absent throughout the class.

With the exception of whales, sharks are the largest living vertebrates. The larger sharks may reach 15 m in length. The dogfish sharks so widely used in zoologic laboratories rarely exceed 1 m.

Sharks, skates, and rays—subclass Elasmobranchii

The elasmobranchs are all carnivores that track their prey using their lateral line system and large olfactory organs. Vision is not well-developed.

Fertilization is internal (a curiously "modern" feature in an ancient group), and many sharks have evolved elaborate reproductive modes; some are live bearers with gestation periods up to 2 years—the longest of any vertebrate.

According to Nelson (1976) there are five living orders of elasmobranchs. All of the more notorious shark members belong to the order Lamniformes (requiem sharks, hammerhead sharks, whale sharks, mackerel sharks, nurse sharks, and sand sharks). The order Squaliformes includes the dogfish shark, familiar to generations of comparative anatomy students. The skates and several groups of rays (sawfish rays, electric rays, stingrays, eagle rays, manta rays, and devil rays) belong to the order Rajiformes.

Much has been written about the propensities of sharks to eat humans, both by those exaggerating their ferocious nature and by those seeking to write them off as harmless. It is true, as the latter group of writers argue, that sharks are by nature timid and cautious. But it also is a fact that certain of them are dangerous to humans and deserve respect. There are numerous authenticated cases of shark attacks by *Carcharodon,* the great white shark (commonly reaching 6 m and often larger); the mako shark *Isurus;* the tiger shark

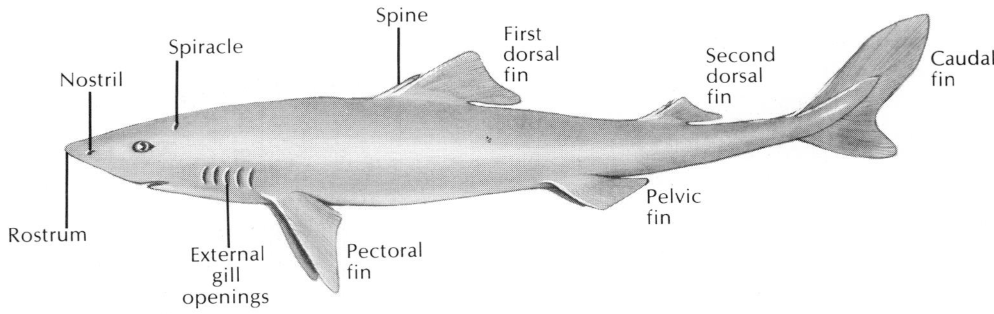

FIG. 23-5 Dogfish shark *Squalus acanthias.* Subclass Elasmobranchii.

Galeocerdo; and the hammerhead shark *Sphyrna.* More shark casualties have been reported from the tropical and temperate waters of the Australian region than from any other. During the Second World War there were several reports of mass shark attacks on the victims of ship sinkings in tropical waters.

Outside of North America, sharks and skates are much used for food. They make up about 1% of the present market for fish. In the United States discs of meat cut from skate "wings" are frequently sold in restaurants as scallops; the flavor and texture are similar to those of scallops, which the customers innocently suppose they are eating. There is a small market for shark leather, which has greater tensile strength and lasting qualities than mammalian leather.

Characteristics

1. **Body fusiform,** with a **heterocercal** caudal fin (Fig. 23-11); paired pectoral and pelvic fins, two dorsal median fins; pelvic fins in male modified for **"claspers"**; fin rays present
2. **Mouth ventral; two olfactory sacs that do not break into the mouth cavity;** jaws present
3. Skin with **placoid** scales (Fig. 23-6) and **mucous glands;** modified placoid scales for teeth
4. **Endoskeleton entirely cartilaginous;** notochord persistent; vertebrae complete and separate; appendicular, girdle, and visceral skeletons present
5. Digestive system with a J-shaped stomach and intestine with a spiral valve; liver, gallbladder, and pancreas present
6. Circulatory system of several pairs of aortic arches; dorsal and ventral aorta, capillary and venous systems, hepatic portal and renal portal systems; two-chambered heart
7. Respiration by means of five to seven pairs of gills with separate and exposed gill slits; **no operculum**
8. No swim bladder
9. Brain of two olfactory lobes, two cerebral hemispheres, two optic lobes, a cerebellum, and a medulla oblongata; ten pairs of cranial nerves
10. Sexes separate; gonads paired; reproductive ducts open into cloaca; oviparous, ovoviviparous, or viviparous; direct development; fertilization internal
11. Kidneys of mesonephros (opisthonephros) type

Structure and natural history

Although sharks to most people have a sinister appearance and fearsome reputation, they are at the same time among the most gracefully streamlined of all fishes. The body of a shark such as a dogfish shark (Fig. 23-5) is fusiform (spindle-shaped). In front of the ventral mouth is a pointed **rostrum;** at the posterior end the vertebral column turns up to form the **heterocercal** tail. The fins consist of the paired **pectoral** and **pelvic** fins supported by appendicular skeletons, two median **dorsal** fins (each with a spine in *Squalus*), and a median **caudal** fin. A median **anal** fin is present in the smooth dogfish *(Mustelus).* In the male the medial part of the pelvic fin is modified to form a **clasper,** which is used in copulation. The paired **nostrils** (blind pouches) are ventral and anterior to the mouth. The lateral eyes are lidless, and behind each eye is a spiracle (remnant of the first gill slit). Five gill slits are found anterior to each pectoral fin. The tough, leathery skin is covered with **placoid scales** (Fig. 23-6), which are modified anteriorly to form replaceable rows of teeth in both jaws. Placoid scales in fact consist of dentin enclosed by an enamel-like substance, and they very much resemble the teeth of other vertebrates.

Sharks are well-equipped for their predatory life. Their vision is less acute than in most bony fishes, but

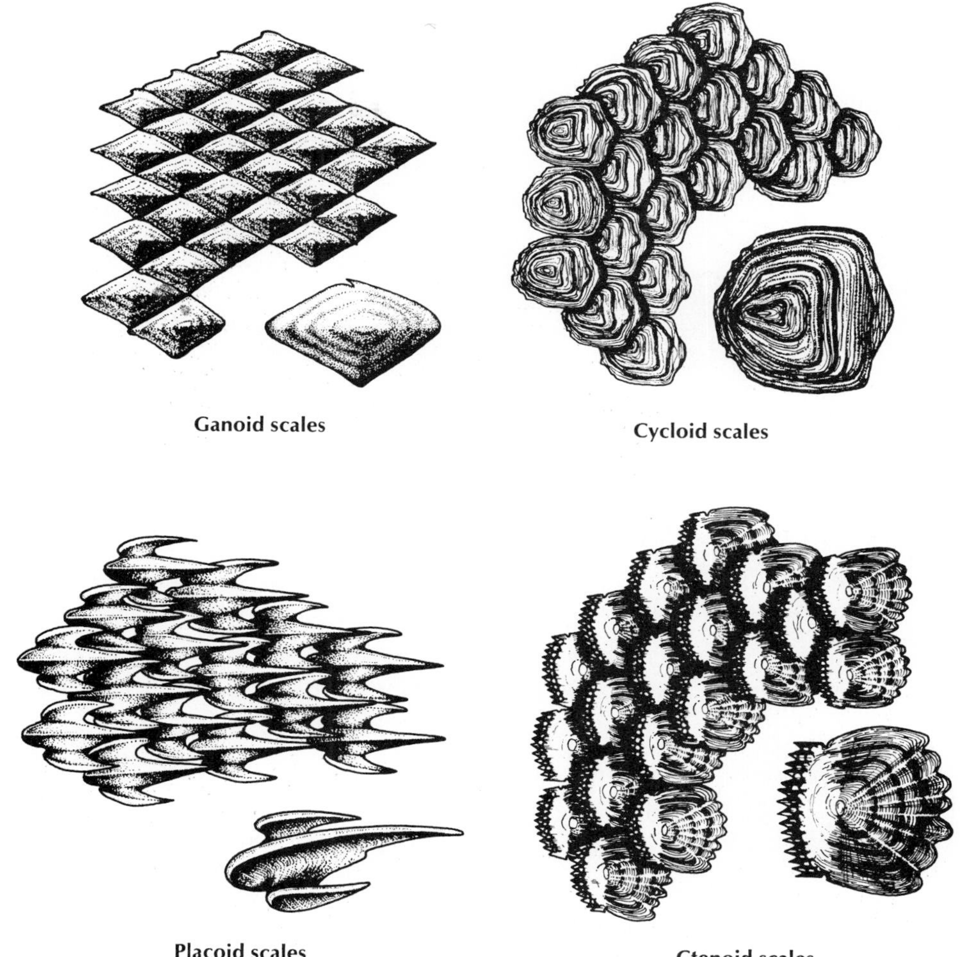

FIG. 23-6 Types of fish scales. Diamond-shaped ganoid scales, present in primitive bony fishes such as the gar, are composed of layers of silvery enamel (ganoin) on the upper surface and bone on the lower. Placoid scales are small, conical toothlike structures characteristic of the chondrichthyes. Modern bony fishes have either cycloid or ctenoid scales. These are thin and flexible and are arranged in overlapping rows.

this is more than compensated for by a keen sense of smell used to guide them to food. A well-developed **lateral line system** serves as a "distance touch" in water for detecting and locating objects and moving animals (predators, prey, and social partners). It is composed of a canal system extending along the side of the body and over the head. The canal opens at intervals to the surface. Inside are special receptor organs **(neuromasts)** that are extremely sensitive to vibrations and currents in the water. Sharks also can detect and aim attacks at prey buried in the sand by sensing the bioelectric fields that surround all animals.

Internally the cartilaginous skeleton is made up of a **chondrocranium,** which houses the brain and auditory organs and partially surrounds the eyes and olfactory organs; a vertebral column; a visceral skeleton; and an appendicular skeleton. The jaws are suspended from the chondrocranium by ligaments and cartilages. Both the upper and the lower jaws are provided with many sharp, triangular teeth that, when lost, are re-

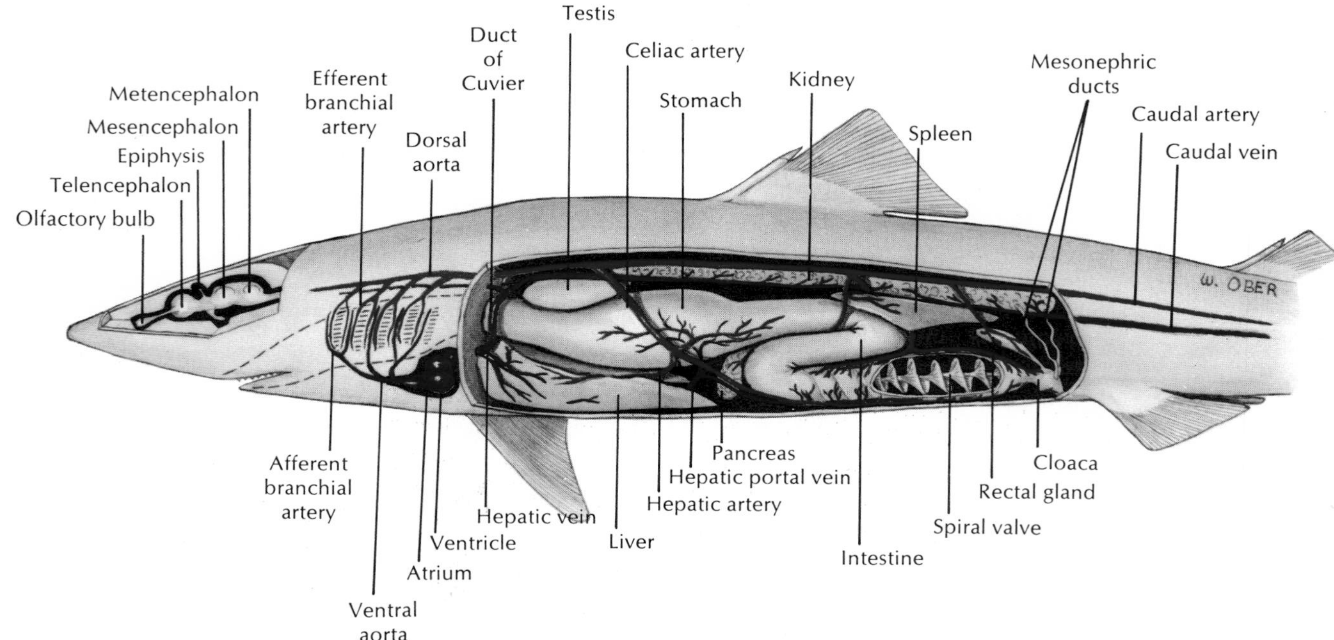

FIG. 23-7 Internal anatomy of dogfish shark *Squalus acanthias.*

placed by other rows of teeth. Teeth serve to grasp the prey, which is usually swallowed whole. The muscles are segmentally arranged and are especially useful in the undulations of swimming.

The mouth cavity opens into the large **pharynx,** which contains openings to the separate gill slits and spiracles. A short, wide esophagus runs to the J-shaped stomach. A **liver** and **pancreas** open into the short, straight **intestine,** which contains the unique **spiral valve** that delays the passage of food and increases the absorptive surface (Fig. 23-7). Attached to the short rectum is the **rectal gland,** which secretes a colorless fluid containing a high concentration of sodium chloride. It assists the kidney in regulating the salt concentration of the blood. The chambers of the **heart** are arranged in tandem formation (Fig. 23-7), and the circulatory system is basically the same as that of the embryonic vertebrate and of the ammocoetes.

The **mesonephric kidneys** are two long, slender organs above the coelom; they are drained by the **wolffian ducts,** which open into a single urogenital sinus at the **cloaca.** The wolffian ducts also carry the sperm from the testes of the male, which uses a clasper to deposit the sperm in the female oviduct. The müllerian duct, or oviduct (paired), carries the eggs from the **ovary** and coelom and is modified into a **uterus** in which a primitive placenta may attach the embryo shark until it is born. Such a relationship is actually **viviparous reproduction;** other species simply retain the developing egg in the uterus without attachment to the mother's wall **(ovoviviparous reproduction).** Still others lay large, yolky eggs immediately after fertilization **(oviparous reproduction).** Some sharks and rays deposit their fertilized eggs in a horny capsule called the "mermaid's purse," which is attached by tendrils to seaweed. Later the young shark emerges from this "cradle."

The embryonic brain is the basic tripartite vertebrate brain—forebrain, midbrain, and hindbrain. These three parts develop into five subdivisions or regions—telencephalon, diencephalon, mesencephalon, metencephalon, and myelencephalon—each with certain functions. There are ten pairs of cranial nerves that are distributed largely to the head regions. Surrounding the spinal cord are the neural arches of the vertebrae. Along the spinal cord a pair of spinal nerves, with united dorsal and ventral roots, is distributed to each body segment.

Elasmobranchs have developed an interesting solution to the physiologic problem of living in a hyperos-

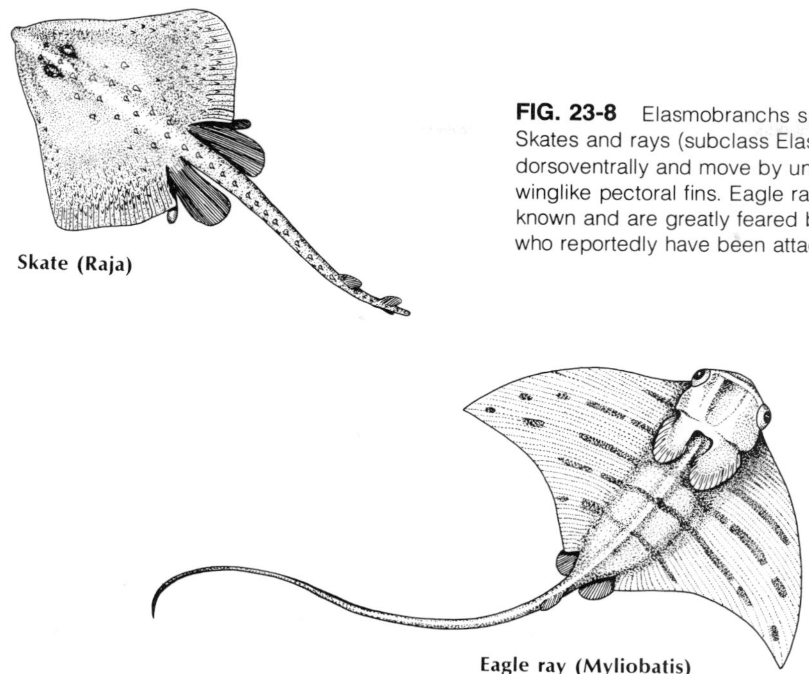

FIG. 23-8 Elasmobranchs specialized for life on the sea floor. Skates and rays (subclass Elasmobranchii) are flattened dorsoventrally and move by undulations of greatly expanded winglike pectoral fins. Eagle rays are among the largest fishes known and are greatly feared by pearl divers in the tropics who reportedly have been attacked and eaten by them.

motic medium. Like the ancestors of the bony fishes, those of sharks and their kin lived in fresh water, only returning to the sea in the Triassic period (about 230 million years ago) after becoming thoroughly adapted to freshwater existence. Their body fluid salt concentration was now much lower than the surrounding seawater; to prevent water from being drawn out of the body osmotically, the elasmobranchs have retained nitrogenous wastes, especially urea and trimethylamine oxide, in the blood. These solutes, combined with the blood salts, raised the blood solute concentration to slightly exceed that of seawater, eliminating an osmotic inequality between their bodies and the surrounding seawater.

The skates and rays (Fig. 23-8) are specialized for bottom dwelling. In these the pectoral fins are greatly enlarged and are used like wings in swimming. The gill openings are on the underside of the head, but the large spiracles are on top. Water for breathing is taken in through these spiracles to prevent clogging the gills, for the mouth is often buried in sand. Their teeth are adapted for crushing their prey—molluscs, crustaceans, and an occasional small fish. Two members of this group are of special interest—the stingrays and the electric rays. In the stingrays the caudal and dorsal fins have disappeared and the tail is slender and whiplike. The tail is armed with one or more saw-edged spines, which can inflict very dangerous wounds. Such wounds may heal slowly and leave complications. Electric rays (Fig. 23-24) have smooth, naked skins and have certain dorsal muscles modified into powerful electric organs, which can give severe shocks and stun their prey. Stingrays also have electric organs in the tail.

Chimaeras—subclass Holocephali

The members of this small group, distinguished by such suggestive names as ratfish (Fig. 23-9), rabbitfish, spookfish, and ghostfish, are remnants of an aberrant line that diverged from the elasmobranchs at least 300 million years ago (Carboniferous or Devonian period). Fossil chimaeras (ky-meer′uz) were first found in the Jurassic period, reached their zenith in the Cretaceous and early Tertiary periods (120 million to 50 million years ago), and have declined ever since.

Today there are only about 25 species extant. Anatomically they present an odd mixture of sharklike and bony fish–like features. Instead of a toothed mouth, their jaws bear large flat plates. The upper jaw is completely fused to the cranium, a most unusual development in fishes. Their food is seaweed, molluscs, echinoderms, crustaceans, and fishes—all in all a surprisingly mixed diet for such a specialized grinding dentition. Chimaeras are not commercial species and are seldom caught. Despite their grotesque shape, they are beautifully colored with a pearly iridescence and have vivid emerald-green eyes that must be seen to be believed.

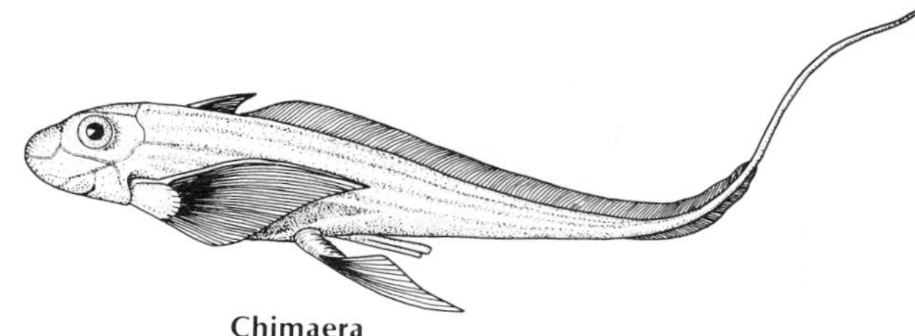

FIG. 23-9 Chimaera, or ratfish, of North American west coast. This species is one of the most handsome of chimaeras, which tend toward bizarre appearances. Subclass Holocephali.

BONY FISHES—CLASS OSTEICHTHYES (TELEOSTOMI)

In no other major animal group do we see better examples of adaptive radiation than among the bony fishes. Their adaptations have fitted them for every aquatic habitat except the most completely inhospitable. Body form alone is indicative of this diversity. Some have fusiform, streamlined bodies and other adaptations for reducing friction. Predaceous, pelagic fish have trim, elongate bodies, powerful tail fins, and other mechanical advantages for swift pursuit. Sluggish bottom-feeding forms have flattened bodies for movement and concealment on the ocean floor. The elongate body of the eel is an adaptation for wriggling through mud and reeds and into holes and crevices. Some, such as pipefishes, are so whiplike that they are easily mistaken for filaments of marine algae waving in the current (Fig. 23-22). Many other grotesque body forms are obviously cryptic or mimetic adaptations for concealment from predators or as predators. Such few examples cannot begin to express the amazing array of physiologic and anatomic specializations for defense and offense, food-gathering, navigation, and reproduction in the diverse aquatic habitats to which bony fishes have adapted themselves. More of these adaptations are described in the pages that follow.

Characteristics

1. **Skeleton more or less bony,** representing the primitive skeleton; vertebrae numerous; notocord may persist in part; **tail usually homocercal**
2. Skin with mucous glands and with embedded dermal scales of three types: **ganoid, cycloid,** or **ctenoid;** some without scales; no placoid scales
3. Fins both median and paired, with **fin rays of cartilage or bone**
4. **Mouth terminal** with many teeth (some toothless); jaws present; olfactory sacs paired and may or may not open into mouth
5. Respiration by gills supported by bony gill arches and covered by a **common operculum**
6. **Swim bladder** often present with or without duct connected to pharynx
7. Circulation consisting of a two-chambered heart, arterial and venous systems, and four pairs of aortic arches; blood of nucleated red cells
8. Nervous system of a brain with small olfactory lobes and cerebrum; large optic lobes and cerebellum; ten pairs of cranial nerves
9. Sexes separate; gonads paired; fertilization usually external; larval forms may differ greatly from adults

Bony fishes vary greatly in size. Some of the minnows are less than 2 cm long; other forms may exceed 3 m in length. The swordfish is one of the largest and may attain a length of 4 m. Most fishes, however, fall between 2 and 30 cm in length.

Classification

Class Osteichthyes—bony fishes. Three subclasses, 69 orders.

Subclass Crossopterygii—lobe-finned fishes. Three extinct orders, one living order (Coelacanthimorpha) containing one species, *Latimeria chalumnae*.

Subclass Dipneusti—lungfishes. Six extinct orders, two living orders containing three genera: *Neoceratodus, Lepidosiren,* and *Protopterus*.

Subclass Actinopterygii—ray-finned fishes. Three superorders and 59 orders.

Superorder Chondrostei (kon-dros′tee-i) (Gr., *chondros,* cartilage, + *osteon,* bone)—**primitive ray-finned fishes.** Ten extinct orders; two living orders containing the bichir *(Polypterus),* sturgeons, and paddlefish.

Superorder Holostei (ho-los′tee-i) (Gr., *holos,* entire, + *osteon,* bone)—**intermediate ray-finned fishes.** Four extinct orders; two living orders containing the bowfin *(Amia)* and gars *(Lepidosteus)*.

Superorder Teleostei (tel′e-os′tee-i) (Gr., *teleos,* complete, + *osteon,* bone)—**climax bony fishes.** Body covered with thin scales without bony layer (cycloid or ctenoid) or scaleless; dermal and chondral parts of skull closely united; caudal fin mostly homocercal; mouth terminal; notochord a mere

FIG. 23-10 *Osteolepis,* primitive rhipidistian (crossopterygian) fish of middle Devonian times. This fish must be considered in direct line of descent between fishes and amphibians because its type of skull was similar to that of primitive land vertebrates, and its lobe fin was of a pattern that could serve as the beginning of a tetrapod limb. This type of fin consisted of median axial bones, with small bones radiating out from median ones. Some of its bones can be homologized with limb bones of tetrapods, such as humerus or femur and the ulna-radius or tibia-fibula elements. *Osteolepis* was covered by primitive cosmoid scales, not found in existing fishes. These scales were of rhombic shape and consisted of basal bony layers, with a spongy layer of blood vessels covered with cosmine (dentin preferred) and enamel. (Modified from Romer, A. S. 1966. Vertebrate paleontology. Chicago, University of Chicago Press.)

vestige; swim bladder mainly a hydrostatic organ and usually not opened to the esophagus; endoskeleton mostly bony. According to Nelson (1976), there are ten extinct orders and 31 living orders. The latter is comprised of 415 families, and approximately 18,000 living species, representing 96% of all living fishes. Seven of the larger orders are as follows:

Anguilliformes—21 families; freshwater eels, moray eels, conger eels, snipe eels.

Salmoniformes—23 families; pikes, whitefish, salmon, trout, smelts, deep-sea luminescent fishes.

Cypriniformes—25 families; suckers, minnows, carp, electric eels.

Siluriformes—30 families; catfish.

Atheriniformes—15 families; flying fish, medakas, killifish, licebearers.

Scorpaeniformes—20 families; rockfish, searobins, greenlings, sculpins, poachers.

Perciformes—146 families; barracudas, mullets, perch, darters, sunfishes, grunters, croakers, Moorish idols, damsel fish, viviparous perch, wrasses, parrot fish, trumpeters, sand perch, stargazers, blennies, wolf fish, eel pouts, mackerels, tunas, swordfish.

Origin, evolution, and diversity

The Osteichthyes are divided into three clearly distinct groups: Crossopterygii (lobe-finned fishes), Dipneusti (lungfishes), and Actinopterygii (ray-finned fishes). What are their origins? As pointed out earlier in this chapter, it seems safest to steer a middle course through current paleontologic debate and assign equal rank to all three groups within the class Osteichthyes. In other words, it is impossible to decide which one of these three groups, if any, might have served as ancestral stock for the other two. It is apparent from the fossil evidence that all three groups were distinct in the Devonian period, some 400 million years ago. They are believed to have descended from an acanthodian of the Silurian period, perhaps a creature similar to the fossil acanthodian illustrated in Fig. 22-5 (p. 482).

Lobe-finned fishes—subclass Crossopterygii

Crossopterygii has had the least spectacular evolutionary radiation of all the Osteichthyes. It is by far the most important in an evolutionary sense because the tetrapods arose from one or more of their ancient members. The crossopterygians had nostrils that opened into the mouth (choanae), lungs as well as gills, and paired **lobed fins.** They first appeared in Devonian times, a capricious period of alternating droughts and floods when their lungs would have been a decided asset, if not absolutely essential, for their survival. They used their strong lobed fins as four legs to scuttle from one disappearing swamp to another that offered more promise for a continuing aquatic existence.

The crossopterygians are divided broadly into two groups. The **rhipidistians** (Fig. 23-10) appeared in the Devonian period, flourished in the late Paleozoic era, and then disappeared. This is the stock from which the amphibians have descended (Fig. 23-1). Among the primitive characteristics of the rhipidistians were the fusiform shape, two dorsal fins, and a heterocercal tail (Fig. 23-11). The paired fins bore sharp resemblances to a tetrapod limb, for they consisted of a basal arrangement of median or axial bones, with other bones

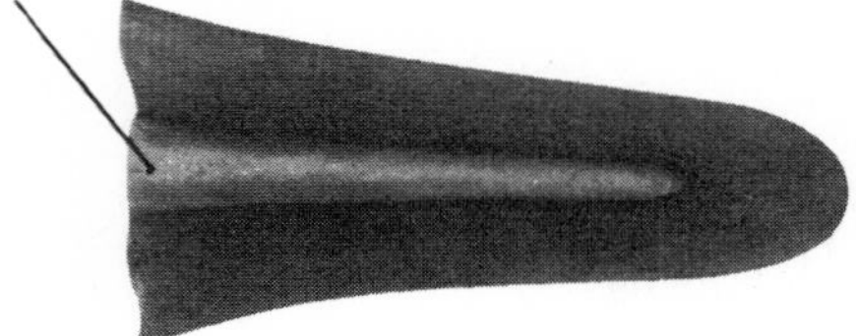

Protocercal (some larval fishes)

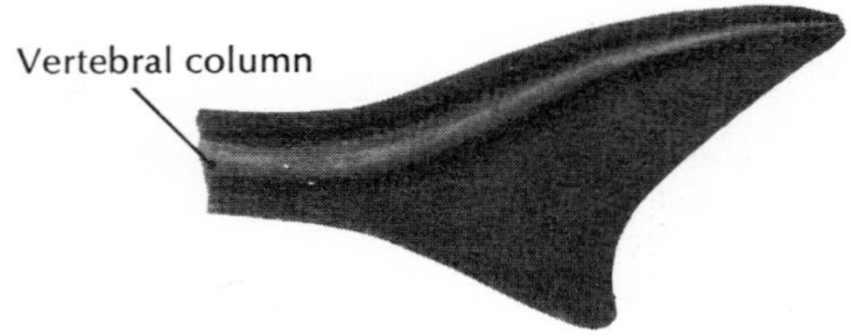

Heterocercal
(many primitive fishes, sturgeon, sharks)
(most ancestral type)

Hypocercal
(some primitive fishes, reptilian ichthyosaur)

Diphycercal
(lungfishes, crossopterygians, Polypterus)

Homocercal (teleost fishes)

FIG. 23-11 Types of caudal fins among fishes. Some functional correlations may be seen among these different types. Heterocercal tail, for example, is found in fishes without swim bladders, for it tends to counteract gravity while swimming.

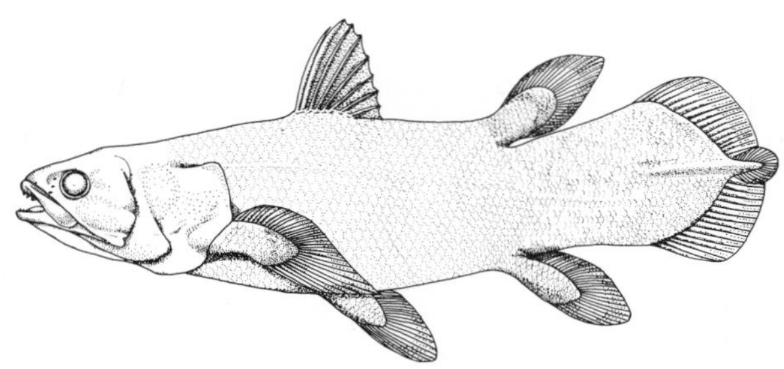

FIG. 23-12 Coelacanth *Latimeria chalumnae*. This surviving marine relic of the crossopterygians that flourished some 350 million years ago has fleshy-based ("lobed") fins with which its ancestors used to pull themselves across land from pond to pond. (Courtesy Vancouver Public Aquarium, British Columbia.)

radiating from these median ones. Some of the proximal bones seem to correspond with the three chief bones of the tetrapod limb. The scales of these primitive fish were of the **cosmoid** type, a thick complex scale of dentinlike cosmine, enamel, and vascular pulp cavities. This type of scale is not found in modern fishes but has been replaced by the bony **cycloid** type (Fig. 23-6).

The other group of crossopterygians was the **coelacanths.** These also arose in the Devonian period, radiated somewhat, and reached their evolutionary peak in the Mesozoic era. At the end of the Mesozoic era they nearly disappeared but left one remarkable surviving species, the living coelacanth *Latimeria chalumnae* (Fig. 23-12). Since the last coelacanths were believed to have become extinct 70 million years ago, the astonishment of the scientific world can be imagined when the remains of a coelacanth were found on a dredge off the coast of South Africa in 1938. An intensive search was begun in the Comoro Islands area near Madagascar where, it was learned, native Comoran fisherman occasionally caught them with hand lines at great depths. By the end of 1976, more than 80 specimens had been caught, many in excellent condition, although none has been kept alive after capture.

The "modern" marine coelacanth is a descendant of the Devonian freshwater stock. The tail is of the **diphycercal** type (Fig. 23-11) but possesses a small lobe between the upper and lower caudal lobes, producing a three-pronged structure (Fig. 23-12). Coelacanths also show some degenerative features, such as more cartilaginous parts and a swim bladder that

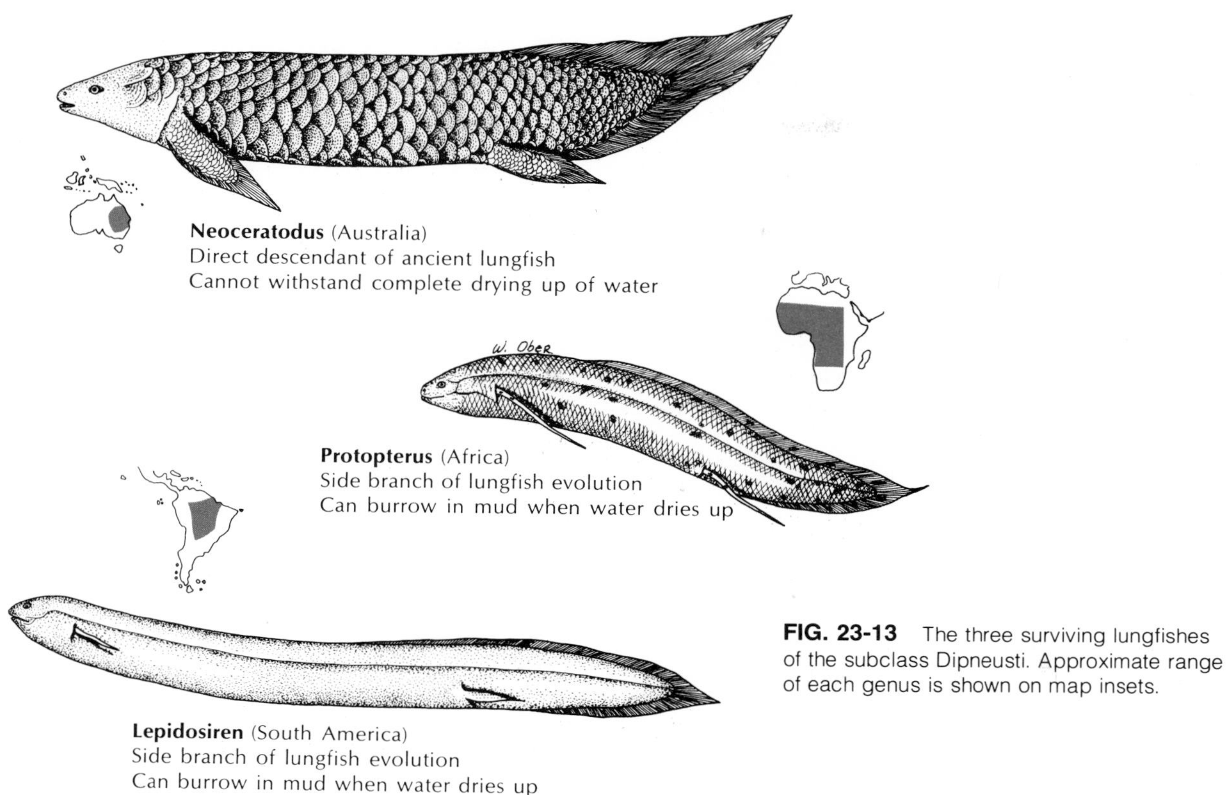

FIG. 23-13 The three surviving lungfishes of the subclass Dipneusti. Approximate range of each genus is shown on map insets.

was either calcified or else persisted as a mere vestige. They also lack the internal nostril so characteristic of crossopterygians, but this is probably a secondary loss after the adoption of a deep-sea existence; obviously neither nostrils nor functional lungs has any relevance for such a life habit.

Lungfishes—subclass Dipneusti

The lungfishes are another relic group of fishes represented today by only three genera (Fig. 23-13). These resemble the lobe-finned fishes in having lobe-shaped paired fins and lungs. However, all lungfishes, extinct or living, differ from the crossopterygians in several significant skeletal features, including the totally different origin of the internal nostrils.

Of the three surviving genera of lungfishes, the least specialized is *Neoceratodus,* the living Australian lungfish, which may attain a length of 1.5 m. This lungfish is able to survive in stagnant, oxygen-poor water by coming to the surface and gulping air into its single lung, but it cannot live out of water. The South American lungfish *Lepidosiren* and the African lungfish *Protopterus* are evolutionary side branches of the Dipneusti, and they can live out of water for long periods of time. *Protopterus* lives in African streams and rivers that run completely dry during the dry season, with their mud beds baked hard by the hot tropical sun. The fish burrows down at the approach of the dry season and secretes a copious slime that is mixed with mud to form a hard cocoon in which it estivates until the rains return (Fig. 23-14).

Ray-finned fishes—subclass Actinopterygii

This huge assemblage contains all the familiar bony fishes. The fossil record reveals that the group had its beginnings in Devonian freshwater lakes and streams. The earliest actinopterygians were small fishes with large eyes and extended mouths. Their tails were **heterocercal** (Fig. 23-11). They had a single dorsal fin and a single anal fin; paired fins were represented by the anterior pectoral fins and the posterior pelvic

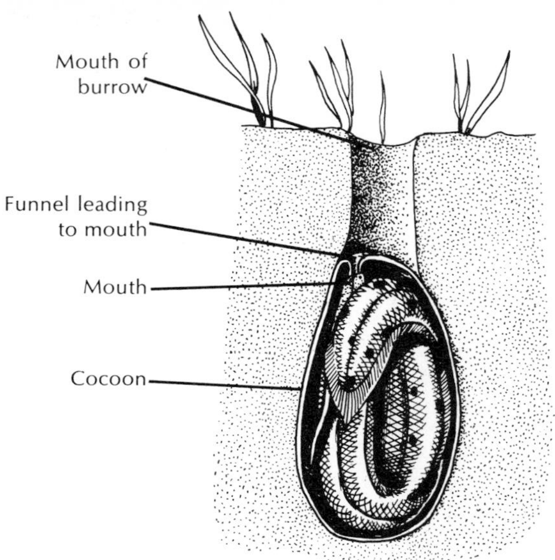

FIG. 23-14 Estivating African lungfish *Protopterus.* As water disappears during dry season, *Protopterus* digs a burrow, coils tightly, and secretes mucus from skin and lips. Combined with mud, the mucus dries into a stiff, papery cocoon. A hollow tube of mucus extruded from the mouth forms a direct opening for air. Estivation may last several months, during which nitrogenous wastes gradually accumulate in body tissues. When rains return, rising water level softens the cocoon, and *Protopterus* breaks out quickly, croaks several times with apparent pleasure, and swims off. Subclass Dipneusti.

FIG. 23-15 White sturgeon *Acipenser.* Sturgeons, largest of all freshwater fishes, were once abundant in North America, but their size and very slow growth rate made them vulnerable to fishermen who destroyed them as nuisances. The largest North American sturgeons were in British Columbia's Fraser River system. Some caught early in this century reached 540 kg (1,200 pounds), but sturgeons of this size will probably never again be seen on this continent. Even larger sturgeons are still collected in Russia, where the female's eggs (roe) are converted into gourmet caviar. Superorder Chondrostei.

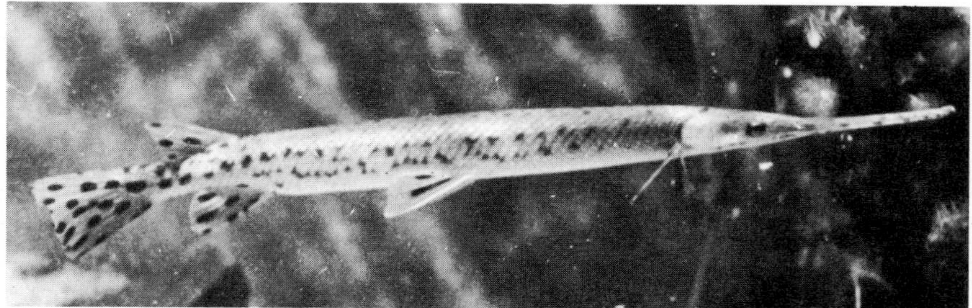

FIG. 23-16 Long-nosed garpike *Lepidosteus osseus.* This and other garpikes thrive in the rivers of the southern United States. They are voracious and solitary feeders that subsist on crayfish and small fishes. On sunny days they may often be seen floating on the river surface looking like drifting pieces of wood. Superorder Holostei.

fins. Their skeletons were largely bone. Their trunks and tails were encased in an armor of heavy, rhombic **ganoid scales** (Fig. 23-6). Most of these early fishes had functional lungs, as did all the Devonian fishes, but they lacked internal nostrils. All had gills (five pairs or less) and spiracles.

These early actinopterygians belonged to the order Palaeonisciformes (pay′lee-o-nis-sif-for′meez). One common genus in the fossil record was *Cheirolepis,* a generalized type that had some resemblance to certain acanthodians (Fig. 22-5, p. 482). The palaeoniscids were distinctly different from their freshwater contemporaries, the crossopterygians and the lungfishes, with which they shared the Devonian swamps and rivers. The palaeoniscids were rare in the Devonian period, became a more secure group in the Carboniferous and Permian periods, but nearly disappeared in the Triassic period when their descendants returned to the sea.

In their evolution, the actinopterygians have passed through three stages. The most primitive group includes the Chondrostei, represented today by the freshwater and marine sturgeons (Fig. 23-15) and the bichir, *Polypterus,* of African rivers (Fig. 23-1). *Polypterus* is an interesting relic with a lunglike swim bladder and many other primitive characteristics; it resembles an ancient palaeoniscid more than any other living descendant. There is no satisfactory explanation for the survival to the present of certain fishes such as this one and the coelacanth *Latimeria* when all of their kin perished millions of years ago.

A second actinopterygian group is the Holostei. There were several lines of descent within this group, which flourished during the Triassic and Jurassic periods. They declined toward the end of the Mesozoic as their successors, the teleosts, crowded them out. But they left two surviving lines, the bowfin, *Amia* (Fig. 23-1), of shallow, weedy waters of the Great Lakes and Mississippi valley, and the garpikes (or simply gars) of eastern North America (Fig. 23-16).

The third group is the Teleostei, the modern bony fishes (Fig. 23-17). Diversity appeared early in teleost evolution, foreshadowing the truly incredible variety of body forms among teleosts today. The skeleton of primitive fishes was largely ossified, but this condition regressed to a partly cartilaginous state among many of the Chondrostei and Holostei. Teleosts, however, have an internal skeleton almost completely ossified like the primitive members. The dermal investing bones of the skull (dermatocranium) and the chondrocranium (endocranium) around the brain and sense organs form a closer union among the teleosts than they did in the primitive bony fishes. Other evolutionary changes among the teleosts were the movement of the pelvic fins forward to the head and thoracic region, the transformation of the lungs of primitive forms into swim bladders (Fig. 23-17) with hydrostatic functions and without ducts, the changing of the heterocercal tail of primitive fishes and of the intermediate superorders into a homocercal form (Fig. 23-11), and the development of the thin **cycloid** and **ctenoid scales** (Fig. 23-6) from the thick ganoid type of early fishes. Among other changes were the loss of the spiracles and the development of stout spines in the fins, especially in the pectoral, dorsal, and anal fins.

STRUCTURAL AND FUNCTIONAL ADAPTATIONS OF FISHES

Locomotion in water

To the human eye, fishes appear capable of swimming at extremely high speeds. But our judgment is unconsciously tempered by our own experience that water is a highly resistant medium to move through. Most fishes, such as a trout or a minnow, can swim maximally about 10 body lengths per second, obviously an impressive performance by human standards. Yet when these speeds are translated into kilometers per hour it means that a 30-cm (1-foot) trout can swim only about 10.4 km (6.5 miles) per hour. The larger the fish, the faster it can swim. A 60-cm salmon can sprint 22.5 km (14 miles) per hour and a 1.2-m barracuda, the fastest fish measured, is capable of 43 km (27 miles) per hour. Fish can swim this fast only for very brief periods during moments of stress; cruising speeds are much lower.

The propulsive mechanism of a fish is its trunk and tail musculature. The axial, locomotory musculature is composed of zigzag muscle bands (myotomes) that on the surface take the shape of a W lying on its side (Fig. 23-18). Internally the muscle bands are deflected forward and backward in a complex fashion that apparently promotes efficiency of movement. The muscles are bound to broad sheets of tough connective tissue, which in turn tie to the highly flexible vertebral column.

The way fishes swim is seen best in a relatively primitive fish such as a shark. When swimming, the shark body assumes the form of a sine wave. Waves

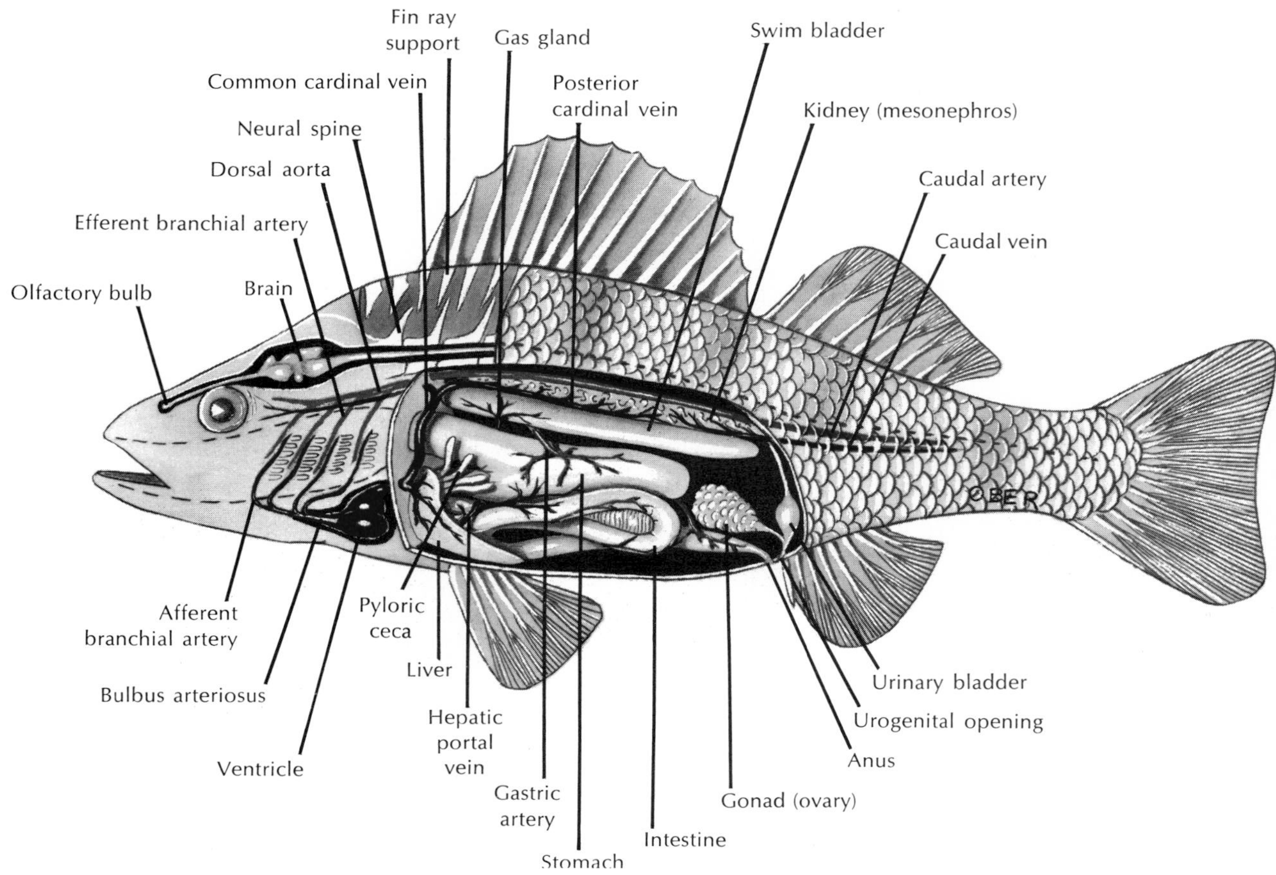

FIG. 23-17 Internal anatomy of the yellow perch *Perca flavescens*, a freshwater teleost.

of contraction begin on one side of the body at the front and proceed to the tail. When this wave has moved some distance, another wave is initiated at the front on the opposite side of the body. The process continues, with waves of contraction moving posteriorly, alternating from one side to the other. In higher bony fishes, the sweeping movement of the tail assumes a greater role.

Swimming is possible only because the density and noncompressibility of water offers great purchase for forward thrust. As a medium for locomotion, water offers another advantage: since the density of water is only slightly less than that of protoplasm, most aquatic animals are almost perfectly supported and need expend no energy overcoming the force of gravity. Consequently, swimming is actually the most economic form of animal locomotion. For example, the energetic cost per kilogram of body weight of traveling 1 km is 0.39 kcal for a salmon (swimming), 1.45 for a gull (flying), and 5.43 for a ground squirrel (walking). However, the low-energy cost of swimming by fish is by no means fully understood. Relatively simple calculations show that a fish moves through water with only about one-tenth the drag of a rigid model of the fish's body. The energy required to propel a submarine is many times greater than that consumed by a whale of similar size and moving at the same speed.

Aquatic mammals and fishes create virtually no turbulence, a feat that humans with twentieth-century ingenuity are a long way from matching. The secret lies in the way aquatic animals bend their bodies and fins (or flukes) to swim and in the textural properties of the body surface. It has recently been shown, for example, that the slimy surface of a fish reduces water

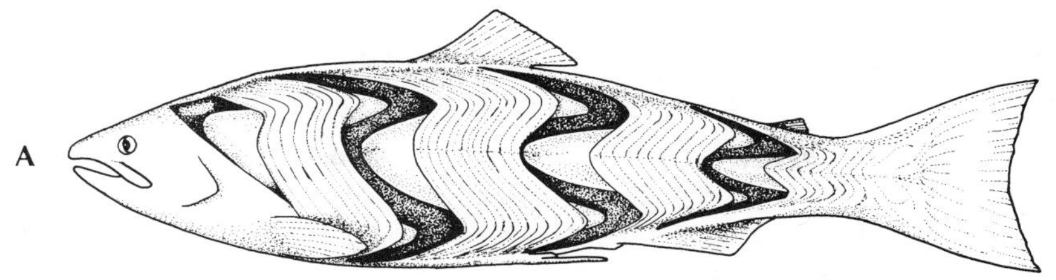

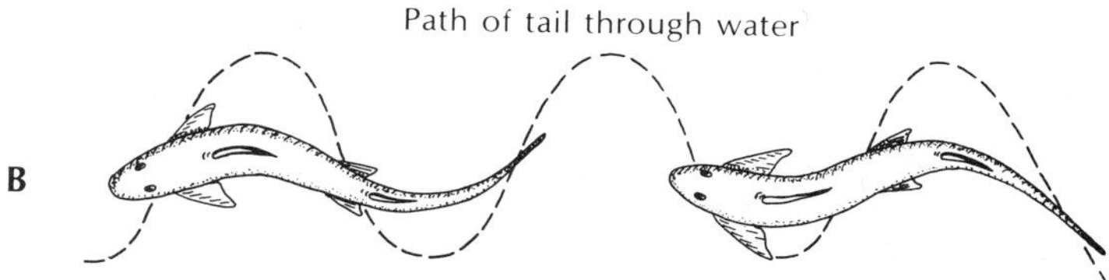

FIG. 23-18 **A,** Trunk musculature of a salmon. Segmental myotomes are W-shaped when viewed from the surface. Musculature has been dissected away in four places to show internal anterior and posterior deflections of myotomes that improve muscular efficiency for swimming. **B,** Motion of swimming fish. Noncompressible water must be pushed aside by the forward motion of the head, driven by the snakelike stroke of the body. (**A,** After Greene from Romer, A. 1970, The vertebrate body, ed. 4, Philadelphia, W. B. Saunders Co.; **B,** modified from Marshall, P. T., and G. M. Hughes, 1967, the physiology of mammals and other vertebrates, New York, Cambridge University Press.)

friction by at least 66%. Understanding the energetics of swimming remains part of the unfinished business of biology.

Neutral buoyancy and the swim bladder

All fishes are slightly heavier than water because their skeletons and other tissues contain heavy elements that are present only in trace amounts in natural waters. To keep from sinking, sharks must always keep moving forward in the water. The asymmetric (heterocercal) tail of a shark provides the necessary tail lift as it sweeps to and fro in the water and the broad head and flat pectoral fins (Fig. 23-5) act as planes to provide head lift. Sharks are also aided in their buoyancy problem by having very large livers containing a special fatty hydrocarbon called **squalene** that has a density of only 0.86. The liver thus acts like a large sack of buoyant oil that helps to compensate for the shark's heavy body.

By far the most efficient flotation device is a gas-filled space. The **swim bladder** serves this purpose in the bony fishes. It arose from the paired lungs of the primitive Devonian bony fishes. Lungs were probably a ubiquitous feature of the Devonian freshwater bony fishes when, as we have seen, the alternating wet and dry climate probably made such an accessory respiratory structure essential for life. The primitive lung may have been similar to the lungs found in the existing *Polypterus,* the chondrostean fish of tropical Africa that so resemble the palaeoniscid ancestors of modern bony fishes (Fig. 23-19). Swim bladders are present in all pelagic bony fishes but are absent in most bottom dwellers, such as flounders and sculpins.

By adjusting the volume of gas in the swim bladder, a fish can achieve neutral buoyancy and remain suspended indefinitely at any depth with no muscular effort. There are severe technical problems, however. If the fish descends to a greater depth, the swim bladder gas is compressed so that the fish becomes heavier and tends to sink. Gas must be added to the bladder to establish a new equilibrium buoyancy. If the fish swims up, the gas in the bladder expands, making the fish lighter. Unless gas is removed, the fish will rise with ever-increasing speed while the bladder continues to expand, until it pops helplessly out of the water. (This is a very real hazard for divers in helmeted diving suits

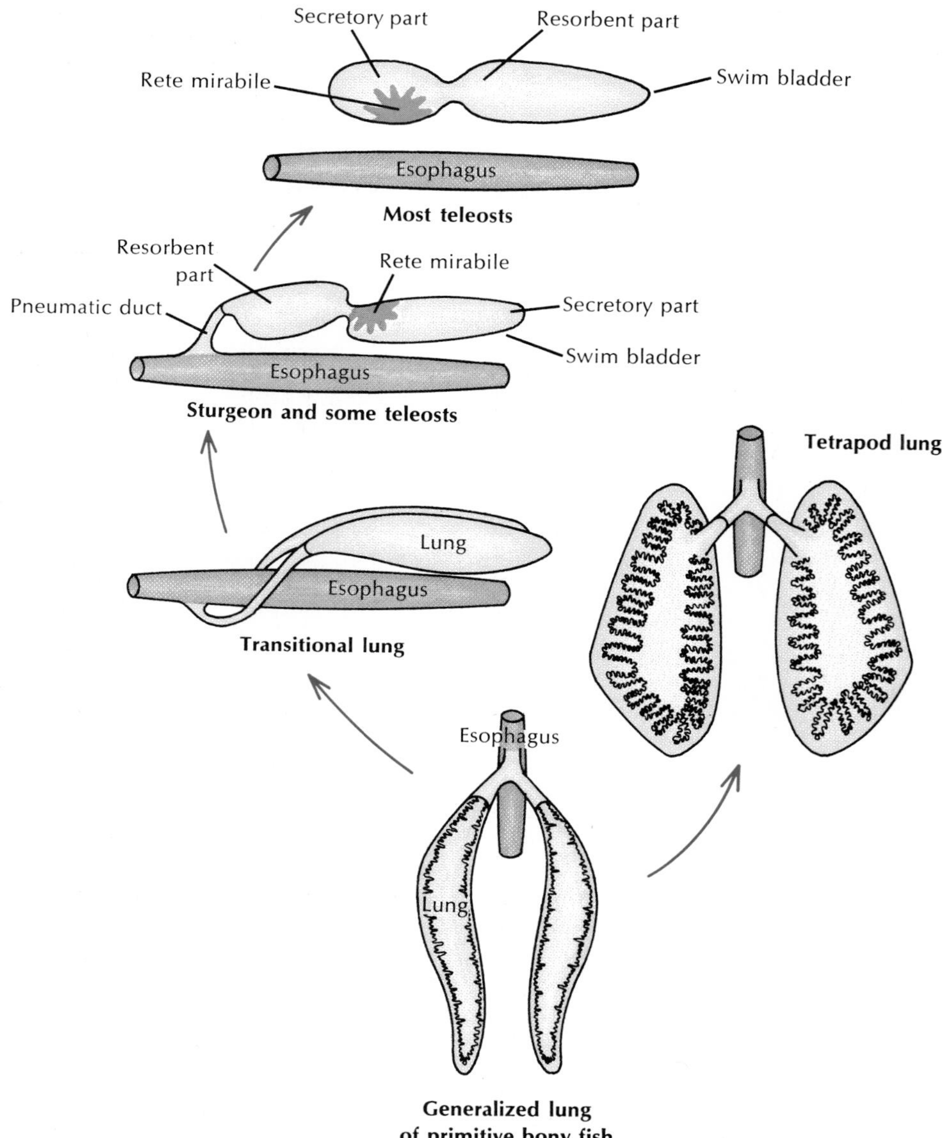

FIG. 23-19 Evolution of lung and swim bladder. Most primitive fishes were provided with lungs, adaptations for the oxygen-depleted environments that existed during the evolution of the Osteichthyes. The lung originated as diverticulum of the foregut. From this early, generalized lung two lines of evolution occurred. One led to the swim bladder of modern teleost fishes. Various transitional stages show that the swim bladder shifted to a dorsal position above the esophagus, becoming a buoyancy organ. The duct has been lost in most teleosts, and the swim bladder, with specialized gas secretion and reabsorption areas, is served by an independent blood supply. Second line of evolution has led to the tetrapod lung found in land forms. There has been extensive internal folding, but no radical change in lung position.

who must carefully adjust the air pressure to prevent overinflation or underinflation of their suits.)

There are two ways fishes adjust gas volume in the swim bladder. The less specialized fishes (trout, for example) have a **pneumatic duct** that connects the swim bladder to the esophagus (Fig. 23-19); these forms must come to the surface and gulp air to charge the bladder and obviously are restricted to relatively shallow depths. More specialized teleosts have lost the pneumatic duct (upper diagram in Fig. 23-19). Gas exchange depends on two highly specialized areas: a **gas gland** that secretes gas into the bladder and a **resorptive area,** or "oval," that can remove gas from the bladder. The gas gland contains a remarkable network of blood vessels **(rete mirabile)** arranged so that a vast number of arteries and veins in a tight bundle run in opposite directions to each other. This is called **countercurrent flow,** an arrangement that makes possible a tremendous multiplication of gas concentration inside the swim bladder. The amazing effectiveness of this device is exemplified by a fish living at a depth of 2,400 m (8,000 feet). To keep the bladder inflated, the gas inside (mostly oxygen, but also variable amounts of nitrogen, carbon dioxide, carbon monoxide, and argon) must have a pressure exceeding 240 atm, much greater than the pressure in a fully charged steel gas cylinder. Yet the oxygen pressure in the fish's blood cannot exceed one-fifth atmosphere—equal to the oxygen pressure at the sea surface.

Respiration

Fish gills are composed of thin filaments covered with a thin epidermal membrane that is folded repeatedly into platelike **lamellae** (Fig. 23-20). These are richly supplied with blood vessels. The gills are located inside the pharyngeal cavity and are covered with a movable flap, the **operculum.** This arrangement provides excellent protection to the delicate gill filaments, streamlines the body, and makes possible a pumping system for moving water through the mouth, across the gills, and out the operculum. Instead of opercular flaps as in bony fishes, the elasmobranchs have a series of **gill slits** out of which the water flows. In both elasmobranchs and bony fishes the branchial mechanism is arranged to pump water continuously and smoothly over the gills, even though to an observer it appears that fish breathing is pulsatile. The flow of water is opposite to the direction of blood flow (countercurrent flow), the best arrangement for extracting the greatest possible amount of oxygen from the water. Some bony fishes can remove as much as 85% of the oxygen from the water passing over their gills. Very active fishes, such as herring and mackerel, can obtain sufficient water for their high oxygen demands only by continually swimming forward to force water into the open mouth and across the gills. Such fish will be asphyxiated if placed in an aquarium that restricts free swimming movements, even though the water is saturated with oxygen.

A surprising number of fishes can live out of water for varying lengths of time by breathing air. Several devices are employed by different fishes. We have already described the lungs of the lungfishes, *Polypterus,* and the extinct crossopterygians. Freshwater eels often make overland excursions during rainy weather, using the skin as a major respiratory surface. The bowfin, *Amia,* has both gills and a lunglike swim bladder. At low temperatures it uses only its gills, but as the temperature and the fish's activity increase, it breathes mostly air with its swim bladder. The electric eel has degenerate gills and must supplement gill respiration by gulping air through its vascular mouth cavity. One of the best air breathers of all is the Indian climbing perch, which spends most of its time on land near the water's edge breathing air through special air chambers above the much-reduced gills.

Osmotic regulation

Fresh water is an extremely dilute medium with a salt concentration (0.001 to 0.005 gram moles per liter [M]) much below that of the blood of freshwater fishes (0.2 to 0.3 M). Water therefore tends to enter their bodies osmotically and salt is lost by diffusion outward. Although the scaled and mucus-covered body surface is almost totally impermeable to water, water gain and salt loss do occur across the thin membranes of the gills. Freshwater fishes are **hyperosmotic regulators** that have several defenses against these problems (Fig. 23-21). First, the excess water is pumped out by the **mesonephric** kidney (p. 699), which is capable of forming a very dilute urine. Second, special **salt-absorbing cells** located in the gill epithelium are capable of actively moving salt ions, principally sodium and chloride, from the water to the blood. This, together with salt present in the fish's food, replaces diffusive salt loss. These mechanisms are so efficient that a freshwater fish devotes only a small part of its total energy expenditure to keeping itself in osmotic balance.

Marine bony fishes are **hypoosmotic regulators** that

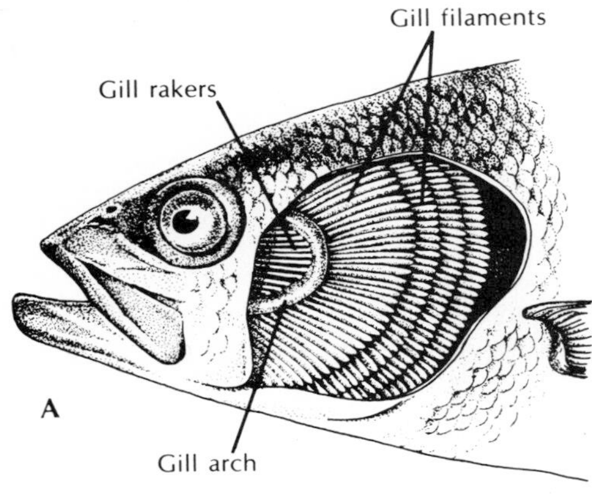

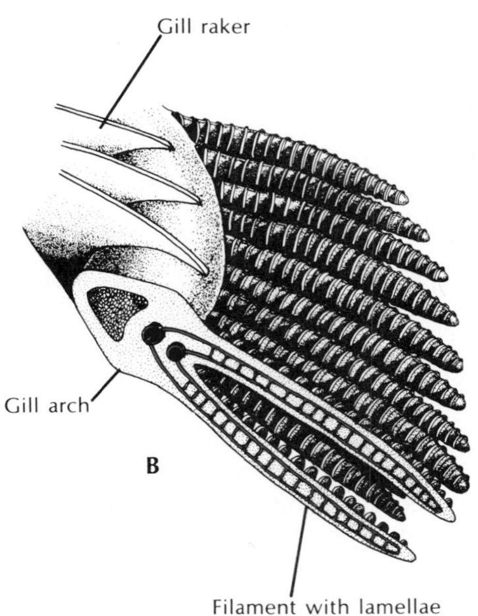

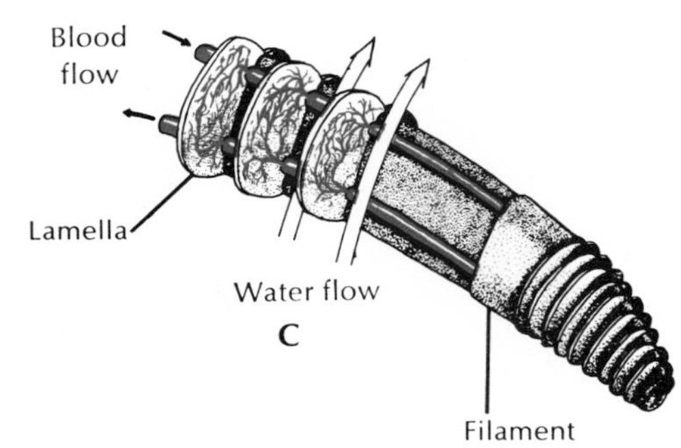

FIG. 23-20 Gills of fishes. Bony, protective flap covering the gills (operculum) has been removed, **A,** to reveal branchial chamber containing the gills. Four gill arches are on each side, each bearing numerous filaments. A portion of gill arch, **B,** shows gill rakers that project forward to strain out food and debris, and gill filaments that project to the rear. A single gill filament, **C,** is dissected to show the blood capillaries within the platelike lamellae. Direction of water flow *(large arrows)* is opposite the direction of blood flow *(small arrows)*.

encounter a completely different set of problems. Having a much lower blood salt concentration (0.3 to 0.4 M) than the seawater around them (about 1 M), they tend to lose water and gain salt. The marine teleost fish quite literally risks drying out, much like a desert mammal deprived of water. Again, marine bony fishes, like their freshwater counterparts, have evolved an appropriate set of defenses (Fig. 23-21). To compensate for water loss, the marine teleost drinks seawater. Although this behavior obviously brings needed water into the body, it is unfortunately accompanied by a great deal of unneeded salt. Unwanted salt is disposed of in two ways: (1) the major sea salt ions (sodium, chloride, and potassium) are carried by the blood to the gills where they are secreted outward by special **salt-secretory cells;** and (2) the remaining ions, mostly the divalent ions (magnesium, sulfate, and calcium), are left in the intestine and voided with the feces. However, a small but significant fraction of these residual divalent salts in the intestine, some 10% to 40% of the total, penetrates the intestinal mucosa and enters the bloodstream. These ions are excreted by the kidney. Unlike the freshwater fish kidney, which forms its urine by the usual filtration-reabsorption sequence typical of most vertebrate kidneys (pp. 701-705), the marine fish's kidney excretes divalent ions by tubular secretion. Since very little if any filtrate is formed, the glomeruli have lost their importance and disappeared altogether in some marine teleosts. The pipefishes and the goosefish, shown in Figs. 23-22 and 23-23, are good examples of "aglomerular" marine fishes.

Most bony fishes are restricted to either a freshwater or a seawater habitat. However some 10% of all teleosts can pass back and forth with ease between both habitats. Examples of these **euryhaline fishes** (Gr., *eurys,* broad, + *hals,* salt), which must have highly

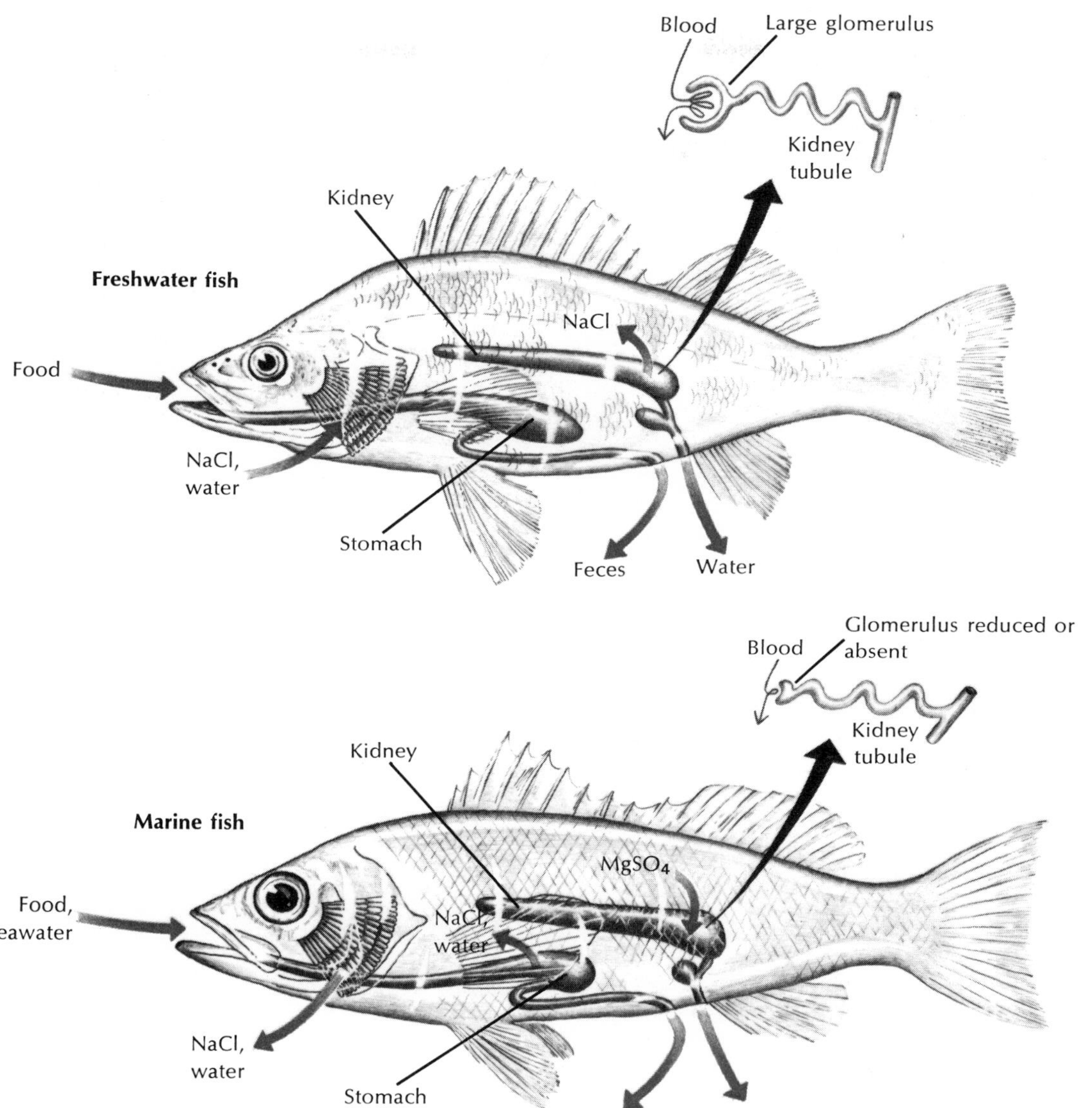

FIG. 23-21 Osmotic regulation in freshwater and marine bony fishes. Freshwater fish maintains osmotic and ionic balance in its dilute environment by actively absorbing sodium chloride across gills (some salt enters with food). To flush out excess water that constantly enters the body, the glomerular kidney produces a dilute urine by reabsorbing sodium chloride. Marine fishes must drink seawater to replace water lost osmotically to its salty environment. Sodium chloride and water are absorbed from the stomach. Excess sodium chloride is secreted outward by the gills. Divalent sea salts, mostly magnesium sulfate, are eliminated with the feces and secreted by the tubular kidney. (Modified from Webster, D., and M. Webster, 1974. Comparative vertebrate morphology. New York, Academic Press, Inc.)

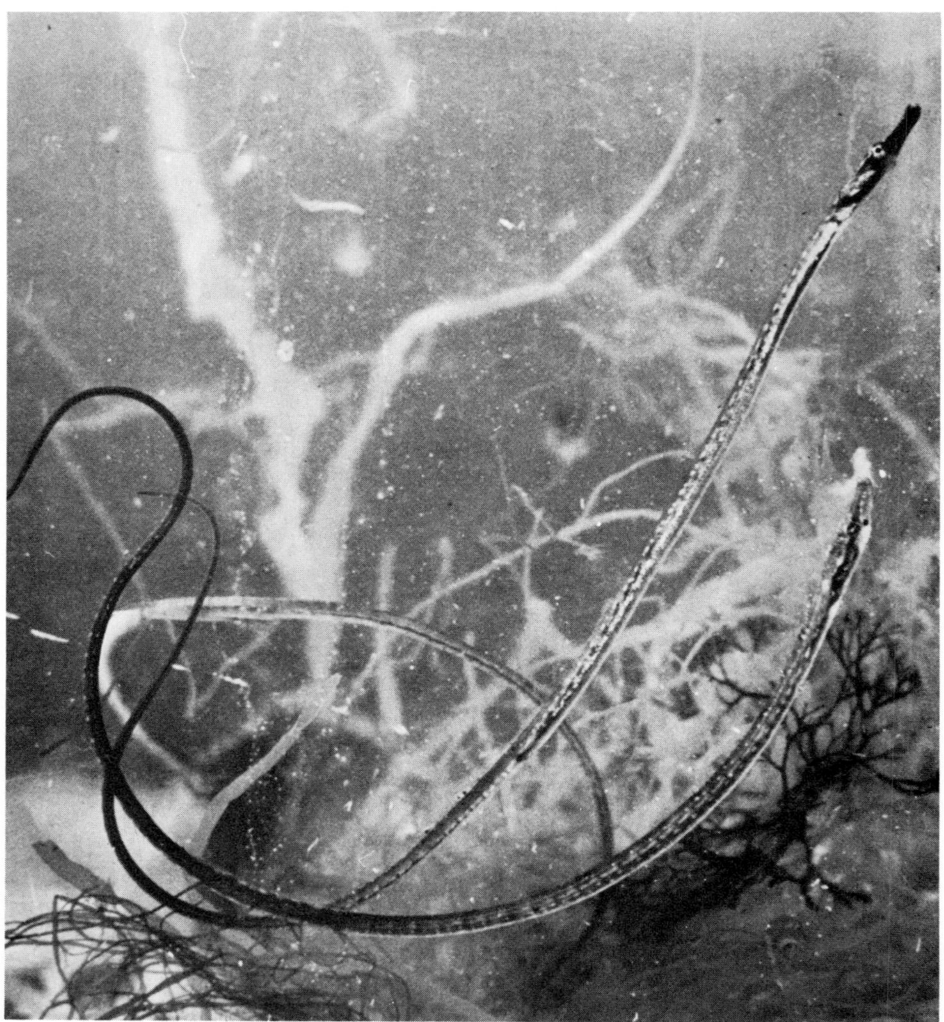

FIG. 23-22 Mimicry in pipefishes *Entelurus aequoreus*. Their shape, coloration (which they change to match the background), and gentle swaying movements combine to make them almost impossible to see in their natural habitat of seaweed, eelgrass and hydroids. Most pipefishes are only a few centimeters in length, although one species of the North American west coast may reach 45 cm. Pipefishes are also interesting because the male incubates the eggs in a special abdominal brood pouch; after mating and fertilization, the eggs are transferred there by the female. (Courtesy B. Tallmark.)

adaptable osmoregulatory mechanisms, are salmon, steelhead trout, many flounders and sculpins, killifish, sticklebacks, and eels. Fishes that can tolerate only very narrow ranges of salt concentration—most freshwater and marine fishes—are said to be **stenohaline** (Gr., *stenos,* narrow, + *hals,* salt). Those fishes that migrate from the sea to spawn in freshwater are **anadromous** (Gr., *ana,* up, + *dromos,* a running), such as salmon, shad, and marine lampreys. Freshwater forms that swim to the sea to spawn are **catadromous** (Gr., *kata,* down, + *dromos,* a running), such as the freshwater eel *Anguilla*.

Feeding behavior

For any fish, feeding is one of the main concerns of day-to-day living. Although many a luckless angler

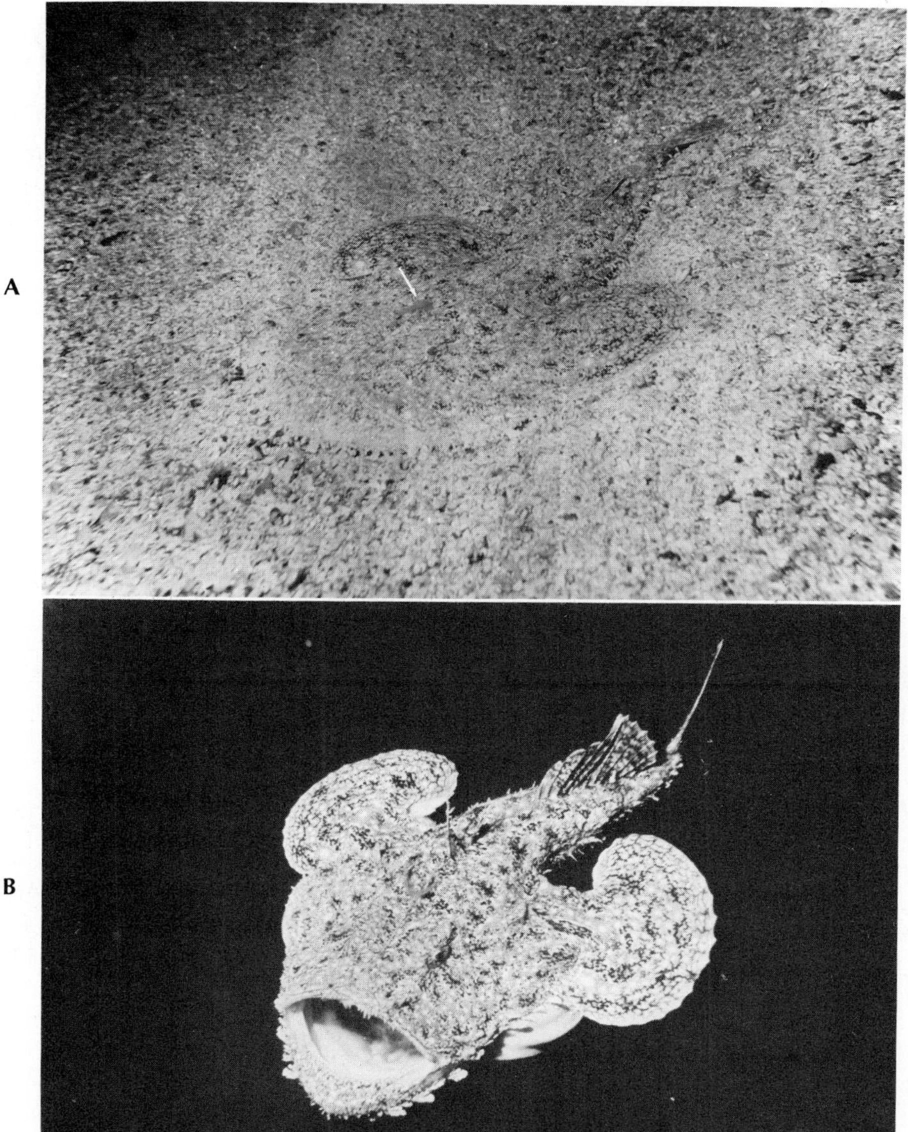

FIG. 23-23 A fish that fishes. **A,** Bedded down in the ocean bottom, beautifully concealed by protective coloration and fringed skin that breaks up the body outline, a goosefish, or angler, *Lophius piscatorius,* awaits its meal. Above its head swings a modified dorsal fin spine on the end of which is a fleshy tentacle *(arrow)* that contracts and expands in a convincing wormlike manner. When a fish approaches the alluring bait, the huge mouth opens suddenly, creating a strong current that sweeps the prey inside: in a split second all is over. Prodded out of its resting place by a skin diver, the goosefish, **B,** reveals its true appearance. Photographed off the Norwegian coast at 25 m depth. (Courtesy T. Lundälv.)

would swear otherwise, the fact is that a fish devotes more time and energy to eating, or searching for food to eat, than to anything else. Throughout the long evolution of fishes, there has been unrelenting selective pressure for those adaptations that enable a fish to come out on the better end of the eat-or-be-eaten contest. Certainly the most far-reaching single event was the evolution of jaws. Their possessors were freed from a mud-grubbing or parasitic existence and could adopt a predatory mode of life. Improved means of capturing larger prey demanded stronger muscles, more agile movement, better balance, and improved special senses. More than any other aspect of its life habit, feeding behavior shapes the fish.

Most fishes are **carnivores** that prey on a myriad of animal foods from zooplankton and insect larvae to large vertebrates. Some deep-sea fishes are capable of eating victims nearly twice their own size—an adaptation for life in a world where meals are necessarily infrequent. Fishes cannot masticate their food as we can because doing so would block the current of water across the gills. Those that do grind their food use powerful pharyngeal teeth in the throat. Most carnivores, however, almost invariably swallow their prey whole, using sharp-pointed teeth in the jaws and on the roof of the mouth to seize their prey. The incompressibility of water makes the task even easier for many large-mouthed predators. When the mouth is opened, a negative pressure is created that sweeps the victim inside.

A second group of fishes are **herbivores** that eat flowering plants, algae, and grasses. Although the plant eaters are relatively few in number, they are crucial intermediates in the food chain, especially in freshwater rivers, lakes, and ponds that contain very little plankton.

The **plankton feeders** that crop the abundant microorganisms of the sea form a third and diverse group of fishes ranging from fish larvae to basking sharks. However, the most characteristic group of plankton feeders are the herringlike fishes (menhaden, herring, anchovies, capelin, pilchards, and others) that are for the most part **pelagic** (that is, open-sea surface dwellers) and travel in large schools. Both phytoplankton and the smaller zooplankton are strained from the water with a sievelike device, the gill rakers (Fig. 23-20). Because plankton feeders are the most abundant of all fishes, they are important food for numerous larger but less abundant carnivores. Certain freshwater fishes, especially members of the large cichlid family, also depend on plankton for food.

A fourth group of fishes are **omnivores** that feed on both plant and animal food. Finally there are the **scavengers** that feed on organic debris (detritus) and the **parasites** that suck the body fluids of other fishes.

Digestion in most fishes follows the vertebrate plan. Apart from several fishes that lack stomachs altogether, the food proceeds from stomach to tubular intestine, which tends to be short in carnivores but may be extremely long and coiled in herbivorous forms. In the carp, for example, the intestine may be nine times the body length, an adaptation for the lengthy digestion required for plant carbohydrates. In carnivores, some protein digestion may be initiated in the acid medium of the stomach but the principal function of the stomach is to store the often large and infrequent meals while awaiting reception by the intestine.

Digestion and absorption proceed simultaneously in the intestine. A curious feature of ray-finned fishes, especially the teleosts, is the presence of numerous **pyloric ceca** (Fig. 23-17) found in no other vertebrate group. Their primary function appears to be fat absorption, although all classes of enzymes (protein-, carbohydrate-, and fat-splitting) are secreted there. They number from two or three to several hundred in some advanced teleost species.

Coloration and concealment

Tropical coral reef fishes bear some of the most resplendent hues and strikingly brilliant color patterns in the animal kingdom. Viewed against the coral reef background of invertebrate and plant life that create a riot of color, the vivid markings and coloration of reef fishes attract relatively little attention. Their coloration is not always for concealment, however, since tropical bony fishes tend toward vivid coloration even in areas of dull and somber backgrounds. Although conspicuous in these circumstances, they are protected by alertness and agility or by their poisonous flesh. In this instance the coloration is an advertisement, warning would-be predators that they should seek their meal elsewhere.

Outside of coral reefs and other littoral habitats of the tropical seas, fishes, like most other animals, characteristically bear colors and patterns that serve to conceal them from enemies. Freshwater fishes wear subdued shades of green, brown, or blue above, grading to silver and yellow-white below. This is **obliterative**

shading. Seen from above against its normal background of water and stream bottom, the fish becomes almost invisible. Seen from beneath as it might be viewed by an aquatic predator, the pale belly of the fish blends with the water surface and sky above. The obliterative coloration is frequently enhanced by blotches, spots, and bars—conflicting patterns that, like the camouflaged ships of the Second World War, tend to break up the outline of the body. Some fishes, such as many flatfish species, can change their color to harmonize with the patterns of their background.

Fish colors are chiefly the result of pigment within **chromatophores** in the dermal layer of the skin. The pigments are red, orange, yellow, and black and can be blended to produce other shades. Pigment dispersion within the branched chromatophores is controlled by the autonomic nervous system. Fish also bear **guanine** in their skin, a purine compound that gives many fishes their silvery appearance.

Another form of concealment is protective resemblance, or **mimicry.** Pipefishes, for example, are of the shape and color of the seaweeds among which they live (Fig. 23-22). Pipefishes even sway slowly like seaweeds in a gentle current. There are many other examples of protective body form that appear to turn their owners into fragmented seaweed, or a leaf, or floating debris, or a weed-covered rock on the bottom. The goosefish, or angler, with its obliterative coloration and numerous fringes and branched appendages of skin, not only becomes nearly invisible when bedded down on the ocean floor but sways a tempting bait above its huge mouth to attract prey (Fig. 23-23).

Electric fishes

The ability to produce strong electric shocks is confined to two groups of vertebrates: elasmobranchs (for example, electric ray) and teleosts (for example, electric eel, electric catfish). Best known and studied is the famous electric eel, *Electrophorus,* of the Amazon and Orinoco river systems of South America. This large, sluggish creature contains powerful electric organs that in 1- to 2.5-m adults can produce paralyzing discharges exceeding 600 volts, quite ample to stun or kill its prey or to discourage potential enemies including large mammals. People who have accidentally met the electric eel under the latter's own terms report it is about as pleasant as contacting an uninsulated high-voltage wire. It is not necessary for the eel to actually touch its victims since the shocks stretch out in an electric field for several meters around the fish. Even small eels less than one-third meter long can discharge pulses exceeding 200 volts.

The source of this impressive performance is the pair of electric organs lying on either side of the vertebral column and extending almost the entire length of the fish. Together they make up about 40% of the body weight. Each organ is composed of longitudinal columns of 6,000 to 10,000 thin, waferlike plates (electroplaxes) stacked one on another like a long cylinder of coins. There are about 60 such columns in each organ. Each plate, or electroplax, is a modified muscle cell innervated by a nerve fiber and has very special electric properties. One side of the plate is a nervous layer that will depolarize when stimulated; the opposite side is a nutritive layer that remains inactive. All the nervous sides of the plates face the tail of the animal. When the eel chooses to discharge its organ, motor impulses are sent out to the electric organs from a special neural center in the brain. These impulses travel to the numerous plates in a highly synchronized way so that all of the thousands of plates discharge simultaneously. Each plate develops a potential charge of about 150 millivolts and because the plates are arranged in series, the individual voltages summate like batteries arranged in series. Thus a high-voltage current flows from the tail to the head of the eel and completes the circuit in the water surrounding the eel. Electric eels must develop high voltages to overcome the high resistance of the fresh water in which they live.

Electric fishes living in the sea, such as the electric ray *Torpedo* (Fig. 23-24), live in a low-resistance medium where high voltages are not required. They generate high amperages instead. In *Torpedo* each electric organ contains some 2,000 columns made up of about 1,000 electroplaxes arranged in vertical stacks. With fewer plates per column, the voltage produced is relatively low (about 50 volts), but the power output may be as much as 6 kilowatts! It is easy to understand why, when seasoned commercial fishermen happen to pick up one of these animals in their trawl, they leave it strictly alone until it is dead and safe to handle.

Migrations

Eel. For centuries naturalists had been puzzled about the life history of the freshwater eel *Anguilla* (an-gwil′la), a common and commercially important species of coastal streams of the North Atlantic. Each fall,

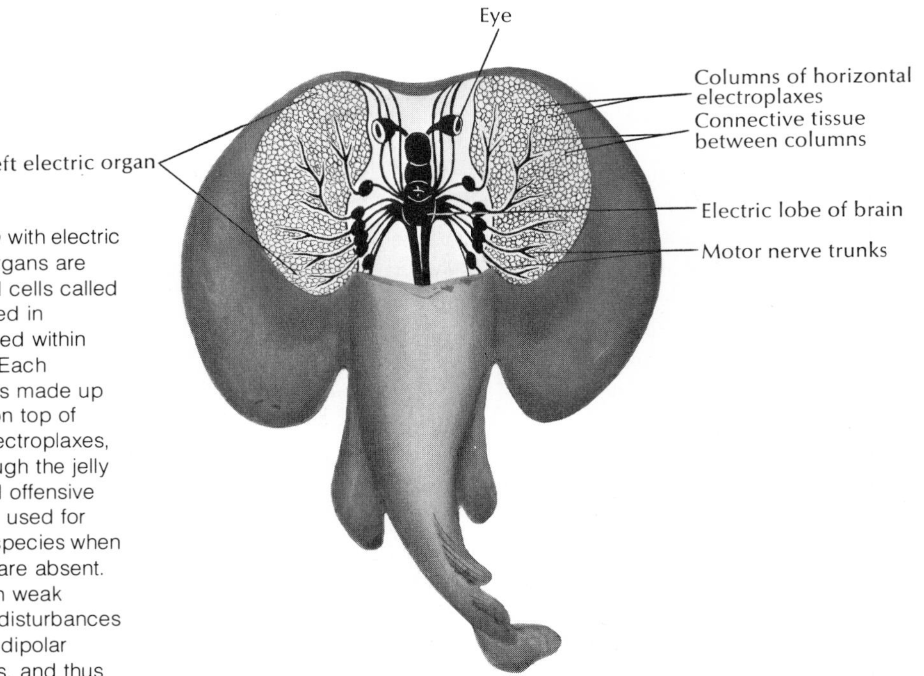

FIG. 23-24 Electric ray *Torpedo* with electric organs uncovered from above. Organs are built up of disclike, multinucleated cells called electroplaxes, which are embedded in jellylike substance and are enclosed within connective-tissue compartments. Each vertical column in an electric ray is made up of a stack of electroplaxes piled on top of each other. Nerve fibers run to electroplaxes, and blood capillaries course through the jelly layer. In addition to defensive and offensive purposes, electric organs may be used for recognition among members of a species when other methods of communication are absent. Some fishes (especially those with weak electric organs) are able to detect disturbances made by an object in the electric dipolar field they form around their bodies, and thus keep informed about their environment.

large numbers of eels were seen swimming down the rivers toward the sea, but no adults ever returned. And each spring countless numbers of young eels, called ''elvers'' (Fig. 23-25), each about the size of a wooden matchstick, appeared in the coastal rivers and began swimming upstream. Beyond the assumption that eels must spawn somewhere at sea, the location of their breeding grounds was totally unknown.

The first clue was provided by two Italian scientists, Grassi and Calandruccio, who in 1896 reported that elvers were not in fact larval eels; rather they were relatively advanced juveniles. The true larval eels, the Italians discovered, were tiny, leaf-shaped, completely transparent creatures that bore absolutely no resemblance to an eel. They had been called **leptocephali** by early naturalists who never suspected their true identity. In 1905 Johann Schmidt, supported by the Danish government, began a systematic study of eel biology that he continued until his death in 1933. With the cooperation of captains of commercial vessels plying the Atlantic, thousands of the leptocephali were caught in different areas of the Atlantic with the plankton nets Schmidt supplied them. By noting where in the ocean larvae in different stages of development were captured, Schmidt and his colleagues eventually reconstructed the spawning migrations.

When the adult eels leave the coastal rivers of Europe and North America, they swim steadily and apparently at great depth for 1 to 2 months until they reach the Sargasso Sea, a vast area of warm oceanic water southeast of Bermuda (Fig. 23-25). Here, at depths of 300 m or more, the eels spawn and die. The minute larvae then begin an incredible journey back to the coastal rivers of Europe. Drifting with the Gulf Stream and preyed on constantly by numerous predators, they reach the middle of the Atlantic after 2 years. By the end of the third year they reach the coastal waters of Europe where the leptocephali metamorphose into elvers, with an unmistakable eel-like body form (Fig. 23-25). Here the males and females part company; the males remain in the brackish waters of coastal rivers and estuaries while the females continue up the rivers, often penetrating hundreds of miles upstream. After 8 to 15 years of growth, the females, now 1 m or more long, return to the sea to join the smaller males; both return to the ancestral breeding grounds thousands of miles away to complete the life cycle.

Schmidt found that the American eel *(Anguilla rostrata)* could be distinguished from the European eel *(A. vulgaris)* because it had fewer vertebrae—an average of 107 in the American eel as compared to an average 114 in the European species. Since the Amer-

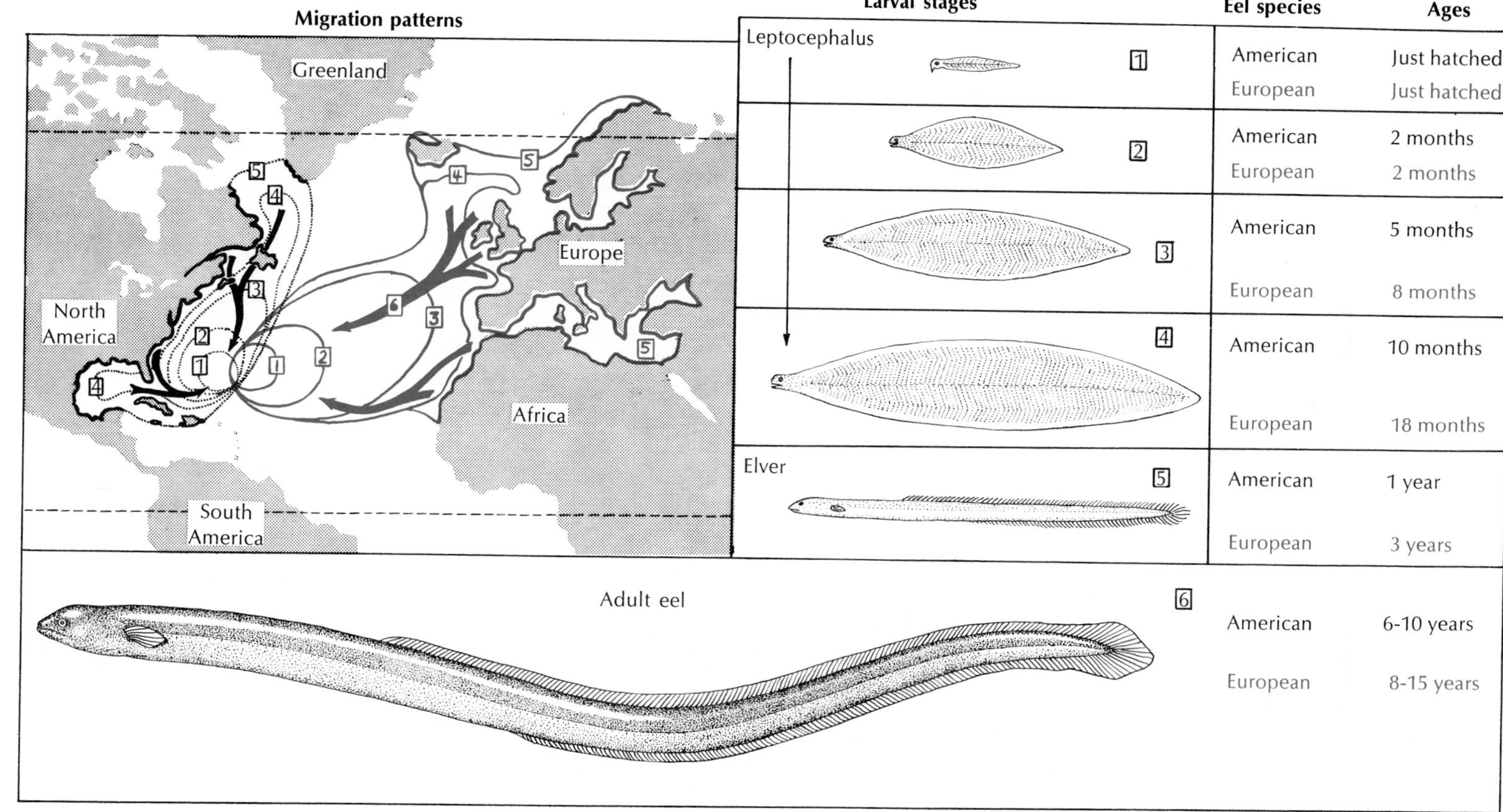

FIG. 23-25 Life histories of the European eel *(Anguilla anguilla)* and the American eel *(Anguilla rostrata)*. *Red,* Migration patterns of European species. *Black,* Migration patterns of American species. Boxed numbers refer to stages of development. Note that the American eel completes its larval metamorphosis and sea journey in 1 year. It requires nearly 3 years for the European eel to complete its much longer journey.

ican eel is much closer to the North American coast line, it requires only about 8 months to make the journey.

Homing salmon. The life history of salmon is nearly as remarkable as that of the eel and certainly has received far more popular attention. Salmon are **anadromous;** that is, they spend their adult lives at sea but return to freshwater to spawn. The Atlantic salmon *(Salmo salar)* and the Pacific salmon (six species of the genus *Oncorhynchus* [on-ko-rink'us]) have this practice, but there are important differences among the seven species. The Atlantic salmon (as well as the closely related steelhead trout) make upstream spawning runs year after year. The six Pacific salmon species (king, sockeye, silver, humpback, chum, and Japanese masu) each make a single spawning run, after which they die.

The virtually infallible homing instinct of the Pacific species is legend: after migrating downstream as a smolt, a sockeye salmon ranges many hundreds of miles over the Pacific for nearly 4 years, grows to 2 to 5 kg in weight, and then returns unerringly to spawn in the headwaters of its parent stream.

Many years ago Canadian biologists marked and released nearly a half million young sockeyes born in a tributary of British Columbia's Fraser River. Eleven thousand of these were recovered 4 years later in the same parent stream; not one was discovered to have strayed into any of the dozens of other Fraser River tributaries.

Experiments have shown that homing salmon are guided upstream by the characteristic odor of their parent stream. The salmon are apparently imprinted with the stream's odor while they are still unhatched embryos, since if the eggs are flown from the parent stream to another stream miles away, the adults still return to the parent stream after their residence at sea. The odor compound is a volatile organic substance, but its exact chemical nature remains unidentified; the embryonic salmon are probably conditioned to a mosaic of compounds released by the characteristic vegetation and soil in the watershed of the parent stream.

How do salmon find their way to the mouth of the river from the trackless miles of the open ocean? We know that they are not capable of distinguishing the parent stream odor while still hundreds of miles at sea. Recent experiments suggest that some migrating fish, like birds, are guided by celestial cues (stars or azimuth position of the sun). To do this the salmon would require time-keeping abilities, certainly not an unreasonable possibility in view of the widespread presence of "biologic clocks" in many organisms from the simplest to the most advanced.

Reproduction and growth

In a group as diverse as the fishes, it is no surprise to find extraordinary variations on the basic theme of sexual reproduction. Fortunately, most fishes favor a simple theme, and it is with these that we will begin our discussion.

The vast majority of fishes are **dioecious,** with **external fertilization and external development** of the eggs and embryos **(oviparous).** The **testes** of the male are elongate, whitish organs divided into lobules that contain cysts of maturing sperm cells. Each cyst contains germ cells in the same stage of development, in contrast to the familiar pattern of higher vertebrates in which the seminiferous tubules at any point contain germ cells in all stages of development. The lobules open into a spermatic duct that runs into a urogenital sinus. The **ovaries** may run the length of the abdominal cavity and are made up of many ovarian follicles supported by connective tissue. The eggs develop within individual follicles. When the eggs are mature they are shed, either directly into the body cavity or into oviducts. In some teleosts, particularly salmon and trout, the oviducts are reduced or absent.

Most fishes spawn at certain times or seasons within a restricted range of temperature. Temperature is critical both for successful spawning and for the survival of the sensitive eggs and young. Many fish, perhaps most, spawn in the spring or early summer when the water temperature is rising. Others, such as cod, many flatfishes, salmon, and certain trout species, spawn in fall or winter, often when water temperatures are at their lowest. In polar water, fishes may live out their existence at temperatures between 0° and −2° C.

Many marine fishes have what appear to be extraordinarily wasteful modes of reproduction. Males and females come together in great schools and, without mating, release vast numbers of germ cells into the water to drift with the current. Female cod release 4 to 6 million eggs at a single spawning. Less than one in a million will survive the numerous perils of the ocean to reach reproductive maturity.

On the whole, freshwater fishes dispose of their eggs

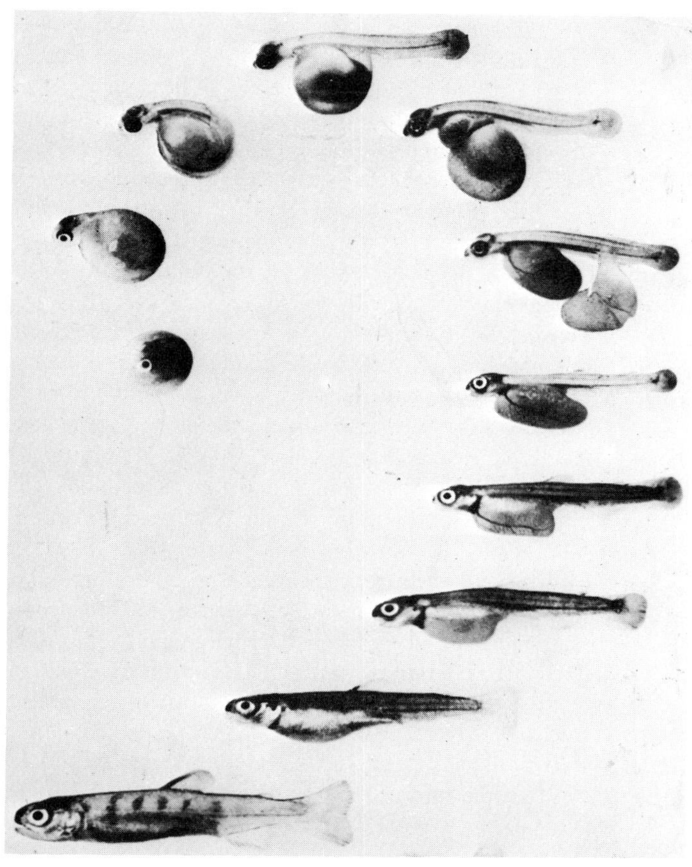

FIG. 23-26 Development of salmon from egg to fingerling. (Courtesy H. Kelly, U.S. Fish and Wildlife Service.)

in a more conservative manner. Some bury their eggs, some attach them to vegetation, and some deposit them in nests. Unlike the tiny, buoyant, transparent eggs of pelagic marine teleosts, those of freshwater species are typically yolky and nonbuoyant. Furthermore, rather elaborate preliminaries to mating are the rule rather then the exception for freshwater fishes. One sex or the other selects a site, prepares a nest, guards it, engages in elaborate courtship ritual to attract a mate, mates, and covers the eggs. Some species guard the nest until the larvae hatch. Hundreds rather than millions of eggs are laid; a trout spawns about a thousand eggs per pound of body weight. It may be taken as axiomatic that fecundity is indirectly related to survival of the young. Whether spawning is profligate or thrifty, natural selection has provided that enough young will reach maturity to perpetuate the species.

There are two other principal modes of reproduction in fishes: (1) **internal fertilization with external development** and (2) **internal fertilization with internal development.** To the second of these two modes belong the **live-bearers.** Several quite unrelated groups have live-bearing members; for example, the ever-popular guppies and mollies, the economically valuable redfish, the mosquito fish *Gambusia,* and all members of the surfperch family Embiotocidae. In one surfperch, *Cymatogaster aggregatus,* male fishes may be retained in the ovary until they are sexually mature! The guppy, however, is a more typical example. The eggs are fertilized within the ovarian follicles where they develop for a period. The follicles then rupture, releasing the embryos into the cavity of the ovary, where they complete their gestation, nourished by their own yolk and possibly by secretions from the follicles.

Live-bearers actually comprise two groups: **ovoviviparous** forms (all of the examples given already), in which the eggs are retained and nourished by egg yolk within the mother, and **viviparous** forms, in

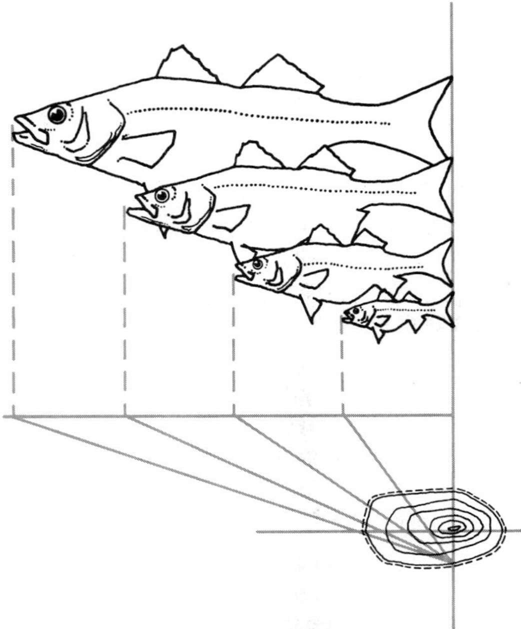

FIG. 23-27 Scale growth. Fish scales disclose seasonal changes in growth rate. Growth is interrupted during winter, producing year marks (annuli). Each year's increment in scale growth is a ratio to the annual increase in body length. Otoliths (ear stones) and certain bones can also be used in some species to determine age and growth rate.

which the young develop some type of placental attachment with the follicle, ovarian wall, or specialized uterus. True viviparity is known only in certain sharks, such as hammerhead and blue sharks, and a few teleost species, such as the dwarf top-minnow *Heterandria* and the four-eyed fishes of the genus *Anableps*. Many authorities consider the distinction between ovoviviparity and viviparity to be a rather artificial one because one can find examples among fishes in which the degree of placental development ranges from almost zero to complete. Furthermore, in some live-bearing sharks the placenta develops only after several months of an ovoviviparous existence.

Soon after the eggs of oviparous species are laid, they take up water and harden. Cleavage follows, and the blastoderm is formed, sitting astride a relatively enormous yolk mass. Soon the yolk mass is enclosed by the developing blastoderm, which now begins to assume a fishlike shape. The fish hatches as a larva that may be very different in appearance from the adult. The eyes and segmented muscles (myotomes) are well-formed, but most conspicuous is the semitransparent, globular mass of yolk, so large that larval movement is nearly impossible (Fig. 23-26). Not until the yolk is totally absorbed and the mouth and digestive tract formed does the larva begin searching for its own food (Fig. 36-1, p. 818). After a period of growth, the larva undergoes a **metamorphosis,** especially dramatic in many marine species, for example, the freshwater eel described earlier (Fig. 23-25). Body shape is refashioned, fin and color patterns change, and the animal becomes a juvenile bearing the unmistakable definitive body form of its species.

Growth is temperature-dependent. Consequently, fish living in temperate regions grow rapidly in summer when temperatures are high and food is abundant but nearly stop growing in winter. Seasonal growth is reflected as annual rings in the scales (Fig. 23-27), a distinctive record of convenience to fishery biologists who wish to determine a fish's age. Unlike birds and mammals, which reach a definitive adult size, most fishes after attaining reproductive maturity continue to grow for as long as they live. This is probably a selective advantage for the species, since the larger the fish the more germ cells it produces, and the greater its contribution to future generations.

Annotated references

Selected general readings

Curtis, B. 1949. The life story of the fish. New York, Harcourt, Brace & Co., Inc. (Dover Publications edition, 1961). *A nontechnical account.*

Harden Jones, F. R. 1968. Fish migration. London, Edward Arnold Ltd. *Scholarly review of fish migration and orientation.*

Hardisty, M. W., and I. C. Potter. 1971-1972. The biology of lampreys, vols. 1 and 2. New York, Academic Press, Inc. *Thorough presentation of systematics, life histories, ecology, behavior, physiology, and economic impact of this small but biologically interesting group.*

Hoar, W. S., and D. J. Randall (eds.). 1969-1972. Fish physiology, vols. 1 to 6. New York, Academic Press, Inc. *This series represents the most complete and authoritative treatise on the functional biology of fishes published. Technical and detailed.*

Lanham, U. 1962. The fishes. New York, Columbia University Press. *A concise account of the evolution, structure, and function of fishes.*

Lineaweaver, T. H., III, and R. H. Backus. 1970. The natural history of sharks. Philadelphia, J. B. Lippincott Co. *One of the best of many books dealing with these intriguing animals.*

Marshall, N. B. 1970. The life of fishes. New York, Universe Books. *Probably the best general biology of fishes.*

Marshall, N. B. 1971. Explorations in the life of fishes. Cambridge, Harvard University Press. *Excellent general biology of fishes. Fairly technical.*

Nelson, J. S. 1976. Fishes of the world. New York, John Wiley & Sons, Inc. *A modern classification of all major groups of fishes.*

Netboy, A. 1974. The salmon: their fight for survival. Boston, Houghton Mifflin Co. *Thorough and readable treatise of the biology of the magnificent but embattled salmon.*

Norman, J. R. 1963. A history of fishes, ed. 2 (revised by P. H. Greenwood). New York, Hill & Wang, Inc. *A revision of a famous treatise that covers nearly all aspects of fish study.*

Rounsefell, G. A. 1975. Ecology, utilization, and management of marine fishes. St. Louis, The C. V. Mosby Co. *Comprehensive textbook of marine fishery science.*

Schultz, L. P., and E. M. Stern. 1948. The ways of fishes. New York, D. Van Nostrand Co., Inc. *A somewhat popularized discussion of the behavior of fish. Especially good for the beginning student in zoology.*

Thomson, K. S. 1969. The biology of lobe-finned fishes. Biol. Rev. **44:**91-154.

Young, J. Z. 1963. The life of vertebrates, ed. 2. New York, Oxford University Press. *Many chapters in this excellent treatise are devoted to the structure, evolution, and adaptive radiation of fish.*

Selected *Scientific American* articles

Applegate, V. C., and J. W. Moffett. 1955. The sea lamprey. **192:**36-41 (Apr.). *Life history and early control measures are described.*

Brett, J. R. 1965. The swimming energetics of salmon. **213:** 80-85 (Aug.). *Describes studies on the remarkable efficiency of swimming by fish.*

Carey, F. G. 1973. Fishes with warm bodies. **228:**36-44 (Feb.). *Some tuna and mackerel shark species employ a circulatory heat exchanger to conserve body heat and increase swimming power.*

Gilbert, P. W. 1962. The behavior of sharks. **207:**60-68 (July).

Gray, J. 1957. How fishes swim. **197:**48-54 (Aug.). *This British biologist's calculations of fish swimming efficiency have been referred to as "Gray's paradox."*

Grundfest, H. 1960. Electric fishes. **203:**115-124 (Oct.).

Hasler, A. D., and J. A. Larsen. 1955. The homing salmon. **193:**72-76 (Aug.). *Describe experiments that indicate salmon locate parent stream by using sense of smell.*

Isaacs, J. D., and R. A. Schwartzlose. 1975. Active animals of the deep-sea floor. **233:**84-91 (Oct.). *Large fishes and other scavengers live on dead animals that fall to the ocean floor.*

Jensen, D. 1966. The hagfish. **214:**82-90 (Feb.).

Johansen, K. 1968. Air-breathing fishes. **219:**102-111 (Oct.). *Not all fishes are water breathers. Many remarkable alternatives for air breathing evolved in fishes; several are described in this article.*

Leggett, W. C. 1973. The migrations of the shad. **228:**92-98 (Mar.).

Lissmann, H. W. 1963. Electric location by fishes. **208:**50-59 (Mar.). *Certain fishes generate weak electric fields to sense their environment and locate prey.*

Lühling, K. H. 1963. The archer fish. **209:**100-108 (July.). *This small fish of southeast Asia spouts a stream of water to down insects it sights on plants above the water.*

McCosker, J. E. 1977. Flashlight fishes. **236:**106-114 (Mar.). *Fishes of the family Anomalopidae use their living light to communicate, attract prey, avoid predators, and see.*

Millot, J. 1955. The coelacanth. **193:**34-39 (Dec.). *Its biology and evolutionary relationships are discussed.*

Ruud, J. T. 1965. The ice fish. **213:**108-114 (Nov.). *A family of transparent Antarctic fishes possess neither red blood cells nor hemoglobin. The physiologic consequences are described.*

Scholander, P. F. 1957. The wonderful net. **196:**96-107 (Apr.). *Describes the arrangement of blood vessels in fish and other vertebrates to form countercurrent exchange systems.*

Shaw, E. 1962. The schooling of fishes. **206:**128-138 (June).

Todd, J. H. 1971. The chemical languages of fishes. **224:** 98-108 (May). *Many fishes emit and sense chemical signals that guide much of their behavior. Experiments with catfishes are described.*

CHAPTER 24

THE AMPHIBIANS

Phylum Chordata
Class Amphibia

A male green treefrog *Hyla cinerea* announces his location in a South Carolina marsh to any attentive females. The ringing *queenk-queenk-queenk* of his call has earned him the local name of "bellfrog."

Photograph by C. P. Hickman, Jr.

The chorus of frogs beside a pond on a spring evening heralds one of nature's dramatic events. Masses of frog eggs soon hatch into limbless, gill-breathing, fish-like tadpole larvae. Warmed by the late spring sun, they feed and grow. Then, almost imperceptibly, a remarkable transformation unfolds. Hind legs appear and gradually lengthen. The tail shortens. Larval teeth are lost, and the gills are replaced by lungs. The eyes develop lids. Forelegs emerge. In a matter of weeks the aquatic tadpole has completed its metamorphosis to an adult frog.

The early members of the class Amphibia (am-fib′e-a) (Gr., *amphi,* both or double, + *bios,* life), of which our chorusing frogs are among the more vociferous modern descendants, originated not in weeks but over millions of years by a lengthy series of almost imperceptible alterations that gradually fitted the vertebrate body plan for life on land. The origin of land vertebrates is no less a remarkable feat for this fact—a feat that incidentally would have a poor chance of succeeding today because well-established competitors make it impossible for a poorly adapted transitional form to gain a foothold.

Even now after some 350 million years of evolution, the amphibians are not completely land adapted; they are quasiterrestrial, hovering between aquatic and land environments. This double life is expressed in their name. Structurally they are between fishes and reptiles. Although adapted for a terrestrial existence, few can stray far from moist conditions. Many, however, have developed devices for keeping their eggs out of open water where the larvae would be exposed to enemies.

Approximately 2,500 species of amphibians are grouped into three living orders: the newts and salamanders (order Urodela), least specialized and most aquatic of all amphibians; the frogs and toads (order Salientia), largest and most successful group of amphibians and closest to the stock from which the higher tetrapods (animals with four legs) descended; and the highly specialized, secretive, earthworm-like tropical caecilians (order Gymnophiona).

THE MOVEMENT ONTO LAND

The movement from water to land is perhaps the most dramatic event in animal evolution, since it involves the invasion of a habitat that in many respects is less suitable for life. The origin of life was conceived in water, animals are mostly water in composition, and all cellular activities proceed in water. Nevertheless, animals eventually moved onto dry land, carrying their watery composition with them. To survive and maintain this fluid matrix, various structural, functional, and behavioral changes had to evolve. Considering that almost every system in the body required some modification, it is remarkable that all vertebrates are basically alike in fundamental structural and functional pattern: whether aquatic or terrestrial, vertebrates are obviously descendants of the same evolutionary limb.

Amphibians were not the first to move onto land. Insects made the transition earlier and plants much earlier still. The pulmonate snails were experimenting with land as a suitable place to live about the same time the early amphibians were. Yet of all these, the amphibian story is of particular interest because their descendants became the most successful and advanced animals on earth.

Physical contrast between aquatic and land habitats

Beyond the obvious difference in water content of aquatic and terrestrial habitats, there are several sharp differences between the two environments of significance to animals attempting to move from water to land.

Greater oxygen content of air. Air contains at least 20 times more oxygen than water. Air has about 210 ml of oxygen per liter; water contains 3 to 9 ml per liter, depending on temperature, presence of other solutes, and degree of saturation. Furthermore the diffusion rate of oxygen is low in water. Consequently aquatic animals must expend far more effort extracting oxygen from water than land animals expend removing oxygen from air.

Greater density of water. Water is about 1,000 times denser than air and about 100 times more viscous. Although water is a much more resistant medium to move through, its high density, only a little less than that of animal protoplasm, buoys up the body. One of the major problems encountered by land animals was the need to develop strong limbs and remodel the skeleton to support their bodies in air.

Constancy of temperature in water. Natural bodies of water, containing a medium with tremendous thermal capacity, experience little fluctuation in temperature. The temperature of the oceans remains almost constant day after day. In contrast, both the range and the fluctuation in temperature are acute on land. Its harsh cycles of freezing, thawing, drying, and flooding, often in unpredictable sequence, present severe thermal problems to terrestrial animals.

Variety of land habitats. The variety of cover and shelter on land were great inducements for its colonization. The rich offerings of terrestrial habitats include coniferous and temperate forests, tropical forests, grasslands, deserts, mountains, oceanic islands, and polar regions. Even so, earth's hydrosphere (oceans, seas, lakes, rivers, and ice sheets), though offering a less diverse range of habitats, contains the greatest number and variety of living things on earth.

Opportunities for breeding on land. The provision of safe shelter for the protection of vulnerable eggs and young is much more readily acccomplished on land than in water habitats.

EVOLUTION OF AMPHIBIANS

The amphibians are descended from the crossopterygians, the lobe-finned fishes. Unlike a fish, which is buoyed and wetted by its medium and supplied with dissolved oxygen, a terrestrial animal must support its own weight, resist drying and rapid temperature change, and extract oxygen from air. Both structural and functional modifications of the fish body plan were required to accomplish these ends.

Appearance of lungs. The lobe-finned fishes flourished in freshwater lakes and streams of the Devonian period, a time of alternating droughts and floods. During dry periods, pools and streams began to dry up, water became foul, and the dissolved oxygen disappeared. The only fishes that could survive such conditions were those that were able to utilize the abundance of oxygen in the air above them. Gills were unsuitable because in air the filaments collapse together into clumps that soon dry out. Virtually all the survivors of this period had a kind of lung that developed as an outgrowth of the pharynx. It was a relatively simple matter to enhance the efficiency of this air-filled cavity by improving its vascularity with a rich capillary network and by supplying it with arterial blood from the last (sixth) pair of aortic arches. Oxygenated blood was returned directly to the heart by a pulmonary vein to form a complete pulmonary circuit. This was the origin of the **double circulation** characteristic of all tetrapods: a **systemic** circulation, which serves the body, and a **pulmonary** circulation, which serves the lungs.

Although it was an important evolutionary development, the simple, saccular amphibian lung is not especially efficient. All amphibians supplement pulmonary respiration with gas exchange across their moist skin. Furthermore, the amphibians fill their lungs the same way their lobe-finned ancestors filled theirs, by gulping a large bubble of air into the mouth, then forcing it into the lung by compression of the mouth cavity. Unlike the higher tetrapods, they cannot draw air into their lungs by expanding the chest cavity.

Limbs for travel on land. The evolution of limbs was also a product of difficult times during the Devonian period. When pools dried up altogether, fishes were forced to move to another pool that still contained water. The crossopterygians were equipped for the task, having strong lobed fins, used originally as swimming stabilizers, that could be adapted as paddles to lever their way across land in search of water. The pectoral fins were especially well-developed in these fishes. These had a series of skeletal elements in the fins and pectoral girdle that clearly foreshadowed the pentadactyl limb of tetrapods. We should note that the development of strong fins, and later, limbs, did not happen so that fish could colonize land but *to permit them to find water and continue living like fish*. Land travel was simply and paradoxically a means for survival in water. But the evolution of lungs and limbs were fortunate and essential specializations that preadapted vertebrates for life on land.

Earliest amphibians

All evidence points to the lobe-finned fishes as ancestors of the modern amphibians. The lobe-fins, abundant and successful in the Devonian period, possessed lungs and strong, mobile fins. Their skull and tooth structure was similar to that of the earliest known amphibians, the Labyrinthodontia, a distinct salamander-like group of the late Devonian period.

A representative of this group was a 350 million–year–old fossil called *Ichthyostega* (Fig. 24-1). *Ichthyostega* possessed several new adaptations that equipped it for life on land. It had jointed, pentadactyl limbs for crawling on land, a more advanced ear structure for picking up airborne sounds, a foreshortening of the skull, and a lengthening of the snout that announced improved olfactory powers for detecting dilute airborne odors. Yet *Ichthyostega* was still fishlike in re-

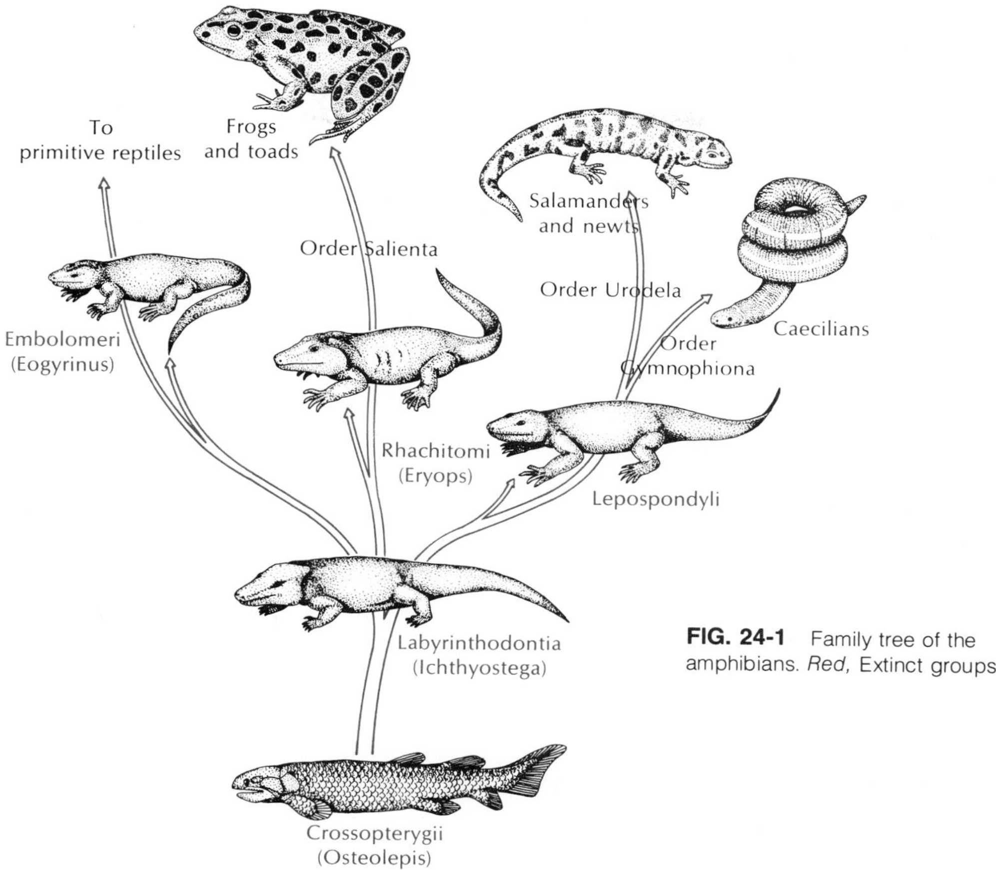

FIG. 24-1 Family tree of the amphibians. *Red,* Extinct groups.

taining a fish tail complete with fin rays and in having opercular (gill) bones.

The capricious Devonian period was followed by the Carboniferous period, characterized by a warm, wet climate during which mosses and large ferns grew in profusion on a swampy landscape. Conditions were ideal for the amphibians. They radiated quickly into a great variety of species, feeding on the abundance of insects, insect larvae, and aquatic invertebrates available: this was the Age of Amphibians.

With water everywhere, however, there was little selective pressure to encourage movement onto land, and many amphibians actually improved their adaptations for living in water. Their bodies became flatter for moving about in shallow water. Many of the urodeles (newts and salamanders), which may have descended from the lepospondyls (Fig. 24-1), developed weak limbs. The tail became better developed as a swimming organ. Even the anurans (frogs and toads), which are the most terrestrial of all amphibians, developed specialized hind limbs with webbed feet better suited for swimming than for movement on land. All groups of amphibians use their porous skin as an accessory breathing organ. This specialization was encouraged by the swampy surroundings of the Carboniferous period but presented serious desiccation problems for life on land.

Amphibians' contribution to vertebrate evolution

Amphibians have met the problems of independent life on land only halfway. To be sure, they made several important contributions to the transition that required the evolution of their own descendants, the reptiles, to complete. Of crucial importance were the change from gill to lung breathing and the development of limbs for locomotion on land. Amphibians also show strengthening changes in the skeleton so that the

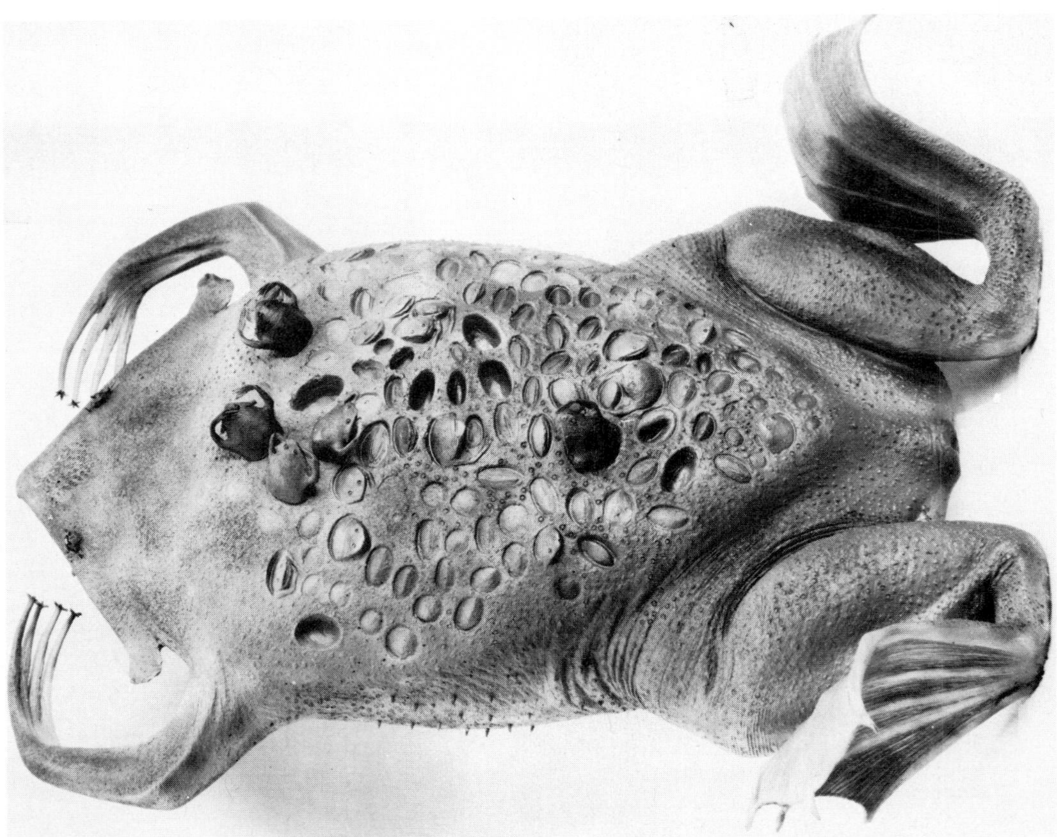

FIG. 24-2 Female Surinam toad, carrying young on her back. As eggs are laid, male assists in positioning them on rough back skin of female. The skin swells, enclosing eggs. The approximately 60 young pass through the tadpole stage beneath the skin and emerge as small frogs. Surinam "toad," actually a frog, is found mainly in Amazon and Orinoco river systems of equatorial South America. (Courtesy American Museum of Natural History.)

body can be supported on land. A start was also made toward shifting special sense priorities from the lateral line system of fish to the senses of smell and hearing. For this, both the olfactory epithelium and the ear required redesigning to improve sensitivities to airborne odors and sounds.

Despite these modifications, the amphibians are basically aquatic animals. They are ectothermic; that is, their body temperature is determined by and varies with the environmental temperature. Their skin is thin, moist, and unprotected from desiccation in air. An intact frog loses water nearly as rapidly as a skinless frog. Most important, the amphibians remain chained to the aquatic environment by their mode of reproduction. Eggs are shed directly into the water or laid in moist surroundings and are externally fertilized (with very few exceptions). The larvae that hatch typically pass through an aquatic tadpole stage.

Many amphibians have developed ingenious devices for laying their eggs elsewhere to give their young protection and a better chance for life. They may lay eggs under logs or rocks, in the moist forest floor, in flooded tree holes, in pockets on the mother's back (Fig. 24-2), or in folds of the body wall. One species of Australian frog even broods its young in its stomach.

However, it remained for the reptiles to complete the conquest of land with the development of a shelled (amniotic) egg, which finally freed the vertebrates from a reproductive attachment to the aquatic environment. With the appearance of reptiles at the end of the Paleozoic period, the halcyon era for the amphibians began to fade. The reptiles captured rule of both

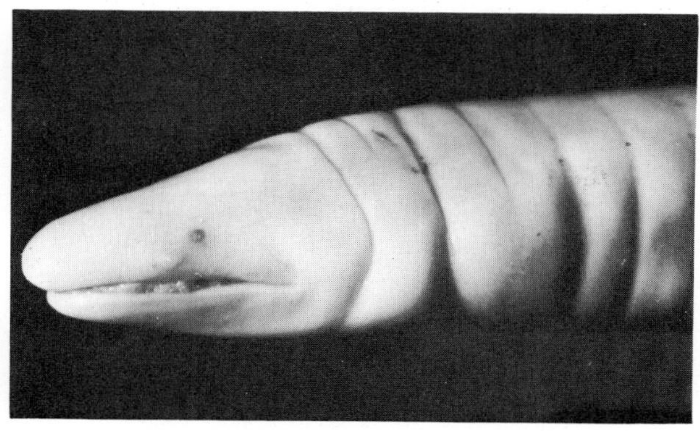

FIG. 24-3 Head and anterior region of caecilian. Order Gymnophiona (Apoda). These legless and wormlike amphibians may reach lengths of 55 cm and diameters of 2 cm. Their body folds give them the appearance of a segmented worm. They have many sharply pointed teeth, a pair of tiny eyes mostly hidden beneath the skin, and a small tentacle between the eye and nostril; some forms have embedded mesodermal scales. (Courtesy General Biological Supply House, Inc., Chicago.)

water and land and removed most amphibians from both environments. From the survivors have descended the three modern orders of amphibians.

Characteristics

1. Skeleton mostly bony, with varying number of vertebrae; ribs present in some, absent in others; notochord does not persist; **exoskeleton absent**
2. Body forms vary greatly from an elongated trunk with distinct head, neck, and tail to a compact, depressed body with fused head and trunk and no intervening neck
3. **Limbs, usually four (tetrapod),** although some are legless; forelimbs of some much smaller than hind limbs, in others all limbs small and inadequate; **webbed feet often present;** no true nails or claws
4. **Skin smooth and moist with many glands,** some of which are poisonous; pigment cells (chromatophores) common, of considerable variety, and in a few capable of undergoing color and pattern changes in accordance with different backgrounds; **no scales,** except concealed dermal ones in some
5. Mouth usually large with small teeth in upper or both jaws; **two nostrils open into anterior part of mouth cavity**
6. Respiration by gills, lungs, skin, and pharyngeal region either separately or in combination; external gills in the larval form and may persist throughout life in some
7. **Circulation with three-chambered heart,** two auricles and one ventricle, and a double circulation through the heart; skin abundantly supplied with blood vessels
8. Ectothermic
9. Excretory system of paired mesonephric kidneys; urea main nitrogenous waste
10. Ten pairs of cranial nerves
11. Separate sexes; fertilization external or internal; predominantly oviparous, rarely ovoviviparous; metamorphosis usually present; **mesolecithal eggs with jelly-like membrane coverings**

Classification of living orders

Order Gymnophiona (jim'no-fy'o-na) (Gr., *gymnos,* naked, + *ophioneos,* of a snake) **(Apoda)—caecilians.** Body wormlike; limbs and limb girdle absent; mesodermal scales may be present in skin; tail short or absent; tropical; one family, 17 genera, approximately 160 species.

Order Urodela (yu'ro-dee'la) (Gr., *oura,* tail, + *delos,* visible) **(Caudata)—salamanders, newts.** Body with head, trunk, and tail; no scales; usually two pairs of equal limbs; eight families, 51 genera, approximately 300 species.

Order Salientia (say'lee-ench'e-a) (L., *saliens,* leaping, + *-ia,* pl. suffix) **(Anura)—frogs, toads.** Head and trunk fused; no tail; no scales; two pairs of limbs; mouth large; lungs; 10 vertebrae, including urostyle; 13 families, 100 genera, approximately 2,000 species.

STRUCTURE AND NATURAL HISTORY

Caecilians—order Gymnophiona (Apoda)

The little-known order Gymnophiona contains some 160 species of burrowing, wormlike creatures commonly called **caecilians** (Fig. 24-1). They are distributed in tropical forests of South America (their principal home), Africa, and southeast Asia. They are characterized by their long, slender body, small scales in the skin, many vertebrae, long ribs, no limbs, and terminal anus. The eyes are small and most species are totally blind as adults (Fig. 24-3). Replacing eyes, which would be useless for a subterranean existence, are special sensory tentacles on the snout. Because they are almost all burrowing forms, they are seldom seen by humans. Their food is mostly worms and small invertebrates, which they find underground. Fertilization is internal, and the male is provided with a protrusible cloaca by which he copulates with the female. The eggs are usually deposited in moist ground near the water; the larvae may be aquatic or the complete lar-

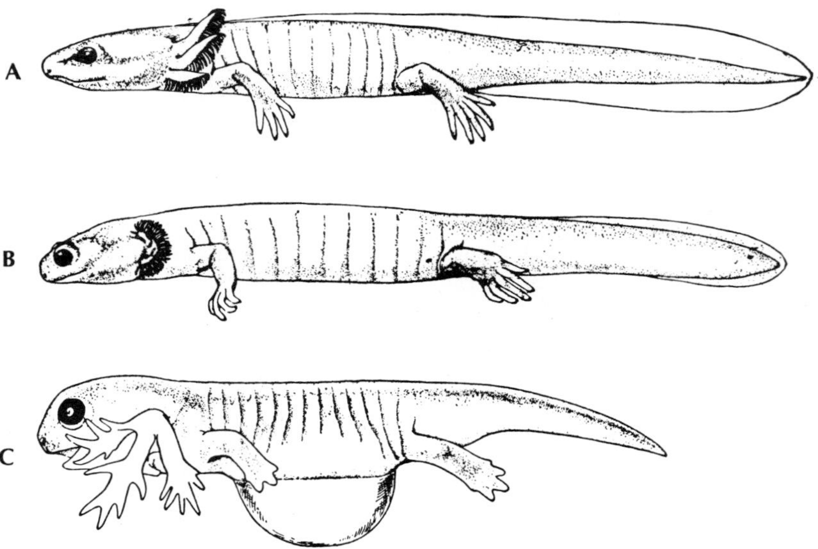

FIG. 24-4 Principal types of salamander larvae. **A,** Pond-type larva *Ambystoma paroticum,* with well-developed gills; **B,** mountain brook–type larva *Dicamptodon ensatus,* with reduced gills; **C,** terrestrial-type larva *Plethodon vandykei,* with large gills but lacking filaments. (From Nobel, G. K. 1931. The biology of the Amphibia. New York, McGraw-Hill Book Co.)

val development may occur in the egg. In some species the eggs are carefully guarded in folds of the body during their development.

Salamanders and newts—order Urodela (Caudata)

As the name of the order suggests, the order Urodela consists of tailed amphibians—the salamanders and newts. This compact, natural group is the least specialized of all the amphibians. Although urodeles are found in almost all temperate and tropical regions of the world, most species occur in North America. Urodeles are typically small; most of the common North American salamanders are less than 15 cm long. Some aquatic forms are considerably longer, and the carnivorous Japanese giant salamander may exceed 1.5 m in length.

Urodeles have primitive limbs set at right angles to the body with forelimbs and hindlimbs of approximately equal size. In some the limbs are rudimentary. One group, the sirens, with minute forelimbs and no hindlimbs at all, is so different from other urodeles that some authorities place them in a completely separate order, Trachystomata.

Salamanders and newts prey on worms, small arthropods, and small molluscs. Most eat only things that are moving. Since their food is rich in proteins, they do not usually store in their bodies great quantities of fat or glycogen. Like all amphibians they are ectotherms and have a low metabolic rate.

Breeding behavior. Some urodeles are wholly aquatic throughout their life cycle, but most are terrestrial, living in moist places under stones and rotten logs, usually not far from water. They do not show as great a diversity of breeding habits as do frogs and toads. The eggs of most salamanders are fertilized internally, usually after the female picks up a packet of sperm **(spermatophore)** that previously has been deposited by the male on a leaf or stick. Aquatic species lay their eggs in clusters or stringy masses in the water. Terrestrial species deposit eggs in small grapelike clusters under logs or excavations in soft earth. Unlike frogs and toads that hatch into fishlike tadpole larvae, the embryos of urodeles hatch from their eggs resembling their parents. The larvae undergo metamorphosis in the course of development, but it is not nearly so revolutionary a change as is the metamorphosis of frog and toad tadpoles to the adult body form.

FIG. 24-5 Common salamanders of the family Plethodontidae. **A,** Red-backed salamander *Plethodon cinereus* has dorsal reddish stripe with gray to black sides. **B,** Two-lined salamander *Eurycea bislineata* is yellow to brown with two dorsolateral black stripes. **C,** Long-tailed salamander *Eurycea longicauda* is yellow to orange with black spots that form vertical stripes on sides of tail. **D,** Slimy salamander *Plethodon glutinosus* is black with white spots. (Photographs by F. M. Hickman.)

Respiration. All salamanders hatch with gills that are morphologically specialized for the life habit of the developing larvae (Fig. 24-4). Larvae that develop in quiet, warm ponds that are often low in dissolved oxygen have well-developed filamentous gills with a large surface area for gas exchange. Salamander larvae from cool, rapid mountain brooks where oxygen is plentiful have reduced gills. Finally, the larvae of terrestrial salamanders have large gills lacking filaments that would collapse and dry out in air. Larval gills are lost in all salamanders except the aquatic forms or those that fail to undergo a complete metamorphosis. Lungs, the characteristic respiratory organ of terrestrial vertebrates, replace the larval gills in most adult amphibians.

Yet some salamanders have dispensed with lungs altogether and thus bear the distinction of being the only vertebrates to have neither lungs nor gills. Members of the large family Plethodontidae, a group containing most of the familiar North American salamanders (Fig. 24-5), are completely lungless, and some members of other urodele families exhibit reductions in lung development. In all amphibians the skin contains extensive vascular nets that serve in varying degrees for the respiratory exchange of oxygen and carbon dioxide. In lungless salamanders the efficiency of cutaneous respiration is increased by the penetration of a capillary network into the epidermis or by the thinning of the epidermis over superficial dermal capillaries. Cutaneous respiration is supplemented by the pumping of air in and out of the mouth where the respiratory gases exchange across the vascularized

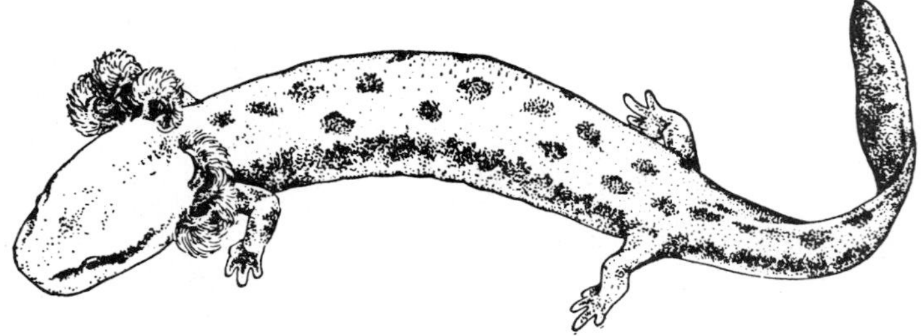

FIG. 24-6 The mudpuppy *Necturus maculosus*—an example of a paedomorphic species. Gills are retained in the breeding adult. (From Nobel, G. K. 1931. The biology of the Amphibia. New York, McGraw-Hill Book Co.)

membranes of the buccal (mouth) cavity (buccopharyngeal breathing). Lungless plethodontid salamanders are believed to have originated in swift streams of the Appalachian mountains, where the water is so cool and well-oxygenated that cutaneous respiration alone was sufficient for life.

Paedomorphosis. Whereas most urodeles complete their development to the definitive adult body form by metamorphosis, there are some species that retain their gills and other larval characteristics even after becoming sexually mature. This condition is called paedomorphosis (Gr., *pais,* child, + *morphe,* form). In effect, ancestral juvenile features have been displaced to the adult stage of descendants. Some are **permanent larvae,** a genetically fixed condition in which the developing tissues fail to respond to the thyroid hormone that, in other amphibians, stimulates metamorphosis. This condition is called **obligatory paedomorphosis.**

Examples of permanent larvae are mud puppies of the genus *Necturus* (Fig. 24-6), which live on bottoms of ponds and lakes and keep their external gills throughout life, and the congo eel *(Amphiuma means)* of the southeastern United States, which with its useless, rudimentary legs superficially resembles an eel more than an amphibian.

There are other species of salamanders that become sexually mature and breed in the larval state, but, unlike the permanent larvae, they may metamorphose to adults if environmental conditions change. This is called **facultative paedomorphosis.** Examples are species of the genus *Ambystoma* and of the genus *Triturus.*

The American axolotl *Ambystoma tigrinum,* widely distributed over Mexico and the southwestern United States, remains in the aquatic, gill-breathing, and fully reproductive larval form unless the water begins to dry up; then it metamorphoses to an adult, loses its gills, develops lungs, and assumes the appearance of an ordinary salamander. In these forms, paedomorphosis is a handy adaptation for remaining aquatic in an environment where water is plentiful. This particular kind of paedomorphosis is often called **neotony** (Gr., *neos,* young, + *teineu,* to extend) because it is achieved by delaying somatic development, but not sexual maturation.

Axolotls can be made to metamorphose by treating them with the thyroid hormone, thyroxin. Thyroxin is essential for normal metamorphosis in all amphibians. Recent research suggests that for some reason the pituitary gland fails to become fully active in neotenous forms and does not release thyrotropin, which is required to stimulate the production of thyroxin by the thyroid gland.

Frogs and toads, order Salientia (Anura)

The more than 2,000 species of frogs and toads that comprise the order Salientia are the most familiar and most successful of amphibians. Frogs and toads are highly specialized for a jumping mode of locomotion, as suggested by the preferred name of the order, Salientia, meaning leaping. The alternate order name, Anura (Gr., *an-,* without, + *oura,* tail), refers to another obvious group characteristic, the absence of tails as adults (although all pass through a tailed larval stage during development).

The Salientia are further distinguished from the Uro-

FIG. 24-7 Some common North American frogs. **A,** Bullfrog *Rana catesbeiana* is the largest of all American frogs. **B,** Green frog *Rana clamitans* is next to bullfrog in size. Its body is usually green, especially around the jaws; it has dark bars on the sides of the legs. **C,** Leopard frog *Rana pipiens* has light-colored dorsolateral ridges and irregular spots. **D,** Spring peepers *Hyla crucifer,* the darlings of warm spring nights, when their characteristic peeping is so often heard. They are small (2 to 3 cm) and light brown, with an X mark on the back. (Courtesy C. Alender.)

dela by their larvae and a dramatic metamorphosis during development. The eggs of most frogs hatch into a tadpole (''polliwog'') stage, with a long, finned tail, both internal and external gills, no legs, specialized mouthparts for herbivorous feeding (salamander larvae, in distinction, are carnivorous), and a highly specialized internal anatomy. They look and act altogether differently from adult frogs. The metamorphosis of the frog tadpole to the adult frog is thus a striking transformation. Neoteny and paedomorphosis are never exhibited in frogs and toads, as they are among salamanders.

The Salientia are an old order—fossil frogs are known from the Jurassic period, 150 million years ago—and today they are a secure and successful group. Frogs and toads occupy a great variety of habitats, despite their aquatic mode of reproduction and water-permeable skin, which prevents them from wandering too far afield from sources of water, and their ectothermy, which bars them from polar and subarctic habitats.

Notwithstanding their success as a distinct group, the frogs and toads are really a specialized side branch of amphibian evolution, and, despite their popularity for education purposes—approximately 20 million are used each year in the United States alone—they are not good representatives of the vertebrate body plan. The primitive and unspecialized salamander would be a much superior choice for the zoology laboratory were not frogs so readily available.

Frogs and toads are divided into 12 families. The best known frog families in North America are Ranidae, containing most of our familiar frogs, and Hylidae, the tree frogs (Fig. 24-7). True toads, belonging to the family Bufonidae, have short legs, stout bodies, and thick skins usually with prominent warts. However, the term ''toad'' is used rather loosely to refer to more or less terrestrial members of several other families.

The largest anuran is the west African *Gigantorana goliath,* which is more than 30 cm long from tip of nose to anus (Fig. 24-8). This giant eats animals as

FIG. 24-8 *Gigantorana goliath* of West Africa, the world's largest frog. This specimen weighed 3.3 kg (approximately 7½ pounds). (Courtesy American Museum of Natural History.)

big as rats and ducks. The smallest frog recorded is *Phyllobates limbatus,* which is only approximately 1 cm long. This tiny frog, which is more than covered by a dime, is found in Cuba. The largest American frog, the bullfrog *Rana catesbeiana* (Fig. 24-7, *A*), reaches a head and body length of 20 cm.

Habitats and distribution. Probably the most abundant and successful of frogs are the 200 to 300 species of the genus *Rana,* found all over the temperate and tropical regions of the world except in New Zealand, the oceanic islands, and southern South America. They are usually found near water, although some, such as the wood frog *Rana sylvatica,* spend most of their time on damp forest floors, often some distance from the nearest water. The wood frog probably returns to pools only for breeding in early spring. The larger bullfrogs *R. catesbeiana* and green frogs *R. clamitans* (Fig. 24-7, *A* and *B*) are nearly always found in or near permanent water or swampy regions. The leopard frog *R. pipiens* (Fig. 24-7, *C*) has a wider variety of habitats and, with all its subspecies and phases, is perhaps the most widespread of all the North American frogs. This is the species most commonly used in biology laboratories and for classic electrophysiologic research. It has been found in some form in nearly every state, though it is sparingly represented along the extreme western part of the Pacific coast. It also extends far into northern Canada and as far south as Panama.

Within the range of any species of frogs, they are often restricted to certain habitats (for instance, to certain streams or pools) and may be absent or scarce in similar habitats of the range. The pickerel frog *(R. palustris)* is especially noteworthy this way, for it is known to be abundant only in certain localized regions.

Most of the larger frogs are solitary in their habits except during the breeding season. During the breeding period most of them, especially the males, are very noisy. Each male usually takes possession of a particular perch, where he may remain for hours or even days, trying to attract a female to that spot. At times frogs are mainly silent, and their presence is not detected until they are disturbed. When they enter the water, they dart about swiftly and reach the bottom of the pool, where they kick up a cloud of muddy water. In swimming, they hold the forelimbs near the body and kick backward with the webbed hind limbs, which propel them forward. When they come to the surface to breathe, only the head and foreparts are exposed, and since they usually take advantage of any protective vegetation, they are difficult to see.

During the winter months most frogs hibernate in the soft mud of the bottom of pools and streams. The wood frog hibernates under stones, logs, and stumps in the forest area. Naturally their life processes are at a very low ebb during their hibernation period, and such energy as they need is derived from the glycogen and fat stored in their bodies during the spring and summer months.

Adult frogs have numerous enemies, such as snakes, aquatic birds, turtles, raccoons, humans, and many others; only a few tadpoles survive to maturity. Although usually defenseless, in the tropics and subtropics many frogs and toads are aggressive, jumping and biting at predators. Some defend themselves by

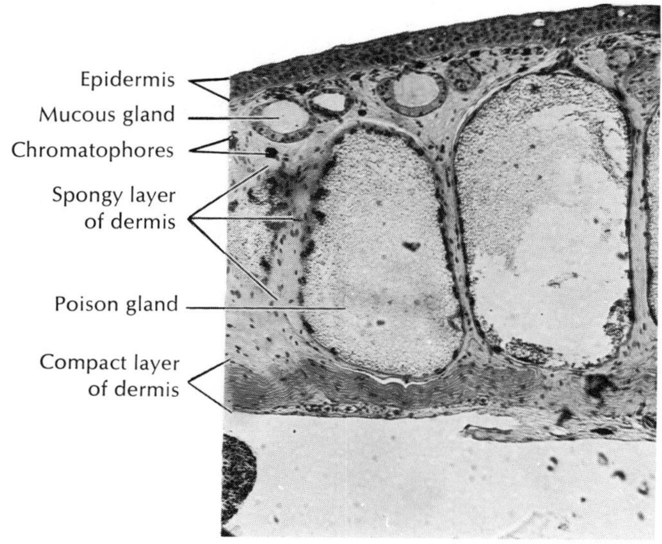

FIG. 24-9 Histologic section of frog skin. Stratified epidermis is seen as dark layer at surface with thicker dermis below. Note small mucous glands and large poison glands in dermal layer. (Photograph by F. M. Hickman.)

feigning death. Most anurans can blow up their lungs so that they are difficult to swallow. When disturbed along the margin of a pond or brook, a frog will often remain quite still; when it thinks it is detected, it will jump, not always into the water where enemies may be lurking but into grassy cover on the bank. When held in the hand a frog may cease its struggles for an instant to put its captor off guard and then leap violently, at the same time voiding its urine. Their best protection is their ability to leap and their use of poison glands. Bullfrogs in captivity will not hesitate to snap at tormenters and are capable of inflicting painful bites.

Integument and coloration. The skin of the frog is thin and moist and is attached loosely to the body only at certain points. Histologically the skin is made up of two layers—an outer stratified **epidermis** and an inner spongy **dermis** (Fig. 24-9). The epidermis consists of a somewhat horny outer layer of epithelium, which is periodically shed, and an inner layer of columnar cells, from which new cells are formed to replace those that are lost. The frog molts a number of times during its active months. In the process the outer layer of skin is split down the back and is worked off as one piece. The dermis is made up mostly of glands, pigment cells, and connective tissue. On its outer portion are the glands—small **mucous glands,** which secrete mucus for keeping the skin moist, and the larger **poison glands,** which secrete a whitish fluid that is highly irritating to enemies. The poison of *Dendrobates,* a South American frog, is used by Indian tribes to poison the points of their arrows. This and the poisons of other dendrobatid frogs are the most lethal animal secretions known, drop for drop more poisonous even than the venoms of sea snakes or any of the most poisonous arachnids.

Skin color in the frog is produced by pigment granules scattered through the epidermis and by special pigment cells, **chromatophores,** located in the dermis (Figs. 24-9 and 24-10). Types of chromatophores include **guanophores,** which contain white crystals; **melanophores** with black and brown pigment; and **lipophores** (xanthophores) with red and yellow pigment. Of these the melanophores are most important for adjusting coloration to blend with the background; some frogs and toads possess remarkably good powers of camouflage (Fig. 24-11).

The degree of dispersion of the dark brown pigment granules of melanin within the melanophores is controlled by melanophore-stimulating hormone (MSH) of the intermediate lobe of the pituitary gland (see Fig. 33-5, p. 760). The release of MSH, which causes pigment dispersion, and thus darkening of the animal, is in turn controlled by the pattern of illumination on the retina of the eyes. When the animal is on a light background, both dorsal and ventral portions of the retina are illuminated by reflected light from the surface and by direct illumination from above. This inhibits the release of MSH, the pigment becomes concentrated in the centers of the melanophores, and the animal becomes pale in color. But on a dark, light-absorbing

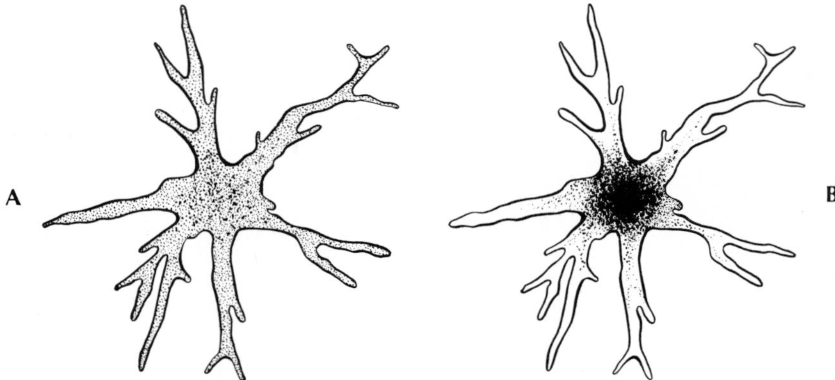

FIG. 24-10 Pigment cells (chromatophores). **A,** Pigment dispersed. **B,** Pigment concentrated. Pigment cell does not contract or expand; color effects are produced by streaming of cytoplasm, carrying pigment granules into cell branches for maximum color effect or to center of cell for minimum effect. Chromatophore of cephalopod, however, does change shape by muscular contraction. Control over dispersal or concentration of pigment is mostly through light stimuli to eye. Melanophore-stimulating hormone is known to influence activity of chromatophore (see text). Types of chromatophores depend on pigments they bear, for example, melanophores (brownish black), xanthophores (yellow or red), and guanophores (white).

background, where only the ventral retina is illuminated, MSH is released from the pituitary and causes the melanophore pigment to disperse, darkening the skin. Both responses are concealing in their effect. Other factors, especially temperature, humidity, and activity, also influence the skin coloration.

Skeletal and muscular systems. In amphibians, as in their fish ancestors, the well-developed **endoskeleton** of bone and cartilage provides a framework for the muscles in movement and protection for the viscera and nervous systems. But movement onto land and the necessity of transforming paddlelike fins into tetrapod legs capable of supporting the body's weight introduced a new set of stress and leverage problems. The changes are most noticeable in the anurans in which the entire musculoskeletal system is specialized for jumping and swimming by simultaneous extensor thrusts of the hind limbs.

AXIAL SKELETON. In amphibians, the vertebral column assumes a new role as a support from which the abdomen is slung and to which the limbs are attached. Since amphibians move with limbs instead of swimming with serial contractions of myotomic trunk musculature, the vertebral column has lost much of the original flexibility characteristic of fishes. Rather it has become a rigid frame for transmitting force from the hind limbs to the body. The anurans are further specialized by an extreme shortening of the body. Typical frogs have only nine trunk vertebrae and a rodlike **urostyle** (Fig. 24-12), which represents several fused caudal vertebrae. The primitive, limbless caecilians, which obviously have not shared these specializations for tetrapod locomotion, may have as many as 200 vertebrae.

The frog skull is also vastly altered as compared to its crossopterygian ancestors; it is much lighter in weight and more flattened in profile and contains fewer bones and less ossification. The front part of the skull, wherein are located the nose, eyes, and brain, is better developed, whereas the back of the skull, which contained the gill apparatus of fishes, is much reduced. Lightening of the skull was essential to mobility on land, and the other changes fitted the frog for its improved special senses and means for feeding and breathing.

APPENDICULAR SKELETON. The **pectoral girdle,** consisting of **suprascapula, scapula, clavicle,** and **coracoid,** serves as support for the forelimbs (Fig. 24-12). The forelimbs in frogs are used mainly to absorb the weight during landing after a jump. The pelvic girdle of frogs is much specialized, consisting of two **innominate** bones, each representing the fusion of a long

FIG. 24-11 Cryptic coloration of the tree frog *(Hyla).* Instead of the excellent protective markings borne by this harmless North American species, several of the dangerously poisonous South American tree frogs display vivid coloration that warns predators to keep away. All tree frogs have suction pads at ends of fingers and toes, enabling them to grip vertical surfaces. (Photograph by L. L. Rue, III.)

ilium, a posterior **ischium,** and a ventral **pubis** (Fig. 24-12). The pattern of bones and muscles in the limbs is the typical tetrapod type. There are three main joints in each limb (hip, knee, and ankle; or shoulder, elbow, and wrist). The hand or foot is basically a five-rayed form with several joints in each of the digits. It is a repetitive system that can be plausibly derived from the bone structure of the crossopterygian fin, which is distinctly suggestive of the amphibian limb; it is not difficult to imagine how selective pressures through millions of years remodeled the former into the latter.

MUSCULATURE. The muscles of the limbs are presumably derived from the radial muscles that moved the fins of fishes up and down, but the muscular arrangement has become so complex in the tetrapod limb that it is no longer possible to see parallels between this and fin musculature. Muscles are usually arranged in antagonistic groups so that movement is effected by the contraction of one group and the relaxation of the opposite. Muscles may be classified by their action.

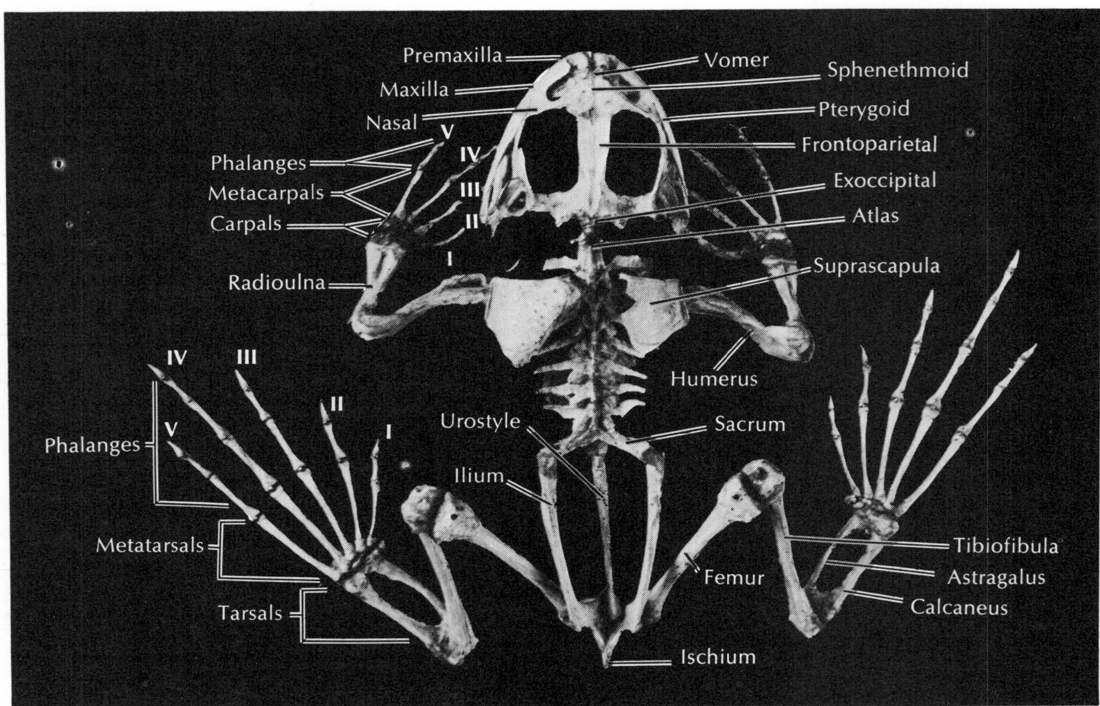

FIG. 24-12 Dorsal view of frog skeleton. (Photograph by F. M. Hickman.)

abductor Moves the part away from the median axis of the body **(deltoid)**
adductor Moves the part toward the median axis of the body **(adductor magnus)**
flexor Bends one part on another part **(biceps brachii)**
extensor Straightens out a part **(triceps brachii)**
depressor Lowers a part **(sternohyoid)**
levator Elevates a part **(masseter)**
rotator Produces a rotary movement **(gluteus)**

Despite the complexity of tetrapod limb musculature, it is possible to recognize two great groups of muscles on any limb: an anterior and ventral group that pulls the limb forward and toward the midline (protraction and adduction), and a second set of posterior and dorsal muscles that serves to draw the limb back and away from the body (retraction and abduction).

The trunk musculature, which in fishes is segmentally organized into powerful muscular bands (myotomes) for locomotion by lateral flexion, was much modified during amphibian evolution. The dorsal (epaxial) muscles are arranged to support the head and brace the vertebral column. The ventral (hypaxial) muscles of the belly are more developed in amphibians than in fishes since they must support the viscera in air without the buoying assistance of water. Anteriorly the hypaxial muscles form the hyoid musculature of the throat, used to raise and lower the floor of the mouth during breathing.

Respiration and vocalization. Amphibians use three respiratory surfaces for gas exchange: the skin (cutaneous breathing), the mouth (buccal breathing), and the lungs. These surfaces are used to varying degrees by different groups and under different environmental conditions. The lungless salamanders of the family Plethodontidae breathe principally through the highly vascular skin (90% to 95% of total gas exchange) and to a much lesser extent through the vascular membranes of the mouth (5% to 10% of total gas exchange). The more terrestrial amphibians such as frogs and toads show a much greater dependence on lung breathing, since cutaneous breathing, simple and direct as it obviously is, suffers from two disadvantages: (1) the skin must be kept thin and moist for gas exchange and consequently is too delicate for a wholly aerial life, and (2) the amount of gas exchange across

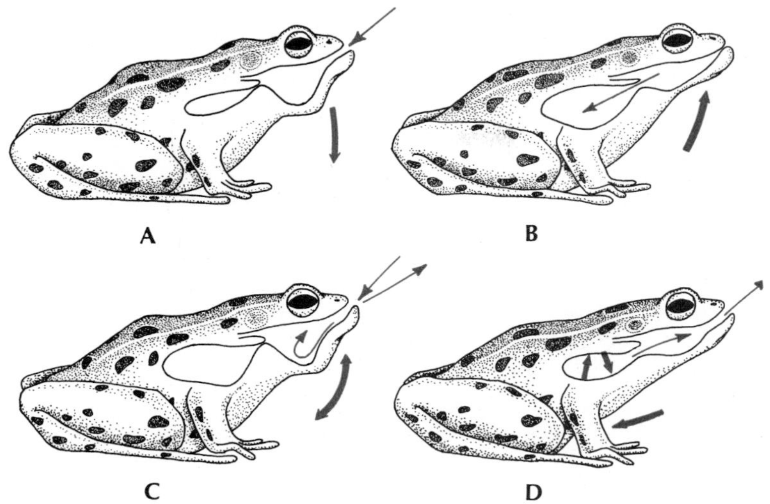

FIG. 24-13 Breathing in frog. The frog, a positive-pressure breather, fills its lungs by forcing air into them. **A,** Floor of mouth is lowered, drawing air in through nostrils. **B,** With nostrils closed and glottis open, the frog forces air into its lungs by elevating floor of mouth. **C,** Mouth cavity rhythmically ventilates for a period. **D,** Lungs are emptied by contraction of body wall musculature and by elastic recoil of lungs. (Modified from Gordon, M. S., and others. 1968. Animal function: principles and adaptations, New York, The Macmillan Co.)

the skin is mostly constant and cannot be varied to match changing demands of the body for more or less oxygen. Nevertheless the skin continues to serve as an important supplementary avenue for gas exchange in anurans, especially during hibernation in winter. Even under normal conditions when lung breathing predominates, most of the carbon dioxide is lost across the skin while most of the oxygen is taken up across the lungs.

The lungs are supplied by pulmonary arteries (derived from the sixth aortic arches) and blood is returned directly to the left atrium by the pulmonary veins. Frog lungs are ovoid, elastic sacs with their inner surfaces divided into a network of septa that are in turn subdivided into small terminal air chambers called **alveoli.** The alveoli of the frog lung are much larger than those of more advanced vertebrates, and consequently the frog lung has a smaller relative surface available for gas exchange: the respiratory surface of the common *Rana pipiens* is about 20 cm^2 per cubic centimeter of air contained, compared to 300 cm^2 for humans. But the problem in lung evolution was not the development of a good internal vascular surface, but rather the problem of moving air. A frog is a positive-pressure breather that fills its lungs by forcing air into them; this contrasts with the negative-pressure system of all the higher vertebrates. The sequence and explanation of breathing in a frog are shown in Fig. 24-13. One can easily follow this sequence in a living frog at rest: rhythmic throat movements of mouth breathing may continue some time before flank movements indicate that the lungs are being emptied.

Both male and female frogs have **vocal cords,** but those of the male are much better developed. They are located in the **larynx,** or voice box. Sound is produced by passing air back and forth over the vocal cords between the lungs and a large pair of sacs (vocal pouches) in the floor of the mouth. The latter also serve as effective resonators in the male. The chief function of the voice is to attract mates. Most species utter characteristic sounds that identify them. Nearly everyone is familiar with the welcome springtime calls of the spring peeper that produces a high-pitched sound surprisingly strident for such a tiny frog. Another sound familiar to residents of the more southern United States is the resonant ''jug-o-rum'' call of the bullfrog. The bass notes of the green frog are banjolike, and those of the leopard frog are long and guttural. Many frog sounds are now available on phonograph records.

Circulation. As in fishes, circulation in amphibians is a closed system of arteries and veins serving a vast peripheral network of capillaries through which blood is forced by the action of a single pressure pump, the heart.

The principal changes in circuitry involve the shift from gill to lung breathing. With the elimination of gills a major obstacle to blood flow was removed from the arterial circuit. But two new problems arose. The first was to provide a blood circuit to the lungs. This was solved as we have seen by converting the sixth aortic arch into pulmonary arteries to serve the lungs and by developing new pulmonary veins for returning oxygenated blood to the heart. The second and evidently more difficult evolutionary problem was to sep-

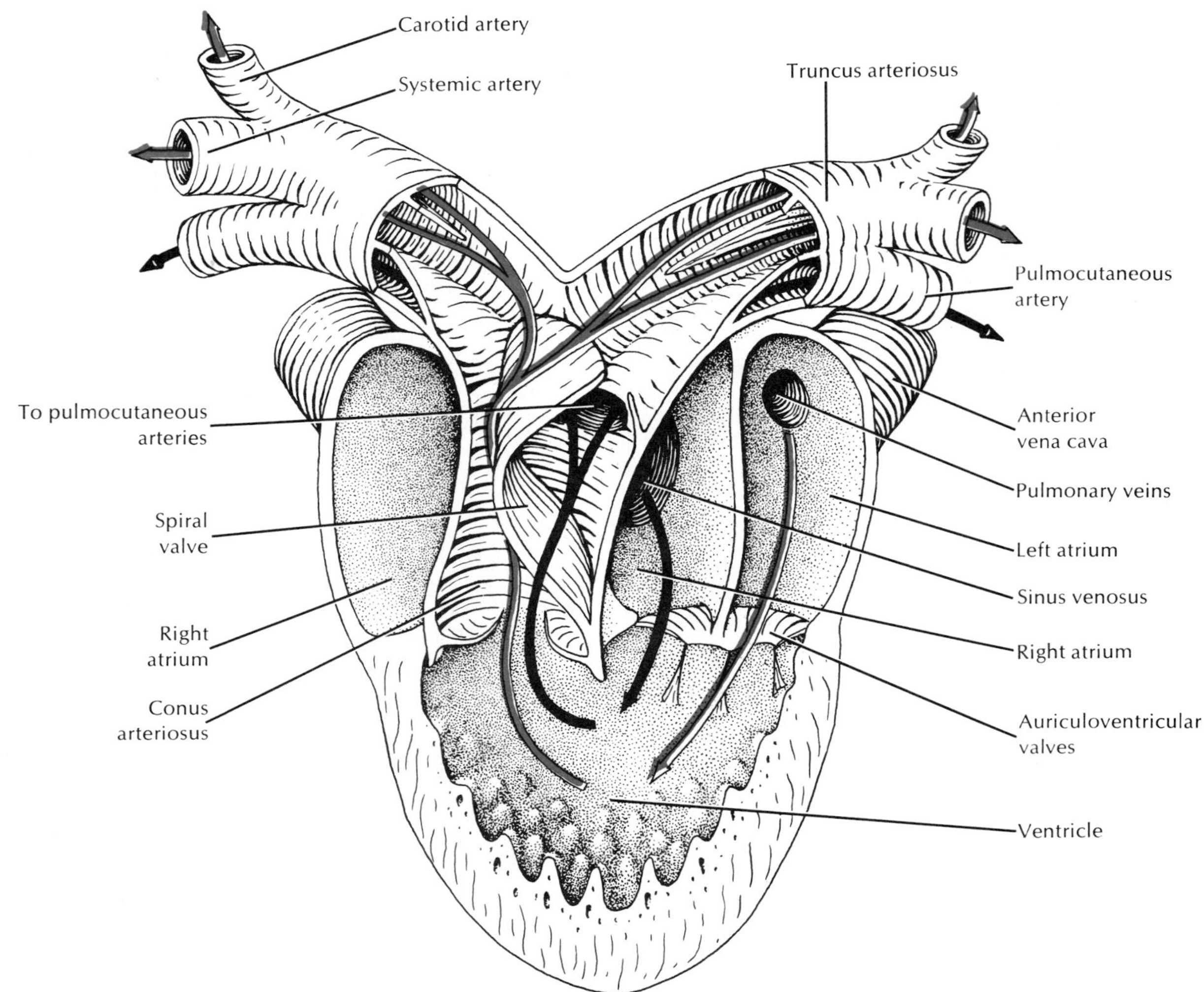

FIG. 24-14 Structure of the frog heart. *Red arrows,* oxygenated blood; *black arrows,* deoxygenated blood.

arate the new pulmonary circulation from the rest of the body's circulation in such a way that oxygenated blood from the lungs would be selectively sent to the body and deoxygenated venous return from the body would be selectively sent to the lungs. In effect this meant creating a double circulation consisting of separate pulmonary and systemic circuits. This was eventually solved by the development of a partition down the center of the heart, creating a double pump, one for each circuit. Amphibians and reptiles have made the separation to variable degrees, but the task was completed by the birds and mammals, which have a completely divided heart of two atria and two ventricles.

The frog heart (Fig. 24-14) has two separate atria and a single undivided ventricle. Blood from the body (systemic circuit) first enters a large receiving chamber, the **sinus venosus,** which forces it into the **right atrium.** The **left atrium** receives freshly oxygenated blood from the lungs. Up to this point the deoxygenated blood from the body and oxygenated blood from the lungs are separated. But now both atria contract almost simultaneously, driving both right and left atrial blood into the single undivided **ventricle.** We should expect that complete admixture of the two circuits would happen here. In fact there is evidence that in at least some amphibians they remain mostly separated, so that when the ventricle contracts, oxygenated pulmonary blood is sent to the systemic circuit and deoxygenated systemic blood is sent to the pulmonary circuit. The **spiral valve** in the **conus arteriosus** (Fig. 24-14) may play an important role in maintaining selective distribution. The matter is controversial and has defied complete analysis despite the application of advanced techniques using radiopaque media and high-speed cineradiography.

Feeding and digestion. Frogs are carnivorous like most other adult amphibians and diet on insects, spiders, worms, slugs, snails, millipedes, or nearly anything else that moves and is small enough to swallow whole. They snap at moving prey with their protrusible tongue, which is attached to the front of the mouth and is free behind. The free end of the tongue is highly glandular and produces a sticky secretion, which adheres to the prey. The teeth on the premaxillae, maxillae, and vomers are used to prevent escape of prey, not for biting or chewing. The digestive tract is relatively short in adult amphibians, a characteristic of most carnivores, and it produces a variety of enzymes for breaking down proteins, carbohydrates, and fats.

The larval stages of anurans (tadpoles) are usually herbivorous, feeding on pond algae and other vegetable matter; they have a relatively long digestive tract, since their bulky food must be submitted to time-consuming fermentation before useful products can be absorbed.

Nervous system and special senses. The frog nervous system is of the basic vertebrate plan that was established in the early jawless fishes. It is commonly divided into (1) the **central nervous system,** consisting of the brain and spinal cord; (2) the **peripheral nervous system,** consisting of the cranial and spinal nerves; and (3) the **autonomic nervous system,** composed of a chain of special ganglia on each side of the spinal column.

The three fundamental parts of the brain—forebrain (telencephalon), concerned with the sense of smell; midbrain (mesencephalon), concerned with vision; and hindbrain (rhombencephalon), concerned with hearing and balance—have undergone dramatic developmental trends as the vertebrates moved onto land and improved their environmental awareness. In general there is increasing cephalization with emphasis on information-

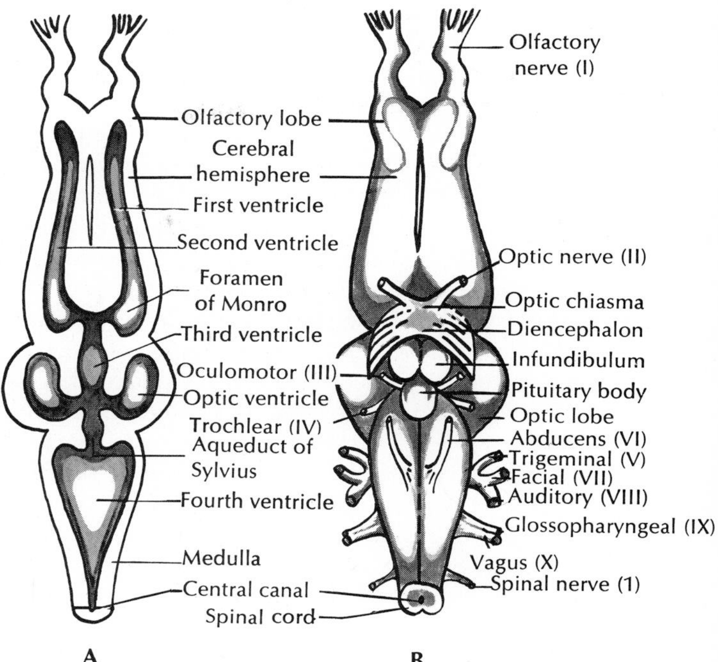

FIG. 24-15 Brain of frog. **A,** Longitudinal section showing dorsal view of ventricles. **B,** Ventral view of brain and cranial nerves.

processing by the brain and a corresponding loss of independence of the spinal ganglia, which are capable of only stereotyped reflexive behavior. Nonetheless a headless frog preserves an amazing degree of purposive and highly coordinated behavior. With only the spinal cord intact, it maintains normal body posture and can with purposive accuracy raise its leg to wipe from its skin a piece of filter paper soaked in dilute acetic acid. It will even use the opposite leg if the closer leg is held.

The forebrain is divided into the **telencephalon** and **diencephalon** (Fig. 24-15). The telencephalon contains the olfactory center, which has assumed much increased importance for the detection of dilute airborne odors on land. The sense of smell is in fact one of the dominant special senses in frogs. The remainder of the telencephalon is the cerebrum, which is of little importance in amphibians and provides no hint of the magnificent development it is destined to attain in the higher mammals. The diencephalon consists of **thalamus, hypothalamus,** and **posterior pituitary,** which are important relay channels and regulative centers. The thalamus and hypothalamus are primitive integrative areas concerned with thirst, hunger, sexual drive, pleasure, and pain.

The most complex integrative activities of the frog brain are centered in the midbrain. The **optic lobe** (tectum) is a dorsal enlargement that integrates sensory information from the eyes as well as from other senses.

The hindbrain is divided into an anterior **metencephalon,** or cerebellum, and a posterior **myelencephalon,** or medulla. The cerebellum is concerned with equilibrium and movement coordination and is not well-developed in amphibians, which stick close to the ground and are not noted for dexterity of movement. The cerebellum becomes vastly developed in the fast-moving birds and mammals. The medulla is really the enlarged anterior end of the spinal cord through which pass all sensory neurons except those of vision and smell. Here are located centers for auditory reflexes, respiration, swallowing, and vasomotor control.

The evolution of a semiterrestrial life for the amphibians has necessitated a reordering of sensory receptor priorities on land. The pressure-sensitive lateral-line (acousticolateral) system of fishes remains only in the aquatic larvae of amphibians and in a few strictly aquatic adult amphibian species. This system of course can serve no useful purpose on land, since it was designed to detect and localize objects in water by reflected pressure waves. Instead the task of detecting airborne sounds devolved on the ear.

The **ear** of a frog is by higher vertebrate standards a primitive structure: a middle ear closed externally by a large **tympanic membrane** (eardrum) and containing a **stapes** (columella) that transmits vibrations to the inner ear. The latter contains the **utricle,** from which arise the semicircular canals, and a **saccule** bearing a diverticulum, the **lagena.** The lagena is partly covered with a **tectorial membrane** that in its fine structure is not unlike that of the much more advanced mammalian cochlea. In most frogs this structure is sensitive to low-frequency sound energy not greater than 4,000 Hz (cycles per second); in the bullfrog the main frequency response is in the 100 to 200 Hz range, which matches the energy of the male frog's low-pitched call. Although it has not been measured, we would expect the ear of the spring peeper to be sensitive to the high-frequency calls this species makes.

In most amphibians, the **eye** is the dominant special sense. It is basically of the fish type, but the lens is located farther from the cornea and is provided with protractor muscles for limited accommodation (focusing on near or distant objects). The **retina** contains both **rods** and **cones,** providing frogs with color vision. The **iris** contains well-developed circular and radial muscles and can rapidly expand or contract the aperture (pupil) to adjust to changing illumination. The upper lid of the eye is fixed, but the lower is folded into a transparent **nictitating membrane** capable of moving across the eye surface. In all, frogs and toads possess good vision, a fact of crucial importance to animals that rely on quick escape to avoid their numerous predators.

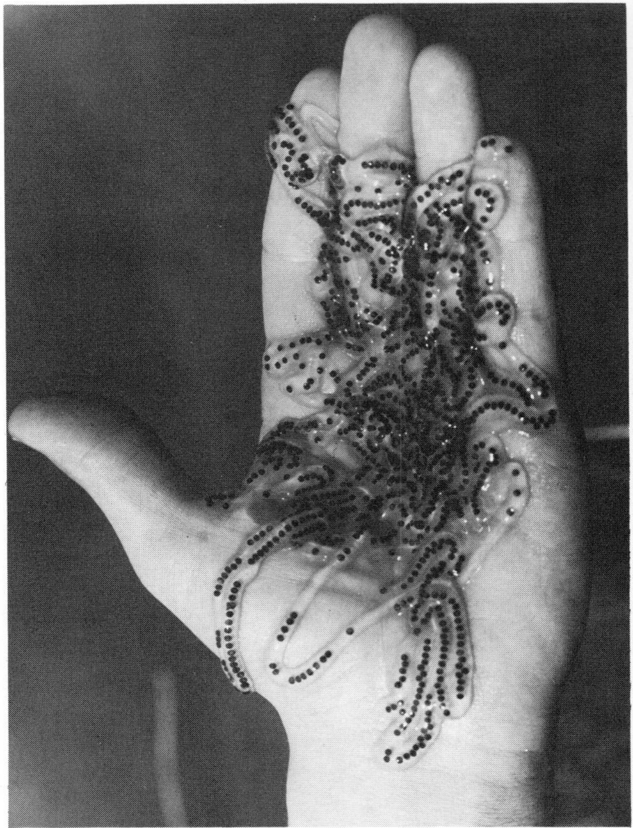

FIG. 24-16 Eggs of American toad. Toads lay eggs in strings; frogs lay eggs in clusters. (Photograph by L. L. Rue, III.)

Other sensory receptors include tactile and chemical receptors in the skin, taste buds on the tongue and palate, and a well-developed olfactory epithelium lining the nasal cavity.

Reproduction and life cycle. Frogs and toads are cold-blooded (ectothermic) animals; their distribution

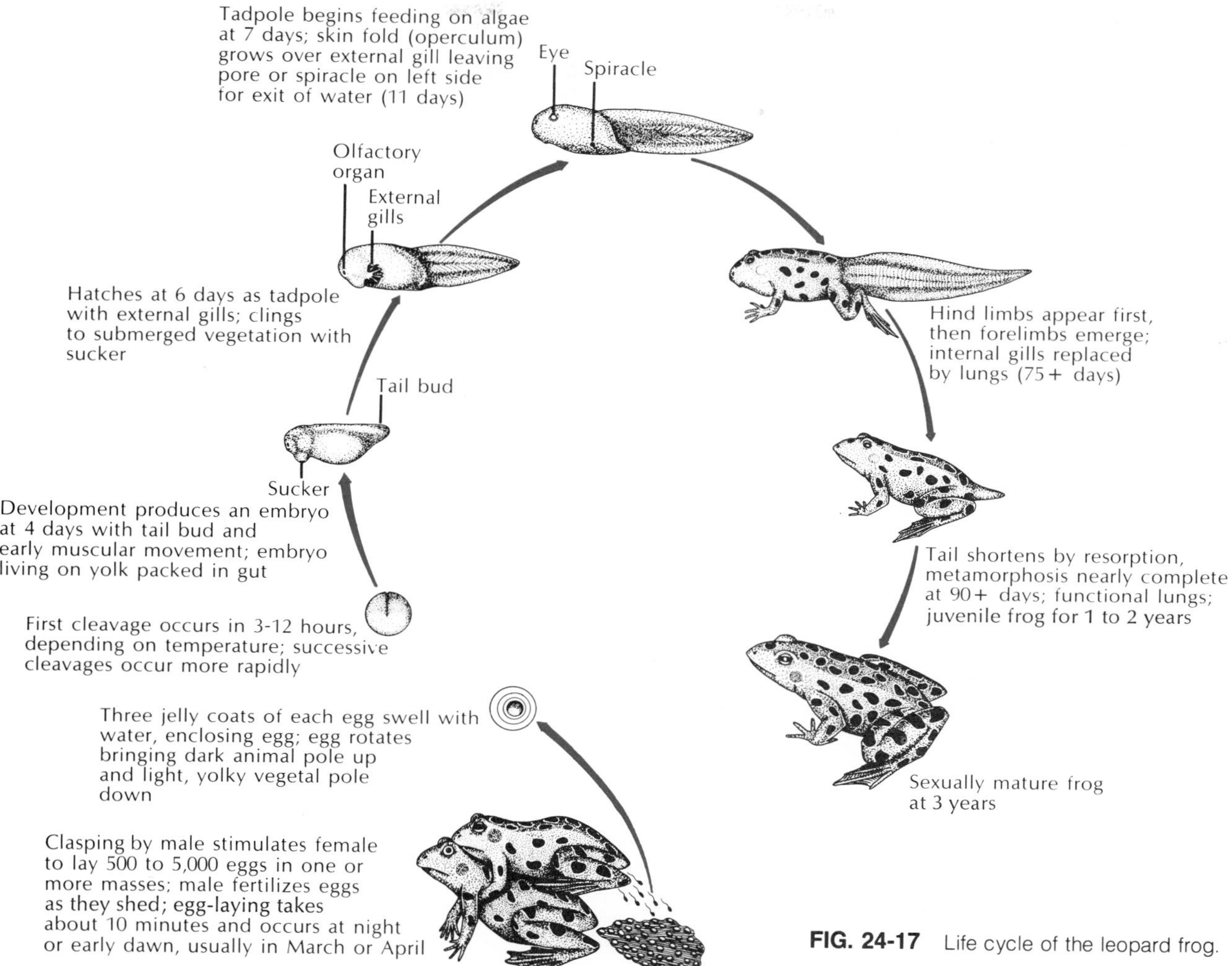

FIG. 24-17 Life cycle of the leopard frog.

and activities are therefore controlled by seasonal changes and climatic conditions. Their activities are restricted to the warmer seasons of the year, when they breed, feed, and grow. During the winter months in colder climates they spend their time in hibernation.

The time of spring emergence varies with different species. One of their first interests after leaving their dormant period is breeding. At this time the males croak and call vociferously to attract females. The breeding season usually extends for several weeks. When their eggs are ripe, the females enter the water and are mounted and clasped by the males in the process called **amplexus.** The male holds the female by pressing the nuptial pads of his thumbs against her breast just back of her forelegs. As the female lays her eggs, the male discharges his seminal fluid containing the sperm over the eggs. The sperm penetrate the jelly layers of the eggs, assisted by **sperm lysins,** which dissolve the membranes. When the first spermatozoon touches the surface of the egg proper, dramatic changes called the **cortical reaction** suddenly occur in the egg cytoplasm. A **fertilization membrane** that prevents additional sperm from entering the egg is immediately formed. Thus only the first sperm penetrates the egg surface **(monospermy)** and is drawn into the egg interior where it swells into a **male pronucleus.** This soon fuses with the **female** (egg) **pronucleus** to form a zygote nucleus with the diploid number of chromosomes. After fertilization the jelly layers absorb water and swell (Fig. 24-16). Eggs are laid in great masses, which may include several thousands in the leopard frog. The egg masses are usually anchored to vegetation or debris by the sticky jelly layers around the egg. Not all the eggs have a chance to develop, for some may not be fertilized and others are eaten by turtles, insects, and other enemies.

Development of the fertilized egg (zygote) begins about 3 hours after fertilization (Fig. 24-17). The process involves cleavage, or segmentation of the egg. Cleavage occurs more rapidly at the black, or animal, pole, where there is more protoplasm and less yolk. By collecting masses of frog eggs early in the morning, it is possible to find eggs in various stages of development, such as 2-, 4-, and 8-cell stages. Generally one finds them still further along in development.

Eggs usually hatch into tadpoles within a period of 6 to 9 days, depending on the temperature (Fig. 24-17). At the time of hatching, the tadpole has a distinct head and body with a compressed tail. The mouth is located on the ventral side of the head and is provided with horny jaws for scraping off vegetation from objects for food. Behind the mouth is a ventral adhesive disc for clinging to objects. In front of the mouth are two deep pits, which later develop into the nostrils. Swellings are found on each side of the head, and these later become external gills. There are finally three pairs of external gills, which are later replaced by three pairs of internal gills within the gill slits. On the left side of the neck region is an opening, the **spiracle,** through which water flows after entering the mouth and passing the internal gills. The hind legs appear first, while the forelimbs are hidden for a time by the folds of the operculum. During metamorphosis the tail is resorbed, the intestine becomes much shorter, the mouth undergoes a transformation into the adult condition, lungs develop, and the gills are resorbed. The leopard frog usually completes its metamorphosis within a year or less; the bullfrog takes 2 or 3 years to complete the process.

Migration of frogs and toads is correlated with their breeding habits. Males usually return to a pond or stream in advance of the females, whom they then attract by their calls. Some salamanders are also known to have a strong homing instinct; guided by olfactory cues, they return year after year to the same pool for reproduction. The initial stimulus for migration in many cases is attributable to a seasonal cycle in the gonads plus hormonal changes that increase their sensitivity to temperature and humidity changes.

Annotated references

Selected general readings

Barbour, R. W. 1971. Amphibians and reptiles of Kentucky. Lexington, Ky., University of Kentucky Press. *Although primarily concerned with a state survey of amphibians and reptiles, this volume is additionally concerned with general distribution of species and contains a number of excellent photographs.*

Cochran, D. M., and C. J. Goin. 1970. The new field book of reptiles and amphibians. New York, G. P. Putnam's Sons. *Probably the most useful field guide available for reptiles and amphibians. Most species descriptions are concise and informative.*

Gans, C. 1974. Biomechanics: an approach to vertebrate biology. Philadelphia, J. B. Lippincott Co. *The final section of this selective treatment is an interesting analysis of the functional anatomy of frog breathing and vocalization.*

Goin, C. J., and O. B. Goin. 1971. Introduction to herpetology, ed. 2. San Francisco, W. H. Freeman & Co. Publishers. *A basic introductory text for the study of amphibians and reptiles.*

Mertens, R. 1960. The world of amphibians and reptiles. London, George G. Harrap. *(Translation) Excellent source of information, beautifully illustrated.*

Moore, J. (ed.). 1964. Physiology of the Amphibia. New York, Academic Press, Inc. *A valuable reference on the Amphibia for the advanced student.*

Noble, G. K. 1931. Biology of the Amphibia. New York, McGraw-Hill Book Co. *An invaluable book despite its age, containing information to be found nowhere else.*

Oliver, J. A. 1955. The natural history of North American amphibians and reptiles. New York, D. Van Nostrand Co. *A good introduction to classification, distribution, and behavior of reptiles and amphibians.*

Porter, G. 1967. The world of the frog and the toad. Philadelphia, J. B. Lippincott Co. *Superbly illustrated natural history.*

Stahl, B. J. 1974. Vertebrate history: problems in evolution. New York, McGraw-Hill Book Co. *A contemporary treatment of vertebrate paleontology. Good account of amphibian beginnings.*

Taylor, D. H., and S. J. Guttman (eds.). 1977. The reproductive biology of amphibians. Society for the study of amphibians and reptiles. New York, Plenum Press. *Collection of 15 symposium papers.*

Twitty, V. C. 1966. Of scientists and salamanders. San Francisco, W. H. Freeman & Co. Publishers. *Delightfully written account of the author's personal study of the experimental biology of salamanders and newts.*

Wright, A. H., and A. A. Wright. 1949. Handbook of frogs of the United States and Canada, ed. 3. Ithaca, N.Y., Comstock Publishing Co. *A volume that has, with time, become a standard reference work for the study of frogs.*

CHAPTER 25

THE REPTILES

Phylum Chordata
Class Reptilia

An eastern cottonmouth, or water moccasin, *Agkistrodon piscivorus piscivorus,* flicks its tongue to "smell" its surroundings. Scent particles trapped on the tongue's surface are transferred to Jacobson's organ, an olfactory organ positioned in the roof of the mouth. Note the heat-sensitive pit organ just anterior to the eye.

Photograph by C. P. Hickman, Jr.

Reptilia (Rep-til′e-a) (L., *repere,* to creep) are the first truly terrestrial vertebrates. With some 6,000 species (about 300 species in the continental United States and Canada) occupying a great variety of aquatic and terrestrial habitats, they are clearly a successful group. Nevertheless, reptiles are perhaps remembered best for what they once were, rather than for what they presently are. The Age of Dinosaurs, which lasted 100 million years encompassing the Jurassic and Cretaceous periods of the Mesozoic era, saw the appearance of a great radiation of reptiles, many of huge stature and awesome appearance, that completely dominated life on land. Then they suddenly declined.

Out of the dozen or so principal groups of reptiles that evolved, four remain today. The most successful of these are the lizards and snakes of the order Squamata. A second group is the crocodilians; having survived for 200 million years, they may finally be made extinct by humans. To a third group belong the turtles of the order Testudines, an ancient group that has somehow survived and remained mostly unchanged from its early reptile ancestors. The last group is a relic stock represented today by a sole survivor, the tuatara of New Zealand.

Reptiles are easily distinguished from amphibians by several adaptations that permit them to live in arid regions and in the sea—habitats barred to amphibians by their reproductive requirements. Reptiles have a dry, scaly skin, almost free of glands, that resists desiccation. Most important, reptiles lay their eggs on land; amphibians must lay their eggs in fresh water or in moist places. This seemingly simple difference was, in fact, a remarkable evolutionary achievement that was to have a profound impact on subsequent vertebrate evolution. To totally abandon an aquatic life, reptiles evolved a sophisticated internally fertilized egg containing a complete set of life-support systems. This **shelled egg** (known also as an "amniotic" egg because of the membranous amnion that encloses the embryo) could be laid on dry land. Within, the embryo floats and develops in an aquatic environment. It is provided with a yolk sac containing its food supply; another membrane, the allantois, serves as a surface for gas exchange through the calcareous or parchmentlike shell; provision is made in it for storing toxic wastes that accumulate during development. The early reptiles that developed this egg must certainly have enjoyed an immediate advantage over the amphibians. They could now hide their eggs in a protected situation away from water—and away from the numerous creatures that fed freely on the eggs provided by amphibians each spring. With the evolution of this ultimate adaptation, conquest of land by the vertebrates was now possible.

ORIGIN AND ADAPTIVE RADIATION OF REPTILES

Biologists generally agree that reptiles arose from labyrinthodont amphibians sometime before the Permian period, which began about 280 million years ago. The oldest stem reptiles belonged to the order Cotylosaura (Fig. 25-1). Forming a transition between the labyrinthodont amphibians and stem reptiles is a lizardlike, partly aquatic animal, approximately 0.5 m long, called *Seymouria*. It was found in 220 million-year-old Permian strata in Texas. It is considered a "stem" reptile since it shows characteristics of both groups and thus appears to have been a transition form. Despite its reptilian skeletal features, *Seymouria* was probably more amphibian than reptile, since there is evidence that it possessed a lateral-line system, typical of amphibians that have an aquatic larval stage. In other words, it is unlikely that *Seymouria* laid amniotic eggs as did the true terrestrial reptiles to follow. Nevertheless, *Seymouria* represents a nearly perfect transition form. From *Seymouria,* or a closely related form, arose the stem reptiles (cotylosaurs), showing more definite reptilian characteristics.

The adaptive radiation of reptiles, especially pronounced in the Triassic period (which followed the Permian period), corresponded with the appearance of new ecologic niches. These were provided by the climatic and geologic changes that were taking place at that time, such as a variable climate from hot to cold, mountain building and terrain transformations, and a varied assortment of plant life.

The Mesozoic era was the age of the great ruling reptiles. Then suddenly they disappeared near the close of the Cretaceous period some 65 to 80 million years ago. What caused their demise? Many changes were occurring during the Cretaceous: modern flowering plants were spreading rapidly, as were the aggressive and intelligent mammals. In general, the modern fauna and flora as we know them today were becoming well-established, and the dinosaurs were not sufficiently adaptable to survive. Their extinction probably resulted from the combined effect of climatic

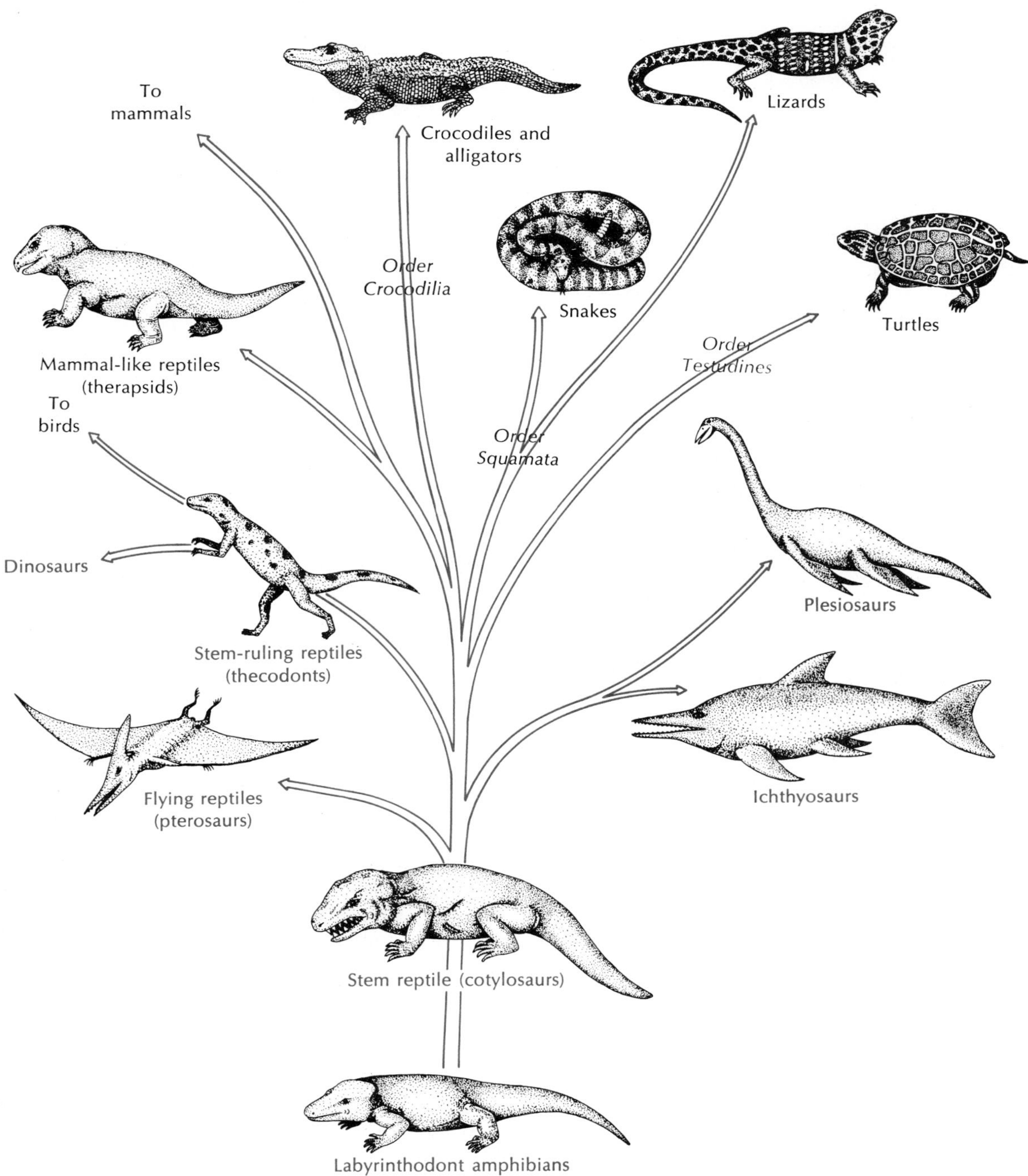

FIG. 25-1 Family tree of the reptiles. Transition from certain labyrinthodont amphibians to reptiles occurred in Carboniferous period to Mesozoic times. This transition was effected by development of the amniote egg, which made land existence possible, although this egg may well have developed before the earliest reptiles had ventured far on land. Explosive adaptation by reptiles may partly have resulted from a variety of ecologic niches into which they could move. Fossil record shows that lines arising from stem reptiles led to ichthyosaurs, plesiosaurs, and stem-ruling reptiles. Some of these returned to the sea. Later radiations led to mammal-like reptiles, turtles, flying reptiles, birds, dinosaurs, and so on. Of this great assemblage, the only reptiles now in existence belong to four orders (Testudines, Crocodilia, Squamata, and Rhynchocephalia). The Rhynchocephalia (not shown in this diagram) is represented by only one living species, the tuatara *(Sphenodon)* of New Zealand. How are the mighty fallen!

and ecologic factors, excessive specialization, and low reproductive potential. But this is speculation, and debate continues among paleontologists. Why did some reptiles survive against the fierce competition of the mammals? Turtles had their protective shells, snakes and lizards evolved in habitats of dense forests and rocks where they could meet the competition of any tetrapod; and crocodiles, because of their size, stealth, and aggressiveness, had few enemies in their aquatic habitats.

Characteristics

1. Body variable in shape, compact in some, elongated in others; **body covered with an exoskeleton of horny epidermal scales** with the addition sometimes of bony dermal plates; **integument with few glands**
2. **Limbs paired, usually with five toes,** and adapted for climbing, running, or paddling; absent in snakes and some lizards
3. Skeleton well-ossified; ribs with sternum forming a complete thoracic basket; **skull with one occipital condyle**
4. Respiration by lungs; **no gills;** cloaca used for respiration by some; branchial arches in embryonic life
5. **Three-chambered heart; crocodiles with four-chambered heart;** usually one pair of aortic arches
6. Ectothermic; some lizards and snakes behaviorally thermoregulate
7. **Kidney a metanephros (paired); uric acid main nitrogenous waste**
8. Nervous system with the optic lobes on the dorsal side of brain; **12 pairs of cranial nerves** in addition to nervus terminalis
9. Sexes separate; **fertilization internal**
10. **Amniotic eggs covered with calcareous or leathery shells;** extraembryonic membranes including the **amnion, chorion, yolk sac,** and **allantois** present during embryonic life

Classification of living reptiles

Order Testudines (tes-tu'din-eez) (L., *testudo,* tortoise) **(Chelonia)**—turtles (250 species). Body in a bony case of dermal plates with dorsal carapace and ventral plastron; jaws without teeth but with horny sheaths; quadrate immovable; vertebrae and ribs fused to shell; anus a longitudinal slit.

Suborder Cryptodira (krip'to-dy'ra) (Gr., *kryptos,* hidden, + *deirē,* neck). Hidden-necked turtles; head withdrawn into shell by flexing neck ventrally. Ten families: Chelydridae (snapping turtles), Platysternidae (big-headed turtle), Emydidae (common terrapins), Kinosternidae (musk and mud turtles), Testudinidae (tortoises), Cheloniidae (sea turtles), Trionychidae (soft-shelled turtles), Dermochelyidae (leatherback turtle), Dermatemydidae (river turtles), Carettochelyoidea (pitted-shelled turtles).

Suborder Pleurodira (ploor'o-dy'ra) (Gr., *pleura,* side, + *deirē,* neck). Side-necked turtles; head withdrawn into shell by bending neck sideways. Two families: Chelidae (snake-necked turtles), Pelomedusidae (pelomedusid turtles).

Order Squamata (squa-ma'ta) (L., *squamatus,* scaly, + *-ata,* characterized by)—snakes (2,700 species), lizards (3,000 species), amphisbaenids (130 species). Skin of horny epidermal scales or plates, which is shed; teeth attached to jaws; quadrate freely movable; vertebrae usually concave in front; anus a transverse slit.

Suborder Sauria (sawr'e-a) (Gr., *sauros,* lizard) **(Lacertilia)—lizards.** Body slender, usually with four limbs; rami of lower jaw fused; eyelids movable; copulatory organs paired. Twenty families; the major ones are Gekkonidae (geckos), Iguanidae (New World lizards), Agamidae (Old World lizards), Chamaeliontidae (chameleons), Lacertidae (Old World lizards), Scincidae (skinks), Helodermatidae (poisonous lizards), Anguidae (plated lizards).

Suborder Serpentes (sur-pen'tes) (L., *serpere,* to creep) **(Ophidia)—snakes.** Body elongated; limbs and ear openings absent; mandibles jointed anteriorly by ligaments; eyes lidless and immovable; tongue bifid and protrusible; teeth conical and on jaws and roof of mouth. Eleven families; the major ones are Leptotyphlopidae (blind snakes), Boidae (boas and pythons), Colubridae (common snakes), Elapidae (poisonous fixed-fang snakes), Viperidae (vipers and pit vipers).

Suborder Amphisbaenia (am'fis-bee'nee-a) (L., *amphis,* on both sides, + *baina,* to walk)—**worm lizards.** Body elongate and of nearly uniform diameter; short tail; no legs (except one genus with short front legs); limb girdles vestigial; eyes hidden beneath skin; only one lung. Two families: Trogonophidae (ovoviviparous worm lizards), Amphisbaenidae (oviparous worm lizards).

Order Crocodilia (croc'o-dil'e-a) (L., *crocodilus,* crocodile, + *-ia,* pl. suffix) **(Loricata).** Four-chambered heart; vertebrae usually concave in front; forelimbs usually with five digits, hind limbs with four digits; quadrate immovable; anus a longitudinal slit. Two families: Crocodylidae (crocodiles, alligators, caimans), Gavialidae (gavial) (25 species).

Order Rhynchocephalia (rin'ko-se-fay'le-a) (Gr., *rhynchos,* snout, + *kephalē,* head). Vertebrae biconcave; quadrate immovable; parietal eye fairly well developed and easily seen; anus a transverse slit. *Sphenodon* only species existing.

How reptiles show advancements over amphibians

Reptiles have tough, dry, scaly skin offering protection against desiccation and physical injury. The skin consists of a thin **epidermis,** shed periodically, and a much thicker, well-developed **dermis.** The dermis is provided with **chromatophores,** the color-bearing cells that give many lizards and snakes their colorful hues. It is also the layer that, unfortunately for their bearers, is converted into alligator and snakeskin leather, so esteemed for expensive pocketbooks and shoes. The characteristic **scales** of reptiles are mostly

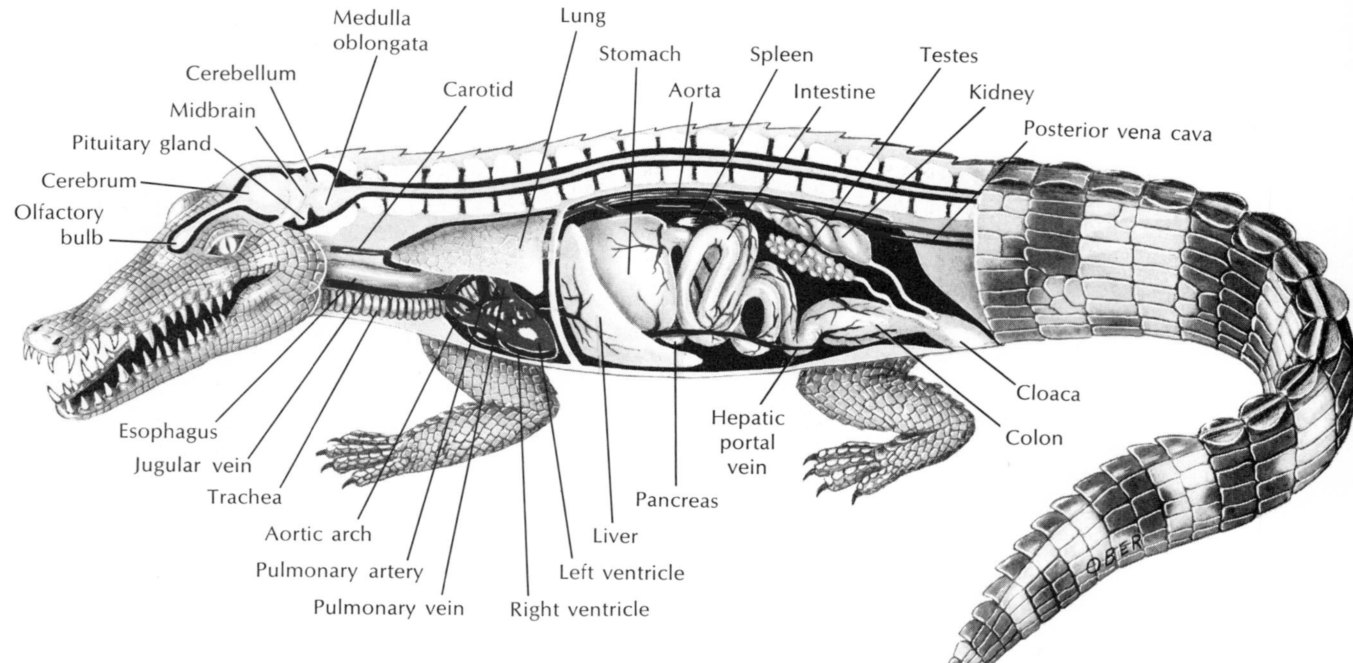

FIG. 25-2 Internal structures of male crocodile.

derived from the epidermis and thus are not homologous to fish scales, which are bony, dermal structures. In some reptiles, such as alligators, the scales remain throughout life, growing gradually to replace wear. In others, such as snakes and lizards, new scales grow beneath the old, which are then shed at intervals. In snakes the old skin (epidermis and scales) is turned inside out when discarded; lizards split out of the old skin leaving it mostly intact and right side out, or it may slough off in pieces.

The shelled egg of reptiles contains food and protective membranes for supporting embryonic development on dry land. The great significance of this adaptation was described earlier in this chapter. The shelled egg is illustrated and described in some detail on p. 826. Amphibian eggs have a gelatinous covering and must be protected from drying. The appearance of the shelled egg marked a great division between amphibians and reptiles and, probably more than any other adaptation, contributed to the decline of amphibians and the ascendance of reptiles.

Reptiles have some form of copulatory organ, permitting internal fertilization. Internal fertilization is obviously a requirement for a shelled egg, since the sperm must reach the egg before the egg is enclosed. Sperm from the paired testes are carried by the vasa deferentia to the copulatory organ, which is an evagination of the cloacal wall. The female system consists of paired ovaries and oviducts. The glandular walls of the oviducts secrete albumin and shells for the large eggs.

Reptiles have a more efficient circulatory system and higher blood pressures than amphibians. In all reptiles the right atrium, which receives unoxygenated blood from the body, is completely partitioned from the left atrium, which receives oxygenated blood from the lungs. In the crocodilians there are two completely separated ventricles as well (Fig. 25-2); in other reptiles the ventricle is incompletely separated. Even in those reptiles with incomplete separation of the ventricles, flow patterns within the heart prevent admixture of pulmonary (oxygenated) and systemic (unoxygenated) blood; all reptiles therefore have two functionally separate circulations.

Reptile lungs are better developed than those of amphibians. Reptiles depend almost exclusively on the lungs for gas exchange, supplemented by pharyngeal membrane respiration in some of the aquatic turtles. The lungs have a larger respiratory surface in reptiles than amphibians, and air is *sucked* into the lungs, as in higher vertebrates, rather than *forced* in by mouth muscles, as in the amphibians. Cutaneous respiration, so important to most amphibians, has been completely abandoned by the reptiles.

The reptilian kidneys are of the advanced metanephros type with their own passageways (ureters) to the exterior. The kidneys are very efficient in producing small volumes of urine, thus conserving precious water. Nitrogenous wastes are excreted as uric acid, rather than urea or ammonia. Uric acid has a low solubility and precipitates out of solution readily; as a result, the urine of many reptiles is a semisolid paste.

All reptiles, except the limbless members, have better body support than the amphibians and more efficiently designed limbs for travel on land. Many of the dinosaurs walked on powerful hindlimbs alone.

The reptilian nervous system is considerably more advanced than the amphibian. Although the reptile's brain is small, the **cerebrum** is increased in size relative to the rest of the brain. The crocodilians have the first true cerebral cortex (neopallium). Central nervous system connections are more advanced, permitting complex kinds of behavior unknown in the amphibians.

Distinctive structures of the reptilian body

In contrast to the soft, naked body of amphibians, reptiles have developed an **exoskeleton** that is largely waterproof. This exoskeleton consists of dead horny scales formed of keratin, a protein substance found in the epidermal layers of the skin. Reptilian scales, unlike fish scales, are derived from the dermis and represent a kind of bone. They may be flat and fitted together like a mosaic or widely separated; in some cases they overlap like roof shingles. Beneath each scale is a dermal vascular papilla, which supplies nutrients to the scale. In some the outer layer of the scales is continually shed in small bits or else is sloughed off periodically in one piece. Turtles, however, add new layers of keratin under the old layers of the platelike scutes, which are modified scales.

The **skeleton** of the early reptiles (cotylosaurs) was similar to that of the labyrinthodont amphibians. Most changes in later reptiles involved a loss of skull elements (by fusion or otherwise) and adaptations for better locomotion. The nostrils of an alligator or crocodile are on the dorsal side of the head, and the internal nares are at the back of the throat, which can be closed off by a fold. Thus the reptile can breathe while holding its prey submerged. The reptilian skull has developed a more flexible joint with the vertebral column, and more efficient girdles were evolved for supporting the body. The toes are provided with claws. Their peglike teeth vary somewhat but do not show the differentiation of the mammal; they may be set in sockets (thecodont) or fused to the surface of the bone (acrodont). In many the teeth are formed in two rows along the edges of the premaxillae in the upper jaw and along the dentaries of the lower jaw. Teeth may also be present on some of the palate bones. They are absent in turtles, in which only a horny beak is present.

The **body cavity** of reptiles is mostly divided into sacs by mesenteries, ligaments, and peritoneal folds. The heart is always enclosed in a pericardial sac (Fig. 25-2). Among the turtles the lungs lie outside the peritoneal cavity. Lizards have a posthepatic septum that divides the peritoneal cavity into two divisions, and crocodiles have a similar one that contains muscle and may function in respiration. This partition, however, is not homologous to the mammalian diaphragm.

Most reptiles are carnivorous and their **digestive system** is adapted for such a diet. All are provided with a tongue. This is large, fleshy, and broad in crocodiles and turtles. In crocodiles a tongue fold with a similar fold of the palate can separate the air passage from the food passage. In some reptiles (chameleon) the tongue is highly protrusible and is used in catching prey. When not in use, the anterior part of such a tongue is telescoped into the posterior portion.

Buccal glands vary a great deal. In snakes the upper labial glands may be modified into poison glands. The stomach is usually spindle-shaped and contains gastric glands. Pebbles or gastroliths are found in some reptiles (crocodiles) for grinding the food. A short duodenum receives the ducts from the liver and pancreas. The walls of the midgut are thrown into folds, but there are few glands. The **cloaca** (Fig. 25-2), which receives the rectum, ureters, and reproductive ducts, may be complicated. A urinary bladder is found in *Sphenodon*, turtles, and most lizards and opens into the ventral wall of the cloaca.

The **excretory system** is made up of the paired elongated or compact metanephric kidneys composed of glomerular nephrons. Ureters carry the urine (fluid in turtles and crocodiles; semisolid with insoluble urates in the others) to the cloaca.

All reptiles have **lungs** for breathing. The glottis

behind the tongue is closed by special muscles. The larynx contains arytenoid and cricoid cartilages but no thyroid cartilages. The trachea is provided with semicircular cartilage rings to prevent collapse. Lungs are mainly simple sacs in *Sphenodon* and snakes (in which one is reduced), but in turtles and crocodiles they are divided into irregular chambers, with the alveoli connected to a branched series of bronchial tubes. In chameleons long, hollow processes of the lungs pass posteriorly among the viscera and represent forerunners of the air sacs of birds. Air is drawn into the lungs by the movements of the ribs and, in crocodiles, by the muscular diaphragm.

Since there is no branchial (gill) circulation in reptiles, on each side the fifth aortic arch is lost, the third becomes the carotid arch, the fourth the systemic arch, and the sixth the pulmonary arteries. The systemic arches are paired, in contrast to the single one of birds and mammals. In crocodiles there are two completely separated ventricles; in the other groups the ventricle is incompletely separated. Crocodiles are thus the first animals with a **four-chambered heart** (Fig. 25-2). The conus and truncus arteriosus are absent. Venous blood is returned to the sinus venosus of the heart through the paired precaval and the single postcaval veins. Renal and hepatic portal systems are present. The blood contains oval, nucleated corpuscles that are smaller than those of amphibians.

The male **reproductive system** consists of paired elongated testes that are connected to the vasa deferentia (the wolffian or mesonephric ducts) (Fig. 25-2). The latter carry sperm to the copulatory organ, which is an evagination of the cloacal wall and is used for internal fertilization. These copulatory organs are single in crocodiles and turtles but are paired in lizards and snakes, in which they are called **hemipenes.** The female system is made up of large paired ovaries, and the eggs are carried to the cloaca by oviducts, which are provided with funnel-shaped ostia. The glandular walls of the oviducts secrete albumin and shells for the large amniotic eggs. The embryonic membranes (amnion, yolk sac, allantois, and chorion), first appear in reptiles and are used for the nourishment, protection, and respiration of the embryo.

The reptilian **nervous system,** though more advanced than the amphibian, is small—never exceeding 1% of the body weight. The cerebrum is enlarged, and there has been a transfer of some nervous integrative activities forward to the forebrain. There are 12 pairs of cranial nerves (amphibians have ten) in addition to the nervus terminalis. A better developed peripheral nervous system is associated with the more efficient movement of reptiles.

Sense organs vary among the different reptilian groups, but in general they are well-developed. All reptiles have a middle and inner ear, and crocodiles have an outer one as well. The middle ear contains the ear ossicle (stapes) and communicates with the pharynx by the eustachian tube. The sense of hearing, however, is poorly developed in most. The lateral-line system is entirely lost. Jacobson's organ, a unique sense organ, is a separate part of the nasal sac and communicates with the mouth; it is especially well-developed in snakes and lizards. It is innervated by a branch of the olfactory nerve and is used in smelling the food in the mouth cavity. In snakes, environmental odors trapped on the flicking tongue are carried into the mouth and to Jacobson's organ. This organ in some form is found in other groups, including the amphibians.

CHARACTERISTICS AND NATURAL HISTORY OF REPTILIAN ORDERS

Turtles—order Testudines (Chelonia)

The turtles are an ancient group that have plodded on from the Triassic period to the present with very little change in their early basic morphology. They are enclosed in shells consisting of a dorsal **carapace** and a ventral **plastron.** Clumsy and unlikely as they appear to be within their protective shells, they are nonetheless a varied and successful group that seems able to accommodate to human presence. The shell is so much a part of the animal that it is built in with the thoracic vertebrae and ribs. The head and appendages can be retracted into this shell for protection. No sternum is found in these forms, and their jaws lack teeth but are covered by a sharp, horny cutting surface. The nasal opening is single. They have lungs, although some aquatic forms gain enough oxygen while submerged by pumping water in and out of the pharynx. All turtles are more tolerant of anoxia than any other reptilian order.

On their toes are horny claws for digging in the sand, where they lay their eggs. Some of the marine forms have paddle-shaped limbs for swimming. Fertilization is internal by means of a cloacal penis on

FIG. 25-3 Two giant tortoises *Geochelone elephantopus* of the Galápagos Islands plod across the crater floor of an extinct volcano. These are part of several remnant populations, all that remain of an estimated half-million tortoises that inhabitated the islands before they were destroyed by nineteenth-century mariners for food. (Photograph by C. P. Hickman, Jr.)

the ventral wall of the male cloaca. All turtles are oviparous.

The terms "turtle," "tortoise," and "terrapin" are applied variously to different members of this order; they are all correctly called turtles. The term "tortoise" is frequently given to land turtles, especially the large forms.

The great marine turtles, buoyed by their aquatic environment, may reach 2 m in length and 725 kg in weight. One is the leatherback. The green turtle, so named because of its greenish body fat, may exceed 360 kg, although most individuals of this economically valuable and heavily exploited species seldom live long enough to reach anything approaching this size. Some land tortoises may weigh several hundred kilograms, such as the giant tortoises of the Galápagos Islands that so intrigued Darwin during his visit there in 1835. Most tortoises are rather slow moving; an hour of determined trudging carries a large Galápagos tortoise approximately 300 m (Fig. 25-3). Their low metabolism probably explains their longevity, for some are believed to live more than 150 years.

One of the most familiar turtles in eastern and central United States is the box tortoise *(Terrapene carolina)*. It is about 15 cm long. It has a high arched carapace, with the front and rear margins curled up. The plastron is hinged, forming two movable parts that can be pulled up against the upper shell so tightly the animal is totally enclosed. The color markings vary, but usually the shell is a dark brown color with irregular yellow spots. Box tortoises are found in woods and fields and sometimes in marshes. Despite their slow

FIG. 25-4 A female snapping turtle *Chelydra serpentina* at her nest. This individual was interrupted while laying and moved aside, revealing the eggs. The flask-shaped nest is dug with the hind legs, 20 to 30 eggs are laid, and the nest is then filled and concealed. The hatchlings emerge in the fall after a 55- to 125-day incubation, depending on environmental temperature. (Photograph by L. L. Rue, III.)

movements, marked individuals have often been found long distances from points of release. They are omnivorous but are especially fond of fruits and berries and grow fat during the strawberry season. They are long-lived and a few have been known to reach 100 years.

Snapping turtles *(Chelydra serpentina)* are found in nearly every pond or lake in eastern and central North America. They grow to be 30 to 35 cm in length and 9 to 18 kg in weight. Ferocious and short-tempered, they are often referred to as the "tigers of the pond." They are entirely carnivorous, living on fish, frogs, waterfowl, or almost anything that comes within reach of their powerful jaws. They are wholly aquatic and come ashore only to lay their eggs (Fig. 25-4).

Lizards and snakes—order Squamata

The lizards and snakes of this order are the most recent products of reptile evolution and by all odds the most successful, comprising approximately 95% of all known living reptiles. The modern lizards began their adaptive radiation during the Cretaceous period when the great dinosaurs were at the climax of their dominance of land. The immediate success of lizards was probably caused by a versatile and highly mobile jaw apparatus, which allowed them to capture large, struggling prey. The snakes appeared during the late Cretaceous period. They probably evolved from burrowing lizards, although the early fossil record of this group is poor. The flexible jaw apparatus of lizards was refined even further in snakes; in these the jaw and skull bones

FIG. 25-5 Marine iguanas *Amblyrhynchus cristatus* of the Galápagos Islands bask in morning sunshine. This is the only marine lizard in the world. It has special salt-removing glands in the eye orbits and long claws that enable it to cling to the bottom while feeding on seaweed, its exclusive diet. It may dive to depths exceeding 10 m (35 feet) and remain submerged more than 30 minutes. (Photograph by C. P. Hickman, Jr.)

are so loosely connected that they can swallow prey several times their own diameter.

The order Squamata is divided into suborders Saura (Lacertilia), the lizards; Serpentes (Ophidia), the snakes; and Amphisbaenidae, the worm lizards. In addition to the difference in jaw articulation and structure, lizards have movable eyelids, external ear openings, and legs (usually); snakes lack these characteristics. Structural features are even more reduced in the secretive worm lizards, a mostly limbless group having both eyes and ears completely hidden beneath the skin.

Lizards—suborder Saura

The lizards are an extremely diversified group, including terrestrial, burrowing, aquatic, arboreal, and aerial members. Among the more familiar groups in this varied suborder are the **geckos,** small, agile, mostly nocturnal forms with adhesive toe pads that enable them to walk upside down and on vertical surfaces; the **iguanas,** New World lizards, often brightly colored with ornamental crests, frills, and throat fans, a group that includes the remarkable marine iguana of the Galápagos Islands (Fig. 25-5); the **skinks,** alert and active lizards with elongate bodies and reduced limbs; and the **chameleons,** a group of arboreal lizards, mostly of Africa and Madagascar. These entertaining creatures catch insects with the sticky-tipped tongue that can be flicked accurately and rapidly to a distance greater than their own body length.

The radiation of lizards into so many different kinds of habitats has been accompanied by a great variety of specialized adaptations. *Draco,* a lizard inhabiting India, is able to volplane from tree to tree because of skin extensions on the side. A few lizards, such as the glass lizards, are limbless (Fig. 25-6).

Many lizards live in the world's hot and arid regions, aided by several adaptations for desert life. Since their skin lacks glands, water loss by this avenue is much reduced. They produce a semisolid urine with

FIG. 25-6 A glass lizard, *Ophisaupis* sp., of the southeastern United States. This legless lizard feels stiff and brittle to the touch and has an extremely long, fragile tail that readily fractures when the animal is struck or seized. Most specimens, such as this one, have only a partly regenerated tip to replace a much longer tail previously lost. Glass lizards can be readily distinguished from snakes by the deep, flexible groove running along each side of body (barely perceptible in photograph). They feed on worms, insects, spiders, birds' eggs, and small reptiles. (Photograph by L. L. Rue, III.)

a high content of crystalline uric acid. This is an excellent adaptation for conserving water and is found in other groups living successfully in arid habitats (birds, insects, and pulmonate snails). Some, such as the Gila monster of the southwestern United States deserts, store fat in their tails, which they draw on during drought to provide both energy and metabolic water (Fig. 25-7).

Especially interesting are the techniques desert lizards use to maintain a relatively constant body temperature, using what physiologists term "behavioral thermoregulation." Lizards, like other reptiles, are primarily ectothermic; if a lizard is placed under constant temperature conditions in the laboratory, its body temperature soon becomes indistinguishable from that of its surroundings. But in its natural environment where surrounding temperatures vary widely, lizards modulate their body temperature by exploiting hour-to-hour changes in thermal flux from the sun. In the early morning they emerge from their burrows and bask in the sun with their bodies flattened to absorb heat. As the day warms they turn to face the sun, to reduce the body area exposed, and raise their bodies from the hot substrate. In the hottest part of the day they may retreat to their burrows. Later they emerge to bask as the sun sinks lower and the air temperature drops.

These behavioral patterns help to maintain a relatively steady body temperature of 36° to 39° C while the air temperature is varying between 29° and 44° C. Some lizards can tolerate intense midday heat without shelter. The desert iguana of the southwestern United States prefers a body temperature of 42° C when active and can tolerate a rise to 47° C, a temperature that is lethal to all birds and mammals and most other lizards. The term "cold-blooded" clearly does not apply to these animals!

Snakes—suborder Serpentes

Snakes are entirely limbless and lack both the pectoral and pelvic girdles (the latter persistent as vestiges in pythons). The numerous vertebrae of snakes, shorter and wider than those of tetrapods, promote quick lateral undulations through grass and over rough terrain. The ribs increase rigidity of the vertebral column, providing more resistance to lateral stresses. The elevation of the neural spine gives the numerous muscles more leverage. Snakes also differ from lizards in having no movable eyelids (their eyes are permanently covered with a third eyelid). This, together with a lack of eyeball mobility, gives snakes their cold, unblinking stare that most people find so unnerving. Snakes have no external ear and are, in fact, totally deaf, although they are sensitive to low-frequency vibrations conducted through the ground.

Snakes need a unique set of special senses to hunt down their prey. In addition to being deaf, most snakes

FIG. 25-7 Gila monster *Heloderma suspectum* of the southwestern United States desert regions and the congeneric Mexican beaded lizard are the only venomous lizards known. These brightly colored, clumsy-looking lizards feed principally on birds' eggs, nestling birds and mammals, and insects. Unlike poisonous snakes, the gila monster secretes venom from glands in the lower jaw. The bite is painful to humans but is seldom fatal. (Courtesy American Museum of Natural History.)

have relatively poor vision, with the tree-living snakes of the tropical forest being a conspicuous exception. In fact, the latter possess excellent binocular vision that helps them track prey through the branches where scent trails would be impossible to follow. But most snakes live on the ground and rely on chemical senses to hunt food. In addition to the usual olfactory areas in the nose, which are not well-developed, there are **Jacobson's organs,** a pair of pitlike organs in the roof of the mouth. These are lined with an olfactory epithelium and are richly innervated. The forked tongue, flicking through the air, picks up scent particles and conveys them to the mouth; the tongue is then drawn past Jacobson's organ or the tips of the forked tongue are inserted directly into the organs. Information is then transmitted to the brain where scents are identified.

The 2,700 species of snakes are not the only limbless "tetrapods"; they are joined by some 160 caecilians, 130 amphisbaenians, and a few true lizards. All evolved from forms that lost their appendages when they took up a burrowing existence. Snakes, however, have radiated into terrestrial, aquatic, and arboreal niches and clearly are the most successful limbless reptiles.

Like lizards, snakes bear rows of scales that overlap like shingles on a roof. They are keeled in some snakes, smooth in others. The ventral surface of nearly all snakes is covered by **scutes,** wide, transverse scales with free-trailing edges.

Four patterns of movement are recognized in snakes. The most typical is **lateral undulation.** Movement follows an S-shaped path, with the snake propelling itself by exerting lateral force against surface irregularities. The snake seems to "flow," since the moving loops appear stationary with respect to the ground. Lateral undulatory movement is fast and efficient under most but not all circumstances. **Concertina movement** enables a snake to move in a narrow passage, as

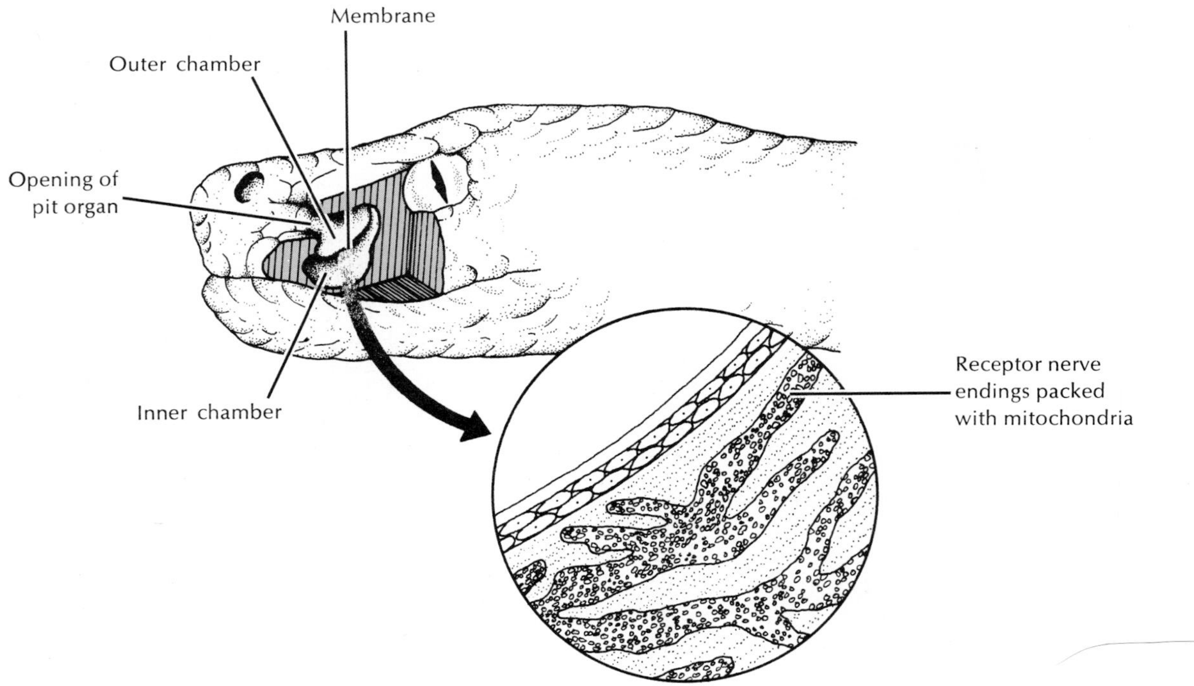

FIG. 25-8 Pit organ of rattlesnake, a pit viper. Cutaway shows location of deep membrane that divides pit into inner and outer chambers. Heat-sensitive nerve endings are concentrated in the membrane.

when climbing a tree by using the irregular channels in the bark. The snake extends forward while bracing S-shaped loops against the sides of the channel. To advance in a straight line as when stalking prey, many snakes use **rectilinear movement,** a form of locomotion that utilizes the large abdominal scutes. Two or three zones of scutes make frictional contact with the ground, while the rest of the body is moved forward. This effective but slow movement depends on very loose, flexible skin and special muscle and bone structure. **Sidewinding** is a fourth form of movement that enables desert vipers to move with surprising speed across loose, sandy surfaces with minimal surface contact. The sidewinder rattlesnake moves by throwing its body forward in loops with its body lying at an angle of about 60 degrees to its direction of travel.

Snakes of the subfamily Crotalinae within the family Viperidae are called **pit vipers** because of special heat-sensitive pits on their heads, between the nostrils and the eyes (Fig. 25-8). All of the best-known North American poisonous snakes are pit vipers, such as the several species of rattlesnakes, the water moccasin, and the copperhead (Fig. 25-9). The pits are supplied with a dense packing of free nerve endings from the fifth cranial nerve. They are exceedingly sensitive to radiant energy (long-waved infrared) and can distinguish temperature differences smaller than 0.2° C from a radiating surface. Pit vipers use the pits to track warm-blooded prey and to aim strikes, which they can make as effectively in total darkness as in daylight.

All pit vipers have a pair of teeth on the maxillary bones modified as fangs. These lie in a membrane sheath when the mouth is closed. When the viper strikes, a special muscle and bone lever system erects the fangs when the mouth opens (Fig. 25-10). The fangs are driven into the prey by the thrust, and venom is injected into the wound along a groove in the fangs.

A pit viper immediately releases its prey after the bite and follows it until it is paralyzed or dies. Then the snake swallows the prey whole. Approximately 1,500 people are bitten and 45 die from pit vipers each year in the United States.

The tropical and subtropical countries are the homes of most species of snakes, both of the venomous and

FIG. 25-9 The copperhead snake *Ancistrodon contortrix,* one of the best-known of American pit vipers, is distributed through the eastern and central United States. Its distinctive brown to copper coloration blends with the dead-leaf ground cover of hardwood forests, making their detection by humans difficult. It is a smaller snake than either the water moccasin or timber rattlesnake, and its bite, though painful, is seldom fatal. As with any snake bite, the danger is much greater to children than to adults. Copperheads feed on frogs, small birds and mammals, and large insects. They are ovoviviparous, bearing six to nine live young in each brood. (Photograph by C. P. Hickman, Jr.)

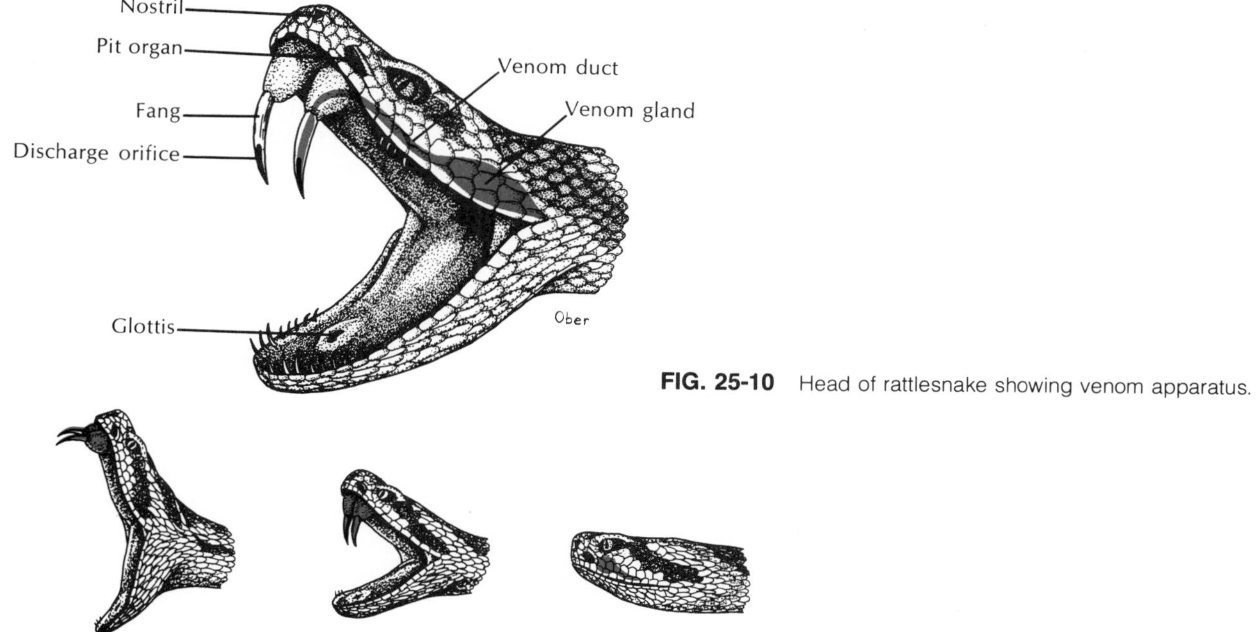

FIG. 25-10 Head of rattlesnake showing venom apparatus.

FIG. 25-11 Female king snake *Lampropeltis getulus,* guarding her eggs. King snakes are highly beneficial constrictors, feeding on rodents and on poisonous snakes. Family Colubridae. (Photograph by L. L. Rue, III.)

nonvenomous varieties. Even there, less than one-third of snakes are venomous; the nonvenomous snakes kill their prey by constriction or by biting and swallowing. Their diet tends to be restricted, many feeding principally on rodents, whereas others feed on fishes, frogs, and insects. Some African, Indian, and neotropical snakes have become specialized as egg eaters.

Poisonous snakes are usually divided into three groups based on the type of fangs. The vipers (family Viperidae) have tubular or grooved fangs at the front of the mouth; the group includes the American pit vipers previously mentioned and the Old World true vipers, which lack facial heat-sensing pits. Among the latter are the common European adder and the African puff adder. A second family of poisonous snakes (family Elapidae) has short, permanently erect fangs so that the venom must be injected by chewing. In this group are the cobras, mambas, coral snakes, and kraits. The highly poisonous sea snakes are placed in a third family (Hydrophiidae).

The saliva of all harmless snakes possesses limited toxic qualities, and it is logical that evolution should have stressed this toxic tendency. There are two types of snake venom. One type acts mainly on the nervous systems (neurotoxic), affecting the optic nerves (causing blindness) or the phrenic nerve of the diaphragm (causing paralysis of respiration). The other type is hemolytic; that is, it breaks down the red blood corpuscles and blood vessels and produces extensive extravasation of blood into the tissue spaces. Many venoms have both neurotoxic and hemolytic properties. The toxicity of a venom is determined by the minimal lethal dose on laboratory animals. By this standard the venoms of the Australian tiger snake and some of the sea snakes appear to be the most deadly of snake poisons. However, several larger snakes are more dangerous. The aggressive king cobra, which may exceed 5.5 m in length, is the largest and probably the most dangerous of all poisonous snakes. In India, where snakes come in constant contact with people, cobra bites cause more than 10,000 deaths each year.

Most snakes are **oviparous** species that lay their shelled, elliptic eggs soon after fertilization (Fig. 25-11). They are laid beneath rotten logs, under rocks, or in holes dug in the ground. Most of the remainder, including all the American pit vipers, except the tropi-

FIG. 25-12 Timber rattlesnake *Crotalus horridus horridus* with newly born young. All pit vipers in the United States are ovoviviparous, giving birth to well-formed young fully capable of capturing their own food. No snake species feeds its young. Timber rattlesnakes once were common in eastern hardwood forests; now they are restricted mostly to second-growth timbered terrain where their rodent food abounds. (Photograph by L. L. Rue, III.)

cal bushmaster, are **ovoviviparous,** giving birth to well-formed young (Fig. 25-12). A very few snakes are **viviparous;** a primitive placenta forms, permitting the exchange of materials between the embryonic and maternal bloodstreams.

Worm lizards—suborder Amphisbaenia

The somewhat inappropriate common name "worm lizards" describes a group of highly specialized, burrowing forms that are neither worms nor true lizards, but certainly are related to the latter. The name of the suborder literally means "walk on both sides" (or both ends), in reference to their peculiar ability to move backward nearly as effectively as forward. They have elongate, cylindric bodies of nearly uniform diameter, and most lack any trace of external limbs. The soft skin is divided into numerous rings, and these rings, combined with the absence of visible eyes and ears (both are hidden under the skin), make the amphisbaenians look like earthworms. The resemblance, though superficial, is the kind of structural convergence that not infrequently occurs when two totally unrelated groups come to occupy similar niches. The amphisbaenians have an extensive distribution in South America and tropical Africa. In the United States, one species, *Rhineura florida,* is found in Florida where it is known as the "graveyard snake."

Crocodiles and alligators, order Crocodilia

The modern crocodiles are the largest living reptiles. They are what remains of a once abundant group in the Jurassic and Cretaceous periods. Having managed to survive virtually unchanged for some 160 million years, the modern crocodilians face a forbidding, and perhaps short, future in a world dominated by humans.

Crocodiles have relatively long slender snouts; alligators have short and broader snouts. With their powerful jaws and sharp teeth, they are formidable antagonists. The members of the group that attack humans are found mainly in Africa and Asia. The estuarine crocodile *(Crocodylus porosus)* found in southern Asia grows to a great size and is very much feared. It is swift and aggressive and will eat any bird or mammal it can drag from the shore to water, where the prey is violently torn to pieces. These crocodiles are known to attack animals as large as cattle, deer, and people.

Alligators are usually less aggressive than crocodiles. They are almost unique among reptiles in being able to make definite sounds. The male alligator can give loud bellows in the mating season. Vocal sacs are found on each side of the throat and are inflated when he calls. In the United States, *Alligator mis-*

FIG. 25-13 American crocodilians. The American alligator *Alligator mississipiensis,* is found along the southeastern United States coast from central Texas to the Atlantic. The American crocodile *Crocodylus acutus,* now limited to the Everglades National Park in Florida, is in danger of extinction. Enlargement of heads shows easily recognized differences between the two genera. The crocodile has a more slender snout, and the fourth tooth from the front of the lower jaw fits *outside* the upper jaw and is visible when the mouth is closed. It is not true, as commonly believed, that there is a difference in the way the jaws are hinged in the two species. Order Crocodilia.

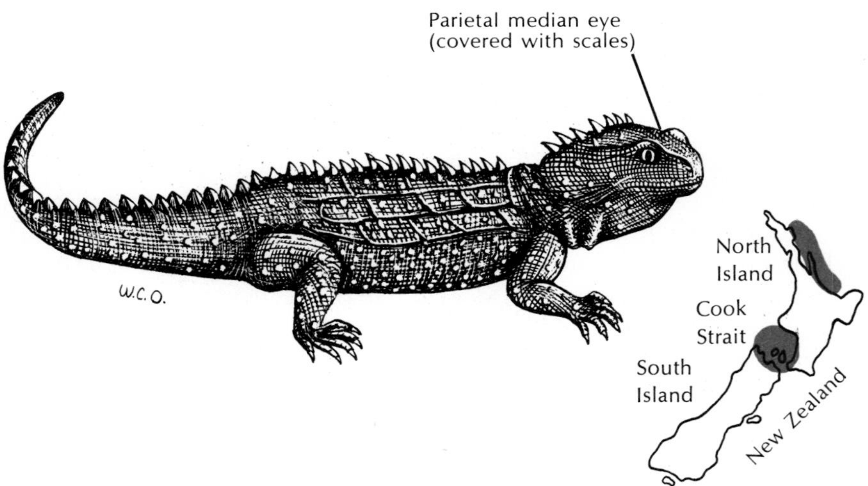

FIG. 25-14 Tuatara *Sphenodon punctatum*—the only living representative of order Rhynchocephalia. This "living fossil" reptile has a well-developed parietal "eye" with retina and lens on top of head. The eye is covered with scales and is considered nonfunctional but may have been an important sense organ in early reptiles. The tuatara is found only in New Zealand.

sissipiensis is the only species of alligator; *Crocodylus acutus,* restricted to extreme southern Florida, is the only species of crocodile (Fig. 25-13).

Alligators and crocodiles are oviparous. Usually from 20 to 50 eggs are laid in a mass of dead vegetation. The eggs are about 7 cm long. The penis of the male is an outgrowth of the ventral wall of the cloaca.

The tuatara—order Rhynchocephalia

Rhynchocephalia is represented by a single living species, the tuatara *(Sphenodon punctatum)* of New Zealand (Fig. 25-14). This animal is the sole survivor of a group of primitive reptiles that otherwise became extinct 100 million years ago. The tuatara was once widespread on the North Island of New Zealand but is now restricted to islets of Cook Strait and off the northern coast of North Island where, under protection of the New Zealand government, it may recover. It is a lizardlike form 0.6 m long or less that lives in burrows often shared with petrels. They are slow-growing animals with a long life; one is recorded to have lived 77 years.

The tuatara has captured the interest of biologists because of its numerous primitive features that are almost identical to those of Mesozoic fossils 200 million years old. These include a primitive skull structure of a type found in early Permian reptiles that were ancestors to the modern lizards. It also bears a well-developed parietal eye, complete with evidences of a retina (Fig. 25-14) and a complete palate. It lacks a copulatory organ. A specialized feature is the teeth, which are fused to the edge of the jaws, rather than being set in sockets. *Sphenodon* represents one of the slowest rates of evolution known among the vertebrates.

Annotated references

Selected general readings

Anderson, P. 1966. The reptiles of Missouri. Columbia, Mo., University of Missouri Press. *A good state survey of reptiles.*

Barbour, R. W. 1971. Amphibians and reptiles of Kentucky. Lexington, Ky., University of Kentucky Press. *Although this volume is primarily concerned with a state survey of amphibians and reptiles, it is in addition concerned with general distribution of species and contains a number of excellent photographs.*

Bellairs, A. 1970. The life of reptiles, vols. 1 and 2. New York, Universe Books. *An accurate, well-written treatise that is primarily concerned with anatomy and evolution. Perhaps a bit advanced for the beginning student.*

Blair, W. F., A. P. Blair, P. Brodkorb, F. R. Cagle, and G. A. Moore. 1968. Vertebrates of the United States, ed. 2. New York, McGraw-Hill Book Co. *An excellent source book that lists and discusses every vertebrate known to occur in the United States.*

Bucherl, W., E. E. Buckey, and V. Deulofeu (eds.). 1968. Venomous animals and their venoms, vols. 1 to 3. New York, Academic Press, Inc.

Carr, A. 1963. The reptiles. New York, Life Nature Library. *There is a wealth of reptile lore in this superbly illustrated volume.*

Cochran, D. M., and C. J. Goin. 1970. The new field book of reptiles and amphibians. New York, G. P. Putnam's Sons. *A useful field guide. Most species descriptions are concise, well written, and clear.*

Conant, R. 1975. A field guide to reptiles and amphibians of eastern and central North America, ed. 2. Boston, Houghton Mifflin Co. *An extremely useful pocket-sized field guide.*

Ernst, C. H., and R. W. Barbour. 1972. Turtles of the United States. Lexington, University of Kentucky Press. *Authoritative monograph on identification, distribution, and detailed life histories of all turtles of the United States.*

Gans, C. (ed.). 1969. Biology of the Reptilia, vols. 1 to 3. New York, Academic Press, Inc. *A work for the advanced student or researcher. The series consists of morphologic, embryologic and physiologic, and ecologic and behavioral portions. Specialists contribute chapters on topics that correspond to their research interests.*

Gans, C. 1974. Biomechanics: an approach to vertebrate biology. Philadelphia, J. B. Lippincott Co. *A selective treatment with sections on locomotion in snakes and amphisbaenians.*

Goin, C. J., and O. B. Goin. 1971. Introduction to herpetology, ed. 2. San Francisco, W. H. Freeman & Co. *A basic introductory text for the study of amphibians and reptiles.*

Klauber, L. M. 1956. Rattlesnakes, vols. 1 and 2. Berkeley, University of California Press. *An excellent monograph on this group of reptiles.*

Minton, S. A., Jr., and M. R. Minton. 1969. Venomous reptiles. New York, Charles Scribner's Sons. *A semipopular account about poisonous snakes and the role they have played in human culture.*

Schmidt, K. P., and R. F. Inger. 1957. Living reptiles of the world. Garden City, N.Y., Hanover House. *A magnificent volume with excellent illustrations of the various families of living reptiles.*

Smith, H. M. 1946. Handbook of lizards of the United States and Canada. Ithaca, N.Y., Comstock Publishing Co. *A standard reference for those interested in the natural history and taxonomy of lizards.*

Wright, A. H., and A. A. Wright. 1957. Handbook of snakes of the United States and Canada, vols. 1 and 2. Ithaca, N.Y., Comstock Publishing Co. *These two volumes are indispensable and are probably the best available on snakes of the United States and Canada.*

Selected *Scientific American* articles

Bakker, R. T. 1975. Dinosaur renaissance. **232:**58-78 (Apr.). *Interesting presentation of evidence that dinosaurs were "warm-blooded" (endothermic).*

Bogert, C. M. 1959. How reptiles regulate body temperature. **200:**105-120 (Apr.). *Many reptiles, especially lizards, achieve a remarkable degree of temperature regulation by behavioral responses.*

Carr, A. 1965. The navigation of the green turtle. **212:**78-86 (May). *These marine turtles make regular migrations over vast ocean routes.*

Gamow, R. T., and J. E. Harris. 1973. The infrared receptors of snakes. **228:**94-100 (May). *The anatomy and physiology of the remarkable "sixth sense" of snakes is described.*

Gans, C. 1970. How snakes move. **222:**82-96 (June).

Minton, S. A., Jr. 1957. Snakebite. **196:**114-122 (Jan.). *A survey of the world's most poisonous snakes and their venoms.*

Pooley, A. C., and C. Gans. 1976. The Nile crocodile. **234:** 114-124 (Apr.). *Social behavior of the largest living reptile.*

Riper, W. Van. 1953. How a rattlesnake strikes. **189:**100-102 (Oct.). *The rattlesnake strike is examined with high-speed photography.*

CHAPTER 26

THE BIRDS

Phylum Chordata
Class Aves

The eyes of this red-tailed hawk, with visual acuity eight times better than ours, gather more detailed information about the environment than all the other senses combined.

Photograph by C. P. Hickman, Jr.

FIG. 26-1 *Archaeopteryx,* the 150-million-year-old ancestor of modern birds. **A,** Cast of the second and most nearly perfect fossil of *Archaeopteryx,* which was discovered in 1877 in a Bavarian stone quarry. **B,** Reconstruction of *Archaeopteryx.* (**A,** Courtesy of the American Museum of Natural History; **B,** drawn by Sheila Ford.)

Of the higher vertebrates, the birds, class Aves (ay′veez) (L., pl. of *avis,* bird), are the most studied, the most observable, the most melodious, and, many think, the most beautiful. With 8,600 species distributed over nearly the entire earth, birds far outnumber all other vertebrates except the fishes. Birds are found in forests and deserts, in mountains and prairies, and on all the oceans. Four species have even visited the North Pole, and one, a skua, was seen at the South Pole. Some birds live in total blackness in caves, finding their way about by echolocation, and others dive to depths greater than 45 m to prey on aquatic life.

The single unique feature that distinguishes birds from all other animals is that they have feathers. If an animal has feathers, it is a bird; if it lacks feathers, it is not a bird. No other vertebrate group bears such an easily recognized and foolproof identification tag.

There is great uniformity of structure among birds. Despite some 130 million years of evolution during which they proliferated and adapted themselves to specialized ways of life, no one has difficulty recognizing a bird as a bird. In addition to feathers, all birds have forelimbs modified into wings (although they may not be used for flight); all have hindlimbs adapted for walking, swimming, or perching; all have horny beaks; and all lay eggs. Probably the reason for this great structural and functional uniformity is that birds evolved into flying machines. This fact greatly restricts diversity, so much more evident in other vertebrate classes. For example, birds do not begin to approach the diversity seen in their warm-blooded evolutionary peers, the mammals, a group that includes forms as unalike as a whale, a porcupine, a bat, and a giraffe.

Birds share with mammals the highest organ system development in the animal kingdom. But a bird's entire anatomy is designed around flight and its perfection. An airborne life for a large vertebrate is a highly demanding evolutionary challenge. A bird, must, of course, have wings for support and propulsion. Bones must be light and hollow yet serve as a rigid airframe. The respiratory system must be incredibly efficient to meet the intense metabolic demands of flight and serve also as a thermoregulatory device to maintain a constant body temperature. A bird must have a rapid and efficient digestion to process an energy-rich diet, it must have a high metabolic rate, and it must have a high-pressure circulatory system. Above all, birds must have a finely tuned nervous system and acute senses, especially superb vision, to handle the complex problems of headfirst, high-velocity flight over the landscape.

ORIGIN AND RELATIONSHIPS

About 150 million years ago, a flying animal drowned and settled to the bottom of a tropical freshwater lake in what is now Bavaria. It was rapidly covered with a fine silt and eventually fossilized. There it remained until discovered in 1861 by a workman splitting slate in a limestone quarry. The fossil was about the size of a crow, with a skull not unlike that of modern birds except that the beaklike jaws bore bony teeth set in sockets like those of reptiles (Fig. 26-1). The skeleton was decidedly reptilian with a long bony tail, clawed fingers, and abdominal ribs. It might have been classified as a reptile except that it carried the unmistakable imprint of **feathers**—those marvels of biologic engineering that only birds possess. The finding was dramatic because it proved beyond reasonable doubt that birds had evolved from reptiles.

Archaeopteryx (ar-kee-op′ter-ix, meaning "ancient wing"), as the fossil was named, was an especially fortunate discovery because the fossil record of birds is disappointingly meager. The bones of birds are lightweight and quickly disintegrate, so that only under the most favorable conditions will they fossilize. Nevertheless, there are certain localities where bird fossils are relatively abundant. One of these is the famous Rancho La Brea tarpits in Los Angeles where in one pit alone were found 30,000 fossil birds representing 81 species. By 1952 over 780 different fossil species had been recorded. Although most of these are relatively recent fossils, enough intermediate forms are known to provide a reasonable picture of bird evolution from the Jurassic period, when *Archaeopteryx* lived, to recent times (Fig. 26-2). Two well-known fossil birds in particular deserve mention. One was *Ichthyornis* (ik-thee-or′nis), a small ternlike sea bird that lived during the Cretaceous period along the shores of North America's inland sea about 100 million years ago, 50 million years after *Archaeopteryx*. The other was *Hesperornis*, a flightless, loonlike diving bird (Fig. 26-2). Both were essentially modern birds in almost every way. By the close of the Cretaceous period, about 63 million years ago, the characteristics of modern birds had been thoroughly molded. There remained only the emergence and proliferation of the modern orders of birds. Hundreds of thousands of bird species have appeared and nearly as many have disappeared, following *Archaeopteryx* to extinction. Only a minute fraction of these nameless species have been discovered as fossils.

Most paleontologists agree that the ancestors of both birds and dinosaurs were derived from a stem group of reptiles called thecodonts. Birds probably evolved from a single ancestor and thus have a monophyletic origin. However, existing birds are divided into two groups: (1) **ratite** (rat′ite) (L., *ratitus,* marked like a raft, from *ratis,* raft), the flightless ostrichlike birds that have a flat sternum with poorly developed pectoral muscles, and (2) **carinate** (L., *carina,* keel), the flying birds that have a keeled sternum on which the powerful flight muscles insert. This division originated from the view that the flightless birds (such as ostrich, emu, kiwi, rhea) represented a separate line of descent that never attained flight. This idea is now completely rejected. The flightless birds are descended from a flying ancestor but lost the use of their wings, which became unnecessary for their mode of life. Flightless forms are ground-living birds that can out-

FIG. 26-2 Family tree of birds showing probable lines of descent and relationship. Thirteen of the most familiar of the 27 recognized living orders of birds are pictured. Two extinct orders, represented by *Archaeopteryx* and *Hesperornis*, are also pictured.

run predators, or they live where few carnivorous enemies are found.

Ancient and modern birds are divided into two subclasses. (A complete classification of the orders of living birds is found on pp. 602 to 604.)

Subclass Archaeornithes (ar'ke-or'ni-theez) (Gr., *archaios,* ancient, + *ornis, ornithos,* bird). Fossil birds. This included *Archaeopteryx* and possibly one or two other genera.
Subclass Neornithes (ne-or'ni-theez) (Gr., *neos,* new, + *ornis,* bird). Modern birds are placed in this group. Some extinct species with teeth are also included here because of their likeness to modern forms.

It may seem paradoxic that birds, with their agile, warm-blooded, colorful, and melodious way of life, should have descended from lethargic, heterothermic, and silent reptiles. Yet the numerous anatomic affinities of the two groups is abundant evidence of close kinship and led the great English zoologist Thomas Henry Huxley to call birds merely "glorified reptiles." This unflattering description has never pleased bird lovers, who answer, "But how wondrously glorified!"

Characteristics

1. Body usually spindle-shaped, with four divisions: head, neck, trunk, and tail; **neck disproportionately long** for balancing and food-gathering
2. Limbs paired, with the **forelimbs usually adapted for flying;** posterior pair variously adapted for perching, walking, and swimming; foot with four toes (chiefly)
3. Epidermal **exoskeleton of feathers** and **leg scales;** thin integument of epidermis and dermis; no sweat glands; oil or preen gland at root of tail; **pinna of ear rudimentary**
4. **Skeleton fully ossified with air cavities or sacs;** skull bones fused with **one occipital condyle;** jaws covered with **horny beaks;** small ribs; vertebrae tend to fuse, especially the terminal ones; sternum well-developed with keel or reduced with no keel; **no teeth**
5. Nervous system well developed, with brain and 12 pairs of cranial nerves
6. Circulatory system of **four-chambered heart,** with the **right aortic arch persisting;** reduced renal portal system; nucleated red blood cells
7. **Endothermic**
8. Respiration by slightly expansible lungs, with thin **air sacs** among the visceral organs and skeleton; **syrinx (voice box) near junction of trachea and bronchi**
9. Excretory system by metanephric kidney; ureters open into cloaca; **no bladder;** semisolid urine; uric acid main nitrogenous waste
10. Sexes separate; testes paired, with the vas deferens opening into the cloaca; **females with left ovary and oviduct;** copulatory organ in ducks, geese, ratites, and a few others
11. Fertilization internal; **amniotic eggs with much yolk** and **hard calcareous shells;** embryonic membranes in egg during development; **incubation external;** young active at hatching **(precocial)** or helpless and naked **(altricial)**

STRUCTURAL AND FUNCTIONAL ADAPTATIONS OF BIRDS FOR FLIGHT

Just as an airplane must be designed and built to rigid aerodynamic specifications if it is to fly, so too must birds meet stringent structural requirements if they are to stay airborne. All the special adaptations found in flying birds come down to two things: more power and less weight. Flight by humans became possible when they developed an internal combustion engine and learned how to reduce the weight-to-power ratio to a critical point. Birds did this millions of years ago. But birds must do much more than fly. They must feed themselves and convert food into high-energy fuel; they must escape predators; they must be able to repair their own injuries; they must be able to air-condition themselves when overheated and heat themselves when too cool; and perhaps most important of all, they must reproduce themselves.

Feathers

Feathers, more than any other single feature, distinguish a bird. A feather is almost weightless, yet possesses incredible toughness and tensile strength. A typical feather consists of a hollow **quill,** or calamus, thrust into the skin, and a **shaft,** or rachis, which is a continuation of the quill and bears numerous **barbs** (Fig. 26-3). The barbs are arranged in closely parallel fashion and spread diagonally outward from both sides of the central shaft, to form a flat, expansive, webbed surface, the **vane.** There may be several hundred barbs in each web. If the feather is examined with a microscope, each barb will appear like a miniature replica of the feather, with numerous parallel filaments, called **barbules,** set in each side of the barb, and spreading laterally from it. There may be 600 barbules on each side of a barb, which adds up to something over a million barbules for the feather. The barbules of one barb overlap the barbules of a neighboring barb in a herringbone pattern and are held together with great tenacity by tiny hooks. Should two adjoining barbs

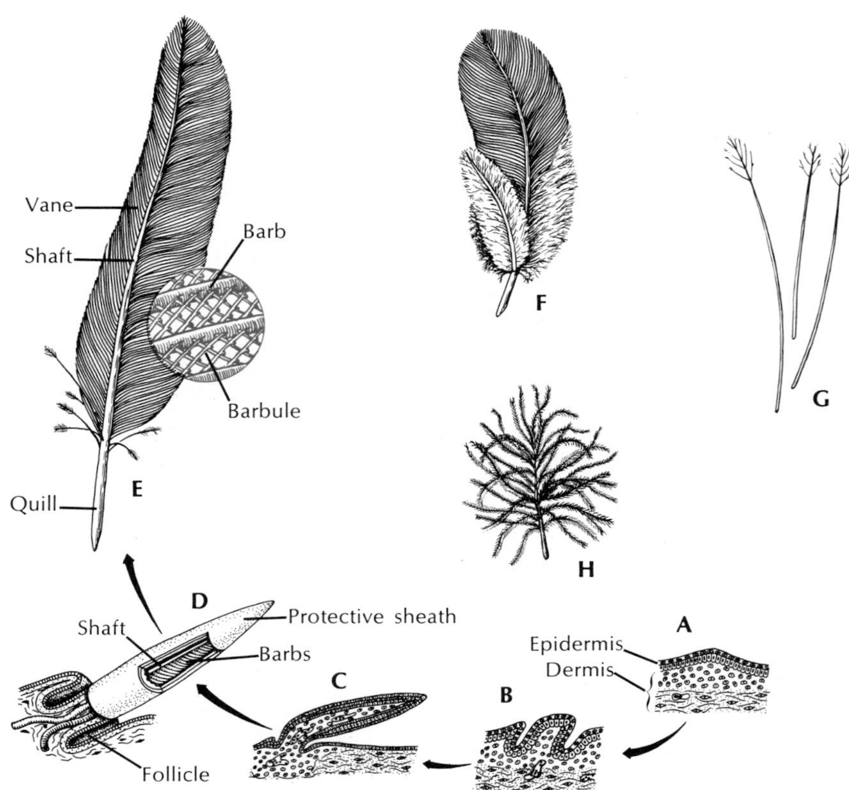

FIG. 26-3 Types of bird feathers and their development. **A** to **E,** Successive stages in the development of a vaned, or contour, feather. **E.** Growth occurs within a protective sheath that splits open when growth is complete, allowing mature feather to spread flat. **F** to **H,** Other feather varieties, including a pheasant vane feather with aftershaft, **F,** filoplumes, **G,** and down feather, **H.** (Principally after Welty, J. C., 1975. The life of birds, ed. 2. Philadelphia, W. B. Saunders Co.)

become separated—and considerable force is needed to pull the vane apart—they are instantly zipped together again by drawing the feather through one's fingertips. The bird, of course, does it with its bill, and much of a bird's time is occupied with preening to keep its feathers in perfect condition.

Types of feathers. There are different types of bird feathers for serving different functions. **Contour feathers** (Fig. 26-3, *E*) give the bird its outward form and are the type we have already described. Contour feathers that extend beyond the body and are used in flight are called **flight feathers. Down feathers** (Fig. 26-3, *H*) are soft tufts hidden beneath the contour feathers. They are soft because their barbules lack hooks. They are especially abundant on the breast and abdomen of water birds and on the young of game birds and function principally to conserve heat. **Filoplume feathers** (Fig. 26-3, *G*) are hairlike, degenerate feathers; each is a weak shaft with a tuft of short barbs at the tip. They are the ''hairs'' of a plucked fowl. They have no known function. The rictal bristles around the mouths of flycatchers and whippoorwills are probably modified filoplumes. A fourth type of highly modified feather, called **powder-down feathers,** are found on herons, bitterns, hawks, and parrots. Their tips disintegrate as they grow, releasing a talclike powder that helps to waterproof the feathers and give them metallic luster.

Origin and development. Feathers are epidermal structures that evolved from the reptilian scale; indeed, a developing feather closely resembles a reptile scale when growth is just beginning. One can imagine that

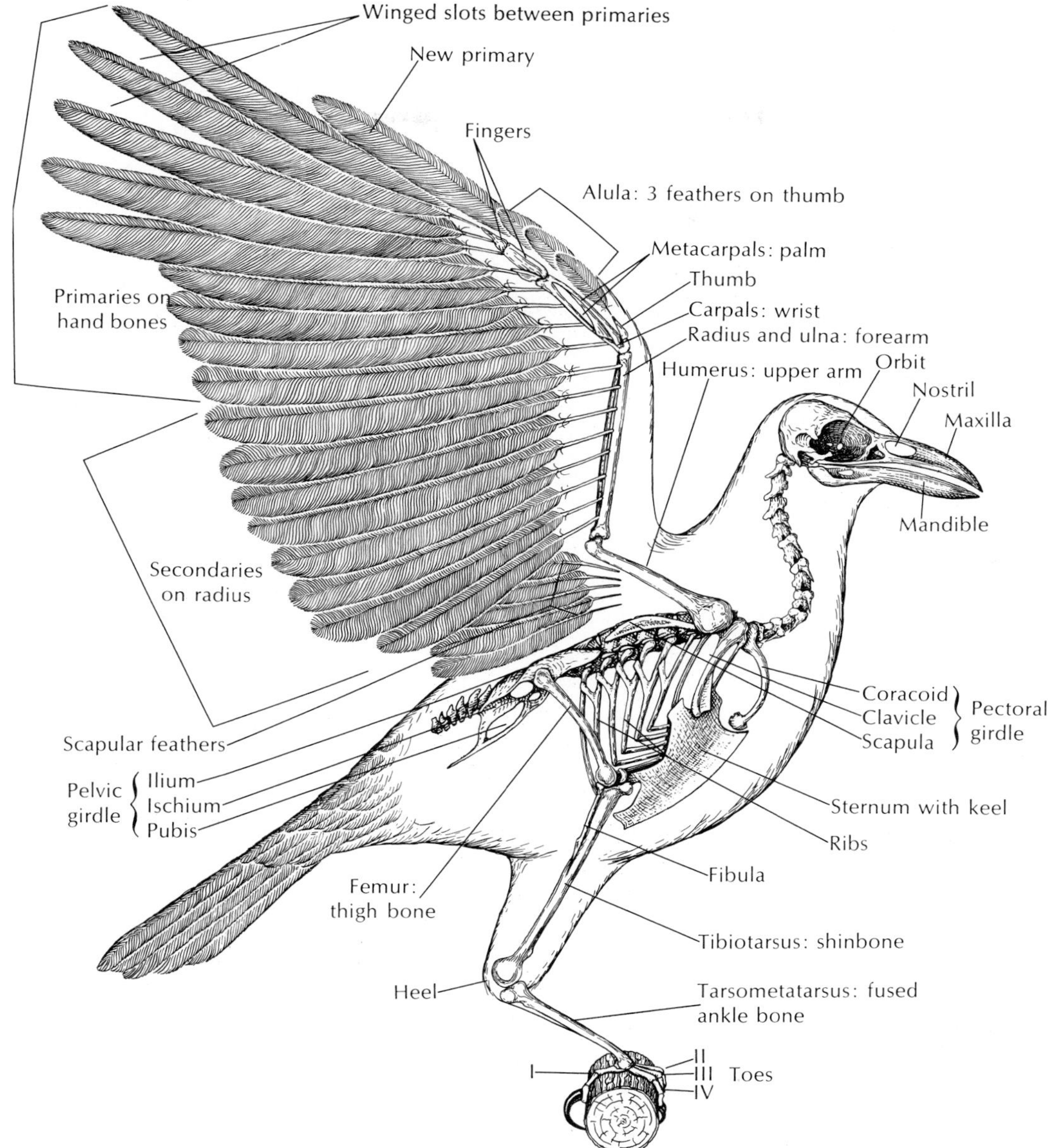

FIG. 26-4 Skeleton of a crow showing position of the flight feathers.

in its evolution, the scale elongated and its edges frayed outward until it became the complex feather of birds. Strangely enough, though modern birds possess both scales (especially on their feet) and feathers, no intermediate stage between the two has been discovered on either fossil or living forms.

Like a reptile's scale, a feather grows from a dermal papilla that pushes up against the overlying epidermis (Fig. 26-3, *A*). However, instead of flattening like a scale, the feather rolls into a cylinder or feather bud and is covered with epidermis. This feather bud sinks in slightly at its base and comes to lie in a feather follicle from which the feather will protrude. A layer of keratin is produced around the cylinder or bud and

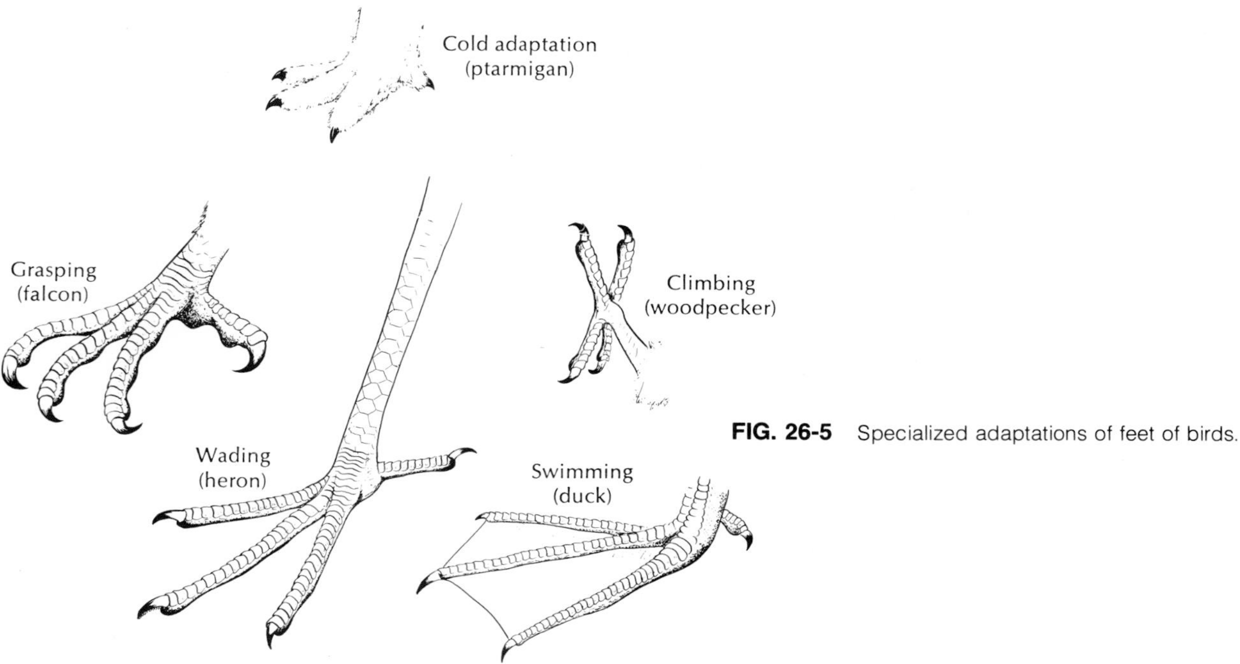

FIG. 26-5 Specialized adaptations of feet of birds.

encloses the pulp cavity of blood vessels. This surface layer of keratin splits away from the deeper layer to form a sheath. The deeper layer now becomes frayed distally to form parallel ridges, the median one of which grows large to form the shaft (contour feathers) and the others the barbs. Then the sheath bursts, and the barbs spread flat to form the vane (Fig. 26-3, *A* to *E*).

The pulp cavity of the quill dries up when growth is finished, leaving it hollow, with openings (umbilici) at its two ends. If the feather is to be a down feather, the sheath bursts and releases the barbs without the formation of a shaft or vane. Pigments (lipochromes and melanins) are added to the epidermal cells during growth in the follicle.

Molting. When fully grown, a feather, like mammalian hair, is a dead structure. The shedding, or molting, of feathers is a highly orderly process. Except in penguins, which molt all at once, feathers are discarded gradually to avoid the appearance of bare spots. Flight and tail feathers are lost in exact pairs, one from each side, so that balance is maintained (Fig. 26-20). Replacements emerge before the next pair is lost and most birds can continue to fly unimpaired during the molting period; only ducks and geese are completely grounded during the molt. Nearly all birds molt at least once a year, usually in late summer after the nesting season. Many birds also undergo a second partial or complete molt just before breeding season, to equip them with their breeding finery, so important for courtship display.

Movement and integration

Skeleton

One of the major adaptations that allows a bird to fly is its light skeleton (Fig. 26-4). Bones are phenomenally light, delicate, and laced with air cavities, yet they are strong. The skeleton of a frigate bird with a 2.1 m (7-foot) wingspan weighs only 114 grams (4 ounces), less than the weight of all its feathers. A pigeon skull weighs only 0.21% of its body weight; the skull of a rat by comparison weighs 1.25%. The bird skull is mostly fused into one piece. The braincase and orbits are large to accommodate a bulging brain and the large eyes needed for quick motor coordination and superior vision.

The anterior skull bones are elongated to form a beak. The lower mandible is a complex of several bones that hinge on two small movable bones, the quadrates. This provides a double-jointed action that permits the mouth to open widely. The upper mandible, consisting of premaxillae, maxillae, and other

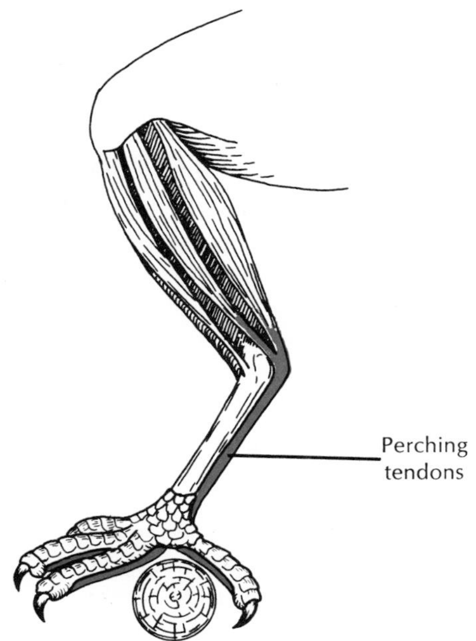

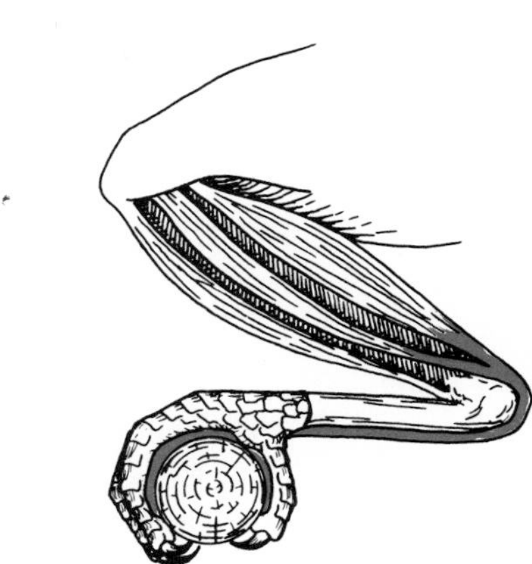

FIG. 26-6 Perching mechanism of a bird. When the bird settles on a branch, the tendons automatically tighten, closing the toes around the perch.

bones, is usually fused to the forehead, but in many birds—parrots, for instance—the upper jaw is hinged also. This adaptation allows greater flexibility of the beak in food manipulation and provides insect-catching species with a wider gap for successful feeding on the wing.

The bones of the pelvis (ilium, ischium, and pubis on each side) are fused with the lumbar and sacral vertebrae to form the **synsacrum,** which bears on each side a socket **(acetabulum)** for the head of the femur. Each leg is made up of the **femur,** the **tibiotarsus** (formed by the fusion of the tibia and proximal tarsals), the **tarsometatarsus** (formed by the fusion of the distal tarsals and metatarsals), and the four **toes** (typically three in front and one behind). A sesamoid bone (the patella) is found at the knee joint. There are two to five phalanges in each toe. Woodpeckers and others have two front and two hind toes. A few birds have only three toes, and the ostrich has two toes, unequal in size. Bird feet show a wide range of adaptations for walking, climbing, seizing, swimming, wading, and so on (Fig. 26-5).

The trunk is a rigid airframe, mainly because of the fusion of the vertebrae and fusion of the ribs with the vertebrae and sternum. Special processes called **uncinate processes** form an additional brace by passing posteriorly from one rib over the one behind. This rigidity affords a firm point of attachment for the wings. To assist in this support and rigidity, the pectoral girdle of scapula, coracoid, and furcula (the latter also called the clavicle, or wishbone) are more or less firmly united and joined to the sternum. Where scapula and coracoid unite is a hollow depression, the glenoid cavity, into which the chief wing bone, the **humerus,** fits as a ball-and-socket joint. The neck of eight to 24 cervical vertebrae is extraordinarily flexible. In all flying birds the sternum (breast bone) is highly modified into a thin, flat keel for the insertion of large wing muscles. The forelimbs, or wing appendages, are the most highly modified of the paired appendages. Each consists of a **humerus,** a **radius** and **ulna,** two **carpals,** and three **digits** (II, III, and IV). The other carpals are fused to the three metacarpals to form two long bones, the **carpometacarpus.** Of the digits, the middle one is longest and carries the large flight feathers. It consists of two phalanges. The second (alula) and fourth digits usually have only one phalanx (Fig. 26-4).

Muscular system

The muscles of birds have become highly adapted for flight. The locomotor muscles of the wings are relatively massive. The largest of these is the **pectoralis,** which depresses the wing in flight. Its antagonist is the **supracoracoideus** muscle, which raises the wing. Surprisingly perhaps, this latter muscle is not located on the backbone (anyone who has been served the back of the chicken knows it offers little meat) but is positioned under the pectoralis on the breast. It is attached by a tendon to the upper side of the humerus of the wing so that it pulls from below by an ingenious "rope and pulley" arrangement. Both of these muscles are anchored to the keel. With the main muscle mass thus placed low in the body, aerodynamic stability is improved.

Leg muscles are not so highly modified as are flight muscles, since birds use their legs to stand, walk, and run much as their reptilian ancestors did. The main muscle mass is located in the thigh, surrounding the femur, and a smaller mass lies over the tibiotarsus (shank, or "drumstick"). Strong, but thin, tendons extend downward through sleevelike sheaths to the toes. Consequently the feet are nearly devoid of muscles, thus explaining the thin, delicate appearance of the bird leg. This arrangement places the main muscle mass near the bird's center of gravity, and at the same time confers great agility to the slender, lightweight feet. And since the feet are made up mostly of bone, tendon, and tough, scaly skin, they are highly resistant to damage from freezing. When a bird perches on a branch, an ingenious toe-locking mechanism (Fig. 26-6) is activated, which prevents a bird from falling off its perch when asleep. The same mechanism causes the talons of a hawk or owl to automatically sink deeply into its victim, as the legs bend under the impact of the strike. The powerful grip of a bird of prey was described by L. Brown (1970): "When an eagle grips in earnest, one's hand becomes numb, and it is quite impossible to tear it free, or to loosen the grip of the eagle's toes with the other hand. One just has to wait till the bird relents, and while waiting one has ample time to realize that an animal such as a rabbit would be quickly paralyzed, unable to draw breath, and perhaps pierced through and through by the talons in such a clutch."

Birds have lost the long reptilian tail, still fully evident in *Archaeopteryx,* and have substituted a pincushion-like muscle mound into which the tail feathers are rooted. It contains a bewildering array of tiny muscles, as many as 1,000 in some species, which control the crucial tail feathers. But the most complex muscular system of all is found in the neck of birds; the thin and stringy muscles, elaborately interwoven and subdivided, provide the bird's neck with the ultimate in vertebrate flexibility.

Nervous system and special senses

A bird's nervous and sensory systems accurately reflect the complex problems of flight and a highly visible existence in which it must gather food, mate, defend territory, incubate and rear young, and at the same time correctly distinguish friend from foe. The brain of a bird has well-developed **cerebral hemispheres, cerebellum,** and **midbrain tectum** (optic lobes). The **cerebral cortex**—the portion in mammals that becomes the chief coordinating center—is thin, unfissured, and poorly developed in birds. But the core of the cerebrum, the **corpus striatum,** has enlarged into the principal integrative center of the brain, where are controlled such activities as eating, singing, flying, and all the complex instinctive reproductive activities. Relatively intelligent birds, such as crows and parrots, have larger cerebral hemispheres than do less intelligent birds, such as chickens and pigeons. The **cerebellum** is a crucial coordinating center where muscle-position sense, equilibrium sense, and visual cues are all assembled and used to coordinate movement and balance. The **optic lobes,** laterally bulging structures of the midbrain, form a visual-association apparatus comparable to the visual cortex of mammals.

Except in flightless birds and in ducks, the senses of smell and taste are poorly developed in birds. This lack, however, is more than compensated by good hearing and superb vision, the keenest in the animal kingdom. As in mammals, the bird ear consists of three regions: (1) the **external ear,** a sound-conducting canal extending to the **ear drum,** (2) a **middle ear,** containing a rodlike **columella** that transmits vibrations to the (3) **inner ear,** where the organ of hearing, the **cochlea,** is located. The bird cochlea is much shorter than the coiled mammalian cochlea, yet birds can hear roughly the same range of sound frequencies as humans. Actually the bird ear far surpasses that of humans in capacity to distinguish differences in intensities and to respond to rapid fluctuations in pitch.

The bird eye resembles those of other vertebrates in gross structure but is relatively larger, less spherical, and almost immobile; instead of turning their eyes, birds turn their heads with their long and flexible necks to scan the visual field. The light-sensitive **retina**

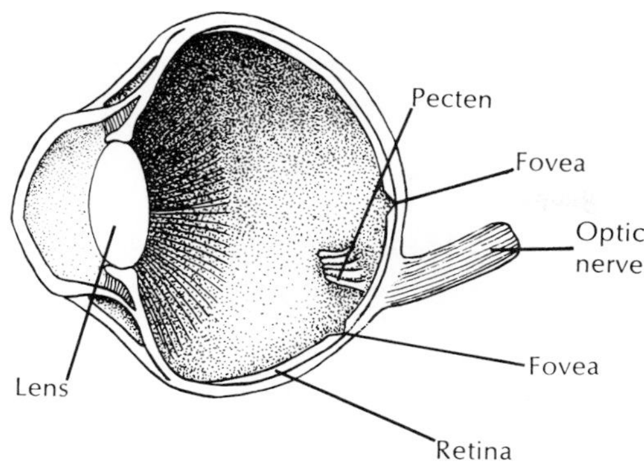

FIG. 26-7 Hawk eye has all the structural components of the mammalian eye, plus a peculiar pleated structure, the pecten, believed to provide nourishment to the retina. The extraordinary vision of the hawk, which has a visual acuity eight times better than that of humans, is attributed to the extreme density of cone cells in the foveae: 1.5 million per fovea compared to 0.2 million for humans. Each hawk eye has two foveae as opposed to one in humans, meaning that each hawk eye focuses on two objects simultaneously —the better to see its next meal!

(Fig. 26-7) is elaborately equipped with rods (for dim light vision) and cones (for color vision). Cones predominate in day birds and rods are more numerous in nocturnal birds. A distinctive feature of the bird eye is the **pecten,** a highly vascularized organ attached to the retina near the optic nerve and jutting out into the vitreous humor (Fig. 26-7). It is believed to supply the retina, which is devoid of blood vessels, with oxygen and nutrients.

The position of a bird's eyes in its head is correlated with its life habits. Vegetarians that must avoid predators have eyes placed laterally to give a wide view of the world; predaceous birds such as hawks and owls have eyes directed to the front. The **fovea,** or region of keenest vision on the retina, is placed (in birds of prey and some others) in a deep pit, which makes it necessary for the bird to focus exactly on the source. Many birds, moreover, have two sensitive spots (foveae) on the retina (Fig. 26-7)—the central one for sharp monocular views and the posterior one for binocular vision. Woodcocks can probably see binocularly both forward and backward. Bitterns, in their freezing stance of bill pointing up, can also see binocularly. The visual acuity of a hawk is believed to be eight times that of a human (enabling it to clearly see a crouching rabbit more than a mile away), and an owl's ability to see in dim light is more than 10 times that of the human eye (Fig. 26-8). Birds have good color vision, especially toward the red end of the spectrum.

FIG. 26-8 Great horned owl, *Bubo virginianus virginianus* (order Strigiformes) with grey squirrel after a successful forage. Owls (order Strigiformes) possess eyes incredibly sensitive to light, enabling them to see prey in light one-hundredth to one-tenth the intensity required by humans (0.000,000,73 footcandle). But this particular ultrasensitivity is traded off for relatively poor visual acuity, narrow visual field, and weak accommodation (ability to focus on near objects). (Photograph by L. L. Rue, III.)

Food, feeding, and digestion

Birds have evolved along with food resources in nearly every environment on earth. In their early evolution, most birds were carnivorous, dieting principally on insects. Insects were well-established on the

FIG. 26-9 A male broad-tailed hummingbird *Selasphorus platycerus* drinks nectar from a flower. This is one of about 50 species of hummingbirds distributed in the western United States and Mexico. While nectar satisfies the intense energy demands of these small birds, they must supplement their diet with insects and spiders to provide the protein required for growth. (Photograph by L. L. Rue, III.)

earth's surface in both variety and numbers long before birds made their appearance, and they presented an enormously valuable food resource only partly exploited by amphibians and reptiles. With the advantage of flight, birds could hunt insects on the wing as well as carry their assault to insect refuges mostly inaccessible to their earthbound tetrapod peers. Today, there is a bird to hunt nearly every insect; they probe the soil, search the bark, scrutinize every leaf and twig, and drill into insect galleries hidden in tree trunks.

Other animal foods — worms, molluscs, crustaceans, fish, frogs, reptiles, mammals, as well as other birds — all found their way into the diet of birds. Fruit- and seed-eating birds are more recent additions, appearing during the Miocene epoch 10 to 15 million years ago. A very large group, nearly one-fifth of all birds, feed on nectar (Fig. 26-9). Thus, as a group, birds eat almost anything, and this is one important reason for their success. Some are omnivores, showing little specialization, while others have become adapted to highly restricted diets. Each extreme offers its own advantage. The opportunistic diet of an omnivore (often termed a **euryphagous,** or "wide-eating" species) allows it to eat whatever is seasonally abundant. However, it must compete with numerous other omnivores for the same broad spectrum of food. The specialist, on the other hand, by refining feeding adaptations, may eliminate all competition for a particular food item. Such a **stenophagous** ("narrow-eating") species has the pantry to itself — but at a price. Should its food speciality be reduced or destroyed for some

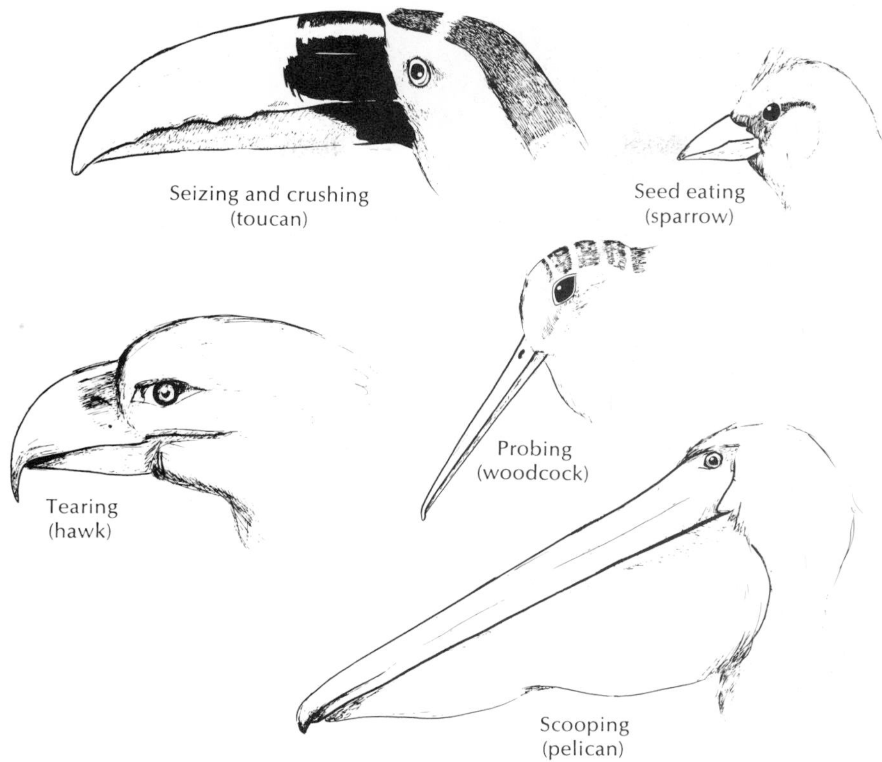

FIG. 26-10 Some bills of birds showing variety of adaptations.

reason (disease, adverse climate, and the like), its very survival as a species may be jeopardized. In fact, few birds cling to a single feeding habit throughout their lives. More typically, a bird may be stenophagous during one season or one period of its life and euryphagous at other times.

The beaks of birds are strongly adapted to specialized food habits—from generalized types, such as the strong, pointed beaks of crows, to grotesque, highly specialized ones in flamingoes, hornbills, and toucans (Fig. 26-10). The beak of a woodpecker is a straight, hard, chisel-like device. Anchored to a tree trunk with its tail serving as a brace, the woodpecker delivers powerful, rapid blows to build nests or expose the burrows of wood-boring insects (Fig. 26-11). It then uses its long, flexible, barbed tongue to seek out insects in their galleries. The woodpecker's skull is especially thick to absorb shock.

How much do birds eat? By a peculiar twist of reality, the commonplace "to eat like a bird" is supposed to signify a diminutive appetite. Yet birds, because of their intense metabolism, are voracious feeders. Small birds eat relatively more than large birds because their metabolic rate is greater. This happens because the oxygen consumption increases only about three-fourths as rapidly as body weight. For example, the resting metabolic rate (oxygen consumed per gram of body weight) of a hummingbird is 12 times that of a pigeon and 25 times that of a chicken. A 3 g hummingbird may eat 100% of its body weight in food each day, an 11 g blue tit about 30%, and a 1,880 g domestic chicken, 3.4%. Obviously food consumption also depends on water content since water has no nutritive value. A 57 g Bohemian waxwing was estimated to eat 170 g of watery *Cotoneaster* berries in one day—three times its body weight! Seed-eaters of equivalent size might eat only 8 g of dry seeds per day.

Birds rapidly process their food with efficient digestive equipment. A shrike can digest a mouse in 3 hours and berries will pass completely through the digestive tract of a thrush in just 30 minutes. Furthermore, birds utilize a very high percentage of the food

FIG. 26-11 Adaptations for chopping holes in wood. This yellow-shafted flicker, *Colaptes auratus,* in common with other woodpeckers, has a heavy skull, chisel-edged beak, strong feet with two opposing toes (instead of the more common one), and stiff tail feathers that serve as a third, bracing leg. The black "mustache" mark beneath the eye of this male is the only visible difference between the sexes. Order Piciformes. (Photograph by L. L. Rue, III.)

they eat. There are no teeth in the mouth, and the poorly developed salivary glands mainly secrete mucus for lubricating the food and the slender, horn-covered tongue. There are few taste buds, although all birds can taste to some extent. Hummingbirds and some others have sticky tongues, and woodpeckers have tongues that are barbed at the end. From the short **pharynx** a relatively long, muscular, elastic **esophagus** extends to the **stomach.** In many birds there is an enlargement **(crop)** at the lower end of the esophagus that serves as a storage chamber (Fig. 26-12).

In pigeons, doves, and some parrots the crop not only stores food but produces milk by the breakdown of epithelial cells of the lining. This "bird milk" is regurgitated by both male and female into the mouth of the young squabs. It has a much higher fat content than cow's milk.

The stomach proper consists of a **proventriculus,** which secretes gastric juice, and the muscular **gizzard,** which is lined with horny plates that serve as millstones for grinding the food. To assist in the grinding process, birds swallow coarse, gritty objects or pebbles, which lodge in the gizzard. Certain birds of prey such as owls form pellets of indigestible materials, for example, bones and fur, in the proventriculus and eject them through the mouth. At the junction of the intestine with the rectum there are paired **ceca,** which may be well-developed in some birds. Two **bile ducts** from the **gallbladder** or liver and two or three **pancreatic ducts** empty into the duodenum, or first part of the intestine. The **liver** is relatively large and bilobed. The terminal part of the digestive system is the **cloaca,** which also receives the genital ducts and ureters; in young birds the dorsal wall of the cloaca bears the **bursa of Fabricius,** which functions to produce immune bodies.

Body temperature regulation

Birds and mammals are **homeothermic.** They are the only animals that generally maintain a nearly constant body temperature, independent of large variations in environmental temperature. They are also **endothermic,** that is, their metabolism is the source of body heat. This distinguishes them from the reptiles,

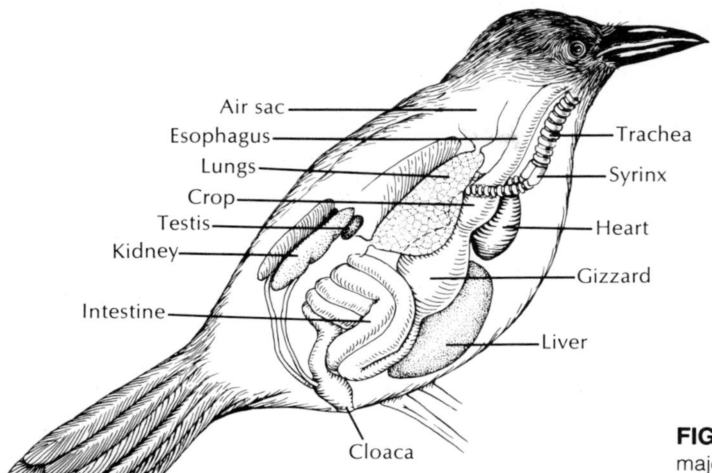

FIG. 26-12 Internal anatomy of a male bird. Locations of major body organs are shown.

many of which are able to elevate their body temperature above that of their surroundings, using an external heat source, principally the sun. Such reptiles are **ectothermic.**

The body temperature of birds is high, ranging between 40° and 42° C as compared to 36° to 39° C for mammals. Some thrushes operate at 43.5° C (110.5° F), a temperature well past the lethal limit for most mammals and only about 2.5° C below the upper lethal temperature for these birds. Small birds tend to have rather more variable body temperatures than do large birds; for example, that of a house wren may fluctuate 8° C over 24 hours.

Body temperature is a balance between metabolic heat production and heat loss through physical means. If a bird begins to overheat, heat loss is accelerated by dilating blood vessels in the skin (to increase radiant heat loss) and by increasing the breathing rate (to increase evaporative cooling). Many species pant in extreme heat. In cold weather, birds ruffle their feathers to form a blanket of warm, insulative air next to the skin. They vasoconstrict peripheral blood vessels to reduce radiant heat loss. And if these physical mechanisms are not sufficient to prevent a drop in body temperature in very cold weather, a bird will shiver. As in mammals, this muscular movement creates needed heat. But it also increases food and oxygen consumption. A sparrow will consume twice as much oxygen and eat twice as much food at 0° C (32° F) as at 37° C (98° F). What person living in a northern climate has not marveled at the ability of the tiny chickadee, a minute furnace of cheerful activity, to survive direct exposure to the coldest winter weather?

Circulation, respiration, and excretion

Circulatory system

The general plan of bird circulation is not greatly different from that of mammals. The four-chambered heart is large, with strong ventricular walls; thus birds share with mammals a complete separation of the respiratory and systemic circulations. However, the right aortic arch, instead of the left as in the mammals, leads to the dorsal aorta. The two jugular veins in the neck are connected by a cross vein, an adaptation for shunting the blood from one jugular to the other as the head is turned around. The brachial and pectoral arteries to the wings and breast are unusually large.

The heartbeat is extremely fast, and, as in mammals, there is an inverse relationship between heart rate and body weight. For example, a turkey has a heart rate at rest of about 93 beats per minute, a chicken has 250 beats per minute, and a black-capped chickadee has a rate of 500 beats per minute when asleep, which may increase to a phenomenal 1,000 beats per minute during exercise. Blood pressure in birds is roughly equivalent to that of mammals of similar size.

Bird's blood contains nucleated, biconvex red corpuscles that are somewhat larger than those of mam-

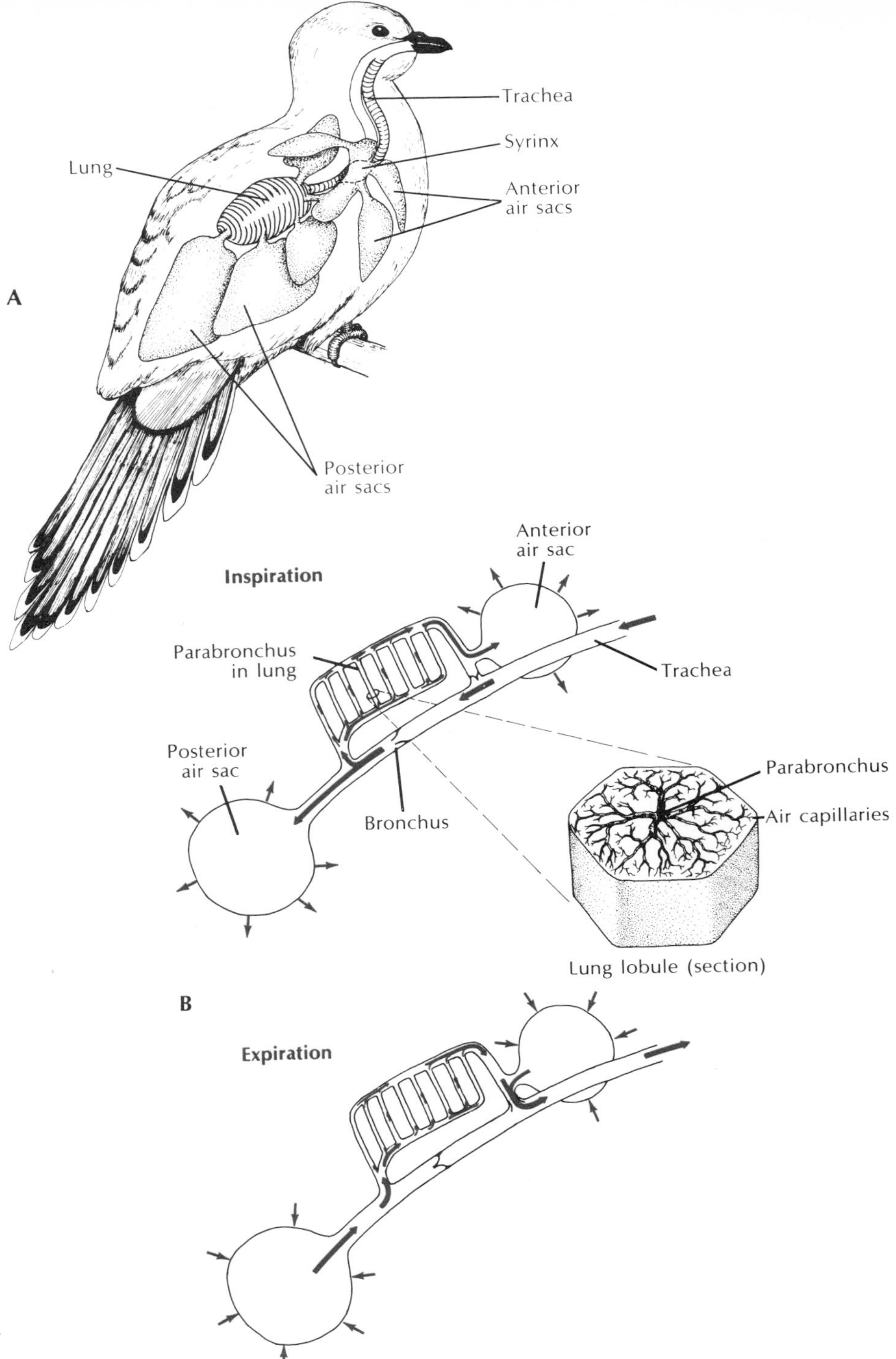

FIG. 26-13 Respiratory system of a bird. **A,** Lungs and air sacs. Only one side of bilateral air sac system is shown. **B,** Direction of air movement through lungs and air sacs of a bird. During inspiration, part of the fresh air passes through the lungs; the rest enters a posterior air sac. During expiration, fresh air stored in posterior air sacs enters the lungs, while stale air in the anterior air sacs is moved directly out. Enlargement of a lung lobule shows numerous air capillary channels that are surrounded by blood capillaries (not shown).

mals. The phagocytes, or mobile ameboid cells, of the blood are unusually active and efficient in birds in the repair of wounds and in destroying microbes.

Respiratory system

The respiratory system of birds differs radically from the lungs of reptiles and mammals and is marvelously adapted for meeting the high metabolic demands of flight. The lungs, which are relatively inexpansible because of their direct attachment to the body wall, are filled with numerous tiny **air capillaries** instead of alveoli of the mammalian type. Really unique, however, is the extensive system of nine interconnecting **air sacs** that are located in pairs in the thorax and abdomen and are even extended by tiny tubes into the centers of the long bones (Fig. 26-13). The air sacs are connected to the lungs in such a way that perhaps 75% of the inspired air bypasses the lungs and flows directly into the air sacs, which serve as reservoirs for fresh air. On expiration, some of this fully oxygenated air is shunted through the lung, while the rest passes directly out. The advantage of such a system is obvious—the lungs receive fresh air during both inspiration and expiration. Rather than locating the respiratory exchange surface deep within blind sacs, which are difficult to ventilate as in mammals, birds pass a continuous stream of fully oxygenated air through a system of richly vascularized air capillaries (Fig. 26-13, *B*). Although many details of the bird's respiratory system are not yet understood, it is clearly the most efficient of any vertebrate system.

In addition to performing its principal respiratory function, the air sac system helps to cool the bird during vigorous exercise. A pigeon, for example, produces about 27 times more heat when flying than when at rest. The air sacs have numerous diverticula that extend inside the larger pneumatic bones of the pectoral and pelvic girdles, wings, and legs. Because they contain warmed air, they provide considerable buoyancy to the bird.

Excretory system

The relatively large paired metanephric kidneys are attached to the dorsal wall in a depression against the sacral vertebrae and pelvis. Urine passes by way of **ureters** to the **cloaca.** There is no urinary bladder. The kidney is composed of many thousands of **nephrons,** each consisting of a renal corpuscle and a nephric tubule. Urine is formed in the usual way by glomerular filtration followed by selective modification of the filtrate in the tubule (the details of this sequence are found on pp. 701 to 705).

The urine of birds differs from that of mammals in having a high concentration of uric acid, rather than urea. This is a strikingly useful adaptation for birds, many of which function on a water-poor economy, eating dry seeds, and drinking little or no water. Uric acid can be excreted using far less water than is required for urea because of uric acid's low solubility. A bird can excrete 1 g of uric acid in only 1.5 to 3 ml of water, whereas a mammal requires 60 ml of water to excrete 1 g of urea. The concentration of uric acid occurs almost entirely in the cloaca, where it is combined with fecal material, and the water is reabsorbed to form a white paste. Thus despite having kidneys that are much less effective in true concentrative ability than mammalian kidneys, birds can excrete uric acid nearly 3,000 times more concentrated than that in the blood. Even the most effective mammalian kidney—those of certain desert rodents—can excrete urea only about 25 times the plasma concentration. This paradox is explained by the insoluble nature of uric acid, which allows it to crystallize out of solution without contributing to the osmotic pressure of the urine.

Marine birds (also marine turtles) have evolved a unique solution for excreting the large loads of salt eaten with their food and in the seawater they drink. Seawater contains about 3% salt and is three times saltier than a bird's body fluids. Yet the bird kidney cannot concentrate salt in urine above about 0.3%. The problem is solved by special **salt glands,** one located above each eye (Fig. 26-14). These glands are capable of excreting a highly concentrated solution of sodium chloride—up to twice the concentration of sea water. The salt solution runs out the internal or external nostrils, giving gulls, petrels, and other sea birds a perpetual runny nose. The development of the salt gland in some birds depends on how much salt the bird takes in its diet. For example, a race of mallard ducks living a semimarine life in Greenland have salt glands 10 times larger than those of ordinary freshwater mallards.

Flight

What prompted the evolution of flight in birds, the ability to rise free of earth-bound concerns, as almost every human being has dreamed of doing? Much as

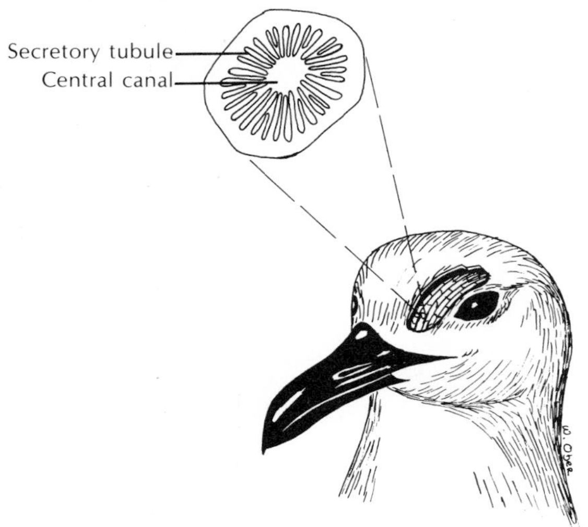

FIG. 26-14 Salt glands of a marine bird (gull). One salt gland is located above each eye. Each gland consists of several lobes arranged in parallel. One lobe is shown in cross section, much enlarged. Salt is secreted into many radially arranged tubules, then flows into a central canal that leads into the nose.

we may envy the birds in their conquest of the air, we also recognize that evolution of flight was the pragmatic result of complex adaptive pressures; certainly the forerunners of birds did not take up flight just to enjoy a new experience. The air was a relatively unexploited habitat stocked with flying insect food. Flight also offered escape from terrestrial predators and opportunity to travel rapidly and widely to establish new breeding areas and to benefit from year-round favorable climate by migrating north and south with the seasons.

Birds unquestionably evolved from reptilian ancestors, but not directly from the membrane-winged flying reptiles (pterodactyls) as one might innocently suppose. They appear to have descended instead from a small group of thecodont reptiles (Pseudosuchia) that gave rise to both dinosaurs and birds. The fossil evidence is far too meager to give us a recorded history of the origins of bird flight. We can only theorize that the proavian reptiles passed through some sequence of swift running, flying leaps, tree climbing, parachute gliding—all leading toward an arboreal existence. Many adaptations of birds, such as the perching mechanism and active climbing habits, indicate such an arboreal apprenticeship. Natural selection favored the swifter and more agile, and eventually wings became strong enough to support the bird in air. One thing seems certain: feathers were an absolute requirement for flight and the evolution of flight and of feathers must have progressed together. There is absolutely no support for the idea that bird ancestors were originally membrane-winged flyers, like bats, that later developed feathers.

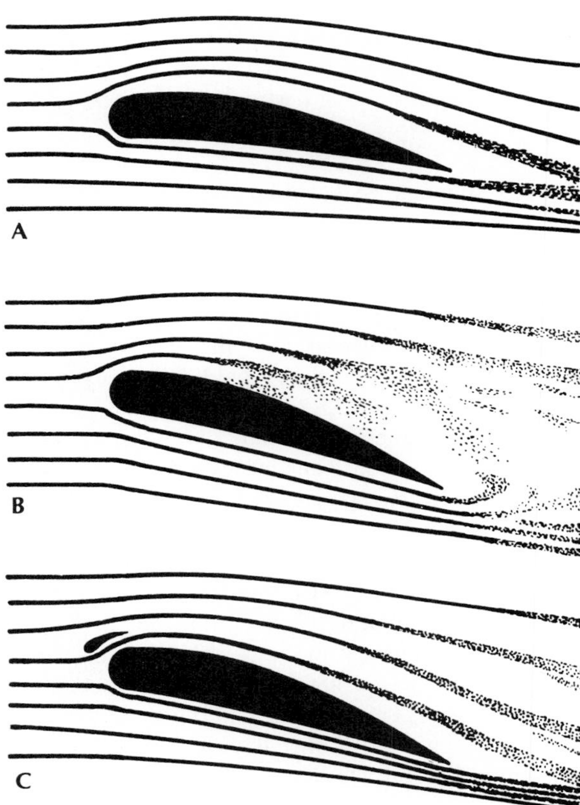

FIG. 26-15 Air patterns formed by airfoil, or wing, moving from right to left. **A,** Normal flight with low angle of attack. As air moves smoothly over the wing, areas of low pressure on upper wing surface and of high pressure on lower wing surface create lift. **B,** Appearance of lift-destroying turbulence on upper wing surface when angle of attack becomes too great. Stalling occurs. **C,** Prevention of stalling by directing a layer of rapidly moving air over upper surface with a wing-slot. (From Welty, J. C. 1975. The life of birds, ed. 2. Philadelphia, W. B. Saunders Co.)

The bird wing as a lift device

Bird flight, especially the familiar flapping flight of birds, is complex. Despite careful analysis by conventional aerodynamic techniques and high-speed photography, there is much about it that is not understood.

FIG. 26-16 Elliptic wing of the black-billed magpie *Pica pica* in slow flight. Note well-developed alula and separation (slotting) between primaries, typical of wings adapted for high maneuverability and low-speed flight. This member of the crow family is a familiar feature of the western North American landscape. Order Passeriformes. (Photograph by C. G. Hampson.)

Nevertheless, the bird wing is an airfoil that is subject to recognized laws of aerodynamics. It is adapted for high lift at low speeds, and, not surprisingly perhaps, it resembles the wings of early low-speed aircraft. The bird wing is streamlined in cross section, with a slightly concave lower surface (cambered) and with small tight-fitting feathers where the leading edge meets the air (Fig. 26-15). Air slips efficiently over the wing, creating lift with minimum drag. Some lift is produced by positive pressure against the undersurface of the wing. But on the upper side, where the airstream must travel farther and faster over the convex surface, a negative pressure is created that provides more than two-thirds of the total lift.

The lift-to-drag ratio of an airfoil is determined by the angle of tilt (angle of attack) and the airspeed (Fig. 26-15). A wing carrying a given load can pass through the air at high speed and small angle of attack, or at low speed and larger angle of attack. But as speed decreases, a point is reached where the angle of attack becomes too steep, turbulence appears on the upper surface, lift is destroyed, and stalling occurs. Stalling can be delayed or prevented by placing a **wing slot** along the leading edge so that a layer of rapidly moving air is directed across the upper wing surface. Wing slots were, and still are, used in aircraft when traveling at low speed. In birds, two kinds of wing slots have developed: (1) the **alula,** or group of small feathers on the thumb (Figs. 26-4 and 26-16), provides a midwing slot, and (2) **slotting between primary feathers** provides a wingtip slot. In many songbirds, these together provide stall-preventing slots for nearly the entire outer (and aerodynamically most important) half of the wing.

Basic forms of bird wings

Bird wings vary in size and form because the successful exploitation of different habitats has imposed special aerodynamic requirements. The following four types of bird wings are easily recognized (Savile, 1957).

Elliptic wings. Birds that must maneuver in forested

FIG. 26-17 High-speed wing of arctic tern *Sterna paradisaea.* Sweepback, taper, and absence of wing slotting adapts tern for high speeds during its incredibly long migrations, from the Arctic to the Antarctic and back each year, with some logging as much as 24,000 miles. This female was photographed beside its nest on an island off the east coast of Sweden. Arctic terns nest so close to water's edge that nests are often flooded during storms. Order Charadriformes. (Photograph by B. Tallmark.)

habitats such as sparrows, warblers, doves, woodpeckers, and magpies (Fig. 26-16) have elliptic wings. This type has a **low aspect ratio** (ratio of length to width). The outline of a sparrow wing is almost identical to that of the British Spitfire fighter plane of Second World War fame—also a highly maneuverable flyer. Elliptic wings are highly slotted between the primary feathers (Fig. 26-16); this is an adaptation that helps to prevent stalling during sharp turns, low-speed flight, and frequent landing and takeoff. Each separated primary feather behaves as a narrow wing with a high angle of attack, providing high lift at low speed. The high maneuverability of the elliptic wing is exemplified by the tiny chickadee, which, if frightened, can change course within 0.03 second.

High-speed wings. Birds that feed on the wing, such as swallows, hummingbirds, and swifts, or that make long migrations, such as plovers, sandpipers, and terns (Fig. 26-17), have wings that sweep back and taper to a slender tip. They are rather flat in section, have a moderately high aspect ratio, and lack the wing-tip slotting characteristic of the preceding group. Sweepback and wide separation of the wing tips reduces "tip vortex," a drag-creating turbulence that tends to develop at wing tips. The fastest birds alive, such as falcons and sandpipers clocked at 360 and 175 km per hour (224 and 109 miles per hour), respectively, belong to this group.

Soaring wing. The oceanic soaring birds have **high-aspect-ratio** wings resembling those of sailplanes. This group includes albatrosses (Fig. 26-18), frigatebirds, gannets, and gulls. Such long, narrow wings lack wing slots and are adapted for high speed, high lift, and dynamic soaring. They have the highest aerodynamic efficiency of all wings but are less maneuverable than the wide, slotted wings of land soarers. Dynamic soarers have learned how to exploit the highly reliable sea winds, using adjacent air currents of different velocities.

High-lift wing. Vultures, hawks, eagles, owls (Fig. 26-19), and ospreys (Fig. 26-20)—predators that carry heavy loads—have wings with slotting, alulas, and pro-

FIG. 26-18 Soaring wing of waved albatross *Diomedea irrorata*. The long, high-aspect wing, like that of a sailplane, has great aerodynamic efficiency and lift but is relatively weak and lacking in maneuverability. Order Procellariiformes. (Photograph by C. P. Hickman, Jr.)

FIG. 26-19 High-lift wing of snowy owl *Nyctea scandiaca*. Wing is deeply emarginated with square slots between primary feathers, a highly efficient pattern for carrying heavy loads (prey) at low speed. Broad tail provides purchase on the air for quick maneuvering and assists in load-carrying. Order Strigiformes. (Photograph by C. G. Hampson.)

nounced camber, all of which promote high lift at low speed. Many of these birds are land soarers, where broad, slotted wings provide the sensitive response and maneuverability required for static soaring in the capricious air currents over land.

MIGRATION AND NAVIGATION

Perhaps it was inevitable that birds, having mastered the art of flight, should use this power to make the long and arduous seasonal migrations that have captured the wonder and curiosity of humans. The term **migration** refers to the regular, extensive, seasonal movements birds make between their summer breeding regions and their wintering regions. The chief advantage seems obvious: it enables birds to live in an optimal climate all the time, where abundant and unfailing sources of food are available to sustain their intense metabolism. Migrations also provide optimal conditions for rearing young when demands for food are especially great. Broods are largest in the far north where the long summer days and the abundance of

FIG. 26-20 Osprey *Pandion haliaetus* landing on nest. Note alulas *(top arrows)* and new primary feathers *(side arrows)*. Feathers are molted in sequence in exact pairs so that balance is maintained during flight. These fish-eating birds have suffered a severe population decline in the United States in recent years because of illegal hunting and poor nesting success. Pesticides concentrated in fish eaten by ospreys causes eggshells to thin and burst during incubation. Order Falconiformes. (Photograph by B. Tallmark.)

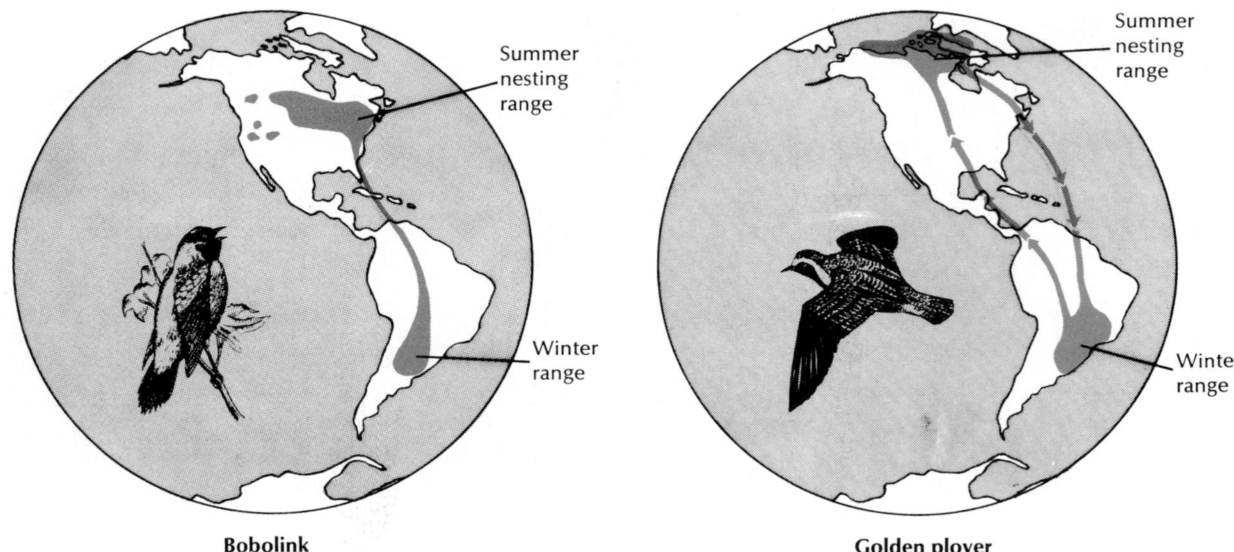

FIG. 26-21 Migrations of the bobolink and golden plover. The bobolink commutes 22,500 km (14,000 miles) each year between nesting sites in North America and its wintering range in Argentina, a phenomenal feat for such a small bird. Although the breeding range has extended to colonies in western areas, these birds take no shortcuts but adhere to the ancestral eastern seaboard route. The golden plover flies a loop migration, striking out across the Atlantic in its southward autumnal migration but returning in the spring via Central America and the Mississippi valley because ecologic conditions are more favorable at that time.

insects combine to provide parents with ample food-gathering opportunity. Predators are relatively rare in the north and the brief once-a-year appearance of vulnerable young birds does not encourage the build-up of predator populations. Migration also vastly increases the amount of space available for breeding and reduces aggressive territorial behavior. And of course migration favors homeostasis by allowing birds to avoid climatic extremes.

Origin of migration

Migration has become so firmly established in the behavior of birds that it has long since become a hereditary instinct. But what prompted the origin of migration? Undoubtedly, it was ecologic pressures. We must suppose that birds, like other animals, will move only when compelled by some necessity. One theory of migration suggests that birds spread over the northern hemisphere when the latter was warm and food conditions were favorable all through the year. When the glacial era came, it forced the birds to go south for survival. When the ice age receded, they came back in the spring, only to be forced south again in winter because of the sharp establishment of the winter and summer seasons. This led in time to the firm fixing of the habit. Another theory centers around the view that the ancestral home of birds was in the tropics and some went north to avoid congestion and competition during the breeding season. After raising their young, they then returned.

Migration routes

Most migratory birds have well-established routes trending north and south. And since most birds (and other animals) live in the northern hemisphere, where most of the earth's land mass is concentrated, most birds are south-in-winter and north-in-summer migrants. Of the 4,000 or more species of migrant birds (a little less than half the total bird species), most breed in the more northern latitudes of the hemisphere; the percentage of migrants in Canada is far higher than is the percentage of migrants in Mexico, for example. Some use different routes in the fall and spring (Fig. 26-21). Some, especially certain aquatic species, complete their migratory routes in a very short time. Others, however, make the trip in a leisurely manner, often stopping here and there to feed. Some of the warblers are known to take 50 to 60 days to migrate from their winter quarters in Central America to their summer breeding areas in Canada.

Not all members of a species perform their migrations at the same time; there is a great deal of straggling so that some members do not reach the summer breeding grounds until after others are well along with their nesting. Some birds, such as the robin, are "weather migrants" that move northward as the winter retreats. Their arrival at summer breeding grounds is variable and dependent on weather. Others, such as purple martins and catbirds, are "instinctive migrants" that are little influenced by year to year weather changes. Records kept of catbirds in a certain eastern state reveal that the birds arrived in the particular locality about the middle of April and did not vary more than a day in a period of 5 years. Many observations also revealed that the same individual bird returns not only to the same locality but also to the same territory that it occupied in previous seasons.

Many of the smaller species migrate at night and feed by day; others migrate chiefly in the daytime; and many swimming and wading birds, either by day or night. The height at which they fly varies greatly. Although some passerines (perching birds) have been observed as high as 6,400 m (nearly 4 miles) with radar height-finders, the great majority (90%) fly below 1,500 m. Migrants tend to fly higher over water than land and higher at night than during the day.

Many birds are known to follow landmarks, such as rivers and coastlines; but others do not hesitate to fly directly over large bodies of water in their routes. Some birds have very wide migration lanes, and others, such as certain sandpipers, are restricted to very narrow ones, keeping well to the coastlines because of their food requirements.

Some species are known for their long-distance migrations. The arctic tern (Fig. 26-17), greatest globe spanner of all, breeds north of the Arctic Circle and in winter is found in the Antarctic regions, 18,000 km away. This species is also known to take a circuitous route in migrations from North America, passing over to the coastlines of Europe and Africa and then to their winter quarters. Other birds that breed in Alaska follow a more direct line down the Pacific coast of North and South America.

Many small songbirds also make great migration treks (Fig. 26-21). Africa is a favorite wintering ground for European birds, and many fly there from Central Asia as well.

Stimulus for migration

Although the Chinese have known for centuries that they could make caged songbirds sing in winter by artificially lengthening their days with candlelight, it was not until 1929 that W. Rowan in Canada performed pioneering experiments showing that the seasonal change in day length is the principal timing factor for bird reproduction and migration. Subsequently many other researchers have shown unequivocally that lengthening days in winter and spring stimulate the development of the gonads and the accumulation of fat—both important internal changes that predispose birds to migrate northward. There is evidence that increasing daylength stimulates the anterior lobe of the pituitary into activity. The release of pituitary gonadotropic hormones in turn set in motion a complex series of physiologic and behavioral changes, leading to gonadal growth, fat deposition, migration, courtship and mating behavior, and care of the young.

Direction-finding in migration

Numerous experiments suggest that birds navigate chiefly by sight. Birds recognize topographic landmarks and follow familiar migratory routes—a behavior assisted by flock migration where navigational resources and experience of older birds can be pooled. But in addition to visual navigation, birds make use of a variety of orientation cues at their disposal. Birds have an innate time sense, a built-in clock of great accuracy; they have an innate sense of direction; and very recent work adds much credence to an old, much debated theory that birds can detect and navigate by the earth's field of gravity. All of these resources are inborn and instinctive, although a bird's navigational abilities may improve with experience.

Recently experiments by German ornithologists G. Kramer and E. Sauer and American ornithologist S. Emlen have demonstrated convincingly that birds can navigate by celestial cues—the sun by day and the stars by night. Using special circular cages, Kramer concluded that birds possessed a built-in time sense that enabled them to maintain compass direction by referring to the sun, regardless of time of day. This is called **sun-azimuth orientation** *(azimuth,* compass-

FIG. 26-22 Flamingos *Phoeniconaias minor* on Lake Nakuru in Kenya, Africa, feeding on blue-green algae. There may be 2 million flamingos on the lake at times. They remove an estimated 70,000 tons of algae from this productive, alkaline lake each year. The lake is threatened by pollution from rapidly growing human population along its shores. Order Ciconiiformes. (Photograph by B. Tallmark.)

bearing of the sun). Sauer's and Emlen's ingenious planetarium experiments strongly suggest that some birds, probably many, learn to use the constellations to navigate at night.

Some of the remarkable feats of bird navigation still defy rational explanation. Most birds undoubtedly use a combination of environmental and innate cues to migrate. Migration is a rigorous undertaking; the target is often small, and natural selection relentlessly prunes off errors in migration, leaving only the best navigators to propagate the species.

REPRODUCTION AND BEHAVIOR

The adage asserts that "birds of a feather flock together." Birds are indeed highly social creatures. Especially during the breeding seasons, sea birds gather in often enormous colonies to nest and rear young. Land birds, with some conspicuous exceptions (such as starlings and rooks), tend to be less gregarious than sea birds during breeding and seek isolation for rearing their brood. But these same species that covet separation from their kind during breeding may aggregate for migration or feeding (Fig. 26-22). Togetherness offers advantages: mutual protection from enemies, greater ease in finding mates, less opportunity for individual straying during migration, and mass huddling for protection against low night temperatures during migration. Certain species may use highly organized cooperative behavior to feed, as do pelicans (Fig. 26-23). At no time are the highly organized social interactions of birds more evident than during the breeding season, as they stake out territorial claims, select mates, build nests, incubate and hatch their eggs, and rear their young.

Reproductive system

In the male the paired **testes** and accessory ducts are similar to those in many other forms. From the **testes** the **vasa deferentia** run to the cloaca. Before being discharged, the sperm are stored in the **seminal vesicle,** the enlarged distal end of the vas deferens. This seminal vesicle may become so large with stored sperm during the breeding season that it causes a cloacal pro-

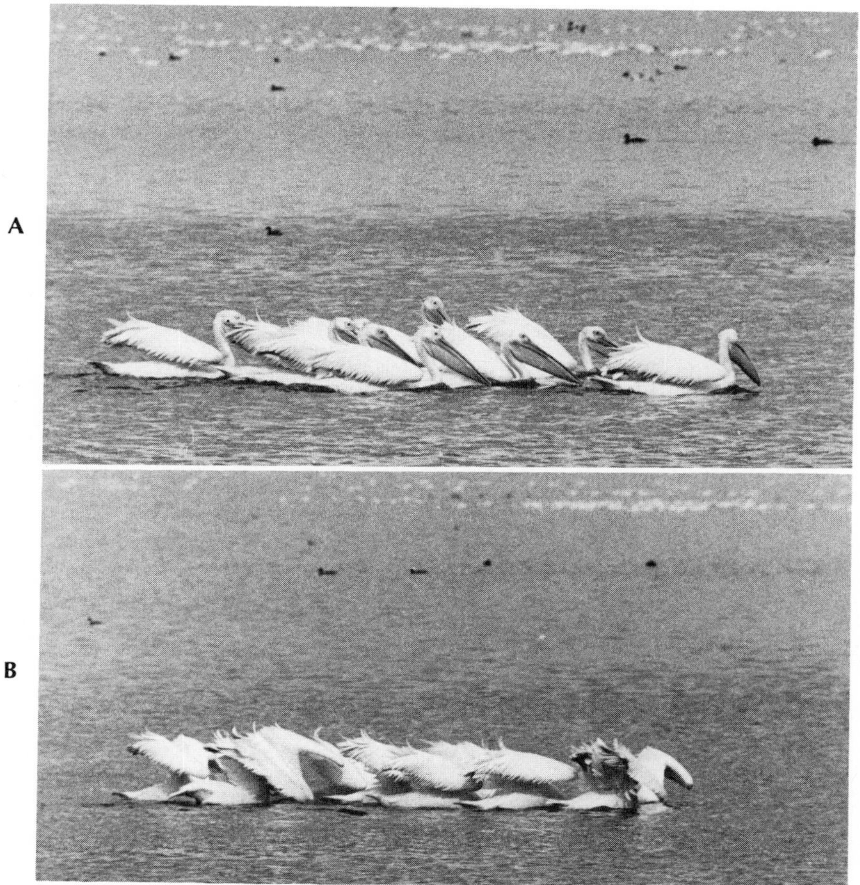

FIG. 26-23 Cooperative feeding behavior by the white pelican *Pelecanus onocrotalus*. **A,** Pelicans on Lake Nakuru, East Africa, form a horseshoe to drive fish together. **B,** Then they plunge simultaneously to scoop up fishes in their huge bills. The pictures were taken 2 seconds apart. Pelicans were attracted to the lake in mid-1960s to feed on fish *(Tilapia grahami)* introduced to control malaria by eating mosquito larvae. Order Pelecaniformes. (Photograph by B. Tallmark.)

tuberance. The high body temperature, which tends to inhibit spermatogenesis in the testes, is probably counteracted by the cooling effect of the abdominal air sacs. The testes of birds undergo a great enlargement at the breeding season, as much as 300 fold, and then shrink to tiny bodies afterward. Some birds, including ducks and geese, have a large, well-developed **copulatory organ** (penis), provided with a groove on its dorsal side for the transfer of sperm. However, in the more advanced birds, copulation is a matter of bringing the cloacal surfaces into contact, usually while the male stands on the back of the female (Fig. 26-24). Some swifts copulate in flight.

In the female of most birds, only the left ovary and oviduct develop; those on the right dwindle to vestigial structures. The ovary is close to the left kidney. Eggs discharged from the ovary are picked up by the expanded end of the oviduct, the **ostium** (Fig. 26-25). The oviduct runs posteriorly to the cloaca. While the eggs are passing down the oviduct, **albumin,** or egg white, from special glands is added to them; farther down the oviduct, the shell membrane, shell, and shell pigments are also secreted about the egg. Fertilization takes place in the upper oviduct several hours before the layers of albumin, shell membranes, and shell are added. Sperm remain alive in the female oviduct for

FIG. 26-24 Culmination of courtship activities is copulation, shown here in a pair of waved albatrosses *Diomedea irrorata.* Order Procellariiformes. (Photograph by C. P. Hickman, Jr.)

many days after a single mating. Hen eggs show good fertility for 5 or 6 days after mating, but then fertility drops rapidly. However, the occasional egg will be fertile as long as 30 days after separation of the hen from the rooster.

Selection of territories and mates

A territory on which to raise a brood is selected in the spring by the male, who jealously guards it against all other males of the same species. The male sings a great deal to help him establish territorial ownership and to announce his presence to females. The female apparently wanders from one territory to another until she settles down with a male. How large a territory a pair takes over depends on location, abundance of food, natural barriers, and the like. In the case of robins a house may serve as the dividing line between two adjacent domains, and each pair will usually stay close to the lawn on its particular side of the house. When members of another species trespass, they are usually ignored; competition is greatest among the members of the same species. Song sparrows and mocking birds, however, try to keep off members of other species as well as their own. Birds may also defend their territories against other species because of environmental limitations and changes, where competition for food or other factors between different species (usually closely related) may occur.

Elaborate courtship rituals precede mating in many birds, such as the prairie chicken, sage grouse, bowerbirds, great crested grebes, and colonial nesting seabirds (Figs. 26-26 and 26-27). Songbirds have simpler rituals, consisting mostly of male displays and songs.

Nesting and care of young

To produce offspring, all birds lay eggs that must be incubated by one or both parents.

Most birds build some form of nest in which to rear their young. These nests vary from depressions on the ground (Fig. 26-28) to huge and elaborate affairs. Some birds simply lay their eggs on the bare ground or rocks and make no pretense of nest-building. Some of the most striking nests are the pendant nests constructed by orioles, the neat lichen-covered nests of

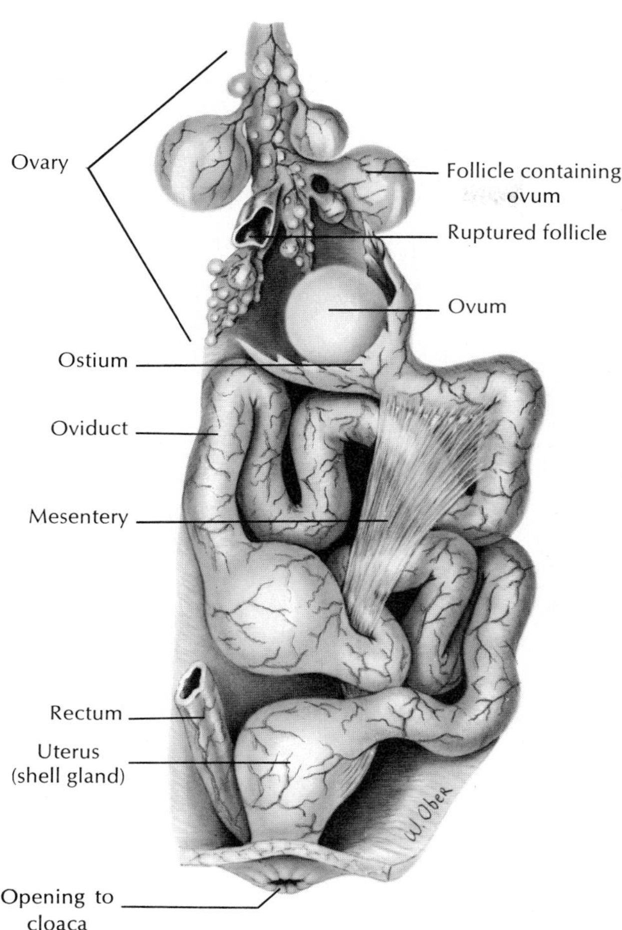

FIG. 26-25 Reproductive system of female bird.

hummingbirds and flycatchers, the chimney-shaped mud nests of cliff swallows, the floating nests of red-necked grebes, and the huge brush pile nests of Australian brush turkeys. Most birds take considerable pains to conceal their nests from enemies. Woodpeckers, chickadees, bluebirds, and many others place their nests in tree hollows or other cavities; kingfishers excavate tunnels in the banks of streams for their nests; and birds of prey build high in lofty trees or on inaccessible cliffs. A few birds such as the American cowbird and the European cuckoo build no nests at all but simply lay their eggs in the nests of birds smaller than themselves. When the eggs hatch, the young are taken care of by their foster parents.

Most American songbirds lay from three to six eggs, but the number of eggs laid in a clutch varies from one or two (some hawks and pigeons) to 18 or 20 (quail). Some birds are determinate layers and lay only a fixed number (clutch) of eggs in a season. If any of the eggs of a set are removed, the deficit is not made up by additional laying (herring gull). Indeterminate layers, however, will continue to lay additional eggs for a long time if some of the first-laid eggs are continually removed (flickers, ducks, domestic poultry). Most birds are probably determinate layers. Many birds such as songbirds lay an egg a day until the clutch is completed; others stagger their egg-laying and lay every other day or so. Domestic geese usually lay every other day, which is probably the pattern for the large birds of prey. The parasitic European cuckoo tends to lay eggs similar in size and pattern to those of the host birds.

The eggs of most songbirds require approximately 14 days for hatching; those of ducks and geese require at least twice that long. Most of the duties of incubation fall to the female, although in many instances

FIG. 26-26 A pair of Galápagos blue-footed boobies *Sula nebouxii* in mutual courtship display. The male at right is sky pointing; the female at left is parading. These and other vivid, stereotyped displays are required to maintain reciprocal stimulation and cooperative behavior during courting, mating, nesting, and caring for the young. (Photograph by C. P. Hickman, Jr.)

both parents share the task, and occasionally only the male performs this work.

Nesting success is very low with many birds, especially in altricial species (see later). Surveys vary among investigators, but one investigation of 170 altricial bird nests reported only 21% as producing at least one young. Of the many causes of nesting failures, predation by snakes, skunks, chipmunks, blue jays, crows, and others is by far the chief factor. Birds of prey probably have a much higher percentage of reproductive success.

When birds hatch, they are of two types: **precocial** and **altricial.** The precocial young, such as quail, fowl, ducks, and most water birds, are covered with down when hatched and can run or swim as soon as their plumage is dry (Fig. 26-29). The altricial ones, on the other hand, are naked and helpless at birth and remain in the nest for a week or more. The young of both types require care from the parents for some time after hatching. They must be fed, guarded, and protected against rain and the sun. The parents of altricial species must carry food to their young almost constantly, for most young birds will eat more than their weight each day. This enormous food consumption explains the rapid growth of the young and their quick exit from the nest (Fig. 26-30). The food of the young, depending on the species, includes worms, insects, seeds, and fruit. Pigeons are peculiar in feeding their young "pigeon milk," the sloughed-off epithelial lining of the crop.

Many birds, such as eagles, Canada geese, albatrosses, and some of the songbirds, are known to mate for life. Others mate only for the rearing of a single brood. There are also cases in which one female mates with several males **(polyandry),** as illustrated by the European cuckoo; in other cases one male mates with several females **(polygyny),** such as the ostrich. In most bird populations there are usually many sexually mature individuals that have no mates at all.

FIG. 26-27 Wild turkey *Meleagris gallapavo* displaying. This species was once abundant over the entire eastern United States. It was disappearing from heavy hunting pressure until a few remnant flocks in Pennsylvania were given full protection, enabling the population to increase and spread. Turkey hunting is now a popular regulated sport in most eastern states. Many would agree with Benjamin Franklin that this handsome, vigorous, and completely native bird would make a more fitting national emblem than the bald eagle, a fish-eating scavenger found also in Canada and Asia. Order Galliformes. (Photograph by L. L. Rue, III.)

BIRD POPULATIONS

Bird populations, like those of other animal groups, vary in size from year to year. Snowy owls (Fig. 26-19), for example, are subject to population cycles that closely follow cycles in their food crop, mainly rodents. Voles, mice, and lemmings in the north have a fairly regular 4-year cycle of abundance; at population peaks, predator populations of foxes, weasels, buzzards, as well as snowy owls, increase because there is abundant food for rearing their young. After a crash in the rodent population, snowy owls move south, seeking alternate food supplies. They occasionally appear in large numbers in southern Canada and northern United States, where their total absence

FIG. 26-28 Blue-footed booby *Sula nebouxii* with two chicks in the nest, a simple depression in the ground. Smaller chick is mostly hidden behind the larger, which is receiving most of the food. The young feed by taking food directly from the crop of the parent. Order Pelecaniformes. (Photograph by C. P. Hickman, Jr.)

FIG. 26-29 Newly hatched chicks of the ring-necked pheasant *Phasianus colchicus torquatus*. These precocial chicks are alert, strong, covered with down, and able to feed themselves. Order Galliformes. (Photograph by L. L. Rue, III.)

FIG. 26-30 A robin *Turdus migratorius* feeds its altricial young. Hatched naked, blind, and helpless, they have phenomenal appetites and develop rapidly. These young have opened their eyes and have learned to gape toward the parent. The parents will supply all the food until the young are nearly adult size. Order Passeriformes. (Photograph by L. L. Rue, III.)

of fear of humans makes them easy targets for thoughtless hunters.

The activities of humans frequently bring about spectacular changes in bird distribution. Both starlings (Fig. 26-31) and house sparrows have been accidentally or deliberately introduced into numerous countries, where they have become the two most abundant bird species on earth with the exception of the domestic fowl. Humans are also responsible for the extinction of many bird species. The dodo became extinct in 1681; since then about 85 other species have followed. Many died naturally, victims of changes in their habitat or competition with better-adapted species. But several have been hunted to extinction, among them the passenger pigeon, which only a century ago darkened the skies over North America in incredible numbers, estimated in the billions (Fig. 26-32).

Hunters kill millions of game birds annually as well as many nongame birds that happen to make convenient targets. An even greater number of game birds die indirectly as the result of eating lead pellets (which they mistake for seeds) or from the crippling effects of embedded pellets. One survey in Wisconsin revealed that the average hunter required 36 shots to down one goose. Of the Canada geese that survived the barrage to fly as far south as the Mississippi Valley, 44% contained embedded lead shot.

But the most destructive effects on birds are usually

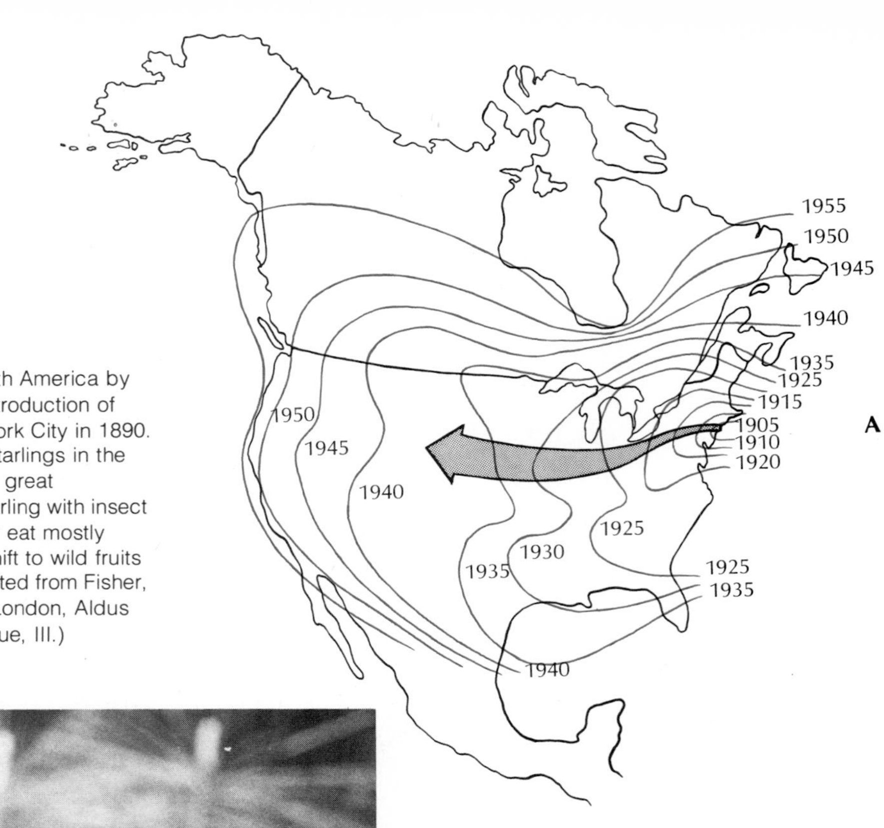

FIG. 26-31 **A,** Colonization of North America by starlings *Sturnus vulgaris* after the introduction of 120 birds into Central Park in New York City in 1890. There are now perhaps 100 million starlings in the United States alone, testimony to the great reproductive potential of birds. **B,** Starling with insect larva. Starlings are omnivorous. They eat mostly insects in spring and summer and shift to wild fruits in fall. Order Passeriformes. (**A,** Adapted from Fisher, J., and R. T. Peterson. 1971. Birds. London, Aldus Books Ltd. **B,** Photograph by L. L. Rue, III.)

FIG. 26-32 "Martha," the last passenger pigeon *Ectopistes migratorius* died in the Cincinnati Zoo in 1914. Passenger pigeons traveled and roosted in huge masses, with nesting areas exceeding 80 sq km (about 30 square miles). In 1869, hunters shot 7.5 million pigeons from a single nesting site. Hunted relentlessly for market, the population eventually dropped too low to sustain colonial breeding. The senior author of this text, when he was a boy, saw Martha some years before she died. (Courtesy Smithsonian Institution, Washington, D.C.)

FIG. 26-33 A female nighthawk *Chordeiles minor* broods two 1-day-old young, their down barely visible under her breast. Nighthawks originally nested on beaches and in old stumps but with the spread of civilization in the United States they discovered building roofs to be ideal nesting sites, convenient to their insect food that was attracted in multitudes to city lights. Now they seldom nest elsewhere. Their nasal "peent" is a familiar summer evening sound above city streets. Order Caprimulgiformes. (Photograph by C. P. Hickman, Jr.)

unintentional. The draining of marshes—more than 99% of Iowa's once-extensive marshland is now farmland—has destroyed waterfowl nesting. Deforestation has likewise had great impact on tree-nesting species. The vertical appendages of civilization, such as television towers, monuments, tall buildings, and electric transmission towers and lines, take a fearful toll during bird migration in bad weather. Tree-spraying programs have virtually eradicated songbirds from certain areas.

Most birds, through their impressive reproductive potential, can replace in numbers those that become victims to human activities. Someone has calculated that a single pair of robins, producing two broods of four young a season, would leave 19,500,000 descendants in 10 years, should all survive at least that long. But although many birds have accommodated to the heavy-handed influence of humans on their environment (Fig. 26-33) and some, like robins, house sparrows, and starlings, even thrive on it, most birds find the changes adverse, and to some species it is lethal.

Classification of living orders

Class Aves (birds) is made up of about 27 orders of living birds and a few fossil orders. More than 8,600 species and many subspecies have been described. Probably only a relatively few species remain to be discovered and named, but many subspecies are added yearly. With their powers of flight and wide distribution, most species of birds are more easily detected than are many animals. Only those that are solitary, shy, and restricted to remote regions have a chance of remaining undiscovered for any length of time. Altogether, the species are grouped into 170 families. Of the 27 recognized orders, 20 are represented by North American species.

The first four orders in the following list make up the group of **ratite,** or flightless, birds; the remainder are the **carinate,** or flying, birds.

Order Struthioniformes (stroo'thi-on-i-for'meez) (LL., *struthio,* ostrich, + L., *forma,* form)—**ostriches.** The ostrich *(Struthio camelus)* is the largest of living birds, with some specimens being 2.4 m tall and weighing 135 kg. These birds cannot fly. The feet are provided with only two toes of unequal size covered with pads, which enable the birds to travel rapidly through sandy country. The ostrich is found in the desert country of Africa and Arabia. There are four species.

Order Rheiformes (re'i-for'meez) (Greek mythology, *Rhea,* mother of Zeus, + form)—**rheas.** These flightless birds are restricted to South America and are often called the American ostrich.

Order Casuariiformes (kazh'u-ar'ee-i-for'meez) (NL., *Casuarius,* type genus, + form)—**cassowaries, emus.** This is a group of flightless birds found in Australia, New Guinea, and a few other islands. Some specimens may reach a height of 1.5 m.

Order Apterygiformes (ap'te-rij'i-for'meez) (Gr., *a,* not, + *pteryx,* wing, + form)—**kiwis.** Kiwis are flightless birds about the size of the domestic fowl, found only in New Zealand. They all belong to the genus *Apteryx,* of which there are three species. Only the merest vestige of a wing is present. The egg is extremely large for the size of the bird.

Order Tinamiformes (tin-am'i-for'meez) (NL., *Tinamus,* type genus, + form)—**tinamous.** These are flying birds found in South America and Mexico. They resemble the ruffed grouse and are classed as game birds. There are more than 60 species in this order.

Order Sphenisciformes (sfe-nis'i-for'meez) (Gr., *sphēniskos,* dim. of *sphen,* wedge, from the shortness of the wings, + form)—**penguins.** Penguins are found in the southern seas, from Antarctica to the Galápagos Islands. Although carinate birds, they use their wings as paddles rather than for flight. They are all web-footed, marine swimmers. The largest penguin is the emperor penguin *(Aptenodytes forsteri)* of the Antarctic, which breeds in enormous rookeries on the shores of that region.

Order Gaviiformes (gay'vee-i-for'meez) (L., *gavia,* bird, probably sea mew, + form)—**loons.** Remarkable swimmers and divers with short legs and heavy body. They live exclusively on fish and small aquatic forms. The familiar great northern diver *(Gavia immer)* is found mainly in northern waters of North America and Eurasia.

Order Podicipediformes (pod'i-si-ped'i-for'meez) (L., *podex,* rump, + *pes, pedis,* foot)—**grebes.** The pied-billed grebe *(Podilymbus podiceps)* is a familiar example of this order. Grebes are most common in old ponds where there are extensive growths of cattails, rushes, and water flags, from which they build their raftlike floating nests. Worldwide distribution.

Order Procellariiformes (pro-sel-lar'ee-i-for'meez) (L., *procella,* tempest, + form)—**albatrosses, petrels, fulmars, shearwaters.** In wingspan (more than 3.6 m in some), albatrosses are the largest of flying birds. *Diomedea* is a common genus of albatrosses. Worldwide distribution.

Order Pelecaniformes (pel'e-can-i-for'meez) (Gr., *pelekan,* pelican, + form)—**pelicans, cormorants, gannets, boobies, etc.** These are fish-eaters with throat pouch and all four toes of each foot included within the web. Worldwide distribution, especially in the tropics.

Order Ciconiiformes (si-ko'nee-i-for'meez) (L., *ciconia,* stork, + form)—**herons, bitterns, storks, ibises, spoonbills, flamingos.** These are long-necked, long-legged, mostly colonial waders. A familiar eastern North American representative is the great blue heron *(Ardea herodias),* which frequents marshes and ponds. Worldwide distribution.

Order Anseriformes (an'ser-i-for'meez) (L., *anser,* goose, + form)—**swans, geese, ducks.** The members of this order have broad bills with filtering ridges at their margins, the foot web is restricted to the front toes, and they have a long breastbone with a low keel. The common domestic mallard duck is *Anas platyrhynchos.* Worldwide distribution.

Order Falconiformes (fal'ko-ni-for'meez) (LL., *falco,* falcon, + form)—**eagles, hawks, vultures, falcons, condors, buzzards.** These are the great diurnal birds of prey. All are strong fliers with keen vision. Worldwide distribution.

Order Galliformes (gal'li-for'meez) (L., *gallus,* cock, + form)—**quail, grouse, pheasants, ptarmigan, turkeys, domestic fowl.** These are henlike vegetarians with strong beaks and heavy feet. Some of the most desirable game birds are in this order. The bobwhite quail *(Colinus virginianus)* is found all over the eastern half of the United States. The ruffed grouse *(Bonasa umbellus),* or partridge, is found in about the same region, but in the woods instead of the open pastures and grain fields, which the bobwhite frequents. Worldwide distribution.

Order Gruiformes (groo'i-for'meez) (L., *grus,* crane, + form)—**cranes, rails, coots, gallinules.** These are prairie and marsh breeders. Worldwide distribution.

Order Charadriiformes (ka-rad'ree-i-for'meez) (NL., *Charadrius,* genus of plovers, + form)—**gulls, oyster catchers, plovers, sandpipers, terns, woodcocks, turnstones, lapwings, snipe, avocets, phalaropes, skuas, skimmers, auks, puffins.** All are shorebirds. They are strong fliers and are usually colonial. Worldwide distribution.

Order Columbiformes (co-lum'bi-for'meez) (L., *columba,* dove, + form)—**pigeons, doves.** All have short necks, short legs, and a short, slender bill. Worldwide distribution.

Order Psittaciformes (sit'ta-si-for'meez) (L., *psittacus,* parrot, + form)—**parrots, parakeets.** These have the upper mandible hinged and movable. Pantropical distribution.

Order Cuculiformes (ku-koo'li-for'meez) (L., *cuculus,* cuckoo, + form)—**cuckoos, roadrunners.** The common cuckoo *(Cuculus canorus)* of Europe lays its eggs in the nests of smaller birds, which rear the young cuckoos. The American cuckoos, black-billed and yellow-billed, rear their own young. Worldwide distribution.

Order Strigiformes (strij'i-for'meez) (L., *strix,* screech owl, + form)—**owls.** Owls are nocturnal predators and have probably the keenest eyes and ears in the animal kingdom. Worldwide distribution.

Order Caprimulgiformes (kap'ri-mul'ji-for'meez) (L., *caprimulgus,* goatsucker, + form)—**goatsuckers, nighthawks, poorwills.** The birds of this group are most active at night and in twilight. They have small, weak legs, wide mouths, and short, delicate bills. The mouth is fringed with bristles in most species. The whippoorwills *(Antrostomus vociferus)* are common in the woods of the Eastern states, and the nighthawk *(Chordeiles minor)* is often seen and heard in the evening flying around city buildings. Worldwide distribution.

Order Apodiformes (up-pod'i-for'meez) (Gr., *apous,* sandmartin; footless, + form)—**swifts, hummingbirds.** The swifts get their name from their speed on the wing. The familiar chimney swift *(Chaetura pelagica)* fastens its nest in chimneys by means of saliva. A swift found in China *(Collocalia)* builds a nest of saliva that is used by the Chinese for soup-making. Most species of hummingbirds are found in the tropics, but there are 14 species in the United States, of which only one, the ruby-throated hummingbird, is found in the eastern part of the country. Most hummingbirds live on nectar, which they suck up with their highly adaptable tongue, although some catch insects also. Worldwide distribtuion.

Order Coliiformes (ka-ly'i-for'meez) (Gr., *kolios,* green woodpecker, + form)—**mousebirds.** These are small birds of uncertain relationship. Restricted to southern Africa.

Order Trogoniformes (tro-gon'i-for'meez) (Gr., *trōgon,* gnawing, + form)—**trogons.** Richly colored birds. Pantropical distribution.

Order Coraciiformes (ka-ray'see-i-for'meez or kor'uh-sigh' uh-for' meez) (NL. Coracii from Gr. *korakias,* a kind of chough [akin to *korax,* raven or crow], + form)—**kingfishers, hornbills, etc.** These birds have strong, prominent bills and colorful plumage. In the eastern half of the United States the belted kingfisher *(Megaceryle alcyon)* is common among most waterways of any size. It makes a nest in a burrow in a high bank or cliff along a water course. Worldwide distribution.

Order Piciformes (pis'i-for'meez) (L., *picus,* woodpecker, + form)—**woodpeckers, toucans, puffbirds, honeyguides.** Woodpeckers are adapted for climbing, with stiff tail feathers and toes with sharp claws. Two of the toes extend forward and two backward. There are many species of woodpeckers in North America, the more common of which are the flickers and the downy, hairy, red-bellied, redheaded, and yellow-bellied woodpeckers. The largest is the pileated woodpecker, which is usually found in deep and remote woods. Worldwide distribution.

Order Passeriformes (pas'er-i-for'meez) (L., *passer,* sparrow, + form)—**perching songbirds.** This is the largest order of birds, containing 56 families and 60% of all birds. Most have a highly developed syrinx. Their feet are adapted for perching on thin stems and twigs. The young are altricial. To this order belong many birds with beautiful songs such as the skylark, nightingale, hermit thrush, mockingbird, meadow lark, robin, and hosts of others. Others of this order, such as the swallow, magpie, starling, crow, raven, jay, nuthatch, and creeper, have no songs worthy of the name. Worldwide distribution.

Annotated references
Selected general readings

Armstrong, E. A. 1963. A study of bird song. Oxford, Oxford University Press. (Dover publication, 1973.) *Comprehensive analysis of bird song.*

Bent, A. C. 1919. Life histories of North American birds. Washington, D.C., U.S. National Museum Bulletins. *More than a score of monographs in this outstanding series provide a wealth of information on birds. A classic that has never been superseded.*

Brown, L. 1970. Eagles. New York, Arco Publishing Co.

Farner, D. S., and J. R. King (eds.). 1971-1975. Avian biology, vols. 1 to 5. New York, Academic Press, Inc. *This advanced treatise composed of contributed chapters is a successor to Marshall's Biology and Comparative Physiology of Birds.*

Fisher, J., and R. T. Peterson. 1971. Birds. An introduction to general ornithology. London, Aldus Books. *Beautifully illustrated general biology of birds by two famous ornithologists.*

Gilliard, E. T. 1958. Living birds of the world. New York, Doubleday & Co., Inc. *A superb book of birds with illustrations, many in color.*

Griffin, D. R. 1964. Bird migration. Garden City, N.Y., The Natural History Press. *Interesting semipopular account of this subject; less detailed treatment but more engaging in style than Matthews' book.*

Headstrom, R. 1949. Birds' nests. New York, Ives Washburn, Inc. *Good photographs of nests with descriptions. Nests are grouped according to patterns of construction.*

Johnsgard, P. A. 1975. Waterfowl of North America. Bloomington, Ind., Indiana University Press. *Thorough treatment of ducks and geese.*

Krutch, J. W., and P. S. Eriksson. 1962. A treasury of birdlore. New York, Paul S. Eriksson, Inc. *Excerpts from the writings of famous ornithologists.*

Lillie, F. R. 1942. On the development of feathers. Biol. Rev. **17**:247-266.

Lorenz, K. Z. 1952. King Solomon's ring. New York, Thomas Y. Crowell Co. *Bird behavior described by a great ethologist.*

Marshall, A. J. 1960-1961. Biology and comparative physiology of birds, vols. 1 and 2. New York, Academic Press, Inc. *A thorough, extensive, and technical treatment of bird physiology by different contributors.*

Matthews, G. V. T. 1968. Bird migration, ed. 2. Cambridge, The University Press. *An authoritative and comprehensive resume of this subject leading to the author's sun-arc hypothesis for bird navigation.*

Newton, A. 1896. A dictionary of birds. London, A. & C. Black. *Despite its age, this famous work contains a wealth of valuable material on many aspects of bird anatomy, physiology, and general biology.*

Peterson, R. T. 1947. Field guide to the birds, ed. 2. Boston, Houghton Mifflin Co. *This popular field guide has a more complete text than the Robbins guide but is not so conveniently arranged. There is a companion volume for western birds.*

Pettingill, O. S., Jr. 1947. Silent wings. Madison, Wisconsin Society for Ornithology, Inc. *An account of the extinction of the passenger pigeon.*

Regal, P. J. 1975. The evolutionary origin of feathers. Q. Rev. Biol. **50**:35-66. *Reviews the problems of deriving a branched feather from an elongate scale.*

Robbins, C. S., B. Brunn, H. Zim, and A. Singer. 1966. Birds of North America: a guide to field identification. New York, Golden Press. *Excellent full-color illustrations of all species, with range maps on opposite pages.*

Rue, L. L., III. 1974. Game birds of North America. New York, Harper & Row, Publishers. *Comprehensive descrip-*

tion of 75 game birds with fine bird paintings by nature artist Douglas Allen.

Savile, D. B. O. 1957. Adaptive evolution in the avian wing. Evolution **11:**212-224. *The author presents a useful classification of bird wings into four flight types according to habitat and function.*

Sparks, J., and T. Soper. 1970. Owls: their natural and unnatural history. New York, Taplinger Publishing Co., Inc. *An engaging, well-illustrated account of owls in fact and fancy.*

Thielcke, G. A. 1976. Bird sounds. Ann Arbor, The University of Michigan Press. *Readable yet comprehensive treatment of bird song.*

Tinbergen, N. 1953. Social behavior in animals. New York, John Wiley & Sons, Inc. *Good introduction to bird behavior.*

Tinbergen, N. 1960. The herring gull's world. New York, Basic Books, Inc., Publishers. *A superb account of bird behavior by a renowned ethologist.*

Wallace, G. J., and H. D. Mahan. 1975. An introduction to ornithology, ed. 3. New York, The Macmillan Co. *College ornithology text.*

Welty, J. C. 1975. The life of birds, ed. 2. Philadelphia, W. B. Saunders Co. *Among the best of the ornithology texts; lucid style and excellent illustrations.*

Selected *Scientific American* articles

Cone, C. D. Jr. 1962. The soaring flight of birds. **206:**130-140 (Apr.). *The nature of air currents and patterns of soaring flight are analyzed.*

Eklund, C. R. 1964. The Antarctic skua. **210:**94-100 (Feb.). *Discusses the biology of this large, aggressive, and cold-adapted bird.*

Emlen, J. E., and R. L. Penny. 1966. The navigation of penguins. **215:**104-113 (Oct.). *Penguins depend on an innate biologic clock and the sun's direction to guide them across hundreds of miles of featureless Antarctic landscape.*

Emlen, S. T. 1975. The stellar-orientation system of a migratory bird. **233:**102-111 (Aug.). *Describes fascinating research with indigo buntings, revealing their ability to orient by the stars.*

Greenewalt, C. H. 1969. How birds sing. **221:**126-139 (Nov.). *The mechanism of bird song is quite different from that of musical instruments or the human voice.*

Keeton, W. T. 1974. The mystery of pigeon homing. **231:**96-107 (Dec.). *Birds use several compass systems in navigation.*

Nicoli, J. 1974. Mimicry in parasitic birds. **231:**92-98 (Oct.).

Peakall, D. B. 1970. Pesticides and the reproduction of birds. **222:**72-78 (Apr.). *Pesticides threaten the survival of several species of birds of prey. The reasons are explained.*

Pennycuick, C. J. 1973. The soaring flight of vultures. **229:**102-109 (Dec.). *Patterns of soaring flight are studied with the aid of a powered glider.*

Sauer, E. G. F. 1958. Celestial navigation by birds. **199:**42-47 (Aug.). *The author describes his ingenious planetarium experiments that demonstrate that migratory birds navigate by the stars.*

Schmidt-Nielsen, K. 1959. Salt glands. **200:**109-116 (Jan.). *The anatomy and physiology of this special salt-excretory organ of marine birds is described.*

Schmidt-Nielsen, K. 1971. How birds breathe. **225:**72-79 (Dec.).

Sladen, W. J. L. 1957. Penguins. **197:**44-51 (Dec.). *These interesting birds have several adaptations that fit them for life in a harsh environment.*

Stettner, L. J., and K. A. Matyniak. 1968. The brain of birds. **218:**64-76 (June).

Tickell, W. L. N. 1970. The great albatrosses. **223:**84-93 (Nov.). *Describes the reproductive behavior and movements of the largest of oceanic birds.*

Tucker, V. A. 1969. The energetics of bird flight. **220:**70-78 (May).

Weis-Fogh, T. 1975. Unusual mechanisms for the generation of lift in flying animals. **233:**80-86 (Nov.). *A study of hovering flight.*

Welty, C. 1955. Birds as flying machines. **192:**88-96 (Mar.). *A description of the remarkable adaptations that fit birds for flight.*

CHAPTER 27
THE MAMMALS

Phylum Chordata
Class Mammalia

A coyote on the western prairie. Though shot, trapped, and poisoned at every opportunity by ranchers who consider coyotes pests, despite their important role in rodent control, and hunted relentlessly for sport, this adaptable carnivore maintains a vigorous population in the West.

Photograph by C. G. Hampson, University of Alberta.

Mammalia (mam-may-′lee-a) (L., *mamma,* breast), with their highly developed nervous system and numerous ingenious adaptations, occupy almost every environment on earth that will support life. Despite their relatively small numbers (4,500 species as compared to 8,600 species of birds, approximately 20,000 species of fishes, and 800,000 species of insects), they are overall the most biologically successful group in the animal kingdom, with the possible exception of the insects. Many potentialities that dwelled more or less latently in other vertebrates are highly developed in mammals. Mammals are exceedingly diverse in size, shape, form, and function. They range in size from the diminutive pigmy shrew, having a body length of less than 4 cm and weight of only a few grams, to whales, which exceed 100 tons in weight.

Yet despite their adaptability, and in some instances because of it, mammals have been influenced by the heavy-handed presence of humans more than any other group of animals. Humans have domesticated numerous mammals for food and clothing, as beasts of burden and as pets. They use millions of mammals each year in biomedical research. They have introduced alien mammals into new habitats, occasionally with benign results, but more frequently with unexpected disaster. Although history provides us with numerous warnings, we continue to overcrop valuable wild stocks of mammals. The valuable whale industry threatens itself with total collapse by exterminating its own resource—a classic example of self-destruction in the modern world, in which each segment of the industry is intent only on reaping all it can today as though tomorrow's supply were of no concern whatever. In some cases destruction of a valuable mammalian resource has been deliberate, such as the officially sanctioned (and tragically successful) policy during the Indian wars of exterminating the bison to drive the Plains Indians to starvation. Although commercial hunting by humans has declined, the ever-increasing human population with the accompanying destruction of wild habitats has harassed and disfigured the mammalian fauna. We are becoming increasingly aware that our presence on this planet as the most powerful product of organic evolution makes us totally responsible for the character of our natural environment. Aware that our welfare has been, and continues to be, closely related to that of the other mammals, it is clearly in our interest to preserve the natural environment of which all mammals, ourselves included, are a part. We need to remember that nature can do without humans, but humans cannot exist without nature.

ORIGIN AND RELATIONSHIPS

Long before the great dinosaurs had reached the peak of their evolutionary success, a group of late Paleozoic reptiles called the **pelycosaurs** appeared and flourished over a period of 40 million years and then nearly disappeared. The pelycosaurs were clearly reptilian, with sprawling gait and undifferentiated teeth. But their descendants, the **therapsids,** were a group of reptiles of the Triassic (early Mesozoic) period with so many mammal-like characteristics that they are considered to be the direct ancestors of the living mammals (Fig. 27-1).

The evolution of the therapsids and their descendants was accompanied by several structural changes that brought them ever closer to full mammalian status. The clumsy limbs of the reptile that stuck out laterally were replaced by straight legs held close to the body, which provided speed and efficiency for hunting. Since reptilian stability was sacrificed by thus raising the animal from the ground, the muscular coordination center of the brain, the cerebellum, took on a greatly expanded role. Among the many changes in the bony structure of the head was the separation of air and food passages in the mouth. This enabled the animal to hold prey in its mouth while breathing. It also made possible prolonged chewing and some predigestion of the food. At some point the premammals acquired warm-bloodedness and those two most characteristic of all mammalian identification tags: hair and mammary glands. Although the fossil sequence provides a good record of skeletal evolution, it is largely silent on the evolution of hair, glands, and warm-bloodedness, and of course, until these characteristics evolved, the premammals were not mammals. Despite the difficulty of selecting a sharp reptile-mammal boundary, it is certain that several groups of premammals had achieved full mammalian status by the end of the Triassic period, some 200 million years ago.

Most of the living mammals belong to the subclass Theria and have descended from a common ancestor of the Jurassic period, some 150 million years ago. However, the monotremes (subclass Prototheria), the egg-laying mammals of Australia, Tasmania, and New Guinea, are so different from the others and possess

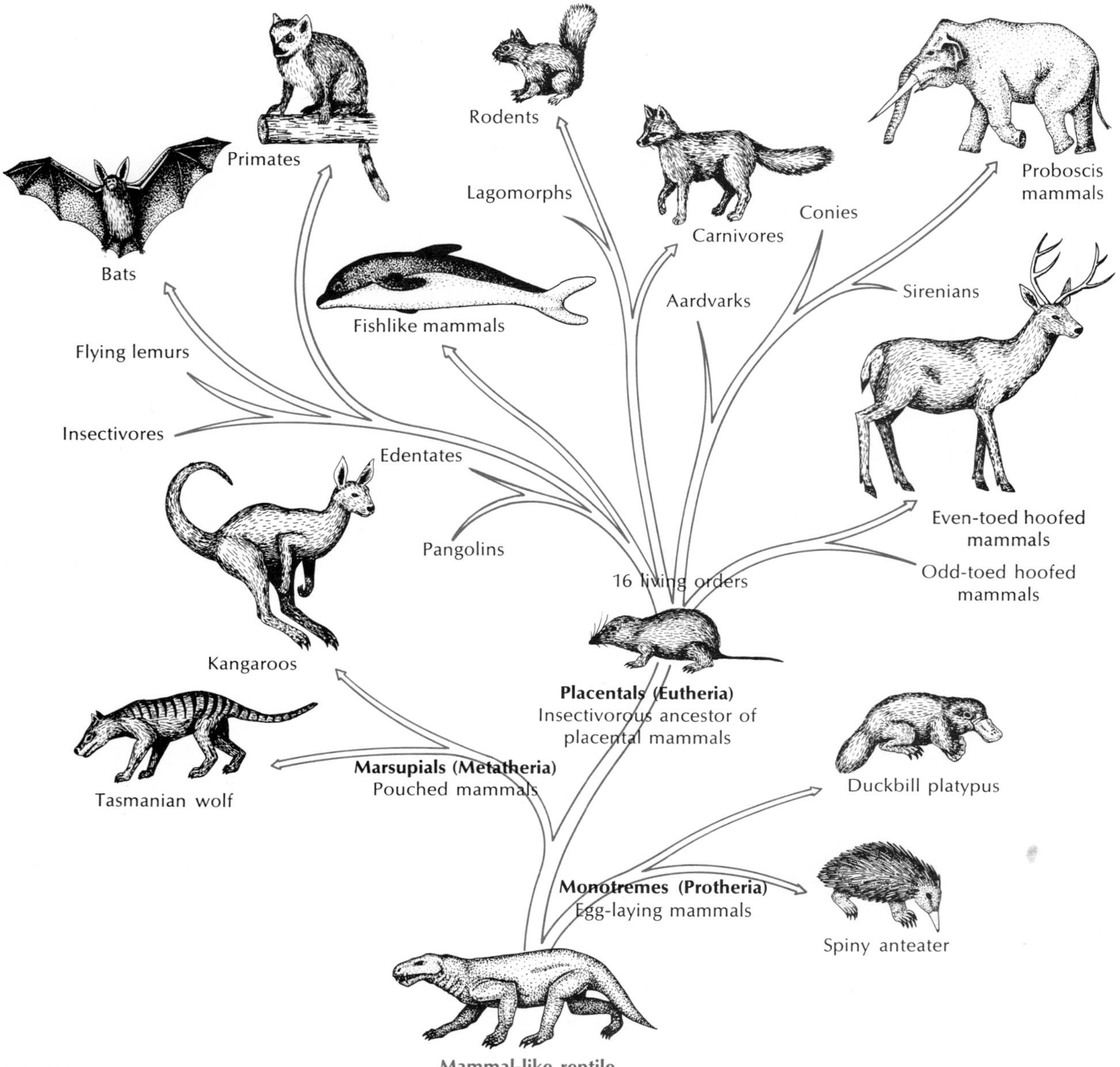

FIG. 27-1 A family tree of the mammals.

so many reptilian characters that they are believed to have descended from an entirely different mammal-like reptile. The separation of Prototheria and Theria probably occurred some 50 million years earlier in the Triassic period. The geologic record during the following Jurassic and Cretaceous periods is fragmentary, in large part because the mammals of these periods were small creatures the size of a rat or smaller, with fragile bones that fossilized only under the most ideal circumstances.

When the dinosaurs vanished near the beginning of the Cenozoic era, the mammals suddenly expanded. This is partly attributable to the numerous ecologic niches vacated by the reptiles, into which the mammals could move as their divergent adaptations fitted them. There were other reasons for their success.

Mammals were agile, warm-blooded, and insulated with hair; they had developed placental reproduction and suckled their young, thus dispensing with vulnerable eggs and nests; and they were more intelligent than any other animal alive. During the Eocene and Oligocene epochs of the Tertiary period (55 to 30 million years ago), the mammals flourished and reached their peak. In terms of number of species, this was the golden age of mammals. They have declined in numbers ever since; only 932 of the 2,864 known mammalian genera (33%) are still living. However, extinction of species within a group is a natural and expected consequence of changing conditions and does not necessarily portend group extinction. Rather than disappearing, mammals are a secure group that dominate the land environment as thoroughly now as they did 50 million years ago.

Class Mammalia is divided into two subclasses as follows: subclass **Prototheria** includes the monotremes, or egg-laying mammals. Subclass **Theria** includes two infraclasses, the **Metatheria,** with one order (marsupials), and the **Eutheria,** with the rest of the mammalian orders, all of which are placental mammals. A complete classification is found on pp. 637 to 640.

Characteristics

Since mammals and birds both evolved from reptiles, we can expect to find, and do find, many structural similarities among the three groups. It is in fact much easier to point to numerous resemblances between the mammals and the reptiles from which they descended, than to point to characteristics that are unique and diagnostic for mammals. Hair is the most obvious mammalian characteristic, although it is vastly reduced in some (such as whales) and although reptilian scales, from which hair is derived, may persist (such as on tails of the rat and beaver). A second unique characteristic of mammals is the method of nourishing their young through milk-secreting glands; reptiles have nothing remotely similar. Although less obvious, several important differences are present in cranial and jaw structure and jaw articulation. Mammals have **diphyodont teeth** (milk teeth replaced by a permanent set of teeth) rather than reptilian **polyphyodont teeth** (successive sets of teeth). But the single most important factor contributing to the success of mammals is the remarkable development of the **neocerebrum,** permitting a level of adaptive behavior, learning, curiosity, and intellectual activity far beyond the capacity of any reptile.

We may summarize the mammalian characteristics as follows:

1. **Body covered with hair,** but reduced in some
2. **Integument with sweat, sebaceous,** and **mammary glands**
3. Skeletal features of skull with **two occipital condyles, seven cervical vertebrae** (usually), and often an elongated tail
4. Mouth with **diphyodont teeth** on both jaws
5. **Movable eyelids** and **fleshy external ears**
6. Four limbs (reduced or absent in some) adapted for many forms of locomotion
7. Circulatory system of a four-chambered heart, **persistent left aorta,** and **nonnucleated, biconcave red blood corpuscles**
8. Respiratory system of lungs and a voice box
9. **Muscular partition (diaphragm) between thorax and abdomen**
10. Excretory system of metanephros kidneys and ureters that usually open into a bladder
11. Nervous systems of a well-developed brain and 12 pairs of cranial nerves
12. Endothermic and homeothermic
13. Cloaca present only in monotremes
14. Separate sexes; reproductive organs of a penis, **testes (usually in a scrotum),** ovaries, oviducts, and vagina
15. Internal fertilization; **eggs develop in a uterus** with **placental attachment** (except in monotremes); **fetal membranes (amnion, chorion, allantois)**
16. Young nourished by **milk from mammary glands**

STRUCTURAL AND FUNCTIONAL ADAPTATIONS OF MAMMALS

Integument and derivatives

The mammalian skin and its modifications especially distinguish mammals as a group. As the interface between the animal and its environment, the skin is strongly molded by the animal's way of life. In general the skin is thicker in mammals than in other classes of vertebrates, although it is made up of the two typical divisions—epidermis and dermis (see Fig. 28-1, *B*, p. 647). Among the mammals the dermis becomes much thicker than the epidermis. The epidermis varies in thickness. It is relatively thin where it is well protected by hair, but in places subject to much contact and use, such as the palms or soles, its outer layers become thick and cornified with keratin.

Hair

Hair is especially characteristic in mammals, although humans are not very hairy creatures, and in

FIG. 27-2 Beaver *Castor canadensis* (order Rodentia) standing on its dam. Although coarse guard hair is wet, the thick layer of fine underfur is nearly dry. Beavers feed mostly on the inner bark of higher, more tender branches of aspen, willow, and birch. They are equipped with powerful jaws and chisel-sharp incisors for felling trees, which are used both for food and for building material for dams. Beavers are valuable conservationists and under protection are recovering rapidly from near-extermination during the nineteenth century. They are regularly moved from lowlands, where they are a nuisance to farmers, to mountain areas where their dams control floods and create marshes for waterfowl. As ponds silt up, they provide rich soil for vegetation, eventually becoming meadows and deciduous forests. (Photograph by L. L. Rue, III.)

FIG. 27-3 Snowshoe, or varying, hare *Lepus americanus* (order Lagomorpha), in brown summer coat, **A,** and white winter coat, **B.** In winter, extra hair growth on the hind feet broadens the animal's support on snow. Snowshoe hares are common residents of the taiga (northern coniferous forests) and are an important food for lynxes, foxes, and other carnivores. Population fluctuations of hares and their predators are closely related. (Photograph by L. L. Rue, III.)

the whales hair is reduced to only a few sensory bristles on the snout. The hair follicle from which a hair grows is an epidermal structure even though it lies mostly in the dermis and subdermal (subcutaneous) tissues (Fig. 28-1, *B*). The hair grows continuously by rapid proliferation of cells in the base of the follicle. As the hair shaft is pushed upward, new cells are carried away from their source of nourishment and turn into the same dense type of keratin **(hard keratin)** that constitutes nails, claws, hooves, and feathers. On a weight basis, hair is by far the strongest material in the body. It has a tensile strength comparable to rolled aluminum, which is nearly twice as strong, weight for weight, as the strongest bone.

A hair is more than a strand of keratin. It consists of three layers: the medulla or pith in the center of the hair, the cortex with pigment granules next to the medulla, and the outer cuticle composed of imbricated scales. The hair of different mammals shows a considerable range of structure. It may be deficient in cortex, such as the brittle hair of deer, or it may be deficient in medulla, such as the hollow, air-filled hairs of the wolverine, so favored by northerners for trimming the hoods of parkas because it resists frost accumulation. The hairs of rabbits and some others are scaled to interlock when pressed together. Curly hair, such as that of sheep, grows from curved follicles.

Each hair follicle is provided with a small strip of muscle that, when contracted, pulls the hair upright. These erector muscles are under the control of the sympathetic nervous system. During certain emotional states (fear and excitement), many animals erect their hair, particularly that on the neck and between the shoulders, thus increasing the apparent size of the body. In humans, contraction of the erector muscles after excitement or cold stimulation causes the hair to stand up and the skin to dimple in above the muscle attachment. The result is "gooseflesh."

Mammals characteristically have two kinds of hair forming the **pelage:** (1) dense and soft **underhair** for insulation and (2) coarse and longer **guard hair** for protection against wear and to provide coloration. The underhair traps a layer of insulating air; in aquatic animals such as the fur seal, otter, and beaver, it is so dense that it is almost impossible to wet it. In water the guard hairs wet and mat down over the underhair, forming a protective blanket (Fig. 27-2). A quick shake when the animal emerges flings off the water and leaves the outer guard hair almost dry.

Molting and coloration. When a hair reaches a certain length, it stops growing. In rare instances, such as the mane of a horse, it may persist as a mature hair throughout the life of the animal. Normally however, it remains in the follicle only until a new growth starts, whereupon it falls out. In humans hair is shed and replaced throughout life. But in most mammals, there are periodic molts of the entire coat. In the simplest cases, such as in foxes and seals, the coat is shed once each year, during the summer months. In the fox, molting begins on the legs and hindquarters and progresses forward, with the new coat appearing as soon as the old is lost.

Most mammals have two annual molts, one in the spring and one in the fall. The summer coat is always much thinner than the winter and is usually a different color. Several of the northern mustelid carnivores (such as the weasel) have white winter coats and colored summer coats. It was once believed that the white winter pelage of arctic animals served to conserve body heat by reducing radiation loss, but recent research has shown that dark and white pelages radiate heat equally. The winter white of arctic animals is simply camouflage in a land of snow. The varying hare of North America has three annual molts: the white winter coat is replaced by a brownish gray summer coat, and this is replaced in autumn by a grayer coat, which is soon shed to reveal the winter white coat beneath (Fig. 27-3).

Outside of the arctic, most mammals wear somber colors for protective purposes. Often the species is marked with "salt-and-pepper" coloration or a disruptive pattern that helps to make it inconspicuous in its natural surroundings. Examples are the leopard's spots, the stripes of the tiger, and the spots of fawns. Zoologists have long wondered what adaptive purpose, if any, is served by the clearly defined black and white pattern of zebras (Fig. 27-16). Although the zebra would appear to be a conspicuous target for predators, naturalists in Africa report that its stripes tend to blur the outline of the animal when it moves, making it difficult to distinguish from the background and from the more uniformly colored species of the African plains. Other mammals, for example, skunks, advertise their presence with conspicuous warning coloration (Fig. 27-4).

An interesting aspect of color is the pair of rump patches of the pronghorn antelope, which are composed of long white hairs erected by special muscles.

FIG. 27-4 Striped skunk, *Mephitis mephitis* (order Carnivora), in warning display. When threatened, a skunk first stamps the ground with its forefeet, then raises its tail with tip hanging down. If further alarmed, it raises the tip of its tail and accurately squirts an irritating and foul-smelling yellowish fluid from a pair of anal glands in the direction of its tormentor. The conspicuous black and white coloration of the skunk is a warning to aggressors of the unpleasant consequences of an attack. (Photograph by L. L. Rue, III.)

When alarmed, the animal can flash these patches in a manner visible for a long distance. They may be used as a warning signal to other members of the herd. The well-known "flag" of the Virginia whitetailed deer serves a similar purpose (Fig. 27-5).

Mammalian coloration is principally caused by pigmentation in the hair, although in a few, bare surfaces of skin may be found with bright hues, such as in the cheeks and ischial callosities of the mandrill (a species of baboon), which may be attributable to pigment or to blood capillaries in the skin. At least two types of chromatophores are found in mammals—melanophores (black and brown pigment) and xanthophores (red and yellow pigment). Pigment granules may also lie outside the regular pigment cells. Although the color of hair may fade to some extent, any noticeable change in the color of a mammal's fur coat must be brought about by molting.

Albinism, or a lack of pigment, may happen in most kinds of mammals and is not to be confused with leucism (Gr., *leukes,* white), the white coloring of arctic mammals and birds. Albinos have red eyes and pink skin, whereas arctic mammals in their white winter coats have dark eyes and often dark-colored ear tips, noses, and tail tips. **Melanism,** or an excess of black pigment, is also occasionally seen among animals.

Derivatives of hair. The hair of mammals has been

FIG. 27-5 Alarmed white-tailed deer *Odocoileus virginianus* (order Artiodactyla) lifts its tail to expose conspicuous white underside. This mild alarm signal serves two functions: (1) it silently warns other deer of impending danger, and (2) it communicates to the predator that it has been seen and thus would be wasting its effort to chase an alerted prey. (Photograph by L. L. Rue, III.)

A

B

FIG. 27-6 Nine-banded armadillo *Dasypus novemcinctus* (order Edendata) in exploring, **A,** and defensive, **B,** positions. Pelvic and pectoral bony shields are separated by nine bony movable bands. This species is abundant in South America and has invaded the southern United States where it is prospering on a diet of insects, scorpions, and worms. Reproduction is unique. Mating occurs in the summer. The single egg is immediately fertilized but does not implant in the uterus for 3 to 4 months. Then it suddenly splits into four cells, which implant separately and develop into four genetically identical embryos. Quadruplets are born in the spring and soon become independent. (Photographs by L. L. Rue, III.)

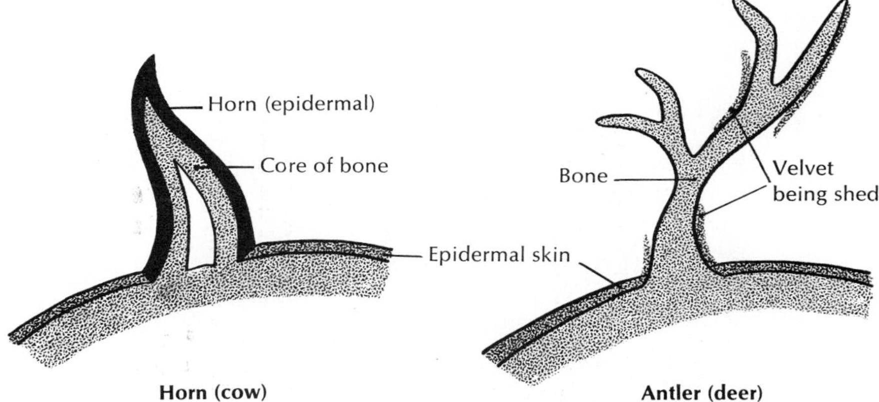

FIG. 27-7 Chief differences between horns and antlers. Bone, a dermal (mesoderm) derivative, forms the basic part of each type, and when epidermal velvet with its hair is shed, bone forms all of antlers. Antlers are shed annually (in winter) when zone of constriction below burr appears near the skull. Horns do not branch and are not shed. Injection of testosterone will prevent shedding of antlers.

modified to serve many purposes. The bristles of hogs, vibrissae on the snouts of most mammals, and the spines of porcupines and their kin are examples.

Vibrissae, commonly and incorrectly called ''whiskers,'' are really sensory hairs that provide an additional special sense to many mammals. The bulb at the base of each follicle is provided with a large sensory nerve. The slightest movement of a vibrissa generates impulses in the nerve endings that travel to a special sensory area in the brain. The vibrissae are especially long in nocturnal and burrowing animals. In seals they apparently serve as a ''distance touch'' sensitive to pressure waves and turbulence in the water caused by objects or passing fish. Vision is of little use to seals hunting in turbid water in which they are frequently found, and investigators have noted that blind seals remain just as fat and healthy as normal seals.

Porcupines, hedgehogs, the echidna, and a few other mammals have developed an effective and dangerous spiny armor; the spines of the common North American porcupine break off at the bases when struck and, aided by backward-pointing hooks on the tips, work deeply into their victim. To assist slow learners like dogs in understanding what they are dealing with, porcupines rattle the spines and prominently display the white markings on the quills toward their tormentors.

Of quite different origin is the armadillo's shell. The scales are small bones of dermal origin and are covered with a tough, horny epidermis. As revealed in the photograph of an armadillo in its defensive posture (Fig. 27-6, *B*), hair grows out between the scales and on the unscaled underside of the body.

Horns and antlers

Three kinds of horns or hornlike substances are found in mammals (Fig. 27-7).

True horns found in ruminants, for example, sheep and cattle, are hollow sheaths of keratinized epidermis that embrace a core of bone arising from the skull. Horns are not normally shed, are not branched (although they may be greatly curved), and are found in both sexes. The horns of North American pronghorn antelope are unique in that they are shed each year after the breeding season. But unlike the shedding of deer antlers, the new horn replaces the old by growing up inside and pushing the outer sheath off.

Antlers of the deer family occur in males only. They are entirely bone when mature. During their annual growth, antlers develop beneath a covering of highly vascular soft skin called **''velvet.''** When growth of the antlers is complete just prior to the breeding season, the blood vessels constrict and the stag tears off the velvet by rubbing the antlers against trees (Fig. 27-8). The antlers are dropped after the breeding season. New buds appear a few months later to herald the next set of antlers. For several years each new pair of antlers is larger and more elaborate than

FIG. 27-8 Bull moose *Alces alces* (order Artiodactyla) shedding velvet. Antlers begin their growth each spring, stimulated by gonadotropin from the pituitary and progressively rising level of testosterone from the testes. When growth is complete in the late summer, vessels at base (burr) of antlers constrict. The skin, called velvet, dies, shrivels, and sloughs off, revealing bony antlers beneath. In late winter after the breeding season, bone is resorbed at the base, weakening the joint, and the antlers drop off. Moose are solitary in their habits; bulls travel alone, whereas cows travel accompanied by calves. Because of the large amount of vegetable food needed to exist, winter is frequently a struggle for survival. Moose, called "elk" in Eurasia, are so abundant in Scandinavia that they are game cropped and marketed commercially. (Photograph by L. L. Rue, III.)

was the previous set. The annual growth of antlers places a strain on the mineral metabolism since during the growing season a large moose or elk must accumulate 50 or more pounds of calcium salts from its vegetable diet.

The **rhinoceros horn** is the third kind of horn. Hairlike horny fibers arise from dermal papillae and are cemented together to form a horn (Fig. 27-9).

Glands

Of all vertebrates, mammals have the greatest variety of integumentary glands. Most fall into one of four classes: sweat, scent, sebaceous, and mammary. All are derivatives of the epidermis.

Sweat glands are simple, tubular, highly coiled glands that occur over much of the body in most mammals. They are present in no other vertebrates. Two

FIG. 27-9 White rhinoceros *Diceros simus* (order Perissodactyla) of east African savanna. Unlike true horns, rhinoceros horn is a dense structure of closely packed fibers, derived from hair. In Asia, where the horn is believed to have aphrodisiac qualities, all three species of Asian rhinoceroses (Indian, Sumatran, Javan) have been hunted close to extinction. Only about 50 Java rhinoceroses remain. The two African species (black and white) have been slaughtered for the same reason but are presently under protection in some areas. Rhinoceroses are grazers (white rhinoceros) and browsers (black rhinoceros) with poor eyesight and a wary, nervous disposition. (Photograph by B. Tallmark.)

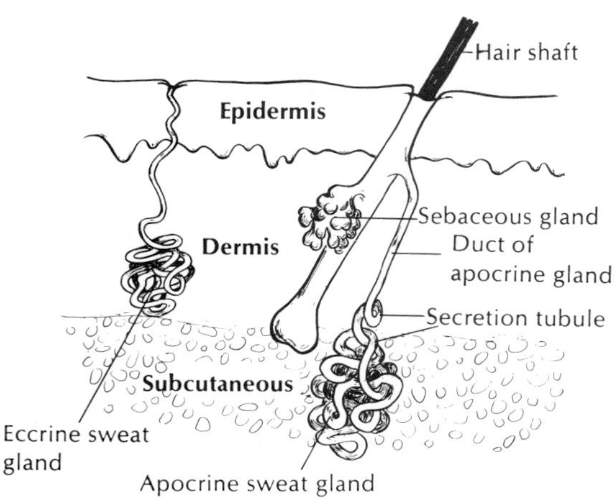

FIG. 27-10 Sweat glands (eccrine and apocrine). Phylogenetically, apocrine glands are older. Eccrine glands, best developed in primates, play an important role in temperature regulation. Most glands in dogs, pigs, cows, horses, and others are apocrine, but these have declined in humans, along with hair, for they develop from follicular epithelium. Apocrine glands are not involved in temperature regulation, but their odorous secretions play a part in sexual attraction.

FIG. 27-11 Olfactory exploration during courtship in thirteen-lined ground squirrels *Citellus tridecemlineatus* (order Rodentia, family Sciuridae). Scent is by far the most important means of communication in mammals. Each produces an odor characteristic of the species; the same species may vary its scent for different communicative purposes. Anal glands, as in these two ground squirrels, are especially common. This species lives on the prairies of western Canada and United States and feeds on grains, other vegetation, and insects. (Photograph by C. G. Hampson.)

kinds of sweat glands may be distinguished, eccrine and apocrine (Fig. 27-10). **Eccrine glands** secrete a watery sweat that functions mainly in temperature regulation (evaporative cooling). They occur in hairless regions, especially the foot pads, in most mammals, although in horses, some apes, and humans they are scattered all over the body. They are much reduced or absent in rodents, rabbits, whales, and others. Dogs are now known to have sweat glands all over the body. In human beings, racial differences are pronounced. Blacks, who have more sweat glands than whites, can tolerate warmer temperatures.

Apocrine glands, the second type of sweat gland, are larger than eccrine glands and have longer and more winding ducts. Their secretory coil is in the subdermis. They always open into the follicle of a hair or where a hair has been. Phylogenetically they are much older than the eccrine gland and are found in all mammals, some of which have only this kind of gland. Women have twice as many apocrine glands as men. They develop about the time of sexual puberty and are restricted (in the human species) to the axillae, mons pubis, breasts, external auditory canals, prepuce, scrotum, and a few other places. Their secretion is not watery like ordinary sweat (eccrine gland) but is a milky, whitish or yellow secretion that dries on the skin to form a plasticlike film. Only the tip of the secretory cell is destroyed in the process of secretion. Their secretion is not involved in heat regulation, but their activity is known to be correlated with certain aspects of the sexual cycle, among other possible functions.

Scent glands are present in nearly all mammals. Their location and functions vary greatly. They are used in communication with members of the same species (Fig. 27-11), to mark territorial boundaries, for warning, or for defense. Scent-producing glands are located in orbital, metatarsal, and interdigital regions (deer); behind the eyes and on the cheek (pika, woodchuck); in preputial regions on the penis (muskrats, beavers, many canines); at the base of the tail (wolves and foxes); at the back of the head (dromedary); and in the anal region (skunks, minks, weasels). These last, the most odoriferous of all glands, open by ducts into the anus; their secretions can be discharged forcefully for several feet (Fig. 27-4). During the mating season many mammals give off strong scents for attracting the opposite sex. Humans are also endowed with scent glands. But civilization has taught us to dislike our own scent, a concern that has stimulated a lucrative deodorant industry to produce an endless output of soaps and odor-masking compounds.

Sebaceous glands (Fig. 27-10) are intimately associated with hair follicles, although some are free and open directly onto the surface. They are classified as **holocrine glands** because the cellular lining of the gland itself is discharged in the secretory process and must be renewed for further secretion. These gland cells become distended with a fatty accumulation, then die and are expelled as a greasy mixture called **sebum** into the hair follicle. Called a "polite fat" because it does not turn rancid, it serves as a dressing to keep the skin and hair pliable and glossy. Most mammals have sebaceous glands all over the body; in humans they are most numerous in the scalp and on the face.

Mammary glands, which provide the name for mammals, are probably modified apocrine glands, although recent studies suggest that they may have derived from sebaceous glands. Whatever their evolutionary origin, they occur on all female mammals and on most, if not all, male mammals; on the latter they are inactive and often covered by hair. They develop by the thickening of the epidermis to form a milk line along each side of the abdomen in the embryo. On certain parts of these lines the mammae appear, while the intervening parts of the ridge disappear.

In the human female, the mammary glands begin at puberty to increase in size because of fat accumulation and reach their maximum development in about the twentieth year. The breasts (or mammae) undergo additional development during pregnancy. In other mammals, the breasts are swollen only periodically when they are distended with milk during pregnancy and subsequent nursing of the young. The outlets of the gland are by elevated nipples (absent in monotremes). The glands are located on the thorax of primates, bats, and a few others but on the abdomen or inguinal region in other mammals. Nipples vary in number from two in the human being, horse, bat, and others to as many as 24 in the koala bear, a marsupial. The number is not always constant in the same species.

Milk varies in composition. Human milk is composed of 1% to 2% protein, 3% to 5% fat, 6% to 8% carbohydrate, 0.2% salts, and the remainder water. Cow's milk has more protein, about the same amount of fat (average 4%), and less carbohydrate. In marine mammals (whales and seals) and arctic mammals (polar bears and caribou), where rapid growth of the young is important for species survival, the milk may contain 30% to 40% fat.

Food and feeding

Mammals have exploited an enormous variety of food sources; some require highly specialized diets, whereas others are opportunistic feeders that thrive on diversified diets. In all, food habits and physical structure are inextricably linked together. A mammal's adaptations for attack and defense and its specialization for finding, capturing, reducing, swallowing, and digesting food all determine a mammal's shape and habits.

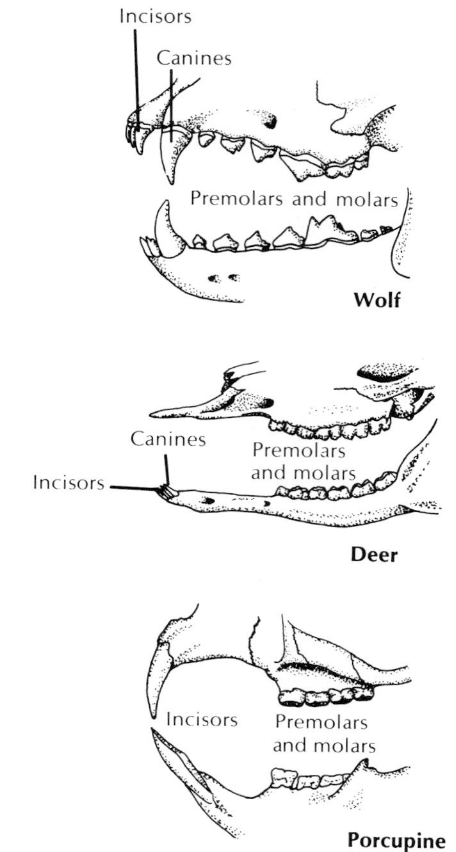

FIG. 27-12 Adaptations of mammal tooth patterns for different kinds of diet. Sharp canines of wolf are designed for stabbing, and premolars and molars are for cutting rather than grinding. Browsing deer has predominantly grinding teeth; lower incisors and canines bite against a horny pad in the upper jaw. Porcupine has no canines; self-sharpening incisors are used for gnawing. (Modified from Carrington, R. 1968. The mammals, New York Life Nature Library.)

Teeth

Perhaps more than any other single physical characteristic, teeth reveal the life-style of a mammal (Fig. 27-12). It has been claimed that, if all mammals except humans were extinct and represented only by fossil teeth, we could still construct a classification as correct as the one we have now, which is based on all anatomic features. All mammals have teeth, except

FIG. 27-13 Malocclusion in the woodchuck *Marmota monax* (order Rodentia, family Sciuridae). Teeth of rodents are two pairs of deeply implanted incisors that grow throughout life to keep pace with wear. Should incisors not meet correctly, continued growth prevents feeding and the animal starves to death, as did this unfortunate one. Natural selection often works harshly to remove genetic defects! Woodchucks (also called "groundhogs") are hibernators. (Photograph by L. L. Rue, III.)

certain whales, monotremes, and anteaters, and their modifications are correlated with what the mammal eats.

Typically, mammals have a **diphyodont** dentition, that is, two sets of teeth: a set of deciduous, or milk, teeth that are replaced by a set of permanent teeth. In any given species, mammalian teeth are modified to perform specialized tasks such as cutting, nipping, gnawing, seizing, tearing, grinding, and chewing. Teeth differentiated in this manner in the individual are called **heterodont,** in contrast to the uniform, **homodont** dentition characteristic of lower vertebrates.

Usually four types of teeth are recognized. **Incisors,** with simple crowns and slightly sharp edges, are mainly for snipping or biting; **canines,** with long conic crowns, are specialized for piercing; **premolars,** with compressed crowns and one or two cusps, are suited for shearing and slicing; and **molars,** with large bodies and variable cusp arrangement, are for crushing and mastication. Molars always belong to the permanent set.

Carnivorous animals have teeth with sharp edges for tearing and piercing. They have well-developed canines, but some of the molars are poorly developed. In the herbivores the canines are suppressed, whereas the molars are broad, with enamel ridges for grinding. Such teeth are also usually high-crowned, in contrast to the low-crowned teeth of carnivores.

The incisors of rodents have enamel only on the anterior surface so that the softer dentin behind wears away faster, resulting in chisel-shaped teeth that are always sharp. Moreover, rodent incisors grow throughout life and must be worn away to keep pace with growth. The failure of two opposing incisors to meet results in serious consequences to the animal (Fig. 27-13).

The **tusks** of the elephant and the wild boar are modifications of teeth. The elephant tusk is a modified upper incisor and may be present in both males and females; in the wild boar the tusk is a modified canine present only in the male. Both are formidable weapons.

The number and arrangement of permanent teeth

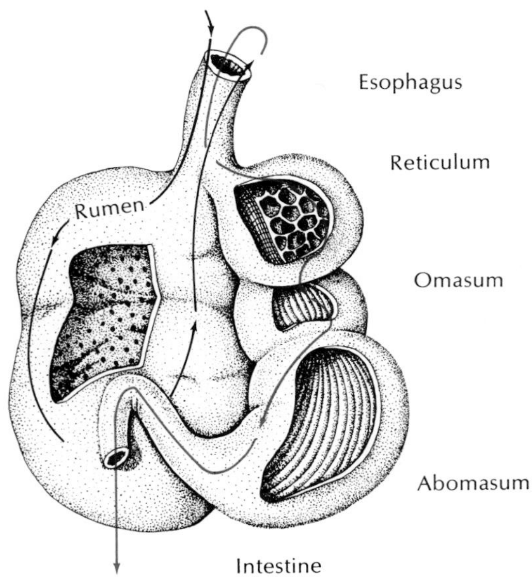

FIG. 27-14 Ruminant's stomach. Food passes first to rumen (sometimes via reticulum) and then is returned to mouth for chewing (chewing the cud, or rumination) *(black arrow).* After reswallowing, food passes to reticulum, omasum, and abomasum for final digestion *(red arrow).* See text for further explanation.

are expressed by a **dental formula.** The figures above the horizontal line represent the number of incisors, canines, premolars, and molars on half of the upper jaw; the figures below the line indicate the corresponding teeth in half of the lower jaw.

Human	Dog
2-1-2-3	3-1-4-2
2-1-2-3	3-1-4-3

Feeding types

On the basis of food habits, animals may be divided into herbivores, carnivores, omnivores, and insectivores.

Herbivorous animals that feed on grasses and other vegetation form two main groups: **browsers** or **grazers,** such as the ungulates (horses, swine, deer, antelope, cattle, sheep, and goats), and the **gnawers** and **nibblers,** such as the rodents and rabbits.

Herbivorous mammals have a number of interesting adaptations for dealing with their massive diet of plant food. Cellulose, the structural carbohydrate of plants, is a potentially nutritious foodstuff, comprised of long chains of glucose. However, the glucose molecules in cellulose are linked by a type of chemical bond that few enzymes can attack. No vertebrates synthesize cellulose-splitting enzymes. Instead the herbivorous vertebrates harbor a microflora of anaerobic bacteria in huge fermentation chambers in the gut. These bacteria break down the cellulose, releasing a variety of fatty acids, sugars, and starches that the host animal can absorb and utilize.

In some herbivores, such as horse and rabbit, the gut has a capacious sidepocket, or diverticulum, called a **cecum,** which serves as a fermentation chamber and absorptive area. Hares and rabbits often eat their fecal pellets, giving the food a second pass through the fermenting action of the intestinal bacteria.

The **ruminants** (cattle, bison, buffalo, goats, antelopes, sheep, deer, giraffe, and okapis) have a huge **four-chambered stomach** (Fig. 27-14). When a ruminant feeds, grass passes down the esophagus to the **rumen,** where it is broken down by the rich microflora and then is formed into small balls of cud. At its leisure the ruminant returns the cud to its mouth where it is deliberately chewed at length to crush the fiber. Swallowed again, the food returns to the rumen where it is partly digested by the cellulolytic bacteria. The pulp then passes to the **reticulum,** then to the **omasum,** and finally to the **abomasum** (''true'' stomach) where proteolytic enzymes are secreted and digestion is completed. Herbivores in general have large and long digestive tracts and must eat a large amount of plant food to survive. A large African elephant weighing 6 tons must consume between 300 and 400 pounds of rough fodder each day to obtain sufficient nourishment for life.

Carnivorous mammals feed mainly on herbivores. This group includes foxes, weasels, cats, dogs, wolverines, otters (Fig. 27-15), lions, and tigers. Carnivores are well equipped with biting and piercing teeth and powerful clawed limbs for killing their prey. Since their protein diet is much more easily digested than is the woody food of herbivores, their digestive tract is shorter and the cecum is small or absent. Carnivores eat separate meals and have much more leisure time for play and exploration.

In general, carnivores lead more active—and by human standards more interesting—lives than do the herbivores. Since a carnivore must find and catch its prey, there is a premium on intelligence; many carnivores, the cats for example, are noted for their stealth and cunning in hunting prey. Although evolution seems

FIG. 27-15 North American otter *Lutra canadensis* (order Carnivora) with a largemouth bass. Otters are excellent divers, often swimming along the bottom of watercourses with serpentine body movements, hunting for crayfish, fishes, frogs, and molluscs. They also hunt from the bank, diving with accuracy when prey is sighted. (Photograph by L. L. Rue, III).

FIG. 27-16 Zebras *Equus burchelli* (order Perissodactyla) at waterhole in East Africa. Animal in foreground is on alert for predatory lions while others drink. If danger is spotted, it will emit a short bark to set the herd running. Zebras travel in large herds of family units, each consisting of a stallion, several mares, and their foals. (Photograph by C. G. Hampson.)

FIG. 27-17 Olive baboons *Papio anubis* (order Primates, suborder Anthropoidea) feeding on large ants. Baboons are omnivorous; they eat seeds, fruits, bulbs, grasses, small mammals and birds, eggs, and insects. Here baboons have torn open an ant nest and pick up disturbed ants as they emerge. Baboons live in troops of variable size, which may include several adult males. They search for food during the day and sleep in trees at night. (Photograph by B. Tallmark.)

FIG. 27-18 Coyote *Canis latrans* (order Carnivora) leaping to break snow crust, while hunting for mice, voles, and shrews. These small mammals, which comprise the principal food of coyotes during the difficult winter months, live on the frozen ground surface, protected by the snow cover above. The resilient coyote is still rather common in North America, despite relentless harassment by humans. (Photograph by C. G. Hampson.)

to have favored the carnivores, their very success has lead to a selection of herbivores capable of either defending themselves or of detecting and escaping carnivores. Thus, for the herbivores there has been a premium on keen senses and agility (Fig. 27-16). Some herbivores, however, survive by virtue of their sheer size (for example, elephants) or by defensive group behavior (for example, muskoxen).

Humans have changed the rules in the carnivore-herbivore contest. Carnivores, despite their intelligence, have suffered much from human presence and have been virtually exterminated in some areas. Herbivores, on the other hand, especially the rodents with their potent reproductive potential, have consistently defeated the most ingenious efforts to banish them from the environment. Indeed the problem of rodent pests in agriculture has been intensified; humans have removed carnivores, which served as the herbivores' natural population control, but have not been able to devise a suitable substitute.

Omnivorous mammals live on both plant food and animals. Examples are pigs, raccoons, rats, bears, hu-

FIG. 27-19 American pika *Ochotona princeps* (order Lagomorpha) making hay. This member of the rabbit order lives in colonies in the Rocky Mountains where it excavates extensive tunnel systems among boulders and rocky ledges. They do not hibernate but prepare for the winter by storing grasses, twigs, thistles, and gooseberries in a haystack in special storerooms after carefully drying them first under the autumn sun. (Photograph by L. L. Rue, III.)

mans, and most other primates (Fig. 27-17). Many carnivorous forms also eat fruits, berries, and grasses when hard pressed. The fox, which usually feeds on mice, small rodents, and birds, will eat frozen apples, beechnuts, and corn when its normal sources are scarce.

Insectivorous mammals are those that subsist chiefly on insects and grubs. Examples are moles, shrews, and most bats. The insectivorous category is not a well-distinguished one however, because many omnivores, carnivores, and even some herbivores will eat insects on occasion, for example, bears, raccoons, mice, baboons (Fig. 27-17), and ground squirrels.

For most mammals, searching for and eating food occupies most of their active life. Seasonal changes in food supplies are considerable in temperate zones. Living may be easy in the summer when food is abundant, but in winter many carnivores must range far and wide to eke out a narrow existence (Fig. 27-18). Some migrate to regions where food is more abundant. Others hibernate and sleep the winter months away. But there are many provident mammals that build up food stores during periods of plenty. This habit is most pronounced in many of our rodents, such as squirrels, chipmunks, gophers, and certain mice. All the tree squirrels—red, fox, and gray—collect nuts, conifer seeds, and fungi and bury these in caches for winter use. Often each item is hidden in a different place (scatter hoarding) and scent-marked to assist relocation in the future. The chipmunk is one of the greatest providers, for it spends the autumn months in collecting nuts and seeds. Some of its caches may exceed a bushel. Pikas lay in large hoards of grass and thistles to carry them over the winter (Fig. 27-19).

Body weight and food consumption

The relationship between body size and metabolic rate has been discussed in relation to food consumption of birds (p. 581) and will be treated again in Chapter 36 (p. 836). The smaller the animal, the greater is its metabolic rate and the more it must eat relative to its body size. This happens because the metabolic rate of an animal—and therefore the amount of food it must eat to sustain this metabolic rate—varies in rough proportion to the surface area, rather than to the body weight. Surface area is proportional to a 0.7 power of body weight. Putting it another way, the amount of food a mammal (or bird) eats is proportional to a 0.7 power of its body weight. This means that as the size of animals gets smaller, their metabolic rate

FIG. 27-20 The masked (common) shrew *Sorex cincereus* (order Insectivora) feeding on a deer mouse it has killed. Note small size of shrew relative to much larger mouse. This tiny but fierce mammal, with a prodigious appetite for insects, mice, snails, and worms, spends most of its time underground and so is seldom seen by humans. Shrews are a primitive group believed to closely resemble the insectivorous ancestors of placental mammals. (Photograph by C. G. Hampson.)

(usually measured as oxygen consumption per gram of body weight) becomes more intense. A 3-gram mouse will consume *per gram* five times more food than does a 10-kg dog and about 30 times more food than does a 50,000-kg elephant. One can easily see why small mammals (shrews, bats, mice) must spend much more time hunting and eating food than do large mammals. The smallest shrews weighing only 2 g may eat more than their body weight each day and will starve to death in a few hours if deprived of food (Fig. 27-20). In contrast, a large carnivore can remain fat and healthy with only one meal every few days. The mountain lion is known to kill an average of one deer a week, although it will kill more frequently when game is abundant.

Body temperature regulation

Mammals share with birds the ability to regulate their body temperature physiologically. Both groups are commonly called "warm-blooded" animals to distinguish them from all other animals, which are "cold-blooded." It is true that the body temperature of mammals is usually (though not always) warmer than the ambient temperature, but a "cold-blooded" animal is not necessarily cold. Tropical fishes and insects, and reptiles basking in the sun, may have body temperatures equaling or surpassing those of mammals. Moreover, many "warm-blooded" mammals hibernate, allowing their body temperature to approach the freezing point of water. Thus the terms "warm-blooded" and "cold-blooded" are hopelessly subjective and nonspecific but are so firmly entrenched in our vocabulary that most biologists find it easier to accept the usage than try to change people.

The term **homeothermic** (constant body temperature) is frequently applied to mammals and birds, and **poikilothermic** (variable body temperature) to all other animals. Again, these terms offer difficulties, since, for example, deep sea fishes live out their lives in an environment having no perceptible temperature change, yet few would argue that they are homeotherms. Furthermore, among the homeothermic birds and mammals there are many that allow their body temperature to fluctuate diurnally or, as with hibernators, seasonally.

For these reasons, many physiologists prefer to call birds and mammals **endothermic,** that is, they maintain a high body temperature by *internal* (metabolic) heat production. Animals that elevate their body temperature above that of their surroundings, using an *external* heat source (solar radiation), are termed **ectothermic.** Many lizards fall into this category. Unfortunately, none of the existing terminology can be applied without qualification to large groups of animals. Nevertheless, it is convenient and usually correct to say that mammals are both endothermic and homeothermic, keeping in mind that the latter term is imprecise in some situations.

Homeothermy has allowed mammals to stabilize their internal temperature so that biochemical processes and nervous function can proceed at steady high levels of activity. Thus they can remain active in winter and exploit habitats unavailable to the poikilotherms. Most mammals have body temperatures between 36° and 38° C (somewhat lower than that of birds, which range between 40° and 42° C). This constant temperature is maintained by a delicate balance between heat pro-

A

B

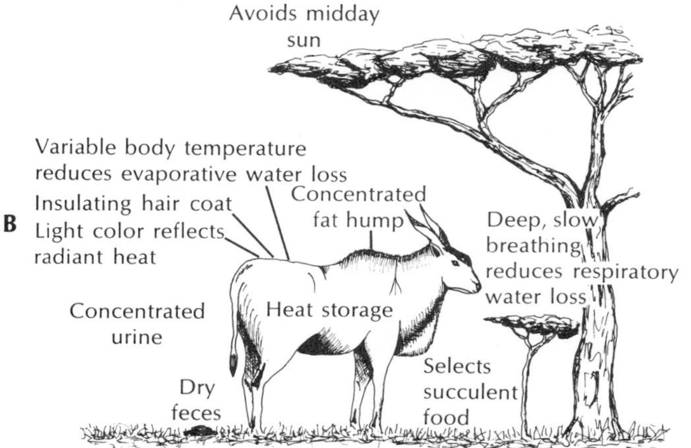

FIG. 27-21 The common eland *Taurotragus oryx* (order Artiodactyla), inhabitant of the arid, open savanna of central Africa. It is one of 72 species of African antelopes that occupy a variety of habitats that include open savannas, bush savannas, marshes, and flooded grasslands. Special food preferences reduce competition between different species. The drawing shows physiologic and behavioral adaptations of the eland for maintaining a constant body temperature in a hot environment. See text for explanation. (Photograph by C. G. Hampson; drawing from Tallmark, B. 1972. Fauna och Flora [Stockholm] **67:**163-175; after Taylor, C. P. 1969. Sci. Am. **220:**88-95.)

FIG. 27-22 African elephant bull *Loxodonta africana* (order Proboscidea) showing large ears, which serve as thermal windows. The ears of a large elephant measure 2 m long and 1.5 m across and are fanned back and forth to accelerate heat loss. The blood may cool as much as 5 centigrade degrees as it circulates through. Elephants also bathe themselves and wallow in mud when overheated. Note cattle egrets *Bulbulcus ibis* on elephant, which feed on insects routed from the brush by the elephant's movements. (Photograph by B. Tallmark.)

duction and heat loss — not a simple matter when mammals are constantly alternating between periods of rest and bursts of activity. Heat is produced by the animals's metabolism, which includes the oxidation of foodstuffs, basal cellular metabolism, and muscular contraction. Heat is lost by radiation and conduction to a cooler environment and by the evaporation of water. The mammal can control both processes of heat production and heat loss within rather wide limits. If it becomes too cool, it can increase heat production by increasing muscular activity (exercising or shivering) and by decreasing heat loss (increasing its insulation). If the mammal becomes too warm, it decreases heat production and increases heat loss. We will examine these processes in the examples that follow.

Adaptations for hot environments

Despite the harsh conditions of deserts — intense heat during the day, cold at night, and scarcity of water, vegetation, and cover — many kinds of animals live there successfully. The smaller desert mammals are mostly fossorial (fitted for digging burrows) and nocturnal. The lower temperature and higher humidity of burrows helps to reduce water loss by evaporation. Water loss is replaced by free water in their food, or by drinking water if it is available. Water is also formed in the cells by the metabolic oxidation of foods. This gain from **oxidation water,** as it is called, can be very significant, since water is not always available for drinking. In fact some desert mammals, such as kangaroo rats and gerbils of Old World deserts, can, if necessary, derive all the water they need from their dry food, thus drinking no water at all. Such animals can produce a highly concentrated urine and form nearly solid feces.

The large desert ungulates obviously cannot escape the desert heat by living in burrows. Animals such as camels and the desert antelopes (gazelle, oryx, and eland) possess a number of adaptations for coping with heat and dehydration. Those of the eland are shown in Fig. 27-21. The mechanisms for controlling water loss and preventing overheating are closely linked. The eland, like other desert antelopes, has a glossy, pallid color that reflects direct sunlight and the fur itself is an excellent insulation that works to keep heat out. The fur is not uniformly distributed over the body, however. Beneath the animal and on the axillae, groin, and scrotum or mammary glands, the pelage is very thin. These areas are provided with a rich capillary network and serve as thermal ''windows'' from which heat can be lost from the blood by convection and conduction. Heat is also lost by convection from the horns, which are well vascularized. The large ears of many mammals living in warm areas serve a similar purpose as heat radiators (such as those of the jackrabbit of the American Southwest, and the African elephant) (Fig. 27-22).

Fat tissue of the eland, an essential food reserve, is concentrated in a single hump on the back, instead of being uniformly distributed under the skin where it would impair heat loss by radiation. The eland avoids evaporative water loss—the only device an animal has for cooling itself when the environmental temperature is higher than that of the body—by permitting its body temperature to vary. At night the body cools down to less than 34° C by radiating heat to the cool surroundings. During the day the body temperature slowly rises to more than 41° C as the body stores heat. Only then must the eland prevent further rise through evaporative cooling by sweating and panting. The large body of the eland serves as a ''heat sink,'' thus conserving water and also reducing heat input, as the rising body temperature approaches that of the hot environment. The eland also conserves water by concentrating its urine and forming dry feces. All these adaptations are also found developed to a similar or even greater degree in camels, the most perfectly adapted of all large desert mammals.

Adaptations for cold environments

In cold environments, mammals use two major mechanisms to maintain homeothermy, which are (1) **decreased conductance,** that is, reduction of heat loss by increasing the effectiveness of the insulation and (2) **increased heat production.**

The excellent insulation of the thick pelage of Arctic animals is familiar to everyone. All mammals living in cold regions of the earth increase the thickness of their fur in winter, some by as much as 50%. As described earlier, the thick underhair is the major insulating layer, whereas the longer and more visible guard hair serves as protection against wear and for protective coloration. But the body extremities (legs, tail, ears, nose) of arctic mammals cannot be insulated as well as can the thorax. To prevent these parts from becoming major avenues of heat loss, they are allowed to cool to low temperatures, often approaching the freezing point. As warm arterial blood passes into a leg, heat is shunted directly from artery to vein and carried back to the core of the body. Without such a device, the blood would lose its heat through the poorly-insulated distal regions of the leg; cold blood would then return directly to the core where it would have to be reheated at a heavy energy cost. A consequence of this **peripheral heat exchange system** is that the legs and feet must operate at low temperatures. The temperatures of the feet of the arctic fox and barren-ground caribou are held just above the freezing point; in fact, the temperature may be below 0° C in the footpads and hooves. To keep feet supple and flexible at such low temperatures, the fats in the extremities have very low melting points, perhaps 30° C, lower than the ordinary body fats. Furthermore, the nerves serving the legs continue to conduct impulses at temperatures far below those that cause ordinary nerves to cold-block.

In severely cold conditions all mammals can produce more heat by augmented **muscular activity,** through exercise or shivering. We are all familiar with the effectiveness of both activities. A human can increase body heat production as much as eighteenfold within 12 minutes by violent shivering, when maximally stressed by cold. Many other mammals can doubtless do as well when under great cold stress. Another source of heat is the increased oxidation of foodstuffs, especially **brown fat stores.** This mechanism is called **nonshivering thermogenesis.**

Small mammals the size of lemmings, voles, and mice meet the challenge of cold environments in a different way. Their very smallness is a disadvantage because the ratio of surface area to body volume is greatest in the smallest mammals. In effect, they have a relatively much greater surface area exposed for heat loss than do large mammals. Moreover, small mammals cannot insulate themselves as well as can large mammals because there is an obvious practical limit to how much pelage a mouse, for example, can carry before it becomes an immobile bundle of fur. Consequently, these forms have successfully exploited the excellent insulating qualities of snow by living under

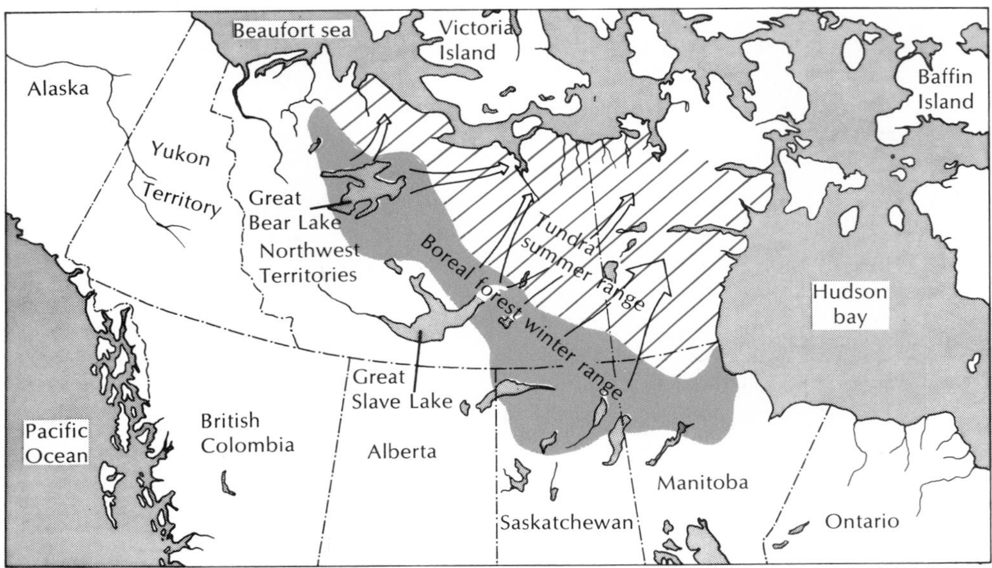

FIG. 27-23 Summer and winter ranges of the barren-ground caribou of Canada. The principal spring migration routes are indicated by arrows; routes vary considerably from year to year. (Adapted from Kelsall, J. P. 1968. The migratory barren-ground caribou of Canada. Ottawa, Canadian Wildlife Service, Queens Printers.)

it in runways on the forest floor, where, incidentally, their food is also located. In this **subnivean environment** the temperature seldom drops below −5° C, even though the air above may fall to −50° C. The snow insulation decreases thermal conductance from small mammals in the same way that pelage does for large mammals. Living beneath the snow is really a kind of avoidance response to cold.

Two additional ways that mammals may survive low temperatures are migration and hibernation.

Migration

Migration is a much more difficult undertaking for mammals than for birds; not surprisingly, few mammals make regular seasonal migrations, preferring instead to center their activities in a defined and limited home range. Nevertheless, there are some striking examples of mammalian migrations.

More migrators are found in North America than on any other continent. The barren-ground caribou of Canada undergo direct and puposeful mass migrations spanning 160 to 1,100 km (100 to 700 miles) twice annually (Figs. 27-23 and 27-24). From winter ranges in the boreal forests (taiga) they migrate rapidly in late winter and spring to calving ranges on the barren grounds (tundra). The calves are born in mid-June. Harassed by warble and nostril flies that bore into their flesh, they move southward in July and August, feeding little along the way. In September they reach the forest, feeding there almost continuously on low ground vegetation. Mating (rut) occurs in October.

The caribou have suffered a drastic decline in numbers. Since primitive times when there were several million, they dropped to less than 200,000 in 1958. The decline has been caused by excessive hunting by humans, poor calf crops, and destruction of the vulnerable forested wintering areas by fires accidentally started by humans during recent exploration and exploitation activities in the north. However, the population has increased slowly since 1958 under protection (Kelsall, 1968).

The plains bison, before its deliberate near-extinction by white people made huge circular migrations to separate summer and winter ranges.

The longest mammal migrations of all are made by the oceanic seals and whales. One of the most remarkable migrations is that of the fur seal, which breeds on the Pribilof Islands about 300 km off the coast of Alaska and north of the Aleutian Islands. From wintering grounds off southern California the females journey 4,800 km (3,000 miles) across open ocean, arriving at the Pribilofs in the spring where they congregate in

FIG. 27-24 Migrating barren-ground caribou *Ranger tarandus groenlandicus* (order Artiodactyla). **A,** Caribou moving southeast in autumn toward forest winter range. Two males in foreground in autumn pelage and velvet-covered antlers. Adult females in background. Cows and younger animals normally lead the herd; males follow. Caribou feed while moving, grazing on low ground vegetation, principally though not exclusively lichens. **B,** Herd of caribou crossing lake in winter. Trails and beds in snow are evident in foreground. (**A,** Courtesy D. Thomas, Canadian Wildlife Service, Ottawa; **B,** from Kelsall, J. P. 1968. The migratory barren-ground caribou of Canada. Ottawa, Canadian Wildlife Service, Queens Printers.)

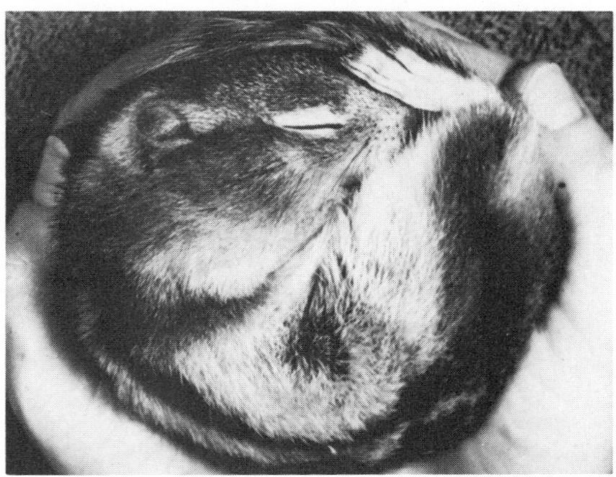

FIG. 27-25 Hibernating golden-mantled ground squirrel *Citellus lateralis* (order Rodentia) of the Rocky Mountains. Like other hibernators, this species rolls into a tight ball with clenched front paws under its chin, eyes and mouth closed tight, ears folded back, and tail wrapped over its head. Body feels decidedly cold to touch and is so rigid it can be rolled or tossed like a ball. Experiments with ground squirrels have revealed that onset and end of hibernation is determined by an internal rhythm; even under conditions of constant temperature and light and with unlimited food available, they hibernate at the normal season. (Photograph by C. G. Hampson.)

enormous numbers. The young are born within a few hours or days after arrival of the cows. Then the bulls, having already arrived and established territories, collect harems of cows, with which they mate and then guard with vigilance. After the calves have been nursed for about 3 months, cows and juveniles leave for their long migration southward. The bulls do not follow but remain in the Gulf of Alaska during the winter.

Although we might expect the only winged mammals, bats, to use their gift to migrate, few of them do. Most spend the winter in hibernation. The four species of American bats that do migrate—the red bat, the silvery-haired bat, the hoary bat, and the Brazilian free-tailed bat—spend their summers in the northern or western states and their winters in the south.

Hibernation

Many small and medium-sized mammals in north-temperate regions solve the problem of winter scarcity of food and low temperature by entering a prolonged and controlled state of dormancy. True hibernators, such as ground squirrels, woodchucks, marmots, and jumping mice, prepare for hibernation by building up large amounts of body fat. Some, such as the marmot, also lay in stores of food in their dens. Entry into hibernation is gradual. After a series of "test drops" during which body temperature drops a few degrees and then returns to normal, the animal cools to within a degree or less of the ambient temperature. Metabolism decreases to a fraction of normal. In the ground squirrel (Fig. 27-25), for example, the respiratory rate drops from a normal rate of 200 per minute to four or five per minute, and the heart rate from 150 to five. In most, body temperature is monitored so that if it drops dangerously close to the freezing point, the animal will awaken. Even under stable temperature conditions, the hibernator awakens at irregular intervals to eliminate wastes and then goes back to sleep. During arousal, the hibernator shivers violently and employs nonshivering thermogenesis to produce heat.

Some mammals, such as bears, badgers, raccoons, and opossums, enter a state of prolonged sleep in winter with little or no drop in body temperature. This is not true hibernation. Bears of the northern forest den-up for several months. Heart rate may drop from 40 to 10 beats per minute, but body temperature remains normal, and the bear is awakened if sufficiently disturbed. One intrepid but reckless biologist learned how lightly a bear sleeps when he crawled into one's den and attempted to measure its rectal temperature with a thermometer!

Flight and echolocation

Mammals have not exploited the skies to the same extent that they have the terrestrial and aquatic environments. However, many mammals scamper about in trees with amazing agility; some can glide from tree to tree, and one group, the bats, is capable of full flight. Gliding and flying evolved independently in several groups of mammals, including the marsupials, rodents, flying lemurs, and bats. And anyone who has watched a gibbon perform in a zoo realizes there is something akin to flight in this primate, too. Among the arboreal squirrels, all of which are nimble acrobats, by far the most efficient is the flying squirrel (Fig. 27-26). These forms actually glide rather than fly, using the gliding skin that extends out from the sides of the body.

Bats, the only group of flying mammals, are nocturnal insectivores and thus occupy a niche left vacant by birds (Fig. 27-27). Their outstanding success is attributed to two things: flight and the capacity to navigate by echolocation. Together these adaptations en-

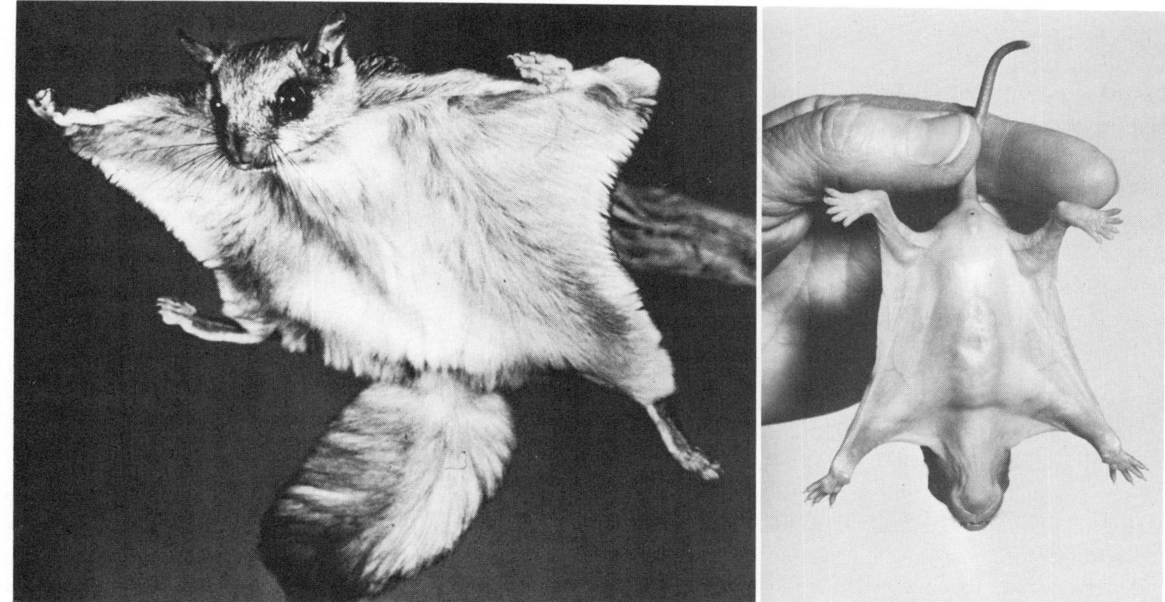

FIG. 27-26 Flying squirrels *Glaucomys sabrinus* (order Rodentia, family Sciuridae). **A,** In full flight. Area of undersurface is nearly trebled when gliding skin is spread. Glides of 40 to 50 m are possible; good maneuverability during flight is achieved by adjusting the position of the gliding skin with special muscles. Flying squirrels are nocturnal and have superb night vision. They feed on nuts, seeds, and insects and hoard food for winter. They do not hibernate, but activity is much reduced in winter. **B,** A 3-week-old youngster reflexly spreads gliding skin when picked up. (Photograph by C. G. Hampson.)

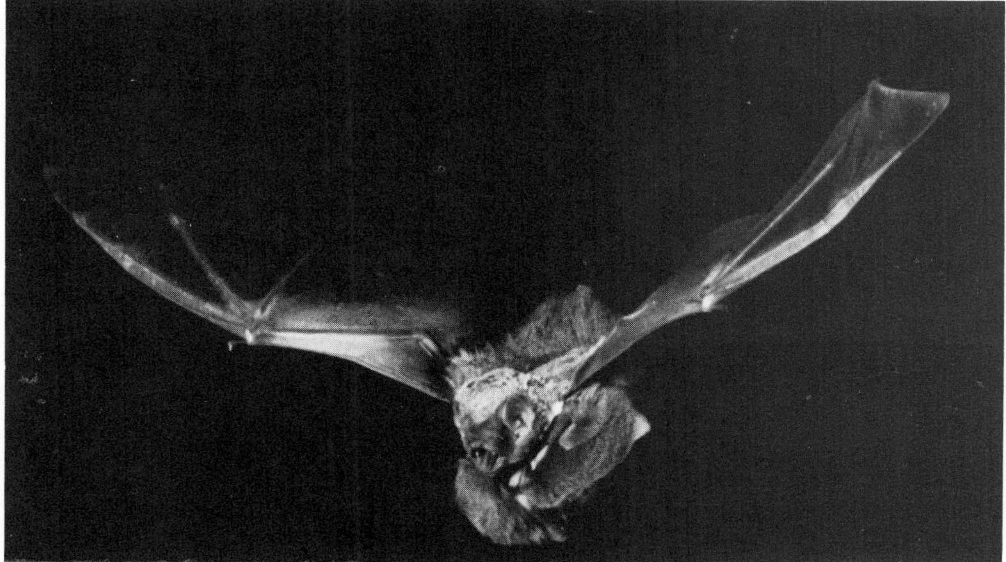

FIG. 27-27 Red bat *Lasiurus borealis* (order Chiroptera) in flight with four young. This species is unusual in giving birth to three or four young; most bats have one or two. The mother carries the young until their combined weight may exceed her own weight; then they are left in the roost, usually a tree. Mortality among the young is rather high. These medium-sized bats are distributed over all the eastern and southern United States. They are strong fliers, migrating northward in spring and southward in autumn. (Photograph by L. L. Rue, III.)

able bats to fly and avoid obstacles in absolute darkness, to locate and catch insects with precision, and to find their way deep into caves (another habitat largely ignored by both mammals and birds) where they sleep away the daytime hours.

Most research has been concentrated on members of the family Vespertilionidae, to which most of our common North American bats belong. When they are in flight, these bats emit short pulses 5 to 10 msec in duration in a narrow directed beam from the mouth. Each pulse is frequency modulated, that is, it is highest at the beginning, up to 100,000 Hertz (Hz, cycles per second) and drops to perhaps 30,000 Hz at the end. Sounds of this frequency are ultrasonic to the human ear, which has an upper limit of about 20,000 Hz. The pulses are produced at a rate of 30 to 40 a second, increasing to perhaps 50 a second as the bat nears an object. Furthermore, the pulses are spaced so that the echo of each is received before the next pulse is emitted, an adaptation that prevents jamming. Since the transmission-to-reception time decreases as the bat approaches an object, it can increase the pulse frequency to obtain more information about the object. The pulse length is also shortened as it nears the object.

The external ears of bats are large, like hearing trumpets, and shaped variously in different species. Less is known about the bat's inner ear, but it obviously is capable of receiving the ultrasonic sounds emitted. Bat navigation is so refined that biologists believe the bat builds up a mental image of its surroundings from echo scanning that is virtually as complete as the visual image from eyes of diurnal animals.

Bats have undergone some adaptive radiation, yet for reasons not fully understood, all are nocturnal, even the fruit-eating bats that use vision and olfaction to find their food instead of sonar. The tropics and subtropics have many nectar-feeding bats that are important pollinators for a wide variety of chiropterophilous ("bat-loving") plants. The flowers of these plants open at night, are white or light in color, and emit a musky batlike odor that the nectar-feeding bats find attractive.

The famed, tropical vampire bat is provided with razor-sharp incisors used to shave away the epidermis of its prey to expose underlying capillaries. After infusing an anticoagulant to keep the blood flowing, it laps up its meal and stores it in a specially modified stomach. It is said that dogs can hear an approaching vampire's sonar and thus awaken and escape.

Reproduction

Fertilization is always internal in mammals. All mammals, except monotremes, which lay eggs, are **viviparous:** the embryo develops in a uterus and is nourished by an intimate connection between embryo and mother, the **placenta.** Most mammals have definite mating seasons, usually in the winter or spring and timed to coincide with the most favorable time of the year for rearing the young after birth. Many male mammals are capable of fertile copulation at any time, but the female mating function is restricted to a periodic cycle, known as the **estrous cycle.** The female receives the male only during a relatively brief period known as **estrus,** or heat (Fig. 27-28).

The estrous cycle is divided into stages marked by characteristic changes in the ovary, uterus, and vagina. **Proestrus,** or period of preparation, when new ovarian follicles grow is followed by estrus, when mating occurs. Almost simultaneously the ovarian follicles burst, releasing the eggs **(ovulation),** which are then fertilized. Implantation of the fertilized egg and **pregnancy** follow. However, should mating and fertilization not occur, estrus is followed by **metestrus,** a period of repair. This stage is followed by **diestrus,** during which the uterus becomes small and anemic. The cycle then repeats itself, beginning with proestrus.

How often females are in heat varies greatly among the different mammals. Those animals that have only a single estrus during the breeding season are called **monestrous;** those that have a recurrence of estrus during the breeding season are called **polyestrous.** Dogs, foxes, and bats belong to the first group; field mice and squirrels are all polyestrous, as are many mammals living in the more tropical regions of the earth. The Old World monkeys and humans have a somewhat different cycle in which the postovulation period is terminated by **menstruation,** during which the lining of the uterus (endometrium) collapses and is discharged with some blood. This is called a **menstrual cycle** and is described in Chapter 35.

Gestation, or period of pregnancy, also varies greatly among the mammals. Mice and rats have a gestation period of approximately 21 days; rabbits and hares, 30 to 36 days; cats and dogs, 60 days; cows, 280 days; and elephants, 22 months. The marsupials (opossum) have a very short gestation period of 13 days; at the end of that time the tiny young leave the vaginal orifice and make their way to the marsupial pouch where they attach themselves to nipples (Fig.

FIG. 27-28 African lions *Panthera leo* (order Carnivora, family Felidae) mating. Lions breed at any season, although predominantly in spring and summer. During the short period a female is receptive, she may mate repeatedly. Three or four cubs are born after gestation of 100 days. Once the mother introduces the cubs into the pride, they are treated with affection by both adult males and females. Cubs go through an 18- to 24-month apprenticeship learning how to hunt and then are frequently driven from the pride to manage for themselves. (Photograph by L. L. Rue, III.)

FIG. 27-29 Baby opossums *Didelphis marsupialis* (order Marsupialia, family Didelphidae) about 2 weeks old in mother's pouch. After the mating in spring and a brief 13-day gestation, bee-sized embryos with enlarged forelegs and sharp hooked claws emerge from cloaca and squirm unaided to pouch and take nipples in their mouths. The pouch is provided with 13 nipples, much less than the 20 to 30 embryos born. The first 13 born are the lucky ones, the remainder are lost. Young remain in the pouch 3 months and then are weaned by autumn and leave the family group. In opossums, as in other marsupials, the young remain in the uterus for such a brief time that only a rudimentary placenta is formed. (Photograph by L. L. Rue, III.)

27-29). Here they remain for more than 2 months before emerging.

The number of young produced by mammals in a season depends on many factors. Usually the larger the animal, the smaller the number of young in a litter. Perhaps one of the greatest factors involved is the number of enemies a species has. Small rodents can reveal astonishing fecundity. Lemmings in captivity may produce as many as 16 litters per year, each with four to eight young. The daughters begin to mate at an

A

B

C

FIG. 27-30 Birth of a white-tailed fawn, *Odocoileus virginianus*. **A,** The fawn's forelegs protrude from the birth canal, and birth quickly follows. **B,** Part of the amnion still clings to the fawn's wet body. **C,** After 13 minutes the fawn attempts to stand. Young of all the members of the order Artiodactyla are precocial; within minutes after birth they are alert to their surroundings and most can walk within an hour. (Photographs by L. L. Rue, III.)

age of 25 days and have their first litters 20 days later. Six weeks later, the granddaughters are reproducing. Most carnivores have but one litter of three to five young a year. Large mammals, such as elephants and horses, usually have only one young.

The condition of the young at birth also varies. Those young born with hair and open eyes and ability to move around are called **precocial** (ungulates and jackrabbits) (Fig. 27-30); those that are naked, blind, and helpless (carnivores and rodents) are known as **altricial.** Most mammals exhibit a great deal of parental care and instruction (young learning by imitation) and will fight fiercely in defense of their young. When disturbed, they will often carry their young to more secure places.

Territory and home range

Virtually all mammals, with aquatic mammals as perhaps the only exception, have **territories**—areas from which individuals of the *same* species are excluded. In fact, most wild mammals, like many people, are basically unfriendly to their own kind, especially so to their own sex during the breeding season. If the mammal dwells in a burrow or den, this forms the center of its territory. If it has no fixed address, the territory is marked out, usually with the highly developed scent glands described earlier in this chapter. Territories vary greatly in size of course, depending on the size of the animal and its feeding habits. The grizzly bear has a territory of several square miles that it guards zealously against all other grizzlies.

FIG. 27-31 Family of prairie dogs *Cynomys ludovicianus* (order Rodentia). These highly social prairie dwellers are plant eaters that comprise an important source of food to many animals. They live in elaborate tunnel systems so closely interwoven that they form "towns" of as many as 1,000 individuals. Towns are subdivided into wards, in turn divided into coteries, the basic family unit, containing one or two adult males, several females and their litters. Although prairie dogs display ownership of burrows with territorial calls, they are friendly with inhabitants of adjacent burrows. The name "prairie dogs" derives from the sharp, doglike bark it makes when danger threatens. Western cattle and sheep ranchers have nearly eradicated prairie dogs in some areas by mass poisoning programs. (Photograph by L. L. Rue, III.)

Mammals usually use natural features of their surroundings in staking their claims. These are marked with secretions from the scent glands, or by urinating or defecating. When an intruder knowingly enters another's marked territory, it is immediately placed at a psychologic disadvantage. Should a challenge follow, the intruder almost invariably breaks off the encounter in a submissive display characteristic for the species. An interesting exception to the strong territorial nature of most mammals is the prairie dog, which lives in large, friendly communities called prairie-dog "towns" (Fig. 27-31). When a new litter has been reared, the adults relinquish the old home to the young and move to the edge of the community to establish a new home. Such a practice is totally antithetic to the behavior of most mammals, which drive off the young when they are self-sufficient.

The **home range** of a mammal is a much larger foraging area surrounding a defended territory. Home ranges are not defended in the same way a territory is; home ranges may in fact overlap, producing a neutral zone used by the owners of several territories, for seeking food.

Mammal populations

A population of animals includes all the animals of a species that interbreed. By this defintion we note that there may be several distinct populations of the same species in a biome (an ecologic entity of plants and animals in an area), but they do not interbreed because they are separated by topographic or climatic barriers, or for some other reason.

All mammals live in communities, each composed of numerous populations of different animal and plant species. Each species is affected by the activities of other species and by the changes, especially climatic, that occur. Thus populations are always changing in size. Populations of small mammals are lowest before the breeding season, and greatest just after the addition of new members. Beyond these expected changes in

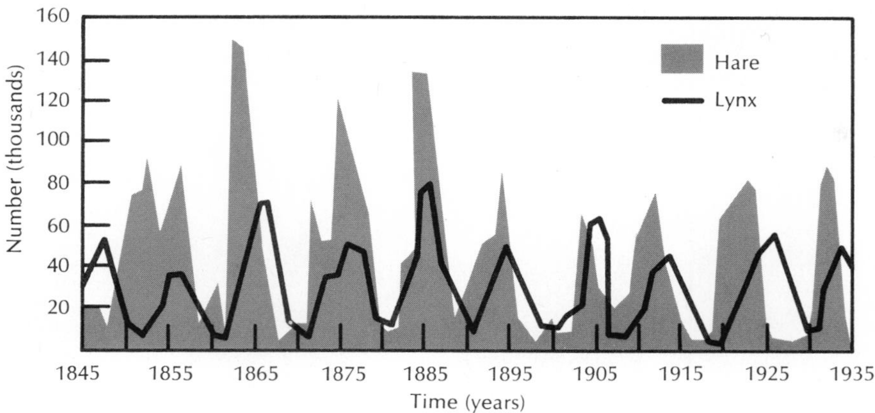

FIG. 27-32 Changes in population of varying hare and lynx in Canada as indicated by pelts received by the Hudson's Bay Company. The abundance of lynx (predator) follows that of the hare (prey). (After Odum, E. P. 1959. Fundamentals of ecology. Philadelphia, W. B. Saunders Co.)

population size, animal populations may fluctuate from other causes. Irregular fluctuations are commonly produced by variations in food supply or by disease. These are **density-independent** causes, since they affect a population whether it is crowded or dispersed. However, the most spectacular fluctuations are **density dependent;** that is, they are correlated with population crowding. Cycles of abundance are common among many rodent species. The population peaks and mass migrations of the Scandinavian and arctic North American lemmings are well-known. The growth of lemming populations in arctic Canada is described by F. Bruemmer:

After a population crash one sees few signs of lemmings. There may be only one to every 10 acres. The next year, they are evidently more numerous; their runways snake beneath the tundra vegetation, and frequent piles of rice-sized droppings indicate the lemmings fare well. The third year one sees them everywhere. The fourth year, usually the peak year of their cycle, the populations explode. Now more than 150 lemmings may inhabit each acre of land and they honeycomb it with as many as 4,000 burrows. Males meet frequently and fight instantly. Males pursue females and mate after a brief but ardent courtship. Everywhere one hears the squeak and chitter of the excited, irritable, crowded animals. At such times they may spill over the land in manic migrations.*

*From Bruemmer, F. 1973. The Arctic. Montreal, Infocor, Ltd., p. 154.

Having devastated the vegetation by tunneling and grazing, they begin long, mass migrations to find new undamaged habitats for food and space. They swim across streams and small lakes as they go, but cannot distinguish these from large lakes and rivers and the sea, in which they drown. Since lemmings are the main diet of many carnivorous mammals and birds, any change in lemming population density affects all their predators as well.

The varying hare (snowshoe rabbit) of North America shows 10-year cycles in abundance. The well-known fecundity of rabbits enables them to produce litters of three or four young up to five times per year. The density may increase to 4,000 hares competing for food in each square mile of northern forest. Predators (owls, minks, foxes, and especially lynxes) also increase (Fig. 27-32). Then the population crashes precipitously, for reasons that have long been a puzzle to scientists. Rabbits die in great numbers, not from lack of food or from an epidemic disease (as was once believed), but evidently from some density-dependent psychogenic cause. As crowding increases, hares become more aggressive, show signs of fear and defense, and stop breeding. The entire population reveals symptoms of pituitary-adrenal gland exhaustion, an endocrine imbalance called "shock disease" that leads to death. But there is much about these dramatic crashes that is not understood. Whatever the causes, population crashes that follow superabundance, though harsh, are clearly advantageous to the species, for the vegetation is allowed to recover, thus providing the survivors a much better chance for successful breeding.

HUMANS AND MAMMALS

Some 10,000 years ago, at the time people developed agricultural methods, they also began the domestication of mammals. Dogs were certainly among the first to be domesticated, probably entering voluntarily into their human dependence. The dog is an extremely adaptable and genetically plastic species derived from wolves. Much less genetically variable and certainly less social than dogs is the domestic cat, probably derived from an African race of wildcat. Wildcats look like oversized domestic cats and are still widespread in Africa and Eurasia. The domestication of cattle, buffalos, sheep, and pigs probably came much later. It is believed that the beasts of burden—horses, camels, oxen, and llamas—probably were subdued by early nomadic peoples. It is of interest that certain domestic species no longer exist as wild animals; for example, the one-humped Arabian camel and the llama and the alpaca of South America. All of the truly domestic animals breed in captivity and have become totaly dependent on humans; many have been molded by selective breeding to yield characteristics that are desirable for our purposes.

Some mammals hold special positions as "domestic" animals. The elephant has never been truly domesticated because it will not breed in captivity. In Asia, adults are captured and submit to a life of toil with astonishing docility. The reindeer of northern Scandinavia are domesticated only in the sense that they are "owned" by nomadic peoples who continue to follow them in their seasonal migrations. The eland of Africa (Fig. 27-21) is presently undergoing experimental domestication in several places. It is placid and gentle, immune to native diseases, and produces an excellent meat.

Finally, we should not leave the subject of domestication without mentioning the albino rat, a domesticated brown rat. It has been suggested that the gentle nature of the albino, which has contributed so much to medical and psychologic research, is the result of a small defect in the amygdaloid nucleus of the brain. Fierce and intractable wild rats can be converted to docile, easily handled, tame rats by destroying a small part of the amygdala in an operation.

In the introduction to this chapter we alluded to the senseless human exploitation of the great marine whales, the largest animals that have ever lived. Despite 50 years of careful scientific study that has shown conclusively that several species of whales are now in real danger of total extinction, ambitious hunting continues by the two major whaling nations, the Soviet Union and Japan.

The whale tragedy is just one example of our inability to reconcile progress with the preservation of wildlife, a dilemma that is explored in the final chapter of this book. The extermination of a species for commercial gain is so totally indefensible that no debate is required. Once a species is extinct, no amount of scientific or technical ingenuity will bring it back. What has taken millions of years to evolve can be destroyed in a decade of thoughtless exploitation. Many people are concerned with the awesome impact we have on wildlife, and there is more determination today to reverse a regrettable trend than ever before. If given half a chance, mammals will usually make spectacular recoveries from human depredations, as have the sea otter and the saiga antelope, both once in danger of extinction and now numerous. Paradoxically, in Africa where conservationists wage a seesaw battle with opposing interests, it appears that the commercial gain of tourism and game cropping will do more to save the fauna than the outraged concern of naturalists.

Mammals can, of course, be enemies of human beings. Rodents and rabbits are capable of inflicting staggering damage to growing crops and stored food (Fig. 27-33). We have provided an inviting forage for rodents with our agriculture, and convenienced them further by removing their natural predators. Rodents also carry various diseases. Bubonic plague and typhus are carried by house rats. Tularemia, or rabbit fever, is transmitted to humans by the wood tick carried by rabbits, woodchucks, muskrats, and other rodents. Rocky Mountain spotted fever is carried by ticks on ground squirrels and dogs. *Trichina* worms and tapeworms are acquired by humans through hogs, cattle, and other mammals.

Classification of living orders

The classification given here recognizes 18 living orders of mammals, although some mammalogists remove the

FIG. 27-33 Brown rat *Rattus norvegicus* (order Rodentia, family Murridae). Originally from the tropical forests of Asia, this species and the less pugnacious tree-living black rat have spread all over the world. Living all too successfully beside human habitations, the brown rat not only causes great damage to food stores but also spreads disease, including bubonic plague (a disease carried by infected fleas that greatly influenced human history in medieval Europe), typhus, infectious jaundice, *Salmonella* food poisoning, and rabies. (Photograph by L. L. Rue, III.)

aquatic carnivores (seals, sea lions, and walruses) from the order Carnivora and place them in a distinct order Pinnipedia. Fourteen extinct orders are not included in this classification.

Subclass Prototheria (pro'to-thir'e-a) (Gr., *prōtos,* first, + *thēr,* wild animal). The egg-laying mammals.

Order Monotremata (mon'o-tre'ma-tah) (Gr., *monos,* single, + *trēma,* hole)—**egg-laying mammals: duckbill platypus, spiny anteaters.** The representatives of this order are from Australia, Tasmania, and New Guinea. The most noted member of the order is the duckbill platypus *(Ornithorhynchus anatinus).* The spiny anteater, or echidna *(Tachyglossus),* has a long, narrow snout adapted for feeding on ants, its chief food. Monotremes represent the only order that is oviparous, and there is no known group of extinct mammals from which they can be derived. Their fossils date from the Pleistocene epoch.

Subclass Theria (thir'e-a) (Gr., *ther,* wild animal)

Infraclass Metatheria (met'a-thir'e-a) (Gr., *meta,* after, + *ther,* wild animal). The marsupial mammals.

Order Marsupialia (mar-su'pe-ay'le-a) (Gr., *marsypion,* little pouch)—**pouched mammals: opossums, kangaroos, koalas, Tasmanian wolf, wombats, bandicoots, numbats, and others.** These are primitive mammals characterized by an abdominal pouch, the **marsupium,** in which they rear their young. Although the young are nourished in the uterus for a short time, there is rarely a placenta present. This order has many representatives; only the opossum is found in the Americas, but the order is the dominant group of mammals in Australia.

Infraclass Eutheria (yu-thir'e-a) (Gr., *eu,* true, + *thēr,* wild animal). The placental mammals.

Order Insectivora (in-sec-tiv'o-ra) (L., *insectum,* an insect, + *vorare,* to devour)—**insect-eating mammals: shrews, hedgehogs, tenrecs, moles.** The principal food of animals in this order is insects. The most primitive of placental mammals, they are widely distributed over the world except Australia and New Zealand. Placental mammals and marsupials are believed to have arisen independently from common ancestors during the Cretaceous period, but in time the placentals became dominant in most parts of the world because of their superior intelligence. Insectivora are small, sharp-snouted animals that spend a great part of their lives underground. The shrews are the smallest of the group; some of them are the smallest mammals known.

Order Chiroptera (ky-rop'ter-a) (Gr., *cheir,* hand, + *pteron,* wing)—**bats.** The wings of bats, the only true flying mammals, are modified forelimbs in which the second to fifth digits are elongated to support a thin integumental membrane for flying. The first digit (thumb) is short with a claw. There are many families and species of bats the world over. The common North American forms are the little brown bat *(Myotis),* the free-tailed bat *(Tadarida),* which lives in the Carlsbad Caverns, and the large brown bat *(Eptesicus).* In the Old World tropics the "flying foxes" *(Pteropus)* are the largest bats, with a wingspread of 1.2 to 1.5 m; they live chiefly on fruits.

Order Dermoptera (der-mop'tera) (Gr., *derma,* skin, + *pteron,* wing)—**flying lemurs.** These are related to the true bats and consist of the single genus *Galeopithecus.* They are found in the Malay peninsula in the East Indies. They are not lemurs (which are primates) and cannot fly in the strict sense of the word, but glide like flying squirrels.

Order Carnivora (car-niv'o-ra) (L., *caro,* flesh, + *vorare,* to devour)—**flesh-eating mammals: dogs, wolves, cats, bears, weasels.** To this extensive order belong some of the most intelligent and strongest of animals. They all have predatory habits, and their teeth are especially adapted for tearing flesh. They are divided among two suborders: Fissipedia, whose feet contain toes, and Pinnipedia, with limbs modified for aquatic life.

Suborder Fissipedia (fi-zi-peed'e-a) (L., *fissus,* cleft, + *ped,* foot). Consists of the well-known carnivores—wolves, tigers, dogs, cats, foxes, weasels, skunks, and many others. They vary in size from certain tiny weasels to the mammoth Alaskan bear and Bengal tiger. They are distributed all over the world except in the Australian and antarctic regions, where there are no native forms. This suborder is divided into certain familiar families, among which are **Canidae** (the dog family), consisting of dogs, wolves, foxes, and coyotes; **Felidae** (the cat family), whose members include the domestic cats, tigers, lions, cougars, and lynxes; **Ursidae** (the bear family), made up of bears; and **Mustelidae** (the fur-bearing family), containing the martens, skunks, weasels, otters, badgers, minks, and wolverines.

Suborder Pinnipedia (pi-ni-peed'e-a) (L., *pinna,* feather, + *ped,* foot). Includes the aquatic carnivores, sea lions, seals, sea elephants, and walruses. Their limbs have been modified as flippers for swimming. They are all saltwater forms, and their food is mostly fish.

Order Tubulidentata (tu'byu-li-den-ta'ta) (L., *tubulus,* tube, + *dens,* tooth)—**aardvark.** The aardvark is the Dutch name for earth pig, a peculiar animal with a piglike body found in Africa. The order is represented by a single species.

Order Rodentia (ro-den'che-a) (L., *rodere,* to gnaw)—**gnawing mammals: squirrels, rats, woodchucks.** The rodents are the most numerous of all mammals. Most of them are small. They are found on all continents and many of the large islands. They have no canine teeth, but their chisel-like incisors (never more than four) grow continually. Their basic adaptive feature, therefore, is gnawing. This adaptation has been largely responsible for the evolution of a very active and diversified group. Some rodents are useful for their fur. The beaver *(Castor),* the largest rodent in the United States, has a valuable pelt. Many rodents are utilized as food by carnivores and by humans. The common families of this order are **Sciuridae** (squirrels and woodchucks), **Muridae** (rats and house mice), **Castoridae** (beavers), **Erethizontidae** (porcupines), **Geomyidae** (pocket gophers), and **Cricetidae** (hamsters, deer mice, gerbils, voles, lemmings).

Order Pholidota (fol'i-do'ta) (Gr., *pholis,* horny scale)—**pangolins.** In this order there is one genus *(Manis)* with eight species. They are an odd group of animals whose body is covered with overlapping horny scales that have arisen from fused bundles of hair. Their home is in tropical Asia and Africa.

Order Lagomorpha (lag'o-mor'fa) (Gr., *lagos,* hare, + *morphe,* form)—**rabbits, hares, pikas.** The chief difference between this order and Rodentia is the presence of four upper incisors, one pair of which is small and the other large, with enamel on the posterior as well as anterior surface of the tooth.

Order Edentata (ee'den-ta'ta) (L., *edentatus,* toothless)—**toothless mammals: sloths, anteaters, armadillos.** These forms are either toothless or else have degenerate peglike teeth without enamel. Most of them live in South and Central America, although the nine-banded armadillo *(Dasypus novemcinctus)* occurs in the southern United States. The sloths are very sluggish animals that have the queer habit of hanging upside down on branches. The hairs of sloths have tiny pits in which green algae grow and render them invisible against a background of mosses and lichens. The extinct ground sloths were represented by some members as large as small elephants.

Order Cetacea (see-tay'she-a) (L., *cetus,* whale)—**fishlike mammals: whales, dolphins, porpoises.** This order is well adapted for aquatic life. Their anterior limbs are modified into broad flippers; the posterior limbs are absent. Some have a fleshy dorsal fin, and the tail is divided into transverse fleshy flukes. The nostrils are represented by a single or double blowhole on top of the head. They have no hair except a few on the muzzle, no skin glands except the mammary and those of the eye, no external ear, and small eyes. The order is divided into two suborders.

Suborder Odontoceti (o-don-te-see'tie) (Gr., *odont,* tooth, + *cete,* whales)—**toothed whales.** Made up of toothed members and represented by the sperm whales, porpoises, and dolphins. The killer whale *(Orcinus)* does not hesitate to attack the larger whales and is destructive to seal rookeries. The sperm whale *(Physeter)* is the source of sperm oil, which is obtained from the head. A peculiar substance, ambergris, is sometimes formed in its stomach and is used in perfumes. Porpoises and dolphins, which also belong to this suborder, are only about 2 m long and feed mainly on gregarious fish.

Suborder Mysticeti (mis-te-see'tie) (Gr., *mystikētos,* whale)—**whalebone whales.** Includes many species, some of which are the largest animals that have ever lived. Instead of teeth, they have a peculiar straining device of whalebone (baleen) attached to the palate, used to filter microscopic animals (plankton), out of the water. The largest of the whales is the blue whale *(Balaenoptera),* which may grow 30 m long and weigh 125 tons.

Order Proboscidea (pro'ba-sid'e-a) (Gr., *proboskis,* elephant's trunk, from *pro,* before, + *boskein,* to feed)—**proboscis mammals: elephants.** These are the largest of living land animals. The two upper incisors are elongated as tusks, and the molar teeth are well developed. There are two species of elephants: the Indian *(Elephas maximus),* with relatively small ears, and the African *(Loxodonta africana),* with large ears. The Asiatic or Indian elephant has long been domesticated and is trained to do heavy work. The taming of the

African elephant is more difficult but was extensively done by the ancient Carthaginians and Romans, who employed them in their armies. Barnum's famous "Jumbo" was an African elephant.

Order Hyracoidea (hy'ra-coi'de-a) (Gr., *hyrax,* shrew)—**hyraxes (conies).** Conies are herbivores that are restricted to Africa and Syria. They have some resemblance to short-eared rabbits but have teeth like rhinoceroses, with hoofs on their toes and pads on their feet. They have four toes on the front and three toes on the back feet.

Order Sirenia (sy-re'ne-a) (Gr., *seiren,* sea nymph)—**sea cows (manatees).** Sea cows, or manatees, are large, clumsy aquatic animals. They have a blunt muzzle covered with coarse bristles, the only hairs these queer animals possess. They have no hind limbs, and their forelimbs are modified into swimming flippers. The tail is broad with flukes but is not divided. They live in the bays and rivers along the coasts of tropical and subtropical seas. There are only two genera living at present: *Trichechus,* found in the rivers of Florida, West Indies, Brazil, and Africa, and *Halicore,* the dugong of India and Australia.

Order Perissodactyla (pe-ris'so-dak'ti-la) (Gr., *perissos,* odd, + *dactylos, toe*)—**odd-toed hoofed mammals: horses, asses, zebras, tapirs, rhinoceroses.** The odd-toed hoofed mammals have an odd number (one or three) of toes, each with a cornified hoof. Both the Perissodactyla and the Artiodactyla are often referred to as **ungulates** (L., *ungula,* hoof) or hoofed mammals, with teeth adapted for chewing. The horse family (Equidae), which also includes asses and zebras, has only one functional toe. Tapirs have a short proboscis formed from the upper lip and nose. The rhinoceros *(Rhinoceros)* includes several species found in Africa and southeastern Asia. All are herbivorous.

Order Artiodactyla (ar'te-o-dak'ti-la) (Gr., *artios,* even, + *daktylos,* toe)—**even-toed hoofed mammals: swine, camels, deer, hippopotamuses, antelopes, cattle, sheep, goats.** Most of these ungulates have two toes, although the hippopotamus and some others have four. Each toe is sheathed in a cornified hoof. Many, such as the cow, deer, and sheep, have horns. Many are ruminants, that is, animals that chew the cud. Like Perissodactyla, they are strictly herbivorous. The group is divided into nine living families and many extinct ones and includes some of the most valuable domestic animals. This extensive order is commonly divided into three suborders: the **Suina** (pigs, peccaries, hippopotamuses), the **Tylopoda** (camels), and the **Ruminantia** (deer, giraffes, sheep, cattle, etc.).

Order Primates (pry-may'teez) (L., *prima,* first)—**highest mammals: e.g., lemurs, monkeys, apes, humans.** This order stands first in the animal kingdom in brain development, possessing especially large cerebral hemispheres. Most of the species are arboreal, apparently derived from tree-dwelling insectivores. The primates represent the end product of a line that branched off early from other mammals and have retained many primitive characteristics. It is believed that their tree-dwelling habits of agility in capturing food or avoiding enemies were largely responsible for their advances in brain structure. As a group, they are generalized, with five digits (usually provided with flat nails) on both forelimbs and hind limbs. All have their bodies covered with hair except humans. Forelimbs are often adapted for grasping, as are the hind limbs sometimes. The group is singularly lacking in claws, scales, horns, and hoofs. There are two suborders.

Suborder Prosimii (pro-sim'ee-i) (Gr., *pro,* before, + *simia,* ape)—**lemurs, tree shrews, tarsiers, lorises, pottos.** These are primitive arboreal primates, with their second toe provided with a claw and a long nonprehensile tail. They look like a cross between squirrels and monkeys. They are found in the forests of Madagascar, Africa, the Malay peninsula and the Philippines. Their food is both plants and small animals.

Suborder Anthropoidea (an'thro-poi'de-a) (Gr., *anthropos,* man)—**monkeys, gibbons, apes, humans.** There are three superfamilies:

Superfamily Ceboidea (se-boi'de-a) (Gr., *kebos,* long-tailed monkey) **(Platyrhinii).** These are New World monkeys, characterized by the broad flat nasal septum, nonopposable thumb, prehensile tail, and the absence of ischial callosities and cheek pouches. Familiar members of this superfamily are the capuchin monkey *(Cebus)* of the organ grinder, the spider monkey *(Ateles),* and the howler monkey *(Alouatta).*

Superfamily Cercopithecoidea (sur'ko-pith'e-koi'de-a) (Gr., *kerkos,* tail, + *pithekos,* monkey) **(Catarrhinii).** These Old World monkeys have the external nares close together, and many have internal cheek pouches. They never have prehensile tails, there are calloused ischial tuberosities on their buttocks, and their thumbs are opposable. Examples are the savage mandrill *(Cynocephalus),* the rhesus monkey *(Macacus)* widely used in biologic investigation, and the proboscis monkey *(Nasalis).*

Superfamily Hominoidea (hom'i-noi'de-a) (L., *homo, hominis,* man). The higher (anthropoid) apes and humans make up this superfamily. Their chief characteristics are lack of a tail and lack of cheek pouches. There are two families: Pongidae and Hominidae. The Pongidae family includes the higher apes, gibbon *(Hylobates),* orangutan *(Simia),* chimpanzee *(Pan),* and gorilla *(Gorilla).* The other family, Hominidae, is represented by a single living species *(Homo sapiens),* modern human beings. Humans differ from the members of family Pongidae in being more erect, in having shorter arms and larger thumbs, and in having lighter jaws with smaller front teeth. Most of the apes also have much more prominent ridges over the eyes. Many human differences from the anthropoid apes are associated with higher human intelligence, human speech centers in the brain, and the fact that human beings are no longer arboreal animals.

Annotated references

Selected general readings

Blair, W. F., A. P. Blair, P. Brodkorr, F. R. Cagle, and G. A. Moore. 1968. Vertebrates of the United States, ed. 2. New York, McGraw-Hill Book Co. *Taxonomic keys are illustrated and identify all vertebrates down to species.*

Bruemmer, F. 1973. The Arctic. Montreal, Infocor, Ltd.

Burt, W. H., and R. P. Grossenheider. 1976. A field guide to the mammals, ed. 3. Boston, Houghton Mifflin Co. *Full color illustrations and range maps.*

Carrington, R. 1975. The mammals. New York, Life Nature Library, Time Inc. *Well-written, beautifully illustrated semipopular treatment.*

Davis, E. E., and F. B. Golley. 1964. Principles of mammalogy. New York, Reinhold Publishing Corp. *This introductory text deals with classification, adaptations, evolution, distribution, populations, and behavior of mammals.*

Eimerl, S., and I. DeVore. 1965. The primates. New York, Life Nature Library, Time Inc. *Imaginatively written and beautifully illustrated semipopular treatment. Highly recommended.*

Griffen, D. R. 1958. Listening in the dark. New Haven, Conn., Yale University Press. *A classic on bats.*

Gunderson, H. L. 1976. Mammalogy. New York, McGraw-Hill Book Co. *Comprehensive text emphasizing adaptations.*

Hall, E. R., and K. R. Kelson. 1959. The mammals of North America, vols. 1 and 2. New York, The Ronald Press Co. *Full descriptions of species and subspecies with distribution maps. Taxonomic keys, records, and revealing line drawings of skull characteristics are included in this authoritative work.*

Kelsall, J. D. 1968. The migratory barren-ground caribou of Canada. Ottawa, Canadian Wildlife Service, Queen's Printers.

van Lawick-Goodall, J., and H. van Lawick. 1971. Innocent killers. Boston, Houghton Mifflin Co. *An engrossing description of the biology of hyenas, jackals, and wild dogs by the author of the acclaimed field study of chimpanzees called* In the Shadow of Man *and her wildlife photographer husband.*

Matthews, L. H. 1969. The life of mammals. London, Weidenfeld & Nicolson. *A well-written general account of mammals—what they are and what they do.*

Mech, D. L. 1970. The wolf: the ecology and behavior of an endangered species. Garden City, N.Y., The Natural History Press. *A thorough, illustrated account; detailed, yet interestingly written.*

Reader's Digest Association. 1970. The living world of animals. London, The Reader's Digest Association. *A fine account of animals, stressing their adaptations and ecology. Fine illustrations; highly recommended.*

Richards, S. A. 1973. Temperature regulation. London, Wykeham Publications. *Clearly written primer on the subject.*

Sadleir, R. M. F. S. 1969. The ecology of reproduction in wild and domestic mammals. London, Methuen & Co. *Wealth of information on mammalian reproductive physiology.*

Sanderson, I. T. 1955. Living mammals of the world. New York, Garden City Books (Hanover House). *A beautiful and informative work of many photographs and concise text material. It is a delight to any zoologist regardless of specialized interest.*

Schaller, G. B. 1963. The mountain gorilla. Chicago, University of Chicago Press. *A thorough and definitive study of the ecology and behavior of this fascinating primate.*

Scheffer, V. B. 1958. Seals, sea lions, and walruses. A review of the Pinnipedia. Stanford, Calif., Stanford University Press. *Treats the group from the standpoint of their evolution, characteristics, and classification. Many fine photographs and a good bibliography are included.*

Scheffer, V. B. 1969. The year of the whale. New York, Charles Scribner's Sons. *Beautifully written and engaging combination of fiction and fact about a baby sperm whale's first year of life.*

Schmidt-Nielsen, K. 1964. Desert animals. New York, Oxford University Press. *Deals with the problems desert-dwelling forms (including people) must meet and solve to survive.*

Seton, E. T. 1925-1928. Lives of game animals, vols. 1 to 4. New York, Doubleday, Doran & Co. *A classic.*

Vaughn, T. A. 1972. Mammalogy. Philadelphia, W. B. Saunders Co. *This basic text takes a systematic approach to the subject.*

Walker, E. P. 1968. Mammals of the world, vols. 1 to 3. Baltimore, The Johns Hopkins University Press. *The only single compendium of information on all known and living mammalian genera. A valuable reference work.*

Young, J. Z. 1957. The life of mammals. New York, Oxford University Press. *This is a well-known work about mammals, especially their anatomy, histology, physiology, and embryology. Classification is not treated.*

Selected *Scientific American* articles

Barnett, S. A. 1967. Rats. **216:**78-85 (Jan.). *Different species of rats are compared and their social behavior described.*

Bartholomew, G. A., and J. W. Hudson. 1961. Desert ground squirrels. **205:**107-116 (Nov.). *Two species living in California's Mojave Desert have developed interesting adaptations for survival in desert heat and aridity.*

Dawson, T. J. 1977. Kangaroos. **237:**78-89 (Aug.). *These marsupials have several adaptations fitting them for life on the Australian grasslands.*

Flyger, V., and M. R. Townsend. 1968. The migration of polar bears. **218:**108-116 (Feb.). *The habitat, biology, and wide migrations of these large arctic carnivores are described.*

Griffin, D. R. 1958. More about bat "radar." **199:**40-44 (July).

Irving, L. 1966. Adaptations to cold. **214:**94-101 (Jan.). *The homeothermic birds and mammals have evolved several adaptations that permit them to survive in cold environments.*

King, J. A. 1959. The social behavior of prairie dogs. **201:** 128-140 (Oct.).

Kooyman, G. L. 1969. The Weddell seal. **221:**100-106 (Aug.). *This Antarctic seal swims for miles under shelf ice on one breath of air, returning unerringly to its breathing hole.*

McVay, S. 1966. The last of the great whales. **215:**13-21 (Aug.). *Most of the 12 commercially hunted whale species have been nearly exterminated. The indifference and unrestricted fishing that has characterized the whaling industry is recounted in this article.*

Modell, W. 1969. Horns and antlers. **220:**114-122 (April). *Their differences, growth, and structure are described.*

Montagna, W. 1965. The skin. **212:**56-66 (Feb.). *The structure and diverse functions of mammalian skin are described.*

Mrosovsky, N. 1968. The adjustable brain of hibernators. **218:**110-118 (Mar.). *Hibernation is preceded by a remarkable series of changes and a resetting of the hypothalamic thermostat.*

Myers, J. H., and C. J. Krebs. 1974. Population cycles in rodents. **230:**38-46 (June). *Population fluctuations appear to be associated with periodic changes in genetic makeup.*

Mykytowycz, R. 1968. Territorial marking by rabbits. **218:** 116-126 (May). *Describes the use of odor-producing glands by Australian colonial rabbits to mark territories.*

Pearson, O. P. 1954. Shrews. **191:**66-70 (Aug.). *This group of tiny, voracious, burrow-dwelling, and elusive forms are among the most fascinating of mammals.*

Pruitt, W. O., Jr. 1960. Animals in the snow. **202:**60-68 (Jan.). *Describes the adaptations of the many homeotherms that live where snow persists for more than half the year.*

Schmidt-Nielsen, K. 1959. The physiology of the camel. **201:**140-151 (Dec.).

Schmidt-Nielsen, K., and B. Schmidt-Nielsen. 1953. The desert rat. **189:**73-78 (July). *The kangaroo rat possesses several adaptations that enable it to live in hot, arid regions, eating only dry food and drinking no water at all.*

Scholander, P. F. 1963. The master switch of life. **209:**92-106 (Dec.). *Diving vertebrates are obviously specialized for making prolonged dives without breathing. One of the most important adaptations is the capacity to redistribute the circulation.*

Taylor, C. R. 1969. The eland and the oryx. **220:**88-95 (Jan.). *These large African antelopes thrive in desert or near-desert regions without drinking water. Their adaptations for heat and aridity are described.*

PART THREE

ACTIVITY OF LIFE

Burchell's zebra with nursing young. (Photograph by L. L. Rue, III.)

Motility, nutrition, respiration, internal transport, excretion, irritability, and integration are necessary activities of animal life. These and the metabolic processes that provide energy for them are functional activities inherent in the simplest to the most complex animals. Animals have evolved various ways to perform these common tasks. Not all animals have specialized systems for carrying out specific functions; specialization of organ systems seems to be correlated with increase in body size and overall complexity. Higher forms exercise finer control over physiologic processes and exhibit greater internal constancy. But regardless of the degree of specialization, all animals possess integrative systems for coordinating internal activities and for relating the animal's behavior to its external environment.

CHAPTER 28
SUPPORT, PROTECTION, AND MOVEMENT

Skeletal muscle of frog. In this teased preparation, motor nerve fibers *(center)* leading to myoneural junctions on the muscle fibers are plainly visible. Cross striations, actually Z-lines between successive functional units (sarcomeres) give muscle fibers a segmented appearance (Original magnification ×5,200.)

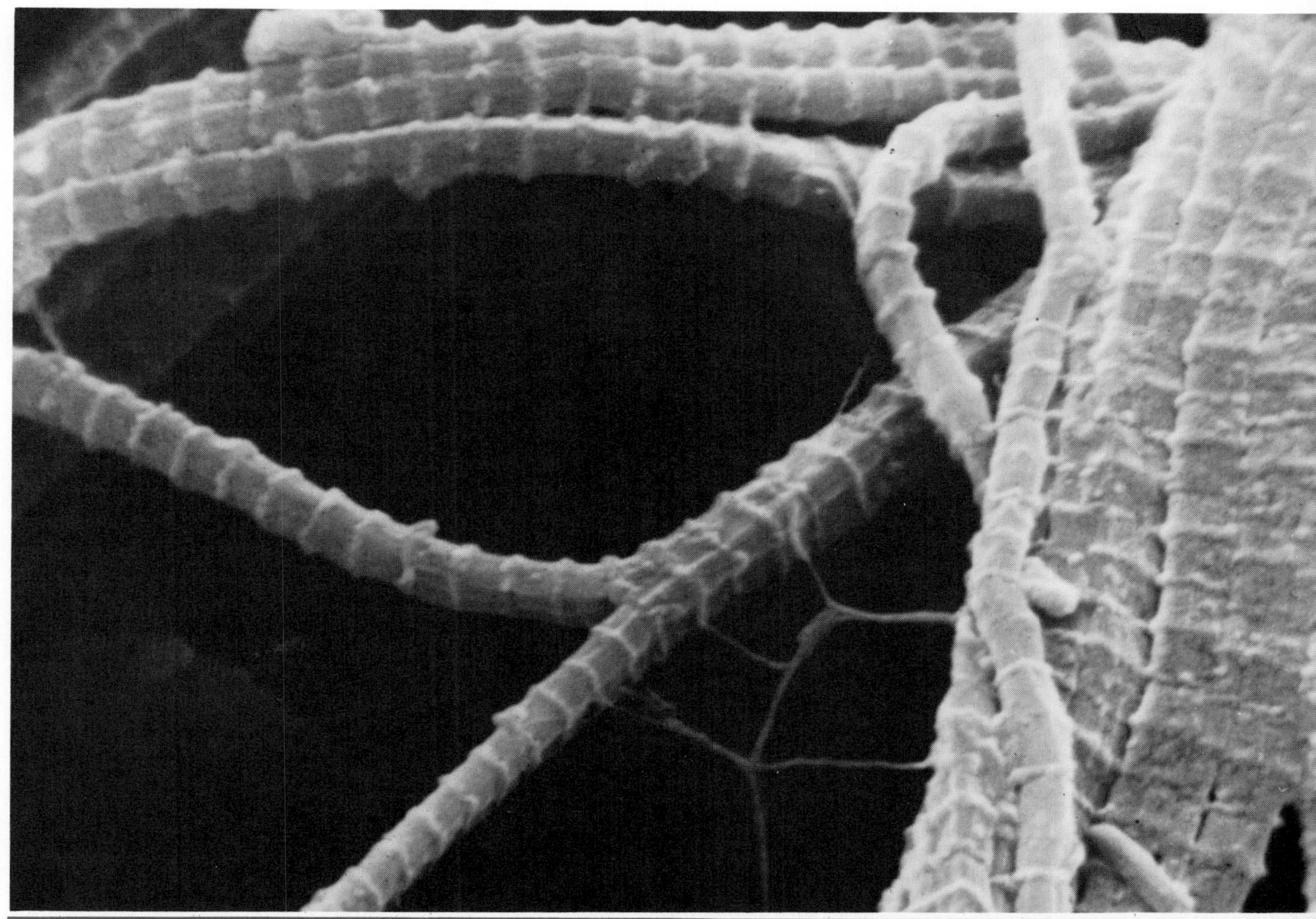

Courtesy P. P. C. Graziadei.

INTEGUMENT AMONG VARIOUS GROUPS OF ANIMALS

The integument is the outer covering of the body, a protective wrapping that includes the skin and all structures that are derived from, or associated with, the skin, such as hair, setae, scales, feathers, and horns. In most animals it is tough and pliable, providing mechanical protection against abrasion and puncture and forming an effective barrier against the invasion of bacteria. It provides moisture-proofing against fluid loss or gain. The skin protects the underlying cells against the damaging action of the ultraviolet rays of the sun. But in addition to being a protective cover, the skin serves a variety of important regulatory functions. For example, in warm-blooded (homeothermic) animals, it is vitally concerned with temperature regulation, since most of the body's heat is lost through the skin. The skin contains sensory receptors that provide essential information about the immediate environment. It has excretory functions and, in some forms, respiratory functions as well. Through skin pigmentation the organism can make itself more, or less, conspicuous. Skin secretions can make the animal attractive or repugnant or provide olfactory cues that influence behavioral interactions between individuals.

Invertebrate integument

Many protozoans have only the delicate cell or plasma membranes for external coverings; others, such as *Paramecium,* have developed a protective pellicle. Most multicellular invertebrates, however, have more complex tissue coverings. The principal covering is a single-layered **epidermis.** Some invertebrates have added a secreted noncellular **cuticle** over the epidermis for additional protection; some groups, such as many parasitic worms, have only a thick resistant cuticle and lack a cellular epidermis.

The molluscan epidermis is delicate and soft and contains mucous glands, some of which secrete the calcium carbonate of the shell. The cephalopod has developed a more complex integument, consisting of a cuticle, a simple epidermis, a layer of connective tissue, a layer of reflecting cells (iridocytes), and, finally, another thicker layer of connective tissue.

Arthropods have the most complex of invertebrate integuments, providing not only protection but also skeletal support. The development of a firm exoskeleton and jointed appendages suitable for the attachment of muscles has been a key feature in the extraordinary evolutionary success of this largest of animal groups. The arthropod integument consists of a single-layered **epidermis** (also called more precisely **hypodermis**), which secretes a complex cuticle of two zones (Fig. 28-1, *A*). The inner zone, the **procuticle (endocuticle),** is composed of protein and chitin. Chitin is a polysaccharide resembling plant cellulose that is laid down in molecular sheets like veneers of plywood, thus providing great strength to the procuticle layer. The outer zone of cuticle, lying on the external surface above the procuticle, is the thin **epicuticle.** The epicuticle is a nonchitinous complex of proteins and lipids, providing a protective, moisture-proof barrier to the integument.

The arthropod cuticle may remain as a tough but soft and flexible layer, or it may be hardened by one of two processes. In the decapod crustaceans—for example, crabs and lobsters—the cuticle is stiffened by **calcification,** the deposition of calcium carbonate. In insects hardening is achieved by a process called **sclerotization,** or **tanning,** in which the protein molecules of the chitin form stabilizing cross-linkages. Arthropod chitin is one of the toughest materials synthesized by animals; it is strongly resistant to pressure and tearing and can withstand boiling in concentrated alkali, yet it is light, having a specific weight of only 1.3.

When arthropods molt, the epidermal cells first divide by mitosis. Enzymes secreted by the epidermis dissolve most of the procuticle; the digested materials are then absorbed and consequently not lost to the body. Then, in the space beneath the old cuticle, a new epicuticle and procuticle are formed. After the old cuticle is shed, the new cuticle is thickened and calcified or sclerotized.

Vertebrate integument

The basic plan of the vertebrate integument, as exemplified by human skin (Fig. 28-1, *B*), includes a thin, outer stratified epithelial layer, the **epidermis,** derived from ectoderm and an inner, thicker layer, the **dermis,** or true skin, which is of mesodermal origin.

Although the epidermis is thin and appears simple in structure, it gives rise to most derivatives of the integument, such as hair, feathers, claws, and hooves. The dermis, containing blood vessels, collagenous fibers, nerves, pigment cells, fat cells, and fibroblasts, functions to support, cushion, and nourish its overlying partner, which is devoid of blood vessels.

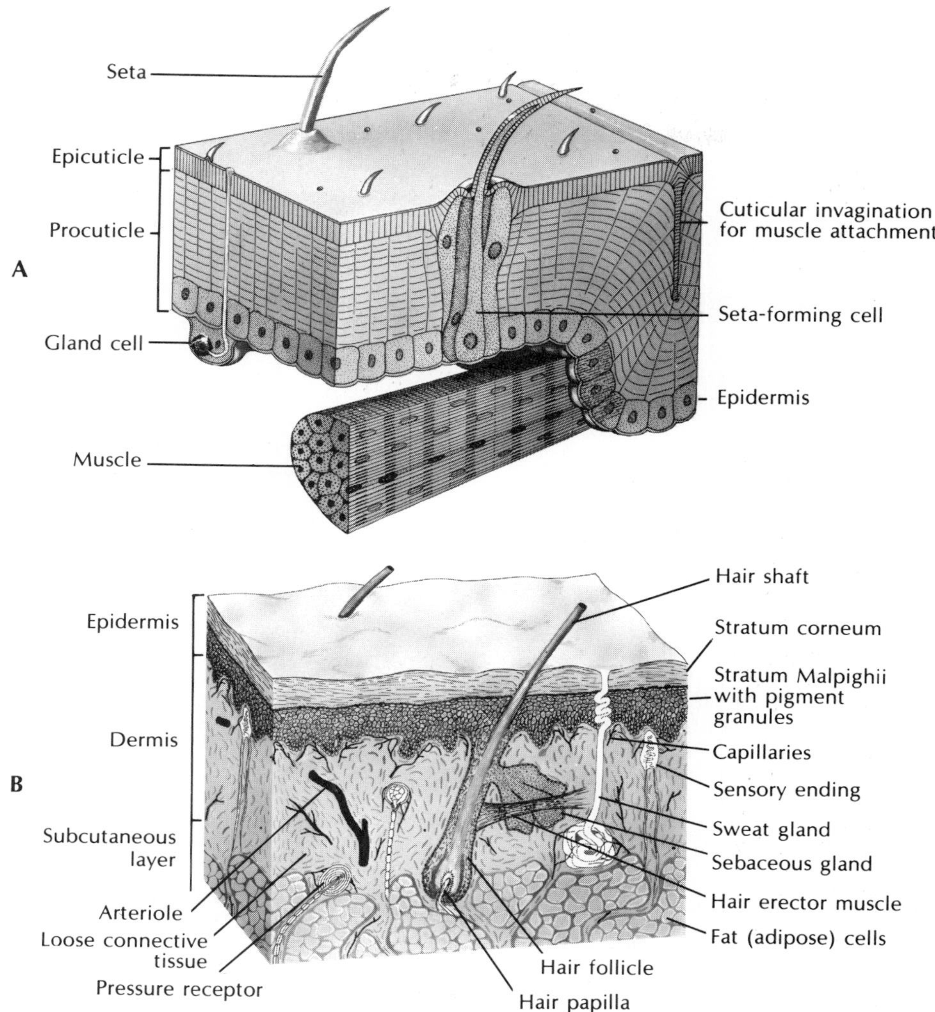

FIG. 28-1 **A,** Structure of insect integument. This reconstruction shows a block of integument drawn at a point where the cuticle invaginates to provide an exoskeletal muscle attachment. **B,** Structure of human skin.

The **epidermis** consists usually of several layers of cells. The basal part is made up of columnar cells that undergo frequent mitosis to renew the layers that lie above. As the outer layers of cells are displaced upward by new generations of cells beneath, an exceedingly tough, fibrous scleroprotein called **keratin,** accumulates in the interior of the cells. Gradually, keratin replaces all metabolically active cytoplasm. The cell dies and is eventually shed, lifeless and scalelike. Such is the origin of dandruff. This process is called **keratinization,** and the cell, thus transformed, is said to be **"cornified."** Cornified cells, highly resistant to abrasion and water diffusion, comprise the **stratum corneum** (Fig. 28-1, *B*). This epidermal layer becomes especially thick in areas exposed to persistent pressure or friction, such as calluses and the human palms and soles.

In fishes and amphibians, the epidermis is provided with mucous glands. These secrete a protective mucous covering for the epidermis, providing waterproofing and significantly reducing surface friction. With the exception of the terrestrial toads, cornified

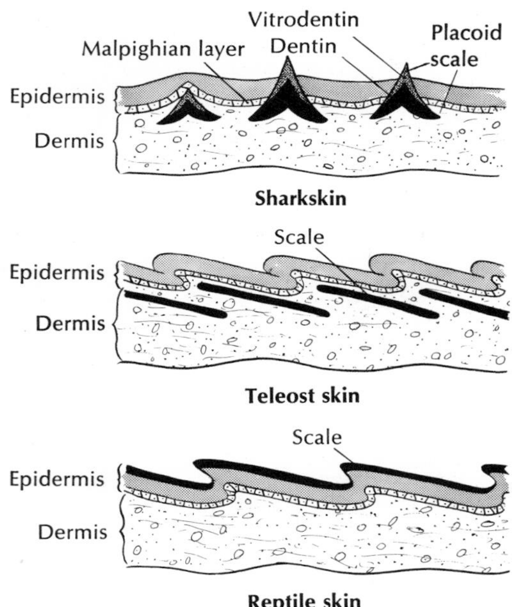

FIG. 28-2 Integument in classes Chondrichthyes, Osteichthyes, and Reptilia showing different types of scales. Placoid scales of sharks are derived from the dermis and have given rise to teeth in all higher vertebrates. Teleost fishes have bony scales from the dermis, and reptiles have horny scales from the epidermis. Only dermal scales are retained throughout life; epidermal scales are shed.

epithelium in these two groups is poorly developed. Poison glands are found in some species; and many deep-sea fishes are provided with luminescent organs, which are, in fact, epidermal glands.

In reptiles, natural selection first begins to exploit the adaptive possibilities of the remarkable protein keratin. The reptilian epidermal scale that develops from keratin is a much lighter and more flexible structure than the bony, dermal scale of fishes (Fig. 28-2), yet it provides excellent protection from abrasion and desiccation. Scales may be overlapping structures, as in snakes and some lizards, or developed into plates, as in turtles and crocodilians. Birds dedicate keratin to new uses. Feathers, beaks, and claws, as well as scales, are all epidermal structures composed of dense keratin.

Mammalian epidermal derivatives include that most diagnostic of all mammalian characteristics, hair, described in Chapter 27. Mammals also have integumentary glands of great diversity, far more than any other vertebrate group. Glands, such as sweat glands, oil glands, and scent glands, are all epidermal derivatives, although many extend deep into the dermis when fully developed. As in the reptiles and birds, the keratin is put to many specialized uses. In addition to hair, it forms the overlying sheath of horn and is the principal component of hoofs, claws, and nails. A claw is shaped to cover the sides, top, and tip of a terminal joint; a nail is flattened and covers the dorsal surface of the distal phalanx; and a hoof extends across the end of the digit and covers the plantar surface also. In the horse the hoof, which is developed from the claw of one toe, is the only part of the foot touching the ground. Other hoofed animals may have spongy pads or other parts of the foot on which to walk.

The **dermis,** as already mentioned, mainly serves a supportive role for the epidermis. Nevertheless, true bony structures, where they occur in the integument, are always dermal derivatives. Heavy bony plates were common in primitive ostracoderms and placoderms but occur in no living fishes. Scales of contemporary fishes are bony dermal structures that have evolved from the bony armor of the Paleozoic fishes, but are much smaller and more flexible. Though of dermal origin, fish scales are intimately associated with the thin, overlying epidermis; in some species the scales protrude through the epidermis, but typically the epidermis forms a continuous sheath that is reflected under the overlapping scales (Fig. 28-2). Dermal bone also forms the flat bones of the skull and gives rise to antlers, which are outgrowths of dermal frontal bone.

Color

Nearly all vertebrates are colored. The colors may be subdued and cryptic when serving protective functions, or vivid and dramatic when serving as important recognition marks. Integumentary color is usually caused by **pigments** but in some vertebrates, especially birds, colors are produced instead by the physical structure of the surface tissue that reflects certain light wavelengths and eliminates others. Colors produced this way are called **structural color,** and they are responsible for the most beautifully iridescent and metallic hues to be found in the animal kingdom. Many butterflies and beetles, and a few fishes, thus share with birds the distinction of being the earth's most resplendent animals. Certain structural colors of feathers are caused by minute air-filled spaces or pores that reflect white light (white feathers) or some portion of the spectrum (for example, Tyndall blue-colored feathers). Iridescent colors are produced when light passes through several layers of thin, transparent film on the feather barbules and is diffracted into its component parts, much as a film of oil on a puddle breaks up sunlight into a rainbowlike spectrum.

Pigment color is caused by **chromatophores,** specialized cells usually located in the dermis, but occasionally in the epidermis. Pigments are large molecules that reflect light rays. In fishes, amphibians, and reptiles the expansion and contraction of pigments within the chromatophores are under hormonal or nervous control (p. 539). When expanded, the pigment is visible; when contracted, the pigment concentrates into a central aggregate too small to be visible. The animal then pales, or assumes the color of a deeper pigment normally masked by the overlying chromatophores. Some reptiles and fishes can shift their color and color patterns in remarkable ways.

Melanophores contain the pigment **melanin.** This is by far the most widespread animal pigment and is responsible for the various earth-colored shades—black, browns, and grays—that most animals wear. Yellow and red colors are often caused by **carotenoid** pigments, which are frequently contained within special pigment cells called **xanthophores.** Most vertebrates are incapable of synthesizing their own carotenoid pigments but must obtain them directly or indirectly from plants. Two entirely different classes of pigments called ommochromes and pterdines are usually responsible for the yellow pigments of molluscs and arthropods. Green colors are rare; when they occur they are usually produced by a combination of structural and pigment colors. **Iridophores,** a third type of chromatophore, contains crystals of guanine or some other purine, rather than pigment. They produce a silvery or metallic effect by reflecting light.

By vertebrate standards, the mammals are a somber-colored group. Most mammals are more or less color-blind, a deficiency that is doubtless connected with the lack of bright colors in the group. Exceptions are the brilliantly colored skin patches of some baboons and mandrills. Significantly, the primates have color vision and thus can appreciate such eye-catching ornaments. The muted colors of mammals are caused by melanin, which is deposited in growing hair by dermal melanophores.

Sunburning and tanning

In general, humans lack the special body coverings that protect other land vertebrates from the damaging action of ultraviolet rays of the sun; we must depend on thickening of the outer layer of the epidermis (stratum corneum) and on epidermal pigmentation for protection from the sun's spectrum. Sunburn and suntanning are caused largely by exposure to the ultraviolet area of the sun's spectrum (wavelength 300 to 390 nm). This spectral band acts almost entirely on epidermis; very little penetrates to the dermis beneath. The ultraviolet rays photochemically decompose nucleoproteins within the nuclei of cells of the deeper layer of the epidermis. Blood vessels then enlarge and other tissue changes occur, producing the red coloration of sunburn. Light skins suntan through the formation in the deeper epidermis of the pigment melanin and by "pigment darkening," that is, the photooxidative blackening of bleached pigment already present in the epidermis. Regrettably, Americans pay dearly for their sun worship with some 300,000 cases of skin cancer each year.

SKELETAL SYSTEMS

Skeletons are supportive systems that provide rigidity to the body, surfaces for muscle attachment, and protection for vulnerable body organs. The familiar bone of the vertebrate skeleton is only one of several kinds of supportive and connective tissues, serving various binding and weight-bearing functions, which are discussed in this section.

Exoskeleton and endoskeleton

Although animal supportive and protective structures take many forms, there are two principal types of skeletons: the **exoskeleton,** typical of molluscs and arthropods, and the **endoskeleton,** characteristic of echinoderms and vertebrates. The invertebrate exoskeleton is mainly protective in function and may take the form of shells, spicules, and calcareous or chitinous plates. It may be rigid, as in molluscs, or jointed and movable as in arthropods. Unlike the endoskeleton, which grows with the animal, the exoskeleton is often a limiting coat of armor, which must be periodically shed (molted) to make way for an enlarged replacement. Some invertebrate exoskeletons, such as the shells of snails and bivalves, grow with the animal. Vertebrates, too, have traces of exoskeleton that serve to remind us of our invertebrate heritage. These are, for example, scales and plates of fishes, fingernails and claws, hair, feathers, and other keratinized integumentary structures.

The vertebrate endoskeleton is formed inside the

body and is composed of bone and cartilage surrounded by soft tissues. It not only supports and protects, but it is also the major body reservoir for calcium and phosphorus. In the higher vertebrates the red blood cells and certain white blood cells are formed in the bone marrow.

Cartilage

Cartilage and bone are the characteristic vertebrate supportive tissues. The **notochord,** the semirigid axial rod of protochordates and vertebrate larvae and embryos, is also a primitive vertebrate supportive tissue. Except in the most primitive chordates, for example, amphioxus and the cyclostomes, the notochord is surrounded or replaced by the backbone during embryonic development. The notochord is composed of large, vacuolated cells and is surrounded by layers of elastic and fibrous sheaths. It is a stiffening device, preserving body shape during locomotion (p. 474).

Vertebrate cartilage is the major skeletal element of primitive vertebrates. Cyclostomes and elasmobranchs have purely cartilaginous skeletons. In contrast, higher vertebrates have principally bony skeletons as adults, with some cartilage interspersed. Cartilage is a soft, pliable, characteristically deep-lying tissue. Unlike connective tissue, which is quite variable in form, cartilage is basically the same wherever it is found. The basic form, **hyaline cartilage** (Fig. 6-10, *B,* p. 110), has a clear, glassy appearance. It is composed of cartilage cells **(chondrocytes)** surrounded by firm complex protein gel interlaced with a meshwork of collagenous fibers. Blood vessels are virtually absent. In addition to forming the cartilagenous skeleton of the primitive vertebrates and that of all vertebrate embryos, hyaline cartilage makes up the articulating surfaces of many bone joints of higher adult vertebrates and the supporting tracheal, laryngeal, and bronchial rings. The basic cartilage has several variants. Among these is **calcified cartilage,** where calcium salt deposits produce a bone-like structure. **Fibrocartilage,** resembling connective tissue, and **elastic cartilage,** containing many elastic fibers, are other variations of basic hyaline cartilage found among the vertebrates.

Bone

Bone differs from other connective and supportive tissues by having significant deposits of inorganic calcium salts laid down in an extracellular matrix. Its structural organization is such that bone has nearly the tensile strength of cast iron, yet is only one-third as heavy. Most bones develop from cartilage **(endochondral bone)** by a complex replacement of embryonic cartilage with bone tissue. A second type of bone is **membrane** (or **dermal) bone** that develops directly from sheets of embryonic cells. In higher vertebrates membrane bone is restricted to bones of the face and cranium; the remainder of the skeleton is endochondral bone. Despite differences in origin, endochondral bone and membrane bone are not distinguishable histologically. As mentioned earlier, the dermal scales and plates of fishes are formed from membrane bone.

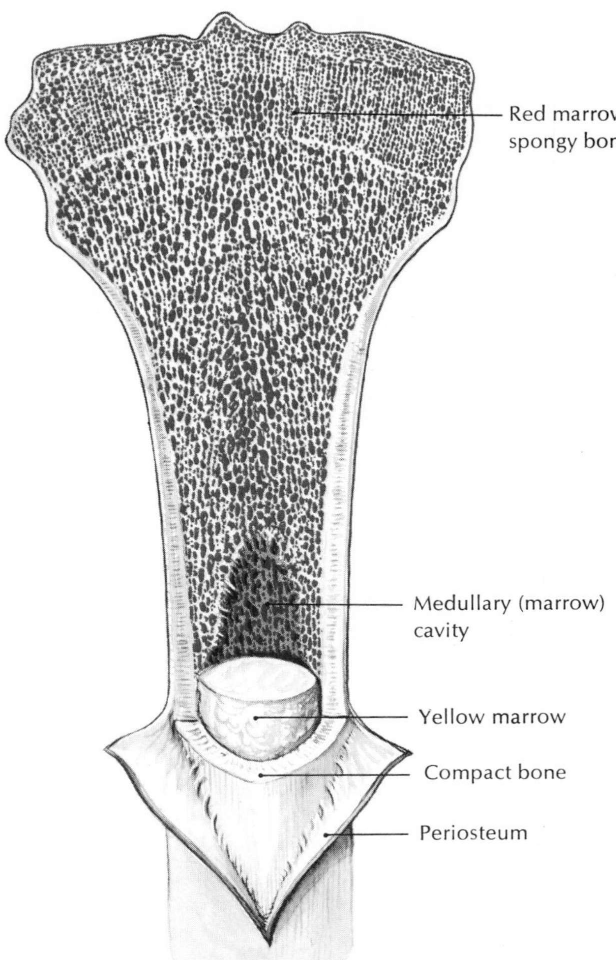

FIG. 28-3 Cutaway section of a long bone, showing appearance of spongy and compact bone. (From Anthony, C. P., and N. J. Kolthoff. 1975. Textbook of anatomy and physiology, ed. 9. St. Louis, The C. V. Mosby Co.)

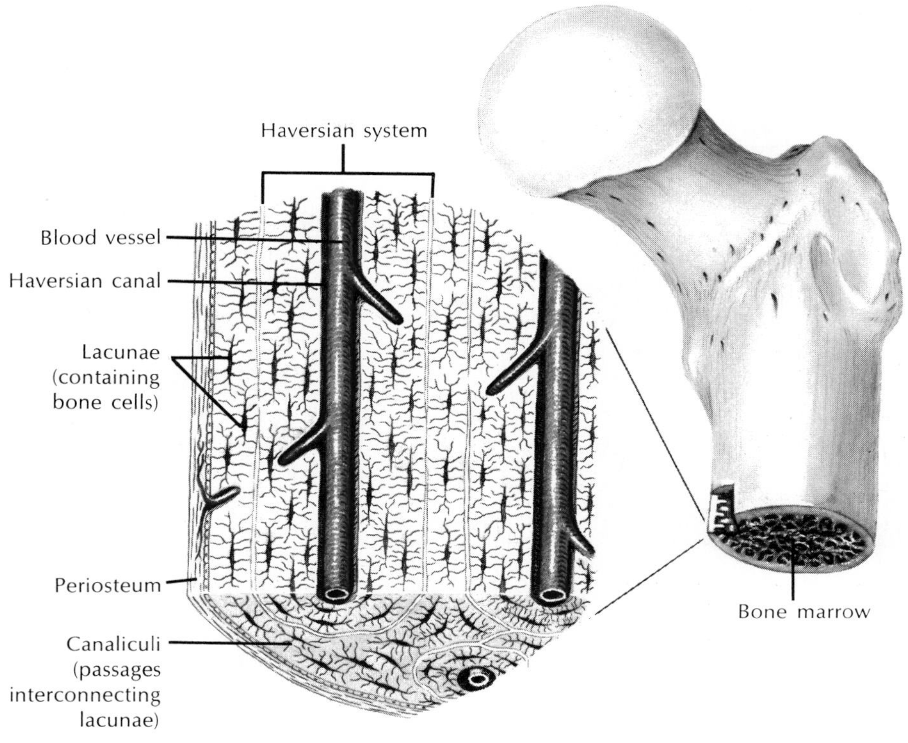

FIG. 28-4 Structure of bone, showing the dense calcified matrix and bone cells arranged into haversian systems. Bone cells are entrapped within the cell-like lacunae, but receive nutrients from the circulatory system via tiny canaliculi that interlace the calcified matrix. Bone cells were known as osteoblasts when they were building bone, but in mature bone (shown here) they become resting osteocytes. Bone is covered with a compact connective tissue called "periosteum."

Two kinds of bone structure are distinguishable—**spongy** (or **cancellous**) and **compact** (Fig. 28-3). Spongy bone consists of an open, interlacing framework of bony tissue, oriented to give maximum strength under the normal stresses and strains that the bone receives. Compact bone is dense, appearing absolutely solid to the naked eye. Both structural kinds of bone are found in the typical long bones of the body such as the humerus (upper arm bone) (Fig. 28-3).

Microscopic structure of bone

Compact bone is composed of a calcified bone matrix arranged in concentric rings. The rings contain cavities **(lacunae)** filled with bone cells **(osteocytes)** that are interconnected by many minute passages **(canaliculi).** These serve to distribute nutrients throughout the bone. This entire organization of lacunae and canaliculi is arranged into an elongated cylinder called a **haversian system** (Fig. 28-4). Bone consists of bundles of haversian systems cemented together and interconnected with blood vessels.

Bone growth is a complex restructuring process, involving both its destruction internally by bone-resorbing cells **(osteoclasts)** and its deposition externally by bone-building cells **(osteoblasts).** Both processes occur simultaneously so that the marrow cavity inside grows larger by bone resorption while new bone is laid down outside by bone deposition. Bone growth responds to several hormones, in particular **parathyroid hormone,** which stimulates bone resorption, and **calcitonin,** which inhibits bone resorption. These two hormones, together with a vitamin D derivative called 1,25-DHCC, are responsible for maintaining a constant level of calcium in the blood (p. 767).

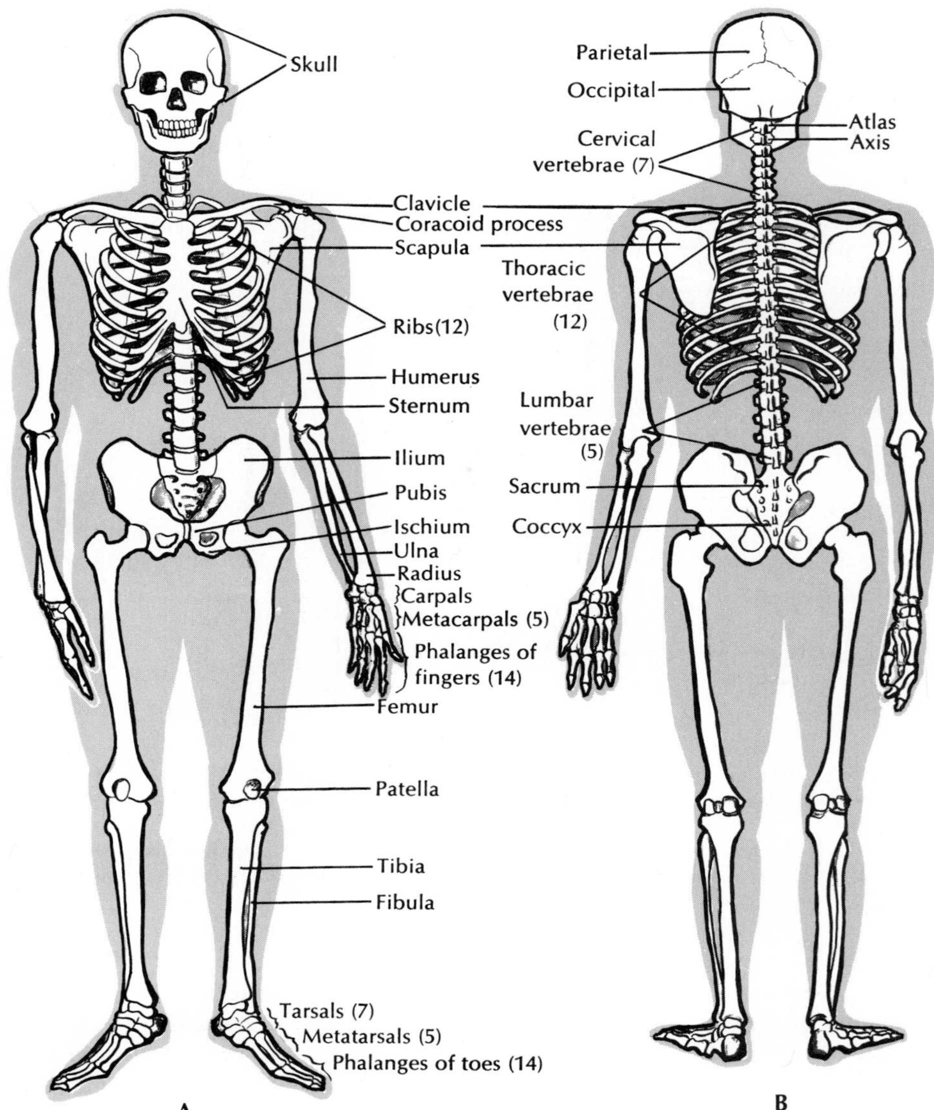

FIG. 28-5 Human skeleton. **A,** Ventral view. **B,** Dorsal view. Numbers in parentheses indicate number of bones in that unit. In comparison with other mammals, the human skeleton is a patchwork of primitive and specialized parts. Erect posture brought about by specialized changes in legs and pelvis enabled primitive arrangement of arms and hands (arboreal adaptation of ancestors of humans) to be used for manipulation of tools. Development of skull and brain followed as a consequence of the premium that natural selection put on dexterity, better senses, and ability to appraise the environment.

Plan of the vertebrate skeleton

The vertebrate skeleton is composed of two main divisions: the **axial skeleton,** which includes the skull, vertebral column, sternum, and ribs and the **appendicular skeleton,** which includes the limbs and the pectoral and pelvic girdles (Fig. 28-5). Not surprisingly the skeleton has undergone extensive remodeling in the course of vertebrate evolution. The move from water to land forced dramatic changes in body form. With cephalization, that is, the concentration of brain, sense organs, and food-gathering and respiratory apparatus in the head, the skull became the most intricate portion of the skeleton. The lower vertebrates have a larger number of skull bones than the more advanced vertebrates. Some fish have 180 skull bones; amphibians and reptiles, 50 to 95; and mammals, 35 or fewer. We have 29.

The numerous elements of the head skeleton belong to three structural and functional components: (1) the **neurocranium,** which is the original, primitive skull that encloses the brain and sense organs, (2) the **splanchnocranium,** which supports the respiratory and food-gathering equipment (gill apparatus and jaws), and (3) the **dermocranium,** which arose from ancient, superficial head armor and serves to protect deep-lying delicate tissues. In the lower vertebrates these three skull components are more or less separated from each other; in higher forms they are all fused together or incorporated into a single unit—the vertebrate skull. There is a basic plan of homology in the skull elements of vertebrates from fishes to human beings; evolution has meant reduction in numbers of bones through loss and fusion in accordance with size and functional changes.

The vertebral column is the main stiffening axis of the postcranial skeleton. In fishes it serves much the same function as the notochord from which it is derived; that is, it provides points for muscle attachment and prevents telescoping of the body during muscle contraction. Since fish musculature is similar throughout the trunk and tail, fish vertebrae are differentiated only into trunk and caudal vertebrae.

With the evolution of tetrapods, the vertebral column becomes structurally adapted to withstand new regional stresses transmitted to the column by the two paired appendages. In the higher tetrapods, the vertebrae are differentiated into **cervical** (neck), **thoracic** (chest), **lumbar** (back), **sacral** (pelvic), and **caudal** (tail) vertebrae. In birds and also in human beings the caudal vertebrae are reduced in number and size, and the sacral vertebrae are fused. The number of vertebrae varies among the different animals. The python seems to lead the list with 435. There are 33 in the human child, but in the adult five are fused to form the **sacrum** and four to form the **coccyx.** Besides the sacrum and coccyx, we have seven cervical, 12 thoracic, and five lumbar vertebrae (Fig. 28-5). The number of cervical vertebrae (seven) is constant in nearly all mammals.

The first two cervical vertebrae, the **atlas** and the **axis,** are modified to support the skull and permit pivotal movements. The atlas bears the globe of the head much as the mythologic Atlas bore the earth on his shoulders. The axis, the second vertebra, permits the head to turn from side to side.

Ribs are long or short skeletal structures that articulate medially with vertebrae and extend into the body wall. Primitive forms have a pair of ribs for every vertebra; they serve as stiffening elements in the connective tissue septa that separate the muscle segments and thus improve the effectiveness of muscle contractions. Many fishes have both dorsal and ventral ribs, and some have numerous riblike intermuscular bones as well—all of which increase the difficulty and reduce the pleasure of our eating certain kinds of fish. Higher vertebrates have a reduced number of ribs, and some, such as the familiar leopard frog, have no ribs at all. Others, such as elasmobranchs and some amphibians, have very short ribs. Human beings have 12 pairs of ribs, but approximately one person in 20 has a thirteenth pair. In mammals the ribs together form the thoracic basket, which supports the chest wall and prevents collapse.

Most vertebrates, fishes included, have paired appendages. All fishes except the agnathans have thin pectoral and pelvic fins that are supported by the pectoral and pelvic girdles, respectively. Forms above the fishes (except snakes and limbless lizards) have two pairs of **pentadactyl** (five-toed) limbs, also supported by girdles. The pentadactyl limb is similar in all tetrapods, alive and extinct; even when highly modified for various modes of life, the elements are rather easily homologized.

Modifications of the basic pentadactyl limb for life in different environments involve the distal elements much more frequently than the proximal, and it is far more common for bones to be lost or fused than for new ones to be added. The elongation of the third toe of the horse, which facilitates running, is described on p. 886. The bird wing is a good example of distal modification. The bird embryo bears 13 distinct wrist and hand bones (carpals and metacarpals), which are reduced to three in the adult. Most of the finger bones (phalanges) are lost, leaving four bones in three digits (Fig. 26-4, p. 575). The proximal bones (humerus, radius, and ulna), however, are little modified in the bird wing.

In nearly all tetrapods the pelvic girdle is firmly attached to the axial skeleton since the greatest locomotory forces transmitted to the body come from the

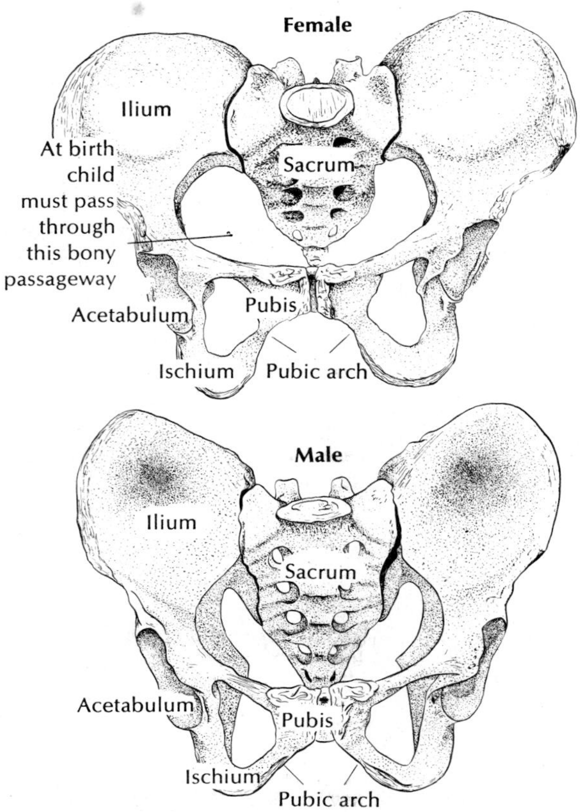

FIG. 28-6 The chief difference between male and female skeletons is the structure of the pelvis. The female pelvis has less depth with broader, less sloping ilia, a more circular bony ring (pelvic canal), a wider and more rounded pubic arch, and a shorter and wider sacrum. Most structures of the female pelvis are correlated with childbearing functions. In the evolution of the human skeleton, the pelvis has changed more than any other part because it has to support the weight of the erect body. (Anterior view.)

hind limbs. The pectoral girdle, however, is much more loosely attached to the axial skeleton, providing the forelimbs with greater freedom for manipulative movements.

The human pectoral girdle is made up of two scapulae and two clavicles; the arm is made up of humerus, ulna, radius, eight carpals, five metacarpals, and 14 phalanges. The pelvic girdle (Fig. 28-6) consists of three fused bones—ilium, ischium, and pubis; the leg is made up of femur, patella, tibia, fibula, seven tarsals, five metatarsals, and 14 phalanges. Each bone of the leg has its counterpart in the arm with the exception of the patella. This kind of correspondence between anterior and posterior parts is called **serial homology.**

ANIMAL MOVEMENT

Movement is a unique characteristic of animals. Plants may show movement, but this usually results from changes in turgor pressure or growth rather than from specialized contractile proteins as in animals. Animal movement occurs in many forms, ranging from barely discernible streaming of cytoplasm, the swelling of mitochondria, or movement of the mitotic spindle during cell division, to obvious movements of powerful striated muscles of vertebrates.

Recently it has become evident that virtually all animal movement depends on a single fundamental mechanism: **contractile proteins,** which can change their form to elongate or contract. This contractile machinery is always composed of ultrafine fibrils—fine filaments, striated fibrils, or tubular fibrils (microtubules)—arranged to contract when powered by **ATP** (adenosine triphosphate). By far the most important protein contractile system is the **actomyosin system,** composed of two proteins, **actin** and **myosin.** This is an almost universal biomechanical system found in protozoans through vertebrates; it performs a long list of diverse functional roles. In this section we will examine the three principal kinds of animal movement: ameboid, ciliary, and muscle.

Ameboid movement

Ameboid movement is a form of movement especially characteristic of the freshwater amebas and other sarcodine protozoans; it is also found in many wandering cells of higher animals, such as white blood cells, embryonic mesenchyme, and numerous other mobile cells that move through the tissue spaces. Ameboid cells constantly change their shape by sending out and withdrawing **pseudopodia** (false feet) from any point on the cell surface. Such cells are surrounded by a delicate, highly flexible membrane called **plasmalemma** (Fig. 8-3, p. 139). Beneath this lies a nongranular layer, the **hyaline ectoplasm,** which encloses the **granular ectoplasm.** Optical studies of an ameba in movement suggest that the outer layer of cytoplasm **(ectoplasm)** actively contracts in the **fountain zone** at the tip of the pseudopod to pull a central core of rather rigid endoplasm forward. The latter is then converted into ectoplasm, which slips posteriorly under

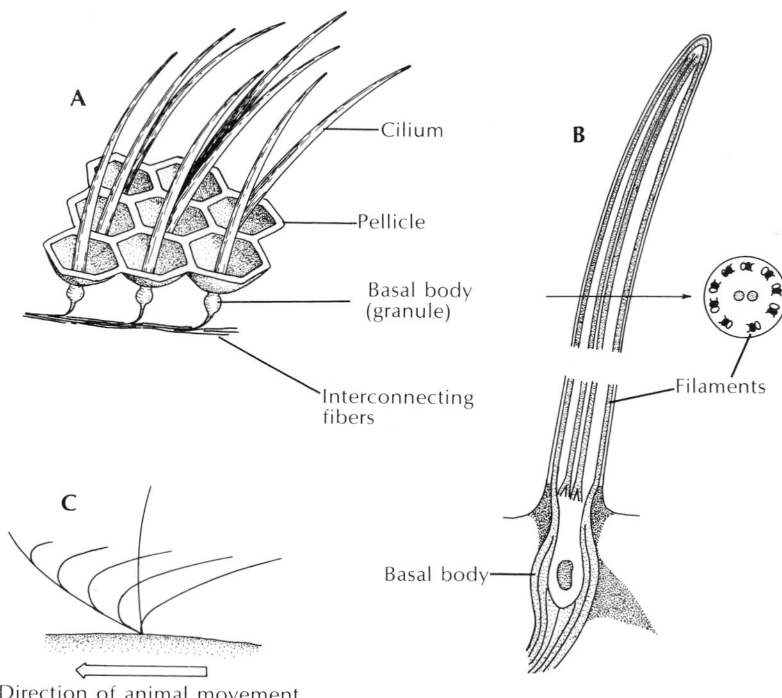

FIG. 28-7 Ciliary structure and movement. **A,** Section of the pellicle of *Paramecium,* showing arrangement of cilia, basal granules, and interconnection fibers (so-called neuromotor system). **B,** Structure of a cilium as revealed by the electron microscope. **C,** Sequence of movements of a cilium of *Paramecium.* Power stroke is to the right and recovery to the left. (**B,** Modified from Rhodin, J., and T. Dalhamn. 1956. Z. Zellforsch. **44:**345.)

the plasmalemma and joins the endoplasm at the rear to begin another cycle. There are other theories of ameboid movement, in particular one that favors the posterior end of the animal as the locus of contraction. According to this theory, an ameba is pushed rather than pulled forward. Although no completely satisfactory analysis exists, it is certain that ameboid movement is based on the same fundamental contractile system that powers vertebrate muscles: an actomyosin machinery driven by ATP.

Ciliary movement

Cilia are minute hairlike motile processes that extend from the surfaces of the cells of many animals (Fig. 28-7). They are a particularly distinctive feature of ciliate protozoans but, except for the nematodes, in which cilia are absent, and the arthropods, in which they are rare, cilia are found in all major groups of animals. Cilia perform many roles, either in moving small animals (such as protozoans) through their aquatic environment or in propelling fluids and materials across the epithelial surfaces of larger animals. Cilia play prominent roles in filter feeding as described in Chapter 31. In some animals, cilia may be modified to form undulating membranes (such as in *Trypanosoma,* p. 149) or grouped in macrocilia each consisting of 2,000 to 3,000 cilia bundled together (such as in ctenophores, p. 212). Cilia are of remarkably uniform diameter (0.1 to 0.5 μm) wherever they are found. The electron microscope has shown that each cilium contains a peripheral circle of nine double filaments and an additional two filaments in the center (Fig. 28-7). (Exceptions to the 9 + 2 arrangement have been noted; certain sperm tails have but one central fibril.)

A **flagellum** is a whiplike structure, longer than a cilium, and usually present singly at one end of a cell. Flagella are found in members of flagellate protozoans, in animal spermatozoa, and in sponges.

According to one currently favored theory of ciliary movement, the fibrils behave as "sliding filaments" that move past one another much like the sliding filaments of vertebrate skeletal muscle that is described in the next section. During contraction, fibrils on the concave side slide outward past fibrils on the convex side to increase curvature of the cilium; during the recovery stroke, fibrils on the opposite side slide outward to bring the cilium back to its starting position. For such a system to work, the fibrils must be interconnected by molecular bridges, which in fact cannot be seen with the electron microscope.

Cilia contract in a highly coordinated way, the rhythmic waves of contraction moving across a ciliated epithelium like windwaves across a field of grain. The columns of cilia are coordinated by an interconnected fiber system through which the excitation wave passes with each stroke.

Muscular movement

Contractile tissue reaches its highest development in muscle cells called **fibers.** Although muscle fibers themselves can only shorten, they can be arranged in so many different configurations and combinations that almost any movement is possible.

Types of vertebrate muscle

Vertebrate muscle is broadly classified on the basis of the appearance of muscle cells (fibers) when viewed with a light microscope (Fig. 6-12, p. 111). **Striated muscle** appears transversely striped (striated), with alternating dark and light bands (Fig. 6-13, p. 111). We can recognize two types of striated muscle: **skeletal** and **cardiac muscle.** A third kind of vertebrate muscle is **smooth** (or visceral) **muscle,** which lacks the characteristic alternating bands of the striated type.

Skeletal muscle is typically organized into sturdy, compact bundles or bands. It is called skeletal muscle because it is attached to skeletal elements and is responsible for movements of the trunk, appendages, the respiratory organs, eyes, mouthparts, and so on. Skeletal muscle fibers are extremely long, cylindric, multinucleate cells, which may reach from one end of the muscle to the other. They are packed into bundles called **fascicles,** which are enclosed by tough connective tissue. The fascicles are in turn grouped into a discrete **muscle** surrounded by a thin connective tissue layer. Most skeletal muscles taper at their ends, where they connect by tendons to bones. Other muscles, such as the ventral abdominal muscles, are flattened sheets.

In most fishes, amphibians, and, to some extent, reptiles, there is a segmented organization of muscles alternating with the vertebrae. The skeletal muscles of higher vertebrates, by splitting, by fusion, and by shifting, have developed into specialized muscles best suited for manipulating the jointed appendages that have evolved for locomotion on land. Skeletal muscle contracts powerfully and quickly but fatigues more rapidly than does smooth muscle. Skeletal muscle is sometimes called **voluntary muscle** because it is stimulated by motor fibers and is under conscious cerebral control.

Smooth muscle lacks the striations typical of skeletal muscle (Fig. 6-12, p. 111). The cells are long, tapering strands, each containing a single nucleus. Smooth muscle cells are organized into sheets of muscle circling the walls of the alimentary canal, blood vessels, respiratory passages, and urinary and genital ducts. Smooth muscle is typically slow acting. It is under the control of the autonomic nervous system; thus, unlike skeletal muscle, its contractions are involuntary and unconscious. The principal functions of smooth muscles are to push the contents of a tube, such as the intestine, along its way by active contractions or to regulate the diameter of a tube, such as a blood vessel, by sustained contraction.

Cardiac muscle, the seemingly tireless muscle of the vertebrate heart, combines certain characteristics of both skeletal and smooth muscle. It is fast-acting and striated like skeletal muscle, but contraction is under involuntary autonomic control like smooth muscle. Actually, the autonomic nerves serving the heart can only speed up or slow down the rate of contraction; the heartbeat originates within specialized cardiac muscle, and the heart will continue to beat even after all autonomic nerves are severed. Until very recently, cardiac muscle was believed to be one large unseparated mass **(syncytium)** of branching, interconnected fibers. Histologists, their understanding vastly increased by the electron microscope, now consider cardiac muscle to be comprised of closely opposed, but separate, uninucleate cell fibers.

Types of invertebrate muscle

Smooth and striated muscles are also characteristic of invertebrate animals, but there are many variations of both types and even instances where the structural and functional features of vertebrate smooth and striated muscle are combined in the invertebrates. Striated muscle appears in invertebrate groups as diverse as the primitive cnidarians and the advanced arthropods. The thickest muscle fibers known—about 3 mm in diameter and 6 cm long, and easily seen with the unaided eye—are those of giant barnacles and Alaska king crabs living along the Pacific coast of North America. These cells are so large that they can be readily cannulated for physiologic studies and are understandably popular with muscle physiologists.

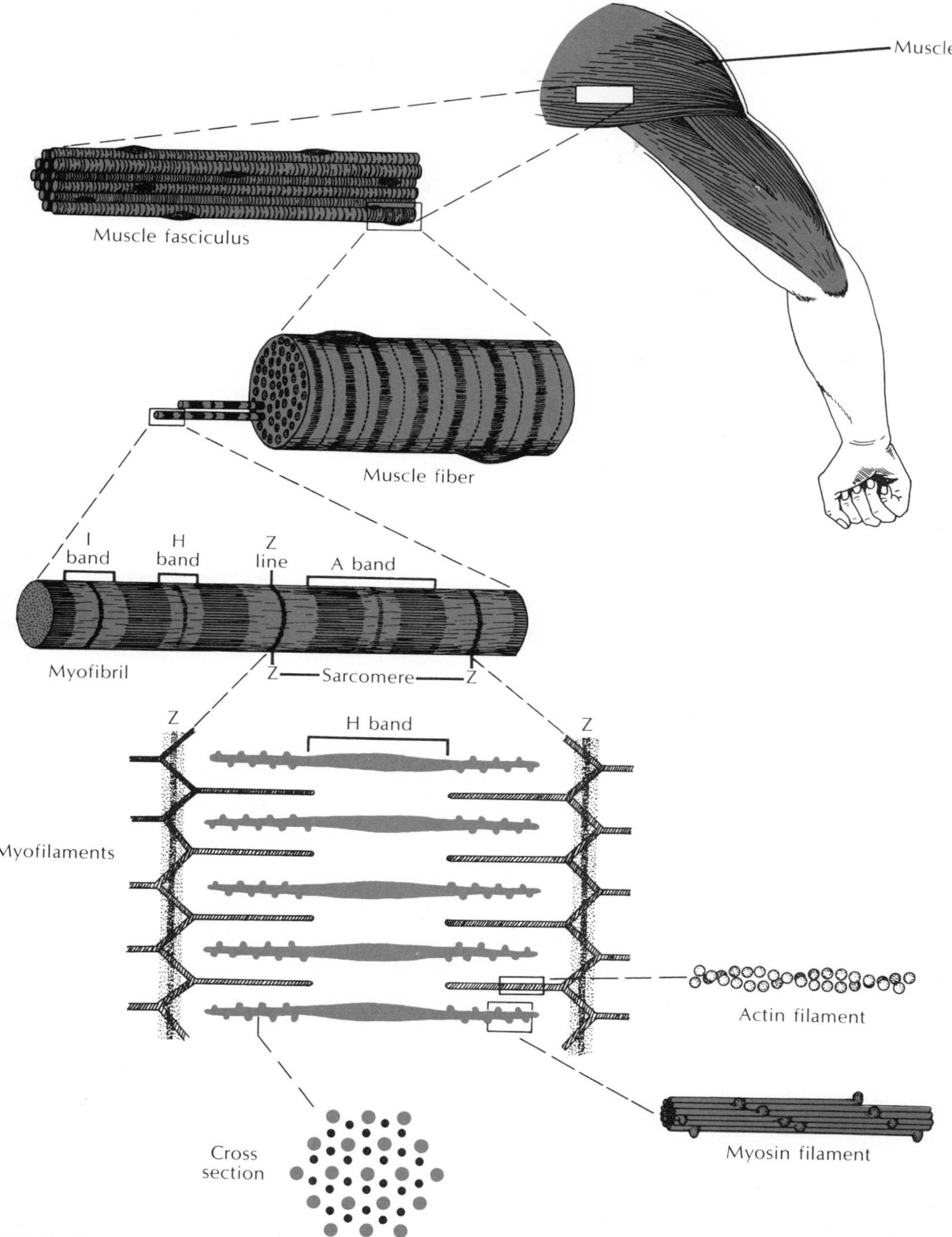

FIG. 28-8 Organization of vertebrate skeletal muscle from gross to molecular level. Actin (thin) and myosin (thick) filaments are enlarged to show supposed shapes of individual molecules and probable positioning of the cross-bridges (shown as knobs) on myosin molecules, which serve to link thick and thin filaments during contraction. Cross section shows that each thick filament is surrounded by six thin filaments and that each thin filament is surrounded by three thick filaments. (Modified from Bloom, W., and D. W. Fawcett. 1975. A textbook of histology. Philadelphia, W. B. Saunders Co.)

It is not possible in this short space to describe adequately the tremendous diversity of muscle structure and function in the vast assemblage of invertebrates. We will mention only two functional extremes.

Bivalve mollusc muscles contain fibers of two types. One kind can contract rapidly, enabling the bivalve to snap shut its valves when disturbed. Scallops use these "fast" muscle fibers to swim in their awkward manner (Fig. 13-17, p. 285). The second muscle type is capable of slow, long-lasting contractions. Using these fibers, a bivalve can keep its valves tightly shut for days or even months. Obviously these are no ordinary muscle fibers! It has been discovered that such retractor muscles use very little metabolic energy and receive remarkably few nerve impulses to maintain the activated state. The contracted state has been likened to a "catch mechanism" involving some kind of stable cross-linkage between the contractile proteins within the fiber. However, despite considerable research, no completely satisfactory explanation for this retractor mechanism exists.

Insect flight muscles are virtually the functional antithesis of the slow, holding muscles of bivalves. The wings of some of the small flies operate at frequencies greater than 1,000 per second. The so-called **fibrillar muscles,** which contract at these incredible frequencies—far greater than even the most active of vertebrate muscles—show unique characteristics. They have very limited extensibility; that is, the wing leverage system is arranged so that the muscles shorten hardly at all during each downbeat of the wings. Furthermore, the muscles and wings operate as a rapidly oscillating system in an elastic thorax (Fig. 17-10, p. 381). Since the muscles rebound elastically during flight, they receive impulses only periodically rather than one impulse per contraction; one reinforcement impulse for every 20 or 30 contractions is enough to keep the system active.

Structure of striated muscle

In recent years the electron microscope and advanced biochemical methods have been focused on the fine structure and function of the striated muscle fiber. These efforts have been so successful that more has been learned of muscle physiology in the last 20 years than in the previous century. The discussion that follows will be limited to the striated muscle, since its physiology is presently much better understood than is that of smooth muscle.

As we earlier pointed out, striated muscle is so named because of the periodic bands, plainly visible under the light microscope, which pass across the widths of the muscle cells. Each cell, or **fiber,** contains numerous **myofibrils** packed together and invested by the cell membrane, the **sarcolemma** (Figs. 28-8 and 28-12). Also present in each fiber are several hundred nuclei usually located along the edge of the fiber, numerous mitochondria (sometimes called **sarcosomes**), a network of tubules called the **sarcoplasmic reticulum** (to be discussed later), and other cell inclusions typical of any living cell. Most of the fiber, however, is packed with the unique **myofibrils,** each 1 to 2 μm* in diameter.

The characteristic banding of the muscle fiber represents the fine structure of the myofibrils that make up the fiber. Alternating light- and dark-staining bands are called the **I bands** and **A bands,** respectively (Fig. 28-8). The functional unit of the myofibril, the **sarcomere,** extends between successive Z lines. The myofibril is actually an aggregate of much smaller parallel units called **myofilaments.** These are of two kinds—thick filaments, 110 Å* in diameter composed of the protein **myosin,** and thin filaments, 50 Å in diameter composed of the protein **actin** (Fig. 28-8). These are the actual contractile proteins of muscle. The thick myosin filaments are confined to the A band region. The thin actin filaments are located mainly in the light I bands but extend some distance into the A band as well. In the relaxed muscle, they do not quite meet in the center of the A band. The Z line is a dense protein, different from either actin or myosin, which serves as the attachment plane for the thin filaments and keeps them in register. These relationships are diagrammed in Fig. 28-8.

Contraction of striated muscle

The thick and thin filaments are spatially arranged in a highly symmetric pattern, so that each thick filament is surrounded by six thin filaments; conversely, each thin filament lies among three thick filaments (see cross section at bottom of Fig. 28-8). The two kinds of filaments are linked together by molecular bridges which, it is believed, extend outward from the thick filaments to hook onto active sites on the thin filaments. During contraction the cross-bridges swing rap-

*1 m = 10^6 micrometers (μm) = 10^{10} angstroms (Å).

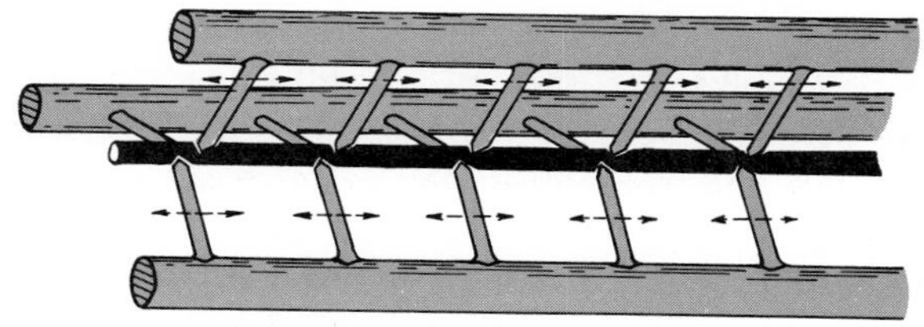

FIG. 28-9 Ratchetlike action of cross-bridges between thick and thin filaments of skeletal muscle fibers. Cross-bridges swing from site to site, pulling thin filaments past the thick. Each thick filament is actually surrounded by six thin filaments and is linked by six sets of cross-bridges. For simplicity, this diagram shows only one set of cross-bridges on each thick filament.

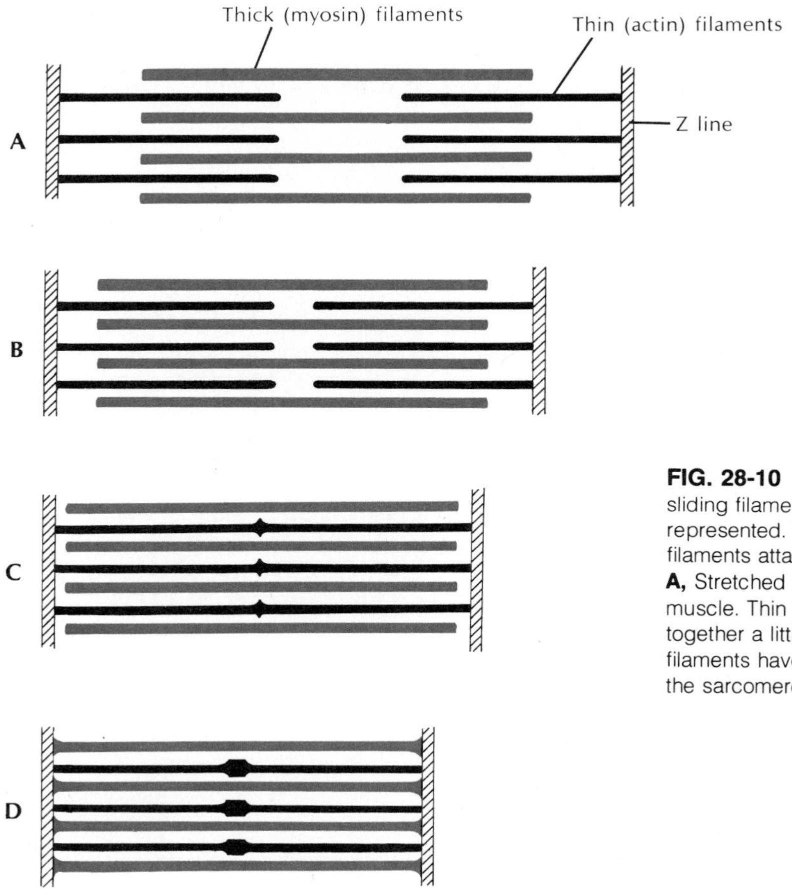

Thick (myosin) filaments
Thin (actin) filaments
Z line
A
B
C
D

FIG. 28-10 Striated muscle contraction according to sliding filament theory. One contractile unit, a sarcomere, is represented. Thick filaments are composed of myosin; thin filaments attached to the Z line are composed of actin. **A,** Stretched muscle. **B,** Resting length. **C,** Contracted muscle. Thin filaments have met in the center and crumpled together a little. **D,** Strongly contracted muscle. Thick filaments have crumpled against the Z lines at either end of the sarcomere.

idly back and forth, alternately attaching and releasing the active sites in succession in a kind of ratchet action (Fig. 28-9).

In 1950, the English physiologists A. F. Huxley and H. E. Huxley independently proposed a **sliding filament model** to explain striated muscle contraction. Their model proposed that the thick and thin filaments slide past one another. Both kinds of filament maintain their original length, but the thin actin filaments now extend farther into the A band as shown in Fig. 28-10. As contraction continues, the Z lines are drawn closer together. During very strong contraction the thin filaments touch and crumple in the center of the A band. Striated muscle contracts so rapidly that each cross-bridge may attach and release 50 to 100 times per second.

The contractile machinery has been most thoroughly studied in mammals, but recent comparative studies indicate a remarkable uniformity of the sliding-filament mechanism throughout the animal kingdom. Even the contractile proteins myosin and actin are biochemically similar in all animals. The actomyosin contractile system evidently appeared very early in animal evolution and proved so flawless that no significant changes occurred thereafter.

Energy for contraction. Muscles perform work when they contract and, of course, require energy to do so. Resting muscles use little energy but consume large amounts during vigorous exercise. Muscles use only 20% of the energy value of food molecules when contracting; the remainder is released as heat. This is a rapid source of body heat as everyone knows; exercising is the quickest way to warm up when one is cold.

The immediate source of energy for muscular contraction is ATP. When muscle is stimulated to contract, the energy released by ATP powers the ratchet-like mechanism between actin and myosin, causing the filaments to telescope.

Although the ATP stored in muscle supplies the immediate energy for contraction, the supply is limited and quickly exhausted. However, muscle contains a much larger energy storage form, **creatine phosphate,** which can rapidly transfer energy for the resynthesis of ATP. Eventually even this reserve is used up and must be restored by the breakdown of carbohydrate. Carbohydrate is available from two sources—from **glycogen** stored in the muscle and from **glucose** entering the muscle from the bloodstream. If muscular contraction is not too vigorous or too prolonged, glucose can be completely oxidized to carbon dioxide and water by **aerobic glycolysis.** But during prolonged or heavy exercise, the blood flow to the muscles, although greatly increased above the resting level, is not sufficient to supply oxygen as rapidly as required for the complete oxidation of glucose. When this happens, the contractile machinery receives its energy largely by **anaerobic glycolysis,** a process that does not require oxygen (p. 92). The presence of this anaerobic pathway, although not nearly as efficient as the aerobic one, is of great importance; without it, all forms of heavy muscular exertion such as running would be impossible,

During anaerobic glycolysis, glucose is degraded to lactic acid with the release of energy. This is used to resynthesize creatine phosphate, which, in turn, passes the energy to ADP for the resynthesis of ATP. Lactic acid accumulates in the muscle and diffuses rapidly into the general circulation. If the muscular exertion continues, the build up of lactic acid causes enzyme inhibition and fatigue. Thus the anaerobic pathway is a self-limiting one, since continued heavy exertion leads to exhaustion. The muscles incur an **oxygen debt** because the accumulated lactic acid must be oxidized by extra oxygen. After the period of exertion, oxygen consumption remains elevated until all of the lactic acid has been oxidized, or resynthesized to glucose.

To summarize, the sequence of chemical sources of energy can be expressed in abridged form as follows:

$$\text{ATP} \rightleftharpoons \text{ADP} + H_3PO_4 + \text{Energy for contraction}$$

$$\text{Creatine phosphate} \rightleftharpoons \text{Creatine} + H_3PO_4 + \text{Energy for resynthesis of ATP (anaerobic)}$$

$$\text{Glucose} \xrightleftharpoons{\text{(anaerobic)}} \text{Lactic acid} + \text{Energy for resynthesis of creatine phosphate}$$

$$\text{Glucose} + O_2 \xrightarrow{\text{(aerobic)}} CO_2 + H_2O + \text{Energy for resynthesis of creatine phosphate}$$

Stimulation of contraction. Skeletal muscle must, of course, be stimulated to contract. If the nerve supply to a muscle is severed, the muscle will **atrophy,** or waste away. Skeletal muscle fibers are arranged in groups of approximately 100, each group under the control of a single motor nerve fiber. Such a group is called a **motor unit.** As the nerve fiber approaches the muscle fibers, it splays out into many slender terminal branches. Each branch attaches to a muscle fiber by a special structure, called a **synapse,** or **myoneural junction** (Fig. 28-11). At the synapse is a tiny gap, or cleft, that thinly separates nerve fiber and muscle fiber. In the synapse is stored a chemical, **acetylcholine,** which is released when a nerve impulse reaches the synapse. This substance is a chemical mediator that diffuses across the narrow junction and acts on the muscle fiber membrane to generate an electric depolarization. The potential spreads rapidly through the muscle fiber, causing it to contract. Thus the synapse is a special chemical bridge that couples together the electric potentials of nerve and muscle fibers.

Coupling of excitation and contraction. For a long time physiologists were puzzled as to how the electric

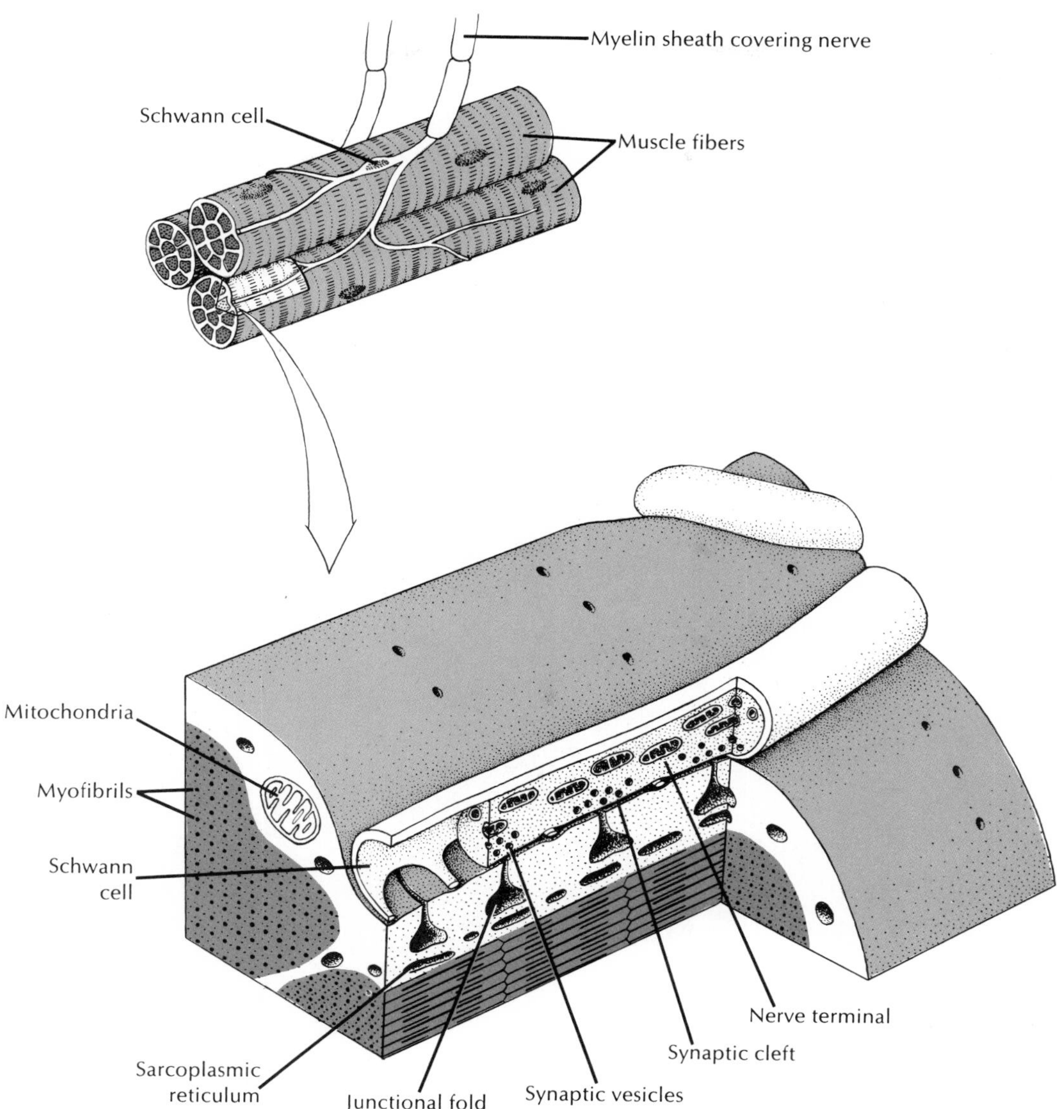

FIG. 28-11 Nerve-muscle synapse (myoneural junction). Slender terminations of motor nerve extend along depressions in muscle fiber. Terminations are covered by extension of a Schwann cell but lack the insulating myelin sheath. The nerve terminal contains synaptic vesicles packed with molecules of acetylcholine. When a nerve impulse arrives at a synapse, vesicles fuse with the terminal membrane, releasing acetylcholine into the synaptic cleft. These transmitter molecules diffuse rapidly across the cleft and bind with receptors on the muscle fiber membrane. Here, biochemical changes trigger a muscle impulse.

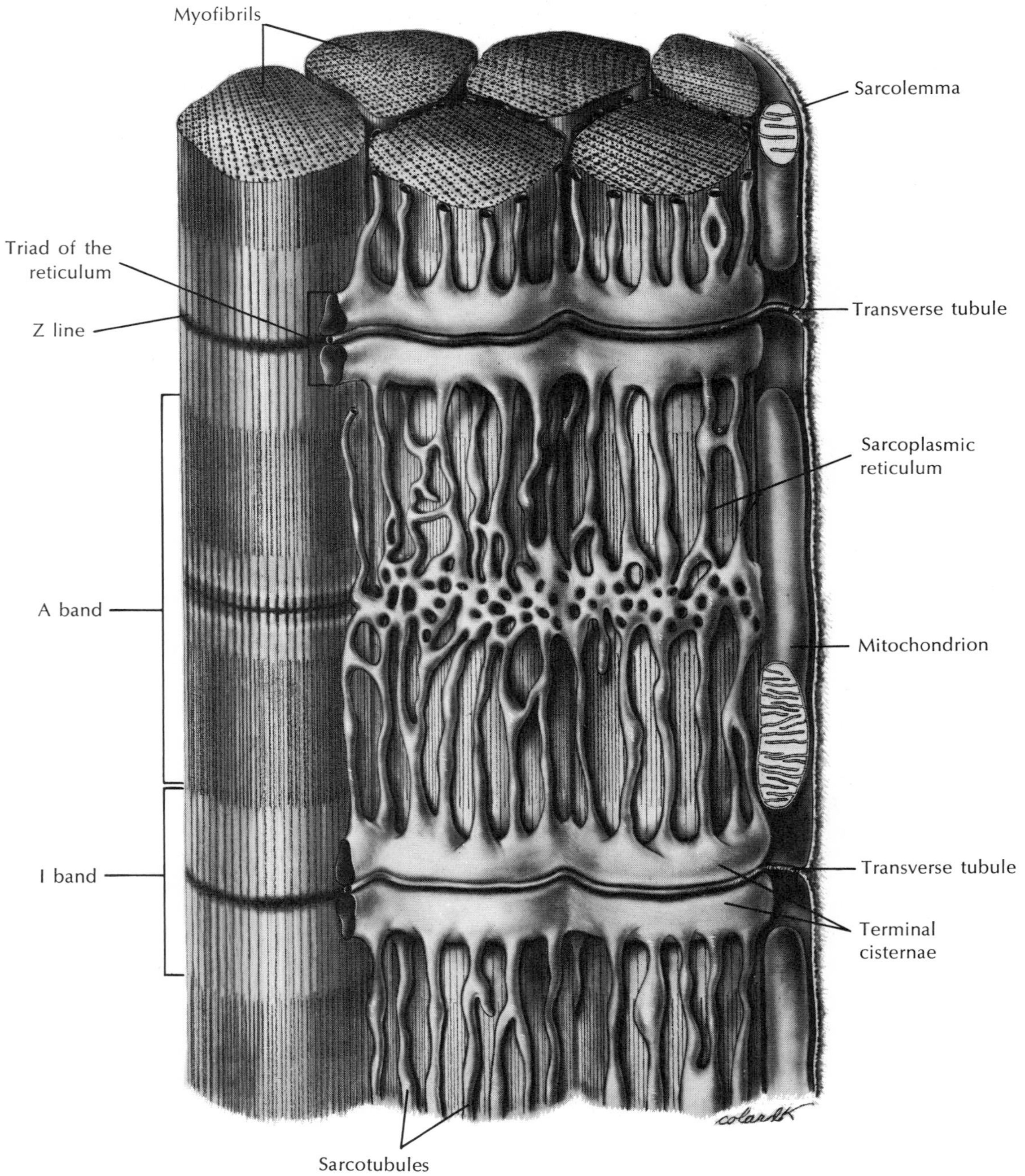

FIG. 28-12 Three-dimensional representation of vertebrate striated muscle showing distribution of sarcoplasmic reticulum and connecting transverse tubules (T system). In the frog muscle shown here, the transverse tubules are positioned at the Z regions where they serve to conduct electrical depolarizations and energy-rich supplies to the myofibrils via the sarcoplasmic reticulum. (From Bloom, W., and D. W. Fawcett. 1975. A textbook of histology. Philadelphia, W. B. Saunders Co.)

potential at the myoneural junction could spread quickly enough through the fiber to cause simultaneous contraction of all the densely packed filaments within. Recently it was discovered that vertebrate skeletal muscle contains an elaborate communication system that performs just this function. This is the endoplasmic reticulum (called the **sarcoplasmic reticulum** in muscle). As shown in the three-dimensional diagram (Fig. 28-12), the sarcoplasmic reticulum is a system of fluid-filled channels running parallel to the myofilaments and coming in close contact at the Z lines to **transverse tubules** (T system), which communicate with the sarcolemma that surrounds the fiber. The sarcoplasmic reticulum and T system are ideally arranged for speeding the electric depolarization from the myoneural junction to the myofilament within. It also serves as a distribution network for glucose, oxygen, minerals, and other supplies needed for muscle contraction.

Annotated references

Selected general readings

See also general references for Part Three, p. 793.

Andrews, W., and C. P. Hickman. 1974. Histology of the vertebrates. St. Louis, The C. V. Mosby Co. *Comparative treatment.*

Bagnara, J. T., and M. E. Hadley. 1973. Chromatophores and color change. Englewood Cliffs, N.J., Prentice-Hall, Inc.

Bendall, J. R. 1969. Muscles, molecules and movement. New York, American Elsevier Publishing Co., Inc. *Contains a wealth of well-organized information on muscle structure and physiology, pitched at the advanced undergraduate level.*

Bloom, W., and D. W. Fawcett. 1975. A textbook of histology, ed. 10. Philadelphia, W. B. Saunders Co. *Authoritative human histology text.*

Bourne, G. H. (ed.). 1972-1977. The structure and function of muscle. vols. 1 to 4. New York, Academic Press, Inc.

Cott, H. B. 1957. Adaptive coloration in animals. London, Methuen & Co., Ltd.

Goodrich, E. S. 1930. Studies on the structure and development of vertebrates. London, Macmillan Co. (Dover reprint 1958.) *A classic of comparative anatomy.*

Hildebrand, M. 1974. Analysis of vertebrate structure. New York, John Wiley & Sons, Inc. *Basic comparative anatomy text with especially helpful discussion of locomotor mechanisms.*

Maderson, P. F. A. (ed.). 1972. The vertebrate integument. Symposium. Am. Zool. **12:**12-171.

Montagna, W., and P. F. Parakkal. 1974. The structure and function of the skin, ed. 3. New York, Academic Press, Inc.

Patt, D. I., and G. R. Patt. 1969. Comparative vertebrate histology. New York, Harper & Row, Publishers.

Romer, A. S., and T. S. Parsons. 1977. The vertebrate body. Philadelphia, W. B. Saunders Co.

Webster, D., and M. Webster. 1974. Comparative vertebrate morphology. New York, Academic Press, Inc.

Zihlman, A., and D. Cramer. 1976. Human locomotion. Nat. Hist. **85:**64-69. *Traces evolutionary changes in the human pelvis.*

Selected *Scientific American* articles

Chapman, C. B., and J. H. Mitchell, 1965. The physiology of exercise. **212:**88-96 (May). *Muscular activity mobilizes several adaptive nervous, respiratory, circulatory, and metabolic responses.*

Cohen, C. 1975. The protein switch of muscle contraction. **233:**36-45 (Nov.). *The roles of calcium and regulatory proteins in muscle contraction are described.*

Hayashi, T. 1961. How cells move. **205:**184-204 (Sept.). *The actomyosin contractile mechanism and its ubiquitous presence throughout the animal kingdom is described.*

Hoyle, G. 1958. The leap of the grasshopper. **198:**30-35 (Jan.). *The powerful muscle system of the grasshopper's hindleg can propel the animal 20 times its body length.*

Hoyle, G. 1970. How is muscle turned on and off? **222:**84-93 (Apr.). *Calcium plays a crucial role in muscle contraction.*

Huxley, H. E. 1965. The mechanism of muscular contraction. **213:**18-27 (Dec.). *The sliding filament theory is decribed.*

Lester, H. A. 1977. The response to acetylcholene. **236:** 106-118 (Feb.). *Action of acetylcholine on cell receptors at the myoneural junction is described.*

McLean, F. C. 1955. Bone **192:**84-91 (Feb.). *Structure and physiology of bone.*

Merton, P. A. 1972. How we control the contraction of our muscles. **226:**30-37 (May). *Voluntary movements of skeletal muscle are controlled by a sensitive feedback mechanism.*

Murray, J. M., and A. Weber. 1974. The cooperative action of muscle proteins. **230:**58-71 (Feb.). *Describes the interactions of muscle proteins in contraction.*

Satir, P. 1974. How cilia move. **231:**44-52 (Oct.).

Smith, D. S. 1965. The flight muscles of insects. **212:**76-88 (June). *Although structurally similar to vertebrate skeletal muscle, the flight muscle of some insects can contract hundreds of times per second.*

CHAPTER 29

INTERNAL FLUIDS

Circulation, immunity, and gas exchange

Human red blood cells entrapped in fibrin clot. Clotting is initiated after tissue damage by the disintegration of platelets in the blood, leading to a complex series of intravascular reactions that end with the conversion of a plasma protein, fibrinogen, into long, tough, insoluble polymers of fibrin. Fibrin and entangled erythrocytes form the blood clot, which arrests bleeding. An aggregation of platelets probably underlies the raised mass of fibrin in center. (Original magnification ×5,180.)

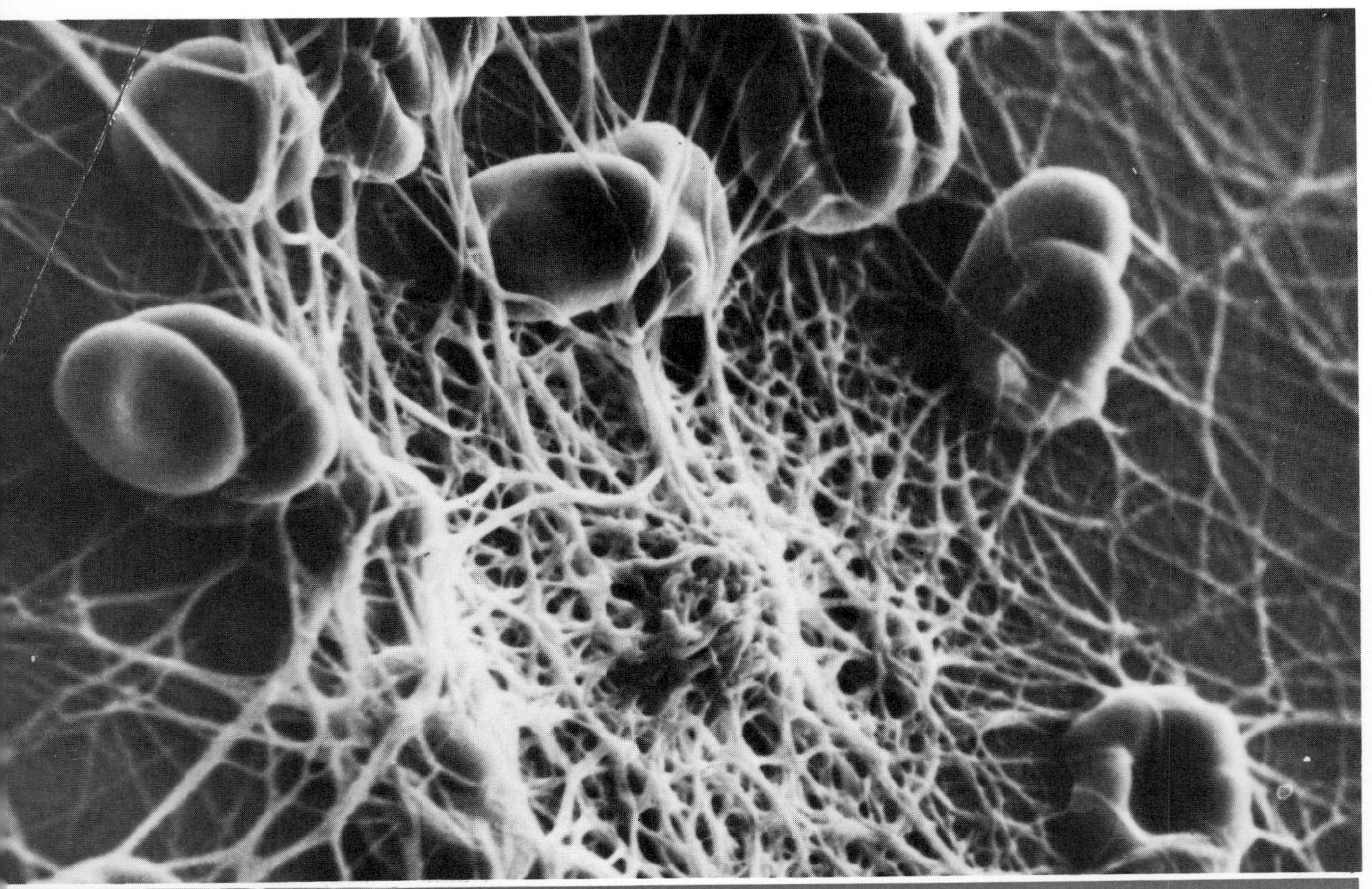

Courtesy N. F. Rodman.

Single-celled organisms live a contact existence with their environment. Nutrients and oxygen are obtained, and wastes are released, directly across the cell surface. These animals are so small that no special internal transport system, beyond the normal streaming movements of the cytoplasm, is required. Even some primitive multicellular forms such as sponges, cnidarians, and flatworms have such a simple internal organization and low rate of metabolism that no circulatory system is needed. Most of the more advanced multicellular organisms, because of their size, activity, and complexity, require a specialized circulatory, or vascular, system to transport nutrients and respiratory gases to and from all tissues of the body.

In addition to serving these primary transport needs, circulatory systems have acquired additional functions. Hormones are moved about, finding their way to target organs where they assist the nervous system to integrate body function. Water, electrolytes, and the many other constituents of the body fluids are distributed and exchanged between different organs and tissues. The body's defenses against microbial invasion are centered in the vascular system. The warm-blooded birds and mammals depend heavily on the blood circulation to conserve or dissipate heat as required for the maintenance of constant body temperature.

INTERNAL FLUID ENVIRONMENT

The body fluid of a single-celled animal is the cellular cytoplasm, a fluid substance in which the various membrane systems and organelles of the cell are suspended. In multicellular animals the body fluids are divided into two main phases, the **intracellular** and the **extracellular.** The intracellular phase (also called intracellular fluid) is the fluid inside all the body's cells. The extracellular phase (or fluid) is the fluid outside and surrounding the cells (Fig. 29-1, *A*). Thus the cells, the sites of the body's crucial metabolic activities, are bathed by their own aqueous environment, the extracellular fluid, which buffers them from the often harsh physical and chemical changes occurring outside the body.

In animals having closed circulatory systems (vertebrates, annelids, and a few other invertebrate groups), the extracellular fluid is further divided into blood **plasma** and **interstitial** fluid (Fig. 29-1, *A*). The blood plasma is contained within the blood vessels, while the interstitial fluid, or tissue fluid as it is sometimes called, occupies the space immediately around the cells. Nutrients and gases passing between the vascular plasma and the cells must traverse this narrow fluid separation. The interstitial fluid is constantly formed from the plasma by filtration through the capillary walls.

Composition of the body fluids

All these fluid spaces—plasma, interstitial, and intracellular—differ from one another in solute composition, but all have one feature in common—they are mostly water. Despite their firm appearance, animals are 70% to 90% water. Humans, for example, are about 70% water by weight: of this, 50% is cell water, 15% is interstitial fluid water, and the remaining 5% is in the blood plasma. As Fig. 29-1, *A*, shows, it is the plasma space that serves as the pathway of exchange between the cells of the body and the outside world. This exchange of respiratory gases, nutrients, and wastes is accomplished by specialized organs (kidney, lung, gill, alimentary canal), as well as by the integument.

The body fluids contain many inorganic and organic substances in solution. Principal among these are the inorganic electrolytes and proteins. Fig. 29-1, *B*, shows that **sodium, chloride,** and **bicarbonate** are the chief extracellular electrolytes, whereas **potassium, magnesium, phosphate,** and **proteins** are the major intracellular electrolytes. These differences are dramatic; they are always maintained despite the continuous flow of materials into and out of the cells of the body. The two subdivisions of the extracellular fluid—plasma and interstitial fluid—have similar compositions except that the plasma has more proteins, which are too large to filter through the capillary wall into the interstitial fluid.

Composition of blood

Among the lower invertebrates that lack a circulatory system (such as flatworms and cnidarians), it is not possible to distinguish a true "blood." These forms possess a clear, watery tissue fluid containing some primitive phagocytic cells, a little protein, and a mixture of salts similar to sea water. All invertebrates with closed circulatory systems maintain a clear separation between blood contained within blood vessels, and tissue (interstitial) fluid surrounding the vessels.

In vertebrates, blood is a complex liquid tissue com-

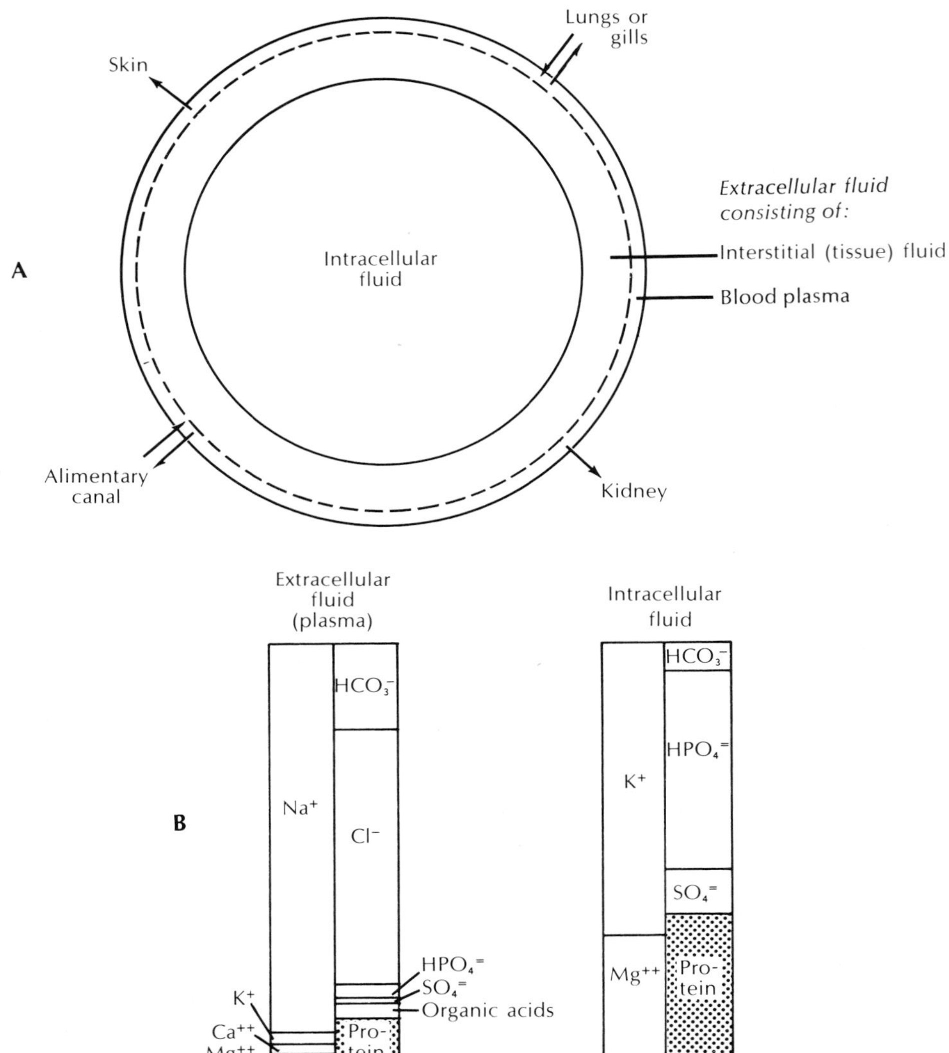

FIG. 29-1 Fluid compartments of body. **A,** All body cells can be represented as belonging to a single large fluid compartment that is completely surrounded and protected by extracellular fluid. This fluid is subdivided into plasma and interstitial fluid. All exchanges with the environment occur across the plasma compartment. **B,** Electrolyte composition of extracellular and intracellular fluids. Total equivalent concentration of each major constituent is shown. Equal amounts of anions (negatively charged ions) and cations (positively charged ions) are in each fluid compartment. Note that sodium and chloride, major plasma electrolytes, are virtually absent from intracellular fluid (actually present in very low concentration). Note the much higher concentration of protein inside the cells.

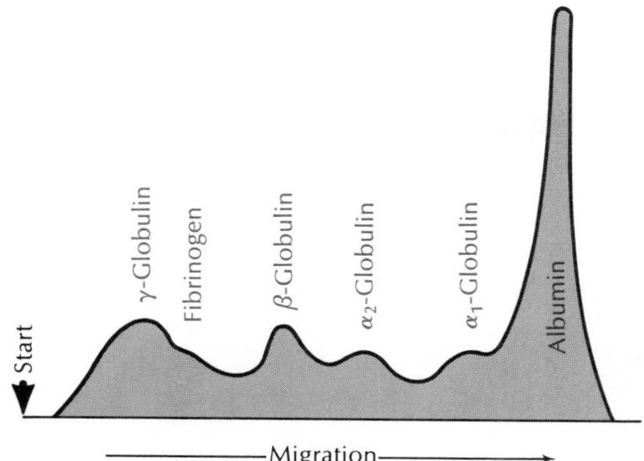

FIG. 29-2 Electrophoretic pattern of major plasma proteins. When blood plasma is placed in an electric field under the right pH conditions, proteins in solution will migrate at different rates according to their net charge. The profile is called a Tiselius pattern, after the Swedish biochemist who developed the technique.

posed of plasma and formed elements, mostly corpuscles, suspended in the plasma. When the red blood corpuscles and other formed elements are spun down in a centrifuge, the blood is found to be about 55% plasma and 45% formed elements.

The composition of mammalian blood is as follows:

Plasma

1. Water 90%
2. Dissolved solids, consisting of the plasma proteins (albumin, globulins, fibrinogen), glucose, amino acids, electrolytes, various enzymes, antibodies, hormones, metabolic wastes, and traces of many other organic and inorganic materials
3. Dissolved gases, especially oxygen, carbon dioxide, and nitrogen

Formed elements (Fig. 29-4)

1. Red blood corpuscles (erythrocytes), for the transport of oxygen and carbon dioxide
2. White blood corpuscles (leukocytes), serving as scavengers and as immunizing agents
3. Platelets (thrombocytes), functioning in blood coagulation

The plasma proteins are a diverse group of large and small proteins that perform numerous functions. They may be separated by classic electrophoretic techniques into six major groups (Fig. 29-2) although more modern chromatographic and immunoelectrophoretic methods show that there are probably hundreds of individual proteins in the plasma. The major protein groups are (1) **albumin,** the most abundant plasma protein, which constitutes 60% of the total; (2) the **globulins** (α_1, α_2, β, and γ), a diverse group of high-molecular weight proteins (35% of total) that includes immunoglobulins and various metal-binding proteins; and (3) **fibrinogen,** a very large protein that functions in blood coagulation.

Red blood cells, or **erythrocytes,** are present in enormous numbers in the blood, about 5.4 million per cubic millimeter (mm^3) in an adult man and 4.8 million per cubic millimeter in women. They are formed continuously from large nucleated **erythroblasts** in the red bone marrow. Here, hemoglobin is synthesized, and the cells divide several times. In mammals the nucleus shrinks during development to a small remnant and eventually disappears altogether. Almost all other characteristics of a typical cell also are lost: ribosomes, mitochondria, and most enzyme systems. What is left is a biconcave disc consisting of a baglike membrane, the **stroma,** packed with about 280 million molecules of the blood-transporting pigment **hemoglobin.** About 33% of the erythrocyte by weight is hemoglobin. The biconcave shape (Fig. 29-3, *A*) is a mammalian innovation that provides a much larger surface for gas diffusion than would a flat or spherical shape. All other vertebrates have nucleated erthrocytes that are usually ellipsoidal, rather than round, discs (Fig. 29-3, *B*).

The erythrocyte enters the circulation for an average life-span of about 4 months. During this time it may journey 700 miles, squeezing repeatedly through the capillaries, which are sometimes so narrow that the erythrocyte must bend to get through. At last it fragments and is quickly engulfed by large scavenger cells called **macrophages** located in the liver, bone marrow, and spleen. The iron from the hemoglobin is salvaged to be used again; the rest of the heme is converted to **bilirubin,** a bile pigment. It is estimated that 10 million erythrocytes are born, and another 10 million destroyed every second in the human body.

The white blood cells, or **leukocytes,** form a wandering system of protection for the body. In adults they number only about 7,500 per cubic millimeter, a ratio of 1 white cell to 700 red cells. There are several kinds of white blood cells: **granulocytes** (subdivided into neutrophils, basophils, and eosinophils), **lymphocytes,** and **monocytes** (Fig. 29-4). All have the capacity to pass through the wall of capillaries and wander by ameboid movement through the tissue

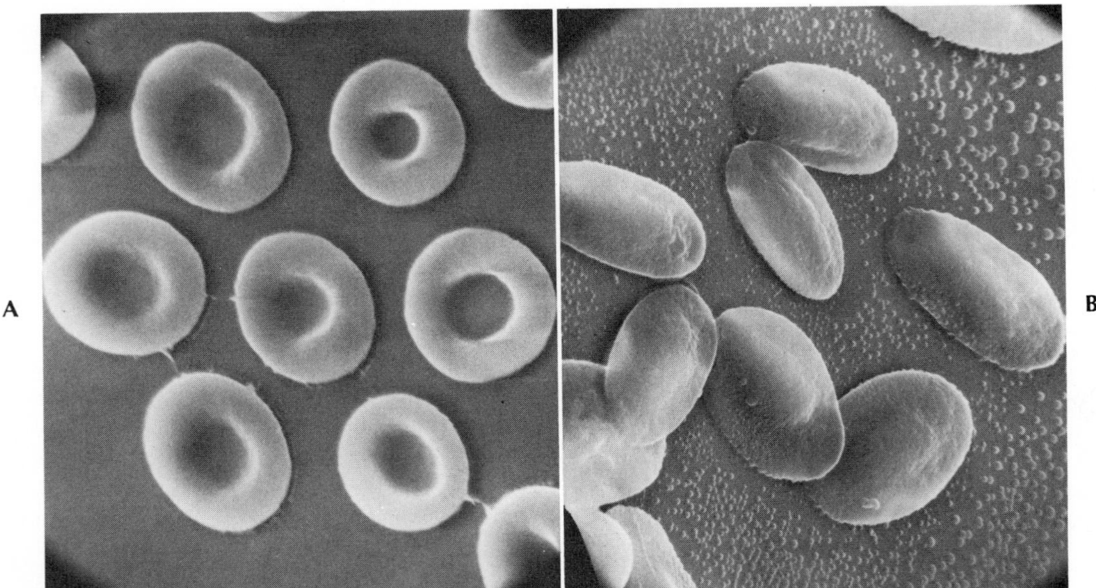

FIG. 29-3 Mammalian and amphibian red blood corpuscles. The erythrocytes of a gerbil, **A,** are biconcave discs containing hemoglobin and surrounded by a tough stroma. The frog erythrocytes, **B,** are convex discs, each containing a nucleus, which is plainly visible in the scanning electron micrograph as a bulge in the center of each cell. (Mammalian erythrocytes, ×6,300; frog erythrocytes, ×2,400.) (Courtesy P. P. C. Graziadei, Florida State University.)

spaces. Monocytes and granulocytes have great power to engulf and digest bacteria and other foreign particulate matter, a process called **phagocytosis.** They also clean up and digest the debris of the body's own tissues, such as fragments of worn-out red blood cells, blood clots, or the remains of wound and disease repair. Lymphocytes are especially important in producing gamma globulins, which act as immune bodies **(antibodies)** that destroy or neutralize toxic molecules **(antigens).** The defense functions of the white blood cells are discussed later in this chapter.

The platelets, or thrombocytes, are minute, colorless bodies about one-third the diameter of red blood cells. They initiate the coagulation of blood. When blood spills from a vessel, as in a wound, the platelets rapidly disintegrate to release factors that start the formation of a clot. Platelets also readily clump together and plug torn vessels by entangling red blood cells (see photograph, p. 664).

CIRCULATION

The circulatory system of vertebrates is made up of a system of tubes, the **blood vessels,** and a propulsive organ, the **heart.** This is a **closed circulation** because the circulating medium, the **blood,** is confined to vessels throughout its journey from the heart to the tissues and back again. Many invertebrates have an **open circulation;** the blood is pumped from the heart into blood vessels that open into tissue spaces. The blood circulates freely in direct contact with the cells and then reenters open blood vessels to be propelled forward again. In invertebrates having open circulatory systems, there is no clear separation of the extracellular fluid into plasma and interstitial fluids, as there must be in closed systems. Closed systems are more suitable for large and active animals because the blood can be moved rapidly to the tissues needing it. In addition, flow to various organs can be readjusted to meet changing needs by varying the diameters of the blood vessels.

The closed circulatory system of vertebrates works along with the **lymphatic system.** This is a fluid ''pick-up'' system. It recollects tissue fluid (lymph) that has been squeezed out through the walls of the capillaries and returns it to the blood circulation. In a sense ''closed'' circulatory systems are not absolutely closed because fluid is constantly leaking out into the tissue

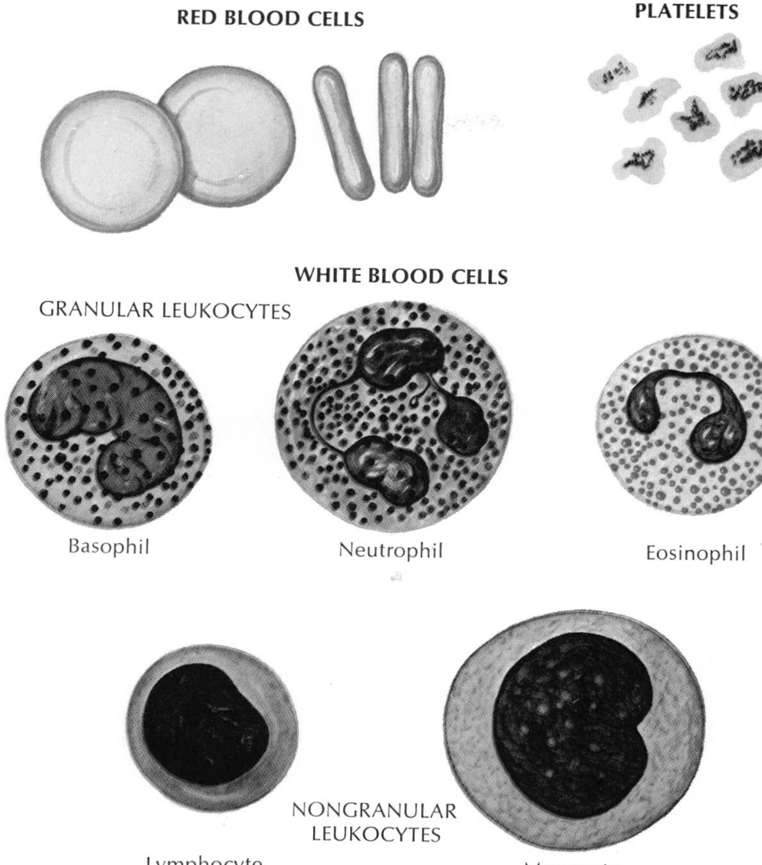

FIG. 29-4 Formed elements of human blood. Hemoglobin-containing red blood cells of humans and other mammals lack nuclei, but those of all lower vertebrates have nuclei. Various leukocytes provide a wandering system of protection for the body. Platelets participate in the body's clotting mechanism. (From Anthony, C. P., and N. J. Kolthoff, 1975. Textbook of anatomy and physiology, ed. 9. St. Louis, The C. V. Mosby Co.)

spaces. However, this leakage is but a small fraction of the total blood flow.

Although it seems obvious to us today that blood flows in a circuit, the first correct description of blood flow by the English physician William Harvey initially received vigorous opposition when published in 1628. Centuries before, Galen had taught that air enters the heart from the windpipe and that blood was able to pass from one ventricle to the other through "pores" in the interventricular septum. He also believed that blood first flowed out of the heart in all vessels, arteries and veins alike, and then returned to the heart by these same vessels—an idea of ebb and flow of the blood.

Even though there was almost nothing right about this theory, it was still doggedly trusted at the time of Harvey's publication. Harvey's conclusions were based on sound experimental evidence. He made use of a variety of animals for his experiments, including a little snake found in English meadows. By tying ligatures on arteries, he noticed that the region between the heart and ligature swelled up. When veins were tied off, the swelling occurred beyond the ligature. When blood vessels were cut, blood flowed in arteries from the cut end nearest the heart; the reverse happaned in veins. By means of such experiments, Harvey worked out a correct scheme of blood circulation, even though he could not see the capillaries that connected the arterial and venous flows.

Plan of the circulatory system

All vertebrate vascular systems have certain features in common. A **heart** pumps the blood into **arteries** that branch and narrow into **arterioles** and then into a vast system of **capillaries.** Blood leaving the capillaries enters **venules** and then **veins** that return the

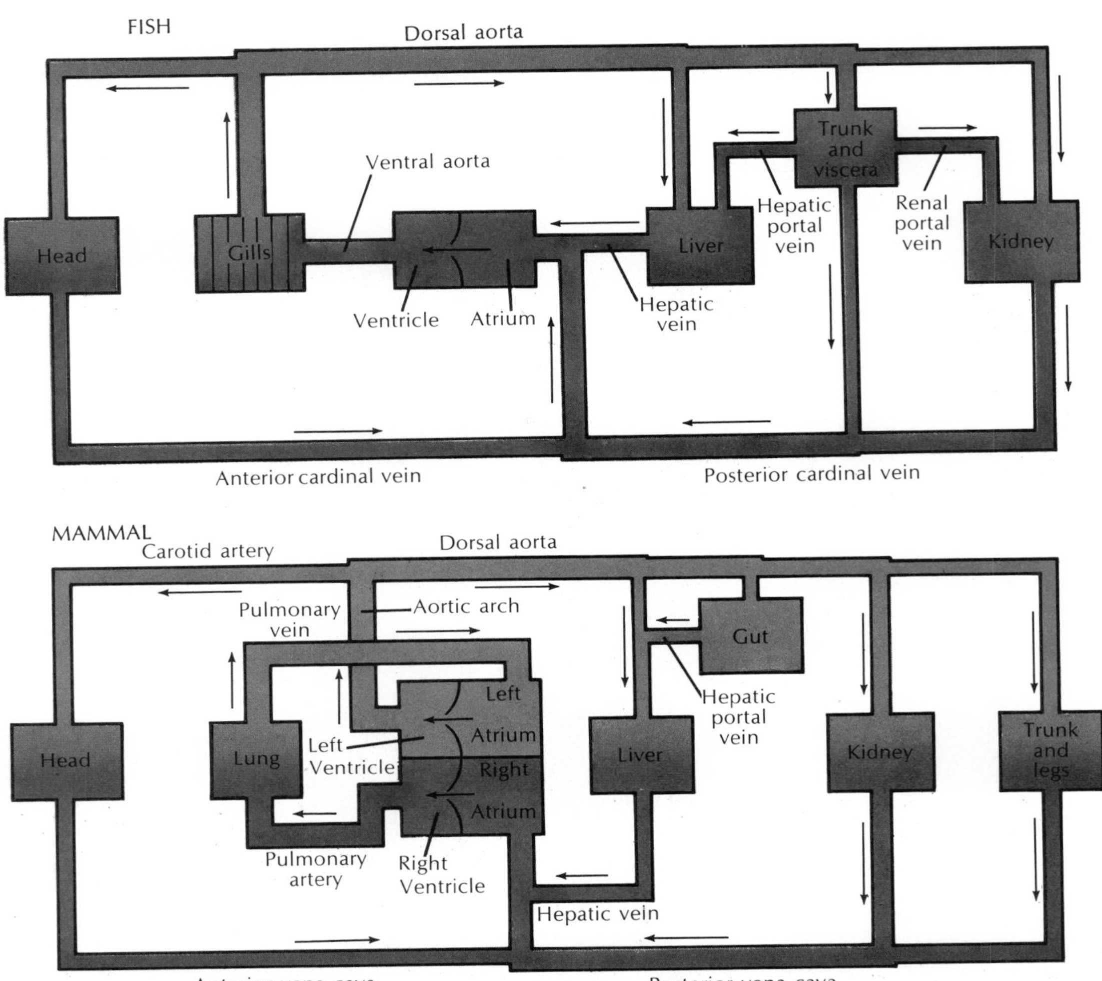

FIG. 29-5 Plan of circulatory system of fishes *(above)* and mammals *(below)*. *Red,* oxygenated blood; *dark red,* deoxygenated blood.

blood to the heart. Fig. 29-5 compares the circulatory systems of gill-breathing (fishes) and lung-breathing (mammals) vertebrates. The principal differences in circulation involve the heart in the transformation from gill to lung breathing. The fish heart contains two main chambers, the **atrium** (or **auricle**) and the **ventricle.** Although there are also two subsidiary chambers, the **sinus venosus** and **conus arteriosus** (not shown in Fig. 29-5), we still refer to the fish heart as a "two-chambered" heart. Blood makes a single circuit through the fish's vascular system; it is pumped from the heart to the gills, where it is oxygenated, and then flows into the dorsal aorta to be distributed to the body organs. After passing through the capillaries of the body organs and musculature, it returns by veins to the heart. In this circuit the heart must provide sufficient pressure to push the blood through two sequential capillary systems, one in the gills and the other in the organ tissues. The principal disadvantage of the single-circuit system is that the gill capillaries offer so much resistance to blood flow that the pressure drops considerably before entering the dorsal aorta. This

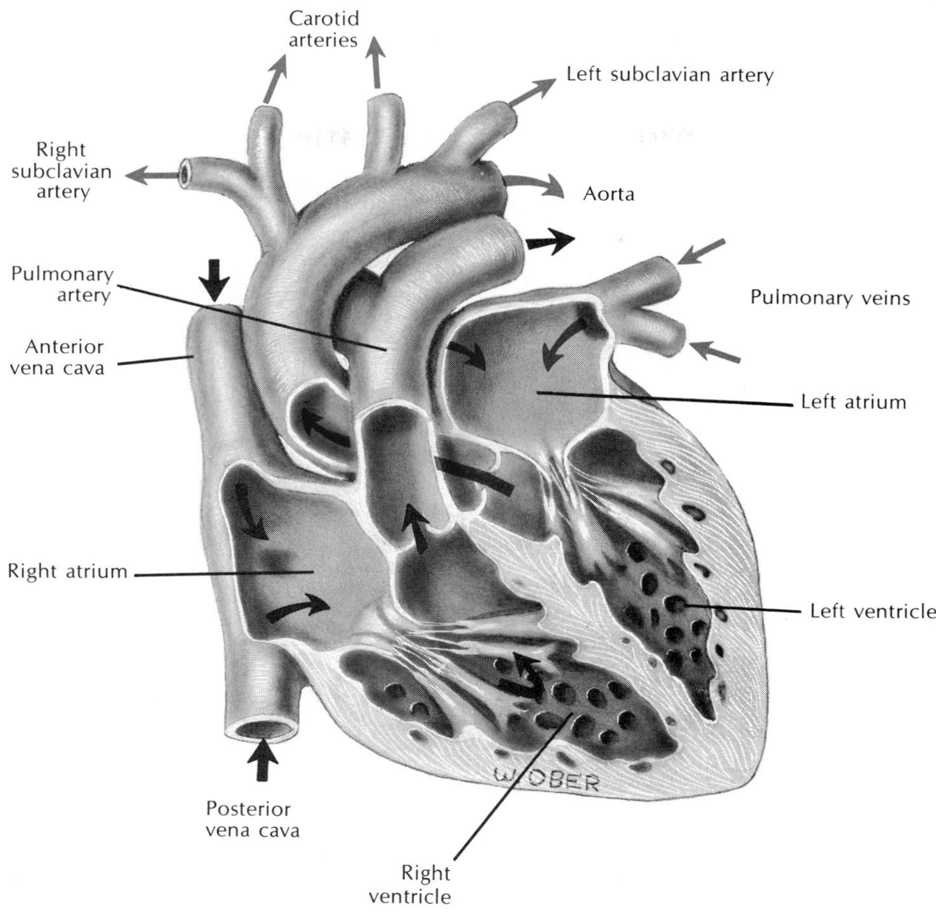

FIG. 29-6 Human heart. Deoxygenated blood *(black arrows)* enters right side of heart and is pumped to the lungs. Oxygenated blood *(red arrows)* returning from the lungs enters left side of the heart and is pumped to the body.

system can never provide high and continuous blood pressure to the body organs.

Evolving land forms with lungs and their need for highly efficient blood delivery had to solve this problem by introducing a **double** circulation. One **systemic** circuit with its own pump provides oxygenated blood to the capillary beds of the body organs; another **pulmonary** circuit with its own pump sends deoxygenated blood to the lungs. Rather than actually developing two separate hearts, the existing two-chambered heart was divided down the center into four chambers—really two two-chambered hearts lying side-by-side.

Needless to say such a great change in the vertebrate circulatory plan, involving not only the heart but the attendant plumbing as well, took many millions of years to evolve (p. 543). The partial division of the atrium and ventricle began with the ancestors of present-day lungfishes. Amphibians accomplished the complete separation of the atrium, but the ventricle is still undivided in this group. In some reptiles the ventricle is completely divided, and the four-chambered heart appears for the first time. All birds and mammals have the four-chambered heart and two separate circuits—one through the lungs (pulmonary) and the other through the body (systemic). The course of the blood through this double circuit is shown in Fig. 29-5.

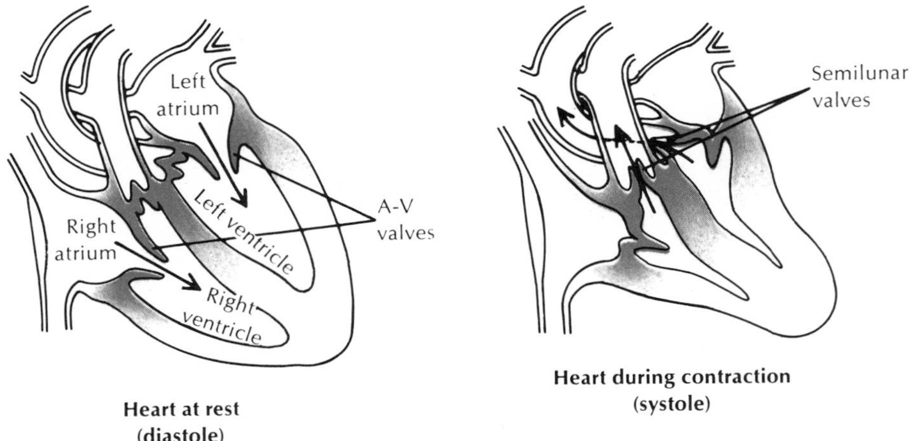

FIG. 29-7 Human heart in systole and diastole.

The heart

The vertebrate heart is a muscular organ located in the thorax and covered by a tough, fibrous sac, the **pericardium** (Fig. 29-6). As we have seen, the higher vertebrates have a four-chambered heart. Each half consists of a thin-walled atrium and a thick-walled ventricle. Heart (cardiac) muscle is a unique type of muscle found nowhere else in the body. It resembles striated muscle, but the cells are branched, and dense end-to-end attachments between the cells are called intercalated discs (Fig. 6-12, p. 111).

There are four sets of valves. **Atrioventricular valves** (A-V valves) separate the cavities of the atrium and ventricle in each half of the heart. These permit blood to flow from atrium to ventricle but prevent backflow. Where the great arteries, the **pulmonary** from the right ventricle and the **aorta** from the left ventricle, leave the heart, **semilunar valves** prevent backflow (Fig. 29-7).

The contraction of the heart is called **systole** (sis′to-lee), and the relaxation, **diastole** (dy-as′to-lee) (Fig. 29-7). The rate of the heartbeat depends on age, sex, and especially exercise. Exercise may increase the **cardiac output** (volume of blood forced from either ventricle each minute) more than fivefold. Both the heart **rate** and the **stroke volume** increase. Heart rates among vertebrates vary with the general level of metabolism and the body size. The cold-blooded codfish has a heart rate of about 30 beats per minute; a warm-blooded rabbit of about the same weight has a rate of 200 beats per minute. Small animals have higher heart rates than do large animals. The heart rate in an elephant is 25 beats per minute; in a human, 70 per minute; in a cat, 125 per minute, in a mouse, 400 per minute, and in the tiny 4-gram shrew, the smallest mammal, the heart rate approaches a prodigious 800 beats per minute. We must marvel that the shrew's heart can sustain this frantic pace throughout this animal's life, brief as it is.

The only rest a heart enjoys is the short interval between contractions. The mammalian heart does an amazing amount of work during a lifetime. Someone has calculated that the heart of an aged human, approaching the end of life, has beat some 2.5 billion times and pumped 300,000 tons of blood!

Excitation of the heart. The heartbeat originates in a specialized muscle tissue, called the **sinoatrial node,** located in the right atrium near the entrance of the caval veins (Fig. 29-8). This tissue serves as the **pacemaker** of the heart. The contraction spreads across the two atria to the **atrioventricular (A-V) node.** From this point the electric activity is conducted very rapidly to the apex of the ventricle through specialized fibers (bundle of His and Purkinje fiber system) and then spreads more slowly up the walls of the ventricles. This arrangement allows the contraction to begin at the apex or "tip" of the ventricles and spread upward to squeeze out the blood in the most efficient way; it ensures that both ventricles will contract simultaneously. Although the vertebrate heart can beat spon-

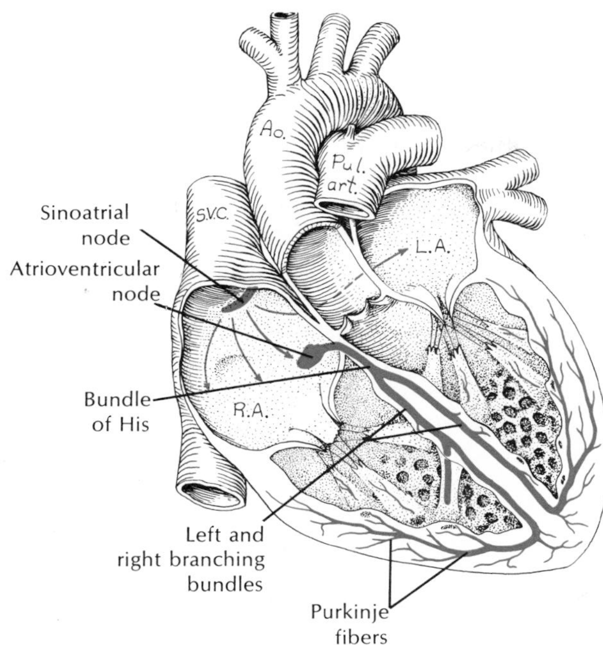

FIG. 29-8 Neuromuscular mechanisms controlling beat of the human heart. Arrows indicate spread of excitation from the sinoatrial node (S-A node), across the atria, to the A-V node (atrioventricular node). Wave of excitation is then conducted very rapidly to ventricular muscle over the specialized bundle of His and Purkinje fiber system.

taneously (the excised fish or amphibian heart will beat for hours in a balanced salt solution), the heart rate is normally under nervous control. The control (cardiac) center is located in the medulla and sends out two sets of motor nerves. Impulses sent along one set, the **vagus** (parasympathetic) nerves, apply a brake action to the heart rate, and impulses sent along the other set, the **accelerator** (sympathetic) nerves, speed it up. Both sets of nerves terminate in the sinoatrial node, thus guiding the activity of the pacemaker. The cardiac center in turn receives sensory information about a variety of stimuli. Pressure receptors (sensitive to blood pressure) and chemical receptors (sensitive to carbon dioxide and pH) are located at strategic points in the vascular system. This information is used by the cardiac center to increase or reduce the heart rate and cardiac output in response to activity or changes in body position. The heart is thus controlled by a series of feedback mechanisms that keep its activity constantly attuned to body needs.

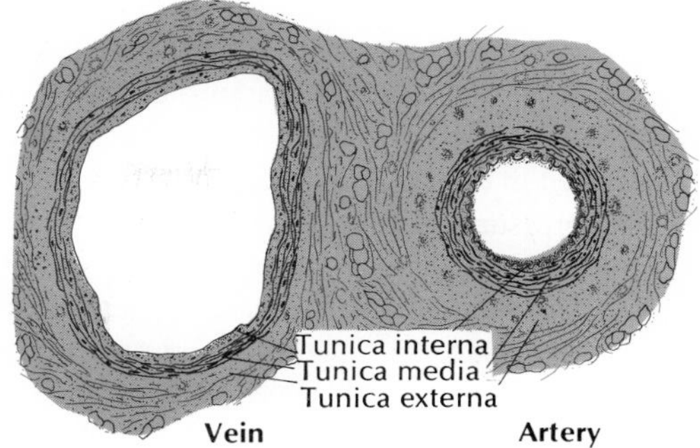

FIG. 29-9 Cross section of vein and corresponding artery.

Coronary circulation

It is no surprise that an organ as active as the heart needs a very good blood supply of its own. The heart muscle of the frog and other amphibians is so thoroughly channeled with spaces between the muscle fibers that sufficient oxygenated blood is squeezed through by the heart's own pumping action. In birds and mammals, however, the heart muscle is very thick and has such a high rate of metabolism that it must have its own vascular **(coronary)** circulation. The coronary arteries break up into an extensive capillary network surrounding the muscle fibers and provide them with oxygen and nutrients. Heart muscle has an extremely high oxygen demand, removing 80% of the oxygen from the blood, in contrast to most other body tissues, which remove only about 30%.

Arteries

All vessels leaving the heart are called arteries whether they carry oxygenated blood (aorta) or deoxygenated blood (pulmonary artery). To withstand high, pounding pressures, arteries are invested with layers of both elastic and tough, inelastic connective tissue fibers. The elasticity of the arteries allows them to yield to the surge of blood leaving the heart during systole and then to squeeze down on the fluid column during diastole. This smooths out the blood pressure. Thus the arterial pressure in humans varies only between a high of 120 mm Hg (systole) and a low of 80 mm Hg (diastole), rather than dropping to zero during diastole as we might expect in a fluid system with an intermittent pump.

As the arteries branch and narrow into **arterioles,**

the walls become mostly smooth muscle (Fig. 29-9). Contraction of this muscle narrows the arterioles and reduces the flow of blood. The arterioles thus control the blood flow to body organs, diverting it to those tissues that need it most. The blood must be given a hydrostatic pressure sufficient to overcome the resistance of the narrow passages through which the blood must flow. Consequently large animals tend to have higher blood pressure than do small animals.

Blood pressure was first measured in 1733 by Stephen Hales, an English clergyman with unusual inventiveness and curiosity. He tied his mare "to have been killed as unfit for service" on her back and exposed the femoral artery. This he cannulated with a brass tube, connecting it to a tall glass tube with the windpipe of a goose. The use of the windpipe was both imaginative and practical; it gave the apparatus flexibility "to avoid inconveniences that might arise if the mare struggled." The blood rose 8 feet in the glass tube and bobbed up and down with the systolic and diastolic beats of the heart. The weight of the 8-foot column of blood was equal to the blood pressure. We now express this as the height of a column of mercury, which is 13.6 times heavier than water. Hale's figures, expressed in millimeters of mercury, indicate he measured a blood pressure of 180 to 200 mm Hg, about normal for a horse. Today, blood pressure can be measured with great accuracy by a sensitive pressure transducer; the electronic signal from this instrument is displayed on a graphic recorder.

Capillaries

The Italian Marcello Malpighi was the first to describe the capillaries in 1661, thus confirming the existence of the minute links between the arterial and venous systems that Harvey knew must be there but could not see. Malpighi studied the capillaries of the living frog's lung, which incidentally is still one of the simplest and most vivid preparations for demonstrating capillary blood flow.

The capillaries are present in enormous numbers, forming extensive networks in nearly all tissues. In muscles there are more than 2,000 per square millimeter (1,250,000 per square inch), but not all are open at once. Indeed, perhaps less than 1% are open in resting skeletal muscle. But when the muscle is active, all the capillaries may open to bring oxygen and nutients to the working muscle fibers and to carry away metabolic wastes.

Capillaries are extremely narrow, averaging less than 10 μm in diameter in mammals, which is hardly any wider than the red blood cells that must pass through them. Their walls are formed of a single layer of thin **endothelial** cells, held together by a delicate basement membrane and connective tissue fibers. Capillaries have a built-in leakiness that allows water and most dissolved substances in the blood plasma to filter through into the interstitial space. The capillary wall is **selectively permeable,** however, which means that it filters some dissolved materials and retains others. In this case the plasma proteins, which are the largest dissolved molecules in the plasma, are held back. These proteins, especially the albumins, contribute an **osmotic pressure** of about 25 mm Hg in mammals (Fig. 29-10). Although small, this protein osmotic pressure is of great importance to fluid balance in the tissues. At the arteriole end of the capillaries the blood pressure is about 40 mm Hg (in humans). This **filtration pressure** forces water and dissolved materials through the capillary endothelium into the tissue space where they circulate freely around the cells. As the blood proceeds through the narrow

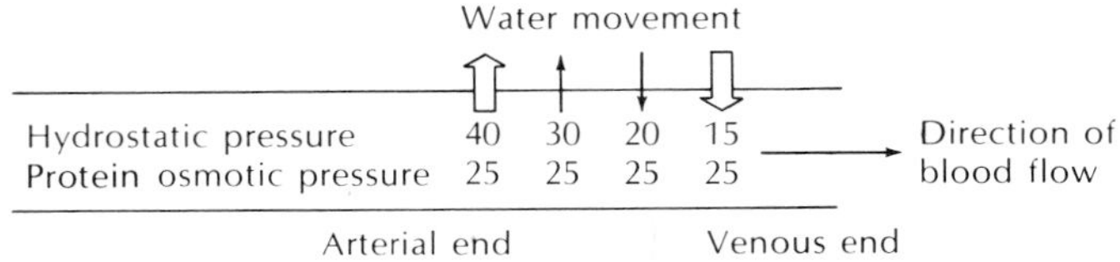

FIG. 29-10 Fluid movement across the wall of a capillary. At arterial end of the capillary, hydrostatic (blood) pressure exceeds protein osmotic pressure contributed by the plasma proteins, and a plasma filtrate (shown as "water movement") is forced out. At venous end, protein osmotic pressure exceeds the hydrostatic pressure, and fluid is drawn back in. In this way plasma nutrients are carried out into the interstitial space where they can enter cells, and metabolic end products from the cells are drawn back into the plasma and carried away.

capillary, the blood pressure drops steadily to perhaps 15 mm Hg. At this point the hydrostatic pressure is less than the osmotic pressure of the plasma proteins, still about 25 mm Hg. Water now is drawn back into the capillaries.

Thus it is the balance between hydrostatic pressure and protein osmotic pressure that determines the direction of capillary fluid shift. Normally water is forced out of the capillary at the arteriole end, where hydrostatic pressure exceeds osmotic pressure, and drawn back into the capillary at the venule end where osmotic pressure exceeds hydrostatic pressure. Any fluid left behind is picked up and removed by the **lymph capillaries.**

Veins

The venules and veins into which the capillary blood drains for its return journey to the heart are thinner walled, less elastic, and of considerably larger diameter than their corresponding arteries and arterioles (Fig. 29-9). Blood pressure in the venous system is low, from about 10 mm Hg where capillaries drain into venules to about zero in the right atrium. Because pressure is so low, the venous return is assisted by one-way valves in the veins, by muscles surrounding the veins, and by the rhythmic pumping action of the lungs. If it were not for these mechanisms, the blood might pool in the lower extremities of a standing animal—a very real problem for people who must stand for long periods. The veins that lift blood from the extremities to the heart contain valves that serve to divide the long column of blood into segments. When the muscles around the veins contract, as in even slight activity, the blood column is squeezed upward and cannot slip back because of the valves. The well-known risk of fainting while standing at stiff attention in hot weather can usually be prevented by deliberately pumping the leg muscles. The negative pressure created in the thorax by the inspiratory movement of the lungs also speeds the venous return by sucking the blood up the large vena cava into the heart.

Lymphatic system

The lymphatic system is an extensive network of thin-walled vessels that is separate from the circulatory system. The system arises as blind-ended lymph capillaries in most tissues of the body. These unite to form larger and larger lymph vessels, which finally drain into veins in the lower neck. Gasparo Aselli, an Italian anatomist, first discovered the nature of lacteals, the lymph capillaries of the intestine, in 1623. In a dog that had recently been fed and cut open, he noticed white cordlike bodies in the mesenteries of the intestine that he first mistook for nerves. When he pricked these cords with a scalpel, a milky fluid gushed out. It is now known that this fluid is largely fat that is carried after digestion to the thoracic duct. The thoracic duct and its relations to the lacteals were discovered by the Frenchman Jean Pecquet in 1647. These vessels are part of the complete lymphatic system demonstrated almost simultaneously but independently by O. Rudbeck in Sweden (1651) and T. Bartholin in Denmark (1653), using dogs and executed criminals.

The lymphatic system (Fig. 29-11) is an accessory drainage system for the body. As we have seen, the blood pressure in the arteriole end of the capillaries forces a plasma filtrate through the capillary walls and into the interstitial space. This **tissue fluid** bathing the cells is a clear, nearly colorless liquid. Tissue fluid and plasma are nearly identical except that tissue fluid contains very little protein, which was screened out as the plasma was squeezed through the capillary walls. Most of the tissue fluid returns to the vascular system at the venous end of the capillaries by the capillary fluid-shift mechanism described earlier. Usually, however, outflow from the capillaries slightly exceeds backflow. This difference is gathered up and returned to the circulatory system by lymphatic vessels. Tissue fluid is called **lymph** as soon as it enters the lymph vessels. Lymph flow is very low, a minute fraction of the blood flow.

Located at strategic intervals along the lymph vessels are **lymph nodes** (Fig. 29-11) that have several defense-related functions. They are effective filters that remove foreign particles, especially bacteria, that might otherwise enter the general circulation. They are also germinal centers for immune cells, which are essential components of the body's defense mechanisms.

Hemostasis: prevention of blood loss

It is essential that animals have ways of preventing the rapid loss of body fluids after an injury. Since blood is flowing and is under considerable hydrostatic pressure, it is especially vulnerable to hemorrhagic loss.

When a vessel is damaged, smooth muscle in the wall contracts, which causes the vessel lumen to nar-

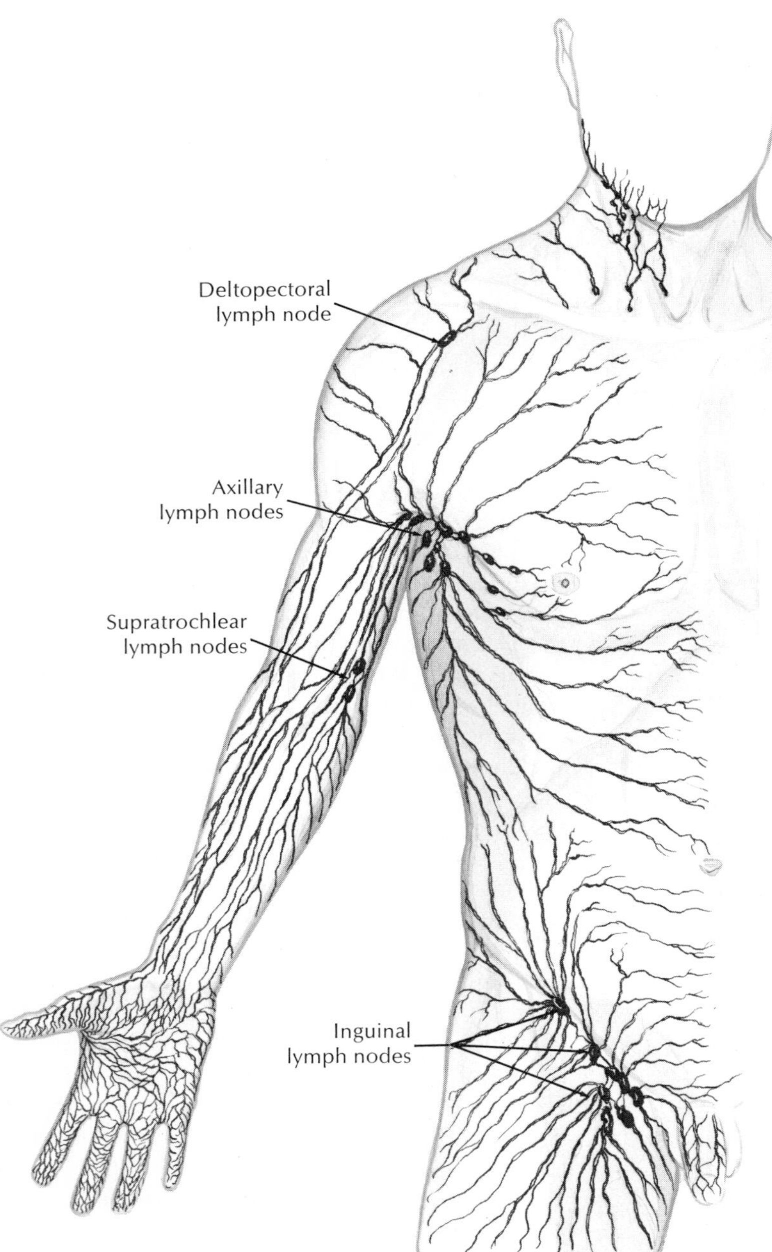

FIG. 29-11 Lymph drainage of the human right front trunk. Superficial lymphatics and lymph nodes are shown. (From Anthony, C. P., and N. J. Kolthoff. 1975. Textbook of anatomy and physiology, ed. 9. St. Louis, The C. V. Mosby Co.)

row, sometimes so strongly that blood flow is completely stopped. This is a primitive but highly effective means of preventing hemorrhage used by invertebrates and vertebrates alike. Beyond this first defense against blood loss, all vertebrates, as well as some of the larger, active invertebrates with high blood pressures, have special cellular elements and proteins in the blood that are capable of forming plugs, or clots, at the injury site.

In higher vertebrates **blood coagulation** is the dominant hemostatic defense. Blood clots form as a tangled network of fibers from one of the plasma proteins, **fibrinogen** (see scanning electron micrograph of a fibrin clot on p. 664). The transformation of fibrinogen into a **fibrin** meshwork that entangles blood cells to form a gel-like clot is catalyzed by the enzyme **thrombin.** Thrombin is normally absent from the blood and only appears when vessels are damaged.

In this process, the blood **platelets** (Fig. 29-4) play a dominant role. Platelets are minute, colorless, incomplete cells lacking nuclei that are present in large numbers in the blood. When the surface of a vessel is damaged, platelets rapidly adhere to the injured surface and release a phospholipid substance. This material, in the presence of calcium, initiates a catalytic sequence resulting in the conversion of the plasma protein prothrombin to thrombin. We may summarize the stages in the formation of fibrin as follows:

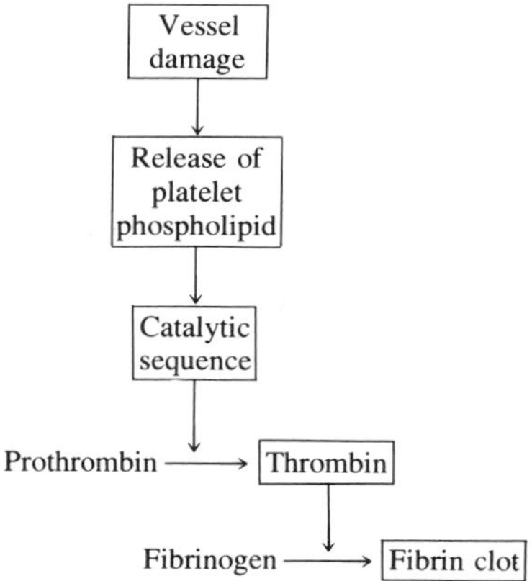

The catalytic sequence in this scheme is an unexpectedly complex series of plasma protein factors, each normally inactive until activated by a previous factor in the sequence. The sequence behaves like a "cascade" with each reactant in the sequence leading to a large increase in the amount of the next reactant. At least 13 different plasma coagulation factors have been recognized. A deficiency of a single factor can delay or prevent the clotting process. Why has such a complex clotting mechanism evolved? Probably it is necessary to provide a fail-safe system capable of responding to any kind of internal or external hemorrhage that might occur, and yet a system that cannot be activated into forming dangerous intravascular clots unless injury has occurred.

Several kinds of clotting abnormalities in man are known. Of these, **hemophilia** is perhaps best known. Hemophilia is a condition characterized by the failure of the blood to clot, so that even insignificant wounds can cause continuous severe bleeding. Called the "disease of kings," it once ran through the royal families of Europe, notably those of Queen Victoria of England and Alfonso XIII, the last king of Spain. Hemophilia is caused by an inherited lack of antihemophilic factor. The disorder is transmitted through females, but almost invariably appears only in males. The disease is by no means limited to royalty, for some 20,000 North Americans have moderate to severe forms of the disease.

IMMUNE SYSTEM

Immunity is all the physiologic mechanisms that enable animals to defend themselves against foreign invaders, such as bacteria, viruses, fungi, and parasites. All multicellular animals have defenses against microbes that would otherwise quickly swamp their cellular machinery. Both bacteria and viruses multiply in abundance when in a favorable medium, and cells and tissue fluids with their nutrients provide an ideal culture medium for microbes.

Every animal's first line of defense is the integument that is exposed to the external environment. Few microbes can penetrate this barrier when it is intact. Many animals increase the effectiveness of the integument by secreting mucus or other glandular materials that contain chemical substances that are toxic to bacteria. Bacteria that adhere to sticky mucus are simply swept away as old mucus is replaced with new. Other animals that lack surface mucous secretions harbor a "friendly" microbial flora on the integument that suppresses the growth of more virulent microorganisms.

Despite such physical and chemical defenses, microbes do frequently breach the external barrier through small breaks and injuries; some viruses even gain entry through intact integument. Once inside, the body's second line of defense, the immune system, responds to the assault. The lower invertebrates appear to have a single but effective nonspecific immune system: **ameboid phagocytic cells** that move through the tissues and circulating fluids, recognizing and engulfing (phagocytizing) foreign material when it is encountered. By migrating into injured areas, these wandering cells also initiate tissue repair. In more advanced invertebrates, phagocytic cells become organized into distinct reticular networks through which the blood filters on its way to the heart. Such struc-

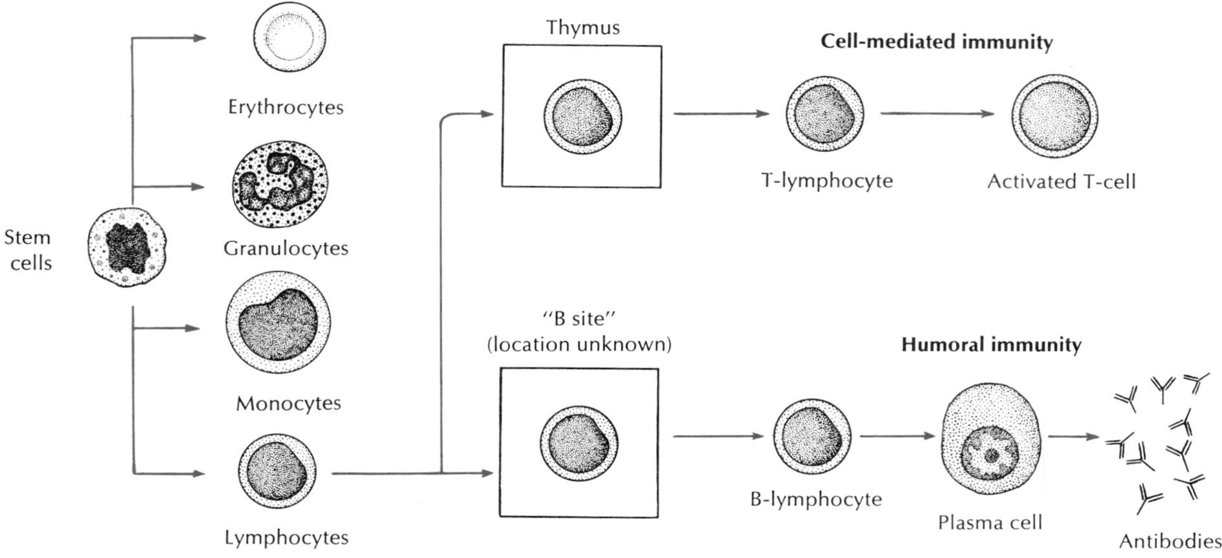

FIG. 29-12 Differentiation of lymphocytes in mammals for cell-mediated and humoral immunity. Stem cells from the embryonic yolk sac migrate to the bone marrow where they divide and produce many kinds of blood cells. Some lymphocytes then travel to the thymus where they are transformed into T-cells which can attack antigens. Other lymphocytes are induced to become B-cells at an unidentified site. When stimulated by an antigen, B-cells produce a clone of antibody-producing plasma cells. (Modified from Cooper, M. D., and A. R. Lawton, III. 1974. Sci. Am. **231**:58-72 [Nov.].)

tures are not unlike the more highly organized lymph nodes of birds and mammals.

In the vertebrates the immune system becomes far more elaborate. In addition to the nonspecific phagocytic cell system inherited from the invertebrates, the vertebrates have evolved a **specific** immune system that forms antibodies in response to invading microbes. This system depends on prior exposure to the foreign body, then recognition and defensive responses on subsequent exposure. We will consider both systems—nonspecific and specific—in this section.

Nonspecific defense mechanisms

Nonspecific defenses are those that do not depend on previous exposure for response to invaders. They recognize a foreign body as foreign the first time it breaches the body's barricades. The basic response to injury is **inflammation,** a sequence of events that leads to destruction of invading microbes and tissue repair. Immediately following the invasion, histamine is released from damaged cells, causing local capillaries to dilate. Blood flow is increased, bringing **neutrophils.** and later, **monocytes** to the inflamed area, where they accumulate in enormous numbers. Neutrophils immediately set about engulfing bacteria, viruses, and foreign debris. Once engulfed, a bacterium is quickly killed and digested with a variety of powerful hydrolytic enzymes. The battle is not all one-sided, however. Should the infection be large, or the bacteria especially virulent, the white blood cells may themselves be killed in large numbers by bacterial toxins and accumulate as pus in the infection site. Occasionally an abscess, or walled-off bag of pus, is formed that will not spontaneously heal. However, the vast majority of minute invasions that are always occurring and of which we are seldom aware are quickly and efficiently disposed of. Once the infection is destroyed, the tissue repairs by cellular regeneration.

Some years ago, a substance named **interferon** was discovered that has proved to be a potent natural inhibitor of virus multiplication in cells. It is a low–molecular weight protein that is rapidly produced by cells in response to viral infection. Unlike antibodies, which we will consider in the following section, interferon is a nonspecific defense. Once it is produced, the cell becomes resistant to many kinds of viruses.

Most importantly, interferon acts to destroy viruses inside the infected cell. Antibodies cannot pass the cell's plasma membrane and thus are ineffective against intracellular viral invasions.

Specific immune mechanisms

Specific immunity is produced in the body following exposure to a specific foreign substance or antigen. Two separate immune systems are involved. One is **cell-mediated immunity,** which defends against viruses and fungi, and the other is **humoral immunity,** which protects the body against bacteria through the production of antibodies (Fig. 29-12). Both kinds of immunity are centered on **lymphocytes,** a small unpigmented type of white blood cell. All lymphocytes are derived from the same blood-forming tissue in bone marrow, but later they develop into two separate lines, having different life histories and becoming separate immune systems.

Cell-mediated immunity

Cell-mediated immunity is conferred by a population of lymphocytes called **T cells,** so termed because they mature in the **thymus,** a gland that in mammals lies in the chest under the sternum. When a viral or fungal invasion occurs, some of the T cells become sensitized to the invading antigen. They enlarge, divide, travel to the invasion site, and there release specific chemicals that eliminate the foreign cells. Other chemicals released by activated T cells are chemotactic factors (chemical attractants) that enlist the aid of neutrophils and monocytes in phagocytizing the invaders. T cells therefore perform two functions: they specifically destroy invading antigens themselves, and they enhance the effectiveness of the body's nonspecific defense system of phagocytic cells. Unlike the humoral system of immunity that we will consider next, the T cells (sensitized lymphocytes) have no permanent address, but constantly circulate in the blood and lymph. This increases the probability that they will encounter an invader right at its site of entry.

Humoral immunity

Humoral immunity is mediated by a separate population of lymphocytes, called **B cells,** that produce antibodies. (''Humoral'' is an old term pertaining to body fluids; in this context it refers to the transport of antibodies in the body fluids.) Unlike T cells, the precursors of B cells do not enter the thymus after their release from bone marrow, but, in mammals, they pass into some as yet unidentified tissue where they are induced to become antibody-producing B cells. In birds, this tissue is the bursa of Fabricius (thus the name B cell), a small lymphoid organ attached to the intestine near the cloaca. When the B cell completes its maturation, it becomes a **plasma cell,** the most active antibody producer (Fig. 29-12).

Plasma cells are an animal's major defense system against bacteria (the T cells combat viruses, fungi, and parasites, but few bacteria). They do this by elaborating enormous numbers of **antibodies,** some 2,000 molecules per second for a life span of several days. An antibody is a specialized protein capable of reacting specifically with an **antigen.** Almost any foreign cell or molecule having a molecular weight of at least 10,000 can act as an antigen; a single bacterium or even a protein bacterial toxin may have several antigenic sites on its surface. Each of these sites can induce the manufacture of a highly specific antibody. When released into the bloodstream, the antibody combines with its corresponding antigen, and no other, in a precise lock-and-key fit.

Antibodies belong to a class of proteins called **immunoglobulins** (or gamma globulins). There are five subclasses of antibodies of which IgG is the most common (the Ig prefix stands for immunoglobulin). An antibody is made of four polypeptide chains, paired in such a way that the molecule consists of two long (''heavy'') chains and two short (''light'') chains in a Y-shaped configuration. Each antibody has two identical antigen-binding sites, one at the end of each arm of the Y (Fig. 29-13). The two arms are joined at the stem of the Y by a hinge that allows the angle of the arms to vary; this flexibility permits the antibody to cross-link two antigenic sites having variable spacing between them.

Antibodies do not themselves destroy antigens. Antibodies *identify* the antigen and then, by attaching to it, activate the attack by a group of plasma enzymes collectively called **complement.** The 11 proteins of the complement system circulate in the plasma all the time in an inactive state. When activated, complement binds to antigen-antibody complexes at special sites (Fig. 29-13), thus coating the foreign substance (antigen) and making it much easier for phagocytes to engulf the foreigner. Complement can also directly destroy microbes without the assistance of phagocytes by

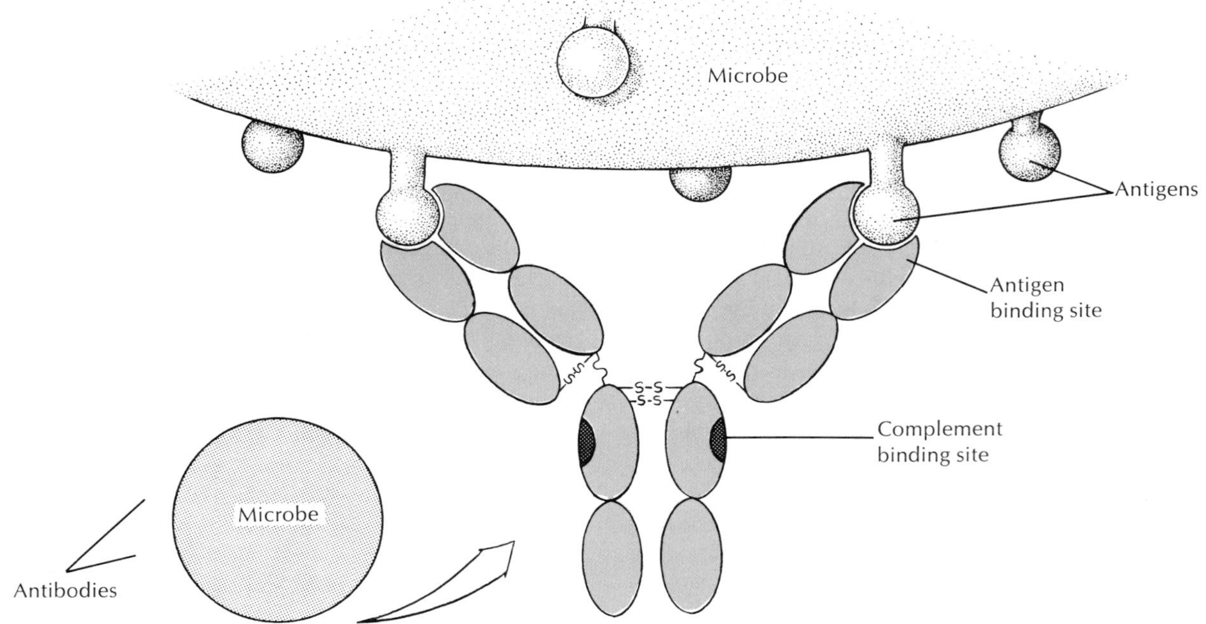

FIG. 29-13 Antibody shown coupled to two antigenic sites on a microbe. The most common type of antibody, immunoglobulin G (IgG), consists of four polypeptide chains—two long, central chains and two, short peripheral chains—arranged in a Y-shaped configuration. The two long chains are held together by disulfide bonds (—S—S—). Each antibody couples through two receptors, which are antigen-binding sites having geometric and chemical specificity for the antigens. Antibodies, together with attached plasma enzymes called complement, modify the surface of the invading microorganisms (lower left), making it easier for white blood cells to engulf and destroy it.

forming lesions on the microbes' lipid-membrane surface.

Having introduced all the main characters and stated their roles in the body's specific defense system, let us summarize their interactions during an infection. When a foreign antigen enters the body for the first time, it is immediately attacked by phagocytes (neutrophils and monocytes) and by T cells (sensitized lymphocytes). At the same time, the antigen triggers antibody synthesis by B cells. For this to happen, the antigen must be transported, usually by lymph vessels, to lymph nodes, the spleen, or other lymphoid cell aggregates where the B cells are located. There the antigen stimulates a small number of B cells to divide and mature rapidly into plasma cells, which begin producing antibodies against the antigen. It usually requires several days for the plasma cells to build up to full-scale antibody production. In doing so, a fraction of the undifferentiated B cells become a "memory bank" for the antigen, ready to respond immediately should the same antigen appear a second time.

In the meantime, while the B cells are "tooling up" for antibody production, the phagocytes are fighting the infection. If they are immediately successful, the antibodies may never enter combat. But if the infection persists for several days, new antibodies activate complement, which directly destroys the microbes and amplifies phagocytic activity. Whether or not the infection is short-lived, or is more persistent, the specific immune system has been alerted, and any subsequent invasion by the antigen will be met with much greater force. The animal now bears an **active immunity** against the antigen.

Until recently, suffering a disease or infection was the only way to obtain an active immunity. Now it may be achieved safely from the injection of vaccines (living or weakened microbes) or microbial deriva-

TABLE 29-1 Major blood groups

Blood type	Genotype	Antigens on red corpuscles	Antibodies in serum	Can give blood to	Can receive blood from	Frequency in United States (%)		
						Whites	Blacks	Chinese
O	OO	None	Anti-A and anti-B	All	O	45	38	46
A	AA, AO	A	Anti-B	A, AB	O, A	41	27	28
B	BB, BO	B	Anti-A	B, AB	O, B	10	21	23
AB	AB	AB	None	AB	All	4	4	13

tives, such as a bacterial toxin. The purpose is to provoke an antibody response, which will leave the antibody-synthesizing machinery poised to respond immediately to a future invasion by that organism.

Passive immunity is gained by directly injecting preformed antibodies from one person or animal to another person. This is the kind of protective immunity passed from mother to fetus across the placenta. Passive immunity is short-lived because the antibodies are soon destroyed and cannot be naturally replaced because the antibody-synthesizing machinery has not been set in motion.

Transplantation and blood antigens

Histocompatibility antigens

When a piece of skin or other tissue is grafted from one individual to another, the transplant seldom survives. Within a few days the graft becomes inflamed, eventually dies, and drops away: it has been rejected by the recipient's tissues. The basis of such a rejection is protein molecules on the graft called **histocompatibility antigens,** which are recognized as foreign by the recipient. Histocompatibility antigens are simply surface proteins present on every cell that give the animal bearing those cells its own chemical identity. No two animals, other than identical twins, have the same antigens; each animal is chemically unique.

Why do histocompatibility antigens exist? Since tissue transplants are a human invention and not a normal event in nature, such antigens do not exist to frustrate graft attempts. It is believed that they are a part of the immunologic defense that protects an animal's cells from its own immune system. In some way, these antigens are recognized as "self" by the developing immune system during embryonic life; thus sensitization to one's own cells does not normally occur.

Blood group antigens

ABO blood types. Blood differs chemically from person to person, and when two different (incompatible) blood types are mixed, **agglutination** of blood cells (clumping together) results. The basis of these chemical differences is naturally occurring antigens on the membranes of red blood cells. The best known of these inherited immune systems is the ABO blood group. The antigens A and B are inherited as dominant genes. Thus as shown in Table 29-1, an individual with, for example, genes AA or AO develops A antigen (blood type A). The presence of a B gene produces B antigens (blood type B), and for the genotype AB both A and B antigens develop on the erythrocytes (blood type AB).

There is an odd feature about the ABO system. Normally we would expect that a type A individual would develop antibodies against type B blood only if B cells were introduced into the body. In fact, type A persons always have anti-B antibodies in their blood, even without the prior exposure to type B blood. Similarly type B individuals carry anti-A antibodies. Type AB blood has neither anti-A nor anti-B antibodies (since if it did, it would destroy its own blood cells), and type O blood has both anti-A and anti-B antibodies.

We see then that the blood group names identify their *antigen* content. Persons with type O blood are called universal donors because, lacking antigens, their blood can be infused into a person with any blood type. Even though it contains anti-A and anti-B antibodies, these are so diluted during transfusion that they do not react with A or B antigens in a recipient's blood. In practice, however, clinicians insist on matching blood types to prevent any possibility of incompatibility.

Rh factor. It is difficult to think of another area of physiology as totally linked with the name of a single researcher as blood grouping is with the name of Karl

Landsteiner. This Austrian—later American—physician discovered the ABO blood groups in 1900. The great importance of his work became abundantly clear during World War I when blood transfusion was first attempted on a large scale. In 1927 Landsteiner in collaboration with Philip Levine in the United States discovered the MN blood group, also present in all people. This classification is not important in transfusions but may be crucial in determining relationship in paternity cases.

In 1940, 10 years after receiving the Nobel Prize in recognition of contributions that were more than adequate for the lifetime of any scientist, Landsteiner made still another famous discovery. This new group was called the Rh factor, named after the Rhesus monkey, in which it was first found. About 85% of individuals have the factor (positive) and the other 15% do not (negative). He also found that Rh-positive and Rh-negative bloods are incompatible; shock and even death may follow their mixing when Rh-positive blood is introduced into an Rh-negative person who has been sensitized by an earlier transfusion of Rh-positive blood.

The Rh factor is inherited as a dominant gene; this accounts for a peculiar and often fatal form of anemia of newborn infants called **erythroblastosis fetalis.** Although the fetal and maternal bloods are separated by the placenta, this separation is not perfect. Some admixture of fetal and maternal bloods usually occurs, especially right after birth when the placenta ("afterbirth") separates from the uterine wall. This admixture of blood, normally of no consequence, can be serious *if* the father is Rh positive, the mother Rh negative, and the fetus Rh positive (by inheriting the factor from the father). The fetal blood, containing the Rh antigen, can stimulate the formation of Rh-positive antibodies in the blood of the mother. The mother is permanently immunized against the Rh factor. During the second pregnancy these antibodies may diffuse back into the fetal circulation and produce agglutination and destruction of the fetal red blood cells. Because the mother is usually sensitized at the end of the first pregnancy, subsequent babies will be more severely threatened than is the first.

Erythroblastosis fetalis can now be prevented by giving an Rh-negative mother anti-Rh antibodies just after the birth of her first child. These antibodies remain long enough to neutralize any Rh-positive fetal blood cells that may enter her circulation, thus preventing her own antibody machinery from being stimulated to produce the Rh-positive antibodies. Active, permanent immunity is blocked.

GAS EXCHANGE

The energy bound up in food must be released by **respiration,** or aerobic energy metabolism. In this cellular process, described in some detail in Chapter 5, oxygen is consumed by the body cells and carbon dioxide is released. In this chapter we will examine **gas exchange:** those activities that exchange oxygen and carbon dioxide between the organism and its environment.

Small aquatic animals such as the one-celled protozoans obtain what oxygen they need by direct diffusion from the environment. Carbon dioxide, the gaseous waste of metabolism, is also lost by diffusion to the environment. Such a simple solution to the problem of gas exchange is really only possible for very small animals (less than 1 mm in diameter) or those having very low rates of metabolism.

As animals became larger and evolved a waterproof covering, specialized devices such as lungs and gills developed that greatly increased the effective surface for gas exchange. But because gases diffuse so slowly through protoplasm, a circulatory system was necessary to distribute the gases to and from the deep tissues of the body.

Even these adaptations were inadequate for advanced animals, with their high rates of cellular respiration. The solubility of oxygen in the blood plasma is so low that plasma alone could not carry enough to satisfy metabolic demands. When special oxygen-transporting blood proteins such as hemoglobin evolved, the oxygen-carrying capacity of the blood was greatly increased. Thus, in higher animals, gas exchange is a two-phase process consisting of (1) **ventilation** (or **breathing**), the exchange of gases between body fluids and the environment via a special respiratory organ (lung or gill); and (2) **blood gas transport,** the movement of gases from the respiratory organ to the cells, often assisted by special transport proteins.

Problems of aquatic and aerial breathing

How an animal respires is largely determined by the nature of its environment. The two great arenas of animal evolution—water and land—are vastly different in their physical characteristics. The most obvious difference is that air contains far more oxygen—at least 20 times more—than does water. Atmospheric air contains oxygen (about 21%), nitrogen (about 79%), carbon dioxide (0.03%), a variable amount of water vapor, and very small amounts of inert gases (helium, argon, neon, and others). These gases are variably soluble in water. The amount of oxygen dissolved depends on the concentration of oxygen in the air and on the water temperature. Water at 5° C fully saturated with air contains about 9 ml of oxygen per liter. (Note that by comparison, air contains about 210 ml of oxygen per liter.) The solubility of oxygen in water decreases as the temperature rises. For example, water at 15° C contains about 7 ml of oxygen per liter, and at 35° C, only 5 ml of oxygen per liter. The relatively low concentration of oxygen dissolved in water is the greatest respiratory problem facing aquatic animals. Unfortunately it is not the only one. Oxygen diffuses much more slowly in water than in air, and water is much denser and more viscous than air. All of this means that successful aquatic animals must have evolved very efficient ways of removing oxygen from water. Yet even the most advanced fishes with highly efficient gills and pumping mechanisms may use as much as 20% of their energy just extracting oxygen from water. By comparison, a mammal uses only 1% to 2% of its resting metabolism to breathe.

It is essential that respiratory surfaces be kept thin and always wet to allow diffusion of gases between the environment and the underlying circulation. This is hardly a problem for aquatic animals, immersed as they are in water, but it is a very real problem for air breathers. To keep the respiratory membranes moist and protected from injury, air breathers have in general developed invaginations of the body surface and then added pumping mechanisms to move air in and out. The lung is the best example of a successful solution to breathing on land. In general, **evaginations** of the body surface, such as gills are most suitable for aquatic respiration and **invaginations,** such as lungs, are best for air breathing. We will now consider the specific kinds of respiratory organs employed by animals.

Cutaneous respiration

Protozoa, sponges, cnidarians, and many worms respire by direct diffusion of gases between the organism and the environment. We have noted that this kind of **integumentary respiration** is not adequate when the mass of living protoplasm exceeds about 1 mm in diameter. But by greatly increasing the surface of the body relative to the mass, many multicellular animals respire in this way. Integumentary respiration frequently supplements gill or lung breathing in larger animals such as amphibians and fishes. For example, an eel can exchange 60% of its oxygen and carbon dioxide through its highly vascular skin. During their winter hibernation, frogs exchange all their respiratory gases through the skin while submerged in ponds or springs.

Gills

Gills are unquestionably the most effective respiratory device for life in water. Gills may be simple **external** extensions of the body surface, such as the **dermal branchiae** of starfish or the **branchial tufts** of marine worms and aquatic amphibians. Most efficient are the **internal** gills of fishes (described on p. 515). Fish gills are thin filamentous structures, richly supplied with blood vessels arranged so that blood flow is opposite to the flow of water across the gills. This arrangement, called **countercurrent flow,** provides for the greatest possible extraction of oxygen from water. Water flows over the gills in a steady stream, pushed and pulled by an efficient branchial pump, and often assisted by the fish's forward movement through the water (Fig. 23-20, p. 516).

Lungs

Gills are unsuitable for life in air because when removed from the buoying water medium, the gill filaments collapse and stick together; a fish out of water rapidly asphyxiates despite the abundance of oxygen around it. Consequently air-breathing vertebrates possess lungs, highly vascularized internal cavities. Lungs of a sort are found in certain invertebrates (pulmonate snails, scorpions, some spiders) but these structures cannot be ventilated and consequently are not very efficient.

Lungs that can be ventilated efficiently are characteristic of the terrestrial vertebrates. The most primitive vertebrate lungs are those of lungfishes (Dipneusti), which use them to supplement, or even replace, gill respiration during periods of drought. Although of simple construction, the lungfish lung (similar to the "generalized lung" in Fig. 23-19, p. 514) is supplied with a capillary network in its largely unfurrowed

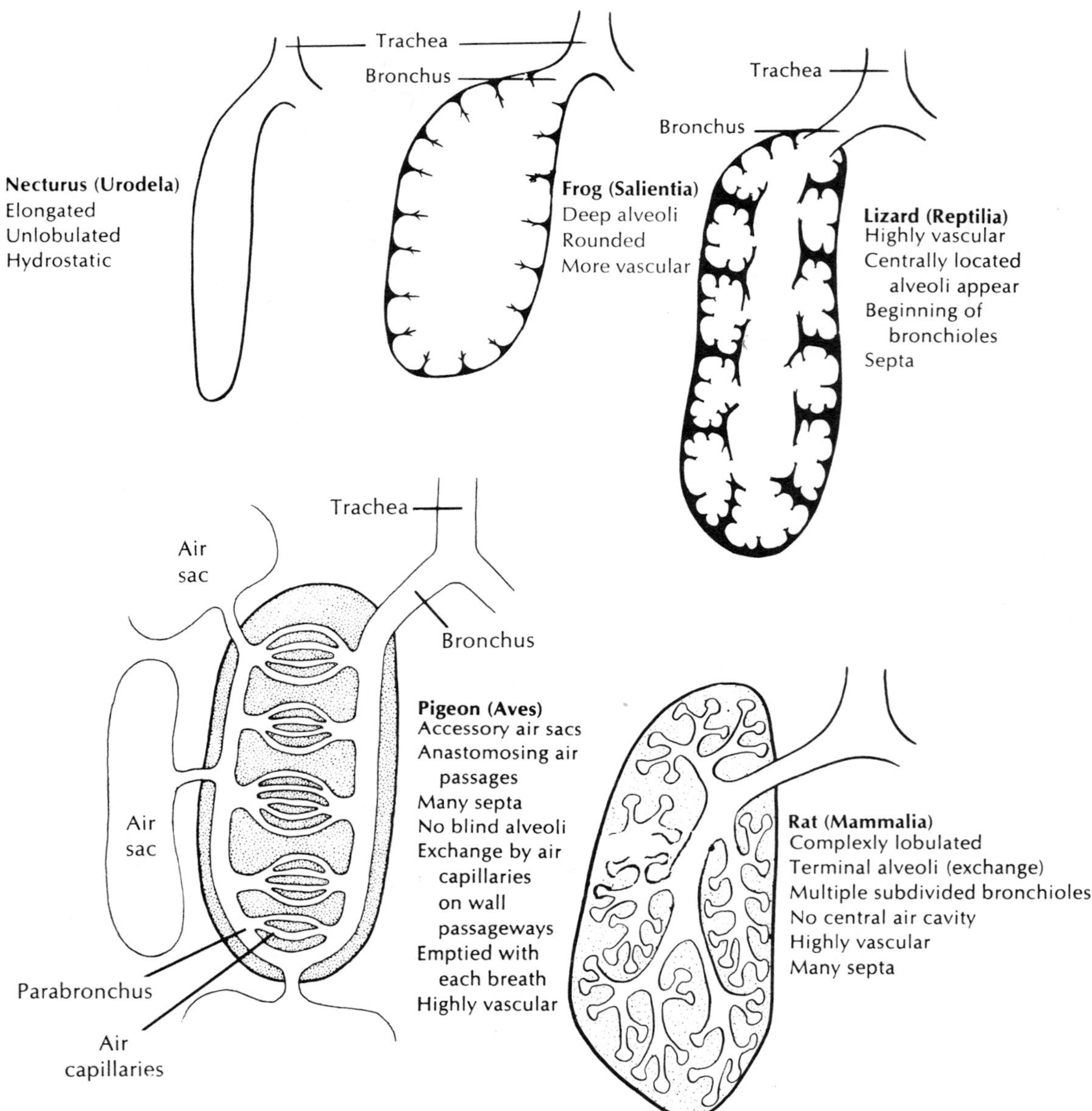

FIG. 29-14 Internal structures of lungs among vertebrate groups. In general, evolutionary trend has been from simple sacs with little exchange surface between blood and air spaces to complex, lobulated structures of complex divisions and extensive exchange surfaces.

walls, a tubelike connection to the pharynx, and a primitive ventilating system for moving air in and out of the lung. Amphibians also have simple baglike lungs, whereas in higher forms the inner surface area is vastly increased by numerous lobulations and folds (Fig. 29-14). This increase is greatest in the mammalian lung, which is complexly divided into many millions of small sacs **(alveoli),** each served by a rich vascular network. It has been estimated that the human lungs have a total surface area of from 50 to 90 m^2—50 times the area of the skin surface—and contain 1,000 miles of capillaries.

Moving air into and out of lungs has been an evolutionary design problem that was, of course, solved, although one wonders if an imaginative biologic engineer, given the proper resources, could not have come up with a better design. Unlike the efficient one-way flow of water across fish gills, air must enter and exit a lung at the same point. Furthermore, a tube of some length—the bronchi, trachea, and mouth cavity—connects the lungs to the outside. This is a "dead-air space" containing a volume of air that shuttles back and forth with each breath, adding to the difficulty of properly ventilating the lungs. In fact, lung ventilation is so inefficient that in normal breathing only about one-sixth of the air in the lungs is replenished with each inspiration.

In one group of vertebrates, the birds, lung efficiency is vastly improved by adding an extensive system of air sacs (Fig. 29-14) that serve as air reservoirs during ventilation. On inspiration, some 75% of the incoming air bypasses the lungs to enter the air sacs. At expiration, some of this fresh air passes directly through the lung passages. Thus the air capillaries receive nearly fresh air during both inspiration and expiration (see Fig. 26-13, p. 584). The beautifully designed bird lung is the result of selective pressures during the evolution of flight and its high metabolic demands.

Frogs force air into the lungs by first lowering the floor of the mouth to draw air into the mouth through the external nares (nostrils); then, by closing the nares and raising the floor of the mouth, the air is driven into the lungs. Much of the time, however, frogs rhythmically ventilate only the mouth cavity, which serves as a kind of auxiliary "lung" (see Fig. 24-13, p. 543). Amphibians therefore employ a **positive pressure** action to fill their lungs, unlike most reptiles, birds, and mammals, which breathe by sucking air into the lungs (**negative pressure** action).

Tracheae

Insects and certain other terrestrial arthropods (centipedes, millipedes, and some spiders) have a highly specialized type of respiratory system; in many respects it is the simplest, most direct, and most efficient respiratory system found in active animals. It consists of a system of tubes **(tracheae)** that branch repeatedly and extend to all parts of the body. The smallest end channels **(air capillaries),** less than 1 μm in diameter, sink into the plasma membranes of the body cells. Oxygen enters the tracheal system through valvelike openings **(spiracles)** on each side of the body and diffuses directly to all cells of the body. Carbon dioxide diffuses out in the opposite direction. Some insects can ventilate the tracheal system with body movements; the familiar telescoping movement of the abdomen of a foraging bee on a warm summer day is an example. The tracheal system is simple because blood is not needed to transport the respiratory gases; the cells have a direct pipeline to the outside.

Mammalian respiration

In mammals the respiratory system is made up of the following: the nostrils (external nares); the **nasal chamber,** lined with mucus-secreting epithelium; the **posterior nares,** which connect to the **pharynx,** where the pathways of digestion and respiration cross; the **epiglottis,** a flap that folds over the **glottis** (the opening to the larynx) to prevent food from going the wrong way in swallowing; the **larynx,** or voice box; the **trachea,** or windpipe; and the two **bronchi,** one to each lung (Fig. 29-15). Within the lungs each bronchus divides and subdivides into smaller tubes **(bronchioles)** that lead to the air sacs **(alveoli).** The walls of the alveoli are thin and moist to facilitate the exchange of gases between the air sacs and the adjacent blood capillaries (Fig. 29-16). Air passageways are lined with mucus-secreting ciliated epithelium and play an important role in conditioning the air before it reaches the alveoli. There are partial cartilage rings in the walls of the trachea, bronchi, and even some of the bronchioles to prevent those structures from collapsing.

In its passage to the air sacs the air undergoes three important changes: (1) it is filtered free from most dust and other foreign substances, (2) it is warmed to body temperature, and (3) it is saturated with moisture.

The lungs consist of a great deal of elastic connective tissue and some muscle. They are covered by a thin layer of tough epithelium known as the **visceral pleura.** A similar layer, the **parietal pleura,** lines the inner surface of the walls of the chest (Fig. 29-15). The two layers of the pleura are in contact and slide over one another as the lungs expand and contract. The "space" between the pleura, called the **pleural cavity,** contains a partial vacuum. Actually, no real pleural space exists; the two pleura rub together, lubricated by lymph. The chest cavity is bounded by the spine, ribs, and breastbone, and floored by the **dia-**

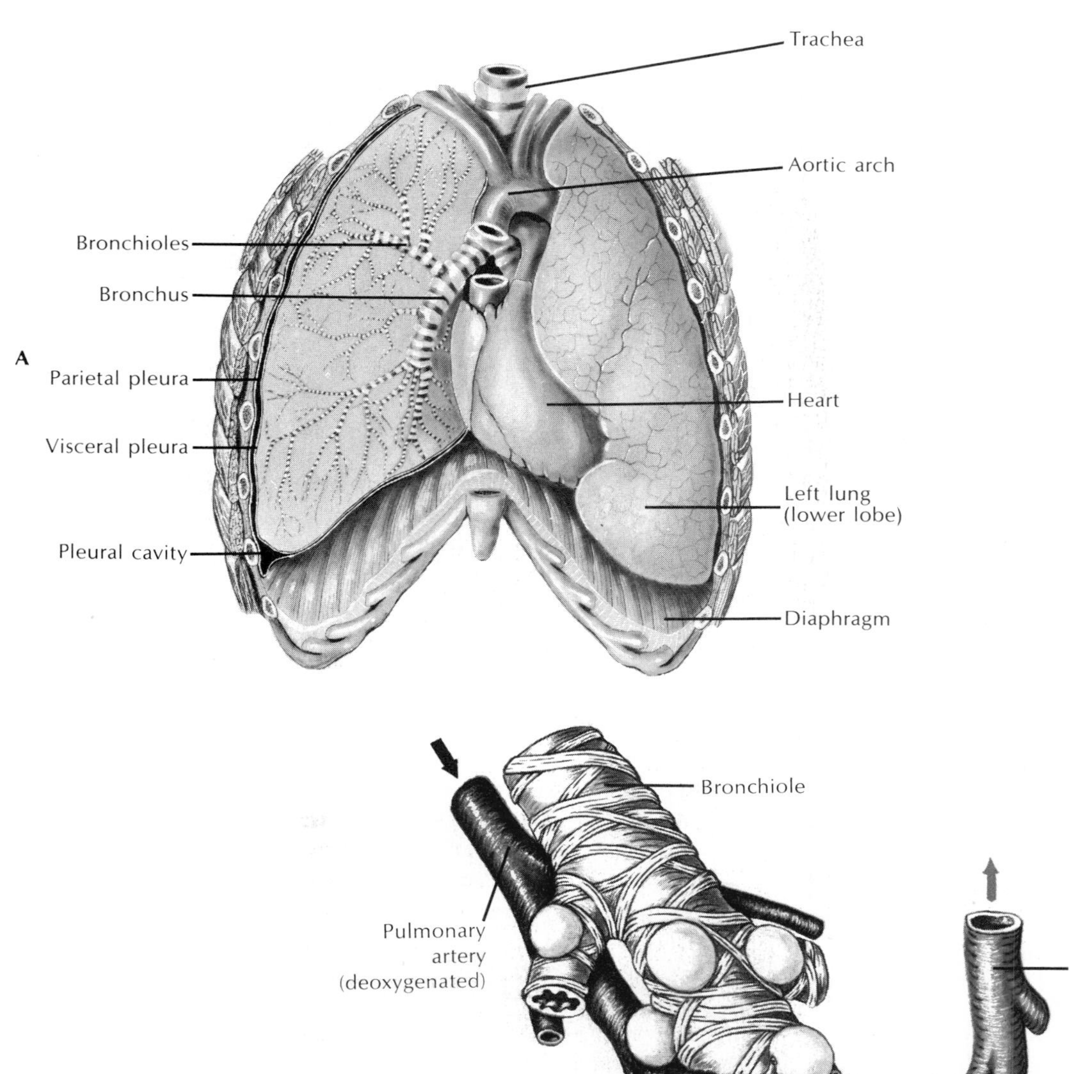

FIG. 29-15 A, Human lungs. **B,** Terminal portion of lung, showing bronchiole and air sacs with their blood supply. Arrows show direction of blood flow. (Approximately ×90.)

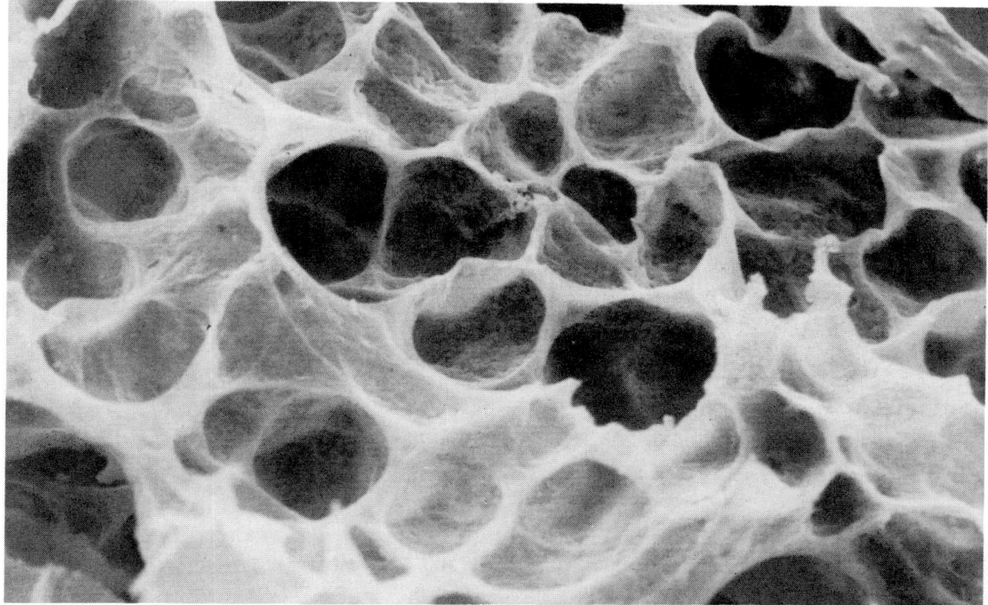

FIG. 29-16 Appearance of inside of terminal bronchus of rat lung showing alveoli and alveolar ducts leading to deeper alveoli. This scanning electron micrograph was made after the inflated lung was quick-frozen in liquid nitrogen to preserve the appearance of the living lung. (×180.) (From Kuhn, C., III, and E. H. Finke. 1972. J. Ultrastructural Res. **38:**161-173.)

phragm, a dome-shaped, muscular partition between chest cavity and abdomen.

Mechanism of breathing

The chest cavity is an air-tight chamber. In **inspiration** the ribs are elevated, the diaphragm is contracted and flattened, and the chest cavity is enlarged. The resultant increase in volume of chest cavity and lungs causes the air pressure in the lungs to fall below atmospheric pressure; air rushes in through the air passageways to equalize the pressure. **Expiration** is a less active process than inspiration. When their muscles relax, the ribs and diaphragm return to their original position and the chest cavity size decreases. The elastic lungs then contract and force the air out.

Control of breathing

Respiration must adjust itself to the varying needs of the body for oxygen. Respiration is normally involuntary and automatic but may come under voluntary control. The rhythmic inspiratory and expiratory movements are controlled by a nervous mechanism centered in the **medulla oblongata** of the brain (Fig. 29-17). By placing tiny electrodes in various parts of the medulla of experimental animals, neurophysiologists located separate **inspiratory** and **expiratory neurons** that act reciprocally to stimulate the inspiratory and expiratory muscles of the diaphragm and rib cage (intercostal) muscles. The rate of breathing is determined by the amount of carbon dioxide in the blood: a slight rise in the blood carbon dioxide stimulates respiration; a fall will decrease breathing.

Composition of inspired, expired, and alveolar airs

The composition of expired and alveolar airs is not identical. Air in the alveoli contains less oxygen and more carbon dioxide than does the air that leaves the lungs. Inspired air has the composition of atmospheric air. Expired air is really a mixture of alveolar and inspired airs. The variations in the three kinds of air are shown in Table 29-2. The water given off in expired air depends on the relative humidity of the external air and the activity of the person. At ordinary room temperature and with a relative humidity of about 50%, an individual in performing light work will lose about 350 ml of water from the lungs each day.

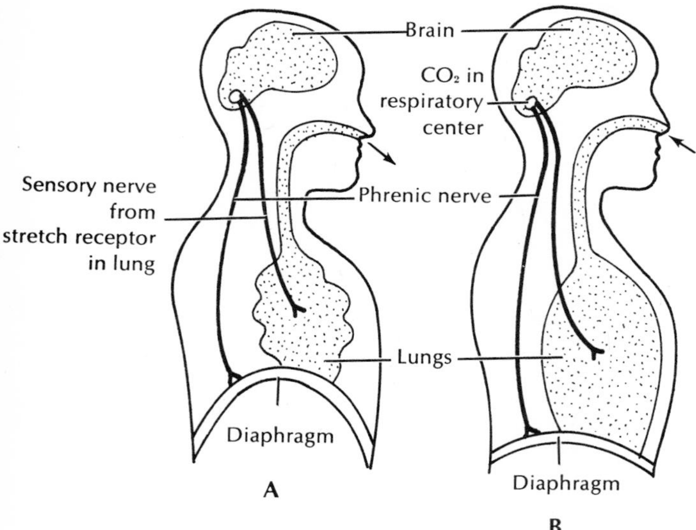

FIG. 29-17 Mechanism of breathing. In inspiration, **B,** carbon dioxide in blood stimulates respiratory center to send impulses by way of phrenic nerves to diaphragm, which, with elevation of ribs, produces inhalation of air. In **A,** impulses from stretch receptors in lungs inhibit respiratory center and exhalation occurs. (See text for further explanation.)

Gaseous exchange in lungs

The diffusion of gases, both in internal as well as external respiration, takes place in accordance with the laws of physical diffusion; that is, the gases pass from regions of high pressure to those of low pressure. The pressure of a gas refers to the partial pressure that that gas exerts in a mixture of gases. If the atmospheric pressure at sea level is equivalent to 760 mm Hg, the partial pressure of oxygen will be 21% (percentage of oxygen in air) of 760, or 159 mm Hg. The partial pressure of oxygen in the lung alveoli is greater (100 mm Hg pressure) than it is in venous blood of lung capillaries (40 mm Hg pressure) (Fig. 29-18). Oxygen then naturally diffuses into the capillaries. In a similar manner the carbon dioxide in the blood of the lung capillaries has a higher concentration (46 mm Hg) than has this same gas in the lung alveoli (40 mm Hg), so that carbon dioxide diffuses from the blood into the alveoli.

In the tissues respiratory gases also move according to their concentration gradients (Fig. 29-18). Here the concentration of oxygen in the blood (100 mm Hg pressure) is greater than in the tissues (0 to 30 mm Hg pressure), and the carbon dioxide concentration in the tissues (45 to 68 mm Hg pressure) is greater than that in blood (40 mm Hg pressure). The gases in each case will go from a high to a low concentration.

TABLE 29-2 Variation in respired air

	Inspired air (vol %)	Expired air (vol %)	Alveolar air (vol %)
Oxygen	20.96	16	14.0
Carbon dioxide	0.04	4	5.5
Nitrogen	79.00	80	80.5

Transport of gases in blood

In some invertebrates the respiratory gases are simply carried dissolved in the body fluids. However, the solubility of oxygen is so low in water that this method is only adequate for animals having low rates of metabolism. For example, only about 1% of our oxygen requirement can be transported in this way. Consequently in just about all the advanced invertebrates and the vertebrates, nearly all the oxygen and a significant amount of the carbon dioxide are transported by special colored proteins, or **respiratory pigments,** in the blood. In most animals (all vertebrates) these respiratory pigments are packaged into blood corpuscles. This is necessary because if this amount of respiratory pigment were free in blood, the blood would have the viscosity of syrup and would barely flow through the blood vessels, if it flowed at all.

Respiratory pigments. The two most widespread respiratory pigments are **hemoglobin,** a red, iron-containing protein present in all vertebrates and many invertebrates, and **hemocyanin,** a blue, copper-containing protein present in the crustaceans and cephalopod molluscs. Among other pigments is **chlorocruorin** (klor-a-kroo′o-rin), a green-colored, iron-containing pigment found in four families of polychaete tube worms. Its structure and oxygen-carrying capacity are very similar to those of hemoglobin, but it is carried free in the plasma rather than being enclosed in blood corpuscles. Some polychaete worms have both chlorocruorin and hemoglobin present in their blood. **Hemerythrin** is a red pigment found in some polychaete worms. Although it contains iron, this metal is not present in a heme group (despite the name of the pigment!), and its oxygen-carrying capacity is poor.

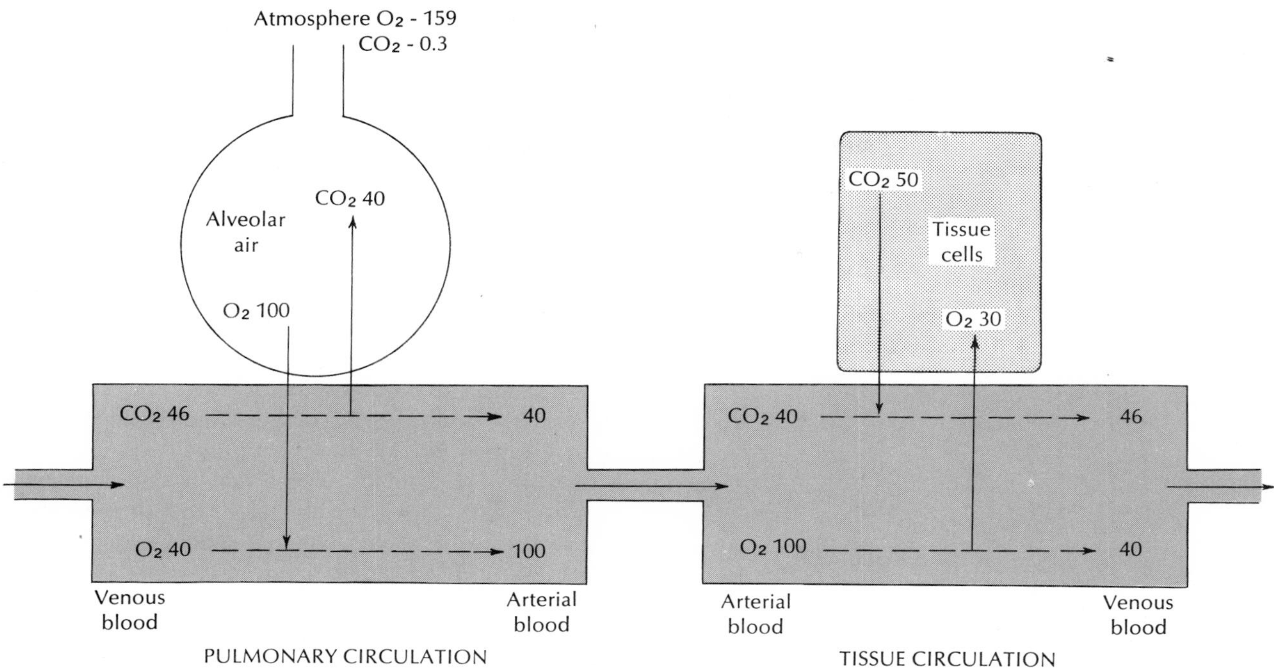

FIG. 29-18 Exchange of respiratory gases in lungs and tissue cells. Numbers present partial pressures in millimeters of mercury.

Hemoglobin and oxygen transport. Hemoglobin is a complex protein. Each molecule is made up of 5% **heme,** an iron-containing compound giving the red color to blood (Fig. 29-19), and 95% **globin,** a colorless protein. The heme portion of the hemoglobin has a great affinity for oxygen; each gram of hemoglobin (there are about 15 g of hemoglobin in each 100 ml of human blood) can carry a maximum of approximately 1.3 ml of oxygen; each 100 ml of fully oxygenated blood contains about 20 ml of oxygen. Of course, for hemoglobin to be of value to the body it must hold oxygen in a loose, reversible chemical combination so that it can be released to the tissues. The actual amount of oxygen bound to hemoglobin depends on the oxygen partial pressure surrounding the blood corpuscles, a relationship expressed in the oxygen dissociation curve in Fig. 29-20. When the oxygen tension is high, as it is in the capillaries of the lung alveoli, hemoglobin becomes almost fully saturated to form oxyhemoglobin. Then, when the oxygenated blood leaves the lung and is distributed to the systemic capillaries in the body tissues, it enters regions of low oxygen partial pressure because oxygen is continuously consumed by cellular oxidative processes. The oxyhemoglobin now releases its bound oxygen, which diffuses into the cells. As the oxygen dissociation curve shows (Fig. 29-20), the lower the surrounding oxygen tension, the greater the quantity of oxygen released. This is an important characteristic because it allows more oxygen to be released to those tissues that need it most (have the lowest oxygen pressure).

Another characteristic facilitating the release of oxygen to the tissues is the sensitivity of oxyhemoglobin to carbon dioxide. Carbon dioxide shifts the oxygen dissociation curve to the right (Fig. 29-20, *B*), a phenomenon that has been called the **Bohr effect** after the Danish scientist who first described it. Therefore as carbon dioxide enters the blood from the respiring tissues, it encourages the release of additional oxygen from the hemoglobin. The opposite event occurs in the lungs; as carbon dioxide diffuses from the venous blood into the alveolar space, the oxygen dissociation curve shifts back to the left, allowing more oxygen to be loaded onto the hemoglobin.

Unfortunately for human beings and other higher

FIG. 29-19 Chemical structure of heme, an iron porphyrin compound composed of four pyrrole rings joined together with methane groups. There are four heme units in each hemoglobin molecule, each capable of carrying one oxygen molecule; thus each hemoglobin molecule when fully loaded is transporting four oxygen molecules.

animals, hemoglobin has even a greater affinity for carbon monoxide (CO) than it has for oxygen—in fact, the affinity is about 200 times greater for carbon monoxide than for oxygen. Carbon monoxide is becoming an atmospheric contaminant of ever-increasing proportions as the world's population and industrialization continue rapidly upward. This odorless and invisible gas displaces oxygen from hemoglobin to form a stable compound called **carboxyhemoglobin.** Air containing only 0.2% carbon monoxide may be fatal. Children and small animals are poisoned more rapidly than adults because of their higher respiratory rate.

Transport of carbon dioxide. The same blood that transports oxygen to the tissues from the lungs must carry carbon dioxide back to the lungs on its return trip. However, unlike oxygen that is transported almost exclusively in combination with hemoglobin, carbon dioxide is transported in three major forms.

1. Most of the carbon dioxide, about 67%, is converted in the red blood cells into bicarbonate and hydrogen ions, by undergoing the following series of reactions:

$$CO_2 + H_2O \rightleftharpoons \underset{\textbf{Carbonic acid}}{H_2CO_3}$$

This reaction would normally proceed very slowly, but an enzyme in the red blood cells, **carbonic anhy-**

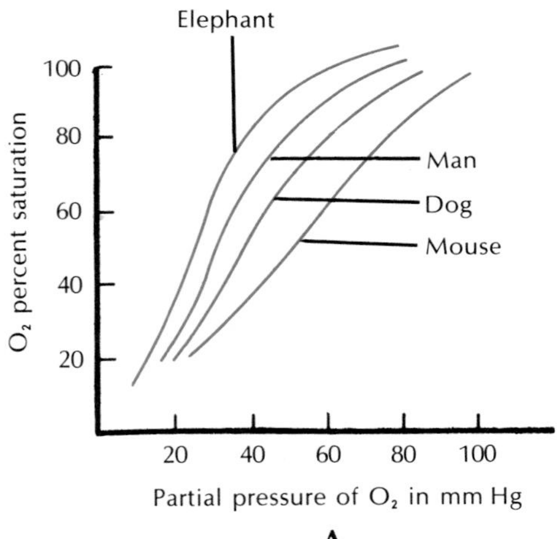

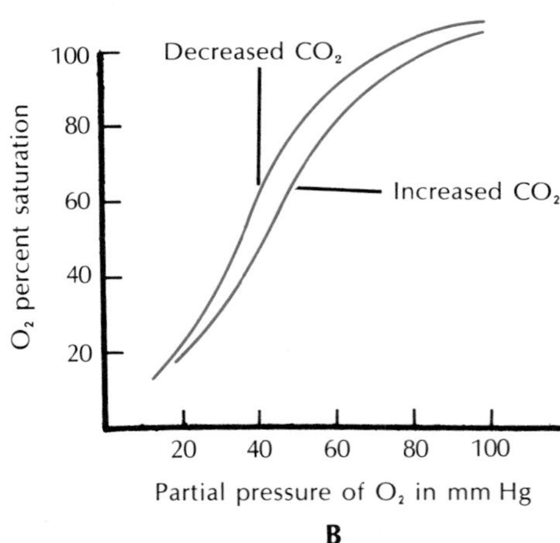

FIG. 29-20 Oxygen dissociation curves. Curves show how the amount of oxygen bound to hemoglobin (oxyhemoglobin) is related to oxygen pressure. **A,** Small animals have blood that gives up oxygen more readily than does the blood of large animals. **B,** Oxyhemoglobin is sensitive to carbon dioxide pressure; as carbon dioxide enters blood from the tissues, it shifts the curve to the right, decreasing affinity of hemoglobin for oxygen.

drase, catalyzes the reaction to proceed almost instantly. As soon as carbonic acid forms, it instantly and almost completely ionizes as follows:

$$\underset{\text{Carbonic acid}}{H_2CO_3} \rightleftharpoons \underset{\text{Bicarbonate ion}}{HCO_3^-} + \underset{\text{Hydrogen ion}}{H^+}$$

The hydrogen ion is buffered by several buffer systems in the blood, the most important of which is hemoglobin; these prevent a severe drop in blood pH. The bicarbonate ion remains in solution in the plasma and red blood cell water, since unlike carbon dioxide, bicarbonate is extremely soluble.

2. Another fraction of the carbon dioxide, about 25%, combines reversibly with hemoglobin. It is carried to the lungs, where the hemoglobin releases it in exchange for oxygen.

3. A third fraction of the carbon dioxide, about 8%, is carried as the physically dissolved gas in the plasma and red blood cells.

Annotated references

Selected general readings

See also the general references for Part Three, p. 793.

Chapman, G. 1967. The body fluids and their function. Institute of Biology's Studies in Biology, no. 8. New York, St. Martin's Press. *Brief, comparative treatment of body fluids, their transport, and regulation.*

Edney, E. G. 1957. The water relations of terrestrial arthropods. New York, Cambridge University Press.

Graubard, M. 1964. Circulation and respiration. The evolution of an idea. New York, Harcourt, Brace & World, Inc. *Historic development of blood flow concepts. Selected writings from Aristotle, Galen, Vesalius, Fabricius, Harvey, Malpighi, Boyle, and others.*

Krogh, A. 1929. The anatomy and physiology of capillaries. New Haven, Conn., Yale University Press. *Classic comparative treatment.*

Marchalonis, J. J. 1977. Immunity in evolution. Cambridge, Harvard University Press. *Well-written treatment of the evolutionary origins and development of vertebrate immunity.*

Satchell, G. H. 1971. Circulation in fishes. Cambridge, England, Cambridge University Press.

Schmidt-Nielsen, K. 1972. How animals work. Cambridge, The University Press. *This paperback deals especially with respiration, temperature regulation, and energy cost of locomotion.*

Snively, W. D., Jr. 1960. Sea within: the story of our body fluid. Philadelphia, J. B. Lippincott Co. *A popular, interesting treatise on the "interior sea" and its importance in our bodies in health and disease.*

Selected *Scientific American* articles

Adolph, E. F. 1967. The heart's pacemaker. **216:**32-37 (Mar.).

Capra, J. D., and A. B. Edmundson. 1977. The antibody combining site. **236:**50-59 (Jan.)

Clarke, C. A. 1968. The prevention of "rhesus" babies. **217:**46-52 (Nov.).

Clements, J. A. 1962. Surface tension in the lungs. **207:**120-130 (Dec.). *The surface of the lungs is coated with a surfactant that lowers surface tension and prevents collapse of the lungs.*

Comroe, J. H., Jr. 1966. The lung. **214:**56-68 (Feb.). *Physiology of the human lung.*

Cooper, M. D., and A. R. Lawton III. 1974. The development of the immune system. **231:**58-72 (Nov.) *Excellent review of the origin of cell-mediated and humoral immunity.*

Cunningham, B. A. 1977. The structure and function of histocompatibility antigens. **237:**96-107 (Oct.). *These are proteins on the cell surface that cause rejection of grafted tissue.*

Fenn, W. O. 1960. The mechanism of breathing. **202:**138-148 (Jan.). *The human respiratory system is described.*

Fox, H. M. 1950. Blood pigments. **182:**20-22 (Mar.). *Compares the characteristics of the major respiratory pigments: hemoglobin, hemocyanin, and chlorocruorin.*

Hock, R. J. 1970. The physiology of high altitude. **222:**52-62 (Feb.). *People living at high altitudes adapt physiologically to chronic low oxygen.*

Laki, K. 1962. The clotting of fibrinogen. **206:**60-66 (Mar.).

Mayerson, H. S. 1963. The lymphatic system. **208:**80-90 (June). *The lymphatics are a crucial "second" circulatory system that picks up fluids leaking from the bloodstream.*

McKusick, V. A. 1965. The royal hemophilia. **213:**88-95 (Aug.). *Story of a defective gene that plagued European royalty for three generations.*

Raff, M. C. 1976. Cell-surface immunology. **234:**30-39 (May). *Recent studies with antibodies and sensitized cells have provided new concepts of cell membrane structure.*

Warren, J. V. 1974. The physiology of the giraffe. **231:**96-105 (Nov.). *The neck is so long that special circulatory adaptations are required to supply the head with blood.*

Wiggers, C. J. 1957. The heart. **196:**74-87 (May). *The structure and physiology of a remarkable organ.*

Williams, C. M. 1953. Insect breathing. **188:**28-32 (Feb.). *How the tracheal system of insects is constructed and functions.*

Wolf, A. V. 1958. Body water. **199:**125-132 (Nov.).

Wood, J. E. 1968. The venous system. **218:**86-96 (Jan.).

Zucker, M. B. 1961. Blood platelets. **204:**58-64 (Feb.). *These minute elements of the blood plasma play important roles in stopping hemorrhage.*

Zweifach, B. W. 1959. The microcirculation of the blood. **200:**54-60 (Jan.). *Describes the anatomy and function of the capillary bed that serves the body's tissues.*

CHAPTER 30

INTERNAL FLUIDS

Homeostasis and excretion

Scanning electron micrograph of longitudinal section of a kidney tubule of *Anoplogaster cornuta,* a marine teleost fish. Note the numerous cilia in the tubular lumen (canal) that drive the formative urine downstream (Original magnification ×2,200.)

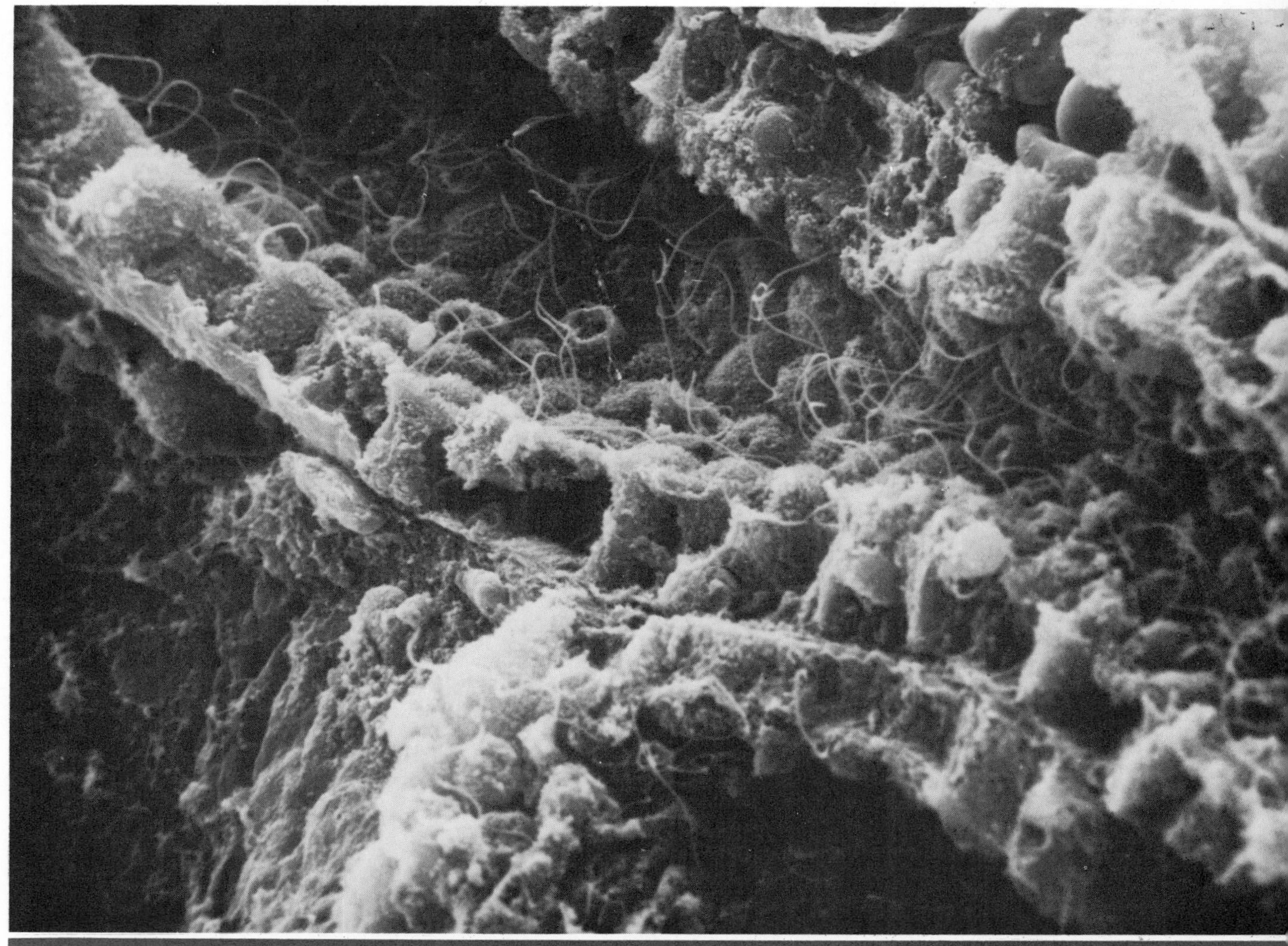

Courtesy G. H. Dobbs, III.

At the beginning of the preceding chapter we described the double-layered environment of the body's cells: the extracellular fluid that immediately surrounds the cells, and the external environment of the outside world. The life-supporting metabolic activities that occur within the body's cells can proceed only as long as they are bathed by a protective extracellular fluid of relatively constant composition. Yet there are many activities that threaten to throw the system out of balance. It is apparent that body fluid composition can be altered either by metabolic events occurring within the cells and tissues or by events occurring across the surface of the body. In other words, a living system is "open at both ends."

On the inside, metabolic activities within the cell require a steady supply of materials, and these activities turn out a continuous flow of products and wastes. On the outside, materials are constantly being exchanged between the plasma and the external environment. Water, which makes up about two-thirds of the body weight of animals, is always entering and leaving the body. Water is also formed within the cells as a by-product of oxidative processes. Ionized inorganic and organic salts are continually moving between the cells and the body fluids and also between the animal and its environment. Protein is constantly being formed, transported, and broken down again within the tissues, yielding nitrogenous wastes that must be excreted.

Obviously, body composition is a dynamic rather than a static thing. It is often described as operating as a **dynamic steady state.** This means that constancy of composition is maintained despite the continuous shifting of components within the system. This kind of internal regulation is called **homeostasis.**

Homeostasis is maintained by the coordinated activities of numerous body systems, such as the nervous system, endocrine system, and especially, the organs that serve as sites of exchange with the external environment. The last includes the kidneys, lungs or gills, alimentary canal, and skin. Through these organs oxygen, foodstuffs, minerals, and other body fluid constituents enter; here water is exchanged and metabolic wastes are eliminated.

The kidney is the chief regulator of the body fluids. It is popularly regarded strictly as an organ of excretion that serves to rid the body of assorted metabolic wastes. But in fact it is as much a regulatory organ as an excretory organ. It is responsible for individually monitoring and regulating the concentrations of body water, of the salt ions, acids, and bases in the blood, and of other major and minor body fluid constituents. In its task of fine-tuning the composition of the internal environment, the kidney is assisted by the other organs of exchange such as the lungs, skin, and digestive tract, as well as by many internal mechanisms.

Several other specialized structures have evolved among the vertebrates that assist in body fluid regulation in various environments; for example, the salt-secreting cells of fish gills and the salt glands of birds and reptiles.

OSMOTIC AND IONIC REGULATION
How aquatic animals meet problems of salt and water balance
Marine invertebrates

Most marine invertebrates are in osmotic equilibrium with their seawater environment. They have body surfaces that are permeable to salts and water so that their body fluid concentration rises or falls in conformity with changes in concentrations of seawater. Because such animals are incapable of regulating their body fluid osmotic pressure, they are referred to as **osmotic conformers.** Invertebrates living in the open sea are seldom exposed to osmotic fluctuations because the ocean is a highly stable environment. Oceanic invertebrates have, in fact, very limited abilities to withstand osmotic change. If they should be exposed to dilute seawater, they die quickly because their body cells cannot tolerate dilution and are helpless to prevent it. These animals are restricted to living in a narrow salinity range and are said to be **stenohaline** (Gr., *stenos,* narrow, + *hals,* salt). An example is the marine spider crab, represented in Fig. 30-1.

Conditions along the coasts and in estuaries and river mouths are much less constant than those of the open ocean. Here animals must be able to withstand large, and often abrupt, salinity changes as the tides move in and out and mix with fresh water draining from rivers. These animals are referred to as **euryhaline** (Gr., *eurys,* broad, + *hals,* salt), meaning that they can survive a wide range of salinity change. Most coastal invertebrates also show varying powers of **osmotic regulation.** For example, the brackish-water shore crab can resist body fluid dilution by dilute (brackish) seawater (Fig. 30-1). Although the body fluid concentration falls, it does so less rapidly than the fall in seawater concentration. This crab is a **hyper-**

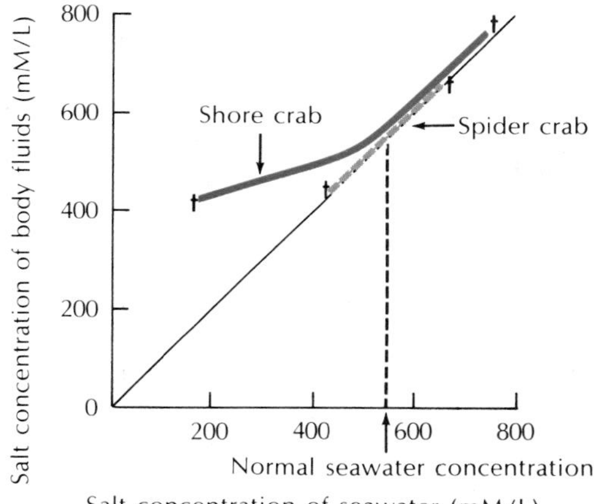

FIG. 30-1 Salt concentration of body fluids of two crabs as affected by variations in the seawater concentration. The 45-degree line represents equal concentration between body fluids and seawater. Since the spider crab cannot regulate its body-fluid salt concentration, it conforms to whatever changes happen in the external seawater environment. The shore crab, however, can regulate osmotic concentration of its body fluids to some degree because in dilute seawater the shore crab can hold its body-fluid concentration above the seawater concentration. For example, when the seawater is 200 mM per liter, the shore crab's body fluids are about 430 mM per liter. Crosses at ends of lines indicate tolerance limits of each species.

osmotic regulator because in a dilute environment it can maintain the concentration of its blood above that of the surrounding water.

What is the advantage of hyperosmotic regulation over osmotic conformity, and how is this regulation accomplished? The advantage is that by regulating against excessive dilution, thus protecting the body cells from extreme changes, these crabs can successfully live in the physically unstable but biologically rich coastal environment. Their powers of regulation are limited however, since if the water is highly diluted, their regulation fails and they die.

To understand how the brackish-water shore crab and other coastal invertebrates achieve hyperosmotic regulation, let us examine the problems they face.

First, the salt concentration of the internal fluids is greater than in the dilute seawater outside. This causes a steady osmotic influx of water. As with the membrane osmometer placed in a sugar solution (p. 77), water diffuses inward because it is more concentrated outside than inside. The shore crab is not nearly as permeable as a membrane osmometer—most of its shelled body surface is in fact almost impermeable to water—but the thin respiratory surfaces of the gills are highly permeable. Obviously the crab cannot insulate its gills with an impermeable hide and still breathe. The problem is solved by removing the excess water through the action of the kidney (the antennal gland located in the crab's thorax).

The second problem is salt loss. Again, because the animal is saltier than its environment, it cannot avoid loss of ions by outward diffusion across the gills. Salt is also lost in the urine. This problem is solved by special salt-absorbing cells in the gills that can actively remove ions from the dilute seawater and move them into the blood, thus maintaining the internal osmotic concentration. This is an **active transport** process that requires energy because ions must be transported against a concentration gradient, that is, from a lower salt concentration in the dilute seawater to an already higher one in the blood.

Invasion of fresh water

Some 400 million years ago, during the Silurian and Lower Devonian periods, the major groups of jawed fishes began to penetrate into brackish-water estuaries and then gradually into freshwater rivers. Before them lay a new unexploited habitat already stocked with food, in the form of insects and other invertebrates, which had preceded them into fresh water. However, the advantages of this new habitat were traded off for a tough physiologic challenge: the necessity of developing effective osmotic regulation.

Freshwater animals must keep the salt concentration of their body fluids higher than that of the water. Water therefore enters their bodies osmotically and salt is lost by diffusion outward. Their problems are similar to those of the brackish-water shore crab, but more severe and unremitting. Fresh water is much more dilute than are coastal estuaries, and there is no retreat, no salty sanctuary into which the freshwater animal can retire for osmotic relief. The animal must, and has, become a permanent and highly efficient hyperosmotic regulator.

The scaled and mucus-covered body surface of a fish is about as waterproof as any flexible surface can be. The water that inevitably enters across the gills is pumped out by the kidney. Even though the kidney is able to make a very dilute urine some salt is lost; this is replaced by salt in food and by active absorption of

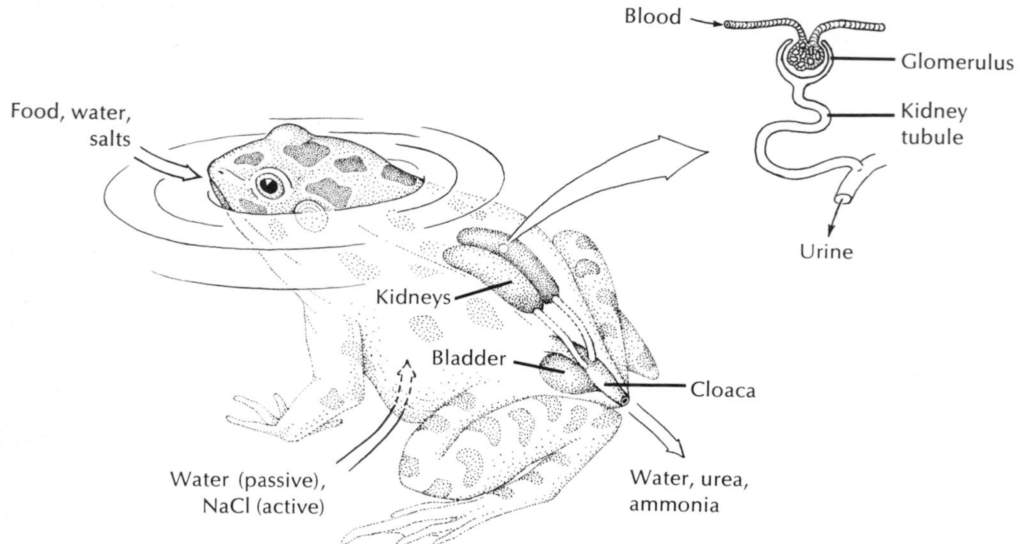

FIG. 30-2 Water and solute exchange in a frog. Water enters the highly permeable skin and is excreted by the kidney. The skin also actively transports ions (sodium chloride) from the environment. The kidney forms a dilute urine by reabsorbing most of the sodium chloride that is filtered. Urine flows into the urinary bladder, where, during temporary storage, nearly all of the remaining sodium chloride is removed and returned to the blood. (Adapted from Webster, D., and H. Webster. 1974. Comparative vertebrate morphology. New York, Academic Press, Inc.)

ions (primarily sodium and chloride) across the gills. The bony fishes that inhabit lakes and streams today are so well adapted to their dilute surroundings that they need expend very little energy to regulate themselves osmotically. Osmotic regulation in fishes is described on p. 515 and illustrated in Fig. 23-21.

Crayfish, aquatic insect larvae, mussels, and other freshwater animals are also hyperosmotic regulators and face the same hazards as freshwater fishes; they tend to gain too much water and lose too much salt. Not surprisingly, all of these forms solved these problems in the same direct way that fishes did. They excrete the excess water as urine, and they actively absorb salt from the water by some salt-transporting mechanism on the body surface.

Amphibians, when they are living in water, also must compensate for salt loss by absorbing salt from the water (Fig. 30-2). They use their skin for this purpose. Physiologists learned some years ago that pieces of frog skin will continue to actively transport sodium and chloride for hours when removed and placed in a specially balanced salt solution. Fortunately for biologists, but unfortunately for frogs, these animals are so easily collected and maintained in the laboratory that frog skin has become a favorite membrane system for studies of ion-transport phenomena.

Marine fishes

The great families of bony fishes that inhabit the seas maintain the salt concentration of their body fluids at about one-third that of seawater (body fluids = 0.3 to 0.4 molar; seawater = 1 molar). Obviously they are osmotic regulators. Bony fishes living in the oceans today are descendants of earlier freshwater bony fishes that moved back into the sea during the Triassic period about 200 million years ago. The return to their ancestral sea was probably prompted by unfavorable climatic conditions on land and the deterioration of freshwater habitats, but we can only guess at the reasons. During the many millions of years that the freshwater fishes were adapting themselves so well to their environment, they established a body fluid concentration equivalent to about one-third that of seawater, thus setting the pattern for all the vertebrates that were to evolve later, whether aquatic, terrestrial, or aerial. The ionic composition of vertebrate body fluid is remarkably like dilute seawater too, a fact that is doubtless related to their marine heritage.

When some of the freshwater bony fishes of the Triassic period ventured back to the sea, they encountered a new set of problems. Having a much lower internal osmotic concentration than the seawater around them, they lost water and gained salt. Indeed marine bony fishes quite literally risk drying out, much like a desert mammal deprived of water. The way marine bony fishes regulate osmotically is described on p. 516 and illustrated in Fig. 23-21. In brief, to compensate for water loss, the marine teleost drinks seawater. This is absorbed from the intestine, and the major sea salt, sodium chloride, is carried by the blood to the gills, where specialized salt-secreting cells transport it back into the surrounding sea. The ions remaining in the intestinal residue, especially magnesium, sulfate, and calcium, are voided with the feces or excreted by the kidney. In this roundabout way, marine fishes rid themselves of the excess sea salts they have drunk, resulting in a net gain of water, which replaces the water lost by osmosis. Samuel Taylor Coleridge's ancient mariner, surrounded by "water, water, everywhere, nor any drop to drink," would doubtless have been tormented even more had he known of the marine fishes' simple solution for thirst. A marine fish carefully regulates the amount of seawater it drinks, consuming only enough to replace water loss and no more.

The cartilaginous sharks and rays (elasmobranchs) solve their water balance problems in a completely different way. This primitive group is almost totally marine. The salt composition of shark's blood is similar to that of the bony fishes, but it also contains a large content of organic compounds, especially urea and trimethylamine oxide. Urea is, of course, a metabolic waste that most animals quickly excrete in the urine. The shark kidney, however, conserves urea, causing it to accumulate in the blood. The blood urea, added to the usual blood electrolytes, raises the blood osmotic pressure to slightly exceed that of seawater. In this way the sharks and their kin turn an otherwise useless waste material into an asset, eliminating the osmotic problem encountered by the marine bony fishes.

How terrestrial animals maintain salt and water balance

The problems of living in an aquatic environment seem small indeed compared to the problems of life on land. Remembering that our bodies are mostly water, that all metabolic activities proceed in water; and that the origins of life itself were conceived in water, it seems obvious that animals were meant to stay in water. Yet many animals, like the plants preceding them, moved onto land, carrying their watery composition with them. Once on land, the terrestrial animals continued their adaptive radiation, undaunted by the threat of dessication, until they became abundant even in some of the most arid parts of the earth. Terrestrial animals lose water by evaporation from the lungs and body surface, by excretion in the urine, and by elimination in the feces. Such losses are replaced by water in the food, by drinking water if it is available, and by forming **metabolic water** in the cells by the oxidation of foodstuffs, especially fats. (The oxidation of fuel hydrogens to form water is described on p. 87.) In some desert rodents, metabolic water may constitute most of the animals' water gain.

TABLE 30-1 Water balance in humans and kangaroo rat, a desert rodent*

	Human (%)	Kangaroo rat (%)
Gains		
Drinking	48	0
Free water in food	40	10
Metabolic water	12	90
Losses		
Urine	60	25
Evaporation (lungs and skin)	34	70
Feces	6	5

*Partly from Schmidt-Nielsen, K. 1972. How animals work, Cambridge, Cambridge University Press.

Particularly revealing is a comparison (Table 30-1) of water balance in humans, a nondesert mammal that drinks water, with that of the kangaroo rat, a desert rodent that may drink no water at all. The kangaroo rat gains all its water from its food (90% as metabolic water derived from the oxidation of foodstuffs and 10% as free moisture in the food). Even though humans eat foods with a much higher water content than the dry seeds that comprise much of the kangaroo rat's diet, we must drink half their total water requirement.

The excretion of wastes presents a special problem in water conservation. The primary end-product of protein catabolism is ammonia, a highly toxic material. Fishes can easily excrete ammonia across their gills, since there is an abundance of water to wash it away. The terrestrial insects, reptiles, and birds have no convenient way to rid themselves of toxic ammonia; instead, they convert it into uric acid, a nontoxic, almost insoluble compound. This enables them to excrete a semisolid urine with little water loss. The use of uric acid has another important benefit. The reptiles and birds lay shelled (amniotic) eggs enclosing the embryos, their stores of food and water, and whatever wastes accumulate during development. By converting ammonia to uric acid, the developing embryo's waste can be precipitated into solid crystals, which are stored harmlessly within the egg until hatching.

Marine birds and turtles have evolved a unique solution for excreting the large loads of salt eaten with their food. Located above each eye is a special **salt gland** capable of excreting a highly concentrated solution of sodium chloride—up to twice the concentration of seawater. In birds the salt solution runs out the nares (p. 585). Marine turtles and lizards shed their salt gland secretion as salty tears. Salt glands are important accessory organs of salt excretion to these animals, since their kidneys cannot produce a concentrated urine, as can the mammalian kidney.

Invertebrate excretory structures

In such a large and varied group as the invertebrates it is hardly surprising that there is a great variety of morphologic structures serving as excretory organs. Many protozoans and some freshwater sponges have special excretory organelles called contractile vacuoles. The more advanced invertebrates have excretory organs that are basically tubular structures that form urine by first producing an ultrafiltrate or fluid secretion of the blood. This enters the proximal end of the tubule and is modified continuously as it flows down the tubule. The final product is urine.

Contractile vacuoles

These tiny spherical intracellular vacuoles of protozoans and freshwater sponges are not true excretory organs, since ammonia and other nitrogenous wastes of metabolism readily leave the cell by direct diffusion across the cell membrane into the surrounding water. The contractile vacuole is really an organ of water balance. Because the cytoplasm of freshwater protozoans is considerably saltier than their freshwater environment, they tend to draw water into themselves by osmosis. In *Paramecium* this excess water is collected by minute canals within the cytoplasm and conveyed to the contractile vacuole (Fig. 8-24, p. 158). This grows larger as water accumulates within it. Finally the vacuole is emptied through a pore on the surface, and the cycle is rhythmically repeated. Although the contractile vacuole has been carefully studied, it is not yet known how this system is able to pump out pure water while retaining valuable salts within the animal. Contractile vacuoles are common in freshwater protozoans but rare or absent from marine protozoans, which are isosmotic with seawater and consequently neither lose nor gain too much water.

Nephridia

The nephridium is the most common type of invertebrate excretory organ. All nephridia are tubular structures, but there are large differences in degree of complexity. One of the simplest arrangements is the flame-cell system (or **protonephridia**) of many acoelomate and pseudocoelomate animals.

In *Planaria* and other flatworms the protonephridium takes the form of two highly branched systems of tubules distributed throughout the body (Fig. 11-4, p. 222). Fluid enters the system through specialized "flame" cells, moves slowly into and down the tubules, and is excreted through pores that open at intervals on the body surface. It is believed that fluid containing wastes enters the flame bulb from the surrounding tissues by **pinocytosis** (cell drinking). In this process fluid-filled vesicles are formed just under the outer cell surface. The vesicles are transported across the cell and set free into the ciliated lumen of the flame cell. Here, the rhythmic beat of the ciliary tuft (the "flame") creates a negative fluid pressure that drives the fluid into the tubular portion of the system. It is probable that as the fluid passes down the tubules, the tubular epithelial cells add certain waste materials to the tubular fluid (secretion) and withdraw valuable materials from it (reabsorption) to complete the formation of urine.

The flame-cell system, like the contractile vacuole of protozoans, is primarily a water balance system, since it is developed best in free-living freshwater forms. Branched flame-cell systems are typical of acoelomate and pseudocoelomate invertebrates that lack circulatory systems. Since there is no circulation to carry wastes to a compact excretory organ such as the kidney of higher invertebrates and vertebrates, the flame-cell system must be distributed to reach the cells directly.

The protonephridium just described is a **"closed"** system, that is, the urine is formed from a fluid that must first enter the tubule by being transported across the flame cells. Another type of nephridium, typical of coelomate invertebrate phyla, such as the annelids, molluscs, sipunculids, brachiopods, and phoronids, is the **open,** or "true," nephridium. In the earthworm *Lumbricus* there are paired nephridia in every segment of the body, except the first three and the last one (Fig. 14-6, p. 309). Each nephridium occupies parts of two successive segments. In the earthworm each nephridium is a tiny, self-contained "kidney" that independently drains to the outside through pores **(nephridiopores)** in the body wall.

Coelomic fluid containing wastes to be excreted is swept into a ciliated, funnel-like opening **(nephrostome)** of the nephridium and carried through a long, twisted tubule of increasing diameter. It then enters a bladder and is finally expelled through a nephridiopore to the outside. The nephridial tubule is surrounded by an extensive network of blood vessels. Solutes, especially sodium and chloride, are reabsorbed from the formative urine during its passage through the tubule. The addition of the blood vascular network to the annelid nephridium makes it a much more versatile and effective system than the flame cell system. However, the basic process of urine formation is the same: fluid flows continuously through a tubule while materials are added here and taken away there, until urine is formed.

Arthropod kidneys

The **antennal glands** of crustaceans form a single, paired tubular structure located in the ventral part of the head. The structure and function of antennal glands are described on p. 356. These excretory devices are an advanced design of the basic nephridial organ. However, they lack open nephrostomes. Instead, a protein-free filtrate of the blood (ultrafiltrate) is formed in the end sac by the hydrostatic pressure of the blood. In the tubular portion of the gland, the filtrate is modified by the selective reabsorption of certain salts and the active secretion of others. Thus, crustaceans have excretory organs that are basically vertebrate-like in the functional sequence of urine formation.

Insects and spiders have a unique excretory system consisting of **malpighian tubules** that operate in conjunction with specialized glands in the wall of the rectum (Fig. 17-16, p. 386). The thin, elastic, blind malpighian tubules are closed and lack an arterial supply. Consequently urine formation cannot be initiated by blood ultrafiltration as in the crustaceans and vertebrates. Instead salt ions, largely potassium, are actively secreted into the tubules. This primary secretion of ions creates an osmotic drag that pulls water, solutes, and waste materials into the tubule. The fluid, or "urine," then drains from the tubules into the intestine, where specialized rectal glands actively reabsorb most of the potassium and water, leaving behind wastes such as uric acid. This unique excretory system is ideally suited for life in dry environments. We must assume that it has contributed to the great success of this most abundant and widespread group of land animals.

VERTEBRATE KIDNEY

The ancestral vertebrate kidney is believed to have extended the length of the coelomic cavity and to have been made up of segmentally arranged uriniferous tubules. Each tubule opened at one end into the coelom by a nephrostome and at the other end into a common archinephric duct. Such a kidney has been called an **archinephros,** or **holonephros,** and is found in the embryos of hagfishes (Fig. 30-3).

Kidneys of higher vertebrates developed from this primitive plan. Embryologic evidence indicates that there are three generations of kidneys: **pronephros, mesonephros,** and **metanephros.** In all vertebrate embryos, the pronephros is the first and most primitive kidney to appear, as its name implies. It becomes the persistent kidney of adult hagfishes. In all other vertebrates it degenerates during development and is replaced by a more centrally located and more structurally advanced kidney, the mesonephros. The mesonephros becomes the persistent kidney of adult fishes and amphibians. But in the developing embryos of amniotes (reptiles, birds, and mammals) the mesonephros is replaced in turn by the metanephros. The metanephros develops behind the mesonephros and is structurally and functionally the most advanced of the three kidney types. Thus three kidneys are formed in succession, each more advanced and each located more caudally than its predecessor.

Vertebrate kidney function

The kidneys of humans and other vertebrates play a critical role in the body's economy. As vital organs their failure means death; in this respect they are neither more not less important than are the heart, lungs, or liver. The kidney is part of many interlocking mechanisms that maintain **homeostasis**—con-

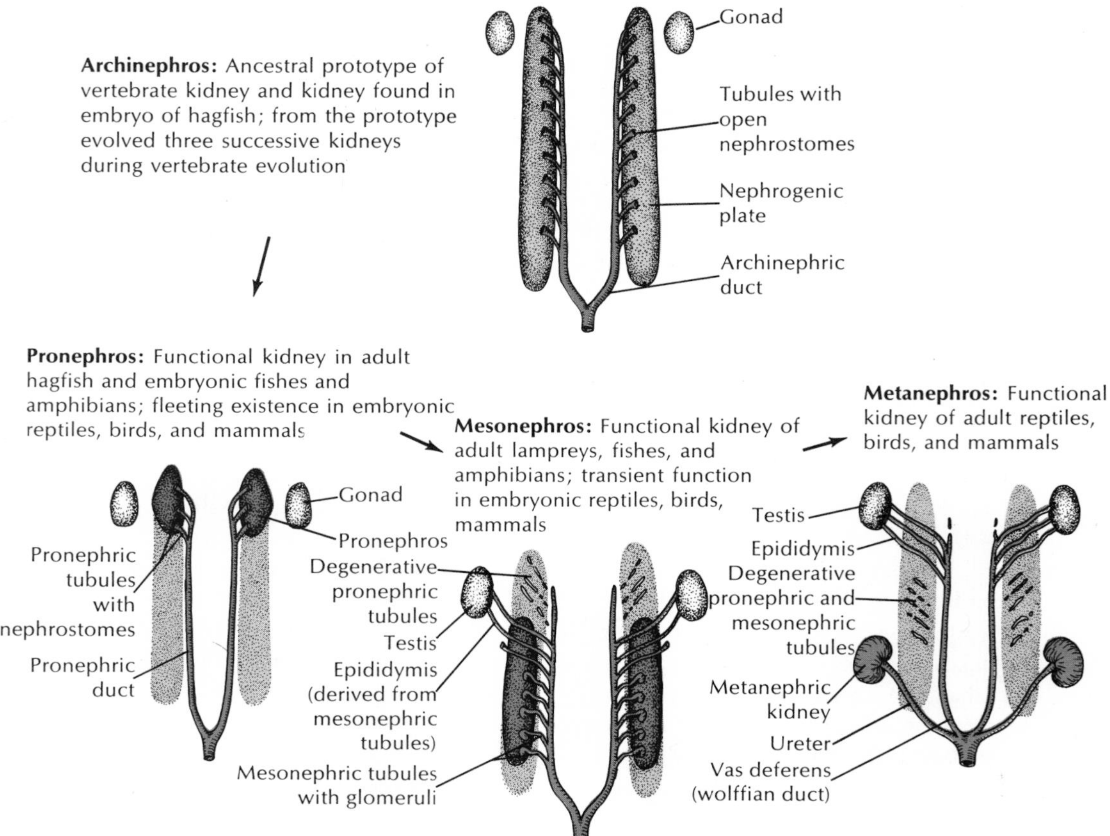

FIG. 30-3 Evolution of male vertebrate kidney from archinephric prototype. *Red,* Functional structures. *Light red,* Degenerative or undeveloped parts.

stancy of the internal environment. However, the kidney's share in this regulatory council is an especially large one. It must, and does, individually monitor and regulate most of the major constituents of the blood and several minor constituents as well. In addition it silently labors to remove a variety of potentially harmful substances that animals deliberately or unconsciously eat, drink, or inhale.

Perhaps even more remarkable than the job the kidney does is the way in which it does it. These small organs, which in humans weigh less than 0.5% of the body's weight, receive nearly 25% of the total cardiac output, amounting to about 2,000 liters of blood per day. This vast blood flow is channeled to approximately 2 million nephrons, which comprise the bulk of the two human kidneys. Each nephron is a tiny excretory unit consisting of a pressure filter **(glomerulus)** and a long **nephric tubule.** Urine formation begins in the glomerulus where an ultrafiltrate of the blood is squeezed into the nephric tubule by the hydrostatic blood pressure. The ultrafiltrate then flows steadily down the twisted tubule. During its travel some substances are added to, and others are subtracted from, the ultrafiltrate. The final product is urine.

All mammalian kidneys are paired structures that lie embedded in fat, anchored against the dorsal abdominal wall. Each of two ureters, 25 to 30 cm (10 to 12 inches) long in humans, extends to the dorsal surface of the **urinary bladder.** Urine is discharged from the bladder by way of the single **urethra** (Fig. 30-4). In the male the urethra is the terminal portion of the reproductive system as well as of the excretory system. In the female the urethra is solely excretory in function, opening to the outside just anterior to the vagina.

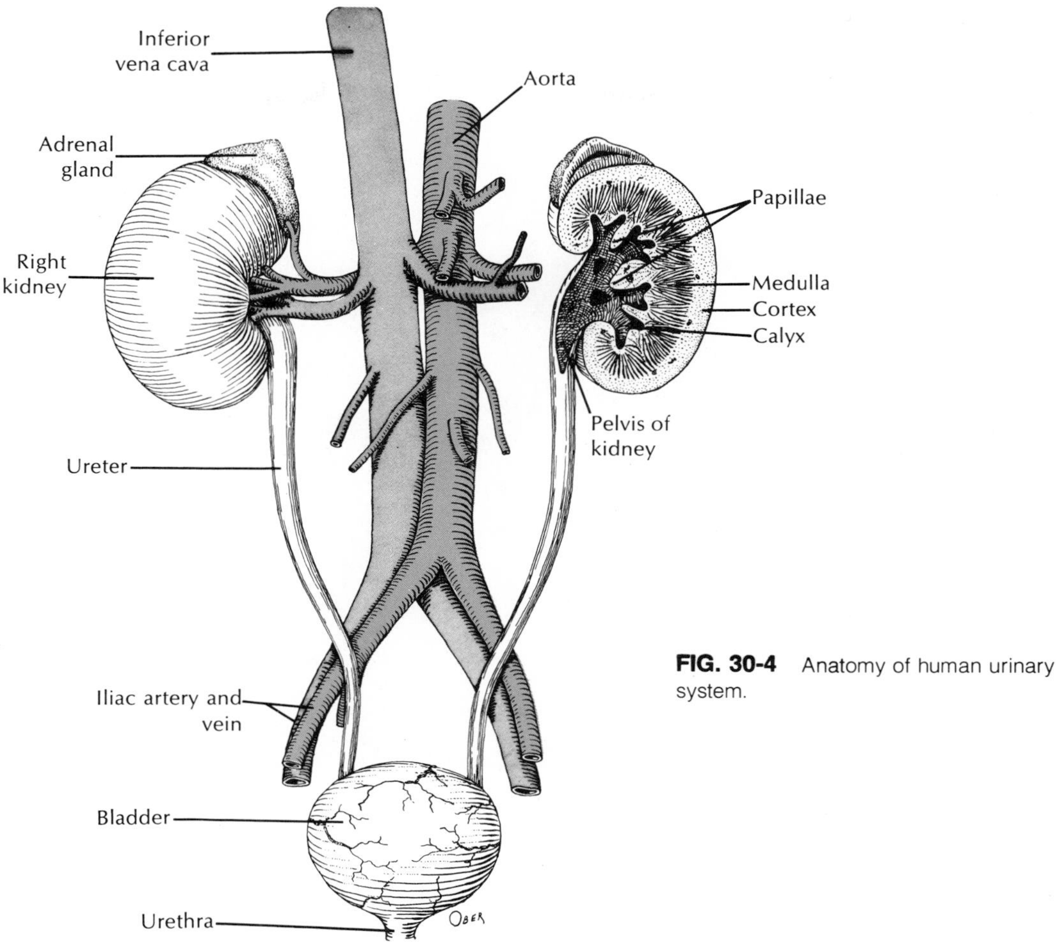

FIG. 30-4 Anatomy of human urinary system.

Since each of the thousands of nephrons in the kidney forms urine independently, each is, in a way, a tiny, self-contained kidney that produces a miniscule amount of urine—perhaps only a few nanoliters per hour. This amount, multiplied by the number of nephrons in the kidney, produces the total urine flow. The kidney is an "in parallel" system of independent units. However, as we will see later, these "independent" nephrons actually work together to create large osmotic gradients in the kidney medulla. This makes it possible for the mammalian kidney to concentrate urine well above the salt concentration of the blood.

As indicated before, the nephron, with its pressure filter and tubule, is intimately associated with the blood circulation (Fig. 30-5). Blood from the aorta is delivered to the kidney by way of the large **renal artery,** which breaks up into a branching system of smaller arteries (Fig. 30-6). The arterial blood flows to each nephron through an **afferent arteriole** to the **glomerulus** (glo-mer'yoo-lus), which is a tuft of blood capillaries enclosed within a thin, cuplike **Bowman's capsule.** Blood leaves the glomerulus via the **efferent arteriole.** This vessel immediately breaks up again into an extensive system of capillaries, the **peritubular capillaries,** which completely surround the nephric tubules. Finally, the blood from these many capillaries is collected by veins that unite to form the **renal vein.** This vein returns the blood to the vena cava.

Glomerular filtration

Let us now return to the glomerulus, where the process of urine formation begins. The glomerulus acts as a specialized mechanical filter in which a protein-free filtrate resembling plasma is driven by the blood pres-

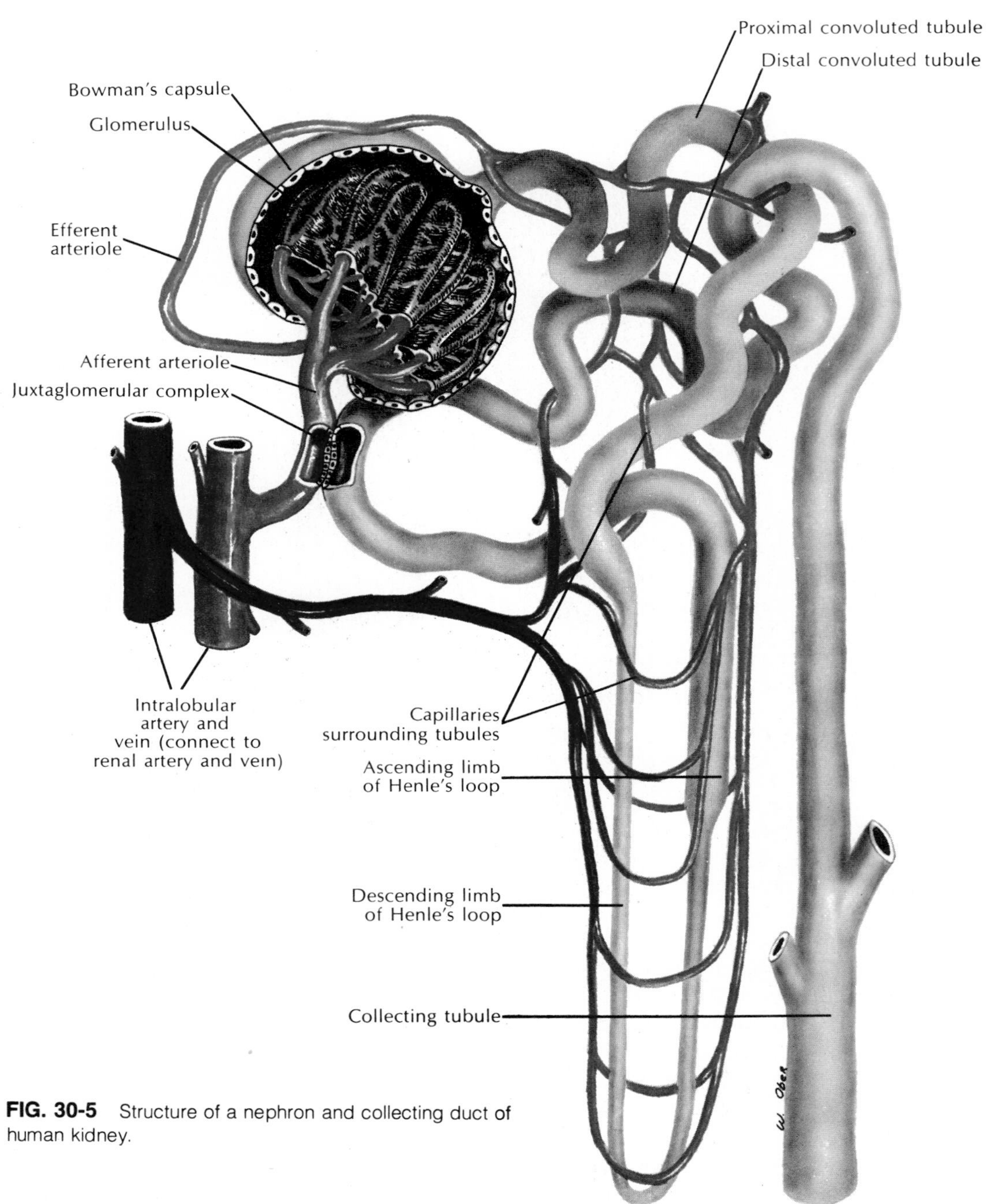

FIG. 30-5 Structure of a nephron and collecting duct of human kidney.

FIG. 30-6 Renal blood flow patterns in the injected kidney of a dog. In this technique, silicone rubber is infused into the kidney, which is then removed from the anesthetized dog and cleared so that only the circulation remains visible. Note the branching system of arteries and arterioles serving the numerous pealike glomeruli scattered through the kidney. The smallest vessels are the peritubular capillaries that surround the nephric tubules (the tubules themselves are not visible). (×25.) (Courtesy M. D. MacFarlane.)

sure across the capillary walls and into the fluid-filled space of Bowman's capsule. As shown in Fig. 30-7, the net filtration pressure is the difference between the blood pressure in the glomerular capillaries, believed to be about 45 mm Hg, and the opposing colloid osmotic and hydrostatic backpressures. Most important of these negative pressures is the colloid (protein) osmotic pressure, which is created because the proteins are too large to pass the glomerular membrane. The unequal distribution of protein causes the water concentration of ultrafiltrate in Bowman's capsule. The osmotic gradient created, about 25 mm Hg, opposes filtration. Though small, the net filtration pressure of 10 mm Hg is sufficient to force the ultra-

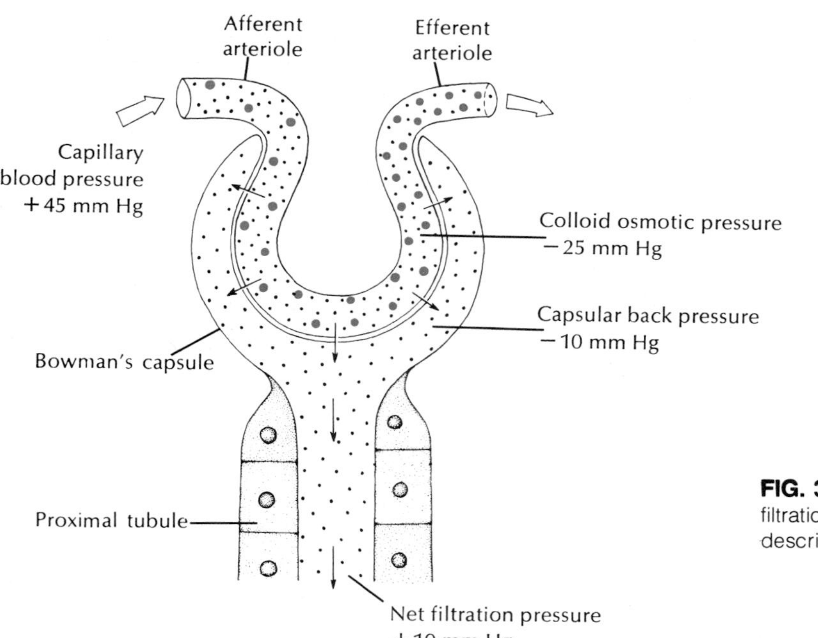

FIG. 30-7 Pressures determining the net filtration pressure in a glomerulus. See p. 703 for description.

filtrate that is formed down the nephric tubule.

The nephric tubule consists of several segments. The first segment, the **proximal convoluted tubule,** leads into a long, thin-walled, hairpin loop called the **loop of Henle** (Fig. 30-5). This loop drops deep into the medulla of the kidney and then returns to the cortex to join the third segment, the **distal convoluted tubule.** The collecting duct empties into the kidney **pelvis,** a cavity that collects the urine before it passes into the **ureter,** on its way to the **urinary bladder** (Fig. 30-4).

Tubular modification of the formative urine

The ultrafiltrate that enters this complex tubular system must undergo extensive modification before it becomes urine. About 200 liters of filtrate are formed each day by the average person's kidneys. Obviously the loss of this volume of body water, not to mention the many other valuable materials present in the filtrate, cannot be tolerated. How does tubular action convert the plasma filtrate into urine?

Two general processes are involved, **tubular reabsorption** and **tubular secretion.** Since the nephric tubules are at all points in close contact with the peritubular capillaries, materials can be transferred from the tubular lumen to the capillary blood plasma (tubular reabsorption) or from the blood plasma to the tubular lumen (tubular secretion).

Tubular reabsorption. The plasma contains a great variety of ions and molecules. With the exception of the plasma proteins, most of which are too large to pass the glomerular filter, all the plasma components are filtered, and most are reabsorbed. Some vital materials, such as glucose and amino acids, are completely reabsorbed. Others, such as sodium, chloride, and most other minerals, are strongly or weakly reabsorbed, depending on the body's need to conserve each mineral. Much of this reabsorption is by **active transport,** in which cellular energy is used to transport materials from the tubular fluid, across the cell, and into the peritubular blood that will return them to the general circulation.

For most substances there is an upper limit to the amount of substance that can be reabsorbed. This upper limit is termed the **transport maximum** for that substance. For example, glucose is normally completely reabsorbed by the kidney because the transport maximum for the glucose reabsorptive mechanism is poised well above the amount of glucose normally present in the plasma filtrate (about 100 mg per 100 ml of filtrate). If the plasma glucose level rises above normal, a condition called hyperglycemia, the concentration in the filtrate will rise accordingly and a greater amount of glucose will be presented to the proximal tubule for reabsorption. As the level rises, eventually a point is reached (about 300 mg per 100 ml of plasma) where the reabsorptive capacity of the tubular cells is saturated. If the plasma level continues to rise, glucose will begin to appear in the urine (glycosuria). This condition happens in the untreated disease diabetes mellitus.

Unlike glucose, which normally never appears in the urine, most of the mineral ions are excreted in the urine in variable amounts. Their excretion is regulated. The reabsorption of sodium, the dominant cation in the plasma, illustrates the flexibility of the reabsorption process. About 600 g of sodium are filtered by the human kidney every 24 hours. Nearly all of this is reabsorbed, but the exact amount is precisely matched to sodium intake. With a normal sodium intake of 4 g per day, the kidney excretes 4 g and reabsorbs 596 g each day. A person on a low-salt diet of 0.3 g of sodium per day still maintains salt balance because only 0.3 g escapes reabsorption. On a very high salt intake, the kidney can excrete up to about 10 g of sodium per day, but not more. It may seem odd that the maximum sodium excretion is only 10 g per day when about 600 g are filtered. One may logically ask: "Why can't more sodium be excreted by simply allowing more to escape tubular reabsorption?" The answer is that the filtration-reabsorption sequence has a built-in restriction in flexibility. Sodium (and other ions) are reabsorbed in both the proximal and distal portions of the convoluted tubule. Some 85% of the salt and water is reabsorbed in the proximal tubule; this is an **obligatory reabsorption** because it is governed entirely by physical processes (the osmotic pressure of the solutes) and cannot be controlled physiologically. In the distal tubule, however, sodium reabsorption is controlled by **aldosterone,** a steroid hormone of the adrenal cortex. This is called **facultative reabsorption,** meaning the reabsorption can be adjusted physiologically according to need. We can say that proximal reabsorption is involuntary and distal reabsorption is voluntary, although of course we are not aware of the adjustments the kidney is performing on our behalf. The flexibility of distal reabsorption varies considerably in different animals: it is restricted in humans but is very broad in many rodents. These differences have appeared because of selective pressures during evolution that fitted rodents for dry environments where they must conserve water and at the same time excrete considerable sodium.

Tubular secretion. In addition to reabsorbing large amounts of materials from the plasma filtrate, the kidney tubules are able to secrete certain substances into the tubular fluid. This process, which is the reverse of tubular reabsorption, enables the kidney to build up the urine concentrations of materials to be excreted, such as hydrogen and potassium ions, drugs, and various foreign organic materials. The distal tubule is the site of most tubular secretion.

In the kidney of bony marine fishes, reptiles, and birds, tubular secretion is a much more highly developed process than it is in mammalian kidneys. Marine bony fishes actively secrete large amounts of magnesium and sulfate, which are by-products of their mode of osmotic regulation, described earlier (p. 516). Reptiles and birds excrete uric acid instead of urea as their major nitrogenous waste. This material is actively secreted by the tubular epithelium. Since uric acid is nearly insoluble, it forms crystals in the urine and requires little water for excretion. Thus the excretion of uric acid is an important adaptation for water conservation.

Water excretion. The total osmotic pressure of the blood is carefully regulated by the kidney. When fluid intake is high, the kidney excretes a dilute urine, saving salts and excreting water. When fluid intake is low, the kidney conserves water by forming a concentrated urine. A dehydrated person can concentrate urine to approximately four times the molar concentration of blood.

The capacity of the kidney of mammals and some birds to produce a concentrated urine involves the loop of Henle, the long hairpin loop between the proximal and distal tubules that extends into the renal medulla. Although the loop of Henle was believed to be the locus for urine concentration, the mechanism has only recently been satisfactorily explained. The loops of Henle constitute a **countercurrent multiplier system.** Flow is in opposite directions in the two limbs, hence the name "countercurrent."

The functional characteristics of this system are as follows. Sodium chloride is actively transported out of the ascending limb and into the surrounding tissue fluid (Fig. 30-8). The ascending limb is relatively imperme-

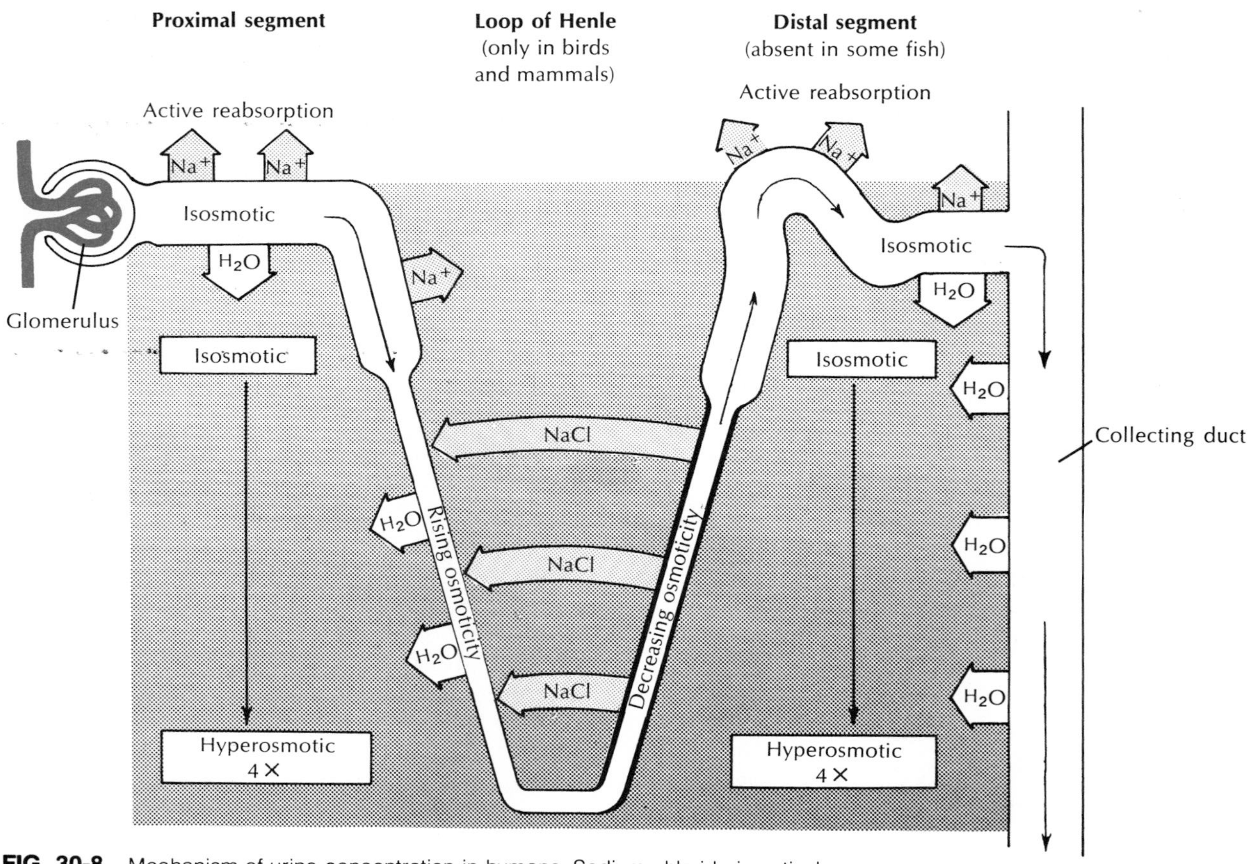

FIG. 30-8 Mechanism of urine concentration in humans. Sodium chloride is actively pumped from the ascending limb of loop of Henle and passively reenters the descending limb, building up the osmotic concentration to approximately four times that of blood. This creates an osmotic gradient for the controlled reabsorption of water from the collecting duct. Many rodent and desert mammals have relatively longer loops of Henle and can produce much more concentrated urine than humans.

able to both sodium chloride and water, so there is minimal passive reentry of sodium chloride or water into this limb. However, the descending limb, which does not transport sodium chloride, is permeable to sodium chloride and water. Consequently the sodium chloride that has been actively pumped out the ascending limb tends to passively enter the descending limb from the tissue fluid. By cycling sodium between the two opposing limbs, the concentration of urine becomes multiplied in the bottom of the loop. A tissue-fluid osmotic concentration is established that is greatest at the bottom of the loop deep in the medulla and lowest at the top of the loop in the cortex.

The actual concentrating of the urine, however, does not occur in the loops of Henle, but in the collecting ducts that lie parallel to the loops. As the urine flows down the collecting duct into regions of increasing sodium concentration, water is osmotically withdrawn from the urine. The amount of water saved and the final concentration of the urine depend on the permeability of the walls of the collecting duct. This is controlled by the **antidiuretic hormone** (ADH, also called vasopressin), which is released by the posterior pituitary gland (neurohypophysis). The release of this hormone is governed in turn by special receptors in the brain that constantly sense the osmotic pressure of the blood. When the blood osmotic pressure drops, as during dehydration, the secretion of ADH is increased. ADH increases the permeability of the collecting duct, probably by expanding pore size in the duct membrane. The result is that water diffuses out of the collecting duct into the surrounding interstitial fluid, and the urine becomes more concentrated. Given this sequence of events for dehydration, we can readily imagine the response of this system to overhydration. ADH secretion decreases, the collecting ducts become relatively impermeable, and urine flow is high.

The varying ability of different mammals to form a concentrated urine is closely correlated with the length of the loops of Henle. The beaver, which has no need to conserve water in its aquatic environment, has short loops and can concentrate its urine to only approximately twice the osmotic concentration of its blood plasma. Humans with relatively longer loops can concentrate urine to 4.2 times the osmotic concentration of the blood. As we would anticipate, desert mammals have much greater urine concentrating powers. The camel can produce a urine 8 times the plasma osmotic concentration, the gerbil 14 times, and the Australian hopping mouse 22 times. In the last creature, the greatest urine concentrator of all, the loops of Henle extend to the tip of a long renal papilla that pushes out into the mouth of the ureter.

Annotated references

Selected general readings

See also general references for Part Three, p. 793.

Brooks, S. M. 1973. Basic facts of body water and ions, ed. 3. New York, Springer Publishing Co., Inc. *Excellent elementary account.*

Deetjen, P., J. W. Boylan, and K. Kramer. 1975. Physiology of the kidney and of water balance. New York, Springer-Verlag Inc. *A primer designed for medical students; a good, balanced treatment of mammalian kidney function.*

Edney, E. B. 1957. The water relations of terrestrial arthropods. New York, Cambridge University Press.

Fraser, E. A. 1950. The development of the vertebrate excretory system. Biol. Rev. **25:**159-187.

Harvey, R. J. 1974. The kidneys and the internal environment. London, Chapman and Hall. *Stresses basic physical and chemical principles that underlie kidney function, especially dissociation and acid-base balance.*

Lockwood, A. P. M. 1963. Animal body fluids and their regulation. Cambridge, Mass., Harvard University Press. *This concise volume deals with the physiology of body fluid regulation in both invertebrates and vertebrates.*

Potts, W. T. W., and G. Parry. 1964. Osmotic and ionic regulation in animals. New York, The Macmillan Co. *An excellent treatise on concepts of homeostatic mechanisms, with special emphasis on osmoregulation and fluid balance*

Riegel, J. A. 1972. Comparative physiology of renal excretion. New York, Hafner Publishing Co., Inc. *Useful review of animal excretion, beginning with vertebrates and ending with the simplest invertebrate systems and a helpful closing chapter on theory of fluid movement.*

Smith, H. W. 1953. From fish to philosopher. Boston, Little, Brown & Co. *Vertebrate kidney evolution.*

Selected *Scientific American* articles

Smith, H. W. 1953. The kidney. **188:**40-48 (Jan.). *The structure, physiology, and evolution of the vertebrate kidney.*

Solomon, A. K. 1962. Pumps in the living cell. **207:**100-108 (Aug.). *Active transport processes in kidney tubules are described.*

CHAPTER 31
DIGESTION AND NUTRITION

Scanning electron micrograph of epithelium of frog stomach. Epithelial cells, resembling cobblestone paving, cover the stomach surface and line gastric pits in which they secrete hydrochloric acid and pepsinogen, which flows out onto the surface. The stomach epithelium is protected from self-digestion by mucous secretions and rapid cellular renewal (Original magnification ×500.)

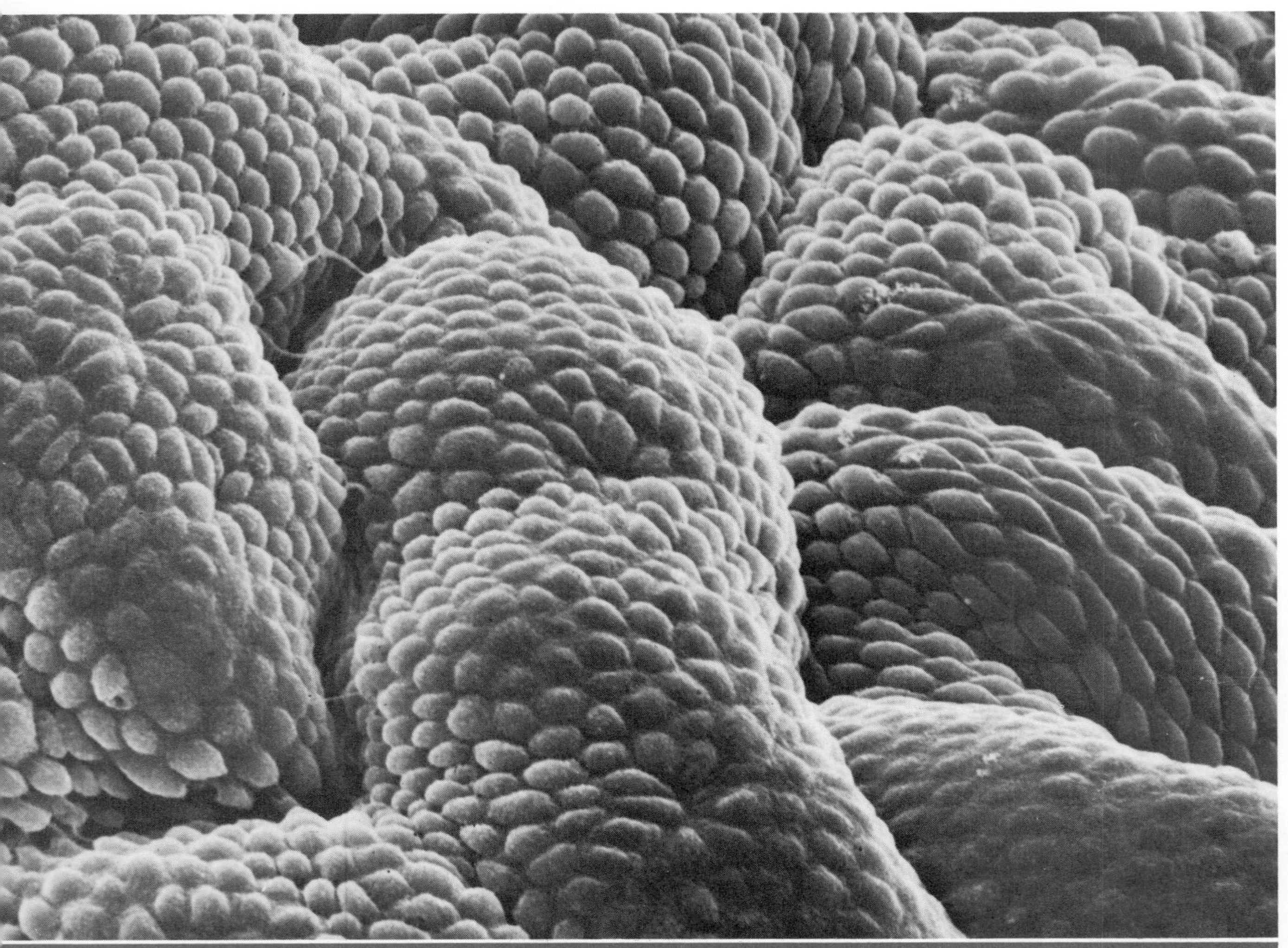

Courtesy P. P. C. Graziadei.

All organisms require energy to maintain their highly ordered and complex structure. This energy is chemical energy that is released by transforming complex compounds acquired from the organisms's environment into simpler ones. Obviously, if living organisms must depend on the breakdown of complex foodstuffs to build and maintain their own complexity, these foodstuffs must somehow be synthesized in the first place. Most of the energy for this synthesis is provided by the powerful radiations of the sun, the one great source of energy reaching our planet, which is otherwise a virtually isolated system.

Elsewhere in this text we discuss energy flow through communities (Chapter 41), nutritive requirements and energy release in living organisms (Chapter 5), and dietary habits of birds and mammals (Chapters 26 and 27). Before proceeding, it may be helpful to draw together some of these discussions in a brief summary.

Organisms capable of capturing the sun's energy by the process of photosynthesis are, of course, the green plants. Green plants are **autotrophic organisms** capable of synthesizing all the essential organic compounds needed for life. Autotrophic organisms need only inorganic compounds absorbed from their surroundings to provide the raw materials for synthesis and growth. Most autotrophic organisms are the chlorophyll-bearing **phototrophs,** although some, the chemosynthetic bacteria, are **chemotrophs,** gaining energy from inorganic chemical reactions.

Almost all animals are **heterotrophic organisms** that depend on already synthesized organic compounds for their nutritional needs. Animals, with their limited capacities to perform organic synthesis, must feed on plants and other animals to obtain the materials they will use for growth, maintenance, and the reproduction of their kind. The foods of animals, usually the complex tissues of other organisms, can seldom be utilized directly. Food is usually too bulky to be absorbed by the body cells and may contain material of no nutritional value. Consequently, food must be broken down, or digested, into soluble molecules sufficiently small to be utilized. One important difference, then, between autotrophs and heterotrophs is that the latter must have digestive systems.

Animals may be divided into a number of categories on the basis of dietary habits. **Herbivorous** animals feed mainly on plant life. **Carnivorous** animals feed mainly on herbivores and other carnivores. **Omnivorous** forms eat both plants and animals. A fourth category is sometimes distinguished, the **insectivorous** animals, which are those birds and mammals that subsist chiefly on insects.

The ingestion of foods and their simplification by digestion are only initial steps in nutrition. Foods reduced by digestion to soluble, molecular form are **absorbed** into the circulatory system and are **transported** to the tissues of the body. There they are **assimilated** into the protoplasm of the cells. Oxygen is also transported by the blood to the tissues, where food products are **oxidized,** or burned, to yield energy and heat. Much food is not immediately utilized but is **stored** for future use. Then the wastes produced by oxidation must be **excreted.** Food products unsuitable for digestion are rejected by the digestive system and are **egested** in the form of feces.

The sum total of all these nutritional processes is called **metabolism.** Metabolism includes both constructive and tearing-down processes. When molecular fragments are formed into larger molecules or when substances are built into new tissues or are stored for later use, such processes are called **synthetic,** or **anabolic, reactions.** The breaking down of complex materials into simpler ones, usually for the release of energy, is called a **degradative,** or **catabolic, reaction.** Both processes occur simultaneously in all living cells.

FEEDING MECHANISMS

Only a few animals can absorb nutrients directly from their external environment. Blood parasites and intestinal parasites may derive all their nourishment as primary organic molecules by surface absorption; some aquatic invertebrates may soak up part of their nutritional needs directly from simple organic molecules dissolved in the water. For most animals, however, working for their meals is the main business of living, and the specializations that have evolved for food procurement are almost as numerous as are the species of animals. In this brief discussion we will consider some of the major food-gathering devices.

Feeding on particulate matter

Drifting microscopic particles fill the upper hundred meters of the ocean. Most of this uncountable multitude is **plankton,** plant and animal microorganisms too small to do anything but drift with the ocean's cur-

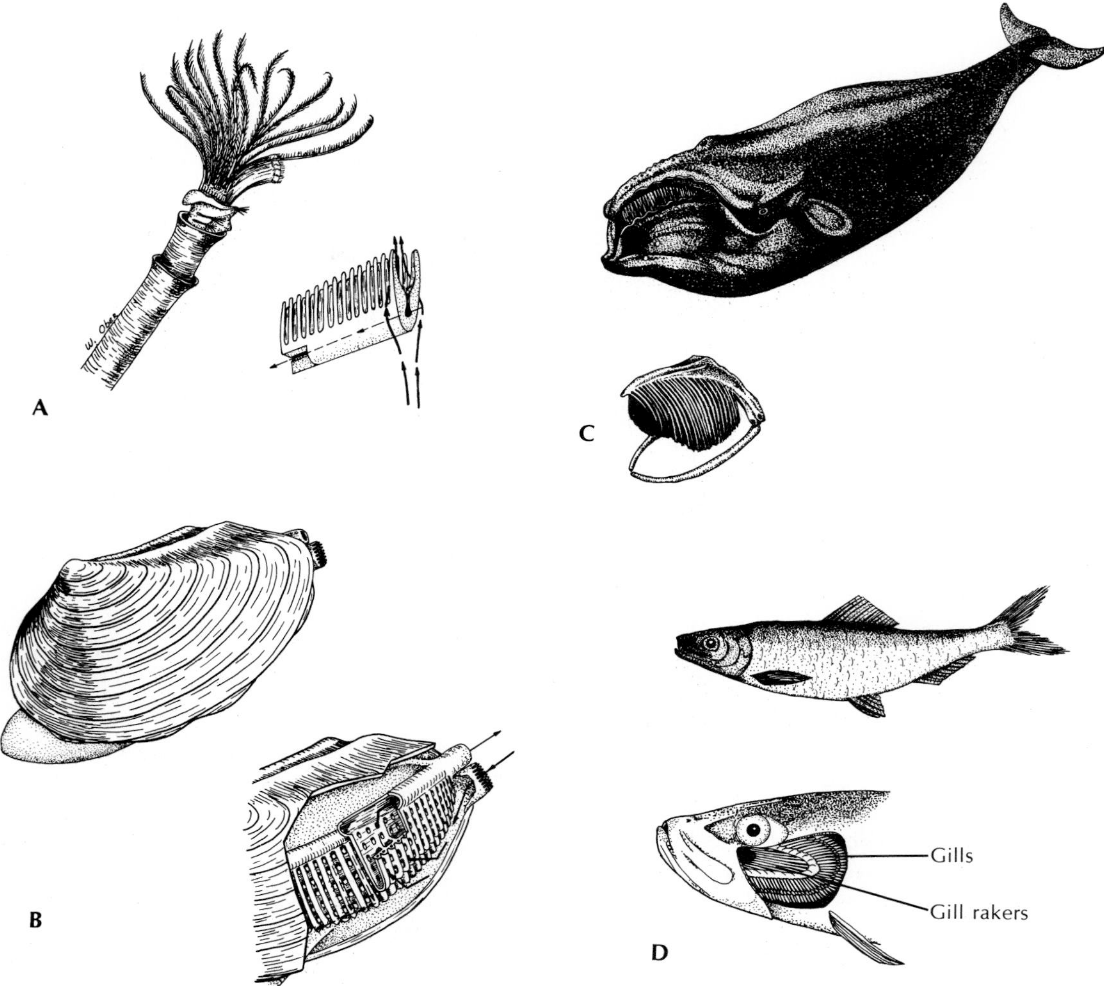

FIG. 31-1 Some filter feeders and their feeding mechanisms. **A,** Marine fan worm (class Polychaeta, phylum Annelida). Fan worms have a crown of tentacles. Numerous cilia on the edges of the tentacles draw water *(solid arrows)* between pinnules where food particles are entrapped in mucus; particles are then carried down a "gutter" in the center of the tentacle to the mouth *(broken arrows).* **B,** Marine clam (class Pelecypoda, phylum Mollusca). Bivalve molluscs use their gills as feeding devices as well as for respiration. Water currents created by cilia on the gills carry food particles into the inhalant siphon and between slits in the gills where they are entangled in a mucous sheet covering the gill surface. Particles are then transported by ciliated food grooves to the mouth (not shown). Arrows indicate direction of water movement. **C,** Right whale (class Mammalia, phylum Chordata). Whalebone whales filter out plankton, principally large copepods called "krill," with whalebone, or baleen. Water enters the swimming whale's open mouth by the force of the animal's forward motion and is strained out through the more than 300 horny baleen plates that hang down like a curtain from the roof of the mouth. Krill and other plankters caught in the baleen are periodically wiped off with the huge tongue and swallowed. **D,** Herring (class Osteichthyes, phylum Chordata). Herring and other filter-feeding fishes use gill rakers which project forward from the gill bars into the pharyngeal cavity to strain off plankters. Herring swim almost constantly, forcing water and suspended food into the mouth; food is strained out by the gill rakers, and the water passes out the gill openings.

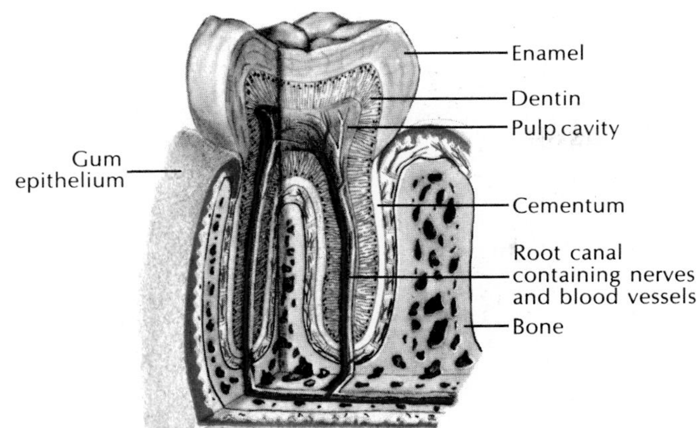

FIG. 31-2 Structure of human molar tooth. Tooth is built of three layers of calcified tissue-covering: enamel, which is 98% mineral and is the hardest material in the body; dentin, which composes the mass of the tooth, and is about 75% mineral; cementum, which forms a thin covering over the dentin in the root of the tooth, and is very similar to dense bone in composition. Pulp cavity contains loose connective tissue, blood vessels, nerves, and tooth-building cells. Roots of the tooth are anchored to the wall of the socket by a fibrous connective tissue layer called the "periodontal membrane." (After Netter, F. H. 1959. The Ciba collection of medical illustrations, vol. 3. Summit, N.J., Ciba Pharmaceutical Products, Inc.)

rents. The rest is organic debris, the disintegrating remains of dead plants and animals. Altogether this oceanic swarm of plankton forms the richest life domain on earth. It is preyed on by numerous larger animals, invertebrates and vertebrates, using a variety of feeding mechanisms. Some protozoans, such as the ameboid sarcodines, ingest particulate food by phagocytosis. The animal, stimulated by the proximity of food, pushes out armlike extensions of the plasmalemma (cell membrane) and engulfs the particle into a food vacuole, in which it is digested. Other protozoans have specialized openings, called cytostomes, through which the food passes to be enclosed in a food vacuole.

By far the most important method to have evolved for feeding on small organisms is **filter feeding** (Fig. 31-1). It is a primitive but immensely successful and widely employed mechanism. The majority of filter feeders use ciliated surfaces to produce currents that draw drifting food particles into their mouths. Most filter-feeding invertebrates, such as the tube-dwelling worms and bivalve molluscs, entrap the particulate food in mucous sheets that convey the food into the digestive tract. Filter feeding is characteristic of a sessile way of life, the ciliary currents serving to bring the food to the immobile or slow-moving animal.

However, active feeders such as tiny copepod crustaceans and herring are also filter feeders, as are immense baleen whales. The vital importance of one component of the plankton, the diatoms, in supporting a great pyramid of filter-feeding animals is stressed by N. J. Berrill*: "A humpback whale . . . needs a ton of herring in its stomach to feel comfortably full—as many as five thousand individual fish. Each herring, in turn, may well have 6,000 or 7,000 small crustaceans in its own stomach, each of which contains as many as 130,000 diatoms. In other words, some 400 billion yellow-green diatoms sustain a single medium-sized whale for a few hours at most." Filter feeding utilizes the abundance and extravagance of life in the sea.

Filter feeders are as a rule nonselective and omnivorous. Sessile filter feeders take what they can get, having only the options of continuing or ceasing to filter. Active filter feeders, however, such as fishes and baleen (whalebone) whales, are much more selective in their feeding.

*Berrill, N. J. 1958. You and the universe, New York, Dodd, Mead & Co.

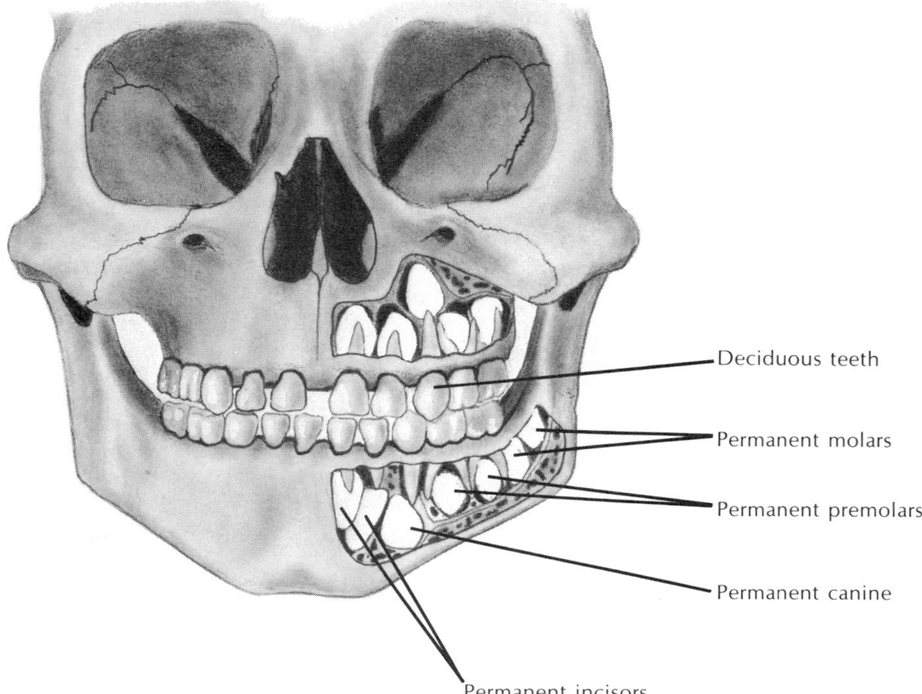

FIG. 31-3 Human deciduous and permanent teeth. Partly dissected skull of a 5-year-old child, showing milk (deciduous) teeth and permanent teeth. Milk teeth begin to erupt at 6 months and are gradually replaced by the permanent teeth beginning at about 6 years of age. There are 20 deciduous teeth, 5 on each side of each jaw, and 32 permanent teeth, 8 on each side of each jaw. These 8 are arranged as follows: 2 incisors, 1 canine (also called cuspid), 2 premolars (bicuspids), 3 molars. The last molar, known as the wisdom tooth, erupts between ages of 17 and 25 or not at all. Upper permanent molars are not seen in this frontal view. (Modified from Arey, L. B. 1965. Developmental anatomy. Philadelphia, W. B. Saunders Co.)

Feeding on food masses

Some of the most interesting animal adaptations are those that have evolved for procuring and manipulating solid food. These, and the animals bearing them, are partly shaped by what the animal eats.

Predators must be able to locate prey, capture it, hold it, and swallow it. Most animals use teeth for this purpose (Fig. 31-2). Although teeth are variable in size, shape, and arrangement, vertebrates as different as fishes and mammals sometimes have remarkably similar tooth arrangements for seizing the prey and cutting it into pieces small enough to swallow.

Mammals characteristically have four different types of teeth, each adapted for specific functions. **Incisors** are for biting and cutting; **canines** are designed for seizing, piercing, and tearing; **premolars** and **molars,** at the back of the jaws, are for grinding and crushing. This basic pattern is well illustrated in human dentition (Fig. 31-3) but is often greatly modified in animals having specialized food habits (Fig. 27-12, p. 618). Herbivores have suppressed canines but have well-developed molars with enamel ridges for grinding. Such teeth are usually high crowned, in contrast to the low-crowned teeth of carnivores. The well-developed, self-sharpening incisors of rodents grow throughout life and must be worn away by gnawing to keep pace with growth (Fig. 27-13, p. 619). Some teeth have become so highly modified that they are no longer useful for biting or chewing food. An elephant's tusk is a modified upper incisor used for defense, attack, and rooting, and the male wild boar has modified canines used as weapons.

Many carnivores among the fishes, amphibians, and reptiles swallow their prey whole. Snakes and some fishes can swallow enormous meals. This, together with the absence of limbs, is associated with some striking feeding adaptations in these groups—recurved teeth for seizing and holding the prey, and distensible jaws and stomachs to accommodate their large and infrequent meals.

Teeth are not vertebrate innovations; biting, scraping, and gnawing devices are common in the invertebrates. Insects, for example, have three pairs of appendages on their heads that serve variously as jaws, teeth, chisels, tongues, or sucking tubes. Usually the first pair is crushing teeth; the second, grasping jaws; and the third, a probing and tasting tongue.

Herbivorous, or plant-eating, animals, whether vertebrate or invertebrate, have evolved special devices for crushing and cutting plant material. Despite its abundance on earth, the woody cellulose that encloses plant cells is to many animals an indigestible and useless material. Cellulase, the enzyme that digests cellulose, is secreted by only a very few herbivorous invertebrates, such as shipworms, certain crustaceans, silverfish, and a few species of wood-boring beetles. All other herbivores, invertebrate or vertebrate, make use of symbiotic intestinal microorganisms to digest cellulose, once it is ground up. One highly specialized mammalian group, the ruminants, is described on p. 620. Certain invertebrates such as snails have rasplike, scraping mouthparts. Insects such as locusts have grinding and cutting mandibles; herbivorous mammals such as horses and cattle use wide, corrugated molars for grinding. All these mechanisms serve to disrupt the tough cellulose cell wall to accelerate its digestion by intestinal microorganisms, as well as to release the cell contents for direct enzymatic breakdown.

Feeding on fluids

Fluid feeding is especially characteristic of parasites but is certainly practiced among free-living forms as well. Most internal parasites (endoparasites) simply absorb the nutrient surrounding them, unwittingly provided by the host. External parasites (ectoparasites) such as leeches, lampreys, parasitic crustaceans, and insects use a variety of efficient piercing and sucking mouthparts to feed on blood or other body fluid. Unfortunately for humans and other warm-blooded animals, the ubiquitous mosquito excels in its blood-sucking habit. Alighting gently, the mosquito sets about puncturing its prey with an array of six needlelike feeding stylets concealed within a long scaly structure, the labium. One stylet is the gutterlike labrum through which the blood is sucked. Another, the hypopharynx, is used to inject an anticoagulant saliva (responsible for the irritating itch that follows the "bite," and serving as vector for microorganisms causing malaria, yellow fever, encephalitis, and other diseases). The remaining stylets (paired mandibles and maxillae) are used to cut and saw into the skin with rapid back-and-forth movements. The mosquito mouthparts are shown spread out in Fig. 17-13, *B* (p. 383). It is of little comfort that only the female of the species dines on blood (the males feed only on sugary fluids, such as flower nectar). Far less troublesome to people are the free-living butterflies, moths, and aphids that suck up plant fluids with long, tubelike mouthparts.

DIGESTION

In the process of digestion, which means literally "carrying asunder," organic foods are mechanically and chemically broken down into small units for absorption. Animal foods vary enormously in composition. Even though food solids consist principally of carbohydrates, proteins, and fats, the very components that make up the body of the consumer, these components must nevertheless be reduced to their simplest molecular units before they can be utilized. Each animal reassembles some of these digested and absorbed units into organic compounds of the animal's own unique pattern. Cannibals enjoy no special metabolic benefit from eating their own kind; they digest their victims just as thoroughly as they do food of another species!

The digestive tract is actually an extension of the outside environment into, or through, the animal. Since most animals eat all manner of organic and inorganic materials, the gut's lining must be something like the protective skin; yet it must be permeable so that foodstuffs can be absorbed. Once in the gut the foods are digested and absorbed as they are slowly moved through it.

Movement is either by **cilia** or by **musculature.** In general, the filter feeders that use cilia to feed, such as bivalve molluscs, also use cilia to propel the food through the gut. Animals feeding on bulky foods rely on well-developed gut musculature. As a rule the gut is lined with two opposing layers of muscle—an outer longitudinal layer, in which the smooth muscle fibers run parallel with the length of the gut, and an inner cir-

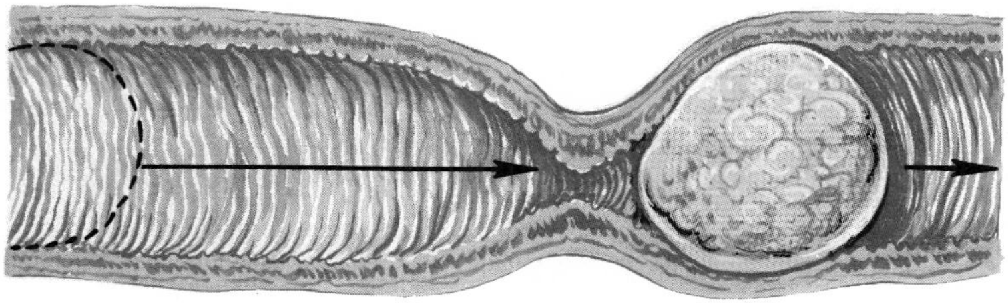

FIG. 31-4 Peristalsis. Food is pushed along before a wave of circular muscle contraction. (From Schottelius, B. A., and D. D. Schottelius. 1978. A textbook of physiology, ed. 18. St. Louis, The C. V. Mosby Co.)

cular layer, in which the muscle fibers embrace the circumference of the gut. This arrangement is ideal for mixing and propelling foods. The most characteristic gut movement is **peristalsis** (Fig. 31-4). In this movement a wave of circular muscle contraction sweeps down the gut for some distance, pushing the food along before it. The peristaltic waves may start at any point and move for variable distances. Also characteristic of the gut are **segmentation** movements that divide and mix the food.

Intracellular versus extracellular digestion

Humans and other vertebrates and the higher invertebrates digest their food **extracellularly** by secreting digestive juices into the intestinal lumen. There, foodstuffs are enzymatically split into molecular units small enough to be selectively absorbed by the intestinal epithelium, transported by the circulation, and utilized by all body cells. Digestion, then, occurs outside the body's tissues.

Intracellular digestion is a primitive process typical of the lower invertebrates. This type of digestion is best illustrated by the single-celled protozoans, which capture food particles by phagocytosis, enclose these particles within food vacuoles, and then digest them. Obviously the big limitation to intracellular digestion is that only small particles of food can be handled. Nevertheless, many multicellular invertebrates practice intracellular digestion. Intracellular digestion is typical of filter-feeding marine animals such as brachiopods, rotifers, bivalves, and cephalochordates, as well as the cnidarians and flatworms. In all of these forms the food particle, phagocytized by the cell, is enclosed within a membrane as a food vacuole (Fig. 31-5). Digestive enzymes are then added. The products of digestion, the simple sugars, amino acids, and other molecules, are absorbed into the cell cytoplasm where they may be utilized or may be transferred to other cells. The inevitable food wastes are extruded from the cell.

It is believed that cellular enzymes are packaged into membrane-bound vacuoles called **lysosomes** (Fig. 31-5). These somehow join with, and discharge their enzymes into, the food vacuoles. All cells seem to contain lysosomes, even those of higher animals that do not practice intracellular digestion. Lysosomes have been called ''suicide-bags'' because they rupture spontaneously in dying or useless cells, digesting the cell contents, which are phagocytized by scavenger cells. Lysosomes also play a role in the lives of healthy cells in cleaning up residues left by growth processes.

It is probable that the obvious limitations of intracellular digestion were responsible for shaping the evolution of extracellular digestion. **Extracellular** digestion offers several advantages: bulky foods may be ingested; the digestive tract can be smaller, more specialized, and more efficient; and food wastes are more easily discarded. Only with extracellular digestion could the enormous variation in feeding methods of the higher animals have evolved.

Vertebrate digestion

The vertebrate digestive plan is similar to that of the higher invertebrates. Both have a highly differentiated alimentary canal with devices for increasing the surface area, such as increased length, inside folds, and diverticula. The more primitive fishes (lampreys and sharks) have longitudinal or spiral folds in their intestines. Higher vertebrates have developed elaborate folds and small fingerlike projections **(villi).** Also, the electron microscope reveals that each cell lining the intestinal cavity is bordered by hundreds of short, delicate processes called **microvilli** (Fig. 31-6). These processes, together with larger villi and intestinal folds, may increase the internal surface of the intestine more than a million times compared to a smooth cylinder

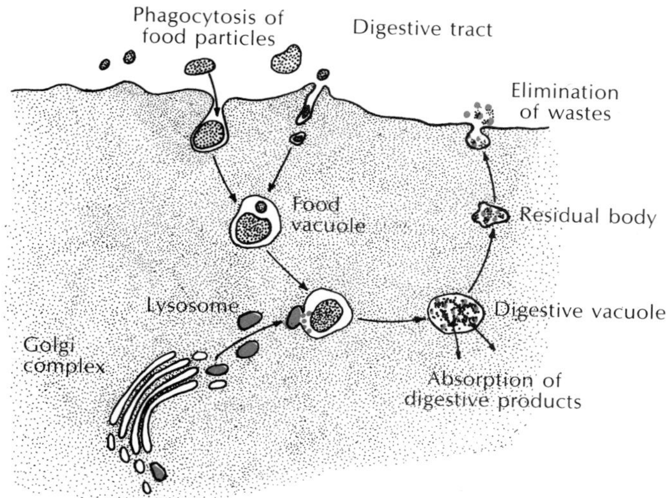

FIG. 31-5 Intracellular digestion. Lysosomes containing digestive enzymes (lysozymes) are produced within the cell, possibly by the Golgi complex. Lysosomes fuse with food vacuoles and release enzymes that digest the enclosed food. Usable products of digestion are absorbed into the cytoplasm and indigestible wastes are expelled to the outside.

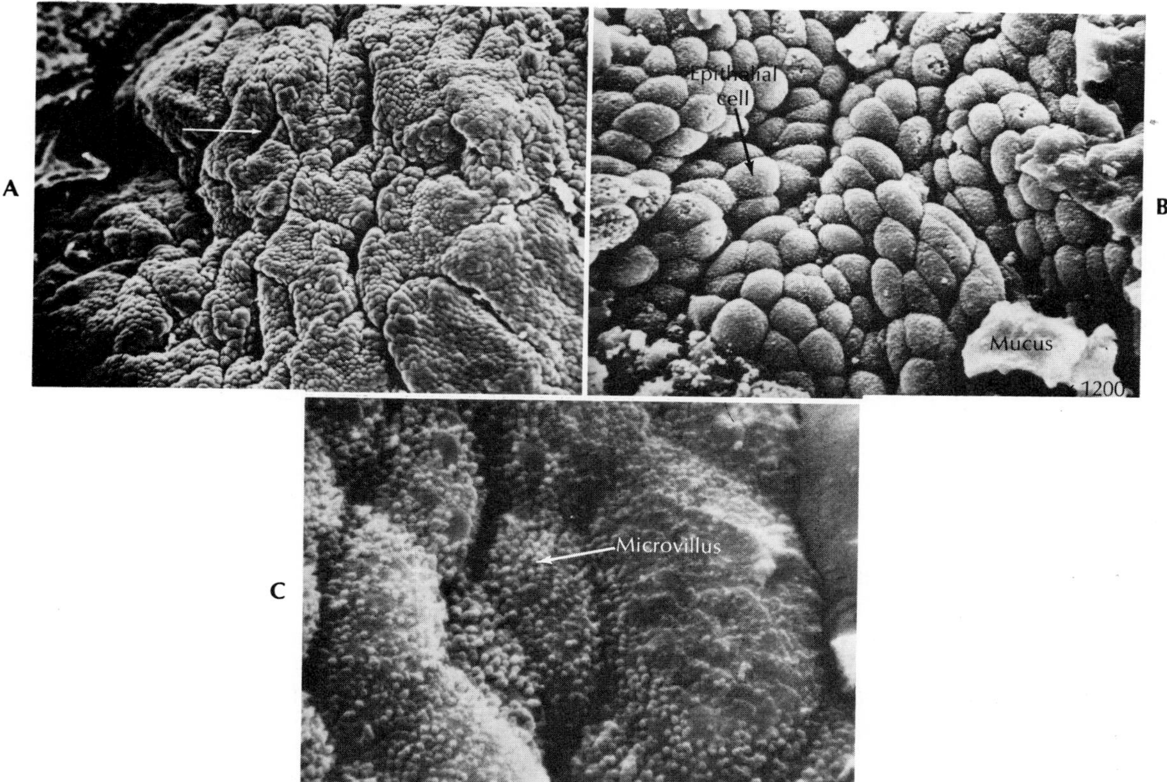

FIG. 31-6 Microscopic structure of human stomach. **A,** At low magnification (×230) the epithelial cells look like paving stones. Numerous openings to gastric glands are evident; one is indicated by white arrow. **B,** A higher magnification (×1,200) shows individual epithelial cells. Microvilli give cell surfaces a fuzzy appearance. Degenerating epithelial cells can be seen at left and at top. **C,** At much higher magnification (×6,500) the individual microvilli are evident on the rounded surfaces of epithelial cells. These scanning electron micrographs were made after the tissue was preserved in a fixative, vacuum dehydrated, and shadowed with a thin lyer of gold-palladium. (From Pfeiffer, C. J. 1970. J. Ultrastructural Res. **33:**252.)

of the same diameter. The absorption of food molecules is enormously facilitated as a result.

Digestion in the human, an omnivore

Anatomy of the human digestive system. The structural plan of the human digestive system is shown in Fig. 31-7. The **mouth** is provided with teeth and a tongue for grasping, masticating, manipulating, and swallowing the food. Three pairs of salivary glands lubricate the food and, in human beings at least, perform limited carbohydrate digestion. In mammals two sets of teeth are formed during life—the temporary, or "milk," teeth (also called deciduous teeth) and the permanent teeth (Fig. 31-3).

The **pharynx** is the throat cavity that serves for the passage of food. It is actually a complex reception chamber, receiving openings from (1) the nasal cavity, (2) the mouth, (3) the middle ear by way of two eustachian tubes, (4) the esophagus, and (5) the trachea via the glottis.

The **esophagus** is a muscular tube connecting the pharynx and stomach. It opens into the stomach at the cardiac sphincter.

The **stomach** is an enlargement of the gut between the esophagus and intestine. The human stomach is divided into the **cardiac** region (adjacent to esophagus), **fundus** (central region), and **pyloric antrum** (region adjacent to intestine). The stomach is principally a storage organ. However, because the stomach secretes a very strong acid (hydrochloric acid) and several enzymes, it also functions to sterilize the food by destroying bacteria, as well as to initiate digestion.

The **small intestine** is the principal digestive and absorptive area of the gut. It is divided grossly into three regions—**duodenum, jejunum,** and **ileum.** Two large digestive glands, the **liver** and **pancreas,** empty into the duodenum by the **common bile duct.**

The **large intestine (colon)** is divided into **ascending, transverse,** and **descending** portions, with the posterior end terminating in the **rectum** and **anus.** At the junction of the large and small intestines is the **colic cecum** and its vestigial **vermiform appendix.** The large intestine lacks villi but contains glands for lubrication.

The stomach, small intestine, and large intestine are all suspended by **mesenteries,** thin sheets of tissue that are modified from the **peritoneum,** or lining, of the coelom and the abdominal organs. Organs such as the liver, spleen, and pancreas are also held in place by mesenteries, which carry blood and lymph vessels as well as nerves to the various abdominal organs.

Action of digestive enzymes. We have already pointed out that digestion involves both mechanical and chemical alterations of food. Mechanical processes of cutting and grinding by teeth and muscular mixing by the intestinal muscles are important in digestion. However, the reduction of foods to small absorbable units relies principally on chemical breakdown by **enzymes,** the highly specific organic catalysts discussed earlier (pp. 79 to 83).

It is well to state that although digestive enzymes are probably the best known and most studied of all enzymes, they represent but a small fraction of the numerous, perhaps thousands, of enzymes that ultimately regulate all processes in the body. The digestive enzymes are **hydrolytic** enzymes **(hydrolases),** so called because food molecules are split by the process of **hydrolysis,** that is, the breaking of a chemical bond by adding the components of water across it:

$$R-R + H_2O \xrightarrow[\text{enzyme}]{\text{Digestive}} R-OH + H-R$$

In this general enzymatic reaction, R—R represents a food molecule that is split into two products, R—OH and R—H. Usually these reaction products must in turn be split repeatedly before the original molecule has been reduced to its numerous subunits. Proteins, for example, are composed of hundreds, or even thousands, of interlinked amino acids, which must be completely separated before the individual amino acids can be absorbed. Similarly, carbohydrates must be reduced to simple sugars. Fats (lipids) are reduced to molecules of glycerol and fatty acids, although some fats, unlike proteins and carbohydrates, may be absorbed without first being completely hydrolyzed. There are specific enzymes for each class of organic compounds. These enzymes are located in various regions of the alimentary canal in a sort of "enzyme chain," in which one enzyme may complete what another has started, the product moving along posteriorly for still further hydrolysis.

Digestion in the mouth. In the mouth, food is broken down mechanically by the teeth and is moistened with saliva from the salivary glands. In addition to **mucin,** which helps to lubricate the food for swallowing, saliva contains the enzyme **amylase.** Salivary amylase is a carbohydrate-splitting enzyme that begins

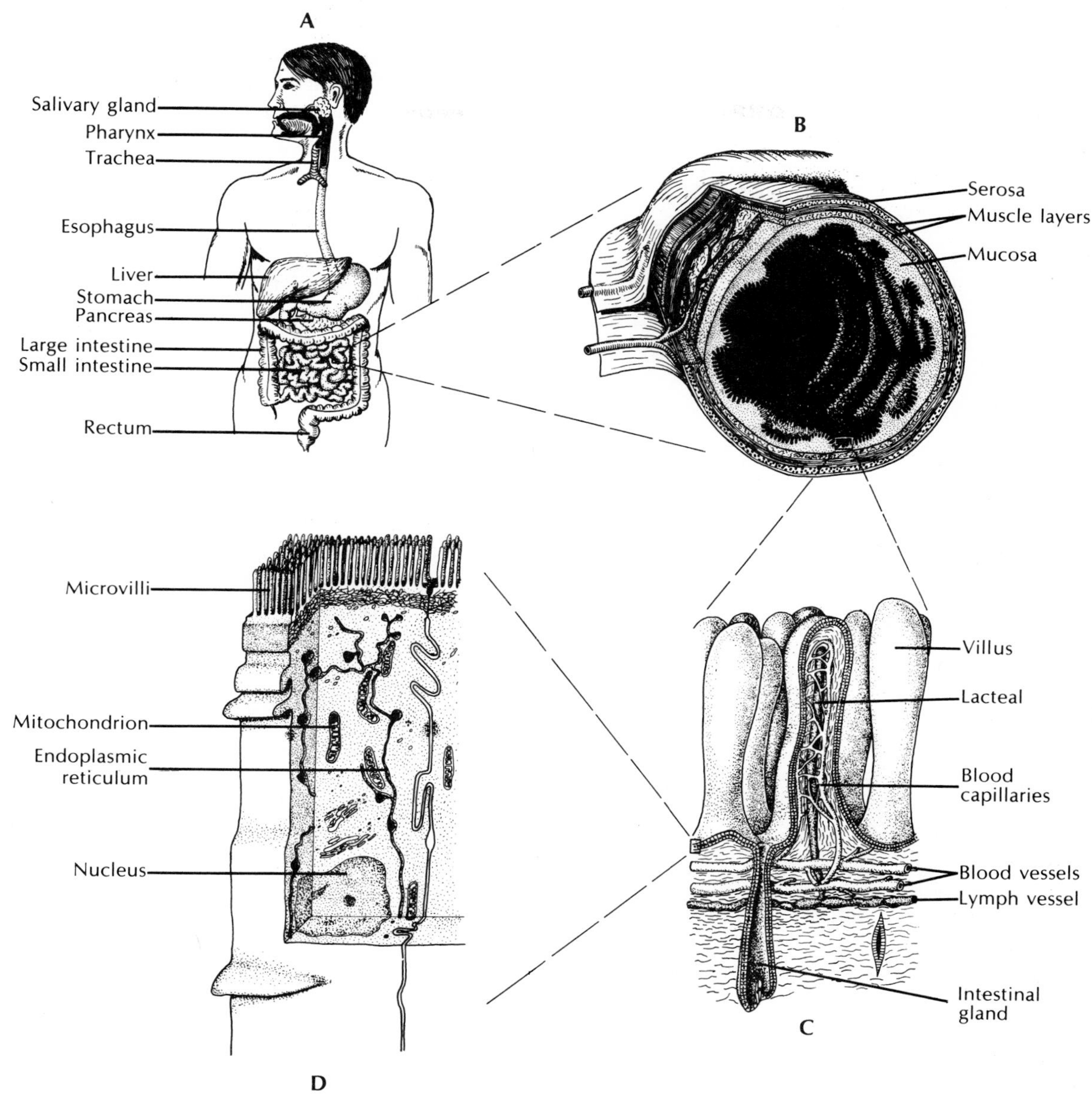

FIG. 31-7 **A,** Human digestive system. **B,** Portion of small intestine. **C,** Portion of mucosa lining of intestine, showing fingerlike villi. **D,** Optical section of single lining cell, as shown by electron microscope.

the hydrolysis of plant and animal starches (Fig. 31-8). Starches, as we have seen previously (p. 33), are long polymers of glucose. Salivary amylase does not completely hydrolyze starch, but breaks it down mostly into 2-glucose fragments called **maltose.** Some free glucose as well as longer fragments of starch are also produced. When the food mass (bolus) is swallowed, salivary amylase continues to act for some time, digesting perhaps half of the starch before the enzyme is inactivated by the acid environment of the stomach.

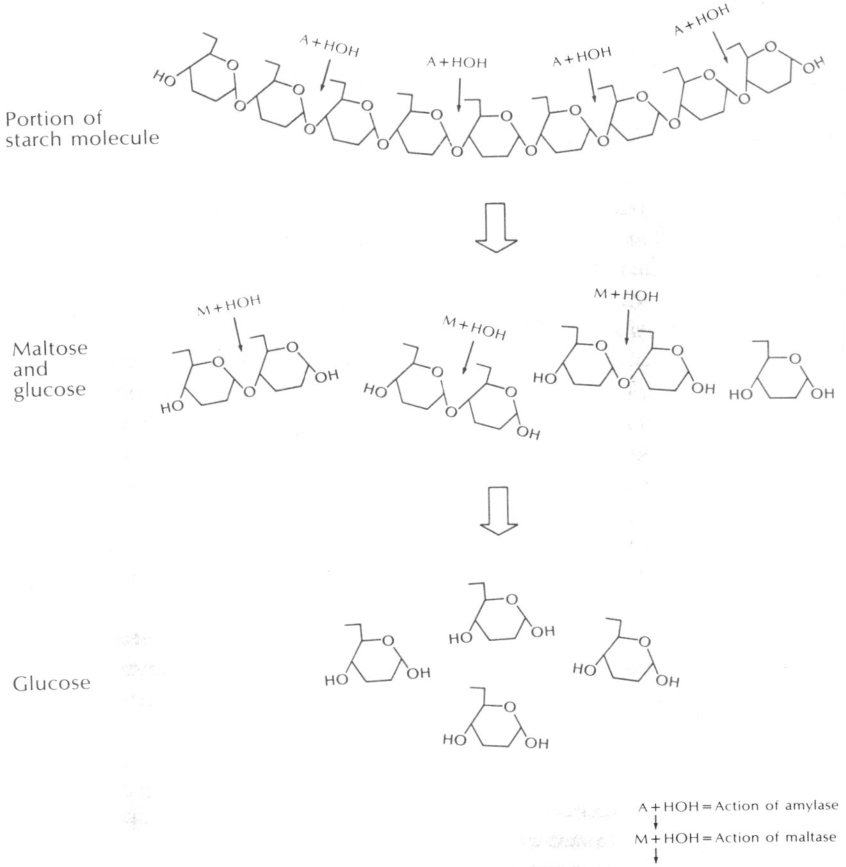

FIG. 31-8 Digestion (hydrolysis) of starch. Long chains of glucose molecules, linked together through oxygen, are first cleaved into disaccharide residues (maltose) by the salivary enzyme amylase. Some glucose may also be split off at the ends of starch chains. The intestinal enzyme maltase then completes the hydrolysis by cleaving the maltose molecules into glucose. A molecule of water is inserted into each enzymatically split bond.

Further starch digestion resumes beyond the stomach when pancreatic amylase is added to the intestinal contents (p. 719).

Swallowing. Swallowing is a reflex process involving both voluntary and involuntary components. Swallowing begins with the tongue pushing the moistened food bolus toward the pharynx. The nasal cavity is reflexly closed by raising the soft palate. As the food slides into the pharynx, the epiglottis is tipped down over the windpipe, nearly closing it. Some particles of food may enter the opening of the windpipe, but are prevented from going further by contraction of laryngeal muscles. Once in the esophagus, the bolus is forced smoothly toward the stomach by peristaltic contraction of the esophageal muscles.

Digestion in the stomach. When food reaches the stomach, the **cardiac sphincter** opens reflexly to allow entry of the food, then closes to prevent regurgitation into the esophagus. The stomach is a combination storage, mixing, digestion, and release center. Gentle peristaltic waves pass over the filled stomach at the rate of about three each minute; churning is most vigorous at the intestinal end where food is steadily released into the duodenum.

About 2 liters of **gastric juice** are secreted each day by deep, tubular glands in the stomach wall. Two types of cells line these glands: (1) **chief cells** secrete an enzyme precursor called **pepsinogen** and (2) **parietal cells** secrete **hydrochloric acid.** Pepsinogen is an inactive form of enzyme that is converted into the active enzyme **pepsin** by hydrochloric acid and by other pepsin already present in the stomach. Pepsin is a **protease** (protein-splitting enzyme) that acts only in an acid medium—pH 1.6 to 2.4. It is a highly specific enzyme that splits large proteins by preferentially breaking down certain peptide bonds scattered along the peptide chain of the protein molecule. Although pepsin, because of its specificity, cannot completely degrade proteins, it effectively breaks them up into a number of small polypeptides. Protein digestion is completed in the intestine by other proteases that can together split all peptide bonds.

Rennin is another enzyme found in the stomachs of the suckling newborn of many mammals, although not of humans. Rennin is a milk-curdling enzyme that transforms the proteins of milk into a finely flocculent form that is more readily attacked by pepsin. Rennin extracted from the stomachs of calves is used in cheesemaking. Human infants, lacking rennin, digest milk proteins with acidic pepsin, the same as adults do.

The unique ability of the stomach to secrete a strong acid at a concentration some 4 million times that found in the blood is still an unsolved question in biology, in part because it has not been possible to collect pure parietal cell secretion or to isolate these cells in culture for study. It is well known that acid solutions can readily destroy organic matter. Since the stomach contains not only a strong acid but also a powerful proteolytic enzyme, it seems remarkable that the stomach mucosa is not digested by its own secretions. That it is not is because of another protective gastric secretion **mucin,** a highly viscous organic compound that coats and protects the mucosa from both chemical and mechanical injury. We should note that despite the popular misconception that "acid stomach" is unhealthy, a notion carefully nourished by the makers of patent medicine, stomach acidity is normal and essential. Sometimes, however, the protective mucous coating fails, allowing the gastric juices to begin digesting the stomach. The result is a peptic ulcer.

The secretion of the gastric juices is intermittent. Although a small volume of gastric juice is secreted continuously, even during prolonged periods of starvation, secretion is normally increased by the sight and smell of food, by the presence of food in the stomach, and by emotional states such as anger and hostility. The most unique and classic investigation in the field of digestion was made by U.S. Army surgeon William Beaumont, during the years 1825 to 1833. His subject was a young, hard-living French Canadian voyageur, named Alexis St. Martin, who in 1822 had accidentally shot himelf in the abdomen with a musket, the blast "blowing off integuments and muscles of the size of a man's hand, fracturing and carrying away the anterior half of the sixth rib, fracturing the fifth, lacerating the lower portion of the left lobe of the lung and the diaphragm, and perforating the stomach." Miraculously, the wound healed, but a permanent opening, or fistula, was formed which permitted Beaumont to see directly into the stomach. St. Martin became a permanent, although temperamental, patient in Beaumont's care, which included food and housing. Over a period of 8 years, Beaumont was able to observe and record how the lining of the stomach changed under different psychic and physiologic conditions, how foods changed during digestion, the effect of emotional states on stomach motility, and many other facts about the digestive processes of his famous patient.

Digestion in the small intestine. The major part of digestion occurs in the small intestine. Three secretions are poured into this region—**pancreatic juice, intestinal juice,** and **bile.** All of these secretions have a high bicarbonate content, especially the pancreatic juice, which effectively neutralizes the gastric acid, raising the pH of the liquefied food mass, now called **chyme,** from 1.5 to 7 as it enters the duodenum. This change in pH is essential because all the intestinal enzymes are effective only in a neutral or slightly alkaline medium.

About 2 liters of **pancreatic juice** are secreted each day. The pancreatic juice contains bicarbonate to neutralize the hydrochloric acid and several enzymes of major importance in digestion. Two powerful proteases, **trypsin** and **chymotrypsin,** are secreted in inactive form as **trypsinogen** and **chymotrypsinogen.** Trypsinogen is activated in the duodenum by **enterokinase,** an enzyme present in the intestinal juice. Chymotrypsinogen is activated by trypsin. These two proteases continue the enzymatic digestion of proteins begun by pepsin, which is now inactivated by the alkalinity of the intestine. Trypsin and chymotrypsin, like pepsin, are highly specific proteases that split apart peptide bonds deep inside the protein molecule. Pancreatic juice also contains **carboxypeptidase,** which splits amino acids off the ends of polypeptides; **pancreatic lipase,** which hydrolyzes fats into fatty acids and glycerol; and **pancreatic amylase,** which is a starch-splitting enzyme identical to salivary amylase in its action.

Intestinal juice from the glands of the mucosal lining furnishes several enzymes. **Aminopeptidase** splits off terminal amino acids from polypeptides; its action is similar to the pancreatic enzyme carboxypeptidase. Three other enzymes are present that complete the hydrolysis of carbohydrates: **maltase** converts maltose to glucose (Fig. 31-8); **sucrase** splits sucrose to glucose and fructose; and **lactase** breaks down lactose (milk sugar) into glucose and galactose.

Bile is secreted by the cells of the **liver** into the **bile duct,** which drains into the upper intestine (duodenum). Between meals the bile is collected into the **gallbladder,** an expansible storage sac that releases the bile when stimulated by the presence of fatty food in the duodenum. Bile contains no enzymes. It is made up of water, bile salts, and pigments. The bile salts (sodium taurocholate and sodium glycocholate) are essential for the complete absorption of fats, which, because of their tendency to remain in large, water-resistant globules, are especially resistant to enzymatic digestion. **Bile salts** reduce the surface tension of fats, so that they are broken up into small droplets by the churning movements of the intestine. This greatly increases the total surface exposure of fat particles, giving the fat-splitting lipases a chance to reduce them. The characteristic golden yellow of bile is produced by the **bile pigments** that are breakdown products of hemoglobin from worn-out red blood cells. The bile pigments also color the feces.

It is well to emphasize the great versatility of the liver. Bile production is only one of the liver's many functions, which include the following: storehouse for glycogen, production center for the plasma proteins, site of protein synthesis and detoxification of protein wastes, destruction of worn-out red blood cells, center for metabolism of fat and carbohydrate, and many others.

Digestion in the large intestine. The liquefied material, now called **chyle,** reaching the large intestine, or **colon,** is low in nutrients, since most important food materials have already been absorbed into the bloodstream from the small intestine. The main function of the colon is the absorption of water and some minerals from the intestinal chyle that enters. In removing more than half the water from the chyle, the colon forms semisolid feces, consisting of undigested food residue, bile pigments, secreted heavy metals, and bacteria. The feces are eliminated from the rectum by the process of **defecation,** a coordinated muscular action that is part voluntary and part involuntary.

The colon contains enormous numbers of bacteria that enter the sterile colon of the newborn infant early in life. In the adult about one-third of the dry weight of feces is bacteria; these include both harmless bacilli as well as cocci that can cause serious illness if they should escape into the abdomen or bloodstream. Normally the body's defenses prevent invasion of such bacteria. The bacteria break down organic wastes in the feces and provide some nutritional benefit by synthesizing vitamin K and small quantities of some of the B vitamins, which are absorbed by the body.

Absorption. Most digested foodstuffs are absorbed from the small intestine, where the numerous finger-shaped **villi** provide an enormous surface area through which materials can pass into the circulation. Little food is absorbed in the stomach because digestion is still incomplete and because of the limited surface exposure. Some materials, however, such as drugs and alcohol are absorbed in part there, which explains their rapid action.

The villi (Fig. 31-7) contain a network of blood and lymph capillaries. The absorbable food products (amino acids, simple sugars, fatty acids, glycerol, and triglycerides as well as minerals, vitamins, and water) pass first across the epithelial cells of the intestinal mucosa and then into either the blood capillaries or the lymph vessels (lacteal). The capillaries pick up nearly all of the absorbed sugars and amino acids. Absorbed fats, however, enter the lacteals.

Carbohydrates are absorbed almost exclusively as simple sugars (for example, glucose, fructose, and galactose) because the intestine is virtually impermeable to polysaccharides. Proteins, too, are absorbed principally as their subunits, amino acids, although it is believed that very small amounts of small proteins or protein fragments may sometimes be absorbed. Simple sugars and amino acids are transferred across the intestinal epithelium by both passive and active processes.

Immediately after a meal these materials are in such high concentration in the gut that they readily diffuse into the blood, where their concentration is initially lower. However, if absorption were passive only, we would expect transfer to cease as soon as the concentrations of a substance became equal on both sides of the intestinal epithelium. This would leave much valuable foodstuff to be lost in the feces. In fact, very little is lost because passive transfer is supplemented by an **active transport** mechanism located in the epithelial cells, which pick up the food molecules and transfer them into the blood. Materials are thus moved **against** their concentration gradient, a process that requires the expenditure of energy. Although not all food products are actively transported, those which are, such as glucose, galactose, and most of the amino acids, are handled by transport mechanisms that are specific for each kind of molecule.

As already described, fat droplets are emulsified by bile salts and then digested by pancreatic lipase. Triglycerides are thus broken down into fatty acids and monoglycerides, which are absorbed by simple diffusion. However, free fatty acids never enter the blood. Instead, during their passage through the intestinal epithelial cells, the fatty acids are resynthesized into triglycerides that accumulate as droplets in the endoplasmic reticulum. The droplets grow larger and larger, then pass out of the cells and into the lacteals. From the lacteals, the fat droplets enter the lymph system and eventually get into the blood by way of the thoracic duct. After a fatty meal, the presence of numerous fat droplets in the blood imparts a milky appearance to the blood plasma.

NUTRITIONAL REQUIREMENTS

The food of animals must include **carbohydrates, proteins, fats, water, mineral salts,** and **vitamins.** Carbohydrates and fats are required as fuels for energy demands of the body and for the synthesis of various substances and structures. Proteins, or actually the amino acids of which they are composed, are needed for the synthesis of the body's specific proteins and other nitrogen-containing compounds. Water is required as the solvent for the body's chemistry and as the major component of all the body fluids. The inorganic salts are required as the anions and cations of body fluids and tissues and form important structural and physiologic components throughout the body. The vitamins are accessory food factors that are frequently built into the structure of many of the enzymes of the body.

All animals require these broad classes of nutrients, although there are differences in the amounts and kinds of food required. The student should note that of the basic food classes listed, some nutrients are used principally as fuels (carbohydrates and lipids), whereas others are required principally as structural and functional components (proteins, minerals, and vitamins). Any of the basic foods (proteins, carbohydrates, fats) can serve as fuel to supply energy requirements, but, conversely, no animal can thrive on fuels alone. A **balanced diet** must satisfy all metabolic requirements of the body—requirements for energy, growth, maintenance, reproduction, and physiologic regulation. The recognition many years ago that many human diseases and those of domesticated animals were caused by, or associated with, dietary deficiencies led biologists to search for specific nutrients that would prevent such diseases. These studies eventually yielded a list of **"essential" nutrients** for human beings and other animal species studied. The essential nutrients are those that are needed for normal growth and maintenance and that **must** be supplied in the diet. In other words, it is "essential" that these nutrients be in the diet because the animal cannot synthesize them from other dietary constituents. Nearly 30 organic compounds (amino acids, fatty acids, and vitamins) and 21 elements have been established as essential for human beings. (Table 31-1). Considering that the body contains thousands of different organic compounds, the list in Table 31-1 is remarkably short. Animal cells have marvelous powers of synthesis enabling them to build compounds of enormous variety and complexity from a small, select group of raw materials.

Carbohydrates are widely consumed because they are more abundant and cheaper than proteins or lipids. In the average diet of Americans and Canadians, about 50% of the total calories (energy content) comes from carbohydrates and 40% comes from lipids. Proteins, essential as they are for structural needs, supply only a little more than 10% of the total calories of the average North American's diet. Actually, many animals can subsist on diets devoid of carbohydrates, provided sufficient total calories and the essential nutrients are present. Eskimos, before the decline of their native culture, lived on a diet that was high in fat and protein and very low in carbohydrate.

Lipids are needed principally to provide energy. However, at least three fatty acids are essential for human beings because they cannot be synthesized (Table 31-1). Certain fatty acids are involved in cell-membrane structure and function and are precursors for the hormonelike prostaglandins (p. 833). In recent

TABLE 31-1 Human nutrient requirements*

Established as essential			Probably essential
Amino acids	Vitamins	Minerals	Tyrosine
Phenylalanine	Water-soluble	Calcium	Histidine
Lysine	Thiamine (B_1)	Phosphorus	Polyunsaturated fatty acids
Isoleucine	Riboflavin (B_2)	Sulfur	
Leucine	Niacin	Potassium	
Valine	Pyridoxine (B_6)	Chlorine	
Methionine	Pantothenic acid	Sodium	
Cystine	Folacin	Magnesium	
Tryptophan	Vitamin B_{12}	Iron	
Threonine	Biotin	Fluorine	
Fatty acids	Choline	Zinc	
Arachidonic	Ascorbic acid (C)	Copper	
Linoleic	Fat-soluble	Silicon	
Linolenic	Vitamins A, D, E, and K	Vanadium	
		Tin	
		Nickel	
		Selenium	
		Manganese	
		Iodine	
		Molybdenum	
		Chromium	
		Cobalt	

*Adapted from Scrimshaw, N. S., and V. R. Young. 1976. The requirements of human nutrition. Sci. Am. **235:**50-64 (Sept.).

years much interest and research have been devoted to lipids in our diets because of the association between fatty diets and the disease **arteriosclerosis** (hardening and narrowing of the arteries). The matter is complex, but evidence suggests that arteriosclerosis may occur when the diet is high in saturated lipids but low in polyunsaturated lipds. For unknown reasons such diets, which are typical of middle-class and affluent North Americans, promote a high blood level of cholesterol, which may deposit in platelike formations in the lining of the major arteries. For this reason the polyunsaturated fatty acids are often considered essential nutrients for human beings. Generally speaking, animal fat is more saturated, whereas fat from plants is more unsaturated.

Proteins are expensive foods and are restricted in the diet. Proteins, of course, are themselves not the essential nutrients, but rather they contain essential amino acids. Of the 20 amino acids commonly found in proteins, nine and possibly 11 are essential to humans (Table 31-1). The rest can be synthesized. One must keep in mind that the terms "essential" and "nonessential" relate only to dietary requirements and to which amino acids can and cannot be synthesized by the body. All 20 amino acids are essential for the various cellular functions of the body. In fact, some of the so-called nonessential amino acids participate in more crucial metabolic activities than the essential amino acids.

In general, animal proteins have more of the essential amino acids than do proteins of plant origin. A human adult would require about 67 g of unfortified whole wheat bread each day to meet the amino acid requirement, but only 19 g of beefsteak to meet these same requirements. One reason that many plants have lower protein scores than animal foods is that plant proteins are often low in certain essential amino acids. All nine of the essential amino acids must be present simultaneously in the diet for protein synthesis. If one or more is missing, the utilization of the other amino acids will be reduced proportionately; they cannot be stored and are broken down for energy. Thus heavy reliance on a single plant source will inevitably lead to protein deficiency. This problem can be corrected if two kinds of plant proteins having complementary strengths in essential amino acids are ingested together. For example, when wheat flour, which is poor only in lysine, is mixed with a legume (beans), which is a good source of lysine but deficient in methionine and cystine, a balanced protein results. Each plant complements the other by having adequate amounts of those amino acids that are deficient in the other.

Because animal proteins are so nutritious, they are

in great demand by all countries. North Americans eat far more animal proteins than do Asians and Africans; on the average a North American eats 61 g of animal protein a day, supplemented by milk and milk products, eggs, cereals, legumes, vegetables, and nuts. In the Middle East, the individual consumption of animal protein is 14 g, in Africa 11 g, and in Asia 8 g.

Undernourishment and malnourishment are the world's major health problems today, afflicting an eighth of the human population. Growing children and pregnant and lactating women are especially vulnerable to the devastating effects of malnutrition. Cell proliferation and growth in the human brain are most rapid in the terminal months of pregnancy and the first year after birth. Adequate protein for neuron development is a requirement during this critical time to prevent neurologic dysfunction. The brains of children who died of protein malnutrition during the first year of life have 15% to 20% fewer brain cells than those of normal children. Malnourished children who survive this period are permanently brain damaged and cannot be helped by later corrective treatment.

Two different types of severe food deficiency are recognized: **marasmus,** general undernourishment from a diet low in both calories and protein, and **kwashiorkor,** protein malnourishment from a diet adequate in calories but deficient in protein. Marasmus is common in infants weaned too early and placed on low-calorie–low-protein diets; these children are listless, and their bodies waste away. Kwashiorkor is a West African word describing a disease a child gets when it is displaced from the breast by a newborn sibling. It is characterized by retarded growth, anemia, weak muscles, bloated body with typical pot belly, acute diarrhea, susceptibility to infection, and high mortality. Ten million children the world over are seriously undernourished or malnourished and 1 million children die of hunger each year in India alone.

The major cause of the world's precarious food situation is recent rapid population growth. The world population was 2 billion in 1930, reached 3 billion in 1960, is 4 billion today, and probably will be 6 billion in just 20 years more. Eighty million people are added each year; the equivalent of the U.S. population is added to the world every 30 months. Thus the search for new ways to increase food production and distribution takes on a desperate urgency. Despite the optimism stimulated by the "green revolution" of the 1960's, with the introduction of new high-yielding wheat and rice strains, it is apparent today that more time, more work, and more money are needed to realize its promise. It was fitting that the 1970 Nobel Peace Prize should go to Dr. Norman Borlaug, who developed several of the new wheat varieties. However, Dr. Borlaug himself emphasizes that despite our best efforts to develop more productive grains, mass famine seems inevitable unless the human population is stabilized.

Vitamins. Vitamins are organic compounds that are required in small amounts in the diet for specific cellular functions. They are not sources of energy, but are often associated with the activity of important enzymes that have vital metabolic roles. Plants and many microorganisms synthesize all the organic compounds they need; animals, however, have lost certain synthetic abilities during their long evolution and depend ultimately on plants to supply these compounds. Vitamins, therefore, represent synthetic gaps in the metabolic machinery of animals. We have seen that several amino acids are also dietary essentials. These are not considered vitamins, however, because they usually enter into the actual **structure** of tissues or proteinaceous tissue secretions. Vitamins, on the other hand, are essential **functional** components of enzyme catalytic systems. Nevertheless, the distinction between certain vitamins (A, D, E, and K) and other dietary essentials not classified as vitamins is not so clear as it was once believed to be.

Vitamins are usually classified as fat-soluble (soluble in fat solvents such as ether) or water-soluble. The water-soluble ones include the B complex and vitamin C. The family of B vitamins, so grouped because the original B vitamin was subsequently found to consist of several distinct molecules, tends to be found together in nature. Almost all animals, vertebrate and invertebrate, require the B vitamins; they are "universal" vitamins. The dietary need for vitamin C and the lipid-soluble vitamins A, D, E, and K tends to be restricted to the vertebrates, although some are required by certain invertebrates.

LIPID-SOLUBLE VITAMINS. **Vitamin A** exists in two forms, as A_1 and A_2 which differ only slightly in chemical structure. Vitamin A consists of half of a β-carotene molecule; carotenes are a class of organic compounds synthesized only by plants. Animals convert the plant carotenoid to vitamin A; so many animal foods, especially butter, eggs, and milk, serve as sources of the vitamin precursor, as do green leafy and yellow vegetables. Vitamin A functions in visual photochemistry and is also necessary for normal cell growth. A deficiency causes night blindness and growth retardation.

Vitamin D is derived from eggs, fish oils, and beef fat. The two forms of vitamin D are vitamin D_2 (ergocalciferol) and vitamin D_3 (cholecalciferol). The latter is synthesized in the skin by irradiation with ultraviolet rays of sunlight. Vitamin D is converted in the liver and kidneys to a hormonal substance (1,25-DHCC) that promotes absorption of calcium in the intestine. A deficiency of vitamin D causes rickets, a disease characterized by defective bone formation.

Vitamin E is a poorly understood group of tocopherols found in grains, meat, milk, and green leafy vegetables. It is necessary for the integrity of biologic membranes. Deficiency causes sterility, defective embryonic growth, muscular weakness, and anemia. Its current popularity with food faddists as a panacea for nearly all ailments is largely wishful thinking.

Vitamin K is necessary for the synthesis of prothrombin in the liver. Its deficiency causes failure of blood to clot. It is scarce in the normal diet but is synthesized by bacteria in the colon.

B-COMPLEX (WATER-SOLUBLE) VITAMINS. **Thiamine (vitamin B_1)** is an essential coenzyme in pyruvate metabolism. Thiamine is synthesized by plants and is found in beans, grain, yeast, and roots. It is also present in eggs and lean meat. A deficiency causes beriberi, nervous system degeneration, and cessation of growth.

Riboflavin (vitamin B_2) is a part of the oxidative flavoprotein coenzyme (FAD). It is present in many foods such as green beans, eggs, liver, and milk. A deficiency, rare in humans, causes nonspecific digestive disturbances and certain kinds of dermatitis.

Nicotinic acid (niacin) is converted in the body to hydrogen donor-acceptor coenzymes (NAD and NADP). It is found in green leaves, egg yolk, wheat germ, and liver. Oxidative metabolism is depressed with its deficiency, producing a complex of symptoms called pellagra.

Pyridoxine (vitamin B_6) is a coenzyme that functions in amino acid metabolism. It is widely distributed in foods, especially meat, eggs, yeast, and cereals. A dietary deficiency is rare in humans but causes a variety of symptoms in experimental animals (dermatitis, growth retardation, anemia).

Pantothenic acid is part of coenzyme A that functions in carbohydrate and lipid metabolism. It is present in many foods and a deficiency is not known in humans. In experimental animals its lack causes dermatitis, growth retardation, graying of hair, and other abnormalities.

Folic acid is essential for growth and blood-cell formation. It is found in green leaves, soybeans, yeast, and egg yolk. A deficiency causes anemia and kidney hemorrhage.

Biotin (vitamin H) functions in fatty-acid oxidation and synthesis. It is found in liver, egg yolk, meat, and many other sources. A deficiency is very rare in humans. In experimental animals a lack causes dermatitis.

Cyanocobalamin (vitamin B_{12}) is essential for red blood cell formation. It is a unique vitamin in that it is synthesized only by bacteria. Chief sources are milk, egg yolk, liver, and oysters. A deficiency causes pernicious anemia.

VITAMIN C. Vitamin C is necessary for the formation of intercellular material. It is found in citrus fruits, tomatoes, and other fresh vegetables. A deficiency causes scurvy, perhaps the most famous deficiency disease of humans. Scurvy was the scourge of sailors and soldiers who subsisted for long periods on dried meats and grains. Scurvy is characterized by capillary bleeding and defective wound healing.

Annotated references

Selected general readings

See also general references for Part Three, p. 793.

Brooks, F. P. 1970. Control of gastrointestinal function. New York, The Macmillan Co. *A concise, well-illustrated summary of human digestive physiology.*

Chrispeels, M. J., and D. Sadava. 1977. Plants, food, and people. San Francisco, W. H. Freeman and Co. *Well-written overview of the human nutrition problem, plants and agriculture, and the green revolution.*

Hegstead, D. M. 1978. Protein-calorie malnutrition. Am. Sci. **66**:61-65. *Its causes are still poorly understood.*

Jennings, J. B. 1973. Feeding, digestion and assimilation in animals, ed. 2. New York, St. Martin's Press, Inc. *A gen-*

eral, comparative approach. Excellent account of feeding mechanisms in animals.

McDonald, P., R. A. Edwards, and J. F. D. Breenhalgh. 1973. Animal nutrition. New York, Hafner Publishing Co., Inc. *Thorough treatment of farm-animal nutrition.*

Morton, G. 1967. Guts. The form and function of the digestive system. Institute of Biology's Studies in Biology, no. 7, New York, St. Martin's Press, Inc.

Selected *Scientific American* articles

Davenport, H. W. 1972. Why the stomach does not digest itself. **226:**86-93 (Jan.).

Jones, J. C. 1978. The feeding behavior of mosquitoes. **238:** 138-148 (June).

Kretchmer, N. 1972. Lactose and lactase. **227:**70-78 (Oct.). *Most adult mammals, humans included, lack lactase, the enzyme that breaks down lactose, or milk sugar.*

Mayer, J. 1976. The dimensions of human hunger. **235:**40-49 (Sept.) *Excellent description of the worsening world nutritional situation.*

Neurath, H. 1964. Protein-digesting enzymes. **211:**68-79 (Dec.). *Describes studies of enzyme structure and how these digestive enzymes exert their catalytic effect.*

Rogers, T. A. 1958. The metabolism of ruminants. **198:**34-38 (Feb.). *How cellulose is digested in the four-chambered stomach of cows and other ruminants.*

Scrimshaw, N. S., and V. R. Young. 1976. The requirements of human nutrition. **235:**50-64 (Sept.). *Describes the difficulties of arriving at recommended energy and nutrient allowances for humans.*

Spain, D. M. 1966. Atherosclerosis. **215:**48-56 (Aug.). *How dietary factors contribute to increasing prevalence of this disease.*

Trowell, H. C. 1954. Kwashiorkor. **191:**46-50 (Dec.). *The most severe and widespread nutritional disease known to medical science is described.*

Young, V. R., and N. S. Schrimshaw. 1971. The physiology of starvation. **225:**14-21 (Oct.). *Studies with human volunteers have helped to clarify the body's nutritional needs.*

CHAPTER 32

NERVOUS COORDINATION

Nervous system and sense organs

Scanning electron micrograph of a scallop eye peering out from among folds of the mantle. Scallops *(Pecten)* bear dozens of eyes, resembling small, bright blue pearls, clearly visible when the valves of the shell are open. Each eye is a remarkably advanced structure with cornea, cellular lens, a double retina, and a tapetum, or reflective layer, containing crystals of guanine. (Original magnification ×275.)

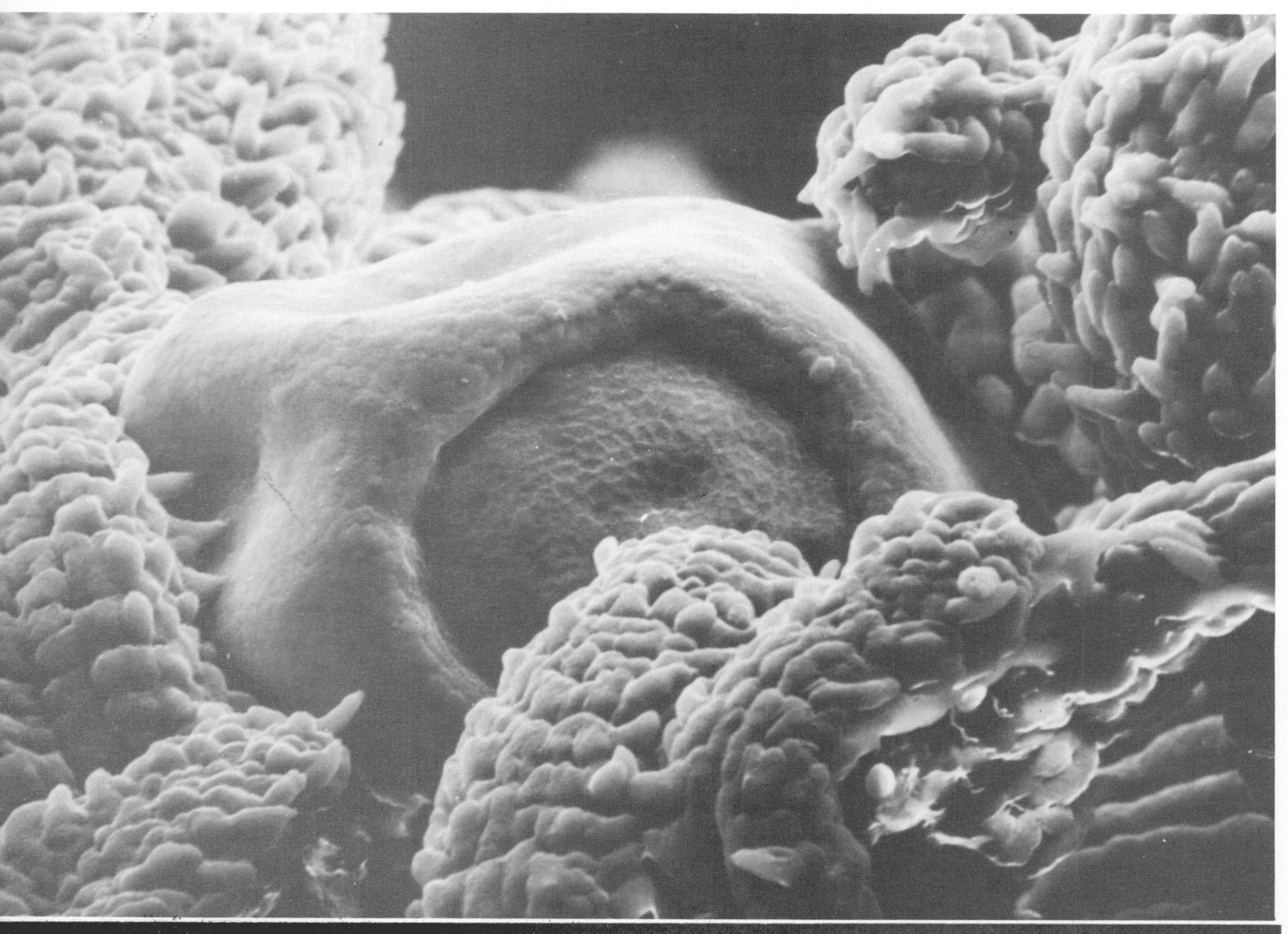

Courtesy P. P. C. Graziadei.

The nervous system originated in a fundamental property of protoplasm—irritability. Each cell responds to stimulation in a manner characteristic of that type of cell. But certain cells are highly specialized for receiving stimuli and for conducting impulses to various parts of the body. Through evolutionary changes, these cells have become organized into the most complex of all body systems: the nervous system. The endocrine system is also important in coordination and interacts continuously with the nervous system in the control of body functions. Functionally, the nervous system differs from the endocrine system in its capacity to monitor external as well as internal changes and to respond immediately to such changes. Nervous responses are measured in milliseconds, whereas the fastest endocrine responses are measured in seconds.

The evolution of the nervous system has been correlated with the development of bilateral symmetry and cephalization. Along with this development, animals acquired exteroceptors and associated ganglia. The basic activity of the nervous system is to code sensory information arising internally or externally and to transmit it to regions of the central nervous system where it is processed into appropriate action. This action may be any of several types, such as simple reflexes, automatic behavior patterns, conscious perception, or learning processes.

NERVOUS SYSTEMS OF INVERTEBRATES

Protozoans are mostly lacking nervous systems as such. They depend on primitive membrane irritability and its conduction across the cell surface to respond to stimuli. Nevertheless there are instances of remarkable neural development in certain protozoans.

The metazoans show a progressive increase in nervous system complexity that we believe recapitulates to some extent the evolution of the nervous system. The cnidarians have a **nerve net** containing bipolar and multipolar cells (protoneurons). These may be separated from each other by synaptic junctions, but they form an extensive network that is found in and under the ectoderm over all the body. An impulse starting in one part of this net will be conducted in all directions, since the synapses do not restrict transmission to one-way movement, as they do in higher animals. There are no differentiated sensory, motor, or connector components in the strict meaning of those terms. Branches of the nerve net connect to receptors in the epidermis and to epitheliomuscular cells. Most responses tend to be generalized, yet many are astonishingly complex for so simple a nervous system (such as the swimming anemone in Fig. 10-21, p. 205). Such a type of nervous system is retained among higher animals in the form of nerve plexuses, which coordinate such generalized movements as intestinal peristalsis.

Flatworms are provided with two anterior **ganglia** of nerve cells from which two main nerve trunks run posteriorly, with lateral branches extending to the various parts of the body (Fig. 11-4, p. 222). This is the true beginning of a differentiation into a **peripheral nervous system** (a communications network extending to all parts of the body) and a **central nervous system,** which coordinates everything. It is also the first appearance of the **linear** type of nervous system, which is more developed in higher invertebrates. These have a more centralized nervous system, with the two longitudinal nerve cords fused (although still recognizable) and many ganglia present. The annelids have a well-developed nervous system consisting of distinctive **afferent** (sensory) and **efferent** (motor) neurons (Fig. 14-8, p. 311). At the anterior end, the ventral nerve cord divides and passes upward around the digestive tract to join the bi-lobed brain. In each segment the double nerve cord bears a double ganglion, each with two pairs of nerves. Arthropods have a system similar to that of earthworms, except that the ganglia are larger, and the sense organs are better developed.

Molluscs have a system of three pairs of ganglia: one pair near the mouth, another pair at the base of the foot, and one pair in the viscera. The ganglia are joined by connectives. The molluscs also have a number of sense organs; they are especially well developed in the cephalopods. Among the echinoderms the nervous system is radially arranged (Fig. 20-12, p. 456).

The nerve cord in all invertebrates is ventral to the alimentary canal and lacks internal cavities. This arrangement is in pronounced contrast to the nerve cord of vertebrates, which is dorsal to the digestive system, single, and hollow.

NERVOUS SYSTEM OF VERTEBRATES

Most vertebrates have a brain much larger than the spinal cord. In lower vertebrates this difference is not significant, but higher in the vertebrate kingdom the brain increases in size, reaching its maximum in mammals, especially humans. Along with this enlargement has come an increase in complexity, bringing better patterns of coordination, integration, and intelligence. The nervous system is commonly divided into central and peripheral parts. The **central nervous system** is housed within the skull and vertebral column and is concerned with integrative activity. The **peripheral nervous system** consists of nerve cells or extensions of nerve cells that lie outside the skull and vertebral column. It is a communications system for the conduction of sensory (input) information from, and motor (output) information to all parts of the body.

Neuron—functional unit of the nervous system

The vertebrate nervous system is composed of **neurons** and **glial cells.** Although neurons are the basic units of the nervous system, they are outnumbered ten to one by the glial cells, which physically support the neurons and are believed to nourish them metabolically (p. 736).

The neuron is a cell body with all its processes. Although neurons assume many shapes, depending on their function and location, a "typical" kind is shown diagrammatically in Fig. 32-1. From the nucleated cell body extends an **axon,** which carries impulses *away* from the cell. Typically several branching **dendrites** extend from the cell body. These carry impulses *toward* the cell body. Axons usually are much longer than dendrites and are often called **nerve fibers.** They are usually covered with a soft, white lipid-containing material called **myelin.** This insulating material is laid down in concentric rings by specialized **Schwann cells** to form a **myelin sheath.** This is enclosed by an outer membrane called the **neurolemma.**

Neurons are commonly divided into **afferent,** or sensory, neurons; **efferent,** or motor, neurons; and **association** neurons (interneurons). Afferent and efferent neurons lie mostly outside the skull and vertebral column; association neurons, which in man comprise 99% of all the nerve cells in the body, lie entirely within the central nervous system. Afferent neurons are connected to receptors. When these respond to some environmental change, they generate **action potentials** in the afferent neurons, which are carried into the central nervous system. Here the impulses may be perceived as conscious sensation. The impulses also

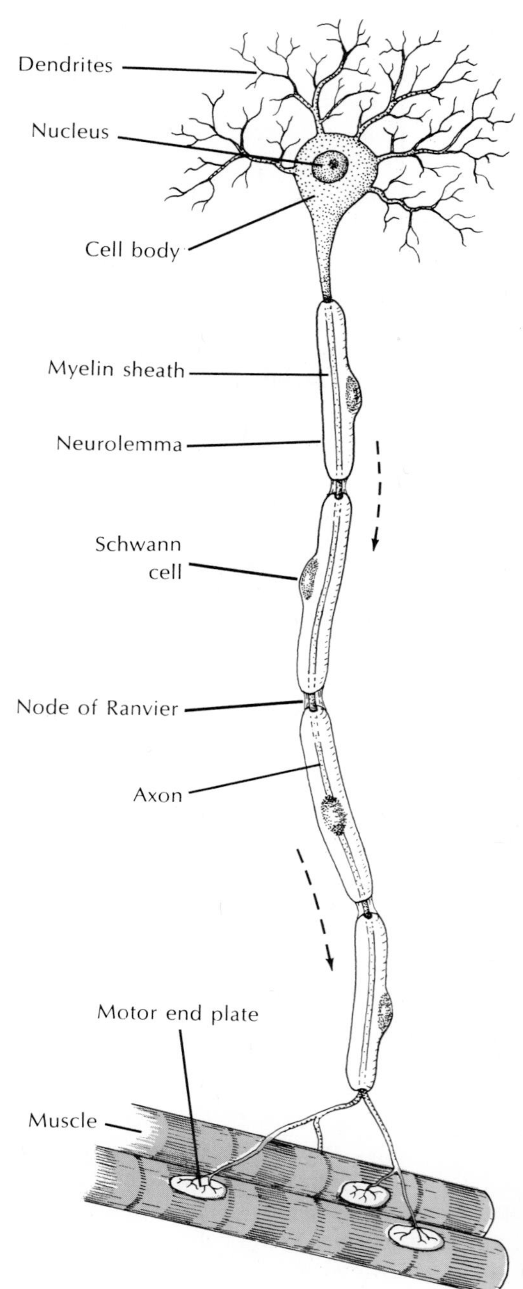

FIG. 32-1 Structure of a motor (efferent) neuron.

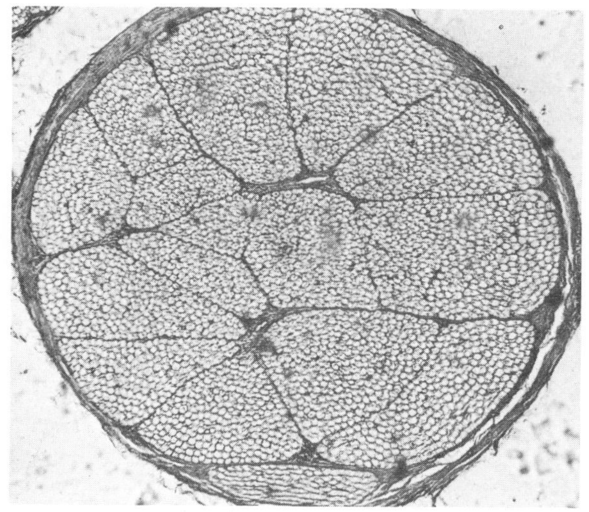

FIG. 32-2 Cross section of nerve showing cut ends of nerve fibers (small white circles). Such a trunk may contain thousands of fibers. Both afferent and efferent fibers are present.

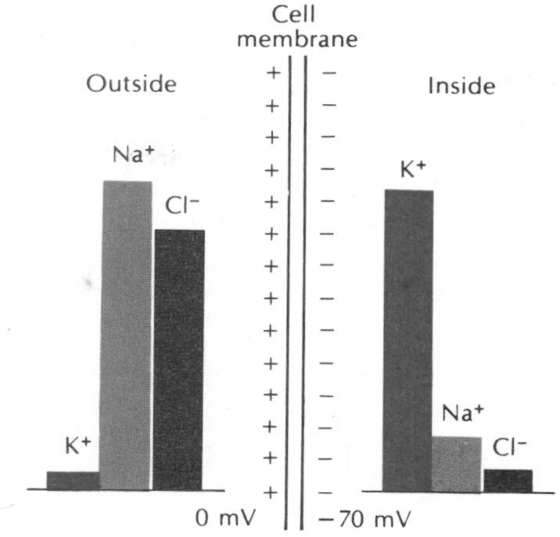

FIG. 32-3 Ionic composition inside and outside a resting nerve cell. An active sodium pump located in the cell membrane drives sodium to the outside, keeping its concentration low inside. Potassium concentration is high inside, and although the membrane is "leaky" to potassium, this ion is held inside by the repelling positive charge outside the membrane.

move to efferent neurons, which carry them out by the peripheral system to **effectors** (muscles or glands).

Nerves (not to be confused with neurons) are actually made up of many nerve processes—axons or dendrites or both—bound together with connective tissue (Fig. 32-2). The cell bodies of these nerve processes are located either in ganglia or somewhere in the central nervous system (brain or spinal cord).

Nature of the nerve impulse

The nerve impulse is the chemical-electrical message of nerves, the common functional denominator of all nervous system activity. Despite the incredible complexity of the nervous system of advanced animals, nerve impulses are basically alike in all nerve fibers and in all animals. It is an *all-or-none* phenomenon; either the fiber is conducting an impulse, or it is not. Because all impulses are alike, the only way a nerve fiber can vary its effect on the tissue it innervates is by changing the **frequency** of impulse conduction. Frequency change is the language of a nerve fiber. A fiber may conduct no impulses at all, or a very few per second up to a maximum approaching 1,000 per second. The higher the frequency (or rate) of conduction, the greater is the level of excitation. The same general rule also applies to sense organs; that is, the more a sense organ is excited by a stimulus, the greater is the frequency of impulses sent out over the axons of the sensory nerves.

Resting potential. To understand what happens when an impulse is conducted down a fiber, we need to know something about the resting, undisturbed fiber. Nerve cell membranes, like all cell membranes, have special permeability properties that create ionic imbalances. Sodium and chloride predominate on the outside of the membrane, whereas potassium ions are more common inside (Fig. 32-3). These differences are quite dramatic; there is about 10 times more sodium outside than in and 25 to 30 times more potassium inside than out. However, the nerve cell membrane is 50 to 70 times more permeable to potassium than to sodium. The result is that potassium ions tend to leak outward, moving down their concentration gradient. This movement of positively charged ions outward creates a **diffusion potential,** with the outside of the membrane positive with respect to the inside.

If the membrane is highly permeable to potassium, why doesn't all of the potassium escape from the cell, allowing the potential to disappear? This does not happen because the potential difference created by outward movement of potassium begins to influence this movement. Since potassium ions are positively charged, they are attracted by the negatively charged inside membrane and repelled by the positively charged ions

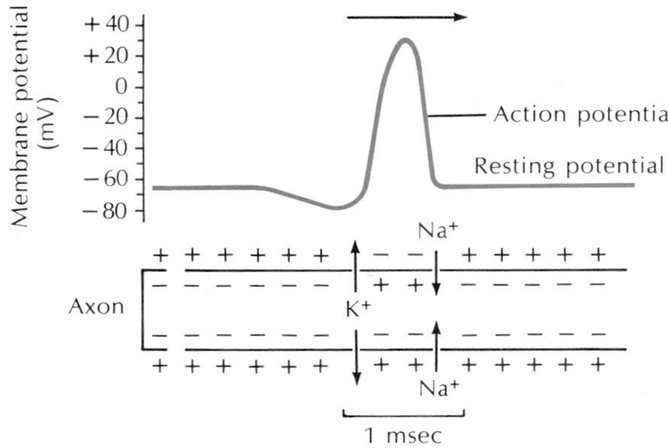

FIG. 32-4 Action potential of nerve impulse. The electric event, moving from left to right, is associated with rapid changes in membrane permeability to sodium and potassium ions. When the impulse arrives at a point, sodium ions suddenly rush in, making the axon positive inside and negative outside. Then the sodium holes close and potassium holes open up. This restores the normal negative resting potential.

outside the membrane. As potassium flows outward the electric force across the membrane becomes large enough to prevent any further net outward movement of potassium. This membrane potential is called an **equilibrium potential** because it is a permanent bioelectric potential produced by a balance between a concentration force favoring the outward flow of potassium and an electric force that opposes it.

Thus, this is the origin of the **resting transmembrane potential,** which is positive outside and negative inside. It is created, as we have seen, by two important characteristics of the living nerve cell: (1) the potassium concentration is much greater inside the cell than outside, and (2) the cell membrane is far more permeable to potassium than to sodium. The resting potential is usually about −70 mV (millivolts), with the inside of the membrane negative to the outside.

Action potential. The nerve impulse is a rapidly moving change in electric potential called the **action potential** (Fig. 32-4). It is a very rapid and brief depolarization of the axon membrane; in fact, not only is the resting potential abolished, but in most nerve fibers the potential actually reverses for an instant so that the outside becomes negative as compared to the inside. Then, as the action potential moves ahead, the membrane returns to its normal resting potential ready to conduct another impulse. The entire event occupies approximately a millisecond. Perhaps the most significant property of the nerve impulse is that it is **self-propagating;** that is, once started the impulse moves ahead automatically, much like the burning of a fuse.

What causes the reversal of polarity in the cell membrane during passage of an action potential? We have seen that the resting potential depends on the high membrane permeability to potassium, some 50 to 70 times greater than the permeability to sodium. When the action potential arrives at a given point, the permeability of the membrane to potassium and sodium is markedly changed. The membrane suddenly becomes approximately 600 times more permeable to sodium, whereas potassium permeability changes very little. Sodium rushes in.

Actually only an extremely small amount of sodium traverses the membrane in that instant—less than one-millionth of the sodium outside—but this brief shift of positive ions inward causes the membrane potential to disappear, even reverse. An electric "hole" is created. Potassium, finding its electric barrier gone, begins to move out, and then, as the action potential passes on, the membrane quickly regains its resting properties. It becomes once again practically impermeable to sodium, and the outward movement of potassium is checked.

The rising phase of the action potential is associated with the rapid influx (inward movement) of sodium (Fig. 32-4). When the action potential reaches its peak, the sodium permeability is restored to normal, and potassium permeability briefly increases above the resting level. This causes the action potential to drop rapidly toward the resting membrane level.

Sodium pump. The resting cell membrane has a very low permeability to sodium. Nevertheless, some sodium ions leak across, even in the resting condition. When the axon is active, sodium flows inward with each passing impulse, and although the amount is very small, it is obvious that the ionic gradient would eventually disappear if the sodium ions were not moved back out again. This is done by a **sodium pump** located in the axon plasma membrane. Although no one has ever actually seen the "pump," we do know quite a bit about it because it has been the object of intense biochemical and biophysical studies. The sodium pump is an active transport device capable of combining with sodium on the inside surface of the membrane, then moving to the outside surface where the sodium is released. It is probably composed of

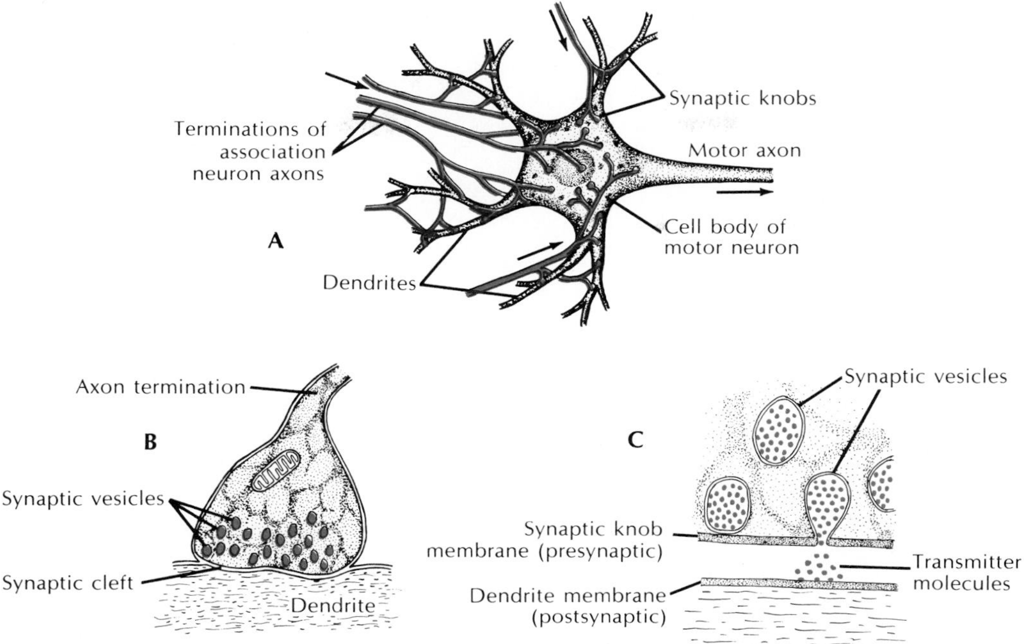

FIG. 32-5 Transmission of impulses across nerve synapses. **A,** Cell body of a motor nerve is shown covered with the terminations of association neurons. Each termination ends in a synaptic knob; thousands of synaptic knobs may be on a single nerve cell body and its dendrites. **B,** Synaptic knob enlarged 60 times more than **A.** An impulse traveling down the axon will cause some synaptic vesicles to move down to the synaptic cleft and rupture, releasing transmitter molecules into the cleft. **C,** Synaptic cleft as it might appear under a high-resolution electron microscope. Transmitter molecules from a ruptured synaptic vesicle move quickly across the gap to produce an electric potential change in the postsynaptic membrane.

phosphate-containing protein molecules. The sodium pump requires energy, since it is moving sodium ''uphill'' against the sodium electric and concentration gradient. This energy is supplied by ATP through cellular metabolic processes. There is evidence that in some cells sodium transport outward is linked to potassium transport inward; the same carrier molecule may act as a two-way shuttle, carrying ions on both trips across the membrane. This kind of pump is called a sodium-potassium pump. Active transport was discussed earlier on p. 79.

Synapses—junction points between nerves

A synapse is a break at the end of a nerve axon where it connects to the dendrites or cell body of the next neuron. Neurons bringing impulses toward synapses are called **presynaptic neurons;** those carrying impulses away are **postsynaptic neurons.** At the synapse, the membranes are separated by a narrow gap, the **synaptic cleft,** having a very uniform width of approximately 20 μm. The synapse is of great functional importance because it acts as a one-way valve that allows nerve impulses to move in one direction only. It is also part of the decision-making equipment of the central nervous system because it is here that information is modulated from one nerve to the next.

The axon of most nerves divides at its end into many branches, each of which bears a synaptic knob that sits on the dendrites or cell body of the next nerve (Figs. 32-5 and 32-6). The axon terminations of several nerves may almost cover a nerve cell body and its dendrites with thousands of synaptic clefts. Because a single impulse coming down a nerve axon splays out into the many branches and synaptic endings on the next nerve cell, many impulses therefore converge at the cell body at one instant.

There are two kinds of synapses, excitatory and

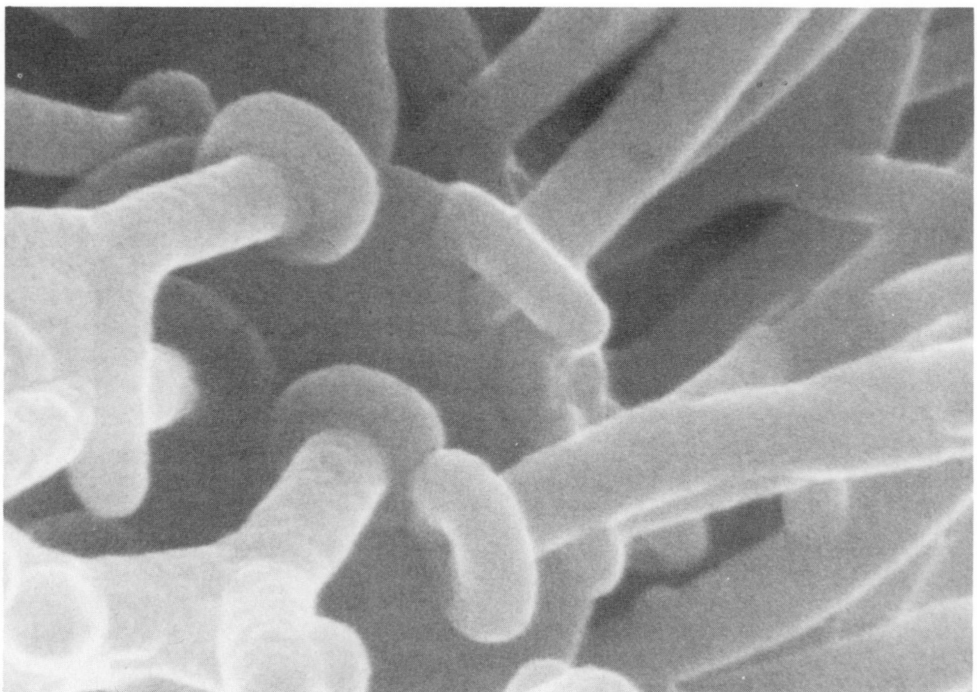

FIG. 32-6 Synaptic knobs in the abdominal ganglion of the marine nudibranch mollusc *Aplysia*. The knobs are strongly attached to rounded surface in left center, which is probably the receptive surface of a postsynaptic neuron. Note firm attachment point of knob in upper center. Note also dividing fibers at left, which form direct connections between fibers. (Scanning electron micrograph, ×6,000.) (From Lewis, E. R., T. E. Everhart, and Y. Y. Zeevi. 1969. Science **165:**1140-1143.)

inhibitory. Both act as chemical bridges in the transmission line. The electron microscope shows that each synaptic knob contains numerous membrane-limited synaptic vesicles (Fig. 32-5, *B*). These are filled with molecules of chemical transmitter. The two most common transmitters are acetylcholine and norepinephrine. Other synaptic transmitters are known, such as γ-aminobutyric acid (GABA) and serotonin, but thus far it has not been possible to determine which of the various transmitters are excitatory and which are inhibitory in the mammalian central nervous system.

When an impulse arrives at the knob, it induces some of the vesicles to move to the base of the knob and release their contents of transmitter molecules into the synaptic cleft (Fig. 32-5, *C*). These move rapidly across the narrow cleft to combine with the postsynaptic membrane, producing a small potential that may be excitatory or inhibitory. Of the thousands of synapses on any neuron, perhaps hundreds are active at one moment, or within a very brief time span. It is the net effect of all the excitatory and inhibitory inputs that determines whether the postsynaptic cell fires or remains silent.

Components of the reflex arc

Neurons work in groups called **reflex arcs** (Fig. 32-7). There must be at least two neurons in a reflex arc, but usually there are more. The parts of a typical reflex arc consist of (1) a **receptor,** a sense organ in the skin, muscle, or other organ; (2) an **afferent,** or sensory, neuron, which carries the impulse toward the central nervous system; (3) a **nerve center,** where synaptic junctions are made between the sensory neurons and the association neurons; (4) the **efferent,** or motor, neuron, which makes synaptic junction with the asso-

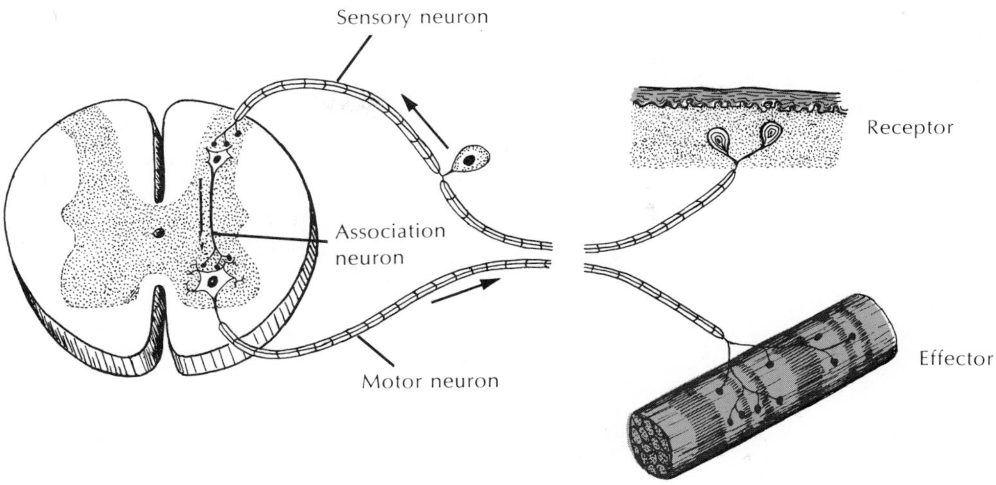

FIG. 32-7 Reflex arc. Impulse generated in the receptor is conducted over an afferent (sensory) nerve to the spinal cord, relayed by an association neuron to an efferent (motor) nerve cell body and by the efferent axon to an effector.

ciation neuron and carries impulses out from the central nervous system; and (5) the **effector,** by which the animal responds to its environmental changes. Examples of effectors are muscles, glands, cilia, nematocysts of cnidarians, electric organs of fishes, and chromatophores.

A reflex arc at its simplest consists of only two neurons—a sensory (afferent) neuron and a motor (efferent) neuron. Usually, however, association neurons are interposed (Fig. 32-7). Association neurons may connect afferent and efferent neurons on the same side of the spinal cord, or on opposite sides or connect them on different levels of the spinal cord, either on the same or opposite sides. In almost any reflex act a number of reflex arcs are involved. For instance, a single afferent neuron may make synaptic junctions with many efferent neurons. In a similar way an efferent neuron may receive impulses from many afferent neurons. In this latter case the efferent neuron is referred to as the **final common path.**

A **reflex act** is the response to a stimulus acting over a reflex arc. It is **involuntary** and may involve the cerebrospinal or the autonomic nervous divisions of the nervous system. Many of the vital processes of the body such as control of breathing, heartbeat, diameter of blood vessels, sweat gland secretion, and others are reflex actions. Some reflex acts are inherited and innate; others are acquired through learning processes (conditioned).

ORGANIZATION OF THE VERTEBRATE NERVOUS SYSTEM

The basic plan of the vertebrate nervous system is a dorsal longitudinal hollow nerve cord that runs from head to tail. During early embryonic development, the central nervous system begins as an ectodermal **neural groove,** which by folding and enlarging becomes a long, hollow, **neural tube.** The cephalic end enlarges into the brain vesicles, and the rest becomes the spinal cord. The spinal nerves (31 pairs in humans) have a dual origin. The **spinal ganglia** (dorsal root ganglia in Fig. 32-8), containing the sensory neurons, differentiate from specialized cells, called **neural crest cells,** that pinch off from the edges of the neural groove as it closes to form a tube. The ventral roots contain motor fibers that originate in the spinal cord. Both dorsal (sensory) and ventral (motor) roots meet some distance beyond the cord to form a mixed **spinal nerve** (Fig. 32-8).

Central nervous system

The central nervous system is composed of the spinal cord and brain.

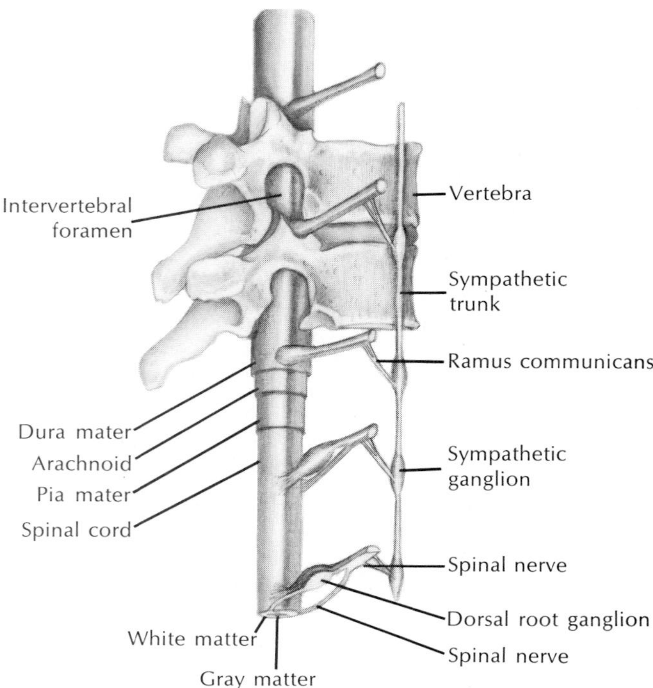

FIG. 32-8 Spinal cord and meninges with relation to spinal nerves, sympathetic system, and vertebrae. Three coats of meninges have been partly cut away to expose spinal cord. Only two vertebrae are shown in position.

Spinal cord

The cord is enclosed by the vertebral canal and additionally protected by three layers, collectively called the **meninges** (men-in'jeez). The three layers are a tough outer **dura mater,** a thin spider web–like **arachnoid,** and a delicate innermost sheath, the **pia mater** (Fig. 32-8). Between the arachnoid and the pia mater is a space containing **cerebrospinal fluid,** a secreted fluid forming a protective cushion and thermal insulation for the cord. The meninges and cerebrospinal fluid blanket are continuous with those covering the brain.

In cross section the cord shows two zones. An inner H-shaped zone of **gray matter** is made up of association neurons and the cell bodies of motor neurons. The outer zone of **white matter** contains nerve bundles of axons and dendrites linking different levels of the cord with each other and with the brain. The fibers are bundled into **ascending tracts** carrying impulses to the brain and **descending tracts** carrying impulses away from the brain. The sensory (ascending) tracts are located mainly in the dorsal part of the cord; the motor (descending) tracts are found ventrally and laterally in the cord. Fibers also cross over from one side of the cord to the other, with the sensory fibers crossing at a higher level than the motor fibers. Although the different tracts cannot be distinguished in a sectioned cord, even with a microscope, their position is known from painstaking mapping experiments.

Brain

Unlike the spinal cord, which has changed little in structure during vertebrate evolution, the brain has changed dramatically. From the primitive linear brain of fishes and amphibians, it has expanded into the deeply fissured, enormously complex brain of mammals. It reaches its highest level in the human brain, which contains some 100 billion cells. The ratio between the weights of the brain and spinal cord affords a fair criterion of an animal's intelligence. In fishes and amphibians this ratio is about 1:1, in humans the ratio is 55:1—in other words, our brain is 55 times heavier than our spinal cord. Although our brain is not the largest (the elephant's brain is four times heavier) nor the most convoluted (that of the porpoise is even more wrinkled), it is by all odds the best. Indeed the human brain—the seat of mind, sensation, and thought—is the most complex structure known to humans. It has no parallel in the living or nonliving world. This "great ravelled knot," as the British physiologist Sir Charles Sherrington called the human brain, may in fact be so complex that it will never be able to understand its own function.

The primitive three-part brain is made up of prosencephalon, mesencephalon, and rhombencephalon (forebrain, midbrain, and hindbrain) (Fig. 32-9; Table 32-1). The prosencephalon and rhombencephalon each divide again to form the five-part brain characteristic of the adults of all vertebrates. The five-part brain includes the telencephalon, diencephalon, mesencephalon, metencephalon, and myelencephalon (Table 32-1). From these divisions the different functional brain structures arise.

The impressive evolutionary improvement of the vertebrate brain has accompanied the increased powers of locomotion and greater environmental awareness of the more advanced vertebrates. In the primitive vertebrate brain each of the three parts was concerned with one or more special senses: the prosencephalon

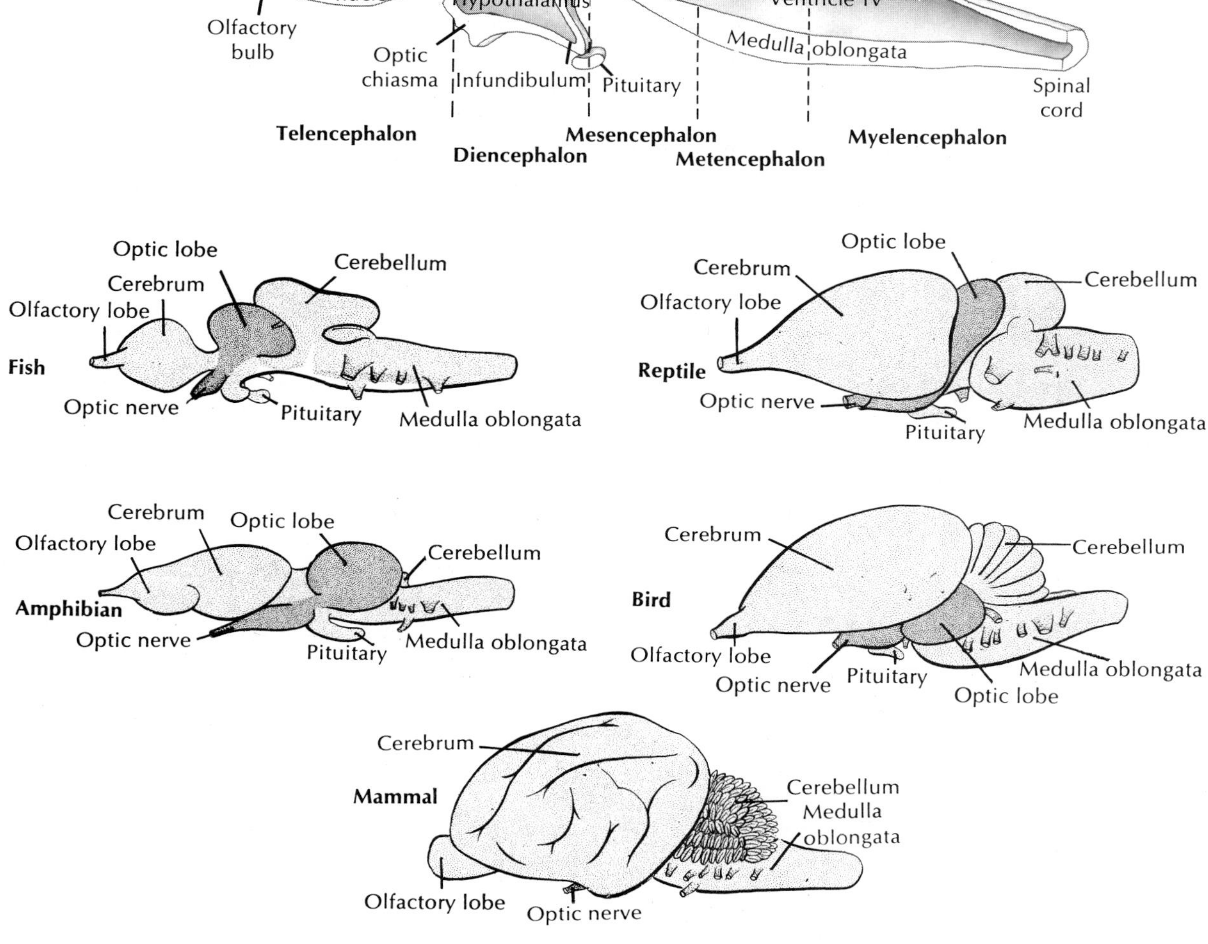

FIG. 32-9 General topography *(top drawing)* and comparative structure of vertebrate brains, showing the principal brain divisions and their development in different vertebrate groups. *(Top, Modified from Romer, A. S. 1949. The vertebrate body, Philadelphia, W. B. Saunders Co.)*

with the sense of smell, the mesencephalon with vision, and the rhombencephalon with hearing and balance. These primitive but very fundamental concerns of the brain have been in some instances amplified, and in others reduced or overshadowed, during continued evolution as sensory priorities were shaped by the animal's habitat and way of life.

The brain is made up of both white and gray matter, with the gray matter on the outside (in contrast to the spinal cord in which the gray matter is inside). The gray matter of the brain is mostly in the convoluted **cortex.** In the deeper white matter of the brain, myelinated bundles of nerve fibers connect the cortex with lower centers of the brain and spinal cord or connect one part of the cortex with another. Also in deeper portions of the brain are clusters of nerve cell bodies

TABLE 32-1 Divisions of the vertebrate brain

Embryonic vesicles		Main components in adults
Prosencephalon (forebrain)	Telencephalon	Olfactory bulbs Cerebrum Lateral ventricles
	Diencephalon	Thalamus Hypothalamus Infundibulum Third ventricle
Mesencephalon (midbrain)	Mesencephalon	Optic lobes (tectum) Cerebral peduncles Red nucleus Aqueduct of Sylvius
Rhombencephalon (hindbrain)	Metencephalon	Cerebellum Pons Fourth ventricle
	Myelencephalon	Medulla

called nuclei (not to be confused with the nuclei of cells) that provide synaptic junctions between the neurons of higher centers and those of lower centers.

The nonnervous elements of the nervous system are the neuroglia (''nerve-glue'') cells. They greatly outnumber the neurons and play indispensable service roles in the functioning of neurons. They bind together the nervous tissue proper, support it metabolically, and serve in the regenerative processes that follow injury or disease. In general, they are necessary to the well-being of neurons. Recently however, some neuroscientists have attributed important information-processing functions to the neuroglia.

The **medulla,** the most posterior division of the brain, is really a conical continuation of the spinal cord. The medulla and the more anterior midbrain constitute the ''brainstem,'' an area in which numerous vital, and largely subconscious, activities are controlled, such as heartbeat, respiration, vasomotor tone, and swallowing. It contains the roots of all the cranial nerves except the first four and is traversed by many sensory and motor fiber tracts. Although it is small in size and largely hidden from view by the much enlarged ''higher'' centers, it is in fact the most vital brain area; whereas damage to higher centers may result in severely debilitating loss of sensory or motor function, damage to the brainstem usually results in death.

The **pons,** between the medulla and the midbrain, is made up of a thick bundle of fibers that carries impulses from one side of the cerebellum to the other.

The **cerebellum,** lying above the medulla, is concerned with equilibrium, posture, and movement. Its development is directly correlated with the animal's mode of locomotion, agility of limb movement, and balance. It is usually weakly developed in amphibians and reptiles, which are relatively clumsy forms that stick close to the ground, and are well developed in the more agile bony fishes. It reaches its apogee in birds and mammals, in which it is greatly expanded and folded. The cerebellum does not initiate movements but operates as a precision error-control center, or servomechanism, that programs a movement initiated somewhere else, such as in the motor cortex. Primates, and especially human beings, which possess a manual dexterity far surpassing that of other animals, have the most complex cerebellum of all, since hand and finger movements may involve the simultaneous contraction and relaxation of hundreds of individual muscles.

Between the medulla and diencephalon is the **midbrain.** This is the anterior portion of the brainstem. The white matter of the midbrain consists of ascending and descending tracts that go to the thalamus and cerebrum. On the upper side of the midbrain are the rounded **optic lobes,** serving as centers for visual and auditory reflexes. The midbrain has undergone little evolutionary change in size among vertebrates but has changed in function. It mediates the most complex behavior of fishes and amphibians. Such integrative functions were gradually assumed by the forebrain in higher vertebrates, and in mammals the midbrain is mainly a reflex center for eye muscles and a relay and analysis center for auditory information.

The **thalamus,** above the midbrain, contains masses of gray matter surrounded by the cerebral hemispheres on each side. This is the relay center for the sensory tracts from the spinal cord. Centers for the sensations of pain, temperature, and touch are supposedly located in the thalamus. In the **hypothalamus** are ''housekeeping'' centers that regulate body temperature, water balance, sleep, appetite, and a few other body functions. The hypothalamus also has neurosecretory cells that produce pituitary-regulating neurohormones.

The anterior region of the brain, the **cerebrum,** can be divided into two anatomically distinct areas: the **paleocortex** and the **neocortex.** As its name implies, the paleocortex is the ancient telencephalon. Originally concerned with smell, it became well developed in the advanced fishes and early terrestrial vertebrates, which depend on this special sense. In mammals, and

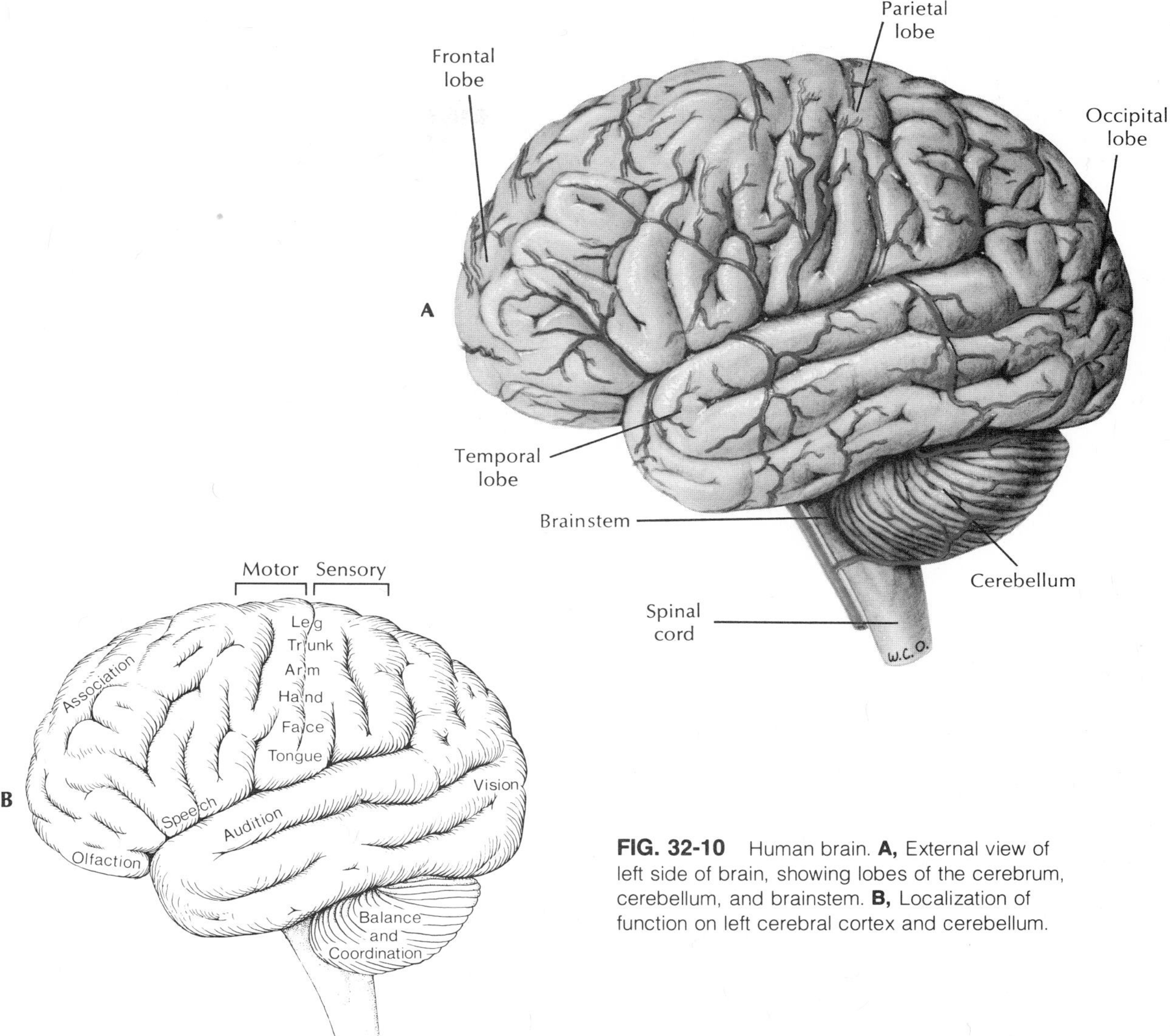

FIG. 32-10 Human brain. **A,** External view of left side of brain, showing lobes of the cerebrum, cerebellum, and brainstem. **B,** Localization of function on left cerebral cortex and cerebellum.

especially in primates, the paleocortex is a deep-lying area called the rhinencephalon (''nose brain''), which actually has little to do with the sense of smell. Instead it seems to have acquired a variety of ill-defined functions concerned with consciousness, the expression of emotions, and sex. Together with a portion of the midbrain it is often called the **limbic system.**

Though a late arrival in vertebrate evolution, the neocortex completely overshadows the paleocortex and has become so expanded that it envelopes the diencephalon and midbrain (Fig. 32-10). Almost all the integrative activities primitively assigned to the midbrain were transferred to the neocortex, or **cerebral cortex** as it is usually called.

Thus in mammals, and especially in human beings, there are two brains, a primitive and an advanced, that mediate quite separate functions. The deep, primitive brain—all of the brain but the cerebral cortex—governs the numerous vital functions that are removed from conscious control: respiration, blood pressure, heart rate, hunger, thirst, temperature balance, salt balance, sexual drive, and basic (sometimes irrational)

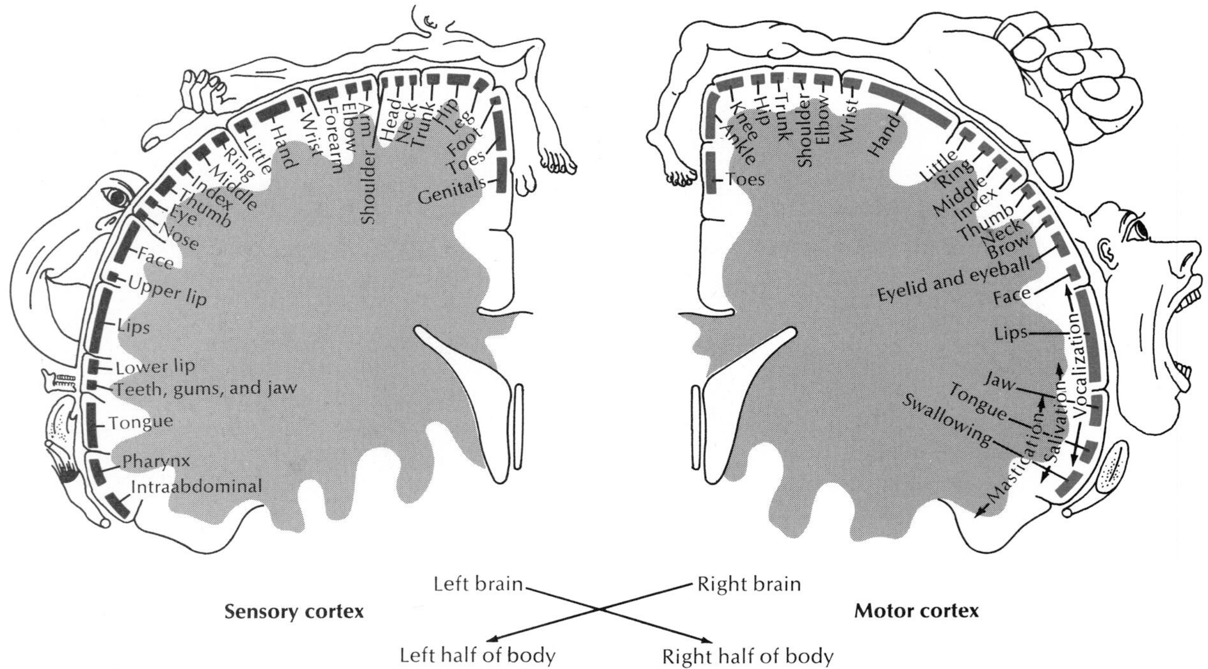

FIG. 32-11 Arrangement of sensory and motor cortices. The location of sensory terminations from different parts of the body are shown at left; the sensory pattern is repeated as a mirror image on the right side. The origin of descending motor pathways to different parts of the body is shown at right. This pattern is repeated on the left brain. The motor cortex lies in front of the sensory (Fig. 32-10, *B*), so the two are not superimposed. Note that ascending and descending tracts from each hemisphere connect to the opposite side of the body.

emotion. It is also a complex endocrine gland that regulates the body's subservient endocrine system. This primitive brain is the *unconscious* mind. The other brain is the "new" brain, the cerebral cortex, wherein are seated intellect and reason. This governing brain is the *conscious* mind.

The brain, of course, operates as a whole, and both the primitive and the advanced brains are intimately interconnected. Unconscious disturbances are communicated to the conscious brain, and the conscious brain may have powerful effects on the unconscious. Memory appears to transcend all brain levels rather than being a property of any particular part of the brain as was once thought.

The cerebral cortex is incompletely divided into two hemispheres by a deep longitudinal fissure. The right and left hemispheres are bridged through the **corpus callosum,** a neural mass lying between and connecting both hemispheres. Via the corpus callosum the two hemispheres are able to transfer information and coordinate mental activities.

Until recently it was thought that one hemisphere, almost always the left, becomes functionally dominant over the other during childhood. This concept is now recognized as misleading. We know now that the left and right brain hemispheres are specialized for entirely different functions: the left brain hemisphere (controlling the right side of the body) for language development, mathematical and learning capabilities, and sequential thought processes; and the right brain hemisphere (controlling the left side of the body) for spatial, musical, artistic, intuitive, and perceptual activities. It has long been known that even extensive damage to the right hemisphere may cause varying degrees of left-sided paralysis but has little effect on intellect. Conversely damage to the left hemisphere usually has

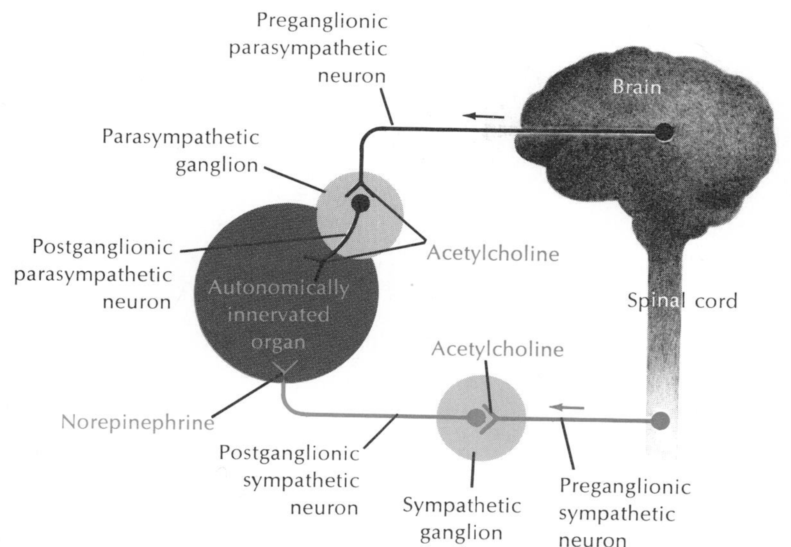

FIG. 32-12 Parasympathetic and sympathetic divisions of the autonomic nervous system, showing location of ganglia. Short preganglionic fibers of sympathetic division *(red)* pass from spinal cord to sympathetic ganglion, which lies close to the spinal cord; long postganglionic fibers continue to effector organ. In contrast, parasympathetic ganglion *(black)* lies in wall of effector organ, and the preganglionic fiber is much longer than the postganglionic. The sympathetic transmitter substance between the postganglionic fiber and effector organ is norepinephrine; at the equivalent parasympathetic ending the transmitter is acetylcholine.

disastrous effects on intellect. Since these differences in brain symmetry and function exist at birth, they appear to be inborn rather than the result of developmental or environmental effects as previously thought. Most people are right-handed (left hemisphere), but handedness is determined separately since it is not always related to left cerebral speech dominance.

It has been possible to localize function in the cerebrum by direct stimulation of exposed brains of people and experimental animals, by postmortem examination of people suffering from various lesions, and by surgical removal of specific brain areas in experimental animals. The cortex contains discrete motor and sensory areas (Figs. 32-10 and 32-11) as well as large "silent" regions, called **association areas,** concerned with memory, judgment, reasoning, and other integrative functions. These regions are not directly connected to sense organs or muscles.

Peripheral nervous system

The peripheral nervous system, the nerve processes connecting the central nervous system to receptors and effectors, can be broadly subdivided into afferent and efferent components. As shown in the following outline, the efferent system is considerably more complex, consisting of a somatic nervous system and an autonomic nervous system.

I. Afferent system (sensory)
II. Efferent system (motor)
   A. Somatic nervous system
   B. Autonomic nervous system
      1. Sympathetic nervous system
      2. Parasympathetic nervous system

Afferent system

Afferent (sensory) neurons carry signals from receptors in the periphery of the body to the central nervous system. The afferent neuron of a reflex arc (Fig. 32-7) is representative of all afferent pathways in the peripheral nervous system. One long nerve process extends from the cell body in the dorsal root ganglion just outside the spinal cord to innervate receptors; another process passes from the cell body into the central nervous system where it connects with other neurons.

Efferent system

Somatic nervous system. Nerve fibers of the somatic division of the peripheral nervous system pass from the brain or the spinal cord to skeletal muscle fibers. These are often called **motor neurons** because, when active, they always cause contraction of muscle cells. Motor neurons in the somatic nervous system may be activated by local reflexes or by higher brain centers. Whether they originate in the brain or in the spinal cord, motor neurons pass without interruption (that is, without synapses) directly to skeletal muscle. They release acetylcholine as the transmitter substance at the nerve endings.

Autonomic nervous system. The autonomic nerves govern the involuntary functions of the body that do not ordinarily affect consciousness. Thus, most people cannot by volition stimulate or inhibit their action. Autonomic nerves control the movements of the alimentary canal and heart, the contraction of the smooth muscle of the blood vessels, urinary bladder, iris of eye, and others, and the secretions of various glands.

Subdivisions of the autonomic system are the **parasympathetic** and the **sympathetic** systems (Fig. 32-12). Most organs in the body are innervated by both sympathetic and parasympathetic fibers, and their actions are antagonistic (Fig. 32-13). If one speeds up an activity, the other will slow it down. However, neither kind of nerve is exclusively excitatory or inhibitory. For example, parasympathetic fibers inhibit heartbeat but will excite peristaltic movements of the intestine; sympathetic fibers will increase heartbeat but slow down peristaltic movement.

The **parasympathetic** system consists of motor nerves, some of which emerge from the brain by certain cranial nerves and others from the pelvic region of the spinal cord by certain spinal nerves. Parasympathetic fibers *excite* the stomach and intestine, urinary bladder, sex organs, bronchi, constrictor of iris, salivary glands, and coronary arteries. Other parasympathetic fibers *inhibit* the heart, intestinal sphincters, sex organs, and sphincter of the urinary bladder.

In the **sympathetic** division the nerve cell bodies are located in the thoracic and upper lumbar areas of the spinal cord. Their fibers pass out through the ventral roots of the spinal nerves, separate from these, and go to the sympathetic ganglia, which are paired and form a chain on each side of the spinal column. From these ganglia some of the fibers run through spinal nerves to the limbs and body wall, where they innervate the blood vessels of the skin, the smooth muscles of the hair, the sweat glands, and others; and some run to the abdominal organs as the splanchnic nerves. Sympathetic fibers *excite* the heart, blood vessels, sphincters of the intestines, urinary bladder, dilator muscles of the iris, and others. They *inhibit* the stomach, intestine, bronchial muscles, and coronary arterioles.

All preganglionic fibers, whether sympathetic or parasympathetic, release **acetylcholine** at the synapse for stimulating the ganglion cells. The terminations of the parasympathetic and sympathetic nervous systems release different types of chemical transmitter substances (Fig. 32-12). The parasympathetic fibers release **acetylcholine** at their endings whereas the sympathetic fibers release **norepinephrine** (also called noradrenaline). These chemical substances produce characteristic physiologic reactions. Since there is some physiologic overlapping of sympathetic and parasympathetic fibers, it is now customary to describe nerve fibers as either **adrenergic** (norepinephrine effect) or **cholinergic** (acetylcholine effect).

SENSE ORGANS

Animals require a constant inflow of information from the environment to regulate their lives. Sense organs are specialized receptors designed for detecting environmental status and change. An animal's sense organs are its first level of environmental perception; they are data input channels for the brain.

A **stimulus** is some form of energy—electric, mechanical, chemical, or radiant. The task of the sense organ is to transform the energy form of the stimulus it receives into nerve action potentials (impulses), the common language of the nervous system. In a very real sense, then, sense organs are **biologic transducers.** A microphone, for example, is an artificial transducer that converts mechanical (sound) energy into electric energy. And like the microphone that is sensitive only to sound, sense organs are, as a rule, quite specific for one kind, or **modality,** of stimulus energy. Thus eyes respond only to light, ears to sound, pressure receptors to pressure, and chemoreceptors to chemical molecules. But, again, all of these different forms of energy are converted into nerve action potentials.

Since all nerve impulses are qualitatively alike, how do animals perceive and distinguish the different **sensations** of varying stimuli? The answer is that the real perception of sensation is done in localized regions of the brain, where each sense organ has its own hookup. Action potentials arriving at a particular sensory area of the brain can be interpreted in only one way. This is why pressure on the eye causes us to see "stars" or other visual patterns; the mechanical distortion of the eye initiates impulses in the optic nerve fibers that are perceived as light sensations. Although the operation probably could never be done, the deliberate surgical

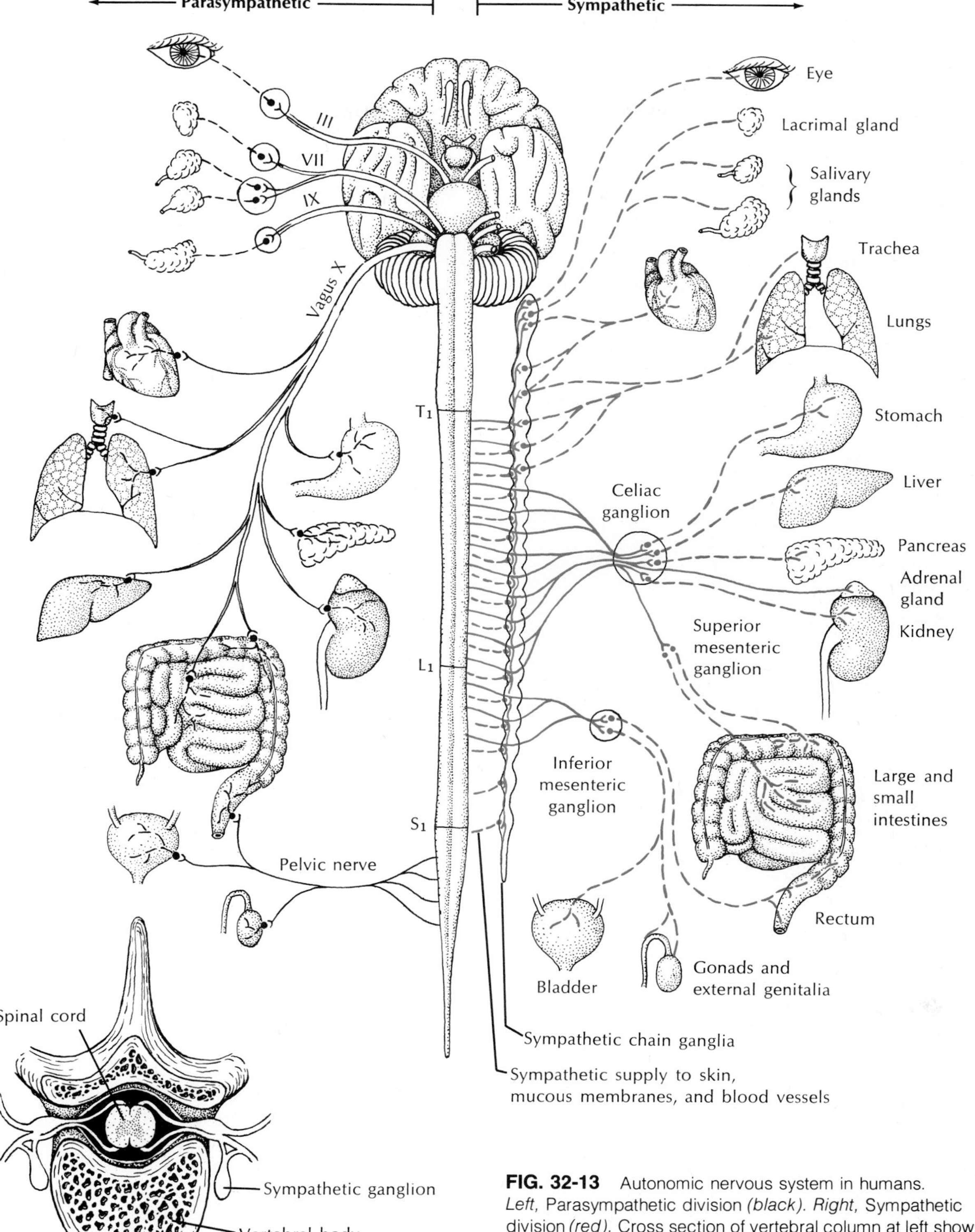

FIG. 32-13 Autonomic nervous system in humans. *Left,* Parasympathetic division *(black). Right,* Sympathetic division *(red).* Cross section of vertebral column at left shows position of sympathetic ganglia.

switching of optic and auditory nerves would cause the recipient to quite literally see thunder and hear lightning!

Classification of receptors

Receptors are classified on the basis of their location. Those near the external surface are called **exteroceptors** and are stimulated by changes in the external environment. Internal parts of the body are provided with **interoceptors,** which pick up stimuli from the internal organs. Muscles, tendons, and joints have **proprioceptors,** which are sensitive to changes in the tension of muscles and provide the organism with a sense of position.

Another way of classifying receptors is on the basis of the energy form used to stimulate them, such as **chemical, mechanical, photo,** or **thermal.**

A

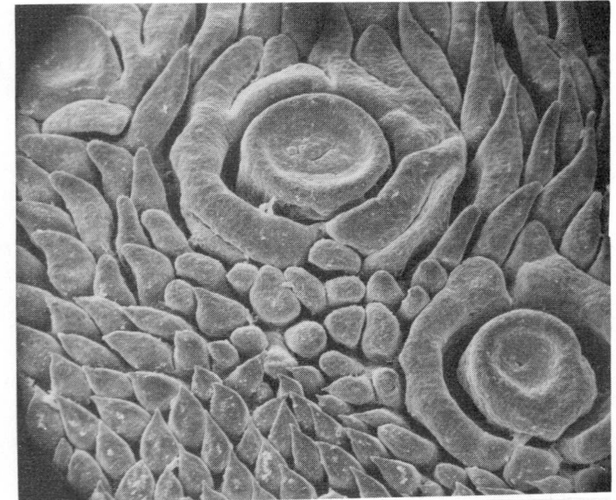

B

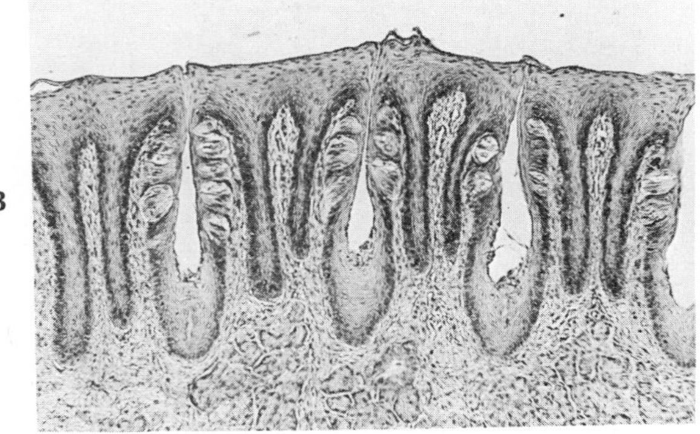

FIG. 32-14 Taste buds. **A,** Scanning electron micrograph of circumvallate papillae on surface of tongue of a young dog. Taste buds (not visible) are located in walls of circular trench surrounding papillae. The numerous filiform papillae surrounding the two circumvallate papillae lack taste buds. (×55.) **B,** Light micrograph of section of rabbit's tongue. Taste buds are little oval bodies on sides of slitlike recesses. (×400.) (**A,** Courtesy P. P. A. Graziadei; **B,** courtesy J. B. Bamberger.)

Chemoreception

Chemoreception is the most primitive and most universal sense in the animal kingdom. It probably guides the behavior of animals more than any other sense. The most primitive animals, protozoans, use **contact chemical receptors** to locate food and adequately oxygenated water and to avoid harmful substances. These receptors elicit a simple trial-and-error behavior, called **chemotaxis.** More advanced animals have specialized **distance chemical receptors.** These are often developed to a truly amazing degree of sensitivity and are responsible for complex behavioral activities. Distance chemoreception, usually referred to as sense of smell, or olfactory sense, guides feeding behavior, location and selection of sexual mates, territorial and trail marking, and alarm reactions of numerous animals.

Many animals produce species-specific odors, called **pheromones,** which comprise a highly developed chemical language. Pheromones are a diverse group of organic compounds released by epithelial glands that serve either to initiate specific patterns of behavior, such as attracting mates or marking trails (releaser pheromones), or to trigger some internal physiologic change such as metamorphosis (primer pheromones). Insects have a variety of chemoreceptors on the body surface for sensing specific pheromones as well as other nonspecific odors.

In all vertebrates and in insects as well, the senses of **taste** and **smell** are clearly distinguishable. Although there are similarities between taste and smell receptors, in general the sense of taste is more restricted in response and is less sensitive than the sense of smell. Taste and smell centers are also located in different parts of the brain.

In higher forms, **taste buds** are found on the tongue and other parts of the mouth cavity (Fig. 32-14). A taste bud consists of a few sensitive cells surrounded by supporting cells and is provided with a small external pore through which the slender tips of the sensory cells project. The basal ends of the sensory cells contact nerve endings from cranial nerves. Taste bud cells in vertebrates have a short life of about 10 days and are continually being replaced. Taste buds are more numerous in ruminants (mammals that chew the cud) than in human beings. They tend to degenerate with age; a child has more buds widely distributed over the mouth.

The four basic taste sensations of human beings—sour, salt, bitter, and sweet—are each attributable to a different kind of taste bud. The tastes for salt and sweet are found mainly at the tip of the tongue, bitter at the base of the tongue, and sour along the sides of the tongue. Of these, bitter taste is by far the most sensitive, since it protects the body against some toxic substances, many of which are bitter.

Sense organs of **smell** (olfaction) are found in a specialized epithelium located either in the nasal cavity (terrestrial vertebrates) or in pouches on the snout (aquatic vertebrates). The sense of smell is much more complex than that of taste. There are millions of olfactory receptor cells in the nasal epithelium, and as many as a thousand of these may converge on a single neuron. This allows great summation and vastly improves sensitivity. Some people can detect many thousands of different odors, and it is obvious that many other vertebrates can easily outdo us. Gases must be dissolved in a fluid to be smelled; therefore the nasal cavity must be moist. The sensory cells with projecting hairs are scattered singly through the olfactory epithelium. Their basal ends are connected to fibers of the olfactory cranial nerve that runs to one of the olfactory lobes. The sensitivity to certain odors approaches the theoretical maximum for the chemical sense. The human nose can detect $^1/_{25}$ millionth of 1 mg of mercaptan, the odoriferous principle of the skunk. This averages out to about 1 molecule per sensory ending.

Since taste and smell are stimulated by chemicals in solution, their sensations may be confused. The taste of food is dependent to a great extent on odors that reach the olfactory membrane through the throat. All the various "tastes" other than the four basic ones (sweet, sour, bitter, salt) are really the result of the flavors' reaching the sense of smell in this manner. The sense of smell is the least understood sense. Of the numerous theories that have been proposed, the favored ones today postulate some kind of **physical interaction** between the odor molecule and a protein receptor site on a cell membrane. This interaction somehow alters membrane permeability and leads to depolarization in the receptor cell, which triggers a nerve impulse. One theory (J. E. Amoore) proposes that odor molecules have specific stereochemical shapes and that the range of detectable odors is attributable to differences in the way the molecule smelled fits the receptor site.

Mechanoreception

Mechanoreceptors are sensitive to quantitative forces such as touch, pressure, stretching, sound, and gravity. Many receptors in and on the body constantly monitor information about conditions within the body (muscle position, body equilibrium, blood pressure, pain, and others) and conditions in the environment (sound and other vibrations such as water currents).

Touch and pain

Although superficial touch receptors are distributed over all the body, they tend to be concentrated in the few areas especially important in exploring and interpreting the environment. In most animals these areas are on the face and limb extremities. Of the more than half a million separate sensitive spots on the human body's surface, most are found on the lips, tongue, and fingertips. Many touch receptors are bare nerve-fiber terminals, but there is an assortment of other kinds of receptors of varying shapes and sizes. Each hair follicle is crowded with receptors that are sensitive to touch.

The sensation of deep touch and pressure is registered by relatively large receptors called **pacinian corpuscles.** They are common in deep layers of skin (Fig. 28-1, *B*, p. 647), in connective tissue surrounding muscles and tendons, and in the abdominal mesenteries. Each corpuscle, easily visible to the naked eye, is built of numerous layers like an onion. Any kind of mechanical deformation of the pacinian corpuscles is converted into nerve action potentials that are sent to sensory areas of the brain.

Pain receptors are relatively unspecialized nerve fiber endings that respond to a variety of stimuli that signal possible or real tissue damage. It is still uncertain whether pain fibers respond directly to injury or indirectly to some substance, such as histamine, which is released by damaged cells.

Lateral line system of fishes

The lateral line is a distant touch reception system for detecting wave vibrations and currents in water.

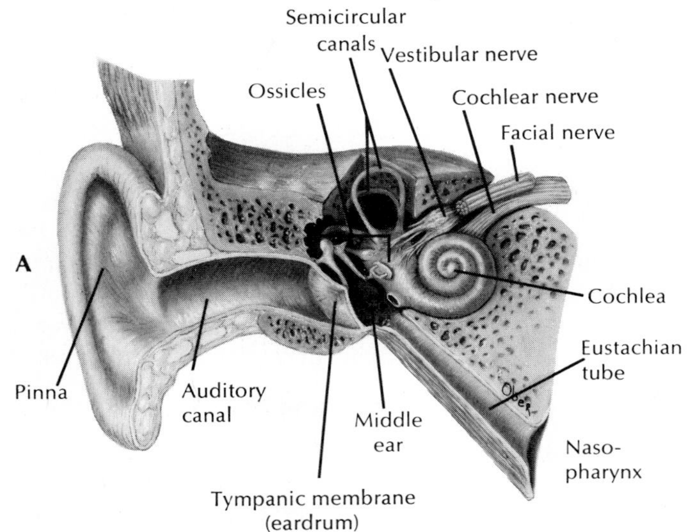

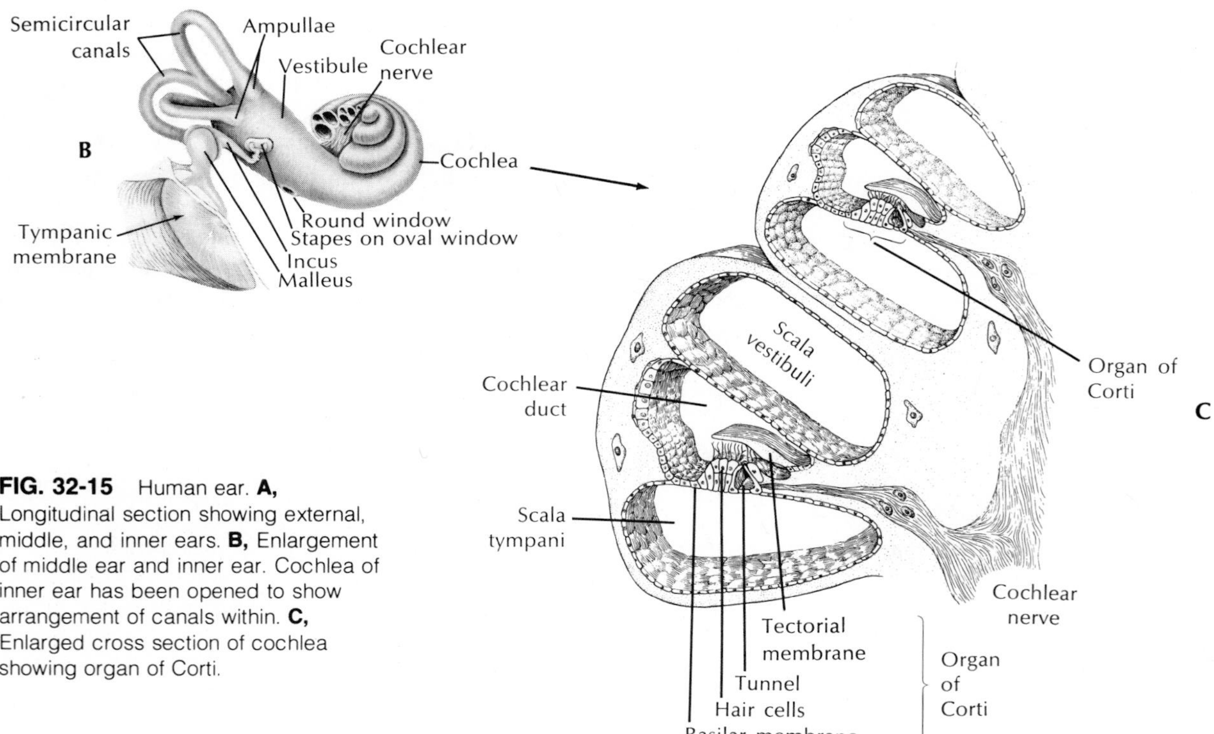

FIG. 32-15 Human ear. **A,** Longitudinal section showing external, middle, and inner ears. **B,** Enlargement of middle ear and inner ear. Cochlea of inner ear has been opened to show arrangement of canals within. **C,** Enlarged cross section of cochlea showing organ of Corti.

The receptor cells, called **neuromasts,** are located free on the body surface in primitive fishes and aquatic amphibians, but in the advanced fishes they are located within canals running beneath the epidermis; at intervals these open to the surface. Each neuromast is a collection of hair cells with the sensory hairs embedded in a gelatinous, wedge-shaped mass, known as a **cupula.** This projects into the center of the lateral line canal so that it will bend in response to any disturbance of water on the body surface. The lateral line system is one of the principal sensory systems that guide fishes in their movements.

Hearing

The ear is a specialized receptor for detecting sound waves in the surrounding air (Fig. 32-15). Another sense, equilibrium, is also associated with the ears of all vertebrate animals. Among the invertebrates, only certain insects have true sound receptors.

In its evolution the vertebrate ear was at first associated more with equilibrium than with hearing. Hearing sense is found only in the internal ear, which is the only part of the ear present in many of the lower vertebrates; the middle and the external ears were added in later evolutionary developments. The internal ear is considered to be a development of part of the lateral line system of fishes. Some fishes apparently can transmit sound from their swim bladders by the weberian ossicles (series of small bones) to some part of the inner ear, since they lack a cochlea.

The ear found in higher vertebrates is made up of three parts: (1) the **inner ear,** which contains the essential organs of hearing and equilibrium and is present in all vertebrates; (2) the **middle ear,** an air-filled

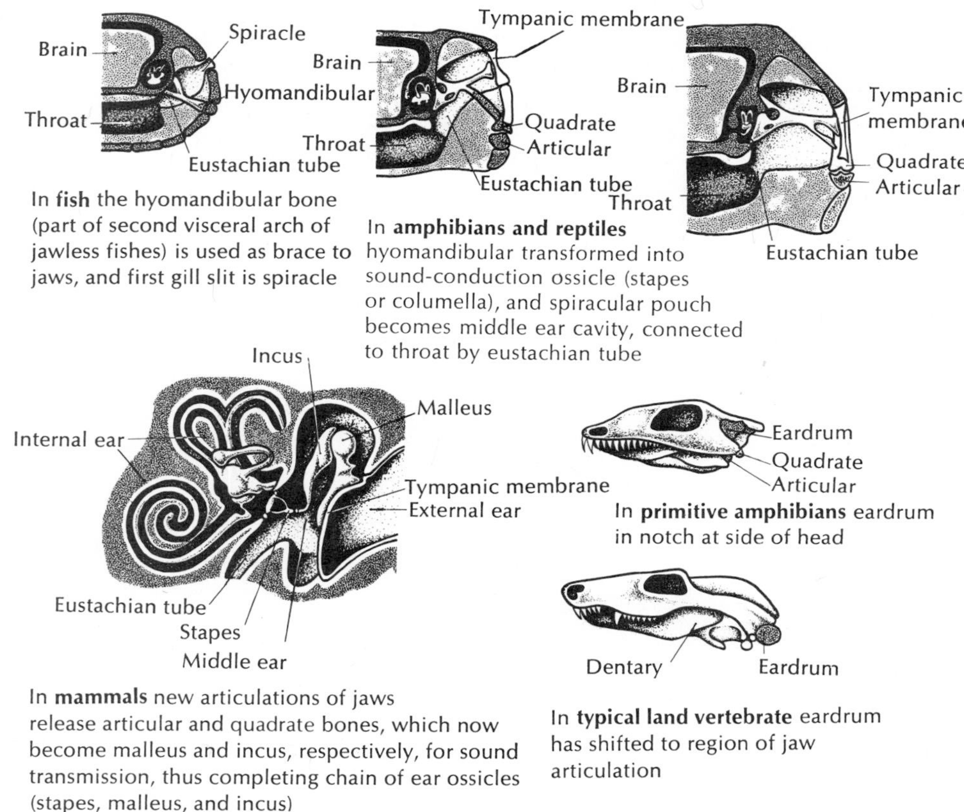

FIG. 32-16 Evolution of middle ear and auditory ossicles. Diagrammatic sections are made through otic region at level of ear and hind end of jaw. (Redrawn from Romer, A. S. 1959. The vertebrate story. Chicago, University of Chicago Press.)

chamber with one or more ossicles for conducting sound waves to the inner ear, present in amphibians and higher vertebrates only (Fig. 32-16); and (3) the **outer ear,** which collects the sound waves and conducts them to the tympanic membrane lying next to the middle ear, present only in reptiles, birds, and mammals. It is most highly developed in the mammals.

Outer ear. The outer, or external, ear of higher vertebrates is made up of two parts: (1) the **pinna,** or skin-covered flap of elastic cartilage and muscles, and (2) the **auditory canal** (Fig. 32-15, *A*). In many mammals, such as the rabbit and cat, the pinna is freely movable and is effective in collecting sound waves. The auditory canal condenses the waves and passes them to the tympanic membrane. The walls of the auditory canal are lined with hair and wax-secreting glands as a protection against the entrance of foreign objects.

Middle ear. The middle ear is separated from the external ear by the **tympanic membrane** (eardrum), which consists of a tightly stretched connective membrane. Within the air-filled middle ear a remarkable chain of three tiny bones—**malleus** (hammer), **incus** (anvil), and **stapes** (stirrup)—conduct the sound waves across the middle ear (Fig. 32-15, *B*). This bridge of bones is so arranged that the force of sound waves pushing against the eardrum is amplified as many as 90 times where the stapes contacts the oval window of the inner ear. Muscles attached to the middle ear bones contract when the ear receives very loud noises, thus protecting the inner ear from damage. However, these muscles cannot contract quickly enough to protect the inner ear from the damaging effects of a sudden blast. The middle ear communicates with the pharynx by means of the eustachian tube, which acts as a safety device to equalize pressure on both sides of the eardrum.

Inner ear. The inner ear consists essentially of two labyrinths, one within the other. The inner one is called the **membranous labyrinth** and is a closed ectodermal sac filled with a fluid, **endolymph.** The part involved with hearing, the **cochlea** is coiled like a snail's shell, making two and a half turns in the human ear (Fig. 32-15, *B*). Surrounding the membranous labyrinth is the **bony labyrinth,** which is a hollowed-out part of the temporal bone and conforms to the shape and contours of the membranous labyrinth. **Perilymph,** a fluid similar to endolymph, is found in the space between the two labyrinths.

The cochlea is divided into three longitudinal canals that are separated from each other by thin membranes (Fig. 32-15, *B* and *C*). These canals become progressively smaller from the base of the cochlea to the apex. One of these canals is called the **vestibular canal** (scala vestibuli); its base is closed by the oval window. The **tympanic canal** (scala tympani), which is in communication with the vestibular canal at the tip of the cochlea, has its base closed by the round window. Between these two canals is the **cochlear duct,** which contains the organ of hearing, the **organ of Corti** (Fig. 32-15, *C*). The latter organ is made up of fine rows of hair cells that run lengthwise from the base to the tip of the cochlea. There are at least 24,000 of these hair cells in the human ear, each cell with many hairs projecting into the endolymph of the cochlear canal and each connected with neurons of the auditory nerve. The hair cells rest on the **basilar membrane,** which separates the tympanic canal and cochlear duct, and are covered over by the **tectorial membrane** found directly above them. These relationships are shown diagrammatically in Fig. 32-17.

Sound waves picked up by the external ear are transmitted through the auditory canal to the eardrum, which is caused to vibrate. These vibrations are conducted by the chain of ear bones to the oval window, which transmits the vibrations to the fluid in the vestibular and tympanic canals. The vibrations of the endolymph cause the basilar membrane with its hair cells to vibrate, so that the latter are bent against the tectorial membrane. This stimulation of the hair cells causes them to initiate nerve action potentials in the fibers of the auditory nerve, with which they are connected.

According to the **place theory** of pitch discrimination, when sound waves strike the inner ear the basilar membrane is set in vibration by a traveling wave of displacement, which increases in amplitude from the oval window toward the apex of the cochlea. This displacement wave reaches a maximum at the region of the basilar membrane where the natural frequency of the membrane corresponds to the sound frequency. Here, the membrane vibrates with such ease that the energy of the traveling wave is completely dissipated. Hair cells in that region will be stimulated, setting up action potentials in fibers of the auditory nerve.

Those impulses that are carried by certain fibers of the auditory nerve are interpreted by the hearing center as particular tones. The **loudness** of a tone depends

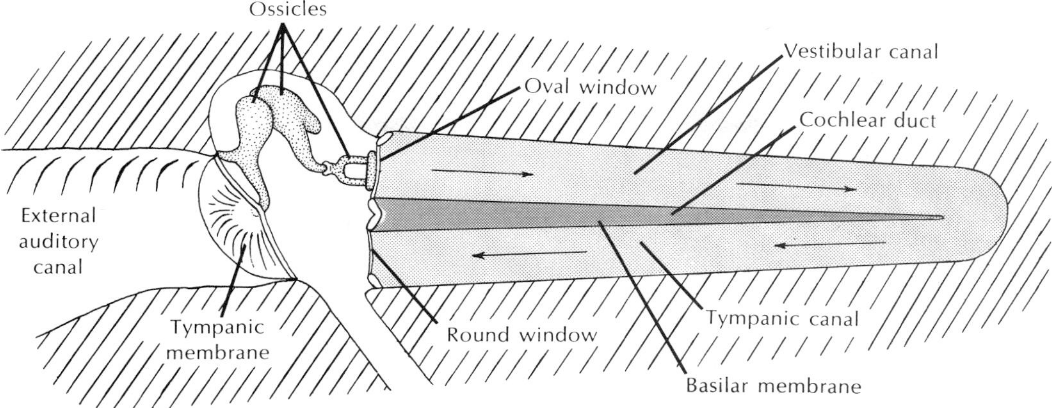

FIG. 32-17 Mammalian ear as it would appear with cochlea stretched out. Sound waves transmitted to the oval window produce vibration waves that travel down the basilar membrane. High-frequency vibrations cause the membrane to resonate at the end near the oval window before dying out; low-frequency tones travel farther down the basilar membrane.

on the number of hair cells stimulated, whereas the **timbre,** or quality, of a tone is produced by the pattern of the hair cells stimulated by sympathetic vibration. This latter characteristic of tone enables one to distinguish among different human voices and different musical instruments, even though the notes in each case may be of the same pitch and loudness.

Sense of equilibrium

Closely connected to the inner ear and forming a part of it are two small sacs, the **saccule** and **utricle,** and three **semicircular canals.** Like the cochlea, the sacs and canals are filled with endolymph. They are concerned with the sense of balance and rotation. They are well developed in all vertebrates, and in some lower forms they represent about all there is of the internal ear, for the cochlea is absent in fishes. They are innervated by the nonacoustic branch of the auditory nerve.

The utricle and saccule are hollow sacs lined with sensitive hairs on which are deposited a mass of minute calcium carbonate crystals called **otoconia.** In bony fishes, these crystals are formed into compact stonelike structures, the **otoliths.** Similar stony accretions are found within statocysts, the balance organs of many invertebrates.

Although the anatomic nature of these static balance organs varies in different groups, they all function in the same basic way: the weight of the stony accretion presses on the hair cells to give information about the position of the head (or entire body) relative to the force of gravity. As the head is tilted in one direction or another, different groups of hair cells are stimulated; conveyed to the brain, this information is interpreted with reference to position.

The semicircular canals of vertebrates are designed to detect changes in movement: acceleration or deceleration. The three semicircular canals are at right angles to each other, one in each plane of space (Fig. 32-15, *B*). They are filled with fluid, and at the opening of each canal into the utricle there is a bulblike enlargement, the **ampulla,** which contains hair cells but no otoconia. Whenever the fluid moves, these hair cells are stimulated. Rotating the head will cause a lag, because of inertia, in certain of these ampullae. This lag produces consciousness of movement. Since the three canals of each internal ear are in different planes, any kind of movement will stimulate at least one of the ampullae.

Vision

Light-sensitive receptors are called **photoreceptors.** These receptors range all the way from simple light-sensitive cells scattered randomly on the body surface of the lowest invertebrates (dermal light sense) to the exquisitely developed vertebrate eye. Although dermal

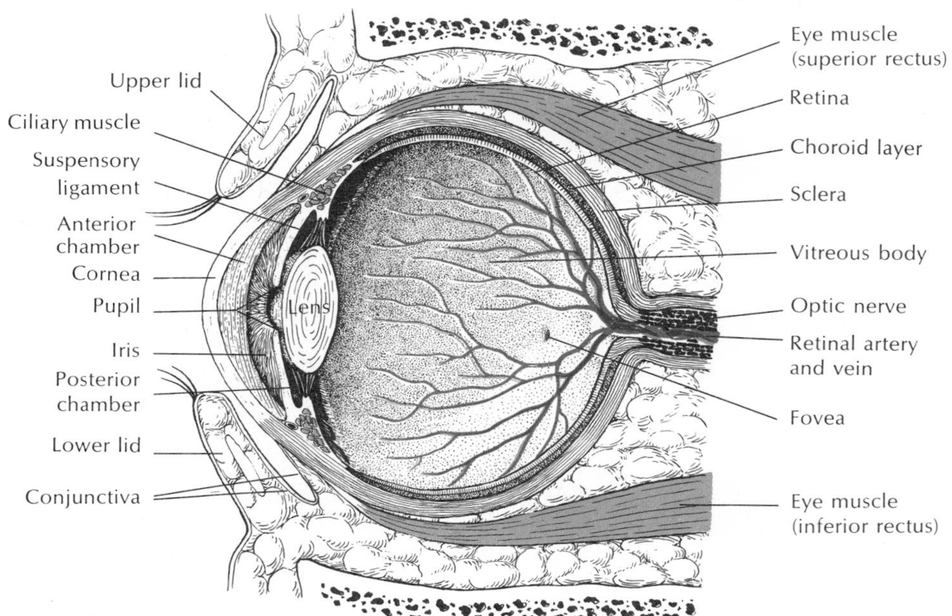

FIG. 32-18 Section through human eye.

light receptors contain little photochemical substance and are far less sensitive than optic receptors, they are important in locomotory orientation, pigment distribution in chromatophores, photoperiodic adjustment of reproductive cycles, and other behavioral changes in many lower invertebrates.

The arthropods have **compound** eyes composed of many independent visual units called **ommatidia.** The eye of a bee contains about 15,000 of these units, each of which views a separate narrow sector of the visual field. Such eyes form a mosaic of images from the separate units. The compound eye probably does not produce a very distinct image of the visual field, but it is extremely well suited to detect motion, as anyone knows who has tried to swat a fly.

The vertebrate eye is built like a camera—or rather we should say a camera is modeled somewhat after the vertebrate eye. It contains a light-tight chamber with a lens system in front that focuses an image of the visual field on a light-sensitive surface (the retina) in back (Fig. 32-18). Because eyes and cameras are based on the same laws of optics, we can wear glasses to correct optic defects in our eyes. But here the similarity between eye and camera ends. The human eye is actually replete with optic shortcomings; projected on the retina of the normal eye are more colored fringe halos, apparitions, and distortions than would be produced by even the cheapest camera lens. Yet the human brain corrects for this "poor design" so completely that we perceive a perfect image of the visual field. It is in the retina and the optic center of the brain that the marvel of vertebrate vision can be understood.

The spherical eyeball is built of three layers: (1) a tough outer white **sclerotic coat** (sclera) serving for support and protection, (2) the middle **choroid coat** containing blood vessels for nourishment, and (3) the light-sensitive **retinal** coat (Fig. 32-18). The **cornea** is a transparent modification of the sclera. A circular curtain, the **iris,** regulates the size of the light opening, the **pupil.** Just behind the iris is the **lens,** a transparent, elastic ball that bends the rays and focuses them on the retina. In land vertebrates the cornea actually does most of the bending of light rays, while the lens adjusts the focus for near and far objects. Between the cornea and the lens is the outer chamber filled with the watery **aqueous humor;** between the lens and the retina is the much larger inner chamber, filled with the viscous **vitreous humor.** Surrounding the margin of the lens and holding it in place is the **suspensory ligament.** This, together with the **ciliary muscle,** a

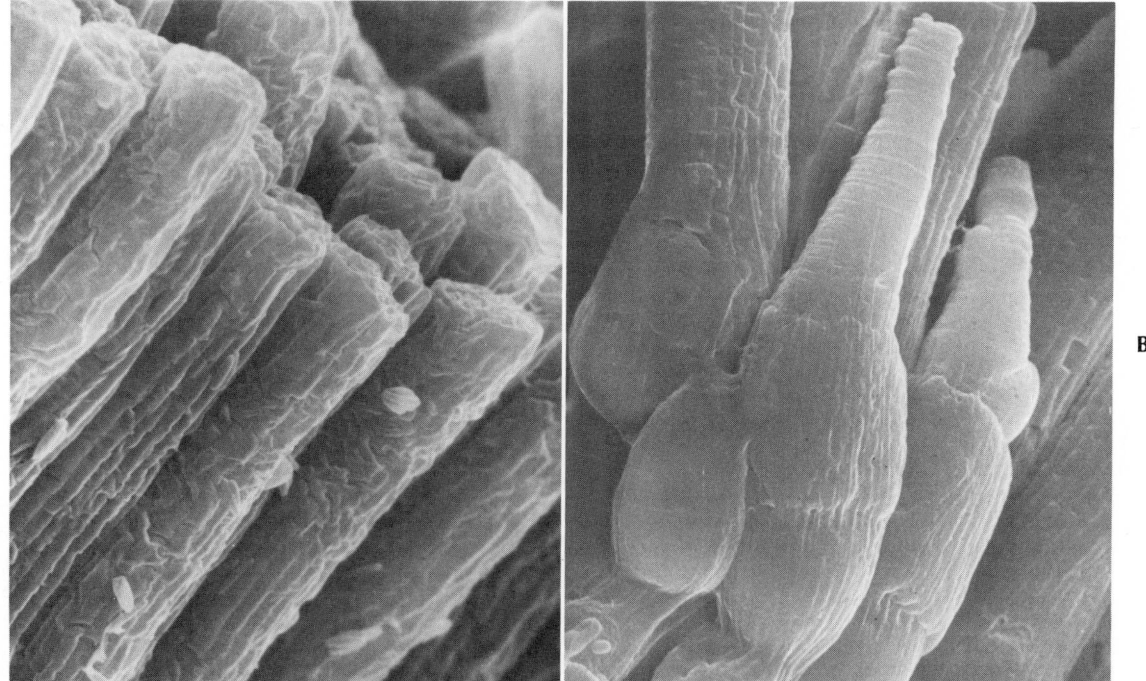

FIG. 32-19 Rods and cones of vertebrate retina. **A,** Outer segments of rods of bullfrog eye. Dendrites of nerve fibers pass up vertical fissures in rods. **B,** Cones and rods of mud puppy *(Necturus)* eye. Cone in center consists of conic outer segment and bulb-shaped inner segment giving rise to nerve fiber at base. Cones are much less sensitive to light than rods, and function in color vision. (Scanning electron micrographs, ×5,000.) (**A,** Courtesy E. R. Lewis; **B,** from Lewis, E. R., Y. Y. Zeevi, and F. S. Werblin. 1969. Brain Res. **15:**559-562.)

ring of radiating muscle fibers attached to the suspensory ligament, makes possible the stretching and relaxing of the lens for close or distant vision (accommodation).

The **retina** is composed of photoreceptors, the **rods** and **cones** (Fig. 32-19). Approximately 125 million rods and 7 million cones are present in each human eye. Cones are primarily concerned with color vision in ample light; rods, with colorless vision in dim light. The retina is actually made up of three sets of neurons in series with each other: (1) photoreceptors (rods and cones), (2) intermediate neurons, and (3) ganglionic neurons whose axons form the optic nerve.

The **fovea centralis,** the region of keenest vision, is located in the center of the retina, in direct line with the center of the lens and cornea. It contains only cones. The acuity of an animal's eyes depends on the density of cones in the fovea. The human fovea and that of a lion contain about 150,000 cones per square millimeter. But many water and field birds have up to 1 million cones per square millimeter. Their eyes are as good as human eyes would be if aided by eight-power binoculars.

Only rods occur at the peripheral parts of the retina. We can see better at night by looking out of the corners of our eyes because the rods, adapted for high sensitivity with dim light, are brought into use.

Chemistry of vision

Each rod contains a light-sensitive pigment known as **rhodopsin.** Each rhodopsin molecule consists of a large, colorless protein, **opsin,** and a small carotenoid molecule, **retinal** (formerly called retinene), a derivative of vitamin A. When a quantum of light strikes a rod and is absorbed by the rhodopsin molecule, the latter undergoes a chemical bleaching process that causes it to split into separate opsin and retinal molecules. This change triggers the discharge of a nerve action potential, which is relayed from the receptor cell to the optic center of the brain. Rhodopsin is then enzymatically resynthesized so that it can respond to a subsequent light signal. The amount of intact rhodopsin in the retina depends on the intensity of light reaching the eye. The dark-adapted eye contains much rhodopsin and is very sensitive to weak light. Conversely most of the rhodopsin is broken down in the light-adapted eye. It takes about half an hour for the light-adapted eye to accommodate to darkness, while the rhodopsin level is gradually built up. The remarkable ability of the eye to dark- and light-adapt vastly increases the versatility of the eye; it enables us to see by starlight and also by the noonday sun, 10 billion times brighter.

The light-sensitive pigments of cones are called **iodopsins.** They are similar to rhodopsin, containing **retinal** combined with a special protein, **cone opsin.** Cones function to perceive color and require 50 to 100 times more light for stimulation than do rods. Consequently night vision is almost totally rod vision; this is why the landscape illuminated by moonlight appears in shades of black and white only. Unlike humans, who have both day and night vision, some vertebrates specialize for one or the other. Strictly nocturnal animals such as bats and owls have pure rod retinas. Purely diurnal forms such as the common gray squirrel and some birds have only cones. They are, of course, virtually blind at night.

Color vision

How does the eye see colors? According to the trichromatic theory of color vision, there are three different types of cones, each with its characteristic pigment, that react most strongly to red, green, and violet light. Colors are perceived by comparing the levels of excitation of the three different kinds of cones. This comparison is made both in nerve circuits in the retina and in the visual cortex of the brain. Color vision is present in some members of all vertebrate groups. Bony fishes and birds have particularly good color vision. Surprisingly, most mammals are color blind; exceptions are primates and a very few other species, such as squirrels.

Annotated references

Selected general readings

See also general references for Part Three, p. 793.

Bachelard, H. S. 1974. Brain biochemistry. New York, John Wiley & Sons, Inc. *Brief but concentrated account of brain chemistry (neurotransmitter substances, hormones, enzymes, drug effects) with helpful diagrams.*

Bullock, T. H., and G. A. Horridge. 1965. Structure and function in the nervous system of invertebrates, Vols. 1 and 2. San Francisco, W. H. Freeman & Co., Publishers. *An excellent summary of nervous integration in invertebrates.*

Burton, M. 1972. The sixth sense of animals. New York, Taplinger Publishing Co., Inc. *Wide-ranging comparative treatment at undergraduate level.*

Droscher, V. B. 1969. The magic of the senses. New York, E. P. Dutton & Co., Inc. *Readable undergraduate level comparative account of the senses, ending with consideration of animal migration and navigation.*

Eccles, J. C. 1957. The physiology of nerve cells. Baltimore, The Johns Hopkins Press. *A summary of certain concepts of nervous integration, including the architecture of the neuron and the transmitter substances of the central nervous system*

Eccles, J. C. 1977. The understanding of the brain, ed. 2. New York, McGraw-Hill Book Co.

Mellon, D., Jr. 1968. The physiology of sense organs. San Francisco, W. H. Freeman & Co., Publishers. *Advanced treatment of sense organ physiology.*

Milne, L., and M. Milne. 1972. The senses of animals and men. New York, Atheneum Publishers. *Beautifully written account of animal senses by these prolific authors. Undergraduate level.*

Mountcastle, V. B. 1974. Medical physiology, vol. 2, ed. 13. St. Louis, The C. V. Mosby Co. *The section on the nervous system is especially well written in this outstanding medical physiology book. Advanced level.*

Nathan, P. 1969. The nervous system. Philadelphia, J. B. Lippincott Co. *Highly readable account of the nervous system and special senses of higher vertebrates.*

Sherrington, C. S. 1947. The integrative action of the nervous system, rev. ed. New Haven, Conn., Yale University Press. *A classic work on the structural and functional plan of the nervous system.*

Selected *Scientific American* articles

Amoore, J. E., J. W. Johnston, Jr., and M. Rubin. 1964. The stereochemical theory of odor. **210:**42-49 (Feb.).

Axelrod, J. 1974. Neurotransmitters. **230:**58-71 (June).

Baker, P. F. 1966. The nerve axon. **214:**75-82 (Mar.). *Describes techniques used to study nerve impulse conduction.*

Dowling, J. E. 1966. Night blindness. **215:**78-84 (Oct.). *The importance of vitamin A in vision.*

Eccles, J. 1965. The synapse. **212:**56-66 (Jan.). *A famous neurophysiologist explains how nerve impulses are transmitted from cell to cell.*

Heimer, L. 1971. Pathways in the brain. **225:**48-60 (July). *A new staining technique has vastly improved studies of neural pathways and connections in the nervous system.*

Heller, H. C., L. I. Crawshaw, and H. T. Hammel. 1978. The thermostat of vertebrate animals. **239:**102-113 (Aug.). *Describes the thermoregulatory functions of the hypothalamus.*

Hendricks, S. B. 1968. How light interacts with living matter. **219:**174-186 (Sept.). *Special pigments mediate the interaction of light with matter in photosynthesis, vision, and photoperiodism.*

Hodgson, E. S. 1961. Taste receptors. **204:**135-144 (May). *The mechanism of taste reception is studied with blowflies.*

Jerison, H. J. 1976. Paleoneurology and the evolution of mind. **234:**90-101 (Jan). *The relationship between brain size and body size during animal evolution provides insight into intelligence.*

Katz, B. 1961. How cells communicate. **205:**209-220 (Sept.). *Nervous communication and nature of the nerve impulse.*

Kennedy, D. 1967. Small systems of nerve cells. **216:**44-52 (May). *The simple nervous systems of invertebrates facilitate studies of nervous integration.*

Land, E. H. 1977. The retinex theory of color vision. **237:** 108-128 (Dec.). *The author proposes that the visual system (retina and cortex) may establish color by comparing the lightnesses of areas on three channels.*

Llinás, R. R. 1975. The cortex of the cerebellum. **232:** 56-71 (Jan.). *The pattern of neuronal connections in the cerebellum can now be related to its role in motor coordination.*

Luria, A. R. 1970. The functional organization of the brain. **222:**66-78 (Mar.).

Melzack, R. 1961. The perception of pain. **204:**41-49 (Feb.). *Pain is greatly modified by experience and "state of mind."*

Miller, W. H., F. Ratliff, and H. K. Hartline. 1961. How cells receive stimuli. **205:**222-238 (Sept.). *Structure and functional properties of receptor cells.*

Nathanson, J. A., and P. Greengard. 1977. "Second messengers" in the brain. **237:**108-119 (Aug.). *A survey of the various neurotransmitters of the vertebrate nervous system and how their chemical messages are translated into physiologic actions.*

Olton, D. S. 1977. Spatial memory. **236:**82-98 (June). *Remembering where it has been is located in an animal's brain region called the hippocampus.*

Ross, J. 1976. The resources of binocular perception. **234:** 80-86. (Mar.). *Binocular vision gives us information about an object even before we are conscious of seeing it.*

Rushton, W. A. H. 1975. Visual pigments and color blindness. **232:**64-74 (Mar.). *There are three types of cone cell, each with one of three visual pigments; color blindness may be caused by lacking one of the pigments.*

Stent, G. S. 1972. Cellular communication. **227:**43-51 (Sept.). *Information processing and communication is discussed with particular emphasis on vision.*

Werblin, F. S. 1973. The control of sensitivity in the retina. **228:**70-79 (Jan.) *Recent studies of neuron interactions in the retina help to explain its versatility over widely ranging light conditions.*

Young, R. W. 1970. Visual cells. **223:**80-91 (Oct.). *How rods and cones renew themselves.*

CHAPTER 33

CHEMICAL COORDINATION

Endocrine system

Photomicrograph of a section of the human thyroid gland showing the cystlike follicles that make up the gland. Each follicle is lined with simple epithelium and is filled with a gelatinous material called colloid. Thyroid hormone is manufactured in the epithelium and is stored in the colloid. (Original magnification ×350.)

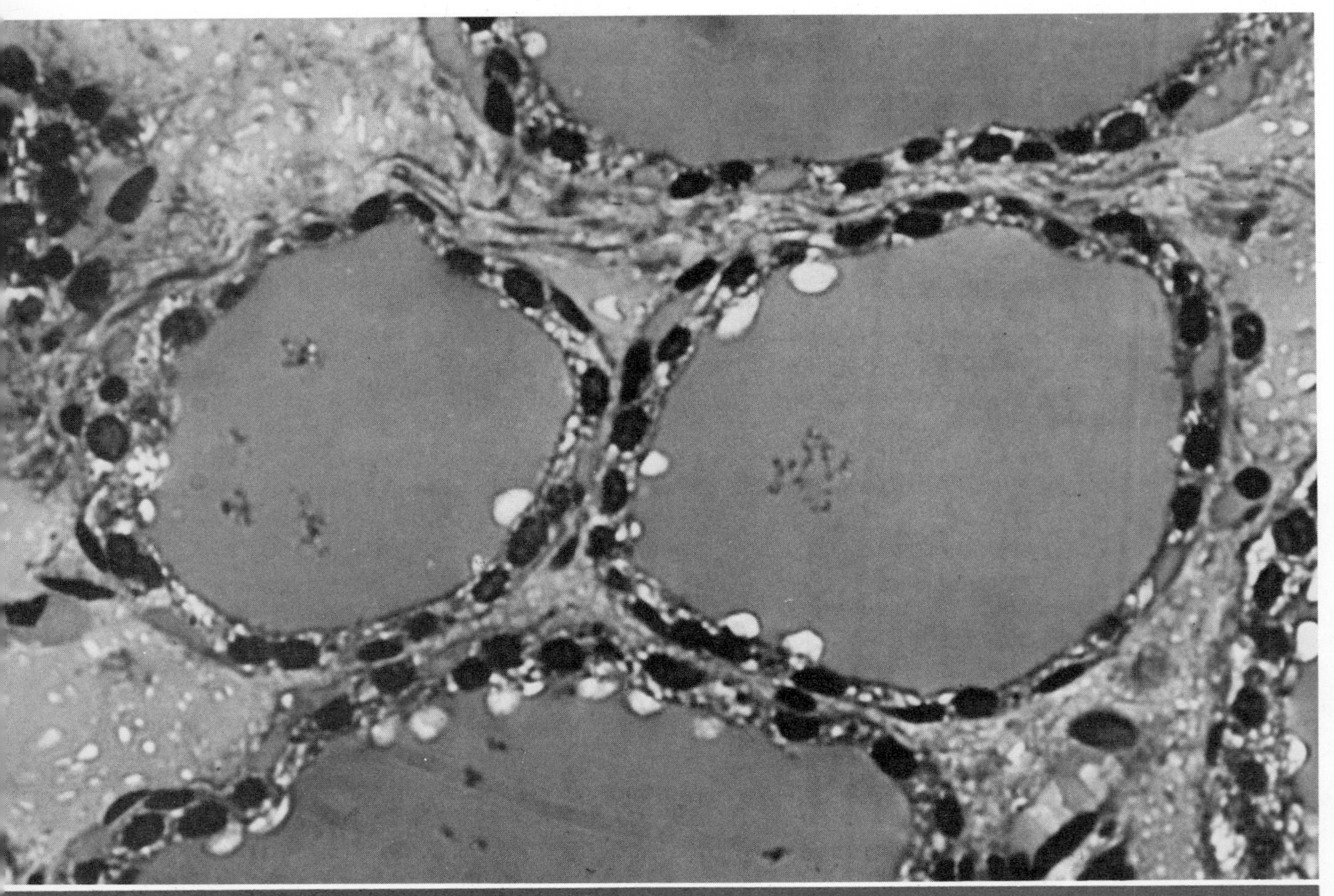

From Schneeberg, N. G. 1970. Essentials of Clinical Endocrinology, St. Louis, The C. V. Mosby Co.

HORMONAL INTEGRATION

The endocrine system is the second great integrative system controlling the body's activities. Endocrine glands secrete **hormones** (Gr., *hormon,* to excite), chemical compounds that are transported by the blood to some part of the body where they initiate definite physiologic responses.

Endocrine glands are small, well-vascularized ductless glands composed of groups of cells arranged in cords or plates. Since the endocrine glands have no ducts, their only connection with the rest of the body is by the bloodstream; they must capture their raw materials from the blood and secrete their finished hormonal products into it. Consequently it is not surprising that the endocrine glands receive enormous blood flows. The thyroid gland is said to have the highest blood flow per unit of tissue weight of any organ in the body. **Exocrine glands,** in contrast, are provided with ducts for discharging their secretions onto a free surface. Examples of exocrine glands are sweat glands and sebaceous glands of skin, salivary glands, and the various enzyme-secreting glands lining the wall of the stomach and intestine.

Since the blood in which hormones are borne is contiguous with the tissue fluid bathing the cells, hormones diffuse into every tissue space in the body. This is quite unlike the discrete action of the nervous system with its network of cablelike nerve fibers that selectively send messages to specific points.

The ubiquitous distribution of hormones makes it possible for certain hormones, such as the growth hormone of the pituitary gland, to affect most, if not all, cells during specific stages of cellular differentiation. However, the circulatory distribution of hormones everywhere creates a potential communications jam, since most hormones must produce highly specific responses at the right time and in the right cells. Confusion is avoided by the specificity of the recipient cells themselves. Each cell will respond only to certain hormones, and to these few only in a highly specific way. This is possible because a cell's receptivity to hormones depends on specialized hormone receptor sites on the cell surface or on certain intracellular components. For example, only certain stomach cells respond to the hormone gastrin, even though gastrin penetrates throughout the body. These particular cells of the stomach bear receptors that selectively bind gastrin; other cells simply ignore the hormone's presence because they lack gastrin-specific receptors. Cells that respond to a specific hormone are called **target-organ cells** for that hormone.

Compared to the nervous system, the endocrine system is slow-acting because of the time required for a hormone to reach the appropriate tissue, cross the capillary endothelium, and diffuse through tissue fluid to, and sometimes into, cells. Thus the minimum response time is a matter of seconds and may be much longer. Furthermore, hormone responses in general are much longer lasting than those under nervous control. Where a sustained effect is required, as in many metabolic and growth processes, or where some concentration or secretion rate must be maintained at a particular level, we expect to find endocrine control.

However, the nervous and endocrine systems really function as a single, united system. There is no sharp separation between the two. As we shall see, the nervous system is itself an endocrine organ that controls most endocrine function. Conversely, several hormones act on the nervous system and may markedly affect many kinds of animal behavior.

Endocrinology is a comparatively young division of animal physiology. Its birthdate is usually given as 1902, the year two English physiologists, W. H. Bayliss and E. H. Starling, demonstrated the action of an internal secretion. They were interested in determining how the pancreas secreted its digestive juice into the small intestine at the proper time of the digestive process. In an anesthetized dog they tied off a section of the small intestine beyond the duodenum (the part of the intestine next to the stomach) and removed all nerves leading to this tied-off loop, but left its blood vessels intact. Bayliss and Starling found that the injection of hydrochloric acid into the blood serving this intestinal loop had no effect on the secretion of pancreatic juice, but when they introduced 0.4% hydrochloric acid directly inside the intestinal loop, a pronounced flow of pancreatic juice into the duodenum occurred through the pancreatic duct. When they scraped off some of the mucous membrane lining of the intestine and mixed it with acid, they found the injection of this extract into the blood caused an abundant flow of pancreatic juice.

They concluded that when the partly digested and slightly acid food from the stomach arrives in the small intestine, the hydrochloric acid reacts with something in the mucous lining to produce an internal secretion, or chemical messenger, which is conveyed by the bloodstream to the pancreas, causing it to secrete pan-

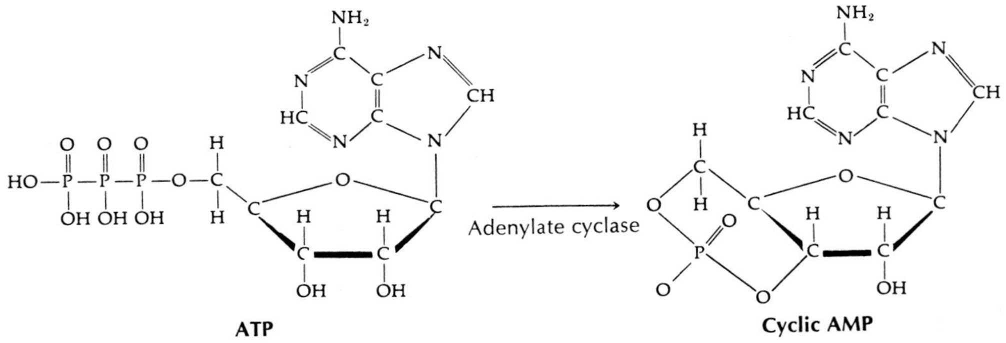

FIG. 33-1 Synthesis of cyclic AMP from ATP.

creatic digestive juices. They called this messenger **secretin.** In a 1905 Croonian lecture at the Royal College of Physicians, Starling first used the word "hormone," a general term to describe all such chemical messengers, since he correctly surmised that secretin was only the first of many hormones that remained to be described.

How hormone secretion rates are controlled

Hormones influence cell functions by altering the rates of a large range of biochemical processes. Some affect enzyme activity and thus alter cellular metabolism, some change membrane permeability, some regulate the synthesis of cellular proteins, and some stimulate the release of hormones from other endocrine glands. Since these are all dynamic processes that must adapt to changing metabolic demands, they must be regulated, not merely activated, by the appropriate hormones. This is achieved by varying hormone output from the gland. However, the concentration of hormone in the plasma depends on two factors: its rate of secretion and the rate at which it is inactivated and removed from the circulation. Consequently an endocrine gland requires information about the hormone level in the plasma in order to control its secretion.

Many hormones, especially those of the pituitary gland, are controlled by **negative feedback systems** that operate between the glands secreting the hormones and the target cells. A feedback pattern is one in which the output is constantly compared with a set point. For example, adrenocorticotropic hormone (ACTH), secreted by the pituitary, stimulates the adrenal gland (the target cells) to secrete cortisol. As the cortisol level in the plasma rises, it acts on the pituitary gland to inhibit the release of ACTH. This is a closed-loop system in which any deviation from the set point (a specific plasma level of cortisol) leads to corrective action in the opposite direction. Such a system is highly effective in preventing extreme oscillations in hormone output.

All hormones are low-level signals. Even when an endocrine gland is secreting maximally, the hormone is so greatly diluted by the large volume of blood it enters that its plasma concentration seldom exceeds 10^{-9} M (one part in a billion). Since hormones have far-reaching and often powerful influences on cells, it is evident that their effects are vastly amplified at the cellular level.

Mechanisms of hormone action

How do hormones exert their effects? The reader can readily appreciate that it is much easier to observe the physiologic effect of a hormone than to determine what the hormone does to produce the effect. Although we have known for years that insulin lowers the blood glucose level, we are still uncertain *how* insulin does this.

There appear to be two basic mechanisms that mediate the actions of hormones. Both mechanisms work through specific receptor molecules on or in the target cells.

1. **Activation of cyclic AMP, the "second messenger."** Cyclic AMP is a nucleotide related in structure to ATP, the "high-energy" molecule of all cellu-

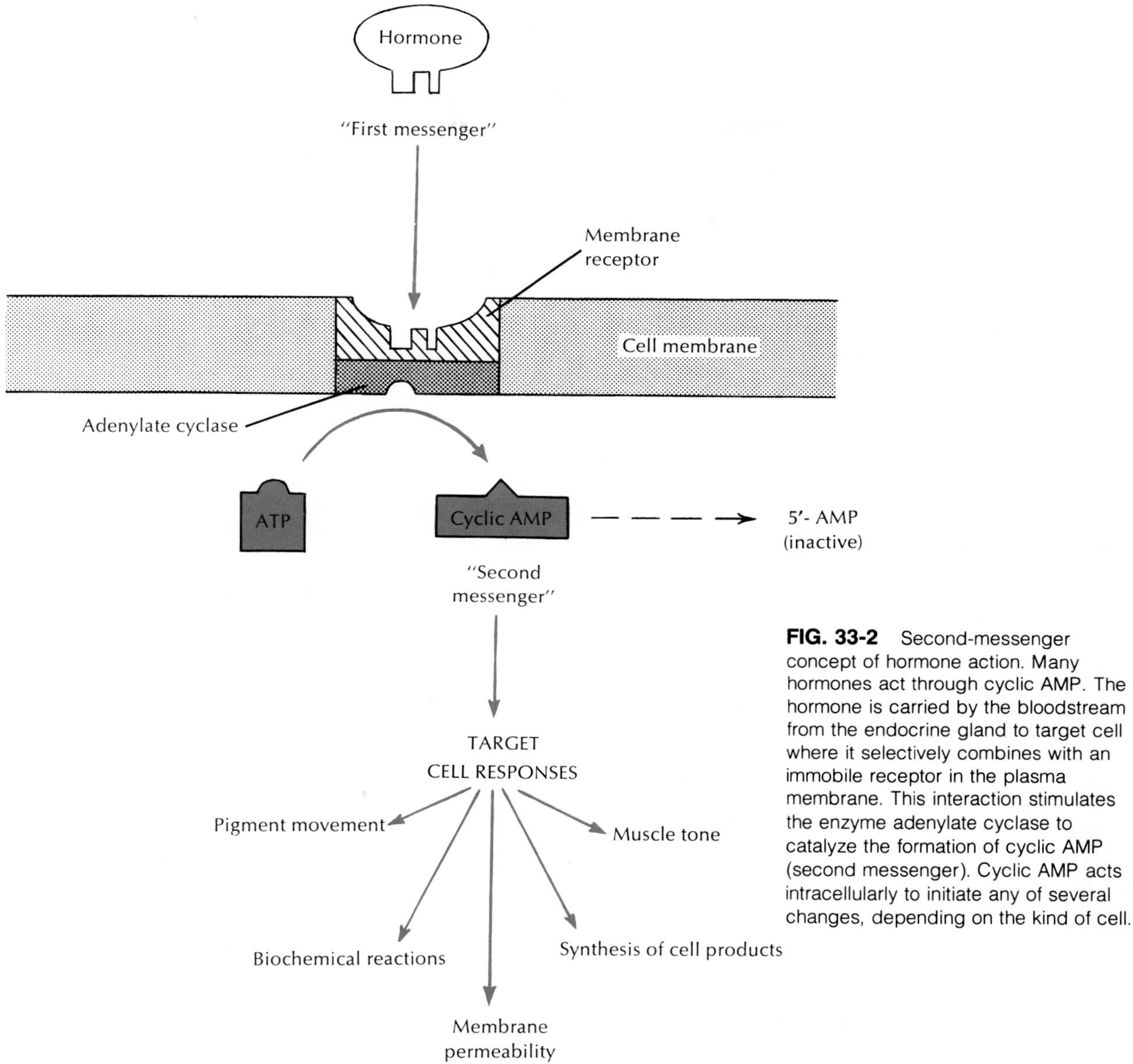

FIG. 33-2 Second-messenger concept of hormone action. Many hormones act through cyclic AMP. The hormone is carried by the bloodstream from the endocrine gland to target cell where it selectively combines with an immobile receptor in the plasma membrane. This interaction stimulates the enzyme adenylate cyclase to catalyze the formation of cyclic AMP (second messenger). Cyclic AMP acts intracellularly to initiate any of several changes, depending on the kind of cell.

lar energy flow (Fig. 33-1). Cyclic AMP is formed by the action of an enzyme, adenylate cyclase, located in the cell membrane (Fig. 33-2). When a hormone (the "first messenger") arrives at its target cell, it binds to a receptor site on the membrane. Adenylate cyclase is activated by the binding and converts some of the ATP in the cytoplasm to cyclic AMP. The cyclic AMP thus generated acts as a "second messenger" that relays the hormone's message to the cell's biochemical machinery, where it alters (usually stimulates) some cellular process. Since many molecules of cyclic AMP may be manufactured by a single bound hormone molecule, the message is amplified, perhaps many thousands of times.

Cyclic AMP mediates the actions of many hormones, including glucagon, epinephrine, ACTH, thyrotropic hormone (TSH), melanophore-stimulating hormone (MSH), and vasopressin. With the exception

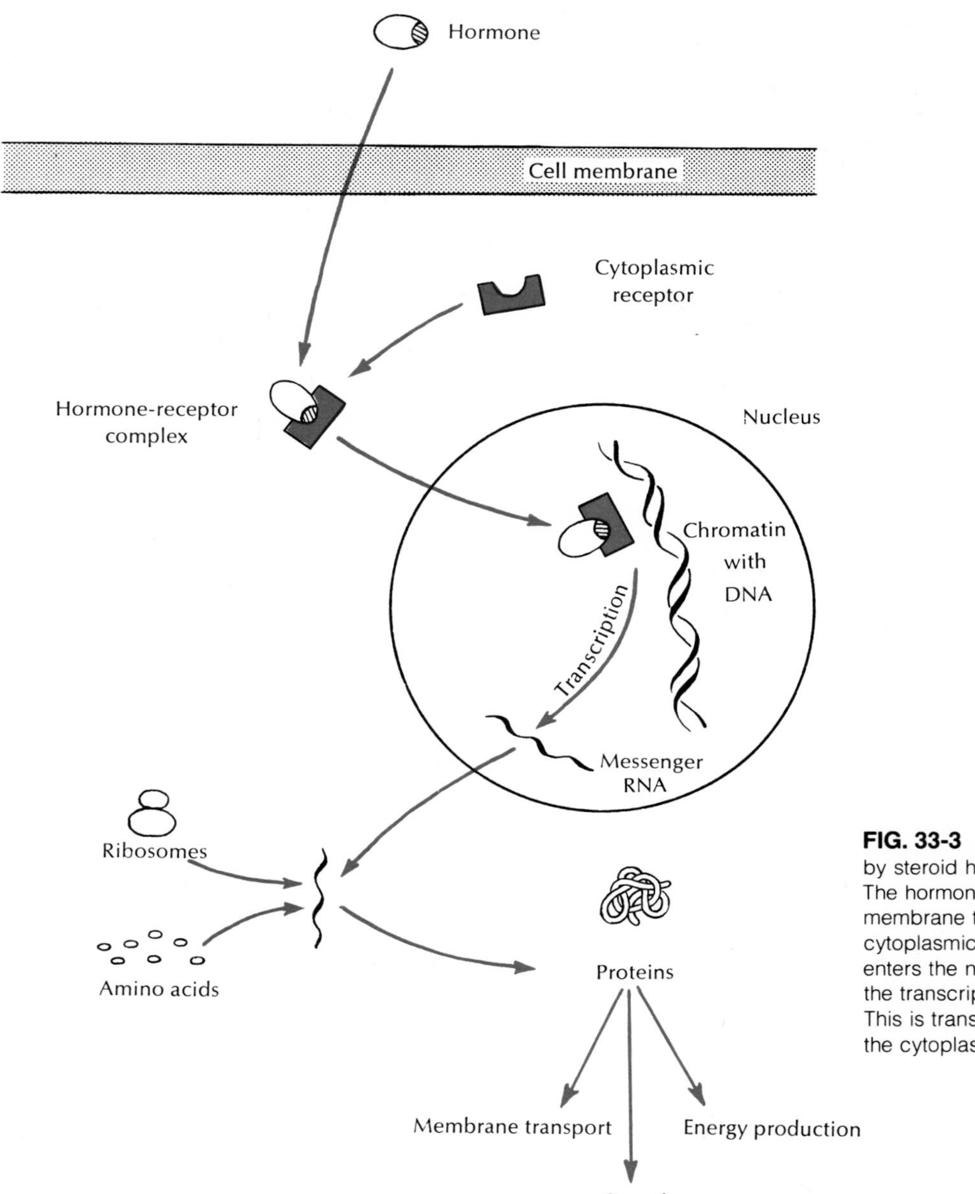

FIG. 33-3 Concept of gene regulation by steroid hormones and thyroxine. The hormone penetrates the cell membrane to combine with a mobile cytoplasmic receptor. The complex enters the nucleus where it stimulates the transcription of messenger RNA. This is translated to specific proteins in the cytoplasm.

of epinephrine, these are all peptides—small proteins but much too large to penetrate the cell membrane. Consequently all act *indirectly* by operating through an immobile receptor on the cell surface.

2. **Induction of protein synthesis.** Several hormones, including all of the steroids (for example, estrogen, testosterone, and aldosterone), stimulate the synthesis of enzymes and other proteins by regulating gene expression. Steroids are small molecules that freely diffuse into cells. Once inside, the steroids bind selectively to *cytoplasmic* receptor molecules found only in the target cells. The hormone-receptor complex then diffuses into the nucleus where it binds directly to certain proteins (nonhistones) in the chromosomes. As a result, gene transcription is increased, and messenger RNA molecules are synthesized on spe-

cific sequences of DNA. Moving from the nucleus into the cytoplasm, the newly formed messenger RNA initiates the formation of new proteins, thus setting in motion the hormone's observed effect (Fig. 33-3). Thyroxine and the insect molting hormone ecdysone [ek′duh-sone] are also believed to act through this mechanism.

As compared to hormones acting through the cyclic AMP mechanism, steroids have a *direct* effect on protein synthesis because they complex with a mobile receptor and move with it into the nucleus to couple with chromosomal proteins.

INVERTEBRATE HORMONES

Over the last 40 years physiologists have shown that the invertebrates have endocrine integrative systems that approach the complexity of the vertebrate endocrine system. Not surprisingly, however, there are few, if any, homologies between invertebrate and vertebrate hormones. Invertebrates have different functional systems, different growth patterns, and different reproductive processes from those of vertebrates, and have been separated from them phylogenetically for a vast span of time. Most studies have been concentrated in the huge phylum Arthropoda, especially the

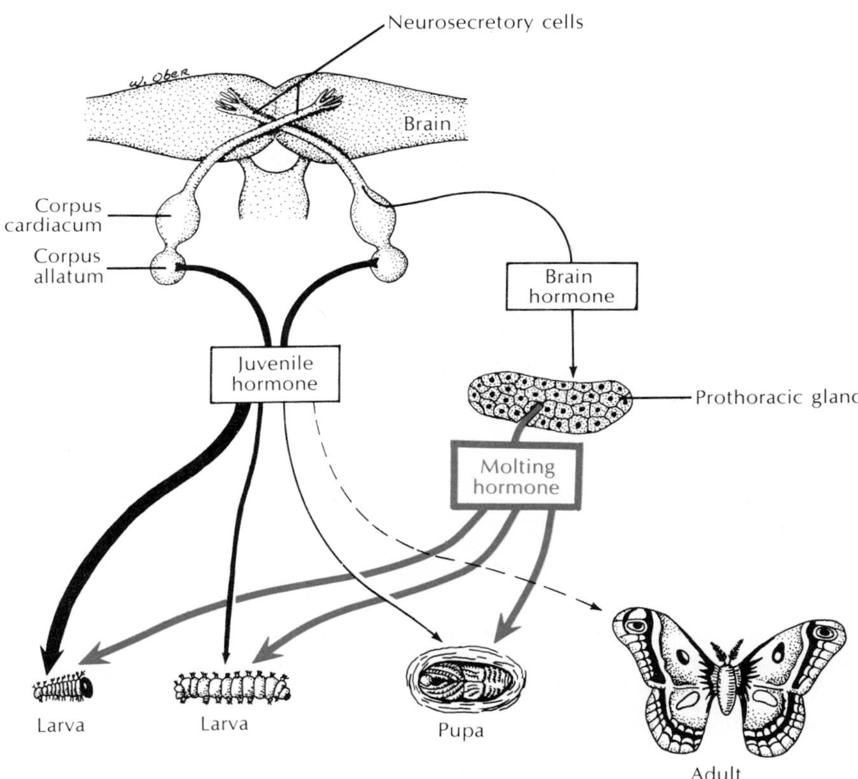

FIG. 33-4 Endocrine control of molting in a butterfly. Butterflies mate in the spring or summer, and eggs soon hatch into the first of several larval stages (called instars). After the final larval molt, the last and largest larva (caterpillar) spins a cocoon in which it pupates. The pupa, or chrysalis, overwinters and an adult emerges in the spring to start a new generation. Two hormones interact to control molting and pupation. The molting hormone, produced by the prothoracic gland and stimulated by a separate brain hormone, favors molting and the formation of adult structures. These effects are inhibited, however, by the juvenile hormone, produced by the corpora allata. Juvenile hormone output declines with successive molts, and the larva undergoes adult differentiation.

insects and crustaceans. However, recent research has revealed hormonal systems in most of the other invertebrate phyla too.

In many invertebrate groups, the principal source of hormones is neurosecretory cells, specialized nerve cells capable of synthesizing and secreting hormones. Their products, called **neurosecretions,** or **neurosecretory hormones,** are discharged directly into the circulation. Neurosecretion is in fact a widespread phenomenon in the animal kingdom. It is a very ancient physiologic activity, and because it serves as a crucial link between the nervous and endocrine systems, we believe that hormones first evolved as nerve-cell secretions. Later nonnervous endocrine glands appeared, especially among the vertebrates, but remained chemically linked to the nervous system by the neurosecretory hormones.

Neurosecretions are known to influence growth, asexual reproduction, and regeneration of hydra (phylum Cnidaria). Neurosecretory hormones also regulate regeneration, training, reproduction, and other aspects of flatworm physiology. We have known for some time that molluscs have neurosecretory hormones, especially among the gastropods and pelecypods. In the polychaete annelids, amputation of a portion of the worm body causes neurosecretory cells in the cerebral ganglia to secrete hormones that trigger regeneration. The cerebral ganglia of young worms produce a ''juvenile hormone'' that has a braking effect on metamorphosis; if the brain is removed, the worms prematurely become sexually mature.

The chromatophores (pigment cells) of shrimp and crabs are controlled by hormones from the **sinus gland** in the eyestalk or in regions close to the brain. Many crustaceans are capable of remarkably beautiful color patterns that change adaptively in relation to their environment; these changes are governed by an elaborate system of endocrine glands and hormones.

Growth and metamorphosis of arthropods are under endocrine control. As described earlier (p. 390), growth of an arthropod is a series of steps in which the rigid, nonexpansible exoskeleton is periodically discarded and replaced with a new larger one. This process is especially dramatic in insects. In the type of development called **holometabolous,** seen in many insect orders (for example, butterflies, moths, ants, bees, wasps, and beetles), there is a series of wormlike larval stages, each requiring the formation of a new exoskeleton; each stage ends with a molt. The last larval stage enters a state of quiescence (pupa) during which the internal tissues are dissolved and rearranged into adult structures **(metamorphosis).** Finally the transformed adult emerges.

Insect physiologists have discovered that molting and metamorphosis are controlled by the interaction of two hormones, one favoring growth and the differentiation of adult structures, the other favoring the retention of larval structures. These two hormones are the **molting hormone** (also referred to as **ecdysone**), produced by the prothoracic glands, and the **juvenile hormone,** produced by the corpora allata (Fig. 33-4). The structures of both hormones have recently been determined. It required extraction from 1,000 kg (about 1 ton) of silkworm pupae to show that the molting hormone is a steroid.

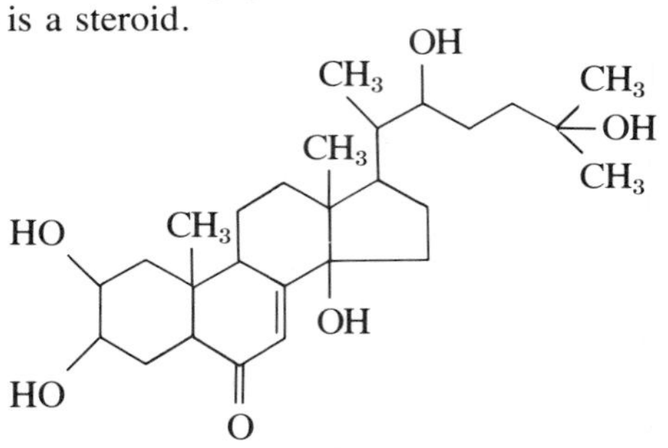

Molting hormone (α-ecdysone) of silkworm

The juvenile hormone has an entirely different structure:

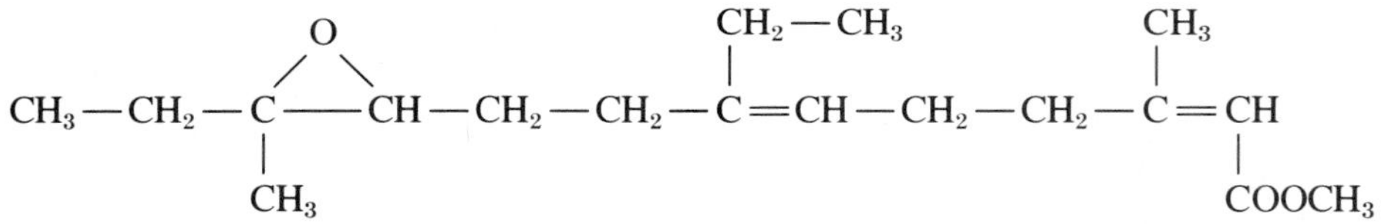

Juvenile hormone of silkworm

The molting hormone is under the control of a neurosecretory hormone from the brain, called **brain hormone** (or ecdysiotropin). At intervals during larval growth, brain hormone is released into the blood and stimulates the release of molting hormone. Molting hormone appears to act directly on the chromosomes to activate genes that promote changes leading to a molt. The molting hormone favors the formation of a pupa and the development of adult structures. It is held in check, however, by the juvenile hormone, which favors the development of larval characteristics. During larval life the juvenile hormone predominates, and each molt yields another larger larva. Finally, the output of juvenile hormone decreases and the final pupal molt occurs.

Chemists have synthesized several potent analogs of the juvenile hormone, which hold great promise as insecticides. Minute quantities of these synthetic analogs induce abnormal final molts, or prolong or block larval development. Unlike the usual chemical insecticides, they are highly specific and do not contaminate the environment.

The pattern of endocrine regulation in crustaceans shows certain parallels to that in insects. Some crustaceans, such as the crayfish, reach a definite adult size after a final molt; others, such as the lobster, molt continuously and keep growing ever larger until death. Molting is controlled by a neurosecretory hormone, called **molt-inhibiting hormone,** produced by the X-organ in the eyestalk, and by a **molting hormone** (ecdysone) produced by the Y-organ located in the thorax. During intermolt, the molt-inhibiting hormone inhibits activity of the Y-organ. But at intervals, just before the molt, the output of molt-inhibiting hormone drops; this permits the Y-organ to secrete molting hormone, which stimulates molting.

VERTEBRATE ENDOCRINE ORGANS

In the remainder of this chapter we will describe some of the best understood and most important of the vertebrate hormones. The hormones of reproduction are discussed in Chapter 35. Space does not permit us to deal with all the hormones and hormone-like substances that have been discovered. The mammalian hormonal mechanisms are the best understood, since laboratory mammals and humans have always been the objects of the most intensive research. Research with the lower vertebrates has revealed that all vertebrates share similar endocrine organs. All vertebrates have a pituitary gland, for example, and all have thyroid glands, adrenal glands (or the special cells of which they are composed), and gonads. Still there are some important differences in the functional roles hormones of these glands play among the different vertebrates, as we will seek to point out.

Hormones of the pituitary gland and hypothalamus

The pituitary gland, or **hypophysis,** is a small gland (0.5 g in humans) lying in a well-protected position between the roof of the mouth and the floor of the brain (Fig. 33-5). It is a two-part gland having a double embryologic origin. The **anterior pituitary** (adenohypophysis) is derived embryologically from the roof of the mouth. The **posterior pituitary** (neurohypophysis) arises from a ventral portion of the brain, the **hypothalamus,** and is connected to it by a stalk, the **infundibulum** (Fig. 33-5, *A*). Although the anterior pituitary lacks any *anatomic* connection to the brain, it is nonetheless *functionally* connected to it by a special portal circulatory system (Fig. 33-5, *B*).

Because of the strategic importance of the pituitary in influencing most of the hormonal activities in the body, the pituitary has been called the body's "master gland." This designation is misleading, however, since most of the pituitary hormones are themselves regulated by neurosecretory hormones from the hypothalamus, as well as by hormones from the target glands they stimulate.

The anterior pituitary consists of an **anterior lobe** (pars distalis) and an **intermediate lobe** (pars intermedia) as shown in Fig. 33-5. The anterior lobe, despite its minute dimensions, produces at least six protein hormones. All but one of these six are **tropic hormones,** that is, they regulate other endocrine glands (Fig. 33-6 and Table 33-1). The **thyrotropic hormone** (TSH) regulates the production of thyroid hormones by the thyroid gland. The **adrenocorticotropic hormone** (ACTH) stimulates the adrenal cortex. Two of the tropic hormones are commonly called **gonadotropins** because they act on the gonads (ovary of the female, testis of the male). These are the **follicle-stimulating hormone** (FSH) and the **luteinizing hormone** (LH). The fifth tropic hormone is **prolactin,** which stimulates milk production by the female mammary glands and has a variety of other effects in the lower vertebrates. The functions of the two gonadotropins

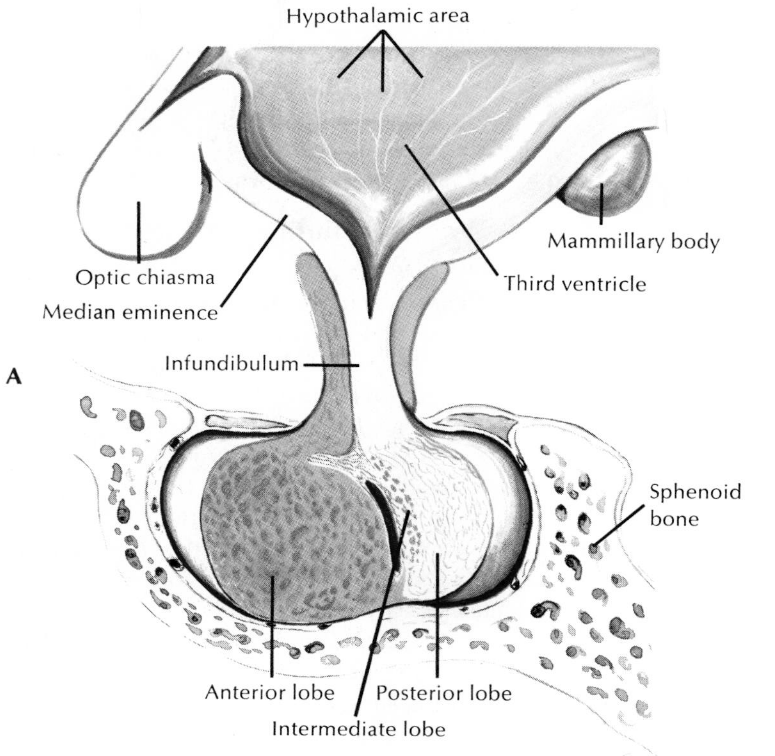

FIG. 33-5 Lateral view of the structure of human pituitary gland and its relationship to hypothalamus. **A,** Posterior lobe is connected directly to hypothalamus by neurosecretory fibers. Anterior lobe is indirectly connected to hypothalamus by a special portal circulation. **B,** Neurosecretory fibers end in the median eminence, in contact with a portal circulation that conveys hormones to the anterior pituitary. (**A** from Schottelius, B. A., and D. D. Schottelius. 1978. Textbook of physiology, ed. 18, St. Louis, The C. V. Mosby Co.)

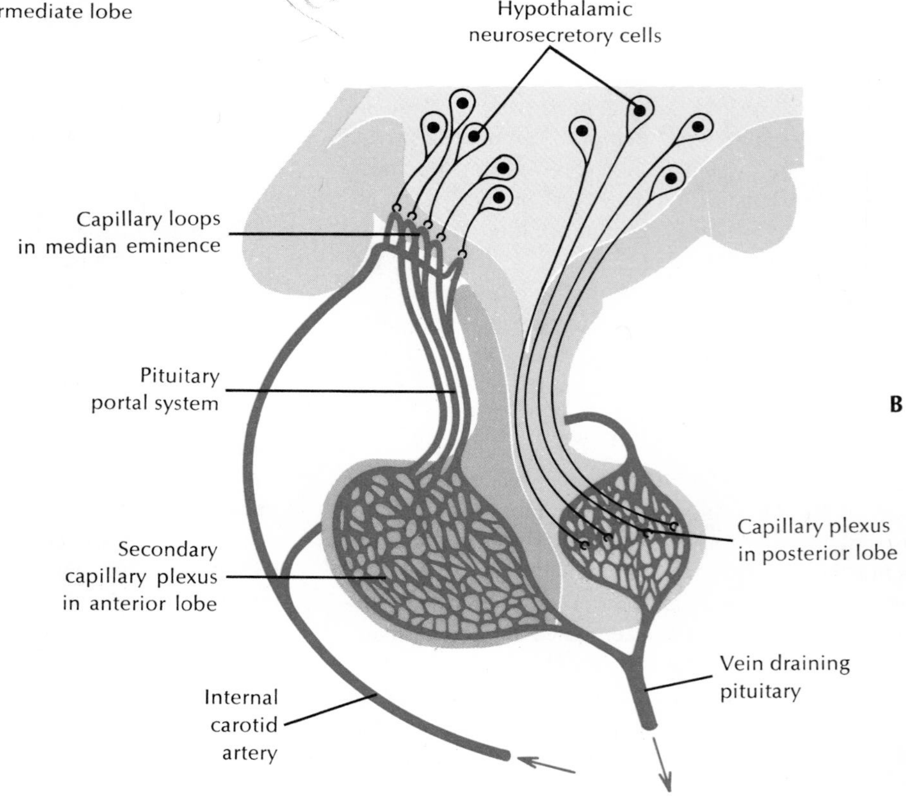

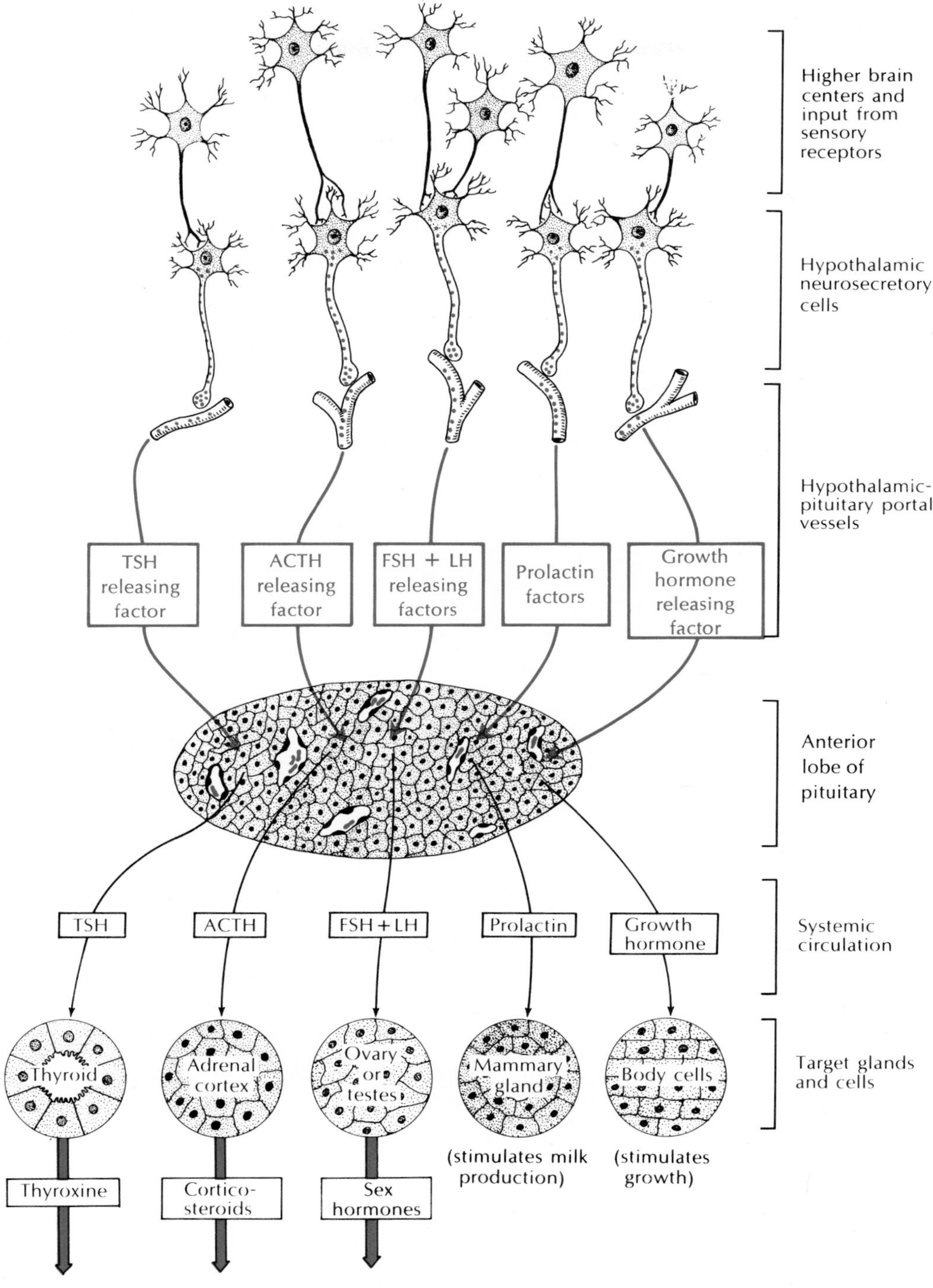

FIG. 33-6 Relationships of hypothalamic, pituitary, and target-gland hormones.

and prolactin are discussed in Chapter 35 in connection with the hormonal control of reproduction.

The sixth hormone of the anterior lobe is the **growth hormone** (also called somatotropic hormone). This hormone performs a vital role in governing body growth through its stimulatory effect on cellular mitosis and protein synthesis, especially in new tissue of young animals. If produced in excess, the growth hormone causes gigantism. A deficiency of hormone in the child or young animal causes dwarfism. Growth hormone and prolactin are so closely related chemically that growth hormone contains considerable prolactin activity; that is, it tends to stimulate milk production (like prolactin) as well as promote growth.

The intermediate lobe of the pituitary (Fig. 33-5) produces **intermedin** (also called melanophore-stimulating hormone [MSH]), which controls the dispersion of melanin within the melanophores of lower vertebrates. Its action in amphibians is described in Chapter 24. In higher vertebrates, intermedin appears to perform no important physiologic role, even though it will cause darkening of the skin in humans if injected into the circulation. Intermedin and ACTH are chemically very similar and ACTH will also cause skin darkening when it is secreted in abnormally large amounts.

As pointed out before, the pituitary gland is not the top director of the body's system of endocrine glands, as endocrinologists once believed. The pituitary serves higher masters, the neurosecretory centers of the hypothalamus; and the hypothalamus is itself under the ultimate control of the brain. The hypothalamus contains groups of neurosecretory cells, which are specialized giant nerve cells. Polypeptide hormones are manufactured in the cell bodies and then travel down the nerve fibers to their endings, where the hormones are stored until released into the blood. The discharge of neurosecretory hormones may occur when a nerve impulse travels down the same neurosecretory fiber (these specialized nerve cells can, in most cases, still perform their original impulse-conducting function) or the release may be activated by ordinary fibers lying alongside them.

Both the anterior and posterior lobes of the pituitary are under hypothalamic control, but in different ways. Neurosecretory fibers serving the posterior lobe pass down the infundibular stalk and into the posterior lobe, ending in proximity to blood capillaries, into which the hormones enter when released. In a sense the posterior lobe is not a true endocrine gland but a storage-release center for hormones manufactured entirely in the hypothalamus.

The relationship of the anterior lobe to the hypothalamus is quite different. Neurosecretory fibers do not pass to the anterior lobe but end some distance above it, in the **median eminence,** at the base of the infundibular stalk (Fig. 33-5). Neurosecretory hormones released here enter a capillary network and complete their journey to the anterior lobe via a short, but crucial, pituitary portal system. These hormones are called **releasing factors** because they govern the release of anterior lobe hormones (Table 33-1). There appear to be releasing factors for all pituitary tropic hormones. Three of the releasing factors have recently been isolated and characterized chemically. They are TRF (thyrotropin-releasing factor, a tripeptide); LRF (luteinizing hormone–releasing factor, a decapeptide); and somatostatin (growth hormone–inhibiting factor, a tetradecapeptide). The structure of TRF is as follows:

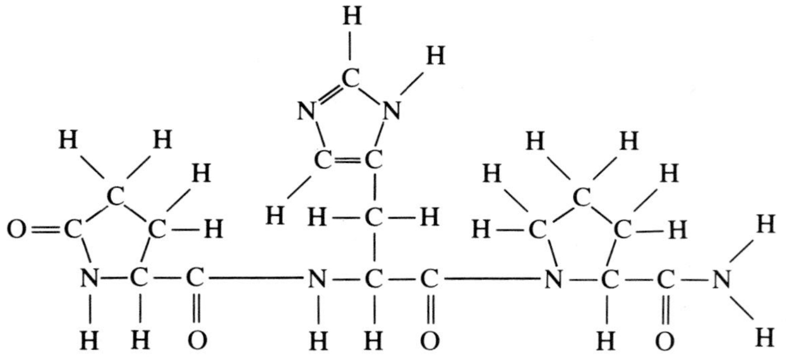

```
                            H
                            |         H
                            C        /
                         N//  \\N
     H     H                \   /              H     H
      \   /    H            C=C             H   \   /    H
        C     /           /   |              \   C     /
O=C /   \ C—H      H  H—C—H        H—C /   \ C—H       H
    \       /                 |                 \       /        /
     N—C—C————————N—C—C————————N—C—C—N
     |   |   ||           |   |   ||            |   ||    \
     H   H   O            H   H   O             H   O      H
```

Pyroglutamic acid–histidine–proline NH_2
(thyrotropin releasing factor)

TABLE 33-1 Hormones of the vertebrate pituitary—chemical nature and actions

	Hormone	Chemical nature	Action
Adenohypophysis			
Anterior lobe	Adrenocorticotropin (ACTH)	Polypeptide	Stimulates adrenal cortex to secrete steroid hormones
	Thyrotropin (TSH)	Glycoprotein	Stimulates thyroid to secrete thyroid hormones
	Gonadotropins (two)		
	1. Follicle-stimulating hormone (FSH)	Glycoprotein	Stimulates gamete production and secretion of sex hormones
	2. Luteinizing hormone (LH, ICSH)	Glycoprotein	Stimulates sex hormone secretion and ovulation
	Prolactin (LTH)	Protein	Stimulates mammary gland growth and secretion in mammals; various reproductive and non-reproductive functions in lower vertebrates
	Growth hormone (GH)	Protein	Stimulates growth
Intermediate lobe	Intermedin (MSH)	Polypeptide	Pigment dispersion in melanophores
Neurohypophysis			
Posterior lobe	Vasopressin (ADH)	Octapeptide	Antidiuretic effect on kidney
	Oxytocin	Octapeptide	Stimulates milk ejection and uterine contraction
	Vasotocin	Octapeptide	Antidiuretic activity
	Isotocin and three others in lower vertebrates	Octapeptides	Functions unknown
Median eminence	Thyrotropin-releasing factor (TRF)	Probably all polypeptides	Control release of anterior lobe hormones
	Corticotropin-releasing factor (CRF)		
	Follicle-stimulating hormone–releasing factor (FRF)		
	Luteinizing hormone–releasing factor (LRF)		
	Prolactin-releasing factor (PRF)		
	Prolactin-inhibiting factor (PIF)		
	Growth hormone–releasing factor (GRF)		
	Somatostatin		

The hormones of the **posterior lobe** are chemically similar polypeptides consisting of eight amino acids (referred to as octapeptides). All vertebrates except the most primitive fishes secrete two posterior lobe octapeptides. However, their chemical structure has changed slightly in the course of evolution. The two posterior lobe hormones secreted, for example, by fishes are not identical to those secreted by mammals. Altogether, seven different posterior lobe hormones have been identified from the different vertebrate groups (Table 33-1).

Vasotocin is found in all vertebrate classes except mammals. It is a water-balance hormone in amphibians, especially toads, in which it acts to conserve water by (1) increasing permeability of the skin (to promote water reabsorption from the environment), (2) stimulating reabsorption from the urinary bladder, and (3) decreasing urine flow. The action of vasotocin

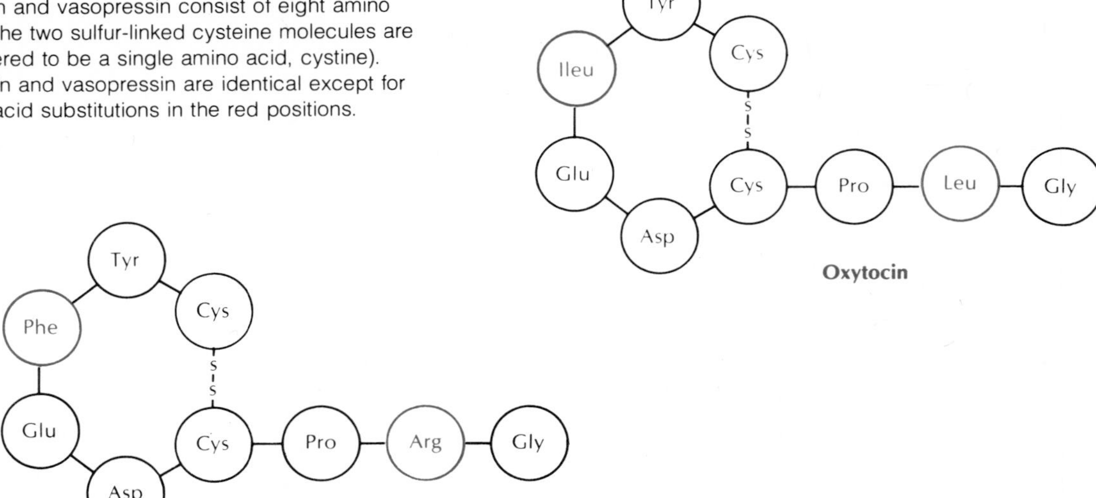

FIG. 33-7 Human posterior lobe hormones. Both oxytocin and vasopressin consist of eight amino acids (the two sulfur-linked cysteine molecules are considered to be a single amino acid, cystine). Oxytocin and vasopressin are identical except for amino acid substitutions in the red positions.

is best understood in amphibians, but it appears to play some water-conserving role in birds and reptiles as well.

The two mammalian posterior-lobe hormones are **oxytocin** and **vasopressin** (Fig. 33-7). They are formed, as we have seen, in the cell bodies of neurosecretory cells in the hypothalamus and are then transported down the nerve cell axons to the posterior lobe. These hormones are among the fastest-acting hormones in the body, since they are capable of producing a response within seconds of their release from the posterior lobe.

Oxytocin has two important specialized reproductive functions in adult female mammals. It causes contraction of uterine smooth muscles during parturition (birth of the young). Physicians sometimes use oxytocin clinically to induce labor and facilitate delivery and to prevent uterine hemorrhage after birth. The second and most important action of oxytocin is that of milk ejection by the mammary glands in response to suckling. Though present, oxytocin has no known function in the male.

Vasopressin, the second posterior lobe hormone, acts on the kidney to restrict urine flow, as already described on p. 000. It is therefore often referred to as the **antidiuretic hormone** (ADH). Vasopressin has a second, weaker effect of increasing the blood pressure through its generalized constrictor effect on the smooth muscles of the arterioles. Although the name "vasopressin" unfortunately suggests that the vasoconstrictor action is the hormone's major effect, it is probably of little physiologic importance, except perhaps to help sustain the blood pressure during a severe hemorrhage.

Hormones of metabolism

Many hormones act to adjust the delicate balance of metabolic activities in the body. Metabolism includes the **anabolic** activities of tissue synthesis, building up of energy reserves, and maintenance of tissue organization and the **catabolic** activities of energy release and tissue destruction. Such activities are mediated almost entirely by enzymes. The numerous enzymatic reactions proceeding within cells are complex, but each step in a sequence is in large part self-regulating, as long as the equilibrium between substrate, enzyme, and product remains stable. However, hormones may alter the activity of crucial enzymes in a metabolic process, thus accelerating or inhibiting the entire process. We must emphasize that hormones never initiate enzymatic processes. They simply alter their rate, speeding them up or slowing them down.

The most important hormones of metabolism are those of the thyroid, parathyroid, and adrenal glands and of the pancreas.

Thyroid hormones

The two thyroid hormones **thyroxine** and **triiodothyronine** are secreted by the thyroid gland. This largest of endocrine glands is located in the neck region of all vertebrates; in many animals, including humans, it is a bilobed structure. The thyroid is made up of thousands of tiny spheres, called **follicles;** each follicle is composed of a single layer of epithelial cells enclosing a hollow, fluid-filled center. This fluid contains stored thyroid hormone that is released into the bloodstream as it is needed.

One of the unique characteristics of the thyroid is its high concentration of iodine; in most animals this single gland contains well over half the body store of iodine. The epithelial cells actively trap iodine from the blood and combine it with the amino acid tyrosine, creating the two thyroid hormones. Each molecule of thyroxine contains four atoms of iodine, as indicated by the following structural formula:

$$\begin{array}{ccccccccc} & I & & I & & & & & \\ HO- & \langle\bigcirc\rangle & -O- & \langle\bigcirc\rangle & -CH_2 & - & CH & - & COOH \\ & I & & I & & & | & & \\ & & & & & & NH_2 & & \end{array}$$

Thyroxine

Triiodothyronine is identical to thyroxine except that it has three instead of four iodine atoms. Thyroxine is formed in much greater amounts than triiodothyronine, but both hormones have two important similar effects. One is to promote the normal growth and development of growing animals. The other is to stimulate the metabolic rate.

The thyroid hormones promote growth by stimulating protein synthesis through their effect on gene transcription. Undersecretion of thyroid hormone dramatically impairs growth, especially of the nervous system. The human **cretin,** a mentally retarded dwarf, is the tragic product of thyroid malfunction from a very early age. Conversely the oversecretion of thyroid hormones causes precocious development, particularly in lower vertebrates. In one of the earliest demonstrations of hormone action, Gudernatsch, in 1912, induced precocious metamorphosis of frog tadpoles by feeding them bits of horse thyroid. The tadpoles quickly resorbed their tails, grew limbs, changed from gill to lung respiration, and became froglets about one-third normal size. The result of a similar experiment is shown in Fig. 33-8.

The control of oxygen consumption and heat production in birds and mammals is the best-known action of the thyroid hormones. The thyroid enables warm-blooded animals to adapt to cold by increasing their heat production. Responding to directives from the hypothalamus, the body's ''thermostat,'' the thyroid gland, releases more thyroxine when the blood temperature begins to fall. Thyroxine stimulates the turnover of ATP, causing cells to produce more heat and store less chemical energy as ATP. Cells burn more fuel when stimulated by thyroid hormone and this is why cold-adapted animals eat more food than do warm-adapted ones, even though they are doing no more work; the food is being converted directly to heat, thus keeping the body warm.

The synthesis and release of thyroxine and triiodothyronine are governed by **thyrotropic hormone** (TSH) from the anterior pituitary gland (Fig. 33-6). Thyrotropic hormone controls the thyroid through a **negative feedback mechanism.** If the thyroxine level in the blood falls, more thyrotropic hormone is released. Should the thyroxine level rise too high, less thyrotropic hormone is released. This sensitive feedback mechanism normally keeps the blood thyroxine level very steady, but certain neural stimuli, as might arise from exposure to cold, can directly increase the release of thyrotropic hormone.

Some years ago, a condition called **goiter** was common among people living in the Great Lakes region of the United States and Canada, as well as in other parts of the earth, such as the Swiss Alps. Goiter is an enlargement of the thyroid gland caused by a deficiency of iodine in the food and water. In striving to produce thyroid hormone with not enough iodine available, the gland hypertrophies, sometimes so much that the entire neck region becomes swollen. Goiter is seldom seen today because of the widespread use of iodized salt.

Hormonal regulation of calcium metabolism

Closely associated with the thyroid gland, and often buried within it, are the parathyroid glands. These tiny glands occur as two pairs in humans but vary in number and position in other vertebrates. They were dis-

FIG. 33-8 Precocious metamorphosis of frog tadpoles *(Rana pipiens)* caused by thyroid hormone treatment. When frog tadpoles had developed hindlimb buds, small amounts of thyroxine were added to aquarium water. In only 3 weeks tadpoles metamorphosed to normal, but miniature, adults about one-third the size of the mother. (From Turner, C. D., and J. T. Bagnara. 1971. General endocrinology, ed. 5, Philadelphia, W. B. Saunders Co.)

covered at the end of the nineteenth century when the fatal effects of "thyroidectomy" were traced to the unknowing removal of the parathyroid glands as well as the thyroid gland. Removal of the parathyroid glands causes the blood calcium to drop rapidly. This leads to a serious increase in nervous system excitability, severe muscular spasms and tetany, and finally death.

The parathyroid glands are vitally concerned with the maintenance of the normal level of calcium in the blood. Actually three hormones are involved: **parathyroid hormone** (parathormone), produced by the parathyroid glands; **calcitonin,** produced by special "C cells" within the thyroid gland (not the same follicle cells that synthesize thyroxine); and a hormonal metabolite of vitamin D. These hormones have cooperative actions that stabilize both the calcium and phosphorus levels in the blood through their action on bone and the intestine.

Bone is a densely packed storehouse of these elements, containing about 98% of the body calcium and 80% of the phosphorus. Although bone is second only to teeth as the most durable material in the body, as evidenced by the survival of fossil bones for millions of years, it is in a state of constant turnover in the living body. Bone-building cells **(osteoblasts)** withdraw calcium and phosphorus (as phosphate) from the blood and deposit them in a complex crystalline form around previously formed organic fibers (Fig. 33-9). Bone-resorbing cells **(osteoclasts),** present in the same bone, tear down bone by engulfing it and releasing the calcium and phosphate into the blood.

These seemingly conflicting activities allow bone to constantly remodel itself, especially in the growing

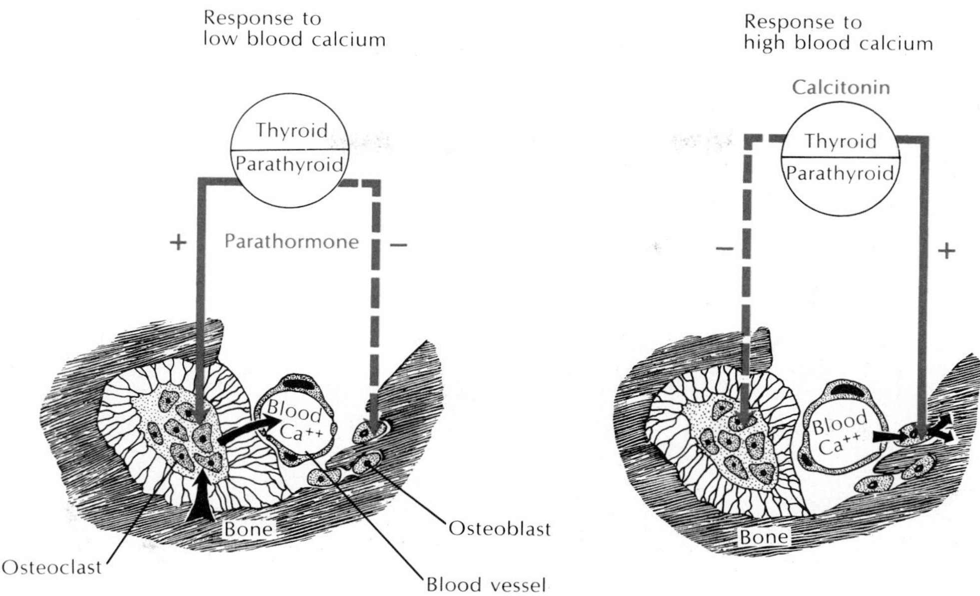

FIG. 33-9 Action of parathormone and calcitonin on calcium resorption and deposition in bone. When blood calcium is low *(left)*, parathyroid gland secretes parathormone, which stimulates large bone-destroying osteoclasts to resorb calcium. Bone-building osteoblasts are inhibited. Calcium and phosphate (not shown) enter the blood and restore the blood calcium level to normal. When blood calcium rises above normal *(right)*, thyroid gland secretes calcitonin that inhibits osteoclasts. Osteoblasts then remove calcium (and phosphate) from the blood and use it to build new bone. Calcitonin may directly stimulate osteoblastic activity. A third hormone, derived from Vitamin D, is necessary for calcium deposition in bones.

animal, for structural improvements to counter new mechanical stresses on the body. Second, they provide a vast and accessible reservoir of minerals that can be withdrawn as the body needs them for its general cellular requirements.

If the blood calcium should drop slightly, the parathyroid gland increases its output of parathormone. This stimulates the osteoclasts to destroy adjacent bone, thus releasing calcium and phosphate into the bloodstream and returning the blood calcium level to normal. Parathormone also acts on the kidney to decrease the excretion of calcium and this, of course, also helps to increase the blood calcium level. Should the calcium in the blood rise above the normal level, the parathyroid gland decreases its output of parathormone. In addition, the thyroid gland is stimulated to release calcitonin. These relationships are shown in Fig. 33-9.

Although the action of this second recently discovered (1962) hormone is not yet clear, evidence suggests that it inhibits bone resorption by the osteoclasts. Calcitonin thus protects the body against a dangerous rise in the blood calcium level, just as parathormone protects it from a dangerous fall in blood calcium. The two act together to smooth out oscillations in blood calcium.

The third hormone is a derivative of vitamin D_3 (cholecalciferol). Unlike other vitamins, most of which are derived exclusively from the diet, vitamin D either may be acquired in the diet or synthesized in the skin by irradiation with ultraviolet light from the sun. It is then converted in the liver and kidney to a hormonal substance, 1,25-dihydroxycholecalciferol, also known as 1,25-DHCC. This steroid is necessary for calcium absorption from the gut. It also promotes mineralization of bone by inducing the formation of calcium-transporting carrier protein.

A deficiency of vitamin D causes **rickets,** a disease characterized by low blood calcium and weak, poorly calcified bones that tend to bend under postural and

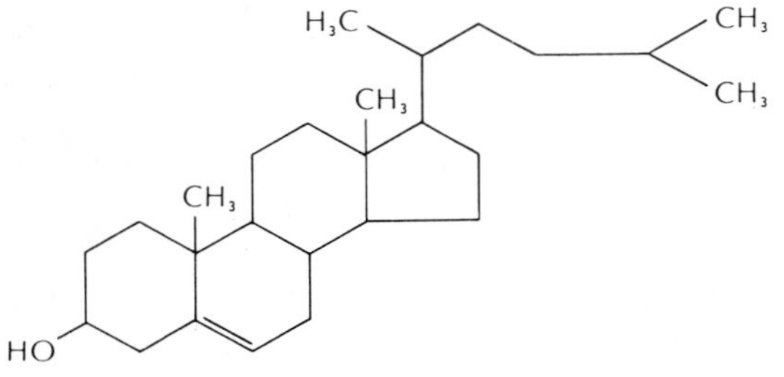

FIG. 33-10 Hormones of the adrenal cortex. Cortisol (a glucocorticoid) and aldosterone (a mineralocorticoid) are two of the many steroid hormones synthesized from cholesterol in the adrenal cortex.

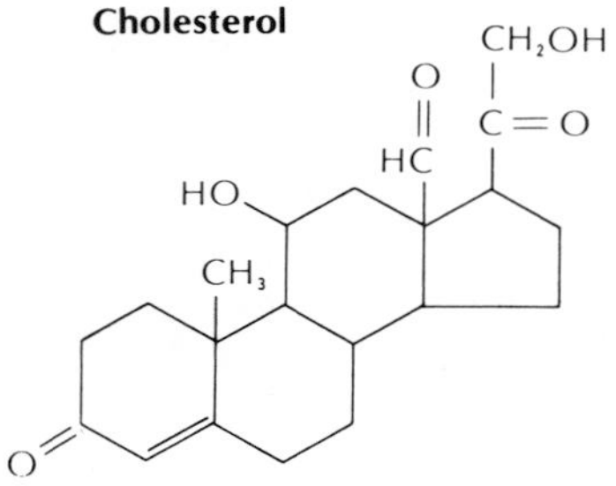

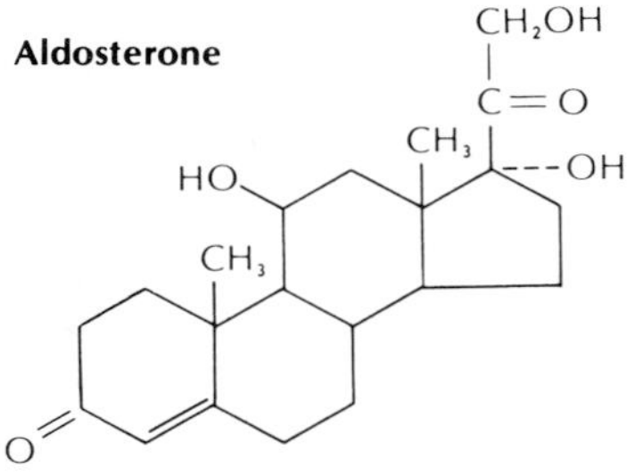

gravitational stresses. Rickets has been called a disease of northern winters, when sunlight is minimal. All children in northern latitudes require supplements of vitamin D in winter. Rickets was once common in the smoke-darkened cities of England and Europe.

Adrenocortical hormones

The vertebrate adrenal gland is a double gland consisting of two very different kinds of tissue: **interrenal** tissue, called **cortex** in mammals, and **chromaffin** tissue, called **medulla** in mammals. The mammalian terminology of cortex (meaning "bark") and medulla (meaning "core") arose because in this group of vertebrates the interrenal tissue completely surrounds the chromaffin tissue like a cover. Although in the lower vertebrates the interrenal and chromaffin tissues are also usually separated, the mammalian terms "cortex" and "medulla" are so firmly fixed in our vocabulary that we commonly use them for all vertebrates instead of the more correct terms "interrenal" and "chromaffin."

Biochemists have found that the adrenal cortex contains at least 30 different compounds, all of them closely related lipoid compounds known as steroids. Only a few of these compounds, however, are true steroid **hormones;** most are various intermediates in the synthesis of steroid hormones from **cholesterol** (Fig. 33-10). The corticosteroid hormones are commonly classified into three groups, according to their function:

1. **Glucocorticoids,** such as **cortisol** (Fig. 33-10) and **corticosterone,** have a number of important effects concerned with food metabolism and inflammation. They cause the conversion of nonglucose compounds, particularly amino acids and fats, into glucose. This process, called **gluconeogenesis,** is an extremely important process that serves to maintain adequate amounts of glucose in the blood when glucose is not being provided in the diet by converting the body's energy reserves of fat and protein to glucose. Cortisol, cortisone, and corticosterone are also **anti-inflammatory.** Because several human diseases are inflammatory diseases (for example, allergies, hypersensitivity, arthritis), these corticosteroids have important medical applications. They must be used with great care, however, since if administered in excess, they may suppress the body's normal repair processes and lower resistance to infectious agents.

2. **Mineralocorticoids,** the second group of corticosteroids, are those that regulate salt balance. **Aldosterone** (Fig. 33-10) and **deoxycorticosterone** are the most important steroids of this group. They promote the tubular reabsorption of sodium and chloride and the tubular excretion of potassium by the kidney. Since sodium usually is in short supply in the diet, and potassium in excess, it is obvious that the mineralocorticoids play vital roles in preserving the correct balance of blood electrolytes. We may also note that the mineralocorticoids **oppose** the anti-inflammatory effect of cortisol and cortisone. In other words, they promote the **inflammatory** defense of the body to various noxious stimuli. Although these opposing actions of the corticosteroids seem self-defeating, they actually are not. They are necessary to maintain readiness of

the body's defenses for any stress or disease threat, yet prevent these defenses from becoming so powerful that they turn against the body's own tissues.

3. **Sex hormones** (such as testosterone, estrogen, progesterone) are produced primarily by the ovaries and testes (pp. 810 to 814). The adrenal cortex is also a minor source of certain steroids that mimic the action of testosterone. These sex hormone–like secretions are of little physiologic significance, except in certain human diseased states.

The synthesis and secretion of the glucocorticoids, such as cortisol are controlled by the **adrenocorticotropic hormone** (ACTH) of the anterior pituitary (Fig. 33-6). As with pituitary control of the thyroid, a negative feedback relationship exists between ACTH and the adrenal cortex: a rise in the level of cortisol suppresses the output of ACTH; a fall in the blood steroid level increases ACTH output. Unlike the glucocorticoids, which are under pituitary control, the mineralocorticoids (especially aldosterone) are regulated by electrolyte levels in the blood.

Adrenal medulla hormones

The adrenal medulla secretes two structurally similar hormones, **epinephrine** (adrenaline) and **norepinephrine** (noradrenaline). Their structures are as follows:

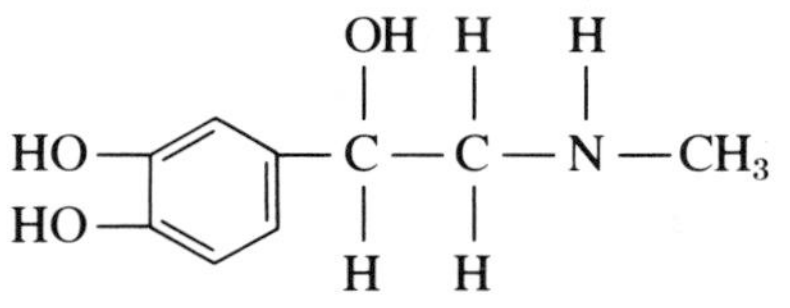

```
          OH  H   H
          |   |   |
HO—[ring]—C — C — N — CH3
HO—       |   |
          H   H
```

Epinephrine

```
          OH  H
          |   |
HO—[ring]—C — C — NH2
HO—       |   |
          H   H
```

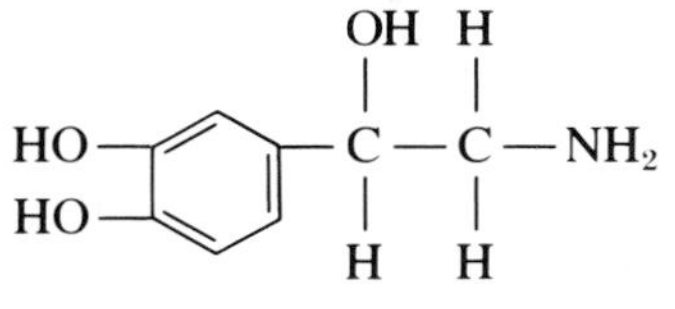

Norepinephrine

Both are derived from catechol,

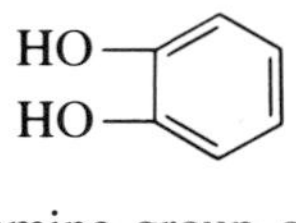

```
HO—[ring]
HO—
```

and each bears an amine group on a two carbon side chains. Consequently both belong to a class of compounds called **catecholamines.**

Norepinephrine is also released at the endings of sympathetic nerve fibers throughout the body, where it serves as a "transmitter" substance to carry neural signals across the gap that separates the fiber and the organ it innervates. The adrenal medulla has the same embryologic origin that sympathetic nerves have; in many respects the adrenal medulla is nothing more than a giant sympathetic nerve ending.

It is not surprising then that the adrenal medulla hormones have the same general effects on the body that the sympathetic nervous system has. These effects center around emergency functions of the body, such as fear, rage, fight, and flight, although they have important integrative functions in more peaceful times as well. We are all familiar with the increased heart rate, tightening of the stomach, dry mouth, trembling muscles, general feeling of anxiety, and increased awareness that attend sudden fright or other strong emotional states. These effects are attributable both to the rapid release into the blood of epinephrine from the adrenal medulla and to increased activity of the sympathetic nervous system.

Epinephrine and norepinephrine have many other effects that we are not so aware of, including constriction of the arterioles (which, together with the increased heart rate, increases the blood pressure), mobilization of liver glycogen to release glucose for energy, increased oxygen consumption and heat production, hastening of blood coagulation, and inhibition of the gastrointestinal tract. All of these changes in one way or another tune up the body for emergencies.

Insulin from the islet cells of the pancreas

The pancreas is both an exocrine and an endocrine organ. The **exocrine** portion produces pancreatic juice, a mixture of digestive enzymes that is conveyed by ducts to the digestive tract. Scattered among the extensive exocrine portion of the pancreas are numerous small islets of tissue, called **islets of Langerhans.** These are the **endocrine** portion of the gland. The islets are without ducts and secrete their hormones directly into blood vessels that extend throughout the pancreas.

Two polypeptide hormones are secreted by different cell types within the islets: **insulin,** produced by the **beta cells,** and **glucagon,** produced by the **alpha cells.** Insulin and glucagon have antagonistic actions of great importance in the metabolism of carbohydrates and fats. Insulin is essential for the utilization of blood glucose by cells, especially skeletal muscle cells.

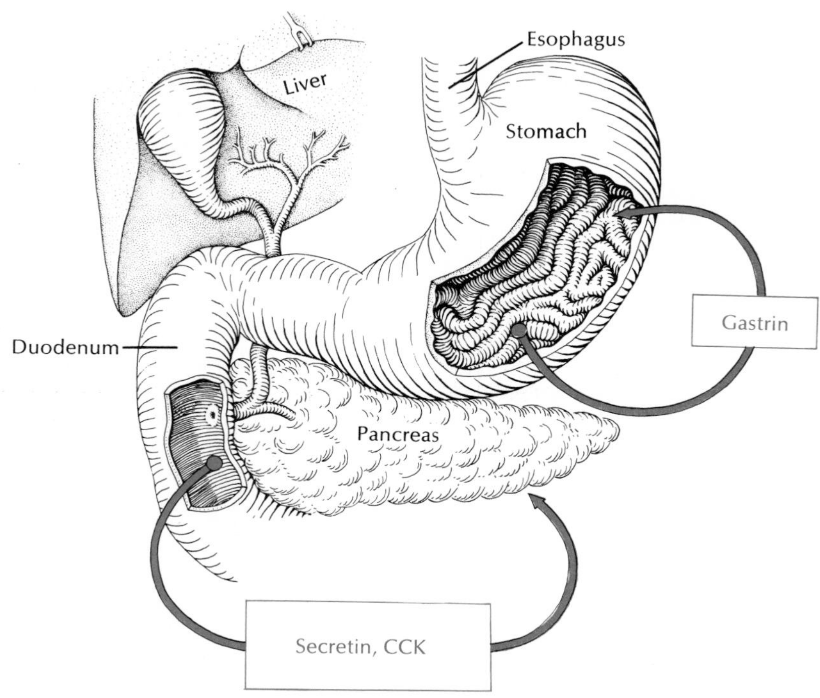

FIG. 33-11 Three hormones of digestion. Arrows show source and target of three gastrointestinal hormones.

Insulin somehow allows glucose in the blood to be transported into the cells. Without insulin, the blood glucose levels rise (hyperglycemia) and sugar appears in the urine. Insulin also promotes the uptake of amino acids by skeletal muscle and inhibits the mobilization of fats in adipose tissue.

Failure of the pancreas to produce enough insulin causes the disease **diabetes mellitus** that afflicts 2% of the population. It is attended by serious alterations in carbohydrate, lipid, protein, salt, and water metabolism, which, if left untreated, may cause death.

The first extraction of insulin in 1921 by two Canadians, Frederick Banting and Charles Best, was one of the most dramatic and important events in the history of medicine. Many years earlier, two German scientists, Von Mering and Minkowski, discovered that surgical removal of the pancreas of dogs invariably caused severe symptoms of diabetes resulting in the animal's death within a few weeks. Many attempts were then made to isolate the diabetes preventive factor, but all failed because powerful protein-splitting digestive enzymes in the exocrine portion of the pancreas destroyed the hormone during extraction procedures. Following a hunch, Banting in collaboration with Best and his physiology professor J. J. R. Macleod tied off the pancreatic ducts of several dogs. This caused the exocrine portion of the gland with its hormone-destroying enzyme to degenerate, but left the islets' tissues healthy, since they were independently served by their own blood supply. Banting and Best then successfully extracted insulin from these glands. Injected into another dog, the insulin immediately lowered the blood sugar level. Their experiment paved the way for the commercial extraction of insulin from slaughterhouse animals. It meant that millions of diabetic persons, previously doomed to invalidism or death, could now look forward to nearly normal lives.

Glucagon, the second hormone of the pancreas, has several effects on carbohydrates and fat metabolism that are opposite to the effects of insulin. For example, glucagon raises the blood glucose level, whereas insulin lowers it. Glucagon and insulin do not have the same effects in all vertebrates, and in some, glucagon is lacking altogether. Glucagon is another example of a hormone that operates through the cyclic AMP second-messenger system.

Hormones of digestion

Gastrointestinal function is coordinated by a family of hormones produced by endocrine cells scattered throughout the gut. Though together they constitute the largest endocrine organ in the body, they have long been neglected by endocrinologists because it has been impossible to apply the classic method of surgical removal of a gland, followed by examination of the effect of its absence. However, by the mid 1970s seven gastrointestinal hormones had been chemically purified or defined.

Of these, we will discuss three of the best understood (Fig. 33-11).

Gastrin is a small polypeptide hormone produced in the mucosa of the pyloric portion of the stomach. When food enters the stomach, gastrin stimulates the secretion of hydrochloric acid by the stomach wall. Gastrin is an unusual hormone in that it exerts its action on the same organ from which it is secreted. Two other hormones of digestion are **secretin** and **cholecystokinin-pancreozymin (CCK).** Both are polypeptide hormones secreted by the intestinal mucosa in response to the entrance of acid and food into the duodenum from the stomach; both stimulate the secretions of pancreatic juice, but their effects differ somewhat. Secretin, the first of all hormones to be discovered (p. 753), stimulates a pancreatic secretion rich in bicarbonate that rapidly neutralizes stomach acid. CCK stimulates the pancreas to release an enzyme-rich secretion.

Annotated references
Selected general readings

See also general references for Part Three, p. 793.

Barrington, E. J. 1975. An introduction to general and comparative endocrinology. Oxford, Oxford University Press.

Bentley, P. J. 1971. Endocrines and osmoregulation: a comparative account of the regulation of water and salt in vertebrates. Vol. 1 of Zoophysiology and Ecology. Berlin, Springer-Verlag. *Environmentally oriented, advanced-level treatment of this subject by an authority in the area.*

Bentley, P. J. 1976. Comparative vertebrate endocrinology. Cambridge, Cambridge University Press. *A comprehensive, up-to-date, and clearly written account that relates endocrine function to the animals' ecology and evolutionary background.*

Highnam, K. C., and L. Hill. 1969. The comparative endocrinology of the invertebrates. New York, American Elsevier Publishing Co., Inc. *Clearly written, well-illustrated comparative account, emphasizing hormonal action.*

Holmes, R. L., and J. N. Ball. 1974. The pituitary gland: a comparative account. New York, Cambridge University Press. *Advanced comprehensive survey.*

Tombes, A. S. 1970. An introduction to invertebrate endocrinology. New York, Academic Press, Inc.

Turner, C. L., and J. T. Bagnara. 1976. General endocrinology, ed. 6. Philadelphia, W. B. Saunders Co. *Popular undergraduate-level text.*

Selected *Scientific American* articles

Gillie, R. B. 1971. Endemic goiter. **224:**92-101 (June). *In some parts of the world, people still suffer from this thyroid disorder caused by insufficient iodine in the diet.*

Guillemin, R., and R. Burgus. 1972. The hormones of the hypothalamus. **227:**24-33 (Nov.). *The functions of two of the brain's "releasing factors" are described.*

Levine, R., and M. S. Goldstein. 1958. The action of insulin. **198:**99-106 (May). *Describes insulin's role in promoting the passage of sugar across cell membranes.*

Li, C. H. 1963. The ACTH molecule. **209:**46-53 (July). *Its composition and what it does.*

Loomis, W. F. 1970. Rickets. **223:**76-91 (Dec.). *The role of calciferol (vitamin D), a sunlight-dependent hormone precursor, in preventing this oldest of air-pollution diseases.*

McEwen, B. S. 1976. Interactions between hormones and nerve tissue. **235:**48-58 (July). *Steroid hormones secreted by gonads and adrenal cortex can be traced to target cells in brain.*

Nathanson, J. A., and P. Greengard. 1977. "Second messengers" in the brain. **237:**108-119 (Aug.). *Analogous messenger mechanisms underlie communication in the endocrine and nervous systems.*

O'Malley, B. W., and W. T. Schrader. 1976. The receptors of steroid hormones. **234:**32-43 (Feb.). *The action of steroid hormones is mediated by receptor molecules that reside only in the target organs.*

Pastan, I. 1972. Cyclic AMP. **227:**97-105 (Aug.). *The hormonal "second messenger" and some of the roles it plays.*

Rasmussen, H., and M. M. Pechet. 1970. Calcitonin. **223:**42-50 (Oct.). *Discovery and action of this calcium-regulating hormone.*

Wilkins, L. 1960. The thyroid gland. **202:**119-129 (Mar.). *How the gland is constructed, what it produces, and how it is controlled.*

Williams, C. M. 1958. The juvenile hormone. **198:**67-74 (Feb.). *This insect hormone prevents metamorphosis of the larva into a pupa until growth is complete.*

CHAPTER 34
ANIMAL BEHAVIOR

A dominant male chacma baboon *Papio ursinus* is groomed by a male social inferior. Dominance hierarchies are common among social animals. Social position is established by aggressive encounters, but once ranking within a group is resolved it serves to reduce social tensions.

Photograph by L. L. Rue, III.

For as long as people have walked the earth, their lives have been touched by, indeed interwoven with, the lives of other animals. They hunted and fished for them, domesticated them, ate them and were eaten by them, made pets of them, revered them, hated and feared them, immortalized them in art, song, and verse, fought them, and loved them. The very survival of ancient peoples depended on knowledge of wild animals. To stalk them they had to know the ways of their quarry. As the hunting society of primitive people gave way to agricultural civilization, an awareness was retained of the interrelationship with other animals.

This awareness is still evident today. Zoos attract more visitors than ever before; wildlife television shows are increasingly popular; game-watching safaris to Africa constitute a thriving enterprise; and millions of pet animals share the cities with us—more than 700,000 pet dogs live in New York City alone. Although ethology, the science of animal behavior, is a young discipline of zoology, people have always been behaviorists. Only recently, however, have we begun to understand how animal societies work.

ETHOLOGY AND THE STUDY OF INSTINCT

In 1973, the Nobel Prize in Physiology and Medicine was awarded to three pioneering zoologists, Karl von Frisch, Konrad Lorenz, and Niko Tinbergen. The citation stated that these three were the principal architects of the new science of ethology. It was the first time any contributor to the behavioral sciences was so honored, and it meant that the discipline of animal behavior, which really takes its roots from the work of Charles Darwin, had arrived.

Ethology, meaning literally "character study," was first used in the late eighteenth century to signify the interpretation of character through the study of gesture. It was an appropriate term for the new field of animal behavior, which had as its objective the study of motor patterns, that is, gestures of animals, with the anticipation that such study would reveal the true characters of animals just as the interpretation of human gestures might reveal the true characters of people. This decidedly restricted interpretation of ethology as a purely descriptive study of the "habits" of animals was considerably modified during ethology's epoch-making period, 1935 to 1950. Today we may define ethology as a discipline that involves the study of the total repertoire of the innate and the learned behavior that animals employ to resolve the problems of survival and reproduction.

One aim of ethologists has always been to describe the behavior of an animal in its **natural habitat.** From the beginning ethologists have been naturalists. Their laboratory has been the out-of-doors and their experiments have been observational ones of animals in their natural surroundings, with nature providing the variables. Ethologists recognized that it makes no more sense to try to study the natural behavior of an animal divorced from its natural surroundings than it does to try to interpret a structural adaptation of an animal apart from the function it serves.

With infinite patience Lorenz, Tinbergen, and their colleagues watched and catalogued the activities and vocalizations of animals during feeding, courtship, and nest building, as well as seemingly insignificant behavioral movements such as head scratching, stretching postures, and turning and shaking movements. Thus these studies concentrated largely on innate motor patterns used for communication within a species.

One of the great contributions of Lorenz and Tinbergen was to demonstrate that behavioral traits are measurable entities like anatomic or physiologic traits. This was to become the central theme of ethology: behavioral traits can be isolated and measured, and they have evolutionary histories. They showed that behavior is not the wavering, transient, unpredictable phenomenon often depicted by earlier writers. In short, behavior is genetically mediated. It is apparent that, if behavior is determined by genes in the same way that genes determine morphologic and physiologic characters (ethologists marshaled abundant evidence showing that it is), then behavior evolves and is adaptive. Thus modern behavioral study is founded on the recognition that the Darwinian view of evolution holds for behavioral traits, as well as for anatomic and functional characters.

STEREOTYPED BEHAVIOR

The European ethologists, through step-by-step analysis of the detailed behavior of animals in nature, especially of fishes and birds, focused on the invariant components of behavior. From such detailed studies emerged several concepts that were first made available to large numbers of North American readers in Niko Tinbergen's influential book, *The Study of Instinct* (1951).

The basic concepts of ethologic theory can be ap-

FIG. 34-1 Egg-rolling movement of the greylag goose, *Anser anser*. (From Lorenz, K., and N. Tinbergen. 1938. Zeit. Tierpsychol. **2:**1-29.)

proached by considering the egg-rolling movement of the greylag goose, described by Lorenz and Tinbergen in 1938 (Fig. 34-1). If a greylag goose is presented with an egg a short distance from her nest, she reaches out with her bill until her lower beak has contacted the egg. Then, extending her bill just beyond the egg, she contracts her neck, pulling the egg carefully toward the nest with the underside of her bill. Since the egg naturally has a tendency to wander off course on the uneven terrain, the goose compensates for this by moving the bill from one side to the other, correcting the course of the egg until she has it in her nest.

If we remove the egg after the goose has begun her retrieval, the goose continues the egg-rolling behavior as though the egg were still there. The side-to-side movements cease, but the tucking movements continue as she slowly pulls her neck back in a straight line toward the nest with the invisible egg. If on another occasion the egg being retrieved slips completely away and rolls down the slope of the nest mound, the goose continues her retrieval movements in vacuo until she is again settled comfortably on her nest. Then, seeing that the egg has not been retrieved, she begins the egg-rolling pattern all over again.

Fixed action patterns

The egg-rolling movement of greylag geese is an example of a **fixed action pattern.** It is a motor act that is highly stereotyped and mostly invariable in its performance. Once the sequence is begun, it is completed whether or not the egg is actually being retrieved. The goose does not have to learn the movement; it is an innate, or instinctive, skill. The goose behaves as though an internal program had been set in motion and must run its course before switching off.

Actually the egg-rolling movement can be separated into two components. The fixed action pattern represents the unvarying, straight retrieval movements that continue even in the absence of a stimulus (the egg). However, the lateral corrective movements that the goose makes to steer the egg past obstructions are the taxis components of the retrieval. A **taxis** (Gr., arrange) is an orientation movement that is guided by external stimulation, in this case the changing tactile stimulation on the beak provided by the wandering egg. Fixed action patterns usually occur with orienting movements (taxes) superimposed on them.

Releasers (sign stimuli)

Much of an animal's behavior is triggered by a few key environmental stimuli. Such key signals, or **sign stimuli,** constitute only a small fraction of the total environmental information available. The egg placed outside the nest of the greylag goose is the specific stimulus that releases egg-retrieval behavior. However, at no time do releasers trigger a fixed action pattern

FIG. 34-2 An oyster catcher, *Haematopus ostralegus* attempts to roll a giant egg model into its nest while ignoring its own egg. (From Tinbergen, N. 1951. The study of instinct. Oxford, Oxford University Press.)

in all members of a species. The animal must be in the proper motivational state. Only a *female* greylag goose *in a nest* responds to a wayward egg outside the nest. But, once she is properly motivated and in the ''right'' surroundings, the greylag goose is not terribly discriminating about what she retrieves. Lorenz and Tinbergen showed that almost any smooth and rounded object placed outside the nest would trigger the egg-retrieval behavior of this species; even a small toy dog and a large yellow balloon were dutifully retrieved.

Sometimes exaggerated sign stimuli release an exaggerated response. In one famous experiment, an oyster catcher tried mostly unsuccessfully to retrieve and incubate an egg four times the normal size while ignoring its own egg (Fig. 34-2). Such an exceptionally effective sign stimulus is called a **supernormal releaser.**

Because sign stimuli are relevant to the animal's survival and reproduction, they are characteristically simple cues, a fact that reduces the chances of making a mistake or misunderstanding the cue. In every case the response is highly predictable. For example, the alarm call of adult herring gulls releases a crouching freeze response in the chicks. Female frogs respond only to the courtship calls of males of their own species. Night-flying moths take evasive maneuvers or drop to the ground when they hear the ultrasonic cries of bats that feed on them; no other sound releases this response. Ethologists have described hundreds of such examples of releasers.

Innate releasing mechanism

The sign stimulus, whether a sound, color, appropriate structure, odor, or movement, releases a highly predictable response. This strict correlation between stimulus and appropriate response led ethologists to conclude that the nervous system is organized into discrete ''centers'' capable of filtering the sensory input, correctly selecting the releaser stimuli, and releasing an appropriate motor response. Lorenz called this organization the **innate releasing mechanism (IRM).** An IRM was thought of as an aggregation of nerve cells that, once triggered by the key stimulus, would direct motor neurons to fire in a programmed manner, causing the animal to perform a stereotyped fixed action pattern. Lorenz used an analogy to illustrate how an IRM operates: the sign stimulus is a key that unlocks only one of many possible doors; once the door (the IRM) is opened, the specific response inside is released.

More controversy surrounds the concept of the IRM than perhaps any other ethologic concept. Physiologists are reluctant to believe that ''centers'' as visualized by Lorenz actually exist. The central nervous system operates as an integrated whole, rather than as a collection of nerve aggregates, each prewired to perform one and only one motor function. Thus the IRM concept has been altered somewhat, but the idea seems to be fundamentally correct; brain ''centers'' as such probably do not exist, but the brain *works* as though they do. If not interpreted too strictly, the IRM is a useful tool to ethologists for understanding the sequence and coordination of stereotyped behavior.

Drive concept

Animals are active because of internal motivation. It has been customary to refer to changes in responsiveness as **drives.** Thus animals show hunger drives, thirst drives, courtship drives, mating drives, migration drives, and so on. Used in this context, a drive refers to a motivational change that results in the performance of certain usually predictable behavior. A hungry animal searches for food in a manner characteristic of its species. The hunting behavior can be thought of as a search for a releasing situation that permits the discharge of prey-capturing behavior.

Ethologists recognize three stages in completing a stereotyped behavior.

1. **Appetitive behavior.** Appetitive behavior is the specific search for a releasing situation; it places an animal in a position to achieve a biologic goal. A hungry fox becomes restless and begins a characteristic hunting behavior that is adapted to the situation; it

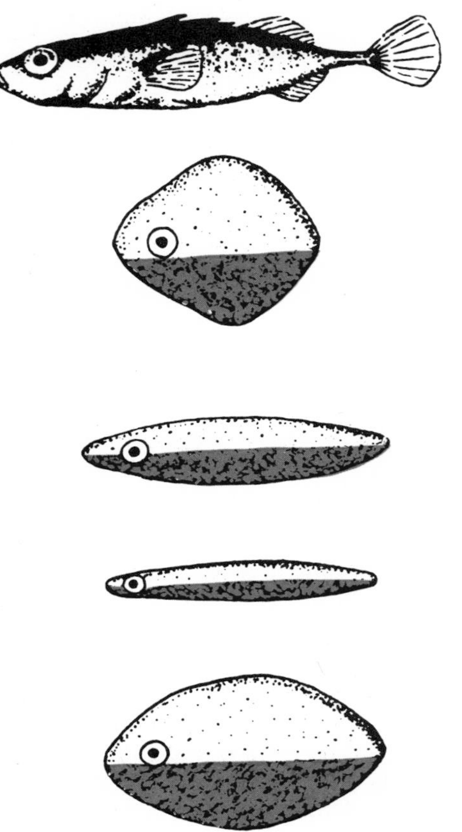

FIG. 34-3 Stickleback models. The carefully made model of the stickleback, at top, without a red belly is attacked much less frequently than the four simple red-bellied models. (From Tinbergen, N. 1951. The study of instinct. Oxford, Oxford University Press.)

knows the terrain, the detours to make, the habits of potential quarry, and the location of a farmer's chicken yard. In lower animals, appetitive behavior is more predictable and less subject to variation from previous learning experience.

2. **Consummatory behavior.** The biologic goal of appetitive behavior is the consummatory act. For the hungry fox, catching and eating a chicken is the climax of the hunger drive. Location of a mate by a songbird following a search (appetitive behavior) releases courtship behavior (consummatory act). A gravid tortoise locates a suitable nest site and lays her eggs. A blowfly's search for food ceases when it has discovered and eaten enough to comfortably fill the foregut. In each case, once the goal is achieved, the goal-directed activity ceases.

3. **Quiescence.** Following the consummatory act, appetitive behavior does not immediately resume. This is the period of quiescence. The sated animal does not hunt. Later, when the animal gets hungry, a new round of appetitive behavior, consummatory act, and quiescence follows.

Although use of the term *drive* is a convenient way to describe a positive motivation toward achieving some biologic goal, the word is increasingly disfavored. Critics contend that it is vague and nonexplanatory and is used to embrace a variety of distantly related phenomena, having different neurophysiologic bases. Yet, the term persists in the absence of any completely acceptable substitute for expressing in a word the idea of motivation toward some biologic goal.

Instinct and learning

The invariable and predictable nature of stereotyped behavior suggests that it is inherited, or innate, behavior. Many kinds of unpracticed stereotyped behavior appear suddenly in animals and are indistinguishable from similar behavior performed by older, experienced individuals. Newly emerged moths and butterflies fly perfectly as soon as their wings are dry. Orb-weaving spiders "know" how to build their webs without practice and without having watched other spiders build theirs. Newly hatched gull chicks crouch in the nest the first time they hear the alarm call of the parent.

To the observer, these behaviors have a mechanical, programmed appearance. This is even more evident when animals act out behaviors released in inappropriate situations. For example, a nesting male stickleback fish vigorously attacks a plump lump of wax with a red underside placed in its tank by experimenters. The stickleback reacts to this unfishlike model as if it were another male stickleback with a red belly intruding into its territory. A carefully made model that closely resembles a stickleback but lacks the red belly is ignored (Fig. 34-3). Male English robins furiously attack a bundle of red feathers placed in their territory but ignore a stuffed juvenile robin without the red feathers (Fig. 34-4). Male red-winged blackbirds attempt to copulate with a crude model consisting of rump and tail feathers of a female, if the tail is elevated in a precopulatory position.

In these and other examples of stereotyped behavior

FIG. 34-4 Two models of the English robin. The bundle of red feathers, at right, is attacked by male robins, whereas the stuffed juvenile bird without a red breast is ignored. (From Tinbergen, N. 1951. The study of instinct. Oxford, Oxford University Press; after Lack, D. 1943. The life of the robin. London, Cambridge University Press.)

released by simple sign stimuli, it is difficult to understand how the animal could have learned the response. These "built-in" or innate behavioral patterns are released in complete form and without practice the first time that an animal of certain age and motivational state encounters the stimulus. To such behaviors the term **instinct,** meaning "driven from within," is applied. It is easy to understand why instinctive behavior is an important adaptation for survival, especially to lower forms that never know their parents. They must be equipped to respond immediately and correctly to their world as soon as they hatch. It is also evident that more advanced animals with longer lives and with parental care or other opportunities for social interactions have the opportunity to improve or change their behavior by learning.

Learning is the modification of behavior by experience. As a rule, behavioral modifications through learning are adaptive, since they improve the animal's fitness. The crouching behavior of gull chicks previously mentioned provides an excellent illustration of the adaptive value of learning. If the parent calls an alarm, the newly hatched chick crouches down—obviously a useful defensive response to overflying predators. After 2 or 3 days, the chick responds to the alarm call by running away before crouching. Later, it runs and hides in a chosen site before crouching.

As the chick grows stronger and more mobile, it also becomes more discriminating in its visual response to moving objects overhead. After several such experiences, it begins to lose its general fear of overflying birds. In other words, its sensitivity to this particular stimulus declines. This very simple kind of learning is called **habituation,** the reduction or elimination of a response in the absence of any reward or punishment. For the gull chick, habituation to overflying objects is clearly appropriate, since running and hiding from every passing shadow consume time and energy better applied to more productive activities.

However, gull chicks do not come to ignore *all* birds flying overhead. Absolute habituation in this instance would be even less appropriate than no habituation at all, since the birds of prey are genuine predators of gull chicks. Chicks that fail to hide from an overflying hawk are less apt to survive than chicks that have retained a healthy fear of this predator.

Lorenz and Tinbergen discovered that chicks discriminate predatory birds from harmless songbirds and ducks on the basis of shape, especially the length of the head and neck. When chicks were presented with a silhouette having a hawklike shape and a short neck such that the head protruded only slightly in front of the wings, they crouched in alarm. Long-necked silhouettes were ignored or aroused only mild interest (Fig. 34-5). In another experiment, a model was built having symmetric wings and with head and tail shaped so that either the front or the rear of the dummy could be regarded as the head or as the tail (Fig. 34-6). When sailed to the left it resembled a goose and caused no alarm; when sailed to the right, however, the model resembled a hawk and did cause alarm. Obviously both the direction of motion and the shape are important in recognition.

Sometimes learning is demonstrated in surprising and amusing ways. Tinbergen describes one such incident. "In order to sail our models, which crossed a meadow where the birds were feeding or resting, at a height of about 10 yards along a wire, running from one tree to another 50 yards away, either Lorenz or I had to climb a tree and mount the dummy we wanted to test out. One family of geese (which also reacted to some of our dummies) very soon associated tree-climbing humans with something dreadful to come, and

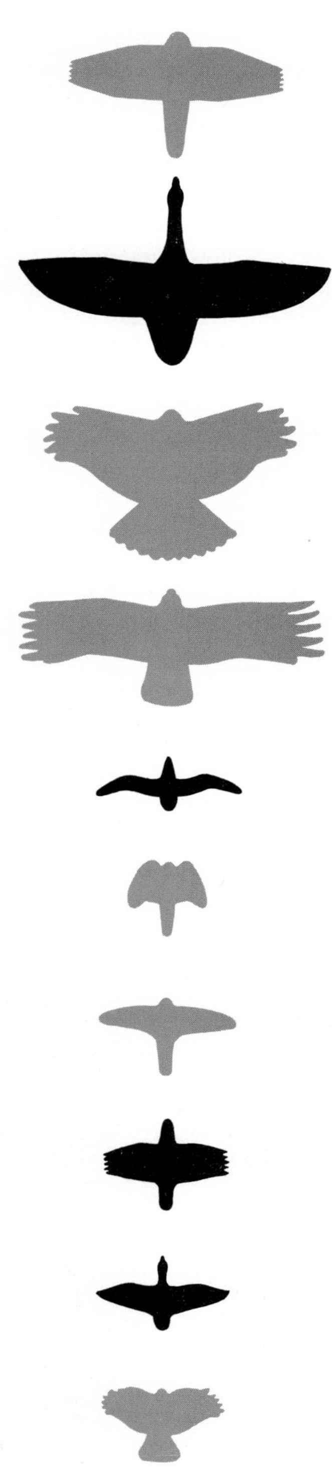

promptly called the alarm and walked off when one of us went up.''*

These observations with models might suggest incorrectly that gull chicks (as well as pheasant and turkey chicks that react similarly) have an innate ability to distinguish short-necked predators from harmless birds having longer necks. But, in fact, subsequent experiments demonstrated that newly hatched chicks, which are at first alarmed by anything that passes overhead—even a falling leaf—gradually become habituated to familiar objects. The chick learns that song and shore birds are common and harmless features of its world. But they never become accustomed to short-necked predators because these are seldom seen. It is the unfamiliar that arouses fear.

The alarm response of herring gull chicks is an example of a simple behavior that becomes altered as the result of experience. As they grow, the chicks store information about their world and become increasingly selective in their alarm response. Animals are internally programmed to respond automatically to certain stimuli; however, this response requires adjustment from the environment to become efficient as the animal grows.

What can we say about the role of genes and that of environment in shaping instinctive and learned behavior? We tread into delicate territory because behaviorists do not all agree on just what constitutes innate behavior. The problem concerns the effects that the environment has on the individual during its development. The idea that behavior must be fixed *either* by heredity *or* by environment has had a long history, resulting in what is termed the **nature versus nurture controversy.** It has proved to be a useless debate because *any* characteristic of an individual is determined by an interaction between information supplied in the genetic blueprint (nature) and by the environment (nurture). Let us examine once again our definitions of instinct and learning and then try to explain how gene-environment interactions contribute to each.

*Tinbergen, N. 1961. The herring gull's world. New York, Basic Books, Inc., p. 216.

FIG. 34-5 Models used by Lorenz and Tinbergen for the study of predator reactions in young fowl. Young gull chicks crouch in alarm when hawk silhouettes *(in red)* pass overhead but ignore shapes of harmless birds. (From Tinbergen, N. 1951. The study of instinct. Oxford, Oxford University Press.)

We defined an instinctive behavior as one that emerges in complete form the first time the animal reacts to the appropriate stimulus. It must have a genetic foundation because without genes there can be no behavior. Specifically the animal's genotype contains instructions that result in the construction of a specific neural organization, which permits certain types of behavior. We are not saying that an instinctive behavioral attribute is determined solely by information contained in specific chromosomal loci. The genetic code is an information-generating device that depends on an environment that supplies materials and provides order for embryonic development. A genotype remains just a genotype if the developing organism cannot obtain the substances required to form tissues, organs, and nervous system.

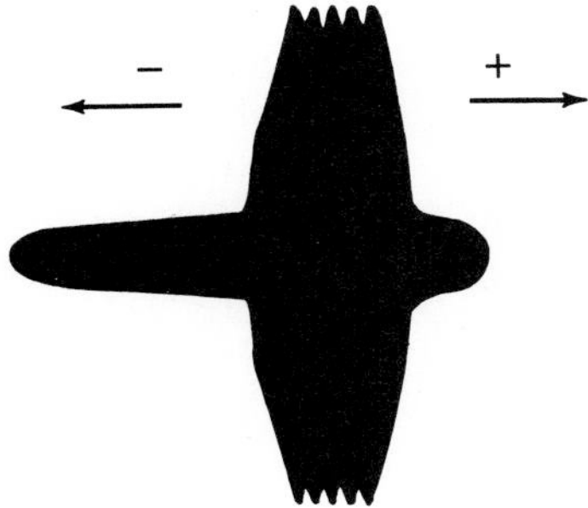

FIG. 34-6 Model that drew positive responses from gull chicks when sailed to the right but none when sailed to the left. (From Tinbergen, N. 1961. The herring gull's world: a study of the social behavior of birds. New York, Basic Books, Inc., Publishers.)

Learning depends on experience encountered by the organism as it interacts with its environment. It also depends on internal programming because the things an organism learns to do best are determined by the genetic blueprint. The nervous system must be designed to facilitate the acquisition of learning at specific stages in the organism's development. In other words, through its genetically determined development, the brain possesses properties enabling it to "anticipate" its eventual use in the modification of behavior. Learned behavior, like instinctive behavior, contains both genetic and environmental components.

One kind of learned behavior that clearly illustrates the interaction of heredity and environment is **imprinting.** As soon as a newly hatched gosling or duckling is strong enough to walk, it follows its mother away from the nest. After it has followed the mother for some time, it follows no other animal (Fig. 34-7). But, if the eggs are hatched in an incubator or if the mother is separated from the eggs as they hatch, the goslings follow the first large moving object they see. As they grow, the young geese prefer the artificial "mother" to anything else, including their true mother. The goslings are said to be imprinted on the artificial mother.

Imprinting was observed at least as early as the first

FIG. 34-7 Canada goose *Branta canadensis* with her imprinted young. (Photograph by L. L. Rue, III.)

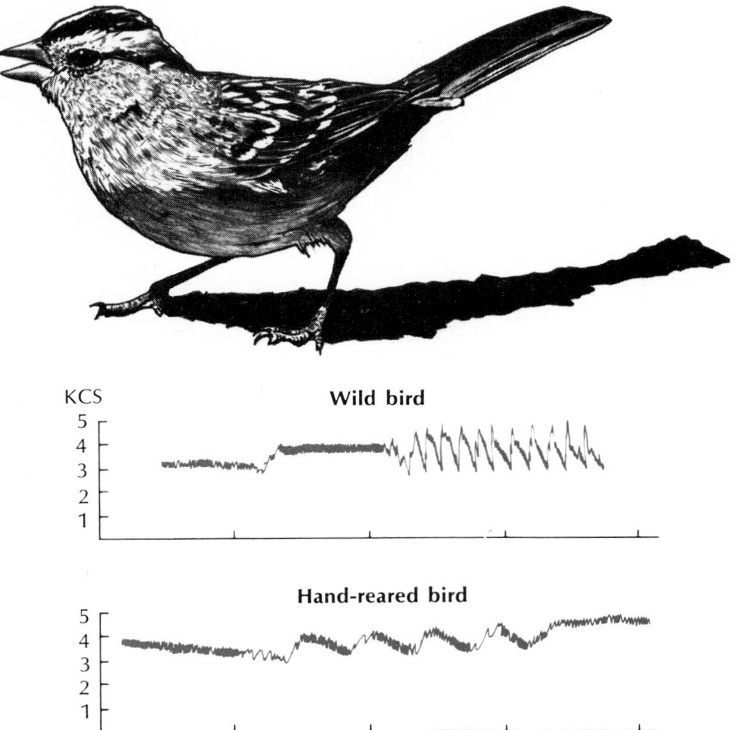

FIG. 34-8 Sound spectrograms of songs of white-crowned sparrows, *Zonotrichia leucophrys. Above,* natural song of wild bird; *below,* abnormal song of isolated bird. (From Alcock, J. 1975. Animal behavior: an evolutionary approach. Sunderland, Mass., Sinauer Associates, Inc.; after M. Konishi, personal communication.)

century A.D., when the Roman naturalist Pliny the Elder wrote of "a goose which followed Lacydes as faithfully as a dog." Konrad Lorenz first studied the imprinting phenomenon objectively and systematically. When Lorenz hand-reared goslings they formed an immediate and permanent attachment to him and waddled after him wherever he went. They could no longer be induced to follow their own mother or another human being. Lorenz found that the imprinting period is confined to a brief *sensitive* period in the individual's early life, and that once established the imprinted bond is retained for life.

What imprinting shows is that the goose or duck brain (or the brain of numerous other birds and mammals that show imprinting-like behavior) is designed to accommodate the imprinting experience. The animal's genotype is provided with an internal template that permits the animal to recognize its mother soon after hatching. Natural selection favors the evolution of animals having a brain structure that imprints in this way because following the mother and obeying her commands is important for survival. The fact that a gosling can be made to imprint to a mechanical toy duck or a human being under artificial conditions is a cost to the system that can be tolerated; the disadvantages of the system's simplicity are outweighed by the advantages of its reliability.

Let us cite one final example to complete our consideration of instinct and learning. The males of many species of birds have characteristic territorial songs that identify the singers to other birds and announce territorial rights to other males of that species. Like many other songbirds, the male white-crowned sparrow must learn the song of its species by hearing the song of its father. If the sparrow is hand-reared in acoustic isolation in the laboratory, it develops an abnormal song (Fig. 34-8). But if the isolated bird is allowed to hear recordings of normal white-crowned sparrow songs during a critical period of 10 to 50 days after hatching, it learns to sing normally. It even imitates the local dialect it hears.

It might appear from this that song characteristic is determined by learning alone. However, if during the critical learning period, the isolated male white-crowned sparrow is played a recording of another species, even a closely related one, it does not learn the song. It learns only the song appropriate to its own species. Thus, although the song must be learned, the brain has been programmed in its development to recognize and learn vocalizations produced by males of its species alone. The sparrow *learns* by example but its attention has been *innately* narrowed to focus on the appropriate example. Learning the wrong song would result in behavioral chaos and natural selection quickly eliminates those genotypes that permit such mistakes to occur.

SOCIAL BEHAVIOR

When we think of "social" animals we are apt to think of highly structured honeybee colonies, herds of antelope grazing on the African plains, schools of herring, or flocks of starlings. But social behavior is by no means limited to such obvious examples of animals of the same species living together in which individuals influence one another.

In the broad sense, any kind of interaction resulting from the response of one animal to another *of the same species* represents social behavior. Even a pair of rival males squaring off for a fight over the possession of a female is a social interaction, although our perceptual bias as people might encourage us to label it antisocial. Thus social aggregations are only one kind of social behavior, and indeed not all animal aggregations are social.

Clouds of moths attracted to a light at night, barnacles attracted to a common float, or trout gathering in the coolest pool of a stream are animal groupings responding to *environmental* signals. Social aggregations, on the other hand, depend on *animal* signals. They remain together and do things together by influencing one another.

Of course, not all animals showing sociality are social to the same degree. All sexually reproducing species must at least cooperate enough to achieve fertilization; but among most mammals, breeding is about the only adult sociality to occur. Alternatively, swans, geese, albatrosses, and beavers, to name just a few, form strong monogamous bonds that last a lifetime. Whether or not adult sociality is strongly or weakly developed, the most persistent social bonds usually form between mother and young, and these, for birds and mammals, usually terminate at fledging or weaning.

Advantages of sociality

Living together may be beneficial in many ways. Each species profits in its own particular way; what confers adaptive value to one species may not for another. One obvious benefit for social aggregations is defense, both passive and active, from predators. Musk-oxen that form a passive defensive circle when threatened by a wolf pack are much less vulnerable than an individual facing the wolves alone.

As an example of active defense, a breeding colony of gulls, alerted by the alarm calls of a few, attack predators en masse; this is certain to discourage a predator more effectively than individual attacks. The members of a prairie dog town, though divided into social units called coteries, cooperate by warning each other with a special bark when danger threatens. Thus every individual in a social organization benefits from the eyes, ears, and noses of all other members of the group.

Predators may also be distracted by the confusion effect created by large numbers of prey grouped together. A fish that can chase down and capture a lone crustacean may be unable to concentrate on a single individual in a large aggregation. Predators may even be frightened by a large aggregate of prey, which they would eat if encountered singly. One ethologist describes the behavior of a captive seal into whose tank a school of anchovies was released. "On previous occasions, when small numbers of the fish were released, the seal rapidly caught and devoured them. The larger number of anchovies, which formed a dense school whose outlines approximated the shape of an animal as large as a seal, not only inhibited pursuit but also apparently frightened the seal out of its pool. Though anecdotal, the account does illustrate the power of masses."*

Sociality offers several benefits to animal reproduction. Social grouping helps to synchronize reproductive behavior through the mutual stimulation that individuals have on one another. In colonial birds the sounds and displays of courting individuals set in motion prereproductive endocrine changes in other individuals of the colony. Because there is more social stimulation, large colonies of gulls produce more young per nest than do small colonies. In colonial species, a population size less than a certain minimum

*Klopfer, P. H. 1974. An introduction to animal behavior, ed. 2, Englewood Cliffs, N.J., Prentice-Hall, Inc., p. 157.

results in inadequate social stimulation and reproductive failure. This effect has been suggested as the cause of extinction of the passenger pigeon once heavy hunting had greatly thinned the population (p. 601).

Another benefit of social grouping to reproductive performance is that it facilitates encounters between males and females. For solitary animals, finding prospective mates may consume much time and energy. Should the population become depleted because of overhunting, habitat destruction, or severe weather, male-female encounters become increasingly rare. For example, on the Galápagos Islands, hunting by humans had reduced the giant tortoise population on one island (Hood) to a point at which the few survivors seldom, if ever, met. Lichens grew on the females' backs because there were no males to scrub them off during mating! Research personnel saved the tortoise from inevitable extinction by collecting them together in a pen, where they began to reproduce.

Still another advantage of sociality to reproduction is care of the young. This is especially well developed in the vertebrates and social insects. Parental care, with direct provisioning of the young and food sharing, increases survival of the brood. Social living also provides opportunities for individuals to give aid to and share food with young other than their own. Such interactions in a social network have resulted in some intricate cooperative behavior among parents, their young, and their kin.

Of the many other advantages of social organization noted by ethologists, we may mention only a few in our brief treatment: cooperation in hunting for food; huddling for mutual protection from severe weather; opportunities for division of labor, especially well developed in the social insects with their caste systems; and the potential for learning and transmitting useful information through the society.

Observers of a seminatural colony of macaques in Japan recount an interesting example of passage of tradition in a society. The macaques were provisioned with sweet potatoes and wheat at a feeding station on the beach of an island colony. One day a young female named Imo was observed washing the sand off a sweet potato in sea water. The behavior was quickly imitated by Imo's playmates and later by Imo's mother. Still later when the young members of the troop became mothers they waded into the sea to wash their potatoes; their offspring imitated them without hesitation. The tradition was firmly established in the troop.

Some years later Imo, an adult, discovered that she could separate wheat from sand by tossing a handful of sandy wheat in the water; allowing the sand to sink, she would scoop up the floating wheat to eat. Again, within a few years, wheat-sifting became a tradition in the troop.

Imo's peers and social inferiors copied her innovations most readily. The adult males, her superiors in the social hierarchy, would not adopt the practice but continued laboriously to pick wet sand grains off their sweet potatoes and scour the beach for single grains of wheat.

If social living offers so many benefits, why haven't all animals through natural selection become social? The answer is that a solitary existence offers its own set of advantages. In the diverse array of ecologic situations in nature, species extract their own optimal ways of life. Species that survive by camouflage from potential predators benefit from a solitary existence for a different reason, their requirement for a large supply of prey. Thus there is no overriding adaptive advantage to sociality that inevitably selects against the solitary way of life. It depends on the ecologic situation.

Aggression and dominance

Many animal species are social because of the numerous benefits that sociality offers. This requires cooperation. At the same time animals, like goverments, tend to look out for their own best interests. In short, they are in competition with one another because of limitations in the common resources that all require for life. Animals may compete for food, water, sexual mates, or shelter, when such requirements are limited in quantity and therefore worth fighting over.

Much of what animals do to resolve competition is called **aggression,** which we may define as an offensive physical action, or threat, to force another to abandon something. Many ethologists consider aggression to be part of a somewhat more inclusive interaction called **agonistic** (Gr., contest) **behavior,** referring to any activity related to fighting—whether aggression, defense, submission, or retreat.

Contrary to the widely held opinion that aggressive behavior aims at the destruction or at least defeat of an opponent, most aggressive encounters involve more saber rattling than real combat. Many species possess dangerous weapons such as teeth, beaks, claws, or horns, which could easily inflict serious injury or even

FIG. 34-9 Male Masai giraffes, *Giraffa camelopardalis* fight for social dominance. Such fights are largely symbolic, seldom resulting in injury. (Photograph by L. L. Rue, III.)

kill an opponent; but such well-armed animals seldom use their weapons against *members of their own species*. However, no such inhibiting mechanism prevents their use in killing prey or in defense against attack from another species. A coyote that uses its teeth effectively in killing a prairie dog would never bite a rival coyote in an aggressive encounter over a mate or territory. A giraffe uses its short horns to fight rivals but uses its far more lethal hoofs in defense against predators (Fig. 34-9). Therefore, when biologists speak of aggression, they are usually (though not always) referring to encounters between conspecifics.

Animal aggression within the species seldom results in injury or death because animals have evolved many symbolic aggressive displays that carry mutually understood meanings. Fights over mates, food, or territory become ritualized tournaments rather than bloody, no-holds-barred battles. When fiddler crabs spar for territory, their large claws usually are only slightly opened. Even in the most intense fighting when the claws are used, the crabs grasp each other in a way that prevents reciprocal injury. Rival male poisonous snakes engage in stylized bouts by winding themselves together; each attempts to butt the other's head with its own until one becomes so fatigued that it retreats. The rivals never bite each other. Many species of fishes contest territorial boundaries with lateral display threats, the males puffing themselves up to look as threatening as possible. The encounter is usually settled when either animal perceives itself to be obviously inferior, folds up its fins, and swims off.

Thus animals fight as though programmed by rules that prevent serious injury. Fights between rival bighorn rams are spectacular to watch, and the sound of clashing horns may be heard for hundreds of yards (Fig. 34-10). But the skull is so well protected by the massive horns that injury occurs only by accident and death is almost unheard of.

The loser in such encounters may simply run away or may signal defeat by a specialized subordination ritual. If it becomes evident to him that he is going to lose anyway, he is better off communicating his submission as quickly as possible and avoiding the cost of a real thrashing. Such submissive displays that signal the end of a fight may be almost the opposite of threat displays (Fig. 34-11). Charles Darwin, whose book *Expression of the Emotions in Man and Animals* (1873) founded the science of ethology, described the seemingly opposite nature of threat and appeasement displays as the "principle of antithesis." The principle remains accepted by ethologists today.

Why doesn't the victor of an aggressive contest kill its opponent? A defeated wolf or dog presents its vulnerable neck to the victor as a sign of complete submission. Although the dominant wolf could easily kill the defeated foe and thus remove a competitor, it never does so. The display of submission has effectively inhibited further aggression by the winner. The best explanation for aggressive restraint is that the winner has little to gain by continuing the fight. Superiority is already assured. By continuing the aggression he merely endangers himself, since a defeated opponent fighting for his life might inflict a wound. It is not difficult to see how natural selection would favor genes

FIG. 34-10 Male bighorn sheep, *Ovis canadensis* fight for social dominance during the breeding season. (Photograph by L. L. Rue, III.)

that induce aggressive restraint. Aggression that is inappropriate runs counter to the maximization of individual fitness. It is maladaptive and consequently is selected against.

The winner of an aggressive competition is dominant to the loser, the subordinant. For the victor, dominance means enhanced access to all the contested resources that contribute to reproductive success: food, mates, territory, and so on. In a social species, dominance interactions often take the form of a **dominance hierarchy.** One animal at the top wins encounters with all other members in the social group; the second in rank wins all but those with the top-ranking individual.

Such a simple, linear hierarchy was first observed in chicken societies by Schjelderup-Ebbe, who called the hierarchy a ''peck-order.'' Once social ranking is established, actual pecking diminishes and is replaced by threats, bluffs, and bows. Top hens and cocks get unquestioned access to feed and water, dusting areas, and the roost. The system works because it reduces the social tensions that would constantly surface if animals had to fight all the time over social position.

Still, life for the subordinates in any social order is apt to be hard. They are the expendables of the social group. They almost never get a chance to reproduce, and when times get difficult they are the first to die. During times of food scarcity, the death of the weaker members helps to protect the resource for the stronger members. Rather than sharing food, the population excess is sacrificed. For the species, it is better for the weaker members to starve before the resources dwindle to crisis levels, rather than share the resource, thus denuding the countryside and risking ultimate starvation of the entire population.

Territoriality

Territorial ownership is another facet of sociality in animal populations. Just as an animal requires food for growth and maintenance and a mate with which to reproduce, it must also have space in which to carry out these activities. Since space, which an animal requires to satisfy its biologic needs, is limited, it must somehow be divided among those that require it.

Territorial defense has been observed in numerous animals: insects, crustaceans, other invertebrates, fishes, amphibians, lizards, birds, and mammals (in-

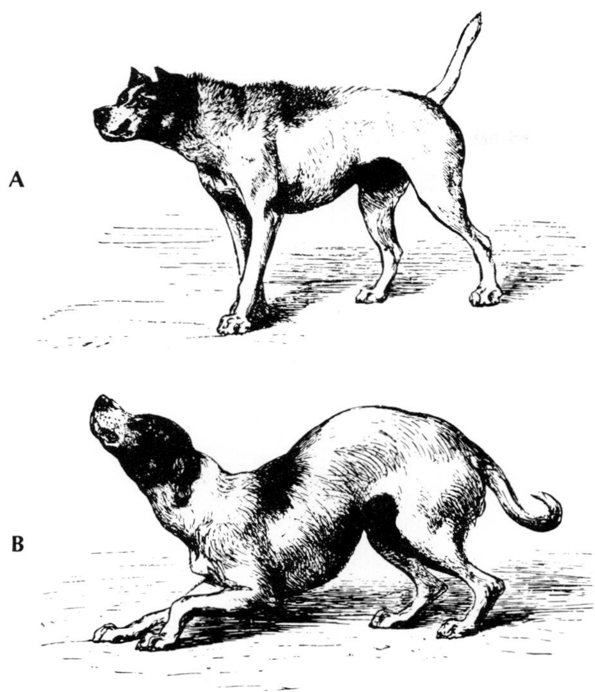

FIG. 34-11 Darwin's principle of antithesis as exemplified by the postures of dogs. In **A,** a dog approaches another dog with hostile, aggressive intentions. In **B,** the same dog is humble and conciliatory. The signals of aggressive display have been reversed. (From Darwin, C. 1873. Expression of the emotions in man and animals. New York, D. Appleton & Co.)

cluding humans). A **territory** is a fixed area from which intruders of the same species are excluded. This involves announcing ownership, defending the area from intruders, and spending long periods of time on the site being conspicuous. Like every other competitive endeavor, there are both costs and advantages.

Holding exclusive rights to a piece of property is beneficial when it ensures access to limited resources, *unless* the territory boundaries cannot be maintained with a minimum of effort. The presumed benefits of a territory are, in fact, numerous: uncontested access to a foraging area; enhanced attractiveness to females, thus reducing the problems of pair-bonding, mating, and rearing the young; reduced disease transmission; reduced vulnerability to predators. But the advantages of holding a territory begin to wane if the individual must spend most of the time in boundary disputes with neighbors.

For territoriality to work, the resources must be defendable, and, in fact, they usually are because natural selection relentlessly disposes of individuals with hopelessly unrealistic territorial appetites. Thus a territory must be large enough to provide the resources needed, but not so large that it cannot be economically defended. In general this is true.

Territory size tends to be flexible. When populations are low, territories are large. When resources, especially food, are abundant, territories tend to shrink, because no benefit is derived from defending large ones. On the other hand, in times of population crowding, defended territories may be abandoned altogether because the sheer numbers of individuals make boundary defense impossible.

Sometimes territories move with the individual. This **individual distance,** as it is termed, is characteristic of the species. It can be observed as the spacing between swallows or pigeons on a wire, in gulls lined up on the beach, or in people queued up for a bus.

Most of the time and energy required for territoriality are expended when the territory is first established. Once the boundaries are located they tend to be respected, and aggressive behavior diminishes as territorial neighbors come to recognize each other. Indeed, neighbors may look so peaceful that an observer who was not present when the territories were established may conclude (incorrectly) that the animals are not territorial. A "beachmaster" sea lion (that is, a dominant male with a harem) seldom quarrels with neighbors who have their own territories to defend. However, he must be on constant vigilance against bachelor bulls who challenge the beachmaster for harem privileges.

Of all vertebrate classes, birds are the most conspicuously territorial. Most male songbirds establish territories in the early spring and defend these vigorously against all males of the same species during spring and summer when mating and nesting are at their height. A male song sparrow, for example, has a territory of approximately three-fourths of an acre. In any given area, the number of song sparrows remains approximately the same year after year. The population remains stable because the young occupy territories of adults that die or are killed. Any surplus in the song sparrow population is excluded from territories and thus not able to mate or nest.

Sea birds such as gulls, gannets, boobies, and albatrosses form colonies that are divided into very small

FIG. 34-12 Sage grouse cock, *Centrocercus urophasianus* displays before four hens, which he has attracted to his "lek." (Photograph by L. L. Rue, III.)

territories just large enough for nesting. These birds' territories cannot include their fishing grounds since they all forage in the sea where the food is always shifting in location and is shared by all.

Several species of North American grouse form specialized, small territories called **leks** (Fig. 34-12). Each lek functions exclusively for mating. Females are attracted to the leks where the males engage in gaudy displays. The females tend to mate preferentially with males occupying the central regions of the lekking grounds. Thus the males fight over these preferred mating territories; the most aggressive and ornamental males mate with most of the females, so there is an intense sexual selection for ornamentation and vivid displays. Once the female has mated, she leaves the lekking grounds and raises her brood alone.

Territorial behavior is not so characteristic of mammals as it is of birds. Mammals are less mobile than birds, and this makes it more difficult for them to patrol a territory for trespassers. Instead, many mammals have **home ranges.** A home range is the total area an individual traverses in its activities. It is not an exclusive defended preserve but overlaps with the home ranges of other individuals of the same species.

For example, the home ranges of baboon troops overlap extensively, although a small part of each range becomes the recognized territory of each troop for its exclusive use. Home ranges may shift considerably with the seasons. A baboon troop may have to shift to a new range during the dry season to obtain water and better grass. Elephants, before their movements were restricted by humans, made long seasonal migrations across the African savanna to new feeding ranges. However, the home ranges established for each season were remarkably consistent in size.

Not surprisingly, size of the home range increases with body size of mammals, since larger animals require more foraging area to satisfy their energy requirements. Accordingly, the home range may be one-fourth of an acre for a field mouse, 100 acres for a deer, 60 square miles for a grizzly bear, and 1,500 square miles for an African hunting dog. In general, carnivores require more foraging space than herbivores, because any given area supports more plant food than animal food.

Communication

Social animals, like people, must be able to communicate with each other. Only through communication can one animal influence the behavior of another. Compared to the enormous communicative potential of human speech, nonhuman communication is severely restricted. Whereas human communication is based mainly, though by no means exclusively, on sounds, animals may communicate by sounds, scents, touch, and movement. Indeed any sensory channel may be used, and in this sense animal communication has richness and variety.

However, unlike our language, which is composed of words with definite meanings that may be rearranged to generate an almost infinite array of new meanings and images, animal communication consists of a limited repertoire of signals. Typically, each signal conveys one and only one message. These messages cannot be divided or rearranged to construct new kinds of information. A single message from the sender may, however, contain several bits of relevant information for the receiver.

The song of a cricket announces to an unfertilized female the species of the sender (males of different species have different songs), his sex (only males sing), his location (source of the song), and his social status (only a male able to defend the area around its burrow sings from one location). This is all crucial information to the female and accomplishes a biologic goal. But there is no way for the male to alter his song to provide additional information concerning food, predators, or habitat, which might improve his mate's chances of survival and thus enhance his own fitness.

The limitations of communication are especially evident in the invertebrates and lower vertebrates. Signals are characteristically stereotyped, and the responses highly predictable and constant throughout the species. This does not mean that such communication is always lacking in intensity and versatility, however. Of the two contrasting examples that follow, mate attraction in silkworms illustrates an extreme case of stereotyped, single-message communication that has evolved to serve a single biologic goal: mating. Meanwhile among the same group—the insects—we find one of the most sophisticated and complex of all nonhuman communication systems, the symbolic language of bees.

Chemical sex attraction in moths

Virgin female silkworm moths have special glands that produce a chemical sex attractant to which the males are sensitive. Adult males smell with their large bushy antennae, covered with thousands of sensory hairs that function as receptors. Most of these receptors are sensitive to the chemical attractant (a complex alcohol called bombykol, from the name of the silkworm *Bombyx mori*) and to nothing else.

To attract the male, the female merely sits quietly and emits a minute amount of bombykol, which is carried downwind. When a few molecules reach the male's antennae, he is stimulated to fly upwind in search of the female. His search is at first random, but, when by chance he approaches within a few hundred yards of the female, he encounters a concentration gradient of the attractant. Guided by the gradient, he flies toward the female, finds her, and copulates with her.

In this example of chemical communication the attractant bombykol, which is really a pheromone, serves as a signal to bring the sexes together. Its effectiveness is assured because natural selection favors the evolution of males with antennal receptors sensitive enough to detect the attractant at great distances (several miles). Males with a genotype that produces a less sensitive sensory system fail to locate a female and thus are reproductively eliminated from the population.

Language of the bees

Honeybees are able to communicate the location of food resources when these sources are too distant to be located easily by individual bees. Communication is done by dances, which are mainly of two forms. The form having the most communicative richness is the **waggle dance** (Fig. 34-13). Bees most commonly execute these dances when a forager has returned from a rich source, carrying either nectar in her stomach or pollen grains packed in basketlike spaces formed by hairs on her legs. The waggle dance is roughly in the pattern of a figure-of-eight made against the vertical surface of the comb. One cycle of the dance consists of three components: (1) a circle with a diameter about three times the length of the bee, (2) a straight run while waggling the abdomen from side to side 13 to 15 times per second, and (3) another circle, turning in the opposite direction from the first. This dance is repeated many times with the circling alternating clockwise and counterclockwise.

The straight, waggle run is the important information component of the dance. Waggle dances are performed almost always in clear weather, and the direction of the straight run is related to the position of the sun. If the forager has located food directly toward the sun, she will make her waggle run straight upward over the vertical surface of the comb. If food was located 60° to the right of the sun, her waggle run is 60° to the right of vertical. We see then that the waggle run points at the same angle relative to the vertical as the food is located relative to the sun.

Distance information about the food source is also coded into bee dances. If the food is close to the hive (less than 50 m), the forager employs a simpler dance called the **round dance.** The forager simply turns a

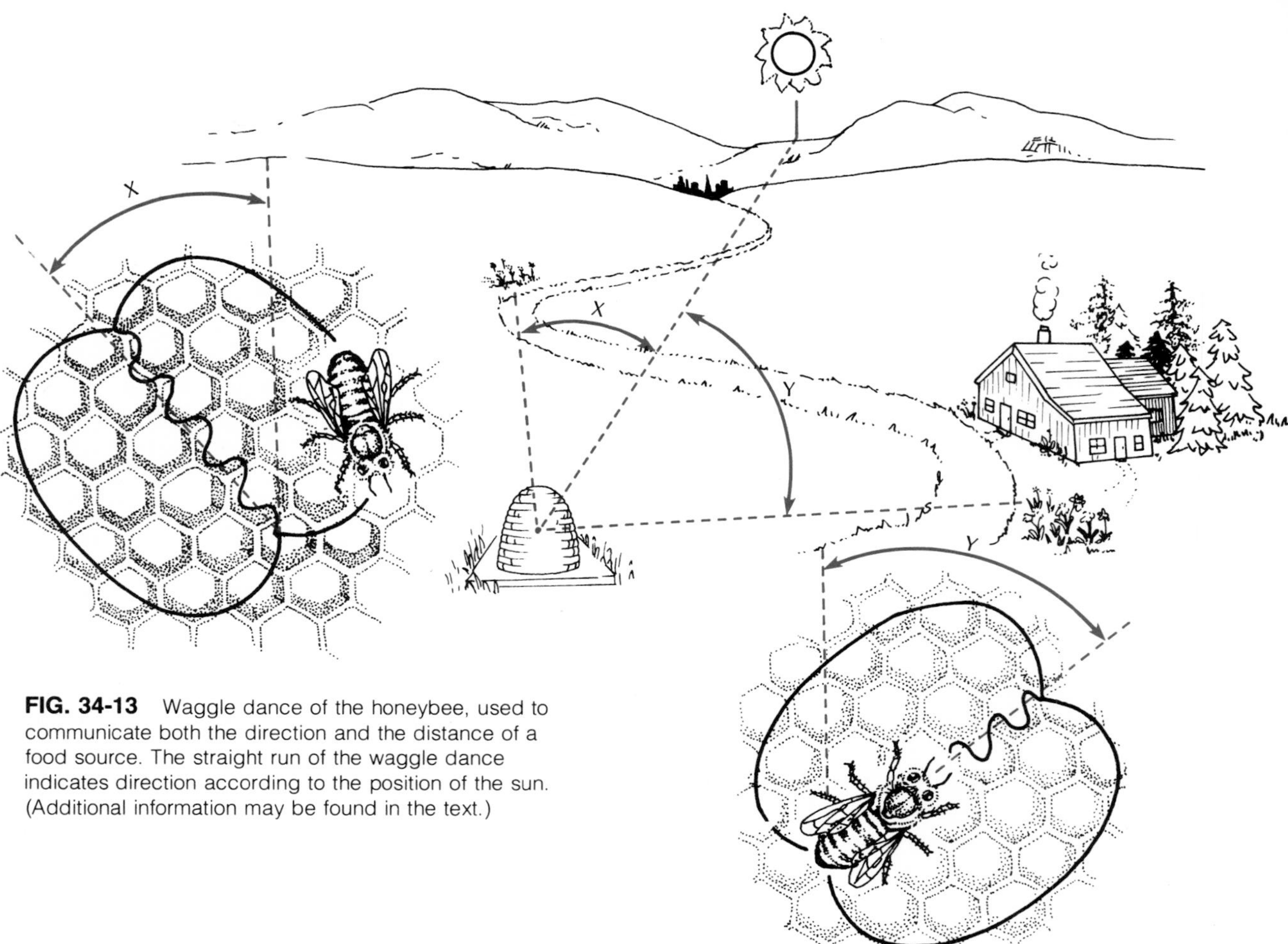

FIG. 34-13 Waggle dance of the honeybee, used to communicate both the direction and the distance of a food source. The straight run of the waggle dance indicates direction according to the position of the sun. (Additional information may be found in the text.)

complete clockwise circle, then turns and completes a counterclockwise circle, a performance that is repeated many times. Other workers cluster around the scout and become stimulated by the dance as well as by the odor of nectar and pollen grains from flowers she has visited. The recruits then fly out and search in all directions but do not stray far. The round dance carries the message that food is to be found in the vicinity of the hive.

If the food source is farther away, the round dances become waggle dances, which provide both distance and direction information. The tempo of the waggle dance is inversely related to the food distance. If the food is about 100 m away, each figure-of-eight cycle lasts about 1.25 seconds; if 1,000 m away, it lasts about 3 seconds; and if about 8 km away (5 miles), it lasts 8 seconds. When food is plentiful, the bees may not dance at all. But when food is scarce, the dancing becomes intense, and the other workers cluster about the returning scouts and follow them through the dance patterns.

The significance of the bee dances was discovered in the 1920s by the German zoologist Karl von Frisch, one of the recipients of the 1973 Nobel Prize. Despite detailed and extensive experiments by von Frisch that supported his original interpretations of the bee dances, the experiments have been criticized, especially by the American biologist A. Wenner, who suggested that the correlation between dance symbolism and food location is accidental. He argued that foraging bees bring back odors characteristic of the food source, and that recruits are stimulated by the dances to go search for flowers bearing those odors. Few biologists were prepared to accept this interpretation, but Wenner's beneficial skepticism stimulated more rigorously controlled experiments (Gould, 1976), which established more conclusively than ever before that the bee dances communicated both distance and direction information. The dances of bees are among the true wonders of the natural world.

Communication by displays

A display is a kind of behavior or series of behaviors that serves a communicative purpose. The release of sex attractant by the female moth and the dances of bees that we have just described are examples of displays; so are the alarm calls of herring gulls, the song of the white-crowned sparrow, the courtship dance of the sage grouse, and the "eyespots" on the hind wings of certain moths that are quickly exposed to startle potential predators. Of course, just about anything an animal does communicates *something* to other animals that see, hear, or smell it. A true display, on the other hand, is a behavior pattern that has been modified through evolution to make it increasingly effective in serving a communicative function. This process is called **ritualization.** Through ritualization, simple movements or traits become more intensive, conspicuous, or precise, and their original undifferentiated function acquires signal value. The result of such intensification is to reduce the possibility of misunderstanding.

The elaborate pair-bonding behavior of the blue-footed boobies, illustrated in Fig. 26-26 (p. 596), exemplifies this point. These displays are performed with maximum intensity when the birds come together after a period of separation. The male at right in the illustration is sky-pointing: the head and tail are pointed skyward, and the wings are swiveled forward in a seemingly impossible position to display their glossy upper surfaces to the female. This is accompanied by a high, piping whistle. The female at left, for her part, is parading. She goose-steps with exaggerated slow deliberation, lifting each brilliant blue foot in turn, as if holding it aloft momentarily for the male to admire. Such highly personalized displays, performed with droll solemnity, appear comical, even inane to the observer. Indeed the boobies, whose name is derived from the Spanish word "bobo" meaning clown, presumably were so designated for their amusing antics.

Needless to say, the birds' amusement plays no part in the ceremonies. The exaggerated nature of the displays ensures that the message is not missed or misunderstood. Such displays are essential to establish and maintain a strong pair bond between male and female. This requirement also explains the repetitious nature of the displays that follow one another throughout courtship and until egg laying. These repeated displays serve to maintain a state of mutual stimulation between male and female, ensuring the degree of cooperation necessary for copulation and subsequent incubation and care of the young. A sexually aroused male has little success with an uninterested female.

Do animals have mental experiences?

Most people, while aware of the enormous capabilities of the human brain that separate us from other animal species, take for granted that animals have feelings and awareness analogous to our own. As we watch animals interacting in behavior that resembles human behavior, we may conclude more or less intuitively that animals have humanlike mental experiences. However, most behavioral scientists have deep-seated objections to even considering such a possibility, believing that because mental experiences in animals cannot be observed, measured, or verified, it is consequently meaningless to suggest that they may exist. Strict behaviorists, who argue that even human mental experiences are unique qualities that cannot be objectively defined, rebel against the notion of animal awareness as anthropomorphic, that is, ascribing human characteristics to other species. Yet, as D. R. Griffin argues in his perceptive book *The Question of Animal Awareness,* "it is actually no more anthropomorphic . . . to postulate mental experiences in another species than to compare its bony structure, nervous system, or antibodies with our own." Griffin points to a curious duality in behavioral studies: rats, pigeons, and monkeys are used as models in behavioral investigations on the implicit assumption that principles revealed from such studies are applicable to human behavior, yet when the question of mental experiences arises, behaviorists reject evolutionary continuity between humans and other animals.

The reluctance of many behaviorists to acknowledge the existence of even simple mental experiences among animals stems in large part from the enormous difference in versatility and complexity that separates human language from any other known animal communication system. One criterion that we mentioned earlier as supposedly unique for human language is that words with discrete meanings and syllables, many of which are meaningless in themselves, can be arranged into almost endless combinations to generate a vast array of new meanings.

Until recently, it was accepted that this characteristic was absent from all nonhuman communication. However, in the early 1970s, a female chimpanzee named Washoe was taught to employ *and combine* gestures, using words from the American Sign Language for the deaf, much as people use spoken words. The discovery that manual gestures and expressive motions were much more appropriate than vocalizations in communicating with apes was a major breakthrough in behavioral research. Since Washoe, sign-language studies have been extended to other chimpanzees and to gorillas; several have acquired "vocabularies" of several hundred reliable signs, some invented by the apes themselves.

Clearly, one problem in assessing the versatility of animal communication is understanding what sensory channel an animal is using. The signals may be visual displays, odors, vocalizations, tactile vibrations, or electric currents (as, for example, among certain fishes). Even more difficult is establishing two-way communication between animals and human beings, since the investigator must translate meanings into symbols the animal can understand. Furthermore, people are poor social partners for most animals. Nevertheless, when the appropriate communication channel was used, the remarkable success with Washoe did happen.

The animal behaviorist Irven DeVore has reported how choosing the proper channel for dialogue can have more than academic interest.*

One day on the savanna I was away from my truck watching a baboon troop when a young juvenile came and picked up my binoculars. I knew if the glasses disappeared into the troop they'd be lost, so I grabbed them back. The juvenile screamed. Immediately every adult male in the troop rushed at me—I realized what a cornered leopard must feel like. The truck was 30 or 40 feet away. I had to face the males. I started smacking my lips very loudly, a gesture that says as strongly as a baboon can, "I mean you no harm." The males came charging up, growling, snarling, showing their teeth. Right in front of me they halted, cocked their heads to one side—and started lip-smacking back to me. They lip-smacked. I lip-smacked, "I mean you no harm." "I mean *you* no harm." It was, in retrospect, a marvelous conversation. But while my lips talked baboon, my feet edged me toward the truck until I could leap inside and close the door.

*From DeVore, Irven. 1972. The marvels of animal behavior. Washington, D.C., National Geographic Society, p. 408.

Despite the recent breakthrough in two-way communication with apes, the strong reductionist traditions of the behavioral sciences tend to cripple attempts to discover whether or not animals possess mental awareness and are conscious of what they are doing. Human thinking has been closely identified with human language. Since people think in words and animals do not, the argument goes, animals cannot have thoughts. Even ignoring serious objections to the implication that all human mental experiences, feelings, and emotions are inseparable from language, it seems more likely that animals and people share certain properties of mental awareness.

The question of mental awareness among animals is not easily approached experimentally and may not even be answered in the foreseeable future. Whether or not we believe animals know what they are doing, they certainly are capable of highly organized instinctive behavior in the absence of any intelligent appreciation of its purpose. Sometimes instinctive behavior misfires, and such incidents often serve to emphasize its stereotyped nature. The following excerpt contains a perfect example of the automatic release of inappropriate behavior in a gannet colony.

A male of an old pair flew into his nest. Normally he would bite his mate on the head with some violence and then go through a long and complicated meeting ceremony, an ecstatic display confined to members of a pair. Unfortunately, the female had caught her lower mandible in a loop of fish netting that was firmly anchored in the structure of the nest. Every time she tried to raise her head to perform the meeting ceremony with her mate she merely succeeded in opening her upper mandible whilst the lower remained fixed in the netting. So she apparently threatened the male with widely gaping beak and he immediately responded by attacking her. With each attack she lowered and turned away her bill (the way in which a female gannet appeases an aggressive male). At once the male stopped biting her and she again turned to greet him but simply repeated the beak-opening and drew another attack. And so it went on despite the fact that these two birds had been mated for years and that the netting, the cause of all the trouble, was clearly visible.*

*Nelson, B. 1968. Galápagos: islands of birds. London, Longmans, Green & Co., Ltd., p. 71.

Despite such limitations to the adaptiveness of instinctive behavior, it obviously functions beautifully most of the time. In recognizing that reasoning and insight are not required for effective, highly organized behavior, we should not conclude that lower animals are little more than nonintelligent machines acting out their lives like so many robots. Although the gannet in this example lacked the "intelligence" to free his mate by purposefully disentangling her beak from the netting, he was capable of appropriately analyzing the thousands of strategic choices he must make during his lifetime: how to find and hold a mate, where to build a nest site and how to defend it, how to locate evasive marine food, and what to do when the environment changes. All of this and more requires endless behavioral adjustments to new situations. Conceivably, this might be accomplished by a machine, but only by one of staggering complexity.

Annotated references

Selected general readings

Alcock, J. 1975. Animal behavior: an evolutionary approach. Sunderland, Mass., Sinauer Associates, Inc. *Well-written and well-illustrated discussion of the genetics, physiology, ecology, and history of behavior in an evolutionary perspective.*

Barash, D. P. 1977. Sociobiology and behavior. New York, Elsevier North-Holland, Inc. *The evolution of social behavior. Clear explanation of how behavior is organized by natural selection.*

Bermant, G. (ed.) 1973. Perspectives on animal behavior. Glenview, Ill., Scott, Foresman & Co. *Collection of 10 original essays by 12 authors. Intermediate level.*

Bramblett, C. A. 1976. Patterns of primate behavior. Palo Alto, Calif., Mayfield Publishing Co. *Behavior of primates with useful field descriptions of 15 best-known primates.*

Brown, J. L. 1975. The evolution of behavior. New York, W. W. Norton & Co., Inc. *Advanced text emphasizing social behavior.*

Carthy, J. D. 1958. An introduction to the behavior of invertebrates. New York, The Macmillan Co. *In this work the author stresses the functions of the sensory patterns of invertebrates and their reactions to the various categories of stimuli.*

Darling, F. F. 1937. A herd of red deer. A study in animal behavior. Oxford, Oxford University Press. *A classic contribution to animal behavior representing a careful and penetrating analysis of the movement, population, reproduction, and other aspects of an interesting group of animals.*

Eibl-Eibesfeldt, I. 1975. Ethology, the biology of behavior, ed. 2. New York, Holt, Rinehart & Winston, Inc. *Excellent general ethology text in the European tradition.*

Eisner, T., and E. O. Wilson (ed.). 1955-1975. Animal behavior. San Francisco, W. H. Freeman & Co., Publishers. *Collection of articles from Scientific American with introductions by Eisner and Wilson.*

Evans, R. I. 1975. Konrad Lorenz: the man and his ideas. New York, Harcourt Brace Jovanovich, Inc. *Dialogue with one of the world's great ethologists, together with a critique of his work, reprints of four important papers, and bibliography.*

Fox, M. W. 1974. Concepts in ethology. Minneapolis, University of Minnesota Press. *Principles of ethology, with many examples from the canids.*

Frings, H., and M. Frings. 1977. Animal communication, ed. 2. Norman, Okla., University of Oklahoma Press. *Comparative intermediate level treatment.*

von Frisch, K. 1971. Bees: their vision, chemical senses, and language, rev. ed. Ithaca, N.Y., Cornell University Press.

Gould, J. L. 1976. The dance-language controversy. Q. Rev. Biol. **51:**211-244.

Griffin, D. R. 1976. The question of animal awareness. New York, The Rockefeller University Press. *An important and provocative book about a controversial question.*

Hahn, E. 1978. Look who's talking. New York, Thomas Y. Crowell Co. *A delightful account of the efforts people have made and are making to communicate with animals.*

Hailman, J. P. 1977. Optical signals: animal communication and light. Bloomington, Indiana University Press. *Advanced monograph on the characteristics of visual signals used in communication.*

Hinde, R. A. 1970. Animal behavior: a synthesis of ethology and comparative psychology, ed. 2. New York, McGraw-Hill Book Co. *Advanced text.*

Klopfer, P. H. 1974. An introduction to animal behavior: ethology's first century, ed. 2. Englewood Cliffs, N.J., Prentice-Hall, Inc. *Historical approach.*

Lorenz, K. Z. 1952. King Solomon's ring. New York, Thomas Y. Crowell Co., Inc. *One of the most delightful books ever written about the behavior of animals.*

National Geographic Society. 1972. The marvels of animal behavior. Washington, D.C., National Geographic Society. *Superbly illustrated collection of essays by leading animal behaviorists.*

Nelson, B. 1968. Galápagos: islands of birds. London, Longmans, Green & Co., Ltd.

Scott, J. P. 1972. Animal behavior, ed. 2. Chicago, University of Chicago Press. *Comprehensive introductory text.*

Sebeok, T. A. (ed.). 1977. How animals communicate. Bloomington, Indiana University Press. *Important collection of essays dealing with the evolution and mechanics of communication in animals.*

Thorpe, W. H. 1974. Animal nature and human nature. Garden City, N.Y., Anchor Press. *Comparison of animal and human behavior; especially useful sections on animal communication.*

Tinbergen, N. 1951. The study of instinct. New York, Oxford University Press, Inc. *Classic introduction to ethology.*

Tinbergen, N. 1961. The herring gull's world: a study of the social behaviour of birds. New York, Basic Books, Inc. *One of the most thorough studies of instinctive behavior ever made. Beautifully written.*

Tinbergen, N. 1965. Animal behavior. New York, Time-Life Books. *Popularized but still highly informative account. Fine illustrations.*

Tinbergen, N. 1966. Social behavior in animals with special reference to vertebrates, ed. 2. New York, Halsted Press.

Wilson, E. O. 1975. Sociobiology: the new synthesis. Cambridge, Mass., Harvard University Press. *An important synthesis of facts and theories on the biologic basis of social behavior.*

Selected *Scientific American* articles

Many important articles not listed below are included in the *Scientific American* anthology edited by Eisner and Wilson.

Burgess, J. W. 1976. Social spiders. **234:**100-106 (Mar.). *Contrary to the usual pattern of solitary behavior among spiders, some species are gregarious and build large communal webs.*

Bertram, B. C. R. 1975. The social system of lions. **232:** 54-65 (May). *The social behavior of lions in their natural habitat.*

Eaton, G. G. 1976. The social order of Japanese macaques. **235:**96-106 (Oct.). *Long-term observations of a confined troop of macaques show that the biologic component of their behavior is much modified by the social component.*

Esch, H. 1967. The evolution of bee language. **216:**96-104 (Apr.).

Hess, E. H. 1972. "Imprinting" in a natural laboratory. **227:**24-31 (Aug.). *Even before the egg hatches, a duckling and its parent interact in ways that will strengthen the juvenile-to-parent bond.*

Hölldobler, B. 1971. Communication between ants and their guests. **224:**86-93 (Mar.). *Some parasites learn the language of their ant hosts to gain free food and shelter.*

Pooley, A. C., and C. Gans. 1976. The Nile crocodile. **234:** 114-124 (Apr.). *The remarkable social behavior of this large reptile includes parental protection of the young.*

Todd, J. H. 1971. The chemical language of fishes. **224:**98-108 (May). *Catfishes produce pheromones to label loses and winners of hierarchic fights.*

Topoff, H. R. 1972. The social behavior of army ants. **227:** 70-79 (Nov.). *A complex social organization is described.*

Watts, C. R., and A. W. Stokes. 1971. The social order of turkeys. **224:**112-118 (June). *Describes a rigidly stratified society.*

Wilson, E. O. 1972. Animal communication. **227:**52-60 (Sept.). *The languages of animals and humans are compared.*

REFERENCES TO PART THREE

The books listed in the following selection are mainly textbooks covering wide areas of physiology. They vary considerably in depth and in the level of background in biology and chemistry required of the reader for a full understanding, as indicated in the annotations.

Barrington, E. J. W. 1968. The chemical basis of physiological regulation. Glenview, Ill., Scott, Foresman & Co. *A selection of topics in comparative physiology, with emphasis on experimental approach. Clearly written.*

Beck, W. S. 1971. Human design: molecular, cellular, and systematic physiology. New York, Harcourt Brace Jovanovich, Inc. *Well-integrated treatment of human anatomy and physiology.*

Florey, E. 1966. General and comparative physiology. Philadelphia, W. B. Saunders Co. *A detailed graduate-level text, numerous illustrations, good invertebrate-vertebrate balance.*

Folk, G. E. 1974. Textbook of environmental physiology, ed. 2. Philadelphia, Lea & Febiger. *Stresses responses of mammals to heat, cold, light, and atmospheric pressure.*

Gordon, M. S. (ed.). 1977. Animal function: principles and adaptations, ed. 3. New York, The Macmillan Co. *Graduate-level vertebrate physiology.*

Guyton, A. C. 1976. Textbook of medical physiology, ed. 5. Philadelphia, W. B. Saunders Co. *A detailed but readable treatment of medical physiology.*

Hill, R. W. 1976. Comparative physiology of animals: an environmental approach. New York, Harper & Row, Publishers. *Highly selective treatment, stressing environmental interactions.*

Hoar, W. S. 1975. General and comparative physiology, ed. 2. Englewood Cliffs, N.J., Prentice-Hall, Inc. *The best-balanced of college comparative physiology texts.*

Prosser, C. L. (ed.). 1973. Comparative animal physiology, ed. 3. Philadelphia, W. B. Saunders Co. *Advanced treatise.*

Schmidt-Nielsen, K. 1975. Animal physiology: adaptation and environment. New York, Cambridge University Press. *Interestingly written selective treatment of comparative physiology, emphasizing physiologic adaptations to the environment.*

Schottelius, B. A., and D. D. Schottelius. 1978. Textbook of physiology, ed. 18. St. Louis, The C. V. Mosby Co. *Intermediate-level college human physiology text. Clearly written and illustrated.*

Vander, A. J., J. H. Sherman, and D. S. Luciano. 1975. Human physiology: the mechanisms of body function, ed. 2. New York, McGraw-Hill Book Co. *Excellent intermediate level human physiology text.*

Vertebrate structures and functions. Readings from *Scientific American* with introductions by N. K. Wessels. 1974. San Francisco, W. H. Freeman & Co., Publishers. *Excellent collection of articles on vertebrate structure and physiology with helpful introductions. Highly recommended supplementary reading.*

Wilson, J. A. 1972. Principles of animal physiology. New York, The Macmillan Co. *Clearly written comparative physiology at the advanced undergraduate level.*

PART FOUR
CONTINUITY AND EVOLUTION OF ANIMAL LIFE

Charles Darwin in 1881, 1 year before his death. (Courtesy American Museum of Natural History.)

Since all living things are mortal, species perpetuation is a central activity in the lives of animals. To comprehend the continuity of life, it is necessary to understand how hereditary units are transmitted from parent to offspring, how such units direct the structural and functional differentiation of a fertilized egg into an adult organism, and how the capacity for evolutionary diversity has been built into the hereditary mechanism.

Organic evolution is the process through which existing animals and plants have attained their diversity of form and behavior by gradual, continuous change from previously existing forms. Evolution comes about by natural selection acting on the inheritable variations of a population. It may be considered a descent in the course of time with modifications. It is an ongoing living process and is taking place today.

The following chapters deal with reproduction, development, and the principles of inheritance, with the evidences for and nature of the evolutionary process that comprises the framework of biology, and with the remarkable and recent divergence of the hominoid line, which gave rise to human beings.

CHAPTER 35

THE REPRODUCTIVE PROCESS

A female, broad-tailed hummingbird, *Selasphorus platycercus,* feeds her young. All birds are oviparous—they lay eggs, then feed the hatchlings until they are self-sufficient. Oviparity is one of three basic reproductive patterns employed by animals to nourish their young.

Photograph by L. L. Rue, III.

All living organisms are capable of giving rise to new organisms similar to themselves (see Principle 10, p. 11). If we admit that all living things are mortal, that every organism is endowed with a life-span that must eventually terminate, we must also acknowledge the indispensability of reproduction. Without reproduction, the evolution of a species is impossible. Like Samuel Butler who concluded that a chicken is just an egg's way of making another egg, many biologists consider the ability to reproduce to be the ultimate objective of all life processes.

The word "reproduction" implies replication, and it is true that biologic reproduction almost always yields a reasonable facsimile of the parent unit. However, sexual reproduction, practiced by the majority of animals, produces the *diversity* needed for survival in a world of constant change. At least for multicellular animals sexual reproduction offers enormous advantages over asexual reproduction, as we strive to point out later.

Reproduction must have developed very early in animal evolution. The process, whether sexual or asexual, embodies a basic pattern: (1) the conversion of raw materials from the environment into the offspring, or sex cells that develop into offspring of a similar constitution and (2) the transmission of a hereditary pattern or code from the parents. The code, of course, is in the deoxyribonucleic acid (DNA) of the genes. The chemical nature of DNA and the genetic code are described in Chapters 2 and 37.

NATURE OF THE REPRODUCTIVE PROCESS

The two fundamental types of reproduction are sexual and asexual. In **asexual** reproduction there is only one parent, and there are no special reproductive organs or cells. Each organism is capable of producing identical copies of itself as soon as it becomes an adult. The production of copies is simple and direct and typically rapid. **Sexual** reproduction involves two parents as a rule, each of which contributes special **sex cells (gametes)** that in union develop into a new individual. The **zygote** formed from this union receives genetic material from *both* parents and accordingly is different from both. The combination of genes produces a genetically unique individual, still bearing the characteristics of the species but also bearing traits that make it different from its parents.

Sexual reproduction, by recombining the parental characters, tends to multiply variations and makes possible a richer and more diversified evolution. Asexual reproduction can only produce carbon copies and must await mutations to introduce variation into the line. This would seem to explain why asexual reproduction is restricted mostly to unicellular forms, which can multiply rapidly enough to offset the disadvantages of relentless replication of identical products.

Of course, in those asexual organisms such as molds and bacteria that are haploid (bear only one set of genes), mutations are immediately expressed and evolution can proceed quickly. In sexual animals, on the other hand, a gene mutation is seldom expressed immediately since it is masked by its normal partner on the homologous chromosome. (Homologous chromosomes are those that pair during mitosis and have genes controlling the same characteristics.) There is only a remote chance that both members of a gene pair will mutate in the same way at the same moment.

Asexual reproduction

As previously pointed out, asexual reproduction is found only among the simpler forms of life, such as protozoans, cnidarians, bryozoans, bacteria, and a few others. It is absent among the higher invertebrates (molluscs and arthropods) and all vertebrates. Even in the phyla in which it occurs, most of the members employ the sexual method of reproduction as well. In these groups, asexual reproduction ensures rapid increase in numbers when the differentiation of the organism has not advanced to the point of forming highly specialized gametes.

The forms of asexual reproduction are fission, budding (both internal and external), fragmentation, and sporulation. **Fission** is common among protozoans and, to a limited extent, metazoans. In this method the body of the parent is divided into two approximately equal parts, each of which grows into an individual similar to the parent. Fission may be either transverse or longitudinal. **Budding** is an unequal division of the organism. The new individual arises as an outgrowth (bud) from the parent. This bud develops organs like those of the parent and then usually detaches itself. If the bud is formed on the surface of the parent, it is an external bud; but in some cases internal buds, or **gemmules,** are produced. Gemmules are collections of many cells surrounded by a dense covering in the body wall. When the body of the parent disintegrates, each gemmule gives rise to a new individual. External budding is common in the hydra and internal budding in the freshwater sponges. Bryozoans also have a form of internal bud called statoblast. **Fragmentation** is a method in which an organism breaks into two or more parts, each capable of becoming a complete animal. This method is found among the Platyhelminthes, Nemertinea, and Echinodermata. **Sporulation** is a method of multiple fission in which many cells are formed and enclosed together in a cyst-like structure. Sporulation occurs in a number of protozoan forms.

Sexual reproduction

Sexual reproduction is the general rule in the animal kingdom. There are usually, although not invariably, two parents, each of which produces special sex cells, called **gametes.** Only one sex, the female, can produce offspring, and this must depend on the intervention of the male. There are two kinds of gametes: the **ovum** (egg), produced by the female, and the **spermatozoon** (sperm), produced by the male. Ova are nonmotile, are produced in relatively small numbers, and may contain a large amount of yolk, the stored food material that sustains early development of the new individual. Sperm are motile, are produced in enormous numbers, and are small. The union of egg and sperm is called **fertilization,** and the resulting cell is known as a **zygote,** which develops into a new individual.

There are two fundamental events that distinguish sexual from asexual reproduction: meiosis and fertilization. **Meiosis** is a distinctive type of gamete-producing nuclear division, actually a double division, that differs from ordinary cell division or mitosis (unfortunately the terms mitosis and meiosis are confusingly similar) in a very important way. The chromosomes split once, but the cell divides *twice,* producing four cells, each with *half* the original number of chromosomes. Meiosis is then followed by fertilization in which two gametes, each containing half the original number of chromosomes, are combined to restore the normal chromosome content of the species. The new cell **(zygote)** contains genetic material from both gametes. It is not difficult to understand why meiosis is required; if ordinary cells from each parent fused to form a zygote, the chromosome number would be doubled with each generation. Meiosis, followed by fertilization, keeps the chromosome number stable.

The zygote now begins to divide and grow by mitosis. The new individual has equal numbers of chromosomes from each parent and accordingly is different from each; it is a unique individual bearing a random assortment of parental characters. This is the great strength of sexual reproduction, the "master adaptation" that keeps introducing new varieties into the population. *Neither meiosis nor fertilization ever occurs in asexual reproduction.*

Sexual reproduction appears in certain protozoans, although most protozoans reproduce asexually by cell division. When sexual reproduction is found, it may involve the formation of male and female gametes that unite to form a zygote or it may involve two mature sexual individuals that join together to exchange nuclear material or to merge cytoplasm. In some cases it is difficult to distinguish sex, for although two parents are involved, they cannot be designated as male and female.

The male-female distinction is clearly evident in the Metazoa. Organs that produce the germ cells are known as **gonads.** The gonad that produces the sperm is called the **testis** (Fig. 35-1) and that which forms the egg, the **ovary** (Fig. 35-2). The gonads represent the **primary sex organs,** the only sex organs found in certain groups of animals. Most metazoans, however, have various **accessory sex organs.** In the primary sex organs the sex cells undergo many complicated changes during their development, the details of which are described in a later section. At present we will distinguish typical biparental reproduction from two alternatives: parthenogenesis and hermaphroditism.

Biparental reproduction

Biparental reproduction is the common and familiar method of sexual reproduction, involving separate and distinct male and female individuals. Each has its own reproductive system and produces only one kind of sex cell, spermatozoon or ovum, but never both. Nearly all vertebrates and many invertebrates have separate sexes, and such a condition is called **dioecious.**

There are instances where animals attain sexual maturity in the larval body form. This is called **paedomorphosis** (Gr., *pais,* child, + *morphē,* form) or paedogenesis. Examples of this genetically fixed condition are the mud puppy *Necturus* and the congo eel *Amphiuma.* Both are permanent larval urodele amphibians. **Neoteny** (Gr., *neos,* young, + *teinen,* to extend) is a form of paedomorphosis found among certain salamanders. One of the best examples is the tiger salamander *(Ambystoma tigrinum)* that in certain parts of its range is found to mate in a larval (axolotl) form. However, unlike true (or "obligatory") paedomorphosis, the neotenous condition is not genetically fixed, and such larvae may transform into adults under certain environmental conditions (p. 536).

Alternatives to biparental reproduction

Parthenogenesis. Parthenogenesis is the development of an egg without the participation of a spermatozoon. Spontaneous, or natural, parthenogenesis is known to occur in rotifers, plant lice, some crustaceans and insects, and several species of desert lizards. Usually several generations of parthenogenetic reproduction alternate with biparental reproduction in which the egg is fertilized. In some cases, however (some platyhelminths, rotifers, and certain wasps) parthenogenesis appears to be the only form of reproduction. The queen bee is fertilized only once by a male (drone) or sometimes by more than one drone. She stores the sperm in her seminal receptacles, and as she lays her eggs, she can either fertilize the eggs or allow them to pass unfertilized. The fertilized eggs become females (queens or workers); the unfertilized eggs become males (drones).

Very rarely, mammalian eggs will spontaneously start developing into embryos without fertilization. In certain strains of mice, such embryos will develop into fetuses before dying. The best parthogenetic development among the higher vertebrates has been found in turkeys in which certain strains, selected to develop without sperm, grow to reproducing adults. From time to time claims arise that spontaneous parthenogenetic development to term has occurred in humans. A British investigation of about 100 cases where the mother denied having had intercourse revealed that in nearly every case the child possessed characters not present in the mother.

Artificial parthenogenesis was discovered in 1900. Eggs that normally are fertilized can be artificially induced to develop without fertilization or the presence of sperm. The agents employed are dilute organic acids, hypertonic salt solutions, and mechanical pricking with a needle. Eggs of certain invertebrates, such as those of the sea urchin, were first used but later, vertebrate eggs were successfully induced to develop without fertilization. Many frogs of both sexes were developed beyond metamorphosis and some to adults. Rarely do such forms complete development and they are often smaller than normal ones. However, several parthenogenetically activated mammalian eggs (mouse and rabbit) have been carried to the blastocyst stage, and there are claims of live births of fully normal rabbits from activated eggs that were returned to the oviduct for further development. However, thus far it has not been possible to repeat these experiments made in the late 1930s by G. Pincus.

Hermaphroditism. Animals that have both male and female organs in the same individual are called hermaphrodites, and the condition is called **hermaphroditism.** The term derives from a combination of the names of the Greek god Hermes and goddess Aphrodite.

In contrast to the dioecious state of separate sexes, hermaphrodites are **monoecious.** Many lower animals (flatworms and hydra) are hermaphroditic. Most avoid self-fertilization by exchanging germ cells with each other. For example, although the earthworm bears both male and female organs, its eggs are fertilized by the copulating mate and vice versa. Another way of preventing self-fertilization is by developing the eggs and sperm at different times.

Several species of teleost fishes exhibit hermaphroditism. Some produce eggs and sperm from different areas of the same gonad and may be self-fertilizing. Certain tropical sea basses known as Bahama groupers *(Petrometopon cruentatum)* begin life as females and later change into males. The gonad is considered in these forms as a compound organ with female, male, and a combination of male and female tissue but with a delayed timing in the appearance of the sperm after the ova degenerate or disappear.

ANATOMY OF REPRODUCTIVE SYSTEMS

The basic plan of the reproductive systems is similar in all animals. However, many structural differences are found among the accessory sex organs, depending on the habits of the animals, methods of fertilizing their eggs, care of the young, and other considerations. Many invertebrates have reproductive systems as complex as those of vertebrates, for example, flatworms, snails, earthworms, and others. There are often complicated accessory sex organs such as reproductive ducts, penis, seminal vesicles, yolk glands, uterus, seminal receptacles, and genital chambers.

In vertebrate animals the reproductive and excretory

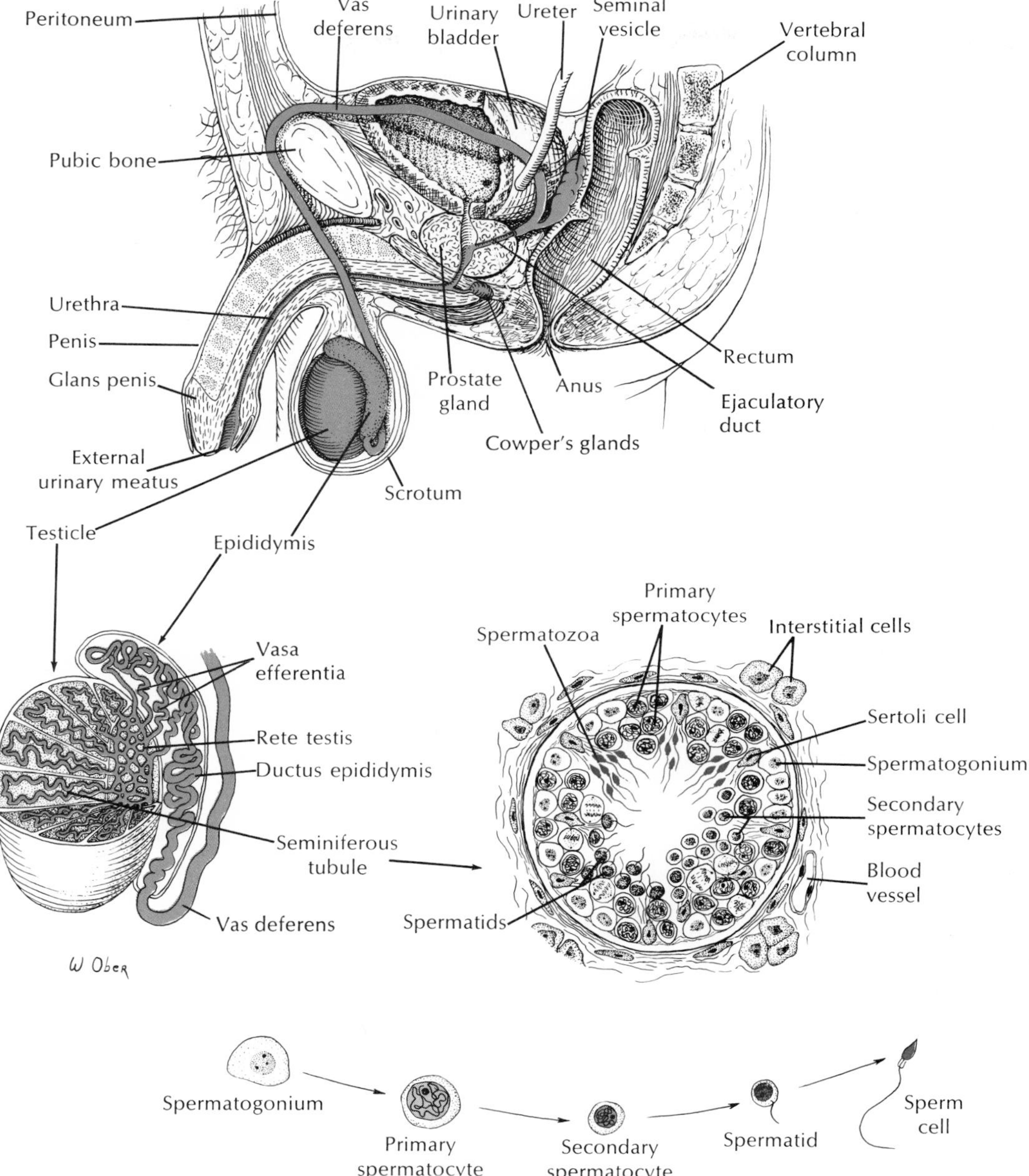

FIG. 35-1 Human male reproductive sytem. *Top*, Median section of male pelvis. *Center left*, Section of left testis. *Center right*, Cross section of one seminiferous tubule, showing different stages in spermatogenesis. *Bottom*, Sequential stages in spermatogenesis.

systems are often referred to as the **urinogenital system** because of their close association, particularly striking during embryonic development, and in the use of common ducts. The male excretory and reproductive systems usually have a more intimate connection than have those of the female. In those forms (some fishes and amphibians) that have an opisthonephric kidney, the **wolffian duct** that drains the kidney also serves as the sperm duct. In male reptiles, birds, and mammals in which there is a metanephric kidney with its own independent duct **(ureter)** to carry away waste, the wolffian duct is exclusively a sperm duct **(vas deferens).** In all these forms, with the exception of mammals higher than monotremes, the ducts open into a **cloaca.** In higher mammals lacking a cloaca, the urinogenital system has an opening separate from the anal opening. The **oviduct** of the female is an independent duct that, however, does open into the cloaca in forms that have a cloaca.

The plan of the reproductive system in vertebrates includes (1) **gonads** that produce the sperm and eggs; (2) **ducts** to transport the gametes; (3) **special organs** for transferring and receiving gametes; (4) **accessory glands** (exocrine and endocrine) to provide secretions necessary for the reproductive process; and (5) **organs for storage** before and after fertilization. This plan is modified among the various vertebrates, and some of the items may be lacking altogether.

Human reproductive systems

Male reproductive system

The male reproductive system (Fig. 35-1) includes testes, vasa efferentia, vasa deferentia, penis, and glands.

Testes (testicles). The testes are paired and are responsible for the production and development of the sperm. Each testis is made up of about 500 **seminiferous tubules,** which produce the sperm, and the **interstitial tissue,** lying among the tubules, which produces the male sex hormone (testosterone). The two testes are housed in the scrotal sac, which hangs down as an appendage of the body. This arrangement keeps the testes at a temperature slightly lower than that of the body. For some reason sperm will not form at body temperatures in most mammals, although they are able to do so in elephants and in birds, which have body temperatures two to three centigrade degrees higher than those of mammals. In some mammals (many rodents) the testes are retained within the body cavity except during the breeding season, when they descend through the inguinal canals into the scrotal sacs.

Vasa efferentia. Vasa efferentia are small tubes connecting the seminiferous tubules to a coiled **vas epididymis** (one for each testis), which serves for the storage of the sperm.

Vasa deferentia. Each vas deferens is a continuation of the epididymis and runs to the urethra, where it joins its counterpart from the other testis. From this point the urethra serves to carry both sperm and urinary products.

Penis. The penis is the external intromittent organ through which the urethra runs. The penis contains erectile tissue, spaces that become engorged with blood during sexual arousal.

Glands. There are at least three pairs of exocrine glands (those with ducts) that open into the reproductive channels. Fluid secreted by these glands furnishes food to the sperm, lubricates the passageways of the sperm, and counteracts the acidity of the urine so that the sperm will not be harmed. The first of these glands is the **seminal vesicle,** which opens into each vas deferens before it meets the urethra. Next are the **prostate glands,** which are really a single fused gland in man; it secretes into the urethra. Near the base of the penis lies the third pair of glands, **Cowper's glands,** which also discharge into the urethra. The secretions of these glands form a part of the seminal discharge.

Female reproductive system

The female reproductive system (Fig. 35-2) includes ovaries, oviducts, uterus, vagina, and vulva.

Ovaries. The ovaries are paired and are contained within the abdominal cavity, where they are held in position by ligaments. Each ovary is about as large as an almond and contains many thousands of developing eggs (ova). Each egg develops within a graafian follicle that enlarges and finally ruptures to release the mature egg (Fig. 35-2). During the fertile period of the woman about 13 eggs mature each year, and usually the ovaries alternate in releasing an egg. Since the female is fertile for only some 30 years, only about 400 of some 400,000 developing eggs in the ovaries of a young woman will have a chance to reach maturity; the others degenerate during her reproductive life and are absorbed.

Oviducts (fallopian tubes). These egg-carrying

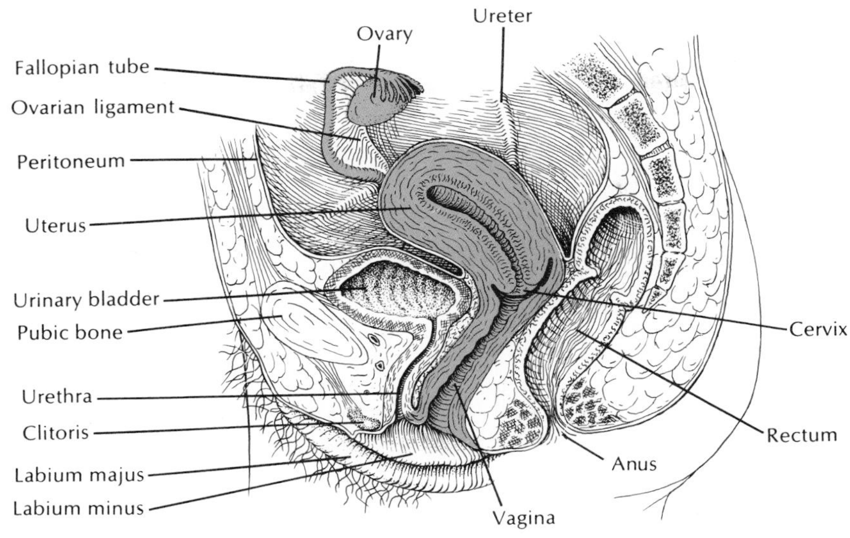

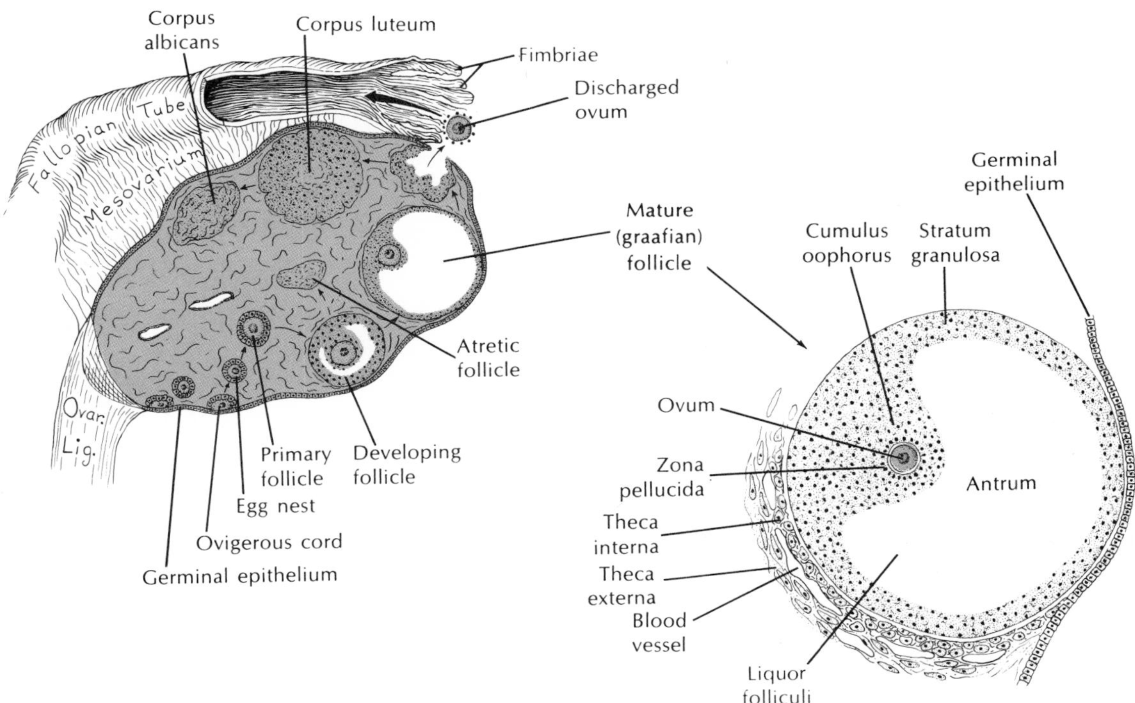

FIG. 35-2 Human female reproductive system. *Top,* Median section through female pelvis. *Left,* Section of ovary showing progressive differentiation of a follicle, ovulation, and formation of a corpus luteum. *Bottom right,* Section of mature follicle.

TABLE 35-1 Chief organ homologies of male and female reproductive systems

Male	Indifferent stage of embryo	Female
Testis	Genital ridge	Ovary
Vas deferens	Wolffian duct	Vestigial
Epididymis	Wolffian body	Vestigial
Appendix of testis	Müllerian duct	Uterus, vagina, fallopian tube
Penis Glans penis	Genital tubercle	Clitoris Glans clitoridis
Anal surface of penis	Genital folds	Labia minora
Scrotum	Genital swellings	Labia majora

tubes are not closely attached to the ovaries but have funnel-shaped ostia for receiving the eggs when they emerge from the ovary. The oviduct is lined with cilia for propelling the egg in its course. The two ducts open into the upper corners of the uterus, or womb.

Uterus. The uterus (womb) is specialized for housing the embryo during the 9 months of its intrauterine existence. It is provided with thick muscular walls, many blood vessels, and a specialized lining—the **endometrium.** The uterus varies with different mammals. It was originally paired but tends to fuse in higher forms.

Vagina. This large muscular tube runs from the uterus to the outside of the body. It is adapted for receiving the male's penis and for serving as the birth canal during expulsion of the fetus from the uterus. Where the vagina and the uterus meet, the uterus projects down into the vagina to form the **cervix.**

Vulva. The vulva refers to the external genitalia and includes two folds of skin covered with hair, the **labia majora;** a smaller pair of folds within the labia majora, the **labia minora;** a small erectile organ, the **clitoris,** at the anterior junction of the labia minora; and a fleshy elevation above the labia majora, the **mons veneris.** The opening into the vagina is the vestibule that is normally closed in the virgin state by a membrane, the **hymen.**

Homology of sex organs

Although the genetic sex of an animal is determined at the time of fertilization, it is not until much later in embryonic development—several weeks for human embryos—that distinct male or female characters become evident. At first the gonads are totally undifferentiated. Also present are two undifferentiated duct systems. If the embryo is a genetic female, one of these duct systems (müllerian ducts), under the influence of female sex hormones, will develop into oviducts, uterus, and vagina; the other duct system remains rudimentary. If the embryo is a genetic male, male sex hormones will cause the alternate duct system (wolffian ducts) to give rise to the duct system of the testes, and the other set becomes rudimentary. Thus for every structure in either system, there is a homologous one in the other. Some of the chief homologues of the male and female reproductive systems are given in Table 35-1.

FORMATION OF REPRODUCTIVE CELLS

The animal body has two obvious contrasting components: the **somatic cells,** which die with the individual; and the **germ cells,** some of which may contribute to the formation of a zygote and thereby to a new generation. The primordial germ cells are set aside at the beginning of embryonic development, usually in the endoderm, and migrate to the gonads (to be discussed further). The germ cells develop into eggs and sperm—nothing else. Other cells of the gonads are somatic cells. They cannot themselves form eggs or sperm, but they are necessary aids in the gametogenesis of the germ cells.

Origin and migration of germ cells

The actual tissue from which the gonads arise appears in early development as a pair of ridges, or pouches, growing into the coelom from the dorsal coelomic lining on each side of the gut near the anterior end of the mesonephros.

Surprisingly perhaps, the primordial ancestors of the germ cells do not arise in the developing gonad but in the yolk-sac endoderm. From studies with frogs and toads, it has been possible to trace the germ-cell line back to the fertilized egg, in which a localized area of "germinal cytoplasm" can be identified in the vegetal pole of the uncleaved egg mass. This material can be followed through subsequent cell divisions of the embryo until it becomes situated in primitive sex cells located deep in the endoderm of the yolk sac. From here they migrate to the developing gonads.

The primitive germ cells are larger than somatic cells and have large round nuclei with prominent nucleoli. They can thus be distinguished from other cells in the embryo. In the human embryo, they wander by ameboid movement from the yolk sac first into the gut mesentery, then to the kidney, and finally into the genital ridges. In some vertebrates, especially in birds, the primitive sex cells travel passively through the bloodstream to the developing gonad. There is some evidence that the gonads release a chemical attractant that helps the germ cells to home in on the gonad. If, for example, the early gonad is removed from its normal position and implanted in some distant site, the primordial germ cells will still migrate to it.

The primordial germ cells are the future stock of gametes for the animal. Once in the gonad they begin to divide by mitosis, increasing their numbers from a few dozen to several thousand.

At first the gonad is sexually indifferent. Recently it has been determined that the indifferent gonad has an inherent tendency to become an ovary. If the embryo is to become a male, male-determining genes on the sex chromosomes direct the primordial germ cells to grow into the central portion (medulla) of the gonad where, eventually, the seminiferous tubules of the testis develop. The inner walls of these tubules are lined with cells that have descended from the primordial germ cells. As the animal approaches sexual maturity, these develop into mature sperm.

If the animal is a genetic female, the *absence* of male-determining genes allows the gonad to follow its inherent tendency to become an ovary. In this alternative, the medullary tissue degenerates and the cortical (outer) tissue forms the substance of the ovary. The primordial germ cells, together with somatic cells in this area, differentiate into ovarian follicles with eggs (ova) (Fig. 35-3). We will return to, and expand on, this subject of sex determination in Chapter 37 (p. 854).

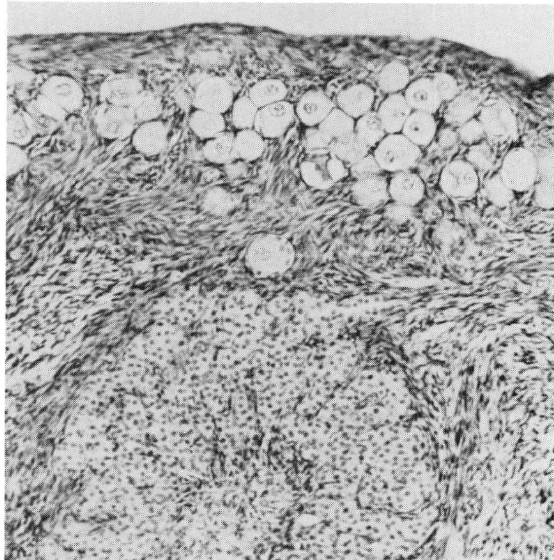

FIG. 35-3 Section of cat ovary showing corpus luteum (large body in lower half of photograph) and primordial follicles containing oogonia (above) (Courtesy J. W. Bamberger.)

Meiosis: maturation division of germ cells

In ordinary cell division, or mitosis, each of the two daughter cells receives exactly the same number and kind of chromosomes. All body (somatic) cells have two sets of chromosomes, one of paternal and the other of maternal origin. The members of such a pair are called **homologous chromosomes.** Mitosis is described in Chapter 4 and is illustrated in Fig. 4-11 (p. 70).

Earlier, we commented that meiosis was one of the fundamental distinguishing events of sexual reproduction. Meiosis is a special kind of maturation division for gametes that *separates* the double set of chromosomes (Fig. 35-4). The result is that mature gametes (eggs and sperm) have only *one* member of each homologous chromosome pair, or a **haploid** (n) number of chromosomes. In humans the zygotes and all body cells normally have the **diploid** number (2n), or 46 chromosomes; the gametes have the haploid number (n), or 23.

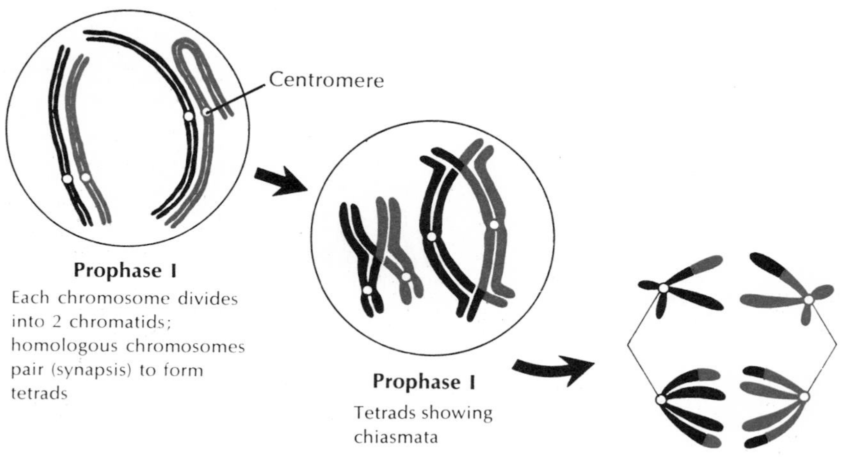

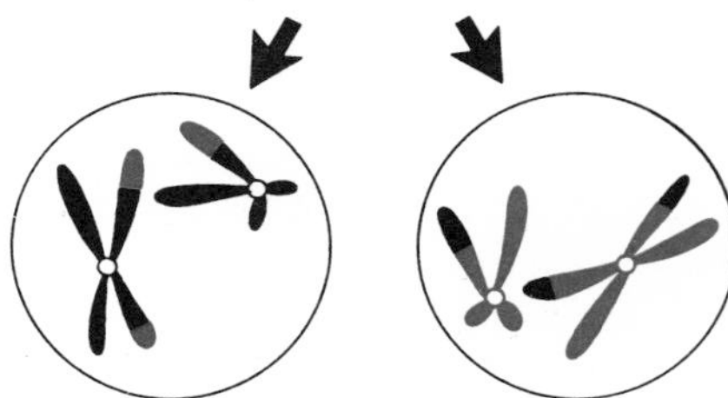

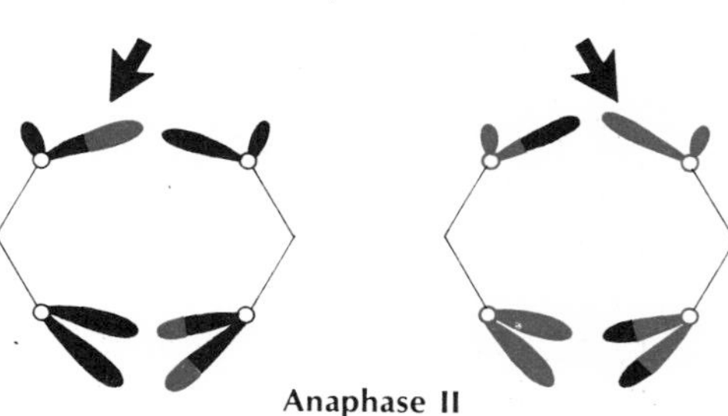

FIG. 35-4 Meiosis in a sex cell with two pairs of chromosomes.

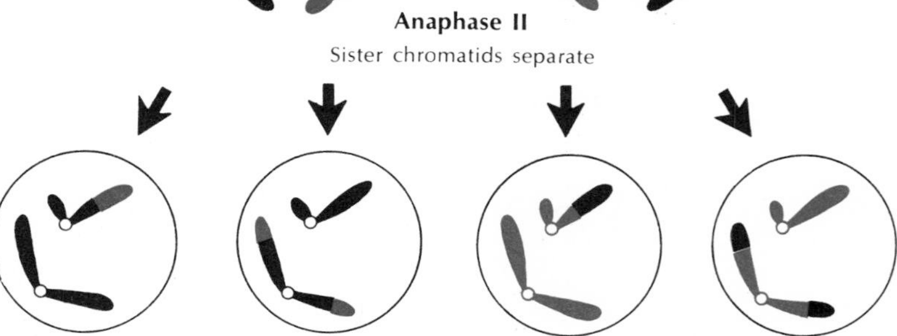

It is helpful to think of meiosis as a modified sequence of two ordinary (mitotic) cell divisions in which the chromosomes divide only once. Each meiotic division is similar to mitosis in its morphologic changes; each passes through the same four stages—prophase, metaphase, anaphase, and telophase—found in mitosis. However, preparation for the first meiotic division (prophase) is much more elaborate than the corresponding stage in mitosis. The two members of each pair of chromosomes come into side-by-side contact **(synapsis).** Since each chromosome has already divided into two daughter chromatids, a **tetrad** of two pairs of two chromatids is formed. The most important event in meiosis occurs when the chromatids of one tetrad exchange parts of chromatids. This phenomenon, called **crossing-over,** is clearly shown in Fig. 35-4. Crossing-over is important between a pair of homologous chromosomes (four chromatids, two from each chromosome) because the hereditary material is redistributed among the four chromatids of one tetrad. The chromosomes exchange equivalent sections bearing the same genes, so each chromatid contains a full set of genes. But the genes are in new combinations.

As prophase continues, the tetrads begin to split apart revealing **chiasmata,** the connection points where crossing-over has occurred. There may be one or more chiasmata present in each tetrad, depending on the number of times crossing-over occurs. Toward the end of prophase the chromosomes shorten and thicken and are ready to enter into the first meiotic division.

At anaphase of the first meiotic division, the tetrads divide and separate, each forming two pairs of chromatids called **dyads,** which migrate to opposite poles of the cell. Cell division follows, and each resulting daughter cell thus contains half the number of chromosomes (dyads) characteristic of the diploid chromosomes of the organism.

The second meiotic division results in a separation and distribution of the sister chromatids of each chromosome to opposite poles and thus involves no reduction in the number of chromosomes. Each chromatid of the original tetrad exists in a separate nucleus. Four cells (gametes) are formed, each containing one complete, haploid set of chromosomes.

The result of the two meiotic divisions is the formation of four cells, each of which has the haploid number of chromosomes or one of each kind of chromosome of the homologous pairs that started meiosis. The first meiotic division is often called a **reduction division** because the number of separate centromeres is reduced by one-half in the two daughter cells, and the second meiotic division is called an **equational division** because the number of centromeres in the daughter cells remains the same as in the parent cell. However, the exchange of chromatid segments (crossing-over) and the fact that there is actually no reduction in total number of chromosomes during the first meiotic division have caused many cytologists to consider the terms "reduction" and "equational" confusing and obsolete.

Stages in meiotic division

Following is a more detailed description of the stages in meiosis.

Prophase I. Most of the unique features of meiosis occur at the beginning of the first meiotic division. The prophase, or the first stage of meiosis, has five substages: leptotene, zygotene, pachytene, diplotene, and diakinesis.

1. **Leptotene.** In this stage the chromosomes (diploid in number) appear as long, thin, threadlike structures and resemble strings of beads because of granules (chromomeres). Each pair of homologous chromosomes is identical as to size, position of centromere, and other characteristics. Each of these early prophase chromosomes is already divided into a pair of indistinguishable sister chromatids. DNA replication is complete.
2. **Zygotene.** This stage involves the pairing of homologous chromosomes **(synapsis)** to form bivalent chromosome units. This process does not occur in mitosis. It results in paired units corresponding to the haploid number.
3. **Pachytene.** The pairing of the chromosomes is completed, and the chromosomes undergo longitudinal contraction so that each bivalent is shorter and thicker, and the two chromosomes of each bivalent become twisted about one another.
4. **Diplotene.** In this stage the homologous chromosomes of each bivalent are now visibly double. Since each homologous chromosome consists of two sister chromatids, each chromosome pair, or bivalent, will show four chromatids, or a **tetrad.** The centromeres remain unsplit at this time. Since longitudinal separation is incomplete, the homologues of the bivalent remain in contact at various points, producing a characteristic X configuration called **chiasma** (Gr., *chiasma,* cross; pl., chiasmata) (Fig. 35-4). Each chiasma represents a region at which two nonsister chromatids are undergoing an exchange of parts (crossing-over). Thus two chromatids of the tetrads are structurally reorganized so that each is made up of an original and an exchanged component. The other two chromatids of the tetrad remain in their original form. Chiasmata are not found in meiosis in which there is no crossing-over, as in the male *Drosophila.*

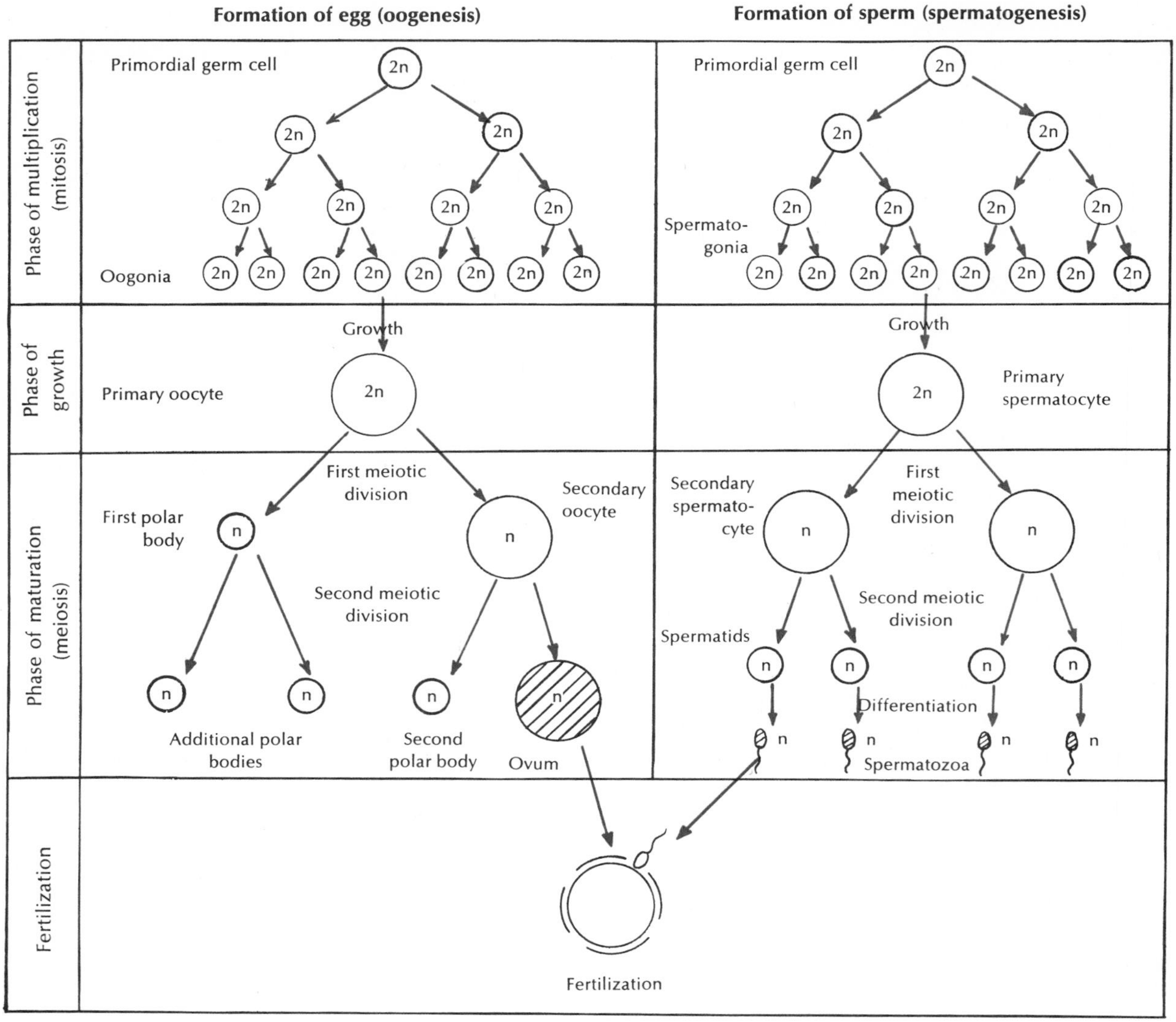

FIG. 35-5 Gametogenesis compared in eggs and sperm.

5. **Diakinesis.** This stage is characterized by a maximum contraction of the chromosomes and a further separation of the homologous chromosomes, although the chromatids remain connected by the chiasmata. At the same time the nucleolus begins to disappear, and the nuclear membrane breaks down.

At the end of the first prophase of meiosis homologous chromosomes have paired, exchanged chromatid segments and have started their longitudinal separation.

Metaphase I. This phase begins when the nuclear membrane disappears, and the spindle is formed. Each homologous chromosome (homologue) has its centromere (kinetochore) and the bivalent chromosomes (tetrads) line up on the equatorial plate, with the centromeres of the two homologues directed toward opposite poles.

Anaphase I. In this stage each homologue of a pair, with its daughter chromosomes united by their centromeres, moves to its respective pole, each centromere taking half of the bivalent with it. Thus whole chromosomes are separated in anaphase, and each of the two resulting cells of the first meiotic division has a haploid number of chromosomes.

Telophase I and interphase. A nuclear membrane may be re-formed around the chromosomes, which often persist in a condensed form; they may become uncoiled; or no membrane may be formed at all and the chromosomes may enter directly into the second meiotic division. The interphase may not exist at all.

Prophase II and metaphase II. A short prophase in which a new spindle starts forming marks the beginning of the second meiotic division. The chromosomes become arranged on the equatorial plate. This is followed by the division of the centromeres for the first time and the longitudinal separation of the sister chromatids of each chromosome. Although these two chromatids are identical in their formation, they differ in those segments that have been exchanged by crossing-over.

Anaphase II and telophase II. The sister chromatids, now called chromosomes, move to their respective poles, and each of the two daughter nuclei has a complete set (genome) that corresponds to the haploid number. In the telophase the cytoplasm divides, and the chromosomes become longer and less visible. A nuclear membrane is then formed around each nucleus.

Gametogenesis

The series of transformations that results in the formation of mature gametes (germ cells) is called gametogenesis.

Although the same essential processes are involved in the maturation of both sperm and eggs, there are some minor differences. Gametogenesis in the testis is called **spermatogenesis** and in the ovary it is called **oogenesis.**

Spermatogenesis (Figs. 35-1 and 35-5)

The walls of the seminiferous tubules contain the differentiating sex cells arranged in a stratified layer five to eight cells deep. The outermost layers contain **spermatogonia** (Fig. 35-1), which have increased in number by ordinary mitosis. Each spermatogonium increases in size and becomes a **primary spermatocyte.** Each primary spermatocyte then undergoes the first meiotic division, as described before, to become two **secondary spermatocytes.**

Each secondary spermatocyte now enters the second meiotic division, without the intervention of a resting period. The resulting cells are called **spermatids,** and each contains the haploid number (23) of chromosomes. A spermatid may have all maternal, all paternal, or both maternal and paternal chromosomes in varying proportions. Without further divisions the spermatids are transformed into mature sperm by losing a great deal of cytoplasm, by condensing the nucleus into a head, and by forming a whiplike tail (Fig. 35-1).

One can see by following the divisions of meiosis that each primary spermatocyte gives rise to four functional sperm, each with the haploid number of chromosomes (Fig. 35-5).

Oogenesis (Figs. 35-2 and 35-5)

The early germ cells in the ovary are called **oogonia,** which increase in number by ordinary mitosis (Fig. 35-3). Each oogonium contains the diploid number of chromosomes. In the human being, after puberty, typically one of these oogonia develops each menstrual month into a functional egg. After the oogonia cease to increase in number, they grow in size and become **primary oocytes.** Before the first meiotic division, the chromosomes in each primary oocyte meet in pairs, paternal and maternal homologues, just as in spermatogenesis. When the first maturation division occurs, the cytoplasm is divided unequally. One of the two daughter cells, the **secondary oocyte,** is large and receives most of the cytoplasm; the other is very small and is called the **first polar body.** Each of these daughter cells, however, has received half the nuclear material or chromosomes.

In the second meiotic division, the secondary oocyte divides into a large **ootid** and a small polar body. If the first polar body also divides in this division, which sometimes happens, there will be three polar bodies and one ootid. The ootid grows into a functional **ovum;** the cytoplasm-starved polar bodies are nonfunctional and disintegrate. This obviously undemocratic hoarding of all accumulated food reserves by just one of four otherwise genetically equal cells is of course a device for avoiding a decrease in ovum size, once all the nutrients have been packaged inside at the end of the growth phase. Thus the mature ovum has the haploid number of chromosomes, the same as the sperm does. However, each primary oocyte gives rise to only **one** functional gamete instead of four as in spermatogenesis (Fig. 35-5).

Gametes and their specializations

Eggs

During **oogenesis,** the egg becomes a highly specialized, very large cell containing condensed food reserves for subsequent growth. It is of interest that the prolonged growth and enormous accumulation of food reserves in the oocyte occur **before** the meiotic, or maturation, divisions begin. (This contrasts with spermatogenesis in which differentiation of the mature sperm occurs only **after** the meiotic divisions.) When the maturation divisions do occur, once the growth phase of the oocyte is complete, they are, as already described, highly unequal: the two meiotic divisions produce one very large mature ovum and two or three polar bodies.

In most vertebrates, the egg does not actually complete all the meiotic divisions before fertilization occurs. The general rule is that the egg completes the first meiotic division and proceeds to the metaphase stage of the second meiotic division, at which point further progress stops. The second meiotic division is completed and the second polar body extruded only if the egg is activated by fertilization.

The most obvious feature of egg maturation is the deposition of yolk. Yolk, usually stored as granules or more organized platelets, is no definite chemical substance but may be lipid or protein or both. In insects and vertebrates, all having more or less yolky eggs, the yolk may be synthesized within the egg from raw materials supplied by the surrounding follicle cells, or preformed lipid or protein yolk may be transferred by pinocytosis from follicle cells to the oocyte. In the latter, the follicle cells are more important in regulating egg growth than factors within the egg itself.

The result of the enormous accumulation of yolk granules and other nutrients (glycogen and lipid droplets) is that an egg grows well beyond the normal limits that force ordinary body (somatic) cells to divide. A young frog oocyte 50 μm in diameter, for example, grows to 1,500 μm in diameter when mature after 3 years of growth in the ovary, and the volume has increased by a factor of 27,000. Bird eggs attain even greater absolute size; a hen egg will increase 200 times in volume in only the last 6 to 14 days of rapid growth preceding ovulation.

Thus eggs are remarkable exceptions to the otherwise universal rule that organisms are composed of relatively minute cellular units. This creates a surface area–to–cell volume ratio problem, since everything that enters and leaves the ovum (nutrients, respiratory gases, wastes, and so on) must pass through the cell membrane. As the egg becomes larger, the available surface per unit of cytoplasmic volume (mass) becomes smaller. As we would anticipate, the metabolic rate of the egg gradually diminishes until, when mature, the ovum is in a sort of suspended animation awaiting fertilization. However, large size is not the only factor leading to quiescence. There is increasing evidence that enzyme and nucleic acid inhibitors, which directly repress metabolic and synthetic activity in the egg, appear toward the end of maturation.

Spermatozoa

Sperm in animals show a greater diversity of form than do ova. A typical spermatozoon is made up of a head, a middle piece, and an elongated tail for locomotion (Fig. 35-6). The head consists of the nucleus containing the chromosomes for heredity and an **acrosome** believed to contain an enzyme that assists in egg penetration. The total length of the human sperm is 50 to 70 μm. Some toads have sperm that exceed 2 mm (2,000 μm) in length. Most sperm, however, are microscopic in size.

The acrosome is a distinctive feature of nearly all the Metazoa (exceptions are the teleost fishes and certain invertebrates). In many species, both invertebrate and vertebrate, the acrosome contains egg-membrane lysins that serve to clear an entrance path through the membranes that form a barrier around the egg. In mammals at least, the lysin is the enzyme hyaluronidase. A striking feature of many invertebrate spermatozoa is the acrosome filament, an extension of varying length in different species, that projects suddenly from the sperm head when the latter first contacts the egg membrane. It serves to penetrate into the egg cytoplasm and is the initial event of fertilization.

The number of sperm in all animals is vastly in excess of the eggs of corresponding females. The number of eggs produced is related to the chances of the young to hatch and reach maturity. This explains the enormous number of eggs produced by certain fishes, as compared with the small number produced by mammals.

HORMONES OF REPRODUCTION

The male and female gonads are endocrine glands as well as gamete-forming glands. Reproduction is a

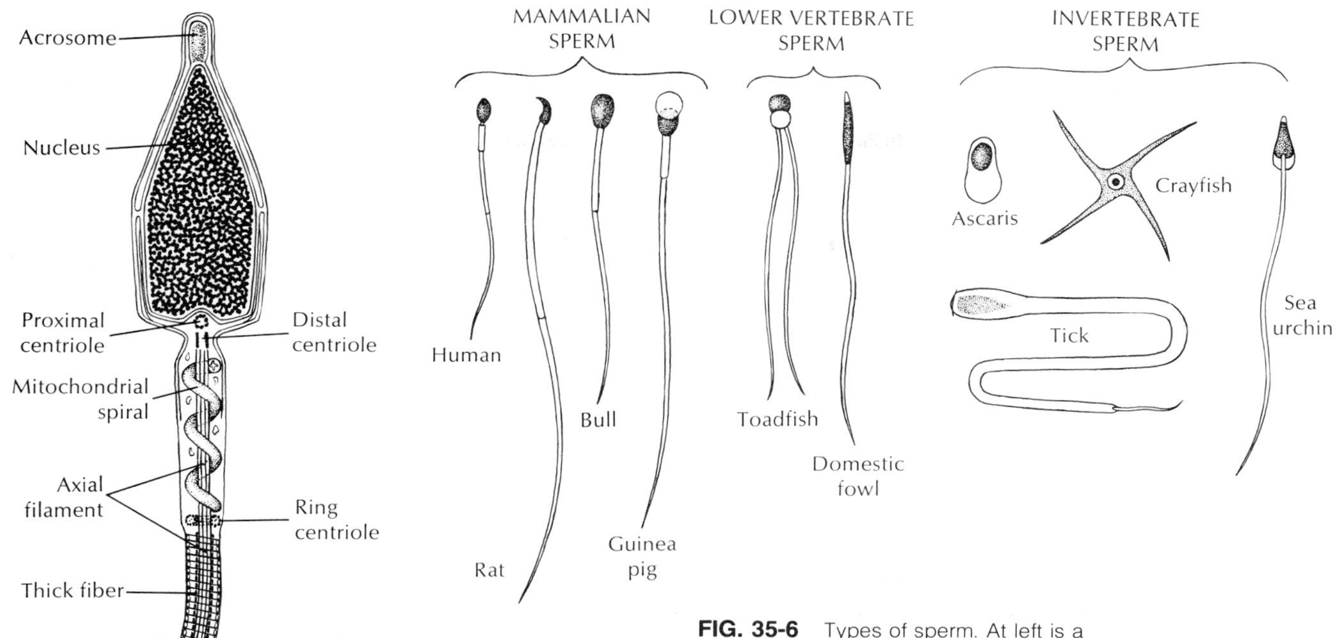

FIG. 35-6 Types of sperm. At left is a semidiagrammatic enlargement of the anterior end of a human spermatazoon.

complex process requiring the coordinated action of many hormones, especially in the female. Although the principal features of reproductive endocrinology are understood, the recent search for effective and safe birth control devices has revealed some disturbing gaps in our knowledge. To cite a single but telling example, no one fully understands how the popular birth control pill works.

The female reproductive activities of almost all animals are cyclic. Usually breeding cycles are seasonal and coordinated so that the young are born at a time of year when conditions for growth are most favorable. Mammals have two major reproductive patterns, the **estrous cycle** (characteristic of most mammals) and the **menstrual cycle** (characteristic of primates only). These two cyclic patterns differ in two ways. First, in the estrous cycle, but not in the menstrual cycle, the female is receptive to the male; that is, she is in ''heat'' only at restricted periods of the year. Second, the menstrual cycle, but not the estrous cycle, ends with the collapse and sloughing of the uterine lining (endometrium). In the estrous animal each cycle ends with the uterine lining simply reverting to its original state, without the bleeding characteristic of the menstrual cycle. (The bleeding that occurs in dogs and cattle when they are in heat is not caused by sloughing of the uterine lining but by red blood cells passing through the wall of the vagina.)

But these differences are really minor variations on a basic theme. The hormonal regulation of the reproductive cycles is so much alike for all mammals that a great deal of what we know about the human reproductive hormones has come from laboratory studies of the ubiquitous white rat, a rodent having an estrous cycle. Differences become more apparent when we study the lower vertebrates, but even fishes share most of the reproductive hormones found in humans. Since our discussion of this vast area must be brief, we will restrict our consideration to humans.

The ovaries produce two kinds of steroid **sex hormones—estrogens** and **progesterone** (Fig. 35-7). Estrogens (principally estradiol) are responsible for the development of the female accessory sex structure (uterus, oviducts, and vagina) and the female secondary sex characters, such as breast development and the characteristic bone growth, fat deposition, and hair distribution of the female. Progesterone is responsible for preparing the uterus to receive the developing em-

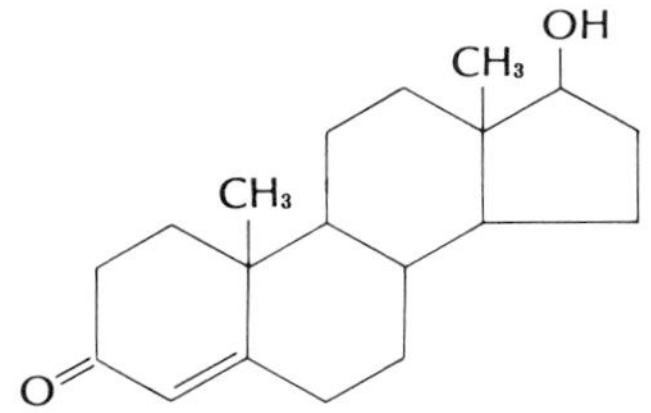

Testosterone

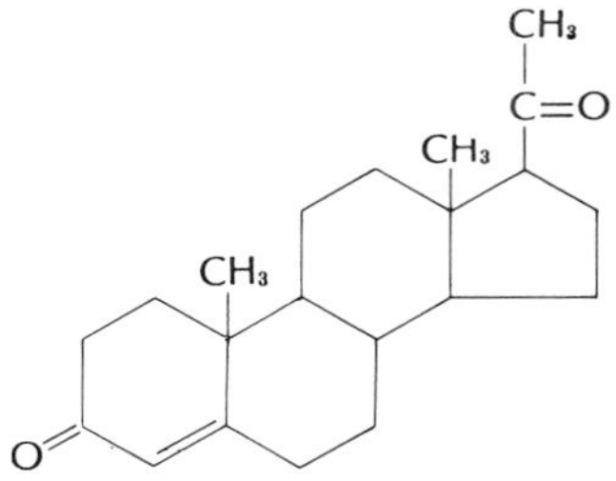

Progesterone

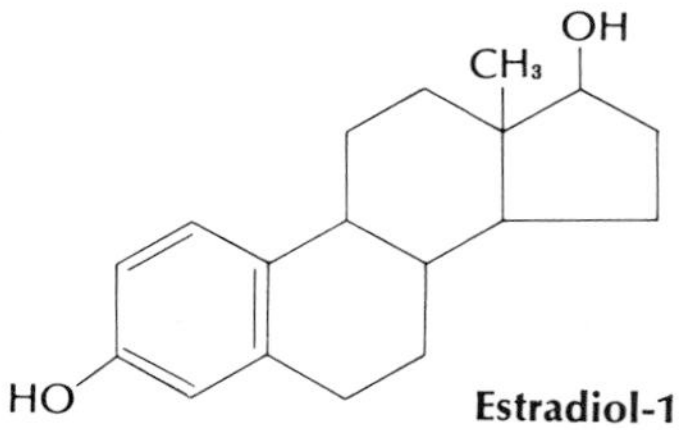

Estradiol-17β

FIG. 35-7 These three sex hormones all show the basic four-ring steroid structure. The female sex hormone estradiol-17β (an estrogen) is a C_{18} (18-carbon) steroid with an aromatic A ring (first ring to left). The male sex hormone testosterone is a C_{19} steroid with a carbonyl group (C=O) on the A ring. The female pregnancy hormone progesterone is a C_{21} steroid, also bearing a carbonyl group on the A ring.

bryo. These hormones are controlled by the pituitary **gonadotropins,** FSH (follicle-stimulating hormone), and LH (luteinizing hormone) (Fig. 35-8; see also Fig. 33-6, p. 761).

The menstrual cycle begins with the release into the bloodstream of FSH from the anterior pituitary (Fig. 35-8). Reaching the ovaries, it stimulates the growth of one of the several thousand follicles present in each ovary. The follicle swells as the egg matures, until it bursts, releasing the egg onto the surface of the ovary. This event, called **ovulation,** normally occurs on about the fourteenth day of the cycle.

Now follows the most critical period of the cycle, for unless the mature egg is fertilized within a few hours it will die. During this period, the egg is swept into an oviduct (fallopian tube) and begins its journey toward the uterus, pushed along by the numerous cilia that line the oviduct walls. If intercourse occurs at this time, the sperm will traverse the uterus and find their way into the oviducts, where one may meet and fertilize the egg. The developing embryo continues down the oviduct, enters the uterine cavity, where it dwells for a day or two, and then implants in the prepared uterine endometrium. This is the beginning of pregnancy.

Let us now examine the intricate series of events that occurs both before and after the beginning of pregnancy. We have seen that the pituitary hormone FSH begins the reproductive cycle by stimulating the growth of at least one of the ovarian follicles. As the follicle enlarges, it releases estrogens that prepare the uterine lining (endometrium) for reception of the embryo.

The rise in blood estrogen is sensed by the pituitary, which responds by stopping the production of FSH. Estrogen also encourages the production of the second pituitary hormone, LH. Ovulation now occurs. LH causes the cells lining the ruptured follicles to proliferate rapidly, filling the cavity with a characteristic spongy, yellowish body called a **corpus luteum** (Fig. 35-3). The corpus luteum, responding to the continued stimulation of LH, manufactures **progesterone** in addition to estradiol. Progesterone, as its name suggests, stimulates the uterus to undergo the final maturation changes that prepare it for gestation.

The uterus is thus fully ready to house and nourish the embryo by the time the latter settles out onto the uterine surface, usually about 7 days after ovulation. If fertilization has *not* occurred, the corpus luteum disappears, and its hormones are no longer secreted.

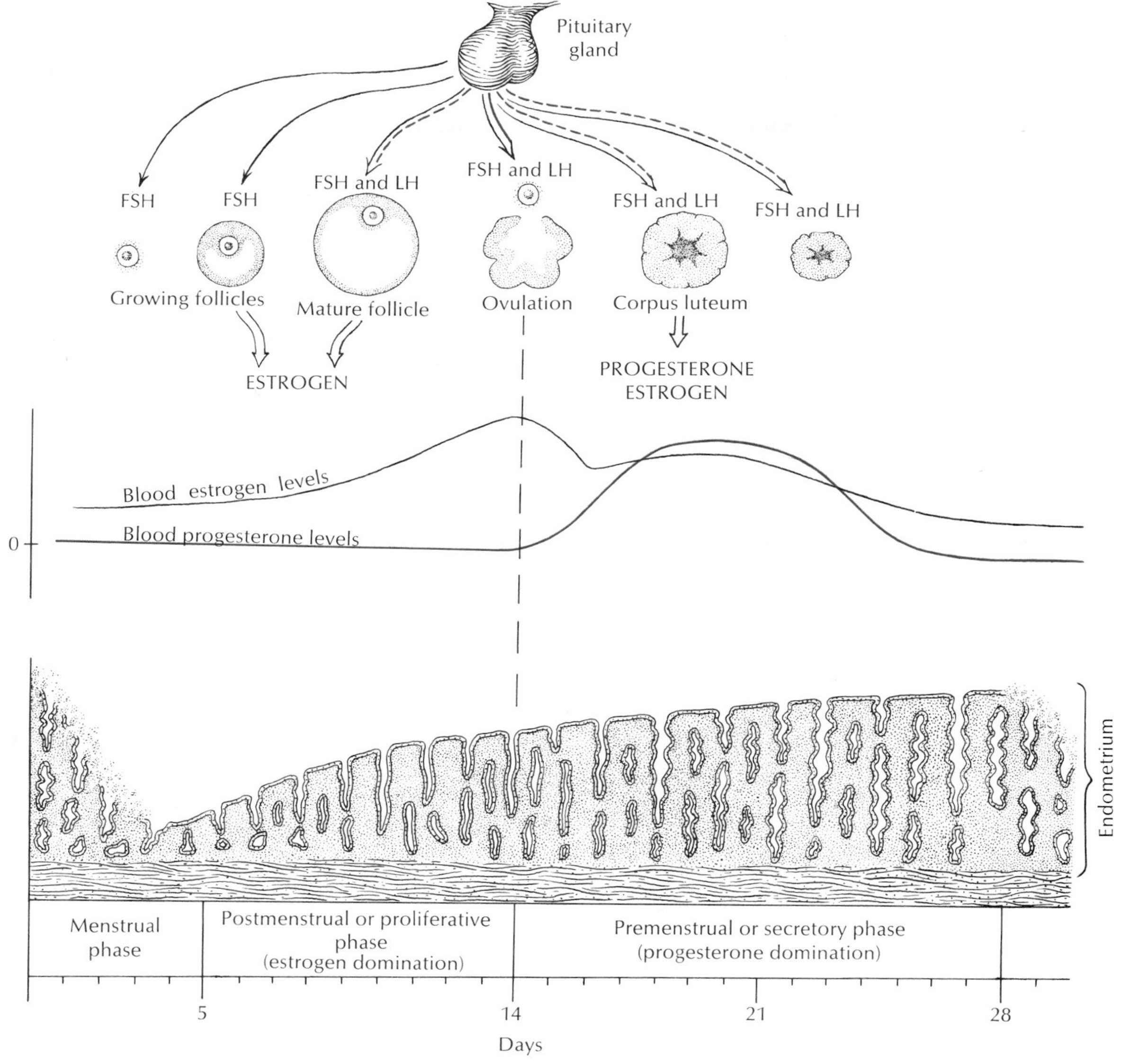

FIG. 35-8 Menstrual cycle in the human female, showing hormonal-ovarian-uterine relationships. The development and eventual collapse of the uterine endometrium is determined by the blood levels of the sex hormones estrogen (black) and progesterone (red). The secretion of the sex hormones (estrogen from the follicle; progesterone from the corpus luteum) is controlled by the interplay of hormones from the pituitary gland.

Since the uterine endometrium depends on progesterone and estrogen for its maintenance, their disappearance causes the endometrial lining to dehydrate and slough off, producing the menstrual discharge. However, if the egg has been fertilized and has implanted, the corpus luteum continues to supply the essential sex hormones needed to maintain the mature uterine endometrium. During the first few weeks of pregnancy the developing placenta itself begins to produce the sex hormones progesterone and estrogen and soon replaces the corpus luteum in this function.

As pregnancy advances, progesterone and estrogen

prepare the breasts for milk production. The actual secretion and release of milk after birth (lactation) is the result of two other hormones, **prolactin** and **oxytocin.** Milk is not secreted during pregnancy because the placental sex hormones inhibit the release of prolactin by the pituitary. The placenta, like the corpus luteum that preceded it, thus becomes a special endocrine gland of pregnancy. After delivery, many mammals eat the placenta (afterbirth), a behavior that serves to remove telltale evidence of a birth from potential predators.

The male sex hormone **testosterone** (Fig. 35-7) is manufactured by the **interstitial cells** of the testes. Testosterone is necessary for the growth and development of the male accessory sex structures (penis, sperm ducts, glands), for development of secondary male sex characters (hair distribution, voice quality, bone and muscle growth), and for male sexual behavior. The same pituitary hormones that regulate the female reproductive cycle, FSH and LH, are also produced in the male, where they guide the growth of the testes and its testosterone secretion.

DIVERSITY OF REPRODUCTIVE PATTERNS

Animals may be divided into three classes on the basis of the methods they employ to nourish their young. Those animals that lay their eggs outside the body for development are called **oviparous.** In such cases the eggs may be fertilized inside or outside the body. Some animals retain their eggs in the body (in the oviduct) while they develop, but the embryo derives its sole nourishment from the egg and not from the mother. These are called **ovoviviparous.** In the third type the egg develops in the uterus, but the embryo early in its development forms an intimate relationship with the walls of the uterus and derives its nourishment directly from food furnished by the mother. Such a type is called **viviparous.** In both of the last two types the young are born alive. Examples of oviparity are found among many invertebrates and vertebrates. All birds are of this type. Ovoviviparity is common among certain fishes, lizards, and a few of the snakes. Viviparity is confined mostly to the mammals and some elasmobranchs. These patterns are described in more detail for fishes on p. 524.

The structural, physiologic, and behavioral aspects of sexual reproduction show an incredible diversity, exceeding that of any other system. The diversity is expressed in the methods for fertilization, kinds of habitats, structure of the reproductive systems, prenatal care of the young, and seasonal and physiologic changes in animals. Bats mate in the fall and the sperm is stored in the female till the following spring before fertilization occurs. The queen honeybee stores up enough sperm from the drone on one nuptial flight to fertilize all the eggs she lays during her lifetime of several years. Salmon spend most of their lives in the sea but spawn far up inland rivers in fresh water. Eels grow to maturity in fresh water streams but migrate to the sea to spawn. Many animals provide nests of various sorts to take care of the young (birds and some fishes). Others have cases for the eggs (insects and spiders). Some female animals carry the eggs attached to the body or to the appendages (crayfish and certain amphibians). Brood pouches for the eggs are provided by such forms as the mussel, the sea horse (a fish), and many others. Mammals are retained in the uterus of the female during early development and later are nourished by the milk from the mammary glands.

These few examples merely begin to express the curious and fascinating range of reproductive adaptations among animals. Evolution seems to have been most imaginative when it was exploiting the adaptive possibilities of sexuality. There is fascinating literature on the subject; some suggested readings are included in the references that follow.

Annotated references

Selected general readings

Asdell, S. A. 1964. Patterns of mammalian reproduction, ed. 2. Ithaca, N.Y., Comstock Publishing Associates.

Austin, C. R., and R. V. Short (eds.). 1972-1976. Reproduction in mammals, vols. 1 to 6. Cambridge, Cambridge University Press. *Highly readable and well-illustrated paperback series dealing with germ cells, fertilization, embryonic and fetal development, reproductive hormones and patterns, the artificial control of reproduction and the evolution of reproduction.*

Carr, D. E. 1970. The sexes. Garden City, N.Y., Doubleday & Co. *Deals with the evolution of sex from protozoans to humans, sex taboos, birth control, and population problems. Written in a popular and frequently amusing style but containing a wealth of information.*

Clegg, A. G., and P. C. Clegg. 1970. Hormones, cells and organisms. Atlantic Highlands, N.J., Humanities Press, Inc.

McCary, J. L. 1973. Human sexuality. New York, D. Van Nostrand Co. *Elementary college text.*

Odell, W. D., and D. L. Moyer. 1971. Physiology of reproduction. St. Louis, The C. V. Mosby Co. *Technical.*

Parkes, A. S. (ed.). 1956. Marshall's physiology of reproduction, ed. 3, vols. 1 to 3. London, Longman Group Ltd.

Perry, J. S. 1971. The ovarian cycle of mammals. Edinburgh, Oliver & Boyd.

Sadleir, R. M. 1973. The reproduction of vertebrates. New York, Academic Press, Inc. *Comparative treatment of reproductive patterns.*

Swanson, H. D. 1974. Human reproduction: biology and social change. New York, Oxford University Press. *Well-written college text dealing with the biologic, psychologic, sociologic, and ethical aspects of the subject.*

Van Tienhoven, A. 1968. Reproductive physiology of vertebrates. Philadelphia, W. B. Saunders Co.

Wickler, W. 1972. The sexual cycle: the social behavior of animals and men. Garden City, N.Y., Doubleday & Co. *A wide-ranging and interesting exploration of sexual behavior in animals, and the origins of human ethics and morals.*

Wood, C. 1969. Human fertility: threat and promise. World of Science Library. New York, Funk & Wagnalls. (In England and Canada: Sex and fertility. 1969. London, Thames & Hudson.) *Succinct, beautifully illustrated account of human reproduction, birth control, population problems, and human destiny. Highly recommended.*

Selected *Scientific American* articles

Berelson, B., and R. Freedman. 1964. A study in fertility control. **210**:29-37 (May). *Family planning program in Taiwan.*

Csapo, A. 1958. Progesterone. **198**:40-46 (Apr.). *The role of progesterone in pregnancy.*

Dahlberg, G. 1951. An explanation of twins. **184**:48-51 (Jan.). *Twinning and multiple births in humans and other mammals.*

Gordon, M. J. 1958. The control of sex. **199**:87-94 (Nov.). *Studies with rabbit sperm cells separated in an electric field.*

Jaffe, F. S. 1973. Public policy on fertility control. **229**:17-23 (July). *Official policy and public attitude toward contraception have undergone substantial changes in the last few years.*

Mittwoch, U. 1963. Sex differences in cells. **209**:54-62 (July). *The recognition of sex indicators in cell nuclei has speeded studies of sex determination in the human species.*

Tietze, C., and S. Lewit. 1977. Legal abortions. **236**:21-27 (Jan.). *This age-old and controversial method of fertility control is now adopted in many countries.*

CHAPTER 36
PRINCIPLES OF DEVELOPMENT

Three-day-old hatchling red-shouldered hawk, *Buteo lineatus,* and two unhatched eggs. The embryos of birds and reptiles develop within shelled eggs, each containing an elaborate system of membranous sacs serving specialized life-support functions. The chicks of hawks, like the young of many bird species, are relatively helpless at hatching and depend on parental care until old enough to fly.

Photograph by L. L. Rue, III

Growth and differentiation are fundamental characteristics of life (see Principles 21 and 22, p. 14). In Chapter 6 a brief survey of early embryonic development was given: types of eggs, fertilization and formation of the zygote, cleavage, and early morphogenesis. In this chapter we will first consider another dimension of development: the coordinating and regulating processes that guide the destiny of a growing embryo. The latter part of the chapter is devoted to the embryology of amniotes, especially human beings, and the special adaptations that support development.

The phenomenon of development is a remarkable, and in many ways awesome, process. How is it possible that a tiny, spherical fertilized human egg, scarcely visible to the naked eye, can unfold into a fully formed, unique person, consisting of thousands of billions of cells, each cell performing a predestined functional or structural role? How is this marvelous unfolding controlled? Obviously all the information needed is contained within the egg, principally in the genes of the egg's nucleus. The fabric of genes is deoxyribonucleic acid (DNA). Thus all development originates from the structure of the nuclear DNA molecules and in the egg cytoplasm surrounding the nucleus. But knowing where the blueprint for development resides is very different from understanding how this control system guides the conversion of a fertilized egg into a fully differentiated animal. This remains a major—many consider it *the* major—unsolved problem of biology. It has stimulated a vast amount of research on the processes and phenomena involved; some early, and in many cases tentative, answers have emerged.

EARLY THEORIES: PREFORMATION VERSUS EPIGENESIS

Early scientists and lay people alike speculated at length about the mystery of development long before the process was submitted to modern techniques of biochemistry, molecular biology, tissue culture, and electron microscopy. An early and persistent idea was that the young animal was preformed in the egg and that development was simply a matter of unfolding what was already there. Some claimed they could actually see a miniature of the adult in the egg or the sperm. Even the more cautious argued that all the parts of the embryo were in the egg, ready to unfold, but so small and transparent they could not be seen. The **preformation theory** was strongly advocated by most seventeenth- and eighteenth-century naturalist-philosophers.

However, William Harvey, the great physiologist of blood circulation fame, could not accept the preformation notion. His famous dictum *ex ovo omnia* (all from the egg) was an important concept at a time when creative ideas in biology were still rare. The preformation theory received its death blow in 1759 when the German embryologist Caspar Friedrich Wolff clearly showed that in the earliest developmental stages of the chick, there was no embryo, only an undifferentiated granular material that became arranged into layers. These continued to thicken in some areas, to become thinner in others, to fold, and to segment, until the body of the embryo appeared. Wolff called this **epigenesis** (origin upon or after), an idea that the fertilized egg contains building material only, somehow assembled by an unknown directing force.

Current ideas of development are essentially epigenetic in concept, although we know far more about what directs the growth and differentiation. We also realize that a bit of the old preformation idea is still with us, since some materials in germ cells (the nucleic acids) are predestined to guide development and are, in a restricted sense, preformed.

DIRECT AND INDIRECT DEVELOPMENT

Even before fertilization, the egg is programmed to follow a special developmental course. The eggs of marine invertebrates and those of freshwater vertebrates typically develop into feeding larval forms that look and behave in totally different manner from the mature adult. In such **indirect** development, the larva represents an adaptive detour in development that enables the animal to exploit microplankton or other food supplies and thus enhance its chances for a successful **metamorphosis** to the juvenile-adult body form. The eggs of animals having indirect development usually contain rather limited food reserves and are programmed for rapid differentiation into tiny, free-swimming larvae that must find their own food to sustain further growth (Fig. 36-1).

The eggs of mammals, birds, and reptiles show **direct** development; that is, the egg develops directly into a juvenile without passing through a highly specialized larval phase. In direct development, either the egg contains a vast amount of stored food, as do the huge, yolky bird and reptile eggs, or the early embryo develops a nourishing, parasitic relationship with the maternal body, as do the small, nearly yolkless eggs of mammals. Direct development is more leisure-

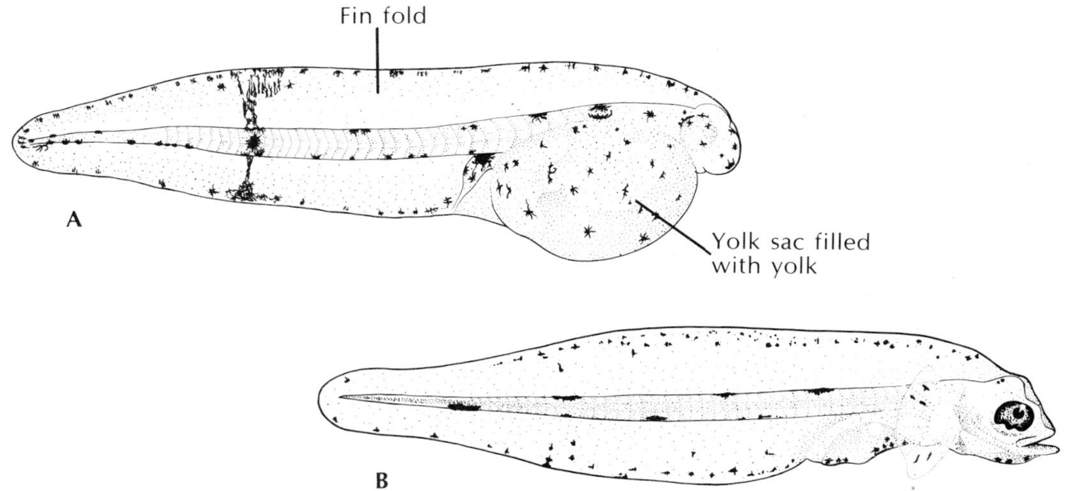

FIG. 36-1 Fish embryos showing yolk sac. **A,** The 1-day-old larva of a marine flounder has a large yolk sac. **B,** By the time the 10-day-old larva has developed a mouth and primitive digestive tract, its yolk supply has been exhausted. It must now catch its own food to survive and continue growing. (From Hickman, C. P., Jr. 1959. The larval development of the sand sole, *Psettichthys melanostictus,* Washington State Fisheries Research Papers **2**:38-47.)

ly than indirect. The eggs are programmed to develop first into extensive sheets of unspecialized cells and then secondarily to differentiate into tissues and organs of a juvenile.

FERTILIZATION

Fertilization is the union of male and female gametes to form a **zygote.** This process accomplishes two things: it activates development, and it provides for the recombination of paternal and maternal genes. Thus it restores the original diploid number of chromosomes characteristic of the species.

For a species to survive, it must ensure that fertilization will occur and that enough progeny will survive to continue the population. Many marine fishes simply set their eggs and sperm adrift in the ocean and rely on the random swimming movements of sperm to make chance encounters with eggs. Even though an egg is a large target for a sperm, the enormous dispersing effect of the ocean, the short life span of the gametes (usually just a few minutes for fish gametes), and the limited range of the tiny sperm all conspire against an egg and a sperm coming together. Accordingly each male releases countless millions of sperm at spawning. The odds against fertilization are further reduced by coordinating the time and place of spawning of both parents.

Ensuring that some eggs are fertilized, however, is not enough. The ocean is a perilous environment for a developing fish, and most never make it to maturity. Thus, the females produce huge numbers of eggs. The common gray cod of the North American east coast regularly spawns 4 to 6 million eggs, of which only two or three eggs on the average will reach maturity. Fishes and other vertebrates that provide more protection to their young produce fewer eggs than do the oceanic marine fishes. The chances of the eggs and sperm meeting is also increased by courtship and mating procedures and the simultaneous shedding of the gametes in a nest or closely circumscribed area.

Internal fertilization, characteristic of the sharks and rays as well as reptiles, birds, and mammals, avoids dispersion of the gametes and protects them. However, even with internal fertilization, vast numbers of sperm must be released by the male into the female tract. Furthermore, the events of ovulation and insemination must be closely synchronized, and the gametes must remain viable for several hours to accomplish fertilization. Sperm may have to travel a considerable distance to reach the egg in the female genital tract, many parts of which may be hostile to sperm. Experiments with rabbits have shown that of the approximately 10 million sperm released into the female vagina, only about 100 reach the site of fertilization.

Oocyte maturation

During oogenesis, described in the preceding chapter, the egg accumulates yolk reserves in preparation for future growth. The nucleus also grows rapidly in size during egg maturation, although not as much as the cell as a whole. It becomes bloated with nuclear sap and so changed in appearance that it was often given another name in older literature, the **germinal vesicle.**

Large amounts of both DNA and RNA accumulate during oogenesis. Early in the oogenesis of large amphibian eggs, the chromosomes become vastly expanded by thin loops thrown out laterally from the chromosomal axis; because of their fuzzy appearance, the chromosomes at this time are called **lampbrush chromosomes** (Fig. 37-14, p. 866). It is believed that ribosomal RNA is being rapidly synthesized on DNA templates as each chromomere puffs out in lampbrush loops, exposing the double helix of DNA. Messenger RNA is also undergoing intense synthesis by the nucleolus during oocyte growth. The messenger RNA subsequently controls the synthesis of protein in the oocyte. In addition, a considerable quantity of DNA appears in the cytoplasm where it becomes bound to mitochondria and yolk platelets. Its function is unknown. Toward the end of oocyte maturation, this intense nucleic acid and protein synthesis gradually winds down. The lampbrush chromosomes contract and migrate to just beneath the egg surface, in preparation for the maturation divisions. The ribosomal RNA and proteins mix with the egg cytoplasm.

The oocyte is now poised for the maturation (meiotic) divisions, which rid it of excess chromosomal material (in the polar bodies, described in Chapter 35) (p. 809) and ready it for fertilization. A vast amount of synthetic activity has preceded this stage, all of which has packed the egg with reserve materials required for subsequent development.

In many vertebrates, the primary oocyte proceeds with the maturation divisions only after it has approached its ultimate size. Even then, however, the divisions typically are interrupted by one or two periods of arrested development before maturation is completed. Human eggs, for example, begin their first meiotic divisions in the ovaries of the prenatal fetus. At birth, all the eggs of the human female infant are developmentally arrested in the diplotene stage of first meiotic prophase. This has been referred to as the "dictyate" stage; the egg is still a primary oocyte, and thus it remains until it is ovulated, a delay of perhaps of 12 to 45 years! The first meiotic division is finally completed just before ovulation, and the second meiotic division occurs after the egg is fertilized.

Fertilization and activation

We began this section with the statement that fertilization accomplishes two ends: (1) activation of the quiescent egg to restore metabolic activity and start cleavage, and (2) insertion of a male haploid nucleus into the cytoplasm. However, we have already seen that the eggs of many species can be artificially induced to develop without sperm fertilization and that some species exhibit natural parthenogenesis (such as rotifers, aphids, bees, and wasps [p. 800]). Obviously the paternal genome is not essential for activation, although in the majority of species, artificial parthenogenesis can seldom be taken past the late cleavage or blastula stage. The artificially activated egg may begin what appears to be normal development, but cleavage soon becomes abnormal and further growth ceases.

One reason that induced parthenogenesis is seldom successful is that the sperm normally contributes the **centriole** needed to form a normal mitotic spindle. For some reason the egg centriole disappears during the maturation divisions, or at least ceases to function. If either natural or artificial parthenogenesis is to succeed, the egg centriole must be rendered capable of division. This is why in most artificial parthenogenesis experiments a so-called second factor of biologic origin (such as blood or tissue fraction) is introduced into the egg cytoplasm. Exactly what this second factor contributes and whether it produces a new self-replicating centriole are not known.

Activation is a dramatic event. In sea urchin eggs, the contact of the spermatozoon with the egg surface sets off almost instantaneous changes in the egg cortex. At the point of contact, a **fertilization cone** appears into which the sperm head is later drawn. From this point, a visible change travels wavelike across the egg surface, causing immediate elevation of a **fertilization membrane** (Fig. 36-2). At the same time **cortical granules,** which form a layer beneath the plasma membrane, explode, releasing materials that fuse together to build up a new egg surface, the **hyaline layer.** This lies between the inner plasma membrane and the outer fertilization membrane.

The **cortical reaction,** as these changes are called, is a crucial event in development. It seems to produce a complete molecular reorganization of the egg cortex.

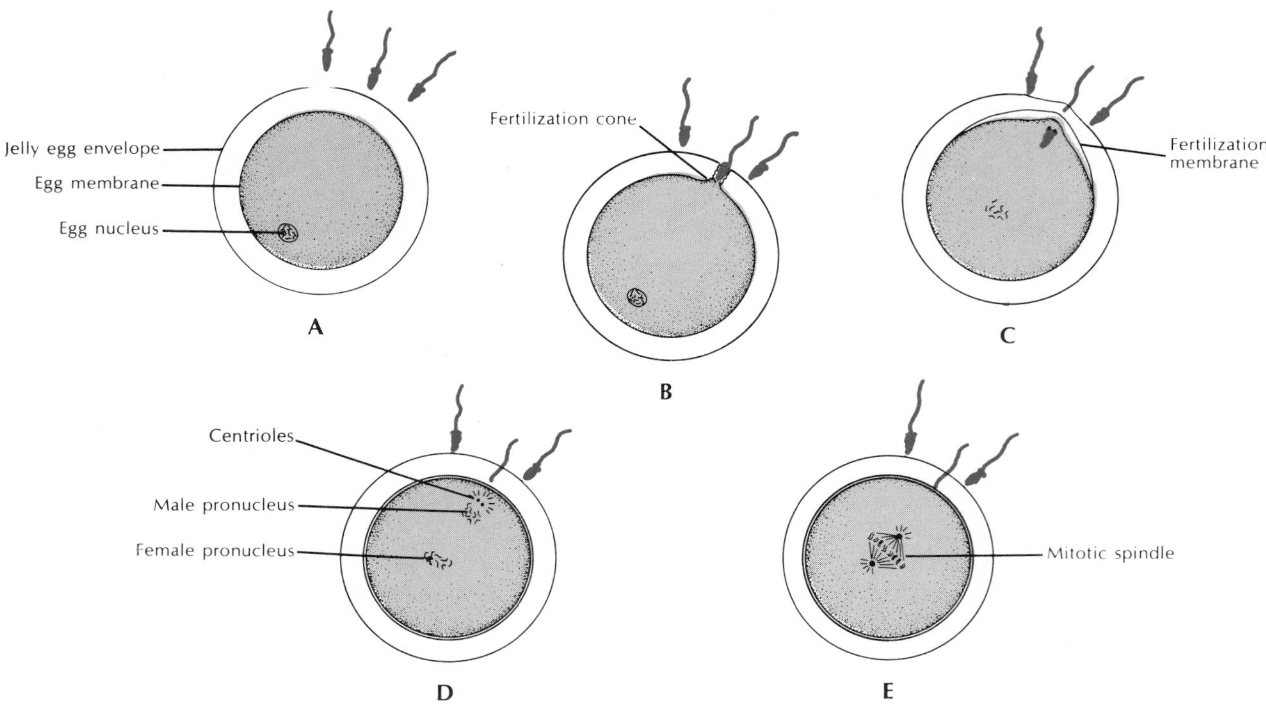

FIG. 36-2 Fertilization of an egg. **A,** Many sperm swim to egg. **B,** The first sperm to penetrate protective jelly envelope and contact egg membrane causes fertilization cone to rise and engulf sperm head. **C,** The fertilization membrane begins to form at the site of penetration and spreads around entire egg, preventing entrance of additional sperm. **D,** Male and female pronuclei approach one another, lose their nuclear membranes, swell, and fuse. **E,** Mitotic spindle forms, signaling creation of a zygote and heralding first cleavage of new embryo.

It also serves to remove one or more inhibitors that have blocked the energy-yielding systems and protein synthesis in the egg and have kept the egg in its quiescent, suspended-animation state. Almost immediately after the cortical reaction, polyribosomes form from the enormous supply of monoribosomes stored in the cytoplasm and begin producing protein. Normal metabolic activity is restored. The reduction divisions, if not completed, are brought to completion. The male and female pronuclei then fuse, and the zygote enters into cleavage.

CLEAVAGE AND EARLY DEVELOPMENT

During cleavage, the zygote divides repeatedly to convert the large, unwieldy cytoplasmic mass into a large number of small, maneuverable cells (called **blastomeres**) clustered together like a mass of soap bubbles. There is no growth during this period, only subdivision of mass, which continues until normal cell size and nucleocytoplasmic ratios are attained. At the end of cleavage the zygote has been divided into many hundreds or thousands of cells (about 1,000 in polychaete worms, 9,000 in amphioxus, and 700,000 in frogs). There is a rapid increase in DNA content during cleavage as the number of nuclei and the amount of DNA are doubled with each division. Apart from this, there is little change in chemical composition or displacement of constituent parts of the egg cytoplasm during cleavage. **Polarity,** that is, a polar axis, is present in the egg, and this establishes the direction of cleavage and subsequent differentiation of the embryo. Usually cleavage is very regular although

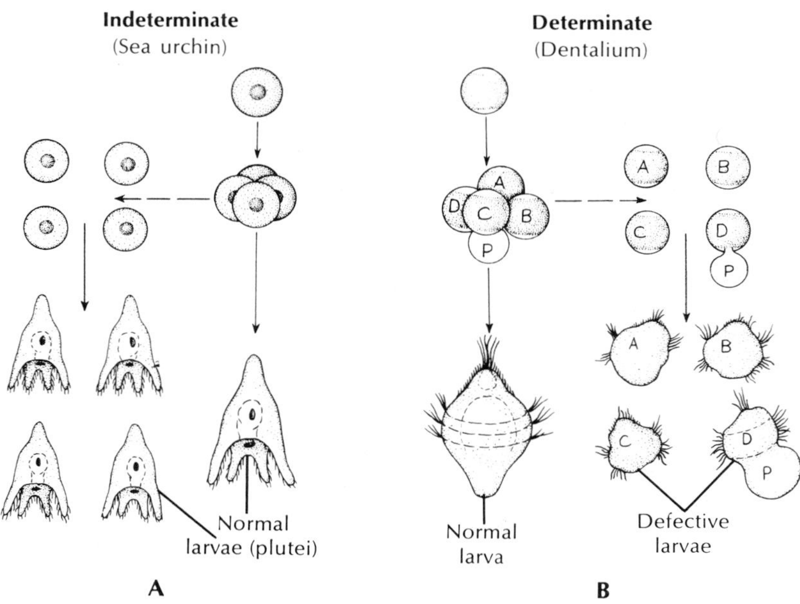

FIG. 36-3 Indeterminate and determinate (mosaic) cleavage. **A,** Indeterminate cleavage. Each of the early blastomeres (such as that of the sea urchin) when separated from the others will develop into a small pluteus larva. **B,** Determinate (mosaic) cleavage. In the mollusc (such as *Dentalium*), when the blastomeres are separated, each will give rise to only a part of an embryo. The larger size of one of the defective larvae is the result of the formation of a polar lobe *(P)* composed of a clear cytoplasm of the vegetal pole, which this blastomere alone receives.

enormously affected by the quantity of yolk present and whether cleavage is radial or spiral, as described earlier (p. 104).

Determinate and indeterminate cleavage

In many species, especially those showing radial cleavage (Deuterostomia), early blastomeres, if separated from each other, will each give rise to a whole larva. This is called **indeterminate cleavage,** meaning that the first blastomeres to form are equipotent; there has been no segregation into potentially different histogenetic regions. Thus if a sea urchin embryo at the 4-cell stage is placed in calcium-free seawater and gently shaken, the blastomeres will fall apart. Replaced in normal seawater, each will subsequently develop into a complete, though small, pluteus larva fully capable of growing into an adult sea urchin (Fig. 36-3). Early blastomeres of frog and rabbit embryos also can be separated and can yield complete embryos. Indeterminate cleavage is also called **regulative.** The reason is that each of the first four or eight blastomeres is capable of regulating its developmental fate to produce a portion of a larva, if it develops in the company of other blastomeres, or a whole larva, if it is forced to develop alone.

The eggs of many invertebrates having spiral cleavage lack this early versatility. If the early blastomeres of a mollusc, annelid, flatworm, or ascidian are separated, each will give rise only to a part of an embryo in accord with their original fate. This is called **determinate** (or mosaic) **cleavage.** Such blastomeres appear to have a fixed informational content as soon as cleavage begins (Fig. 36-3). The terms ''determinate'' and ''indeterminate'' can be used only in a provisional sense, since there are many eggs that do not fit clearly into either category.

The explanation behind these two types of cleavage seems to lie in the extent to which early blastomeres depend exclusively on information segregated by the early divisions (determinate cleavage) or whether they

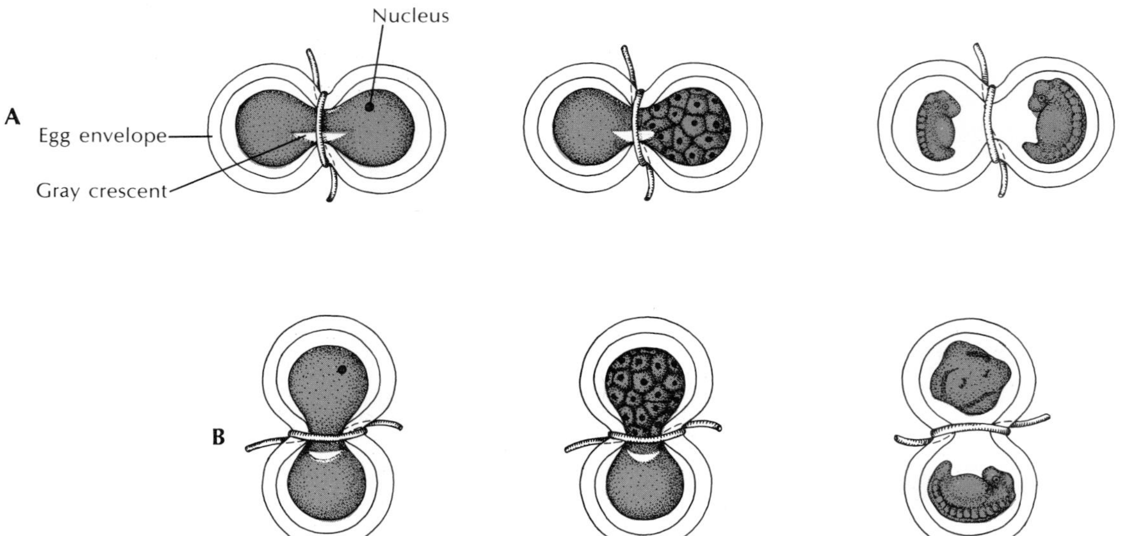

FIG. 36-4 Spemann's delayed nucleation experiments. Two kinds of experiments were performed. **A,** Hair ligature was used to partly divide an uncleaved fertilized newt egg. Both sides contained part of the gray crescent. Nucleated side alone cleaved until a descendant nucleus crossed over the cytoplasmic bridge. Then both sides completed cleavage and formed two complete embryos. **B,** Hair ligature was placed so that the nucleus and gray crescent were completely separated. Side lacking the gray crescent became an unorganized piece of belly tissue; other side developed normally.

are able to fall back on a supply of information from the nucleus (indeterminate cleavage). In the determinate cleavage type, the information-containing material is localized so early that each blastomere receives a portion that is qualitatively unique. With such mosaic (determinate) eggs it is often possible to map out the fates of specific areas on the egg surface that are known to be presumptive for specific structures, such as germ layers, notochord, and nervous system.

From the evidence of identical twins, humans apparently have the indeterminative type of cleavage, for identical twins come from the same zygote. Of course, the totipotency (the capacity to develop into a complete embryo) of the blastomeres in indeterminate cleavage is strictly limited. After the first three or four cleavages, the fate of the blastomeres becomes fixed.

Significance of the cortex

Early cleavage proceeds independently of nuclear genetic information, guided instead by information deposited in the egg during its maturation. It was once believed that the visible particulate material in the cytoplasm had determinative properties. However, it was soon discovered that if the egg is strongly centrifuged so that everything inside—nucleus, mitochondria, lipid droplets, yolk, and other inclusions—is thoroughly displaced, the embryo still develops perfectly. If sea urchin eggs are examined by electron microscope after being centrifuged for 5 minutes at several thousand times the force of gravity, the only thing not affected is the plasma membrane and a gel-like layer just beneath the plasma membrane **(plasmagel layer).** Yet development proceeds normally.

This and similar experiments show conclusively that the plasmagel (or cortical) layer of the egg contains an invisible but dynamic organization that determines the pattern of cleavage. Cortical organization is at first labile (especially so in indeterminate eggs) but soon becomes regionally fixed and irreversible. Thus, as cleavage progresses, the cortex becomes segregated into territories having specific determinative properties. This explains why different blastomeres bear different cytodifferentiation properties.

Delayed nucleation experiments

Another kind of experiment that demonstrates the importance of specific cortical regions of the egg was first carried out many years ago by Hans Spemann, a German embryologist. Spemann put ligatures of human hair around newt eggs (amphibian eggs similar to frog eggs) just as they were about to divide, constricting them until they were almost, but not quite, separated into two halves (Fig. 36-4). The nucleus lay in one-half of the partially divided egg; the other side was anucleate, containing only cytoplasm. The egg then completed its first cleavage division on the side containing the nucleus; the anucleate side remained undivided. Eventually, when the nucleated side had divided into about 16 cells, one of the cleavage nuclei would wander across the narrow cytoplasmic bridge to the anucleate side. Immediately this side began to divide.

With both halves of the embryo containing nuclei, Spemann drew the ligature tight, separating the two halves of the embryo. He then watched their development. Usually two complete embryos resulted. Although the one embryo possessed only one-sixteenth the original nuclear material, and the other contained fifteen-sixteenths, they both developed normally. The one-sixteenth embryo was initially smaller, but caught up by about 140 days. This proves that every nucleus of the 16-cell embryo contains a complete set of genes; all are equivalent.

Sometimes, however, Spemann observed that the nucleated half of the embryo developed only into an abnormal ball of "belly" tissue, although the half that received the delayed nucleus developed normally. Why should the more generously endowed fifteen-sixteenths embryo fail to develop and the small one-sixteenth embryo live? The explanation, Spemann discovered, depended on the position of the **gray crescent,** a crescent-shaped, pigment-free area on the egg surface. In amphibian eggs the gray crescent forms at the moment of fertilization and determines the plane of bilateral symmetry of the future animal. If one-half of the constricted embryo lacked any part of the gray crescent, it would not develop.

Obviously, then, there must be cytoplasmic inequalities involved. The egg cortex in the area of the gray crescent contains substances that are essential for normal development. Since all the nuclei of the 16-cell embryo are equivalent, each capable of supporting full development, it is clear that the cytoplasmic environment is crucial to nuclear expression. The nuclei are all alike, but the cytoplasm (or cortex) throughout the embryo is not all alike. In some way chemically different regions of the egg, created during the early growth of the egg (oogenesis), are segregated out into specific cells during early cleavage. Thus, although all nuclei have the same information content, cytoplasmic substances surrounding the nucleus determine what part of the genome will be expressed and when (see Principle 24, p. 15).

PROBLEM OF DIFFERENTIATION

The egg cortex contains the primary guiding force for early development (cleavage and blastula formation), and synthetic activities during this period are supported by nucleic acids and reserves laid down in the egg cytoplasm *before* fertilization. However, continued differentiation past cleavage or the blastula stage requires the action of genetic information in the nuclear chromosomes. There is considerable evidence to show that this is so. If the maternal nucleus is removed from an activated but unfertilized frog egg, the anucleate egg will cleave more or less normally but arrest at the outset of gastrulation, when nuclear information is required.

Hybridization experiments have also been carried out in which the maternal nucleus is removed from the egg by microsurgery, just after the egg has been fertilized, but before the pronuclei have time to fuse. Such a zygote, called an **andromerogone,** contains *maternal* cytoplasm and a *paternal* (sperm) nucleus. The egg will cleave to the blastula stage normally; whether it will gastrulate depends on the compatibility of the sperm with the maternal cytoplasm. If the egg was fertilized with a sperm of the same species, it will develop into a haploid (but otherwise normal) tadpole. If it was fertilized with a sperm of a closely related species, it may develop to the neurula stage before development stops. If fertilized with a distantly related species, development will arrest in the blastula or early gastrula stage. These differences are explained as varying degrees of immunologic incompatibilities that appear when the paternal (sperm) genes begin coding for proteins in the embryo. The more foreign the sperm, the greater the incompatibility and the earlier the stage of arrest.

Finally, the most direct evidence that nuclear genes become active at or shortly before gastrulation is that the production of messenger RNA increases sharply at this time. This can be demonstrated with the use of radioactively labeled RNA precursors.

Gene expression

Gastrulation is a critical time of orderly and integrated cell movements. By folding processes (invagination and evagination) that segregate single epithelial layers and by splitting processes (delamination) that segregate multiple layers, the three prospective germ layers—ectoderm, mesoderm, and endoderm—are formed. This is followed by the rapid differentiation of germ layers into rudimentary, and later functional, tissues and organs.

As cells differentiate, they obviously use only a part of the instructions their nuclei contain. Cells that are differentiating into a thyroid gland are not concerned with that part of the genome that codes for a striated muscle, for example. The unneeded genes are in some way switched off. Are the unused genes destroyed? Recent **nuclear transplantation** experiments strongly indicate that they are not. In 1952, R. Briggs and T. J. King in the United States developed a technique for surgically extracting the nucleus from an unfertilized frog egg, or inactivating it with ultraviolet light. With a micropipette they introduced a substitute nucleus taken from the cell of a frog embryo. Employing this technique, J. B. Gurdon and R. Laskey in England showed that if the substitute nucleus were taken from a blastula cell, it supported development of the egg to the tadpole stage in 80% of the transplants. Nuclei from older embryos did not support development as well, but even the nucleus from a fully differentiated intestinal or skin cell from a tadpole supported normal development in *some* cases. These experiments are of great significance because they demonstrate that the nucleus of a differentiated cell can be forced backward from its specialized state and once again make available all its genetic information.

The basic problem is **gene expression.** What determines that a particular blastomere of, say, a 100-cell embryo, will differentiate into muscle or skin or thyroid gland? Presumably the set of genes responsible for the development of a thyroid gland will be set in motion by the chemical environment found *only* in the region of the future thyroid gland. But how can such a unique chemical environment be created unless some *previous* genic action made the thyroid region different from the rest of the body? And even this earlier gene action must have been expressed in a unique chemical environment, or else thyroid glands would grow all over the body.

This kind of argument quickly takes one back to the fertilized egg itself. If genes are the same in all nuclei of the early embryo, then the only way differences can develop is through some interaction between these nuclei and the surrounding cytoplasm. We have already seen that the basic polarity of the egg and the organizing qualities of the egg cortex provide an early opportunity for such interactions.

Attention is now turned to the nuclear DNA, from which genes are made. It is clear that not all genes are active all the time. In fact, for any cell at any given time, only a very small part of the genetic information present is being used.

Let us briefly summarize what is presently known of the transmission of genetic information. Genetic information is coded in the sequence of nucleotides in DNA molecules (see Principle 18, p. 13). DNA serves as a template for the synthesis of messenger RNA in the nucleus. Messenger RNA then migrates out through nuclear pores into the cytoplasm, where it attaches to a ribosome. Here the messenger RNA serves as a template for the synthesis of specific proteins (see Principle 19, p. 13). In this way cytoplasmic proteins are formed that may be specific for that cell.

Evidence to date suggests that at the beginning of development most nuclear genes are inactive. Only small amounts of messenger RNA are being produced on the DNA templates. As development proceeds, new cytoplasmic proteins appear, indicating that more genic DNA is producing messenger RNA. Evidently different genes are activated (or "derepressed") in different parts of the young embryo, and this differential gene activity is responsible for embryonic differentiation.

What is the mechanism by which genes are repressed and then derepressed at specific times during development? We presently do not know. Whatever the mechanism is, it seems certain that the kinds of cytoplasm present in different cells determine what genes come into action. Nucleocytoplasmic interactions form the basis of the organized differentiation of tissues that characterizes animal development.

Embryonic inductors and organizers

Embryonic induction is a widespread phenomenon in development. It is the capacity of one tissue to evoke a specific developmental response in another. The classic experiments were reported in 1924 by H. Spemann and O. Mangold. When a piece of dorsal blastopore lip from a salamander gastrula is transplanted into a ventral or lateral position of another salamander gastrula, it invaginates and develops a notochord and

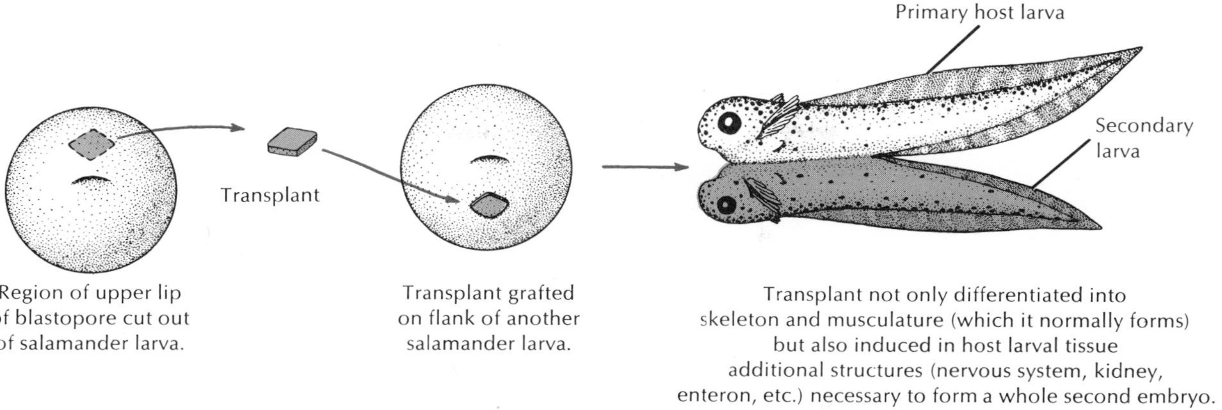

FIG. 36-5 Spemann's famous organizer experiment.

muscle somites. It also induces the *host* ectoderm to form a neural tube. Eventually a whole system of organs develops where the graft was placed, and this grows into a nearly complete secondary embryo (Fig. 36-5). This creature is composed partly of grafted tissue and partly of induced host tissue.

It was soon found that *only* grafts from the dorsal lip of the blastopore were capable of inducing the formation of a complete or nearly complete secondary embryo. This area corresponds to the presumptive areas of notochord, muscle somites, and prechordal plate. It was also found that only ectoderm of the host could be induced to develop a nervous system in the graft and that the reactive ability is greatest at the early gastrula stage and declines as the recipient embryo gets older.

Spemann called the dorsal lip area the **primary organizer** because it was the only region capable of inducing the development of a complete embryo in the host. This region can be traced back to the gray crescent of the undivided fertilized egg, a cortical area that, as we have already seen, is crucial in directing early development of amphibian embryos (Fig. 36-4). Many other examples of embryonic induction have been discovered, both in amphibians, a favorite for study, as well as in other vertebrate and invertebrate species.

Efforts to discover the chemical nature of the inductor have met with much less success. Embryologists were dismayed to find that a great variety of denatured animal materials could cause inductive responses. Obviously the dorsal lip graft did not "organize" a particular differentiation; rather it evoked a response that was already part of the induced tissue's total developmental capacity and repertoire. The inductor appears to play a permissive rather than an instructive role by providing a favorable environment in which the stimulated tissue releases a preprogrammed response.

In normal development, inductors operate by cell-to-cell contact. Since one tissue must actually touch another, inductors, unlike hormones, are contact chemical signals. Usually a tissue that has differentiated acts as an inductor for an adjacent undifferentiated tissue. Timing is important. Once a primary inductor sets in motion a specific developmental pattern in one tissue, numerous secondary inductions follow. What emerges is a sequential pattern of development involving not only inductors but cell movement, changes in adhesive properties of cells, and cell proliferation. This is why developmental patterns are so difficult to unravel. There is no master control panel that directs development, but rather a sequence of local patterns in which one step in development is a subunit of another.

The chemical nature of inductors is still unknown. Despite the enormous research effort that has been devoted to induction, the phenomenon has remained so resistant to experimental unveiling that embryologists find themselves with little more insight into the question than they had half a century ago. Nevertheless, in showing that one step in development is a necessary requisite for the next, Hans Spemann's

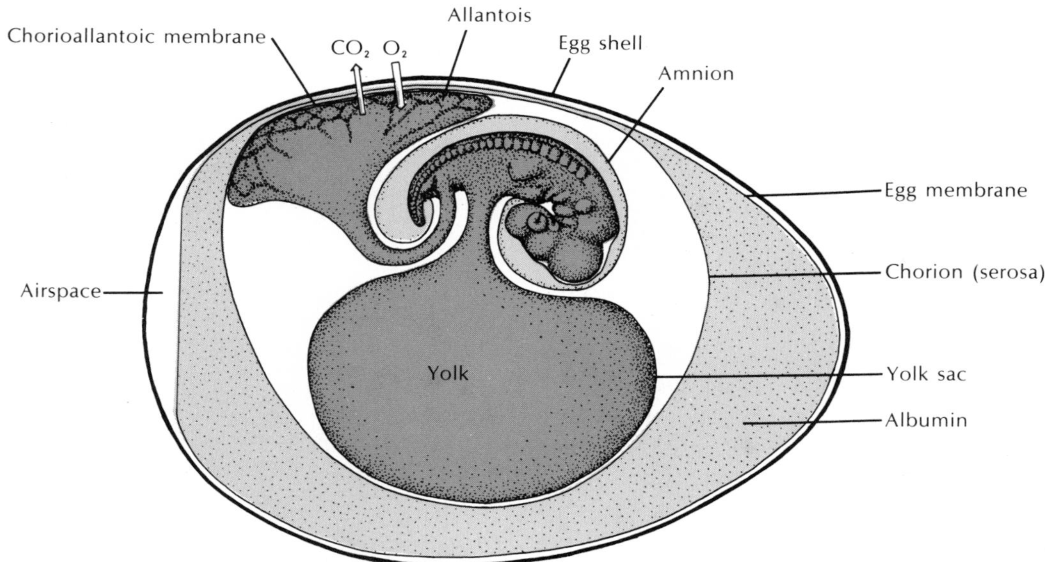

FIG. 36-6 Amniotic egg at early stage of development showing a chick embryo and its extraembryonic membranes. Porous shell allows gaseous exchange of oxygen and carbon dioxide. Circulatory channels from embryo's body to allantois and yolk sac are not shown.

induction experiments were among the most significant events in experimental embryology.

AMNIOTES AND THE AMNIOTIC EGG

Reptiles, birds, and mammals form a natural grouping of vertebrates distinguished by having an amniotic egg. They are called **amniotes,** meaning that they develop an amnion, one of the extraembryonic membranes that make the development of these forms unique among animals.

As rapidly growing living organisms, embryos have the same basic animal requirements as adults—food, oxygen, and disposal of wastes. For the embryos of marine invertebrates that show indirect development, gas exchange is a simple matter of direct diffusion. Food can be acquired as soon as the embryo develops a mouth and begins feeding on plankton. All eggs of aquatic animals are provided with just enough stored yolk to allow growth to this critical stage. Beyond this point, the embryo (now called a free-swimming **larva**) is on its own.

Yolk enclosed in a membranous **yolk sac** is a conspicuous feature of all fish embryos (Fig. 36-1). The yolk is gradually used up as the embryo grows; the yolk sac shrinks and finally is enclosed within the body of the embryo. The mass of yolk is an **extraembryonic** structure, since it is not really a part of the embryo proper, and the yolk sac is an **extraembryonic membrane.** Bird and reptile eggs are also provided with large amounts of yolk to support early development. In birds (direct development), the yolk reaches relatively massive proportions, since it must nourish a baby bird to a much more advanced stage of growth at hatching than a larval fish.

In abandoning an aquatic life for a land existence the first terrestrial animals had to evolve a sophisticated egg containing a complete life-support system. Thus appeared the **amniotic egg,** equipped to protect and support the growth of embryos on dry land. In addition to the **yolk sac** containing the nourishing yolk are three other membranous sacs—**amnion, chorion,** and **allantois.** All are referred to as extraembryonic membranes because, again, they are accessory structures that develop beyond the embryonic body and are discarded when the embryo hatches. (We should note that since the amnion is only one of four extraembryonic membranes, the ''amniotic'' egg could just as well be called a ''chorionic'' or an ''allantoic'' egg.)

The **amnion** is a fluid-filled bag that encloses the embryo and provides a private aquarium for development (Fig. 36-6). Floating freely in this aquatic environment, the embryo is fully protected from shocks and adhesions. The evolution of this structure, from which the amniotic egg takes its name, was crucial to the successful habitation of land.

The **allantois,** another component in the support system for embryos of land animals, is a bag that grows out of the hindgut of the embryo (Fig. 36-6). It collects the wastes of metabolism (mostly uric acid). At hatching, the young animal breaks its connection with the allantois and leaves it and its refuse behind in the shell.

The **chorion** (also called **serosa**) is an outermost extraembryonic membrane that completely encloses the rest of the embryonic system. It lies just beneath the shell (Fig. 36-6). As the embryo grows and its need for oxygen increases, the allantois and chorion fuse to form a **chorioallantoic membrane.** This double membrane is provided with a rich vascular network, connected to the embryonic circulation. Lying just beneath the porous shell, the vascular chorioallantoic membrane serves as a provisional lung across which oxygen and carbon dioxide can freely exchange. And although nature did not plan for it, the chorioallantoic membrane of the chicken egg has been used extensively by generations of experimental embryologists as a place to culture chick and mammalian tissues.

The great importance of the amniotic egg to the establishment of a land existence cannot be overemphasized. Amphibians must return to water to lay their eggs. But the reptiles, even before they took to land, developed the amniotic egg with its self-contained aquatic environment enclosed by a tough outer shell. Protected from drying out and provided with yolk for nourishment, such eggs could be laid on dry land, far from water. Reptiles were thus freed from aquatic life and could become the first true terrestrial tetrapods.

Incidentally, the sexual act itself comes from the requirement that the egg be fertilized *before* the egg shell is wrapped around it, if it is to develop. Thus the male must introduce the sperm into the female tract so that the sperm can reach the egg before it passes to that part of the oviduct where the shell is secreted. Hence, as one biologist puts it, it is the egg shell, and not the devil, that deserves the blame for the happy event we know as sex.

HUMAN DEVELOPMENT

The amniotic egg, for all its virtues, has one basic flaw: placed neatly in a nest, it makes fine food for other animals. It was left for the mammals to evolve the best solution for early development: allow the embryo to grow within the protective confines of the mother's body. This has resulted in important modifications in the development of mammals as compared with other vertebrates. The earliest mammals, descended from early reptiles, were egg layers. Even today the most primitive mammals, the monotremes (for example, duckbill platypus, spiny anteater), lay large yolky eggs that closely resemble bird eggs. In the marsupials (pouched mammals such as the opossum and kangaroo), the embryos develop for a time within the mother's uterus. But the embryo does not "take root" in the uterine wall, as do the embryos of the more advanced **placental mammals,** and consequently it receives little nourishment from the mother. The young of marsupials are therefore born immature and are sheltered and nourished in a pouch of the abdominal wall.

All other mammals, the placentalians, nourish their young in the uterus by means of a **placenta.**

Early development of the human embryo

The eggs of all placental mammals, though relatively enormous on the cellular scale of things, are small by egg standards. The human egg is about 0.1 mm in diameter, barely visible to the unaided eye. It contains very little yolk. After fertilization in the mouth (ampulla) of the oviduct, the cleaving egg begins a leisurely 5-day journey down the oviduct toward the uterus, propelled by a combination of ciliary action (especially in the ampullary region) and muscular peristalsis. Cleavage is very slow: 24 hours for the first cleavage and 10 to 12 hours for each subsequent cleavage. By comparison, frog eggs cleave once every hour. Cleavage produces a small ball of 20 to 30 cells (called the morula) within which a fluid-filled cavity appears, the **segmentation cavity** (Fig. 36-7). This is comparable to the blastocoel of a frog's egg. The embryo is now called a **blastocyst.**

At this point, development of the mammalian embryo departs radically from that of lower vertebrates. A mass of cells, called the **inner cell mass,** develops on one side of the peripheral cell, or **trophoblast,** layer (Fig. 36-7, *A*). The inner cell mass will form the embryo, while the surrounding trophoblast will form the

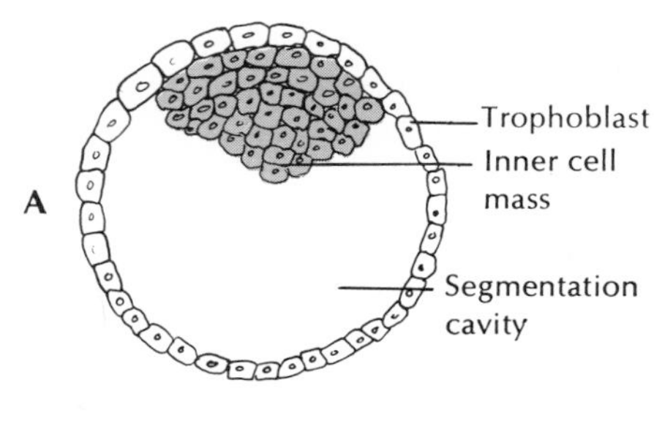

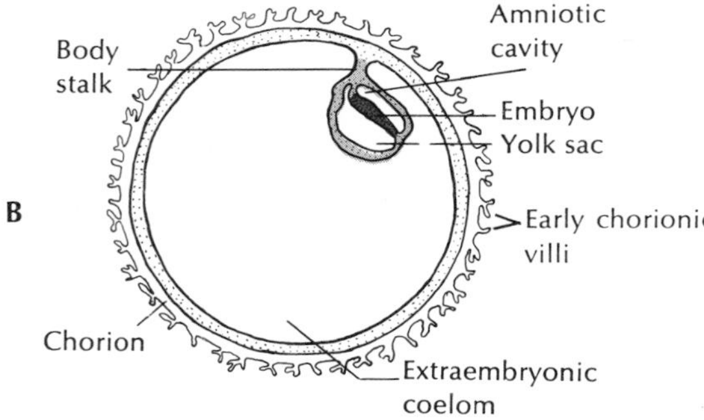

FIG. 36-7 Early development of human embryo. **A,** Blastocyst at about 7 days. Trophoblast gives rise to chorion, which later becomes part of placenta; inner cell mass becomes embryo. **B,** Blastocyst at about 15 days. Note amniotic cavity and yolk sac; primary villi are developing from chorion to begin establishment of placenta.

placenta (Fig. 36-7, *B*). When the blastocyst is about 6 days old and is composed of about 100 cells, it contacts and implants into the uterine endometrium. Very little is known of the forces involved in implantation, or why, incidentally, the intrauterine birth control devices are so effective in preventing successful implantation. On contact, the trophoblast cells proliferate rapidly and produce enzymes that break down the epithelium of the uterine endometrium. This allows the blastocyst to sink into the endometrium. By the eleventh or twelfth day the blastocyst, now totally buried, has eroded through the walls of capillaries and small arterioles; this releases a pool of blood that bathes the embryo. At first, the minute embryo derives what nourishment it requires by direct diffusion from the surrounding blood. But very soon, a remarkable fetal-maternal structure, the **placenta,** develops to assume these exchange tasks.

The placenta

The placenta is a marvel of biologic engineering. Serving as a provisional lung, intestine, and kidney for the embryo, it performs elaborate selective activities without ever allowing the maternal and fetal bloods to intermix. (Very small amounts of fetal blood may regularly escape into the maternal system without causing harm.) The placenta permits the entry of foodstuffs, hormones, vitamins, and oxygen and the exit of carbon dioxide and metabolic wastes. Its action is highly selective, for it allows some materials to enter that are chemically quite similar to others that are rejected.

The two circulations are physically separated at the placenta by an exceedingly thin membrane only 2 μm thick, across which materials are transferred by diffusive interchange. The transfer occurs across thousands of tiny fingerlike projections, called **chorionic villi,** which develop from the original chorion membrane (Figs. 36-7 and 36-8). These projections sink like roots into the uterine endometrium after the embryo implants. As development proceeds and embryonic demands for food and gas exchange increase, the great proliferation of villi in the placenta vastly increases its total surface area. Although the human placenta at term measures only about 18 cm (7 inches) across, its total absorbing surface is about 13 square meters—50 times the surface area of the skin of the newborn infant.

Since the mammalian embryo is protected and nourished by the placenta, what becomes of the various embryonic membranes of the amniotic egg whose functions are no longer required? Surprisingly perhaps, all of these special membranes are still present, although they may be serving new functions. The yolk sac is retained, empty and purposeless, a vestige of our distant past (Fig. 36-8). Perhaps evolution has not had enough time to discard it. The amnion remains unchanged, a protective water jacket in which the embryo weightlessly floats. The remaining two extraembryonic membranes, the allantois and chorion, have been totally redesigned. The allantois is no longer needed as a urinary bladder. Instead it becomes the stalk, or **umbilical cord,** that links the embryo physically and functionally with the placenta. The chorion, the outermost membrane, forms most of the placenta itself.

One of the most intriguing questions the placenta presents is why it is not rejected by the mother's tissues. The placenta is a uniquely successful foreign transplant, or **allograft.** Since the placenta is an embryonic structure, containing both paternal and maternal antigens, we should expect it to be rejected by

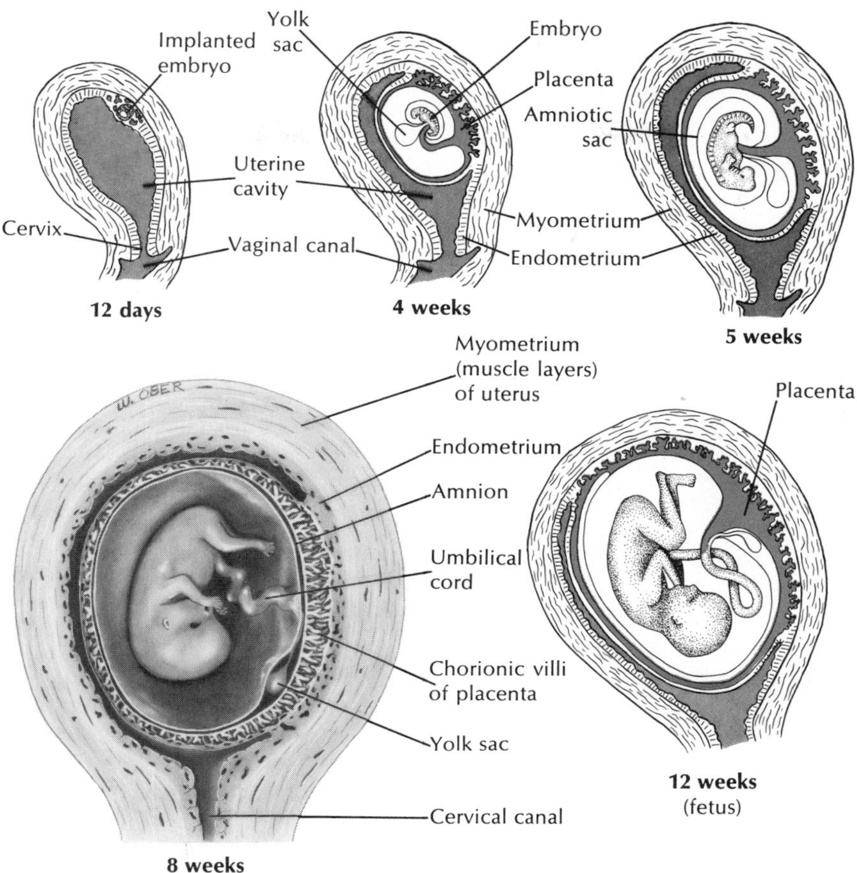

FIG. 36-8 Development of human embryo from 12 days to 12 weeks. The 8-week-old embryo and uterus are drawn to life size. As the embryo grows, placenta develops into a disclike structure attached to one side of the uterus. Note vestigial yolk sac.

the uterine tissues, just as a piece of a child's skin will be rejected by the child's mother should a surgeon attempt a grafting transplant. The placenta in some way circumvents the normal rejection phenomenon, a matter of the greatest interest to immunologists seeking ways to transplant tissues and organs successfully.

Development of systems and organs

Pregnancy may be divided into four phases: the first phase of 6 or 7 days is the period of cleavage and blastocyst formation and ends when the blastocyst implants in the uterus. The second phase of about 2 weeks is the period of gastrulation and formation of the neural plate. The third phase, called the **embryonic period,** is a crucial and sensitive period of primary organ system differentiation. This phase ends at about the eighth week of pregnancy. The last phase, known as the **fetal period,** is characterized by rapid growth, proportional changes in body parts, and final preparation for birth.

During gastrulation the three germ layers are formed. These differentiate, as we have seen, first into primordial cell masses and then into specific organs and tissues. During this process, cells become increasingly committed to specific directions of differentiation.

Derivatives of ectoderm

The brain, spinal cord, and nearly all the outer epithelial structures of the body develop from the primi-

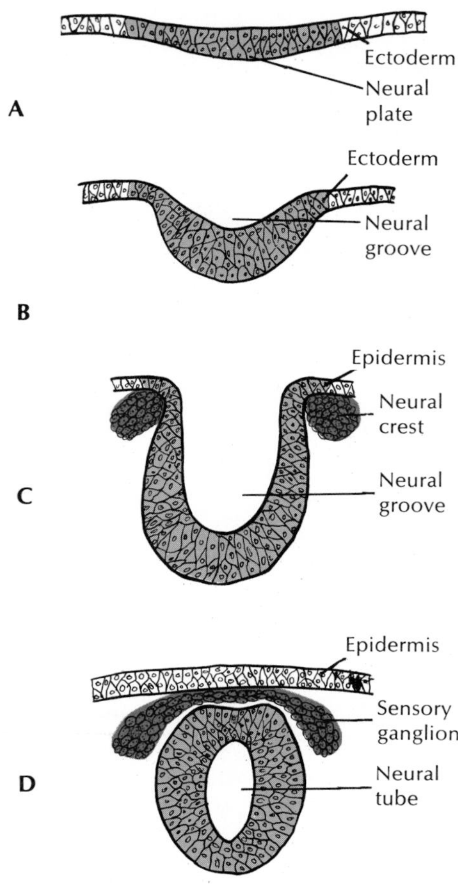

FIG. 36-9 Development of neural tube and neural crest from neural plate ectoderm (cross section).

tive ectoderm. They are among the earliest organs to appear. Just above the notochord, the **ectoderm** thickens to form a **neural plate** (Fig. 36-9). The edges of this plate rise up, fold, and join together at the top to create an elongated, hollow **neural tube.** The neural tube gives rise to most of the nervous system: anteriorly it enlarges and differentiates into the brain, cranial nerves, and eyes; posteriorly it forms the spinal cord and spinal motor nerves. Sensory nerves arise from special **neural crest** cells pinched off from the neural tube before it closes.

How are the billions of nerve axons in the body formed? What directs their growth? Biologists were intrigued with these questions that seemed to have no easy solutions. Since a single nerve axon may be more than a meter in length (for example, motor nerves running from the spinal cord to the feet), it seemed impossible that a single cell could spin out so far. It was suggested that nerve fibers grew from a series of preformed protoplasmic bridges along its route. The answer had to await the development of one of the most powerful tools available to biologists, the cell-culture technique. In 1907 embryologist Ross G. Harrison discovered that he could culture living neuroblasts (embryonic nerve cells) for weeks outside the body by placing them in a drop of frog lymph hung from the underside of a cover slip. Watching nerves grow for periods of days, he saw that each nerve fiber was the outgrowth of a single cell. As the fibers extended outward, materials for growth flowed down the axon center to the growing tip, where they were incorporated into new protoplasm.

The tissue-culture technique is now used extensively by scientists in all fields of active biomedical research, not just by embryologists. The great impact of the technique has been felt only in recent years. Harrison was twice considered for the Nobel Prize (1917 and 1933), but he failed ever to receive the award because, ironically, the tissue culture method was then believed to be "of rather limited value."

The second question—what directs nerve growth—has taken longer to unravel. An idea held well into the 1940s was that nerve growth is a random, diffuse process. It was believed that the nervous system developed as an equipotential network, or blank slate, that later would be shaped by usage into a functional system. The nervous system just seemed too incredibly complex for one to imagine that nerve fibers could find their way selectively to predetermined destinations. Yet it appears that this is exactly what they do! Recent work indicates that each of the billions of nerve cell axons acquires a chemical identification tag that in some way directs it along a correct path. Many years ago Ross Harrison observed that a growing nerve axon terminates in a "growth cone," from which extend numerous tiny threadlike processes (Fig. 36-10). These are constantly reaching out, testing the environment in all directions, to guide the nerve chemically to its proper destination. This chemical guidepost system, which must, of course, be genetically directed, is just one example of the amazing precision that characterizes the entire process of differentiation.

Derivatives of endoderm

In the frog embryo the primitive gut makes its appearance during gastrulation with the formation of an

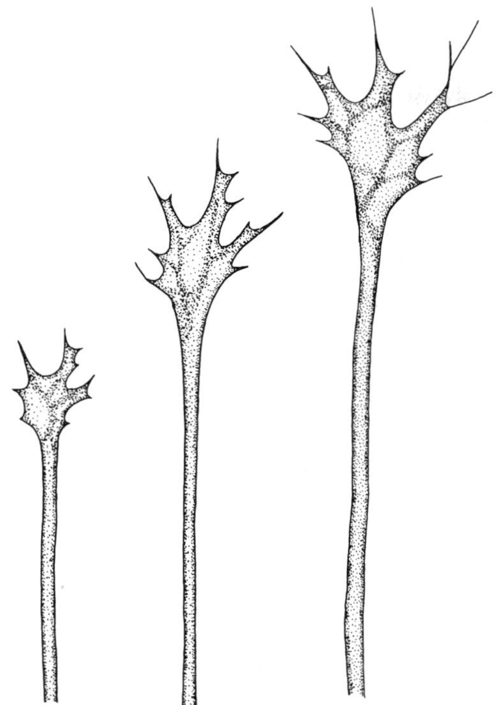

FIG. 36-10 Growth cone at the growing tip of a nerve axon. Materials for growth flow up axon to growth cone from which numerous threadlike pseudopodial processes extend. These appear to serve as a pioneering guidance system for the developing axon.

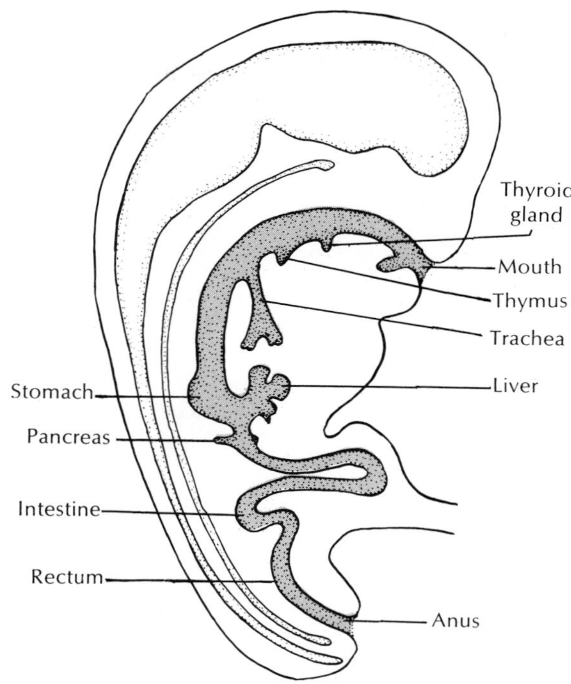

FIG. 36-11 Derivatives of alimentary canal.

internal cavity, the **archenteron.** From this simple endodermal cavity develop the lining of the digestive tract, lining of the pharynx and lungs, most of the liver and pancreas, the thyroid and parathyroid glands, and the thymus.

The **alimentary canal** is early folded off from the yolk sac by the growth and folding of the body wall (Fig. 36-11). The ends of the tube open to the exterior and are lined with ectoderm, whereas the rest of the tube is lined with endoderm. The **lungs, liver,** and **pancreas** arise from the foregut.

Among the most intriguing derivatives of the digestive tract are the pharyngeal (gill) arches and pouches, which make their appearance in the early embryonic stages of all vertebrates (Fig. 36-12). In fishes, the gill arches develop into gills and supportive structures and serve as respiratory organs. When the early vertebrates moved onto land, gills were unsuitable for aerial respiration and were replaced by lungs.

Why then do gill arches persist in the embryos of terrestrial vertebrates? Certainly not for the convenience of biologists who use these and other embryonic structures to reconstruct lines of vertebrate descent (see p. 887 and Principle 7, p. 11). Even though the gill arches serve no respiratory function in either the embryos or adults of terrestrial vertebrates, they remain as necessary primordia for a great variety of other structures. For example, the first arch and its endoderm-lined pouch (the space between adjacent arches) form the upper and lower jaws and inner ear of higher vertebrates. The second, third, and fourth gill pouches contribute to the tonsils, parathyroid gland, and thymus. We can understand then why gill arches and other fishlike structures appear in early mammalian embryos. Their original function has been abandoned, but the structures are retained for new purposes. It is the great conservatism of early embryonic development that has so conveniently provided us with a telescoped evolutionary history.

Derivatives of mesoderm

The intermediate germ layer, the mesoderm, forms the vertebrate skeletal, muscular, and circulatory struc-

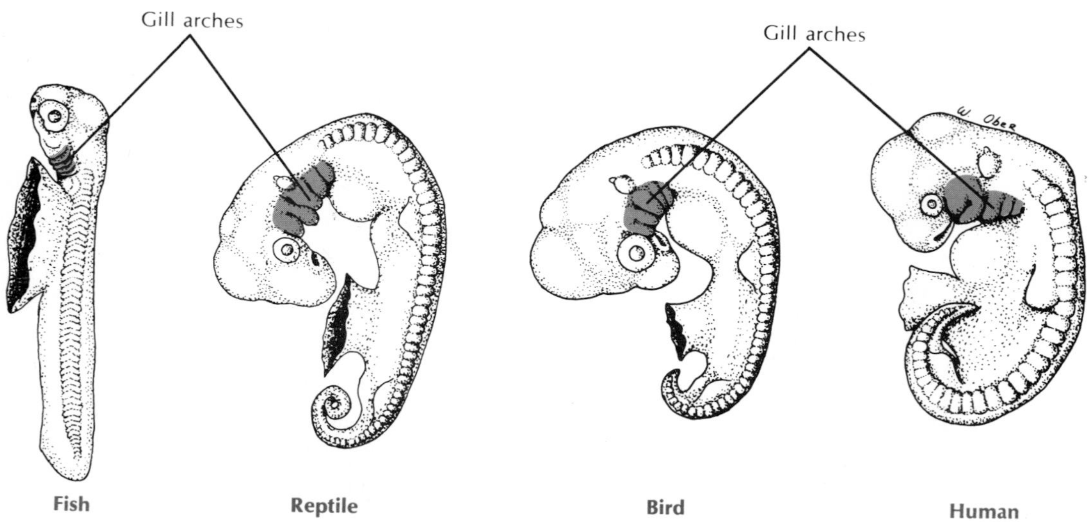

FIG. 36-12 Comparison of gill arches of different vertebrate embryos at equivalent stages of development.

tures and the kidney. As vertebrates have increased in size and complexity, the mesodermally derived supportive, movement, and transport structures make up an ever greater proportion of the body bulk.

Most **muscles** arise from the mesoderm along each side of the spinal cord (Fig. 36-13). This mesoderm divides into a linear series of somites (38 in humans) that by splitting, fusion, and migration become the muscles of the body and axial parts of the skeleton. The **limbs** begin as buds from the side of the body. Projections of the limb buds develop into digits.

Although the primitive mesoderm appears after the ectoderm and endoderm, it gives rise to the first functional organ, the embryonic **heart.** Guided by the underlying endoderm, clusters of precardiac mesodermal cells move ameba-like into a central position between the underlying primitive gut and the overlying neural tube. Here the heart is established, first as a single, thin tube.

Even while the cells group together, the first twitchings are evident. In the chick embryo—a favorite and nearly ideal animal for experimental embryology studies—the primitive heart begins to beat on the second day of the 21-day incubation period; it begins beating before any true blood vessels have formed and before there is any blood to pump. As the ventricle primordium develops, the spontaneous cellular twitchings become coordinated into a feeble, but rhythmic, beat. Then, as the atrium develops behind the ventricle, followed by the sinus venosus behind the atrium, the heart rate quickens. Each new heart chamber has an intrinsic beat faster than its predecessor. Finally, the **sinoatrial node** develops in the sinus venosus and takes command of the entire heartbeat. This becomes the heart's **pacemaker.** As the heart builds up a strong and efficient beat, vascular channels open within the embryo and across the yolk. Within the vessels are the first primitive blood cells, suspended in plasma. The early development of the heart and circulation is crucial to continued embryonic development because without a circulation the embryo could not obtain materials for growth. Food is absorbed from the yolk and carried to the embryonic body; oxygen is delivered to all the tissues, and carbon dioxide and other wastes are carried away. The embryo is totally dependent on extraembryonic support systems, and the circulation is the vital link between them.

Fetal circulation

The fetal circulation is another engineering marvel. During fetal development the placenta must function for the lungs, since these, of course, are nonfunctional in the fetus, which has no access to air. Thus the blood must be diverted to the placenta from the fetal lungs, and done so in such a way that instantaneous conversion to the adult pattern of circulation is allowed when the newborn infant takes its first breath. There are

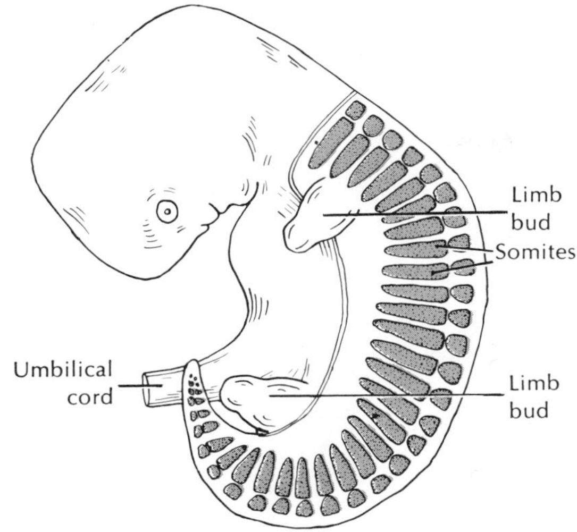

FIG. 36-13 Mammalian embryo showing muscle somites.

several parts of the solution, as explained below and illustrated in Fig. 36-14.

The interchange of nutrients and respiratory gases occurs via the **umbilical cord,** which is composed of one umbilical vein and two umbilical arteries. The umbilical vein, which carries oxygenated blood from the placenta, passes through the umbilical cord and then to the liver, through which the blood passes via a special embryonic vessel, the **ductus venosus.** This vessel joins the inferior vena cava, which enters the right atrium. At this point in the normal adult pattern, the venous blood would be pumped into the right ventricle and from there to the pulmonary artery and lungs. However, the collapsed fetal lungs cannot possibly accept the total volume of blood entering the right side of the heart. A partial solution to this problem is the provision of a large connection, the **ductus arteriosus,** between the pulmonary artery and the aorta into which the blood is diverted. Not all the venous blood follows this route, however, because a second diversionary passage, a hole in the interatrial septum called the **foramen ovale,** allows about half the blood to pass directly from the right to the left atrium. Thus, both ventricles receive approximately equal amounts of blood to pump, as they do in the postnatal condition, even though their usual connection, the pulmonary circuit, is not yet operating. The fetal circuit is completed by two umbilical arteries from the iliac arteries that return blood to the placenta.

At birth, the placental bloodstream is cut off, and the pulmonary circulation opens as the infant draws the first lung-expanding breath. The system is ingeniously prepared for this sudden event. The muscular walls of the ductus arteriosus are particularly sensitive to the blood-oxygen tension. As soon as the blood oxygen rises with the onset of breathing, the ductus arteriosus closes. The other diversionary opening, the foramen ovale, is provided with a flap valve that shuts automatically at birth as the blood pressure in the more powerful left side of the heart rises above that of the right side. In time, all traces of the foramen ovale are obliterated, and the ductus arteriosus and the umbilical vessels become reduced to fibrous cords.

Birth (parturition)

Hormonal changes preceding birth

During pregnancy, the placenta gradually takes over most of the functions of regulating growth and development of the uterus and the fetus. As an endocrine gland, it secretes estradiol and progesterone, hormones that are secreted by the ovaries and corpus luteum in the early periods of pregnancy (p. 812). The placenta also produces **chorionic gonadotropin,** a hormone that now assumes the role of the LH and FSH pituitary hormones. These cease their secretions about the second month of pregnancy. Chorionic gonadotropin maintains the corpus luteum so that the latter may continue secreting progesterone and estradiol necessary for the integrity of the placenta. In the later stages of pregnancy the placenta becomes a totally independent endocrine organ, requiring support from neither the corpus luteum nor the pituitary gland.

What stimulates birth? Why does pregnancy not continue indefinitely? What factors produce the onset of labor (the rhythmic contractions of the uterus)? Thus far we have only tentative answers to these questions. It has long been known that estrogens stimulate uterine contractions, whereas progesterone, also secreted by the placenta, blocks uterine activity. Just before birth there is a very sharp increase in the plasma level of estrogens and a decrease in progesterone. This appears to remove the ''progesterone block'' that keeps the uterus quiescent throughout pregnancy. Thus labor contractions can proceed.

Recently much more information on the control of birth has emerged, involving **prostaglandins,** a group of hormonelike long-chain fatty acids. Prostaglandins

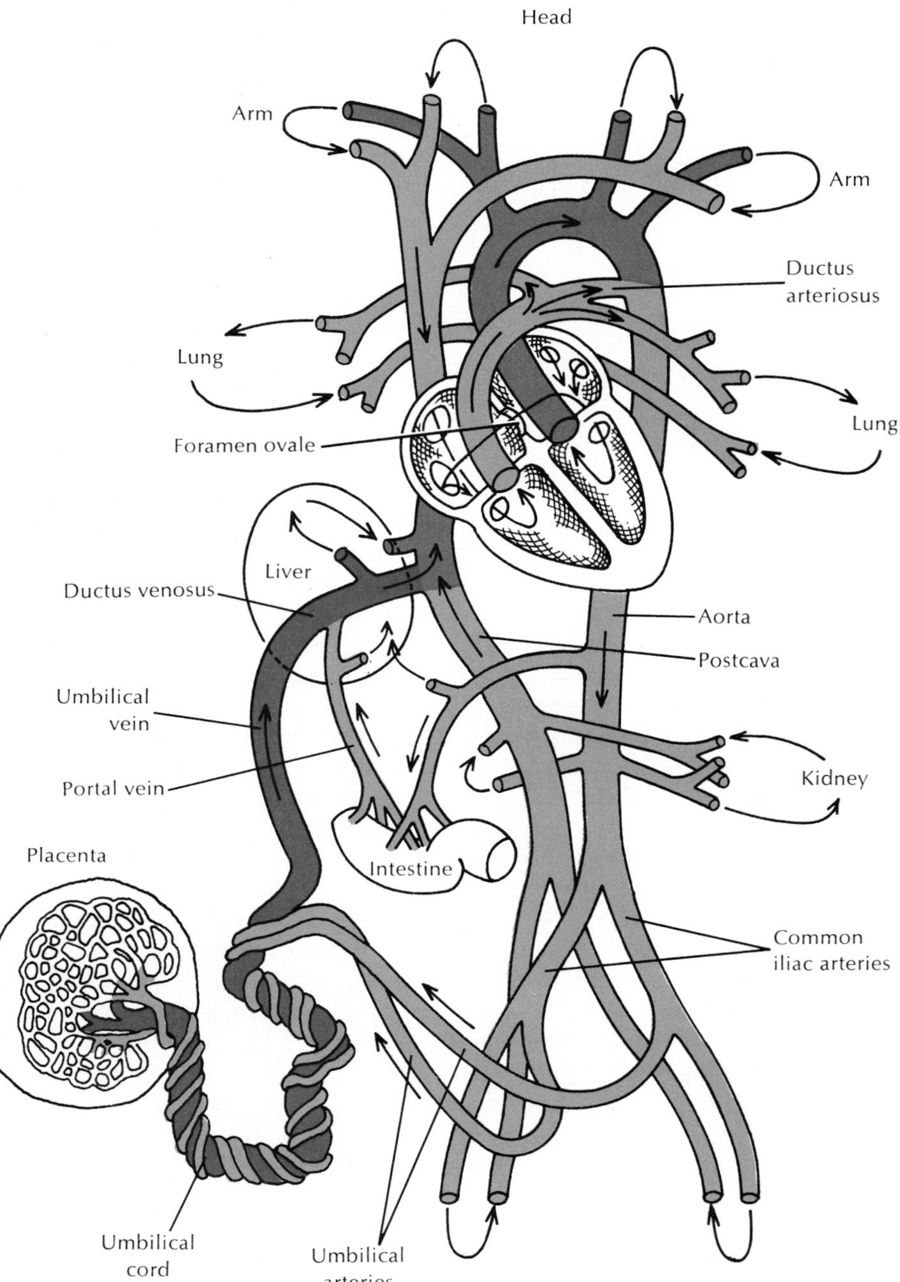

FIG. 36-14 Fetal circulation. Blood carrying food and oxygen from placenta is brought by umbilical vein to liver, is shunted through ductus venosus to vena cava where it mixes with deoxygenated blood returning from fetal tissues, and then is carried to right atrium. Two fetal shortcuts bypass nonfunctioning lungs. Some blood goes directly from right atrium to left atrium by an opening, the foramen ovale, and so enters systemic circulation. Some blood passes from pulmonary trunk into aorta by a temporary duct, the ductus arteriosus. At birth, placental exchange ceases; ductus venosus, foramen ovale, umbilical arteries, and ductus arteriosus all close; and regular postnatal pulmonary and systemic circuits take over.

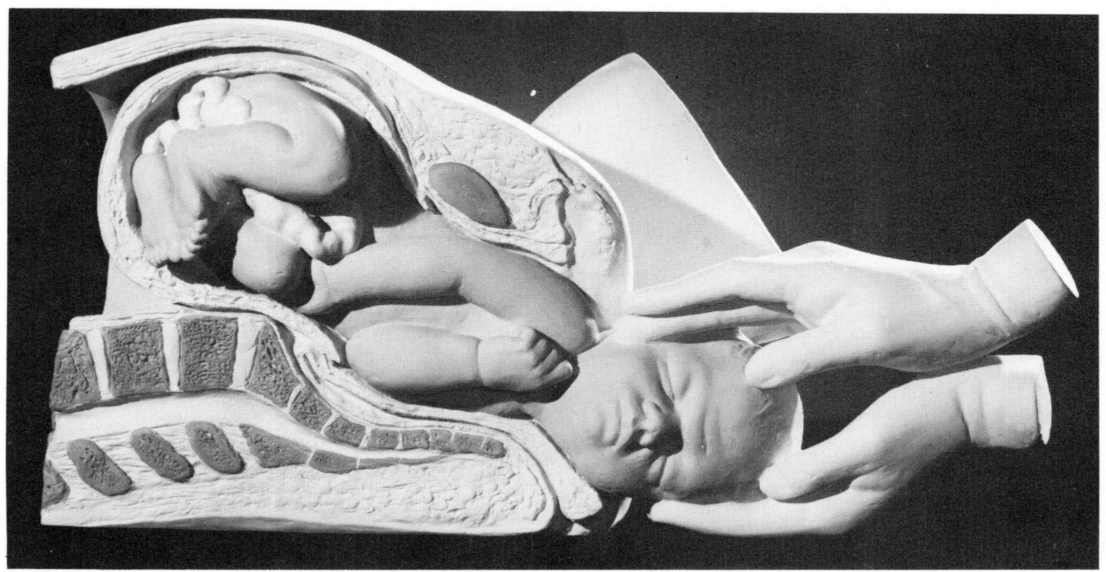

FIG. 36-15 Birth of a baby. (Courtesy American Museum of Natural History.)

have complex actions on virtually all body tissues and organs. They are especially powerful in stimulating contractions and, when used pharmacologically, are very effective in inducing labor or abortion. There is increasing evidence from studies on sheep (which, like human females, have an endocrine placenta during pregnancy) that the fetus itself controls the onset of labor. In some way not yet understood, the fetus, when it reaches normal term weight, releases an endocrine signal that stimulates the placenta to produce prostaglandins and thus initiates labor.

Stages in childbirth

The first major signal that birth is imminent is the so-called labor pains, caused by the rhythmic contractions of the uterine musculature. These are usually slight at first and occur at intervals of 15 to 30 minutes. They gradually become more intense, longer in duration, and more frequent. Some women compare the contractions to menstrual cramps. To some they are painful; to others they are not. Labor contractions may last anywhere from 4 to 24 hours, usually longer with the first child. The squeezing action of these contractions changes the position of the baby, usually so that the head presses against the cervix.

Childbirth occurs in three stages. In the first stage the neck (cervix), or opening of the uterus into the vagina, is enlarged by the pressure of the baby in its bag of amniotic fluid, which may be ruptured at this time. In the second stage the baby is forced out of the uterus and through the vagina to the outside (Fig. 36-15). In the third stage the placenta, or afterbirth, is expelled from the mother's body, usually within 10 minutes after the baby is born.

Multiple births

Many mammals give birth to more than one offspring at a time or to a litter, each member of which has come from a separate egg. Most higher mammals, however, have only one offspring at a time, although occasionally they may have plural young. The armadillo *(Dasypus)* is almost unique among mammals in giving birth to four young at one time—all of the same sex, either male or female, and all derived from one zygote.

Human twins may come from one zygote (identical twins) or two zygotes (nonidentical, or fraternal, twins). Triplets, quadruplets, and quintuplets may include a pair of identical twins. The other babies in such multiple births usually come from separate zygotes. Fraternal twins do not resemble each other more than other children born separately in the same family, but identical twins are, of course, strikingly alike and always of the same sex. Embryologically, each member of fraternal twins has its own placenta, chorion, and amnion. Usually (but not always) identical twins share the same chorion and the same placenta, but each has its own amnion. Sometimes identical twins fail to separate completely and form Siamese twins, in which the organs of one may be a mirror image of the organs of the other. The frequency of twin births in comparison to single births is approximately 1 in 86, that of triplets approximately 1 in 86^2, and that of quadruplets approximately 1 in 86^3.

LONGEVITY, AGING, AND DEATH

We will end our consideration of development with one aspect that some might wish to ignore: the inevitability of death. Each animal has a characteristic life span. The allotted lifetime of a human being would appear to be the biblical 3 score and 10 years. Even today, despite the achievements of modern medicine, the average life span is still about 70 years. Life spans among other animals vary enormously, and it is a common understanding that large animals live longer than small ones. The Indian elephant may reach 70 years; a hippopotamus or rhinoceros, 50 years; a horse, 45 years; a black bear or gorilla, 25 years; a beaver, 20 years; dogs and cats, 15 years; a gray squirrel, 10 years; a chipmunk, 5 years; a house mouse, 2 years; and a shrew, 1 year. Among birds, some parrots and owls may exceed 60 years (the record appears to be a 68-year-old eagle owl). Cranes and ostriches may live 40 years. However, most of our familiar songbirds are fortunate indeed to reach 10 years, even under the benign conditions of captivity.

Reptiles are noted for longevity, and centenarians are common among giant Galápagos tortoises. However, they are probably the only animals that can routinely outlive people. We live far longer—perhaps three times longer—than mammals of our body size should. Our exceptional longevity has been attributed to **neoteny**—the retention of juvenile characteristics in the adult. Compared to any other primate, we are born as helpless embryos; our brains develop slowly, continuing to grow at fetal rates long after birth; our bones ossify late; and we reach maturity only after an extended childhood. The postponement of maturity and onset of our senility have been attributed to our large brain—gestation is shortened and postnatal growth extended so that birth occurs when the brain is only one-fourth its ultimate size.

The nineteenth-century physiologist Max Rubner had a theory of aging that has received recent renewed support. He concluded that all mammals in their lifetimes use up about the same number of calories per unit of body weight. A mouse that lives for 2 years lives its life far more intensively than an elephant that lives for 70 years. This relationship is explained by **scaling theory,** an interpretation of the relationship among body shape, body size, metabolic rate, and lifetime. In most animals, total oxygen consumption scales only about three-fourths as fast as body weight. This means that among mammals—the group that reveals this relationship most clearly—small mammals generate far more heat in proportion to their body weight than large ones. Many physiologic properties scale at the same rate to body weight. Both heart rate and breathing rate increases about three-fourths as fast as body weight. Lifetimes also scale at about the same rate. This means that the hearts of all mammals, regardless of size, beat about the same number of times —800 million—in their allotted life span. Because the breathing rate is about one-fourth the heart rate, a mammal will breathe about 200 million times. Lifetime is scaled to life's pace, and all mammals, should they live out their "natural" lives, will survive about the same length of *biologic* time. Again, we deviate dramatically from our mammalian kin: our heart beats three times more often, and we take three times as many breaths as the average mammal of comparable body size.

The outward signs of aging are obvious, but the basic process remains obscure. Most theories attribute aging to progressively defective cellular function that follows an orderly program of senescence encoded in the cell's genes. Since consideration of all contemporary theories of cellular aging are well beyond the scope of this book, we will mention only certain well-documented observations. In one approach to the study of aging, structural cells called fibroblasts are removed from a piece of tissue, transferred to a culture dish containing a nutrient medium, and allowed to divide. After a few divisions, a small number of cells are subcultured in a new dish. After several subculturings, a point is reached when the cells fail to divide, and they die. The number of divisions is accepted as a measure of the life span of the cell type. Human fibroblasts taken from embryonic tissue will undergo approximately 50 population doublings before they die; those taken from a young adult are capable of about 30 doublings; and those taken from older individuals will divide about 20 times (Hayflick, 1968). The declining potential for cell division has been taken as evidence of progressive cellular deterioration with increasing age of the cell donor.

Cells from short-lived species have less potential for cell division than do those from a long-lived animal. Fibroblasts from the Galápagos tortoise have been observed to undergo as many as 114 doublings and live far longer in culture than human fibroblasts, a not unexpected finding in view of this creature's life span, which may exceed 150 years.

As cells age, defective proteins, especially inactive enzymes, begin to accumulate in the cells. This is probably caused by biochemical mistakes arising from the mutation of genes that code for protein synthesis. There is much more visible damage in chromosomal structure in cells from aged than from young animals.

We ought not to leave the impression that all aging is cellular. The extracellular fibrous protein **collagen,** which accounts for 40% of all body protein, has long been known to undergo degenerative molecular cross-linking with increasing age. Collagen is a nonrenewable protein produced early in life. As collagen ages it begins to interfere with the molecular traffic between cells and the extracellular fluid that bathes them. Because collagen is a major component of tendons, ligaments, bone, and the lining of blood vessels, its deteriorative hardening causes senescent alterations in these structures.

Why do animals age at all? While the cause of senescence is poorly understood, its value as a deteriorative process that inevitably leads to death is obvious: it prevents disastrous overpopulation. All organisms have impressive reproductive potential. For example, a pair of house flies in a single summer could produce 19×10^{19} offspring and descendants, if all reproduce and live at least until autumn. Under these circumstances the earth would disappear beneath a thick blanket of flies. Another more fundamental and often overlooked reason is that death is essential to natural selection. Natural selection is equated with reproductive success (see p. 9). But individuals cannot remain as reproductive or postreproductive adults indefinitely because they would compete with their own offspring for limited resources. No offspring could survive in such a world. The species would cease to adapt to environmental change (and consequently would cease to evolve) because selection for favorable variations would end. Consequently there has been a selection for life spans that terminate in postreproductive years. Senescence is only one of several agencies that lead to death—disease, starvation, predation, and adverse climate all take their toll —but death itself is the agency that no animal can escape.

Parenthetically, we must admit that unicellular organisms may have what is in one sense "eternal life": the parent cell does not die; it divides to produce two or more progeny and continues to do so for as long as conditions compatible with its survival persist. Nevertheless, unicellular organisms are as subject to natural selection as are multicellular ones. As environmental conditions change, only those organisms within the range of variation in the population that have the characteristics allowing survival will transmit those characteristics to their progeny.

Annotated references

Selected general readings

Austin, C. R. 1965. Fertilization. Englewood Cliffs, N.J., Prentice-Hall, Inc. *Cytology, physiology, and behavior of fertilization.*

Balinsky, B. I. 1975. An introduction to embryology, ed. 4. Philadelphia, W. B. Saunders Co. *One of the best animal embryology texts at the advanced undergraduate level. Stresses mechanisms of development.*

Berrill, N. J., and G. Karp. 1976. Development. New York, McGraw-Hill Book Co. *Balanced developmental biology text, embracing the entire scope of this complex field, yet readable and well ordered.*

Bonner, J. T. 1974. On development. Cambridge, Mass., Harvard University Press. *General but thoughtful overview of development.*

Brachet, J. 1974. Introduction to molecular embryology. London, The English Universities Press Ltd. *Molecular approach.*

Conklin, E. G. 1929. Problems of development. Am. Naturalist **63:**5-36. *Fascinating classic essay by a great American embryologist.*

DeBeer, G. R. 1951. Embryos and ancestors. London, Oxford University Press.

Ebert, J. D., and I. M. Sussex. 1970. Interacting systems in development, ed. 2. New York, Holt, Rinehart & Winston, Inc. *Clearly written and balanced undergraduate level paperback.*

Finch, G. E., and L. Hayflick (eds.). 1977. Handbook of the biology of aging. New York, Van Nostrand Reinhold

Co. *A reference work encompassing all aspects of aging from molecular to whole animal.*

Gurdon, J. B. 1974. The control of gene expression in animal development. Cambridge, Mass., Harvard University Press. *A thin but technical book.*

Kohn, R. R. 1971. Principles of mammalian aging. Englewood Cliffs, N.J., Prentice-Hall, Inc.

Markert, C. L., and H. Ursprung. 1971. Developmental genetics. Englewood Cliffs, N.J., Prentice-Hall, Inc. *Clearly written consideration of the question of gene expression and its control.*

Nelsen, O. E. 1953. Comparative embryology of the vertebrates. New York, McGraw-Hill Book Co. *This treatise is a comprehensive study of the comparative morphology of the vertebrates and protochordates. Contains a wealth of comparative information difficult to find in other texts.*

Oppenheimer, J. M. 1967. Essays in the history of embryology and biology. Cambridge, Mass., M. I. T. Press. *A collection of influential writings by distinguished developmental biologists.*

Rockstein, M. (ed.). 1974. Theoretical aspects of aging. New York, Academic Press, Inc. *Contributed chapters deal with theories of aging.*

Sacher, G. A. 1978. Longevity and aging in vertebrate evolution. Bioscience **28**(8):497-501. *Argues against the idea that "senescence genes" are responsible for aging.*

Spemann, H. 1938. Embryonic development and induction. New Haven, Conn., Yale University Press. *An authoritative and classic work by a pioneer in this aspect of embryology.*

Spratt, N. T., Jr. 1971. Developmental biology. Belmont, Calif., Wadsworth Publishing Co., Inc. *College text.*

Trinkaus, J. P. 1969. Cells into organs. Englewood Cliffs, N.J., Prentice-Hall, Inc. *Stresses cellular differentiation.*

Selected *Scientific American* articles

Allen, R. D. 1959. The moment of fertilization. **201:**124-134 (July). *Studies with sea urchin eggs have revealed the complexity of this critical biologic event.*

Bryant, P. J., S. V. Bryant, and V. French. 1977. Biological regeneration and pattern formation. **237:**66-81 (July). *Studies of regeneration reveal much about basic principles of development.*

Ebert, J. D. 1959. The first heartbeats. **201:**87-96 (Mar.). *The formation and early function of the heart is studied with chick embryos.*

Edwards, R. G. 1966. Mammalian eggs in the laboratory. **215:**72-81 (Aug.).

Edwards, R. G., and R. E. Fowler. 1970. Human embryos in the laboratory. **223:**44-54 (Dec.). *This and the 1966 article by Edwards deal with recent successes in culturing human and other mammalian eggs and embryos for observation.*

Epel, D. 1977. The program of fertilization. **237:**128-138 (Nov.). *Describes the fusion of egg and sperm and the initial events of development.*

Fischberg, M., and A. W. Blackler. 1961. How cells specialize. **205:**124-140 (Sept.). *Early steps in differentiation of the embryo are programmed into the egg before fertilization.*

Gray, G. W. 1957. "The organizer." **197:**79-88 (Nov.). *The history of embryologic studies on the organizing qualities of the blastopore lip of amphibian embryos. The famous experiments of Spemann and others are recounted.*

Gurdon, J. B. 1968. Transplanted nuclei and cell differentiation. **219:**24-35 (Dec.). *An extension of the work first successfully done by Briggs and King.*

Hadorn, E. 1968. Transdetermination of cells. **219:**110-120 (Nov.). *How cells of a fruit fly larva can change predestined fate by being transplanted into an adult fly.*

Hayflick, L. 1968. Human cells and aging. **218:**32-37 (Mar.).

Hinton, H. E. 1970. Insect eggshells. **223:**84-91 (Aug.). *Complexity revealed with scanning electron microscope.*

Jacobson, M., and R. K. Hunt. 1973. The origins of nerve-cell specificity. **228:**26-35 (Feb.). *How nerves find direction during growth.*

Moscona, A. A. 1961. How cells associate. **205:**142-162 (Sept.). *Describes the forces that promote cellular aggregation and binding.*

Patterson, P. H., D. D. Potter, and E. J. Furshpan. 1978. The chemical differentiation of nerve cells. **239:**50-59 (July).

Reynolds, S. R. M. 1952. The umbilical cord. **187:**70-74 (July). *Some facts about its performance.*

Singer, M. 1958. The regeneration of body parts. **199:**79-88 (Oct.). *Studies on the remarkable ability of amphibians to regrow lost limbs.*

Taylor, T. G. 1970. How an eggshell is made. **222:**88-95 (March). *How the hen mobilizes body stores of calcium to build eggshells.*

Wessells, N. K. 1971. How living cells change shape. **225:**76-82 (Oct.). *How microtubules and microfilaments make cell movement possible, so important in development.*

Wessells, N. K., and W. J. Rutter. 1969. Phases in cell differentiation. **220:**36-44 (Mar.). *Cultivation of embryonic pancreas tissues reveals stages of specialization.*

CHAPTER 37
PRINCIPLES OF INHERITANCE

A sea lion *(Zalophus wollebaeki)* and her nursing pup. Each generation hands on to the next a genetic blueprint that guides orderly development and preserves the fidelity of reproduction.

Photograph by C. P. Hickman, Jr.

MEANING OF HEREDITY

Heredity establishes the continuity of life forms. Although offspring and parents in a particular generation may look different, there is nonetheless a basic sameness that runs from generation to generation for any species of plant or animal. In other words, "like begets like." Yet children are not precise replicas of their parents. Some of their characteristics show resemblances to one or both parents, but they also demonstrate many traits not found in either parent. What is actually inherited by an offspring from its parents is a certain type of germinal organization **(genes)** that, under the influence of environmental factors, guides the orderly sequence of differentiation of the fertilized egg into a human being, bearing the unique physical characteristics as we see them. Each generation hands on to the next the instructions required for maintaining continuity of life (see Principle 13, p. 12).

The gene is the unit entity of inheritance, the germinal basis for every characteristic that appears in an organism. In a sense, the primary function of an organism is to reproduce genes. The organism is a device, a vehicle for the transfer of genes from one generation to its descendants. It is a repository in which a portion of the gene "pool" of the population has been temporarily entrusted.

The study of what genes are and how they work is the science of genetics. It is a science that deals with the underlying causes of *resemblance,* as seen in the remarkable fidelity of reproduction, and of *variation,* which is the working material for organic evolution. Genetics has shown that all living forms use the same information storage, transfer, and translation system, and thus it has provided an explanation for both the stability of all life and for its probable descent from a common ancestral form. This is one of the most important unifying concepts of biology.

We will approach the subject in a historical sequence, since this leads us from the general to the specific and enables us to comprehend certain principles that govern inheritance. This places us in a better position to understand the mechanisms of inheritance and the molecular organization of the genetic material, topics that are taken up in later sections of this chapter.

MENDEL AND THE LAWS OF INHERITANCE

Mendel's investigations

The first person to understand the quantitative principles of heredity was Gregor Johann Mendel (1822-1884) (Fig. 37-1), who lived with the Augustinian monks at Brünn (Brno), Moravia. At that time Brünn was a part of Austria, but now it is in the central part of Czechoslovakia. While conducting breeding experiments in a small monastery garden from 1856 to 1864, he examined with great care the progeny of many thousands of plants. He worked out in elegant simplicity the laws governing the transmission of characters from parent to offspring. His discoveries, published in 1886, were of great potential significance, coming, as they did, just after Darwin's publication of *Origin of Species.* Yet these discoveries remained unappreciated and forgotten until 1900—some 35 years after the completion of the work and 16 years after Mendel's death.

Mendel's classic observations were based on the garden pea because it had been produced in pure strains by careful selection by gardeners over a long period of time. For example, some varieties were definitely dwarf and others were tall. A second reason for selecting peas was that they were self-fertilizing, but also capable of cross-fertilization. To simplify his problem he chose single characters and characters that were sharply contrasted. Mere quantitative and intermediate characters were carefully avoided.* Mendel selected seven pairs of these contrasting characters: tall plants, dwarf plants; smooth seeds, wrinkled seeds; green cotyledons, yellow cotyledons; inflated pods, constricted pods; yellow pods, green pods; axial position of flowers, terminal position of flowers; and transparent seed coats, brown seed coats (Fig. 37-1).

Mendel crossed plants having one of these characters with others having the contrasting characteristic. He did this by removing the stamens from a flower to prevent self-fertilization and then placing on the stigma of this flower the pollen from the flower of the plant that had the contrasting character. He also prevented the experimental flowers from being pollinated from other sources such as wind and insects. When the cross-fertilized flower bore seeds, he noted the kind of plants (hybrids) that were produced from the planted seeds. Subsequently he crossed these hybrids among themselves to see what would happen.

*In not reporting conflicting findings, which must surely have arisen as they do in any original research, Mendel has been accused of "cooking" his results. The chances are, however, that he carefully avoided ambiguous material to strengthen his central message, which we still regard as an exemplary achievement in experimental analysis.

FIG. 37-1 Seven experiments of Gregor Mendel. (Portrait courtesy American Museum of Natural History.)

Mendel's first law

In one of Mendel's original experiments, he pollinated pure-line tall plants with the pollen of pure-line dwarf plants. He found that all the progeny, the first filial generation (F_1), were tall, just as tall as the tall parents of the cross. The reciprocal cross—dwarf plants pollinated with tall plants—gave the same result. This always happened no matter which way the cross was made. Obviously, this kind of inheritance was not a blending of two characteristics, since none of the progeny was of intermediate size.

Next Mendel selfed (self-fertilized) the tall F_1 plants and raised several hundred progeny, the second filial (F_2) generation. This time, *both* tall and dwarf plants appeared. Again, there was no blending (no plants of intermediate size), but the appearance of dwarf plants from all tall parental plants was surprising. The dwarf characteristic, present in the grandparents but not in the parents, had reappeared. When he counted the actual number of tall and dwarf plants in the F_2 generation, he discovered that there were almost exactly three times more tall plants than dwarf ones (787 tall and 277 dwarf) which was close to a ratio of 3:1 (exactly 2.84:1).

Mendel then repeated this experiment for the six other contrasting characters that he had chosen, and in every case he obtained ratios very close to 3:1 (Fig. 37-1). At this point it must have been clear to Mendel that he was dealing with hereditary determinants for the contrasting characters that did not blend when brought together. Even though the dwarf characteristic disappeared in the F_1 generation, it reappeared fully expressed in the F_2 generation. He realized that the F_1 generation plants carried determinants (which he called "factors") of both tall and dwarf parents, even though only the tall characteristic was expressed in the F_1 generation.

Mendel called the tall factor **dominant** and the short **recessive.** Similarly, the other pairs of characters that he studied showed dominance and recessiveness. Thus when plants with yellow unripe pods were crossed with green unripe pods, the hybrids all contained yellow pods. In the F_2 generation the expected ratio of 3 yellow to 1 green was obtained. Whenever a dominant factor (gene) is present, the recessive one cannot produce an effect. The recessive factor will show up *only* when both factors are recessive, or, in other words, a pure condition.

Genetic terminology

Before proceeding further, let us introduce some genetic terminology. In representing his crosses Mendel used letters as symbols: for dominant characters he employed capitals and for recessives, the corresponding lower-case letters. Thus the factors for pure tall plants might be represented by **TT,** the pure recessive by **tt,** and the mix, or **hybrid,** of the two plants by **Tt.** This type of cross is called a **monohybrid cross** because it involves only one pair of contrasting characters. When the gametes are united in any cross, a zygote is formed. The zygote bears the complete genetic constitution of the organism. In the example of tall and dwarf plants, the pure tall plants **(TT)** produce only **T** gametes; the pure dwarf plants **(tt)** produce only **t** gametes. When these gametes unite, the resulting zygote is **Tt** and thus is a hybrid, or **heterozygote,** since the factors **T** and **t** are not alike. On the other hand, the pure tall plants **(TT)** and pure dwarf plants **(tt)** are **homozygotes,** meaning that the factors, or genes, are alike.

In the cross between tall and dwarf plants there are two types of *visible* characters—tall and dwarf. These are called **phenotypes.** On the basis of genetic formulas there are three *hereditary* types—**TT, Tt,** and **tt.** These are called **genotypes.** A genotype is a gene combination **(TT, Tt,** or **tt),** and the phenotype is the appearance of the organism (tall or dwarf).

When a gene exists in more than one form, the different forms are called **alleles** (or **allelomorphs**). **T** and **t** are alleles for the gene for plant height. The allele **T** is dominant and the allele **t** is recessive. The position of a gene on a chromosome is called a **locus** (pl., loci).

In diagram form, one of Mendel's original crosses (tall plant and dwarf plant) could be represented in this manner:

	(tall)			(dwarf)
Parents	**TT**		×	**tt**
Gametes	all **T**			all **t**
F_1		**Tt** (all tall)		
Crossing hybrids	**Tt**		×	**Tt**
Gametes	**T, t**			**T, t**
F_2 genotypes	**TT**	**Tt**	**tT**	**tt**
F_2 phenotypes	(tall)	(tall)	(tall)	(dwarf)

It is convenient in most Mendelian crosses to use the checkerboard method devised by Punnett for representing the various combinations resulting from a cross. In the F_2 cross the following scheme would apply:

Sperm \ Eggs	½ **T**	½ **t**
½ **T**	¼ **TT** (homozygous tall)	¼ **Tt** (hybrid tall)
½ **t**	¼ **Tt** (hybrid tall)	¼ **tt** (homozygous dwarf)

Ratio: 3 tall to 1 dwarf.

The next step was an important one, because it enabled Mendel to test his hypothesis that every plant contained nonblending factors from both parents. He self-fertilized the plants in the F_2 generation; that is, the stigma of a flower was fertilized by the pollen of the same flower. The results showed that self-pollinated F_2 dwarf plants produced only dwarf plants, whereas one-third of the F_2 tall plants produced tall and the other two-thirds produced both tall and dwarf in the ratio of 3:1, just as the F_1 plants had done.

F_2 plants: tall
- ¼ **TT** $\xrightarrow{\text{selfed}}$ all **TT** (homozygous tall)
- ½ **Tt** $\xrightarrow{\text{selfed}}$ 1 **TT**:2 **Tt**:1 **tt** (3 tall: 1 dwarf)

dwarf ¼ **tt** $\xrightarrow{\text{selfed}}$ all **tt** (homozygous dwarf)

This experiment showed that the dwarf plants were pure because they at all times gave rise to short plants when self-pollinated; the tall plants contained both pure tall and hybrid tall. It also demonstrated that, although the dwarf character disappeared in the F_1 plants, which were all tall, the character for dwarfness appeared in the F_2 plants.

Mendel reasoned that the factors for tallness and dwarfness were units that did not blend when they were together. The F_1 generation (the first generation of hybrids, or first filial generation) contained both of these units or factors, but when these plants formed their germ cells, the factors separated so that each germ cell had only one factor. In a pure plant both factors were alike; in a hybrid they were different. He concluded that individual germ cells were always pure with respect to a pair of contrasting factors, even though the germ cells were formed from hybrids in which the contrasting characters were mixed.

This idea formed the basis for his first principle, the **law of segregation,** which states that whenever two factors are brought together, they segregate into separate gametes (see Principle 14, p. 12). Either one of the pair of genes of the parent passes with equal frequency to the gametes.

Mendel's great contribution was his quantitative approach to inheritance. This really marks the birth of genetics, since before Mendel people thought that traits were blended like mixing together two colors of paint, a notion that unfortunately still lingers in the minds of many. If this were true, variability would be lost in hybridization between individuals. With particulate inheritance, on the other hand, different variations are retained and can be shuffled about and resorted like blocks.

Testcross

The dominant characters in the offspring of a cross are all of the same phenotypes whether they are homozygous or heterozygous. For instance, in Mendel's experiment of tall and dwarf characters, it is impossible to determine the genetic constitution of the tall plants of the F_2 generation by mere inspection of the tall plants. Three-fourths of this generation are tall, but which of them are heterozygotes?

As Mendel reasoned, the test is to cross the questionable individuals with pure recessives. If the tall plant is homozygous, all the plants in such a testcross are tall, thus:

Parents	**TT** (tall) × **tt** (dwarf)
Offspring	**Tt** (hybrid tall)

If, on the other hand, the tall plant is heterozygous, half of the offspring are tall and half dwarf, thus:

Parents	**Tt** × **tt**
Offspring	**Tt** (tall) or **tt** (dwarf)

The testcross is often used in modern genetics for the analysis of the genetic constitution of the offspring, as well as for a quick way to make desirable homozygous stocks of animals and plants.

Intermediate inheritance

A cross that always distinguishes the heterozygotes from the pure dominants is exemplified by the four-

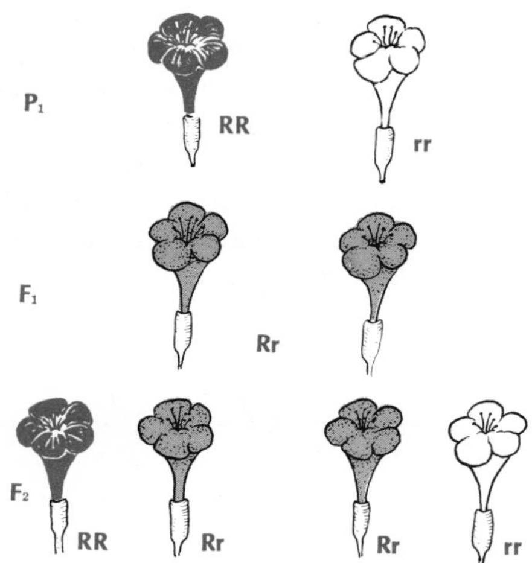

FIG. 37-2 Cross between red and white four-o'clock flowers. Red and white are homozygous; pink is heterozygous.

o'clock flower *(Mirabilis)* (Fig. 37-2). Whenever a red-flowered variety is crossed with a white-flowered variety, the hybrid (F_1), instead of being red or white according to whichever is dominant, is actually intermediate between the two and is pink. Thus the homozygotes are either red or white, but the heterozygotes are pink. The testcross is therefore unnecessary to determine the nature of the genotype or the individuals.

In the F_2 generation, when pink flowers are crossed with pink flowers, one-fourth are red, one-half pink, and one-fourth white. This cross may be represented in this fashion:

	(red flower)		(white flower)	
Parents	**RR**	×	**rr**	
Gametes	all **R**		all **r**	
F_1		**Rr** (all pink)		
Crossing hybrids	**Rr**	×	**Rr**	
Gametes	**R, r**		**R, r**	
F_2 genotypes	**RR**	**Rr**	**rR**	**rr**
F_2 phenotypes	(red)	(pink)	(pink)	(white)

In this kind of cross neither of the genes shows complete dominance. Therefore the heterozygote is indeed a blending of both red and white characters. It is easy to see how observations of this type would encourage the notion of the blending theory of inheritance. However, in the cross of red and white four-o'clock flowers, *only* the hybrid is a phenotypic blend; the homozygous strains breed true to the parental phenotypes (red or white).

Mendel's second law

Thus far we have considered crosses involving alleles of a single gene (monohybrid cross). Mendel also carried out experiments on peas that differed from each other by two or more genes, that is, experiments involving two or more phenotypic characters.

Mendel had already established that tall plants were dominant to dwarf. He also noted that crosses between plants bearing yellow cotyledons and plants bearing green cotyledons produced plants with yellow cotyledons in the F_1 generation; therefore yellow was dominant to green. The next step was to make a cross between plants differing in these two characteristics. When a tall plant with yellow cotyledons **(AABB)** was crossed with a dwarf plant with green cotyledons **(aabb)**, the F_1 plants were tall and yellow, as expected **(AaBb)**.

Parents	**AABB** (tall, yellow)	×	**aabb** (dwarf, green)
Gametes	all **AB**		all **ab**
F_1		**AaBb** (tall, yellow)	

When the F_1 hybrids were crossed with each other, the result was four different types (phenotypes) of plants in a ratio of 9:3:3:1. They were:

9 tall with yellow cotyledons
3 tall with green cotyledons
3 dwarf with yellow cotyledons
1 dwarf with green cotyledons

Mendel already knew that a cross between two plants bearing a single pair of alleles of the genotype **Aa** would yield a 3:1 ratio. Similarly a cross between two plants with the genotypes **Bb** would yield the same 3:1 ratio. If we examine *only* the tall and dwarf phenotypes in the outcome of the dihybrid experiment, they total up to 12 tall and 4 dwarf, giving a ratio of 3:1. Likewise, there is a total of 12 plants with yellow cotyledons and 4 plants with green—again a 3:1 ratio. Thus the monohybrid ratio prevails for both traits when they are considered independently. The 9:3:3:1 ratio is nothing more than two 3:1 ratios combined:

$$3:1 \times 3:1 = 9:3:3:1$$

The F_2 genotypes and phenotypes are as follows:

Genotype	Class	Phenotype
1 **AABB** 2 **AaBB** 2 **AABb** 4 **AaBb**	9 **A-B-**	9 tall yellow
1 **AAbb** 2 **Aabb**	3 **A-bb**	3 tall green
1 **aaBB** 2 **aaBb**	3 **aaB-**	3 dwarf yellow
1 **aabb**	1 **aabb**	1 dwarf green

The results of this experiment show that the segregation of alleles for plant height is entirely independent of the segregation of alleles for cotyledon color. Neither has any influence on the other. This is the basis of Mendel's second law, the **law of independent assortment,** which states: whenever two or more pairs of contrasting characters are brought together in a hybrid, the alleles of different pairs segregate independently of one another (see Principle 14, p. 12).

The dihybrid experiment is shown in the Punnett square below:

When Mendel worked out the ratios for his various crosses, they were approximations and not certainties. In his 3:1 ratio of tall and short plants, for instance, the resulting phenotypes did not come out exactly 3 tall to 1 short. All genetic experiments are based on probability; that is, the outcome of the events is uncertain, and there is an element of chance in the final results. Probabilities are expressed in fractions; they are always a number between 0 and 1.

We may define probability as:

$$\text{probability (p)} = \frac{\text{number of times an event happens}}{\text{total number of trials or possibilities for the event to happen}}$$

For example, the probability (p) of a coin falling heads when tossed is ½ because the coin has two sides. The probability of rolling a three on a die is 1/6 because the die has six sides.

The probability of independent events occurring together (ordered events) involves the **product rule,** which is simply the product of their individual probabilities. When two coins are tossed together, the

	(tall, yellow)		(dwarf, green)
Parents	**AABB**	×	**aabb**
Gametes	all **AB**		all **ab**
F_1		**AaBb** (hybrid tall, hybrid yellow)	
Crossing hybrids	**AaBb**	×	**AaBb**
Gametes	**AB, Ab, aB, ab**		**AB, Ab, aB, ab**
F_2		(see checkerboard)	

	AB	**Ab**	**aB**	**ab**
AB	**AABB** pure tall pure yellow	**AABb** pure tall hybrid yellow	**AaBB** hybrid tall pure yellow	**AaBb** hybrid tall hybrid yellow
Ab	**AABb** pure tall hybrid yellow	**AAbb** pure tall pure green	**AaBb** hybrid tall hybrid yellow	**Aabb** hybrid tall pure green
aB	**AaBB** hybrid tall pure yellow	**AaBb** hybrid tall hybrid yellow	**aaBB** pure dwarf pure yellow	**aaBb** pure dwarf hybrid yellow
ab	**AaBb** hybrid tall hybrid yellow	**Aabb** hybrid tall pure green	**aaBb** pure dwarf hybrid yellow	**aabb** pure dwarf pure green

Ratio: 9 tall yellow to 3 tall green; 3 dwarf yellow to 1 dwarf green.

probability of getting two heads is ½ × ½ = ¼, or 1 chance in 4. Or, the probability of rolling two threes simultaneously with two dice is

$$p \text{ (two threes)} = 1/6 \times 1/6 \times 1/36$$

Probabilities are predictions, and the more the coins are tossed, or the dice rolled, the more closely the number of favorable cases approaches the number predicted by the p value.

We can use the product rule to predict the ratios of inheritance in monohybrid or dihybrid (or larger) crosses, if the genes sort independently in the gametes (as they did in all of Mendel's experiments). In other words, the mechanism of placing **A** into a gamete is independent of the mechanism of putting **a** into a gamete. Therefore, in a monohybrid cross the probability that a sperm carries the dominant is ½ and the same applies to an egg. In a dihybrid cross involving **Aa** and **Bb,** the same thing applies: the probability of any gene appearing in a gamete is ½. Now we can apply the product rule to an F_1 plant **AaBb** to determine the frequency of each kind of gamete:

p of gamete being **AB** = ½ × ½ = ¼
p of gamete being **Ab** = ½ × ½ = ¼
p of gamete being **aB** = ½ × ½ = ¼
p of gamete being **ab** = ½ × ½ = ¼

From this point we can easily derive the probabilities for the genotype in each box of the Punnett square on p. 845.

p of plant being **AABB** = ¼ × ¼ = 1/16
p of plant being **AABb** = ¼ × ¼ = 1/16

and so on for all sixteen boxes. Collecting all similar phenotypes together we get:

Tall, yellow cotyledon	9/16 = 9
Tall, green cotyledon	3/16 = 3
Dwarf, yellow cotyledon	3/16 = 3
Dwarf, green cotyledon	1/16 = 1

Thus we have the 9:3:3:1 ratio, derived by the product rule. In fact, one quickly learns by experience how to determine the ratios of phenotypes without using either the Punnett squares or the product rule. In a dihybrid (9:3:3:1 ratio), for instance, those phenotypes that make up the dominants of each gene are 9/16 of the whole F_2 generation; each of the 3/16 phenotypes consists of one dominant and one recessive; and the 1/16 phenotype consists of two recessives.

The F_2 ratios in any cross involving more than one pair of contrasting pairs can be found by combining the ratios in the cross of one pair of alleles. Thus the number of genotypes are $(3)^n$ and the proportion of phenotypes $(3:1)^n$ when one allele is dominant and the other recessive. For example, let us suppose that in a cross of two pairs of alleles the phenotypes are in the ratio of $(3:1)^2$, or 9:3:3:1. The genotypes in such a cross are $(3)^2$, or 9. If three pairs of characters are involved (trihybrid cross), the proportions, or ratios, of the phenotypes are then $(3:1)^3$, or 27:9:9:9:3:3:3:1. The genotypes are $(3)^3$, or 27. Obviously, as the number of gene pairs increases, the number of phenotypes and genotypes rises steeply.

Multiple alleles

Earlier we defined alleles as the alternate forms of a gene. Many dissimilar alleles may occupy the same gene locus on a chromosome but not, of course, all at one time. Thus more than two alternative characters may affect the same character. An example is the set of multiple alleles that affect coat color in rabbits. The different alleles are C (normal color), c^{ch} (chinchilla color), c^h (Himalayan color), and c (albino). The four alleles fall into a dominance series with C dominant over everything. The dominant allele is always written to the left and the recessive to the right:

Cc^h = normal color
$c^{ch}c^h$ = chinchilla color
c^hc = Himalayan color

Multiple alleles arise through mutations at the gene locus over long periods of time. Any gene may mutate in several different ways (p. 872) if given time, and thus can give rise to slightly different genes or alleles at the same locus. It is thought that most genes now present in an organism at any locus are mutated alleles of an original gene.

Gene interaction

The types of crosses previously described are simple in that the characters involved result from the action of a single gene, but many cases are known in which the characters are the result of two or more genes. Mendel probably did not appreciate the real significance of the genotype, as contrasted with the visible character—the phenotype. We now know that many different genotypes may be expressed as a single phenotype.

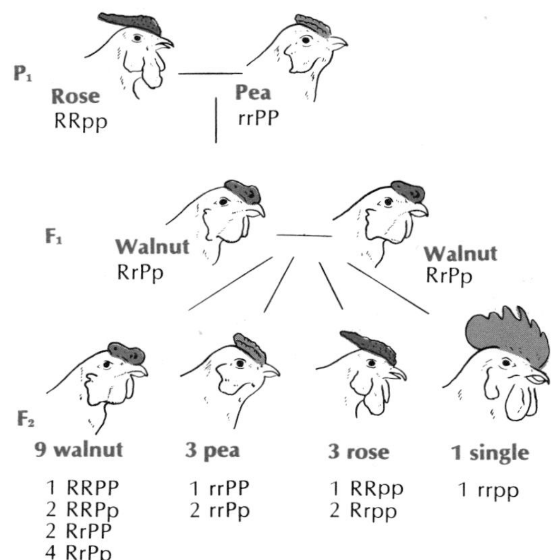

FIG. 37-3 Heredity of comb forms in chickens.

Also many genes have more than a single effect. A gene for eye color, for instance, may be the ultimate cause for eye color, yet at the same time it may be responsible for influencing the development of other characters as well. Some of these special forms of inheritance are described in the following discussions on supplementary factors, epistasis, and multiple gene inheritance.

Supplementary factors

The variety of comb forms found in chickens illustrates the interaction of supplementary genes (Fig. 37-3). The common forms of comb are rose, pea, walnut, and single. Of these, the pea comb and the rose comb are dominant to the single comb. For example, when a pea comb is crossed with a single comb, all the F_1 offspring are pea and the F_2 offspring show a ratio of 3 pea to 1 single. When the two dominants, pea and rose, are crossed with each other, an entirely new kind of comb, walnut, is found in the F_1 generation. Each of these genes supplements the other in the production of a kind of comb different from each of the dominants. In the F_2 generation the ratio is 9 walnut to 3 rose to 3 pea to 1 single. The walnut comb cannot thus be considered a unit character but is merely the phenotype's expression of pea and rose when they act together.

Inspection of the ratio reveals that two genes are involved. If **P** represents the gene for the pea comb and **p** represents its recessive allele and if **R** represents the gene for the rose comb and **r** its recessive allele, then the pea comb formula would be **rrPP;** the formula for rose comb would be **RRpp.** Any individual having both dominant alleles has a walnut comb. When no dominant allele is present, the comb is single. The cross may be diagramed as follows:

	(pea comb)		(rose comb)
Parents	**rrPP**	×	**RRpp**
Gametes	all **rP**		all **Rp**
F_1	**RrPp**	×	**RrPp**
	(walnut)		(walnut)
Gametes	**RP, rP, Rp, rp**		**RP, rP, Rp, rp**

By the checkerboard method, the F_2 generation shows 9 walnut, 3 pea, 3 rose, and 1 single. Genotypes with the combinations of **PR** give walnut phenotypes; those with **P,** pea; those with **R,** rose; and those lacking in both **P** and **R,** single.

Epistasis

When a gene at one locus affects the expression of a gene at another locus acting on the same trait, the first is said to be epistatic to the second. In rabbits the gene for pigment (of any kind) is epistatic to that for agouti, a barred pigment pattern of the fur. The gene for pigment has two alleles, **A** and **a.** If the allele **A** is present, the rabbit is pigmented and the color depends on other genes. If the individual is homozygous recessive, **aa,** the rabbit is albino, no matter what other genes for pigment may be present.

The expression of the second gene, **B** for agouti, depends on the presence or absence of the **A** gene. This interaction can be summarized as follows:

	(agouti)		(albino)
Parents	**AABB**	×	**aabb**
Gametes	all **AB**		all **ab**
F_1	**AaBb**	×	**AaBb**
	(agouti)		(agouti)
Gametes	**AB, Ab, aB, ab**		**AB, Ab, aB, ab**

The F_2 genotypes and phenotypes are as follows:

Genotypes		Phenotype
1 **AABB** 2 **AaBB** 2 **AABb** 4 **AaBb**	9 **A-B-**	9 agouti

Genotypes	Classes	Phenotypes
1 **AAbb** 2 **Aabb**	3 **A-bb**	3 black
1 **aaBB** 2 **aaBb**	3 **aaB-**	4 albino
1 **aabb**	1 **aabb**	

The F_2 phenotypic ratio is 9:3:4. This happens because the last two phenotypic classes are both albino and are combined. The dominant allele **A** must be present for any color to appear. Also if only the recessive gene for agouti is present (homozygous **bb**), the animal's color becomes black, an intermediate between agouti and albino.

Epistasis is not the same as dominance, with which it might be confused. Dominance is the phenotypic expression of one member of a pair of alleles of a *single* gene. In epistasis the alleles of *two* genes on separate loci are involved, with one gene affecting the phenotypic expression of the other gene.

Multiple gene inheritance

Whenever several sets of alleles produce a cumulative effect on the same character, they are called **multiple genes,** or **polygenic** factors. Several characteristics in humans are influenced by multiple genes. In such cases the characters, instead of being sharply marked, show continuous variation between two extremes. This is sometimes called **blending,** or **quantitative inheritance.** In this kind of inheritance the children are often more or less intermediate between the two parents.

One illustration of such a type is the degree of pigmentation in crosses between the black and white human races. The cumulative genes in such crosses have a quantitative expression. Many genes are believed to be involved in skin pigmentation, but we will simplify our explanation by assuming that there are only two pairs of independently assorting genes. Thus, a person with very dark pigment has two genes for pigmentation on separate chromosomes **(AABB).** Each dominant allele contributes one unit of pigment. A person with very light pigment has alleles **(aabb)** that contribute no color. (Freckles that often appear in the skin of very light people represent pigment contributed by entirely separate genes.) The offspring of very dark and very light parents would have an intermediate skin color **(AaBb).**

The children of parents having intermediate skin color show a range of skin color, depending on the number of genes for pigmentation they inherit. Their skin color ranges from very dark **(AABB),** through dark (**AABb** or **AaBB**), intermediate (**AAbb** or **AaBb** or **aaBB**), light (**Aabb** or **aaBb**), to very light (**aabb**). It is thus possible for parents heterozygous for skin color to produce children with darker or lighter colors than themselves.

The relationships can be seen in the following diagram:

	(very dark)		(very light)
Parents	**AABB**	×	**aabb**
Gametes	**AB**		**ab**
F_1	**AaBb**	×	**AaBb**
	(intermediate)		(intermediate)
Gametes	**AB, Ab, aB, ab**		**AB, Ab, aB, ab**

The F_2 generation shows this ratio:

1 **AABB**	very dark
2 **AaBB**	dark
2 **AABb**	dark
4 **AaBb**	intermediate
1 **AAbb**	intermediate
2 **Aabb**	light
1 **aaBB**	intermediate
2 **aaBb**	light
1 **aabb**	very light

Collecting the phenotypes, the ratio becomes: 1 very dark:4 dark:6 intermediate:4 light:1 very light.

When there are many genes involved in the production of traits, this situation may be represented in graphs as distribution curves, usually bell-shaped. A graph for height indicates that the height of most people is near the average height, indicated by the high point of such a bell-shaped curve and that few people are much shorter or taller than average. If one locus is much more important than others, the alleles at this locus might split the graph into two or three bell-shaped curves representing different genotypes. Because we never find this kind of graph for the height of human populations, we know that humans do not have a principle for tall and dwarf lines like that of garden peas.

Environmental effects: penetrance and expressivity

Penetrance refers to the percentage frequency with which a gene manifests a phenotypic effect. If a dominant gene or a recessive gene in a homozygous state always produces a detectable effect, it is said to have

complete penetrance. If dominant or homozygous recessive genes fail to show phenotypic expression in every case, it is called **incomplete** or **reduced penetrance.** Environmental factors may be responsible for the degree of penetrance because some genes may be more sensitive to such influences than others. The genotype responsible for diabetes mellitus, for instance, may be present, but because of reduced penetrance, the disease does not always occur. All of Mendel's experiments apparently had 100% penetrance.

The phenotypic variation in the expression of a gene is known as **expressivity.** For instance, a heritable allergy may cause more severe symptoms in one person than in another. Environmental factors may cause different degrees in the appearance of a phenotype. Temperature affects the expression of the genes for dark-colored fur in Siamese cats and Himalayan rabbits. Normally the tail, feet, ears, and nose—the areas that have a cooler body temperature—are dark, but under warm conditions there may be less than normal darkening, whereas in colder environmental temperatures there may be some darkening of the entire animal. Other genes in the hereditary constitution may also modify the expression of a trait. What is inherited is a certain genotype, but how it is expressed phenotypically is determined by the environment and other factors.

Genes that have more than one effect are called **pleiotropic.** Most genes may have multiple effects. Even those genes that produce visible effects probably have numerous physiologic effects not detected by the geneticist. The recessive gene in fruit flies that produces (in a homozygous condition) vestigial wings also affects other traits such as the halteres (balancers), bristles, reproductive organs, and length of life.

Lethal genes

A lethal gene is one that, when present in a homozygous condition, causes the death of the offspring. It has been known for a long time that the yellow strain of the house mouse *(Mus musculus)* is heterozygous, with the allele for yellow coat dominant to the allele for wild-type gray coat. When two yellow mice are bred together, some of the progeny are yellow and some are gray, always in a ratio of 2:1. This is a marked departure from the expected Mendelian 3:1 ratio consisting of 1 pure yellow **(YY),** 2 hybrid yellow **(Yy),** and 1 pure gray **(yy).** The explanation is that the homozygous yellow always dies as an embryo, because the **YY** genotype is lethal. The peculiar feature of this inheritance is that the **Y** allele is *dominant* in its effect on coat color (the **Yy** hybrid is always yellow, never gray) but is *recessive* in its lethality because it must be homozygous **(YY)** to express its lethal effect. This is a good example of a pleiotropic effect, since the gene affects the phenotype of the mouse in two different ways.

CHROMOSOMAL THEORY OF INHERITANCE

Cytogenetics

Mendel knew nothing of genes and chromosomes. To explain the phenotypic ratios he observed in his garden pea experiments, Mendel proposed a purely abstract, intellectual model, based on the segregation and random assortment of invisible "factors." Until 1900, the infant science of genetics was mostly dormant, existing mainly in the minds of a handful of European and North American biologists. In contrast, the vigorous discipline of cytology, wherein lay the key to understanding the physical basis of inheritance, progressed rapidly throughout the nineteenth century. One after another, crucial concepts of cell organization were unveiled: nuclear division (1844); presence of chromosomes (1848) and their longitudinal separation into daughter chromosomes during mitosis (1880); constancy of chromosome number in a species (1885); and significance of meiosis (1891).

When the discovery of Mendel's laws was announced in 1900 (a date that may be considered the beginning of modern genetics), the time was ripe to demonstrate the parallelism that existed between these fundamental laws of inheritance and the cytologic behavior of the chromosomes. In a series of experiments by Boveri in Germany, and Sutton, McClung, and Wilson in the United States, it was shown that the hereditary laws of Mendel could be explained by the behavior of chromosomes at meiosis and fertilization. This led to an analysis of the chromosome structure and the idea of the gene as the physical basis of hereditary traits. The outstanding work of Thomas Hunt Morgan and his colleagues on the fruit fly *(Drosophila)* led to the mapping of chromosomes in which the location of genes was more or less definitely determined. Out of this work developed the new hybrid science of **cytogenetics.**

Physical structure of chromosomes

The structure of chromosomes is described briefly in Chapter 4 (p. 68). Let us now consider the cytologic properties of chromosomes in more detail. The genetic material in nondividing cells resides in **chromatin,** a diffuse, irregular material dispersed throughout the nucleus. Prior to cell division, the chromatin condenses into deeply staining, wormlike bodies of varying length and shape, the chromosomes. They vary enormously in size in different species, but are of predictable size and shape for a particular species. Even within a single organism, however, the chromosomes may vary in size from tissue to tissue. Some chromosomes have the **centromere** or **primary constriction** (point of spindle-fiber attachment) at the midpoint, with equal arms on each side (metacentric); others have the centromere closer to one end, with unequal arms (acrocentric) or at the very end of the chromosome (telocentric).

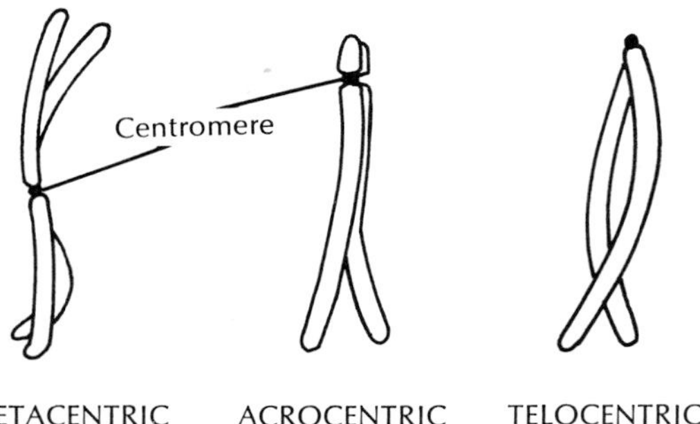

Staining differences are evident on different parts of the chromosome during prophase of cell division. Condensed and deeply staining regions are called **heterochromatin,** in contrast to the less densely stained remainder of the chromosome, called **euchromatin.** These regions are constant for any chromosome. Heterochromatin regions may not have functional genes, although DNA is present there and may code for microtubules. Investigations have thus far failed to reveal the function of heterochromatin, but there is the suggestion that it represents genetic polymorphism that is expressed in different blood groups, tissue enzymes, and so on.

Landmarks on chromosomes, other than the **centromere,** are one or more **secondary constrictions,** where the nucleoli are located, and **chromomeres,** densely staining enlargements that are conspicuous features of many chromosomes during early prophase of meiosis I. Chromosomes look like numerous small beads strung together on a string; this gave rise to the erroneous idea that each bead is a gene.

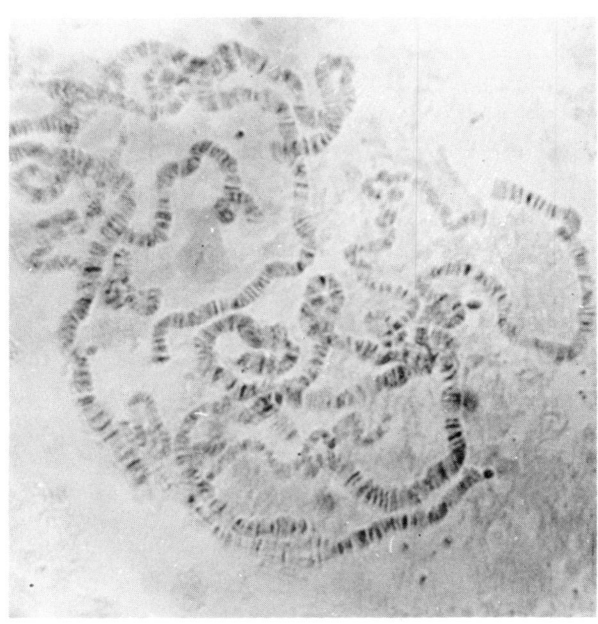

FIG. 37-4 Chromosomes from a salivary gland cell of larval fruit fly *Drosophila.* These are among the largest chromosomes found in animal cells. Bands of nucleoproteins may be loci of genes. Such chromosomes are sometimes called polytene because they appear to be made up of many chromonemata. These chromosomes are not confined to salivary glands but are also known to occur in other organs, such as gut and malpighian tubules of most dipteran insects. Technique for their study is simply to crush salivary glands in drop of acetocarmine, between cover glass and slide, so that chromosomes are set free from the nuclei and are spread out as shown in the photograph. (Courtesy General Biological Supply House, Inc., Chicago.)

Giant chromosomes

Details about the structure of chromosomes have been difficult to obtain because of their small size in most animals. Since most chromosomes are only a few microns in length, and each one bears thousands of genes, there was little hope of discovering the real nature of the physical units of heredity. About 1934 T. Painter of the University of Texas and two German investigators, E. Heitz and H. Bauer, independently discovered in the salivary glands of the larvae of *Drosophilia* and other flies chromosomes many times

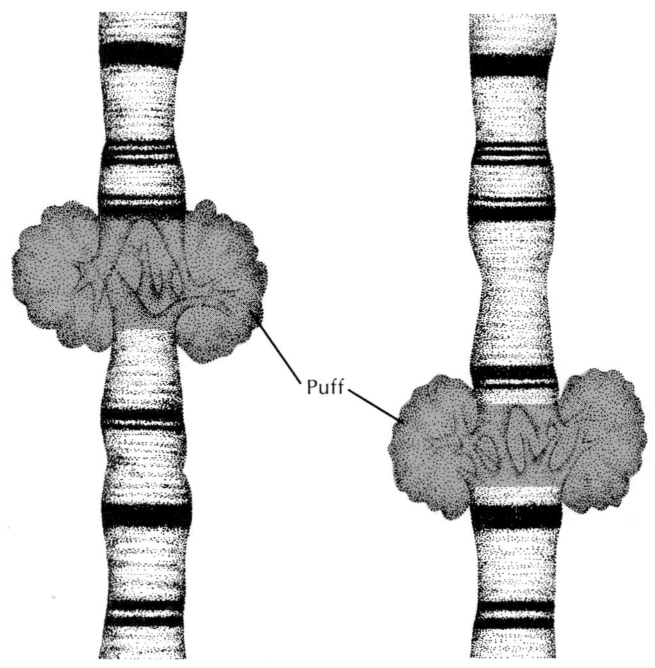

FIG. 37-5 Puffing in one of the bands of a salivary gland chromosome of a midge larva *(Chironomus).* Swelling, or puff, indicates activity in a region where RNA (and perhaps some DNA) is being produced, and it may include single bands or a group of adjacent ones. Puffs always include the same bands that occur in a definite sequence during development of larva.

larger than those of the ordinary somatic or germinal chromosomes of these forms. Actually, these giant chromosomes had been discovered as early as 1881 by the Italian cytologist Balbiani in the larval forms of the midge fly *Chironomus,* but he did not recognize them as chromosomes. Their rediscovery marked a new era in the development of cytogenetics.

The salivary glands of larval flies are a pair of club-shaped bodies attached to the pharynx. Each gland is composed of only about 100 unusually large cells. Salivary tissue grows by an increase in cell size and not by an increase in cell number. The giant chromosomes of these cells are elongated, ribbonlike bodies about 100 to 200 times longer than the ordinary chromosome (Fig. 37-4). In some flies they lie separated from each other; in *Drosophila* they are attached to a dark, heterochromatic mass called the **chromocenter.**

What are these chromosomes, and what do they show? The chromosomes are somatic prophase chromosomes with the homologous chromosomes closely paired throughout their length. One of their most striking characteristics is the transverse bands that appear along their length. Another feature is the number of **chromonemata** (central threads) they possess. In the ordinary somatic chromosome there may be only one or two of these gene strings, but in the salivary gland chromosomes there may be between 512 and 1,024 *(Drosophila).* This indicates that the chromonemata have divided many times without being accompanied by the division of the whole chromosome; hence, they are often called **polytene** chromosomes. A polytene chromosome is a typical mitotic chromosome that has uncoiled and undergone many repeated duplications, which have remained together in the same nucleus (karyokinesis without cytokinesis).

The transverse bands are made up of chromatic granules, the chromomeres. These bands result from the lateral apposition of the chromomeres on the adjacent fibrils or chromonemata. More than 6,000 of these bands have been found on the three large chromosomes of *Drosophilia.* The bands contain much DNA, and each may be considered the equivalent of the conceptual gene. In the regions between the bands there is little DNA. Another aspect of giant chromosomes is the "puffs," which are local and reversible enlargements in the bands (Fig. 37-5). Each "puff" may be the result of the unfolding or uncoiling of the chromosomes in a band. The puffing may be large (Balbiani's rings) or small. The size of the puff is an indication of gene activity. Evidence seems to indicate that RNA synthesis is produced at the puff, where

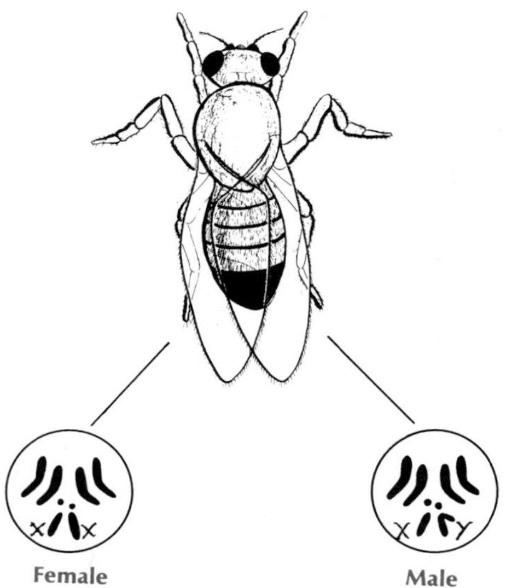

FIG. 37-6 Fruit fly *(Drosophila melanogaster)* and diploid set of chromosomes of each sex. Sex chromosomes X and Y are marked. Male fruit flies have three black bands on the abdomen, which is rounded at tip; females have five black bands, with pointed abdomen. Many genetic concepts have been developed from extensive investigations on this now classic animal.

it makes a complementary copy of a DNA strand. The RNA messenger is then carried to the ribosome, where it serves for synthesis of proteins. (A discussion of the coding for protein synthesis is found on p. 864.)

The even larger lampbrush chromosomes found in amphibian oocytes and some invertebrates are characterized by loops extending laterally, giving the appearance of a brush. These chromosomes appear to be composed of two chromatids that form loops (gene loci) when they are active but are coiled up within a chromomere when at rest.

Constancy of chromosome number

Every somatic cell in a given organism contains the same number of chromosomes. Somatic chromosomes of diploid organisms are found in pairs, with the members of each pair being alike in size, in position of spindle attachment, and in bearing genes relating to the same hereditary characters. In each homologous pair of chromosomes, one has come from the father and the other from the mother. The lowest diploid number in the cell of any organism is two, which is found in certain roundworms; the largest number (300 or more) is found in some protozoa. In most forms the number is between 12 and 40, the most common diploid number being 24. The fruit fly has a diploid number of four (Fig. 37-6), and humans have 46 (Fig. 37-7).

Obviously many forms wholly unrelated have the same number of chromosomes, and so the number is without significance. What matters is the nature of the genes on the chromosomes. Some chromosomes are as small as 0.25 μm in length, and (with the exception of giant chromosomes) some are as long as 50 μm. Lampbrush chromosomes may be as much as 800 μm in length. Within the haploid set of chromosomes of most species there are considerable differences in size and shape. In humans, most chromosomes are 4 to 6 μm in length, although some range up to 12 μm in length.

Mature germ cells have only half as many chromosomes as somatic cells because the meiotic divisions have reduced the chromosomes into haploid sets (pp. 805 to 809).

Significance of meiosis

Because it is so important, let us review once again the significance of meiosis. As previously noted, meiosis is a double cell division in which the chromosomes replicate only once. When the paired, homologous chromosomes separate during the first meiotic (''reductional'') division, the homologous genes must also separate, one gene going to each of the germ cells produced. Thus at the end of the maturation process each mature gamete (egg or sperm) contains one gene of every pair or a single set of every kind of gene (haploid number), instead of two genes for each character as in somatic cells (diploid number). The haploid number of chromosomes found in the germ cells does not consist of just any half of the diploid somatic number but must include one of the members of each homologous pair of chromosomes. It is a matter of chance in the reductional division whether the paternal chromosome of a homologous pair goes to one daughter cell or the other, and the same is true of the maternal chromosome.

However, meiosis does more than merely reduce the chromosome number to the haploid condition so that when gametes join at fertilization, the correct somatic chromosome number is restored. During prophase of the first meiotic division, when the homologous chromosomes are paired, chromatid breaks occur, and there is a physical exchange of reciprocal chromosomal parts. This phenomenon is called cross-

A

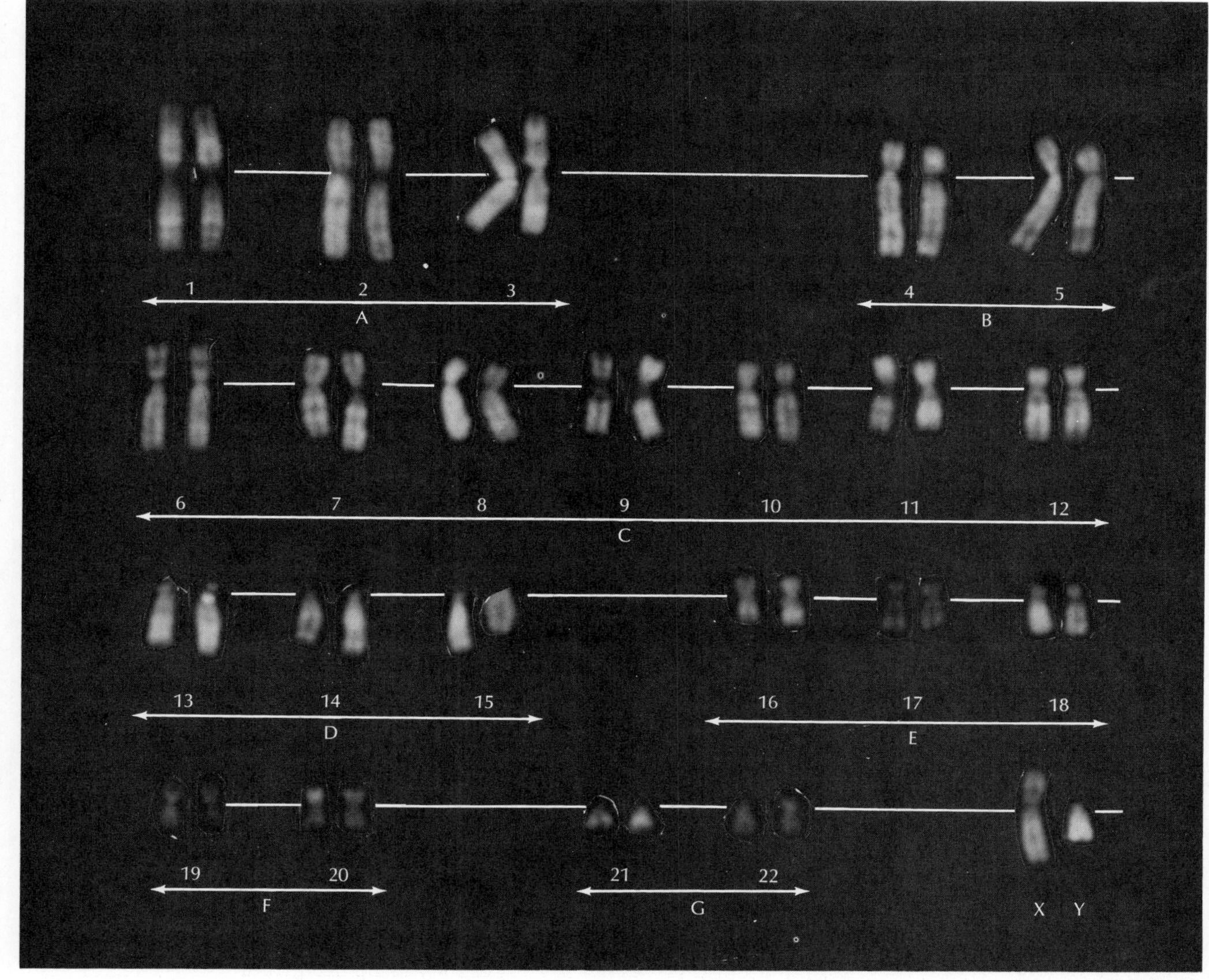

Continued.

FIG. 37-7 Human male **(A)** and human female **(B)** chromosomes. The 46 chromosomes (diploid number) are arranged according to a standard pattern (karyotype). All of the human chromosomes can be identified because they differ in size, centromere position, and banding patterns produced by specific staining techniques. (Courtesy L. R. Emmons, Washington and Lee University.)

ing-over, and, it will be recalled, the chiasmata are its visible manifestation (Fig. 35-4, p. 806). Crossing-over generates **recombinate** chromosomes, that is, chromosomes bearing genotypes that are different from the parental combinations. Therefore, specific arrays of genes on a chromosome do not always go together as a package, a point we will return to later in this chapter (p. 859).

The independent assortment (segregation) of chromosomes during meiosis is a completely random process that produces many new recombinations of chromosomes in the gametes formed. In the fruit fly, which has 4 pairs of chromosomes, there are 2^4, or 16, different kinds of sperm or kinds of eggs. In humans with 23 pairs of chromosomes, the number of possible combinatons increases to 2^{23}, or approximately 8.4 million. The number of variations is further increased by chromosomal crossing-over during meiosis, which

B

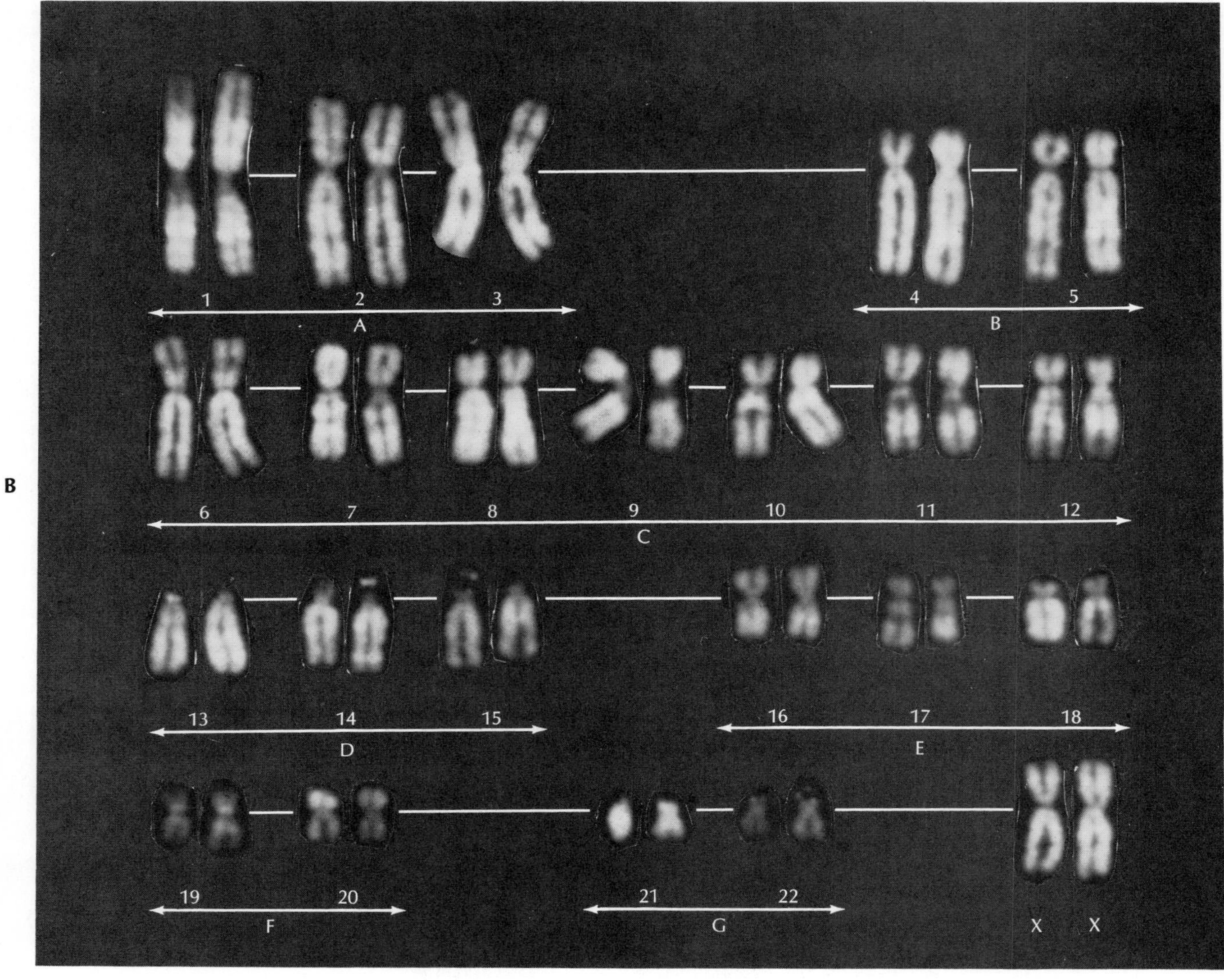

FIG. 37-7, cont'd For legend see p. 853.

allows recombinaton of linked genes between homologous chromosomes. Furthermore, the random fusion of gametes from *both* parents adds yet another source of variation. All of these factors enrich biologic variation, which, in turn, makes a diversified evolution possible.

Sex determination

Before the importance of the chromosomes in heredity was realized in the early 1900s, the determination of sex was totally unknown. Speculation produced several incredible theories, for example, that the two testicles of the male contained different types of semen, one begetting males, the other females. It is not difficult to imagine the abuse and mutilation of domestic animals that occurred when attempts were made to alter the sex ratios of herds. Another theory asserted that sex of the offspring was determined by the more heavily sexed parent. An especially masculine father should produce sons, an effeminate father only daughters. Such ideas were not testable and have lingered until recently.

The first really scientific clue to the determination of sex came in 1891 when Henking noticed that the sperm of a heteropteran insect were of two types, produced in equal numbers. One type carried 11 chromosomes; the other carried the same 11 plus a densely staining body that contrasted sharply with the chromosomes. Unable to determine its true nature, he labeled it "X" for unknown. Eleven years later McClung, working with grasshoppers, realized that the sex of the embryo was determined by whichever of the two types of sperm fertilized the egg. He incorrectly concluded that the sperm carrying the odd "X" chromosome was male-determining, and the sperm lacking this chromosome was female-determining. Actually, in this species the reverse is the case, and this error was soon corrected by others. But McClung nevertheless was the first to establish the chromosomal basis of sex determination. Later, the sex-determining X chromosomes came to be called **sex chromosomes** to distinguish them from the **autosomes** that carry other genes, which determine other characteristics.

The particular type of sex determination discovered by McClung is often called the XX-XO type, which indicates that the females have two X chromosomes and the males only one X chromosome (the O represents its absence). The XX-XO method of sex determination can be depicted as follows:

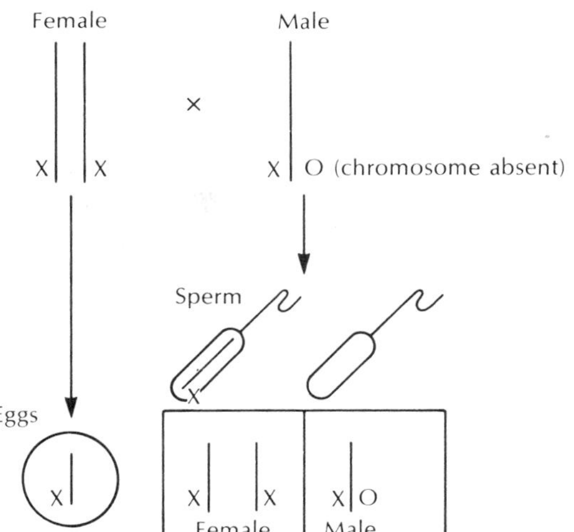

Later, other types of sex determination were discovered. In humans and many other forms there are the same number of chromosomes in each sex; however, the sex chromosomes (XX) are alike in the female but are heteromorphic, or unlike (XY), in the male. Hence the human egg contains 22 autosomes + one X chromosome; the sperm are of two kinds: half carry 22 autosomes + one X, and half bear 22 autosomes + one Y. The Y chromosome is much smaller than the X. At fertilization, when two X chromosomes come together, the offspring are female; when XY, they are male. The XX-XY kind of determination is depicted as follows:

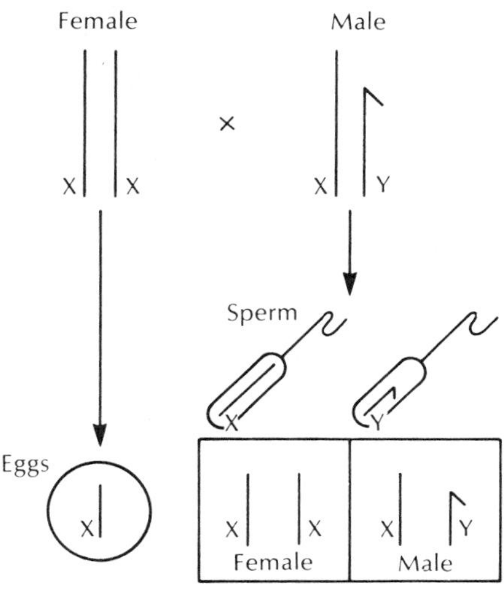

A third type of sex determination is found in birds, moths, butterflies, and some fishes in which the male is the homogametic sex with two X chromosomes and the female is heterogametic with an X and a Y chromosome. This is just the reverse of that found in humans.

How the sex-determining mechanism works

From the foregoing discussion, we see that there are really two common chromosomal mechanisms of sex determination: XX-XO and XX-XY. In the grasshopper and other animals with the XX-XO mechanism, sex is determined by the balance between the number of X chromosomes and autosomes present. The autosomes bear male-determining genes whereas the X chromosome bears female-determining genes. An organism with two X chromosomes is a female because the two Xs, with their female-determining genes, outweigh the male-determining genes of the autosomes. If only one X is present, the balance is tipped in favor of the autosomes, and the organism is a male. The sex of

Drosophila is determined by this mechanism, even though *Drosophila* has a Y chromosome (Fig. 37-6), because the Y has no effect on sex determination (the Y is essential for male fertility in this animal, however).

Until recently it was thought that sex determination in mammals was a balance mechanism like that of *Drosophila*. However, we now know that the short arm of the Y chromosome contains male-determining genes, and for an embryo to develop as a male the short arm of the Y chromosome must be present.

In humans, why is the X chromosome so much larger than the Y? These two chromosomes are certainly not genetic equals. The X chromosome contains several genes that have nothing to do with sex; in fact only one gene is known for certain to have a direct bearing on sex determination. The rest specify the synthesis of various metabolic enzymes, blood-clotting factors, and factors for normal color vision—all completely unrelated to sex. These are called **sex-linked genes** because they are on the X chromosome. Their pattern of inheritance is unusual, and it will be discussed later.

The Y chromosome is small and apparently carries no sex-linked genes. It consists of repetitious DNA base sequences of no known functional significance. The Y appears to contain just one important gene, the male-determining gene, located on the short arm of the chromosome.

Since the X chromosome carries several genes that are essential for normal cellular function and the Y chromosome carries none, every animal must have *at least* one X chromosome. And of course the XX-XY system assures that they will: the female has two, the male one. To keep the balance between X-chromosome genes and autosome genes the same in the female as in the male, one of the female's X chromosomes is always inactivated. This happens very early in development when the embryo consists of only a few cells, according to the widely accepted Lyon hypothesis (after Mary Lyon, who discovered this and other aspects of human sex determination). At this time it is strictly a matter of chance whether the maternal or paternal X chromosome is the one inactivated in each cell. All descendants of that cell, however, retain the same inactivated chromosome. Consequently, the adult female is a patchwork of tissues expressing the genes of one or the other X chromosome, never both. This is an odd feature that normally is of no consequence in the human female, but there is one dramatic example of its effect.

Tortoise-shell (calico, tabby) cats, which are spotted black and yellow, are *always* females; the males are either all black or all yellow. This happens because the genes for fur color are located on the X chromosome, that is, they are sex-linked. A male, having only one X chromosome, will be either black or yellow. But a female has two X chromosomes and, if she is heterozygous for fur color, her fur will be a mosaic of yellow and black patches. Each patch represents the descendants of an early embryonic skin cell having one operating and one inactive X chromosome.

Expression of phenotypic sex

Genetic sex is just one aspect of sex determination. Having an XX or an XY genetic constitution does not in itself produce a female or male. The embryo is at first totally indifferent sexually, despite its genetic sex. Even the gonad is sexually indifferent and bipotential, since it can differentiate into either a testis or an ovary. Gonad differentiation was discussed in Chapter 35 (p. 805), and we pointed out there that the primordial gonad has an inherent tendency to become an ovary, and the rest of the body to become female, *unless* the male-determining gene on the Y chromosomes redirects gonadal differentiation into a testis.

This brings us to the subject of sex hormones and their important role in sex determination. Until the gonads begin producing sex hormones (androgens by the testes, estrogens by the ovaries), the gonads are remarkably labile. Just how labile has been demonstrated with fish and amphibian larvae, by exposing genetic males to estrogens and genetic females to androgens. This treatment completely reverses the functional sex; the genetic males, now functional females, produce fully fertile eggs! They are female behaviorally and morphologically—indeed in every way except genetically. Likewise, the genetic females that have been exposed to testosterone at an early age become phenotypic males that produce fertile sperm. Obviously in these forms, genetic sex merely points to a developmental direction that, of course, is normally followed.

In mammals, gonad differentiation is more predetermined, and treatment with hormones of the opposite sex can never cause an ovary to become a testis nor a testis an ovary. There is a good reason for this.

Mammals are viviparous, the fetus developing in the uterus where it is exposed to female sex hormones manufactured by the mother's ovaries and by its own placenta. If the mammalian gonad were as plastic as that of fishes, mammals would give birth only to females. Nevertheless, sex determination in mammals, as in any vertebrate, is a complicated sequence of events that spreads throughout the body—and can, unfortunately, go wrong at more than one stage during differentiation.

The first step in creating a functional male, for example, is determination of genetic sex; in the mammalian male, this means the formation of an XY zygote at fertilization. Next, a gene or genes on the Y chromosome direct the indifferent gonad to become a testis. The testis soon begins synthesizing androgens, principally testosterone, that causes male-specific organs to differentiate: prostate, seminal vesicles, and penis. In the female, estrogens have a parallel effect in inducing female organs to develop: fallopian tubes, uterus, and vagina. The important point is that hormone production is controlled by the structure of the gonad and not *directly* by its genotype.

The sex hormones are also responsible for the development of the secondary sex characters—physical differences between the sexes that serve to attract opposite sexes to each other, facilitate copulation, and provide protection and nutrition for the young. In the human species, these changes are most dramatic at puberty, at which time sex hormone production by the ovaries and testes increases significantly. A young girl's oviducts, uterus, and vagina all enlarge; changes in bone structure cause the waist to narrow and the pelvis to broaden. Breast development and deposition of fat under the skin at this time both contribute to the characteristic feminine figure. In the male at puberty, increased testosterone production stimulates beard growth, lowering of the voice, and increased muscular development.

Finally, in all vertebrates, the sex hormones, by influencing brain function, determine behavioral sex, especially the distinctive male-female differences in sexual desire, sexual advertising, courtship behavior, territoriality, and care of the young.

Sex-linked inheritance

It has long been known that the inheritance of some characters depends on the sex of the parent carrying the gene and the sex of the offspring. An example is red-green color blindness in humans in which red and green colors are indistinguishable to varying degrees. Color-blind men greatly outnumber color-blind women. When color blindness does appear in women, their fathers are color blind. Furthermore, if a woman with normal vision who is a carrier of color blindness bears sons, half of them are likely to be color blind, regardless of whether the father had normal or affected vision. How are these observations explained?

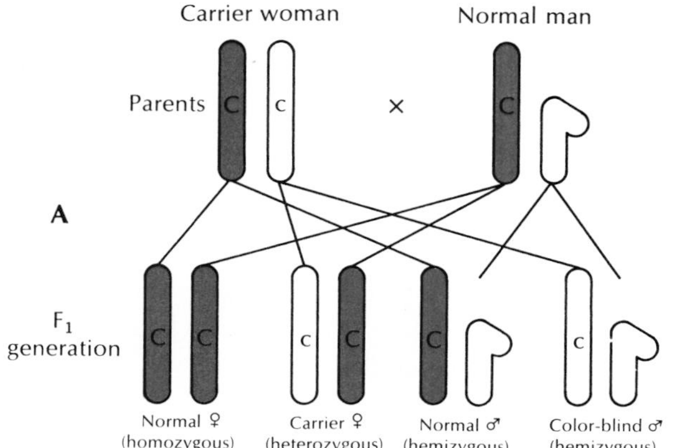

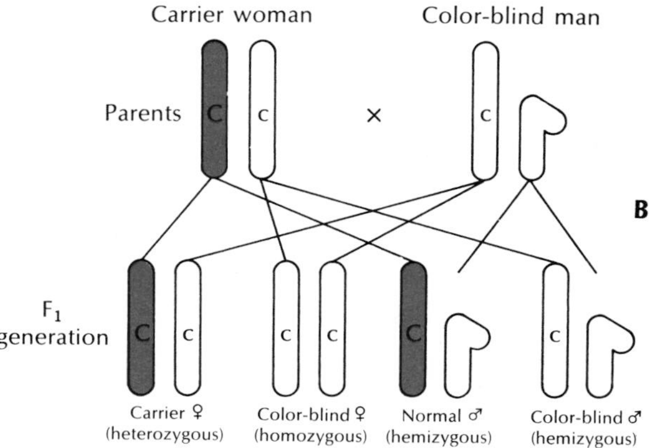

FIG. 37-8 Sex-linked inheritance of red-green color blindness in humans. The gene determining normal color vision is carried on the X chromosome. The Y chromosome carries no gene for color vision. **A,** Carrier mother and normal father produce color blindness in one-half of their sons but in none of their daughters. **B,** Half of both sons and daughters of carrier mother and color-blind father are color blind.

The color-blindness defect is a recessive trait carried on the X chromosome that is visibly expressed either when both genes are defective in the female or when only one defective gene is present in the male. The inheritance pattern is shown in Fig. 37-8. When the mother is a carrier and the father is normal, half of the sons but none of the daughters are color blind. However, if the father is color blind, half of the sons *and* half of the daughters are color blind. It is easy to understand then why the defect is much more prevalent in males: a single sex-linked recessive gene in the male has a visible effect. What would be the outcome of a mating between a homozygous normal woman and a color-blind man?

Another example of a sex-linked character was discovered by Morgan in *Drosophila*. The normal eye color of this fly is red, but mutations for white eyes do occur. The genes for eye color are known to be carried in the X chromosome. If a white-eyed male and a red-eyed female are crossed, all the F_1 offspring are red eyed, because this trait is dominant (Fig. 37-9). If these F_1 offspring are interbred, all the females of F_2 have red eyes, half the males have red eyes, and the other half have white eyes. No white-eyed females are found in this generation; only the males have the recessive character (white eyes). The gene for being

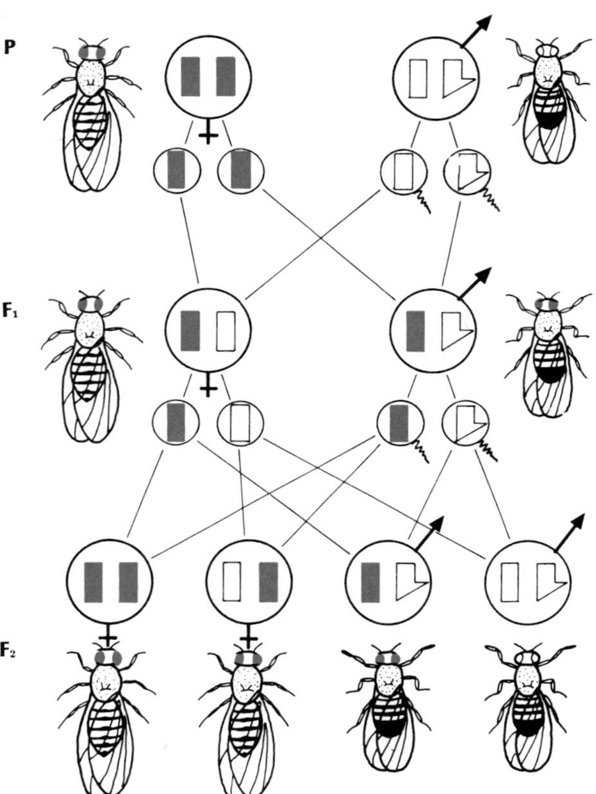

FIG. 37-9 Sex determination and sex-linked inheritance of eye color in fruit fly *(Drosophila)*. Normal red eye color is dominant to white eye color. If a homozygous red-eyed female and a white-eyed male are mated, all F_1 flies are red eyed. When F_1 flies are intercrossed, F_2 yields approximately 1 homozygous red-eyed female and 1 heterozygous red-eyed female to 1 red-eyed male and 1 white-eyed male. Genes for red eyes and white eyes are carried by sex (X) chromosomes; Y carries no genes for eye color.

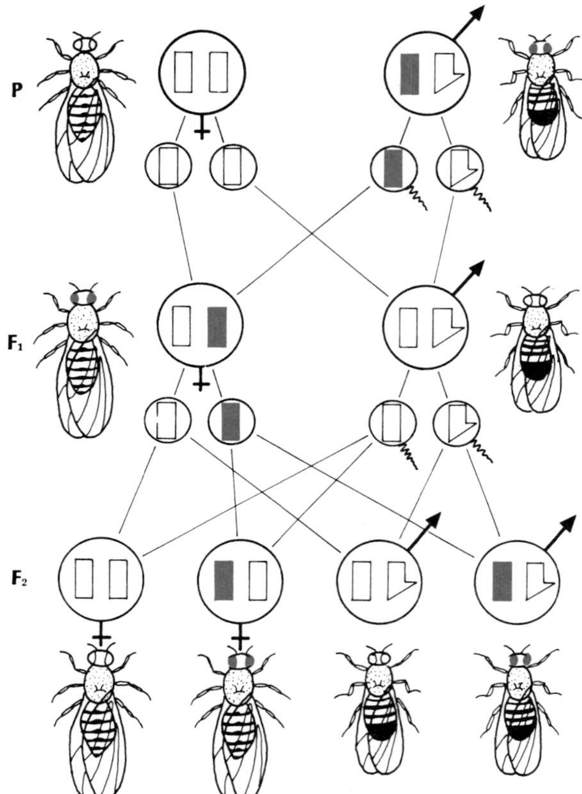

FIG. 37-10 In cross of a homozygous white-eyed female and a heterozygous red-eyed male (reciprocal cross of Fig. 37-9), F_1 consists of white-eyed males and red-eyed females. In the F_2, there are equal numbers of red-eyed and white-eyed females and red-eyed and white-eyed males.

white eyed is recessive and should appear in a homozygous condition. However, since the male has only one X chromosome (the Y does not carry a gene for eye color), white eyes appear whenever the X chromosome carries the gene for this trait. If the reciprocal cross is made in which the females are white eyed and the males red eyed, all the F_1 females are red eyed and all the males are white eyed (Fig. 37-10). This is called **crisscross inheritance** because the phenotypes of the parents are exchanged in the offspring. If these F_1 offspring are interbred, the F_2 generation shows equal numbers of red-eyed and white-eyed males and females.

If the allele for red eyed is represented by **R,** white eyed by **r,** the female sex chromosome by X, and the male sex chromosome lacking a gene for eye color by Y, the following diagrams show the inheritance of this eye color:

Parents	X^RX^R (red ♀)	×	X^rY (white ♂)	
F_1	X^RX^r (red ♀)	×	X^RY (red ♂)	
Gametes	X^R, X^r		X^R, Y	
F_2 genotypes	X^RX^R	X^RY	X^rY	X^rX^R
F_2 phenotypes	(red ♀)	(red ♂)	(white ♂)	(red ♀)

RECIPROCAL CROSS

Parents	X^rX^r (white ♀)	×	X^RY (red ♂)	
F_1	X^rX^R (red ♀)	×	X^rY (white ♂)	
Gametes	X^r, X^R		X^r, Y	
F_2 genotypes	X^rX^r	X^rY	X^RX^r	X^RY
F_2 phenotypes	(white ♀)	(white ♂)	(red ♀)	(red ♂)

Autosomal linkage and crossing-over

Linkage

Since Mendel's laws were rediscovered in 1900, it became apparent that, contrary to Mendel's second law, not all factors segregate independently. Indeed, many traits are inherited together. Since the number of chromosomes in any organism is relatively small compared to the number of traits, each chromosome must contain many genes. All genes present on a chromosome are **linked.** Linkage simply means that the genes are on the same chromosome, and all genes present on homologous chromosomes belong to the same linkage groups. Therefore there should be as many linkage groups as there are chromosome pairs.

In *Drosophila,* in which this principle has been worked out most extensively, there are four linkage groups that correspond to the four pairs of chromosomes found in these fruit flies. Small chromosomes have small linkage groups, and large chromosomes have large groups. More than 500 genes have been mapped in the fruit fly, and they all are distributed among the four pairs of chromosomes.

Let us see how the Mendelian ratios can be altered by linkage, as illustrated by one of Morgan's experiments on *Drosophila.* In fruit fly genetics, the normal allele of any gene is called the **wild type,** since that allele is the most widespread in the wild state. It is usually dominant over its sister alleles, which are considered mutations of the normal wild allele. Morgan made a cross between wild-type fruit flies with normal body and normal wings and flies bearing two recessive mutant characters of black body and vestigial wings.

As expected the F_1 generation of this cross was phenotypically the wild type, confirming that the alleles for black body and vestigial wing were recessive.

Morgan than made a testcross to learn more about the genotype of the F_1 generation. This was done by breeding back the F_1 hybrid generation (AaBb) to the double recessive flies of the parental generation (aabb). With independent assortment we should expect four different phenotypes in approximately equal numbers:

Phenotype	Expected ratio	Numbers obtained
Wild type	1	586 (46%)
Normal body, vestigial wing	1	106 (8%)
Black body, normal wing	1	111 (9%)
Black body, vestigial wing	1	465 (37%)

Instead Morgan obtained an excess of **parental** types (wild type and double-recessive type) and a deficiency of the two gene recombinations (called **recombinant** types). In a testcross such as this, linkage is indicated if the proportion of parental types exceeds 50%. This cross yielded 83% parental types and 17% recombinant types. Morgan concluded that the wild type and black-vestigial type had entered the dihybrid cross together and stayed together, or linked.

Crossing-over

Linkage, however, is usually only partial. In the experiment just described, the parental forms totaled 83% rather than 100% as expected if linkage were complete. The fact that some recombinant types appear means that the linked genes have indeed separated, in this experiment 17% of the time. Separation of genes located on the same chromosome occurs because of **crossing-over.**

As described earlier, during the protracted prophase

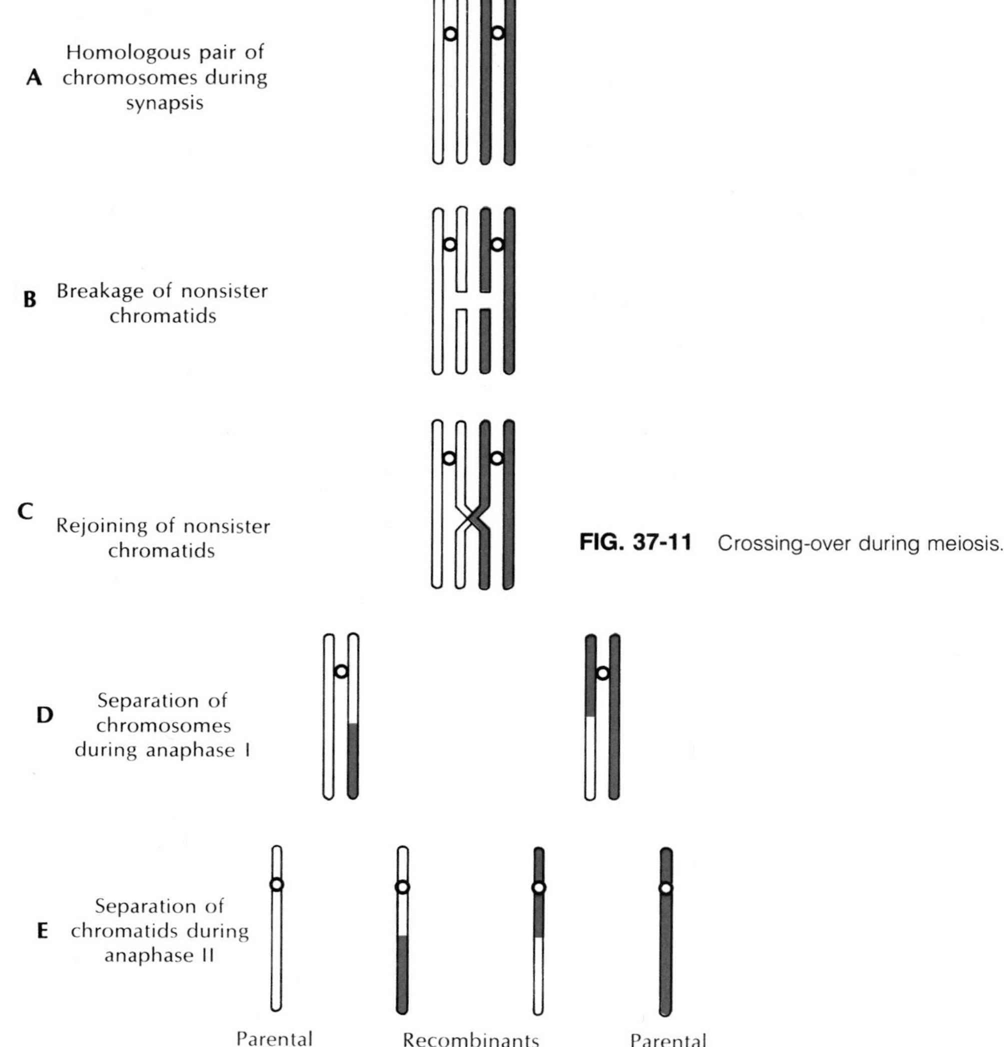

FIG. 37-11 Crossing-over during meiosis.

of the first meiotic division, homologous chromosomes break and exchange equivalent portions; genes "cross over" from one chromosome to its homologue, and vice versa (Fig. 37-11). Each chromosome consists of two sister chromatids held together by means of a synaptonemal complex. Breaks and exchanges occur at corresponding points on nonsister chromatids. (Breaks and exchanges also occur between sister chromatids but have no genetic significance because sister chromatids are identical.) Crossing-over then is a means for exchanging genes between homologous chromosomes and as such greatly increases the amount of genetic recombination. The frequency of crossing-over varies with the species, but usually at least one and often several crossovers occur each time chromosomes pair.

Gene mapping

Crossing-over makes possible the construction of chromosome maps and provides proof that the genes lie in a linear order on the chromosomes. Crossing-over does not occur randomly throughout the length of the chromosome. The greater the distance is between genes, the greater is the probability that a crossover occurs between them. Two genes located at opposite ends of the chromosome are separated almost every time a

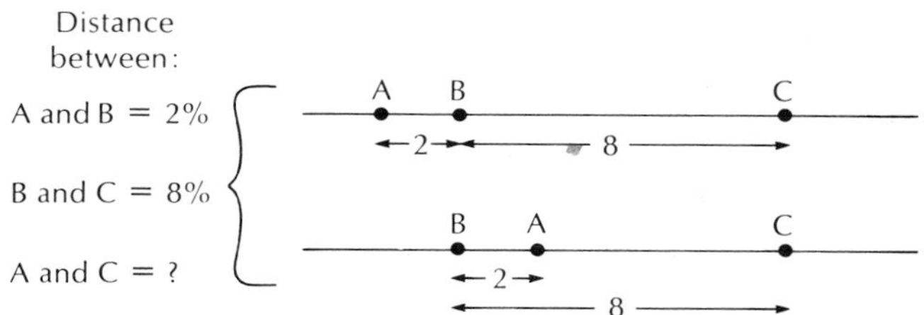

FIG. 37-12 Tentative arrangement of three genes on a chromosome. Distances are given in linkage map units with each unit defined as equal to 1% frequency of recombination. The third cross, *A* and *C*, is necessary to resolve which of the two possible arrangements proposed by the first two crosses is the correct one.

break occurs; if they are located close together, they are separated only when the chromosome chances to break between them. Therefore the *frequency of recombination* indicates the relative position of the genes.

In our example of the cross between wild-type and black-vestigial–type flies, the frequency of recombination (percentage of offspring that are recombinants) is 17%. By itself, this value does not indicate the location of the two genes involved. But, if a third gene is added, their arrangement can be determined by making three crosses. Let us take a hypothetical example of three genes (**A, B, C**) on the same chromosome (Fig. 37-12).

In the determination of their comparative linear position on the chromosome, we first need to find the crossing-over value between any two of these genes. If **A** and **B** have a crossing-over rate of 2% and **B** and **C** of 8%, then the crossing-over percentage between **A** and **C** should be either the sum (2 + 8) or the difference (8 − 2). If it is 10%, **B** lies between **A** and **C;** if 6%, **A** is between **B** and **C.** By laborious genetic experiments for many years, the famed chromosome maps in *Drosophila* were worked out in this manner (Fig. 37-13). More recent cytologic investigations on the giant chromosomes present in the salivary glands of fruit fly larvae tend to prove the correctness of the linear order if not the actual position of the genes on the chromosomes.

Chromosomal aberrations

Structural and numerical deviations from the norm that affect many genes at once are called chromosomal aberrations. They are sometimes called chromosomal mutations, but most cytogeneticists prefer to use the term mutation to refer to qualitative changes within a gene; gene mutations will be discussed on p. 872.

Despite the incredible precision of meiosis, chromosomal aberrations do occur, and they are more common than one might think. They are responsible for great economic benefit in agriculture. Unfortunately, they are also responsible for many human genetic malformations. It is estimated that five out of every 1,000 humans are born with *serious* genetic defects, attributable to chromosomal anomalies. An even greater number of chromosomal defects are aborted spontaneously, far more than ever reach term.

Changes in chromosome numbers are called **euploidy** when there is the addition or deletion of whole chromosome sets and **aneuploidy** when a single chromosome is added or subtracted from a diploid set. The most common kind of euploidy is **polyploidy,** the carrying of one or more additional sets of chromosomes. Such aberrations are much more common in plants than animals. Animals are much less tolerant of chromosomal aberrations because sex determination requires a delicate balance between the numbers of sex chromosomes and autosomes. Many domestic plant species are polyploid (cotton, wheat, apples, oats, tobacco, and others), and perhaps one-third of flowering plants are believed to have originated in this manner. Horticulturists favor polyploids and often try to develope them because they have more intensely colored flowers and more vigorous vegetative growth.

Aneuploidy is usually caused by nondisjunctional separation of chromosomes during meiosis. If a pair of chromosomes fails to separate during the first or second meiotic divisions, both members go to one pole and none to the other. This results in one gamete having n − 1 number of chromosomes and another having n + 1 number of chromosomes. If the n − 1 gamete is fertilized by a normal n gamete, the result is a **monosomic** animal. Survival is rare because the lack of one chromosome gives an uneven balance of genetic instructions. **Trisomy,** the result of the fusion of a normal n gamete and an n + 1 gamete, is much more common, and several kinds of trisomic conditions are known in humans. Perhaps the most familiar is **trisomy 21,** often called **mongolism.** As the name indicates, it involves an extra chromosome 21 combined with the

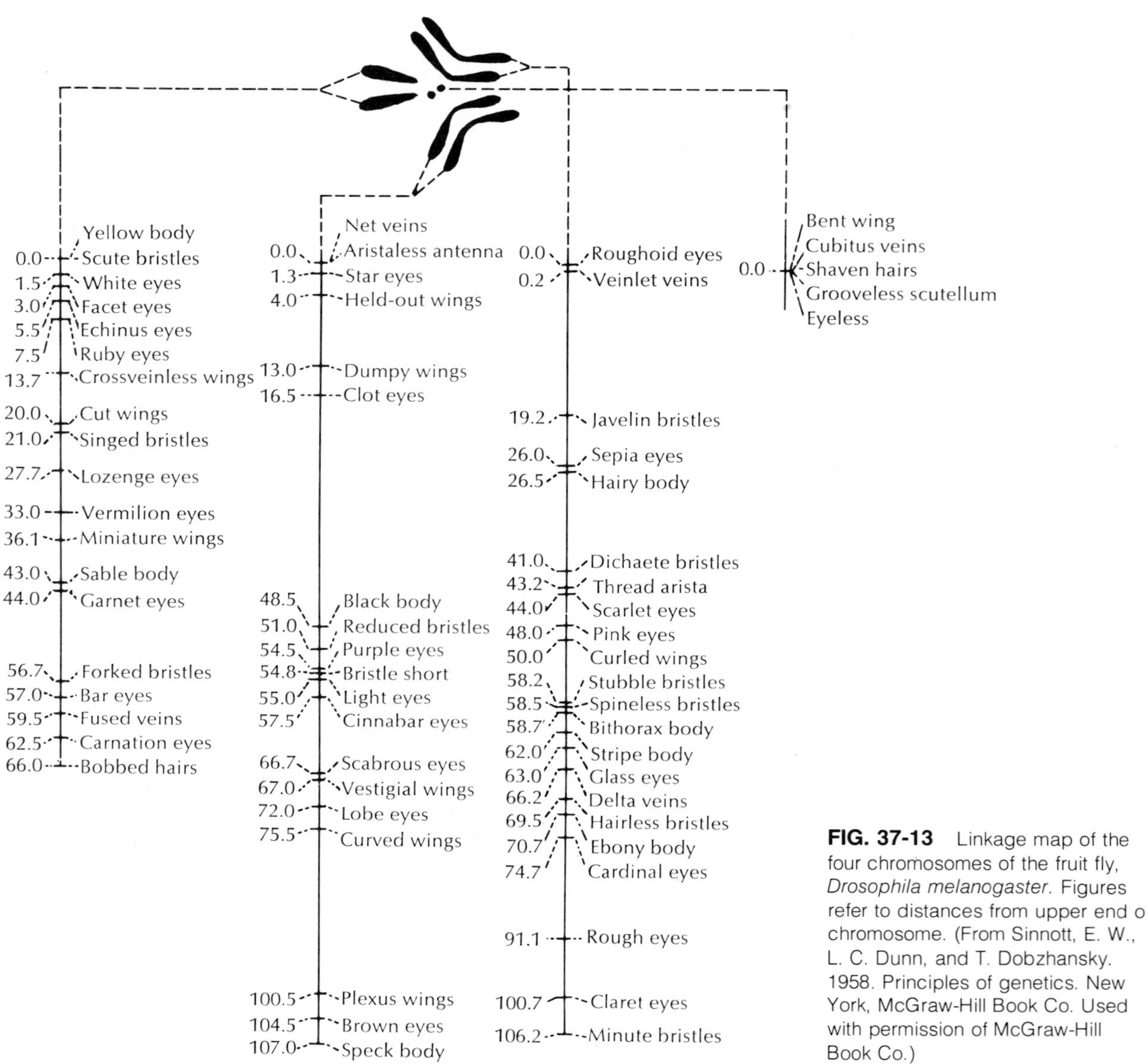

FIG. 37-13 Linkage map of the four chromosomes of the fruit fly, *Drosophila melanogaster.* Figures refer to distances from upper end of chromosome. (From Sinnott, E. W., L. C. Dunn, and T. Dobzhansky. 1958. Principles of genetics. New York, McGraw-Hill Book Co. Used with permission of McGraw-Hill Book Co.)

chromosome pair 21, and it is caused by nondisjunction of that pair during meiosis. It occurs spontaneously, and there seldom is any family history of the abnormality. However, the risk of its appearance rises dramatically with increasing age of the mother: it occurs 40 times as often in women over 40 years old than among women between the ages of 20 and 30.

Structural aberrations involve whole sets of genes within a chromosome. A portion of a chromosome may be reversed, placing the linear arrangement of genes in reverse order (inversion); nonhomologous chromosomes may exchange sections (translocation); entire blocks of genes may be lost (deletion); or a section of chromosome may attach to a normal chromosome (duplication). These are all structural changes that often produce phenotypic changes that are inherited in the normal Mendelian manner. Not all such alterations are deleterious. Duplications, though rare, are important for evolution because they supply additional genetic information that may assume new functions.

GENE THEORY
The gene concept

The term "gene" (Gr., *genos,* descent) was coined by W. Johannsen in 1909 to refer to the hereditary factors of Mendel (1865). Both cytologic and genetic studies showed that genes, though of unknown chemical nature, were the fundamental units of inheritance. They were regarded as indivisible units of the chromosomes on which they were located. The American *Drosophila* geneticist T. H. Morgan, in the early part of this century, was fond of referring to the supposed linear arrangement of genes on chromosomes as "beads on a string." The metaphor was a useful one at the time but has proved to be only partially true. It is correct that genes are arranged linearly on chromosomes but it is not correct that they behave like indivisible beads. Studies with multiple mutant alleles demonstrated that alleles were in fact divisible by recombination; that is, *portions* of a gene are separable, and they have a fine structure.

As a result of new insights developed after 1950 on gene structure and function, the gene emerged as a **unit of function,** called a **cistron.** A cistron is by no means the smallest divisible unit of a gene because mutations may occur within a cistron, but it is the smallest *functional* region on a chromosome. In general, a "gene" is synonymous with "cistron," although a classic Mendelian gene may consist of more than one cistron. The reader may already detect a semantic problem here and may wonder if introducing the synonym "cistron" for gene is helpful. To the geneticist it is helpful because it replaces the unitary concept of the physical gene with the concept of an operational gene composed of one or more functional components, each of which embraces an array of mutant sites. Out of this, the gene concept emerges more or less intact, as a unit that behaves in a Mendelian fashion.

As the chief functional unit of genetic material, genes determine the basic architecture of every cell, the nature and life of the cell, the specific protein syntheses, the enzyme formation, the self-reproduction of the cell, and, directly or indirectly, the entire metabolic function of the cell. By their property to mutate, to be assorted and shuffled around in different combinations, genes have become the basis for our modern interpretation of evolution. Genes are molecular patterns that can maintain their identities for many generations, can be self-duplicated in each generation, and can control cell processes by allowing their specificities to be copied.

One gene–one enzyme hypothesis

Since genes act to produce different phenotypes, we may infer that their action follows the scheme: gene → gene product → phenotypic expression. Further, we may suspect that the gene product is usually a protein, because proteins, acting as enzymes, antibodies, hormones, and structural elements throughout the body, are the single most important group of biomolecules. In the late 1930s, a few geneticists began to narrow their attention to one probable gene product in particular: enzymes. Enzymes were known to be involved in every aspect of life. Cellular activities cannot proceed without them because every cellular biosynthetic pathway entails the action of a specific enzyme at every step.

The first clear, well-documented study to link genes and enzymes was carried out on the common bread mold *Neurospora* by Beadle and Tatum in the early 1940s. This organism was ideally suited to a study of gene function for several reasons: these molds are much simpler to handle than fruit flies, they grow readily in well-defined chemical media, and they are haploid organisms that are consequently unencumbered with dominance relationships. Furthermore, mutations were readily induced by irradiation with ultraviolet light. Ultraviolet-light–induced mutants, grown and tested in specific nutrient media, were found to have single-gene mutations that were inherited in accord with Mendelian principles of segregation. Each mutant strain was shown to be defective in one enzyme, which prevented that strain from synthesizing one or more complex molecules. Putting it another way, the ability to synthesize a particular molecule was controlled by a single gene.

From these experiments Beadle and Tatum set forth an important and exciting formulation: **one gene produces one enzyme** (see Principle 20, p. 14). For this work Beadle and Tatum were awarded the Nobel Prize in 1958. The new hypothesis was soon consolidated by the research of others who studied other biosynthetic pathways. We now know of hundreds of inherited disorders, including dozens of human hereditary diseases, that are caused by single mutant genes that result in the loss of a specific essential enzyme. Thus the one gene–one enzyme hypothesis has proved to be of enormous value as a stimulus to a biochemical approach to gene function. Today we recognize that while most genes direct the synthesis of enzymes, others code for polypeptide chains, antibodies, hormones, and various kinds of RNA.

Protein synthesis

Enzymes are proteins composed of long chains of amino acids linked by peptide bonds. (See pp. 32 to 36 for a description of protein structure.) The sequence of amino acids in a polypeptide chain is highly ordered, and it is this sequence, together with the spiraling and complex folding of the chain, that provides each kind of protein with a unique identity. The loss, misplacement, or substitution of a single amino acid will alter this structure and may interfere with, or even destroy, a protein's function. We should expect, then, that the primary structure of a protein—that is, its amino acid sequence—is a direct reflection of the linear structure of the gene. Again mutations, which geneticists have employed in many ingenious ways to reveal the nature of gene function, were used to demonstrate the anticipated **collinearity** between a gene and its corresponding polypeptide. This correspondence may seem obvious but it was very difficult to prove and occupied the best efforts of many biochemical geneticists for more than a decade. Collinearity was finally verified by Yanofsky and his colleagues in 1966. Out of this painstaking genetic dissection emerged one of the important concepts of biology: genes determine the primary sequence of amino acids in specific proteins (see Principles 17 to 19, p. 13).

STORAGE AND TRANSMISSION OF GENETIC INFORMATION

What a gene can do is intimately associated with its chemical structure. Let us now consider the molecular structure and function of genes.

Nucleic acids: molecular basis of inheritance

It has been known for many years that genes and chromosomes are made up chiefly of nucleoproteins. These are macromolecules composed of nucleic acids, special basic proteins called histones, and some complex residual proteins. Of particular interest are the **nucleic acids.** Their structure is described in some detail in Chapter 2 (p. 36) and the reader is encouraged to review that section at this time. In summary, a nucleic acid is chemically made up of **nucleotides,** which in turn are composed of a purine or pyrimidine base, a sugar, and phosphoric acid.

On the basis of the kind of sugar (deoxyribose or ribose), the nucleic acids are divided into two main groups: **deoxyribonucleic acid (DNA)** and **ribonucleic acid (RNA).** DNA occurs in the nucleus, where it is the major structural component of genes (see Principle 17, p. 13); RNA is found throughout the cell, being especially abundant in nucleoli and in the cytoplasm. Thus a nucleic-acid molecule is made up of many nucleotides joined to form long chains (Fig. 2-17, p. 37). The purine units are adenine and guanine; the pyrimidines are cytosine, thymine, and uracil. Five kinds of nucleotides are recognized on the basis of these purines and pyrimidines: (1) adenine-sugar-phosphate, (2) guanine-sugar-phosphate, (3) cytosine-sugar-phosphate, (4) thymine-sugar-phosphate, and (5) uracil-sugar-phosphate. The DNA molecule has the first four of these nucleotides (Fig. 2-18, p. 838); the RNA has the first three and the last one. Although the phosphate-sugar part of the long chain of nucleotides is regular, the purine or pyrimidine base attached to the sugar is not always the same. The order of these bases varies from one section to another of the nucleic acid molecule. Depending on the proportion and sequence of the nucleotides, there is an almost unlimited variety of nucleic acids.

The positive identification of the nucleic acids as the repository of genetic information came in the early 1950s after many decades of research. Although DNA was first isolated by Miescher in 1869 (who called it nuclein), just 4 years after Mendel published his findings, it was not until 1944 that O. T. Avery and his associates implicated DNA as the genetic material in bacteria. However, many geneticists were reluctant to accept these conclusions, arguing that Avery's findings were a special case having little relevance to inheritance in other organisms. Then, in 1952, Hershey and Chase established beyond all reasonable doubt that DNA was the heritable material. This launched an intensive activity in laboratories all over the world, culminating in the momentous discovery of the structure of DNA by Watson and Crick at Cambridge University in 1953. A lively and absorbing personal account of the discovery is given in James Watson's book *The Double Helix*.

The double helix

The Watson-Crick model of DNA structure had to suggest plausible answers to such problems as (1) how specific directions are transmitted from one generation to another, (2) how DNA could control protein synthesis, and (3) how the DNA molecule could duplicate itself. Classic genetics and cytology had shown how a cell divides to form two cells and how each cell receives a set of chromosomes with their genes identical in structure to the preexisting set. But nothing was known about how a chemical substance could carry out the specifications required by the genetic substance of a gene. The elegance of the Watson-Crick hypothesis lies in the perfect manner in which it fits the data and in the way that it can be tested.

Wilkins succeeded in getting very sharp x-ray diffraction patterns that revealed three major periodic spacings in crystalline DNA. These periodicities of 3.4, 20, and 34 Å were interpreted by Watson and Crick as the space distance between successive nucleotides in the DNA chains, the width of the chain, and the distance between successive turns of the helix, respectively. The x-ray diffraction photograph, with certain limitations, also gave indications of the spatial arrangement of some of the atoms within the large molecule. Watson and Crick came up with the idea that the molecule is bipartite, with an overall helical configuration. Accordingly, their model showed that the DNA molecule looks like two interlocked coils. Each "coil" consists of a sugar-phosphate backbone held together by phosphodiester bonds. The two interlocked coils are held together by hydrogen bonds between the purine and pyrimidine bases that form the core of the double helix.

Base pairing. The two DNA strands run in opposite directions, that is, they are **antiparallel.** This is evident from an examination of Fig. 2-18 (p. 38). The two strands are also **complementary**—the sequence of bases along one strand specifies the sequence of bases along the other strand. Each strand could thus serve as a **template** for the production of its complement.

From examination of their original wire model of DNA, Watson and Crick proposed a base-pairing rule. Adenine at one point on a strand must bond with thymine at the corresponding point on the other strand. Similarly, guanine must bond with cytosine. This happens because only these pairs are of proper size and arrangement of hydrogens and acceptor atoms for efficient bonding.

The purine bases adenine and guanine (A and G) are large, whereas the pyrimidine bases thymine and cytosine (T and C) are small. To fit into the structure of the DNA molecule, each pair must consist of one large and one small base; thus the small bases T and C, T and T, or C and C could not bond in pairs because they would not meet in the middle. The big bases (A and G) could not bond in any combination because they would produce a bulge in the helix. Neither can A and C nor G and T (although in these pairs one is large and one is small) form bond mates because their hydrogen bonds would not pair off properly. Thus there is only one correct way for the bases to bond: A with T and G with C. This is shown in the circled cross sections to the right of the DNA molecule in Fig. 2-18 (p. 38).

Although all DNA molecules have the same general pattern, each one is unique because of the varying sequence of bases. This sequence spells out the genetic instructions that determine the sequence of amino acids in a protein organized by the gene (see Principle 17, p. 13). This genetic message, or "code," is written in a language of only four symbols (adenine, guanine, thymine, and cytosine) and is translated into protein amino acid sequences (see Principle 18, p. 13). The coding will be discussed subsequently.

DNA replication

Every time the cell divides, the structure of the nuclear DNA must be precisely copied in the daughter cells. The wonder of the DNA double helix model was that its structure suggested how it could be replicated: each existing strand of the DNA molecule serves as a template for the synthesis of a new strand. This is the essential requirement of a genetic molecule, which was not satisfied by other models (such as single or triple helix) proposed before the Watson-Crick model. The two strands of the helix could unwind to expose the nucleotide bases. The double helix then separates, perhaps like a zipper, and new nucleotides are incorporated to synthesize new strands. Each of the four kinds of nucleotides consists of one of the DNA bases, one sugar unit, and three phosphate groups. Only one of the three phosphate groups is used in making the backbone of the new strand; the other two are necessary to provide energy for synthesis.

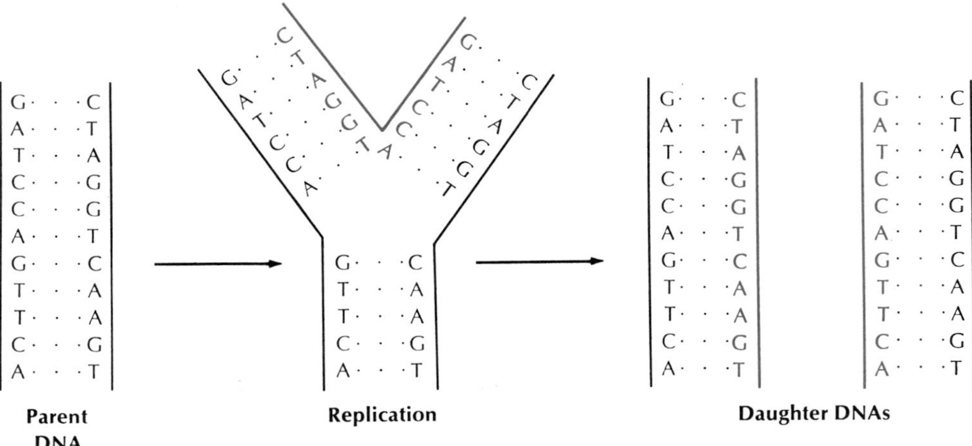

Because of the strict complementarity (base-pairing) rule, each exposed base will pair only with its complementary base. Thus each of the two strands will reform a double helix identical with the parental molecule. Note that each daughter molecule will contain one parental strand and one new strand. The transcription process has been clearly visualized in electron micrographs prepared by O. L. Miller and his associates (Fig. 37-14).

A large protein enzyme, **DNA polymerase,** is essential for replication. The enzyme does not determine which of the four bases will be added to the new chain (base selection is determined by complementation with the base sequence in the parent strand) but it does hold the reacting molecules in position and vastly accelerates the reaction.

If replication proceeds as predicted, a replication "fork" should be visible where the double helix is "unzipping" during replication. Forks have in fact been seen in the circular DNA molecules from bacteria. How the fork is formed is still unclear. Nor is it yet known how the two intertwined strands of DNA untwist during replication. They may rotate during the process, but it is inconceivable that the entire chromosomal DNA molecule, millions of base pairs in length, flips about within the chromosome. When the hydrogen bonds holding base pairs together are broken during replication, the two strands move apart from one another, even though in the double helical form they appear to be interlocked.

DNA coding by base sequence

We saw earlier that the gene has a linear structure that is collinear with the sequence of amino acids in a protein. Since DNA is the genetic material and is composed of a linear sequence of base pairs, an obvious extension of the Watson-Crick model is that the sequence of base pairs in DNA codes for, and is collinear with, the sequence of amino acids in a protein (see Principle 18, p. 13). The coding hypothesis had to account for the way a string of four different bases —a four-letter alphabet—could dictate the sequence of 20 different amino acids.

In the coding procedure, obviously there cannot be a 1:1 correlation between four bases and 20 amino acids. If the coding unit (often called a word, or **codon**) consists of two bases, only 16 words (4^2) can be formed, which cannot account for 20 amino acids. Therefore, the protein code must consist of at least three bases or three letters because 64 possible words (4^3) can be formed by four bases when taken as triplets. This means that there must be a considerable redundancy of triplets (codons), since DNA codes for just 20 amino acids. Later work by Crick confirmed that nearly all of

FIG. 37-14 Genes in action. This remarkable electron micrograph shows DNA transcribing ribosomal RNA molecules in the oocyte nucleoli of the spotted newt *Diemictylus (Triturus) viridescens.* These are tandemly repeated genes separated by spacer (nontranscribing) segments of DNA. The newest and shortest ribosomal RNA transcripts are close to the initiation site on the gene. When completed molecules reach the termination site, they are freed to move out of the nucleus into the cytoplasm, where they form ribosomes. Enormous quantities of ribosomes are synthesized in maturing amphibian eggs. (×25,000.) (Courtesy O. L. Miller, Jr. and D. L. Beatty, University of Virginia.)

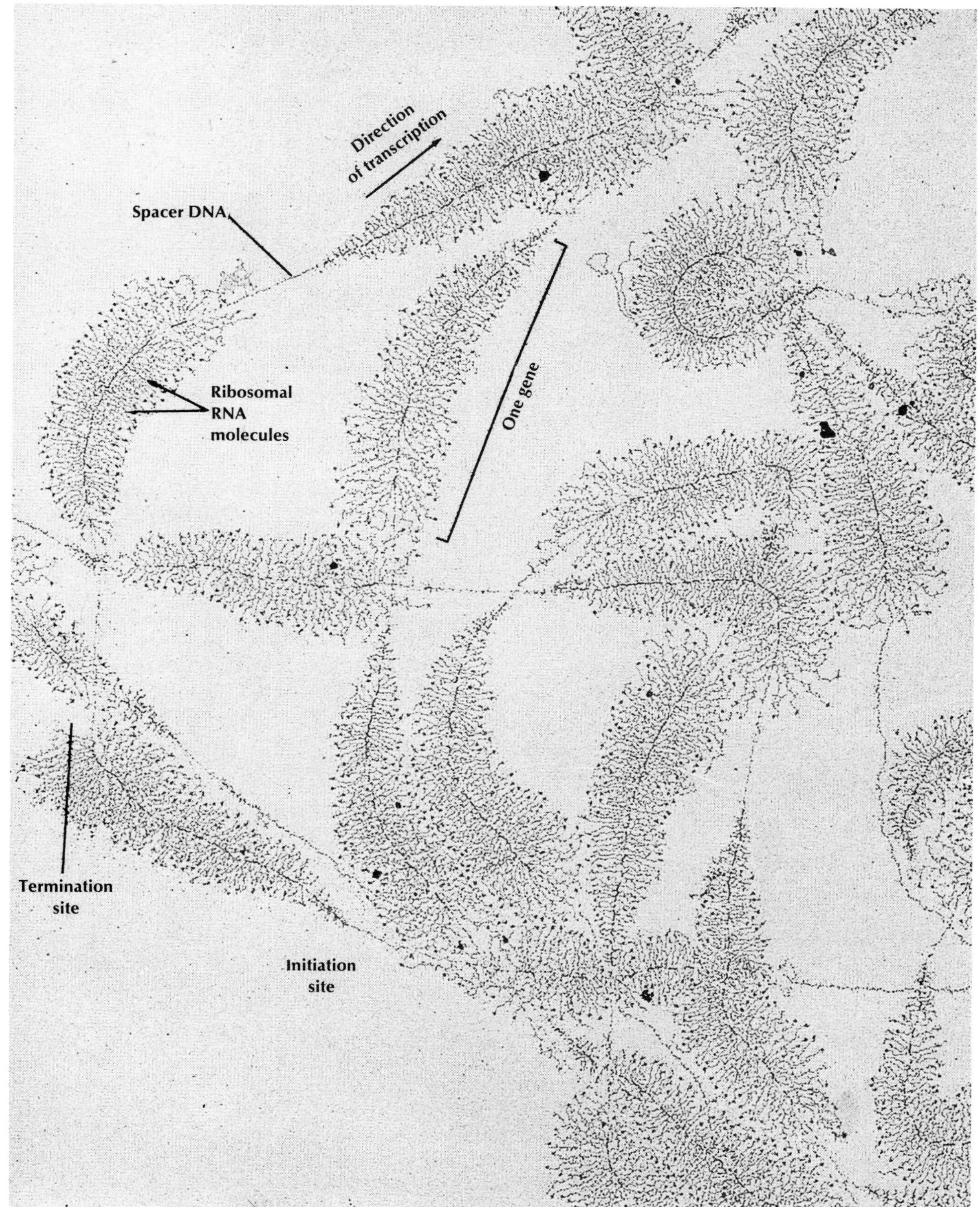

FIG. 37-14 For legend see opposite page.

TABLE 37-1 The genetic code: codons (code triplets) between messenger RNA and specific amino acids

Codons	Amino acid
GCU, GCC, GCA, GCG	Alanine
CGU, CGC, CGA, CGG, AGA	Arginine
AAU, AAC	Asparagine
GAU, GAC	Aspartic acid
UGU, UGC	Cysteine
GAA, GAG	Glutamic acid
CAA, CAG	Glutamine
GGU, GGC, GGA, GGG	Glycine
CAU, CAC	Histidine
AUU, AUC, AUA	Isoleucine
CUU, CUC, CUA, CUG, UUA, UUG	Leucine
AAA, AAG	Lysine
AUG	Methionine; initiation of message
UUU, UUC	Phenylalanine
CCU, CCC, CCA, CCG	Proline
AGU, AGC, UCU, UCC, UCA, UCG	Serine
ACU, ACC, ACA, ACG	Threonine
UGG	Tryptophan
UAU, UAC	Tyrosine
GUU, GUC, GUA, GUG	Valine
UAA, UAG, UGA	Termination of message of one gene

the amino acids are specified by more than one code triplet (Table 37-1).

Transcription and the role of messenger RNA

Information is coded in nuclear DNA, whereas protein synthesis occurs in the cytoplasm. It is obvious that an intermediary is required to bridge the two regions. This intermediary is another nucleic acid called **messenger RNA** (see Principle 19, p. 13). Recall from Chapter 2 (p. 39) that RNA (ribonucleic acid) differs from DNA in three important ways: (1) it is *single* stranded and not a double helix, (2) it has ribose sugar in its nucleotides instead of deoxyribose, and (3) it has the pyrimidine uracil (U) instead of thymine (T). Despite these differences, RNA is constructed very much like a single strand of DNA.

In eukaryotes, messenger RNA is believed to be **transcribed** directly from DNA in the nucleus, with each of the many messenger RNAs being determined by a gene or a particular segment of DNA. In this process of making a complementary copy of one strand or gene of DNA in the formation of messenger RNA, an enzyme, **RNA polymerase,** is needed. The messenger RNA contains a sequence of bases that complements the bases in one of the two DNA strands just as the DNA strands complement each other. Thus, A in the coding DNA strand is replaced by U in messenger RNA; C is replaced by G; G is replaced by C; and T is replaced by A. Only one of the two chains is used as the template for RNA synthesis because only one bears the AUG codon that initiates a message (Table 37-1). The reason why only one strand of the double-stranded DNA is a "coding strand" is that messenger RNA otherwise would always be formed in complementary pairs, and enzymes also would be synthesized in complementary pairs. In other words, two different enzymes would be produced for every DNA coding sequence instead of one. This would certainly lead to metabolic chaos.

Translation: final stage in information transfer

The translation process takes place on **ribosomes,** granular structures composed of protein and nucleic acid (see Principle 18, p. 13). The messenger RNA molecules fix themselves to the ribosomes to form a messenger RNA–ribosome complex. Since only a short section of a messenger RNA molecule is in contact with a single ribosome, the messenger RNA usually fixes itself to several ribosomes at once. This arrangement, called a polyribosome, allows several proteins of the same kind to be synthesized at once, one on each ribosome of the polyribosome (Fig. 37-15).

The assembly of proteins on the messenger RNA–ribosome complex requires the action of another kind of RNA called **transfer RNA.** The transfer RNA molecules collect the free amino acids from the cytoplasm and deliver them to the messenger RNA–ribosome complex, where they are assembled into a protein. There is a special transfer RNA molecule for every amino acid. Furthermore each transfer RNA is accompanied by a special **activating enzyme.** These enzymes are necessary to sort out and attach the correct amino acid to each transfer RNA by a process called **loading.**

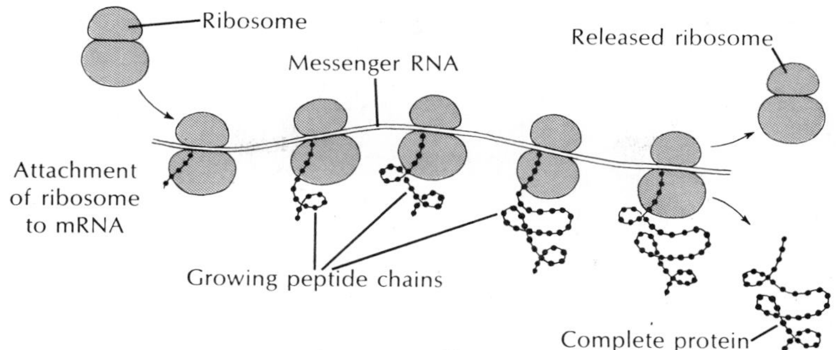

FIG. 37-15 How the protein chain is formed. As ribosomes move along messenger RNA, the amino acids are added stepwise to form the polypeptide chain.

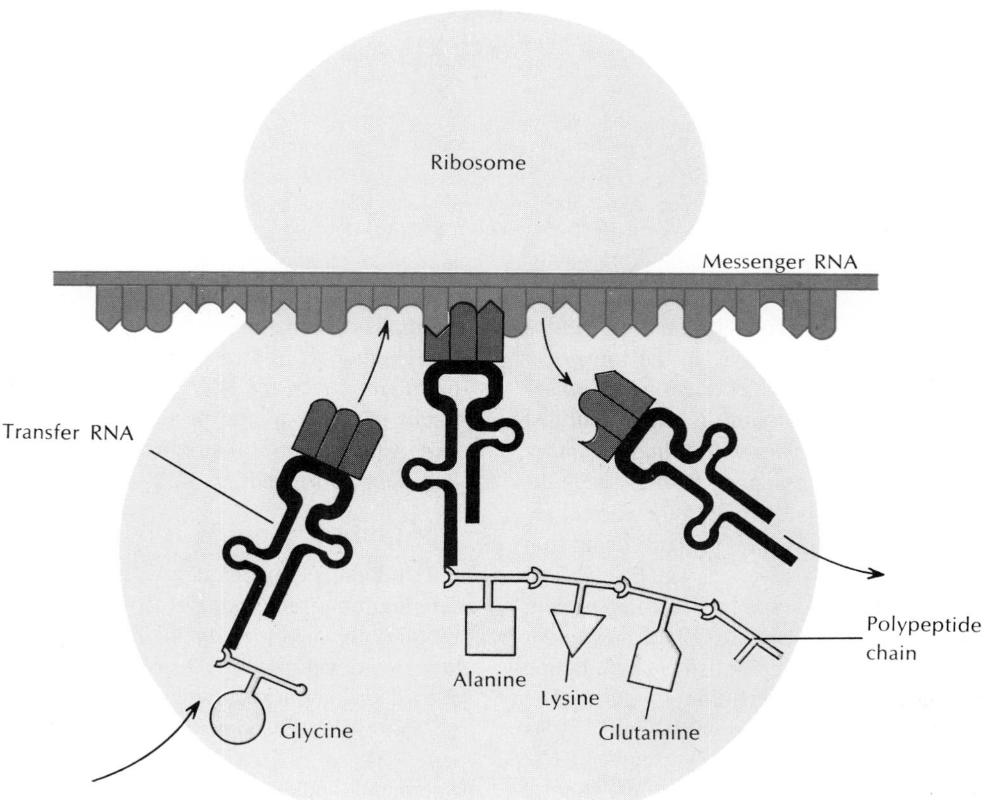

FIG. 37-16 Formation of a polypeptide chain on messenger RNA. As the ribosome moves down the messenger RNA molecule, transfer RNA molecules with attached amino acids enter the ribosome *(left)*. Amino acids are joined together into a polypeptide chain, and transfer RNA molecules leave the ribosome *(right)*.

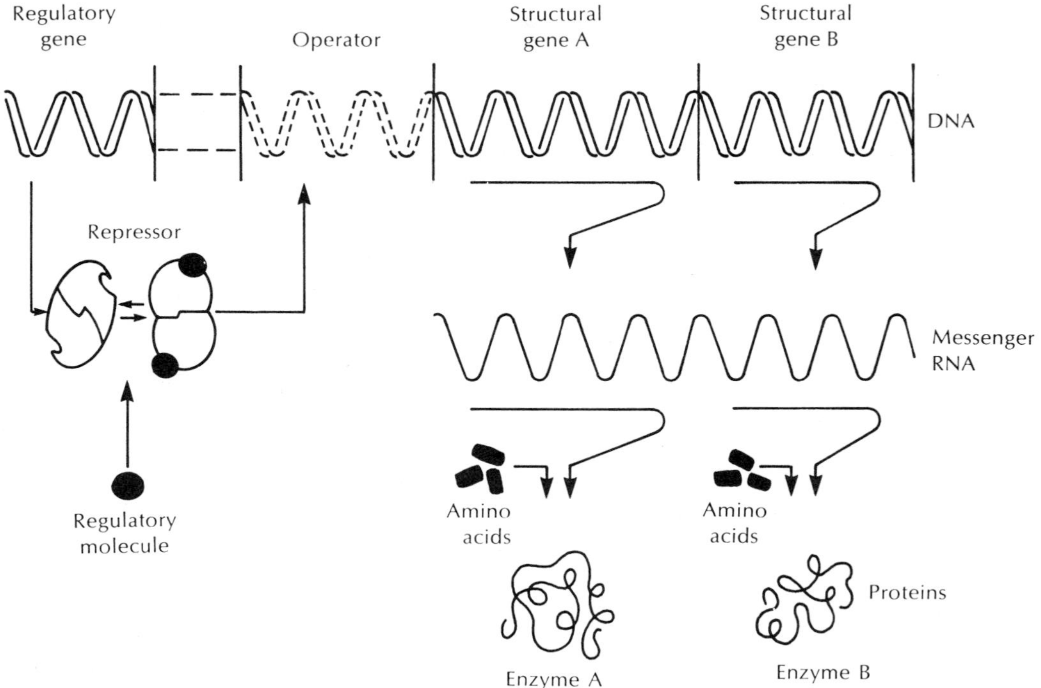

FIG. 37-17 Regulation of gene function. Regulator gene acts by way of repressor on the operator. Repressor, in combination with regulatory molecules (hormones or products of cellular metabolism), binds with and inhibits operator, thus preventing structural genes from producing messenger RNA. (Modified from Monod, J. 1966. Endocrinology **78:**412-425.)

The transfer RNAs are surprisingly large molecules that are folded in a complicated way in the form of a cloverleaf (Fig. 37-16). On this cloverleaf a special sequence of three bases (the anticodon) is exposed in just the right way to form base pairs with complementary bases (the codon) in the messenger RNA. The anticodon on the transfer RNA is the key to the correct sequencing of amino acids in the protein being assembled.

For example, alanine is assembled into a protein when it is signaled by the codon GCG in a messenger RNA. This translation is accomplished by alanine transfer RNA in which the anticodon is CGC. The alanine transfer RNA is first loaded with alanine by its activating enzyme. The loaded alanine transfer RNA enters the messenger RNA–ribosome complex where it fits precisely into the right place on the messenger RNA strand. Then the next loaded transfer RNA specified by the messenger RNA code (glycine transfer RNA, for example) enters the messenger RNA–ribosome complex and attaches itself beside the alanine transfer RNA. The two amino acids are united with a peptide bond (with the assistance of a molecule of ATP), and the alanine transfer RNA falls off. The process continues stepwise as the protein chain is built (Fig. 37-15). A protein of 500 amino acids can be assembled in less than 30 seconds.

Regulation of gene function

In the preceding chapter we saw how the orderly differentiation of an organism from fertilized egg to adult requires the involvement of genetic material at every stage of development. Developmental biologists have provided convincing evidence that every cell in a developing embryo is genetically equivalent (see Principle 24, p. 15). This was first demonstrated by Spemann's delayed nucleation experiments (p. 822) and more recently confirmed with the elegant nuclear transplantation experiments of Gurdon and his associates in England (p. 824). Thus it is clear that as tissues differentiate, they use only a part of the genetic instructions present in every cell. Certain genes express themselves only at certain times and not at other times. Indeed, there is reason to believe that in a particular cell or tissue, most of the genes are inactive at any given moment. The problem in development is to explain how, if every cell has a full gene complement, certain genes are "turned on" and produce proteins that are required for a particular developmental stage while the other genes remain silent.

Actually, whereas the developmental process brings the question of gene activation clearly into focus, gene regulation is necessary throughout an organism's life existence. The cellular enzyme systems that control all functional processes obviously require genetic regulation because enzymes have powerful effects even in minute amounts. Enzyme synthesis must be responsive to the influences of supply and demand.

Operon model for gene regulation

The mechanism of gene transcription and translation that we have described in this chapter explains how the code, carried on DNA molecules of the nucleus, is transcribed into definite proteins synthesized in the cytoplasm. It does not explain how genes are turned on as their products are needed by the cell; nor does it explain why certain enzymes are not formed when they are not needed.

In one group of organisms, the bacteria, we now have a clear understanding of how genetic material is utilized to regulate the synthesis of proteins. The **operon concept,** as it is called, was proposed by two French scientists, J. Monod and F. Jacob, in 1961. By 1965, when Monod and Jacob were awarded the Nobel Prize for their research, the model had been strongly supported by additional genetic and biochemical evidence from several other laboratories around the world.

In this model, the basic unit carrying the genetic code for protein synthesis is the **structural gene,** whose actions are regulated in a complex way (Fig. 37-17). Adjacent to the structural gene is a special segment of DNA called the **operator.** Together, the structural gene and operator comprise a genetic unit called the **operon.** The operon may contain a single structural gene or several structural genes of related function. The operator is in turn negatively controlled by a **regulator gene** that can produce a protein molecule called a **repressor.** The repressor may bind with and so inhibit the operator. The operator, then, is the receptor site for the repressor. While repressed, the operator prevents the structural gene or genes from forming messenger RNA.

For the operon to resume functioning, that is, form messenger RNA required for protein synthesis, the operator must be **derepressed.** One way that this can be accomplished is for the repressor molecule to bind with and be inactivated by other materials in the cell. This renders the repressor incapable of turning the operator off. The regulatory molecule could be a hormone, or it could be some material synthesized within the cell by the enzymes produced by the operon. For example, if there is a high concentration of a particular enzyme in the cell, this high concentration can act as a feedback through the repressor to block the action of the operator so that the structural gene can no longer produce the enzymes. Alternatively, the repressor may be changed to an inactive form by a lower-than-normal concentration of the enzyme.

This important **operon hypothesis,** which has been described in a much-abbreviated form, seeks to explain the mechanism for repressing the synthesis of enzymes when they are not needed and for inducing them when they are needed. Since it is based almost entirely on work with bacteria and viruses, it remains to be seen whether or not it also applies to higher forms of life.

SOURCES OF PHENOTYPIC VARIATION

We will conclude this chapter by considering the creative force of evolution: biologic variation. Without variability among individuals, there could be no continued adaptation to a changing environment and no evolution.

There are actually several sources of variability, some of which we have already described. The independent assortment of chromosomes during meiosis is a random process that creates new chromosomal recombinations in the gametes. In addition, chromosomal crossing-over during meiosis allows recombination of linked genes between homologous chromosomes, further increasing variability. Finally, the random fusion of gametes from both parents produces still another source of variation.

Thus sexual reproduction multiplies variation and provides the diversity and plasticity necessary for response to environmental change. Sexual reproduction with its sequence of gene segregation and recombination, generation after generation, is as the geneticist T. Dobzhansky has said, the "master adaptation which makes all other evolutionary adaptations more readily accessible.''

Although sexual reproduction reshuffles and amplifies whatever genetic diversity exists in the population, there must be ways to generate *new* genetic material. This happens through gene mutations and chromosomal aberrations.

Earlier (p. 861) we discussed the importance of chromosomal aberrations that involve structural changes affecting many genes at one time. These often produce phenotypic changes that are inherited in the normal Mendelian manner. Gene mutations, however, are the ultimate source of variability in all living systems; they are the subject of the remainder of this discussion.

Gene mutations

Gene mutations are chemicophysical changes in genes resulting in a visible alteration of the original character. Although the actual mutation cannot be visually detected under the microscope, all gene mutations are believed to be changes of nucleotides in chromosomal DNA. A mutation may involve a codon substitution, the deletion of one or more bases from a codon, or the insertion of additional bases into the DNA chain. Most mutations are ''point'' mutations involving a single base pair.

Mutations are random because they are unpredictable and unrelated to the needs of the organism. Once a gene is mutated, it faithfully reproduces its new self just as it did before it was mutated. *Many, perhaps nearly all, mutant genes are actually harmful because they replace adaptive genes that have evolved and served the organism through its long evolution.* Sometimes, however, mutations are advantageous. These are of great significance to evolution because they furnish new possibilities upon which natural selection works. Natural selection determines which new genes have survival merit; the environment imposes a screening process that passes the fit and eliminates the unfit.

When an allele of a gene is mutated into the new allele it tends to be recessive and its effects are normally masked by its partner allele. Only in the homozygous condition can such mutant genes express their phenotypic effect. Thus a population carries a reservoir of mutant recessive genes, some of which are lethal but seldom expressed. Inbreeding encourages the formation of homozygotes and increases the probability of recessive mutants appearing.

Most mutations are destined for a brief existence. There are cases, however, in which mutations may be harmful to one animal under one set of environmental conditions and helpful under a different set. Should the environment change at the same time that a favorable mutation appears, there could be a new adaptation beneficial to the species. The changing environment of earth has provided numerous opportunities for new gene combinations and mutations, as evidenced by the great diversity of animal life today.

Frequency of mutations

Although mutation is thought of as a completely random process, there is increasing evidence that some genes are more stable than others. In other words, different mutation rates prevail at different loci. Nevertheless it is possible to estimate average spontaneous mutation rates for different organisms.

All genes are extremely stable. In the well-studied fruit fly *Drosophila* there is approximately 1 detectable mutation per 10,000 loci (rate of 0.01% per locus per generation). The rate for humans is 1 per 10,000 to 1 per 100,000 loci per generation. If we accept the latter, more conservative figure, then a single normal allele is expected to go through 100,000 generations before it is mutated. However, since human chromosomes contain 100,000 loci, each person carries approximately one new mutation. Similarly, each spermatozoan produced contains, on the average, one mutant allele.

Since most mutations are deleterious, these statistics are anything but cheerful figures. Fortunately, most genes are recessive and are not detected by natural selection until by chance they have increased enough in frequency for homozygotes to be produced. At this point most are eliminated, since the zygote or individual carrying the mutants dies.

All animals carry large numbers of lethal and semilethal alleles in their gene pools. In human populations this has been called the ''genetic load'' (Muller), or load of hidden mutations. By genetic recombination these continue to surface to produce mutant individuals less ''fit'' (in the Darwinian sense) than the ''wild-type'' genotype.

Annotated references

Selected general readings

Beadle, G., and M. Beadle. 1966. The language of life. New York, Doubleday & Co., Inc. *Interesting and lucid account of genetics.*

Dobzhansky, T. 1951. Genetics and the origin of species, ed. 3. New York, Columbia University Press. *Important, advanced treatment of genetics.*

Dunn, L. C. 1965. A short history of genetics. New York, McGraw-Hill Book Co. *Engaging account of the development of the science of genetics.*

Goodenough, U. 1978. Genetics, ed. 2. New York, Holt, Rinehart and Winston.

Hexter, W., and H. T. Yost, Jr. 1976. The science of genetics. Englewood Cliffs, N.J., Prentice-Hall, Inc. *Introductory text.*

Jenkins, J. B. 1975. Genetics. Boston, Houghton Mifflin Co. *Comprehensive introductory text.*

Mayr, E. 1970. Population, species, and evolution. Cambridge, Mass., Harvard University Press. *This excellent reference is an abridgement of the author's* Animal Species and Evolution.

Mettler, L., and T. G. Gregg. 1969. Population, genetics and evolution. Englewood Cliffs, N.J., Prentice-Hall, Inc. *Sound elementary text.*

Suzuki, D. T., and A. J. F. Griffiths. 1976. An introduction to genetic analysis. San Francisco, W. H. Freeman and Co. *Concepts and experimental basis of genetics presented in a historic sequence.*

Swanson, C. P., T. Merz, and W. J. Young. 1967. Cytogenetics. Englewood Cliffs, N.J., Prentice-Hall, Inc. *Good source of information on the chromosomal theory of inheritance.*

Watson, J. D. 1969. The double helix. New York, Atheneum Publishers. *An enlightening and entertaining history of events leading to an understanding of the genetic code.*

Watson, J. D. 1976. Molecular biology of the gene, ed. 3. New York, The Benjamin Co., Inc. *Concise, well-illustrated popular text.*

Whitehouse, H. L. K. 1973. Towards an understanding of the mechanism of heredity, ed. 3. London, Edward Arnold Ltd. *Excellent treatment of both classic and molecular genetics.*

Selected *Scientific American* articles

Baern, A. G. 1956. The chemistry of hereditary disease. **195:**126-136 (Dec.).

Beermann, W., and U. Clever. 1964. Chromosome puffs. **210:**50-58 (Apr.).

Benzer, S. 1962. Fine structure of the gene. **206:**70-84 (Jan.).

Britten, R. J., and D. E. Kohne. 1970. Repeated segments of DNA. **222:**24-31 (Apr.).

Brown, D. D. 1973. The isolation of genes. **229:**20-29 (Aug.).

Clark, B. F. C., and Marcker, K. A. 1968. How proteins start. **218:**36-42 (Jan.).

Cohen, S. N. 1975. The manipulation of genes. **233:**24-33 (July). *It is now possible to transfer the DNA of one species into an unrelated species.*

Crick, F. H. C. 1962. The genetic code. **207:**66-74 (Oct.).

Crick, F. H. C. 1966. The genetic code: III. **215:**55-63 (Oct.).

Hurwitz, J., and J. J. Furth. 1962. Messenger RNA. **206:**41-49 (Feb.).

Ingram, V. M. 1958. How do genes act? **198:**68-74 (Jan.).

Kornberg, A. 1968. The synthesis of DNA. **219:**64-78 (Oct.).

Maniatis, T., and M. Ptashne. 1976. A DNA operator-repressor system. **234:**64-76 (Jan.). *Describes how the operon system works in a bacterial virus.*

McKusick, V. A. 1971. The mapping of human chromosomes. **224:**104-112 (Apr.).

Miller, O. L., Jr. 1973. The visualization of genes in action. **228:**34-42 (Mar.).

Mirsky, A. E. 1968. The discovery of DNA. **218:**78-88 (June).

Nirenberg, M. W. 1963. The genetic code: II. **208:**80-94 (Mar.).

Nomura, M. 1969. Ribosomes. **221:**28-35 (Oct.).

Ptashne, M., and W. Gilbert. 1970. Genetic repressors. **222:**36-44 (June).

Rich, A. 1963. Polyribosomes. **209:**44-53 (Dec.).

Rich, A., and S. H. Kim. 1978. The three-dimensional structure of transfer RNA. **238:**52-62 (Jan.).

Sager, R. 1965. Genes outside the chromosomes, **212:**71-79 (Jan.).

Sinsheimer, R. L. 1962. Single-stranded DNA. **207:**109-116 (July).

Stein, G. S., J. S. Stein, and L. J. Kleinsmith. 1975. Chromosomal proteins and gene regulation. **232:**46-57 (Feb.). *The role of histone and nonhistone proteins in turning genes off and on is explained.*

Temin, H. 1972. RNA-directed DNA synthesis. **226:**25-33 (Jan.).

Thomasz, A. 1969. Cellular factors in genetic transformation. **220:**38-44 (Jan.).

Yanofsky, C. 1967. Gene structure and protein structure. **216:**80-95 (May).

CHAPTER 38

ORGANIC EVOLUTION

A Galápagos ground finch *Geospiza conirostris* examines its reflection in a camera lens. This is one of 13 species of finches that evolved from a single ancestor that colonized the remote Galápagos Islands long ago. The little finches were an inspiration to Charles Darwin in the development of his theory of natural selection that revolutionized biology.

Photograph by C. P. Hickman, Jr.

The singularity of earth is not that it harbors life, for life surely must exist elsewhere in the universe. It is that life on earth is so enormously diverse and pervasive. Even from space the ubiquity of life is evident in the kaleidoscopic patterns of grasslands, forests, lakes, and croplands stretching across the earth's surface, their colors changing with the seasons.

Each ecosystem is biologically and physically distinct. Each is occupied by a great variety of organisms, living interdependently and manifesting every conceivable kind of adaptation to their surroundings. Nutrients are withdrawn and again released; energy captured by plants flows through the system bringing order out of disorder in apparent defiance of the second law of thermodynamics; organisms die and are replaced by offspring, a renewal that threads its descent faithfully through generations of ancestors. It is a drama of incalculable complexity that has been and is now unfolding before our eyes. Incapable at present of fully understanding how such intricacy is perpetuated year after year, we say simply that the balance of nature is preserved.

Yet despite the evident permanence of the natural world, change characterizes all things on earth and in the universe. Countless kinds of animals and plants have flourished and disappeared, leaving behind an imperfect fossil record of their former existence. The earth itself bears its own record of change, transformed as it is by processes that have occurred over a vast span of time and still are at work today. These changes are irreversible, and for life at least they appear in the broad sense to be directional and progressive, as primitive organisms of the young earth have yielded to the advanced, complex creatures of the present. We call this historical process of descent with modification "organic evolution" (see Principle 4, p. 10).

An understanding of what a species is and of the evolutionary process that brought it to the place it is today underlies all biologic knowledge. Charles Darwin's great contribution was that he provided a logical explanation for evolutionary change—natural selection. Natural selection, through its effect on reproductive success, determines the genetic composition of the population and the appearance of individual adaptations.

In our treatment of the principle of organic evolution in this chapter, we consider Darwin's theory of natural selection and the evidences for evolution, the concept of species and how they change with time, and the question of how new species arise. The evolutionary concept and the mechanism of natural selection was summarized in Chapter 1 (p. 8), and the reader should find it helpful to reread that overview before undertaking the more detailed account here.

DEVELOPMENT OF THE IDEA OF ORGANIC EVOLUTION

Evolution is no longer a subject for debate among biologists; as an event it is known and accepted, even though the forces that determine the course of evolution are not fully understood. Of course, it was not always thus. Prior to the eighteenth century, much of the speculation on the origin of species rested on myth and superstition rather than on observation. Nevertheless, long before this time, there were those who were thinking about and attempting to interpret the order of nature.

Some of the early Greek philosophers, notably Xenophanes, Empedocles, and Aristotle, developed the germ of the idea of change and natural selection within the restrictions of a belief in spontaneous generation. They recognized fossils as evidence of a former life that they thought had been destroyed by some natural catastrophe. Living in a spirit of intellectual inquiry, the Greeks failed to establish an evolutionary concept probably because of their limited experience with the natural world.

With the gradual decline of ancient science, beginning well before the rise of Christianity, debate on evolution virtually ended. The opportunity for fresh thinking became even more restricted as the biblical account of earth's creation became accepted as a tenet of faith. The year 4004 B.C. was fixed by Archbishop James Ussher (midseventeenth century) as the time of true creation of life. In this atmosphere evolutionary views were considered rebellious and heretical.

Still, some speculation continued. The French naturalist Buffon (1707-1788) stressed the influence of environment on the modifications of animal type. He also extended the age of earth to 70,000 years from 6,000 years. Another French zoologist, Jean Baptiste de Lamarck (1744-1829), elaborated a theory of the evolutionary ascent of animals on a scale or ladder. He was the first biologist to make a convincing case for the idea that fossils were the remains of extinct animals and not, as some argued, stones molded in chance imitations of life. However, his explanation of the method of evolutionary change was less inspired.

Lamarck supposed that the use and disuse of organs

by animals striving to adapt to their environment brought about changes that were passed on by inheritance. In other words, an animal's *need* for a structure encouraged its development. The giraffe, according to Lamarck, developed its long neck as the result of constant stretching for food for many generations. The explanation for the limbless snake was that legs were a handicap in moving through dense vegetation; legs were not used and consequently were lost. A strong man who had developed his muscles by exercise was supposed to pass this strength on to his sons. Unfortunately, despite the appeal of Lamarck's theory of **inheritance of acquired characteristics,** no critical evidence has been produced to support it.

While eighteenth- and early nineteenth-century zoologists wrestled with conflicting concepts of organic evolution, geologists were marshaling much sound evidence for the physical evolution of the earth's crust. The geologist Sir Charles Lyell (1797-1875) in his *Principles of Geology* (1830-1833) stated that the forces that produced changes in the earth's surface in the past are the same as those that operate at present. He called his theory ''uniformitarianism.'' He reasoned that if the earth is *now* affected by volcanic activity, by wind and flowing water and frost, and by mountain-building and faulting, it was affected by these same forces, operating in a ''uniform''—or similar—manner, in the past. As one writer of the time stated, ''no vestige of a beginning—no prospect of an end.''

Lyell was able to show that such forces, acting over long periods of time, could account for all observed changes, including the formation of fossil-bearing rocks. His familiarity with fossils, with the natural history of contemporary marine and freshwater animals, and with sedimentary rocks led him to conclude that the earth's age must be reckoned in millions of years rather than in thousands.

Charles Darwin (1809-1882) was thus not the first to propose the idea of evolution nor even to suggest a mechanism for its action. His predecessors had already established an intellectual climate that made a theory of evolution possible, if not inevitable. Darwin himself later acknowledged that the theory of natural selection had been suggested by others before him. But so forcefully did Darwin present his ideas and his array of carefully collected scientific data that no one since has been able to challenge his preeminence in this field. As the noted English biologist Sir Julian Huxley wrote, ''Charles Darwin effected the greatest of all revolutions in human thought, greater than Einstein's or Freud's or even Newton's, by simultaneously establishing the fact and discovering the mechanism of organic evolution.''*

DARWIN'S GREAT VOYAGE OF DISCOVERY

''After having been twice driven back by heavy south-western gales, Her Majesty's ship *Beagle,* a ten-gun brig, under the command of Captain Fitz Roy, R.N., sailed from Devonport on the 27th of December, 1831.'' Thus began Charles Darwin's account of the historic 5-year voyage of the *Beagle* around the world (Fig. 38-1). Darwin, not quite 23 years old, had been invited to serve as naturalist, without pay, on the *Beagle,* a small vessel only 90 feet in length, which was about to depart on a second extensive surveying voyage to South America and the Pacific. It was the beginning of one of the most important voyages of the nineteenth century.

Charles Darwin (Fig. 38-2) was born on February 12, 1809, on the same day that Abraham Lincoln was born in a Kentucky log cabin. Unlike Lincoln, Darwin was born with position and security. His grandfather, Dr. Erasmus Darwin, naturalist, physician, and poet, had earlier achieved recognition by clearly stating the evolutionary theory of the inheritance of acquired characteristics, later embellished by Lamarck. His grandfather on his mother's side, Josiah Wedgwood, was founder of the famous Wedgwood pottery works. Assured of a financial inheritance sufficient to support himself comfortably all his life, Charles Darwin's name might well have been lost in obscurity as have numerous others in similar positions. Indeed, his youth promised the possibility of just such a fate.

He was indifferent to schooling. The boyish pursuits of collecting minerals, insects, bird's eggs, and shells later gave way to a passion for hunting and riding. At age 16, Charles entered medical school but failed, unable to stand the sight of pain. His concerned father then decreed that Charles should become a clergyman and sent him to Cambridge where, as Charles later said, he largely wasted 3 years. Nevertheless he finished tenth in his class and graduated shortly before joining the *Beagle* expedition. His father, who con-

*Huxley, J. In Bowman, R. I. (ed.). 1966, The Galapagos, Berkeley, Calif. University of California Press.

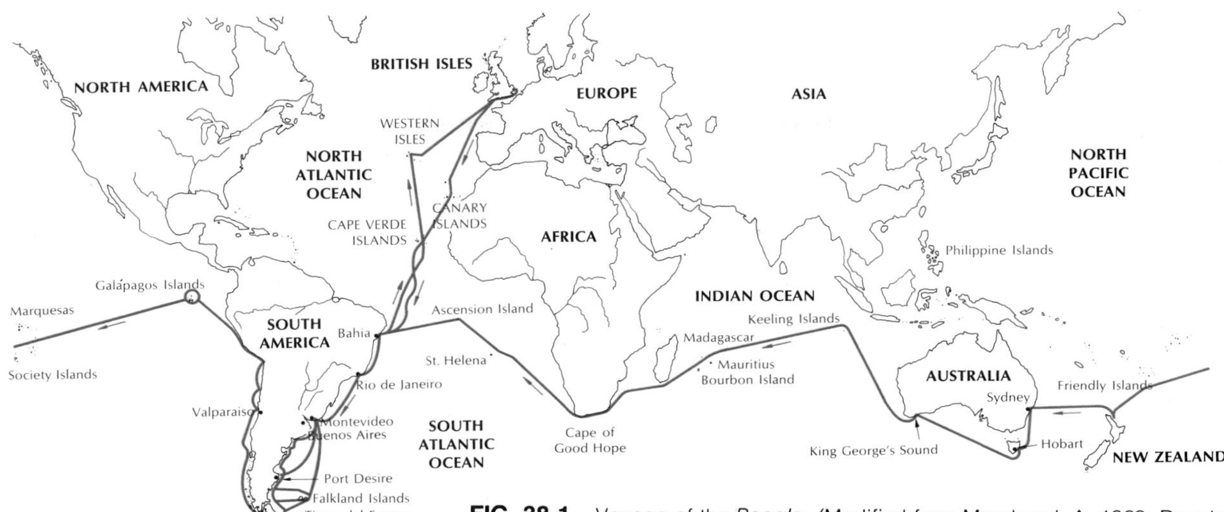

FIG. 38-1 Voyage of the *Beagle.* (Modified from Moorhead, A. 1969. Darwin and the *Beagle.* New York; Harper & Row, Publishers.)

sidered the *Beagle* voyage the final seal to his son's fate as a professional idler, was finally persuaded to give his consent to the venture.

During the 5-year voyage (1831-1836), Darwin endured almost constant seasickness and the erratic companionship of the authoritarian Captain Fitz Roy. But his youthful physical strength and early training as a naturalist equipped him for his work. The *Beagle* made many stops along the harbors and coasts of South America and adjacent regions. Darwin made extensive collections and observations on the fauna and flora of these regions. He unearthed numerous fossils of animals long since extinct and noted the resemblance between fossils of the South American pampas and the known fossils of North America. In the Andes he encountered seashells embedded in rocks at 4,000 m (about 13,000 feet). He experienced a severe earthquake and watched the mountain torrents that relentlessly wore away the earth. He read and was impressed by the writings of Charles Lyell. Reflecting at length on activities that have shaped the earth over vast periods of time, he began to realize that he was witnessing evolution.

Finally in mid-September of 1835, the *Beagle* arrived at the Galápagos Islands, a volcanic archipelago straddling the equator 600 miles west of the coast of Ecuador (Fig. 38-3). The fame of the islands stems from their infinite strangeness. They are unlike any other islands on earth. Some visitors today are struck with an impression of awe and wonder; others with a sense of depression and dejection. Darwin described them thus:

> These islands at a distance have sloping uniform outline, excepting where broken by sundry paps and hillocks; the whole black Lava, completely covered by small leafless brushwood and low trees. The fragments of Lava where most porous, are reddish like cinders; the stunted trees show little signs of life. The black rocks heated by the rays of the Vertical sun, like a stove, give to the air a close and sultry feeling. The plants also smell unpleasantly. The country was compared to what one might imagine the cultivated parts of the Infernal regions to be.

Circled by capricious currents, surrounded by shores of twisted lava, bearing skeletal brushwood baked by the equatorial sun, almost devoid of lush tropic vegetation, inhabited by strange reptiles and by convicts stranded by the Ecuadorian government, the islands indeed had few admirers among mariners. By the middle of the seventeenth century, the islands were already known to the Spaniards as "Las Islas Galápagos"—the tortoise islands. The giant tortoises, used for food first by buccaneers and later by American and British whalers, sealers, and ships of war, were the islands' principal attraction. At the time of Darwin's visit, the tortoises were already being heavily exploited.

A

B

FIG. 38-2 Charles Darwin and H.M.S. *Beagle*. **A,** Darwin in 1840, 4 years after the *Beagle* returned to England. **B,** H.M.S. *Beagle* in the Straits of Magellan. (**A,** from a watercolor by George Richmond; **A** and **B,** courtesy American Museum of Natural History.)

FIG. 38-3 The Galápagos Islands. (Photograph by C. P. Hickman, Jr.)

During the *Beagle's* 5-week visit to the Galápagos, Darwin began to develop his views of the evolution of life on earth. After returning to England Darwin wrote:

> The natural history of these islands is eminently curious, and well deserves attention. Most of the organic productions are aboriginal creations, found nowhere else; there is even a difference between inhabitants of the different islands, yet all show a marked relationship with those of America, though separated from that continent by an open space of ocean, between 500 and 600 miles in width. The archipelago is a little world within itself . . .

Darwin's original observations on the giant tortoises, the marine iguanas that feed on seaweed, and the family of drab finches that had evolved into several species, each with distinct beak and feeding adaptations, all contributed to the turning point in Darwin's thinking.

Darwin was struck by the fact that, although the Galápagos Islands and the Cape Verde Islands (visited earlier in this voyage of the *Beagle*) were similar in climate and topography, their fauna and flora were altogether different. He recognized that Galápagos plants and animals were related to those of the South American mainland, yet differed from them sometimes in curious ways. Each island often contained a unique species of a particular group of animals that was nonetheless related to forms on other islands. In short, Galápagos life must have originated in continental South America and then undergone modification in the various environmental conditions of the separate islands. He concluded that living forms were neither divinely created nor immutable; they were, in fact, the products of evolution. Although Darwin devoted only a few pages to Galápagos animals and plants in his monumental *The Origin of Species,* published more than two decades after visiting the islands, his observations on the unique character of the animals and plants were, in his own words, the "origin of all my views."

On October 2, 1836, the *Beagle* landed in England. Most of Darwin's extensive collections had long since preceded him to England, as had most of his notebooks and diaries kept during the 5-year cruise. Three years after the *Beagle's* return to England, Darwin's journal was published. It was an instant success and required two additional printings within the first year. In later versions, Darwin made extensive changes and abbreviated the original ponderous title typical of nineteenth century books to simply *The Voyage of the Beagle*. The fascinating account of his observations written in a simple, appealing style has made the book one of the most lasting and popular travel books of all time.

Curiously, the main product of Darwin's voyage, his theory of evolution, did not appear in print for more than 20 years after the *Beagle's* return. Darwin began gathering together the evidence for evolution almost as soon as the voyage ended. He carried out an extensive correspondence with botanists and zoologists and followed progress in biology and related fields.

In 1838 he "happened to read for amusement" an essay on populations by T. R. Malthus, who stated that animal and plant populations, including human populations, tend to increase beyond the capacity of the environment to support them. Darwin had already been gathering information on the artificial selection of animals under domestication by humans. After reading Malthus' article, Darwin realized that a process of selection in nature, a "struggle for existence" because of overpopulation, could be a powerful force for evolution of wild species.

He allowed the idea to develop in his mind until it was presented in an essay in 1844—still unpublished. Finally in 1856 he began to pull together his voluminous data into a work on the origin of species. He expected to write four volumes, a "very big" book, "as perfect as I can make it." However, his plans were to take an unexpected turn.

In 1858, he received a manuscript from Alfred Russel Wallace (1823-1913), an English naturalist in Malaya with whom he had been corresponding. Darwin was stunned to find that in a few pages Wallace summarized the main points of the natural selection theory that Darwin had been working on for two decades. Rather than withhold his own work in favor of Wallace as he was inclined to do, Darwin was persuaded by two close friends, the geologist Lyell and the botanist Hooker, to publish his views in a brief statement that would appear together with Wallace's paper in the Journal of the Linnaean Society. Portions of both papers were read before an unimpressed audience on July 1, 1858.

For the next year, Darwin worked urgently to prepare an "abstract" of the planned four-volume work. This was published in November, 1859, with the title *On the Origin of Species by Means of Natural Selection or the Preservation of Favoured Races in the Struggle for Life*. The 1,250 copies of the first printing were sold the first day; the book instantly generated a storm that has never completely abated. His views were to have extraordinary consequences on scientific and religious thought and remain among the greatest intellectual achievements of all time.

Once Darwin's excessive caution had been swept away by the publication of *The Origin of Species*, he entered an incredibly productive period of evolutionary thinking for the next 23 years, producing book after book. He died April 19, 1882, and was buried in Westminster Abbey. The little *Beagle* had already disappeared, having been retired in 1870 and presumably broken up for scrap.*

The biologic fame of the Galápagos Islands that contributed so much to Darwin's thinking had been secured by his first book, *The Voyage of the Beagle*. The unusual flora and fauna of the Galápagos continued to excite biologists, and numerous expeditions visited the islands. Unfortunately the famous tortoises and once-great herds of fur seals were relentlessly hunted throughout the nineteenth century and finally were almost exterminated. At the same time the natural balance of the island communities was seriously damaged by the accidental or purposeful introduction of black rats, dogs, cats, pigs, goats, and donkeys, all of which have become feral on most of the islands and continue to menace the native animals and plants today.

Rising concern over these events eventually culminated in the establishment of the Charles Darwin Research Station at Academy Bay, Santa Cruz Island, in 1964. The entire archipelago has now been proclaimed a national park by Ecuador, and legislation protecting the wildlife has been passed. Although the native flora and fauna were seriously disfigured by human activities and many dangers remain, there is renewed hope that these unique islands are finally receiving the protection they deserve.

*The history of the famous H.M.S. *Beagle* and its elusive fate are described by K. S. Thomson, 1975. American Scientist **63**(6):664.

EVIDENCES FOR EVOLUTION
Reconstructing the past
Fossils and their formation

The strongest and most direct evidence for evolution is the fossil record of the past. Fossils and the sediments containing them provide dim and imperfect views of an ancient life (Fig. 38-4). The complete record of the past is always beyond our reach since many groups left no fossils. Indeed, were the past by some miracle to be suddenly laid out for our scrutiny, the relationships of organisms would be incalculably more complex than they are now. Yet, incomplete as the record is, biologists rely on the discoveries of new fossils and the continued study of existing fossils to interpret phylogeny and relationships of both plant and animal life. It would be difficult indeed to make sense out of the evolutionary patterns or classification of organisms without the support of the fossil record. The documentary evidence for evolution as a general process, the progressive changes in life from one geologic era to another, the past distribution of lands and seas, and the environmental conditions of the past (paleoecology) are all dependent on what fossils teach us.

A fossil may be defined as the remains of past life uncovered from the crust of the earth. It refers not only to complete remains (mammoths and amber insects), actual hard parts (teeth and bones), and petrified skeletal parts that are infiltrated with silica or other minerals by water seepage (ostracoderms and molluscs), but also to molds, casts, impressions, and fossil excrement (coprolites).

The fossil record is biased because preservation is selective. Vertebrate skeletal parts and invertebrates with shells or other hard structures have left the best record (Fig. 38-5). It is unlikely that jellyfish, worms, caterpillars, and such, fossilize, but now and then a rare chance discovery such as the Burgess shale deposits of British Columbia and the Precambrian fossil bed of South Australia reveals an enormous amount of

A

Continued.

FIG. 38-4 Reconstruction of the appearance of *Corythosaurus,* a dinosaur from the upper Cretaceous of North America, approximately 90 million years ago. **A,** Portion of skeleton as discovered and partially worked out of rocks in New Mexico quarry. **B,** Skeleton, 30 feet in length, on slab. **C,** Drawing prepared from skeleton. **D,** Artist's reconstruction of living animal. (**A** and **B** courtesy American Museum of Natural History; **C** and **D** from Colbert, E. H. 1969. Evolution of the vertebrates, ed. 2. New York, John Wiley & Sons, Inc.)

FIG. 38-4, cont'd For legend see p. 881.

FIG. 38-5 This bit of Green River shale from Farson, Wyoming, bears the impression of a "double-armored herring" *(Knightia),* approximately 9 cm long, which swam there during the Eocene epoch, approximately 55 million years ago.

information about soft-bodied organisms. Certain regions have apparently provided ideal conditions for fossil formation, for example, the tar pits of Rancho La Brea in Hancock Park, Los Angeles; the great dinosaur beds of Alberta, Canada, and Jensen, Utah; the Olduvai Gorge of South Africa; and many others.

A common method of fossil formation is the quick burial of animals under water-borne sediments. Rapid burial is usually important because it slows or prevents decomposition by oxidation, solution, and bacterial action.

Most fossils are laid down in deposits that become stratified. The numerous layers of rock exposed in riverside cliffs or roadbanks were formed mainly by the accumulation of sand and mud at the bottoms of seas or lakes. If left undisturbed, which is rare, the older strata are the deeper ones; however, the layers are usually tilted or folded or show faults (cracks). Often old deposits exposed by erosion are later covered with new deposits in a different plane. Sedimentary rock such as limestone may be exposed to tremendous pressures or heat during mountain building and may be metamorphosed into rocks such as crystalline quartzite, slate, or marble. Fossils are often destroyed during these processes.

Since various fossils are correlated with certain strata, they often serve as a means of identifying the strata of different regions. Certain widespread marine invertebrate fossils such as various Foraminifera and echinoderms are such good indicators of specific geologic periods that they are called "index" or "guide" fossils.

Geologic time

Long before the age of the earth was known, geologists began dividing its history into a table of succeeding events, using as a basis the accessible deposits and correlations from sedimentary rock. Time was divided into eons, eras, periods, and epochs. These are shown on the end papers inside the front and back covers of this book. Time during the last eon (Phanerozoic) is expressed in eras (Cenozoic), periods (Tertiary), epochs (Paleocene), and sometimes smaller divisions of an epoch.

As far as the fossil record is concerned, the recorded history of life begins near the base of the Cambrian period of the Paleozoic era approximately 600 million years B.P. (before present). The period before the Cambrian is called the Precambrian era. Although the Precambrian era occupies 85% of all geologic time, it is a puzzling era because of the lack of macrofossils. It has received much less attention than later eras partly because of the absence of fossils and partly because oil, which provides the commercial incentive for much geologic work, seldom exists in Precambrian formations. There are, however, evidences for life in the Precambrian era: well-preserved bacteria and algae, as well as casts of jellyfish, sponge spicules, soft corals, segmented flatworms, and worm trails. Most, but not all, are microfossils.

There are two explanations for the lack of conventional fossils beyond the Cambrian-Precambrian "barrier": (1) most rocks older than 600 million years have undergone so much metamorphosis or distortion that any fossils that might have been present were destroyed, and (2) animal species with hard parts and shells evolved after the early Cambrian period. We have already seen (Chapter 3) that life originated perhaps 3 billion years ago and that major phyla of marine invertebrates were well established and had diverged to a considerable extent before the Cambrian period. It is probable therefore that toward the end of the Precambrian era, the seas contained a wealth of greatly diverse life of which we can never catch more than a glimpse.

How are fossils and rocks dated? The succession of geologic events was well established long before there was any reliable knowledge of the absolute time scale. Paleontologists knew, for example, that dinosaurs and ichthyosaurs were animals of the Mesozoic era, that amphibians originated in the Devonian period, and that saber-toothed tigers and mastodons roamed the forests of the Pleistocene epoch. At the end of the nineteenth century, the earth was thought to be 100 million years old, and the geologic time scale was adjusted accordingly. But with no measure of the absolute passage of time, figures of 1 billion or 10 billion years could just as well have been accepted.

The three principal methods now used for dating geologic formations are based on the radioactive decay of naturally occurring elements into other elements. This proceeds independently of pressure and temperature changes and therefore is not affected by often violent earth-building activities.

One method, potassium-argon dating, depends on the decay of potassium-40 (^{40}K) to argon-40 (^{40}A) (12%) and calcium-40 (^{40}Ca) (88%).

$$^{40}K \xrightarrow{\text{electron capture}} {}^{40}A \qquad ^{40}K \xrightarrow{\beta\text{-emission}} {}^{40}Ca$$

The half-life of ^{40}K is 1.3 billion years. The age of the rock is determined by measuring both the residual potassium content and the ^{40}A content (calcium is unreliable). The method is technically difficult, and certain kinds of rocks cannot be dated because the gaseous argon escapes; however, the procedure is now routinely used in geology laboratories.

The uranium-thorium-lead method uses three isotopes of uranium and thorium that decay to lead at different rates (0.7 to 13.9 billion years) through a complex series of radioactive steps. The age of the rock is determined by measuring the accumulation of stable lead (^{204}Pb).

A third method of dating, the rubidium-strontium method, is complicated and restricted to rubidium-rich minerals.

The well-known carbon 14 (^{14}C) dating method deserves mention, although the short half-life of ^{14}C restricts its use to quite recent events. Although its accuracy is limited to approximately 40,000 years, it is especially useful for archeologic studies. This method is based on the production of radioactive ^{14}C (half-life of approximately 5,570 years) in the upper atmosphere by bombardment of nitrogen 14 (^{14}N) with cosmic radiation. The radioactive ^{14}C enters the tissue of living animals and plants, and an equilibrium is established between atmospheric ^{14}C and ^{14}C in the organism. At death, ^{14}C exchange with the atmosphere stops. In 5,570 years only half of the original ^{14}C remains in the preserved fossil. Its age is found by comparing the ^{14}C content of the fossil with that of living organisms.

As with all radioactive methods, there are complications. One is that the amount of ^{14}C in the atmosphere is disturbed by fossil-fuel burning (decreases ^{14}C) and thermonuclear explosions (increases ^{14}C). Corrections must be applied for these disturbances.

There are many uncertainties in radioactive dating, and the techniques require sophisticated instrumentation and great care in methodology. Not all rocks or fossils can be dated. But, if two or more different methods provide concordant answers, the age can be accepted as reliable since it is most improbable that different isotopes would leach out of the sample at the same rate.

Evolution of the horse

The fossil record provides no more convincing or more complete evolutionary line of descent than that of the horse. The evolution of this form extends back to the Eocene epoch some 60 million years ago, and much of it took place in North America. This record would at first seem to indicate a straight-line evolution, but actually the history of the horse family is made up of many lineages, that is, descent from many lines. The phylogeny of the horse is extensively branched,

FIG. 38-6 Evolution of the horse family, showing the transition from browsing to grazing and evolution of the front leg. Only a few of the many lines of horse evolution are shown. See text for further explanation.

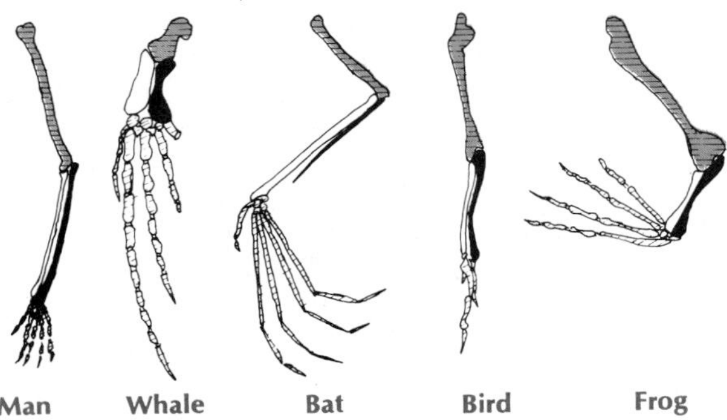

FIG. 38-7 Forelimbs of five vertebrates to show skeletal homologies. *Red,* humerus; *white,* radius; *black,* ulna; *white striated,* wrist and phalanges. Most generalized or primitive limb is that of humans—the feature that has been a primary factor in human evolution because of its wide adaptability. Various types of limbs have been structurally modified for adaptations to particular functions.

with most of the branches now extinct. There were millions of years when little change occurred; there were other epochs when changes took place relatively rapidly.

No real change in the feet occurred during the Eocene epoch, but at least three types of feet developed later and were found in different groups during the late Cenozoic era. Only one of these three types is found today.

There was extensive adaptive radiation throughout the horse's evolutionary history. In the evolution of the horse the morphologic changes of the limbs and teeth were of primary importance, along with a progressive increase in size of most of the types in the direct line of descent.

The earliest member of the horse phylogeny is considered to be *Hyracotherium,* also known as *Eohippus,* the dawn horse (Fig. 38-6). It was rodentlike in appearance and about the size of a small dog. Its forefeet had four digits, and the hind legs had three. Most of the weight was borne on pads, making the feet well adapted for travel on the soft earth of tropical North America. The teeth had short crowns and long roots and were specialized for browsing on shrubs.

As the tropical conditions of central North America later gave way to open, grassy plains, the horse gradually developed feet better suited for rapid travel on dry land and grinding teeth for grazing. This transition began with *Mesohippus,* which flourished in the Oligocene epoch some 40 million years ago. *Mesohippus,* though still a browser, was larger than *Eohippus* and had three toes on each foot. The middle toe was larger and better developed than the others. *Mesohippus* was a very common and successful animal in North America, as evidenced by the numerous fossil remains discovered.

Mesohippus gave rise to several lineages, one of which, *Merychippus,* was the direct ancestor of modern horses. *Merychippus* lived during the Miocene epoch, some 25 million years ago, when grassy prairies were becoming widespread all over the earth under the influence of an increasingly cooler and drier climate. Its teeth were high-crowned, with complex crests and ridges for grinding; this development made possible the complete switch to a grazing diet. It was still three-toed, but the lateral toes were high above the ground so that the weight of the body was thrown on the middle toe.

Pliohippus, a descendant of *Merychippus,* persisted into the Pleistocene epoch, some 2 million years ago. *Pliohippus* discarded the side toes (although splint bones remained to attest to their former existence) and became a single-toed horse. It gave rise to *Equus,* the modern horse, of which there are some 60 domesticated breeds, and to zebras and asses. During the Pleistocene Ice Age, the *Equus* genus spread from North America into Eurasia, Africa, and South America.

The sequel to the horse evolution story is strange. Having pursued a remarkable evolution on the North American continent for 60 million years, the horse became extinct in both Americas toward the end of the Pleistocene epoch. The cause of its demise is unknown, whether by excessive hunting by early Indians, infectious disease, predation, or some climatic catastrophe. It is certain that no horses were present when white people first came to the Americas. However, horses still existed in Europe and Asia and were reintroduced into America from Europe by the early Spanish colonists.

The development of this remarkable animal, which has had an enormous influence on the course of human history, was closely associated with the climate and ecologic changes of its habitat. As the habitat changed, so did the horse. It became a grazer instead of browser and developed the body size and foot structure conducive for fleetness in rapid travel over a firm, prairie surface. It is an extraordinary example of evolutionary adaptation to changing ecologic opportunity (see Principle 5, p. 10).

Natural system of classification

Biologists long before Darwin discovered that, despite the great diversity of life, it was possible to arrange living things into a logical system of classification. The "natural system of classification" that originated with Aristotle and later was formalized by Linnaeus is based on the degree of similarity of morphologic characters.

For example, the domestic cat is clearly related to the wild cats and jungle cats of the Old World, and all are included in the same genus, *Felis*. This genus shares an obvious relationship with the "big cats" (lions, tigers, leopards, and jaguars of the genus *Panthera),* as well as lynxes and bobcats (genus *Lynx*); so all are combined into the same family Felidae. All members of this family share a number of traits that set them apart from all other families, such as a round head, 30 teeth, and digitigrade feet with retractile claws. At the same time, the cat family is clearly related as a group to a number of other families (for example, the dog family, bear family, raccoon family, weasel and otter family, and hyaena family). All of these bear a distinctive anatomy and way of life that characterizes them as flesh-eaters; they are of the order Carnivora.

Thus animals can be grouped together with respect to certain combinations of traits. This fact makes possible the classification of the approximately 2 million species of animals and strongly suggests that groups of animals are interrelated. For centuries it remained a puzzle why the system should work so well. The theory of evolution provided a solution: similarities exist because of descent from a common ancestor. True genetic relationship is reflected in the similarity of morphologic traits (see Principle 6, p. 10).

Comparative anatomy: homologous resemblances

The concepts of homology and analogy were contrasted in Chapter 6 (p. 117). **Homology** refers to similarity in structure and development of organs because of common ancestry and similar genetic basis. **Analogy** refers to resemblance in function but not in structure or development. The wings of a robin and a moth are analogous because both are used in flight, but they are quite different in structure and origin. The robin's wing is, however, homologous to the arm of a human because they are both pentadactyl (five-toed) structures that share a common embryologic origin; they are homologous despite the different functions they perform.

The similar position and embryologic origin of the pentadactyl limb (Fig. 38-7) provide only one of several striking examples of homology among the vertebrates. Mammals characteristically have seven cervical vertebrae in the neck. Mouse, human, elephant, and giraffe—even the porpoise, which has lost the capacity to turn its head—have seven cervical vertebrae, an indication that all have descended from an ancestral mammalian stock that bore this number. Evolutionary theory provides the most rational explanation for the fact that animals differing so much in size and mode of life have seven neck vertebrae, as well as numerous other skeletal similarities.

Vertebrate examples could be extended indefinitely. The skulls of various mammals are unitary structures composed of several bones and obviously constructed from the same pattern. The vertebrate brain presents a common pattern; from fish to philosopher it is constructed of olfactory lobes, cerebral hemispheres, optic lobes, cerebellum, and medulla (p. 734). All vertebrates share the same array of endocrine glands. Comparative anatomy courses are based on fundamental similarities—homologies—in every vertebrate organ system.

Comparative embryology

If homologies exist in the comparative anatomy of adults, we might expect—and do find—impressive embryologic homologies, since adult structures are attained by embryonic development. The early embryos of fishes, amphibians, reptiles, birds, and mammals look very similar, and all share several features in common, such as gill slits, aortic arches, notochord neural tube, and postanal tail (see Principle 7, p. 11).

The vertebrate gill (pharyngeal) arches serve as a single example (Fig. 36-12, p. 832). In fishes the embryonic gill arches later serve as respiratory organs. In the adults of terrestrial vertebrates the gill arches disappear altogether or become modified beyond recognition. The appearance of gill arches in the embryos of these terrestrial vertebrates and other pronounced similarities between the embryos of fishes and higher vertebrates are a strong indication of a common vertebrate ancestry.

Embryonic development is thus a record, although a considerably modified one, of evolutionary history. Biologists of the last century were so impressed by embryonic similarities between widely separated vertebrate groups that they used embryonic development as important evidence in reconstructing lines of evolutionary descent within the animal kingdom.

Comparative biochemistry

Common ancestry can now be demonstrated just as forcefully by homologous biochemical compounds as by homologous anatomic structures. The chemical composition of living things, based as it is on nucleic acids, proteins, carbohydrates, and fats, is itself evidence of the kinship of all life. Despite the diversity of form and function of animals and plants, numerous chemical compounds perform virtually identical metabolic roles in all. The citric acid (Krebs) cycle that releases electrons from carbon compounds; the cytochrome system that passes electrons to molecular oxygen; the ornithine cycle that synthesizes urea from surplus ammonia; the metabolic pathways that oxidize fatty acids; these and many more are identical in a wide variety of species of plants and animals (see Principle 3, p. 10).

It has been possible to trace the evolution of at least one protein, cytochrome c, to the origin of eukaryotic cells more than 1 billion years ago. Cytochrome c is a small enzyme composed of 104 amino acid units. As recently noted, the amino acid sequences from various species are different; furthermore, just how much they differ corresponds well with the distance that any two species are separated on the phylogenetic tree. Elaborate pedigrees constructed entirely from sequence diversion agree remarkably well with those obtained by classic morphology and embryology. Such studies are of great importance because they carry evolutionary study much closer to the understanding of the genetic variation that underlies all evolutionary adaptation.

NATURAL SELECTION

Darwin's theory of natural selection

Darwin's concept of natural selection is based on a few simple principles supported by an abundance of facts (see Principle 4, p. 10).

1. **All organisms show variation.** No two individuals are exactly alike. They differ in size, color, physiology, behavior, and other ways. Darwin realized that much of this variation is inherited even though he did not understand how. The genetic basis of heredity was to be discovered many years later. Darwin pointed out that humans have taken advantage of inherited variation to produce useful new breeds and races of livestock and plants. He believed that, if selection is possible under human control, it can also be produced by agencies operating in nature. He reasoned that natural selection can have the same effect as artificial selection.

2. **In nature all organisms produce more offspring than can survive.** In every generation the young are more numerous than the parents. Darwin calculated that, even in a slow-breeding species such as the elephant, a single pair breeding from ages 30 to 90 and having only 6 young in this span could give rise to 19 million elephants in 750 years. Why, he asked, are there not more elephants?

3. **Accordingly there is a struggle for survival.** If more individuals are born than can possibly survive, there must be a severe struggle for existence among them. Darwin wrote in *The Origin of Species,* "It is the doctrine of Malthus applied with manifold force to the whole animal and vegetable kingdoms." Competition for food, shelter, and space becomes increasingly severe as overpopulation develops.

4. **Some individuals of a species have a better chance for survival than others.** Because individuals of a species differ, some are favored in the struggle for survival; others are handicapped and eliminated.

5. **The result is natural selection.** Out of the struggle for existence there results the survival of the fittest.* Under natural selection, individuals bearing favorable variations survive and have a chance to breed and transmit their characteristics to their offspring. The less fit die without reproducing. Natural selection is simply the differential survival or reproduction of favored variants. The process continues with each succeeding generation so that organisms gradually become better adapted to their environment. Should the en-

*The popular phrase "survival of the fittest" was not originated by Darwin but was coined a few years earlier by the British philosopher Herbert Spencer, who anticipated some of Darwin's principles of evolution. Unfortunately the phrase later came to be coupled with unbridled aggression and violence in a bloody, competitive world. In fact, natural selection operates through many tame characteristics of living things; fighting prowess is only one of several means toward successful reproductive advantage.

vironment change, there must also be a change in those characters that have survival value or the species is eliminated. Reproduction is what really counts in natural selection.

6. **Through natural selection new species originate.** The differential reproduction of variants can gradually transform species and result in the long-term improvement of types. According to Darwin, when different parts of an animal or plant population are confronted with slightly different environments, each diverges from the other. In time they differ enough from each other to form separate species. In this way two or more species may arise from a single ancestral species. Through adaptation to a changed environment, a group of animals may also diverge enough from their ancestors to become a different species.

Appraisal of Darwin's theory

Darwin's theory of evolution was supported by an overwhelming amount of evidence that had never before been accumulated. He convincingly demonstrated evolution and then provided a logical explanation for it. Through the well-chosen body of evidence so lucidly expounded in *The Origin of Species,* he made evolution understandable. Nevertheless, despite its elegance it is hardly surprising that parts of his theory have been modernized because of more recent biologic knowledge. When Darwin wrote *The Origin of Species,* nothing was known about the inheritance of variation. It was not even known that sexual reproduction involved the combination of a single sperm with a single egg. Thus one always marvels that Darwin's deductions and conclusions were so sound in view of the scientific ignorance that prevailed at the time.

Two weaknesses in his theory centered around the concepts of variation and its inheritance. Darwin made little distinction between variations induced by the environment (physical or chemical) and those that involve alterations of the germ plasm or the chromosomal material. It is now known that many of the types of variations Darwin stressed are noninheritable. Only variations arising from changes in the genes (mutations) are inherited and furnish the material on which natural selection can act.

In 1868 Gregor Johann Mendel published his theories on inheritance in a journal of the Society of Natural Science of Brünn. Darwin presumably never read the work (although it was found in Darwin's extensive library after his death), and, if there were others who knew of it, they failed to bring it to his attention. It is interesting to speculate on how the genetic implications of Mendel's study would have influenced Darwin's thinking had Mendel written to him of his results.

It is altogether possible that Darwin could not have seen the relationship of the hereditary mechanisms to gradual changes and continuous varieties that represent the hub of Darwinian evolution. It took others years to establish the relationship of genetics to Darwin's natural selection theory. Darwin did not point out the cumulative tradition of human evolution, that is, the capacity of the human race to transmit experience from generation to generation, now called our "cultural evolution."

The first step in modernizing Darwin's theory was the demonstration that the operative units in inheritance are self-reproducing, nonblending genes. Next there was the discovery of chromosomes in which genes were found to be precisely located in linkage groups. Biologists could then understand how the ultimate source of variation — mutations — could be inherited in the usual Mendelian fashion.

Darwin also did not appreciate the real significance of geographic isolation in the differentiation of new species. Biologists now agree that isolation of some kind is an essential element in the formation of new species. In rare instances it is true that a mutation or chance hybridization may produce a group of organisms that are infertile with the parent stock; accordingly a new species may arise in a single step without isolation. Otherwise geographic or ecologic isolation provides the opportunity for populations to develop their own unique characters independently of the parent stock.

What happens to characters when they are nonadaptive or indifferent? Variations that are neither useful nor harmful are not affected by natural selection and may be transmitted to succeeding generations as fluctuating variations. This explains many variations that have no significance from an evolutionary viewpoint.

Genetic variation within species

Biologic variation exists within all natural populations. We can verify this fact simply by measuring a single character in several dozen specimens selected from a local population: tail length in mice, length of a leg segment in grasshoppers, number of gill rakers in sunfish, number of peas in pods, height of adult males of the human species. When the values are graphed with respect to frequency distribution, they tend to approximate a normal, or bell-shaped, probability curve. Most individuals fall near the mean, fewer fall above or below the mean, and the extremes make up the "tails" of the frequency curve with increasing rarity. The larger the population sample, the closer the frequencies resemble a normal curve. Sometimes significant deviations from normality are discovered in populations and such instances may reveal important facts about variations within the population.

As explained in earlier chapters, biparental reproduction and Mendelian assortment furnish a means of shuffling genes into various combinations. Some of those combinations that work best have a better chance to be retained in the population. For instance, biparental reproduction makes it possible for two favorable mutant genes (one from each of two individuals) to come together within the same individual, but within an asexual population the two favorable genes would be together only when both mutations occurred in the same individual. Sexual reproduction thus speeds the rate of evolution and better enables the organisms to respond to changing environments.

Gene pool. All of the alleles of all genes of a population are referred to collectively as the **gene pool.** The interbreeding population is the visible manifestation of the gene pool, which thus has continuity through successive generations. The gene pool of large populations must be enormous, for at observed mutation rates many mutant alleles can be expected at all gene loci. Let us suppose that there are two alleles present, **A** and **a.** Among the individuals of the population there are three possible genotypes: **AA, Aa,** and **aa.** When there are three alleles present, **A, a,** and *a*, there are six possible genotypes: **AA, Aa, A***a*, **aa, a***a*, and *aa.* With four alleles present there are 16 possible genotypes. Increasing the number of alleles at a locus increases the possible genotypes enormously.

Role of mutations. Changes in uniparental populations occur by the addition or elimination of a mutation; in biparental populations the mutant gene may combine with all existing combinations and thus double the types. With only 10 alleles at each of 100 loci, the number of mating combinations is 10^{100}. When an organism has thousands of pairs of genes, the amount of diversity is staggering. Even though many genes are found together on a single chromosome and tend to stay together in inheritance, this linkage is often broken by crossing-over.

If no new mutations occurred, the shuffling of the old genes would produce an inconceivably great number of combinations. But this is not the whole story because genes exert different influences in the presence of other genes. Gene A may act differently in the presence of gene B from what it does in the presence of gene C (epistasis, p. 847). The diversity produced by this interaction and the addition of new mutations now and then add to the complication of population genetics.

If this diversity is possible in a single population, suppose two different populations with different gene pools should mix by interbreeding. It is easy to see that many more combinations of genes and their phenotypic expression would occur. This means that populations have enormous possibilities for variation.

Variation and natural selection. What does variation signify for evolution? As already stated, genetic variation produced in whatever manner furnishes the material on which natural selection works to produce evolution. Natural selection does this by favoring beneficial variations and eliminating those that are either neutral or harmful to the organism. Selective advantages of this type represent a very slow process, but on a geologic time scale they can bring about striking evolutionary changes represented by the various taxonomic units (species, genera, and so forth), adaptive radiation groups, and the various kinds of adaptations.

The fact must be stressed, however, that natural selection works on the whole animal and not on a single hereditary characteristic. The organism that possesses the most beneficial combination of characteristics or "hand of cards" is going to be selected over one not so favored. This concept may help to explain some of those puzzling instances in which an animal may have certain characteristics of no advantage or that are actually harmful, but in the overall picture it has a winning combination. Thus, in a population, pools of variations are created on which natural selection can work to produce evolutionary change.

Genetic equilibrium

In the human population, brown eyes are dominant to blue, curly hair is dominant to straight, and Roman nose is dominant to straight nose. Why hasn't the dominant gene gradually supplanted the recessive one in each instance so that we are all brown-eyed, curly-haired, and Roman-nosed? It is a common belief that a character dependent on a dominant gene increases in proportion because of its dominance. This is not the case, for there is a tendency in *large* populations for genes to remain in equilibrium generation after generation. A dominant gene does not change in frequency with respect to its recessive allele(s).

This principle is based on a basic law of population genetics called the **Hardy-Weinberg equilibrium.** According to this law, gene frequencies and genotype ratios in large biparental populations reach an equilibrium in one generation and remain constant thereafter unless disturbed by new mutations, by natural selection, or by genetic drift (chance). The rule does not operate in small populations (see Principle 15, p. 12).

A rare gene, according to this principle, does not disappear merely because it is rare. That is why certain rare traits, such as albinism and cystic fibrosis, persist for endless generations. Variation is retained even though evolutionary processes are not in active operation. Whatever changes occur in a population—gene flow from other populations, mutations, natural selection, and migration—involve the establishment of a new equilibrium with respect to the gene pool, and this new balance is maintained until upset by disturbing factors.

The Hardy-Weinberg law is a logical consequence of Mendel's first law of segregation and expresses the tendency toward equilibrium inherent in Mendelian heredity.

Let us select for our example a population having a single locus bearing just two alleles **T** and **t.** The phenotypic expression of this gene might be, for example, the ability to taste a chemical compound called phenylthiocarbamide. Individuals in the population will be of three genotypes for this locus, **TT, Tt** (both tasters), and **tt** (nontasters). In a sample of 100 individuals, let us suppose we have determined that there are 20 of **TT** genotype, 40 of **Tt** genotype, and 40 of **tt** genotype. We could then set up a table showing the allelic frequencies as follows (remember that every individual has two loci for the gene and thus two alleles):

Genotype	Number of individuals	Number of T alleles	Number of t alleles
TT	20	40	
Tt	40	40	40
tt	40		80
Total	100	80	120

Of the 200 alleles, the proportion of the **T** allele is 80/200 = 0.4 (40%); and the proportion of the **t** allele is 120/200 = 0.6 (60%). It is customary in presenting this equilibrium to use "*p*" and "*q*" to represent the two allele frequencies. The dominant gene is represented by *p*, the recessive by *q*. Thus:

$$p = \text{frequency of } \mathbf{T} = 0.4$$
$$q = \text{frequency of } \mathbf{t} = 0.6$$
$$\text{therefore} \quad p + q = 1$$

Now, having calculated allele frequencies in the sample, let us determine whether these frequencies will change spontaneously in a new generation of a population. Assuming the mating is random (and this is important; all mating combinations of genotypes must be equally probable), each individual will contribute an equal number of gametes to the "common pool" from which the next generation is formed. This being the case, the frequencies of gametes in the "pool" will be proportional to the allele frequencies in the sample. That is, 40% of the gametes will be **T**, and 60% will be **t** (ratio of 0.4:0.6). Both eggs and sperm will, of course, show the same frequencies. The next generation is formed as follows:

Sperm \ Eggs	**TT** = 0.2	**Tt** = 0.4	**tt** = 0.4
TT = 0.2	.04 **TT**	.04 **TT** .04 **Tt**	.08 **Tt**
Tt = 0.4	.04 **TT** .04 **Tt**	.04 **TT** .08 **Tt** .04 **tt**	.08 **Tt** .08 **tt**
tt = 0.4	.08 **Tt**	.08 **Tt** .08 **tt**	.16 **tt**

Collecting the genotypes, we have:

$$\text{frequency of } \mathbf{TT} = .16$$
$$\text{frequency of } \mathbf{Tt} = .48$$
$$\text{frequency of } \mathbf{tt} = .36$$

Next, we determine the values of *p* and *q* from the randomly mated population:

$$\mathbf{T}\ (p) = \frac{.16 + \frac{1}{2}\ (.48)}{1} = 0.4$$

$$\mathbf{t}\ (q) = \frac{.36 + \frac{1}{2}\ (.48)}{1} = 0.6$$

The new generation bears exactly the same genotype frequencies as the parent population! Note that there has been no increase in the frequency of the dominant gene **T.** Thus, **in a freely interbreeding, sexually reproducing population, the frequency of each allele remains constant generation after generation.** The more mathematically minded reader will recognize that the genotype frequencies **TT, Tt,** and **tt** are actually a binomial expansion of $(p + q)^2$:

$$(p + q)^2 = p^2 + 2pq + q^2 = 1$$

This arithmetic approach is a very convenient way to calculate the frequency of genotypes in a population. For example, albinism in humans is caused by a recessive allele **a.** Only one person in 20,000 is an albino, and this individual must be homozygous **(aa)** for the recessive allele. Obviously there are many carriers in the population with normal pigmentation, that is, people who are heterozygous **(Aa)** for albinism. What is their frequency? Assuming again that mating is random (a questionable assumption for albinos, but one that we will accept for our example), genotype distribution will be $p^2 =$ **AA,** $2pq =$ **Aa,** and $q^2 =$ **aa.** Only the frequency of genotype **aa** is known with certainty: 1/20,000.
Therefore:

$$q^2 = 1/20{,}000$$

$$q = \sqrt{1/20{,}000} = \frac{1}{141}$$

$$p = 1 - q = \frac{140}{141}$$

The frequency of carriers is

$$\mathbf{Aa} = 2pq = 2 \times \frac{140}{141} \times \frac{1}{141} = \frac{1}{70}$$

One person in every 70 is a carrier! Even though a recessive trait may be rare, it is amazing how common a recessive allele may be in a population. There is a message here for anyone proposing a eugenics program designed to free a population of a "bad" gene. It is practically impossible. Since only the homozygous recessives reveal the phenotype to be artificially selected against (by sterilization, for example), the gene continues to surface from the heterozygous carriers. For a recessive allele present in two of every 100 persons (but homozygous in only one in 10,000 persons), it would require 50 generations of complete selection against the homozygotes just to reduce its frequency to one in 100 persons.

Processes of evolution: how genetic equilibrium is upset

The Hardy-Weinberg law shows that populations do not change—and thus do not evolve—as long as certain conditions are met: (1) very large population, (2) random mating, (3) no mutations, (4) no selection, and (5) no migration. We already know that these conditions cannot all prevail indefinitely and that the Hardy-Weinberg equilibrium can be disturbed by mutations, genetic drift, natural selection, and migration. Earlier (p. 872) we discussed mutations, the ultimate source of variability in all populations. We will now look at other factors that cause gene frequencies to change and thus lead to evolution (see Principle 16, p. 13).

Genetic drift

Genetic drift (Wright, 1969) refers to changes in gene frequency resulting from purely random sampling fluctuations. By such means a new mutant gene may be able to spread through a small population until it becomes homozygous in all the organisms of a population (random fixation), or it may be lost altogether from a population (random extinction). Such a condition naturally would upset the gene frequency equilibrium.

How does the principle apply? Let us suppose a few individuals at random became isolated from a large general population. This could happen by some freakish accident of physical conditions, such as a flood carrying a small group of field mice to a remote habitat or a disease epidemic wiping out most of a population. In the general population individuals would be represented by both homozygotes, **TT,** for example, and heterozygotes, **Tt.** It might be possible for the small, isolated group to be made up only of **TT** individuals and the **t** gene would be lost altogether. Also, when only a small number of offspring are produced, certain genes may, by chance, be included in the germ cells, and others may not be represented. It is possible in this way for heterozygous genes to become homozygous. The new group may in time have gene pools different from the ancestral population.

We should note also that most breeding populations of animals are small. A natural barrier such as a stream may be effective in separating two breeding populations. Thus chance could result in the presence or absence of genes without being directed at first by natural selection.

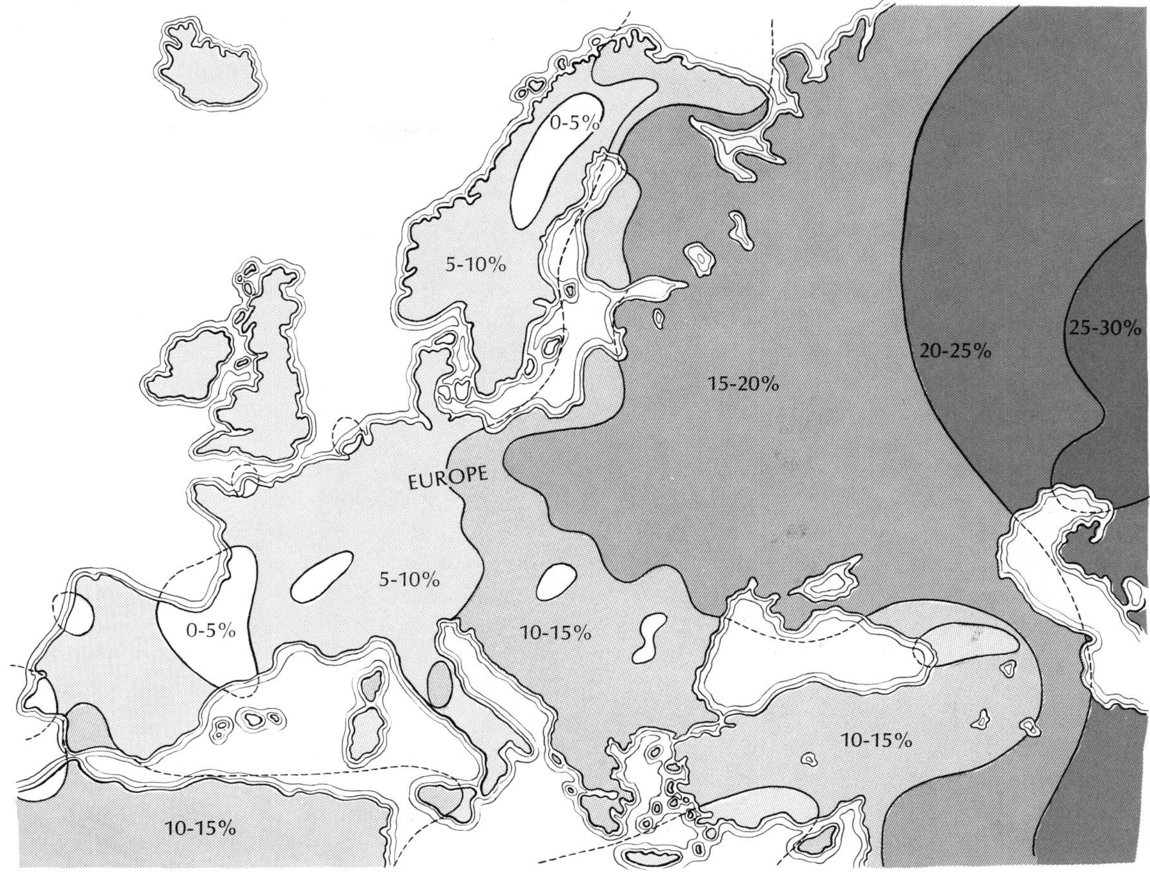

FIG. 38-8 Frequencies of the blood-type allele B among humans in Europe. The allele is common in the east and rare in the west. The allele may have arisen as a new mutant gene in a small population of people in Asia that were, or became, isolated from their parent population eastward where the allele did not exist. By random genetic drift, the new allele may have spread through the founding population. The spread of the allele westward began when the founding population lost its isolation and intermixed with people in the west. (Adapted from Mourant, A. E. 1954. The distribution of human blood. Toronto, Ryerson Press.)

Genetic drift appears to be the cause of the uneven distribution of the ABO blood groups among human populations in different parts of the world (Fig. 38-8). These differences could result from small migrant groups, which were atypical of their parent populations, settling in new areas. American Indians have a high frequency of type O blood, whereas type B is very rare. This presumably arose from the founding population of Asians that migrated into North America bearing this gene combination. If this is true, the migrants were an atypical sample of Asians (in whom type O blood is uncommon).

There are many who deny the importance of genetic drift, but it is generally agreed that, in bisexually reproducing species, evolution proceeds more rapidly when a population is broken up into isolated or partially isolated breeding communities and that the smaller the population the greater is the importance of genetic drift.

Natural selection

We have already stated that natural selection is the principal guiding force in evolution, and we have alluded to its mechanism. We will now more firmly define what it does. *Natural selection guides evolution by sorting out new adaptive combinations from a population gene pool derived from mutations, recombination, and other sources of genetic variation acting over many generations.*

Let us examine this definition. Sexually reproducing populations are composed of many individuals, and each individual has a different genotype. These differences have arisen by mutations, by genetic recombination, and perhaps by genetic drift, nonrandom mating, or migration. Suppose that one of these genotypes is better than any of the others, that is, the animal bearing it is *better adapted* to the environment than are other animals having different genotypes. What basis have we for concluding that a particular genotype is better, that is, has greater **adaptive value** than others? The basis is reproductive success. Most organisms contribute progeny to the next generation. The genotype that contributes the most progeny, and therefore the most genotypes to the next generation, has the greatest adaptive value and chance of success. If this differential reproduction continues systematically generation after generation, the genotype will increase in frequency while others decline in frequency. Natural selection, then, is the difference in gene pools from one generation to the next. Changing gene frequencies is the whole basis of evolution.

We have stressed, and we will stress once more, that natural selection works on the whole organism—the entire phenotype—and not on individual genes. The winning genotype may well contain genes that have detrimental effects on the individual but are preserved because other genes are strongly beneficial.

Polymorphism. Up to this point we have considered a population in terms of gene frequencies. We will now extend our consideration to actual examples of how natural selection works. Many species of plants and animals are represented in nature by two or more clearly distinguishable kinds of individuals, a condition called **polymorphism** (many forms). Polymorphic variation may involve color differences such as blue and white forms of the snow goose or the black and gray forms of the gray squirrel, or some other morphologic or physiologic character may be involved. It does *not* refer to seasonal variations in pelage or plumage (for example, the gray coat in summer of the varying hare changing to white in winter), for such seasonal alterations affect all members of the population alike.

An example of **stable polymorphism** is provided by a European species of ladybird beetle. Individuals show two color phases, red with black spots and black with red spots. Both forms interbreed freely. In summer the black form, behaving as a Mendelian dominant, increases and by fall outnumbers the red form. In winter, however, the red form recovers and by spring outnumbers the black form. The recessive red form is in some way favored and selected for during the cold winter months.

A classic example of **transient polymorphism** is industrial melanism (dark pigmentation) in the peppered moth of England (Fig. 38-9). Before 1850 the peppered moth was always white with black speckling in the wings and body. In 1850 a mutant black form of the species appeared. It became increasingly common, reaching frequencies of 95% in Manchester and other heavily industrialized areas by 1900. The peppered moth, like many similar species, normally rests in exposed places, depending on its cryptic coloration for protection. The mottled pattern of the normal white form blends perfectly with the lichen-covered tree trunks. With increasing industrialization, the soot and grime from thousands of chimneys killed the lichens and darkened the bark of trees for miles around centers such as Manchester. (This part of England was known as the "Black Country.") Against a dark background the white moth is conspicuous to predatory birds, whereas the mutant black form is camouflaged. The result was rapid natural selection: the easier-to-see white form was preferentially selected by birds, whereas the melanic form was subjected to far less predation. Selection pressure thus tended to eliminate the white form while favoring the genes that contributed to black wings and body.

Of special interest is the rapidity of the change. Rather than requiring thousands of years, the change occurred in less than 100 years. White moths still survived in nonpolluted areas of England, and with the recent institution of pollution-control programs the white forms are beginning to reappear in the woods around cities. Industrial melanism is a dramatic example of shifting gene frequencies as selection pressures change with a changing environment.

Another way to describe the selective quality of the genes that determine color is in terms of their **fitness.** Fitness describes the way individuals differ in their reproductive success because of differences in the genes that they carry. In this case the genes for the mutant black form increased the fitness of the black

FIG. 38-9 Light and melanic forms of the peppered moth *(Biston betularia)* on a lichen-covered tree in an unpolluted countryside **(A)** and on a soot-covered oak near industrial Birmingham, England **(B).** The dark melanic coloration that appeared in 1850 is controlled by a dominant allele. (Photographs by H. B. D. Kettlewell.)

moths in England's "Black Country." This same trait, however, confers low fitness in nonpolluted countryside areas. On the other hand, genes for the white form confer high fitness in nonpolluted countryside areas but low fitness near industrial centers.

Polymorphism clearly has adaptive value to species exposed to different environmental conditions. It is widespread throughout the animal kingdom. Examples are blood types of humans and other animals, albinism in many animals, silver foxes in litters of gray foxes, and rufous and gray phases of screech owls in the same brood. Frequently, polymorphic forms have been mistaken for separate species. If exposed to a persistent environmental shift that favors reproductive success by one or the other of separate polymorphic forms, one may become dominant over a particular part of the animals's range. This is but a short step to the establishment of a well-defined subspecies.

EVOLUTION OF NEW SPECIES

Previously we have described how populations change over periods of time because of genetic variation. Mutations are the fundamental source of genetic variation, further enhanced by sexual interchange and recombination. Natural selection acting on these variations is the force that produces evolutionary change and maintains the adaptive well-being of populations. In this section we consider how new species arise. The process by which one form becomes genetically isolated from another is called **speciation.**

Species concept

Although a definition of species is needed before we can theorize about species formation, biologists do not agree on a single rigid definition that applies in all cases. Carolus Linnaeus, the great Swedish naturalist who founded our system of classification, saw plant and animal species as fixed entities, static immutable units subject only to minor and unimportant variations (p. 120). This is the traditional concept of species: namely, a group within which individuals closely resemble one another but that is clearly distinguishable from any other group. The limitation of the fixed entity definition is biologic variation. Many species are made up of individuals that can be arranged into completely intergrading series so that sometimes the gaps between species can be detected only with the greatest difficulty, and it becomes almost impossible to decide where two species are to be separated. Yet species are realities in nature. We observe and collect them, study them, sample their populations, and describe their behavior.

The difficulty with the traditional system is that classification is based on the animals' appearance alone (size, color, length of various parts, and so on). Many other criteria can be and presently are being used: behavioral differences, reproductive characteristics, genetic composition, and ecologic niche. The single property that maintains the integrity of a species more than any other property, especially among biparental organisms, is **interbreeding.** The members of a species can interbreed freely with each other, produce fertile offspring, and share a common genetic pool. Interbreeding of different species is usually either physically impossible or produces sterile offspring. There are, of course, exceptions, but they are rare. Species are thus usually considered genetically closed systems, whereas **races** within a species are open systems and can exchange genes. *A species therefore may be defined as a group of organisms of interbreeding natural populations that is reproductively isolated from other groups and that shares common gene pools* (Dobzhansky, 1951; Mayr, 1970).

Speciation

Geographic isolation

The definition of species states that gene pools of different species are isolated from each other. With the exception of occasional hybridization, gene flow between species does not occur. Isolation, then, is a crucial factor in evolution. Unless some individuals of a population can be segregated from the parent population for many generations, new adaptive variations that may arise become lost through interbreeding with the parent population. Geographic isolation permits a unique sample of the population gene pool to be "pinched off" or segregated so that diversification can occur.

Let us imagine a single interbreeding population of mammals that is split into two isolated populations by some geographic change—the uplifting of a mountain barrier, the sinking and flooding of a geologic fault, or a climatic change that creates a hostile ecologic barrier such as a desert. Gene flow between the two isolated populations is no longer possible. The populations are almost certainly different from each other even when first isolated, since just by chance alone one population contains alleles not present in the other.

With the passage of time genetic recombination accentuates the difference, especially if the populations are small. Mutations that appear in the separated populations are certain to be different. Furthermore the climatic conditions of the separated regions are different; one may be warm and moist, the other cool and dry. Natural selection acting on the isolated gene pools favors those mutations and recombinations that best adapt the populations to their respective environments, unique food supplies, and new predator-prey relationships.

The two populations continue to diverge morphologically, physiologically, and behaviorally until distinct geographic races are formed. After some indefinite time span, perhaps only a few thousand years, but more likely a period of hundreds of thousands or even millions of years, evolutionary diversification progresses to the point that two races are reproductively isolated. If the barrier separating them fails (erosion of mountains, shifting of river flow, reestablishment of a hospitable ecologic bridge), the two populations again intermix. However, they may no longer interbreed, in which case they are distinct species.

Speciation under these conditions is by no means inevitable. It happens only under the most ideal conditions: small population (and gene pool) size, fortunate mutations, and the right combination of selection pressures. The chances are good, in fact, that one or both populations isolated in this way become extinct. But over the vast span of time that organic evolution has had to work, the special conditions required for speciation have been repeated numerous times, indeed millions of times.

Adaptive radiation on islands

Many times in the earth's history, a single parental population has given rise not just to one or two new species but to an entire family of species. The rapid multiplication of related species, each with their unique specializations that fit them for particular ecologic niches, is called **adaptive radiation.**

Young islands are especially ideal habitats for rapid evolution. If they are formed by volcanos that rose from a platform on the ocean floor, as were the Galápagos Islands, they are at first devoid of life. In time they are colonized by plants and animals from the continent or from other islands. These newcomers arrive to an especially productive situation for evolution because of the abundance of new opportunities.

On the crowded mainland almost every ecologic niche is already occupied. Every animal is specialized and adapted for a particular way of life, and all food sources are exploited by one species or another. Although variations continue to appear within a population, those offspring in each generation most like their parents are most likely to survive, because they are best suited for that particular set of environmental conditions. Competition between different species is too keen on the mainland, where evolution has been proceeding for a very long time, for many new evolutionary experiments to succeed.

On a young island, however, new arrivals find new ecologic opportunities and no competitors. Those that survive the sea or air voyage and the landfall may be able to become established, multiply, and spread out. They are in a new land that in all probability differs ecologically from their original home. New variations may serve some of the descendants of the colonizers in establishing themselves in new niches. Offspring that differ slightly from their parents may find that these differences enable them to exploit alternate food sources as yet unutilized by other animals. They thus flourish and produce offspring, some of which bear similar characteristics. With each passing generation these animals become increasingly successful at this alternate way of life, at the same time becoming increasingly different from their ancestors. The outcome is a new genetic blueprint, in short, a new species. Equally well, immigrants to an island may fail to become established if the limited variety that is provided by a small group of colonizers does not happen to fit the new environment.

On archipelagos, such as the Galápagos Islands, isolation plays an extremely important evolutionary role. Not only is the entire archipelago isolated from the continent, but each island is geographically isolated from the others by the sea, the most inhospitable environment for land animals. Moreover, each island is different from every other to a greater or lesser extent in its physical, climatic, and biotic characteristics. As colonizing animals find their way to the different islands, they are presented with an environmental challenge that is unique for each island. Moreover, each new arrival carries with it only a small fraction, a biased sample, of the population's gene pool. This further stimulates diversification, already encouraged by isolation, new ecologic opportunities, and lack of competition. Archipelagos, more than any other place on the earth's surface, offer the raw materials and opportunities for rapid evolutionary changes of great magnitude.

Darwin's finches

Let us consider the evolution of the family of 13 famous Galápagos finch species (Fig. 38-10). Their fame rests not on their beauty—they are, in fact, inconspicuous, dull-colored, and rather unmusical in song—but on the enormous impact they had in molding Darwin's theory of natural selection. Darwin noticed that the Galápagos finches (the name "Darwin's finches" was popularized in the 1940s by the British ornithologist David Lack) are clearly related to each other, but that each species differs from the others in several respects, especially in size and shape of the beak and in feeding habits. Darwin reasoned that, if the finches were specially created, it would require the strangest kind of coincidence for 13 similar finches

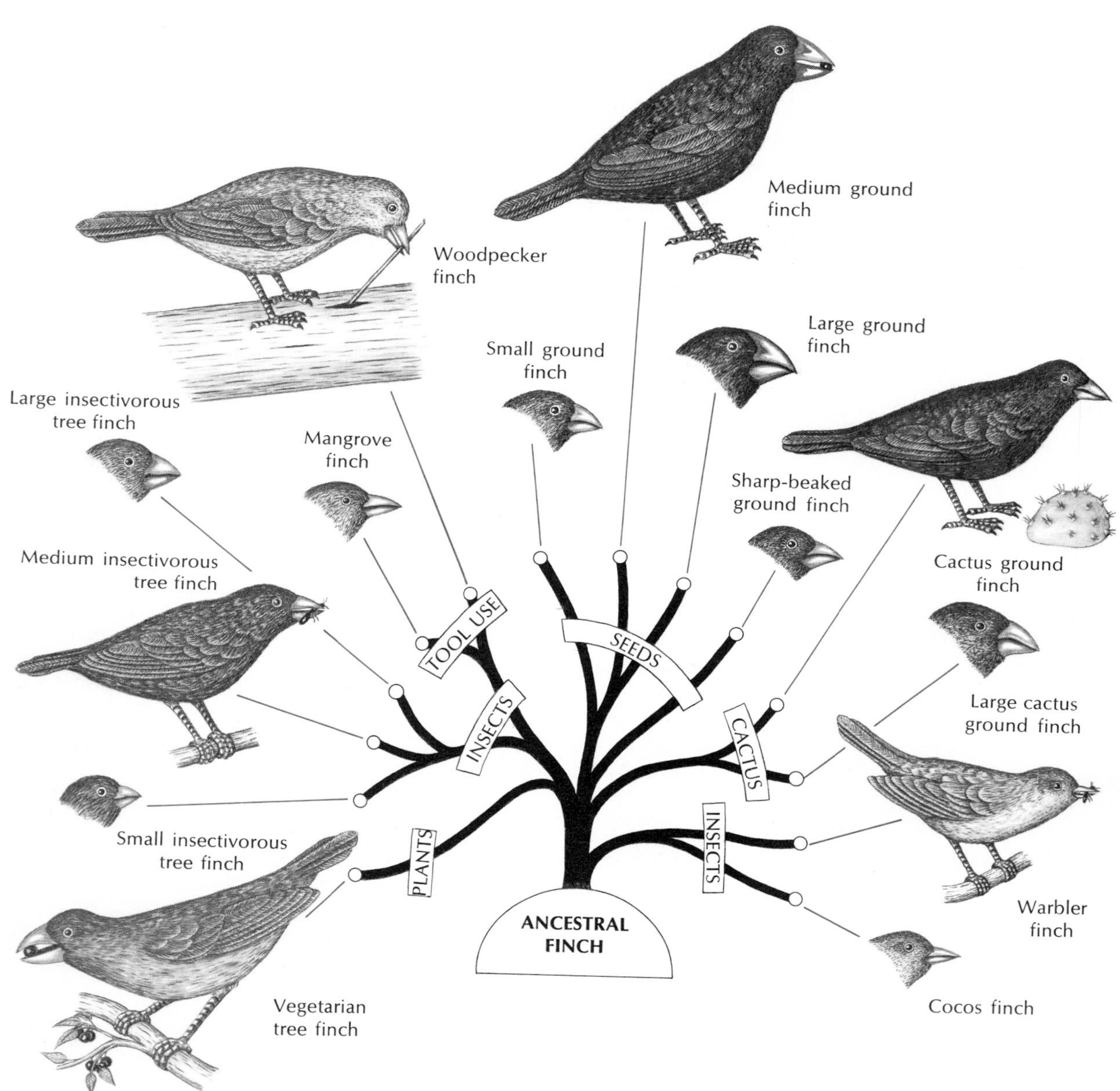

FIG. 38-10 Adaptive radiation in Darwin's finches.

to be created on the Galápagos Islands and nowhere else.

The 13 finches are grouped into several genera according to beak shape and feeding habits. The most clearly finchlike are the six species of **ground finches** *(Geospiza)*. They feed mainly on seeds on the ground and are equipped with bills of different strengths and sizes for handling seeds varying in size and hardness. One, the cactus ground finch, feeds mostly on cactus; its slightly curved beak is adapted for probing into the yellow cactus flowers.

The three **tree finches** *(Camarhynchus)* have parrotlike beaks, live in trees, and excavate twigs and branches for concealed insects. The single **warbler finch** *(Certhidea olivacea)* is a delicate warblerlike finch that feeds exclusively on insects with its slender bill. The **vegetarian tree finch** *(Platyspiza crassirostris),* the largest Galápagos finch, feeds on fruits and soft seeds with its short, thick beak.

The two most remarkable finches are the **woodpecker finch** and the **mangrove finch** *(Cactospiza)*. Lacking the long, protrusive tongue of true woodpeckers, they often use small cactus spines or twigs as tools to probe for insect larvae deep in holes in trunks or branches of trees. This is one of the very few recorded cases of tool use among animals below humans and apes.

Darwin's finches are an excellent example of how new forms may originate from a single colonizing ancestor. Lacking competitors from other land birds, they underwent adaptive radiation, evolving along various lines to occupy available niches that on the mainland would have been denied to them. They thus assumed the characteristics of mainland families as diverse and unfinchlike as warblers and woodpeckers.

A fourteenth finch is found on isolated Cocos Island, far to the north of the Galápagos archipelago. It is similar in appearance to the Galápagos finches and almost certainly has descended from the same ancestral stock. Since the Cocos Island is a single small island with no opportunities for isolation of finch populations, there has been no species multiplication.

Allopatric and sympatric speciation

The Cocos Island example emphasizes the fundamental requirement for geographic isolation in speciation. In its absence, there is no other reasonable way for gene pools to become separated long enough for genomic differences to accumulate. This is often called the **allopatric theory of speciation,** meaning that species can originate only in populations that occupy different communities and therefore do not interbreed.

Is there any chance or any mechanism, then, for **sympatric** speciation to occur? That is, can new species originate within a single population without geographic isolation? Most biologists think not, while admitting that it may occur as a highly improbable event.

One clear case of sympatric speciation is the artificial creation of new species by hybridization. In 1924 Karpechenko crossed the radish *(Raphanus sativus)* with the cabbage *(Brassica oleracea),* each having 18 chromosomes. The 18 chromosome hybrids (9 chromosomes from each parent) were sterile, as hybrids usually are, but by chance some of the hybrids doubled their chromosome number, becoming tetraploid. This event is called spontaneous autoallopolyploidy.

The tetraploid plant with 36 chromosomes, bearing characteristics of both the radish and the cabbage, proved to be fertile. A new sympatric "instant species" was formed called *Raphanobrassica*. It could not be crossed with either of the parents and was reproductively isolated. Unfortunately the new species is of no practical use because it has roots like a cabbage and leaves like a radish.

Spontaneous allopolyploidy has probably occurred as a natural event for the production of some of our food plants such as the bread wheats, potatoes, and raspberries. Although it may be a significant factor in plant evolution, it is probably a rare process in that of animals.

Annotated references

Selected general readings

Bodmer, W. F., and L. L. Cavalli-Sforza. 1976. Genetics, evolution and man. San Francisco, W. H. Freeman and Co. Publishers.

Colp, R., Jr. 1977. The illness of Charles Darwin. Chicago, University of Chicago Press. *The author argues that Darwin's notorious chronic illness was caused by psychic stress from his work on evolution, a thesis not widely accepted.*

Darwin, C. 1839. The voyage of the Beagle. New York, Doubleday Co., Inc. (1962). *This Natural History Library edition contains an excellent introduction by Leonard Engel.*

Darwin, C. 1859. On the origin of species by means of natural selection, or the preservation of favoured races in the struggle for life. London, John Murray. *There were five subsequent editions by the author. The recent Mentor Book edition is introduced by Sir Julian Huxley.*

Dawkins, R. 1976. The selfish gene. New York, Oxford University Press. *Provocative and controversial approach to evolutionary theory.*

Dillon, L. S. 1978. Evolution: concepts and consequences. St. Louis, The C. V. Mosby Co. *Comprehensive college text.*

Dobzhansky, T. 1951. Genetics and the origin of species, ed. 3. New York, Columbia University Press. *Important though advanced treatise.*

Dobzhansky, T. 1970. Genetics of the evolutionary process. New York, Columbia University Press.

Dobzhansky, T., F. J. Ayala, G. L. Stebbins, and J. W. Valentine. 1977. Evolution. San Francisco, W. H. Freeman and Co. Publishers. *Comprehensive but somewhat uneven treatment.*

Dobzhansky, T., and others. 1967. Evolutionary biology, New York, Plenum Publishing Corp. *A series of volumes with chapters by contributors on evolution, appearing annually.*

Eaton, T. H., Jr. 1970. Evolution. New York, W. W. Norton & Co., Inc. *Well-written account with an ecologic approach.*

Grant, V. 1963. The origin of adaptations. New York, Columbia University Press.

Grant, V. 1977. Organismic evolution. San Francisco, W. H. Freeman and Co. Publishers *Compressed treatment of evolutionary principles.*

Lack, D. 1947. Darwin's finches: an essay on the general biological theory of evolution. New York, Cambridge University Press. *A classic study of adaptive radiation in the famous Galápagos finches.*

Mayr, E. 1970. Population, species and evolution. Cambridge, Mass., Harvard University Press. *This excellent reference is an abridgement of the author's* Animal Species and Evolution.

Mayr, E. 1972. The nature of the Darwinian revolution. Science **176:**981-989.

Mayr, E. 1976. Evolution and the diversity of life: selected essays. Cambridge, Mass., Harvard University Press. *Excellent.*

Merrill, D. J. 1962. Evolution and genetics. New York, Holt, Rinehart & Winston, Inc. *Elementary introduction, strong on the evidences for evolution.*

Mettler, L., and T. G. Gregg. 1969. Population, genetics and evolution. Englewood Cliffs, N.J., Prentice-Hall, Inc. *Balanced introductory treatment.*

Moody, P. A. 1970. Introduction to evolution, ed. 3. New York, Harper & Row, Publishers. *A popular introductory text.*

Salthe, S. N. 1972. Evolutionary biology. New York, Holt, Rinehart & Winston, Inc. *Excellent, clearly developed account of evolutionary theory.*

Stebbins, G. L. 1971. Processes of organic evolution. Englewood Cliffs, N.J., Prentice-Hall, Inc. *Solid introductory text.*

Thornton, I. 1971. Darwin's Islands. A natural history of the Galápagos. Garden City, N.Y., Natural History Press. *Best general account of the Galápagos Islands, describing island history, animals, evolution, and future outlook.*

Volpe, E. P. 1977. Understanding evolution, ed. 3. Dubuque, Iowa, William C. Brown Co., Publishers. *Succinct, clearly developed paperback.*

Wright, S. 1969. Evolution and genetics of populations, vol. 2. Chicago, University of Chicago Press.

Selected *Scientific American* articles

In addition to the articles listed below, the Sept. 1978 issue is devoted exclusively to organic evolution, with articles by Ernst Mayr, F. J. Ayala, J. W. Schopf, J. Maynard Smith, and others.

Barghoorn, E. S. 1971. The oldest fossils. **244:**30-42 (May).

Bishop, J. A., and L. M. Cook. 1975. Moths, melanism and clean air. **232:**90-99 (Jan.). *The evolution of industrial melanism in moths is described.*

Cavalli-Sforza, L. L. 1969. "Genetic drift" in an Italian population. **221:**30-37 (Aug.).

Clarke, B. 1975. The causes of biological diversity. **233:** 50-60 (Aug.). *Argues that diversity is maintained by, as well as is the basis of, natural selection.*

Cole, Fay-Cooper. 1959. A witness at the Scopes trial. **200:** 120-130 (Jan.).

Darlington, C. D. 1959. The origins of Darwinism. **200:** 60-66 (May).

Eiseley, L. C. 1959. Alfred Russel Wallace. **200:**70-84 (Feb.).

Evans, H. E., and R. E. Matthews. 1975. The sand wasps of Australia. **233:**108-115 (Dec.). *The genus* Bembix *is remarkably diverse, apparently because they were able to fill available ecologic niches.*

Kettlewell, H. B. D. 1959. Darwin's missing evidence. **200:** 48-53 (Mar.).

Kurtén, B. 1969. Continental drift and evolution. **220:**54-64 (Mar.).

Lack, D. 1953. Darwin's finches. **188:**66-72 (Apr.).

Newell, N. D. 1972. The evolution of reefs. **226:**54-65 (June). *Major events in the earth's history are reflected in changes in tropical reef communities.*

Wilson, E. O. (ed.) 1950-1974. Ecology, evolution, and population biology. San Francisco, W. H. Freeman and Co. Publishers. *Collection of Scientific American articles with introductions by E. O. Wilson.*

CHAPTER 39

HUMAN EVOLUTION

Fossil humans are revealed to us by two kinds of relics: (1) bone evidences, usually rare and often in poor states of preservation, and (2) products of human handiwork, often numerous and well preserved. These skeletons of North American Indians are in good condition because they are recent—less than 1,000 years old. Far older hominid fossils of the crucial Pliocene and Pleistocene epochs are virtually never found in such ideal conditions.

Photograph by Jack Zehrt.

While the question of human origins has preoccupied human thoughts for centuries and has formed the basis of many of the world's religions,the case for human antiquity really takes its roots in Darwin's 1859 publication of *The Origin of Species*. Darwin did not include the human species among the evidences used to support his theory of natural selection, but few missed the implication: humans have evolved slowly from lower forms of life. It was an idea repugnant to the Victorian world because many felt it degrading to be related, even distantly, to an ape. Actually, neither Charles Darwin nor Thomas H. Huxley, who vigorously supported and defended Darwin's theory, argued that humans were descended *directly* from apes, but rather that apes are the closest relatives of humans. It was an unfortunate misunderstanding, since it was thought that acceptance of the theory of evolution carried with it an obligation to believe that we are descended from gorillas or chimpanzees. This confusion greatly delayed broad acceptance of evolutionary theory by the general public.

Today, this storm of controversy seems quaint if not ridiculous to most people but it helps if we remember that this era was a turning point in the history of human knowlege, especially in our attitudes toward *Homo sapiens* as an animal species and to our place in nature. Furthermore, there was at that time virtually no fossil evidence that linked humans with apelike creatures. Darwin and Huxley rested their case mostly on cogent anatomic comparisons between humans and apes. The close resemblance of many features of human anatomy to those of apes could be explained only by assuming a common origin. Most differences in anatomy related to the upright human posture.

The search for fossils, especially for a "missing link" that would prove a connection between apes and people, began slowly but was spurred by the unearthing of two excellent skeletons of the Neanderthal man in the 1880s and then by the discovery of the famous Java man *(Homo erectus)* by Eugene Dubois in 1891. The most spectacular discoveries, however, have been made in the last three decades in Africa, especially in the 1970s. There is great excitement now among paleontologists and anthropologists, although it still is not possible to reconstruct fully the descent of genus *Homo* from the existing fossil evidence (which is still disappointingly meager compared to that of many nonprimate mammals).

We are no longer searching for a mythical "missing link" between living apes and humans. Both descended from a common primate ancestor in the past and, though closely related much as cousin is to cousin, they have evolved along independent lineages. Today's refined methods of investigation and modern dating techniques have established beyond reasonable doubt the authenticity and antiquity of the fossil evidence. Misunderstandings, disputes, and contradictions have been cleared up; elaborate hoaxes exposed; and doubt and ridicule gradually silenced. Nevertheless, old ideas die hard, and reluctance to accept the fact that people are mutable creatures like other animals still lingers among some people in the United States and Canada.

We will begin our story by looking beyond apes and monkeys to a more distant past when mammalian forms emerged bearing primate characteristics; then we will investigate the hominid evolutionary record leading to modern human beings.

OUR CLOSEST RELATIVES—PRIMATES

Humans are primates, a fact that even the nonevolutionist Linnaeus correctly recognized. (The classification of order Primates, the most advanced of mammalian orders is found on p. 640.) Order Primates is divided into two suborders—Prosimii and Anthropoidea. The prosimians include the more primitive primates, such as the tree shrews, lemurs, lorises, and tarsiers (Fig. 39-1). The more advanced anthropoids are separated into three superfamilies, the Ceboidea, or New World monkeys (Fig. 39-2), with prehensile tail and nonopposable thumb; the Cercopithecoidea, or Old World monkeys (Fig. 39-3), with opposable thumb and nonprehensile tail; and Hominoidea, the great apes and humans. The hominoids are in turn divided into two families—Pongidae and Hominidae. The pongids (great apes) include the gibbons (three species), gorilla (one species), orangutan (one species), and chimpanzees (two species) (Fig. 39-4). The hominids include one living species, *Homo sapiens*. The reader should beware of the potential confusion with the terms **hominoid,** which encompasses the great apes and humans, and **hominid,** which refers only to humanlike types, both living and extinct.

The basic adaptive features that have guided the evolution of this great mammalian order culminating in humans have revolved mainly around the arboreal habits of the group. Humans abandoned the brachiat-

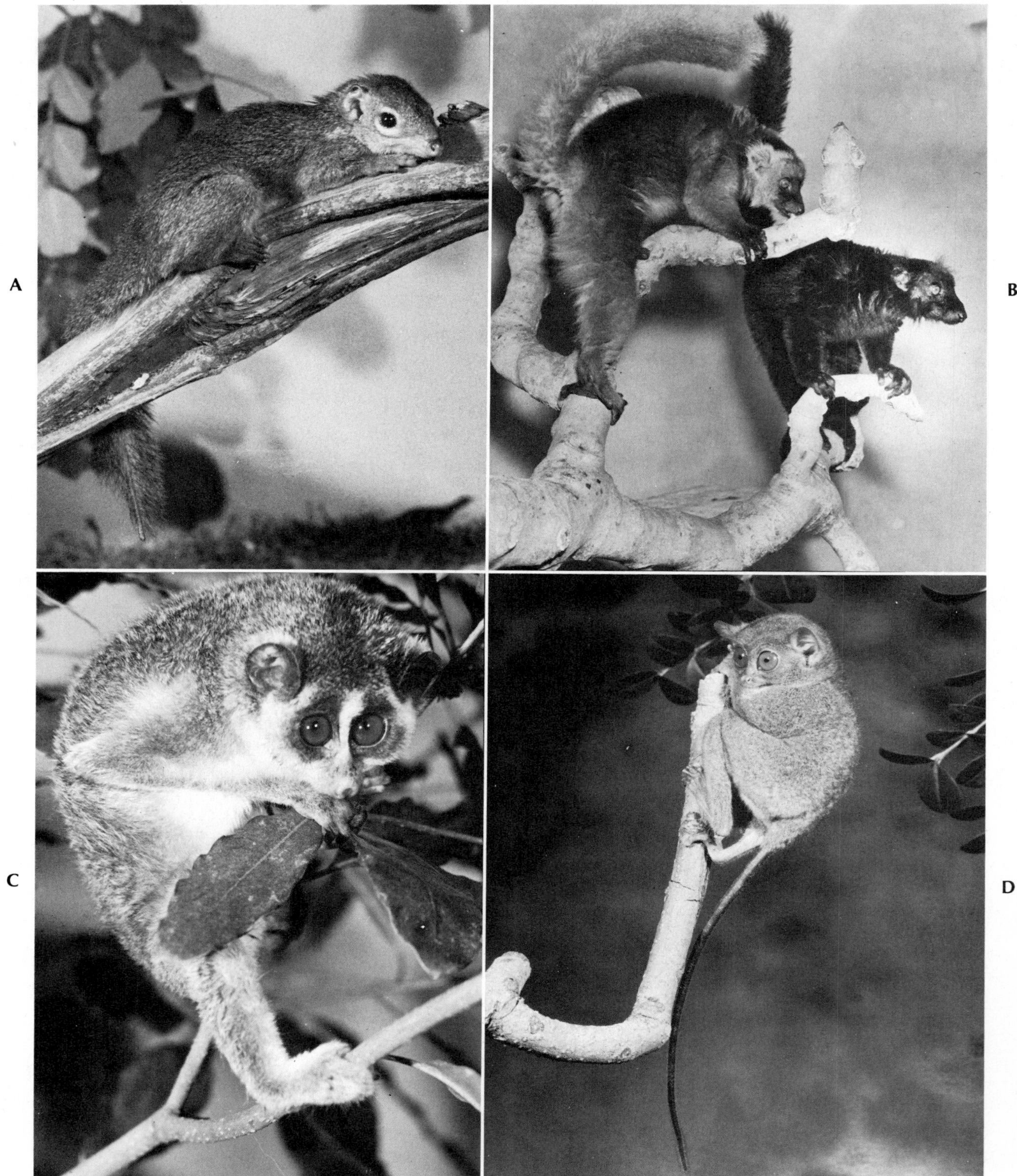

FIG. 39-1 Representative prosimians. **A,** Common tree shrew *Tupaia glis* of tropical rain forests and montane forests of southeast Asia. **B,** Black lemur *Lemur macaco* of Madagascar and Comoro islands. **C,** Slender loris *Loris tardigradus* of southern India and Sri Lanka (Ceylon). **D,** Mindanao tarsier *Tarsius syrichta carbonarius* of Mindanao Island in the Philippines. (Courtesy of the San Diego Zoo; **A, B,** and **C** by Ron Garrison.)

FIG. 39-2 Representative New World monkeys. **A,** Red howler *Alouatta seniculus* of tropical rain forests and deciduous forests throughout South America. **B,** Squirrel monkey *Saimiri sciureus* of the tropical rain forests of central and northern South America. Note the round, widely-separated nostrils, the flat appearance of the faces, and the prehensile tails. New World monkeys are completely arboreal, reveal little sexual dimorphism, and lack the ischial callosities (sitting pads) of Old World monkeys. (Courtesy of the San Diego Zoo; **A** by Ron Garrison.)

ing habit (using the arms for swinging) and took up a terrestrial existence, but there are anatomic structures, such as the limbs, hands, and feet, which plainly indicate the basic ancestral traits for an arboreal life. The free rotation of limbs in their sockets, the movable digits on all four limbs, and the opposable thumbs are all modifications for grasping branches and swinging through trees. Omnivorous food habits have produced a characteristic dentition.

Along with skeletal specializations, there are many correlated features that have evolved in primates, such as muscular, neural, and sensory adaptations superior to those in other animals. Better vision and development of other sense organs, as well as the proper coordination of limb and finger muscles, have meant an enlargement of appropriate regions of the brain. Precise timing and judgments of distance were necessary concomitants of an arboreal life and required a larger cerebral cortex. Concurrent enlargement of the brain led to a level of intelligence and alertness of mind characteristic of the higher primates.

Early in the Cenozoic era the basic radiation of primitive placental mammals began and continued until more than a score of orders arose and took over the niches formerly occupied by reptiles. The most primitive order was the Insectivora, familiar examples of which are the present terrestrial moles and shrews. But some members of this order, the tree shrews, took to living in trees. Living tree shrews in Asia thus may be considered a transition between the primitive insectivore ancestors and the primates. They look superficially like squirrels (Fig. 39-1, *A*), but their body structure bears characters that foreshadow the primates. They have a first digit that can be opposed to the others,

FIG. 39-3 Representative Old World monkeys. **A,** Mandrill *Mandrillus* sphinx of high forested regions of central Africa. **B,** Kikuyu colobus *Colobus polykomos kikuyuensis* and baby of rain and montane forests of central Africa. **C,** Hamadryas (sacred) baboon *Papio hamadryas* of extreme east-central Africa. **D,** Olive baboon *Papio anubia* and baby of central Africa. Note the comma-shaped, thinly-separated nostrils, downward-pointing noses, and absence of prehensile tails. Old World monkeys are of robust form, may be either arboreal or ground dwelling, reveal distinct sexual dimorphism, and possess ischial callosities. (**A, B,** and **C,** Courtesy of the San Diego Zoo; **A** and **C,** by Ron Garrison; **D,** by L. L. Rue, III.)

A

B

C

D

FIG. 39-4 Anthropoid apes (pongids). **A,** White-handed gibbon *Hylobates lar* of rain forests, semideciduous forests, and montane forests of southeast Asia. **B,** Bornean orangutan *Pongo pygmaeus* of the rain forests of Borneo. **C,** Lowland gorilla *Gorilla gorilla gorilla* of rain forests and equatorial Africa. **D,** Chimpanzee *Pan troglodytes* of western Africa. (Courtesy of the San Diego Zoo; **A** and **D** by Ron Garrison.)

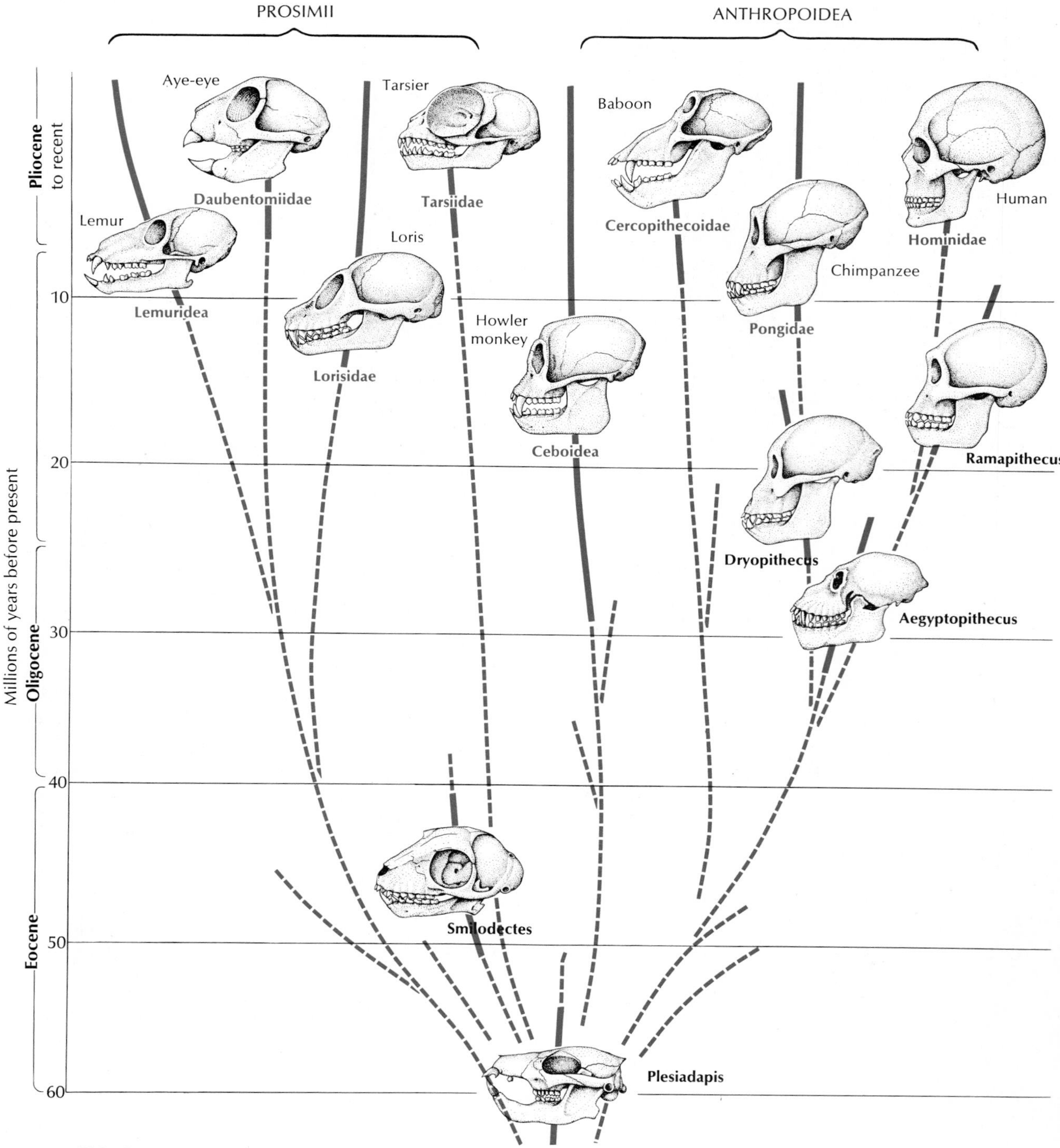

FIG. 39-5 Phylogeny of the primates, showing evolutionary relationships. Extinct forms are indicated by their generic names. Extant groups are indicated by the family name (superfamily in the case of Ceboidea) and common name of the example.

and a relatively large brain in which the visual portion is especially well-developed, an adaptation for an arboreal life. Zoologists today tend to include the tree shrews among the primates, although some place them in the order Insectivora. Whether or not the Paleocene (70 million years B.P.) tree shrews are actually ancestors of primates as most authorities believe, they indicate how a transition form might have looked.

The earliest primates were prosimians, small nimble animals adapted for living in trees. Although they possessed good muscular coordination and sense organs, their intelligence was low; intelligence among modern primates is a later development in their evolution. Besides the tree shrews, the Prosimii include the lemurs (mostly in Madagascar), the tarsiers of the East Indies, and the lorises of Asia and Africa (Fig. 39-1). Some modern prosimians have many primitive characteristics, such as eyes on the side, long snouts, and long tails (lemurs); but in the tarsiers the large eyes have moved forward and the muzzle is short. Their larger brain and better-developed placenta also place the tarsiers nearer the anthropoids than the lemurs (Fig. 39-5). Although tarsiers have all but vanished, many Paleocene and Eocene fossil genera have been found. Certain basic adaptations developed by primates, such as the opposability of the thumb and great toe, stereoscopic vision, and the ability to judge distance, fitted them for a tree-dwelling existence.

The earliest fossils of monkeys and apes have been found in late Eocene and early Oligocene deposits, 40 million years B.P. By far the most important source of these early fossils is the Fayum Depression—a hot, dry desert about 60 miles southwest of Cairo, Egypt. In the Oligocene epoch, however, Fayum was a tropic rain forest, laced with rivers, on the edge of the Mediterranean, which extended much farther inland than it does today. Here a relatively abundant fossil record indicates that several genera of both monkeys and apes, all very small but clearly monkeylike and not prosimian-like, scampered among the trees along the swamp-lined rivers. Thus, no later than the Oligocene epoch, both monkeys and apes arose independently from prosimian stocks.

Some of these monkeys migrated in some way to South America, perhaps by rafting on felled trees or masses of vegetation, where their isolated descendants gave rise to the New World monkeys. Although it seems impossible that any primate could survive the journey from Africa to South America today, the two continents were much closer together 40 million years ago than they are today (see maps of continental drift on p. 930). The New World monkeys (superfamily Ceboidea) are more primitive and are usually smaller than Old World monkeys, although they are similar in appearance. Their nostrils are far apart, and they have long prehensile tails (Fig. 39-2). The thumb is only slightly opposable to the other fingers. Because the New World monkeys split off early from the prosimians and are not closely related to Old World monkeys despite many anatomic and behavioral resemblances, they are not important to human evolution.

The Old World monkeys have their nostrils close together, and they do not use their tails as prehensile organs (Fig. 39-3). Like the New World monkeys, they are four-footed in terrestrial locomotion, walking on their palms and soles. Most are arboreal, but the short-tailed baboons show a tendency for ground dwelling.

The Old World monkeys cannot be considered ancestral to the anthropoid apes or humans. The primitive, hominoid apes evolved independently from prosimian ancestors. Many fossils of these small, monkey-sized animals have been found in Miocene and Pliocene deposits. Although apelike, they lack certain specializations of modern apes. One early form of particular interest, because it may represent an early lineage that founded both ape and human lines of descent, is *Aegyptopithecus* (Fig. 39-5), the "dawn ape". This spaniel-sized form is known from several late Oligocene (30 million years B.P.) fossil remains, including a nearly complete skull with cranium, upper and lower jaws, and virtually all teeth. The dentition is distinctly apelike but the skull is monkeylike. This fact, together with limb bones discovered at Fayum in 1977, suggests that the animal was a quadruped. Many paleontologists believe that *Aegyptopithecus* is a transition form leading to the hominids.

ANCESTORS OF APES AND HUMANS

In the Miocene epoch, beginning about 25 million years B.P., several new and significant characters enter the story. Lines of descent become clearer (though by no means complete) because the fossils are more abundant and not so old. The earth was becoming cooler, and this trend continued into the Pliocene epoch, at which time polar ice caps were beginning to form. Winters were longer and summers were shorter in many parts of the world. Tropic and subtropic forests gave way to grasslands over vast areas of North America, Europe, and Asia. In this setting the anthropoid apes (pongids) began an adaptive radiation that lasted well into the Pliocene epoch. Of the numerous pongid fossils from the Miocene and Pliocene epochs, one in particular revealed an intriguing blend of monkey, ape, and human traits. This was *Dryopithecus* (Gr., *drys,* tree, + *pithecus,* ape), a diverse genus that embraces several African, European, and Indian species (Fig. 39-5). One species, *D. africanus,* resembled chimpanzees and is believed to be in a direct line of descent leading to modern chimpanzees. Another much larger species *(D. major)* is probably a predecessor of the gorilla. Still another form having obvious affinities to the chimpanzee was *Proconsul,* discovered at Lake Victoria in Africa in 1930. *Proconsul's* position in ape evolution has been long disputed, but most authorities now agree that *Proconsul* is not a distinct genus, but an African member of the widespread genus *Dryopithecus. Dryopithecus* appeared in the late Oligocene, some 30 million years B.P. and flourished until about 10 million years B.P.

The confusing diversity of Miocene apes has now been collected as a subfamily of pongids called the *dryopithecines.* They clearly are ancestral to the pongids and probably to the hominids as well, although the exact lineages remain to be worked out. As a group they were essentially arboreal fruit-eaters of woodland savannas. They were probably good brachiators, with arms and legs of about the same length. On the ground, they were quadrupedal, but most of them walked on their knuckles, much as chimpanzees do today. Their large size (one was as large as the modern gorilla) and intelligence enabled them to exploit both arboreal and terrestrial environments in relative safety.

The pongids today (gibbons, orangutan, chimpanzees, and gorilla) have longer arms than legs, a long trunk compared to lower limbs, curved legs with the knees turned outward, large canines, laterally compressed dental arch (not rounded), long protruding face, and a brain about one-third as large as ours. Some modern apes have largely abandoned the arboreal way of life. Chimpanzees and gorillas are quite at home on the ground. A more or less erect posture is characteristic of many of them, although they may use their long arms for support while walking. More abundant food on the ground may have induced them to come down out of the trees.

Despite the large gap that separates humans from apes, there is abundant evidence for common ancestry that becomes ever more convincing. By using a variety of new biochemical techniques to estimate the "genetic distance" between humans and chimpanzees, we find that their macromolecules are almost all alike. The genetic distance between them is in fact so close that their structural genes are virtually identical. Yet chimpanzees are obviously very different biologically from people. This may be explained by mutations that affect the *expression* of genes. Small changes in the timing of activation or in the level of activity of single genes, it has been suggested, could vastly influence embryonic development and lead to the large behavioral and anatomic difference between humans and chimpanzees. Gene regulation is a poorly understood but exciting frontier in evolutionary study.

First hominids

At some point, the hominids separated from the apes and underwent a radiation of their own. Giving up an arboreal life was a necessary prelude to their amazing evolution. The climatic changes during the Miocene and Pliocene epochs that greatly reduced the forests also provided an impetus to an existence on the open savannas. The upright carriage of the human body, of no advantage for tree dwelling, can be advantageous in open country. Because there are more dangerous predators on the ground than in trees, selection pressures would have fostered longer and stronger hind legs for faster upright running. Standing up also provides a better view over tall grasses and bushes. And perhaps most important, it frees the hands for using tools. We should note, however, that some paleontologists, notably S. Washburn, argue that tool and weapon use *preceded* and promoted walking on two legs.

Thus, during the Miocene and Pliocene epochs, emerging hominids slowly evolved toward an upright posture. This important transition was really an enormous leap because it required extensive redesigning of the skeleton and muscle attachments. Unfortunately we have only a glimpse of how it may have proceeded.

A late Miocene fossil fragment first discovered in 1932 by G. E. Lewis, and long ignored by paleontologists, was *Ramapithecus brevirostris,* dated at about 12 to 14 million years B.P. (Figs. 39-5 and 39-6) Since *Ramapithecus* was found in the Siwalik Hills of India, Lewis named his discovery after the Hindu epic hero Rama. Interest in this form was rekindled when in 1961

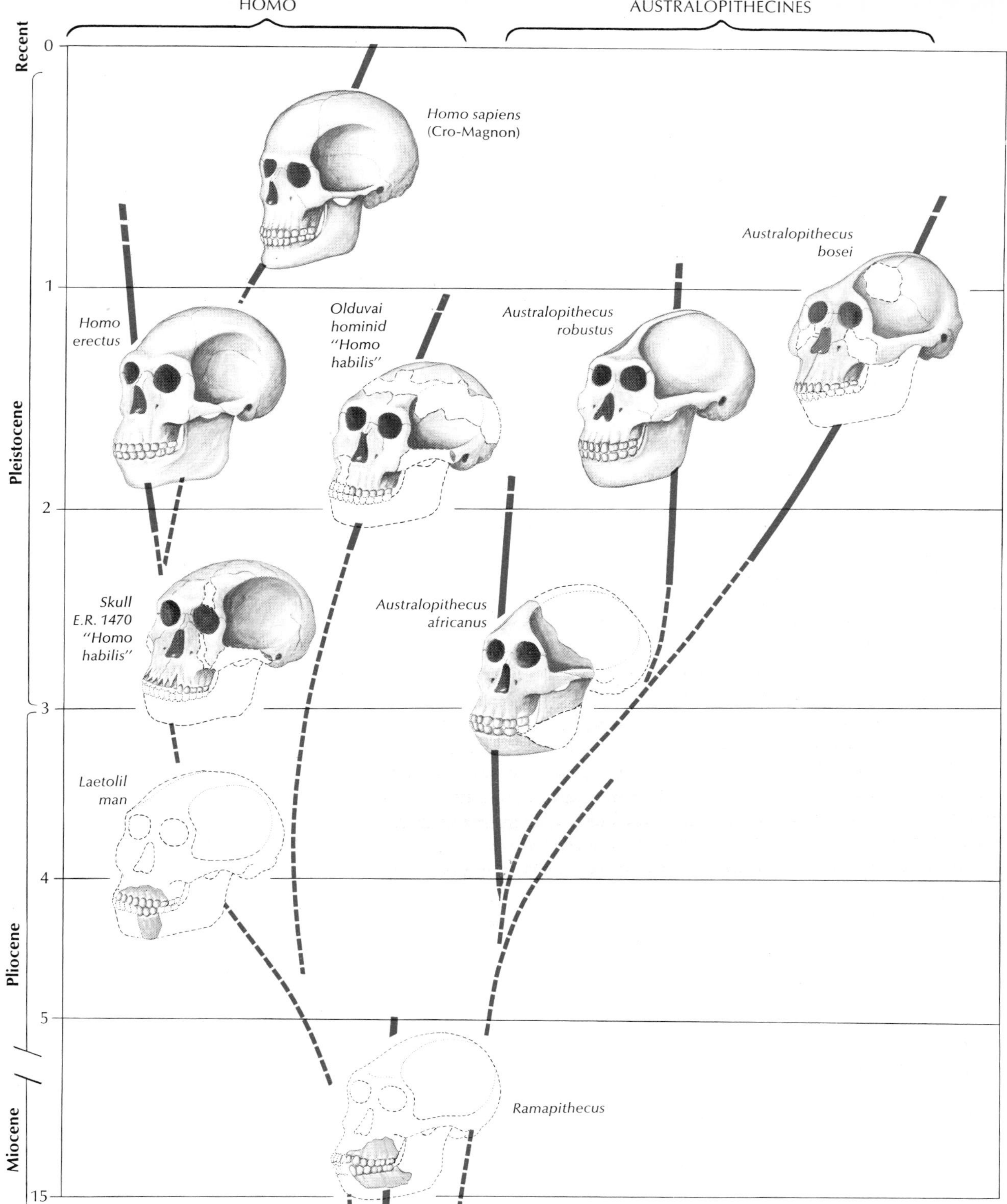

FIG. 39-6 Phylogeny of the hominids. Relationships as indicated by broken lines are highly tentative. Solid lines indicate intervals when forms are known to have lived. Shown is a four-lineage hypothesis in which the stem form *Ramapithecus* is antecedent to four separate lines of descent: (1) *Australopithecus africanus*, (2) *A. robustus*, (3) Olduvai hominid *(Homo habilis)*, and (4) Laetolil-Skull 1470 *(H. habilis)*, which led to *Homo sapiens*. Many paleoanthropologists contest the validity of the species *Homo habilis*, believing instead that all late Pliocene and early Pleistocene hominids should be grouped under the genus *Australopithecus*. According to these more conservative views, *Homo erectus* and *Homo sapiens* have descended from *Australopithecus*.

the husband-and-wife team of Louis and Mary Leakey discovered more fragments of *Ramapithecus* in South Africa that, pieced together with Lewis' find, definitely established the human character of the jaw. Other discoveries followed, and by 1977 some two dozen separate jaw and tooth fossils of *Ramapithecus* had been described. The teeth of *Ramapithecus* are low in form, rather than high as in apes, and the dental arch is conspicuously hominid in form. All in all, *Ramapithecus* appears to be the most appropriate fossil representing an early stage in the hominid line. Until a better candidate is unearthed, we may consider *Ramapithecus* ancestral to humans. Still, *Ramapithecus* is known only from tooth and jaw fragments. Lacking hip or leg bones or vertebrae, or even a complete cranium, it is impossible to say whether this creature stood upright like humans or crouched forward like apes.

Australopith hominids—the near humans

After *Ramapithecus,* there is an early Pliocene gap in our family history from 8 million to 4 million years B.P., for which virtually no hominid fossils, other than a few tantalizing fragments, have been discovered. Yet the early Pliocene epoch was an eventful one in which the hominid line began to branch into different routes. One in particular gave rise to the first certain bipedal genus *Australopithecus*. The fossil skull of an immature anthropoid was discovered in 1924 at Tuang, South Africa, by Raymond Dart, an anatomy professor at a Johannesburg University. He named his find *Australopithecus* (L., *auster,* south, + Gr., *pithecus,* ape) *africanus,* "the southern African ape," and he was certain from the beginning that it was in the direct lineage to humans. The skull was beautifully preserved but mostly enclosed in a rock-hard mixture of sand, soil, and pebbles (breccia). Dart picked away at his find for several years, at last fully revealing the gracile skull of a juvenile, aged 5 or 6 years, containing the endocast of a brain and a fully developed set of milk teeth (Fig. 39-7).

Additional fossils of this and related types, including skeletons that were nearly complete, have been found by a number of investigators since that time. Two much more robust forms of *Australopithecus* are now recognized, *A. robustus* and *A. boisei*. The gracile *A. africanus* was a contemporary of *A. robustus* in the late Pliocene epoch. *A. robustus* in turn is thought to have given rise to the superrobust *A. boisei* that inhabited eastern and southern Africa for perhaps a million years before becoming extinct in the mid-Pleistocene epoch, about 600,000 years B.P. (Fig. 39-6).

Although Africa seems to be the basic home of this group, similar fossils have been found as far away as Java. The volume of their brain casts varied from about 450 to 600 cc and overlapped the range of the chimpanzee and gorilla. Certain bones, such as those in the pelvis, were hominid rather than pongid. They walked in an erect or semierect position, as attested to by the shape of the leg and foot bones. Their dentition was more human than apelike. They may be regarded as the earliest known type with distinctly humanlike characteristics.

EMERGENCE OF HOMO THE TRUE HUMAN

Perhaps more than any other place on earth, Olduvai Gorge in Tanzania deserves the distinction of being the most dramatic source of information about prehistoric humans. Here Louis and Mary Leakey, later joined by their son Richard, worked over a period that spans four decades to record the geologic, animal, and human history of the region. The stratigraphy of Olduvai is clear, so that absolute dating is possible. There is an abundance of both animal fossils and primitive human tools.

Perhaps the most exciting find at Olduvai was Mary Leakey's discovery (1959) of a nearly intact skull, lacking the jaw, of *Australopithecus boisei*. It was originally described as *Zinjanthropus boisei,* the Olduvai Gorge man (Fig. 39-8). This young male has been dated as 1.7 million years old, but more recent finds by Richard Leakey at Lake Rudolf in Kenya have pushed the age of *A. boisei* back to nearly 3 million years B.P.

Homo habilis—primitive human or advanced ape?

In the early 1960s a second skull, together with stone tools, was found at Olduvai close to the site where *A. boisei* had been discovered. The Leakeys awarded their find with the binomen *Homo habilis,* meaning "able man," because of their conviction that this man was the first user of primitive stone tools. The short and slender *Homo habilis* had a brain capacity estimated at 657 cc. This is minute by modern human stan-

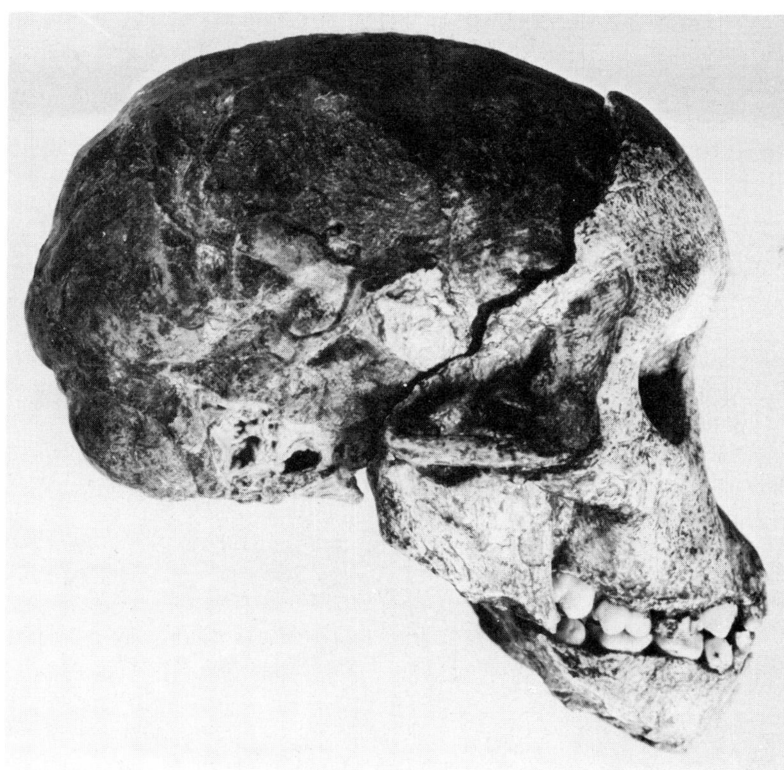

FIG. 39-7 Skull and cranial endocast of the southern African hominid, *Australopithecus africanus.* (From Dillon, L. S. 1978. Evolution. St. Louis, The C. V. Mosby Co.)

dards (average 1,350 cc) but is significantly larger than the australopith range of 430 to 530 cc. *Homo habilis* has been found associated with tools in more recent strata at Olduvai, indicating that it persisted for nearly a million years, well into the Pleistocene epoch. Opinion is sharply divided on the relationship of *Homo habilis* to *Australopithecus*. Some, including the Leakeys, believe that *Homo habilis* is in the direct lineage to *Homo sapiens*. Others, however, believe that the name bearer *Homo habilis* is not significantly different from specimens of *Australopithecus,* and consequently is an invalid name.

Skull 1470

Further confounding the uncertain descent of hominids was the discovery in 1972 by Richard Leakey of a remarkable skull known by its registration number as Skull 1470—"1470 man." It is at least 1.6 million years old and may be more than 2.5 million years old. Skull 1470 had a cranial capacity of 775 cc and resembled the skull of *Homo erectus* (discussed further) in several significant ways. This startling find appeared to prove that *Homo* was a contemporary of *Australopithecus,* as the Leakeys had claimed for some time. An even older hominid fossil find announced by Mary Leakey late in 1975 has been firmly potassium-argon dated at 3.35 to 3.75 million years B.P. This find—jaws and teeth of 11 distinct individuals—was discovered at a site 40 km south of Olduvai called Laetolil. The fossils closely resemble the jaw and teeth of Skull 1470 (Fig. 39-6).

Thus, if Skull 1470 and the Laetolil fossils are of the genus *Homo,* as the Leakeys and others believe, it means that *Homo's* origins have been pushed back to more than 3.5 million years B.P. The Leakeys contend that *Homo* and *Australopithecus* broke off from a common ancestral line some 4 million years ago B.P. *Homo* coexisted with *Australopithecus* for 3 million years, but unlike the latter, which was probably a vegetarian, *Homo* used tools made from stones and bones to kill animals for food. *Homo* was, of course, fully erect and bipedal.

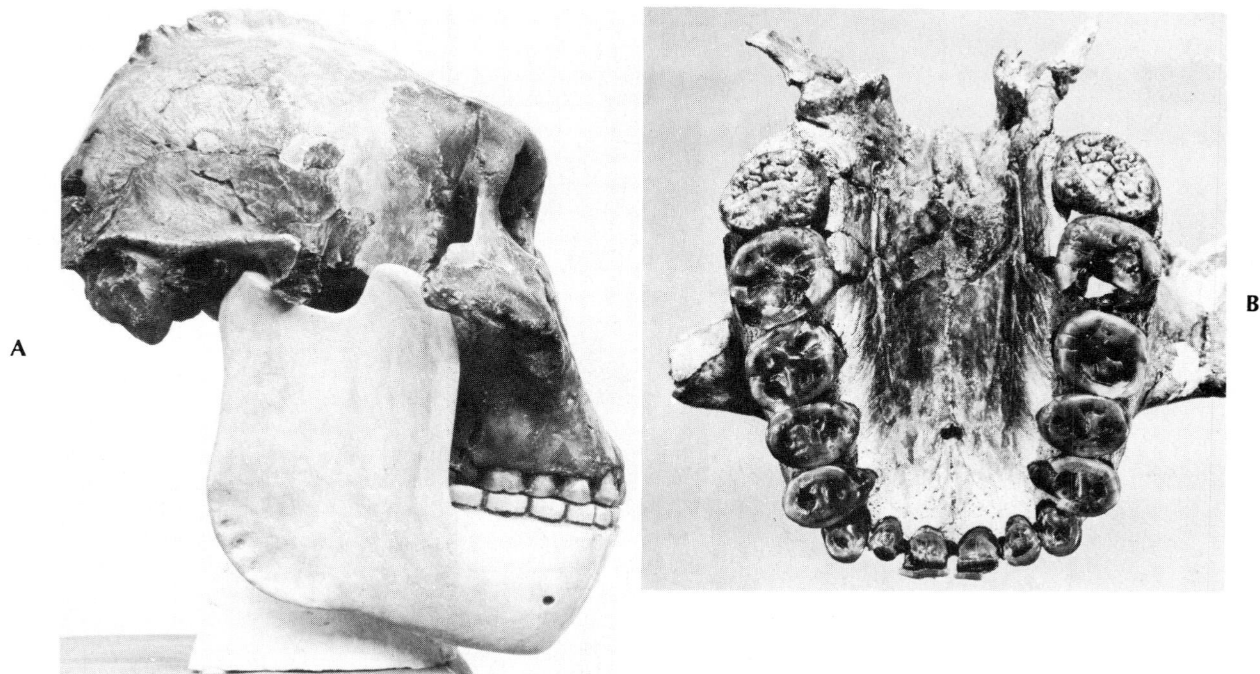

FIG. 39-8 Olduvai Gorge man. **A,** In many features this Tanzanian skull shows a strong resemblance to the autralopithecines from South Africa; the lower jaw has been restored. **B,** The palate and upper dental arch of this form, which was originally described under the name *Zinjanthropus boisei*. (Photographs by R. Klomfass; from Dillon, L. S. 1978. Evolution. St. Louis, The C. V. Mosby Co.)

Homo erectus—Pleistocene hominids

Opinion still wavers on whether or not to award *Homo habilis* full membership in the "true human" club. Let us turn now to *Homo erectus,* who was uncontestably a true human being. The story of the discovery and eventual respectability of *Homo erectus* begins with Dutch anatomist Eugene Dubois' famous fossil find in eastern Java in 1891. Dubois named his find *Pithecanthropus erectus* ("erect ape man"). Dubois' ape man, consisting of only a skull cap and thigh bone, was so much older than anything else so humanlike discovered up to that time that no one knew how to interpret it. Dubois himself vacillated over its relationship to apes and hominids, and finally, piqued by public suspicion of its authenticity, he locked up his specimens and refused to let anyone see them.

In 1927 Davidson Black, a Canadian anatomist teaching in a Peking university, announced the discovery near Peking of a skull that closely resembled the Java man (Fig. 39-9). Continued excavation at the site after Black's death in 1933 eventually revealed 14 skulls, some in excellent state of preservation, together with several lower jaws and other skeletal parts. Several crudely worked tools were also found and, most important, ash heaps indicating the use of fire. Black's Peking man and Dubois' Java man are now recognized as separate races of the same species *Homo erectus*.

Homo erectus was a large hominid standing 150 to 170 cm (5 to 5.5 feet) tall. It had a low but distinct forehead, strong browridges, and no chin (Fig. 39-10). Both jaws were fairly massive and bore large teeth. The cranial capacity averaged about 1,000 cc. The pelvic and leg bones indicate that *H. erectus* had long, straight legs and was an excellent walker.

How old is *Homo erectus?* A 1976 discovery by Richard Leakey of a complete *Homo erectus* skull has been reliably dated at 1.5 to 1.8 million years B.P. This is much older than either the Java or Peking hominids are thought to be (about 0.5 to 0.7 million years B.P.,

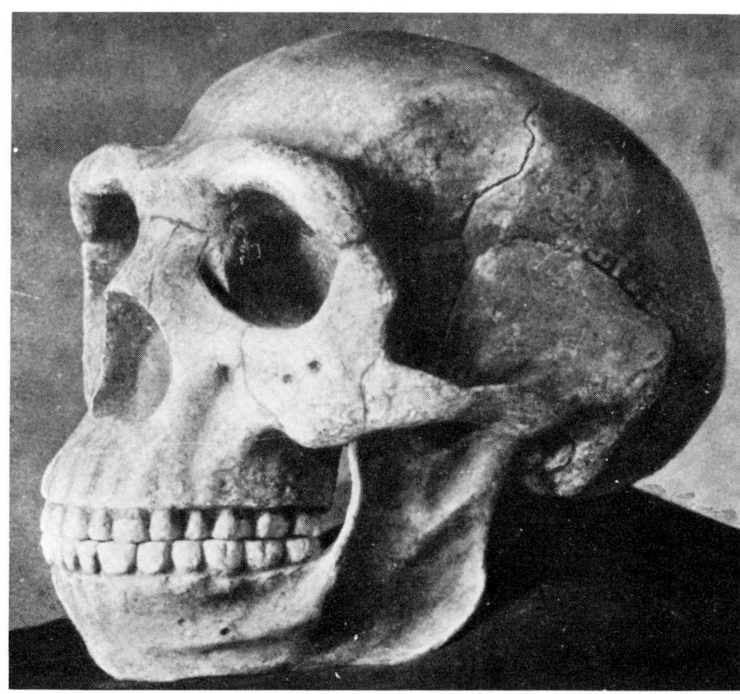

FIG. 39-9 Peking man. Pictured is a cast of the best preserved of 14 skulls discovered at a site near Peking, China. (From Dillon, L. S. 1978. Evolution. St. Louis, The C. V. Mosby Co.)

although neither has been accurately dated). *Homo erectus* lived until about 300,000 years ago, thus spanning a period of about 1.5 million years.

Homo erectus occupied much of the Old World as indicated by fossil finds in eastern and northern Africa and several deposits in Europe as well as in China and Java. *Homo erectus* successfully hunted large animals, butchered them using stone and bone tools, and cooked them at campsites, often in cave chambers. Fire was probably also used for warmth. The site near Peking, China has yielded the remains of some 60 species of animals used by *H. erectus* as food: small rodents, bats, bears, horses, deer, and even rhinoceroses and elephants. However, *H. erectus,* who was really a vegetarian turned meat eater, varied its diet with many vegetable foods. Some authorities believe as much as 75% of the diet was vegetarian.

Homo erectus was a social species that lived in tribes of 20 to 50 people. The tribes interbred and migrated across the tropic Old World and even into temperate regions where they survived severe winters by wrapping themselves in animal skins and using fire for warmth. They often used caves for shelter but there are clear indications that they also built crude wooden shelters. Their brain size and brain configuration suggest that they communicated by speech. All in all, *Homo erectus* pursued a complex culture. At the time of their disappearance about 300,000 years B.P., they had become dependent on cooperative behavior and social control, accompanied by the beginnings of moral and ethical codes.

Homo sapiens—modern hominids

Swanscombe and Steinhelm skulls

Between *Homo erectus* and the establishment of the modern polytypic species of *Homo sapiens* (''wise man''), hominid evolution has threaded a complex course. The earliest fossils showing clear-cut *Homo sapiens* features are the Swanscombe hominid from southeastern England and the Steinhelm hominid from near Stuttgart, West Germany. These skulls are 250,000 to 350,000 years old. They had prominent browridges and a brain capacity of 1200 cc (Steinhelm) to 1300 cc (Swanscombe), which places them well within the range of variation of modern humans (1,000 to 2,000 cc). These skulls are of particular interest because they are intermediate in position between *Homo erectus* and modern *Homo sapiens*. Com-

FIG. 39-10 Artist's concept of the Java man. (Courtesy of the artist, Zdeněk Burian; from Dillon, L. S., 1978. Evolution. St. Louis, The C. V. Mosby Co.)

pared to *H. erectus,* they have larger brains; decreased face, jaw, and tooth sizes; less robust browridges; and less thick skulls. Indeed, the major evolutionary changes from *Homo erectus* on are confined to the head. *Homo sapiens* was becoming ever more intelligent, as manifested not only by increasing brain volume but by increasing complexity of social culture, communication, and environmental manipulation.

Neanderthals

The matching of stone tools with fossil evidence shows that a large number of human subcultures developed during the Pleistocene epoch. These became increasingly varied and tangled as the Neanderthals, modern human's immediate predecessors, emerged some 100,000 years ago. The find from which the Neanderthals' name was derived was made in the Neander Valley near Düsseldorf, West Germany in 1856. New discoveries have been made repeatedly since then, and the Neanderthals are perhaps the best-documented fossil humans. These European fossil finds have been associated with a rich array of tools belonging to a paleolithic (Mousterian) culture. It is characterized by a particular pattern of flaking in a wide variety of blades, borers, spear points, axes, scrapers, and numerous specialized tools. The Neanderthals used such tools for hunting, killing and butchering, food processing, hide preparation, woodworking, and tool making. Their populations dominated the Old World from Europe to Africa and Asia in late Pleistocene times. The species was not a homogeneous one, but varied from place to place in response to local condi-

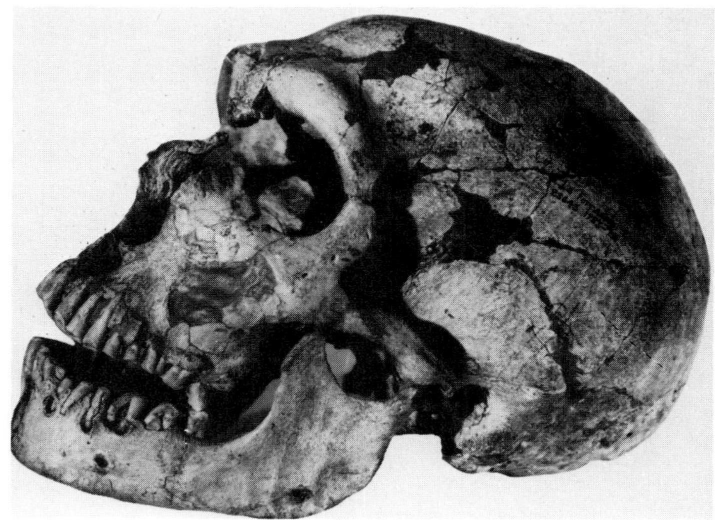

FIG. 39-11 European Neanderthal skull. This "classic" form is the La Ferrassie skull from a site near Burgue, France. (Courtesy Musée de l'Homme; from Dillon, L. S. 1978. Evolution. St. Louis, The C. V. Mosby Co.)

tions and the isolation of populations one from another.

The "classic" Neanderthals that ranged widely in Europe and North Africa had a large skull, comparable to that of modern Europeans, but the brain was probably less convoluted. The face was relatively massive with forward projecting jaws, heavy browridges, and weakly formed chin (Fig. 39-11). However, in the Middle East other so-called "progressive" Neanderthals were becoming more like modern humans. These modern traits—less massive browridges and a more rounded skull in the back—were first revealed in the Mount Carmel "man" (probably a woman) discovered in Israel in 1931. Then in 1957 several superb skulls, even complete skeletons, were dug out of a cave at Shanidar in northern Iraq (Fig. 39-12). These and other discoveries indicate that the Neanderthals were extremely varied. This resulted from portions of the total gene pool becoming isolated for thousands of years at a time in different areas of the Old World. As a whole, however, the gene pool was slowly drifting toward modern humans.

Cro-Magnons

During the last interglacial period about 30,000 to 40,000 years ago, the Neanderthal race was replaced in Europe rather suddenly by the Cro-Magnon race, which emerged from an unknown source (Fig. 39-13). They may be descendants of early, generalized Neanderthals of Asia and may have been responsible for the eventual extermination of the Neanderthals. Like the Neanderthals, the Cro-Magnons were not homogeneous but rather a mixture of people who showed considerable physical variations in different localities. They had a far superior culture (Perigordian) and left artistic paintings and carvings in their caves. They are considered ancestors of modern humans and represent the modern type of human. They were about 183 cm (6 feet) tall, had a high forehead but no supraorbital ridges, a rather prominent chin, and a brain capacity as large as (or larger than) present-day humans. Their physical characteristics are matched today by the Basques in northern Spain and certain Swedes in southern Sweden. Attempts to discern the characteristics of present-day races in early populations of *Homo sapiens* have not been successful. It is not known whether they were white, black, or brown.

The evolutionary course of modern humans in the last 30,000 to 40,000 years has exhibited the same pattern of divergence and extinction demonstrated by our forebears and that of other organisms. The essential characteristics of human phylogeny, such as a superior brain and wide adaptability, can all be attributed to the strictly quantitative effects of mutations that could have happened at any evolutionary level. In other words, we are the outcome of the basic forces of evolution that have directed the evolution of every organism from the time life first originated. Mutation, isolation, population factors, genetic drift, and natural selection

A

B

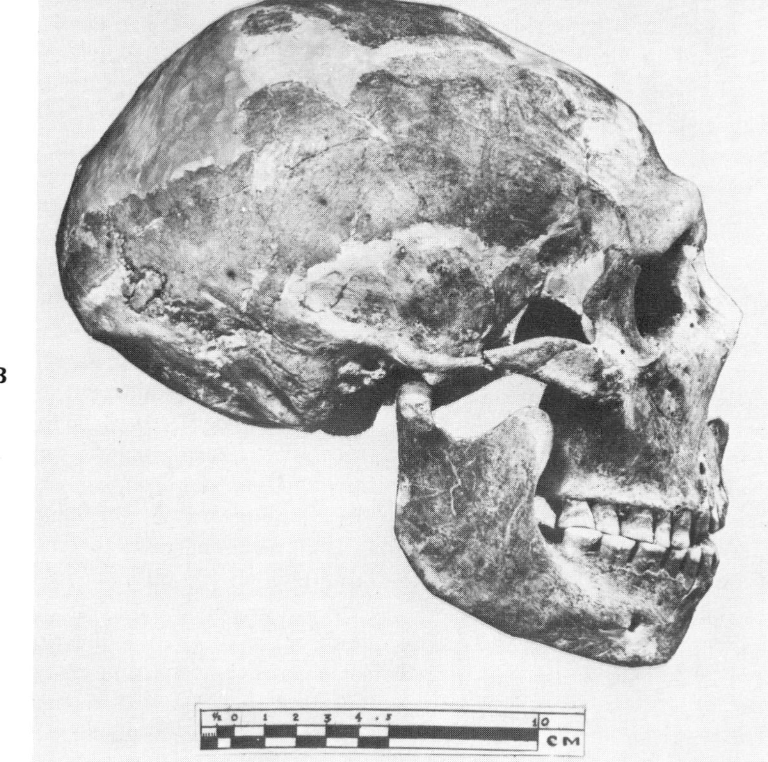

FIG. 39-12 The Shanidar I skull. **A,** The skull still partly enclosed in the rocky matrix in which it was found, from the Shanidar cave in northern Iraq. **B,** The skull after restoration. This well-preserved specimen, about 46,000 years old, is from a Neanderthal man, who died at about 40 years of age. (Courtesy R. S. Solecki, Columbia University, New York.)

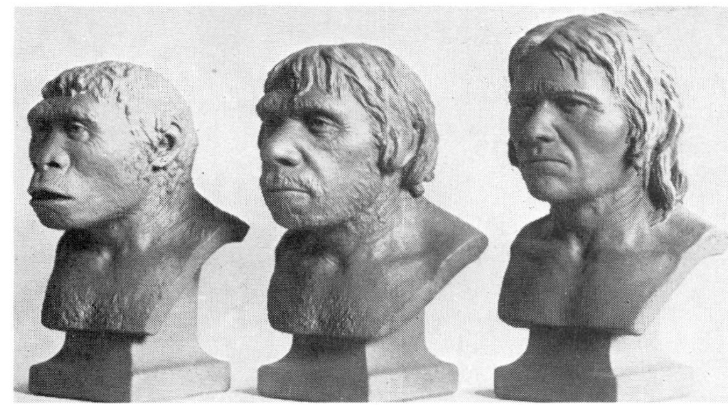

FIG. 39-13 Restoration of prehistoric hominids. A Cro-Magnon *(right)* is compared to a Neanderthal *(center)* and a Java man *(left)*. (Courtesy J. M. McGregor.)

—all general processes of evolutionary progress—have operated for us the same as for other animals.

OUR UNIQUE POSITION

We are animals and the products of evolution. However, we have what no other animal has—a psychosocial evolution, or a directional cultural pattern that involves a constant feedback between past and future experience. Although human evolution has become increasingly cultural as opposed to genetic, we are still subjected to the same biologic forces and principles that affect the lives and evolution of other animals. All animals are adapted to their environment, but through culture, we have found ways to surround ourselves with an environment of our own making.

We are a definite species. Our population is commonly divided into races or populations that are genetically distinguished from others. A so-called race has certain genes or gene combinations that may be more or less unique, although races grade into each other and do not have definite boundaries. Pure races are nonexistent and there is no fixed number of races. Races are adaptations to local conditions. It is believed that as primitive peoples spread over geographic areas, they became adapted to certain regions, and natural selection stamped on them certain distinguishing features. Races at present are losing their biologic significance because human adaptation to environment is becoming largely cultural. Rapid mobility and communication have shrunk the size of our planet so that isolation of races seldom exists. Genes can shift through human populations now with amazing speed and racial intermixtures are far more common than formerly. Since all races have more or less unique potentialities, hybrid vigor could operate within racial interbreeding just as it does for other animals.

We are the only animal that knows how to make and use tools effectively. Perhaps this capability more than any other has been responsible for our dominant position in the animal kingdom. We are able to expend enormous energies in manipulating the environment to suit ourselves, often resulting in destructive consequences to other living forms and not infrequently to ourselves.

Another unique characteristic is our capacity for conceptual thought. We have a symbolic language of wide and specific expression. With words we can carve concepts out of experience. This has resulted in cumulative experience that can be transmitted from one generation to another. In other animals, transmission never spans more than one generation. We owe much to our arboreal ancestry. This promoted binocular vision, a fine visuotactile discrimination, and manipulative skills in the use of our hands. If a horse (with one toe instead of five fingers) had human intellect and culture, could it accomplish what humans have? Compared to any other animal, we are truly extraordinary products of evolution.

Annotated references

Selected general readings

Brace, C. L. 1967. The stages of human evolution: human and cultural origins. Englewood Cliffs, N.J., Prentice-Hall, Inc.

Brace, C. L., and J. Metress (eds.). 1973. Man in evolutionary perspective. New York, John Wiley & Sons, Inc. *Readings in physical anthropology with excellent explanatory introductions to each selection.*

Buettner-Janusch, J. 1973. Physical anthropology: a perspective. New York, John Wiley & Sons, Inc. *Lucid presentation of human evolution.*

Clark, W. E. le Gros. 1967. Man-apes or ape-men? The story of discoveries in Africa. New York, Holt, Rinehart and Winston, Inc. *Describes australopith discoveries.*

Dillon, L. S. 1978. Evolution: concepts and consequences, ed. 2. St. Louis, The C. V. Mosby Co. *College text with comprehensive treatment of human evolution in final chapters.*

Edey, M. A. 1972. The emergence of man: the missing link. New York, Time-Life Books. *Survey of the australopiths.*

Eimerl, S., I. DeVore, and editors of Time-Life Books. 1974. The primates. Life Nature Library. New York, Time-Life Books. *Superbly illustrated natural history of our nearest relatives.*

Howell, F. C., and editors of Time-Life Books. 1973. Early man. Life Nature Library. New York, Time-Life Books. *Beautifully written and illustrated survey of human evolution. Highly recommended.*

King, M. C., and A. C. Wilson. 1975. Evolution at two levels in humans and chimpanzees. Science **188:**107-116. *Describes recent work on measuring the genetic distance between humans and chimpanzees.*

Klein, R. G. 1977. The ecology of early man in southern Africa. Science **197:**115-126. *Traces human ecology through 3 million years.*

McHenry, H. M. 1975. Fossils and the mosaic nature of human evolution. Science **190:**425-431. *Provides support for an early proposal that hominids became bipedal at least 3 million years* B.P., *well before the brain began to enlarge.*

Miller, D. A. 1977. Evolution of primate chromosomes. Science **198:**1116-1124. *Gene mapping among the pongids suggests that the gorilla, not the chimpanzee, may be human's closest relative.*

Pfeiffer, J. E. 1972. The emergence of man. New York, Harper & Row, Publishers. *Well-written anthropology.*

Pilbeam, D. 1972. The ascent of man: an introduction to human evolution. New York, The Macmillan Co. *Survey of fossils from Ramapithecus to modern humans.*

Poirier, F. E. 1977. Fossil evidence: the human evolutionary journey, ed 2. St. Louis, The C. V. Mosby Co.

Simons, E. L. 1972. Primate evolution: an introduction to man's place in nature. New York, The Macmillan Co.

Solecki, R. S. 1972. Shanidar: the first flower people. New York, Alfred A. Knopf, Inc. *Describes the exciting discovery and excavation of the Iraq Neanderthals.*

Tuller, R. M. 1977. The human species: its nature, evolution, and ecology. New York, McGraw-Hill Book Co. *Clearly presented, well-illustrated treatment.*

Von Koenigswald, G. H. R. 1971. The evolution of man. Ann Arbor, The University of Michigan Press. *Concise, rather technical account.*

Selected *Scientific American* articles

Clark, J. D. 1958. Early man in Africa. **199:**76-83 (July).

Dobzhansky, T. 1960. The present evolution of man. **203:**206-217 (Sept.).

Eckhardt, R. B. 1972. Population genetics and human origin. **226:**94-103 (Jan.).

Eiseley, L. C. 1953. Fossil man. **189:**65-72 (Dec.).

Holloway, R. L. 1974. The casts of fossil hominid brains. **231:**106-115 (July). *Molds made from fossil hominid skulls indicate the human brain began enlargement some 3 million years ago.*

Howells, W. W. 1960. The distribution of man. **203:**112-127 (Sept.)

Howells, W. W. 1966. Homo erectus. **215:**46-53 (Nov.).

Issac, G. 1978. The food-sharing behavior of protohuman hominids. **238:**90-108 (Apr.).

Leakey, L. S. B. 1954. Olduvai Gorge. **190:**66-71 (Jan.).

MacNeish, R. S. 1971. Early man in the Andes. **224:**36-46 (Apr.) *Humans have lived in Peru for at least 22,000 years.*

Sahlins, M. D. 1960. The origin of society. **203:**76-86 (Sept.).

Simons, E. L. 1964. The early relatives of man. **211:**50-62 (July).

Simons, E. L. 1967. The earliest apes. **217:**28-35 (Dec.).

Simons, E. L. 1977. Ramapithecus. **236:**28-35 (May).

Smith, P. E. L. 1976. Stone-age man on the Nile. **235:**30-38 (Aug.). *Hunters and gatherers lived along the Nile for thousands of years before the Pharaohs.*

Walker, A., and R. E. F. Leakey. 1978. The hominids of East Turkana. **239:**54-66 (Aug.).

Washburn, S. L. 1960. Tools and human evolution. **203:**63-75 (Sept.).

Washburn, S. L. 1978. The evolution of man. **239:**194-208 (Sept.). *A personal evaluation of human evolution based on recent fossil finds.*

PART FIVE

THE ANIMAL AND ITS ENVIRONMENT

Canada geese flushed from a cornfield. (Photograph by L. L. Rue, III.)

Life was created and shaped by environmental forces. No organism is for an instant free from the requirements of the surroundings in which it lives. The environment therefore is the totality of all extrinsic factors, both physical and biotic, that in any way affect the life and behavior of an organism.

In the following chapters we deal with the relationships of animals to the world around them and with the special human impact on the creatures that share the biosphere with us.

CHAPTER 40

THE BIOSPHERE AND ANIMAL DISTRIBUTION

A pronghorn antelope buck *(Antilocapra americana)* of the grasslands biome, one of the principal terrestrial biomes within the biosphere.

Photograph by Irene Vandermolen, Leonard Rue Enterprises.

THE BIOSPHERE

All life is confined to a thin veneer of the earth called the **biosphere.** From the first remarkable photographs of earth taken from the Apollo spacecraft, revealing a beautiful blue and white globe lying against the limitless backdrop of space, viewers were struck and perhaps humbled by our isolation and insignificance in the enormity of the universe. The phrase ''spaceship earth'' became a part of the vocabulary, and the realization evolved that all the resources we will ever have for sustaining life are restricted to a thin layer of land and sea and a narrow veil of atmosphere above it. We could better appreciate just how thin the biosphere is if we could shrink the earth and all of its dimensions to a 1-m sphere. We would no longer perceive vertical dimensions on the earth's surface. The highest mountains would fail to penetrate a thin coat of paint applied to our shrunken earth; a fingernail's scratch on the surface would exceed the depth of the ocean's deepest trenches.

Fitness of the earth's environment

In Chapter 1 we commented that astronomers have estimated conservatively that in the vastness of the universe there are millions of stars similar to our sun that have planetary systems. The astronomer Harlow Shapley believes that in our galaxy alone there are *at least* 100,000 planets having conditions suitable for life. If he is correct, we can no longer maintain that only the earth harbors life. Nevertheless, there are so many requirements for life that only a small number of planets can fulfill the special conditions that would permit evolution of life at all similar to that on earth.

First, a planet suitable for life must receive a steady supply of light and heat from its sun for many billions of years. This means that its orbit must be nearly circular.

Second, water must be present on the planet to permit the evolution of complex biochemical systems based on carbon. Earth is indeed a watery planet, since water covers 71% of its surface. Viewed from space, water, not land, is earth's most conspicuous surface feature. Fanciful systems of life based on ammonia and silicates rather than water and carbon have been suggested, but such speculations are totally hypothetical and in any case would require environmental conditions vastly different from those on earth.

Third, the temperature must be suitable, meaning practically within the range of $-50°$ to $+100°$ C, as it is over most of the earth's surface. Life at temperatures above 100° C is impossible because biopolymers based on carbon and water would be rapidly hydrolyzed. Temperatures much below the freezing point of water prevent growth of organisms by slowing chemical processes, although some inactive organisms may survive storage in liquid nitrogen ($-195°$ C) or even in liquid helium ($-269°$ C).

Fourth, all life requires a suitable array of major and minor elements. Oxygen, carbon, hydrogen, and nitrogen form 95% of protoplasm and thus dominate the composition of life on earth. These four are supported by seven other major elements (phosphorus, calcium, potassium, sulfur, sodium, chlorine, and magnesium) and a large number of minor or ''trace'' elements. Perhaps 46 elements in all are found in protoplasm; many are essential for life, whereas others are present in protoplasm only because they exist in the environment with which the organism interacts. Not all planets possess elements in the same proportions, even in our solar system.

There are many other properties of earth that make it an especially fit environment for life. It is large enough to have a surface density that permits molecules to collect and align properly. Protoplasm is in a colloidal state, an intermediate between conditions too solid to allow change and conditions too fluid or gaseous to permit molecular organization. The earth's gravity is strong enough to hold an extensive gaseous atmosphere but not so strong that more than a trace of free hydrogen remains. Another consideration, especially important for life on earth today, is the oxygen-ozone atmospheric screen that absorbs lethal ultraviolet radiation from the sun. So effective is this absorption that rays with wavelengths shorter than 283 nm (nanometer = 10^{-12} m) fail to reach the earth's surface.

In his classic book *The Fitness of the Environment,* published in 1913, the biochemist L. J. Henderson maintained that earth possesses ''the best of all possible environments for life.'' We may agree; however, we realize that the surface of any body that is the size and age of earth in a similar orbit revolving about a similar sun and having a similar elemental composition should also have an excellent environment for life.

As was described in Chapter 3, life's origins depended on the formation of a few basic molecules composed of atoms that exist on all stars and planets in the universe. The same laws of physics and chemistry that operated during life's origins on earth must also apply throughout the universe. Therefore the sequence of events that led to life on earth may also have occurred on other planets in our galaxy and on planets in other galaxies as well.

Even so, there is no reason to conclude that all other earthlike planets in the universe have undergone evolutions that produced DNA, chlorophyll, plants, flowers, fungi, molluscs, fishes, frogs, and people. The manner in which living organisms on earth capture and exchange energy, move, respond, reproduce, and grow is not an inevitable pattern of life. There is no objective evidence that the evolutionary process follows a predetermined pattern or is channeled and pointed to some inevitable goal.

When we realize that advanced mammals such as the human species possess an enormous number of inherited characteristics that have appeared in response to changing adaptive pressures during a long period of evolution, it is obvious that humans are only one of an almost infinite variety of structural and intellectual combinations that might have appeared. We who read this are the improbable product of a long ancestry of organisms shaped and molded by heredity and environment. If we could start all over again at the dawn of life on a primitive earth, it is most unlikely that the result of 3 billion years of evolution would be the same. Life may well exist on other earthlike planets, but the kind of life on each planet is surely unique.

The organism and its environment share a reciprocal relationship. The environment is fit for the organism, and the organism is fitted to the environment and adapts to its changes. As an open system, an animal is forever receiving and giving off materials and energy. The building materials for life are obtained from the physical environment, either directly by producers such as green plants or indirectly by consumers that return inorganic substances to the environment by excretion or by the decay and disintegration of their bodies.

The living form is a transient link that is built up out of environmental materials, which are then returned to the environment to be used again in the re-creation of new life. Life, death, decay, and re-creation have been the cycle of existence since life began.

In this continuous interchange between organism and environment, both are altered in the process, and a favorable relationship is preserved. The environment of earth, with its living and nonliving components, is not a static entity but has undergone an evolution in every way as dramatic as the evolution of the animal kingdom. It is still changing today, more rapidly than ever before, under the heavy-handed influence of humans.

The primitive earth of 3.5 billion years ago, barren, stormy, and volcanic with a reducing atmosphere of ammonia, methane, and water, was wonderfully fit for the prebiotic syntheses that led to life's beginnings. Yet, it was totally unsuited, indeed lethal, for the kinds of living organisms that inhabit the earth today, just as early forms of life could not survive in our present environment. The appearance of free oxygen in the atmosphere, produced largely if not almost entirely by life, is an example of the reciprocity between organism and environment. Although oxygen was at first poisonous to early forms of life, its gradual accumulation over the ages from photosynthesis forced protective biochemical alterations to appear that led eventually to complete dependence on oxygen for life.

Earth's biosphere and the organisms in it have evolved together. Again, as living organisms adapt and evolve, they act on and produce changes in their environment; in so doing they must themselves change.

Subdivisions of the biosphere

The biosphere may be conveniently divided into three major subdivisions—**hydrosphere, lithosphere,** and **atmosphere.** The hydrosphere refers to the aquatic portions of the biosphere, the streams, rivers, ponds, oceans, and wherever else water may be found. The lithosphere is made up of the crust of the earth, especially the solid portions such as rocks. Surrounding the other two subdivisions is the atmosphere, which forms a gaseous envelope. Plants and animals require and obtain inorganic components from each of these subdivisions. From the hydrosphere, they get water that makes up about 75% of living material. The lithosphere furnishes the essential minerals and chemicals, whereas the atmosphere supplies oxygen, nitrogen, and carbon dioxide. These inorganic substances are needed in all living organisms.

Hydrosphere

About 71% of the earth's surface is water, which also forms part of the lithosphere and atmosphere. Not only is water the most abundant constituent of protoplasm, but it is also the source of hydrogen that is so fundamental to the metabolic reactions of all living substance. Water also serves as one of the sources of oxygen in the bodies of organisms.

Water is distributed over the earth by a global hydrologic cycle that consists of three principal phases: evaporation, precipitation, and runoff. Evaporation occurs from oceans, from land, and from other water surfaces. Some five-sixths of the evaporation is from the ocean; the quantity of water that evaporates from its surface is equivalent to a depth of about 1 m annually. One important aspect of the global water balance is that more water is evaporated from the ocean than is returned to it by precipitation. Ocean evaporation therefore provides much of the rainfall that supports land ecosystems.

On land, more water is received via precipitation than is lost by evaporation from the surface and by plant transpiration. The excess is intercepted by living organisms, especially plants, runs off the land surface into streams, or filters into the ground-water system (aquifers). Precipitation is not evenly distributed among the continents; South America with its enormous tropical rain forests receives twice as much precipitation per unit of area as do other continents, and evaporation and runoff are correspondingly high.

Runoff from land produces erosion and wears away the surfaces of continents. This leveling-off process, however, is offset by the geologic uplift, thus bringing marine sediments above sea level. Living things also carry water through their bodies in addition to what they need themselves and in doing so speed up the return of water to the atmosphere. The metabolism of organisms thus accelerates the cycle of water and may profoundly influence weather conditions, not only locally, but also over extensive areas, such as the rank vegetation of jungles.

Displacement of water in the oceans is also cyclic. The sun-warmed waters of the tropic seas expand and rise, whereas the cold polar waters contract, sink, and flow slowly along the bottom toward the equator. The spinning of the earth eastward creates a force—the Coriolis force—that deflects the winds and currents, causing the sea current to turn clockwise in the northern hemisphere and counterclockwise in the southern hemisphere. Ocean currents, such as the Gulf Stream, have enormous influences on climatic conditions in all parts of the world.

As pointed out in Chapter 2, the very nature of water itself is of great significance to life. It has great chemical stability, it is the most versatile of solvents, and it stores large amounts of thermal energy. Because of its unique density behavior, water reaches maximum density at 4° C; therefore ice is lighter than liquid water and floats, enabling life to continue beneath the ice cover. Water also has high surface tension and low viscosity, both properties important to life.

In the long overall picture, the cyclic changes of ice ages, of which at least four great glacial periods have occurred in the last million years, involve the advance or retreat of polar ice. Warm interglacial periods have actually freed the poles from ice. (Amphibian fossils of the Quaternary period have been discovered in the Antarctic.) Melting of polar ice during the present warm trend in temperature has gradually raised the level of ocean water and made possible a steady advance of biota toward the poles.

Lithosphere

The lithosphere is the ultimate source of all the mineral elements required by living organisms. Below the loose soil and subsoil is the solid bedrock of sedimentary rock, such as limestone and sandstone, resting on a thicker base of igneous and metamorphic rocks. Below this layer is a thicker stratum composed mostly of basaltic rock. These two layers form the so-called crust of the earth. As far as life is concerned, only the more superficial part of the lithosphere is involved, although the tectonic movements of folding and breaking of the bedrock plus the rise of molten lava in volcanos may bring additional minerals to the surface.

Plate tectonic theory states that the lithosphere is divided into seven rocky plates about 100 km (60 miles) thick that float on a partially melted layer of rock, the upper mantle. The plates drift like rafts across the earth's surface at an average drift rate of 6.6 cm per year. The crust of the earth shows striking changes over long periods of time. Uplifting and buckling are always occurring, resulting in mountain building and shifting of land masses. Mountains produce profound changes in climate conditions, such as the unequal distribution of thermal energy and moisture. Moisture-laden clouds not able to pass mountain barriers may dump their contents on one side alone so that this inequality results in favorable rainfall with fertility on one side and desert on the other. These factors influence the distribution of both vegetation and animals, which must adapt to these different conditions to survive.

Billions of tons of dissolved inorganic and organic matter, as well as much undissolved matter, are carried into the oceans each year—the result of weathering processes on land. Many important nutrients are irretrievably lost to animal and plant life. They are not returned from the ocean but must be derived from further weathering of parent materials. Of the major nutrients, phosphorus is in shortest supply, and insufficient phosphorus most commonly limits biologic productivity both on land and in fresh water. Some valuable minerals are replaced by the decay of animals and plants and in the long run by the upheaval of the sea floor to form new land. This last is a slow process and life could be greatly affected by the gradual decline of certain key minerals. Many ecologists have been concerned about this decline, especially because of human influence in the misuse of certain resources.

Atmosphere

The earth's atmosphere is commonly subdivided into three great strata—**troposphere, stratosphere,** and **ionosphere.** The troposphere, the layer nearest the earth, extends upward to about 8 km above the poles and 15 km above the equator. All life is confined to this layer, as are haze, dust, clouds, and air currents. Air masses warmed in the tropics rise, those cooled at the poles sink, and all are shifted laterally by the rotation of the earth. These effects produce enormous weather systems that profoundly influence life on the earth's surface.

Above the troposphere is the stratosphere, which extends to about 50 km above the earth's surface. In the lower stratosphere is the oxygen-ozone atmospheric screen layer mentioned earlier; it is concentrated mostly between 20 and 25 km. Above the stratosphere is the ionosphere, the outer reaches of which extend some 3,500 km from the earth. Air particles in this extremely rarified atmosphere are largely ionized by solar radiation. Temperatures are cold (to −90° C) in the lower ionosphere but rise with increasing altitude and may reach 1,650° C in the upper region during periods of high solar activity.

The gases present in the lower atmosphere (troposphere) are (by volume) nitrogen, 78%; oxygen, 21%; argon, 0.93%; carbon dioxide, 0.03%; variable amounts of water vapor; and trace amounts of neon, helium, methane, krypton, ozone, and xenon. Also present are variable amounts of pollutants, the most important of which are carbon monoxide; sulfur oxides; hydrocarbons; nitrogen oxides; and particulate dust, ash, and spray. Because the atmospheric gases have mass, they are attracted by the earth's gravitational field and thus exert a pressure on the surface of the earth. At sea level, the normal atmospheric pressure is sufficient to support a column of mercury 760 mm in height. Each of the gases in air contributes to the total pressure in direct proportion to its percent composition (Dalton's law of partial pressure). Thus the pressure of oxygen in dry air at sea level is 21% of 760 mm Hg, or 159 mm Hg; that of nitrogen is 78% of 760 mm Hg, or 593 mm Hg; and that of carbon dioxide is 0.03% of 760, or 0.23 mm Hg.

The presence of water vapor in air also contributes to the total gas pressure. To calculate the partial pressure of gases in air that has not been dried before measurement, it is necessary to first subtract the water vapor pressure from the total atmospheric pressure. For example, the water vapor pressure of atmospheric air at 23° that is saturated with water vapor is 21 mm Hg. For oxygen, $760 - 21 = 739$ mm Hg $\times 0.21 = 155$ mm Hg. The corrected partial pressures of nitrogen and carbon dioxide are calculated similarly.

Atmospheric oxygen has originated almost entirely from plant photosynthesis. As discussed in Chapter 3, the primitive earth contained a reducing atmosphere devoid of oxygen. When oxygen-evolving photosynthesis appeared about 3 billion years ago, oxygen gradually began to accumulate in the atmosphere. It is believed that by the mid-Paleozoic era, some 400 million years B.P., the oxygen concentration had reached its present level of about 21%. Since then oxygen consumption by animals and plants has approximately equaled oxygen production. The present surplus of free oxygen in the atmosphere resulted from the fossilization of plants before they could decay or be consumed by animals. As these vast stores of fossil fuels are burned by our industrialized civilization, the oxygen surplus that accumulated over the ages could conceivably be depleted. Fortunately, this is unlikely for two reasons: (1) most of the total fossilized carbon is deposited in noncombustible shales and rocks, and (2) the oxygen reserves in the atmosphere and in the oceans are so enormous that the supply could last thousands of years, even if all photosynthetic replenishment suddenly were to cease. However, there is concern that the rapid input of carbon dioxide into the atmosphere from the industrial burning of fossil fuels may significantly affect the earth's heat budget.

With the exception of anaerobic animals (which can carry on oxidative processes without free oxygen), oxygen is a necessary prerequisite for respiration in animals. Its major function in cellular metabolism is its role as the final hydrogen acceptor, resulting in the formation of water. Oxygen also dissolves in water, a fact that makes possible aquatic life. Each of the atmospheric gases dissolves in water according to its partial pressure, its solubility, and the temperature of the water. The solubility coefficient, defined as the volume of gas dissolved in one volume of water exposed to gas at one atmospheric pressure is, at 0° C, 0.0486 for oxygen, 0.0235 for nitrogen, and 1.704 for carbon dioxide. Thus oxygen is a little more than twice as soluble as nitrogen, and carbon dioxide is about 30 times more soluble than oxygen. At sea level and at 0° C, water exposed to atmospheric air will dissolve 21% of 0.0486, or 0.0102 ml of oxygen per milliliter of water (10.2 ml/liter). As the water temperature increases, gas solubility decreases. At 10° C, water will dissolve 8.02 ml of oxygen per liter and at 20° C, 6.57 ml per liter, a reduction of about 40%. Therefore, the amount of oxygen in water is only one-twentieth to one-fortieth of that present in air. Furthermore, rising water temperature decreases the oxygen content of the water while at the same time the metabolic demand of aquatic animals for oxygen is increasing. These physical characteristics present a special problem for the extraction of oxygen from water and have been a severely limiting factor in both the evolution and distribution of aquatic animals. Lakes at high altitudes, for example those in the high Andes, lack fish altogether because the partial pressure of oxygen is too low to support life.

Carbon dioxide plays a unique role in the ecology of animals, although it makes up a small percentage of atmospheric air. It is a product of animal respiration, alcoholic fermentation, decay of organic matter, the action of acids on carbonates in water, crustal outgassing from volcanoes, and the combustion of carbonaceous fuels (such as coal, oil, and natural gas). Carbon dioxide is removed from the atmosphere by plant photosynthesis and by precipitation as carbonates in the ocean. The latter is a major reservoir of biospheric carbon dioxide. The circulation of carbon dioxide in the carbon cycle will be described in the following chapter (p. 953).

Because of carbon dioxide's high solubility in water, much of the free carbon dioxide produced by animal respiration and by agricultural and industrial activities is rapidly passed into lakes and the sea. Here it enters into chemical combination with water to form carbonic acid (H_2CO_3). This dissociates to form bicarbonate (HCO_3), or "half-bound" carbon dioxide, which in turn combines with available limestone to form carbonate (CO_3), or "fixed" carbon dioxide. These compounds serve both as important buffers to keep the hydrogen ion concentration of natural waters near neutrality and as nutrients for aquatic life.

Nitrogen is a chemically inert gas, but nitrogen is one of the chief constituents of living matter. Atmospheric nitrogen is the ultimate source. It may be fixed as nitrites or nitrates by electric discharges and later washed to earth by rain or snow. In the nitrogen cycle, described on p. 954, nitrogen-fixing bacteria living symbiotically with legumes form nitrites and nitrates by the reactions **ammonification** and **nitrification** and add them to the soil where plants can get their nitrogen supplies. Animals get their nitrogen from eating plants or other animals.

The atmosphere has a low degree of buoyancy and cannot be used as a permanent habitat by organisms. It is used as a passageway by those forms that are specialized to use it as a medium. Most of the life found within it is restricted to its lower boundary.

ANIMAL DISTRIBUTION (ZOOGEOGRAPHY)

The ecology of dispersal

The study of zoogeography tries to explain why animals are found where they are, their patterns of dispersal, and the factors responsible for their dispersal. With the exception of the human species, which is able to live nearly anywhere on earth, and those creatures such as the house mouse and the cockroach which share human habitations, the different kinds of animals occupy limited habitats on earth. It is not always easy to explain why animals are distributed as they are, since similar habitats on separate continents may be occupied by quite different kinds of animals. A particular species may be absent from a region that supports similar animals for any of three reasons: (1) there may be barriers that prevent it from getting there; (2) having gotten there, it may be unable to adapt to the new habitat or compete successfully with resident species; or (3) having once arrived and adapted, it has subsequently evolved into a distinct species.

Thus there are always good reasons why animals are found where they are (or are not found where one thinks they ought to be), and it is the task of the zoogeographer to discover what these reasons are. Usually this means going back into the past. The fossil record plainly shows that animals once flourished in regions from which they are presently absent. Extinction has played a major role, but many groups left descendants that migrated to other regions and survived. For example, camels originated in North America, where their fossils are found, but spread to Eurasia by way of Alaska (true camels) and to South America (llamas) during the Pleistocene epoch. Then they became extinct in North America at the close of the Ice Age. Thus the past history of an animal species or its ancestor must be known before one can understand why it is where it is. The surface of the earth is undergoing constant change. Many areas that are now land were once covered with seas; fertile plains may be claimed by advancing desert; impassable mountain barriers may arise where none existed before; or inhospitable ice fields may retreat before a warmer climate to be replaced by forests. Geologic change has been responsible for much of the alteration in animal (and plant) distribution and has been a powerful influence in shaping organic evolution.

By **dispersion,** animals spread into new localities from their place of origin. Changes in environment may force animals to shift their domiciles, although most great groups have spread to gain favorable conditions, not to avoid unfavorable ones. The reproductive rate of animals is so great that there is continuous pressure on individuals to move into new areas, thus reducing the competition for food, shelter, and space to live and breed. This **population pressure** serves to expand the range of the species and maintain healthy populations. Survival of the species may depend on successful dispersion to new areas when the old habitat ceases to be habitable because of geographic and ecologic changes through time. Dispersion followed by isolation is also a biologic prerequisite for the formation of new species because, as we have pointed out earlier (p. 896), geographic isolation of a founding colony from the parent population permits new adaptive variations to arise.

Dispersal movements involve emigration from one region and immigration into another. Dispersal is a *one-way,* outward movement that must be distinguished from migration, a *periodic* movement back and forth between two localities, such as the seasonal migration of birds. Dispersing animals may move **actively** under their own power whether they are creepers, climbers, runners, hoppers, or flyers; or they may be **passively** dispersed by wind and air currents; by floating or rafting on rivers, lakes, or the sea; or by hitching rides on other animals. The chief deterrents to the spread of animals are barriers of various kinds, such as mountain chains, cliffs, canyons, sand dunes, lava flows, rivers, and, most effective of all, oceans. However, that which comprises a barrier to one animal may serve as an avenue of dispersal for another. A river may obstruct the dispersal of a gray squirrel but facilitate the movement of a beaver. In addition to the physical barriers listed are climatic barriers, such as deserts, and biotic barriers, such as competition between species, predators, changes in vegetation, and absence of suitable food. Deserts are especially important barriers to most animals since deserts have an inhospitable climate, which is too hot and too dry for most species, and are lacking in vegetation for cover, in food, and in suitable places to live. Deserts are of course the chosen habitats for a variety of animals bearing appropriate desert adaptations; to these animals, nondesert habitats are barriers to dispersal.

Pathways of dispersal

The oceans are the strongest, most stable, and most effective of all barriers to land animals. Of principal interest to the zoogeographer is how living organisms have been able to disperse from one continental land mass to another and to oceanic islands, separated as they are from one another by the inhospitable sea.

Continental drift theory

The recent emergence of plate tectonic theory in the earth sciences—the concept that the earth's surface is composed of an interlocking mosaic of shifting rocky plates—has revived interest in an old and previously controversial theory: continental drift. In 1912 the German meteorologist Alfred Wegener proposed that the earth's continents had been drifting like rafts following the breakup of a single great landmass called Pangaea (''all land''). According to recent workers who have considerably revised Wegener's dating, this occurred some 200 million years ago. Two great supercontinents were formed: a northern Laurasia and a southern Gondwana, separated from each other by the Tethys Sea (Fig. 40-1). At the end of the Jurassic period, some 135 million years ago the supercontinents began to fragment and drift apart. Laurasia split into North America, most of Eurasia, and Greenland. Gondwana split into South America, Africa, Madagascar, Arabia, India, Australia, and Antarctica. This

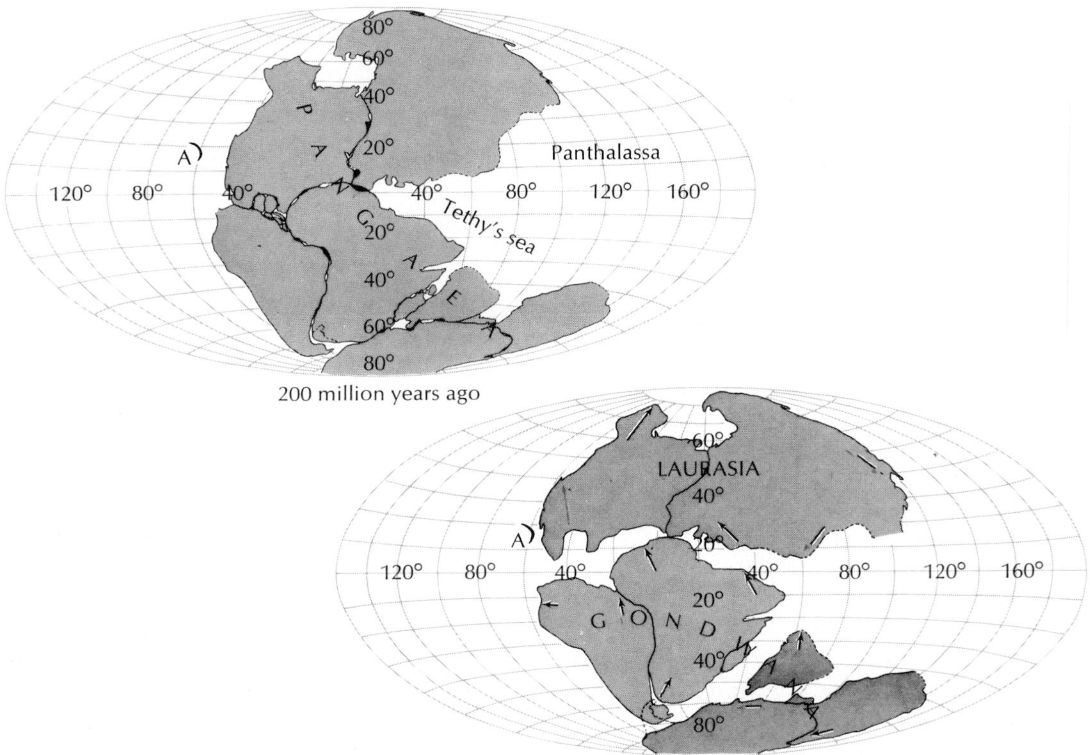

200 million years ago

180 million years ago

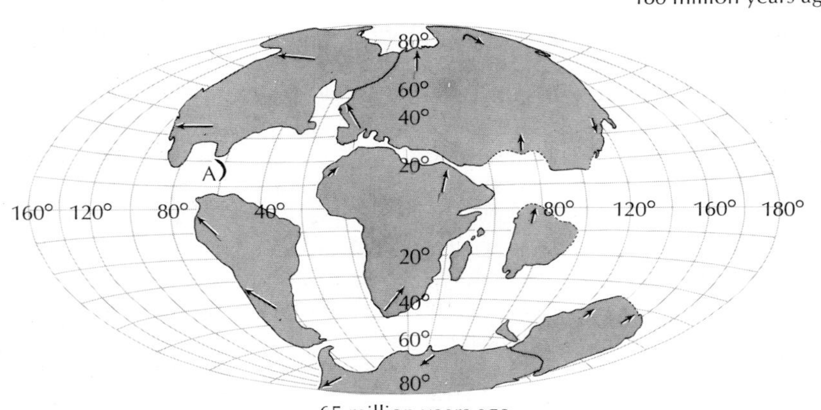

65 million years ago

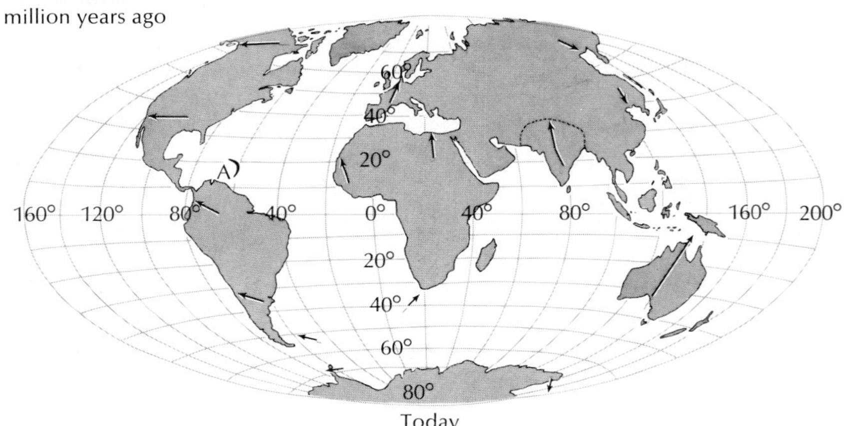

Today

FIG. 40-1 Hypothesized drift of continents over the past 200 million years from an original single land mass to their present positions. The universal land mass Pangaea first separated into two supercontinents (Laurasia and Gondwana). These later broke up into smaller continents. The black arrows indicate vector movements of the continents. The black crescent labeled *A* is a modern geographic reference point representing the Antilles arc in the West Indies. (Modified from Dietz, R. S., and J. C. Holden. 1970. Sci. Am. **223:**30-41 [Oct.].)

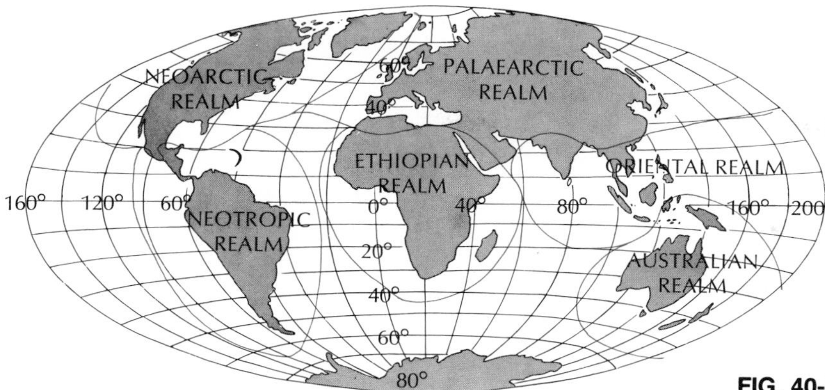

FIG. 40-2 The six major faunal realms of the earth.

theory is supported by the appearance of fit between the continents, by recent airborne paleomagnetic surveys, by seismographic studies, by the presence of midoceanic ridges where the tectonic plates are born, and by a wealth of biologic data.

Continental drift helps to explain the similarity of invertebrate fossils in Africa and South America, as well as certain similarities in present-day fauna at the same latitudes on the two continents. However, the continents have been separated for all of the Cenozoic era and probably for much of the Mesozoic era as well, much too long to explain the dispersal of modern organisms. Continental drift theory is, however, enormously useful in explaining interconnections between flora and fauna of the past. Structural geologists now believe that the fragmentation of the giant land mass Pangaea beginning some 200 million years ago is only the last of several earlier episodes, during which a giant universal continent dispersed and then coalesced again. Thus the continents evidently have never had fixed positions on the earth's surface but have always been on the move. Such geologic changes have been very important in the evolution of the major animal taxa.

Land bridges

Since the continents have been separated for at least 100 million years, some other pathway of dispersal must have been responsible for the present distribution of modern animal groups. Temporary land bridges between continents that may exist for only a short geologic time serve as important dispersal routes for land animals. One well-established and important land bridge that no longer exists connected Asia and North America across the Bering Strait. Today a land bridge connects North and South America at the isthmus of Panama, but from the mid-Eocene epoch (50 million B.P.) to the mid-Pliocene epoch (5 million B.P.), the two continents were completely separated by a water barrier. During this long period, all of the major groups of placental mammals were evolving in distinctive directions on each continent. When the land bridge was reestablished in the Pliocene epoch, a tide of mammals began to flow in both directions. For a period both continents gained in mammal diversity, but direct competition for similar ecologic niches soon led to the extinction of large numbers of mammals on both continents. North American invaders such as deer, foxes, bears, raccoons, tapirs, llamas, mastadons, and antelopes displaced many South American residents occupying similar niches, and today nearly half of the South American mammals are descendants of recent North American invaders. Only a few South American invaders survived in North America: porcupines, armadillos and opossums. Several other South American groups, including the giant ground sloths and the giant armadillos, entered North America but subsequently died out.

Routes of dispersal are of three principal kinds. A **corridor** is a land bridge that allows free movement of animals in both directions. The isthmus of Panama and isthmus of Suez were both of this type before they were crossed by man-made canals. A **filter route** is one that allows only specially adapted animals to pass; it may be a land bridge that has an unfavorable climate or is narrow or has many competitors. A **sweepstakes route** is one that only a very few species may pass, and these usually by chance. Remote oceanic islands are reached by a relatively few species, and usually after lengthy delays. The more distant the island from a continent, the fewer the species are that reach it, and the more disharmonious are the resulting fauna and flora. Some groups such as mammals, amphibians, and freshwater fishes that are well represented on continents are often scarce or absent from certain oceanic islands.

Major faunal realms

On the basis of animal distribution, it is possible to recognize the world as divided into several distinct realms of animal distribution, separated one from another by topographic and climatic barriers. Philip L. Sclater (1858) first proposed a region system based on bird families. Later (1876), Alfred Russel Wallace, codiscoverer with Charles Darwin of the principle of natural selection, modified Sclater's pattern and applied it to vertebrates in general. Wallace recognized six land realms based on animal ranges and, although later scientists slightly modified his biogeographic boundaries, the wallacean regions are the ones we recognize today (Fig. 40-2). Great oceanic barriers serve as longitudinal dividers between the Nearctic and Palaearctic realms in the north and the Neotropic and Australian realms in the south. The great northern land mass is divided by a subtropic, warm temperate dry belt into Palaearctic, Ethiopian, and Oriental realms. Following is a brief description of each realm and the unique and distinctive forms of animal life each possesses.

Nearctic. This region includes North America as far south as southern Mexico. Typical animals are the wolf, bear, caribou, mountain goat, beaver, elk, bison, lynx, bald eagle, and red-tailed hawk.

Palaearctic. This realm consists of Europe, Asia north of the Himalaya Mountains, Afghanistan, Iran, and North Africa. Its animals include the tiger, wild boar, camel, and hedgehog.

Australian. This realm includes Australia, New Zealand, New Guinea, and certain adjacent islands. Some of the most primitive mammals are found here, such as the monotremes (duckbill) and marsupials (kangaroo and Tasmanian wolf), but few placental mammals. Most of the birds are also different from those of other realms, such as the cassowary, emu, and brush turkey. The primitive lizard *Sphenodon* is found in New Zealand.

Neotropical. This realm includes South and Central America, part of Mexico, and the West Indies. Among its many animals are the llama, sloth, New World monkey, armadillo, anteater, vampire bat, anaconda, toucan, and rhea.

Ethiopian. This realm is made up of Africa south of the Sahara desert, Madagascar, and Arabia. It is the home of higher apes, elephant, rhinoceros, lion, zebra, antelope, ostrich, secretary bird, and lungfish.

Oriental. This region includes Asia south of the Himalaya Mountains, India, Sri Lanka (Ceylon), Malay Peninsula, Southern China, Borneo, Sumatra, Java, and the Philippines. Its characteristic animals are the tiger, Indian elephant, certain apes, pheasant, jungle fowl, and king cobra.

THE DIVERSITY OF BIOMES

The six faunal realms just described are large areas embracing continental land masses, each containing a characteristic and distinctive animal composition. Each of these zoologic regions is divisible into smaller zones based on regional differences in animals and plants living there. One approach to the geographic subdivision of realms is the system of **life-zones.** Life-zones are transcontinental belts running east and west, each bearing a characteristic type of vegetation and fauna adapted for a particular latitudinal climate. Although the life-zone system was used extensively for decades, and still is occasionally, many of the recognized zones embraced too many different kinds of habitats to be useful.

A more workable approach is to classify plant forms into natural biotic units called **biomes.** A biome is a distinctive combination of plants and animals of the important "climax" species, especially the dominant plants (primary producers) that form a stable, self-maintaining community. Biomes are based mainly on predominant plant formations, but since animals depend on plants, each biome supports a characteristic fauna.

Each biome is distinctive and sometimes there are abrupt boundaries between adjacent biomes, especially when there is a sudden change in soil type or drainage. More often, however, biomes grade into one another over broad areas. Such transition areas are called **ecotones.** The ecotone may contain interspersed species of both biomes or the communities of one biome may exist as more or less isolated outposts inside the territory of another biome. Such areas are especially interesting because the variety and density of plant and animal life is often greater in ecotones than in the uninvaded biome.

Terrestrial biomes

Terrestrial biomes are more varied than are those of the water because there are more variable conditions on land. Physical differences in the air are expressed in such factors as humidity, temperature, pressure, and winds to which air-dwelling forms must adapt themselves, as well as types of soils and vegetation. Variations from profuse rainfall to none at all; topographic differences of mountains, plains, hills, and valleys; climatic differences from arctic conditions to those of the tropics; temperature variations from those in hot deserts to those of high altitudes and polar zones; and air and sunlight differences from those of daily variations to great storms—all these factors have influenced animal life and have been responsible for directing its evolutionary development.

Land forms have become specially adapted for living in the soil (subterranean), on the open ground, on the forest floor, in vegetation, and in the air. Although more species of animals live on land than in water, there are fewer phyla represented among terrestrial forms. The chief land organisms are the mammals, birds, reptiles, amphibians, worms, protozoans, and arthropods.

The principal terrestrial biomes are temperate deciduous forest, temperate coniferous forest, tropical forest, grassland, tundra, and desert. In this brief survey, we will refer especially to the biomes of North America and will consider predominant features of each.

Temperate deciduous forest

The temperate deciduous forest, best-developed in eastern North America, encompasses several forest types that change gradually from the northeast to the south. Deciduous, broad-leaved trees such as oak, maple, and beech that shed their leaves in winter predominate. Seasonal aspects are better defined in this biome than in any other. The relatively dense forests in summer form a closed canopy that creates deep shade below. Consequently there has been a selection for understory plants that grow rapidly in the spring and flower early before the canopy develops. The mean annual precipitation is relatively high (75 to 125 cm, or 30 to 50 in) and rain falls periodically throughout the year. The heavy exploitation of the deciduous forests of North America began in the seventeenth century and reached a peak in the nineteenth century. Logging removed nearly all of the once-magnificent stands of temperate hardwoods. With the opening of the prairie for agriculture, many eastern farms were abandoned and allowed to return gradually to deciduous forests.

Among characteristic fauna of these forests are the deer, fox, bear, beaver, squirrel, flying squirrel, raccoon, skunk, wildcat, rattlesnake, and copperhead. There are also various songbirds, birds of prey, and amphibians. Insects and other invertebrates are common, since decaying logs afford excellent shelters for them.

Coniferous forest

In North America the coniferous forests form a broad, continuous, continent-wide belt stretching across Canada and Alaska, and south through the Rocky Mountains into Mexico. It is dominated by evergreens—pine, fir, spruce, and cedar—which are adapted to withstand freezing and take full advantage of short summer growing seasons. The conical-shaped trees with their flexible branches shed snow loads easily. The northern area is the **boreal** (northern) **forest,** often referred to as **taiga** (a Russian word, pronounced "tie-ga"). The taiga is dominated by white and black spruce, balsam, subalpine fir, larch, and birch. In the central region of North America, the taiga merges into **lake forest,** dominated by white pine, red pine, and eastern hemlock. However, most of this forest was destroyed by logging and is replaced by shrubby wasteland that characterizes much of Michigan, Wisconsin, and Minnesota today. The large **southern pine forests** occupy much of the southeastern United States.

Mammals of the boreal and lake coniferous forests are deer, moose, elk, snowshoe hare, several furbearers, and a variety of rodents. They are adapted physiologically or behaviorally for long, cold, snowy winters. Common birds are chickadees, nuthatches, warblers, and jays. One bird, the red crossbill, has a beak highly specialized for picking seeds from cones. Mosquitoes and flies are pests to both animals and humans in this biome. Southern coniferous forests lack

FIG. 40-3 Air plants (bromeliads) in a tropical forest. The pineapple-like bromeliads and other epiphytes (plants that live on the surface of other plants) are common in tropical America. They tolerate lack of soil and a precarious water supply. Large bromeliads are often little ecosystems in themselves, supporting a variety of small, interdependent plant and animal life. (Photograph by L. L. Rue, III.)

many mammals found in the north, but they have more snakes, lizards, and amphibians.

Tropical forest

The worldwide equatorial belt of tropical forests are areas of high rainfall, high humidity, relatively high and constant temperatures, and little seasonal variation in day length. These conditions have nurtured luxurious, uninterrupted growth that reaches its greatest intensity in the rain forests. In sharp contrast to temperate deciduous forests, dominated as they are by a relatively few tree species, tropical forests contain thousands of species, none of which is dominant. A single hectare typically contains 50 to 70 tree species as compared to 10 to 20 tree species in an equivalent area in the eastern United States. Climbing plants and epiphytes are common among the trunks and limbs (Fig. 40-3). A distinctive feature of tropical forests is the stratification of life into six, and occasionally as many as eight, feeding strata.

Insectivorous birds and bats occupy the air above the canopy; below it birds, fruit bats, and mammals feed on leaves and fruit. In the middle zones are arboreal mammals (such as monkeys and tree sloths), numerous birds, insectivorous bats, insects, and amphibians. A distinct group of animals range up and down the trunks, feeding from all strata. On the ground are large mammals lacking climbing ability, such as the large rodents of South America (for example, capabara, paca, and agouti). Finally, a mixed group of

FIG. 40-4 Sonoran desert of Arizona and central Mexico has a diversified vegetation. Conspicuous here are giant seguaro cactus, creosote bushes, and bur sage. (Photograph by L. L. Rue, III.)

small insectivorous, carnivorous, and herbivorous animals search the litter and lower tree trunks for food. The tropical forests, especially the enormous expanse centered in the Amazon basin, are the most seriously threatened of forest ecosystems. Large areas are being cleared for agriculture, but because of low soil fertility, farms are soon abandoned. Once eradicated from large areas, tropical species cannot recolonize.

Grassland

The most extensive prairie biome in the world is in North America. In the early nineteenth century this sea of grass, broken only by occasional woodlands of pine, juniper, and oak, supported vast herds of bison, totaling perhaps 60 million animals. This biome—both its plant and animal components—has been almost totally destroyed by humans. By 1888 less than 1,000 bison remained. Virtually all of the major native grasses have been replaced by cultivated grains in croplands and by alien species of grasses in grazing lands.

The original grasslands were controlled by the climate. Rainfall is too scant to support trees but is sufficient to prevent the formation of deserts. Although the dominant herbivore, the bison, is gone, several typical prairie herbivores remain, for example, jackrabbits, prairie dogs, ground squirrels, and antelope. Mammalian predators include coyotes, ferrets, and badgers, although, of these, only coyotes are common. Wolves that once followed the bison herds have disappeared with their prey.

Tundra

The tundra is characteristic of severe, cold climatic regions, especially the treeless arctic regions and high mountain tops. Plant life must adapt itself to a short growing season of about 60 days and to a soil that remains frozen for most of the year. Most tundra regions are covered with bogs, marshes, ponds, and a spongy mat of decayed vegetation, although high tundras may be covered only with lichens and grasses. Despite the

FIG. 40-5 **A,** Running-water habitats such as this brook rapids contain many organisms having adaptations for maintaining their position in the current. **B,** A water penny, larva of the riffle beetle *Psephenus*. This flat larva is 3.5 to 6 mm long and is adapted for clinging to lower surfaces of stones in swift brooks. The adult is a somewhat flattened, blackish beetle, 4 to 6 mm long, that may be found in the water or in bordering vegetation. (Photographs by F. M. Hickman.)

thin soil and short growing season, the vegetation of dwarf woody plants, grasses, sedges, and lichens may be quite profuse. The plants of the alpine tundra of high mountains, such as the Rockies and Sierra Nevadas, may differ from the arctic tundra in some respects. Characteristic animals of the arctic tundra are the lemming, caribou, musk ox, arctic fox, arctic hare, ptarmigan, and (during the summer) many migratory birds.

Desert

Deserts are extremely arid regions where rainfall is low (less than 25 cm a year), and water evaporation is high. The largest desert in the world is the huge Sahara-Arabian-Gobi desert, stretching from west Africa to central Asia and occupying 8% of the earth's nonpolar land mass. In large portions of this area, it may rain only every few years. The deserts of western North America and Mexico are often called "near desert" because rainfall is substantially greater than on the great desert of Africa and Asia (Fig. 40-4). Desert plants, such as various shrubs and cacti, have reduced foliage and other adaptations for conserving water. Many large desert animals have developed remarkable anatomic and physiologic adaptations for keeping cool and conserving water (p. 626). Most smaller animals avoid the most severe desert conditions by living in burrows. Many are nocturnal. Mammals found there include the white-tailed deer, peccary, cottontail, jackrabbit, kangaroo rat, pocket mouse, ground squirrel, badger, gray fox, skunk, and the like. Birds typical to desert life are the roadrunner, cactus wren, turkey vulture, cactus woodpecker, burrowing owl, Gambel's quail, raven, hummingbird, and flicker. Reptiles are numerous, such as the horned lizard, Gila monster, race runner, collared lizard, chuckwalla, coral snake, rattlesnake, and bull snake. A few species of toads are also common. Arthropods include a great variety of scorpions, spiders, centipedes, and insects.

Freshwater biomes

Freshwater biomes include the running-water, or **lotic,** habitats, and the standing-water, or **lentic,** habitats. Freshwater streams are running-water habitats. Standing-water habitats include ponds and lakes.

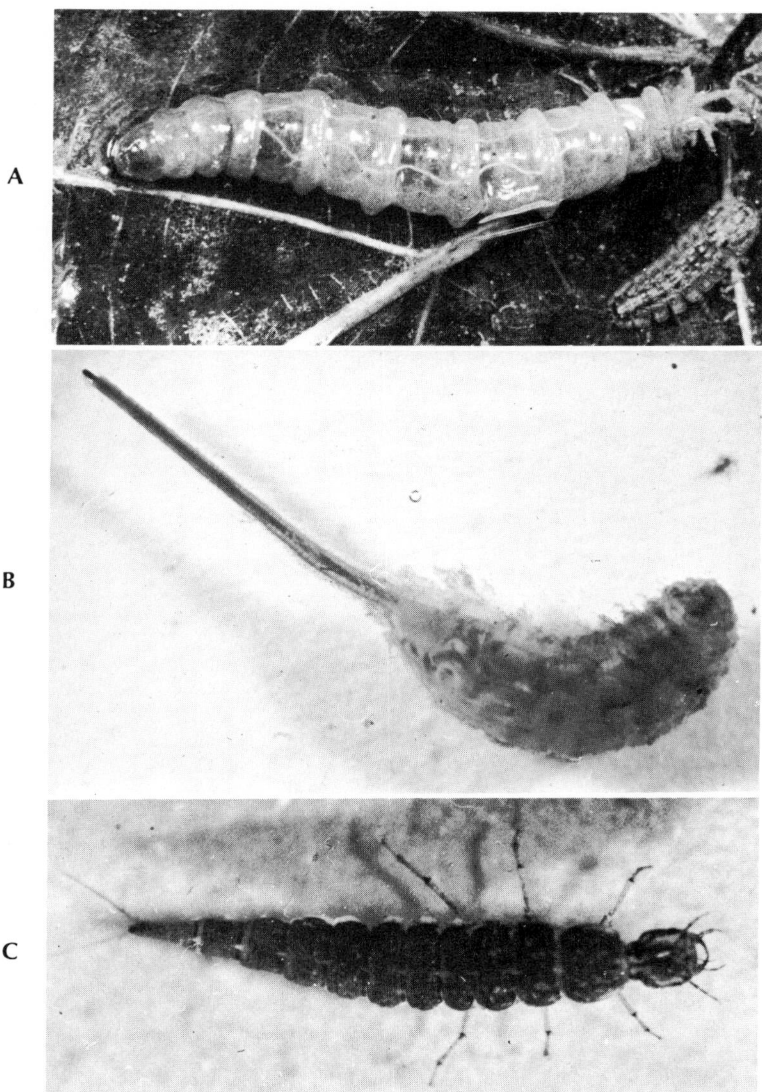

FIG. 40-6 Pond water larvae. **A,** Crane fly larva found in decaying debris along bank. Beside it is small crustacean *Asellus.* Adult crane fly resembles oversize mosquito with extralong legs. **B,** Rat-tailed maggot *Eristalis* gets its air by means of its caudal respiratory tube ("rat-tail"), which is in sections like telescope and can be extended to four times the length of body. The adult fly resembles a bee. **C,** Larval form of *Dytiscus,* the giant diving beetle (Fig. 17-8, *C,* p. 379). Both larvae, called water tigers, and adults are predaceous. These were all found in tiny midwestern woodland pond in early March. (Photographs by F. M. Hickman.)

Running-water habitats

Freshwater streams are lotic habitats that follow a gradient from mountain brooks to streams and rivers. From its origin at a spring, pond, or mountain lake, a permanent stream flows downward at a rate determined by the vertical face, volume of flow, and streambed volume. As the stream enlarges at lower altitudes, the current diminishes, the temperature tends to rise, oxygen content tends to drop, and bottom silting increases. Lotic habitats have a high primary productivity, as much as 30 times that of standing water habitats, because flowing water continually brings in nutrients and carries away wastes. Oxygen is usually at or near saturation for the existing temperature in the swifter portions of a stream because of the constant contact of moving water with the atmosphere.

All animals that occur in streams are adapted for maintaining position in the unremitting current (Fig. 40-5). Many of the forms found in such habitats have organs of attachment, such as suckers and modified appendages, streamlined body shapes for reducing surface resistance, or stages adapted for creeping under stones.

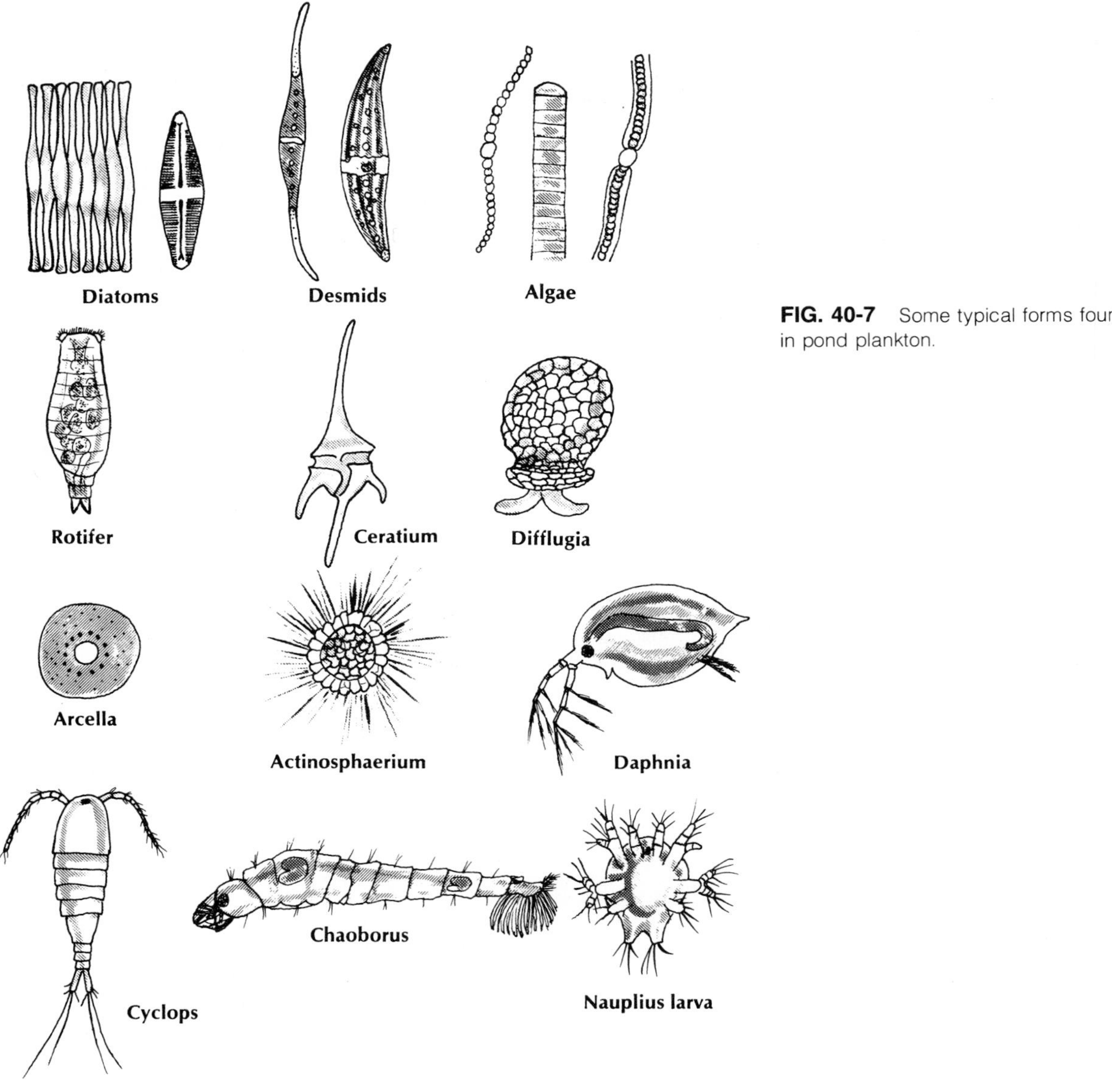

FIG. 40-7 Some typical forms found in pond plankton.

Ponds

Unlike streams, ponds have feeble currents or none at all. They vary, depending on their age and location. Most ponds contain a great deal of vegetation that tends to increase with the age of the pond. Many of them have very little open water in the center, for the vegetation, both rooted and floating types, has largely taken over. As ponds fill up, the higher plants become progressively more common. The bottoms of ponds vary from sandy and rocky (young ponds) to deep and mucky (old ponds). The water varies in depth from a few centimeters to a meter or two, although some may be deeper. Ponds are too shallow to be stratified, for the force of the wind is usually sufficient to keep the entire mass of water in circulation. Because of this, the gases (oxygen and carbon dioxide) are uniformly distributed through the water, and the temperature is fairly uniform.

Animal communities of ponds are usually similar to those of bays in larger bodies of water (lakes). The

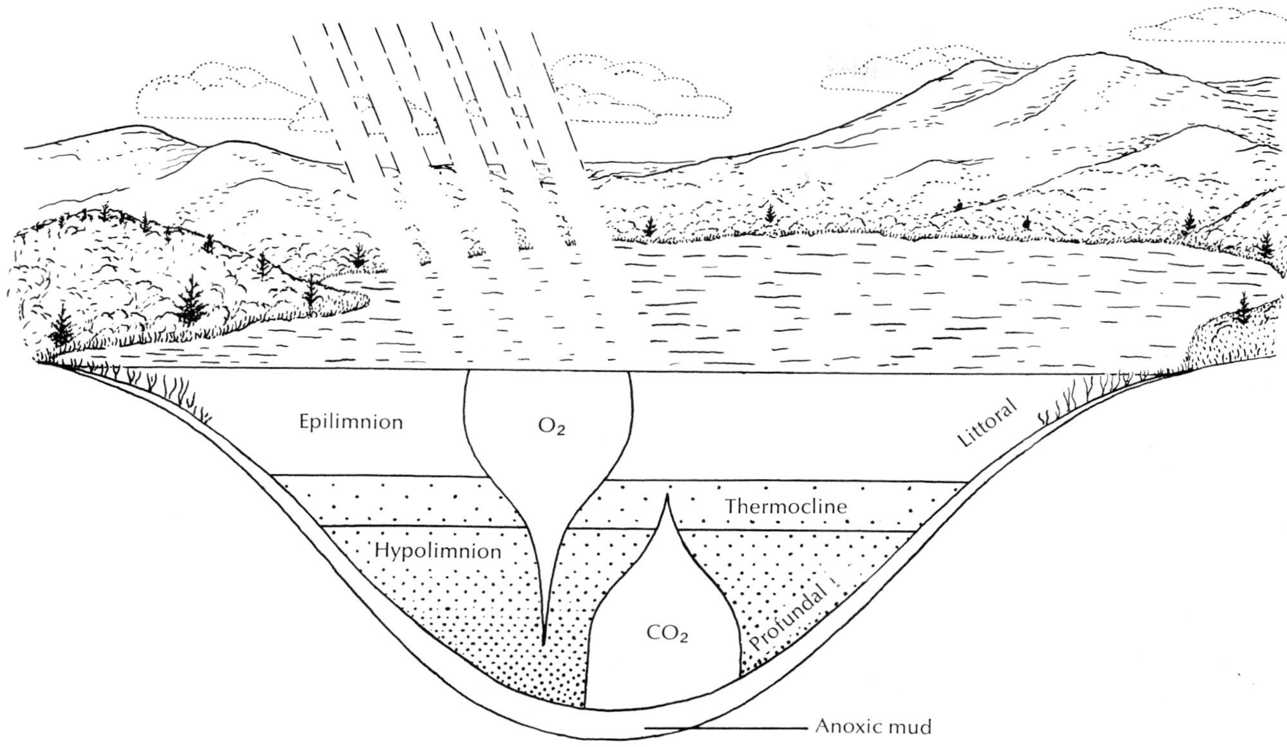

FIG. 40-8 Stratification of a lake in summer. Warm surface water is separated from cold deep water by the thermocline. Oxygen is abundant in the well-illuminated, wind-stirred epilimnion, but scanty or absent at the bottom. Carbon dioxide may be abundant at the bottom where respiration and organic decomposition predominate.

large amount of vegetation and plant decomposition products affords an excellent habitat for many forms. Among the common forms found are varieties of snails and mussels, larvae of flies (Fig. 40-6), beetles, caddis flies, dragonflies, many kinds of crustaceans and midges, and many species of frogs. The community of animals that live on the bottom or among the submerged vegetation is called **benthos.** Many swimming forms, called **nekton,** are also found in the larger ponds. Because of the abundance of food, many birds are usually found in and around ponds. These include herons, killdeers, ducks, grebes, and blackbirds. Most ponds also have plankton composed of microscopic plants and animals, such as protozoans, crustaceans, worms, rotifers, diatoms, and algae (Fig. 40-7). Plankton floats on or near the surface and is shifted about passively by the winds.

Those forms that are found in ponds ordinarily require less oxygen than those living in streams or rapids.

Lakes

The distinction between lakes and ponds is not sharply defined. Lakes are usually distinguished from ponds by having continuous and permanent water in their centers and by having some sandy shores. The bottoms of lakes depend to a great extent on exposure to winds. Where waves are common, the bottoms are usually sandy, but protected areas may contain a great deal of deposited bottom muck. The surface of water in proportion to the total volume of water is less in a lake than in a pond, for lakes are deeper. In large lakes, such as the Great Lakes of North America, the depth rarely exceeds 500 m; in moderate-sized lakes the depth is much less. Oxygen is scarcer in the deeper regions where there is little circulation. Light penetration depends on the sediment in water. Most of the light is absorbed by the first meter of surface water, and little penetrates beyond a few meters.

Waters of many northern, temperate lakes undergo seasonal changes in temperature that cause them to become stratified (Fig. 40-8). After the ice melts in the spring, the sun warms the surface water. When the temperature reaches 4° C, the temperature at which water attains its greatest density, the strong winds of spring mix the water from top to bottom until it is uniformly 4° C. This is the **spring overturn,** an important event that brings nutrients up from the bottom and carries oxygen down from the top. As the surface water continues to warm in early summer, its density decreases. Eventually, the warmer and lighter surface water ceases to mix with the colder and heavier bottom water, even during periods of strong winds. Stratification has occurred, and the warm surface water, the **epilimnion,** remains separated from the deeper water, the **hypolimnion,** for the rest of the summer. The area of separation is a temperature barrier called the **thermocline,** a zone where the temperature drops steeply. Above the thermocline, the surface water continues to warm during the summer. Below the thermocline, the water remains still and cold, often only a degree or two warmer than it was in winter.

The decomposition of organic material at the bottom may deplete the hypolimnion of oxygen during the summer and increase the concentration of carbon dioxide and methane. This summer stagnation period requires special adaptations by the animals living there: some are anaerobic (bacteria), some tolerate extended oxygen debts (worms), some make nightly excursions into the epilimnion to obtain oxygen (certain insect larvae), and some encyst and wait out the stagnation period in a dormant condition (certain crustaceans).

In the autumn, colder weather cools the surface, causing the water to sink. The thermocline begins to disappear, and soon the water temperature is uniformly cold (4° C) from top to bottom. Winds again stir the lake throughout, recharging it with nutrients and oxygen. This is the **fall overturn.** In winter the surface water cools, becomes lighter, and may freeze, while the bottom water remains at 4° to 5° C. If the lake is covered with ice and snow, there may be a winter stagnation period during which the oxygen concentration gradually falls as a result of consumption by animals and decomposition of organic matter.

Several habitats are found in lakes. In the terrigenous bottom habitat where the water is shallow and vegetation is absent, snails and mayfly nymphs might be found. In the cove habitat, which has a great deal of vegetation—submerged, emergent, and floating—one might find bryozoans, small crustaceans, snails, and some insect larvae; in open water one might find plankton and nekton. On the bottom (benthic habitat) are animals that require little light and oxygen, such as a few annelids, bivalves, and midge larvae.

Marine biomes

By almost any measure, the oceans represent by far the largest portion of the earth's biosphere. They cover 71% of the earth's surface to an average depth of 3.75 km (2.3 miles), with their greatest depths reaching to more than 11.5 km (7.2 miles) below sea level. Life occurs throughout this enormous volume, while on land, life is confined to within a few meters above or below the land's surface. The marine world is a relatively uniform one as compared to land, and in many respects it is less demanding on life forms. However, the evident monotony of the ocean's surface belies the variety of life below. The oceans are the cradle of life, and this is reflected in the variety of organisms living there—more than 200,000 species of plants and animals. The vast majority of these forms, about 98%, live on the sea bed **(benthic);** only 2% live freely in the open ocean **(pelagic).** Of the benthic forms, most occur in the intertidal zone or shallow depths of the tropical oceans. Less than 1% live in the deep ocean below 2,000 m.

The two principal determinants of such uneven distribution of biomass in the ocean are nutrient availability and light penetration into the water. The most productive areas are concentrated along continental margins and a few open areas where the waters are enriched by organic nutrients (principally phosphates and nitrates) and organic debris lifted by upwelling currents into the sunlighted, or **photic,** zone. Photosynthetic activity is limited to the depth of the photic zone, usually less than 150 m, although in especially clear waters it may extend to more than 300 m. Below the photic zone, the lack of both nutrients and light prevent the growth of phytoplankton, which forms the base of the food pyramid. In the zone of perpetual darkness all life depends on the "rain" of organic debris from the photic zone above.

In addition to the principal division of the marine habitat into benthic (bottom) and pelagic (open water), we can distinguish several zones based on depth and distance from shore (Fig. 40-9). The **epipelagic** zone is equivalent to the photic zone and reaches down to 100 to 200 m. Below this is the **mesopelagic** zone, extending down to 1,000 m. Only shorter (blue) wave-

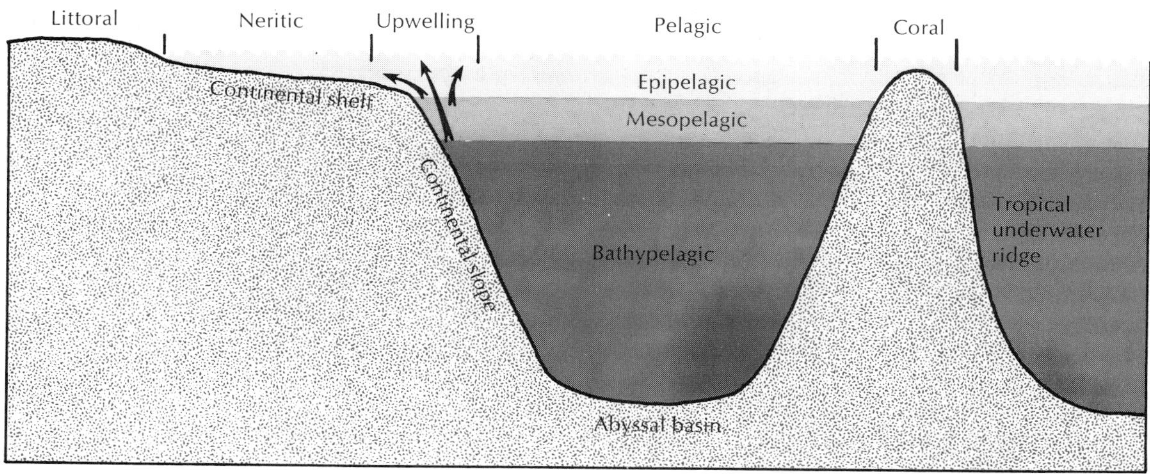

FIG. 40-9 Major marine zones and biomes.

lengths of light penetrate into this zone, too little to support effective photosynthesis. This is followed by the **bathypelagic** zone (1,000 m to the bottom). The total biomass may be in excess of 1,000 g/m^2 in the inshore photic zone. At 200 m the biomass drops to about 50 g/m^2 and the deeper one goes, the sparser life becomes. At 1,000 m the biomass is 1 g/m^2, and in the abyssal depths of the ocean it is only 0.1 g/m^2. These figures emphasize the importance of the inshore region, even though it makes up less than 2.5% of the total oceanic area.

An important faunistic boundary is the permanent thermocline in which the water temperature drops from the mixed surface temperature to the cold water of the deep ocean. The thermocline begins at about 200 m and extends between 500 to 1000 m. Below 2,000 m the water temperature averages about 3.3° C over 60% of the earth. Unlike the seasonal thermocline of fresh-water lakes, considered earlier, the permanent thermocline of the ocean is unaffected by climatic changes on the surface. Nutrients captured by the phytoplankton biomass tend to sink below the thermocline, becoming unavailable to surface productivity unless flushed upward by upwelling.

The life of the ocean is divided into biomes, each with its own distinctive life forms (Fig. 40-9). Five major biomes are the (1) **littoral,** or tidal zone, where sea and land meet; (2) **neritic,** or shallow-water zone, surrounding the continents, and extending out to a depth of about 200 m; (3) **upwelling,** where currents along coastal boundaries carry nutrient-rich water into the photic zone; (4) **pelagic,** the vast open sea; and (5) **coral reefs,** built by carbonate-secreting organisms where tropical land masses protrude into the photic zone. Coral-reef formation was described in Chapter 10, so will not be included in the brief survey of marine biomes that follows.

Littoral biome

The littoral biome is biologically the richest and, paradoxically, physically the harshest of all biomes. At its margins it is subjected to pounding surf (as much as 15,000 kg/cm^2), sun, wind, rain, extreme temperature fluctuations, erosion, and sedimentation. These influences have created numerous distinctive habitats, often in proximity to one another, each harboring a specialized fauna. Animals living here must be firmly attached and strongly protected. Barnacles, snails, chitons, limpets, and mussels are shielded by their shells and either grow permanently to the rocky surface or maintain position with powerful suction organs. Some species of sea urchins bore recesses into the rocks to become permanently imprisoned in their stony cups as they grow. On sandy and muddy shores, animals possess adaptations for burrowing and surviving a shifting matrix.

The littoral biome of temperate regions of the earth embraces three zones. The upper **spray zone,** moistened by saltwater spray above the high-tide mark, supports a few green algae and lichens that are grazed by periwinkle snails (*Littorina* sp.). Small spray-zone barnacles (*Chthalamus* sp.) also survive here. The middle **tidal zone** is typically covered with brown and red algae, especially the common brown alga (''seaweed'') with its distinctive floats (*Fucus* sp.). Barnacles (*Balanus* sp.), limpets, chitons, snails, and other herbivores are common in this region. The lower **subtidal zone,** which is always submerged, supports forests of brown algae (*Laminaria* sp.) and a great variety of smaller algae, sea urchins, starfishes, anemones, molluscs, and fishes.

Estuaries are also part of the littoral biome. An **estuary** is a semienclosed transition zone where fresh water flows into the sea. Depending on the substrate, currents, climate, and other physical conditions, an estuary may take the form of a mud or sand flat, tidal marsh, or mangrove swamp. Salinity is variable because the amount of fresh water entering the estuary varies throughout the year. Temperature extremes are often severe. Despite their physical harshness, estuaries are typically rich in nutrients and consequently support a diverse fauna. Most estuarine organisms are benthic, either burrowing into or attaching to the bottom. Conspicuous features of estuaries are oyster communities, around which a host of symbiotic organisms associate. Motile inhabitants of estuaries are chiefly fishes and crustaceans. All estuarine animals must tolerate fluctuations in salinity, which is always changing with the ebb and flow of the tide and variable discharge of fresh water from rivers and streams. The different ways animals meet this problem is discussed in Chapter 30.

Neritic biome

The neritic biome overlies the continental shelf, an area usually defined as lying above the 200 m contour. Although it occupies only 7.5% of the total oceanic area, it ranks high in importance because here are concentrated most of the world's fisheries. The neritic zone is more productive than the open sea because upwelling at the shelf edge together with the greater turbulence of continental shelf waters brings nutrients into all strata of the water. Furthermore, the continental shelf benefits from nutrients washed into the sea by streams and rivers. Sinking nutrients, rather than disappearing into the ocean's depths, fall on the shelf floor where they stimulate a rich growth of seaweed, marine grasses, and other plants. The increased plant growth in turn supports a diverse animal life.

Upwelling biome

The upwelling biome supports much of the productivity of the neritic biome as well as a rich fishery of its own. Areas of upwelling, though small in area and restricted to a few regions of coastal currents, are important sources of nutrient renewal for the surface photic zone of the oceans. In such regions deep, low-temperature water in which nutrients have accumulated well up to slowly replace surface water swept away by prevailing winds.

Among the most important upwelling regions is the Peru Current. Until recently, the Peruvian anchovy (or anchoveta) comprised 22% of all the fish caught in the world. The total catch of all the fish in the United States in 1969 was only 20% of the Peruvian anchovy catch! Anchovies are converted into fish meal, which is exported as livestock feed, and fish oil, which is used in margarine, paint, and other products. The base of this food chain is principally photosynthetic colonial diatoms, which convert upwelling inorganic nutrients into energy-rich organic compounds. The diatoms are harvested directly by crustaceans and by the anchovies, which have gill rakers adapted for filter-feeding. Predation on the anchovies is intense. "Guano" birds (principally cormorants and gannets) that feed almost exclusively on anchovies have over the centuries deposited vast accumulations of droppings (guano) along the coastal deserts of Peru. Tuna, sea lions, and squid also feast on the anchovies. In recent years humans have become the greatest predator of all. Following more than a decade of intense commercial fishing, the Peruvian anchovy fishery suffered a collapse in 1972 from which it has not yet recovered. There is concern that this great industry will experience the same fate that befell the California sardine fishery and the Japanese herring fishery, both fisheries of upwelling biomes. These intensively harvested fisheries fell below points of spontaneous recovery and appear to have disappeared forever.

Pelagic biome

The pelagic biome beyond the continental shelves occupies 90% of the total oceanic area. Despite its vastness, it is relatively impoverished biologically because the constant rain of sinking plankton and detritus from the photic zone carries nutrients below the permanent thermocline and into the bathypelagic zone where they are immobilized. Nitrates and phosphates, the most important limiting nutrients, are commonly 100 times more concentrated in the neritic than in the epipelagic zone. Productivity is high only in areas where nutrients are replenished by current convergences (Arctic and Antarctic seas) and upwelling. The polar areas are in fact enormously productive. Phytoplankton, the base of the food chain, is eaten by euphausiid shrimp (krill), which are harvested by blue and fin whales. Before their egregious overexploitation by humans, the whales are estimated to have consumed 77 million tons of Antarctic krill per year, far more than the entire catch of all the fishes, crustaceans, and molluscs caught by all the world's fishing fleets in any one year! Because the baleen whales have been nearly destroyed by hunting, the Antarctic krill now remain unharvested.

Far below the epipelagic zone exists the abyssal habitat, characterized by enormous pressure, perpetual darkness, and a constant temperature near 0° C. It has remained a world unknown to humans until recently, when baited cameras, bathyscaphs, and deep-water trawls have been lowered to view and sample the ocean bottom. Benthic forms of the sea floor survive on that meager portion of the gentle rain of organic debris from above that escapes consumption by organisms in the water column. Despite the unusual properties of the habitat, sea anemones, sea urchins, crustaceans, and polychaete worms—indeed nearly all major invertebrate groups—exist here as well as fishes.

Annotated references

Selected general readings

Andrewartha, H. G., and L. C. Birch. 1954. The distribution and abundance of animals. Chicago, University of Chicago Press. *An analysis of animal populations and the factors that influence their abundance and distribution.*

Birch, L. C. 1954. The role of weather in determining the distribution and abundance of animals. Cold Spring Harbor Symposium **22:**203-218.

Cushing, D. H., and J. J. Walsh (eds.). 1976. The ecology of the seas. Philadelphia, W. B. Saunders Co. *Fourteen selections by specialists, with excellent sectional introductions by the editors. An important reference in marine ecology.*

Darlington, P. J. 1957. Zoogeography: the geographical distribution of animals. New York, John Wiley & Sons, Inc.

Elton, C. S. 1958. The ecology of invasions by animals and plants. New York, John Wiley & Sons, Inc. *A classic on animal dispersion.*

Engel, L., and the editors of Time-Life Books. 1972. The sea. Life Nature Library. New York, Time-Life Books. *This handsomely illustrated volume provides a popularized yet balanced treatment of the sea and its biota.*

Henderson, L. J. 1913. The fitness of the environment. New York, The Macmillan Co. *Explains how conditions on our planet made life possible.*

Hubbs, C. L. 1958. Zoogeography. Washington, D.C., American Association for the Advancement of Science. *This includes the papers presented at two symposia at Stanford and Indianapolis. Part I deals with the Origin and Affinities of the Land and Freshwater Fauna of Western North America, and Part II deals with the Geographic Distribution of Contemporary Organisms.*

Jarman, C. 1972. Atlas of animal migration. New York, Doubleday & Co. *Maps and diagrams show how the various groups migrate.*

Kendeigh, S. C. 1974. Ecology. Englewood Cliffs, N.J., Prentice-Hall, Inc. *The last section of this textbook deals with geographic ecology.*

McCan, T. T. 1973. Ponds and lakes. New York, Crane, Russak & Co., Inc. *The life and ecology of ponds and lakes and how they are studied.*

McConnaughey, B. H. 1978. Introduction to marine biology, ed. 3. St. Louis, The C. V. Mosby Co. *Unified presentation of the marine environment and its life.*

McNaughton, S. J., and L. L. Wolf. 1973. General ecology. New York, Holt, Rinehart and Winston, Inc. *This well-written text offers an especially comprehensive treatment of biomes.*

Simpson, G. G. 1953. Evolution and geography. Eugene, Ore., State System of Higher Education. *An excellent appraisal of the main concepts of animal distribution.*

Udvardy, M. D. F. 1969. Dynamic zoogeography. New York, Van Nostrand Reinhold Co. *Textbook stressing ecology and mechanisms of dispersal of land animals.*

Selected *Scientific American* articles

The biosphere. 1970. **223:**44 (Sept.). *This special issue includes articles on the various cycles—energy, water, oxygen, nitrogen, and mineral—and also on human food and energy and materials production as processes in the biosphere.*

Bullard, E. 1969. The origin of the oceans. **221:**66-75 (Sept.).

Dietz, R. S., and J. C. Holden. 1970. The breakup of Pangaea. **223:**30-41 (Oct.). *The sequence of continental drift since the early Mesozoic era is mapped.*

Fairbridge, R. W. 1960. The changing level of the sea. **202:** 70-79 (May).

Gautier, T. N. 1955. The ionosphere. **193:**123-138 (Sept.).

Gregg, M. C. 1973. The microstructure of the ocean. **228:** 64-77 (Feb.).

Hallam, A. 1975. Alfred Wegener and the hypothesis of continental drift. **232:**88-97 (Feb.). *Wegener's theory that anticipated the modern concept of continental drift remained unaccepted for some 50 years by the scientific community.*

Idyll, C. P. 1973. The anchovy crisis. **228:**22-29 (June). *The Peruvian anchovy fishery and its recent collapse are described.*

Kurtén, B. 1969. Continental drift and evolution. **220:**54-64 (Mar.). *The breakup of early supercontinents in the past helps to explain the diversification of mammals.*

Landsberg, H. E. 1953. The origin of the atmosphere. **189:** 82-85 (Aug.).

Milne, L. J., and M. Milne. 1978. Insects of the water surface. **238:**134-142 (Apr.). *The surface of ponds creates a special environment for four types of insects.*

Welty, C. 1957. The geography of birds. **197:**118-126 (July). *Despite their mobility, most birds are confined to restricted zoogeographic regions.*

Woodwell, G. M. 1978. The carbon dioxide question. **238:** 34-43 (Jan.). *Human activities are increasing the carbon dioxide content of the atmosphere.*

CHAPTER 41
ANIMAL ECOLOGY

A community of acorn barnacles (*Balanus* sp.) in the intertidal zone of a northern Atlantic rocky coast. When again covered by the tide, these cirripedian crustaceans will open and begin sweeping plankton from the water with their specialized feathery legs.

Photograph by C. P. Hickman, Jr.

Life is confined largely to the interfaces between land, air, and water. It does not penetrate very far below the land surface, very high into the atmosphere, or very abundantly into the depths of the ocean. In the preceding chapter we defined where life does occur in the biosphere. Living things within the biosphere may be examined at several different levels of organization.

The highest level of organization is called the **ecosystem,** literally the ''environment-system.'' It includes both living (biotic) and nonliving (abiotic) components of the total environment. Other levels of organization are parts or fractions of the ecosystem. Consequently, the ecosystem may be considered the basic functional unit of ecologic study.

An ecosystem, as defined by the worker studying it, may be large such as a grassland, forest, lake, or cropland, or it may be more restricted as, for example, a river bank or tree hole. Whatever their sizes and differences in physical and biologic structures, all ecosystems function in much the same manner: the sun's energy is fixed by plants and then transferred to consumers and decomposers. The substrate provides nutrients that are cycled and recycled through the various living components of the ecosystem. There is always some flow of energy and nutrients between different ecosystems, so that all are interrelated within the earth's biosphere.

The next level of organization is the **community,** an assemblage of living organisms sharing the same environment and having a certain distinctive unity (Fig. 41-1). Communities comprise the living elements of an ecosystem. Like ecosystems, communities may be large or small, ranging from the coniferous forest community that may span a continent to the inhabitants of a rotting log community or the community of microorganisms living in the large intestine of a human. The elements of a community are closely interdependent.

The **population,** the next lower level of organization, is an interbreeding group of organisms of the *same species* sharing a particular space. Every community is composed of several populations, including those of plants, animals, and microorganisms. Energy and nutrients flow through a population. Its size is regulated by its relationships to other populations in the community and by the abiotic characteristics of the ecosystem in which it is found. Because members of a population are interbreeding, they share a gene pool and thus are a distinct genetic unit.

The lowest level of organization within the biosphere, in ecologic terms, is the **organism** itself. It is the living expression of the species. Each organism responds to its environment (see Principle 27, p. 15). The effect of the environment on all members of a species is called natural selection; the results of natural selection determine the evolution of the species, that is, its success or failure in response to changes in the environment. Although an organism itself does not evolve, its fate, when taken together with the fates of all other members of its species, is a factor in the evolution of the species.

As the unit on which natural selection acts, the organism reflects the response of the population to environmental change. The ecologist who is to understand why animals are distributed as they are must examine the varied mechanisms that animals use to compensate for environmental stresses and alterations.

Ecologists have, in fact, become increasingly interested in the physiologic and behavioral mechanisms of animals. Both are intrinsic to the animal-habitat interrelationship. For example, the success of certain warm-blooded species under extreme temperature conditions such as in the Arctic or in a desert depends on near-perfect balance between heat production and heat loss and between appropriate insulation and special heat exchangers. Other species succeed in these situations by escaping the most extreme conditions by migration, hibernation, or torpidity. Insects, fishes, and other poikilotherms (animals having variable body temperature) compensate for temperature change by altering biochemical and cellular processes involving enzymes, lipid organization, and the neuroendocrine system. Thus the physiologic capacities with which the animal is endowed permit it to live under changing and often adverse environmental conditions. Physiologic studies are necessary to answer the ''how'' questions of ecolgy.

The animal's behavioral responses are also part of the animal-habitat interaction and are of interest to the ecologist. Behavior is involved in obtaining food, finding shelter, escaping enemies and unfavorable environments, finding a mate, courting, and caring for the young. As pointed out earlier (p. 773), genetically determined mechanisms such as behavioral repertoires are acted on by natural selection. Those that improve adaptability to the environment assist in survival and the evolution of the species (see Principle 5, p. 10).

The term **ecology,** coined in the last century by the German zoologist Ernst Haeckel, is derived from the

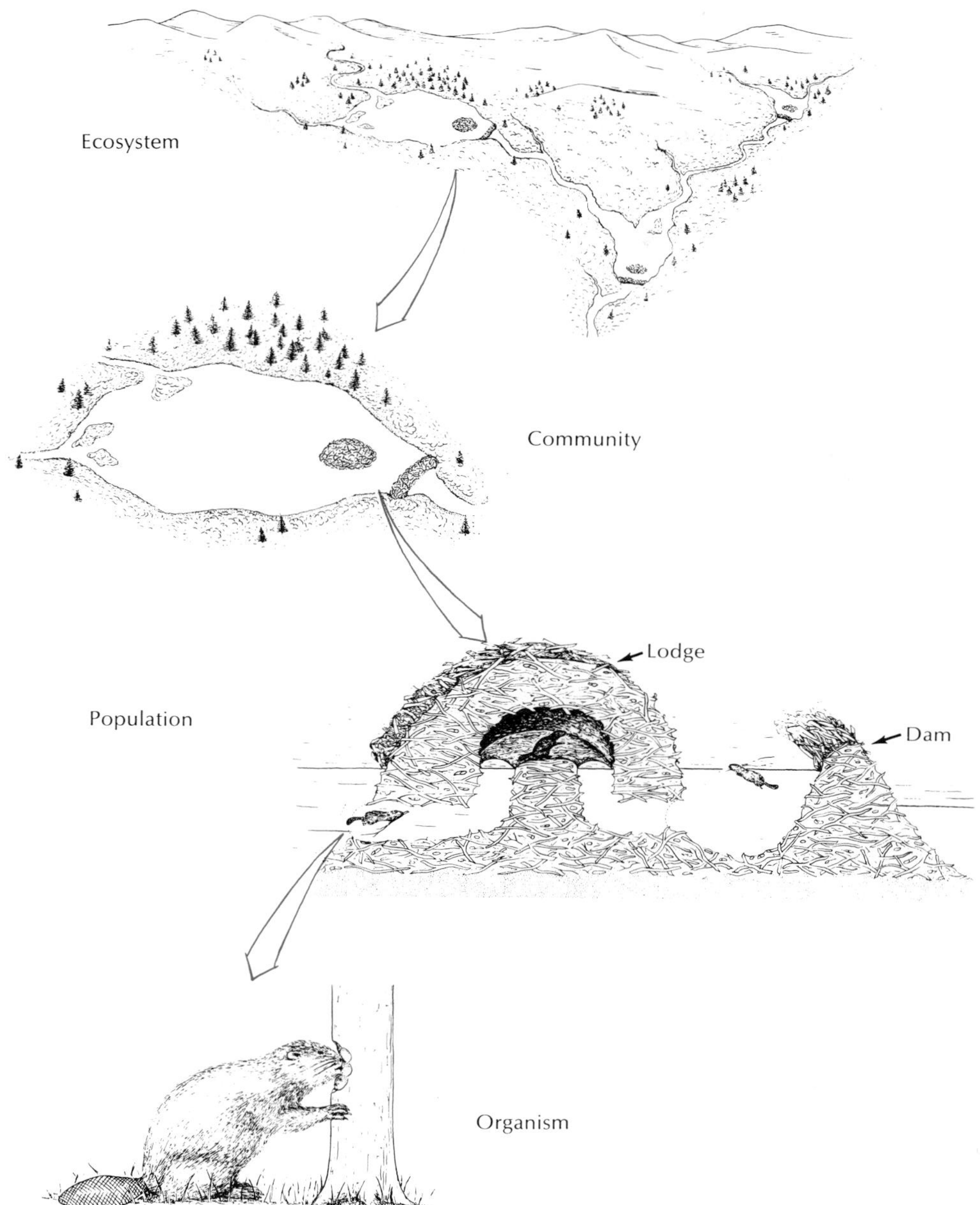

FIG. 41-1 Relationship between ecosystem, community, population, and organism.

Greek *oikos,* meaning "house" or "place to live." Haeckel called ecology the "relation of the animal to its organic as well as inorganic environment." Although we no longer restrict ecology to animals alone, Haeckel's definition is still basically sound.

The term **environment** can mean a number of things. It is frequently used in reference to the organism's immediate surroundings, but it is not always clear whether it is meant to include the living, as well as the nonliving, surroundings. To the ecologist it certainly includes both. Ultimately the environment consists of everything in the universe external to the organism. Since ecology is really biology of the environment, it obviously is a broad and far-flung field of study. In a sense, there is very little that is *not* ecology.

The ecologist may choose to focus on any level of organization within the biosphere. **Ecosystem analysis** is largely interdisciplinary, incorporating physics, chemistry, and other sciences to assist in the comprehension of the role of the environment in determining the distribution and abundance of organisms. **Community ecology** is similar to ecosystem analysis but is of more restricted scope. It is possible to focus on the interactions of a few species and to study energy transfers in detail. **Population biology** stresses genetics, evolution, seasonal changes, and other phenomena within a single species. Some ecologists study the organism itself **(organismic biology)** to see how it responds to the environment, hour by hour, day by day; such studies have become physiologic and behavioral. All contribute to ecologic understanding. None stands alone.

ECOSYSTEM ECOLOGY

The ecosystem is a complex conceptual unit composed of abiotic and biotic elements. The abiotic component is the nonliving physical and chemical environment. It can be characterized by its physical parameters, such as temperature, moisture, light, and altitude, and by its chemical features, which include various essential nutrients. These characteristics determine the basic nature of the ecosystem.

The biotic component—the populations of plants, animals, and microorganisms that form the communities of the ecosystem—may be categorized into producers, consumers, and decomposers. The **producers,** mostly green plants, are **autotrophs** that utilize the energy of the sun to synthesize sugars from carbon dioxide by photosynthesis. This energy is made available to the **consumers** and **decomposers.** They are the **heterotrophs** that exploit the self-nourishing autotrophs by converting organic compounds of plants into compounds required for their own growth and activity. In this section we consider the flow of energy through the ecosystem, which involves the concepts of productivity and the food chain, biogeochemical cycling, and limiting factors within the environment.

Solar radiation and photosynthesis

All life depends on the energy of the sun (see Principle 2, p. 9). The sun releases electromagnetic energy produced by the nuclear transmutation of hydrogen to helium. Solar radiation received at the earth's surface extends from wavelengths of approximately 280 to 13,500 nm. Ultraviolet radiation with wavelengths of less than 280 nm is cut off sharply by the ozone layer in the upper atmosphere (Fig. 41-2). Radiation with wavelengths greater than 1,050 nm, comprising approximately 45% of the total energy, is long-wave infrared radiation that heats the atmosphere, warms the earth, and produces currents of air and water. The most important part of solar radiation lies between wavelengths of 310 and 1,050 nm; this is the portion that we call **light** because of its effect on the human retina. It is also the range that controls all important photobiologic processes, including photosynthesis, photochemical effects, phototropism (orientation of plants toward light), and animal vision.

The flow of energy through the ecosystem begins with **photosynthesis.** Energy enters the ecosystem as visible light; it is utilized by plants to produce adenosine triphosphate (ATP) and to synthesize carbon compounds for themselves and for the animals that feed on them. Although certain details of photosynthesis remain uncertain, the major events can be outlined.

Light striking a green plant is absorbed by chlorophyll, sending low-energy electrons into a higher energy level. These excited electrons drop back to a ground state in approximately 10^{-7} seconds, but in this brief interval their energy is channeled into a sequence of energy-yielding reactions. Part of the energy is used to synthesize ATP; the remainder causes the reduction of pyridine nucleotides (NADP). Both ATP and reduced NADP are then used to synthesize sugars from carbon dioxide and water. Photosynthesis in the individual leaf begins at low light intensity and at first increases linearly; that is, the rate of photosynthesis increases as light intensity increases until it reaches a maximum. The leaf achieves its highest rate of photosynthesis at only approximately one-tenth the intensity of full sunlight.

The amount of solar energy that reaches our earth's atmosphere is estimated at 15.3×10^8 g-cal/m²/yr. Much of this energy is dissipated by dust particles or is used in the evaporation of water. Only a very small fraction is used in the photosynthetic conversion of carbon dioxide to carbohydrates. Calculated on an annual or growing season basis, the photosynthetic efficiency of land areas is approximately 0.3% and of the ocean approximately 0.13%. These estimates are very low because they are based on the total energy

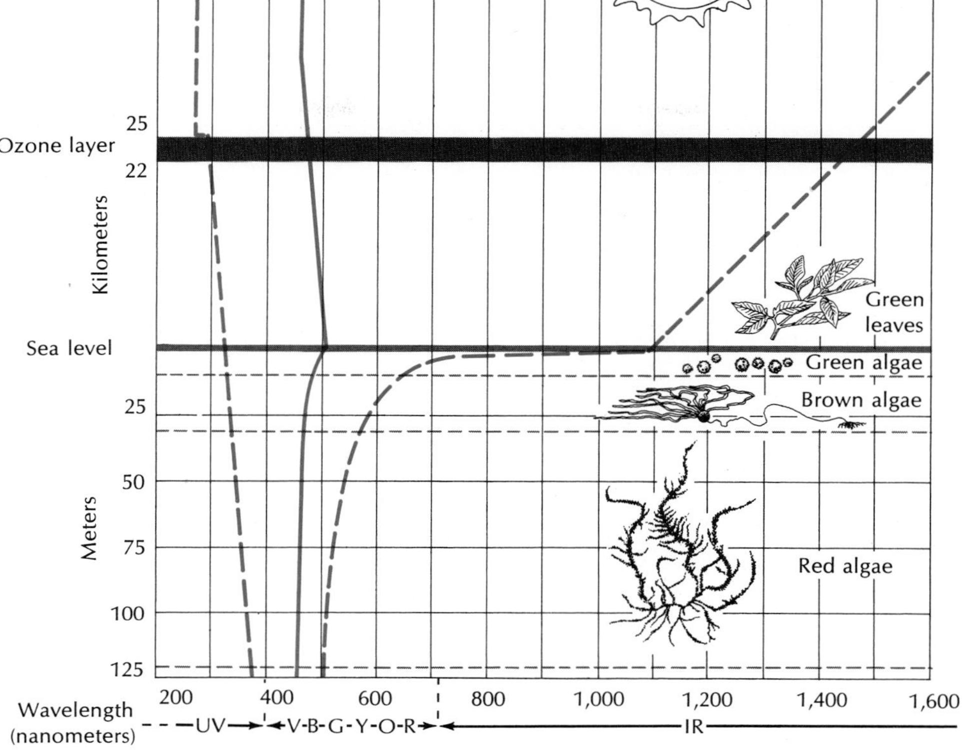

FIG. 41-2 Narrowing of spectrum of sunlight by atmospheric absorption and by absorption in sea water. The solid red line from top to bottom represents wavelengths of maximum intensity. Broken red lines locate the wavelength boundaries within which 90% of the solar energy is concentrated. (Modified from Wald, G., 1959. Life and light. Scientific American, San Francisco, W. H. Freeman & Co.)

available for the year rather than on the growing season alone. During brief intervals of very active growth, plants may store a maximum of 19% of the available light energy.

Production and the food chain

The energy accumulated by plants in photosynthesis is called **production.** Because it is the first step in the input of energy into the ecosystem, the *rate* of energy storage by plants is known as **primary productivity.** The total rate of energy storage, the **gross productivity,** is not entirely available for growth because plants also use energy for maintenance and reproduction. When this energy consumption, or plant **respiration,** is subtracted from the gross productivity, the **net primary productivity** remains. Plant growth results in the accumulation of plant **biomass.** Biomass is expressed as the *weight* of dry organic matter per unit of area and thus differs from productivity, which is the *rate* at which organic matter is formed by photosynthesis.

The level of productivity of different ecosystems depends on the availability of nutrients and the limitations of temperature levels and moisture availability. Highly productive ecosystems are flood-plain forests, swamps and marshes, estuaries, coral reefs, and certain crop ecosystems (for example, rice and sugar cane). In such systems net production can exceed 3,000 $g/m^2/yr$ of dry organic matter. Less productive (1,000 to 3,000 $g/m^2/yr$) are most temperate forests, most agricultural crops, lakes and streams, and grasslands. Least productive (50 to 200 $g/m^2/yr$) are tundra and alpine regions, deserts, and the open ocean. Extreme desert, rock, and ice regions have virtually zero productivity.

The net productivity of plants is the energy that supports all the rest of life on earth. Plants are eaten

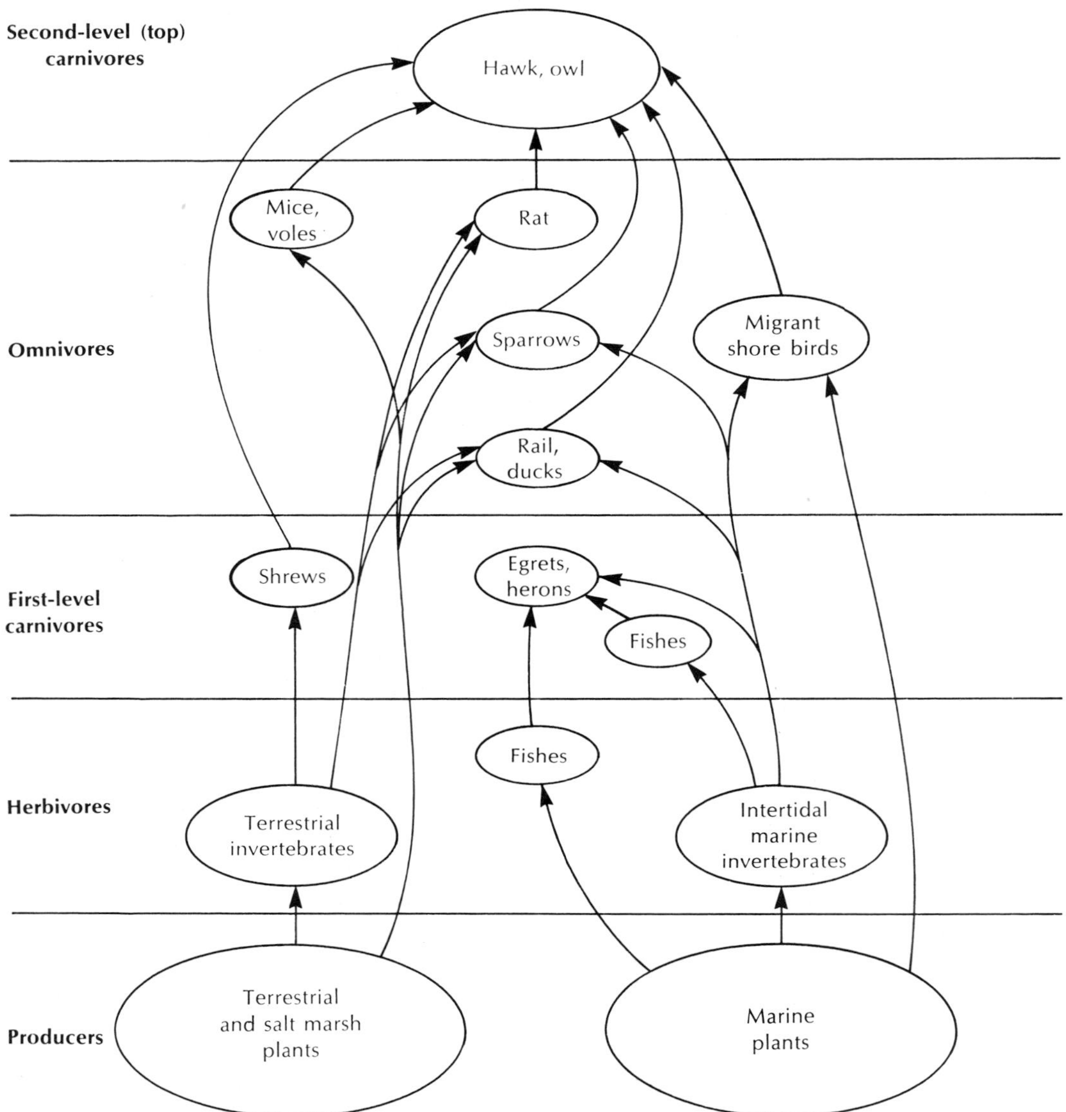

FIG. 41-3 A midwinter food web in a *Salicornia* salt marsh of the San Francisco Bay area. (Modified from Smith, R. L. 1974. Ecology and field biology, ed. 2, New York, Harper & Row, Publishers; after Johnston, R. F. 1956. Wilson Bull. **68:**91.)

by consumers, which are themselves consumed by other consumers, and so on in a series of steps called the **food chain.** Food chains are descriptions of the way energy flows through the ecosystem. A diagram of a food chain shows arrows leading from one species to another, meaning that the first species is food for the second. But the first may be food for several other organisms as well. Seldom, in fact, does one organism live exclusively on another. Although some food chains are simple and short—for example, the one in which the whale feeds mainly on plankton—it is more common for several food chains to be interwoven into a complex **food web** (Fig. 41-3).

Despite their complexity, food chains tend to follow a pattern. Green plants, the base of the food chain, are eaten by grazing **herbivores,** which can convert the energy stored in plants into animal tissue. Herbivores may be eaten by small **carnivores** and these by large carnivores. There may be two or three, sometimes even four, levels of carnivores. At the end of the chain are the **top carnivores** that, lacking predators, die and decompose, replenishing the soil with nutrients for plants, which start the chain.

There are numerous examples of food chains. In the forest, for instance, there are many small insects (primary consumers) that feed on plants (producers). A smaller number of spiders and carnivorous insects (secondary consumers) prey on small insects; still fewer birds (tertiary consumers) live on the spiders and carnivorous insects; and finally one or two hawks (quaternary or top consumers) prey on the other birds.

The decomposers have traditionally been considered the final step in the herbivore-carnivore food chain, since they reduce organic matter into nutrients that become available again to the producers. Ecologists now recognize that the decomposers comprise their own distinct **detritus food chain,** consisting of **detritus feeders,** such as earthworms, mites, millipedes, crabs, aquatic worms, and molluscs, and **microorganisms,** such as bacteria and fungi. Dead organic matter such as fallen leaves or dead animals is decomposed and utilized by fungi, bacteria, and protozoa. Detritus feeders then eat the microorganisms as well as much of the dead organic matter directly. The detritus feeders are in turn eaten by small carnivores; the detritus food chain thus leads up into herbivore-carnivore food chains.

Ecologic pyramids

At each transfer within the food chain, 80% to 90% of the available energy is lost as heat. This limits the number of steps in the chain to four or five. Therefore the number of top consumers that can be supported by a given biomass of plants depends on the length of the chain.

Humans, who occupy a position at the end of the chain, may eat the grain that fixes the sun's energy; this very short chain represents an efficient use of the potential energy. Humans also may eat beef from animals that eat grass that fixes the sun's energy; the addition of a step in the chain decreases the available energy by an order of ten. In other words, it requires ten times as much plant biomass to feed humans as meat eaters than to feed humans as grain eaters. Let us consider the person who eats the bass that eats the sunfish that eats the zooplankton that eats the phytoplankton that fixes the sun's energy. The ten-fold loss of energy that occurs at each level in this five-step chain explains why bass do not form a very large part of the human diet. In this particular food chain, for a person to gain a pound by eating bass, the pond must produce 5 tons of phytoplankton biomass.

These figures need to be considered as we look to the sea for food. The productivity of the oceans is, in fact, very low and largely limited to regions of upwelling where nutrients are brought up and made available to the phytoplankton producers (p. 942). Such areas occupy only approximately 1% of the ocean. The rest is a watery desert.

Ocean fisheries supply 18% of the world's protein, but most of this is used to supplement livestock and poultry feed. If we remember the rule of ten-to-one loss in energy with each transfer of material, then the use of fish as food for livestock rather than as food for people is a poor use of a valuable resource in a protein-deficient world. Of the fishes that we do eat, the preference is for species such as flounder, tuna, and halibut, which are three or four steps up the food chain. Every 125 g of tuna requires 1 metric ton of phytoplankton food. If we are to derive greater benefit from the oceans as a food source in the future, we must eat more of the less palatable fishes that are lower in the food chain.

When we examine the food chain in terms of biomass at each level, it is apparent that we can construct **ecologic pyramids** of numbers or of biomass. A pyramid of numbers (Fig. 41-4, *A*), also known as an **Eltonian pyramid** (after the British ecologist Charles Elton, who first devised the scheme), depicts the number of individual organisms that is transferred between each level in the food chain. Although providing a vivid impression of the great difference in numbers of organisms involved in each step of the chain, a pyramid of numbers does not indicate the actual weight of organisms at each level.

More instructive are pyramids of biomass (Fig. 41-4, *B*), which depict the total bulk of organisms. Such pyramids usually slope upward because mass and energy are lost at each transfer. However, in some aquatic ecosystems in which the producers are the algae that have short life spans and rapid turnover rates, the pyramid is inverted. This happens because the algae can tolerate heavy exploitation by the zooplankton

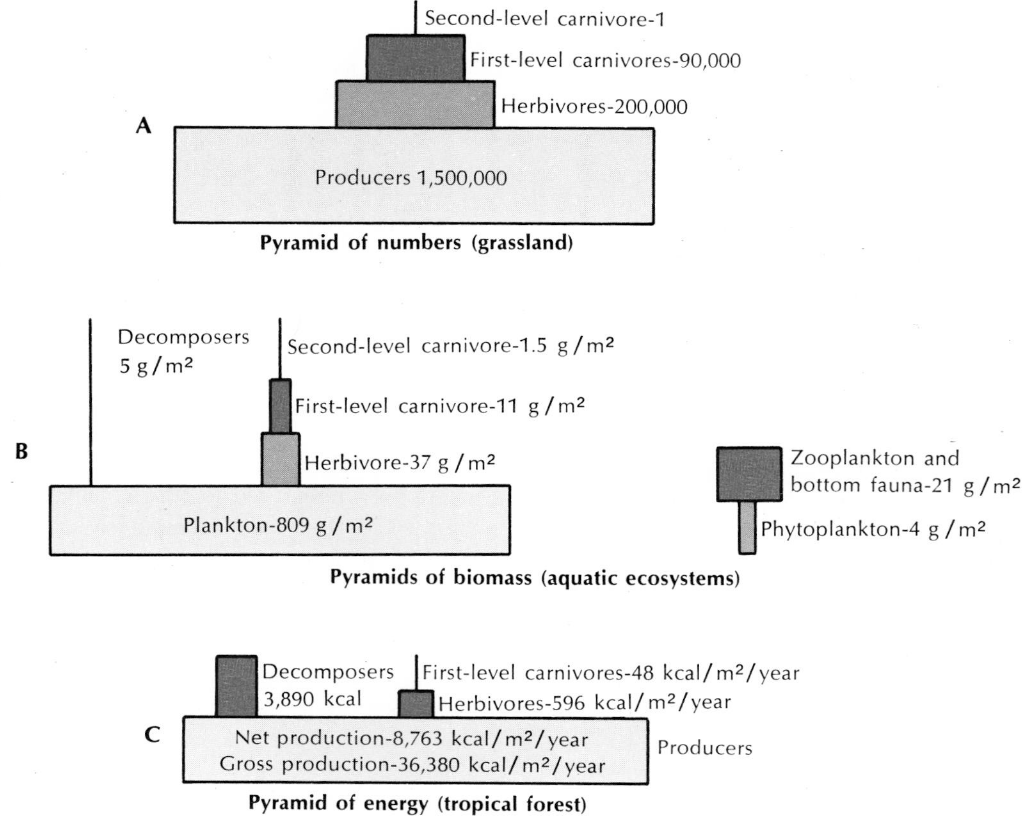

FIG. 41-4 Ecologic pyramids of numbers, biomass, and energy. Pyramids are generalized, since area within each level is not scaled proportionally to quantitative differences in units given. (Redrawn from Smith, R. L. 1974. Ecology and field biology, ed. 2, New York, Harper & Row, Publishers; and Odum, E. P. 1971. Fundamentals of ecology, ed. 3, Philadelphia, W. B. Saunders Co.)

consumers. Therefore the base of the pyramid is smaller than the biomass it supports.

A third type of pyramid is the pyramid of energy, which shows rate of energy flow between levels (Fig. 41-4, *C*). An energy pyramid is never inverted because less energy is transferred from each level than was put into it. A pyramid of energy gives the best overall picture of community structure because each level reveals its true importance in the community regardless of its biomass.

Nutrient cycles

All of the elements essential for life are derived from the environment, where they are present in the air, soil, rocks, and water. When plants and animals die and their bodies decay or when organic substances are burned or oxidized, the elements and inorganic compounds essential for life processes, which we refer to as nutrients, are released and returned to the environment. Decomposers fulfill an essential role in this process by feeding on plant and animal remains and on fecal material. The result is that nutrients flow in a perpetual cycle between the biotic and abiotic components of the biosphere.

Nutrient cycles are often called **biogeochemical cycles** because they involve exchanges between living organisms (bio-) and the rocks, air, and water of the earth's crust (geo-). Geochemistry is the discipline that deals with the chemical composition of the earth and the exchange of elements therein.

Nutrient and energy cycles are closely interrelated since both influence the abundance of organisms in an

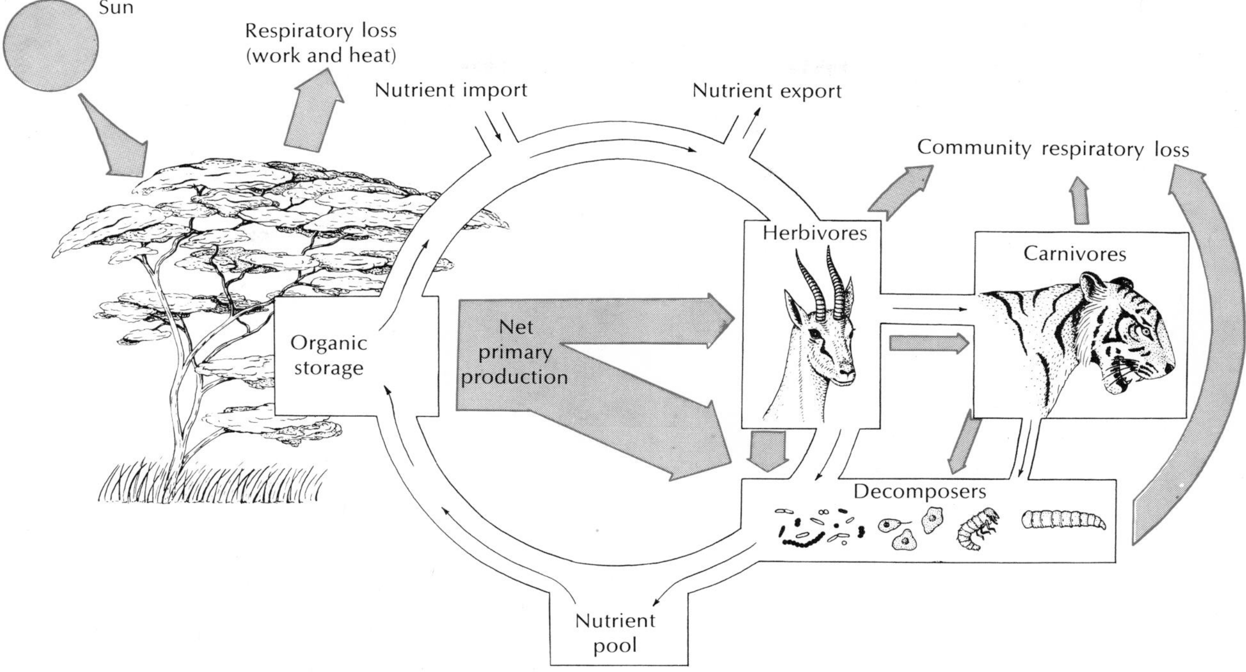

FIG. 41-5 Nutrient cycles and energy flow in a terrestrial ecosystem. Note that nutrients are recycled, whereas energy flow *(red)* is one way. (Modified from Smith, R. L. 1974. Ecology and field biology, ed. 2. New York, Harper & Row, Publishers.)

ecosystem. However, unlike nutrients, which recirculate, energy flow follows one direction; it does not follow a cycle because it is lost as heat as it is used. The continuous input of energy from the sun keeps nutrients flowing and the ecosystem functioning. This interrelationship is depicted in Fig. 41-5. Among the most important biogeochemical cycles are those of carbon and nitrogen.

Carbon cycle

Carbon is the basic constituent of organic compounds and living tissue and is required by plants for the fixation of energy photosynthesis. It is hardly necessary to emphasize the total dependence of life on the availability of this element. Carbon circulates between carbon dioxide (CO_2) gas in the atmosphere and living organisms through assimilation and respiration; it is also withdrawn into long-term reserves of fossil fuel deposits (humus and peat and finally coal and oil) (Fig. 41-6).

The cycling of carbon parallels and is linked to the flow of energy that begins with the fixation of energy during photosynthetic production. Plants synthesize glucose, a 6-carbon compound, from carbon dioxide that is withdrawn from the atmosphere; they then use this sugar to build higher carbohydrates, especially cellulose, the structural carbohydrate of plants. Plants require 1.6 kg of carbon dioxide from the atmosphere for each kilogram of cellulose produced. The concentration of carbon dioxide in the atmosphere today is only 0.03% of air, in contrast to the relative abundance of atmospheric oxygen, 21% of air.

Two aspects of the low concentration of carbon dioxide require emphasis. The first is that the availability of carbon dioxide limits energy fixation by plants. Physiologists have shown that, if the atmospheric carbon dioxide is increased by 10%, plant photosynthesis increases 5% to 8%.

The second point is that there has been a tremendous increase in the consumption of fossil fuels by humans in the last two decades. More than 10 billion tons of carbon dioxide enter the atmosphere each year from human industrial and agricultural activities; of this 75% comes from burning fossil fuels and 25% from

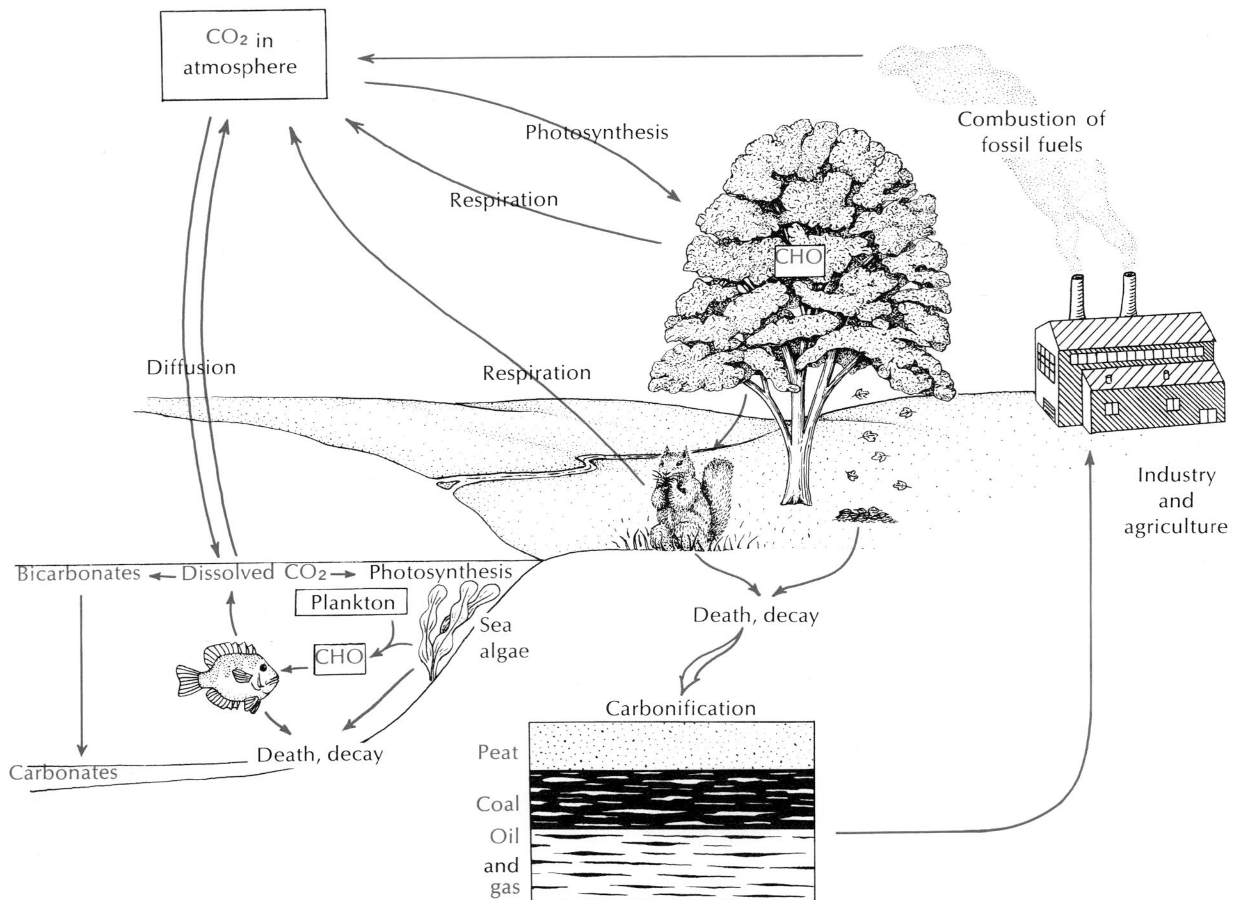

FIG. 41-6 Carbon cycle showing circulation of carbon as CO_2 gas between living components of environment and long-term storage as fossil fuels.

the release of carbon dioxide from the soil because of frequent plowing (a surprising and unappreciated effect of present agricultural practice). However, some authorities argue that the input of carbon dioxide into the atmosphere by human activities is not really chemical pollution, since the increase is removed by photosynthesis or by precipitation as carbonates in the ocean. There is concern, however, that even small increases in carbon dioxide, which traps radiated heat, may increase the temperature of the earth's biosphere.

Nitrogen cycle

Like carbon, nitrogen (N_2) is also a basic and essential constituent of living material, particularly in the amino acids, which are the building blocks of protein. Despite the high concentration of nitrogen in the atmosphere (79% of air), it is almost totally unavailable to life forms in its gaseous state. The most important contribution of the nitrogen cycle is converting nitrogen into a chemical form that living organisms can use. This conversion is called **nitrogen fixation.** Some atmospheric nitrogen is fixed by lightning, which produces ammonia and nitrates that are carried to earth by rain and snow. But at least ten times as much nitrogen is biologically fixed by bacteria and by blue-green algae (Fig. 41-7).

Most important in terrestrial systems are bacteria associated with legumes (members of the pea family) whose nitrogen contribution to soil enrichment is well known. The old agricultural practice of allowing fields

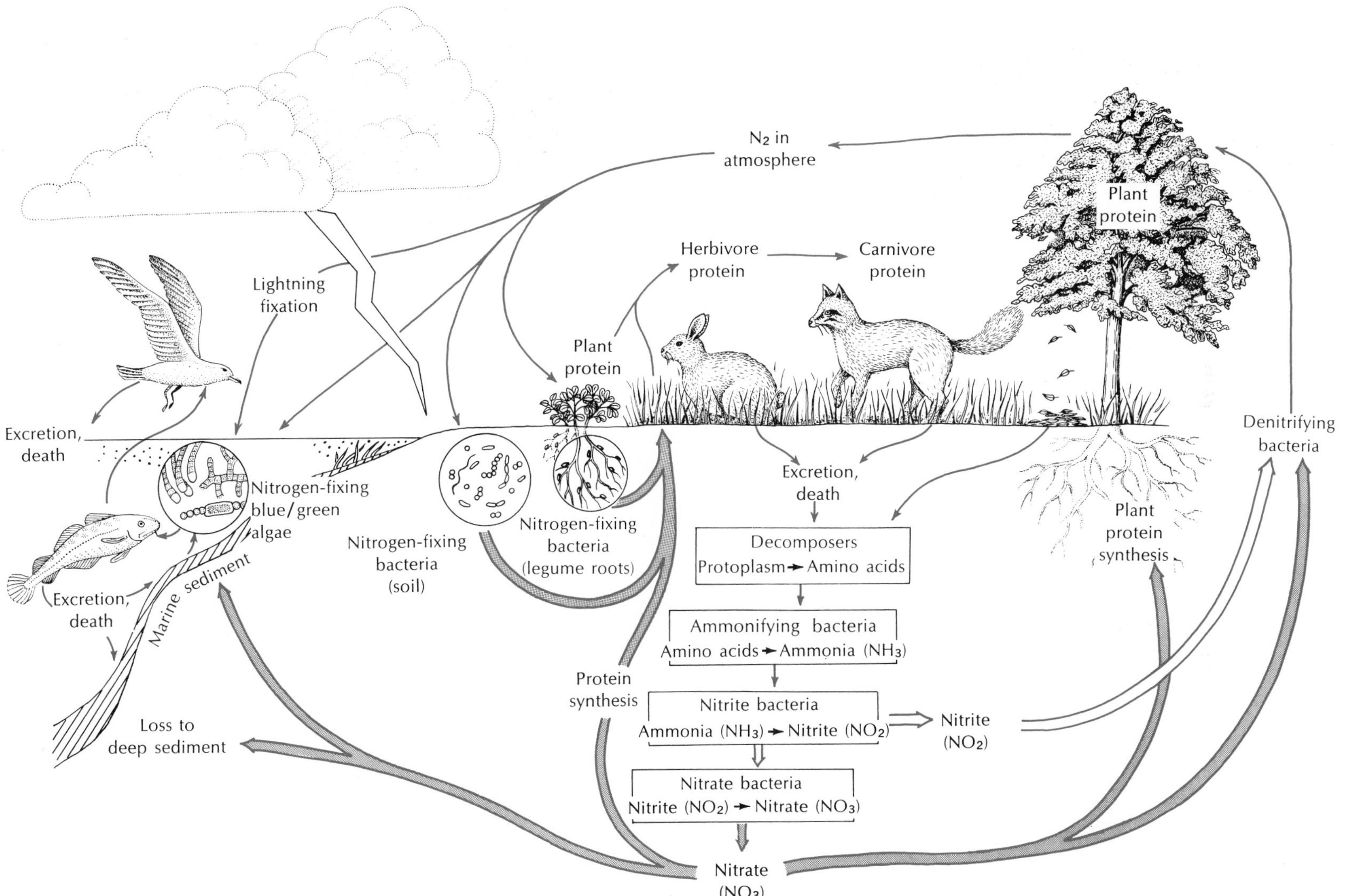

FIG. 41-7 Nitrogen cycle showing circulation of nitrogen between organisms and through environment *(red)*. Microorganisms responsible for key conversions are indicated in circles and boxes.

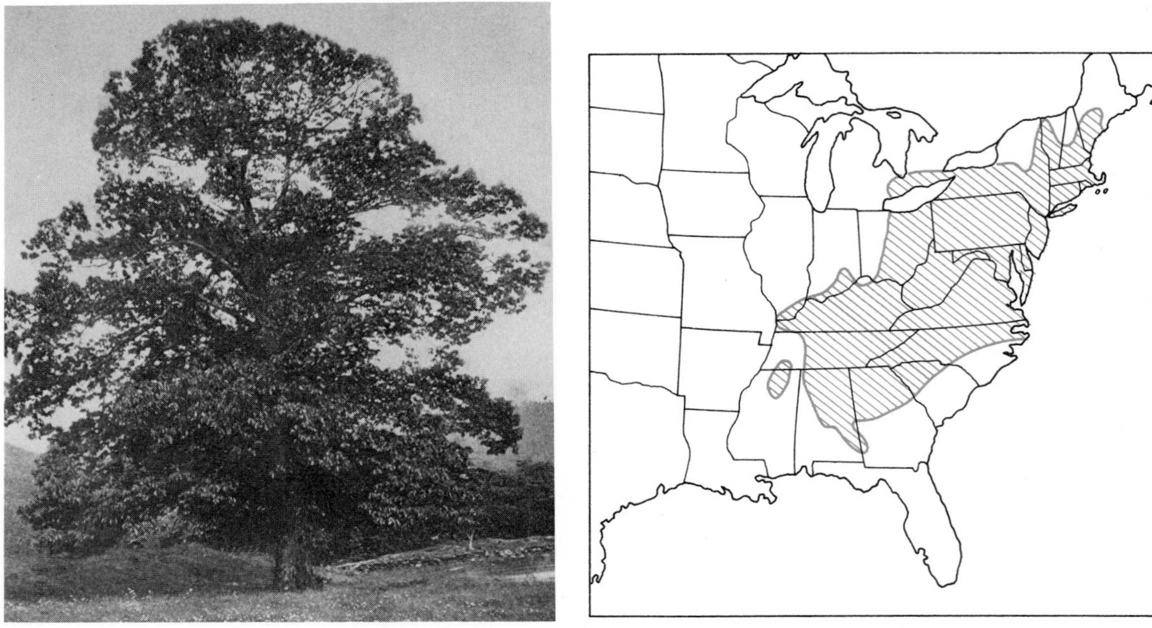

FIG. 41-8 American chestnut tree, *Castanea dentata,* and its original distribution in North America. Photograph taken circa 1900. (Map from Krebs, C. J. 1972. Ecology, New York, Harper & Row, Publishers.)

to lie fallow (without crops) for a season every few years enabled nitrogen fixers to replenish the available nitrogen. Aerobic rhizobia bacteria produce nodules on the roots of legumes in which molecular nitrogen is converted to ammonia and nitrates, which plants can use to build protein. Plant proteins are transferred to consumers that build their own proteins from the amino acids supplied. Plants' and animals' waste products (urea and excreta) and the ultimate decomposition of the organisms themselves provide for the return of organic nitrogen to the substrate. Then decomposers break down proteins, freeing ammonia, which is utilized by bacteria in a series of steps as shown in Fig. 41-7. Finally nitrate (NO_3) is produced, which may be utilized once again by plants or may be degraded to inorganic nitrogen by denitrifying bacteria, and is returned to the atmosphere. Nitrates also are carried by runoff to streams and lakes and eventually to the sea. A cycle similar to this terrestrial cycle occurs in aquatic ecosystems except that there is a steady loss of nitrogen to deep-sea sediments.

The nitrogen cycle is a near-perfect self-regulating cycle in which nitrogen losses in one phase are balanced by nitrogen gains in another. There is little absolute change in nitrogen in the biosphere as a whole. However, human activities have caused steady losses of soil nitrogen by slowing natural addition of organic nitrogen through current agricultural practices and by the harvesting of timber. The latter causes an especially heavy nitrogen outflow, both from timber removal and from soil disturbance.

COMMUNITIES

Communities represent the most tangible concept in ecology. We are all familiar with the differences between forest, grassland, desert, and salt marsh, and we have little trouble picturing, at least in a general way, the kinds of plant and animal communities associated with each of these ecosystems. Communities comprise the *biotic* portion of the ecosystem; each consists of a certain combination of species that forms a functional unit. Although communites are sometimes difficult to define because the assemblages of species within similar communities are not always the same, communities do exist, and they all possess a number of attributes. It is beyond the scope of this book to discuss all the aspects of community form and function. In this section we examine three important principles that operate in community organization: ecologic dominance, ecologic niche, and ecologic succession.

Ecologic dominance

Biologic communities are typically dominated by a single species or limited group of species that determines the nature of the local environment. All other species in that community must adapt to conditions created by the dominants. Communities often are named for the dominant species, for example, black spruce forest, beech-maple forest, oyster community, and coral reef. It is not always easy to specify just what constitutes a dominant species. It may be the most numerous, the largest, or the most productive; or it may in some other manner exert the greatest influence on the rest of the community. In a woodland, a tree obviously means more to the community than a poison ivy plant (even though a casual visitor may carry away a more lasting impression of the poison ivy).

Dominant species achieve their status by occupying space that might otherwise be occupied by other species. When a dominant species is eliminated for some reason, the community changes. The American chestnut once dominated large regions of eastern United States where it comprised more than 40% of the overstory trees in climax deciduous forests. After the invasion in 1900 of the chestnut blight (a fungus), which was apparently brought into New York City on nursery stock from Asia, the chestnut was eliminated from its entire range within a few years (Fig. 41-8). The chestnut's position was replaced by oak, hickory, beech, and red maple, and chestnut-oak forests became oak and oak-hickory forests. Tree squirrel populations that relied on chestnuts for their major food source decreased to a fraction of their former abundance.

The appearance of the forests changed. To the subdominant species, whose specialized environmental requirements were attuned to the particular conditions of a chestnut forest, the change required adjustments that some could not meet. The shift in dominance produced changes throughout the community.

Ecologic niche

An animal's position in the environment is characterized by more than the habitat in which we expect to find it; it also has a "profession," a special role in life that distinguishes it from all other species. This is its **niche,** which Elton defined as the animal's place in the biotic environment, that is, what it does and its relation to its food and to its enemies. The niche concept has now been broadened to include an animal's position in time and space as well. The expression "every animal had its niche" conveys the idea rather well (see Principle 27, p. 15).

It is an accepted rule that no two species can occupy the same niche at the same time. If they did, they would be in direct competition for exactly the same food and space. Should this happen, one species would have to diverge by natural selection into a different niche or face extinction.

Closely related species that live close together are called **sympatric** (meaning literally "same country") species, as opposed to **allopatric** (different country) species that occupy different geographic regions. Sympatric species might be expected to have similar niches and be in danger of direct competition. This can be avoided in a number of ways, as the following examples illustrate.

In parts of eastern and southern Africa the two species of rhinoceros, the black and the white, are sympatric: they share the same habitat. However, the black rhinoceros is a browser feeding on leaves and woody plants, whereas the white rhinoceros is a grazer eating grasses and herbs. They do not compete for the same food and consequently occupy distinct ecologic niches.

The three species of boobies of the Galápagos Islands—the white, blue-footed, and red-footed—offer a second example. All three are plunge divers that feed on the same kinds of marine fishes frequenting the ocean around the islands. The blue-footed booby, however, always fishes close to the shore; the white booby flies a mile or two from land to fish; and the red-footed booby makes long hunting forays many miles from shore. Again, although sympatric, they do not compete for the same food. The three species have divided up the food resources so that a different portion of the sea becomes the undisputed hunting territory of each species.

The niche concept is a fundamental one to the biologist for it explains why animals avoid endless struggles with other animals. A species and its niche are reflections of the same thing: a unique way of life. A species is master of its own niche and thus is not in direct competition with similar species, even though competition was used to decide the boundaries of the

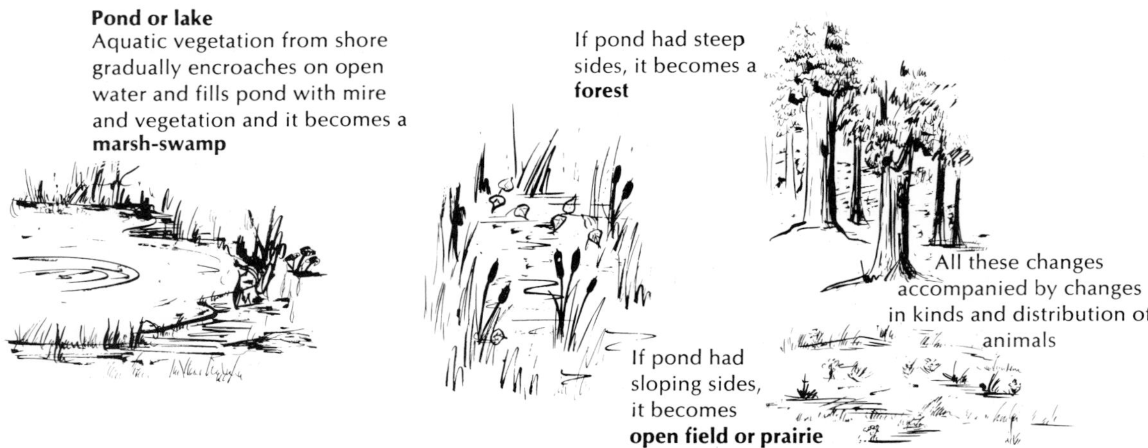

FIG. 41-9 Ecologic succession in a pond or lake.

niche in the first place. This concept was discussed in Chapter 38 in which we considered how speciation occurs.

Ecologic succession

Communities are not static but are continually changing, a process called **ecologic succession.** Succession is the change in biotic and abiotic composition of a site. Over a period of years a site would have a series of communities, each succeeding, and being succeeded by, other communities as the site ages. Succession begins on a site following a geologic or other disturbance (for example, recession of a glacier, erosion, fire, flood, logging, or abandonment of cultivated ground). Perhaps abandoned cropland is the most familiar example of succession. Grasses and weeds quickly appear and are succeeded by bushes such as sumac, locust, blackberries, and hawthorns. Later oak, maple, hickory, and pine trees dominate the abandoned field.

One community is succeeded by another until a fairly stable end product is attained. Such a sequence of communities is called a **sere** and involves early pioneer communities, transient communities, and finally a **climax community,** which is more or less balanced with its environment.

A small lake begins as a clear body of water with a sandy bottom and shores more or less free from vegetation. As soil is washed into the lake by the surrounding streams, mud and vegetable muck gradually replace the sandy bottom. Vegetation grows up along the sides of the lake and begins a slow migration into the lake, resulting in a bog or marsh. The first plant life is aquatic or semiaquatic, consisting of filamentous algae on the surface and later of rooted plants, such as *Elodea,* bulrushes, and cattails. As the water recedes and the shore becomes firm, the marshy plants are succeeded by shrubs and trees, such as alders and larches and, later, beeches and maples. Eventually the lake may be replaced by a forest, especially if its sides and slopes are steep; if the sides have gentle slopes, a grassy region may replace the site of the lake. The terminal forest or grassland is a climax community (Fig. 41-9).

In its beginning a lake may contain fish that use the gravelly or sandy bottoms for spawning. When the bottoms are replaced by muck, these fishes will be replaced by others that spawn in aquatic vegetation. Eventually, no fishes may be able to live in the habitat; but other forms, such as snails, crayfish, many kinds of insects, and birds, are able to live in the swampy, boggy community. As the community becomes a forest or grassland, there will be other successions of animal life.

In general, communities tend to go from a state of instability to one of stability, or climax. The term ''stability,'' however, must be used in a relative sense, for changes are inevitable.

POPULATIONS

As defined earlier, a population is a group of organisms belonging to the same species that share a particular space. Whether the population is gray squirrels or deer in an eastern woods or bluegill sunfish in a farm fishpond, it bears a number of attributes unique to the group. A population shares a common gene pool; it has a certain density, birth rate, death rate, age ratio, and reproductive potential; and it grows and differentiates much like the individual organism of which it is composed.

Natural balance within populations

One characteristic of populations that we take for granted is their basic stability. We expect to see approximately the same number of robins and starlings on our lawns each year. Fireflies, nighthawks, and whippoorwills are an accepted and appreciated background of summer evenings; they always return each year in approximately the same numbers, and their predictability is comforting. Of course, there are good years and bad years for game animals, and the buzz of cicadas is more pervasive during some summers than during others, but these fluctuations are seldom violent. The populations of plants and animals around us remain in balance, provided that we do not interfere.

Yet with few exceptions animals have reproductive potentials far beyond that required for replacement of their numbers. Insects lay thousands of eggs, field mice can produce as many as 17 litters of four to seven young each year, and a single female codfish may spawn 6 million eggs per season. It has been said that the descendants of a pair of flies, if unchecked, would weigh more than the earth in a few years. The potential for rapid growth of a population, the so-called biotic potential rate, is large but never proceeds unchecked indefinitely because of the limitations of the environment. What keeps populations in check? How is the "balance of nature" explained?

The growth of a population in a limited space and with a limited input of nutrients is described by a **population growth curve** that is sigmoid in shape (Fig. 41-10). In the beginning, with ample space and abundant food, the starting population breeds and grows as fast as its reproductive potential allows. The curve turns steeply upward as more and more animals join the reproductive population; the increase is exponential. But, as space gets crowded and food begins to vanish, the curve straightens out and then flattens as the upper population limit is reached. The reason why the rate of growth of the population becomes zero is, in a word, competition. Growth is suppressed by competition for space and competition for the limited input of food.

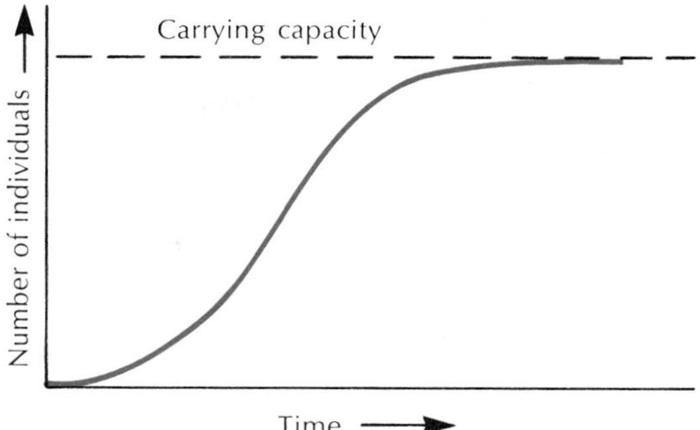

FIG. 41-10 Population growth curve.

The simple sigmoid growth curve is representative of many plant and animal populations, both in nature and in the laboratory. It can be easily reproduced in the laboratory with cultures of yeast or paramecia—indeed, with any of a great variety of microorganisms or animals. In nature the curve represents the annual plankton blooms of ponds and lakes, eruptions of insect pests, and growth of weeds in old fields. When populations first approach their growth limit, they tend to overshoot into a temporarily unlimited environment and then decline and oscillate about some lower limit. The level of final stabilization, about which the population fluctuates, is the **carrying capacity** of the environment.

For example, when sheep were introduced on the island of Tasmania in approximately 1800, their growth was represented by a sigmoid curve with a small overshoot, followed by mild oscillations around a final population size of 1,700,000 sheep (Fig. 41-11, *A*). A similar pattern but with larger fluctuations was recorded for a population of ring-necked pheasants introduced on an island in Ontario, Canada (Fig. 41-11, *B*). Population overshoots are especially characteristic of populations of higher animals because their life histories are long and complicated and because, after the population already has consumed the resources of the environment, reproduction continues for a time.

The growth of human population was slow for a very long period of time. For most of our evolutionary history, humans have been hunter-gatherers, who depended on and were limited by the natural productivity of the environment. With the development of agriculture, the carrying capacity of the environment increased, and the population grew steadily from 5 million around 8000 B.C., when agriculture was introduced, to 16 million around 4000 B.C. Despite terrible famines, disease, and war, which took their toll, the population reached 500 million by 1650. With the coming of the Industrial Revolution in Europe and England in the eighteenth century, followed by a medical revolution, discovery of new lands for colonization, and better agricultural practices, the carrying capacity of the earth for people increased dramatically. The population doubled to 1 billion around 1850. It doubled again to 2 billion by 1930 and to 4 billion by 1976. Thus the growth has been exponential and remains high (Fig. 41-11, *C*).

However, recent surveys provide hope that the world population growth is slackening. In just 5 years (1970 to 1975) the annual growth rate decreased from 1.9% to 1.64%. At 1.64%, it will take 42 years for the world population to double rather than 36 years at the higher annual growth rate figure. The decrease is credited to better family planning programs, but increasing deaths from starvation also are a contributing factor.

Unlike other animals that cannot change and increase the carrying capacity of their environment, we are unique in finding ways. Unfortunately, when the carrying capacity is increased, we respond as would any animal by increasing our population. Because the earth's resources are finite, the time will inevitably come when the carrying capacity can be extended no further. The question is whether we will be able to anticipate this limit and check our population growth in time or whether we will overshoot the resources and experience a catastrophic decline as do most other higher animals introduced into a new environment. Indeed, for millions of Third World people time has already run out. The rapid growth of population is not something that thinking people view with equanimity.

How population growth is curbed

What determines the number of animals in a natural population? For a laboratory culture of animals the answer seems fairly clear. They grow until they reach the carrying capacity of the environment, at which point growth is suppressed by competition for the limited resources of space and food. Thus the forces that limit growth arise from within the population; they are **density-dependent** mechanisms caused by crowding.

Food and space limitations are obviously density-dependent factors and it is not difficult to understand how these same limitations, as well as other natural hazards, could act on wild populations in a density-dependent way. As crowding increases, the quest for food becomes increasingly critical. The competition that results may affect the population in either of two ways. All members may scramble for what is left until the resource is subdivided into so many small parts that none of the population gets enough for self-maintenance. Mass starvation follows with only the strongest or luckiest individuals surviving. This **scramble type** of competition, which is typical of certain herbivore populations, is wasteful and produces large oscillations in population density.

Or competition may be of the **contest type.** Socially dominant individuals, or those with defended territories denied to others, claim portions of the food resource for themselves and prevent the unsuccessful from having any part of it. These unfortunates starve or in their weakened condition become victims of disease or predation, but the remainder stay healthy. Population numbers remain high and wild oscillations in density are prevented.

Crowding brings forth other agencies of death that operate to keep populations in check. High density encourages **disease** in both humans and animals. Occasionally epidemics sweep through crowded populations as did the terrible bubonic plague (Black Death) through the crowded, dirty cities of Europe in the fourteenth century. **Predation** increases when prey become abundant and easy to catch. **Shelter** for suitable nest sites and as refuge from bad weather becomes less available with crowding.

The continuous competition for food and space to live brings forth yet another density-dependent force: **stress.** Although the physiology of stress is not thoroughly understood, there is ample evidence that when certain natural populations become crowded, such as populations of lemmings, voles, and snowshoe hares, a neuroendocrine imbalance involving the pituitary and adrenal glands appears. Growth is suppressed. Reproduction fails, and individuals become irritable and aggressive. Under such conditions snowshoe hares suffer from a lethal "shock disease" that results in a population crash.

Overcrowding may also cause **emigration** away from the birth area. Overcrowded mice increase their locomotor activity and begin to explore new areas. In the case of lemmings, overpopulation produces the famous mass "marches" recorded at intervals in Scandi-

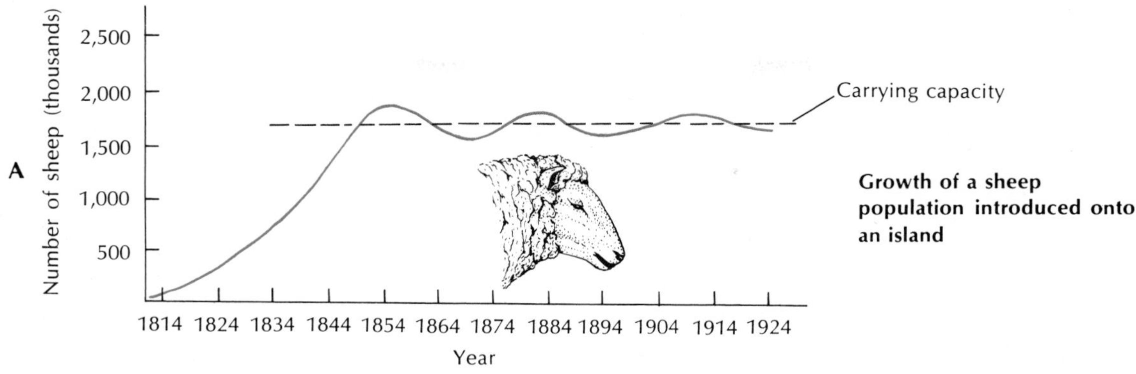

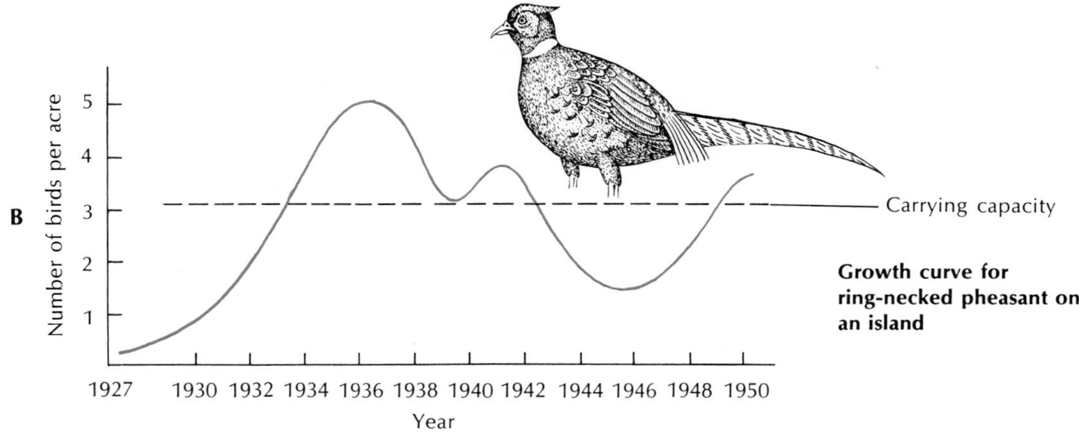

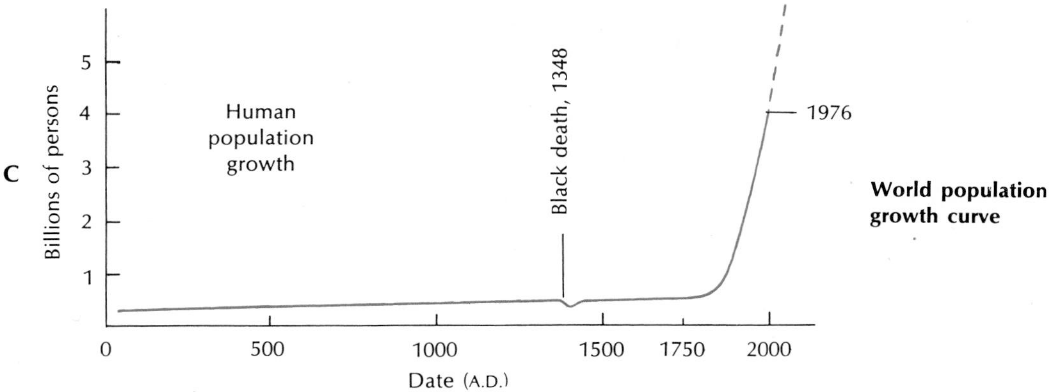

FIG. 41-11 Growth curves for sheep, ring-necked pheasant, and human populations.

navia. When songbird populations become overcrowded the less successful become ''vagabond'' birds that are forced out into less preferred habitats. In fact, emigration is one major force resulting in the colonization of new habitats. The rapid dispersal of the European starling throughout the United States and southern Canada, following its introduction into New York City in 1890, is an excellent example (p. 600).

Not all of the forces that limit growth are density-dependent. Extreme and rapid changes in the weather, or unusually cold, hot, wet, or dry weather, is a **density-independent** hazard of varying severity to animal populations. Local insect populations are sometimes pushed to the point of extinction by a severe winter. Hurricanes and volcanic eruptions may destroy entire populations. Hailstorms have been known to kill most of the young of wading bird populations. Prairie grass and forest fires kill everything unable to escape.

Vertebrate populations are more resistant than invertebrate populations to catastrophic weather changes but are influenced nonetheless. Populations of Virginia white-tailed deer in the northern parts of its range are much more affected by unusually severe winters than by hunting pressure. One study of deer populations over a period of 70 years in the Adirondack Mountains of New York revealed that during heavy snow years fawn mortality was exceptionally high. Heavy snow covered low-growing food and made travel so difficult that more energy was consumed in searching for scarce food than was gained in eating it. During three successive severe winters (1969 to 1971) five out of six fawns died, and the deer population dropped to one-third of its former density.

Thus population numbers are influenced by factors generated within the population and by forces from without, and no single mechanism can explain how population growth is curbed. The evidence, however, indicates that most natural populations are commonly controlled by density-dependent forces. Humans are the greatest force of all. By altering animal habitats we change the balances that set the old limits to animal numbers. Our activities can increase or exterminate whole populations of animals.

Annotated references

Selected general readings

Allee, W. C., A. E. Emerson, O. Park, T. Park, and K. P. Schmidt. 1949. Principles of animal ecology. Philadelphia, W. B. Saunders Co. *A valuable comprehensive text and reference, although dated.*

Andrewartha, H. G. 1970. Introduction to the study of animal populations. Chicago, University of Chicago Press. *A summary of the classic 1954 text by Andrewartha and Birch.*

Andrewartha, H. G., and L. C. Birch. 1954. The distribution and abundance of animals. Chicago, The University of Chicago Press. *A stimulating and in many respects pioneering book in population ecology. Though some concepts presented are now considered controversial, the book remains a valuable source of information.*

Bates, M. 1960. The forest and the sea. New York, Random House, Inc. *Excellent general approach to ecology. Highly recommended.*

Boughey, A. S. 1971. Fundamental ecology. Scranton, Pa., Intext Educational Publishers. *Elementary text of ecologic principles.*

Boughey, A. S. 1971. Man and the environment. New York, The Macmillan Co. *Well-written combination of ecology, human evolution, and pollution.*

Clapham, W. B. 1973. Natural ecosystems. New York, The Macmillan Co. *The ecosystem and how people have altered the environment.*

Colinvaux, P. A. 1973. Introduction to ecology. New York, John Wiley & Sons, Inc. *Highly readable and interesting approach to general ecology. Stresses population ecology.*

Ehrlich, P. R., and A. H. Ehrlich. 1972. Population, resources, environment, ed. 2. San Francisco, W. H. Freeman & Co., Publishers. *Well-written source book on the subjects in title.*

Elton, C. E. 1942. Voles, mice and lemmings. Oxford, Clarendon Press. *A classic on animal populations.*

Fuller, W. A., and J. C. Holmes. 1972. The life of the far north. New York, McGraw-Hill Book Co. *Interesting, beautifully illustrated, authoritative treatment of the Arctic. Other excellent books in the* Our Living World of Nature *series published by McGraw-Hill include forest, African plains, and seashore.*

Henderson, L. J. 1913. The fitness of the environment (reprint). Boston, Beacon Press. *A classic that should still be read.*

Kendeigh, S. C. 1974. Ecology with special reference to animals and man. Englewood Cliffs, N.J., Prentice-Hall, Inc. *A sound presentation of basic principles, with emphasis on animal ecology.*

Krebs, C. J. 1978. Ecology: experimental analysis of distribution and abundance, ed. 2, New York, Harper & Row, Publishers. *Important treatment of population ecology.*

Leopold, A. 1949. A Sand County almanac. New York, Oxford University Press, Inc. *A beautiful account by a great conservationist; now a favorite of the ecology movement.*

MacArthur, R. H. 1972. Geographical ecology. New York, Harper & Row, Publishers. *One of the most important and influential books of recent years.*

Miller, G. T., Jr. 1971. Energy, kinetics and life. Belmont, Calif., Wadsworth Publishing Co. Inc. *Fine general ecology text.*

Odum, E. P. 1971. Fundamentals of ecology, ed. 3. Philadelphia, W. B. Saunders Co. *One of the most popular general ecology texts.*

Phillipson, J. 1966. Ecological energetics. New York, St. Martin's Press, Inc. *Brief, clear treatment of the subject.*

Scientific American. 1970. The biosphere. San Francisco, W. H. Freeman & Co., Publishers. *Excellent collection of articles on the ecosystem.*

Smith, R. L. 1974. Ecology and field biology. New York, Harper & Row, Publishers. *Comprehensive, clearly written, well-illustrated general ecology text. An excellent summary follows each chapter.*

Selected *Scientific American* articles

Bell, R. H. V. 1971. A grazing ecosystem in the Serengeti. **225:**86-93 (July). *The migration of grazers across Tanzania is synchronized with the availability of specific tissues of grasses.*

Bolin, B. 1970. The carbon cycle. **223:**124-132 (Sept.).

Bormann, F. H., and G. E. Likens. 1970. The nutrient cycles of an ecosystem. **223:**92-101 (Oct.).

Brill, W. J. 1977. Biological nitrogen fixation. **236:**68-81 (Mar.). *Only a few bacteria and algae can fix atmospheric nitrogen into ammonia. A hypothetical fixation sequence and its control is described.*

Cloud, P., and A. Gibor, 1970. The oxygen cycle. **223:**110-123 (Sept.).

Cole, LaMont C. 1958. The ecosphere. **198:**83-92 (Apr.).

Deevey, E. S., Jr. 1970. Mineral cycles. **223:**148-158 (Sept.).

Delwiche, C. C. 1970. The nitrogen cycle. **223:**136-146 (Sept.).

Gates, D. M. 1971. The flow of energy in the biosphere. **224:**88-100 (Sept.). *Plants capture only a fraction of the solar energy that falls on the earth, but this fraction maintains all life.*

Goze, J. R., R. T. Holmes, G. E. Likens, and F. H. Bormann. 1978. The flow of energy in a forest ecosystem. **238:**92-102 (Mar.).

Janick, J., C. H. Noller, and C. L. Rhykerd. 1976. The cycles of plant and animal nutrition. **235:**74-86 (Sept.). *Energy and nutrients for human consumption are processed by chains of organisms.*

Myers, J. H., and C. J. Krebs. 1974. Population cycles in rodents. **230:**38-46 (June). *The cyclic rise and fall in rodent populations are associated with genetic changes in the population.*

Peixoto, J. P., and M. A. Kettoni. 1973. The control of the water cycle. **228:**46-61 (Apr.).

Penman, H. L. 1970. The water cycle. **223:**98-108 (Sept.).

Richards, P. W. 1973. The tropical rain forest. **229:**58-67 (Dec.). *One of the oldest ecosystems is giving way to the activities of humans.*

Wynne-Edwards, V. C. 1964. Population control in animals. **211:**68-74 (Aug.).

CHAPTER 42

ANIMALS IN THE HUMAN ENVIRONMENT

"The still hunt," by James Henry Moser. This painting was commissioned for the Smithsonian Institution and was first displayed at the 1888 Cincinnati Exhibition as part of a crusade to save the bison from extinction. Although vast herds like this no longer existed in 1888—the bison was already an endangered species—many who had crossed the prairies in the first half of the nineteenth century attested to the painting's authenticity. Market hunters such as the one portrayed not infrequently shot more than 100 bison from a single position before moving on.

Courtesy of the Mammoth Museum, Yellowstone National Park.

Prologue

In 1846, 23-year-old Francis Parkman, guided by a desire to study American Indians and haunted by images of a western wilderness yet unseen, began a journey westward on the Oregon Trail. Near the Platte River in what is now Nebraska, he described the prairie scene that unfolded before him while riding back to camp after a furious buffalo chase.*

> The face of the country was dotted far and wide with countless hundreds of buffalo. They trooped along in files and columns, bulls, cows, and calves, on the green faces of the declivities in front. They scrambled away over the hills to the right and left; and far off, the pale swells in the extreme distance were dotted with innumerable specks. Sometimes I surprised shaggy old bulls grazing alone, or sleeping behind the ridges I ascended. They would leap up at my approach, stare stupidly at me through their tangled manes, and then gallop heavily away. The antelope were very numerous, and as they are always bold when in the neighborhood of buffalo, they would approach to look at me, gaze intently with their great round eyes, then suddenly leap aside, and stretch lightly away over the prairie, as swiftly as a race horse. Squalid, ruffian-like wolves sneaked through the hollows and sandy ravines. Several times I passed through villages of prairie dogs, who sat, each at the mouth of his burrow, holding his paws before him in a supplicating attitude, and yelping away most vehemently, whisking his little tail with every squeaking cry he uttered. Prairie dogs are not fastidious in their choice of companions; various long checkered snakes were sunning themselves in the midst of the village, and demure little gray owls, with a large white ring around each eye, were perched side by side with the rightful inhabitants. The prairie teemed with life.

When the first British settlements were founded in Virginia and Massachusetts, there could not have been less than 60 million bison, or buffalo *(Bison bison),* roaming the American wilderness. The two recognized subspecies, *B.b. bison,* the plains bison, and the somewhat larger and darker wood bison, *B.b. athabascae,* extended from northwestern Canada to Mexico and from the Rocky Mountains to the Atlantic seaboard from New York to Florida (Fig. 42-1). On the great plains, where their domain was most secure, the bison made enormous annual migrations in vast herds numbering millions of animals. In the spring they moved northward to graze on new grass that had sprung up behind the retreating snow; in autumn they returned southward. Late eighteenth- and early nineteenth-century observers recorded moving herds of bison covering areas of several square miles so densely that nowhere was the ground itself visible. Because they were nomadic, bison did not overgraze the land despite their incredible numbers. There was also ample herbage to feed the numerous antelopes and prairie dogs observed by Parkman. The bison population itself easily sustained losses inflicted from predation by their natural enemies—the wolves, grizzly bears, and pumas—and even the hunting pressure by Indians had very little effect on their numbers.

To the Plains Indians the bison were providers of virtually every necessity of life. The high-quality meat was eaten fresh or was dried, pounded, and stored as pemmican for future use. The skins provided homes, clothing, moccasins, and warm robes. The rawhide was turned into saddles, ropes, and quivers; and the sinews became bowstrings. Spoons and other utensils were carved out of the horns; hooves were reduced to glue; and the ribs became scrapers and sled runners.

The bison culture of the Plains Indians reached full flower in the seventeenth and eighteenth centuries after the Indians acquired horses, which had been introduced to North America by the Spaniards at the end of the sixteenth century. With their new mobility, hunting became easier, and the tribes could follow the bison migrations, carrying all their belonging with them. When rifles were introduced to the Indians by white traders along the Missouri, Platte, and Arkansas Rivers, hunting pressure increased substantially, but even this made few inroads on the total bison population because the Indian population was low.

In eastern America the rapid westward advance of pioneers was accompanied by decimation of the bison. By 1810 no bison remained east of the Mississippi. Their extinction clearly portended the ultimate fate of the plains bison. In 1846, when Parkman traveled the Oregon Trail, the bison were restricted to the western plains and mountains, although there were still probably 30 million left. The final act in the great bison slaughter began in the mid-nineteenth century. Congress in 1864 authorized the building of the Union Pacific Railroad across the heart of the continent, directly through the bison range. Thousands of hunters moved over the plains, killing bison to feed the huge construction gangs. As the railroad penetrated westward, "buffalo" robes and tongues, the latter considered a delicacy in the East, were collected and shipped

*Parkman, F. 1931. The Oregon Trail. New York, Farrar and Rinehart.

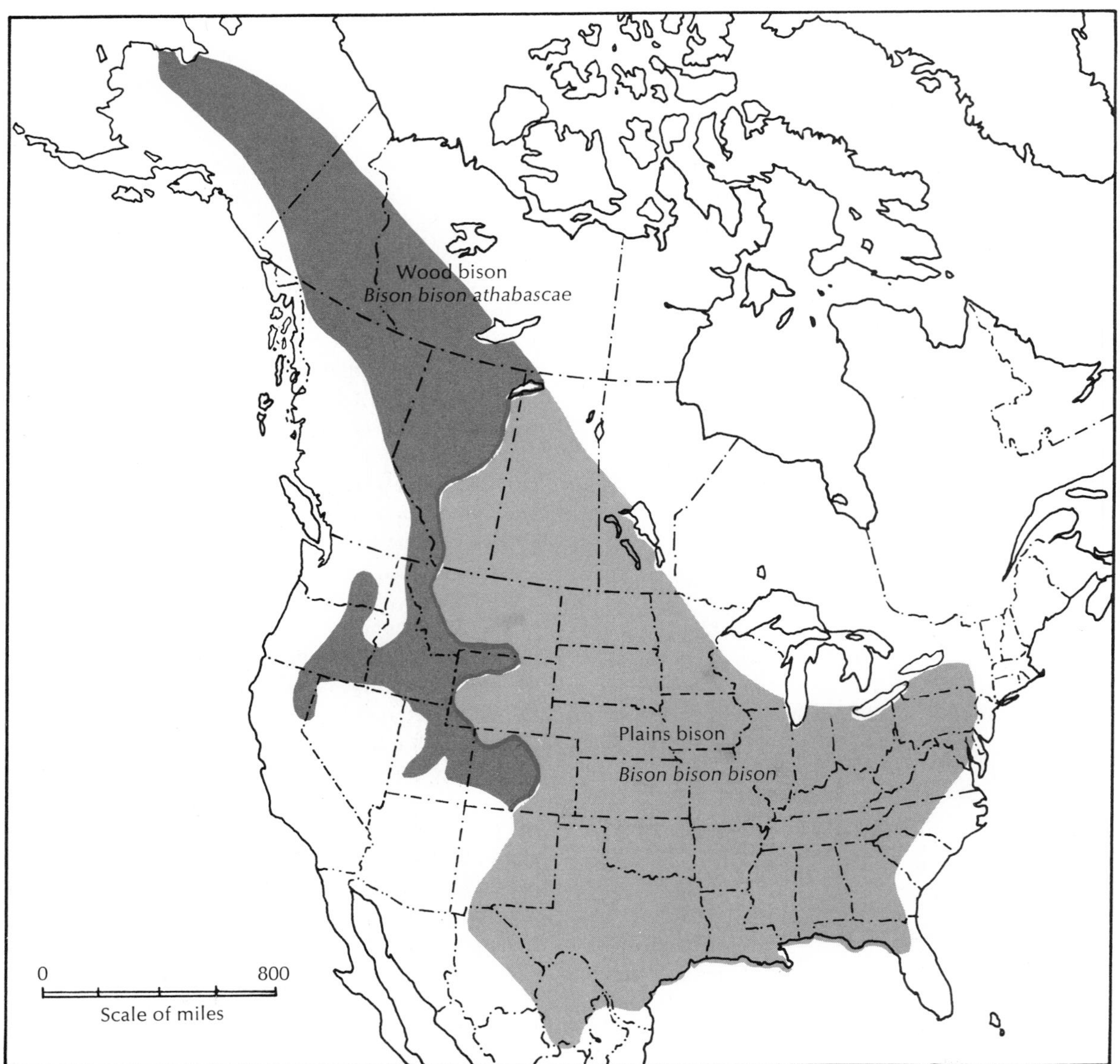

FIG. 42-1 Distribution of American bison before the arrival of Europeans to North America. (From Hall, E. R., and K. R. Kelson. 1959. The mammals of North America. New York, The Ronald Press Co.)

east by rail. The carcasses were left to rot on the ground. Hunting pressure reached its zenith in the early 1870s when some 4 million bison were killed each year.

Contributing to the slaughter were thousands of soldiers instructed to kill bison in order to starve out the Plains Indians. By 1883 the annihilation was virtually complete, except for a single herd of about 10,000 animals in North Dakota. In September of that year, hunters were sent to North Dakota, and by posting all available water holes, they destroyed the last free herd of bison (Fig. 42-2).

Those who protested the destruction were ignored by politicians, who believed it to be a justifiable means to deprive the Plains Indians of a way of life. Fortunately, small groups of semidomesticated bison still existed on

FIG. 42-2 Huge stacks of skulls and other bones representing approximately 25,000 bison await shipment beside a Saskatchewan railway siding in 1890. For a brief period following the bison slaughter, bones collected from the prairies were shipped east to be ground into fertilizer. (From Hewitt, C. G. 1921. The conservation of wildlife of Canada. New York, Scribner's.)

a few private estates. A census taken in 1889 estimated the total remnant population throughout the United States and Canada at 541 animals. Following the senseless North Dakota slaughter of 1883, strong voices of protest from a few influential individuals aroused public interest in the bisons' plight. In Yellowstone National Park, a herd of 20 bison that were constantly threatened by poachers were given full protection in 1890. Some farmers succeeded in building up small breeding herds from remnant groups they had been protecting.

By the turn of the century the bison population was beginning to recover. The American Bison Society, formed in 1905, received both federal government backing and powerful private support in its mission to reestablish the American bison as a viable species secure from extinction. The National Bison Range was established in Montana in 1908, and other reserves in both the United States and Canada followed. Shortly after the turn of the century, a large herd of wood bison was discovered near Lake Athabasca in western Canada. The 17,300-square-mile Wood Buffalo National Park in Alberta and Northwest Territories was established in 1922. By 1967, there were 23,000 bison in North America, including plains bison, wood bison, hybrids of the two subspecies, and zoo captives. Most exist as wards of humans.

The destruction of the American bison was a tragic episode in American history that becomes only slightly more comprehensible when viewed in the context of the exploitive spirit of the times. Within a few decades 60 million large animals of a single species were destroyed, a holocaust of unprecedented magnitude. The herds were vast, probably the largest aggregation of land animals on earth, and few persons could foresee their fate. The same mentality resulted in the annihilation of the passenger pigeon. Fortunately, unlike the passenger pigeon, the bison were able to flourish in captivity and could recover from remnant populations. It was inevitable that the bison would disappear from the plains, since roaming herds of half-ton animals were incompatible with a West given over to agriculture. We cannot deny that the human species must exploit nature to live, but the destruction of the bison was cataclysmic in scope, senseless in direction, and erosive of human values. Their fate is a sobering commentary on the turbulent relationship between wildlife and advancing human civilization.

It is of interest to compare the great prairies that "teemed with life" when Francis Parkman saw them in 1846 with their appearance at the turn of the century just a few years after the arrival of homesteaders. Bison were replaced by longhorn cattle and sheep, and the ranges were overstocked. Sheep, with their close-

FIG. 42-3 The "black blizzards" of the 1930s moved millions of tons of soil off the prairies, darkening the skies to the Atlantic seaboard. Thousands of "dusted out" families streamed westward to California, a migration later portrayed in John Steinbeck's novel *The Grapes of Wrath.* This photograph was taken in 1936 in Gregory County, South Dakota. (Courtesy United States Department of Agriculture, Soil Conservation Service, Washington, D.C.)

cropping teeth, were especially destructive; within a few years, the highly palatable prairie grasses, legumes, and forbs disappeared. These species were replaced by wheat grasses, bromegrasses, and annual weeds—rapidly converting the prairie into low-grade pasture. With continued grazing, mesquite appeared, and the topsoil thinned. Boomtime "sod-busting" for growing wheat accelerated soil deterioration. No longer bound by grass roots, the soil was lifted by wind and washed away by rain. The American naturalist and conservationist William Temple Hornaby wrote in 1908 of the Montana prairie, where "the awful sheep herds have gone over it, like swarms of locusts, and now the earth looks scalped and bald and lifeless. Today it is almost as bare of cattle as of buffaloes, and it will be years in recovering from the fatal passage of sheep."

Conditions were ideal for the creation of the Dust Bowl, requiring only an extended drought, which occurred in the 1930s. From Texas to Canada the "black blizzards" swept across an exhausted land, joining economic depression in bringing misery to prairie farmers (Fig. 42-3). A single storm on May 12, 1934 lifted an estimated 300 million tons of soil from Texas, Oklahoma, Colorado, and Kansas and deposited it eastward in amounts up to 100 tons per square mile.

Today the American prairie is a changed land. Soil conservation practices introduced in the late 1930s eventually transformed the prairies into the most productive agricultural region of the world, dominated by a monoculture of cereal grains (Fig. 42-4). Except for pests, few animals exist in the uniformly cultivated fields, but certain areas are still important refuges for wildlife. One of these is the broad Platte River valley in Nebraska, through which Francis Parkman traveled in 1846. In south-central Nebraska, where the Platte River curves eastward through an enormous bordering marshy area, migratory birds are attracted in great numbers: hundreds of thousands of ducks, countless songbirds, and most of the mid-continent population of

FIG. 42-4 Harvesting wheat in the North American grain belt. Sound soil conservation practices instituted after the Dust Bowl have transformed the prairies into the world's most productive cropland. Today, the average North American farmer, using irrigation, fertilizers, machines, and a considerable input of fossil fuel energy, produces enough to feed himself and 50 others. (Photograph taken near Regina, Saskatchewan, by L. L. Rue, III.).

sandhill cranes and white-fronted geese. Over the years, however, the marshy wetlands have been gradually drained for cropland until only 20% of the original area remains for wildlife.

The Platte River itself is seriously threatened by an unremitting demand for water for irrigation and other human needs. The yearly flow today is only 16% of what it was before the Nebraska Territory became a state. If the additional water developments that are planned for the Platte River are all carried out, the river will be dry within 35 years. Attempts by the U.S. Fish and Wildlife Service to set aside a portion of the wetlands as a permanent refuge for wildlife have thus far been defeated, largely by suspicious landowners who fear they will be displaced from their land. Meanwhile, cattle raisers, farmers, and industry vigorously compete for the remaining water, a dispute in which the wildlife has no voice.

Little remains of the vast climax grasslands that occupied the heart of North America only 150 years ago. No other ecosystem on this continent has been as thoroughly transformed from its natural state. It emphasizes our capacity as humans to reshape the face of nature and manipulate the environment to suit our own needs. This capacity is both the outcome of our biologic success as a species and the origin of our ecologic problems.

THE ENVIRONMENT AND HUMAN EVOLUTION

Humans as hunters

Hunting is an ancient human activity. For most of the period that true humans (genus *Homo*) have walked the earth, their survival depended on the success of the hunt. Only in the last evolutionary instant of human existence on earth have humans of necessity replaced hunting with work. However, the ancient hunting urge, basically a male pursuit, is abundantly manifested today in big-game hunting, fox-hunting, game-bird hunting, angling, and numerous other expressions of sport hunting that have survived cultural suppression.

While humans today stand fully revealed as predators, we have our origins in an apelike lineage called the dryopithecines, which thrived on a diet of fruit rather than meat. Human evolution, treated in Chapter 39, was a sequence of fundamental changes in anatomy, behavior, and intelligence from which humans emerged as hunters. Our early dryopithecine ancestors were creatures of the forest. They were probably arboreal, but were capable of adapting in varying degrees to foraging on the fringes of the forest and in patches of open terrain within the forest. Earlier we indicated that in the late Miocene and early Pliocene epochs, climatic changes over the earth produced major environmental transformations in North America and Eurasia. Tropical and subtropical forests were replaced by temperate forests and great expanses of grasslands. At this time the semiterrestrial ground-feeding *Ramapithecus* appeared in the hominid lineage and moved into the forest-grassland edges as well as onto the expanding savanna.

The dentition of *Ramapithecus* differed in several important respects from its pongid contemporaries. The early hominid incisors were relatively small and unsuited for defense or threat displays. The sides of the molar crowns were more nearly parallel than are those of any pongids. The proportions of the lower and upper jaws and the muscles of mastication indicated an adaptation for grinding and chewing small hard objects such as nuts, seeds, stems, roots, and grains. Furthermore, the rapid wear patterns of the teeth of fossil hominids suggest such a vegetarian diet.

During the Pliocene epoch the grasslands continued their expansion. Hominids of this period gradually left the dwindling forests altogether to become creatures of the open terrain. As they did so, they consumed more meat and fewer plant foods because plant foods were much scarcer on the savanna than in the forests. Manual dexterity and excellent eye-hand coordination, together with superior intelligence and upright posture, were advantages that enabled the hominid lineage to surpass the rest of the wildlife. They became increasingly dependent on animal food, ranging from insects, tortoises, lizards, and rodents to carrion abandoned by predators on the savanna.

We can speculate that carrion feeding and the first use of tools—sharpened pebbles used to tear away parts of hoofed animals killed by predators—appeared together. The effectiveness of such tools were demonstrated by the late Louis Leakey, who skinned and butchered an antelope in less than 20 minutes using a pebble tool found at Olduvai. Weapons, too, doubtless were needed, since lions and hyenas are not expected to relinquish their kills voluntarily. Hominids were now aggressors. Using tools and weapons of their own invention, they became hunters, fully equipped for a predatory existence.

These changes occurred slowly, since some 10 million years separate *Ramapithecus*, the first true hominid, from *Homo erectus*, the first indisputable big-game hunter. The beginnings of social organization, as well as the use of tools, appeared among the Pleistocene australopithecines that lived in open habitats 2 to 3 million years B.P. A new social order, especially the family unit, developed to maximize food-foraging potentials. Division of labor between the sexes was required: men ranged widely to hunt game and collect kills abandoned by lions, while women cared for the young, gathered plant foods, and guarded the home site. Family coherence was strengthened by the length of human childhood and associated necessary parental care. The use of language, which presumably existed in at least rudimentary form some 3 million years ago, also reinforced the family bond and facilitated intergroup communication. With individuals united into families and families into tribes and clans, cooperative hunting behavior was employed, making early humans far more formidable adversaries of wildlife. The stage was readied for humans to use their intelligence to exploit a new way of life on the open savanna. Gradually, almost imperceptibly, early hominids acquired Biblical dominion "over the fish of the sea, and over the fowl of the air, and over the cattle, and over all the earth, and over every creeping thing that creepeth upon the earth" (Genesis 1).

There is no evidence that the small-brained australopithecines used any but the smallest stone tools. Nor had they discovered the use of fire. The large-brained *Homo erectus*, who shared much of the Pleistocene epoch with the australopithecines, took a much larger step toward environmental conquest. Wooden clubs and spears entered the arsenal of hunting weapons, and there is evidence that bolas were employed about 400,000 years ago to capture antelopes. Even more significant, *Homo erectus* learned the use of fire, perhaps as early as 800,000 years ago. With fire they warmed themselves, smoked animals from caves, lighted the darkness to ward off real and imagined terrors of the night, and cooked and tenderized their meat. There is also evidence that Pleistocene humans set fire to the vegetation to concentrate their prey, perhaps stampeding them over cliffs or into swamps where they would be trapped. They also must have discovered that the new growth that sprang up following grass and forest fires attracted wildlife and eased the task of finding it. Humans were thus beginning to modify the environment for their own ends. They were also conducting cooperative big-game drives that were having an increasing impact on wildlife.

Modern humans' immediate predecessors, the Neanderthals that appeared about 100,000 years ago, left a relatively rich array of fossils, artifacts, and evidences of cultural activities. Already the world's most formidable predators, they depended entirely on their skill as hunters for their survival. Mass slaughters were not uncommon. At one site at the foot of a high rock cliff in southern France were discovered the bones of at least 40,000 wild horses in a deposit up to 245 cm (8 feet) thick. These were presumably stampeded over the cliff by Neanderthal hunters who set grass fires. The refinement of mass harvesting techniques of this kind coincided with a human population explosion that brought the total world population to perhaps 5 million about 20,000 years ago. Large mammal herds could not sustain their numbers against mass slaughter techniques employed by nomadic peoples who followed the herds wherever they went. Animal populations went into serious decline. Magic hunting rituals and polychrome cave art of primitive hunters are thought to be attempts to restore animals to their former abundance as well as to achieve spiritual domination over their quarry.

For countless millennia humans had drawn all their sustenance from the animal kingdom. If the hunter outwitted the quarry, the family could eat; if he failed, they starved. Animals that shared the hunter's world occupied his thoughts and plans and were the center of all his activities. Although he killed the animals ruthlessly, he nevertheless respected them, felt a kinship with them, and believed himself closely related to them. But as successful hunters, humans were putting themselves out of the hunting business.

Impact of agriculture

Agriculture, the domestication and cultivation of plants, especially cereal crops, arose independently in the Near East and Central America as the hunter-gatherer economies began to fail. The first cultivated cereals, perhaps wheat and barley, were probably selected from natural hybrids of wild species. Sites along the Nile Valley in Egypt provide evidence that milled grain was used for food as early as 15,000 years ago. The domestication of animals soon followed. Sheep were domesticated in Iraq at least 11,000 years ago and evidences of domesticated goats were discovered in 10,000-year-old Palestinian sites. Cattle (8,000 B.P.) and pigs (5,000 to 6,000 B.P.) soon followed. Cattle became perhaps the earliest beasts of burden, used for pulling carts, turning water wheels, and carrying people, as well as providing milk and meat.

The advent of agriculture was nothing less than a revolution. No subsequent human event—not the great plagues nor the Industrial Revolution nor the devastating wars of humankind—has had a greater permanent impact on the course of human history. The hunter-gatherer economy that had persisted for more than 2 millions years was brought rapidly to a close, although certain hunting populations persisted, some to the present century (such as Australian aborigines, pygmies of the Congo rain forest, and North American Eskimos). A nomadic economy was replaced by more or less permanent villages and communities centered in agricultural areas. Civilization, defined as a state of social culture that is organized around cities and that practices agriculture, was born. The most important derivative of the new agricultural technology was a vast increase in the carrying capacity of the human environment. A given area of agricultural land could support hundreds of times more people than it could during the early hunter economy. The human population began to soar. Agricultural production continued to diversify, and more and more land was turned to the production of crops and domesticated animals.

As limits to growth were reached, new lands were conquered and exploited. The human species now emerged as a powerful reshaper of the earth. Forests were cut or burned, irrigation systems were built, hillsides were terraced, and grasslands were converted to barren wastelands by overgrazing. Landscapes have suffered the greatest ravages where civilization has existed the longest. The vegetation of the Mediterranean was largely destroyed 2,500 years ago. Most of the pastureland of Spain, Italy, Greece, and Yugoslavia was severely damaged by erosion, which was caused by grazing of domestic animals, especially goats. Before the birth of Christ large sections of northern Africa and the Middle East were so heavily overcropped that they became deserts or wastelands, largely abandoned by both humans and animals. Great civilizations that once flourished in Mesopotamia (Iraq), Palestine, Greece, Italy, and China fell victim to exploitive overgrazing.

Despite the calamities that had befallen the Mediterranean arena, until about 300 years ago most of the earth's biosphere was virtually unmarked by human activity. In the Americas, the forests, lakes, prairies, and rivers lay wild and natural and rich in native fauna and flora. Most of northern Europe and Asia were dominated by virgin forests, and the soil remained untouched. The environments of Australia, New Zealand, the East Indies, and the Pacific islands were yet to feel the violent impact of civilization's arrival. The seas were clean and abounded in marine life undisturbed by human beings.

This was the scene that the Swedish zoologist-conservationist K. Curray-Lindahl calls "the last idyll." It was a heritage of incalculable richness for future generations. At that point, 300 years ago, history provided abundant warnings of the bitter harvest to be reaped from unbridled human exploitation of nature. Tragically, they were lessons unheeded. Human population was about to begin an unprecedented growth. Stimulated by the Industrial Revolution, the discovery of abundant fossil fuels for energy, the colonization of new lands, improved agricultural technology, and medical advances, the human population doubled between the years 1650 and 1850, reaching 1 billion people. It doubled again to 2 billion people by 1930; in 1976 it stood at 4 billion people. Now, with each passing day, the problem of feeding so many people intensifies. The population of the human species has reached plague proportions.

Today very little of the earth's biosphere resembles its original pristine state. Numerous animal species have been destroyed by human activities in the last three centuries and many others are threatened with extinction. The present ecologic crisis is, writes the American biologist W. H. Murdy, a "crisis in human evolution."* We must, he argues, participate in our own evolution, using our knowledge to ensure our survival and the enhancement of human cultural values in an environment of which all living things are a part. We alone have the enormous responsibility of preserving that which is not yet destroyed by countering the destructive forces within us. Our options are rapidly diminishing.

*Murdy, W. H. 1975. Anthropocentrism: a modern version. Science **187:**1168-72.

WILDLIFE AND PEOPLE
Pitting wildlife against progress

In June 1978, the United States Supreme Court ruled that construction on the $120 million Tillico Dam nearing completion in Tennessee must be stopped to save from certain extinction an endangered species of fish, the snail darter. The snail darter, discovered only in 1973, requires shallow fast-flowing water for spawning, a critical habitat that could be destroyed by the dam's reservoir.

The ruling did nothing to dissipate polarization of public opinion that had already developed over the issue. Proponents of the dam maintained that it would provide badly needed power, improve downstream flood control, and create hundreds of miles of shoreline for recreation. Tennessee already has 85 to 90 species of darters of interest only to ichthyologists, say proponents, and one species would never be missed. Their viewpoint was summed up by Senator Jake Garn of Utah, who stated, "I do not believe that any animal, no matter how worthless, ought to be allowed to halt any project, no matter how valuable."

Preservationists, rather than defending the snail darter's right to survive, questioned the necessity of the dam itself. They argued that the 16,000 acres of productive cropland to be inundated was of much greater value than the benefits of the dam. As with all dams, the life of this one would not exceed a few decades because of silting within the reservoir, whereas the inundated land would be lost forever. Furthermore, 200 historic and archeologic sites, including the old Cherokee village of Tenasi, from which the state takes its name, would be lost. Finally, opponents pointed out, the flood control benefits would cut the flood crest by only 2 inches at Chattanooga, the nearest city. Their viewpoint was perhaps best expressed by U.S. Secretary of the Interior Cecil D. Andrus, who said, "Common sense dictates that we shouldn't dam every river, stream, or creek in America. Let me suggest that we may have developed the best hydroelectric sites, that we have built some of the best reclamation projects, and that having done this, the law of diminishing returns requires that we proceed with every caution."

The emotionally charged snail darter–Tillico dam issue clearly brought into focus the continuing confrontation between human beings and their environment—a conflict that has no easy resolution and from which there can emerge no winners. It pits wildlife against "progress," intangible, long-term benefits of nature against the tangible, short-term benefits of public-works projects, principles against practicality. That a small fish could quite literally bring to a halt an enormously expensive construction project astounded Congress, delighted preservationists, and left the average citizen incredulous.

The source of the snail darter's reprieve from extinction, however temporary it may turn out to be, was the Endangered Species Act, a surprisingly powerful piece of legislation passed by Congress in 1966 and expanded and reinforced in 1973. Congress declared in 1966 that "one of the unfortunate consequences of growth and development . . . has been the extermination of some native fish and wildlife . . . and serious losses in other species of native wild animals." It pledged through the Act to conserve and protect native fish and wildlife threatened with extinction. The most significant section of the act prohibits federal agencies from jeopardizing endangered species or habitats that have been designated as critical. This legislation was a landmark in wildlife protection because it was the first time that a United States domestic law was not enacted mainly to promote human, rather than wildlife, interests. In the past, wildlife reserves or seasons closed to hunting, trapping, and fishing were usually established to perpetuate animals for sport, future market use, or some other nonaltruistic motive.

Extinction

The Endangered Species Act grew out of rapidly rising public concern for the destructive effects of civilization on wildlife. One measure of human impact is the extinction rate of animals. Since 1600, when reasonably accurate record-keeping began, approximately 225 species of vertebrate animals are known to have become extinct (the true extinction total is far higher, but includes numerous invertebrates and tropical animals unknown to biologists). The rate of extinction has increased significantly between 1600 and 1900: 21 extinctions in the seventeenth century, 36 in the eighteenth century, and 84 in the nineteenth century. Toward the end of the nineteenth century the number of extinctions increased alarmingly, and although conservation efforts instituted after 1920 stabilized the rate somewhat, 85 more species had disappeared forever by 1974. Furthermore, numerous animal species are today quite literally on the verge of extinction: these are the approximately 175 critically endangered species and subspecies recognized by the U.S. Fish and Wildlife Service and by the International Union for the Conservation of Nature and Natural Resources.*

In North America, the rapid transition from wilderness to a highly industrialized society has had an especially severe impact on native wildlife. Sixteen mammals, eight birds, and nine fishes are known to have vanished in the continental United States alone. If animals from Hawaii and Puerto Rico are included, the list rises to 62 birds, mammals, and fishes that have become extinct since 1600, most of them in the twentieth century. The victims include, among mammals, the great sea mink (last seen in 1860), Merriam's elk (1906), and six subspecies of North American wolf. Extinct birds include the great auk (1833), the Labrador duck (1875), the passenger pigeon (1914), the Carolina parakeet (1914), and the heath hen (1932).

Extinction is, of course, a natural event in organic evolution, as we have discussed earlier in this book (especially in Chapter 38). Sometimes individual species disappear gradually as their populations wane in competition with better-adapted forms: at other times whole taxa become extinct within a relatively brief geologic period, as for example the demise of all the dinosaurs at the end of the Cretaceous period (70 million years B.P.) or the concurrent extinction of many large mammals and ground-nesting birds at the end of the Pleistocene epoch of the Cenozoic era (11,000 years B.P.). Extinctions may be associated with global or regional climatic changes that adversely affect vegetation and habitat or with the invasion of exotic animals that outcompete the endemic fauna; or, in the case

*The International Union for the Conservation of Nature and Natural Resources of Lausanne, Switzerland periodically publishes a "Red Data Book" listing threatened wildlife. The recognized categories are (1) critically endangered species that are in immediate danger of extinction, such as the California condor, the whooping crane, and the blue whale, (2) vulnerable or threatened species that are still abundant but are depleted and are under threat, for example, the grizzly bear, the California bighorn, and the prairie falcon, (3) rare species, which comprise small populations and, though not threatened, are subject to risk, for example the Galápagos tortoise and the Komodo dragon. This and other such lists published by federal agencies of different nations are unfortunately replete with statistical confusion arising from differences of opinion as to when species or subspecies should be added or removed from the lists, and to which category each belongs.

of recent extinctions, they may have been caused by human predation.

One theory of the massive Pleistocene extinction that by no means has been discredited proposes that early populations of *Homo sapiens,* using weapons, fire, and ingenious hunting techniques, exterminated vulnerable large mammals and ground-nesting birds. Within a relatively brief period, giant bison, wild horses, long-necked camels, ground sloths, mammoths, and mastodons disappeared from the American plains. In Eurasia, mammoths, wooly rhinoceroses, Irish elk, and cave bears became extinct. It has been possible to trace the path of early human immigrations into Africa and Asia, the Americas, Australia, and New Zealand and to note an associated drastic reduction in certain groups of Pleistocene fauna. Nevertheless, humans were still relatively rare creatures during the Pleistocene epoch. Their weapons were primitive. Many question the capacity of early humans to exterminate big game animals, when eighteenth- and nineteenth-century American Indians, using horses and firearms, had little noticeable effect on the populations of plains bison.

Although we cannot fully assess the impact of early humans on animal extinction, the spread of civilization across the earth has unmistakably been linked to the demise of many animal species. In historic times, waves of extinction attended the arrival of explorers and colonizers to each new continent, island, or archipelago. At first, most extinctions were caused by direct hunting, usually for food but often for commercial products such as furs and ornamental feathers. The development of the musket was responsible for the sharp rise in bird extinctions during the late eighteenth and early nineteenth centuries. Later, various indirect causes of extinction assumed prominence: habitat destruction, exotic animal introductions, and pest and predator control.

Causes of extinction today

Habitat alteration

The destruction of animal habitat is today the major threat to survival of presently endangered species. When habitats are destroyed—whether by logging, drainage of marshes, damming of rivers, urban development, or agriculture—animals must move and adapt to other habitats or die. Although it has been difficult to educate the public about the importance of habitat preservation—many still believe that conservation is primarily the protection of animals against wanton killing—nearly everyone is aware that the destruction of an ecosystem by strip-mining, clean-cut logging, or subdivision development adversely affects animals.

Habitat encroachment by humans often has subtle and far-flung effects that are not so readily apparent. Development and land exploitation more frequently than not cause habitat fragmentation. As patches of habitat become smaller and more isolated from each other, many animals that require large home ranges find themselves restricted to areas that are too small for their food, shelter, and reproductive requirements. Animals in isolated habitats of restricted size become much more vulnerable to nearby human activities, such as pesticide use, air pollution, noise, and human intrusion. Not infrequently the exploitation of a small area may have widespread ramifications to wildlife in undisturbed areas. A pond or stream may be the only source of water to wildlife for miles around; if it is destroyed by pollution or diverted for irrigation, large numbers of animals are adversely affected.

Migratory animals are especially vulnerable to habitat alteration. The habitat of ducks, geese, and other migratory waterfowl embraces the northern breeding grounds, southern wintering grounds, and all the areas they visit in between. All portions are necessary for their survival. Migratory fishes, such as salmon and steelhead trout, if barred from their spawning grounds by a dam or polluted stretch of river, will be destroyed just as effectively as if they were decimated by overfishing.

Yet another dimension of habitat alteration is successional changes that follow the clearing or burning of a mature forest. Species requiring stable, mature forest wilderness (such as grizzly bears, wolverines, bighorn sheep, caribou, and ivory-billed woodpeckers) decrease in numbers following a severe forest fire; whereas the luxurious new plant growth that follows a fire, fertilized by nutrients in the ash, encourages successional species such as deer, elk, rabbits, grouse, and quail (Fig. 42-5). Not all forest fires are bad, despite the admonitions of Smokey the Bear. Periodic forest fires stimulate forest diversity, recirculate nutrients, and control diseases that tend to flourish in overmature forest stands. Some species such as Kirtland's warbler depend on fires for their survival. This endangered species requires 8- to 20-year-old jack pines for nesting. Jack pine cones remain tightly sealed by resins, opening only during a fire to release their seeds. Fire sup-

FIG. 42-5 Regrowth after a forest fire provides excellent habitat for game species such as deer, elk, bear, grouse, and quail. Controlled fires are now considered an important forest management tool for removing diseased trees and opening land to new growth. (Photograph of Tillamook Burn in Oregon, by L. L. Rue, III.)

pression in the Kirtland warbler's Michigan habitat resulted in uniform stands of trees too old for the birds' narrow habitat requirement. In 1964, when the cause of the warbler's demise was discovered, the United States Forest Service instituted periodic controlled burns to encourage the species' recovery.

The rapid destruction of the tropical forests has alarming implications for both humans and animals. Some 40% of the tropics of the world has been destroyed during the past 150 years, and virtually all the rest will be cleared by the year 2000. Some 400 million people make their living by slash-and-burn techniques, growing crops for a year or two, then moving on. Within 25 years, when the human population is expected to have at least doubled in size, there will be no tropical forest left. Enormous economic disruption and widespread starvation of truly frightening proportions seem inevitable. Most tropical countries, beset as they are by overwhelming human problems arising from explosive population growth, have neither the resources nor the motivation to study and protect the natural communities on which they depend. The immense biologic wealth of the American tropics will be largely destroyed before it is ever appreciated or understood by humans. There are probably more than a million unnamed plants and animals in the tropics; many, perhaps most, will become extinct within the next 25 years.

Hunting

Until relatively recently uncontrolled hunting for meat and oil and for commercial animal products (furs, hides, ornamental feathers, horns, and tusks) was the most serious menace to wildlife. Historically, the marine mammals, especially the fur-bearers (fur seals, harp seals, and sea otters) were subjected to egregious

FIG. 42-6 Starving white-tailed deer. Hunting is necessary to crop deer populations that otherwise quickly proliferate beyond the capacity of the land to support them in winter. Humans have replaced natural predators. (Photograph by L. L. Rue, III.)

mass slaughters wherever they were discovered around the globe. The sea otter suffered 170 years of unrelenting persecution along northern Pacific coasts until, in 1910, it was considered extinct. Belatedly, the animal was granted international protection, a move that encountered no objections from the fur hunters who found no profit in hunting for any remnant colonies that might still exist. Fortunately, some had survived and were first sighted in 1938 off Bixby Creek on the California coast. Other colonies were soon discovered in Alaska and elsewhere. Slowly, under strict protection, the species is recovering.

The northern fur seal was nearly exterminated by a century and a half of sealing on the Pribilof Islands by Russian, American, Canadian, and British sealers. In 1870 the United States attempted to limit harvesting, but Canadian and British sealers continued indiscriminate and wasteful shooting in the open sea. Finally in 1911, the United States arranged an international treaty that stopped open-sea hunting. From a low of about 125,000 seals, the population has regained its former size of more than 2 million animals. Presently, some 30,000 fur seals are collected annually by licensed sealers, operating in a rational, well-managed industry. The baby–harp-seal industry along the coast of Labrador and in the Gulf of St. Lawrence is also well managed, but the bloody method of harvesting has sparked an outcry from animal lovers and protectionists that may well bring an end to the persecution. It must be remembered that seal harvesting is supported by human demand for fur coats, not by the sealers' putative desire for butchery.

The vigorous public denunciation of harp-seal hunting stems in part from the ''Bambi factor''—our separation of animals into ''good'' and ''bad'' based on human illusions of imaginary animal characteristics. Baby harp seals, with their big, dark, liquid eyes, look lovable and innocent; it seems inconceivable that any-

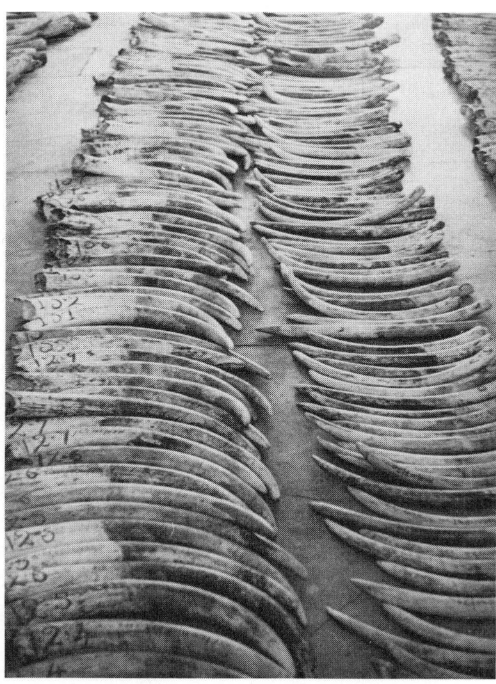

FIG. 42-7 Elephant tusks taken from poachers by Kenya conservation authorities. These are only a few from an enormous warehouse in Mombasa containing thousands of huge tusks, representing, of course, the slaughter of thousands of elephants. (Photograph by F. M. Hickman.)

one could club them to death by the thousands and skin them for their fur—or, for that matter, any other reason. Yet snakes, coyotes, and weasles, which are just as innocent (though perhaps not so lovable), seldom benefit from public outcries over their persecution. Nevertheless, as long as an irrational public is concerned only about furry creatures with warm brown eyes, conservation organizations will use them as a selling point. The conservation movement, however, is firmly committed to the preservation of the entire pyramid of life.

Perhaps the saddest and least comprehensible slaughter of animal life today is the exploitation of the great whales. It is a true war of extermination, and the two major whaling nations, the Soviet Union and Japan, are fully aware of it. Despite rapidly growing international pressure, which has placed the whaling industry on the defensive, it appears that whaling will cease only when whale populations are driven so low that hunting becomes unprofitable. The whaling fleets will disappear, their crews will disperse, and substitutes will be discovered for the raw materials derived from whales. Then, if it is not too late, the whales may recover to their former magnificence.

Present-day game hunting and angling are, as we mentioned earlier, atavistic displays of an ancient hunting urge. ''There is in the passion for hunting,'' wrote Charles Dickens, ''something deeply implanted in the human breast.'' Sport hunting is a well-managed industry in the United States and Canada, probably the best in the world. None of the 74 bird species and 35 mammal species hunted legally in the United States is either threatened or endangered. Nevertheless, much of the urban public associates hunting with the wanton and repugnant killing of beautiful forest creatures. The confrontation between the nation's 21 million hunters and the numerous antihunting organizations is not new, but the polarization of viewpoints is more intense than ever before. However one views hunting emotionally, the morality of killing animals and the arguments about cruelty are side issues to conservation, often obscuring more important issues.

Two inescapable facts are that (1) hunting interests in North America have been a principal force in acquiring land for wildlife refuges and migratory bird sanctuaries and (2) until humans are prepared to reestablish wolves, coyotes, mountain lions, lynxes, and other natural predators into animal habitats, controlled hunting is necessary to manage game populations (Fig. 42-6). It should be added, however, that because hunters killed the carnivores and raptors in the first place (because these feed on the hunter's game), hunters themselves are responsible for the present predator-prey imbalance. It is unlikely that the hunter-nonhunter controversy will disappear. As long as hunters continue to shatter the peacefulness of autumn, endanger hikers, and annoy farmers, the American public's tolerance for hunting will continue to decline.

In other parts of the world, game hunting still is a major threat to numerous species such as Bengal tigers, leopards, cheetahs, rhinoceroses, crocodiles, and elephants (Fig. 42-7). Animal poaching in game preserves is increasingly prevalent as the rising human population of Africa and Asia places enormous strains on available food supplies. How long can national parks and game preserves be protected for the enjoyment of tourists, while natives adjacent to the preserves starve to death? Here, as elsewhere, it is human problems that conflict with wildlife.

Feral domestic and introduced animals

Humans have introduced numerous species into new habitats all over the world, where they affect the indigenous fauna by preying on them (cats, dogs, mongooses) or competing with them for food and habitat (rabbits, rats, mice, cattle, goats, sheep, pigs). Many introductions are deliberate, others are accidental, and both are facilitated by the mobility of humans. In 1929 a noted tropical entomologist, while on a rice ship from Trinidad to Manila, amused himself by listing every kind of animal on board, from cockroaches and fleas to rats and pet animals. He found no less than 41 species, mostly insects. When he opened his luggage in Manila, out walked several beetles.

Of course, not every animal introduced to a new land manages to establish itself, but of those that do, nearly all change the balance of the ecosystem in some way. Many have inflicted an enormous cost to agriculture alone—the fire ant, the boll weavil, and the gypsy moth are just a few latter-day examples in the United States—and are responsible for much, if not most, of the heavy use of pesticides on crops and forests. The demise of both the American elm and the chestnut (p. 957) was caused by accidental introduction of diseases carried by insect vectors.

The natural balance of island communities is especially vulnerable to damage by introduced animals. In the absence of natural checks that keep their populations regulated on the continent, introduced species often flourish at the expense of the native island fauna, many of which have evolved in the absence of mammalian predators and consequently lack protective avoidance behavior. Flightless birds are especially vulnerable. A classic example is the dodo, a very large flightless member of the pigeon family, that inhabited the island of Mauritius. Its fate was sealed by the first Dutch colonists who arrived in 1644 with pigs and dogs. Both the colonists and the introduced animals preyed on the dodos, annihilating the last of them in 1693. The dodo thus bears the distinction of being the first extinct animal whose demise could be fully documented.

The disfigurement of the native fauna and flora of the Galápagos Islands by introduced animals was mentioned earlier (p. 880). No archipelago has suffered worse than the Hawaiian Islands, where the whole of the native lowland forest community has been destroyed by introduced plants and animals. Sixteen species of birds became extinct between 1837 and 1945, half of them members of the extraordinary honeycreeper family. Only 14 species of honeycreepers survive; nearly all are endangered. Although some of the original forest remains in Hawaii, especially in upland habitats where cultivation is impractical, all is overrun by feral domestic animals, rats, and a host of other introduced vertebrate and invertebrate pests. It is but a matter of time before what remains of the marvelous indigenous fauna and flora of Hawaii has either vanished forever or is disfigured beyond recognition.

The unique Pleistocene fauna of New Zealand that still flourished when the Polynesians arrived in the ninth century A.D. was dominated by large ratite (flightless) birds (21 species of moas and four species of kiwis) and had a rich representation of other birds, reptiles, and amphibians. The only land mammals were two bats and perhaps a rodent. Today, largely as a result of the depredations of introduced animals, all 27 moa species are extinct. Three kiwi species still survive under strict protection. Nearly one-third of the original fauna has vanished in a thousand years of human habitation. Approximately 60 species of birds and mammals have been introduced (Fig. 42-8), of which about half have become problem animals. While conservation programs are vigorously supported in New Zealand, it will never be possible to reconstitute the original fauna and flora.

Predator and pest control

Predators and pests that compete with humans for the same food have long been relegated to the category of "bad" living things. Among mammalian predators, wolves, coyotes, bobcats, grizzly bears, and cougars kill livestock and game, and in doing so have engendered the wrath of sheepherders and hunters. In the United States and Canada, millions of taxpayer dollars have been spent to shoot, poison, and trap predators to benefit a relatively small but vociferous group. As a result, cougars, grizzly bears, and wolves have virtually disappeared from western grazing lands. The adaptable coyote, however, has been able to maintain a relatively healthy population despite vigorous control programs. Biologists had long argued that elimination of all predators by mass poisoning programs was poor management, since a very large part of the coyotes' diet consists of rodents, which compete with sheep for forage. In 1972, yielding finally to citizen protests over mass poisoning programs—stimulated in part by such polemics as Jack Olsen's book, *Slaughter the Ani-*

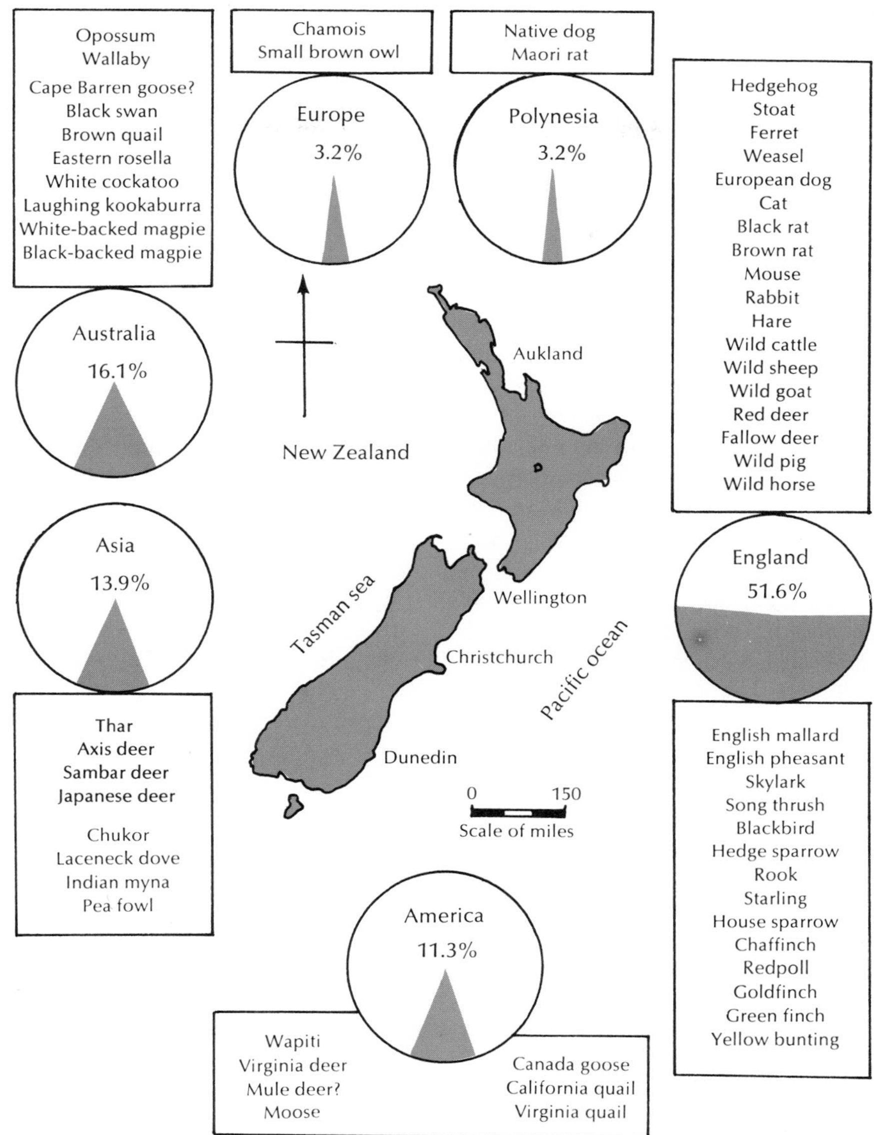

FIG. 42-8 Introduced birds and mammals that have established populations in New Zealand, with countries of origin. (From Elton, C. S. 1958. The ecology of invasions by animals and plants. New York, John Wiley & Sons Inc; after Wodzicki, K. A. 1950. Bull. D.S.I.R., New Zealand **98:**1-255.)

mals, Poison the Earth—the United States Fish and Wildlife Service banned all poisoning on public lands. Contrary to fears expressed by the sheep industry, the ban neither resulted in a dramatic increase in coyotes nor was responsible for decline in the sheep industry, which was already waning for economic reasons.

Animal traffic

The pet trade is big business in North America, England, and Europe. Between 1967 and 1974 American imports of exotic mammals, birds, and reptiles increased six-fold, reaching nearly 4 million annually in the mid-1970s. More than 100 million fishes, mostly tropical, are imported each year for the 20 million Americans that own aquariums. An astonishing variety of endangered species are sold illegally as pets, especially in the United States. The list includes monkeys of various species, cheetahs, leopards of three species, ocelots, numerous species of endangered and threatened birds, lizards, snakes, and fishes. The result is that the exotic pet trade, a curious phenomenon that seems to be associated with an affluent public's avidity for new kinds of gratification, is rapidly assuming a prominent position among the causes of extinction.

The collection and shipment of animals from Southeast Asia and Central and South America by careless hunters and callous dealers are extraordinarily wasteful —only about one animal in ten survives the numerous hazards and gross mistreatment that occurs between capture and purchase by the ultimate owner. Most collecting is done by indigenous peasant hunters who know little about live-capture techniques. Mortality is high at the time of capture. Many species are collected by shooting the mother in order to capture the baby; not infrequently several females are killed to secure a single uninjured infant. The rarer the animal, the greater is the price it will fetch, and the more enterprising are its human predators. Thus, an animal's rarity often promotes those elements that were responsible for its demise in the first place. Of those wild animals that survive the perilous trip to American homes, most die within a few weeks from improper care. Most of the rest are disposed of as people become bored with them or discover they cannot cope with them.

Some 60% of all wild mammals imported into the United States are destined for biomedical research, and of these, the great majority are primates. They are used for the preparation of vaccines, for physiologic and behavioral experiments, and for defense-related research. The global demand for primates for research is of such proportions, estimated at 200,000 animals per year, that many species of both New World and Old World monkeys have been pushed close to extinction. Fifty-four of the 149 species of primates (36%) are severely threatened today. The demand by zoos and circuses for primates, especially for the pongids (chimpanzee, orangutan, gorilla, and gibbon) also contributes to the drain, although in general, zoos are far less damaging to wildlife than the general public imagines. For example, in 1974 zoos imported into the United States only 4% of the wild mammals and 2% of the wild birds; the pet trade and biomedical research consumed the rest. Furthermore, nearly all of the larger zoos breed animals in captivity; zoos trade animals with each other and in some instances have been able to rebuild stocks of severely endangered species.

To relieve the stress that biomedical research places on native primate populations, several American users are beginning breeding programs for rhesus monkeys and other primates. However, demand is certain to outpace the production of home-breeding programs for some years to come. India, the major supplier of rhesus monkeys, recently (1978) banned further export to the United States because of an apparent breach of conditions of export (a few of the 12,000 Indian monkeys reaching the United States each year had found their way into weapons-related radiation experiments, a use specifically prohibited by agreement). The ban may seriously impede biomedical research for a period but should at least stimulate domestic breeding programs. Biomedical researchers can only welcome any move that will reduce their contribution to the world wildlife crisis.

THE ENVIRONMENTAL MOVEMENT

In the 1850's Chief Seattle of the Dunwanish tribe in Puget Sound admonished president Franklin Pierce in a moving and visionary letter:

> The great Chief in Washington sends word that he wishes to buy our land. How can you buy or sell the sky—the warmth of the land? The idea is strange to us. Yet we do not own the freshness of the air or the sparkle of the water. How can you buy them from us? Every part of this earth is sacred to my people. Every shiny pine needle, every sandy shore, every mist in the dark woods, every clearing and humming insect is holy in the memory and experience of my people:
>
> We know that the white man does not understand our ways. One portion of the land is the same to him as the next, for he is a stranger who comes in the night and takes from the land whatever he needs. The earth is not his brother but his enemy, and when he has conquered it he moves on. He leaves his father's graves, and his children's birthright is forgotten.

The American Indians, harmoniously and inextricably linked with the creatures around them and in spiritual kinship with nature, could not comprehend the white settlers' reckless and indifferent exploitation of land. The Indians venerated nature: the white settlers abused it. We may never fully understand why the ethics of these two peoples diverged as far as they did, but it is possible to reconstruct historical events that led to the white settlers' attitude—an attitude that threatens our very survival today.

We have already traced the beginnings of human society. Early human beings were merely a part of wildlife's ecologic web. People exploited nature for food and shelter as did the animals around them, but as long as people were fundamentally hunters and gatherers, their environmental requirements were small, and they remained in harmony with their surroundings. As civilization developed, people began to draw away from nature, spinning an increasingly artificial world around themselves. To satisy their needs, new and more effective ways were developed to obtain them from the environment. When this caused habitat destruction and diminishing returns, the change was regarded as an inevitable and acceptable consequence of a chosen way of life. Agriculture, as we have already seen, vastly accelerated both habitat alteration and wildlife extermination, since wild animals now became competitors with humans for their food. Changes often occurred so slowly that people were unaware of how altered their environment had become.

At the time the first settlers came to America, much of the wildlife of England and Europe belonged to the privileged classes. Only the nobility hunted deer and grouse; the peasants hunted rabbits. Such nonegalitarian restraints were in large part responsible for the presence of wild boar and other large mammals in densely settled parts of Europe for hundreds of years.

In America, natural resources and wildlife appeared inexhaustible. From the first, settlers encountered deer, turkey, salmon, and sturgeon in abundance. No aristocracy restricted their right to hunt them; wildlife belonged to everyone. Thus began the "great American experiment," founded on the philosophy that every individual was entitled to any resources that could be gathered through personal initiative. It was an exploitive and destructive ethic, but no one viewed it that way. Living was hard for the settler despite the plenitude of natural resources. It was the pioneer family pitted against trees, wild animals, and hostile Indians—not conditions expected to foster a conservation ethic.

Even before the pioneers had reached the Mississippi River, attitudes were changing along the eastern seaboard. As the wildlife disappeared, eastern colonists began enacting laws to protect game and fish. Poorly enforced and relatively ineffective at first, such efforts nevertheless reflected changing public attitudes toward wildlife. Gradually and almost grudgingly, the federal government began to assume leadership in environmental legislation.

The establishment of Yellowstone National Park in 1872 marked the real beginning of the environmental movement in the United States. Environmental wounds were becoming obvious to everyone, and by the turn of the century, an increasing number of determined Americans began to prod a reluctant Congress to act on environmental legislation. Public sentiment for animals in jeopardy had been quickened by the loss of the passenger pigeon and the near miss with the bison, the latter becoming a symbol of vanishing America.

At a time when pollution, energy and economic crises, and burgeoning social problems all besiege the nation's conscience, most people are apt to respond indifferently to the passing of a few remote species of animals. If one has never seen a whooping crane or a blue whale—and never expects to see one or any other animal on the endangered species list—of what possible concern is their extinction? And is not the sacrifice of a few insignificant animals and plants necessary for progress? Such myopic views are but affirmations of a philosophy at least as old as civilization: that the value of plants and animals is determined by their benefit to humans. This is an anthropocentric position that sees living things merely as instruments to human survival. It ignores the interrelatedness of all living components of the environment while subscribing to the notion that because humans are the highest animal, they alone can make correct decisions about the exploitation of nature. According to this view, some animals and plants are more valuable to humans than others. Such a narrowly conceived attitude is rejected by modern ecologic theory, which reaffirms the value of all living things in nature because all are parts of an interrelated whole.

Extinction, therefore, is an extremely serious result of environmental abuse because it is irreversible; an extinct animal or plant can never be recreated. Extinction removes a functional part from the web of life, and both diversity and stability of the ecosystem are reduced. Vanishing wildlife is an "early warning signal" of environmental malaise, and when we see the list of endangered animals lengthen year after year, we cannot remain indifferent.

Many people have not remained indifferent. With the establishment of Yellowstone National Park the conservation movement almost at once began to gather momentum. In 1886 Canada created Glacier National Park in British Columbia, followed in 1887 by magnificent Banff National Park in the Canadian Rockies. Australia and New Zealand followed suit with their own national parks in the next decade; soon national parks were springing up all over the world. Admittedly, during the early decades of the environmental movement, national parks were created to preserve scenic beauty; the preservation of wildlife was incidental. But as the surrounding land became emptied of wildlife, parks acquired status as sanctuaries. And while Americans and Canadians pondered the new idea of establishing parks just for animals, large game preserves were already appearing in Africa. Today, the 386 national wildlife refuges scattered among all but one state in the United States and the hundreds of national parks and game preserves all over the world are manifestations of an environmental movement that has matured far beyond the most fanciful visions of the founders of Yellowstone National Park.

Conservation is much more than trying to fence off beautiful national parks and wildlife sanctuaries from an otherwise deteriorating biosphere. Most of all, it requires a common understanding among all humanity that our fate is inextricably linked to the living world at large. It is a worldwide problem of enormous complexity. In the final analysis, anything that we do now or in the future becomes futile unless the growth of the human population can be stemmed. That is the ultimate problem.

Many believe that we are on a threshold of a new awakening, that we will be able to bridle our self-indulgence, our extraordinary propensity to isolate ourselves from nature, and then reaffirm both the meaningfulness of the human experience and our kinship with all living things.

Annotated references

Selected general readings

The environmental movement has spawned an avalanche of literature, some of which has been given over to emotional condemnation of just about every aspect of human interaction with wildlife, and some of which is concerned principally with pollution. We have not included these kinds of treatment in the following selection.

Bates, M. 1960. The forest and the sea. New York, Random House, Inc. *A lucid, thoughtful treatment of nature's economy and human intervention.*

Borland, H. 1975. The history of wildlife in America. Washington, D.C., National Wildlife Federation. *Beautifully illustrated overview with an appendix containing a summary of United States wildlife habitats.*

Curry-Lindahl, K. 1972. Let them live. New York, William Morrow & Co., Inc. *A continent-by-continent survey of animals threatened with extinction.*

Domalain, J. Y. 1977. Confessions of an animal trafficker. Natural History **86**(5):54-67. *A vivid and disturbing description of the wasteful exotic pet trade.*

Elton, C. S. 1958. The ecology of invasions by animals and plants. New York, John Wiley & Sons, Inc. *Numerous examples of the often disastrous consequences of purposeful and accidental introductions.*

Fairfax, S. K. 1978. A disaster in the environmental movement. Science **194:**743-748. *The author charges that the National Environmental Policy Act of 1969 has wasted the resources of environmentalists and has stifled citizen participation in making decisions relating to environmental protection.*

Guggisberg, C. A. W. 1970. Man and wildlife. New York, Arco Publishing Co., Inc. *Well-written historical account of human exploitation of nature. A useful worldwide survey of national parks and wildlife sanctuaries is included.*

Hill, D. O. 1975. Vanishing giants. Audubon **77**(1):56-107. *This beautifully illustrated article includes a factual survey of the major whale species.*

H. R. H. The Prince Philip, Duke of Edinburgh, and J. Fisher. 1970. Wildlife crisis. New York, Cowles Book Co., Inc. *A global view of endangered wildlife.*

Matthews, L. H. 1975. Man and wildlife. New York, St. Martin's Press, Inc. *Excellent account of human-wildlife conflicts.*

Milne, L. J., and M. Milne. 1971. The cougar doesn't live here any more. Englewood Cliffs, N.J., Prentice-Hall, Inc. *Human impact on the global environment of animals.*

Mitchell, J. G. 1976. Fear and loathing in wolf country. Audubon **78**(3):20-39. *Examines conflicting perspectives of wolves by game hunters and preservationists in Minnesota and Alaska.*

Murdy, W. H. 1975. Anthropocentrism: a modern version. Science **187:**1168-72.

National Wildlife Federation. 1974. Endangered species. National Wildlife, vol. 12, no. 3. *This special issue considers several aspects of endangered species.*

Nietschmann, B. 1977. The Bambi factor. Natural History **86**(7):84-87. *Why there are "good" and "bad" animals.*

Scheffer, V. B. 1974. A voice for wildlife. New York, Charles Scribner's Sons. *Examines the causes of confrontation between people and wildlife, considers solutions, and takes a look at the future wildlife ethic. A thoughtful and well-developed book.*

Street, P. 1970. Wildlife preservation. Chicago, Henry Regnery Co. *Problems of preservation of mammals, birds, and reptiles are described through case histories.*

Trefethen, J. B. 1975. An American crusade for wildlife. New York, Winchester Press. *Excellent history of the conservation movement in the United States.*

Uetz, G., and D. L. Johnson. 1974. Breaking the web. Environment **16**(10):31-39. *The impact of humans on animal extinction.*

Vogt, B. 1978. Now, the river is dying. National Wildlife **16**(4):4-11. *Battle for the water of the Platte River in Nebraska reflects a vastly larger problem throughout the American west.*

Watt, K. E. F., and others. 1977. The unsteady mix: environmental problems, growth and culture. Honolulu, The University Press of Hawaii. *Economic, social, and technological aspects of environmental problems.*

Westman, W. E. 1977. How much are nature's services worth? Science **197:**960-964. *Measuring social benefits of ecosystem functioning is difficult and controversial but important for public understanding.*

Selected *Scientific American* articles

Calder, W. A., III. 1978. The kiwi. **239:**132-142 (July). *In filling ecologic niches left vacant by mammals, this New Zealand bird has evolved several mammalian characteristics.*

Photograph by Sally Mann.

"If I were to name the three most precious resources of life, I should say books, friends, and nature; and the greatest of these, at least the most constant and always at hand, is nature. Nature we have always with us, an inexhaustible storehouse of that which moves the heart, appeals to the mind, and fires the imagination—health to the body, a stimulus to the intellect, and joy to the soul."

John Burroughs
"Leaf and Tendril," 1908

GLOSSARY

aboral (ab-o′rəl) (L., *ab,* from, + *os,* mouth). A region opposite the mouth.
acanthor (ə-kan′thor) (Gr., *akantha,* spine or thorn, + *-or*). First larval form of acanthocephalans in the intermediate host.
acclimatization (ə-klī′mə-də-zā′shən) (L., *ad,* to, + Gr., *klima,* climate). Gradual physiologic adaptation in response to relatively long-lasting environmental changes.
acetabulum (as′ə-tab′ū-lum) (L., a little saucer for vinegar). True sucker, especially in flukes and leeches. The socket in the hip bone that receives the thigh bone.
acinus, pl. **acini** (as′ə-nəs, as′ə-ni) (L., grape). A small lobe of a compound gland or a saclike cavity at the termination of a passage.
acoelomate (a-sēl′ə-māt′) (Gr., *a,* not, + *koilōma,* cavity). Without a coelom, as in flatworms and proboscis worms.
acontium (ə-kän′chē-əm), pl. **acontia** (Gr., *akontion,* dart). Thread bearing nematocysts located on mesentery of sea anemone.
actin (Gr., *aktis,* ray). A protein in the contractile tissue that forms the thin myofilaments of striated muscle.
adaptation (L., *adaptatus,* fitted). Adjustment to environment by an organism so that it becomes more fit for existence.
adductor (ə-duk′tər) (L., *ad,* to, + *ducere,* to lead). A muscle that draws a part toward a median axis, or a muscle that draws the two valves of a mollusc shell together.
adenine (ad′nēn, ad′ə-nēn) (Gr., *adēn,* gland, + *-ine,* suffix). A purine base; component of nucleotides and nucleic acids.
adenosine (ə-den′ə-sēn) **(di-, tri-) phosphate** (ADP and ATP). A nucleotide composed of adenine, ribose, sugar, and two (ADP) or three (ATP) phosphate units; ATP is an energy-rich compound that, with ADP, serves as a phosphate bond–energy transfer system in cells.
adipose (ad′ə-pōs) (L., *adeps,* fat). Fatty tissue; fatty.
adrenaline (ə-dren′ə-lən) (L., *ad,* to, + *renalis,* pertaining to kidneys). A hormone produced by the adrenal, or suprarenal, gland; epinephrine.
adsorption (ad-sorp′shən) (L., *ad,* to, + *sorbere,* to suck in). The adhesion of molecules to solid bodies.
aerobic (a-rō′bik) (Gr., *aēr,* air, + *bios,* life). Dependent on oxygen for respiration.
afferent (af′ə-rənt) (L., *ad,* to, + *ferre,* to bear). Adj. Leading or bearing toward some organ, for example, nerves conducting impulses toward the brain, or blood vessels carrying blood toward an organ; opposed to efferent.
aggression (L., *aggredi,* attack). A primary instinct usually associated with emotional states; an offensive behavior action.
agonistic behavior (Gr., *agōnistēs,* combatant). Any behavior related to aggression, such as threat, attack, fighting, as well as defensive responses to aggression.
alate (ā′lāt) (L., *alatus,* wing). Winged.
allantois (ə-lan′tois) (Gr., *allas,* sausage, + *eidos,* form). One of the extraembryonic membranes of the amniotes that functions in respiration and excretion in birds and reptiles and plays an important role in the development of the placenta in most mammals.
allele (ə-lēl′) (Gr., *allēlōn,* of one another). One of a pair, or series, of genes that are alternative to each other in heredity and are situated at the same locus in homologous chromosomes.
allograft (a′lō-graft) (Gr., *allos,* other, + graft). A piece of tissue or an organ transferred from one individual to another individual of the same species, but not identical twins; homograft.
allometry (ə-lom′ə-trē) (Gr., *allos,* other, + *metry,* measure). Relative growth of a part in relation to the whole organism.
allopatric (Gr., *allos,* other, + *patra,* native land). In separate and mutually exclusive geographic regions.
alpha helix (Gr., *alpha,* first, + *helix,* spiral). Literally the first spiral arrangement of the genetic DNA molecule; regular coiled arrangement of polypeptide chain in proteins; secondary structure of proteins.
altricial (al-tri′shəl) (L., *alcatrices,* nourishers). Referring to young animals; especially birds having the young hatched in an immature, dependent condition.

---

bat / āpe / ärmadillo / herring / fēmale / finch / līce / crocodile / crōw / duck / ūnicorn / ə indicates unaccented vowel sound "uh" as in mammal, fishes, cardinal, heron, vulture / stress as in bi-ol′o-gy, bi′o-log′i-c

alula (al′yə-lə) (L., dim. of *ala,* wing). The first digit or thumb of a bird's wing, much reduced in size.

alveolus (al-vē′ə-ləs) (L., dim. of *alveus,* cavity, hollow). A small cavity or pit, such as a microscopic air sac of the lungs, terminal part of an alveolar gland, or bony socket of a tooth.

ambulacra (am′byə-lak′rə) (L., *ambulare,* to walk). In echinoderms, radiating grooves where podia of water-vascular system project to outside.

amebocyte (ə-mē′bə-sīt) (Gr., *amoibē,* change, + *kytos,* hollow vessel). Any free body cell capable of movement by pseudopodia; certain types of blood cells and tissue cells.

ameboid (ə-mē′boid) (Gr., *amoibē,* change, + *-oid,* like). Ameba-like in putting forth pseudopodia.

amictic (ə-mik′tic) (Gr., *a,* without, + *miktos,* mixed or blended). Pertaining to female rotifers, which produce only diploid eggs that cannot be fertilized, or to the eggs produced by such females.

amino acid (ə-mē′nō) (amine, an organic compound). An organic acid with an amino radical (NH_2). Makes up the structure of proteins.

amitosis (ā′mī-tō′səs) (Gr., *a,* not, + *mitos,* thread). A form of cell division in which mitotic nuclear changes do not occur; cleavage without separation of daughter chromosomes.

amnion (am′nē-än) (Gr., *amnion,* membrane around the fetus). The innermost of the extraembryonic membranes forming a fluid-filled sac around the embryo in amniotes.

amniote (am′nē-ōt′). Having an amnion; as a noun, an animal that develops an amnion in embryonic life, that is, reptiles, birds, and mammals.

amphiblastula (am′fə-blas′chə-lə) (Gr., *amphi,* on both sides, + *blastos,* germ, + L., *-ula,* small). Free-swimming larval stage of certain marine sponges; blastula-like, but with only the cells of the animal pole flagellated; those of the vegetal pole unflagellated.

amphid (am′fəd) (Gr., *amphidea,* anything that is bound around). One of a pair of anterior sense organs in certain nematodes.

amplexus (am-plek′səs) (L., embrace). The copulatory embrace of frogs or toads.

amylase (am′ə-lās′) (L., *amylum,* starch, + *ase,* suffix meaning enzyme). An enzyme that breaks down starch into smaller units.

anadromous (an-ad′rə-məs) (Gr., *anadromos,* running upward). Refers to those fishes that migrate up streams from the sea to spawn.

anaerobic (an′ə-rō′bik) (Gr., *an,* not, + *aēr,* air, + *bios,* life). Not dependent on oxygen for respiration.

analogy (L., *analogus,* ratio). Similarity of function but not in origin. (Adj.: **analogous.**)

anamorphic (a′nə-mor′fik) (Gr., *ana,* again, + *morphē,* form). Gradually changing form to one of greater complexity. In arthropods, involves the acquisition of additional body segments during development.

anastomosis (ə-nas′tə-mō′səs) (Gr., *ana,* again, + *stoma,* mouth). A union of two or more arteries, veins, or fibers to form a branching network.

androgen (an′drə-jən) (Gr., *anēr, andros,* man, + *genēs,* born). Any of a group of vertebrate male sex hormones.

Angstrom (after Ångström, Swedish physicist). A unit of one ten-millionth of a millimeter (one ten-thousandth of a micrometer); it is represented by the symbol Å.

anhydrase (an-hī′drās) (Gr., *an,* not, + *hydōr,* water, + *ase,* enzyme suffix). An enzyme involved in the removal of water from a compound. Carbonic anhydrase promotes the conversion of carbonic acid into water and carbon dioxide.

anlage (än′lä-gə) (Ger., *anlage,* laying-out, foundation). Rudimentary form; primordium.

antenna (L., *antenna, antemna,* sail yard). A sensory appendage on the head of arthropods, or one of the second pair of the two such pairs of structures in crustaceans (second antennae); first pair are antennules (first antennae).

anterior (L., comparative of *ante,* before). The head end of an organism, or (as adj.) toward that end.

anticodon (an′tī-kō′don). A sequence of three nucleotides in transfer RNA that is complementary to a codon in messenger RNA.

aperture (ap′ər-chər) (L., *apertura,* from *aperire,* to uncover). An opening; the opening into the first whorl of a gastropod shell.

apical (ap′ə-kl) (L., *apex,* tip). Pertaining to the tip or apex.

apopyle (ap′-ə-pīl) (Gr., *apo,* away from, + *pylē,* gate). In sponges, opening from the radial canal into the spongocoel.

appendicular (L., *ad,* to, + *pendare,* to hang). Pertaining to appendages; pertaining to vermiform appendix.

arboreal (är-bōr′ē-al) (L., *arbor,* tree). Living in trees.

archenteron (ärk-en′tə-rän′) (Gr., *archē,* beginning, + *enteron,* gut). The main cavity of an embryo in the gastrula stage; it is lined with endoderm and represents the future digestive cavity.

archeocytes (ärk′ē-ō-sites) (Gr., *archaios,* beginning, + *kytos,* hollow vessel). Ameboid cells of varied function in sponges.

artiodactyl (är′ti-o-dak′təl) (Gr., *artios,* even, + *daktylos,* toe). One of an order or suborder of mammals with two or four digits on each foot.

ascon (Gr., *askos,* bladder). Simplest form of sponges, with canals leading directly from the outside to the interior.

asexual. Without distinct sexual organs; not involving gametes.

assimilation (L., *assimilatio,* bringing into conformity). Absorption and building up of digested nutriments into complex organic protoplasmic materials.

atoke (ā′-tōk) (Gr., *a,* without, + *tokos,* offspring). Anterior part of a marine polychaete, as distinct from the posterior part (epitoke) during the breeding season.

ATP. In biochemistry, an ester of adenosine and triphosphoric acid. Adenosine triphosphate.

atrium (ā′trē-əm) (L., vestibule). One of the chambers of the heart; the tympanic cavity of the ear; the large cavity containing the pharynx in tunicates and cephalochordates.

auricle (aw′ri-kl) (L., *auricula,* dim. of *auris,* ear). One of the less muscular chambers of the heart; atrium; the external ear, or pinna; any earlike lobe or process.

autosome (aw'tō-sōm) (Gr., *autos,* self, + *sōma,* body). Any chromosome that is not a sex chromosome.

autotomy (aw-täd'ə-mē) (Gr., *autos,* self, + *tomas,* a cutting). The automatic breaking off of a part of the body.

autotroph (aw'tō-trōf) (Gr., *autos,* self, + *trophos,* feeder). An organism that makes its organic nutrients from inorganic raw materials.

autotrophic nutrition (Gr., *autos,* self, + *trophia,* denoting nutrition). Nutrition characterized by the ability to utilize simple inorganic substances for the synthesis of more complex organic compounds, as in green plants and some bacteria and true fungi.

avicularium (L., *avicula,* small bird, + *aria,* like or connected with). Modified zooid that is attached to the surface of the major zooid in Ectoprocta and resembles a bird's beak.

axial (L., *axis,* axle). Relating to the axis, or stem; on or along the axis.

axolotl (ak'sə-lot'l) (Nahuatl, *atl,* water, + *xolotl,* doll, servant, spirit). Larval stage of any of several species of the genus *Ambystoma* (such as *A. tigrinum*) exhibiting neotenic reproduction.

axoneme (ak'sə-nēm) (Gr., *axon,* an axis, + *nēma,* thread). Portion of a flagellum or cilium proceeding from the kinetosome and consisting of a cylinder of nine pairs of fibrils surrounding a single pair.

axopodium (ak'sə-pō'di-um) (Gr., *axon,* an axis, + *podion,* small foot). Long, slender, more or less permanent pseudopodium found in certain sarcodine protozoa. (Also **axopod.**)

basal body. Specialized granule in a cell, found basal to a flagellum or cilium, identical in structure to a centriole; also called kinetosome or blepharoplast.

benthos (ben'thäs) (Gr., depth of the sea). Those organisms that live along the bottom of seas and lakes. (Adj., **benthic.**)

biogenesis (bī'ō-jen'ə-səs) (Gr., *bios,* life, + *genesis,* birth). The doctrine that life originates only from preexisting life.

bioluminescence. Method of light production by living organisms in which usually certain proteins (luciferins), in the presence of oxygen and an enzyme (luciferase), are converted to oxyluciferins with the liberation of light.

biomass (Gr., *bios,* life, + *maza,* lump or mass). The weight of total living organisms or of a species population per unit of area.

biome (bī'ōm) (Gr., *bios,* life, + *ōma,* abstract group suffix). Complex of plant and animal communities characterized by climatic and soil conditions; the largest ecologic unit.

biosphere (Gr., *bios,* life, + *sphaira,* globe). That part of Earth containing living organisms.

bipinnaria (L., *bi-,* double, + *pinna,* wing, + *-aria,* like or connected with). Free-swimming, ciliated, bilateral larva of the asteroid echinoderms; develops into the brachiolaria larva.

blastocoel (blas'tō-sēl') (Gr., *blastos,* germ, + *koilos,* hollow). Cavity of the blastula.

blastomere (Gr., *blastos,* germ, + *meros,* part). An early cleavage cell.

blastopore (Gr., *blastos,* germ, + *poros,* passage, pore). External opening of the archenteron in the gastrula.

blastula (Gr., *blastos,* germ, + L., *-ula,* diminutive). Early embryonic stage of many animals; consists of a hollow mass of cells.

blepharoplast (blə-fā'rə-plast) (Gr., *blepharon,* eyelid, + *plastos,* formed). A centriole from which arises the axoneme of a flagellum. Also called a basal body or kinetosome.

B.P. Before the present.

brachial (brāk'ē-əl) (L., *brachium,* forearm). Referring to the forearm.

brachiolaria (brak'ē-ō-lār'ē-ə) (L., *brachiola,* little arm, + *-aria,* pertaining to). This asteroid larva develops from the bipinnaria larva and has three preoral holdfast processes.

branchial (brank'ē-əl) (Gr., *branchia,* gills). Referring to gills.

buccal (buk'əl) (L., *bucca,* cheek). Referring to the mouth cavity.

buffer Any substance or chemical compound that tends to keep pH constant when acids or bases are added.

carapace (kar'ə-pās) (F., fr. Sp., *carapacho*). Shieldlike plate covering the cephalothorax of certain crustaceans; dorsal part of the shell of a turtle.

carbohydrate (L., *carbo,* charcoal, + Gr., *hydōr,* water). Compounds of carbon, hydrogen, and oxygen; aldehyde or ketone derivatives of polyhydric alcohols, with hydrogen and oxygen atoms attached in a 2 to 1 ratio.

carboxyl (kär-bäk'səl) (carbon + oxygen + *-yl,* chemical radical suffix). The acid group of organic molecules—COOH.

carnivore (kär'nə-vōr') (L., *carnivorus,* fleshing-eating). One of the flesh-eating mammals of the order Carnivora.

carotene (kar'ə-tēn) (L., *carota,* carrot, + *-ene,* unsaturated straight-chain hydrocarbons). A red, orange, or yellow pigment belonging to the group of carotenoids; precursor of vitamin A.

cartilage (L., *cartilago;* akin to L., *cratis,* wickerwork). A translucent elastic tissue that makes up most of the skeleton of embryos and very young vertebrates; in higher forms much of it is converted into bone.

caste (kast) (L., *castus,* pure, separated). One of the polymorphic forms within an insect society, each caste having its specific duties, as queen, worker, soldier, and so on.

catadromous (kə-tad'rə-məs) (Gr., *kata,* down, + *dromos,* a running). Refers to those fishes that migrate from fresh water to the ocean to spawn.

catalyst (kad'ə-ləst) (Gr., *kata,* down, + *lysis,* a loosening). A substance that accelerates a chemical reaction but does not become a part of the end product.

cecum, caecum (sē'kəm) (L., *caecus,* blind). A blind pouch at the beginning of the large intestine; any similar pouch.

cellulose (sel'ū-lōs) (L., *cella,* small room). Chief polysaccharide constituent of the cell wall of green plants and some fungi; an insoluble carbohydrate $(C_6H_{10}O_5)_x$ that is converted into glucose by hydrolysis.

centriole (sen'trē-ol) (Gr., *kēntron,* center of a circle, + L., *-ola,* small). A minute cytoplasmic organelle usually found in the centrosome and considered to be the active division center of the cell; organizes spindle fibers during mitosis and meiosis.

centrolecithal (sen'tro-les'ə-thəl) (Gr., *kentron,* center, + *lekithos,* yolk, + Eng., *-al,* adj.). Pertaining to an insect egg with the yolk concentrated in the center.

centromere (sen'trə-mir) (Gr., *kentron,* center, + *meros,* part). A small body or constriction on the chromosome to which a spindle fiber attaches during mitosis or meiosis.

cephalization (sef'ə-li-zā'shən) (Gr., *kephale,* head). The process by which specialization, particularly of the sensory organs and appendages, became localized in the head end of animals.

cephalothorax (sef'ə-lä-thō'raks) (Gr., *kephale,* head, + thorax). A body division found in many Arachnida and higher Crustacea, in which the head and some or all of the thoracic segments are fused.

cercaria (ser-kār'ē-ə) (Gr., *kerkos,* tail, + L., *-aria,* like or connected with). Tadpolelike larva of trematodes (fluke).

chelicera (kə-lis'ə-rə), pl. **chelicerae** (Gr., *chēlē,* claw, + *keras,* horn). One of a pair of pincerlike head appendages on the members of the subphylum Chelicerata.

chelipeds (kēl'ə-peds) (Gr., *chēlē,* claw, + L., *pes, pedis,* foot). First pair of legs in most decapod crustaceans; specialized for seizing and crushing.

chiasma (kī-az'mə), pl. **chiasmata** (Gr., cross). An intersection or crossing, as of nerves; an exchange of partners in meiosis.

chitin (kī'tən) (Fr., *chitine,* fr. Gr., *chitōn,* tunic). A horny substance that forms part of the cuticle of arthropods and is found sparingly in certain other invertebrates; a nitrogenous polysaccharide, insoluble in water, alcohol, dilute acids, and digestive juices of most animals.

chloragogue cells (klōr'ə-gog) (Gr., *chlōros,* light green, + *agōgos,* a leading, a guide). Modified peritoneal cells, greenish or brownish, clustered around the digestive tract of certain annelids; apparently they aid in elimination of nitrogenous wastes and in food transport.

chlorocruorin (klō'rō-kroo'ə-rən) (Gr., *chlōros,* light green, + L., *cruor,* blood). A greenish iron-containing respiratory pigment dissolved in the blood plasma of certain marine polychaetes.

chlorophyll (klō'ro-fil) (Gr., *chlōros,* light green, + *phyllōn,* leaf). Green pigment found in plants and in some animals; necessary for photosynthesis.

choanocyte (kō-an'ō-sīt) (Gr., *choanē,* funnel, + *kytos,* hollow vessel). One of the flagellate collar cells that line cavities and canals of sponges.

cholinergic (kōl'-i-nər'jik) (Gr., *chōle,* bile, + *ergon,* work). Type of nerve fiber that releases acetylcholine from axon terminal.

chorion (kō'rē-on) (Gr., *chorion,* skin). The outer of the double membrane that surrounds the embryo of reptiles, birds, and mammals; in mammals it helps form the placenta.

chromatid (krō'mə-tid) (Gr., *chromato-,* fr. *chrōma,* color, + L., *-id,* feminine stem for particle of specified kind). A half chromosome between early prophase and metaphase in mitosis; a half chromosome between synapsis and second metaphase in meiosis. At the anaphase stage each chromatid is known as a daughter chromosome.

chromatin (kro'mə-tin) (Gr., *chrōma,* color). The nucleoprotein material of a chromosome; regarded as the physical basis of heredity.

chromatophore (krō-mat'ə-fōr) (Gr., *chrōma,* color, + *phorein,* to bear). Pigment cell, usually in the dermis, in which the pigment usually can be dispersed or concentrated. (Sometimes used to refer to chloroplast or chromoplast.)

chromomere (krō'mō-mir) (Gr., *chrōma,* color, + *meros,* part). One of the chromatin granules of characteristic size on the chromosome; may be identical to a gene or a cluster of genes.

chromonema (krō-mə-nē'mə) (Gr., *chrōma,* color, + *nema,* thread). A convoluted thread in prophase of mitosis or the central thread in a chromosome.

chromosome (krō'mə-sōm) (Gr., *chrōma,* color, + *sōma,* body). One of the DNA bodies that arises from the nuclear network during mitosis, splits longitudinally, and carries the linear sequence of genes.

chrysalis (kris'ə-lis) (L., from Gr., *chrysos,* gold). The pupal stage of a butterfly.

cilium, pl. **cilia** (sil'i-əm) (L., *cilium,* eyelid). A hairlike, vibratile organelle process found on many animal cells. Cilia may be used in moving particles along the cell surface or, in ciliate protozoans, for locomotion.

cinclides (sing'klid-əs), sing. **cinclis** (sing'kləs) (Gr., *kinklis,* latticed gate or partition). Small pores in the external body wall of sea anemones for extrusion of acontia.

circadian (sər'kə-dē'-ən) (L., *circa,* around, + *dies,* day). Occurring at a period of approximately 24 hours.

cirrus (sir'əs) (L., curl). A hairlike tuft on an insect appendage; locomotor organelle of fused cilia; male copulatory organ of some invertebrates.

cleavage (OE., *cleofan,* to cut). Process of nuclear and cell division in animal zygote.

climax (klī'maks) (Gr., *klimax,* ladder). A state of dynamic equilibrium; a culmination of the succession in the like forms of a community.

cline (klīn) (Gr., *klinein,* slope, bend). A pattern of gradual genetic change in a population according to its geographic range.

clitellum (klī-tel'əm) (L., *clitellae,* packsaddle). Thickened saddlelike portion of certain midbody segments of many oligochaetes and leeches.

cloaca (klō-ā'kə) (L., sewer). Posterior chamber of digestive tract in many vertebrates, receiving feces and urogenital products. In certain invertebrates, a terminal portion of digestive tract that serves also as respiratory, excretory, or reproductive duct.

clone (klōn) (Gr., *klōn,* twig). All descendants derived by asexual reproduction from a single individual.

cnidoblast (nī'dō-blast) (Gr., *knidē,* nettle, + *blastos,* germ). Modified interstitial cell that holds the nematocyst.

cnidocil (nī'dō-sil) (Gr., *knidē, nettle,* + L., *cilium,* hair). Triggerlike spine on nematocyst.

coacervate (kō'ə-sər'vət) (L., *coacervatus,* to heap up). An aggregate of colloidal droplets held together by electrostatic forces.

cochlea (kōk'lēə) (L., snail, fr. Gr., *kochlos,* a shellfish). A tubular cavity of the inner ear containing the essential organs of hearing; occurs in crocodiles, birds, and mammals; spirally coiled in mammals.

codon (kō'dän) (L., code, + on). A sequence of three adjacent nucleotides that code for one amino acid.

coelenteron (sē-len'tər-on) (Gr., *koilos,* hollow, + *enteron,* intestine). Internal cavity of a cnidarian; gastrovascular cavity; archenteron.

coelom (sē'lōm) (Gr., *koilōma,* cavity). The body cavity in triploblastic animals, lined with mesodermal peritoneum.

coelomocyte (sē-lō'mə-cīt) (Gr., *koilōma,* cavity, + *kytos,* hollow vessel). Another name for amebocyte; primitive or undifferentiated cell of the coelom and the water-vascular system.

coelomoduct (sē-lō'mə-dukt) (Gr., *koilos,* hollow, + L., *ductus,* a leading). A duct that carries gametes or excretory products (or both) from the coelom to the exterior.

coenecium, coenoecium (sə-nēs(h)'ē-um) (Gr., *koinos,* common, + *oikion,* house). The common secreted investment of an ectoproct colony; may be chitinous, gelatinous, or calcareous.

coenzyme (kō-en'zīm) (L., prefix, *co-,* with, + Gr., *enzymos,* leavened, from *en,* in, + *zymē,* leaven). A required substance in the activation of an enzyme; a prosthetic or nonprotein constituent of an enzyme.

collenchyme (käl'ən-kīm) (Gr., *kolla,* glue, + *enchyma,* infusion). A gelatinous mesenchyme containing undifferentiated cells; found in cnidarians and ctenophores.

colloblast (käl'ə-blast) (Gr., *kolla,* glue, + *blastos,* germ). A glue-secreting cell on the tentacles of ctenophores.

colloid (kä'loid) (Gr., *kolla,* glue, + *eidos,* form). A two-phase system in which particles of one phase are dispersed in the second phase.

comb plate. One of the plates of fused cilia that are arranged in rows for ctenophore locomotion.

commensalism (kə-men'səl-iz'əm) (L., *cum,* together with, + *mensa,* table). A symbiotic relationship in which one individual benefits and the other is unharmed.

community (L., *communitas,* community, fellowship). An assemblage of organisms that are associated in a common environment and interact with each other in a self-sustaining and self-regulating relation.

conjugation (kon'jū-gā'shun) (L., *conjugare,* to yoke together). Temporary union of two ciliate protozoans while they are exchanging chromatin material and undergoing nuclear phenomena resulting in binary fission.

conspecific (L., *con,* together, + species). A member of the same species.

contractile vacuole. A clear fluid-filled cell vacuole in protozoans and a few lower metazoans; takes up water and releases it to the outside in a cyclic manner, for osmoregulation and some excretion.

copulation (kop'ū-lā'shun) (F., fr. L., *copulare,* to couple). Sexual union to facilitate the reception of sperm by the female.

corium (kō'rē-əm) (L., leather). The deep layer of the skin, dermis.

cornea (kor'nē-ə) (L., *corneus,* horny). The outer transparent coat of the eye.

cortex (kor'teks) (L., bark). The outer layer of a structure.

crista (kris'ta), pl. **cristae** (L., crest). A crest or ridge on a body organ or organelle.

cryptobiotic (Gr., *kryptos,* hidden, + *biōticus,* pertaining to life). Living in concealment; refers to insects and other animals that live in secluded situations, such as underground or in wood; also tardigrades and some nematodes, rotifers, and others that survive harsh environmental conditions by assuming for a time a state of very low metabolism.

ctenoid scales (ten'oid) (Gr., *kteis, ktenos,* comb). Thin, overlapping dermal scales of the more advanced fishes; posterior margins are serrate, or comblike.

cuticle (kū'ti-kəl) (L., *cutis,* skin). A protective, noncellular, organic layer secreted by the external epithelium (hypodermis) of many invertebrates. In higher animals, the term refers to the epidermis or outer skin.

cycloid scales (sī'-kloid) (Gr., *kyklos,* circle). Thin, overlapping dermal scales of the more primitive fishes; posterior margins are smooth.

cydippid larva (sī-dip'pid) (Gr., *kydippe,* mythologic Athenian maiden). Free-swimming larva of most ctenophores; superficially similar to the adult.

cystid (sis'tid) (Gr., *kystis,* bladder). In an ectoproct, the dead secreted outer parts plus the adherent underlying living layers.

cytochrome (sī'tō-krōm) (Gr., *kytos,* hollow vessel, + *chrōma,* color). One of several iron-containing pigments that serve as hydrogen carriers in aerobic respiration.

cytokinesis (sī'tō-kin-ē'sis) (Gr., *kytos,* hollow, + *kinesis,* movement). Changes that occur in the cytoplasm during cell division.

cytopharynx (Gr., *kytos,* hollow vessel, + *pharynx,* throat). Short tubular gullet in ciliate protozoans.

cytoplasm (sī'tō-plasm) (Gr., *kytos,* hollow, + *plasma,* mold). The living matter of the cell, excluding the nucleus.

cytosol (sīt'ō-sol) (Gr., *kytos,* hollow vessel, + L., *sol,* fr. *solutus,* to loosen). Unstructured portion of the cytoplasm in which the organelles are bathed.

cytosome (sī'tə-sōm) (Gr., *kytos,* hollow vessel, + *sōma,* body). The cell body inside the plasma membrane.

cytostome (sī'tə-stōm) (Gr., *kytos,* hollow vessel, + *stoma,* mouth). The cell mouth in ciliate protozoans.

deme (dēm) (Gr., populace). A local population of closely related animals.

demography (də-mäg'grə-fē) (Gr., *demos,* people, + *grāphy*). The properties of the rate of growth and the age structure of populations.

deoxyribose (dē-ok'sē-rī'bōs) (*deoxy,* loss of oxygen, +

ribose, pentose sugar). A 5-carbon sugar having 1 oxygen atom less than ribose; a component of deoxyribonucleic acid (DNA).

dermal (Gr., *derma,* skin). Pertaining to the skin; cutaneous.

dermis. The inner, sensitive mesodermal layer of skin; corium.

desmosome (des'mə-sōm) (Gr., *desmos,* bond, + *soma,* body). Buttonlike plaque serving as an intercellular connection.

determinate cleavage. The type of cleavage, usually spiral, in which the fate of the early blastomeres can be foretold; mosaic cleavage.

detritus (də-trī'tus) (L., that which is rubbed or worn away). Any fine particulate debris of organic or inorganic origin.

Deuterostomia (du'də-rō-stō'mē-ə) (Gr., *deuteros,* second, secondary, + *stoma,* mouth). A group of higher phyla in which cleavage is indeterminate and primitively radial. The endomesoderm is enterocoleous, and the mouth is derived away from the blastopore. Includes Echinodermata, Chordata, and a number of minor phyla. Compare with Protostomia.

diapause (dī'ə-pawz) (Gr., *diapausis,* pause). A period of arrested development in the life cycle of insects and certain other animals, during which physiologic activity is very low and they are highly resistant to unfavorable external conditions.

diffusion (L., *diffusus,* dispersion). The movement of particles or molecules from area of high concentration of the particles or molecules to area of lower concentration.

digitigrade (dij'ə-də-grād') (L., *digitus,* finger, toe, + *gradus,* step, degree). Walking on the digits with the posterior part of the foot raised; compare **plantigrade.**

dihybrid (Gr., *dis,* twice, + L., *hibrida,* mixed offspring). A hybrid whose parents differ in two distinct characters; an offspring having different alleles at two different loci; for example, **AaBb.**

dimorphism (dī-mor'fizm) (Gr., *di,* two, + *morphē,* form). Existence within a species of two distinct forms according to color, sex, size, organ structure, and the like. Occurrence of two kinds of zooids in a colonial organism.

dioecious (dī-ē'shəs) (Gr., *di-,* two + *oikos,* house). Having male and female organs in separate individuals.

diphycercal (dif'i-ser'kəl) (Gr., *diphyēs,* twofold, + *kerkos,* tail). A tail that tapers to a point, as in lungfishes; vertical column extends to tip without upturning.

diphyodont (di'fi-ə-dänt) (Gr., *diphyēs,* twofold, + *odous,* tooth) Having deciduous and permanent sets of teeth successively.

dipleurula (dī-ploor'ū-lə) (Gr., *di-,* two, + *pleura,* rib, side, + L., *-ula,* small). A hypothetical, simple ancestral form of echinoderms; elongated and bilaterally symmetric, with three pairs of coelomic sacs.

diploid (dip'loid) (Gr., *diploos,* double, + *eidos,* form). Having the somatic (double, or 2n) number of chromosomes or twice the number characteristic of a gamete of a given species.

disulfide. A compound of two atoms of sulfur combined with a chemical element or radical.

DNA. Abbreviation for deoxyribonucleic acid, a high molecular weight substance on the chromosomes that codes for all hereditary information.

dominance hierarchy. A social ranking, formed through agonistic behavior, in which individuals are associated with each other so that some have greater access to resources than do others.

dorsal (dor'səl) (L., *dorsum,* back). Toward the back, or upper surface, of an animal.

DPN. Abbreviation for diphosphopyridine nucleotide, a hydrogen carrier in respiration; now called NAD, nicotinamide adenine dinucleotide.

drive. A state of activity directed toward satisfying a specific need.

dyad (dī'əd) (Gr., *dyas,* two). One of the groups of two chromosomes formed by the division of a tetrad.

ecdysis (ek'dəh-sis) (Gr., *ekdysis,* a stripping, escape). Shedding of outer cuticular layer; molting, as in insects or crustaceans.

ecologic niche. The role of an organism in a community; its unique way of life and its relationship to other biotic and abiotic factors.

ecology (Gr., *oikos,* house, + *logos,* discourse). That part of biology that deals with the relationship between organisms and their surroundings.

ecosystem (ek'ō-sis-təm) (eco[logy], from Gr., *oikos,* house, + system). An ecologic unit consisting of both the biotic communities and the nonliving (abiotic) environment, which interact to produce a stable system.

ecotone (ek'ō-tōn) (eco[logy], from Gr., *oikos,* home, + *tonos,* stress). The transition zone between two adjacent communities.

ectoderm (ek'tō-derm) (Gr., *ektos,* outside, + *derma,* skin). Outer layer of cells of an early embryo (gastrula stage); one of the germ layers, also sometimes used to include tissues derived from ectoderm.

ectoplasm (ec'tō-plazm) (Gr., *ektos,* outside, + *plasma,* form). The cortex of a cell or that part of cytoplasm just under the cell surface; contrasts with endoplasm.

ectothermic (ek'tō-therm'ic) (Gr., *ectos,* external, + *thermē,* heat). Having a variable body temperature derived from heat acquired from the environment; contrasts with **endothermic.**

effector (L., *efficere,* bring to pass). An organ, tissue, or cell that becomes active in response to stimulation.

efferent (ef'ə-rənt) (L., *ex,* out, + *ferre,* to bear). Adj. Leading or conveying away from some organ, for example, nerve impulses conducted away from the brain, or blood conveyed away from an organ; as opposed to **afferent.**

embryogenesis (em'brē-ō-jen'ə-səs) (Gr., *embryon,* embryo, + *genesis,* origin). The origin and development of the embryo; embryogeny.

emigrate (L., *emigrare,* to move out). To move *from* one area to another to take up residence; Opposite of immigrate.

emulsion (ə-məl'shən) (L., *emulsus,* milked out). A colloidal system in which both phases are liquids.

endemic (en-dem′ik) (Gr., *en-*, in, + *demos*, populace). Peculiar to a certain region or country; native to a restricted area; not introduced.

endergonic (en-dər-gän′ik) (Gr., *endon*, within, + *ergon*, work). Used in reference to a chemical reaction that requires energy.

endocrine (en′də-krən) (Gr., *endon*, within, + *krinein*, to separate). Refers to a gland that is without a duct and that releases its product directly into the blood or lymph.

endocytosis (en′dō-sī-tō′sis) (Gr., *endon*, within, + *kytos*, hollow vessel). The engulfment of matter by phagocytosis and of macromolecules by pinocytosis.

endoderm (en′də-dərm) (Gr., *endon*, within, + *derma*, skin). Innermost germ layer of an embryo, forming the primitive gut; also may refer to tissues derived from endoderm.

endometrium (en′də-mē′tre′-əm) (Gr., *endon*, within, + *mētra*, womb). The mucous membrane lining the uterus.

endoplasm (en′də-pla-zəm) (Gr., *endon*, within, + *plasma*, mold or form). That portion of cytoplasm that immediately surrounds the nucleus.

endoplasmic reticulum The cytoplasmic double membrane with ribosomes (rough) or without ribosomes (smooth).

endoskeleton (Gr., *endon*, within, + *skeletos*, hard). An internal skeleton or supporting framework, as opposed to **exoskeleton.**

endostyle (en′də-stīl) (Gr., *endon*, within, + *stylos*, a pillar). Ciliated groove(s) in the floor of the pharynx of tunicates, cephalochordates, and larval cyclostomes, used for accumulating and moving food particles to the stomach.

endothermic (en′də-therm′ic) (Gr., *endon*, within, + *thermē*, heat). Having a body temperature determined by heat derived from the animal's own oxidative metabolism; contrasts with ectothermic.

enterocoel (en′tər-ō-sēl′) (Gr., *enteron*, gut, + *koilos*, hollow). A type of coelom that is formed by the outpouching of a mesodermal sac from the endoderm of the primitive gut.

enterocoelic mesoderm formation. Embryonic formation of mesoderm by a pouchlike outfolding from the archenteron, which then expands and obliterates the blastocoel, thus forming a large cavity, the coelom, lined with mesoderm.

enterocoelomate (en′ter-ō-sēl′ō-māte) (Gr., *enteron*, gut, + *koilōma*, cavity, + Eng., *-ate*, state of). An animal having an enterocoel, such as an echinoderm or a vertebrate.

enteron (en′tə-rän) (Gr., intestine). The digestive cavity.

entozoic (en-tə-zō′ic) (Gr., *entos*, within, + *zoon*, animal). Living within another animal; internally parasitic (chiefly parasitic worms).

enzyme (en′zīm) (Gr., *enzymos*, leavened, from *en*, in, + *zyme*, leaven). A protein substance, produced by living cells, that is capable of speeding up specific chemical transformations, such as hydrolysis, oxidation, or reduction, but is unaltered itself in the process; a biologic catalyst.

ephyra (ef′ə-rə) (Gr., *Ephyra*, Greek city). Refers to castle-like appearance. Stage in development of Scyphozoa.

epidermis (ep′ə-dər′məs) (Gr., *epi*, upon, + *derma*, skin). The outer, nonvascular layer of skin of ectodermal origin; in invertebrates, a single layer of ectodermal epithelium.

epididymis (ep′ə-did′ə-məs) (Gr., *epi*, over, + *didymos*, testicle). That part of the sperm duct that is coiled and lying near the testis.

epigenesis (ep′ə-jen′ə-sis) (Gr., *epi*, over, + *genesis*, birth). The embryologic (and generally accepted) view that an embryo is a new creation that develops and differentiates step by step from an initial stage; the progressive production of new parts that were nonexistent as such in the original zygote.

epigenetics (ep′ə-jə-net′iks) (Gr., *epi*, over, + *genesis*, birth). Study of those mechanisms by which the genes produce phenotypic effects.

epilimnion (ep-ə-lim′nē-än) (Gr., *epi*, upon, + *limnē*, lake). Upper water that overlies the thermocline of a lake or sea.

epimorphic (ep′ə-mor′fik) (Gr., *epi*, upon, + *morphē*, form). Having the same form in successive stages of growth; pertains especially to insects and crustaceans in which the juvenile hatching from the egg is morphologically similar to the adult.

epithelium (ep′i-thē′lē-um) (Gr., *epi*, upon, + *thele*, nipple). A cellular tissue covering a free surface or lining a tube or cavity.

epitoke (ep′i-tōk) (Gr., *epitokos*, fruitful). Posterior part of a marine polychaete when swollen with developing gonads during the breeding season.

estrus (es′trəs) (Gr., *oistros*, a gadfly). Heat or rut, especially of the female during ovulation.

estuary (es′chə-we′rē) (L., *aestuarium*, estuary). An arm of the sea where the tide meets the current of a freshwater stream.

ethology (e-thäl′-ə-jē) (Gr., *ethos*, character, + *logos*, discourse). The study of animal behavior in natural environments.

eukaryotic, eucaryotic (ū′-ka-rē-ot′ik) (Gr., *eu*, good, true, + *karyon*, nut, kernel). Containing a visible nucleus or nuclei.

euryhaline (yū′-rə-hā′-līn) (Gr., *eurys*, broad, + *hals*, salt). Able to tolerate wide ranges of saltwater concentrations.

eurytopic (yū-rə-täp′ik) (Gr., *eurys*, broad, + *topos*, place). Refers to an organism with a wide distribution range.

eutely (yū′te-lē) (Gr., *euteia*, thrift). Condition of a body composed of a constant number of cells or nuclei in all adult members of a species, as in nematodes, rotifers, acanthocephalans, and others.

evagination (ē-vaj′ə-nā′shən) (L., *e*, out, + *vagina*, sheath). An outpocketing from a hollow structure.

evolution (L., *evolvere*, to unfold). Organic evolution is any genetic change in organisms, or more strictly a change in gene frequency, from generation to generation.

exergonic (ek′sər-gän′ik) (Gr., *exo*, outside of, + *ergon*, work). Referring to an energy-yielding reaction.

exocrine (ek′sə-krən) (Gr., *exo*, outside, + *krinein*, to separate). A type of gland that releases its secretion through a duct; as opposed to endocrine.

exoskeleton (ek'sō-skel'ə-tən) (Gr., *exo,* without, + *skeletos,* hard). A hard supporting structure secreted by ectoderm or epidermis; external, as opposed to **endoskeleton.**

exteroceptor (ek'stər-ō-sep'tər) (L., *exter,* outward, + *capere,* to take). A sense organ excited by stimuli from the external world.

FAD. Abbreviation for flavine adenine dinucleotide, a hydrogen acceptor in the respiratory chain.

fatty acid. Any of a series of saturated acids having the general formula $C_nH_{2n}O_2$; occurs in natural fats of animals and plants.

fiber, fibril (L., *fibra,* thread). These two terms are often confused. Fiber is a fiberlike cell or a strand of protoplasmic material produced or secreted by a cell and lying outside the cell. Fibril is a strand of protoplasm produced by a cell and lying within the cell.

filter feeding. Any feeding process by which food (usually in fine particles) is filtered from water in which it is suspended.

fission (L., *fissio,* a splitting). Asexual reproduction by a division of the body into two or more parts.

flagellum (flə-jel'əm) (L., a whip). Whiplike organelle of locomotion.

flame bulb. Specialized hollow excretory structure of one or several small cells containing a tuft of cilia (the "flame") and situated at the end of a minute tubule; connected tubules ultimately open to the outside. (See **solenocyte, protonephridium.**)

fluke (O.E., *flōc,* flatflsh). A member of class Trematoda.

food vacuole. A digestive organelle in the cell.

fovea (fō'vē-ə) (L., a small pit). A small pit or depression; especially the fovea centralis, a small rodless pit in the retina of some vertebrates, a point of acute vision.

free energy. The energy available for doing work in a chemical system.

gamete (ga'mēt, gə-mēt') (Gr., *gamos,* marriage). A mature haploid sex cell, either male or female.

ganoid scales (ga'noid) (Gr., *ganos,* brightness). Thick, bony, rhombic scales of some primitive bony fishes; not overlapping.

gastrodermis (gas'tro-dər'mis) (Gr., *gastēr,* stomach, + *derma,* skin). Lining of the digestive cavity of cnidarians.

gastrovascular cavity (Gr., *gaster,* stomach, + L., *vasculum,* small vessel). Body cavity in certain lower invertebrates that functions in both digestion and circulation and has a single opening serving as both mouth and anus.

gastrula (gas'trə-lə) (Gr., *gastēr,* stomach, + L., *ula,* dim.). Embryonic stage, usually cap or sac shaped, with walls of two layers of cells surrounding a cavity (archenteron) with one opening (blastopore).

gastrulation (gas'trə-lā'shən) (Gr., *gaster,* belly). Process by which an early metazoan embryo becomes a gastrula, acquiring first two and then three layers of cells.

gel (jel) (from gelatin, from L., *gelare,* to freeze). That state of a colloidal system in which the solid particles form the continuous phase and the fluid medium the discontinuous phase.

gemmule (jem'yūl) (L., *gemma,* bud, + *ula,* dim). Asexual, cystlike reproductive unit in freshwater sponges; formed in summer or autumn and capable of overwintering.

gene (Gr., *genos,* descent). The part of a chromosome that is the hereditary determiner and is transmitted from one generation to another. It occupies a fixed chromosomal locus and can best be defined only in a physiologic or operational sense.

gene pool. A collection of all of the alleles of all of the genes in a population.

genetic drift. Change in gene frequencies by chance processes in the evolutionary process of animals. In small populations, one allele may drift to fixation, becoming the only representative of that gene locus.

genome (jē'nōm) (Gr., *genos,* offspring, + L., *-oma,* abstract group). The total number of genes in a haploid set of chromosomes.

genotype (jēn'ō-tīp) (Gr., *genos,* offspring, + *typos,* form). The genetic constitution, expressed and latent, of an organism; the total set of genes present in the cells of an organism; as opposed to **phenotype.**

genus (jē-nus), pl. **genera** (L., *genus,* race). A taxonomic rank between family and species.

germ layer. In the animal embryo, one of three basic layers (ectoderm, endoderm, mesoderm) from which the various organs and tissues arise in the multicellular animal.

germ plasm. The germ cells of an organism, as distinct from the somatoplasm; the hereditary material (genes) of the germ cells.

gestation (je-stā'shən) (L., *gestare,* to bear). The period in which offspring are carried in the uterus.

glochidium (glō-kid'-e-əm) (Gr., *glochis,* point, + *-idion,* diminutive). Bivalved larval stage of freshwater mussels.

glomerulus (glä-mer'u-ləs) (L., *glomus,* ball). A ball-shaped network of capillaries projecting into a renal corpuscle in a kidney. Also, a small spongy mass of tissue in the proboscis of hemichordates, presumed to have an excretory function. Also, a concentration of nerve fibers situated in the olfactory bulb.

glycogen (glī'kə-jən) (Gr., *glykys,* sweet, + *genēs,* produced). A polysaccharide constituting the principal form in which carbohydrate is stored in animals; animal starch.

glycolysis (glī-kol'i-sis) (Gr., *glykys,* sweet, + *lyein,* to loosen). Enzymatic breakdown of glucose (especially) or glycogen into pyruvate or lactate with release of energy.

gnathobase (nāth'ə-bās') (Gr., *gnathos,* jaw, + base). A median basic process on certain appendages in some arthropods, usually for biting or crushing food.

Golgi complex (gōl'jē) (after Golgi, Italian histologist). An organelle in cells that serves as a collecting and packaging center for secretory products.

gonad (gō'nad) (Gr., *gonos,* a primary sex gland). A sex gland (ovary in the female and testis in the male).

gonangium (gō-nan'jē-əm) (Gr., *gonos,* seed, + *angeion,* dim. of vessel). Reproductive zooid of hydroid colony (Cnidaria).

gonoduct (Gr., *gonos,* offspring, + duct). Duct leading from a gonad to the exterior.

gonophore (gän′ə-fōr) (Gr., *gonos,* offspring, + *phorein,* to bear). A reproducing zooid of a hydroid colony representing the medusa stage, but differing from a medusa in remaining attached.

gonopore (gän′ə-pōr) (Gr., *gonos,* offspring, seed, + *poros,* an opening). A genital pore found in many invertebrates.

gregarious (L., *grex,* herd). Living in groups or flocks.

guanine (gwä′nēn) (Sp. fr. Quechura, *huanu,* dung). A white crystalline purine base, $C_5H_5N_5O$, occurring in various animal tissues and in guano and other animal excrements; one of the nitrogenous bases in DNA and RNA.

gynandromorph (ji-nan′drə-mawrf) (Gr., *gyn,* female, + *andr,* male, + *morph,* form). A bisexual form with the characteristics of both sexes; bisexual mosaic.

habitat (L., *habitare,* to dwell). The place where an organism normally lives or where individuals of a population live.

habituation. A kind of learning in which continued exposure to the same stimulus produces diminishing responses.

halter (hal′tər), pl. **halteres** (hal-ti′rēz) (Gr., *halter,* leap). In Diptera, small club-shaped structure on each side of the metathorax representing the hindwings; believed to be sense organs for balancing; also called balancer.

haploid (Gr., *haploos,* single). The reduced, or *n,* number of chromosomes, typical of gametes, as opposed to the diploid, or 2*n,* number found in somatic cells. In certain lower phyla, some mature animals have a haploid number of chromosomes.

hectocotylus (hek-tə-kät′ə-ləs) (Gr., *hekaton,* hundred, + *kotylē,* cup). Specialized, and sometimes autonomous, arm that serves as a male copulatory organ in cephalopods.

hemal system (hē′məl) (Gr., *haima,* blood). System of small vessels in echinoderms; function unknown.

hemerythrin (hē′mə-rith′rin) (Gr., *haima,* blood, + *erythros,* red). A red iron-containing respiratory pigment found in the blood of some polychaetes, sipunculids, priapulids, and brachiopods.

hemimetabolous (he′mi-mə-ta′bə-ləs) (Gr., *hēmi-,* half, + *metabolē,* change). Refers to gradual metamorphosis during development of insects, without a pupal stage.

hemoglobin (Gr., *haima,* blood, + L., *globulus,* globule). An iron-containing respiratory pigment occurring in vertebrate red blood cells and in blood plasma of many invertebrates; a compound of an iron porphyrin heme and a protein globin.

hepatic (hə-pat′ic) (Gr., *hēpatikos,* of the liver). Pertaining to the liver.

herbivore ([h]ərb′ə-vōr′) (L., *herba,* green crop, + *vorare,* to devour). Any organism subsisting on plants. Adj. **herbivorous.**

hermaphrodite (hə(r)-maf′rə-dīt) (Gr., *hermaphroditos,* containing both sexes; from Greek mythology, Hermaphroditos, son of Hermes and Aphrodite). An organism with both male and female functional reproductive organs. **Hermaphroditism** may refer to an aberration in unisexual animals; **monoecism** implies that this is the normal condition for the species.

hermatypic (hər-mə-ti′pik). Relating to reef-forming corals.

heterocercal (het′ər-o-sər′kəl) (Gr., *heteros,* other, + *kerkos,* tail). In some fishes, a tail with the upper lobe larger than the lower and the end of the vertebral column somewhat upturned in the upper lobe; for example, as in sharks.

heterotroph (hət′ə-rō-träf) (Gr., *heteros,* another, + *trophos,* feeder). An organism that obtains both organic and inorganic raw materials from the environment in order to live; includes most animals and those plants that do not carry on photosynthesis.

heterozygote (het′ə-rō-zī′gōt) (Gr., *heteros,* another, +*zygōtos,* yoked). An organism in which the pair of alleles for a trait is composed of different genes (usually dominant and recessive); derived from a zygote formed by the union of gametes of dissimilar genetic constitution.

hibernation (L., *hibernus,* wintry). The act of passing the winter in a resting state.

histone (hi′stōn) (Gr., *histos,* tissue). Any of several simple proteins found in cell nuclei and complexed at one time or another with DNA. Histones yield a high proportion of basic amino acids on hydrolysis.

holoblastic cleavage (Gr., *holos,* whole, + *blastos,* germ). Complete and approximately equal division of cells in early embryo. Found in mammals, *Amphioxus,* and many aquatic invertebrates that have eggs with a small amount of yolk.

holometabolous (hō′lō-mə-ta′bə-ləs) (Gr., *holo,* complete, + *metabolē,* change). Refers to complete metamorphosis during development of insects, with a pupal stage.

holophytic nutrition (hōl′ō-fit′ik) (Gr., *holo,* whole, + *phyt,* plant). Occurs in green plants and certain protozoans and involves synthesis of carbohydrates from carbon dioxide and water in the presence of light, chlorophyll, and certain enzymes.

holozoic nutrition (Gr., *holos,* whole, + *zoikos,* of animals). That type of nutrition that involves ingestion of liquid or solid organic food particles.

home range. The area over which an animal ranges in its activities. Unlike territories, home ranges are not defended.

homeostasis (hō′mē-ō-stā′sis) (Gr., *homeo,* similar, + *stasis,* state or standing). Maintenance of an internal steady state by means of self-regulation.

homeothermic (hō-mē-ō-thər′mik) (Gr., *homeo,* alike, + *thermē,* heat). Having a nearly uniform body temperature, regulated independently of the environmental temperature; "warm-blooded."

hominid (häm′ə-nid) (L., *homo, homonis,* man). A member of the family Hominidae, represented by one living species, *Homo sapiens,* and its closest fossil relatives.

hominoid (häm′ə-noid). Relating to the Hominoidea, a superfamily of primates to which the great apes and humans are assigned.

homocercal (hō′mə-ser′kal) (Gr., *homos,* same, common, + *kerkos,* tail). A tail with the upper and lower lobes symmetric and the vertebral column ending near the middle of the base, as in most teleost fishes.

homograft. See **allograft.**

homology (hō-mäl′ə-jē) (Gr., *homologos,* agreeing). Similarity of parts or organs of different organisms caused by similar embryonic origin and evolutionary development from a corresponding part in some remote ancestor. Also, correspondence in structure of different parts of the same individual. May also refer to a matching pair of chromosomes. (Adj. **homologous.**)

homozygote (hō-mə-zī'gōt) (Gr., *homos,* same, + *zygotos,* yoked). An organism in which the pair of alleles for a trait is composed of the same genes (either dominant or recessive but not both). Adj. **homozygous.**

honeydew. A sweet secretion produced by aphids, leafhoppers, and phyllas that is the principal food of some ants; also eaten by bees and wasps.

humoral (hū'mər-əl) (L., *humor,* a fluid). Pertaining to an endocrine secretion. Also pertaining to a normal functioning fluid of the body, as the blood or lymph.

hyaline (hī'ə-lən) (Gr., *hyalos,* glass). Adj., glassy, translucent. Noun, a clear, glassy, structureless material occurring, for example, in cartilage, vitreous body, mucin, and glycogen.

hydranth (hī'dranth) (Gr., *hydōr,* water, + *anthos,* flower). Nutritive zooid of hydroid colony.

hydroid. The polyp form of cnidarian as distinguished from the medusa form. Any cnidarian of the class Hydrozoa, order Hydroida.

hydrolysis (Gr., *hydōr,* water, + *lysis,* a loosening). The decomposition of a chemical compound by the addition of water; the splitting of a molecule into its groupings so that the split products acquire hydrogen and hydroxyl groups.

hydrostatic skeleton. A mass of fluid or plastic parenchyma enclosed within a muscular wall to provide the support necessary for antagonistic muscle action; for example, parenchyma in acoelomates and perivisceral fluids in pseudocoelomates serve as hydrostatic skeletons.

hydroxyl (hydrogen + oxygen + *yl*). Containing an OH^- group, a negatively charged ion formed by alkalies in water.

hypertonic (Gr., *hyper,* over, + *tonos,* tension). Refers to a solution whose osmotic pressure is greater than that of another solution with which it is compared; contains a greater concentration of dissolved particles and gains water through a semipermeable membrane from a solution containing particles. Hyperosmotic; as opposed to **hypotonic.**

hypertrophy (hī-pər'trə-fē) (Gr., *hyper,* over, + *trophē,* nourishment). Abnormal increase in size of a part or organ.

hypodermis (hī'pə-dər'məs) (Gr., *hypo,* under, + L., *dermis,* skin). The cellular layer lying beneath and secreting the cuticle of annelids, arthropods, and certain other invertebrates.

hypolimnion (hī-pō-lim'nē-än) (Gr., *hypo,* under, + *limnē,* lake). The water below the thermocline of a lake.

hypothalamus (hī-pō-thal'ə-məs) (Gr., *hypo,* under, + *thalamos,* inner chamber). A ventral part of the forebrain beneath the thalamus; one of the centers of the autonomic nervous system.

hypotonic (Gr., *hypo,* under, + *tonos,* tension). Refers to a solution whose osmotic pressure is less than that of another solution with which it is compared or taken as standard; contains a lesser concentration of dissolved particles and loses water during osmosis. Hypo-osmotic; as opposed to **hypertonic.**

imago (ə mā'gō). The adult and sexually mature insect.

indeterminate cleavage. A type of early cleavage in which the fate of the blastomeres is not predetermined as to tissues or organs; for example, in echinoderms and vertebrates.

indigenous (ən-dij'ə-nəs) (L., *indigena,* native). Pertains to organisms that are native to a particular region; not introduced.

inductor (in-duk'ter) (L., *inducere,* to introduce, lead in). In embryology, a tissue or organ that causes the differentiation of another tissue or organ.

inquiline (in'kwə-līn) (L., *inquilinus,* lodger). A relation in which a socially parasitic species lives in the abode of its host insect.

instar (inz'tär) (L., form). Stage in the life of an insect or other arthropod between molts.

instinct (L., *instinctus,* impelled). Genetically programmed behavior; unlearned, stereotyped behavior.

integument (ən-teg'ū-mənt) (L., *integumentum,* covering). An external covering or enveloping layer.

interstitial (in-tər-sti'shəl) (L., *inter,* among, + *sistere,* to stand). Situated in the interstices or spaces between body cells or organs.

introvert (L., *intro,* inward, + *vertere,* to turn). The anterior narrow portion that can be withdrawn (introverted) into the trunk of a sipunculid worm.

invagination (in-vaj'ə-nā'shən) (L., *in,* in, + *vagina,* sheath). An infolding of a layer of tissue to form a saclike structure.

irritability (L., *irritare,* to provoke). A general property of all organisms involving the ability to respond to stimuli or changes in the environment.

isolecithal (ī'sə-les'ə-thəl) (Gr., *isos,* equal, + *lekithos,* yolk, + *-al*). Pertaining to a zygote (or ovum) with yolk evenly distributed. Homolecithal.

isotonic (Gr., *isotonos, isos,* equal, + *tonikos,* tension). Said of solutions having the same or equal osmotic pressure; isosmotic.

isotope (Gr., *isos,* equal, + *topos,* place). One of several different forms (species) of a chemical element, differing from each other in atomic mass but not in atomic number.

keratin (ker'ə-tən) (Gr., *kera,* horn, + *-in,* suffix of proteins). A scleroprotein found in epidermal tissues and modified into hard structures such as horns, hair, and nails.

kinetosome (kin-et'ə-sōm) (Gr., *kinētos,* moving, + *sōma,* body). The self-duplicating granule at the base of the flagellum or cilium; similar to centriole.

kinin (kī'nin) (Gr., *kinein,* to move, + *-in,* suffix of hormones). A type of local hormone that is released near its site of origin; also called parahormone or tissue hormone.

kwashiorkor (native name in Ghana). Malnutrition caused by diet high in carbohydrate and extremely low in protein.

labium (lā'bē-əm) (L., a lip). The lower lip of the insect formed by fusion of the second pair of maxillae.

labrum (lā'brəm) (L., a lip). The upper lip of insects and crustaceans situated above or in front of the mandibles; also refers to the outer lip of a gastropod shell.

labyrinthodont (lab′ə-rin′thə-dänt) (Gr., *labyrinthos,* labyrinth, + *odous, odontos,* tooth). A group of fossil stem amphibians from which most amphibians later arose. They date from the late Paleozoic period.

lacteal (lak′tē-əl) (L., *lacteus,* of milk). One of the lymph vessels in the villus of the intestine. (Adj., relating to milk.)

lacuna (lə-kū′nə), pl. **lacunae** (L., pit, cavity). A sinus; a space between cells; a cavity in cartilage or bone.

lagena (lə-jē′nə) (L., large flask). Portion of the primitive ear in which sound is translated into nerve impulses; evolutionary beginning of cochlea.

lamella (lə-mel′ə) (L., dim. of *lamina,* plate.). One of the two plates forming a gill in a bivalve mollusc. One of the thin layers of bone laid concentrically around a haversian canal. Any thin, platelike structure.

larva (lär′və), pl. **larvae** (L., a ghost). An immature stage that is quite different from the adult.

lek (lek) (Sw., play, game). An area where animals assemble for communal courtship display and mating.

lemniscus (lem-nis′kəs) (L., ribbon). One of a pair of internal projections of the epidermis from the neck region of Acanthocephala, which functions in fluid control in the protrusion and invagination of the proboscis.

lentic (len′tik) (L., *lentus,* slow). Of or relating to standing water such as swamp, pond, or lake.

leukocyte (lū′kə-sīt′) (Gr., *leukos,* white, + *kytos,* hollow vessel). A common type of white blood cell with beaded nucleus.

lipase (lī′pās) (Gr., *lipos,* fat, + *-ase,* enzyme suffix). An enzyme that accelerates the hydrolysis or synthesis of fats.

lipid, lipoid (li′pid) (Gr., *lipos,* fat). Certain fatlike substances that often contain other groups, such as phosphoric acid; lipids combine with proteins and carbohydrates to form principal structured components of cells.

lithosphere (lith′ə-sfir) (Gr., *lithos,* rock, + *sphaira,* ball). The rocky component of the earth's surface layers.

littoral (lit′ə-rəl) (L., *litoralis,* seashore). Adj., Pertaining to the shore. Noun, that portion of the sea floor from the shore to the continental shelf; often used to refer only to the portion of the shore between the high and low tide levels, the intertidal; in lakes, the shallow part from the shore to the lakeward limit of aquatic plants.

locus (lō′kəs), pl. **loci** (lō-sī) (L., place). Position of a gene in a chromosome.

lophophore (lōf′ə-fōr) (Gr., *lophos,* crest, + *phoros,* bearing). Tentacle-bearing ridge or arm that is an extension of the coelomic cavity in lophophorate animals (ectoprocts, brachiopods, and phoronids).

lotic (lō′tik) (L., *lotus,* action of washing or bathing). Of or pertaining to running water, such as a brook or river.

lumen (lū′mən) (L., light). The cavity of a tube or organ.

lysosome (lī′sə-sōm) (Gr., *lysis,* loosing, + *soma,* body). Intracellular organelle consisting of a membrane enclosing several digestive enzymes that are released when the lysosome ruptures.

macromolecule. A very large molecule, such as a protein or a starch.

macronucleus (ma′krō-nū-klē-əs) (Gr., *makros,* long, large, + *nucleus,* kernel). The larger of the two kinds of nuclei in ciliate protozoa; controls all cell functions except reproduction.

madreporite (ma′drə-pōr′īt) (Fr., *madrépore,* reef-building coral, + *-ite,* suffix for some body parts). Sievelike structure that is the intake for the water-vascular system of echinoderms.

malacostracan (mal′ə-käs′trə-kən) (Gr., *malako,* soft, + *ostracon,* shell). Any member of the crustacean subclass Malacostraca, which includes both aquatic and terrestrial forms of crabs, lobsters, shrimps, pill bugs, sand fleas, and others.

Malpighian tubules (mal-pig′ ē-ən) (Marcello Malpighi, Italian anatomist, 1628-1694). Blind tubules opening into the hindgut of nearly all insects and some myriapods and arachnids, and functioning primarily as excretory organs.

mantle. Soft extension of the body wall in certain invertebrates, for example, brachiopods and molluscs, in which it usually secretes a shell; thin body wall of tunicates.

marasmus (mə-raz′məs) (Gr., *marasmos,* to waste away). Malnutrition, especially of infants, caused by a diet deficient in both calories and protein.

marsupial (mär-sū′pē-əl) (Gr., *marsypion,* little pouch). One of the pouched mammals of the subclass Metatheria.

matrix (mā′triks) (L., *mater,* mother). The intercellular substance of a tissue or that part of a tissue into which an organ or process is set.

maturation (L., *maturus,* ripe). The process of ripening; the final stages in the preparation of gametes for fertilization.

maxilla (mak-sil′ə) (L., dim. of *mala,* jaw). One of the upper jawbones in vertebrates; one of the head appendages in arthropods.

maxillipeds (mak-sil′ə-ped) (L., *maxilla,* jaw, + *-ped,* foot). Legs that have become specialized for feeding, derived from appendages on somites just posterior to the maxillary somite in crustaceans.

medulla (mə-dul′ə) (L., marrow). The inner portion of an organ in contrast to the cortex or outer portion; hindbrain.

medusa (mə-dū′sə) (Gr. mythology, female monster with snake-entwined hair). A jellyfish, or the free-swimming stage in the life cycle of cnidarians.

meiosis (mī-ō′səs) (Gr., fr. *meioun,* to make small). The nuclear changes that occur in the last two divisions in the formation of the mature egg or sperm, by means of which the chromosomes are reduced from the diploid to the haploid number.

melanin (mel′ə-nin) (Gr., *melas,* black). Black or dark brown pigment found in plant or animal structures.

menopause (men′ō-pawz) (Gr., *men,* month, + *pauein,* to cease). In the human female that time of life when ovulation ceases; cessation of the menstrual cycle.

menstruation (men′stroo-ā′shən) (L., *menstrua,* the menses, fr. *mensis,* month). The discharge of blood and uterine tissue from the vagina at the end of a menstrual cycle.

meroblastic (mer-ə-blas′tik) (Gr., *meros,* part, + *blastos,* germ). Pertaining to cleavage occurring in zygotes having a large amount of yolk at the vegetal pole; cleavage restricted to a small area on the surface of the egg.

mesenchyme (me'zn-kīm) (Gr., *mesos,* middle, + *enchyma,* infusion). Embryonic connective tissue; irregular or amebocytic cells often embedded in gelatinous matrix.

mesoderm (me'zə-dərm) (Gr., *mesos,* middle, + *derma,* skin). The third germ layer, formed in the gastrula between the ectoderm and endoderm; gives rise to connective tissues, muscle, urogenital and vascular systems, and the peritoneum.

mesoglea (mez'ō-glē'ə) (Gr., *mesos,* middle, + *gloios,* glutinous substance). The layer of jellylike or cement material between the epidermis and gastrodermis in cnidarians and ctenophores; also may refer to jellylike matrix between epithelial layers in sponges.

mesonephros (me-zō-nef'rōs) (Gr., *mesos,* middle, + *nephros,* kidney). The middle of three pairs of embryonic renal organs in vertebrates. Functional kidney of fishes and amphibians; its collecting duct is a wolffian duct. (Adj. **mesonephric.**)

messenger RNA (mRNA). A form of ribonucleic acid that carries genetic information from the gene to the ribosome, where it determines the order of amino acids as a polypeptide is formed.

metabolism (Gr., *metabolē,* change). A group of processes that includes nutrition, production of energy (respiration), and synthesis of protoplasm; the sum of the constructive (anabolic) and destructive (catabolic) processes.

metacercaria (mə'tə-sər-ka'rē-ə) (Gr., *meta,* after, + *kerkos,* tail, + L., *aria,* connected with). Fluke larva (cercaria) that has lost its tail and become encysted in an intermediate aquatic host.

metamere (met'ə-mir) (Gr., *meta,* after, + *meros,* part). A repeated unit of structure; a somite, or segment.

metamerism (mə-ta'mə-ri'zəm) (Gr., *meta,* after, + *meros,* part). Condition of being made up of serially repeated parts (metameres); serial segmentation.

metamorphosis (Gr., *meta,* after, + *morphē,* form, + *-osis,* state of). Sharp change in form during postembryonic development; for example, tadpole to frog or larval insect to adult.

metanephridium (me'tə-nə-fri'di-əm) (Gr., *meta,* after, + *nephros,* kidney). A type of tubular nephridium with the inner open end draining the coelom and the outer open end discharging to the exterior.

metanephros (me'tə-ne'fräs) (Gr., *meta,* between, after, + *nephros,* kidney). Embryonic renal organs of vertebrates arising behind the mesonephros; the functional kidney of reptiles, birds, and mammals. It is drained by a ureter.

micrometer (μm) (mī'kräm'me-tər) (Gr., neuter of *mikros,* small, + meter). One one-thousandth of a millimeter; about 1/25,000 of an inch. Formerly designated micron (μ).

micronucleus. A small nucleus found in ciliate protozoa; controls the reproductive functions of these organisms.

microtubule (Gr., *mikros,* small, + L., *tubule,* pipe). A long, tubular cytoskeletal element with an outside diameter of 200 to 270 Å. Microtubules influence cell shape and play important roles during cell division.

microvillus (Gr., *mikros,* small, + L., *villus,* shaggy hair). Narrow, cylindric cytoplasmic projection from epithelial cells; microvilli form the brush border of several types of epithelial cells.

mictic (mik'tik) (Gr., *miktos,* mixed or blended). Pertaining to haploid egg of rotifers or the females that lay such eggs.

miracidium (mīr'ə-sid'ē-əm) (Gr., *meirakidion,* youthful person). A minute ciliated larval stage in the life of flukes.

mitochondrion (mīd'ə-kän'drē-ən) (Gr., *mitos,* a thread, + *chondrion,* dim. of *chondros,* corn, grain). An organelle in the cell in which aerobic metabolism takes place.

mitosis (mī-tō'səs) (Gr., *mitos,* thread, + *-osis,* state of). Cell division in which there is an equal qualitative and quantitative division of the chromosomal material between the two resulting nuclei; ordinary cell division (indirect).

monoecious (mə-nē'shəs) (Gr., *monos,* single, + *oikos,* house). Having both male and female gonads in the same organism; hermaphroditic.

monohybrid (Gr., *monos,* single, + L., *hybrida,* mongrel). A hybrid offspring of parents different in one specified character.

monomer (mä'nə-mər) (Gr., *monos,* single, + *meros,* part). A molecule of simple structure, but capable of linking with others to form polymers.

monophyletic (mä'nə-phī-le'tik) (Gr., *mono,* single, + *phyletikos,* pertaining to a phylum). Referring to a taxon whose units all evolved from a single parent stock; as opposed to **polyphyletic.**

monosaccharide (mä'nə-sa'kə-rīd) (Gr., *monos,* one, + *sakcharon,* sugar, from Sanskrit *sarkarā,* gravel, sugar). A simple sugar that cannot be decomposed into smaller sugar molecules; the most common are pentoses (such as ribose) and hexoses (such as glucose).

morphogenesis (mor'fə-je'nə-səs) (Gr., *morphē,* form, + *genesis,* origin). Development of the architectural features of organisms; formation and differentiation of tissues and organs.

morphology (Gr., *morphē,* form, + L., *logia,* study, from Gr. *logos,* word). The science of structure. Includes cytology, or the study of cell structure; histology, or the study of tissue structure; and anatomy, or the study of gross structure.

morula (mär'u-lə) (L., *morum,* mulberry, + *ula,* diminutive). Solid group of cells in early stage of segmentation.

mosaic cleavage. Type characterized by independent differentiation of each part of the embryo; determinate cleavage.

Müller's larva. Free-swimming ciliated larva that resembles a modified ctenophore, characteristic of certain marine polyclad turbellarians.

mutation (myū-tā'shən) (L., *mutare,* to change). A stable and abrupt change of a gene; the heritable modification of a character.

mutualism (myū'chə-wə-li'zəm) (L., *mutuus,* lent, borrowed, reciprocal). A type of symbiosis in which two different species derive benefit from their association.

myofibril (Gr., *mys,* muscle, mouse, + L., dim. of *fibra,* fiber). A contractile filament within muscle or muscle fiber.

myoneme (mī′ə-nēm′) (Gr., *mys,* muscle, mouse, + *nēma,* thread). Long contractile fibril in certain protozoans.

myosin (mī′ə-sin) (Gr., *mys,* muscle, mouse). A large protein of contractile tissue that forms the thick myofilaments of striated muscle. During contraction it combines with actin to form actomyosin.

myotome (mī′ə-tōm′) (Gr., *mys,* muscle, mouse, + *tomos,* cutting). A voluntary muscle segment in cephalochordates and vertebrates; that part of a somite destined to form muscles; the muscle group innervated by a single spinal nerve.

nacre (nā′kər) (F., mother-of-pearl). Innermost lustrous layer of mollusc shell, secreted by mantle epithelium. (Adj. **nacreous.**)

NAD. Abbreviation of nicotinamide adenine dinucleotide; see **DPN.**

naiad (nā′əd) (Gr., *naias,* a water nymph). An aquatic, gill-breathing nymph.

nares (na′rēz), sing. **naris** (L., nostrils). Openings into the nasal cavity, both internally and externally, in the head of a vertebrate.

natural selection. A nonrandom reproduction of genotypes that results in the survival of those best adapted to their environment and elimination of those less well adapted; leads to evolutionary change.

nauplius (naw′plē-əs) (L., a kind of shellfish). A free-swimming microscopic larval stage of certain crustaceans, with three pairs of appendages (antennules, antennae, and mandibles) and median eye. Characteristic of ostracods, copepods, barnacles, and some others.

nekton (nek′tən) (Gr., neuter of *nēktos,* swimming). Term for actively swimming organisms, essentially independent of wave and current action. Compare with **plankton.**

nematocyst (ne-mad′ə-sist′) (Gr., *nēma,* thread, + *kystis,* bladder). Stinging organoid of cnidarians.

neoteny (nē′ə-tē′nē, nē-ot′ə-nē) (Gr., *neos,* new, + *teinein,* to extend). The attainment of sexual maturity in the larval condition. Also the retention of larval characters into adulthood.

nephridium (nə-frid′ē-əm) (Gr., *nephridios,* of the kidney). One of the segmentally arranged, paired excretory tubules of many invertebrates, notably the annelids. In a broad sense, any tubule specialized for excretion and/or osmoregulation, with an external opening and with or without an internal opening.

nephron (ne′frän) (Gr., *nephros,* kidney). Functional unit of kidney structure of reptiles, birds, and mammals, consisting of a Bowman's capsule body, its glomerulus, and the attached uriniferous tubule.

neurosecretory cell (nu′rō-sə-krēd′ə-rē). Any cell (neuron) of the nervous system that produces a hormone.

nitrogen fixation (Gr., *nitron,* soda, + *gen,* producing). Oxidation of ammonia to nitrites and of nitrites to nitrates, as by action of bacteria.

notochord (nōd′ə-kord′) (Gr., *nōtos,* back, + *chorda,* cord). An elongated cellular cord, enclosed in a sheath, which forms the primitive axial skeleton of chordate embryos and adult cephalochordates.

nucleic acid (nu-klē′ik) (L., *nucleus,* kernel). One of a class of molecules composed of joined nucleotides; chief types are deoxyribonucleic acid (DNA), found in cell nuclei (chromosomes), and ribonucleic acid (RNA), found both in cell nuclei (chromosomes and nucleoli) and in cytoplasmic ribosomes.

nucleolus (nu-klē′ə-ləs) (dim. of *nucleus*). A deeply staining body within the nucleus of a cell and containing RNA.

nucleoplasm (nu′klē-ə-plazm′) (L., *nucleus,* kernel, + Gr., *plasma,* mold). Protoplasm of nucleus, as distinguished from cytoplasm.

nucleoprotein. A molecule composed of nucleic acid and protein; occurs in the nucleus and cytoplasm of all cells.

nucleotide (nū′klē-ə-tīd). A molecule consisting of a phosphate, a 5-carbon sugar (ribose or deoxyribose), and a purine or a pyrimidine; the purines are adenine and guanine, and the pyrimidines are cytosine, thymine, and uracil.

nuptial flight (nəp′shəl). The mating flight of insects, especially that of the queen with a male or males.

nymph (L., *nympha,* nymph, bride). An immature stage (following hatching) of a hemimetabolic insect that lacks a pupal stage.

ocellus (ō-sel′əs) (L., dim. of *oculus,* eye). A simple eye or eyespot in many types of invertebrates.

ommatidium (ä′mə-tid′ē-əm) (Gr., *omma,* eye, + *idium,* small). One of the optic units of the compound eye of arthropods and molluscs.

oncomiracidium (än′kō-mīr′ə-sid′ē-əm) (Gr., *onkos,* barb, hook, + *meirakidion,* youthful person). A ciliated larva of a monogenetic trematode.

ontogeny (än-tä′jə-nē) (Gr., *ontos,* being, + *-geneia,* act of being born, fr. *genēs,* born). The life history of an individual from egg to senescence.

ooecium (ō-ēs′ē-əm) (Gr., *ōion,* egg, + *oikos,* house, + L., *-ium,* from Ger., *-ion,* diminutive). Brood pouch; compartment for developing embryos in ectoprocts.

oogonium (ō′ə-gōn′ē-əm) (Gr., *ōion,* egg, + *gonos,* offspring). A cell that, by continued division, gives rise to oocytes; an ovum in a primary follicle immediately before the beginning of maturation.

operculum (ō-per′kyə-ləm) (L., cover). The gill cover in bony fishes; horny plate in some snails.

operon (äp′ə-rän). A genetic unit consisting of a cluster of genes that are under the control of an operator and a repressor.

ophthalmic (äf-thal′mik) (Gr., *ophthalmos,* an eye). Pertaining to the eye.

opisthaptor (ä′pəs-thap′tər) (Gr., *opisthen,* behind, + *haptein,* to fasten). Posterior attachment organ of a monogenetic trematode.

organelle (Gr., *organon,* tool, organ, + L., *ella,* diminutive). Specialized part of a cell; literally, a small organ that performs functions analogous to organs of multicellular animals.

organoid (or-gə-noid′) (Gr., *organon,* tool, organ, + *eidos,* form). A morphologically differentiated part of a cell, such as a mitochondrion or Golgi body.

osmoregulation. Maintenance of proper internal salt and water concentrations in a cell or in the body of a living organism; active regulation of internal osmotic pressure.

osmosis (oz-mō′sis) (Gr., *ōsmos,* act of pushing, impulse). The flow of solvent (usually water) through a semipermeable membrane.

osphradium (äs-frā′dē-əm) (Gr., *osphradion,* small bouquet, dim. of *osphra,* smell). A sense organ in aquatic snails and bivalves that tests incoming water.

ossicles (L., *ossiculum,* small bone). Small separate pieces of echinoderm endoskeleton; tiny bones of middle ear of vertebrates.

ostium (L., door). An opening.

otolith (ōd′əl-ith′) (Gr., *ous, otos,* ear, + *lithos,* stone). Calcareous concretions in the membranous labyrinth of the inner ear of lower vertebrates or in the auditory organ of certain invertebrates.

oviparity (ō′və-pa′rəd-ē) (L., *ovum,* egg, + *parere,* to bring forth). Reproduction in which eggs are released by the female; development of offspring occurs outside the maternal body. (Adj. **oviparous** [ō-vip′ə-rəs]).

ovipositor (ō′və-päz′əd-ər) (L., *ovum,* egg, + *positus,* fr. *ponere,* to place). In many female insects, a structure at the posterior end of the abdomen for laying eggs.

ovoviviparity (ō′vo-vī-və-par′əd-ē) (L., *ovum,* egg, + *vivere,* to live, + *parere,* to bring forth). Reproduction in which eggs develop within the maternal body but without additional nourishment from the parent and hatch within the parent or immediately after laying. (Adj. **ovoviviparous** [ō′vo-vī-vip′ə-rəs]).

oxidation (äk′sə-dā′shən) (Fr., *oxider,* to oxidize, fr. Gr., *oxys,* sharp, + *-ation*). The loss of an electron by an atom or molecule; addition of oxygen chemically to a substance.

oxidative phosphorylation (ok′sə-dād′iv fäs′fər-i-lā′shən). The conversion of inorganic phosphate to energy-rich phosphate of ATP.

paedogenesis (pē-dō-jen′ə-sis) (Gr., *pais,* child, + *genēs,* born). Reproduction by immature or larval animals caused by acceleration of maturation. Progenesis.

paedomorphosis (pē-dō-mor′fə-səs) (Gr., *pais,* child, + *morphē,* form). Displacement of ancestral juvenile features to later stages of the ontogeny of descendants.

pair bond. An affiliation between an adult male and an adult female for reproduction. Characteristic of monogamous species.

papilla (pə-pil′ə) (L., nipple). A small nipplelike projection. A vascular process that nourishes the root of a hair, feather, or developing tooth.

parabiosis (pa′rə-bī-ō′sis) (Gr., *para,* beside, + *biosis,* mode of life). The fusion of two individuals, resulting in mutual physiologic intimacy.

parapodium (pa′rə-pō′dē-əm) (Gr., *para,* subsidiary, + *pous, podos,* foot). One of the paired flat, lateral processes on each side of most segments in polychaete annelids; variously modified for locomotion, respiration, or feeding.

parasitism (par′ə-sīd′iz-əm) (Gr., *parasitos,* from *para,* beside, + *sitos,* food). The condition of an organism living in or on another organism at whose expense the parasite is maintained; destructive symbiosis.

parasympathetic (par′ə-sim-pə-thed′ik) (Gr., *para,* beside, + *sympathes,* sympathetic, from *syn,* with, + *pathos,* feeling). One of the subdivisions of the autonomic nervous system, whose fibers originate in the brain and in anterior and posterior parts of the spinal cord.

parenchyma (pə-ren′kə-mə) (Gr., anything poured in beside). In lower animals, a spongy mass of vacuolated mesenchyme cells filling spaces between viscera, muscles, or epithelia; the specialized tissue of an organ as distinguished from the supporting connective tissue.

parthenogenesis (pär′thə-nō-gen′ə-sis) (Gr., *parthenos,* virgin, + L. fr. Gr., *genesis,* origin). Unisexual reproduction involving the production of young by females not fertilized by males; common in rotifers, cladocerans, aphids, bees, ants, and wasps. A parthenogenetic egg may be diploid or haploid.

pathogenic (path′ə-jen′ik) (Gr., *pathos,* disease, + *gennan,* to produce). Producing or capable of producing disease.

peck order. A hierarchy of social privilege in a flock of birds.

pecten (L., comb). A pigmented, vascular, and comblike process that projects into the vitreous humor from the retina at point of entrance of the optic nerve in the eyes of all birds and many reptiles.

pectoral (pek′tə-rəl) (L., *pectoralis,* fr. *pectus,* the breast). Of or pertaining to the breast or chest, to the pectoral girdle, or to a pair of horny shields of the plastron of certain turtles.

pedicel (ped′ə-sel) (L., *pediculus,* little foot). A small or short stalk or stem. In insects, the second segment of an antenna or the waist of an ant.

pedicellaria (ped′ə-sə-lar′ē-ə) (L., *pediculus,* little foot, + *-aria,* like or connected with). One of many minute pincerlike organs on the surface of certain echinoderms.

pedipalps (ped′ə-palps′) (L., *pes, pedis,* foot, + *palpus,* stroking, caress). Second pair of appendages of arachnids.

pedogenesis. See **paedogenesis.**

peduncle (pē′dən-kəl) (L., *pedunculus,* dim. of *pes,* foot). A stalk; a band of white matter joining different parts of the brain.

pelagic (pə-laj′ik) (Gr., *pelagos,* the open sea). Pertaining to the open ocean.

pellicle (pel′ə-kəl) (L., *pellicula,* dim. of *pellis,* skin). Thin, translucent, secreted envelope covering many protozoans.

pentadactyl (pen-tə-dak′təl) (Gr., *pente,* five, + *daktylos,* finger). With five digits, or five fingerlike parts, to the hand or foot.

peptidase (pep′tə-dās) (Gr., *peptein,* to digest, + *-ase,* enzyme suffix). An enzyme that breaks down simple peptides, releasing amino acids.

peptide bond. A bond that binds amino acids together into a polypeptide chain, formed by removing an OH from the carboxyl group of one amino acid and an H from the amino group of another to form an amide group —CO—NH—.

periostracum (pe-rē-äs'trə-kəm) (Gr., *peri,* around, + *ostrakon,* shell). Outer horny layer of a mollusc shell.

periproct (per'ə-präkt) (Gr., *peri,* around, + *prōktos,* anus). Region of aboral plates around the anus of echinoids.

perisarc (per'ə-särk) (Gr., *peri,* around, + *sarx,* flesh). Sheath covering the stalk and branches of a hydroid.

perissodactyl (pə-ris'ə-dak'təl) (Gr., *perissos,* odd, + *dactylos,* finger, toe). Pertaining to an order of ungulate mammals with an odd number of digits on each foot.

peristalsis (per'ə-stal'səs) (Gr., *peristaltikos,* compressing around). The series of alternate relaxations and contractions that serve to force food through the alimentary canal.

peristomium (per'ə-stō'mē-əm) (Gr., *peri,* around, + *stoma,* mouth). Foremost true segment of an annelid; it bears the mouth.

peritoneum (per'ə-tə-nē'əm) (Gr., *peritonaios,* stretched across). The membrane that lines the coelom and is reflected over the coelomic viscera.

pH (*p*otential of *h*ydrogen). A symbol of the relative concentration of hydrogen ions in a solution; pH values are from 0 to 14, and the lower the value, the more acid or hydrogen ions in the solution. Equal to the negative logarithm of the hydrogen ion concentration.

phagocyte (fag'ə-sīt) (Gr., *phagein,* to eat, + *kytos,* hollow vessel). Any cell that engulfs and devours microorganisms or other foreign particles.

phagocytosis (fag'ə-sī-tō'səs) (Gr., *phagein,* to eat, + *kytos,* hollow). The engulfment of a foreign particle by a phagocyte.

pharynx (far'inks), pl. **pharynges** (Gr.). The part of the digestive tract between the mouth cavity and the esophagus that, in vertebrates, is common to both digestive and respiratory tracts. In cephalochordates, the gill slits open from it.

phasmid (faz'mid) (Gr., *phasma,* apparition, phantom, + *-id*). One of a pair of glands or sensory structures found in the posterior end of certain nematodes.

phenotype (fē'nə-tīp') (Gr., *phainein,* to show). The visible characters of an organism; opposed to genotype of the hereditary constitution.

pheromone (fer'ə-mōn) (Gr., *phorein,* to carry). Substance released by one organism that influences the behavior or physiology of another organism; ectohormone.

phosphagen (fäs'fə-jən) (phosphate + *gen*). A term for creatine-phosphate and arginine-phosphate, which store and may be sources of high-energy phosphate bonds.

phosphatide (fäs'fə-tīd') (phosphate + *-ide*). A lipid with phosphorus, such as lecithin. A complex phosphoric ester lipid, such as lecithin, found in all cells. Phospholipid.

phosphorylation (fäs'fə-rə-lā'shən). The addition of a phosphate group, such as H_2PO_3, to a compound.

photosynthesis (fōd'ō-sin'thə-sis) (Gr., *phōs,* light, + *synthesis,* action or putting together). The synthesis of carbohydrates from carbon dioxide and water in chlorophyll-containing cells exposed to light.

phototropism (fō-tä'trō-piz'm) (Gr., *phōs,* light, + *trope,* turn). A tropism in which light is the orienting stimulus. An involuntary tendency for an organism to turn toward (positive) or away from (negative) light.

phylogeny (fī-läj'ə-nē) (Gr., *phylon,* tribe, race, + *geneia,* origin). The origin and development of any taxon, or the evolutionary history of its development.

phylum (fī'ləm), pl. **phyla** (NL. fr. Gr., *phylon,* race, tribe). A chief category of taxonomic classification into which are grouped organisms of common descent that share a fundamental pattern of organization.

pilidium (pī-lid'ē-əm) (Gr., *pilidion,* dim. of *pilos,* felt cap). Free-swimming hat-shaped larva of nemertine worms.

pinna (pin'ə) (L., feather, sharp point). The external ear. Also a feather, wing, fin, or similar part.

pinocytosis (pin'o-cī-tō'sis, pīn'o-cī-to'sis) (Gr., *pinein,* to drink, + *kytos,* hollow vessel, + *-osis,* condition). Taking up of fluid by living cells; cell-drinking.

placenta (plə-sen'tə) (L., flat cake; Gr., *plakous,* fr. Gr., *plax, plakos,* anything flat and broad). The vascular structure, embryonic and maternal, through which the embryo and fetus are nourished while in the uterus.

placoid scale (pla'koid) (Gr., *plax, plakos,* tablet, plate). Type of scale found in cartilaginous fishes, with basal plate of dentine embedded in the skin and a backward-pointing spine tipped with enamel.

plankton (plank'tən) (Gr., neuter of *planktos,* wandering). The passively floating animal and plant life of a body of water. Compare with **nekton.**

plantigrade (plan'tə-grād') (L., *planta,* sole, + *gradus,* step, degree). Pertaining to animals that walk on the whole surface of the foot (for example, humans and bears). Compare with **digitigrade.**

planula (plan'yə-lə) (NL., dim. of L., *planus,* flat). Free-swimming, ciliated larval type of cnidarians; usually flattened and ovoid, with an outer layer of ectodermal cells and an inner mass of endodermal cells.

plasma membrane (plaz'mə) (Gr., *plasma,* a form, mold). A living, external, limiting, protoplasmic structure that functions to regulate exchange of nutrients across the cell surface.

pleiotropic (plī'ə-trä'pik) (Gr., *pleiōn,* more, + *tropos,* to turn). Pertaining to a gene producing more than one effect; affecting multiple phenotypic characteristics.

pleopod (plē'ə-päd') (Gr., *plein,* to sail, + *pous, podos,* foot). One of the appendages on the abdomen of a crustacean.

pleura (plu'rə) (Gr., side, rib). The membrane that lines each half of the thorax and covers the lungs.

plexus (plek'səs) (L., network, braid). A network, especially of nerves or of blood vessels.

pluteus (plū'dē-əs), pl. **plutei** (L., *pluteus,* movable shed, reading desk). Echinoid larva with elongated processes like the supports of a desk; originally called "painter's easel larva."

poikilothermic (poi-ki'lə-thər'mik) (Gr., *poikilos,* variable, + thermal). Pertaining to animals whose body temperature is variable and fluctuates with that of the environment; cold-blooded. Compare with **ectothermic.**

polarization (L., *polaris,* polar, + Gr., *-iz-,* make). The arrangement of positive electric charges on one side of a surface membrane and negative electric charges on the other side (in nerves and muscles).

polymer (pä′lə-mər) (Gr., *polys,* many, + *meros,* part). A chemical compound composed of repeated structural units called monomers.

polymerization (pə-lim′ər-ə-zā′shən). The process of forming a polymer or polymeric compound.

polymorphism (pä′lē-mor′fi-zəm) (Gr., *polys,* many, + *morphe,* form). The presence in a species of more than one type of individual.

polynucleotide (*poly* + nucleotide). A nucleotide of many mononucleotides combined.

polyp (päl′əp) (Fr., *polype,* octopus, fr. L., *polypus,* many-footed). The sessile stage in the life cycle of cnidarians.

polypeptide (pä-lē-pep′tīd) (Gr., *polys,* many, + *peptein,* to digest). A molecule consisting of a few to many joined amino acids, not as complex or with as many amino acids as a protein.

polyphyletic (pä′lē-fī-led′ik) (Gr., *polys,* many, + *phylon,* tribe). Derived from more than one ancestral source; as opposed to **monophyletic.**

polyphyodont (pä′lē-fī′ə-dänt) (Gr., *polyphyes,* manifold, + *odous,* tooth). Having several sets of teeth in succession.

polyploid (pä′lə-ploid′) (Gr., *polys,* many, + *ploidy,* number of chromosomes). Characterized by a chromosome number that is greater than two full sets of homologous chromosomes.

polysaccharide (pä′lē-sak′ə-rid, -rīd) (Gr., *polys,* many, + *sakcharon,* sugar, fr. Sanskrit *śarkarā,* gravel, sugar). A carbohydrate composed of many monosaccharide units; for example, glycogen, starch, and cellulose.

polysome (polyribosome) (Gr., *polys,* many, + *soma,* body). Two or more ribosomes connected by a molecule of messenger RNA.

pongid (pän′jəd) (L., *Pongo,* type genus of orangutan). Of or relating to the primate family Pongidae, comprising the anthropoid apes (gorillas, chimpanzees, gibbons, orangutans).

population (L., *populus,* people). A group of organisms of the same species inhabiting a specified geographic locality.

portal system (L., *porta,* gate). System of large veins beginning and ending with a bed of capillaries; for example, hepatic-portal and renal-portal systems in vertebrates.

preadaptation. The possession of a condition not necessarily adapted to the ancestral environment but that predisposes an organism for survival in some other environment.

prebiotic synthesis. The chemical synthesis that occurred before the emergence of life.

precocial (prē-kō′shəl) (L., *praecoquere,* to ripen beforehand). Referring (especially) to birds whose young are covered with down and are able to run about when newly hatched.

predaceous, predacious (prē-dā′shəs) (L., *praeda,* prey, plunder). Living by killing and consuming other animals, predatory.

predator (pred′ə-tər) (L., *praeda,* plunder, prey). An organism that preys on other organisms for its food.

prehensile (prē-hen′səl) (L., *prehendere,* to seize). Adapted for grasping.

primate (prī′māt) (L., *primus,* first). Any mammal of the order Primates, which includes the tarsiers, lemurs, marmosets, monkeys, apes, and humans.

primitive (L., *primus,* first). Primordial; ancient; little evolved; said of species closely approximating their early ancestral types.

proboscis (prō-bäs′əs) (Gr., *pro,* before, + *boskein,* feed). A snout or trunk; tubular sucking or feeding organ with mouth at the end as in planarians, leeches, and insects; the sensory and defensive organ at the anterior end of certain invertebrates.

producers (L., *producere,* to bring forth). Organisms, such as plants, able to produce their own food from inorganic substances.

progesterone (prō-jes′tə-rōn′) (L., *pro,* before, + *gestare,* to carry). Hormone secreted by the corpus luteum and the placenta; prepares the uterus for the fertilized egg and maintains the capacity of the uterus to hold the embryo and fetus.

proglottid (prō-gläd′əd) (Gr., *proglōttis,* tongue tip, from *pro,* before, + *glōtta,* tongue, + *-id,* suffix). A segment of a tapeworm.

prokaryotic, procaryotic (pro-kar′ē-ot′ik) (Gr., *pro-,* before, + *karyon,* kernel, nut). Not having a visible nucleus or nuclei. Prokaryotic cells were more primitive than eukaryotic cells and persist today in the bacteria and blue-green algae.

pronephros (prō-nef′rəs) (Gr., *pro-,* before, + *nephros,* kidney). Most anterior of three pairs of embryonic renal organs of vertebrates; functional only in adult hagfishes and larval fishes and amphibians; vestigial in mammalian embryos. (Adj. **pronephric.**).

proprioceptor (prō′prē-ə-sep′tər) (L., *proprius,* own, particular, + receptor). Sensory receptor located deep within the tissues, especially muscles, tendons, and joints, that is responsive to changes in muscle stretch, body position, and movement.

prosimian (prō-sim′ē-ən) (Gr., *pro,* before, + L., *simia,* ape). Any member of a group of primitive, arboreal primates; lemurs, tarsiers, lorises, and so on.

prosoma (prō-sōm′ə) (Gr., *pro,* before, + *sōma,* body). Anterior part of an invertebrate in which primitive segmentation is not visible; fused head and thorax of arthropod; cephalothorax.

prosopyle (präs′ə-pīl) (Gr., *prosō, forward,* + *pylē,* gate). Connections between the incurrent and radial canals in some sponges.

prostaglandins (prös′tə-glan′dəns). A family of fatty acid tissue hormones, originally discovered in semen, known to have powerful effects on smooth muscle, nerves, circulation, and reproductive organs.

prostomium (prō-stō'mē-əm) (Gr., *pro,* before, + *stoma,* mouth). In most annelids and some molluscs, that part of the head located in front of the mouth.

protease (prō'tē-ās) (Gr., *protein* + *-ase,* enzyme). An enzyme that digests proteins; includes proteinases and peptidases.

protein (prō'tēn, prō'tē-ən) (Gr., *protein,* fr. *proteios,* primary). A macromolecule of carbon, hydrogen, oxygen, and nitrogen and sometimes sulfur and phosphorus; composed of chains of amino acids joined by peptide bonds; present in all cells.

prothrombin (pro-thräm'bən) (Gr., *pro,* before, + *thrombos,* clot). A constituent of blood plasma that is changed to thrombin by a catalytic sequence that includes thromboplastin, calcium, and plasma globulins; involved in blood clotting.

protist (prō'tist) (Gr., *protos,* first). A member of the kingdom Protista, generally considered to include the unicellular eukaryotic organisms (protozoans and eukaryotic algae).

protonephridium (prō'tō-nə-frid'ē-əm) (Gr., *protos,* first, + *nephros,* kidney). Primitive osmoregulatory or excretory organ consisting of tubule with terminating flame bulb or solenocyte; the unit of a flame bulb system.

protoplasm (prō'tə-plazm) (Gr., *protos,* first + *plasma,* form). Organized living substance; cytoplasm and karyoplasm of the cell.

Protostomia (prō'də-stō'mē-ə) (Gr., *protos,* first, + *stoma,* mouth). A group of higher phyla in which cleavage is determinate and primitively spiral, the endomesoderm is derived from the 4d blastomere, and the mouth from or near the blastopore. Includes the Annelida, Arthropoda, Mollusca, and a number of minor phyla. Compare with Deuterostomia.

proventriculus (prō'ven-trik'yə-ləs) (L., *pro-,* before, + *ventriculum,* ventricle). In birds, the glandular stomach between the crop and gizzard. In insects, a muscular dilation of the foregut armed internally with chitinous teeth.

proximal (L., *proximus,* nearest). Situated toward or near the point of attachment; opposite of **distal,** distant.

psammolittoral (sam'ə-lid'-rəl) (Gr., *psammos,* sand, + L., *litoralis,* seashore). Pertaining to the intertidal areas of sandy beaches or the intertidal biota of such regions.

psammon (sa'män) (Gr., *psammos,* sand). Microfauna and microflora inhabiting the interstices between grains of sand of sandy beaches; the psammolittoral biota.

pseudocoel (sū'dō-sēl) (Gr., *pseudēs,* false, + *koilōma,* cavity). A body cavity not lined with peritoneum and not a part of the blood or digestive systems.

pseudopodium (sū'də-pō'dē-əm) (Gr., *pseudēs,* false, + *podion,* small foot, + *eidos,* form). A temporary cytoplasmic protrusion extended out from a protozoan or ameboid cell and serving for locomotion or for taking up food.

puff. The pattern of swelling of specific bands or gene loci on giant chromosomes during the larval and imaginal stages of flies.

pupa (pyū'pə) (L., girl, doll, puppet). Inactive quiescent stage of the holometabolous insects. It follows the larval stage and precedes the adult stage.

purine (pyū'rēn) (L., *purus,* pure, + *urina,* urine). Organic base with carbon and nitrogen atoms in two interlocking rings. The parent substance of adenine, guanine, and other naturally occurring bases.

pyrimidine (pī-rim'ə-dēn) (alter. of pyridine, fr. Gr., *pyr,* fire, + *-id,* adjective suffix, + *ine*). An organic base composed of a single ring of carbon and nitrogen atoms; parent substance of several bases found in nucleic acids.

queen. In entomology, the single fully-developed female in a colony of social insects such as bees, ants, and termites, distinguished from workers, unproductive females, and soldiers.

radial cleavage. Type in which early cleavage planes are symmetric to the polar axis, each blastomere of one tier lying directly above the corresponding blastomere of the next layer; indeterminate cleavage.

radula (ra'jə-lə) (L., scraper). Rasping tongue of certain molluscs.

recombinant DNA. DNA technique in which restriction enzymes are used to divide DNA molecules at particular sequences, thus allowing a hybrid molecule to be formed by joining a genetic segment from one organism to a similarly cut segment from another organism; a technique for joining DNA molecules of unrelated organisms.

redia (rē'dē-ə), pl. **rediae** (rē'dē-ē) (Redi, Italian biologist). A larval stage in the life cycle of flukes; it is produced by a sporocyst larva, and in turn gives rise to many cercariae.

releaser (L., *relaxare,* to unloose). Simple stimulus that elicits an innate behavior pattern.

replication (L., *replicatio,* a folding back). In genetics, the duplication of one or more DNA molecules from the preexisting molecule.

respiration (L., *respiratio,* breathing). Gaseous interchange between an organism and its surrounding medium. In the cell, the release of energy by the oxidation of food molecules.

rete mirabile (rē'tē mə-rab'ə-lē) (L., wonderful net). A network of small blood vessels so arranged that the incoming blood runs countercurrent to the outgoing blood and thus makes possible efficient exchange between the two bloodstreams. Such a mechanism serves to maintain the high concentration of gases in the fish swim bladder.

rhabdoid (rab'doid) (Gr., *rhabdos,* rod). Rodlike structures in the cells of the epidermis or underlying parenchyma in certain turbellarians. They are discharged in mucous secretions.

rheoreceptor (rē'ə-rē-cep'tər) (Gr., *rheos,* a flowing, + receptor). A sensory organ of aquatic animals that responds to water current.

rhopalium (rō-pā'lē-əm) (NL., fr. Gr., *rhopalon,* a club). One of the marginal, club-shaped sense organs of certain jellyfishes; tentaculocyst.

rhynchocoel (ring'kō-sēl) (Gr., *rhynchos,* snout, + *koilos,* hollow). In nemertines, the dorsal tubular cavity that contains the inverted proboscis. It has no opening to the outside.

ribosome (rī'bō-sōm). A small organelle composed of protein and ribonucleic acid. May be free in the cytoplasm or attached to the membranes of the endoplasmic reticulum; thought to function in protein synthesis.

ritualization. In ethology, the evolutionary modification, usually intensification, of a behavior pattern to serve communication.

RNA. Ribonucleic acid, of which there are several different kinds, such as messenger RNA, ribosomal RNA, and transfer RNA.

rostrum (räs'trəm) (L., ship's beak). A snoutlike projection on the head.

sagittal (saj'ə-dəl) (L., *sagitta,* arrow). Pertaining to the median anteroposterior plane that divides a bilaterally symmetric organism into right and left halves.

saprobe (sa'prōb) (Gr., *sapros,* rotten, + *bios,* life). Any organism living on dead or decaying plant or animal life.

saprophagous (sə-präf'ə-gəs) (Gr., *sapros,* rotten, + *-phagos,* fr. *phagein,* to eat). Feeding on decaying matter; saprobic; saprozoic.

saprophyte (sap'rə-fīt) (Gr., *sapros,* rotten, + *phyton,* plant). A plant living on dead or decaying organic matter.

saprozoic nutrition (sap-rə-zō'ik) (Gr., *sapros,* rotten, + *zoion,* animal). Nutrition of an animal by absorption of dissolved salts and simple organic nutrients from surrounding medium; also refers to feeding on decaying matter.

sarcolemma (sär'kə-lem'ə) (Gr., *sarx,* flesh, + *lemma,* rind). The thin noncellular sheath that encloses a striated muscle fiber.

schizocoel (skiz'ō-sēl) (Gr., *schizo,* fr. *schizein,* to split, + *koilōma,* cavity). A body cavity formed by the splitting of the embryonic mesoderm. Noun **schizocoelomate** refers to an animal with a schizocoel, such as an arthropod or mollusc. Adj. **schizocoelous.**

schizocoelous mesoderm formation (skiz'ō-sēl-ləs). Embryonic formation of the mesoderm as cords of cells between ectoderm and endoderm; splitting of these cords results in the coelomic space.

schizogony (skə-zä'gə-nē) (Gr., *schizein,* to split, + *gonos,* seed). Multiple asexual fission.

sclerite (skle'rīt) (Gr., *sklēros,* hard). A hard chitinous or calcareous plate or spicule; one of the plates making up the exoskeleton of arthropods, especially of insects.

scleroblast (skler'ō-blast) (Gr., *sklēros,* hard, + *blastos,* germ). An amebocyte specialized to secrete a spicule, found in sponges.

sclerotic (skle-räd'ik) (Gr., *sklēros,* hard). Pertaining to the tough outer coat of the eyeball.

sclerotization (skli'rə-tə-zā'shən). Process of hardening of the cuticle of arthropods by the formation of stabilizing cross-linkages in the amino acid chains of the protein.

scolex (skō'leks) (Gr., *skōlēx,* worm, grub). The holdfast, or so-called head, of a tapeworm; it bears suckers and, in some, hooks, and from it new proglottids are budded off.

scrotum (skrō'təm) (L., bag). The pouch that contains the testes in most mammals.

scyphistoma (sī-fis'tə-mə) (Gr., *skyphos,* cup, + *stoma,* mouth). A stage in the development of scyphozoan jellyfish just after the larva becomes attached.

seminiferous (sem-ə-nif'-rəs) (L., *semen,* semen, + *ferre,* to bear). Pertains to the tubules that produce or carry semen in the testes.

semipermeable (L., *semi,* half, + *permeabilis,* capable of being passed through). Permeable to small particles, such as water and certain inorganic ions, but not to larger molecules.

septum, pl. **septa** (L., fence). A wall between two cavities.

sere (sēr) (L., *series,* row, kind, fr. *serere,* to join). The sequence of communities that develop in a given situation, from pioneer to terminal climax during ecologic succession.

serosa (sə-rō'sə) (NL., fr. L., *serum,* serum). The outer embryonic membrane of birds and reptiles. Chorion.

serotonin (sir'ə-tōn'ən) (L., *serum,* serum). A phenolic amine, found in the serum of clotted blood and in many other tissues, that possesses several poorly understood metabolic, vascular, and neural functions; 5-hydroxytryptamine.

serum (sir'əm) (L., whey, serum). The liquid that separates from the blood after coagulation; blood plasma from which fibrinogen has been removed. Also, the clear portion of a biologic fluid separated from its particulate elements.

sessile (ses'əl) (L., *sessilis,* low, dwarf, fr. *sedere, sessum,* to sit). Attached at the base; sedentary; fixed to one spot, not able to move about.

seta, pl. **setae** (sēd'ə, sē'tē) (L., bristle). A needlelike chitinous structure of the integument of annelids and related forms.

siliceous (sə-li'shəs) (L., *silex,* flint). Containing silica.

simian (sim'ē-ən) (L., *simia,* ape). Pertaining to monkeys or apes.

sinus (sī'nəs) (L., curve) A cavity or space in tissues or in bone.

siphonoglyph (sī-fän'ə-glif') (Gr., *siphōn,* reed, tube, siphon, + *glyphē,* carving). Ciliated furrow in the gullet of sea anemones.

solenocyte (sō-len'ə-sīt) (Gr., *solēn,* pipe, + *kytos,* hollow vessel). Special type of flame bulb in which the bulb bears a flagellum instead of a tuft of cilia. See **flame bulb, protonephridium.**

soma (sō'mə) (Gr., body). The whole of an organism except the germ cells (germ plasm).

somatic (sō-mad'ik) (Gr., *sōma,* body). Refers to the body; for example, somatic cells in contrast to germ cells.

somatoplasm (sō'mə-də-pla'zəm) (Gr., *sōma,* body, + *plasma,* anything formed). The living matter that makes up the mass of the body as distinguished from germ plasm, which makes up the reproductive cells. The protoplasm of body cells.

somite (sō'mīt) (Gr., *sōma,* body). One of the blocklike masses of mesoderm arranged segmentally (metamerically) in a longitudinal series beside the neural tube of the embryo; metamere.

speciation (spē'sē-ā'shən) (L., *species,* kind). The evolutionary process by which new species arise; the process by which variations become fixed.

species (spē'shez, spē'sēz) sing. and pl. (L., particular kind). A group of interbreeding individuals of common ancestry that are reproductively isolated from all other such groups; a taxonomic unit ranking below a genus and designated by a binomial consisting of its genus and the species name.

spermatheca (spər'mə-thē'kə) (Gr., *sperma,* seed, + *thēkē,* a case). A sac in the female reproductive organs for the reception and storage of sperm.

spermatogonium (spər'mad-ə-gō'nē-əm) (Gr., *sperma,* seed, + *gonē,* offspring). Precursor of mature male reproductive cell; gives rise directly to a spermatocyte.

spermatophore (spər-mad'ə-for') (Gr., *sperma, spermatos,* seed, + *phorein,* to bear). Capsule or packet enclosing sperm, produced by males of several invertebrate groups and a few vertebrates.

sphincter (sfingk'tər) (Gr., *sphinkter,* band, sphincter, from *sphingein,* to bind tight). A ring-shaped muscle capable of closing a tubular opening by constriction.

spicule (spi'kyūl) (L., dim. of *spica,* point). One of the minute calcareous or siliceous skeletal bodies found in sponges, radiolarians, soft corals, and sea cucumbers.

spiracle (spi'rə-kəl) (L., *spiraculum,* fr. *spirare,* to breathe). External opening of a trachea in arthropods. One of a pair of openings on the head of elasmobranchs for passage of water. Exhalent aperture of tadpole gill chamber.

spiral cleavage. A type of early embryonic cleavage in which cleavage planes are diagonal to the polar axis, and unequal cells are produced by the alternate clockwise and counterclockwise cleavage around the axis of polarity; determinate cleavage.

spongin (spən'jin) (L., *spongia,* sponge). Fibrous, scleroprotein material making up the skeletal network of horny sponges.

spongocoel (spən'jō-sēl') (Gr., *spongos,* sponge, + *koilos,* hollow). Central cavity in sponges.

sporocyst (spō'rə-sist) (Gr., *sporos,* seed, + *kystis,* pouch). A larval stage in the life cycle of flukes; it originates from a miracidium.

sporozoite (spō'rə-zō'īt) (Gr., *sporos,* seed, + *zōon,* animal, + *-ite,* suffix for body part). A stage in the life history of many sporozoan Protozoa; released from spores.

statoblast (stad'ə-blast) (Gr., *statos,* standing, fixed, + *blastos,* germ). Biconvex capsule containing germinative cells and produced by most freshwater ectoprocts by asexual budding. Under favorable conditions it germinates to give rise to new zooid.

statocyst (Gr., *statos,* standing, + *kystis,* bladder). Sense organ of equilibrium; a fluid-filled cellular cyst containing one or more granules (statoliths) used to sense direction of gravity.

statolith (Gr., *statos,* standing, + *lithos,* stone). Small calcareous body resting on tufts of cilia in the statocyst.

stenohaline (sten-ə-hā'līn, -lən) (Gr., *stenos,* narrow, + *hals,* salt). Pertaining to aquatic organisms that have restricted tolerance to changes in environmental saltwater concentration.

stenotopic (sten-ə-tä'pik) (Gr., *stenos,* narrow, + *topos,* place). Refers to an organism with a narrow range of adaptability to environmental change; having a restricted geographic distribution.

stereogastrula (ste'rē-ə-gas'trə-lə) (Gr., *stereos,* solid, + *gastēr,* stomach, + L., *-ula,* diminutive). A solid type of gastrula, such as the planula of cnidarians.

sterol (ste'rōl) **steroid** (ste'roid) (Gr., *stereos,* solid, + L., *-ol,* fr. *oleum,* oil). One of a class of organic compounds containing a molecular skeleton of four fused carbon rings; sterols include cholesterol, sex hormones, adrenocortical hormones, and vitamin D.

stigma (Gr., *stigma,* mark, tattoo mark). Eyespot in certain protozoans. Spiracle of certain terrestrial arthropods.

stolon (stō'lən) (L., *stolō, stolonis,* a shoot, or sucker of a plant). A rootlike extension of the body wall that gives rise to buds that may develop into new zooids, thus forming a compound animal in which the zooids remain united by the stolon. Found in some colonial anthozoans, hydrozoans, ectoprocts, and ascidians.

stoma (stō'mə) (Gr., mouth). A mouthlike opening.

stomochord (stō'mə-kord) (Gr., *stoma,* mouth, + *chordē,* cord). Anterior evagination of the dorsal wall of the buccal cavity into the proboscis of hemichordates; the buccal diverticulum.

strobila (strō'bə-lə) (Gr., *strobilē,* lint plug like a pine cone [*strobilos*]). A stage in the development of the scyphozoan jellyfish.

sycon (Gr., *sykon,* fig). A type of canal system in certain sponges. Sometimes called syconoid.

symbiosis (sim'bī-ōs'əs, sim'bē-ōs'əs) (Gr., *syn,* with, + *bios,* life). The living together of two different species in an intimate relationship; includes mutualism, commensalism, and parasitism.

sympatry (sim'pə-trē) (Gr., *syn,* with, + *patra,* native land). Having the same or overlapping regions of distribution. (Adj. **sympatric.**)

synapse (si'naps, si-naps') (Gr., *synapsis,* contact, union). The place at which a nerve impulse passes between neuron processes, typically from an axon of one nerve cell to a dendrite of another nerve cell.

syncytium (sin-sish'e-əm) (Gr., *syn,* with, + *kytos,* hollow vessel). A mass of protoplasm containing many nuclei and not divided into cells.

syngamy (sin'ga'mē) (Gr., *syn,* with, + *gamos,* marriage). Sexual reproduction by the union of gametes in fertilization.

syrinx (sir'inks) (Gr., shepherd's pipe). The vocal organ of birds located at the base of the trachea.

tactile (tak'til) (L., *tactilis,* able to be touched, fr. *tangere,* to touch). Pertaining to touch.

tagma, pl. **tagmata** (Gr., *tagma,* arrangement, order, row). A compound body section of an arthropod resulting from embryonic fusion of two or more segments; for example head, thorax, abdomen.

tagmatization, tagmosis. Organization of the arthropod body into tagmata.

taiga (tī'gä) (Russ.). Habitat zone characterized by large tracts of coniferous forests, long, cold winters, and short summers; most typical in Canada and Siberia.

taxis (taks'əs) (Gr., *taxis,* arrangement). An orientation movement by a (usually) simple organism in response to an environmental stimulus.

tegument (teg'yə-ment) (L., *tegumentum,* fr. *tegere,* to cover). An **integument;** external covering in cestodes and trematodes, formerly thought to be a cuticle.

telencephalon (tel'en-sef'ə-lon) (Gr., *telos,* end, + *encephalon,* brain). The most anterior vesicle of the brain; the anteriormost subdivision of the prosencephalon that becomes the cerebrum and associated structures.

teleology (tel'ē-äl'ə-jē) (Gr., *telos,* end, + L., *-logia,* study of, fr. Gr., *logos,* word). The philosophic view that natural events are goal-directed and are preordained; as opposed to scientific view of mechanical determinism.

telolecithal (te-lō-les'ə-thəl) (Gr., *telos,* end, + *lekithos,* yolk, + *-al*). Having the yolk concentrated at one end of an egg.

telson (tel'sən). Posterior projection of the last body segment in many crustaceans.

template (tem'plət). A pattern or mold guiding the formation of a duplicate; often used with reference to gene duplication.

tentaculocyst (ten-tak'u-lō-sist) (L., *tentaculum,* feeler, + Gr., *kystis,* pouch). One of the sense organs along the margin of medusae; a rhopalium.

territory (L., *territorium,* fr. *terra,* earth). A restricted area preempted by an animal or pair of animals, usually for breeding purposes, and guarded from other individuals of the same species.

test. A shell or hardened outer covering.

tetrad (te'trad) (Gr., *tetras,* four). Group of four chromatids formed by synapsis and resulting from the splitting of paired homologous chromosomes.

tetrapods (te'trə-päds) (Gr., *tetras,* four, + *pous,* foot). Four-footed vertebrates; the group includes amphibians, reptiles, birds, and mammals.

therapsid (thə-rap'sid) (Gr., *theraps,* an attendant). Extinct Mesozoic mammal-like reptile from which true mammals evolved.

thermocline (thər'mō-klīn) (Gr., *thermē,* heat, + *klinein,* to swerve). Layer of water separating upper warmer and lighter water from lower colder and heavier water in a lake or sea; a stratum of abrupt change in water temperature.

tornaria (tor-na'rē-ə) (L., *tornare,* to turn). A free-swimming larva of enteropneusts that rotates as it swims; resembles somewhat the bipinnaria larva of echinoderms.

torsion (LL., *torsio,* fr. L., *torquere,* to twist). A twisting phenomenon in gastropod development that alters the position of the visceral and pallial organs by 180 degrees.

trachea (trā'kē-ə) (ML., windpipe, trachea, fr. Gr. [*artēria*] *tracheia,* rough [artery]). The windpipe; any of the air tubes of insects.

transcription. Formation of messenger RNA from the coded DNA.

transfer RNA (tRNA). A form of RNA of about 70 nucleotides, which are adapter molecules in the synthesis of proteins. A specific amino acid molecule is carried by transfer RNA to a ribosome–messenger RNA complex for incorporation into a polypeptide.

translation (L., a transferring). The process in which the genetic information present in messenger RNA is used to direct the order of specific amino acids during protein synthesis.

trichocyst (trik'ə-sist) (Gr., *thrix,* hair, + *kystis,* bladder). Saclike protrusible organelle in the ectoplasm of ciliates, which discharges as a threadlike weapon of defense.

triploblastic (trip'lō-blas'tik) (Gr., *triploos,* triple, + *blastos,* germ). Pertaining to metazoans in which the embryo has three primary germ layers—ectoderm, mesoderm, and endoderm.

trochophore (trōk'ə-fōr) (Gr., *trochos,* wheel, + *pherein,* to bear). A free-swimming ciliated marine larva characteristic of many molluscs, certain bryozoans, brachiopods, and marine worms; an ovoid or piriform body with preoral circlet of cilia and sometimes a secondary circlet behind the mouth.

trophallaxis (trōf'ə-lak'səs) (Gr., *trophē,* food, + *allaxis,* barter, exchange). Exchange of food between young and adults, especially among those of certain social insects.

trophoblast (trōf'ə-blast) (Gr., *trephein,* to nourish, + *blastos,* germ). Outer ectodermal nutritive layer of blastodermic vesicle; in mammals it is part of the chorion and attaches to the uterine wall.

trophozoite (trōf'ə-zō'īt) (Gr., *trophē,* food, + *zōon,* animal). Adult stage in the life cycle of a sporozoan in which it is actively absorbing nourishment from the host.

tube feet. Numerous small, muscular, fluid-filled tubes projecting from body of echinoderms; part of water-vascular system; used in locomotion, clinging, food-handling, and respiration.

tundra (tən'drə) (Russ., from Lapp, *tundar,* hill). Terrestrial habitat zone, between taiga in south and polar region in north; characterized by absence of trees, short growing season, and mostly frozen soil during much of the year.

tunic (L., *tunica,* tunic, coat). In tunicates, a cuticular, cellulose-containing covering of the body secreted by the underlying body wall.

turbellarian (tər'bə-lar'ə-an) (L., *turbellae,* a stir or tumult). Free-living flatworm of phylum Platyhelminthes.

typhlosole (tif'lə-sōl') (Gr., *typhlos,* blind, + *sōlēn,* channel, pipe). A longitudinal fold projecting into the intestine in certain invertebrates such as the earthworm.

umbilical (L., *umbilicus,* navel). Refers to the navel, or umbilical, cord.

umbo (əm'bō), pl. **umbones** (əm-bō'nēz) (L., boss of a shield). One of the prominences on either side of the hinge region in a bivalve mollusc shell; the ''beak'' of a brachiopod shell.

ungulate (ən′gyə-lət) (L., *ungula,* hoof). Hoofed; any hoofed mammal.

urethra (yə-rē′thrə) (Gr., *ourethra,* urethra). The tube from the urinary bladder to the exterior in both sexes.

uriniferous tubule (yu′rə-nif′rəs) (L., *urina,* urine, + *ferre,* to bear). One of the tubules in the kidney extending from a malpighian body to the collecting tubule.

utricle (yū′trə-kəl) (L., *utriculus,* little bag). That part of the inner ear containing the receptors for dynamic body balance; the semicircular canals lead from and to the utricle.

vacuole (vak′yə-wōl′) (L., *vacuus,* empty, + Fr., *-ole,* diminutive). A membrane-bounded, fluid-filled space in a cell.

valence (vā′ləns) (L., *valere,* to have power). Degree of combining power of an element as expressed by the number of atoms of hydrogen (or its equivalent) that that element can hold (if negative) or displace in a reaction (if positive). The oxidation state of an element in a compound. The number of electrons gained, shared, or lost by an atom when forming a bond with one or more other atoms.

valve. One of the two shells of a typical bivalve mollusc or brachiopod. Also, a structure that temporarily closes a passage or orifice or permits passage in one direction only.

vector (L., a bearer, carrier, fr. *vehere, vectum,* to carry). An animal, usually an insect, that carries and transmits pathogenic microorganisms from one host to another host.

veliger (vēl′ə-jər, vel-) (L., sail-bearing). Larval form of certain molluscs; develops from the trochophore and has the beginning of a foot, mantle, shell, and so on.

velum (vē′ləm) (L., *veil,* sail). A membrane on the subumbrella surface of jellyfish of class Hydrozoa.

vestige (ves′tij) (L., *vestigium,* footprint). A rudimentary organ that may have been well-developed in some ancestor or in the embryo.

villus (vil′əs), pl. **villi** (L., tuft of hair). A small, fingerlike, vascular process on the wall of the small intestine; one of the branching, vascular processes on the embryonic portion of the placenta.

virus (vī′rəs) (L., slimy liquid, poison). A submicroscopic noncellular particle composed of a nucleoprotein core and a protein shell; parasitic and will grow and reproduce in a host cell.

viscera (vis′ər-ə), sing. **viscus** (L., pl. of *viscus,* internal organ). Internal organs in the body cavity.

vitalism (L., *vita,* life). The view that natural processes are controlled by supernatural forces and cannot be explained through the laws of physics and chemistry alone; as opposed to mechanism.

vitamin (L., *vita,* life, + *amine,* fr. former supposed chemical origin). An organic substance required in small amounts as a catalyst for normal metabolic function; must be supplied in the diet or by intestinal flora because the organism cannot synthesize it.

vitelline membrane (və-tel′ən, vī′təl-ən) (L., *vitellus,* yolk of an egg). The noncellular membrane that encloses the egg cell.

viviparity (vī′və-par′ə-dē) (L., *vivus,* alive, + *parere,* to bring forth). Reproduction in which eggs develop within the female body, with nutritional aid of maternal parent as in therian mammals, many reptiles, and some fishes; offspring are born as juveniles. Adj. **viviparous** (vī-vip′ə-rəs).

water-vascular system. System of fluid-filled closed tubes and ducts peculiar to echinoderms; used to move tentacles and tube feet that serve variously for clinging, food-handling, locomotion, and respiration.

zoecium, zooecium (zō-ē′shē-əm) (Gr., *zōon,* animal, + *oikos,* house). Cuticular sheath or shell of Ectoprocta.

zoochlorella (zō′ə-klōr-el′ə) (Gr., *zōion,* life, + *Chlorella*). Any of various minute green algae (usually *Chlorella*) that live symbiotically within the cytoplasm of some protozoans and other invertebrates.

zooid (zō-oid) (Gr., *zōion,* life). An individual member of a colony of animals, such as colonial cnidarians and ectoprocts.

zygote (Gr., *zygōtos,* yoked). The fertilized egg.

APPENDIX

DEVELOPMENT OF ZOOLOGY

Certain key discoveries have greatly influenced progress in the study of zoology. This appendix presents some of the major landmarks in the development of biology and the individuals whose names are commonly associated with them. It is very difficult to appraise the historic development of any field of study. One investigator often gets the credit for an important discovery when many others should share in the prestige. No one individual has a monopoly on ideas; advances in science are built on the results of many causes and the work of many minds.

In this brief outline, the student may be able to see some of the relationships that exist between one discovery and another. The discoveries are not completely isolated as they may sometimes appear. One may note also that fundamental discoveries in a particular branch of biology tend to be grouped fairly close together chronologically because that particular interest may have dominated the thought of biologic investigators at that time.

The reader may wish to locate important discoveries in a given area without reading the entire list. Such an effort may be made easier by use of the following list of key symbols. Any such system of classification suffers some limitations, however. Many basic discoveries are not easily categorized because they have implications in wide areas. The list of categories must be short, or it becomes too cumbersome. Nevertheless, we hope that the classification will be useful.

□ Taxonomy, systematics, and environmental fields (behavior, ecology, and the like)
○ Tissue-and organ-level anatomy and physiology
△ Cell biology
* Developmental biology
‡ Genetics
× Evolution and paleontology
† Of *particular* medical significance

ORIGINS OF BASIC CONCEPTS AND KEY DISCOVERIES IN ZOOLOGY

384-322 B.C.: *Aristotle. The foundation of zoology as a science.*

Although this great pioneer zoologist and philosopher cannot be appraised by modern standards, there is scarcely a major subdivision of zoology to which he did not make some contribution. He was a true scientist, for he emphasized observational and experimental methods. Despite his lack of scientific background, he was one of the greatest scientists of all time.

○ **130-200** A.D.: *Galen. Development of anatomy and physiology.*

This Roman investigator has been praised for his clear concept of scientific methods and blamed for passing down to others for centuries certain glaring errors. His influence was so great that for centuries after his period students considered him the final authority on anatomic and physiologic subjects.

1347: *William of Occam. Occam's razor.*

This principle of logic has received its name from the fact that it is supposed to cut out unnecessary and irrelevant hypotheses in the explanation of phenomena. The gist of the principle is that, of several possible explanations, the one that is simplest, has the fewest assumptions, and is most consistent with the data at hand is the most probable one.

○ **1543:** *Vesalius, Andreas. First modern interpretation of anatomic structures.*

With his insight into fundamental structure, Vesalius ushered in the dawn of modern biologic investigation. Many aspects of his interpretation of anatomy are just now beginning to be appreciated.

○ **1616-1628:** *Harvey, William. First accurate description of blood circulation.*

Harvey's classic demonstration of blood circulation was the key experiment that laid the foundation of modern physiology. He explained bodily processes in physical terms, cleared away much of the mental rubbish of mystic interpretation, and gave an auspicious start to experimental physiology.

○ **1627:** *Aselli, G. First demonstration of lacteal vessels.*

This discovery, coming at the same time as Harvey's great work, supplemented the discovery of circulation.

○ **1649:** *Descartes, René. Early concept of reflex action.*

Descartes postulated the idea that impulses originating at the receptors of the body were carried to the central nervous system where they activated muscles and glands by what he called "reflection."

∗ **1651:** *Harvey, William. Aphorism of Harvey: Omne vivum ex ovo (all life from the egg).*

Although Harvey's work as an embryologist is overshadowed by his demonstration of blood circulation, his *De Generatione Animalium,* published in 1651, contains many sound observations on embryologic processes.

○ **1652:** *Bartholin, Thomas. Discovery of lymphatic system.*

The significance of the thoracic duct in its relation to the circulation was determined in this investigation.

○ **1658:** *Swammerdam, Jan. Description of red blood corpuscles.*

This discovery, together with his observations on the valves of the lymphatics and the alterations in shape of muscles during contraction, represented early advancements in the microscopic study of bodily structures.

○ **1660:** *Malpighi, Marcello. Demonstration of capillary circulation.*

By demonstrating the capillaries in the lung of a frog, Malpighi was able to complete the scheme of blood circulation because Harvey never saw capillaries and thus never included them in his description.

△ **1665:** *Hooke, Robert. Discovery of the cell.*

Hooke's investigations were made with cork and the term "cell" fits cork much better than it does animal cells, but by tradition the misnomer has stuck.

△ **1672:** *de Graaf, R. Description of ovarian follicles.*

De Graaf's name is given to the mature ovarian follicle, but he believed that the follicles were the actual ova, an error later corrected by von Baer.

△□ **1675-1680:** *van Leeuwenhoek, Anthony. Discovery of protozoa.*

The discoveries of this eccentric Dutch microscopist revealed a whole new world of biology.

□ **1693:** *Ray, J. Concept of species.*

Although Ray's work on classification was later overshadowed by Linnaeus, Ray was really the first to make the species concept apply to a particular kind of organism and to point out the variations that exist among the members of a species.

○ **1733:** *Hales, Stephen. First measurement of blood pressure.*

This was further proof that the bodily processes could be measured quantitatively—more than a century after Harvey's momentous demonstration.

○ **1734:** *Borelli, A. Analysis of fish locomotion.*

The nature of fish propulsion has been the subject of many investigations from Aristotle down to the present. Borelli was really the first to demonstrate that the vibration of the tail, and not the fins, is the chief propulsive agency in fish movement. Fresh insights into this problem have been furnished in recent years by J. Gray, R. Bainbridge, E. Kramer, and J. R. Nursall.

∗ **1745:** *Bonnet, Charles. Discovery of natural parthenogenesis.*

Although somewhat unusual in nature, this phenomenon has yielded much information about meiosis and other cytologic problems.

× **1745:** *Maupertuis, P. M. Early concept of evolutionary process.*

Although Maupertuis' ideas are speculative and not based on experimental observation, he foretold many of the concepts of variation and natural selection that Darwin was to demonstrate with convincing evidence.

□ **1758:** *Linnaeus, Carolus. Development of binomial nomenclature system of taxonomy.*

So important is this work in taxonomy that 1758 is regarded as the starting point in the determination of the generic and specific names of animals. Besides the value of his binomial system, Linnaeus gave taxonomists a valuable working model of conciseness and clearness that has never been surpassed.

∗ **1759:** *Wolff, Kaspar F. Embryologic theory of epigenesis.*

This embryologist, the greatest before von Baer, did much to overthrow the preformation theory then in vogue and, despite many shortcomings, laid the basis for the modern interpretation of embryology.

○ **1760:** *Hunter, John. Development of comparative investigations of animal structure.*

This vigorous eighteenth-century anatomist gave a powerful impetus not only to anatomic observations but also to the establishment of natural history museums.

□ **1763:** *Adanson, M. Concept of empiric taxonomy.*

This botanist proposed a scheme of classification that grouped individuals into taxa according to shared characters. A species would have the maximum number of shared characters according to this scheme. The concept has been revived recently by exponents of numerical taxonomy. It lacks the evaluation of the evolutionary concept and has been criticized on this account.

‡ **1763:** *Kölreuter, J. G. Discovery of quantitative inheritance (multiple genes).*

Kölreuter, a pioneer in plant hybridization, found that certain plant hybrids had characters more or less intermediate between the parents in the F_1 generation, but in the F_2 there were many gradations from one extreme to the other. An explanation was not forthcoming until after Mendel's laws were discovered.

□ **1768-1779:** *Cook, James. Influence of geographic exploration on biologic development.*

This famous sea captain made possible a greater range of biologic knowledge because of the able naturalists whom he took on his voyages of discovery.

△ **1772:** *Priestley, J., and J. Ingenhousz. Concept of photosynthesis.*

These investigators first pointed out some of the major aspects of this important phenomenon, such as the use of light energy for converting carbon dioxide and water into released oxygen and the retention of carbon.

△ **1774:** *Priestley, J. Discovery of oxygen.*

The discovery of this element is of great biologic interest because it helped in determining the nature of oxidation and the exact role of respiration in organisms.

△ **1778:** *Lavoisier, Antoine L. Nature of animal respiration demonstrated.*

A basis for the chemical interpretation of the life process was given a great impetus by the careful quantitative studies of the changes during breathing made by this great investigator. His work also meant the final overthrow of the mystic phlogiston theory that had held sway for so long.

□ **1781:** *Abildgaard, P. First experimental life cycle of a tapeworm.*

Life cycles of parasites may be very complicated, involving several hosts. That a parasitic worm could require more than one host was a revolutionary concept in Abildgaard's day and was widely disbelieved until the work of Küchenmeister over 60 years later.

△ **1781:** *Fontana, F. Description of the nucleolus.*

Fontana discovered this organelle in the slime from

the skin of an eel, and until relatively recently its function has been a subject of controversy. We now know that ribosomal RNA is synthesized there.

× **1791:** *Smith, William. Correlation between fossils and geologic strata.*

By observing that certain types of fossils were peculiar to particular strata, Smith was able to work out a method for estimating geologic age. He laid the basis of stratigraphic geology.

○ **1792:** *Galvani, L. Animal electricity.*

The lively controversy between Galvani and Volta over the twitching of frog legs led to extensive investigation of precise methods of measuring the various electric phenomena of animals.

× **1796:** *Cuvier, Georges. Development of vertebrate paleontology.*

Cuvier compared the structure of fossil forms with that of living ones and concluded that there had been a succession of organisms that had become extinct and were succeeded by the creation of new ones. To account for this extinction, Cuvier held to the theory of catastrophism, or the simultaneous extinction of animal populations by natural cataclysms.

× **1801:** *de Lamarck, J. B. Evolutionary concept of use and disuse.*

de Lamarck gave the first clear-cut expression of a theory to account for organic evolution. His assumption was that acquired characters were inherited; modern evolutionists have refuted this part of the theory.

○ **1802:** *Young, T. Theory of trichromatic color vision.*

Young's theory suggested that the retina contained three kinds of light-sensitive substances, each having a maximum sensitivity in a different region of the spectrum and each being transmitted separately to the brain. The three substances combined produced the colors of the environment. The three pigments responsible are located in three kinds of cones. Young's theory has been modified in certain details by other investigators.

○ **1811:** *Bell, C., and F. Magendie. Discovery of the functions of dorsal and ventral roots of spinal nerves.*

This demonstration was a starting point for an anatomic and functional investigation of the most complex system in the body.

∗ **1817:** *Pander, C. H. First description of three germ layers.*

The description of the three germ layers was first made on the chick, and later the concept was extended by von Baer to include all vertebrates.

∗ **1824:** *Prévost, P., and J. B. A. Dumas. Cell division first described.*

The significance of their description of the cleavage of the frog egg was unappreciated because the cell theory had not yet been developed.

∗ **1827:** *von Baer, Karl. Discovery of mammalian ovum.*

The very tiny ova of mammals escaped de Graaf's eyes, but von Baer brought mammalian reproduction into line with that of other animals by detecting the ova and their true relation to the follicles.

△ **1828:** *Brown, Robert. Brownian movement first described.*

This interesting phenomenon is characteristic of living protoplasm and sheds some light on the structure of protoplasm.

□ **1828:** *Thompson, J. V. Nature of plankton.*

Thompson's collections of these small forms with a tow net, together with his published descriptions, are the first records of the vast community of planktonic animals. He was also the first to work out the true nature of barnacles.

△ **1828:** *Wöhler, F. First to synthesize an organic compound.*

Wöhler succeeded in making urea (a compound formed in the body) from the inorganic substance of ammonium cyanate, thus $NH_4OCN \rightarrow NH_2CONH_2$. This success in producing an organic substance synthetically was the stimulus that resulted in the preparation of thousands of compounds by others.

∗ **1830:** *Amici, G. B. Discovery of fertilization in plants.*

Amici was able to demonstrate the tube given off by the pollen grain and to follow it to the micropyle of the ovule through the style of the ovary. Later it was established that a sperm nucleus of the pollen makes contact and union with the nucleus of the egg.

∗ × **1830:** *von Baer, Karl. Biogenetic law formulated.*

Von Baer's conception of this law was conservative and sounder in its implication than has been the case with many other biologists (Haeckel, for instance), for von Baer stated that embryos of higher and lower forms resemble each other more the earlier they are compared in their development, and *not* that the embryos of higher forms resemble the adults of lower organisms.

× **1830:** *Lyell, C. Modern concept of geology.*

The influence of this concept not only did away with the catastrophic theory but also gave a logical interpretation of fossil life and the correlation between the formation of rock strata and the animal life that existed at the time these formations were laid down.

△ **1831:** *Brown, Robert. First description of cell nucleus.*

Others had seen nuclei, but Brown was the first to name the structure and to regard the nucleus as a general phenomenon. This description was an important preliminary to the formulation of the cell theory a few years later, for Schleiden acknowledged the importance of the nucleus in the development of the cell concept.

○ **1833:** *Hall, M. Concept of reflex action.*

Hall described the method by which a stimulus can produce a response independently of sensation or volition and coined the term "reflex action." It remained for the outstanding work of the great Sir Charles Sherrington in the twentieth century to explain much of the complex nature of reflexes.

○ **1833:** *Purkinje, J. E. Discovery of sweat glands.*

This discovery opened up a new field of investigation into problems of the skin and its structure that have not yet been fully resolved.

†○ **1835:** *Bassi, Agostino. First demonstration of a microorganism as an infective agent.*

Bassi's discovery that a certain disease of silkworms was caused by a small fungus represents the beginning of the germ theory of disease that was to prove so fruitful in the hands of Pasteur and other able investigators.

△ **1835:** *Dujardin, Felix. Description of living matter (protoplasm).*

Dujardin associated the jellylike substance that he found in protozoa and that he called "sarcode" with the life process. This substance was later called "protoplasm," and the sarcode idea may be considered a significant landmark in the development of the protoplasm concept.

†○ **1835:** *Owen, Richard. Discovery of Trichinella.*

This versatile investigator is chiefly remembered for his researches in anatomy, but his discovery of this very common parasite in humans is an important landmark in the history of parasitology.

△○ **1837:** *Berzelius, J. J. Formulation of the catalytic concept.*

Berzelius called substances that caused chemical conversions by their mere presence catalytic in contrast to analytic; this term described the real chemical affinity between substances in a chemical reaction.

△○ **1838:** *von Liebig, Justus. Foundation of biochemistry.*

The idea that vital activity could be explained by chemicophysical factors was an important concept for biologic investigators studying the nature of life.

△ **1838-1839:** *Schleiden, M. J., and T. Schwann. Formulation of the cell theory.*

The cell doctrine, with its basic idea that all plants and animals were made up of similar units, represents one of the truly great landmarks in biologic progress. Many biologists believe the work of Schleiden and Schwann has been rated much higher than it deserves in the light of what others had done to develop the cell concept. R. Dutrochet (1824) and H. von Mohl (1831) described all tissues as being composed of cells.

△ **1839:** *Mulder, G. J. Concept of the nature of proteins.*

Mulder proposed the name "protéine," later "protein," for the basic constituents of protoplasmic materials.

△ **1839-1846:** *Purkinje, J. E., and Hugo von Mohl. Concept of protoplasm established.*

Purkinje proposed the name "protoplasm" for living matter, and von Mohl did extensive work on its nature, but it remained for Max Schultze (1861) to give a clear-cut concept of the relations of protoplasm to cells and its essential unity in all organisms.

□ **1839:** *Verhulst, P. F. Logistic theory of population growth.*

According to this theory, animal populations have a slow initial growth rate that gradually speeds up until it reaches a maximum and then slows down to a state of equilibrium. By plotting the logarithm of the total number of individuals against time, an S-shaped curve results that is somewhat similar for all populations.

□○ **1840:** *von Liebig, J., and F. F. Blackman. The law of minimum requirements.*

Von Liebig interpreted plant growth as being dependent on essential requirements that are present in minimum quantity; some substances, even in minute quantities, were needed for normal growth. Later Blackman discovered that the rate of photosynthesis is restricted by the factor that operates at a limiting intensity (for example, light or temperature). The concept has been modified in various ways since it was formulated. We now know that a factor interaction other than the minimum ones plays a part.

× **1840:** *Miller, H. Appraisal of geologic formation, Old Red Sandstone.*

These Devonian deposits in Scotland and parts of England represent one of the most important vertebrate-bearing sediments ever discovered. From them much knowledge about early vertebrates, such as ostracoderms, placoderms, and bony fishes, has been obtained.

○ **1840:** *Müller, J. Theory of specific nerve energies.*

This theory states that the kind of sensation experienced depends on the nature of the sense organ with which the stimulated nerve is connected. The optic nerve, for instance, conveys the impression of vision however it is stimulated.

○ **1842:** *Bowman, William. Histologic structure of the nephron (kidney unit).*

Bowman's accurate description of the nephron afforded physiologists an opportunity to attack the problem of how the kidney separates waste from the blood, a problem not yet fully solved.

∗ **1842:** *Steenstrup, J. J. S. Alternation of generations described.*

Metagenesis, or alternation of sexual and asexual reproduction in the life cycle, exists in many animals and plants. The concept had been introduced before this time by L. A. de Chamisso (1819).

× **1843:** *Owen, Richard. Concepts of homology and analogy.*

Homology, as commonly understood, refers to similarity in embryonic origin and development, whereas analogy is the likeness between two organs in their functioning. Owen's concept of homology was merely that of the same organ in different animals under all varieties of form and function.

○ **1844:** *Ludwig, C. Filtration theory of renal excretion.*

About the same time that Bowman described the malpighian corpuscle, Ludwig showed that the corpuscle functions as a passive filter and that the filtrate that passes through it from the blood carries into the urinary tubules the waste products that are concentrated by the resorption of water as the filtrate moves down the tubules. Other investigators, Cushny, Starling, and Richards, have confirmed the theory by actual demonstration.

○△ **1845:** *von Helmholtz, H., and J. R. Mayer. The formulation of the law of conservation of energy.*

This landmark in human thinking showed that in any system, living or nonliving, the power to perform mechanical work always tends to decrease unless energy is added from without. Physiologic investigation could now advance on the theory that the living organism is an energy machine and obeys the laws of the science of energetics.

‡ **1848:** *Hofmeister, W. Discovery of chromosomes.*

This investigator made sketches of bodies, later known to be chromosomes, in the nuclei of pollen mother cells *(Tradescantia)*. Schneider further described these elements in 1873, and Waldeyer named them in 1888.

□ **1848:** *von Siebold, C. T. E. Establishment of the status of protozoa.*

Von Siebold emphasized the unicellular nature of protozoa, fitted them into the recently developed cell theory, and established them as the basic phylum of the animal kingdom.

○ **1851:** *Bernard, Claude. Discovery of the vasomotor system.*

Bernard showed how the amount of blood distributed to the various tissues by the small arterioles was regulated by vasomotor nerves of the sympathetic nervous system, as he demonstrated in the ear of a white rabbit.

∗ **1851:** *Waller, A. V. Importance of the nucleus in regeneration.*

When nerve fibers are cut, the parts of the fibers peripheral to the cut degenerate in a characteristic fashion. This wallerian degeneration enables one to trace the course of fibers through the nervous system. In the regeneration process the nerve cell body (with its nucleus) is necessary for the downgrowth of the fiber from the proximal segment.

○ **1852:** *von Helmholtz, Herman. Determination of rate of nervous impulse.*

This landmark in nerve physiology showed that the difficult phenomena of nervous activity could be expressed numerically.

○△ **1852:** *von Kölliker, Albrecht. Establishment of histology as a science.*

Many histologic structures were described with marvelous insight by this great investigator, starting with the publication of the first text in histology *(Handbuch der Gewebelehre)*.

○ **1852:** *Stannius, H. F. Stannius' experiment on the heart.*

By tying a ligature as a constriction between the sinus venosus and the atrium in the frog and also one around the atrioventricular groove, Stannius was able to demonstrate that the muscle tissues of the atria and ventricles have independent and spontaneous rhythms. His observations also indicated that the sinus is the pacemaker of the heartbeat.

†□ **1852:** *Küchenmeister, G. F. H. Relationship of ''bladder worms'' to adult tapeworms.*

This was the first demonstration that ''bladder worms'' (cysticerci) were the juveniles of taeniid tapeworms. It was followed rapidly by numerous other life-cycle studies of cestodes.

∗ **1854:** *Newport, G. Discovery of the entrance of spermatozoon into a frog's egg.*

This was a significant step in cellular embryology, although its real meaning was not revealed until the concept of fertilization as the union of two pronuclei was formulated about 20 years later (Hertwig, 1875).

†○ **1855:** *Addison, T. Discovery of adrenal disease.*

The importance of the adrenals in maintaining normal body functions was shown when Addison described a syndrome of general disorders associated with the pathology of this gland.

× **1856:** *Discovery of Neanderthal fossil hominid (Homo sapiens).*

Many specimens of this type of fossil hominid have been discovered (Europe, Asia, Africa) since the first one was found near Düsseldorf, Germany. Their culture was Mousterian (100,000 to 40,000 years B.P.), and although of short stature their cranial capacity was as large as or larger than that of modern humans.

○ **1857:** *Bernard, Claude. Discovery of formation of glycogen by the liver.*

Bernard's demonstration that the liver forms glycogen from substances brought to it by the blood showed that the body can build up complex substances as well as tear them down.

□ **1858:** *Sclater, P. L. Distribution of animals on basis of zoologic regions.*

This was the first serious attempt to study the geographic distribution of organisms, a study that eventually led to the present science of zoogeography. A. R. Wallace worked along similar lines.

∗△ **1858:** *Virchow, Rudolf. Aphorism of Virchow: Omnis cellula e cellula (every cell from a cell).*

△ **1858:** *Virchow, Rudolf. Formulation of the concept of disease from the viewpoint of cell structure.*

He laid the basis of modern pathology by stressing the role of the cell in diseased tissue.

× **1859:** *Darwin, Charles. The concept of natural selection as a factor in evolution.*

Although Darwin did not originate the concept of organic evolution, no one has been more influential in the development of evolutionary thought. The publication of *On The Origin of Species* represents the greatest single landmark in the history of biology.

○ **1860:** *Bernard, Claude. Concept of the constancy of the internal environment (homeostasis).*

Bernard's realization that there are mechanisms in living organisms that tend to maintain internal conditions within relatively narrow limits, despite external change, is one of the most valuable generalizations in physiology.

∗ **1860:** *Pasteur, Louis. Aphorism of Pasteur: Omne vivum e vivo (every living thing from the living).*

∗ **1860:** *Pasteur, Louis. Refutation of spontaneous generation.*

Pasteur's experiment with the open S-shaped flask proved conclusively that fermentation or putrefaction resulted from microbes, thus ending the long-standing controversy regarding spontaneous generation.

□ **1860:** *Wallace, A. R. The Wallace line of faunal delimitation.*

As originally proposed by Wallace, there was a sharp boundary between the Australian and Oriental faunal regions so that a geographic line drawn between certain islands of the Malay Archipelago, through the Makassar Strait between Borneo and Celebes, and between the Philippines and the Sanghir Islands separated two distinct and contrasting zoologic regions. On one side Australian forms predominated; on the other, Oriental ones. The validity of this division has been questioned by zoogeographers in the light of more extensive knowledge of the faunas of the two regions.

○ **1861:** *Claparéde, E. Discovery of giant nerve axons of annelids.*

These large nerve fibers were first found and described by Ehrenberg (1836) in the Crustacea. As new microscopic techniques were developed, many investigations have been made on the structure and functions of these unique nerve cells. Their chief function is in escape mechanisms involving widespread and synchronous muscular contractions.

△ **1861:** *Graham, Thomas. Colloidal states of matter.*

This work on colloidal solutions has resulted in one of the most fruitful concepts regarding protoplasmic systems.

□ **1862:** *Bates, H. W. Concept of mimicry.*

Mimicry refers to the advantageous resemblance of one species to another for protection against predators. It involves palatable or edible species that imitate dangerous or inedible species that enjoy immunity because of warning coloration against potential enemies.

□ **1864:** *Haeckel, E. Modern zoologic classification.*

The broad features of zoologic classification as we know it today were outlined by Haeckel and others, especially R. R. Lankester, in the third quarter of the nineteenth century. B. Hatschek is another zoologist who deserves much credit for the modern scheme, which is constantly undergoing revision. Other schemes of classification in the early nineteenth century did much to resolve the difficult problem of classification, such as those of Cuvier, de Lamarck, Leuckart, Ehrenberg, Vogt, Gegenbauer, and Schimkevitch. More recently, L. H. Hyman performed invaluable service in the arrangement of animal taxonomy.

△ **1864:** *Schultze, Max. Protoplasmic bridges between cells.*

The connection of one cell to another by means of protoplasmic bridges has been demonstrated in both plants and animals. Modern techniques, including electron microscopy, have shown that many cells communicate by gap junctions, which have channels capable of passing molecules of up to 1,000 daltons in molecular weight.

△ **1865:** *Kekulé, F. A. Concept of the benzene ring.*

The concept of the benzene ring established the later development of synthetic chemistry; it also aided in understanding the basic structure of life that is so dependent on the capacity of carbon atoms to form rings and chains.

‡ **1866:** *Haeckel, E. Nuclear control of inheritance.*

At the time the hypothesis was formed, the view that the nucleus transmitted the inheritance of an animal had no evidence to substantiate it.

□ **1866:** *Kovalevski, A. Taxonomic position of tunicates.*

The great Russian embryologist showed in the early stages of development the similarity between amphioxus and the tunicates and how the latter could be considered a branch of the phylum Chordata.

‡ **1866:** *Mendel, Gregor. Formulation of the first two laws of heredity.*

These first clear-cut statements of inheritance made possible an analysis of hereditary patterns with mathematic precision. Rediscovered in 1900, his paper confirmed the experimental data geneticists had found at that time. Since that time Mendelian genetics has been considered the chief cornerstone of hereditary investigation.

○ **1866:** *Schultze, Max. Histologic analysis of the retina.*

Schultze's fundamental discovery that the retina contained two types of visual cells—rods and cones—helped to explain the differences in physiologic vision of high and low intensities of light.

×✳ **1867:** *Kovalevski, A. Germ layers of invertebrates.*

The concept of the primary germ layers laid down by Pander and von Baer was extended to invertebrates by this investigator. He found that the same three germ layers arose in the same fashion as those in vertebrates. Thus an important embryologic unity was established for the animal kingdom.

○ **1869:** *Langerhans, P. Discovery of islet cells in pancreas.*

These islets were first found in the rabbit and since their discovery they have been one of the most investigated tissues in animals. The theory that they have come from the transformation of pancreatic acinus cells has now given way to the theory that they have originated from the embryonic tubules of the pancreatic duct.

△ **1869:** *Miescher, F. Isolation of nucleoprotein.*

From pus cells, a pathologic product, Miescher was able to demonstrate in the nuclei certain phosphorus-rich substances, nucleic acids, which are bound to proteins to form nucleoproteins. In recent years these complex molecules have been the focal point of significant biochemical investigations on the chemical properties of genes, with the wider implications of a better understanding of growth, heredity, and evolution.

1871: *Quetelet, L. A. Foundations of biometry.*

The applications of statistics to biologic problems is called biometry. By working out the distribution curve of the height of soldiers, Quetelet showed biologists how the systematic study of the relationships of numeric data could become a powerful tool for analyzing data in evolution, genetics, and other biologic fields.

□ **1872-1876:** *Challenger expedition.*

Not only did this expedition establish the science of oceanography, but a vast amount of material was collected that greatly extended the knowledge of the variety and range of animal life.

□ **1872:** *Dohrn, Anton. Establishment of Naples Biological Station.*

The establishment of this famous station marked the first of the great biologic stations, and it rapidly attained international importance as a center of biologic investigation.

○ **1872:** *Ludwig, C., and E. F. W. Pflüger. Gas exchange of the blood.*

By means of the mercurial blood pump, these investigators separated the gases from the blood and thereby threw much light on the nature of gaseous exchange and the place where oxidation occurred (in the tissues).

‡ **1873:** *Schneider, Anton. Description of nuclear filaments (chromosomes).*

In his description of cell division, Schneider showed nuclear structures that he termed "nuclear filaments," a further description of what are now known as chromosomes.

×□ **1874:** *Haeckel, E. H. Taxonomic position of phylum Chordata.*

The great German evolutionist based many of his conclusions on the work of the Russian embryologist Kovalevski, who in 1866 showed that the tunicates as well as amphioxus had vertebrate affinities.

× **1874:** *Haeckel, E. H. Gastrea theory of metazoan ancestry.*

According to this theory, the hypothetic ancestor of all the Metazoa consisted of two layers (ectoderm and endoderm) similar to the gastrula stage in embryonic development, and the endoderm arose as an invagination of the blastula. Thus the diploblastic stage of ontogeny was to be considered as the repetition of this ancestral form. This theory in modified form has had wide acceptance.

‡∗ **1875:** *Hertwig, O. Concept of fertilization as the conjugation of two sex cells.*

The fusion of the pronuclei of the two gametes in the process of fertilization paved the way for the concept that the nuclei contained the hereditary factors and that both maternal and paternal factors are brought together in the zygote.

△ **1875:** *Strasburger, Eduard. Description of "indirect" (mitotic) cell division.*

The accurate description of the processes of cell division that Strasburger made in plants represents a great pioneer work in the rapid development of cytology during the last quarter of the nineteenth century.

△ **1876:** *Pasteur, Louis. The Pasteur effect.*

In his studies of fermentation processes in winemaking, Pasteur discovered that cells consumed much more glucose and produced much more lactic acid in the absence of oxygen than in its presence. It is now known that the effect occurs in many different kinds of cells and is a consequence of the lower energy yield of glycolysis compared to oxidative phosphorylation.

○ **1877:** *Ehrlich, P. Discovery of mast cells.*

Mast cells are granular cells of connective tissue and are associated with inflammatory processes and new growth. They are known also to be the site of the concentration and release of histamine in hypersensitive reactions.

○ **1877:** *Pfeffer, W. Concept of osmosis and osmotic pressure.*

Pfeffer's experiments on osmotic pressure and the determination of the pressure in different concentrations laid the foundation for an understanding of a general phenomenon in all organisms.

○ **1878:** *Balfour, F. M. Relationship of the adrenal medulla to the sympathetic nervous system.*

By showing that the adrenal medulla has the same origin as the sympathetic nervous system, Balfour really laid the foundation for the interesting concept of the similarity in the action of epinephrine and sympathetic nervous mediation.

○△ **1878:** *Kühne, W. Nature of enzymes.*

The study of the action of chemical catalysts has steadily increased with biologic advancement and now represents one of the most fundamental aspects of biochemistry.

△ **1879:** *Flemming, W. Chromatin described and named.*

Flemming shares with a few others (especially Strasburger, 1875) the description of the details of mitotic cell division. The part of the nucleus that stains deeply he called **chromatin** (colored), which gives rise to the chromosomes.

∗**1879:** *Fol, Hermann. Penetration of ovum by a spermatozoon described.*

Fol was the first to describe a thin conelike body extending outward from the egg to meet the sperm. Compare Dan's acrosome reaction (1954).

△ **1879:** *Kossel, A. Isolation of nucleoprotein.*

Nucleoproteins were isolated in the heads of fish sperm; they make up the major part of chromatin. They are combinations of proteins with nucleic acids, and this study was one of the first of the investigations that interests biochemists at the present time.

Nobel Laureate (1910).

† **1880:** *Laveran, C. L. A. Protozoa as pathogenic agents.*

This French investigator first demonstrated that the causative organism for malaria is a protozoan. This discovery led to other investigations that revealed the role the protozoans play in causing diseases such as sleeping sickness and kala-azar. It remained for Sir Ronald Ross to discover the role of mosquitoes in spreading malaria.

Nobel Laureate (1907).

○ **1880:** *Ringer, S. Influence of blood ions on heart contraction.*

The pioneer work of this investigator determined the inorganic ions necessary for contraction of frog hearts and made possible an evaluation of heart metabolism and the replacement of body fluids.

△ **1882:** *Flemming, W. First accurate counts of nuclear filaments (chromosomes) made.*

△ **1882:** *Flemming, W. Mitosis and spireme named.*

†△ **1882:** *Metchnikoff, Élie. Role of phagocytosis in immunity.*

The theory that microbes are ingested and destroyed by certain white corpuscles (phagocytes) shares with the theory of chemical bodies (antibodies) the chief explanation for the body's natural immunity. Metchnikoff first studied phagocytosis in the cells of starfish and water fleas *(Daphnia).*

Nobel Laureate (1908).

△ **1882:** *Strasburger, E. Cytoplasm and nucleoplasm named.*

□ **1882-1924:** *The Albatross of the Fish Commission.*

The *Albatross,* under the direction of the U.S. Fish Commission, was second only to the famed *Challenger* in advancing scientific knowledge about oceanography.

○△ **1883:** *Golgi, Camillo, and R. Cajal. Silver nitrate technique for nervous elements.*

The development and refinement of this technique gave a completely new picture of the intricate relationships of neurons. Modifications of this method have given valuable information concerning the cellular element—the Golgi apparatus.

Nobel Laureates (1906).

∗ **1883:** *Hertwig, O. Origin of term "mesenchyme."*

This important tissue may arise embryonically from all three germ layers, but chiefly from the mesoderm. It is a protoplasmic network whose meshes are filled with a fluid intercellular substance. Mesenchyme gives rise to a great variety of tissues, and both its cells and intercellular substance may be variously modified; its most common derivative is connective tissue.

†□ **1883:** *Leuckart, R., and A. P. Thomas. Life history of sheep liver flukes.*

This discovery represents the first time a complete life cycle was worked out for a digenetic trematode. Working independently, these investigators found that the trematode used a snail intermediate host.

‡ **1883:** *Roux, W. Allocation of hereditary functions to chromosomes.*

This theory could not have been much more than a guess when Roux made it, but how fruitful was the idea in the light of the enormous amount of evidence since accumulated!

△ **1884:** *Flemming, W., E. Strasburger, and E. van Beneden. Demonstration that nuclear filaments (chromosomes) double in number by longitudinal division.*

This concept represented a further step in understanding the precise process in indirect cell division.

∗ **1884:** *Kollman, J. Concept of neoteny.*

Neoteny refers to the retention of larval characters and retardation of somatic growth after the gonads have become sexually mature. It was first described in the axolotl larval form of the tiger salamander *Ambystoma tigrinum.*

○ **1884:** *Rubner, Max. Quantitative determinations of the energy value of foods.*

Although Liebig and others had estimated calorie values of foods, the investigations of Rubner put their determinations on a sound basis. His work made possible a scientific explanation for metabolism and a basis for the study of comparative nutrition.

△ **1884:** *Strasburger, E. Prophase, metaphase, and anaphase named.*

‡ **1885:** *Hertwig, O., and E. Strasburger. Concept of the nucleus as the basis of heredity.*

The development of this idea occurred before Mendel's laws of heredity were rediscovered in 1900, but it anticipates the important role the nucleus with its chromosomes was to assume in hereditary transmission.

‡ **1885:** *Rabl, Karl. Concept of the individuality of the chromosomes.*

The view that the chromosomes retain their individuality through all stages of the cell cycle is accepted by all cytologists, and much evidence has accumulated in proof of the theory. However, it has been virtually impossible to demonstrate the chromosome's individuality through all stages.

✳ **1885:** *Roux, W. Mosaic theory of development.*

In the early development of the frog's egg, Roux showed that the determinants for differentiation were segregated in the early cleavage stages and that each cell or groups of cells would form only certain parts of the developing embryo (mosaic or determinate development). Later, other investigators showed that in many forms blastomeres, separated early, would give rise to whole embryos (indeterminate development).

✳‡ **1885:** *Weismann, August. Formulation of germ plasm theory.*

Weismann's great theory of the germ plasm stresses the idea that there are two types of protoplasm—germ plasm, which gives rise to the reproductive cells or gametes, and somatoplasm, which furnishes all the other cells. Germ plasm, according to the theory, is continuous from generation to generation, whereas the somatoplasm dies with each generation and does not influence the germ plasm.

1886: *Establishment of Woods Hole Biological Station.*

This station is by all odds the greatest center of its kind in the world. Many of the biologists in America have studied there, and the station has increasingly attracted investigators from many other countries as well. The influence of Woods Hole on the progress of biology cannot be overestimated.

△ **1886:** *MacMunn, C. A. Discovery of cytochrome.*

This iron-bearing compound was rediscovered in 1925 by D. Keilin and has been demonstrated in most types of cells. Its importance in cellular oxidation and the metabolic pathway has made possible a workable theory of cell respiration and has stimulated the study of intracellular localization of enzymes and how they behave in metabolic processes.

△ **1887:** *Fischer, Emil. Structural patterns of proteins.*

The importance of proteins in biologic systems has made their study the central theme of much modern biochemical work.

Nobel Laureate (1902).

✳ **1887:** *Haeckel, E. H. Concept of organic form and symmetry.*

Symmetry refers to the spatial relations and arrangements of parts in such a way as to form geometric designs. Although many others before Haeckel's time had studied and described types of animal form, Haeckel has given us our present concepts of organic symmetry, as revealed in his monograph on radiolarians collected on the *Challenger* expedition.

‡✳ **1887:** *Weismann, August. Prediction of the reduction division.*

Weismann formulated the hypothesis that the separation of undivided whole chromosomes must take place in one of the maturation divisions (reduction division), or else the number of chromosomes would double in each generation *(reductio ad absurdum).*

‡✳ **1887:** *van Beneden, E. Demonstration of chromosome reduction during maturation.*

Weismann predicted this important event before it was actually demonstrated. Only by this method can the constancy of chromosome numbers be maintained.

‡ **1888:** *Waldeyer, W. Chromosome named.*

‡ **1889:** *Hertwig, R., and E. Maupas. True nature of conjugation in paramecium.*

The process of conjugation was described (even by van Leeuwenhoek) many times and its sexual significance interpreted, but these two investigators independently showed the details of pregamic divisions and the mutual exchange of the micronuclei during the process.

†○ **1889:** *von Mering, J., and O. Minkowski. Effect of pancreatectomy.*

The classic experiment of removing the pancreas stimulated research that led to the isolation of the pancreatic hormone insulin by Banting (1922).

† **1890:** *Smith, Theobald. Role of an arthropod in disease transmission.*

The transmission of the sporozoan *Babesia,* which is the active agent in causing Texas cattle fever, by the tick *Boophilus,* represented the first demonstration of the role of an arthropod as a vector of disease.

✳ **1891:** *Driesch, H. Discovery of totipotent cleavage.*

The discovery that each of the first several blastomeres, if separated from each other in the early cleavage of the sea urchin embryo, would develop into a complete embryo stimulated investigation on totipotent and other types of development.

× **1890-1891:** *Dubois, Eugene. Discovery of the fossil human Pithecanthropus erectus (now Homo erectus).*

Although not the first fossil human to be found, the Java man represents one of the first significant primitive humans that have been discovered.

†□ **1892:** *Ivanovski, D. Discovery of viruses.*

Ivanovski found that there was a disease-producing agent much smaller than bacteria. This discovery was the start of the many investigations on the nature of viruses.

× **1893:** *Dollo, I. Concept of the irreversibility of evolution.*

In general, the overall evolutionary process is one-way and irreversible insofar as a whole complex genetic system is concerned, although back mutations and restricted variations may occur over and over.

○ **1893:** *His, W. Anatomy and physiology of the atrioventricular node and bundle.*

The specialized conducting tissue of the heart has given rise to many investigations, not the least of which were those of His, whose name was given to the intricate system of branches that are reflected over the inner surface of the ventricles.

∗**1894:** *Driesch, H. Constancy of nuclear potentiality.*

Driesch's view was that all nuclei of an organism were equipotential but that the activity of nuclei varied with different cells in accordance with the differentiation of tissues. That all genetic factors are present in all cells is supported by the constancy of DNA for each set of chromosomes and the similarity of histone proteins in the different somatic cells of an organism. This theory, however, is being challenged in the much investigated problems of differentiation and growth, and there has been some evidence to show that nuclear potentialities do vary with different stages in development.

□ **1894:** *Merriam, C. H. Concept of life zones in North America.*

This scheme is based on temperature criteria and the importance of temperature in the distribution of plants and animals. According to this concept, animals and plants are restricted in their northward distribution by the total quantity of heat during the season of growth and reproduction, and their southward distribution is restricted by the mean temperature during the hottest part of the year.

□ **1894:** *Morgan, C. Lloyd. Concept of animal behavior.*

The modern interpretation of comparative psychology really dates from certain basic principles laid down by this psychologist. Among these principles was the one in which he stated that the actions of an animal should be interpreted in terms of the simplest mental processes (Morgan's canon).

○ **1894:** *Oliver, G., and E. A. Sharpey-Schafer. Demonstration of the action of a hormone.*

The first recorded action of a specific hormone was the demonstration of the effect of an extract of the suprarenal (adrenal) gland on blood vessels and muscle contraction.

†□ **1895:** *Bruce, D. Life cycle of protozoan blood parasite (Trypanosoma).*

The relation of this parasite to the tsetse fly and to wild and domestic animal infection in Africa is an early demonstration of the role of arthropods as vectors of disease.

†○ **1895:** *Roentgen, W. Discovery of x-rays.*

This great discovery was quickly followed by its application in the interpretation of bodily structures and processes and represents one of the greatest tools in biologic research.

Nobel Laureate (1901).

× **1896:** *Baldwin, J. M. Baldwin evolutionary effect.*

It is the belief that genetic selection of genotypes will be channeled or canalized in the same direction as the adaptive modifications that were formerly nonhereditary. It is now understood that the *capacity* to respond to environmental conditions is itself hereditary.

× **1896:** *Russian Hydrographic Survey. Biology of Lake Baikal.*

This lake in Siberia is more than 800 km long, 80 km wide, and has an extreme depth of more than a mile (the deepest lake in the world). Its unique fauna is a striking example of evolutionary results from long-continued isolation. Up to 100% of the species in certain groups are endemic (found nowhere else). This remarkable fauna represents the survival of ancient freshwater animals that have become extinct in surrounding areas.

○ **1897:** *Abel, J. J., and A. C. Crawford. Isolation of the first hormone (epinephrine).*

The purification and isolation of one of the active principles of the suprarenal medulla led to its chemical nature and naming by J. Takamine (1901) and its synthesis by F. Stolz (1904).

○ **1897:** *Braun, F. Invention of the cathode-ray oscillograph.*

This instrument has been one of the most useful tools ever invented for measuring electric events in excitable tissues. Erlanger and Gasser (1922) later combined an amplifier with the oscillograph to record nerve action potentials.

Nobel Laureate (1909).

△○ **1897:** *Buchner, E. Discovery of zymase.*

Buchner's discovery that an enzyme (a nonliving substance) manufactured by yeast cells was responsible for fermentation resolved many problems that had baffled Pasteur and other investigators. Zymase is now known to consist of a number of enzymes.

Nobel Laureate (1907).

× **1897:** *Canadian Geological Survey. Dinosaur fauna of Alberta, Canada.*

In the rich fossil beds along the Red Deer River in Alberta there was found the fauna of the Upper Cretaceous time, and a whole new revelation of the dinosaur world has been made from the study of these fossils.

†○ **1897:** *Eijkman, C. Discovery of the cause of a dietary deficiency disease.*

Eijkman's pioneer work on the causes of beriberi led to the isolation of the antineuritic vitamin (thiamine). This work may be called the key discovery that resulted in the development of the important vitamin concept.

Nobel Laureate (1929).

○ **1897:** *Huot, A. Discovery of the aglomerular fish kidney.*

This discovery in the angler fish *(Lophius)* and later in the toadfish and others proved to renal physiologists that renal tubules of kidneys could secrete as well as reabsorb substances.

†□ **1897:** *Ross, Ronald. Life history of malarial parasite (Plasmodium).*

This notable achievement represents a great landmark in the field of parasitology and the climax of the work of many investigators on the problem.

Nobel Laureate (1902).

△○ **1897:** *Sherrington, C. S. Concept of the synapse in the nervous system.*

If the nervous system is composed of discrete units, or neurons, functional connections must exist between these units. Sherrington showed how individual nerve cells could exert integrative influences on other nerve cells by graded excitatory or inhibitory synaptic actions. The electron microscope has in recent years added much to a knowledge of the synaptic structure.

Nobel Laureate (1932).

△ **1898:** *Benda, C., and C. Golgi. Discovery of mitochondria and the Golgi apparatus.*

These interesting cytoplasmic inclusions were actually seen by various observers before this date, but they were both named in 1898, and the real study of them began at this time. W. Flemming and R. Altmann first demonstrated mitochondria. The Golgi apparatus was demonstrated by V. St. George (1867) and G. Platner (1855), but Camillo Golgi, with his silver nitrate impregnation method, gave the first clear description of the apparatus in nerve cells. Mitochondria are now known to play an important role in cell respiration and energy production.

× **1898:** *Osborn, H. F. Concept of adaptive radiation in evolution.*

This concept states that, starting from a common ancestral type, many different forms of evolutionary adaptations may occur. In this way evolutionary divergence can take place, and the occupation of a variety of ecologic niches is made possible, according to the adaptive nature of the invading species.

△ **1899:** *Hardy, W. B. Appraisal of conventional fixation methods.*

This investigation showed that the common preservatives used in killing and fixing cells produced either fibrous networks or fine emulsions and that the method of fixation determined which of these states is produced (artifacts).

The structures found in stained cells needed confirmation by other methods because the fibrillar, reticular, and other appearances of protoplasm are artifacts from the type of fixation and staining employed.

× **1900:** *Chamberlain, T. C. Theory of the freshwater origin of vertebrates.*

The evidence for this theory as first proposed was based mainly on the discovery of early vertebrate fossils in sediments thought to be of freshwater origin, such as Old Red Sandstone. The fossil evidence has been reevaluated by recent investigators, who interpret it to support a marine origin of the vertebrates.

‡ **1900:** *Correns, K. E., E. von Tschermak-Seysenegg, and Hugo De Vries. Rediscovery of Mendel's laws of heredity.*

In their genetic experiments on plants these three investigators independently obtained results similar to Mendel's, and in their survey of the literature found that Mendel had published his now famous laws in 1866. A few years later, W. Bateson and others found that the same laws applied to animals.

× **1900:** *Andrews, C. F. Discovery of fossil beds in the Fayum Depression region of Egypt.*

Here Andrews and others discovered numerous Eocene and Oligocene fossils of protomonkeys believed to be ancestral to Old World monkeys and thus in the lineage leading to hominids.

†○ **1900:** *Landsteiner, Karl. Discovery of blood groups.*

This fundamental discovery made possible successful blood transfusions and initiated a tremendous amount of work on the biochemistry of blood.

Nobel Laureate (1930).

✶ **1900:** *Loeb, J. Discovery of artificial parthenogenesis.*

The possibility of getting eggs that normally undergo fertilization to develop by chemical and mechanical methods has been accomplished in a number of different animals from the sea urchin and frog eggs (Loeb, 1900) to the rabbit egg (Pincus, 1936). The phenomenon has been a useful tool in studying the biology of fertilization.

‡ **1901:** *Montgomery, T. H. Homologous pairing of maternal and paternal chromosomes in the zygote.*

Montgomery showed that in synapsis before reduction division, each pair is made up of a maternal and a paternal chromosome. This phenomenon, verified a year later by W. S. Sutton, is of fundamental importance in the segregation of hereditary factors (genes).

×‡ **1901:** *De Vries, Hugo. Mutation theory of evolution.*

De Vries concluded from his study of the evening primrose *Oenothera lamarckiana* that new characters appear suddenly and are inheritable. Although the variations in *Oenothera* were probably not mutations at all, since many of them represented hybrid combinations, the evidence for the theory from other sources has steadily mounted until now the theory affords the most plausible explanation for evolutionary progress.

× **1902:** *Kropotkin, P. Mutual aid as a factor in evolution.*

This concept was not new by any means, but Kropotkin elaborated on the importance of social life at all levels of animal life in the survival patterns of the evolutionary process.

‡ **1902:** *McClung, C. E. Discovery of sex chromosomes.*

The discovery in the grasshopper that a certain chromosome (X) had a synaptic mate (Y) different in appearance or else lacked a mate altogether gave rise to the theory that certain chromosomes determined sex. H. Henking actually discovered the X chromosome in 1891.

○ **1903:** *Bayliss, W. M., and E. H. Starling. Action of the hormone secretin.*

The demonstration of the action of secretin, a hormone released from the mucosa of the stomach, marked the real birth of the science of endocrinology. It was the first unequivocal proof that physiologic functions could be chemically integrated without participation of the nervous system.

‡ **1903:** *Boveri, T., and W. S. Sutton. Parallelism between chromosome behavior and Mendelian segregation.*

This theory states that synaptic mates in meiosis correspond to the Mendelian alternative characters and that the formula of character inheritance of Mendel could be explained by the behavior of the chromosomes during maturation. This is therefore a cytologic demonstration of Mendelian genetics.

‡ **1903:** *Sutton, W. S. Constitution of the diploid group of chromosomes.*

Sutton related Mendelian transmission of inherited characters with cytology by showing that the diploid group of chromosomes is made up of two chromosomes of each recognizable size, one member of which is paternal and the other maternal in origin.

○ **1904:** *Cannon, W. B. Mechanics of digestion by x-rays.*

The clever application of x-rays to a study of the movements and other aspects of the digestive system has revealed an enormous amount of information on the physiology of the alimentary canal. Cannon first used this technique in 1898.

○ **1904:** *Carlson, A. J. Pacemaking activity of neurogenic hearts.*

Heartbeats may originate in muscle (myogenic hearts) or in ganglion cells (neurogenic hearts). Carlson showed that in the arthropod *Limulus* the pacemaker was located in certain ganglia on the dorsal surface of the heart and that experimental alterations of these ganglia by temperature or other means altered the heart rate. Neurogenic hearts are restricted to certain invertebrates.

□ **1904:** *Jennings, H. S. Behavior patterns in Protozoa.*

The careful investigations of this lifelong student of the behavior of these lower organisms led to concepts such as the trial-and-error behavior and many of our important beliefs about the various forms of tropisms and taxes.

□ **1904:** *Nuttall, G. H. F. Serologic relationships of animals.*

This method of determining animal relationships is striking evidence of evolution. It has been used in recent years to establish the taxonomic position of animals whose classification has not been determined by other methods.

○ **1905:** *Haldane, J. B. S., and J. G. Priestley. Role of carbon dioxide in the regulation of breathing.*

By their clever technique of obtaining samples of air from the lung alveoli, these investigators showed how the constancy of carbon dioxide concentration in the alveoli and its relation to the concentration in the blood was the chief regulator of the mechanism of respiration.

△ **1905:** *Zsigmondy, R. Application of ultracentrifuge to colloids.*

This has made possible a study of colloidal particles and the finer details of protoplasmic systems.

Nobel Laureate (1925).

‡ **1906:** *Bateson, W., and R. C. Punnett. Discovery of linkage of hereditary units.*

Although Bateson and Punnett first discovered linkage in sweet peas, it was Morgan and associates who correctly interpreted this great genetic concept. All seven pairs of Mendel's alternative characters were in separate chromosomes, a fact that simplified the problem.

○ **1906:** *Einthoven, W. Mechanism of the electrocardiogram.*

The invention of the string galvanometer (1903) by Einthoven supplied a precise tool for measuring the bioelectric activity of the heart and quickly led to the electrocardiogram, which gives accurate information about disturbances of the heart's rhythm.

Nobel Laureate (1924).

○ **1906:** *Hopkins, F. G. Analysis of dietary deficiency.*

Hopkins tried to explain dietary deficiency by a biochemical investigation of the lack of essential amino acids in the diet, an approach that has led to many important investigations in nutritional requirements.

Nobel Laureate (1929).

△ **1906:** *Tswett, M. Principle of chromatography.*

This is the separation of chemical components in a mixture by differential migration of materials according to structural properties within a special porous sorptive medium. It was refined and became a widely used method 30 or 40 years later (Martin and Synge, 1941). The technique was first used by Tswett to separate pigments of plants.

† **1906-1913:** *Gorgas, W. C. Control of malaria and yellow fever.*

By intelligent and industrious application of the knowledge that mosquitoes carry malaria and yellow fever, Gorgas saved 71,000 lives and made possible the successful construction of the Panama Canal.

✳ **1907:** *Boveri, T. Qualitative differences of chromosomes.*

Boveri showed in his classic experiment with sea urchin eggs that chromosomes have qualitatively different effects on development. He found that only those cells developed into larvae that had one of each kind of chromosome; those cells that did not have representatives of each kind of chromosome failed to develop.

× **1907:** *Discovery of the Heidelberg fossil human (Homo heidelbergensis, now H. erectus heidelbergensis).*

This fossil consisted of a lower jaw with all its teeth. The jaw shows many simian characteristics, but the teeth show patterns of primitive humans. This type is supposed to have existed during mid-Pleistocene times and is more or less intermediate between Java and modern humans.

○ **1907:** *Hopkins, F. G. Relationship of lactic acid to muscular contraction.*

Hopkins showed that, after being formed in muscular contraction, a part of the lactic acid is oxidized to furnish energy for the resynthesis of the remaining lactic acid into glycogen. This discovery did much to clarify part of the metabolic reactions involved in the complicated process of muscular contraction.

Nobel Laureate (1929).

○ **1907:** *Keith, A., and M. J. Flack. Discovery of the sinoauricular (S-A) node.*

The precursor of this node is the sinus tissue of the primitive heart of cold-blooded animals. In mammals this node is embedded in the muscle of the right auricle near the openings of the superior and inferior venae cavae and initiates the beat and sets the pace (hence, pacemaker) for the mammalian heart.

The atrioventricular (A-V) node, which lies in the septum of the atria near the A-V valves, was discovered in 1906 by Tawara.

∗ **1907:** *Wilson, H. V. Reorganization of sponge cells.*

In this classic experiment, Wilson showed that the disaggregation of sponges, by squeezing them through fine silk bolting cloth so that they are separated into minute cell clumps, resulted in the surviving cells coming together and organizing themselves into small sponges when they were in seawater.

‡ **1908:** *Garrod, A. E. Discovery that gene products are proteins.*

Certain hereditary disorders are caused by enzyme deficiencies wherein the mutations of a single gene may be responsible for controlling the specificity of a particular enzyme in certain disorders. Garrod's work was largely ignored until 1940.

×‡ **1908:** *Hardy, G. H., and W. Weinberg. Hardy-Weinberg population formula.*

This important theorem states that, in the absence of factors (such as mutation and selection) causing change in gene frequency, the frequency of a particular gene in any large population will reach an equilibrium in one generation and thereafter will remain stable regardless of whether the genes are dominant or recessive. Its mathematic expression forms the basis for the calculations of population genetics.

○△ **1909:** *Arrhenius, S., and S. P. L. Sörensen. Determination of hydrogen ion concentration (pH).*

The sensitivity of most biologic systems to acid and alkaline conditions has made pH values of the utmost importance in biologic research.

Arrhenius, Nobel Laureate (1903).

∗ **1909:** *Bataillon, E. Discovery of the pseudogamy concept.*

Pseudogamy is the activation of an unfertilized egg by a sperm without the participation of the male chromosomes of the sperm in the hereditary pattern of the resulting embryo. The sperm may be from a male of the same species or from a different species.

‡ **1909:** *Castle, W. E., and J. C. Philips. The inviolability of germ cells to somatic cell influences.*

That germ cells are relatively free from somatic cell influences was shown by the substitution of a black guinea pig ovary in a white guinea pig that gave rise to black offspring when mated to a black male.

‡ **1909:** *Janssens, F. A. Chiasmatype theory.*

When homologous chromosomes are paired before the reduction division, they or their chromatids form visible crosslike figures or chiasmata, which Janssens interpreted as the visible exchange of parts of two homologous chromatids, although he could not actually prove this point. This phenomenon of crossing-over is the key to the genetic mapping of chromosomes so extensively worked out by Morgan and his school with *Drosophila*.

‡ **1909:** *Johannsen, W. Gene, genotype, and phenotype named.*

× **1909:** *Johannsen, W. Limitations of natural selection on pure lines.*

This investigator found that when a hereditary group of characters becomes genetically homogeneous, natural selection cannot change the genetic constitution with regard to these characters. He showed that selection could not itself create new genotypes but was effective only in isolating genotypes already present in the group; it therefore could not effect evolutionary changes directly.

†□ **1909:** *Nicolle, C. J. H. Body louse as vector of typhus fever.*

The demonstration that typhus fever was transmitted from patient to patient by the bite of the body louse paved the way for the control of this dreaded epidemic disorder by using DDT to delouse populations.

Nobel Laureate (1928).

○ **1910:** *Dale, H. H. Nature of histamine.*

Dale and colleagues found that an extract from ergot had the properties of histamine (β-imidazolyl ethylamine), which can be produced synthetically by splitting off carbon dioxide from the amino acid histidine. It is also a constituent of all tissue cells, from which it may be released by injuries or other causes. The pronounced effect of histamine in dilating small blood vessels, contracting smooth muscle, and stimulating glands has caused it to be associated with many physiologic phenomena, such as anaphylaxis, shock, and allergies.

Nobel Laureate (1936).

† **1910:** *Ehrlich, Paul. Chemotherapy in the treatment of disease.*

The discovery of salvarsan as a cure for syphilis represents the first great discovery in this field. Another was the dye sulfanilamide, discovered by Domagk in 1935.

Nobel Laureate (1908).

□ **1910:** *Heinroth, O. Concept of imprinting as a type of behavior.*

This is a special type of learning that is demonstrated by birds (and possibly other animals). It is based on the fact that in many species of birds the hatchlings are attracted to the first large object they see and thereafter will follow that object to the exclusion of all others. Although the behavior pattern appears to be strongly fixed, there is some doubt about its irreversibility.

†‡ **1910:** *Herrick, J. B. Discovery of sickle cell anemia.*

In this inherited condition, red blood cells have an abnormal sickle shape and do not function normally. Investigation has shown that the S hemoglobin found in this type of anemia differs from the normal A hemoglobin in the protein component but not in the heme component. Glutamic acid, one of the 300 amino acids in the molecule of normal hemoglobin, is replaced by valine in the sickle cell hemoglobin. The inheritance is supposed to involve a single gene.

‡ **1910:** *Morgan, T. H. Discovery of sex linkage.*

Morgan and colleagues discovered that the results of a cross between a white-eyed male and a red-eyed female in *Drosophila* were different from those obtained from the reciprocal cross of a red-eyed male with a white-eyed female. This was a crucial experiment, for it showed for the first time that how a trait behaved in heredity depended on the sex of the parent, in contrast to most Mendelian characters, which behave genetically the same way whether introduced by a male or female parent.

‡ **1910-1920:** *Morgan, T. H. Establishment of the theory of the gene.*

The extensive work of Morgan and associates on the localization of hereditary factors (by genetic experiments) on the chromosomes of the fruit fly *(Drosophila)* represents some of the most significant work ever performed in the field of heredity.

Nobel Laureate (1933).

□ **1910:** *Murray, J., and J. Hjort. Deep-sea expedition of the Michael Sars.*

Of the many expeditions for exploring the depths of the oceans, the *Michael Sars* expedition, made in the North Atlantic regions, must rank among the foremost. The expedition yielded an immense amount of information about deep-sea animals, as well as many important concepts regarding the ecologic pattern of animal distribution in the sea.

□ **1910:** *Pavlov, I. P. Concept of the conditioned reflex.*

The idea that acquired reflexes play an important role in the nervous reaction patterns of animals has greatly influenced the development of modern psychology.

Nobel Laureate (1904).

* **1911:** *Child, C. M. Axial gradient theory.*

This theory attempts to explain the pattern of metabolism from the standpoint of localized regional differences along the axes of organisms. The differences in the metabolic rate of different areas have made possible an understanding of certain aspects of regeneration, development, and growth.

× **1911:** *Cuénot, L. The preadaptation concept.*

Preadaptation refers to a morphologic or physiologic character that has been selected for (arisen) in one environment, but which is coincidentally adaptive in a new environment. This favorable conjunction of characters and suitable environment is considered to be an important factor in progressive evolution (opportunistic evolution).

† **1911:** *Funk, C. Vitamin hypothesis.*

Vitamin deficiency diseases refer to those diseases in which effects can be definitely traced to the lack of some essential constituent of the diet. Thus beriberi is caused by an insufficient amount of thiamine, scurvy by a lack of vitamin C, and so on.

* **1911:** *Harvey, E. B. Cortical changes in the egg during fertilization.*

In the mature egg, cortical granules gather at the surface of the egg, but on activation of the egg these granules, beginning at the point of sperm contact, disappear in a wavelike manner around the egg. These granules are supposed to release material during their breakdown that helps form the ensuing fertilization membrane.

△ **1911:** *Rutherford, E. Concept of the atomic nucleus.*

This cornerstone of modern physics must be of equal interest to the biologist in the light of the rapid advances of molecular biology. Rutherford also discovered (1920) the proton, one of the charged particles in the atomic nucleus.

Nobel Laureate (1908).

× **1911:** *Walcott, C. D. Discovery of Burgess shale fossils.*

The discovery of a great assemblage of beautifully preserved invertebrates in the Burgess shale of British Columbia and their careful study by an American paleontologist represent a landmark in the fossil record of invertebrates. These fossils date from the middle Cambrian age and include the striking *Aysheaia,* which has a resemblance to the extant *Peripatus.*

△○ **1912:** *Carrel, A. Technique of tissue culture.*

The culturing of living tissues in vitro, that is, outside of the body, has given biologists an important tool for studying tissue structure and growth. Although Carrel developed the technique, Ross Harrison introduced tissue culture in 1907 when he found that parts of living tissues in suitable media under suitable conditions could live and multiply.

Nobel Laureate (1912).

○∗**1912:** *Gudernatsch, J. F. Role of the thyroid gland in the metamorphosis of frogs.*

This investigator found that the removal of the thyroid gland of tadpoles prevented metamorphosis into frogs, and also that the feeding of thyroid extracts to tadpoles induced precocious metamorphosis. In 1919 W. W. Swingle also showed that the presence and absence of inorganic iodine would produce the same results.

△ **1912:** *Kite, G. L. Micrurgic study of cell structures.*

The use of micromanipulators for the microdissection of living cells has greatly enriched our knowledge of the finer microscopic details of protoplasm, chromosomes, cell division, and many other phenomena of cells. The method has been developed by many workers, such as H. D. Schmidt (1859), M. A. Barber (1904), and W. Seifriz (1921), but R. Chambers has perhaps done more to refine its use and to employ it in experimental cell research.

□ **1912:** *Wegener, A. L. Concept of continental drift theory.*

This theory postulates that the continents were originally joined together in one or two large masses that gradually broke up during geologic time, and the fragments drifted apart to form the current land masses. The theory has been revived recently on the basis of surveys of paleomagnetism, seisomgraphic studies, tectonic plate studies, and a wealth of biologic evidence, for example, the finding of an amphibian fossil in Antarctic regions.

△ **1913:** *Michaelis, L., and M. Menton. Enzyme-substrate complex.*

On theoretic and mathematic grounds these investigators showed that an enzyme forms an intermediate compound with its substrate (the enzyme-substrate complex) that subsequently decomposes to release the free enzyme and the reaction products. D. Keilin of Cambridge University and B. Chance of the University of Pennsylvania later demonstrated the presence of such a complex by color changes and measurements of its rate of formation and breakdown that agree with theoretic predictions.

× **1913:** *Reck, H. Discovery of Olduvai Gorge fossil deposits.*

This region in East Africa has yielded an immense amount of early mammalian fossils as well as the tools of Stone Age humans, such as stone axes. Among the interesting fossils discovered were elephants with lower jaw tusks, horses with three toes, the odd ungulate (chalicothere) with claws on the toes, and, more recently, several hominid fossils (see Leakey, Mary D., 1959).

□ **1913:** *Shelford, V. E. Law of ecologic tolerance.*

This law states that the potential success of an organism in a specific environment depends on how it can adjust within the range of its toleration to the various factors to which the organism is exposed.

‡ **1913:** *Sturtevant, A. H. Formation of first chromosome map.*

By the method of crossover percentages, it has been possible to locate the genes in their relative positions on chromosomes—one of the most fruitful discoveries in genetics, for it led to the extensive mapping of the chromosomes in *Drosophila.*

○ **1913:** *Tashiro, S. Metabolic activity of propagated nerve impulse.*

The detection of slight increases in carbon dioxide production in stimulated nerves, as compared with inactive ones, was evidence that conduction in nerves is a chemical change. Later (1926), A. V. Hill was able to measure the heat given off during the passage of an impulse. Increased oxygen consumption and other metabolic changes have also been measured in excited nerves.

○ **1914:** *Kendall, E. C. Isolation of thyroxine.*

The isolation of thyroxine in crystalline form was a landmark in endocrinology. Its artificial synthesis was done by Harington in 1927.

∗ **1914:** *Lillie, F. R. Role of fertilizin in fertilization.*

According to this theory, the jelly coat of eggs contains a substance, fertilizin, now known to be an acid mucopolysaccharide, which combines with the antifertilizin on the surface of sperm and causes the sperm to clump together.

‡ **1914:** *Shull, G. H. Concept of heterosis.*

When two standardized strains or races are crossed, the resulting hybrid generation may be noticeably superior to both parents as shown by greater vigor, vitality, and resistance to unfavorable environmental conditions. First worked out in corn (maize), such hybrid vigor may also be manifested by other kinds of hybrids. Although its exact nature is still obscure, the phenomenon may be caused by the bringing together in the hybrid of many dominant genes of growth and vigor that were scattered among the two inbred parents, or it may be caused by the complementary reinforcing action of genes when brought together.

‡ **1916:** *Bridges, C. B. Discovery of nondisjunction.*

Bridges explained an aberrant genetic result by a suggested formula that later he was able to confirm by cytologic examination. A pair of chromosomes failed to disjoin at the reduction division so that both chromosomes passed into the same cell. It was definite proof that genes are located on the chromosomes.

○ **1916:** *Lillie, F. R. Theory of freemartin.*

A female calf is sexually abnormal when it is born as a twin to a normal male, and this had been a baffling problem for centuries. Lillie demonstrated in convincing fashion that hormones from the earlier developing gonads of the male circulate into the blood of the female and alter the sex differentiation of the latter. As a result the gonads of the freemartin never reach maturity and so she remains sterile.

○ **1917:** *Kopeč, S. Demonstration of hormonal factors in invertebrate physiology.*

This investigation showed that the brain was necessary for insect metamorphosis, for when the brain was removed from the last instar larva of a certain moth, pupation failed to occur; when the brain was grafted into the abdomen, pupation was resumed.

× **1917:** *Broili, F. Discovery of amphibian-reptilian fossil Seymouria.*

This interesting fossil found near Seymour, Texas, has characteristics of both amphibians and reptiles and thus throws some light on the relations between these two vertebrate classes.

□ **1917:** *Grinnell, J. Concept of the ecologic niche.*

The niche is the role of an animal in relation to all of the resources in its environment, both physical and biologic. C. Elton has done much to develop the concept.

○ **1918:** *Krogh, A. Regulation of the motor mechanism of capillaries.*

The mechanism of the differential distribution of blood to the various tissues has posed many problems from the days of Harvey and Malpighi, but Krogh showed that capillaries were not merely passive in this distribution but that they had the power to contract or dilate actively, according to the needs of the tissues. Both nervous and chemical controls are involved.

Nobel Laureate (1920).

□ **1918:** *Loeb, J. Forced movements, tropisms, and animal conduct.*

Regarded as the founder of a mechanist "school," Loeb opposed the generally anthropomorphic and teleologic interpretations of animal behavior. While Loeb carried his mechanistic doctrine to an extreme, he did inspire more rigorous, experimental approaches in animal behavior studies.

○ **1918:** *Starling, E. H. The law of the heart.*

Within physiologic limits, the more the ventricles are filled with incoming blood, the greater is the force of their contraction at systole. This is an adaptive mechanism for supplying more blood to tissues when it is needed. It is now recognized that only the hearts of lower vertebrates obey Starling's law; the intact heart of mammals is regulated principally by nervous and hormonal controls.

□ **1918:** *Szymanski, J. S. Demonstration of time-measuring mechanism of animals.*

Szymanski showed that animals had some means of measuring time independently of such physical factors as light and temperature, for he discovered that 24-hour activity patterns were synchronized with the day-night cycle when animals were kept in constant darkness and temperature. This work has led to the concept (demonstrated by many investigations) that animals have some kind of internal clock whereby they can measure certain cycles independent of external factors. The mechanism in most cases is still unexplained.

‡ **1918:** *Vavilov, N. I. Biologic centers of origin as reservoirs of desirable genes.*

The Russian botanist and plant geographer stressed the importance of tracing strains of cultivated plants to the locale of their original cultivation, where inferior plants (by present standards) may contain valuable genes already selected by natural selection. Such a pool of genes, he maintained, could by selection and intercrossing afford genetic banks for constructing new and superior genotypes.

1919: *Aston, F. W. Discovery of isotopes.*

Radioactive isotopes have proved of the utmost value in biologic research because of the possibility of tracing the course of various elements in living organisms.

Nobel Laureate (1922).

△ **1920:** *Herzog, R. O., and W. Jancke. Development of x-ray diffractometry.*

When a parallel beam of x-rays is passed through crystallized biologic material, the rays are spread and an image of the diffracted pattern is recorded (by rings, spots, and the like). By measurements and mathematic calculations information about structure can be obtained.

□ **1920:** *Howard, H. E. Territorial patterns of bird behavior.*

In many species a mating pair of birds establishes and defends a specific territory against others of the same species. Usually the male asserts his claim by singing at points close to the boundaries of his staked-out claim.

△ **1921:** *Hopkins, F. G. Isolation of glutathione.*

The discovery of this sulfur-containing compound gave a great impetus to the study of the complicated nature of cellular oxidation and metabolism.

Nobel Laureate (1929).

○ **1921:** *Langley, J. N. Concept of the functional autonomic nervous system.*

Langley's concept of the functional aspects of the autonomic system dealt mainly with the mammalian type, but the fundamental principles of the system have been applied with modification to other groups as well. Two divisions—sympathetic and parasympathetic—are recognized in the functional interpretation of excitation and inhibition in the antagonistic nature of the two divisions.

○ **1921:** *Loewi, O., and H. H. Dale. Isolation of acetylcholine.*

This key demonstration led to the neurohumoral concept of the transmission of nerve impulses to muscles.

Nobel Laureates (1936).

○ **1921:** *Richards, A. N. Collection and analysis of glomerular filtrate of the kidney.*

This experiment was direct evidence of the role of the glomeruli as mechanical filters of cell-free and protein-free fluid from the blood and was striking confirmation of the Ludwig-Cushny theory of kidney excretion.

∗ **1921:** *Spemann, Hans. Organizer concept in embryology.*

Certain parts of the developing embryo known as organizers induce specific developmental patterns in responding tissue. While many aspects of induction remain obscure, Spemann's work was important in showing that development is a sequence of programmed stages and that each stage is necessary for the next.

Nobel Laureate (1935).

†○ **1921, 1922:** *Banting, F., C. H. Best, and J. J. R. Macleod. Extraction of insulin.*

The great success of this hormone in relieving a distressful disease, diabetes mellitus, and the dramatic way in which active extracts were obtained have made the isolation of this hormone the best known in the field of endocrinology.

Banting and Macleod, Nobel Laureates (1923).

○ **1922:** *Erlanger, J., and H. S. Gasser. Differential conduction of nerve impulses.*

By using the cathode-ray oscillograph, these investigators found that there were several different types of mammalian nerve fibers that could be distinguished structurally and that had different rates of conducting nervous impulses according to the thickness of the nerve sheaths (most rapid in the thicker ones).

Nobel Laureates (1944).

○ **1922:** *Kopeč, S. Concept of neurosecretory systems.*

This concept has emerged from the work of many investigators. R. Cajalin (1899) discovered that the nerve fibers to the neurohypophysis are extensions of the neurons of the hypothalamus. Kopeč found that the substance responsible for metamorphosis in the larva of moths originated in the brain where certain cerebral ganglia served as glands of internal secretion. Kopeč's pioneering studies were largely ignored by vertebrate endocrinologists until the 1950s, when through the studies of B. Scharrer, E. Sharrer, B. Hanström, G. W. Harris, and others, it was recognized that vertebrates possess a neurosecretory system (hypothalamo-hypophyseal) analagous to invertebrate systems.

○ **1922:** *Schiefferdecker, P. Distinction between eccrine and apocrine sweat glands.*

In mammals certain large sweat glands that develop in connection with the hair follicles in localized regions respond to stresses, such as fear, pain, and sex. This concept has stimulated investigation on the histology and physiology of the glandular activity of the skin. Apocrine glands were described as early as 1846 by W. E. Horner, an English investigator.

□ **1922:** *Schjelderup-Ebbe, T. Social dominance-subordinance hierarchies.*

This observer found certain types of social hierarchies among birds in which higher ranking individuals could peck those of lower rank without being pecked in return. Those of the first rank dominated those of the second rank, who dominated those of the third, and so on. Such an organization, once formed, may be permanent.

□ **1922:** *Schmidt, Johannes. Life history of the freshwater eel.*

The long, patient work of this oceanographer in solving the mystery of eel migration from the freshwater streams of Europe to their spawning grounds in the Sargasso Sea near Bermuda represents one of the most romantic achievements in natural history.

△ **1923:** *Hevesy, G. First isotopic tracer method.*

Tracer methodology has proved especially useful in biochemistry and physiology. For instance, it has been possible by the use of these labeled units to determine the fate of the particular molecule in all steps of a metabolic process and the nature of many enzymatic reactions. The exact locations of many elements in the body have been traced by this method.

Nobel Laureate (1943).

△ **1923:** *Warburg, Otto. Manometric methods for studying metabolism of living cells.*

The Warburg apparatus has been useful in measuring gas exchange and other metabolic processes of living tissues. It proved of great value in the study of enzymatic reactions in living systems and was a standard tool in many biochemical laboratories for many years.

Nobel Laureate (1931).

○ **1924:** *Cleveland, L. R. Symbiotic relationships between termites and intestinal flagellates.*

This study was made on one of the most remarkable examples of mutualism known in the animal kingdom. Equally important were the observations this investigator and others found in the symbiosis between the wood-roach *(Cryptocercus)* and its intestinal protozoa.

△ **1924:** *Feulgen, R. Test for nucleoprotein.*

This microchemical test is widely used by cytologists and biochemists to demonstrate the presence of DNA (deoxyribonucleic acid), one of the two major types of nucleic acids.

○ **1924:** *Houssay, B. A. Role of the pituitary gland in regulation of carbohydrate metabolism.*

This investigator showed that when a dog had been made diabetic by pancreatectomy, the resulting hyperglycemia and glucosuria could be abolished by removing the anterior pituitary gland.

Nobel Laureate (1947).

* ○ **1925:** *Baltzer, F. Sex determination in Bonellia.*

This was a classic discovery of the influence of environmental factors on sex determination in the echiurid worm *Bonellia*. It showed that sex determination in some animals was not necessarily chromosomal, and the juveniles could be ambipotent with respect to sex.

○ **1925:** *Barcroft, J. Function of the spleen.*

Barcroft and associates showed that the spleen served as a blood reservoir that in time of stress adds new corpuscles to the circulation. The spleen reservoir is especially important in hemorrhage and shock.

× **1925:** *Dart, Raymond. Discovery of Australopithecus africanus.*

This important fossil, once referred to as the "missing link," bore many human characteristics but had a brain capacity only slightly greater than the higher apes. Most authorities place this hominid on or near the main branch of human ancestry.

△ **1925:** *Mast, S. O. Nature of ameboid movement.*

By studying the reversible sol-gel transformation in the protoplasm of an ameba. Mast not only gave a logical interpretation of ameboid movement but also initiated many fruitful concepts about the contractile nature of protoplasmic gel systems, such as the furrowing movements (cytokinesis) in cell divisions.

†○ **1925:** *Minot, G. R., M. W. P. Murphy, and G. H. Whipple. Liver treatment of pernicious anemia.*

That the feeding of raw liver had a pronounced effect in the treatment of pernicious anemia (a serious blood disorder) was discovered by these investigators. Much later investigation by numerous workers has led to some understanding of the antipernicious factor or vitamin B_{12}, whose chemical name is cyanocobalamin. This complex vitamin contains, among other components, porphyrin that has a cobalt atom instead of iron or magnesium at its center.

Nobel Laureates (1934).

□ **1925:** *Rowan, W. Photoperiodism hypothesis of bird migration.*

Rowan demonstrated that birds subjected to artificially increased day length in winter increased the size of their gonads and showed a striking tendency to migrate out of season. Other workers, stimulated by Rowan's experiments, showed that naturally increasing day length in spring acts through the brain neurosecretory system to set in motion the migratory disposition in birds.

△ **1926:** *Fujii, K. Finer analysis of the chromosome.*

With the development of the smear and squash techniques, it was possible with the light microscope to demonstrate the internal structure of a chromosome, which formerly was described as a rod-shaped body. The newer version emphasizes a coiled filament (chromonema) that runs through the matrix of the chromosome and bears the genes. In certain stages of cell division two such threads are spirally coiled around each other so compactly that they appear as one thread.

○ **1926:** *Hill, A. V. Measurement of heat production in nerve.*

By applying the principle of the thermocouple, which Helmholtz had used to detect the heat of a contracting muscle, Hill and other workers were able to measure the different phases of heat release, such as initial heat and recovery heat.

Nobel Laureate (1922).

△ **1926:** *Sumner, J. B. Isolation of enzyme urease.*

The isolation of the first enzyme in crystalline form was a key discovery to be followed by others that have helped unravel the complex nature of these important biologic substances.

Nobel Laureate (1946).

○ **1927:** *Bozler, E. Analysis of nerve net components.*

Bozler's demonstration that the nerve net of cnidarians was made up of separate cells and contained synaptic junctions resolved the old problem of whether or not the plexus in this group of animals was an actual network.

□ **1927:** *Coghill, G. E. Innate behavior patterns of Amphibia.*

Coghill's studies on the origin and growth of the behavior patterns of salamanders by following the sequence of the emergence of coordinated movements and nervous connections through all stages of embryonic development have represented one of the most fruitful investigations in animal behavior. He showed how broad, general movements preceded local reflexes and how the probable phylogenetic appearance of behavior patterns originated.

○ **1927:** *Eggleton, P., G. P. Eggleton, C. H. Fiske, and Y. Subbarow. Role of phosphagen (phosphocreatine) in muscular contraction.*

The demonstration that phosphagen is broken down during muscular contraction and then is resynthesized during recovery gave an entirely new concept of the initial energy necessary for the contraction process. Confirmation of this discovery received a great impetus from the discovery of E. Lungsgaard (1930) that muscles poisoned with monoiodoacetic acid, which inhibits the production of lactic acid from glycogen, would still contract and that the amount of phosphocreatine broken down was proportional to the energy liberated. It is now known that phosphocreatine is a high-energy phosphate storage compound, readily donating its phosphate group to ADP to form ATP.

○ **1927:** *Heymans, C. Role of carotid and aortic reflexes in respiratory control.*

The carotid sinus and aortic areas contain pressoreceptors and chemoreceptors, the former responding to mechanical stimulation, such as blood pressure, and the latter to oxygen lack. When the pressoreceptors are stimulated, respiration is inhibited; when the chemoreceptors are stimulated, the respiratory rate is increased. (A much more potent stimulator of respiration, however, is the effect of carbon dioxide on the respiratory center of the brain.)

Nobel Laureate (1938).

‡ **1927:** *Muller, H. J. Artificial induction of mutations.*

By subjecting fruit flies *(Drosophila)* to mild doses of x-rays, Muller found that the rate of mutation could be increased 150 times over the normal rate.

Nobel Laureate (1946).

× **1927:** *Stensiö, E. A. Appraisal of the Cephalaspida (ostracoderm) fish fossil.*

The replacement of amphioxus as a prototype of vertebrate ancestry by the ammocoetes lamprey larva, currently of great interest, has to a great extent been the result of this careful fossil reconstruction. It is generally believed that living Agnatha (lamprey and hagfish) are descended from these ancient forms.

× **1928:** *Garstang, W. Theory of the ascidian ancestry of chordates.*

According to this theory, primitive chordates were sessile, filter-feeding marine organisms very similar to present-day ascidians. The actively swimming prevertebrate was considered a later stage in chordate evolution. The tadpole ascidian larva, with its basic organization of a vertebrate, had evolved within the group by progressive evolution and by neoteny became sexually mature, ceased to metamorphose into a sessile, mature ascidian, and through adaptation to freshwater conditions became the true vertebrate.

‡△ **1928:** *Griffith, F. Discovery of the transforming principle (DNA) in bacteria (genetic transduction).*

By injecting living nonencapsulated bacteria and dead encapsulated bacteria of the *Pneumococcus* strain into mice, it was found that the former acquired the ability to grow a capsule and that this ability was transmitted to succeeding generations. This active agent or transforming principle (from the encapsulated type) was isolated by other workers later and was found to consist of DNA. This is excellent evidence that the gene involved is the nucleic acid deoxyribonucleic acid (DNA).

F. Sanfelice (1893) had actually found the same principle when he discovered that nonpathogenic bacilli grown in a culture medium containing the metabolic products of true tetanus bacilli would also produce toxins and would do so for many generations.

○ **1928:** *Koller, G., and E. B. Perkins. Hormonal control of color changes in crustaceans.*

These investigators found out independently that the chromatophores of crustaceans were regulated by a substance that originated in the eyestalk and was carried by the blood. Before this time, the common belief was that nerves served as the principal control.

△○ **1928:** *Wieland, H., and A. Windaus. Structure of the cholesterol molecule.*

Sterol chemistry has been one of the chief focal points in the investigation of such biologic products as vitamins, sex hormones, and cortisone. Cholesterol is the precursor of bile acids and steroid hormones in animals.

Nobel Laureates (1927, 1928).

○ **1929:** *Berger, Hans. Demonstration of brain waves.*

The science of electroencephalography, or the electric recording of brain activity, is in its infancy because of the complexity of the subject, but much has been revealed about both the healthy and the diseased brain by this technique.

○ **1929:** *Butenandt, A., and E. A. Doisy. Isolation of estrone.*

This discovery was the first isolation of a sex hormone and was arrived at independently by these two investigators. Estrone was found to be the urinary and transformed product of estradiol, the actual hormone. The male hormone testosterone was synthesized by Butenandt and L. Ruzicka in 1931. The second female hormone, progesterone, was isolated from the corpora lutea of sow ovaries in 1934.

Doisy, Nobel Laureate (1943).

○ **1929:** *Castle, W. B. Discovery of the antianemic factor.*

Castle and associates showed that the gastric juice contained an enzymelike substance (intrinsic factor) that reacts with a dietary factor (extrinsic) to produce the antianemic principle. The latter is stored in the liver of healthy individuals and is drawn on for the maintenance of activity in bone marrow (erythropoietic tissue), that is, the formation of red blood cells. If the intrinsic factor is missing from gastric juice, pernicious anemia occurs. At present, authorities believe that vitamin B_{12} is both the extrinsic factor as well as the antianemic principle (erythrocyte-maturing factor).

† **1929:** *Fleming, A. Discovery of penicillin.*

The chance discovery of this drug from molds and its development by H. Florey a few years later gave us the first of a notable line of antibiotics that have revolutionized medicine.

Nobel Laureate (1945).

○ **1929:** *Heymans, C. Discovery of the role of the carotid sinus and bodies in regulating the respiratory center and arterial blood pressure.*

Heymans found that the carotid sinus contains pressure-sensitive receptors (baroreceptors) that, acting through a nervous reflex to the medulla, help to keep the blood pressure stabilized. He also discovered oxygen-sensitive chemoreceptors in the carotid and aortic bodies. These are the origin of a reflex that responds to a drop in blood oxygen tension by increasing respiratory rate and blood pressure.

Nobel Laureate (1938).

△ **1929:** *Lohmann, K., C. Fiske, and Y. Subbarow. Discovery of ATP.*

The discovery of ATP (adenosine triphosphate) culminated a long search for the direct source of energy in biochemical reactions of many varieties, such as muscular contraction, vitamin action, and many enzymatic systems.

× **1930:** *Fisher, R. A. Statistical analysis of evolutionary variations.*

With Sewall Wright and J. B. S. Haldane, Fisher has analyzed mathematically the interrelationships of the factors of mutation rates, population sizes, selection values, and others in the evolutionary process. Although many of their theories are in the empiric stage, evolutionists in general agree that they have great significance in evolutionary interpretation.

1930: *Lawrence, E. O. Invention of the cyclotron.*

The importance to biologic research of artificial radioactive isotopes, the synthesis of which is made possible by this invention, cannot be overestimated.

Nobel Laureate (1939).

○ **1930:** *Northrop, J. H. Crystallization of the enzymes pepsin and trypsin.*

This was a further step in the elucidation of the nature of enzymes that form the core of biochemical processes.

Nobel Laureate (1946).

△ **1931:** *Lewis, W. H. Concept of pinocytosis.*

This refers to a discontinuous process of fluid engulfment by cells, in contrast to diffusion. Many types apparently exist, but often the process involves membranous pseudopodia that enclose droplets of surrounding fluid. These vesicles are then pinched off and sucked into the interior of the cell. The phenomenon may also be concerned with the uptake and transport of substances by cells.

‡ **1931:** *Stern, C., H. Creighton, and B. McClintock. Cytologic demonstration of crossing-over.*

Proof that crossing-over in genes is correlated with exchange of material by homologous chromosomes was independently proved by Stern in *Drosophila* and by Creighton and McClintock in corn. By using crosses of strains that had homologous chromosomes distinguishable individually, it was definitely demonstrated cytologically that genetic crossing-over was accompanied by chromosomal exchange.

○□ **1932:** *Bethe, A. Concept of the ectohormone (pheromone).*

These substances are secreted to the outside of the body by an organism where they affect the physiology or behavior of another individual of the same species. They have been demonstrated in insects, where they may function in trail-making, sex attraction, development control, and the like. (Gr., *pherein,* to carry, + *hormōn,* exciting, stirring up.)

× **1932:** *Danish scientific expedition. Discovery of fossil amphibians (ichthyostegids).*

These fossils were found in the upper Devonian sediments in east Greenland and appear to be intermediate between advanced crossopterygians *(Osteolepis)* and early amphibians. They are the oldest known forms that can be considered amphibians. Many of their characters show primitive amphibian conditions.

× **1932:** *Lewis, G. Edward. Discovery of Ramapithecus fossil.*

Ramapithecus brevirostris was a small humanlike ape of the Miocene and the earliest known hominid. Since Lewis' initial find in India, other fossil fragments of this genus have been discovered in East Africa, Greece, Turkey, and Hungary.

×‡ **1932:** *Wright, S. Genetic drift as a factor in evolution.*

In small populations the Hardy-Weinberg formula of gene frequency may not apply because of "sampling error." If some genotypes are by chance underrepresented in matings in a small population, gene frequencies in the population may change.

△ **1933:** *Collander, R., and H. Bärlund. Measurement of cell permeability.*

Their quantitative measurements of cell membrane permeability to nonelectrolytes of varying molecular size and lipid solubility made possible testable hypotheses and profoundly influenced conceptions of membrane structure.

○ **1933:** *Goldblatt, M., and U. S. von Euler. Discovery of prostaglandins.*

These fatty acid derivative compounds have been isolated from many mammalian tissues (such as seminal plasma, pancreas, seminal vesicle, brain, and kidney). They have a variety of hormonelike and pharmacologic actions, such as stimulation of smooth muscle contractions and relaxation, lowering blood pressure, inhibition of enzymes and hormones, and so on.

× **1933, 1938:** *Haldane, J. B. S., and A. I. Oparin. Heterotroph theory of the origin of life.*

This theory is based on the idea that life was generated from nonliving matter under the conditions that existed before the appearance of life and that have not been duplicated since. The theory stresses the idea that living systems at present make it impossible for any incipient life to gain a foothold as primordial life was able to do.

‡ **1933:** *Painter, T. S., E. Heitz, and H. Bauer. Rediscovery of giant salivary chromosomes.*

These interesting chromosomes were first described by Balbiani in 1881, but their true significance was not realized until these investigators rediscovered them. It has been possible in a large measure to establish the chromosome theory of inheritance by comparing the actual cytologic chromosome maps of salivary chromosomes with the linkage maps obtained by genetic experimentation.

○ **1933:** *Wald, G. Discovery of vitamin A in the retina.*

The discovery that vitamin A is a part of the visual purple molecule of the rods not only gave a better understanding of an important vitamin but also showed how night blindness can occur whenever there is a deficiency of this vitamin in the diet.

Nobel Laureate (1967).

△ **1934:** *Bensley, R. R., and N. L. Hoerr. Isolation and analysis of mitochondria.*

This demonstration suggested an explanation for the behavior and possible functions of these mysterious bodies that intrigued cytologists for a generation, and much has been learned about them in recent years.

○ **1934:** *Dam, H., and E. A. Doisy. Identification of vitamin K.*

The isolation and synthesis of this vitamin are important not merely because of the vitamin's practical value in certain forms of hemorrhage but also because of the light they throw on the physiologic mechanism of blood clotting.

Nobel Laureates (1943).

△ **1934-1935:** *Danielli, J. F., and H. Dawson. Concept of the cell (plasma) membrane.*

Danielli proposed a hypothesis that cell membranes consist of two layers of lipid molecules surrounded on the inner and outer surfaces by a layer of protein molecules. The electron microscope reveals the plasma membrane of a thickness of 75 to 100 Å, consisting of two dark (protein) membranes (25 to 30 Å thick) separated by a light (lipid) interval membrane of 25 to 30 Å thickness. Recent studies confirm the Danielli-Dawson model, while providing a greater dynamic role of the peripheral proteins.

○ **1934:** *Wigglesworth, V. B. Role of the corpus allatum gland in insect metamorphosis.*

This small gland lies close to the brain of an insect, and it has been shown that during the larval stage this gland secretes a juvenile hormone that causes the larval characters to be retained. Metamorphosis and maturation occur as the gland secretes less and less of the hormone. Removal of the gland causes the larva to undergo precocious metamorphosis; grafting the gland into a mature larva will cause the latter to grow into a giant larval form. The gland was first described by A. Nabert in 1913.

○ **1935:** *Hanström, B. Discovery of the x-organ in crustaceans.*

This organ, together with the related sinus gland, constitutes an anatomic complex that has proved of great interest in understanding crustacean endocrinology. The neurosecretory cells in the x-organ (part of the brain) produce a molt-inhibiting hormone that is stored in the sinus gland of the eyestalk; while a molting hormone is produced in a y-organ. The interrelations of these two hormones control the molting process.

○ **1935-1936:** *Kendall, E. C., and P. S. Hench. Discovery of cortisone.*

Kendall had first isolated from the adrenal glands this substance that he called compound E. Its final stages were prepared by Hench later and involved a long tedious chemical process. A hormone that controls the synthesis and release of cortisone and similar steroids was isolated in 1943 from the pituitary and was called ACTH (adrenocorticotropic hormone).

Nobel Laureates (1950).

△ **1935:** *Stanley, W. M. Isolation of a virus in crystalline form.*

This achievement of isolating a virus (tobacco mosaic) is noteworthy not only in giving information about these small agencies responsible for many diseases but also in affording much speculation on the differences between the living and the nonliving.

Nobel Laureate (1946).

□ **1935:** *Tansley, A. G. Concept of the ecosystem.*

The relatively young science of ecology has added many new terms, but the ecosystem is considered one of the basic functional units in ecology, for it best encompasses the environmental relations of organisms in their entirety. It includes both biotic and abiotic factors. The concept under different terminology had been used by others in the early development of ecology.

‡ **1935:** *Timofeeff-Ressovsky, N. W. Target theory of induction of gene mutations.*

Timofeeff-Ressovsky's discovery that mutation can be induced in a gene if a single electron is detached by high-energy radiation gave rise to one of the two prevailing theories of how radiation affects mutation rate.

‡ **1936:** *Demerec, M., and M. E. Hoover. Correspondence between salivary gland chromosome bands and normal chromosome maps.*

By means of three stocks of *Drosophila,* each with a different deficiency at one end of the X chromosome, these investigators were able to find the approximate location on the giant chromosomes of the same genes found on chromosome maps constructed by the percent of crossing-over.

○ **1936:** *Young, J. Z. Demonstration of giant fibers in squid.*

These giant fibers are formed by the fusion of the axons of many neurons whose cell bodies are found in a ganglion near the head. Each fiber is really a tube, up to 1 mm in diameter, consisting of an external sheath filled with liquid axoplasm. These giant fibers control the contraction of the characteristic mantle that surrounds these animals. Because of their size, they have been a very valuable system in neurophysiology.

‡ **1937:** *Blakeslee, A. F. Artificial production of polyploidy.*

By applying the drug colchicine to dividing cells, Blakeslee found that cell division in plants is blocked after the chromosomes have divided (metaphase), and thus the cell has double the normal number of chromosomes. When applied to hybrid plants, this principle makes possible the production of new, polyploid plants.

△ **1937:** *Findlay, G. W. M., and F. O. MacCullum. Discovery of interferon.*

These protein substances, produced by cells in response to viruses and other foreign substances, selectively inhibit virus replication and are one of the body's important defenses against viral infection. They also may play a role in regulation of the immune response.

△ **1937:** *Krebs, H. A. Citric acid (tricarboxylic acid) cycle.*

In a brilliant example of reasoning and deduction, Krebs showed the existence of a cycle of reactions central to aerobic energy production. In the cycle, two carbon fragments from carbohydrate, fatty acids, or amino acids are oxidized to carbon dioxide. The electrons derived from the oxidation reactions in the cycle are passed through a further series of reactions (electron transport system), in which energy is captured in ATP, and finally to oxygen as the final electron acceptor to produce water.

Nobel Laureate (1953).

× **1937:** *Sonneborn, T. M. Discovery of mating types in paramecium.*

Sonneborn's discovery that only individuals of complementary physiologic classes (mating types) would conjugate opened up a new era of protozoan investigation that bids to shed new light on the problems of species concept and evolution.

○ **1937:** *Werle, E., W. Gotze, and A. Keppler. Discovery of kinins.*

Kinins are local hormones produced in blood or tissues that bring about dilation of blood vessels and other changes. They are also found in wasp venom. They have no connection with special glands, are peptides in nature, are very evanescent, perform their functions rapidly, and are then quickly inactivated by enzymes. They also have powerful effects on smooth muscle wherever it is found.

△○ **1937:** *König, P., and A. Tiselius. Development of electrophoresis.*

Electrophoresis is an extremely valuable method of separating proteins in a solution and, after numerous refinements, is currently in wide use. It depends on the differential migration of proteins on an inert carrier when an electric field is applied.

□ **1938:** *Remane, A. Discovery of the new phylum Gnathostomulida.*

This marine phylum was first described by P. Ax in 1956, and its taxonomic position is still under appraisement. The members of the phylum are small wormlike forms (about 0.5 mm long) and show great diversity among the different species. One of their major characteristics is their complicated jaws provided with teeth. They live in sandy substrata and are worldwide in distribution. They show some relationships to the Turbellaria, Gastrotricha, and Rotifera.

○ **1938:** *Schoenheimer, R. Use of radioactive isotopes to demonstrate synthesis of bodily constituents.*

By labeling amino acids, fats, carbohydrates, and the like with radioactive isotopes, it was possible to show how these were incorporated into the various constituents of the body. Such experiments demonstrated that parts of the cell were constantly being synthesized and broken down and that the body must be considered a dynamic equilibrium.

□ **1938:** *Skinner, B. F. Measurement of motivation in animal behavior.*

Skinner worked out a technique for measuring the rewarding effect of a stimulus, or the effects of learning on voluntary behavior. His experimental animals (rats) were placed in a special box (Skinner's box) containing a lever that the animal could manipulate. When the rat presses the lever, small pellets of food may or may not be released, according to the experimental conditions.

○△ **1938:** *Svedberg, T. Development of ultracentrifuge.*

In biologic and medical investigation this instrument has been widely used for the purification of substances, the determination of particle sizes in colloidal systems, the relative densities of materials in living cells, the production of abnormal development, and the study of many problems concerned with electrolytes.

Nobel Laureate (1926).

○ **1939:** *Brown, F. A., Jr., and O. Cunningham. Demonstration of molt-inhibiting hormone in eyestalk of crustaceans.*

Although C. Zeleny (1905) and others had shown that eyestalk removal shortened the intermolt period in crustaceans, Brown and Cunningham were the first to present evidence to explain the effect as being caused by a molt-inhibiting hormone present in the sinus gland.

× **1939:** *Discovery of coelacanth fishes.*

The collection of a living specimen of this ancient fish *(Latimeria),* followed later by other specimens, has brought about a complete reappraisal of this "living fossil" with reference to its ancestry of the amphibians and land forms.

✳ **1939:** *Hörstadius, S. Analysis of the basic pattern of regulative and mosaic eggs in development.*

The masterful work of this investigator has done much to resolve the differences in the early development of regulative eggs (in which each of the early blastomeres can give rise to a whole embryo) and mosaic eggs (in which isolated blastomeres produce only fragments of an embryo). Regulative eggs were shown to have two kinds of substances and both were necessary in proper ratios to produce normal embryos. Each of the early blastomeres has this proper ratio and thus can develop into a complete embryo; in mosaic eggs the regulative power is restricted to a much earlier time scale in development (before cleavage), and thus each isolated blastomere will give rise only to a fragment.

×□ **1939:** *Huxley, J. Concept of the cline in evolutionary variation.*

This concept refers to the gradual and continuous variation in character over an extensive area because of adjustments to changing conditions. This idea of character gradients has proved a very fruitful one in the analysis of the mechanism of evolutionary processes, for such a variability helps to explain the initial stages in the transformation of species.

✳ **1939:** *Pincus, G. Artificial parthenogenesis of the mammalian egg.*

This experiment of producing a normal, fatherless rabbit showed that Loeb's classic method could be made to apply to the eggs of the highest group of animals, and that the primary physiologic process of fertilization is the activation of the egg.

†‡ **1940:** *Landsteiner, Karl, and A. S. Wiener. Discovery of Rh-blood factor.*

Not only was a knowledge of the Rh factor of importance in solving a fatal infant's disease, but it has also yielded a great deal of information about relationships of human races.

Landsteiner, Nobel Laureate (1930).

△‡ **1941:** *Beadle, G. W., and E. L. Tatum. Biochemical mutation.*

By subjecting the bread mold *Neurospora* to x-ray irradiation, they found that genes responsible for the synthesis of certain vitamins and amino acids were inactivated (mutated) so that a strain of this mold carrying the mutant genes could no longer grow unless these particular vitamins and amino acids were added to the medium on which the mold was growing. This outstanding discovery has revealed as never before the precise way in which a single gene controls the specificity of a particular enzyme and has greatly stimulated similar research on other simple forms of life such as bacteria and viruses.

Nobel Laureates (1958).

○ **1941:** *Cori, C. F., and G. T. Cori. Lactic acid metabolic cycle.*

The regeneration of muscle glycogen reserves in mammals involves the passage of lactic acid from the muscles through the blood to the liver, the conversion of lactic acid there to glycogen, the production of blood glucose from the liver glycogen, and the synthesis of muscle glycogen from the blood glucose.

Nobel Laureates (1947).

○ **1941:** *von Szent-Györgyi, Albert. Role of ATP in muscular contraction.*

The demonstration showing that muscles get their energy for contraction from ATP (adenosine triphosphate) has done much to explain many aspects of the puzzling problem of muscle physiology.

Nobel Laureate (1937).

△○ **1941:** *Martin, A. J. P., and R. L. M. Synge. Development of partition chromatography.*

This is a powerful method for separation of chemically similar substances in mixtures. It depends on differential solubility of a compound in stationary and moving phases of solvents. Variants of the method include column, paper, and thin-layer chromatography.

✳ **1942:** *McClean, D., and I. M. Rowlands. Discovery of the enzyme hyaluronidase in mammalian sperm.*

This enzyme dissolves the cement substance of the follicle cells that surround the mammalian egg and facilitates the passage of the sperm to the egg. This discovery not only aided in resolving some of the difficult problems of the fertilization process but also offers a logical explanation of cases of infertility in which too few sperm may not carry enough of the enzyme to afford a passage through the inhibiting follicle cells.

△ **1943:** *Claude, A. Isolation of cell constituents.*

By differential centrifugation, Claude found it possible to separate, in relatively pure form, particulate components, such as mitochondria, microsomes, and nuclei. These investigations led immediately to a more precise knowledge of the chemical nature of these cell constituents and aided the elucidation of the structure and physiology of the mitochondria—one of the great triumphs in the biochemistry of the cell.

Nobel Laureate (1974).

✳ **1943:** *Holtfreter, J. Tissue synthesis from dissociated cells.*

By dissociating the cells of embryonic tissues of amphibians (by dissolving with enzymes or other agents the intercellular cement that holds the cells together) and heaping them in a mass, he found that the cells in time coalesced and formed the type of tissue from which they had come. This is an application to vertebrates of the discovery of Wilson with sponge cells. It showed that, by some mechanism, cells could recognize other cells of the same type.

‡ **1943:** *Sonneborn, T. M. Extranuclear inheritance.*

The view that in paramecia cytoplasmic determiners (plasmagenes) that are self-reproducing and capable of mutation can produce genetic variability has thrown additional light on the role of the cytoplasm in hereditary patterns.

△‡ **1944:** *Avery, O. T. C., C. M. MacLeod, and M. McCarty. Agent responsible for bacterial transformation.*

These workers were able to show that the bacterial transformation of nonencapsulated bacteria to encapsulated cells was really attributable to the DNA fraction of debris from disrupted encapsulated cells to which the nonencapsulated cells were exposed. This key demonstration showed for the first time that nucleic acids and not proteins were the basic material of heredity. However, geneticists remained skeptical until Avery's conclusions were proved correct nearly a decade later.

△ **1945:** *Cori, Carl. Hormone influence on enzyme activity.*

The delicate balance that insulin and the diabetogenic hormone of the pituitary exercise over the activity of the enzyme hexokinase in carbohydrate metabolism has opened up a whole new field of the regulative action of hormones on enzymes.

Nobel Laureate (1947).

□ **1945:** *Griffin, D., and R. Galambos. Development of the concept of echolocation.*

Echolocation refers to a type of perception of objects at a distance by which echoes of sound are reflected back from obstacles and detected acoustically. These investigators found that bats generated their own ultrasonic sounds that were reflected back to their own ears so that they were able to avoid obstacles in their flight without the aid of vision. Their work climaxed an interesting series of experiments inaugurated as early as 1793 by Spallanzani, who believed that bats avoided obstacles in the dark by reflection of sound waves to their ears. Others who laid the groundwork for the novel concept were C. Jurine (1794), who proved that ears were the all-important organs in perception; H. S. Maxim (1912), who advanced the idea that the bat made use of sounds of low frequency inaudible to human ears; and H. Hartridge (1920), who proposed the hypothesis that bats emitted sounds of high frequencies and short wavelengths (ultrasonic sounds).

△ **1945:** *Lipmann, F. Discovery of coenzyme A.*

The discovery of this important catalyst made possible a better understanding of the breaking down of fatty acid chains and furnished important further knowledge of the reactions in the tricarboxylic acid cycle.

Nobel Laureate (1953).

△ **1945:** *Porter, K. R. Description of the endoplasmic reticulum.*

The endoplasmic reticulum is a very complex cytoplasmic structure consisting of a lacelike network of irregular anastomosing tubules and vesicular expansions within the cytoplasmic matrix. Associated with the reticulum complex are small dense granules of ribonucleoprotein and other granules known as microsomes that are fragments of the endoplasmic reticulum. The reticulum complex plays an important role in the synthesis of proteins.

△ **1946:** *Auerbach, C., and J. M. Robson. The chemical production of mutations.*

Besides radiation effects (Muller, 1927), many inorganic and organic compounds (such as mustard gas, oils, alkaloids, and phenols) are now known to have mutagenic effects such as chromosome breakage. It is believed that such agents disrupt the nucleic acid metabolism responsible for protein synthesis and reduplication.

‡ **1946:** *Lederberg, J., and E. L. Tatum. Sexual recombination in bacteria.*

These investigators found that two different strains of bacteria *(Escherichia coli)* could undergo conjugation and exchange genetic material, thereby producing a hereditary strain with characteristics of the two parent strains. W. Hayes (1952) found that recombination still occurred after one parent strain was killed.

Nobel Laureates (1958).

× **1946:** *Libby, W. F. Radiocarbon dating of fossils.*

The radiocarbon age determination is based on the fact that carbon 14 in the dead organism disintegrates at the rate of one-half in 5,560 years, one-half of the remainder in the next 5,560 years, and so on. This is based on the assumption that the isotope is mixed equally through all living matter and that the cosmic rays (which form the isotopes) have not varied much in periods of many thousands of years. The limitation of the method is around 40,000 years.

Nobel Laureate (1960).

× **1946:** *White, E. I. Discovery of the primitive chordate fossil Jamoytius.*

The discovery of this fossil in the freshwater deposits of Silurian rock in Scotland throws some light on the early ancestry of vertebrates, for this form seems to be intermediate between amphioxus or ammocoetes larva (of lampreys) and the oldest known vertebrates, the ostracoderms. Morphologically, *Jamoytius* represents the most primitive vertebrate yet discovered. It could well serve as an ancestor of such forms as amphioxus and the jawless ostracoderms.

○ **1947:** *Holtz, P. Discovery of norepinephrine (noradrenaline).*

This hormone (vasoconstriction effects) has since been found in most vertebrates and shares with epinephrine (metabolic effects) the functions of the chromaffin part of the adrenal gland.

× **1947:** *Sprigg, R. C. Discovery of Precambrian fossil bed.*

The discovery of a rich deposit of Precambrian fossils in the Ediacara Hills of South Australia has been of particular interest because the scarcity of such fossils in the past has given rise to vague and uncertain explanations about Precambrian life. It was all the more remarkable that the fossils discovered were those of soft-bodied forms, such as jellyfish, soft corals, and segmented worms, including the amazing *Spriggina,* which shows relationship to the trilobites. The fossils are prearthropod but not preannelid.

△ **1947:** *von Szent-Györgyi, Albert. Concept of the contracticle substance actomyosin.*

This protein complex consists of the two components, actin and myosin, and is considered the source of muscular contraction when triggered by ATP. Neither actin nor myosin singly will contract.

Nobel Laureate (1937).

□ **1948:** *von Frisch, Karl. Communication mechanisms of honeybees.*

After many years of patient work, von Frisch deciphered the meaning of bee "dances," behavior patterns of individual bees returning to the hive. The dances communicate information to the other bees regarding location of food and water, location and suitability of sites for a new hive, and other information.

Nobel Laureate (1973).

□ **1948:** *Hess, W. R. Localization of instinctive impulse patterns in the brain.*

By inserting electrodes through the skull, fixing them in position, and allowing such holders to heal in place, it was possible to study the brain of an animal in its ordinary activities. When the rat could automatically and at will stimulate itself by pressing a lever, it did so frequently when the electrode was inserted in the hypothalamus region of the brain, indicating a pleasure center. In this way, by placing electrodes at different centers, rats can be made to gratify such drives as thirst, sex, and hunger.

Nobel Laureate (1949).

△ **1948:** *Hogeboom, G. H., W. C. Schneider, and G. E. Palade. Separation of mitochondria from the cell.*

This was an important discovery in unraveling the amazing enzymatic activity of the mitochondria in the tricarboxylic acid cycle. The role mitochondria play in the energy transfer of the cell has earned for these rod-shaped bodies the appellation of the ''powerhouse'' of the cell.

Palade, Nobel Laureate (1974).

□ **1948:** *Johnson, M. W. Relation between echo-sounding and the deep-scattering layer of marine waters.*

The development of a sound transmitter and a receiver coupled with a timing mechanism for recording the time between an outgoing sound impulse and the echo of its return has made possible an accurate method for determing depths in the ocean. By means of this device a deep-scattering layer far above the floor of the ocean was discovered that scattered the sound waves and sent back echoes. This scattering layer tends to rise toward the surface at night and sink to a depth of many hundred meters by day. Johnson saw a striking parallelism between the shifting of this scattering layer and the diurnal vertical migration of plankton or pelagic animals.

△ **1949:** *de Duve, C. Discovery of the lysosome organelle.*

These small particles were first identified chemically, and later (1955) morphologically with the electron microscope. They are supposed to provide the enzymes for digestion of materials taken into the cell by pinocytosis and phagocytosis and for the digestion of the cell's own cytoplasm. When the cell dies, their enzymes are also released and digest the cell (autolysis).

Nobel Laureate (1974).

† **1949:** *Enders, J. F., F. C. Robbins, and T. H. Weller. Cell culture of animal viruses.*

These investigators found that poliomyelitis virus could be grown in ordinary tissue cultures of nonnervous tissue instead of being restricted to host systems of laboratory animals or embryonated chick eggs.

Nobel Laureates (1954).

○ **1949:** *von Euler, U. S. Role of norepinephrine as transmitter.*

Von Euler isolated and identified norepinephrine as the neurotransmitter in the sympathetic nervous system (1940). Later he isolated and characterized norepinephrine storage granules in nerves and described how the substance was taken up, stored, and released by them.

Nobel Laureate (1970).

‡ **1949:** *Pauling, L. Genic control of protein structure.*

Pauling and his colleagues demonstrated a direct connection between specific chemical differences in protein molecules and alterations in genotypes. Making use of the hemoglobin of patients with sickle cell anemia (which is caused by a homozygous condition of an abnormal gene), he was able by the method of electrophoresis to show a considerable difference in the behavior of this hemoglobin in an electric field compared with that from a heterozygote or from a normal person.

Nobel Laureate (1954).

○ **1949:** *Selye, H. Concept of the stress syndrome.*

In 1937 Selye began his experiments, which led to what he called the ''general adaptation syndrome'' that involved the chain reactions of many hormones, such as cortisone and ACTH, in meeting stress conditions faced by an organism. Whenever the stress experience exceeds the limitations of these body defenses, serious degenerative disorders may result.

△ **1950:** *Callan, H. G., and S. G. Tomlin. The bilamellar organization of the nuclear envelope.*

By means of the electron microscope, these investigators were the first to show that the nuclear envelope was composed of two membranes that are interrupted by discontinuities.

△ **1950:** *Caspersson, T., and J. Brachet. Biosynthesis of proteins.*

Investigations to solve the vital problem of protein synthesis from free amino acids have been under way since Emil Fischer's (1887) outstanding work on protein structure. Many competent research workers, such as Bergmann, Lipmann, and Schoenheimer, have made contributions to an understanding of protein synthesis, but Caspersson and Brachet were the first to point out the significant role of ribonucleic acid (RNA) in the process—and most biochemical investigations on the problem since that time have been directed along this line.

△ **1950:** *Chargaff, E. Base composition of DNA.*

The discovery that the amount of purines was equal to the amount of pyrimidines in DNA, the amount of adenine was equal to that of thymine, and the amount of cytosine was equal to that of guanine paved the way for the DNA model of Watson and Crick. The two major functions of DNA are replication and hereditary information storage.

○ **1950:** *Simpson, M., and C. H. Li. Coordination of hormones for balanced development of a tissue.*

One hormone may control the size of an organ or tissue and a different hormone may be responsible for its maturation. Thus the growth hormone of the adenohypophysis may cause a bone to grow in length, but thyroxine from the thyroid is necessary for a fully differentiated bone.

□ **1950:** *Lorenz, K. Development of ethology.*

Lorenz is regarded as the founder of ethology, a biologic approach to analyzing behavior, emphasizing a comparative method and stressing innate factors in behavior development. His classic studies on the greylag goose first came to the attention of English-speaking scientists during the 1950s, though the work itself commenced in the 1930s.

Nobel Laureate (1973).

‡ **1951:** *Lewis, E. B. Concept of pseudoallelism in genetics.*

On the basis of a series of mutations in *Drosophila,* Lewis concluded that certain adjacent loci were closely linked, affected the same trait, and probably arose from a common ancestral gene instead of being a single locus with multiple genes.

‡ **1952:** *Beermann, W. Concept of the chromosomal puff.*

Puffs are local and reversible enlargements in the bands or loci of giant chromosomes. It is believed that puffs are related to the activation of genes and the consequent synthesis of RNA. Puffiing may indicate a correlation between hormonal action and larval development in insect metamorphosis, in which the phenomenon was first discovered.

✶ **1952:** *Briggs, R., and T. J. King. Demonstration of possible differentiated nuclear genotypes.*

The belief that all cells of a particular organism have the same genetic endowment has been questioned as the result of the work of these investigators, who transplanted nuclei of different ages and sources from blastulas and early gastrulas into enucleated zygotes and observed variable and abnormal development.

△ **1952:** *Chase, M., and A. D. Hershey. DNA as the basis of gene structure.*

By using radioactive isotopes, they showed that when a bacterium is infected by a bacterial virus only the viral DNA enters the host cell, with the protein coat remaining behind. Compare with H. Fraenkel-Conrat and R. C. Williams (1955), who were able to separate the protein coat from the RNA core in tobacco mosaic virus.

□ **1952:** *Kramer, G. Orientation of birds to positional changes of sun.*

This discovery showed that birds (starlings and pigeons) can be trained to find food in accordance with the position of the sun. He found that the general orientation of the birds shifted at a rate (when exposed to a constant artificial sun) that could be predicted on the basis of the birds' correcting for the normal rotation of the earth. Birds were able to orient themselves in a definite direction with reference to the sun, whether the light of the sun reached them directly or was reflected by mirrors. They were capable also of finding food at any time of day, thus indicating an ability to compensate for the sun's motion across the sky.

△ **1952:** *Palade, G. E. Analysis of the finer structure of the mitochondrion.*

The important role of mitochondria in the enzymatic systems and cellular metabolism has focused much investigation on the structure of these cytoplasmic inclusions. Each mitochondrion is bounded by two membranes; the outer is smooth and the inner is thrown into small folds or cristae that project into a homogenous matrix in the interior. Some modifications of this pattern are found.

Nobel Laureate (1974).

△ **1952:** *Zinder, N., and J. Lederberg. Discovery of the transduction principle.*

Transduction is the transfer of DNA from one bacterial cell to another by means of a phage. It occurs when an infective phage picks up from its disintegrated host a small fragment of the host's DNA and carries it to a new host where it becomes a part of the genetic equipment of the new bacterial cell.

Lederberg, Nobel Laureate (1958).

‡△ **1953:** *Crick, F. H. C., J. D. Watson, and M. H. F. Wilkins. Chemical structure of DNA.*

Crick and Watson formulated the hypothesis that

the DNA molecule was made up of two chains twisted around each other in a helic structure with pairs of nitrogenous bases projecting toward each other—adenine opposite thymine and guanine opposite cytosine. Genes are considered to be segments of these molecules, with the sequence of bases coding for the amino acids in a protein. Each of the complementary strands acts as a model or template to form a new strand before cell division. The hypothesis has been widely accepted and is now amply confirmed.

Nobel Laureates (1962).

△ **1953:** *Lwoff, A. Concept of the prophage.*

When a bacterial virus enters a bacterial cell, it may enter a vegetative state and produce more phages, or it may become a property of the bacterial cell and become incorporated into the genetic material of its host. The bacterial cell can then reproduce itself and the phage through many generations. The term "prophage" refers to the hereditary ability to produce bacteriophages under such conditions.

Nobel Laureate (1965).

△ **1953:** *Palade, G. E. Description of cytoplasmic ribosomes.*

The description of ribonucleic acid–rich granules (usually on the endoplasmic reticulum) represents one of the key links in the unraveling of the mechanism of protein synthesis. The ribosomes (about 250 Å in diameter) serve as the site of protein synthesis; the number of ribosomes (polysomes) involved in the synthesis of a protein depends on the length of messenger RNA and the protein being synthesized.

Nobel Laureate (1974).

× **1953:** *Urey, H., and S. Miller. Demonstration of the possible primordia of life.*

By exposing a mixture of water vapor, ammonia, methane, and hydrogen gas to electric discharge (to simulate lightning) for several days, these investigators found that several complex organic substances, such as the amino acids glycine and alanine, were formed when the water vapor was condensed into water. This demonstration offered a very plausible theory to explain how the early beginnings of life substances could have started by the formation of organic substances from inorganic ones.

∗**1954:** *Dan, J. C. Acrosome reaction.*

In echinoderms, annelids, and molluscs Dan has shown that the acrosome region of the spermatozoon forms a filament and releases an unknown substance at the time of fertilization. Evidence seems to indicate that the filament is associated with the formation of the fertilization cone. Other observers had described similar filaments before Dan made his detailed descriptions. The filament (1 to 75 μm long) may play an important role in the entrance of the sperm into the cytoplasm.

○ **1954:** *Du Vigneaud, V. Synthesis of pituitary hormones.*

This investigator isolated the posterior pituitary hormones, oxytocin and vasopressin. Both were found to be polypeptides of amino acids, and oxytocin was the first polypeptide hormone to be produced artificially. Oxytocin contracts the uterus during childbirth and releases the mother's milk; vasopressin raises blood pressure and decreases urine production.

○ **1954:** *Katz, B. Impulse transmission at nerve junction.*

The active transmitter substance was found to be acetylcholine, which is stored at the nerve endings in quantum packets and is released in the passage of nerve impulses.

Nobel Laureate (1970).

○ **1954:** *Huxley, H. E., A. F. Huxley, and J. Hanson. Theory of muscular contraction.*

By means of electron microscopic studies and x-ray diffraction studies, these investigators showed that the proteins actin and myosin were found as separate filaments that apparently produced contraction by a sliding reaction in the presence of ATP. The concept has been widely accepted.

A. F. Huxley, Nobel Laureate (1963).

△ **1954:** *Sanger, F. Structure of the insulin molecule.*

Insulin is the important hormone used in the treatment of diabetes. It was the first protein for which a complete amino acid sequence was known. The molecule was found to be made up of 17 different amino acids in 51 amino acid units. Although one of the smallest proteins, its formula contains 777 atoms.

Nobel Laureate (1958).

△ **1955:** *Fraenkel-Conrat, H., and R. C. Williams. Analysis of the chemical nature of a virus.*

In tobacco mosaic virus these workers were able to separate the protein, which makes up the outer cylinder of the virus, from the nucleic acid, the inner core of the cylinder. Neither the protein fraction nor the nucleic acid by itself was able to grow or infect tobacco, but when the two fractions were recombined, the resulting particles behaved like the original virus.

△ **1955-1957:** *Kornberg, A., and S. Ochoa. Synthesis of nucleic acids outside of cells (in vitro).*

By mixing the enzyme polymerase, extracted from the bacterium *Escherichia coli,* with a mixture of nucleotides and a tiny amount of DNA, Kornberg was able to produce synthetic DNA. Ochoa obtained the synthesis of RNA in a similar manner by using the enzyme polynucleotide phosphorylase from the bacterium *Azotobacter vinelandii*. This significant work reveals more insight into the mechanism of nucleic acid duplication in the cell.

Nobel Laureates (1959).

× **1955:** *Kettlewell, H. B. D. Natural selection in action: industrial melanism in moths.*

Using field techniques, Kettlewell was able to demonstrate a selective advantage of dark (melanic) mutants in a barklike cryptic moth species in industrial areas of Great Britain, where the tree trunks had been darkened by soot. This demonstration provided an evolutionary explanation for the increasing incidence of melanism in cryptic moths that has been noted throughout the industrialized regions of the world and constitutes a classic example of the effects of a strong selective pressure.

○ **1956:** *von Bekesy, G. The traveling wave theory of hearing.*

Helmholtz (1868) had proposed the resonance theory of hearing on the basis that each cross fiber of the basilar membrane, which increases in width from the base to the apex of the cochlea, resonates at a different frequency; von Bekesy showed that a traveling wave of vibration is set up in the basilar membrane and reaches a maximal vibration in that part of the membrane appropriate for that frequency.

Nobel Laureate (1961).

△ **1956:** *Borsook, H., and P. C. Zamecnik. Site of protein synthesis.*

By injecting radioactive amino acids into an animal, they found that the ribosome of the endoplasmic reticulum is the place where proteins are formed.

‡ **1956:** *Ingram, V. M. Nature of a mutation.*

By tracing the change in one amino acid unit out of more than 300 units that make up the protein hemoglobin, Ingram was able to pinpoint the difference between normal hemoglobin and the mutant form of hemoglobin that causes sickle cell anemia.

○ **1956:** *Peart, W. S., and D. F. Elliot. Isolation of angiotensin.*

Ever since Volhard in 1928 suggested that a substance in the kidney might be responsible for certain cases of hypertension, investigators have been trying to identify this substance. Among the landmarks in the development of the concept were the Goldblatt clamp (an artificial constriction of the renal artery) that caused something in the kidney to elevate blood pressure; the discovery of the enzyme renin by Page and others; the action of this enzyme on a blood protein (renin substrate) to form an inactive substance (angiotensin I); and finally the conversion of the inactive form into the active angiotensin II by means of a converting enzyme.

△ **1956:** *Sutherland, E. W., and T. W. Rall. Discovery of cyclic AMP.*

An intercellular mediating agent (cyclic adenosine-3′,5′-monophosphate, or cyclic AMP) is found in all living animal tissues, and changes in cyclic AMP levels (by effects of hormones) cause the hormones to produce different target effects depending on the type of cell in which they are found. Cyclic AMP is formed by a metabolic reaction in which the enzyme adenyl cyclase converts ATP to cyclic AMP.

Sutherland, Nobel Laureate (1971).

‡ **1956:** *Tjio, J. H., and A. Levan. Revision of human chromosome count.*

The time-honored number of chromosomes in humans, 48 (diploid), was found by careful cytologic technique to be 46 instead.

‡△ **1957:** *Benzer, S. Concept of the cistron.*

The newer view of the gene has greatly changed the classic concept of the gene as an entity that controlled mutation, hereditary recombination, and function. There are actually several subunits, of which the cistron, the unit of function, is the largest. Many units of recombination may be found in a single cistron. The mutation unit (muton) is variable but usually consists of two to five nucleotides.

△ **1957:** *Calvin, M. Chemical pathways in photosynthesis.*

By using radioactive carbon 14, Calvin and colleagues were able to analyze step by step the incorporation of carbon dioxide and the identity of each intermediate product involved in the formation of carbohydrates and proteins by plants.

Nobel Laureate (1961).

△ **1957:** *Holley, R. W. The role of transfer RNA in protein synthesis.*

Nucleotides of transfer RNAs differ from each other only in their bases. Holley also devised methods that precisely established the transfer RNAs that were used in the transfer of certain amino acids to the site of protein synthesis.

Nobel Laureate (1968).

□ **1957:** *Ivanov, A. V. Analysis of phylum Pogonophora (beard worms).*

Specimens of this phylum were collected in 1900 and represent the most recent phylum to be discovered and evaluated in the animal kingdom. Collections have been made in the waters of Indonesia, the Okhotsk Sea, the Bering Sea, and the Pacific Ocean. They are found mostly in the abyssal depths. Thought originally to belong to the Deuterostomia division, more recent evidence indicates they are protostomes. At present 80 species divided into two orders have been described.

△ **1957:** *Perutz, M. F., and J. C. Kendrew. Structure of hemoglobin.*

The mapping of a complex globular protein molecule of 600 amino acids and 10,000 atoms arranged in a three-dimensional pattern represented one of the great triumphs in biochemistry. Myoglobin of muscle, which acts as a storehouse for oxygen and contains only one heme group instead of four (hemoglobin), was found to contain 150 amino acids.

Nobel Laureates (1962).

□ **1957:** *Sauer, F. Celestial navigation by birds.*

By subjecting Old World warblers to various synthetic night skies of star settings in a planetarium, Sauer was able to demonstrate that the birds made use of the stars to guide them in their migrations.

△‡ **1957:** *Taylor, J. H., P. S. Woods, and W. L. Hughes. Application of the tracer method to organization and replication of chromosomes.*

By using radioactive materials as markers, these investigators were able to show that each new chromosome consists of one-half of old material and one-half of newly synthesized substances. This was confirmation of the Crick-Watson model of the nucleic acid molecule that is supposed to divide or unwind into two single threads, and each half reduplicates itself to form a complete double strand.

○ **1958:** *Lerner, A. B. Discovery of melatonin in the pineal gland.*

Lerner and associates discovered that the production of melatonin in the pineal gland is increased in darkness and decreased in the light. Activity of the hormone appears important in regulation of gonadal functions affected by photoperiod, and its discovery represented a breakthrough in understanding the function of the pineal gland.

△ **1958:** *Meselson, M., and F. W. Stahl. Confirmation "in vivo" of the duplicating mechanism in DNA.*

This was really confirmation of the self-copying of DNA in accordance with Watson and Crick's scheme of the structure of DNA. These investigators found that after producing a culture of bacterial cells labeled with heavy nitrogen 15 and then transferring these bacteria to a cultural medium of light nitrogen 14, the resulting bacteria had a DNA density intermediate between heavy and light, as would be expected on the basis of the Watson-Crick hypothesis.

○ **1959:** *Burnet, F. M. Clonal selection theory of immunity.*

One of the most puzzling aspects of the acquired immune response has been the specificity of the antigen-antibody reaction. Burnet suggested that we have a large variety of antibody-producing cells, each present in such a small number that its product cannot be detected. When exposed to a foreign substance (antigen), a clone of cells that can make antibody to that antigen is "selected" and multiplies rapidly, thus increasing antibody to a detectable level. A corollary to the theory is that antigens present when the organism is embryonic or very young are recognized as "self" and the immune system does not respond. This was confirmed by P. Medawar.

Burnet and Medawar, Nobel Laureates (1960).

‡ **1959:** *Ford, C. E., P. A. Jacobs, and J. H. Tjio. Chromosomal basis of sex determination in humans.*

By discovering that certain genetic defects were associated with an abnormal somatic chromosomal constitution, it was possible to determine that male-determining genes in humans were located on the Y chromosome. Thus a combination of XXY (47 instead of the normal 46 diploid number) produced sterile males (Klinefelter's syndrome), and those with XO combinations (45 diploid number) gave rise to Turner's disease or immature females.

× **1959:** *Leakey, Mary D. Discovery of Australopithecus (Zinjanthropus) boisei fossil hominid.*

This fossil is a robust form within the small-brained, large-jawed genus *Australopithecus,* first discovered by Dart (1925). Its status is at present controversial: some believe it to be a species distinct from *A. africanus,* while others believe that *A. africanus* and *A. boisei* are females and males within a single, polytypic species.

†‡ **1959:** *LeJeune, J., M. Gautier, and R. Turpin. Abnormal chromosome pattern in humans.*

The discovery of the presence of an extra chromosome (autosome) in the tissue cultures obtained from mongoloid children has aroused much interest in the human cytogenetic pattern in its relation to disease states. This was the first clear-cut case of such an etiologic mechanism in the explanation of a disease and has stimulated many investigations of clinical interest along similar lines. Such a condition of an extra chromosome is called **trisomy;** a condition of one less chromosome is called **monosomy.**

○ **1959:** *Butenandt, A. F. J. Chemical identification of a pheromone.*

Butenandt and associates chemically analyzed the first pheromone—the sex attractant substance of silk moths *(Bombyx mori)*. Named bombykol, it is a doubly unsaturated fatty alcohol with 16 carbon atoms. The chemical nature of numerous other pheromones has since been determined.

○ **1959-1960:** *Yalow, R., and S. A. Berson. Development of radioimmunoassay.*

The development of this technique made it possible for the first time to measure minute quantities of a hormone directly in a complex mixture of proteins such as serum. A radioactively labeled hormone is mixed with a specific antibody to the hormone, together with a sample of the unknown. The labeled compound competes with the unlabeled in the formation of the antigen-antibody complex, and the amount of unknown is obtained by comparison with a standard curve prepared with known amounts of unlabeled hormone. The technique was originally described for insulin, but it has been rapidly extended to many other hormones and has revolutionized endocrinology.

Yalow, Nobel Laureate (1977).

□ **1960:** *Tinbergen, L. Development of the concept of specific searching images in predators.*

Tinbergen's studies of songbird predation in Dutch pinewoods laid the foundation for a field becoming known as behavioral ecology. Current interest focuses on foraging strategies of predators.

Nobel Laureate (1973).

△ **1960:** *Hurwitz, J., A. Stevens, and S. Weiss. Enzymatic synthesis of messenger RNA.*

The exciting knowledge of the coding system of DNA and its translation to protein synthesis (ribosomes) was further elucidated when it was discovered that an enzyme, RNA polymerase, was responsible for the synthesis of RNA from a template pattern of DNA.

△ **1960:** *Jacob, F., and J. Monod. The operon hypothesis.*

The operon hypothesis is a postulated model of how enzyme synthesis is regulated in the cell. The model proposes that refinement in regulation involves an inducible system for allowing structural genes to synthesize needed enzymes and a repressible system that cuts off the synthesis of unneeded enzymes.

Nobel Laureates (1965).

△ **1960:** *Strell, M., and R. B. Woodward. Synthesis of chlorophyll a.*

Strell and Woodward with the aid of many co-workers finally solved this problem, which had been the goal of organic chemists for many generations.

Woodward, Nobel Laureate (1965).

△ **1961:** *Hurwitz, J., A. Stevens, and S. B. Weiss. Confirmation of messenger RNA.*

Messenger RNA transcribes directly the genetic message of the nuclear DNA and moves to the cytoplasm, where it becomes associated with a number of ribosomes or submicroscopic particles containing protein and nonspecific structural RNA. Here the messenger RNA molecules serve as templates against which amino acids are arranged in the sequence corresponding to the coded instructions carried by messenger RNA.

‡△ **1961:** *Jacob, F., and J. Monod. The role of messenger RNA in the genetic code.*

The transmission of information from the DNA code in the genes to the ribosomes represents an important step in the unraveling of the genetic code, and these investigators proposed certain deductions for confirmation of the hypothesis, which in general has been established by many researchers.

Nobel Laureates (1965).

○ **1961:** *Miller, J. F. A. Function of the thymus gland.*

Long known as a transitory organ that persists during the early growth period of animals, the thymus is now recognized as a vital processing center in the development of certain lymphocytes (T cells). These cells are very important in cell-mediated immunity.

‡△ **1961:** *Nirenberg, M. W., and J. H. Matthaei. Deciphering the genetic code.*

By adding a synthetic RNA composed entirely of uracil nucleotides to a mixture of amino acids, these investigators obtained a polypeptide made up solely of a single amino acid, phenylalanine. On the basis of a triplet code, it was concluded that the RNA code word for phenylalanine was UUU and its DNA complement was AAA. This was the beginning of coding the various amino acids and represents a key demonstration for understanding the genetic code.

Nirenberg, Nobel Laureate (1968).

△ **1961:** *Mitchell, P. The chemiosmotic-coupling hypothesis of ATP formation.*

Although it is known that the oxidation of food molecules in the cell results in net synthesis of ATP from ADP and inorganic phosphate, the molecular mechanism to drive the reaction is still unclear. Present evidence supports the chemiosmotic hypothesis, which suggests that the energy derived from electron transport serves to pump hydrogen ions across the inner mitochondrial membrane, actively setting up an electrochemical gradient. The gradient of hydrogen ions is then coupled with an ATPase complex in the inner membrane, providing the free energy to form the high-energy phosphate bond.

○ **1962:** *Copp, H. Discovery of calcitonin.*

This peptide hormone is produced by the ultimobranchial glands (lower vertebrates) or their embryologic derivatives, the parafollicular cells of the thyroid gland (mammals). It lowers the blood calcium and increases calcium uptake by bone cells, thus antagonizing the action of parathormone.

△ **1962:** *Perry, R. P. Cellular sites of synthesis of RNA.*

By using different labeled RNA precursors and other cytochemical techniques, this investigator concluded that messenger RNA and transfer RNA synthesis are localized in the chromosomes of the cell; the ribosomal RNA of greater molecular weight is produced in the nucleolus.

× **1964:** *Hamilton, W. D. Concepts of kin selection and inclusive fitness.*

The problem of neuter castes in social insects had destroyed Lamarckian and baffled Darwinian evolutionary explanations. Hamilton, however, demonstrated that the peculiar mode of reproduction in most social insects having neuter castes (haplodiploidy, or males haploid and females diploid) resulted in a situation in which sisters would share more common genes with one another than with their own progeny. Hence, helping to rear "kin" (sisters) could be more adaptive from the point of view of one's genes (inclusive fitness) than rearing one's own young (individual fitness). These ideas have since been extended to social behaviors of many species—vertebrate as well as invertebrate—and provide a fundamental core of the emerging field of sociobiology.

× **1964:** *Hoyer, B. H., B. J. McCarthy, and E. T. Bolton. Phylogeny and DNA sequence.*

These investigators presented evidence that certain homologies exist among the polynucleotide sequences in the DNA sequence of such different forms as fishes and humans. Such sequences may represent genes that have been retained with little change throughout vertebrate history. Possible phenotypic expressions may be bilateral symmetry, notochord, hemoglobin, and so on.

△‡ **1966:** *Khorana, H. G. Proof of code assignments in the genetic code.*

By using alternating codons (CUC and UCU) in an artificial RNA chain, he was able to synthesize a polypeptide of alternating amino acids (leucine and serine) for which these codons respectively stood.

Nobel Laureate (1968).

△‡ **1967:** *Ptashne, M. Isolation of first repressor.*

Repressors are protein substances supposedly formed by regulatory genes and function by preventing a structural gene from making its product when not needed by the cells.

□ **1967:** *MacArthur, R. H., and E. O. Wilson. Theoretic ecology.*

Mathematic models in conjunction with field studies form the basis of theoretic ecology. This is well-shown in the *Theory of Island Biogeography* by MacArthur and Wilson. By the methods found in their book and the investigations of many others, it has been possible to determine the equilibrium of species, the number of extinctions, and other factors on islands. This viewpoint has served as a revelation in ecologic studies.

□ **1968:** *Goodall, J. van Lawick. Behavior of free-living chimpanzees.*

A long-term, complete study of primate social behavior, was done in the field, providing an impetus and stimulus for a large number of similar studies on other primates.

△ **1970:** *Temin, H. M., and D. Baltimore. Demonstration of DNA synthesis from an RNA template.*

Many viruses carry RNA rather than DNA as a genetic material, and it was not known how the RNA virus could enter the host cell and induce the host to synthesize more RNA virus particles. Temin and Baltimore independently found an RNA-dependent DNA polymerase in RNA viruses; thus, the RNA virus reproduces by entering a host cell and producing a DNA copy from the RNA template; then the host cell machinery makes new virus particles from the DNA template.

Nobel Laureates (1975).

○△ **1970:** *Edelman, G. M., and R. R. Porter. The structure of gamma globulin.*

By using myeloma tumors (which contain pure immunoglobulin proteins) the investigators were able to work out after many years a complete analysis of the large gamma globulin molecule, which is made up of 1,320 amino acids and 19,996 atoms, with a molecular weight of 150,000.

Nobel Laureates (1972).

○ **1972:** *Woodward, R. B., and A. Eschenmoser. The synthesis of vitamin B_{12}.*

Vitamin B_{12} is the last vitamin to be synthesized. This complex molecule is not made up of polymers; new methods of organic chemistry were required before it could be synthesized. It contains the metal ion cobalt.

△ **1973:** *Cohen, S. N., and H. W. Boyer. Recombinant DNA.*

This was the first demonstration that restriction enzymes could be used to transfer gene sequences from the genome of one organism into that of another.

□ **1973:** *The Endangered Species Act.*

This omnibus legislation, created in 1966 and greatly strengthened in 1973, is the first United States domestic law concerned exclusively with wildlife and enacted for purely altruistic motives. Its strength resides in that portion of the act specifically prohibiting federally funded projects from jeopardizing endangered species or their habitats. Some 400 species of plants and animals have been placed on the endangered list.

△ **1975:** *Miller, J., and P. Lu. Alteration of amino acid sequence.*

By inserting known amino acids to replace naturally occurring amino acids in the lac repressor protein, these investigators have been able to determine which amino acids are necessary for the various functions of the lac repressor.

△ **1975:** *Sanger, J. W. Chromosome migration on spindle fibers.*

The proteins (actin and myosin) that contract muscles were found to be involved in the migration of chromosomes along the spindle fibers. Actin bundles are present on chromosomal spindle fibers.

† **1977:** *Eradication of smallpox.*

The global eradication of variola major, the virulent form of smallpox, was achieved by making preventive immunization available to underdeveloped areas through a World Health Organization campaign.

BOOKS AND PUBLICATIONS THAT HAVE GREATLY INFLUENCED DEVELOPMENT OF ZOOLOGY

Aristotle. 336-323 B.C. De anima, Historia animalium, De partibus animalium, *and* De generatione animalium. *These biologic works of the Greek thinker have exerted an enormous influence on biologic thinking for centuries.*

Vesalius, Andreas. 1543. De fabrica corporis humani. *This work is the foundation of modern anatomy and represents a break with the Galen tradition. His representations of some anatomic subjects, such as muscles, have never been surpassed. Moreover, he treated anatomy as a living whole, a viewpoint present-day anatomists are beginning to copy.*

Fabricius of Aquapendente. 1600-1621. De formato foetu *and* De formatione ovi pulli. *This was the first illustrated work on embryology and may be said to be the beginning of the modern study of development.*

Harvey, William. 1628. Anatomic dissertation concerning the motion of the heart and blood. *This great work represents one of the first accurate explanations in physical terms of an important physiologic process. It initiated an experimental method of observation that gave an impetus to research in all fields of biology.*

Descartes, René. 1637. Discourse on method. *This philosophic essay gave a great stimulus to a mechanistic interpretation of biologic phenomena.*

Buffon, Georges. 1749-1804. Histoire naturelle. *This extensive work of many volumes collected together natural history facts in a popular and pleasing style. It had a great influence in stimulating a study of nature. Many eminent biologic thinkers, such as Erasmus Darwin and Lamarck, were influenced by its generalizations, which here and there suggest an idea of evolution in a crude form.*

Linnaeus, Carolus. 1758. Systema naturae. *In this work there is laid the basis for the classification of animals and plants. With few modifications, the taxonomic principles outlined therein have been universally adopted by biologists.*

Wolff, Caspar Friedrich. 1759. Theoria generationis. *The theory of epigenesis was here set forth for the first time in opposition to the preformation theory of development so widely held up to the time of Wolff's work.*

von Haller, Albrecht. 1760. Elementa physiologiae. *An extensive summary of various aspects of physiology that greatly influenced physiologic thinking for many years. Some of the basic concepts therein laid down are still considered valid, especially those on the nervous system.*

Malthus, Thomas R. 1798. Essay on population. *This work stimulated evolutionary thinking among such people as Darwin and Wallace.*

de Lamarck, Jean Baptiste. 1809. Philosophie zoologique. *This publication was of great importance in focusing the attention of biologists on the problem of the role of the environment as a factor in evolution. Lamarck's belief that all species came from other species represented one of the first clear-cut statements on the mutability of species, even though his theory of use and disuse has not been accepted by most biologists.*

Cuvier, Georges. 1817. Le règne animal. *A comprehensive biologic work that dealt with classification and a comparative study of animal structures. Its plates are still of value, but the general plan of the work was marred by a disbelief in evolution and a faith in the doctrine of geologic catastrophes. The book, however, exerted an enormous influence on contemporary zoologic thought.*

von Baer, Karl Ernst. 1828-1837. Entwickelungsgeschichte der Thiere. *In this important work are laid down the fundamental principles of germ layer formation and the similarity of corresponding stages in the development of embryos that have proved to be the foundation studies of modern embryology.*

Audubon, John J. 1828-1838. The birds of America. *The greatest of all ornithologic works, it has served as the model for all monographs dealing with a specific group of animals. The plates, the work of a master artist, have seldom been surpassed.*

Lyell, Charles. 1830-1833. Principles of geology. *From a biologic viewpoint this great work exerted a profound influence on biologic thinking, for it did away with the theory of catastrophism and prepared the way for an evolutionary interpretation of fossils and the forms that arose from them.*

Beaumont, William. 1833. Experiments and observations on the gastric juice and the physiology of digestion. *In this classic work the observations Beaumont made on various functions of the stomach and digestion were accurate and thorough. The work paved the way for the brilliant investigations of Pavlov, Cannon, and Carlson of later generations.*

Müller, Johannes. 1834-1840. Handbook of physiology. *The principles set down in this work by perhaps the greatest of all physiologists have set the pattern for the development of the science of physiology.*

Darwin, Charles. 1839. Journal of researches (Voyage of the *Beagle*). *This book reveals the training and development of the naturalist and the material that led to the formulation of Darwin's concept of organic evolution.*

Schwann, Theodor. 1839. Mikroskopische Untersuchungen über die Uebereinstimmung in der Struktur und dem Wachstum der Thiere und Pflanzen. *The basic principles concerning the cell doctrine are laid down in this classic work.*

Kölliker, Albrecht. 1852. Mikroskopische Anatomie. *This was the first textbook in histology and contains contributions of the greatest importance in this field. Many of the histologic descriptions Kölliker made have never needed correction. In many of his biologic views he was far ahead of his time.*

Maury, Matthew F. 1855. The physical geography of the sea. *This work has often been called the first textbook on oceanography. This pioneer treatise stressed the integration of such knowledge as was then available about tides, winds, currents, depths, circulation, and such matters. Maury's work represents a real starting point in the fascinating study of the oceans and has had a great influence in stimulating investigations in this field.*

Virchow, Rudolf. 1858. Die Cellularpathologie. *In this work Virchow made the first clear distinction between normal and diseased tissues and demonstrated the real nature of pathologic cells. The work also represents the death knell to the old humoral pathology, which had held sway for so long.*

Darwin, Charles. 1859. On the origin of species. *One of the most influential books ever published in biology. Although built around the theme that natural selection is the most important factor in evolution, the great influence of the book has been attributable to the great array of evolutionary evidence it presented. It also stimulated constructive thinking on a subject that had been vague and confusing before Darwin's time.*

Marsh, George P. 1864. Man and nature: physical geography as modified by human action. *A work that had an early and important influence on the conservation movement in America.*

Mendel, G. 1866. Versuche über Pflanzenhybriden. *Careful, controlled pollination technique and statistical analysis gave a scientific explanation that has influenced all geneticists after the "rediscovery" in 1900 of Mendel's classic paper on the two basic laws of inheritance.*

Owen, Richard. 1866. Anatomy and physiology of the vertebrates. *This work contains an enormous amount of personal observation on the structure and physiology of animals, and some of the basic concepts of structure and function, such as homologue and analogue, are here defined for the first time.*

Brehm, Alfred E. 1869. Tierleben. *The many editions of this work over many years have indicated its importance as a general natural history.*

Bronn, Heinrich G. (ed.). 1873 to present. Klassen und Ordnungen des Tier-Reichs. *This great work is made up of exhaustive treatises on the various groups of animals by numerous authorities. Its growth extends over many years, and it is one of the most valuable works ever published in zoology.*

Balfour, Francis M. 1880. Comparative embryology. *This is a comprehensive summary of embryologic work on both vertebrates and invertebrates up to the time it was published. It is often considered the beginning of modern embryology.*

Semper, Karl. 1881. Animal life as affected by the natural conditions of existence. *This work first pointed out the modern ecologic point of view and laid the basis for many ecologic concepts of existence that have proved important in the further development of this field of study.*

Bütschli, Otto. 1889. Protozoen (Bronn's Klassen und Ordnungen des Tier-Reichs). *This monograph has been of the utmost importance to students of Protozoa. No other work on a like scale has ever been produced in this field of study.*

von Hertwig, Richard. 1892. Lehrbuch der Zoologie. *This text has proved to be an invaluable source of material for many generations of zoologists. Its illustrations have been widely used in many other textbooks.*

Weismann, August. 1892. Das Keimplasma. *Weismann predicted from purely theoretic considerations the necessity of meiosis or reduction of the chromosomes in the germ cell cycle, a postulate that was quickly confirmed cytologically by others.*

Hertwig, Oskar. 1893. Zelle und Gewebe. *In this work a clear distinction is made between histology as the science of tissues and cytology as the science of cell structure and function. Cytology as a study in its own right really dates from this time.*

Korschelt, E., and K. Heider. 1893. Lehrbuch der vergleichende Entwicklungsgeschichte der wirbellosen Thiere, 4 vols. *A treatise that has been a valuable tool for all workers in the difficult field of invertebrate embryology.*

Wilson, Edmund B. 1896. The cell in development and heredity. *This and subsequent editions represented the most outstanding work of its kind in the English language. Its influence in directing the development of cytogenetics cannot be overestimated, and in summarizing the many investigations in cytology, the book has served as one of the most useful tools in the field.*

Pavlov, Ivan. 1897. Le travail des glandes digestives. *This work marks a great landmark in the study of the digestive system, for it describes many of the now classic experiments that Pavlov conducted, such as the gastric pouch technique and the rate of gastric secretions.*

De Vries, Hugo. 1901. Die Mutationstheorie. *The belief that evolution is the result of sudden changes or mutations is advanced by one who is commonly credited with the initiation of this line of investigation into the causes of evolution.*

Sherrington, Charles. 1906. The integrative action of the nervous system. *The basic concepts of neurophysiology laid down in this book have been little altered since its publication. Much of the work done in this field has served to confirm the nervous mechanism he here outlines.*

Garrod, Archibald. 1909. Inborn errors of metabolism. *This pioneer book showed that certain congenital diseases were caused by defective genes that failed to produce the proper enzymes for normal functioning. It laid the basis for biochemical genetics, which later received a great impetus from the work of Beadle and Tatum.*

Henderson, Lawrence J. 1913. The fitness of the environment. *This book has pointed out in a specific way the reciprocity that exists between living and nonliving nature and how organic matter is fitted to the inorganic environment. It has exerted a considerable influence on the study of ecologic aspects of adaptation.*

Shelford, Victor E. 1913. Animal communities in temperate America. *This work was a pioneer in the field of biotic community ecology and has exerted a great influence on ecologic study.*

Bayliss, William M. 1915. Principles of general physiology. *If a classic book must meet the requirements of masterly analysis and synthesis of what is known in a particular discipline, then this great work must be called one.*

Wegener, A. 1915. The origin of continents and oceans. *In the first (German) edition of this book, Wegener developed the idea of continental drift–involving the movement of continental masses away from a central common area of origin to their present dispersed positions on the globe. Wegener's interpretation of continental history was out of favor for many decades but was substantiated by geophysical work (plate tectonics) in the 1960s and 1970s. Biogeographers now invoke wegenerian hypotheses to explain distribution of many groups of plants and animals.*

Matthew, W. D. 1915. Climate and evolution. *Matthew, in contrast to Wegener, viewed the positions of continents as permanent. He explained distribution of plants and animals by dispersal between continents across land bridges, such as the Bering bridge between Asia and Alaska and the Panamanian bridge between Central and South America. Matthew's ideas dominated biogeographic thinking until the revival of Wegener's theories in the 1960s and 1970s.*

Morgan, T. H., A. H. Sturtevant, C. B. Bridges, and H. J. Muller. 1915. The mechanism of Mendelian heredity. *This book gave an analysis and synthesis of Mendelian inheritance as formulated from the epoch-making investigations of the authors. This classic work will always stand as a cornerstone of our modern interpretation of heredity.*

Doflein, F. 1916. Lehrbuch der Protozoenkunde, ed. 6 (revised by E. Reichenow, 1949). *A standard treatise on Protozoa. Its many editions have proved helpful to all workers in this field.*

Thompson, D. W. 1917. Growth and form. *This pioneer work deals with the problems of growth and form in relation to physical and mathematic principles. It has thrown much light on these difficult subjects.*

Kukenthal, W., and T. Krumbach. 1923. Handbuch der Zoologie. *An extensive modern treatise on zoology that covers all phyla. The work has been an invaluable tool for all zoologists who are interested in the study of a particular group.*

Fisher, R. A. 1930. Genetical basis of natural selection. *This work has exerted an enormous influence on the newer synthesis of evolutionary mechanisms so much in vogue at present.*

Dobzhansky, T. 1937. Genetics and the origin of species. *The vast change in the explanation of the mechanism of evolution, which emerged about 1930, is well-analyzed in this work by a master evolutionist. Other syntheses of this new biologic approach to the evolutionary problems have appeared since this work was published, but none of them has surpassed the clarity and fine integration of Dobzhansky's work.*

Spemann, Hans. 1938. Embryonic development and induction. *In this work the author summarizes his pioneer investigations that have proved so fruitful in experimental embryology.*

Hyman, L. H. 1940. The invertebrates: Protozoa through Ctenophora. 1951. Platyhelminthes and Rhynchocoela, The acoelomate bilateria. 1951. Acanthocephala, Aschelminthes, and Entoprocta, the pseudocoelomate bilateria. 1955. Echinodermata, the coelomate bilateria. 1959. Smaller coelomate groups: Chaetognatha, Hemichordata, Pogonophora, Phoronida, Ectoprocta, Brachiopoda, Sipunculida, The coelomate bilateria. 1967. Mollusca I. *This series of volumes is a monumental work by a single author and is the only such coverage of the invertebrates in the English language. It is a benchmark in completeness, accuracy, and incisive analysis. It is an absolutely essential reference for all teachers and investigators in invertebrate zoology.*

Schrödinger, Erwin. 1945. What is life? *The emphasis placed on the physical explanation of life gave a new point of view of biologic phenomena so well expressed in the current molecular biologic revolution.*

Grassé, P. P. (ed.). 1948. Traité de zoologie. *This is a series of many treatises by various specialists on both invertebrates and vertebrates. Since it is relatively recent and comprehensive, the work is valuable to all students who desire detailed information on the various groups.*

Allee, W. C., A. E. Emerson, O. Park, T. Park, and K. P. Schmidt. 1949. Principles of animal ecology. *Ecology has become a major field of biologic study, and the basic principles laid down in this comprehensive work are a landmark in the field.*

Tinbergen, N. 1951. The study of instinct. *The classic statement of the approaches and theoretic interpretations of early ethologists, especially K. Lorenz, is presented. This work brought ethology to the attention of American behavioral scientists, generating both controversy and collaboration.*

Blum, H. F. 1951. Time's arrow and evolution. *In this thought-provoking book, Blum explores the relation between the second law of thermodynamics (time's arrow) and organic evolution and recognizes that mutation and natural selection have been restricted to certain channels in accordance with the law, even though these two factors appear to controvert the principle of pointing the direction of events in time.*

Simpson, G. G. 1953. The major features of evolution. *In a comprehensive synthesis of modern evolutionary theory, the author draws on evidence from paleontology, population genetics, and systematics. This book, along with Simpson's earlier* Tempo and mode in evolution, *have greatly influenced our thinking about the mechanisms of evolution within the framework of natural selection.*

Crick, Francis H. C., and James D. Watson. 1953. Genetic implications of the structure of deoxyribonucleic acid. *The solution of the structure of the DNA molecule has served as the cornerstone for an explanation of genetic replication and control of the cell's attributes and functions.*

Burnet, F. M. 1959. The clonal selection theory of acquired immunity. *This was an important new theory to explain the specificity of the acquired immune response and has to a great extent replaced the template theory formulated by Haurowitz in 1930.*

Carson, R. L. 1962. Silent spring. *Though a polemic addressed to a lay audience, this book was highly valuable in increasing public awareness of the danger to the ecosystem of indiscriminate use of pesticides.*

Hinde, R. A. 1970. Animal behaviour: a synthesis of ethology and comparative psychology. *The most ambitious attempt to date to meld the concepts developed by students of behavior from biology and psychology during some 50 years of independent efforts.*

Macarthur, R. H. 1972. Geographical ecology. *One of the most important theoretic ecology books to have appeared in the 1970s.*

Wilson, E. O. 1975. Sociobiology: the new synthesis. *The rapidly developing fields of behavioral biology, behavioral ecology, and population genetics are reviewed. Wilson attempts to provide the basis for a new biologic science, utilizing concepts developed in these heretofore disparate fields, devoted to analyzing social behaviors at all phylogenetic levels.*

INDEX

C

D

G

H

J

K

L

N

T

ORIGIN OF LIFE AND GEOLOGIC TIME TABLE

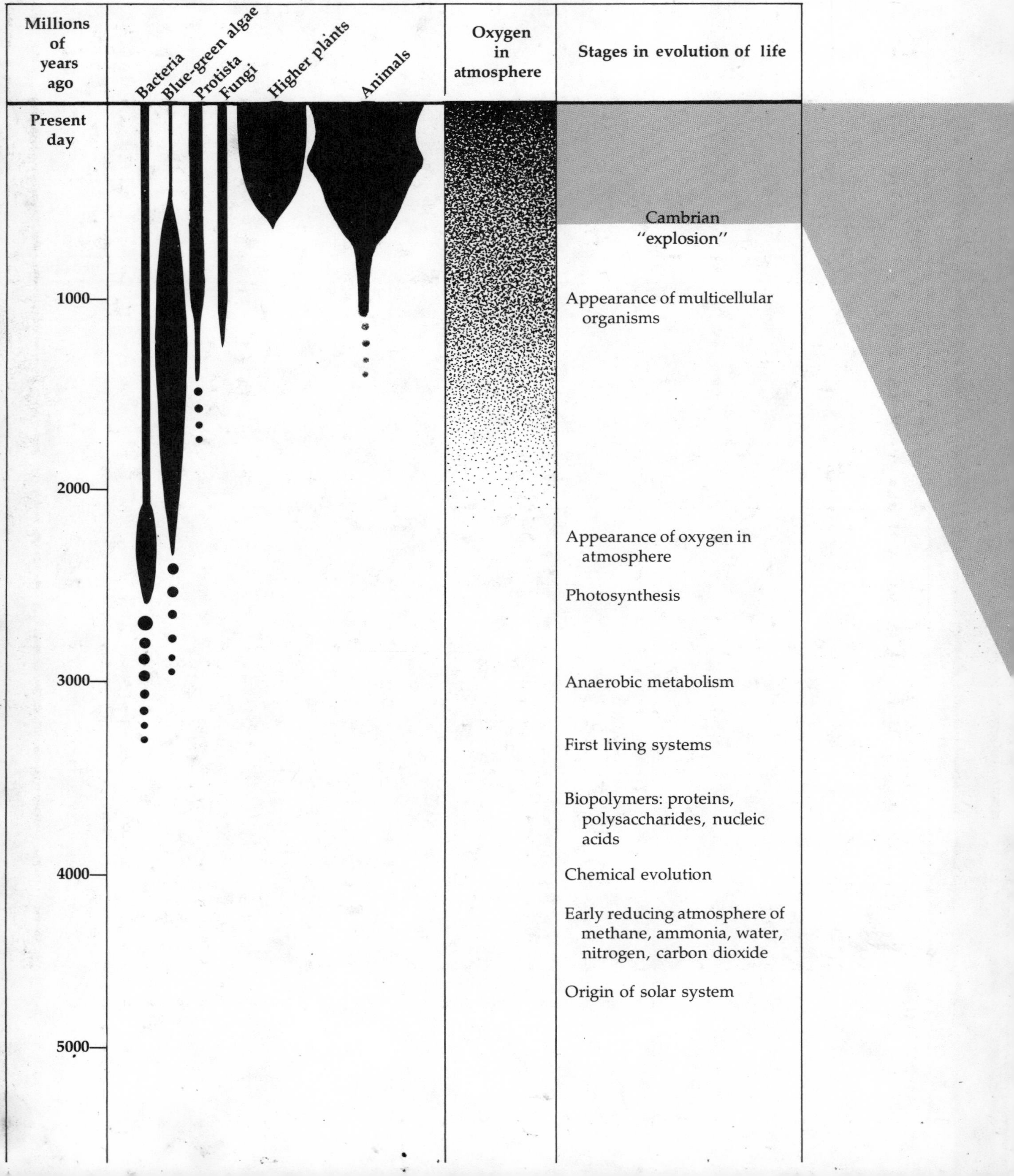

Bacteria
Blue-green algae
Protista
Fungi
Higher plants
Animals

Millions of years ago	Oxygen in atmosphere	Stages in evolution of life
Present day		
		Cambrian "explosion"
1000		Appearance of multicellular organisms
2000		
		Appearance of oxygen in atmosphere
		Photosynthesis
3000		Anaerobic metabolism
		First living systems
		Biopolymers: proteins, polysaccharides, nucleic acids
4000		Chemical evolution
		Early reducing atmosphere of methane, ammonia, water, nitrogen, carbon dioxide
		Origin of solar system
5000		